ENGINEERING THERMODYNAMICS

(Principles and Practices)

"Dedicated to my wife, Mrs. Manjit Kaur, for her endless patience, continuous moral support and encouragement. Her cheerful compliments and inspiring words lifted my spirits on countless occassions. She has enriched my life and my gratitute to her is boundless"

ENGINEERING THERMODYNAMICS

(Principles and Practices)

by

DR. D.S. KUMAR

B.Sc. Engg. (Hons.), M.Sc. Engg, Ph.D, FIE

Formerly :

Professor Mech. Engg, Punjab Engg. College, Chandigarh,
Director Academic Affairs, Punjab Technical Univ., Jalandhar,
Principal S.U.S. College of Engg. & Technology, Tangori (Mohali) and
Advisor, Manav Rachna Educational Institutions, Faridabad

S.K. KATARIA & SONS®

Publisher of Engineering Books

4885/109, Prakash Mahal, Dr. Subhash Bhargav Lane,
Opposite Delhi Medical Association, Daryaganj, New Delhi–110002
Phone: +91-11-43551243; Mobile: +91-9871775858, 7678425209
e-mail: katariabook@gmail.com, katariabooks@yahoo.com
Website: www.skkatariaandsons.com

(e-books, subscriptions and bulk buying also available on our website)

Published by:

S.K. Kataria & Sons®
4885/109, Prakash Mahal, Dr. Subhash Bhargav Lane,
Opposite Delhi Medical Association, Daryaganj, New Delhi–110002
Phone: +91-11-43551243; Mobile: +91-9871775858, 7678425209
e-mail: katariabook@gmail.com, katariabooks@yahoo.com

First Edition : June, 2012; Reprint (edited) : June, 2015; Reprint: 2026

Price : ₹ 895/-

Laser Type Setting:
PLUS COMPUTERS, Delhi

Printed at:
New A.S. Offset Printers, Delhi

Preface

The subject of Engineering Thermodynamics is a core course taught to the students of all disciplines of engineering in almost all the technical institutes and universities. The subject has a wide scope and its applications are extensive. This treatise on Principles and Practices in ENGINEERING THERMODYNAMICS presents an exhaustive and extensive exposition of the following topics which are included in the latest syllabi of courses related to Heat Engines, Applied Thermodynamics and, Thermal Science and Engineering.

- Thermodynamic system, property, state and equilibrium; heat and work interactions; gas laws; First law of thermodynamics and its applications to closed and open systems.
- Second law of thermodynamics; heat engine, refrigerator and heat pump; concept of entropy, reversibility and availability.
- Steam and its properties; various boilers, mountings and accessories; draught and performance criterion for boilers; fuels and combustion, stoichiometry of combustion reactions.
- Steam power cycles; simple and compound steam engines; flow through nozzles; steam turbines and condensers.
- Air standard cycles; internal combustion engines; recriprocating and rotary compressors; gas turbines and jet propulsion.
- Basics of heat transfer; refrigeration and psychrometrics.
- Thermodynamics relations, Maxwell relations and *Tds* equations.

The author had taught these topics for several years at Punjab Engg College, Chandigarh and this monograph is essentially based on the lectures delivered by him. From the experience gained through useful class discussions and feed back, these class notes were revised to improve their clarity and necessary explanatory notes were added during each teaching semester.The subject matter has thus been thoroughly tested in the class room and found suitable.

The text presents an amalgam of numerous ideas and is the outcome of the distillation of the available text books written by eminent authors and many of them being simply outstanding. The author has, however, adopted his own arrangement of the subject and method of presentation. This aspect is clearly reflected by the following distinctive features incorporated in this book :

- Precise and logical presentation of the basic concepts and principles
- Text written in simple, lucid and easily understandable language
- Full use of line diagrams made to supplement the text and to explain a particular phenomenon as clearly as possible
- Step-wise solutions provided to a wide variety of problems of standard comparable to those set for engineering degree, AMIE, GATE and Engineering Service Examinations
- Salient points provided at the end of each chapter; this is to help the students to comprehend the subject matter prior to sitting for any test or examination
- Inclusion of sections on objective questions keeping in view the recent trend of such questions being asked in the various university and competitive examinations

The author strongly believes that this treatise on engineering thermodynamics will contribute effectively to the existing literature in the field, and will be a useful text for the students of engineering curriculum.

Acknowledgments: On completion of the book Engineering Thermodynamics, I wish to acknowledge my thanks and indebtedness to all those who have been associated with this work directly/indirectly. Acknowledgment is due and is hereby made to all the authors whose works have been freely consulted in the preparation of this text. I owe an enormous debt of gratitute to my colleagues and students for much valued assistance in the form of discussion and feed back. A word of thanks is due to the publishers too for their considerable patience and cooperation throughout. Above all, I wish to place on record my sincerest thanks and apologies to my near and dear ones who willingly endured certain hardships as a result of my pre-occupation with this work.

My effort in writing this book will be amply rewarded if the readers send their constructive suggestions and objective criticism with a view to enhance the usefulness of the book. Further, I would gratefully acknowledge the contribution of the readers by way of bringing to my attention the errors and printing mistakes that might have inadvertently crept into the book.

D.S. Kumar

MRIU, Faridabad
May, 2012

Contents

Chapter 1

Basic Concepts and Definitions

Learning objectives : After study of the subject matter presented in this chapter, the reader will acquire knowledge about :

- *microscopic and macroscopic perspectives of thermodynamics*
- *system, boundary and surroundings; open, closed and isolated systems; control volume and control surface*
- *thermodynamic equilibrium*
- *state, path, process and cycle*
- *reversible and irreversible processes*

1.1. THERMODYNAMICS: WHAT IS IT?

Thermodynamics: heat-force-action

Thermodynamics is the fascinating branch of science which deals with energy transfer and its effect on the state or condition of a system. Essentially, thermodynamics pertains to the study of :

• Interaction of system and surroundings; it relates the changes which the system undergoes to the influences to which it is put.

• Energy and its transformation; energy inter-conversions in the form of heat and work. James Joule had proved with his well-known experiments with churner that mechanical work can be converted into heat energy. The credit for using heat by converting into mechanical work goes to James Watt who produced the first steam engine and paved the way for industrial revolution.

• Relationship between heat, work and physical properties such as pressure, volume and temperature of the working substance employed to obtain energy conversion.

• Feasibility of a process and the concept of equilibrium processes.

Thermodynamics has been excellently defined as the science of three "Es" namely energy, entropy and equilibrium.

The laws, principles and concepts of thermodynamics are important and indispensable tools in the innovation, design, development and improvement of engineering processes, equipment and devices which deal with effective utilization of energy. Notable applications of engineering thermodynamics in the field of energy technology are:

• power producing devices, *e.g.*, internal combustion engines and gas turbines, steam and nuclear power plants.

• power consuming devices, *e.g.*, fans, blowers and compressors, refrigeration and air-conditioning plants.

• chemical process plants and direct energy conversion devices.

A large number of processes in various fields such as agriculture, textiles, dairy, drugs and pharmaceutical industry are also governed by thermodynamic principles.

Thermodynamics is essentially based upon experimental results and observations of common experience; there is no mathematical proof to the zeroth, 1st and 2nd laws of thermodynamics. The evidence of their validity stems from the fact that not in a single case these principles have been violated. The laws, however, lay down the general restrictions within which energy transformations are observed to occur.

- **Zeroth law** deals with thermal equilibrium, relates to the concept of equality of temperature and provides a means for measuring temperature.
- **First law** pertains to the conservation of energy (*i.e.* energy can neither be created nor destroyed) and introduces the concept of internal energy.
- **Second law** relates the direction of flow of heat, dictates limits on the conversion of heat into work and provides a yardstick to evaluate the performance of various processes. It also tells us whether a particular change is feasible or not and introduces the concept of entropy. It will be shown later that *first law denies the possibility of creating or destroying energy, the second law denies the possibility of utilizing energy in a particular way.*
- Third law defines the absolute zero of entropy.

Carnot (1792-1832), Joule (1818-1889), Kelvin (1824-1907) and Clausius (1822-1888) were the main scientists who developed and formalized thermodynamics as a science in the nineteenth century. Later, Gibbs (1839-1903) developed and broadened the scope of thermodynamics to such an extent that its principles could be applied to analyse almost any physical and chemical phenomenon.

1.2. MACROSCOPIC AND MICROSCOPIC APPROACH : CLASSICAL AND STATISTICAL THERMODYNAMICS

There are two approaches for investigating the behaviour of a system. In the ***macroscopic approach*** which is followed in classical thermodynamics

- structure of the matter is not considered; no attention is focussed on the behaviour of individual particles constituting the matter.
- the volumes considered are very large compared to molecular dimensions and the system is regarded as continuum devoid of any voids and cavities.
- study is made of overall effect of several molecules; the behaviour and activities of the molecules are averaged , *i.e.,* their effect integrated.

Only a few variables are needed to describe the state or condition of matter under consideration and these time-averaged characteristics or variables can be perceived by our senses and measured directly by instruments. For example, in the piston-cylinder assembly of an IC engine, the volume occupied by the gas for each position of the piston can be determined by measuring the cylinder diameter and the piston travel. The pressure exerted by the gas and its temperature can be measured by means of a pressure gauge and a thermocouple respectively. The state or condition of the system (gas in the cylinder) can then be completely described by specifying the measured values of pressure, temperature and volume. These few large scale characteristics/properties, known as ***macroscopic co-ordinates,*** are considered sufficient to provide complete description of the system under consideration at a particular instant. Further, it is possible to set-up or deduce many relations among their time-averaged characteristics. However, the particular form of relationship cannot be explained. The macroscopic approach, however, offers results of sufficient accuracy and validity.

Statistical thermodynamics adopts ***microscopic approach*** in which the matter is considered to be comprised of a large number of tiny particles called molecules (1 cm^3 of a monoatomic gas at atmospheric pressure and temperature contains 6×10^5 atoms) which move randomly in chaotic fashion. At a particular instant each particle has a definite position, velocity and energy and these characteristics change very frequently due to collision between the particles. The overall behaviour of the matter is then predicted by statistically averaging the behaviour of individual particles. The salient aspects of microscopic approach may be summed up as:

(*i*) necessity of complete knowledge of the structure of matter,

(*ii*) requirement of a large number of variables for complete specification of the state of matter,

(*iii*) easy and precise measurement is not possible of the variables used to describe the state of matter.

Microscopic view does help to explain certain phenomenon which cannot be analysed by macroscopic approach. However, the microscopic approach is rather complex, cumbersome and time consuming. The macroscopic approach is more practical and consequently the overwhelming majority of thermodynamic analysis is made by it. Engineering thermodynamic analysis is macroscopic and not microscopic.

Macroscopic approach	Microscopic approach
1. Attention is focussed on a certain quantity of matter without taking into account the events occuring at molecular level.	1. Matter constituting the system is considered to comprise a large number of discrete particles called molecules. These molecules have different velocities and energies, and these parameters constantly change with time.
2. Analysis is concerned with gross or overall behaviour of the system, and this approach is adopted in the study of classical thermodynamics.	2. A knowledge of the structure of matter is essential in analysing the behaviour of the system, and this approch is adopted in the study of statistical thermodynamics.
3. A few properties are needed to describe the system.	3. Large number of variables are needed to describe the system.
4. The properties like pressure and temperature etc needed to describe the system can be easily measured, and felt by our senses.	4. The properties like velocity, momentum and kinetic energy which describe the behaviour of the molecules can neither be felt by our senses nor easily measured by instruments.
5. The properties of the system are their average values.	5. The properties are defined for each molecule individually.
6. The macroscopic approach requires simple mathematical formulae for analysing the system.	6. Number of molecules is very large and as such the microscopic approach requires advanced statistical and mathematical methods to explain any change in the system.

Note : The relationship between macroscopic and microscopic point of view lies in the fact that macroscopic properties are in fact the average properties of a large number of microscopic characteristics. Obviously, when both the methods are applied to a practical system, they give the same result.

Thermodynamics, however, never explains how a process is executed, the time rate at which the process is in progress and time of completion. Such investigations are made through kinetic theory of gases, heat and mass transfer and chemical kinetics. Again, thermodynamics fails when the system is at high vacuum, *i.e.*, it has less number of particles.

1.3. THERMODYNAMIC SYSTEM, SURROUNDINGS AND BOUNDARY

A ***thermodynamic system*** represents a prescribed and identifiable (fixed) quantity of matter under consideration to analyse a problem; to study the changes in its properties due to exchange of energy in the form of heat and work. The system may be a quantity of steam, a mixture of vapour and gas or a piston-cylinder assembly of an IC engine and its contents.

For the description of a thermodynamic system, some of the following quantities need to be specified:

(*i*) quantity as well as composition of the matter

(*ii*) measurable properties such as pressure, temperature and volume of the system

(*iii*) energy of the system

The combination of matter and space, external to the system, that may be influenced by changes in the system is called *surroundings* or *environment*. The thermodynamic system and surroundings are separated by an envelope called *boundary* of the system. The boundary represents the limit of the system and may be either real or imaginary, and may change shape, volume, position and/or orientation relative to the observer. For example, an elastic balloon which is initially spherical in shape may get deformed into a cylindrical shape or some other geometrical shape or it may get squeezed to a reduced volume during a certain process. As such the boundary of the gas enclosed in the balloon would not retain the same shape and size. Further, the boundary may be diathermal or adiabatic depending upon whether it allows or not exchange of energy in the form of heat. The walls or boundaries which do not allow heat transfer to take place across them are named *adiabatic*. In contrast, the walls that do allow heat interaction across them are called *diathermic*.

The surroundings can also be regarded as a system, and hence thermodynamics is largely concerned with interaction between systems.

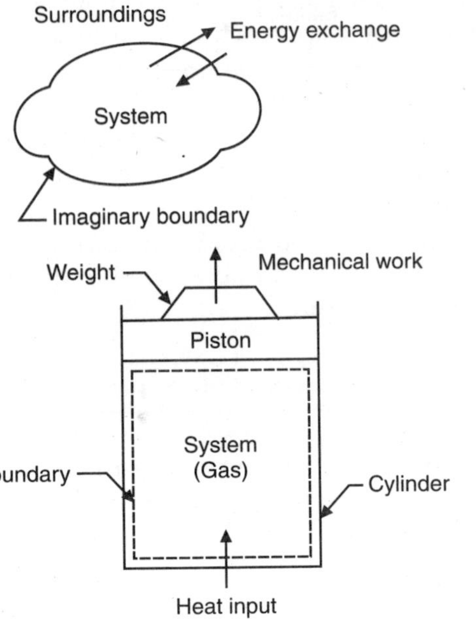

Fig. 1.1. Representation of system, boundary and surroundings

Thermodynamic systems can be classified into :

(*a*) Closed system : A closed system can exchange energy in the form of heat and work with its environment but there is no mass transfer across the system boundary. The mass within the system remains the same and constant, though its volume can change against a flexible boundary. Further, the physical nature and chemical composition of the mass may change. Thus a liquid may evaporate, a gas may condense or a chemical reaction may occur between two or more components of the system.

Examples of closed system are :

(*i*) *Cylinder fitted with a movable piston* : The gas can receive or reject heat, expand or contract according to whether the piston is moved outward or inward but no matter (gas) crosses the system boundary.

(*ii*) *Bomb calorimeter* : Here electric energy crosses the boundary to cause a spark between electrodes and initiate combustion. There is heat transfer across the boundary after combustion; however there is no mass transfer either way.

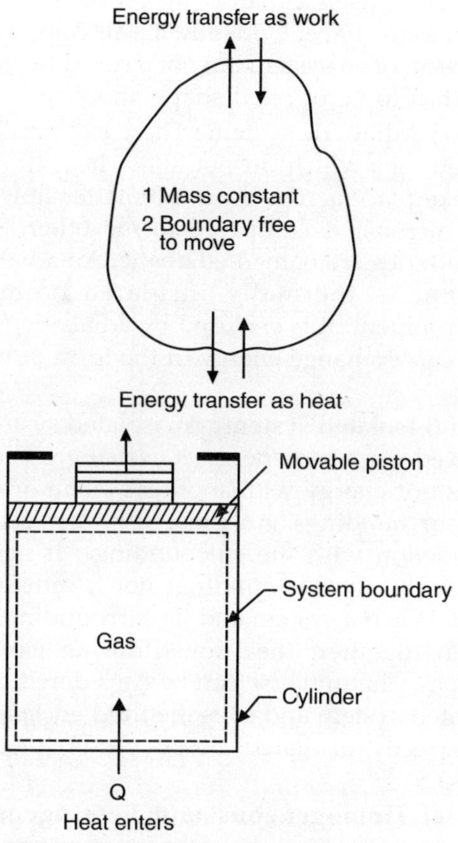

Fig. 1.2. Closed system

(*iii*) *Motor-car battery* : No mass transfer (neglecting issuing of the gases at the battery poles) occurs across the boundary enclosing the battery.

(*iv*) *Pressure cooker, Kitchen refrigerator, Ice cream freezer* (manual crank operation) etc.

(*b*) Open system : An open system has mass exchange with the surroundings along with transfer of energy in the form of heat and work. The mass within the system does not necessarily remain constant; it may change depending upon mass inflow and mass outflow.

Examples of an open system are :

(*i*) *Water wheel* : It is a device that converts potential energy of water into mechanical work. Water enters the wheel from head race side and leaves it to the tail race from the other end, and as such the mass crosses the system (wheel) boundary. The work output due to rotation of the wheel also crosses the system boundary.

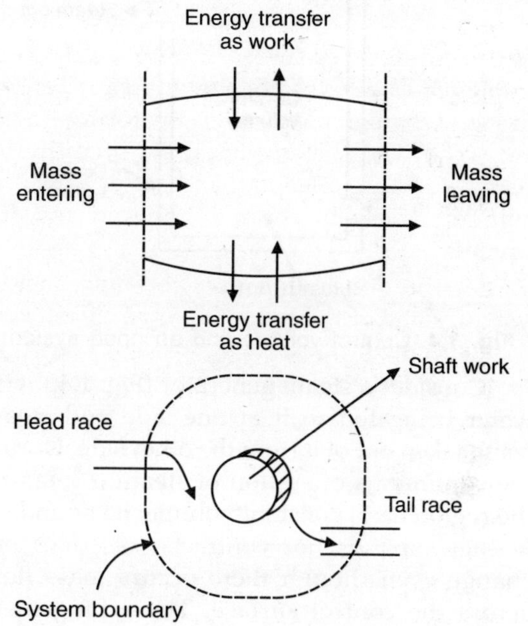

Fig. 1.3. Open system

(*ii*) *Motor car engine* : The engine initially draws charge (mixture of air and petrol) and finally exhausts the spent-up gases to the surrounding atmosphere; the mass flow occurs across the system (engine) boundary.

(*iii*) *Steam generator (Boiler)* : Boiler is a device which converts water from liquid to vapour phase. Here contents of the system change; water flows into and steam flows out of the system.

Most of the engineering devices are open systems involving flow of fluids through them.

Control volume and control surface

A *control region* or *control volume* is defined as any region in space which is separated from its environment by a control surface which may be either physical or imaginary. The control system is normally taken to be fixed in shape, position and orientation relative to the observer (Note that a thermodynamic system has a boundary which can change its shape and position). Heat and work interactions are present across the control surface and matter flows continuously in and out of the control volume.

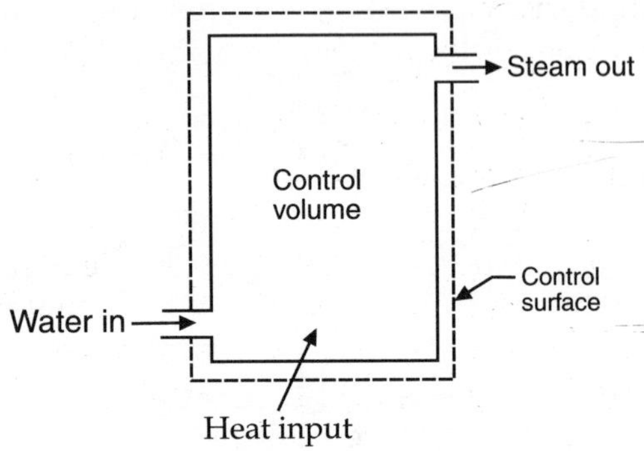

Fig. 1.4. Control volume and an open system

Consider a steam generator (Fig. 1.4.) with water being fed to it at one side and steam being taken out of it from the other side. During the continuous operation of steam formation, the region has a constant volume; its boundary neither expands nor contracts, *i.e.*, does not change even though there occurs mass flow across the control surface. The arrangement represents an open system also as mass crosses the system boundary.

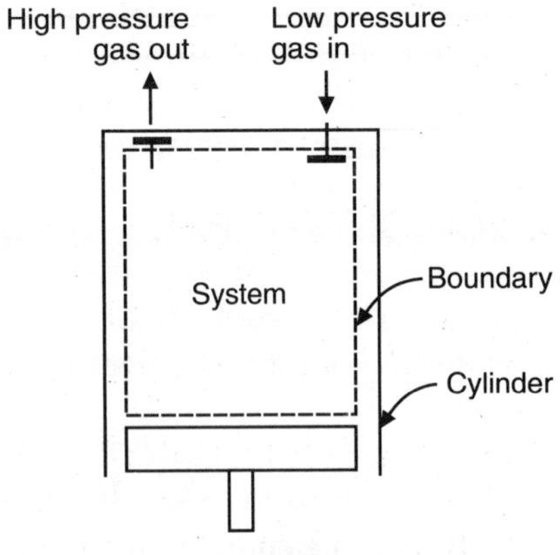

Fig. 1.5. An open system but not a control volume

Refer to the piston-cylinder assembly (Fig. 1.5.) of a reciprocating compressor. When the piston moves downward, the atmospheric air is breathed in through the suction valve; and the compressed air is discharged out through the delivery valve during upward movement of the piston. Obviously the mass flow crosses the system boundary. Furthermore, the boundary itself expands or contracts, *i.e.*, it changes its shape. Evidently the air compressor forms an open system but not a control volume that has to be of fixed shape and size.

(*c*) Adiabatic system : There exist walls or boundaries which do not allow heat transfer to take place across them; no matter how hot one member is compared to the other. Such boundaries are named adiabatic. An adiabatic system is thermally insulated from its environment; it is enclosed by adiabatic walls and can exchange energy in the form of work only.

(*d*) Isolated system : An isolated system is of fixed mass and energy; it exchanges neither mass nor energy with another system or with the surroundings. An isolated system has no interaction with the surroundings; it neither influences the surroundings nor is influenced by it. When a system and its surroundings are taken together, they constitute an isolated system. The universe can be considered as an isolated system and so is the fluid enclosed in a perfectly insulated closed vessel (thermos flask).

(*e*) Homogeneous and heterogeneous systems : Based on phase change, a system may be classified as a homogeneous system and a heterogeneous system. A phase represents a quantity of matter that is uniform throughout in physical structure and in chemical composition. Physical uniformity implies that the matter is all gas, or all liquid or all solid. Uniformity of chemical composition means that the chemical composition does not vary from one part of the system to another. An iron piece, a liquid contained in a vessel, a gas enclosed within a container, and a mixture of gases represent one-phase systems. A system consisting of liquid and gas is a two-phase system of a liquid phase and a gaseous phase. Likewise a mixture of solid, liquid and gas constitutes a three-phase system.

A system consisting of a single phase is called homogeneous system. The examples are:

(*i*) ice, water, dry saturated steam
(*ii*) mixture of ammonia in water
(*iii*) mixture of air and water vapour

(*iv*) water plus nitric acid

A system whose mass content is non-uniform throughout, *i.e.*, it consists of more than one-phase is called heterogeneous system. The examples are:

(*i*) mixture of ice and water

(*ii*) mixture of non-miscible liquids (water and mercury)

(*iii*) water plus gasoline

(*iv*) wet steam (vapours in contact with liquid being evaporated)

1.4. THERMODYNAMIC PROPERTY: INTENSIVE AND EXTENSIVE PROPERTIES

A thermodynamic property refers to the characteristics which can be used to describe the condition or state of a system, *e.g.*, temperature, pressure, chemical composition, colour, volume, energy etc. The salient aspects of a thermodynamic property are:

(*i*) it is a measurable characteristic describing a system, and helps to distinguish one system from another,

(*ii*) it has a definite unique value when the system is in a particular state,

(*iii*) it is dependent only on the state of the system; it does not depend on the path or route the system follows to attain that particular state,

(*iv*) its differential is exact.

If P is a thermodynamic property with dP representing its differential change, then the integral between initial state 1 to final state 2 of the system will have only value given by

$$\int_1^2 dP = (P_2 - P_1) \qquad ...(1.1)$$

Since a thermodynamic property is a function of the state of a system, it is referred to as a **point function** or a **state function**.

There are two kinds of thermodynamic properties namely intensive and extensive.

(*a*) ***Intensive property*** is independent of the extent or mass of the system. Its value remains the same whether one considers the whole system or only a part of it. Examples are pressure, temperature, density, composition, viscosity, thermal conductivity, electrical potential etc.

(*b*) ***Extensive property*** depends on mass or extent of the system ; its value depends on how big a portion of the system is being considered. Examples include energy, enthalpy, entropy, volume etc.

Let a system with value of a certain characteristic/property P be divided into a number of parts. If P_1, P_2, P_3 etc denote the value of that property for various parts of the system, then

For an intensive property :

$$P = P_1 = P_2 = P_3 = ...$$

For an extensive property :

$$P = P_1 + P_2 + P_3 + ...$$

(*c*) ***Specific property*** **:** An extensive property expressed per unit mass of the system. Examples include specific volume, specific energy, specific entropy, specific enthalpy etc.

$$\text{specific volume } v = V/m$$

$$\text{specific energy } e = E/m$$

where V, E and m are volume, energy and mass of the system respectively.

Like extensive properties, specific properties too are additive.

(*d*) ***Molar property*** **:** Ratio of an extensive property to the mole number (or the property per unit mole). For example

$$\text{Molar vol } v = \frac{\text{total volume}}{\text{mole number}} = \frac{V}{n} \text{ m}^3/\text{mol}$$

Check for a Property : A given expression $dP = M\,dx + N\,dy$ would be a thermodynamic property if its differential is exact , *i.e.*, if

$$\left[\frac{\partial M}{\partial y}\right]_x = \left[\frac{\partial N}{\partial x}\right]_y \qquad ...(1.2)$$

EXAMPLE 1.1

Discuss whether the following quantities represent the system properties or not.

(i) $p\,dv + v\,dp$ (ii) $p\,dv$ (iii) $v\,dp$

where p is the pressure and v is specific volume.

Solution : The exercise resolves to determining whether the given differential is an exact differential or not. A differential is taken to be exact if its integration between two end states can be carried out without requiring any information on the variation of variables.

(*i*) $p\,dv + v\,dp = d(pv)$: Its integration $\int d(pv) = pv$ has been obtained without requiring any additional information as to how p and v are related. The quantity is thus an exact differential and hence a property of a system.

(*ii*) $p\,dv$: p is a function of v and they are connected by a line path on p-v plane. The integration $\int p\,dv$ can be determined only if a relationship between variation of p and v is known. Thus $p\,dv$ is not an exact differential and hence not a property.

(*iii*) $v\,dp$: It is not an exact differential (on the reasoning stated above at (*ii*) and hence not a property.

EXAMPLE 1.2

If $pv = RT$, determine which of the following quantities are thermodynamic properties:

$$(i)\ \frac{v}{T}dp + \frac{p}{T}dv \qquad (ii)\ \frac{dT}{T} + \frac{p}{T}dv$$

In these expressions p is pressure, v is specific volume, T is absolute temperature and R is a gas constant.

Solution : A quanity $dP = M\,dx + N\,dy$ would be an exact differential, and hence a thermodynamic property if

$$\left(\frac{\partial M}{\partial y}\right)_x = \left(\frac{\partial N}{\partial x}\right)_y$$

(*i*) Comparing the given identity with $M\,dx + N\,dy$

$$M = \frac{v}{T} = \frac{R}{p}$$

for a perfect gas, and $x = p$

$$N = \frac{p}{T} = \frac{R}{v}$$

for a perfect gas, and $y = v$

$$\left(\frac{\partial M}{\partial y}\right)_x = \left\{\frac{\partial}{\partial v}\left(\frac{R}{p}\right)\right\}_p = 0$$

and $$\left(\frac{\partial N}{\partial x}\right)_y = \left\{\frac{\partial}{\partial p}\left(\frac{R}{v}\right)\right\}_v = 0$$

Since $\left(\frac{\partial M}{\partial y}\right)_x = \left(\frac{\partial N}{\partial x}\right)_y$; the given expression is an exact differential and hence a property.

(*ii*) Comparing the given identity with $M\,dx + N\,dy$

$$M = \frac{1}{T} \text{ and } x = T$$

$$N = \frac{p}{T} = \frac{R}{v}$$

for a perfect gas and $y = v$

$$\left(\frac{\partial M}{\partial y}\right)_x = \left\{\frac{\partial}{\partial v}\left(\frac{1}{T}\right)\right\}_T = 0$$

and $$\left(\frac{\partial N}{\partial x}\right)_y = \left\{\frac{\partial}{\partial T}\left(\frac{R}{v}\right)\right\}_v = 0$$

Since $\left(\frac{\partial M}{\partial y}\right)_x = \left(\frac{\partial N}{\partial x}\right)_y$, the given expression is an exact differential and hence a property.

1.5. EQUILIBRIUM: MECHANICAL, CHEMICAL, THERMAL AND THERMODYNAMIC

Equilibrium is the concept associated with the absence of tendency for spontaneous change in the value of any macroscopic property of the system when it is isolated from its surroundings.

(*a*) *Mechanical equilibrium* (Equality of pressure) : Condition or state in which there is no unbalanced force within the system and nor at its boundaries. Mechanical equilibrium implies uniformity of pressure, *i.e.*, there is only one value of pressure for the entire system. If pressure gradients exist within the system, internal changes (diffusion) take place to wipe out the unbalance and attain the state of mechanical equilibrium. Similarly, if the weight on the piston from the piston-cylinder assembly (Fig. 1.2) is removed, the system boundary will expand until the mechanical equilibrium is attained. Mention may also be made of still water contained in a stationary tank. This system too is considered to be in mechanical equilibrium even though the pressure increases as one moves downward from the free water surface.

(*b*) *Chemical equilibrium* (Equality of chemical potential) : A system in mechanical equilibrium may undergo a spontaneous change of internal structure due to chemical reaction or diffusion (transfer of matter). For example , there will be spontaneous change in the properties of a mixture of oxygen and gasoline once it is ignited. Chemical equilibrium represents that condition or state of the system when all chemical reactions in it have ceased and there is no mass diffusion.

(*c*) *Thermal equilibrium* (Equality of temperature): Condition or state in which the temperature of the system is uniform.

Let a system in mechanical and chemical equilibrium be separated from its surroundings by a diathermic wall (heat flow is allowed). If the surroundings are at a temperature higher than that of the system, then heat will flow towards the system and with the passage of time it will become hotter than what it was before. Subsequently both will attain the same temperature and the heat interaction will cease. When such a stage is reached, *i.e.*, there is no change taking place in the condition of system and surroundings, they are said to be in thermal equilibrium with each other. A system is thus said to be in thermal equilibrium if there exists uniformity of temperature throughout the system or between the system and its surroundings.

It is to be emphasised that the fluids comprising the systems in thermal contact are not reactive chemically. For example, the thermal contact between acid and water, although both at the same temperature, would result in rise in temperature due to the release of heat of dilution, and would not conform to the definition of thermal equilibrium.

(*d*) *Thermodynamic equilibrium*: A system which is simultaneously in a state of mechanical equilibrium, thermal equilibrium and chemical equilibrium is said to be in a state of thermodynamic equilibrium.

1.6. STATE, PATH, PROCESS AND CYCLE

Consider a system constituted by gas enclosed in the piston cylinder assembly of a reciprocating machine. Corresponding to position of the piston at any instant, the condition of the system will be prescribed by pressure, volume and temperature of the gas. When all such properties have a definite value, the system is said to exist at a definite state. *State is thus condition of the system identified by thermodynamic properties*. With properties as co-ordinates, the state of the system in thermodynamic equilibrium can be represented by a point. (Fig. 1.6.)

When the piston moves outwards, the properties of the system change (pressure decreases and volume increases). Any such operation in which properties of the system change is called a *change of state*. The locus of the series of states through which a system passes in going from initial state to its final state constitutes the *path*. A complete specification of the path is referred to as a *process*.

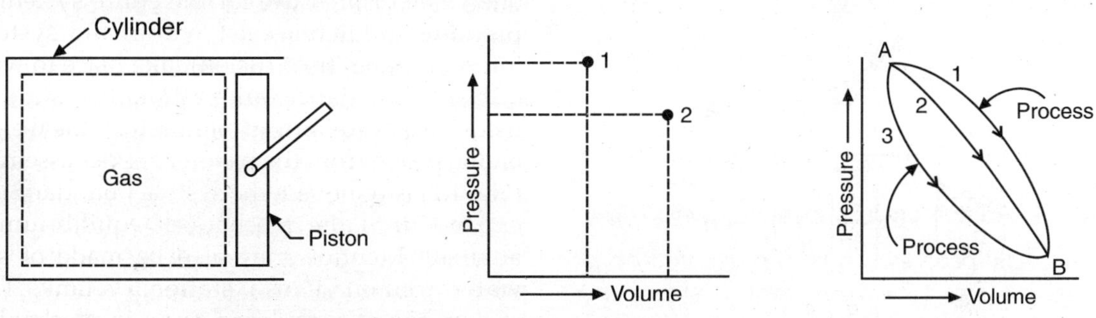

Fig. 1.6. Representation of state, a process and system paths

Apparently *a thermodynamic process represents a transition in which a system changes from one state to another.* A process is simply a change in the state of the system and there are infinite ways for a system to change from one state to another state. Reference Fig. 1.6., *A* and *B* represent the terminal states of the piston and the paths *A*1*B*, *A*2*B*, and *A*3*B* traced by the system from state *A* to state *B* represent the processes. Certain processes are identified by special names, *e.g.*, an isochoric process is a constant volume process, an isobaric process is a constant pressure process, and an isothermal process is a constant temperature process.

When a system in a given state undergoes through a series of processes such that the final state is identical with the initial state, a *cyclic process* or a *cycle* is said to have been executed. An essential feature of the cycle is that the intital condition of the system is restored after the processes. The change in the value of any property of the system for a cyclic process is zero. Mathematically

$$\oint dP = 0 \qquad ...(1.3.)$$

where *P* is any property and symbol $\oint$ represents integration around a cycle.

Note: In an IC engine, the condition of the exhaust gases is not the same as the condition of the charge sucked in. As such, these engines execute only a mechanical cycle and not a thermodynamic cycle.

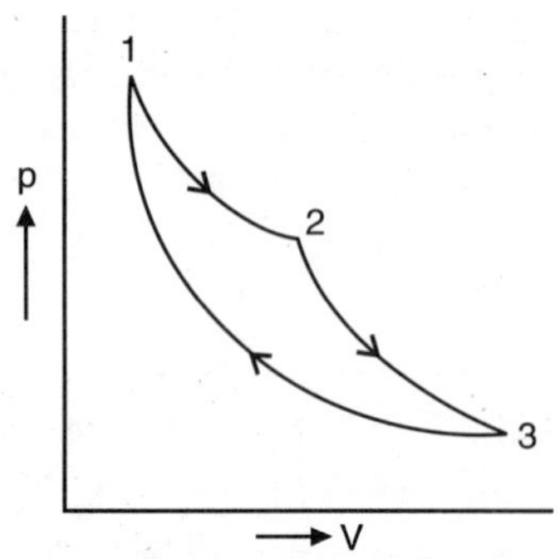

Fig. 1.7. A three process thermodynamic cycle

A cyclic process can be represented on *p*-*v* diagram (Fig. 1.7.) if the various states and paths are known. The cycle 1-2-3-1 comprises three processes 1-2, 2-3 and 3-1.

1.7. QUASI-STATIC OR QUASI-EQUILIBRIUM PROCESS

Some unbalanced potential must exist either within the system or between a system and its surroundings to promote the change of state during a thermodynamic process. Consider a system of gas contained in a cylinder fitted with a piston upon which are placed many very small pieces of weights. The upward force exerted by the gas just balances the weights on the piston. The system is initially in equilibrium state identified by pressure p_1, volume v_1 and temperature T_1. When these weights are removed slowly, one at a time, the unbalanced potential is infinitesimally small. The piston will slowly move upwards and at any particular instant of piston travel, the system would be almost close to state of equilibrium. The departure of the state of system from the thermodynamic equilibrium state will be infinitesimally small. Every state passed by the system will be an equilibrium state. The locus of a series of such equilibrium states is called a ***quasi-static*** or ***quasi-equilibrium*** process and can be represented graphically as a continuous line on a state diagram. A quasi-static process is thus a succession of equilibrium states.

A quasi-static process can be viewed as a sufficiently slow process which allows the system to adjust itself internally so that properties in one part of the system do not change any faster than those at other parts. Since quasi-static processes are considered extensively during interaction of system and surroundings it has been suggested that the name *thermostatics* would be more appropriate than thermodynamics.

In contrast, if all the weights are removed suddenly, the unbalanced potential would be finite. The piston will jump up, strike the stoppers and finally come to equilibrium state 2 at pressure p_2, volume v_2, and temperature T_2 after many oscillations. However, the intermediate states, through which the gas (system) passes are indeterminate and can not be represented on the state diagram. The dotted line 1-2 (Fig. 1.8) indicates the general direction

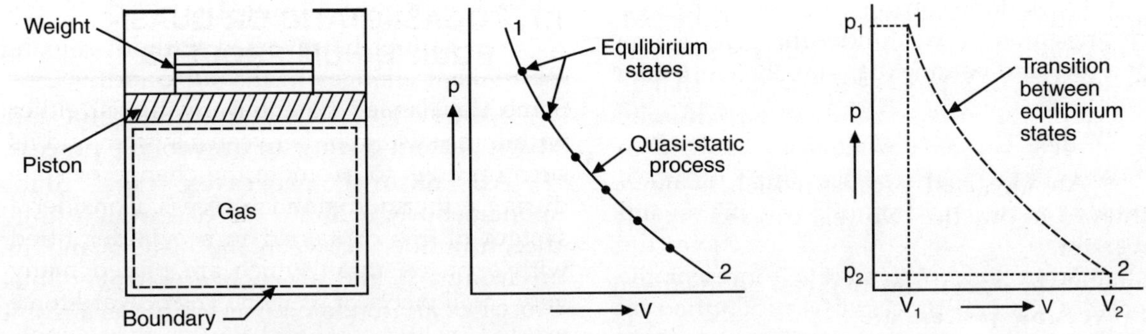

Fig. 1.8. Equilibrium states and Quasi-static process

of change between the terminal equilibrium states. Such a process is a non-equilibrium or a non-quasistatic process. Most of the real processes are dynamic and non-quasistatic, even though they may be in equilibrium at the terminal (initial and final) states.

1.8. REVERSIBLE AND IRREVERSIBLE PROCESSES

A thermodynamic process is ***reversible*** if the system passes through a continuous series of equilibrium states. An equilibrium state of the system can be located on any diagram drawn for properties (such as pressure-volume diagram) at any instant during the process. The reversible process is then shown by a line drawn through the points representing the equilibrium states. Fig. 1.9. shows a reversible process on A-B plot whose path is 1-2. If the process is reversed, path 2-1 will be followed and that will restore the system as well as surroundings back to their respective initial states.

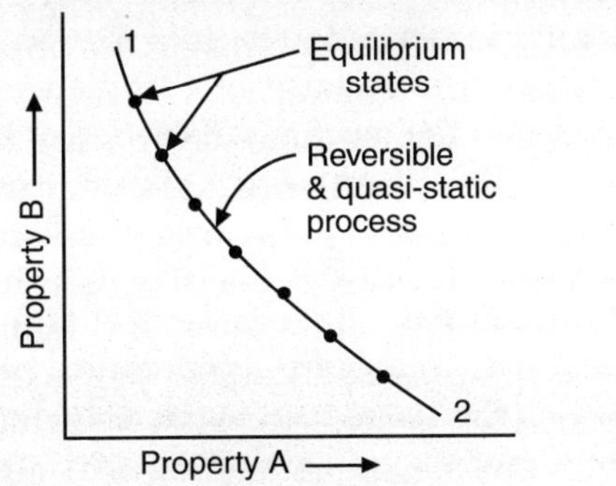

Fig. 1.9. Quastatic and reversible process

The following conditions need to the satisfied for a process to be reversible:

• There should be no friction; solid or fluid

• The heat exchange to or from the system, if any, should be only through infinitely small temperature difference

• The process should be quasi-static, it should proceed at infinitely slow speed. For this the pressure difference between the system and surroundings must be infinitely small.

Salient Features

• A quasistatic process which can be carried out in the reverse direction along the same path

• It leaves no trace of its occurrence in the system and surroundings. When a process is reversibly done, no after effects or changes are evident either in the system or in the surroundings.

• It can proceed in either direction without violating the second law of thermodynamics.

• No wastage of energy accompanies a reversible process; a system going through a reversible process results in maximum efficiency.

• A reversible process yields the maximum work in engines, and requires minimum work in devices such as refrigerators, pumps and compressors.

• Degree of reversibility can be considered as a measure of the perfection which can be attained in any thermodynamic transaction. As such, a reversible process is used to set the criterion for a standard efficiency.

• Work done during a reversible process is represented by area under the process line. For a pressure-volume diagram, the work done equals $\int_1^2 p\,dv$.

• An idealised process which is never achieved in practice. No real process is truly reversible.

Some of the operations that can be idealised as reversible process are:

• Motion without friction

• Restricted and controlled expansion or compression

• Isothermal and frictionless adiabatic process

• Elastic stretching of a solid

• Restrained discharge of a battery

• Electric circuit with zero resistance

• Polarization and mangetisation effects; and electrolysis

A process is ***irreversible*** if the system passes through a sequence of non-equilibrium states. During such a process, the properties of the system donot have a unique value and accordingly cannot be located on any property diagram. When an irreversible process is made to proceed in the backward direction, the original state of the system is not restored. (Fig 1.10). An irreversible process cannot be represented by a full line on diagram of properties since the properties are definite only at the initial and final states.

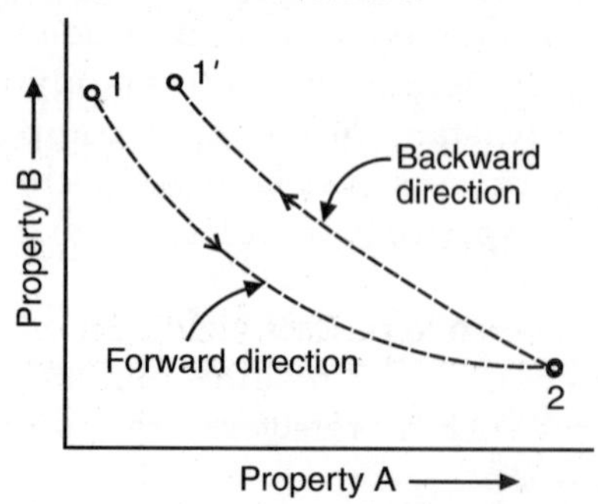

Fig. 1.10. Irreversible process

An irreversible process is identified by the following characteristics:

• it can be carried out in one direction only,

• it occurs at a finite rate,

• it cannot be reversed without causing permanent changes in the surroundings,

• the system is never in equilibrium state at any instant during an irreversible process.

All natural processes take place spontaneously at finite speeds through finite discontinuities between the system and its environment and as such are irreversible. Processes are rendered irreversible as a result of degradation of energy by such factors as friction, turbulence, diffusion, throttling and combustion etc.

Some notable examples of irreversible process are:

• Spontaneous chemical reaction

• Viscous flow; fluid flow with friction

• Inelastic (plastic) deformation and hysteresis effects

• Electric circuit with resistance

• Diffusion of gases; mixing of dissimilar gases

• Energy transfer as heat with finite temperature differences

• Free expansion and throttling process

EXAMPLE 1.3

(a) Distinguish between

(i) Mechanical and thermal irreversibility

(ii) Internal and external irreversibility

(b) How the following aspects render a process irreversible?

(i) frictional resistance (ii) free expansion and (iii) stirring of liquid by means of a paddle wheel.

Solution : (*a*) Mechanical irreversibility is generally associated with fluid friction between the molecules because of turbulence at high speeds, and the mechanical friction at bearings, between the piston and cylinder walls, and between the atmosphere and rotating parts of a mechanical device. Thermal irreversibility, on the other hand, is because of energy transfer as heat due to a finite temperature difference between the parts of a system, or between a system and its environment (surroundings).

Internal irreversibility is confined to the working substance itself. External irreversibility is external to the working substance (if mechanical) and between the system and its surroundings (if thermal). While analysing a thermodynamic system, the boundaries are quite often so chosen that most of the causes of irreversibility are excluded. The system will then experience no irreversible effects within the system boundaries and will be called internally reversible system.

(*b*) (*i*) The work is spent in overcoming the frictional resistance and that gets converted into heat. The heat thus generated cannot be reconverted completely into work (second law of thermodynamics) and accordingly the stock of the available energy is reduced. This degradation of energy by frictional resistance renders the process irreversible.

(*ii*) In a free expansion process, a finite pressure difference occurs that transforms into kinetic energy and is dissipated into random motion. Further, no work is done as the process is unresisted. Both these factors lead to irreversibility.

(*iii*) When the fluid is stirred by means of a shaft and paddle wheel, there is work transfer to the fluid. This work is absorbed locally through turbulent eddies and fluid friction. It is not possible under any circumstances to make the eddies unite and drive the paddle wheel.

1.9. WORKING SUBSTANCE

A working substance refers to the fluid employed in thermodynamic devices to serve as a medium for the transport of energy between the system and its surroundings. A fluid, which may be gas, vapour, liquid or any non-chemically reactive mixture of these constitutes the matter normally considered in engineering thermodynamics. Fluids can be conveniently made to absorb, store and release energy. Further, they can be easily transported from place to place through pipes.

• Combustion products of hydrocarbons obtained by igniting air-fuel mixture form the working substance in petrol and diesel engines, gas turbines and jet propulsion.

• Water vapour is the working substance in steam power plants. The water is vaporised to steam in the boiler, the steam is fed to the turbine which is set into rotary motion and mechanical work is produced. The spent-up steam (steam leaving the turbine) is condensed to liquid water and is returned to the boiler.

• Refrigerants (NH_3, SO_2, Freon-12 etc.) are the working substances used in refrigeration, air-conditioners and ice-making plants.

A working substance may exist in solid, liquid and gaseous phase or in a combination of these phases. Further it may exist in different conditions, *i.e.*, at different pressure, temperature and density levels. Steam and refrigerants change their phase during processes occurring within the system and are called **vapours**. The substance such as air and combustion products of hydrocarbons in air are well above their critical temperature. In these substances, no change in phase takes place and these are called **gases**.

A **pure substance** is homogeneous, and has the same chemical and physical compositions. Examples of a pure substance are:

(*i*) Gases: nitrogen; superheated steam

(*ii*) Gas mixtures: air; oxygen plus gaseous fuels

(*iii*) Combination of liquid and vapour of the same matter: water plus steam; liquid oxygen plus vaporised oxygen.

The mixture of liquid air and gaseous air is not a pure substance.

The state of a pure substance in equilibrium is completely defined if any two independent intensive properties are known. This is known as the **two-property rule** for a pure substance.

1.10. UNITS AND DIMENSIONS

The remaining part of this introductory chapter deals with the

(*i*) the properties which are significant when making a study of a thermodynamic system, and

(*ii*) the units in which these properties are expressed.

Mass, force and weight

Mass is an indication of the quantity of matter present within a system. The more mass means more matter.

Force is an external agent which tends to change the speed or direction of a system. A force is applied whenever the system needs to be accelerated or decelerated.

Weight is a force which the system exerts due to gravitational acceleration. Weight is actually the force with which the system is attracted towards the earth. Where there is no gravity, there is no weight and this condition of the system is known as weightlessness. The weight of a system varies from place to place but mass remains constant. The mass of a given collection of matter cannot be altered by a change of location, a deformation of shape, a change of temperature, and through chemical reactions which the matter may undergo.

Almost all the major countries of the world use SI units (System International units) in which the fundamental units are metre (m) for measuring length, kilogram (kg) for measuring mass, and second (s) for measuring time. A complete dimensional system can be built up with these three dimensions.

Newton's second law of motion states that force is proportional to the product of mass and acceleration. That is

$$F \propto (m) \times (a) \qquad F = \frac{ma}{g_c}$$

where m is mass, a is acceleration and g_c is proportionality constant. The SI unit of force is newton (N) which represents the force required to accelerate 1 kg mass with an acceleration of 1 m/s². Thus

$$1\text{ N} = \frac{1\text{ kg} \times 1\text{ m/s}^2}{g_c}$$

$$\text{and } g_c = \frac{1\text{ kg m}}{\text{Ns}^2}$$

The constant of proportionality becomes unity for defining unit of force as newton and is accordingly dropped. Such a practice corresponds to writing Newton's second law as

$$F = ma \quad \text{and} \quad W = mg \qquad ...(1.4)$$

The weight (W) of system equals the product of mass (m) and local gravitational acceleration. The mass of a system remains constant whereas weight varies with change in value of gravitational acceleration from one place to another. The value of g at sea level is 9.8066 m/s² and is generally taken as 9.81 m/s².

EXAMPLE 1.4

Calculate the force exerted by a piston of 0.5 kg mass on the connecting rod when it is accelerated at the rate of 1500 m/s².

Solution : From Newton's law $F = ma$

$$= 0.5 \times 1500$$

$$= 750\text{ kgm/s}^2 = \mathbf{750\ N}$$

EXAMPLE 1.5

A metallic piece weighs 75.3. N at a location where acceleration due to gravity is 9.75 m/s². What is the mass of the metallic piece ? Proceed to calculate reading on the spring balance if the weighing is done at the moon where gravitational acceleration measures 1.67 m/s.

Solution : From Newton's law, $W = mg$

∴ Mass of metallic piece,

$$m = \frac{W}{g} = \frac{75.3}{9.75} = 7.72\text{ kg}$$

The mass of the metallic piece remains the same regardless of its location. The weight, however, changes with a change in gravitational acceleration. The weight of the metallic piece at the moon will be,

$$W_{moon} = 7.72 \times 1.67 = 12.89\text{ N}$$

Spring balance indicates the gravitational force, *i.e.*, weight of the body. Accordingly spring balance would read **12.89 N.**

EXAMPLE 1.6

A mass of 1 kg is weighed by a beam balance at a location where g = 9.35 m/s². What will be the reading of the beam balance?

The same mass is next weighed by a spring balance which was calibrated at sea level where g = 9.81 m/s². What will be the indicated weight ? If the calibration had been made at a location where g = 9.35 m/s², what would be the indicated weight and the weight in standard newton ?

Solution : (*i*) A beam balance gives only the mass which remains same regardless of its location. Hence, when 1 kg mass is weighed by a beam balance at a location where $g = 9.35$ m/s^2, it will give a reading of 1 kg mass only.

(*ii*) A spring balance indicates the gravitational force, *i.e.*, the weight of a body. The same mass will give different readings on the spring balance at different locations.

At the place of calibration,

$$W_1 = mg_1 = 1 \times 9.81 = 9.81 \text{ N}$$

At any other place, $W_2 = mg_2$

$$\therefore \; W_2 = W_1 \frac{g_2}{g_1} = 9.81 \times \frac{9.35}{9.81} = \mathbf{9.35 \; N}$$

(*iii*) If the calibration were done at g_1 = 9.35 m/s^2 and measurement was also at g_2 = 9.35 m/s^2, then indicated weight will be

$$W_2 = W_1 \frac{g_2}{g_1} = 9.81 \times \frac{9.35}{9.35} = \mathbf{9.81 \; N}$$

(*iv*) Standard newton is the weight indicated by the spring balance at any location with the scale always calibrated at $g = 9.81$ m/s^2

$$\text{Standard newton} = 9.81 \times \frac{9.35}{9.81} = \mathbf{9.35 \; N}$$

1.11. SPECIFIC WEIGHT, MASS DENSITY AND SPECIFIC VOLUME

(*a*) **Specific weight** (*w*) of a substance is its weight per unit volume

$$w = \frac{W}{V} \qquad ...(1.5a)$$

where *W* is the weight of the substance occupying volume *V*. The weight of the substance is the force with which it is attracted to the centre of earth. It is product of mass and local gravitational acceleration, *i.e.*, $W = mg$. Since weight is expressed in Newton, the unit of measurement of specific weight is N/m^3. In terms of fundamental units, the dimensional formula of specific weight is $\left[\frac{F}{L^3}\right]$ or $\left[\frac{M}{L^2T^2}\right]$.

(*b*) **Density** (ρ) is a measure of the amount of working substance contained in a given volume and is defined as the mass per unit volume.

$$\rho = \frac{m}{V} \qquad ...(1.5b)$$

where *m* is the mass of the substance having volume *V*. The volume is a measure of the space occupied and *m* is a measure of the ability of the particles of the substance to resist acceleration. The units of density correspond to those of mass and volume. The dimensional formula of density in fundamental units is $\left[\frac{M}{L^3}\right]$ or $\left[\frac{FT^2}{L^4}\right]$ and the corresponding units are kg/m^3 or Ns^2/m^4.

Relations 1.5 *a*, *b* are valid only when the medium (substance) fills the given volume completely without any blank space, *i.e.*, the fluid is a continuum. For a non-homogeneous fluid, these relations give average specific weight and density. To determine the absolute value of *w* and ρ at any point, the volume is regarded as tending to zero and the limit of the corresponding ratio is calculated

$$w = \underset{V \to 0}{Lt} \frac{W}{V} = \frac{dW}{dV}$$

$$\rho = \underset{V \to 0}{Lt} \frac{m}{V} = \frac{dm}{dV} \qquad ...(1.6)$$

The weight *W* and the mass *m* of a substance are related to each other by the expression $W = mg$. Dividing this expression throughout by volume *V* of the working substance, we obtain

$$\frac{W}{V} = \frac{mg}{V} \quad \text{or} \quad w = \rho g \qquad ...(1.7)$$

Equation 1.7 reveals that specific weight *w* changes with location depending upon gravitational pull.

(*c*) **Specific volume** (*v*) represents the volume per unit mass of the substance; specific volume is the reciprocal of mass density

$$v = \frac{V}{m}; \quad v = \frac{1}{\rho} \qquad ...(1.8)$$

The specific volume has the units m^3/kg.

The concept of specific volume is found to be practically more useful in the study of flow of gases and vapours.

Note :

(*i*) Density is mass per unit volume, while specific volume is volume per unit mass.

(*ii*) The density is an invariant quantity whereas the specific weight may vary with location on the earth.

(*iii*) A specific quantity does not depend on the extent (mass) of the system and as such the specific volume, density and specific weight are intensive properties of a thermodynamic system.

(*iv*) The specific volume and density of a working medium are sometimes expressed on the basis of mole rather than mass. A mole of a substance has a mass numerically equal to molecular weight of the substance. 1 kg mol of oxygen has a mass of 32 kg and 1 kg mol of nitrogen has a mass of 28 kg.

The molar specific volume is designated by v and has units of m^3/kg mol.

(*d*) Specific gravity (*s*) refers to the ratio of specific weight (or mass density) of a fluid to the specific weight (or mass density) of a standard fluid. For liquids the standard fluid is water at 4°C, and for gases the standard fluid is taken either air at 0°C or hydrogen at the same temperature. Specific gravity is dimensionless and has no units.

A statement that specific gravity of mercury is 13.6 implies that mercury is 13.6 times heavier than water.

EXAMPLE 1.7

A fan delivers 4 m³ of air having a mass equal to 5 kg. Calculate weight, specific weight, density and specific volume of the air being delivered.

Solution : From Newton's law : $W = mg$

$$= 5 \times 9.81 = \mathbf{49.05\ N}$$

Specific weight is a measure of weight per unit volume

$$\therefore \text{Specific weight } w = \frac{49.05}{4} = \mathbf{12.26\ N/m^3}$$

Mass density is related to specific weight by the relation : $w = \rho g$

$$\therefore \text{Mass density } \rho = \frac{12.26}{9.81} = \mathbf{1.249\ kg/m^3}$$

Alternatively, density ρ = mass per unit volume

$$= \frac{5}{4} = 1.25 \text{ kg/m}^3$$

Specific volume is the reciprocal of density

∴ Specific volume

$$v = \frac{1}{\rho} = \frac{1}{1.25} = \mathbf{0.8\ m^3/kg}$$

1.12. WORK, ENERGY AND POWER

Work is done when the point of application of force moves in the direction of force. The amount of work equals the product of force and displacement in the direction of force.

Work done = force × distance; $W = FL$

where F is the force, W is the work done and L is the distance moved. The unit of work is joule (J) which is defined as the work done when a force of one newton is exerted through a distance of one metre in the direction of the force.

$$1 \text{ J} = 1 \text{ Nm}$$

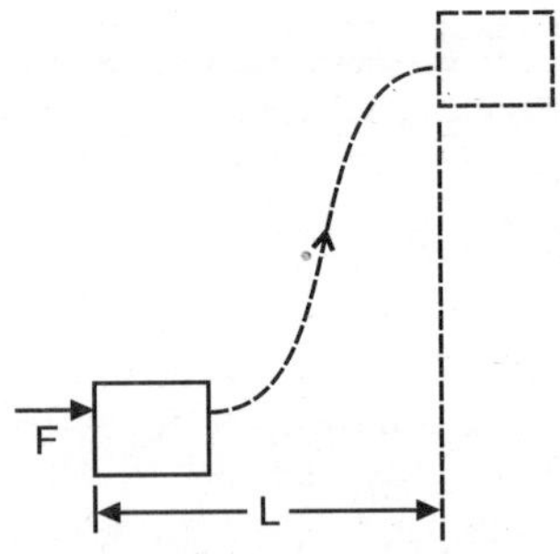

Fig. 1.11. Concept of work as used in mechanics

Energy is defined as the capacity to produce a change from existing conditions, *i.e.*, capacity to exert force through a distance and do work. Energy cannot be seen; however its presence can be felt by properties of the system. When energy is added to or subtracted from the system, there occurs a change in one or more properties of the system. The energy of a system may take the form of (*i*) *Stored energy, i.e.*, energy contained within the system boundaries. Examples are kinetic energy, potential energy and internal energy.

(*ii*) *Energy in transit, i.e.*, the energy which crosses the system boundaries. Examples are heat energy, mechanical energy and electrical energy.

Whatever may be the form of the energy, its unit is Joule (J) or kilo joule (kJ). The energy per unit mass is the specific energy, the unit of which is kJ/kg. The industrial unit of energy is kilowatt hour (kWh) which is equivalent to work of one kilowatt for one hour.

$$1 \text{ kWh} = 3.6 \times 10^6 \text{ J} = 3600 \text{ kJ}$$

Power is the rate of work or energy transfer. The unit of power is watt (W), kilowatt (kW) or megawatt (MW).

$$1 \text{ W} = 1 \text{ J/s} = 1 \text{ Nm/s}$$

$$1 \text{ kW} = 10^3 \text{ W and } 1 \text{ MW} = 10^6 \text{ W}$$

Potential energy and kinetic energy

Potential energy is the energy possessed by a system due to its position or elevation; it represents the ability to do work against the body weight in lifting it from the surface of earth.

Consider a mass m kilogram elevated through a distance z metres above the surface of earth, then

Weight of the body (system) = mg

Potential energy = work done against gravity

= weight × height raised

$$= mg\,z \qquad ...(1.9)$$

Potential energy of a system is taken zero when its mass rests upon the earth.

When the height of mass changes from elevation z_1 to z_2, the change in potential energy (ΔPE) equals the work done on the system in raising it from lower elevation to higher elevation and is given by

$$\Delta PE = mg(z_2 - z_1) \qquad ...(1.10)$$

Kinetic energy is the energy possessed by a system by virtue of its motion. Consider a system of constant mass m which moves through a distance dx in the direction of force acting upon it.

Elementary work done = force × elemental distance moved

$$= \left(m\frac{dV}{dt}\right) \times dx$$

$$= mV\,dV \qquad \left(\because \frac{dx}{dt} = V\right)$$

where V is the instantaneous velocity in the direction of distance moved. Then,

Kinetic energy gained by the body = total work done

$$= m\int V\,dV$$

If the system experiences a change in velocity from V_1 to V_2, then,

Change in kinetic energy,

$$\Delta KE = m\int_{V_1}^{V_2} V\,dV$$

$$= \frac{1}{2}m\left(V_2^2 - V_1^2\right) \qquad ...(1.11)$$

If kinetic energy of the system at rest with respect to the earth is arbitrarily assigned a zero value, then

Kinetic energy,

$$KE = \frac{1}{2}mV^2 \qquad ...(1.12)$$

where V is the velocity of the system relative to the earth.

Both potential and kinetic energies are external forms of energy since they do not depend upon the composition of the fluid comprising the system. Further, the work done in accelerating a system (*KE*) or raising it (*PE*) gets stored and can be later recovered by reversing the direction of process.

The sum of the potential energy and kinetic energy of a system is called **mechanical energy** of the system.

EXAMPLE 1.8

A box of mass 80 kg is moving at a speed of 10 m/s on a ropeway. If the box is 40 m above the ground level, estimate the potential energy and kinetic energy of the box.

Solution : Potential energy = mgh

$= 80 \times 9.81 \times 40$

$= 41392$ J = **41.39 kJ**

$$\text{Kinetic energy} = \frac{1}{2} mV^2 = \frac{1}{2} \times 80 \times 10^2$$

$$= 4000 \text{ J} = \mathbf{4 \text{ kJ}}$$

EXAMPLE 1.9

A vehicle accelerates a glider of 125 kg mass from rest to a speed of 50 km/hr. Make calculations for the work done on the glider by the vehicle. What change would occur in the kinetic energy of the glider if subsequently its velocity reduces to 20 km/hr on the application of brakes ?

Solution : Initial velocity of glider, $V_1 = 0$ m/s

Velocity of glider after acceleration

$$V_2 = 50 \text{ km/hr}$$

$$= \frac{50 \times 10^3}{3600} = 13.89 \text{ m/s}$$

The work done on the glider in accelerating it from velocity V_1 to velocity V_2 equals the change in kinetic energy of the glider.

∴ Work done on the glider,

$$W = \frac{1}{2} m\left(V_2^2 - V_1^2\right)$$

$$= \frac{1}{2} \times 125 \times 13.89^2 = \mathbf{12058 \text{ J}}$$

Upon application of brakes,

Velocity of glider,

$$V_3 = 20 \text{ km/hr}$$

$$= \frac{20 \times 10^3}{3600} = 5.56 \text{ m/s}$$

Change in kinetic energy,

$$\Delta\text{KE} = \frac{1}{2} m\left(V_3^2 - V_2^2\right)$$

$$= \frac{1}{2} \times 125 \, (5.56^2 - 13.89^2)$$

$$= \mathbf{10126 \text{ J}}$$

EXAMPLE 1.10

What will be the kinetic energy in kWh of an aeroplane which has a mass of 30 tons and is travelling at 1000 km/hr speed ? If this plane is made to nose vertically upwards at this speed with power off, calculate the vertical distance through which the plane will move. Take $g = 9.80$ m/s²

Solution : Speed of the plane,

$$V = 1000 \text{ km/hr}$$

$$= \frac{1000 \times 10^3}{3600} = 277.78 \text{ m/s}$$

Kinetic energy of the plane

$$= \frac{1}{2} mV^2 = \frac{1}{2} \times (30 \times 10^3) \times (277.78)^2$$

$$= 77161743 \text{ J} = 77162 \text{ kJ}$$

1 kWh equals 3600 kJ

∴ Kinetic energy of the plane

$$= \frac{77162}{3600} = \mathbf{21.43 \text{ kWh}}$$

When the power is off and the plane is made to move vertically upwards, the kinetic energy gets converted to potential energy. Thus

$$77161743 = mgh$$

∴ Vertical distance through which the plane moves,

$$h = \frac{77161743}{\left(1000 \times 10^3\right) \times 9.80} = \mathbf{7.87 \text{ m}}$$

EXAMPLE 1.11

An artificial satellite of mass 500 kg is moving towards the moon. Make calculations for the kinetic and potential energies.

(i) relative to the earth when 50 km from launching and travelling at 2500 km/hr. Take earth's gravitational field equal to 7.9 m/s².

(ii) relative to the moon when travelling at the same velocity and 50 km from its destination where 1 kg mass has a weight of 3 N.

Solution : Speed of the satellite,

$$V = 2500 \text{ km/hr}$$

$$= \frac{2500 \times 10^3}{3600}$$

$$= 694.45 \text{ m/s}$$

(*i*) Kinetic energy

$$= \frac{1}{2} \times 500 \times (694.45)^2$$

$$= 120.6 \times 10^6 \text{ J}$$

$$= \mathbf{120.6 \text{ MJ}}$$

Potential energy

$$= mgh = 500 \times 7.9 \times (50 \times 10^3)$$

$$= 197.5 \times 10^6 \text{ J} = \mathbf{197.5 \text{ MJ}}$$

(*ii*) 1 kg mass weighs 3N and hence 500 kg will weigh (relative to the moon) as 1500 N

Potential energy

$$= mgh = 1500 \times (50 \times 10^3)$$
$$= 75 \times 10^6 \text{ J} = \mathbf{75\ MJ}$$

Kinetic energy

$$= \frac{1}{2} mV^2 = \frac{1}{2} \times 500 \times (694.45)^2$$
$$= 120.6 \times 10^6 = \mathbf{120.6\ MJ}$$

It is to be noted that kinetic energy is not dependent on gravitational field of the moon.

1.13. PRESSURE

A fluid element or mass is essentially acted upon by two categories of forces: body forces and surface forces. *Body forces* on fluid elements are caused by agencies such as gravitational, electric or magnetic fields. The magnitude of these forces is proportional to the mass of the fluid. *Surface forces* represent the action of the surrounding fluid on the element under consideration through direct contact. A force acting on portion dA of a surface can be resolved into two components; one normal and other tangential to the area dA.

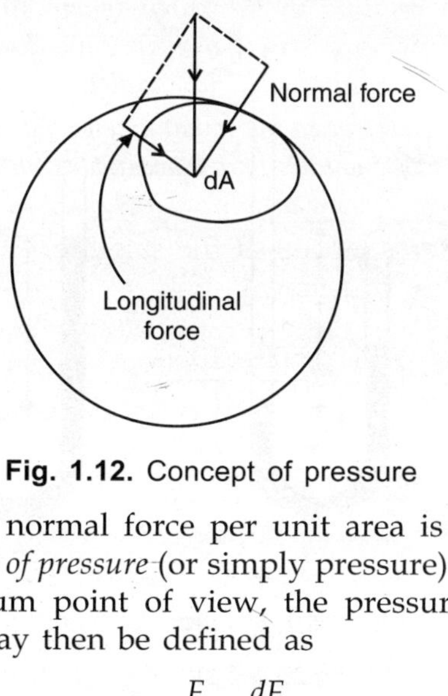

Fig. 1.12. Concept of pressure

The normal force per unit area is called *intensity of pressure* (or simply pressure). From continuum point of view, the pressure at a point may then be defined as

$$p = \underset{A \to 0}{Lt} \frac{F}{A} = \frac{dF}{dA} \qquad \text{...(1.13)}$$

where dF is the force acting normal to the infinitesimal area dA. If the total normal force F acts uniformly over the entire area A, then $p = F/A$. Pressure has the dimensions of $[FL^{-2}]$ and is usually expressed in N/m^2 or pascal (Pa).

$$1 \text{ pascal (Pa)} = 1 \text{ N/m}^2$$

i.e., pascal represents the force of one newton acting on an area of 1 m^2.

The unit of pascal is very small. Engineers express pressure in kilopascal (kPa) and Megapascal (MPa)

$$1 \text{ kPa} = 10^3 \text{ Pa}; 1\text{MPa} = 10^6 \text{ Pa}$$

Some other commonly used units of pressure measurement are:

$$1 \text{ bar} = 10^5 \text{ N/m}^2 = 10^5 \text{ Pa}$$

1 standard atmosphere (atm)

$$= 1.01325 \text{ bar} = 101325 \text{ Pa}$$
$$= 760 \text{ mm of mercury} = 760 \text{ Torr}$$

The pressure exerted by 1 mm column of mercury is named as Torr. Usually pressure below atmospheric pressure is expressed in Torr

$$750 \text{ mm of mercury} = 1 \text{ bar} = 100 \text{ kPa}$$

The following terms are generally associated with measurement of pressure.

(*i*) *Atmospheric pressure* (p_{at}) : This is the pressure exerted by the envelope of air surrounding the earth's surface. The standard atmospheric pressure is defined as the pressure produced by a column of mercury 760 mm high; the specific gravity of mercury being 13.5951 kg/m^3 and the acceleration due to gravity being 9.80665 m/s^2. From hydrostatic equation

$$p_{at} = \rho g h$$
$$= (13.5951 \times 1000) \times 9.80665 \times \frac{760}{1000}$$
$$= 101325 \text{ N/m}^2$$
$$= 1.01325 \times 10^5 \text{ Pa} = 1.01325 \text{ bar}$$

(*ii*) *Absolute pressure* (p_{abs}) : Pressure has been defined as the force per unit area due to interaction of fluid particles amongst themselves. A zero pressure intensity will occur when molecular momentum is zero. Such a situation can occur only when there is a perfect vacuum, *i.e.*, a vanishingly small population of gas molecules or of molecular velocity. Pressure intensity measured from this state of vacuum or zero pressure is called absolute pressure.

(*iii*) *Gauge pressure* (p_g) *and vacuum* (p_{vac}) : Instruments and gauges used to measure fluid pressure generally measure the difference between the unknown pressure p and the existing atmospheric pressure p_{at} as shown in Fig. 1.13.

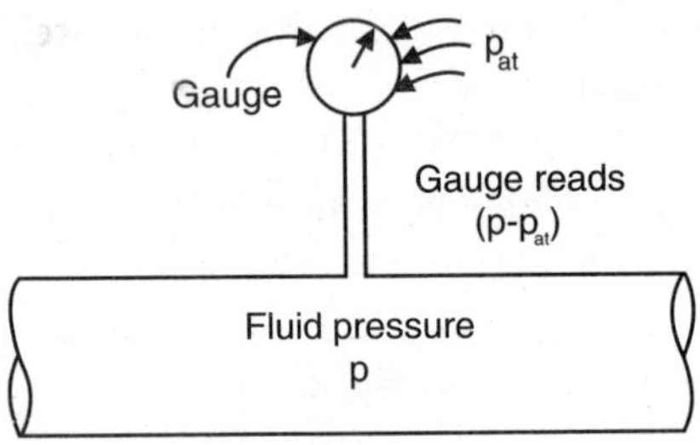

Fig. 1.13. Gauge pressure

When the unknown pressure is more than atmospheric pressure, the pressure recorded by the instrument is called *gauge pressure*. A pressure reading below the atmospheric pressure is known as *vacuum, rarefaction or negative pressure*. Actual absolute pressure is then the algebraic sum of the gauge indication and the atmospheric pressure

$$p_{abs} = p_{at} + p_g$$
$$p_{abs} = p_{at} - p_{vac} \quad ...(1.14)$$

Relationship between these pressure terms has been illustrated in Fig. 1.14.

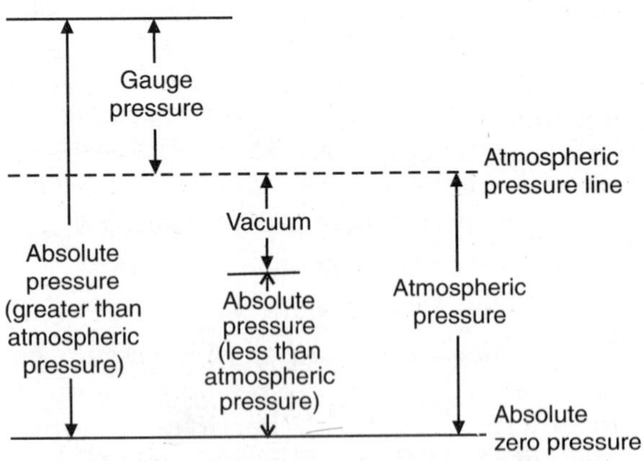

Fig. 1.14. Relationship between absolute, gauge, and atmospheric pressure

Manometers measure pressure by balancing a column of liquid against the pressure to be measured. Height of the column so balanced is noted and converted to the desired pressure units. Manometers may be vertical, inclined, open, differential or compound. Choice of any type depends on its sensitivity of measurement, ease of operation and magnitude of pressure being measured. Manometers can be used to measure gauge, differential, atmospheric and absolute pressure.

The simplest and useful pressure measuring device consists of a transparent tube bent in the form of letter U and filled with a manometric liquid whose density in known.

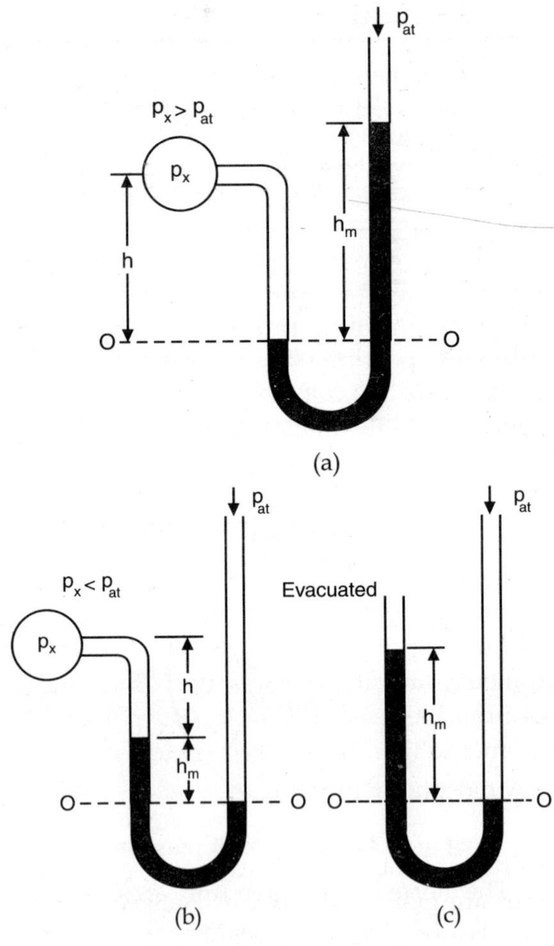

Fig. 1.15. U-tube manometers

Fig. 1.15.(*a*): One limb of the U-tube in open to the atmosphere and the other is connected to a vessel having pressure higher than the atmospheric pressure. ($p_x > p_{at}$). Due to greater pressure p_x in the vessel, the manometric liquid is forced down in the left limb of the U-tube and there is corresponding rise of manometric liquid in the right limb. Under equilibirum conditions, the pressure along a horizontal line 0-0 is same for either limb of the manometer. Then according to the principle of hydrostatics;

$$p_x + \rho\, g\, h = p_{at} + \rho_m\, g\, h_m$$

where p_{at} is the atmospheric pressure, ρ_m and ρ are the densities of the manometric liquid and the liquid in the vessel. If ρ is insignificantly small as compared to ρ_m, then

$$p_x = p_{at} + \rho_m\, g\, h_m \quad ...(1.15)$$

Fig. 1.15.(*b*). One limb of the U-tube is open to the atmosphere and the other is connected to a vessel having pressure less than that of the atmosphere ($p_x < p_{at}$). Due to negative pressure p_x in the vessel, the manometric liquid is sucked upwards in the left limb and there is corresponding fall of manometric liquid in the right limb. Pressure in the two limbs at the same level 0-0 are equal and, therefore, in accordance with hydrostatic equation,

$$p_x + \rho\, g\, h + \rho_m\, g\, h_m = p_{at}$$

Neglecting the density of the liquid in the vessel as compared to that of manometric liquid

$$p_x - p_{at} = -\rho_m\, g\, h_m \quad ...(1.16)$$

Fig. 1.15(*c*): The U-tube is closed at one end and the other end communicates with atmospheric pressure. This gauge is known as barometer, and it measures the atmospheric pressure in terms of height h_m of a column of manometric liquid, usually mercury.

Atmospheric pressure

$$p_{at} = \rho_m\, g\, h_m \quad ...(1.17)$$

Pressures higher than two or three atmospheres are measured with mechanical gauges of the bourdon tube or diaphragm type. In these gauges, the fluid pressure is applied to a hollow tube, moveable diaphragm or bellows. The deflection thus obtained is transmitted, through suitable mechanism, to a needle which indicates pressure on a pre-calibrated dial.

1.14. ATMOSPHERIC PRESSURE AND TEMPERATURE

Standard temperature and pressure (STP) : Under standard temperature and pressure, the temperature is taken as 15°C and the pressure is taken as 1.01325 bar (101.325 kN/m^2 or 760 mm of Hg).

Normal temperature and pressure (NTP) : Under normal temperature and pressure conditions, the temperature is taken as 0°C and pressure is taken as 1.01325 bar.

EXAMPLE 1.12

Calculate the pressure difference in N/m^2 which corresponds to one centimeter difference in liquid-surface level in a manometer when the liquid is (i) water (ii) mercury. The densities of the two liquids may be taken as 1000 kg/m^3 and 13568 kg/m^3 respectively.

Solution : Pressure difference

$$\Delta p = \rho_m\, g\, h_m$$

(*i*) For water :

$$\Delta p = 1000 \times 9.81 \times 0.01 = \mathbf{98.1\ N/m^2}$$

(*ii*) For mercury :

$$\Delta p = 13568 \times 9.81 \times 0.01 = \mathbf{1331\ N/m^2}$$

EXAMPLE 1.13

Determine the pressure difference indicated by a U-tube mercury (density = 13600 kg/m^3) manometer if the difference in the level of mercury in the two limbs is 52.5 mm. How would this level difference be affected if the mercury is replaced by carbon tetrachloride whose specific gravity is 1.5. ?

Solution : Indicated pressure difference

$$\Delta p = \rho_m\, g\, h_m = 13600 \times 9.81 \times \frac{52.5}{1000} = \mathbf{7004.34\ N/m^2}$$

Again, $\Delta p = \rho\, g\, h$ for carbon tetrachloride as the manometric liquid.

$$7004.34 = (1.5 \times 1000) \times 9.81 \times h$$

$$h = \mathbf{0.476\ m}$$

∴ Difference in level with carbon tetrachloride

= 476 mm

EXAMPLE 1.14

A turbine is supplied with steam at a pressure of 20 bar gauge. After expansion in the turbine the steam passes to a condenser which is maintained at a vacuum of 250 mm of mercury by means of pumps. Express the inlet and exhaust steam pressures in N/m² and kPa.

The barometer reads 750 mm of mercury.

Solution :

Atmospheric pressure

$$p_{at} = 750 \text{ mm of mercury}$$
$$= (13.6 \times 1000) \times 9.81 \times \frac{750}{1000}$$
$$= 1 \times 10^5 \text{ N/m}^2 = 1 \text{ bar}$$

(*i*) At inlet to turbine:

gauge pressure = 20 bar

absolute pressure = p_{abs}

$$= p_{at} + p_g = 1 + 20 = 21 \text{ bar}$$
$$= 21 \times 10^5 \text{ N/m}^2$$
$$= 21 \times 10^5 \text{ Pa} = \mathbf{2100 \text{ kPa}}$$

(*ii*) At condenser :

vacuum pressure = 250 mm of mercury

absolute pressure,

$$p_{abs} = p_{at} - p_{vac} = 750 - 250$$
$$= 500 \text{ mm of mercury}$$
$$= (13.6 \times 1000) \times 9.81 \times \frac{500}{1000}$$
$$= 66708 \text{ N/m}^2$$
$$= 66708 \text{ Pa} = \mathbf{66.708 \text{ kPa}}$$

EXAMPLE 1.15

A vessel of cylindrical shape, 40 cm diameter and 60 cm height, contains 3 kg gas. The pressure gauge mounted on the vessel indicates 600 mm of mercury. If the barometer reading is 750 mm of mercury, calculate the absolute pressure of gas. Also determine the density and specific volume of the gas.

Solution : Absolute pressure

$$p_{abs} = p_{at} + p_g$$
$$= 750 + 600 = 1350 \text{ mm of Hg}$$
$$= 1300 \times 9.81 \times \frac{1350}{1000}$$
$$= 180112 \text{ N/m}^2$$
$$= 180.112 \text{ kPa} = \mathbf{1.80 \text{ bar}}$$

$$\text{Volume of vessel} = \frac{\pi}{4} d^2 h$$
$$= \frac{\pi}{4} (0.4)^2 \times 0.6 = 0.07536 \text{ m}^3$$

Density ρ is defined as mass per unit volums.

∴ Density of gas

$$\rho = \frac{\text{mass}}{\text{volume}}$$
$$= \frac{3}{0.07536} = \mathbf{39.81 \text{ kg/m}^3}$$

Specific volume v is the reciprocal of mass density.

∴ Specific volume

$$v = \frac{1}{\rho} = \frac{1}{39.81} = \mathbf{0.0251 \text{ m}^3/\text{kg}}$$

EXAMPLE 1.16

A basic barometer can be used as an altitude measuring device in aeroplanes. The ground control reports a barometric reading of 750 mm of mercury while the pilot's reading is 690 mm of mercury. Estimate the altitude of the plane from the ground level if the average specific weight of air is 11.97 N/m³.

Solution : Pressure is related to a fluid column by the expression $p = \rho g h$

∴ Pressure at the ground level

$$= 13600 \times 9.81 \times \frac{750}{1000}$$
$$= 100062 \text{ N/m}^2$$

Pressure at plane level

$$= 13600 \times 9.81 \times \frac{690}{1000}$$
$$= 92057 \text{ N/m}^2$$

∴ Difference of pressure at the ground and plane level

$$= 100062 - 92057 = 8005 \text{ N/m}^2$$

In terms of altitude Z of plane, this difference is pressure is given by

$$\Delta P = \rho g Z = w Z$$

where w is specific weight of air

$$8005 = 11.97 Z \ ; \quad Z = \frac{8005}{11.97} = 668.75$$

Thus the altitude of the plane from the ground level is 668.75 m.

EXAMPLE 1.17

A U-tube mercury manometer has been employed to measure the pressure of gas flowing through a pipe line. One limb of the manometer communicates with the gas pressure and the other limb is open to atmosphere. If the difference in level of mercury in the two limbs is 40 cm, determine the gas pressure and express it in kPa, bar and atmosphere. The barometer reads 750 mm of mercury.

Solution : Refer Fig. 1.16 for the arrangement between manometer and the gas pipe line.

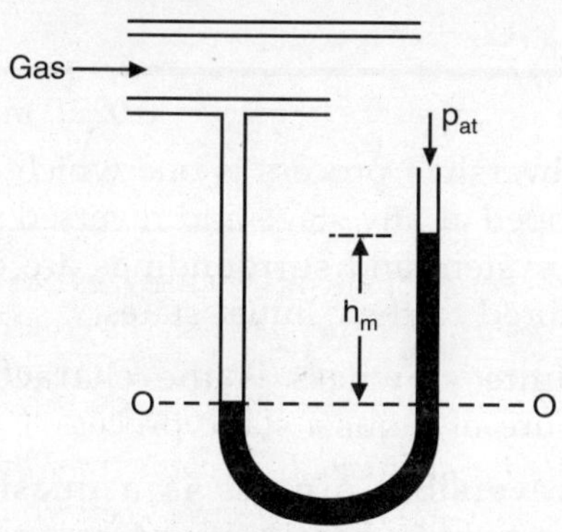

Fig. 1.16.

Atmospheric pressure, p_{at} = 750 mm of Hg

$$= 13600 \times 9.81 \times \frac{750}{1000}$$

$$= 1 \times 10^5 \text{ N/m}^2$$

Neglecting density of gas compared to that of manometric liquid (mercury), the hydrostatic equation gives

$$p = p_{at} + \text{pressure due to mercury column } h_m$$

$$= p_{at} + \rho_m g h_m$$

$$= 1 \times 10^5 + \left(13600 \times 9.81 \times \frac{40}{100}\right)$$

$$= 153366 \text{ N/m}^2$$

$$= 153.4 \text{ kPa} = 1.534 \text{ bar}$$

$$= \frac{1.534}{1.01325} = \mathbf{1.514 \text{ atmosphere}}$$

EXAMPLE 1.18

A U-tube mercury manometer with one arm open to atmosphere is used to measure the pressure of steam flowing through a pipe. The level of mercury in the open-ended arm stands 95 mm higher than that in the arm connected to steam pipe. Some of the steam condenses in the arm in communication with the steam pipe. If the height of this column is 32 mm, make calculations for the absolute pressure of steam.

Take atmospheric pressure equal to 760 mm of mercury.

Solution : Refer Fig 1.17 for the arrangement between manometer and the steam pipe.

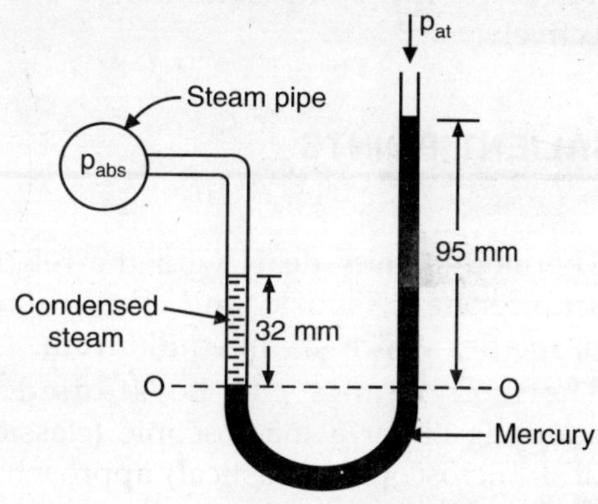

Fig. 1.17.

Under equilibrium conditions, the pressure along the horizontal line 0 – 0 is same for either arm of the manometer. Then from the hydrostatic equation

$$p_{abs} + \rho_w g h_w = p_{at} + \rho_m g h_m$$

$$\text{or } p_{abs} + (1000 \times 9.81 \times 0.032)$$

$$= (13600 \times 9.81 \times 0.76) + (13600 \times 9.81 \times 0.095)$$

$$\text{or } p_{abs} + 313.92 = 101396 + 12674.5$$

∴ Absolute pressure of steam flowing through the pipe,

$$p_{abs} = 101396 + 12674.5 - 313.92$$

$$= 113757 \text{ N/m}^2 = 113.76 \text{ kPa}$$

$$= \mathbf{1.137 \text{ bar}}$$

EXAMPLE 1.19

A large chamber is separated into two compartments which are maintained at different pressures (Fig. 1.18). Pressure gauge A reads 180 kPa, and pressure gauge B reads 120 kPa. If the barometric pressure is 100 kPa, determine the absolute pressure existing in the compartments and the reading of gauge C.

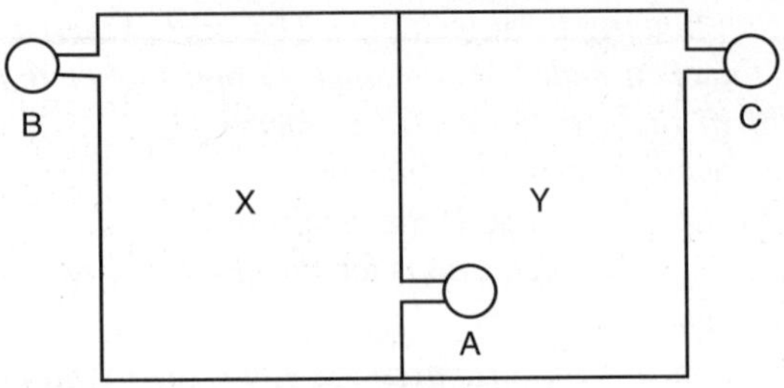

Fig. 1.18.

Solution : Let x kPa and y kPa be the absolute pressures in the compartments X and Y respectively.

Barometric pressure = 100 kPa

Reading on gauge

$$A = x - y; \quad 180 = x - y \qquad (i)$$

Reading on gauge

$$B = x - 100; \quad 120 = x - 100 \qquad (ii)$$

From equations (i) and (ii)

$$x = \mathbf{220\ kPa}$$

and $$y = \mathbf{40\ kPa}$$

Reading on gauge

$$C = y - 100$$
$$= 40 - 100 = \mathbf{-60\ kPa}$$

SALIENT POINTS

1. Thermodynamics deals with the relation among heat, work and the system properties which are in equilibrium.
2. Thermodynamics can be studied by adopting either a macroscopic (classical) or a microscopic (statistical) approach.
3. System represents a finite quantity of matter or a prescribed region of space. A system may be open, close or isolated.
4. The envelope surrounding the system is called boundary. The boundary may change its shape or size.
5. Every thing external to the system is called surroundings or environments. A system and its environment together constitute an isolated system.
6. Property defines the state or condition of a system. Intensive properties donot depend on the mass of the system. Extensive properties depend on the mass of the system. All the specific properties are intensive properties.
7. When a path is completely specified, then the change of state is called a process.
8. A process represents transformation of the system from one fixed state to another fixed state.
9. A reversible process is one which can be stopped at any stage and reversed so that the system and surroundings are exactly restored to their initial states.
10. Infinite slowness is the characteristic feature of a quasi static process.
11. A reversible process is a quasi-static process, but a quasi-static process need not be a reversible process.
12. Any process or series of processes whose end states are identical is called a cycle.
13. A system is said to be in thermodynamic equilibrium, if it is simultaneously in the state of mechanical, thermal and chemical equilibrium.
14. Phase refers to the quantity of matter which is homogeneous throughout in chemical composition and physical structure.

    A homogeneous system is of single phase where as a heterogeneous system consists of two or more phases.
15. The types of matter normally considered in engineering thermodynamics are: Gas, vapour, liquid or any non-chemically reactive mixture of these.

REVIEW QUESTIONS

A. Conceptual and conventional questions.

1. Define thermodynamics and state its scope in the field of energy technology.
2. How does classical thermodynamics differ from statistical thermodynamics ?
3. Explain the concept of macroscopic and microscopic view point as applied to the study of thermodynamics.
4. Name the various laws of thermodynamics and state briefly the purpose served by each law.
5. Is it possible to prove the laws of thermodynamics, like any theorem in mathematics ?
6. What is the relationship between a system and its environment when the system is :
   (*i*) adiabatic, (*ii*) isolated ?
7. Specify, giving reasons whether the following systems are open, closed or isolated when in operation
   (*i*) Kitchen refrigerator
   (*ii*) Scooter engine
   (*iii*) Manually operated ice-cream freezer
   (*iv*) Liquid cooling system of an automobile
   (*v*) Radio, switched on
   (*vi*) Universe
   (*vii*) Thermometer surrounded by a medium at high temperature.
   Identify all mass and energy inflows and outflows across the system boundaries.
8. Define control volume and control surface. How does control volume differ from an open system.
   Comment on the validity of the following system:
   "An open system defined for a fixed region and control volume are synonymous."
9. What do you understand by the term phase as applied to a system ? Differentiate between homogeneous and heterogeneous systems.
10. State the difference between extensive, intensive and specific properties of a thermodynamic system.
11. State, giving brief justification, whether the following properties of a system are intensive or extensive:
    Pressure, temperature, mass, density, volume, specific volume, elevation (altitude), surface area, kinetic energy, velocity, electric charge.
12. Name some properties which are relevant to thermodynamics. Why shape, colour and odour are irrelevant in thermodynamics ?
13. Define property and mention its main characteristics in relation to a cyclic process.
14. Explain the terms state, path, process and cycle.
15. What is meant by thermodynamic equilibrium ? How does it differ from thermal equilibirum ?
16. Does a contact between acid and water at the same temperature conform to thermal equilibrium ?
17. Define a quasistatic process and state its salient characteristics.
18. Why would the name "thermostatics" be more appropriate than thermodynamics as applied to the science of thermal engineering ?
19. State the conditions for a process to be reversible.
20. What types of matter are normally considered in engineering thermodynamics ?
21. What is the concept of continuum ? How density and pressure are defined using this concept ?
22. Differentiate between absolute and gauge pressure.
23. A body of fixed mass is weighed at the poles ($g = 9.825 \text{ m/s}^2$) by a spring balance which was calibrated at standard gravitational acceleration ($g = 9.806 \text{ m/s}^2$). If the spring balance reads 200 N, make calculations for the mass of the body.
    [Ans. 20.356 kg]
24. A system has a mass of 4 kg and it is moving at a velocity of 50 m/s and height 100 m above the sea level. Determine its potential energy and kinetic energy relative to the earth.
    [Ans. 3.924 kJ, 5 kJ]
25. A forced draught fan supplies air to the furnace of a boiler at draught of 40 mm of water. If the barometer reads 760 mm of mercury, make calculations for the absolute pressure of air supply in kPa.
    [Ans. 101.715 kPa]
26. A vessel of cylindrical shape, 50 cm diameter and 75 cm height, contains 4 kg of a gas. The pressure gauge mounted on the vessel indicates 620 mm of mercury above atmosphere. If the barometer reading is 760 mm of mercury, calculate the absolute pressure of gas in bar. Also determine the density and specific volume of the gas.
    [Ans. 1.841 bar, 27.21 kg/m^3, 0.03675 m^3/kg]

27. The gauge pressure in a steam boiler is 30 bars, the rarefaction in the condenser as indicated by a vacuum gauge is 0.95 bar. Find the absolute pressure in the steam boiler and condenser if atmospheric pressure of air is 1.01 bar. Express these pressures in kN/m^2.
[Ans. 3101 kN/m^2, 6 kN/m^2]

28. The air supply to an internal combustion engine is metered by observing the pressure drop across an orifice in the air supply line to the engine. The pressure drop is measured by means of a manometer containing paraffin having a density of 800 kg/m^3. Express the difference of level of mercury of 20 cm in manometer in (*i*) bar, (*ii*) metres of air. Take density of air as 1.23 kg/m^3.
[Ans. 0.015696 bar, 130.08 m]

29. The absolute pressure in a tank as determined by a U-tube manometer is stated to be 1.25 bar. The difference in the liquid level in the two limbs in 80 cm. Taking barometer pressure equal to 750 mm of Hg, determine the density of liquid used in the manometer.
[Ans. 3190 kg/m^3]

30. Measurements of pressure at the base and top of a mountain are 74 cm and 60 cm of mercury respectively. Work out the height of mountain if air has a mass density of 1.22 kg/m^3.
[Ans. 1560.4 m]

B. Fill in the blanks with appropriate word/words

1. A system is separated from its surroundings by a .... which may be .... or ....
2. The system and its surroundings taken together constitute an .... system.
3. The .... property of a system remains the same whether one considers the whole system or a part of it.
4. The state is the condition of a system identified by its ....
5. A quasi-static process is one in which all the states through which the system passes are very close to ....
6. A reversible process leaves no trace of its occurrence in the .... and ....
7. The fluid employed in thermodynamic devices for the transport of energy between the system and surroundings is called the ....
8. Any process or series of processes whose end states are identical is called a ....
9. A .... is homogeneous, and has the same chemical and physical compositions.
10. A system consisting of a single phase is called .... system.
11. A system is said to be in .... if it is simultaneously in the state of mechanical, thermal and chemical equilibrium.

Answers

1. boundary, rigid or flexible 2. isolated
3. intensive 4. thermodynamic properties
5. equilibrium state 6. system, surroundings
7. working substance 8. cycle 9. pure substance
10. homogeneous 11. thermodynamic equilibrium

C. State true or false. If false, write the correct statement

(*i*) The laws of thermodynamics are based on logical reasonings and experimental observations but have no mathematical support.
(*ii*) In a microscopic study of thermodynamics, a limited number of thermodynamic variables are required to describe the state of the system.
(*iii*) A system may enter into an interaction with its surroundings via mass and energy transfer.
(*iv*) A system need not remain in equilibrium in all states during a quasi-static process.
(*v*) An ice cube placed into a thermos flask, with its cap tightened, constitutes an isolated system.
(*vi*) All properties are state or point functions.
(*vii*) Intensive properties per unit mass are extensive properties.
(*viii*) Any change from one state to another is called a path.
(*ix*) A process with identical end states is called a cycle.
(*x*) A pure substance is an equilibrium mixture of solid, liquid and vapour.

Answers

(*i*) True (*ii*) False; matter is considered to be comprised of a large number of tiny particles called molecules (*iii*) True (*iv*) False; need to remain necessarily in equilibrium and any departure is at most infinitesimal (*v*) True (*vi*) True (*vii*) False; extensive properties per unit mass are intensive properties (*vii*) False; is called a process (*ix*) True (*x*) True

D. Multiple choice questions.

1. The study of thermodynamics provides answer to the followings, except
   (*a*) whether a process is feasible or not
   (*b*) how much energy is required for a process to take place
   (*c*) rate or speed with which a process occurs
   (*d*) extent to which a reaction/process takes place

2. Choose the false statement in the context of macroscopic approach followed in classical thermodynamics :
   (*a*) considers gross behaviour of matter
   (*b*) requires no hypothesis on the detailed structure of matter
   (*c*) predicts the behaviour of individual fluid particles by statistical averaging
   (*d*) deals with characteristics which can be perceived by senses and measured directly

3. Which aspect is not true concerning microscopic study of thermodynamics ?
   (*a*) a knowledge of the structure of matter is essential
   (*b*) a limited number of variables/properties are needed to describe the state of matter
   (*c*) the values of these variables cannot be measured
   (*d*) statistical averaging is adopted to predict the behaviour of individual fluid particles

4. A thermodynamic system refers to
   (*a*) any defined region in space
   (*b*) a specified mass in fluid flow
   (*c*) a specified region of constant volume
   (*d*) a prescribed and identifiable quantity of matter

5. A closed thermodynamic system manifests when
   (*a*) matter is not allowed to cross the boundary, but energy transfer does occur across the boundary
   (*b*) there is transfer of both mass and energy across the system boundaries
   (*c*) there is only transfer of mass, but no heat and work energy are transferred
   (*d*) there is absolutely no interaction of the system with surroundings across its boundaries

6. In an open thermodynamic system
   (*a*) mass content of the system under consideration remains same
   (*b*) transfer of mass and/or energy takes place
   (*c*) there is only mass transfer even though there may not be any exchange of energy with the system environment
   (*d*) the system exchanges energy with the surroundings in the form of heat energy only

7. A thermodynamic system is referred to be an isolated system when there is transfer of .... across the system boundaries
   (*a*) only mass
   (*b*) only energy
   (*c*) both mass and energy
   (*d*) neither mass nor energy

8. A system and its environment put together constitute
   (*a*) an adiabatic system
   (*b*) an isolated system
   (*c*) a segregated system
   (*d*) a homogeneous system

9. Choose the open thermodynamic system
   (*a*) manual ice cream freezer
   (*b*) centrifugal pump
   (*c*) pressure cooker
   (*d*) automobile storage battery

10. Which one of the following represents a closed system ?
    (*a*) bomb calorimeter
    (*b*) steam generator
    (*c*) universe
    (*d*) exhaust stroke of an IC engine

11. Identify the heterogeneous system
    (*a*) atmospheric air
    (*b*) mixture of hydrogen and oxygen
    (*c*) cooling fluid in a radiator
    (*d*) mixture of ice, water and steam

12. Which aspect is not true regarding thermodynamic property of a system ?
    (*a*) a charteristic that describes the condition of the system
    (*b*) a variable whose value depends upon the particular path in going from one state to another
    (*c*) an intensive property is independent of the extent or mass of the system
    (*d*) an extensive property can be made intensive, if it is divided by mass and is called specific property

13. The value of an extensive property is essentially dependent on
   (*a*) mass or extent of the system
   (*b*) interaction of system with its surroundings
   (*c*) path followed by the system in going from one state to another
   (*d*) nature of boundaries, rigid or flexible
14. Which one of the properties given below is an intensive property of the system ?
   (*a*) composition (*b*) volume
   (*c*) kinetic energy (*d*) entropy
15. All of the following are intensive properties of a system, except
   (*a*) viscosity (*b*) temperature
   (*c*) density (*d*) potential energy
16. Choose the specific property of a thermodynamic system
   (*a*) viscosity (*b*) density
   (*c*) pressure (*d*) temperature
17. Which one of the following quantities presents the property of a system ?
   (*a*) $\int p\,dv$ (*b*) $\int v\,dp$
   (*c*) $\int (p\,dv + v\,dp)$
18. Which amongst the followings is not a property of the system ?
   (*a*) pressure
   (*b*) composition
   (*c*) thermal conductivity
   (*d*) heat
19. Spot the odd one out
   (*a*) thermal conductivity
   (*b*) kinetic energy
   (*c*) work
   (*d*) pressure
20. Choose the correct statement
   (*a*) temperature is an extensive property
   (*b*) mass remains same in an open system
   (*c*) the system boundaries may expand or collapse
   (*d*) an isolated system allows exchange of energy in the form of heat only
21. Identify the wrong statement
   (*a*) the laws of thermodynamics cannot be derived mathematically
   (*b*) the quantity of matter constituting a system remains constant
   (*c*) the kinetic and potential energies possessed by a system can be converted into heat
   (*d*) the system and its surroundings taken together constitute an isolated system
22. Which of the followings is not valid in the context of a quasi-static process ?
   (*a*) pressure is uniform and remains practically constant
   (*b*) friction is negligible and the process proceeds at infinitesimally small speed
   (*c*) process takes place due to finite unbalanced potential within the system or between it and the surroundings
   (*d*) process can be represented graphically as a continuous line on a state diagram
23. All of the following statements are true for a quasi-static process except
   (*a*) proceeds very slowly
   (*b*) can proceed in either direction
   (*c*) need not be a reversible process
   (*d*) occurs when the driving force is finite but very small

Answers

1. (*c*)	2. (*c*)	3. (*b*)	4. (*d*)	5. (*a*)
6. (*c*)	7. (*d*)	8. (*b*)	9. (*b*)	10. (*a*)
11. (*d*)	12. (*b*)	13. (*a*)	14. (*a*)	15. (*d*)
16. (*b*)	17. (*c*)	18. (*d*)	19. (*c*)	20. (*c*)
21. (*b*)	22. (*c*)	23. (*c*)		

Chapter **2**

Temperature and Zeroth Law of Thermodynamics

Learning objectives : A study of the subject matter covered in this chapter will enable the reader to understand :

- *zeroth law of thermodynamics and its significance*
- *thermometry ; thermometric property and thermometric substances*
- *temperature scales ; Centigrade, Kelvin and International*
- *perfect gas thermometer, resistance thermometer and thermo-couple*

Temperature may be defined as the

(*i*) degree of hotness and coldness of a body or an environment measured on a definite scale;

(*ii*) driving force or potential causing the flow of energy as heat;

(*iii*) measure of the mean kinetic energy of the molecules of a system. A change in the temperature of the system accounts for the change in the molecular motion and hence the kinetic energy of the molecules.

(*iv*) parameter which determines whether or not a system is in thermal equilibrium with another system.

Temperature is an intensive property independent of the size and mass of the system.

2.1. EQUALITY OF TEMPERATURE

Two systems are equal in temperature if no change occurs in any property when they are brought into communication.

Consider two systems, perfectly insulated from the surroundings and at different temperatures. When they are brought into physical contact, the energy in the form of heat will flow from the hot system to the cold system, and there would occur changes in their physical properties such as length, electrical resistance etc. After a sufficient long time of contact, no further change occurs and thermal equilibrium is established. The two systems are then said to have the same temperature.

2.2. ZEROTH LAW OF THERMODYNAMICS

When two systems have equality of temperature (thermal equilibrium) to a third system, they in turn have equality of temperature (thermal equilibrium) with each other.

Consider two systems S_1 and S_3 which are in thermal equilibrium with each other. Accordingly no change occurs in their physical properties when they are brought into good contact. It is assumed that the systems S_1 and S_3 donot react each other chemically or electrically. Likewise, let the system S_3 be separately equal in temperature to another system S_2 so that no change occurs when they communicate with each other. It is then an observed fact that the systems S_1 and S_2 will not undergo any physical changes when brought into contact. The systems S_1 and S_2

are then taken to be in thermal equilibrium, *i.e.*, equal in temperature to each other.

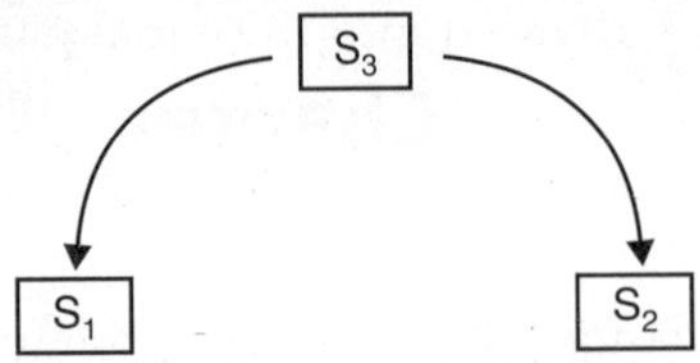

Fig. 2.1. Concept of Zeroth law

The Zeroth law was developed by R.H. Fowler in 1931, and the first and the second laws already existed at that time. However, during the development of first and second laws, the basics of the postulate suggested by Fowler were assumed to be true without stating it so. Accordingly for the logical development of the subject of thermal science, it was thought appropriate to call the law proposed by Fowler as the Zeroth law of thermodynamics so that it precedes the first and second laws.

The Zeroth law provides the basis for the measurement of temperature of a system. The temperature measuring device (thermometer) is brought into contact with a system A which is at a known thermal level (the reference temperature) and the two are allowed to reach a state of thermal equilibrium. Some measurable characteristic of the thermometer, called the thermometric property, is then recorded. The thermometer is subsequently placed in thermal contact with another system B whose temperature is needed to be measured. When thermal equilibrium gets established between the thermometer and the system B, the thermometric property of the thermometer is again noted. If the two thermometric property readings (one measured while in contact with system A and the other measured while in contact with system B) are identical, then the systems A and B are in thermal equilibrium with each other, *i.e.*, there is equality of temperature between the two systems.

2.3. THERMOMETRY

Thermometry may be defined as the act of measuring temperature with accuracy and precision. Temperature measurement depends upon the establishment of thermodynamic equilibrium between the system and the device used to measure the temperature, *e.g.*, a thermometer bulb or thermocouple wires. The sensing element of the device has certain physical characteristics which change with temperature and this effect is taken as a measure of temperature. A property or characteristic which changes in value as a function of temperature is called ***thermometric property***; and the corresponding substance is known as ***thermometric substance***.

The physical characteristics which are so used could be:

(*i*) a change in dimensions, *i.e.*, expansion or contraction of material in the form of the solid, liquid or gas (mercury-in-glass thermometer, bimetallic thermometer, gas thermometer).

(*ii*) a change in electrical resistance of metals and semiconductors (resistance thermometers and thermistors)

(*iii*) a thermo-electric emf for two different metals and alloys joined together (thermo-couples)

(*iv*) a change in the intensity and colour of emitted radiations (radiation thermometers).

(*v*) fusion of materials when exposed to temperature (pyrometric cone).

Calibration is then achieved through comparison with established standards. The temperature scales serve to define temperature in terms of observable characteristics of materials. Any one of the following mathematical relationship between temperature t and the thermometric property x can be utilized to establish a temperature scale.

$$t = Ax$$

..... Direct proportionality

$$t = Ax + B$$

..... Linear relationship

$$t = \alpha x^2 + \beta x + \gamma$$

..... Quadratic relationship

$$t = A' + B'x + C'x^2 + D'x^3$$

..... Polynomial relationship

The constants appearing in these relations can be determined from any two fixed thermometric points, *e.g.*, the ice point (freezing point) and steam point (boiling point) of water. A fixed point refers to an easily reproducible state of an arbitrarily chosen standard system.

2.4. TEMPERATURE SCALES

A quantitative measure of the temperature of a system requires reference to some datum plane or reference condition, and the establishment of a suitable temperature unit. Many temperature scales and reference points have been proposed; the important ones are listed below:

2.4.1. Centigrade and Fahrenheit scales

On both these scales, the ice point (equilibrium temperature of ice and air saturated pure water) and the steam point (equilibrium temperature of pure water and water vapour) at one standard atmospheric pressure are used as fixed points.

Let these temperatures be represented by t_i and t_s and the corresponding thermometric properties be x_i and x_s; the suffix *i* and *s* refer to ice point and steam point respectively. Considering linear correlation, $t = a\,x + b$, between the temperature *t* and thermometric property *x*, we obtain

$$t_i = a\,x_i + b \quad \text{and} \quad t_s = a\,x_s + b$$

$$\text{or} \quad a = \frac{t_s - t_i}{x_s - x_i} \quad \text{and} \quad b = t_i - \frac{t_s - t_i}{x_s - x_i}x_i$$

$$\therefore \quad t = \frac{t_s - t_i}{x_s - x_i}x + t_i - \frac{t_s - t_i}{x_s - x_i}x_i$$

$$= t_i + (t_s - t_i) \times \frac{x - x_i}{x_s - x_i} \qquad \text{...(2.1)}$$

The centrigrade scale, abbreviated °C, assigns 0°C to ice point and 100°C to steam point and the interval between these points is divided into 100 equal parts. Thus in centigrade scale t_i = 0°C and $(t_s - t_i)$ = 100°C and accordingly equation 2.1 takes the form:

$$t_C = 100\frac{x - x_i}{x_s - x_i} \qquad \text{...(2.2)}$$

The corresponding values on Fahrenheit scale, abbreviated °F, are 32°F and 212°F with the interval divided into 180 equal parts. That gives :

$$t_F = 32 + 180\frac{x - x_i}{x_s - x_i} \qquad \text{...(2.3)}$$

From equations 2.2. and 2.3, the following correlation can be established between temperatures measured on Centigrade and Fahrenheit scales :

$$\frac{t_C}{100} = \frac{t_F - 32}{180}$$

$$\text{or} \quad t_C = \frac{5}{9}(t_F - 32) \qquad \text{...(2.4)}$$

The use of ice point and steam point as fixed points is considered unsatisfactory because:

(*i*) it is difficult to achieve and maintain equilibrium between pure ice and ice and air saturated water,

(*ii*) the steam point is extremely sensitive to changes in pressure.

2.4.2. Kelvin and Rankine absolute scales

Thermodynamically there is a condition of no molecular activity and hence no heat content in a body. The temperature at this condition is the lowest temperature possible and is referred to as ***absolute zero***. On Kelvin and Rankine scales the absolute temperature is hypothetically placed at – 273.16°C and – 459.67°F.

2.4.3. Thermodynamic scale

The efficiency of an ideal engine operating on the Carnot cycle between any two temperatures is given by

$$\eta = \frac{T_1 - T_2}{T_1} = \frac{Q_1 - Q_2}{Q_1} \qquad \text{...(2.5)}$$

where Q_1 represents the heat absorbed at temperature T_1 and Q_2 represents the heat rejected at temperature T_2. This identity may be rewritten as

$$\frac{T_1}{T_2} = \frac{Q_1}{Q_2} \qquad \text{...(2.6)}$$

The above expression forms the basis of a thermodynamic temperature scale devised by Lord Kelvin. This expression states that *"the ratio of any two temperatures on the Kelvin or thermodynamic scale is equal to the ratio of the heat absorbed to the heat rejected by a Carnot engine operating between those temperatures."*

The Kelvin temperature scale is completely defined if a number is selected to describe the temperature of a chosen fixed point. At the tenth conference of Weights and Measures held in 1954, the fixed point was chosen as the **triple point of water**; the triple point refers to the state at which all the three phases - solid, liquid and gas co-exist simultaneously. The triple point of water can be easily reproduced in the laboratory and has been found to be superior compared to the ice point or the steam point for standardisation purposes. The triple point was arbitrarily assigned a value of 273.16.

With triple point of water as the single fixed point, the direct proportionality relationship ($t = ax$) between the absolute temperature T and the thermometric property x can be written as

$$\frac{T}{T_{tp}} = \frac{x}{x_{tp}}; \quad T = T_{tp}\frac{x}{x_{tp}} \qquad ...(2.7)$$

where T_{tp} is the triple point of water on absolute scale and x_{tp} is the value of thermometric property at this point. It is to be noted that:

• The Kelvin scale has a value of 273.16 at 0°C and a value of 373.16 at 100°C; it is sometimes called the absolute Celsius scale and is denoted by K. There is an equivalent Fahrenheit scale, called the Rankine scale R where the triple point of water is 491.69° R and the ice point is 491.67° R. The Kelvin and Rankine scales are related as

$$\frac{T_R}{T_K} = \frac{491.69}{273.16} = \frac{180}{100} = \frac{9}{5}$$

• Kelvin scale is obtained by sliding each value on the Celsius scale upward by 273.16 degree, *i.e.* $T_K = t_C + 273.16$. Likewise in Fahrenheit and Rankine scale : $T_R = t_F + 459.67$

• In engineering applications the above correlations are approximated as

$$T_K = t_C + 273$$
$$T_R = t_F + 460 \qquad ...(2.8)$$

In SI units temperature is denoted simply by K without the degree symbol.

• Each division on both the Kelvin and Celsius scale is equal.

2.4.4. Perfect gas thermometer

The thermometric properties of all the materials do not change linearly with temperature. As such, the different thermometers making use of different thermometric substances, would give different temperature readings of a system except at the fixed points at which calibration has been done. Obviously to have reliable and reproducible value of temperatures, it becomes necessary to devise a temperature scale which should not depend on the nature of thermometric substance. This aspect leads to the establishment of an ***ideal gas temperature scale.***

The behaviour of an ideal gas furnishes the basis for developing the thermodynamic scale. An ideal gas is one which behaves according to the equation

$$p\,v = RT \qquad ...(2.9)$$

where p is the absolute pressure, v is the specific volume, R is the gas constant for a particular gas and T is the Kelvin or thermodynamic temperature. This equation is only an approximation to the actual behaviour of gases. However, all gases and vapours tend to behave like an ideal gas at sufficiently low pressure (in the limit when pressure approaches zero).

Equation 2.9 may take the form

$$T = T_{ref}\left(\frac{p}{p_{ref}}\right) \text{ at constant volume}$$

$$T = T_{ref}\left(\frac{v}{v_{ref}}\right) \text{ at constant pressure}$$

$$...(2.10)$$

Temperature on thermodynamic scale can thus be determined either

(*a*) by observing the change in pressure of a constant volume of ideal gas (constant volume thermometer); or

(*b*) by observing the change in volume of a fixed quantity of the ideal gas at constant pressure (constant pressure thermometer).

Refer Fig 2.2 which depicts the schematics of a constant volume gas thermometer. The thermometer consists of a

- bulb which encloses a fixed mass of an ideal gas
- capillary tube
- U-tube manometer with flexible bend

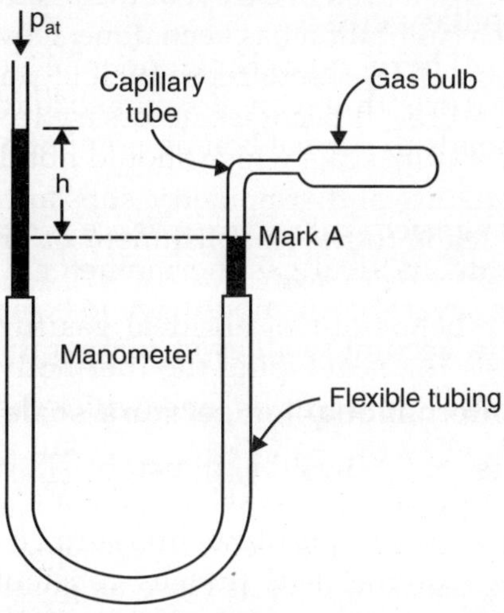

Fig. 2.2. Constant volume gas thermometer

The capillary tube connects the gas bulb to one limb of the manometer; the other limb being open to atmosphere. The flexible bend of the manometer helps to raise or lower the limb open to atmosphere so that meniscus of mercury can be made to stand at fixed mark A in the limb communicating with the bulb. This is to ensure that volume of the gas in the bulb is maintained at constant value. During operation, the bulb communicates with a constant temperature bath. There is exchange of heat between the bath and the bulb until both attain the state of thermal equilibrium (equality of temperature). Due to this heat transfer to the bulb the gas enclosed within the bulb, expands and that pushes the mercury downwards in the right limb of the manometer. The flexible tubing is then adjusted so that mercury level again attains the position of fixed mark A. The difference in the level of mercury in the two limbs is recorded, and absolute pressure p of gas is determined from the hydrostatic equation.

$$p = p_{at} + \rho g h \qquad \text{...(2.11)}$$

The various steps in the operation of gas thermometer are outlined below:

(*i*) The bulb is filled with a certain mass of gas such that initially the gas has a pressure p_{tp} = 1000 mm of mercury absolute when the bulb interacts with a constant temperature bath maintained at the triple point of water, T_{tp}. Subsequently the bulb is exposed to the system whose temperature T is to be measured. Let the corresponding absolute gas pressure be p mm of mercury. The ratio (p/p_{tp}) is determined and designated as $(p/p_{tp})_{1000}$.

(*ii*) A certain amount of gas is withdrawn from the bulb so that at the triple point of water , the gas has now an absolute pressure of 800 mm of mercury. The bulb is next again brought into direct contact with the system, the corresponding gas pressure is noted and ratio $(p/p_{tp})_{800}$ is determined.

(*iii*) The procedure is repeated; a series of measurements with different amount of gas in the bulb are made and a set of (p/p_{tp}) values is obtained.

Experimental observations indicate that for all gases such as hydrogen , helium , oxygen and nitrogen etc., the values of pressure ratio (p/p_{tp}) plotted against p_{tp} are straight lines. Further, when these straight lines are extrapolated, they converge and meet at a point. The actual temperature T of the system is then worked out as

$$\frac{T}{T_{tp}} = \underset{p_{tp} \to 0}{Lt} \frac{p}{p_{tp}} \qquad \text{...(2.12)}$$

Since the value T_{tp} = 273.16 is fixed, the absolute temperature T of the system can be worked out.

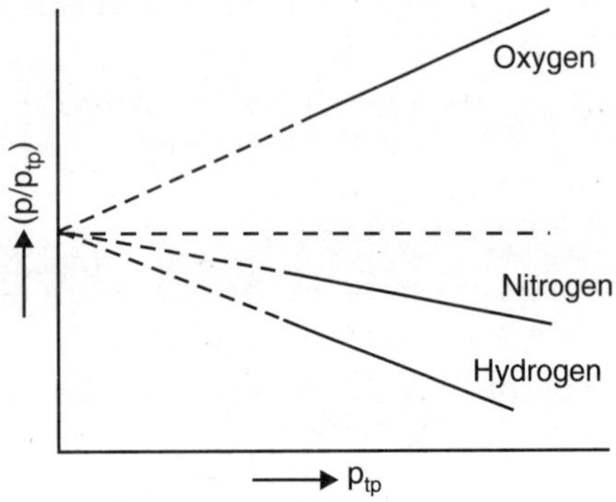

Fig. 2.3. (p/p_{tp}) vs p_{tp}

If the unknown temperature T in equation 2.12 corresponds to the steam point T_s then

$$\frac{T_s}{T_{tp}} = \underset{p_{tp}\to 0}{Lt}\left(p_s/p_{tp}\right)$$

Since there is a negligible temperature difference between the triple point and ice point, (*i.e.* 273.16 – 273.15 = 0.01°C), the triple point can be taken as ice point. Then

$$\frac{T_s}{T_i} = \underset{p_i\to 0}{Lt}\left(p_s/p_i\right)$$

When the values of (p_s/p_i) are plotted against p_i and the resulting line is extrapolated, we get;

$$\frac{p_s}{p_i} = 1.366 \text{ at } p_i = 0$$

Thus, $$\frac{T_s}{T_i} = \underset{p_i\to 0}{Lt}\left(p_s/p_i\right) = 1.366 \quad \text{...(2.13)}$$

The Kelvin scale has the same numerical interval as the Celsius scale.

$$T_s - T_i = 100 \quad \text{...(2.14)}$$

Equations 2.12 and 2.13 define Kelvin or absolute centigrade scale and are independent of the properties of the gas enclosed in the bulb.

In a constant pressure thermometer, the temperature – volume relationship will be of the form

$$\frac{T}{T_{tp}} = \underset{V_{tp}\to 0}{Lt}\left(V/V_{tp}\right) \quad \text{...(2.15)}$$

where V_{tp} is the volume of the gas at the triple point of water T_{tp} and V is the volume of the gas at the system temperature T.

The following are the practical difficulties associated with the use of ideal gas thermometer:

(*i*) The actual gases do not obey the gas laws exactly and, therefore, to obtain a physically reasonable temperature scale, the deviation of the gas from ideal gas behaviour is observed and corrections made thereof.

(*ii*) Corrections are needed to be applied for the change in volume of the bulb due to variations of temperature, internal pressure and capillary etc.

(*iii*) There is a limit on the lowest temperature that can be measured. This corresponds to normal boiling point of helium gas.

(*iv*) Greater skill and experience is required to operate an ideal gas thermometer. This is because several corrections have to be applied to obtain accurate values of temperature.

2.4.5. International temperature scale

This scale has been established and adopted to provide an experimental basis for the calibration of specific thermometers to indicate temperatures as close as possible to the Kelvin thermodynamic scale. The international temperature scale is based upon a number of fixed and reproducible equilibrium points to which numerical values are assigned (Table 2.1).

For interpolation between these primary fixed points and numerous other secondary fixed points, a thermometer is specified with a formula relating the thermometric property to the temperature. The method of evaluating constants in the correlation is also specified. For example :

(*a*) From – 182.97 to 0°C (oxygen point to ice point): A platinum resistance thermometer of standard design is used, and the temperature t is defined by a polynomial of the form,

Table 2.1. Fundamental fixed points and their numerical values at 1.01325 N/m²

	Fixed point	Temperature 0°C
(i)	Oxygen point (temperature of equilibrium between liquid oxygen and its vapour)	– 182.97
(ii)	Ice point (temperature of equilibrium between ice and saturated water)	0
(iii)	Steam point (temperature of equilibrium between liquid water and its vapour)	100
(iv)	Sulphur point (temperature of equilibrium between liquid sulphur and its vapour)	444.6
(v)	Silver point (temperature of equilibrium between solid and liquid silver)	960.8
(vi)	Gold point (temperature of equilibrium between solid and liquid gold)	1063.0

$$R_t = R_0[1 + \alpha t + \beta t^2 + \gamma(t - 100)t^3] \quad ...(2.16)$$

where R_t is the resistance at temperature t in °C, R_0 is the resistance at 0°C, and constants α, β, γ are determined from measured values of R_t at steam, sulphur and oxygen points respectively.

(*b*) From 0°C to 630.5°C (ice point to freezing point of antimony). It is also based on the platinum resistance thermometer, and the temperature t is defined by

$$R_t = R_0(1 + \alpha t + \beta t^2) \quad ...(2.17)$$

where constants α and β are as specified at (a) above.

(*c*) From 630.5°C to 1063.0°C (Freezing point of antimony to gold point). It is based on standard platinum versus platinum rhodium alloy thermocouple and the temperature t is defined by the three term equation

$$E = a + bt + ct^2 \quad ...(2.18)$$

where E is the emf of the thermocouple when the reference junction is at zero degree Celsius and the other junction is at a temperature t degree Celsius. The constants a, b and c are determined by calibration at the freezing point of antimony and the silver and gold points.

For measuring temperatures higher than the gold point, the intensity of radiation of any convenient wavelength is compared with the intensity of radiation of the same wavelength emitted by a perfectly black body at gold point. The temperature is then determined by applying Planck's law of thermal radiation.

The international temperature scale can be easily and rapidly used to calibrate scientific and industrial instruments used for temperature measurement.

EXAMPLE 2.1.

Two thermometers, one Centigrade and other Fahrenheit, are immersed in a fluid. After the thermometers reach equilibrium with the fluid, it is noted that both the thermometers indicate the same numerical value. Find that identical numerical value shown by the thermometers. What would be the corresponding temperature of the fluid expressed in degrees Kelvin and degrees Rankine ?

Solution: The Fahrenheit and Celsius (centigrade) temperature scales are related by

$$t_C = \frac{5}{9}(t_F - 32)$$

For $\quad t_C = t_F = x$

$$x = \frac{5}{9}(x - 32); \quad x = -40$$

$$\therefore\ t_C = -40°C \text{ and } t_F = -40°F$$

On absolute scales;

$$T_K = t_C + 273.15$$
$$= -40 + 273.15 = 233.15 \text{ K}$$

$T_R = t_F + 459.67$

$= -40 + 459.67 = 419.67°R$

Alternatively:

$t_C = T_K - 273.15$

and $t_F = T_R - 459.67$

Hence for $t_C = t_F$

$T_K - 273.15 = T_R - 459.67$ ...(*i*)

The absolute temperature in Kelvin scale and Rankine scale are related by

$T_R = 1.8\ T_K$ ...(*ii*)

From expressions (*i*) and (*ii*)

$T_K - 273.15 = 1.8\ T_K - 459.67$

or $T_K = \dfrac{459.67 - 273.15}{1.8 - 1}$

$= 233.15$ K

or $t_C = 233.15 - 273.15$

$= -40°C$

$T_R = 233.15 \times 1.8$

$= 419.67°R$

or $t_F = 419.67 - 459.67$

$= -40°F$

EXAMPLE 2.2.

A new temperature scale in degrees N is desired with freezing point at 100°N and the boiling point at 400°N. Establish a correlation between degrees Celsius and degrees N. What would be the absolute temperature at 0°N ?

Solution : Let the correlating expression be of the form

$$t = a\,x + b$$

where a and b are constants

Substituting the values of ice point and steam point in this correlation, we get

$$100 = a\,x_i + b \quad \text{and} \quad 400 = a\,x_s + b$$

Upon solution:

$$a = \frac{300}{x_s - x_i} \text{ and } b = 100 - \frac{300}{x_s - x_i} x_i$$

$$\therefore\ t_N = \frac{300}{x_s - x_i} x + 100 - \frac{300}{x_s - x_i} x_i$$

$$= \frac{x - x_i}{x_s - x_i} 300 + 100 \qquad ...(i)$$

$$\text{Also } t_C = \frac{x - x_i}{x_s - x_i} \times 100 \qquad ...(ii)$$

From expressions (*i*) and (*ii*) :

$$t_N = 3t_C + 100$$

When working in absolute units

$$\frac{T_s}{T_i} = \underset{p_i \to 0}{Lt} \left(\frac{p_s}{p_i}\right) = 1.366$$

$T_s - T_i = 400 - 100 = 300$

$\therefore\ 1.366\ T_i - T_i = 300$

or $T_i = \dfrac{300}{0.366} = 820°N$

and $T_s = 820 \times 1.366 = 1120°N$

Also $T_N = (t_N - 100) + 820 = t_N + 720$

At $t_N = 0$

$\therefore\ T_N = 0 + 720 = 720°N$

EXAMPLE 2.3.

The temperature t on a thermometric scale is defined in terms of a property P by the relation,

$t = a \log_e P + b$, where a and b are constants.

The temperatures of the ice point and steam point are assigned the numbers 32 and 212 respectively. Experiment gives values of P of 1.86 and 6.81 at the ice point and steam point respectively. Evaluate the temperature corresponding to a reading of P = 2.50 on the thermometer.

Solution : Substituting the values for ice point and steam point in the given correlation, we obtain

$$32 = a \log_e 1.86 + b$$

$$212 = a \log_e 6.81 + b$$

Upon solution: $a = 138.7$ and $b = -54.07$

Therefore, the given identity can be rewritten as

$$t = 138.7 \log_e P - 54.07$$

When the thermometer indicates $P = 2.50$,

$$t = 138.7 \log_e 2.50 - 54.07 = \mathbf{73°}$$

EXAMPLE 2.4.

The readings t_A and t_B of two thermometers A and B on the Celsius scale agree at the ice point and the steam point, and are assumed to be related by the expression

$$t_A = l + m\,t_B + n\,t_B^2$$

where l, m and n are constants. When both thermometers are immersed in a well stirred oil bath, A reads 51°C and B reads 50°C. Determine the reading on thermometer A when thermometer B reads 25°C. Which thermometer can be taken to be correct ?

Solution : At ice point : $t_A = t_B = 0°C$

$$\therefore \quad 0 = l + m \times 0 + n \times (0)^2; \quad l = 0$$

At steam point :

$$t_A = t_B = 100°C$$

$$\therefore \quad 100 = 0 + m \times 100 + n \times (100)^2$$
$$= 100\, m + (100)^2\, n$$

or $\quad m + 100\, n = 1 \qquad ...(i)$

Also when

$$t_A = 51° \text{ C},$$
$$t_B = 50° \text{ C and, therefore,}$$
$$51 = 0 + m \times 50 + n \times (50)^2$$
$$= 50\, m + (50)^2\, n \qquad ...(ii)$$

or $\quad m + 50\, n = \dfrac{51}{50}$

Solution of equations (*i*) and (*ii*) gives

$$m = 1.04 \quad \text{and} \quad n = -0.0004$$

The given identity may then be rewritten as

$$t_A = 1.04\, t_B - 0.0004\, t_B^2$$

When $t_B = 25°C$, we will have

$$t_A = 1.04 \times 25 - 0.0004 \times 25^2$$
$$= 25.75°C$$

Both the thermometers agree at ice and steam point; at any other point they read different. If A is presumed to be correct, B becomes faulty and vice-versa. From the data available, it is rather difficult to say which thermometer is correct. However, it can be predicted that

(*i*) one of the thermometers may be correct; or

(*ii*) both are faulty.

EXAMPLE 2.5.

Consider a particular Celsius scale assigned the value of 0°C′ to the steam point and 100°C′ to the ice point.

(a) Using an ideal gas as the thermometric medium, set up a relationship between 0°C′ and the pressure for a constant volume thermometer. Proceed to derive the correlation between °C and °C′. At what temperature are the °C and °C′ scales numerically equal.

(b) What is the numerical value of the absolute zero for the particular scale ? What is 200 K in °C′ ?

Solution : Let temperature be related to pressure by a linear law of the form

$t = a\,p + b$ where a and b are constants

Further let p_s and p_i be the pressures in the constant volume thermometer at steam point and ice point respectively. Then for the particular Celsius scale for which $t_s = 0°C'$ and $t_i = 100°C'$,

$$100 = a\, p_i + b \quad \text{and} \quad 0 = a\, p_s + b$$

That gives :

$$a = \frac{100}{p_i - p_s}$$

and
$$b = 100 - 100\frac{p_i}{p_i - p_s}$$

$$\therefore \quad t_{c'} = \frac{100}{p_i - p_s}p + 100 - 100\frac{p_i}{p_i - p_s}$$
$$= 100 - 100\,\frac{p - p_i}{p_s - p_i}$$

or
$$\frac{p - p_i}{p_s - p_i} = \frac{100 - t_{c'}}{100} \qquad ...(i)$$

Similarly substituting 0°C for t_i and 100°C for t_s on the normal Celsius scale, we would obtain

$$t_c = \frac{100}{p_s - p_i}(p - p_i)$$

or
$$\frac{p - p_i}{p_s - p_i} = \frac{t_c}{100} \qquad ...(ii)$$

From expressions (*i*) and (*ii*)

$$\frac{t_c}{100} = \frac{100 - t_{c'}}{100}$$

The two centigrade scales are related by $t_c = 100 - t_{c'}$

when $\quad t_c = t_{c'} = t$ (say)

$$t = 100 - t\,; \quad \mathbf{t = 50°}$$

At 50°, both the scales will have the same numerical value

(*b*) At absolute zero, $t_c = -273°C$

$$\therefore \quad -273 = 100 - t_{c'}; \quad \mathbf{t_{c'} = 373°}$$

At 200 K; $t_c = 200 - 273$

$$= -73°C \text{ and therefore,}$$
$$t_{c'} = 100 - t_c = 100 - (-73) = \mathbf{173°}$$

EXAMPLE 2.6.

It is proposed to devise a new scale with value 5 assigned to ice point and 20 to the steam point. The pressure of an ideal gas at constant volume is to be used as thermometric medium.

(a) Set up a linear relationship between the pressure and the temperature (in °N) on the new scale. What is the Kelvin absolute zero in this scale ?

(b) Derive a relationship between °N and °K.

(c) If the pressure at the steam point was 12 atm, what is the pressure at 0°N and what is temperature in °N when the pressure is 2 atm ?

Solution : Let temperature t be related to pressure p of the constant volume thermometer by a linear law of the form

$$t = a\,p + b$$

where a and b are constants.

Further, let p_s and p_i be the pressures in the constant volume thermometer at steam and ice points respectively. Then for the new scale for which t_s = 20°N and t_i = 5°N,

$$20 = a\,p_s + b \text{ and } 5 = a\,p_i + b$$

That gives : $a = \dfrac{15}{p_s - p_i}$

and $b = 5 - \dfrac{15}{p_s - p_i}\,p_i$

$$\therefore\ t_N = \frac{15}{p_s - p_i}\,p + 5 - \frac{15}{p_s - p_i}\,p_i$$

$$= 5 + 15\,\frac{p - p_i}{p_s - p_i}$$

or $$\frac{p - p_i}{p_s - p_i} = \frac{t_N - 5}{15} \quad ...(i)$$

Similarly substituting 0°C for t_i and 100°C for t_s on the Celsius scale, we would obtain

$$t_C = \frac{100}{p_s - p_i}(p - p_i)$$

or $$\frac{p - p_i}{p_s - p_i} = \frac{t_C}{100} \quad ...(ii)$$

From expressions (*i*) and (*ii*),

$$\frac{t_C}{100} = \frac{t_N - 5}{15}$$

or $$\frac{t_C}{20} = \frac{t_N - 5}{3}$$

At absolute zero, t_C = – 273°C

$$\frac{-273}{20} = \frac{t_N - 5}{3}$$

$$t_N = \frac{3}{20}(-273) + 5 = \mathbf{-35.95}$$

(*b*) Writing (T_K – 273) for t_C where subscript K stands for temperature on absolute scale, we get

$$\frac{T_K - 273}{20} = \frac{t_N - 5}{3}$$

$$\therefore\ T_K = 273 + \frac{20}{3}(t_N - 5)$$

$$= 273\left[1 + \frac{20}{273\times 3}t_N - \frac{20}{273\times 3}\times 5\right]$$

$$= 273\,[1 + 0.0244\,t_N - 0.122]$$

$$= 273\,(0.878 + 0.244\,t_N)$$

(*c*) For a constant volume thermometer, pressure is proportional to temperature being measured. That is $P = f\,T_K$. The proportionality constant f can be determined from the condition that when T_K = 373 K, P = 12 atm

$$\therefore\quad f = \frac{12}{373}\ \text{atm/K}$$

and $$P = \frac{12}{373} \times 273\,(0.878 + 0.0244\,t_N)$$

$$= 8.78\,(0.878 + 0.0244\,t_N)$$

When $t_N = 0$

$$P = 8.78 \times 0.878 = \mathbf{7.71\ atm}$$

When P = 2 atm

$$2 = 8.78\,(0.878 + 0.0244\,t_N)$$

$$= 7.71 + 0.2142\,t_N$$

or $$t_N = \frac{2 - 7.71}{0.2142} = -\,26.66°\text{N}$$

EXAMPLE 2.7.

How does a resistance thermometer measure temperature ?

A platinum resistance thermometer has a resistance of 150Ω at 100°C. When exposed to high temperature gas, the resistance increases to 300 Ω. Determine the temperature of hot gas. It may be assumed that at 0°C, the temperature coefficient of resistance for platinum is 0.0039 per degree celcius.

Solution : The electrical resistance of most metals increases with temperature and this forms the basis of temperature measurement by a resistance thermometer. An adequate approximation of temperature-resistance relationship is

$$R_t = R_o(1 + \alpha t + \beta t^2)$$

where R_t is the resistance at any temperature t °C, R_o is the resistance at 0°C, and α and β are constants depending on the material. The constants R_o, α and β are determined from measured values of resistance at the ice, steam and sulphur points respectively.

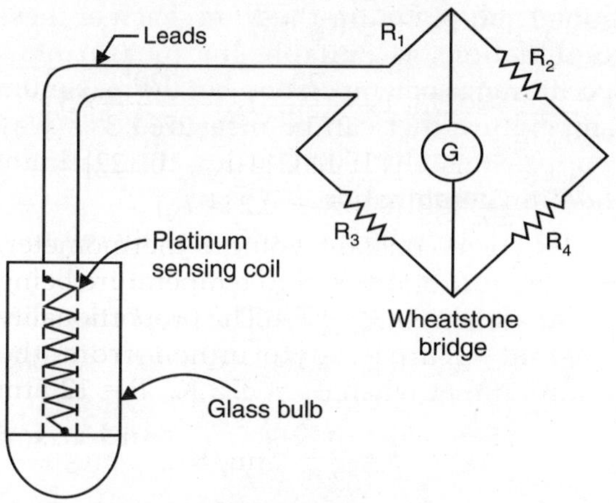

Fig. 2.4. Resistance thermometer

Over a limited temperature around 0°C, the following linear relationship is equally valid.

$$R = R_o\,(1 + \alpha\,\theta)$$

where α is the temperature coefficient of resistance per degree centigrade and θ is the temperature relative to 0°C.

If a change in temperature from θ_1 to θ_2 is considered, then

$$R_1 = R_o\,(1 + \alpha\,\theta_1)$$

and $R_2 = R_o(1 + \alpha\,\theta_2)$

Rearrangement gives

$$\theta_2 = \theta_1 + \frac{R_2 - R_1}{\alpha R_o} \quad ...(i)$$

The linear relation implies that changes in resistance are directly proportional to change in temperature.

The thermometer comprises a resistance element, suitable electrical leads and a resistance measuring device. The resistance element is usually in the form of a coil of very fine platinum, nickle or copper wound non-conductively onto a ceramic former and enclosed in a pyrex glass tube. Leads are taken out of the thermometer and the change in resistance is measured by Wheatstone bridge circuit.

(*b*) Inserting the appropriate values in relation (*i*) given above

$$\theta_2 = 100 + \frac{300 - 150}{0.0039 \times 100} = 484.61°C$$

Thus the hot gas has a temperature equal to 484.61°C.

EXAMPLE 2.8.

A resistance thermometer has been used to measure the temperature of the copper windings of a motor. When the motor is not running, the resistance of the motor is found to be 80 ohms at 30°C. Subsequently the motor is run at full load for some time and then switched off. Immediately thereafter the resistance is measured and found to increase to 95 ohms. Determine the temperature of the windings when the motor is running at full load. The variation of resistance with temperature conforms to the relation :

$$R = R_o\,(1 + 0.004t)$$

where R_o is the resistance at 0°C.

Solution : When the motor is not running

$$80 = R_o(1 + 0.004 \times 30)$$
$$= R_o(1 + 0.12)$$
$$= 1.12\,R_o \quad ...(i)$$

When the motor is stopped after running it for some time at full load

$$95 = R_o(1 + 0.004\,t) \quad ...(ii)$$

From expressions (*i*) and (*ii*)

$$\frac{95}{80} = \frac{R_o\,(1 + 0.004t)}{1.12\,R_o}$$

$$1 + 0.004t = \frac{95}{80} \times 1.12 = 1.33$$

∴ Temperature of the windings when the motor is running,

$$t = \frac{1.33 - 1}{0.004} = \mathbf{82.5°C}$$

EXAMPLE 2.9.

State the principle of working of a thermocouple in the act of the temperature measurement.

The following correlation conforms to temperature-emf relation for a thermocouple whose reference junction is maintained at 0°C

$$\in = 0.25\, t - 6 \times 10^{-4}\, t^2 \;\; mV$$

where t is the temperature of hot junction on a gas thermometer scale. Determine the temperature recorded by thermocouple at a location where the gas thermometer reads 80°C. The milli voltmeter used for measuring the emf produced has been calibrated at ice point and steam point.

Solution : A thermocouple consists of two dissimilar electrical conductors joined at two ends to form a closed loop. When a temperature difference exists between the junctions, an emf is set up. The size of emf depends upon the temperature difference and the materials involved. For a particular thermocouple with one end (cold junction) maintained at ice point, the emf produced is taken as an index of the temperature of the other end (hot junction).

The thermocouple is calibrated by measuring the thermal emf at various known temperatures, the reference (cold) junction being kept at 0°C. For most of the thermocouples, the results of such measurements can be represented by the following equation:

$$\in = At + \frac{1}{2}Bt^2 + \frac{1}{3}Ct^3$$

where t is the temperature in °C.

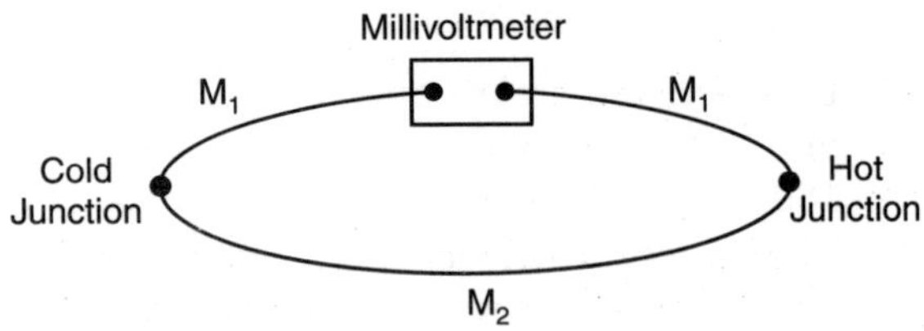

Fig. 2.5. Thermocouple

The most common metal combinations are iron-constanton, copper-constanton, chromium-alumel and platinum-rhodium. Each of these combinations is suitable for measuring a specific range of temperature but the maximum temperature that can be measured this way is approximately 1500°C using the platinum rhodium combination.

(*b*) At ice point; t = 0°C and therefore

$$\in = 0 \text{ mV}$$

At steam point; t = 100°C and therefore

$$\in = 0.25 \times 100 - 6 \times 10^{-4} \times (100)^2 = 19 \text{ mV}$$

At 80°C: $\in = 0.25 \times 80 - 6 \times 10^{-4} \times (80)^2 = 16.16$ mV

When the gas thermometer reads 80°C, the thermocouple will read

$$= \frac{100}{19} \times 16.16 \quad \text{or} \quad \mathbf{85.05°C}$$

SALIENT POINTS

1. Temperature of a system is that property which determines the ability of the system to transfer energy in the form of heat to any other system in thermal contact.
2. Two systems would be equal in temperature if no change occurs in any property when they are brought into communication.
3. Two systems which are in thermal equilibrium with a third system are in thermal equilibrium with each other. This is the essence of Zeroth law of thermo-dynamics.
4. The Zeroth law, like other thermodynamic laws, is based on experience and does not have any mathematical proof.
5. The Zeroth law provides the basis for temperature measurements.
6. Temperature scales are defined by assigning numbers to the ice point, to the steam point and to the equal spaced marks between them.
7. The ice point is the equilibrium temperature of ice with water at standard atmospheric pressure and it is assigned a value of 0°C on Celsius scale (also called centigrade scale) and 32°F on Fahrenheit scale.

   The steam point is the equilibrium temperature of pure water with its own vapour at standard atmospheric pressure and it is assigned a value of 100°C on

Celsius scale and 212°F on Fahrenheit scale.

Between the steam and ice points, there are 100 equal parts on Celsius scale and 180 equal parts on Fahrenheit scale.

8. The temperature measured on the Celsius and Fahrenheit scales are correlated by the expression

$$\frac{t_C}{100} = \frac{t_F - 32}{180} \quad \text{or} \quad t_C = \frac{5}{90}(t_F - 32)$$

9. The Kelvin temperature scale has been defined with reference to a single point—the triple point of water which is assigned a value of 0.01°C.

At the triple point of water, the ice, the liquid water and water vapour coexist in equilibrium state.

10. The ideal gas temperature scale does not depend on the nature of thermometric material.

11. The International Practical Temperature Scale (IPTS-68) was adopted in the year 1968 by the international committee on weights and measures, and is based on a number of easily reproducible fixed points together with the methods of interpolating within them.

REVIEW QUESTIONS

A. Conceptual and conventional questions.

1. State the concept of temperature and equality of temperature.
2. Define and explain the Zeroth law of thermodynamics. Why it is so called ?
3. Name and define the law that forms the basis for temperature measurement.
4. List any five physical properties of matter which can be used for the measurement of temperature.
5. What do you understand by thermometric property and thermometric substance ?
6. Establish a correlation between Centigrade and Fahrenheit temperature scales.
7. What is meant by absolute scale of temperature ? How is it established ?
8. How does Celsius temperature scale differ from absolute Kelvin scale ?
9. What is meant by fixed points of a thermometric scale ?
10. Point out the significance of international temperature scale.
11. State the principle of working of a gas thermometer. How a constant volume gas thermometer is used to establish the absolute temperature scale ? What corrections need to be applied for correct estimate of temperature by such thermometers ?
12. A Centigrade and a Fahrenheit thermometers are placed in a tank containing a fluid. After the thermometers reach thermal equilibrium with the fluid, it is observed that reading in the Fahrenheit thermometer is numerically twice that on the Centigrade thermometer. Calculate the actual temperature in degrees Kelvin.

[Ans. 433 K**]**

13. A constant volume gas thermometer containing helium gives a reading of gas pressure p of 1000 mm and 1360 mm of mercury at the ice point and steam point respectively. Express the gas thermometer centigrade temperature t_c in terms of gas pressure. Also determine the atmospheric temperature if the thermometer registers 1050 mm when left standing in the atmosphere.

[Ans. t = 0.278 p - 278; 13.9°C**]**

14. Given that the ice point is 0°C and 32°F and that the steam point is 100°C and 212°F, set up a correlation between the Celsius and Fahrenheit scales. Determine the temperature at which the Celsius and Fahrenheit scales have the same numerical value. Is there any temperature at which the Kelvin and Rankine scales have the same numerical value ?

[Ans. - 40°; - 0.75°**]**

15. An adequate approximation of temperature-resistance relationship applicable to a resistance thermometer is given by $R_t = R_o(1 + \alpha t)$ where R_t and R_o are values of resistance at temperature t °C and 0°C respectively, and α is the thermal coefficient of resistance. The thermometer is calibrated by immersion in boiling water (100°C) and boiling sulphur (445°C). The thermometer indicates the resistance values at these temperatures to be 14.5 ohms and 29 ohms respectively. Make

calculations for the temperature of a fluid for which the thermometer resistance is 23.5 ohms.

16. A certain thermometer is calibrated using ice and steam points designating them as being temperatures of 0° C and 100°C respectively. The thermodynamic function chosen to establish the scale is $t = a \log_e x + b$; instead of the linear scale $t = ax + b$. Determine the constants a and b in terms of x_s and x_i. Further proceed to show that the new scale is given by

$$t = 100 \frac{\log_e (x / x_i)}{\log_e (x_s / x_i)}$$

17. When the reference junction of a thermocouple is kept at the ice point and the test junction is the Celsius temperature t, the emf $\in$ of the thermocouple is given by the equation

$$\in = a t + b t^2$$

where a = 0.20 mV/deg and $b = -50 \times 10^{-4}$ mV/deg^2. Compute the emf when t = –100°C, 200°C, 400°C and 500°C and draw a graph of $\in$ against t in this range.

[Ans. –25 mV, 20 mV, 0 mV, –25mV**]**

18. The resistance of a platinum wire is found to be 11 ohms at the ice point, 15.247 ohms at the steam point and 28.887 ohms at the sulphur point (425°C). Find the constants A and B in the equation

$$R = R_0(1 + At + Bt^2)$$

and plot R against t in the range 0 to 600°C.

[Ans. $A = 0.01234$, $B = 2.3517 \times 10^{-5}$**]**

B. Fill in the blanks with appropriate word/words.

1. Temperature is an .... property independent of the size and mass of the system.
2. The .... law provides the basis for the measurement of temperature of a system.
3. Temperature measurement depends upon the establishment of .... between the system and the device used to measure the temperature.
4. Thermodynamically the condition of no molecular activity occurs at .... which defines the lowest temperature possible.
5. On Kelvin temperature scale, the triple point of water has been arbitrarily assigned a value of ....
6. Each division on both the .... and .... scale is equal.
7. The .... is based upon a number of fixed and reproducible equilibrium points to which numerical values are assigned.
8. The ideal gas temperature scale does not depend on nature of ....
9. Two blocks which are at different states are brought into contact with each other and allowed to reach a final state of thermal equilibrium. The final temperature is specified by the ....

Answers

1. intensive 2. third 3. thermodynamic equilibrium 4. absolute zero 5. 273.16 6. Kelvin and Celcius 7. International temperature scale 8. thermometric material 9. third law

C. Multiple choice questions.

1. Which statement corresponds to Zeroth law of thermodynamics ?
   (*a*) work can be completely converted into heat
   (*b*) two systems in thermal equilibrium with a third system are in thermal equilibrium with each other
   (*c*) there can not be transfer of heat from a system at low temperature to another system at high temperature
   (*d*) it is practically impossible to attain a zero degree absolute temperature
2. Zeroth law of thermodynamics states that :
   (*a*) two thermodynamic systems are always in thermal equilibrium with each other
   (*b*) if two systems are in thermal equilibrium, then the third system will also be in thermal equilibrium.
   (*c*) two systems not in thermal equilibrium with a third system are also not in thermal equilibrium with each other.
   (*d*) When two systems are in thermal equilibrium with a third system, they are in thermal equiulibrium with each other
3. Zeroth law of thermodynamics forms the basis of .... measurement
   (*a*) pressure (*b*) temperature
   (*c*) heat exchange (*d*) work

Answers

1. (*b*) 2. (*d*) 3. (*b*)

Chapter 3

Transient Energies: Work and Heat

Learning objectives : A study of the subject matter included in this chapter will enable the reader to have an understanding of :

- *thermodynamic perspective of work done*
- *displacement work in a quasi-static process*
- *works other than mechanical work*
- *heat interaction between system and surroundings, specific heat and heat capacity*
- *comparison between work and heat interactions*

Thermal science is essentially concerned with energy in the form of mechanical work and heat, it deals with their transfer across the system boundaries and their mutual interconversions. These energies manifest at the system boundary only during any interaction between the system and surroundings, and lose their identity as soon as they enter or leave a system. Attention has been focussed in this chapter on the transient concept of work and heat interactions, and the associated property changes of the system.

3.1. WORK AS DEFINED IN MECHANICS

Work is done when the point of application of a force moves in the direction of force. The amount of work done equals the product of force and the resultant displacement in the direction of force.

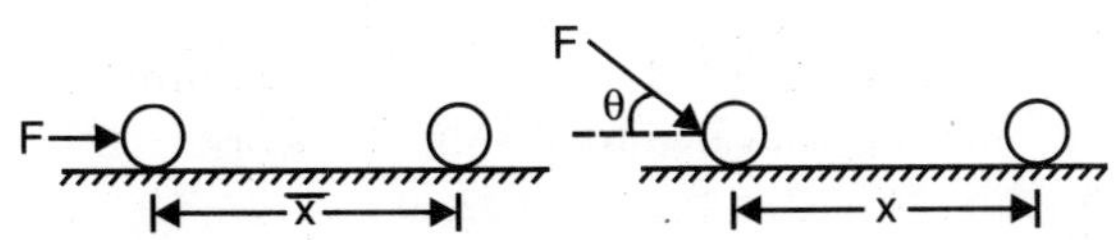

Fig. 3.1.

With reference to Fig 3.1, F is the force acting on a body and the body moves a distance x in the direction of force. Then the work done on the body is

Work done = force × distance ; $W = Fx$

Most of the time, the force and the displacement are not collinear. Work done by the force is then expressed as the component of force in the direction of motion and the distance moved. With reference to Fig. 3.1(b), consider the force F to be inclined at an angle θ with the direction of motion. Then

Component of force in the direction of motion = $F \cos \theta$

Work done = $F \cos \theta \times x$

When $\theta = 90°$, $\cos \theta = 0$ and accordingly no work will be done. Hence if the force and displacement are at right angles, then the work done will be zero.

In case the force changes with displacement, then the work can be obtained by the expression

$$W = \int_1^2 F(x)\,dx$$

3.2. WORK AS DEFINED IN THERMODYNAMICS

The thermodynamic definition of work relates to the concept of system and its surroundings.

(*i*) Work refers to the energy transferred (without transfer of mass and with the help of a mechanism) because of a property difference, other than temperature, that exists between the system and its surroundings.

Work is the energy transfer by reasons other than temperature difference.

(*ii*) Work refers to the interaction between system and surroundings whose sole effect, external to the system, could be reduced to the lifting of mass against the gravitational force.

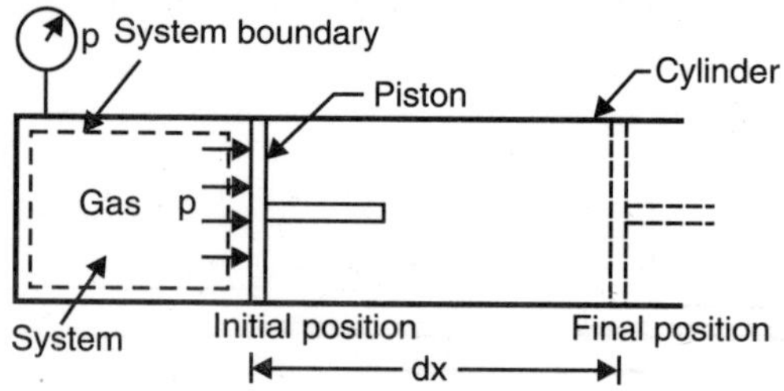

Fig. 3.2. Piston cylinder arrangement

Consider a system formed by a certain amount of gas contained in a piston cylinder arrangement. Let the piston move outward through a small distance dx ($dx \to 0$) during a small interval dt. Since the piston moves through a small distance, the pressure p acting on the face of the piston can be assumed to remain constant. A single pressure gauge will then indicate the system pressure p, which also prevails over the entire system boundary. The infinitesimal work done by the system is

$$\delta W = \text{force} \times \text{distance moved}$$
$$= (\text{pressure} \times \text{area of piston}) \times \text{distance moved}$$
$$= p\,A\,dx \qquad \text{...(3.1)}$$

It is to be noted that $A\,dx$ represents the infinitesimal volume dV swept by the piston.

$$\therefore \quad \delta W = p \times (A\,dx) = p\,dV \qquad \text{...(3.2)}$$

If the piston moves through a finite distance, the work done by the piston can be evaluated by integrating equation 3.2 between the initial and final states.

Reference Fig. 3.3. the expansion of the gas in the piston cylinder assembly can be reduced to the raising of a mass m against gravity by manipulating surroundings external to the system.

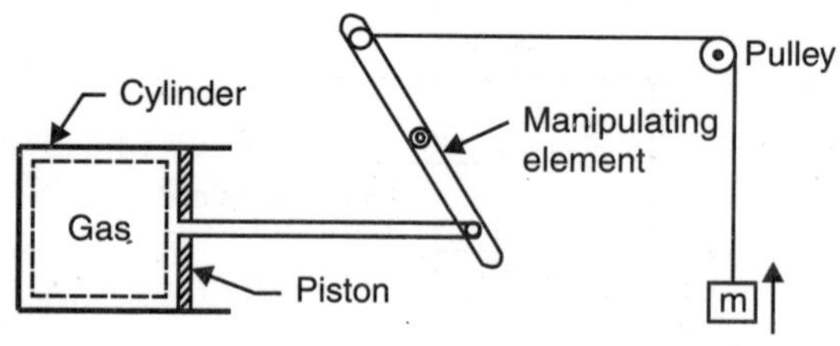

Fig. 3.3. Work done equivalent to raising of mass

Consider an electrical storage battery (system) whose terminals are connected to an electrical resistance coil through a switch. The circuit external to the battery constitutes the surroundings. When the switch is closed, the current flows through the resistance coil, and the resistance (surroundings) become warmer and the charge of the battery (system) decreases. Obviously there has been interaction between the system and the surroundings. According to mechanics, this interaction cannot be classified as work because there has been no action of force through a distance or of a torque through an angle.

However, as per thermodynamic concept, the battery (system) does work as the electrical energy crosses the system boundary because of potential gradient (not temperature). Further the external resistance can be replaced by an ideal frictionless motor - pulley arrangement which can wind a string and thereby raise the suspended mass. The sole effect, external to the system, is the raising of a mass. As such interaction of battery with resistance coil is a work.

3.3. DISPLACEMENT WORK : p dV

Refer to equation 3.2 resulting from expansion of gas in a piston - cylinder assembly as discussed above in section 3.2. The expansion of the system (gas) by outward movement of the piston can be represented on p-V diagram like 1-2 (Fig. 3.5.). The shaded area represents the work done due to small movement of the piston and the total work done by the gas during expansion process is

$$W = \int_1^2 p\,dV \qquad \text{...(3.3)}$$

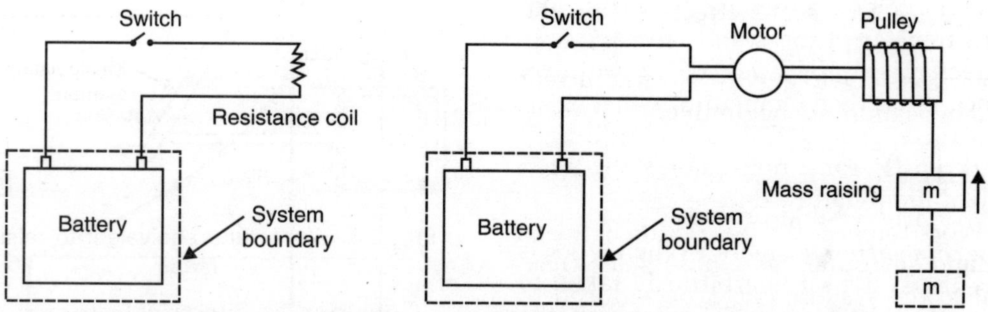

Fig. 3.4. Equivalence of current and work interaction between the system and surroundings

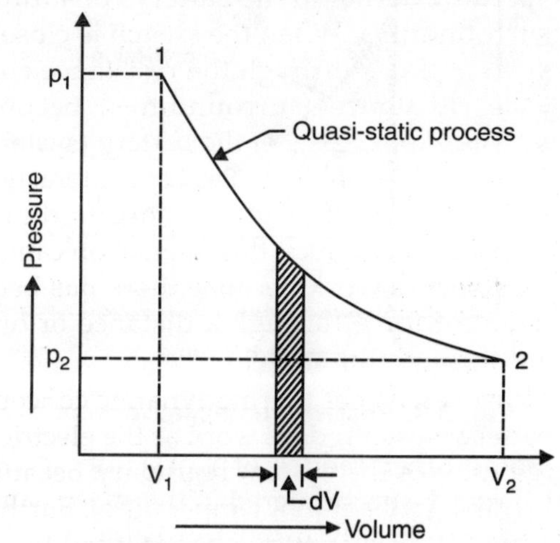

Fig. 3.5. Quasi-static p dV work

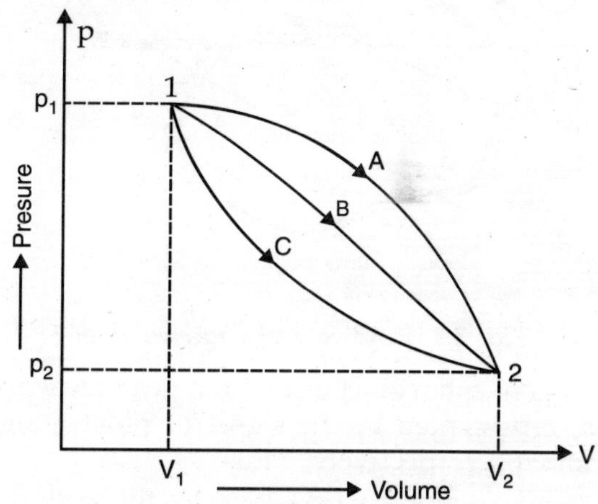

Fig. 3.6. Work-a path function

Inspection of the p-V plot (Fig 3.5) shows that mere specifying the end states 1 and 2 does not determine the area (or work); the nature of the curve needs to be known also. The curve may be arched upwards or it may sag downwards, and the area under the curve will vary accordingly. For the same initial and final states, the work done by the system in following the paths A, B and C is different. Obviously work is a path function and not a point function. Accordingly the work transfer across the system boundaries is not classified as a thermodynamic property.

The integral $\int_1^2 p\,dV$ can be evaluated, and hence computations of work done made, only if we know the relation between p and V or the path of the process.

Salient aspects of work:

• Work is energy in transit and can be identified only when the process is being executed. Work does not exist either before or after the interaction.

• Work is boundary phenomenon; it is an interaction across the boundary. It will be wrong to say that work is possessed or stored by the system. It is the transferred energy that can be stored indefinitely by the system and not work.

• Work is something which happens to a system but it is not the characteristic of the system.

Equation 3.3 for work done is valid only if

– the system is closed.

– the process is quasi-static, *i.e*, frictionless and reversible.

– there are no viscous effects within the system.

– the effects due to gravity, electric current and magnetisation are negligible.

$W \neq \int p\,dV$ for a non-quasi-static and irreversible process.

Sign convention for work : When work is done by the system it is arbitrarily taken to be positive, and when work is done on the system it is taken to be negative.

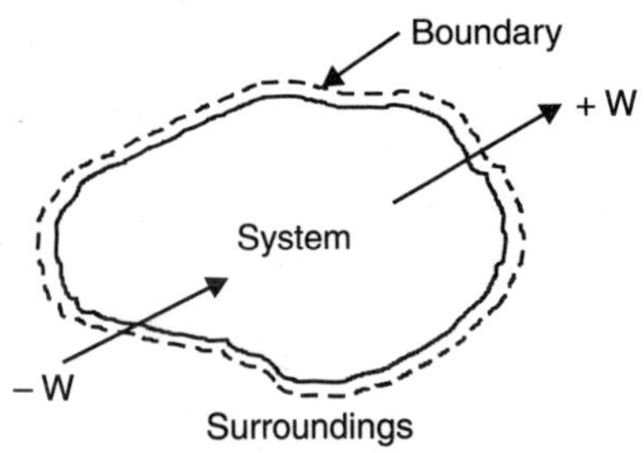

Fig. 3.7. Positive and negative work

Conventionally positive and negative work are represented by the signs for plus(+) and minus(–) respectively. Thus

+ W denotes work done by the system

– W denotes work done on the system.

When a system does positive work, its surroundings do an equal amount of negative work. Conversely when the surroundings do positive work, the system does an equal amount of negative work. Thus in any process

$$W_{system} + W_{surroundings} = 0 \qquad ...(3.4)$$

The numerical value of area under p-V curve is not affected whether the gas expands from state 1 (high pressure) to state 2 (low pressure) or is being compressed from state 2 to state 1. The signs of work done will, however, be different. It can be recognised by adopting the convention that

(a) Positive work occurs if one proceeds clockwise round the closed area being measured.

(b) Work is negative if one proceeds in anticlockwise direction.

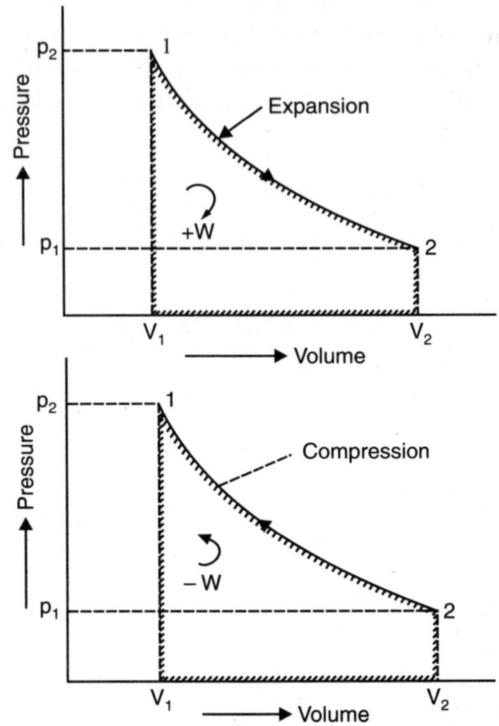

Fig. 3.8. Positive and negative work

Some other modes of work interactions which may be encountered in thermodynamic analysis are briefly discussed below:

• ***Stretching of a wire:*** When a wire of cross-sectional area A is subjected to an external force F, its length changes from l to $(l + dl)$ and the infinitesimal amount of work done is

$$\delta W = -\,F\,dl$$

The -ve sign stems from the fact that work must be done on the system (wire) to bring about a change in its length.

From mechanics of solids,

Stress induced in the wire $\sigma = \dfrac{F}{A}$

strain for a differential change in length

$$d\epsilon = \frac{dl}{l}$$

Then $\quad \delta W = -\,(\sigma A) \times l\,d\epsilon$

and $\quad W = -\,Al \int \sigma\,d\epsilon$

To integrate the above relations, a knowledge of the relationship between stress σ and strain ∈ is necessary. If extension of the wire is limited to within the elastic limits, we have

$$\text{Modulus of elasticity} = \frac{\text{stress}}{\text{strain}}; \quad E = \frac{\sigma}{\in}$$

Therefore for a finite change of length (or strain)

$$W = -AlE\int \in d\in$$

$$= -\frac{1}{2}AEl\,(\in_2^2 - \in_1^2) \quad \text{...(3.5)}$$

Equation 3.5 is essentially valid when the system is perfectly homogeneous and the extension or compression process is reversible.

• ***Surface film.*** A film on the surface of a liquid is acted upon by the force of surface tension which tends to make the surface area of the liquid minimum. Accordingly the free liquid surface behaves like a stretched membrane.

With reference to Fig. 3.9, one side of the liquid film held between a wire frame is moved with the help of a sliding wire.

Force applied on the liquid film,

$$F = 2l\sigma$$

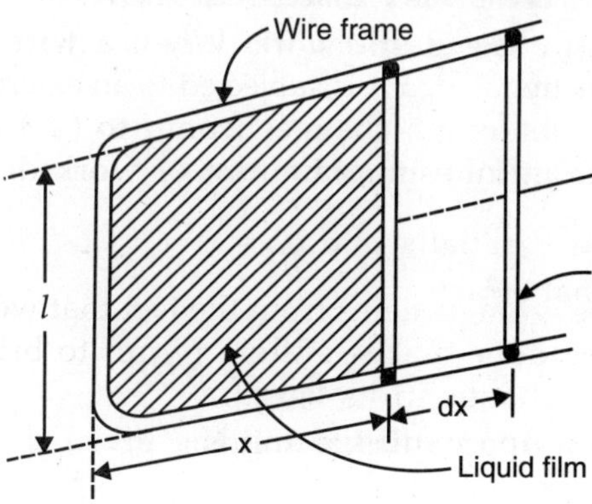

Fig. 3.9.

The factor 2 stems from the fact that two film surfaces act on the wire. Then work done,

$$W = -\int_{x_1}^{x_2} F\,dx = -\int_{x_1}^{x_2} 2l\sigma\,dx$$

The factor $2l\,dx$ represents a change in the surface area dA of the liquid film. Then

$$W = -\int_{x_1}^{x_2} 2l\sigma\,dx$$

$$= -\int_{A_1}^{A_2} \sigma\,dA = -\sigma(A_2 - A_1) \quad \text{...(3.6)}$$

• ***Electrical work.*** Consider a resistance coil (system) through which current I flows; this current flow is given by $I = dC/d\tau$ where dC is the charge in coulombs that crosses the system boundary in time $d\tau$.

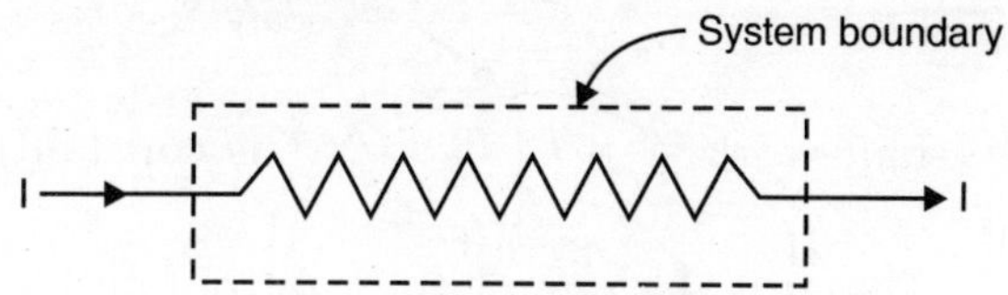

Fig. 3.10.

Electrical work done by flow of current through electrolytic cell/conductor/ resistor etc. is given by

$$\delta W = -E\,dC = -EI\,d\tau$$

$$\text{or} \quad W = -\int_1^2 EI\,d\tau \quad \text{...(3.7)}$$

where E is the potential difference.

• ***Magnetic work.*** Work done in the magnetisation of a paramagnetic substance is given by

$$W = -\int_1^2 \mu_o\,H\,d(V\,dM) \quad \text{...(3.8)}$$

where H is the strength of the magnetic field, M is the magnetisation per unit volume, V is the volume and μ_o is the permeability of free space.

• ***Shaft work.*** Consider a shaft of radius r fixed at one end and with a tangential force F applied at the free end.

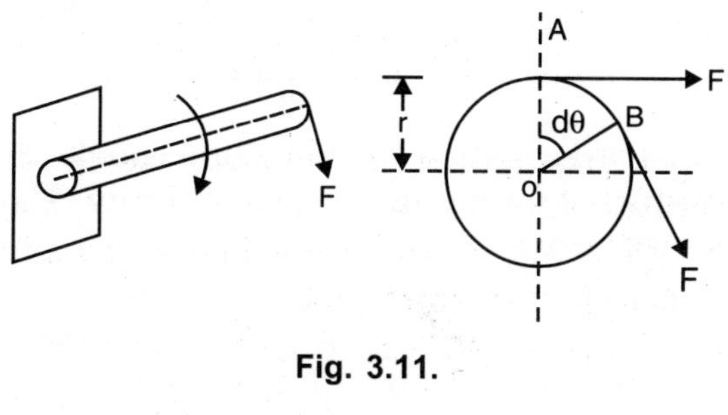

Fig. 3.11.

The shaft is then subjected to torque T which equals the product of force and the shaft radius. That is

Torque $T = F \times r$

If F is in newtons (N) and r is in metres (m), then the unit of torque will be Nm. As a result of this torque, the shaft undergoes an angular displacement $d\theta$. Then

linear displacement (length of arc AB)

$$= r\,d\theta$$

$$\text{work done } \delta W = \text{force} \times \text{distance moved} = F \times r\,d\theta = T d\theta$$

$$\therefore \text{ shaft work } W = T\int_1^2 d\theta \quad (\because T \text{ is constant})$$

Power is the rate of work output and as such

$$\text{shaft power} = T\int_1^2 \frac{d\theta}{d\tau} = T\omega \qquad \text{...(3.9)}$$

where $\omega = d\theta/d\tau$ is the angular velocity (in radians) of the shaft. If the shaft turns N revolutions per minute, then

$$\omega = \frac{2\pi N}{60} \text{ radians/sec}$$

Engineering importance of shaft work lies in the fact that most power plants deliver their output in the form of shaft rotation.

EXAMPLE 3.1

A paddle wheel used for mixing and stirring of fluids turns 600 revolutions per minute when 2.5 Nm torque is applied to it. Make calculation for the power transmitted to a liquid through the wheel.

Solution : Angular velocity

$$\omega = \frac{2\pi N}{60} = \frac{2\pi \times 600}{60} = 62.8 \text{ rad/s}$$

Power transmitted by the shaft equals the product of torque and angular velocity. That is

$$P = T\omega$$
$$= 2.5 \times 62.8 = 150 \text{ Nm/s} = \mathbf{150 \text{ W}}$$

EXAMPLE 3.2

An electric potential difference of 220 V is maintained across a 450 ohm resistor for a period of 8 minutes. Estimate the electrical work done and power.

Solution : From ohm's law,

$$\text{Current } I = \frac{E}{R} = \frac{220}{450} = 0.489 \text{ amperes}$$

Work done

$$W = E\,I\,\tau = 220 \times 0.489 \times (8 \times 60)$$
$$= 51638.4 \text{ J} \simeq \mathbf{51.64 \text{ kJ}}$$

This is the work done on the system and as such is negative.

Power is the rate at which the work is supplied. That is

$$\text{Power} = \frac{W}{\tau}$$
$$= EI = 220 \times 0.489 = \mathbf{107.58 \text{ W}}$$

EXAMPLE 3.3

A steel rod of 10 mm in diameter and 500 mm in length extends by 10 mm when a tensile force is applied to it in a direction parallel to the length of the rod. Taking Young's modulus of steel as 200 GN/m^2, calculate stress and force applied on the rod and the work done by the force.

Solution : The elastic work done by the rod is given by

$$W = -\frac{1}{2} A E l\,(\epsilon_2^2 - \epsilon_1^2)$$

Now : initial strain $\epsilon_1 = 0$

final strain

$$\epsilon_2 = \frac{dl}{l} = \frac{10}{500} = 0.02$$

Then upon substituting the given data, we obtain

$$W = -\frac{1}{2} \times \left[\frac{\pi}{4}(0.01)^2\right] \times (200 \times 10^9) \times 0.5 \times 0.02^2$$
$$= -1570 \text{ Nm}$$
$$= -1570 \text{ J} = \mathbf{-1.57 \text{ kJ}}$$

The negative sign is indicative of the fact that work is done on the rod when it elongates

Stress $\sigma = E \times \in$

$= (200 \times 10^9) \times 0.02$

$= \mathbf{4 \times 10^9\ N/m^2}$

Force $F = \sigma \times A$

$= (4 \times 10^9) \times \frac{\pi}{4}(0.01)^2$

$= \mathbf{0.314 \times 10^6\ N}$

EXAMPLE 3.4

A gas undergoes a reversible non-flow process according to the relation p = (– 3V + 15) where V is the volume in m³ and p is the pressure in bar. Determine the work done when the volume changes from 3 to 6 m³.

Solution : $p = (-3V + 15)$ bar

$= (-3V + 15) \times 10^5\ N/m^2$

Work done

$$W_{1\text{-}2} = \int p\,dV$$

$$= \int_{V_1}^{V_2} (-3V + 15) \times 10^5\,dV$$

$$= \left| -3\frac{V^2}{2} + 15V \right|_3^6 \times 10^5$$

$$= \left[\frac{-3}{2}(6^2 - 3^2) + 15(6-3) \right] \times 10^5$$

$$= (-40.5 + 45) \times 10^5$$

$$= \mathbf{40.5 \times 10^5\ Nm\ (or\ J)}$$

EXAMPLE 3.5

A non-flow reversible process occurs for which pressure and volume are correlated by the expression p = (V² + 6/V) where p is in bar and V is in m³. What amount of work will be done when volume changes from 2 to 4 m³?

Solution : $p = \left(V^2 + \frac{6}{V}\right)$ bar

$= \left(V^2 + \frac{6}{V}\right) \times 10^5\ N/m^2$

Work done, $W_{1\text{-}2} = \int p\,dV$

$$= \int_{V_1}^{V_2} \left(V^2 + \frac{6}{V}\right) \times 10^5\,dV$$

$$= 10^5 \left[\frac{V^3}{3} + 6\log_e V \right]_2^4$$

$$= 10^5 \left[\frac{4^3 - 2^3}{3} + 6\log_e \frac{4}{2} \right]$$

$$= (18.67 + 4.16) \times 10^5$$

$$= \mathbf{22.83 \times 10^5\ Nm}\ \text{(or J)}$$

EXAMPLE 3.6

Consider a gas enclosed in a piston-cylinder arrangement. The gas is initially at 150 kPa and occupies a volume of 0.03 m³. The gas is now heated until the volume of the gas increases to 0.1 m³. Calculate the work done by the gas if volume of the gas is inversely proportional to the pressure.

Solution : The path followed by the system is $V \propto \frac{1}{p}$. That is

$$pV = \text{constant } C = p_1V_1 = p_2V_2$$

or $p = \frac{C}{V}$

Work done,

$$W_{1\text{-}2} = \int p\,dV = \int_{V1}^{V2} \frac{C}{V}\,dV$$

$$= C\log_e \frac{V_2}{V_1} = p_1V_1 \log_e \frac{V_2}{V_1}$$

$$= 150 \times 10^3 \times 0.03 \log_e \frac{0.1}{0.03}$$

$$= 5.41 \times 10^3\ \text{Nm (or J)} = \mathbf{5.41\ kJ}$$

EXAMPLE 3.7

The pressure-volume correlation for a non-flow reversible (quasi-static) process is given by

p = (8 – 4V) bar, where V is in m³

If 150 kJ of work is supplied to the system, determine the final pressure and volume of the system. Take initial volume = 0.6 m³

Solution : $p = (8 - 4V)$ bar

$= (8 - 4V) \times 10^5\ N/m^2$

Work done,

$$W_{1\text{-}2} = \int p\,dV = \int_{V_1}^{V_2} (8 - 4V) \times 10^5\,dV$$

$$= 10^5 \times \left| 8V - 2V^2 \right|_{0.6}^{V_2}$$

$$= 10^5 \left[8(V_2 - 0.6) - 2(V_2^2 - 0.6^2) \right]$$

$$= -\,10^5\,(2V_2^2 - 8V_2 + 4.08)\ \text{J}$$

Since work has been supplied to the system, it is negative and, therefore,

$$-\,150 \times 10^3 = -\,10^5\,(2V_2^2 - 8V_2 + 4.08)$$

or $2V_2^2 - 8V_2 + 2.58 = 0$

Solution of this quadratic equation gives

Final volume, V_2 = **0.707 m³**

Final pressure,

$$p_2 = 8 - 4V_2$$
$$= 8 - 4 \times 0.707 = \mathbf{5.172\ bar}$$

EXAMPLE 3.8

A spherical balloon of 1m diameter contains a gas at 200 kPa and 300 K. The gas inside the balloon is heated until the pressure reaches 500 kPa. During the process of heating, the pressure of gas inside the balloon is proportional to the diameter of the balloon. Make calculations for the work done by the gas inside the balloon.

Solution : p_1 = 200 kPa,

D_1 = 1 m and T = 300 K

$p \propto D$ or $p = C\,D$

where C is a constant

$200 = C \times 1\,;\ C = 200$ kPa/m

Now, $p_2 = CD_2$

$500 = 200\,D_2$ or $D_2 = 2.5$ m

$$V = \frac{\pi}{6} D^3 \quad \text{or} \quad dV = \frac{\pi}{2} D^2\,dD$$

Work done,

$$W_{1\text{-}2} = \int_1^2 p\,dV = \int_1^2 CD \times \frac{\pi}{2} D^2\,dD$$

$$= \frac{\pi C}{8}\left(D_2^4 - D_1^4\right)$$

$$= \frac{\pi \times 200}{8}\left(2.5^4 - 1^4\right) \simeq \mathbf{2988\ kJ}$$

EXAMPLE 3.9

A fluid at a pressure of 3 bar and with specific volume of 0.18 m³/kg is contained in a cylinder behind a piston. The fluid expands reversibly to a pressure of 0.6 bar according to the law $p = \dfrac{C}{v^2}$ where C is a constant. Calcualte the work done by the fluid on the piston.

Solution : Refer Fig. 3.12 for the pressure-specific volume diagram for the fluid undergoing expansion.

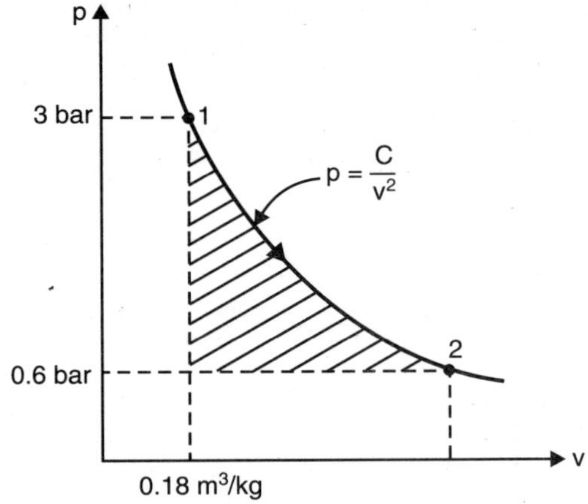

Fig. 3.12.

The work done by the fluid (system) on the piston (surroundings) equals the shaded area and is given by

$$W = \int_1^2 p\,dv = C\int_1^2 \frac{dv}{v^2}$$

$$= C\left| -\frac{1}{v} \right|_1^2 = C\left(\frac{1}{v_1} - \frac{1}{v_2} \right) \quad ...(i)$$

Given : $C = p\,v^2 = p_1\,v_1^2 = 3 \times (0.18)^2$
$= 0.0972$ bar (m³/kg)²

and $v_2 = \sqrt{\dfrac{C}{p_2}} = \sqrt{\dfrac{0.0972}{0.6}} = 0.402$ m³

Upon substituion of values of C, v_1 and v_2 in expression (*i*), we get

work done by the fluid,

$$W = 0.0972 \times 10^5 \left[\frac{1}{0.18} - \frac{1}{0.402} \right]$$

$= 29840$ Nm/kg = **29.84** kJ/kg

EXAMPLE 3.10

One kg of a certain fluid is contained in a piston-cylinder assembly at a pressure of 10 bar. The fluid is allowed to expand reversibly behind the piston until the volume becomes two-times of its initial value. During the expansion process, the relation between pressure and volume is given by pv^2 = constant. The fluid is then cooled reversibly at constant pressure until the piston regains its original position. Finally, the fluid is supplied heat reversibly with the piston firmly locked in position and the fluid pressure rises to initial value of 10 bar. If the fluid has an initial volume of 0.05 m³, make calculations for the net work done by the fluid.

Solution : Refer Fig. 3.13 for the pressure-volume diagram for the three different processes undergone by the fluid. The net work done by the fluid is given by the hatched area and it equals

$$W_{net} = W_{1-2} + W_{2-3} + W_{3-1}$$

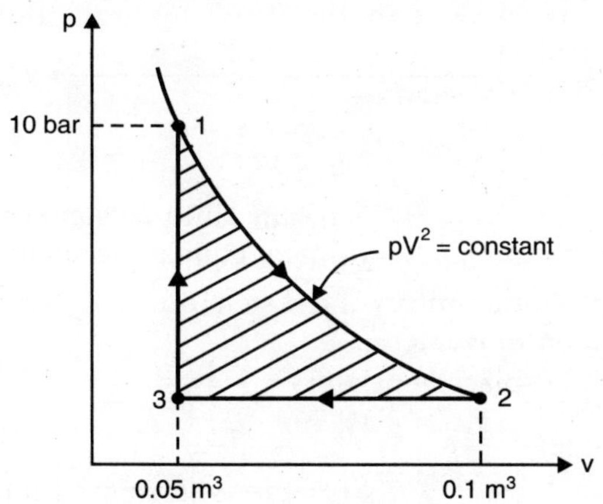

Fig. 3.13.

(*i*) $$W_{1-2} = \int_1^2 p\,dv = C\int_1^2 \frac{dv}{v^2}$$

$$= C\left|-\frac{1}{v}\right|_1^2 = C\left(\frac{1}{v_1} - \frac{1}{v_2}\right)$$

Given : $C = p\,v^2 = p_1\,v_1^2$
$= 10 \times (0.05)^2$
$= 0.025$ bar m⁶

$$\therefore \quad W_{1-2} = 0.025 \times 10^5 \left(\frac{1}{0.05} - \frac{1}{0.1}\right)$$
$$= 25000 \text{ Nm}$$

(*ii*) During process 2 – 3, the pressure remains constant and the work done is given by

$$W_{2-3} = p_2(V_3 - V_2)$$

From the relation $p\,V^2$ = constant

$$p_2 = p_1\left(\frac{V_1}{V_2}\right)^2$$

$$= 10 \times \left(\frac{0.05}{0.1}\right)^2 = 2.5 \text{ bar}$$

$$\therefore \quad W_{2-3} = 2.5 \times 10^5\,(0.05 - 0.1)$$
$$= -\,12500 \text{ Nm}$$

The negative sign implies that during the constant pressure cooling process, the work is done on the fluid.

(*iii*) Since the piston is locked in position during the process 3 – 1 of heat supply to the fluid,

$$W_{3-1} = 0$$

∴ Net work done by the fluid is

$$W_{net} = 25000 - 12500 + 0$$
$$= 12500 \text{ Nm} = \mathbf{12.5\ kJ}$$

EXAMPLE 3.11

A cylinder with a frictionless piston contains 0.1 m³ of gas at 200 kPa pressure. The piston is connected to a coil spring which exerts a force proportional to the displacement from its equilibrium position. The gas is heated until the volume is doubled at which pressure is 500 kPa. Determine the work done by the gas.

Take atmospheric pressure equal to 100 kPa.

Solution : Force balance at any position of the pistion gives :

$$pA = p_a\,A + K\,x$$

where displacement x = volume/area

$$= p_a\,A + \frac{KV}{A}$$

$$\therefore \quad p = p_a + \frac{K}{A^2}\,V$$

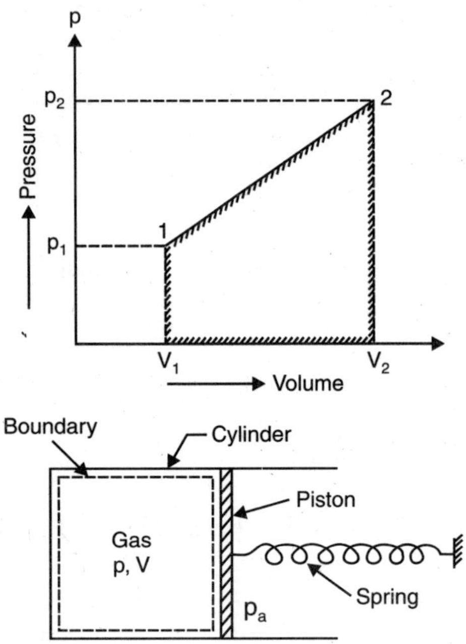

Fig. 3.14.

Work done,

$$W = \int p\, dV = \int_1^2 \left\{ p_a + \frac{K}{A^2} V \right\} dV$$

$$W = p_a (V_2 - V_1) + \frac{K}{2A^2} (V_2^2 - V_1^2)$$

$$W = p_a (V_2 - V_1) + \frac{K}{2A^2} [(V_2 - V_1)(V_2 + V_1)]$$

$$W = (V_2 - V_1) \left[p_a + \frac{K}{2A^2} (V_2 + V_1) \right]$$

$$= (V_2 - V_1) \left[p_a + \frac{p_2 - p_a}{2} + \frac{p_1 - p_a}{2} \right]$$

$$= (V_2 - V_1) \left[\frac{p_1 + p_2}{2} \right]$$

which equals the area under the line 1-2

Given: $p_1 = 200$ kPa, $p_2 = 500$ kPa,

$V_1 = 0.1 \text{m}^3$; $V_2 = 2\ V_1 = 0.2\ \text{m}^3$

$\therefore$ Work done by the gas

$$= (0.2 - 0.1) \left[\frac{200 + 500}{2} \right]$$

$$= \mathbf{35\ kJ}$$

EXAMPLE 3.12

A fluid contained in a horizontal cylinder fitted with a frictionless leak proof piston is continuously agitated by a stirrer passing through the cylinder cover. The diameter of the cylinder is 50 cm and the piston is held against the fluid due to atmospheric pressure equal to 100 kPa. The stirrer turns 8000 revolutions with an average torque of 1.5 Nm. If the piston slowly moves outwards by 60 cm, determine the net work transfer to the system.

Solution :

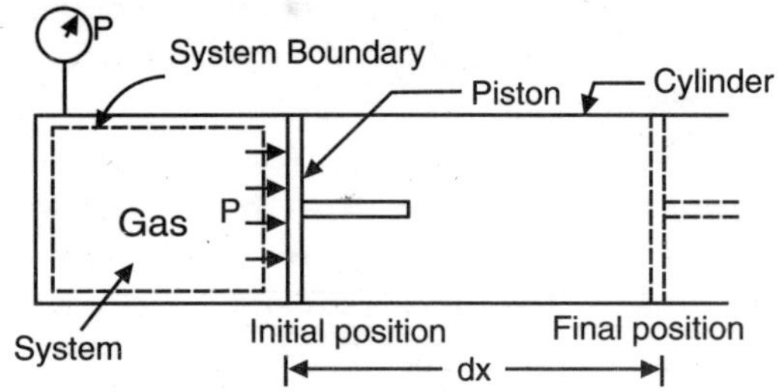

Fig. 3.15.

Work done by the stirrer upon the fluid system,

Paddle work,

$$W_p = T\omega = T \times 2\pi N$$

$$= 1.5 \times 2\pi \times 8000$$

$$= 75360 \text{ Nm (or J)} = 75.36 \text{ kJ}$$

This is a negative work for the fluid system.

Work done by the system due to outward piston movement,

Displacement work,

$$W_d = p\, dV = p\, A l$$

$$= 100 \times 10^5 \times \left(\frac{\pi}{4} (0.5)^2 \times 0.6 \right)$$

$$= 11.775 \text{ Nm (or J)} = 11.775 \text{ kJ}$$

$\therefore$ Net work transfer to the system,

$$W_n = W_p + W_d$$

$$= -75.36 + 11.775 = \mathbf{-63.585\ kJ}$$

EXAMPLE 3.13

An inelastic flexible balloon originally folded completely has been inflated to a volume of 0.75 m^3 by compressed helium gas withdrawn from a storage bottle. Sketch the system before and after the process, and calculate the work done by the *system comprising the gas initially in the bottle. Given that the balloon is light, requires no stretching and that the barometer reads 750 mm of mercury.*

Solution : Refer Fig. 3.16 for the sketch of the system before and after the process of inflating a balloon. The firm line B_1 coincides with the inner surface of the storage bottle and indicates the system boundary before the process. At the end of the process, the boundary also encloses 0.75 m^3 content of the balloon and has been shown by the dotted line B_2.

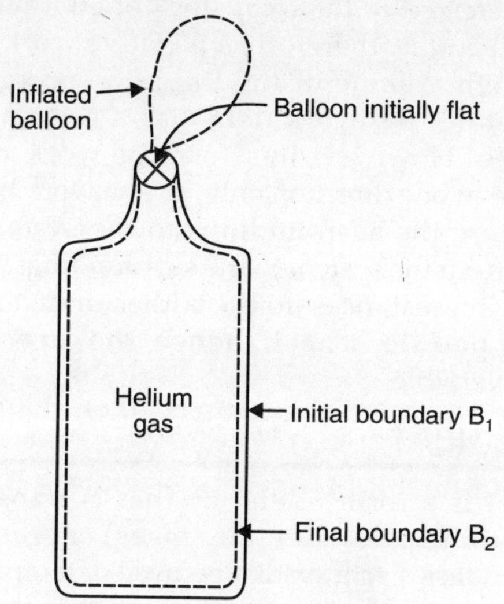

Fig. 3.16. Filling a balloon from bottle

Atmospheric pressure,

$$p_{at} = 760 \text{ mm of Hg} = 101325 \text{ N/m}^2$$

Displacement work,

$$W_d = \underset{\text{balloon}}{\int p\,dV} + \underset{\text{bottle}}{\int p\,dV}$$

The bottle is rigid and as such no work is involved in it. Further since the balloon fabric is light, inelastic and unstressed, the pressure in the balloon can be assume to be atmospheric at all times. Therefore,

$$W_d = \int p\,dV = p_{at}\,(V_2 - V_1)$$
$$= 101325 \times (0.75 - 0)$$
$$= 75994 \text{ Nm (or J)} = \mathbf{75.994\ kJ}$$

This work is done by the system and as such is positive. The atmosphere would do an equivalent amount of negative work on the system.

EXAMPLE 3.14

Determine the work done by air which enters an evacuated bottle from the atmosphere when the valve is opened. The atmospheric pressure is 1.01325 bar and 0.5 m^3 of air measured at atmospheric conditions enters the bottle. Sketch the system before and after the process.

Solution :

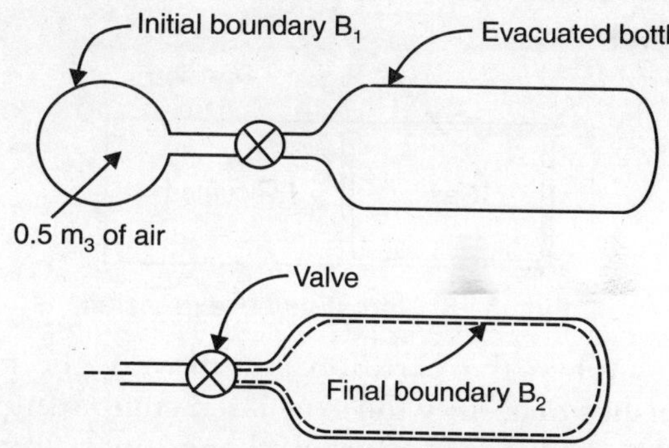

Fig. 3.17. Filling a bottle from atmosphere

Refer Fig. 3.17 which shows the system arrangement with initial (B_1) and final (B_2) system boundaries. The displacement work done by the system is given by

$$W_d = \underset{\text{free-air boundary}}{\int p\,dV} + \underset{\text{bottle}}{\int p\,dV}$$

No work is done by the part of the boundary in contact with the bottle as the bottle is rigid and the boundary does not move. Work is done only by the moving part of the boundary external to the bottle, and the pressure over this moving part is uniform at 1.01325 bar.

$$\therefore\ W_d = \underset{\text{free-air boundary}}{\int p\,dV} = 1.01325 \times 10^5 \times (0 - 0.5)$$
$$= -\,50662.5 \text{ Nm (or Joule)}$$
$$= \mathbf{-\,506.625\ kJ}$$

Since the free air boundary is contracting, the work done by the system is negative. The surroundings would do an equivalent amount of positive work at the boundary.

3.4. CASES OF W ≠ $\int p\,dV$

3.4.1. Unrestrained (or Free) expansion

The expansion of gas against a vacuum is called free expansion; essentially it is expansion which is not restrained by an opposing force.

Consider an insulated and rigid container divided into two compartments by a partition wall. One compartment contains an ideal gas and the other compartment is completely evacuated.

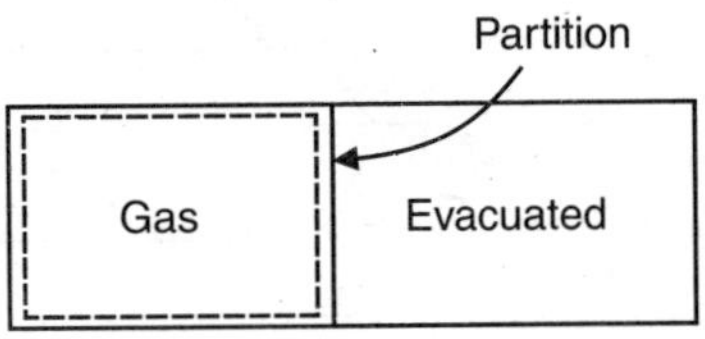

Fig. 3.18. Unrestrained expansion

When the partition is removed, the gas rushes to occupy the evacuated compartment also and there occurs a change in volume. Even though the initial and final states may be in equilibrium, yet the process is not quasi-static and reversible. The vacuum does not offer any resistance to the expansion process. The process is unrestricted, does not pass through equilibrium states and cannot be represented on diagram of properties. Though this free expansion process involves a change of volume yet the work transfer is zero.

3.4.2. Paddle work : dv = 0 yet W ≠ 0

Consider a fluid system being stirred by a paddle wheel. The wheel is inserted into the fluid system, communicates with a mass m hung from a rope which passes over a frictionless pulley. When the mass moves downward, the paddle wheel rotates. The differential work transfer equals decrease in potential energy of the mass and is given by

$$\delta W = T\,d\theta = mg\,dz$$

where T is the torque transmitted by the shaft in rotating through an angle $d\theta$.

Total work transfer,

$$W = \int T\,d\theta = \int mg\,dz$$

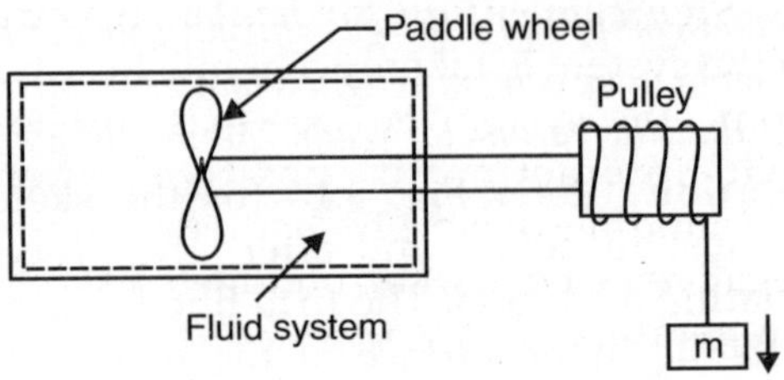

Fig. 3.19. Paddle wheel work

The paddle work arises from the resistance offered by the fluid to the paddle rotation. The energy associated with paddle work is transferred to the total stock of energy in the fluid and is, therefore, a negative work. Even though volume of the fluid system remains constant, the work done has a certain finite value. However, this form of work can be done in one direction only, *i.e.* the work transfer is from the surroundings to the system and not from the system to the surroundings. Under no circumstances do the eddies unite to drive the paddle wheel, hence the process is irreversible.

3.5. HEAT

Heat is a form of energy that is transferred (without transfer of mass) across the boundaries of a system because of temperature difference between the system and its surroundings, and is always from a higher temperature to a lower temperature. No heat interaction is possible between the system and its surroundings if both are in thermal equilibrium, *i.e.*, at the same temperature. Two conditions, namely temperature difference and communication, must be fulfilled simultaneously for heat interaction between two systems to occur; absence of either requirement means absence of heat interaction.

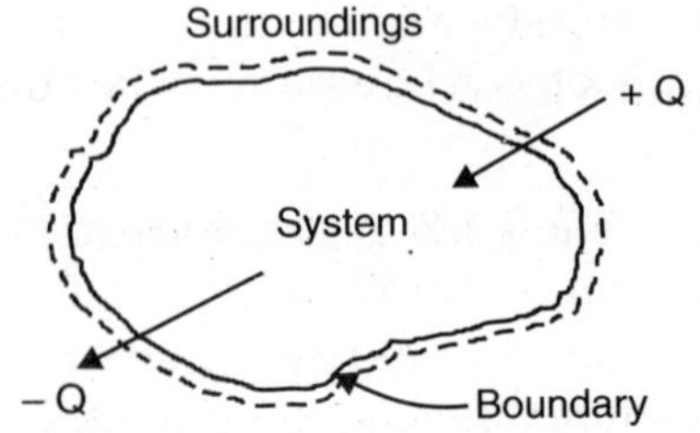

Fig. 3.20. Positive and negative heat

Sign convention for heat : Heat transfer to the system is taken +ve while heat transfer from the system is –ve. This implies that heat interaction is positive when the surroundings have a higher temperature, and negative when the surroundings have a lower temperature.

Comparison of heat and work

Similarities :

• Both heat and work are path functions; they are not properties of the system and their differentials are not exact.

• They represent energy crossing the system boundary; and hence they are a boundary phenomenon.

• Both represent transient phenomenon; these energy interactions occur only when a system undergoes change of state. They don't exist either before or after the interaction.

• Both are something which happen to a system but they are not the characteristics of a system.

• Concepts of heat and work are associated not with a 'store' but with a 'process'.

• Work and heat are inexact differentials and their change cannot be written as the difference of their values at the end ststes. That is

$$\int_1^2 \delta Q \neq Q_2 - Q_1 \quad \text{but } {}_1Q_2 \quad \text{or} \quad Q_{1-2}$$

$$\int_1^2 \delta W \neq W_2 - W_1 \quad \text{but } {}_1W_2 \quad \text{or} \quad W_{1-2}$$

This indicates the change in Q or W as and when a path 1 – 2 has been followed.

The operator δ is used to denote inexact differentials, while operator d is used to denote exact differentials.

Dissimilarities :

• There cannot be work transfer in a stable system. No such restriction exists for heat interaction.

• Heat is energy interaction due to temperature difference only; work is by reasons other than temperature difference.

• Sole effect external to the system could be reduced to the raising of mass against gravitational force in case of work. However, during heat transfer other effects are also observed.

• Heat is a low grade energy whereas work is a high grade energy.

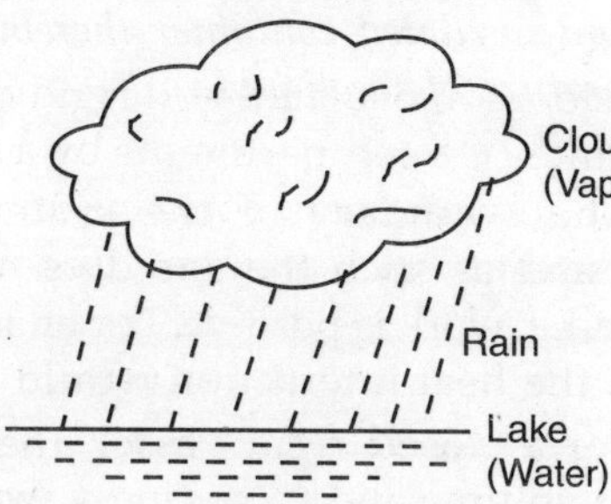

Fig. 3.21. Analogy between rain and work/heat transfer

Both heat and work may be likened to speech between two persons. Though the effect of speech may remain for ever yet the communication comes to an end with last word spoken. Further, heat and work correspond to rain which falls from the clouds to the lake (Fig. 3.21). Rain refers to the distinct form of water in falling from clouds to lake. The rain is stored neither in the cloud nor in the lake. The rain becomes water as soon as it reaches the lake. It is the water (not rain) that is stored in the lake.

EXAMPLE 3.15

State the direction of the heat transfer Q and work transfer W for each of the following processes. The systems to be considered are printed in bold italics.

(a) ***Gas*** *in an insulated cylinder expands as the piston is slowly moved outwards*

(b) A closed rigid vessel containing ***steam*** *at a temperature of 200°C is left standing in an atmosphere which is at 20°C temperature.*

(c) A ***sealed calorimeter*** *containing a mixture of air and fuel is immersed in a water tank. The mixture is ignited by an infinitesimal spark of negligible energy. Consider the calorimeter to be rigid and (i) insulated and (ii) non-insulated.*

(d) The ***air*** *in a tyre and connecting pump. The pump plunger is pushed down, forcing air*

into the tyre. The tyre, pump walls and connecting tube can be thought of to be non-conducting.

*(e) The **water** and **water vapour** in a rigid metallic container. The container is set on a stove, and there occurs an increase in both pressure and temperature of the contents of the container.*

*(f) A **viscous liquid** is stirred by a paddle wheel in an insulated container thereby raising the temperature of the liquid.*

Solution :

(*a*) The boundary of the system (gas) expands and as such the gas does work on the piston, *i.e.* work is positive. For an insulated cylinder, the heat interaction would be zero.

(*b*) For a closed rigid vessel, there is no change in volume and accordingly work done is zero. Since the steam is at a temperature higher than that of the surrounding atmosphere, the heat is rejected to the atmosphere, *i.e.*, heat interaction is negative.

(*c*) When the calorimeter is insulated no heat flow can occur across it, however, in case of non-insulated calorimeter heat will flow out of it and by convention it is negative. Assuming the calorimeter walls to be rigid, the work transfer is zero (neglecting electrical energy).

(*d*), (*e*), (*f*) Adopting the reasoning outlined above, the signs of work and heat interactions for these cases are given below:

d : *W* is negative and *Q* is zero

e : *W* is zero and *Q* is positive

f : *W* is negative and *Q* is zero

3.6. SPECIFIC HEAT AND HEAT CAPACITY

Whether a system has received or rejected heat is indicated by a change in its temperature; the temperature grows with the receipt of heat and diminishes with heat rejection. The quantity of heat received or rejected by a system during the process of heating or cooling is measured in terms of the product of mass, specific heat and temperature changes,

$$Q = m\,c\,(t_2 - t_1) \qquad \text{... (3.10)}$$

where Q is the heat gained by the system in kJ; m is the mass of the substance in kg; $(t_2 - t_1)$ is the temperature rise in degree centigrade or kelvin; and c is an experimental factor called *specific heat* and expressed in kJ/kg K.

The specific heat may be defined as the quantity of heat that enters or leaves a unit mass of the substance when it experiences unit degree change in temperature.

The product ($m \times c$) is separately termed as *heat capacity* or *water equivalent* of the system.

It has been predicted from experimental results that the value of specific heat depends upon

- substance comprising the system;
- condition under which the change of state occurs; and
- state of the system at the time of heat interaction.

The specific heat for liquids and solids varies primarily with temperature and does not differ much with nature of the process. In case of gases, specific heat is also dependent on the path of the thermodynamic process which may be carried out isochorically (at constant volume), isobarically (at constant pressure) or polytropically. Accordingly the specific heats for liquids and solids are quoted without any mention of the process involved in heat exchange. But for gases, it is the usual practice to quote two specific heats namely specific heat at constant pressure and specific heat at constant volume.

The *specific heat at constant volume* (c_v) is defined as the amount of heat added at constant volume to change the temperature of unit mass of the gas by 1 degree. Likewise for heat addition at constant pressure, the quantity of heat required for raising the temperature of unit mass of the gas through 1 degree temperature difference is called *specific heat at constant pressure* (c_p). These specific heats are properties of the gas, and are really not constant but vary both with pressure and temperature. However, for moderate changes in pressure and temperature, they are assumed constant.

For water: c = 4.186 kJ/kg-deg

For air : c_p = 1.005 kJ/kg-deg

c_v = 0.718 kJ/kg-deg

The ratio $\frac{c_p}{c_v}$ is given the symbol γ and is called adiabatic exponent. For air $\gamma = 1.4$

Mass, volume and molar specific heats

Depending upon selected unit of substance, a distinction is made amongst the following forms of specific heats:

(*i*) Mass specific heat (c) related to 1 kg mass of gas and expressed in kJ/kg-deg.

(*ii*) Volume specific heat (c') related to 1 m^3 of gas in standard physical conditions, *i.e.* at a pressure of 101325 N/m^2 and a temperature of 0°C and expressed in kJ/m^3-deg

(*iii*) Molar specific heat (c_{mol}) related to 1 kilo mole of gas, and measured in units of kJ/kilo mole-deg

The mass, volume and molar specific heats are correlated by the expression

$$c = c'v_o = c_{mol}/M \qquad ...(3.11)$$

where v_o is the specific volume at standard thermodynamic conditions and M is the molecular mass.

EXAMPLE 3.16

A mild steel piece of 5 kg mass is heated to 115°C and dropped into an insulated vessel containing 30 kg of water at 20°C. Due to heat flow from the hot mild steel piece, the rise in temperature of water is measured to be 3.5°C. Make calculations for the specific heat of mild steel.

Solution : Final temperature of water - steel piece system, $t_f = 20 + 3.5 = 23.5$°C

From energy balance,

Heat lost by mild steel piece

= Heat gained by water

$$m_s\, c_s\, (t_{s1} - t_{s2}) = m_w\, c_w\, (t_{w2} - t_{w1})$$

$$5\, c_s\, (115 - 23.5) = 30 \times 4.186\, (23.5 - 20)$$

$$(\because t_{s2} = t_{w2} = t_f)$$

$\therefore$ Specific heat of mild steel,

$$c_s = \frac{30 \times 4.186 \times 3.5}{5 \times 91.5} = \mathbf{0.96\ kJ/kg\ K}$$

EXAMPLE 3.17

During a particular process, the specific heat of the working fluid comprising a system is given by the relation

$$c = (0.2 + 0.002\ T)\ kJ/kg\ K$$

where T is the absolute temperature in degrees Kelvin.

What amount of heat is required to be supplied to this fluid system to raise its temperature from 300 K to 400 K ? Also determine the mean value of specific heat.

Solution : The total heat interaction for change in temperature from T_1 to T_2 can be calculated by

$$Q = m\int_{T_1}^{T_2} c\, dT$$

$$= 2\int_{300}^{400} (0.2 + 0.002\, T)\, dT$$

$$= 2\left|0.2\,T + \frac{0.002\,T^2}{2}\right|_{300}^{400}$$

$$= 2\left[0.2\,(400 - 300) + 0.001\,(400^2 - 300^2)\right]$$

$$= \mathbf{180\ kJ}$$

The mean specific heat is given by

$$Q = m\ c_m\ (T_2 - T_1)$$

$$180 = 2\ c_m\ (400 - 300)$$

$$\therefore \quad c_m = \frac{180}{2(400 - 300)} = \mathbf{0.9\ kJ/kg\ K}$$

EXAMPLE 3.18

An automobile vehicle of 1500 kg mass is running at a speed of 60 km/hr. The brakes are applied and the vehicle is brought to rest. Calculate the rise in the temperature of the brakes if their mass is 15 kg.

Take specific heat of the material of brakes = 0.46 kJ/kg K

Solution : Speed of the vehicle = 60 km/hr

$$= \frac{60 \times 1000}{3600} = 16.67\ \text{m/s}$$

Kinetic energy of the vehicle

$$= \frac{1}{2}\, mV^2 = \frac{1}{2} \times 1500 \times 16.67^2$$

$$= 208417\ \text{J} = \mathbf{208.417\ kJ}$$

When the vehicle is brought to rest by the application of brakes, its kinetic energy is transformed into heat energy which is taken up by the brakes. That is

$$208.417 = m\, c\, \Delta t = 15 \times 0.46 \times \Delta t$$

∴ Rise in temperature of brakes,

$$\Delta t = \frac{208.417}{15 \times 0.46} = \mathbf{32.3°C}$$

EXAMPLE 3.19

Find the rise in temperature of water when it falls down from a height of 100 m. It may be presumed that all the heat generated stays in water.

Take specific heat of water

= 4186 J/Kg K

Solution : Consider unit mass of water.

Decrease in potential energy of water as it falls downward through 100 m height is given by

$$\Delta (PE) = m\,g\,h$$
$$= 1 \times 9.81 \times 100 = 981 \text{ J}$$

This decrease in potential energy appears as heat energy that raises the temperature of water. That is

$$981 = m\,c\,\Delta t = 1 \times 4186 \times \Delta t$$

∴ Rise in temperature of water,

$$\Delta t = \frac{981}{1 \times 4186} = \mathbf{0.234°C}$$

EXAMPLE 3.20

It is proposed to heat 2 litres of water initially at 15°C to a temperature of 80°C. What time will be required to achieve this objective if 0.5 kW electric heater has been supplied for this purpose.

It may be presumed that the system does not exchange any heat with the surroundings.

Take specific heat of water

= 4.187 kJ/kg K

Solution : Mass of water, m = 2 litres

$= 0.002 \text{ m}^3 = 0.002 \times 1000 = 2 \text{ kg}$

($\because \rho$ for water $= 1000 \text{ kg/m}^3$)

Heat to be supplied to water $= m\,c\,\Delta t$

$= 2 \times 4.187 \times (80 - 15)$

$= 544.31 \text{ kJ}$

Power of heater

$= 0.5 \text{ kW} = 0.5 \text{ kJ/s}$

Let t be the time in seconds for which the heater must operate.

Then heat supplied by the heater equals $0.5\,t$ kJ.

$$\therefore \quad 544.31 = 0.5 \times t$$

$$t = \frac{544.31}{0.5} = 1088.62 \text{ s} = \mathbf{18.14 \text{ min}}$$

SALIENT POINTS

1. Heat and work are energy interactions which bring about changes in the properties of the system.
2. Work is said to be done by a system if the sole effect on things external to the system can be reduced to the raising of a weight.
3. Heat is a form of energy that is transferred across the system boundaries by virtue of temperature difference.
4. Work done by the system is taken +ve, and the workd done on the system is taken – ve.
5. Heat flow into the system is taken +ve, and the heat flow out of the system is taken –ve.
6. The salient characteristics of heat and work are:
   (*i*) energy in transit; can be identified only during the course of interaction
   (*ii*) its differential is not exact; hence a path function and not a property of the system
   (*iii*) not stored in a system.
7. Heat like work cannot be stored by a system. Accordingly it is wrong to say total heat or heat content of a system.
8. The unit of work is Nm or Joule. The rate at which work is done by or upon a system is called power. The unit of power is J/s or Watt.
9. Apart from displacement work $\int p\,dV$, the other forms of work which get involved during interaction are:
   - Elastic work
   - Stretching of liquid film

- Shaft work
- Electrical work, polarization and magnetization

10. The specific heat of a substance is defined as the amount of energy required in the form of heat to change the temperature of a unit mass of the substance by one degree

$$C = \frac{Q}{m\,dT} \text{ J/kg K}$$

11. For gases, two specific heats are quoted namely specific heat at constant pressure (c_p) and specific heat at constant volume (c_v).

For air : $c_p = 1.005$ kJ/kg K
and $c_v = 0.718$ kJ/kg K

The ratio c_p/c_v is given the symbol γ and is called adiabatic exponent. For air $\gamma = 1.4$.

12. The mass, volume and molar specific heats are correlated by the expression

$$c = c' v_o = \frac{c_{mol}}{M}$$

where v_o is the specific volume at standard thermodynamic conditions and M is the molecular mass.

REVIEW QUESTIONS

A. Conceptual and conventional questions.

1. Define thermodynamic work.
2. What is the significance of the term $\int p\,dV$ in a non-flow system ?
3. Show that work is a function of a process.
4. What does the area under the curve of a process on p-v diagram represent ?
5. In what respects are the heat and work interactions (*i*) similar, and (*ii*) dissimilar ?
6. How do the terms energy content and energy transfer differ from each other ?
7. Is it correct to say that heat content of the system is 5 kJ ?
8. Why heat (or work) transfer across the system boundaries is not classified as a thermodynamic property ?
9. Define specific heat ? Why gases have two specific heats ?
10. Set up an expression for the displacement work done during any arbitrary process.
11. Is it possible that $W \neq 0$ even if $dV = 0$? If so give an example.
12. What are +ve and –ve work (or heat) interactions ?
13. Is differential of heat (or work) exact ?
14. Consider an electrical storage battery discharging through a resister. Apply thermodynamic definition of work and show that the effect external to the system can be reduced to the raising of a weight.
15. Air initially at 75 kPa pressure, 1000 K temperature and occupying a volume of 0.12 m^3 is compressed isothermally until the volume is halved and subsequently it undergoes further compression at constant pressure till the volume is halved again. Sketch the processes on *p-V* diagram and make calculations for the work done.
[Ans. –10.74 kJ**]**
16. A gas system, comprised by a piston and cylinder, undergoes a change of state such that the product of pressure and volume remains constant. If the process begins at a pressure 300 kPa and a volume 0.015 m^3 and proceeds until the pressure falls to half its initial value, determine the magnitude and direction of work transfer.
[Ans. + 3119 Nm**]**
17. A non-flow reversible process occurs for which pressure and volume are correlated by the expression $pV = 150$, where V is the volume in m^3 and p is the pressure in bar. Make calculations for the work done on or by the system as pressure increases from 10 bar to 100 bar. Indicate the nature of process, whether expansion or compression.
[Ans. –34.5 MJ, Compression**]**
18. A gas contained in a piston cylinder arrangement expands from 0.75 m^3 volume to 1.25 m^3 volume whilst the pressure remains constant at 200 kPa. If the gaseous system receives 80 kJ of work from a paddle wheel, determine the net work done by the system.
[Ans. 20 KJ**]**
19. An ideal gas requires 1150 kJ/kg of heat to raise its temperature from 20°C to 100°C, when heated at constant pressure. When heat is

supplied to the same gas at constant volume, the heat requirement is 825 kJ for the same temperature range. Determine specific heat at constant pressure, specific heat at constant volume and the adiabatic exponent.

[Ans. 14.375 kJ/kg K, 10.31 kJ/kg K, 1.394**]**

20. A fluid contained in a horizontal cylinder fitted with a frictionless leak-proof piston is continuously agitated by a stirrer passing through the cylinder cover. During stirring operation for 10 minutes, the piston slowly moves outwards through 20 cm distance against a pressure of 1 bar. The stirrer turns 600 revolutions and the net work done on the fluid during the process is estimated to be 2kJ. If the cylinder diameter is 40 cm, estimate the torque in the driving shaft and the shaft power output of the motor.

[Ans. 0.1197 Nm, 0.00752 kW**]**

B. Fill in the blanks with appropriate word/words.

1. Heat and work are .... interactions which bring about a change in the .... of the system
2. $W \neq \int p\,dV$ for .... and .... process.
3. Both heat and work are .... functions; they are not .... of the system and their differentials are not ....
4. Thermodynamic work is the energy transfer for reasons other than ....
5. Heat is energy in .... and can be identified only when the .... is being executed.
6. The expression $\int_1^2 p\,dV$ gives the measure of work done during .... process.
7. The specific heat for .... and .... does not differ much with nature of the process.
8. The product of mass and specific heat of a substance is termed as .... or ....
9. The term .... may be defined as the quantity of heat that enters or leaves a unit mass of the substance when it experiences unit degree change in temperature.
10. A tank containing 5 kg of a certain gas is supplied 240 kJ of heat and that causes increase in temperature from 15°C to 60°C. The specific heat of the gas would be ....

Answers

1. energy, properties 2. non-quasistatic, irreversible process 3. path, properties, exact 4. temperature difference 5. transit, process 6. non-flow reversible 7. solids, liquids 8. heat capacity, water equivalent 9. specific heat 10. 1.067 kJ/kg K

C. Multiple choice questions.

1. Identify the false statement in the context of heat and work
   (*a*) they exist only during interaction between the system and the boundary
   (*b*) are path functions and exact differential
   (*c*) may or may not occur simultaneously
   (*d*) occur due to some unbalanced potential which tends to promote a change of state
2. As differentials, heat and work would be described mathematically as
   (*a*) inexact (*b*) exact
   (*c*) discontinuity (*d*) point function
3. Heat interaction between a system and its surroundings
   (*a*) represents energy in transit
   (*b*) does not depend on the choice of the system
   (*c*) can be identified after the completion of process
   (*d*) is the property of the system and its differential is exact.
4. Identify the wrong statement concerning thermodynamic work
   (*a*) $W = \int p\,dv$ for a reversible process only
   (*b*) $W = 0$ if $dv = 0$
   (*c*) $W \neq \int p\,dv$ for a free expansion process
   (*d*) is a path function and inexact differential
5. The work done equals $\int p\,dv$ for
   (*a*) an irreversible process
   (*b*) unrestricted expansion
   (*c*) non-flow, quasi-static process
   (*d*) the arrangement where work is done on the gas as the paddle wheel turns by a falling weight

Answers

1. (*b*) 2. (*a*) 3. (*a*) 4. (*a*) 5. (*c*)

Chapter **4**

Ideal and Real Gases

Learning objectives : Attention has been focussed in this chapter to know about the following aspects of gases :

- *ideal and perfect gas*
- *basic gas laws: Boyle's law and Charles' law*
- *characteristic gas equation*
- *Avogadro's law and universal gas constant*
- *van der Waal's equation and reduced equation of state*
- *compressibility factor*

4.1. CONCEPT OF AN IDEAL GAS

An ideal (or perfect) gas has no intermolecular forces of attraction or repulsion, does not change its phase during a thermodynamic process, and obeys a set of common rules governing change of its properties. There exists a relationship between pressure p, specific volume v and temperature T of a substance and is given by

$$\Phi\,(p, v, T) = 0 \quad \text{or} \quad p = f\,(v, T) \qquad ...(4.1)$$

When a substance is in a gaseous state, it experiences an appreciable change in specific volume with change of pressure and temperature, and as such equation 4.1 is applied widely for gases. Under special conditions ($p \to 0$), even the real gases behave in a similar manner. That is

$$p\,v = (RT)_{p \to 0} \quad \text{or} \quad p = \left(\frac{RT}{v}\right)_{p \to 0} \qquad ...(4.2)$$

The state of zero pressure of a real gas is called *ideal state* and this identical behaviour of a real gas at high temperature and low pressure is called *ideal behaviour*. At reduced pressure and elevated temperature, the gas molecules lie far apart and molecular force of attraction between them tends to be small. Gases deviate from ideal behaviour at high pressure and low temperature , *i.e.* at high densities.

An ideal gas follows both the equation of state $p\,v = RT$ and the relationships $u = f_1\,(T)$ and $h = f_2\,(T)$, *i.e.* the internal energy u and enthalpy h are functions of temperature alone. For real gases, specific heats are found to vary with temperature. However, over reasonably small ranges of temperature, this variation is negligibly small. This leads to the concept of the *perfect gas* which is ideal and has constant values of specific heats, c_p and c_v.

4.2. BASIC GAS LAWS

4.2.1. Boyle's law

If the state of a gas changes at constant temperature, then the volume of a given mass of perfect gas is inversely proportional to the absolute pressure.

$$V \propto \left(\frac{1}{p}\right) \quad \text{or} \quad pV = \text{constant} \qquad ...(4.3)$$

Thus the product of absolute pressure and volume of a given quantity of gas is constant when the temperature remains constant. Boyle's law is essentially valid only at very low pressure or at moderately high temperature.

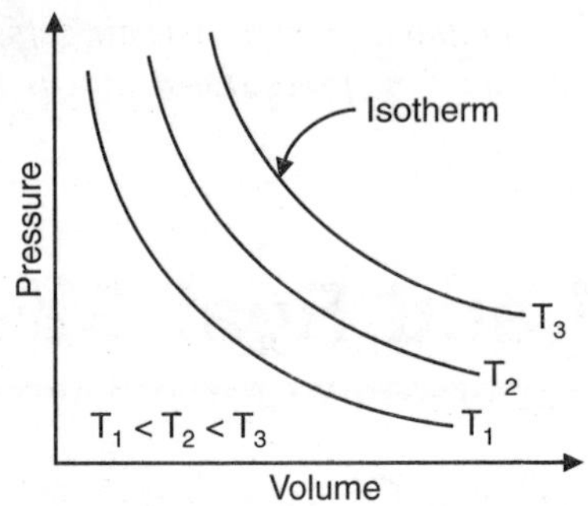

Fig. 4.1. Boyle's law on P-V plot

If plotted on p-V diagram, Boyle's law represents a rectangular hyperbolic curve. Each curve corresponds to a different temperature, and is called *isotherm*. For any two points on the curve

$$p_1 V_1 = p_2 V_2$$

The process occuring at constant temperature is known as **isothermal process.**

EXAMPLE 4.1

At temperature t = 15°C and pressure p = 0.98 bar, the specific volume of a gas equals 0.45 m³/kg. Subsequently the pressure drops to p_2 = 0.6 bar while the temperature remains constant. Make calculations for the density of gas under the changed conditions.

Solution : Since temperature remains constant, Boyle's law applies and therefore,

$$p_1 v_1 = p_2 v_2 \; ; \quad v_2 = \frac{p_1 v_1}{p_2}$$

$$\therefore \qquad v_2 = \frac{0.98 \times 0.45}{0.6} = 0.735 \text{ m}^3/\text{kg}$$

Density is the reciprocal of specific volume and therefore,

$$\rho_2 = \frac{1}{v_2} = \frac{1}{0.735} = \mathbf{1.36\ kg/m^3}$$

4.2.2. Charles' law

If the state of a perfect gas changes at constant pressure, then the volume of a given mass of perfect gas varies directly as absolute temperature. Thus,

$V \propto T$ when p = remains constant

$$\text{or} \quad \frac{V}{T} = \text{constant} \qquad \text{...(4.4)}$$

It was found by Gay-Lussac and Regnault that at constant pressure, the change in volume of any perfect gas corresponding to a unit degree temperature change is given by 1/273 of its volume at 0°C.

Let V_0 = Volume of the gas at 0°C

V_t = Volume of the gas at t °C

When pressure remains constant, then

$$V_t = V_0 + V_0 \times \frac{t}{273} = V_0\left(1 + \frac{t}{273}\right) \text{...(4.5)}$$

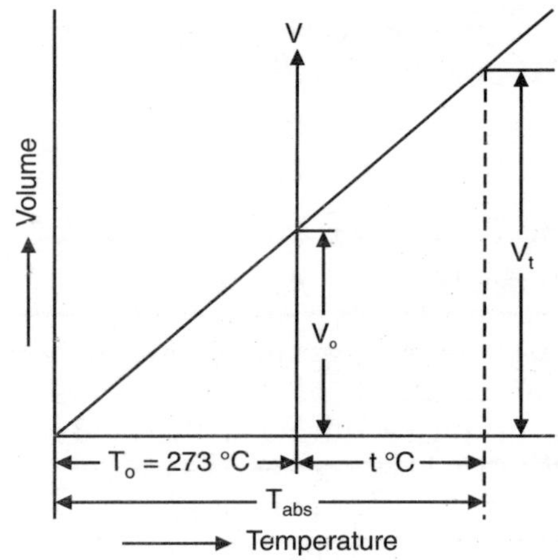

Fig. 4.2. Charles law on V-T diagram

At temperature t = – 273°C, the volume V_t of the gas will be zero. This is taken to be point of absolute zero temperature. Further,

$$V_t = V_0\left(\frac{273 + t}{273}\right) = \frac{V_0 T_t}{T_0}$$

$$\frac{V_t}{V_0} = \frac{T_t}{T_0} \; ; \quad \text{or} \quad \frac{V_t}{T_t} = \frac{V_0}{T_0} \qquad \text{...(4.6)}$$

Charles' law has also been stated as follows :

"The absolute pressure of a perfect gas varies directly with absolute temperature if the volume of the gas is maintained constant during the process."

That is : $p \propto T$

$$\frac{p}{T} = \text{constant at constant volume ...(4.7)}$$

EXAMPLE 4.2

0.75 kg of an ideal gas at 15°C temperature is contained in a cylinder of 0.5 m³ capacity. The gas is heated at constant pressure until it attains a temperature of 150° C. Determine the density of gas when it has been heated.

Solution : $T_1 = 15°C = 273 + 15 = 288$ K

$T_2 = 273 + 150 = 423$ K

$$v_1 = \frac{V}{m} = \frac{0.5}{0.75} = 0.667 \text{ m}^3/\text{kg}$$

Since pressure remains constant, the Charles' law applies and therefore,

$$\frac{v_1}{T_1} = \frac{v_2}{T_2}$$

$$v_2 = v_1 \frac{T_2}{T_1} = \frac{0.667 \times 423}{288} = 0.98 \text{ m}^3/\text{kg}$$

Density is reciprocal of specific volume and, therefore

$$\rho_2 = \frac{1}{v_2} = \frac{1}{0.98} = \mathbf{1.02 \text{ kg/m}^3}$$

EXAMPLE 4.3

The pressure in a car tyre was checked at 15°C temperature and a reading of 1.75 bar was indicated by the pressure gauge. How would the gauge reading be effected if during running the temperature rises to 40°C? It may be presumed that air behaves as an ideal gas and the tyre does not stretch. The atmospheric pressure is equivalent to 1 bar.

Solution : $p_1 = 1.75 + 1 = 2.75$ bar absolute

$T_1 = 273 + 15 = 298$ K

$T_2 = 273 + 40 = 313$ K

Since the volume of air in the tyre remains constant, Charles' law applies and therefore,

$$\frac{p_1}{T_1} = \frac{p_2}{T_2} \quad ; \quad p_2 = p_1 \frac{T_2}{T_1}$$

$$\therefore \quad p_2 = 2.75 \times \frac{313}{298} = 2.888 \text{ bar}$$

Pressure gauge reading at 40°C would be $= 2.888 - 1 =$ **1.888 bar**

EXAMPLE 4.4

Cold air at 300 K is made to enter the air heater through a 50 cm diameter pipe. The air gets heated to 1050 K at constant pressure in the air heater. Make calculations for the diameter of outlet pipe which is to carry the heated air away from the heater. It may be presumed that the velocity of the air remains the same in both the pipes.

Solution : Since heating process is being done at constant pressure, Charles' law applies and therefore,

$$\frac{V_1}{T_1} = \frac{V_2}{T_2} \text{ or } \frac{V_2}{V_1} = \frac{T_2}{T_1}$$

The volume flow equals the product of cross-sectional area of pipe and the velocity of air.

$$\therefore \quad \frac{(\pi/4)\, d_2^2 \times \text{velocity of air}}{(\pi/4)\, d_1^2 \times \text{velocity of air}} = \frac{T_2}{T_1}$$

$$\text{or} \quad d_2 = d_1 \left(\frac{T_2}{T_1}\right)^{1/2}$$

$$= 0.5 \times (1050/300)^{1/2}$$

$$= 0.935 \text{ m} = \mathbf{93.5 \text{ cm}}$$

EXAMPLE 4.5

A cylinder containing gas at 27°C is kept at constant pressure by a gas-tight dead weight piston which is 15 cm from the head of cylinder. The gas is supplied heat, and its temperature increases to 65°C. Make calculations for the distance moved by the piston.

Solution : Since the heat is being supplied at constant pressure, Charles' law applies and, therefore,

$$\frac{V_1}{T_1} = \frac{V_2}{T_2} \quad ; \quad \frac{V_2}{V_1} = \frac{T_2}{T_1}$$

Let l_1 be the distance of piston from head of the cylinder before heat is supplied, and l_2 be the corresponding distance after heat is supplied. Then

$$\frac{\text{area of cylinder} \times l_2}{\text{area of cylinder} \times l_1} = \frac{T_2}{T_1}$$

$$\therefore \quad l_2 = l_1 \times \frac{T_2}{T_1}$$

$$= 0.15 \times \left(\frac{273+65}{273+27}\right)$$

$$= 0.169 \text{ m} = \textbf{16.9 cm}$$

4.3. CHARACTERISTIC GAS EQUATION (Equation of State of an Ideal Gas)

The characteristic gas equation represents the relation between p, v, and T and can be set up by applying Boyle's law and Charles' law to a perfect gas undergoing a change of state.

Consider a unit mass of an ideal gas that passes from state p_1, v_1, T_1 to another state identified by p_2, v_2, T_2.

Let this change be

(*i*) first at constant pressure p_1 to some intermediate volume v_i and

(*ii*) then at constant temperature T_1 to final volume v_2

Process 1 - *i* is at constant pressure and, therefore, state of the gas changes in conformity with Charles' law. That is

$$\frac{v_1}{T_1} = \frac{v_i}{T_i}; \quad v_i = v_1\frac{T_i}{T_1} \qquad \ldots(a)$$

Process *i* - 2 is at constant temperature and, therefore, change in the state of the gas is according to Boyle's law. That is

$$p_i v_i = p_2 v_2; \quad v_i = p_2\frac{v_2}{p_i} \qquad \ldots(b)$$

From expressions (*a*) and (*b*),

$$v_1\frac{T_i}{T_1} = p_2\frac{v_2}{p_i}$$

Since $T_i = T_2$ and $p_i = p_1$, we get

$$v_1\frac{T_2}{T_1} = p_2\frac{v_2}{p_1}, \quad \text{or} \quad \frac{p_1 v_1}{T_1} = \frac{p_2 v_2}{T_2}$$

$$\therefore \quad \frac{pv}{T} = R \text{ (a constant)} \qquad \ldots(4.8)$$

The identity $p\,v = RT$ is the equation of state or the characteristic gas equation for a unit mass of gas. Since $v = V/m$ where V is the volume of the gas and m is its mass, equation 4.8 can be rewritten as

$$pV = mRT \qquad \ldots(4.9)$$

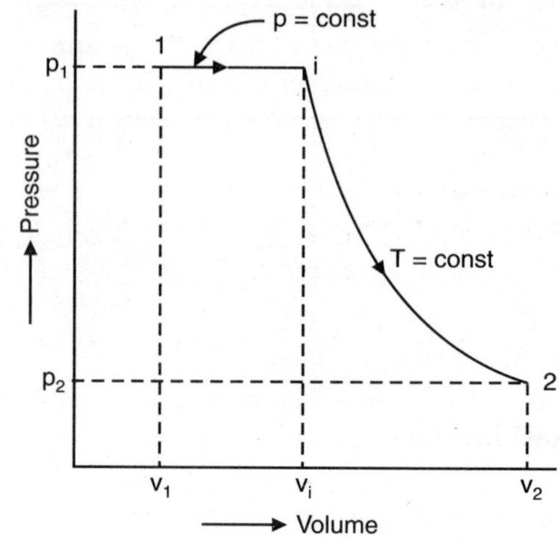

Fig. 4.3. Characteristic gas equation

The above correlation is the equation of state for any arbitrary amount of gas. This equation is essentially true only for a perfect gas. In practice no known gas is perfect and this gas equation causes an error of about 3% at large pressure ranges. However, at room temperature and pressure, this gas equation can be adopted even for real gases of low molecular masses.

The constant R has a particular value for each gas and its units depend on the units of pressure volume, mass and temperature.

In SI units: p is in N/m^2, V is in m^3, m is in kg and T is in degrees kelvin. Units of gas constant R then work out to be

$$R = \left[\frac{N}{m^2}\right] \times [m^3] \times \left[\frac{1}{Kg}\right] \times \left[\frac{1}{K}\right]$$

$$= \text{Nm/kg K} = \text{J/kg K}$$

The gas constant R has thus units of energy per unit mass and unit temperature difference. It may also be said that R represents the amount of work done at constant pressure when unit mass of gas expands reversibly due to temperature rise of one degree kelvin.

The gas constant R, though constant for a particular gas, has different values for different gases. Table 4.1 gives the values of gas constant R for gases most frequent in use.

Table 4.1.

Gas	Molecular mass	Gas constant J/kg K	Gas constant × molecular mass J/kg mole K
Air	28.97	287	
Hydrogen	2	4124	
Nitrogen	28	297	Approximately 8314 for all gases
Oxygen	32	260	
Helium	4	2077	
Carbon dioxide	44	189	

EXAMPLE 4.6

Determine the mass of nitrogen present in a vessel of 2.5 m^3 capacity. The pressure and temperature gauges mounted on the vessel indicate gas to be at 80 bar pressure and 25°C temperature. For nitrogen gas constant R = 297 J/kg deg.

Solution : $p = 80$ bar $= 80 \times 10^5$ N/m^2

$T = (25 + 273) = 298$ K

Invoking characteristic gas equation,

$$pV = mRT$$

$$80 \times 10^5 \times 2.5 = m \times 297 \times 298$$

∴ Mass of nitrogen,

$$m = \frac{80 \times 10^5 \times 2.5}{297 \times 298}$$

$$= \mathbf{225.97\ kg}$$

EXAMPLE 4.7

A high altitude chamber, the volume of which is 30 m^3, is put into operation by reducing the pressure from 1.013 bar to 0.35 bar and temperature from 27°C to 5°C. How many kg of air must be removed from the chamber during the process ? Express this mass as volume measured at 1.013 bar and 27°C. Take R = 287 J/kg K for air.

Solution : Invoking characteristic gas equation, $PV = mRT$

Initial mass of air,

$$m_1 = \frac{p_1V_1}{RT_1} = \frac{(1.013 \times 10^5) \times 30}{287\,(273 + 27)}$$

$$= 35.296 \text{ kg}$$

Final mass of air,

$$m_2 = \frac{p_2V_2}{RT_2} = \frac{0.35 \times 10^5 \times 30}{287\,(273 + 5)}$$

$$= 13.160 \text{ kg}$$

∴ Mass of air removed during the process,

$m_1 - m_2 = 35.296 - 13.160 =$ **22.136 kg**

Volume of this mass at 1.013 bar and 27°C is given by,

$$V = \frac{mRT}{p} = \frac{22.136 \times 287 \times (273 + 27)}{1.013 \times 10^5}$$

$$= \mathbf{18.815\ m^3}$$

EXAMPLE 4.8

A spherical vessel of 1.5 m diameter containing air at 315 K is evacuated until the vacuum inside the vessel is 730 mm of mercury; the evacuation process is carried out at constant temperature. Determine the mass of air pumped out. Proceed to calculate the pressure in the tank if subsequently the air is cooled to 275 K. For air R = 287 J/Kg K and take atmospheric pressure equal to 760 mm of mercury.

Solution : Volume of the tank,

$$V = \frac{4}{3}\pi r^3 = \frac{4}{3}\pi\left(\frac{1.5}{2}\right)^3 = 1.766 \text{ m}^3$$

Before evacuation :

$p_1 = 760$ mm of mercury

$= 1.01325$ bar

$V_1 = 1.766$ m^3 and $T_1 = 315$ K

$$m_1 = \frac{p_1V_1}{RT_1} = \frac{1.01325 \times 10^5 \times 1.766}{287 \times 315}$$

$$= 1.9793 \text{ kg}$$

After evacuation :

$$p_2 = 730 \text{ mm of mercury (vacuum)}$$
$$= 760 - 730$$
$$= 30 \text{ mm of mercury (absolute)}$$
$$= \frac{30}{760} \times 1.01325 = 0.034 \text{ bar}$$
$$V_2 = 1.766 \text{ m}^3 \text{ and } T_2 = 3.5 \text{ K}$$
$$m_2 = \frac{p_2 V_2}{RT_2} = \frac{0.034 \times 10^5 \times 1.766}{287 \times 315}$$
$$= 0.0664 \text{ kg}$$

∴ Mass of air pumped out

$$= m_1 - m_2$$
$$= 1.9793 - 0.0664 = \mathbf{1.9129} \text{ kg}$$

(*ii*) Since the cooling is at constant volume, Charles law applies, and we can write

$$\frac{p_2}{T_2} = \frac{p_3}{T_3}$$
$$\therefore \quad p_3 = \frac{T_3}{T_2} p_2 = \frac{275}{315} \times 30$$
$$= \mathbf{26.2} \text{ mm of mercury.}$$

4.4. AVOGADRO'S LAW AND UNIVERSAL GAS CONSTANT

Under identical conditions of temperature and pressure, equal volumes of all gases have same number of molecules.

Consider two different ideal gases having the same volume, same pressure and same temperature. Further, let their molecular masses be denoted by M_1 and M_2 respectively. In accordance with Avogadro's law, each gas will contain the same number of molecules, say n and so the masses of the gases would be

$$m_1 \propto n\, M_1 = k\, n\, M_1$$
$$m_2 \propto n\, M_2 = k\, n\, M_2 \quad \text{...(4.10)}$$

where k is the proportionality constant

If ρ_1 and ρ_2 are the densities of the gases, then

$$m_1 = \rho_1 V \text{ and } m_2 = \rho_2 V$$

where $V_1 = V_2 = V$ is the volume of each gas.

When the values of masses are substituted in equation (4.10), we get

$$\frac{\rho_1 V}{\rho_2 V} = \frac{k\, n\, M_1}{k\, n\, M_2}; \quad \frac{\rho_1}{\rho_2} = \frac{M_1}{M_2} \quad \text{...(4.11)}$$

i.e. the densities of the gases are directly proportional to their molecular masses at the same temperature and pressure.

Substituting $\rho = 1/v$ in equation (4.11), we obtain,

$$\frac{v_2}{v_1} = \frac{M_1}{M_2} \quad \text{...(4.12)}$$

i.e. the specific volumes of the gases are inversely proportional to their molecular masses at the same temperature and pressure.

Equation 4.12 may be recast as

$$v_1 M_1 = v_2 M_2\ ;\ v\, M = \text{constant} \quad \text{...(4.13)}$$

i.e. the product of molecular mass and specific volume is constant for all ideal gases under identical conditions of pressure and temperature.

The quantity $v\,M$ is called *molar volume*; it represents the volume of one kilo mole (kg mole) of ideal gas. One *kilo mole* is the quantity of a substance whose mass in kilograms is numerically equal to its molecular mass.

- 1 kg mole of nitrogen (molecular mass M = 28) has a mass of 28 kg
- 1 kg mole of oxygen (molecular mass M = 32) has a mass of 32 kg
- 1 kg mole of air (apparent molecular mass M = 29) has a mass of 29 kg

At standard conditions of t = 0°C = 273.15 K and p = 101325 N/m^2, the volume of one kilo mole of all gases is equal to 22.4135 m^3. This 22.4135 m^3 of volume is known as *mole*. Mole, therefore, is a bigger unit of volume.

$$\text{Molar volume} = v_o M = 22.4135 \text{ m}^3/\text{kg mole} \quad \text{...(4.14)}$$

The above equation helps to find the specific volume and density of all gases under standard conditions.

Specific volume,

$$v_o = \frac{22.4135}{M} \text{ m}^3/\text{kg}$$

Density,

$$\rho_o = \frac{M}{22.4135} \text{ kg/m}^3$$

For molar volume of a gas, the characteristic gas equation can be written as

$$p\,V_{mol} = M\,R\,T = R_{mol}\,T$$

where R_{mol} = (gas constant R × molecular mass M) is same for all gases and is called the *molar gas constant* or *universal gas constant*; its value does not depend on the type of gas. Under standard conditions of

$$p = p_o = 101325 \text{ N/m}^2$$
$$T = T_0 = 273.15 \text{ K}$$
$$V_{mol} = 22.4135 \text{ m}^3/\text{kg mole}$$

the universal gas constant turns out to be equal to

$$R_{mol} = \frac{101325 \times 22.4135}{273.15}$$
$$= 8314.31 \text{ J/kg mole K} \quad ...(4.15)$$

The universal equation of state, related to 1 kg mole of gas takes the form

$$p\,V_{mol} = 8314.31\,T \simeq 8314\,T \quad ...(4.16)$$

This equation is most general one for all gases, because it relates three ideal gas laws (Boyle's, Charles' and Avogadro's), and includes the universal gas constant which is independent of the nature of gas.

The universal gas constant helps to evaluate the gas constant for any gas if its molecular mass M is known.

$$R = \frac{8314}{M} \text{ J/kg K}$$

EXAMPLE 4.9

A pressure bottle stores 5 m³ of an inert gas at pressure 10 MPa and temperature 300 K. Make calculations for the mass, density, specific volume and molar volume of the gas. Take molecular mass of gas = 28.

Solution : Characteristic gas constant,

$$R = \frac{R_o}{M} = \frac{8314}{28} = 296.93 \text{ J/kg K}$$

Invoking the gas equation, $pV = mRT$

$$\text{mass of gas, } m = \frac{pV}{RT}$$
$$= \frac{10 \times 10^6 \times 5}{296.93 \times 300} = \mathbf{561.3\ kg}$$

density of gas, ρ = mass per unit volume

$$= \frac{561.3}{6} = \mathbf{112.26\ kg/m^3}$$

Specific volume is the reciprocal of density

∴ specific volume,

$$v = \frac{1}{\rho} = \frac{1}{112.26} = \mathbf{0.0089\ m^3/kg}$$

Molar volume represents the volume of 1 kg mole of the gas, *i.e.*, volume corresponding to mass equal to its molecular mass.

Molar volume,

$$V_{mol} = \frac{MRT}{p}$$
$$= \frac{28 \times 296.93 \times 300}{10 \times 10^6} = \mathbf{0.2496\ m^3}$$

EXAMPLE 4.10

Two spheres, each of capacity 2 m³, are connected by a pipe with a valve inserted in between. When the valve lies in the closed position one sphere contains oxygen at 50 kPa and 320 K and the other contains oxygen at 45 kPa and 290 K. Subsequently the valve is opened and the entire system is allowed to attain the equilibrium conditions. At this state, the final temperature is noted to be 300 K. Presuming that the volume of the connecting pipe is negligible, determine the final pressure of the composite system.

Solution : Gas constant R for oxygen (molecular mass 32)

$$R = \frac{R_o}{M} = \frac{8314}{32} = 259.8 \text{ J/kg K}$$

Let suffix 1 and 2 refer to state of oxygen in sphere 1 and sphere 2 respectively. The mass of oxygen in each sphere can be worked out by using the ideal gas equation, $pV = mRT$

$$m_1 = \frac{p_1 V_1}{RT_1} = \frac{50 \times 10^3 \times 2}{259.8 \times 320} = 1.203 \text{ kg}$$

$$m_2 = \frac{p_2 V_2}{RT_2} = \frac{45 \times 10^3 \times 2}{259.8 \times 290} = 1.155 \text{ kg}$$

Then at the equilibrium state :

$$m = m_1 + m_2$$
$$= 1.203 + 1.155 = 2.358 \text{ kg}$$
$$V = V_1 + V_2$$
$$= 2 + 2 = 4 \text{ m}^3 \text{ and } T = 300 \text{ K}$$

Again using the ideal gas equation

$$p = \frac{mRT}{V} = \frac{2.358 \times 259.8 \times 300}{4}$$

$$= 45945.63 \text{ N/m}^3 \simeq \mathbf{45.94} \text{ kPa}$$

EXAMPLE 4.11

Nitrogen is to be stored at 15 MPa and 300 K in a steel cylinder of 0.5 m³ capacity. The cylinder is protected against excessive pressure by a fusible plug which would melt and allow the gas to escape if the temperature rises too high. Make calculations for the mass of oxygen which the cylinder will hold at the designed condition. Also determine the temperature at which the fusible plug must melt in order to limit the pressure in the cylinder to 16 MPa.

Solution : Gas constant for nitrogen (molecular mass = 28)

$$R = \frac{R_o}{M} = \frac{8314}{28} = 296.93 \text{ J/kg K}$$

(*a*) Mass of nitrogen at the designed conditions can be worked out by using the characteristic gas equation. That is

$$m = \frac{pV}{RT}$$

$$= \frac{(15 \times 10^6) \times 0.5}{296.93 \times 300} = \mathbf{84.2} \text{ kg}$$

(*b*) When the fusible plug is about to melt,

$p_2 = 16$ MP$_a$

$V_2 = 0.5$ m^3 and $m_2 = 84.2$ kg

Then from characteristic equation,

$$T_2 = \frac{p_2 V_2}{m_2 R}$$

$$= \frac{(16 \times 10^6) \times 0.5}{84.2 \times 296.93} = 320 \text{ K}$$

Alternatively: $\frac{p_1}{T_1} = \frac{p_2}{T_2}$ as the volume of cylinder remains constant

$$T_2 = \frac{p_2}{p_1} T_1 = \frac{16}{15} \times 300 = 320 \text{ K}$$

$$\therefore \quad t = 320 - 273 = \mathbf{47°C}$$

EXAMPLE 4.12

Air at 20°C and 7 bar is contained in a rigid tank of 0.57 m³ volume capacity. The tank is provided with a valve that opens at a pressure of 8.5 bar and remains open until the the pressure drops to 8.15 bar. A fire occurs and that causes the valve to operate once in the manner described. Determine:

(a) air temperature just before the valve opens,

(b) mass of air lost due to fire.

It may be presumed that temperature of the air remains constant during discharge and air in the tank behaves as an ideal gas.

Solution : Let suffix 1 represent the initial conditions and suffix 2 represent the conditions before the valve opens. For these conditions

$T_1 = 293$ K ; $p_1 = 7$ bar

and $p_2 = 8.5$ bar

Since volume remains constant, Charles law applies and we have

$$\frac{p_1}{T_1} = \frac{p_2}{T_2}\ ; \quad \frac{7}{293} = \frac{8.5}{T_2}$$

∴ Temperature of air just before the valve opens,

$$T_2 = \frac{8.5 \times 293}{7} = \mathbf{355.78} \text{ K or } \mathbf{82.78°C}$$

(*b*) Let suffix 3 denote the conditions at the time of valve closing. Then using the characteristic gas equation

$$m_2 = \frac{p_2 V_2}{RT_2}$$

$$= \frac{8.5 \times 10^5 \times 0.57}{287 \times 355.78} = 4.745 \text{ kg}$$

During discharge of air, the volume and temperature remain constant. That gives

$$m_3 = \frac{p_3 V_3}{RT_3}$$

$$= \frac{8.15 \times 10^5 \times 0.57}{287 \times 355.78} = 4.549 \text{ kg}$$

Mass of air that leaks out

$$= m_2 - m_3$$

$$= 4.745 - 4.549 = \mathbf{0.196 \text{ kg}}$$

EXAMPLE 4.13

A rigid pipe of 2.5 cm diameter and 6 m length has been pressurised with dry air. At that instant, the temperature and gauge pressure of air in the pipe are observed to be 40°C and 6 bar. Some times later, the corresponding readings are 30°C and 5 bar. Is there a leakage in the pipe ? If so, make calculations for the mass of air that leaks through the fitting provided in the pipeline. Take atmosphere pressure equal to 1 bar

Solution : Let suffix 1 refer to the initial state when the pipe has been pressurised.

$$V_1 = \frac{\pi}{4} d^2 l$$

$$= \frac{\pi}{4}(0.025)^2 \times 6 = 0.00294 \text{ m}^3$$

$$p_1 = 6 + 1 = 7 \text{ bar}$$

and $T_1 = 273 + 40 = 313$ K

Using ideal gas equation, initial mass of

$$\text{air } m_1 = \frac{p_1 V_1}{RT_1} = \frac{7 \times 10^5 \times 0.00294}{287 \times 313} = 0.0229 \text{ kg}$$

Let suffix 2 denote the state when the pressure and temperature readings have been taken sometimes later.

$$V_2 = V_1 = 0.00294 \text{ m}^3$$

($\because$ pipe is rigid and so volume remains constant)

$$p_2 = 5 + 1 = 6 \text{ bar}$$

and $T_2 = 273 + 30 = 303$ K

Then, final mass of air,

$$m_2 = \frac{p_2 V_2}{RT_2}$$

$$= \frac{6 \times 10^5 \times 0.00294}{287 \times 303} = 0.02028 \text{ kg}$$

Since the initial mass is greater than the final mass, the air leaks out and the mass of air leakage is

$$m_1 - m_2 = 0.0229 - 0.02028 = \mathbf{0.00262} \text{ kg}$$

EXAMPLE 4.14

A spherical balloon of 8 m radius is floating in atmosphere at 100 kPa pressure and 300 K temperature. Determine the mass and number of moles of air displaced by the balloon. If hydrogen gas fills the balloon under same conditions of temperature and pressure, calculate the mass and number of moles of hydrogen. Molecular masses of air and hydrogen are stated to be 28.97 and 2 respectively.

Solution : Volume of balloon $= \frac{4}{3}\pi r^3$

$$= \frac{4}{3}\pi \times 8^3 = 2143.6 \text{ m}^3$$

The balloon would displace the air volume equivalent to its own volume.

$\therefore$ Volume of air displaced, $V = 2143.6$ m^3

Gas constant for air,

$$R = \frac{8314}{28.97} = 287 \text{ J/kg K}$$

Mass of air, $m = \frac{pV}{RT}$

$$= \frac{100 \times 10^3 \times 2143.6}{287 \times 300} = 2489.7 \text{ kg}$$

The mass of air equals (number of moles × molecular mass)

$$\therefore \text{Moles of air displaced} = \frac{2489.7}{28.97} = 85.94$$

Under identical conditions of pressure and temperature, equal volumes of all gases contain the same number of moles. Therefore, the moles of hydrogen will be equal to moles of air.

Moles of hydrogen = 85.94

Mass of hydrogen = 85.94 × 2

= 171.88 kg

Alternatively : Gas constant for hydrogen,

$$R = \frac{8314}{2} = 4157 \text{ J/kg K}$$

Mass of hydrogen, $m = \frac{pV}{RT}$

$$= \frac{100 \times 10^3 \times 2143.6}{4157 \times 300} = 171.88 \text{ kg}$$

$$\text{Moles of hydrogen} = \frac{171.88}{2} = \mathbf{85.94}$$

EXAMPLE 4.15

Consider a spherical balloon of 5 cm radius filled with hydrogen at 300 K and atmospheric pressure.

The surrounding air is at 20°C and the barometer reads 75 cm of mercury. Determine the payload that can be lifted with the aid of this balloon.

Solution : Volume of balloon

$$= \frac{4}{3}\pi r^3 = \frac{4}{3}\pi \times 5^3 = 523.34 \text{ m}^3$$

Pressure of hydrogen in balloon

= atmospheric pressure = 75 cm of mercury

$$= \frac{75}{76} \times 1.01325 \simeq 1 \text{ bar}$$

Gas constant for hydrogen,

$$R = \frac{8314}{2} = 4157 \text{ J/kg K}$$

Mass of hydrogen filling the balloon,

$$m_1 = \frac{pV}{RT}$$

$$= \frac{1 \times 10^5 \times 523.34}{4157 \times 300} = 41.96 \text{ kg}$$

Volume of air displaced equals the volume of balloon

∴ Mass of air displaced,

$$m_2 = \frac{pV}{RT}$$

$$= \frac{1 \times 10^5 \times 523.34}{287 \times 293} = 622.35 \text{ kg}$$

The load which can be lifted by the balloon due to buoyant force

= 622.35 – 41.96 = **580.39 kg**

EXAMPLE 4.16

The heat produced by combustion of fuel in a bomb calorimeter is determined by using oxygen taken from a cylinder of 0.006 m³ capacity at an absolute pressure 12.5 MPa and a temperature 300 K. Make calculations for the number of charges that can be made if the volume of bomb calorimeter is 0.0004 m³ and the oxygen in the calorimeter is at 300 K temperature and 2 MPa pressure.

Solution : Characteristic gas constant for oxygen, $R = \frac{8314}{32} = 259.8$ J/kg K

Invoking characteristic gas equation :

$$pV = mRT$$

Mass of oxygen in the cylinder,

$$m_1 = \frac{pV}{RT}$$

$$= \frac{12.5 \times 10^6 \times 0.006}{259.8 \times 300} = 0.9623 \text{ kg}$$

Unconsumed mass of oxygen left in the calorimeter,

$$m_2 = \frac{pV}{RT}$$

$$= \frac{2 \times 10^6 \times 0.006}{259.8 \times 300} = 0.154 \text{ kg}$$

Mass of oxygen consumed $= m_1 - m_2$

= 0.9623 – 0.154 = 0.8083 kg

Mass of oxygen used for charging the calorimeter,

$$m_3 = \frac{pV}{RT}$$

$$= \frac{2 \times 10^6 \times 0.0004}{259.8 \times 300} = 0.01026 \text{ kg}$$

∴ Number of oxygen charges that can be made,

$$= \frac{0.8083}{0.01026} = 78.55 \simeq \mathbf{78}$$

4.5. van der Waal's EQUATION OF STATE

The concept of a perfect gas, that obeys the equation of state $pv = RT$, stems from the kinetic theory of gases that makes the following assumptions about the gas molecules:

- The collision of molecules with one another and with the walls of the container is perfectly elastic.
- There is no molecular attraction between the particles of gas, and that they are in a state of continuous motion.
- The volume occupied by the gas molecules is negligible as compared to the volume of the gas.

Real gases differ from ideal ones due to presence of the intermolecular forces and also to the finite molecular volumes. The van der

Waal's equation of state has been set up by incorporating the following two corrections into the equation of state, $pv = RT$. These modifications take into account the departure of a real gas from an ideal one.

(*i*) Let the equation of state be presented in the form, $v = RT/p$. With increasing pressure, the volume diminishes and if $p \to \infty$, then $v \to 0$. This aspect is in full agreement with the definition of an ideal gas according to which the molecules occupy an infinitely small volume. However in real gases, the molecules occupy a finite volume and an allowance also needs to be made for the volume of voids existing between molecules. As such, the free volume available for molecular motion shall be equal to $(v - b)$, where $b = v_{mol} + v_{void}$.

The quantity b represents the smallest volume to which the gas can be compressed. For each real gas, the quantity b takes a definite numerical value.

(*ii*) Due to the presence of forces of molecular interaction in real gases, there will be a smaller number of impacts with the walls of the container. This may be attributed to the fact that the molecules positioned at the walls are attracted by adjacent molecules inside the container. Consequently, compared to an ideal gas, the pressure exerted by a real gas will be smaller by a quantity Δp. This pressure reduction is directly proportional to the number of molecules affected by molecular interaction, or directly proportional to the square of the density of gas, or inversely proportional to the square of its specific volume.

$$\Delta p = a\,\rho^2 = a/v^2$$

where the proportionality factor a has a definite numerical value for each gas.

When these two corrections are introduced, we obtain the van der Waal's equation of state.

$$\left(p + \frac{a}{v^2}\right)(v - b) = RT \qquad ...(4.17)$$

In this equation,

- the coefficient b is intended to correct for the volume occupied by the gas molecules, and is known as *co-volume*.
- the term $\frac{a}{v^2}$ is a correction that accounts for the molecular forces of attraction; this term is called the *force of cohesion*.

If the volume of one mole of gas is considered, then the van der Waal's equation can be written as

$$\left(p + \frac{a}{v^2}\right)(v - b) = R_o\,T \qquad ...(4.18)$$

where p is in N/m^2, v is in m^3/kg mole, T is in degrees Kelvin, R_o is 8314 J/kg mole K. The units of coefficients a and b are $Nm^4/(kg\ mole)^2$ and m^3/kg mole respectively.

The van der Waal's equation has been found to be sufficiently accurate over a wide range of pressure and temperature, and to wide variety of gases.

4.5.1. Determination of constants a and b

van der Waal's equation can be expanded and written in the form

$$pv^3 - (pb + RT)v^2 + av - ab = 0 \qquad ...(4.19)$$

For given values of p and T, this cubic equation in v will have three roots with the following characteristics:

(*i*) the three roots are real and +ve for a certain range of pressure at low temperatures. With reference to Fig. 4.4, these three values for volume have been indicated as v_1, v_2, and v_3.

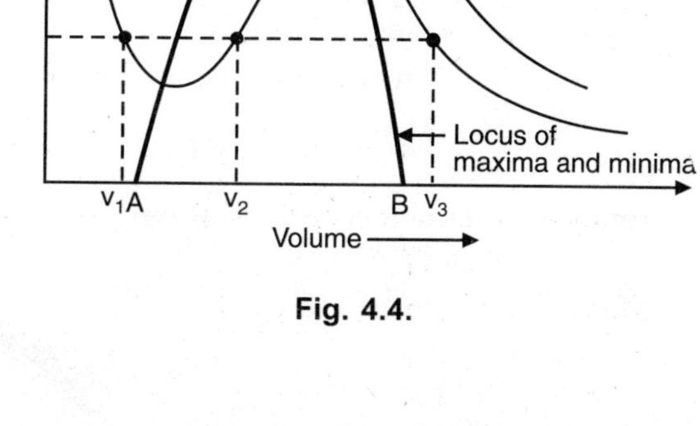

Fig. 4.4.

(*ii*) with rise in temperature, the three real roots approach each other and become equal at the critical point denoted by C.

(*iii*) above critical temperature, there is only one real root for all values of pressure.

If we plot pressure against volume, we get a family of isotherms (contours of constant temperature). The locus of maxima and minima points of the curves has been shown by the thick line ACB. The apex C of this curve is called the **critical point**. The values of pressure, volume and temperature corresponding to the critical point are called the **critical parameters** for a gas and are denoted by the symbols p_c, v_c and T_c respectively. At the critical point, where the roots coincide, the isotherm has zero slope. Further, the slope changes at the critical point , *i.e.*, it exhibits a point of inflection. That is

$$\left(\frac{\partial p}{\partial v}\right)_c = 0 \text{ and } \left(\frac{\partial^2 p}{\partial v^2}\right)_c = 0 \quad ...(4.20)$$

The subscript *c* refers to the values at critical point.

The van der Waal's equation at the critical point is

$$p_c = \frac{RT_c}{v_c - b} - \frac{a}{v_c^2} \quad ...(4.21)$$

$$\left(\frac{\partial p}{\partial v}\right)_c = \frac{-RT_c}{(v_c - b)^2} + \frac{2a}{v_c^3} = 0$$

or $$RT_c = \frac{2a}{v_c^3}(v_c - b)^2 \quad ...(i)$$

Further $$\left(\frac{\partial^2 p}{\partial v^2}\right)_c = \frac{2RT_c}{(v_c - b)^3} - \frac{6a}{v_c^4} = 0$$

or $$RT_c = \frac{3a}{v_c^4}(v_c - b)^3 \quad ...(ii)$$

From expressions (*i*) and (*ii*),

$$\frac{2a}{v_c^3}(v_c - b)^2 = \frac{3a}{v_c^4}(v_c - b)^3$$

or $$2 = \frac{3}{v_c}(v_c - b) = \frac{3v_c - 3b}{v_c}$$

or $$2v_c = 3v_c - 3b; \quad \boldsymbol{v_c = 3b} \quad ...(4.22)$$

Substituting $v_c = 3b$ in expression (*i*)

$$RT_c = \frac{2a}{27b^3}(3b - b)^2$$

$$= \frac{2a \times 4b^2}{27b^3} = \frac{8a}{27b}$$

That gives :

$$\boldsymbol{T_c = \frac{8a}{27bR}} \quad ...(4.23)$$

When the values of v_c and T_c are substituted in the van der Waal's equations, we get

$$p_c = \frac{R(8a/27bR)}{(3b - b)} - \frac{a}{9b^2}$$

$$= \frac{a}{27b^2}$$

That gives :

$$\boldsymbol{p_c = \frac{a}{27b^2}} \quad ....(4.24)$$

Remarks

(*i*) Equations 4.22 to 4.24 can be applied to determine the values of critical parameters v_c, T_c and p_c from known values of van der Waal's constants *a* and *b*

(*ii*) The critical volume v_c is difficult to measure experimentally and accordingly the van der Waals constants are quite often estimated from measured values of p_c and T_c. Rearrangement of equations 4.23 and 4.24 gives

$$T_c = \frac{8a}{27Rb}$$

$$= \frac{8 \times (27b^2 p_c)}{27Rb} = \frac{8b\,p_c}{R}$$

$$\therefore \quad b = \frac{RT_c}{8p_c} \quad ...(4.25)$$

Further $$a = 27b^2 p_c$$

$$= 27\left(\frac{RT_c}{8p_c}\right)^2 p_c$$

$$= \frac{27}{64}\frac{R^2 T_c^2}{p_c} \quad ...(4.26)$$

(*iii*) The parameter $Z = pv/RT$ at critical point for all gases obeying van der Waal's equation of state works out as

$$Z_c = \frac{p_c v_c}{RT_c}$$

$$= \frac{(a/27b^2) \times 3b}{R(8a/27Rb)} = 0.375 \quad ...(4.27)$$

However, the experiments show that this value varies form gas to gas and for most of the gases Z_c lies between 0.23 to 0.31. Apparently Z_c depends upon molecular structure of the gas. Accordingly, the accuracy and utility of van der Waal's equation is limited, *i.e.*, this equation too is not very accurate for predicting the p, v, T behaviour of all gases.

4.6. REDUCED PARAMETERS

The reduced property of a substance is defined as

"The ratio of the property of the substance at the given state to the value of the same property at critical state of the substance"

Thus: Reduced temperature $T_r = \dfrac{T}{T_c}$

Reduced pressure $p_r = \dfrac{p}{p_c}$

Reduced volume $v_r = \dfrac{v}{v_c}$

4.7. REDUCED EQUATION OF STATE

It refers to the van der Waal's equation of state in terms of reduced parameters.

The van der Waal's equation of state is given by

$$\left(p + \frac{a}{v^2}\right)(v - b) = RT$$

and the parameters p, v and T in terms of their reduced values and van der Waal's constants are

$$p = p_c\, p_r = \frac{a}{27b^2}\, p_r$$

$$v = v_c\, v_r = 3b\, v_r$$

$$T = T_c\, T_r = \frac{8a}{27Rb}\, T_r$$

When these values are substituted in van der Waal's equation, we get

$$\left(\frac{a}{27b^2}\, p_r + \frac{a}{9b^2 v_r^2}\right) \times (3b\, v_r - b) = R \times \frac{8a}{27Rb}\, T_r$$

$$\text{or} \quad \frac{a}{27b^2}\left(p_r + \frac{3}{v_r^2}\right) \times b(3v_r - 1) = \frac{8a}{27b}\, T_r$$

$$\frac{a}{27b}\left(p_r + \frac{3}{v_r^2}\right) \times (3v_r - 1) = \frac{8a}{27b}\, T_r$$

$$\text{or} \quad \left(p_r + \frac{3}{v_r^2}\right)(3v_r - 1) = 8T_r \quad ...(4.28)$$

Equation 8.28 is independent of R, a and b which are constants specific to individual gases. Further, the equation contains only reduced parameters and is called the **reduced equation of state**. Again, this equation is true for all fluids which obey the van der Waal's equation of state.

4.8. LAW OF CORRESPONDING STATES

Two gases are said to be in corresponding states:

"If the gases have same values for any two of the three reduced variables, they also have the same value for the third variable"

Denoting reduced variables of the gases by subscript 1 and 2, we have

$$\left(p_{r_1} + \frac{3}{v_{r_1}^2}\right)(3v_{r_1} - 1) = 8T_{r_1}$$

$$\text{and} \quad \left(p_{r_2} + \frac{3}{v_{r_2}^2}\right)(3v_{r_2} - 1) = 8T_{r_2}$$

If $p_{r_1} = p_{r_2}$ and $v_{r_1} = v_{r_2}$, then T_{r_1} must be equal to T_{r_2}. Thus if the two gases have equal values of reduced pressure and reduced volume, then they will have the same value of reduced temperature. This also implies that the gases are considered to be in corresponding states if their pressures, volumes and temperatures are of the same fractions (or multiples) of the critical pressure, volume and temperature of the two gases. Mathematically

$$T_r = f(p_r, v_r) \quad ...(4.29)$$

for all gases and the function is the same. This relation is referred as the **generalised equation** of **state**, and is most accurate in the vicinity of critical point.

4.9. COMPRESSIBILITY FACTOR

The compressibility factor Z is defined as the ratio of the actual volume v of the gas to the value predicated by the ideal gas equation ($pv = RT$) at the same temperature and pressure. That is

$$Z = \frac{\text{actual volume of the gas}}{\text{volume as predicted by the ideal gas equation}}$$

$$= \frac{v}{RT/p} = \frac{pv}{RT} \qquad ...(4.30)$$

The factor Z is dimensionless and becomes unity for an ideal gas at all temperatures and pressures.

The magnitude of Z for a certain gas at a particular temperature and pressure gives an indication of the extent of deviation of the gas from ideal gas behaviour.

The equation $Z = p\,v/RT$ shows that the compressibility factor Z is a function of p and T. Thus the experimental p-v-T data of a pure substance can be fitted into an equation of the form

$$Z = f(p\,,\,T) \qquad ...(4.31)$$

Accordingly, a plot can be made of isotherms (the lines of constant temperature) on the coordinates of p and Z. Such a plot is referred to as the **compressibility chart**. The general shape of such a plot has been depicted in Fig. 4.5.

Such plots show that;

(*i*) A different compressibility chart is needed for each gas.

(*ii*) A general similarity exists in the compressibility chart of different gases; the difference lies in the scales (numerical values).

(*iii*) As the pressure is reduced, the compressibility factor for any gas approaches unity

As $p \to 0\,;\, Z \to 1$

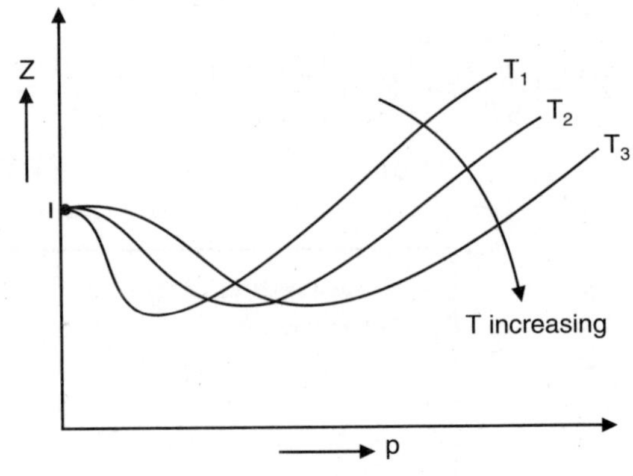

Fig. 4.5.

This aspect stems from the fact that the gases behave more like a perfect gas as the pressure is reduced.

(*iv*) The value of Z for any gas may be more or less than unity depending upon the pressure and temperature of the gas.

For real gases $Z \lesseqgtr 1$. The value of $Z < 1$ means that the actual density is greater, and the value $Z > 1$ means that the actual density is less than that predicted by the ideal equation of state.

(*v*) From the law corresponding states

$$f(p_r\,,\,v_r\,,\,T_r) = 0$$

and the compressibility factor Z can be expressed as

$$Z = \frac{pv}{RT} = \frac{p_c\,p_r}{R}\,\frac{v_c\,v_r}{T_c\,T_r}$$

$$= \frac{p_c\,v_c}{RT_c} \times \frac{p_r\,v_r}{T_r} = Z_c \times \frac{p_r\,v_r}{T_r}$$

The factor Z_c is called the **critical compressibility factor**, and for a gas obeying van der Waal's equation of state

$$Z_c = \frac{p_c\,v_c}{RT_c} = \frac{3}{8}$$

$$\therefore \qquad Z = \frac{3}{8}\,\frac{p_r\,v_r}{T_r}$$

The van der Waal's equation of state expressed in the reduced co-ordinates (Equation 4.28) may then be written as

$$\left[p_r + \frac{27\,p_r^2}{64\,T_r^2\,Z^2}\right]\left[\frac{8\,T_r\,Z}{p_r} - 1\right] = 8\,T_r$$

or

$$Z^3 - \left[\frac{p_r}{8T_r} + 1\right] Z^2 + \left[\frac{27\, p_r}{64\, T_r^2}\right] Z - \frac{27 p_r^2}{512\, T_r^2} = 0 \quad ...(4.32)$$

This is another form of van der Waal's equation and can be expressed as

$$Z = f(p_r, T_r) \quad ...(4.33)$$

This relation is referred to as the **generalised equation of state.**

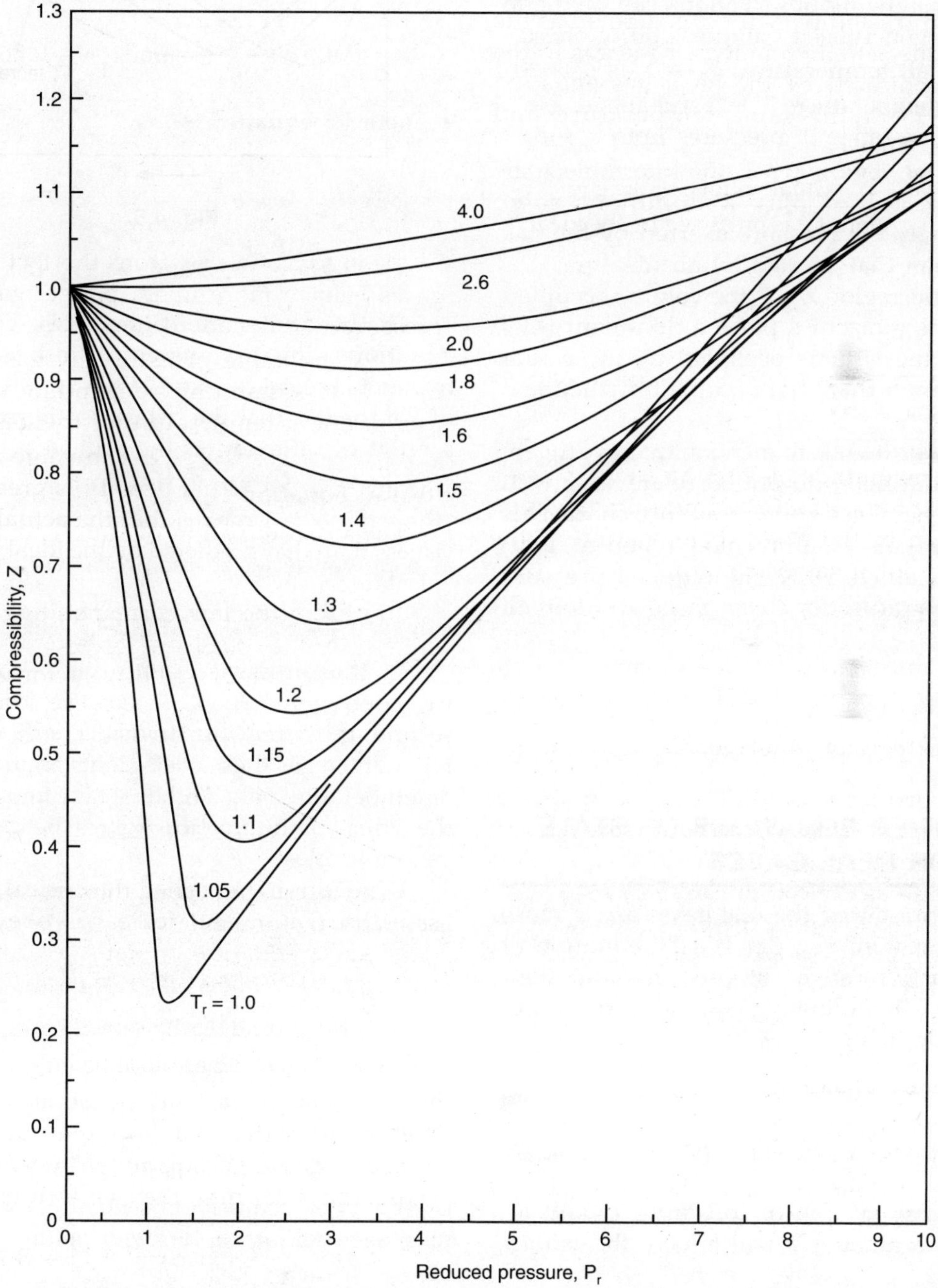

Fig. 4.6. Generalised compressibility chart

If the actual values of Z are plotted against the values of p_r for constant T_r, we get a plot of the form shown in Fig. 4.6. This plot is called the **generalised compressibility chart,** and is same for all gases with accuracy within 5 percent.

The following observations can be made from the generalised compressibility chart:

- At all temperatures $Z \to 1$ as $p_r \to 0$
- At temperatures $T_r = 2$ and above, $Z = 1$ over a wide range of pressures upto $p_r = 5$.
- In the region $Z < 1$, the intermolecular attractive forces play a dominant role. Accordingly the pressure exerted by the gas is less than that exerted by an ideal gas.
- In the region $Z > 1$, the volume occupied by the gas molecules plays a dominant role. As such the volume occupied by the actual gas is greater than that occupied by the ideal gas.

It is worthwhile to mention that hydrogen, helium and neon do not correlate very well on the generalised compressibility chart. This difficulty gets circumvented when at temperatures above 50 K, the reduced pressure and temperature for these gases are defined as

$$p_r = \frac{p}{p_c + 8} \text{ and } T_c = \frac{T}{T_c + 8} \quad \text{...(4.34)}$$

where p is in atmospheres and T is in kelvin.

4.10. OTHER EQUATIONS OF STATE FOR REAL GASES

At high pressures, the real gases are in close agreement with van der Waal's equation of state. For all ranges of pressure and temperatures, the following equations are found to be more true.

- **Bertheolot equation**

$$\left(p + \frac{a}{Tv^2}\right)(v - b) = RT \quad \text{...(4.35)}$$

In terms of critical constants T_c and p_c, the gas parameters a and b have the values

$$a = \frac{27}{64}\frac{R^2T_c^3}{p_c} \quad \text{and} \quad b = \frac{1}{8}\frac{RT_c}{p_c}$$

- **Redlich-Kwong equation**

$$p = \frac{RT}{v - b} - \frac{a}{T^{1/2}\, v(v + b)} \quad \text{...(4.36)}$$

$$a = 0.4278\frac{R^2T_c^{2.5}}{p_c} \quad \text{and} \quad b = 0.0867\frac{RT_c}{p_c}$$

- **Dieterici equation**

$$p = \frac{RT}{v - b}e^{-a/RTv} \quad \text{...(4.37)}$$

$$a = 0.541\frac{R^2T_c^2}{p_c} \quad \text{and} \quad b = 0.135\frac{RT_c}{p_c}$$

In the above mentioned two-constant equations, the parameters a and b can be related to critical constants T_c and p_c by making use of the fact that the critical isotherm shows a point of inflexion at the critical point.

- **Virial equation**

The virial equation of state is expressed as a power series in reciprocal volumes.

$$\frac{pv}{RT} = 1 + \frac{B}{v} + \frac{C}{v^2} + \frac{D}{v^3} + \ldots. \quad \text{...(4.38)}$$

In this pressure explict virial equation, the coefficients B, C, D etc are known as second, third and fourth virial coefficient etc. For a given gas these coefficients are functions of temperature only, and they take into account the effect of interaction forces between the gas molecules

In an alternative form, the virial equation is expressed as a power series in the pressure.

$$\frac{pv}{RT} = 1 + B'p + C'p^2 + D'p^3 + \ldots \quad \text{...(4.39)}$$

In this volume explict virial equation, the factors B', C', D' etc are virial coefficients and these depend on the nature of the gas and temperature only.

The virial coefficients B', C', D' are related to the virial coefficients B, C, D and the necessary relation is derived below:

The pressure explicit virial equation is

$$\frac{pv}{RT} = 1+\frac{B}{v}+\frac{C}{v^2}+\frac{D}{v^3}+....$$

or $$p = \frac{RT}{v}+\frac{BRT}{v^2}+\frac{CRT}{v^3}+\frac{DRT}{v^4}+... \quad ...(i)$$

The volume explict virial equation is

$$\frac{pv}{RT} = 1+B'p+C'p^2+D'p^3+...$$

or $$p = \frac{RT}{v}+\frac{B'pRT}{v}+\frac{C'p^2RT}{v}+\frac{D'p^3RT}{v} \quad ...(ii)$$

When p from expression (*i*) is substituted in expression (*ii*), we get

$$p = \frac{RT}{v}+\frac{B'RT}{v}\left[\frac{RT}{v}+\frac{BRT}{v^2}+\frac{CRT}{v^3}+\frac{DRT}{v^4}+....\right]$$

$$+\frac{C'RT}{v}\left[\frac{RT}{v}+\frac{BRT}{v^2}+\frac{CRT}{v^3}+\frac{DRT}{v^4}+...\right]^2$$

$$+\frac{D'RT}{v}\left[\frac{RT}{v}+\frac{BRT}{v^2}+\frac{CRT}{v^3}+\frac{DRT}{v^4}+...\right]^3 \quad ...(iii)$$

Both the expressions (*i*) and (*iii*) represent the virial equation as power series in reciprocal volumes. We get the following results when the like terms are compared:

Terms of $\frac{1}{v^2}$ give :

$$\frac{BRT}{v^2} = \frac{B'(RT)^2}{v^2}$$

$$\therefore \quad B' = \frac{B}{RT}$$

Terms of $\frac{1}{v^3}$ give :

$$\frac{CRT}{v^3} = \frac{B'B(RT)^2}{v^3}+\frac{C'(RT)^3}{v^3}$$

or $$C = B'BRT + C'(RT)^2$$

$$= \left(\frac{B}{RT}\right)BRT+C'(RT)^2$$

$$= B^2 + C'(RT)^2$$

$$\therefore \quad C' = \frac{C-B^2}{(RT)^2}$$

Terms of $\frac{1}{v^4}$ give :

$$\frac{DRT}{v^4} = \frac{B'C(RT)^2}{v^4}+\frac{2BC'(RT)^3}{v^4}+\frac{D'(RT)^4}{v^4}$$

or $$D = B'C(RT)+2BC'(RT)^2+D'(RT)^3$$

$$= \left(\frac{B}{RT}\right)C(RT)+2B \times\frac{(C-B^2)}{(RT)^2}\times(RT)^2+D'(RT)^3$$

$$= BC+2BC-2B^3+D'(RT)^3$$

$$\therefore \quad D' = \frac{D-3BC+2B^3}{(RT)^3}$$

The coefficients in the virial equations are determined from experimental *p-v-T* data. The equation is valid essentially for gases at low and medium densities.

EXAMPLE 4.17

A vessel of 2.75 m³ capacity contains 5 kg of oxygen at 300 K . Determine the pressure exerted by oxygen using van der Waal's equation. Compare the result obtained if oxygen is treated as ideal gas. The constants appearing in the van der Waal's equation have the values :

a = 139.35 × 10³ Nm⁴/(kg mol)² and

b = 0.0314 m³/kg mol

Solution : Volume occupied by 1 kg mole of oxygen $= \frac{2.75}{5}\times 32 = 17.6\ m^3$

The van der Waal's equation is written as

$$\left(p+\frac{a}{v^2}\right)(v-b) = R_o\,T$$

or $$p = \frac{R_o\,T}{v-b}-\frac{a}{v^2}$$

Substituting the given values in the above relation, we get

$$p = \frac{8314\times 300}{(17.6-0.0314)} - \frac{139.35\times 10^3}{(17.6)^2}$$

$= 141969.2 - 449.86$

$= 142419 \text{ N/m}^2 \simeq$ **1.43 bar**

(*ii*) Gas constant for oxygen,

$$R = \frac{R_o}{M} = \frac{8314}{32} = 259.8 \text{ J/kg K}$$

Using the ideal gas equation, $pV = mRT$

$$p = \frac{mRT}{V} = \frac{5 \times 259.8 \times 300}{2.75}$$

$= 1.417 \times 10^5 \text{ N/m}^2 =$ **1.417** bar

EXAMPLE 4.18

A cylinder of volume capacity 0.1 m³ is filled with 1.35 kg of ammonia at 2 MPa pressure. Using van der Waal's equation of state, determine the temperature at which ammonia exists in the cylinder. The van der Waals constants for ammonia are:

$a = 422.55 \times 10^{-3}\ P_a(m^3/mol)^2$ *and*

$b = 37 \times 10^{-6}\ m^3/mol$

Compare the result obtained if ammonia is treated as ideal gas.

Solution : From Avogadro's law, $m = nM$

moles of ammonia

$$n = \frac{1.35}{17} = 0.0794$$

molar volume

$$v = \frac{0.1}{0.0794} = 1.259 \text{ m}^3/\text{kg mol}$$

$a = 422.55 \times 10^{-3} \text{ Pa(m}^3/\text{mol)}^2$

$= 422.55 \times 10^3 \text{ Nm}^4/(\text{kg mol})^2$

$b = 37 \times 10^{-6} \text{ m}^3/\text{mol}$

$= 37 \times 10^{-3} \text{ m}^3/\text{kg mol}$

The van der Waal's equation of state is written as

$$\left(p + \frac{a}{v^2}\right)(v - b) = R_o T$$

Substituting the values of p, v, a and b in the above relation, we get

$$\left[2 \times 10^6 + \frac{422.55 \times 10^3}{(1.259)^2}\right]\left[1.259 - 37 \times 10^{-3}\right] = 8314\ T$$

or $(2.266 \times 10^6)\ (1.222) = 8314\ T$; $T =$ **333** K

(*b*) Gas constant for ammonia (molecular mass = 17)

$$R = \frac{R_o}{M} = \frac{8314}{17} = 489.06 \text{ J/kg K}$$

Using the ideal gas equation $pv = mRT$

$$T = \frac{pv}{mR} = \frac{(2 \times 10^6) \times 0.1}{1.35 \times 489.06} = \mathbf{302.92} \text{ K}$$

EXAMPLE 4.19

One kg-mol of oxygen at 350 K undergoes a reversible non-flow isothermal expansion and the volume increases from 0.06 m³/kg to 0.15 m³/kg. Using van der Waal's equation of state, calculate the final pressure and the work done during the process. How does this value for work compare with that obtained by using ideal gas equation of state?

Solution : The van der Waal's equation of state is prescribed by the relation

$$\left(p + \frac{a}{v^2}\right)(v - b) = R_o T$$

where p = pressure of gas is N/m², v is molar volume and R_o is universal gas constant. For oxygen, the constants a and b have the values.

$a = 139.35 \times 10^3 \text{ Nm}^4/(\text{kg-mol})^2$

and $b = 0.0314 \text{ m}^3/\text{kg-mol}$

(*a*) $v_1 = 0.06 \times 32 = 1.92 \text{ m}^3/\text{kg-mol}$

$v_2 = 0.15 \times 32 = 4.8 \text{ m}^3/\text{kg-mol}$

Final pressure

$$p_2 = \frac{R_o T}{v_2 - b} - \frac{a}{v_2^2}$$

$$= \frac{8314 \times 350}{4.8 - 0.0314} - \frac{139.35 \times 10^3}{(4.8)^2}$$

$= 610221 - 6048$

$= 604173 \text{ N/m}^2 =$ **6.04 bar**

(*b*) Work done

$$W = \int_1^2 p\,dv = \int_1^2 \left[\left(\frac{R_o T}{v - b}\right) - \frac{a}{v^2}\right] dv$$

$$= R_o T \left|\log_e (v - b)\right|_{v_1}^{v_2} + \left|\frac{a}{v}\right|_{v_1}^{v_2}$$

$$= R_o T \left[\log_e \left(\frac{v_2 - b}{v_1 - b}\right)\right] + a\left[\frac{1}{v_2} - \frac{1}{v_1}\right]$$

$$= 8314 \times 350 \left[\log_e \left(\frac{4.8 - 0.0314}{1.92 - 0.0314} \right) \right]$$

$$+ 139.35 \times 10^3 \left[\frac{1}{4.8} - \frac{1}{1.92} \right]$$

$= 2695269 - 43547$

$= 2651722$ Nm/kg-mol

$\simeq$ **2652** kJ/kg-mol

(*c*) Work done in an isothermal process using ideal gas equations

$$W = \int p\,dv = \int \frac{dv}{v} = R_o T \log_e \left(\frac{v_2}{v_1} \right)$$

$$= 8314 \times 350 \log_e \frac{4.8}{1.92}$$

$= 2666314$ Nm/kg-mol

$\simeq$ **2666** kJ/kg-mol

SALIENT POINTS

1. Boyle's law provides a reciprocal relationship between pressure p and volume of a gas at constant temperature.

$$p \propto \frac{1}{v} \quad \text{or} \quad pv = \text{constant}$$

(at constant temperature)

2. Charles law provides direct relationship

(*a*) between volume and temperature of a gas at constant pressure

$$v \propto T \quad \text{or} \quad \frac{v}{T} = \text{constant}$$

(at constant pressure)

(*b*) between pressure and temperature of a gas at constant volume

$$p \propto T \quad \text{or} \quad \frac{p}{T} = \text{constant}$$

(at constant volume)

3. The ideal equation of a state is given by

$$pv = RT \quad \text{or} \quad pV = mRT$$

where v is the specific volume, V is the total volume and R is the gas constant for the particular gas.

If p is in N/m^2, V is in m^3, m is in kg and T is in degrees kelvin, the units of the gas constant R work out to be

Nm/kg K or J/kg K

4. The gas constant R, though constant for a particular gas, has different values for different gases. For air R = 287 J/kg K.

5. In practice, no gas obeys the ideal gas equation rigidly. It is essentially valid at low pressures and high temperatures where the density is sufficiently low.

6. An ideal gas has no forces of inter molecular attraction. It obeys the law $pv = RT$. The specific heat capacities are not constant but are functions of temperature. A perfect gas obeys the law $pv = RT$ and has constant specific heats.

7. The gas constant multiplied by molecular mass (MR) is a constant for all gases. This constant is called the universal or molar gas constant, and it is equal to

$$R_o = MR = 8314 \text{ J/kg mol K}$$
$$= 8.314 \text{ J/mol K}$$
$$= 8.314 \text{ kJ/kg mol K}$$

8. The value of gas constant R for any gas can be calculated by dividing the universal gas constant by molecular mass of the gas.

9. Avogadro's hypothesis: Equal volumes of different gases at the same pressure and temperature contain equal number of molecules.

10. The mass of 22.41 m^3 of any gas at NTP (atmospheric pressure 1.01325 bar and 0°C) is equal to its molecular mass in kg. For example 22.41 m^3 of hydrogen at NTP has a mass of 2 kg.

11. A 22.41 m^3 of volume is known as a mole. Obviously mole is a bigger unit of volume.

12. The van der Waal's equation of state is prescribed as

$$\left(p + \frac{a}{v^2} \right)(v - b) = RT$$

where a and b are the constants for the gas in question, v represents the volume per unit mass and R is the gas constant.

13. If the volume of one mole is considered, then the van der Waal's equation is written as

$$\left(p+\frac{a}{v^2}\right)(v-b) = R_0T$$

The units of p, v, T, R, a and b are follows
$p \to N/m^2$; $v \to m^3/kg$ mol ; $T \to K$;
$a \to Nm^4/(kg\ mole)^2$; $b \to m^3/kg$ mole and
$R \to 8314$ J/kg mole K.

14. Every equation of state satisfies the following criteria:
    - the slope of an isotherm on a p versus v diagram is negative.
    - it reduces to the ideal gas law, $pv = RT$, as $p \to 0$
    - the critical isotherm exhibits a point of inflexion at the critical point.
15. In terms of van der Waal's constants a and b, the critical parameters are:

$$v_c = 3b\ ; \quad T_c = \frac{8a}{27\,Rb}$$

and $$P_c = \frac{a}{27\,b^2}$$

Alternatively: $a = \dfrac{27\,R^2\,T_c^2}{64\,P_c}$ and $b = \dfrac{RT_c}{8\,P_c}$

16. For all gases which obey the van der Waal's equation of state

$$\frac{P_c v_c}{R\,T_c} = 0.375$$

17. The compressibility factor Z is defined as the ratio of actual volume v to be volume predicted by the gas at the same temperature and pressure. That is

$$Z = \frac{v}{RT/p} = \frac{pv}{RT}$$

For an ideal gas $Z = 1$ at all temperatures and pressures. For a real gas Z differs from unity and the value indicates the extent of deviation from the ideal gas law.

REVIEW QUESTIONS

A. Conceptual and conventional questions.

1. What is an ideal gas ? How does it differ from a perfect gas ?
2. When is the relation $pv = RT$ expected to be applicable ?
3. Enumerate the laws of perfect gases ?
4. What is an equation of state ?
5. State the conditions necessary for a substance to be defined as perfect gas.
6. The specific heat of a gas varies linearly with temperature only. Would the gas obey the relation $pv = RT$?
7. Define the characteristic gas constant. How does it differ from universal gas constant? Write the units for these constants.
8. What are the dimensions of the gas constant and what is its physical significance ?
9. State and explain Avogadro's law. Show that the product of gas constant and molecular mass has the same value for all gases.
10. Define mole. Is it a unit of mass or volume ?
11. Write down the van der Waal's equation of state. How does it differ from the ideal gas equation ?
12. Define co-volume and force of cohesion.
13. What are the units of constants a and b appearing in van der Waal's equation ?
14. Explain the factors which are taken into account by van der Waal in modifying the ideal gas equation.
15. Define the terms reduced pressure and reduced volume. Express the van der Waal's equation of state in terms of reduced parameters.
16. Develop the following form of van der Waal's equation in terms of reduced parameters.

$$\left(p_r+\frac{3}{v_r^2}\right)(3v_r-1) = 8T_r$$

17. State the law of corresponding states.
18. Define compressibility and explain its significance
19. Set up an expression for the van der Waals equation of state in terms of the compressibility factor, reduced temperature and reduced pressure.
20. Determine the mass of air contained in a room measuring 10 m × 6 m × 4 m if the pressure is 1.01325 bar and the temperature is 27°C.

[Ans. 282.44 kg**]**

21. An air bottle of 20 litres capacity used for starting a diesel engine contains air at 0.11 MPa and 20°C. After charging the bottle, the pressure reaches 1.5 MPa and temperature rises to 50°C. Make calculations for mass of air added to the bottle.

22. Determine the density and specific volume of carbon monoxide at a pressure 100 kPa and a temperature of 300 K. For CO, gas constant = 297.4 J/kg K.

[**Ans.** 0.89m^3/kg, 1.125 kg/m^3]

23. A tank of 0.35 m^3 capacity contains H_2S (molecular mass 34) at 300 K . When 2.5 kg of gas is withdrawn, the temperature in the tank becomes 288 K and pressure 10.5 bar absolute. Calculate the mass of gas initially present in the tank. Also determine the initial pressure of gas.

[**Ans.** 7.5 kg, 15.72 bar]

24. A steel cylinder contains oxygen at a pressure of 3 MPa and temperature 300 K. After using a quantity of gas, the pressure drops to 0.6 MPa and temperature becomes 290 K. Determine the mass of oxygen used. It may be presumed that initially 1000 litres of oxygen was put in the cylinder at normal temperature pressure (NTP).
Take the density of oxygen at NTP
= 1.43 gm./litre.

[**Ans.** 1134.14 gm]

25. A vessel of 0.1 m^3 volume contains oxygen at p = 9 bar gauge and temperature 27°C. Subsequently when oxygen was pumped into the vessel, the pressure increased to 105 bar and the temperature became 67°C. If ambient pressure is 1 bar, make calculations for oxygen pumped into the cylinder.

[**Ans.** 10.717 kg]

26. 5 kg of carbondioxide occupies 1.5 m^3 at 300 K. Determine the pressure exerted by CO_2 gas using van der Waal's equation. How this result would compare with the one obtained by treating CO_2 as an ideal gas. The constants a and b appearing in van der Waal's equation have the values a = 3.6285 × 10^5 Nm^4/(kg mol)^2 and b = 0.0423 m^3/kg mol.

[**Ans.** 1.681 bar, 1.89 bar]

B. Fill in the blanks with appropriate word/words.

1. A perfect gas obeys the law .... and has constant ....
2. The gas constant R has units of energy per unit .... and unit ....
3. For an ideal gas, the .... and .... are functions of temperature alone.
4. Molar volume is equal to .... at normal temperature and pressure.
5. The value of gas constant R for any gas can be calculate by dividing the universal gas constant by .... of the gas.
6. In the van der Waal's equation

$$\left(p+\frac{a}{v^2}\right)(v-b)=R_oT$$, the coefficient b is a correction for the volume occupied by .... and is known as ....

7. The ratio of the property of substance at the given state to the value of the same property at the criticial state of the substance is called the ....
8. If the gases have same values for any two of the reduced variables, they also have the same value for the third variable. This statement represents the ....
9. The .... is defined as the ratio of the actual volume of the gas to the value predicted by the ideal gas equation at the same temperature and pressure.

Answers

1. $pv = RT$, specific heats 2. mass, temperature difference 3. internal energy, enthalpy 4. 22.41 m^3 5. molecular mass 6. gas molecules, co-volume 7. reduced property 8. law of corresponding states 9. compressibility factor

C. Multiple choice questions.

1. Boyle's law states that
   (*a*) at constant volume, the pressure of a gas varies directly with temperature
   (*b*) at constant pressure, the volume of a gas varies directly with temperature
   (*c*) under constant temperature conditions, the product of pressure and volume of a given mass of gas is always constant
   (*d*) the volume of a given mass of gas held at constant temperature is directly proportional to its pressure
2. For all perfect gases, the change in volume is 1/273 of original volume at 0°C for every 1°C change in temperature while pressure remains constant. This statement is referred to as
   (*a*) Boyle's law
   (*b*) Charles' law

(c) Gay-Lussac's law
(d) Joule's law

3. The characteristic gas equation $pv = mRT$ is essentially valid for a
(a) real gas (b) perfect gas
(c) monoatomic gas (d) mixture of gases

4. Which one of the following identity is correct ?
(a) $G = MR$ (b) $G = M/R$
(c) $G = R/M$ (d) $G = M^2 R$
where G is universal gas constant. R is characteristic gas constant and M is the molecular weight

5. Which one of the following gases will have the maximum value of gas constant R ?
(a) nitrogen (b) carbon dioxide
(c) sulphur dioxide (d) oxygen

6. Consider the following statements
An ideal gas
1. has no intermolecular forces of attraction or repulsion
2. does not change its phase during a thermodynamic process
3. obeys a set of common rules govering change of its properties

Which of these statements is/are correct ?
(a) 1 and 2 (b) 2 and 3
(c) 3 alone (d) 1, 2 and 3

7. Which aspect is true in the context of Avogadro's hypothesis ?
(a) the product of molecular weight and the characteristic gas constant is same for all gases
(b) the pressure of a gas mixture equals the sum of partial pressures of the gases comprising the mixture
(c) for a gas, the specific heat at constant pressure is greater than the specific heat at constant volume
(d) the molecular weight of all gases occupy the same volume under same conditions of pressure and temperature

8. Molar volume is equal to .... at normal temperature and pressure
(a) 29.27 m^3 (b) 22.4 m^3
(c) 42.7 m^3 (d) 84.8 m^3

9. The ideal gas laws can be applied, with least error, to
(a) boiling water (b) wet steam
(c) dry saturated steam
(d) superheated steam

10. In the van der Waal's equation
$\left(p + \frac{a}{v^2}\right)(v - b) = RT$, the constant b accounts for
(a) the finite volume occupied by the gas molecules
(b) the molecular forces of attraction
(c) the elastic collisions between the gas molecules
(d) the momentum of gas molecules in random fashion

11. In the general gas equation $pv = RT$, van-der-Waal introduced a correction factor a/v^2 which is added to pressure p. This factor represents
(a) mean velocity of the gas molecules
(b) attraction force between molecules
(c) volume occupied by the molecules
(d) effective area of the molecules

12. In the van-der-Waal's equation
$\left(p + \frac{a}{v^2}\right)(v - b) = RT,$ the units of constant a are
(a) Nm^2 (b) N/m^4
(c) Nm^4 (d) None of the above

13. In the van-der-Waal's equation
$\left(p + \frac{a}{v^2}\right)(v - b) = RT,$ a greater value of constant a indicates
(a) higher density of the gas
(b) less force of cohesion
(c) closeness of the gas to the ideal conditions
(d) that the gas can be liquefied easily

Answers

1. (c)	2. (b)	3. (b)	4. (a)	5. (a)
6. (d)	7. (d)	8. (b)	9. (d)	10. (a)
11. (b)	12. (c)	13. (d)		

❑❑❑

Chapter 5

Mixture of Gases

Learning objectives : A study of the subject matter of this chapter on thermodynamic properties of gas mixtures will enable the reader to :

- *explain the concept of mole and mass fraction*
- *define partial presure and explain Daltons law of partial pressures*
- *explain the Amagat Leduc law of partial volumes*
- *set up the relation between partial pressure, mole fraction and volume fraction*
- *set up the relations for gas constant, molecular mass, specific heats and entropy of the gas mixture*

The working fluids employed in most of the heat engines are not homogeneous gases, but are mixture of gases such as air, combustion products. The constituent gases are assumed to be ideal, do not interact with one another chemically and the gas mixture obeys the laws of ideal homogeneous gases. Each gas in the mixture behaves independently, occupies the entire volume of the mixture and exerts its own pressure on the walls of the container.

5.1. MOLE FRACTION AND MASS FRACTION

Consider a homogeneous non-reacting (inert) mixture of ideal gases *a, b, c*... etc at temperature *T*, pressure *p* and occupying a volume *V* as shown in Fig. 5.1. Further let

n_a, n_b, n_c ... = number of moles of individual gases in the mixture

m_a, m_b, m_c ... = masses of individual gases in the mixture

Then for the total mass *m* and the total number of moles *n* we have:

$$m = m_a + m_b + m_c + \ldots \qquad \ldots(5.1)$$

and $$n = n_a + n_b + n_c + \ldots \qquad \ldots(5.2)$$

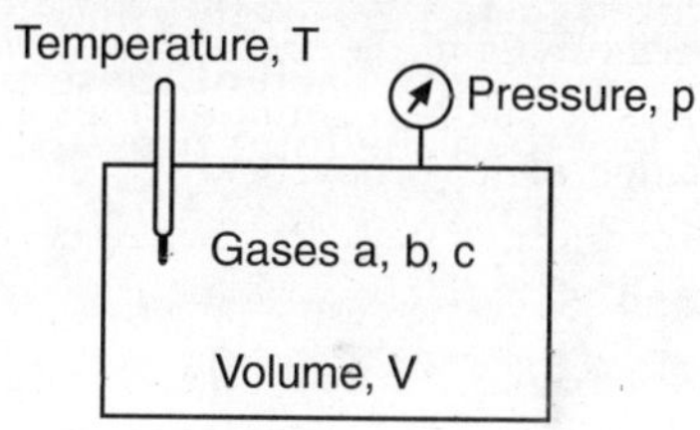

Fig. 5.1.

The **mass fraction** of each component of a mixture is defined as the ratio of the mass of a given component to the mass of the entire mixture.

Upon dividing expression 5.1 by mass *m* of the gas mixture,

$$1 = \frac{m_a}{m} + \frac{m_b}{m} + \frac{m_c}{m} + \ldots$$

$$= x_a + x_b + x_c + \ldots$$

where x_a, x_b, x_c ... are the mass fractions. Obviously ***"the sum of the parts by mass (mass fractions) of the constituents in a mixture equals unity"***

When an analysis of gas mixture is made on the mass basis, it is called a **gravimetric analysis**.

The **mole fraction** of each component of a mixture is defined as the ratio of the moles of each constituent gas to the total number of moles contained in the gas mixture.

Upon dividing expression 5.2 by number of moles n of the gas mixture,

$$1 = \frac{n_a}{n} + \frac{n_b}{n} + \frac{n_c}{n} + \ldots$$
$$= y_a + y_b + y_c + \ldots$$

where y_a, y_b, y_c ... are the mole fractions. Obviously ***the sum of the parts by moles (mole fractions) of the constituents in a mixture equals unity.***

5.2. PARTIAL PRESSURE AND DALTON'S LAW

Consider a homogeneous non-reacting mixture of ideal gases a, b, c ... etc at temperature T, pressure p and occupying a volume V. (Fig. 5.2)

Further, let it be presumed that each constituent of the mixture exists separately at temperature T and volume V, and pressures p_a, p_b, p_c ... exerted by individual gases are measured separately. Each of these pressures would be less than the total pressure p of the mixture.

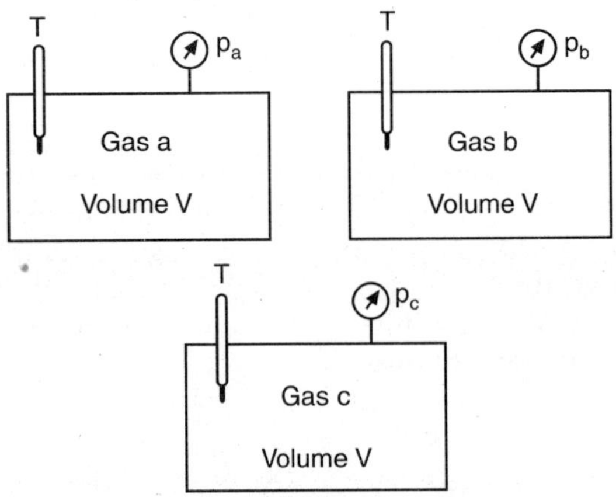

Fig. 5.2.

When the equation of state for an ideal gas is applied to the gas mixture as well as the constituent gases, we have

For the mixture :

$$pV = mRT$$
$$= nMRT = nR_{mol}T \qquad \ldots(5.3)$$

where R_{mol} is the universal gas constant (R_{mol} = 8314 J/kg mole K) and M is the molecular mass

For the constituent gases

$$p_a\ V = n_a\ R_{mol}T$$
$$p_b\ V = n_b\ R_{mol}T$$
$$p_c\ V = n_c\ R_{mol}T \text{ etc.}$$

Upon adding for the components

$$(p_a + p_b + p_c + \ldots)\ V = (n_a + n_b + n_c + \ldots)\ R_{mol}T$$
$$= nR_{mol}T \qquad \ldots(5.4)$$

From expressions 5.3 and 5.4,

$$p = p_a + p_b + p_c + \ldots$$
$$p = \Sigma p_i \qquad \ldots(5.5)$$

where $p_i = \dfrac{n_i R_{mol}T}{V}$ represents the pressure the component i would exert if it alone occupied the volume V at temperature T. This is called the **partial pressure** of the ith component of the gas mixture. Thus

Partial pressure is defined as the pressure which each individual component of a gas mixture would exert if it alone occupied the volume of the mixture at the same temperature.

Further, the equation 5.5 stipulates that total pressure of a mixture of ideal gases is equal to the sum of the partial pressures of the individual gas components of the mixture.

This is known as **Dalton's law of partial pressures.**

The following relations are implicit in the Dalton's law:

$$t = t_a = t_b = t_c$$
$$V = V_a = V_b = V_c$$

and $$m = m_a + m_b + m_c \qquad \ldots(5.6)$$

where t, V and m respectively represent the temperature, volume and mass. In terms of specific volume v,

$$m\ v = m_a\ v_a = m_b\ v_b = m_c\ v_c \qquad \ldots(5.7)$$

Combining expressions 5.6 and 5.7, we may write

$$\frac{m}{mv} = \frac{m_a}{m_a v_a} + \frac{m_b}{m_b\ v_b} + \frac{m_c}{m_c v_c}$$

or $$\frac{1}{v} = \frac{1}{v_a} + \frac{1}{v_b} + \frac{1}{v_c} \qquad \ldots(5.8)$$

The reciprocal of specific volume is density and so we can write

$$\rho = \rho_a + \rho_b + \rho_c \quad ...(5.9)$$

which means that density of the mixture is equal to the sum of the densities of the components.

5.3. AMAGAT-LEDUC LAW OF PARTIAL VOLUMES

The partial (reduced) volume of a gas component represents the volume that this gas component would occupy if its temperature and pressure were equal to these parameters of the gas mixture

The ratio of the partial (reduced volumes) of each constituent gas to the total volume of the gas mixture is called the volume fraction

$$r_a = \frac{V_a}{V}, \quad r_b = \frac{V_b}{V}, \quad r_c = \frac{V_c}{V}$$

where r_a, r_b, r_c ... are volume fractions; V_a, V_b, V_c... are partial (reduced) volumes of each constituent gas; and V is the total volume of the gas mixture.

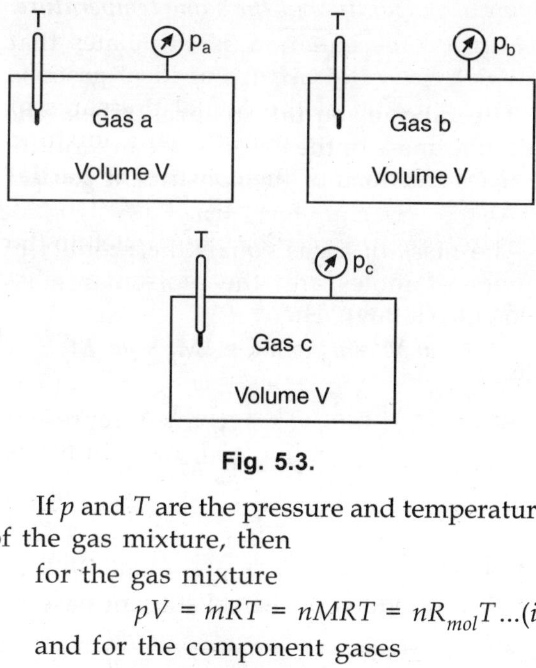

Fig. 5.3.

If p and T are the pressure and temperature of the gas mixture, then

for the gas mixture

$$pV = mRT = nMRT = nR_{mol}T \quad ...(i)$$

and for the component gases

$$p\,V_a = n_a R_{mol} T$$
$$p\,V_b = n_b R_{mol} T$$
$$p\,V_c = n_c R_{mol} T$$

Upon adding for the components

$$p(V_a + V_b + V_c + ...) = (n_a + n_b + n_c + ...)R_{mol}T$$
$$= nR_{mol}T \quad ...(ii)$$

From expressions (*i*) and (*ii*)

$$V = V_a + V_b + V_C + ... \quad ...(5.10)$$

Equation 5.10 states that the total volume of a mixture of ideal gases is equal to the sum of the partial (reduced) volumes of the constituent gases. This is known as **Amagat-Leduc** law of partial volumes.

Upon dividing expression 5.10 by the total volume of the gas mixture,

$$1 = \frac{V_a}{V} + \frac{V_b}{V} + \frac{V_c}{V} + ...$$
$$= r_a + r_b + r_c + \quad ... \quad ...(5.11)$$

Thus the sum of parts by volumes (volume fractions) is equal to unity.

When the analysis of a gas mixture is made on volume basis, it is called **volumetric analysis.**

5.4. RELATION BETWEEN PARTIAL PRESSURE, MOLE FRACTION AND VOLUME FRACTION

The characteristic gas equation can be written as

$$pV = mRT = nR_{mol}T$$

for the gas mixture ...(*i*)

$$p_i V = m_i\,R_i T = n_i\,R_{mol}T$$

for a constituent of the gas mixture ...(*ii*)

From these two identities: $\frac{p_i}{p} = \frac{n_i}{n}$ ...(*iii*)

Further, the gas equation for the reduced or partial volumes of the gas can be written as

$$pV_i = m_i\,R_i T = n_i\,R_{mol}T \quad ...(iv)$$

From identities (*i*) and (*iv*) : $\frac{V_i}{V} = \frac{n_i}{n}$ ...(*v*)

Comparing identities (*iii*) and (*v*)

$$\frac{p_i}{p} = \frac{n_i}{n} = \frac{V_i}{V} \quad ...(5.12)$$

i.e., partial pressure ratio = mole fraction = volume fraction

That is : ***In a mixture of ideal gases, the partial pressure ratio, the mole fraction and the volume fraction are all equal.***

Equation 5.12 suggests that:

$p_i \propto n_i$, *i.e.*, the partial pressure of a particular gas in the gaseous mixture is directly proportional to the number of moles of that gas

$p_i \propto V_i$, *i.e.*, the partial pressure of a particular gas in the gaseous mixture is directly proportional to the partial or reduced volume of that gas.

In terms of masses, the identities (*i*) and (*ii*) give

$$\frac{p_i}{p} = \frac{m_i}{m}\frac{R_i}{R}$$

or $$p_i = p\, x_i \left(\frac{R_i}{R}\right) \quad ...(5.13)$$

That is : the partial pressure p_i of any constituent gas in a mixture is equal to the pressure p of the gas mixture multiplied by mass fraction x_i of the constituent gas and by the ratio R_i/R between the gas constants of the constituent gas and the gas mixture.

5.5. GAS CONSTANT, MOLECULAR MASS AND SPECIFIC HEATS OF THE GAS MIXTURE

Gas constant :

For each constituent of gas mixture contained in a vessel of volume V and at temperature T, we can write

$$p_a V = m_a R_a T$$
$$p_b V = m_b R_b T$$
$$p_c V = m_c R_c T$$

upon adding :

$$(p_a + p_b + p_c)\, V = (m_a R_a + m_b R_b + m_c R_c)T$$

According to Dalton's law, the sum of the partial pressures of the constituent gases gives the total pressure p exerted by the mixture

$$\therefore \quad pV = (m_a R_a + m_b R_b + m_c R_c)\, T \quad ...(i)$$

For the gas mixture, we have :

$$pV = (m_a + m_b + m_c)\, RT$$
$$= mRT \quad ...(ii)$$

Comparison of the identities (*i*) and (*ii*) gives

$$mR = m_a R_a + m_b R_b + m_c R_c$$

or $$R = \frac{m_a}{m} R_a + \frac{m_b}{m} R_b + \frac{m_c}{m} R_c$$
$$= x_a R_a + x_b R_b + x_c R_c$$
$$= \Sigma\, x_i R_i \quad ...(5.14)$$

Recalling that for any gas : $MR = R_{mol}$

The gas constant R for the gas mixture can then be rewritten as

$$R = R_{mol}\left(\frac{x_i}{M_i}\right) \quad ...(5.15)$$

Molecular mass :

Equation 5.15 can be recast as

$$\frac{R_{mol}}{M} = x_a \frac{R_{mol}}{M_a} + x_b \frac{R_{mol}}{M_b} + x_c \frac{R_{mol}}{M_c}$$

or $$\frac{1}{M} = \frac{x_a}{M_a} + \frac{x_b}{M_b} + \frac{x_c}{M_c}$$

or $$M = \frac{1}{\left(\frac{x_a}{M_a}\right) + \left(\frac{x_b}{M_b}\right) + \left(\frac{x_c}{M_c}\right)}$$
$$= \frac{1}{\Sigma \frac{x_i}{M_i}} \quad ...(5.16)$$

The above equation prescribes the molecular mass of the gas mixture in terms of the mass fractions of the constituent gases.

Also $\quad m = m_a + m_b + m_c$

The mass of a gas equals the product of number of moles and the molecular mass (Avogadro's law). Then

$$n\, M = n_a M_a + n_b M_b + n_c M_c$$

or $$M = \frac{n_a}{n} M_a + \frac{n_b}{n} M_b + \frac{n_c}{n} M_c$$
$$= y_a M_a + y_b M_b + y_c M_c$$
$$= \Sigma\, y_i M_i \quad ...(5.17)$$

The above equation prescribes the molecular mass of the gas mixture in terms of the mole fractions of the constituent gases.

Specific heats :

Gibbs—Dalton law states that : The internal energy, enthalpy and entropy of a gaseous mixture are respectively equal to the sums of the internal energies, the enthalpies and the entropies which each component of the gas mixture would have, if each alone occupied the volume of the mixture at the temperature of the mixture.

Mathematically ;

$$m\,u = m_a\,u_a + m_b\,u_b + m_c\,u_c$$

or $$m\,c_v\,dt = m_a\,(c_v)_a\,dt + m_b\,(c_v)_b\,dt + m_c\,(c_v)_c\,dt$$

Therefore, the specific heat at constant volume c_v for the mixture is

$$c_v = \frac{m_a}{m}(c_v)_a + \frac{m_b}{m}(c_v)_b + \frac{m_c}{m}(c_v)_c$$

$$= x_a\,(c_v)_a + x_b\,(c_v)_b + x_c\,(c_v)_c$$

$$= \Sigma\, x_i\,(c_v)_i \qquad ...(5.18)$$

Like wise :

$$m\,h = m_a\,h_a + m_b\,h_b + m_c\,h_c$$

or $$m\,c_p\,dt = m_a\,(c_p)_a\,dt + m_b\,(c_p)_b\,dt + m_c\,(c_p)_c\,dt$$

Therefore, the specific heat at constant pressure c_p for the mixture is

$$c_p = \frac{m_a}{m}(c_p)_a + \frac{m_b}{m}(c_p)_b + \frac{m_c}{m}(c_p)_c$$

$$= x_a\,(c_p)_a + x_b\,(c_p)_b + m_c\,(c_p)_c$$

$$= \Sigma\, x_i\,(c_p)_i \qquad ...(5.19)$$

Entropy of a mixture is given by

$$m\,s = m_a\,s_a + m_b\,s_b + m_c\,s_c$$

or $$s = \frac{m_a}{m}s_a + \frac{m_b}{m}s_b + \frac{m_c}{m}s_c$$

$$= x_a\,s_a + x_b\,s_b + x_c\,s_c$$

Noting that mass fraction of constituent is constant, the change in the entropy of the mixture can be written as

$$ds = x_a\,ds_a + x_b\,ds_b + x_c\,ds_c$$

where ds_a, ds_b, ds_c can be calculated from $T\,ds$ relation.

$$T\,ds = dh - v\,dp = c_p\,dT - v\,dp$$

or $$ds = c_p\frac{dT}{T} - \frac{v}{T}dp$$

$$= c_p\frac{dT}{T} - R\frac{dp}{p}$$

which upon integration gives

$$(ds)_i = \left[c_p \log_e \frac{T_2}{T_1} - R \log_e \frac{p_2}{p_1}\right]_i \qquad ...(5.20)$$

The subscript i refers to ith component of the gas mixture.

5.6. RELATION BETWEEN VOLUMETRIC AND GRAVIMETRIC ANALYSIS

For constituent a of the gas mixture consisting of gases a, b and c

mass fraction

$$x_a = \frac{m_a}{m_a + m_b + m_c}$$

From Avogadro's law $m = nM$ and accordingly we can write

$$\frac{m_a}{m_a + m_b + m_c} = \frac{n_a M_a}{n_a M_a + n_b M_b + n_c M_c}$$

$$= \frac{(n_a/n)\,M_a}{(n_a/n)M_a + (n_b/n)M_b + (n_c/n)M_c}$$

$$= \frac{y_a M_a}{y_a M_a + y_b M_b + y_c M_c}$$

Since for an ideal gas, the mole fraction y is equal to the volume fraction r, we may write

$$\frac{m_a}{m_a + m_b + m_c} = \frac{r_a M_a}{r_a M_a + r_b M_b + r_c M_c}$$

$$= \frac{(V_a/V)\,M_a}{(V_a/V)\,M_a + (V_b/V)M_b + (V_c/V)\,M_c}$$

$$\therefore \quad \frac{m_i}{m} = \frac{(V_i/V)\,M_i}{\sum (V_i/V)\,M_i} \qquad ...(5.21)$$

The above equation helps to make a change from parts by mass (mass fraction) to parts by volume (volume fraction).

EXAMPLE 5.1

A rigid vessel contains 1 kg of nitrogen at 1.8 MPa pressure and 300 K temperature. The pressure is required to be raised to 2.5 MPa and this is achieved by adding oxygen. Determine the mass of oxygen added if the temperature is presumed to remain constant.

Solution : According to Avogadro's law:

$$m = nM$$

∴ Number of moles of nitrogen,

$$n_n = \frac{m_n}{M_n} = \frac{1}{28} = 0.0357$$

Partial pressure of nitrogen, p_n = 1.8 MPa

Partial pressure of oxygen in the mixture of nitrogen and oxygen is given by Dalton's law

$$p_o = p - p_n$$
$$= 2.5 - 1.8 = 0.7 \text{ MPa}$$

The partial pressure of any gas in the mixture is directly proportional to the number of moles of that gas. That is

$$\frac{p_o}{n_o} = \frac{p_n}{n_n}$$

∴ Moles of oxygen,

$$n_o = n_n \times \frac{p_o}{p_n}$$

$$= 0.0357 \times \frac{0.7}{1.8} = 0.01388$$

∴ Mass of oxygen added,

$$m_o = n_o M_o = 0.01388 \times 32$$
$$= \mathbf{0.4443 \ kg}$$

EXAMPLE 5.2

An engine uses mixture of fuel and air consisting of 0.08 kg of octane ($C_8 H_{18}$) and 1.2 kg of air. The measurements indicate that temperature and pressure in the intake maifold are 300 K and 101.3 kPa respectively. Make calculations for the volume of the mixture, and the partial pressure of air in the mixture.

For air, molecular mass = 29

Solution : Let suffix *a* and *f* refer to air and fuel respectively.

Moles of air,

$$n_a = \frac{m_a}{M_a} = \frac{1.2}{29} = 0.0414$$

Molecular mass of fuel,

$$M_f = 8 \times 12 + 18 \times 1 = 114$$

Moles of fuel,

$$n_f = \frac{0.08}{114} = 0.0007017$$

Number of moles for the mixture,

$$n = n_a + n_f$$
$$= 0.0414 + 0.0007017 = 0.0421$$

From the characteristic equation,

Volume of mixture

$$V = \frac{nR_{mol}T}{p}$$

$$= \frac{0.0421 \times 8314 \times 300}{101.3 \times 10^3} = \mathbf{1.0366 \ m^3}$$

(b) Partial pressure of air in the mixture

$$p_a = p\frac{n_a}{n}$$

$$= 101.3 \times \frac{0.0414}{0.0421} = \mathbf{99.616 \ kPa}$$

EXAMPLE 5.3

A closed vessel of 0.40 m³ capacity contained air at 80 kN/m² and 300 K. Hydrogen was added and the pressure in the vessel was thereby raised to 82 kN/m² at the same temperature. Find the final masses of oxygen, nitrogen and hydrogen in the vessel and their respective partial pressures. Air contains 77% of nitrogen on mass basis.

Take gas constant R = 287 J/kg K for air and 4124 J/kg K for hydrogen

Solution : Applying the characteristic gas equation, $pV = mRT$, the mass of air in the vessel is given by

$$m_a = \frac{pV}{RT} = \frac{80 \times 10^3 \times 0.4}{287 \times 300}$$
$$= 0.3717 \text{ kg}$$

∴ Mass of nitrogen,

$$m_n = 0.3717 \times 0.77 = 0.2862 \text{ kg}$$

Mass of oxygen,

$$m_o = 0.3717 \times 0.23 = 0.0855 \text{ kg}$$

Partial pressure of hydrogen,

$$p_h = 82 - 80 = 2 \text{ kN/m}^2$$

Mass of hydrogen,

$$m_h = \frac{p_h V}{R_h T} = \frac{2 \times 10^3 \times 0.4}{4124 \times 300}$$
$$= 0.0006466 \text{ kg}$$

Partial pressure of oxygen,

$$p_o = \frac{m_o R_o T}{V} = \frac{0.0855 \times 287 \times 300}{0.4}$$
$$= 18.404 \text{ mN/m}^2$$

Partial pressure of nitrogen,

$$p_n = 80 - 18.404$$
$$= 61.596 \text{ kN/m}^2$$

EXAMPLE 5.4

Determine the gas constant, density and partial pressures of the components of a gas mixture consisting of 10 mass fractions of air and 1 mass fractions of lighting gas. Take the density of lighting gas ρ = 0.5 kg/m³ at 101325 N/m² pressure and 273 K temperature.

Solution : Let suffix 1 and 2 refer to air and lighting gas respectively.

Gas constant for air, R_1 = 287 J/kg K

Applying the characteristic gas equation, $pv = RT$, the gas constant of lighting gas is

$$R_2 = \frac{pv}{T} = \frac{p}{\rho T} = \frac{101325}{0.5 \times 273}$$

$$= 742.3 \text{ J/kgK}$$

(i) The gas constant for the mixture is given by

$$R_{mix} = x_1 R_1 + x_2 R_2$$

$$= \frac{10}{11} \times 287 + \frac{1}{11} \times 742.3$$

$$= \mathbf{328.39 \text{ J/kg K}}$$

(*ii*) Density of gas mixture, $\rho_{mix} = \dfrac{p}{R_{mix} T}$

$$= \frac{101325}{328.39 \times 273} = \mathbf{1.13 \text{ kg/m}^3}$$

(*iii*) Partial pressure of air,

$$p_1 = p x_1 \frac{R_1}{R_{mix}}$$

$$= 101325 \times \frac{10}{11} \times \frac{287}{328.39}$$

$$= \mathbf{80503.7 \text{ N/m}^2}$$

Partial pressure of lighting gas,

$$p_2 = 101325 \times \frac{1}{11} \times \frac{742.3}{328.39}$$

$$= \mathbf{20821.6 \text{ N/m}^2}$$

Alternatively : From Dalton's law of partial pressures, $p = p_1 + p_2$

∴ Partial pressure of lighting gas,

$$p_2 = p - p_1$$

$$= 101325 - 80503.7$$

$$= 20821.3 \text{ N/m}^2$$

EXAMPLE 5.5

A vessel contains 10 kg of oxygen, 8 kg of nitrogen and 25 kg of carbondioxide at 375 K temperature and 250 kPa pressure. Make calculations for the capacity of the vessel, the partial pressure of each gas present in the vessel, and the total pressure in the vessel when the temperature is raised to 450 K.

Solution :

$$R_{mix} = x_1 R_1 + x_2 R_2 + x_3 R_3$$

$$= \frac{10}{43} \times \frac{8314}{32} + \frac{8}{43} \times \frac{8314}{28} + \frac{25}{43} \times \frac{8314}{44}$$

$$= 60.42 + 55.24 + 109.86$$

$$= 225.52 \text{ J/kg K}$$

$$V_{mix} = \frac{m R_{mix} T}{p}$$

$$= \frac{43 \times 225.52 \times 375}{250 \times 10^3} = 14.546 \text{ m}^3$$

For the fixed mass of gas mixture at constant volume heating,

$$\frac{p_1}{T_1} = \frac{p_2}{T_2}; \quad p_2 = p_1 \frac{T_2}{T_1}$$

$$p_2 = 250 \times \frac{450}{375} = 300 \text{ kPa}$$

The partial pressure of a gas component in a gas mixture is determined from the relation, $p_i = p\, x_i\, R_i / R_{mix}$

The mass fractions of oxygen, nitrogen and carbon dioxide in the gas mixture are :

$$x_1 = \frac{10}{43}; \; x_2 = \frac{8}{43}; \; x_3 = \frac{25}{43}$$

The corresponding gas constants are:

$$R_1 = \frac{8314}{32}; \; R_2 = \frac{8314}{28}$$

$$R_3 = \frac{8314}{44} \text{ J/kg K}$$

$$\therefore \quad P_{O2} = 250 \times \frac{10}{43} \times \frac{8314/32}{225.52}$$

$$= 66.98 \text{ kPa}$$

$$P_{N2} = 250 \times \frac{8}{43} \times \frac{8314/28}{225.52}$$

$$= 61.24 \text{ kPa}$$

$$P_{CO2} = 250 \times \frac{25}{43} \times \frac{8314/44}{225.52}$$

$$= 121.78 \text{ kPa}$$

EXAMPLE 5.6

Consider a mixture of ideal gases comprising 2 kg of nitrogen and 3 kg of carbon dioxide at 250 kPa pressure and 300 K temperature. Make calculations for :

(a) mole fraction, partial pressure and partial volume of each constituent,

(b) molecular mass, gas constant, volume and density of the mixture,

(c) specific heats and adiabatic exponent for the mixture.

Take $\gamma = 1.286$ for CO_2 and $\gamma = 1.4$ for nitrogen.

Solution : From Avogadro's law: $m = nM$

$\therefore$ Number of moles of nitrogen,

$$n_{N_2} = \frac{2}{28} = 0.0714$$

Number of moles of carbon dioxide,

$$n_{CO_2} = \frac{3}{44} = 0.0682$$

Total number of moles

$$= 0.0714 + 0.0682 = 0.1396$$

Mole fraction of nitrogen,

$$y_{N_2} = \frac{0.0714}{0.1396} = 0.5115$$

Mole fraction of carbon dioxide,

$$y_{CO_2} = \frac{0.0682}{0.1396} = 0.4885$$

The partial pressures of the constituent gases are proportional to mole fractions ($p_i = py_i$).

$$p_{N_2} = 250 \times 0.5115 = 127.875 \text{ kPa}$$

$$p_{CO_2} = 250 \times 0.4885 = 122.125 \text{ kPa}$$

The partial volume of each constituent can be worked out by applying the characteristic gas equation ($pV = mRT$)

$$R_{N_2} = \frac{8314}{28} = 296.9 \text{ J/kg K}$$

$$R_{CO_2} = \frac{8314}{44} = 188.9 \text{ J/kg K}$$

Partial volume of nitrogen,

$$V_{N_2} = \frac{2 \times 296.9 \times 300}{250 \times 10^3} = 0.712 \text{ m}^3$$

Partial volume of carbon dioxide,

$$V_{CO_2} = \frac{3 \times 188.9 \times 300}{250 \times 10^3} = 0.680 \text{ m}^3$$

(*b*) Equivalent molecular mass of the mixture,

$$M_{mix} = \Sigma\, y_i\, M_i$$
$$= 0.5115 \times 28 + 0.4885 \times 44$$
$$= 14.322 + 21.494 = 35.816$$

Equivalent gas constant of the mixture

$$R_{mix} = \Sigma\, x_i\, R_i = \frac{2}{5} \times 296.9 + \frac{3}{5} \times 188.9$$
$$= 118.76 + 113.34$$
$$= 232.1 \text{ J/kg K}$$

Alternatively

$$R_{mix} = \frac{R_{mol}}{M_{mix}} = \frac{8314}{35.816}$$
$$= 232.1 \text{ J/kg K}$$

Volume of the mixture,

$$V = \frac{mRT}{p}$$
$$= \frac{5 \times 232.1 \times 300}{250 \times 10^3} = 1.393 \text{ m}^3$$

Density of the mixture,

$$\rho = \frac{m}{V} = \frac{5}{1.393} = 3.589 \text{ kg/m}^3$$

(*c*) From the relations

$$c_p - c_v = R \quad \text{and} \quad \frac{c_p}{c_v} = \gamma$$

we have $c_v = \dfrac{R}{\gamma - 1}$

$$\therefore \quad (c_v)_{N_2} = \frac{296.9}{1.4 - 1} = 742.25 \text{ J/kg K}$$

$$(c_p)_{N_2} = 742.25 \times 1.4 = 1039.15 \text{ J/kg K}$$

$$(c_v)_{CO_2} = \frac{188.9}{1.286 - 1} = 660.49 \text{ J/kg K}$$

$$(c_p)_{CO_2} = 660.49 \times 1.286$$
$$= 849.39 \text{ J/kg K}$$

For the mixture:

$$(c_v)_{mix} = \Sigma x_i\, (c_v)_i$$
$$= \frac{2}{5} \times 742.25 + \frac{3}{5} \times 660.49$$
$$= 296.9 + 396.3 = 693.2 \text{ J/kg K}$$

$$(c_p)_{\text{mix}} = \Sigma x_i\,(c_p)_i$$
$$= \frac{2}{5} \times 1039.15 + \frac{3}{5} \times 849.39$$
$$= 415.66 + 509.63$$
$$= 925.29 \text{ J/kg K}$$

Hence, adiabatic exponent for the mixture,

$$(\gamma)_{\text{mix}} = \frac{(c_p)_{mix}}{(c_v)_{mix}} = \frac{925.29}{693.2} = \mathbf{1.335}$$

EXAMPLE 5.7

A mixture of 1 kg of oxygen and 2 kg of nitrogen occupies 1.2 m³ volume at temperature 300 K. Assuming perfect gas behaviour, determine the following parameters for the gas mixture.

(a) specific volume, (b) pressure, (c) gas constant and (d) molecular mass.

Solution : Mass of the mixture, $m = m_{O_2} + m_{N_2}$

$$= 1 + 2 = 3 \text{ kg}$$

Specific volume represents the volume per unit mass. That is

$$v = \frac{V}{m} = \frac{1.2}{3} = \mathbf{0.4\ m^3/kg}$$

(*b*) Total pressure equals the sum of partial pressure of each component of the mixture

For oxygen : $m_o = 1$ kg

and $R = \frac{8314}{32} = 259.81$ J/kg K

$$p_{O_2} = \frac{mRT}{V} = \frac{1 \times 259.81 \times 300}{1.2}$$
$$= 64952.5 \text{ N/m}^2$$

For nitrogen : $m = 2$ kg and

$$R = \frac{8314}{28}\ 296.93 \text{ J/kg K}$$

$$p_{N_2} = \frac{mRT}{V} = \frac{2 \times 296.93 \times 300}{1.2}$$
$$= 148465 \text{ N/m}^2$$

∴ Total pressure for the mixture

$$= 64952.5 + 148465$$
$$= 213417.5 \text{ N/m}^2$$
$$= \mathbf{213.42\ kN/m^2}$$

(*c*) The gas constant for the mixture is

$$R_{mix} = \frac{\sum m_i R_i}{m} = \Sigma\, x_i\, R_i$$

where x_i denotes the mass fraction

$$= \frac{1}{3} \times 259.81 + \frac{2}{3} \times 296.93$$
$$= \mathbf{284.56\ J/kg\ K}$$

(*d*) The molecular mass for the mixture is

$$M_{mix} = \Sigma y_i\, M_i$$

where y_i represents the mole fraction

From Avogadro's law $n = \frac{m}{M}$, and hence

$$\text{moles of oxygen} = \frac{1}{32} = 0.03125$$

$$\text{moles of nitrogen} = \frac{2}{28} = 0.07143$$

Total number of moles = 0.03125 + 0.07143 = 0.10268

$$\therefore \quad M_{mix} = \frac{0.03125}{0.10268} \times 32 + \frac{0.07143}{0.10268} \times 28$$
$$= 9.739 + 19.478 = \mathbf{29.217}$$

Alternatively, the molecular mass for the mixture can also be worked out as

$$M_{mix} = \frac{R_{mol}}{R_{mix}} = \frac{8314}{284.56} = 29.217$$

EXAMPLE 5.8

A mixture of perfect gases at 300 K has the following composition by volume

N_2 = 60%; O_2 = 15%, and CH_4 = 25%.

If the partial pressure of CH_4 is 0.5 bar, make calculations for the following parameters of the mixture.

(a) partial pressures of N_2 and O_2,

(b) mass proportion of mixtures, and

(c) gas constant and volume per mole of mixture.

Solution : For each component of the mixture, the volume fraction and the mole fraction are equal. That is

$$\frac{V_{N2}}{V} = \frac{n_{N2}}{n}$$

$$\frac{V_{O2}}{V} = \frac{n_{O2}}{n}$$

$$\frac{V_{CH4}}{V} = \frac{n_{CH4}}{n}$$

where n is the total number of moles

∴ mole fraction of nitrogen, y_{N2}= 0.6

mole fraction of oxygen, y_{O2} = 0.15
mole fraction of methane, y_{CH4} = 0.25

(*a*) Using the relation: $p_i = y_i\, p$;

$$0.5 = 0.25\, p$$

∴ Total pressure of the mixture,

$$p = \frac{0.5}{0.25} = 2 \text{ bar}$$

Partial pressure of nitrogen,

$$p_{N2} = y_{N2} \times p = 0.6 \times 2 = \mathbf{1.2\ bar}$$

Partial pressure of oxygen,

$$p_{O2} = y_{O2} \times p = 0.15 \times 2 = \mathbf{0.3\ bar}$$

(*b*) Mass fraction $x_i = \frac{m_i}{m} = \frac{M_i n_i}{M n} = y_i \frac{M_i}{M}$

Equivalent molecular weight for the mixture,

$$M = \Sigma y_i M_i$$
$$= 0.6 \times 28 + 0.15 \times 32 + 0.25 \times 16$$
$$= 16.8 + 4.8 + 4 = 25.6$$

Hence, mass fraction for nitrogen,

$$x_{N2} = \frac{0.6 \times 28}{25.6} = 0.656$$

mass fraction for oxygen,

$$x_{O2} = \frac{0.15 \times 32}{25.6} = 0.1875$$

mass fraction for methane,

$$x_{CH_4} = \frac{0.25 \times 16}{25.6} = 0.562$$

(*c*) Gas constant for the mixture,

$$R = \frac{R_{mol}}{M} = \frac{8314}{25.6} = 324.76 \text{ J/kg K}$$

Volume per mole of mixture

$$= \frac{R_{mol}\, T}{p} = \frac{8314 \times 300}{2 \times 10^5}$$

$$= \mathbf{12.47\ m^3/mol}$$

EXAMPLE 5.9

A mixture 1 mole of carbon dioxide and 3 moles of air is contained in a vessel. If the pressure and temperature are 1 bar and 25 °C respectively, make calculations for the following parameters in relation to the mixture:

(a) the total mass,

(b) the percentage carbon content,

(c) the apparent molecular weight and the gas constant, and

(d) the specific volume.

It may be presumed that volumetric analysis of air by volume is 79% nitrogen and 21% by oxygen.

Solution : We know that the volumetric analysis is identical with molar analysis. That is

$$\frac{n_i}{n} = \frac{V_i}{V}$$

∴ moles of oxygen, n_{O_2} = 0.21 × 3 = 0.63

moles of nitrogen n_{N_2} = 0.79 × 3 = 2.37

Further from the correlation: $m_i = n_i\, M_i$, we have

∴ mass of carbon dioxide,

$$m_{CO_2} = 1 \times 44 = 44 \text{ kg}$$

mass of oxygen,

$$m_{O_2} = 0.63 \times 32 = 20.16 \text{ kg}$$

mass of nitrogen,

$$m_{N_2} = 2.37 \times 28 = 66.36 \text{ kg}$$

Hence total mass of the mixture,

$$m = 44 + 20.16 + 66.36$$
$$= \mathbf{130.52\ kg}$$

(*b*) The carbon has a molecular weight equal to 12. This implies that there are 12 kg of carbon present for every mole of carbon dioxide.

∴ Percentage of carbon in the mixture

$$= \frac{12}{130.52} \times 100 = \mathbf{9.194\%}$$

(*c*) Total number of moles

$$= 1 + 0.63 + 2.37 = 4$$

Apparent molecular weight of the mixture,

$$M_{mix} = \sum y_i M_i$$
$$= \sum \frac{n_i}{n} M_i = \frac{1}{n} \sum n_i M_i$$
$$= \frac{1}{4}(1 \times 44 + 0.63 \times 32 + 2.37 \times 28)$$
$$= \mathbf{32.63}$$

Gas constant for the mixture,

$$R_{mix} = \frac{R_{mol}}{M_{mix}} = \frac{8314}{32.63} = \mathbf{254.8\ J/kg\ K}$$

(*d*) From the characteristic equation,

$$p\,v = RT$$

$$\text{specific volume } v = \frac{254.8 \times 298}{1 \times 10^5}$$

$$= \mathbf{0.759\ m^3/kg}$$

EXAMPLE 5.10

A gas mixture contains hydrogen, nitrogen and carbon monoxide in equal molar proportions. If the components of the mixture are treated as perfect gases, make calculations for

(a) gas constant and ratio of specific heats for the mixture,

(b) work required to compress 1 kg of the mixture in a reversible adiabatic flow through a pressure ratio of 1.5.

Take initial temperature as 290 K and neglect changes in kinetic and potential energy.

Solution : Consider 1 mole of mixtures. Then

$\Sigma n_i = 1$ and

$$n_{H_2} = n_{N_2} = n_{CO} = \frac{1}{3}$$

Equivalent molecular weight of the mixture

$$M = \Sigma\, y_i\, M_i$$

$$= \frac{1}{3} \times 2 + \frac{1}{3} \times 28 + \frac{1}{3} \times 28$$

$$= 19.333$$

Gas constant for the mixture,

$$R = \frac{R_{mol}}{M} = \frac{8314}{19.333}$$

$$= \mathbf{430.04\ J/kg\ K = 0.430\ kJ/kg\ K}$$

The specific heat of an ideal gas at constant pressure is given by

$$c_p = \left(1 + \frac{f}{2}\right)\frac{R_{mol}}{M}$$

where f = degrees of freedom of the molecule. At low temperatures, for monatomic gas f = 3, for diatomic gas f = 5 and for polyatomic gas f = 6.

Hydrogen, nitrogen and carbon monoxide are all diatomic gases. Therefore

$$(c_p)_{H_2} = \frac{7}{2} \times \frac{8.314}{2} = 14.55 \text{ kJ/kg K}$$

$$(c_p)_{N_2} = \frac{7}{2} \times \frac{8.314}{28} = 1.04 \text{ kJ/kg K}$$

$$(c_p)_{CO} = \frac{7}{2} \times \frac{8.314}{28} = 1.04 \text{ kJ/kg K}$$

Now:

$$(c_p)_{mix} = \frac{\sum m_i\, c_{p_i}}{\sum m_i} = \frac{\sum n_i\, M_i\, c_{p_i}}{\sum n_i\, m_i}$$

where

$$\sum n_i\, M_i\, c_{p_i} = \frac{1}{3} \times 2 \times 14.55 + \frac{1}{3} \times 28 \times 1.04 + \frac{1}{3} \times 2 \times 1.04$$

$$= 9.7 + 9.71 + 9.71 = 29.12$$

$$\sum n_i\, M_i = \frac{1}{3} \times 2 + \frac{1}{3} \times 28 + \frac{1}{3} \times 28$$

$$= 19.35$$

$$\text{Hence } (c_p)_{mix} = \frac{29.12}{19.35} = 1.505 \text{ kJ/kg K}$$

$$(c_v)_{mix} = (c_p)_{mix} - R$$

$$= 1.505 - 0.430$$

$$= 1.075 \text{ kJ/Kg K}$$

$$\therefore \text{Ratio of specific values, } \gamma = \frac{1.505}{1.075} = \mathbf{1.4}$$

(*b*)
$$T_2 = T_1 \left(\frac{p_2}{p_1}\right)^{\frac{\gamma-1}{\gamma}}$$

$$= 290 \times (1.5)^{\frac{1.4-1}{1.4}} = 325.66 \text{ K.}$$

Applying steady flow energy equation on unit mass basis,

$$\left(h_1 + \frac{V_1^2}{2} + gz_1\right) + Q = \left(h_2 + \frac{V_2^2}{2} + gz_2\right) + W_s$$

Since the process is adiabatic, Q = 0. Further neglecting change in kinetic and potential energy,

$$W_s = (h_1 - h_2) = (c_p)_{mix}\,(T_1 - T_2)$$

$$= 1.508\ (290 - 325.66)$$

$$= \mathbf{-\ 53.668\ kJ/kg}$$

The negative sign indicates that work has to be supplied to the gas during the compression process.

EXAMPLE 5.11

A certain gaseous fuel has the following composition by volume:

$$H_2 = 7\%;\ CH_4 = 2\%$$
$$CO = 27.6\%\ CO_2 = 4.8\%$$

and remaining nitrogen.

Calculate the apparent molecular weight, the characteristic gas constant and density of the mixture at NTP. If the pressure of the mixture is 1 bar, proceed to workout the partial pressure of the constituent gases.

Solution : Volume of nitrogen

$$= 100 - (7 + 2 + 27.6 + 4.8) = 58.6\%$$

Constituent	*Volume proportion*	*Molecular mass*	*Mass proportion*	*Mass fraction*
Hydrogen	0.07	2	0.14	$\frac{0.14}{26.708} = 0.00524$
Methane	0.02	16	0.32	0.01198
Carbon monoxide	0.276	28	7.728	0.28935
Carbon dioxide	0.048	44	2.112	0.07907
Nitrogen	0.586	28	16.408	0.61435
			26.708	≃ 1.00

The mass and volume proportions are correlated by the expression

$$m_i = \left(\frac{V_i}{V}\right) M_i$$

Then :

The gas constant for the mixture is

$$R_{mix} = \Sigma x_i\ R_i = R_{mol}\ \Sigma\ x_i/M_i$$

$$= 8314\left[\frac{0.00524}{2} + \frac{0.01198}{16} + \frac{0.28935}{28} + \frac{0.07907}{44} + \frac{0.61435}{28}\right]$$

$$= \mathbf{311.28\ J/kg\ K}$$

Apparent molecular mass,

$$M_{mix} = \frac{R_{mol}}{R_{mix}} = \frac{8314}{311.28} = \mathbf{26.71}$$

Density of the mixture at normal temperature pressure (NTP) can be calculated either by applying gas equation for the mixture or from the well known result that molecular volume of a gas at NTP is 22.4 m^3.

Density = mass per unit volume

The mass of 1 mole of the gas mixture will be equal to its molecular mass.

$$\therefore\ \text{Density}\ \rho = \frac{26.71}{22.4} = 1.192\ \text{kg/m}^3$$

or $\quad$ Density $\rho = \frac{p}{RT} = \frac{1.013 \times 10^5}{311.28 \times 273}$

$$= 1.192\ \text{kg/m}^3$$

(*b*) The partial pressure of the constituent gases are proportional to parts by volume or mole fraction ($p_i = p\ V_i/V$).

Since the pressure of the gas is (1 bar), the partial pressure of each constituent gas will be numerically equal to the volume fraction.

Thus partial pressure of hydrogen will be 0.07 bar, of methane 0.02 bar, of carbon monoxide 0.276 bar, of carbon dioxide 0.048 bar and that of nitrogen 0.586 bar.

EXAMPLE 5.12

A gaseous mixture contains by volume 21% of nitrogen, 50% of hydrogen and 29% of carbon dioxide. Make calculations for the gas constant and ratio of specific heats for the mixture. Take:

c_v for N_2 = 0.724; for H_2 = 10.11; and for CO_2 = 0.716 kJ/kg K

(b) A cylinder contains 0.084 m^3 of this mixture at 1 bar and 10°C. The mixture is compressed polytropically to one-fifth of its original volume according to the relationship $pV^{1.2}$ = constant. Determine the work required and also the heat transfer to or from the gas during the compression process.

Solution : Assume 1 mole of mixture containing 0.21 mole of N_2, 0.5 mole of H_2 and 0.29 mole of CO_2.

$\therefore$ Mass of nitrogen = 0.21 × 28 = 5.88 kg

Mass of hydrogen = 0.5 × 2 = 1.00 kg

Mass of carbon dioxide = 0.29 × 44 = 12.76 kg

Total mass of mixture = 5.88 + 1.00 + 12.76 = 19.64 kg

Thus parts by mass of each gas is

$$x_{N_2} = \frac{5.88}{19.64} = 0.2994$$

$$x_{H_2} = \frac{1.00}{19.64} = 0.0509$$

$$x_{CO_2} = \frac{12.76}{19.64} = 0.6497$$

The gas constant for the mixture is

$$R_{mix} = \Sigma\, x_i\, R_i = R_{mol} \sum \frac{x_i}{M_i}$$

$$= 8314 \left[\frac{0.2994}{28} + \frac{0.0509}{2} + \frac{0.6497}{44}\right]$$

$$= 8314\ [0.0107 + 0.0254 + 0.0147]$$

$$= \mathbf{422.35\ J/kg\ K}$$

The specific heat at constant volume for the mixture is

$$(c_v)_{mix} = \Sigma\, x_i\, (c_v)_i$$

$$= 0.2994 \times 0.724 + 0.0509 \times 10.11 + 0.6497 \times 0.716$$

$$= 0.2167 + 0.5147 + 0.4652$$

$$= 1.1966\ kJ/kg\ K$$

Then

$$(c_p)_{mix} = (c_v)_{mix} + R_{mix}$$

$$= 1.1966 + 422.35 \times 10^{-3}$$

$$= 1.6189\ kJ/kg\ K$$

Hence, ratio of specific heats for the mixture,

$$\gamma = \frac{(c_p)_{mix}}{(c_v)_{mix}} = \frac{1.6189}{1.1966} = \mathbf{1.353}$$

(*b*) $T_2 = T_1 \left(\frac{V_1}{V_2}\right)^{n-1}$

$$= 283 \times (5)^{1.2-1} = 390.5\ K$$

mass of mixture,

$$m = \frac{pV}{RT} = \frac{1 \times 10^5 \times 0.084}{422.35 \times 283}$$

$$= 0.0703\ kg$$

$$\text{Work done } W_{1\text{-}2} = \frac{p_1V_1 - p_2V_2}{n-1}$$

$$= \frac{mR\,(T_1 - T_2)}{n-1}$$

$$= \frac{0.0703 \times 422.35\,(283 - 390.5)}{1.2 - 1}$$

$$= -\ 15959\ J = \mathbf{-\ 15.96\ kJ}$$

The negative sign indicates that work has been done on the gas

Change in internal energy,

$$dU = m\ c_v\ (T_2 - T_1)$$

$$= 0.0703 \times 1.196\ (390.5 - 283)$$

$$= 9.04\ kJ$$

Then, from the energy equation for the closed system,

$$Q_{1\text{-}2} = W_{1\text{-}2} + dU = -\ 15.96 + 9.04$$

$$= \mathbf{-\ 6.92\ kJ}$$

The negative sign indicates that heat is transferred from the gas to the surroundings.

EXAMPLE 5.13

During a steady flow adiabatic process nitrogen and hydrogen are mixed in the ratio of 4 kg of hydrogen per kg of nitrogen. The hydrogen gas enters at 200 kPa and 315 K and nitrogen gas at 160 kPa and 525 K. After mixing, the pressure of the mixture is 180 kPa. Determine the final temperature of the mixture, and the net entropy change per kg of the mixture.

Assume specific heat at constant pressure for nitrogen and hydrogen as 1.035 kJ/kg K and 14.35 kJ/kg K respectively.

Solution : From the relation, $R_{mol} = MR$, we have

Gas constant for nitrogen,

$$R_{N_2} = \frac{8314}{28} = 296.92\ J/kg\ K$$

Gas constant for hydrogen,

$$R_{H_2} = \frac{8314}{2} = 4157\ J/kg\ K$$

(*a*) From the energy equation:

$$m_1\ h_1 + m_2\ h_2 = m_3\ h_3$$

Substituting the values of m_2 and m_3 in the above identity, we get

$$m_1h_1 + 4\ m_1h_2 = 5\ m_1h_3$$

or $\quad 0.2\ h_1 + 0.8\ h_2 = h_3$

or $\quad 0.2\ (c_p)_{N_2}\ T_{N_2} + 0.8\ (c_p)_{H_2}\ T_{H_2}$

$$= (c_p)_{mix} \times T_{mix} \quad ...(i)$$

where T_{mix} is the temperature of the mixture.

The specific heat at constant pressure for the mixture is

$$(c_p)_{mix} = \Sigma\, x_i\ (c_p)_i$$

where x_i is the mass fraction of each constituent of the mixture.

$$(c_p)_{mix} = 0.2 \times 1.035 + 0.8 \times 14.35$$
$$= 11.687 \text{ kJ/kg K}$$

Then, from expression (*i*), we have

$$0.2 \times 1.035 \times 525 + 0.8 \times 14.35 \times 315$$
$$= 11.687\ T_{mix}$$

or $\quad 108.675 + 3616.2 = 11.687\ T_{mix}$

∴ Temperature of the mixture,

$$T_{mix} = \frac{108.675 + 3616.2}{11.687} = \mathbf{318.72\ K}$$

(*b*) The entropy change of the mixture equals the sum of entropy changes in the constituent gases. Then

$$(dS)_{mix} = (dS)_{N2} + (dS)_{H2}$$

In terms of pressure and temperature, the entropy change for a gas constituent is

$$dS = c_p \log_e \frac{T_f}{T_1} - R \log_e \frac{p_f}{p_1}$$

where:

T_f is the final temperature of each constituent in the mixture and it equals the temperature of the mixture; T_f = 318.72 K

p_f is the partial pressure of each constituent in the mixture and it is given by

$$p_f = px_i$$

where x_i denotes the mass fraction.

$$(p_f)_{N_2} = 180 \times \frac{0.2/28}{(0.2/28) + (0.8/2)}$$
$$= 180 \times 0.01754 \simeq 3.16 \text{ kPa}$$

$$(p_f)_{H_2} = p - (p_f)_{N_2}$$
$$= 180 - 3.16 = 176.84 \text{ kPa}$$

One kg of mixture consists of 0.2 kg of nitrogen and 0.8 kg of hydrogen

$$(dS)_{N2} = 0.2\left[1.035 \log_e \frac{318.72}{525} - \frac{8314}{28 \times 1000} \log_e \frac{3.16}{160}\right]$$
$$= 0.2\left[-0.5167 - (-1.1653)\right]$$
$$= 0.1279 \text{ kJ/K}$$

$$(dS)_{H2} = 0.8\left[14.35 \log_e \frac{318.72}{315} - \frac{8314}{2 \times 1000} \log_e \frac{176.84}{200}\right]$$
$$= 0.8\left[0.1683 - (-0.5116)\right]$$
$$= 0.5439 \text{ kJ/K}$$

∴ For one kg of mixture

$$(dS)_{mix} = 0.1279 + 0.5439$$
$$= \mathbf{0.6736\ kJ/K}$$

EXAMPLE 5.14

A gaseous mixture contained in a rigid tank of 1 m^3 volume has a molar composition of 25% ethane (C_2H_6), 30% methane (CH_4) and the remainder 45% air. The mixture has a pressure of 1.2 bar and temperature 300 K. It becomes necessary to change the composition such that the final mixture contains 60 mole percent methane and the tank is always held at 300 K. Determine:

(a) mass of methane to be added,

(b) final pressure of the gas mixture.

Solution : Initially the gas mixture has a molar composition of

$$y_{C_2H_6} = 0.25 \qquad y_{CH_4} = 0.30$$
and $\quad y_{air} = 0.45$

The total number of moles in the mixture can be determined by applying the characteristic equation,

$$n = \frac{pV}{R_{mol}T} = \frac{1.2 \times 10^5 \times 1}{8.314 \times 300}$$
$$= 48.11 \text{ mol}$$

where universal gas constant

R_{mol} = 8.314 kJ/kg mol K = 8.314 J/mol K

Then the number of moles of the constituent gases in the mixture are :

$$n_{C_2H_6} = 0.25 \times 48.11 = 12.027$$

$$n_{CH_4} = 0.3 \times 48.11 = 14.433$$

$$n_{air} = 0.45 \times 48.11 \quad 21.649$$

Let n_m be the number of moles of methane present in the tank after a certain mass of methane has been added. Then

$$0.6 = \frac{n_m}{n_m + n_{C_2H_6} + n_{air}}$$

$$= \frac{n_m}{n_m + 12.027 + 21.649}$$

or $\quad 0.6\, n_m + 7.2162 + 12.989 = n_m$

$$\therefore \quad n_m = \frac{7.2162 + 12.989}{0.4} = 50.51$$

Hence moles of methane to the added

$$= 50.51 - 14.433 = 36.077$$

mass of methane to be added

$$= 36.077 \times 16 = 577.2 \text{ gm}$$

$$= 5.77 \text{ kg}$$

(b) Total moles $n = n_{C_2H_6} + n_{CH_4} + n_{air}$

$$= 12.027 + 50.51 + 21.649 = 84.186$$

$$\text{Final pressure } P_f = \frac{n R_{mol} T}{V}$$

$$= \frac{84.186 \times 8.314 \times 300}{1}$$

$$= 209977 \text{ N/m}^2 \simeq \mathbf{2.1 \text{ bar}}$$

EXAMPLE 5.15

A perfect gas mixture contained in a vessel has volumetric composition of 75 percent hydrogen (H_2) and 25 percent carbon monoxide (CO). It is desired to have the mixture in proportion of 55% H_2 and 45% CO, and this is done by removing some of the mixture and adding more CO. Assuming that pressure and temperature remain constant during the process, calculate per mole of mixture the mass of mixture to be removed and mass of CO to be added.

Solution : For the mixture

$$M = \Sigma y_i M_i = \sum \frac{n_i}{n} M_i$$

Since volumetric analysis is identical with molar analysis,

$$M = \sum \frac{V_i}{V} M_i$$

$$= 0.75 \times 2 + 0.25 \times 28 = 8.25$$

Let x and y denote the masses in kg of mixture to be removed and CO to be added. Then

$$\text{moles of mixture removed} = \frac{x}{8.25}$$

$$\text{moles of } CO \text{ added} = \frac{y}{28}$$

moles of H_2 in the mixture removed

$$= 0.75 \times \frac{x}{8.25} = 0.0909\, x$$

moles of H_2 initially in the mixture

$$= 0.75 \times 1 = 0.75$$

$\therefore$ moles of hydrogen remaining in the vessel

$$= 0.75 - 0.0909\, x$$

Sionce 1 mole of new mixture contains 55% H_2 and 45% CO, we have

$$0.75 - 0.0909\, x = 0.55; \qquad x = 2.20$$

$\therefore$ Mass of mixture removed, x = **2.20 kg**

Since the pressure and temperature remain constant, the number of moles in the vessel remain same throughout.

$\therefore$ moles of mixture removed

= moles of CO added

$$\frac{x}{8.25} = \frac{y}{28}$$

or $\quad y = \frac{28}{8.25} \times x = \frac{28}{8.25} \times 2.20$

$$= 7.467$$

Hence, mass of CO added = **7.467 kg**

EXAMPLE 5.16

Two vessels are connected by a valve. One vessel contains 0.6 kg mole of nitrogen gas at 50°C and 15 bar pressure. The other vessel holds 3.5 kg of the same gas at 25°C and 6.5 bar. The valve is opened and the gases are allowed to mix while having heat interactions with the surroundings. If the final equilibrium temperature is 30°C, calculate the final equilibrium pressure and the amount of heat transferred to the surroundings.

(b) Considering the vessels to be perfectly insulated, find the final temperature and pressure

which would have been reached. Take adiabatic exponent $\gamma = 1.4$.

Solution : Refer Fig. 5.4 for the gases contained in the two vessels which are connected by a valve.

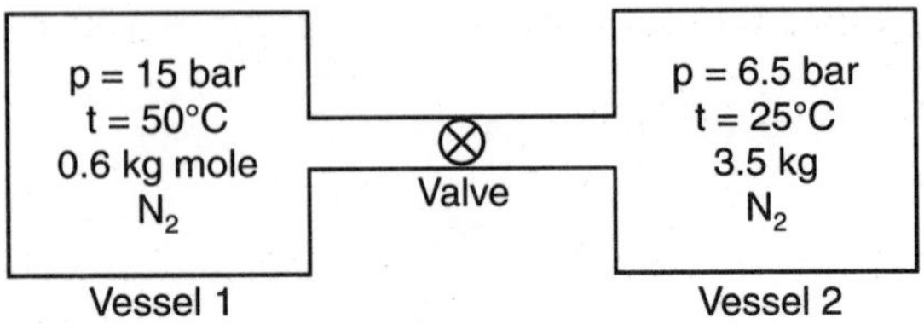

Fig. 5.4.

Characteristic gas constant R for nitrogen,

$$R = \frac{R_{mol}}{M} = \frac{8314}{28} = 296.93 \text{ J/kg K}$$

Mass of nitrogen in vessel 1,

$$m_1 = 0.6 \times 28 = 16.8 \text{ kg}$$

Volume of vessel 1,

$$V_1 = \frac{mRT}{p} = \frac{16.8 \times 296.93 \times 323}{15 \times 10^5}$$

$$= 1.0742 \text{ m}^3$$

Volume of vessel 2,

$$V_2 = \frac{3.5 \times 296.93 \times 298}{6.5 \times 10^5}$$

$$= 0.4764 \text{ m}^3$$

Total volume of both the vessels

$$V = V_1 + V_2$$

$$= 1.0742 + 0.4764 = 1.5506 \text{ m}^3$$

Total mass of nitrogen gas

$$m = m_1 + m_2$$

$$= 16.8 + 3.5 = 20.3 \text{ kg}$$

Final temperature after mixing,

$$T = 30 + 273 = 303 \text{ K}$$

Then from the characteristic gas equation, Final equilibrium pressure,

$$p = \frac{mRT}{V} = \frac{20.3 \times 296.93 \times 303}{1.5506}$$

$$= 1177858 \text{ N/m}^2 = \mathbf{11.78 \text{ bar}}$$

From non-flow energy equation,

$$\delta Q = \delta W + dU$$

Since there is no work transfer, the heat interaction equals the change in internal energy. That is

Heat interaction $Q = U_f - U_i$

Measuring the internal energy above the datum of absolute zero

($U = 0$ kJ/kg at $T = 0$ K), we have

U_i (initial internal energy, *i.e.*, before mixing)

$$= m_1 c_{v_1} T_1 + m_2 c_{v_2} T_2$$

$$c_{v_1} = c_{v_2} = \frac{R}{\gamma - 1} = \frac{296.93}{1.4 - 1}$$

$$= 742 \text{ J/kg K} = 0.742 \text{ kJ/kg K}$$

$$\therefore \quad U_i = 16.8 \times 0.742 \times 323 + 3.5 \times 0.742 \times 298$$

$$= 4026.4 + 773.9 = 4800.3 \text{ kJ}$$

U_f (final internal energy, *i.e.*, after mixing)

$$= m\, c_v T = 20.3 \times 0.742 \times 303$$

$$= 4563.97 \text{ kJ}$$

∴ Heat interaction

$$Q = 4563.97 - 4800.3 = -236.3 \text{ kJ}$$

The negative sign indicates that heat is transferred to the surroundings.

(*b*) Since there is no work transfer and the vessels are perfectly insulated ($\delta Q = 0$, $\delta W = 0$), we have

$$dU = 0 \quad \text{or} \quad U_i = U_f$$

Then, if T_f is the final equilibrium temperature

$$m_1 c_{v_1} T_1 + m_2 c_{v_2} T_2 = m\, c_v\, T_f$$

or
$$T_f = \frac{m_1 T_1 + m_2 T_2}{m} \qquad (\because c_{v_1} = c_{v_2} = c_v)$$

$$= \frac{16.8 \times 323 + 3.5 \times 298}{20.3} = \mathbf{318.7 \text{ K}}$$

Final pressure,

$$p_f = \frac{mRT}{V} = \frac{20.3 \times 296.93 \times 318.7}{1.5506}$$

$$= 1238889 \text{ N/m}^2 = \mathbf{12.39 \text{ bar}}$$

EXAMPLE 5.17

A rigid and insulated tank is divided into two compartments by a partition which is removed to allow the contents to mix and achieve an equilibrium temperature. Before mixing, the following information is known about the gases in the two compartments:

Compartment A

Volume = 1 m^3

Content = air at 1 bar and 300 K

c_p for air = 29 J/mol K

c_v for air = 20.8 J/mol K

Compartment B

Volume = 2 m^3

Content = helium at 5 bar and 1000 K

c_p for helium = 20.8 J/mol K

c_v for helium = 12.48 J/mol K

Determine:

(a) the molar composition of the mixture,

(b) the final temperature and pressure of the mixture,

(c) the change in entropy of helium and air, and

(d) the net entropy change.

Solution : For air in compartment *A*,

$$n_{air} = \frac{pV}{R_{mol}T} = \frac{1 \times 10^5 \times 1}{8.314 \times 300} = 40.09$$

where n_{air} is the number of moles of air.

For helium in compartment *B*,

$$n_{He} = \frac{5 \times 10^5 \times 2}{8.314 \times 1000} = 120.28$$

where n_{He} is the number of moles of helium.

Total number of moles of the mixture,

$$\Sigma n = 40.09 + 120.28 = 160.37$$

∴ Mole fraction of air,

$$y_{air} = \frac{40.09}{160.37} \simeq \mathbf{0.25}$$

Mole fraction for helium,

$$y_{He} = \frac{120.28}{160.37} \simeq \mathbf{0.75}$$

(*b*) If T_f is the final temperature of the mixture, one can write

$$dU = n_{air} (c_v)_{air} (T_f - 300) + n_{He} (c_v)_{He} (T_f - 1000)$$
$$= 40.09 \times 20.8 (T_f - 300) + 120.28 \times 12.48 (T_f - 1000)$$
$$= 2334.96\ T_f - 1751255$$

Since the tank is insulated and rigid, we have

$$\delta Q = 0 \quad \text{and} \quad \delta W = 0$$

Then from the energy equation for a closed system, the change in internal energy $dU = 0$.

$$2334.96\ T_f - 1751255 = 0$$

∴ Final temperature $T_f = \dfrac{1751255}{2334.96} = 750$ K

and final pressure,

$$P_f = \frac{(n_{air} + n_{He})\, R_{mol}\, T_f}{V}$$
$$= \frac{(40.09 + 120.28) \times 8.314 \times 750}{3}$$
$$= 333329 \text{ N/m}^2 \simeq \mathbf{3.33\ bar}$$

(*c*) The partial pressures of the constituent gases in the mixture are proportional to volume or mole fraction ($p_i = p\ y_i$)

∴ Partial pressure of air,

$$p_{air} = P_f\ y_{air} = 3.33 \times 0.25 = \mathbf{0.8325\ bar}$$

Partial pressure of helium,

$$p_{He} = P_f\ y_{He} = 3.33 \times 0.75 = \mathbf{2.4975\ bar}$$

In terms of temperature and pressure changes, the entropy change of a constituent gas is given by

$$ds = c_p \log_e \frac{T_f}{T_1} - R_{mol} \log_e \frac{p}{p_1}$$

Universal gas constant

$$R_{mol} = 8.314 \text{ kJ/k mol K} = 8.314 \text{ J/mol K}$$

$$\therefore\ (ds)_{air} = 29 \log_e \frac{750}{300} - 8.314 \log_e \frac{0.8325}{1}$$
$$= 26.57 + 1.52 = 28.09 \text{ J/mol K}$$

$$(ds)_{He} = 20.8 \log_e \frac{750}{1000} - 8.314 \log_e \frac{2.4975}{5}$$
$$= -\ 5.983 + 5.771$$
$$= -\ 0.212 \text{ J/mol K}$$

Net or total entropy change

$$(dS)_{net} = n_{air}\,(dS)_{air} + n_{He}(dS)_{He}$$
$$= 40.09 \times 28.09 + 120.28 \times (-0.212)$$
$$= 1126.13 - 25.5 = \mathbf{1100.63\ J/K}$$

EXAMPLE 5.18

A vessel is divided into three compartments: (a), (b) and (c) by two partitions

Part (a) contains oxygen and has a volume of 0.1 m³

Part (b) contains nitrogen and has a volume of 0.2 m³

Part (c) contains carbon dioxide and has a volume of 0.05 m³

All three parts are at a pressure of 2 bar and a temperature of 13°C. When the partitions are removed and the gases mix, determine

(i) the final pressure in the vessel and the partial pressure of each gas

(ii) the change of entropy of each constituent. The vessle may be taken as being completely isolated from its surroundings.

Solution : The gases in all the three compartments are at the same pressure 2 bar and temperature 13°C. When these gases are allowed to mix and there are no heat and work interactions, the gaseous mixture will have the same values for these parameters. That is, for the mixture

$$p = 2 \text{ bar} \quad \text{and} \quad t = 13°\text{ C}$$

The partial pressure for a constituent gas of the mixture is $p_i = y_i\,p$, where y_i is the mole fraction which equals the volume fraction.

mole fraction for oxygen

$$= \frac{0.1}{0.35} = 0.2857$$

mole fraction for nitrogen

$$= \frac{0.2}{0.35} = 0.5714$$

mole fraction for carbon dioxide

$$= \frac{0.05}{0.35} = 0.1429$$

$$\therefore \quad p_{O_2} = 0.2857 \times 2 = \mathbf{0.5714} \text{ bar}$$

$$p_{N_2} = 0.5714 \times 2 = \mathbf{1.1428} \text{ bar}$$

and $\quad p_{CO_2} = 0.1429 \times 2 = \mathbf{0.2858}$ bar

(*b*) From the characteristic equation,

$$pV = n\,R_{mol}\,T$$

number of moles of oxygen,

$$n_{O_2} = \frac{2\times10^5 \times 0.1}{8.314\times286} = 8.411 \text{ mol}$$

number of moles of nitrogen,

$$n_{N_2} = \frac{2\times10^5 \times 0.2}{8.314\times286} = 16.822 \text{ mol}$$

number of moles of carbon dioxide,

$$n_{CO_2} = \frac{2\times10^5 \times 0.05}{8.314\times286} = 4.205 \text{ mol}$$

In terms of pressure and temperature changes, the entropy change of a constituent gas is given by

$$ds = c_p \log_e \frac{T_f}{T_1} - R_{mol} \log_e \frac{p}{P_1}$$

$$= -R_{mol} \log_e \frac{p}{p_1} \qquad (\because T_f = T_1)$$

$$\therefore \quad (ds)_{O_2} = -8.314 \log_e \frac{0.5714}{2}$$
$$= 10.416 \text{ J/mol K}$$
$$= 10.416 \times 8.411 = 87.61 \text{ J/K}$$
$$= 0.0871 \text{ kJ/K}$$

$$(ds)_{N_2} = -8.314 \log_e \frac{1.1428}{2}$$
$$= 4.653 \text{ J/mol K}$$
$$= 4.653 \times 16.822 = 78.27 \text{ J/K}$$
$$= 0.07827 \text{ kJ/K}$$

$$(ds)_{CO_2} = -8.314 \log_e \frac{0.2858}{2}$$
$$= 16.176 \text{ J/mol K}$$
$$= 16.176 \times 4.205$$
$$= 68.02 \text{ J/K} = 0.0680 \text{ kJ/K}$$

SALIENT POINTS

1. The sum of the parts of mass (mass fractions) and by moles (mole fractions) of the constituents in a gaseous mixture equal unity. Mathematically

$$x_a + x_b + x_c + ... = 1$$
$$y_a + y_b + y_c + ... = 1$$

2. Dalton's law: The total pressure of a mixture of ideal gases is equal to the sum of partial pressures of the individual gas components of the mixture.
   Partial pressure is defined as the pressure which each individual component of a gas mixture would exert if it alone occupied the volume of the mixture at the same temperature.
3. Density of the gas mixture is equal to the sum of the densities of the components. Mathematically

$$\rho = \rho_a + \rho_b + \rho_c + ...$$
$$\frac{1}{v} = \frac{1}{v_a} + \frac{1}{v_b} + \frac{1}{v_c} + ...$$

   where v is the specific volume
4. Amagat-Leduc law: The total volume of a mixture of ideal gases is equal to the sum of partial (reduced) volumes of the constitutent gases.
5. In a mixture of ideal gases, the partial pressure ratio, the mole fraction and the volume fraction are all equal.
6. $R_{mix} = \sum x_i R_i = R_{mol} \sum \left(\frac{x_i}{M_i}\right)$

$$M_{mix} = \sum \frac{1}{\left(\frac{x_i}{M_i}\right)} = \sum y_i M_i$$
$$(c_v)_{mix} = \sum x_i (c_v)_i$$
$$(c_p)_{mix} = \sum x_i (c_p)_i$$

7. The mass of volume proportions are correlated by the expression

$$m_i = \left(\frac{V_i}{V}\right) M_i$$

8. Gibbs-Dalton law: "The internal energy, enthalpy and entropy of a gaseous mixture are respectively equal to the sums of the internal energies, the enthalpies and the entropies which each component of the gas mixture would have, if each alone occupied the volume of the mixture at the temperature of the mixture".

REVIEW QUESTIONS

1. How does a mixture of ideal gases behave in relation to an ideal gas ?
2. Which two properties are common and additive respectively for a mixture of ideal gases ?
3. State the Dalton's law of partial pressures. On what assumptions this law is based ?
4. What is meant by mass, volume and molar fractions ?
5. Define partial pressure and partial (or reduced) volume of a gas mixture component.
6. Set up an expression for the partial pressure of a gas in the gaseous mixture in terms of mass and volume fraction.
7. Determine the molecular mass, gas constant and density at standard temperature pressure (p = 101325 N/m^2 and t = 0°C) for air. It may be presumed that air contains 21 percent of oxygen and 79 percent of nitrogen in volume fractions.
   [**Ans.** 28.84, 288.28 J/kg K, 1.287 kg/m^3]
8. A perfect gas mixture consists of 3 kg of nitrogen and 5 kg of carbon dioxide at a pressure of 300 kPa and a temperature of 290 K. Calculate the specific heats at constant volume and at constant pressure for the gas mixture. Given
   For nitrogen : c_v = 0.745 kJ/kg K and
   c_p = 1.042 kJ/kg K
   For carbon dioxide : c_v = 0.653 kJ/kg K and
   c_p = 0.842 kJ/kg K
   [**Ans.** 0.6875 kJ/kg K and 0.917 kJ/kg K)
9. A vessel of 0.4 m^3 capacity contains a mixture of air and carbon dixoide. The mixture comprises 5 kg of air, 3 kg of carbon dioxide

and is at 300 K temperature. Determine the gas constant and pressure of the gas mixture, and partial pressures of the components.

[**Ans.** 2560.23 J/kg, 15.02 bar, 10.79 bars and 4.23 bars]

10. A 1.5 kg mixture of two gases in the proportion of 40% and 60% is stored at 20°C in a closed vessel of 5 m^3 capacity. If the gas constant for the constituent gases is 287 J/kg K and 294 kJ/kg K, make calculations for the partial pressures, total pressure and gas constant for the mixture.

[**Ans.** 10.09 kPa, 15.5 kPa, 25.59 kPa, 291 J/kg K]

11. A mixture of perfect gases at a pressure of 3 bar and a temperature of 290 K has the following composition: 3 kg of N_2 and 5 kg of CO_2

Determine (*i*) the mole fraction of each constituent, (*ii*) the equivalent molecular weight of the mixture, (*iii*) the equivalent gas constant of the mixture, (*iv*) the partial pressures and partial volume, (*v*) the volume and density of the mixture.

[**Ans.** (*i*) 0.486, 0.54 (*ii*) 36.23, (*iii*) 229.4 J/kg K, (*iv*) 1.458 bar, 1.52 bar, (*v*) 1.769 m^3, 4.52 kg m^2]

❏❏❏

Chapter 6

First Law of Thermodynamics

Learning objectives : A study of the subject matter included in this chapter will enable the reader to :

- *explain energy conversions and the essence of first law of thermodynamics*
- *state important consequences/corollaries of the first law*
- *define internal energy as a system property*
- *make an analysis of the non-flow processes regarding relationship between state parameters, evaluation of work and heat involved, and calculations for change in internal energy*
- *explain Joule's law: internal energy as a function of absolute temperature only*

6.1. JOULE'S PADDLE-WHEEL EXPERIMENT

A known mass of water is taken into a rigid and well insulated container provided with a paddle wheel. The insulation is provided to prevent any heat interaction with the surroundings. The water (system) is stirred by the paddle wheel which is driven by allowing a known mass m (connected to the paddle wheel by means of chords of negligible mass and frictionless pulley) to fall through a specified height h. The work transferred to the water system is calculated in terms of decrease in potential energy of the falling mass ($\Delta PE = mgh$). Due to this work interaction, the temperature of the water increases. The change in thermal energy of the water system is computed from rise in temperature and heat capacity of the water and the container ($Q = mc\Delta t$).

Joule conducted a number of experiments involving different types of work interactions, and found that the work expended was proportional to increase in thermal energy. That is

$$Q \propto W \quad \text{or} \quad Q = W/J \qquad ...(6.1)$$

where the proportionality constant J was termed as Joule's equivalent or mechanical equivalent of heat. In SI system of units, both heat and work are measured in Joules and as such the proportionality constant J takes a unit value.

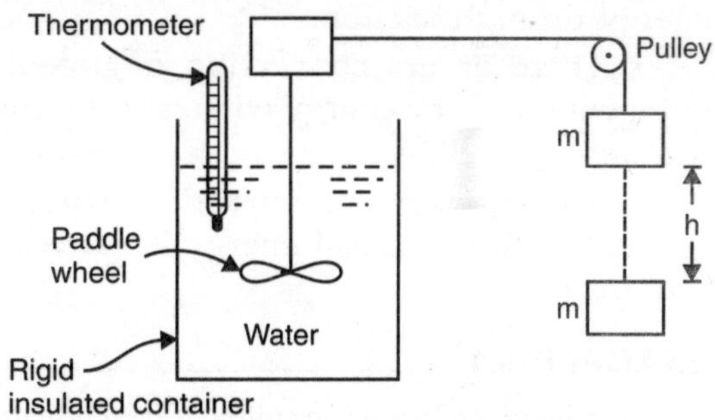

Fig. 6.1. Joules's paddle wheel experiment

6.2. ENERGY CONVERSIONS

Consider water available in the natural or artificial high level water reservoir. When the water falls, its potential energy gets transferred to kinetic energy. The high velocity jet impinges on the blades of a turbine and sets them into rotary motion. The mechanical energy made available at the turbine shaft is utilized to run an electric generator which is coupled to the turbine shaft. The electrical power thus generated, if shorted through a resistance coil produces heat. The different species of energy, though qualitatively different, are simply diverse manifestations of a single fundamental potency namely energy and disappearance of one form gives way to another form.

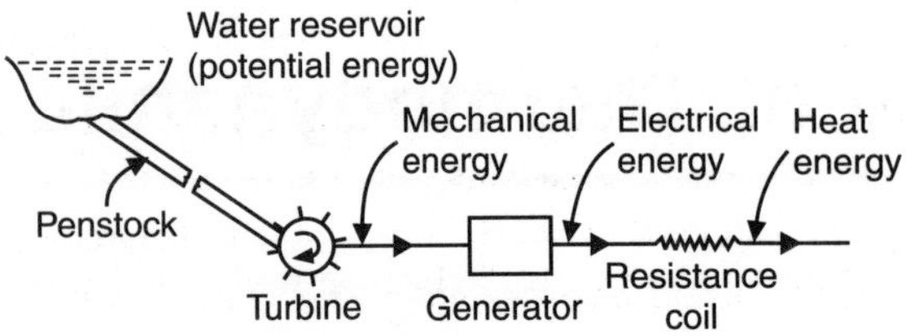

Fig. 6.2. Energy conversions

Some other notable examples of energy transformations are :

- The heat of fuel input to an internal combustion engine can be accounted for as output in the form of mechanical energy at the shaft and heat loss to cooling medium and surroundings etc.
- When a running automotive vehicle is brought to rest by applying brakes, the kinetic energy of the vehicle is converted into heat energy through friction.
- There occurs conversion of potential energy into kinetic energy when a fluid flows through a pipeline.
- During flow of current through a resistance, the electrical energy is converted into heat.

EXAMPLE 6.1

A closed insulated vessel contains 200 kg of water. A paddle wheel immersed in the water is driven at 400 rev/min with an average torque of 500 Nm. If the test run is made for 30 minutes, determine rise in the temperature of water.

Take specific heat of water = 4.186 kJ/kg K

Solution : Total number of revolutions made by the paddle wheel,

$$N = 400 \times 30 = 12000$$

Work supplied to water

$$= 2\pi NT = 2\pi \times 12000 \times 500$$
$$= 37680 \times 10^3 \text{ Nm (or J)}$$

This work will get transferred to heat which will result in raising the temperature of water. Therefore,

$$37680 \times 10^3 = mc\Delta t$$
$$= 200 \times (4.186 \times 10^3) \times \Delta t$$

∴ Rise in temperature of water,

$$\Delta t = \frac{37680 \times 10^3}{200 \times 4.186 \times 10^3} = \mathbf{45°C}$$

EXAMPLE 6.2

A locomotive train of 2 × 10⁶ kg mass is driven by the energy generated by burning coal. The train requires a tractive effort of 40 N per 1000 kg of its dead mass. What distance the train will travel if it consumes 1000 kg of coal and 10 percent of heat generated by burning coal is converted into mechanical work ?

Take calorific value of coal = 32000 kJ/kg

Solution : Heat energy released by burning of coal,

$$= \text{mass of coal} \times \text{calorific value of coal}$$
$$= 1000 \times 32000 = 32 \times 10^6 \text{ kJ}$$

Mechanical energy available by propelling the train

$$= 0.1 \times (32 \times 10^6)$$
$$= 32 \times 10^5 \text{ kJ} \quad ...(i)$$

Force required to move the train (tractive effort)

$$= \frac{40}{1000} \times (2 \times 10^6) = 80 \times 10^3 \text{ N}$$

Let x metre be the distance travelled by the train. Then

Work done = force × distance moved

$$= 80 \times 10^3 \times x \quad \text{Nm (or J)}$$
$$= 80x \text{ kJ} \quad ...(ii)$$

Equating expressions (*i*) and (*ii*),

$$80x = 32 \times 10^5;\ x = 40 \times 10^3 \text{ m} = \mathbf{40\ km}$$

6.3. FIRST LAW OF THERMODYNAMICS

It is a special case of the law of conservation of energy and stipulates that :

– Energy can neither be created nor destroyed; it is always conserved.However, it can change from one kind into another.

– Total energy of an isolated system, in all its forms remains constant.

– All the energy that goes into a system

comes out in some other form. Energy does not vanish and has the ability to be converted into any other form of energy.

– No machine can produce energy without corresponding expenditure of energy. Energy cannot appear from nothing, nor can it convert into nothing.

The first law of thermodynamics cannot be proved mathematically. Its validity stems from the fact that neither it nor any of its consquences have been contradicted.

6.3.1. First law for a cyclic process

A process is cyclic if the initial and final states of the system executing the process are identical.

A system represented by state point 1 undergoes a process 1 – *a* – 2, and comes back to initial state following the path 2 – *b* – 1. All properties of the system are restored, when the initial state is reached. During the execution of these processes :

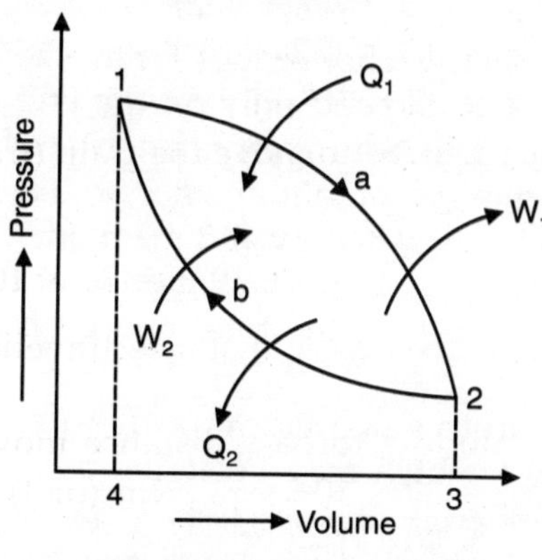

Fig. 6.3. Cyclic process

(*i*) area 1 – *a* – 2 – 3 – 4 – 1 represents the work W_1 done by the system during expansion process 1 – *a* – 2,

(*ii*) area 2 – 3 – 4 – 1 – *b* – 2 represents the work W_2 supplied to the system during compression process 2 – *b* – 1

(*iii*) area 1 – *a* – 2 – *b* – 1 represents the net work $(W_1 - W_2)$ delivered by the system.

Since the system regains its initial state, there is no change in the energy stored by the system. For energy balance and in accordance with the law of conservation of energy, an equivalent amount of heat energy must have been received by the system. Accordingly the first law of thermodynamics for a cyclic process can be mathematically expressed as

$$W_{1-2} = Q_{1-2}; \quad \oint \delta Q = \oint \delta W \quad \text{...(6.2)}$$

i.e., cyclic integral of heat $\oint \delta Q$ *is equal to cyclic integral of work* $\oint \delta W$; both forms of energy being expressed in the same units.

$$\text{or} \quad \oint (\delta Q - \delta W) = 0 \quad \text{...(6.3)}$$

It follows from the above equation that *if a system is taken through a cycle of processes so that it returns to the same state or condition from which it started, the sum of heat and work effects will be zero.*

It is to be noted that though both δQ and δW are path functions, their difference $(\delta Q - \delta W)$ is a point function as the integral $\oint (\delta Q - \delta W)$ is zero.

EXAMPLE 6.3

A system undergoes a cycle composed of four processes. The heat transfers in each process are : 400 kJ, – 365 kJ, – 200 kJ and 250 kJ. The respective work transfers are 140 kJ, 0, – 55 kJ and 0. Is the data consistent with first law of thermodynamics ?

Solution : Net heat interaction,

$$\oint \delta Q = 400 - 365 - 200 + 250$$

$$= 85 \text{ kJ}$$

Net work transfer,

$$\oint \delta W = 140 + 0 - 55 + 0 = 85 \text{ kJ}$$

Since $\oint \delta Q = \oint \delta W$, the given data conforms to the first law of thermodynamics.

EXAMPLE 6.4

The working fluid in an engine executes a cyclic process, and two work interactions are involved : 10 kJ to the working fluid and 30 kJ from the working fluid. The working cycle also involves three heat transfers two of which are 75 kJ to the working fluid and 40 kJ from the working fluid.

Determine the magnitude and direction of the third heat interaction.

Solution : $\oint \delta W = -10 + 30 = 20$ kJ

$$\oint \delta Q = 75 - 40 + x$$

where x is the third heat interaction.

From first law of thermodynamics,

$$\oint \delta Q = \oint \delta W \text{ for a cyclic process}$$

$75 - 40 + x = 20$; $x = $ **- 15 kJ**

The negative sign indicates that heat flows out of the system.

EXAMPLE 6.5

Consider a steam turbine power plant system where the work and heat interactions across the system boundary are :

(i) Power available at the turbine shaft is 600 kW and the feed pump requires 3 kW of work to pump the condensate back into the boiler.

(ii) Steam generation in the boiler requires 2500 kJ/kg of heat, and in the condenser 1800 kJ/kg of heat is rejected to cooling water.

Make calculations for the steam flow rate in kg/hr round the cycle.

Solution : $\oint \delta W = 600 - 3$

$= 597 \text{ kW} = 597 \times 3600 \text{ kJ/hr}$

If m is the steam flow rate in kg/hr, then

$$\oint \delta Q = m\,(2500 - 1800) = 700m \text{ kJ/hr}$$

From the first law of thermodynamics,

$$\oint \delta Q = \oint \delta W;\ 700m = 597 \times 3600$$

$$m = \frac{597 \times 3600}{700} = \mathbf{3070\ kg/hr}$$

EXAMPLE 6.6

A fully charged battery of a car is lying on the shelf of a store at constant temperature. There occurs gradual discharge from the battery and during this act it loses 1000 kJ to the environment. When the battery is subsequently recharged slowly to its initial state, it consumes 0.5 kWh of electricity. Work out the heat transfer during the recharging process. Consider the battery as system.

Solution : Discharging process :

$Q_{1-2} = -1000$ kJ ; $W_{1-2} = 0$

Recharging process :

$W_{1-2} = -0.5$ kWh

$= -0.5 \times 3600 = -1800$ kJ

Applying, first law to the complete discharge and recharge process,

$$\oint \delta Q = \oint \delta W;\ Q_{2-1} - 1000 = -1800 + 0$$

$Q_{2-1} = $ **- 800 kJ**

The battery loses 800 kJ of heat during the recharging process.

6.3.2. Corollaries of first law

The first law of thermodynamics has a number of important consequences which are stated in the form of corollaries.

Corollary 1. (First law for a process) : When a system executes a process, the net heat interaction equals the net work interaction plus change in stored energy.

That is, $\delta Q = \delta W + dE$ ...(6.4)

The energy change has been shown in section 6.4 to depend only on the end states and hence E is a property. Internal energy, kinetic energy, potential energy, electrical energy, chemical energy and magnetic energy are some of the important modes of energy which add to give the total energy E of the system.

The above aspect is referred to as corollary 1 which may be stated as :

"There exists a property of a closed system such that a change in its value during any change of state is given by the difference between the heat supplied and work done"

$$dE = \delta Q - \delta W$$

The property E is called energy of the system. This energy resides within the system; it increases or decreases with change of state.

Corollary 2. (First law for an isolated system) : For an isolated system, both heat and work interactions are absent ($\delta Q = 0$, $\delta W = 0$) and therefore,

$dE = 0$ or $E = \text{constant}$ ...(6.5)

The above identity indicates that the energy of a system remains unchanged if the system is isolated from its surroundings as regards heat and work interactions.

This fact is often referred to as the principle of conservation of energy and may be stated as :

"Energy can neither be created nor destroyed; however, it can be converted from one form to another"

Corollary 3. (Perpetual motion machine of first kind) : A perpetual motion machine of first kind (*PMM*1) is an imaginary device which produces a continuous supply of work without absorbing any energy from the surroundings or from other systems. Such a machine, in effect, creates energy from nothing and violates the first law of thermodynamics. Despite several attempts to create such a mechanism, no device has yet been invented. Corollary 3 may then be stated as,

"A perpetual motion machine of the first kind is impossible"

EXAMPLE 6.7

A system consisting of 2 kg of water is moving at a velocity of 30 m/s at an elevation of 400 m above the sea level. If the value of its internal energy is 100 kJ/kg, determine the total energy of the system.

Solution : Specific energy (energy per unit mass) is given by

$$e = u + \frac{V_1^2}{2} + gz$$

$$= 100 + \left(\frac{30^2}{2}\times 10^{-3}\right) + \left(9.81\times 400\times 10^{-3}\right)$$

$$= 100 + 0.415 + 3.925 = 104.34 \text{ kJ/kg}$$

Total energy,

$$E = m \times e$$

$$= 2 \times 104.34 = 208.68 \text{ kJ}$$

It is to be noted that kinetic and potential energies are very small when compared with the value of internal energy.

6.4. STORED ENERGY—A PROPERTY OF THE SYSTEM

Consider a closed system which changes from state 1 to state 2 by path *A* and returns back to original state 1 by any one of the following paths,

(*i*) along path 2*B*1
(*ii*) along path 2*C*1
(*iii*) along path 2*D*1

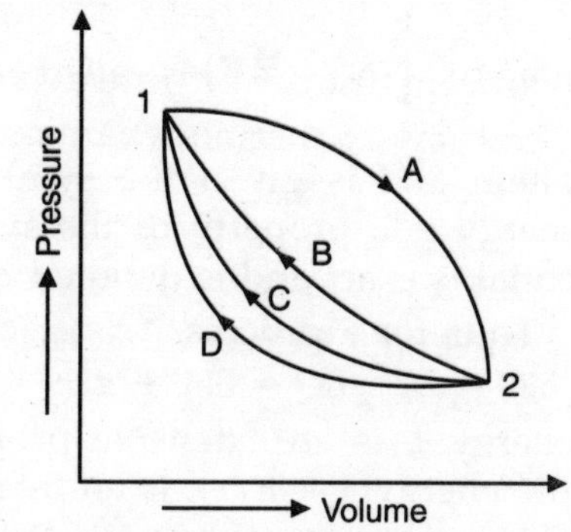

Fig. 6.4. Energy—a Property of the System

Invoking first law for the cyclic process 1 - *A* - 2 - *B*- 1

$$\oint(\delta Q - \delta W) = 0$$

or
$$\int_{1,\text{ via } A}^{2}(\delta Q - \delta W) + \int_{2,\text{ via } B}^{1}(\delta Q - \delta W) = 0 \quad \ldots(i)$$

Likewise for the cycle 1 - *A* - 2 - *C* - 1,

$$\int_{1,\text{ via } A}^{2}(\delta Q - \delta W) + \int_{2,\text{ via } C}^{1}(\delta Q - \delta W) = 0 \quad \ldots(ii)$$

and for the cycle 1 - *A* - 2 - *D* - 1,

$$\int_{1,\text{ via } A}^{2}(\delta Q - \delta W) + \int_{2,\text{ via } D}^{1}(\delta Q - \delta W) = 0 \quad \ldots(iii)$$

Comparing equations (*i*), (*ii*) and (*iii*), we may write :-

$$\int_{2,\text{ via } B}^{1}(\delta Q - \delta W) = \int_{2,\text{ via } C}^{1}(\delta Q - \delta W)$$

$$= \int_{2,\text{ via } D}^{1}(\delta Q - \delta W)$$

Since *B*, *C* and *D* represent arbitrary paths between the state point 2 and state point 1, it can be concluded that the integral $\int_{2}^{1}(\delta Q - \delta W)$

(*i*) remains the same irrespective of the path along which the system proceeds,

(*ii*) is solely dependent on the initial and final states of the system; is a point function and hence a property.

The integral $\int_2^1 (\delta Q - \delta W)$ is called energy of the system and is given the symbol E. Further energy is a property of the system; its differential is exact and is denoted by dE.

Thus for a process,

$$\delta Q - \delta W = dE \; ; \; \delta Q = \delta W + dE \quad ...(6.6)$$

The energy E is an extensive property. The specific energy, $e = E/m$, is an intensive property.

In a reversible non-flow process,

$\delta W = p\,dV$ and therefore,

$$\delta Q = p\,dV + dE \quad ...(6.7)$$

The energy E includes all forms of energy in the system, *i.e.*,

$$E = U + \text{kinetic energy} + \text{potential energy} + \text{chemical energy} + \text{electrical energy} + \text{magnetic energy} + ...$$

where energy U is associated with molecular motion and is stored in the molecules and atomic structure of the substance. In an ideal gas, there are no molecular forces of attraction and repulsion and the energy U depends only on temperature, *i.e.*, $U = f(T)$. Energy of this kind is called *internal energy, intrinsic energy* or *microscopic energy*.

For a closed system, there is no flow of fluid into and out of the system and as such the kinetic energy of the mass of system is not to be considered. There is also no appreciable change in the potential energy of the system mass. Further, the process is quasi-static in nature and the system moves from one equilibrium state to another. Thus no chemical reaction or change in chemical composition is involved and hence no change in chemical energy of the system.

In the absence of kinetic energy, potential energy, chemical energy, electrical and magnetic effects, equation 6.6 can be rewritten as

$$\delta Q = \delta W + dU$$
$$= p\,dV + dU \quad ...(6.8)$$

Equation 6.8 is an analytical expression of the first law of thermodynamics and is often referred to as the basic equation of thermodynamics. It shows that the heat received by a system is utilized to change its internal energy and to make the system perform external work. If the expansion of the system is without any transfer of heat ($\delta Q = 0$), the work will be performed at the cost of internal energy. Further, if the system receives heat without any change in its boundaries ($\delta W = 0$), the entire heat added will go in for increasing the internal energy of the system.

EXAMPLE 6.8

In a general compression process, 2 kJ of mechanical work is supplied to 5 kg of working substance, and 800 J of heat is rejected to the cooling jacket. Calculate the change in specific internal energy.

Solution : From non-flow energy equation,

$$\delta Q = \delta W + dU$$
$$\therefore \quad dU = \delta Q - \delta W$$
$$= -800 - (-2000) = 1200 \text{ J}$$

The heat transfer has been taken negative as it flows out of system. Further work is also negative as this has been supplied to the system.

du = change in internal energy per unit mass

$$= \frac{1}{5} \times 1200 = \mathbf{240\ J/kg}$$

EXAMPLE 6.9

A domestic refrigerator is loaded with fresh food and the door closed. During a certain period, the machine consumes 1.25 kWh of electrical energy in cooling the food and the internal energy of the system decreases by 4500 kJ. Find the magnitude and sign of net heat transfer for the system. The refrigerator and its contents may be considered as the system.

Solution : Work supplied to refrigerator,

$$\delta W = 1.25 \text{ kWh} = 1.25 \times 3600 = 4500 \text{ kJ}$$

This work interaction is negative as work has been supplied to the system.

Change in internal energy,

$dU = -4500$ kJ (since decreases)

From non-flow energy equation,

$$\delta Q = \delta W + dU$$
$$= -4500 + (-4500) = \mathbf{-9000\ kJ}$$

Since this is negative, heat flows out of the system.

EXAMPLE 6.10

A storage battery, having a terminal potential of 12 volts, draws a current of 8 amperes for 2.5 hours. If the stored energy of the battery decreases by 1250 kJ, evaluate the heat interaction across the boundary enveloping the storage battery.

Solution : Work done in 2.5 hours

$$= VI\tau = \frac{12 \times 8 \times 2.5}{1000} \text{ kWh}$$
$$= \frac{12 \times 8 \times 2.5}{1000} \times 3600 = 864 \text{ kJ}$$

This is a positive work as the current flows out of the battery. Further, the battery forms a closed system, and then from the non-flow energy equation,

$$\delta Q = \delta W + dU$$
$$= 864 + (-1250) = \mathbf{386\ kJ}$$

EXAMPLE 6.11

A 0.75 kW motor drives a paddle which stirs for an hour 50 kg of water contained in a thermally insulated rigid tank. Assuming the process to be at constant volume, determine rise in temperature and change in internal energy of water.

Take c for water = 4.186 kJ/kg K

Solution : Paddle work done in one hour

$= 0.75$ kWh

$= 0.75 \times 3600 = 2700$ kJ

This is a –ve work as this has been supplied to the system (water). The heat interaction during the operation is zero as the tank is thermally insulated. Further, the tank is rigid and accordingly the displacement work $\int p\, dv$ is also zero. Then from non-flow energy equation,

$$\delta Q = \left(\int p dv + \text{paddle work}\right) + dU$$

∴ Change in internal energy,

$dU = -$ paddle wheel $= 2700$ kJ

This change in internal energy of water is reflected by an increase dT in its temperature.

Thus,

$$2700 = m\, c\, dT = 50 \times 4.186 \times dT$$
$$\therefore dT = \frac{2700}{50 \times 4.186} = \mathbf{12.90°C}$$

EXAMPLE 6.12

A stationary mass of gas is compressed without friction from an initial state of 0.4 m³ and 0.1 MPa to a final state of 0.2 m³ and 0.1 MPa; the pressure remaining constant during the process. If there is a transfer of 45 kJ of heat from the gas during the process, evaluate change in the internal energy of the gas.

Solution : $Q_{1-2} = -45$ kJ

(–ve because heat is lost by the gas)

$$W_{1-2} = \int p\, dV$$
$$= p(V_2 - V_1)$$
$$= 0.1 \times 10^6 (0.2 - 0.4)$$
$$= -20 \times 10^3 \text{ Nm (or J)} = -20 \text{ kJ}$$

From non-flow energy equation,

$$Q_{1-2} = W_{1-2} + dU$$
$$-45 = -20 + dU; \quad dU = \mathbf{-25\ kJ}$$

The internal energy of the gas decreases by 25 kJ during the compression process.

EXAMPLE 6.13

A closed system having a mass of 50 kg has an initial velocity of 10 m/s. Subsequently its velocity increases to 30 m/s and its elevation also rises by 40 m. During this process, the system receives 30000 J of heat and 4500 J of work. If the system delivers 0.002 kWh of electrical energy, make calculations for a change in the internal energy of the system.

Solution :

$$\Delta PE = mg(h_2 - h_1)$$
$$= 50 \times 9.81 \times 40 = 19620 \text{ J}$$
$$\Delta KE = \frac{1}{2} m\left(V_2^2 - V_1^2\right)$$

$$= \frac{1}{2} \times 50 \times (30^2 - 10^2)$$
$$= 20000 \text{ J}$$

Q_{1-2} = + 30000 J (+ve sign because heat is received by the system)

W_{1-2} = work delivered – work received
$= 0.002 \times 1000 \times 3600 - 4500 = 2700$ J

From non-flow energy equation :

$$Q_{1-2} = W_{1-2} + dE$$

The energy E includes the internal energy U, the kinetic energy and the potential energy.

$$\therefore \quad Q_{1-2} = W_{1-2} + dU + \Delta KE + \Delta PE$$
$$30000 = 2700 + dU + 20000 + 19620$$

Then, change in internal energy,

$$dU = 30000 - 2700 - 20000 - 19620$$
$$= \mathbf{-12320 \text{ J}}$$

The internal energy of the system decreases by 12320 J during the process undergone by the system.

EXAMPLE 6.14

A workpiece of 30 cm diameter is being turned on a lathe and the tangential force on the cutting tool is estimated to be 100 N. When the turning operation is executed for 10 minutes, the increase in the internal energy of the workpiece works out 125 kJ. Make calculations for the heat transfer from the workpiece if the workpiece turns at 180 rev/min.

Solution : Revolutions made by the workpiece in 10 minutes,

$$N = 180 \times 10 = 1800$$

Torque, T = force × radius

$$= 100 \times \frac{0.3}{2} = 15 \text{ Nm}$$

Work done on the workpiece,

$$\delta W = 2\pi NT = 2\pi \times 1800 \times 15$$
$$= 169560 \text{ Nm (or J)} = 169.56 \text{ kJ}$$

This work is negative as it has been on the workpiece (system)

Increase in internal energy, dU = 150 kJ

From first law of thermodynamics (non-flow energy equation)

$$\delta Q = \delta W + dU$$
$$= -169.56 + 125 = \mathbf{-44.56 \text{ kJ}}$$

Since this is negative, heat flows from the system (workpiece).

EXAMPLE 6.15

The work output and heat input involved by a system in process A are 20 kJ and 15 kJ respectively. Another process B between the same end conditions involves a heat input of 10 kJ. Determine the change in internal energy involved and also the work done in process B. Show that if a cycle is formed using processes A and B, the given data is consistent with the first law.

Solution : For process 1 – A – 2

$$\delta W = 20 \text{ kJ and } \delta Q = 15 \text{ kJ}$$

From non-flow energy equation

$$dU = U_2 - U_1$$
$$= \delta Q - \delta W = 15 - 20 = -5 \text{ kJ}$$

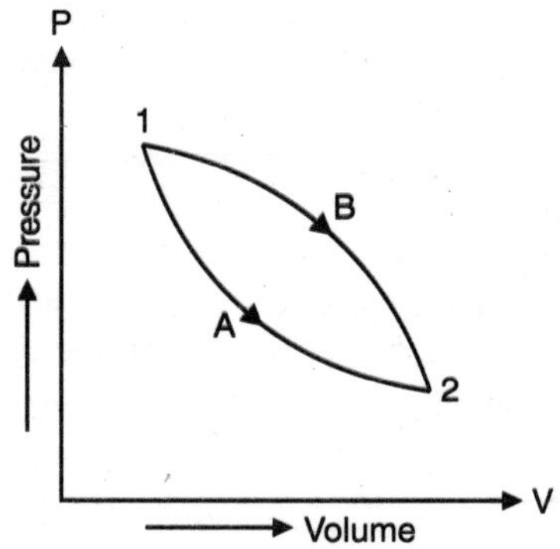

Fig. 6.5.

The negative sign implies that the internal energy of the system has decreased.

For process 1 – B – 2 :

$$\delta Q = 10 \text{ kJ}$$
$$dU = U_2 - U_1$$
$$= -5 \text{ kJ as calculated above}$$
$$\therefore \quad \delta W = \delta Q - dU = 10 - (-5) = 15 \text{ kJ}$$

When the cycle is formed by the processes A and B

$$\oint \delta Q = Q_{1-A-2} + Q_{2-B-1} = 15 - 10 = 5 \text{ kJ}$$
$$\oint \delta W = W_{1-A-2} + W_{2-B-1}$$
$$= 20 + (-15) = 5 \text{ kJ}$$

It is to be noted that :

$$Q_{2-B-1} = -Q_{1-B-2}$$
and $$W_{2-B-1} = -W_{1-B-2}$$

Since $\oint \delta Q = \oint \delta W$, the data conforms to the first law of thermodynamics.

EXAMPLE 6.16

A mixture of powdered coal and air at 10 bar pressure and 300 K temperature is contained in an insulated rigid vessel. When the mixture is ignited, heat energy is liberated which causes a rise in both pressure and temperature; the temperature attains a value of 550 K. Evaluate the increase in energy of the system comprising the vessel and the contents.

(b) Upon removal of insulation, a heat transfer of 50 kJ of heat causes the temperature to drop to the initial value of 300 K. What will be the decrease in energy of the system during the process?

(c) If the initial energy of the system is 30 kJ, write down the energy values after the process (a) and process (b).

Solution : The vessel is rigid; there occurs no change in the volume of the system and therefore, $W_{1-2} = 0$. Further, the system is insulated and accordingly $Q_{1-2} = 0$. Then from non-flow energy equation,

$$Q_{1-2} = W_{1-2} + dU; \quad 0 = 0 + dU$$

$$\therefore \quad dU = 0$$

Even though there has been rise in both pressure and temperature of the system, there will be no change in energy. All that happens is that the chemical energy of the system gets changed to internal energy.

(*b*) $W_{1-2} = 0; \quad Q_{1-2} = -50 \text{ kJ}$

$$Q_{1-2} = W_{1-2} + dE$$

$$= 0 + dE \quad \text{or} \quad dE = -50 \text{ kJ}$$

$\therefore$ Decrease in energy = 50 kJ

(*c*) When insulation is provided; $dE = 0$. That is

final energy = initial energy = 30 kJ

When insulation is removed; $dE = -50$ kJ. That is

$$E_2 - E_1 = -50; \quad E_2 - 30 = -50$$

$$\therefore \quad E_2 = \mathbf{-20 \text{ kJ}}$$

EXAMPLE 6.17

A combustible mixture is enclosed in a 10 cm diameter vertical cylinder fitted with a frictionless piston. The piston exerts a constant pressure of 300 kPa upon the mixture and can move freely. The upper surface of the piston is exposed to the atmosphere. When the mixture is ignited, the combustion products make the piston move outwards in quasi-static manner and the heat flows to the surroundings. After completion of the reaction and with the contents having been reduced to the initial temperature of 300 K, it is found that the piston has moved a distance of 8 cm and the magnitude of heat transfer to the surroundings is 4 kJ. Make calculations for the work done and change in energy of the system.

Solution : Work done under constant pressure conditions,

$$W_{1-2} = \int p\, dV$$

$$= 300 \times 10^3 \times \left[\frac{\pi}{4}(0.1)^2 \times 0.08\right]$$

$$= 188.4 \text{ Nm (or J)} = 0.1884 \text{ kJ}$$

Since this is positive, work is done by the system.

From non-flow energy equation,

$$Q_{1-2} = W_{1-2} + dU; \; -4 = 0.1884 + dU$$

Heat interaction has been taken negative as it flows out of the system.

$$dU = -4 - 0.1884 = -4.1884 \text{ kJ}$$

$\therefore$ Decrease in energy of the system $= \mathbf{4.1884 \text{ kJ}}$

EXAMPLE 6.18

The relationship between pressure and volume in a non-flow process is prescribed by the expression:

$p = \left(\frac{3}{V} + 2\right)$ *where presssure p is in bar and volume V is in m³.*

During the process 1600 kJ of heat is added to the gas and the volume changes form 1.2 m³ to 4 m³. Determine the change in internal energy.

Solution : Work done

$$W_{12} = \int_1^2 p\, dV = \int_1^2 \left(\frac{3}{V} + 2\right) \times 10^5 \times dV$$

$$= 10^5 \left|3\log_e V + 2V\right|_{1.2}^{4}$$

$$= 10^5\left[3\log_e\frac{4}{2} + 2(4-1.2)\right]$$

$= 10^5\ [3 \times 1.203 + 2 \times 2.8]$

$= 9.2116 \times 10^5$ J = **921.16** kJ

From the non-flow energy equation,

$\delta Q = \delta W + dU$, we have

change in internal energy,

$dU = \delta Q - \delta W$

$= 1600 - 921.16 =$ **678.84** kJ

EXAMPLE 6.19

A cylinder fitted with a movable piston contains 0.04 m^3 of air at 10 bar pressure and 400 K temperature. The air expands according to the law $p = \left[\frac{A}{V^2} - \frac{B}{V}\right]$ to a final pressure of 1 bar and 0.2 m^3.

Determine work done, change in internal energy and heat absorbed or rejected during the expansion process.

Take c_v = 0.0718 kJ/kgK

Solution : Substituting the data corresponding to initial and final conditions in the given expression, we may write

$$10 = \frac{A}{0.04^2} - \frac{B}{0.04}$$

and $$1 = \frac{A}{0.2^2} - \frac{B}{0.2}$$

or $10 = 625\,A - 25\,B$

and $1 = 25\,A - 5\,B$

Solving for A and B, we get

$A = 0.01$ and $B = -0.15$

work done

$$W_{1\text{-}2} = \int_1^2 p\,dV = \int_1^2\left(\frac{A}{V^2} - \frac{B}{V}\right)dV$$

$$= A\left|-\frac{1}{V}\right|_1^2 - B\left|\log_e V\right|_1^2$$

$$= A\left[\frac{1}{V_1} - \frac{1}{V_2}\right] - B\log_e\frac{V_2}{V_1}$$

Substituting the values of A, B, V_1 and V_2, we obtain

$$W_{1\text{-}2} = \left[0.01\left\{\frac{1}{0.04} - \frac{1}{0.2}\right\} - (-0.15)\log_e\frac{0.2}{0.04}\right]\times 10^5$$

$= (0.2 + 0.2414) \times 10^5$

$= 0.4414 \times 10^5$ Nm (kJ) = **44.14** kJ

(*b*) From ideal gas equation $pv = mRT$,

$$m = \frac{p_1V_1}{RT_1} = \frac{10\times 10^5 \times 0.04}{287\times 400}$$

$= 0.3484$ kg

$$T_2 = \frac{p_2V_2}{mR} = \frac{1\times 10^5\times 0.2}{0.3484\times 287} = 200 \text{ K}$$

change in internal energy

$dU = mc_v\,dT$

$= 0.3484 \times 0.718 \times (200 - 400) =$ **– 50 kJ**

(*c*) From non-flow energy equation,

$\delta Q = \delta W + dU$,

Heat interaction

$Q_{1\text{-}2} = 44.14 - 50 =$ **– 5.86** kJ

The negative sign indicates loss or rejection of heat.

EXAMPLE 6.20

A closed system undergoes a thermodynamic cycle consisting of four separate and distinct processes. The heat and work transferred in each process are as tabulated below :

Process	*Heat transfer in kJ/min*	*Work transfer in kJ/min*
1 – 2	0	– 10,500
2 – 3	30,000	0
3 – 4	– 3,000	30,000
4 – 1	– 9,000	– 1,500

Show that the data is consistent with the first law of thermodynamics and proceed to evaluate the net work output in kW, thermal efficiency of the cycle and the change in internal energy for each process.

Solution : $\oint \delta Q = 0 + 30000 - 3000 - 9000$

$= 18000$ kJ/min

$$\oint \delta W = -10500 + 0 + 30000 - 1500$$
$$= 18000 \text{ kJ/min}$$

Since $\oint \delta Q = \oint \delta W$, the data is in accordance with the first law.

(*b*) Network output

$$= 18000 \text{ kJ/min} = \frac{18000}{60}$$
$$= 300 \text{ kJ/s} = 300 \text{ kW}$$

(*c*) The ratio of net work done by system to the heat supplied is known as thermal efficiency. Therefore,

$$\text{Thermal efficiency} = \frac{18000}{30000} = 0.6 \text{ or } 60\%$$

(*d*) Changes in internal energy for each of the process can be evaluated by applying the non-flow energy equation,

$$\delta Q = \delta W + dU$$

Process 1–2 :

$$dU = \delta Q - \delta W = 0 - (-10500)$$
$$= 10500 \text{ kJ/min}$$

Process 2–3 :

$$dU = 30000 - 0 = 30000 \text{ kJ/min}$$

Process 3–4 :

$$dU = -3000 - 30000$$
$$= -33000 \text{ kJ/min}$$

Process 4–1 :

$$dU = -9000 - (-1500)$$
$$= -7500 \text{ kJ/min}$$

For the complete cycle

$$\oint dU = 10500 + 30000 - 33000 - 7500 = 0$$

We see that for the complete cycle, the total change in internal energy is zero. This has to be so because in a cyclic process the system returns to the initial state and internal energy depends only on the state point and is independent of path.

EXAMPLE 6.21

A piston and a cylinder machine contains a fluid system, which passes through a complete cycle of four processes. During a cycle, the sum of all heat transfer is – 150 kJ and the system completes 100 cycles per minute. Complete the following table, and calculate the net rate of work output

Process	*Q (kJ/min)*	*W(kJ/min)*	*dU(kJ/min)*
1 – 2	0	2150	....
2 – 3	20000	0	....
3 – 4	– 2000	....	– 36,000
4 – 1	....	....	....

Solution :

Process 1–2 : $\delta Q = \delta W + dU$

$$0 = 2150 + dU$$
$$\therefore \quad dU = -2\,150 \text{ kJ/min}$$

Process 2–3 : $\delta Q = \delta W + dU$

$$20000 = 0 + dU$$
$$\therefore \quad dU = 20000 \text{ kJ/min}$$

Process 3–4 : $\delta Q = \delta W + dU$

$$-2000 = \delta W + (-36\,000)$$
$$\therefore \quad \delta W = 34000 \text{ kJ/min}$$

Given : $\oint \delta Q = -150$ kJ per cycle

The system completes 100 cycles per minute and therefore

$$Q_{1-2} + Q_{2-3} + Q_{3-4} + Q_{4-1}$$
$$= -150 \times 100 = -15000$$

or $0 + 20000 - 2000 + Q_{4-1} = -15000$

Thta gives, $Q_{4-1} = -33000$

The cyclic integral of any property is zero and accordingly $\oint dU = 0$. Thus

$$-2150 + 20000 - 36000 + dU_{41} = 0$$
$$\therefore \quad dU_{41} = 18150 \text{ kJ/min}$$

Again applying the non-flow energy equation to process 4–1,

$$W_{4-1} = Q_{4-1} - dU_{41}$$
$$= -33000 - 18150$$
$$= -51150 \text{ kJ/min}$$

The complete table is then as shown below:

Process	Q (kJ/min)	W (kJ/min)	dU (kJ/min)
1 - 2	0	2150	- 2150
2 - 3	20000	0	20000
3 - 4	- 2000	34000	- 36000
4 - 1	- 33000	- 51150	18150
	ΣQ = - 15000 kJ/min	ΣW = - 15000 kJ/min	ΣdU = 0

(*b*) Work output

$$= -15000 \text{ kJ/min}$$

$$= \frac{-15000}{60} \text{ kJ/s} = -250 \text{ kJ/s}$$

The negative sign implies that work has been done by the surroundings on the given fluid system.

EXAMPLE 6.22

Consider a system taken from state 1 to state 2 (Fig. 6.6.) along the path 1 – a – 2; 100 kJ of heat flows into the system and the system does 40 kJ of work.

(a) Evaluate the heat that flows into the system along the path 1 – b – 2 if it is accompanied by 20 kJ of work transfer from the system.

(b) Calculate the work done and heat exchanged if the system returns to the initial state 1 along the straight path 2–1.

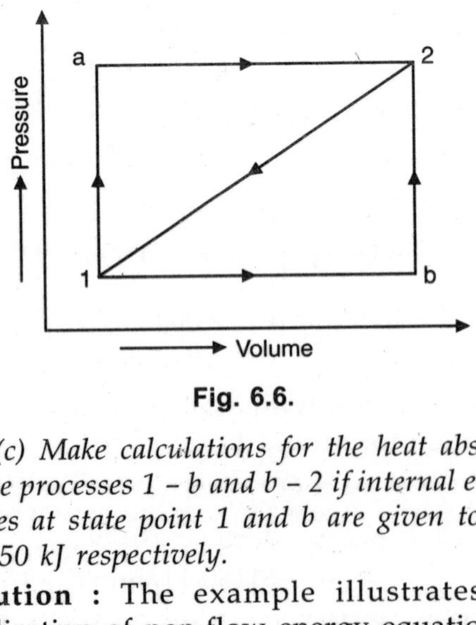

Fig. 6.6.

(c) Make calculations for the heat absorbed in the processes 1 – b and b – 2 if internal energy values at state point 1 and b are given to be 0 and 50 kJ respectively.

Solution : The example illustrates the application of non-flow energy equation,

$\delta Q = \delta W + dU$, applied to a process.

For path 1 - a - 2 :

$$\delta Q = 100 \text{ kJ and } \delta W = 40 \text{ kJ}$$

$$\therefore U_2 - U_1 = Q_{1-a-2} - W_{1-a-2}$$
$$= 100 - 40 = \mathbf{60\ kJ}$$

For path 1 - b - 2 :

$$\delta W = 20 \text{ kJ and } U_2 - U_1 = 60 \text{ kJ}$$

$$\therefore Q_{1-b-2} = W_{1-b-2} + (U_2 - U_1)$$
$$= 20 + 60 = \mathbf{80\ kJ}$$

(*b*) **Straight path 2—1 :** For path 1 - *a* - 2, no work is done during constant volume process 1 - *a* and the entire work of 40 kJ is done during the constant pressure path *a* - 2. Therefore, area under line *a* - 2 = 40 kJ

Similarly, area under line 1 - *b* = 20 kJ

area of rectangle 1 - *a* - 2 - *b*

$$= 40 - 20 = 20 \text{ kJ}$$

area of triangular portion 2 - 1 - *b*

$$= \frac{1}{2} \times 20 = 10 \text{ kJ}$$

area under straight line 2—1 (return path)

= area 2 - 1 - *b* + area under 1 - *b*

= 10 + 20 = 30 kJ

∴ Work done on the system,

$$W_{2-1} = -30 \text{ kJ}$$

$$Q_{2-1} = W_{2-1} + (U_1 - U_2)$$
$$= -30 - 60 = -90 \text{ kJ}$$

Since this is negative, heat flows out of the system

(*c*) For path 1 - *b* - 2

$$\text{Work done} = W_{1-b} + W_{b-2}$$

$$20 = W_{1-b} + 0 \, ; \; W_{1-b} = 20 \text{ kJ}$$

$$Q_{1-b} = W_{1-b} + (U_b - U_1)$$
$$= 20 + (50 - 0) = \mathbf{70\ kJ}$$

Also $$Q_{b-2} = Q_{1-b-2} - Q_{1-b}$$
$$= 80 - 70 = \mathbf{10\ kJ}$$

EXAMPLE 6.23

When a closed system executes a certain non-flow process, the work and heat interactions per degree rise in temperature at each temperature attained are given by

$$\frac{\delta W}{dT} = (4 - 0.08\ T)\ kJ/kg \text{ and } \frac{\delta Q}{dT} = 1.00\ kJ/K$$

Make calculations for the increase or decrease in the internal energy of the system if it is to operate between the temperature limits of 200°C and 400°C.

Solution : Work interaction,

$$W = \int (4 - 0.08\,T)dT$$

$$= \left| 4T - \frac{0.08\,T^2}{2} \right|_{473}^{673}$$

$$= 4\,(673 - 473) - \frac{0.08}{2}(673^2 - 473^2)$$

$$= -\ 8368 \text{ kJ}$$

Since it is negative, work is done on the system.

Heat interaction

$$= \int 1.00\, dT = |1.00\,T|_{473}^{673}$$

$$= 1.00\ (673 - 473) = 200 \text{ kJ}$$

Since it is positive, heat is supplied to the system.

Now, for a closed system :

$$\delta Q = \delta W + dU$$

$$200 = -\ 8368 + dU\ ;\ dU = 8568 \text{ kJ}$$

∴ Change of internal energy,

$(U_2 - U_1)$ = **8568 kJ (increase)**

EXAMPLE 6.24

A piston-cylinder arrangement contains 2 kg of working substance. When the piston moves outward, the temperature of the system falls from 85°C to 30°C and net heat rejected through the cylinder walls amounts to 25 kJ. The specific internal energy and temperature are related by the expression,

$$u = (u_0 + t/1.25)\ kJ$$

where u_0 is constant and t is the temperature in °C. Make calculations for the work done on or by the system.

Solution:

$$u_1 = (u_0 + t_1/1.25)$$
$$= (u_0 + 85/1.25) = (u_0 + 68) \text{ kJ/kg}$$
$$u_2 = (u_0 + t_2/1.25)$$
$$= (u_0 + 30/1.25) = (u_0 + 24) \text{ kJ/kg}$$
$$du = u_2 - u_1$$
$$= (u_0 + 24) - (u_0 + 68)$$
$$= -\ 44 \text{ kJ/kg}$$
$$dU = m\ du = 2 \times (-\ 44) = -\ 88 \text{ kJ}$$

Now, for a closed system :

$$\delta Q = \delta W + dU$$
$$-\ 20 = \delta W + (-\ 88)$$

The heat has been taken negative as it flows out of the system.

$$\delta W = -\ 20 + 88 = \mathbf{68\ kJ}$$

Since δW is positive, work is done by the system.

6.5. NON-FLOW PROCESSES (FIRST LAW AND CLOSED SYSTEMS)

In thermodynamics, we encounter with a number of processes wherein one or another state parameter remains constant. The basic thermodynamic processes used in engineering practice are being analysed here with regard to

(*i*) relationship between state parameters such as pressure, temperature and volume,

(*ii*) evaluation of work and heat involved in the process, and

(*iii*) calculation of the change in internal energy.

6.5.1. Constant volume or isochoric process

A change in the state of a system at constant volume is called *isochoric process*. An isochoric process results when the gas system is heated or cooled in an enclosed space.

During an isochoric process, both the pressure and temperature change. The pressure and temperature increase when heat is supplied to the system, and decrease when heat is rejected by the system. The isochoric process can be depicted on *p* - *V* co-ordinates by a straight line parallel to the axis of pressures.

For the initial state 1:

$$p_1 V_1 = mRT_1 \quad ...(i)$$

For the final state 2 :

$$p_2 V_2 = mRT_2 \quad ...(ii)$$

Dividing (*i*) by (*ii*) and noting that $V_1 = V_2$, we obtain

$$\frac{p_1}{p_2} = \frac{T_1}{T_2} \quad \text{or} \quad \frac{p}{T} = \text{constant}$$

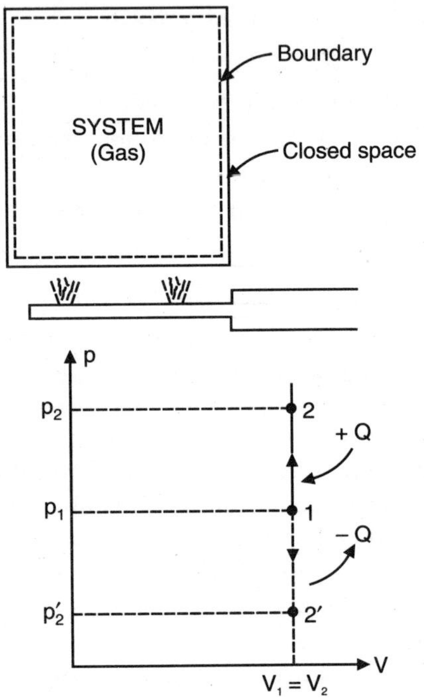

Fig. 6.7. Constant volume process

Thus in an isochoric process, the pressure of gas changes in direct proportion to absolute temperature (Charles' law).

Since there is no expansion of gas ($dV = 0$), no mechanical work is done on or by the system. This also becomes clear from the fact that area under the pressure-volume is zero. From the non-flow energy equation $\delta Q = \delta W + dU$, it follows that since $\delta W = 0$ all the heat is utilized to change the internal energy of the system. Heat added during a constant volume process is given by

$\delta Q = mc_v(T_2 - T_1)$ and therefore,

$dU = \delta Q = mc_v (T_2 - T_1)$

For a unit mass of gas,

$$du = \delta q = c_v dT \; ; \quad c_v = \frac{du}{dT}$$

Accordingly specific heat at constant volume c_v may be defined as the rate of change of specific internal energy with respect to change of temperature when the process is carried out at constant volume. Since c_v has been defined solely in terms of the state properties, it is a state property itself.

EXAMPLE 6.25

A closed vessel contains 3 kg of CO_2 at pressure 70 kPa and temperature 300 K. Heat is supplied to the vessel till the gas attains 140 kPa of pressure. Calculate : (a) final temperature, (b) work done on or by the gas, (c) heat added, and (d) change in internal energy.

For CO_2 : take c_v = 0.65 kJ/kg K

Solution : The closed vessel implies that the change in the state of the system is at constant volume and that the process is isochoric. For such a constant volume process, $\frac{p_1}{T_1} = \frac{p_2}{T_2}$

∴ Final temperature,

$$T_2 = \frac{p_2}{p_1} T_1 = \frac{140}{70} \times 300 = \mathbf{600\ K}$$

(*b*) No volume change takes place and consequently the work done on or by the gas is zero

(*c*) Heat added during a constant volume process is given by

$$Q = mc_v (T_2 - T_1)$$
$$= 3 \times 0.65 \times (600 - 300) = \mathbf{585\ kJ}$$

(*d*) From the non-flow energy equation,

$$dU = \delta Q - \delta W$$
$$= 585 - 0$$
$$= \mathbf{585\ kJ} \text{ (increase)}$$

Alternatively,

$$dU = mc_v (T_2 - T_1)$$
$$= 3 \times 0.65 \times (600 - 300)$$
$$= \mathbf{585\ kJ}$$

EXAMPLE 6.26

An oxygen cylinder of 0.25 m^3 capacity contains oxygen at a pressure of 30 bar and 293 K temperature. The stop valve is opened and a certain *mass of oxygen is used. Measurements show that the oxygen left in the cylinder is at 15 bar pressure and 288 K temperature. Determine the mass of oxygen used.*

After closure of the valve, the oxygen remaining in the cylinder gradually attains its initial temperature of 293 K. Make calculations for the amount of heat transferred through the cylinder walls to the atmosphere.

It may be presumed that oxygen has a density of 1.43 kg/m³ at 0°C and 1.01325 bar, and $\gamma = 1.4$.

Solution : Applying characteristics gas equation, $pv = mRT$, we have

$$R = \frac{V}{m} \times \frac{p}{T} = \frac{p}{\rho T}$$

$$= \frac{1.01325 \times 10^5}{1.43 \times 273}$$

$$= 259.55 \text{ J/kg K}$$

Initial mass of oxygen in the cylinder,

$$m_1 = \frac{p_1 V_1}{R T_1}$$

$$= \frac{30 \times 10^5 \times 0.25}{259.55 \times 293}$$

$$= 9.862 \text{ kg}$$

After the valve is opened and the gas is used, oxygen left in the cylinder

$$m_2 = \frac{p_2 V_2}{R T_2}$$

$$= \frac{15 \times 10^5 \times 0.25}{259.55 \times 288} = 5.017 \text{ kg}$$

$\therefore$ mass of oxygen used $= m_1 - m_2$

$= 9.862 - 5.017 = \mathbf{4.845}$ **kg**

(*b*) From the non-flow energy equation; $\delta Q = \delta W + dU$

The cylinder is rigid (no change in volume takes place) and consequently the work done on or by the gas is zero

$\therefore$ Heat exchange

$$\delta Q = dU = mc_v \, dT$$

$$= m\left(\frac{R}{\gamma - 1}\right) dT$$

$$= 5.017 \times \frac{259.55}{1.4 - 1} \times (293 - 288)$$

$$= 16277 \text{ J} \simeq \mathbf{16.28} \text{ kJ}$$

6.5.2. Constant pressure or isobaric process

A change in the state of system at constant pressure is called *isobaric process*. An isobaric process can be carried out in a piston-cylinder assembly with the piston loaded with a constant dead weight. A perfect gas enclosed in the cylinder and on the underside of the piston is considered as the system. When the gaseous system is supplied heat, there occurs an increase in temperature and volume of the gas, and accordingly the piston is pushed upwards. Ignoring friction between the cylinder and piston, the piston moves upwards slowly and the system passes through a series of equilibrium positions maintaining constant intensity of pressure.

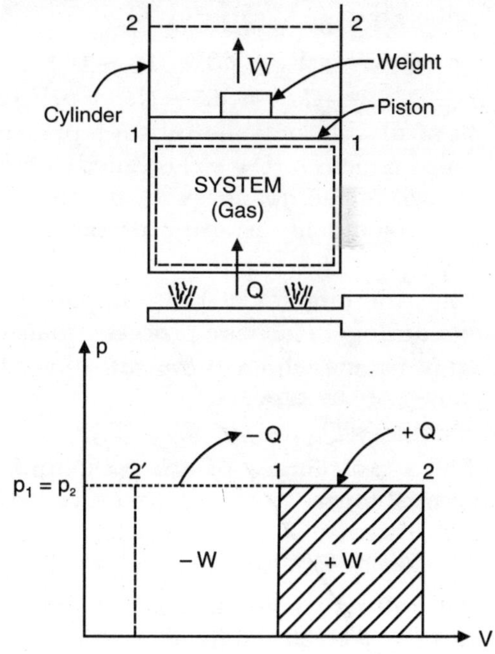

Fig. 6.8. Constant pressure process

During an isobaric process, both the volume and temperature change. The reception of heat causes the temperature of the gas and its volume to increase, and the rejection of heat causes these parameters to diminish. On $p - V$ coordinates, an isobaric process is depicted by a straight line parallel to the axis of volumes.

(*i*) For the initial state:

$$p_1 V_1 = mRT_1 \quad ...(a)$$

For the final state :

$$p_2 V_2 = mRT_2 \quad ...(b)$$

Dividing the expression (*a*) by the expression (*b*) and noting that $p_1 = p_2$, we obtain

$$\frac{V_1}{V_2} = \frac{T_1}{T_2} \quad \text{or} \quad \frac{V}{T} = \text{constant}$$

i.e., in an isobaric process, the volume of a given mass of gas is directly proportional to the absolute temperature (Charles' law).

(*ii*) Mechanical work,

$$W_{1\text{-}2} = \int_1^2 p\,dv = p\,(V_2 - V_1)$$

(*iii*) For a non-flow process :

$$\delta Q = \delta W + dU$$

or $Q_{1\text{-}2} = p(V_2 - V_1) + (U_2 - U_1)$

$= (U_2 + p_2 V_2) - (U_1 + p_1 V_1)$

Sum of internal energy and pressure volume product, *i.e*, $(U + pV)$ is called *enthalpy*. Since each of the quantities U, p and V are properties, enthalpy is also a property.

$\therefore \; Q_{1\text{-}2} = H_2 - H_1$

Thus the amount of heat supplied to a system during an isobaric process equals the difference in enthalpies at the end and at the beginning of the process.

$$H_2 - H_1 = Q_{1\text{-}2} = mc_p\,(T_2 - T_1)$$

For a unit mass of the gas, and in differential form,

$$dh = \delta q = c_p dT \; ; \; c_p = \frac{dh}{dT}$$

Accordingly specific heat at constant pressure c_p may be defined as the rate of change of specific enthalpy with respect to change of temperature when the process is carried out at constant pressure. Since c_p has been defined solely in terms of the state properties, it is a state property itself.

Physical significance of gas constant *R*: During an isobaric process,

specific work done,

$$W_{1\text{-}2} = \int p\,dv = p\,(v_2 - v_1)$$

Since, $p_1 = p_2 = p$

$$p_1 v_1 = RT_1 \quad \text{and} \quad p_2 v_2 = RT_2$$

$$W_{1\text{-}2} = R(T_2 - T_1) \; ; \quad R = \frac{W_{1-2}}{(T_2 - T_1)}$$

Thus the *gas constant is equal to the work of one kilogram of gas in an isobaric process when the temperature changes by one degree.*

Relationship between specific heats and gas constant : Let a unit mass of an ideal gas undergo constant volume process and constant pressure process separately through temperature range from T_1 to T_2.

At constant volume process :

$$q_{1\text{-}2} = c_v\,(T_2 - T_1) \; ; \; w_{1\text{-}2} = 0 \quad \text{and}$$

therefore, change in internal energy

$$= c_v\,(T_2 - T_1)$$

At constant pressure process :

$$q_{1\text{-}2} = c_p\,(T_2 - T_1)$$

$$w_{1\text{-}2} = p(V_2 - V_1) \text{ and}$$

therefore, change in internal energy

$$= c_p\,(T_2 - T_1) - p(V_2 - V_1)$$

For an ideal gas $u = f(T)$ only. Since both processes have been carried out through the same temperature range, the change in internal energy would be same for both cases.

$$\therefore \; c_p\,(T_2 - T_1) - p(V_2 - V_1) = c_v\,(T_2 - T_1)$$

$$c_p\,(T_2 - T_1) - R(T_2 - T_1) = c_v\,(T_2 - T_1)$$

or $c_p - R = c_v \; ; \; c_p - c_v = R$

Alternatively : Consider the system of an ideal gas which satisfies the relations

$$pv = RT \; ; \; c_v = du/dT \text{ and } c_p = dh/dT$$

Enthalpy is the sum of internal energy and pressure volume product. That is

$$h = u + pv = u + RT$$

Upon differentiation :

$$dh = du + R\,dT$$

$$\frac{dh}{dT} = \frac{du}{dT} + R$$

$$c_p = c_v + R \; ; \; c_p - c_v = R$$

Thus the difference between specific heat at constant pressure and specific heat at constant volume for any gas having the equation of state $pv = RT$ is equal to the gas constant R. The value of c_p is greater than c_v because the heat supplied to a gas at constant pressure is utilized for two purposes; for doing the external work and for increasing the internal energy. The ratio c_p/c_v is designated by γ and is called adiabatic index or adiabatic exponent.

From the results $c_p - c_v = R$ and $c_p/c_v = \gamma$, the following useful results can be setup :

$$c_v = \frac{R}{\gamma - 1} \; ; \; c_p = R\frac{\gamma}{\gamma - 1}$$

If M is the molecular mass of the gas, then molar specific heats for the gas are given by

$$c_{p\,mol} = Mc_p \; ; \text{ and } \; c_{v\,mol} = Mc_v$$

Molar specific heat is defined as the quantity of heat required to raise the temperature of one mole (22.4 m^3) of a gas by 1 degree.

Multiplying both sides of the relation, $c_p - c_v = R$, by molecular mass of the gas, we obtain

$$Mc_p - Mc_v = MR$$

But MR = 8314 Joule/kg mole deg
$\simeq$ 8.314 kJ/kg mole deg and hence

$$c_{p\,mol} - c_{v\,mol} \simeq 8.314 \text{ kJ/kg mole deg}$$

Thus for all ideal gases the difference between the molar specific heats at constant pressure and constant volume is the same and is approximately equal to 8.314 kJ/kg mol per degree temperature difference.

EXAMPLE 6.27

An ideal gas requires 1150 kJ/kg of heat to raise its temperature from 20°C to 100°C when heated at constant pressure. When heat is supplied to the same gas at constant volume, the heat requirement is 825 kJ for the same temperature range. Determine the specific heat at constant pressure, specific heat at constant volume, adiabatic exponent, characteristic gas constant and the molecular mass of the gas.

Solution : At constant pressure process,

$$Q = mc_p dT$$

$$1150 = 1 \times c_p(100 - 20)$$

$$c_p = \mathbf{14.375 \text{ kJ/kg K}}$$

At constant volume process, $Q = mc_v dT$

$$825 = 1 \times c_v \; (100 - 20);$$

$$c_v = \mathbf{10.312 \text{ kJ/kg K}}$$

Adiabatic exponent,

$$\gamma = \frac{c_p}{c_v} = \frac{14.375}{10.312} = \mathbf{1.394}$$

Characteristic gas constant,

$$R = c_p - c_v$$
$$= 14.375 - 10.312$$
$$= \mathbf{4.063 \text{ kJ/kg K}}$$

Molecular mass,

$$M = \frac{R_{mol}}{R} = \frac{8.314}{4.063} = \mathbf{2.046}$$

EXAMPLE 6.28

A closed rigid vessel containing 10 kg of oxygen at 290 K is supplied heat until its pressure becomes two-fold that of initial value. Identify the process and calculate the final temperature, change in internal energy and enthalpy, and heat ineraction across the system boundary.

Take c_v = 0.65 kJ/kg K

Solution : Characteristic gas constant,

$$R = \frac{R_{mol}}{\text{molecular mass}}$$

$$= \frac{8314}{32} = 259.8 \text{ J/kg K}$$

Since the vessel is closed and rigid, there will be no change in the volume of gas enclosed in it. As such the situation corresponds to that of an isochoric (constant volume) process. Since volume remains constant, Charles' law applies and therefore,

$$\frac{T_2}{T_1} = \frac{p_2}{p_1}$$

$$T_2 = \frac{p_2}{p_1} \times T_1 = 2 \times 290 = \mathbf{580 \text{ K}}$$

$$dU = mc_v \; (T_2 - T_1)$$
$$= 10 \times 0.65 \times (580 - 290)$$
$$= \mathbf{1885 \text{ kJ}}$$

Now, $c_p = c_v + R$
$= 0.65 + 0.2598 = 0.9098$ kJ/kg K
$dH = mc_p (T_2 - T_1)$
$= 10 \times 0.9098 \times (580 - 290)$
$=$ **2638.4 kJ**

From non-flow energy equation,

$Q_{1-2} = W_{1-2} + dU$

But $W_{1-2} = 0$ as volume remains constant

$\therefore \; Q_{1-2} = 0 + 1885 =$ **1885 kJ**

EXAMPLE 6.29

1.5 kg of Nitrogen contained in a cylinder at pressure 6 bar and temperature 300 K expands three times its original volume. Assuming the expansion process to be isobaric, make calculations for : (i) initial volume, (ii) final temperature, (iii) work done by gas, (iv) heat added and (v) change in internal energy.

For Nitrogen : $c_p = 1.05$ kJ/kg K and $R = 295$ J/kg K

Solution : The initial volume of the gas can be worked out by invoking characteristic gas equation, *i.e.*,

$$V_1 = \frac{mRT_1}{p_1} = \frac{1.5 \times 295 \times 300}{6 \times 10^5} = \mathbf{0.221\ m^3}$$

(*ii*) Final volume,

$V_2 = 3V_1 = 3 \times 0.221 = 0.663$ m^3

For a constant pressure process, Charles law applies, *i.e*, $\frac{V_1}{T_1} = \frac{V_2}{T_2}$

$\therefore$ Final temperature,

$$T_2 = \frac{V_2}{V_1} T_1 = 3 \times 300 = \mathbf{900\ K}$$

(*iii*) Work done,

$W_{1-2} = p(V_2 - V_1)$
$= 6 \times 10^5 [0.663 - 0.221]$
$= 2.652 \times 10^5$ Nm (or J)
$=$ **265.2 kJ**

(*iv*) Heat added,

$Q_{1-2} = mc_p (T_2 - T_1)$
$= 1.5 \times 1.05 (900 - 300) =$ **945 kJ**

(*v*) From non-flow energy equation

$dU = Q_{1-2} - W_{1-2}$
$= 945 - 265.2$
$=$ **679.8 kJ (increase)**

EXAMPLE 6.30

It is desired to compress 10 kg of gas from 1.5 m^3 to 0.3 m^3 at a constant pressure of 15 bar. During this compression process, the temperature rises from 20°C to 150°C and the increase in internal energy is 3250 kJ. Calculate the work done, heat interaction and change in enthalpy during the process. Also workout the average value of specific heat at constant pressure.

Solution : Work done

$W_{1-2} = p (V_2 - V_1)$
$= 15 \times 10^5 (0.3 - 1.5)$
$= - 18 \times 10^5$ Nm (or J)
$=$ **– 1800 kJ**

The negative sign indicates that the work has to be done on the system.

From non-flow energy equation,

$\delta Q = \delta W + dU$
$Q_{1-2} = W_{1-2} + (U_2 - U_1)$
$= - 1800 + 3250 =$ **1450 kJ**

Since it is positive, heat has been supplied to the system.

For a constant pressure process, the change in enthalpy equals the heat supplied.

$\therefore \; dH = Q_{1-2} =$ **1450 kJ**

For unit mass of gas,

$$dh = \frac{1450}{10} = 145 \text{ kJ}$$

$\therefore$ Average value of specific heat at constant pressure,

$$c_p = \frac{dh}{dT} = \frac{145}{(150 - 20)} \simeq \mathbf{1.12\ kJ/kg\ K}$$

EXAMPLE 6.31

At 100 kN/m^2 pressure and 288 K temperature, the mass density of certain gas is 1.875 kg/m^3. Under constant pressure conditions, 1.2 kg of this gas requires 300 kJ of heat to raise its temperature from 15°C to 300°C. Make calculations for:

(i) characteristic gas constant, (ii) specific heat capacity at constant pressure and at constant volume, (iii) change of internal energy and work interaction.

Solution : Applying characteristic gas equation, $pV = mRT$, we have

$$R = \frac{V}{m} \times \frac{p}{T} = \frac{p}{\rho T} \qquad \left(\because \frac{m}{V} = \rho\right)$$

$$R = \frac{100 \times 10^3}{1.875 \times 288} = \mathbf{186.2}\ \text{J/kg K}$$

(ii) At constant pressure process

$$Q = mc_p\, dT$$

$$300 = 1.2 \times c_p \times (300 - 15)$$

$$c_p = 0.8772\ \text{kJ/kg K}$$

Using the relation $c_p - c_v = R$, we get

$$c_v = c_p - R = 0.8772 - 0.1862 = \mathbf{0.692}\ \text{kJ/kg K}$$

(iii) Change in internal energy,

$$dU = mc_v dT = 1.2 \times 0.692\,(300 - 15) = \mathbf{236.66}\ \text{kJ}$$

(iv) From the non-flow energy equation :

$$\delta Q = \delta W + dU$$

∴ Work interation

$$W_{1-2} = Q_{1-2} - dU = 300 - 236.66 = \mathbf{63.34}\ \text{kJ}$$

The positive sign implies that work is done by the gas on the surroundings.

EXAMPLE 6.32

A vertical piston cylinder assembly that initially has a volume of 0.1 m³ is filled with 0.1 kg of nitrogen gas. This is weighted so that the pressure of nitrogen is always maintained at 1.15 bar. Heat transfer is allowed to take place until the volume is reduced to 75 per cent of initial volume. Determine :

(a) initial and final temperatures of nitrogen

(b) magnitude and direction of heat transfer.

Assume the process to be quasi static and take c_v = 0.745 kJ/kg K for nitrogen.

Solution : For nitrogen (molecular mass = 28), the gas constant is

$$R = \frac{R_{mol}}{M} = \frac{8314}{28} = 296.93\ \text{J/kg K}$$

From ideal gas equation $pV = mRT$

$$T_1 = \frac{p_1 V_1}{mR} = \frac{1.15 \times 10^5 \times 0.1}{0.1 \times 296.93} = \mathbf{387.3}\ \text{K}$$

For a constant pressure process, Charles' law applies, *i.e.*,

$$\frac{V_1}{T_1} = \frac{V_2}{T_2}$$

∴ Final temperature

$$T_2 = \frac{V_2}{V_1} \times T_1 = 0.75 \times 387.3 = \mathbf{290.47}\ \text{K}$$

(b) Work done

$$W_{1-2} = p(V_2 - V_1) = 1.15 \times 10^5\,(0.75 \times 0.1 - 0.1) = -\,2875\ \text{Nm (or J)} = -\,2.875\ \text{kJ}$$

Change in internal energy,

$$dU = mc_v\,(T_2 - T_1) = 0.1 \times 0.745\,(290.47 - 387.3) = -\,7.214\ \text{kJ}$$

From the non-flow energy equation, $\delta Q = \delta W + dU$, we have

Heat interaction

$$Q_{1-2} = -\,2.875 - 7.214 = -\,10.089\ \text{kJ}$$

The negative sign shows that the heat is rejected by the system.

EXAMPLE 6.33

A closed system consists of a fluid inside a cylinder fitted with a frictionless piston. When the fluid was stirred by a paddle wheel, 120 kJ of mechanical work was supplied along with 40 kJ of energy in the form of heat. At the same time, the piston moved in such a way that pressure remained constant at 200 kPa and the volume changed from 2 m³ to 4 m³. Make calculations for the change in internal energy and enthalpy of the fluid system.

Solution : Work done by the fluid on the piston,

$$W_{1-2} = p\,(V_2 - V_1) = 200 \times 10^3\,(4 - 2) = 4 \times 10^5\ \text{Nm (or J)} = 400\ \text{kJ}$$

Work done on fluid by paddle wheel

$$= -\,120\ \text{kJ}$$

∴ Net work done by the fluid

$= 400 - 120 = 280$ kJ

From non-flow energy equation,

$\delta Q = \delta W + dU$

Change in internal energy,

$$dU = Q_{1-2} - W_{1-2}$$
$$= 40 - 280 = \mathbf{-240\ kJ}$$

Change in enthalpy equals the sum of change in internal energy and the change in pressure-volume product. That is

$$dH = (U_2 - U_1) + (p_2V_2 - p_1V_1)$$
$$= (U_2 - U_1) + p\,(V_2 - V_1) \qquad (\because p_1 = p_2 = p)$$
$$= -240 + \{200 \times 10^3\,(4 - 2)\} \times \frac{1}{10^3}$$
$$= -240 + 400 = \mathbf{160\ kJ}$$

EXAMPLE 6.34

The following expession gives the specific heat at constant pressure of a gas that undergoes a non-flow quasi-static constant pressure process

$$c_p = \left[2.5 + \frac{60}{T + 40}\right] \text{ kJ/kg K}$$

where T is in °C

The pressure during the process is maintained at 2 bar and the volume changes from 0.3 m³/kg to 0.56 m³/kg. The corresponding change in temperature is from 25°C to 300°C. Determine

(a) work done and heat added,

(b) changes in internal energy and enthalpy.

Solution : Work done during constant pressure non-flow process is

$$W_{1-2} = p(V_2 - V_1) = 2 \times 10^5(0.56 - 0.3)$$
$$= 0.52 \times 10^5 \text{ J/kg}$$
$$= \mathbf{520}\text{ kJ/kg}$$

Heat supplied at constant pressure for unit mass of gas is

$$Q_{1-2} = \int_1^2 c_p\, dT = \int_1^2 \left(2.5 + \frac{60}{T + 40}\right) dT$$
$$= \left|2.5\,T + 60 \log_e (T + 40)\right|_1^2$$
$$= 2.5\,(300 - 25) + 60 \log_e \frac{300 + 40}{25 + 40}$$
$$= 687.5 + 99.26 = \mathbf{786.76}\text{ kJ/kg}$$

(*b*) From non-flow energy equation

$\delta Q = \delta W + dU$

Change in internal energy,

$$dU = 786.76 - 520 = 266.76 \text{ kJ/kg}$$

The enthalpy change during an isobaric (constant pressure) process equals the amount of heat supplied to the system.

∴ Change in enthalpy = **786.76** kJ/kg

EXAMPLE 6.35

A vessel of 2.5 m³ capacity contains 1 kg mole of nitrogen at 100°C. Determine the mass, the pressure and specific volume of the gas.

(b) What would be the final pressure if subsequently the gas is cooled to atmospheric temperature of 25°C?

(c) Proceed to determine the increase in internal energy, the increase in enthalpy, and the heat interaction both in magnitude and direction.

For nitrogen, the ratio of specific heats is 1.4

Solution : 1 kg mole of nitrogen will have mass equal to its molecular weight expressed in kilograms

∴ mass of nitrogen $m = 28$ kg

Gas constant

$$R = \frac{R_{mol}}{M} = \frac{8314}{28} = 296.93 \text{ J/kg K}$$

Assuming nitrogen to be a perfect gas, we have $pV = mRT$ and therefore

$$p = \frac{mRT}{V}$$
$$= \frac{28 \times 296.93 \times 373}{2.5}$$
$$= 1.24 \times 10^6 \text{ N/m}^2 = \mathbf{12.4}\text{ bar}$$

Specific volume

$$v = \frac{V}{m} = \frac{2.5}{28} = \mathbf{0.089}\text{ m}^3\text{/kg}$$

(*b*) The change in the state of the gas occurs at constant volume and for an isochoric process

$$\frac{p_1}{T_1} = \frac{p_2}{T_2}$$

∴ Final pressure

$$p_2 = \frac{T_2}{T_1} p_1 = \frac{298}{373} \times 12.4 = \mathbf{9.907\ bar}$$

(c) From the relations

$$\frac{c_p}{c_v} = \gamma \quad \text{and} \quad c_p - c_v = R, \text{ we have}$$

$$c_v = \frac{R}{\gamma - 1} = \frac{296.93}{0.4}$$
$$= 742.32 \text{ J/kg K} = 0.742 \text{ kJ/kg K}$$
$$c_p = \gamma\, c_v = 1.4 \times 0.742$$
$$= 1.039 \text{ kJ/kg K}$$

For a perfect gas

$$dU = m\, c_v\, dT \quad \text{and} \quad dH = m\, c_p\, dT$$

∴ Change in internal energy

$$dU = 28 \times 0.742 \times (25 - 100)$$
$$= -1558.2 \text{ kJ (decrease)}$$

Change in enthalpy

$$dH = 28 \times 1.039 \times (25 - 100)$$
$$= -2181.9 \text{ kJ (decrease)}$$

Applying the non-flow energy equation for the process

$\delta Q = \delta W + dU$; $\delta W = 0$ as the change in volume is zero

$$\therefore \quad \delta Q = dU = -1558.2 \text{ kJ}$$

The negative sign indicates that the heat transfer is from the gas (system) to the surroundings.

EXAMPLE 6.36

A 25 cm diameter cylinder is fitted with a frictionless piston which is loaded with 1200 N weight. Initially the piston is 30 cm above the bottom position and space beneath the piston contains CO_2 at 290 K temperature. The gas is supplied heat from an electric heater of 1 watt power. Presuming that there is no loss of heat to the surroundings, evaluate the distance through which piston will be pushed outwards during this heating for a period of 1 hour. Take barometric pressure equivalent to 1 bar. For CO_2 :

$$c_p = 820 \text{ J/kg K}$$

Solution : Absolute gas pressure,

$$p = \text{gauge pressure} + \text{atmospheric pressure}$$

$$= \frac{1200}{\frac{\pi}{4}(0.25)^2} + 1 \times 10^5$$

$$= 124458 \text{ N/m}^2$$

Volume of gas,

$$V = \frac{\pi}{4}(0.25)^2 \times 0.3 = 0.0147 \text{ m}^3$$

Characteristic gas constant,

$$R = \frac{8314}{44} \approx 189 \text{ J/kg K}$$

Mass of gas,

$$m = \frac{pV}{RT}$$
$$= \frac{124458 \times 0.0147}{189 \times 290} = 0.0334 \text{ kg.}$$

Heat supplied from the heater

$$= 1 \text{ watt-hr} = 3600 \text{ J}$$

Since the piston carries a dead weight, the situation corresponds to constant pressure, and for a constant pressure process, the heat supplied equals $m\, c_p\, dT$. Therefore,

$$mc_p\,(T_2 - T_1) = 3600$$
$$0.0334 \times 820\,(T_2 - 290) = 3600$$
$$\therefore \quad T_2 = 417.62 \text{ K}$$

Now, $c_p - c_v = R$; $820 - c_v = 189$

$$\therefore \quad c_v = 631 \text{ J/kg K}$$

Change in internal energy,

$$dU = mc_v\,(T_2 - T_1)$$
$$= 0.0334 \times 631 \times (417.62 - 290)$$
$$= 2689.64 \text{ J}$$

From non-flow energy equation,

$$Q_{1\text{-}2} = W_{1\text{-}2} + dU$$
$$3600 = W_{1\text{-}2} + 2689.64$$

∴ Displacement work,

$$W_{1\text{-}2} = 3600 - 2689.64 = 910.36 \text{ J}$$

The work done equals the product of load and the distance by which the piston is raised. That is

$$910.36 = 1250 \times x \quad \therefore \quad x = \mathbf{0.728 \text{ m}}$$

EXAMPLE 6.37

100 litres of hydrogen gas at 300 K temperature and 5 bar pressure is contained in a cylinder fitted with a frictionless piston. The piston carries some dead weight, can move freely and its upper part is exposed to atmospheric pressure. Identify the process.

There is input of paddle work to the gas and its temperature rises to 350 K. Evaluate (i) work done, (ii) change in internal energy and enthalpy, and (iii) paddle work input.

Take $c_p = 14.25$ *kJ/kg K and*

$c_v = 10.1$ *kJ/kg K*

Solution : The dead weight exerts a constant force on the piston, and as such the situation corresponds to displacement work under isobaric conditions.

Gas constant,

$$R = c_p - c_v = 14.25 - 10.10$$
$$= 4.15 \text{ kJ/kg K} = 4150 \text{ J/kg K}$$

As the pressure remains constant, Charles' law applies and therefore,

$$V_2 = \frac{T_2}{T_1} V_1$$
$$= \frac{350}{300} \times (100 \times 10^{-3}) = 0.1167 \text{ m}^3$$

Mass of hydrogen,

$$m = \frac{p_1 V_1}{R T_1}$$
$$= \frac{5 \times 10^5 \times 0.1}{4150 \times 300} = 0.0402 \text{ kg}$$

(*i*) Displacement work,

$$W_{1-2} = \int p dV = p (V_2 - V_1)$$
$$= 5 \times 10^5 (0.1167 - 0.1)$$
$$= 8350 \text{ Nm (or J)} = \mathbf{8.35 \text{ kJ}}$$

(*ii*) Change in internal energy,

$$dU = mc_v (T_2 - T_1)$$
$$= 0.0402 \times 10.1 (350 - 300)$$
$$= \mathbf{20.30 \text{ kJ}}$$

Change in enthalpy,

$$dH = mc_p (T_2 - T_1)$$
$$= 0.0402 \times 14.25 (350 - 300)$$
$$= \mathbf{28.64 \text{ kJ}}$$

(*iii*) Paddle work input $= W_{1-2} + dU$

$$= 8.35 + 20.30 = \mathbf{28.65 \text{ kJ}}$$

6.5.3. Isothermal (constant temperature) process

A change in the state of a system at constant temperature is called *isothermal process*; here changing variables are pressure and volume.

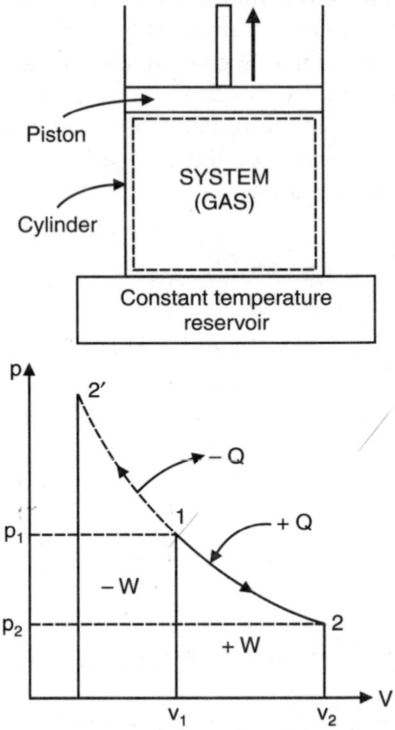

Fig. 6.9. Isothermal (constant temperature) process

An isothermal process can be performed in a piston-cylinder assembly which is surrounded by a constant temperature T reservoir. When the ideal gas inside the cylinder is at temperature T at any instant of time, there exists thermal equilibrium between the system and surroundings. As such there is no heat interaction across the system boundary in either direction. Subsequently when the piston moves outwards, the gas expands and its temperature falls by an amount dT. Heat then flows from the reservoir to the gas until its original temperature is restored. Similar events follow and the process is completed.

(*i*) For the initial state :

$$p_1 V_1 = mRT_1 \quad ...(a)$$

For the final state :

$$p_2 V_2 = mRT_2 \quad ...(b)$$

Upon dividing expression (*a*) by expression (*b*) and noting that $T_1 = T_2$, we obtain

$$p_1 V_1 = p_2 V_2 \quad \text{or} \quad pV = \text{constant}$$

i.e., in an isothermal process the pressure of the gas is inversely proportional to its volume (Boyle's law). In p - V co-ordinates, an isothermal process is plotted according to equation pV = constant. This equation is an equilateral hyperbola and is called an isothermal line.

(*ii*) Work done,

$$W_{1-2} = \int_1^2 p\, dV$$

Since pV = constant C, $p = \frac{C}{V}$

$$\therefore\ W_{1-2} = C\int_1^2 \frac{dV}{V}$$

$$= C \log_e \frac{V_2}{V_1} = p_1 V_1 \log_e \frac{V_2}{V_1}$$

$$= p_1 V_1 \log_e \frac{p_1}{p_2} \quad (\because p_1 V_1 = p_2 V_2)$$

(*iii*) Since in an isothermal process, $T_2 = T_1 = T$, the change in the internal energy process

$dU = mc_v (T_2 - T_1) = 0$. The non-flow energy equation gives :

$$Q_{1-2} = W_{1-2}$$

i.e., in an isothermal process all the heat received by the gas is utilized to accomplish mechanical work. During isothermal expansion heat is received by the working fluid, and in isothermal compression the working fluid rejects heat to the surroundings.

The isothermal curve (pV = constant) on p–V plot is hyperbolic in character and accordingly the isothermal process may be referred to as hyperbolic process.

EXAMPLE 6.38

A rigid and insulated tank of 0.1 m^3 volume contains nitrogen at 325 K and 1.2 bar. An electrial resistor placed inside the tank is connected to a battery and current flows through the resistor till the temperature of the gas is raised to 375 K. Determine the work done and final pressure of gas in the tank.

For nitrogen gas, the molar heat capacity at constant volume is $\frac{5}{2}$ R_{mol} where R_{mol} is the universal gas constant.

Solution : Gas constant for nitrogen

$$R = \frac{R_{mol}}{M} = \frac{8314}{28} = 296.93 \text{ J/kg K}$$

The mass of oxygen can be worked out by applying the characteristic gas equation, *i.e.*,

$$m_1 = \frac{pV}{RT}$$

$$= \frac{1.2 \times 10^5 \times 0.1}{296.93 \times 325} = 0.1243 \text{ Kg}$$

$$c_v = \frac{c_{v\,mol}}{M} = \frac{5}{2}\frac{R_{mol}}{M}$$

$$= \frac{5}{2} \times \frac{8314}{28} = 742.32 \text{ J/kg K}$$

Change in internal energy,

$$dU = m\, c_v\, dT$$

$$= 0.1243 \times 0.742 \times (375 - 325)$$

$$= 4.61 \text{ kJ}$$

Since the tank is insulated, the heat interaction is zero and then from non-flow energy equation, $\delta Q = \delta W + dU$, we have

$$0 = \delta W + 4.61\ ; \quad \delta W = -\ 4.61 \text{ kJ}$$

Since δW is negative, the work is done by battery on the gas.

(*b*) The rigid vessel implies that the change in the state of the gas is at constant volume and the process is isochoric. For such a constant volume process,

$$\frac{p_1}{T_1} = \frac{p_2}{T_2}$$

$\therefore$ Final presure,

$$p_2 = \frac{T_2}{T_1} p_1 = \frac{375}{325} \times 1.2 = \mathbf{1.3846 \text{ bar}}$$

EXAMPLE 6.39

A fluid expands reversibly behind a piston from initial conditions of pressure 600 kPa and volume 0.03 m^3 to a final volume of 0.09 m^3. Presuming isothermal conditions, estimate the work done, change in internal energy, and heat supplied or rejected by the fluid system.

Solution : For an isothermal process, the work done during expansion follows from the relation :

$$W_{1-2} = p_1 V_1 \log_e \frac{V_2}{V_1}$$

$$= \left(600 \times 10^3\right) \times 0.03 \times \log_e \frac{0.09}{0.03}$$

$$= 19775 \text{ Nm (or J)}$$

(*b*) Since there is no change in temperature during an isothermal process, change in internal energy of the system equals zero.

(*c*) From non-flow energy equation;

$$\delta Q = \delta W + dU$$
$$= 19775 + 0$$
$$= 19775 \text{ J} = \mathbf{19.775 \text{ kJ}}$$

The positive sign indicates that heat is given to the system.

EXAMPLE 6.40

Air initially at 60 kPa pressure, 800 K temperature and occupying a volume of 0.1 m³ is compressed isothermally until the volume is halved and subsequently it goes further compression at constant pressure till the volume is halved again. Sketch the process on p - V plot and make calculations for the total work done and total heat interaction for the two processes. Assume ideal gas behaviour for air and take c_p = 1.005 kJ/kg K.

Solution : $V_1 = 0.1$ m³; $V_2 = 0.05$ m³ and $V_3 = 0.025$ m³

Isothermal process 1—2

Since temperature is constant, Boyle's law applies and therefore,

$$p_2 = p_1 \frac{V_1}{V_2} = 0.6 \times \frac{0.1}{0.05} = 1.2 \text{ bar}$$

Constant pressure process

Since the pressure remains constant, Charles' law applies and therefore,

$$T_3 = T_2 \frac{V_3}{V_2} = 800 \times \frac{1}{2} = 400 \text{ K}$$

$$(T_2 = T_1 = 800 \text{ K})$$

From the characteristic gas equation,

$$m = \frac{p_1 V_1}{R T_1}$$

$$= \frac{0.6 \times 10^5 \times 0.1}{287 \times 800} = 0.0261 \text{ kg}$$

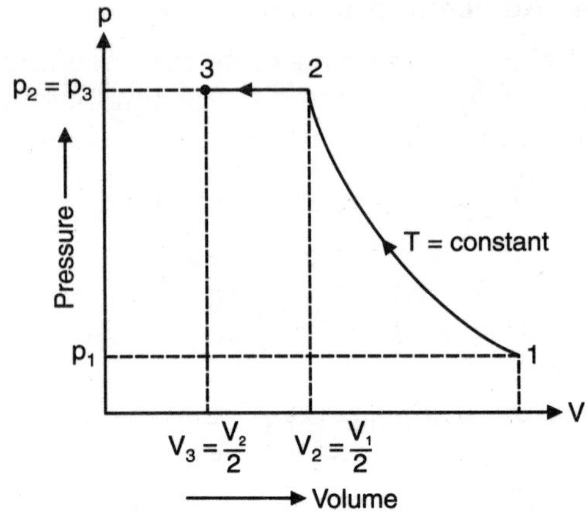

Fig. 6.10.

Work interactions

$$W_{1-2} = p_1 V_1 \log_e \frac{V_2}{V_1}$$

$$= 0.6 \times 10^5 \times 0.1 \times \log_e \left(\frac{1}{2}\right)$$

$$= -4158.9 \text{ Nm (or J)} = -4.159 \text{ kJ}$$

$$W_{2-3} = p_2(V_3 - V_2)$$
$$= 1.2 \times 10^6 (0.025 - 0.05)$$
$$= -3000 \text{ Nm (or J)} = -3 \text{ kJ}$$

Total work done,

$$W_{1-3} = W_{1-2} + W_{2-3}$$
$$= -4.159 - 3 = \mathbf{-7.159 \text{ kJ}}$$

Heat interactions

$$Q_{1-2} = W_{1-2} + dU = W_{1-2} + (U_2 - U_1)$$
$$= W_{1-2} + mc_v (T_2 - T_1)$$
$$= -4.159 \text{ kJ} \quad (dU = 0 \text{ as } T_2 = T_1)$$

$$Q_{2-3} = W_{2-3} + (U_3 - U_2)$$
$$= W_{2-3} + m\, c_v (T_3 - T_2)$$
$$= W_{2-3} + m\, c_v (T_3 - T_1)$$

From the relation $c_p - c_v = R$, we have

$$c_v = c_p - R$$
$$= 1.005 - 0.287 = 0.718 \text{ kJ/Kg K}$$

Then:

$$Q_{2-3} = -3 + 0.0261 \times 0.718 (400 - 800)$$
$$= -10.496 \text{ kJ}$$

$$\therefore Q_{1-3} = Q_{1-2} + Q_{2-3}$$
$$= -4.159 + (-10.496)$$
$$= \mathbf{-14.655 \text{ kJ}}$$

6.5.4. Adiabatic process

A change in the state of the system without heat exchange with the surrounding medium is called an *adiabatic process;* here changing variables are pressure, volume and temperature. An adiabatic process can be carried out by the expansion or compression of gas in a cylinder whose walls are insulated. A reversible adiabatic process is called ***isentropic process***.

• **Governing equation :** Considering unit mass of gas, the non-flow energy equation can be written as

$$dq = \delta w + du = p\,dv + c_v\,dT$$

For an adiabatic operation $dq = 0$ and, therefore,

$$c_v\,dT + p\,dv = 0 \qquad ...(i)$$

Differentiating of gas equation $pv = RT$ gives

$$dT = \frac{p\,dv + v\,dp}{R}$$

Upon substitution of this value of dT in expression (*i*), we get

$$c_v \frac{(p\,dv + v\,dp)}{R} + p\,dv = 0$$

or $c_v(p\,dv + v\,dp) + pR\,dv = 0$

or $c_v\,(p\,dv + v\,dp) + p\,(c_p - c_v)dv = 0$

$(\because R = c_p - c_v)$

or $c_p\,p\,dv + c_v\,v\,dp = 0$

Dividing both sides by $c_v p\,v$,

$$\frac{c_p}{c_v}\frac{dv}{v} + \frac{dp}{p} = 0\,;\quad \gamma\frac{dv}{v} + \frac{dp}{p} = 0$$

Upon integration,

$\gamma \log_e v + \log_e p = \log_e$ constant

or $pv^\gamma = e^{\text{constant}} =$ constant

Hence an adiabatic process follows the law

$$p_1V_1^\gamma = p_2V_2^\gamma = pV^\gamma = \text{constant } C$$

• Work done $W_{1\text{-}2} = \int_1^2 p\,dV$

Since $pV^\gamma =$ constant C; $p = \dfrac{C}{V^\gamma}$

$$\therefore W_{1\text{-}2} = C\int_1^2 \frac{dV}{V^\gamma} = \frac{C}{1-\gamma}\left(V_2^{1-\gamma} - V_1^{1-\gamma}\right)$$

As $C = p_1V_1^\gamma = p_2V_2^\gamma$, we multiply the first term inside the bracket by $p_2V_2^\gamma$ and the second term by $p_1V_1^\gamma$, and obtain

$$W_{1\text{-}2} = \frac{p_2V_2 - p_1V_1}{1-\gamma} = \frac{p_1V_1 - p_2V_2}{\gamma - 1}$$

• Since $\delta Q = 0$ for an adiabatic process, the energy equation for non-flow process takes the form

$$\delta W + dU = 0$$

or $U_1 - U_2 = \dfrac{p_1V_1 - p_2V_2}{\gamma - 1}$

i.e., the work of adiabatic expansion is done at the expense of a reduction in the internal energy.

• From the gas equation $\dfrac{p_1V_1}{T_1} = \dfrac{p_2V_2}{T_2}$ and the adiabatic law $p_1V_1^\gamma = p_2V_2^\gamma$, the following relations can be set up:

$$\frac{T_2}{T_1} = \left(\frac{p_2}{p_1}\right)^{\frac{\gamma-1}{\gamma}} = \left(\frac{V_1}{V_2}\right)^{\gamma-1}$$

EXAMPLE 6.41

2 kg of an ideal gas is compressed adiabatically from pressure 100 kPa and temperature 220 K to a final pressure of 400 kPa. Make calculations for : (a) initial volume, (b) final volume and temperature (c) work performed, (d) heat added to or subtracted from the system, and (e) change in internal energy. It may be presumed that for the given ideal gas

$$c_p = 1 \text{ kJ/kg K} \quad \text{and} \quad c_v = 0.707 \text{ kJ/kg K}$$

Solution : $c_p - c_v = R$

$\therefore$ $R = (1 - 0.707)$

$= 0.293$ kJ/kg K $= 293$ J/kg K

(*a*) Invoking characteristic gas equation,

$$V_1 = \frac{mRT_1}{p_1} = \frac{2 \times 293 \times 220}{100 \times 10^3} = \mathbf{1.29\ m^3}$$

(*b*) Adiabatic index,

$$\gamma = \frac{c_p}{c_v} = \frac{1}{0.707} = 1.414$$

For an adiabatic process: $p_1V_1^\gamma = p_2V_2^\gamma$

∴ Final volume,

$$V_2 = \left(\frac{p_1}{p_2}\right)^{1/\gamma} \times V_1$$

$$= \left(\frac{100}{400}\right)^{1/1.414} \times 1.29 = \mathbf{0.484\ m^3}$$

Again from characteristic gas equation, Final temperature,

$$T_2 = \frac{p_2V_2}{mR}$$

$$= \frac{400 \times 10^3 \times 0.484}{2 \times 293} = \mathbf{330.4\ K}$$

(*c*) Work performed

$$= \frac{p_1V_1 - p_2V_2}{\gamma - 1}$$

$$= \frac{100 \times 10^3 \times 1.29 - 400 \times 10^3 \times 0.484}{1.414 - 1}$$

$$= -156.04 \times 10^3 \text{ Nm (or J)}$$

$$= \mathbf{-156.04\ kJ}$$

The negative sign indicates that work was done on the gas system.

(*d*) Adiabatic compression implies that there is no heat transfer to or from the system.

(*e*) From non-flow energy equation,

$$dU = \delta Q - \delta W = 0 - (-156.04)$$

$$= \mathbf{156.04\ kJ}$$

The positive value of dU implies that internal energy of the gas has increased.

EXAMPLE 6.42

A certain perfect gas is contained in a perfectly insulated piston cylinder arrangement at 100 kPa pressure and 290 K temperature. When the gas undergoes a non-flow reversible compression process to 500 kPa, its temperature rises to 350 K. If the work done on the gas during the compression process is 50 kJ/kg, evaluate the adiabatic exponent, specific heat at constant volume, the characteristic gas constant and molecular mass of the gas.

Solution : Perfect insulation implies no transfer of heat and hence the process is adiabatic.

From the relation,

$$\frac{T_2}{T_1} = \left(\frac{p_2}{p_1}\right)^{\frac{\gamma-1}{\gamma}}$$

$$\frac{\gamma - 1}{\gamma} = \frac{\log_e\left(\frac{T_2}{T_1}\right)}{\log_e\left(\frac{p_2}{p_1}\right)} = \frac{\log_e\left(\frac{350}{290}\right)}{\log_e\left(\frac{500}{100}\right)}$$

$$= \frac{0.1880}{1.6094} = 0.1168$$

$$\gamma - 1 = 0.1168\,\gamma$$

$$\therefore \quad \gamma = \frac{1}{1 - 0.1168} = \mathbf{1.132}$$

From non-flow energy equation,

$$Q_{1\text{-}2} = W_{1\text{-}2} + dU$$

$$0 = -50 + dU \quad \therefore \quad dU = 50 \text{ kJ/kg}$$

Now, $dU = mc_v\,(T_2 - T_1)$

$$50 = 1 \times c_v \times (350 - 290);$$

$$c_v = \mathbf{0.833\ kJ/kg\ K}$$

Characteristic gas constant, $R = c_v\,(\gamma - 1)$

$$= 0.833 \times (1.132 - 1)$$

$$= \mathbf{0.1099\ kJ/kg\ K}$$

$$\text{Molecular mass, } M = \frac{8.314}{0.1099} = \mathbf{75.65}$$

EXAMPLE 6.43

0.5 kg of an ideal gas expands adiabatically until its pressure is halved. During expansion, the gas does 30 kJ of external work and its temperature falls from 500 K to 410 K. Make calculations for the adiabatic exponent and the characteristic gas constant.

Solution : From the relation;

$$\frac{T_2}{T_1} = \left(\frac{p_2}{p_1}\right)^{\frac{\gamma-1}{\gamma}} \text{ and noting that}$$

$$p_2 = \frac{p_1}{2}$$

$$\frac{\gamma - 1}{\gamma} = \frac{\log_e \frac{T_2}{T_1}}{\log_e \frac{p_2}{p_1}} = \frac{\log_e \frac{410}{500}}{\log_e \frac{1}{2}}$$

$$= \frac{-0.1985}{-0.693} = 0.2864$$

or $\gamma - 1 = 0.2864\,\gamma$

$$\therefore \quad \gamma = \frac{1}{1-0.2864} = \mathbf{1.401}$$

For adiabatic expansion process $\delta Q = 0$ and therefore, from non-flow energy equation,

$$\delta Q = \delta W + dU$$

$$W_{1\text{-}2} = -dU = -mc_v(T_2 - T_1)$$

$$30 = -0.5 \times c_v (410 - 500)$$

$$\therefore \quad c_v = 0.667 \text{ kJ/kg K}$$

Gas constant,

$$R = c_v(\gamma - 1) = 0.667 \times (1.4 - 1)$$

$$= 0.267 \text{ kJ/kg K} = \mathbf{267 \text{ J/kg K}}$$

EXAMPLE 6.44

An insulated cylinder of 0.4 m diameter and 0.8 m length contains 10 kg of oxygen. Paddle work is done on the gas to increase its pressure from 3 bar to 6 bar. Determine the change in internal energy, work done on the gas and the change in enthalpy.

c_p = 0.91 kJ/kg K and c_v = 0.64 kJ/kg K

Solution : Volume of cylinder,

$$V = \frac{\pi}{4}d^2 l = \frac{\pi}{4}(0.4)^2 \times 0.8 = 0.10 \text{ m}^3$$

Gas constant,

$$R = c_p - c_v$$

$$= 0.91 - 0.64$$

$$= 0.27 \text{ kJ/kg K} = 270 \text{ J/kg K}$$

Initial state :

$$p_1 = 3 \text{ bar}$$

$$V_1 = 0.1 \text{ m}^3 \quad \text{and} \quad m = 10 \text{ kg}$$

Then from characteristic gas equation,

$$T_1 = \frac{p_1 V_1}{mR} = \frac{3\times 10^5 \times 0.1}{10\times 270} = 11.11 \text{ K}$$

Final state :

$$p_2 = 6 \text{ bar}$$

$$V_2 = 0.1 \text{ m}^3 \quad \text{and} \quad m = 10 \text{ Kg}$$

$$T_2 = \frac{p_2 V_2}{mR} = \frac{6\times 10^5 \times 0.1}{10\times 270} = 22.22 \text{ K}$$

(*i*) Change in internal energy,

$$dU = m\,c_v(T_2 - T_1)$$

$$= 10 \times 0.64\,(22.22 - 11.11)$$

$$= 71.10 \text{ kJ} = 71100 \text{ J}$$

(*ii*) Using non-flow energy equation,

$$\delta Q = \delta W + dU$$

Here $\delta Q = 0$ as the cylinder is insulated

$$\therefore \quad \delta W = -dU = -71.10 \text{ kJ}$$

The negative sign indicates that work is done on the gas.

(*iii*) Enthalpy represents the sum of internal energy and pressure-volume product. That is

$$H = U + pV$$

Change in enthalpy,

$$dH = dU + d(pV)$$

$$= dU + (p_2V_2 - p_1V_1)$$

$$= 71100 + (6 \times 10^5 \times 0.1 - 3 \times 10^5 \times 0.1)$$

$$= 71100 + 30000 = \mathbf{101100 \text{ J}}$$

6.5.5. Polytropic process

Each of the isochoric, isobaric, isothermal and adiabatic processes are quasi-static in operation under ideal working conditions. For all these processes, the relation between pressure and volume can be described by a single equation $pV^n = C$ where n is a positive number. The value of index n remains constant throughout in any particular process. The equation $pV^n = C$ is called polytropic law, and the process in which the state of the system changes in accordance with this correlation is called a polytropic process.

- *Work done,*

$$W_{1\text{-}2} = \int_1^2 p\,dV$$

Since pV^n = constant C, $p = \dfrac{C}{V^n}$

$$\therefore \; W_{1\text{-}2} = C\int_1^2 \frac{dV}{V^n} = \frac{C}{1-n}\left(V_2^{1-n} - V_1^{1-n}\right)$$

As $C = p_1V_1^n = p_1V_2^n$, we multiply the first term inside the bracket by $p_2V_2^n$ and the second term by $p_1V_1^n$, and obtain

$$W_{1\text{-}2} = \frac{p_2V_2 - p_1V_1}{1-n}$$

$$= \frac{p_1V_1 - p_2V_2}{n-1} = \frac{mR(T_1 - T_2)}{n-1}$$

For unit mass of gas

$$w_{1\text{-}2} = \frac{p_1v_1 - p_2v_2}{n-1} = \frac{R(T_1 - T_2)}{n-1}$$

• *Change in internal energy,*

$U_2 - U_1 = mc_v (T_2 - T_1)$

or $u_2 - u_1 = c_v (T_2 - T_1)$ for unit mass of gas

• *Heat interaction* : The heat transfer during a polytropic process can be determined by applying the non-flow energy equation,

$$q_{1\text{-}2} = w_{1\text{-}2} + (u_2 - u_1)$$

$$= \frac{p_1v_1 - p_2v_2}{n-1} + c_v(T_2 - T_1)$$

$$= \frac{p_1v_1 - p_2v_2}{n-1} + \frac{R}{\gamma - 1}(T_2 - T_1)$$

$$= \frac{p_1v_1 - p_2v_2}{n-1} + \frac{p_2v_2 - p_1v_1}{\gamma - 1}$$

$$= \frac{\gamma - n}{\gamma - 1} \times \frac{p_1v_1 - p_2v_2}{n-1}$$

$$= \frac{\gamma - n}{\gamma - 1} \times \text{polytropic work done}$$

Since $\frac{\gamma - n}{\gamma - 1}$ is positive, the sign of work and heat interactions are same.

In a polytropic expansion process, work is done by the working fluid as well as heat is absorbed.

• *Polytropic specific heat* : On the basis of unit mass of the system

$$q_{1\text{-}2} = w_{1\text{-}2} + du$$

$$= \frac{p_1v_1 - p_2v_2}{n-1} + c_v(T_2 - T_1)$$

$$= \frac{R(T_1 - T_2)}{n-1} + c_v(T_2 - T_1)$$

$$= \frac{(c_p - c_v)(T_1 - T_2)}{n-1} + c_v(T_2 - T_1)$$

$$(\because R = c_p - c_v)$$

$$= (T_2 - T_1)\left[c_v - \frac{c_p - c_v}{n-1}\right]$$

$$= \frac{c_p - nc_v}{1-n}(T_2 - T_1)$$

$$= c_n (T_2 - T_1)$$

where $c_n = \frac{c_p - n\,c_v}{1-n}$ is called the polytropic specific heat.

Substituting $\frac{c_p}{c_v} = \gamma$, we get

$$c_n = \frac{\gamma - n}{1-n}c_v = -\left(\frac{\gamma - n}{n-1}c_v\right)$$

c_n is negative when $\frac{\gamma - n}{n-1}$ is positive, *i.e.*, when $\gamma > n > 1$

• *Distinction between different polytropic processes*

(*i*) If $n = 0$, then the identity $pV^n = C$ becomes $pV^0 = C$ or $p = C$ which indicates a constant pressure process.

(*ii*) If $n = 1$, then the identity $pV^n = C$ becomes $pV = C$ which represents a constant temperature or isothermal process.

(*iii*) If $n = \gamma$, then the identity pV^n becomes $pV^\gamma = C$ which conforms to an adiabatic process.

(*iv*) Taking nth root of both sides of tne identity $pV^n = C$, we get

$$p^{1/n}\, V = C^{1/n} = C'$$

where C' is another constant.

Now if $n = \infty$, then $p^{1/n} = p^{1/\infty} = p^0 = 1$

Accordingly, the identity $pV^n = C$ takes the form $V = C$ which indicates a constant volume process. The different polytropic processes commonly encountered in engineering practice have been shown on the $p - V$ plot in Fig. 6.11. Starting from the initial state, the lower right quadrant shows the expansion processes and the upper left quadrant shows the compression processes. Upon differentiation of polytropic law pV^n = constant C, we get

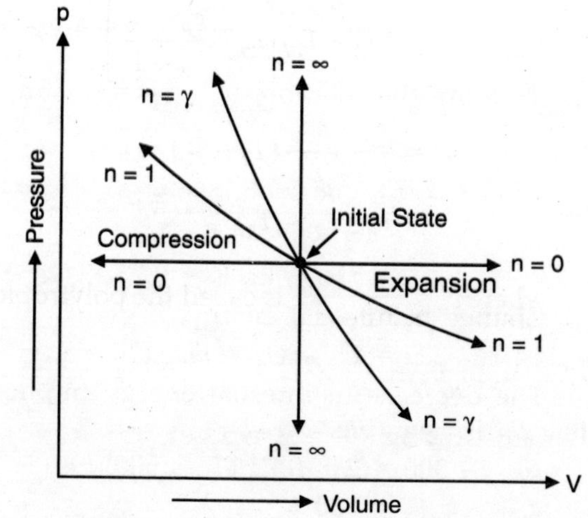

Fig. 6.11. Polytopic process on p-V plot

$$V^n dp + pnV^{n-1}\,dV = 0\ ;\quad \frac{dp}{dV} = -n\frac{p}{V}$$

The slope of the curve thus increases in the negative direction with increase of exponent n.

- *Polytropic index n* : The value of the polytropic index n during change of state 1 to state 2 can be worked out as follows :

For a polytropic process :

$$pV^n = C \text{ or } p_1V_1^{\,n} = p_2V_2^{\,n}$$

Taking logs on both sides,

$$\log_e p_1 + n\log_e V_1 = \log_e p_2 + n\log_e V_2$$

or $n(\log_e V_1 - \log_e V_2) = \log_e p_2 - \log_e p_1$

$$\therefore\ n = \frac{\log_e p_2 - \log_e p_1}{\log_e V_1 - \log_e V_2} = \frac{\log_e(p_2/p_1)}{\log_e(V_1/V_2)}$$

- *p – V – T relations* : From polytropic law and gas equation we have

$$\frac{p_2}{p_1} = \left(\frac{V_1}{V_2}\right)^n \text{ and } \frac{p_2}{p_1} = \frac{V_1}{V_2}\times\frac{T_2}{T_1}$$

Equating the two values of (p_2/p_1), we get

$$\frac{T_2}{T_1} = \frac{(V_1/V_2)^n}{(V_1/V_2)} = \left(\frac{V_1}{V_2}\right)^{n-1}$$

Again $V_1/V_2 = (p_2/p_1)^{1/n}$ and $\dfrac{V_1}{V_2} = \dfrac{p_2}{p_1}\times\dfrac{T_1}{T_2}$

Equating the two values of V_1/V_2,

$$\left(\frac{p_1}{p_2}\right)^{1/n} = \left(\frac{p_2}{p_1}\right)\times\frac{T_1}{T_2}$$

$$\therefore\ \frac{T_2}{T_1} = \frac{(p_2/p_1)}{(p_2/p_1)^{1/n}} = (p_2/p_1)^{1-1/n}$$

$$= \left(\frac{p_2}{p_1}\right)^{\frac{n-1}{n}}$$

EXAMPLE 6.45

3 kg of air kept at an absolute pressure of 100 kPa and temperature of 300 K is compressed polytropically until the pressure and temperature become 1500 kPa and 500 K respectively. Evaluate the polytropic exponent, the final volume, the work of compression and the heat interaction.

Take gas constant R = 287 J/kg K

Solution : From the relation

$$\frac{T_2}{T_1} = \left(\frac{p_2}{p_1}\right)^{\frac{n-1}{n}}$$

$$\frac{n-1}{n} = \frac{\log_e \frac{T_2}{T_1}}{\log_e \frac{p_2}{p_1}}$$

$$= \frac{\log_e \frac{500}{300}}{\log_e \frac{1500}{100}} = \frac{0.5108}{2.708} = 0.1886$$

or $\quad n - 1 = 0.1886\,n\ ;\ n = \dfrac{1}{1-0.1886} = \mathbf{1.23}$

Initial volume,

$$V_1 = \frac{mRT_1}{p_1} = \frac{3\times 287\times 300}{100\times 10^3} = 2.583\ \text{m}^3$$

Final volume,

$$V_2 = \frac{mRT_2}{p_2} = \frac{3\times 287\times 500}{1500\times 10^3} = 0.287\ \text{m}^3$$

Work of compression,

$$W_{1-2} = \frac{p_1V_1 - p_2V_2}{n-1} = \frac{mR(T_1 - T_2)}{n-1}$$

$$= \frac{3 \times 287 \times (300 - 500)}{1.23 - 1}$$

$$= -748696 \text{ Nm (or J)} \simeq -748.7 \text{ kJ}$$

The negative sign indicates that work has to be done on the gas.

$$\text{Now, } c_v = \frac{R}{\gamma - 1} = \frac{287}{1.4 - 1}$$

$$= 7175 \text{ J/kg K} = 0.7175 \text{ kJ/kg K}$$

$$dU = mc_v (T_2 - T_1)$$

$$= 3 \times 0.7175 \times (500 - 300)$$

$$= 430.5 \text{ kJ}$$

From non-flow energy equation,

$$Q_{1-2} = W_{1-2} + dU$$

$$= -748.7 + 430.5 = \mathbf{-318.2 \text{ kJ}}$$

The negative sign indicates that heat is rejected by the gas.

Alternatively:

$$Q_{1-2} = \frac{\gamma - n}{\gamma - 1} \times \text{polytropic work done}$$

$$= \frac{1.4 - 1.23}{1.4 - 1} \times (-748.7) = \mathbf{-318.2 \text{ kJ}}$$

EXAMPLE 6.46

2 kg of an ideal gas occupies a volume of 0.3 m^3 at 10 bar pressure and 500 K temperature. When this gas expands polytropically ($pV^{1.2}$ = C) the internal energy decreases by 300 kJ. Presuming adiabatic exponent γ = 1.4, determine:

(a) specific gas constant,

(b) final temperature, pressure and volume of gas, and

(c) heat and work interactions across the system boundary.

Solution : From characteristic gas equation, $pV = mRT$

Gas constant,

$$R = \frac{pV}{mT}$$

$$= \frac{10 \times 10^5 \times 0.3}{2 \times 500} = \mathbf{300 \text{ J/kg K}}$$

(*b*) From the relation, $(c_p - c_v) = R$ and

$$c_p/c_v = \gamma$$

$$c_v = \frac{R}{\gamma - 1} = \frac{300}{1.4 - 1}$$

$$= 750 \text{ J/kg K} = 0.75 \text{ kJ/kg K}$$

Change in internal energy,

$$dU = U_2 - U_1 = mc_v (T_2 - T_1)$$

The decrease in internal energy implies that dU is negative.

$$-300 = 2 \times 0.75 (T_2 - 500)$$

$$\therefore \quad T_2 = \mathbf{300 \text{ K}}$$

From the relations :

$$\frac{T_2}{T_1} = \left(\frac{p_2}{p_1}\right)^{\frac{n-1}{n}} = \left(\frac{V_1}{V_2}\right)^{n-1}$$

$$p_2 = p_1\left(\frac{T_2}{T_1}\right)^{\frac{n}{n-1}} = 10 \times \left(\frac{300}{500}\right)^{\frac{1.2}{1.2-1}}$$

$$= \mathbf{0.467 \text{ bar}}$$

$$V_2 = V_1\left(\frac{p_1}{p_2}\right)^{1/n}$$

$$= \left(\frac{10}{0.467}\right)^{1/1.2} \times 0.3 = \mathbf{3.85 \text{ m}^3}$$

(*c*) Work done,

$$W_{1-2} = \frac{p_1V_1 - p_2V_2}{n-1} = \frac{mR(T_1 - T_2)}{n-1}$$

$$= \frac{2 \times 300 \times (500 - 300)}{1.2 - 1}$$

$$= 600 \times 10^3 \text{ J} = \mathbf{600 \text{ kJ}}$$

From non-flow energy equation,

$$Q_{1-2} = W_{1-2} + (U_2 - U_1)$$

$$= 600 + (-300) = \mathbf{300 \text{ kJ}}$$

Alternatively:

$$Q_{1-2} = \frac{\gamma - n}{\gamma - 1} \times \text{polytropic work done}$$

$$= \frac{1.4 - 1.2}{1.4 - 1} \times 600 = \mathbf{300 \text{ kJ}}$$

EXAMPLE 6.47

One kg of air at a pressure of 7 bar and a temperature of 360 K undergoes a reversible polytropic process which may be represented by $pV^{1.1}$ *= constant. If the final pressure is 1.4 bar, evaluate (a) final specific volume and temperature, (b) work done and heat transfer. How the work and heat interactions would be affected if the process is irreversible and 15 kJ of work is lost due to internal friction?*

Take R = 287 J/kg K and $\gamma = 1.4$

Solution : Assuming air to be perfect gas,

$$p_1V_1 = mRT_1$$

$$\therefore \quad V_1 = \frac{1 \times 287 \times 360}{7 \times 10^5} = 0.1476 \text{ m}^3/\text{kg}$$

From the given polytropic law,

$$V_2 = V_1\left(\frac{p_1}{p_2}\right)^{1/n} = 0.1476 \times \left(\frac{7}{1.4}\right)^{1/1.1}$$

$$= \mathbf{0.637\ m^3/kg}$$

$$T_2 = \frac{p_2V_2}{mR}$$

$$= \frac{1.4 \times 10^5 \times 0.637}{1 \times 287} = \mathbf{310.73\ K}$$

(*b*) Work done,

$$W_{1\text{-}2} = \frac{p_1V_1 - p_2V_2}{n-1} = \frac{mR(T_1 - T_2)}{n-1}$$

$$= \frac{1 \times 287 \times (360 - 310.73)}{1.1 - 1}$$

$$= 141405 \text{ Nm (or J)} = 141.405 \text{ kJ}$$

Now,

$$c_v = \frac{R}{\gamma - 1} = \frac{287}{1.4 - 1}$$

$$= 717.5 \text{ J/kg K} = 0.7175 \text{ kJ/kg K}$$

Change in internal energy, $dU = mc_v(T_2 - T_1)$

$$= 1 \times 0.7175\ (310.73 - 360)$$

$$= -\ 35.35 \text{ kJ/kg}$$

From non-flow energy equation,

$$\delta Q = \delta W + dU$$

$$Q_{1\text{-}2} = W_{1\text{-}2} + dU = 141.405 + (-\ 35.35)$$

$$= \mathbf{106.055\ kJ/kg}$$

Alternatively: $Q_{1\text{-}2} = \dfrac{\gamma - n}{\gamma - 1} \times \text{polytropic work}$

$$= \frac{1.4 - 1.1}{1.4 - 1} \times 141.405$$

$$= \mathbf{106.053\ kJ/kg}$$

(*c*) The end states are the same, but the process is irreversible with work dissipation in friction.

$\therefore\ dU = -\ 35.35$ kJ/kg since the end states are the same and internal energy is the property or point function.

Net work done

= displacement work - work dissipated in friction

= 141.405 - 15 = 126.405 kJ

and net heat transfer is given by

$$Q_{net} = dU + \text{net work done}$$

$$= -\ 35.35 + 126.405$$

$$= \mathbf{91.055\ kJ/kg}$$

EXAMPLE 6.48

A gaseous mixture of mean molecular mass 44 occupies a volume of 0.1 m^3 *at 1 bar pressure and 340 K temperature. The gas is compressed polytropically according to the law* pV^n *to 0.015* m^3 *volume and 10 bar pressure. Evaluate the amount of work done, change in internal energy and heat added or subtracted during the compression process. For the gas, mean molar specific heat at constant volume = 25 kJ/kg mole K.*

Solution : (*i*) From the polytropic law,

$$p_1V_1^{\,n} = p_2V_2^{\,n}$$

$$n = \frac{\log_e \dfrac{p_2}{p_1}}{\log_e \dfrac{V_1}{V_2}}$$

$$= \frac{\log_e\left(\dfrac{10}{1}\right)}{\log_e\left(\dfrac{0.1}{0.015}\right)} = \frac{2.302}{1.897} = 1.214$$

(*ii*) Work done,

$$W_{1\text{-}2} = \frac{p_1V_1 - p_2V_2}{n-1}$$

$$= \frac{1 \times 10^5 \times 0.1 - 10 \times 10^5 \times 0.015}{1.214 - 1}$$

$= -\ 23365$ Nm (or J)

$= \mathbf{-\ 23.365\ kJ}$

Negative sign implies work supplied to the system.

(*iii*) Change in internal energy,

$$dU = U_2 - U_1 = mc_v\,(T_2 - T_1)$$

$$\frac{T_2}{T_1} = \left(\frac{V_1}{V_2}\right)^{n-1}$$

$$T_2 = 340\left(\frac{0.1}{0.015}\right)^{1.214-1} = 510.26 \text{ K}$$

$$c_v = \frac{\text{molar specific heat}}{\text{molecular mass}}$$

$$= \frac{25}{44} = 0.5682 \text{ kJ/kg K}$$

$$R = \frac{R_{mole}}{\text{molecular mass}}$$

$$= \frac{8314}{44} = 188.95 \text{ J/kg K}$$

Mass of gas,

$$m = \frac{p_1V_1}{RT_1}$$

$$= \frac{1 \times 10^5 \times 0.1}{188.95 \times 340} = 0.1556 \text{ kg}$$

$$\therefore\ dU = 0.1556 \times 0.5682\ (510.26 - 340)$$

$$= 15.053 \text{ kJ/kg}$$

(*iv*) Heat interaction

$$Q_{1\text{-}2} = W_{1-2} + dU$$

$$= -\ 23.365 + 15.053 = \mathbf{-\ 8.317\ kJ}$$

Negative sign implies that heat has been rejected by the system.

EXAMPLE 6.49

During a certain expansion process of a perfect gas, the variation in presure is estimated to conform to the linear relation $p = aV + b$ where a and b are constants and V represents the volume in m^3. The gas has a mass of 0.75 kg, the initial and final pressures are 4 bar and 2 bar, and the corresponding volumes are 0.1 m^3 and 0.2 m^3. Make calculations for the work and heat interactions during the process. Comment on the nature of process.

Solution : The pressure variation for the given expansion process is of the form $p = aV + b$.

The values of constants a and b can be determined from the values of pressure and temperature at the initial and final state points.

$$4 = 0.1a + b \quad \text{and} \quad 2 = 0.2a + b$$

From these expressions:

$$a = -20 \quad \text{and} \quad b = 6$$

The law of expansion becomes

$$p = (-20V + 6) \text{ bar}$$

$$= (-20V + 6) \times 10^5 \text{ N/m}^2$$

(*a*) Work done

$$W_{12} = \int_1^2 p\,dV = \int_1^2 (-20V + 6) \times 10^5\,dV$$

$$= 10^5 \left| \frac{-20V^2}{2} + 6V \right|_{0.1}^{0.2}$$

$$= 10^5\left[-10(0.2^2 - 0.1^2) + 6(0.2 - 0.1)\right]$$

$$= 30000 \text{ Nm} = \mathbf{30\ kJ}$$

(*b*) The given data shows that

$$p_1V_1 = 4 \times 0.1 = 0.4$$

$$p_2V_2 = 2 \times 0.2 = 0.4$$

Thus $p_1V_1 = p_2V_2$ and consequently

$$T_1 = T_2 \quad \text{and} \quad dU = 0$$

Then from non-flow energy equation

$$\delta Q = \delta W + dU, \quad \text{we have}$$

$$= 30 + 0 = \mathbf{30\ kJ}$$

(*c*) Even though $T_1 = T_2$, the given process cannot be treated as isothermal. The relation is true only for the initial and final conditions. Further, the given relation is a straight line path whereas isothermal path is a parabolic curve.

EXAMPLE 6.50

A perfect gas undergoes the following three separate and distinct processes to execute a cycle:

(i) constant volume process during which 80 kJ of heat is supplied to the gas,

(ii) constant pressure process during which 85 kJ of heat is lost to the surroundings and 20 kJ of work is done on it.

(iii) adiabatic process which restores the gas back to its initial state.

Evaluate the work done during adiabatic process and the value of internal energy at all the state points if initially its value is 95 kJ.

Solution : This example illustrates the application of non-flow energy equation and the 1st law applied to a cyclic process.

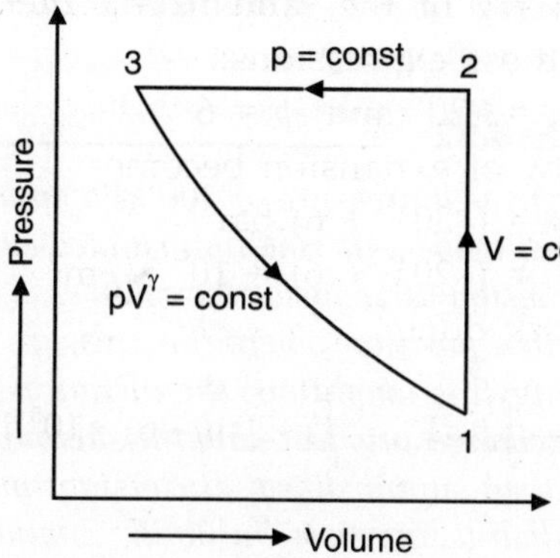

Fig. 6.12.

Process 1–2 :

$$Q_{1-2} = 80 \text{ kJ} ; \; W_{1-2} = 0$$

$$U_2 - U_1 = Q_{1-2} - W_{1-2}$$

$$= 80 - 0 = 80 \text{ kJ}$$

$$\therefore \; U_2 = 80 + 95 = \mathbf{175 \text{ kJ}}$$

Process 2–3 :

$$Q_{2-3} = -85 \text{ kJ} ; \; W_{2-3} = -20 \text{ kJ}$$

$$U_3 - U_2 = Q_{2-3} - W_{2-3}$$

$$= -85 - (-20) = -65 \text{ kJ}$$

$$\therefore \; U_3 = -65 + 175 = \mathbf{110 \text{ kJ}}$$

For the complete cycle :

$\oint \delta Q = \oint \delta W$. That is

$$Q_{1-2} + Q_{2-3} + Q_{3-1} = W_{1-2} + W_{2-3} + W_{3-1}$$

$$80 - 85 + 0 = 0 + (-20) + W_{3-1}$$

$\therefore$ Work done during adiabatic process,

$$W_{3-1} = \mathbf{15 \text{ kJ}}$$

EXAMPLE 6.51

One kg of air at 1 bar and 300 K is compressed adiabatically till its pressure becomes 5 times the original pressure. Subsequently it is expanded at constant pressure and finally cooled at constant volume to return to its original state. Calculate the heat and work interactions, and change in internal energy for each process and for the cycle.

Solution : Refer fig. 6.13 which represents the given processes on *p-V* coordinates.

Let it be presumed that for air:

$$c_p = 1.005 \text{ kJ/kg K}$$

$$c_v = 0.718 \text{ kJ/kg K}$$

Then $\gamma = \dfrac{c_p}{c_v} = \dfrac{1.005}{0.718} = 1.4$

$$R = c_p - c_v = 1.005 - 0.718$$

$$= 0.287 \text{ kJ/kg K} = 287 \text{ J/K}$$

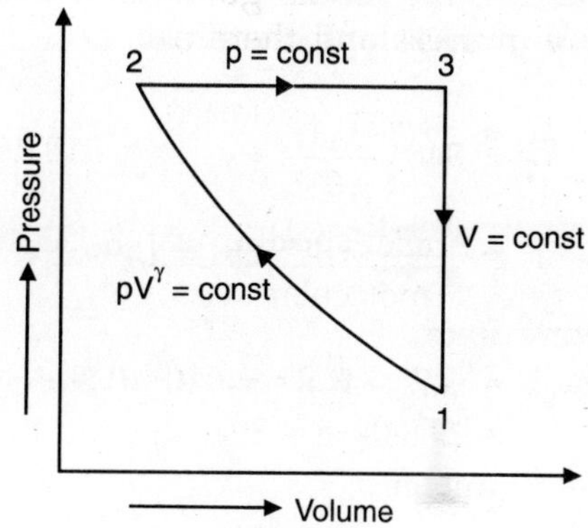

Fig. 6.13.

Process 1-2 (adiabatic)

$$V_1 = \frac{mR\,T_1}{p_1}$$

$$= \frac{1 \times 287 \times 300}{1 \times 10^5} = 0.861 \text{ m}^3$$

$$V_2 = \left(\frac{p_1}{p_2}\right)^{1/\gamma} \times V_1$$

$$= \left(\frac{1}{5}\right)^{1/1.4} \times 0.861 = 0.273 \text{ m}^3$$

$$T_2 = \left(\frac{p_2}{p_1}\right)^{\frac{\gamma-1}{\gamma}} \times T_1$$

$$= (5)^{\frac{1.4-1}{1.4}} \times 300 = 475.36 \text{ K}$$

Work done,

$$W_{1-2} = \frac{p_1 V_1 - p_2 V_2}{\gamma - 1} = \frac{mR\,(T_1 - T_2)}{\gamma - 1}$$

$$= \frac{1 \times 287\,(300 - 475.36)}{1.4 - 1}$$

$$= -125281 \text{ J} = -125.28 \text{ kJ}$$

Heat transfer, $Q_{1-2} = 0$ as the process is adiabatic. Then from non-flow energy equation,

$\delta Q = \delta W + dU$

Change in internal energy,

$dU_{1-2} = -\delta W = 125.28$ kJ

Process 2-3 (Constant pressure)

$V_2 = 0.273$ m^3 ; $V_3 = 0.861$ m^3

$p_3 = p_2 = 5$ bar and $T_2 = 475.36$ K

Charles' law holds good for a constant pressure process and therefore,

$$T_3 = \frac{V_3}{V_2} \times T_2$$

$$= \frac{0.861}{0.273} \times 475.36 = 1499.2 \text{ K}$$

Work done,

$$W_{2-3} = p(V_3 - V_2) = 5 \times 10^5 (0.861 - 0.273)$$
$$= 294000 \text{ J} = 294 \text{ kJ}$$

Heat transfer,

$$Q_{2-3} = m\, c_p (T_3 - T_2)$$
$$= 1 \times 1.005 (1499.2 - 475.36)$$
$$= 1028.96 \text{ kJ}$$

Then from non-flow energy equation,

$\delta Q = \delta W + dU$

$$dU_{2-3} = Q_{2-3} - W_{2-3}$$
$$= 1028.96 - 294 = 734.96 \text{ kJ}$$

Process 3-1 (Constant volume)

$p_3 = 5$ bar ; $V_3 = V_1 = 0.861$ m^3

$T_3 = 1499.2$ K and $T_1 = 300$ K

Work done, $W_{3-1} = 0$ as volume remains constant

Heat transfer

$$Q_{3-1} = m\, c_v (T_1 - T_3)$$
$$= 1 \times 0.718 (300 - 1499.2)$$
$$= -861.02 \text{ kJ}$$

Then from non-flow energy equation,

$\delta Q = \delta W + dU$

$$dU_{3-1} = Q_{3-1} - W_{3-1}$$
$$= -861.02 - 0 = -861.02 \text{ kJ}$$

For the cycle :

Net work done $= W_{1-2} + W_{2-3} + W_{3-1}$

$= -125.28 + 294 + 0 = 168.72$ kJ

Net heat transfer

$= 0 + 1028.96 - 861.02 = 167.94$ kJ

Obviously for a cyclic process $\oint \delta Q = \oint \delta W$

$$\oint dU = 125.28 + 734.96 - 861.02 \simeq 0$$

This has to be so because internal energy is a property of the system and the cyclic integral of a property is zero.

EXAMPLE 6.52

0.2 m^3 of air at a pressure of 400 kPa and 430 K is contained inside a thermally insulated cylinder. A movable piston fitted inside the cylinder is pushed outwards to a position where the pressure is 100 kPa. During this operation, the piston assembly does not exchange any heat with the surroundings. The air is then supplied heat at constant pressure till its enthalpy increases by 60 kJ. Evaluate the net work done by the gaseous system. What should be the index of expansion of a single reversible polytropic process if that is to produce the same work whilst operating between the same initial and final states ?

Solution : *Process* 1–2 : Since the piston-cylinder assembly does not exchange heat with the surroundings, the expansion process 1–2 is isentropic (reversible adiabatic) and for such a process

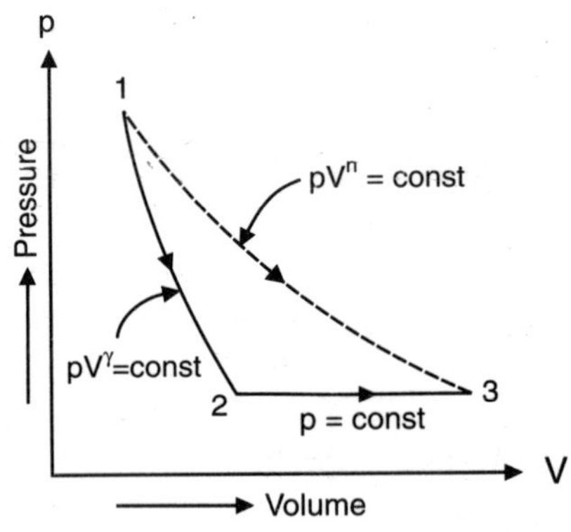

Fig. 6.14.

$$T_2 = T_1 \left(\frac{p_2}{p_1}\right)^{\frac{\gamma-1}{\gamma}}$$

$$= 430 \times \left(\frac{100}{400}\right)^{\frac{1.4-1}{1.4}} = 289.25 \text{ K}$$

$$V_2 = V_1 \left(\frac{p_1}{p_2}\right)^{1/\gamma}$$

$$= 0.2 \times \left(\frac{400}{100}\right)^{\frac{1}{1.4}} = 0.538 \text{ m}^3$$

Mass of gas,

$$m = \frac{p_1V_1}{RT_1} = \frac{400 \times 10^3 \times 0.2}{287 \times 430} = 0.6482 \text{ kg}$$

Work done,

$$W_{1\text{-}2} = \frac{p_1V_1 - p_2V_2}{\gamma - 1} = \frac{mR(T_1 - T_2)}{\gamma - 1}$$

$$= \frac{0.6482 \times 287 \times (430 - 289.25)}{1.4 - 1}$$

$$= 65460 \text{ Nm (or J)} = 65.46 \text{ kJ}$$

Process 2–3 : The heat supplied at constant pressure equals the change in enthalpy

$$\therefore \quad Q_{2\text{-}3} = (H_3 - H_2) = mc_p\,(T_3 - T_2)$$

$$60 = 0.6482 \times 1.005 \times (T_3 - 289.25)$$

$$T_3 = 381.35 \text{ K}$$

Also, Charles' law holds good for a constant pressure process and therefore,

$$V_3 = V_2\frac{T_3}{T_2} = 0.538 \times \frac{381.35}{289.25} = 0.7093 \text{ m}^3$$

Work done,

$$W_{2\text{-}3} = p_2\,(V_3 - V_2)$$

$$= 100 \times 10^3\,(0.7093 - 0.538)$$

$$= 17130 \text{ Nm (or J)} = 17.13 \text{ kJ}$$

Total work done by both the processes,

$$W_{1\text{-}2} + W_{2\text{-}3} = 65.46 + 17.13 = 82.59 \text{ kJ}$$

According to the given condition this work must equal the work done by a hypothetical polytropic process between the same initial and final states. That is

$$82.59 = \frac{p_1V_1 - p_3V_3}{n - 1} = \frac{mR(T_1 - T_3)}{n - 1}$$

$$= \frac{0.6482 \times 287 \times (430 - 381.35)}{n - 1} \times 10^{-3}$$

Solution gives : n = **1.1096**

EXAMPLE 6.53

60 litres of an ideal gas at 290 K and 1 bar is compressed adiabatically to 10 bar. It is then cooled at constant volume and further expanded isothermally so as to reach the condition from where it started. Evaluate : (a) pressure at the end of constant volume cooling, (b) change in internal energy during constant volume process, (c) net work done and heat transfer during the cycle.

Assume c_p = 14.25 kJ/kg K and
c_v = 10.15 kJ/kg

Solution : $\gamma = \frac{c_p}{c_v} = \frac{14.25}{10.15} = 1.404$

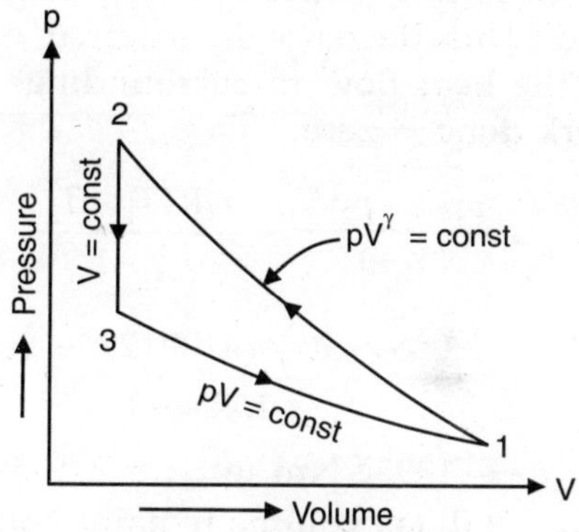

Fig. 6.15.

Characteristic gas constant,

$$R = c_p - c_v = 14.25 - 10.15$$

$$= 4.10 \text{ kJ/kg K}$$

$$V_1 = 60 \text{ litres} = 0.060 \text{ m}^3$$

Considering adiabatic process 1–2,

$$V_2 = \left(\frac{p_1}{p_2}\right)^{\frac{1}{\gamma}} V_1$$

$$= \left(\frac{1}{10}\right)^{\frac{1}{1.404}} \times 0.060 = 0.01164 \text{ m}^3$$

$$T_2 = T_1\left(\frac{p_2}{p_1}\right)^{\frac{\gamma-1}{\gamma}}$$

$$= 290 \times \left(\frac{10}{1}\right)^{\frac{1.404-1}{1.404}} = 562.47 \text{ K}$$

Considering isothermal process, 3–1,

$$p_3 = p_1\frac{V_1}{V_3} = \frac{1 \times 0.06}{0.01164} \qquad (\because V_3 = V_2)$$

$$= \textbf{5.155 bar}$$

(*b*) Mass of gas, $m = \frac{p_1V_1}{RT_1}$

$$= \frac{1 \times 10^5 \times 0.06}{4100 \times 290} = 5.046 \times 10^{-3} \text{ kg}$$

Change in internal energy during constant volume process 2–3,

$$U_3 - U_2 = mc_v (T_3 - T_2)$$
$$= 5.046 \times 10^{-3} \times 10.15\ (290 - 560.47)$$
$$= -13.85 \text{ kJ}$$

During constant volume cooling process, temperature and hence internal energy is reduced. This decrease in internal energy equals the heat flow to surroundings since the work done is zero.

(c) $$W_{1\text{-}2} = \frac{p_1V_1 - p_2V_2}{\gamma - 1} = \frac{mR(T_1 - T_2)}{\gamma - 1}$$

$$= \frac{5.046 \times 10^{-3} \times 4100\,(290 - 562.46)}{1.404 - 1}$$

$$= -13952 \text{ Nm (or J)} = -13.952 \text{ kJ}$$

$W_{2\text{-}3} = 0$ (as volume remains constant)

$$W_{3\text{-}1} = p_3V_3 \log_e \frac{V_1}{V_3} = p_1V_1 \log_e \frac{p_3}{p_1}$$

$$= 1 \times 10^5 \times 0.06 \log_e \frac{5.155}{1}$$

$$= 9840 \text{ Nm (or J)} = 9.84 \text{ kJ}$$

Net work done,

$$W_n = W_{1\text{-}2} + W_{2\text{-}3} + W_{3\text{-}1}$$
$$= -13.952 + 0 + 9.84 = -4.112 \text{ kJ}$$

Negative sign means that work has been done on the system.

For a cyclic process

$$\oint \delta Q = \oint \delta W$$

∴ Heat transfer during the complete cycle

$$= -4.112 \text{ kJ}$$

Negative heat transfer implies that heat has been lost from the system.

EXAMPLE 6.54

0.2 m^3 of an ideal gas at a pressure of 2 MPa and 600 K is expanded isothermally to 5 times the initial volume. It is then cooled to 300 K at constant volume and then compressed back polytropically to its initial state. Determine the net work done and heat transfer during the cycle.

Solution : Considering process 1–2 which is isothermal, *i.e.*, $T_1 = T_2 = 600$ K

$$p_2 = \frac{p_1V_1}{V_2} = \frac{2 \times 10^6 \times 0.2}{(5 \times 0.2)}$$
$$= 400 \times 10^3 \text{ N/m}^2 = 0.4 \text{ MPa}$$

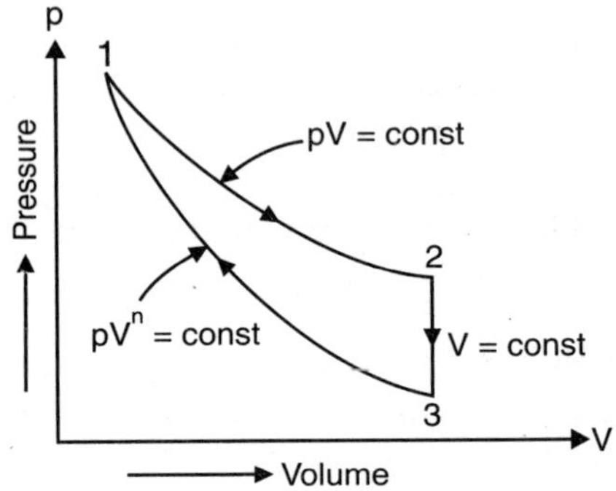

Fig. 6.16.

Work done, $$W_{1\text{-}2} = p_1V_1 \log_e \frac{V_2}{V_1}$$
$$= 2 \times 10^6 \times 0.2 \log_e 5$$
$$= 643775 \text{ J}$$
$$= 643.775 \text{ kJ}$$

Considering process 2–3 which is at constant volume, *i.e.*,

$$V_2 = V_3 = 5 \times 0.2 = 1 \text{m}^3$$

$$p_3 = p_2 \times \frac{T_3}{T_2} = 400 \times 10^3 \times \frac{300}{600}$$
$$= 200 \times 10^3 \text{ N/m}^2 = 0.2 \text{ MPa}$$

$W_{2\text{-}3} = 0$ as volume remains constant.

For the polytropic process 3–1;

$$p_3V_3^n = p_1V_1^n$$

$$n = \frac{\log_e \frac{p_1}{p_3}}{\log_e \frac{V_3}{V_1}} = \frac{\log_e \frac{2}{0.2}}{\log_e \frac{1}{0.2}}$$

$$= \frac{2.3026}{1.6094} = 1.433$$

$$W_{3\text{-}1} = \frac{p_3V_3 - p_1V_1}{n - 1}$$

$$= \frac{0.2 \times 10^6 \times 1 - 2 \times 10^6 \times 0.2}{1.433 - 1}$$

$$= -\ 461894 \text{ J} = -\ 461.894 \text{ kJ}$$

Net work done,

$$W_n = W_{1\text{-}2} + W_{2\text{-}3} + W_{3\text{-}1}$$
$$= 643.774 + 0 + (-\ 461.894)$$
$$= \mathbf{181.88\ kJ}$$

For a cyclic process

$$\oint \delta Q = \oint \delta W$$

∴ Heat transfer during the complete cycle

$$= \mathbf{181.88\ kJ}$$

EXAMPLE 6.55

A system consisting of 1 kg of an ideal gas at 6 bar pressure and 0.01 m³ volume executes a cyclic process comprising the following three distinct operations :

(i) reversible expansion to 0.05 m³ volume and 2 bar pressure, presuming pressure to be a linear function of volume ($p = a + bV$)

(ii) reversible cooling at constant pressure, and

(iii) reversible hyperbolic compression according to law pV = const.

This brings the gas back to the initial conditions.

Sketch the cycle on p – V diagram. Calculate the work done in each process stating whether it is done on or by the system and evaluate the net cyclic work and heat transfer.

Solution : *Process 1–2 (linear law)*

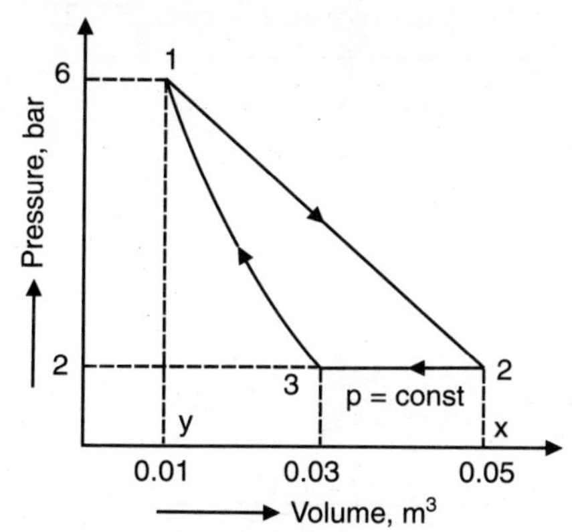

Fig. 6.17.

Let the linear law for the reversible expansion process 1–2 be of the form

$$p = a + bV$$

The values of constants a and b can be determined from the values of pressure and volume at the state points 1 and 2.

$$6 = a + 0.01\ b$$
$$2 = a + 0.05\ b$$

From these expressions

$$b = -\ 100 \text{ and } a = 7$$

∴ Work done during expansion process, 1–2,

$$W_{1-2} = \int_1^2 p dV = \int_1^2 (a + bV) dV$$

$$= 10^5 \int_{0.01}^{0.05} (7 - 100\,V) dV$$

$$= 10^5 \left| 7V - 100 \frac{V^2}{2} \right|_{0.01}^{0.05}$$

$$= 16000 \text{ Nm}$$

Alternatively : $W_{1\text{-}2}$ = area under the process line 1–2

= area of trapezium 1–2–x–y

$$= \frac{\text{sum of parallel lines}}{2} \times \text{base width}$$

$$= \left[\frac{6+2}{2} \times 10^5 \right] \times (0.05 - 0.01)$$

$$= \mathbf{16000\ Nm}$$

This work is done by the system.

Process 2–3 (constant pressure)

$$p_3 = p_2 = 2 \text{ bar}$$

The volume V_3 can be worked out from the hyperbolic compression 3–1. Thus

$$V_3 = \frac{p_1 V_1}{p_3} = \frac{6 \times 0.01}{2} = 0.03 \text{ m}^3$$

$$W_{2\text{-}3} = p_2 (V_3 - V_2) = 2 \times 10^5 (0.03 - 0.05)$$
$$= -\ 4000 \text{ Nm}$$

This is the work done on the system.

Process 3–1 (hyperbolic process)

The compression process 3–1 is isothermal in nature and accordingly

$$W_{3\text{-}1} = p_3 V_3 \log_e \frac{V_1}{V_3}$$

$$= 2 \times 10^5 \times 0.03 \log_e \frac{0.01}{0.03}$$

$$= -6588 \text{ Nm}$$

This is the work done on the system.

Net work done,

$$W_{net} = W_{1-2} + W_{2-3} + W_{3-1}$$
$$= 16000 - 4000 - 6588$$
$$= 5412 \text{ Nm} = \mathbf{5.412 \text{ kJ}}$$

For a cyclic process

$$\oint \delta Q = \oint \delta W$$

∴ Heat transfer during the complete cycle

$$= 5.412 \text{ kJ}$$

EXAMPLE 6.56

A banquet hall of 150 m³ volume is to be used for a formal dinner for 20 persons. Each person occupies 0.075 m³ of space and has an average heat transfer rate of 500 kJ/hr. If the air-conditioning system of the hall fails, determine the increase in internal energy and temperature of air in the hall during the first 15 minutes of failure. Assume that the hall is well-insulated and receives no heat from outside. Take air in the room as the system under consideration.

(b) If the hall and all its contents are considered as a system, how much will be the increase of internal energy of the system?

Solution : Volume of air = volume of room − volume occupied by persons

$$= 150 - 20 \times 0.075 = 148.5 \text{ m}^3$$

Assuming air to be at conditions of 1 bar and 20°C, the mass of air is

$$m = \frac{pV}{RT} = \frac{1 \times 10^5 \times 148.5}{287 \times (273 + 20)}$$

$$= 176.6 \text{ kg}$$

No work is done on or by the air and therefore, from the non-flow energy equation

$$dU = Q_{1-2} = 20 \times 500 = 10000 \text{ kJ/hr}$$

For 15 minutes period,

$$dU = 10000 \times \frac{15}{60} = \mathbf{2500 \text{ kJ}} \text{ (increase)}$$

Now, $dU = mc_v (T_2 - T_1)$

∴ Rise in temperature of air,

$$(T_2 - T_1) = \frac{2500}{0.718 \times 176.6} = \mathbf{19.717°C}$$

The specific heat at constant volume has been assumed to be equal to 0.718 kJ/kg K.

(*b*) When all the contents of the hall are considered as a system, there is no change in internal energy *i.e.*, $dU = 0$. Increase in internal energy of the air due to increase in its temperature is compensated by the decrease in internal energy of the persons present in the hall.

EXAMPLE 6.57

A cylinder closed at both ends is divided into two compartments by a frictionless and freely movable piston; both the piston and cylinder are perfectly insulated regarding heat interactions. One compartment contains air and the other contains nitrogen. Initially both air and nitrogen are at 1 bar pressure, 298 K temperature and 0.02 m³ volume. An electric heater on the side of air compartment heats the air till the volume of nitrogen is reduced to 0.01 m³. Evaluate the final temperature of air and heat supplied to it.

For air c_v = 715 J/kg K, R = 287 J/kg K and for nitrogen γ = 1.4.

Solution :

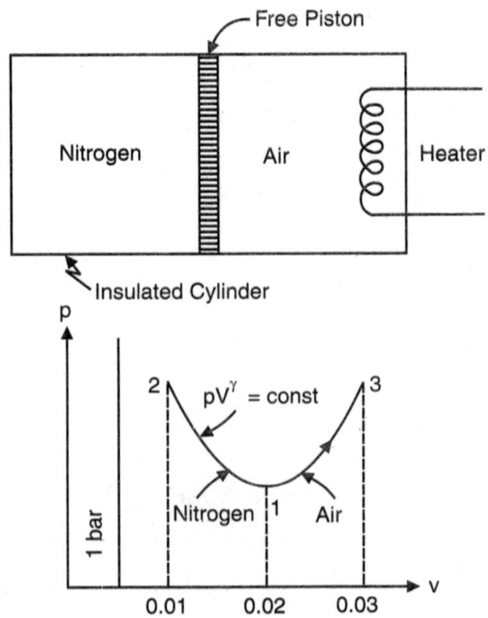

Fig. 6.18.

For nitrogen : Initial volume, $V_1 = 0.02\ m^3$
Final volume, $V_2 = 0.01\ m^3$

The following points need to be taken care of :

- Air expands on heating, pushes the piston and nitrogen gets compressed. The compression of nitrogen is adiabatic as the cylinder walls are thermally insulated.
- Piston moves freely and as such the final pressure in both the compartments will be same.

Invoking the adiabatic law,

$$p_1 V_1^\gamma = p_2 V_2^\gamma$$

$$p_2 = p_1\left(\frac{V_1}{V_2}\right)^\gamma$$

$$= 1\times\left(\frac{0.02}{0.01}\right)^{1.4} = 2.639 \text{ bar} = p_3$$

Work done on nitrogen,

$$W_{1\text{-}2} = \frac{p_1V_1 - p_2V_2}{\gamma - 1}$$

$$= \frac{1\times10^5\times0.02 - 2.639\times10^5\times0.01}{1.4-1}$$

$$= -1597.5 \text{ Nm}$$

For oxygen : Initial volume $V_1 = 0.02\ m^3$
Final volume, V_3 = total volume of cylinder − volume occupied by nitrogen
$= (0.02 + 0.02) - 0.01 = 0.03\ m^3$

Using the gas equation for the process undergone by air,

$$\frac{p_1V_1}{T_1} = \frac{p_3V_3}{T_3};\quad T_3 = \frac{p_3}{p_1}\times\frac{V_3}{V_1}\times T_1$$

$$= \frac{2.639}{1}\times\frac{0.03}{0.02}\times 293 = 1159.84 \text{ K}$$

Mass of air,

$$m = \frac{p_1V_1}{RT_1}$$

$$= \frac{1\times10^5\times0.02}{287\times293} = 0.02378 \text{ kg}$$

Change in internal energy,

$$(U_3 - U_1) = mc_v\,(T_3 - T_1)$$
$$= 0.02378 \times 715 \times (1159.84 - 293)$$
$$= 14739 \text{ Nm}$$

The work done by air equals the work done on nitrogen, and therefore, from non-flow energy equation

Heat supplied to air $= W_{1\text{-}3} + dU$
$= 1597.5 + 14739$
$= 16336.5$ Nm $=$ **163.36 kJ**

EXAMPLE 6.58

A fluid is contained in a cylinder by a spring loaded, frictionless piston so that the pressure in the fluid is a linear function of volume ($p = a + bV$). The internal energy of the fluid in kJ is given by the expression

$U = 32 + 3\,pV$ where p is in kPa and V is in m^3.

The initial and final pressures are 150 kPa and 350 kPa and the corresponding volumes are 0.02 m^3 and 0.05 m^3. Make calculations for the direction and magnitude of work and heat interactions.

Solution : Relation between pressure and volume, $p = a + bV$. The values of constants a and b can be determined from the values of pressure and temperature at the initial and final state points.

$$150 = a + 0.02\,b \quad\text{and}\quad 350 = a + 0.05\,b$$

From these expressions

$$a = 16.67 \text{ and } b = 6666.67$$

The law of expansion becomes

$$p = 16.67 + 6666.67\,V$$

(*a*) Work involved during the process

$$W_{1\text{-}2} = \int_1^2 p\,dV = \int_1^2 (16.67 + 6666.67\,V)\,dV$$

$$= 16.67\,(V_2 - V_1) + \frac{6666.67}{2}(V_2^2 - V_1^2)$$

$$= 16.67\,(0.05 - 0.02) + \frac{6666.67}{2}(0.05^2 - 0.02^2)$$

$$= \mathbf{7.5\ kJ}$$

Since $W_{1\text{-}2}$ is positive, the work has been done by the system; the magnitude being 7.5 kJ.

(*b*) Change in internal energy of the fluid during the process

$$U_2 - U_1 = (32 + 3\,p_2V_2) - (32 + 3\,p_1V_1)$$
$$= 3\,(p_2V_2 - p_1V_1)$$

$= 3\ (350 \times 0.05 - 150 \times 0.02)$
$= \mathbf{43.50\ kJ}$

Applying non-flow energy equation, $\delta Q = \delta W + dU$, the heat interaction is

$Q_{1-2} = 7.5 + 43.50 = \mathbf{50.50\ kJ}$

Since Q_2 is positive, the heat flows into the system; the magnitude being 50.5 kJ.

EXAMPLE 6.59

A piston-cylinder arrangement contains 1.5 kg of an ideal gas at 0.3 m^3 volume and 6 bar pressure. The gas expands polytropically and the expansion process continues till the pressure reduces to 1.2 bar. The internal energy is related to pressure p and volume V by the following identity

$$U = 2.4\ pV + 105$$

where U is in kJ, p is in kPa and V is in m^3. Considering expansion to be frictionless, evaluate the changes in internal energy, heat transfer to or from the system, and work done on or by the system. Take polytropic index n= 1.25.

(b) How the work transfer would be affected if 35 kJ of heat is transferred to the system during the expansion process ? Comment on the difference in work transfer for the two cases.

Solution : From the polytropic law of expansion;

$$V_2 = V_1\left(\frac{p_1}{p_2}\right)^{1/n}$$

$$= 0.3 \times \left(\frac{6}{1.2}\right)^{1/1.25} = 1.087\ \text{m}^3$$

From the given correlation,

$$U_1 = 2.4\ p_1V_1 + 105$$
$$U_2 = 2.4\ p_2V_2 + 105$$

∴ Change in internal energy,

$$U_2 - U_1 = 2.4\ (p_2V_2 - p_1V_1)$$

Given : $p_1 = 6$ bar $= 600$ kPa

$p_2 = 1.2$ bar $= 120$ kPa

$V_1 = 0.3\ \text{m}^3;\quad V_2 = 1.087\ \text{m}^3$

$$U_2 - U_1 = 2.4 \times (120 \times 1.087 - 600 \times 0.3)$$
$$= -\ 118.95\ \text{kJ}$$

The negative sign implies that internal energy of the system has decreased.

For the frictionless polytropic expansion, Work done,

$$W_{1-2} = \frac{p_1V_1 - p_2V_2}{n-1}$$

$$= \frac{6 \times 10^5 \times 0.3 - 1.2 \times 10^5 \times 1.087}{1.25 - 1}$$

$$= 198240\ \text{J} = 198.24\ \text{kJ}$$

The positive work means that work is done by the system.

From non-flow energy equation,

$$Q_{1-2} = W_{1-2} + dU$$
$$= 198.24 + (-\ 118.95) = 79.29\ \text{kJ}$$

Thus 79.29 kJ of heat flows into the system during the process.

(*b*) Since the initial and final conditions are same, the change in internal energy will be same as calculated above.

This stems from the fact that energy changes correspond only to end states.

$U_2 - U_1 = -\ 118.95$ kJ ; $Q_{1-2} = 35$ kJ

$$\therefore\ W_{1-2} = Q_{1-2} - (U_2 - U_1)$$
$$= 35 - (-\ 118.95) = 153.95\ \text{kJ}$$

The difference may be attributed to the fact that :

- Work depends on path as well as type of the process; whether with or without friction.
- Work is not always equal to $\int p\, dV$.

Obviously the process (*b*) involves friction and is an irreversible process.

EXAMPLE 6.60

The piston cylinder arrangement of an engine contains 0.04 m^3 of a gas initially at 24 bar pressure and 2000 K temperature. The gas is allowed to expand to a volume of 0.16 m^3 and a pressure of 4 bar. The variation of specific heats with temperature is of the form

$$c_p = a + 0.0001\ T \text{ and } c_v = b + 0.0001\ T$$

and at 2000 K, the values of c_p and c_v are 1.095 kJ/kg K and 0.795 kJ/ kg K respectively. Evaluate, (a) index of expansion and final temperature, (b) work done by the gas, (c) change in internal energy of the system, (d) heat flow to or from the gas, and (e) polytropic specific heat.

Solution : From the polytropic law,

$$p_1V_1^n = p_2V_2^n$$

$$n = \frac{\log_e \frac{p_1}{p_2}}{\log_e \frac{V_2}{V_1}} = \frac{\log_e \frac{24}{4}}{\log_e \frac{0.16}{0.04}}$$

$$= \frac{1.792}{1.386} = \mathbf{1.293}$$

$$T_2 = T_1\left(\frac{p_2}{p_1}\right)^{\frac{n-1}{n}}$$

$$= 2000 \times \left(\frac{4}{24}\right)^{\frac{1.293-1}{1.293}} = 1332.6 \text{ K}$$

(*b*) Work done,

$$W_{1-2} = \frac{p_1V_1 - p_2V_2}{n-1}$$

$$= \frac{24 \times 10^5 \times 0.04 - 4 \times 10^5 \times 0.16}{1.293 - 1}$$

$$= 109215 \text{ J} = 109.215 \text{ kJ}$$

(*c*) Gas constant at initial temperature of 2000 K

$$R = c_p - c_v = 1.095 - 0.795$$
$$= 0.3 \text{ kJ/kg K}$$

Mass of gas,

$$m = \frac{p_1V_1}{RT_1}$$

$$= \frac{24 \times 10^5 \times 0.04}{(0.3 \times 1000) \times 2000} = 0.16 \text{ kg}$$

At 2000 K

$1.095 = a + 0.0001 \times 2000\,;\ a = 0.895$

$0.795 = b + 0.0001 \times 2000\,;\ b = 0.595$

$$\therefore c_{p\ mean} = 0.895 + 0.0001 \times \frac{(2000 + 1332.6)}{2}$$

$$= 1.0616 \text{ kJ/kg K}$$

$$c_{v\ mean} = 0.595 + 0.0001 \times \frac{(2000 + 1332.6)}{2}$$

$$= 0.7616 \text{ kJ/kg K}$$

Change of internal energy,

$$dU = mc_{v\ mean}\,(T_2 - T_1)$$

$$= 0.16 \times 0.7616\,(1332.6 - 2000)$$
$$= \mathbf{-\ 81.33\ kJ}$$

(*d*) From steady flow energy equation,

$$Q_{1-2} = W_{1-2} + dU$$
$$= 109.215 - 81.33 = \mathbf{27.885\ kJ}$$

Heat interaction with the surroundings could also be evaluated from the relation :

$$Q_{1-2} = \frac{\gamma - n}{\gamma - 1} \times \text{polytropic work done}$$

$$\text{Here, } \gamma = \frac{c_{p\ mean}}{c_{v\ mean}} = \frac{1.0616}{0.7616} = 1.394$$

$$\therefore Q_{1-2} = \frac{1.394 - 1.293}{1.394 - 1} \times 109.215$$
$$= 28 \text{ kJ}$$

(*e*) Polytropic specific heat

$$c_n = c_v \frac{\gamma - n}{1 - n} = 0.7616 \times \frac{1.394 - 1.293}{1 - 1.293}$$

$$= \mathbf{0.2625\ kJ/kg\ K}$$

6.6. JOULE'S LAW

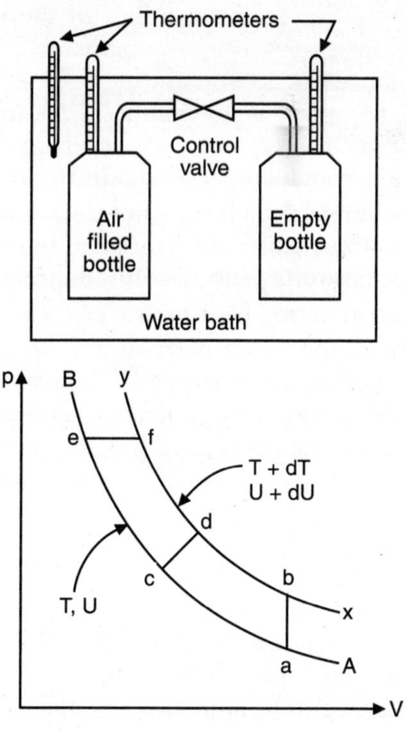

Fig. 6.19. Concept of internal energy (Joule's law)

Internal energy of an ideal gas is a function of absolute temperature only, $u = f(T)$, and is independent of pressure and volume changes.

Joule's experimental set up consists of two bottles, one empty and one filled with air, interconnected by means of a valve. Both the bottles are immersed in a water bath. The bath and the bottles are provided with temperature measuring devices. When the valve is opened, the air rushes to the empty bottle till pressure becomes same in both the bottles. Experimental observations made by Joule showed that both pressure and volume change during the process but there occurs no change in temperature either of the bath or in the bottles.

The system within the boundaries of the two bottles is a closed one and accordingly from the non-flow energy equation

$$\delta Q = \delta W + dU \; ; \quad dU = \delta Q - \delta W$$

Since there is no difference in temperature between the system (bottles) and surroundings (water bath), the heat interaction is zero. Also the flow of air from the filled to the empty bottle constitutes a free expansion process for which work interaction is zero. As such $dU = 0$, *i.e.*, there is no change in the internal energy of the system. Accordingly, the following conclusion was made by Joule :

The internal energy is not dependent on changes of pressure and volume of the system but changes only with absolute temperature.

This is referred to as Joule's Law of internal energy.

An isothermal curve, pV = constant, on pressure-volume diagram also represents a constant internal energy. Fig. 6.19 shows two such curves : curve AB with T = constant and curve XY with $(T + dT)$ = constant. If internal energy along AB is U = constant, then internal energy along XY can be taken as $(U + dU)$.

A shift from curve AB to curve XY can be carried out in numerous ways; three such processes are indicated in Fig. 6.19.

Constant volume process $a-b$; constant temperature process $c-d$; and constant pressure process $e-f$.

In all such processes, other than constant volume process, the heat supplied to a system does mechanical work in addition to increasing its internal energy. Accordingly, it is considered worthwhile to choose constant volume process for making calculations of changes in internal energy of a system. For a constant volume process displacement (mechanical) work is zero and, therefore, the non-flow energy equation can be written as

$$\delta Q = dU \; ; \quad dU = m\, c_v\, dT$$

or $$U_2 - U_1 = m\, c_v\, (T_2 - T_1)$$

which is the desired expression for change in internal energy occuring during any process; reversible or irreversible.

6.7. FREE EXPANSION

The expansion of gas against vacuum is called free or unresisted expansion process.

Consider a thermally insulated rigid tank which is initially divided into two compartments by a membrane. One compartment contains an ideal gas and the other is completely evacuated. When the membrane is removed or punctured, the gas expands and rushes into the evacuated compartment to occupy the entire volume of the container.

The salient aspects of this free expansion process are:

(*i*) The process is highly irreversible since the gas is eddying continuously during the process. The process takes place suddenly and is not quasi-static in nature. Although the initial and final states are in equilibrium, the changes in pressure and volume occuring at intermediate stages cannot be predicted and as such the free expansion process cannot be represented on $p-V$ diagram.

(*ii*) No work is done on or by the system as there is no resistance to be overcome. The integral $\int p\, dV$ has a value for the process but work done is zero ($\delta W = 0$).

(*iii*) Since the system is thermally insulated it does not exchange any heat with the surroundings ($\delta Q = 0$). The process is, therefore, adiabatic but irreversible.

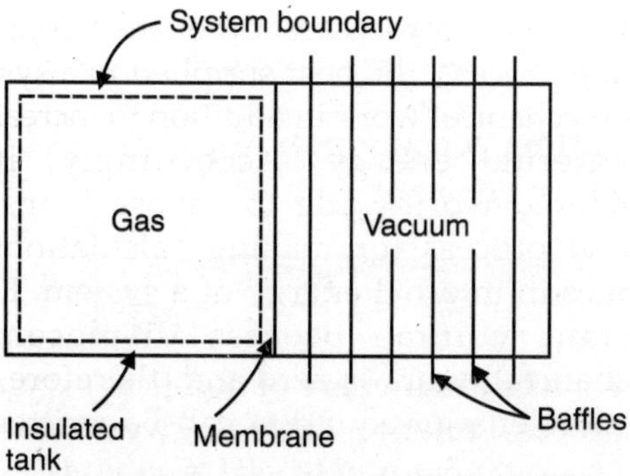

Fig. 6.20. Free expansion process

(*iv*) From the non-flow energy equation, $\delta Q = \delta W + dU$. Since both the work and heat interactions are zero, dU is also zero, *i.e.*, there is no change in internal energy of the gas. For an ideal gas, $U = f(T)$ only, and therefore,

$$dU = 0 = mc_v\,(T_2 - T_1)$$

Obviously there is no change in the temperature of the system, *i.e*, initial temperature is equal to the final temperature. Further, enthalpy change, $dH = mc_p dT$ and since there is no change in temperature, the enthalpy of the system would also remain same.

(*v*) Enthalpy equals the sum of internal energy and pressure volume product

$$H = U + pV = \text{constant}$$

Since internal energy U during the process is constant, the pressure-volume product must also remain constant. As such, the values of pressure and volume change in such a manner that their product remains constant.

(*vi*) It will be misnomer to call the free expansion process as adiabatic process (since $\delta Q = 0$) or isothermal process (since temperature remains constant). Both these processes can be carried out in reversed direction whereas the free expansion process is highly irreversible. It cannot be retraced back without the help of external energy, *i.e*, without an input of work. However, the free expansion process may be called as *constant internal energy* or *constant enthalpy process.*

(*vii*) If a large number of baffles are provided in the evacuated space and these are removed slowly one after the other, a number of intermediate states can be plotted and these will trace out a line of constant internal energy.

SALIENT POINTS

1. Energy can neither be created nor destroyed but only converted from one form to another.
2. For any closed system taken through a cyclic process, the net work delivered to surroundings is equal to the net heat taken from the surroundings. That is $\oint \delta Q = \oint \delta W$.
3. The energy of a system comprises of *PE*, *KE*, internal energy and terms accounting for electricity, magnetism and surface tension.
4. Internal energy is the microscopic energy, *i.e.*, the energy stored in the molecular motions. It is the energy stored in the molecular structure due to heat and work interactions.
5. The change in energy of a system depends on the end states and is independent of the path of process. It is a point function and an exact differential. Thus energy is a property of the system.
6. In a isolated system, the energy of the system remains constant.
7. Consider the non-flow energy equation:
$$\delta Q = \delta W + dU$$
   (*a*) δQ, δW and δU have to be in consistent units
   (*b*) Heat supplied to gas, work done by gas and increase in internal energy are taken as +ve quantities.
   (*c*) Heat received by gas, work done on gas and decrease in internal energy are taken as –ve quantities.
8. A perpetual motion machine of first kind is a hypothetical machine which produces continuous supply of work without absorbing energy from the surroundings. No machine can produce energy without

corresponding expenditure of energy. As such a perpetual motion machine of the first kind is impossible.

9. The internal energy of a gas is function of temperature only (Joule's law) and for any reversible process change in internal energy $(U_2 - U_1)$
$$= m\, c_v\, (T_2 - T_1)$$

10. Enthalpy is the property of the system defined by
$$H = U + pV$$

11. The following useful correlations are helpful in the solution of problems:
$$c_v = \left(\frac{du}{dT}\right)_v \; ; \; c_p = \left(\frac{dh}{dT}\right)_p$$
$$\frac{c_p}{c_v} = \gamma \quad \text{and} \quad (c_p - c_v) = R$$
Further for any polytropic process:
pV^n = constant
$$\frac{T_2}{T_1} = \left(\frac{p_2}{p_1}\right)^{\frac{n-1}{n}} = \left(\frac{V_1}{V_2}\right)^{n-1}$$
For a brief summary of non-flow expansion process, refet to Table A given below:

12. Change in enthalpy and heat interactions are different quantities. For the solitary case of constant pressure, change in enthalpy equals the heat interaction
$$Q_{1\text{-}2} = H_2 - H_1 = m\, c_p\, (T_2 - T_1)$$

13. The major characteristics of free expansion process are : unresisted, irreversible and no work done.

Table A: Summary of non-flow expansion process 1–2

Process	*Work output*	*Heat received*
1. Isochoric (constant volume)	$W_{1\text{-}2} = 0$	$Q_{1\text{-}2} = U_2 - U_1$ $= m\, c_v\, (T_2 - T_1)$
2. Isobaric (constant pressure)	$W_{1\text{-}2} = p\int dV$	$Q_{1\text{-}2} = H_2 - H_1$ $= m\, c_p\, (T_2 - T_1)$
3. Isothermal (constant temperature)	$W_{1\text{-}2} = p_1 V_1 \log_e \frac{V_2}{V_1}$	$Q_{1\text{-}2} = W_{1\text{-}2}$
4. Adiabatic	$W_{1\text{-}2} = \frac{p_1V_1 - p_2V_2}{\gamma - 1}$ $= \frac{mR\,(T_1 - T_2)}{\gamma - 1}$	$Q_{1\text{-}2} = 0$
5. Polytropic (pV^n = const)	$W_{1\text{-}2} = \frac{p_1V_1 - p_2V_2}{n - 1}$ $= \frac{mR\,(T_1 - T_2)}{n - 1}$	$Q_{1\text{-}2} = W_{1\text{-}2} + (U_2 - U_1)$ $= \frac{\gamma - n}{\gamma - 1} \times W_{1\text{-}2}$

REVIEW QUESTIONS

A. Conceptual and conventional questions.

1. What is mechanical equivalent of heat ? Write down its value when heat is expressed in kJ and work in Nm.
2. What is the essence of first law of thermodynamics ? Write down expressions for the first law applied to (*i*) a cycle (*ii*) a process.
3. Define internal energy. Is it a function of state or a process ?
4. What is the change of internal energy in a cyclic process equal to ?

5. Explain the perpetual motion machine of first kind.
6. Name and state the property introduced by the first law of thermodynamics.
7. Show that internal energy is a property of the system.
8. State the difference between standard symbols of E and U.
9. Define enthalpy. How is it related to internal energy ?
10. Define an isochoric/isobaric/isothermal process. How it is represented graphically on $p-V$ diagram.
11. Define the specific heats at constant volume and constant pressure.
12. Account for the existence of two values for the specific heat of a gas, and derive the relation between them and the characteristic gas constant.
13. Define an adiabatic process. Show that for a reversible adiabatic process of a given mass of perfect gas : pV^{γ} = constant.
14. How are the state parameters p, V and T related in a polytropic process ? What are the values of polytropic exponent for isochoric, isobaric, isothermal and adiabatic processes ?
15. How is the polytropic exponent determined and within what limits can it change ?
16. Develop the following expression for the heat transfer from a mass of gas undergoing reversible expansion process obeying the polytropic law, pV^n = constant.

$$Q_{1-2} = \frac{\gamma - n}{\gamma - 1} \times \text{polytropic work done}$$

17. Show that the work done per kilogram of a perfect gas during an adiabatic expansion for which pV^{γ} = constant is given by

$$W_{1-2} = \frac{R(T_1 - T_2)}{\gamma - 1}$$

where T_1 and T_2 are initial and final temperature and R is the characteristic gas constant.
18. Show that when an ideal gas is compressed in a piston-cylinder assembly accordingly to the polytropic law pV^n = constant, the heat rejected, increase in internal energy and work done are in the ratio

$(\gamma - n) : (n - 1) : (\gamma - 1)$
19. Show that polytropic specific heat is given by the expression :

$$c_n = -\, c_v \frac{\gamma - n}{n - 1}$$

What is the significance of its negative sign ?
20. State and explain Joule's law of internal energy for an ideal gas.
21. Point out the salient features of a free expansion process.
22. During the compression stroke of an engine, the work done on the working substance in the engine cylinder is 80 kJ/kg and the heat rejected to the surroundings is 50 kJ/kg. Determine the change in internal energy and state whether it is an increase or a decrease.

[Ans. 30 kJ/kg, increase**]**
23. A closed system passes from state 1 to state 2 while 100 kJ of heat is supplied to the system and 140 kJ of work is obtained. To restore the system to initial state 80 kJ of work need to be supplied to the system. What would be the magnitude and the sign of heat interaction during the return path ? Show the processes on property diagram.

[Ans. - 40 kJ**]**
24. A system executes a cyclic process during which there are four heat transfers as given below:

$$Q_{1-2} = 880 \text{ kJ} \quad Q_{2-3} = -100 \text{ kJ}$$
$$Q_{3-4} = -720 \text{ kJ} \quad \text{and} \quad Q_{4-1} = 200 \text{ kJ}$$

The work done during three of the processes are given as

W_{1-2} = 60 kJ; W_{2-3} = - 40 kJ and W_{3-4} = 80 kJ

What will be the work interaction during the process 4—1 ? Is it done on or by the system?

[Ans. W_{4-1} = 160 kJ**]**
25. The work and heat transfer per degree of temperature change for a system executing a non-flow process are given by

$$\frac{\delta W}{dt} = \frac{1}{7.5} \text{ kJ/°C and } \frac{\delta Q}{dt} = \frac{1}{2.5} \text{ kJ/°C.}$$

Calculate the change in internal energy of the system as its temperature increases from 120°C to 240°C.

[Ans. 32 kJ increase**]**
26. A closed system consists of water contained in a cylinder and being stirred by a paddle wheel. During the process, 35 kJ/hr of work was imparted to the system, and the internal energy increased to 145 kJ from its initial value of 120 kJ during one hour of stirring. Determine

the heat transfer. Is the temperature of the system rising or falling ?

[Ans. - 10 kJ; rising]

27. A perfectly insulated tank contains 25 kg of water which is stirred by a paddle wheel coupled to 1 kW motor. If the stirring continues for half an hour, calculate the rise in temperature of water.

For water, specific heat = 4.186 kJ/kg K

[Ans. 17.20 °C]

28. One kg of air is expanded slowly in a piston-cylinder arrangement from 6 bar and 0.015 m^3 to a final volume of $0.1 m^3$. During the process, heat is exchanged with the surroundings at a rate sufficient to make the process isothermal. Make calculations for final pressure, work done, change in internal energy and heat interaction.

[Ans. 0.9 bar, 170.75 kJ, 0, 170.75 kJ]

29. For an ideal gas, the values of specific heats at constant pressure and volume are stated to be 0.984 kJ/kg K and 0.728 kJ/kg K. Determine the characteristic gas constant and the ratio of specific heats for the gas.

One kg of this gas is heated at constant pressure from 25°C to 200°C. Make calculations for the heat added, work done and change in internal energy. Proceed to determine the pressure and final volume if the initial volume of the gas is 2 m^3.

[Ans. 0.256 kJ/kg K, 1.351, 172.2 kJ, 44.8 kJ, 127.4 kJ, 38.14 kN/m^2 and 3.175 m^3]

30. It is proposed to compress 5 kg of CO_2 from an initial state of 20 bar and 15°C to a final state of 80 bar pressure. If the compression is done in a reversible adiabatic manner, find the final temperature, the work transfer and the change of internal energy.

Treat CO_2 as a perfect gas and take $\gamma = 1.27$

[Ans. 387 K, - 346 kJ, 346 kJ]

31. A stationary system of 2 kg air, initially at pressure 5 bar and temperature 500 K, expands reversibly to pressure 2 bar following the polytropic law $pV^{1.25}$ = constant. Assuming that air behaves as a perfect gas, determine the final temperature, work done and heat transferred.

[Ans. 416 K, 193 kJ, 72 kJ]

32. 12 kg of air at an absolute pressure of 1.5 bar and 300 K is compressed at constant temperature until its volume reduces to 1/4 th of original value. Make calculations for initial and final parameters, work performed and heat transfer.

[Ans. V_1 = 6.9 m^3, V_2 = 1.725 m^3, p_2 = 6 bar, W_{1-2} = - 1435 kJ, Q_{1-2} = - 1435 kJ]

33. A mass of perfect gas at 800 kPa, 600 K is contained in a cylinder. Expansion of the form pV^n = constant takes place until the volume is five times the initial volume and the pressure is 100 kPa. Make calculations for the value of exponent n, and the work and heat transfers per unit mass.

Take $\gamma = 1.4$ and c_p = 14.5 kJ/kg K

[Ans. 1.29, 3194 kJ, 878.6 kJ]

34. A cylinder contains 1.25 m^3 of gas at 80 kPa and 375 K. The gas is compressed polytropically until its volume reduces to one-fourth of its initial volume and the final pressure becomes 480 kPa. Make calculations for the mass of gas, value of index n for compression, change in internal energy, work done and heat transfer to or from the gas.

Take R = 287 J/kg K and $\gamma = 1.4$

[Ans. 0.929, 1.28, 122.29 kJ, - 171.15 kJ, -48.78 kJ]

35. 2 kg of air at 480 kPa pressure and 450 K temperature expands polytropically upto a pressure of 120 kPa. Find the initial volume, final volume and temperature, work done and heat transfer.

Take : $n = 1.3$; c_v = 0.715 kJ/kg K

and R = 0.3 kJ/kg K.

[Ans. 0.5625m^3; 1.634 m^3 and 326 K; 246.4 kJ and 61.6 kJ]

36. 0.15 m^3 of air at 6 bar and 313.6 K is contained in a cylinder fitted with a frictionless piston. The gas is allowed to expand in a quasi-equilibrium manner following polytropic law $pV^{1.3}$ = constant. If the final volume is 1 m^3, find (*i*) final pressure and temperature, (*ii*) workdone, change in internal energy and heat exchange.

[Ans. 0.509 bar, 177.35 K, 130.35 kJ, -97.9 kJ and 32.65 kJ]

37. 0.45 kg of air at 450 K expands isentropically to three times of its original volume and the temperature falls to 290 K. If 50 kJ of work is done by the air during this expansion process, evaluate the adiabatic exponent, gas constant and specific heats at constant pressure and temperature.

[Ans. 1.40, 0.278 kJ/kg K, 0.695 kJ/kg K, 0.973 kJ/kg K]

38. Air initially at 300 kPa pressure and 0.02 m^3 volume expands according to the pressure-volume relationship pV = constant until the pressure reduces to 150 kPa. Subsequently it is compressed at constant pressure to its original volume of 0.02 m^3. Sketch the process on $p-V$ diagram and make calculations for the work interaction per unit mass of air.

[Ans. 1.159 kJ/kg**]**

39. A fixed quantity of an ideal gas can be expanded in two different ways between the same initial and final states :
    (*a*) An expansion process during which the work output is 50 kJ and the heat transfer is 40 kJ. Evaluate the change in internal energy.
    (*b*) An expansion process during which heat received is 30 kJ. Evaluate the work transfer produced.

[Ans. –10 kJ, 40 kJ**]**

40. Consider a stationary fluid system that undergoes through a cycle comprising the following reversible processes :
    $a-b$: isochoric with 250 kJ/kg of internal energy
    $b-c$: adiabatic expansion to original pressure with work output of 80 kJ/kg.
    $c-d$: isobaric compression to original volume with heat rejection of 200 kJ/kg.
    Show that the data is consistent with first law of thermodynamics and determine the net work output and thermal efficiency of the cycle.

[Ans. 50kJ, 20%**]**

B. Fill in the blanks with appropriate word/words.

1. For a quasi-static process, the work output is found to be .... and the work input ....
2. Heat supplied to a system equals the workdone in case of a non-flow process carried out ....
3. For a closed system the difference between heat added to and work done by the system is equal to ....
4. Internal energy is a .... of the system and for an ideal gas it is a function of .... alone.
5. The internal energy of an ideal gas is a function of its absolute temperature only. This statement refers to ....
6. The identity $\delta q = p\,dv + du$ is true for .... process and .... system.
7. The change in the enthalpy of a system equals the heat supplied under constant .... conditions.
8. For an adiabatic process, the pressure and volume are related as pv^γ = constant. The exponent γ represents the ratio of .... and .... and for air its value equals ....
9. The polytropic process pv^n = constant for which the index n equals infinity represents a .... process
10. A perpetual motion machine of first kind is represented by a machine that ....

Answers

1. maximum, minimum 2. isothermally 3. internal energy 4. property, temperature 5. Joule's law 6. reversible, closed 7. pressure 8. specific heat at constant pressure, specific heat at constant volume, 1.4 9. constant volume 10. produces a continuous supply of work without absorbing any energy from the surroundings

C. Multiple choice questions.

1. First law of thermodynamics refers to conservation of
   (*a*) mass (*b*) momentum
   (*c*) energy (*d*) force
2. Which of the following statements does not represent the essence of first law of thermodynamics?
   (*a*) heat input to a system equals the net work done plus change in internal energy
   (*b*) the sum of heat and work effects will be zero if the system undergoes a cycle
   (*c*) heat input cannot be more than the work output
   (*d*) for an isolated system, energy remains constant
3. In the context of a closed non-flow system, the net energy transferred as heat and work equals the change in
   (*a*) enthalpy (*b*) internal energy
   (*c*) entropy (*d*) flow work
4. The internal energy of an ideal gas is a function of its absolute temperature only.
   This statement refers to
   (*a*) Avogadro's law (*b*) Maxwell law
   (*c*) Joule's law (*d*) Regnault's law
5. In a general compression process, 1 kJ of mechanical work is supplied to 2 kg of fluid

and 400 J of heat is rejected to the cooling jacket. The change in specific internal energy would be

(*a*) 700 J (*b*) 350 kJ (*c*) 300 kJ (*d*) 600 kJ

6. A paddle wheel used for stirring a liquid contained in a tank supplied 5000 kJ of work and during the stirring operation the tank lost 1500 kJ of heat to the surroundings. If the tank and liquid are considered as a system, the change in its internal energy will be

(*a*) 1500 kJ (*b*) 3500 kJ
(*c*) 5000 kJ (*d*) 6500 kJ

7. When a system is taken from state point *a* to state point *b* along the path *a-c-b*, 80 kJ of heat flows into the system and it does 30 kJ of work.

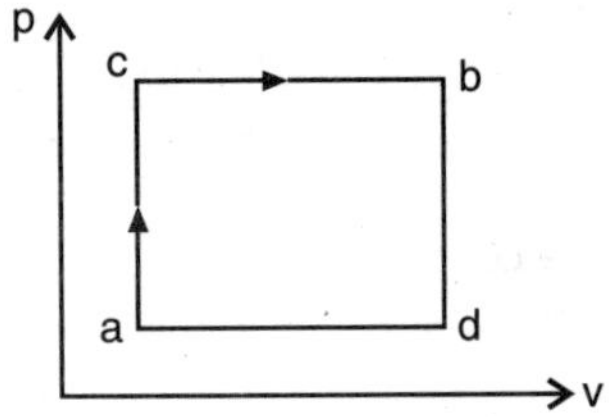

How much heat will flow into the system along path *a-d-b* if the work done by it along this path is 10 kJ ?

(*a*) 20 k (*b*) 40 kJ
(*c*) 60 kJ (*d*) 80 kJ

8. Refer to cycle *abcda* executed by an ideal gas. The net work done during the cycle equals

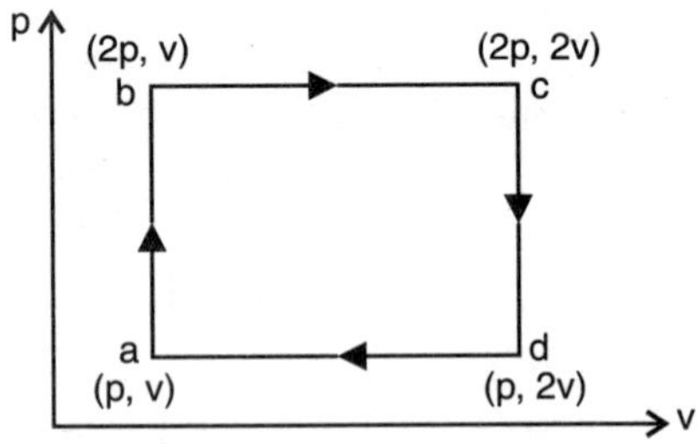

(*a*) zero (*b*) pv
(*c*) $2pv$ (*d*) $4pv$

9. Heat is being supplied to air in a cylinder fitted with a frictionless piston held by a constant weight. The process is

(*a*) isothermal (*b*) adiabatic
(*c*) isobaric (*d*) isochoric

10. An ideal gas of mass *m* in state A expands to state point B via three paths. If Q_1, Q_2 and Q_3 represent the heats absorbed by the gas along these paths, then

(*a*) $Q_1 < Q_2 < Q_3$ (*b*) $Q_1 > Q_2 > Q_3$
(*c*) $Q_1 < Q_2 > Q_3$ (*d*) $Q_1 > Q_2 < Q_3$

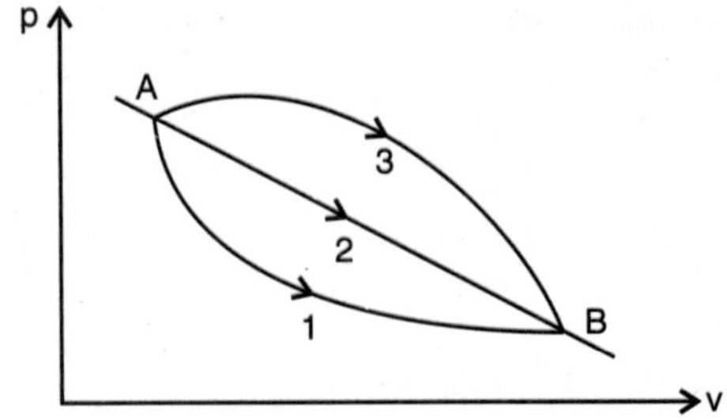

11. Which aspect is true in the context of an isochoric process ?
(*a*) no transfer of mass and energy
(*b*) product of pressure and volume remains constant
(*c*) free expansion takes place
(*d*) work output from the system is zero

12. An ideal gas is made to go from state A to state B in the following two different ways:
(*i*) an isobaric and then an isochoric process
(*ii*) an isochoric and than an isobaric process

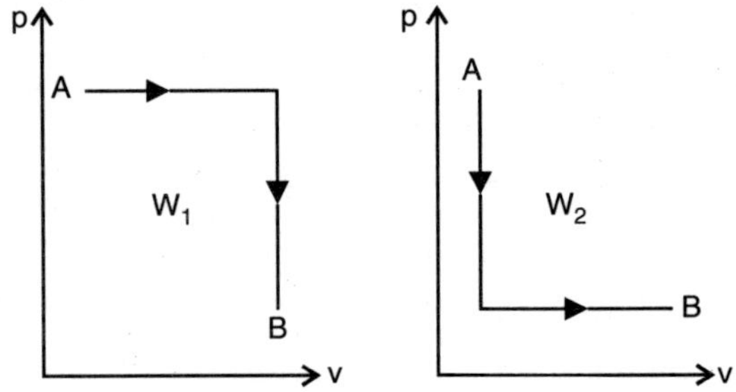

The work done by the gas in two cases are W_1 and W_2 respectively. Then

(*a*) $W_1 = W_2$ (*b*) $W_1 > W_2$
(*c*) $W_1 < W_2$ (*d*) $W_1 = 1/4\ W_2$

13. The change in internal energy of a closed system during a certain reversible process equals the heat transferred if the process takes place at
(*a*) constant volume
(*b*) constant pressure
(*c*) constant temperature
(*d*) such a situation never arises

14. Which parameter remains constant during a reversible isothermal process ?
(*a*) rate of heat exchange
(*b*) internal energy
(*c*) enthalpy
(*d*) entropy

15. Which of the following parameter changes during an isothermal process ?
    (*a*) pressure
    (*b*) temperature
    (*c*) internal energy
    (*d*) (pressure × specific volume)

16. Work output from a system is at the expense of internal energy in a non-flow process carried out
    (*a*) at constant pressure
    (*b*) at constant volume
    (*c*) adiabatically
    (*d*) polytropically

17. During an adiabatic process
    (*a*) temperature of gas is varied but the pressure is kept constant
    (*b*) total kinetic energy of the gas molecules remains constant
    (*c*) gas is compressed at constant temperature and there is no change of entropy
    (*d*) expansion and compression of a gas occur without any exchange of heat with the surroundings

18. For an adiabatic process, the temperature and pressure are related as, T_2/T_1 =
    (*a*) $(p_2/p_1)^{\gamma-1}$ (*b*) $(p_2/p_1)^{1/\gamma-1}$
    (*c*) $(p_2/p_1)^{\frac{\gamma-1}{\gamma}}$ (*d*) $(p_2/p_1)^{\gamma/\gamma-1}$

19. During a general polytropic expansion process characterised by pv^n = constant, the work done is equal to
    (*a*) $p_1v_1 - p_2v_2$ (*b*) $p_1v_1 \log_e \frac{v_2}{v_1}$
    (*c*) $\frac{p_1 v_1 - p_2 v_2}{n-1}$ (*d*) $\frac{p_1 v_1 - p_2 v_2}{n+1}$

20. The heat absorbed or rejected during a polytropic process (pv^n = constant) is
    (*a*) $\frac{\gamma-n}{\gamma-1} \times W$ (*b*) $\frac{\gamma-n}{\gamma+1} \times W$
    (*c*) $\frac{\gamma+1}{\gamma-1} \times W$ (*d*) $\frac{\gamma-n}{n-1} \times W$
    where *W* is the work done during the process.

21. For a closed system undergoing an expansion process, the work output
    (*a*) increases with increase in *n*
    (*b*) increases with decrease in *n*
    (*c*) is maximum when $n = 0$
    (*d*) is independent of *n*
    where *n* is the index of expansion in the polytropic equation pv^n = constant

22. Consider a perfectly insulated container which has two compartments separated by a partition. One compartment contains a perfect gas and the other compartment is in a state of perfect vaccum. Upon removal of the partition, the gas expands and fills the whole space. Such an expansion process is referred to as
    (*a*) adiabatic expansion
    (*b*) isothermal expansion
    (*c*) throttling
    (*d*) free or unrestricted expansion

23. Which aspect does not pertain to a free expansion process ?
    (*a*) no gain or loss of heat
    (*b*) work done is zero
    (*c*) pressure remains constant
    (*d*) no change in the temperature of the system

24. When a gas or vapour is allowed to escape through a minute aperature (i.e. leakage of steam through the safety valve of a pressure cooker), the leakage process is called
    (*a*) constant pressure (*b*) adiabatic
    (*c*) throttling (*d*) isothermal

25. Air is being forced by the bicycle pump into a tyre against a pressure of 4.5 bar. A slow downward movement of the piston can be approximated as
    (*a*) isobaric (*b*) adiabatic process
    (*c*) isothermal process (*d*) throttling process

26. Which parameter can be considered to remain constant if the value of exponent *n* in the polytropic equation pv^n = constant takes a unit value ?
    (*a*) enthalpy (*b*) entropy
    (*c*) internal energy (*d*) pressure or volume

27. A system at a given state undergoes change through the following expansion processes to reach the same final volume
    1. isothermal 2. isobaric
    3. adiabatic (γ = 1.4)
    4. polytropic (n = 1.2)
    The correct ascending order of the work output in these four processes is

(*a*) 1, 2, 4, 3 (*b*) 1, 4, 3, 2
(*c*) 4, 1, 3, 2 (*d*) 4, 1, 2, 3

28. Give a correct match between the processes listed in column A and the value of exponent n, in the relationship pv^n = constant, listed in column B.

Column A	Column B
(*i*) adiabatic	(*a*) $n = 0$
(*ii*) isobaric	(*b*) $n = 1$
(*iii*) isochoric	(*c*) $n = \frac{c_p}{c_v} = \gamma$
(*iv*) isothermal	(*d*) $\pm \infty$

29. Identify the statement that does not represent the concept of the first law of thermodynamics
(*a*) the heat input can not be more than work output
(*b*) the total energy remains constant for an isolated system
(*c*) the sum of all heat and work interactions will be zero if the system undergoes a cyclic process
(*d*) the heat absorbed by a system equals the net work output plus gain in internal energy

30. During a certain reversible process, the change in internal energy of a closed system is stated to be equal to the net heat transferred. Such a condition exists when the process is executed
(*a*) adiabaticy (*b*) isobarically
(*c*) isochorically (*d*) isothermally

31. The contents of a well-insulated tank are heated by a resistor of 25 Ω in which 10 A current is flowing. The tank along with its contents are considered as a thermodynamic system, and the work done by the system and the heat transfer to the system are positive. The rates of heat (Q), work (W) and change in internal energy (dU) during the process in kW are
(*a*) $Q = 0$, $W = -2.5$ and $dU = +2.5$
(*b*) $Q = +2.5$, $W = 0$ and $dU = +2.5$
(*c*) $Q = -2.5$, $W = 0$ and $dU = -2.5$
(*d*) $Q = 0$, $W = 2.5$ and $dU = -2.5$

32. Which of the following statements represents the tenet of perpetual motion machine of first kind ?
(*a*) a machine that has the highest performance
(*b*) a machine that converts input heat completely into work
(*c*) a machine that can transfer heat from a low temperature reservoir to a high temperature reservoir without taking any work input
(*d*) a machine that generates its own driving power and gives the work output

Answers

1. (*c*)	2. (*c*)	3. (*b*)	4. (*c*)	5. (*c*)
6. (*b*)	7. (*c*)	8. (*b*)	9. (*c*)	10. (*a*)
11. (*d*)	12. (*b*)	13. (*a*)	14. (*b*)	15. (*a*)
16. (*c*)	17. (*d*)	18. (*c*)	19. (*c*)	20. (*a*)
21. (*c*)	22. (*d*)	23. (*c*)	24. (*c*)	25. (*c*)
26. (*c*)	27. (*a*)	28. *i-c* ; *ii-a* ; *iii-d* ; *iv-b*		
29. (*a*)	30. (*c*)	31. (*a*)	32. (*d*)	

Chapter **7**

Steady Flow Energy Equation
(Flow Process and Ist Law)

Learning objectives : Attention hes been focussed in this chapter to enable the reader to :

- *explain flow process, control volume and flow work*
- *make distinction between steady and unsteady flow processes*
- *set up the steady flow energy equation*
- *apply the steady flow energy equation to flow processes in turbines, pumps and compressors, boilers and condensers*
- *explain throttling process and its importance in engineering*
- *deal with unsteady flow processors, tank filling and emptying processes*

7.1. FLOW RATE AND CONTINUITY EQUATION

Consider flow of an ideal fluid through a tube. At the inlet section 1−1, the flow characteristics are : tube cross-sectional area A, average fluid density ρ and the mean flow velocity V. The corresponding parameters at the exit section 2−2 are $(A + dA)$, $(\rho + d\rho)$ and $(V + dV)$.

The mass flow entering the tube at section 1−1 during time interval dt is given by $(\rho A V dt)$. During the same time interval a quantity of mass flow $(\rho + d\rho)\,(A + dA)\,(V + dV)dt$ out flows from the exit section. The fluid mass that accumulates between the two sections is given by

$$dm = \rho A V\,dt - (\rho + d\rho)\,(A + dA)\,(V + dV)dt$$

Simplification yields

$$\frac{dm}{dt} = -\,(AV\,d\rho + \rho V\,dA + \rho A\,dV)$$

For a steady flow, the time rate of change of mass is zero. That gives

$$AV\,d\rho + \rho V\,dA + \rho A\,dV = 0$$

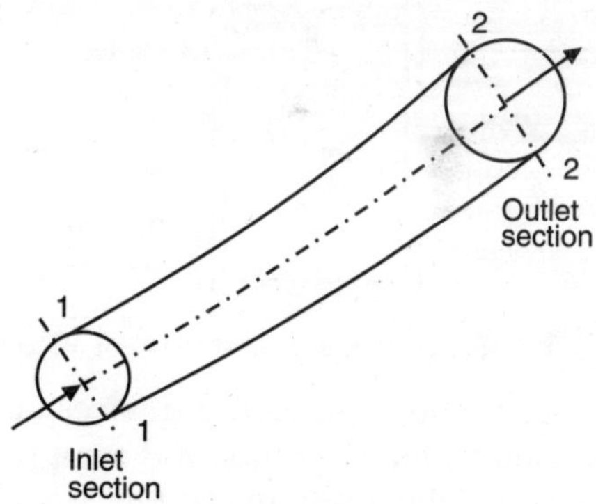

Fig. 7.1. Steady flow through a tube

Dividing throughout by $\rho A V$, one obtains

$$\frac{d\rho}{\rho} + \frac{dA}{A} + \frac{dV}{V} = 0$$

$$\text{or}\quad d(\rho A V) = 0$$

$$\rho A V = \text{constant} \qquad ...(7.1)$$

Evidently the mass of fluid per unit time passing through a section of a tube is constant.

The equation 7.1 is referred to as the continuity equation and it represents the mass conservation. For any two sections 1–1 and 2–2,

$$\dot{m}\text{(mass flow rate)} = \rho_1 A_1 V_1 = \rho_2 A_2 V_2$$

Since specific volume v is the reciprocal of density ρ,

$$\dot{m} = \frac{A_1V_1}{v_1} = \frac{A_2V_2}{v_2}$$

The mass flow rate $\dot{m}$ is the quantity ρAV or AV/v which passes a reference point per unit time ; its value is expressed in kg/s.

7.2. FLOW PROCESS AND CONTROL VOLUME

A *flow process* constitutes an open system through which the working fluid enters and leaves from the surface of a system. In addition there may also be energy interactions in the form of heat and work, internal energy, gravitational potential energy and kinetic energy.

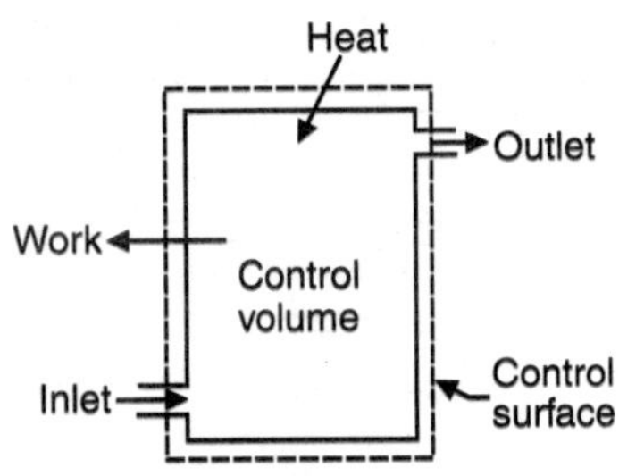

Fig. 7.2. Flow process and control volume

Many of the engineering devices such as engines and turbines, pumps and compressors, boilers and heat exchangers etc. deal with flowing fluids and their analysis is done with the concept of control volume and control surface. An imaginary envelope is supposed to exist around the piece of equipment. This envelope is called *control surface* and the space thus bounded is called *control volume*. In contrast to system boundary, the control surface is fixed in position and does not change its shape or orientation during the flow process. The concept of control volume helps to relate the energy interaction (work and heat) with average change of state of fluid from entrance to exit section.

7.3. FLOW WORK OR FLOW ENERGY

Flow energy or flow work refers to the work which a fluid does when it moves in or out of any passage. Essentially, flow work represents the work required either to push a certain quantity of fluid in or out of the passage against the existing pressure.

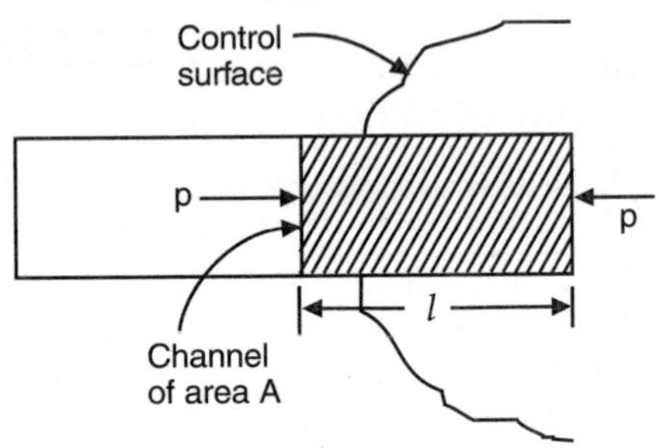

Fig. 7.3. Flow work

Consider a unit mass of fluid about to enter the control volume through a channel of cross-sectional area A. If p is the pressure acting uniformly at the entrance section, the force acting on the element is pA. Let l be the distance through which the fluid mass is pushed in, then

Work done = force × distance moved = pAl

The quantity Al represents the volume of fluid which is about to enter the control volume.

$\therefore$ Flow work per unit mass $= p\,v$

Thus, flow work is expressed in terms of properties of the system; it represents the product of pressure and specific volume of the flowing fluid. Flow work is supplied by a pump located in the surroundings. Flow work is supplied to the plant when mass enters, and to the environment (surroundings) when the mass leaves the plant.

7.4. STEADY AND UNSTEADY FLOW PROCESS

A flow process is said to be *steady* when the fluid parameters at any point of the control volume remain constant with respect to time; the parameters may, however, be different at different cross-sections of the flow passage. This means that quantities like velocity, pressure, temperature etc. are functions only of location and do not vary with time. If any one of these parameters is represented by the symbol P, then mathematically a steady flow is defined as $\frac{\partial P}{\partial t} = 0$, *i.e.*, the time rate of change of fluid characteristics at a position is zero.

A steady flow process is characterised by the following conditions in a control volume

- no change in the mass of fluid in the control volume; mass inflow equals the mass outflow.
- the fluid is uniform in state, composition and velocity both at entrance to and at exit from the control volume.
- the heat and work interactions at the control surface are at a constant rate.
- the state of fluid at any point is same at all times.
- no change in chemical composition of the fluid within the control volume; change in the chemical energy is not involved.

Flow process is *unsteady* when the conditions vary with respect to time. Unsteadiness refers to changing parameters with the passage of time at a position in the control volume. Symbolically $\frac{\partial P}{\partial t} \neq 0$ defines unsteady flow process.

7.5. STEADY FLOW ENERGY EQUATION (SFEE)

Consider flow of fluid through a generalised system (Fig. 7.4). During a small time interval *dt* there occurs a flux (flow) of mass and energy into the fixed control volume; entry is at section 1 and exit occurs at section 2.

The fluid enters the control volume at section 1 with average velocity V_1, pressure p_1, specific volume v_1 and internal energy u_1. The corresponding values at the exit section 2 are V_2, p_2, v_2 and u_2. Further during the fluid flow between the two chosen sections, heat q and mechanical or shaft work w_s may also cross the control surface.

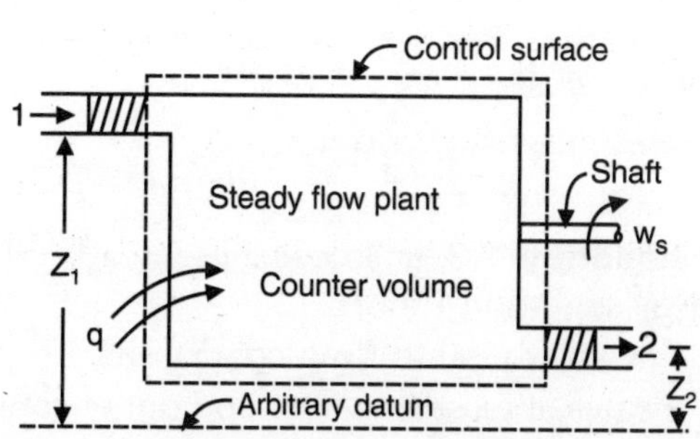

Fig. 7.4. Control volume for first law analysis of a steady flow process

The following species of energy are taken into account while drawing up the energy balance :

(*i*) internal energy stored by the fluid,

(*ii*) kinetic and potential energies,

(*iii*) flow work which is the work required to push the fluid in or out of the control volume.

(*iv*) heat and mechanical (shaft) work which cross the control volume.

Since energy can neither be created nor destroyed, energy balance in the symbolic form may be written as :

$$m_1\left(u_1 + p_1v_1 + \frac{V_1^2}{2} + gz_1\right) + Q = m_2\left(u_2 + p_2v_2 + \frac{V_2^2}{2} + gz_2\right) + W_s \quad \text{...(7.2)}$$

Equation 7.2 is the *general energy equation* and is equally applicable to compressible and incompressible fluids, liquids and gases, ideal and real fluids.

Assuming continuity of mass (which means no accumulation or diminution of matter within the machine) the mass of working substance entering the machine is equal to the mass of the working substance leaving it, *i.e.*,

$$m_1 = m_2 = m. \text{ Then}$$

$$m\left(u_1 + p_1v_1 + \frac{V_1^2}{2} + gz_1\right) + Q = m\left(u_2 + p_2v_2 + \frac{V_2^2}{2} + gz_2\right) + W_s$$

By definition : $h = u + pv$. Therefore,

$$m\left(h_1 + \frac{V_1^2}{2} + gz_1\right) + Q = m\left(h_2 + \frac{V_2^2}{2} + gz_2\right) + W_s \quad ...(7.3)$$

Equation 7.3 is known as steady flow energy equation (SFEE)

For steady state flow conditions,

- rates of mass flow into and out of control volume are equal and constant with respect to time.
- no variation in rates at which heat and work cross the surface.
- state, velocity and elevation of fluid mass entering and leaving the surface do not vary with time.

The steady flow energy equation can be written on the basis of unit mass or on the basis of unit time.

1. SFEE on unit mass basis

$$m\left(h_1 + \frac{V_1^2}{2} + gz_1\right) = m\left(h_2 + \frac{V_2^2}{2} + gz_2\right) + W_s$$

Each term inside the brackets is expressed per unit mass and has units of energy (J/kg).

For unit mass,

$$\left(h_1 + \frac{V_1^2}{2} + gz_1\right) + q = \left(h_2 + \frac{V_2^2}{2} + gz_2\right) + w_s$$

All the terms represent energy flow per unit mass of fluid (J/kg).

Note : All the quantities must be expressed in consistent units.

- If Q is expressed in J, W_s in J, h in J/kg, V in m/s, z in m, g in m/s^2 and m in kg, then

$$m\left(h_1 + \frac{V_1^2}{2} + gz_1\right) + Q = m\left(h_2 + \frac{V_2^2}{2} + gz_2\right) + W_s$$

- If Q is expressed in kJ, W_s in kJ, h in kJ/kg, V in m/s, z in m, g in m/s^2 and m in kg, then

$$m\left(h_1 + \frac{V_1^2}{2\times 10^3} + \frac{gz_1}{10^3}\right) + Q = m\left(h_2 + \frac{V_2^2}{2\times 10^3} + \frac{gz_2}{10^3}\right) + W_s$$

2. SFEE on unit time basis

$$\dot{m}\left(h_1 + \frac{V_1^2}{2} + gz_1\right) + \dot{Q} = \dot{m}\left(h_2 + \frac{V_2^2}{2} + gz_2\right) + \dot{W}_s$$

where all terms represent energy flow per unit time (J/s).

$\dot{m}$ = rate of fluid flow across the surface in kg/s

$\dot{Q}$ = rate of heat transfer across the control surface in J/s

$\dot{W}_s$ = rate of shaft work transfer across the control surface in J/s

and all the terms in brackets have units of J/kg.

When there are a number of streams entering and leaving the control surface, the SFEE can be written in the following form :

$$\dot{Q} - \dot{W}_s = \sum_{\text{outlet}} \dot{m}\left(h + \frac{V^2}{2} + gz\right) - \sum_{\text{inlet}} \dot{m}\left(h + \frac{V^2}{2} + gz\right)$$

SFEE on unit mass basis is usually more convenient when only a single stream of fluid enters and leaves a control volume. When more than one fluid streams is involved, analysis is made by using SFEE on unit time basis.

7.6. ENGINEERING APPLICATIONS OF STEADY FLOW ENERGY EQUATION

The steady flow energy equation applies to flow processes in many of the engineering plants such as mechanical power generation in engines and turbines; heat transfer processes in boilers and condensers; power consumption in pumps and compressors; pipe line flows, and flows through nozzles and diffusers. In certain flow processes, some of the energy terms in SFEE are negligibly small and can be omitted without much error. However, it is appropriate to write the complete equation first and then make suitable changes regarding certain terms which are unnecessary.

7.6.1. Nozzle and diffuser

A nozzle is a device for increasing the velocity of a steadily flowing stream at the expense of its pressure drop. The enthalpy of the fluid decreases and the flow of fluid is accelerated.

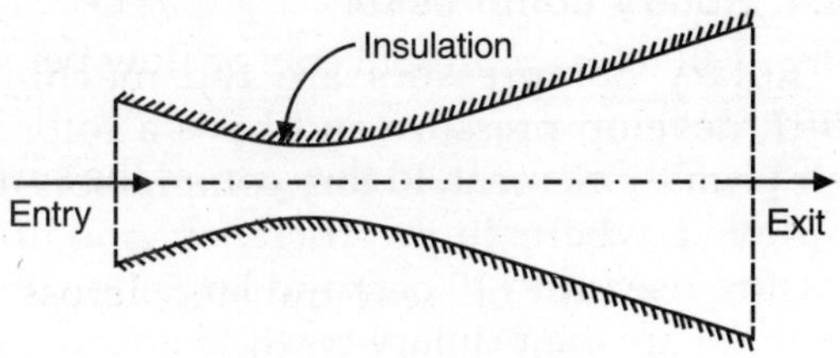

Fig. 7.5. Nozzle

The flow through a nozzle is characterised by the following features:

- no shaft work ; $W_s = 0$
- walls are thermally insulated, *i.e.*, the flow is adiabatic ; $Q = 0$
- change in elevation between the entry and exit point is negligible ;

$$(z_1 - z_2) = 0$$

With these stipulations, the SFEE reduces to

$$\frac{V_2^2 - V_1^2}{2} = (h_1 - h_2)$$

Thus, increase in kinetic energy
= decrease of enthalpy

From continuity of mass flow,

$$m = \frac{A_1 V_1}{v_1} = \frac{A_2 V_2}{v_2}; \; V_1^2 = V_2^2 \left(\frac{A_2 v_1}{A_1 v_2} \right)^2$$

$$\therefore \; V_2^2 - V_2^2 \left(\frac{A_2 v_1}{A_1 v_2} \right)^2 = 2\,(h_1 - h_2)$$

Upon simplification

$$V_2 = \left[\frac{1}{1 - (A_2 v_1 / A_1 v_2)^2} \right]^{1/2} \sqrt{2(h_1 - h_2)}$$

When the inlet velocity or velocity of approach V_1 is small compared to the exit velocity V_2, we get

$$V_2 = \sqrt{2(h_1 - h_2)}$$

If the working fluid is a perfect gas, then

$$pv = RT; \; h = c_p T; \; \frac{T_2}{T_1} = \left(\frac{p_2}{p_1} \right)^{(\gamma - 1)/\gamma}$$

and the flow equations can be expressed as

$$\frac{V_2^2 - V_1^2}{2} = \text{change in kinetic energy}$$
$$= (h_1 - h_2)$$
$$= c_p\,(T_1 - T_2)$$
$$= c_p T_1 \left[1 - \left(\frac{p_2}{p_1} \right)^{(\gamma - 1)/\gamma} \right]$$

A *diffuser* is a passage of varying cross-section that serves to achieve reduction in velocity of the flowing fluid. When the flow is subsonic (velocity is less than the sonic speed) the cross-section of the diffusing passage diverges from entry to exit. However, when the flowing fluid is in the supersonic regime (velocity is more than the sonic value) the cross-section of the diffusing duct decreases down from entry to exit.

7.6.2. Boiler

Boiler is an equipment/apparatus used for the generation of steam. Thermal energy released by combustion of fuel is transferred to water which vaporises and gets converted into steam at the desired pressure and temperature. The steam thus generated is used for :

(*i*) producing mechanical work by expanding it in steam engine or steam turbine,

(*ii*) heating the residential and industrial buildings in cold weather, and

(*iii*) performing certain processes in the sugar mills, chemical and textile industries.

The characteristic features for flow through a boiler are :

- no shaft work ; $W_s = 0$

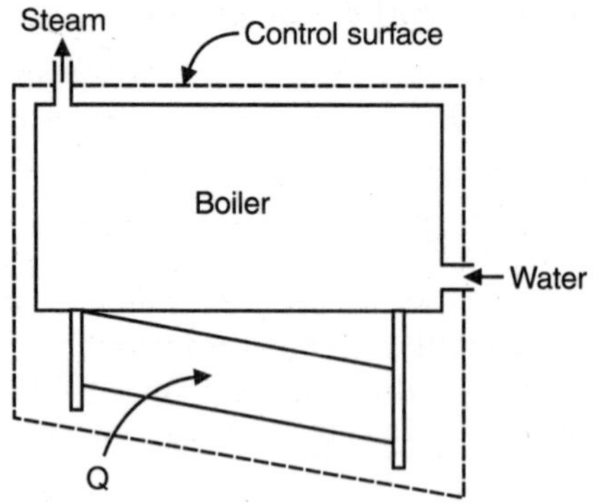

Fig. 7.6. Boiler

• not much changes in kinetic energy of fluid ; $\frac{V_2^2 - V_1^2}{2} = 0$

• elevation change between inlet and exit point is negligible.

With the stipulations, the SFEE reduces to

$$mh_1 + Q = mh_2$$
$$Q = m\,(h_2 - h_1)$$

Apparently the heat energy supplied to the system in a boiler increases the enthalpy or heat content of the system.

7.6.3. Steam or gas turbine

A turbine is a device for obtaining work from a flow of fluid expanding from high pressure to low pressure. The output from the turbine may be used to run a generator and produce electric power.

The characteristic features of flow through a steam or gas turbine are :

• negligible velocity change in the flow of fluid,

• negligible potential energy change,

• no transfer of heat as its walls are thermally insulated.

With these stipulations, the steady flow energy equation becomes,

$$mh_1 = mh_2 + W_s$$
$$\text{or} \quad W_s = m\,(h_1 - h_2)$$

Apparently work is done at the expense of enthalpy.

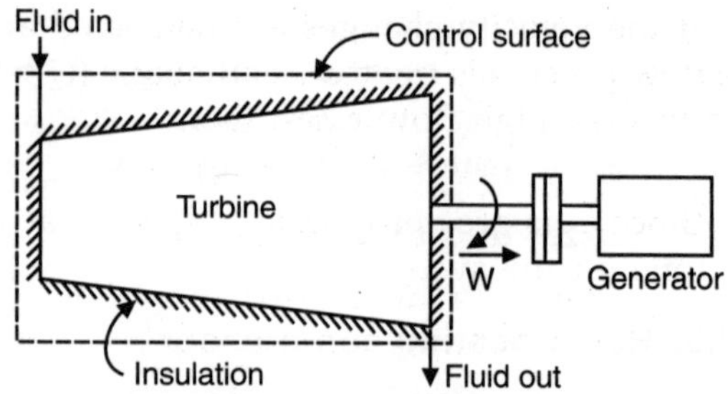

Fig. 7.7. Flow through a turbine

7.6.4. Rotary compressor

Rotary compressors are the machines which develop pressure and have a rotor as their primary element. Rotary compressors are employed where high efficiency, medium pressure rise (upto 10 bar) and large flow rate (50 m^3/s) are the primary considerations. The characteristic features of flow through a rotary compressor are :

• shaft work is done on the system and the fluid is compressed ; W is negative.

• potential energy changes are negligible.

• heat is lost from the surrounding ; Q is negative.

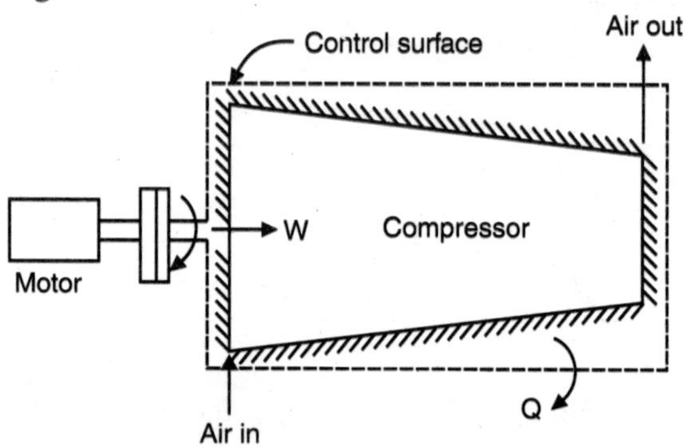

Fig. 7.8. Rotary compressor

With these stipulations, the steady flow energy equation may be written as

$$m\,(h_1 + \frac{V_1^2}{2}) - Q = m\,(h_2 + \frac{V_2^2}{2}) - W_s$$

$$\text{or} \quad W_s = m\,(h_2 + \frac{V_2^2}{2}) - m\,(h_1 + \frac{V_1^2}{2}) + Q$$

If the velocity changes are neglected and the flow process is treated as adiabatic ($Q = 0$) due to very high flow rates, then

$$W_s = m\,(h_2 - h_1)$$

Since h_2 is greater than h_1, W_s is negative or done on the system.

7.6.5. Reciprocating compressor

A reciprocating compressor is a contrivance which is used for increasing the pressure of a fluid (gas or air), and has a piston-cylinder mechanism as the primary element. The unit sucks in definite quantity of fluid, compresses through a required pressure ratio and then delivers the compressed fluid to a receiver.

The receiver is a vessel which is sufficiently large and reduces the flow fluctuations in the control volume.

Reciprocating compressors are used when small quantity of fluid (maximum limit 5 m^3/s) with high pressure (as high as 1000 bar) is required.

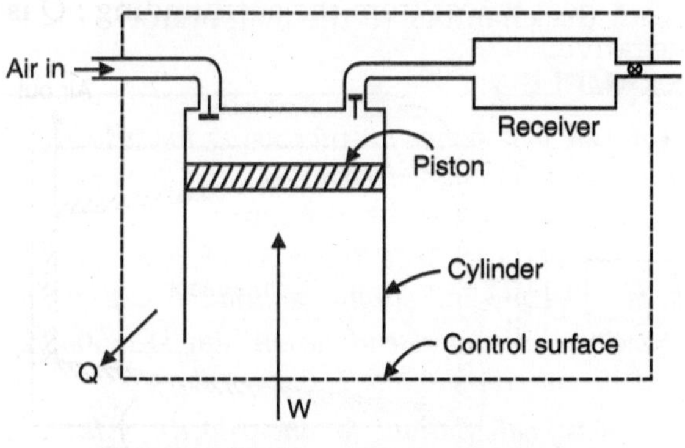

Fig. 7.9. Reciprocating compressor

The characteristic features of steady flow through a reciprocating compressor are :

- potential and kinetic energy changes are negligible.
- shaft work is done on the system and the fluid is compressed ; W_s is negative.
- appreciable heat transfer is involved; heat is lost from the system and Q is negative.

With these stipulations, the steady flow energy equation becomes

$$mh_1 - Q = mh_2 - W_s$$

or $$W_s = Q + m\,(h_2 - h_1)$$

7.6.6. Centrifugal water pump

A pump is a unit that transfers the mechanical energy of a motor or an engine into the potential energy of a liquid. Centrifugal pumps belong to the category of dynamic pressure pumps wherein the pumping of liquids from lower level to higher level is affected by rotary motion of one or more rotating wheels called impellers.

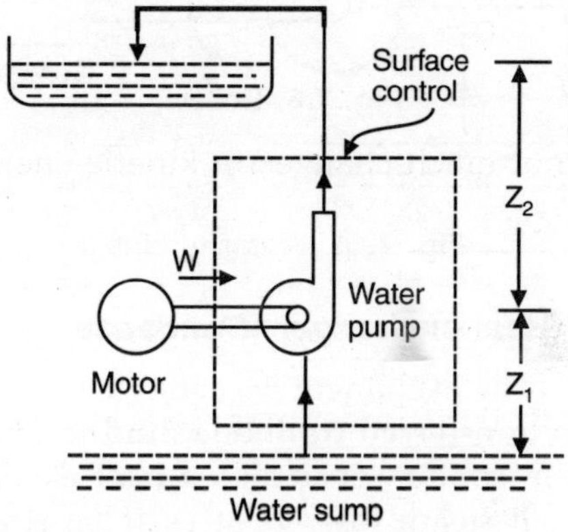

Fig. 7.10. Centrifugal water pump

Here $Q = 0$; $du = 0$ as there is no change in temperature of water.

W is taken as negative because it is done on the system, *i.e.*, work enters the control volume.

The steady flow energy then becomes

$$m\left(p_1 v_1 + \frac{V_1^2}{2} + gz_1\right) = m\left(p_2 v_2 + \frac{V_2^2}{2} + gz_2\right) - W$$

7.6.7. Hydraulic turbine

A hydraulic turbine is a prime mover wherein a gradual change in the momentum of a fluid is utilized to set a mobile unit into motion. The turbine uses the potential and kinetic energy of water and converts it into usable mechanical energy.

Here $Q = 0$; $du = 0$ as there is no change in temperature of water. Work is taken as positive because it is done by the system.

The steady flow energy then becomes

$$m\left(p_1 v_1 + \frac{V_1^2}{2} + gz_1\right) = m\left(p_1 v_1 + \frac{V_2^2}{2} + gz_2\right) + W$$

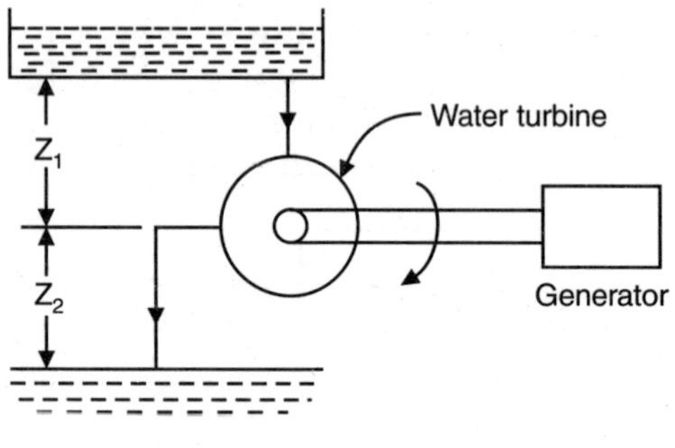

Fig. 7.11. Water turbine

7.6.8. Heat exchanger (Condenser)

A heat exchanger is a device in which heat is transferred from one fluid to another. A steam condenser is also a heat exchanger in which steam loses heat as it passes over the tubes through which water is flowing. Obviously, there are two steady flow streams : one for steam and the other for cooling water. The flow through a heat exchanger is characterised by

– no shaft or mechanical work ; $W_s = 0$.

– no change in kinetic energy and potential energy.

– perfectly insulated system ; no external heat interaction.

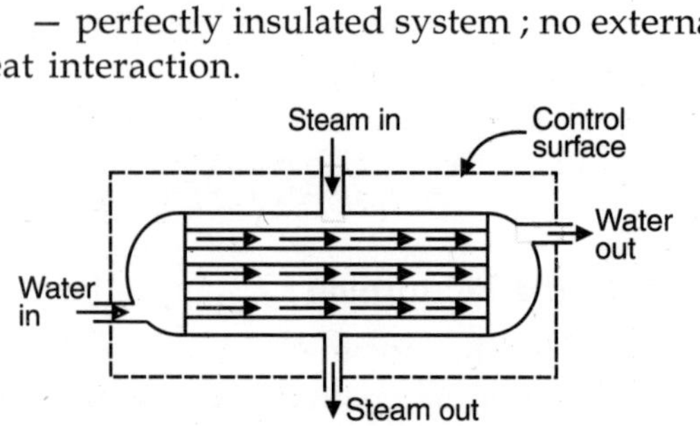

Fig. 7.12. Heat exchanger

From energy balance, we may write heat lost by steam

= heat gained by cooling water

$$m_s (h_1 - h_2) = m_w (h_4 - h_3)$$

where m_s is the mass flow of steam and m_w is the mass flow of the cooling water.

7.6.9. Evaporator

An evaporator is the component of a refrigeration system, and is used to extract heat from the chamber which is to be kept at low temperature. The refrigerating liquid enters the evaporator, absorbs latent heat from the chamber at constant pressure and comes out as a vapour. The characterising features of steady flow through an evaporator are :

– no work is absorbed or supplied.

– no changes in kinetic energy and potential energy.

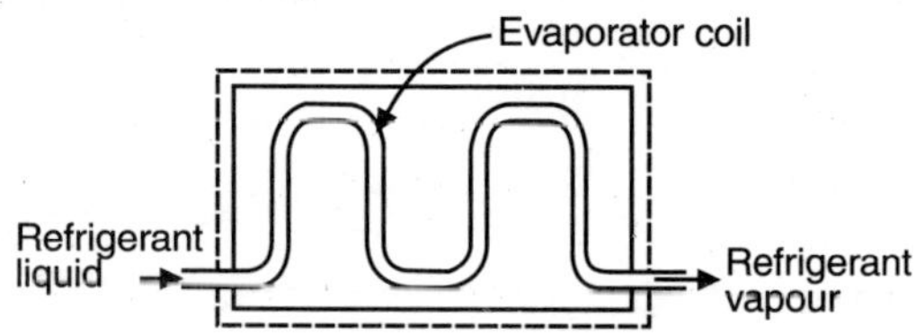

Fig. 7.13. Evaporatior

The steady flow energy then becomes

$$mh_1 + Q = mh_2; \quad Q = m(h_2 - h_1)$$

Q is taken as positive because heat flows from the chamber to the evaporator.

EXAMPLE 7.1

A nozzle is a device for increasing the velocity of a steadily flowing stream. At inlet to a certain nozzle, the fluid parameters are :

Enthalpy = 2850 kJ/kg; velocity = 50 m/s; area = 0.1 m² and specific volume = 0.18 m³/kg.

At the discharge end the enthalpy is 2650 kJ/kg and the specific volume is 0.49 m³/kg.

Make calculations for the velocity of fluid at exit from the nozzle, mass flow rate of fluid, and the exit area of the nozzle.

The nozzle is horizontal and there is negligible heat loss from it.

Solution : Applying steady flow energy equation for unit mass,

$$u_1 + p_1 v_1 + \frac{V_1^2}{2} + gz_1 + q = u_2 + p_2 v_2 + \frac{V_2^2}{2} + gz_2 + w_s$$

With the given assumptions of $z_1 = z_2$ and $q = 0$, and noting that shaft work w_s is zero:

$$\frac{V_2^2 - V_1^2}{2} = (u_1 + p_1v_1) - (u_2 + p_2v_2)$$
$$= (h_1 - h_2)$$
$$= 2850 - 2650$$
$$= 200 \text{ kJ} = 200 \times 10^3 \text{ J}$$
$$V_2^2 = 2 \times 200 \times 10^3 + (50)^2$$
$$= 402500$$

∴ Flow velocity at exit from nozzle,

$$V_2 = \sqrt{(402500)} = \mathbf{634.43 \ m/s}$$

(*b*) From continuity equation,

$$\dot{m} = \frac{A_1V_1}{v_1} = \frac{A_2V_2}{v_2}$$

∴ Mass flow rate,

$$\dot{m} = \frac{A_1V_1}{v_1} = \frac{0.1 \times 50}{0.18} = \mathbf{27.78 \ kg/s}$$

(*c*) $\dot{m} = \dfrac{A_2V_2}{v_2}$; $27.78 = \dfrac{A_2 \times 634.43}{0.49}$

∴ Exit area of nozzle,

$$A_2 = \frac{27.78 \times 0.49}{634.43}$$
$$= 0.02146 \text{ m}^2 = \mathbf{214.6 \ cm^2}$$

EXAMPLE 7.2

A perfect gas flows through a nozzle where it expands in a reversible adiabatic manner. The inlet conditions are 22 bar, 500°C, 38 m/s. At exit the pressure is 2 bar. Determine the exit velocity and exit area if the flow rate is 4 kg/s.

Take R = 190 J/kg K and γ = 1.35

Solution : Applying steady flow energy equation for a unit mass,

$$u_1 + p_1v_1 + \frac{V_1^2}{2} + gz_1 + q$$
$$= u_2 + p_2v_2 + \frac{V_2^2}{2} + gz_2 + w_s$$

During flow through a nozzle there is no shaft work, the walls are thermally insulated and the change in elevation between the entry and exit point is zero. Therefore,

$$u_1 + p_1v_1 + \frac{V_1^2}{2} = u_2 + p_2v_2 + \frac{V_2^2}{2}$$

or
$$\frac{V_2^2 - V_1^2}{2} = (u_1 + p_1v_1) - (u_2 + p_2v_2)$$
$$= h_1 - h_2 = c_p (T_1 - T_2)$$

Now, $c_p = \dfrac{\gamma}{\gamma - 1} R = \dfrac{1.35}{1.35 - 1} \times 190$

$$= 733 \text{ J/kg K}$$

and, $T_2 = T_1\left(\dfrac{p_2}{p_1}\right)^{\frac{\gamma-1}{\gamma}} = 773\left(\dfrac{2}{22}\right)^{\frac{1.35-1}{1.35}}$

$$= 414 \text{ K}$$

∴ $\dfrac{V_2^2 - V_1^2}{2} = 733\ (773 - 414)$

$$= 263147 \text{ J}$$

and, $V_2 = \sqrt{263147 \times 2 + 38^2} = \mathbf{726 \ m/s}$

(*b*) From characteristic gas equation, $pv = RT$, the specific volume of air at exit from the nozzle,

$$v_2 = \frac{RT_2}{p_2} = \frac{190 \times 414}{2 \times 10^5}$$
$$= 0.3933 \text{ m}^3/\text{kg}$$

From continuity considerations, mass flow rate, $\dot{m} = \dfrac{A_2V_2}{v_2}$

∴ Area at nozzle exit,

$$A_2 = \frac{\dot{m}v_2}{V_2} = \frac{4 \times 0.3933}{726}$$
$$= \mathbf{2.167 \times 10^{-3} \ m^2}$$

EXAMPLE 7.3

In a steady flow process, the fluid flows through a machine at the rate of 15 kg/min. Between the entrance and exit of the machine, the relevant data regarding the working fluid is

Parameter	*Inlet*	*Outlet*
Velocity	*5 m/s*	*8 m/s*
Pressure	*100 kPa*	*700 kPa*
Specific volume	*0.45 m³/kg*	*0.125 m³/kg*

The working fluid leaves the machine with internal energy 160 kJ/kg greater than that at entrance and during the process 7200 kJ/min of heat is lost to the surroundings. Assuming entrance and exit pipes to be at the same level, calculate the shaft work and the ratio of inlet pipe diameter to outlet pipe diameter.

Solution : From continuity considerations, the mass flow rate entering equals the mass flow leaving. That is

$$\dot{m} = \frac{A_1V_1}{v_1} = \frac{A_2V_2}{v_2}$$

$$\frac{A_1}{A_2} = \frac{V_2}{V_1} \times \frac{v_1}{v_2} = \frac{8}{5} \times \frac{0.45}{0.125} = 5.76$$

∴ Ratio of pipe diameters,

$$\frac{D_1}{D_2} = \sqrt{5.76} = \mathbf{2.4}$$

(*b*) Applying steady flow energy equation for unit mass (all terms expressed in kJ/kg)

$$u_1 + p_1v_1 + \frac{V_1^2}{2} + gz_1 + q$$

$$= u_2 + p_2v_2 + \frac{V_2^2}{2} + gz_2 + w_s$$

Since, $z_1 = z_2$, we have

$$w_s = (u_1 - u_2) + (p_1v_1 - p_2v_2) + \frac{V_1^2 - V_2^2}{2} + q$$

It is to be noted that:

(*i*) $q = -7200$ kJ/min $= \frac{-7200}{15}$

$= -480$ kJ/kg

The –ve sign with q because heat is lost from the system.

(*ii*) $w_s =$

$$-160 + \left[\frac{100 \times 10^3 \times 0.45 - 700 \times 10^3 \times 0.125}{10^3}\right]$$

$$+ \frac{8^2 - 5^2}{2 \times 10^3} - 480$$

$= -160 - 42.5 + 0.0195 - 480$

$= -682.48$ kJ/kg

$= -(682.48 \times 15)$ kJ/min

$= -\frac{682.48 \times 15}{60} =$ **– 170.62 kJ/s**

The –ve sign implies that work has been supplied to the machine.

EXAMPLE 7.4

In a centrifugal compressor, the suction and delivery pressures are 100 kPa and 550 kPa respectively. The compressor draws 15 m³/min of air which has a specific volume of 0.77 m³/kg. At delivery point, the specific volume is 0.20 m³/kg. The compressor is driven by a 40 kW motor and during passage of air through the compressor, the heat lost to the surroundings is 30 kJ/kg of air. Neglecting changes in the potential and kinetic energy, make calculations for increase in internal energy per kg of air.

Solution : Mass of air,

$$m = \frac{\text{total volume}}{\text{specific volume}}$$

$$= \frac{15}{0.77} = 19.48 \text{ kg/min}$$

$$= \frac{19.48}{60} = 0.3246 \text{ kg/s}$$

Compressor work $= m \times w$

where w is the compressor work in kJ/kg

$40 = 0.3246\, w$

$$w = \frac{40}{0.3246} = 123.23 \text{ kJ/kg}$$

Applying steady flow energy equation per unit mass of air (all terms expressed in kJ/kg)

$$u_1 + p_1v_1 + \frac{V_1^2}{2} + gz_1 + q$$

$$= u_2 + p_2v_2 + \frac{V_2^2}{2} + gz_2 + w$$

Neglecting changes in kinetic energy and potential energy,

$$u_2 - u_1 = (p_1v_1 - p_2v_2) + q - w$$

Substituting $q = -30$ kJ/kg (– ve because heat is lost to surroundings) and $w = -123.23$ kJ (– ve because work is supplied to the system) and expressing each term in kJ/kg

$u_2 - u_1 = (100 \times 0.77 - 550 \times 0.2)$

$+ (-30) - (-123.23)$

$= (77 - 110) - 30 + 123.23$

$=$ **60.23 kJ/kg**

Thus increase in internal energy of air is 60.23 kJ/kg.

EXAMPLE 7.5

Air at 1 bar pressure, 290 K temperature flows steadily at the rate of 120 m³/hr into a compressor where its pressure and temperature are respectively raised to 15 bar and 390 K. During the compression process, the heat transfer from the compresor is 10 percent of work transfer from the machine. Neglecting changes in kinetic energy and potential energy, evaluate the work and heat interactions. Presume that air behaves as a perfect gas.

Solution : Applying steady flow energy equation on time basis (all terms expressed in kJ/s)

$$\dot{m}\left(u_1 + p_1v_1 + \frac{V_1^2}{2} + gz_1\right) + \dot{Q}$$

$$= \dot{m}\left(u_2 + p_2v_2 + \frac{V_2^2}{2} + gz_2\right) + \dot{W}_s$$

Neglecting changes in kinetic energy and potential energy,

$$\dot{Q} - \dot{W}_s = \dot{m}\,[(u_2 + p_2v_2) - (u_1 + p_1v_1)]$$
$$= \dot{m}\,(h_2 - h_1) = \dot{m}c_p\,(T_2 - T_1)$$

Mass flow rate,

$$\dot{m} = \frac{pV}{RT} = \frac{1 \times 10^5 \times 120}{287 \times 290}$$
$$= 144.18 \text{ kg/hr} = 0.04005 \text{ kg/s}$$

$$\therefore\ \dot{Q} - \dot{W}_s = 0.04005 \times 1.005 \times (390 - 290)$$
$$= 4.025 \text{ kJ/s}$$

Given : $\dot{Q} = 10$ percent of $\dot{W}_s = 0.1\dot{W}_s$

$0.1\,\dot{W}_s - \dot{W}_s = 4.025\,;\quad \dot{W}_s = -\,4.47$ kW

$\dot{Q} = $ **– 0.447 kW**

Thus the work transfer to the compressor is 4.47 kW and heat transfer from the compressor is 0.447 kW.

EXAMPLE 7.6

A steam turbine operates under steady flow conditions receiving steam at the following state : pressure 15 bar ; internal energy 2700 kJ/kg ; specific volume 0.17 m³/kg and velocity 100 m/s.

The exhaust of steam from the turbine is at 0.1 bar with internal energy 2175 kJ/kg, specific volume 15 m³/kg and velocity 300 m/s. The intake is 3 meters above the exhaust. The turbine develops 35 kW and heat loss over the surface of turbine is 20 kJ/kg. Determine the steam flow rate through the turbine.

Solution : Applying steady flow energy equation per unit mass of steam (all terms expressed in kJ/kg)

$$u_1 + p_1v_1 + \frac{V_1^2}{2} + gz_1 + q$$

$$= u_2 + p_2v_2 + \frac{V_2^2}{2} + gz_2 + w_s$$

Noting that q is negative as heat is lost from the surface, we get

$$2700 + \frac{15 \times 10^5 \times 0.17}{10^3} + \frac{100^2}{2 \times 10^3} + \frac{9.81 \times 3}{10^3} - 20$$

$$= 2175 + \frac{0.1 \times 10^5 \times 15}{10^3} + \frac{300^2}{2 \times 10^3} + \frac{9.81 \times 0}{10^3} + w_s$$

or $\quad 2700 + 255 + 5 + 0.0294 - 20$

$$= 2175 + 150 + 45 + 0 + w_s$$

∴ work available from the turbine,

$$w_s = 570.0294 \text{ kJ/kg}$$

It may be noted that potential energy changes are negligible and may be omitted without much error.

Power output = $m\,w_s$

where m is the steam flow rate in kg/s

$$35 = m \times 570.0294$$

m = **0.0614 kg/s**

EXAMPLE 7.7

A gasoline engine operates steadily with a work output of 36 kW and the heat transfer from the engine to a stream of cooling water is at the rate of 125 × 10³ kJ/hr. The sum of all other heat transfers to the surrounding atmosphere and the engine foundation is 25 × 10³ kJ/hr. The only stream of fluid other than the cooling water is the stream which enters as fuel-air mixture at the rate of 150 kg/hr. Determine the enthalpy change of this stream in kJ/kg of mixture. Neglect changes in kinetic energy and potential energy.

Solution : Work output, $w_s = 36$ kW

$$= 36 \times 3600 = 129.6 \times 10^3 \text{ kJ/hr}$$

The heat transfer from the engine consists of heat transfer to the cooling water and to the surroundings including the engine foundation.

Heat transfer,

$$\dot{Q} = 125 \times 10^3 + 25 \times 10^3$$
$$= 150 \times 10^3 \text{ kJ/hr}$$

This heat transfer is negative as heat is lost from the system.

Applying steady flow energy equation on time basis,

$$\dot{m}(u_1 + p_1v_1 + \frac{V_1^2}{2} + gz_1) + \dot{Q}$$

$$= \dot{m}(u_2 + p_2v_2 + \frac{V_2^2}{2} + gz_2) + \dot{W}_s$$

Neglecting changes in kinetic and potential energy

$$\dot{Q} - \dot{W}_s = \dot{m}[(u_2 + p_2v_2) - (u_1 + p_1v_1)]$$

$$= \dot{m}(h_2 - h_1)$$

$$\therefore h_2 - h_1 = \frac{\dot{Q} - \dot{W}_s}{\dot{m}}$$

$$= \frac{-150 \times 10^3 - 129.6 \times 10^3}{150}$$

$$= \mathbf{-1864\ kJ}$$

The negative sign implies that enthalpy of the fuel-air mixture has decreased.

EXAMPLE 7.8

Determine the power required to drive a pump which raises the water pressure from 1 bar at entry to 25 bar at exit and delivers 2000 kg/hr of water. Neglect changes in volume, elevation and velocity and assume specific volume of water to be 0.001045 m³/kg.

Solution : The flow through a pump is a steady flow process and, therefore, for unit mass of water

$$u_1 + p_1v_1 + \frac{V_1^2}{2} + gz_1 + q$$

$$= u_2 + p_2v_2 + \frac{V_2^2}{2} + gz_2 + w_s$$

The pumping process is characterised by:

– $V_1 = V_2$; $z_1 = z_2$ and $v_1 = v_2$ (given)

– no heat interaction ; $q = 0$

– no change in internal energy; $u_1 = u_2$ because the fluid (water) does not experience any change in temperature.

With these stipulations, the steady flow energy equation becomes

$$p_1v_1 - p_2v_2 = w_s$$

$$\therefore \quad w_s = v\,(p_1 - p_2)$$

$$= 0.001045 \times (1 \times 10^5 - 25 \times 10^5)$$

$$= -2508\ \text{J/kg}$$

Power required to drive the pump = mw_s

$$= \frac{2000}{3600} \times (-2508)$$

$$= -1393\ \text{J/s} = \mathbf{-1.393\ kW}$$

The –ve sign indicates that work must be done on water.

EXAMPLE 7.9

A centrifugal pump delivers 2750 kg of water per minute from initial pressure of 0.8 bar absolute to a final pressure of 2.8 bar absolute. The suction is 2 m below and the delivery is 5 m above the centre of pump. If the suction and delivery pipes are of 15 cm and 10 cm diameter respectively, make calculations for the power required to run the pump.

Solution : Refer section 7.6.6 for the centrifugal pump installation. Here there is no change in internal energy and no heat interaction. The steady flow energy equation on unit mass basis may then be written as

$$p_1v_1 + \frac{V_1^2}{2} + gz_1 = p_2v_2 + \frac{V_2^2}{2} + gz_2 + w_s$$

or

$$\frac{p_1}{\rho_1} + \frac{V_1^2}{2} + gz_1 = \frac{p_2}{\rho_2} + \frac{V_2^2}{2} + gz_2 + w_s \quad ...(i)$$

From continuity consideration,

$$\dot{m} = \rho_1 A_1 V_1 = \rho_2 A_2 V_2$$

Water is incompressible and does not experience any change in density. For water, density = 1000 kg/m³

$$\frac{2750}{60} = 1000 \times \frac{\pi}{4}(0.15)^2 V_1$$

$$= 1000 \times \frac{\pi}{4}(0.1)^2 V_2$$

$$\therefore \quad V_1 = 2.59\ \text{m/s};\ V_2 = 5.84\ \text{m/s}.$$

Taking suction level as datum for elevation head, $z_1 = 0$ and $z_2 = 2 + 5 = 7$ m

Substituting the relevent values in equation (*i*) we get

$$w_s = \frac{p_1 - p_2}{\rho} + \frac{V_1^2 - V_2^2}{2} + g(z_1 - z_2) \quad ...(\rho_1 = \rho_2 = \rho)$$

$$= \frac{0.8 \times 10^5 - 2.8 \times 10^5}{1000}$$

$$+\frac{2.59^2 - 5.84^2}{2} + 9.81(0-7)$$

$= -\ 200 - 13.70 - 68.67$

$= -\ 282.37$ J/kg

The -ve sign indicates that work is absorbed by the pump. The power required to drive the pump will be

$$= mw_s = \frac{2750}{60} \times (282.37)$$

$= 12942$ J/s = **12.942 kW**

EXAMPLE 7.10

In a conference hall, comfortable temperature conditions are maintained in winter by circulating hot water through a piping system. The water enters the piping system at 3 bar pressure and 50°C temperature (enthalpy = 240 kJ/kg) and leaves at 2.5 bar pressure and 30 °C temperature (enthalpy = 195 kJ/kg). The exit from the piping system is 15 m above the entry. If 30 MJ/hr of heat needs to be supplied to the hall, make calculations for the quantity of water circulated through the pipe per minute. Assume that there are no pumps in the system and that the change in kinetic energy is negligible.

Solution : Applying steady flow energy equation per unit mass of hot water

$$u_1 + p_1v_1 + \frac{V_1^2}{2} + gz_1 + q$$

$$= u_2 + p_2v_2 + \frac{V_2^2}{2} + gz_2 + w_s$$

or $$h_1 + \frac{V_1^2}{2} + gz_1 + q$$

$$= h_2 + \frac{V_2^2}{2} + gz_2 + w_s$$

Neglecting changes in kinetic energy and noting that shaft work, $w_s = 0$ (no pumps),

$$q = (h_2 - h_1) + g(z_2 - z_1)$$

$$= (195 - 240) + \frac{9.81 \times 15}{10^3}$$

$= -\ 44.853$ kJ/kg

The - ve sign indicates that heat is given out by water.

Mass of water to be circulated

$$= \frac{30 \times 10^6}{44.853} \times \frac{1}{60} = \textbf{11.15 kg/min}$$

EXAMPLE 7.11

Steam flows steadily into a condenser and at entry it has an enthalpy of 2050 kJ/kg and its velocity is 500 m/s. The condensate (condensed steam) has an enthalpy of 200 kJ/kg and it leaves with 10 m/s velocity. The exit from the condenser is in line with the inlet. Determine the heat transfer to cooling water per unit mass of steam condensed.

Solution : Applying steady flow energy equation per unit mass of steam flow,

$$u_1 + p_1v_1 + \frac{V_1^2}{2} + gz_1 + q$$

$$= u_2 + p_2v_2 + \frac{V_2^2}{2} + gz_2 + w_s$$

or $$h_1 + \frac{V_1^2}{2} + gz_1 + q = h_2 + \frac{V_2^2}{2} + gz_2 + w_s$$

Noting that shaft work is zero and there is no change in potential energy (intake and outlet are in line),

$$q = (h_2 - h_1) + \frac{V_2^2 - V_1^2}{2}$$

$$= (200 - 2050) + \frac{10^2 - 500^2}{2 \times 10^3}$$

$= \textbf{- 1974.75 kJ/kg}$

The - ve sign implies that heat is rejected by the system (steam).

EXAMPLE 7.12

Water enters a boiler with enthalpy 150 kJ/kg and gets vaporised into steam with enthalpy 2825 kJ/kg. Apply steady flow energy equation to the boiler, make appropriate assumptions and workout the heat transferred per kg of steam generated.

Solution : Refer section 7.6.2

On unit mass basis of steam raised, the steady flow energy equation is given by

$$q = h_2 - h_1$$

$= 2825 - 150 =$ **2675 kJ/kg**

EXAMPLE 7.13

The adiabatic mixing process of saturated water at enthalpy 650 kJ/kg with saturated steam at enthalpy 2350 kJ/kg is carried out at constant pressure. If the mixing ratio of water to steam is 1:8 by mass, make calculations for the enthalpy of mixture. Neglect changes in kinetic and potential energy and assume that there is no work except flow work.

Solution : From mass balance, $m_m = m_w + m_s$

$= 1 + 8 = 9$ kg

The suffix m, w and s refer to mixture, water and steam respectively.

From energy balance :

$$m_w h_w + m_s h_s = m_m h_m$$

The above identity has been written with the following stipulations :

– kinetic and potential energy changes are negligible

– no shaft work ; $w_s = 0$

– no heat interaction ($q = 0$) as the mixing process is adiabatic

$\therefore 1 \times 650 + 8 \times 2350 = 9\, h_m$; $19450 = 9\, h_m$

Thus, enthalpy of mixture,

$$h_m = \frac{19450}{9} = \mathbf{2161\ kJ/kg}$$

EXAMPLE 7.14

A turbojet engine is in flight at 250 m/s in the atmosphere at – 15°C. The engine operates with fuel-air ratio of 0.018 and the fuel used has a calorific value of 44000 kJ/kg. Due to incomplete combustion of fuel, only 95% of energy is liberated by fuel. The engine loses 20 kJ/kg of air, and the measurements indicate that temperature of gases at exit from the nozzle is 600°C and the enthalpy values for air and gases are 250 kJ/kg and 900 kJ/kg respectively. Neglecting potential energy changes for the flow streams, make calculations for the velocity of exhaust jet.

Solution : Refer Fig. 7.14. for the given system

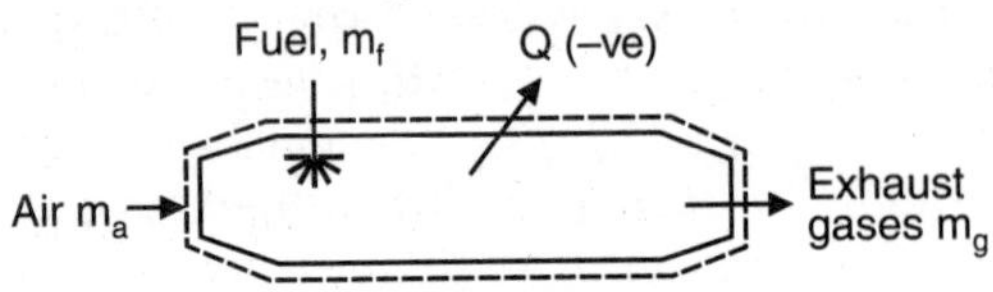

Fig. 7.14. Turbojet in flight

From mass balance : $m_g = m_a + m_f$

$m_g = 1 + 0.018 = 1.018$ kg per kg of air

The suffix g, a and f refer to gas, air and fuel respectively.

From energy balance (neglecting potential energy changes and no shaft work)

$$m_a \left(h_a + \frac{V_a^2}{2}\right) + m_f E_f + Q = m_g \left(h_g + \frac{V_g^2}{2}\right)$$

The term $m_f E_f$ refers to the heat energy liberated by fuel. Expressing each term in kJ, we get

$$1\left(250 + \frac{250^2}{2 \times 10^3}\right) + 0.018 \times 44000 \times 0.95 - 20 = 1.018\left(900 + \frac{V_g^2}{2 \times 10^3}\right)$$

or $281.25 + 752.4 - 20 = 916.2 + 5.09 \times 10^{-4}\, V_g^2$

or $\qquad 5.09 \times 10^{-4}\, V_g^2 = 97.45$

$\therefore$ Velocity of exhaust jet,

$$V_g = \sqrt{\frac{97.45}{5.09 \times 10^{-4}}} = \mathbf{437.55\ m/s}$$

EXAMPLE 7.15

An incompressible metal with density 9400 kg/m³ and specific heat 205 J/kg K extrudes slowly through a die under a pressure of 200 MPa. Evaluate rise in temperature of the metal presuming that there is no cooling of the metal during the extruding process.

Solution : Applying steady flow energy equation per unit mass of the metal

$$u_1 + p_1 v_1 + \frac{V_1^2}{2} + g z_1 + q = u_2 + p_2 v_2 + \frac{V_2^2}{2} + g z_2 + w_s$$

The extrusion process is characterised by:

– no shaft work, $w_s = 0$

– no heat loss to surroundings, $q = 0$

– change in kinetic energy is negligible as the extrusion process is done slowly.

– changes in potential energy are insignificant and can be ignored.

With these stipulations, the SFEE can be written as

$$u_1 + p_1v_1 = u_2 + p_2v_2$$

$$\text{or} \quad u_2 - u_1 = p_1v_1 - p_2v_2$$

Since the metal is incompressible,

$$v_1 = v_2 = \frac{1}{\rho} = \frac{1}{9400} \text{ m}^3/\text{kg}$$

$$\therefore \quad (u_2 - u_1) = \frac{p_1 - p_2}{\rho}$$

$$= \frac{200 \times 10^6 - 0}{9400}$$

$$= 212.766 \text{ J/kg}$$

The atmospheric pressure has been neglected; it being much smaller than the extrusion pressure.

Now, $u_2 - u_1 = c\,(T_2 - T_1)$

$\therefore$ Rise in temperature of metal,

$$(T_2 - T_1) = \frac{21276.6}{205} = \mathbf{103.79°C}$$

EXAMPLE 7.16

A perfect gas has a molar mass of 26 kg/kmol and a value of γ = 1.25. Make calculations for the heat rejected.

(a) when 1 kg mass of the gas is contained in a rigid vessesl at 2.5 bar and 225°C, and is then cooled until the pressure falls to half the initial value;

(b) when 1 kg/s mass of the gas enters a pipline at 300°C, and flows steadily to the end of the pipe where the temperature is 25°C. The changes in velocity of gas in the pipeline are negligible.

Solution : Since the vessel is rigid, the change in the state of the gas occurs at constant volume and accordingly $p_1/T_1 = p_2/T_2$

$\therefore$ Final temperature,

$$T_2 = T_1 \times \frac{p_2}{p_1} = 598 \times \frac{1}{2} = 299 \text{ K}$$

For the given gas (molecular mass 26), the gas constant is

$$R = \frac{R_{mol}}{M} = \frac{8314}{26} = 319.77 \text{ J/kg K}$$

$$c_v = \frac{R}{\gamma - 1} = \frac{319.77}{1.25 - 1}$$

$$= 1279.07 \text{ J/kg K} = 1.279 \text{ kJ/kg K}$$

Change in internal energy,

$$dU = mc_v\, dT$$

$$= 1 \times 1.279 \times (299 - 598)$$

$$= -\,382.42 \text{ kJ}$$

Using non-flow energy equation:

$$\delta Q = \delta W + dU$$

Here $\delta W = 0$ as the vessel is rigid, *i.e.*, no change in volume

$$\therefore \quad \delta Q = 0 + (-\,382.42) = \mathbf{-\,382.42} \text{ kJ}$$

The negative sign implies that heat is rejected by the gas.

(*b*) Applying steady flow energy equation,

$$\dot{m}\left(u_1 + p_1v_1 + \frac{V_1^2}{2} + gz_1\right) + \dot{Q}$$

$$= \dot{m}\left(u_2 + p_2v_2 + \frac{V_2^2}{2} + gz_2\right) + \dot{W}_s$$

Neglecting changes in kinetic energy and potential energy, and noting that no work is done

$$\dot{m}\,(u_1 + p_1v_1) + \dot{Q} = \dot{m}\,(u_2 + p_2v_2)$$

$$\text{or} \quad \dot{Q} = \dot{m}\left[(u_2 + p_2v_2) - (u_1 + p_1v_1)\right]$$

$$= \dot{m}\,(h_2 - h_1) = \dot{m}\,c_p\,(T_2 - T_1)$$

As per calculations made in part (*a*), c_v = 1.279 kg/kg K and γ = 1.25 (given)

$$c_p = \gamma\, c_v = 1.25 \times 1.279$$

$$= 1.5987 \text{ kJ/kg K}$$

$$\therefore \quad \dot{Q} = 1 \times 1.5987 \times (25 - 300)$$

$$= -\,439.64 \text{ kJ/s} = \mathbf{-\,439.64} \text{ kW}$$

The negative sign implies that heat is rejected by the gas.

EXAMPLE 7.17

Air at 290 K temperature passes through a heat exchanger at 30 m/s velocity and its temperature gets raised to 1100 K. Subsequently the heated air enters a turbine with the same velocity and its expansion continues till the temperature drops to 900 K. After exit from the turbine at 45 m/s, further expansion occurs in a nozzle and the temperature falls to 790 K. If mass flow rate of air is 2 kg/s, determine :

(a) rate of heat transfer to air in the heat exchanger,

(b) power output from the turbine, and

(c) velocity at exit from the nozzle.

Assume no heat loss and take

$$c_p = 1.005 \text{ kJ/kg K for air}$$

Solution : Refer Fig 7.15. for the schematic arrangement of the heat exchanger, turbine and nozzle.

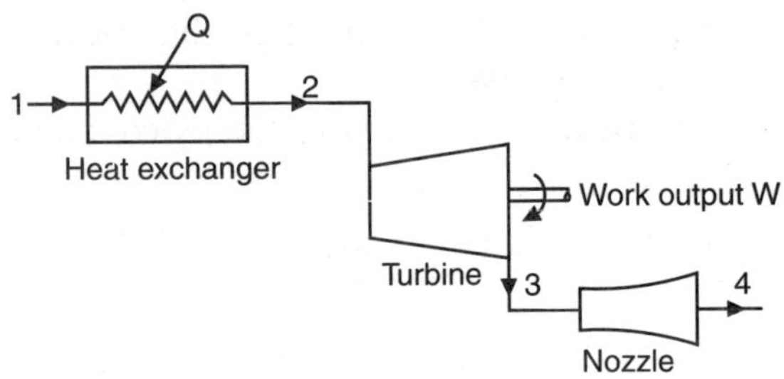

Fig. 7.15.

(*a*) Applying steady flow energy equation for the heat exchanger,

$$m\left(h_1 + \frac{V_1^2}{2} + gz_1\right) + Q = m\left(h_2 + \frac{V_2^2}{2} + gz_2\right) + W_s$$

The characteristic features for flow through heat exchanger are :

- no shaft work; $W_s = 0$

- negligible kinetic and potential energy changes ; $V_1 = V_2$ and $z_1 = z_2$

Accordingly,

$$Q_{1-2} = m(h_2 - h_1) = m\,c_p\,(T_2 - T_1)$$
$$= 2 \times 1.005 \times (1100 - 290)$$
$$= \mathbf{1620} \text{ kJ/s}$$

(*b*) Writing steady flow energy equation for turbine,

$$m\left(h_2 + \frac{V_2^2}{2} + gz_2\right) + Q = m\left(h_3 + \frac{V_3^2}{2} + gz_3\right) + W_s$$

The characteristic features for flow through a turbine are :

- no transfer of heat as walls are insulated, $Q = 0$

- negligible potenital energy change; $z_1 = z_2$

Accordingly

$$m\left(h_2 + \frac{V_2^2}{2}\right) = m\left(h_3 + \frac{V_3^2}{2}\right) + W_s$$

∴ Work output from turbine,

$$W_s = m\left[(h_2 - h_3) + \frac{V_2^2 - V_3^2}{2}\right]$$
$$= m\left[c_p\,(T_2 - T_3) + \frac{V_2^2 - V_3^2}{2}\right]$$
$$= 2\left[1.005 \times (1100 - 900) + \frac{30^2 - 45^2}{2 \times 1000}\right]$$
$$= 2\,[200 - 0.5625] = 398.87 \text{ kJ/s}$$
$$= \mathbf{398.87} \text{ kW}$$

(*c*) Writing steady flow energy equation for nozzle,

$$m\left(h_3 + \frac{V_3^2}{2} + gz_3\right) + Q = m\left(h_4 + \frac{V_4^2}{2} + gz_4\right) + W_s$$

The characteristic features for flow through a nozzle are

- no shaft work; $W_s = 0$

- walls are thermally insulated, *i.e.*, the flow is adiabatic, $Q = 0$

- change in elevation between the entry and exist point is negligible ; $(z_1 - z_2) = 0$

Accordingly

$$\frac{V_4^2}{2} = (h_3 - h_4) + \frac{V_3^2}{2}$$
$$= c_p(T_3 - T_4) + \frac{V_3^2}{2}$$

or $$\frac{V_4^2}{2 \times 1000} = 1.005(900 - 790) + \frac{45^2}{2 \times 1000}$$
$$= 111.0125$$

∴ Velocity at exit from the nozzle,

$$V_4 = \sqrt{111.0125 \times 2000} = \mathbf{471.2}\ \text{m/s}$$

7.7. THROTTLING PROCESS

Throttling is the expansion of fluid from high pressure to low pressure. This process occurs when fluid passes through an obstruction (partially opened valve or a small orifice) placed in the fluid flow passage.

Fig. 7.16. shows the schematics of porous plug experiment performed by Joule and Thomson in 1852. A stream of incompressible fluid (gas) is made to pass steadily through a porous plug placed in an insulated and horizontal pipe. The upstream conditions of pressure p_i and temperature T_i are held constant and the corresponding values at exit are measured. The friction of the narrow passage causes the pressure to drop and accordingly the exit pressure p_e is less than the intake pressure p_i.

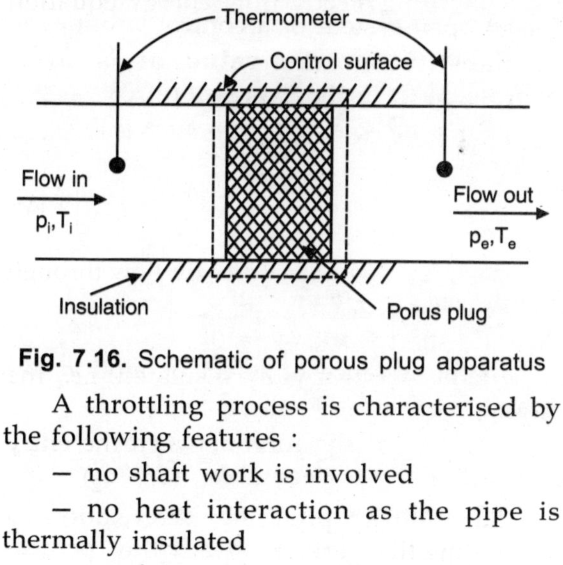

Fig. 7.16. Schematic of porous plug apparatus

A throttling process is characterised by the following features :

– no shaft work is involved

– no heat interaction as the pipe is thermally insulated

– no change in potential energy ($z_1 = z_2$) as the pipe is placed horizontally

– negligible changes in kinetic energy

With these stipulations the steady flow energy equation,

$$h_1 + \frac{V_1^2}{2} + gz_1 + q = h_2 + \frac{V_2^2}{2} + gz_2 + w_s$$

transforms to

$h_1 = h_2$, *i.e.*, enthalpy of fluid remains constant during throttling.

Thus the throttling expansion process is an ***isenthalpic process***. If the fluid undergoing throttling behaves as an ideal gas for which $h = c_pT$, we get

$$c_pT_1 = c_pT_2\ ;\ T_1 = T_2$$

Again for a perfect gas, internal energy is a function of temperature alone and equality of temperature implies that $u_1 = u_2$. Apparently a throttling process takes place at constant enthalpy, constant temperature and constant internal energy.

Throttling is an irreversible process and involves degradation of energy and its dissipation in turbulence.

Joule-Thomson coefficient, inversion point and inversion curve

For real gases, enthalpy is a function of both temperature and pressure. As such even though enthalpy remains constant during throttling, the temperature need not remain the same. Experimental test-runs can be conducted by keeping upstream conditions constant but with different down stream pressures. This is achieved by having porous plugs of different sizes. The exit temperature of the fluid at different exit pressures is measured. Since the upstream pressure and temperature conditions are kept constant, the enthalpy of the fluid for all measured conditions of exit pressure and temperature would be constant. The results are plotted as a constant enthalpy (isenthalpy) curve on $T-p$ diagram. Several enthalpy curves can be obtained by repeating the experiments with several inlet conditions.

The slope of an isenthalpic curve is called the Joule - Thomson coefficient, μ_{JT}. That is

$$\mu_{JT} = \left(\frac{\partial T}{\partial p}\right)_{h=\text{constant}} \qquad ...(7.4)$$

This coefficient may be +ve, –ve or zero. The point on the isenthalpic curve where $\mu_{JT} = 0$ is called the *inversion point*. Thus the inversion point denotes the maximum value of temperature on $T-p$ plot. The locus of all inversion points is called the inversion curve. The Joule - Thomson coefficient is positive on the left side of inversion curve, is zero at the inversion point, and is negative on the right side of inversion curve.

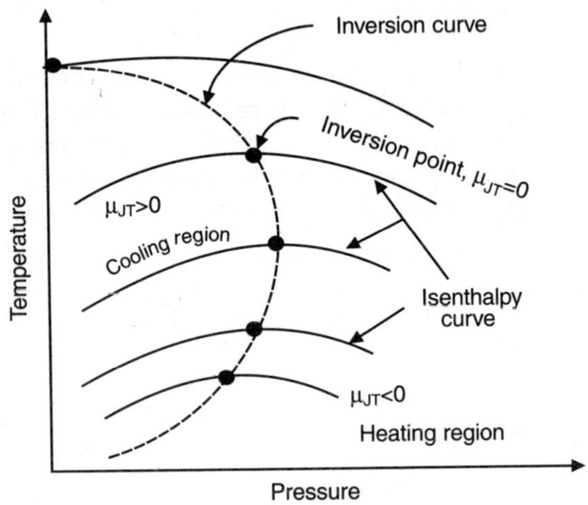

Fig. 7.17. Isenthalpy and inversion curves for a real gas

Throttling is always accompanied by pressure drop, *i.e.*, Δp is –ve. That leads to drop in temperature when μ_{JT} is +ve. Accordingly when a real gas is throttled at the condition such that its state lies to the left of inversion curve, the gas will get cooled. Thus the region of $\mu_{JT} > 0$ represents the region of cooling. Likewise, when μ_{JT} is –ve, the temperature change is +ve and therefore, the throttling of real gas would produce heating effect.

A knowledge of the inversion temperatures and inversion curves of real gases is of considerable importance in the design of refrigeration and liquefaction equipment.

Throttling process is commonly used

- to determine the dryness fraction of steam
- for the liquefaction of gases where μ_{JT} is positive
- to reduce the pressure of the refrigerant before entry into the evaporator of refrigeration plant.

7.8. FLOW WORK AND NON-FLOW WORK

The steady flow equation on unit mass basis is

$$h_1 + \frac{V_1^2}{2} + gz_1 + q = h_2 + \frac{V_2^2}{2} + gz_2 + w_{sf}$$

The suffix *sf* indicates the steady flow process.

In differential form,

$$\delta q - \delta w_{sf} = dh + d(V^2/2) + d(gz)$$

$$= d(u + pv) + d(V^2/2) + d(gz)$$

$$= du + p\,dv + v\,dp + d(V^2/2) + d(gz) \quad ...(i)$$

For a closed system, the reversible heat interaction δq is given by,

$$\delta q = du + p\,dv \quad ...(ii)$$

This also represents the heat interaction for an open system or a control volume.

Substituting the value of δq from expression (*i*) in the expression (*ii*),

$$du + p\,dv - \delta w_{sf} = du + p\,dv + v\,dp + d\left(\frac{V^2}{2}\right) + d\,(gz)$$

$$\text{or} \quad \delta w_{sf} = -v\,dp - d\left(\frac{V^2}{2}\right) - d\,(gz)$$

In the absence of effect of motion and gravity

$$\delta w_{sf} = -v\,dp\ ; \quad w_{sf} = -\int v\,dp \quad ...(7.5)$$

Equation 7.5 prescribes the relation for evaluating the work of a steady flow process.

For the non-flow process :

$$w_{nf} = \int p\,dv \quad ...(7.6)$$

Equations 7.5 and 7.6 for the steady flow work and non-flow work have been drawn on p - v plot in Fig. 7.18.

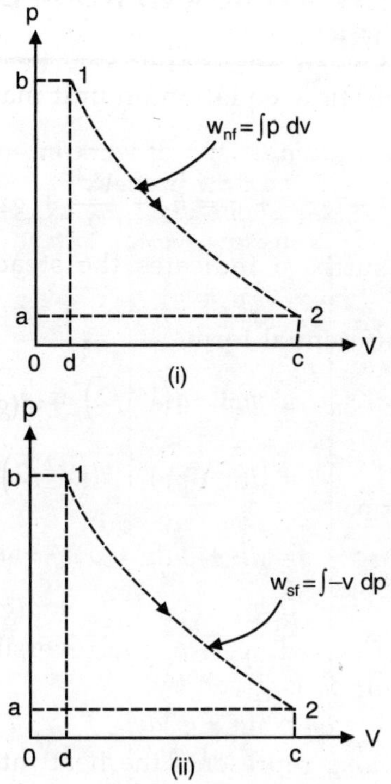

Fig. 7.18. (i) Non-flow work and (ii) Steady flow work

The area $(1-2-a-b)$ represents the work done in a steady flow process. That is

$$\text{area } (1-2-a-b) = -\int v\,dp = w_{sf}$$

From the p - v plots indicated in Fig. 7.18,

$$\text{area } (1-2-a-b) = \text{area } (1-2-c-d) + \text{area } (0-b-1-d) - \text{area } (0-a-2-c)$$

or
$$w_{sf} = \int p\,dv + p_1v_1 - p_2v_2$$

where,

p_1v_1 = flow work brought into the control volume by virtue of incoming fluid

p_2v_2 = flow work leaving the control volume by virtue of outgoing fluid

$\int p\,dv$ = work done within the control volume, *i.e.*, non-flow work.

The sum total of these work terms represents the work done in a steady flow process.

7.9. FLOW WORK PROCESSES

In any steady flow process, the work is prescribed by

$$\delta w_{sf} = -v\,dp - d(KE) - d(PE)$$

7.9.1. Isochoric (constant volume) process: v = const

$$-\int_1^2 v\,dp = -v\,(p_2 - p_1) = v\,(p_1 - p_2)$$

$$\therefore \quad w_{sf} = v\,(p_1 - p_2) + \frac{V_1^2 - V_2^2}{2} + g\,(z_1 - z_2)$$

Neglecting changes in kinetic energy and potential energy,

$$w_{sf} = v\,(p_1 - p_2)$$

7.9.2. Isobaric (constant pressure) process: p = const

$$-\int_1^2 v\,dp = 0$$

$$\therefore \quad w_{sf} = 0 + \frac{V_1^2 - V_2^2}{2} + g\,(z_1 - z_2)$$

Neglecting changes in kinetic energy and potential energy,

$$w_{sf} = 0$$

7.9.3. Isothermal (constant temperature) process : pv = const

$$-\int_1^2 v\,dp = -\int_1^2 \frac{RT_1}{p}dp = RT_1 \log_e \frac{p_1}{p_2}$$

$$\therefore \quad w_{sf} = RT_1 \log_e \frac{p_1}{p_2} + \frac{V_1^2 - V_2^2}{2} + g\,(z_1 - z_2)$$

Neglecting changes in kinetic energy and potential energy,

$$w_{sf} = RT_1 \log_e \frac{p_1}{p_2}$$

$$= RT_1 \log_e \frac{v_2}{v_1}$$

7.9.4. Polytropic process :

$$pv^n = \text{const};\ v = \left(\frac{C}{p}\right)^{1/n}$$

$$-\int_1^2 v\,dp = -\int_1^2 \left(\frac{C}{p}\right)^{1/n} dp$$

$$= -C^{1/n}\left|\frac{p^{-\frac{1}{n}+1}}{\frac{-1}{n}+1}\right|_1^2$$

$$= -\frac{n}{n-1}\left|p^{\frac{n-1}{n}}C^{\frac{1}{n}}\right|_1^2$$

$$= -\frac{n}{n-1}\left|p^{\frac{n-1}{n}}p^{\frac{1}{n}}v\right|_1^2$$

$$= -\frac{n}{n-1}|pv|_1^2$$

$$= \frac{n}{n-1}(p_1v_1 - p_2v_2)$$

$$= \frac{n}{n-1}R(T_1 - T_2)$$

$$\therefore \quad w_{sf} = \frac{n}{n-1}R(T_1 - T_2) + \frac{V_1^2 - V_2^2}{2} + g(z_1 - z_2)$$

Neglecting changes in kinetic energy and potential energy,

$$w_{sf} = \frac{n}{n-1}R(T_1 - T_2)$$

During polytropic expansion $T_1 > T_2$ and so w_{sf} is +ve which implies that work is done by the system.

The work interaction during *reversible adiabatic process* pv^γ = constant can be calculated in the same manner by simply replacing polytropic index n by adiabatic index γ.

$$w_{sf} = \frac{\gamma}{\gamma-1}(p_1v_1 - p_2v_2)$$

$$= \frac{\gamma}{\gamma-1}R(T_1 - T_2)$$

Refer Table 7.1 for comparison of expressions for work done in a non-flow and in a steady flow process undergone by an ideal gas.

Table 7.1. Comparison of work in non-flow and flow process

Process	Non-flow work $\int p\,dv$	Steady flow $-\int v\,dp$
• isochoric ($v = c$)	0	$v(p_1 - p_2)$
• isobaric ($p = c$)	$p(v_2 - v_1)$	0
• isothermal ($t = c$)	$p_1v_1 \log_e \frac{v_2}{v_1}$	$p_1v_1 \log_e \frac{v_2}{v_1}$
• polytropic ($pv^n = c$)	$\frac{p_1v_1 - p_2v_2}{n-1}$	$\frac{n}{n-1}(p_1v_1 - p_2v_2)$
• adiabatic ($pv^\gamma = c$)	$\frac{p_1v_1 - p_2v_2}{\gamma-1}$	$\frac{\gamma}{\gamma-1}(p_1v_1 - p_2v_2)$

7.10. UNSTEADY FLOW PROCESSES

In the steady flow analysis made above, there is neither a change in the energy of control volume nor in the properties of fluid at any section with time. However, in many of thermodynamic processes such as in the filling or evacuation of a tank, the internal energy as well as mass of the tank changes with time. Likewise, the condition of water in the cylinder jacket of an internal combustion engine is time dependent. Such processes are referred to as unsteady flow processes.

7.10.1. Tank filling process

The tank or bottle initially contains fluid having mass m_i at the state p_i, v_i and T_i. The corresponding values after the filling operation are m_f, p_f, v_f and T_f. In addition, there may be heat and work interactions with the surroundings. The work interaction is possible by change in volume of the bottle or by internal electrical/mechanical devices.

Mass of fluid entering $= m_f - m_i$

Energy of entering fluid $= (m_f - m_i)\,e$

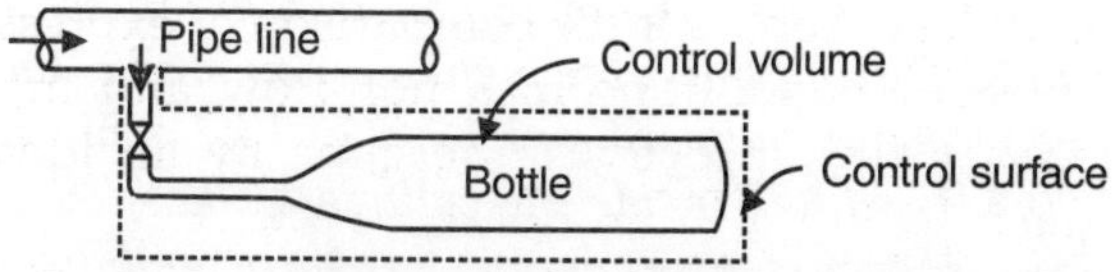

Fig. 7.19. Filling process of a bottle

Energy balance then gives :

$$(m_f - m_i)\, e_p + Q = (m_f u_f - m_i u_i) + W$$

where u_f and u_i denote the final and initial specific internal energy of the fluid mass. In the filling process of a bottle from a large reservoir (pipeline), the properties of the entering fluid stream are essentially constant and so will be the energy e_p accompanying unit mass of fluid at entrance to control volume.

$$e_p = u_p + p_p v_p + \frac{V_p^2}{2} = h_p + \frac{V_p^2}{2}$$

$$\therefore\ (m_f - m_i)\left(h_p + \frac{V_p^2}{2}\right) + Q = (m_f u_f - m_i u_i) + W$$

The suffix p refers to state of fluid in the pipe line.

In case of thermally insulated bottle and in the absence of any work transfer,

$$Q = 0 \quad \text{and} \quad W = 0$$

That gives,

$$(m_f - m_i)\left(h_p + \frac{V_p^2}{2}\right) = m_f u_f - m_i u_i$$

Neglecting kinetic energy of incoming fluid,

$$(m_f - m_i)\, h_p = m_f u_f - m_i u_i$$

If the bottle is initially empty ($m_i = 0$),

$$m_f h_p = m_f u_f\ ; \quad h_p = u_f$$

Thus the specific internal energy of fluid charged into empty insulated bottle is equal to the specific enthalpy of the filling fluid in the charging pipe line.

If the fluid is an ideal gas, then temperature of gas in the bottle after it is charged is given by

$$c_p T_p = c_v T_2\ ; \quad T_2 = \gamma T_p$$

7.10.2. Tank emptying process

The tank emptying process is the reverse of filling process, *i.e.*, there is flow of fluid from the tank (bottle) to the surroundings. The surroundings are much larger than the tank being emptied and so the energy,

$$e_p = h_p + V_p^2/2$$

accompanying unit mass of fluid at exit from the control volume will be constant.

Analogous to filling process, the energy balance can be written as

$$(m_i - m_f)\left(h_p + \frac{V_p^2}{2}\right) + Q = m_i u_i - m_f u_f$$

where h_p and V_p are the specific enthalpy and velocity of leaving fluid.

For no heat transfer and negligible exit velocity

$$(m_i - m_f)\, h_p = m_i u_i - m_f u_f$$

Further if the tank is to be fully emptied ($m_f = 0$)

$$m_i h_p = m_i u_i\ ; \quad h_p = u_i$$

EXAMPLE 7.18

A household gas cylinder initially evacuated is filled by 15 kg gas supply of enthalpy 625 kJ/kg. After filling, the gas in the cylinder has the following parameters :

pressure 10 bar; enthalpy 750 kJ/kg and specific volume 0.0487 m³/kg

Evaluate the heat received by the cylinder from the surroundings.

Solution : The energy balance equation for a filling process is given by

$$(m_f - m_i)\left(h_p + \frac{V_p^2}{2}\right) + Q = m_f u_f - m_i u_i + W$$

The suffix p refers to the state of gas in the intake pipe.

Neglecting kinetic energy of gas in the pipe line and any work interaction

$$(m_f - m_i)\, h_p + Q = m_f u_f - m_i u_i$$

The cylinder is initially completely evacuated, *i.e.*, $m_i = 0$ and therefore

$$m_f\, h_p + Q = m_f u_f$$

Given : $(m_f - m_i) = 15$ kg or $m_f = 15$ kg and noting that $u_f = h_f - p_f v_f$, we get

$$Q = m_f u_f - m_f h_p = m_f\left[(h_f - p_f v_f) - h_p\right]$$

$$= 15\left[\left(750 - \frac{10\times10^5\times0.0487}{10^3}\right) - 625\right]$$

$$= \mathbf{1144.5\ kJ}$$

EXAMPLE 7.19

0.56 m³ of air at 0.2 MPa is contained in a fully insulated rigid tank. The tank communicates through a valve with a pipe line carrying high pressure air at 300 K temperature. The valve is opened and the air is allowed to flow into the tank until the pressure of air in the tank is raised to 1 MPa. Determine the mass of air that enters the tank. Neglect kinetic energy of incoming air. For air take:

$$\gamma = 1.4 \quad and \quad R = 287\ J/kg\ K$$

Solution : The situation corresponds to a filling process for which the energy balance equation is given by

$$(m_f - m_i)\left(h_p + \frac{V_p^2}{2}\right) + Q = m_f u_f - m_i u_i + W$$

With the stipulations of no heat transfer (insulated tank) and work interaction (rigid tank) and neglecting kinetic energy of air in the supply pipe line, the energy equation reduces to

$$(m_f - m_i)\ h_p = m_f u_f - m_i u_i$$

or $(m_f - m_i)\ c_p T_p = m_f c_v T_f - m_i c_v T_i$

From characteristic gas equation, $pV = mRT$

$$m_f T_f = \frac{p_f V}{R}$$

and $m_i T_i = \frac{p_i V}{R} \qquad (V_i = V_f = V)$

$$\therefore\ (m_f - m_i)\ c_p T_p = c_v \frac{V}{R}(p_f - p_i)$$

or $(m_f - m_i) = \frac{1}{T_p}\frac{V}{\gamma R}(p_f - p_i)$

$$= \frac{1}{300}\times\frac{0.56}{1.4\times287}\left(1\times10^6 - 0.2\times10^6\right)$$

$$= \mathbf{3.72\ kg}$$

The mass of air that enters the tank from the supply pipe line is 3.72 kg.

EXAMPLE 7.20

After the completion of exhaust stroke of an IC engine, the piston cylinder assembly remains filled up with 1 × 10⁻⁴ kg of combustion products at 800 K. During the subsequent suction stroke, the piston moves outward and 16 × 10⁻⁴ kg of air at 290 K is sucked inside the cylinder. The suction process occurs at constant pressure and heat interaction is negligible. Evaluate the temperature of gases at the end of suction stroke. For air and gases, take c_p = 1 kJ/kg.

Solution : The energy equation for a filling process is given by

$$(m_f - m_i)\left(h_p + \frac{V_p^2}{2}\right) + Q = m_f u_f - m_i u_i + W$$

The suffix p refers to state of fluid in the intake pipe. With the stipulation of no heat interaction and neglecting kinetic energy of air in the supply pipe line,

$$(m_f - m_i)\ h_p = m_f u_f - m_i u_i + W$$

The suction occurs at constant pressure and hence

$$W = \int_i^f p\,dv = p\,(V_f - V_i)$$

$$= p\,(m_f v_f - m_i v_i)$$

$$= m_f (p_f v_f - p_i v_i)$$

$$\therefore\ (m_f - m_i)h_p = m_f u_f - m_i u_i + m_f(p_f v_f - p_i v_i)$$

$$= m_f(u_f + p_f v_f) - m_i(u_i + p_i v_i)$$

$$= m_f h_f - m_i h_i$$

or $(m_f - m_i)\ c_p T_p = m_f c_p T_f - m_i c_p T_i$

or $(m_f - m_i)\ T_p = m_f T_f - m_i T_i$

Substituting the given data, we get

$$(16\times10^{-4} - 1\times10^{-4})\times290 = 16\times10^{-4}\ T_f - 1\times10^{-4}\times800$$

$$4350\times10^{-4} = 16\times10^{-4}\ T_f - 800\times10^{-4}$$

∴ Temperature of gas at the end of suction stroke,

$$T_f = \frac{(4350 + 800)\times10^{-4}}{16\times10^{-4}} = \mathbf{321.87\ K}$$

EXAMPLE 7.21

An insulated rigid tank contains 8 kg of air at 1.5 bar pressure and 310 K temperature. It is filled with air from a large reservoir at 15 bar and 335 K. If the air behaves as a perfect gas, make calculations for the amount of air added and its temperature.

Solution : The situation corresponds to a filling process for which the energy balance equation is given by

$$(m_f - m_i)\left(h_p + \frac{V_p^2}{2}\right) + Q = m_f u_f - m_i u_i + W$$

The suffix p prefers to state of air in the supply pipe line.

With the stipulations of no heat transfer (insulated tank) and work interactions (rigid tank) and neglecting kinetic energy of air in the supply pipeline, the energy equation reduces to

$$(m_f - m_i)\, h_p = m_f u_f - m_i u_i$$

or $$(m_f - m_i)\; c_p\, T_p = c_v(m_f T_f - m_i T_i)$$

or $$(m_f - m_i)\, T_p = \frac{m_f T_f - m_i T_i}{\gamma} \quad ...(i)$$

Here, $m_i = \frac{p_i V_i}{RT_i}$ and $m_f = \frac{p_f V_f}{RT_f}$

Upon division $$\frac{m_f}{m_i} = \frac{p_f}{p_i}\frac{T_i}{T_f} \quad (\because V_i = V_f)$$

$$= \frac{15}{1.5}\frac{T_i}{T_f} = 10\frac{T_i}{T_f}$$

or $$\frac{m_f}{m_i}\frac{T_f}{T_i} = 10 \quad ...(ii)$$

Substituting this result in expression (*i*)

$$(m_f - m_i)\, T_p = \frac{(10-1)m_i T_i}{\gamma}$$

$$(m_f - 8)\; 335 = \frac{9\times 8\times 310}{1.4};\; m_f = 55.60 \text{ kg}$$

∴ Mass added

= 55.60 – 8 = **47.60 kg**

and, $T_f = 10\frac{m_i T_i}{m_f}$ from expression (*ii*)

$$= \frac{10\times 8\times 310}{55.60} = \mathbf{446.04\ K}$$

EXAMPLE 7.22

1.25 kg of air at 25°C is contained in a storage cylinder of fixed capacity. The cylinder is filled by opening a valve on the cylinder by taking atmospheric air which has the state conditions mentioned below :

Temperature 25°C ; pressure 101.3 kPa ; and specific volume 0.85 m³/kg.

After 30 minutes of intake, the mass of air in the bottle becomes 3.0 kg and the temperature rises to 35°C. Determine the average rate of heat transfer.

Neglect work interaction and kinetic energy of air in supply pipe line and assume the following relation between internal energy u in kJ/kg and temperature t in degree centigrade

$$u = u_0 + 0.148\ t$$

Solution : The energy equation for a filling process is given by

$$(m_f - m_i)\left(h_p + \frac{V_p^2}{2}\right) + Q = m_f u_f - m_i u_i + W$$

The suffix p refers to state of fluid in the intake pipe.

With the stipulation of no work interaction and neglecting kinetic energy of gas in the supply pipe line,

$$(m_f - m_i)\; h_p + Q = m_f u_f - m_i u_i$$

Now, $m_f = 3$ kg ; $m_i = 1.25$ kg

and $m_f - m_i = 1.75$ kg

$$u_f = u_0 + 0.418 \times 35$$
$$= (u_0 + 14.63) \text{ kJ/kg}$$
$$u_i = u_0 + 0.418 \times 25$$
$$= (u_0 + 10.45) \text{ kJ/kg}$$
$$h_p = u_p + p_p v_p$$
$$= (u_0 + 0.418 \times 25) + \frac{101.3\times 10^3 \times 0.85}{10^3}$$
$$= (u_0 + 10.45 + 86.11) \text{ kJ/kg}$$
$$Q = m_f u_f - m_i u_i - (m_f - m_i)\, h_p$$
$$= 3\,(u_0 + 14.63) - 1.25\,(u_0 + 10.45) - 1.75\,(u_0 + 10.45 + 86.11)$$
$$= -\,138.15 \text{ kJ}$$

∴ Average rate of heat transfer,

$$= \frac{-138.15}{30} = \mathbf{-\ 4.605\ kJ/min}$$

EXAMPLE 7.23

An insulated and rigid tank contains 5 cm³ of air at 10 bar and 425 K. The air is then let off to atmosphere through a valve. Determine the work obtainable by utilizing the kinetic energy of the discharge air.

Take c_p = 1 kJ/kg K, c_v = 0.714 kJ/kg K, atmospheric pressure = 1 bar

Solution : The situation corresponds to an emptying process for which the energy balance equation is given by

$$(m_i - m_f)\left(h_p + \frac{V_p^2}{2}\right) + Q = (m_i\, u_i - m_f\, u_f) + W$$

With the stipulations of no heat transfer (insulated tank) and work interaction (rigid tank), the energy equation reduces to

$$(m_i - m_f)\left(h_p + \frac{V_p^2}{2}\right) = (m_i\, u_i - m_f\, u_f)$$

The suffix p refers to state of fluid in the discharge pipe.

$$(m_i - m_f)\,\frac{V_p^2}{2} = m_i\, u_i - m_f\, u_f - (m_i - m_f)\, h_p$$

$$\gamma = \frac{c_p}{c_v} = \frac{1.0}{0.714} \simeq 1.4, \quad p_f = 1 \text{ bar}$$

$$R = c_p - c_v = 1 - 0.714 = 0.286 \text{ kJ/kg K} = 286 \text{ J/kg K}$$

$$\frac{T_i}{T_f} = \left(\frac{p_i}{p_f}\right)^{\frac{\gamma-1}{\gamma}} = \left(\frac{10}{1}\right)^{\frac{1.4-1}{1.4}} = 1.932$$

$$\therefore \quad T_f = \frac{T_i}{1.932} = \frac{425}{1.932} = 220 \text{ K}$$

$$m_i = \frac{p_i V_i}{RT_i} = \frac{10 \times 10^5 \times 5}{286 \times 425} = 41.14 \text{ kg}$$

$$m_f = \frac{p_f V_f}{RT_f} = \frac{1 \times 10^5 \times 5}{286 \times 220} = 7.95 \text{ kg}$$

Now :

$$m_i u_i = m_i\, c_v\, T_i = 41.14 \times 0.714 \times 425 = 12483.93 \text{ kJ}$$

$$m_f u_f = m_f\, c_v\, T_f = 7.95 \times 0.714 \times 220 = 1248.78 \text{ kJ}$$

$$(m_i - m_f)\, c_p\, T_f = (41.14 - 7.95) \times 1.0 \times 220 = 7301.8 \text{ kJ}$$

$$\therefore\ (m_i - m_f)\,\frac{V_p^2}{2} = 12483.93 - 1248.78 - 7301.8 = 3933.35 \text{ kJ}$$

SALIENT POINTS

1. The mass flow rate $\dot{m}$ is the quantity which passes a reference point per unit time; its value is expressed in kg/s
$$\dot{m} = \rho A V = \frac{AV}{v}$$
where V is the flow velocity
For steady flow (no accumulation or diminution of matter within the machine)
$$\dot{m} = \frac{A_1 V_1}{v_1} = \frac{A_2 V_2}{v_2}$$
2. A flow process constitutes an open system through which the working fluid enters and leaves from the surface of the system.
3. Flow work or flow energy refers to the work which a fluid does when it moves in or out of any passage. It represents the product of pressure and specific volume of the flowing fluid.
4. The general energy equation for flow through a machine is written as
$$m_1\left(u_1 + p_1 v_1 + \frac{V_1^2}{2} + g z_1\right) + Q = m_2\left(u_2 + p_2 v_2 + \frac{V_2^2}{2} + g z_2\right) + W_s$$
For steady flow $m_1 = m_2 = m$ and by definition $h = u + p\,v$
Then,
$$m\left(h_1 + \frac{V_1^2}{2} + g z_1\right) + Q = m\left(h_2 + \frac{V_2^2}{2} + g z_2\right) + W_s$$

5. The steady flow energy equation can be written on the basis of unit mass or on the basis of unit time.

On unit mass basis:

$$m\left(h_1 + \frac{V_1^2}{2} + gz_1\right) + Q$$

$$= m\left(h_2 + \frac{V_2^2}{2} + gz_2\right) + W_s$$

- Each term inside the brackets is expressed per unit mass and has units of energy
- Normally : Q and W_s are in kJ
  m is in kg ; h is in kJ/kg
  v is in m/s ; z is in m and
  g is in m/s²

$$m\left(h_1 + \frac{V_1^2}{2\times 10^3} + \frac{gz_1}{10^3}\right) + Q$$

$$= m\left(h_2 + \frac{V_2^2}{2\times 10^3} + \frac{gz_2}{10^3}\right) + W_s$$

On unit time basis:

$$\dot{m}\left(h_1 + \frac{V_1^2}{2\times 10^3} + \frac{gz_1}{10^3}\right) + \dot{Q}$$

$$= \dot{m}\left(h_2 + \frac{V_2^2}{2\times 10^3} + \frac{gz_2}{10^3}\right) + \dot{W}_s$$

where all terms represent energy flow per unit time (kJ/s).

6. Throttling is the expansion of fluid from high pressure to low pressure.
   For an ideal gas, throttling process takes place at constant enthalpy, constant temperature and constant internal energy.
7. The slope of a constant enthalpy line or a p-T diagram at a particular state may have a positive, zero or a negative value. The slope is called Joule-Thomson coefficient, μ and is given by

$$\mu = \left(\frac{\partial T}{\partial p}\right)_h = 0 \text{ for an ideal gas.}$$

   The states where $\mu = 0$ are called inversion states and the locus of these states is called the inversion curve.
8. An unsteady flow manifests when the rates at which mass and energy enter the control volume are not equal to the rates at which mass and energy move out of the control volume.
9. The processes of emptying and filling a vessel represent unsteady flow and are taken as adiabatic.

REVIEW QUESTIONS

1. What is a steady flow process ? State all the assumptions made for such a flow process.
2. Explain the concept of flow work.
3. Write the steady flow energy equation and point out the significance of various terms involved.
4. How does control volume differ from a system?
5. Differentiate between pv and $p\,dv$ and give examples of the inclusion of these terms in energy equation.
6. Make an energy analysis of the following devices:
   steam turbine, rotary compressor, heat exchanger, combustion chamber and centrifugal pump
7. What is the significance of $p\,dv$ and $v\,dp$ areas on $p-v$ diagram ?
8. Show that the work done in a steady flow process is given by $-\int p\,dv$
9. What is a throttling process? Point out its salient aspects.
10. Define Joule–Thomson coefficient, inversion point and inversion curve.
11. Sketch the isenthalpic curves for a real gas and indicate there upon the inversion curve.
12. Write down the energy equation for (*a*) filling of a tank, and (*b*) emptying of tank.
13. Refrigerant Freon-12 enters the condenser of a refrigeration plant with enthalpy 223.75 kJ/kg and leaves with enthalpy 64.60 kJ/kg. Cooling water enters at 15°C and leaves at 20°C. Make calculations for the mass flow rate of cooling water for unit mass of refrigerant.
    [Ans. 7.60 kg**]**

14. Air at 100 kPa and 290 K flows steadily through a compressor at a rate of 5 m^3/s. During the compression process, the pressure and temperature of air are respectively raised to 250 kPa and 400 K. There is also a heat loss of 15 kJ/s to the cooling water. Determine the power required to drive the compressor. Presume that air behaves as a perfect gas and neglect changes in kinetic energy and potential energy.

[Ans. 679.07 kW]

15. A gas turbine receives 60 kg/min of fuel-air mixture at 4 MPa pressure, 1320 K temperature with a velocity of 200 m/s. At exit the mixture has 0.1 MPa pressure and 100 m/s velocity. The turbine is well insulated and the flow process can be assumed adiabatic. Evaluate the power output of the turbine. Neglect changes in elevation and take $c_p = 1.05$ kJ/kg K, $\gamma = 1.4$ and $R = 300$ J/kg K.

[Ans. 918 kW]

16. 0.8 kg/s of air flows through a compressor under steady state conditions. The properties of air at entry are : pressure 1 bar, velocity 10 m/s, specific volume 0.95 m^3/kg and internal energy 30 kJ/kg. The corresponding values at exit are 8 bar, 6 m/s, 0.2 m^3/kg and 124 kJ/kg. The outlet is in-line with the intake. Determine the power input to compressor and the pipe diameter at entry and exit.

[Ans. 175.2 kW, 0.311m, 0.184 m]

17. A steady flow of water at enthalpy 250 kJ/kg enters a section of the heating plant of a building in which there are no pumps. The water leaves the section at enthalpy 200 kJ/kg. The exit pipe is 25 m above the inlet pipe.
Neglecting changes in kinetic energy, make calculations for the heat transfer from the water.

[Ans. 49.755 kJ/kg]

18. The gas leaving the turbine jet engine flows steadily into the jet pipe with enthalpy 960 kJ/kg and velocity 250 m/s relate to the pipe. The exit from the pipe is at enthalpy 860 kJ/kg, and the exhaust is in line with intake. Neglecting heat loss from the system, determine the relative velocity of gas leaving the pipe.

[Ans. 512.35 m/s]

19. A thermodynamic system operates under steady flow conditions, the fluid entering at 2 bar and leaving at 10 bar. The entry velocity is 30 m/s and the exit velocity is 10 m/s. During the process 25 MJ/hr of heat from an external source is supplied and the increase in enthalpy is 5 kJ/kg. The exit point is 20 m above the entry point. Determine flow of work from the system if fluid flow rate is 45 kg/min.

[Ans. 3.35 kW]

20. A centrifugal compressor delivers 16 kg of air per minute at a pressure ratio 7 : 1, the suction pressure being 1 bar. The flow velocity and the specific volume at inlet are 5 m/s and 0.45 m^3/kg. The corresponding values at exit are 8 m/s and 0.15 m^3/kg. When the air flows through the compressor, its enthalpy increases by 150 kJ/kg and the heat loss to the surroundings is 800 kJ/min. Neglecting any change in potential energy, make calculations for the power required to drive the compressor, and the ratio of inlet pipe diameter to outlet pipe diameter.

[Ans. 53.33 kW, 2.19]

21. A vessel of constant volume 0.3 m^3 contains air at 1.5 bar and is connected, via a valve, to a large main carrying air at a temperature of 38 °C and high pressure. The valve is opened allowing air to enter the vessel and raising the pressure therein to 7.5 bar. Assuming the vessel and valve to be thermally insulated, find the mass of air entering the vessel.

[Ans. 1.44 kg]

□□□

Chapter 8

Second Law of Thermodynamics

Learning objectives : A study of the subject matter included in this chapter will enable the reader to have an understanding of :

- *limitations of first law*
- *heat source and heat sink; heat engine and heat pump/refrigerator*
- *thermal efficiency of heat engine, and coefficient of performance of heat pump/refrigerator*
- *Kelvin-Planck and Clausius statements of second law of thermodynamics*
- *Carnot cycle for heat engines and heat pumps/refrigerators*
- *Carnot theorem, thermodynamic temperature scale and Clausius inequality*

8.1. LIMITATIONS OF FIRST LAW

First law of thermodynamics deals with conservation and conversion of energy. It stipulates that when a thermodynamic process is carried out, energy is neither gained nor lost. Energy only transforms from one form into another and the energy balance is maintained. The law, however, fails to state the conditions under which energy conversions are possible. The law presumes that any change of a thermodynamic state can take place in either direction. However, this is not true; particularly in the inter-conversion of heat and work. Processes proceed spontaneously in certain directions but not in opposite directions, even though the reversal of processes does not violate the first law.

For example

(1) Consider a running automobile vehicle stopped by applying brakes. The process changes the kinetic energy of the vehicle into heat and the brakes get heated up. The increase in the internal energy of the brakes is in accordance with the first law. The stopping of the vehicle by friction, however, constitutes an irreversible process. Cooling of brakes to their initial state never puts the vehicle back into motion. Heat in the brakes cannot be converted back into mechanical work even though that act would not violate the principle of energy conversion.

(2) The temperature of liquid contained in a vessel rises when it is churned by paddle work. However, the paddle work can not be restored on cooling the liquid to its original state.

(3) Consider two bodies A and B at different temperatures t_A and t_B ($t_A > t_B$) in thermal contact into an insulated box. Energy in the form of heat will flow from body A to body B because of temperature difference. This energy flow continues till both the bodies attain the same temperature level at some intermediate value between these original temperatures. In conformity with first law, heat lost by body A equals the heat gained by body B. First law would be equally satisfied if the reverse process were to occur, *i.e.*, the energy in the same amount gets transferred from body B to body A so that both bodies regain their initial temperatures. This, however, is never seen to happen.

(4) When a block slides down a rough plane, it becomes warmer. However, the reverse process where the block slides up the plane and becomes cooler is not true even though the first law will still hold good.

(5) Work converts easily and completely into heat. However, there does exist a maximum limit upto which conversion of heat is possible in a heat engine. As a source of energy, work is superior to heat, and a complete conversion of low grade energy (heat) into high grade energy (work) is impossible.

(6) Electric current flowing through a resistor produces heat. Electric current once dissipated as heat cannot be converted back into electricity.

(7) Water flows from a higher level to a lower level, and the reverse is not automatically possible. A mechanical energy from an external source would be required to pump the water back from the lower level to higher level.

(8) Fuel (coal, diesel, petrol) burns with air to form the products of combustion. Fuel once burnt cannot be restored back to its original form.

(9) When hydrogen and oxygen are kept in an isolated system, they produce water on chemical initiation. However, the water left in the isolated system never dissociates into hydrogen and oxygen.

(10) Persons always grow old...............

The examples cited above indicate that:

- first law fixes the exchange rate between heat and work, and places no restrictions on the direction of change.
- processes proceed spontaneously in certain directions, but the reverse is not automatically attainable even though the reversal of the processes does not violate the first law.
- first law provides a necessary but not a sufficient condition for a process to occur, and
- there does exist some directional law which would tell whether a particular process occurs or not. Answer is provided by the second law of thermodynamics.

8.2. THERMAL RESERVOIR

A thermal (heat) reservoir is that part of environment which can exchange heat energy with a system. It has sufficiently large heat capacity and its temperature is not affected by the quantity of heat transferred to or from it. A thermal reservoir is thus characterised by its temperature which remains constant. The changes that do take place in the thermal reservoir as heat enters or leaves are so slow and so small that processes within it are quasi-static. The reservoir which is at high temperature and supplies heat is known as ***heat source***. Examples are a boiler furnace, a combustion chamber and a nuclear reactor etc. The reservoir which is at low temperature and to which heat is transferred is called the ***heat sink***. Atmospheric air, ocean and river etc. constitute the heat sink.

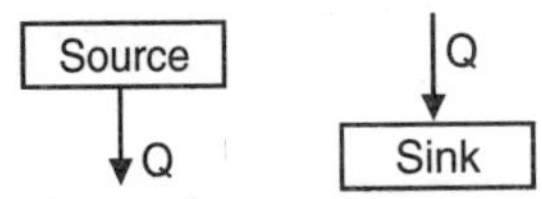

Fig. 8.1. Heat source and heat sink

8.3. HEAT ENGINE, REFRIGERATOR AND HEAT PUMP

A heat engine is a thermodynamic device used for continuous production of work from heat when operating in a cyclic process. Both heat and work interactions take place across the boundary of this cyclically operating device.

A heat engine is characterised by the following features:

– reception of heat Q_1 from a high temperature source at T_1

– partial conversion of heat received to mechanical work W

– rejection of remaining heat Q_2 to a low temperature sink at temperature T_2

– cyclic/continuous operation and

– working substance flowing through the engine.

The performance of any machine is expressed as the ratio of *what we want* to *what we have to pay for*. In the context of an engine, work is obtained at the expense of heat input. Accordingly, the performance of a heat engine is given by net work output to the entire amount of heat supplied to the working medium, and this ratio is called ***thermal efficiency***, η_{th} (Thermal efficiency is a measure of the degree of useful utilization of heat received in a heat engine).

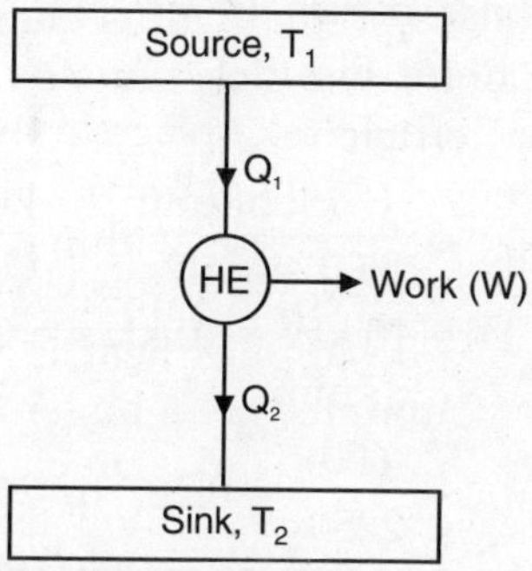

Fig. 8.2. Energy interactions in a heat engine

$$\eta_{th} = \frac{\text{net work output}}{\text{total heat supplied}}$$

Application of the principle of energy conservation (First law) to the heat engine, which undergoes a cycle gives : $W = Q_1 - Q_2$

$$\therefore \quad \eta_{th} = \frac{Q_1 - Q_2}{Q_1} = 1 - \frac{Q_2}{Q_1} \quad ...(8.1)$$

Obviously, thermal efficiency of a heat engine operating between two thermal reservoirs is always less than unity. To increase the thermal efficiency, it is necessary to reduce Q_2 (heat rejected) with Q_1 (heat supplied) remaining constant. Thermal efficiency could be equal to unity if $Q_1 \rightarrow \infty$ and $Q_2 = 0$ which, however, can not be realized in practice.

Refrigerators and heat pumps are *reversed* heat engines. The adjective *reversed* means operating backwards. The direction of heat and work interactions are opposite to that of a heat engine, *i.e.*, work input and heat output. These machines (refrigerators and heat pumps) are used to remove heat from a body at low temperature level and then transfer this heat to another body at high potential of temperature. When the main purpose of the machine is to remove heat from the cooled space, it is called a ***refrigerator***. A refrigerator operates between the temperature of surroundings and a temperature below that of the surroundings. Refrigerators are essentially used to preserve food items and drugs at low temperature.

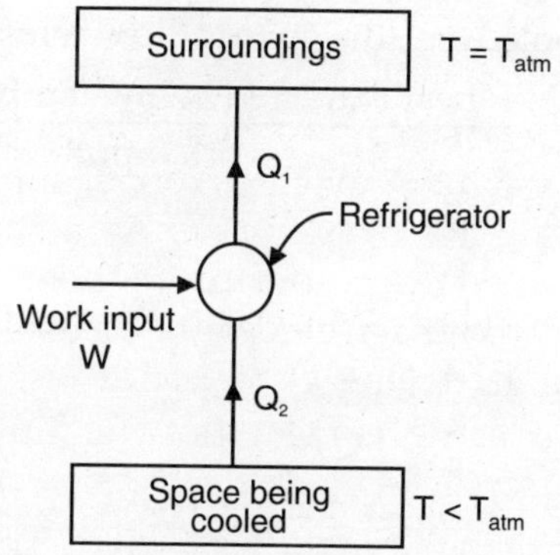

Fig. 8.3. (a) Energy interactions in a refrigerator

The term ***heat pump*** is applied to a machine whose objective is to heat a medium which may already be warmer than its surroundings. A heat pump thus operates between the temperature of the surroundings and a temperature above that of the surroundings. Heat pumps are generally used to keep the rooms warm in winter. The transfer of heat against a reverse temperature gradient in a refrigerator and heat pump is accomplished by supplying energy to the machine. A schematic representation of heat pump and a refrigerator has been shown in Fig. 8.3.

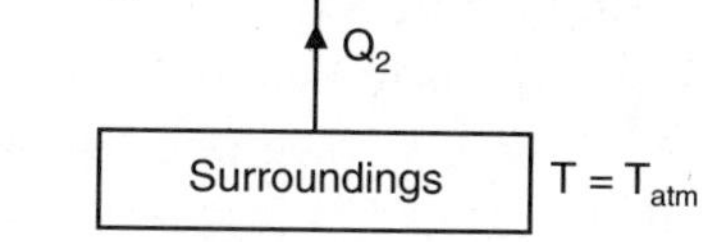

Fig. 8.3. (b) Energy interaction in a heat pump

The performance of a refrigerator and heat pump is expressed in terms of *coefficient of performance* (*COP*) which represents the ratio of desired effect to work input.

$$COP = \frac{\text{desired effect}}{\text{work input}}$$

In a refrigerator, the desired effect is the amount of heat Q_2 extracted from the space being cooled, *i.e.*, the space at low temperature

$$(COP)_{ref} = \frac{\text{heat extracted at low temperature}}{\text{work input}}$$

$$= \frac{Q_2}{W}$$

From energy conservation principle:

$$W = Q_1 - Q_2$$

$$\therefore\ (COP)_{ref} = \frac{Q_2}{Q_1 - Q_2} \quad \text{...(8.2)}$$

For most of the refrigerating machines, the values of COP lie between 3 and 4, and the COP are greatest when temperature differences are least.

In a heat pump, the desired effect is the amount of heat Q_1 supplied to the space being heated.

$$(COP)_{heat\ pump} = \frac{Q_1}{Q_1 - Q_2} \quad \text{...(8.3)}$$

$$= 1 + \frac{Q_2}{Q_1 - Q_2} = 1 + (COP)_{ref}$$

Thus the COP of a machine operating as a heat pump is higher than the COP of the same machine when operating as a refrigerator by unity.

EXAMPLE 8.1

A heat engine produces work equivalent to 80 kW with an efficiency of 40 percent. Determine the heat transfer rate to and from the working fluid.

Solution : Refer Fig. 8.2.

Thermal efficiency,

$$\eta_{th} = \frac{\text{work output}}{\text{heat supplied}} = \frac{W}{Q_1}$$

$\therefore$ Heat supplied to engine,

$$Q_1 = \frac{W}{\eta_{th}} = \frac{80}{0.4} = \mathbf{200\ kW}$$

From the principle of energy conservation (1st law): $Q_1 = W + Q_2$

$\therefore$ Heat rejected by the engine,

$$Q_2 = Q_1 - W = 200 - 80 = \mathbf{120\ kW}$$

EXAMPLE 8.2

A heat engine develops 10 kW power when receiving heat at the rate of 2250 kJ/min. Evaluate the corresponding rate of heat rejection from the engine and its thermal efficiency.

Solution : Refer Fig. 8.2

Thermal efficiency,

$$\eta_{th} = \frac{\text{work output}}{\text{heat supplied}} = \frac{W}{Q_1}$$

Given: $W = 10\ kW = 10\ kJ/s = 600\ kJ/min$ and $Q_1 = 2250\ kJ/min$

$$\therefore \quad \eta_{th} = \frac{600}{2250} = 0.267 \text{ or } 26.7\%$$

From the principle of energy conservation (1st law): $Q_1 = W + Q_2$

$\therefore$ Amount of heat rejected,

$$Q_2 = Q_1 - W$$

$$= 2250 - 600 = \mathbf{1650\ kJ/min}$$

EXAMPLE 8.3

A reversed heat engine absorbs 250 kJ of heat from a low temperature region and has a mechanical input of 100kJ. What would be the heat transferred to the high temperature region ? Also evaluate the coefficient of performance of the reversed engine when working as a refrigerator and as a heat pump.

Solution : Refer Fig. 8.3.

From the principle of energy conservation (1st law):

$$Q_1 = W + Q_2$$

$\therefore$ Heat transferred to the high temperature region,

$$Q_1 = W + Q_2 = 100 + 250 = \mathbf{350\ kJ}$$

$$(COP)_{ref} = \frac{Q_2}{W} = \frac{250}{100} = \mathbf{2.5}$$

$$(COP)_{heat\ pump} = \frac{Q_1}{W} = \frac{350}{100} = \mathbf{3.5}$$

EXAMPLE 8.4

A machine operating as a heat pump extracts heat from the surrounding atmosphere, is driven by a 7.5 kW motor and supplies 2×10^5 kJ/hr heat to a house needed for its heating in winter. Find *coefficient of performance for the heat pump. How this COP will be affected if the objective of the same machine is to cool the house in summer requiring 2×10^5kJ/hr of heat rejection? Comment on the result.*

Solution : Refer Fig. 8.3.

Work input to heat pump,

$$W = 7.5 \text{ kW} = 7.5 \text{ kJ/s} = 0.27 \times 10^5 \text{kJ/hr}$$

From the principle of energy conservation:

$W = Q_1 - Q_2$

∴ Heat extracted from the surrounding atmosphere,

$$Q_2 = Q_1 - W = 2\times10^5 - 0.27\times10^5 = 1.73\times10^5 \text{ kJ/hr}$$

$$(COP)_{heat\ pump} = \frac{\text{heat supplied to the room } (Q_1)}{\text{work input } (W)} = \frac{2\times10^5}{0.27\times10^5} = \mathbf{7.407}$$

$$(COP)_{ref} = \frac{\text{heat extracted from the room}(Q_2)}{\text{work input } (W)} = \frac{1.73\times10^5}{0.27\times10^5} = \mathbf{6.407}$$

Apparently the same machine has two values of its performance depending upon the task being accomplished.

EXAMPLE 8.5

A heat pump requires 36.5 kW for its operation and has coefficient of performance equal to 5.

(a) Determine the magnitude of heat transfer from the working fluid

(b) The heat transfer from the pump is used to heat the water flowing through the radiator of a building. Determine the flow rate of the heated water if its temperature rises by 20°C when flowing through the heat pump. The flow velocity of water may be neglected.

Solution : If Q_1 is the heat rejected at higher temperature, then

$$(COP)_{heat\ pump} = \frac{Q_1}{W}$$

$$Q_1 = W \times COP = 36.5 \times 5 = 182.5 \text{ kW}$$

Q_1 represents the heat transfer from the pump to the water flowing through the radiator.

Applying steady flow energy equation on unit time to radiator,

$$\dot{m}\left(h_1 + \frac{V_1^2}{2} + gz_1\right) + \dot{Q} = \dot{m}\left(h_2 + \frac{V_2^2}{2} + gz_2\right) + \dot{W}_s$$

Neglecting kinetic and potential energy changes and no shaft work, we have

$$\dot{Q} = \dot{m}(h_2 - h_1) = \dot{m}\, c_p\, \Delta T$$

or $182.5 = \dot{m} \times 4.186 \times 20$;

$$\dot{m} = 2.18 \text{ kg/s} = \mathbf{7847.6} \text{ kg/hr}$$

8.3.1. Kelvin-Planck statement

It is impossible to construct an engine that operates in a cycle and produces no effect other than work output and exchange of heat with a single heat reservoir

The statement implies that no heat engine can be developed that receives a certain amount of heat from a high temperature source and converts that into an equivalent amount of work. *i.e,* $W = Q$. The thermal efficiency of such a engine $\eta = W/Q = 1$ or 100 percent.

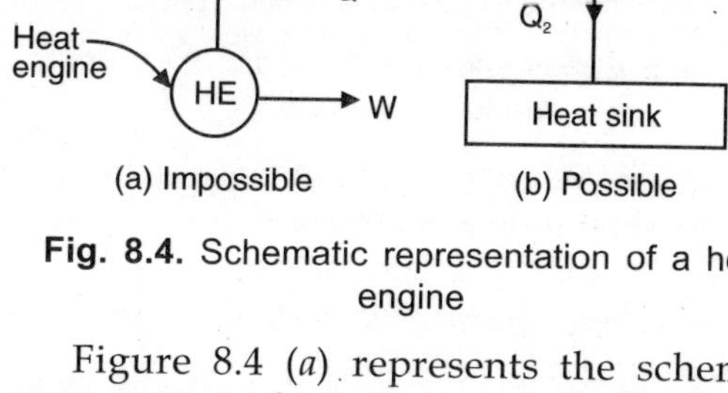

Fig. 8.4. Schematic representation of a heat engine

Figure 8.4 (*a*) represents the schematic arrangement of a heat engine that exchanges heat with a single heat source and is 100 percent efficient. Such a system satisfies the principle of energy conservation (1st law) but violates the Kelvin statement of second law. Obviously Kelvin-Planck statement tells that no heat engine can have thermal efficiency equal to 100 percent.

The only alternative for continuous power output from a heat engine is that a portion of the heat received must be rejected to a heat reservoir at low temperature (heat sink). This engine receives Q_1 units of heat, rejects Q_2 units of heat and converts $(Q_1 - Q_2)$ units of heat into work per cycle. All possible heat engines conform to this representation (Fig. 8.4*b*).

8.3.2. Clausius statement

It is impossible to construct a device that operates in a cycle and produces no effect other than the transfer of heat from a system at low temperature to another system at high temperature

The statement implies that heat cannot flow of itself from a system at low temperature to a system at high temperature. The schematic arrangement that is prohibited by Clausius statement is shown in Fig 8.5(*a*).

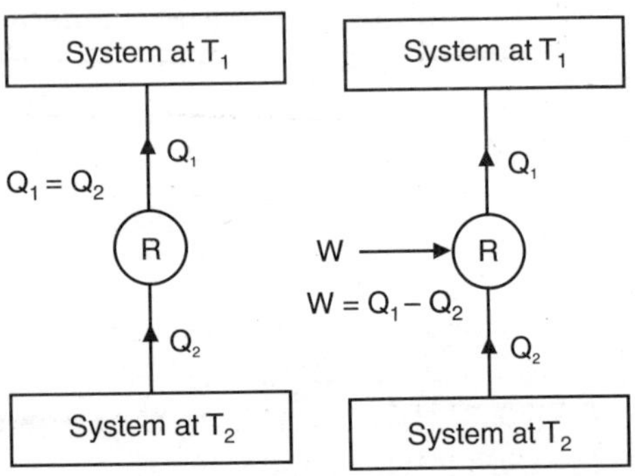

(a) Impossible (b) Impossible

Fig. 8.5. Schematic representation of a heat pump/refrigerator

The coefficient of performance of such an arrangement equals $COP = Q/W = Q/0 = \infty$. Obviously the Clausius statement tells that COP of a heat pump/refrigerator cannot be equal to infinity.

The only alternative for the transfer of heat from low temperature to high temperature level is that some external work must be supplied to the machine as shown in Fig. 8.5(*b*).

Whereas the Kelvin-Planck statement is applied to heat engines, the Clausius statement concerns heat pumps and refrigerators. Both the Kelvin-Planck and Clausius statements are negative statements, they have no mathematical proof. The law is based on experimental observations, and to-date no observation has been made that contradicts the law and this aspect is taken as sufficient evidence of its validity.

8.4. EQUIVALENCE OF KELVIN-PLANCK AND CLAUSIUS STATEMENTS

The Kelvin-Planck and Clausius statements, though worded differently, are interlinked and are complementary to each other. It is impossible to have a device satisfying one statement and violating the other. Any device that violates Clausius statement leads to violation of Kelvin-Planck statement and vice-versa.

8.4.1. Violation of Clausius statement leading to violation of Kelvin-Planck statement

Fig 8.6 (*a*) depicts a refrigerator R that operates in a cycle and transfers Q_2 units of heat from a low temperature reservoir at T_2 to a high temperature reservoir at T_1 without any work input from an external agency (surroundings). This is in violation of the Clausius statement. Indicated alongwith is a heat engine E that too operates in a cycle. This engine takes Q_1 units of heat from the high temperature reservoir, delivers $(Q_1 - Q_2)$ units of work to the surroundings and rejects the remaining Q_2 units of heat to the low temperature reservoir. The engine operates in conformity with the Kelvin-Planck statement.

Fig 8.6 (*b*) illustrates the heat and work interactions for the refrigerator and heat engine when coupled together. This composite system constitutes a device that receives $(Q_1 - Q_2)$ units of heat from the high temperature reservoir and converts it completely into an equivalent amount of work $W = (Q_1 - Q_2)$ without rejecting any heat to the low temperature reservoir. This operation of the composite system is in violation of the Kelvin-Planck statement.

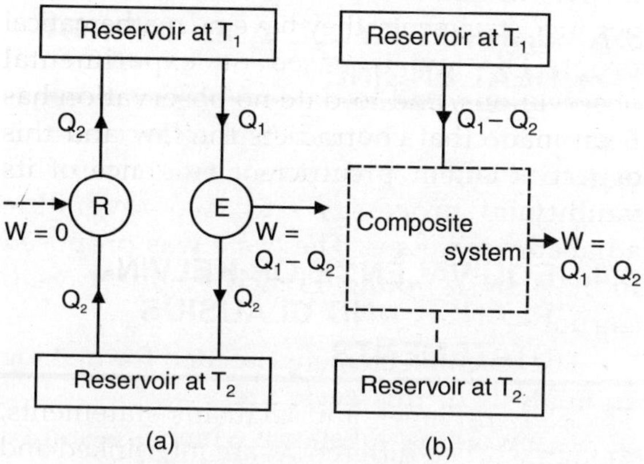

Fig. 8.6. Violation of Clausius statement leads to violation of Kelvin Planck statement

Thus violation of Clausius statement leads to violation of Kelvin-Planck statement also.

8.4.2. Violation of Kelvin-Planck statement leading to violation of Clausius statement

Figure 8.7 (*a*) depicts an engine E which operates from a single heat reservoir at temperature T_1. It receives Q_1 units of heat from this reservoir and converts it completely into an equivalent amount of work $W = Q_1$ without rejecting any heat to the low temperature reservoir at T_2. This is in violation with the Kelvin-Planck statement. Indicated alongwith it is a refrigerator R which extracts Q_2 units of heat from the low temperature reservoir, is supplied with Q_1 units of work from an external agency (surroundings) and supplies $(Q_1 + Q_2)$ units of heat to the high temperature reservoir. The refrigerator operates in conformity with the Clasius statement.

Fig. 8.7(*b*) illustrates the work and heat interactions for the refrigerator and heat engine when coupled together. The output of the engine is utilized to drive the refrigerator. This composite system constitutes a device which transfers heat from the low temperature reservoir to the high temperature reservoir without any work input from an external agency. This is in violation of the Clausius statement.

Thus violation of Kelvin-Planck statement leads to violation of Clausius statement also.

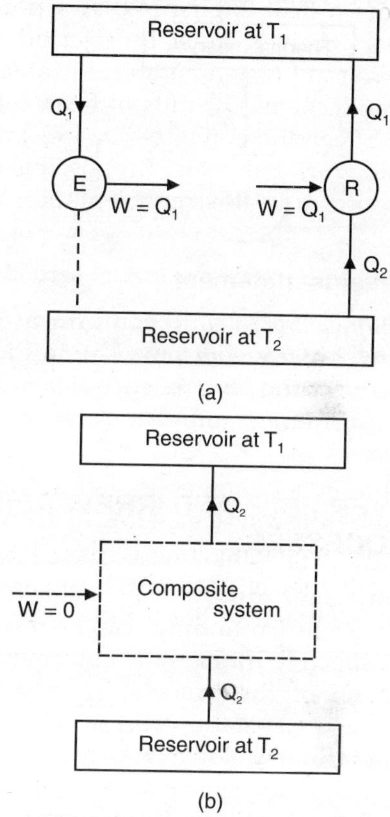

Fig. 8.7. Violation of Kelvin-Planck statement leads to violation of Clausius statement

8.5. PERPETUAL MOTION MACHINE OF SECOND KIND

A practical heat engine exchanges heat with two thermal reservoirs (source and sink) and its thermal efficiency is given by :

$$\eta_{th} = \frac{W_{net}}{Q_1} = \frac{Q_1 - Q_2}{Q_1} = 1 - \frac{Q_2}{Q_1}$$

where Q_1 is the heat supplied from the source, Q_2 is the heat rejected to the sink and W_{net} is the net work delivered to the surroundings.

If $Q_2 = 0$, then $W_{net} = Q_1$ and $\eta_{th} = 100$ percent . That is, if the engine exchanges heat only with one thermal reservoir (Fig 8.8), then the entire heat supplied to it gets converted into an equivalent amount of work and the efficiency becomes 100 percent. Such a heat engine is called a perpetual motion machine of the second kind, abbreviated as *PMM*2.

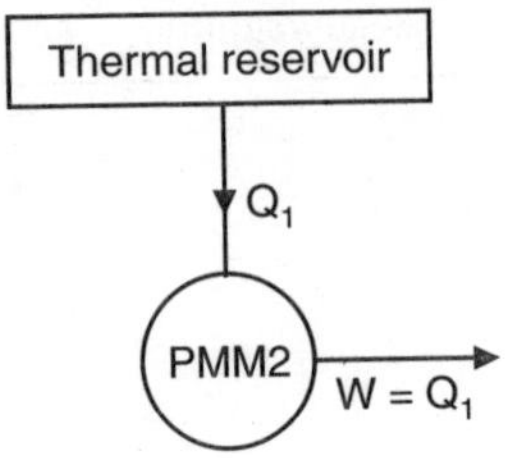

Fig. 8.8. Perpetual motion machine of second kind

The *PMM*2 operates in conformity with the first law, but it violates the Kelvin-Planck statement of second law. As such it is only of hypothetical interest and cannot be realised in practice.

8.6. REVERSIBLE AND IRREVERSIBLE PROCESSES

A thermodynamic process is reversible if both the system and surroundings can be restored to their respective initial states by reversing the direction of the process. A reversible process is a quasi-static process, a process carried out infinitely slowly with infinitesimal gradient with the system passing through a continuous series of equilibrium states. The reversible processes can proceed in either direction without violating the second law of thermodynamics. A system going through a reversible process results in maximum efficiency. In other words, a reversible process yields the maximum work in engines and requires minimum work in devices such as refrigerators, pumps and compressors.

All natural processes take place simultaneously at finite speeds through finite discontinuities between the system and its environment and as such are irreversible. Processes are rendered irreversible as a result of degradation of energy by such factors as friction, turbulence, diffusion, inelasticity and electrical resistance etc. These elements can be reduced but cannot be completely eliminated.

Note : Refer section 1.8 for salient features and notable examples of reversible and irreversible processes.

8.7. CARNOT CYCLE AND CARNOT HEAT ENGINE

A carnot cycle is a hypothetical cycle consisting of four distinct processes : two reversible isothermal processes and two reversible adiabatic processes. The cycle was proposed in 1824 by a young French engineer, Sadi Carnot.

The essential elements needed for making an analysis of this cycle are :-

– a working substance which is assumed to be a perfect gas

– two heat reservoirs; the hot reservoir (heat source) at temperature T_1 and the cold reservoir (heat sink) at temperature T_2

– piston- cylinder arrangement for getting the work out of the working substance. The piston and cylinder walls (excluding the cylinder head) are taken as perfect heat insulators. The cylinder head is imagined to provide alternatively a diathermic cover (perfect heat conductor) and an adiabatic cover (perfect heat insulator).

The piston-cylinder arrangement is shown in Fig 8.9. There is no friction to the movement of piston inside the cylinder.

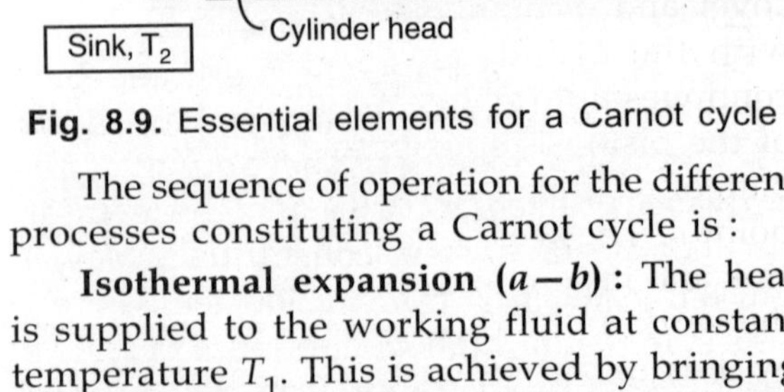

Fig. 8.9. Essential elements for a Carnot cycle

The sequence of operation for the different processes constituting a Carnot cycle is :

Isothermal expansion (*a* – *b*) : The heat is supplied to the working fluid at constant temperature T_1. This is achieved by bringing the heat source in good thermal contact with the cylinder head through diathermic cover. The gas expands isothermally from state point a (p_a, V_a) to state point b (p_b, V_b). The heat supplied equals the work done which is represented by area under the curve a-b on pressure-volume plot and is given by

$$Q_1 = W_{a-b} = p_a V_a \log_e \frac{V_b}{V_a}$$

$$= m\,R\,T_1 \log_e \frac{V_b}{V_a} \qquad ...(8.4)$$

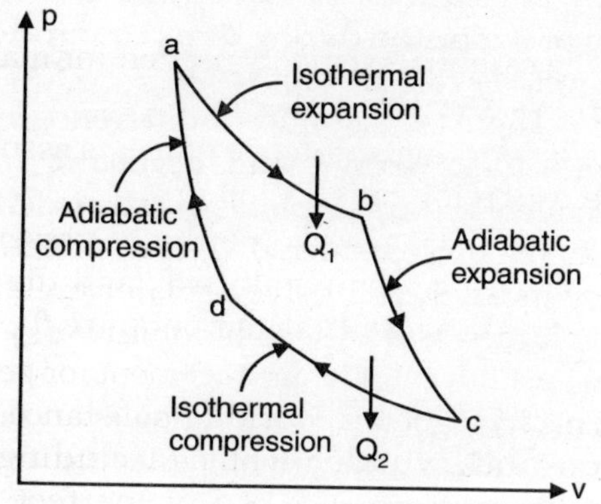

Fig. 8.10. Carnot cycle on P-V plot

Adiabatic expansion $(b-c)$: At the end of isothermal expansion (state point b), the heat source is replaced by adiabatic cover. The expansion continues adiabatically and reversibly upto state point c (p_c, V_c). Work is done by the working fluid at the expense of internal energy and its temperature falls to T_2 at state point c.

Isothermal compression $(c-d)$: After state point c, the piston starts moving inwards and the working fluid is compressed isothermally at temperature T_2. The constant temperature T_2 is maintained by removing the adiabatic cover and bringing the heat sink in contact with the cylinder head. The compression continues upto state point d and this position of the piston is so chosen that it lies on the path of reversible adiabatic drawn from state point a. The working fluid loses heat to the sink and its amount equals the work done on the working fluid. This work is represented by area under the curve c-d and its amount is given by.

$$Q_2 = W_{c-d} = p_c V_c \log_e \frac{V_c}{V_d}$$

$$= m\,R\,T_2 \log_e \frac{V_c}{V_d} \qquad ...(8.5)$$

Adiabatic compression $(d-a)$: At the end of isothermal compression (state point d), the heat sink is removed and is replaced by adiabatic cover. The compression now proceeds adiabatically and reversibly till the working fluid returns back to its initial state point a. Work is done on the working fluid, the internal energy increases and temperature is raised to T_1.

Since all the processes that constitute a Carnot cycle are reversibe, the Carnot cycle is referred to as a reversible cycle. Further, a cyclic heat engine working on the Carnot cycle is called Carnot engine and its thermal efficiency is given by

$$\eta = \frac{\text{net work output}}{\text{heat input}} = \frac{W_{net}}{Q_1}$$

There are no heat interactions along the reversible adiabatic processes b–c and d–a, and application of the first law of thermodynamics for the complete cycle gives

$$\oint \delta W = \oint \delta Q$$

or $\quad W_{net} = Q_1 - Q_2$

$$= mRT_1 \log_e \frac{V_b}{V_a} - mRT_2 \log_e \frac{V_c}{V_d}$$

$$\therefore \quad \eta = \frac{mRT_1 \log_e \frac{V_b}{V_a} - mRT_2 \log_e \frac{V_c}{V_d}}{mRT_1 \log_e \frac{V_b}{V_a}}$$

$$= 1 - \frac{T_2}{T_1} \times \frac{\log_e \frac{V_c}{V_d}}{\log_e \frac{V_b}{V_a}} \qquad ...(8.6)$$

For the adiabatic expansion processes b–c and d–a,

$$\frac{T_b}{T_c} = \left(\frac{V_c}{V_b}\right)^{\gamma - 1} \quad \text{and} \quad \frac{T_a}{T_d} = \left(\frac{V_d}{V_a}\right)^{\gamma - 1}$$

Since $T_b = T_a = T_1$, and $T_c = T_d = T_2$, the above expression give

$$\frac{T_1}{T_2} = \left(\frac{V_c}{V_b}\right)^{\gamma - 1} = \left(\frac{V_d}{V_a}\right)^{\gamma - 1}$$

$$\text{or } \frac{V_c}{V_b} = \frac{V_d}{V_a} \text{ or } \frac{V_c}{V_d} = \frac{V_b}{V_a}$$

Substituting the above relation in equation 8.6, we get

$$\eta = 1 - \frac{T_2}{T_1} = \frac{T_1 - T_2}{T_1} \quad ...(8.7)$$

Following conclusions can be made with respect to efficiency of a Carnot engine :

(1) The efficiency is independent of the working fluid and depends upon the temperatures of source and sink.

(2) If $T_2 = 0$, the engine will have an efficiency of 100 percent. However, that means absence of heat sink which is violation of Kelvin-Planck statement of the second law.

(3) The efficiency is directly proportional to temperature difference (T_1 - T_2) between the source and the sink.

(4) Higher the temperature difference between source and sink, the higher will be the efficiency obtained.

(5) The efficiency increases with an increase in temperature of source and a decrease in temperature of sink.

(6) If $T_1 = T_2$, no work will be done and efficiency will be zero.

Metallurgical considerations and the high cost of temperature resisting materials limit the higher temperature T_1. The lower temperature T_2 is limited by atmospheric or sink conditions.

Carnot cycle is impracticable

Carnot cycle gives the maximum possible thermal efficiency which can be obtained for the given temperature limits. The Carnot engine, however, is a hypothetical device and it is not possible to devise it due to the following reasons :

(*i*) All the four processes have to be reversible. This necessitates that the working fluid must have no internal friction between the fluid particles and no mechanical friction between the piston and cylinder walls.

(*ii*) The heat absorption and rejection have to take place with infinitesimal temperature differences. Accordingly, the rate of energy transfer will be very low and the engine will deliver only infinitesimal power.

(*iii*) For attaining isothermal operation, the piston movement is required to be very slow. However, the piston must move fast for the adiabatic process. A variation in the speed of the piston during different processes of a cycle is rather impossible.

(*iv*) There is insignificant difference in the slopes of isothermal and adiabatic lines. Consequently the p–V plot is greatly extended both in the horizontal and vertical directions. The cylinder then involves greater pressures and volumes, and as such becomes bulky and heavy.

This hypothetical device, however, serves as a yardstick or standard of perfection against which the performance of any practical heat engine can be compared.

8.8. REVERSED HEAT ENGINE (CARNOT HEAT PUMP)

Refrigerators and heat pumps are reversed heat engines. The adjective *reversed* means that the reversible processes constituting a heat engine are individually reversed and carried out in reverse order. When a reversible process is operated backwards, all the energy transfers associated with the process get reversed in direction, but remain the same in magnitude.

Fig. 8.11 shows the p–V plot of a Carnot heat pump (*i.e.* reversed Carnot heat engine). The sequence of operation is :

a–d : Isentropic (reversible adiabatic) expansion of working fluid in the clearance space of the cylinder. The temperature falls from T_1 to T_2.

d–c : Isothermal expansion during which heat Q_2 is absorbed at temperature T_2 from the space being cooled.

c–b : Isentropic compression of working fluid. The temperature rises from T_2 to T_1.

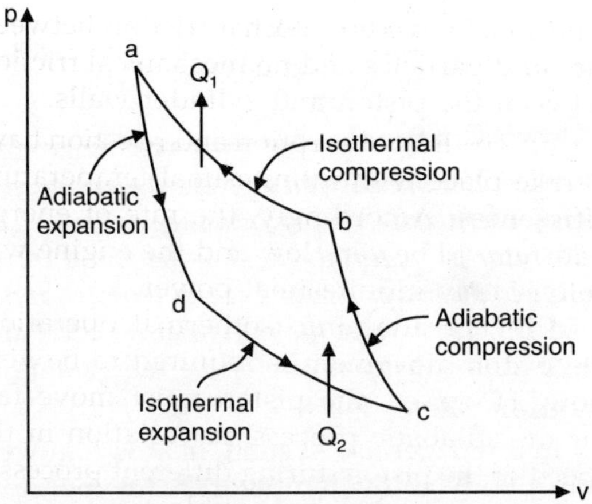

Fig. 8.11. Carnot heat pump

b–a : Isothermal compression of working fluid during which heat Q_1 is rejected to a system at higher temperature.

As outlined above in section 8.7 ,

$$Q_1 = mRT_1 \log_e \frac{V_b}{V_a}$$

$$Q_2 = mRT_2 \log_e \frac{V_c}{V_d}$$

Also $\dfrac{V_c}{V_d} = \dfrac{V_b}{V_a}$

Therefore, for a heat pump

$$(COP)_{heat\ pump} = \frac{Q_1}{Q_1 - Q_2}$$

$$= \frac{mRT_1 \log_e \frac{V_b}{V_a}}{mRT_1 \log_e \frac{V_b}{V_a} - mRT_2 \log_e \frac{V_c}{V_d}}$$

$$= \frac{T_1}{T_1 - T_2} \quad \text{...(8.8)}$$

For a refrigerator,

$$(COP)_{ref} = \frac{Q_2}{Q_1 - Q_2}$$

$$= \frac{mRT_2 \log_e \frac{V_c}{V_d}}{mRT_1 \log_e \frac{V_b}{V_a} - mRT_2 \log_e \frac{V_c}{V_d}}$$

$$= \frac{T_2}{T_1 - T_2} \quad \text{...(8.9)}$$

Although theoretically more efficient , Carnot heat pumps like the Carnot heat engines have not been used in practice. This is because the machine has to run at high speed during portions of the cycle when adiabatic processes are being performed and run very slowly during other portions when isothermal processes are being carried out. The variation of speed during a stroke is not practicable.

EXAMPLE 8.6

A heat engine working on Carnot cycle converts one-fifth of the heat input into work. When the temperature of the sink is reduced by 80°C, the efficiency gets doubled. Make calculations for the temperature of source and sink.

Solution : The Carnot heat engine is a reversible engine and its efficiency is given by :

$$\eta = \frac{T_1 - T_2}{T_1}$$

In the first case :

$$\frac{T_1 - T_2}{T_1} = \frac{1}{5}$$

$$5T_1 - 5T_2 = T_1 ; \quad T_1 = 1.25\ T_2$$

In the second case :

$$\frac{T_1 - (T_2 - 80)}{T_1} = 2 \times \frac{1}{5} = \frac{2}{5}$$

$$5T_1 - 5T_2 + 400 = 2T_1 ; \; 3T_1 + 400 = 5T_2$$

Substituting $T_1 = 1.25\ T_2$ as obtained above,

$$3(1.25T_2) + 400 = 5T_2 ; \quad T_2 = 320 \text{ K}$$

∴ Temperature of sink, T_2 = **320 K**

and temperature of source,

$T_1 = 1.25 \times 320$ = **400 K**

EXAMPLE 8.7

A reversible heat engine delivers 0.6 kW power and rejects heat energy to a reservoir at 300 K at the rate of 24 kJ/min. Make calculations for the engine efficiency and the temperature of the thermal reservoir supplying heat to the engine.

Solution : Refer Fig 8.2.

Work output, W = 0.6 kW = 0.6 kJ/s

Heat rejected, Q_2 = 24 kJ/min = 0.4 kJ/s

From the principle of energy conservation (1st law): $Q_1 = W + Q_2$

∴ Heat supplied to the engine,

$$Q_1 = 0.6 + 0.4 = 1.0 \text{ kJ/s}$$

Thermal efficiency,

$$\eta_{th} = \frac{\text{work output}}{\text{heat supplied}}$$

$$= \frac{0.6}{1.0} = 0.6 = 60\ \%$$

For a reversible heat engine, thermal efficiency is also given by

$$\eta_{th} = \frac{T_1 - T_2}{T_1}\ ;\ 0.6 = \frac{T_1 - 300}{T_1}$$

$$0.6\ T_1 = T_1 - 300$$

or $$T_1 = \frac{300}{0.4} = \mathbf{750\ K}$$

EXAMPLE 8.8

An engine mounted on a ship has a thermal efficiency 80% of that of the corresponding Carnot cycle. The engine receives heat from the sea at 300 K and rejects heat to the atmosphere at 280 K. The work output from the engine is dissipated through an agitator to heat 500 kg of sea water to 355 K. What quantity of heat must be extracted from the sea water to provide the required heating effect ?

Take specific heat of sea water

$$C_p = 4.186 \text{ kJ /kg K}$$

Solution : Efficiency of Carnot cycle,

$$\eta = \frac{T_1 - T_2}{T_1} = \frac{300 - 280}{300} = \frac{20}{300}$$

Thermal efficiency of engine

$$= 0.8 \times \text{Carnot efficiency}$$

$$= 0.8 \times \frac{20}{300} = 0.0533$$

Energy (work) needed for heating the water,

$$mc_p\Delta T = 500\times 4.186\ (355 - 300)$$

$$= 115115 \text{ kJ}$$

In terms of heat and work interactions,

$$\eta = \frac{Q_1 - Q_2}{Q_1} = \frac{W}{Q_1}$$

∴ Heat to be extracted from sea water,

$$Q_1 = \frac{W}{\eta} = \frac{115115}{0.0533} = \mathbf{2159756\ kJ}$$

EXAMPLE 8.9

A refrigerator operates on a reversed Carnot cycle whose coefficient of performance is 5.0. The evaporator is maintained at a temperature of – 6°C and the power required to run the refrigerator is 3.5 kW. Determine the refrigerating effect and the condenser temperature of the refrigerator.

Solution : Refer Fig 8.3.

In a refrigerator system, heat is removed from a body at low temperature (evaporator) and this heat is transferred to another body (condenser) at a high potential of temperature. Further, refrigerator effect is the amount of heat extracted from the body at low temperature (evaporator).

$$T_2 = -6°\text{C} = 273 + (-6) = 267 \text{ K}$$

For a refrigerator working on reversed Carnot cycle,

$$(COP)_{ref} = \frac{T_2}{T_1 - T_2}\ ;\ 5 = \frac{267}{T_1 - 267}$$

∴ Condenser temperature,

$$T_1 = 267 + \frac{267}{5} = \mathbf{320.4\ K}$$

In terms of heat and work interactions,

$$(COP)_{ref} = \frac{Q_2}{Q_1 - Q_2} = \frac{Q_2}{W}$$

∴ Heat extracted from the space being cooled (refrigeration effect),

$$Q_2 = W \times (COP)_{ref}$$

$$= 3.5 \times 5 = \mathbf{17.5\ kW}$$

EXAMPLE 8.10

A domestic food refrigerator maintains a temperature of –10°C whilst the ambient air temperature is – 30°C. The heat leakage into the freezer is estimated to be at the continuous rate of 2 kJ/s. Determine the least power needed to pump out this heat continuously.

Solution : The refrigerator is required to remove heat from the freezer at the same rate at which heat leaks into it. That is

$$Q_2 = 2 \text{ kJ/s}$$

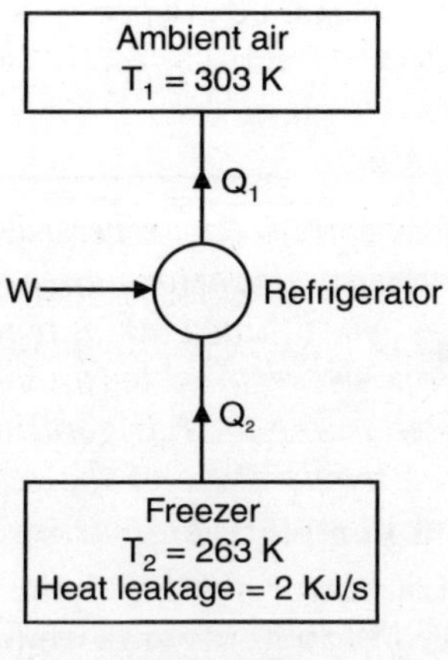

Fig. 8.12.

For minimum power requirement, the refrigerator operates on reversed Carnot cycle. Accordingly

$$COP = \frac{T_2}{T_1 - T_2} \quad \text{...}(i)$$

In terms of heat and work interactions

$$COP = \frac{Q_2}{W} = \frac{Q_2}{Q_1 - Q_2} \quad \text{...}(ii)$$

From expressions (*i*) and (*ii*)

$$\frac{T_1 - T_2}{T_2} = \frac{Q_1 - Q_2}{Q_2} \quad \text{or} \quad \frac{T_1}{T_2} = \frac{Q_1}{Q_2}$$

$$\therefore \quad Q_1 = \frac{T_1}{T_2} \times Q_2$$

$$= \frac{303}{263} \times 2 = 2.304 \text{ kJ/s}$$

and $W = Q_1 - Q_2$

$= 2.304 \times 2 =$ **0.304** kJ/s

Hence the least power required to pump the heat continuosly is 0.304 kW

EXAMPLE 8.11

A refrigerator operating on a reversed Carnot cycle consumes 150 W power in summer when the ambient atmosphere is at 310 K. The heat leakage into the refrigerator through the doors is estimated to be at the continuous rate of 15 W per degree temperature difference between the ambient air and the cold space of the refrigerator. For continuous operation of the refrigerator, determine the temperature at which the cold space is maintained.

Solution : The refrigerator is requried to remove heat from the cold space at which heat leaks into it. That is

$$Q_2 = 15(T_1 - T_2)$$

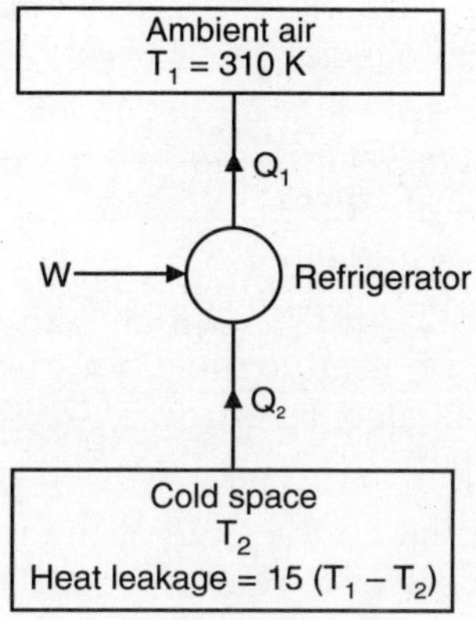

Fig. 8.13.

Since the refrigerator operates on the reversed Carnot cycle, its coefficient of performances can be written as

$$COP = \frac{T_2}{T_1 - T_2} \quad \text{...}(i)$$

In terms of heat and work interactions

$$COP = \frac{Q_2}{W} \quad \text{...}(ii)$$

From expressions (*i*) and (*ii*),

$$\frac{T_2}{T_1 - T_2} = \frac{15(T_1 - T_2)}{150}$$

$$\frac{T_2}{310 - T_2} = \frac{T_1 - T_2}{10} = \frac{310 - T_2}{10}$$

or $(310 - T_2)^2 = 10\ T_2$

or $96100 + T_2^2 - 620\ T_2 = 10\ T_2$

or $T_2^2 - 630\ T_2 + 96100 = 0$

$$T_2 = \frac{630 \pm \sqrt{396900 - 384400}}{2}$$

$$= \frac{630 \pm 111.80}{2}$$

$= 370.9$ K or 259.1 K

The temperature $T_2 = 370.9$ K is greater than the temperature of ambient air and as such is inadmissible. Therefore the cold space has to be maintained at a temperature of **259.1 K.**

EXAMPLE 8.12

It is proposed to design a refrigeration plant for a food store which is to be maintained at -5°C. The ambient temperature is 25°C and the estimated heat transfer from the store is at the rate of 5 kW. If the system operates on reversed Carnot heat engine cycle, determine the performance index and the minimum power required to operate the refrigeration plant.

Solution : Refer Fig 8.3.

$$T_1 = 25°C = 273 + 25 = 298 \text{ K}$$
$$T_2 = -5°C = 273 + (-5) = 268 \text{ K}$$

The performance of a refrigeration system is expressed in terms of coefficient of performance, and the *COP* of the refrigerator working between two thermal reservoirs T_1 (298 K) and T_2 (268 K) is given by

$$(COP)_{ref} = \frac{T_2}{T_1 - T_2} = \frac{268}{298 - 268} = 8.93$$

In terms of heat interactions,

$$(COP)_{ref} = \frac{Q_2}{Q_1 - Q_2}\ ; \quad 8.93 = \frac{5}{Q_1 - 5}$$

or $\quad 8.93\, Q_1 - 44.65 = 5$

$$Q_1 = \frac{49.65}{8.93} = 5.56 \text{ kW}$$

From principle of energy conservation (1st law): $Q_1 = W + Q_2$

∴ Power required to operate the refrigerator,

$$W = Q_1 - Q_2 = 5.56 - 5 = \mathbf{0.56\ kW}$$

A reversible refrigerator requires the minimum power to operate it. Therefore, the minimum power required is 0.56 kW.

EXAMPLE 8.13

A perfect reversed heat engine is used for making ice at – 5°C from water available at 25°C. The temperature of freezing mixture is – 10°C . Calculate the quantity of ice formed per kWh.

For ice : specific heat = 2.1 kJ/kg K and latent heat = 335 kJ/kg.

Solution : The temperature limits for the reversed heat engine are :

$$T_1 = 25°C = 298 \text{ K}$$
and $\quad T_2 = -10°C = 263 \text{ K}$

A perfect reversed heat engine works on Carnot cycle for which *COP* in terms of temperatures is given by

$$COP = \frac{T_2}{T_1 - T_2} = \frac{263}{298 - 263} = 7.514$$

Work input = 1 kWh = 3600 kJ

In terms of heat and work interactions,

$$COP = \frac{Q_2}{Q_1 - Q_2} = \frac{Q_2}{W}$$

∴ Total heat extracted from ice,

$$Q_2 = W \times COP$$
$$= 3600 \times 7.514 = 27050 \text{ kJ/hr}$$

Removal of heat from 1 kg of water at 25°C so that it becomes ice at - 5°C is given by

$$= mc_{pw}\Delta T + mL + mc_{pi}\Delta T$$
$$= 1\times 4.186\ (25 - 0) + 1\times 335 + 1 \times 2.1\ [0 - (-5)]$$
$$= 104.65 + 335 + 10.5$$
$$= 450.15 \text{ kJ}$$

∴ Ice formed per kWh

$$= \frac{27050}{450.15} = \mathbf{60.09\ kg}$$

EXAMPLE 8.14

An engine working on Carnot cycle absorbs Q_1 units of heat from a source at T_1 and rejects Q_2 units of heat to a sink at T_2 . Because of imperfect heat transfer, the maximum (θ_1) and minimum (θ_2) absolute temperatures attained during the cycle are given by

$$\theta_1 = T_1 - KQ_1 \quad \text{and} \quad \theta_2 = T_2 + KQ_2$$

where K is a constant. Show that efficiency of the plant is given by

$$\eta = 1 - \frac{T_2}{T_1 - 2KQ_1}$$

Solution : The engine operates on Carnot cycle and its efficiency is given by

$$\eta = \frac{Q_1 - Q_2}{Q_1} = \frac{\theta_1 - \theta_2}{\theta_1}$$

$$\therefore \quad \frac{Q_2}{Q_1} = \frac{\theta_2}{\theta_1}$$

From the given relations :

$$KQ_1 = T_1 - \theta_1 \quad \text{and} \quad KQ_2 = \theta_2 - T_2$$

$$\therefore \quad \frac{\theta_2 - T_2}{T_1 - \theta_1} = \frac{Q_2}{Q_1} = \frac{\theta_2}{\theta_1}$$

Upon simplification :

$$\theta_2 = \frac{T_2\,\theta_1}{2\theta_1 - T_1} = \frac{T_2\,\theta_1}{2(T_1 - KQ_1) - T_1} = \frac{T_2\,\theta_1}{T_1 - 2KQ_1}$$

Substituting this value of θ_2 in the expression for thermal efficiency

$$\eta = \frac{\theta_1 - \left(\dfrac{T_2\theta_1}{T_1 - 2KQ_1}\right)}{\theta_1} = 1 - \frac{T_2}{T_1 - 2KQ_1}$$

which is the required expression.

EXAMPLE 8.15

A Carnot refrigerator operates between temperature -10°C and 30°C. Determine its coefficient of performance. Subsequently it is desired to change the temperatures so as to make the coefficient of performance exactly equal to 7.0. The amount of increase in the higher temperature is equal to amount of decrease in the lower temperature. Workout the new temperatures.

Solution : T_1 = 30°C = 303 K and

T_2 = - 10°C = 263 K

$$COP = \frac{T_2}{T_1 - T_2} = \frac{263}{303 - 263} = 6.575$$

Let the change in temperature be x. Then

$$COP = 7 = \frac{263 - x}{(303 + x) - (263 - x)} = \frac{263 - x}{40 + 2x}$$

or $280 + 14x = 263 - x$; $x = -1.133$ degree

Hence, T_1 = 303 - 1.133 = **301.867 K** and

T_2 = 263 - (-1.133) = **264.133 K**

EXAMPLE 8.16

Which is more effective way to increase the efficiency of a Carnot heat engine: to increase the source temperature T_1 while the sink temperature T_2 is held constant or to decrease the sink temperature by the same amount while the source temperature is held constant.

How this result would be affected in case of a Carnot heat pump ?

Solution : Let η_1 be the efficiency when the source temperature is increased by dT while the sink temperature is held constant, and

η_2 be the efficiency when the sink temperature is decreased by dT while the source temperature is held constant. Then

$$\eta_1 = \frac{T_1 - T_2}{T_1} = \frac{(T_1 + dT) - T_2}{T_1 + dT} \quad ...(i)$$

$$\eta_2 = \frac{T_1 - T_2}{T_1} = \frac{T_1 - (T_2 - dT)}{T_1} = \frac{(T_1 + dT) - T_2}{T_1} \quad ...(ii)$$

It is to be noted that

- numerators for the expressions (*i*) and (*ii*) are same.
- denominator $(T_1 + dT)$ in expression (*i*) is more than the denominator T_1 in expression.

$\therefore \quad \eta_2 > \eta_1$

Hence, a decrease in the sink temperature while the source temperature is held constant is more effective in increasing the efficiency of a Carnot heat engine.

(*b*) Let $(COP)_1$ be the coefficient of performance of the heat pump when the temperature of the high temperature reservoir is increased by dT while the low temperature reservoir is held at constant temperature, and $(COP)_2$ be the coefficient of performance if the temperature of the low temperature reservoir is decreased by dT while the high temperature reservoir is held at constant temperature. Then

$$(COP)_1 = \frac{T_1}{T_1 - T_2} = \frac{T_1 + dT}{(T_1 + dT) - T_2} \quad ...(iii)$$

$$(COP)_2 = \frac{T_1}{T_1 - T_2} = \frac{T_1}{T_1 - (T_2 - dT)}$$

$$= \frac{T_1}{(T_1 + dT) - T_2} \quad ...(iv)$$

An examination of expressions (*iii*) and (*iv*) shows that $(COP)_1 > (COP)_2$. That is, an increase in the high temperature reservoir while the temperature of low temperature reservoir is held constant is more effective in increasing the coefficient of performance of a heat pump.

EXAMPLE 8.17

A reversible engine works between three heat reservoirs A, B and C. The engine receives equal amount of heat from reservoir A and B at temperatures T_A and T_B respectively and rejects heat to a reservoir C at temperature T_C. If the efficiency of the engine is α times the efficiency of a reversible engine operating between two reservoirs A and C only, show that

$$\frac{T_A}{T_B} = 2(1-\alpha)\frac{T_A}{T_C} + (2\alpha - 1)$$

Solution : Refer Fig 8.14.

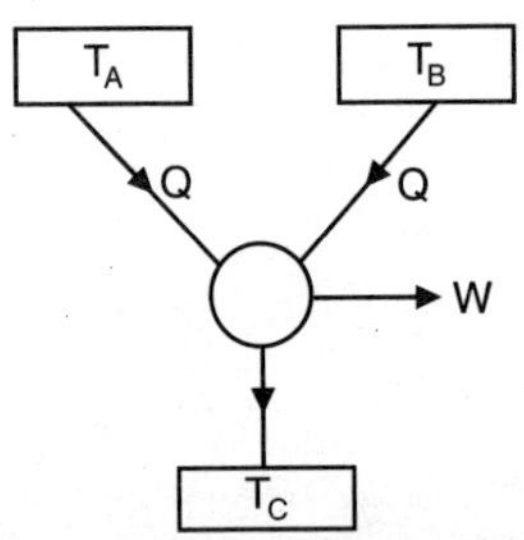

Fig. 8.14.

Let the engine be supplied Q units of heat from each of the thermal reservoirs A and B.

When operating between reservoirs A and C

$$\eta_{th} = \frac{T_A - T_C}{T_A}$$

Since Q units of heat is supplied from reservoir A, the work output will be

$$Q\left(\frac{T_A - T_C}{T_A}\right)$$

Likewise when the operation is between reservoirs B and C, the work output will be

$$Q\left(\frac{T_B - T_C}{T_B}\right)$$

For the given engine,

Total heat input = $2Q$

Total work output

$$= Q\left(\frac{T_A - T_C}{T_A}\right) + Q\left(\frac{T_B - T_C}{T_B}\right)$$

Thermal efficiency

$$= \frac{Q\left(\frac{T_A - T_C}{T_A}\right) + Q\left(\frac{T_B - T_C}{T_B}\right)}{2Q}$$

$$= \frac{1}{2}\left[1 - \frac{T_C}{T_A} + 1 - \frac{T_C}{T_B}\right]$$

$$= \frac{1}{2}\left[2 - \frac{T_C}{T_A} - \frac{T_C}{T_B}\right]$$

As per the given condition,

$$\frac{1}{2}\left[2 - \frac{T_C}{T_A} - \frac{T_C}{T_B}\right] = \alpha\left[\frac{T_A - T_C}{T_A}\right] = \alpha\left(1 - \frac{T_C}{T_A}\right)$$

Multiplying both sides by $2\frac{T_A}{T_C}$, we get

$$2\frac{T_A}{T_C} - 1 - \frac{T_A}{T_B} = 2\alpha\frac{T_A}{T_C} - 2\alpha$$

or
$$\frac{T_A}{T_B} = (2\alpha - 1) + 2\frac{T_A}{T_C} - 2\alpha\frac{T_A}{T_C}$$

$$= (2\alpha - 1) + 2\frac{T_A}{T_C}(1 - \alpha)$$

$$= 2(1 - \alpha)\frac{T_A}{T_C} + (2\alpha - 1)$$

EXAMPLE 8.18

A reversible heat engine in outer space operates between a high temperature reservoir at T_1 and radiating panel at T_2. The radiation heat transfer from the panel is proportional to its area and fourth power of its absolute temperature. Show that for a given work output and the value of T_1, the area of panel will be a minimum when $T_2/T_1 = 3/4$. Also find the corresponding minimum area.

Solution : Refer Fig 8.15.

For a reversible heat engine,

$$\eta = \frac{W}{Q_1} = \frac{Q_1 - Q_2}{Q_1} = \frac{T_1 - T_2}{T_1}$$

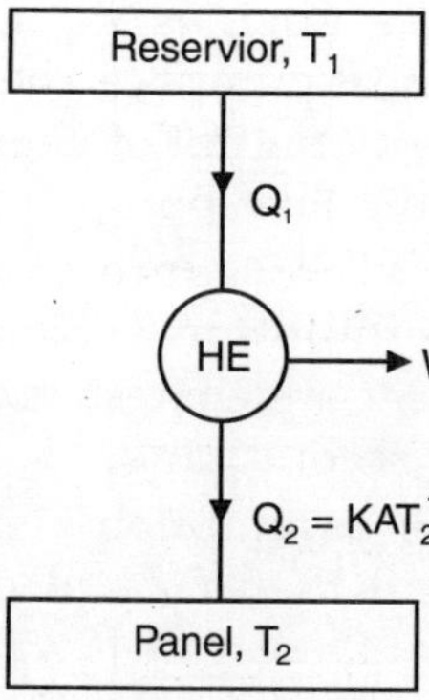

Fig. 8.15.

Accordingly :

$$\frac{Q_1}{T_1} = \frac{Q_2}{T_2}$$

and $$\frac{W}{T_1 - T_2} = \frac{Q_1}{T_1} = \frac{Q_2}{T_2} = \frac{KAT_2^4}{T_2}$$

or $$A = \frac{W}{KT_2^3(T_1 - T_2)}$$

where K is an empirical constant.

For a given W and T_1, the panel area A is minimum when T_2^3 $(T_1 - T_2)$ is maximum. Differentiating it with respect to T_2 and setting the derivative equal to zero we obtain :

$$\frac{d}{dT_2}\left(T_2^3T_1 - T_2^4\right) = 0$$

or $$3T_2^2T_1 - 4T_2^3 = 0$$

That gives :

$$\frac{T_2}{T_1} = \frac{3}{4} = 0.75$$

$$\therefore \; A_{min} = \frac{W}{K(0.75T_1)^3(T_1 - 0.75T_1)}$$

$$= 9.48\frac{W}{KT_1^4}$$

8.9. CARNOT THEOREM

No heat engine operating in a cycle between two given thermal reservoirs, with fixed temperatures, can be more efficient than a reversible engine operating between the same thermal reservoirs

Consider a reversible engine E_A and an irreversible engine E_B operating between the same thermal reservoirs at temperatures T_1 and T_2. For the same quantity of heat Q withdrawn from the high temperature source, the work output from these engines is W_A and W_B respectively. As such the heat given off by the reversible engine is $(Q - W_A)$ and that from irreversible engine is $(Q - W_B)$.

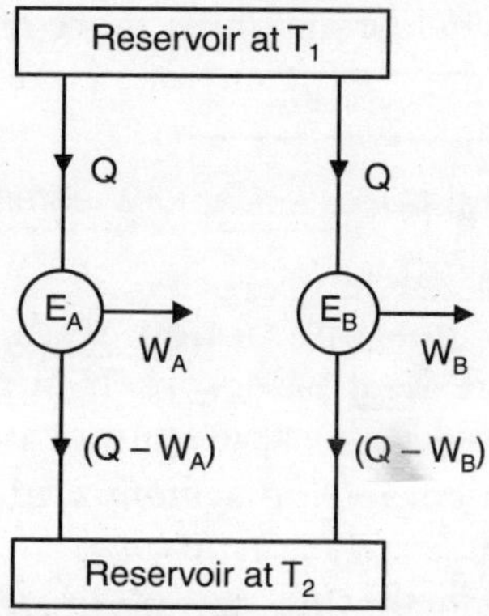

Fig. 8.16. (a) Carnot theorem

Let it be presumed that the irreversible engine E_B is more efficient than the reversible engine E_A. Then $\frac{W_B}{Q} > \frac{W_A}{Q}$; $W_B > W_A$

and $(Q - W_A) < (Q - W_B)$ *i.e,* output from irreversible engine is more than that from reversible engine.

Let the reversible engine E_A now be made to operate as a refrigerator/heat pump ; the irreversible engine continues to operate as an engine.

Since engine E_A is reversible, the magnitudes of heat and work interactions will remain the same but their directions will be reversed. The work required to drive the refrigerator can be withdrawn from the irreversible engine by having a direct coupling between the two.

Fig. 8.16 (*b*) shows the work and heat interactions for the composite system constituted by the reversible engine (now operating as refrigerator) and the irreversible engine.

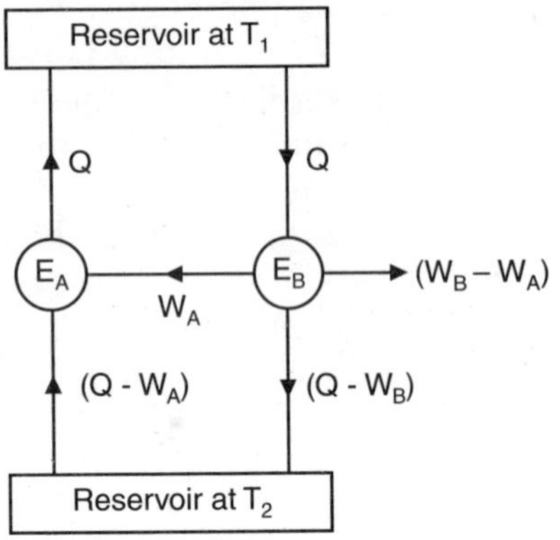

Fig. 8.16. (b) Carnot theorem

The net effect is

– No net interaction with the high temperature heat reservoir. It supplies and recovers back the same amount of heat.

– The composite system withdraws

$(Q - W_A) - (Q - W_B) = (W_B - W_A)$ units of heat from the low temperature reservoir and converts that into equivalent amount of work output.

The combination thus constitutes a perpetual motion of the second kind in violation of the second law. Obviously the assumption that the irreversible engine is more efficient than the reversible engine is wrong. Hence, an irreversible engine cannot have efficiency higher than that from a reversible engine operating between the same thermal reservoirs.

Corollaries of Carnot's Theorem

Corollary 1 : ***All reversible engines operating between the two given thermal reservoirs, with fixed temperature, have the same efficiency.***

Let both the heat engines E_A and E_B depicted in Fig. 8.16 be reversible in operation. Both operate between the same thermal reservoirs at temperatures T_1 and T_2, and draw the same quantity of heat Q from the high temperature source.

Let it be assumed that:

– engine E_A is more efficient than engine E_B

– engine E_B with lower efficiency is reversed to operate as a heat pump/refrigerator.

– both the engines E_A and E_B (now operating as a heat pump) can be coupled and a part of the work output of engine E_A can be used to drive the heat pump.

Adopting a procedure similar to that outlined above in section 8.9, it can be shown that the composite system results in a perpetual motion of the second kind.

Likewise, if assumption is made that E_B is more efficient than E_A and the operating of E_A is reversed then again it will be observed that the assumption violates the second law.

Therefore, neither engine E_A nor engine E_B can be more efficient than the other. Hence all reversible engines operating between the same two thermal reservoirs must have equal efficiencies.

Corollary 2 : ***The efficiency of any reversible heat engine operating between two reservoirs is independent of the nature of working fluid and depends only on the temperature of the reservoirs.***

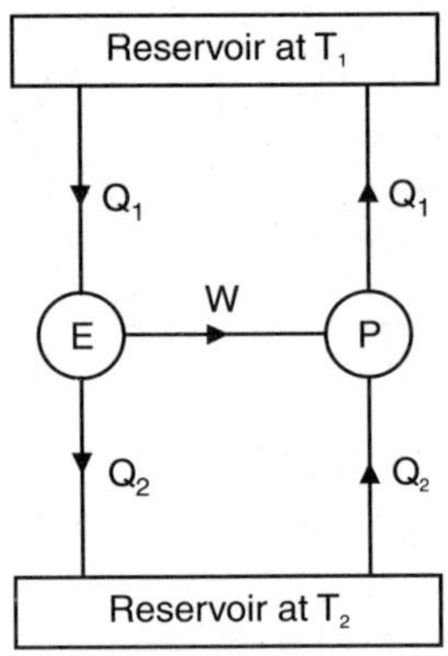

Fig. 8.17. (a)

Let a Carnot engine E and a Carnot pump P operate between two thermal reservoirs at temperatures T_1 and T_2. The engine receives Q_1 units of heat from the reservoir at T_1 and rejects heat Q_2 to the reservoir T_2. The engine is coupled to the heat pump ; the engine output is used to drive the heat pump. Since the Carnot pump is a reversed Carnot engine, its work and heat transfers will be same (as for Carnot engine) but with directions reversed.

The pump then absorbs heat Q_2 from the reservoir at T_2 and rejects heat Q_1 to the reservoir at T_1 From the composite system depicted in Fig 8.17(*a*), it is observed that net output of the engine-heat pump combination is zero.

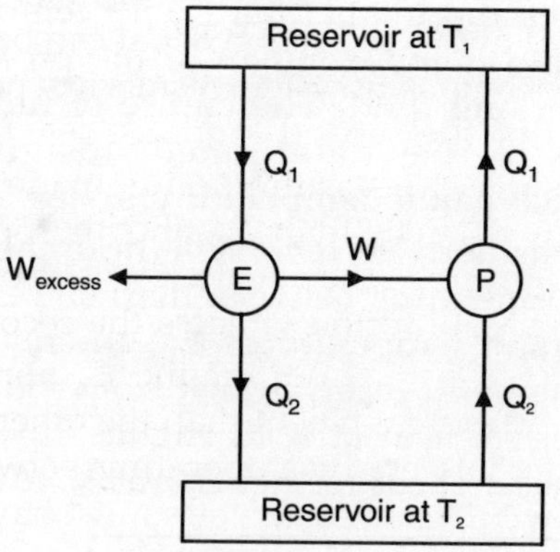

Fig. 8.17. (b)

Let us assume that apart from temperature limits, the efficiency of a machine depends upon some other factor say the working fluid. The engine may then deliver work more than that required by the heat pump. That would lead to the arrangement Fig 8.17(*b*) which indicates that :

– no net interaction with the high temperature reservoir. It supplies and recovers back the same amount of heat.

– the combination essentially works with one reservoir and delivers work.

The combination thus constitutes a perpetual motion of the second kind in iolation of the second law. Hence the assumption that the efficiency of a reversible engine depends on a factor other than the temperature is false.

EXAMPLE 8.19

An inventor claims to have developed an engine that operates between a source at 450 K and a sink at 280 K, and is capable of delivering 0.15 kWh of work for every 1200 kJ of heat received. As a patent officer, would you issue a patent for such an engine ?

Solution : For the engine proposed by the inventor,

Work output, $W = 0.15$ kWh

$= 0.15 \times 3600 = 540$ kJ

Thermal efficiency,

$$\eta_{th} = \frac{\text{work output}}{\text{heat supplied}}$$

$$= \frac{540}{1200} = 0.45 = 45\ \%$$

Thermal efficiency of a reversible engine working between the same temperature limits,

$$\eta_{rev} = \frac{T_1 - T_2}{T_1} = \frac{450 - 280}{450}$$

$$= 0.378 \text{ or } 37.8\ \%$$

Since no engine can be more efficient than a reversible engine, the claim is false and the patent is not to be issued.

EXAMPLE 8.20

An engine manufacturer claims to have developed a heat engine with the following specifications:

Power developed = 75 kW

Fuel burnt = 5 kJ/hr

Heating value of fuel = 75,000 kJ/kg

Temperature limits = 1000 K and 400 K

Is the claim of the manufacturer true or false? Give reasons for your answer.

Solution : For the engine proposed by the manufacturer,

Work done = 75 kW = 75 kJ/s

$= 75 \times 3600 = 27 \times 10^4$ kJ/hr

Heat supplied = mass of fuel × heating value of fuel

$= 5 \times 75000 = 37.5 \times 10^4$ kJ/hr

Thermal efficiency,

$$\eta_{th} = \frac{\text{work done}}{\text{heat supplied}}$$

$$= \frac{27 \times 10^4}{37.5 \times 10^4} = 0.72 \text{ or } 72\%$$

Thermal efficiency of a reversible engine working between the same temperature limits,

$$\eta_{rev} = \frac{T_1 - T_2}{T_1}$$

$$= \frac{1000 - 400}{1000} = 0.6\% = 60\%$$

Since the manufacturer's claim of efficiency is greater than the efficiency of a reversible engine, the claim is false and the engine needs to be redesigned.

8.10. THERMODYNAMIC TEMPERATURE SCALE

A quantitative measure of the temperature of a body requires reference to some datum plane or reference condition and the establishment of a suitable temperature scale. The temperature scale is established by assigning certain numerical values to two easily reproducible states such as the freezing point and the boiling point of water. The temperature is then measured by noting a change in the thermodynamic property of a material. The thermodynamic properties which are so used could be a change in the

– length of mercury column in a capillary tube

– emf generated at the junction of two dissimilar metals

– resistance of a wire

– volume or pressure change of a gas

A temperature scale defined in terms of some physical properties of a material is called the ***empirical scale.***

There exist different relations between the temperature and thermodynamic property of different materials. Accordingly the empirical temperature scale based upon a change in the thermodynamic property of one material does not agree at all points with the scale based upon the same thermodynamic property of another material. Different thermometers tend to give different temperatures except at the fixed points at which they are calibrated. This aspect necessitates that a temperature scale be established which is independent of the nature of thermometric material.

A temperature scale which is independent of the property of thermometric substance is defined as thermodynamic temperature scale.

The thermodynamic temperature scale is established based on the fact that the efficiency of a reversible heat engine does not depend on the nature of working medium but depends only on the temperatures of the two thermal reservoirs between which it operates (Carnot theorem).

$$\eta_{rev} = \phi\,(t_1, t_2) \quad \text{...(8.10)}$$

where ϕ signifies a form of function that connects the temperature with temperature scale and is independent of the property of the working fluid. The nature of function ϕ needs to be determined to give the thermodynamic temperature scale.

Consider two reversible heat engines E_1 and E_2 operating between thermal reservoirs of pairs of temperatures (t_1, t_2) and (t_2, t_3). The engine E_1 receives heat from source at t_1 and rejects heat at t_2 to engine E_2 which in turn rejects heat to the sink at t_3. (Fig 8.18).

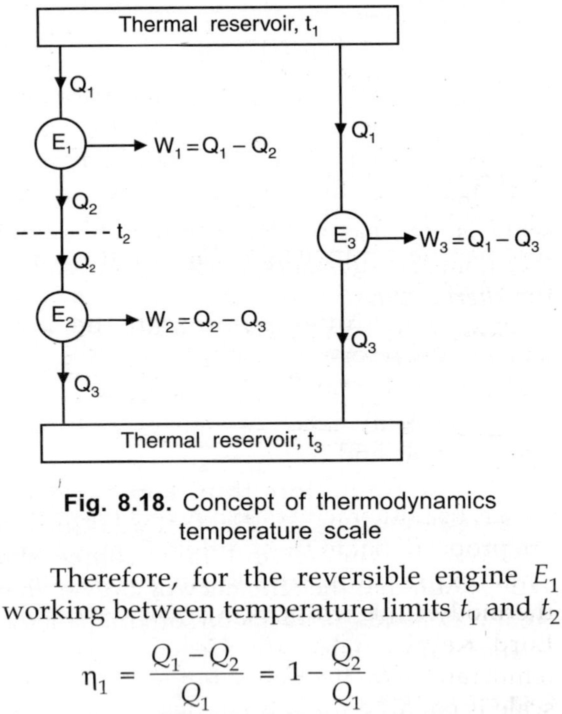

Fig. 8.18. Concept of thermodynamics temperature scale

Therefore, for the reversible engine E_1 working between temperature limits t_1 and t_2

$$\eta_1 = \frac{Q_1 - Q_2}{Q_1} = 1 - \frac{Q_2}{Q_1}$$

$$= 1 - \frac{1}{Q_1/Q_2} = 1 - \frac{1}{f(t_1, t_2)} \quad \text{...(8.11)}$$

i.e., $\dfrac{Q_1}{Q_2} = f\,(t_1, t_2)$ where f is any function.

Likewise for the reversible engine E_2 working between t_2 and t_3,

$$\frac{Q_2}{Q_3} = f\,(t_2, t_3)$$

The engines E_1 and E_2 constitute another heat engine E_3 operating between t_1 and t_3. That engine gives

$$\frac{Q_1}{Q_3} = f(t_1, t_3)$$

Now we may write

$$\frac{Q_1}{Q_2} = \frac{Q_1/Q_3}{Q_2/Q_3}$$

$$f(t_1, t_2) = \frac{f(t_1, t_3)}{f(t_2, t_3)} \quad ...(8.12)$$

The right hand side of the equation must be a function of t_1 and t_2 only. Hence t_3 must cancel out, and the nature of the function f must be of the form

$$\frac{Q_1}{Q_2} = f(t_1, t_2) = \frac{\psi(t_1)}{\psi(t_2)} \quad ...(8.13)$$

where ψ is another function of t. The choice of function ψ depends upon chosen scale of temperature and has infinite variety of forms. If a single form is selected, a temperature scale is defined which is independent of thermometric substance. Such a scale is called the ***thermodynamic temperature scale.***

Kelvin had chosen the form of function as $\psi(t) = T$, and thus equation 8.13 may be written as

$$\frac{Q_1}{Q_2} = f(t_1, t_2) = \frac{T_1}{T_2} \quad ...(8.14)$$

i.e, heat interactions in a reversible engine re proportional to the absolute temperature. This expression forms the basis of a thermodynamic temperature scale devised by Lord Kelvin. Thus the ratio of any two temperatures on the Kelvin or thermodynamic scale is equal to the ratio of heat absorbed to heat rejected by a Carnot engine operating between those temperatures. The correlation suggests that if one of the two temperatures is fixed and assigned a number, then by measuring the heat interactions for the reversible engine, the other temperature can be determined. Denoting such reference temperature by T_0, the value of any other temperature T is fully and uniquely determined by $T = T_0(Q/Q_0)$. The ratio (Q/Q_0) is fixed as a consequence of second law of thermodynamics, and is a function only of the temperature of reservoirs. At present, the fixed (reference temperature) point is taken as the triple point (the state at which solid, liquid and vapour phase are in equillibrium) of water and this has been assigned a numerical value of 273.16 K. Then for a Carnot engine operating between thermal reservoirs at temperatures T and T_t (triple point of water), we may write

$$T = T_t \frac{Q}{Q_t}$$

$$= 273.16 \frac{Q}{Q_t} \quad ...(8.15)$$

Apparently in the Kelvin scale, the amount of heat supplied plays the role of thermodynamic property. The amount of heat supplied changes with temperature just like emf in a thermocouple.

If $Q = 0$ in equation 8.15, then $T = 0$

The thermodynamic temperature scale thus presents an idea of absolute zero.

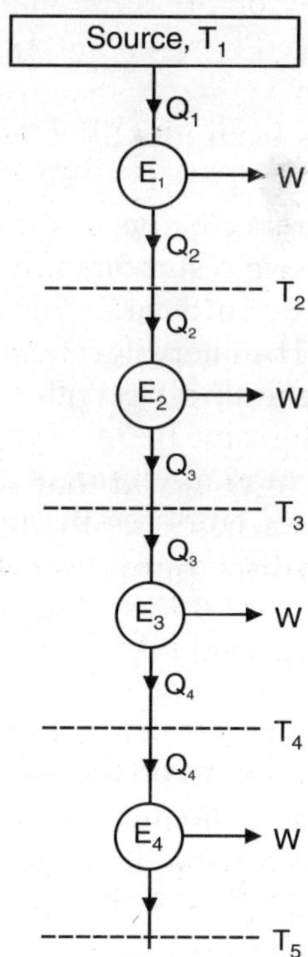

Fig. 8.19. Reversible engines arranged in series

Further consider a series of reversible engines (Fig 8.19) extending from a source at T_1 to lower temperature such that :

– all the reversible engines operate between only two reservoirs,

– each sink is a source for the following engine ; heat rejected from one engine becomes the heat supply to the next engine,

– the temperatures of thermal reservoirs are so chosen that work done by each engine is same.

Thus, $W = Q_1 - Q_2$

$= Q_2 - Q_3 = Q_3 - Q_4 = \text{..........}$

From the definition of Kelvin scale,

$$\frac{Q_1}{Q_2} = \frac{T_1}{T_2} ; \frac{Q_2}{Q_3} = \frac{T_2}{T_3} ; \frac{Q_3}{Q_4} = \frac{T_3}{T_4}$$

That gives :

$$\frac{Q_1}{T_1} = \frac{Q_2}{T_2} = \frac{Q_3}{T_3} = \frac{Q_4}{T_4} = \text{.......} \quad ...(a)$$

Also $W = \eta_1 Q_1 = \eta_2 Q_2 = \eta_3 Q_3 = \text{.....}$

$$= \frac{T_1 - T_2}{T_1} Q_1$$

$$= \frac{T_2 - T_3}{T_2} Q_2 = \frac{T_3 - T_4}{T_3} Q_3 \quad ...(b)$$

It follows from identities (*a*) and (*b*) that

$T_1 - T_2 = T_2 - T_3 = T_3 - T_4$

The difference between the temperatures of the successive reservoirs are the same and can provide intervals or degrees of temperature. The intervals can be made smaller or larger depending upon the requirements of the scale.

It is also to be noted that when the heat rejected approaches zero, the temperature of heat rejection also approaches zero as a limit.

Since the efficiency $\eta = \frac{T_1 - T_2}{T_1}$ of a heat engine cannot be greater than one, the lowest possible temperature on thermodynamic scale is zero. Thus it can be said that a definite zero point exists on the absolute scale but this cannot be attained without a violation of the second law.

Note: If the steam point and the ice point are chosen as fixed reference temperatures, and the number of divisions between these two points is taken as 100, then

$$T_s - T_i = 100 \quad ...(c)$$

A reversible engine working between heat source at temperature of steam point and sink at ice point has an efficiency of 0.268.

$$\eta = \frac{T_s - T_i}{T_s} ; 0.268 = 1 - \frac{T_i}{T_s} \quad ...(d)$$

Upon solution of identities (*c*) and (*d*), we get

$T_s = 373.15$ K and $T_i = 273.15$ K

where K stands for degrees Kelvin. For practical use, these values are approximated to

$T_s = 373$ K and $T_i = 273$ K

Therefore, it follows that : K = °C + 273

8.11. CLAUSIUS INEQUALITY

Recapitulate that

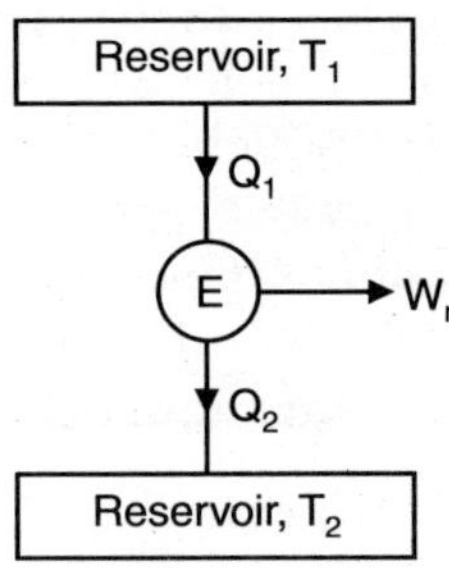

Fig. 8.20. Reversible heat engine

– the efficiency of a reversible heat engine operating within the temperature limits T_1 and T_2 is given by

$$\eta = \frac{Q_1 - Q_2}{Q_1} = \frac{T_1 - T_2}{T_1}$$

$$= 1 - \frac{Q_2}{Q_1} = 1 - \frac{T_2}{T_1}$$

$$\therefore \quad \frac{Q_1}{T_1} = \frac{Q_2}{T_2} \quad \text{or} \quad \frac{Q_1}{T_1} - \frac{Q_2}{T_2} = 0$$

With positive sign convention for heat absorption and negative for heat rejection

$$\frac{Q_1}{T_1} - \frac{(-Q_2)}{T_2} = 0 \quad \text{or} \quad \frac{Q_1}{T_1} + \frac{Q_2}{T_2} = 0$$

or $\sum \frac{Q}{T} = 0$ for a reversible engine. ...(8.16)

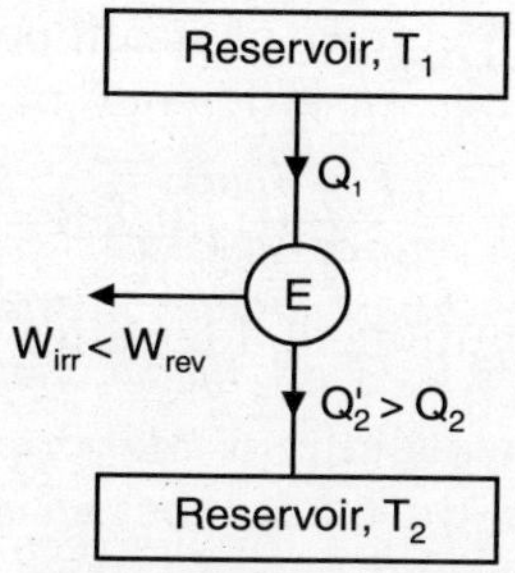

Fig. 8.21. Irreversible heat engine

– the efficiency of an irreversible engine is lower than that of a reversible engine operating within the same temperature limits. Therefore,

$$\frac{Q_1 - Q_2'}{Q_1} < \frac{T_1 - T_2}{T_1}$$

or $$1 - \frac{Q_2'}{Q_1} < 1 - \frac{T_2}{T_1}$$

or $$\frac{Q_1}{T_1} < \frac{Q_2'}{T_2}$$

Taking the sign convention for heat interaction into account, we may write

$$\frac{Q_1}{T_1} - \frac{(-Q_2')}{T_2} < 0 \quad \text{or} \quad \frac{Q_1}{T_1} + \frac{Q_2}{T_2} < 0$$

or $\sum \frac{Q}{T} < 0$ for an irreversible engine ...(8.17)

Fig 8.22 shows a system performing reversible cyclic process 1–2–3–4 on p–V diagram. Let the cycle be cut by a family of closely spaced adiabatic lines similar to *ad*, *bc* , *ef* etc. The elementary cycle *abcd* then consists of two adiabatics *da* and *bc*, and two elementary paths *ab* and *cd*. Since the adiabatics lie close to each other, the distance *ab* tends to be zero and accordingly the temperatures at points *a* and *b* can be thought to be equal. Likewise the points *c* and *d* would also be at another equal temperature. The elementary paths *ab* and *cd* then conform to isothermal conditions and the cycle *abcd* constitutes an elementary Carnot cycle. Likewise *befc* is another elementary Carnot cycle. Thus the given cycle 1–2–3–4 can be imagined to consist of a large number (say *i*) of Carnot cycles.

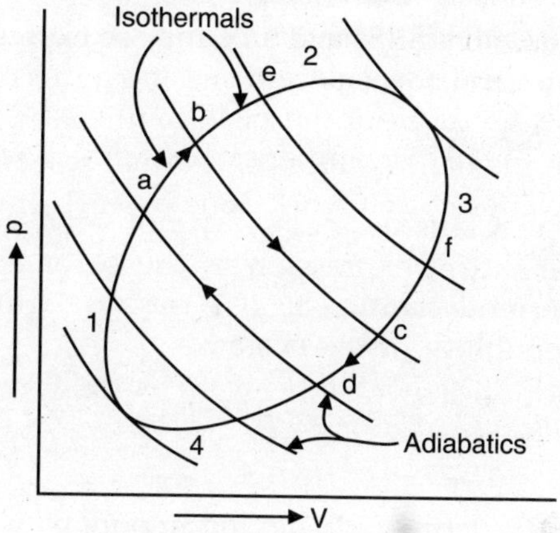

Fig. 8.22. Clausius inequality

For the Carnot cycle *abcd*, the process *bc* represents adiabatic expansion, and for the adjoining Carnot cycle *befc* it represents an equal amount of adiabatic compression. As such for the composite reversible cycle 1–2–3–4, the adiabatics contribute neither towards work nor towards heat interactions. The work output from the engine is as a consequence of net heat transfer during the elementary isothermal processes. Equation 8.16 is valid for each elementary cycle and, therefore, for the reversible cycle 1–2–3–4, we may write

$$\sum \frac{Q_i}{T_i} = 0 \quad \text{or} \quad \oint_R \frac{\delta Q}{T} = 0 \quad ...(8.18)$$

The subscript *R* refers to reversible cycle. Equation 8.18 is called *Clausius theorem* and be stated as :

The algebraic sum of the ratio $\frac{\delta Q}{T}$ *i.e, the heat interaction to the absolute temperature for a reversible heat engine is equal to zero.*

Likewise for an irreversible cycle

$$\sum \frac{Q_i}{T_i} < 0 \quad \text{or} \quad \oint_I \frac{\delta Q}{T} < 0 \quad ...(8.19)$$

The subscript *I* refers to irreversible cycle. Equation 8.19. is called *Clausius inequality* and may be stated as :

The algebraic sum of the ratio $\frac{\delta Q}{T}$ i.e, the heat interaction to the absolute temperature for an irreversible heat engine is less than zero.

Equations 8.18. and 8.19. may be expressed in a general form as

$$\oint \frac{\delta Q}{T} \leq 0 \text{ for any engine.} \qquad ...(8.20)$$

The Clausius inequality not only gives mathematical expression to the second law of thermodynamics, it also throws light on irreversibility of the system.

For an irreversible cycle, we may write

$$\oint \frac{\delta Q}{T} + I = 0$$

where the term I indicates the amount by which the given cycle is irreversible. For a reversible cycle, $I = 0$. Further, a cycle in which the integral $\oint \frac{\delta Q}{T}$ is greater than zero violates the second law and is impossible.

EXAMPLE 8.21

A heat engine is supplied with 2512 kJ/min of heat at 650°C. Heat rejection takes place at 100°C. Specify which of the following heat rejections represent a reversible, irreversible or impossible result:

(i) 867 kJ/min (ii) 1015 kJ/min

(iii) 1494 kJ/min

Solution : $T_1 = 650°C = 650 + 273 = 923$ K

$T_2 = 100 + 273 = 373$ K

$$(i) \quad \oint \frac{\delta Q}{T} = \frac{Q_1}{T_1} - \frac{Q_2}{T_2} = \frac{2512}{923} - \frac{867}{373}$$

$$= 2.721 - 2.324 = 0.347 > 0$$

This cycle is impossible.

$$(ii) \quad \oint \frac{\delta Q}{T} = \frac{Q_1}{T_1} - \frac{Q_2}{T_2}$$

$$= \frac{2512}{927} - \frac{1015}{373}$$

$$= 2.721 - 2.721 = 0$$

This cycle is reversible.

$$(iii) \quad \oint \frac{\delta Q}{T} = \frac{Q_1}{T_1} - \frac{Q_2}{T_2} = \frac{2512}{927} - \frac{1494}{373}$$

$$= 2.721 - 4.005 = -1.284 < 0$$

This cycle is irreversible.

EXAMPLE 8.22

An inventor claims that his engine absorbs 300 kJ of energy from a thermal reservoir at 325 K and delivers 75 kJ of work. The inventor also states that his engine has two heat rejections : 125 kJ to a reservoir at 300 K and 100 kJ to a reservoir at 275 K. Check the validity of his claim.

Solution : The validity of the proposed engine is checked on the basis whether it satisfies or not the first and second laws of thermodynamics. The engine is theoretically feasible if both the laws are satisfied. The device becomes impossible if any of the two laws is violated.

$$\oint \delta Q = 300 - 125 - 100 = 75 \text{ kJ}$$

$$\oint \delta W = 75 \text{ kJ}$$

Since, $\oint \delta Q = \oint \delta W$, the data is in accordance with the first law.

Applying Clausius theorem,

$$\oint \frac{\delta Q}{T} = \frac{Q_1}{T_1} - \frac{Q_2}{T_2} - \frac{Q_3}{T_3}$$

$$= \frac{300}{325} - \frac{125}{300} - \frac{100}{275} = 0.416 > 0$$

The Clausius theorem requires that $\oint \frac{\delta Q}{T} \leq 0$ and as the proposed engine gives $\oint \frac{\delta Q}{T} > 0$, the claim of the inventor is not acceptable.

EXAMPLE 8.23

A household refrigerator absorbs heat at 2°C and rejects heat to the surroundings at 50°C. Its compressor is driven by 3 kW motor and 50 MJ/hr are absorbed at the low temperature. Evaluate the amount of heat rejected per hour and the irreversibility in J/hr.

Solution : $Q_2 = 50$ MJ/hr $= 50000$ kJ/hr

$W = 3$ kW $= 3 \times 3600 = 10800$ kJ/hr

From the principle of energy conservation,

$$Q_1 = W + Q_2$$

$$= 10800 + 50000 = \mathbf{60800 \text{ kJ/hr}}$$

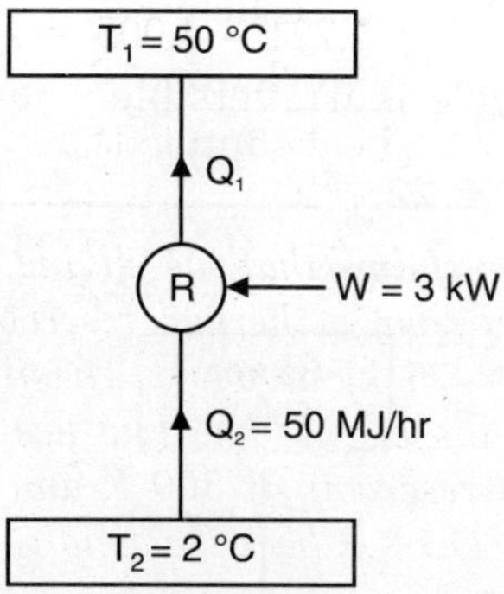

Fig. 8.23.

Thus the refrigerator rejects 60800 kJ/hr of heat

Taking into account the sign convention for heat interaction, the Clausius inequality for an irreversible cycle can be written as

$$\oint \frac{\delta Q}{T} + I = 0$$

where I is irreversibility

or $$I = \frac{Q_1}{T_1} - \frac{Q_2}{T_2}$$

$$= \frac{60800}{50+273} - \frac{50000}{2+273}$$

$$= 188.23 - 181.82$$

$$= \mathbf{6.41\ kJ/hr}$$

EXAMPLE 8.24.

Figure 8.24 shows a system that undergoes a reversible cycle during which it exchanges heat with three thermal reservoirs. Show that

$$\frac{Q_1}{\frac{1}{T_2}-\frac{1}{T_3}} = \frac{Q_2}{\frac{1}{T_3}-\frac{1}{T_1}} = \frac{Q_3}{\frac{1}{T_1}-\frac{1}{T_2}}$$

regardless of the actual process involved.

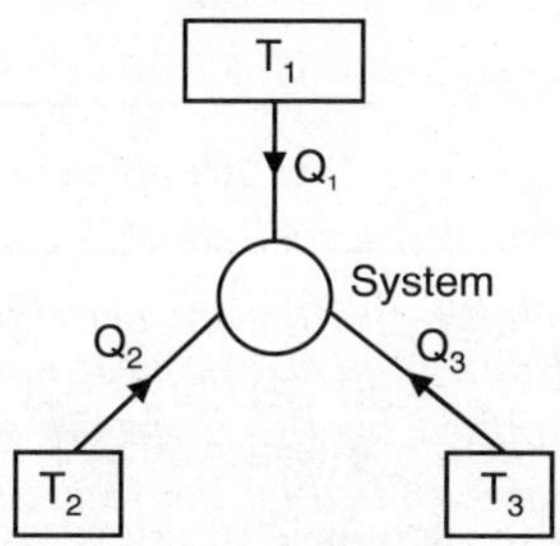

Fig. 8.24.

Solution : From the principle of energy conservation,

$$Q_1 + Q_2 + Q_3 = 0 \qquad ...(i)$$

Since the system executes a cyclic process, the Clausius theorem gives

$$\frac{Q_1}{T_1} + \frac{Q_2}{T_2} + \frac{Q_3}{T_3} = 0 \qquad ...(ii)$$

From expression (*i*) $Q_3 = -(Q_1 + Q_2)$ and substituting this value of Q_3 in expression (*ii*), we obtain

$$\frac{Q_1}{T_1} + \frac{Q_2}{T_2} + \frac{-(Q_1+Q_2)}{T_3} = 0$$

or $$Q_1\left(\frac{1}{T_1}-\frac{1}{T_3}\right) = Q_2\left(\frac{1}{T_3}-\frac{1}{T_2}\right)$$

or $$\frac{Q_1}{\frac{1}{T_2}-\frac{1}{T_3}} = \frac{Q_2}{\frac{1}{T_3}-\frac{1}{T_1}} \qquad ...(a)$$

From expression (*i*) $Q_2 = -(Q_1 + Q_3)$ and substituting this value of Q_2 in expression (*ii*), we obtain

$$\frac{Q_1}{T_1} + \frac{-(Q_1+Q_3)}{T_2} + \frac{Q_3}{T_3} = 0$$

or $$Q_1\left(\frac{1}{T_1}-\frac{1}{T_2}\right) = Q_3\left(\frac{1}{T_2}-\frac{1}{T_3}\right)$$

or $$\frac{Q_1}{\frac{1}{T_2}-\frac{1}{T_3}} = \frac{Q_3}{\frac{1}{T_1}-\frac{1}{T_2}} \qquad ...(b)$$

From identities (*a*) and (*b*),

$$\frac{Q_1}{\frac{1}{T_2}-\frac{1}{T_3}} = \frac{Q_2}{\frac{1}{T_3}-\frac{1}{T_1}} = \frac{Q_3}{\frac{1}{T_1}-\frac{1}{T_2}}$$

which is the required expression.

EXAMPLE 8.25

A reversible heat engine receives heat from two thermal reservoirs maintained at constant temperatures of 750 K and 500 K. The engine develops 100 kW and rejects 3600 kJ/min of heat to a heat sink at 250 K. Determine thermal efficiency of the engine and heat supplied by each thermal reservoir.

Solution : Work output = 100 kW = 100 kJ/s

= 6000 kJ/min

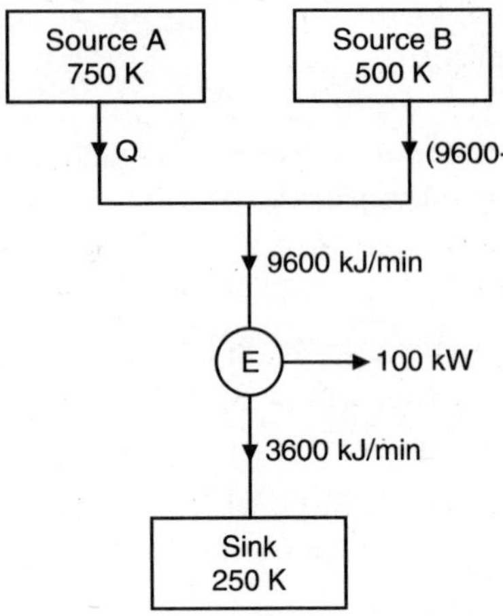

Fig. 8.25.

Application of the first law of thermodynamics to heat engine gives ;

Total heat supplied

$$= 6000 + 3600 = 9600 \text{ kJ/min}$$

$$\eta = \frac{\text{work done}}{\text{heat supplied}}$$

$$= \frac{6000}{9600} = 0.625 \quad \text{or} \quad 62.5\%$$

Let Q be the heat supplied by source A. Then heat supplied by source B is $(9600 - Q)$.

As the engine is reversible, the Clausius theorem gives

$$\oint \frac{\delta Q}{T} = 0 = \frac{Q}{750} + \frac{9600 - Q}{500} - \frac{3600}{250}$$

Multiplying throughout by 750, we get

$$0 = Q + 14400 - 1.5\,Q - 10800$$

$$Q = \frac{14400 - 10800}{0.5} = 7200 \text{ kJ/min}$$

Thus, heat supplied by source at 750 K

= **7200 kJ/min**

and heat supplied by source at 500 K

= (9600 – 7200) = **2400 kJ/min**

EXAMPLE 8.26

A reversible heat engine is supplied 900 kJ of heat from a heat source at 500 K. The engine develops 300 kJ of net work and rejects heat to two heat sinks at 400 K and 300 K. Determine the engine thermal efficiency and magnitude of heat interaction with each of the sink.

Solution : Thermal efficiency,

$$\eta = \frac{\text{work done}}{\text{heat supplied}}$$

$$= \frac{300}{900} = 0.333 \text{ or } 33.3\%$$

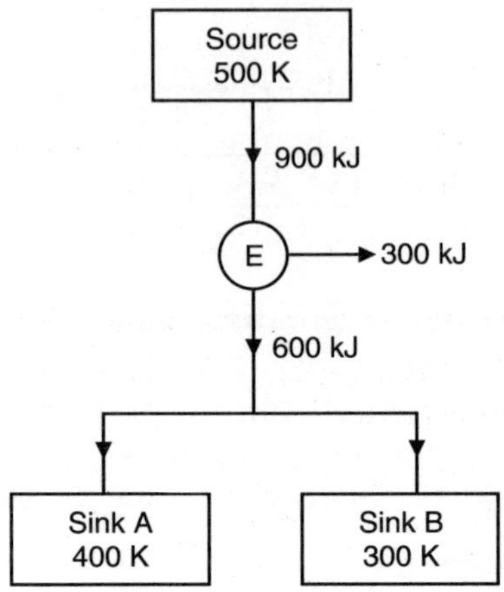

Fig. 8.26.

From energy conservation principle,

Heat rejected = heat supplied – work done

= 900 – 300 = 600 kJ

Let Q be the heat rejected to sink at 400 K. Then heat rejected to sink at 300 K is $(600 – Q)$. As the engine is reversible, the Clausius theorem gives :

$$\oint \frac{\delta Q}{T} = 0 = \frac{900}{500} - \frac{Q}{400} - \frac{600 - Q}{300}$$

Multiplying throughout by 60, we get

$$0 = 108 - 0.15\,Q - 120 + 0.2\,Q$$

$$Q = \frac{120 - 108}{0.05} = 240 \text{ kJ}$$

Thus heat rejected to sink at 400 K = **240 kJ** and heat rejected to sink at 300 K

= (600 – 240) = **360 kJ**

EXAMPLE 8.27

A heat engine working on Carnot cycle absorbs heat from three thermal reservoirs at 1000K, 800 K and 600 K. The engine does 10 kW of net work and rejects 400 kJ/min of heat to a heat sink at 300 K. If the heat supplied by the reservoir at 1000 K is 60% of the heat supplied by the reservoir at 600 K, make calculations for the quantity of heat absorbed by each reservoir.

Solution : Work output = 10 kJ/s

= 10 × 60 = 600 kJ/min

Application of first law of thermodynamics to engine gives,

Total heat supplied

= work done + heat rejected

= 600 + 400

= 1000 kJ/min

Let Q_1 kJ/min be the heat absorbed by the engine from the source at 600 K. Then heat absorbed from the source at 1000 K is 0.6 Q_1

∴ Heat taken from the source at 800 K,

$$Q_2 = 1000 - Q_1 - 0.6\,Q_1 = (1000 - 1.6\,Q_1) \text{ kJ/min}$$

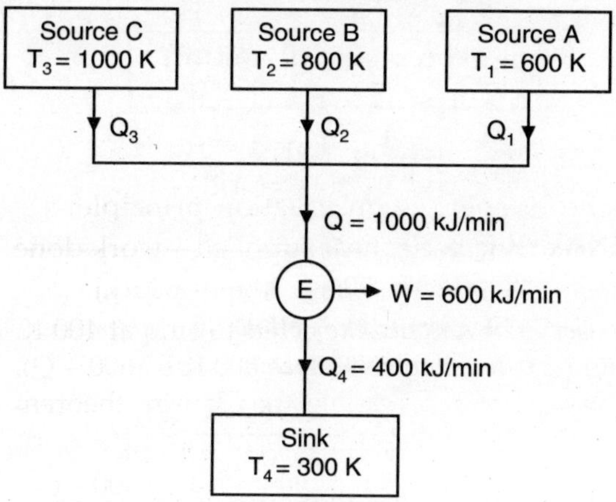

Fig. 8.27.

As the engine is reversible, the Clausius theorem gives:

$$\oint \frac{\delta Q}{T} = 0 = \frac{Q_1}{T_1} + \frac{Q_2}{T_2} + \frac{Q_3}{T_3} - \frac{Q_4}{T_4}$$

$$0 = \frac{Q_1}{600} + \frac{1000 - 1.6Q_1}{800} + \frac{0.6Q_1}{1000} - \frac{400}{300}$$

Upon solution :

Q_1 = 312.5 kJ/min (from source at 600 K)

Q_2 = 1000 – 1.6 × 312.5 = 500 kJ/min (from source at 800 K)

Q_3 = 0.6 Q_2 = 312.5 × 0.6 = 187.5 kJ/min (from source at 1000 K)

EXAMPLE 8.28

A reversible heat engine receives equal quantity of heat from two thermal reservoirs at temperatures T_1 and T_2, and rejects heat to a heat sink at T_3. Presuming that efficiency of this engine is α times the efficiency of a reversible engine absorbing the same amount of heat only from reservoir at T_1 and rejecting heat to sink at T_3, show that

$$\alpha = \frac{1}{2}\left[\frac{T_2 - T_3}{T_1 - T_3} + \frac{T_2}{T_3}\right] \times \frac{T_1}{T_2}$$

Solution : Refer to Fig 8.28 for the schematic arrangement of the two engines

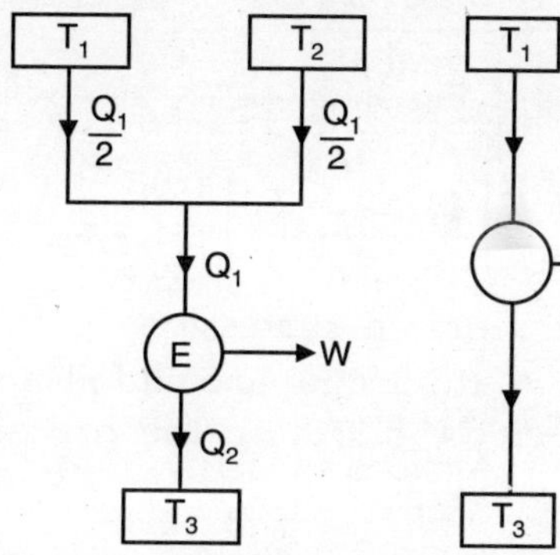

Fig. 8.28.

(*i*) Since the given engine is reversible, the Clausius theorem gives :

$$\oint \frac{\delta Q}{T} = 0 = \frac{Q_1}{2T_1} + \frac{Q_1}{2T_2} - \frac{Q_2}{T_3}$$

If W is the work output from the engine, then from the principle of energy conservation,

$$Q_1 = W + Q_2\,; \quad Q_2 = Q_1 - W$$

$$\text{Then : } 0 = \frac{Q_1}{2T_1} + \frac{Q_1}{2T_2} - \frac{Q_1 - W}{T_3}$$

$$\therefore \quad W = Q_1\left[\frac{2T_1T_2 - T_3T_2 - T_3T_1}{2T_1T_2}\right]$$

Thermal efficiency,

$$\eta_1 = \frac{\text{work output}}{\text{heat supplied}} = \frac{2T_1T_2 - T_3T_2 - T_3T_1}{2T_1T_2}$$

(*ii*) Thermal efficiency of a reversible engine operating between temperature limits T_1 and T_3 is

$$\eta_2 = \frac{T_1 - T_3}{T_1}$$

According to the given conditions

$$\eta_1 = \alpha\, \eta_2$$

$$\therefore \quad \frac{2T_1T_2 - T_3T_2 - T_3T_1}{2T_1T_2} = \alpha \frac{T_1 - T_3}{T_1}$$

Hence

$$\alpha = \frac{T_1}{2T_1T_2}\left[\frac{2T_1T_2 - T_3T_2 - T_3T_1}{T_1 - T_3}\right]$$

$$= \frac{1}{2T_2}\left[\frac{T_1T_2 - T_3T_2}{T_1 - T_3} + \frac{T_1T_2 - T_3T_1}{T_1 - T_3}\right]$$

$$= \frac{1}{2T_2}\left[T_2\left(\frac{T_1 - T_3}{T_1 - T_3}\right) + T_1\left(\frac{T_2 - T_3}{T_1 - T_3}\right)\right]$$

$$= \frac{1}{2}\frac{T_1}{T_2}\left[\frac{T_2}{T_1} + \frac{T_2 - T_3}{T_1 - T_3}\right]$$

which is the required expression.

Solution to this example could also have been obtained by adopting the procedure outlined in Example 8.14

EXAMPLE 8.29

A Carnot engine E_1 operates between temperatures T_1 and T_2, rejects its waste heat at its sink temperature T_2 to Carnot engine E_2 which operates between temperature limits T_2 and T_3. Set up an expression relating the thermal efficiency of engines E_1 and E_2 to that of a Carnot engine E_3 operating directly between temperatures T_1 and T_3.

Solution : Refer Fig 8.29.

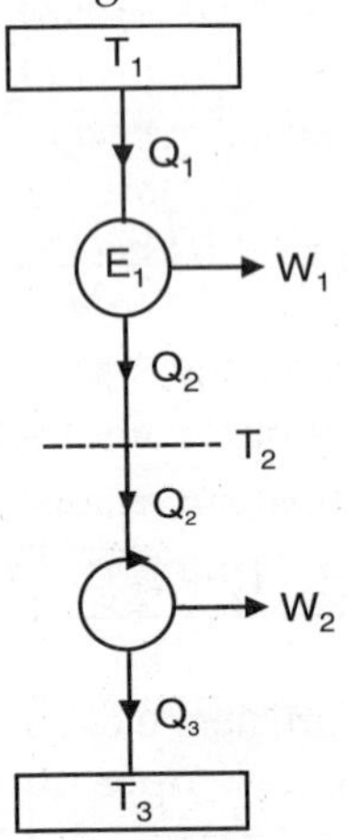

Fig. 8.29.

$$\eta_1 = \frac{T_1 - T_2}{T_1} = 1 - \frac{T_2}{T_1}$$

or $\quad 1 - \eta_1 = \frac{T_2}{T_1}$

Likewise $\quad 1 - \eta_2 = \frac{T_3}{T_2}$

$$\therefore \quad (1 - \eta_1)(1 - \eta_2) = \frac{T_3}{T_1} \qquad ...(i)$$

For the Carnot engine operating directly between the temperature limits T_1 and T_3,

$$\eta_3 = \frac{T_1 - T_3}{T_1} = 1 - \frac{T_3}{T_1}$$

or $\quad 1 - \eta_3 = \frac{T_3}{T_1} \qquad ...(ii)$

From expressions (*i*) and (*ii*),

$$1 - \eta_3 = (1 - \eta_1)(1 - \eta_2)$$
$$= 1 - \eta_2 - \eta_1 + \eta_1\eta_2$$
$$\therefore \eta_3 = \eta_1 + \eta_2 - \eta_1\eta_2$$

EXAMPLE 8.30

Two Carnot refrigerators A and B are arranged in series. Workout the following expression for COP of the composite system

$$COP = \frac{COP_A \times COP_B}{1 + COP_A + COP_B}$$

where COP_A is the coefficient of performance of refrigerator A and COP_B is the coefficient of performance of refrigerator B.

A Carnot refrigerator absorbs 250 kJ of heat from a reservoir at 150 K and rejects heat to a reservoir at 450 K. This heat serves as the energy input to a second Carnot refrigerator which operates between 450 K and 1500 K. Determine the coefficient of performance of (a) the cold refrigerator (b) the hotter refrigerator and (c) the composite system. Check the answer of (c) with the relation established above.

Solution : Refer Fig 8.30

$$COP_A = \frac{Q_3}{Q_2 - Q_3}; \; COP_AQ_2 - COP_AQ_3 = Q_3$$

$$\therefore \quad Q_2 = \frac{1 + COP_A}{COP_A}Q_3 \qquad ...(i)$$

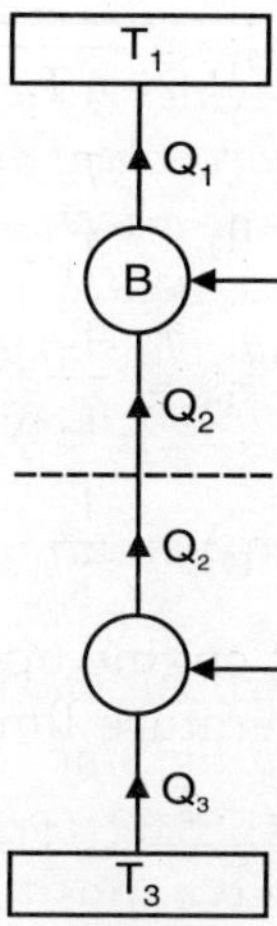

Fig. 8.30.

Likewise

$$Q_1 = \frac{1+COP_B}{COP_B} Q_2 \qquad ...(ii)$$

From (*i*) and (*ii*)

$$\frac{Q_1}{Q_3} = \frac{(1+COP_B)(1+COP_A)}{COP_A\ COP_B}$$

For the Carnot refrigerator working directly between the reservoirs T_3 and T_1,

$$COP = \frac{Q_3}{W_A + W_B}$$

$$= \frac{Q_3}{(Q_2 - Q_3) + (Q_1 - Q_2)}$$

$$= \frac{Q_3}{Q_1 - Q_3} = \frac{1}{\frac{Q_1}{Q_3} - 1}$$

Substituting the value of $\frac{Q_1}{Q_3}$ as obtained above, we get

$$COP = \frac{1}{\frac{(1+COP_A)(1+COP_B)}{COP_A\ COP_B} - 1}$$

$$= \frac{COP_A\ COP_B}{(1+COP_A)(1+COP_B) - COP_A\ COP_B}$$

$$= \frac{COP_A\ COP_B}{1 + COP_A + COP_B}$$

which is the required expression.

(*b*) For the colder refrigerator *A*,

$$COP_A = \frac{Q_3}{Q_2 - Q_3} = \frac{T_3}{T_2 - T_3}\ ;\ \frac{T_3}{T_2} = \frac{Q_3}{Q_2}$$

∴ Heat outflow

$$Q_2 = Q_3 \times \frac{T_2}{T_3} = 250 \times \frac{450}{150} = 750 \text{ kJ}$$

$$COP_A = \frac{250}{750 - 250} = \mathbf{0.50}$$

For the hotter refrigerator *B*,

$$COP_B = \frac{Q_2}{Q_1 - Q_2} = \frac{T_2}{T_1 - T_2}\ ;\ \frac{T_2}{T_1} = \frac{Q_2}{Q_1}$$

∴ Heat outflow

$$Q_1 = Q_2 \times \frac{T_1}{T_2} = 750 \times \frac{1500}{450} = 2500 \text{ kJ}$$

$$COP_B = \frac{750}{2500 - 750} = \mathbf{0.4286}$$

For the overall system,

$$COP = \frac{Q_3}{Q_1 - Q_3} = \frac{250}{2500 - 250} = \mathbf{0.111}$$

Upon checking with the relation established above we get,

$$COP = \frac{0.50 \times 0.4286}{1 + 0.50 + 0.4286} = 0.111$$

EXAMPLE 8.31

A reversible heat engine receives heat from a high temperature reservoir at T_1 K and rejects heat to a low temperature sink at 800 K. A second reversible engine receives the heat rejected by the first engine at 800 K and rejects to a cold reservoir at 280 K. Make calculations for temperature T_1: (i) for equal thermal efficiencies of the two engines, (ii) for the two engines to deliver the same amount of work.

Solution : Refer Fig 8.31 for the schematic arrangement of the composite system.

(*i*) Efficiency of engine E_1 :

$$\eta_1 = \frac{T_1 - T_2}{T_1} = 1 - \frac{T_2}{T_1}$$

Efficiency of engine E_2 :

$$\eta_2 = \frac{T_2 - T_3}{T_2} = 1 - \frac{T_3}{T_2}$$

Since both the engines operate with equal thermal efficiency,

$$1 - \frac{T_2}{T_1} = 1 - \frac{T_3}{T_2}; \quad T_2 = \sqrt{T_1 T_3}$$

Thus the intermediate temperature T_2 is the geometric mean of temperatures T_1 and T_3 if the engines have the same thermal efficiencies.

$$T_1 = \frac{T_2^2}{T_3} = \frac{800 \times 800}{280} = \mathbf{2285.7\ K}$$

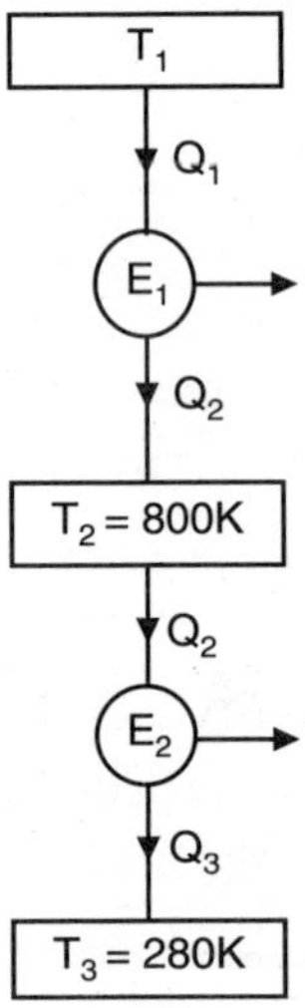

Fig. 8.31.

(*ii*) Work delivered by engines E_1 and E_2 are

$$W_1 = Q_1 - Q_2 \text{ and } W_2 = Q_2 - Q_3$$

Since each engine delivers the same amount of work,

$$Q_1 - Q_2 = Q_2 - Q_3 \quad \text{or} \quad \frac{Q_1}{Q_2} + \frac{Q_3}{Q_2} = 2$$

Further the engines are reversible and,

therefore, $\frac{Q_1}{Q_2} = \frac{T_1}{T_2}$ and $\frac{Q_3}{Q_2} = \frac{T_3}{T_2}$

$$\therefore \frac{T_1}{T_2} + \frac{T_3}{T_2} = 2 \quad \text{or} \quad T_2 = \frac{T_1 + T_3}{2}$$

Thus the intermediate temperature T_2 is the arithmetic mean of temperatures T_1 and T_3 if the engines produce the same amount of work output

$$T_1 = 2T_2 - T_3 = 2 \times 800 - 280 = \mathbf{1320\ K}$$

EXAMPLE 8.32

Two reversible heat engines A and B are arranged in series. Engine A rejects heat directly to engine B. A receives 200 kJ at a temperature of 421°C from the hot source while engine B is in communication with a cold sink at a temperature of 5°C. If the work output of A is twice that of B, find

(a) intermediate temperature between A and B,

(b) efficiency of each engine, and

(c) heat rejected to the sink.

Solution : Refer Fig 8.32 for the schematic arrangement of the composite system.

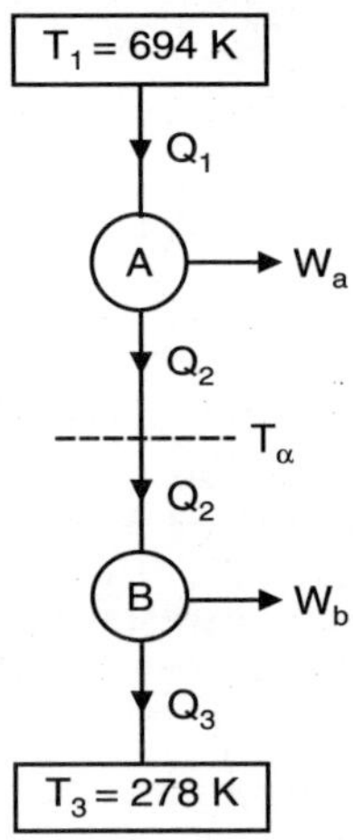

Fig. 8.32.

Works delivered by engines A and B are

$$W_a = Q_1 - Q_2$$

and $$W_b = Q_2 - Q_3$$

Since the work output of engine A is twice that of B,

$$Q_1 - Q_2 = 2(Q_2 - Q_3)$$

or $$Q_1 - Q_2 = 2Q_2 - 2Q_3$$

or $$Q_1 + 2Q_3 = 3Q_2$$

or $$\frac{Q_1}{Q_2} + 2\frac{Q_3}{Q_2} = 3 \qquad \text{...}(i)$$

Further the engines are reversible and therefore

$$\frac{T_1 - T_2}{T_1} = \frac{Q_1 - Q_2}{Q_1}$$

$$\text{and} \quad \frac{T_2 - T_3}{T_2} = \frac{Q_2 - Q_3}{Q_2}$$

The above two expressions can be written as

$$\text{or } 1 - \frac{T_2}{T_1} = 1 - \frac{Q_2}{Q_1} \text{ and } 1 - \frac{T_3}{T_2} = 1 - \frac{Q_3}{Q_2}$$

$$\text{or} \quad \frac{T_2}{T_1} = \frac{Q_2}{Q_1} \text{ and } \frac{T_3}{T_2} = \frac{Q_3}{Q_2} \quad ...(ii)$$

From expressions (*i*) and (*ii*)

$$\frac{T_1}{T_2} + 2\frac{T_3}{T_2} = 3$$

$$\text{or} \quad \frac{694}{T_2} + 2 \times \frac{278}{T_2} = 3$$

or $(694 + 556) = 3T_2$; $T_2 =$ **416.67** K

Thus the intermediate temperature between engines A and B is 416.67 K or 143.67°C

(*b*) Efficiency of engine A,

$$\eta_a = \frac{T_1 - T_2}{T_1} = \frac{694 - 416.67}{694}$$

$= \mathbf{0.3996}$ or 39.96 %

Efficiency of engine B,

$$\eta_a = \frac{T_2 - T_3}{T_2} = \frac{416.67 - 218}{416.67}$$

$= \mathbf{0.3328}$ or 33.28%

(*c*) From expression (*ii*),

$$Q_2 = Q_1 \times \frac{T_2}{T_1} = 200 \times \frac{416.67}{694}$$

$= 120.08$ kJ

$$Q_3 = Q_2 \times \frac{T_3}{T_2} = 120.08 \times \frac{278}{416.67}$$

$= \mathbf{80.12}$ kJ

Thus the heat rejected to the cold sink is **80.12** kJ

EXAMPLE 8.33

Two reversible engines E_1 and E_2 are kept in series between a hot reservoir at temperature T_1 of 600 K and a cold reservoir at a temperature T_2 of 300 K. Engine E_1 receives 500 kJ of heat from the reservoir at T_1. Presuming that both the engines have equal thermal efficiency determine:

(i) the temperature at which heat is rejected by engine E_1 and is received by engine E_2

(ii) the thermal efficiency of each engine

(iii) the work done by engine E_1 and E_2 and

(iv) the heat rejected by engine E_2 to cold reservoir

Solution : Refer Fig. 8.33 for the schematic arrangement of the composite system.

Since the engines are reversible, their thermal efficiencies are

$$\eta_1 = \frac{T_1 - T_0}{T_1} = 1 - \frac{T_0}{T_1}$$

$$\eta_2 = \frac{T_0 - T_2}{T_0} = 1 - \frac{T_2}{T_0}$$

where T_0 is the temperature at which engine E_1 rejects heat to engine E_2. Since both the engines operate with equal thermal efficiency

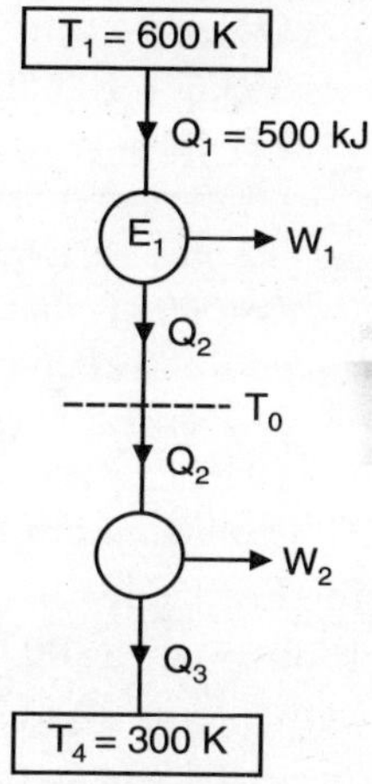

Fig. 8.33.

$$1 - \frac{T_0}{T_1} = 1 - \frac{T_2}{T_0}$$

$$\text{or} \quad \frac{T_0}{T_1} = \frac{T_2}{T_0}$$

i.e., $T_0 = \sqrt{T_1 T_2}$

$= \sqrt{6000 \times 300} = 424.3$ K

(*ii*) $\eta_1 = \dfrac{T_1 - T_0}{T_1} = \dfrac{600 - 424.3}{600}$

$= 0.2928$ or 29.28 %

$$\eta_2 = \frac{T_0 - T_2}{T_0} = \frac{424.3 - 300}{424.3}$$
$$= 0.2928 \text{ or } 29.28\ \%$$

(*iii*) Let W_1 and W_2 be the works performed by engines E_1 and E_2 respectively. Then

$$\eta_1 = \frac{W_1}{Q_1} = \frac{W_1}{500}$$

or $\frac{W_1}{500} = 0.2928$; $W_1 = 146.4$ kJ

and $Q_2 = Q_1 - W_1$
$= 500 - 146.4 = 353.6$ kJ

Likewise for engine E_2

$$\eta_2 = \frac{W_2}{Q_2} = \frac{W_2}{353.6}$$

or $\frac{W_2}{353.6} = 0.2928$; $W_2 = 103.53$ kJ

Heat rejected by engine E_2 to cold reservoir,

$$Q_3 = Q_2 - W_2$$
$$= 353.6 - 103.53 = \mathbf{250.07} \text{ kJ}$$

EXAMPLE 8.34

Three Carnot heat engines are arranged in series. The first engine takes 4000 kJ of heat from a source at 2000 K and delivers 1800 kJ of work ; the second and third engines deliver 1200 kJ and 500 kJ of work respectively. Make calculations for the exhaust temperature of the second and third Carnot engine.

Solution : Refer Fig 8.34.

Total work output = 1800 + 1200 + 500
= 3500 kJ

From principle of energy conservation, the heat rejected to the sink at temperature T_4 is

$$Q_4 = 4000 - 3500 = 500 \text{ kJ}$$

For the overall Carnot engine ,

$$\frac{T_1}{T_4} = \frac{Q_1}{Q_4}$$

∴ Exhaust temperature of the third Carnot engine,

$$T_4 = \frac{Q_4}{Q_1} \times T_1 = \frac{500}{4000} \times 2000 = \mathbf{250\ K}$$

For the composite system with engines E_1 and E_2 in series

$$Q_3 = Q_1 - W_1 - W_2$$
$$= 4000 - 1800 - 1200 = 1000 \text{ kJ}$$

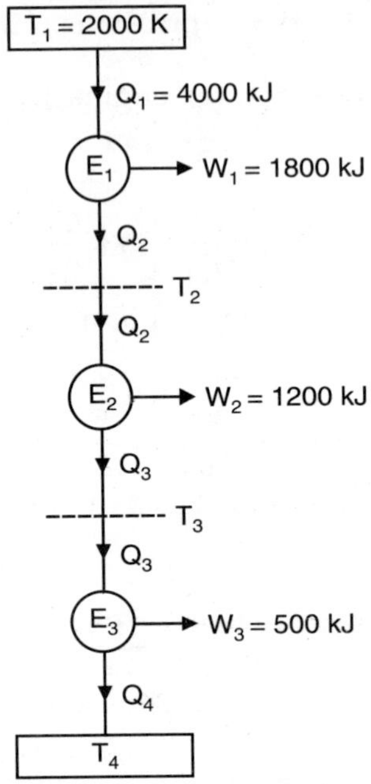

Fig. 8.34.

$$\frac{T_1}{T_3} = \frac{Q_1}{Q_3}$$

$$T_3 = \frac{Q_3}{Q_1} \times T_1 = \frac{1000}{4000} \times 2000 = \mathbf{500\ K}$$

Therefore, exhaust temperature of the second Carnot engine is 500 K.

EXAMPLE 8.35

Three real heat engines have the same thermal efficiency and are connected in series. The first engine absorbs 2400 kJ of heat from a thermal reservoir at 1250 K and the third engine rejects its waste of 300 kJ to a sink at 150 K. Determine the work output from each engine.

Solution : Refer Fig 8.34 For the series arrangement of three engines.

$$Q_1 = 2400 \text{ kJ} \quad \text{and} \quad Q_4 = 300 \text{ kJ}$$

For engine E_1:

$$\eta_1 = \frac{Q_1 - Q_2}{Q_1}$$
$$= 1 - \frac{Q_2}{Q_1} ; \quad \frac{Q_2}{Q_1} = 1 - \eta_1$$

Likewise :

$$1 - \eta_2 = \frac{Q_3}{Q_2} \text{ for engine } E_2 \text{, and}$$

$$1 - \eta_3 = \frac{Q_4}{Q_3} \text{ for engine } E_3$$

Thus,

$$(1-\eta_1)(1-\eta_2)(1-\eta_3) = \frac{Q_2}{Q_1} \times \frac{Q_3}{Q_2} \times \frac{Q_4}{Q_3} = \frac{Q_4}{Q_1}$$

$$\text{or } (1 - \eta)^3 = \frac{Q_4}{Q_1} \qquad (\because \eta_1 = \eta_2 = \eta_3)$$

$$\text{or } (1 - \eta)^3 = \frac{300}{2400} = \frac{1}{8}; \quad \eta = 0.5$$

Thus each engine has a thermal efficiency of 50%

$$\text{Now : } \eta = \frac{W_1}{Q_1} = \frac{W_1}{2400}$$

$$W_1 = 0.5 \times 2400 = \mathbf{1200\ kJ}$$

$W_1 = Q_1 - Q_2$ and therefore

$$Q_2 = 2400 - 1200 = \mathbf{1200\ kJ}$$

$$\text{Again } \eta = \frac{W_2}{Q_2} = \frac{W_2}{1200}$$

$$W_2 = 0.5 \times 1200 = \mathbf{600\ kJ}$$

$W_2 = Q_2 - Q_3$ and therefore

$$Q_3 = 1200 - 600 = \mathbf{600kJ}$$

Further,

$$\therefore \quad \eta = \frac{W_3}{Q_3} = \frac{W_3}{600}$$

$$W_3 = \eta Q_3 = 0.5 \times 600 = \mathbf{300kJ}$$

Check : Total work done

$$= Q_1 - Q_4 = 2400 - 300 = 2100 \text{ kJ}$$

Also, $W = W_1 + W_2 + W_3$

$$= 1200 + 600 + 300 = \mathbf{2100\ kJ}$$

EXAMPLE 8.36

Three Carnot engines E_1, E_2, E_3 operate between temperatures of 1000 K and 300 K. Make calculations for the intermediate temperatures if the work produced by the engines are in the ratio of 4 : 3 : 2.

Solution : Refer Fig 8.35 for the schematic arrangement of the three Carnot engines in series.

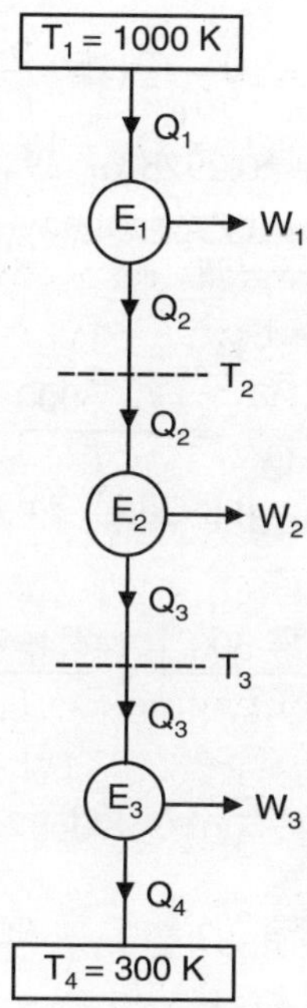

Fig. 8.35.

For engine E_1 :

$$\eta_1 = \frac{W_1}{Q_1} = \frac{T_1 - T_2}{T_1} = \frac{1000 - T_2}{1000}$$

$$Q_1 = W_1 + Q_2$$

$$\therefore \quad \frac{W_1}{W_1 + Q_2} = \frac{1000 - T_2}{1000}$$

$$Q_2 = W_1 \frac{T_2}{1000 - T_2}$$

For engine E_2 :

$$\eta_2 = \frac{W_2}{Q_2} = \frac{T_2 - T_3}{T_2}$$

Substituting the value of Q_2 as obtained above,

$$\eta_2 = \frac{W_2}{W_1}\left(\frac{1000 - T_2}{T_2}\right) = \frac{T_2 - T_3}{T_2}$$

Since,

$$\frac{W_2}{W_1} = \frac{3}{4}, \text{ we get}$$

$$\frac{3}{4}(1000 - T_2) = T_2 - T_3$$

$$\text{or } 750 - \frac{3}{4}T_2 = T_2 - T_3$$

or $$T_2 = \frac{4}{7}(750 + T_3) \qquad ...(i)$$

Also, $$Q_2 = Q_3 + W_2$$

$$\therefore \frac{W_2}{Q_3 + W_2} = \frac{T_2 - T_3}{T_2}; \qquad Q_3 = W_2 \frac{T_3}{T_2 - T_3}$$

For Engine E_3:

$$\eta_3 = \frac{W_3}{Q_3} = \frac{T_3 - 300}{T_3}$$

Substituting the value of Q_3 as outlined above

$$\frac{W_3}{W_2}\left(\frac{T_2 - T_3}{T_3}\right) = \frac{T_3 - 300}{T_3}$$

Since, $\frac{W_3}{W_2} = \frac{2}{3}$, we get

$$\frac{2}{3}(T_2 - T_3) = T_3 - 300 \qquad ...(ii)$$

From expressions (*i*) and (*ii*)

$$\frac{2}{3}\left[\frac{4}{7}(750 + T_3) - T_3\right] = T_3 - 300$$

Solution gives T_3 = 455.80 K

$$\therefore \quad T_2 = \frac{4}{7}(750 + T_3)$$

$$= \frac{4}{7}(750 + 455.80) = \mathbf{689\ K}$$

EXAMPLE 8.37

A heat engine is used to drive a heat pump. The heat transfers from the heat engine and the heat pump are used to heat the water circulating though the radiators of a building. The engine has a thermal efficiency of 27 percent and the coefficient of performance for the heat pump is 4. Determine the ratio of heat transfer to the circulating water to the heat supplied to the heat engine.

Solution : Refer Fig 8.36 for the arrangement of the composite system.

For the engine,

$$\eta = \frac{W_e}{Q_1} = \frac{W_e}{Q_2 + W_e} = 0.27$$

For the pump,

$$COP = \frac{Q_4}{W_p} = \frac{Q_4}{W_e} = 4$$

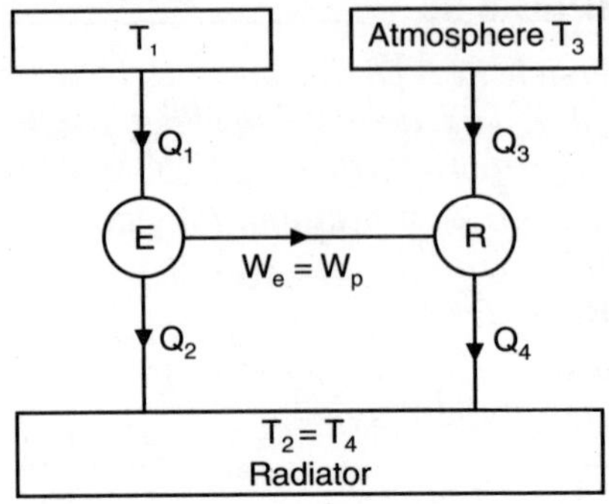

Fig. 8.36.

From the above identities

$$0.27 = \frac{\frac{Q_4}{4}}{Q_2 + \left(\frac{Q_4}{4}\right)}$$

or $$0.27 = \frac{Q_4}{4Q_2 + Q_4}$$

or $$1.08\ Q_2 + 0.27\ Q_4 = Q_4$$

or $$1.08\ Q_2 = 0.73\ Q_4$$

or $$\frac{Q_4}{Q_2} = \frac{1.08}{0.73} = 1.48$$

$$\therefore \frac{Q_2 + Q_4}{Q_2} = 2.48 \qquad ...(a)$$

Heat supplied to the engine,

$$Q_1 = Q_2 + W_e$$

$$= Q_2 + \frac{Q_4}{4} = Q_2 + \frac{1.48 Q_2}{4}$$

$$\therefore \quad Q_1 = 1.37\ Q_2 \qquad ...(b)$$

From expressions (*a*) and (*b*)

$$\frac{Q_2 + Q_4}{Q_1} = \frac{Q_2 + Q_4}{Q_2} \times \frac{Q_2}{Q_1}$$

$$= \frac{Q_2 + Q_4}{Q_2} \times \frac{Q_2}{1.37\ Q_2}$$

$$= \frac{2.48}{1.37} = 1.81$$

Hence,

$$\frac{\text{heat supplied to circulating water}}{\text{heat supplied to engine}} = \mathbf{1.81}$$

EXAMPLE 8.38

A reversible heat engine operates between 875 K and 310 K and drives a reversible refrigerator operating between 310 K and 255 K. The engine receives 2000 kJ of heat and the net work output from the arrangement equals 350 kJ. Make calculations for the cooling effect.

Solution : Refer Fig 8.37 for the arrangement of the composite system.

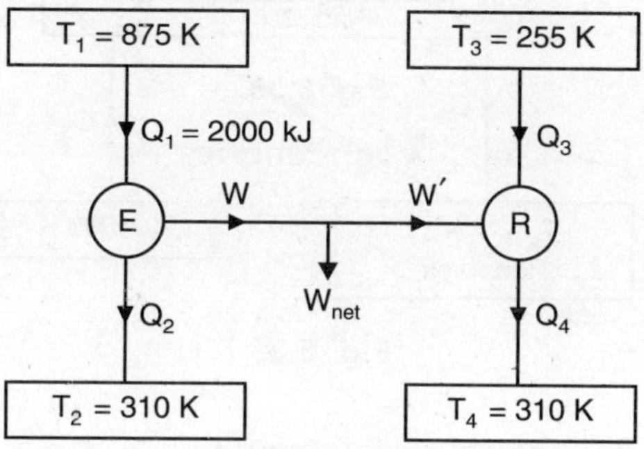

Fig. 8.37.

If W is the work performed by the engine, then

$$\eta = \frac{W}{Q_1} = \frac{W}{2000}$$

Since the engine is reversible and it operates between temperatures T_1 and T_2, the efficiency is also given by

$$\eta = \frac{T_1 - T_2}{T_1} = \frac{875 - 310}{875} = 0.6457$$

$$\therefore \frac{W}{2000} = 0.6457\,; \quad W = 1291.4 \text{ kJ}$$

Let W' represent the work needed to run the refrigerator, then

The net work output of the combined engine and refrigerator is

$$W_{net} = W - W'\,; \qquad W' = W - W_{net}$$

∴ Work required to run the refrigerator,

$$W' = 1291.4 - 350 = 941.4 \text{ kJ}$$

If Q_3 is the cooling effect, *i.e.*, heat extracted by the refrigerator, then coefficient of performance (COP) of the refrigerator is

$$COP = \frac{Q_3}{W'} = \frac{Q_3}{941.4}$$

Since the refrigerator is reversible and it operates between temperatures T_3 and T_4, its COP is also given by

$$COP = \frac{T_3}{T_4 - T_3} = \frac{255}{310 - 255} = 4.636$$

$$\therefore \frac{Q_3}{941.4} = 4.636\,; \quad Q_3 = \mathbf{4364.3} \text{ kJ}$$

The refrigerator thus has a cooling effect equal to 4364.3 kJ

EXAMPLE 8.39

A heat engine operating between two reservoirs at 1000 K and 300 K is used to drive a heat pump which extracts heat from the reservoir at 300 K, at a rate twice that at which the engine rejects heat to it. If the efficiency of the engine is 40% of the maximum possible and the coefficient of performance of the heat pump is 50% of the maximum possible, make calculations for the temperature of the reservoir to which the heat pump rejects heat. Also workout the rate of heat rejection from the heat pump if the rate of supply of heat to the engine is 50 kW.

Solution : Refer Fig 8.38.

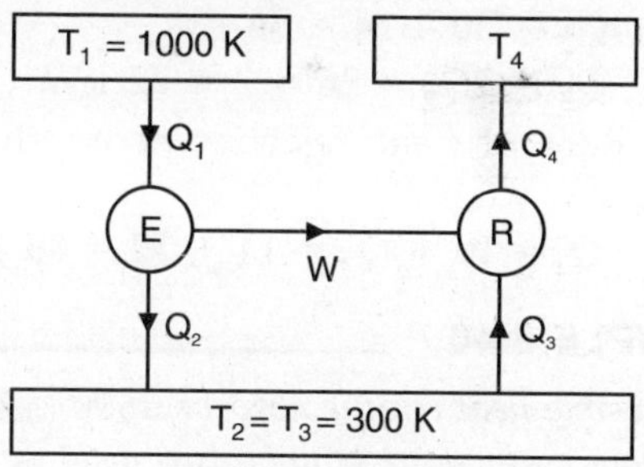

Fig. 8.38.

For engine : $\eta_{max} = \frac{T_1 - T_2}{T_1}$

$$= \frac{1000 - 300}{1000} = 0.7$$

$$\eta_{actual} = 0.4 \times 0.7 = 0.28$$

$$\therefore \frac{W}{Q_1} = 0.28$$

But, $Q_1 = W + Q_2$ (first law)

$$\therefore \frac{W}{W + Q_2} = 0.28$$

$$\text{or} \quad \frac{W}{W + \frac{Q_3}{2}} = 0.28 \quad (\because Q_3 = 2Q_2 \text{ given})$$

That gives : $Q_3 = 5.14$ W

For heat pump :

$$(COP)_{max} = \frac{T_4}{T_4 - T_3} = \frac{T_4}{T_4 - 300}$$

$$(COP)_{actual} = 0.5 \times \frac{T_4}{T_4 - 300}$$

$$\therefore \quad \frac{Q_4}{W} = 0.5 \times \frac{T_4}{T_4 - 300}$$

But, $Q_4 = Q_3 + W$ (1st law)

$= 5.14\ W + W = 6.14\ W$

$$\therefore \quad \frac{6.14\ W}{W} = 0.5 \times \frac{T_4}{T_4 - 300};$$

Solution gives :

T_4 = **326.6 K**

Therefore, the temperature of the reservoir receiving heat from the heat pump is 326.6 K

(*b*) $Q_1 = 50$ kW ; $\frac{W}{Q_1} = 0.28$

$\therefore$ $W = 0.28 \times 50 = 14$kW

$Q_2 = 50 - 14 = 36$ kW

$Q_3 = 2Q_2 = 36 \times 2 = 72$ kW

$\therefore$ Rate of heat rejection from the heat pump,

$Q_4 = W + Q_3 = 14 + 72 =$ **86 kW**

EXAMPLE 8.40

A reversible heat engine absorbs 2500 kJ/cycle of heat from a constant temperature heat source at 2000 K, and rejects some energy as waste heat to a reservoir X. The work output from the engine is used to drive a reversible refrigerator ; its source temperature being 300 K. The heat outflow from the refrigerator is also taken to the reservoir X. If the total heat flow into the reservoir is 3000 kJ/cycle, make calculations for the temperature of reservoir X.

Solution : Refer Fig 8.39.

$Q_2 + Q_4 = 3000$ kJ/cycle

The work developed by the engine equals the work required to run the refrigerator. That is

$Q_1 - Q_2 = W = Q_4 - Q_3$

$\therefore$ $Q_3 = (Q_2 + Q_4) - Q_1$

$= 3000 - 2500$

$= 500$ kJ/cycle.

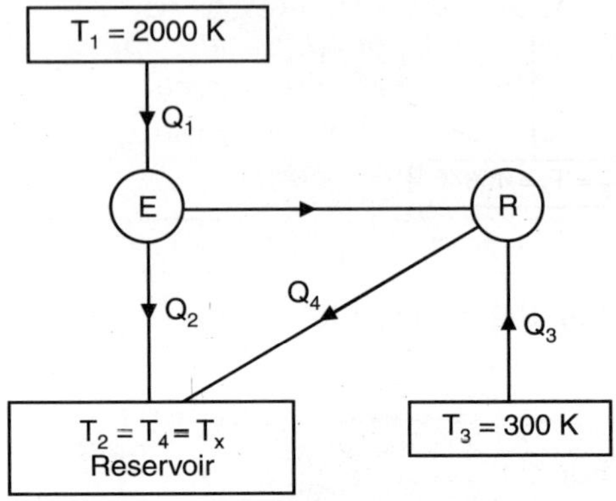

Fig. 8.39.

For the reversible engine $\frac{Q_1}{T_1} = \frac{Q_2}{T_2}$ and

for the reversible refrigerator $\frac{Q_3}{T_3} = \frac{Q_4}{T_4}$.

Since $T_2 = T_4 = T_x$ (say), the summation of these expressions gives

$$\frac{Q_2 + Q_4}{T_x} = \frac{Q_1}{T_1} + \frac{Q_3}{T_3}$$

$$\text{or} \quad \frac{3000}{T_x} = \frac{2500}{2000} + \frac{500}{300} = 2.917$$

$\therefore$ Temperature of the common reservoir X,

$$T_x = \frac{3000}{2.917} = \mathbf{1028.5\ K}$$

EXAMPLE 8.41

A reversible heat engine operates within the higher and lower temperature limits of 1400 K and 400 K respectively. The entire output from this engine is utilized to operate a heat pump. The pump works on reversed Carnot cycle, extracts heat from a reservoir at 300 K and delivers it to the reservoir at 400 K. If 100 kJ/s of net heat is supplied to the reservoir at 400 K, calculate the heat supplied to the engine by the reservoir at 1400 K.

Solution : Refer Fig 8.40 for schematic arrangement of the composite system.

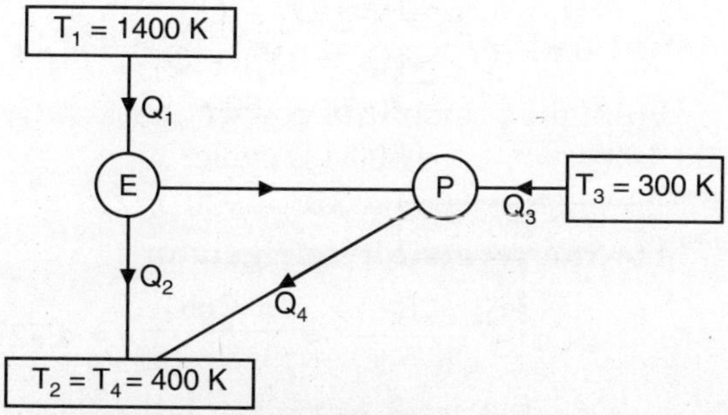

Fig. 8.40.

For the reversible engine :

$$\eta = \frac{Q_1 - Q_2}{Q_1} = \frac{T_1 - T_2}{T_1}; \frac{Q_1}{Q_2} = \frac{T_1}{T_2}$$

For the reversible pump :

$$COP = \frac{Q_4}{Q_4 - Q_3} = \frac{T_4}{T_4 - T_3}; \frac{Q_3}{Q_4} = \frac{T_3}{T_4}$$

Since the entire output of the heat engine is utilised to drive the pump,

$$Q_1 - Q_2 = Q_4 - Q_3$$

$$Q_1 - \frac{T_2}{T_1} Q_1 = Q_4 - Q_4 \frac{T_3}{T_4}$$

or

$$\frac{Q_1}{Q_4} = \frac{T_1}{T_4}\left(\frac{T_4 - T_3}{T_1 - T_2}\right)$$

$$= \frac{1400}{400}\left(\frac{400 - 300}{1400 - 400}\right) = 0.35$$

Heat supplied to the reservoir at 400 K,

$$100 = Q_2 + Q_4 = Q_1 \frac{T_2}{T_1} + \frac{Q_1}{0.35}$$

$$= Q_1 \times \frac{400}{1400} + \frac{Q_1}{0.35} = 3.4284\ Q_1$$

∴ Energy taken from the reservoir at 1400 K,

$$Q_1 = \frac{100}{3.4284} = \mathbf{29.168\ kJ/s}$$

EXAMPLE 8.42

A Carnot heat engine operates between 825 K and 550 K constant temperature thermal reservoirs. Half of the work output from this engine drives a generator and the other half is used to run a reversed Carnot engine that abstracts heat from the 550 K temperature reservoir and rejects that to a thermal reservoir at 715 K. Determine the heat rejected to the reservoir by the reversed engine as a percentage of heat supplied from the 825 K reservoir to the heat engine. Proceed to calculate the heat received per hour if the generator output is 400 kW.

Solution : Refer Fig 8.41 for the schematic arrangement of the composite setup.

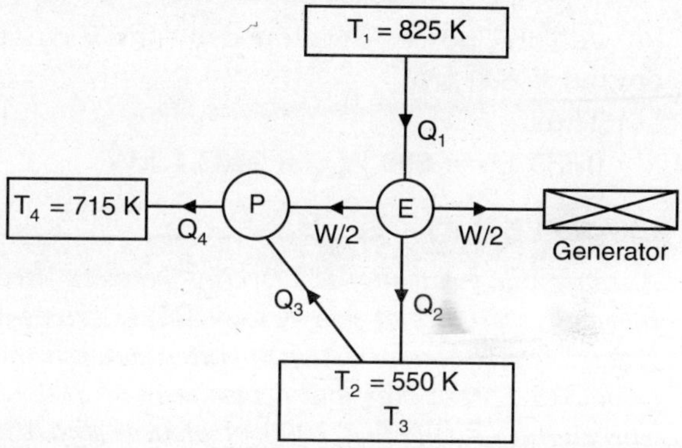

Fig. 8.41.

The Carnot heat engine is a reversible engine and its efficiency is given by

$$\eta = \frac{T_1 - T_2}{T_1} = \frac{825 - 550}{825} = 0.333$$

If Q_1 and Q_2 are the heat interactions for the Carnot engine, then

$$\eta = \frac{Q_1 - Q_2}{Q_1} = \frac{W}{Q_1}$$

∴ Work output from the Carnot engine,

$$W = 0.333\ Q_1$$

Thus, work input to the reversed Carnot engine $= \frac{W}{2} = 0.1665 Q_1$

COP of the reversed heat engine

$$= \frac{T_4}{T_4 - T_3} = \frac{715}{715 - 550} = 4.333$$

In terms of heat and work interactions,

$$COP = \frac{Q_4}{Q_4 - Q_3} = \frac{Q_4}{W_p}$$

where W_p is the work input to the reversed engine.

or $$4.333 = \frac{Q_4}{0.1665\,Q_1}$$

or $$\frac{Q_4}{Q_1} = 0.7215 = \mathbf{72.15\%}$$

(*b*) Power output from the generator $= 400$ kW

∴ Total power developed by the Carnot engine $= 800$ kW

Thus,

$0.333\ Q_1 = 800\ ;\ Q_1 = \mathbf{2402.4\ kW}$

EXAMPLE 8.43

A reversible heat engine working between two thermal reservoirs at 875 K and 315 K drives a reversible refrigerator which operates between the same 315 K reservoir and a reservoir at 260 K. The engine is supplied 2000 kJ of heat and the net work output from the composite system is 350 kJ. Make calculations for the heat transfer to the refrigerator and the net heat interaction with the reservoir at 315 K temperature.

Solution : Refer Fig 8.42.

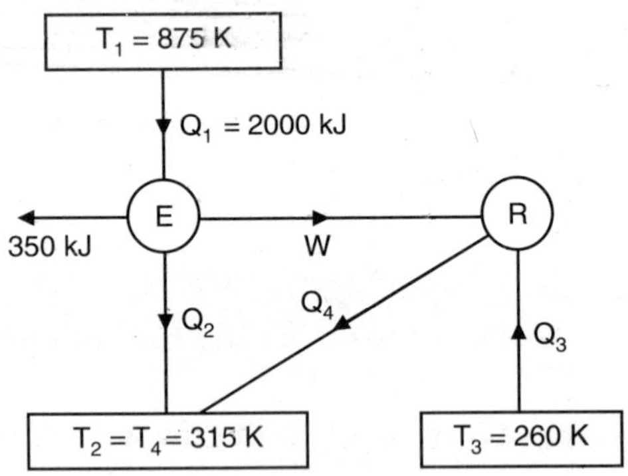

Fig. 8.42.

For the reversible heat engine :

$$\eta_{th} = \frac{T_1 - T_2}{T_1} = \frac{875 - 315}{875} = 0.64 \text{ or } 64\%$$

In terms of heat and work interactions

$$\eta_{th} = \frac{\text{work output}}{\text{heat supplied}}$$

$$0.64 = \frac{W + 350}{2000} \quad \text{or} \quad W = 930 \text{ kJ}$$

From the principle of energy conservation (1st law)

$$Q_2 = 2000 - 350 - 930 = 720 \text{ kJ}$$

For the reversible refrigerator :

$$COP = \frac{T_3}{T_3 - T_4} = \frac{260}{315 - 260} = 4.727$$

In terms of heat and work interactions

$$COP = \frac{Q_3}{W}\ ;\ 4.727 = \frac{Q_3}{930}$$

or $$Q_3 = 4.727 \times 930 = \mathbf{4396\ kJ}$$

From the principle of energy conservation (1st law)

$$Q_4 = Q_3 + W = 4396 + 930 = 5326 \text{ kJ}$$

∴ Net heat transfer to the refrigerator at 315 K,

$$= Q_2 + Q_4 = 720 + 5326 = \mathbf{6046\ kJ}$$

EXAMPLE 8.44

An ice-making plant produces 750 kg of ice from liquid water available at 0°C whilst the ambient atmospheric temperature is 310 K. The task is accomplised by running a refrigerating unit that obtains its motive power from a heat engine. Hot water at 350 K produced by solar heating is used as heat source for the heat engine and the engine uses the ambient atmosphere as sink. Determine:

(a) Power required to run the refrigerating unit,

(b) Ratio of the heat extracted from freezing water to the heat supplied to the heat engine,

(c) Rate at which energy is rejected to atmosphere by both the machines.

It may be presumed that both the devices operate as ideal machine and the enthalpy of fusion of water at 0°C is 333.45 kJ/kg.

Solution : Refer Fig 8.43 for the schematic arrangement of the refrigerator and the heat engine.

Heat extracted from liquid water for making ice,

$$Q_3 = \frac{750 \times 333.45}{3600} = 69.468 \text{ kJ/s}$$

For the reversible refrigerator,

$$COP = \frac{T_3}{T_4 - T_3} = \frac{273}{310 - 273} = 7.378$$

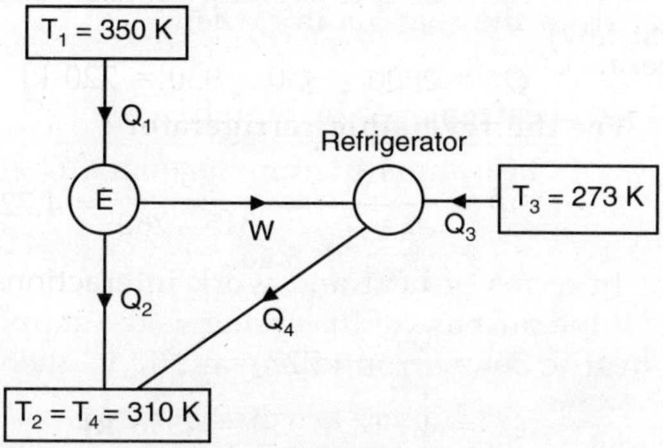

Fig. 8.43.

In terms of heat and work interactions,

$$COP = \frac{Q_3}{W} \; ; \quad 7.378 = \frac{69.468}{W}$$

∴ Power consumption of refrigerating unit,

$$W = \frac{69.468}{7.378} = \mathbf{9.415} \text{ kJ/s}$$

(*b*) For the reversible heat engine,

$$\eta = \frac{T_1 - T_2}{T_1} = \frac{350 - 310}{350} = 0.1143$$

In terms of heat and work interactions,

$$Q_1 = \frac{W}{\eta} = \frac{9.415}{0.1143} = 82.371 \text{ kJ/s}$$

$$\therefore \text{ Ratio } \frac{\text{heat extracted from freezing water}}{\text{heat supplied to heat engine}}$$

$$= \frac{69.468}{82.371} = \mathbf{0.843}$$

(*c*) For the reversible heat engine

$$\frac{Q_1}{Q_2} = \frac{T_1}{T_2}$$

$$Q_2 = \frac{T_2}{T_1} Q_1 = \frac{310}{350} \times 82.371$$

$$= 72.957 \text{ kJ/s}$$

For the reversible refrigerator,

$$\frac{Q_3}{Q_4} = \frac{T_3}{T_4}$$

$$Q_4 = \frac{T_4}{T_3} Q_3 = \frac{310}{273} \times 69.468$$

$$= 78.883 \text{ kJ/s}$$

∴ Heat rejected to ambient atmosphere by both devices,

$$Q_2 + Q_4 = 72.957 + 78.883 = \mathbf{151.84 \text{ kJ/s}}$$

EXAMPLE 8.45

A reversible heat pump is required to maintain a temperature of 0°C in a refrigerator while rejecting heat to the surroundings at 300 K. If the heat removed rate from the refrigerator is 25 kW, determine the coefficient of performance of the machine and the work input required.

Proceed to determine the overall COP of the system if the power required to run the machine is developed by a reversible engine which operates between higher and lower temperature limits of 650 K and 300 K respectively.

Solution : Refer fig. 8.44 for the heat pump alone and the composite system.

The machine operates as a refrigerator and its *COP* is

$$COP = \frac{T_2}{T_1 - T_2} = \frac{273}{300 - 273} = 10.11$$

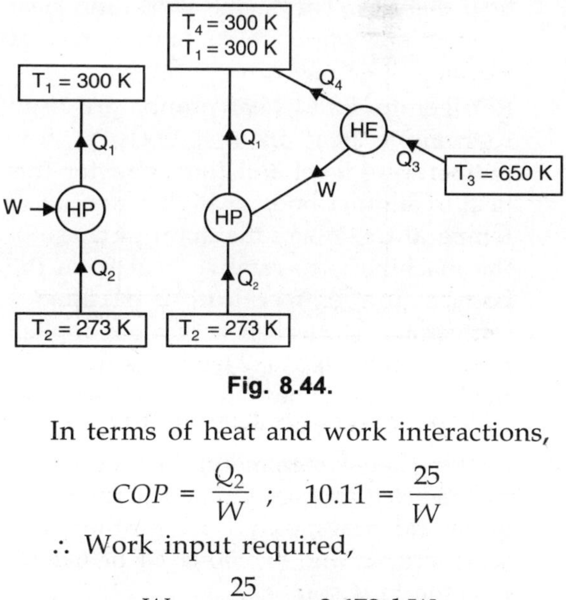

Fig. 8.44.

In terms of heat and work interactions,

$$COP = \frac{Q_2}{W} \; ; \quad 10.11 = \frac{25}{W}$$

∴ Work input required,

$$W = \frac{25}{10.11} = \mathbf{2.473 \text{ kW}}$$

Heat dissipated to surroundings,

$$Q_1 = W + Q_2$$
$$= 2.473 + 25 = \mathbf{27.473} \text{ kW}$$

If the machines is operating as a heat pump, its *COP* will be

$$COP = \frac{Q_1}{W} = \frac{27.473}{2.473} = 11.11$$

(*b*) For the reversible heat engine :

$$T_3 = 650 \text{ K and } T_4 = 300 \text{ K}$$

$$\eta = \frac{T_3 - T_4}{T_3} = \frac{Q_3 - Q_4}{Q_3}$$

or $$1 - \frac{T_4}{T_3} = 1 - \frac{Q_4}{Q_3}$$

or $$\frac{Q_3}{T_3} = \frac{Q_4}{T_4}; \quad \frac{Q_4 + W}{T_3} = \frac{Q_4}{T_4}$$

or $$\frac{Q_4 + 2.473}{650} = \frac{Q_4}{300}; \quad Q_4 = 2.12 \text{ kW}$$

Then : $$Q_3 = Q_4 + W$$
$$= 2.12 + 2.473 = 4.593 \text{ kW}$$

If the desired objective is the removal of heat from the regrigerator, then overall *COP* =

$$\frac{\text{heat removed by the refrigerator } (Q_2)}{\text{heat supplied from the source } (Q_3)}$$

$$= \frac{2.5}{4.593} = \mathbf{5.44}$$

If the purpose of the system is the supply of heat to the surroundings at 300 K, then

$$\text{overall } COP = \frac{Q_1 + Q_4}{Q_3} = \frac{27.473 + 2.12}{4.593} = 6.44$$

It may be noted that

$$(COP)_{heat\ pump} = 1 + (COP)_{refrigerator}$$

SALIENT POINTS

1. A heat engine is a thermodynamic device which produces work from heat supplied to it and has cyclic operation.
2. Refrigerators and heat pumps are reversed heat engines. They have work and heat interactions opposite to that of a heat engine.
3. Refrigerators and heat pumps are used to remove heat from a body at low temperature level and then transfer this heat to another body at high potential of temperature. When the main purpose of the machine is to remove heat from the body at low temperature, it is called a refrigerator. The term heat pump is applied to a machine whose objective is to heat a medium which may already be warmer than the surroundings.
4. **Kelvin-Planck statement:** It is impossible to construct an engine that operates in a cycle and produces no effect other than work output and exchange of heat with a single heat source.

   Clausius statement: It is impossible to construct a device that operates in a cycle and produces no effect other than the transfer of heat from a system at low temperature to another system at high temperature.

   Though worded differently both these statements are interlinked and complimentary to each other. Any device that violates Clausius statement leads to violation of Kelvin-Planck statement and vice versa.
5. A perpetual motion machine of the second kind is a hypothetical machine which converts the entire heat supplied to it into an equivalent amount of work and the efficiency becomes 100 percent.
6. **Carnot cycle:** A hypothetical cycle consisting of two reversible isothermal processes and two reversible adiabatic processes. The thermal efficiency of this cycle is independent of the working fluid and is given by

$$\eta_{carnot} = \frac{T_1 - T_2}{T_1} = 1 - \frac{T_2}{T_1}$$

   where T_1 and T_2 are the temperatures of source and sink respectively.

For high efficiency, heat should be supplied at the highest possible temperature and rejected at the lowest possible temperature.

7. **Corollaries:**

   (*a*) It is impossible to construct an engine operating between two heat reservoirs which will have efficiency higher than that of a reversible engine operating between the same two reservoirs.

   (*b*) All reversible engines operating between the same two heat reservoirs have the same efficiency.

   (*c*) The efficiency of any reversible engine operating between two reservoirs is independent of the nature of the working fluid and depends solely on the temperatures of the reservoirs.

8. Clausius unequality is given by

$$\oint \frac{dQ}{T} = 0 \text{ for a reversible cycle}$$

$$\oint \frac{dQ}{T} < 0 \text{ for a irreversible cycle}$$

For an irreversible cycle, we may write

$$\oint \frac{dQ}{T} + I = 0$$

where the parameter I indicates the amount by which the given cycle is irreversible. For a reversible cycle, $I = 0$.

Further, a cycle in which the integral $\oint \frac{dQ}{T}$ is greater than zero violates the second law and is impossible.

9. A temperature scale defined in terms of some physical properties of a material is called the empirical scale.

   The temperature scale which is independent of the property of thermometric substance is defined as thermodynamic temperature scale.

REVIEW QUESTIONS

A. Conceptual and conventional questions.

1. State the limitations of first law of thermodynamics.
2. What is a thermal energy reservoir ? Define in terms of heat source and heat sink.
3. What features characterise a heat engine ?
4. Define thermal efficiency of a heat engine. Can it be 100 percent?
5. Differentiate between a reversible and a reversed machine.
6. What is a heat pump? How does it differ from a refrigerator?
7. Show that *COP* of a heat pump is greater than *COP* of a refrigerator by unity.
8. State the Kelvin-Planck and Clausius statements of the second law of thermodynamics, and establish the equivalence between them.
9. What is perpetual motion machine of second kind ? Why such a machine cannot be constructed in actual practice?
10. Explain the term reversibility as applied to a thermodynamic process.
11. List suitable examples of reversible and irreversible processes.
12. What main conclusions can be drawn with respect to the efficiency of a Carnot engine?
13. Derive an expression for the efficiency of a Carnot cycle. What is the difficulty in constructing an engine working on Carnot cycle.
14. In a refrigerator, heat is transferred from a lower temperature to a higher temperature. Is this a violation of the second law of thermodynamics? Explain.
15. Is it possible for a heat engine to operate without rejecting any heat to a low temperature reservoir? Explain.
16. In the absence of any friction and other irreversibilities, can a heat engine have an efficiency of 100 percent? Explain.
17. State and prove Carnot theorem.
18. Prove that no heat engine operating in a cycle between two constant temperature heat reservoirs can be more efficient than a reversible engine operating between the same two reservoirs.
19. State the Carnot theorem in the context of a heat pump/refrigerator.
20. Prove that the efficiency of an engine working on a reversible cycle depends only on the

temperature of source and sink and is independent of the working fluid.

21. Comment on the validity of the statement: All reversible engines operating between the same two thermal reservoirs have the same efficiency.
22. What is a thermodynamic temperature scale or absolute temperature scale?
23. Which principle or fact is made use of in establishing the thermodynamic temperature scale?
24. Explain the establishment of a thermodynamic temperature scale. Why is the thermodynamic temperature scale called the absolute temperature scale?
25. How many fixed points or reference points are used in establishing the temperature scale and what are those fixed points?
26. State the difference between the thermodynamic and empirical scales of temperature.
27. Explain the concept of Clausius inequality.
28. State the importance of second law of thermodynamics.
29. A Carnot engine operates between two reservoirs whose difference in temperature is 200°C. If the work output of the engine is 0.5 times the heat rejected, make calculations for the temperature of source and sink and the thermal efficiency of the engine.

    [Ans. 327°C, 127°C, 0.33]
30. An engine operating on a Carnot cycle works within temperature limits of 600 K and 300 K. If the engine receives 2000 kJ of heat, evaluate the work done and thermal efficiency of the engine.

    [Ans. 1000 kJ, 0.5]
31. Determine the power required to run a refrigerator that transfers 2000 kJ/min of heat from a cooled space at 0°C to the surrounding atmosphere at 27°C. The refrigerator operates on reversed Carnot cycle.

    [Ans. 3.29 kW]
32. A reversible heat engine operates between reservoirs at 420 K and 280 K. If output from the engine is 2.5 kJ, determine the efficiency of the engine and its heat interactions with the heat reservoirs.

    Subsequently the engine is reversed and made to operate as heat pump between the same reservoirs. Make calculations for the coefficient of performance of the heat pump and the power input required when the heat transfer rate from the 280 K reservoir is 5 kW.

    [Ans. (*a*) 33.3%, 7.5 kJ, 5kJ; (*b*) 3, 2.5 kW]
33. A reversible heat engine receives heat from two thermal reservoirs at 870 K and 580 K, and rejects 50 kW of heat to a sink at 290 K. If the engine output is 85 kW, make calculations for the engine efficiency and heat supplied by each reservoir.

    [Ans. 62.96%, 105 kW , 30 kW]
34. A reversible heat engine interacts with three thermal reservoirs at 750 K, 650K and 550 K respectively. The engine absorbs 2400 kJ/min of energy as heat from the reservoir at 750 K and does 400 kJ/min of net work. Determine the magnitude and direction of heat interactions of the engine with the other two reservoirs.

    [Ans. 1560 kJ/min and 400 kJ/min]
35. A reversible heat engine operating between thermal reservoirs at 900 K and 300 K is used to drive a reversible refrigerator for which the temperature limits are 300 K and 250 K. The engine absorbs 1800 kJ of energy as heat from the reservoir at 900 K and the net output from the engine-refrigerator system is 360 kJ. Make calculations for the heat extracted from the refrigerator cabinet and the net heat rejected to the reservoir at 300 K.

    [Ans. 5040 kJ, 5640 kJ]
36. A cyclic reversible refrigerating plant extracts heat from a space at low temperature T_1 and also receives heat at a high temperature T_2. The entire heat is then transferred to the environment at an intermediate temperature T_3. Determine the quantity of heat extracted from the refrigerating plant in terms of T_1, T_2, T_3 and Q_2. The quantity Q_2 represents the heat received at the high temperature. The system does not exchange any work with the surroundings.

    $$\left(\textbf{Ans. } Q_2 \times \frac{T_1}{T_3 - T_1} \times \frac{T_2 - T_3}{T_2}\right)$$
37. Two Carnot heat engines operate so that the exhaust of one is the intake of the other; the highest and lowest temperature limits are 1200 K and 400 K. Determine the intermediate temperature if (*i*) both the engines give the same work output (*ii*) both the engines have the same thermal efficiency.

    [Ans. 800 K, 692.82 K]

38. Set up a relationship between the coefficients of performance of a heat pump and a refrigerator operated between the same temperature limits. A Carnot refrigerator operates between 0°C and 30°C. What fraction of the cooling effect is required as the work input?

[Ans. 0.11]

39. A reversible engine operates between temperatures T_1 and T ($T_1 > T$). The energy rejected by this engine is received by a second reversible engine at the same temperature T. The second engine rejects the heat at temperature T_2 ($T_2 < T$). Prove that
(*a*) $T = (T_1 + T_2)/2$ if the engines produce same work output and
(*b*) $T = \sqrt{T_1 T_3}$ if the engines develop the work at the same efficiency.

40. A reversible engine operates between a source at 1200 K and two sinks, one at 400 K and another at 300 K. The heat rejected to both the sinks is same. Determine thermal efficiency of the engine.

[Ans. 71.43%]

41. A heat engine receives 600 kJ of heat from a high temperature source at 1000 K during a cycle. It converts 150 kJ of this heat to net work and rejects the remaining 450 kJ to a low temperature sink at 300 K. Determine if this heat engine violates the second law of thermodynamics on the basis of (*i*) the Carnot principle and (*ii*) the Clausius inequality.

[Ans. The heat engine is in compliance with the Carnot principle and the Clausius inequality].

42. Three Carnot engines *A*, *B*, *C* working between temperatures of 1000 K and 300 K are in a series combination . The work produced by these engines are in the ratio of 5: 4: 3. Make calculations of temperature for the intermediate reservoirs.

[Ans. 708 K, 475 K]

43. A reversible heat engine operates between systems at constant temperatures of 150°C and 10°C. If work output from the engine is 2000 Nm, make calculations for the engine efficiency and the heat interactions between the engine and the thermal reservoirs.

(*b*) Subsequently the engine is reversed and operates as a heat pump between the same thermal reservoirs. Determine the coefficient of performance of the heat pump and the power input required when the heat transfer rate from the reservoir at 10°C is 40 kW.

[Ans. (*a*) 33.1%, 6.04 kJ, 4.04 kJ
(*b*) 3.02, 19.8 kW]

44. A Carnot heat engine absorbs heat from a thermal reservoir at temperature 600 K and rejects heat to a thermal reservoir at temperature T_x. The power available from this engine is utilised to drive a reversed Carnot cycle that takes heat from a reservoir at temperature 300 K and supplies heat to the reservoir at T_x. Make calculations for the temperature T_x if heat supplied to the Carnot heat engine equals the heat taken by the reversed heat engine. Proceed to calculate the performance parameters for both the machines.

[Ans. 400 K, 33.33%, 3]

B. Fill in the blanks with appropriate word/words.

1. The .... law of thermodynamics does not specify the conditions under which conversion of heat into work is possible.
2. The .... law of thermodynamics puts a limit on the conversion of a given amount of heat into work.
3. The processes involved in a Carnot cycle are .... and ....
4. The efficiency of a Carnot cycle is independent of .... and depends upon the temperatures of .... and ....
5. Refrigerators and heat pumps are .... heat engines.
6. The performance of a refrigerator and heat pump is expressed in terms of ....
7. All .... have the same efficiency when working between the same ....
8. A temperature scale which is independent of the property of thermometric substance is defined as ....
9. For a ...., the entire heat supplied to it gets converted into equivalent amount of work and the efficiency becomes .... percent.

Answers

1. first 2. second 3. two aidabatic, two isothermal 4. working substance, source and sink 5. reversed 6. coefficient of performance 7. reversible engines 8. thermodynamic temperature scale 9. perpetual motion of second kind

C. Multiple choice questions.

1. Which is the essence of second law of thermodynamics ?
   (*a*) the whole of heat supplied to a system can be converted into equivalent mechanical work
   (*b*) no engine can be 100 per cent efficient
   (*c*) a refrigerator can reduce the temperature to absolute zero
   (*d*) reversible engines working between the same temperature limits can have different thermal efficiencies

2. "Heat can not be transported from a system at low temperature to another system at high temperature without the aid of external agency". This statement of second law is attributed to
   (*a*) Gay Lussac (*b*) Clausius
   (*c*) Max-Planck (*d*) Joule Thomson

3. In a Carnot cycle, the rejection of heat is
   (*a*) at constant pressure
   (*b*) at constant volume
   (*c*) at constant temperature
   (*d*) partly at constant pressure and partly at constant volume

4. The efficiency of a Carnot cycle engine depends on
   (*a*) speed of the engine
   (*b*) working fluid; whether gas or vapour
   (*c*) temperature at which heat is supplied and that at which it is rejected
   (*d*) temperature of ambient air

5. The Carnot cycle is impracticable because it
   (*a*) requires a perfect gas as the working fluid
   (*b*) necessitates all processes comprising the cycle to be reversible
   (*c*) demands high speeds for its working
   (*d*) involves greater pressures and volumes

6. For a heat engine operating on Carnot cycle, the work output is 1/4th of the heat rejected to the sink. The thermal efficiency of the engine would be
   (*a*) 10% (*b*) 20%
   (*c*) 30% (*d*) 50%

7. A heat engine is supplied with 280 kJ/s of heat at a constant fixed temperature of 520 K and heat rejection takes place at 260 K temperature. If the engine is reversible, the heat rejected would be approximately equal to
   (*a*) 85 kJ/s (*b*) 110 kJ/s
   (*c*) 140 kJ/s (*d*) 155 kJ/s

8. Three engines A, B and C operating on Carnot cycle respectively use air, steam and helium as the working fluid. If all the engines operate with in the same high and low temperature limits, then which engine will have the highest efficiency ?
   (*a*) Engine A (*b*) Engine B
   (*c*) Engine C
   (*d*) All engines will have the same efficiency

9. If the thermal efficiency of a Carnot heat engine is 40 per cent, then coefficient of performance of a refrigerator working within the same temperature limits would be
   (*a*) 1.5 (*b*) 2.5
   (*c*) 3.5 (*d*) 4.5

10. A small change Δt can be affected either in source temperature T_1 or the sink temperature T_2. Which (out of change in T_1 and T_2) will result in higher efficiency ?
    (*a*) increase in T_1
    (*b*) decrease in T_2
    (*c*) efficiency will be equally affected
    (*d*) result will depend upon the amount of change Δt

11. A heat pump working on a reversed Carnot cycle has a *COP* of 5. If it works as a refrigerator taking 1 kW of work input, the refrigeration effect will be
    (*a*) 1 kW (*b*) 2 kW
    (*c*) 3 kW (*d*) 4 kW

12. A refrigeration machine working on reversed Carnot cycle takes out 2 kW per minute of heat from the system while working between the temperature limits of 300 K and 200 K. Then the coefficient of performance and power consumed by the cycle will be respectively
    (*a*) 1 and 1 kW (*b*) 1 and 2 kW
    (*c*) 2 and 1 kW (*d*) 2 and 2 kW

13. Which expression holds good for a cycle of processes reversible or irreversible ?
    (*a*) cyclic $\int \frac{\delta Q}{T} > 0$
    (*b*) cyclic $\int \frac{\delta Q}{T} < 0$
    (*c*) cyclic $\int \frac{\delta Q}{T} = 0$
    (*d*) cyclic $\int \frac{\delta Q}{T} \geq 0$

14. An engine is supplied with 1120 kJ/s of heat and the source and sink are maintained at constant fixed temperatures of 560 K and 280 K respectively. If heat rejection is 840 kJ/s, indicate whether the given cycle is
    (a) reversible
    (b) irreversible
    (c) impossible
    (d) unpredictable; insufficient data
15. In a Carnot cycle, the addition and rejection of heat take place it
    (a) constant pressure
    (b) constant volume
    (c) constant temperature
    (d) constant enthalpy
16. No heat engine can operate by exchanging heat with a single temperature source. This statement refers to the
    (a) Joule's law
    (b) Carnot theorem
    (c) Clausius statement
    (d) Kelvin Planck statement
17. Identify the invalid option in the context of the efficiency of a Carnot heat engine
    (a) is always less than 100 percent
    (b) depends on the temperature of heat source and that of heat sink
    (c) is same for all reversible heat engines provided those operate between the same temperatures of source and sink
    (d) is dependent on whether the working fluid is an ideal gas or a vapour
18. Energy tends to gravitate towards a lower degree of intensity until or unless there does not occur any transformation. This statement is related to which law of thermodynamics?
    (a) zeroth (b) first
    (c) second (d) third
19. Identify the process which can approach reversibility
    (a) throttling
    (b) combustion of fuel
    (c) dissolvation of solids in a solvent
    (d) controlled expansion or compression of gases
    (e) stropping of vehicle by the application of brakes

Answers

1. (b)	2. (b)	3. (c)	4. (c)	5. (d)
6. (b)	7. (c)	8. (d)	9. (a)	10. (b)
11. (d)	12. (c)	13. (d)	14. (b)	15. (c)
16. (d)	17. (d)	18. (c)	19. (d)	

□□□

Chapter 9

Entropy

Learning objectives : A study of the subject matter presented in this chapter will enable the reader to deal with the following aspects :

- *Clausius theorem and concept of entropy*
- *entropy as a state property*
- *entropy principle; entropy for an isolated system*
- *temperature entropy plot and its significance*
- *entropy changes during different thermodynamic processes*

9.1. WHAT IS ENTROPY ?

From Clausius theorem, the cyclic integral of the quantity $\delta Q/T$ for a reversible process is zero. This suggests that the quantity $\delta Q/T$ is a point function and hence a property of the system. This property is called *entropy* (denoted by S) and its change from initial equilibrium state 1 to final equilibrium state 2 during a reversible process is

$$\int_1^2 \left(\frac{\delta Q}{T}\right)_{rev} = \int_1^2 dS = S_2 - S_1 \qquad ...(9.1)$$

Entropy is an extensive property as it depends on the mass of the system.

The unit of entropy is kJ/K while that of specific entropy (entropy per unit mass) is kJ/kg K. The entropy of a system increases with absorption of heat and decreases with heat rejection. For a reversible adiabatic process $\delta Q = 0$ and so will be the entropy change. Thus it may be stated that *entropy is a property of the system that remains constant during a reversible adiabatic process*. The reversible adiabatic process during which entropy remains constant is called ***isentropic process.*** However, an isentropic process need not be an adiabatic process as the entropy during a particular process may be kept constant by heat transfer to or from the system.

Essentialy entropy is a measure of the degree of randomness of the molecules comprising the system. Higher the disorder, greater is the increase in entropy.

9.2. ENTROPY IS A POINT FUNCTION

Consider a system taken from initial state 1 to final state 2 by following the reversible path A. Subsequently the system may be brought back to initial state by following the reversible paths B or C. The paths A and B together constitute a reversible cycle and so do the paths A and C. Then in accordance with Clausius theorem

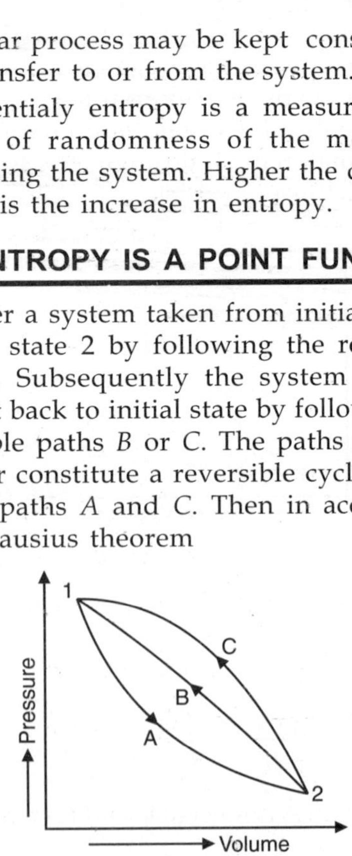

Fig. 9.1. Entropy is a point function

$$\int_{1}^{2\ via\ A} \frac{\delta Q}{T} + \int_{2}^{1\ via\ B} \frac{\delta Q}{T} = 0$$

Also $$\int_{1}^{2\ via\ A} \frac{\delta Q}{T} + \int_{2}^{1\ via\ C} \frac{\delta Q}{T} = 0$$

The above expressions indicate that

$$\int_{2}^{1\ via\ B} \frac{\delta Q}{T} = \int_{2}^{1\ via\ C} \frac{\delta Q}{T} \qquad ...(9.2)$$

Thus :

(*i*) the integral of the quantity $(\delta Q/T)_{rev}$ is path independent and depends only on the end states.

(*ii*) the quantity $(\delta Q/T)_{rev}$ is an exact differential and hence a property of the system.

9.3. ENTROPY CHANGE IN AN IRREVERSIBLE PROCESS

Consider a system that changes from state 1 to state 2 by a reversible process *A*. Subsequently the system is brought back to the original state 1 through an irreversible process *B*.

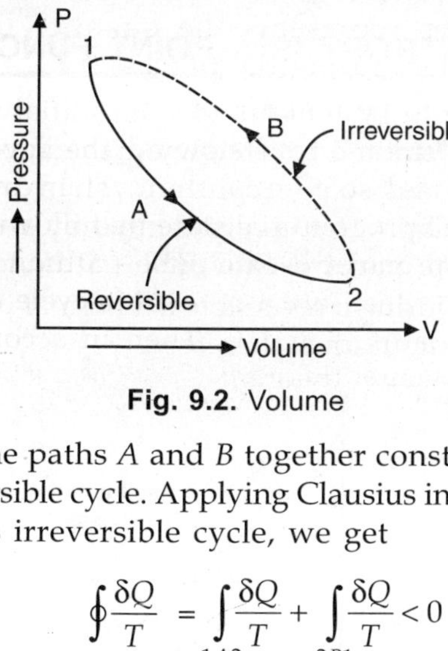

Fig. 9.2. Volume

The paths *A* and *B* together constitute an irreversible cycle. Applying Clausius inequality to this irreversible cycle, we get

$$\oint \frac{\delta Q}{T} = \int_{1A2} \frac{\delta Q}{T} + \int_{2B1} \frac{\delta Q}{T} < 0 \qquad ...(9.3)$$

The path *A* is reversible and for a reversible process, $\int \left(\frac{\delta Q}{T}\right)_{rev} = dS$

Accordingly $\int_{1A2} dS + \int_{2B1} \frac{\delta Q}{T} < 0$

For a reversible path, the limits of integration can be reversed. The above identity can then be rewritten as

$$-\int_{2A1} dS + \int_{2B1} \frac{\delta Q}{T} < 0 \text{ or } dS > \left(\frac{\delta Q}{T}\right)_{irrev} \qquad ...(9.4)$$

Thus the entropy change of a system undergoing an irreversible process is given by $dS > \left(\frac{\delta Q}{T}\right)_{irrev}$. Hence, for the general case, it may be stated that

$$dS \geq \frac{\delta Q}{T} \text{ or } S_2 - S_1 \geq \int_1^2 \frac{\delta Q}{T} \qquad ...(9.5)$$

The equality sign holds good for a reversible process and the inequality sign for an irreversible process.

The equation 9.5 shows that entropy increases during an irreversible process and is conserved (remains constant) during a reversible process. Recollecting that for an adiabatic process (whether reversible or irreversible) $\delta Q = 0$ and, therefore, $dS \geq 0$. This implies that entropy change is zero in a reversible adiabatic process and greater than zero in an irreversible adiabatic process.

9.4. ENTROPY PRINCIPLE

Entropy of an isolated system either increases or, in the limit, remains constant.

An isolated system does not undergo any energy (work or heat) interaction with the surroundings, and the total energy of all possible states remains constant. Therefore, for an isolated system ($\delta Q = 0$),

$$(dS)_{iso} \geq 0 \qquad ...(9.6)$$

For a reversible process :

$(dS)_{iso} = 0$ or S = constant

For an irreversible process :

$(dS)_{iso} > 0$ or entropy increases

Thus it may be stated that entropy of an isolated system can never decrease. It always increases with every irreversible process and remains constant only when the process is reversible. This is known as *principle of entropy increase*.

An isolated system results when any system and its surroundings are enclosed within a single boundary. The isolated system (also referred to as universe) includes every thing which is affected by the process.

Refer to the isolated system, consisting of a system and its surroundings, as shown in Fig. 9.3. If $T_{sys} > T_{sur}$, there occurs irreversible transfer of heat from the system to the surroundings.

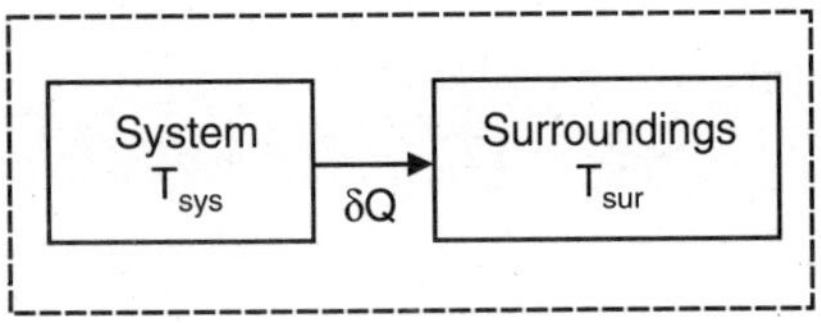

Fig. 9.3. Isolated system and entropy principle

$$(dS)_{sys} = \frac{-\delta Q}{T_{sys}}$$

$$(dS)_{sur} = \frac{\delta Q}{T_{sur}}$$

Therefore, entropy change for the isolated system or surroundings is

$$(dS)_{iso} = (dS)_{sys} + (dS)_{sur}$$

$$= \delta Q\left[\frac{1}{T_{sur}} - \frac{1}{T_{sys}}\right] \quad ...(9.7)$$

Since T_{sur} is less than T_{sys}, we can write $(ds)_{iso} > 0$

That is the entropy of an isolated system increases if the isolated system undergoes an irreversible process. In case $T_{sys} = T_{sur}$, the process would occur reversibly and the entropy will remain constant. Thus we may generalise and state that

$$dS \geq 0 \text{ for an isolated system}$$

If the entropy decreases locally at some region within an isolated system, it must be compensated by greater increase of entropy somewhere within the system. The net effect of an irreversible system would then be an increase in the entropy of the whole system. Since the universe is an isolated system and all real processes are irreversible, the entropy of the universe must be progressively increasing. Thus *it is an age of entropy growth and the entropy of the world tends towards a maximum*.

9.5. TEMPERATURE - ENTROPY PLOT

When a system is taken, reversibly from initial state 1 to final state 2,

$$Q_{rev} = \int_1^2 TdS$$

This implies that the area under a curve for a reversible process plotted on $T-S$ co-ordinates represents the quantity of heat interaction during the process.

It is to be noted that :

– During a reversible adiabatic process, $\delta Q = 0$ and so S = constant. The reversible adiabatic process is represented by a vertical (constant entropy) line on $T-S$ diagram.

– During a reversible isothermal process, T = constant and, therefore,

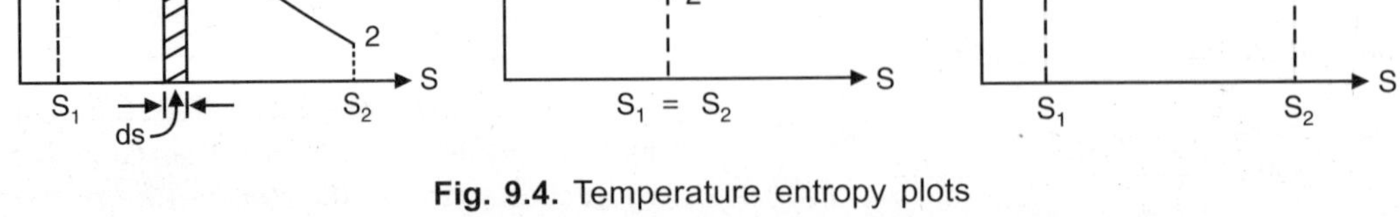

Fig. 9.4. Temperature entropy plots

$$Q_{rev} = T \int_1^2 dS = T\,(S_2 - S_1)$$

The reversible isothermal process is represented by a horizontal (constant temperature) line on $T-S$ diagram.

Recapitulate that a Carnot cycle consists of two reversible isothermals and two reversible adiabatics. Accordingly a Carnot cycle would plot as a rectangle on $T-S$ diagram.

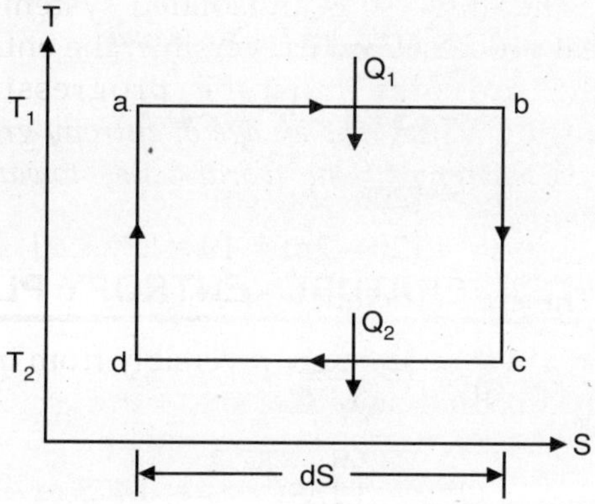

Fig. 9.5. Carnot cycle

Process a – b : Isothermal expansion at constant temperature T_1 during which heat Q_1 is supplied to the system.

$$Q_1 = \text{area under the process line } a-b = T_1 dS$$

Process b – c : Adiabatic expansion during which work is done by the system.

Process c – d : Isothermal compression at constant temperature T_2 during which heat Q_2 is rejected by the system.

$$Q_2 = \text{area under the process line } c-d = T_2\, dS$$

Process d – a : Adiabatic compression during which work is done on the system

Efficiency of Carnot cycle,

$$\eta = \frac{\text{work done per cycle}}{\text{heat supplied per cycle}} = \frac{Q_1 - Q_2}{Q_1} = \frac{T_1\, dS - T_2\, dS}{T_1 dS} = \frac{T_1 - T_2}{T_1} \quad \text{...(9.8)}$$

The area of the rectangle *abcd* represents the work output per cycle and it equals,

$$Q_1 - Q_2 = (T_1 - T_2)\, dS$$

EXAMPLE 9.1

A system comprising 1.5 kg of fluid undergoes a reversible process at 500 K during which it receives 0.75 MJ of heat. Determine the increase in specific entropy of the system. How would this increase in specific entropy of the system for the same initial and final states be affected if: (i) the process had been irreversible (ii) heat supply had been doubled.

Solution : Heat supplied per kg of fluid,

$$\delta q = \frac{0.75}{1.5} = 0.5 \text{ MJ} = 500 \text{ kJ}$$

change in entropy,

$$ds = \frac{\delta q}{T} = \frac{500}{500} = \mathbf{1\ kJ/kg\ K}$$

Entropy, like any other property, is dependent entirely upon the state of the system and is independent of the path or heat supplied to achieve that state. Therefore, even if the process is irreversible and the heat supply is doubled, the change in specific entropy will remain 1 kJ/kg K.

EXAMPLE 9.2

A system comprising 5 kg of fluid changes reversibly from an initial state of specific entropy 4 kJ/kg K and temperature 280 K to a final state of specific entropy 4.25 kJ/kg K and temperature 360 K. Evaluate the heat flow involved if the entropy varies linearly with temperature.

Solution : Refer Fig. 9.6.

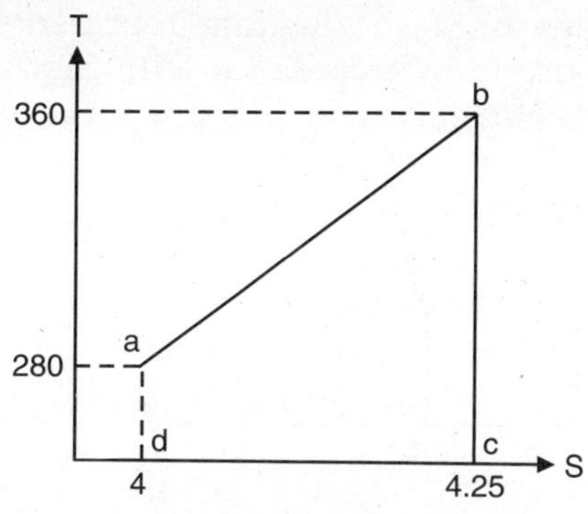

Fig. 9.6.

From the correlation : $\delta q = T\,ds$

Heat flow = area under the line *ab*

$$= \text{base} \times \text{mean ordinate}$$

$$= (4.25 - 4) \times \frac{1}{2}(280 + 360)$$

$$= 80 \text{ kJ/kg}$$

∴ Total heat flow,

$$\delta Q = m\,\delta q = 5 \times 80 = \mathbf{400\ kJ}$$

EXAMPLE 9.3

A heat engine cycle is represented by a circle of 4 cm diameter on T – S plane whose centre lies on 500 K temperature line. The temperature scale is 1 cm = 100 K and entropy scale is 1 cm = 0.1 kJ/K. If the engine completes 10 such cycles per second, determine the power developed by the engine. Proceed to calculate the heat supplied to the engine and the efficiency of engine.

Solution : Refer Fig. 9.7.

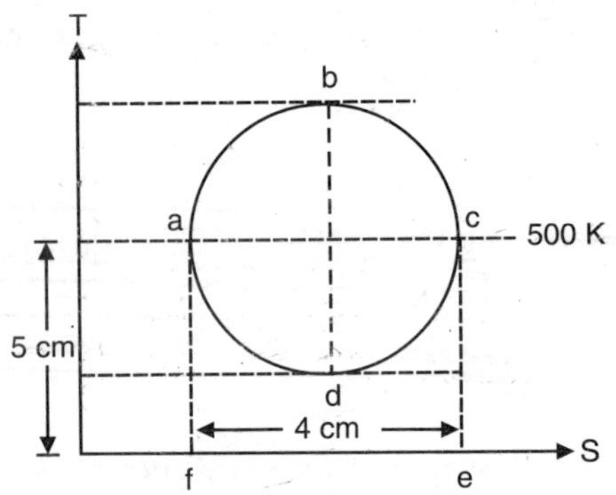

Fig. 9.7.

One cm^2 of area on $T - S$ plane is a measure of

$$= 100 \times 0.1 = 10 \text{ kJ}$$

Area on $T - S$ diagram represents heat transfer and in accordance with first law of thermodynamics

$$\oint \delta Q = \oint \delta W$$

Therefore, the cyclic area on $T - S$ diagram gives the work output

∴ Work output per cycle

$$= \text{area of the circle } abcd$$

$$= \frac{\pi}{4} \times 4^2 = 4\pi \text{ cm}^2$$

$$= 4\pi \times 10 = 40\pi \text{ kJ}$$

Power developed = work done/s

$$= 40\pi \times 10 = 400\pi \text{ kJ/s}$$

$$= \mathbf{1256.6\ kW}$$

Heat supplied, Q_1 = area under *abc*

$$= \text{area } abc + \text{area } acef$$

$$= \frac{1}{2}\left(\frac{\pi}{4} \times 4^2\right) + 5 \times 4$$

$$= (2\pi + 20) \text{ cm}^2$$

$$= (2\pi + 20) \times 10 = 262.8 \text{ kJ/cycle}$$

Heat rejected, Q_2 = area under *cda*

$$= \text{area } acef - \text{area } cda$$

$$= 5 \times 4 - \frac{1}{2}\left(\frac{\pi}{4} \times 4^2\right)$$

$$= (20 - 2\pi) \text{ cm}^2$$

$$= (20 - 2\pi) \times 10 = 137.2 \text{ kJ/cycle}$$

Thermal efficiency of engine,

$$\eta = \frac{Q_1 - Q_2}{Q_1}$$

$$= \frac{262.8 - 137.2}{262.8} = \mathbf{0.478 \text{ or } 47.8\%}$$

EXAMPLE 9.4

Heat flows from a reservoir at 800 K to another reservoir at 250 K. If entropy change of the hot reservoir is – 4 kJ/K, determine the entropy change of the cold reservoir.

Solution : Entropy change is defined as,

$$dS = \frac{\delta Q}{T}$$

∴ Heat lost by the hot reservoir,

$$Q = T\,dS = 800\,(-4) = -3200 \text{ kJ}$$

An equivalent amount of heat is received by the cold reservoir.

∴ Entropy change of cold reservoir,

$$= \frac{3200}{250} = \mathbf{12.8\ kJ/K}$$

EXAMPLE 9.5

Heat flows from a hot reservoir at 800 K to another reservoir at 250 K. If the entropy change of the overall process is 4.25 kJ/K, make calculations for the heat flowing out of the high temperature reservoir.

Solution : Let Q be the heat that flows out of the hot reservoir. An equivalent amount will be the heat received by the cold reservoir.

For the high temperature reservoir, $(dS)_h = \dfrac{-Q}{T_1}$

For the low temperature reservoir, $(dS)_l = \dfrac{Q}{T_2}$

Net entropy change, $(dS)_{net} = (dS)_h + (dS)_l$

$$\therefore \quad 4.25 = \frac{-Q}{800} + \frac{Q}{250}\ ; \ Q = \mathbf{1545.45\ kJ}$$

EXAMPLE 9.6

A lump of steel of mass 8 kg at 1000 K is dropped in 80 kg of oil at 300 K. Make calculations for the entropy change of steel, the oil and the universe. Take specific heats of steel and oil as 0.5 kJ/kg K and 3.5 kJ/kg K, respectively.

Solution : Let T_f be the final equilibrium temperature. Then from the principle of energy conservation,

heat lost by steel = heat gained by oil

$$(m\,c_p\,\Delta T)_{steel} = (m\,c_p\,\Delta T)_{oil}$$

$$8 \times 0.5 \times (1000 - T_f) = 80 \times 3.5 \times (T_f - 300)$$

Solution gives : $T_f = 309.86$ K

Entropy change of steel,

$$(dS)_{steel} = mc_p \log_e \frac{T_f}{T_1}$$

$$= 8 \times 0.5 \times \log_e \frac{309.86}{1000}$$

$$= \mathbf{-\ 4.686\ kJ/K}$$

Entropy change of oil,

$$(dS)_{oil} = mc_p \log_e \frac{T_f}{T_1}$$

$$= 80 \times 3.5 \times \log_e \frac{309.86}{300}$$

$$= \mathbf{9.0547\ kJ/K}$$

$\therefore$ Net change of entropy (*i.e.*, entropy change of universe) is,

$$(dS)_{universe} = (dS)_{steel} + (dS)_{oil}$$

$$= -\ 4.686 + 9.0547$$

$$= \mathbf{4.369\ kJ/K\ (increase)}$$

EXAMPLE 9.7

In a certain heat exchanger, 50 kg of water is heated per minute from 50°C to 110°C by hot gases which enter the heat exchanger at 250°C. If the flow rate of gases is 100 kg/min, estimate the net change of entropy.

c_p (water) = 4.186 kJ/kg K; c_p (gas) = 1 kJ/kg K

Solution : Assuming no loss of heat to surroundings, the principle of energy conservation gives,

heat lost by gases = heat gained by water

$$(m\,c_p\,\Delta T)_{gas} = (mc_p\,\Delta T)_{water}$$

$$100 \times 1.0 \times (250 - t_{g_2}) = 50 \times 4.186\ (110 - 50)$$

$\therefore$ Final temperature of hot gas,

$$t_{g_2} = 124.42°\text{C}$$

Entropy change of gases,

$$(dS)_{gas} = mc_p \log_e \frac{T_{g_2}}{T_{g_1}}$$

$$= 100 \times 1.0 \log_e \frac{124.42 + 273}{250 + 273}$$

$$= -\ 27.46\ \text{kJ/kg}$$

Entropy change of water,

$$(dS)_{water} = mc_p \log_e \frac{T_{w2}}{T_{w1}}$$

$$= 50 \times 4.186 \log_e \frac{110 + 273}{50 + 273}$$

$$= 35.66\ \text{kJ/kg}$$

Net change of entropy,

$$(dS)_{net} = (dS)_{gas} + (dS)_{water}$$

$$= -\ 27.46 + 35.66 = \mathbf{8.2\ kJ/kg}$$

EXAMPLE 9.8

A hot iron forging (specific heat 0.5 kJ/kg K) weighs 30 kg and has a temperature of 500°C. The forging is dropped into 200 kg oil mass (sp heat 2.5 kJ/kg K) at 25°C for quenching. Make calculations for the entropy change of forging, entropy change of oil and entropy change of the composite system. It may be presumed that there is no loss of heat to the surroundings.

Solution : Let suffix f and o represent the forging and oil respectively. From the principle of energy conservation,

heat lost by forging = heat gained by oil

$$m_f c_f (T_{f_1} - T_m) = m_o c_o (T_m - T_{o_1})$$

where T_m is the common temperature of the composite system

T_{f_1} = 500 + 273 = 773 K

T_{o_1} = 25 + 273 = 298 K

30 × 0.5 (773 – T_m) = 200 × 2.5 (T_m – 298)

Solution gives : T_m = 311.83 K

(*i*) Entropy change of forging

$$= m_f c_f \log_e \frac{T_m}{T_{f_1}}$$

$$= 30 \times 0.5 \log_e \frac{311.83}{773}$$

$$= \mathbf{-\ 13.617\ kJ/K}$$

(*ii*) Entropy change of oil

$$= m_o c_o \log_e \frac{T_m}{T_{o_1}}$$

$$= 200 \times 2.5 \log_e \frac{311.83}{298}$$

$$= \mathbf{22.682\ kJ/K}$$

(*iii*) Change in entropy of the system is

$$(dS)_{system} = (dS)_{forging} + (dS)_{oil}$$
$$= -\ 13.617 + 22.682$$
$$= \mathbf{9.065\ kJ/K\ (increase)}$$

EXAMPLE 9.9

5 kg of water at 30°C is mixed with 1 kg of ice at 0°C. The process of mixing is adiabatic and the system is open to atmosphere. Make calculations for the temperature of mixture and the change of entropy for the spontaneous mixing process. Take specific heat of water = 4.187 kJ/kg K and latent heat of ice = 335 kJ/kg.

Solution : Let T_m be the temperature of the mixture after adiabatic mixing.

From the principle of energy conservation,

heat lost by water =heat gained by ice

5 × 4.187 (30 – t_m) = 335 + 4.187 (t_m – 0)

Solution gives : t_m = 11.67°C

(*i*) Change in entropy of water is given by

$(dS)_{water}$ = entropy change due to cooling of water from 30°C to 11.67°C

$$= mc_p \log_e \frac{T_m}{T_1}$$

$$= 5 \times 4.187 \log_e \frac{11.07 + 273}{30 + 273}$$

$$= -\ 1.3004\ \text{kJ/K}$$

(*ii*) Change in entropy of ice is given by

$(dS)_{ice}$ = entropy change due to melting of ice at constant temperature + entropy change due to heating of ice water from 0°C to 11.67°C

$$= \frac{L}{T_1} + mc_p \log_e \frac{T_m}{T_1}$$

$$= \frac{335}{273} + 1 \times 4.187 \log_e \frac{11.67 + 273}{0 + 273}$$

$$= 1.4024\ \text{kJ/K}$$

(*iii*) Change in entropy of the system is

$$(dS)_{system} = (dS)_{water} + (dS)_{ice}$$
$$= -\ 1.3064 + 1.4024$$
$$= \mathbf{0.096\ kJ/K\ (increase)}$$

EXAMPLE 9.10

A block of mass m_a, specific heat c_a and temperature T_a is brought into thermal contact with another block of corresponding parameters m_b, c_b and T_b. Show that the change in entropy in the process of thermal equilibrium is necessarily positive.

Solution : Let it be presumed that initially the block *A* is at a higher temperature, *i.e*, $T_a > T_b$. As such the block *A* will lose heat which will be taken up by the block *B*. Then from the principle of energy conservation,

heat lost by block *A* = heat gained by block *B*

$$-\ m_a c_a dt_a = m_b c_b dt_b$$

The negative sign stems from the fact that dt_a is negative.

Overall entropy change,

$$dS = dS_a + dS_b = \frac{\delta Q_a}{T_a} + \frac{\delta Q_b}{T_b}$$

$$= \frac{m_a c_a dt_a}{T_a} + \frac{m_b c_b dt_b}{T_b}$$

$$= \frac{m_a c_a dt_a}{T_a} + \left(\frac{-\ m_a c_a dt_a}{T_b} \right)$$

$$= m_a c_a \left[\frac{1}{T_a} - \frac{1}{T_b} \right] dt_a$$

The terms $\left(\frac{1}{T_a}-\frac{1}{T_b}\right)$ and (dt_a) are both negative and accordingly $dS > 0$, *i.e.*, overall entropy change is positive.

EXAMPLE 9.11

m_1 kg of water at T_1 is isobarically and adiabatically mixed with m_2 kg of water at T_2 ($T_1 > T_2$). Show that for equal masses of water, the entropy change of the mixture is given by

$$(dS)_{universe} = 2m\, c_p \log_e\left[\frac{T_1+T_2}{2\sqrt{T_1T_2}}\right]$$

and prove that the change is necessarily positive.

Solution : Let T_f be the final temperature of the mixture. Then from the principle of energy conservation,

heat lost by m_1 kg of water

= heat gained by m_2 kg of water

$$m_1 c_p (T_1 - T_f) = m_2 c_p (T_f - T_2)$$

or $$T_f = \frac{m_1T_1 + m_2T_2}{m_1 + m_2}$$

Change in entropy of m_1 kg of water, whose temperature decreases from T_1 to T_f is

$$(dS)_1 = m_1c_p \log_e \frac{T_f}{T_1}$$

$$= m_1c_p \log_e \frac{m_1T_1 + m_2T_2}{(m_1+m_2)T_1}$$

Change in entropy of m_2 kg of water whose temperature increases from T_2 to T_f is

$$(dS)_2 = m_2c_p \log_e \frac{T_f}{T_2}$$

$$= m_2c_p \log_e \frac{m_1T_1 + m_2T_2}{(m_1+m_2)T_2}$$

∴ Change in entropy of universe is

$$(dS)_{universe} = (dS)_1 + (dS)_2$$

$$= m_1c_p \log_e \frac{m_1T_1 + m_2T_2}{(m_1+m_2)T_1} + m_2c_p \log_e \frac{m_1T_1 + m_2T_2}{(m_1+m_2)T_2}$$

If $m_1 = m_2 = m$, then the above expression for change in entropy of universe takes the form

$$(dS)_{universe} = mc_p \log_e \frac{T_1+T_2}{2T_1} + mc_p \log_e \frac{T_1+T_2}{2T_2}$$

$$= mc_p \left[\log_e \frac{T_1+T_2}{2T_1} + \log_e \frac{T_1+T_2}{2T_2}\right]$$

$$= mc_p \log_e \left[\frac{T_1+T_2}{2T_1} \times \frac{T_1+T_2}{2T_2}\right]$$

$$= mc_p \log_e \left[\frac{T_1+T_2}{2\sqrt{T_1T_2}}\right]^2$$

$$= 2mc_p \log_e \frac{T_1+T_2}{2\sqrt{T_1T_2}}$$

which is the required expression.

The term $\frac{T_1+T_2}{2}$ represents the arithmetic mean and this is always greater than geometric mean represented by $\sqrt{T_1T_2}$. Accordingly the term $\log_e \frac{T_1+T_2}{2\sqrt{T_1T_2}}$ is positive and as such the change in entropy is always positive, *i.e*, the entropy of the universe is increasing.

EXAMPLE 9.12

A heat engine operates between two thermal reservoirs; source at temperature T_1 and sink at temperature T_2. If the source and sink are of mass m and specific heat c, set up the following expression for the maximum work output possible

$$W_{max} = m\, c\, (\sqrt{T_1} - \sqrt{T_2})^2$$

Solution : For maximum work output or maximum efficiency, the heat engine must be reversible and for a reversible cycle engine, Clausius equation gives $\oint \frac{\delta Q}{T} = 0$. This implies that entropy change for a reversible cycle is zero.

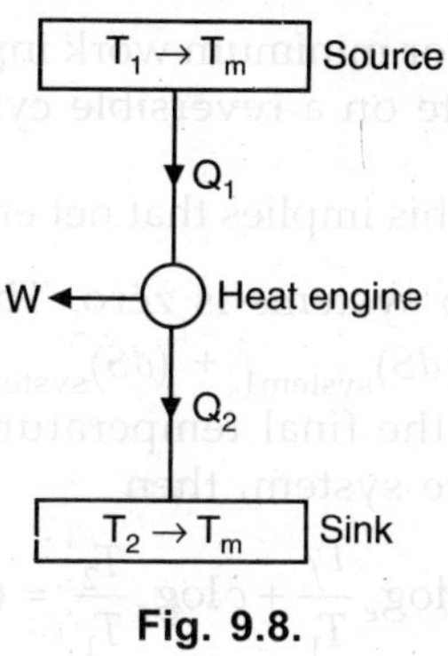

Fig. 9.8.

Let T_m be the common temperature when the working of the engine stops. Then,

$$(dS)_{source} + (dS)_{sink} = 0$$

$$mc \log_e \frac{T_m}{T_1} + mc \log_e \frac{T_m}{T_2} = 0$$

$$mc\left(\log_e \frac{T_m}{T_1} + \log_e \frac{T_m}{T_2}\right) = 0$$

$$mc \log_e \frac{T_m^2}{T_1 T_2} = 0$$

$$\log_e \frac{T_m^2}{T_1 T_2} = 0 = \log_e 1$$

$$\therefore \quad T_m^2 = T_1 T_2 \text{ or } T_m = \sqrt{T_1 T_2}$$

Now, work done

$$= \text{heat supplied} - \text{heat rejected}$$

$$= m\,c\,(T_1 - T_m) - m\,c\,(T_m - T_2)$$

$$\therefore W_{max} = mc\left[T_1 - \sqrt{T_1 T_2}\right] - mc\left[\sqrt{T_1 T_2} - T_2\right]$$

$$= mc\left[T_1 + T_2 - 2\sqrt{T_1 T_2}\right]$$

$$= mc\left(\sqrt{T_1} - \sqrt{T_2}\right)^2$$

EXAMPLE 9.13

A heat pump operates between two identical systems; both being at temperature T_1 to start with. Due to operation of pump, one of the systems gets cooled down to temperature T_2. Show that for this operation of pump, the minimum work required by the heat pump is

$$W_{min} = c\left[\frac{T_1^2}{T_2} + T_2 - 2T_1\right]$$

where c is the specific heat of the two identical systems.

Solution : For minimum work input, the pump is to operate on a reversible cycle for which $\oint \frac{\delta Q}{T} = 0$. This implies that net entropy change for the two systems is zero. That is

$$(dS)_{system 1} + (dS)_{system 2} = 0$$

If T_f is the final temperature of the high temperature system, then

$$c \log_e \frac{T_f}{T_1} + c \log_e \frac{T_2}{T_1} = 0$$

or $$\log_e\left(\frac{T_f}{T_1} \times \frac{T_2}{T_1}\right) = 0 = \log_e 1$$

$$\therefore \quad T_f = \frac{T_1^2}{T_2}$$

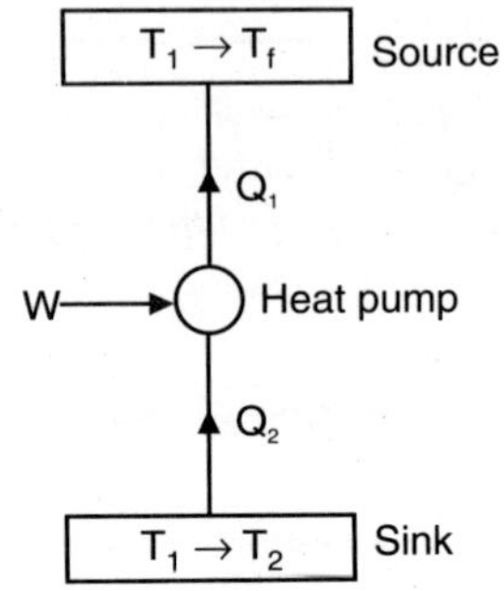

Fig. 9.9.

From the principle of energy conservation,

$$\oint \delta W = \oint \delta Q$$

$\therefore$ Work input = (heat rejected at high temperature) – (heat absorbed at low temperature)

$$= c\int_{T_1}^{T_f} dT - c\int_{T_2}^{T_1} dT$$

$$= c\,(T_f - T_1) - c\,(T_1 - T_2)$$

$$= c\,(T_f + T_2 - 2T_1]$$

Substituting the value of

$T_f = \frac{T_1^2}{T_2}$, we get

Minimum work input,

$$W_{min} = c\left[\frac{T_1^2}{T_2} + T_2 - 2T_1\right]$$

which is the required expression.

EXAMPLE 9.14

Two bodies of equal heat capacities C and temperature T_1 and T_2 form an adiabatically closed system. Determine the final temperature if this system is brought to an equilibrium state: (i) freely and (ii) reversibly. Proceed to calculate the maximum work which can be obtained from this system.

Solution : When the system attains the equilibrium state freely

heat given by body at temperature T_1
= heat gained by body at temperature T_2

$$C(T_1 - T_f) = C(T_f - T_2)$$

or $\quad T_1 - T_f = T_f - T_2\,;\quad T_f = \dfrac{T_1 + T_2}{2}$

(*ii*) When the system comes to equilibrium irreversibly: $(dS)_{total} = 0$

$$\therefore\ C\log_e \frac{T_f}{T_1} + C\log_e \frac{T_f}{T_2} = 0$$

or $\quad \log_e \dfrac{T_f^2}{T_1 T_2} = 0 = \log_e 1$

or $\quad \dfrac{T_f^2}{T_1 T_2} = 1\,;\quad T_f = \sqrt{T_1\,T_2}$

(*iii*) Maximum work,

$$\begin{aligned} W_{max} &= Q_1 - Q_2 \\ &= C(T_1 - T_f) - C(T_f - T_2) \\ &= C[T_1 + T_2 - 2T_f] \\ &= C\left[T_1 + T_2 - 2\sqrt{T_1 T_2}\right] \\ &= C\left[\sqrt{T_1} - \sqrt{T_2}\right]^2 \end{aligned}$$

EXAMPLE 9.15

A system maintained at constant volume is initially at temperature T_1 and a heat reservoir at the lower temperature T_0 is available. Determine the maximum work obtainable as the system is cooled to temperature T_0.

Solution : When the system is cooled to temperature T_0 under constant volume condition,

Entropy change for the system,

$$(dS)_{system} = c_v \log_e \frac{T_0}{T_1}$$

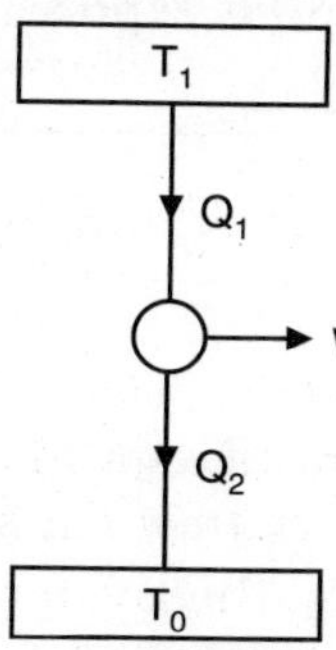

Fig. 9.10.

Entropy change for the reservoir,

$$(dS)_{reservoir} = \frac{Q_2}{T_0} = \frac{Q_1 - W}{T_0}$$

where $Q_1 = c_v(T_1 - T_0)$. That gives

$$(dS)_{reservoir} = \frac{c_v(T_1 - T_0) - W}{T_0}$$

$\therefore$ Entropy production (net entropy change) as a result of cooling

$$(dS)_{net} = c_v \log_e \frac{T_0}{T_1} + \frac{c_v(T_1 - T_0) - W}{T_0}$$

For a possible working arrangement

$$(dS)_{net} \geq 0$$

$$\therefore\ c_v \log_e \frac{T_0}{T_1} + \frac{c_v(T_1 - T_0) - W}{T_0} \geq 0$$

For maximum work obtainable, we have

$$c_v \log_e \frac{T_0}{T_1} + \frac{c_v(T_1 - T_0) - W_{max}}{T_0} = 0$$

or $\quad T_0\,c_v \log_e \dfrac{T_0}{T_1} + c_v(T_1 - T_0) - W_{max} = 0$

That gives :

$$W_{max} = c_v\left[(T_1 - T_0) + T_0 \log_e \frac{T_0}{T_1}\right]$$

EXAMPLE 9.16

The thermal capacity at constant volume of thermodynamic system initially at 300 K is prescribed by the relation:

$C_v = AT^2$ where $A = 0.042$ J/K^3

A low temperature sink is available at 200 K and a heat engine is made to operate between the reservoirs. Determine the maximum amount of work that can be obtained as the system is cooled down to the temperature of the sink.

Solution : Refer Fig. 9.11.

Heat taken from the system,

$$Q_1 = \int C_v\,dT$$

$$= \int_{300}^{200} 0.042\,T^2\,dT = 0.042\left|\frac{T^3}{3}\right|_{300}^{200}$$

$$= \frac{0.042}{3}(200^3 - 300^3)$$

$$= -266 \times 10^3 \text{ J}$$

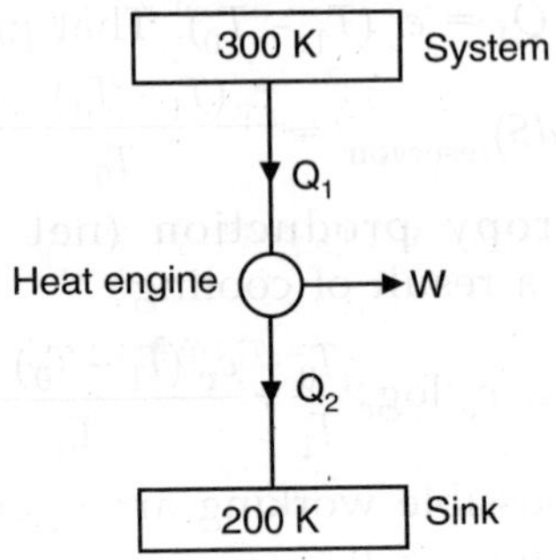

Fig. 9.11.

Change in entropy of the system,

$$(ds)_{\text{system}} = \int C_v \frac{dT}{T}$$

$$= \int 0.042T^2 \times \frac{dT}{T}$$

$$= 0.042 \left| \frac{T^2}{2} \right|_{300}^{200}$$

$$= \frac{0.042}{2}(200^2 - 300^2)$$

$$= -1050 \text{ J/K}$$

Change in entropy of the sink,

$$(ds)_{\text{sink}} = \frac{Q_2}{T_{\text{sink}}}$$

$$= \frac{Q_1 - W}{200} = \frac{266 \times 10^3 - W}{200}$$

Net entropy change of the arrangement,

$$(dS)_{\text{net}} = (ds)_{\text{system}} + (ds)_{\text{sink}}$$

$$= -1050 + \frac{266 \times 10^3 - W}{200}$$

For a possible working arrangement

$$(ds)_{\text{net}} \geq 0$$

$$\therefore \quad -1050 + \frac{266 \times 10^3 - W}{200} \geq 0$$

For maximum work obtainable, we have

$$-1050 + \frac{266 \times 10^3 - W_{\text{max}}}{200} = 0$$

or $-210 \times 10^3 + 266 \times 10^3 - W_{\text{max}} = 0$

$\therefore W_{\text{max}} = (266 - 210) \times 10^3 =$ **56000 J**

EXAMPLE 9.17

With reference to Fig. 9.12, abcd represents a Carnot cycle bounded by two reversible adiabatics and two reversible isotherms at temperatures T_1 and T_2; T_1 being greater than T_2. The oval figure is a reversible cycle whose heat is absorbed at temperature less than or equal to T_1, and rejected at temperature greater than or equal to T_2. Show that efficiency of the oval cycle is less than that of the Carnot cycle.

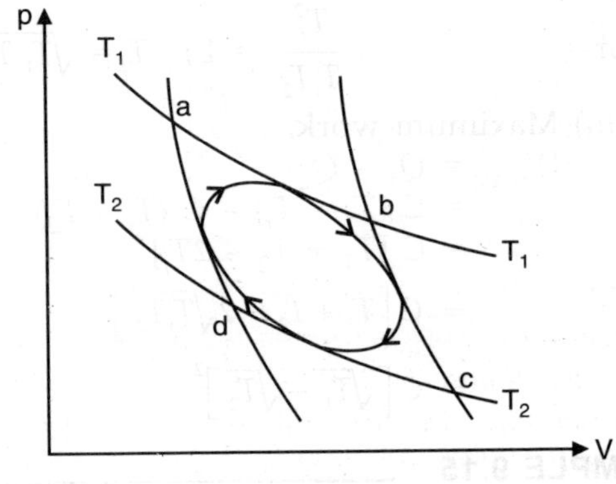

Fig. 9.12.

Solution : Carnot cycle *abcd* operates between temperatures T_1 and T_2 and accordingly its thermal efficiency is given by

$$\eta_{carnot} = \frac{T_1 - T_2}{T_1}$$

The oval cycle receives heat at temperature $\leq T_1$ and rejects at temperature $\geq T_2$ and is a reversible cycle. Entropy change is defined as $dS = \delta Q/T$. Therefore,

For the high temperature reservoir,

$$dS_1 = \frac{-\delta Q_1}{T_1}$$

For the low temperature reservoir,

$$dS_2 = \frac{\delta Q_2}{T_2}$$

$\therefore$ Net entropy change,

$$(dS)_{net} = dS_1 + dS_2 = \frac{-\delta Q_1}{T_1} + \frac{\delta Q_2}{T_2}$$

For a possible cycle, $(dS)_{net} \geq 0$

Hence $\dfrac{-\delta Q_1}{T_1} + \dfrac{\delta Q_2}{T_2} \geq 0$

or $\dfrac{\delta Q_1}{T_1} \leq \dfrac{\delta Q_2}{T_2}$ or $\dfrac{\delta Q_2}{T_2} \geq \dfrac{\delta Q_1}{T_1}$

or $\dfrac{\delta Q_2}{\delta Q_1} \geq \dfrac{T_2}{T_1}$ or $1 - \dfrac{\delta Q_2}{\delta Q_1} \leq 1 - \dfrac{T_2}{T_1}$

or $\dfrac{\delta Q_1 - \delta Q_2}{\delta Q_1} \leq \dfrac{T_1 - T_2}{T_1}$

The term $\dfrac{(\delta Q_1 - \delta Q_2)}{\delta Q_1}$ represents the efficiency of oval cycle where as the term $\dfrac{(T_1 - T_2)}{T_1}$ represents the efficiency of Carnot cycle. Obviously the efficiency of the oval cycle is less than the efficiency of the Carnot cycle.

EXAMPLE 9.18

An inventor claims that he has developed a heat engine which absorbs 1200 kJ and 800 kJ of heat from reservoirs at 800 K and 600 K respectively, and rejects 600 kJ and 200 kJ of heat to reservoirs at 400 K and 300 K. The engine is further stated to give an output equivalent to 1200 kJ. Determine whether the engine suggested by the inventor is theoretically possible.

Solution : Refer Fig. 9.13.

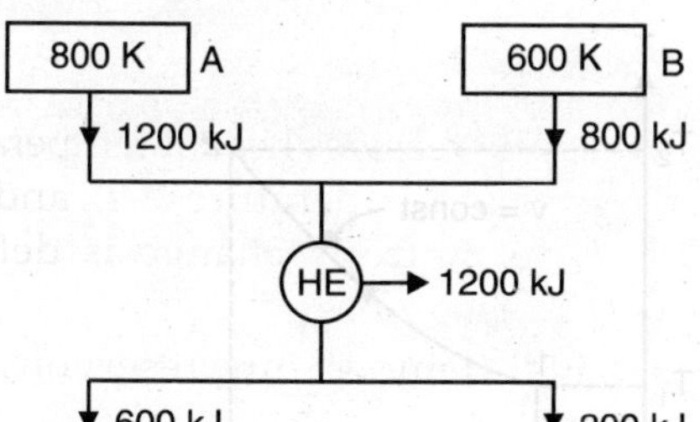

Fig. 9.13.

The entropy change of a system is defined as $dS = \dfrac{\delta Q}{T}$.

Hence entropy change for reservoir A,

$$dS_a = \frac{-1200}{800} = -1.5 \text{ kJ/K}$$

Entropy change for reservoir B,

$$dS_b = \frac{-800}{600} = -1.33 \text{ kJ/K}$$

Entropy change for reservoir C,

$$dS_c = \frac{600}{400} = 1.5 \text{ kJ/K}$$

Entropy change is reservoir D

$$dS_d = \frac{200}{300} = 0.667 \text{ kJ/K}$$

Net entropy change for the arrangement (universe)

$$= dS_a + dS_b + dS_c + dS_d$$
$$= -1.5 - 1.333 + 1.5 + 0.667$$
$$= -0.666 \text{ kJ/K}$$

For any possible cycle $(dS)_{net} \geq 0$. For the proposed engine, the net entropy change is found to be negative and this is not possible. Obviously the engine suggested by the inventor is theoretically not possible.

EXAMPLE 9.19

The connections of a reversible engine to three sources at 400 K, 300 K and 200 K are as shown in Fig. 9.14. The engine draws 1200 kJ from source at 400 K and produces 200 kJ of work.

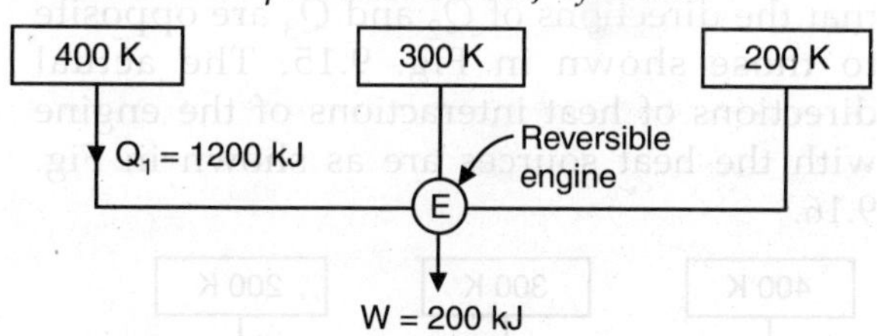

Fig. 9.14.

(a) Determine the amount and directions of heat interactions with the other heat sources.

(b) Make calculations for the entropy changes due to each of the heat interactions with the engine.

(c) How much entropy change occurs for the cycle?

Solution : Assuming the directions of heat interactions as shown in Fig. 9.15 and noting that the engine is reversible.

$$\oint \frac{\delta Q}{T} = 0$$

$$\therefore \quad \frac{Q_1}{T_1} + \frac{Q_2}{T_2} - \frac{Q_3}{T_3} = 0$$

$$\text{or} \quad \frac{1200}{400} + \frac{Q_2}{300} - \frac{Q_3}{200} = 0$$

$$\text{or} \quad 2Q_2 - 3Q_3 = -1800 \quad ...(i)$$

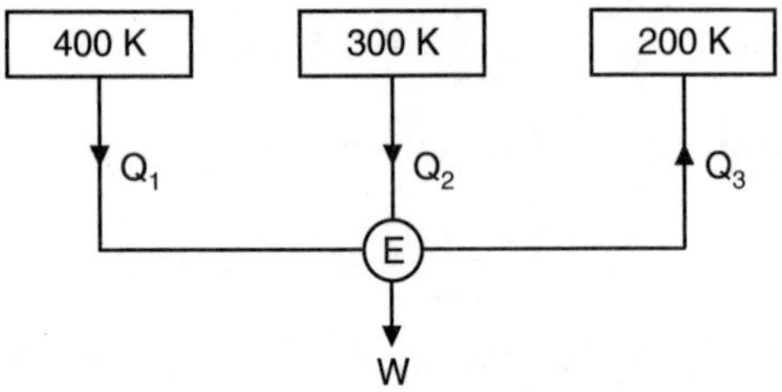

Fig. 9.15.

Also from the law of conservation of energy,

$$Q_1 + Q_2 - Q_3 = W$$
$$1200 + Q_2 - Q_3 = 200$$
$$\text{or} \quad Q_2 = Q_3 - 1000 \quad ...(ii)$$

From expression (*i*) and (*ii*),

$$2(Q_3 - 1000) - 3Q_3 = -1800$$
$$\text{or} \quad 2Q_3 - 2000 - 3Q_3 = -1800$$

That gives :

$$Q_3 = -200 \text{ kJ}$$
$$\text{and} \quad Q_2 = -200 - 1000 = -1200 \text{ kJ}$$

The negative sign with Q_2 and Q_3 indicates that the directions of Q_2 and Q_3 are opposite to those shown in Fig. 9.15. The actual directions of heat interactions of the engine with the heat sources are as shown in Fig. 9.16.

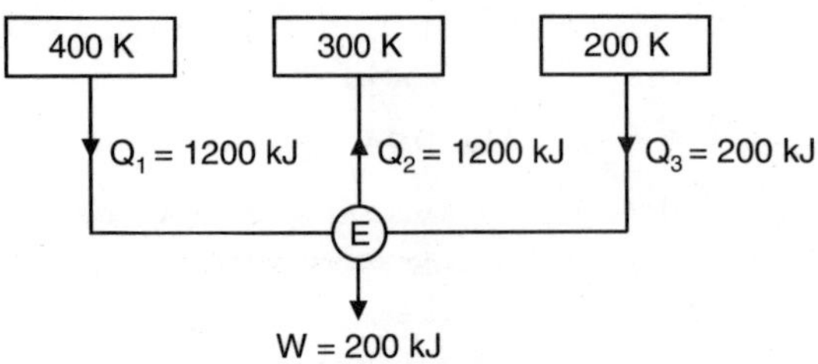

Fig. 9.16.

$$(b) \quad ds_1 = \frac{Q_1}{T_1} = \frac{-1200}{400} = -3 \text{ kJ/K}$$

$$ds_2 = \frac{Q_2}{T_2} = \frac{1200}{300} = 4 \text{ kJ/K}$$

$$ds_3 = \frac{Q_3}{T_3} = \frac{-200}{200} = -1 \text{ kJ/K}$$

(*c*) Entropy change for the cycle,

$$dS = ds_1 + ds_2 + ds_3$$
$$= -3 + 4 - 1 = 0$$

Thus the entropy change for the cycle is zero and this has to be so because entropy is a property.

9.6. ENTROPY CHANGE DURING A PROCESS

In the absence of gravity, motion, electricity and magnetic effects, the first law of thermodynamics can be expressed as

$\delta q = du + \delta w$ for unit mass of substance

Further if the process is reversible :

$$\delta w = p\, dv$$

$$\therefore \quad (\delta q)_{rev} = du + p\, dv$$

From the definition of entropy :

$T\, ds = (\delta q)_{rev}$ and, therefore,

$$T\, ds = du + p\, dv \quad ...(i)$$

Also, $h = u + pv$

$dh = du + p\, dv + v\, dp$ and thus

$$T\, ds = dh - v\, dp \quad ...(ii)$$

Expressions (*i*) and (*ii*) can be used to work out the entropy change during any process ; whether reversible or irreversible.

9.6.1. Constant volume process

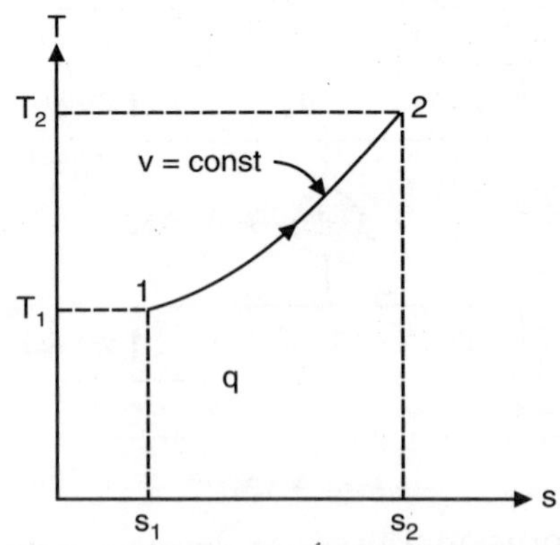

Fig. 9.17. T-s diagram: constant volume process

For unit mass of substance

$$\delta q = c_v\, dT$$

$$\text{or} \qquad \frac{\delta q}{T} = c_v \frac{dT}{T}$$

$$\text{or} \qquad ds = c_v \frac{dT}{T}$$

Integrating both sides, we obtain

$$\int_1^2 ds = c_v \int_1^2 \frac{dT}{T}$$

$$\therefore \; s_2 - s_1 = c_v \log_e \frac{T_2}{T_1} = c_v \log_e \frac{p_2}{p_1} \quad ...(9.9)$$

9.6.2. Constant pressure process

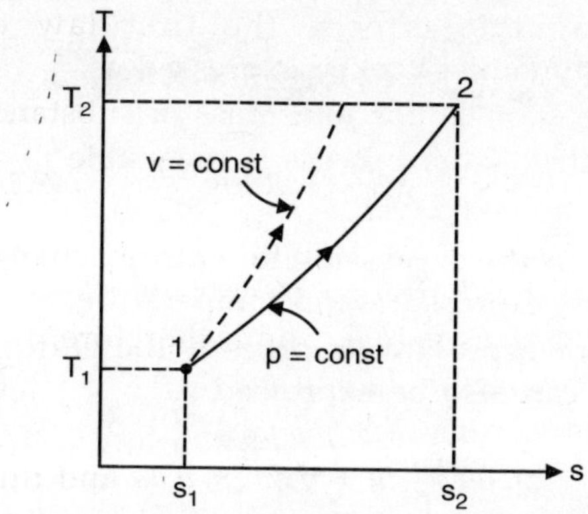

Fig. 9.18. T-s diagram: constant pressure process

For unit mass of substance

$$\delta q = c_p \, dT$$

$$\text{or} \qquad \frac{\delta q}{T} = c_p \frac{dT}{T}$$

$$\text{or} \qquad ds = c_p \frac{dT}{T}$$

Integrating both sides, we get

$$\int_1^2 ds = c_p \int_1^2 \frac{dT}{T}$$

$$\therefore \; s_2 - s_1 = c_p \log_e \frac{T_2}{T_1} = c_p \log_e \frac{v_2}{v_1} \quad ...(9.10)$$

9.6.3. Isothermal process

For unit mass of substance

$$\delta q = du + p \, dv$$

Since the temperature remains constant, $du = 0$ and therefore

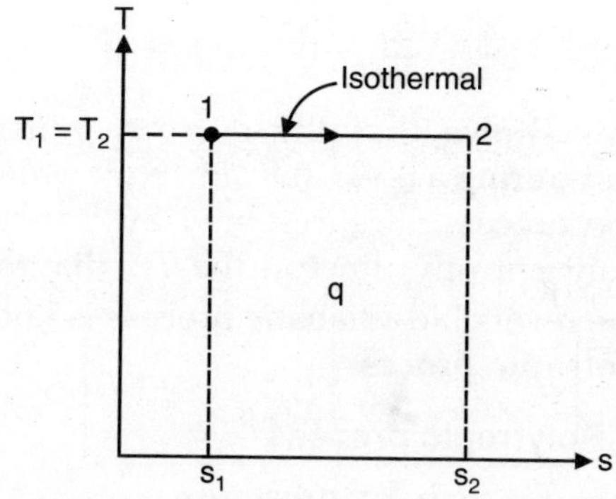

Fig. 9.19. T-s diagram: isothermal process

$$\delta q = p \, dv$$

$$\text{or} \qquad \frac{\delta q}{T} = \frac{p}{T} dv = R \frac{dv}{v}$$

$$\text{or} \qquad ds = R \frac{dv}{v}$$

Integrating both sides, we get

$$\int_1^2 ds = R \int_1^2 \frac{dv}{v}$$

$$\therefore \; s_2 - s_1 = R \log_e \frac{v_2}{v_1} = R \log_e \frac{p_1}{p_2} \quad ...(9.11)$$

The reversible isothermal process is represented by a horizontal (constant temperature) line on the *T–s* diagram.

9.6.4. Reversible adiabatic (isentropic) Process

During an adiabatic process, heat is neither supplied to or rejected by the system.

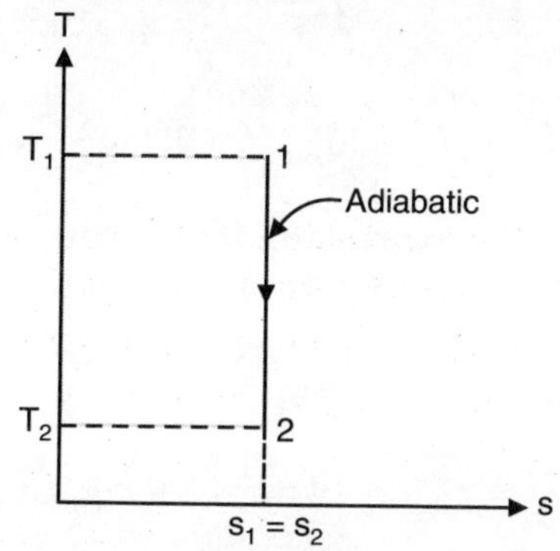

Fig. 9.20. T-s diagram: adiabatic process

$$\delta q = 0$$

or $\quad \dfrac{\delta q}{T} = 0 \;; \quad ds = 0$

This shows that the entropy remains constant during a reversible adiabatic process, and the process is represented by a vertical (constant entropy) line on the T-s diagram.

The reversible adiabatic process is known as **isentropic process.**

9.6.5. Polytropic process

For unit mass of substance

$$\delta q = du + p\,dv = c_v\,dT + p\,dv$$

or $$\frac{\delta q}{T} = c_v\frac{dT}{T} + \frac{p}{T}dv$$

or $$ds = c_v\frac{dT}{T} + R\frac{dv}{v}$$

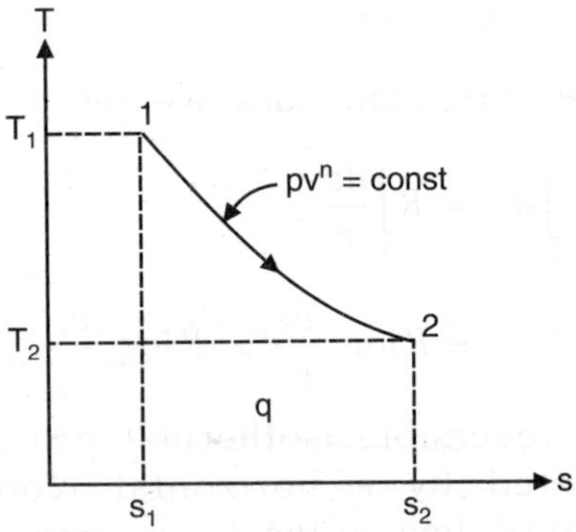

Fig. 9.21. T-s diagram: Polytropic process

Upon integration between the initial and final states

$$\int_1^2 ds = c_v\int_1^2 \frac{dT}{T} + R\int \frac{dv}{v}$$

$$\therefore \quad s_2 - s_1 = c_v \log_e\frac{T_2}{T_1} + R\log_e\frac{v_2}{v_1} \quad ...(9.12)$$

This is the expression for entropy change in terms of temperature and volume and changes during a polytropic process.

- Equation 9.12 can be recast as

$$s_2 - s_1 = c_v \log_e\left(\frac{p_2 v_2}{p_1 v_1}\right) + R\log_e\frac{v_2}{v_1}$$

$$= c_v \log_e\frac{p_2}{p_1} + c_v \log_e\frac{v_2}{v_1} + R\log_e\frac{v_2}{v_1}$$

$$= c_v \log_e\frac{p_2}{p_1} + (R + c_v)\log_e\frac{v_2}{v_1}$$

$$= c_v \log_e\frac{p_2}{p_1} + c_p\log_e\frac{v_2}{v_1} \quad ...(9.13)$$

This is the expression for entropy change in terms of pressure and volume.

- Equation 9.12 can be recast as

$$s_2 - s_1 = c_v \log_e\frac{T_2}{T_1} + R\log_e\left(\frac{T_2}{T_1}\frac{p_1}{p_2}\right)$$

$$s_2 - s_1 = c_v \log_e\frac{T_2}{T_1} + R\log_e\frac{T_2}{T_1} + R\log_e\frac{p_1}{p_2}$$

$$= (c_v + R)\log_e\frac{T_2}{T_1} + R\log_e\frac{p_1}{p_2}$$

$$= c_p \log_e\frac{T_2}{T_1} + R\log_e\frac{p_1}{p_2} \quad ...(9.14)$$

This is the expression for entropy change in terms of pressure and temperature.

- For a polytropic process, the entropy change can also be expressed as

$$ds = \frac{\delta q}{T} = \frac{\gamma - n}{\gamma - 1}\frac{\delta w}{T}$$

$$= \frac{\gamma - n}{\gamma - 1}\frac{p\,dv}{T} = \frac{\gamma - n}{\gamma - 1}R\frac{dv}{v}$$

Upon integration between the initial and final states,

$$s_2 - s_1 = \frac{\gamma - n}{\gamma - 1}R\log_e\frac{v_2}{v_1}$$

$$= (\gamma - n)c_v \log_e\frac{v_2}{v_1}$$

$$= \left(c_p - nc_v\right)\log_e\frac{v_2}{v_1} \quad ...(9.15)$$

This is the expression for entropy change in terms of volume ratio.

From the polytropic law pv^n = constant, we have

$$\frac{v_2}{v_1} = \left(\frac{p_1}{p_2}\right)^{\frac{1}{n}}; \quad \log_e\frac{v_2}{v_1} = \frac{1}{n}\log_e\left(\frac{p_1}{p_2}\right)$$

Substituting this value of volume ratio in equation 9.15. we get

$$s_2 - s_1 = \frac{c_p - nc_v}{n} \log_e \left(\frac{p_1}{p_2} \right) \quad ...(9.16)$$

which is an expression for entropy change in terms of pressure ratio.

Further,

$$\frac{T_2}{T_1} = \left(\frac{p_2}{p_1} \right)^{\frac{n-1}{n}}$$

or $$\log_e \left(\frac{p_1}{p_2} \right) = -\frac{n}{n-1} \log_e \frac{T_2}{T_1}$$

When this value of pressure ratio is substituted in equation 9.16., it transforms to

$$s_2 - s_1 = \frac{c_p - nc_v}{n} \times \left(-\frac{n}{n-1} \log_e \frac{T_2}{T_1} \right)$$

$$= -\frac{c_p - nc_v}{n-1} \log_e \frac{T_2}{T_1}$$

$$= -\frac{\gamma - n}{n-1} c_v \log_e \frac{T_2}{T_1}$$

$$= c_n \log_e \frac{T_2}{T_1} \quad ...(9.17)$$

Since $\gamma > n > 1$, c_n is negative and so the entropy change is positive.

Constant volume and constant pressure lines on T–S diagram

For unit mass of substance, the entropy changes are

$$ds = c_v \frac{dT}{T} \text{ for constant volume process,}$$

and

$$ds = c_p \frac{dT}{T} \text{ for constant pressure process}$$

Therefore, slope of the constant volume line on $T-S$ diagram,

$$\frac{dT}{ds} = \frac{T}{c_v} \quad ...(i)$$

and slope of the constant pressure line on $T-S$ diagram

$$\frac{dT}{ds} = \frac{T}{c_p} \quad ...(ii)$$

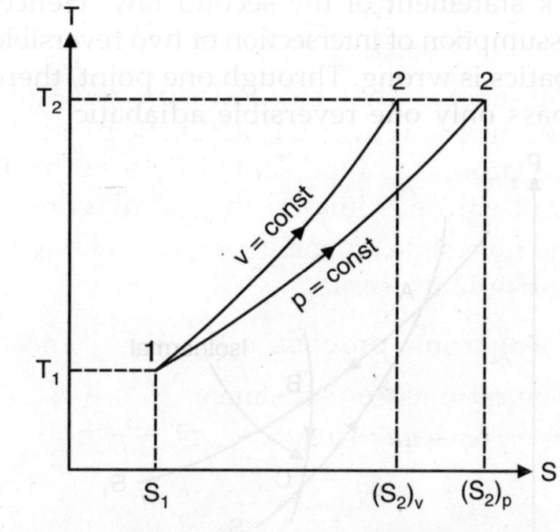

Fig. 9.22. Constant volume and constant pressure lines on T-S diagram

Since $c_p > c_v$, the above expressions indicate that constant volume line on $T-S$ diagram is steeper than constant pressure line working between the same temperature limits. Further since c_p and c_v are constant, $\left(\frac{dT}{ds}\right) \propto T$ both for the constant pressure and constant volume processes.

Can two reversible adiabatic paths intersect each other

For a reversible adiabatic process, heat interaction is zero and so is the change in entropy ($dS = \delta Q/T$). Two constant entropy curves S_1 and S_2 have been depicted on p–v diagram. If we assume that these curves intersect at point A as shown, then at the point of intersection, there will be two values of entropy. However, entropy is a property and point function, and therefore, cannot have two values at any point.

Further, let a reversible isothermal be plotted on the $p-v$ diagram and let it intersect the reversible adiabatics at B and C. The reversible processes AB, BC, CA then constitute a reversible cycle. The included area ABC represents the net work output. However, this work has been obtained from an engine which exchanges heat with a single thermal reservoir during the isothermal process BC (there is no heat intraction during the adiabatics AB and CB). This is in violation of the Kelvin-

Planck statement of the second law. Hence, our assumption of intersection of two reversible adiabatics is wrong. Through one point, there can pass only one reversible adiabatic.

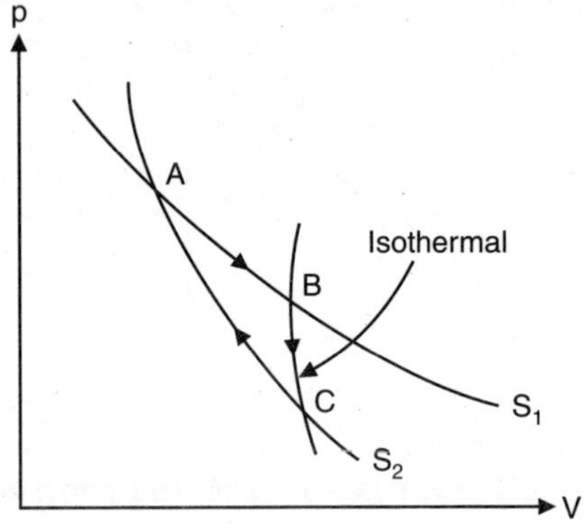

Fig. 9.23. Two reversible adiabatics intersecting each other

EXAMPLE 9.20

A rigid cylinder containing 0.005 m³ of nitrogen (molecular mass 28) at 1 bar and 290 K is heated reversibly until the temperature becomes 360 K. Determine the heat supplied and entropy change. Assume nitrogen to be perfect gas and take γ = 1.4.

Solution : A rigid cylinder implies that the heating of the gas is done at constant volume.

Gas constant,

$$R = \frac{R_{mol}}{\text{molecular mass}}$$

$$= \frac{8314}{28} = 296.93 \text{ J/kg K}$$

$$m = \frac{p_1 v_1}{RT_1} = \frac{1 \times 10^5 \times 0.005}{296.93 \times 290}$$

$$= 5.806 \times 10^{-3} \text{ kg}$$

$$c_v = \frac{R}{\gamma - 1} = \frac{296.93}{1.4 - 1}$$

$$= 742.32 \text{ J/kg K}$$

Heat supplied $= mc_v(T_2 - T_1)$

$= 5.806 \times 10^{-3} \times 742.32\ (360 - 290)$

$= 301.694 \cong 0.302$ kJ

Change in entropy,

$$(S_2 - S_1) = mc_v \log_e \frac{T_2}{T_1}$$

$$= 5.806 \times 10^{-3} \times 742.32 \log_e \frac{360}{290}$$

$$= \mathbf{0.9319 \text{ J/kg K}}$$

EXAMPLE 9.21

A piston-cylinder arrangement contains 0.03 m³ of nitrogen at 1 bar and 290 K. The piston moves inwards and the gas is compressed isothermally and reversibly until the pressure becomes 4 bar. Determine change in entropy and work done. Assume nitrogen to be a perfect gas.

Solution : Gas constant,

$$R = \frac{R_{mol}}{\text{molecular mass}}$$

$$= \frac{8314}{28} = 296.93 \text{ J/kg K}$$

$$m = \frac{p_1 v_1}{RT_1} = \frac{1 \times 10^5 \times 0.03}{296.93 \times 290}$$

$$= 0.03484 \text{ kg}$$

For an isothermal process, the change in entropy is given by

$$(S_2 - S_1) = mR \log_e \frac{v_2}{v_1}$$

$$= mR \log_e \frac{p_1}{p_2}$$

$$= 0.03484 \times 296.93 \log_e \frac{1}{4}$$

$$= -14.341 \text{ J/K} \simeq -0.01434 \text{ kJ/K}$$

Heat interaction,

$$Q = T(S_2 - S_1)$$

$$= 290 \times (-0.01434) = -4.1586 \text{ kJ}$$

During an isothermal process, work done equals the heat interaction

∴ Work done = – 4.1586 kJ

Alternatively, Work done

$$= p_1 v_1 \log_e \frac{v_2}{v_1}$$

$$= p_1 v_1 \log_e \frac{p_1}{p_2}$$

$$= 1 \times 10^5 \times 0.03 \log_e \frac{1}{4}$$

$$= -4158.88 \text{ J} \simeq -4.1589 \text{ kJ}$$

EXAMPLE 9.22

A piston-cylinder arrangement contains 1 kg of air at pressure 1 bar and temperature 300 K. When 10 kJ of work is done on air, its parameters change to 2.5 bar and 310 K. If the piston - cylinder assembly exchanges heat only with a reservoir maintained at 300 K, make calculations for entropy change of the air, entropy change of the reservoir and heat transfer from the air. Comment on the truth or falsehood of the claim.

For air c_p = 1.005 kJ/kg K and R = 0.287 kJ/kg K

Solution : In terms of pressure and temperature, the entropy change is

$$S_2 - S_1 = mc_p \log_e \frac{T_2}{T_1} + mR \log_e \frac{p_1}{p_2}$$

$$= 1 \times 1.005 \log_e \frac{310}{300} + 1 \times 0.287 \log_e \frac{1}{2.5}$$

$$= 0.03295 - 0.26298$$

$$= - 0.23 \text{ kJ/K}$$

$$dU = m\, c_v\, dT = m\, (c_p - R)\, dT$$

$$= 1\, (1.005 - 0.287)\, (310 - 300)$$

$$= 7.18 \text{ kJ}$$

From principle of energy conservation,

$$\delta Q = \delta W - \delta U$$

$$= (- 10) + 7.18 = - 2.82 \text{ kJ}$$

This heat transfer out of the air is received by the reservoir, and therefore, entropy change for the reservoir is

$$(dS)_{reservoir} = \frac{\delta Q}{T} = \frac{2.82}{300} = 0.0094 \text{ kJ/K}$$

Overall entropy change = $(dS)_{air} + (dS)_{reservoir}$

$$= - 0.23 + 0.0094$$

$$= - 0.2206 \text{ kJ/K}$$

But negative entropy change for a composite system (universe) is impossible and as such the claim is false.

EXAMPLE 9.23

One kg of air initially at 7 bar pressure and 360 K temperature expands polytropically ($pv^{1.2}$ = const) until the pressure is reduced to 1.4 bar. Determine : (i) final specific volume and temperature, (ii) change of internal energy, work done and heat interaction, and (iii) change in entropy. Take R = 287 J/kg K and γ = 1.4.

Solution : From characteristic equation, $pv = RT$, for unit mass of air

$$v_1 = \frac{RT_1}{p_1} = \frac{287 \times 360}{7 \times 10^5}$$

$$= 0.1476 \text{ m}^3\text{/kg}$$

$$v_2 = v_1 \left(\frac{p_1}{p_2}\right)^{\frac{1}{n}} = 0.1476 \times \left(\frac{7}{1.4}\right)^{1/1.2}$$

$$= 0.5641 \text{ m}^3\text{/kg}$$

$$T_2 = T_1 \left(\frac{p_2}{p_1}\right)^{\frac{n-1}{n}} = 360 \left(\frac{1.4}{7}\right)^{\frac{1.2-1}{1.2}}$$

$$= 275 \text{ K}$$

Alternatively :

$$T_2 = \frac{p_2 v_2}{R} = \frac{1.4 \times 10^5 \times 0.5641}{287}$$

$$= 275 \text{ K}$$

(*ii*) $du = c_v\, dT = \dfrac{R}{\gamma - 1} dT$

$$= \frac{287}{1.4 - 1}(275 - 360) = - 60987 \text{ J/kg}$$

$$\delta W = \frac{p_1 v_1 - p_2 v_2}{n - 1} = \frac{R(T_1 - T_2)}{n - 1}$$

$$= \frac{287\,(360 - 275)}{1.2 - 1} = 121975 \text{ J/kg}$$

$$\delta q = \delta w + du$$

$$= 121975 - 60987 = 60988 \text{ J/kg}$$

Alternatively, $\delta q = \dfrac{\gamma - n}{\gamma - 1} \times$ work done

$$= \frac{1.4 - 1.2}{1.4 - 01} \times 121975$$

$$\simeq 60988 \text{ J/kg} \simeq 61 \text{ kJ/kg}$$

(*iii*) Change in entropy,

$$s_2 - s_1 = c_v \log_e \frac{T_2}{T_1} + R \log_e \frac{v_2}{v_1}$$

$$= \frac{R}{\gamma - 1} \log_e \frac{T_2}{T_1} + R \log_e \frac{v_2}{v_1}$$

$$\therefore s_2 - s_1 = \frac{287}{1.4-1}\log_e\frac{275}{360} + 287\log_e\frac{0.5641}{0.1476}$$
$$= -193.2 + 384.8$$
$$= 191.6 \text{ J/kg K} = \mathbf{0.1916 \text{ kJ/kg K}}$$

EXAMPLE 9.24

A gas initially at 1.5 bar pressure, 0.15 m³ volume and 300 K was compressed polytropically ($pv^{1.25}$ = constant) to 15 bar pressure. Determine the change in entropy. Also work out the approximate entropy change obtained by dividing the heat transferred by the mean absolute temperature during compression. Take c_p = 1.04 kJ/kg K and c_v = 0.745 kJ/kg K

Solution : $R = c_p - c_v = 1.04 - 0.745$
$= 0.295$ kJ/kg K

$$m = \frac{p_1 v_1}{RT_1}$$
$$= \frac{1.5\times10^5\times0.15}{0.295\times10^3\times300} = 0.254 \text{ kg}$$
$$v_2 = v_1\left(\frac{p_1}{p_2}\right)^{\frac{1}{n}}$$
$$= 0.15\left(\frac{1.5}{15}\right)^{1/1.25} = 0.0238 \text{ m}^3$$
$$s_2 - s_1 = mc_p\log_e\frac{v_2}{v_1} + mc_v\log_e\frac{p_2}{p_1}$$
$$= 0.254\times1.04\log_e\frac{0.0238}{0.15} + 0.254\times0.745\log_e\frac{15}{1.5}$$
$$= -0.4863 + 0.4357$$
$$= \mathbf{-0.0506 \text{ kJ/kg}}$$

$$(b)\ W_{1\text{-}2} = \frac{p_1v_1 - p_2v_2}{n-1}$$
$$= \frac{1.5\times10^5\times0.15 - 15\times10^5\times0.0238}{1.25-1}$$
$$= -52800 \text{ J} = -52.80 \text{ kJ}$$

The negative sign indicates that work is done on the air.

$$T_2 = T_1\left(\frac{p_2}{p_1}\right)^{\frac{n-1}{n}}$$
$$= 300\left(\frac{15}{1.5}\right)^{\frac{1.25-1}{1.25}} = 475.46 \text{ K}$$
$$dU = mc_v(T_2 - T_1)$$
$$= 0.254\times0.745\ (475.46 - 300)$$
$$= 33.20 \text{ kJ}$$

From principle of energy conservation,
$$Q_{1\text{-}2} = W_{1\text{-}2} + dU$$
$$= -52.80 + 33.20 = -19.60 \text{ kJ}$$

The negative sign indicates that heat flows from the air

$$T_{mean} = \frac{(300+475.46)}{2} = 387.73 \text{ K}$$
$$\therefore S_2 - S_1 = \frac{Q_{12}}{T_{mean}}$$
$$= \frac{-19.60}{387.73} = \mathbf{-0.05055 \text{ kJ/K}}$$

EXAMPLE 9.25

A pure substance is initally at 5 bar pressure and 0.25 m³. It undergoes reversible adiabatic compression according to the law $pv^{1.3}$ = constant till pressure becomes 30 bar. Determine:

(a) changes in enthalpy, internal energy and entropy,

(b) heat and work interactions during the process.

Solution : For the reversible adiabatic process,
$$p_1v_1^n = p_2v_2^n$$
$$v_2 = v_1\left(\frac{p_1}{p_2}\right)^{1/n}$$
$$= 0.25\left(\frac{5}{30}\right)^{1/1.3} = 0.063 \text{ m}^3$$

(*a*) For a pure substance undergoing an infinitesimal reversible process
$$du = \delta q - p\,dv = T\,ds - p\,dv$$
$$dh = d(u + pv) = du + p\,dv + v\,dp$$
$$= (T\,ds - p\,dv) + p\,dv + v\,dp$$
$$= T\,ds + v\,dp$$

For the reversible adiabatic process,
$ds = 0$ and accordingly
$$dh = v\,dp$$

$\therefore$ change in enthalpy,

$$(h_2 - h_1) = \int_1^2 v\,dp$$

$$= \int_{p_1}^{p_2} \left(\frac{p_1 v_1^n}{p}\right)^{1/n} dp$$

$$= (p_1\,v_1^n)^{1/n} \int_{p_1}^{p_2} p^{-\frac{1}{n}}\,dp$$

$$= (p_1\,v_1^n)^{1/n} \left|\frac{p^{-\frac{1}{n}+1}}{-\frac{1}{n}+1}\right|_{p_1}^{p_2}$$

$$= (p_1\,v_1^n)^{1/n} \left[p_2^{\frac{n-1}{n}} - p_1^{\frac{n-1}{n}}\right]\frac{n}{n-1}$$

$$= \frac{n\,(p_2 v_2 - p_1 v_1)}{n-1}$$

$$(\because\ p_1 v_1^{\,n} = p_2 v_2^{\,n})$$

$$= \frac{1.3}{1.3-1}\left[30\times10^5\times0.063 - 5\times10^5\times0.25\right]$$

$= 277333$ J $=$ **277.33 kJ**

Noting that :

$$h_2 - h_1 = (u_2 + p_2 v_2) - (u_1 + p_1 v_1)$$
$$= (u_2 - u_1) + (p_2 v_2 - p_1 v_1)$$

or $(u_2 - u_1) = (h_2 - h_1) - (p_2 v_2 - p_1 v_1)$

Now : $p_2 v_2 - p_1 v_1$

$= 30 \times 10^5 \times 0.063 - 5 \times 10^5 \times 0.25$

$= 64000$ J $= 64$ kJ

$\therefore$ Change in internal energy,

$(u_2 - u_1) = 277.33 - 64 =$ **213.33** kJ

Since the compression process is reversilbe adiabatic, change in entropy, $(s_2 - s_1) = 0$

(*b*) Heat interaction $Q_{1-2} = 0$ as the process is adiabatic. From energy equation applied to a closed process,

$$\delta q = \delta w + du$$

$\therefore$ work interaction

$W_{1-2} = -du =$ **– 213.33** kJ

The negative sign indicates that work is done on the pure substance.

EXAMPLE 9.26

One mole of nitrogen contained in a rigid and insulated tank at 300 K is stirred by a paddle wheel which is operated by a 60 W electric motor for 1.5 minutes. The molar specific heat at constant volume for nitrogen is given by

$$c_v = 20 + 5 \times 10^{-3}\,T$$

where c_v is in J/mol K and T is in kelvin, Make calculations for the change in internal energy, final temperature and change in entropy of the gas.

Solution : From the principle of energy conservation for a closed system, we have

$$\delta q = \delta w + du$$

Here δq is zero as the tank is insulated.

$\therefore$ Change in internal energy,

$du = -\delta w = 60 \times (1.5 \times 60) =$ **5400** J

(*b*) The change in internal energy is also given by

$$du = \int_{T_1}^{T_2} c_v\,dT = \int_{T_1}^{T_2} (a + bT)\,dT$$

$$= a\,(T_2 - T_1) + \frac{b}{2}(T_2^2 - T_1^2)$$

$$\therefore\quad 5400 = 20\,(T_2 - 300) + \frac{5\times10^{-3}}{2}(T_2^2 - 300^2)$$

$$= 20\,T_2 - 6000 + 2.5 \times 10^{-3}\,T_2^2 - 225$$

or $\quad 2.5\times10^{-3}\,T_2^2 + 20\,T_2 - 11625 = 0$

Solution of this quadratic equation gives

$$T_2 = \frac{-20 \pm \sqrt{20^2 - 4\times2.5\times10^{-3}\times(-11625)}}{2\times2.5\times10^{-3}}$$

$$= \frac{-20 \pm \sqrt{400 + 116.25}}{5\times10^{-3}} = \frac{-20 \pm 22.72}{5\times10^{-3}}$$

$= 544$ K or $- 8944$ K

The value $- 8844$ K is in admissible and therefore,

Final temperature of gas, $T_2 = 544$ K

(*c*) The entropy change can be calculated as

$$ds = \int\frac{dq}{T} = \int\frac{du + p\,dv}{T}$$

Here $dv = 0$ as the tank is rigid

$$\therefore \quad ds = \int \frac{du}{T} = \int \frac{c_v\, dT}{T}$$

$$= \int (a + bT) \frac{dT}{T}$$

$$= a \log_e \frac{T_2}{T_1} + b\,(T_2 - T_1)$$

$$= 20 \log_e \frac{544}{300} + 5 \times 10^{-3} (544 - 300)$$

$$= \mathbf{13.12}\ \text{J/mol K}$$

EXAMPLE 9.27

Air flows steadily through an insulated pipe and following measurements have been recorded of pressure and temperature at two stations which lie at sufficient distance apart.

	Station 1	*Station 2*
Pressure	*1.5 bar*	*1.0 bar*
Temperature	*350 K*	*290 K*

Considering air to be an ideal gas, determine the flow direction of air.

Solution : For steady flow process, the entropy of the fluid at exit has to be greater than or equal to its entropy at inlet. That is

$$s_{exit} \geq s_{inlet} \quad \text{or} \quad s_{exit} - s_{inlet} \geq 0$$

The following criterion can thus form the basis for determining the flow direction.

- If $s_{exit} - s_{inlet} > 0$; the process is irreversible and it proceeds in the forward direction.
- If $s_{exit} - s_{inlet} = 0$; the process is reversible and feasible.
- If $s_{exit} - s_{inlet} < 0$; the process cannot proceed in the forward direction. However, it is spontaneous in the reverse direction.

Let it be presumed that the air flows from station 1 to station 2. Then in terms of temperature and pressure changes, the entropy change is given by

$$s_2 - s_1 = c_p \log_e \frac{T_2}{T_1} + R \log_e \frac{p_1}{p_2}$$

$$= 1.005 \log_e \frac{290}{350} + 0.287 \log_e \frac{1.5}{1.0}$$

$$= -\ 0.1889 + 0.1164$$

$$= -\ 0.0725 \text{ kJ/K}$$

Since $s_2 < s_1$, the process cannot proceed from station 1 to station 2. Obviously the flow direction of air has to be from station 2 to station 1.

EXAMPLE 9.28

One inventor claims that when 2 kg of air is supplied to a magic tube at 4 bar and 20°C, two equal mass streams at 1 bar are produced; one at - 20°C and the other at 80°C.

Another inventor claims that it is possible to produce two equal mass streams; one at - 40°C and the other at 40°C.

Considering the system to be adiabatic, determine which claim is correct and why ? Assume ambient condition at 0°C.

Solution : For this irreversible process, the claim will be correct if the entropy of the system increases after the process.

Inventor 1 : Inlet condition = 2 kg of air at 4 bar and 20°C.

Total entropy at inlet,

$$S_1 = m\, c_p \log_e \frac{T}{T_0}$$

$$= 2 \times 1.005 \log_e \frac{293}{273}$$

$$= 0.1422 \text{ kJ/K}$$

Exit condition = Two equal mass streams, one at - 20°C and other at 80°C.

Total entropy at exit,

$$S_2 = m_1\, c_p \log_e \frac{T_1}{T_0} + m_2\, c_p \log_e \frac{T_2}{T_0}$$

$$= 1 \times 1.005 \log_e \frac{253}{273} + 1 \times 1.005 \log_e \frac{353}{273}$$

$$= -\ 0.0765 + 0.2582 = 0.1817 \text{ kJ/K}$$

Since $S_2 > S_1$, the claim is acceptable.

Inventor 2 : $S_1 = 0.1422$ kJ/k as calculated above

$$S_2 = 1 \times 1.005 \log_e \frac{233}{273} + 1 \times 1.005 \log_e \frac{313}{273}$$

$$= -\ 0.1592 + 0.1374 = -\ 0.0218 \text{ kJ/K}$$

Since $S_2 < S_1$, the claim is false as it violates the second law of thermodynamics.

EXAMPLE 9.29

In a particular chemical plant, air at 1 bar pressure and 475 K temperature needs to be supplied to a unit at a rate of 8 × 10^{-3} K mol/s. At another unit, air is available at the same flow rate and pressure but has a temperature of 1100 K. The plant engineer devices a heat engine that operates between the hot air as source and ambient air at 300 K as sink, and supplies the air at 475 K for use in the unit requiring that. Calculate the maximum power that can be delivered by the engine. Assume air to be an ideal gas with adiabatic exponent γ = 1.4.

Solution : The temperature of the hot air has to be brought down from temperature T_1 = 1100 K to T_2 = 475 K and the corresponding change in entropy is given by

$$(dS)_{source} = n\,c_p \log_e \frac{T_2}{T_1} + R \log_e \frac{p_1}{p_2}$$

$$= n\frac{\gamma}{\gamma-1} R_{mol} \frac{T_2}{T_1} \qquad (\because p_1 = p_2)$$

$$= 8\times10^{-3} \times \frac{1.4}{1.4-1} \times 8.314 \log_e \frac{475}{1100}$$

$$= -\ 0.1955 \text{ kJ/K}$$

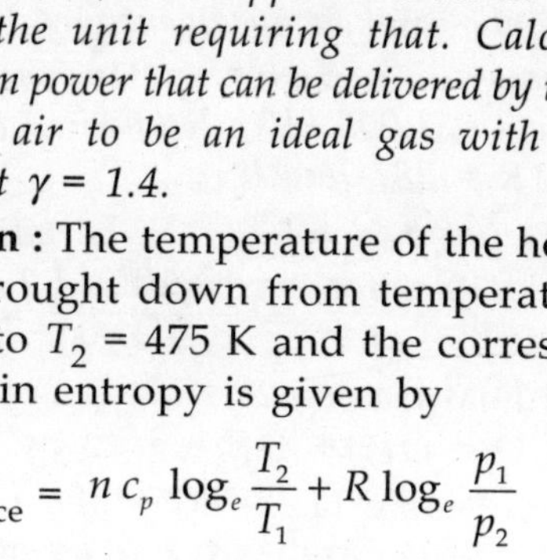

Fig. 9.24.

For any practical arrangement,

$$(ds)_{source} + (ds)_{sink} = 0$$

That gives

$$(ds)_{sink} = 0.1955 \text{ kJ/K}$$

∴ Heat rejected to sink

$$Q_2 = 300 \times 0.1955 = 58.65 \text{ kJ/s}$$

Heat supplied to engine

$$Q_1 = n\,c_p(T_1 - T_2)$$

$$= n\frac{\gamma}{\gamma-1} R_{mol}(T_1 - T_2)$$

$$= 8\times10^{-3}\times\frac{1.4}{1.4-1} \times 8.314(1100 - 475)$$

$$= 145.5 \text{ kJ/s}$$

∴ Power delivered by the engine,

$$W = Q_1 - Q_2$$

$$= 145.5 - 58.65$$

$$= 86.85 \text{ kJ/s} = \mathbf{86.85 \text{ kW}}$$

EXAMPLE 9.30

Show that the approximate change of entropy during a polytropic process equals the quantity of heat transferred divided by the mean absolute temperature.

One kg of air at 320 K is compressed in a cylinder according to the polytropic law $pv^{1.3}$ = const. If the compression ratio is 15, make calculations for the entropy change of air during the compression process stating whether it is an increase or decrease. What would be the percentage error if the entropy change is calculated by dividing the quantity of heat exchanged by the mean absolute temperature during the process ?

Take γ = 1.4 and c_v = 0.718 kJ/kg K

Solution : Refer Fig. 9.25. The curve 1–2 represents the polytropic process from state point 1 to state point 2. The area under the process curve 1–2 on $T-S$ diagram represents the heat transfer during the process. The curve 1–2 usually has a small slope and can be considered to be a straight line (shown dotted).

Heat interaction

= area of trapezium 1–2–3–4

= base × mean ordinate

$$= dS \times \frac{1}{2}(T_1 + T_2)$$

= entropy change × mean temperature during the process

Thus approximate change of entropy can be calculated by dividing the heat transfer by the mean absolute temperature during the process.

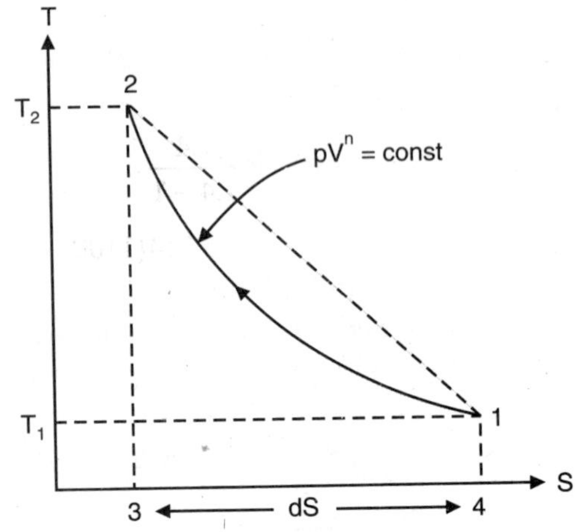

Fig. 9.25.

(*b*) From the polytropic law,

$$\frac{T_2}{T_1} = \left(\frac{v_1}{v_2}\right)^{n-1}$$

$$T_2 = 320 \times (15)^{1.3-1} = 721 \text{ K}$$

$$S_2 - S_1 = -mc_v \frac{\gamma - n}{n-1} \log_e \frac{T_2}{T_1}$$

$$= -1 \times 0.718 \left(\frac{1.4-1.3}{1.3-1}\right) \log_e \frac{721}{320}$$

$$= \mathbf{-\ 0.1944\ kJ/K}$$

The negative sign indicates decrease in entropy. Heat transferred during a polytropic process is given by

$$Q = mc_v \frac{\gamma - n}{n-1}(T_1 - T_2)$$

$$= 1 \times 0.718 \times \frac{1.4-1.3}{1.3-1}(320-721)$$

$$= -\ 95.97 \text{ kJ}$$

Average temperature during the process,

$$T_{av} = \frac{T_1 + T_2}{2} = \frac{320+721}{2} = 520.5 \text{ K}$$

Approximate change of entropy,

$$= \frac{Q}{T_{av}} = \frac{-95.97}{520.5} = -\ 0.1844 \text{ kJ/K}$$

∴ Percentage error

$$= \frac{0.1944 - 0.1844}{0.1944} = 0.0514 \text{ or } \mathbf{5.14\ \%}$$

The approximate value of entropy change is lower, because in the relation $Q = T_{av} \times dS$ actual value of heat transfer is substituted instead of approximate value (*i.e.*, area under the straight line) which is higher.

EXAMPLE 9.31

1 m^3 of air is heated reversibly at constant pressure from 290 K to 580 K, and is then cooled reversibly at constant volume back to initial temperature. If the initial pressure is 1 bar, work out the net heat flow and overall change in entropy. Represent the processes on T – S plot.

Take c_p = 1.005 kJ/kg K and
R = 287 J/kg K

Solution : Mass of air,

$$m = \frac{p_1 v_1}{RT_1} = \frac{1 \times 10^5 \times 1}{287 \times 290} = 1.20 \text{ kg}$$

Total heat transfer,

$$Q = Q_{1\text{-}2} + Q_{2\text{-}3}$$
$$= m\, c_p (T_2 - T_1) + m\, c_v (T_3 - T_2)$$
$$= m\, c_p (T_2 - T_1) + m\, c_v (T_1 - T_2)$$
$$= m\, (T_2 - T_1)\,(c_p - c_v)$$
$$= m\, R\, (T_2 - T_1)$$
$$= 1.20 \times 287 \times (580 - 290)$$
$$= 99876 \text{ J} \simeq 99.88 \text{ kJ}$$

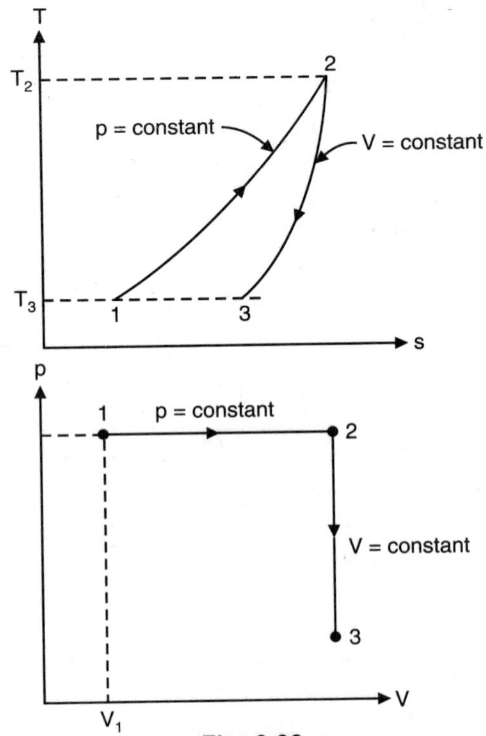

Fig. 9.26.

Entropy change during constant pressure process 1 – 2,

$$S_2 - S_1 = mc_p \log_e \frac{T_2}{T_1}$$

$$= 1.20 \times 1.005 \log_e \frac{580}{290}$$

$$= 0.8359 \text{ kJ/K}$$

Entropy change during constant volume process 2 – 3,

$$S_3 - S_2 = mc_v \log_e \frac{T_3}{T_2}$$

$$= m(c_p - R) \log_e \frac{T_1}{T_2}$$

$$= 1.20 \times (1.005 - 0.287) \log_e \frac{290}{580}$$

$$= -0.5972 \text{ kJ/K}$$

∴ Overall change in entropy

$$= (S_2 - S_1) + (S_3 - S_2)$$

$$= 0.8359 + (-0.5972)$$

$$= \mathbf{0.2387\ kJ/K}$$

EXAMPLE 9.32

A closed system contains air at pressure 1 bar, temperature 290 K and volume 0.02 m³. This system undergoes a thermodynamic cycle consisting of the following three processes in series:

Process 1 – 2 : Constant volume heat addition till pressure becomes 4 bar

Process 2 – 3 : Constant pressure cooling

Process 3 – 1 : Isothermal heating to initial state

Represent the cycle on T – s and p – v plots and evaluate the change in entropy for each process.

Take c_v = 0.718 kJ/kg K and R = 287 J/kg K

Solution : Refer Fig. 9.27.

$$m = \frac{p_1 V_1}{R T_1}$$

$$= \frac{1 \times 10^5 \times 0.02}{287 \times 290}$$

$$= 0.024 \text{ kg}$$

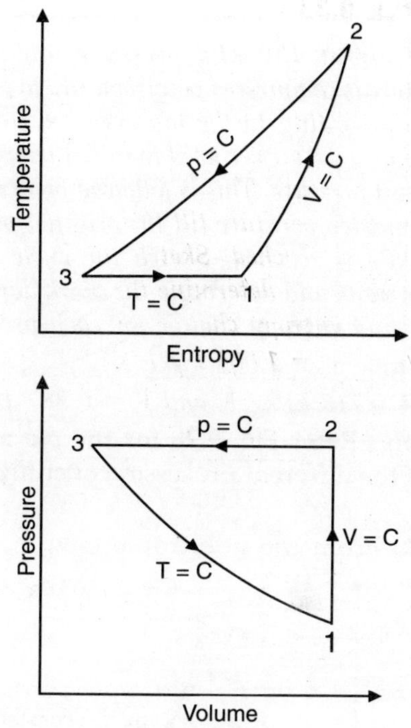

Fig. 9.27.

Process 1 – 2 :

$$T_2 = \frac{p_2}{p_1} T_1 = \frac{4}{1} \times 290 = 1160 \text{ K}$$

$$S_2 - S_1 = m\, c_v \log_e \frac{T_2}{T_1}$$

$$= 0.024 \times 0.718 \log_e \frac{1160}{290}$$

$$= \mathbf{0.02389\ kJ/K}$$

Process 2 – 3 :

$$T_3 = T_1 = 290 \text{ K}$$

$$S_3 - S_2 = m\, c_p \log_e \frac{T_3}{T_2} = m(c_v + R) \log_e \frac{T_1}{T_2}$$

$$= 0.024 \times (0.718 + 0.287) \log_e \frac{290}{1160}$$

$$= \mathbf{-0.0334\ kJ/K}$$

Process 3 – 1 : $p_3 = p_2 = 4$ bar

$$S_1 - S_3 = mR \log_e \frac{p_3}{p_1}$$

$$= 0.024 \times 0.287 \log_e \left(\frac{4}{1}\right)$$

$$= \mathbf{0.0095\ kJ/K}$$

EXAMPLE 9.33

3 kg of air at 150 kPa pressure and 360 K temperature is compressed polytropically to pressure 750 kPa according to the law $pV^{1.2}$ = constant. Subsequently the air is cooled to initial temperature at constant pressure. This is followed by expansion at constant temperature till the original pressure of 150 kPa is reached. Sketch the cycle on p-v and T-s plots and determine the work done, heat transfer and entropy change for each process.

Assume c_p = 1.005 kJ/kg K

c_v = 0.718 kJ/kg K and R = 0.287 J/kg K

Solution : Refer Fig. 9.28 for the *p-v* and *T-s* plots of the different processes constituting the cycle.

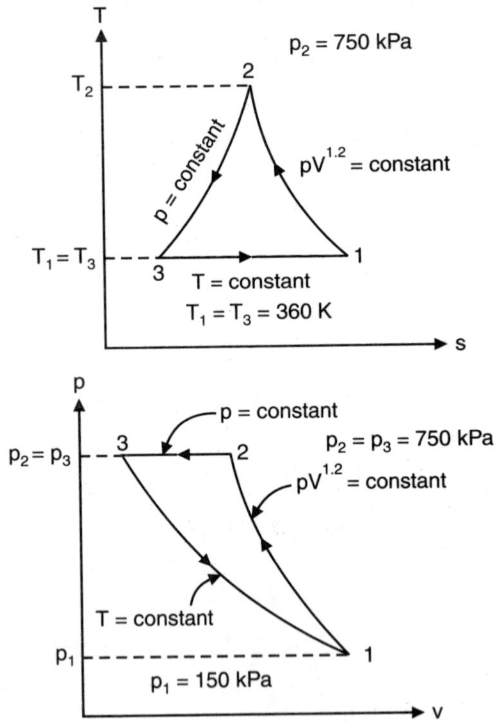

Fig. 9.28.

Process 1 - 2

$$\frac{T_2}{T_1} = \left(\frac{p_2}{p_1}\right)^{\frac{n-1}{n}}$$

$$T_2 = 360 \times \left(\frac{750}{150}\right)^{\frac{1.2-1}{1.2}} = 471 \text{ K}$$

Work done,

$$W_{1-2} = \frac{p_1V_1 - p_2V_2}{n-1} = \frac{mR\,(T_1 - T_2)}{n-1}$$

$$= \frac{3 \times 0.287\,(360 - 471)}{1.2 - 1}$$

$$= \mathbf{-\ 477.85}\ \text{kJ}$$

Heat transfer,

$$Q_{1-2} = W_{1-2} + dU_{1-2}$$
$$= W_{1-2} + m\ c_v(T_2 - T_1)$$
$$= -\ 477.85 + 3 \times 0.718(471 - 360)$$
$$= \mathbf{-\ 238.76}\ \text{kJ}$$

Entropy change,

$$S_2 - S_1 = m\,c_p \log_e \frac{T_2}{T_1} + mR \log_e \frac{p_1}{p_2}$$

$$= 3 \times 1.005 \log_e \frac{471}{360} + 3 \times 0.287 \log_e \frac{150}{750}$$

$$= 0.8095 + (-\ 1.3857) = -\ 0.5762 \text{ kJ/K}$$

Process 2 - 3

Work done,

$$W_{2-3} = p_2(V_3 - V_2)$$
$$= mR(T_3 - T_2)$$
$$= 3 \times 0.287\ (360 - 471)$$
$$= -\ 95.57 \text{ kJ}$$

Heat transfer,

$$Q_{2-3} = m\ c_p\ (T_3 - T_2)$$
$$= 3 \times 1.005\ (360 - 471)$$
$$= -\ 334.66 \text{ kJ}$$

Entropy change,

$$S_3 - S_2 = m\,c_p \log_e \frac{T_3}{T_2}$$

$$= 3 \times 1.005 \log_e \frac{360}{471}$$

$$= -\ 0.8104 \text{ kJ/K}$$

Process 3 - 1

Work done,

$$W_{3-1} = p_3V_3 \log_e \frac{V_1}{V_3} = m\,RT_3 \log_e \frac{p_3}{p_1}$$

$$= 3 \times 0.287 \times 360 \log_e \frac{750}{150}$$

$$= 498.86 \text{ kJ}$$

Heat transfer,

$$Q_{3-1} = W_{3-1} + (U_1 - U_3)$$
$$= W_{3-1} + m\,c_v\,(T_1 - T_3)$$
$$= W_{3-1} = 498.86 \text{ kJ } (\because T_1 = T_3)$$

Entropy change,

$$S_1 - S_3 = m\,R \log_e \frac{p_3}{p_1}$$

$$= 3 \times 0.287 \log_e \frac{750}{150}$$

$$= 1.3857 \text{ kJ/k}$$

Net change of entropy during the cycle,

$$\oint ds = (S_2 - S_1) + (S_3 - S_2) + (S_1 - S_3)$$

$$= -0.5762 - 0.8104 + 1.3857 = 0$$

This has to be because entropy remains constant for a reversible cyclic process.

EXAMPLE 9.34

Show that when a perfect gas changes from a state (p_1, V_1, T_1) to another state (p_2, V_2, T_2), the change in entropy per unit mass is given by

$$s_2 - s_1 = c_v \log_e \frac{p_2}{p_1} + c_p \log_e \frac{V_2}{V_1}$$

(b) Determine the value of index n so that the gain of entropy during the heating of gas at constant volume between temperatures T_1 and T_2 will be the same as that during an expansion according to the law pV^n = constant between the same temperatures. Further, proceed to show that the heat supplied per unit mass of the gas will be the same in each case.

Solution : For unit mass of the gas,

$$\delta q = du + p\,dv = c_v\,dT + p\,dv$$

$$\frac{\delta q}{T} = c_v \frac{dT}{T} + \frac{p}{T} dv$$

or $$ds = c_v \frac{dT}{T} + R\frac{dv}{v} \qquad (\because pv = RT)$$

Upon integration :

$$s_2 - s_1 = c_v \log_e \frac{T_2}{T_1} + R \log_e \frac{V_2}{V_1}$$

The above identity can be recast as

$$s_2 - s_1 = c_v \log_e \left(\frac{p_2 V_2}{p_1 V_1}\right) + R \log_e \frac{V_2}{V_1}$$

$$= c_v \log_e \frac{p_2}{p_1} + c_v \log_e \frac{V_2}{V_1} + R \log_e \frac{V_2}{V_1}$$

$$= c_v \log_e \frac{p_2}{p_1} + (c_v + R) \log_e \frac{V_2}{V_1}$$

$$= c_v \log_e \frac{p_2}{p_1} + c_p \log_e \frac{V_2}{V_1}$$

which is the required expession for entropy change in terms of pressure and volume.

(*b*) Entropy change during constant volume process

$$= c_v \log_e \frac{T_2}{T_1}$$

Entropy change during polytropic (pV^n = const) process,

$$= c_v \frac{\gamma - n}{n-1} \log_e \frac{T_2}{T_1}$$

For the same entropy change during these processes, we have

$$c_v \log_e \frac{T_2}{T_1} = c_v \frac{\gamma - n}{n-1} \log_e \frac{T_2}{T_1}$$

or $$1 = \frac{\gamma - n}{n-1}$$

That gives: $n = \dfrac{\gamma + 1}{2}$

(*c*) For a closed system, the principle of energy conservation gives

$$\delta q = \delta w + du$$

For the constant volume process, $\delta w = 0$

$$\therefore \qquad \delta q = du = c_v\,(T_2 - T_1)$$

For the polytropic (pv^n) = constant), process

$$\delta q = c_v\,(T_2 - T_1) + \frac{p_1 v_1 - p_2 v_2}{n-1}$$

$$= c_v\,(T_2 - T_1) + \frac{R\,(T_1 - T_2)}{n-1}$$

$$= c_v\,(T_2 - T_1) + (c_p - c_v)\frac{T_1 - T_2}{n-1}$$

$$= (T_2 - T_1)\left[c_v - \frac{c_p - c_v}{n-1}\right]$$

Since $c_p = \gamma\, c_v$ and $n = \dfrac{\gamma+1}{2}$, we get

$$\delta q = (T_2 - T_1)\left[c_v - \frac{\gamma c_v - c_v}{\dfrac{\gamma+1}{2} - 1}\right]$$

$$= (T_2 - T_1)\left[c_v - \frac{2c_v(\gamma - 1)}{\gamma - 1}\right]$$

$$= -\, c_v\,(T_2 - T_1)$$

Obviously the magnitude of heat interaction in the given processes is same. These processes have been depicted in Fig. 9.29 on p-v and T-s coordinates.

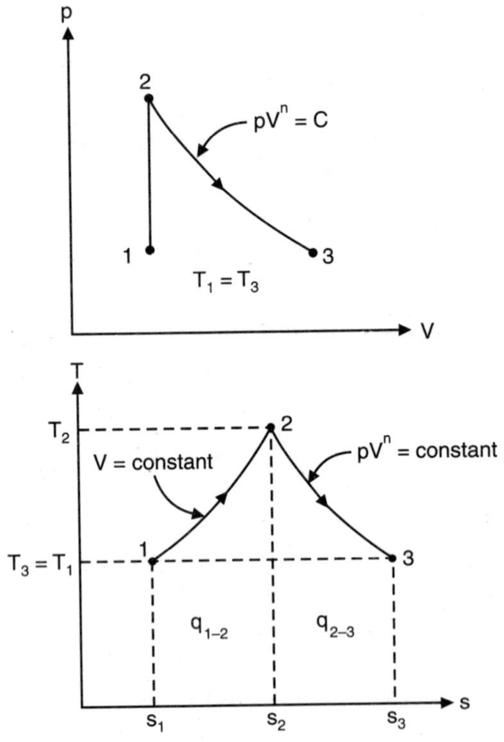

Fig. 9.29.

EXAMPLE 9.35

One kg of water at 273 K is heated to 373 K by bringing it into contact with a heat source at 373 K. Determine the entropy change of water, the entropy change of heat source, and the entropy change of universe.

(b) What will be the entropy change of the universe if the heating of the water is done from 273 to 373 K by first bringing it in contact with a heat source at 323 K and then with a heat source at 373 K ?

(c) Is it possible to heat the water from 273 K to 373 K with almost no change in the entropy of the universe ? Discuss.

Solution : Refer Fig. 9.30 and assume a reversible path between the end states which are in equilibrium.

Entropy change of water,

$$(ds)_{water} = \int mc\,\frac{dT}{T} = mc\log_e\frac{T_2}{T_1}$$

$$= 1 \times 4.186 \times \log_e\frac{373}{273} = 1.3055 \text{ kJ/K}$$

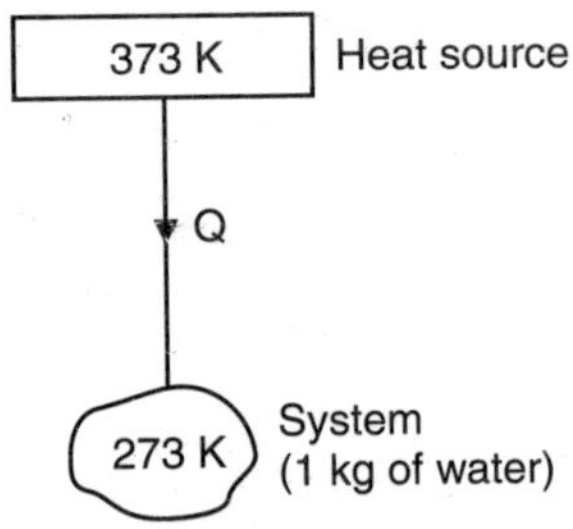

Fig. 9.30.

The temperature of the heat source remains constant irrespective of the heat withdrawn from it. The heat withdrawn from the source equals the heat taken by water to raise its temperature from 273 K to 373 K. That is

$$Q = mc\,dT$$
$$= 1 \times 4.186\,(373 - 273) = 418.6 \text{ kJ}$$

Entropy change of the source,

$$(ds)_{source} = \frac{-Q}{T} = \frac{-418.6}{373} = -\,1.122 \text{ kJ/kg}$$

Then : net entropy change (entropy of the universe),

$$(ds)_{universe} = (ds)_{water} + (ds)_{source}$$
$$= 1.3055 + (-\,1.122)$$
$$= 0.1835 \text{ kJ/K}$$

(*b*) The heating of water is being carried out in two stages : first from 273 to 323 K by bringing it into contact with a heat source at 323 K and then from 323 to 373 K by bringing it contact with a heat source at 373 K.

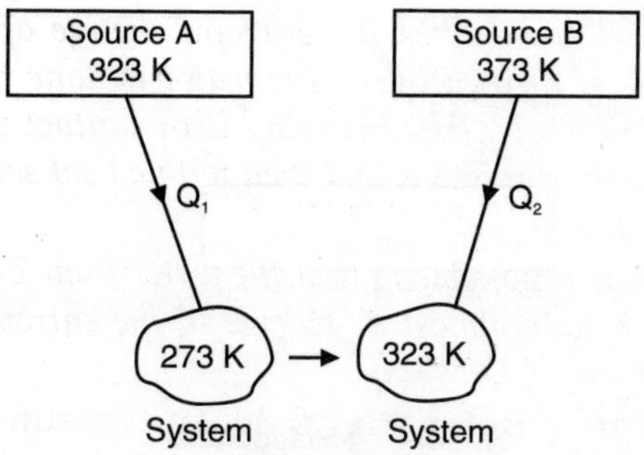

Fig. 9.31.

$$(ds)_{water} = \int_{273}^{323} mc \frac{dT}{T} + \int_{323}^{373} mc \frac{dT}{T}$$

$$= mc\left[\log_e \frac{323}{273} + \log_e \frac{373}{323}\right]$$

$$= 1 \times 4.186\ (0.1680 + 0.1439)$$

$$= 1.3056 \text{ kJ/K}$$

The heat withdrawn from a heat source is given by $mc\ dT$. Then

$$(ds)_{source\ A} = -\frac{1 \times 4.186\ (323 - 273)}{323}$$

$$= -\ 0.648 \text{ kJ/K}$$

$$(ds)_{source\ B} = -\frac{1 \times 4.186\ (373 - 323)}{373}$$

$$= -\ 0.561 \text{ kJ/K}$$

$$\therefore (ds)_{universe} = 1.3056 + (-\ 0.648) + (-\ 0.561)$$

$$= \mathbf{0.0966} \text{ kJ/kg}$$

(*c*) There will be almost no change in the entropy of the universe if the water is heated reversibly. That is possible if the water is heated in infinite steps by bringing it in contact with infinite number of heat sources in succession. Then at any instant, the temperature difference between the water and the heat source in contact will be infinitesimally small and the heating will be done reversibly.

EXAMPLE 9.36

One kg of ice at – 4°C is exposed to the atmosphere which is at 20°C. The ice melts and comes into equilibrium with the atmosphere. Determine the entropy increase of the universe. For ice, take c_p = 2.09 kJ/kg K and latent heat of fusion = 333 kJ/kg

Proceed to calculate the minimum amount of work necessary to convert the water back into ice at – 4°C.

Solution : Heat absorbed by ice from the atmosphere

= heat absorbed in solid phase + latent heat + heat absorbed in liquid phase

$$= 1 \times 2.09\ [0 - (-\ 4)] + 333 + 1 \times 4.186 \times (20 - 0)$$

$$= 8.36 + 333 + 83.72 = 425.08 \text{ kJ}$$

Entropy change of atmosphere

$$(ds)_{atmosphere} = \frac{-Q}{T}$$

$$= \frac{-\ 425.08}{293} = -\ 1.4508 \text{ kJ/K}$$

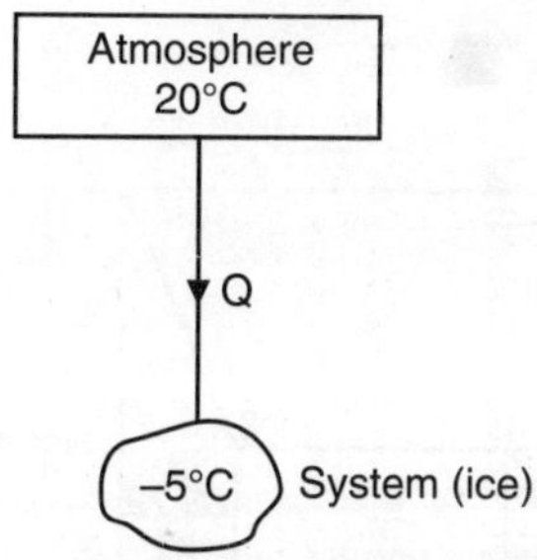

Fig. 9.32.

The ice melts in phases and so will be change in its entropy.

Entropy change of ice as it gets heated from – 4°C to 0°C

$$(ds)_{sys} = \int_{269}^{273} mc_p \frac{dT}{T}$$

$$= 1 \times 2.09 \log_e \frac{273}{269}$$

$$= 0.03085 \text{ kJ/K}$$

Entropy change of ice as it melts at 0°C to become water at 0°C

$$(ds_2)_{sys} = \frac{333}{273} = 1.2198 \text{ kJ/K}$$

Entropy change of water as it gets heated from 0°C to 20°C

$$(ds_3)_{sys} = \int_{273}^{293} m\, c_p \frac{dT}{T}$$

$$= 1 \times 4.186 \log_e \frac{293}{273}$$
$$= 0.2961 \text{ kJ/K}$$

Total entropy change of ice as it melts into water

$$(ds)_{sys} = ds_1 + ds_2 + ds_3$$
$$= 0.03085 + 1.2198 + 0.2961$$
$$= 1.5467 \text{ kJ/K}$$

Entropy change of universe,

$$(ds)_{uni} = (ds)_{atm} + (ds)_{sys}$$
$$= -1.4508 + 1.5467$$
$$= \mathbf{0.0959} \text{ kJ/K (increase)}$$

The temperature-entropy diagram for the system as ice at –4°C melts and subsequently gets converted to water at 20°C is shown in Fig. 9.33.

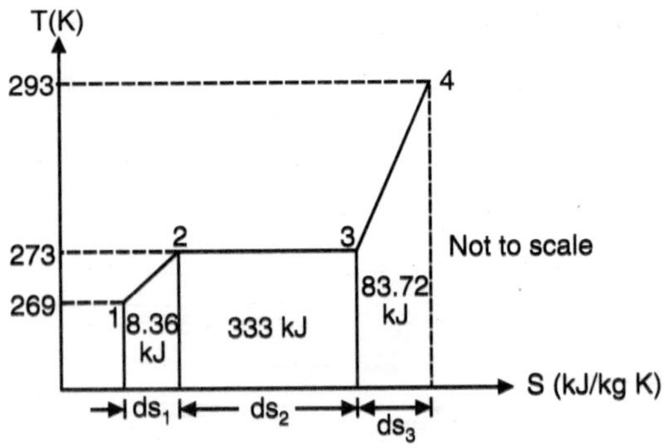

Fig. 9.33.

(*b*) The water at 20°C can be converted into ice at –4°C by removing 425.08 kJ of heat from it. The task can be accomplished by a refrigerator that brings the system back from state point 4 to state point 1. The entropy change for the system will remain same. However, its sign will be negative as the heat is removed from the system.

The refrigerator operates in a cycle and as such the entropy change of the working fluid in the refrigerator will be zero. That is

$$(ds)_{ref} = 0$$

Entropy change of the atmosphere

$$(ds)_{atm} = \frac{Q+W}{T}$$

∴ Net entropy change (entropy change of the universe)

$$(ds)_{uni} = (ds)_{sys} + (ds)_{ref} + (ds)_{atm}$$
$$= (s_1 - s_4) + 0 + \frac{Q+W}{T}$$

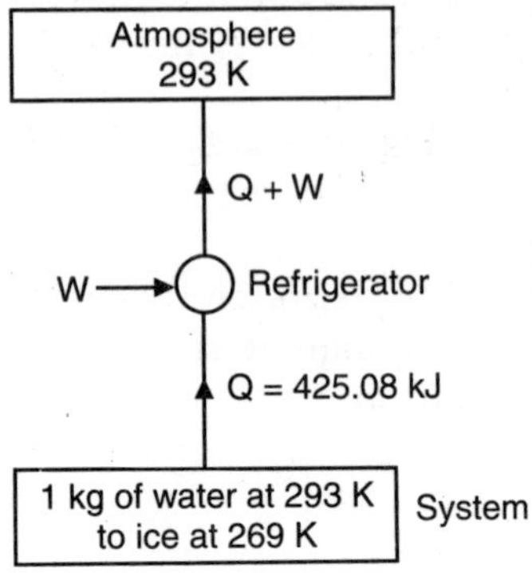

Fig. 9.34.

For a possible working arrangement,

$$(ds)_{uni} \geq 0$$

$$\therefore \left[(s_1 - s_4) + 0 + \frac{Q+W}{T}\right] \geq 0$$

For minimum work input to refrigerator

$$s_1 - s_4 + \frac{Q + W_{min}}{T} = 0$$

$$\text{or } W_{min} = T(s_4 - s_1) - Q$$
$$= 293\ (1.5467) - 425.08$$
$$= \mathbf{28.10} \text{ kJ}$$

9.7. ENTROPY CHANGE WITH VARIABLE SPECIFIC HEAT

Consider that the specific heats of a gas vary with temperature according to the linear relation $c_p = a + kT$ and $c_v = b + kT$ where a, b and k are constants and T is the temperature in degree Kelvin.

For unit mass of gas:

$$\delta q = T\,ds = du + p\,dv$$
$$= c_v\,dT + p\,dv$$

$$\text{or} \quad ds = c_v \frac{dT}{T} + \frac{p\,dv}{T} = c_v \frac{dT}{T} + R\frac{dv}{v}$$
$$(\because pv = RT)$$

Since, $R = c_p - c_v$

$$= (a + kT) - (b + kT) = a - b$$

we may write,

$$ds = (b + kT)\frac{dT}{T} + (a - b)\frac{dv}{v}$$

This represents the differential form of entropy change. Upon integration between the initial (1) and final (2) limits,

$$s_2 - s_1 = b \log_e \frac{T_2}{T_1} + k(T_2 - T_1) + (a - b)\log_e \frac{v_2}{v_1} \quad ...(9.18)$$

Following expressions can be set up for the entropy change by suitable manipulations to this equation:

- $\dfrac{T_2}{T_1} = \left(\dfrac{v_1}{v_2}\right)^{n-1}$

$$\log_e \frac{T_2}{T_1} = (n-1)\log_e \frac{v_1}{v_2} = -(n-1)\log_e \frac{v_2}{v_1}$$

$$\therefore s_2 - s_1 = b \log_e \frac{T_2}{T_1} + k(T_2 - T_1) - \frac{a-b}{n-1}\log_e \frac{T_2}{T_1}$$

$$= \left(b - \frac{a-b}{n-1}\right)\log_e \frac{T_2}{T_1} + k(T_2 - T_1) \quad ...(9.19)$$

This is an expression for entropy change in terms of temperature only

- Entropy change $s_2 - s_1 =$

$$b \log_e \frac{T_2}{T_1} + k(T_2 - T_1) + a \log_e \frac{v_2}{v_1} - b \log_e \frac{v_2}{v_1}$$

$$= a \log_e \frac{v_2}{v_1} + b \log_e \left(\frac{T_2}{T_1} \times \frac{v_1}{v_2}\right) + k(T_2 - T_1)$$

$$= a \log_e \frac{v_2}{v_1} + b \log_e \frac{p_2}{p_1} + k(T_2 - T_1) \quad ...(9.20)$$

This shows entropy change in terms of pressure, volume and temperature.

- Entropy change $s_2 - s_1 =$

$$b \log_e \frac{T_2}{T_1} + k(T_2 - T_1) + a \log_e \frac{v_2}{v_1} - b \log_e \frac{v_2}{v_1}$$

$$= a \log_e \left(\frac{T_2}{T_1} \times \frac{p_1}{p_2}\right) + b \log_e \left(\frac{T_2}{T_1} \times \frac{v_1}{v_2}\right) + k(T_2 - T_1)$$

$$= a \log_e \frac{T_2}{T_1} - a \log_e \frac{p_2}{p_1} + b \log_e \frac{p_2}{p_1} + k(T_2 - T_1)$$

$$= a \log_e \frac{T_2}{T_1} + (b - a)\log_e \frac{p_2}{p_1} + k(T_2 - T_1) \quad ...(9.21)$$

which is an expression for entropy change in terms of pressure and temperature only.

- Setting up the formula $T^b v^{a-b} e^{kT} =$ constant for the adiabatic expansion of gas.

$$ds = (a - b)\frac{dv}{v} + (b + kT)\frac{dT}{T}$$

$$s_2 - s_1 = a \log_e v - b \log_e v + b \log_e T + kT$$

$$= a \log_e v + b \log_e \frac{T}{v} + kT$$

$$= a \log_e v + b \log_e \frac{p}{R} + kT$$

$$= \log_e v^a + \log_e \left(\frac{p}{a-b}\right)^b + kT$$

For adiabatic expansion, $s_2 - s_1 = 0$

That gives : $v^a p^b e^{kT}$ = constant

$p\, v^{a-b} e^{kT}$ = constant

and $T^b v^{a-b} e^{kT}$ = constant

The above expressions can be set up by taking kT on the right side and taking antilog of the resulting expression.

EXAMPLE 9.37

Find the entropy change of 5 kg of a perfect gas whose temperature varies from 150°C to 200°C during a constant volume process. The specific heat varies linearly with absolute temperature and is presented by the relation :

c_v = (0.45 + 0.009 T) kJ/kg K.

Solution : Entropy variation for a constant volume process is given by

$$dS = mc_v \frac{dT}{T}$$

$$= 5 \times (0.45 + 0.009\,T)\frac{dT}{T}$$

Upon integration :

$$s_2 - s_1 = 5 \times 0.45 \int_{T_1}^{T_2} \frac{dT}{T} + 5 \times 0.009 \int_{T_1}^{T_2} dT$$

$$= 2.25 \log_e \left(\frac{T_2}{T_1}\right) + 0.045\,(T_2 - T_1)$$

Given : T_1 = (150 + 273) = 423 K

and T_2 = (200 + 273) = 473 K

$$\therefore \; s_2 - s_1 = 2.25 \log_e \frac{473}{423} + 0.045(473 - 423)$$
$$= 0.2514 + 2.25 = \mathbf{2.5014 \text{ kJ/K}}$$

EXAMPLE 9.38

Determine the entropy of a gas at 30 bar pressure and 800 K temperature. Given that entropy of the gas at 1 bar pressure and 273 K is zero, and the specific heats of the gas vary linearly with absolute temperature according to the following correlations

c_p = *(0.94 + 0.0002 T) kJ/kg K and*

c_v = *(0.65 + 0.0002 T) kJ/kg K*

Solution : From the relation $Tds = du + p\,dv$

$$ds = c_v \frac{dT}{T} + \frac{p}{T}dv = c_v \frac{dT}{T} + R\frac{dv}{v}$$

Upon integration:

$$s_2 - s_1 = \int c_v \frac{dT}{T} + R\log_e \frac{v_2}{v_1}$$
$$= \int c_v \frac{dT}{T} + (c_p - c_v)\log_e \frac{p_1}{p_2} \times \frac{T_2}{T_1}$$
$$= \int \left(\frac{0.65}{T} + 0.0002\right) dT + 0.29\log_e \frac{1}{30} \times \frac{800}{273}$$
$$= 0.65 \log_e \frac{800}{273} + 0.0002(800 - 273)$$
$$+ 0.29 \log_e \frac{1}{30} \times \frac{800}{273}$$
$$= 0.6988 + 0.1054 - 0.6746$$
$$= \mathbf{0.1296 \text{ kJ/kg K}}$$

EXAMPLE 9.39

An insulated vessel of 0.6 m^3 *capacity is divided by a rigid conducting diaphragm into two chambers A and B, each having a capacity of 0.3 m*^{3}*. Chamber A contains air at 1.5 bar pressure and 300 K temperature and the corresponding parameters for air in chamber B are 4.5 bar and 450 K. Calculate: (i) final equilibrium temperature, (ii) final pressure on each side of the diaphragm, and (iii) entropy change of the system.*

For air take c_v = *0.715 kJ/kg K*

Solution : Mass of air,

$$m_a = \frac{p_a v_a}{RT_a}$$
$$= \frac{1.5 \times 10^5 \times 0.3}{287 \times 300} = 0.5226 \text{ kg}$$
$$m_b = \frac{p_b v_b}{RT_b}$$
$$= \frac{4.5 \times 10^5 \times 0.3}{287 \times 450} = 1.0453 \text{ kg}$$

(*i*) Let T_f be the final equilibrium temperature. Since the diaphragm is conducting heat gained by air in chamber A = heat lost by air in chamber B

$$0.5226 \times c_v (T_f - 300) = 1.0453 \times c_v (450 - T_f)$$

Solution gives: $T_f \simeq$ **400 K**

(*ii*) Since the volume of each chamber remains constant, change in pressure will be proportional to absolute temperature (Charles' law)

$$p_a = \frac{1.5 \times 400}{300} = \mathbf{2 \text{ bar}}$$
$$p_b = \frac{4.5 \times 400}{450} = \mathbf{4 \text{ bar}}$$

(*iii*) Entropy change for chamber

$$A = m_a c_v \log_e \frac{T_f}{T_1}$$
$$= 0.5226 \times 0.715 \log_e \frac{400}{300}$$
$$= 0.1075 \text{ kJ/K}$$

Entropy change for chamber

$$B = m_b c_v \log_e \frac{T_f}{T_1}$$
$$= 1.0453 \times 0.715 \log_e \frac{400}{450}$$
$$= -0.0880 \text{ kJ/K}$$

Net change of entropy

$$= 0.1075 - 0.088$$
$$= \mathbf{0.0195 \text{ kJ/K}}$$

SALIENT POINTS

1. For any reversible process, the cyclic integral of the quantity $\delta Q/T$ is zero (Clausius theorem). This suggests that the quantity $\delta Q/T$ is a point function and hence a property of the system. This property is called entropy.
2. Entropy is an extensive property and has units of kJ/K.
3. Entropy of a system increases with addition of heat and decreases with rejection of heat.
4. Entropy is a measure of the degree of randomness of the molecules comprising the system. Higher the disorder, greater is the increase in entropy.
5. **Entropy principle:** Entropy of an isolated system either increases or, in the limit, remains constant.
6. **Entropy changes for a closed system (per kg) :**

   (*i*) *General case:*

   (*a*) $s_2 - s_1 = c_v \log_e \frac{T_2}{T_1} + R \log_e \frac{v_2}{v_1}$

   (in terms of T and v)

   (*b*) $s_2 - s_1 = c_v \log_e \frac{p_2}{p_1} + c_p \log_e \frac{v_2}{v_1}$

   (in terms of p and v)

   (*c*) $s_2 - s_1 = c_p \log_e \frac{T_2}{T_1} + R \log_e \frac{p_1}{p_2}$

   (in terms of p and T)

   (*ii*) Constant volume process :

   $$s_2 - s_1 = c_v \log_e \frac{T_2}{T_1} = c_v \log_e \frac{p_2}{p_1}$$

   (*iii*) Constant pressure process :

   $$s_2 - s_1 = c_p \log_e \frac{T_2}{T_1} = c_p \log_e \frac{v_2}{v_1}$$

   (*iv*) Isothermal process :

   $$s_2 - s_1 = R \log_e \frac{v_2}{v_1} = R \log_e \frac{p_1}{p_2}$$

   (*v*) Adiabatic process : $s_2 - s_1 = 0$

   (*vi*) Polytropic process :

   $$s_2 - s_1 = \frac{n - \gamma}{n - 1} c_v \log_e \frac{T_2}{T_1}$$

7. Constant volume lines on temperature-entropy plot are steeper than the constant pressure lines.
8. Two reversible adiabatic paths cannot intersect each other.
9. The reversible adiabatic process during which entropy remains constant is called isentropic process. However, an isentropic process need not be an adiabatic process as the entropy during a particular process may be kept constant by heat transfer to or from the system.

REVIEW QUESTIONS

A. Conceptual and conventional questions.

1. Under what circumstances does $dS = \frac{dQ}{T}$?
2. Name and define the property that remains constant during a reversible adiabatic process.
3. What are the two requirements for a process to be isentropic ?
4. What expression for heat transfer is equivalent to the term $p\,dv$ for work transfer ?
5. An analysis of a process shows an entropy production of – 0.025 kJ/kg K. What comment can you make on this result ?
6. Comment on the statement: "The entropy of the universe tends to be maximum."
7. Entropy is defined in terms of a reversible process. How can then it be evaluated for an irreversible process ?
8. A system is taken from state 1 to state 2 by following two different processes, a reversible process R and an irreversible process I. Is entropy change of the system same for both processes ? Explain.
9. What does the principle of entropy increase, specify ?

10. Show that whenever a system executes a complete cyclic process, $\oint \frac{\delta Q}{T}$ is less than zero or in the limit is equal to zero. Hence prove that entropy is a property of the system.
11. "There exists a property of a closed system such that a change in its value is equal to $\oint_1^2 \frac{\delta Q}{T}$ for any irreversible process undergone by the system between state 1 and state 2." Prove this statement.
12. How do the values of the integral $\oint_1^2 \frac{\delta Q}{T}$ compare for a reversible process and an irreversible process between the end states ?
13. Define entropy and show that for an irreversible process

$$\int dS > \int \frac{\delta Q}{T}$$

14. Prove that heat absorbed during a process is approximately equal to the change in entropy multiplied by the mean absolute temperature during the process.
15. Sketch the Carnot cycle on the *T-S* diagram, and derive an expression for its efficiency.
16. A closed system undergoes the internally reversible process from temperature T_1 = 200 K, S_1 = 1 kJ/K to temperature T_2 = 600 K and S_2 = 3 kJ/K. What would be the heat interaction during the process if variation is assumed to be linear.

[Ans. 800 kJ]

17. A fluid transfers 2000 kJ of heat to the environment at 300 K. If the entropy change of the fluid is - 5 kJ/K determine the overall entropy change and comment on the nature of the process whether possible, reversible, irreversible.

[Ans. 1.67 kJ/K, possible and irreversible]

18. An electric motor of 5 kW is subjected to a braking test for one hour. The heat generated by frictional forces in the process gets dissipated to the surroundings at 300 K. Determine the resulting increase in entropy.

[Ans. 60 kJ/K]

19. A gear box receives 750 kW power and delivers 745 kW power when operating in a steady state at 325 K. If the surroundings are at 280 K, determine the rate of entropy growth in the gear box and in the surroundings.

[Ans. 0.923 kJ/K-min, 1.0714 kJ/K-min]

20. Air in a cylinder is compressed reversibly and isothermally from 95 kPa, 25°C to 290 kPa. The initial volume is 0.162 m^3. Determine the work input, the heat transfer, and the entropy change of the air being compressed.

[Ans. - 17.2 kJ, - 17.2 kJ, - 0.0576 kJ/K]

21. One kg mole of nitrogen is contained in a rigid vessel of 2.5 m^3 volume at 375 K temperature. Evaluate the change in entropy if the gas cools to 300 K. Take γ = 1.4.

[Ans. 4.637 kJ/K decrease]

22. For a non-flow process, show that entropy change per kg of gas from state point 1 to state point 2 is given by

$$s_2 - s_1 = c_p \log_e \frac{T_2}{T_1} - R \log_e \frac{p_2}{p_1}$$

where the symbols *s*, *T* and *p* represent entropy, temperature and pressure respectively.
23. Show that when a perfect gas changes from state point p_1, v_1, T_1 to another state p_2, v_2, T_2, the increase in entropy per unit mass is given by

$$s_2 - s_1 = c_v \log_e \frac{p_2}{p_1} + c_p \log_e \frac{v_2}{v_1}$$

24. Set up the following expression for change of entropy per kg of a gas during a polytropic expansion process

$$s_2 - s_1 = c_v \frac{\gamma - n}{n-1} \log_e \frac{T_1}{T_2} = c_v \frac{\gamma - n}{n} \log_e \frac{p_1}{p_2}$$

where the symbols have the usual meanings.
25. For a limited temperature range, the specific heats of an ideal gas vary linearly with temperature as follows

$$c_p = a + kT \text{ and } c_v = b + kT$$

where a, b and k are constants. Derive an expression for the change of specific entropy of this gas in terms of initial and final pressures and temperatures.
26. The specific heats of an ideal gas vary linearly with temperature such that

$$c_p = a + kT \text{ and } c_v = b + kT$$

Show that an isentropic process for this gas will have the equation

$$T^b v^{(a-b)} e^{kT} = \text{constant}$$

27. 0.5 m^3 of air at 200°C and 6 bar expands to 1 bar according to the law (*i*) pV = constant (*ii*) $pv^{1.2}$ = constant and (*iii*) adiabatically. Make

calculations for the change in entropy for each of the process.

Take $\gamma = 1.4$ and $c_v = 0.715$ kJ/kg K.

[Ans. 1.137 kJ/K; 0.4723 kJ/K, zero]

28. 0.2 kg of air at 1.5 bar pressure and 300 K temperature is compressed to 15 bar pressure according to the polytropic law $pv^{1.25}$ = const. Make calculations for (*i*) initial and final parameters, (*ii*) work done on or by the air, (*iii*) heat flow to or from the air and (*iv*) change of entropy stating whether it is an increase or decrease.

Take $R = 0.287$ kJ/kg K and $c_v = 0.718$ kJ/kg K

[Ans. $v_1 = 0.1148$ m^3, $v_2 = 0.0182$ m^3, $T_2 = 475.5$ K, $W = -40.25$ kJ, $Q = -15$ kJ and $dS = -0.04$ kJ/K]

29. One kg of an ideal gas of molecular mass 30 is compressed polytropically according to the law $pV^{1.25}$ = constant from a pressure of 1 bar and 60°C to a pressure of 16 bar. Make calculations for (*i*) the temperature at the end of compression, (*ii*) the work done on or by the gas, (*iii*) the heat transfer to or from the gas, and (*iv*) the entropy change during the process. Take $\gamma = 1.38$ and use only calculated values of specific heats.

[Ans. 580 K, – 273.5 kJ, – 93.5 kJ, 0.2102 kJ/K]

B. Fill in the blanks with appropriate word/words.

1. For a reversible process the quantity $ds = \frac{\delta Q}{T}$ is .... of the path of process and represents a .... of the system.
2. In the identity $ds \geq \frac{\delta Q}{T}$, the equality sign stands for a .... process and inequality sign stands for .... process.
3. Entropy of an isolated system either .... or ....
4. For any reversible process, the change in entropy of the .... and .... is zero.
5. The entropy of the system does not change in a ....
6. The increase in entropy is small when heat is added at .... temperature and is greater when addition of heat is made at .... temperture.
7. The slope of constant volume line on *T-s* diagram is .... than that of constant pressure line working between the same temperature limits.
8. When a system is at zero absolute temperature, the entropy of a system is zero. This is the essence of ....

Answers

1. independent, property 2. reversible, irreversible 3. increases, remains constant 4. system, surroundings 5. reversible cycle 6. high, low 7. more 8. third law of thermodynamics

C. Multiple choice questions.

1. In statistical thermodynamics, entropy is defined as
   (*a*) reversible heat transfer
   (*b*) measure of reversibility of a system
   (*c*) degree of randomness
   (*d*) a universal property
2. Zeroth law : temperature :: second law : ....
   (*a*) efficiency (*b*) enthalpy
   (*c*) internal energy (*d*) entropy
3. All of the following statements are correct, except
   (*a*) entropy is a measure of the level of irreversibility associated with a process
   (*b*) during a reversible process, the total entropy of the interacting bodies remains constant
   (*c*) entropy of the universe is continuously decreasing
   (*d*) entropy is a property like pressure, temperature, volume and so on and as such it is a point function
4. Choose the incorrect statement
   (*a*) calculations can be made for the entropy change and not for the absolute value of entropy
   (*b*) the freezing of water is accompanied by an increase in entropy
   (*c*) the amount of heat transfer is represented by the area under a reversible process on a temperature-entropy diagram
   (*d*) the entropy increase is proportional to energy degradation.
5. A system undergoes a state change from 1 to 2. The second law then tells that for the process to be feasible, the entropy change $(s_2 - s_1)$ of the system

(*a*) is positive or zero
(*b*) is negative or zero
(*c*) is zero
(*d*) can be +ve, –ve or zero

6. The relation $ds = \delta Q/T$ for entropy change during a process is valid for
   (*a*) all real processes
   (*b*) only for reversible processes
   (*c*) only for irreversible processes
   (*d*) both for reversible and irreversible processes

7. When a system undergoes a process such that $\int \delta Q/T = 0$ and $\Delta S > 0$, the process is
   (*a*) irreversible adiabatic
   (*b*) reversible adiabatic
   (*c*) isothermal (*d*) isobaric

8. A perfect gas enclosed in a piston-cylinder arrangement executes a reversible adiabatic expansion process. Consider the following statements :
   1. entropy will increase
   2. entropy change will be zero
   3. entropy change of the surroundings will be zero.

   Which of these statements are correct?
   (*a*) 1 and 3 (*b*) 2 alone
   (*c*) 2 and 3 (*d*) 1 alone

9. Which of the followings is a correct statement?
   (*a*) a reversible adiabatic process is an isentropic process
   (*b*) an irreversible adiabatic process is a constant entropy process
   (*c*) entropy decreases during an irreversible adiabatic process
   (*d*) an isentropic process is an adiabatic process

10. The entropy of universe tends to
    (*a*) become zero
    (*b*) remain constant
    (*c*) be maximum
    (*d*) attain a certain finite minimum value

11. For any natural process, the entropy change would be
    (*a*) zero
    (*b*) positive
    (*c*) negative
    (*d*) unpredictable; insufficient data

12. For a fixed quantity of heat added to a system under specific conditions, the entropy change will be maximum in case of
    (*a*) isentropic process (*b*) isobaric process
    (*c*) isothermal process (*d*) isochoric process

13. A system undergoes a change of state $A \rightarrow B$. Then
    1. $S_A = S_B$
    2. $S_B - S_A$ is positive
    3. $S_B - S_A$ is negative
    4. Unpredictable as the given information is incomplete

    Which of the identities given above is/are feasible?
    (*a*) 1 and 2 (*b*) 2 and 3
    (*c*) 1 and 3 (*d*) 4 only

14. Consider an ideal gas enclosed in a piston cylinder arrangement. If the gas is made to undergo reversible adiabatic expansion, then which of the following identities is correct?
    (*a*) the entropy of the gas system increases
    (*b*) the entropy of the surroundings decreases
    (*c*) there is no change in the entropy of the gas system
    (*d*) $(\Delta S)_{sys} + (\Delta S)_{surr} = 0$

15. A system of 100 kg mass undergoes a change of state where upon there occurs a change in its specific entropy from 0.5 kJ/kg K to 0.6 kJ/kg K. Simultaneously there is a decrease in the entropy of its surroundings from 80 kJ/K to 75 kJ/K. The data suggests that the process is
    (*a*) impossible (*b*) reversible
    (*c*) irreversible (*d*) adiabatic

Answers

1. (*c*)	2. (*d*)	3. (*c*)	4. (*b*)	5. (*a*)
6. (*b*)	7. (*a*)	8. (*b*)	9. (*a*)	10. (*c*)
11. (*b*)	12. (*c*)	13. (*a*)	14. (*d*)	15. (*c*)

Chapter **10**

Availability and Irreversibility

Learning objectives : The subject matter of this chapter will enable the reader to deal with :

- *available and non-available energy, and availability*
- *quality of energy and law of degradation of energy*
- *loss of available energy due to heat transfer*
- *availability of non-flow and steady flow systems*
- *Helmholtz and Gibb's functions*
- *effectiveness and irreversibility*
- *absolute entropy and third law of thermodynamics*

10.1. HIGH AND LOW GRADE ENERGY

The second law of thermodynamics puts a limit on the conversion of a given amount of heat into work. Complete conversion of heat into work is not possible. A portion of the heat supplied to any engine must be rejected. This implies that no heat engine can have an efficiency equal to 100 percent. On the contrary, the electrical work can be completely converted to mechanical work without any loss (neglecting friction and turbulence effects). The different sources of energy may thus be divided into two groups :

– *High grade energy*: Energy that can be completely transformed into shaft work without any loss and hence is fully utilizable. Examples are mechanical and electric work; water, wind and tidal power; kinetic energy of jets; animal and manual power.

– *Low grade energy*: Energy of which only a certain portion can be converted into mechanical work. Examples are heat or thermal energy; heat from nuclear fission or fusion; heat from combustion of fuels such as coal, wood, oil etc.

10.2. AVAILABLE AND UNAVAILABLE ENERGY

The portion of thermal energy input in cyclic engine which gets converted into mechanical work is referred to as *available energy.*

The portion of thermal energy which is not utilizable and is rejected to the sink (surroundings) is called *unavailable energy*.

The terms *exergy* and *anergy* are synonymous with available energy and unavailable energy, respectively. Thus

Energy = exergy + anergy

Lower the temperature at which heat is rejected, lesser would be the amount of heat rejected and more would be the work output for a given amount of heat supplied. Lower temperature corresponds to temperature of atmosphere or a large mass of water (ocean). The available work from a system becomes maximum when its state is brought in equilibrium to the state of surroundings. This maximum work is known as *availability* which may be defined as the maximum work available when a system comes to equilibrium with its environment by means of work and heat transfer. Availability is thus a composite property depending on the state of both the system and surroundings.

When the system comes to complete equilibrium with its environment, all intensive and specific properties will be common, and no energy difference will exist to promote further work. Such a state of the system is called *dead state*. The concept of dead state of a system implies that:

(*i*) the system is stable and uniform in composition

(*ii*) its pressure and temperature are equal to that of surroundings.

(*iii*) the system has a zero velocity and minimum potential energy.

The dead state is designated by affixing subscript '0' to the properties.

Any change in the state of the system from the dead state is a measure of the available work that can be obtained from it. Farther the initial state of system from the dead state (either above or below), higher will be the available energy. Available work of the system decreases as its state approaches the dead state.

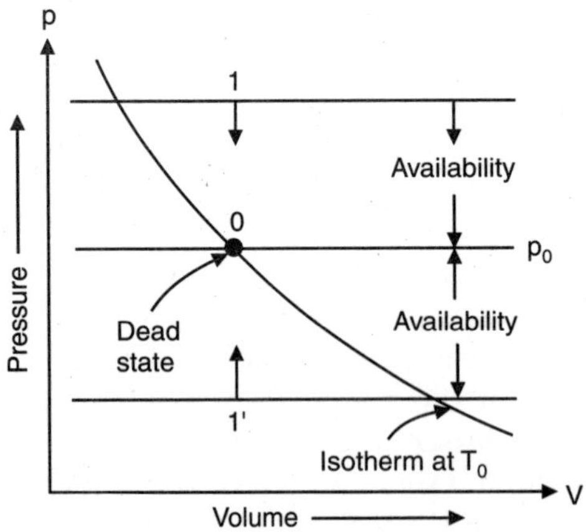

Fig. 10.1. Dead state and availability

When the system is reduced in state to temperature T_0 of the environment, its capacity to do work by utilizing the temperature difference ceases and its available energy becomes zero. Similar situation of zero available energy prevails when pressure and potential heads of the system are reduced to the level of surroundings.

The following two cases arise when considering available and unavailable portions of heat energy :

Case 1 : *Heat is withdrawn at constant temperature*

Fig 10.2. represents a reversible engine that operates between a constant temperature reservoir at temperature T and a sink at temperature T_0. Corresponding to heat Q supplied by the reservoir, the available work W_{max} is given by

$$\eta = \frac{T - T_0}{T} = \frac{W_{max}}{Q}$$

$$\therefore \; W_{max} = Q\left(\frac{T - T_0}{T}\right)$$

$$= Q\left(1 - \frac{T_0}{T}\right)$$

$$= Q - \frac{Q}{T} T_0 = Q - T_0 \, ds \quad ...(10.1)$$

where ds represents the change of entropy of the system during the process of heat supply.

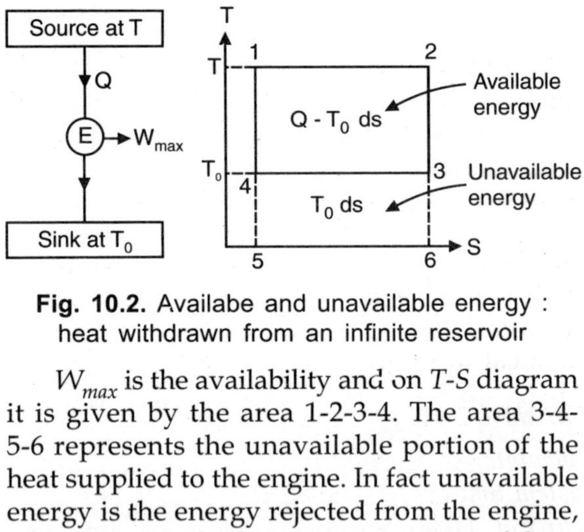

Fig. 10.2. Availabe and unavailable energy : heat withdrawn from an infinite reservoir

W_{max} is the availability and on *T-S* diagram it is given by the area 1-2-3-4. The area 3-4-5-6 represents the unavailable portion of the heat supplied to the engine. In fact unavailable energy is the energy rejected from the engine, and hence represents the portion of heat supplied that can not be converted into work. The unavailable energy equals the product of the lowest temperature of heat rejection and the change of entropy of the system during the process of heat supply (unavailable energy $= T_0 \, ds$)

Case 2 : *Heat is withdrawn at varying temperature*

In case of a finite reservoir, the temperature changes as heat is withdrawn from it, and as such the supply of heat to the engine is at varying temperature. The analysis is then made by breaking the process into a series of infinitesimal Carnot cycles each supplying δQ of heat at the temperature T (different for each cycle) and rejecting heat at the constant temperature T_0. Maximum amount of work (available energy) then equals

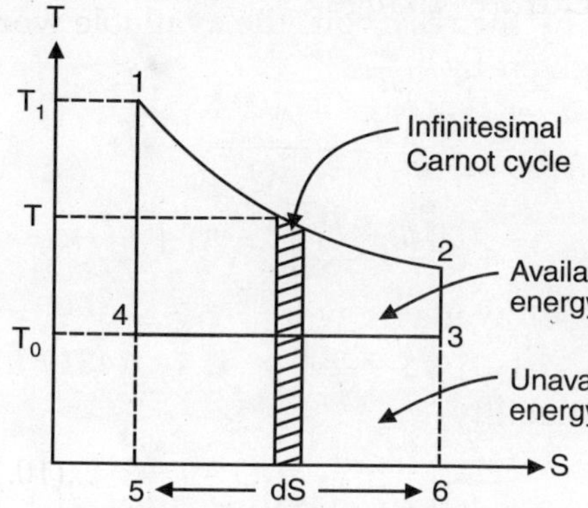

Fig. 10.3. Available and unavailable energy: heat supply at varying temperature

$$W_{max} = \int \left(1 - \frac{T_0}{T}\right) \delta Q$$

$$= \int \delta Q - \int T_0 \frac{\delta Q}{T}$$

$$= \int \delta Q - T_0 \int \frac{\delta Q}{T}$$

$$= Q - T_0\, ds \qquad ...(10.2)$$

It is to be seen that expressions for both the available and unavailable parts are identical in the two cases.

EXAMPLE 10.1

Calculate the availability and unavailability of a system that absorbs 15000 kJ of heat from a heat source at 500 K temperature while the environment is at 290 K temperature.

Solution : Entropy change,

$$dS = \frac{Q}{T} = \frac{15000}{500} = 30 \text{ kJ/K}$$

Unavailable work

$$= T_0\, dS = 290 \times 30 = \mathbf{8700\ kJ}$$

Available work

$$= Q - T_0\, dS$$

$$= 15000 - 290 \times 30 = \mathbf{6300\ kJ}$$

EXAMPLE 10.2

0.2 kg of air initially at 575 K temperature receives 300 kJ of heat reversibly at constant pressure. Determine the available and unavailable energies of the heat added.

Take c_p for air = 1.005 kJ/kg K and temperature of surroundings 300 K.

Solution : Let T_2 be the temperature of air after the addition of heat at constant pressure. Then

$$300 = mc_p\,(T_2 - T_1)$$

$$= 0.2 \times 1.005\,(T_2 - 575)$$

$$= 0.201\, T_2 - 115.57$$

$$\therefore \quad T_2 = \frac{300 + 115.57}{0.201} = 2067.5 \text{ K}$$

Entropy change,

$$dS = mc_p \log_e \frac{T_2}{T_1}$$

$$= 0.2 \times 1.005 \log_e \frac{2067.5}{575}$$

$$= 0.2572 \text{ kJ/K}$$

Unavailable work

$$= T_0\, dS$$

$$= 300 \times 0.2572 = \mathbf{77.16\ kJ}$$

Available work

$$= Q - T_0\, dS$$

$$= 300 - 300 \times 0.2572 = \mathbf{222.84\ kJ}$$

EXAMPLE 10.3

A temperature of 2280 K is attained in a furnace when hot products of combustion are produced by burning fuel in air at atmospheric pressure. Determine the availability of heat in the combustion products. It may be presumed that combution products behave as a perfect gas and the heat withdrawl occurs at constant pressure. Take c_p = 1.00 kJ/kg K for the gas and ambient temperature 300 K.

Solution : Heat transfer per kg of products of combustion,

$$= mc_p \Delta t$$

$$= 1 \times 1.00 \times (2280 - 300)$$

$$= 1980 \text{ kJ/kg}$$

Change in entropy of combustion products as these cool down from 2280 K to 300 K,

$$dS = mc_p \log_e \frac{T_2}{T_1}$$

$$= 1 \times 1.00 \log_e \frac{300}{2280}$$

$$= -2.028 \text{ kJ/K}$$

Availability of heat

$$= Q - T_0 dS$$

$$= 1980 - 300 \times 2.028$$

$$= \mathbf{1371.55 \text{ kJ/kg}}$$

EXAMPLE 10.4

A lump of 800 kg of steel at 1250 K is to be cooled to 500 K. If it is desired to use the steel as source of energy, calculate the available and unavailable energies. Take specific heat of steel as 0.5 kJ/kg K and ambient temperature 300 K.

Solution : Heat transfer from steel,

$$Q = mc_p \Delta t$$

$$= 800 \times 0.5 (1250 - 500)$$

$$= 300000 \text{ kJ} = \mathbf{300 \text{ MJ}}$$

Change in entropy of steel as it cools down from 1250 K to 500 K at constant pressure

$$dS = mc_p \log_e \frac{T_2}{T_1}$$

$$= 800 \times 0.5 \log_e \frac{500}{1250} = -366.52 \text{ kJ/K}$$

Unavailable energy

$$= T_0 dS$$

$$= 300 \times 366.52$$

$$= 109956 \text{ kJ} \simeq \mathbf{110 \text{ MJ}}$$

Available energy

$$= Q - T_0 dS = 300 - 110$$

$$= \mathbf{190 \text{ MJ}}$$

EXAMPLE 10.5

Heat is supplied reversibly from a heat source to a reversible engine and during this process, the temperature of the working fluid increases from 525 K to 875 K. Taking water equivalent as 100 kJ/K and presuming that heat rejection during the cycle takes place at ambient temperature of 290 K, determine the total heat abstracted, availability and the loss of available work.

Solution : Heat abstracted = $m\, c\, dT$

The product of mass of m and specific heat c is called the water equivalent. Then

Heat abstracted,

$$Q = \int_{T_1}^{T_2} 100\, dT = 100(T_2 - T_1)$$

$$= 100(875 - 525) = \mathbf{35000 \text{ kJ}}$$

(*b*) Entropy change,

$$dS = \int_{T_1}^{T_2} mc \frac{dT}{T} = 100 \log_e \frac{T_2}{T_1}$$

$$= 100 \log_e \frac{875}{525} = 51.1 \text{ kJ/K}$$

Loss of available work

$$= T_0\, dS = 290 \times 51.1 = \mathbf{14819 \text{ kJ}}$$

Availability,

$$A = Q - T_0\, dS$$

$$= 35000 - 14819 = \mathbf{20181 \text{ kJ}}$$

10.3. LOSS OF AVAILABLE ENERGY DUE TO HEAT TRANSFER THROUGH A FINITE TEMPERATURE DIFFERENCE

Consider a certain quantity of heat Q transferred from a system at constant temperature T_1 to another system at constant temperature T_2 ($T_1 > T_2$). Before heat is transferred, the energy Q is available at T_1 and the ambient temperature is T_0.

∴ Initial available energy,

$$(AE)_1 = Q\left(1 - \frac{T_0}{T_1}\right)$$

After heat transfer, the energy Q is available at T_2 and again the ambient temperature is T_0

∴ Final available energy,

$$(AE)_2 = Q\left(1 - \frac{T_0}{T_2}\right)$$

Change in available energy

$$= (AE)_1 - (AE)_2$$

$$= Q\left(1 - \frac{T_0}{T_1}\right) - Q\left(1 - \frac{T_0}{T_2}\right)$$

$$= T_0\left[\frac{-Q}{T_1}+\frac{Q}{T_2}\right]$$

$$= T_0\,(dS_1 + dS_2)$$

$$= T_0\,(dS\,)_{net} \qquad ...(10.3)$$

where $dS_1 = -Q/T_1$, $dS_2 = Q/T_2$ and $(dS)_{net}$ is the net change in the entropy of the combination of the two interacting systems. This total entropy change is called entropy of universe or entropy production. Since the heat transfer has been through a finite temperature difference, the process is irreversible, *i.e.*, $(dS)_{net} > 0$ and hence there is loss or decrease of available energy.

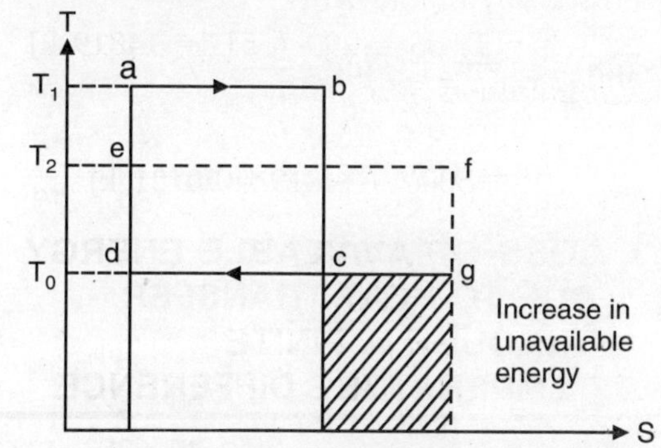

Fig. 10.4. Decrease in available energy due to heat transfer through a finite temperature difference

Reference Fig. 10.4.

(*i*) *abcd* is the power cycle when heat is available at T_1. Area under *cd* then represents the unavailable portion of energy.

(*ii*) *efgd* is the power cycle when heat is available at T_2. Area under *dg* then represents the unavailable energy.

(*iii*) Increase in unavailable energy due to irreversible heat transfer is then represented by the shaded area under *cg*. The increase in unavailable energy equals the decrease in available energy, and is given by the product of the ambient temperature and the net increase in the entropy of the interacting systems.

The above aspects lead us to conclude that :

– Whenever heat is transferred through a finite temperature difference, there is always a loss of available energy.

– Greater the temperature difference $(T_1 - T_2)$, the more net increase in entropy and, therefore, loss of available energy.

– The available energy of a system at a higher temperature is more than that at a lower temperature, and decreases progressively as the temperature falls. Stated in other words, quality of energy of a fluid at higher temperature is superior to that at lower temperature; a high temperature fluid has capacity to do more work under the same identical conditions.

– The concept of available energy provides a useful measure of the quality of energy. Energy is said to be degraded each time it flows through a finite temperature difference. The second law may, therefore, be referred to as *law of degradation of energy.* Whereas the first law tells that energy is always conserved quantity wise, the second law emphasizes that energy always degrades quality wise. Energy is always conserved but its quality is always degraded.

EXAMPLE 10.6

A source at 1000 K is available for transfer of heat at the rate of 100 kW to a system at 500 K. If these temperatures remain constant, determine the entropy production accompanying the heat transfer and the increase of unavailable energy. Take ambient temperature as 300 K.

Solution : Heat transfer

$$Q = 100 \text{ kW} = 100 \text{ kJ/s} = 6000 \text{ kJ/min}$$

Entropy change of high temperature source,

$$dS_1 = -\frac{Q}{T_1} = -\frac{6000}{1000} = -6 \text{ kJ/K min}$$

The negative sign stems from the fact that heat flows out of the high temperature source. The system at 500 K receives the same amount of heat and its entropy change is

$$dS_2 = \frac{6000}{500} = 12 \text{ kJ/K min}$$

∴ Entropy production as a result of heat transfer is,

$$(dS)_{net} = dS_1 + dS_2$$
$$= -6 + 12 = 6 \text{ kJ/K min}$$

Increase of unavailable energy

$= T_0 \times (dS)_{net}$

$= 300 \times 6 =$ **1800 kJ**

EXAMPLE 10.7

During a certain process, a vapour condenses at 700 K and transfers heat to water evaporating at 500 K. The steam thus produced is used in a power cycle which rejects heat to surroundings at 300 K. Make calculations for the fraction of energy that becomes unavailable due to irreversible heat transfer.

Solution : Let Q be the amount of heat transferred. The available energy associated with condensing vapour is

$$= Q\left(1 - \frac{T_0}{T_1}\right) = \frac{Q}{T_1}(T_1 - T_0)$$

Entropy change of condensing vapour,

$$dS_1 = \frac{-Q}{T_1}$$

Entropy change for the evaporating water,

$$dS_2 = \frac{Q}{T_2}$$

∴ Entropy production as a result of heat transfer is,

$$(dS)_{net} = dS_1 + dS_2 = \frac{-Q}{T_1} + \frac{Q}{T_2}$$

Increase of unavailable energy $= T_0 \times (dS)_{net}$

$$= T_0\left(\frac{-Q}{T_1} + \frac{Q}{T_2}\right) = \frac{T_0\,Q}{T_1\,T_2}(T_1 - T_2)$$

∴ Required fraction is

$$= \frac{\dfrac{T_0\,Q}{T_1\,T_2}(T_1 - T_2)}{\dfrac{Q}{T_1}(T_1 - T_0)}$$

$$= \frac{T_0\,(T_1 - T_2)}{T_2\,(T_1 - T_0)} = \frac{300\,(700 - 500)}{500\,(700 - 300)}$$

$= \mathbf{0.3}$

EXAMPLE 10.8

In a counter flow heat exchanger 0.25 kg/s of water is heated from 30°C to 60°C by hot gases which enter at 180°C and leave at 80°C. Make calculations for the mass flow of gases, heat transferred and loss of available energy due to this heat transfer.

Take c_p for gases = 1.08 kJ/kg K and ambient temperature = 27°C.

Solution : From the principle of energy conservation,

heat given by hot gases

= heat gained by water

$(m\,c_p\,\Delta t)_{gas} = (m\,c_p\,\Delta t)_{water}$

= heat transfer Q

$m_g \times 1.08 \times (180 - 80) = 0.25 \times 4.186\,(60 - 30)$

Upon solution : $Q = 31.395$ kJ/s and

$m_g = 0.2907$ kg

$$(dS)_{gases} = m_g\,c_{pg}\,\log_e \frac{T_{2g}}{T_{1g}}$$

$$= 0.2907 \times 1.08 \times \log_e \frac{80 + 273}{180 + 273}$$

$= -\,0.0783$ kJ/K

$$(dS)_{water} = m_w\,c_{pw}\,\frac{T_{2w}}{T_{1w}}$$

$$= 0.25 \times 4.186 \log_e \frac{60 + 273}{30 + 273}$$

$= 0.1092$ kJ/K

∴ Entropy production as a result of heat transfer is

$(dS)_{net} = (dS)_{gases} + (dS)_{water}$

$= -\,0.0783 + 0.1092 = 0.309$ kJ/K

Increase in unavailable energy (or loss of available energy)

$= T_0\,(dS)_{net}$

$= (273 + 27) \times 0.0309 =$ **9.27 kJ**

EXAMPLE 10.9

20 kg of water at 90°C is mixed with 30 kg of water at 30°C, and the pressure remains constant during the mixing operation. Calculate the decrease in available energy. It may be presumed that the surroundings are at 10°C temperature and for water

c_p = 4.18 kJ/kg K

Solution : The available energy of a system of mass m, specific heat c_p and at temperature T is given by

$$AE = m c_p \int_{T_o}^{T} \left(1 - \frac{T_o}{T}\right) dT$$

where T_o is the temperature of surroundings. Then available energy of 20 kg of water at 90°C (363 K)

$$(AE)_1 = 20 \times 4.18 \int_{283}^{363} \left(1 - \frac{283}{T}\right) dT$$

$$= 83.6\left[(363 - 283) - 283 \log_e \frac{363}{283}\right]$$

$$= 798.4 \text{ kJ}$$

Available energy of 30 kg of water at 30°C (303 K),

$$(AE)_2 = 30 \times 4.18 \int_{283}^{303} \left(1 - \frac{283}{T}\right) dT$$

$$= 125.4\left[(303 - 283) - 283 \log_e \frac{303}{283}\right]$$

$$= 85.3 \text{ kJ}$$

Total available energy,

$$(AE)_{total} = 798.4 + 85.3 = 883.7 \text{ kJ}$$

Let t_f be the final temperature after the mixing process. Then from the principle of energy conservation,

heat lost by water at 90° temperature

= heat gained by water at 30°C

$$20 \times 4.18\ (90 - t_f) = 30 \times 4.18\ (t_f - 30)$$

or $20(90 - t_f) = 30(t_f - 30)$

$$t_f = \frac{20 \times 90 + 30 \times 30}{20 + 30}$$

$$= 54°\text{C or } 327 \text{ K}$$

Total mass after mixing = 20 + 30 = 50 kg

Available energy of 50 kg of water at 327 K

$$= 50 \times 4.18 \int_{283}^{327} \left(1 - \frac{283}{T}\right) dT$$

$$= 209\left[(327 - 283) - 283 \log_e \frac{327}{283}\right]$$

$$= 650 \text{ kJ}$$

∴ Decrease in available energy due to mixing

= total energy before mixing – total energy after mixing

= 883.7 – 650 = **233.7** kJ

EXAMPLE 10.10

A system at 450 K receives 225 kJ/s of heat energy from a source at 1500 K, and the temperatures of both the system and source remain constant during the heat transfer process. Represent the process on temperature-entropy diagram and determine :

(a) net change in entropy,

(b) available energy of heat source and system, and

(c) decrease in available energy.

Take atmospheric temperature equal to 300 K.

Solution : Refer Fig. 10.5. for the representation of processes on *T-s* diagram.

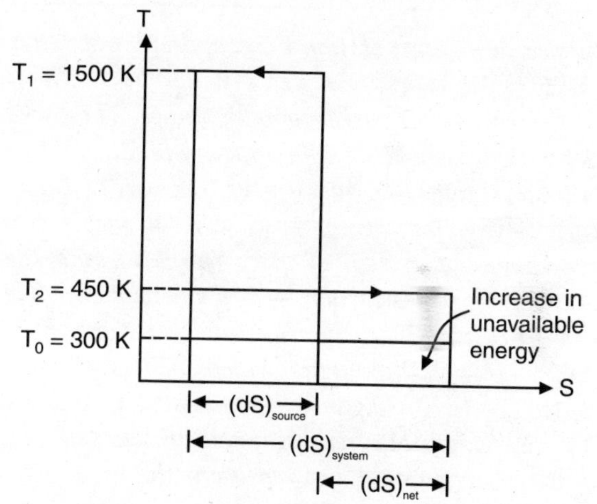

Fig. 10.5.

Entropy change of heat source

$$(dS)_{source} = \frac{Q}{T_1} = \frac{-225}{1500}$$

$$= -0.150 \text{ kJ/s K}$$

Entropy change of system,

$$(dS)_{system} = \frac{Q}{T_2} = \frac{225}{450}$$

$$= 0.5 \text{ kJ/s K}$$

Net change in entropy,

$$(dS)_{net} = (dS)_{source} + (dS)_{system}$$

$$= -0.15 + 0.5$$

$$= \mathbf{0.35 \text{ kJ/s K}}$$

(*b*) Available energy of heat source,

$$A_1 = Q - T_0(dS)_{source}$$
$$= 225 - 300 \times 0.15$$
$$= \mathbf{180} \text{ kJ/s}$$

Available energy of system,

$$A_2 = Q - T_0(dS)_{system}$$
$$= 225 - 300 \times 0.5 = \mathbf{75} \text{ kJ/s}$$

(*c*) Decrease in available energy

$$= A_1 - A_2 = 180 - 75$$
$$= \mathbf{105} \text{ kJ/s}$$

The decrease in available energy may also be determined by using the following relation,

Decrease in available energy

$$= T_0(dS)_{net}$$
$$= 300 \times 0.35 = 105 \text{ kJ/s}$$

EXAMPLE 10.11

In a certain power station the evaporation of water at 250°C is done in a steam boiler by the heat transfer from hot gases from a furnace. The gases get cooled from 1100°C to 550°C and all the heat transfer from the gases goes to water. Make calculations for the increase in total entropy of the combined system of gas and water, and the increase in the unavailable energy due to irreversible heat transfer. Take specific heat of the gas c_p = 1.005 kJ/kg K, the latent heat of vaporisation of water L = 1690 kJ/kg and the ambient temperature T_0 = 27°C. Obtain the results on the basis of 1 kg of water evaporated and represent the process on temperature entropy diagram.

Solution : Refer Fig. 10.6.

To evaporate 1 kg of water at 250°C, the heat to be transferred is 1690 kJ and this comes from m_g kg of gas that gets cooled from 1100°C to 550°C. From principle of energy conservation,

$$m_g \times 1.005\ (1100 - 550) = 1690$$

$$\therefore\ m_g = \frac{1690}{1.005\,(1100 - 550)} = 3.057 \text{ kg}$$

Change in entropy of gas,

$$dS_g = m_g c_{pg} \log_e \frac{T_{2g}}{T_{1g}}$$

$$= 3.057 \times 1.005 \log_e \frac{550 + 273}{1100 + 273}$$

$$= -1.572 \text{ kJ/K}$$

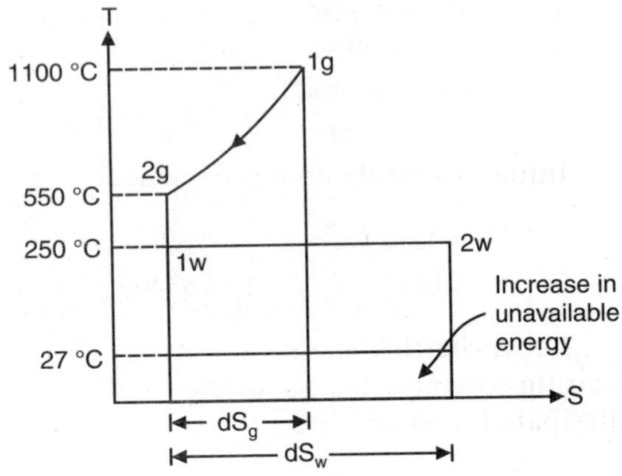

Fig. 10.6.

Change in entropy of water,

$$dS_w = \frac{\text{latent heat}}{T}$$

$$= \frac{1690}{250 + 273} = 3.231 \text{ kJ/K}$$

Entropy production as a result of heat transfer is

$$(dS)_{net} = dS_g + dS_w$$
$$= -1.572 + 3.231 = \mathbf{1.659\ kJ/K}$$

Increase in unavailable energy/or loss of available energy

$$= T_0\ (dS)_{net}$$
$$= (27 + 273) \times 1.659$$
$$= \mathbf{497.7\ kJ/kg\ of\ water}$$

EXAMPLE 10.12

The moment of interia of a flywheel is 0.52 kg m[2] and it rotates at a speed of 3000 rpm in a large heat insulated system, the temperature of which is 15°C. The kinetic energy of the flywheel is dissipated as frictional heat at the shaft bearings which have water equivalent of 7.5 kJ/K. Determine:

(a) temperature of the bearings when the vehicle has come to rest,

(b) greatest possible amount of frictional heat which may be returned to the flywheel as high grade energy, and

(c) final rorational speed of flywheel if it is set in motion with the available energy.

Solution : Initial angular speed of flywheel

$$\omega_1 = \frac{2\pi N_1}{60}$$

$$= \frac{2\pi \times 300}{60} = 314.2 \text{ rad/s}$$

Initial available energy flywheel,

$$E = \frac{1}{2} I\omega^2 = \frac{1}{2} \times 0.52 \times (314.2)^2$$

$$= 25668 \text{ Nm (or J)} = 25.668 \text{ kJ}$$

Let Δt be the rise in temperature of the bearing when the energy of the flywheel gets dissipated as heat. Then

$$mc\ \Delta T = E$$

The product of mass m and specific heat c of the brake material is called water equivalent and its value is stated as 7.5 kJ/K. Then

Rise in temperature of bearing,

$$\Delta t = \frac{25.668}{7.5} = 3.42°\text{C}$$

∴ Final temperature of the bearing

$$= 15 + 3.42 = \mathbf{18.42°C} = \mathbf{291.42\ K}$$

(*b*) Maximum amount of energy which may be returned to flywheel as high grade energy

$$mc \int_{288}^{291.42} \left(1 - \frac{288}{T}\right) dT$$

$$= 7.5\left[(291.42 - 288) - 288 \log_e \frac{291.42}{288}\right]$$

$$= 7.5\ [3.42 - 3.398] = \mathbf{0.165\ kJ}$$

Amount of energy rendered unavailable

$$= 25.668 - 0.165 = 25.503 \text{ kJ}$$

(*c*) Let ω be the angular rotational speed of flywheel when it is set into motion with the available energy. Then

$$\frac{1}{2} I\omega_2^2 = 0.165 \times 10^3$$

or $$\frac{1}{2} \times 0.52\ \omega_2^2 = 0.165 \times 10^3$$

Solution gives : $\omega_2 = 25.19$ rad/s

$$\therefore N_2 = \frac{60 \times \omega_2}{2\pi}$$

$$= \frac{60 \times 25.19}{2\pi} = \mathbf{240.67} \text{ rev/min}$$

EXAMPLE 10.13

A rigid vessel has two compartments; compartment A contains one kg of oxygen at 1 bar and 450 K and the compartment B contains one kg of hydrogen at the same pressure and temperature. The gases are allowed to mix by removing the diaphragm separating them. If the ambient temperature is 300 K, determine the loss in availability.

For oxygen: $c_p = 0.92$ kJ/kg K

$c_v = 0.65$ kJ/kg K

For hydrogen $c_p = 14.32$ kJ/Kg K

$c_v = 10.17$ kJ/Kg K

Solution : For oxygen :

$$R = c_p - c_v = 0.92 - 0.65 = 0.26 \text{ kJ/kg K}$$

$$V = \frac{mRT}{P}$$

$$= \frac{1 \times (0.26 \times 10^3) \times 450}{1 \times 10^5} = 1.17 \text{ m}^3$$

For hydrogen:

$$R = 14.32 - 10.17 = 4.15 \text{ kJ/kg K}$$

$$V = \frac{1 \times (4.15 \times 10^3) \times 450}{1 \times 10^5} = 18.675 \text{ m}^3$$

Upon removal of diaphragm, both oxygen and hydrogen will have a volume equal to

$$V_2 = 1.17 + 18.675 = 19.845 \text{ m}^3$$

Entropy change of oxygen

$$dS_{ox} = \left(m\,R \log_e \frac{V_2}{V_1}\right)_{ox}$$

$$= 1 \times 0.26 \log_e \frac{19.845}{1.17} = 0.736 \text{ kJ/K}$$

Entropy change of hydrogen

$$dS_{hy} = \left(m\,R \log_e \frac{V_2}{V_1}\right)_{hy}$$

$$= 1 \times 4.15 \log_e \frac{19.845}{18.675} = 0.252 \text{ kJ/K}$$

Total change in entropy as a result of mixing is

$$(dS)_{net} = dS_{ox} + dS_{hy}$$
$$= 0.736 + 0.252 = 0.988 \text{ kJ/K}$$

Loss in availability/or gain in unavailable energy

$$= T_0 \times (dS)_{net} = 300 \times 0.988 = \mathbf{296.4\ kJ}$$

10.4. AVAILABILITY OF A NON-FLOW OR CLOSED SYSTEM

Consider a piston-cylinder arrangement (closed system) in which the fluid at p_1, V_1, T_1, expands reversibly to the environmental state with parameters p_0, V_0, T_0. The following energy (work and heat) interactions take place:

– The fluid expands and expansion work W_{exp} is obtained. From the principle of energy conservation, $\delta Q = \delta W + dU$, we get:

$$-Q = W_{exp} + (U_0 - U_1)$$

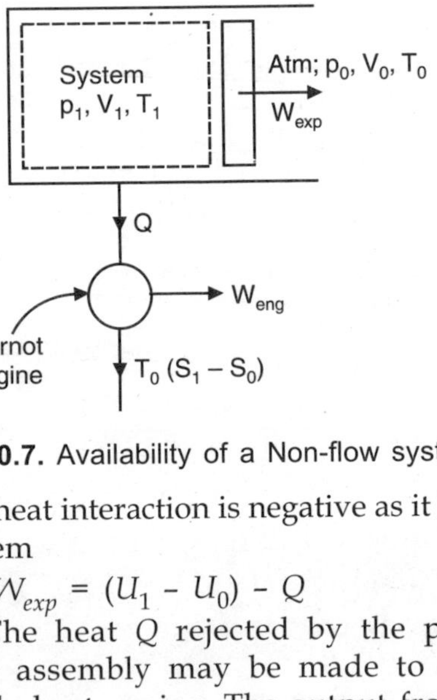

Fig. 10.7. Availability of a Non-flow system

The heat interaction is negative as it leaves the system

$$\therefore \quad W_{exp} = (U_1 - U_0) - Q$$

– The heat Q rejected by the piston-cylinder assembly may be made to run a reversible heat engine. The output from the reversible engine equals

$$W_{eng} = Q\left(1 - \frac{T_0}{T_1}\right) = Q - T_0\,(S_1 - S_0)$$

The sum total of expansion work W_{exp} and the engine work W_{eng} gives ***maximum work*** obtainable from the arrangement.

$$W_{max} = [(U_1 - U_0) - Q] + [Q - T_0\,(S_1 - S_0)]$$
$$= (U_1 - U_0) - T_0\,(S_1 - S_0)$$

The piston moving outwards has to spend a work in pushing the atmosphere against its own pressure. This work, which may be called as the *surrounding work*, is simply dissipated, and as such is not useful. It is given by

$$W_{surr} = p_0(V_0 - V_1)$$

The energy available for work transfer less the work absorbed in moving the environment is called the ***useful work*** or ***net work***.

∴ Maximum available useful work or net work,

$$W_{net} = W_{max} - W_{surr}$$
$$= (U_1 - U_0) - T_0\,(S_1 - S_0) - p_0\,(V_0 - V_1)$$
$$= (U_1 + p_0V_1 - T_0S_1) - (U_0 + p_0V_0 - T_0S_0)$$
$$= A_1 - A_0 \qquad \text{...(10.4)}$$

where $A = (U + p_0V - T_0S)$ is known as ***non-flow availability function***. It is a composite property of the system and surroundings as it consists of three extensive properties of the system (U, V, and S) and two intensive properties of the surroundings (p_0 and T_0).

When the system undergoes a change from state 1 to state 2 without reaching the dead state, then

$$(W_u)_{max} = W_{net} = (A_1 - A_0) - (A_2 - A_0)$$
$$= A_1 - A_2$$

It is worth while to mention that availability is a function of the properties of the surroundings as well as the end states of the closed system.

The changes in entropy for the closed system are calculated from the property relation

$$TdS = dU + p\,dV$$

Upon integration between the initial and final states, we obtain

$$\int_1^2 dS = \int_1^2 \frac{mc_v\,dT}{T} + \int_1^2 \frac{p\,dV}{T}$$
$$= \int_1^2 mc_v\,\frac{dT}{T} + \int_1^2 m\,R\,\frac{dV}{V}$$
$$\left(\because \frac{p}{T} = \frac{mR}{V}\right)$$

or
$$S_2 - S_1 = m\,c_v \log_e \frac{T_2}{T_1} + mR \log_e \frac{V_2}{V_1}$$

EXAMPLE 10.14

One kg of air is contained in a piston cylinder assembly at 10 bar pressure and 500 K temperature. The piston moves outwards and the air expands to 2 bar pressure and 350 K temperature. Make calculations for

(i) the availability in the initial and final states

(ii) the maximum useful work, and

(iii) the irreversibility for the system.

Assume that the system is insulated and the environmental conditions are at 1 bar and 290 K. Further for air :

$R = 0.287$ *kJ/kg K,* $c_v = 0.718$ *kJ/kg K,*

and $c_p = 1.005$ *kJ/kg K*

Solution : (*i*) Availability in the initial state,

$$A_1 = (U_1 - U_0) + p_0(V_1 - V_0) - T_0(S_1 - S_0)$$

Now : $(U_1 - U_0) = mc_v(T_1 - T_0)$

$= 1 \times 0.718\ (500 - 290)$

$= 150.78$ kJ

$$V_1 = \frac{mRT_1}{p_1} = \frac{1 \times 287 \times 500}{10 \times 10^5}$$

$= 0.1435$ m^3

$$V_o = \frac{mRT_o}{p_o} = \frac{1 \times 287 \times 290}{1 \times 10^5}$$

$= 0.8323$ m^3

$\therefore p_0(V_1 - V_0) = 1 \times 10^5\ (0.1435 - 0.8323)$

$= -\ 68880$ Nm (or J)

$= -\ 68.88$ kJ

Further

$$S_o - S_1 = m\, c_v \log_e \frac{T_0}{T_1} + mR \log_e \frac{V_0}{V_1}$$

$$= 1 \times 0.718 \times \log_e \frac{290}{500}$$

$$+ 1 \times 0.287 \times \log_e \frac{0.8323}{0.1435}$$

$= -\ 0.3911 + 0.5045 = 0.1134$ kJ/K

$\therefore T_0(S_1 - S_0) = 290 \times (-\ 0.1134) = -\ 32.88$ kJ

Accordingly :

$A_1 = 150.78 + (-\ 68.88) - (-\ 32.87)$

$=$ **114.77** kJ

Availability in the final state,

$$A_2 = (U_2 - U_0) + p_0(V_2 - V_0) - T_0(S_2 - S_0)$$

Now $(U_2 - U_0) = mc_v\ (T_2 - T_o)$

$= 1 \times 0.718\ (350 - 290)$

$= 43.08$ kJ

$$V_2 = \frac{mRT_2}{p_2} = \frac{1 \times 287 \times 350}{1 \times 10^5} = 0.5022 \text{ m}^3$$

$$V_0 = \frac{mRT_o}{p_o} = \frac{1 \times 287 \times 290}{2 \times 10^5} = 0.8323 \text{ m}^3$$

$\therefore p_0(V_2 - V_0) = 1 \times 10^5(0.5022 - 0.8323)$

$= -\ 33010$ Nm (or J)

$= -\ 33.01$ kJ

Further

$$S_0 - S_2 = mc_v \log_e \frac{T_0}{T_2} + mR \log_e \frac{V_0}{V_2}$$

$$= 1 \times 0.718 \log_e \frac{290}{350}$$

$$+ 1 \times 0.287 \log_e \frac{0.8323}{0.5022}$$

$= -\ 0.135 + 0.145 = 0.01$ kJ/K

$\therefore T_0(S_2 - S_0) = 290 \times (-\ 0.01) = -\ 2.90$ kJ

Accordingly :

$A_2 = 43.08 + (-\ 33.01) - (-\ 2.90)$

$=$ **12.97** kJ

(*ii*) For a non-flow process, the maximum useful work (net work) obtainable is given by the change in availability. That is

$W_{net} = A_1 - A_2$

$= 114.78 - 12.97 =$ **101.81** kJ

(*iii*) The irreversibility is given by

$$I = T_o(dS)_{univ} = T_0\left[(dS)_{sys} + (dS)_{surr}\right]$$

In terms of pressure and temperature changes, the entropy change of the system is given by

$$(dS)_{sys} = m\left[c_p \log_e \frac{T_2}{T_1} - R \log_e \frac{p_2}{p_1}\right]$$

$$= 1\left[1.005 \log_e \frac{350}{500} - 0.287 \log_e \frac{2}{10}\right]$$

$$= 1[-0.358 - (-0.462)] = 0.104 \text{ kJ/K}$$

Since the system is insulated, there will be no heat interaction with the surrounding and accordingly $(dS)_{surr} = 0$

Then, irreversibility

$$I = 290 \times 0.104 = \mathbf{30.16\ kJ}$$

EXAMPLE 10.15

A closed system contains 2 kg of air and during an adiabatic expansion process, there occurs a change in its pressure from 500 kPa to 100 kPa and in its temperature from 350 to 320 K. If the volume doubles during the process, make calculations for the maximum work, the change in availability and the irreversibility.

For air take $c_v = 0.718$ *kJ/kg K and* $R = 0.287$ *kJ/kg K*

The surrounding conditions may be assumed to be 10 kPa and 300 K.

Solution : When a system undergoes a change from state 1 to state 2 during a mass-flow process, the maximum obtainable work is given by

$$W_{max} = (U_1 - U_2) - T_0\,(S_1 - S_2)$$

Now : $(U_1 - U_2) = mc_v(T_1 - T_2)$

$$= 2 \times 0.718\,(350 - 320)$$
$$= 43.08 \text{ kJ}$$

$$(S_2 - S_1) = m\left[c_v \log_e \frac{T_2}{T_1} + R \log_e \frac{V_2}{V_1}\right]$$

$$= 2\left[0.718 \log_e \frac{320}{350} + 0.287 \log_e 2\right]$$

$$= 0.2693 \text{ kJ/K}$$

Then :

$$W_{max} = 43.08 - 300\,(-\,0.2693)$$
$$= \mathbf{123.87}\text{ kJ}$$

(*b*) Change in availability is given by

$$A_1 - A_2 = (U_1 - U_2) - T_0\,(S_1 - S_2) + p_0(V_1 - V_2)$$
$$= W_{max} + p_0(V_1 - V_2)$$

Now : $V_1 = \dfrac{mRT_1}{p_1}$

$$= \frac{2 \times 0.287 \times 350}{500} = 0.4018 \text{ m}^3$$

$$V_2 = 2V_1 = 2 \times 0.4018 = 0.8036 \text{ m}^3$$

$$\therefore \quad A_1 - A_2 = 123.87 + 100(0.4018 - 0.8036)$$
$$= \mathbf{83.69}\text{ kJ}$$

The change in availability is a measure of the maximum useful work, *i.e.*, net work obtainable from the arrangement.

(*c*) Irreversibility I = maximum work − actual work

The actual work can be obtained from the energy equation for the closed process:

$$\delta q = \delta w + du$$

That is,

$$W_{act} = Q - (U_2 - U_1)$$

Since the process is adiabatic, $Q = 0$ and accordingly

$$W_{act} = -\,(U_2 - U_1) = (U_1 - U_2)$$
$$= mc_v\,(T_1 - T_2)$$
$$= 2 \times 0.718\,(250 - 320) = 43.08 \text{ kJ}$$

∴ Irreversibility

$$I = 123.87 - 43.08 = \mathbf{80.79}\text{ kJ}$$

Alternatively :

$$I = (U_1 - U_2) - T_0(S_1 - S_2) - (U_1 - U_2)$$
$$= T_0\,(S_2 - S_1) = T_0(dS)_{net}$$

$$(dS)_{sys} = m\left[c_v \log_e \frac{T_2}{T_1} + R \log_e \frac{V_2}{V_1}\right]$$

$$= 2\left[0.718 \log_e \frac{325}{350} + 0.287 \log_e 2\right]$$

$$= -\,0.2693 \text{ kJ/K}$$

For adiabatic process $(dS)_{surr} = 0$

∴ Irreversibility

$$I = 300 \times 0.2693 = 80.79 \text{ kJ}$$

10.5. AVAILABILITY OF A STEADY FLOW SYSTEM

Consider a steady flow system and let it be assumed that the flowing fluid has the following properties and characteristics :

Internal energy u, specific volume v, specific enthalpy h, pressure p, velocity C and location z.

The properties of the fluid would change when flowing through the system. Let subscript 1 indictate the properties of the system at inlet and subscript 0 be used to designate the fluid parameters at outlet. Further let Q units of heat be rejected by the system and let the system deliver W_s units of work.

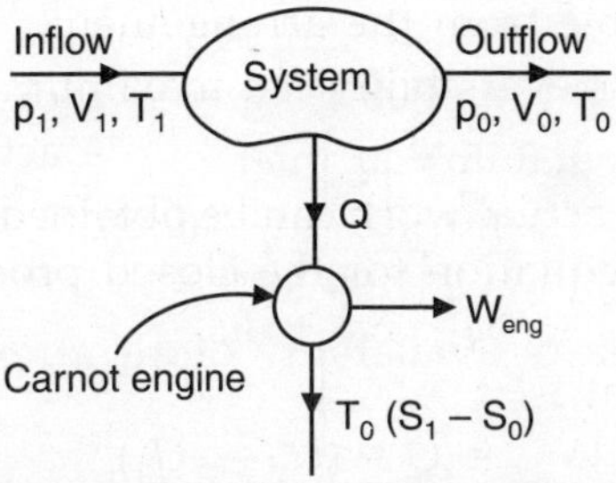

Fig. 10.8. Availability of a steady flow system

The steady flow energy equation may then be written as

$$U_1 + p_1V_1 + \frac{C_1^2}{2} + gz_1 - Q$$

$$= U_0 + p_0V_0 + \frac{C_0^2}{2} + gz_o + W_s$$

Neglecting potential and kinetic energy changes,

$$U_1 + p_1V_1 - Q = U_0 + p_0V_0 + W_s$$

$$H_1 - Q = H_0 + W_s$$

$$\therefore \quad \text{shaft work } W_s = (H_1 - H_0) - Q$$

The heat Q rejected by the system may be made to run a reversible heat engine. The output from this engine equals

$$W_{eng} = Q\left(1 - \frac{T_0}{T_1}\right) = Q - T_0\,(S_1 - S_0)$$

$\therefore$ Maximum available useful work or net work,

$$W_{net} = W_s + W_{eng}$$

$$= (H_1 - H_0) - Q + Q - T_0\,(S_1 - S_0)$$

$$= (H_1 - T_0\,S_1) - (H_0 - T_0\,S_0)$$

$$= B_1 - B_0 \qquad \text{...(10.5)}$$

where $B = (H - T_0S)$ is known as the ***steady flow availability function.*** It is a composite property of system and surroundings involving two extensive properties H and S of the system and one intensive property T_0 of the surroundings. Alternative names for the term B are the Darrieus function and the Keenam function.

When the system changes from state 1 to some intermediate state 2, the change in steady flow availability function is given by

$$dB = (B_1 - B_0) - (B_2 - B_0)$$

$$= B_1 - B_2 \qquad \text{...(10.6)}$$

The changes in entropy for the steady flow can be calculated from the property relation

$$T\,ds = dH - V\,dp$$

Upon integration between the initial and final states, we obtain

$$\int_1^2 dS = \int_1^2 \frac{mc_p\,dT}{T} - \int_1^2 \frac{V\,dp}{T}$$

$$= \int_1^2 mc_p\,\frac{dT}{T} - \int_1^2 mR\,\frac{dp}{p}$$

$$\left(\because \frac{V}{T} = \frac{mR}{p}\right)$$

or $$S_2 - S_1 = m\,c_p\,\log_e \frac{T_2}{T_1} - mR\,\log_e \frac{p_2}{p_1}$$

EXAMPLE 10.16

A single stage air turbine is to operate with an inlet pressure and temperature of 6 bar and 800 K. The outlet pressure and temperature are 1.0 bar and 400 K. During expansion, the turbine loses 25 kJ/kg to the surroundings which are at 1 bar and 300 K. For unit mass flow rate, determine the decrease in availability, the maximum work and the reversibility. For air take

$$c_p = 1.005 \text{ kJ/kg K and } R = 0.287 \text{ kJ/kg K}$$

Solution : Decrease in availability,

$$b_1 - b_2 = (h_1 - h_2) - T_0\,(s_1 - s_2)$$

$$\text{Now : } h_1 - h_2 = c_p\,(T_1 - T_2)$$

$$= 1.005\,(800 - 400)$$

$$= 402 \text{ kJ/kg}$$

$$s_1 - s_2 = R\,\log_e \frac{p_2}{p_1} - c_p\,\log_e \frac{T_2}{T_1}$$

$$= 0.287\,\log_e \frac{1}{6} - 1.005\,\log_e \frac{400}{800}$$

$$= -\,0.514 - (-\,0.472) = -\,0.042 \text{ kJ/kg K}$$

$$\therefore \quad b_1 - b_2 = 402 - 300\,(-\,0.042)$$

$$= 414.6 \text{ kJ/kg}$$

Max work

W_{max} = change in availability
= **414.6 kJ/kg**

From steady flow energy equation,

$Q + h_1 = W + h_2$

$W = (h_1 - h_2) + Q$
$= 1.005\ (800 - 400) + (-20)$
$= 402 - 20 = 382$ kJ/kg

∴ Irreversibility,

$I = W_{max} - W$
$= 414.6 - 382 =$ **32.6 kJ/kg**

Alternatively :

$I = T_0\ [dS_{sys} + dS_{surr}]$

$$= 300\left[1.005 \log_e \frac{400}{800} - 0.287 \log_e \frac{1}{6} + \frac{20}{300}\right]$$

$= 300[-0.6966 + 0.5142 + 0.0667]$
= **32.57 kJ/kg**

EXAMPLE 10.17

The air, in a steady flow, enters the system at pressure 12 bar and temperature 180°C with a velocity of 100 m/s. The corresponding values at exit from the system are 1.5 bar, 20°C and 50m/s. If the surroundings are at 1 bar pressure and 20°C, determine.

(i) reversible work and actual work assuming the process to be adiabatic

(ii) irreversibility and effectiveness of the system.

Make your calculations for unit mass flow and take

$c_p = 1.005$ *kJ/kg K and* $R = 0.287$ *kJ/lg K*

Solution : Availability of air at the inlet

$$= (h_1 - h_0) - T_0(s_1 - s_0) + \frac{V_1^2}{2}$$

Now :

$h_1 - h_0 = c_p(T_1 - T_0)$
$= 1.005\ (453 - 293) = 160.8$ J/kg

In terms of temperature and pressure, the entropy change is given by

$$s_1 - s_0 = c_p \log_e \frac{T_1}{T_0} - R \log_e \frac{p_1}{p_0}$$

$$= 1.005 \log_e \frac{453}{293} - 0.287 \log_e \frac{12}{1.5}$$

$= 0.4378 - 0.5968 = -0.159$ kJ/kg

∴ $T_0(s_1 - s_0) = 293 \times (-0.159) - 46.59$kJ/kg

Further,

$$\frac{V_1^2}{2} = \frac{100^2}{2}$$

$= 5000$ J/kg $= 5$ kJ/kg

∴ Avaiability at inlet

$= 160.8 - (-46.59) + 5$
$= 212.39$ kJ/kg

Similarly availability of air at exit,

$$= (h_2 - h_0) - T_0\,(s_2 - s_0) + \frac{V_2^2}{2}$$

Now : $h_2 - h_0 = c_p\ (T_2 - T_0)$
$= 1.005\ (293 - 293) = 0$

$$s_2 - s_0 = c_p \text{ loge } \frac{T_2}{T_0} - R \log_e \frac{p_2}{p_0}$$

$$= 1.005 \log_e \frac{293}{293} - 0.287 \log_e \frac{1.5}{1}$$

$= 0 - 0.1164 = -0.1164$ kJ/kg K

$$\frac{V_2^2}{2} = \frac{50^2}{2 \times 100} = 1.25 \text{ kJ/kg}$$

∴ Availability of air at exit

$= 0 - 293\ (-0.1164) + 1.25$
$= 35.35$ kJ/kg

∴ Reversible or theoretical work

$W_{rev} = 212.39 - 35.35 = 177.04$ kJ/kg

Actual work output can be calculated by using steady flow energy equation assuming the process to be adiabatic (no heat interaction)

$$W_{act} = (h_1 - h_2) + \frac{V_1^2 - V_2^2}{2}$$

$$= c_p\,(T_1 - T_2) + \frac{V_1^2 - V_2^2}{2}$$

$$= 1.005\,(453 - 293) + \frac{100^2 - 50^2}{2 \times 10^3}$$

$= 160.8 + 3.75 = 164.55$ kJ/kg

(*b*) Irreversibility,

$I = W_{rev} - W_{act}$
$= 177.04 - 164.55 = 12.49$ kJ/kg

$$\text{Effectiveness, } \varepsilon = \frac{W_{act}}{W_{rev}}$$

$$= \frac{164.55}{177.04} = 0.9294 \text{ or } \mathbf{92.94\%}$$

10.6. HELMHOLTZ AND GIBB'S FUNCTIONS

Consider a closed system which is initially and finally at temperature of the environment around it and exchanges heat only with the atmosphere. Then

$$\delta Q = \delta W + dU \qquad \text{... 1st law}$$

$$\delta Q = TdS \qquad \text{... 2nd law}$$

Hence for a non-flow reversible process,

$$W_{max} = U_1 - U_2 + Q$$
$$= (U_1 - U_2) + T_0 (S_2 - S_1)$$
$$= (U_1 - T_0S_1) - (U_2 - T_0S_2)$$

Replacing T_0 by T_1 or T_2 (since the process is at constant temperature), we have

W_{max} (maximum work out put from the system)

$$= (U_1 - T_1S_1) - (U_2 - T_2S_2)$$
$$= A_1 - A_2 \qquad \text{...(10.7)}$$

The term $(U - TS)$ is called the **Helmholtz function** and is defined as the difference between the internal energy and the product of temperature and entropy. Thus if a closed system passes from one state to another state at the same temperature while exchanging heat only with the surrounding atmosphere at that temperature ($T_2 = T_1 = T_0$), the maximum work of the process is equal to the decrease in Helmholtz function of the system.

The Helmholtz function is useful in solution of problems involving chemical and electro-chemical processes which generally occur at constant temperature. However, its usefulness is restricted to non-flow processes only.

In case of flow processes, flow work at entry and flow work at exit need to be included in the expression for useful work. At entry the flow work p_1V_1 is supplied from the surroundings to the system while at exit point the flow work p_2V_2 is supplied by the system to the surroundings. Accordingly it follows that

$$W_{max} = (U_1 - U_2) + T_0 (S_2 - S_1) - (p_2V_2 - p_1V_1)$$
$$= (U_1 + p_1V_1 - T_0S_1) - (U_2 + p_2V_2 - T_0S_2)$$

The term T_0 can be replaced by T_1 or T_2 as the process is at constant temperature of surroundings.

$$W_{max} = (U_1 + p_1V_1 - T_1S_1) - (U_2 + p_2V_2 - T_2S_2)$$
$$= (H_1 - T_1S_1) - (H_2 - T_2S_2)$$
$$= G_1 - G_2 \qquad \text{...(10.8)}$$

The term $(H - TS)$ is called **Gibb's function** and is defined as the difference between enthalpy and product of temperature and entropy. Thus if a substance enters and leaves a steady flow system at the temperature of surrounding atmosphere and exchanges heat only with the atmosphere, the maximum work that can be produced is equal to the decrease in Gibb's function of the substance. The Gibb's function is a property which represents the upper limit for the work done in a steady flow isothermal process during which a system exchanges heat only with the surroundings. Gibb's function is useful in the study of vapour pressure, phase equilibrium and chemical reactions.

The changes of both Helmholtz and Gibb's functions are called free energy *i.e.*, energy that is free to be converted into work. Further, both the Helmholtz and Gibb's functions establish a criterion for thermodynamic equilibrium. At equilibrium, these functions are at their minimum values.

10.7. EFFECTIVENESS AND IRREVERSIBILITY

The concept of availability leads to the following parameters which are helpful in the assessment of the performance of a process or cycle:

Effectiveness represents the fraction of maximum useful work $(W_u)_{max}$ actually utilized and is expressed as the ratio of useful work W_u to the maximum useful work $(W_u)_{max}$.

During cooling or expansion process (*i.e.* work producing), the work produced is at the expense of loss of availability of the system, and the work produced increases the availability of the environment which is the objective.

$$\varepsilon = \frac{W_u}{(W_u)_{max}}$$

$$= \frac{\text{gain of availability of the environment}}{\text{loss of availability of the system}} \quad ...(10.9)$$

All of the work performed by a system W_{exp} is not available for delivery; a certain portion of it has to be spent in pushing out the atmosphere. The work done on the atmosphere equals $p_0 dV$ where p_0 is the atmospheric pressure and dV is the change in volume

$$W_u = W_{exp} - W_{surr} = W_{exp} - p_0\, dV$$

Likewise the maximum useful work is

$$(W_u)_{max} = W_{max} - p_0\, dV$$

In a steady flow system, there is no change in the volume of the system, *i.e.* no work is done on the atmosphere. Therefore,

$(W_u)_{max} = W_{max}$ in a steady flow

Further, for a reversible process,

$$(W_u)_{max} = W_u$$

Accordingly the effectiveness is unity.

(*ii*) for a heating or compression process

$$\epsilon = \frac{\text{loss of availability of system}}{\text{gain of availability of the environment}} \quad ...(10.10)$$

Irreversibility denotes the loss of work due to the fact that all real processes are irreversible, and is defined as the difference between the maximum work output from the system and the expansion work. That is

$$I = \left[(U_1 - T_0S_1) - (U_2 - T_0S_2)\right] - \left[-Q - (U_2 - U_1)\right]$$

$$= T_0\,(S_2 - S_1) + Q$$

Change in entropy of environment due to addition of heat Q at constant atmospheric temperature T_0,

$$(dS)_{surr} = \frac{Q}{T_0}\,; \quad Q = T_0\,(dS)_{surr}$$

Note: The sign convention for heat interaction Q; it is negative for the system and positive for the surroundings.

$$\therefore \quad I = T_0\left[(dS)_{sys} + (dS)_{surr}\right]$$

$$= T_0\,(dS)_{net}$$

$$= T_0(dS)_{univ} \quad ...(10.11)$$

Similarly for the steady flow process,

$$I = \left[(h_1 - T_0S_1) - (h_2 - T_0S_2)\right] - \left[(h_1 - h_2) - Q\right]$$

$$= T_0\,(S_2 - S_1) + Q$$

$$= T_0\,(dS)_{system} + T_0\,(dS)_{surr}$$

$$= T_0\left[(dS)_{sys} + (dS)_{surr}\right]$$

$$= T_0(dS)_{net} = T_0(dS)_{univ} \quad ...(10.12)$$

The same expressions (10.11 and 10.12) for irreversibility apply to both flow and non-flow processes, and it prescribes that irreversibility equals the product of temperature of surroundings and the entropy production. The quantity $T_0(dS)_{univ}$ also represents an increase in unavailable energy (or anergy) and accordingly it may be stated that irreversibility amounts to increase in anergy.

In terms of rate of loss of available energy, the equation 10.11 or 10.12 may be written as

$$I = T_0\,(dS)_{univ} = T_0\,S_{gen} \quad ...(10.13)$$

This is referred to as the Gouy-stodola equation. There would be minimum exergy loss with minimum rate of entropy generation in a thermodynamically efficient process.

EXAMPLE 10.18

One kg of air is heated from 25°C to 50°C by mixing it with a certain quantity of air at 120°C. Assuming mixing process to be adiabatic and neglecting changes in kinetic and potential energies, evaluate the mass of heating air and effectiveness of the heating process. Take c_p = 1.005 kJ/kg K and ambient temperature to be 20°C.

Solution : Let suffix 1, 2, 3 denote the air streams at 25°C, 120°C and 50°C respectively. If m is the mass of heating air, then from the principle of energy conservation,

$$c_p t_1 + m c_p t_2 = (1 + m) c_p t_3$$

or $$m = \frac{t_3 - t_1}{t_2 - t_3}$$

$$= \frac{50 - 25}{120 - 50} = 0.357 \text{ kg}$$

Gain in availability of the air stream which is being heated from $t_1 = 25°C$ ($T_1 = 298$ K) to $t_3 = 50°C$ ($T_3 = 323$ K) taken as system is

$$B_3 - B_1 = (h_3 - h_1) - T_0 (s_3 - s_1)$$

$$= c_p (T_3 - T_1) - T_0 c_p \log_e \frac{T_3}{T_1}$$

$$= 1.005 (323 - 298) - 293 \times 1.005 \log_e \frac{323}{298}$$

$$= 25.12 - 23.72 = 1.4 \text{ kJ/kg}$$

$$= 1 \times 1.4 = 1.4 \text{ kJ}$$

$$(m = 1 \text{ kg})$$

Loss in availability of the air stream which is being cooled from $t_2 = 120°C$ ($T_2 = 393$ K) to $t_3 = 50°C$ ($T_3 = 323$ K) taken as surrounding is

$$(B_2 - B_3) = (h_2 - h_3) - T_0 (s_2 - s_3)$$

$$= c_p (T_2 - T_3) - T_0 c_p \log_e \frac{T_2}{T_3}$$

$$= 1.005 (393 - 323) - 293 \times 1.005 \log_e \frac{393}{323}$$

$$= 70.35 - 57.76 = 12.59 \text{ kJ/kg}$$

$$= 0.357 \times 12.59 = 4.49 \text{ kJ}$$

$$(m = 0.357 \text{ kg})$$

For a heating process,

Effectiveness

$$= \frac{\text{increase in availability of the system}}{\text{loss in availability of the surroundings}}$$

$$= \frac{1.40}{4.49} = \mathbf{0.312} \quad \textbf{or} \quad \mathbf{31.2\%}$$

EXAMPLE 10.19

Water is to be heated at constant pressure from 25°C to 80°C. If the heat source is at a constant temperature of 500°C and the ambient temperature is 20°C, what would be the gain in availability of water and effectiveness of the heating process ? For water take c_p = 4.187 kJ/kg K.

Solution : Gain in availability of water (useful work),

$$= (h_2 - h_1) - T_0 (s_2 - s_1)$$

$$= c_p (t_2 - t_1) - T_0 c_p \log_e \frac{T_2}{T_1}$$

$$= 4.187 (80 - 25)$$

$$- 293 \times 4.187 \log_e \frac{80 + 273}{25 + 273}$$

$$= 22.498 \text{ kJ/Kg}$$

Maximum possible work, *i.e,* loss of availability of surroundings,

= heat supplied × Carnot efficiency

With temperature of source and sink as 500°C and 20°C,

Carnot efficiency

$$= \frac{(500 + 273) - (20 + 273)}{500 + 273} = 0.62$$

∴ Loss of availability of surroundings

$$= c_p (t_2 - t_1) \times \eta_{Carnot}$$

$$= 4.187 (80 - 25) \times 0.62 = 147.78 \text{ kJ}$$

∴ Effectiveness

$$= \frac{\text{useful work}}{\text{maximum possible usefulwork}}$$

$$= \frac{\text{gain in availability of system}}{\text{loss in availability of surroundings}}$$

$$= \frac{22.498}{147.78} = \mathbf{0.157 \text{ or } 15.7\%}$$

10.8. THIRD LAW OF THERMODYNAMICS (NERNST LAW)

The degree of atomic or molecular activity of a substance depends upon its temperature. As absolute zero temperature is approached, the randomness of molecules ceases and the entropy becomes equal to zero. This is the essence of third law of thermodynamics which states that:

At absolute zero temperature, the entropy of all homogeneous crystalline (condensed) substances in a state of equilibrium becomes zero.

When expressed mathematically : $\underset{T \to 0}{dt}\ S = 0$

This implies that at absolute zero temperature, the entropy ceases to be a function of state and approaches a constant value independent of the parameters of the state. However, for many substances like alloys, all amorphous bodies, chemical compounds CO, NO etc., the entropy does not tend to zero as $T \to 0$, but takes a finite positive value. This is because these substances are not found in equilibrium state.

The third law provides an absolute base from which the entropy of each substance can be measured. The entropy relative to this base is referred to as the absolute entropy. Further, this law is helpful in

– measurement of chemical affinity, *i.e.*, the action of chemical forces of the reacting substances.

– explaining the behaviour of solids at very low temperatures.

– analysing the chemical and phase equilibrium.

SALIENT POINTS

1. Availability of a system is the maximum work obtained under ideal conditions which reduce the system to its dead state.
2. The unavailable energy of heat withdrawn from finite as well as infinite reservoir is given by $T_0\, dS$.
3. Availability of a heat source is
$$A = Q - T_0\, dS$$
4. An increase in unavailable energy due to heat transfer through a finite temperature difference is $T_0\,(dS)_{univ}$
5. Energy is said to be degraded each time it is transferred through a finite temperature difference.
6. Availability of a closed system is
$$W_{max} = A - A_0 = (U - T_0 S + p_0 V) - (U_0 - T_0 S_0 + p_0 V_0)$$
7. Availability of a steady flow system is
$$W_{max} = A - A_0 = \left(h - T_0\, S + \frac{C_1^2}{2} + gz_1\right) - \left(h_0 - T_0\, S_0 + \frac{C_0^2}{2} + gz_0\right)$$
8. Non-flow available function
$$A = U + p_0 V - T_0 S$$
differs from the steady flow available function $B = H - T_0 S = U + pV - T_0 S$ in pressure term only.
9. Both A and B are composite properties of the system and surroundings; they involve properties of both the system and surroundings.
Gibb's function $G = (H - TS)$
$= (U + pV - TS)$
is a property of the system only. However, at the dead state (environmental conditions), the properties of the system and surroundings are same and, therefore,
$$A \equiv B \equiv G$$
i.e., these parameters become numerically equal.
10. The concept of entropy is used
—in the correlation of physical data.
—to determine whether a process is reversible, irreversible or impossible.
—to make calculations for heat transfer in a reversible process.
—as a convenient co-ordinate for depicting ideal processes in many machines.
11. Irreversibility of a system is
$$I = T_0 (dS)_{net}$$
12. Effectiveness of a system is defined as the ratio of useful work to maximum useful work.

REVIEW QUESTIONS

A. Conceptual and conventional questions.

1. What is high grade and low grade energy ? Give important examples of each category.
2. Define the terms available energy and unavailable energy.

3. What do you understand by exergy and anergy ?
4. Define dead state.
5. What is meant by availability ?
6. Set up an expression for available energy for an infinite energy source at temperature T when the ambient temperature is T_0.
7. Comment on the statement "Energy is always conserved, but its quality is always degraded."
8. Why is the second law of thermodynamics referred to as the law of degradation of energy?
9. Setup expressions for availabilities of a closed system and a steady flow open system.
10. What are Helmholtz function and Gibb's function ?
11. What is the difference between the Gibb's function and the availability function ? When these two functions are numerically equal ?
12. Explain the concept of the third law of thermodynamics.
13. Define irreversibility. Show that irreversibility of a process is given by the product of the temperature of surroundings and the net entropy change.
14. 0.25 kg of air at constant pressure and a temperature of 250ºC receives 300 kJ of heat reversibly. Determine the available and unavailable energy if the temperature of surroundings is 27ºC. Take c_p = 1.005 kJ/kg K.

    [Ans. 210.4 kJ, 89.60 kJ]
15. 2 kg of water at 40°C is mixed with 3 kg of water at 80°C in a steady flow process. Determine the temperature of the resulting mixture and the entropy change of the mixing process. What would be the unavailable energy with respect to the receiver at 45°C ?

    [Ans. 64°C, 0.0362 kJ/K, 11.51 kJ]
16. 500 kJ of heat at constant temperature of 800 K was removed and supplied to a system at constant temperature of 700 K. Find the net loss of available energy as a result of this irreversible heat transfer. Take ambient temperature as 300 K.

    [Ans. 26.79 kJ]
17. 5 kg of air at 1.5 bar and 27°C is heated reversibly at constant pressure to 227°C. If the lowest available temperature is 20°C, how much is the available energy ?

    Take c_p = 1.005 kJ/kg K.

    [Ans. 59.58 J/kg]
18. A heat source at 900 K transfers heat at the rate of 50 kW to a system at 560 K. Assuming that these temperatures remain constant, make calculations for change in entropy of the source, the entropy production accompanying the heat transfer, original available energy and the available energy after heat transfer.

    [Ans. – 3.333 kJ/K-min ; 2.024 kJ/K-min ; 1866.56 kJ, 1393 kJ/K-min]
19. In a boiler for generating steam, water is evaporated at 520 K, while the combustion gases are cooled from 1350 K to 400 K. Determine the loss in available energy for unit mass of steam generated.

    Take c_p for gases 1.08 kJ/kg K, latent heat of steam 1731 kJ/kg and ambient temperature 300 K.

    [Ans. 556.8 kJ]
20. A mass of 0.25 kg of air in a closed system expands from 2 bar and 60°C to 1 bar and 40°C, while receiving 1.05 kJ of heat from a reservoir at 100°C. The surrounding atmosphere is at 0.95 bar and 27°C. Determine the maximum work. How much of this work would be done on the atmosphere ?

    [Ans. 23.44 kJ, 9.986 kJ]
21. Air expands through a turbine from 500 kPa, and 520°C to 100 kPa and 300°C. During expansion 10 kJ/kg of heat is lost to the surroundings which are at 98 kPa and 20°C. Neglecting the changes in kinetic and potential energies, determine per kg of air (*i*) the irreversibility (*ii*) the decrease in availability and (*iii*) the maximum work.

    For air take c_p = 1.005 kJ/kg K and R = 0.287 kJ/kg K

    [Ans. 49.6 kJ/kg, 260.7 kJ/kg 260.7 kJ/kg]
22. A heat source at constant temperature 1000 K transfers heat uniformly at 100 watt to a system operating in a Carnot engine located in an environment at 300 K. If the source transfers irreversibly to a system at 800 K, find the net increase in entropy and loss in availability.

    [Ans. 0.025 J/K-s, 7.5 J/s]
23. (*a*) For a system of perfect gas, prove that two reversible adiabatic lines on any thermodynamic co-ordinates cannot intersect.

    (*b*) Prove that the constant volume lines on $T-S$ diagram are steeper than constant pressure lines working between the same temperature limits.
24. 10 kg of water at 30°C is mixed with 1 kg of ice at 0°C, the mixing is adiabatic and the system is open to atmosphere at 10°C.

Determine the net increase in entropy and unavailable energy when the system attains common temperature after the spontaneous mixing process. Take specific heat of water and ice = 4.18 kJ/kg K and 2.1 kJ/kg K respectively and latent heat of fusion of ice = 335 kJ/kg.

[Ans. 1.088kJ/K, 307.8K J]

25. In a parallel flow heat exchanger, water is heated from 60°C to 80°C by an oil which enters at 250°C and leaves at 100°C. Determine the loss in availability on the basis of one kg of oil per second.

    Take specific heat of oil = 2.5 kJ/kg K and ambient temperature = 27°C.

    [Ans. 59.9 k]

26. The exhaust gases from a gas turbine are at 950 K and a mass flow rate of 25 kg/s. These gases are used to heat water from 300 K to 425 K. If the rate of water flow is 32.5 kg/s, determine the loss of available energy as a result of heat transfer from the gases (sp. heat 1.1 kJ/kg K) to water (sp. heat 4.187 kJ/kg K). It may be presumed that the lowest available sink temperature is 300 K and the heat transfer process is carried out at constant pressure.

B. Fill in the blanks with appropriate word/words.

1. The portion of thermal energy put in cyclic engine which gets converted into mechanical work is called the ....
2. The availability of a system at any given state is the maximum .... obtainable as the system goes to ....
3. The concept of .... implies that pressure and temperature of the system are equal to that of surroundings.
4. The change in availability of a system is equal to change in .... of the system at constant temperature and pressure.
5. For a non-flow process, the maximum useful work obtainable is given by the change in ....
6. The parameter ($u - Ts$) defined as the difference between internal energy and product of temperature and entropy is called the ....
7. The non-flow availability function for a closed system is expressed as ....
8. The non-flow availability function $A = (U + p_0V - T_0S)$ consists of .... properties of the system and .... properties of the surroundings.
9. The .... law provides an absolute base from which the entropy of each substance can be measured.

Answers

1. available energy 2. useful work, dead state 3. dead state 4. Gibb's function 5. availability 6. Helmholtz function 7. $A = U + p_oV - T_oS$ 8. three extensive and two intensive 9. third law

□□□

Chapter 11

Thermodynamic Relations

Learning objectives : The mathematical treatment of different thermodynamic properties of a system presented in this chapter will enable the reader to:

- *define the reciprocal and cyclic relations*
- *prescribe the property relations and provide a check for thermodynamic property*
- *set up Maxwell's thermodynamic relations and Tds equations*
- *apply Maxwell's relations and Tds equations in deriving the equations for specific heats, heat capacity, entropy and internal energy*
- *define volume expansivity, isothermal compressibility and adiabatic compressibility*

11.1. RECIPROCAL AND CYCLIC RELATIONS

Thermodynamic relations pertain to the relations between different thermodynamic properties of a system. Such relations help to determine changes in certain properties which are rather difficult to measure experimentally. Further, through these relations, the limited experimental data can be used to derive equations of state for different substances which do not undergo chemical reactions or a change in phase.

Before establishing some general thermodynamic relations, it is desirable to review a few principles of partial differential calculus.

Consider the relationship $f(x, y, z) = 0$ among the three variables x, y and z. Then

$$x = f(y, z)$$

or $$dx = \left(\frac{\partial x}{\partial y}\right)_z dy + \left(\frac{\partial x}{\partial z}\right)_y dz \quad ...(11.1)$$

The subscript denotes the variable held constant.

Also $y = f(x, z)$

or $$dy = \left(\frac{\partial y}{\partial x}\right)_z dx + \left(\frac{\partial y}{\partial z}\right)_x dz \quad ...(11.2)$$

When dy from equation 11.2 is substituted in equation 11.1, one obtains

$$dx = \left(\frac{\partial x}{\partial y}\right)_z \left[\left(\frac{\partial y}{\partial x}\right)_z dx + \left(\frac{\partial y}{\partial z}\right)_x dz\right] + \left(\frac{\partial x}{\partial z}\right)_y dz$$

or $$dx = \left(\frac{\partial x}{\partial y}\right)_z \left(\frac{\partial y}{\partial x}\right)_z dx + \left[\left(\frac{\partial x}{\partial y}\right)_z \left(\frac{\partial y}{\partial z}\right)_x + \left(\frac{\partial x}{\partial z}\right)_y\right] dz \quad ...(11.3)$$

x and z are independent variables in the above expression and accordingly

$$\left(\frac{\partial x}{\partial y}\right)_z \left(\frac{\partial y}{\partial x}\right)_z = 1 \quad ...(11.4)$$

and $$\left(\frac{\partial x}{\partial y}\right)_z \left(\frac{\partial y}{\partial z}\right)_x + \left(\frac{\partial x}{\partial z}\right)_y = 0 \quad ...(11.5)$$

Equation 11.4 can be rewritten as

$$\left(\frac{\partial x}{\partial y}\right)_z = \frac{1}{(\partial y/\partial x)_z} \quad ...(11.6)$$

and this is called the ***reciprocal relation.***

Further, the equation 11.5 can be recast as

$$\left(\frac{\partial x}{\partial y}\right)_z \left(\frac{\partial y}{\partial z}\right)_x = -\left(\frac{\partial x}{\partial z}\right)_y = -\frac{1}{\left(\frac{\partial z}{\partial x}\right)_y}$$

$$\text{or} \quad \left(\frac{\partial x}{\partial y}\right)_z \left(\frac{\partial y}{\partial z}\right)_x \left(\frac{\partial z}{\partial x}\right)_y = -1 \quad ...(11.7)$$

Equation 11.7 is called the ***cyclic relation.***

EXAMPLE 11.1

Prove that the ideal gas equation pv = RT satisfies the cyclic relation.

Solution : From the given ideal gas equation

$p = \frac{RT}{v}$, *i.e.*, the pressure changes due to changes in volume and temperature (the gas constant R being constant for a particular gas). The cyclic relation then slipulates that

$$\left(\frac{\partial p}{\partial v}\right)_T \times \left(\frac{\partial v}{\partial T}\right)_p \times \left(\frac{\partial T}{\partial p}\right)_v = -1$$

The three partial derivates can be determined separately and we get

$$\left(\frac{\partial p}{\partial v}\right)_T = RT\frac{\partial}{\partial v}\left(\frac{1}{v}\right) = -\frac{RT}{v^2}$$

$$\left(\frac{\partial v}{\partial T}\right)_p = \frac{R}{p}\frac{\partial}{\partial T}(T) = \frac{R}{p}$$

$$\left(\frac{\partial T}{\partial p}\right)_v = \frac{v}{R}\frac{\partial}{p}(p) = \frac{v}{R}$$

Substituting these values in left hand side of the cyclic relation, we obtain

$$-\frac{RT}{v^2}\times\frac{R}{p}\times\frac{v}{R} = -\frac{RT}{pv} = -1$$

$$(\because pv = RT)$$

and it equals the right hand side. That proves that the given gas equation satisfies the cyclic relation.

11.2. PROPERTY RELATIONS

• For a unit mass of the system, the first and second law of thermodynamics are prescribed by the relations

$$\delta q = \delta w + du = p\,dv + du$$

$$\text{and} \quad \delta q = T\,ds$$

$$\therefore \quad T\,ds = p\,dv + du$$

$$\text{or} \quad du = T\,ds - p\,dv \quad ...(11.8)$$

• The ***enthalpy*** is defined as the sum of internal energy and the flow work (*i.e.* pressure volume product). Thus for a unit mass,

$$h = u + pv$$

$$\begin{aligned} \text{or} \quad dh &= du + p\,dv + v\,dp \\ &= (T\,ds - p\,dv) + p\,dv + v\,dp \\ &= T\,ds + v\,dp \quad ...(11.9) \end{aligned}$$

• The ***Helmholtz free energy*** is defined as the difference between the internal energy and the product ot temperature and entropy. Thus for a unit mass,

$$f = u - Ts$$

$$\begin{aligned} \text{or} \quad df &= du - T\,ds - s\,dT \\ &= (T\,ds - p\,dv) - T\,ds - s\,dT \\ &= -p\,dv - s\,dT \quad ...(11.10) \end{aligned}$$

• The ***Gibbs free energy*** is defined as the difference between enthalpy and the product of temperature and entropy. Thus for a unit mass

$$g = h - Ts$$

$$\begin{aligned} \text{or} \quad dg &= dh - T\,ds - s\,dT \\ &= (T\,ds + v\,dp) - T\,ds - s\,dT \\ &= v\,dp - s\,dT \quad ...(11.11) \end{aligned}$$

Since u, h, f and g are properties of the system, the relations 11.8 to 11.11 are often referred to as the *property relations* (also called the *Gibbs relations*). These relations are applicable to all processes whether reversible or irreversible, and whether in a closed system or in an open system.

11.3. CHECK FOR A PROPERTY

Consider z to be a function of two independent variables x and y. Mathematically

$$z = f(x, y)$$

$$\text{or} \quad dz = \left(\frac{\partial z}{\partial x}\right)_y dx + \left(\frac{\partial z}{\partial y}\right)_x dy \quad ...(11.12)$$

where z is an exact differential , *i.e.*, it represents a thermodynamic property. Let the partial derivative in the above identity be written as

$$\left(\frac{\partial z}{\partial x}\right)_y = M \text{ and } \left(\frac{\partial z}{\partial y}\right)_x = N$$

Equation 11.12 may then be written as

$$dz = M\,dx + N\,dy$$

$$\text{Further : } \frac{\partial M}{\partial y} = \frac{\partial}{\partial y}\left(\frac{\partial z}{\partial x}\right)_y = \frac{\partial^2 z}{\partial x\,\partial y}$$

$$\text{and } \frac{\partial N}{\partial x} = \frac{\partial}{\partial x}\left(\frac{\partial z}{\partial y}\right)_x = \frac{\partial^2 z}{\partial x\,\partial y}$$

$$\text{Thus } \left(\frac{\partial M}{\partial y}\right)_x = \left(\frac{\partial N}{\partial x}\right)_y$$

This leads us to conclude that considering the expression: $dz = M\,dx + N\,dy$

If z represents thermodynamic property of the system, then

$$\left(\frac{\partial M}{\partial y}\right)_x = \left(\frac{\partial N}{\partial x}\right)_y$$

11.4. MAXWELL RELATIONS

Consider the expression : $du = T\,ds - p\,dv$

Here internal energy u is a thermodynamic property and the expression indicates that internal energy varies with entropy s and volume v. That is

$$u = f(s, v)$$

$$\text{or } du = \left(\frac{\partial u}{\partial s}\right)_v ds + \left(\frac{\partial u}{\partial v}\right)_s dv$$

Equating the coefficients of ds and dv, we may write

$$T = \left(\frac{\partial u}{\partial s}\right)_v \text{ and } p = -\left(\frac{\partial u}{\partial v}\right)_s$$

When differentiated, we obtain

$$\left(\frac{\partial T}{\partial v}\right)_s = \frac{\partial^2 u}{\partial v\,\partial s} \text{ and } \left(\frac{\partial p}{\partial s}\right)_v = -\frac{\partial^2 u}{\partial v\,\partial s}$$

$$\therefore \left(\frac{\partial T}{\partial v}\right)_s = -\left(\frac{\partial p}{\partial s}\right)_v \quad ...(11.13)$$

Alternatively : Comparing the given identity with $dz = M\,dx + N\,dy$

$$M = T \qquad N = -p$$
$$x = s \qquad y = v$$

Since z corresponds to u which is a thermodynamic property, the following identity must hold good:

$$\left(\frac{\partial M}{\partial y}\right)_x = \left(\frac{\partial N}{\partial x}\right)_y, \text{ i.e., } \left(\frac{\partial T}{\partial v}\right)_s = -\left(\frac{\partial p}{\partial s}\right)_v$$

- Consider the expression :

$$dh = T\,ds + v\,dp$$

Here enthalpy h is a thermodynamic property and the expression indicates that enthalpy is a function of entropy and pressure. That is

$$h = f(s, p)$$

$$\text{or } dh = \left(\frac{\partial h}{\partial s}\right)_p ds + \left(\frac{\partial h}{\partial p}\right)_s dp$$

Equating the coefficients of ds and dv, we may write

$$T = \left(\frac{\partial h}{\partial s}\right)_p \text{ and } v = \left(\frac{\partial h}{\partial p}\right)_s$$

When differentiated, we obtain

$$\left(\frac{\partial T}{\partial p}\right)_s = \frac{\partial^2 h}{\partial s\,\partial p} \text{ and } \left(\frac{\partial v}{\partial s}\right)_p = \frac{\partial^2 h}{\partial s\,\partial p}$$

$$\therefore \left(\frac{\partial T}{\partial p}\right)_s = \left(\frac{\partial v}{\partial s}\right)_p \quad ...(11.14)$$

Alternatively : Comparing the given identity with $dz = M\,dx + N\,dy$

$$M = T \quad N = v$$
$$x = s \quad y = p$$

Since z corresponds to h which is a thermodynamic property, the following identity must hold good,

$$\left(\frac{\partial M}{\partial y}\right)_x = \left(\frac{\partial N}{\partial x}\right)_y, \text{ i.e., } \left(\frac{\partial T}{\partial p}\right)_s = \left(\frac{\partial v}{\partial s}\right)_p$$

- Consider the expression :

$$df = -p\,dv - s\,dT$$

Here Helmholtz free energy f is a thermodynamic property and the expression indicates that it varies with volume v and temperature T. That is,

$$f = f(v, T)$$

or
$$df = \left(\frac{\partial f}{\partial v}\right)_T dv + \left(\frac{\partial f}{\partial T}\right)_v dT$$

Equating the coefficients of dv and dT, we may write

$$p = -\left(\frac{\partial f}{\partial v}\right)_T \quad \text{and} \quad s = -\left(\frac{\partial f}{\partial T}\right)_v$$

When differentiated, we obtain,

$$\left(\frac{\partial p}{\partial T}\right)_v = -\frac{\partial^2 f}{\partial T\,\partial v}$$

and
$$\left(\frac{\partial s}{\partial v}\right)_T = -\frac{\partial^2 f}{\partial T\,\partial v}$$

$$\therefore \quad \left(\frac{\partial p}{\partial T}\right)_v = \left(\frac{\partial s}{\partial v}\right)_T \qquad \text{...(11.15)}$$

Alternatively : Comparing the given expression with $dz = M\,dx + N\,dy$

$M = -p \qquad N = -s$

$x = v \qquad y = T$

Since z corresponds to f which is a thermodynamic property, the following identity must hold good,

$$\left(\frac{\partial M}{\partial y}\right)_x = \left(\frac{\partial N}{\partial x}\right)_y \quad i.e., \quad -\left(\frac{\partial p}{\partial T}\right)_v = -\left(\frac{\partial s}{\partial p}\right)_T$$

- Consider the expression :

$$dg = v\,dp - s\,dT$$

Here the Gibbs free energy g is a thermodynamic property and the expression indicates that it varies with pressure p and temperature T. That it

$$g = f(p, T)$$

or
$$dg = \left(\frac{\partial g}{\partial p}\right)_T dp + \left(\frac{\partial g}{\partial T}\right)_p dT$$

Equating the coefficients of dp and dT, we may write

$$v = \left(\frac{\partial g}{\partial p}\right)_T \quad \text{and} \quad s = -\left(\frac{\partial g}{\partial T}\right)_p$$

When differentiated, we obtain

$$\left(\frac{\partial v}{\partial T}\right)_p = \frac{\partial^2 g}{\partial p\,\partial T}$$

and
$$\left(\frac{\partial s}{\partial p}\right)_T = -\frac{\partial^2 g}{\partial p\,\partial T}$$

$$\therefore \quad \left(\frac{\partial v}{\partial T}\right)_p = -\left(\frac{\partial s}{\partial p}\right)_T \qquad \text{...(11.16)}$$

Alternatively : Comparing the given expression with $dz = M\,dx + N\,dy$

$m = v \qquad N = -s$

$x = p \qquad y = T$

Since z corresponds to g which is a thermodynamic property, the following identity must hold good.

$$\left(\frac{\partial M}{\partial y}\right)_x = \left(\frac{\partial N}{\partial x}\right)_y \quad i.e., \quad \left(\frac{\partial v}{\partial T}\right)_p = -\left(\frac{\partial s}{\partial p}\right)_T$$

Equations 11.13 to 11.16 are called *Maxwell relations*. They relate the properties p, v, T and s of a system, and provide means for calculating the entropy of a substance from p-v-T relations.

11.5. *T-ds* EQUATIONS

(*i*) Consider variation of internal energy u with temperature T and specific volume v. Then

$$u = f(T, v)$$

or
$$du = \left(\frac{\partial u}{\partial T}\right)_v dT + \left(\frac{\partial u}{\partial v}\right)_T dv$$

$$= c_v\,dT + \left(\frac{\partial u}{\partial v}\right)_T dv \qquad \text{...(a)}$$

Invoking property relation :

$$du = T\,ds - p\,dv \qquad \text{...(b)}$$

Comparing identities (*a*) and (*b*), and equating the coefficients

$$c_v = T\left(\frac{\partial s}{\partial T}\right)_v$$

Introducing total differential of ds in terms of $s = f(T,v)$

$$ds = \left(\frac{\partial s}{\partial T}\right)_v dT + \left(\frac{\partial s}{\partial v}\right)_T dv$$

$$= c_v \frac{dT}{T} + \left(\frac{\partial s}{\partial v}\right)_T dv$$

From Maxwell relation 11.15 :

$$\left(\frac{\partial p}{\partial T}\right)_v = \left(\frac{\partial s}{\partial v}\right)_T$$

$$\therefore \quad ds = c_v \frac{dT}{T} + \left(\frac{\partial p}{\partial T}\right)_v dv \quad ...(11.17)$$

The term $c_v \frac{dT}{T}$ gives change in entropy at constant volume and the term $\left(\frac{\partial p}{\partial T}\right)_v dv$ represents change in entropy at constant temperature.

Equation 11.17 is known as first $T\,ds$ equation and is often recast as

$$T\,ds = c_v\,dT + T\left(\frac{\partial p}{\partial T}\right)_v dv \quad ...(11.18)$$

(*ii*) Consider variation of enthalpy h with temperature T and pressure p. Then

$$h = f(T, p)$$

or
$$dh = \left(\frac{\partial h}{\partial T}\right)_p dT + \left(\frac{\partial h}{\partial p}\right)_T dp$$

$$= c_p\,dT + \left(\frac{\partial h}{\partial v}\right)_T dv \quad ...(c)$$

Invoking property relation :

$$dh = T\,ds + v\,dp \quad ...(d)$$

Comparing identities (*c*) and (*d*), and equating the coefficients,

$$c_p = T\left(\frac{\partial s}{\partial T}\right)_p$$

Introducing total differential of ds in terms of $s = f(T, p)$

$$ds = \left(\frac{\partial s}{\partial T}\right)_p dT + \left(\frac{\partial s}{\partial p}\right)_T dp$$

$$= c_p \frac{dT}{T} + \left(\frac{\partial s}{\partial p}\right)_T dp$$

From Maxwell relation 11.16 :

$$\left(\frac{\partial s}{\partial p}\right)_T = -\left(\frac{\partial v}{\partial T}\right)_p$$

$$\therefore \quad ds = c_p \frac{dT}{T} - \left(\frac{\partial v}{\partial T}\right)_p dp \quad ...(11.19)$$

The term $c_p \frac{dT}{T}$ gives change in entropy at constant pressure and the term $\left(\frac{\partial v}{\partial T}\right)_p dp$ represents change in entropy at constant temperature.

Equation 11.19 is known as second $T\,ds$ equation and is often recast is

$$T\,ds = c_p\,dT - T\left(\frac{\partial v}{\partial T}\right)_p dp \quad ...(11.20)$$

11.6. HEAT CAPACITY RELATIONS

Equating the first and second $T\,ds$ equations, we get

$$c_v\,dT + T\left(\frac{\partial p}{\partial T}\right)_v dv = c_p\,dT - T\left(\frac{\partial v}{\partial T}\right)_p dp$$

or $(c_p - c_v)\,dT$

$$= T\left(\frac{\partial p}{\partial T}\right)_v dv + T\left(\frac{\partial v}{\partial T}\right)_p dp$$

or
$$dT = \frac{T\left(\frac{\partial p}{\partial T}\right)_v dv}{c_p - c_v} + \frac{T\left(\frac{\partial v}{\partial T}\right)_p dp}{c_p - c_v} \quad ...(i)$$

From the fundamental relation $T = f(p, v)$

$$dT = \left(\frac{\partial T}{\partial p}\right)_v dp + \left(\frac{\partial T}{\partial v}\right)_p dv \quad ...(ii)$$

Comparing identities (*i*) and (*ii*), and equating the coefficients

$$\frac{T\left(\frac{\partial p}{\partial T}\right)_v}{c_p - c_v} = \left(\frac{\partial T}{\partial v}\right)_p$$

or $c_p - c_v = T\left(\frac{\partial p}{\partial T}\right)_v \times \left(\frac{\partial v}{\partial T}\right)_p$ ...(*iii*)

Following cyclic relationship holds good among the thermodynamic variables p, T and v

$$\left(\frac{\partial p}{\partial T}\right)_v \times \left(\frac{\partial T}{\partial v}\right)_p \times \left(\frac{\partial v}{\partial p}\right)_T = -1 \quad ...(iv)$$

From identities (*iii*) and (*iv*), we have

$$c_p - c_v = T\left[\frac{-1}{\left(\frac{\partial T}{\partial v}\right)_p \times \left(\frac{\partial v}{\partial p}\right)_T}\right] \times \left(\frac{\partial v}{\partial T}\right)_p$$

$$= -T\left(\frac{\partial v}{\partial T}\right)_p^2 \times \left(\frac{\partial p}{\partial v}\right)_T \quad ...(11.21)$$

The following important conclusions can be drawn from the above thermodynamic relation :

(*i*) For any substance $\left(\frac{\partial p}{\partial v}\right)_T$ is negative, and the term $\left(\frac{\partial v}{\partial T}\right)_p^2$ is always positive. Obviously $(c_p - c_v)$ is positive and accordingly c_p is always greater than c_v.

(*ii*) $c_p \to c_v$ as $T \to 0$. Apparently c_p equals c_v at zero absolute temperature.

(*iii*) For solids and liquids, the volume change with temperature is negligible under constant pressure conditions. That is $\left(\frac{\partial v}{\partial T}\right)_p \to 0$. That gives $c_p - c_v = 0$, *i.e*, $c_p = c_v$. For water, such a situation exists at 4°C where specific volume is minimum or density is maximum.

(*iv*) For an ideal gas: $pv = RT$

$$\left(\frac{\partial v}{\partial T}\right)_p = \frac{R}{p} = \frac{v}{T}$$

$$\left(\frac{\partial p}{\partial v}\right)_T = \frac{\partial}{\partial v}\left(\frac{RT}{v}\right)_p = -\frac{RT}{v^2}$$

$$\therefore \quad c_p - c_v = -T \times \left(\frac{v}{T}\right)^2 \times \left(\frac{-RT}{v^2}\right) = R$$

(*v*) Quite often, it is considered worth while to express equation 11.21 in terms of *volume expansivity* β and *isothermal compressibility* α.

• ***Volume expansivity*** (coefficient of volume expansion) β is defined as the change in volume with change in temperature per unit volume, keeping the pressure constant.

That is $\beta = \frac{1}{v}\left(\frac{\partial v}{\partial T}\right)_p$ ...(11.22)

where the subscript p indicates that the partial differential is to be taken at constant pressure.

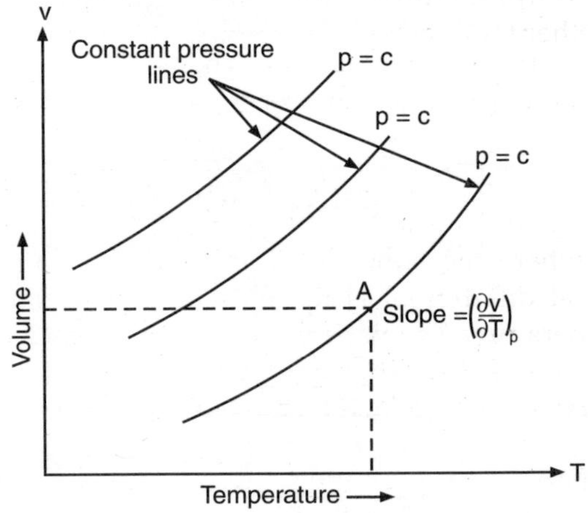

Fig. 11.1. Volume expansivity

With reference to volume-temperature diagram at constant pressure as shown in Fig. 11.1, the slope at any point A is given by $\left(\frac{\partial v}{\partial T}\right)_p$.

This slope divided by volume at that point gives the volume expansivity β.

For a perfect gas : $pv = RT$

or $v = \frac{RT}{p}$ and therefore

$$\left(\frac{\partial v}{\partial T}\right)_p = \frac{R}{P}$$

That gives : $\beta = \frac{1}{v} \times \frac{R}{P} = \frac{1}{v} \times \frac{v}{T} = \frac{1}{T}$

Apparently, the volume expansivity of a perfect gas is a function of temperature only. It varies inversely with absolute temperature and is independent of both pressure and volume.

• ***Isothermal compressibility α_T and adiabatic compressibility α_S***

In thermodynamics, compressibility is a measure of the relative volume change of a fluid in response to a pressure change. That is

$$\alpha = -\frac{1}{v}\left(\frac{\partial v}{\partial p}\right) \qquad ...(11.23)$$

This statement is, however, incomplete because for any fluid system, the magnitude of compressibility depends strongly on whether the process is isothermal or adiabatic. Accordingly we define the isothermal compressibility as

$$\alpha_T = -\frac{1}{v}\left(\frac{\partial v}{\partial p}\right)_T$$

where the subscript T indicates that the partial differential is to be taken at constant temperature. The negative sign implies that a positive pressure increment results in a negative volume increment, *i.e.*, an increase in pressure causes a decrease in volume. Isothermal compressibility thus represents the change in volume with the change in pressure per unit volume, keeping the temperature constant. With reference to pressure-volume diagram at constant temperature as shown in Fig. 11.2, the slope at any point is given by $-\left(\frac{\partial v}{\partial p}\right)_T$. This slope divided by volume at that point gives the isothermal compressibility α_T.

For a perfect gas : $pv = RT$

or $\quad v = \frac{RT}{p}$ and therefore

$$\left(\frac{\partial v}{\partial p}\right)_T = RT\left(-\frac{1}{p^2}\right)$$

That gives :

$$\alpha_T = -\frac{1}{v}\,RT\left(-\frac{1}{p^2}\right)$$

$$= -\frac{1}{v} \times pv\left(-\frac{1}{p^2}\right) = \frac{1}{p}$$

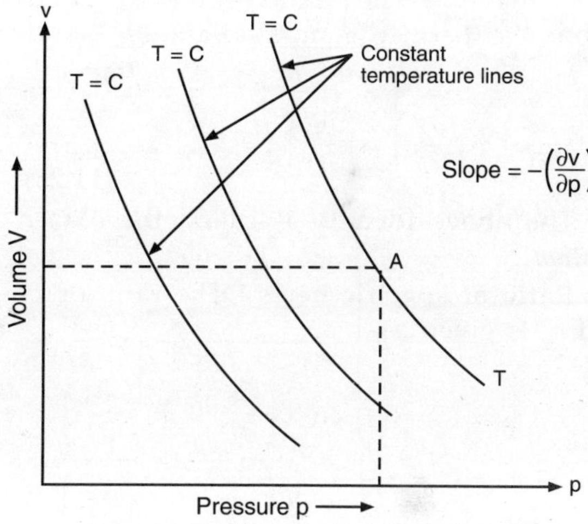

Fig. 11.2. Isothermal compressibility

Apparently the isothermal compressibility of a perfect gas varies inversely with pressure and is independent of both temperature and volume.

The adiabatic compressibility is defined as

$$\alpha_s = -\frac{1}{v}\left(\frac{\partial v}{\partial p}\right)_s$$

where the subscript s indicates that the partial differential is to be taken at constant entropy. The adiabatic compressibility thus represents the change in volume with change in pressure per unit volume when entropy is kept constant.

Considering changes in volume due to changes in pressure and temperature,

$$v = f(p, T)$$

or
$$dv = \left(\frac{\partial v}{\partial T}\right)_p dT + \left(\frac{\partial v}{\partial p}\right)_T dp$$

$$\therefore \quad \frac{dv}{v} = \frac{1}{v}\left(\frac{\partial v}{\partial T}\right)_p dT + \frac{1}{v}\left(\frac{\partial v}{\partial p}\right)_T dp$$

$$= \beta\, dT - \alpha\, dp \qquad ...(11.24)$$

The above correlation expresses volume change in terms of volume expansivity and isothermal compressibility.

Equation 11.21 can be rewritten as

$$c_p - c_v = \frac{\left[\frac{1}{v}\left(\frac{\partial v}{\partial T}\right)_p\right]^2}{-\frac{1}{v}\left(\frac{\partial v}{\partial p}\right)_T} Tv = \frac{\beta^2}{\alpha} Tv \quad ...(11.25)$$

The above identity is called the *Mayer relation.*

Ratio of specific heats : The ratio of c_p and c_v is given as

$$\gamma = \frac{c_p}{c_v} = \frac{T\left(\frac{\partial s}{\partial T}\right)_p}{T\left(\frac{\partial s}{\partial T}\right)_v} = \frac{\left(\frac{\partial s}{\partial T}\right)_p}{\left(\frac{\partial s}{\partial T}\right)_v}$$

From the cyclic correlations

$$\left(\frac{\partial s}{\partial T}\right)_p\left(\frac{\partial T}{\partial p}\right)_s\left(\frac{\partial p}{\partial s}\right)_T = -1$$

and $\left(\frac{\partial s}{\partial T}\right)_v\left(\frac{\partial T}{\partial v}\right)_s\left(\frac{\partial v}{\partial s}\right)_T = -1$, we have

$$\left(\frac{\partial s}{\partial T}\right)_p = -\left(\frac{\partial p}{\partial T}\right)_s\left(\frac{\partial s}{\partial p}\right)_T$$

and $$\left(\frac{\partial s}{\partial T}\right)_v = -\left(\frac{\partial v}{\partial T}\right)_s\left(\frac{\partial s}{\partial v}\right)_T$$

$$\therefore \quad \gamma = \frac{\left(\frac{\partial p}{\partial T}\right)_s\left(\frac{\partial s}{\partial p}\right)_T}{\left(\frac{\partial v}{\partial T}\right)_s\left(\frac{\partial s}{\partial v}\right)_T} = \frac{\left(\frac{\partial p}{\partial v}\right)_s}{\left(\frac{\partial p}{\partial v}\right)_T} \quad ...(11.26)$$

Since γ is greater than 1,

$$\left(\frac{\partial p}{\partial v}\right)_s > \left(\frac{\partial p}{\partial v}\right)_T$$

This indicates that on *p-v* plot, the slope of an isentropic is greater than that of an isotherm (Fig. 11.3).

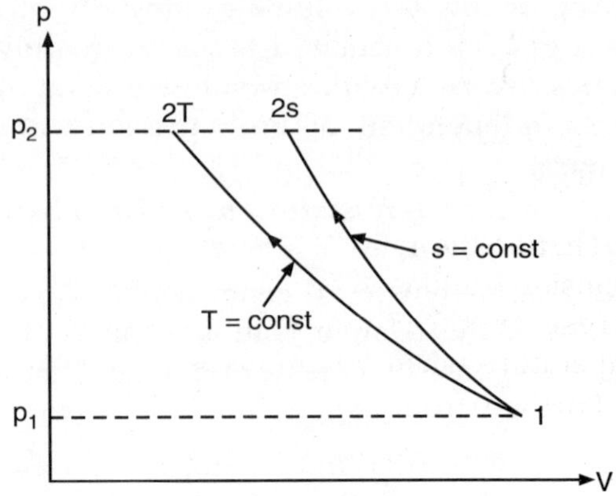

Fig. 11.3.

• ***Volume change in terms of volume expansivity and isothermal compressibility :***

Considering changes in volume due to changes in pressure and temperature,

$$v = f(p, T)$$

or $$dv = \left(\frac{\partial v}{\partial T}\right)_p dT + \left(\frac{\partial v}{\partial p}\right)_T dp$$

$$\therefore \quad \frac{dv}{v} = \frac{1}{v}\left(\frac{\partial v}{\partial T}\right)_p dT + \frac{1}{v}\left(\frac{\partial v}{\partial p}\right)_T dp$$

Since $\beta = \frac{1}{v}\left(\frac{\partial v}{\partial T}\right)_p$ or $\left(\frac{\partial v}{\partial T}\right)_p = \beta v$ ...(*i*)

$$\alpha_T = -\frac{1}{v}\left(\frac{\partial v}{\partial p}\right)_T \text{ or } \left(\frac{\partial v}{\partial p}\right)_T = -\alpha_T v \quad ...(ii)$$

$$\therefore \quad \frac{dv}{v} = \frac{1}{v}(\beta v)\, dT + \frac{1}{v}(-\alpha_T v)\, dp$$
$$= \beta\, dT - \alpha_T\, dp$$

From cyclic property of differentials,

$$\left(\frac{\partial p}{\partial T}\right)_v\left(\frac{\partial T}{\partial v}\right)_p\left(\frac{\partial v}{\partial p}\right)_T = -1$$

Substituting the values of $\left(\frac{\partial v}{\partial T}\right)_p$ and $\left(\frac{\partial v}{\partial p}\right)_T$ from identities (*i*) and (*ii*), we obtain

$$\left(\frac{\partial p}{\partial T}\right)_v \left(\frac{1}{\beta v}\right) (-\alpha_T v) = -1$$

or $$\left(\frac{\partial p}{\partial T}\right)_v = \frac{\beta v}{\alpha_T v} = \frac{\beta}{\alpha_T}$$

That is, the ratio of coefficient of volume expansion β and isothermal compressibility α_T represents the change in pressure with temperature when volume is kept constant.

Further, from expression 11.26

$$\gamma = \frac{(\partial p/\partial v)_s}{(\partial p/\partial v)_T} = \frac{(\partial v/\partial p)_T}{(\partial v/\partial p)_S}$$

$$= \frac{(-1/v)(\partial v/\partial p)_T}{(-1/v)(\partial v/\partial p)_S} = \frac{\alpha_T}{\alpha_S}$$

Apparently the adiabatic exponent γ (the ratio of specific heats c_p/c_v) is equal to the ratio of isothermal compressibility and adiabatic compressibility. Since γ is always greater than unity, the above expression indicates that α_T is always greater than α_S.

11.7. RELATIONS FOR INTERNAL ENERGY AND ENTHALPY

(*a*) When a system undergoes an infinitesimal reversible process, the change in internal energy between two equilibrium states is given by

$$du = T\,ds - p\,dv$$

The first $T\,ds$ equation gives

$$T\,ds = c_v\,dT + T\left(\frac{\partial p}{\partial T}\right)_v dv$$

$$\therefore \quad du = c_v\,dT + T\left(\frac{\partial p}{\partial T}\right)_v dv - p\,dv$$

$$= c_v\,dT + \left[T\left(\frac{\partial p}{\partial T}\right)_v - p\right] dv \quad ...(i)$$

Considering the functional relationship $u = f(T, v)$ which expresses internal energy u as a function of temperature T and volume v.

$$u = f(T, v)$$

or $$du = \left(\frac{\partial u}{\partial T}\right)_v dT + \left(\frac{\partial u}{\partial v}\right)_T dv \quad ...(ii)$$

The equivalence between (*i*) and (*ii*) gives

$$c_v = \left(\frac{\partial u}{\partial T}\right)_v$$

and $$\left(\frac{\partial u}{\partial v}\right)_T = T\left(\frac{\partial p}{\partial T}\right)_v - p \quad ...(11.27)$$

The above identity is called the ***energy equation***.

(*b*) When a system undergoes an infinitesimal reversible process, the change in enthalpy between two equilibrium states is given by

$$dh = T\,ds + v\,dp$$

The second $T\,ds$ equation gives

$$T\,ds = c_p\,dT - T\left(\frac{\partial v}{\partial T}\right)_p dp$$

$$\therefore \quad dh = c_p\,dT - T\left(\frac{\partial v}{\partial T}\right)_p dp + v\,dp$$

$$= c_p\,dT + \left[v - T\left(\frac{\partial v}{\partial T}\right)_p\right] dp \quad ...(iii)$$

Considering the functional relationship $h = f(T, p)$ which expresses enthalpy as a function of temperature T and pressure p.

$$h = f(T, p)$$

or $$dh = \left(\frac{\partial h}{\partial T}\right)_p dT + \left(\frac{\partial h}{\partial p}\right)_T dp \quad ...(iv)$$

The equivalence between (*iii*) and (*iv*) gives

$$c_p = \left(\frac{\partial h}{\partial T}\right)_p$$

and $$\left(\frac{\partial h}{\partial p}\right)_T = v - T\left(\frac{\partial v}{\partial T}\right)_p \quad ...(11.28)$$

EXAMPLE 11.2

Set up the following expression for the specific heats at constant volume and at constant pressure:

$$\left(\frac{\partial c_v}{\partial v}\right)_T = T\left(\frac{\partial^2 p}{\partial T^2}\right)_v$$

and $$\left(\frac{\partial c_p}{\partial p}\right)_T = -T\left(\frac{\partial^2 v}{\partial T^2}\right)_p$$

Further, proceed to show that for a perfect gas obeying $pv = RT$, the specific heats c_p and c_v are function of temperature alone.

Solution : The specific heat at constant volume is

$$c_v = T\left(\frac{\partial s}{\partial T}\right)_v$$

$$\text{or} \quad \left(\frac{\partial c_v}{\partial v}\right)_T = T\frac{\partial^2 s}{\partial T \partial v}$$

From Maxwell relation:

$$\left(\frac{\partial s}{\partial v}\right)_T = \left(\frac{\partial p}{\partial T}\right)_v$$

$$\text{or} \quad \frac{\partial^2 s}{\partial v \partial T} = \left(\frac{\partial^2 p}{\partial T^2}\right)_v$$

$$\therefore \quad \left(\frac{\partial c_v}{\partial v}\right)_T = T\left(\frac{\partial^2 p}{\partial T^2}\right)_v$$

which is the required result.

For the perfect gas: $p = \dfrac{RT}{v}$

$$\text{or} \quad \frac{\partial p}{\partial T} = \frac{R}{v} \quad \text{and} \quad \frac{\partial^2 p}{\partial T^2} = 0$$

$$\text{Hence} \quad \left(\frac{\partial c_v}{\partial v}\right)_T = 0$$

$\therefore$ c_v is independent of volume and is a function of temperature alone.

(*b*) The specific heat at constant pressure is

$$c_p = T\left(\frac{\partial s}{\partial T}\right)_p$$

$$\text{or} \quad \left(\frac{\partial c_p}{\partial p}\right)_T = T\frac{\partial^2 s}{\partial T \partial p}$$

From Maxwell relation:

$$\left(\frac{\partial s}{\partial p}\right)_T = -\left(\frac{\partial v}{\partial T}\right)_p$$

$$\text{or} \quad \frac{\partial^2 s}{\partial p \partial T} = -\left(\frac{\partial^2 v}{\partial T^2}\right)_p$$

$$\therefore \quad \left(\frac{\partial c_p}{\partial p}\right)_T = -\left(\frac{\partial^2 v}{\partial T^2}\right)_p$$

which is the required result.

For a perfect gas, $v = \dfrac{RT}{p}$

$$\text{or} \quad \frac{\partial v}{\partial T} = \frac{R}{P} \quad \text{and} \quad \frac{\partial^2 v}{\partial T^2} = 0$$

$$\text{Hence} \quad \left(\frac{\partial c_p}{\partial p}\right)_T = 0$$

$\therefore$ c_p is independent of pressure and is a function of temperature alone.

EXAMPLE 11.3

Determine the difference in c_p and c_v for water at 300 K for which

coefficient of volume expansion β $= 2 \times 10^{-4}\ K^{-1}$

isothermal compressibility α $= 4.85 \times 10^{-4}\ MPa^{-1}$, and

specific volume $v = 0.001003\ m^3/kg$

(b) What would be the value of $(c_p - c_v)$ for an ideal gas.

Solution : The difference in specific heats for any substance is given by

$$c_p - c_v = \frac{\beta^2}{\alpha} T v$$

Therefore, for liquid water

$$c_p - c_v = \frac{(2 \times 10^{-4})^2}{4.85 \times 10^{-4} \times 10^{-6}} \times 300 \times 0.001003$$

$$= 24.816 \text{ J/kg K}$$

$$= \mathbf{0.0248\ kJ/kg\ K}$$

(*b*) For an ideal gas : $pv = RT$

$$\beta = \frac{1}{v}\left(\frac{\partial v}{\partial T}\right)_p = \frac{p}{RT} \times \frac{\partial}{\partial T}\left(\frac{RT}{p}\right)$$

$$= \frac{p}{RT} \times \frac{R}{p} = \frac{1}{T}$$

$$\alpha = -\frac{1}{v}\left(\frac{\partial v}{\partial p}\right)_T$$

$$= -\left(\frac{p}{RT}\right) \times \frac{\partial}{\partial p}\left(\frac{RT}{p}\right)_T$$

$$= -\left(\frac{p}{RT}\right) \times \left(\frac{-RT}{p^2}\right) = \frac{1}{p}$$

$$\therefore \quad c_p - c_v = \frac{\beta^2}{\alpha} T v = \frac{p}{T^2} \times T v = \frac{pv}{T} = R$$

EXAMPLE 11.4

Show that for a perfect gas, the difference between the specific heats $(c_p - c_v)$ can be expressed as

$$c_p - c_v = \left[p + \left(\frac{\partial u}{\partial v}\right)_T\right]\left(\frac{\partial v}{\partial T}\right)_p$$

$$= pv\beta + v\beta\left(\frac{\partial u}{\partial v}\right)_T$$

where β is the coefficient of volume expansion.

Solution : The property relations for internal energy u and enthalpy h are

$$T\,ds = du + p\,dv$$

and $$T\,ds = dh - v\,dp$$

$$\therefore \quad du + p\,dv = dh - v\,dp \qquad ...(i)$$

From the functional relations $u = f(T, v)$ and $h = f(T, p)$, we have

$$du = \left(\frac{\partial u}{\partial T}\right)_v dT + \left(\frac{\partial u}{\partial v}\right)_T dv$$

$$= c_v\,dT + \left(\frac{\partial u}{\partial v}\right)_T dv$$

and $$dh = \left(\frac{\partial h}{\partial T}\right)_p dT + \left(\frac{\partial h}{\partial p}\right)_T dp$$

$$= c_p\,dT + \left(\frac{\partial h}{\partial p}\right)_T dp$$

Substituting the values of du and dh in expression (i), we get

$$c_v\,dT + \left(\frac{\partial u}{\partial v}\right)_T dv + p\,dv$$

$$= c_p\,dT + \left(\frac{\partial h}{\partial p}\right)_T dp - v\,dp$$

or $$c_v\,dT + \left[p + \left(\frac{\partial u}{\partial v}\right)_T\right] dv$$

$$= c_p\,dT - \left[v - \left(\frac{\partial h}{\partial p}\right)_T\right] dp$$

The above equation is valid for any process. Writing it for the case when $dp = 0$, we have

$$(c_p - c_v)\,(dT)_p = \left[p + \left(\frac{\partial u}{\partial v}\right)_T\right](dv)_p$$

or $$c_p - c_v = \left[p + \left(\frac{\partial u}{\partial v}\right)_T\right]\left[\frac{\partial v}{\partial T}\right]_p$$ which is the required result.

(b) The coefficient of volume expansion β is defined as

$$\beta = \frac{1}{v}\left(\frac{\partial v}{\partial T}\right)_p$$

and accordingly, we can write

$$c_p - c_v = \left[p + \left(\frac{\partial u}{\partial v}\right)_T\right] v\beta$$

$$= pv\beta + v\beta\left(\frac{\partial u}{\partial v}\right)_T$$

EXAMPLE 11.5

Set up a T ds relation in the following form:

$$T\,ds = c_v\,dT + \left(\frac{\beta T}{\alpha}\right) dv$$

where β is the coefficient of volume expansion, α is the isothermal compressibility, and the other symbols have their usual meanings.

Solution : The given relation expresses entropy s as a function of temperature T and volume v. That is

$$s = f(T, v)$$

or $$ds = \left(\frac{\partial s}{\partial T}\right)_v dT + \left(\frac{\partial s}{\partial v}\right)_T dv$$

It is known that:

$$\left(\frac{\partial s}{\partial v}\right)_T = \left(\frac{\partial p}{\partial T}\right)_v \qquad \text{...Maxwell relation}$$

and $$T\left(\frac{\partial s}{\partial T}\right)_v = c_v\,; \quad \left(\frac{\partial s}{\partial T}\right)_v = \frac{c_v}{T}$$

Making these substitutions, we get

$$T\,ds = c_v\,dT + T\left(\frac{\partial p}{\partial T}\right)_v dv \qquad ...(i)$$

The parameters β and α are defined as

$$\beta = \frac{1}{v}\left(\frac{\partial v}{\partial T}\right)_p$$

and $$\alpha = -\frac{1}{v}\left(\frac{\partial v}{\partial p}\right)_T$$

$$\therefore \quad \frac{\beta}{\alpha} = \frac{1}{v}\left(\frac{\partial v}{\partial T}\right)_p \times \left[\frac{-v}{(\partial v/\partial p)_T}\right]$$

$$= -\left(\frac{\partial v}{\partial T}\right)_p \times \left(\frac{\partial p}{\partial v}\right)_T \qquad ...(ii)$$

The cyclic relation between p, v and T gives

$$\left(\frac{\partial p}{\partial v}\right)_T \times \left(\frac{\partial v}{\partial T}\right)_p \times \left(\frac{\partial T}{\partial p}\right)_v = -1$$

or $$-\left(\frac{\partial v}{\partial T}\right)_p \times \left(\frac{\partial p}{\partial v}\right)_T = \frac{1}{(\partial T/\partial p)_v}$$

$$= \left(\frac{\partial p}{\partial T}\right)_v \qquad ...(iii)$$

From relations (*ii*) and (*iii*),

$$\frac{\beta}{\alpha} = \left(\frac{\partial p}{\partial T}\right)_v$$

When the value of $\left(\frac{\partial p}{\partial T}\right)_v$ as obtained above is substituted in expression (*i*), we get

$$T\,ds = c_v\,dT + \frac{\beta T}{\alpha}dv$$

which is the required relation.

EXAMPLE 11.6

Set up a T ds relation in the following form

$$T\,ds = c_p\,dT - \beta\, v\, T\,dp$$

where β is the coefficient of volume expansion. The other symbols have their usual meanings

Solution : The given relation expresses entropy s as a function of temperature T and pressure p. That is

$$s = f(T, p)$$

or $$ds = \left(\frac{\partial s}{\partial T}\right)_p dT + \left(\frac{\partial s}{\partial p}\right)_T dp$$

It is known that :

$$\left(\frac{\partial s}{\partial p}\right)_T = -\left(\frac{\partial v}{\partial T}\right)_p \quad \text{...Maxwell relation}$$

and $$T\left(\frac{\partial s}{\partial T}\right)_p = c_p$$

Making these substitutions, we get

$$T\,ds = c_p\,dT - T\left(\frac{\partial v}{\partial T}\right)_p dp \qquad ...(i)$$

The coefficient of volume expansion is defined as

$$\beta = \frac{1}{v}\left(\frac{\partial v}{\partial T}\right)_p \quad \text{or} \quad \left(\frac{\partial v}{\partial T}\right)_p = \beta\, v$$

When this value of $\left(\frac{\partial v}{\partial T}\right)_p$ is substituted in expression (*i*), we get

$$T\,ds = c_p\,dT - \beta\, v\, T\,d_p$$

which is the required relation.

EXAMPLE 11.7

Set up the following relation for the change in internal energy of a substance from the measured values of temperature and volume

$$du = c_v\,dT + \left(\frac{\beta T}{\alpha} - p\right)dv$$

where β is the coefficient of volume expansion, α is the isothermal compressibility, and the other symbols have their usual meanings.

Proceed therefrom to show that internal energy of an ideal gas is independent of pressure and volume and is a function of temperature only.

Solution : The given relation expresses internal energy u as a function of temperature T and volume v. That is

$$u = f(T, v)$$

or $$du = \left(\frac{\partial u}{\partial T}\right)_v dT + \left(\frac{\partial u}{\partial v}\right)_T dv$$

Since $\left(\frac{\partial u}{\partial T}\right)_v$ defines c_v, we can write

$$du = c_v\,dT + \left(\frac{\partial u}{\partial v}\right)_T dv \qquad ...(i)$$

Consider the property relation :

$$du = T\,ds - p\,dv$$

Keeping T constant, let this relation be divided by dv. That gives

$$\left(\frac{\partial u}{\partial v}\right)_T = T\left(\frac{\partial s}{\partial v}\right)_T - p$$

Substituting the Maxwell relation

$$\left(\frac{\partial s}{\partial v}\right)_T = \left(\frac{\partial p}{\partial T}\right)_v \text{, we get}$$

$$\left(\frac{\partial u}{\partial v}\right)_T = T\left(\frac{\partial p}{\partial T}\right)_v - p \qquad ...(ii)$$

The cyclic relation between p, v and T gives

$$\left(\frac{\partial p}{\partial v}\right)_T \times \left(\frac{\partial v}{\partial T}\right)_p \times \left(\frac{\partial T}{\partial p}\right)_v = -1$$

or $$\left(\frac{\partial p}{\partial T}\right)_v = -\left(\frac{\partial p}{\partial v}\right)_T \times \left(\frac{\partial v}{\partial T}\right)_p \qquad ...(iii)$$

The parameters β and α are defined as

$$\beta = \frac{1}{v}\left(\frac{\partial v}{\partial T}\right)_p$$

and $$\alpha = -\frac{1}{v}\left(\frac{\partial v}{\partial p}\right)_T$$

$$\therefore \quad \frac{\beta}{\alpha} = \frac{1}{v}\left(\frac{\partial v}{\partial T}\right)_p \times \left[\frac{-v}{(\partial v/\partial p)_T}\right]$$

$$= -\left(\frac{\partial v}{\partial T}\right)_p \times \left(\frac{\partial p}{\partial v}\right)_T \qquad ...(iv)$$

Relations (*iii*) and (*iv*) are then combined to give

$$\left(\frac{\partial p}{\partial T}\right)_v = \frac{\beta}{\alpha}$$

When the value of $\left(\frac{\partial p}{\partial T}\right)_v$ as obtained above is substituted in expression (*ii*), we get

$$\left(\frac{\partial u}{\partial v}\right)_T = \frac{\beta T}{\alpha} - p$$

The above correlation is then substituted in expression (*i*) and we obtain

$$du = c_v\, dT + \left(\frac{\beta T}{\alpha} - p\right) dv$$

which is the required relation.

(*b*) For an ideal gas $pv = RT$

$$\beta = \frac{1}{v}\left(\frac{\partial v}{\partial T}\right)_p$$

$$= \frac{p}{RT}\frac{\partial}{\partial T}\left(\frac{RT}{p}\right)_p$$

$$= \frac{p}{RT} \times \frac{R}{p} = \frac{1}{T}$$

$$\alpha = -\frac{1}{v}\left(\frac{\partial v}{\partial p}\right)_T$$

$$= -\frac{p}{RT} \times \frac{\partial}{\partial p}\left(\frac{RT}{p}\right)_T$$

$$= -\frac{p}{RT} \times \left(-\frac{RT}{p^2}\right) = \frac{1}{p}$$

$$\therefore \quad du = c_v\, dT + \left(\frac{Tp}{T} - p\right) dv = c_v\, T$$

Thus the internal energy of an ideal gas is independent of pressure and volume and is a function of temperature only.

EXAMPLE 11.8

Set up the following relation for the change in enthalpy of a substance from the measured values of temperature and pressure

$$dh = c_p\, dT + v\,(1 - \beta T)\, dp$$

where β is the coefficient of volume expansion. The other symbols have their usual meanings.

Proceed therefrom to show that enthalpy of an ideal gas is independent of pressure and volume and depends only on temperature.

Solution : The given relation expresses enthalpy h as a function of temperature T and pressure p. That is

$$h = f\,(T, p)$$

or $$dh = \left(\frac{\partial h}{\partial T}\right)_p dT + \left(\frac{\partial h}{\partial p}\right)_T dp$$

Since $\left(\frac{\partial h}{\partial T}\right)_p$ defines c_p, we can write

$$dh = c_p\, dT + \left(\frac{\partial h}{\partial p}\right)_T dp \qquad ...(i)$$

From the property relation :

$$dh = T\, ds + v\, dp$$

Keeping T constant, let this relation be divided by dp. That gives

$$\left(\frac{\partial h}{\partial p}\right)_T = T\left(\frac{\partial s}{\partial p}\right)_T + v$$

Substituting the Maxwell relation

$$\left(\frac{\partial s}{\partial p}\right)_T = -\left(\frac{\partial v}{\partial T}\right)_p, \text{ we get}$$

$$\left(\frac{\partial h}{\partial p}\right)_T = -T\left(\frac{\partial v}{\partial T}\right)_p + v$$

The coefficient of volume expansion is defined as

$$\beta = \frac{1}{v}\left(\frac{\partial v}{\partial T}\right)_p$$

$$\therefore \quad \left(\frac{\partial h}{\partial p}\right)_T = -\beta v t + v$$

$$= v(1 - \beta T) \qquad ...(ii)$$

When the value of $\left(\frac{\partial h}{\partial p}\right)_T$ as obtained above is substituted in expression (*i*), we get

$$dh = c_p\, dT + v(1 - \beta T)\, dp$$

which is the required expression.

(*b*) For an ideal gas $pv = RT$

$$\beta = \frac{1}{v}\left(\frac{\partial v}{\partial T}\right)_p$$

$$= \frac{p}{RT}\frac{\partial}{\partial T}\left(\frac{RT}{p}\right)_p$$

$$= \frac{p}{RT} \times \frac{R}{p} = \frac{1}{T}$$

$$\therefore \quad dh = c_p\, dT + v\left(1 - T \times \frac{1}{T}\right) = c_p\, dT$$

Thus the enthalpy of an ideal gas is independent of pressure and volume and is a function of temperature only.

EXAMPLE 11.9

Over a certain range of pressures and temperatures, the equation of state of a certain gas is prescribed by the relation

$$v = \frac{RT}{p} - \frac{C}{T^3} \text{ where } C \text{ is a constant}$$

Set up expressions for a change in enthalpy and entropy of the gas. Consider the change to occur under isothermal conditions.

Solution : From the relations

$$dh = T\, ds + v\, dp$$

and $\quad T\, ds = c_p\, dT - T\left(\frac{\partial v}{\partial T}\right)_p dp$; we have

$$dh = c_p\, dT - T\left(\frac{\partial v}{\partial T}\right)_p dp + v\, dp$$

For an isothermal process (T = constant)

$$(dh)_T = \left[v - T\left(\frac{\partial v}{\partial T}\right)_p\right] dp \qquad ...(i)$$

From the given equation of state :

$$v = \frac{RT}{p} - \frac{C}{T^3}$$

$$\left(\frac{\partial v}{\partial T}\right)_p = \frac{R}{p} + \frac{3C}{T^4}$$

and $\left[v - T\left(\frac{\partial v}{\partial T}\right)_p\right] dp$

$$= \left[\frac{RT}{p} - \frac{C}{T^3} - \frac{RT}{p} - \frac{3C}{T^3}\right] dp$$

$$= -\frac{4C}{T^3}\, dp$$

The identity (*i*) may then be written as

$$(dh)_T = -\frac{4C}{T^3}\, dp$$

Upon integration between the state point 1 and state point 2, we have

$$h_2 - h_1 = -\frac{4C}{T^3}(p_2 - p_1)_T$$

$$= \frac{4C}{T^3}(p_1 - p_2)_T$$

(*b*) From the $T\, ds$ relation:

$$T\, ds = c_p\, dT - T\left(\frac{\partial v}{\partial T}\right)_p dp, \text{ we have}$$

$$ds = c_p \frac{dT}{T} - \left(\frac{\partial v}{\partial T}\right)_p dp$$

For an isothermal (constant temperature process),

$$(ds)_T = -\left(\frac{\partial v}{\partial T}\right)_p dp$$

From the given equation of state:

$$v = \frac{RT}{p} - \frac{C}{T^3}$$

$$\left(\frac{\partial v}{\partial T}\right)_p = \frac{R}{p} + \frac{3C}{T^4}$$

$$\therefore \quad (ds)_T = -\left[\frac{R}{p} + \frac{3C}{T^4}\right] dp$$

Upon integration between the state point 1 and state point 2,

$$(s_2 - s_1)_T = -\left[R \log_e \frac{p_2}{p_1} + \frac{3C}{T^4}(p_2 - p_1)_T\right]$$

$$= R \log_e \frac{p_1}{p_2} + \frac{3C}{T^4}(p_1 - p_2)_T$$

EXAMPLE 11.10

Set up the T ds equation in the form

$$T\,ds = c_v \left(\frac{\partial T}{\partial p}\right)_v dp + c_p \left(\frac{\partial T}{\partial v}\right)_p dv$$

Further, proceed to show that this relation can be recast as

$$T\,ds = c_v \frac{\alpha}{\beta} dp + \frac{c_p}{\beta v} dv$$

where β is the coefficient of volume expansion and α is the isothermal compressibility factor.

Solution : Consider variation of entropy s with pressure p and specific volume v. Then

$$s = f(p, v)$$

or
$$ds = \left(\frac{\partial s}{\partial p}\right)_v dp + \left(\frac{\partial s}{\partial v}\right)_p dv$$

$$= \left(\frac{\partial s}{\partial T}\right)_v \left(\frac{\partial T}{\partial p}\right)_v dp + \left(\frac{\partial s}{\partial T}\right)_p \left(\frac{\partial T}{\partial v}\right)_p dv$$

Substituting the relations:

$$\left(\frac{\partial s}{\partial T}\right)_v = \frac{c_v}{T} \quad \text{and} \quad \left(\frac{\partial s}{\partial T}\right)_p = \frac{c_p}{T},$$

we can write

$$ds = \frac{c_v}{T}\left(\frac{\partial T}{\partial p}\right)_v dp + \frac{c_p}{T}\left(\frac{\partial T}{\partial v}\right)_p dv$$

or
$$T\,ds = c_v \left(\frac{\partial T}{\partial p}\right)_v dp + c_p \left(\frac{\partial T}{\partial v}\right)_p dv \quad \text{...(i)}$$

which is the required expression.

This identity is often known as *Third T ds equation.*

(*b*) The parameters β and α are defined as

$$\beta = \frac{1}{v}\left(\frac{\partial v}{\partial T}\right)_p$$

and
$$\alpha = -\frac{1}{v}\left(\frac{\partial v}{\partial p}\right)_T$$

Then: $\left(\frac{\partial v}{\partial T}\right)_p = \beta v$ or $\left(\frac{\partial T}{\partial v}\right)_p = \frac{1}{\beta v}$

From the cyclic relationship

$$\left(\frac{\partial T}{\partial p}\right)_v \times \left(\frac{\partial p}{\partial v}\right)_T \times \left(\frac{\partial v}{\partial T}\right)_p = -1$$

or
$$\left(\frac{\partial T}{\partial p}\right)_v = -\frac{1}{\left(\frac{\partial p}{\partial v}\right)_T \times \left(\frac{\partial v}{\partial T}\right)_p}$$

$$= -\left(\frac{\partial v}{\partial p}\right)_T \Big/ \left(\frac{\partial v}{\partial T}\right)_p = \frac{\alpha}{\beta}$$

Making the substitution for $\left(\frac{\partial T}{\partial p}\right)_v$ and $\left(\frac{\partial T}{\partial v}\right)_p$ in expression (*i*), we get

$$T\,ds = c_v\left(\frac{\alpha}{\beta}\right) dp + c_p\left(\frac{1}{\beta v}\right) dv$$

$$= c_v \frac{\alpha}{\beta} dp + \frac{c_p}{\beta v} dv$$

which is the required result.

EXAMPLE 11.11.

Set up the following relations for a gas which conforms to the van der Waal's equation of state.

$$(a)\ c_p - c_v = \frac{R}{1 - 2a(v-b)^2/RTv^3}$$

(b) $T(v-b)^{R/c_v}$ = *constant for an isentropic process where the symbols have their usual meanings.*

Solution : From generalised relation for c_p and c_v, we have

$$c_p - c_v = T\left(\frac{\partial p}{\partial T}\right)_v \left(\frac{\partial v}{\partial T}\right)_p$$

The van der Waals equation of state is

$$\left(p+\frac{a}{v^2}\right)(v-b) = RT$$

or
$$p = \frac{RT}{v-b} - \frac{a}{v^2}$$

Differentiating p with respect to T at constant volume, we get

$$\left(\frac{\partial p}{\partial T}\right)_v = \frac{R}{v-b}$$

$$\therefore \quad c_p - c_v = \frac{RT}{v-b}\times\left(\frac{\partial v}{\partial T}\right)_p$$

Again when volume v in the van der Waal's equation is differentiated with respect to temperature T treating pressure p as constant, we get

$$\left(p+\frac{a}{v^2}\right)\left(\frac{\partial v}{\partial T}\right)_p + (v-b)\left(\frac{-2a}{v^3}\right)\left(\frac{\partial v}{\partial T}\right)_p = R$$

or
$$\frac{RT}{v-b}\left(\frac{\partial v}{\partial T}\right)_p + (v-b)\left(\frac{-2a}{v^3}\right)\left(\frac{\partial v}{\partial T}\right)_p = R$$

or
$$\left(\frac{\partial v}{\partial T}\right)_p = \frac{R}{\left[\frac{RT}{(v-b)^2} - \frac{2a}{v^3}\right][v-b]}$$

$$= \frac{R(v-b)v^3}{RTv^3 - 2a(v-b)^2}$$

$$\therefore \quad c_p - c_v = \frac{RT}{v-b}\times\frac{R(v-b)v^3}{RTv^3 - 2a(v-b)^2}$$

$$= RT\times\frac{Rv^3}{RTv^3\left[1-2a(v-b)^2/RTv^3\right]}$$

$$= \frac{R}{1-2a(v-b)^2/RTv^3}$$

(*b*) Consider the first $T\,ds$ equation

$$T\,ds = c_v\,dT + T\left(\frac{\partial p}{\partial T}\right)_v dv$$

The van der Waal's equation of state is

$$\left(p+\frac{a}{v^2}\right)(v-b) = RT \quad \text{or} \quad p = \frac{RT}{v-b} - \frac{a}{v^2}$$

Differentiating p with respect to T treating v as constant, we get

$$\left(\frac{\partial p}{\partial T}\right)_v = \frac{R}{v-b}$$

Then
$$T\,ds = c_v\,dT + \frac{RT}{v-b}dv$$

or
$$ds = c_v\frac{dT}{T} + \frac{R}{v-b}dv$$

At constant entropy $ds = 0$ and, therefore

$$c_v\frac{dT}{T} + \frac{R}{v-b}dv = 0$$

or
$$\frac{dT}{T} + \frac{R}{c_v}\frac{dv}{v-b} = 0$$

By integration:

$$\log_e T + \frac{R}{c_v}\log_e(v-b) = \log_e \text{constant}$$

or
$$\log_e\left[T\times(v-b)^{R/c_v}\right] = \log_e \text{constant}$$

or
$$T(v-b)^{R/c_v} = \text{constant}$$

EXAMPLE 11.12

Set up the expressions for changes in internal energy, enthalpy and entropy for a gas which obeys the van der Waal's equation

$$\left(p+\frac{a}{v^2}\right)(v-b) = RT$$

Solution : When a system undergoes an infinitesimal reversible process, the change in internal energy between two equilibrium states is given by

$$du = c_v\,dT + \left[T\left(\frac{\partial p}{\partial T}\right)_v - p\right]dv$$

From the given van der Waal's equation

$$p = \frac{RT}{v-b} - \frac{a}{v^2}$$

and $\left(\frac{\partial p}{\partial T}\right)_v = \frac{R}{v-b} - 0 = \frac{R}{v-b}$

$$\therefore \quad u_2 - u_1 = \int_1^2 du$$

$$= c_v \int_1^2 dT + \int_1^2 \left[\frac{RT}{v-b} - p\right] dv$$

$$= c_v \int_1^2 dT$$

$$+ \int_1^2 \left[\frac{RT}{v-b} - \left(\frac{RT}{v-b} - \frac{a}{v^2}\right)\right] dv$$

$$= c_v \int_1^2 dT + \int_1^2 \frac{a}{v^2} dv$$

$$= c_v (T_2 - T_1) + a\left(\frac{1}{v_1} - \frac{1}{v_2}\right)$$

which is the desired expression for change in internal energy.

Under constant temperature conditions, $c_v (T_2 - T_1) = 0$ and accordingly

$$(u_2 - u_1)_T = a\left(\frac{1}{v_1} - \frac{1}{v_2}\right)$$

(*b*) When a system undergoes an infinitesimal reversible process, the change in enthalpy between two equilibrium states is

$$dh = c_p\, dT + \left[v - T\left(\frac{\partial v}{\partial T}\right)_p\right] dp$$

Under constant temperature conditions, $dT = 0$ and then

$$(dh)_T = \left[v - T\left(\frac{dv}{dT}\right)_p\right](dp)_T \qquad ...(a)$$

Considering the functional relation

$$p = f(v, T)$$

$$dp = \left(\frac{\partial p}{\partial v}\right)_T dv + \left(\frac{\partial p}{\partial T}\right)_v dT$$

or $\quad (dp)_T = \left(\frac{\partial p}{\partial v}\right)_T dv + 0 \quad$ as $\; dT = 0$

Substituting the above value of $(dp)_T$ in identity (*a*), we get

$$(dh)_T = \left[v - T\left(\frac{\partial v}{\partial T}\right)_p\right]\left(\frac{\partial p}{\partial v}\right)_T dv$$

$$= \left[v\left(\frac{\partial p}{\partial v}\right)_T - T\left(\frac{\partial v}{\partial T}\right)_p\left(\frac{\partial p}{\partial v}\right)_T\right] dv \qquad ...(b)$$

Using the cyclic relation for p, v and T, we have

$$\left(\frac{\partial v}{\partial T}\right)_p \left(\frac{\partial T}{\partial p}\right)_v \left(\frac{\partial p}{\partial v}\right)_T = -1$$

or $$\left(\frac{\partial v}{\partial T}\right)_p \left(\frac{\partial p}{\partial v}\right)_T = \frac{-1}{(\partial T/\partial p)_v}$$

$$= -\left(\frac{\partial p}{\partial T}\right)_v$$

Substituting this value in expression (*b*), we get

$$(dh)_T = \left[v\left(\frac{\partial p}{\partial v}\right)_T + T\left(\frac{\partial p}{\partial T}\right)_v\right] dv$$

From van der Waal's equation

$$\left(\frac{\partial p}{\partial v}\right)_T = \frac{\partial}{\partial v}\left[\frac{RT}{v-b} - \frac{a}{v^2}\right]_T$$

$$= -\frac{RT}{(v-b)^2} + \frac{2a}{v^3}$$

$$\left(\frac{\partial p}{\partial T}\right)_v = \frac{\partial}{\partial T}\left[\frac{RT}{v-b} - \frac{a}{v^2}\right]_v = \frac{R}{(v-b)}$$

$$\therefore \quad (dh)_T =$$

$$\left[v\left\{-\frac{RT}{(v-b)^2} + \frac{2a}{v^3}\right\} + T\left\{\frac{R}{v-b}\right\}\right] dv$$

Upon integration between state points 1 and 2,

$$(h_2 - h_1)_T = -RT\int_1^2 \frac{v}{(v-b)^2} dv$$

$$+ 2a\int_1^2 \frac{dv}{v^2} + RT\int_1^2 \frac{dv}{v-b}$$

$$= -RT\left[\log_e \frac{v_2 - b}{v_1 - b} - b\left\{\frac{1}{v_2 - b} - \frac{1}{v_1 - b}\right\}\right]$$

$$- 2a\left(\frac{1}{v_2} - \frac{1}{v_1}\right) + RT\log_e \frac{v_2 - b}{v_1 - b}$$

$$= bRT\left[\frac{1}{v_2 - b} - \frac{1}{v_1 - b}\right] - 2a\left[\frac{1}{v_2} - \frac{1}{v_1}\right]$$

which is the desired expression for change in enthalpy.

(c) Consider the first $T\,ds$ equation,

$$T\,ds = c_v\,dT + T\left(\frac{\partial p}{\partial T}\right)_v dv$$

The van der Waal's equation of state is

$$\left(p + \frac{a}{v^2}\right)(v - b) = RT$$

or

$$p = \frac{RT}{v - b} - \frac{a}{v^2}$$

Then $\left(\frac{\partial p}{\partial T}\right)_v = \frac{\partial}{\partial T}\left[\frac{RT}{v - b} - \frac{a}{v^2}\right]_v = \frac{R}{v - b}$

$$\therefore \quad T\,ds = c_v\,dT + \frac{RT}{v - b}dv$$

or

$$ds = c_v\frac{dT}{T} + \frac{R}{v - b}dv$$

Upon integration between state points 1 and 2

$$s_2 - s_1 = c_v\int_1^2 \frac{dT}{T} + R\int_1^2 \frac{dv}{v - b}$$

$$= c_v \log_e \frac{T_2}{T_1} + R\log_e\left[\frac{v_2 - b}{v_1 - b}\right]$$

which is the desired expression for change in entropy.

EXAMPLE 11.13

Show that for any fluid

(a) $\left(\frac{\partial h}{\partial v}\right)_T = v\left(\frac{\partial p}{\partial v}\right)_T + T\left(\frac{\partial p}{\partial T}\right)_v$

(b) $\left(\frac{\partial h}{\partial p}\right)_T = v - T\left(\frac{\partial v}{\partial T}\right)_p$

Further, proceed to prove that for any pure substance obeying van der Waal's equation, the enthalpy is given by

$$h = \frac{RTb}{v - b} - \frac{2a}{v} + f(T)$$

where f(T) is arbitrary and R, a, b are the constants appearing in van der Waal's equation.

Solution : For unit mass, the enthalpy change is given by

$$dh = T\,ds + v\,dp$$

and from first $T\,ds$ equation

$$T\,ds = c_v\,dT + T\left(\frac{\partial p}{\partial T}\right)_v dv$$

$$\therefore \quad dh = c_v\,dT + T\left(\frac{\partial p}{\partial T}\right)_v dv + v\,dp$$

Under constant temperature conditions $dT = 0$ and therefore

$$(dh)_T = T\left(\frac{\partial p}{\partial T}\right)_v dv + v\,dp$$

or

$$\left(\frac{\partial h}{\partial v}\right)_T = T\left(\frac{\partial p}{\partial T}\right)_v + v\left(\frac{\partial p}{\partial v}\right)_T$$

which is the required result.

(b) $\left(\frac{\partial h}{\partial p}\right)_T = \left(\frac{\partial h}{\partial v}\right)_T \times \left(\frac{\partial v}{\partial p}\right)_T$

$$= \left[T\left(\frac{\partial p}{\partial T}\right)_v + v\left(\frac{\partial p}{\partial v}\right)_T\right] \times \left(\frac{\partial v}{\partial p}\right)_T$$

$$= T\left(\frac{\partial p}{\partial T}\right)_v\left(\frac{\partial v}{\partial p}\right)_T + v$$

From cyclic relation :

$$\left(\frac{\partial p}{\partial T}\right)_v \times \left(\frac{\partial T}{\partial v}\right)_p \times \left(\frac{\partial v}{\partial p}\right)_T = -1$$

or

$$\left(\frac{\partial p}{\partial T}\right)_v \times \left(\frac{\partial v}{\partial p}\right)_T = -\left(\frac{\partial v}{\partial T}\right)_p$$

Making this substitution, we get

$$\left(\frac{\partial h}{\partial p}\right)_T = -T\left(\frac{\partial v}{\partial T}\right)_p + v$$

$$= v - T\left(\frac{\partial v}{\partial T}\right)_p$$

which is the required result.

(c) From van der Waal's equation

$$p = \frac{RT}{v-b} - \frac{a}{v^2}$$

The differentiations with respect to v at constant temperature T, and with respect to T at constant v give

$$\left(\frac{\partial p}{\partial v}\right)_T = -\frac{RT}{(v-b)^2} + \frac{2a}{v^3}$$

and $$\left(\frac{\partial p}{\partial T}\right)_v = \frac{R}{v-b}$$

$$\therefore \quad \left(\frac{\partial h}{\partial v}\right)_T = v\left[-\frac{RT}{(v-b)^2} + \frac{2a}{v^3}\right] + T\left(\frac{R}{v-b}\right)$$

$$= -\frac{RTv}{(v-b)^2} + \frac{2a}{v^2} + \frac{RT}{v-b}$$

$$= \frac{-RTv + RT(v-b)}{(v-b)^2} + \frac{2a}{v^2}$$

$$= -\frac{RTb}{(v-b)^2} + \frac{2a}{v^2}$$

That gives :

$$h = \frac{RTb}{v-b} - 2a + f(T)$$

Obviously h depends on T and v.

EXAMPLE 11.14

The following expressions for the equation of state and specific heat c_p are obeyed by a certain gas:

$$v = \frac{RT}{p} + \alpha T^2$$

and $$c_p = A + BT + Cp$$

where α, A, B, and C are constants. Obtain an expression for the Joule-Thomson coefficient.

Solution : Joule-Thomson coefficient represents the numerical value of the slope of an isenthalpy on T-p diagram at any point and is denoted by u_j. Mathematically,

$$u_j = \left(\frac{\partial T}{\partial p}\right)_h$$

The property relation for dh is :

$$dh = T\,ds + v\,dp$$

From second $T\,ds$ equation, we have

$$T\,ds = c_p\,dT - T\left(\frac{\partial v}{\partial T}\right)_p dp$$

$$\therefore \quad dh = c_p\,dT - T\left(\frac{\partial v}{\partial T}\right)_p dp + v\,dp$$

$$= c_p\,dT - \left[T\left(\frac{\partial v}{\partial T}\right)_p - v\right]dp$$

For a constant enthalpy process $dh = 0$ and therefore,

$$0 = (c_p\,dT)_h - \left[\left\{T\left(\frac{\partial v}{\partial T}\right)_p - v\right\}dp\right]_h$$

or $$\left(\frac{\partial T}{\partial p}\right)_h = u_j = \frac{1}{c_p}\left[T\left(\frac{\partial v}{\partial T}\right)_p - v\right] \quad \text{...(a)}$$

For the given gas:

$$v = \frac{RT}{p} + \alpha T^2$$

$$\left(\frac{\partial v}{\partial T}\right)_p = \frac{R}{p} + 2\alpha T$$

Substuting the given values of c_p, v and $\left(\frac{\partial v}{\partial T}\right)_p$ as calculated above in identity (a), we get

$$u_j = \frac{1}{A + BT + Cp} \times \left[T \times \left(\frac{R}{p} + 2\alpha T\right) - \left(\frac{RT}{p} + \alpha T^2\right)\right]$$

$$= \frac{1}{A + BT + Cp} \times \left[\frac{RT}{p} + 2\alpha T^2 - \frac{RT}{p} - \alpha T^2\right]$$

$$= \frac{1}{A+BT+CP}(\alpha T^2) = \frac{\alpha T^2}{A+BT+Cp}$$

EXAMPLE 11.15

By using Maxwell's relations of thermodynamics, show that Joule-Thomson coefficient μ_j of a gas can be expressed as

$$\mu_j = \left(\frac{\partial T}{\partial p}\right)_h = \frac{T^2}{c_p}\left[\frac{\partial}{\partial T}\left(\frac{v}{T}\right)\right]_p$$

Solution : The Joule-Thomson coefficient μ_j represents the numerical value of the slope of an isenthalpy on T-p diagram at any point and it works out as

$$\mu_j = \frac{1}{c_p}\left[T\left(\frac{\partial v}{\partial T}\right)_p - v\right] \quad \text{(Example 11.14)}$$

$$= \frac{1}{c_p}T^2\left[\frac{1}{T}\left(\frac{\partial v}{\partial T}\right)_p - \frac{v}{T^2}\right]$$

$$= \frac{1}{c_p}T^2\left[\frac{1}{T}\left(\frac{\partial v}{\partial T}\right)_p + v\frac{\partial}{\partial T}\left(\frac{1}{T}\right)\right]$$

$$= \frac{T^2}{c_p}\left[\frac{\partial}{\partial T}\left(\frac{v}{T}\right)_p\right]$$

which is the required result.

EXAMPLE 11.16

Set up an expression for $(\Delta h)_T$ for a substance that obeys the equation of state given by

$$p = \left(\frac{RT}{v} - \frac{a}{v^2}\right)$$

Solution : When a system undergoes an infinitesimal reversible process the change in internal energy between two equilibrium states is given by

$$du = T\,ds - p\,dv$$

The first $T\,ds$ equation gives

$$Tds = c_v\,dT + T\left(\frac{\partial p}{\partial T}\right)_v dv$$

$$\therefore \quad du = c_v\,dT + T\left(\frac{\partial p}{\partial T}\right)_v dv - p\,dv$$

$$= c_v\,dT + \left[T\left(\frac{\partial p}{\partial T}\right)_v - p\right]dv$$

From the given equation

$$\left(\frac{\partial p}{\partial T}\right)_v = \frac{R}{v} - 0 = \frac{R}{v}$$

$$\therefore \quad u_2 - u_1 = \int_1^2 du$$

$$= c_v\int_1^2 dT + \int_1^2\left[\frac{RT}{v} - \left(\frac{RT}{v} - \frac{a}{v^2}\right)\right]dv$$

$$= c_v\int_1^2 dT + \int_1^2 \frac{a}{v^2}\,dv$$

$$= c_v\,(T_2 - T_1) + a\left(\frac{1}{v_1} - \frac{1}{v_2}\right)$$

Under constant temperature conditions, $c_v(T_2 - T_1) = 0$ and accordingly

$$(u_2 - u_1)_T = a\left(\frac{1}{v_1} - \frac{1}{v_2}\right)$$

From the given identity : $pv = RT - \frac{a}{v}$

and under constant temperature conditions.

$$(p_2 v_2 - p_1 v_1)_T = a\left(\frac{1}{v_1} - \frac{1}{v_2}\right)$$

Now

$$(\Delta h)_T = (h_2 - h_1)_T$$

$$= (p_2 v_2 + u_2)_T - (p_1 v_1 + u_1)_T$$

$$= (p_2 v_2 - p_1 v_1)_T + (u_2 - u_1)_T$$

$$= a\left(\frac{1}{v_1} - \frac{1}{v_2}\right) + a\left(\frac{1}{v_1} - \frac{1}{v_2}\right) = 2a\left(\frac{1}{v_1} - \frac{1}{v_2}\right)$$

EXAMPLE 11.17

The enthalpy a pure substance can be expressed as a function of any two of independent variables p, v and T. Using the method of partial derivatives, set up the following expressions for change in enthalpy.

1. $\left(\frac{\partial h}{\partial v}\right)_T = \frac{1}{K}(T\beta - 1)$

2. $\left(\frac{\partial h}{\partial p}\right)_T = v\ (1 - \beta T)$

3. $\left(\frac{\partial h}{\partial p}\right)_v = C_v\ \frac{K}{\beta} + v$

4. $C_v + v\ \frac{\beta}{K}$

where K denotes isothermal compressibility and β is coefficient of volume expansion.

Solution : The isothermal compressibility K and the coefficient of volume expansion β are defined as

$$K = -\ \frac{1}{v}\left(\frac{\partial v}{\partial p}\right)_T$$

and
$$\beta = \frac{1}{v}\ \left(\frac{\partial v}{\partial T}\right)_p$$

That gives :

$$\left(\frac{\partial v}{\partial p}\right)_T = -\ Kv$$

and
$$\left(\frac{\partial v}{\partial T}\right)_p = \beta\ v$$

The cyclic relation for the thermodynamic properties p, T, v gives

$$\left(\frac{\partial p}{\partial T}\right)_v \left(\frac{\partial T}{\partial v}\right)_p \left(\frac{\partial v}{\partial p}\right)_T = -\ 1$$

Using the above expressions, the cyclic relation becomes

$$\left(\frac{\partial p}{\partial T}\right)_v \times \frac{1}{\beta v} \times (-\ Kv) = -\ 1$$

or
$$\left(\frac{\partial p}{\partial T}\right)_v = \frac{\beta}{K} \qquad ...(i)$$

(1) $h = u + pv$ and let $h = h(v, T)$. Differentiating h with respect to v(keeping T constant), we obtain

$$\left(\frac{\partial h}{\partial v}\right)_T = \left[\left(\frac{\partial u}{\partial v}\right)_T + p\right] + v\ \left(\frac{\partial p}{\partial v}\right)_T \qquad ...(ii)$$

Let $u = u(v, T)$

and
$$du = \left(\frac{\partial u}{\partial T}\right)_v dT + \left(\frac{\partial u}{\partial v}\right)_T dv$$

Substituting it in the Tds equation :

$T\ ds = du + p\ dv$, we get

$$T\ ds = \left[\left(\frac{\partial u}{\partial T}\right)_v dT + \left(\frac{\partial u}{\partial v}\right)_T dv\right] + p\ dv$$

or
$$ds = \frac{1}{T}\ \left(\frac{\partial u}{\partial T}\right)_v dT + \frac{1}{T}\ \left[\left(\frac{\partial u}{\partial v}\right)_T + p\right] dv$$

Now, let $s = s(T, v)$

$$ds = \left(\frac{\partial s}{\partial T}\right)_v dT + \left(\frac{\partial s}{\partial v}\right)_T dv$$

Comparing the above two expressions for ds, we can write

$$\left(\frac{\partial s}{\partial T}\right)_v = \frac{1}{T}\ \left(\frac{\partial u}{\partial T}\right)_v$$

and
$$\left(\frac{\partial s}{\partial v}\right)_T = \frac{1}{T}\ \left[p + \left(\frac{\partial u}{\partial v}\right)_T\right]$$

Differentiating the first identity with respect to v and the second with respect to T, we get

$$\left(\frac{\partial^2 s}{\partial s\ \partial T}\right) = \frac{1}{T}\ \frac{\partial^2 u}{\partial v\ \partial T}$$

and
$$\frac{\partial^2 s}{\partial T\ \partial s} = -\ \frac{1}{T^2}\ \left[p + \left(\frac{\partial u}{\partial v}\right)_T\right] + \frac{1}{T}\ \left[\left(\frac{\partial p}{\partial T}\right)_v + \frac{\partial^2 u}{\partial v\ \partial T}\right]$$

Since s is a property, the order of taking derivatives is immaterial. The right hand sides of the above two equations can then be equated and therefore

$$\frac{1}{T}\ \frac{\partial^2 u}{\partial v\ \partial T} = -\ \frac{1}{T^2}\ \left[p + \left(\frac{\partial u}{\partial v}\right)_T\right] + \frac{1}{T}\ \left[\left(\frac{\partial p}{\partial T}\right)_v + \frac{\partial^2 u}{\partial v\ \partial T}\right]$$

Simplification gives :

$$p + \left(\frac{\partial u}{\partial v}\right)_T = T\ \left(\frac{\partial p}{\partial T}\right)_v \qquad ...(iii)$$

Equation (*ii*) can then be written as

$$\left(\frac{\partial h}{\partial v}\right)_T = T\left(\frac{\partial p}{\partial T}\right)_v + v\left(\frac{\partial p}{\partial v}\right)_T$$

Substituting the results from identities at (*i*), we have

$$\left(\frac{\partial h}{\partial v}\right)_T = T\,\frac{\beta}{K} + v\left(\frac{1}{-Kv}\right)$$

$$= \frac{T\beta}{K} - K = \frac{1}{K}\,(T\beta - 1)$$

(2) Let $h = h(p, T)$

and $dh = \left(\frac{\partial h}{\partial T}\right)_p dT + \left(\frac{\partial h}{\partial p}\right)_T dp$

Substituting it in the $T\,ds$ equation :

$T\,ds = dh - v\,dp$, we get

$$T\,ds = \left[\left(\frac{\partial h}{\partial T}\right)_p dT + \left(\frac{\partial h}{\partial p}\right)_T dp\right] - \frac{1}{T}\,v\,dp$$

$$\text{or } ds = \frac{1}{T}\left(\frac{\partial h}{\partial T}\right)_p dT + \frac{1}{T}\left[\left(\frac{\partial h}{\partial p}\right)_T - v\right] dp$$

Now, let $s = s(T, p)$

$$ds = \left(\frac{\partial s}{\partial T}\right)_p dT + \left(\frac{\partial s}{\partial p}\right)_T dp$$

Comparing the above two expression for ds, we can write

$$\left(\frac{\partial s}{\partial T}\right)_p = \frac{1}{T}\left(\frac{\partial h}{\partial T}\right)_p$$

and $\left(\frac{\partial s}{\partial p}\right)_T = \frac{1}{T}\left[\left(\frac{\partial h}{\partial p}\right)_T - v\right]$

Differentiating the first identity with respect to p and the second with respect to T, we have

$$\frac{\partial^2 s}{\partial p\,\partial T} = \frac{1}{T}\,\frac{\partial^2 h}{\partial p\,\partial T}$$

and $\frac{\partial^2 s}{\partial T\,\partial p} = -\frac{1}{T^2}\left[\left(\frac{\partial h}{\partial p}\right)_T - v\right] + \frac{1}{T}\left[\frac{\partial^2 h}{\partial T\,\partial p} - \left(\frac{\partial v}{\partial T}\right)_p\right]$

Since s is a property, the order of taking derivatives is immaterial. The right hand sides of the above two identities can then the equated and therefore

$$\frac{1}{T}\,\frac{\partial^2 h}{\partial p\,\partial T} = -\frac{1}{T^2}\left[\left(\frac{\partial h}{\partial p}\right)_T - v\right] + \frac{1}{T}\left[\frac{\partial^2 h}{\partial T\,\partial p} - \left(\frac{\partial v}{\partial T}\right)_p\right]$$

or $\left(\frac{\partial h}{\partial p}\right)_T - v = -T\left(\frac{\partial v}{\partial T}\right)_p$

Since $\left(\frac{\partial v}{\partial T}\right)_p = \beta\,v$, we can write

$$\left(\frac{\partial h}{\partial p}\right)_T = v - T\beta v = v(1 - \beta T)$$

(3) Taking the derivative of $h = u + pv$ with respect to p keeping v constant,

$$\left(\frac{\partial h}{\partial p}\right)_T = \left(\frac{\partial u}{\partial p}\right)_v + v$$

$$= \left(\frac{\partial u}{\partial T}\right)_v \times \left(\frac{\partial T}{\partial p}\right)_v + v$$

Now : $\left(\frac{\partial p}{\partial T}\right)_v = \frac{\beta}{K}$

as outlined at (*i*) above and $\left(\frac{\partial u}{\partial T}\right)_v = c_v$

Then : $\left(\frac{\partial h}{\partial p}\right)_v = c_v\,\frac{K}{\beta} + v$

(4) Taking the derivative of $h = u + pv$ with respect to T while keeping v constant, we have

$$\left(\frac{\partial h}{\partial T}\right)_v = \left(\frac{\partial u}{\partial T}\right)_v + v\left(\frac{\partial p}{\partial T}\right)_v$$

Now $\left(\frac{\partial u}{\partial T}\right)_v = c_v$ and $\left(\frac{\partial p}{\partial T}\right)_v = \frac{\beta}{K}$ as outlined above at (*i*). Then

$$\left(\frac{\partial h}{\partial T}\right)_v = c_v + v\,\frac{\beta}{K}$$

REVIEW QUESTIONS

1. State the condition for exact differential.
2. Derive Maxwell's equations and state their importance in thermodynamics.
3. Define the terms volume expansivity and isothermal compressibility.
4. The volume expansivity β and isothermal compressibility of a substance are given by

$$\beta = \frac{1}{T} \quad \text{and} \quad \alpha = \frac{1}{p}$$

Set up the corresponding equation of state for the substance. $(pv/T = \text{constant})$

5. From the following property relations $du = T\,ds - p\,dv$ and $dh = T\,ds + v\,dp$ show that for the perfect gas equation

$$\left(\frac{\partial u}{\partial v}\right)_T = 0 \quad \text{and} \quad \left(\frac{\partial h}{\partial p}\right)_T = 0$$

6. Derive the first and second $T\,ds$ equations and set up the expression for the difference in heat capacities c_p and c_v. State the significance of this expression.
7. Derive the following relations for the difference in heat capacities

$$c_p - c_v = -\left(\frac{\partial v}{\partial T}\right)_p^2 \left(\frac{\partial p}{\partial v}\right)_T = \frac{vT\beta^2}{\alpha}$$

where α and β are isothermal compressibility and volume expansivity respectively. The other symbols have their usual meanings.
What conclusions can be drawn from these relations?

8. Show that the ratio of specific heats for a gas is given by

$$\gamma = \frac{c_p}{c_v} = \left(\frac{\partial p}{\partial v}\right)_s \left(\frac{\partial v}{\partial p}\right)_T$$

9. Show that for a gas obeying van der Waal's equation of state
(a) Difference of specific heats is given by

$$c_p - c_v = \frac{R}{1 - \dfrac{2a(v-b)^2}{v^3 RT}}$$

(b) Joule Thomson coefficient is given by

$$u_j = \frac{v}{c_p}\,\frac{2a(v-b)^2 - RTbv^2}{RTv^3 - 2a(v-b)^2}$$

10. Using Maxwel's relation's show that for a pure substance

$$T\,ds = c_p\,dT - T\,v\,\beta\,dp$$
$$= c_v\,dT + T\frac{\beta}{k}dv$$
$$= \frac{k\,c_v\,dp}{\beta} + \frac{c_p}{\beta v}dv$$

where β is the coefficient of cubical expansion, k is the coefficient of compressibility, c_p and c_v are specific heats at constant pressure and constant volume respectively.

11. If $u = f(T, v)$ and $h = (T, p)$, show that

(a) $du = c_v\,dT + \left(\frac{T\beta}{k} - p\right)dv$

(b) $dh = v(1 - \beta T)\,dp$

12. (a) Explain the terms of cubical expansion β and coefficient of compressibility k. Hence, show that

$$\frac{\beta}{k} = \left(\frac{\partial p}{\partial T}\right)_v$$

(b) Using Maxwell's and other equations show that

$$c_p - c_v = \left[p + \left(\frac{\partial u}{\partial v}\right)_T\right]\left(\frac{\partial v}{\partial T}\right)_p = \frac{\beta^2 Tv}{k}$$

13. Set up the following relations

(a) $\left(\frac{\partial u}{\partial v}\right)_T = T\left(\frac{\partial p}{\partial T}\right)_v - p$

(b) $\left(\frac{\partial h}{\partial p}\right)_T = v - T\left(\frac{\partial v}{\partial T}\right)_p$

□□□

Chapter 12

Gas Power Cycles

Learning objectives : The subject matter included in this chapter will enable the reader to:

- *define the concept of air standard cycle and air standard efficiency*
- *explain the different air standard cycles and set up expressions for cycle efficiency and mean effective pressure*
- *compare the performance of otto, diesel and dual cycle on the basis of compression ratio, maximum pressure/temperature and heat addition*

12.1. HEAT ENGINE

A heat engine is a device in which chemical energy of fuel is converted into heat energy which is further utilized to produce mechanical work. Heat engines are broadly classified into external combustion and internal combustion engines.

(*a*) **External combustion engines :** The combustion of fuel takes place outside the cylinder in a furnace. The combustion products thus produced raise the pressure and temperature of another fluid (say water) which gets converted into vapour. These vapours expand in a steam turbine/steam engine and generate mechanical power.

(*b*) **Internal combustion engines :** The combustion of fuel takes place inside a cylinder. The heat energy is liberated by fuel and the combustion products are available at high temperature and pressure. These gases subsequently expand to produce mechanical work.

By the kind of fuel used and the method of ignition adopted, the IC engines are classified as :

(*i*) *Spark ignition (Petrol) engines :* The combustible mixture of air and fuel is drawn into a cylinder and compressed. An electric spark ignites the mixture and energy is liberated by the spontaneous exothermic chemical reaction and the temperature is raised to a high value. The gaseous products expand, work is done on the piston and that turns the engine crank shaft. Subsequently the spent up gases (gases which have done work) are expelled to the atmosphere, and a fresh mixture of fuel and air is sucked in.

(*ii*) *Compression ignition (Diesel) engines :* Air is drawn into the cylinder and is compressed to a high pressure. The compression process raises the temperature of air above the ignition temperature of fuel. Towards the end of compression process, diesel fuel is injected and it quickly ignites. The combustion products expand, work is done and finally the exhaust takes place.

It is to be noted that in both the spark and compression ignition engines, the working fluid is never restored to the initial state. The condition of the exhaust gases is not the same as the condition of the charge sucked in. The spent up gases are expelled to the atmosphere and fresh charge is sucked in during each cycle. As such, these engines execute only a mechanical cycle and not a thermodynamic cycle. A true thermodynamic cycle comprises sequence of operations during which the working fluid undergoes changes in fluid properties but eventually it returns to the initial state.

The mechanical cycles executed by the I C engines can be converted into equivalent thermodynamic cycles if air is taken as the working fluid, and the addition and rejection of heat are due to the processes of heat transfer alone. The thermodynamic cycle with air as the working fluid is called an ***air standard cycle***.

12.2. AIR STANDARD EFFICIENCY

The mechanical cycle of petrol and diesel engines can be modelled into a thermodynamic cycle if the following assumptions are made.

(1) The engine operates in a closed cycle. The cylinder is filled with constant amount of working medium and the same fluid is used repeatedly.

(2) The working fluid is pure air; it behaves as a perfect gas and its properties at any state can be calculated by applying the characteristic gas equation.

(3) The working fluid is homogeneous throughout at all times and no chemical reaction takes place.

(4) The compression and expansion processes are truly adiabatic; no heat is gained or lost.

(5) All the processes are internally reversible; no mechanical or friction loss.

(6) The specific heats, c_p and c_v, and their ratio γ have constant values.

(7) The actual combustion of fuel is replaced by heat supply from a heat source. The source is brought into contact with the cylinder head at appropriate time. The cylinder head is perfect conductor during that operation. Heat is thus added from external heat source and not at the expense of combustion of fuel.

(8) The exhaust process is replaced by an equivalent amount of heat rejection to a heat sink by which the working fluid is restored to the initial state.

Both the source and sink are of infinite capacity so that when the heat is transferred from or to them, their temperatures remain unchanged.

From second law of thermodynamics, only a portion of heat supplied to a system is converted into work and the remainder has to be rejected to an outside sink. If Q_1 and Q_2 represent the heat addition and heat rejection respectively, then assuming no friction and heat losses

$$\text{Work done} = Q_1 - Q_2$$

$$\text{Thermal efficiency} = \frac{\text{work done}}{\text{heat input}}$$

$$= \frac{Q_1 - Q_2}{Q_1} \quad ...(12.1)$$

A thermodynamic cycle working with air as the working fluid is called an ***air standard cycle***, and the thermal efficiency of an air standard cycle is called ***air standard efficiency***.

The approach and concept of ideal air cycle helps to

(*i*) indicate the ultimate performance, *i.e.*, to determine the maximum ideal efficiency of a specific thermodynamic cycle.

(*ii*) study qualitatively the influence of different variables on the performance of an actual engine.

(*iii*) evaluate one engine relative to another.

Apparently an air standard cycle establishes a standard of perfection by which performances of actual engines can be compared.

12.3. PISTON-CYLINDER ARRANGEMENT

The essential elements needed for making an analysis of air standard cycle are :

– a working fluid which is assumed to be a perfect gas.

– two heat reservoirs; the working fluid receives heat from heat source at higher temperature in one part of the cycle and rejects heat to heat sink at lower temperature during another part of the cycle.

– piston-cylinder arrangement for getting work out of the working fluid. The piston and cylinder walls (excluding the cylinder head) are taken as perfect heat insulators. The cylinder head is imagined to provide alternately a diathermic cover (perfect heat conductor) and an adiabatic cover (perfect heat insulator).

The piston-cylinder arrangement is shown in Fig. 12.1. There is no friction to the movement of piston inside the cylinder.

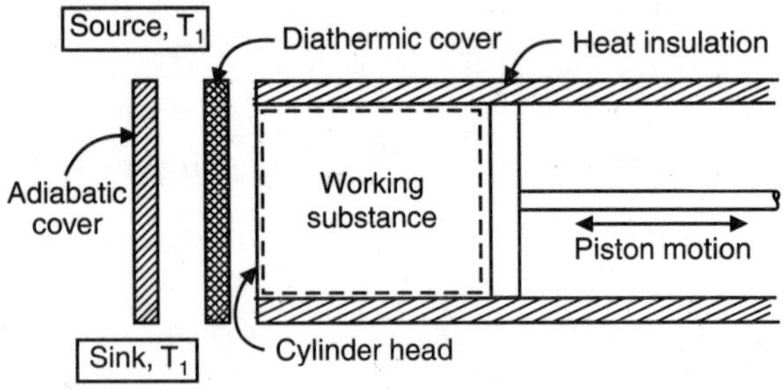

Fig. 12.1. Essential elements for an air standard cycle

With reference to piston-cylinder assembly of an internal combustion engine shown in Fig. 12.2, some of the terms that are common to reciprocating piston engines are defined below:

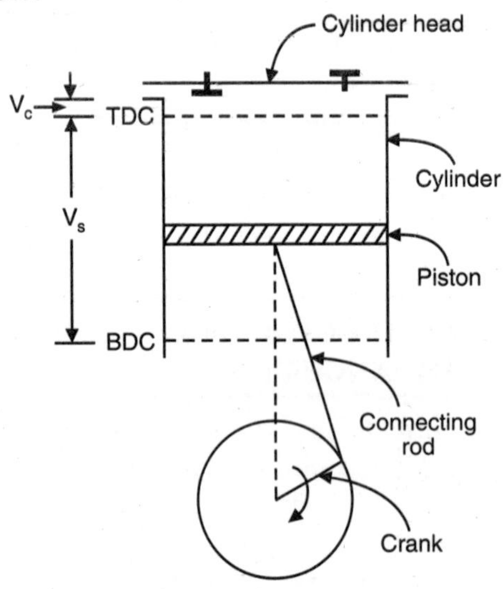

Fig. 12.2.

Dead centres : The extreme positions of the piston in the cylinder are called dead centres. The piston reverses its direction of motion at the dead centres and apparently its velocity is zero. For a vertical configuration, the dead centres are called the top and bottom dead centres. For horizontal disposition of piston-cylinder assembly, the extreme position of the piston towards the cover end and crank end side of the cylinder are referred to as the inner and outer dead centres, respectively.

Stroke length and swept volume : The distance between the top and bottom dead centre is called the piston stroke. The piston stroke equals double the radius of crank, *i.e.*, $l = 2r$.

The volume swept by the piston as it moves from top dead centre to bottom dead centre is called the ***swept volume***.

$$\text{Swept volume } V_S = \frac{\pi}{4} d^2 l$$

where d is the inside diameter of the cylinder; it is called bore.

Clearance space and clearance volume : Theoretically a piston may flush with the cylinder head but in actual practice some clearance is always kept for wear and tear and for the reception of valves. The space between the cylinder head and the piston at top dead centre position is called ***clearance space*** and the corresponding volume is called ***clearance volume***. The clearance volume is usually expressed as a percentage of swept volume.

Clearance volume,

$$V_C = \frac{\pi}{4} d^2 \times \text{clearance space}$$

Cylinder volume and compression ratio : The volume of working fluid in the cylinder, when the piston is at the bottom dead centre is referred to as cylinder volume and it equals the clearance volume plus swept volume.

Cylinder volume

$$= \text{clearance volume} + \text{swept volume}$$

$$= V_C + V_S$$

The ratio of cylinder volume to clearance volume is called the ***compression ratio***.

Compression ratio,

$$r = \frac{\text{cylinder volume}}{\text{clearance volume}}$$

$$= \frac{V_C + V_S}{V_C} = 1 + \frac{V_S}{V_C} \qquad \text{...(12.2)}$$

Higher the compression ratio, better will be the performance of an engine.

Piston speed : The distance travelled by the piston in one minute is called *piston speed.*

Piston speed $= 2\,lN$ m/min ...(12.3)

where l is the stroke length in metre, and N is the speed of the engine in rev/min.

Mean effective pressure : The pressure variation versus volume inside the cylinder of a reciprocating engine is plotted with the help of an engine indicator.

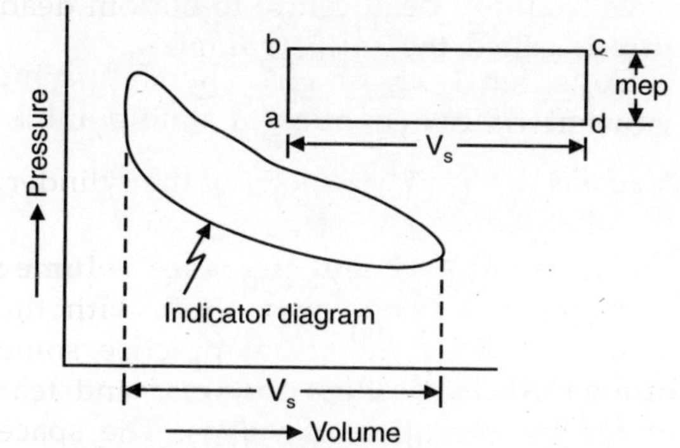

Fig. 12.3.

The resulting contour is closed one and is referred to as indicator diagram. The area enclosed by the contour is a measure of the work output per cycle from the engine.

Mean effective pressure (mep) is defined as the average pressure acting on the piston which will produce the same output as is done by the varying pressure during a cycle.

That is

area of indicator loop

= area of rectangle *abcd*

The height of the rectangle then represents the mean effective pressure.

$$mep = \frac{\text{work done per cycle}}{\text{swept volume}}$$

$$= \frac{\text{area of indicator loop}}{\text{length of loop}} \qquad \text{...(12.4)}$$

Mean effective pressure is usually expressed in bar or kN/m^2.

Mean effective pressure is used as a parameter to compare the performance of reciprocating engines of equal size. An engine that has a large value of *mep* will deliver more net work and will thus perform better.

EXAMPLE 12.1.

An engine, bore 7.5 cm and stroke 10 cm, has a compression ratio of 5 to 1. To increase the compression ratio, 1.25 mm is machined off the cylinder head face. Calculate the new compression ratio.

Solution : Swept volume $V_S = \frac{\pi}{4} d^2 l_S$ and clearance volume $V_C = \frac{\pi}{4} d^2 l_C$ where l_S and l_C represent the stroke length and clearance length respectively and d is the cylinder bore.

Compression ratio,

$$r = \frac{V_C + V_S}{V_C} = 1 + \frac{V_S}{V_C}$$

$$5 = 1 + \frac{\pi/4\, d^2 l_S}{\pi/4\, d^2 l_C} = 1 + \frac{l_S}{l_C}$$

∴ Clearance length,

$$l_C = \frac{l_S}{4} = \frac{10}{4} = 2.5 \text{ cm}$$

Clearance length after machining

$$= 2.5 - 0.125 = 2.375$$

∴ New compression ratio

$$= 1 + \frac{10}{2.375} = \mathbf{5.21}$$

12.4. CARNOT CYCLE

The Carnot cycle is operated with two isothermals and two adiabatic operations.

The four stages of operation of a Carnot cycle have been shown plotted on p–V and T–s diagram in Fig. 12.4. The sequence of operations is

1. Isothermal expansion 1 – 2
2. Adiabatic expansion 2 – 3
3. Isothermal compression 3 – 4
4. Adiabatic compression 4 – 1

These operations and salient aspects of Carnot cycle have already been explained in chapter 8. The cycle has the highest possible efficiency given by

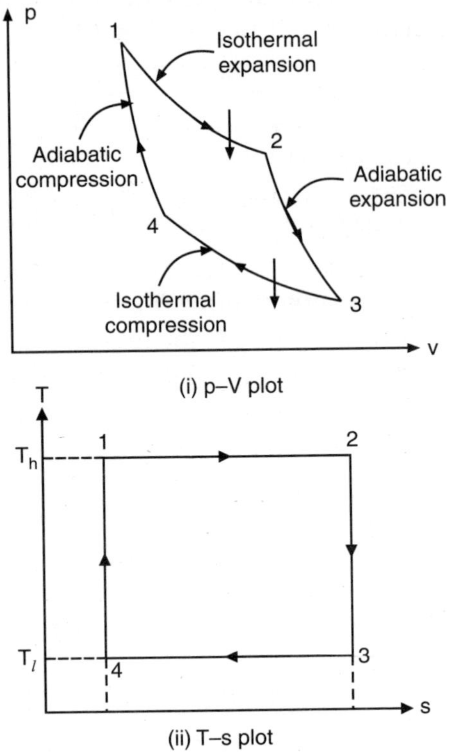

Fig. 12.4. Carnot cycle

$$\eta_{carnot} = \frac{T_h - T_l}{T_h} \quad ...(12.5)$$

where T_h and T_l denote the highest and lowest temperature respectively in the operation of the cycle.

The Carnot cycle is only of theoretical importance and is unsuitable (for the reasons already discussed) as a basic cycle for actual reciprocating engines. The cycle is not seriously considered as a device for producing power. The cycle, however, permits the designer to arrive at maximum efficiency that can be obtained from any heat engine under the given conditions.

EXAMPLE 12.2

The following data pertains to an air standard Carnot cycle:

Minimum temperature in the cycle = 290 K

Minimum pressure in the cycle = 100 kN/m²

Pressure after isothermal compression = 350 kN/m²

Pressure after isentropic compression = 1050 kN/m²

Determine the efficiency and mean effective pressure for the cycle. Proceed to calculate the power output if the engine makes 2 cycles/sec.

Solution : Refer Fig. 12.4. The subscripts denote the respective state points in the cycle.

Given : $p_1 = 1050$ kN/m²; $p_3 = 100$ kN/m²

and $p_4 = 350$ kN/m²

$T_3 = T_4 = T_l = 290$ K

Consider 1 kg of air. Then applying characteristic gas equation at state point 3,

$$V_3 = \frac{mRT_3}{p_3} = \frac{1 \times 287 \times 290}{100 \times 10^3} = 0.832 \text{ m}^3$$

For the isothermal compression process 3 – 4:

$$V_4 = \frac{p_3}{p_4} \times V_3 = \frac{100}{350} \times 0.832 = 0.238 \text{ m}^3$$

For the isentropic compression process 4 – 1 :

$$V_1 = \left(\frac{p_4}{p_1}\right)^{\frac{1}{\gamma}} \times V_4 = \left(\frac{350}{1050}\right)^{\frac{1}{1.4}} \times 0.238 = 0.109 \text{ m}^3$$

$$T_1 = T_4 \left(\frac{p_1}{p_4}\right)^{\frac{\gamma - 1}{\gamma}} = 290 \times \left(\frac{1050}{350}\right)^{\frac{1.4-1}{1.4}} = 397 \text{ K}$$

The temperature $T_1 = T_2 = 397$ K is the maximum temperature in the cycle. Efficiency of air standard Carnot cycle is

$$\eta = \frac{T_h - T_l}{T_h}$$

where T_h and T_l are the higest and lowest operating temperastures of the cycle

$$\therefore \quad \eta = \frac{397 - 290}{397} = \mathbf{0.269} \text{ or } \mathbf{26.9}\ \%$$

(*b*) Work done per cycle for the Carnot cycle is

$$= mR(T_h - T_l) \log_e r$$

where $r = \frac{V_2}{V_1}$

From the isentropic processes

$$\frac{T_h}{T_l} = \left(\frac{V_4}{V_1}\right)^{\gamma-1} = \left(\frac{V_3}{V_2}\right)^{\gamma-1}$$

$$\therefore \quad \frac{V_4}{V_1} = \frac{V_3}{V_2}$$

$$\text{or} \quad \frac{V_2}{V_1} = \frac{V_3}{V_4} = \frac{0.832}{0.238} \simeq 3.496$$

∴ Work done per cycle

$$= 1 \times 287(397 - 290) \text{ loge } 3.496$$

$$= 38436 \text{ Nm} = 38.436 \text{ kNm}$$

(*c*) Swept volume

$$= V_3 - V_1$$

$$= 0.832 - 0.109 = 0.723 \text{ m}^3$$

Mean effective pressure,

$$mep = \frac{\text{work done per cycle}}{\text{swept volume per stroke}}$$

$$= \frac{38436}{0.723}$$

$$= 53162 \text{ N/m}^2 = 53.132 \text{ kN/m}^2$$

(*d*) Work done = 38436 Nm/cycle

Since the engine makes two cycles per second,

Work done/sec

$$= 38436 \times 2$$

$$= 76872 \text{ Nm/s (or J/s)}$$

$$= 76.872 \text{ kJ/s} = \mathbf{76.872\ kW}$$

EXAMPLE 12.3

In an ideal engine operating on the Carnot cycle, the ratio of isentropic compression is 6 and isothermal expansion is 1.5. If the maximum pressure and temperature are limited to 20 bar and 700 K, make calculations for the following parameters :

(a) temperature and pressure at main points of the cycle,

(b) change in entropy during isothermal expansion,

(c) thermal efficiency and mean effective pressure of the cycle, and

(d) theoretical power if the engine makes 180 cycles per minute.

It may be assumed that air has a volume of 0.21 m³ at the beginning of isothermal expansion.

Solution : Refer Fig. 12.4. The subscripts denote the respective state points in the cycle.

Given : p_1 = 20 bar; T_1 = 700 K and $V_1 = 0.21 \text{ m}^3$

$$\frac{V_4}{V_1} = 6 \quad \text{and} \quad \frac{V_2}{V_1} = 1.5$$

For the isothermal expansion process 1 – 2 :

$$p_2 = p_1 \frac{V_1}{V_2} = 20 \times \frac{1}{1.5} = 13.33 \text{ bar}$$

For the isentropic exapansion process 2 – 3 :

$$p_3 = p_2 \left(\frac{V_2}{V_3}\right)^{\gamma}$$

Again, from isentropic processes 4 – 1 and 2 – 3,

$$\frac{T_h}{T_l} = \left(\frac{V_4}{V_1}\right)^{\gamma-1} = \left(\frac{V_3}{V_2}\right)^{\gamma-1}$$

$$\therefore \quad \frac{V_4}{V_1} = \frac{V_3}{V_2} \quad \text{or} \quad \frac{V_1}{V_4} = \frac{V_2}{V_3}$$

and accordingly

$$p_3 = p_2 \left(\frac{V_1}{V_4}\right)^{\gamma} = 13.33 \times \left(\frac{1}{6}\right)^{1.4}$$

$$= 1.08 \text{ bar}$$

$$\frac{T_1}{T_4} = \left(\frac{V_4}{V_1}\right)^{\gamma-1} = (6)^{1.4-1} = 2.05$$

$$\frac{p_1}{p_4} = \left(\frac{V_4}{V_1}\right)^{\gamma} = (6)^{1.4} = 12.29$$

$$\therefore \quad T_4 = \frac{700}{2.05} = 341.5 \text{ K} \quad \text{and}$$

$$p_4 = \frac{20}{12.29} = 1.63 \text{ bar}$$

Thus : $p_1 = 20$ bar; $p_2 = 13.33$ bar
$p_3 = 1.08$ bar and $p_4 = 1.63$ bar
$T_1 = T_2 = 700$ K
and $T_3 = T_4 = 341.5$ K

(*b*) Change in entropy during isothermal expansion,

$$S_2 - S_1 = mR \log_e \left(\frac{V_2}{V_1}\right)$$

$$= \frac{p_1 V_1}{T_1} \log_e \frac{V_2}{V_1}$$

$$= \frac{20 \times 10^5 \times 0.21}{700} \log_e 1.5$$

$= 243.3$ J/k $=$ **0.2433** kJ/K

(*c*) Heat supplied, $Q_1 = T_1 (S_2 - S_1)$
$= 700 \times 0.2433 = 170.31$ kJ

Heat rejected, $Q_2 = T_4 (S_3 - S_4)$
$= 341.5 \times 0.2433 = 83.09$ kJ

It is to be noted that decrease in entropy during heat rejection equals increase in entropy during heat addition.

Work done $= Q_1 - Q_2 = 170.31 - 83.09$
$= 87.22$ kJ

$$\text{Efficiency, } \eta = \frac{\text{work done}}{\text{heat supplied}}$$

$$= \frac{87.22}{170.31} = \mathbf{0.512} \text{ or } \mathbf{51.2}\ \%$$

Now: $\frac{V_4}{V_1} = 6$; $V_4 = 6 \times 0.21 = 1.26$ m^3

$\frac{V_2}{V_1} = 1.5$; $V_2 = 1.5 \times 0.21 = 0.315$ m^3

Since $\frac{V_4}{V_1} = \frac{V_3}{V_2}$, we have :

$$V_3 = V_2 \times \frac{V_4}{V_1} = 0.315 \times 6 = 1.89 \text{ m}^3$$

∴ Swept volume,
$V_s = V_3 - V_1$
$= 1.89 - 0.21 = 1.68$ m^3

Mean effective pressure,

$$mep = \frac{\text{work done}}{\text{swept volume}}$$

$$= \frac{87.22 \times 10^3}{1.68}$$

$=$ **51917** N/m^2 $=$ **0.519** bar

(*d*) Work done $= 87.22 \times 10^3$ Nm/cycle

Since the engine makes 180 cycles per minute,

$$\text{work done/sec} = 87.22 \times 10^3 \times \frac{180}{60}$$

$= 261660$ Nm/s (or J/s)
$=$ **261.66** kJ/s $=$ **261.66** kW

EXAMPLE 12.4

For an ideal engine operating on Carnot cycle with a perfect gas as the working fluid, the lower temperature has a fixed value and so is the ratio of the greatest to the least volume fixed at x : 1. However, the volume compression ratio r of the reversible adiabatic compression is variable. Show that for maximum work obtainable from the cycle, the following identity holds good

$$(\gamma - 1) \log_e \frac{x}{r} + \frac{1}{r^{\gamma - 1}} = 1$$

Solution : With reference to Fig. 12.4,

$$\frac{V_3}{V_1} = x \text{ and } \frac{V_4}{V_1} = r$$

Considering unit mass of the working fluid
Work done = heat supplied – heat rejected

$$= R T_h \log_e \frac{V_2}{V_1} - R T_l \log_e \frac{V_3}{V_4}$$

Since the cycle of operation is limited by two pairs of parallel curves (two isothermals and two adiabatics),

$$\text{expansion ratio } \frac{V_2}{V_1} = \text{compression ratio } \frac{V_3}{V_4}$$

$$\text{Also } \frac{V_3}{V_4} = \frac{V_3}{V_1} \times \frac{V_1}{V_4} = x \times \frac{1}{r} = \frac{x}{r}$$

∴ work done

$$= R\,T_h \log_e \frac{x}{r} - RT_l \log_e \frac{x}{r}$$

$$= R\,(T_h - T_l) \log_e \frac{x}{r}$$

$$= RT_l\left(\frac{T_h}{T_l} - 1\right)\log_e \frac{x}{r}$$

From isentropic compression process :

$$\frac{T_h}{T_l} = \left(\frac{V_4}{V_1}\right)^{\gamma-1} = r^{\gamma-1}$$

∴ work done,

$$W = R\,T_l\left[r^{\gamma-1} - 1\right]\log_e \frac{x}{r}$$

Differentiating W with respect to r (holding T_l and x as constant) and equating the differential to zero, we get

$$\frac{dW}{dr} = R\,T_l\left[\left\{r^{\gamma-1} - 1\right\}\left\{\frac{r}{x}\times(-xr^{-2})\right\}\right.$$

$$\left. + \log_e \frac{x}{r}\left\{(\gamma-1)\,r^{\gamma-2}\right\}\right] = 0$$

or $$\left(r^{\gamma-1} - 1\right)\left(-\frac{1}{r}\right) + (\gamma-1)\,r^{\gamma-2}\log_e \frac{x}{r} = 0$$

or $$-r^{\gamma-2} + \frac{1}{r} + (\gamma-1)\,r^{\gamma-2}\log_e \frac{x}{r} = 0$$

or $$r^{\gamma-2}\left[-1 + \frac{1}{r\times r^{\gamma-2}} + (\gamma-1)\log_e \frac{x}{r}\right] = 0$$

or $$-1 + \frac{1}{r^{\gamma-1}} + (\gamma-1)\log_e \frac{x}{r} = 0$$

or $$(\gamma-1)\log_e \frac{x}{r} + \frac{1}{r^{\gamma-1}} = 1$$

which is the required result.

12.5. OTTO (CONSTANT VOLUME) CYCLE

The cycle was presented by Beau De Rochas (1862) and was successfully applied by a German scientist Nikolous A Otto to produce a successful four stroke cycle engine in 1876, that was far superior to any IC engine previously built. The thermodynamic cycle is operated with isochoric (constant volume) heat addition, and consists of two adiabatics processes and two constant volume changes.

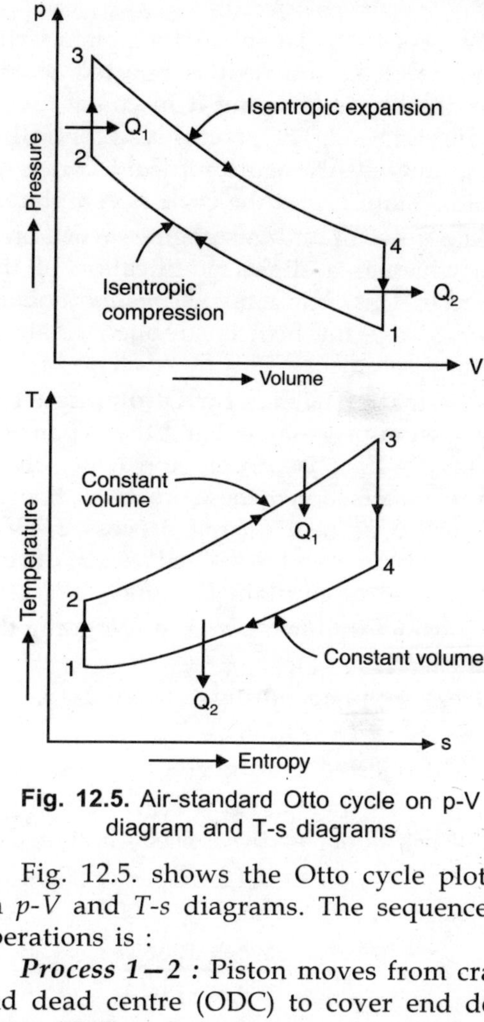

Fig. 12.5. Air-standard Otto cycle on p-V diagram and T-s diagrams

Fig. 12.5. shows the Otto cycle plotted on p-V and T-s diagrams. The sequence of operations is :

Process 1 – 2 : Piston moves from crank end dead centre (ODC) to cover end dead centre (IDC) and an ideal gas with initial state 1 is compressed isentropically to state point 2, through compression ratio, $r = V_1/V_2$

Process 2 – 3 : The piston is momentarily at rest at IDC and heat is added to the working fluid at constant volume from an external heat source which is brought into contact with the cylinder head. The pressure rises and the ratio $\alpha = p_3/p_2$ is called the ***explosion ratio.***

Process 3 – 4 : The increased high pressure exerts a greater amount of force on the piston and pushes it towards the ODC. Expansion of working fluid takes place isentropically and work is done by the system. The volume ratio V_4/V_3 is called ***isentropic expansion ratio***.

Process 4−1 : The piston is momentarily at rest at ODC and heat is rejected to the external sink by bringing it in contact with the cylinder head. The process is so controlled that ultimately the working fluid comes to its initial state 1 and the cycle is completed.

Many petrol and gas engines work on a cycle which is a slight modification of the Otto cycle. This cycle is called ***constant volume cycle*** because the heat is supplied to air at constant volume.

Thermal efficiency for Otto cycle : The compression process 1−2 and the expansion process 3−4 are isentropic and hence there is no heat interaction in these processes. Energy is absorbed as heat during process 2−3 at constant volume, and is rejected as heat during process 4−1 at constant volume.

Consider unit mass of air undergoing the cyclic change,

Heat supplied during process 2−3,

$$Q_1 = c_v (T_3 - T_2)$$

Heat rejected during process 4−1,

$$Q_2 = c_v (T_4 - T_1)$$

$$\text{Work done} = \text{heat supplied} - \text{heat rejected}$$
$$= c_v (T_3 - T_2) - c_v (T_4 - T_1)$$

Thermal efficiency,

$$\eta = \frac{\text{work done}}{\text{heat supplied}}$$

$$= \frac{c_v (T_3 - T_2) - c_v (T_4 - T_1)}{c_v (T_3 - T_2)}$$

Since c_v remains constant and is independent of temperature,

$$\eta = \frac{(T_3 - T_2) - (T_4 - T_1)}{T_3 - T_2}$$

$$= 1 - \frac{T_4 - T_1}{T_3 - T_2} \qquad \text{...(12.6)}$$

The expression for thermal efficiency in terms of compression ratio r and adiabatic exponent γ can be set up by expressing all the temperatures in terms of lowest temperature T_1 in the cycle.

At point 2 : $V_2 = \dfrac{V_1}{r}$

where r is the compression ratio

$$p_2 = p_1 \left(\frac{V_1}{V_2}\right)^{\gamma} = p_1 r^{\gamma}$$

$$T_2 = T_1 \left(\frac{V_1}{V_2}\right)^{\gamma - 1} = T_1 \, r^{\gamma - 1}$$

At point 3 : $V_3 = V_2 = \dfrac{V_1}{r}$

$$p_3 = p_2 \, \alpha = p_1 \, \alpha \, r^{\gamma}$$

where α is the explosion pressure ratio

$$T_3 = T_2 \frac{p_3}{p_2}$$
$$= T_2 \, \alpha = T_1 \, \alpha \, r^{\gamma - 1}$$

At point 4 : $V_4 = V_1$

$$p_4 = p_3 \left(\frac{V_3}{V_4}\right)^{\gamma} = p_3 \left(\frac{V_2}{V_1}\right)^{\gamma}$$

$$= \frac{p_3}{r^{\gamma}} = \frac{p_1 \, r^{\gamma} \, \alpha}{r^{\gamma}} = p_1 \, \alpha$$

$$T_4 = T_3 \left(\frac{V_3}{V_4}\right)^{\gamma - 1}$$

$$= T_1 \, \alpha \, r^{\gamma - 1} \times \frac{1}{r^{\gamma - 1}} = T_1 \, \alpha$$

When these temperature values are substituted in expression 12.6, we get

$$\eta = 1 - \frac{T_4 - T_1}{T_3 - T_2}$$

$$= 1 - \frac{T_1 \alpha - T_1}{T_1 \alpha r^{\gamma - 1} - T_1 r^{\gamma - 1}}$$

$$= 1 - \frac{T_1 (\alpha - 1)}{T_1 r^{\gamma - 1} (\alpha - 1)}$$

$$= 1 - \frac{1}{r^{\gamma - 1}} \qquad \text{...(12.7)}$$

Thus thermal efficiency of an Otto cycle engine depends upon compression ratio r and the adiabatic exponent γ, *i.e.*, on the nature of working medium. The thermal efficiency increases with both r and γ. In actual engines working on Otto cycle, the compression ratio varies from 5 to 8 depending upon the quality of fuel. At compression ratios higher than this, the temperature after combustion becomes high and that may lead to spontaneous and uncontrolled combustion of fuel in the cylinder. The phenomenon of uncontrolled combustion in petrol engines is called ***detonation*** and it leads to poor engine efficiency and in structural damage of engine parts.

Fig. 12.6. shows the variation of air standard efficiency of Otto cycle with compression ratio. The curve tends to become rather flat at higher compression ratios which means that though the efficiency is still increasing, the rate of increase starts diminishing.

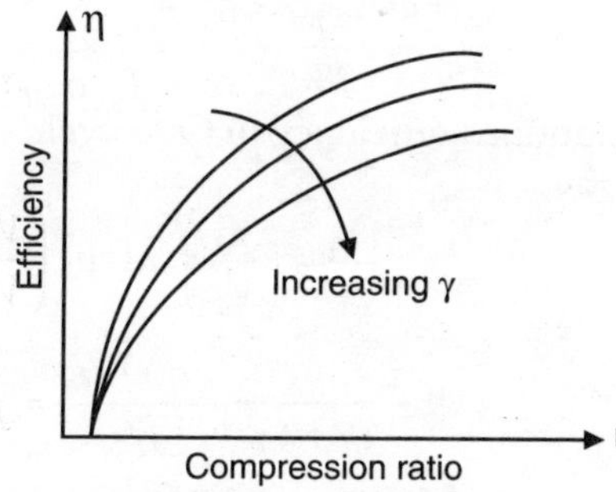

Fig. 12.6. Efficiency of Otto cycle versus compresssion ratio

Condition for maximum work : The ambient conditions determine the minimum temperature T_1 while the maximum temperature T_3 is determined by the design conditions and metallurgical considerations of the piston and cylinder.

For unit mass of air,

Work output,

$$W = c_v (T_3 - T_2) - c_v (T_4 - T_1)$$

$$\frac{W}{c_v} = T_3 - T_2 - T_4 + T_1 \qquad ...(i)$$

From adiabatic compression process 1–2 and adiabatic expansion process 3–4, we have

$$T_2 = T_1 \left(\frac{V_1}{V_2}\right)^{\gamma-1} = T_1\, r^{\gamma-1}$$

$$T_4 = T_3 \left(\frac{V_3}{V_4}\right)^{\gamma-1}$$

$$= T_3 \left(\frac{V_2}{V_1}\right)^{\gamma-1} = \frac{T_3}{r^{\gamma-1}}$$

$$= \frac{T_3}{T_2/T_1} = \frac{T_1 T_3}{T_2}$$

Substituting $T_4 = \frac{T_1 T_3}{T_2}$ in expression (i), we get

$$\frac{W}{c_v} = T_3 - T_2 - \frac{T_1 T_3}{T_2} + T_1$$

The intermediate temperature T_2 for maximum work output can be obtained by differentiating the above expression with respect to T_2 and setting the derivative equal to zero. That is

$$\frac{1}{c_v}\frac{dW}{dT_2} = -1 + \frac{T_1 T_3}{T_2^2}$$

$$= 0 \text{ for maximum work}$$

$$\therefore \quad T_2^2 = T_1 T_3$$

$$\text{or} \quad T_2 = \sqrt{T_1 T_3} \qquad ...(12.8)$$

Similarly :

$$\frac{W}{c_v} = T_3 - \frac{T_1 T_3}{T_4} - T_4 + T_1$$

$$\frac{1}{c_v}\frac{dW}{dT_4} = \frac{T_1 T_3}{T_4^2} - 1$$

$$= 0 \text{ for maximum work}$$

$$\text{or} \quad T_4 = \sqrt{T_1 T_3} \qquad ...(12.9)$$

Thus for maximization of work,

$T_2 = T_4 = \sqrt{T_1 T_3}$, *i.e.*, the intermediate temperatures T_2 and T_4 must be equal.

The corresponding maximum work is

$$W_{max} = c_v (T_3 - T_2 - T_4 + T_1)$$

$$= c_v \left(T_3 - \sqrt{T_1 T_3} - \sqrt{T_1 T_3} + T_1\right)$$

$$= c_v \left(T_3 + T_1 - 2\sqrt{T_1 T_3}\right) \qquad ...(12.10)$$

Mean effective pressure : With reference to Fig. 12.5, the net work done per cycle for unit mass of air is

Work done = heat supplied – heat rejected

$$= c_v (T_3 - T_2) - c_v (T_4 - T_1)$$

Swept volume

$$= (V_1 - V_2) = V_1\left(1 - \frac{V_2}{V_1}\right)$$

$$= \frac{RT_1}{p_1}\left(1 - \frac{1}{r}\right) = \frac{RT_1}{p_1\, r}(r-1)$$

By definition of mean effective pressure

$$mep = \frac{\text{work done per cycle}}{\text{swept volume}}$$

$$= \frac{c_v(T_3 - T_2) - c_v(T_4 - T_1)}{\frac{RT_1}{p_1 r}(r-1)}$$

$$= \frac{c_v}{R}\frac{p_1\, r}{(r-1)}\left[\frac{(T_3 - T_2) - (T_4 - T_1)}{T_1}\right] \quad ...(12.11)$$

The expression for mean effective pressure in terms of compression ratio r, explosion ratio α, induction pressure p_1 and adiabatic exponent γ can be set up by expressing all the temperatures in terms of lowest temperature T_1 in the cycle.

As outlined above :

$$T_2 = T_1\, r^{\gamma-1};\; T_3 = T_1\, \alpha\, r^{\gamma-1}$$

$$T_4 = T_1\, \alpha$$

When these temperature values are substituted in expression 12.11,

we get : mep

$$= \frac{c_v}{R}\frac{p_1\, r}{(r-1)}\left[\frac{T_1 \alpha r^{\gamma-1} - T_1 r^{\gamma-1} - (T_1\alpha - T_1)}{T_1}\right]$$

$$= \frac{c_v}{R}\frac{p_1\, r}{(r-1)}\left[\left(r^{\gamma-1} - 1\right)(\alpha - 1)\right]$$

But, $\frac{c_v}{R} = \frac{1}{\gamma - 1}$

$$\therefore\; mep = \frac{p_1\, r}{(r-1)(\gamma-1)}\left[\left(r^{\gamma-1} - 1\right)(\alpha - 1)\right] \quad ...(12.12)$$

which is the required expression of mean effective pressure for an Otto cycle.

EXAMPLE 12.5

The following data pertains to Rolls Royce Merline engine :

Cylinder bore (dia) = 13.7 cm; Stroke length = 13.0 cm ; Clearance volume = 280 cm³

Express clearance as a percentage of swept volume and calculate the air standard efficiency of the engine operating on four-stroke cycle.

Solution : Swept volume,

$$V_S = \frac{\pi}{4} d^2\, l = \frac{\pi}{4}(13.7)^2 \times 13$$

$$= 1915.4 \text{ cm}^3$$

Clearance volume,

$$V_C = 280 \text{ cm}^3$$

$$\therefore \quad \frac{V_C}{V_S} = \frac{280}{1915.4}$$

$$= 0.1462 \text{ or } \mathbf{14.62\ \%}$$

Compression ratio,

$$r = \frac{V_C + V_S}{V_C}$$

$$= \frac{280 + 1915.4}{280} = 7.84$$

Air standard efficiency of Otto cycle engine is given by,

$$\eta = 1 - \frac{1}{r^{\gamma-1}}$$

$$= 1 - \frac{1}{(7.84)^{1.4-1}}$$

$$= 0.562 \text{ or } \mathbf{56.2\ \%}$$

EXAMPLE 12.6

In an automobile device working on Otto cycle, energy generated per cycle is twice as much rejected through the exhaust. Calculate the thermal efficiency and compression ratio. Consider the working fluid as an ideal gas with $\gamma = 1.4$

Solution : Efficiency, $\eta = \frac{Q_1 - Q_2}{Q_1} = 1 - \frac{Q_2}{Q_1}$

Since $Q_1 = 2\, Q_2$, we get

$$\eta = 1 - \frac{1}{2} = 0.5 \text{ or } \mathbf{50\%}$$

In terms of compression ratio r,

$$\eta = 1 - \frac{1}{r^{\gamma-1}}; \quad 0.5 = 1 - \frac{1}{r^{1.4-1}}$$

$$\text{or} \quad r = \left(\frac{1}{1 - 0.5}\right)^{1/0.4} = \mathbf{5.66}$$

EXAMPLE 12.7

An Otto cycle takes in air at 1 bar and 15°C. The compression ratio is 8 to 1 and 2000 kJ/kg of energy is released to air in each cycle. To what value must the compression ratio be raised to increase the net work per cycle by 20 percent ?

Solution : $\eta = 1 - \dfrac{1}{r^{\gamma-1}}$

$$= 1 - \frac{1}{8^{1.4-1}} = 0.5647$$

$$W_{net} = \eta \times \text{heat supplied}$$
$$= 0.5647 \times 2000 = 1129.4 \text{ kJ/kg}$$

For the new cycle,

$$W_{net} = 1.2 \times 1129.4 = 1355.28 \text{ kJ/kg}$$

$$\eta = \frac{W_{net}}{\text{heat supplied}}$$
$$= \frac{1355.28}{2000} = 0.6776$$

In terms of compression ratio,

$$\eta = 1 - \frac{1}{r^{\gamma-1}} \; ; 0.6676 = 1 - \frac{1}{r^{1.4-1}}$$

or $$r = \left(\frac{1}{1-0.6776}\right)^{1/0.4} = \mathbf{16.95}$$

It is to be noted that for modest increase in net work output, the compression ratio has to be increased by a large value.

EXAMPLE 12.8

An air standard Otto cycle is designed to operate with the following data :

Maximum cycle pressure and temperature : 5 MPa and 2250 K

Minimum cycle pressure and temperature : 0.1 MPa and 300 K

Determine the net work output per unit mass of working fluid and the thermal efficiency.

Solution : Refer Fig. 12.7.

$$\text{Compression ratio, } r = \frac{V_1}{V_2} = \frac{T_1}{p_1} \times \frac{p_2}{T_2}$$

For constant volume heat addition process,

$$\frac{p_2}{T_2} = \frac{p_3}{T_3}$$

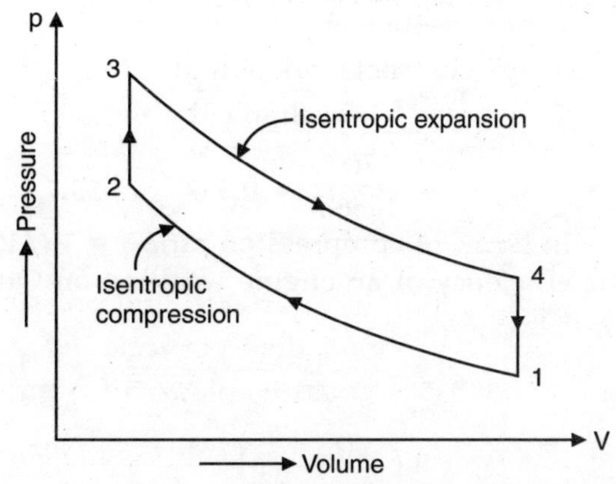

Fig. 12.7.

$$\therefore \quad r = \frac{T_1}{p_1} \times \frac{p_3}{T_3} = \frac{300}{0.1} \times \frac{5}{2250} = 6.67$$

$$\eta = 1 - \frac{1}{r^{\gamma-1}} = 1 - \frac{1}{6.67^{0.4}}$$
$$= 0.532 \quad \text{or} \quad \mathbf{53.2\ \%}$$

From adiabatic compression process,

$$T_2 = T_1 \left(\frac{V_1}{V_2}\right)^{\gamma-1}$$
$$= 300 \times (6.67)^{0.4} = 640.87 \text{ K}$$

Heat supplied per unit mass of working fluid,

$$Q_1 = c_v\,(T_3 - T_2)$$
$$= 0.718\,(2250 - 640.87)$$
$$= 1155.35 \text{ kJ/kg}$$

Work done, $W = \eta \times Q_1$

$$= 0.532 \times 1155.35$$
$$= \mathbf{614.65\ kJ/kg}$$

EXAMPLE 12.9.

In an air standard Otto cycle engine, the temperature at the end of compression stroke is 650 K and the maximum cycle temperature is 2400 kJ/kg. If the engine delivers 700 kJ/kg of net work, find the thermal efficiency and compression ratio of the engine.

Solution : Refer Fig. 12.7

Heat added,

$$Q_1 = c_v\,(T_3 - T_2)$$
$$= 0.718\,(2400 - 650)$$
$$= 1256.\ 5 \text{ kJ/kg}$$

Thermal efficiency,

$$\eta = \frac{\text{net work output}}{\text{heat supplied}}$$

$$= \frac{700}{1256.5} = 0.557 \text{ or } \mathbf{55.7\ \%}.$$

In terms of compression ratio $r = V_1/V_2$, the efficiency of an engine working on Otto cycle is

$$\eta = 1 - \frac{1}{r^{\gamma-1}}; \quad 0.557 = 1 - \frac{1}{r^{1.4-1}}$$

$$\therefore \quad r = \left(\frac{1}{1-0.557}\right)^{1/0.4}$$

$$= (2.257)^{2.5} = \mathbf{7.65}$$

EXAMPLE 12.10.

In an air standard Otto cycle, the ambient conditions determine the minimum temperature while the maximum temperature is determined by the design conditions and metallurgical considerations of the piston and cylinder. For fixed value of minimum temperature T_1 and maximum temperature T_3, show that for maximum work output, the compression ratio should have the value

$$r = \left(\frac{T_3}{T_1}\right)^{\frac{1}{2(\gamma-1)}} = \left(\frac{T_3}{T_1}\right)^{1.25} \quad \text{if } \gamma = 1.4$$

Solution : Refer Fig. 12.8.

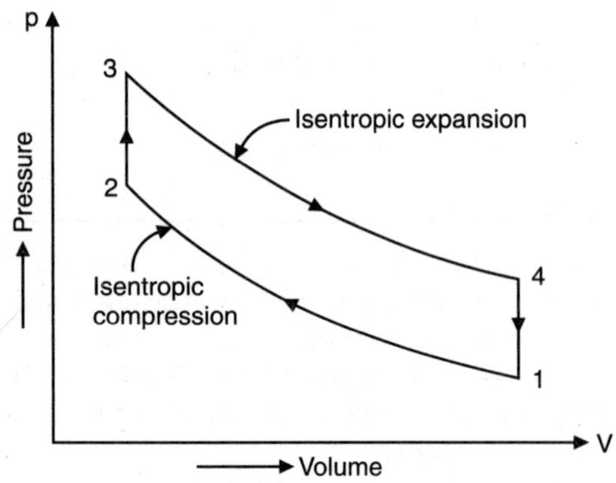

Fig. 12.8.

Work output,

$$W = c_v (T_3 - T_2) - c_v (T_4 - T_1)$$

From adiabatic compression process 1—2,

$$T_2 = T_1 \left(\frac{V_1}{V_2}\right)^{\gamma-1} = T_1\ r^{\gamma-1}$$

From adiabatic expansion process 3—4,

$$T_4 = T_3 \left(\frac{V_3}{V_4}\right)^{\gamma-1}$$

$$= T_3 \left(\frac{V_2}{V_1}\right)^{\gamma-1} = T_3\ r^{1-\gamma}$$

$$\therefore W = c_v \left(T_3 - T_1 r^{\gamma-1}\right) - c_v \left(T_3 r^{1-\gamma} - T_1\right)$$

The compression ratio r for maximum work output can be obtained by differentiating the above expression with respect to r and setting the derivative equal to zero. That is

$$\frac{dW}{dr} = -c_v T_1 (\gamma-1) r^{\gamma-2} - c_v T_3 (1-\gamma) r^{-\gamma}$$

$$= 0 \quad \text{for maximum work}$$

$$\therefore \quad -T_1 (\gamma-1) r^{\gamma-2} - T_3 (1-\gamma) r^{-\gamma} = 0$$

$$\text{or} \quad \frac{r^{\gamma-2}}{r^{-\gamma}} = \frac{T_3}{T_1}$$

$$\text{or} \quad r^{2(\gamma-1)} = \frac{T_3}{T_1}$$

$$\therefore \quad r = \left(\frac{T_3}{T_1}\right)^{\frac{1}{2(\gamma-1)}}$$

$$= \left(\frac{T_3}{T_1}\right)^{\frac{1}{2(1.4-1)}} = \left(\frac{T_3}{T_1}\right)^{1.25}$$

which are the required results.

EXAMPLE 12.11.

The upper and lower temperature limits for an Otto cycle are 1500 K and 300 K respectively. What compression ratio is required to develop maximum work? Estimate the maximum theoretical power developed by an engine working on this cycle when the air flow rated is 0.35 kg/min.

Solution : Refer Fig. 12.8

The conditions for maximum work output are

$$r = \left(\frac{T_3}{T_1}\right)^{\frac{1}{2(\gamma-1)}}$$

$$= \left(\frac{1500}{300}\right)^{\frac{1}{2\times 0.4}} = 7.476$$

$$T_2 = T_4 = \sqrt{T_1 T_3}$$

$$= \sqrt{300 \times 1500} = 670.82 \text{ K}$$

Work done per kg mass of air

= heat supplied – heat rejected

$= c_v (T_3 - T_2) - c_v (T_4 - T_1)$

$= 0.718 (1500 - 670.82) - 0.718 (670.82 - 300)$

$= 329.1$ kJ/kg

∴ Maximum power developed

$$= 329.1 \times \frac{0.35}{60}$$

= 1.92 kJ/s = **1.92 kW**

EXAMPLE 12.12.

An engine working on Otto cycle has a compression ratio of 6 : 1. If the specific heat at constant volume increases by 1 percent, determine the corresponding percentage change in thermal efficiency.

Solution : In terms of compression ratio, the efficiency of Otto cycle is given by

$$\eta = 1 - \frac{1}{r^{\gamma-1}}$$

$$= 1 - \frac{1}{r^{R/c_v}} = 1 - r^{-R/c_v}$$

or $\quad 1 - \eta = r^{-R/c_v}$

Taking logs: $\log_e (1 - \eta) = \dfrac{-R}{c_v} \log_e r$

Upon differentiating η with respect to c_v, we get

$$\frac{-1}{1-\eta} d\eta = \frac{R}{c_v^2} \log_e r \, dc_v$$

$$= \frac{c_v (\gamma - 1)}{c_v^2} \log_e r \, dc_v$$

or $\quad d\eta = -(1 - \eta)(\gamma - 1) \log_e r \dfrac{dc_v}{c_v}$

$$\therefore \quad \frac{d\eta}{\eta} = -\left(\frac{1-\eta}{\eta}\right)(\gamma - 1) \log_e r \frac{dc_v}{c_v}$$

Now, $\quad \eta = 1 - \dfrac{1}{r^{\gamma-1}} = 1 - \dfrac{1}{6^{0.4}} = 0.5116$

$$\therefore \frac{d\eta}{\eta} = -\left(\frac{1 - 0.5116}{0.5116}\right) \times 0.4 \log_e 6 \times \left(\frac{1}{100}\right)$$

$= -0.006842$ or **– 0.6842 %**

That is the efficiency of Otto cycle decreases by 0.6842 % with increase in specific heat at constant volume by 1%.

EXAMPLE 12.13

Consider an air standard Otto cycle that has a heat addition of 2800 kJ/kg of air, a compression ratio of 8, and a pressure and temperature at the beginning of compression process of 1 bar, 300 K. Determine :

(a) the maximum pressure and temperature in the cycle

(b) the thermal efficiency and (c) the mean effective pressure.

For air: c_p = 1.005 kJ/kg K,

c_v = 0.718 kJ/kg K and R = 287 kJ/kg K

Solution : Refer Fig. 12.9.

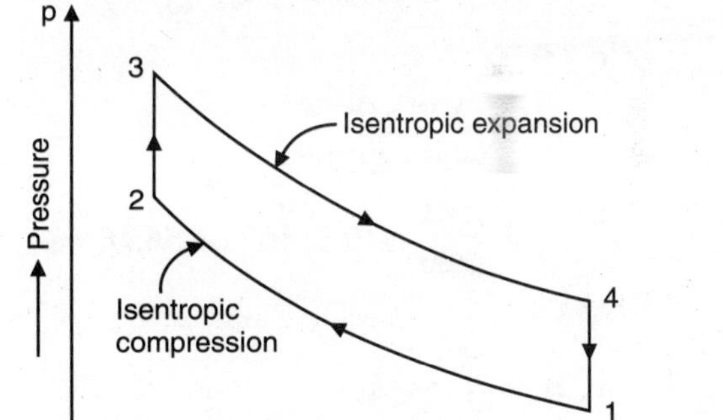

Fig. 12.9.

(*a*) Considering adiabatic compression process 1–2,

$$p_2 = p_1 \left(\frac{V_1}{V_2}\right)^{\gamma} = 1 \times (8)^{1.4} = 18.38 \text{ bar}$$

$$T_2 = T_1 \left(\frac{V_1}{V_2}\right)^{\gamma-1}$$

$$= 300 \times (8)^{0.4} = 689.2 \text{ K}$$

·Heat added, $Q_1 = c_v\,(T_3 - T_2)$ for unit mass

$$2800 = 0.718\,(T_3 - 689.2)$$

$$\therefore \quad T_3 = \frac{2800}{0.718} + 689.2 = \mathbf{4588.92\ K}$$

Considering constant volume process 2—3,

$$p_3 = p_2\,\frac{T_3}{T_2}$$

$$= 18.38 \times \frac{4588.92}{689.2} = \mathbf{122.38\ bar}$$

(*b*) Considering adiabatic expansion process 3—4,

$$\frac{T_3}{T_4} = \left(\frac{V_4}{V_3}\right)^{\gamma-1} = \left(\frac{V_1}{V_2}\right)^{\gamma-1} = 8^{0.4}$$

$$T_4 = \frac{4588.92}{8^{0.4}} = 1997.8\ \text{K}$$

Heat rejected,

$$Q_2 = c_v\,(T_4 - T_1)$$
$$= 0.718\,(1997.8 - 300)$$
$$= 1219\ \text{kJ/kg}$$

Work done

= heat added - heat rejected

= 2800 - 1219 = 1581 kJ/kg

Efficiency,

$$\eta = \frac{\text{work done}}{\text{heat supplied}}$$

$$= \frac{1581}{2800} = 0.5646 \text{ or } \mathbf{56.46\ \%}$$

Check :

$$\eta = 1 - \frac{1}{r^{\gamma-1}}$$

$$= 1 - \frac{1}{8^{0.4}} = 0.5647 \text{ or } 56.47\ \%$$

(*c*) From characteristics gas equation:

$$p_1 v_1 = RT_1$$

$$v_1 = \frac{RT_1}{p_1} = \frac{287 \times 300}{1 \times 10^5}$$

$$= 0.861\ \text{m}^3/\text{kg}$$

$$v_2 = \frac{v_1}{8} = \frac{0.861}{8} = 0.1076\ \text{m}^3/\text{kg}$$

Stroke volume $= v_1 - v_2 = 0.861 - 0.1076$

$= 0.7534\ \text{m}^3/\text{kg}$

$$mep = \frac{\text{work done per cycle}}{\text{swept volume}}$$

$$= \frac{1581}{0.7534}$$

$= 2099\ \text{kN/m}^2 = \mathbf{20.99\ bar}$

EXAMPLE 12.14.

In an air standard Otto cycle, the compression ratio is 6 and the isentropic compression begins at 100 kPa and 373 K. Heat is added at constant volume until the pressure reaches a maximum value of 3500 kPa.

(a) Find the pressure and temperature at the salient points of the cycle, and the amount of work done per cycle.

(b) How the air standard efficiency of this cycle compares with that of a Carnot cycle operating with in the same extreme temperature limits ? How do you explain the difference between the two values ?

(c) If the engine has a relative efficiency of 50%, determine the fuel consumption per kWh. Assume that the fuel used has a calorific value of 41860 kJ/kg.

Solution : Refer Fig. 12.10. The subscripts denote the respective state points in the cycle.

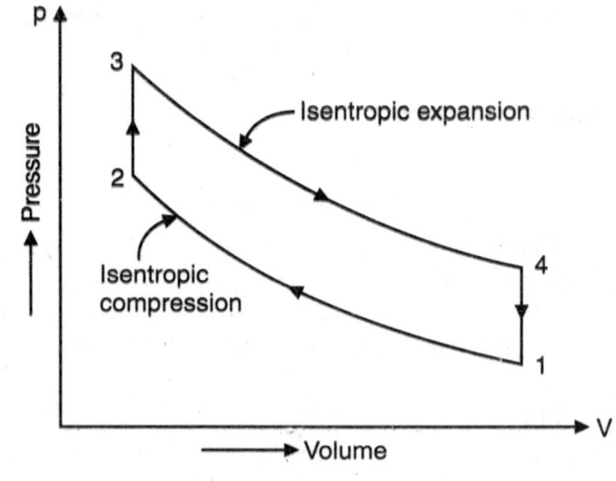

Fig. 12.10.

Given : $r = \dfrac{V_1}{V_2} = 6\ ;\quad P_1 = 100\ \text{kPa}$

$T_1 = 373\ \text{K}\ ;$ and $P_3 = 3500\ \text{kPa}$

(*a*) Consider 1 kg of air

Applying characteristic equation at state point 1, we get

$$V_1 = \frac{mRT_1}{p_1}$$

$$= \frac{1 \times 287 \times 373}{100 \times 10^3} = 1.0705 \text{ m}^3$$

Process 1 – 2 : Since the compression ratio is given equal to 6,

$$V_2 = \frac{V_1}{6} = \frac{1.0705}{6} = 0.1784 \text{ m}^3$$

Further, $p_2 = p_1 \left(\frac{V_1}{V_2}\right)^{\gamma}$

$= 100 \times (6)^{1.4} = 1228.6$ kPa

$$T_2 = T_1 \left(\frac{V_1}{V_2}\right)^{\gamma - 1}$$

$= 373 \times (6)^{1.4 - 1} = 763.78$ K

Process 2 – 3 : $V_3 = V_2 = 0.1784$ m³ and $p_3 = 3500$ kPa

$$T_3 = \frac{p_3}{p_2} \times T_2$$

$$= \frac{3500}{1228.6} \times 763.78 = 2175.83 \text{ K}$$

Process 3 – 4 : $V_4 = V_1 = 1.0705$ m³

$$p_4 = p_3 \left(\frac{V_3}{V_4}\right)^{\gamma}$$

$$= 3500 \times \left(\frac{0.1784}{1.0705}\right)^{1.4} = 284.8 \text{ kPa}$$

Process 4 – 1 : The rejection of heat takes place at constant volume and therefore

$$T_4 = T_1 \frac{p_4}{p_1} = 373 \times \frac{284.8}{100} = 1062.3 \text{ K}$$

Heat supplied during constant volume process 2 – 3

$Q_1 = m\ c_v\ (T_3 - T_2)$

$= 1 \times 0.718\ (2175.83 - 763.78)$

$=$ **1013.85** kJ

Heat rejected during constant volume process 4 – 1,

$Q_2 = m\ c_v\ (T_4 - T_1)$

$= 1 \times 0.718\ (1062.3 - 373)$

$= 494.92$ kJ

Work done,

$W = Q_1 - Q_2 = 1013.85 - 494.92$

$=$ **518.93** kJ

(*b*) Cycle efficiency,

$$\eta = \frac{\text{work done}}{\text{heat supplied}}$$

$$= \frac{518.93}{1013.85} = \mathbf{0.5118} \text{ or } \mathbf{51.18}\ \%$$

Alternatively, the efficiency of Otto cycle can also be calculated by employing the relation.

$$\eta = 1 - \frac{1}{r^{\gamma - 1}}$$

$$= 1 - \frac{1}{6^{1.4 - 1}} = 0.512 \text{ or } \mathbf{51.2}\ \%$$

The efficiency of a corresponding Carnot cycle is

$$\eta_{carnot} = \frac{T_h - T_l}{T_h}$$

$$= \frac{2175.83 - 373}{2175.83}$$

$= 0.8286$ or **82.86** %

Thus we see that efficiency of the Carnot cycle is greater than that of the given Otto cycle. This has to be so because in Carnot cycle the entire addition and rejection of heat takes place at the maximum and minimum temperatures of the cycle.

(*c*) Actual efficiency

$=$ air standard η × relative η

$= 0.512 \times 0.5 = 0.256$

1 kWh ≡ 3600 kJ and therefore

Heat supplied per kWh

$$= \frac{3600}{0.256} = 14062.5 \text{ kJ}$$

∴ Fuel consumed per kWh

$$= \frac{14062.5}{41860} = \mathbf{0.336}$$

EXAMPLE 12.15.

The following operations comprise a certain air standard cycle :

(i) compression from 1 bar and 313 K through volume compression ratio 6 : 1

(ii) heat addition at constant volume with peak pressure reaching 50 bar

(iii) expansion to initial volume, and

(iv) heat rejection at constant volume.

Sketch the cycle on p – v and T– s plots. If compression and expansion follow the polytropic law $pV^{1.25}$ = constant, determine the mean effective pressure, heat added per kg of charge, and heat interaction between the working fluid and cylinder walls. State the direction of heat flow.

Solution : Refer Fig. 12.11 for the $p - v$ and T–s plots of the given cycle. The subscripts denote the respective state points in the cycle.

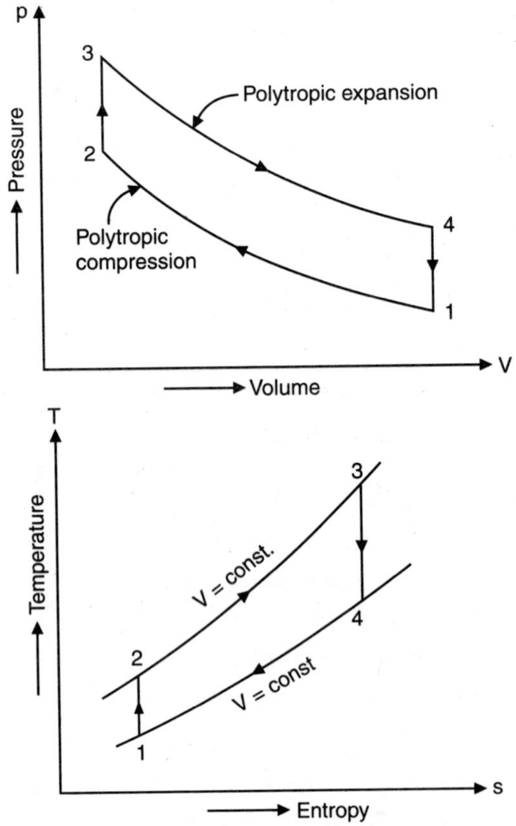

Fig. 12.11.

From polytropic expansion process 1 – 2:

$$p_2 = p_1\left(\frac{V_1}{V_2}\right)^n$$

$$= 1 \times (6)^{1.25} = 9.39 \text{ bar}$$

$$T_2 = T_1\left(\frac{V_1}{V_2}\right)^{n-1}$$

$$= 313 \times (6)^{1.25 - 1} = 489.87 \text{ K}$$

From constant volume process 2 – 3

$$T_3 = \frac{p_3}{p_2} \times T_2$$

$$= \frac{50}{9.39} \times 489.87 = 2608.47 \text{ K}$$

From polytropic expansion process 3 – 4

$$p_4 = p_3\left(\frac{V_3}{V_4}\right)^n$$

$$= 50 \times \left(\frac{1}{6}\right)^{1.25} = 5.32 \text{ bar}$$

From constant volume process 4 – 1

$$T_4 = \frac{p_4}{p_1} \times T_1$$

$$= \frac{5.32}{1} \times 313 = 1665.16 \text{ K}$$

(*a*) Work done during expansion process 3 – 4

$$W_{exp} = \frac{p_3V_3 - p_4V_4}{n-1}$$

Work done during compression process 1 – 2,

$$W_{comp} = \frac{p_1V_1 - p_2V_2}{n-1}$$

∴ Net work done per cycle

$$= \frac{(p_3V_3 - p_4V_4) + (p_1V_1 - p_2V_2)}{n-1}$$

Swept volume, $V_s = V_1 - V_2$

Mean effective pressure,

$$mep = \frac{\text{work done}}{\text{swept volume}}$$

$$= \frac{(p_3V_3 - p_4V_4) + (p_1V_1 - p_2V_2)}{(n-1)(V_1 - V_2)}$$

Substituting the values, we get

$$mep = \frac{(50 \times 1 - 5.32 \times 6) + (1 \times 6 - 9.39 \times 1)}{(1.25 - 1)(6 - 1)}$$

$$= \mathbf{11.752} \text{ bar}$$

The above calculations have, been made by assuming unit clearance volume, *i.e.*, V_2 = unit volume and then V_1 = 6 units

(*b*) Heat input per kg of charge,

$$Q_1 = c_v(T_3 - T_2)$$
$$= 0.718(2608.47 - 489.87)$$
$$= \mathbf{1521.15\ kJ}$$

$$W_{comp} = \frac{p_1V_1 - p_2V_2}{n-1} = \frac{R(T_1 - T_2)}{n-1}$$

$$= \frac{0.287}{1.25-1}(313 - 489.87)$$

$$= -203 \text{ kJ/kg}$$

The negative sign implies that work has been done on the working fluid.

Heat exchange during compression

$$= \frac{\gamma - n}{n-1} \times W_{comp}$$

$$= \frac{1.4 - 1.25}{1.25-1} \times (-203)$$

$$= -76.12 \text{ kJ/kg}$$

The negative sign implies that heat has been rejected to the surroundings.

$$W_{exp} = \frac{p_3V_3 - p_4V_4}{n-1} = \frac{R(T_3 - T_4)}{n-1}$$

$$= \frac{0.287}{1.25-1}(2608.47 - 1665.16)$$

$$= 1082.92 \text{ kJ/kg}$$

Heat exchange during expansion

$$= \frac{\gamma - n}{n-1} \times W_{exp}$$

$$= \frac{1.4 - 1.25}{1.25-1} \times 1082.92$$

$$= 406.09 \text{ kJ/kg}$$

The positive sign implies that during expansion, heat has been added to the working substance.

EXAMPLE 12.16.

The compression curve on the pressure-volume diagram of a gas engine working on Otto cycle follows the polytropic law $pV^{1.3}$ = constant. At two points on this curve at 1/4th stroke and 3/4th stroke, the pressures are measured to be 140 kN/m² and 360 kN/m². Determine the compression ratio of the engine. Proceed to calculate the gas consumption per kW hour if relative efficiency is 0.4 and the gas has calorific value of 18840 kJ/m³.

Solution : Refer Fig. 12.12.

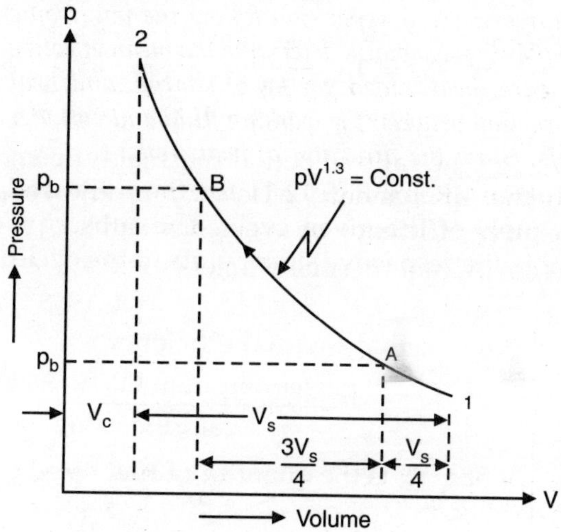

Fig. 12.12.

Volume at point *A*,

$$V_a = V_C + V_S - \frac{V_S}{4}$$
$$= V_C + 0.75\ V_S$$

Volume at point *B*,

$$V_b = V_C + V_S - \frac{3}{4}V_S$$
$$= V_C + 0.25\ V_S$$

Using the given polytropic law

$$p_a\ V_a^{1.3} = p_b\ V_b^{1.3}$$

$$\frac{V_a}{V_b} = \left(\frac{p_a}{p_b}\right)^{1/1.3}$$

or $$\frac{V_C + 0.75\ V_S}{V_C + 0.25\ V_S} = \left(\frac{360}{140}\right)^{1/1.3} = 2.067$$

or $V_C + 0.75\ V_S = 2.067\ V_C + 0.5167\ V_S$

$$\therefore \quad \frac{V_S}{V_C} = 4.5735$$

(*a*) Compression ratio,

$$r = \frac{V_1}{V_2} = \frac{V_C + V_S}{V_C}$$

$$= 1 + \frac{V_S}{V_C} = 1 + 4.5735 = \mathbf{5.5735}$$

In terms of compression ratio, the air standard efficiency of an engine working on Otto cycle is given by

$$\eta_{Otto} = 1 - \frac{1}{r^{\gamma-1}} = 1 - \frac{1}{(5.5735)^{0.4}}$$

$$= 1 - \frac{1}{1.988}$$

$$= 1 - 0.503 = 0.497 \text{ or } \mathbf{49.7\ \%}$$

(*b*) The ratio of actual thermal efficiency to the air standard efficiency is known as relative efficiency or efficiency ratio.

∴ Actual thermal efficiency = 0.497 × 0.4 = 0.1988

Now, actual thermal efficiency

$$= \frac{\text{power output in kW} \times 3600}{m_f \times \text{calorific value}}$$

where m_f is the quantity of fuel used per kW-hour

$$0.1988 = \frac{1 \times 3600}{m_f \times 18840}$$

∴ m_f = **0.961 m³/ kW-hr**

The fuel used per kW-hr is called specific fuel consumption.

12.6. DIESEL (CONSTANT PRESSURE) CYCLE

The Diesel cycle was conceived and developed by Rudolph Diesel in 1893 to the point where he was able to begin commercial production of the corresponding diesel engine in his own factory. Fig. 12.13 shows the diesel cycle plotted on $p-V$ and $T-s$ diagrams.

The sequence of operations is

Process 1—2 : Piston moves from crank end dead centre (ODC) to cover end dead centre (IDC) and air with initial state 1 is compressed isentropically to state point 2 through compression ratio $r = V_1/V_2$.

Process 2—3 : Heat is added to compressed air at constant pressure from an external source which is brought into contact with the cylinder head. The heat supply is stopped at point 3 which is called the cut off point, and the volume ratio $\rho = V_3/V_2$ is called ***cut off ratio*** or ***isobaric expansion ratio***.

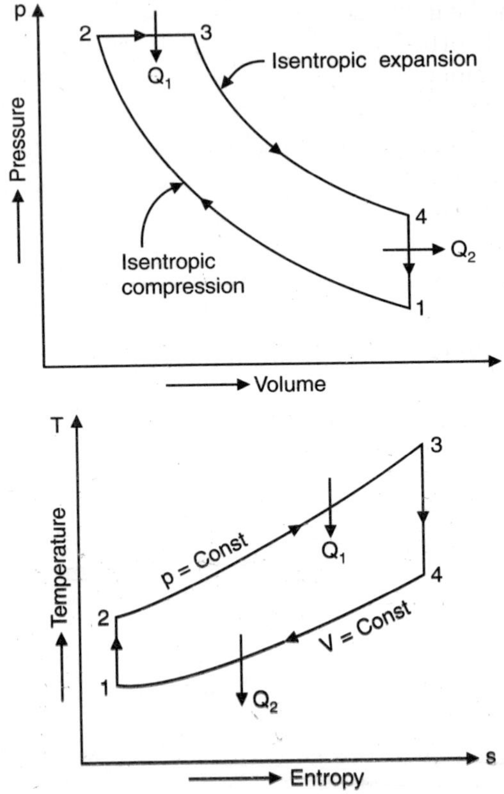

Fig. 12.13. Air-standard diesel cycle on p-V diagram and T-s diagrams

Process 3—4 : The increased high pressure exerts a greater amount of force on the piston and pushes it towards the ODC. Expansion of working fluid (air) takes place isentropically and work is done by the system. The volume ratio V_4/V_3 is called ***isentropic expansion ratio.***

Process 4—1 : The piston is momentarily at rest at ODC and heat is rejected to the external sink by bringing it in contact with the cylinder head. The process is so controlled that ultimately the air comes to its initial state 1 and the cycle is completed.

The low speed diesel engines work on a cycle which is slight modification of the Diesel cycle. This thermodynamic cycle is called constant pressure cycle because heat is supplied to the air at constant pressure.

Thermal efficiency for diesel cycle : The compression process 1–2 and the expansion process 3–4 are isentropic and hence there is no heat interaction in these processes. Energy is absorbed as heat during process 2–3 at constant pressure, and is rejected as heat during process 4–1 at constant volume.

Consider unit mass of air undergoing the cyclic change.

Heat supplied during process 2–3,

$$Q_1 = c_p\,(T_3 - T_2)$$

Heat rejected during process 4–1,

$$Q_2 = c_v\,(T_4 - T_1)$$

Work done = heat supplied – heat rejected

$$= c_p\,(T_3 - T_2) - c_v\,(T_4 - T_1)$$

Thermal efficiency,

$$\eta = \frac{\text{work done}}{\text{heat supplied}}$$

$$= \frac{c_p\,(T_3 - T_2) - c_v\,(T_4 - T_1)}{c_p\,(T_3 - T_2)}$$

$$= 1 - \frac{1}{\gamma}\frac{(T_4 - T_1)}{(T_3 - T_2)} \qquad \text{...(12.13)}$$

The expression for thermal efficiency in terms of compression ratio r, cut off ratio ρ and adiabatic exponent γ can be set up by expressing all the temperatures in terms of lowest temperature T_1 in the cycle.

At point 2 :

$$V_2 = \frac{V_1}{r}$$

where r is the compression ratio

$$p_2 = p_1\left(\frac{V_1}{V_2}\right)^{\gamma} = p_1\,r^{\gamma}$$

$$T_2 = T_1\left(\frac{V_1}{V_2}\right)^{\gamma-1} = T_1\,r^{\gamma-1}$$

At point 3 : $p_3 = p_2 = p_1\,r^{\gamma}$

$$V_3 = V_2\,\rho$$

where ρ is the cut off ratio

$$T_3 = T_2\frac{V_3}{V_2} = T_2\,\rho = T_1\,r^{\gamma-1}\,\rho$$

At point 4 :

$$V_4 = V_1$$

$$p_4 = p_3\left(\frac{V_3}{V_4}\right)^{\gamma} = p_3\left(\frac{V_3}{V_1}\right)^{\gamma}$$

$$= p_3\left(\frac{V_3/V_2}{V_1/V_2}\right)^{\gamma} = p_3\,(\rho/r)^{\gamma}$$

$$T_4 = T_3\left(\frac{V_3}{V_4}\right)^{\gamma-1} = T_3\left(\frac{\rho}{r}\right)^{\gamma-1}$$

$$= \frac{T_1\,r^{\gamma-1}\,\rho\,\rho^{\gamma-1}}{r^{\gamma-1}} = T_1\,\rho^{\gamma}$$

When these temperature values are substituted in expression 12.13, we get

$$\eta = 1 - \frac{1}{\gamma}\left[\frac{T_1\,\rho^{\gamma} - T_1}{T_1\,r^{\gamma-1}\rho - T_1\,r^{\gamma-1}}\right]$$

$$= 1 - \frac{1}{\gamma}\left[\frac{T_1\left(\rho^{\gamma} - 1\right)}{T_1\,r^{\gamma-1}(\rho - 1)}\right]$$

$$= 1 - \frac{1}{r^{\gamma-1}}\left[\frac{\rho^{\gamma} - 1}{\gamma(\rho - 1)}\right] \qquad \text{...(12.14)}$$

Apparently the efficiency of diesel cycle depends upon the compression ratio (r) and cut off ratio (ρ) and hence upon the quantity of heat supplied. Fig. 12.14 shows the air standard efficiency of diesel cycle for various cut off ratio.

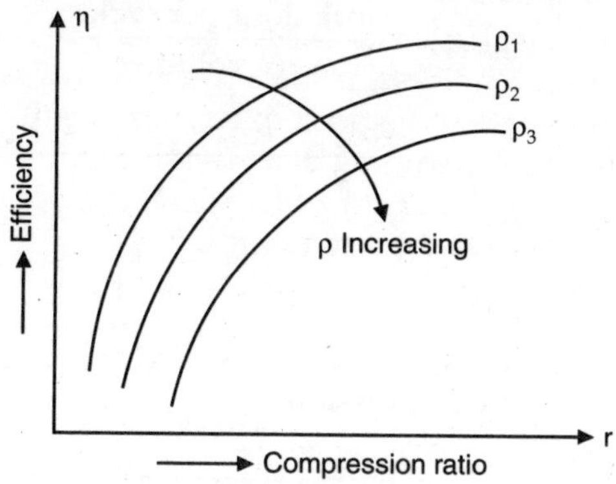

Fig. 12.14.

Further examination of factor K reveals that with an increase in the cut off ratio ρ, the value of the factor $K = \left[\left(\rho^{\gamma} - 1\right)/\gamma\,(\rho - 1)\right]$ increases. That implies that for a diesel engine at constant compression ratio, the efficiency would increase with decrease in ρ and in the limit $\rho \to 1$, the efficiency would become $1 - \frac{1}{r^{\gamma-1}}$. Since the factor K is always greater than unity, the diesel cycle is always less efficient than a corresponding Otto cycle having the same compression ratio. This aspect is, however, of little importance because the diesel engines operate on much higher compression ratio, (14 to 18) compared to those for spark ignition engines operating on Otto cycle. High compression ratios for diesel engines are must not only for high efficiency but also to prevent diesel knock; a phenomenon which leads to uncontrolled and rapid combustion in diesel engines.

Mean effective pressure : With reference to Fig. 12.13, the work done per cycle for unit mass of air is

$$\text{Work done} = \text{heat supplied} - \text{heat rejected} = c_p (T_3 - T_2) - c_v (T_4 - T_1)$$

Swept volume

$$= (V_1 - V_2) = V_1 \left(1 - \frac{V_2}{V_1}\right)$$

$$= \frac{R\,T_1}{p_1}\left(1 - \frac{1}{r}\right) = \frac{R\,T_1}{p_1\,r}(r-1)$$

By definition of mean effective pressure

$$mep = \frac{\text{work done per cycle}}{\text{swept volume}}$$

$$= \frac{c_p (T_3 - T_2) - c_v (T_4 - T_1)}{R\,T_1 / p_1\,r \times (r-1)}$$

$$= \frac{c_v}{R}\frac{p_1\,r}{(r-1)}\left[\frac{\gamma (T_3 - T_2) - (T_4 - T_1)}{T_1}\right] \quad \text{...(12.15)}$$

The expression for mean effective pressure in terms of compression ratio r, cut off ratio ρ induction pressure p_1 and adiabatic exponent γ can be set up by expressing all the temperatures in terms of lowest temperature T_1 in the cycle.

As outlined above :

$$T_2 = T_1\,r^{\gamma-1}; \quad T_3 = T_1\,\rho\,r^{\gamma-1}$$
$$T_4 = T_1\,\rho^{\gamma}$$

When these temperature values are substituted in equation 12.14,we get :

$$mep = \frac{c_v}{R}\frac{p_1\,r}{(r-1)}\left[\frac{\gamma\left(T_1\,\rho\,r^{\gamma-1} - T_1\,r^{\gamma-1}\right) - \left(T_1\,\rho^{\gamma} - T_1\right)}{T_1}\right]$$

$$= \frac{c_v}{R}\frac{p_1\,r}{(r-1)}\left[\gamma\,r^{\gamma-1}(\rho-1) - \left(\rho^{\gamma}-1\right)\right]$$

$$\text{But } \frac{c_v}{R} = \frac{1}{\gamma-1}$$

$$\therefore\ mep = \frac{p_1\,r}{(r-1)(\gamma-1)} \times \left[\gamma\,r^{\gamma-1}(\rho-1) - \left(\rho^{\gamma}-1\right)\right] \quad \text{...(12.16)}$$

which is the required expression of mean effective pressure for a diesel cycle.

EXAMPLE 12.17.

The following particulars refer to the German Mercedes 190 D car operating on four-stroke in-line diesel engine cycle :-

Compression ratio 21 : 1 and expansion ratio 10.5 : 1. Determine the cut off ratio and air standard efficiency.

Solution : Refer Fig. 12.15.

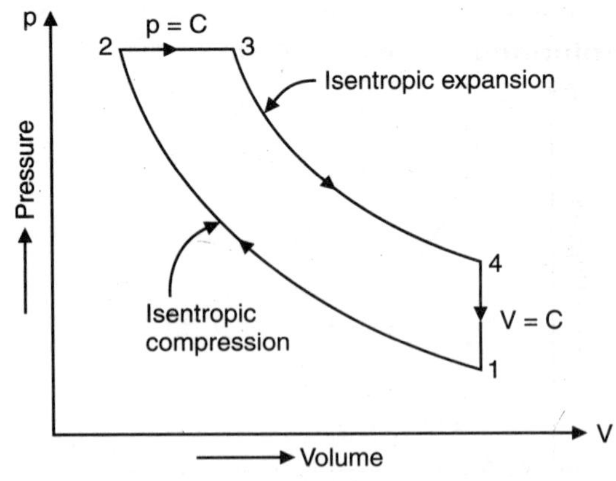

Fig. 12.15.

$$\text{Given : } \frac{V_1}{V_2} = 21 \text{ and } \frac{V_4}{V_3} = 10.5$$

$$\text{Cut off ratio, } \rho = \frac{V_3}{V_2} = \frac{V_3}{V_4} \times \frac{V_4}{V_2}$$

$$= \frac{V_3}{V_4} \times \frac{V_1}{V_2} \quad (\because V_1 = V_4)$$

$$= \frac{1}{10.5} \times 21 = 2$$

In terms of compression ratio r and cut off ratio ρ, the efficiency of a diesel engine cycle is given by

$$\eta = 1 - \frac{1}{r^{\gamma-1}}\left[\frac{\rho^\gamma - 1}{\gamma(\rho-1)}\right]$$

$$= 1 - \frac{1}{21^{0.4}}\left[\frac{2^{1.4} - 1}{1.4\,(2-1)}\right]$$

$$= 1 - \frac{1}{3.38}\left[\frac{2.639 - 1}{1.4}\right]$$

$$= 1 - 0.346 = 0.654 \text{ or } \mathbf{65.4\ \%}$$

EXAMPLE 12.18.

The following data pertains to a compression ignition engine working on air standard diesel cycle :

Cylinder bore (dia) = 15 cm

Stroke length = 25 cm

Clearance volume = 400 cm³

Calculate the air standard efficiency of the engine if fuel injection takes place at constant pressure for 5% of the stroke. How this efficiency value will be affected if the fuel supply continues upto 8% of the stroke?

Solution : Refer Fig. 12.16.

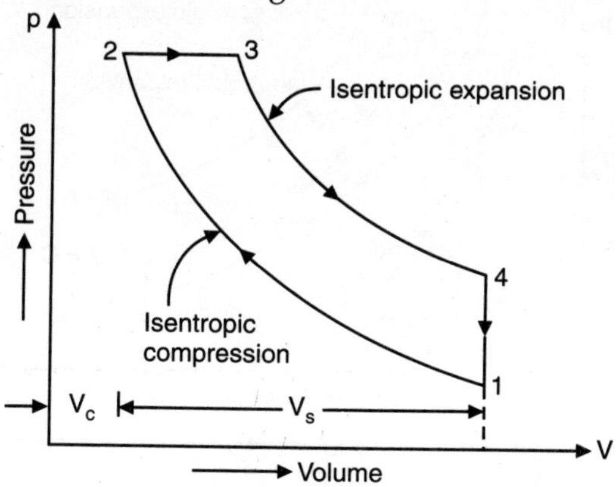

Fig. 12.16.

Swept volume,

$$V_S = \frac{\pi}{4} d^2 l$$

$$= \frac{\pi}{4}(15)^2 \times 25 = 4416 \text{ cm}^3$$

Compression ratio,

$$r = 1 + \frac{V_S}{V_C}$$

$$= 1 + \frac{4416}{400} = 10.04$$

Volume at the point of cut off,

$$V_3 = V_2 + \frac{5}{100} \times V_S$$

or $$V_3 = 400 + \frac{5}{100} \times 4416 = 620.8 \text{ cm}^3$$

$$(\because V_2 = V_C)$$

Cut off ratio,

$$\rho = \frac{V_3}{V_2} = \frac{620.8}{400} = 1.552$$

In terms of compression ratio r and cut off ratio ρ, the efficiency of a diesel engine cycle is given by

$$\eta = 1 - \frac{1}{r^{\gamma-1}}\left[\frac{\rho^\gamma - 1}{\gamma(\rho-1)}\right]$$

$$= 1 - \frac{1}{12.04^{0.4}}\left[\frac{1.552^{1.4} - 1}{1.4\,(1.552-1)}\right]$$

$$\eta = 1 - \frac{1}{2.705}\left[\frac{1.85 - 1}{0.773}\right]$$

$$= 0.5935 \text{ or } 59.35\ \%$$

When the fuel supply is cut off at 8% of stroke

$$V_3 = 400 + \frac{8}{100} \times 4416 = 753.28 \text{ cm}^3$$

$$\rho = \frac{V_3}{V_2} = \frac{753.28}{400} = 1.883$$

$$\eta = 1 - \frac{1}{12.04^{0.4}}\left[\frac{1.883^{1.4} - 1}{1.4\,(1.883-1)}\right]$$

$$= 1 - \frac{1}{2.705}\left[\frac{2.425 - 1}{1.236}\right]$$

$$= 0.574 \text{ or } \mathbf{57.4}\ \%$$

∴ Percentage loss in efficiency

$$= 59.35 - 57.4 \approx \mathbf{2\%}$$

EXAMPLE 12.19.

A heat engine draws in air at 1 bar and 300 K. Measurements indicate maximum pressure and temperature values equivalent to 70 bar and 2000 K respectively. Determine the air standard efficiency if the engine works on diesel cycle.

How the air standard efficiency of this cycle compares with that of a Carnot cycle working within the same maximum and minimum temperature limits ? How do you explain the difference between the two values?

Solution : Refer Fig. 12.17.

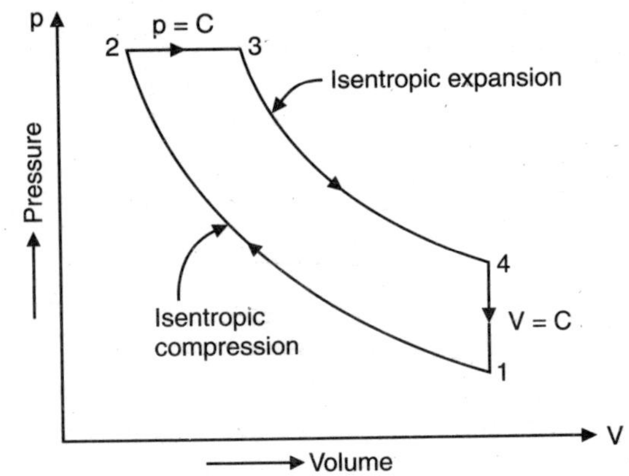

Fig. 12.17.

$$p_1 = 1 \text{ bar};\ p_2 = p_3 = 70 \text{ bar}$$
$$T_3 = 2000 \text{ K}$$

From isentropic compression process 1–2,

$$\left(\frac{V_1}{V_2}\right)^{\gamma} = \frac{p_2}{p_1} = 70$$

Compression ratio,

$$r = \frac{V_1}{V_2} = 70^{1/1.4} = 20.77$$

$$T_2 = T_1\left(\frac{V_1}{V_2}\right)^{\gamma-1}$$
$$= 300 \times (20.77)^{0.4} = 1009.5 \text{ K}$$

From the constant pressure heat addition process 2–3, we have

Cut off ratio,

$$\rho = \frac{V_3}{V_2} = \frac{T_3}{T_2} = \frac{2000}{1009.5} = 1.981$$

In terms of compression ratio r and cut off ratio ρ, the thermal efficiency of an engine working on diesel cycle is given by

$$\eta = 1 - \frac{1}{r^{\gamma-1}}\left[\frac{\rho^{\gamma}-1}{\gamma(\rho-1)}\right]$$
$$= 1 - \frac{1}{20.77^{0.4}}\left[\frac{1.981^{1.4}-1}{1.4\,(1.981-1)}\right]$$
$$= 0.653 \text{ or } 65.3 \%$$

The efficiency of a corresponding Carnot cycle is

$$\eta_{Carnot} = \frac{T_h - T_l}{T_h}$$
$$= \frac{2000-300}{2000} = 0.85 \quad \text{or } 85\%$$

It is observed that the efficiency of the Carnot cycle is greater than that of the given diesel cycle. This has to be so because in Carnot cycle, the entire addition and rejection of heat takes place at the maximum and minimum cycle temperatures, and this is not the case in diesel cycle.

EXAMPLE 12.20.

In an air standard diesel cycle, the compression ratio is 15 and the fluid properties at the beginning of compression are 100 kPa and 300 K. For a peak temperature of 1600 K, calculate (a) the percentage of stroke at which cut off occurs (b) the cycle efficiency and (c) the work output per kg of air.

Solution : Refer Fig. 12.18.

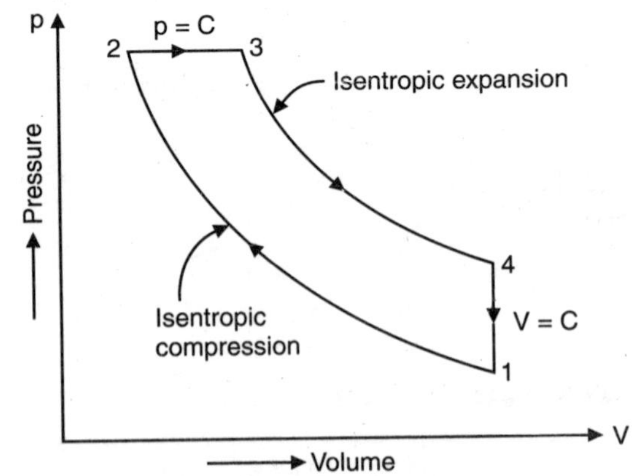

Fig. 12.18.

(*a*) Considering adiabatic compression process1–2,

$$T_2 = T_1 \left(\frac{v_1}{v_2}\right)^{\gamma - 1}$$

$$= 300 \times (15)^{1.4 - 1} = 886 \text{ K}$$

From constant pressure process 2–3,
Cut off ratio,

$$\rho = \frac{v_3}{v_2} = \frac{T_3}{T_2}$$

$$= \frac{1600}{886} = 1.806$$

$$v_1 = \frac{R\,T_1}{p_1} = \frac{287 \times 300}{100 \times 1000} = 0.861 \text{ m}^3/\text{kg}$$

$$v_2 = \frac{v_1}{15} = \frac{0.861}{15} = 0.0574 \text{ m}^3/\text{kg}$$

$$v_3 = 1.806 \times 0.0574$$

$$= 0.1036 \text{ m}^3/\text{kg}$$

$\therefore$ % age cut off

$$= \frac{v_3 - v_2}{v_1 - v_2} = \frac{0.1036 - 0.0574}{0.861 - 0.0574}$$

$$= 0.0576 \text{ or } \mathbf{5.67\ \%}$$

(*b*) In terms of compression ratio r and cut off ratio ρ, the thermal efficiency of a diesel engine cycle is given by,

$$\eta = 1 - \frac{1}{r^{\gamma-1}}\left[\frac{\rho^{\gamma} - 1}{\gamma(\rho - 1)}\right]$$

$$= 1 - \frac{1}{15^{0.4}}\left[\frac{1.806^{1.4} - 1}{1.4\,(1.806 - 1)}\right]$$

$$= 1 - \frac{1}{2.954}\left[\frac{2.288 - 1}{1.1284}\right]$$

$$= 0.614 \text{ or } \mathbf{61.4\ \%}$$

(*c*) Heat supplied during constant pressure process 2–3,

$$Q_1 = c_p\,(T_3 - T_2)$$

$$= 1.005\,(1600 - 886)$$

$$= 717.57 \text{ kJ/kg}$$

Net work output

$$= \eta \times Q_1$$

$$= 0.614 \times 717.57 = \mathbf{440.59\ kJ/kg}$$

EXAMPLE 12.21.

An engine working on diesel cycle has a compression ratio of 15 and fuel supply is cut off at 8% of stroke. If the engine has a relative efficiency of 50%, determine the fuel consumption per kW-hr. Assume that the fuel has a calorific value of 42000 kJ/kg.

Solution : Refer Fig. 12.19.

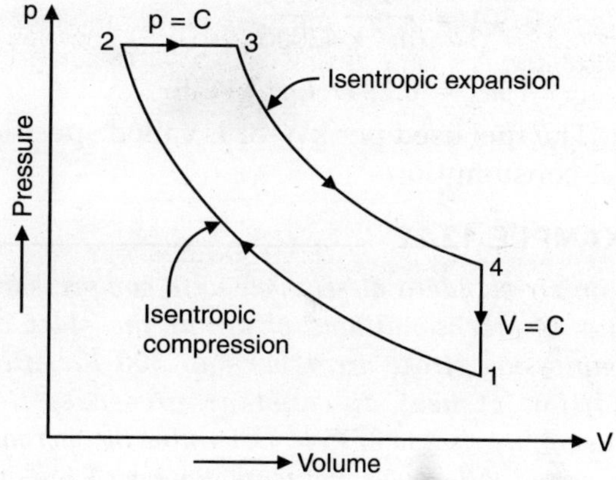

Fig. 12.19.

Volume at cut off,

$$V_3 = V_2 + 0.08\,(V_1 - V_2)$$

$$\text{Cut off ratio, } \rho = \frac{V_3}{V_2}$$

$$= 1 + 0.08\left(\frac{V_1}{V_2} - 1\right)$$

$$= 1 + 0.08\,(15 - 1)$$

$$= 2.12$$

In terms of compression ratio r and cut off ratio ρ, the efficiency of a diesel cycle engine is given by

$$\eta = 1 - \frac{1}{r^{\gamma-1}}\left[\frac{\rho^{\gamma} - 1}{\gamma(\rho - 1)}\right]$$

$$= 1 - \frac{1}{15^{0.4}}\left[\frac{2.12^{1.4} - 1}{1.4\,(2.12 - 1)}\right]$$

$$= 1 - \frac{1}{2.954}\left[\frac{2.863 - 1}{1.568}\right]$$

$$= 1 - 0.398 = 0.602 \text{ or } 60.2\ \%$$

The ratio of actual thermal efficiency to the air standard efficiency is known as relative efficiency or efficiency ratio.

$\therefore$ Actual thermal efficiency

$$= 0.602 \times 0.5 = 0.301$$

Now, actual thermal efficiency

$$= \frac{\text{power output in kW} \times 3600}{m_f \times \text{calorific value}}$$

where m_f is the mass of fuel used per kW-hr

$$0.301 = \frac{1 \times 3600}{m_f \times 42000}$$

$$\therefore \quad m_f = \mathbf{0.2847\ kg/\ kW\text{-}hr}$$

The fuel used per kW-hr is called specific fuel consumption.

EXAMPLE 12.22

In an air-standard diesel cycle with compression ratio 14, the conditions of air at the start of compression stroke are 1 bar and 300 K. After addition of heat at constant pressure, the temperature rises to 2775 K. Determine the thermal efficiency of the cycle, net work done per kg of air and the mean effective pressure.

Solution : Refer Fig. 12.20.

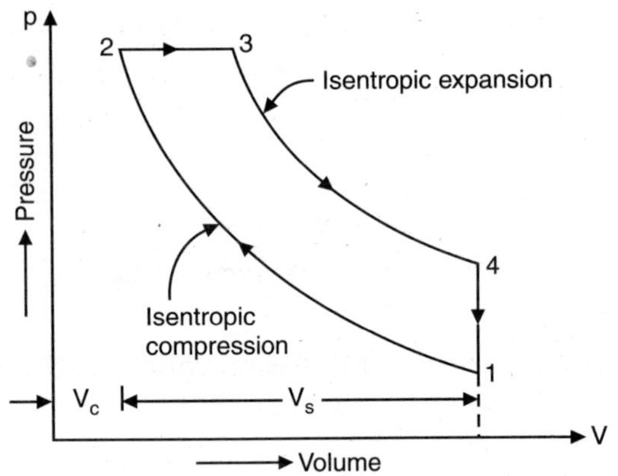

Fig. 12.20.

From adiabatic compression process 1 – 2,

$$T_2 = T_1 \left(\frac{v_1}{v_2}\right)^{\gamma - 1}$$

$$= 300 \times (14)^{0.4} = 862 \text{ K}$$

From constant pressure heat addition process 2 - 3,

$$\text{Cut off ratio, } \rho = \frac{v_3}{v_2}$$

$$= \frac{T_3}{T_2} = \frac{2775}{862} = 3.22$$

In terms of compression ratio r and cut off ratio ρ, thermal efficiency of an engine working on diesel cycle is given by

$$\eta = 1 - \frac{1}{r^{\gamma-1}}\left[\frac{\rho^\gamma - 1}{\gamma(\rho - 1)}\right]$$

$$= 1 - \frac{1}{14^{0.4}}\left[\frac{3.22^{1.4} - 1}{1.4(3.22 - 1)}\right]$$

$$= 1 - \frac{1}{2.874}\left[\frac{5.14 - 1}{3.108}\right]$$

$$= 0.536 \text{ or } \mathbf{53.6\ \%}$$

(*b*) Heat supplied $= c_p (T_3 - T_2)$

$$= 1.005 (2775 - 862)$$

$$= 1922.56 \text{ kJ/kg}$$

Work done = efficiency × heat supplied

$$= 0.536 \times 1922.56$$

$$= \mathbf{1030.49\ kJ/kg}$$

(*c*) From characteristic gas equation,

$$v_1 = \frac{R T_1}{p_1}$$

$$= \frac{287 \times 300}{1 \times 10^5} = 0.861 \text{ m}^3/\text{kg}$$

$$v_2 = \frac{0.861}{14} = 0.0615 \text{ m}^3/\text{kg}$$

Swept volume $= v_1 - v_2$

$$= 0.861 - 0.0615$$

$$= 0.7995 \text{ m}^3/\text{kg}$$

Mean effective pressure

$$= \frac{\text{work done}}{\text{swept volume}} = \frac{1030.49}{0.7995}$$

$$= 1288.92 \text{ kN/m}^2 \approx \mathbf{10.89\ bar}$$

EXAMPLE 12.23.

In a diesel cycle, air a 0.1 MPa and 300 K is compressed adiabatically until the pressure rises to 5 MPa. If 700 kJ/kg of energy in the form of heat is supplied at constant pressure, determine the compression ratio, cut off ratio, thermal efficiency and mean effective pressure.

Solution : Refer Fig. 12.21.

(*a*) From adiabatic compression process

$$p_1 v_1^\gamma = p_2 v_2^\gamma$$

$$\frac{v_1}{v_2} = \left(\frac{p_2}{p_1}\right)^{\frac{1}{\gamma}}$$

or $$r = \left(\frac{5}{0.1}\right)^{\frac{1}{1.4}} = 16.35$$

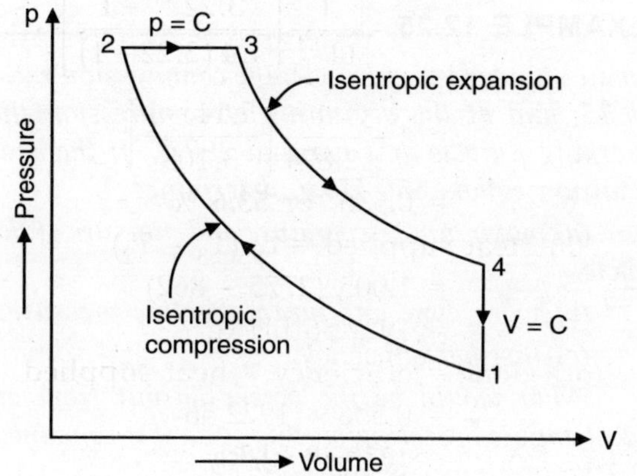

Fig. 12.21.

$$T_2 = T_1\left(\frac{v_1}{v_2}\right)^{\gamma-1}$$

$$= 300 \times (16.35)^{1.4-1} = 917.3 \text{ K}$$

Heat supplied,

$$Q_1 = c_p\,(T_3 - T_2)$$

$$700 = 1.005\,(T_3 - 917.3)$$

$$\therefore \quad T_3 = \frac{700}{1.005} + 917.3 = 1613.82 \text{ K}$$

Cut off ratio,

$$\rho = \frac{v_3}{v_2} = \frac{T_3}{T_2} = \frac{1613.82}{917.3} = 1.759$$

(*b*) In terms of ccmpression ratio r and cut off ratio ρ, the thermal efficiency of an air standard diesel cycle is given by

$$\eta = 1 - \frac{1}{r^{\gamma-1}}\left[\frac{\rho^{\gamma}-1}{\gamma\,(\rho-1)}\right]$$

$$= 1 - \frac{1}{16.35^{0.4}}\left[\frac{1.759^{1.4}-1}{1.4\,(1.759-1)}\right]$$

$$= 1 - \frac{1}{3.0578}\left[\frac{2.2048-1}{1.0626}\right]$$

$$= 1 - 0.3708 = 0.6292 \text{ or } \mathbf{62.92\ \%}$$

(*c*) Work done

= thermal efficiency × heat supplied

= 0.6292 × 700 = 440.44 kJ/kg

From characteristic gas equation,

$$p_1 v_1 = R\,T_1$$

$$v_1 = \frac{R\,T_1}{p_1} = \frac{287 \times 30}{0.1 \times 10^6}$$

$$= 0.861 \text{ m}^3/\text{kg}$$

$$v_2 = \frac{0.861}{16.35} = 0.05266 \text{ m}^3/\text{kg}$$

Swept volume

$$= v_1 - v_2$$

$$= 0.861 - 0.05266$$

$$= 0.8083 \text{ m}^3/\text{kg}$$

Mean effective pressure

$$= \frac{\text{work done}}{\text{swept volume}} = \frac{440.44}{0.8083}$$

$$= 544.89 \text{ kN/m}^2 \simeq \mathbf{0.545\ MPa}$$

EXAMPLE 12.24.

A four stroke engine working on diesel cycle has a piston diameter of 25 cm, a stroke of 40 cm and a clearance volume of 1200 cc. The fuel injection takes place for 5 % of stroke. If the induction pressure corresponds to 1 bar and engine turns 5 rev/sec, find the air standard efficiency, mean effective pressure and power developed.

Solution : Refer Fig. 12.22.

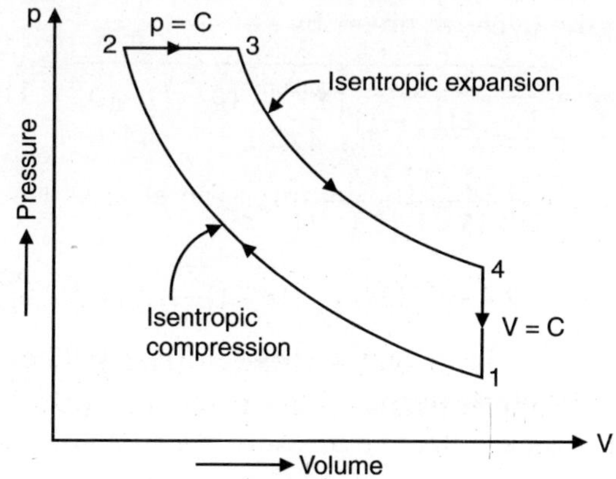

Fig. 12.22.

Swept volume,

$$V_S = \frac{\pi}{4}\,d^2\,l$$

$$= \frac{\pi}{4}(25)^2 \times 40 = 19625 \text{ cm}^3$$

Clearance volume, $V_C = 1200 \text{ cm}^3$

Compression ratio,

$$r = \frac{V_C + V_S}{V_C}$$

$$= \frac{1200 + 19625}{1200} = 17.35$$

Volume at the point of cut off,

$$V_3 = V_2 + 0.05\ V_S$$

$$= 1200 + 0.05 \times 19625$$

$$= 2181.25 \text{ cm}^3$$

Cut off ratio,

$$\rho = \frac{V_3}{V_2} = \frac{2181.25}{1200} = 1.818$$

In terms of compression ratio r and cut off ratio ρ, the efficiency of a diesel engine cycle is given by

$$\eta = 1 - \frac{1}{r^{\gamma-1}}\left[\frac{\rho^\gamma - 1}{\gamma(\rho - 1)}\right]$$

$$= 1 - \frac{1}{17.35^{0.4}}\left[\frac{1.818^{1.4} - 1}{1.4(1.818 - 1)}\right]$$

$$= 1 - \frac{1}{3.13}\left[\frac{2.31 - 1}{1.145}\right]$$

$$= 1 - 0.365 = 0.635 \text{ or } \mathbf{63.5\ \%}$$

(*b*) The mean effective pressure of a diesel engine cycle is given by

$$mep = \frac{p_1\, r}{(r-1)(\gamma-1)}\left[\gamma\, r^{\gamma-1}(\rho - 1) - (\rho^\gamma - 1)\right]$$

$$= \frac{1 \times 17.35}{(17.35-1)(1.4-1)} \times$$

$$\left[1.4 \times 17.35^{0.4}(1.818 - 1) - (1.818^{1.4} - 1)\right]$$

$$= 2.653\,(3.584 - 1.31) = \mathbf{6.03\ bar}$$

(*c*) Work output

= mean effective pressure × swept volume

$$= 6.03 \times 10^5 \times (19625 \times 10^{-6})$$

$$= 11834 \text{ Nm/cycle}$$

$$= 11.834 \text{ kJ/cycle}$$

For a 4-stroke engine, the cycle is completed in two strokes and accordingly the number of effective working cycles equal half the number of revolutions/second.

∴ Power developed

$$= 11.834 \times \frac{5}{2}$$

$$= 29.58 \text{ kJ/s} = \mathbf{29.58\ kW}$$

EXAMPLE 12.25.

In air standard diesel cycle, the compression ratio is 15, and at the beginning of compression the working fluid is at 1 bar and 290 K. If the heat addition equals 850 kJ/kg, determine:

(a) maximum temperature and pressure of the cycle,

(b) work done and mean effective pressure,

(c) thermal efficiency.

What would be the power output from an ideal engine operating on this cycle if it consumes 0.2 kg of air per second ?

Solution : Refer Fig. 12.23. The subscripts denote the respective state points in the cycle.

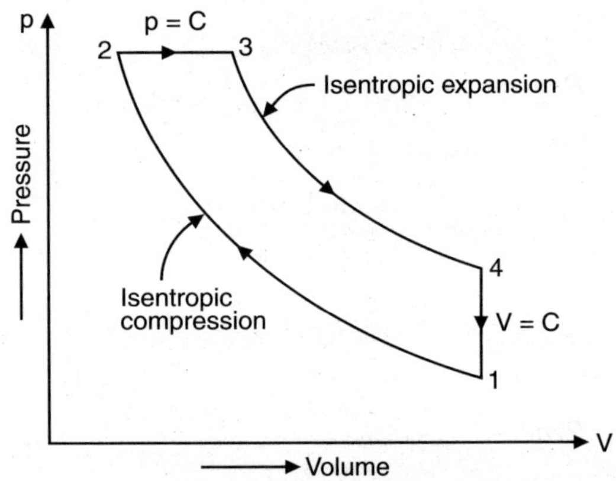

Fig. 12.23.

Given : $\frac{V_1}{V_2} = 15$

$T_1 = 290$ K and $P_1 = 1$ bar

Applying characteristic equation to 1 kg of air at state point 1, we get

$$V_1 = \frac{mRT_1}{p_1}$$

$$= \frac{1 \times 287 \times 290}{1 \times 10^5} = 0.832 \text{ m}^3$$

Process 1 – 2 : Since the compression ratio is given equal to 15,

$$V_2 = \frac{V_1}{15} = \frac{0.832}{15} = 0.0555 \text{ m}^3$$

$$p_2 = p_1\left(\frac{V_1}{V_2}\right)^{\gamma}$$

$$= 1 \times (15)^{1.4} = 44.31 \text{ bar}$$

$$T_2 = T_1\left(\frac{V_1}{V_2}\right)^{\gamma-1}$$

$$= 290 \times (15)^{1.4-1} = 856.72 \text{ K}$$

Process 2 – 3 :

$$p_3 = p_2 = 44.31 \text{ bar}$$

Heat added

$$Q_1 = mc_p\,(T_3 - T_2)$$

or $$850 = 1 \times 1.005(T_3 - 856.4);$$

$$T_3 = 1702.17 \text{ K}$$

$$V_3 = \frac{T_3}{T_2} \times V_2 = \frac{1702.17}{856.4} \times 0.0555$$

$$= 0.1103 \text{ m}^3$$

Process 3 – 4 :

$$V_4 = V_1 = 0.832 \text{ m}^3$$

$$T_4 = T_3 \times \left(\frac{V_3}{V_4}\right)^{\gamma-1}$$

$$= 1702.17 \times \left(\frac{0.1103}{0.832}\right)^{1.4-1}$$

$$= 758.55 \text{ K}$$

Process 4 – 1 :

$$P_4 = p_1 \times \frac{T_4}{T_1}$$

$$= 1 \times \frac{758.55}{290} = 2.615 \text{ bar}$$

(*a*) The maximum temperature and pressure in the cycle are :

Maximum temperature,

$$T_3 = \mathbf{1702.17} \text{ K}$$

Maximum pressure,

$$p_2 = 44.31 \text{ bar}$$

(*b*) Heat rejected,

$$Q_2 = mc_v(T_4 - T_1)$$

$$= 1 \times 0.718\ (758.55 - 290)$$

$$= 336.41 \text{ kJ}$$

Work done,

$$W = \text{heat supplied} - \text{heat rejected}$$

$$= 850 - 336.41 = 513.59 \text{ kJ}$$

Mean effective pressure,

$$mep = \frac{\text{work done}}{\text{swept volume}} = \frac{\text{work done}}{V_1 - V_2}$$

$$= \frac{513.59}{0.832 - 0.0555}$$

$$= 661.84 \text{ kN/m}^2 \simeq \mathbf{6.62} \text{ bar}$$

cycle efficiency,

$$\eta = \frac{\text{work done}}{\text{heat supplied}} = \frac{513.59}{850}$$

$$= \mathbf{0.604} \text{ or } \mathbf{60.4}\ \%$$

(*c*) Work done = 513.59 kJ/kg

$$= 513.59 \times 0.2$$

$$= 102.718 \text{ kJ/s} = \mathbf{102.72} \text{ kW}$$

EXAMPLE 12.26.

The volume ratio of compression and expansion for a diesel engine as measured from an indicator diagram are 15.3 and 7.5 respectively. The pressure and temperature at the beginning of the compression are 1 bar and 27°C.

Assuming an ideal engine, determine the mean effective pressure, the ratio of maximum pressure to mean effective pressure and cycle efficiency.

Also find the fuel consumption per kWh if the indicated thermal efficiency is 0.5 of ideal efficiency, mechanical efficiency is 0.8 and the calorific value of oil is 42000 kJ/kg.

For air $c_p = 1.005$ *kJ/kg K*

$c_v = 0.718$ *kJ/kg K and* $\gamma = 1.4$.

Solution : Refer Fig. 12.24. The subscripts denote the respective state points in the cycle.

Given : $\dfrac{V_1}{V_2} = 15.3$; $\dfrac{V_4}{V_3} = 7.5$

$T_1 = 300$ K and $p_1 = 1$ bar

Process 1 – 2 is adiabatic compression process and therefore

$$p_2 = p_1\left(\frac{V_1}{V_2}\right)^{\gamma}$$

$$= 1 \times (15.3)^{1.4} = 45.56 \text{ bar}$$

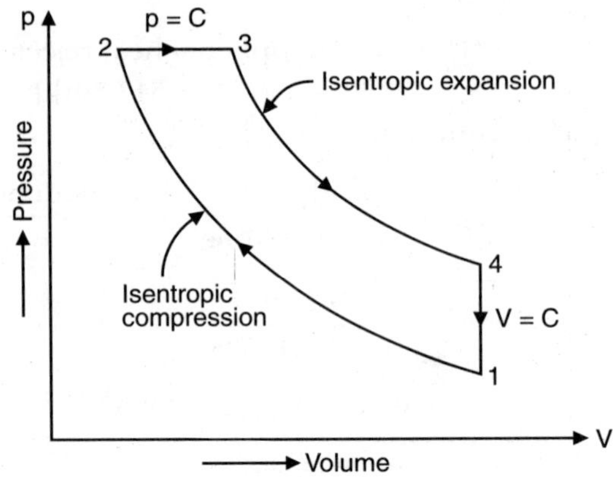

Fig. 12.24.

This is the maximum pressure in the cycle.

$$T_2 = T_1\left(\frac{V_1}{V_2}\right)^{\gamma-1}$$

$$= 300 \times (15.3)^{1.4-1} = 893.3 \text{ K}$$

Process 2–3 is a constant pressure process and hence

$$p_3 = p_2 = 45.56 \text{ bar}$$

$$\frac{V_2}{T_2} = \frac{V_3}{T_3}$$

$$\therefore \quad T_3 = \frac{V_3}{V_2}\times T_2 = \left(\frac{V_3}{V_4}\times\frac{V_4}{V_2}\right)\times T_2$$

$$= \left(\frac{V_3}{V_4}\times\frac{V_1}{V_2}\right)\times T_2 \quad (\because V_4 = V_1)$$

$$= \left(\frac{1}{7.5}\times 15.3\right)\times 893.3$$

$$= 1822.3 \text{ K}$$

Process 3 – 4 is an adiabatic expansion process and thus

$$T_4 = T_3\left(\frac{V_3}{V_4}\right)^{\gamma-1}$$

$$= 1822.3\times\left(\frac{1}{7.5}\right)^{1.4-1} = 813.98 \text{ K}$$

(*a*) Assuming that volume of working fluid (air) at state point 1 is 1 m^3, the mass of air involved in the process is

$$m = \frac{p_1 V_1}{R T_1}$$

$$= \frac{1\times 10^5\times 1}{287\times 300} = 1.161 \text{ kg}$$

Heat supplied,

$$Q_1 = m\, c_p\, (T_3 - T_2)$$
$$= 1.161 \times 1.005 \times (1822.3 - 893.3)$$
$$= 1083.96 \text{ kJ}$$

Heat rejected,

$$Q_2 = m\, c_v\, (T_4 - T_1)$$
$$= 1.161 \times 0.718 \times (813.98 - 300)$$
$$= 428.45 \text{ kJ}$$

Work done = heat supplied – heat rejected

$$= 1083.96 - 428.45 = 655.51 \text{ kJ}$$

Swept volume

$$= V_1 - V_2 = V_1 - \frac{V_1}{15.3} = \frac{14.3\, V_1}{15.3}$$

$$= \frac{14.3\times 1}{15.3} = 0.935 \text{ m}^3$$

∴ Mean effective pressure, *mep*

$$= \frac{\text{work done}}{\text{swept volume}}$$

$$= \frac{655.31}{0.935}$$

$$= 700.87 \text{ kN/m}^2 \simeq 7 \text{ bar}$$

Ratio of maximum to mean effective pressure,

$$= \frac{45.56}{7} = \mathbf{6.508}$$

Cycle efficiency,

$$\eta = \frac{\text{work done}}{\text{heat supplied}}$$

$$= \frac{655.51}{1083.96} = \mathbf{0.6047 \text{ or } 60.47\%}$$

(*b*) Indicated thermal, η

$$= 0.5 \times \text{ideal } \eta$$
$$= 0.5 \times 0.6047 = 0.3024$$

Brake thermal η

$$= \text{indicated thermal } \eta \times \text{mechanical } \eta$$
$$= 0.3024 \times 0.8 = 0.242$$

1 kWh ≡ 3600 kJ and therefore

Heat supplied per kWh

$$= \frac{3600}{0.242} = 14876 \text{ kJ}$$

∴ Fuel consumption per kWh

$$= \frac{14876}{42000} = \mathbf{0.354}$$

Alternatively : Brake thermal efficiency

$$\eta = \frac{\text{Brake power}}{(m_f/3600) \times CV}$$

$$= \frac{1}{(m_f/3600) \times 42000}$$

$$\text{or } 0.242 = \frac{3600}{42000\, m_f}$$

$$\text{or} \quad m_f = \frac{3600}{0.242 \times 42000}$$

$$= \mathbf{0.354\ kg/kWh}$$

EXAMPLE 12.27.

Two engines are to operate on Otto and diesel cycles with the following data :

maximum temperature = 1500 K

exhaust temperature = 700 K

ambient conditions = 1 bar and 300 K

For these engines make a comparison between their compression ratio, maximum pressure and cycle efficiency.

Solution : Refer Fig. 12.25 for the p–V plots for the engines operating on Otto cycle and diesel cycle.

(a) Consider Otto cycle : T_3 = 1500 K and T_4 = 700 K

$$p_1 = 1 \text{ bar and } T_1 = 300 \text{ K}$$

Compression ratio $r = \frac{V_1}{V_2}$ and maximum pressure = p_3

From isentropic expansion process 3–4,

$$\frac{T_3}{T_4} = \left(\frac{V_4}{V_3}\right)^{\gamma-1}$$

$$\text{or} \quad \frac{V_4}{V_3} = \left(\frac{T_3}{T_4}\right)^{\frac{1}{\gamma-1}} = \left(\frac{1500}{700}\right)^{\frac{1}{1.4-1}} = 6.72$$

$$\frac{V_4}{V_3} = \frac{V_1}{V_2} = r \text{ compression ratio.}$$

Therefore, the compression ratio for the Otto cycle is 6.72.

From isentropic compression process 1–2,

$$T_2 = T_1\left(\frac{V_1}{V_2}\right)^{\gamma-1}$$

$$= 300 \times (6.72)^{1.4-1} = 642.8 \text{ K}$$

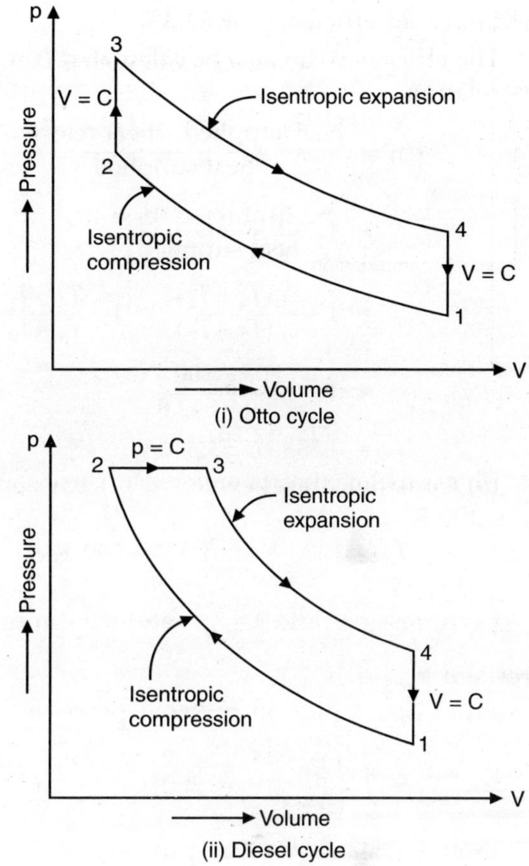

Fig. 12.25.

$$p_2 = p_1\left(\frac{V_1}{V_2}\right)^{\gamma}$$

$$= 1 \times (6.72)^{1.4} = 14.4 \text{ bar}$$

From constant volume process 2–3,

$$p_3 = p_2 \times \frac{T_3}{T_2}$$

$$= 14.4 \times \frac{1500}{642.8} = 33.6 \text{ bar}$$

Therefore, the maximum pressure for the Otto cycle is **33.6** bar

$$\text{Cycle } \eta = 1 - \frac{1}{r^{\gamma-1}}$$

$$= 1 - \frac{1}{(6.72)^{1.4-1}}$$

$$= 0.533 \text{ or } 53.3\%$$

The engine operating on Otto cycle will thus have an efficiency of **53.3%**

The efficiency can also be calculated from the relation

$$\eta = \frac{\text{heat supplied - heat refected}}{\text{heat supplied}}$$

$$= 1 - \frac{\text{heat rejected}}{\text{heat supplied}}$$

$$= 1 - \frac{c_v(T_4 - T_1)}{c_v(T_3 - T_2)} = 1 - \frac{T_4 - T_1}{T_3 - T_2}$$

$$= 1 - \frac{700 - 300}{1500 - 642.8}$$

$$= 0.533 \text{ or } 53.3\ \%$$

(*b*) Consider diesel cycle : p_1 = 1 bar and T_1 = 300 K

$$T_3 = 1500 \text{ K and } T_4 = 700 \text{ K}$$

Compression ratio $r = \dfrac{V_1}{V_2}$ and maximum pressure = $p_2 = p_3$

From isentropic compression process,

$$T_2 = T_1\left(\frac{V_1}{V_2}\right)^{\gamma-1} = 300\left(\frac{V_1}{V_2}\right)^{0.4} \quad ...(i)$$

From constant pressure process,

$$T_2 = T_3 \times \frac{V_2}{V_3} = 1500 \times \left(\frac{V_2}{V_3}\right) \quad ...(ii)$$

$$\therefore\ 300\left(\frac{V_1}{V_2}\right)^{0.4} = 1500 \times \frac{V_2}{V_3}$$

or $$\left(\frac{V_1}{V_2}\right)^{0.4} \times \frac{V_3}{V_2} = 5 \quad ...(iii)$$

From isentropic expansion process 3 – 4,

$$\frac{T_4}{T_3} = \left(\frac{V_3}{V_4}\right)^{\gamma-1}$$

or $$\frac{V_3}{V_4} = \left(\frac{T_4}{T_3}\right)^{\frac{1}{\gamma-1}}$$

$$= \left(\frac{700}{1500}\right)^{\frac{1}{1.4-1}} = 0.1488 \quad ...(iv)$$

Now $$\frac{V_3}{V_2} = \frac{V_1}{V_2} \times \frac{V_3}{V_1}$$

$$= \frac{V_1}{V_2} \times \frac{V_3}{V_4} = 0.1488\,\frac{V_1}{V_2}$$

Substituting the value of $\dfrac{V_3}{V_2}$ in identity (*iii*) we get

$$\left(\frac{V_1}{V_2}\right)^{0.4} \times 0.1488\,\frac{V_1}{V_2} = 5$$

or $$\left(\frac{V_1}{V_2}\right)^{1.4} = \frac{5}{0.1488} = 33.6$$

$$\therefore\ \frac{V_1}{V_2} = (33.6)^{1/1.4} = 12.3$$

Thus the compression ratio for the diesal engine is **12.3.**

Again considering the isentropic compression process,

$$T_2 = T_1\left(\frac{V_1}{V_2}\right)^{\gamma-1}$$

$$= 300 \times (12.3)^{0.4} = 818.62 \text{ K}$$

$$p_2 = p_1\left(\frac{V_1}{V_2}\right)^{\gamma}$$

$$= 1 \times (12.3)^{1.4} = 33.6 \text{ bar}$$

Thus the maximum pressure for the Diesel cycle is **33.6** bar.

The efficiency of diesel cycle can be calculated by using the relation

$$\eta = \frac{\text{heat supplied - heat rejected}}{\text{heat supplied}}$$

$$= 1 - \frac{\text{heat rejected}}{\text{heat supplied}}$$

$$= 1 - \frac{c_v\,(T_4 - T_1)}{c_p(T_3 - T_2)} = 1 - \frac{T_4 - T_1}{\gamma(T_3 - T_2)}$$

$$= 1 - \frac{700 - 300}{1.4(1500 - 818.62)}$$

$$= 1 - 0.419 = 0.581 \text{ or } 58.1\%$$

The engine operating on the diesel cycle will thus have an efficiency of **58.1**%

The comparison between the desired parameters for the two cycles (Otto/diesel) has been shown below

Ratio of compression ratios

$$= \frac{6.72}{12.3} = 0.546$$

Ratio of maximum pressures

$$= \frac{33.6}{33.6} = 1.0$$

Ratio of cycle efficiency

$$= \frac{0.533}{0.581} = 0.917$$

12.7. DUAL COMBUSTION (MIXED OR COMPOSITE) CYCLE

This is a cycle in which the addition of heat is partly at constant volume and partly at constant pressure.

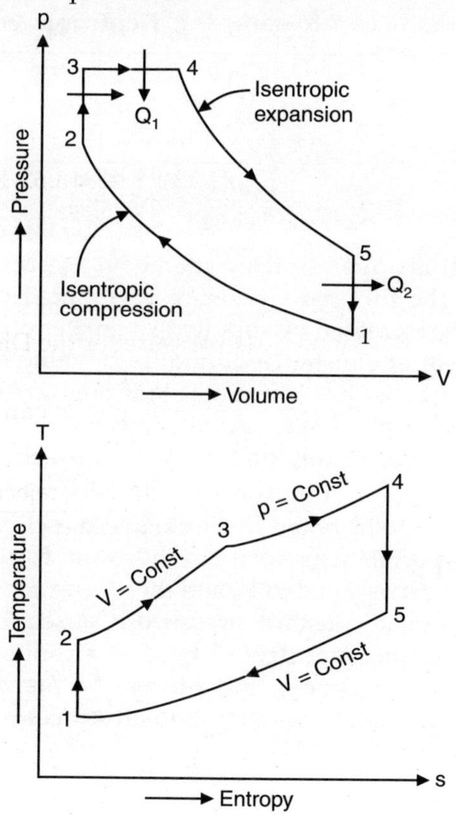

Fig. 12.26. Pressure-volume and temperature-entropy diagrams for air standard cycle

Fig. 12.26 shows the dual cycle plotted on p–V and $T-s$ diagrams. The sequence of operations is :

Process 1–2 : Piston moves from crank end dead centre (ODC) to cover end dead centre (IDC) and air with initial state 1 is compressed isentropically to state point 2 through compression ratio $r = V_1/V_2$

Process 2–3 : The piston is momentarily at rest at IDC and heat is added to air at constant volume from an external heat source which is brought into contact with the cylinder head. The pressure rises and the ratio $\alpha = p_3/p_2$ is called the ***explosion ratio.***

Process 3–4 : Addition of heat at constant pressure during which preliminary expansion of air takes place. The heat supply is stopped at point 4 which is called the cut off point. The ratio $\rho = V_4/V_3$ is called the ***cut off ratio*** or ***isobaric expansion ratio***.

Process 4–5 : Expansion of air takes place isentropically and the work is done by the system. The volume ratio V_5/V_4 is called the ***isentropic expansion ratio***.

Process 5–1 : The piston is momentarily at rest at ODC and heat is rejected to the external sink by bringing it in contact with the cylinder head. The process is so controlled that ultimately the initial parameters are restored and the cycle is completed.

The high speed diesel engines work on a cycle which is slight modification of the dual cycle. This thermodynamic cycle is often referred to as the ***limited pressure cycle.***

Thermal efficiency for dual cycle : The compression process 1–2 and the expansion process 4–5 are isentropic and hence there is no heat interaction in these processes. Energy is absorbed as heat during process 2–3 at constant volume and during process 3–4 at constant pressure, and is rejected as heat during process 5–1 at constant volume.

Consider unit mass of air undergoing the cyclic change.

Heat added, Q_1 = heat added during process 2–3 + heat added during process 3–4

$$= c_v (T_3 - T_2) + c_p (T_4 - T_3)$$

Heat rejected, Q_2 = heat rejected during constant volume process 5–1

$$= c_v (T_5 - T_1)$$

Work done = heat supplied – heat rejected

$$= c_v (T_3 - T_2) + c_p (T_4 - T_3) - c_v (T_5 - T_1)$$

Thermal efficiency,

$$\eta = \frac{\text{work done}}{\text{heat supplied}}$$

$$= \frac{c_v (T_3 - T_2) + c_p (T_4 - T_3) - c_v (T_5 - T_1)}{c_v (T_3 - T_2) + c_p (T_4 - T_3)}$$

$$= 1 - \frac{(T_5 - T_1)}{(T_3 - T_2) + \gamma(T_4 - T_3)} \quad ...(12.17)$$

The expression for thermal efficiency in terms of compression ratio $r = V_1/V_2$, explosion ratio $\alpha = p_3/p_2$, cut off ratio $\rho = V_4/V_3$ and adiabatic exponent γ can be worked out by expressing all the temperatures in terms of lowest temperature T_1 in the cycle.

At point 2 :

$$V_2 = \frac{V_1}{r} \; ; \; p_2 = p_1 \left(\frac{V_1}{V_2}\right)^{\gamma} = p_1 r^{\gamma}$$

$$T_2 = T_1 \left(\frac{V_1}{V_2}\right)^{\gamma - 1} = T_1 r^{\gamma - 1}$$

At point 3 :

$$V_3 = V_2 = \frac{V_1}{r}$$

$$p_3 = p_2 \alpha = p_1 \alpha r^{\gamma}$$

$$T_3 = T_2 \frac{p_3}{p_2} = T_2 \alpha = T_1 \alpha r^{\gamma - 1}$$

At point 4 :

$$V_4 = V_3 \rho \; ; \; p_4 = p_3 = p_1 \alpha r^{\gamma}$$

$$T_4 = T_3 \frac{V_4}{V_3} = T_3 \rho = T \alpha \rho r^{\gamma - 1}$$

At point 5 :

$$V_5 = V_1$$

$$p_5 = p_4 \left(\frac{V_4}{V_5}\right)^{\gamma}$$

$$= p_4 \left(\frac{V_4}{V_1}\right)^{\gamma} = p_3 \left(\frac{V_4 / V_3}{V_1 / V_3}\right)^{\gamma}$$

$$= p_3 \left(\frac{V_4 / V_3}{V_1 / V_2}\right)^{\gamma} = p_3 \left(\frac{\rho}{r}\right)^{\gamma}$$

$$T_5 = T_4 \left(\frac{V_4}{V_5}\right)^{\gamma - 1} = T_4 \left(\frac{\rho}{r}\right)^{\gamma - 1}$$

$$= \frac{T_1 \alpha \rho r^{\gamma - 1} \times \rho^{\gamma - 1}}{r^{\gamma - 1}} = T_1 \alpha \rho^{\gamma}$$

When these temperature values are substituted in expression 12.17, we get

$$\eta = 1 - \frac{T_1 \alpha \rho^{\gamma} - T_1}{\left(T_1 \alpha r^{\gamma - 1} - T_1 r^{\gamma - 1}\right) + \gamma \left(T_1 \alpha \rho r^{\gamma - 1} - T_1 \alpha r^{\gamma - 1}\right)}$$

$$= 1 - \frac{1}{r^{\gamma - 1}} \left[\frac{T_1 (\alpha \rho^{\gamma} - 1)}{T_1 (\alpha - 1) + \gamma T_1 (\alpha \rho - \alpha)}\right]$$

$$= 1 - \frac{1}{r^{\gamma - 1}} \left[\frac{\alpha \rho^{\gamma} - 1}{(\alpha - 1) + \gamma \alpha (\rho - 1)}\right] \quad ...(12.18)$$

It is apparent from the above expression that the thermal efficiency of a mixed cycle can be increased by supplying a greater portion of heat at constant volume (high value of α) and smaller portion at constant pressure (low value of ρ). In the actual high speed diesel engines operating on this cycle, it is achieved by early fuel injection and an early cut off.

It is to be noted that Otto and diesel cycles are special cases of the dual cycle. If $\rho = 1$, V_4 becomes equal to V_3 and there is no addition of heat at constant pressure. Consequently the entire quantity of heat is supplied at constant volume, and obviously the cycle becomes the Otto cycle. Substituting $\rho = 1$ in equation 12.18, We get

$$\eta = 1 - \frac{1}{r^{\gamma - 1}}$$

which is the same as for Otto cycle.

Similarly if $\alpha = 1$, the heat addition is only at constant pressure with the result that the cycle becomes diesel cycle. Substituting $\alpha = 1$ in equation 12.18, we get

$$\eta = 1 - \frac{1}{r^{\gamma-1}}\left[\frac{\rho^{\gamma}-1}{\gamma(\rho-1)}\right]$$

which is the same as for diesel cycle.

Mean effective pressure : With reference to Fig. 12.26 , the work done per cycle per unit mass of air is

Work done =heat supplied – heat rejected

$$= c_v (T_3 - T_2) + c_p (T_4 - T_3) - c_v (T_5 - T_1)$$

Swept volume

$$= (V_1 - V_2) = V_1(1 - V_2/V_1)$$

$$= \frac{RT_1}{p_1}\left(1-\frac{1}{r}\right) = \frac{RT_1}{p_1 r}(r-1)$$

By definition of mean effective pressure,

$$mep = \frac{\text{work done per cycle}}{\text{swept volume}}$$

$$= \frac{c_v (T_3 - T_2) + c_p (T_4 - T_3) - c_v (T_5 - T_1)}{RT_1 / p_1 r \times (r-1)}$$

$$= \frac{c_v}{R}\frac{p_1 r}{(r-1)}\left[\frac{(T_3-T_2)+\gamma(T_4-T_3)-(T_5-T_1)}{T_1}\right] \quad ...(12.19)$$

The expression for mean effective pressure in terms of compression ratio r, explosion ratio α, cut off ratio ρ, induction pressure p_1 and adiabatic exponent γ can be set up by expressing all the temperatures in terms of lowest temperature T_1 in the cycle.

As outlined above :

$$T_2 = T_1 r^{\gamma-1}; \quad T_3 = T_1 \alpha r^{\gamma-1}$$

$$T_4 = T_1 \alpha \rho r^{\gamma-1}; \quad T_5 = T_1 \alpha \rho^{\gamma}$$

When these temperature values are substituted in equation 12.19, we get

$$mep = \frac{c_v}{R}\frac{p_1 r}{(r-1)}\left[\frac{\begin{array}{r}\left(T_1\alpha r^{\gamma-1} - T_1 r^{\gamma-1}\right)\\ +\gamma\left(T_1\alpha\rho r^{\gamma-1} - T_1\alpha r^{\gamma-1}\right)\\ -\left(T_1\alpha\rho^{\gamma} - T_1\right)\end{array}}{T_1}\right]$$

$$= \frac{c_v}{R}\frac{p_1 r}{(r-1)}\left[\begin{array}{r}(\alpha-1)r^{\gamma-1} + \gamma\alpha r^{\gamma-1}(\rho-1)\\ -\left(\alpha\rho^{\gamma}-1\right)\end{array}\right]$$

But $\dfrac{c_v}{R} = \dfrac{1}{\gamma-1}$

$\therefore$ mep =

$$= \frac{p_1 r}{(r-1)(\gamma-1)}\left[\begin{array}{r}(\alpha-1)r^{\gamma-1} + \gamma\alpha r^{\gamma-1}(\rho-1)\\ -\left(\alpha\rho^{\gamma}-1\right)\end{array}\right] \quad ...(12.20)$$

which is the required expression of mean effective pressure for limited pressure (dual) cycle.

As pointed out earlier, the dual cycle becomes an Otto cycle for $\rho = 1$ and a diesel cycle for $\alpha = 1$

For $\rho = 1$, the equation 12.20 reduces to

$$mep = \frac{p_1 r}{(r-1)(\gamma-1)}\left[(\alpha-1)r^{\gamma-1} - (\alpha-1)\right]$$

$$= \frac{p_1 r}{(r-1)(\gamma-1)}\left[\left(r^{\gamma-1}-1\right)(\alpha-1)\right]$$

which is the same as that of mep for Otto cycle derived earlier (Equation 12.12).

For α =1, the equation 12.20 , reduces to

$$mep = \frac{p_1 r}{(r-1)(\gamma-1)}\left[\gamma r^{\gamma-1}(\rho-1) - \left(\rho^{\gamma}-1\right)\right]$$

which is the same as that of mep for diesel cycle derived earlier (Equation 12.16).

EXAMPLE 12.28.

A compression ignition engine working on dual cycle has a cylinder diameter of 20 cm and stroke 30 cm. At the start of compression process, the cylinder is full of air at 1 bar and 300 K. The engine operates with compression ratio of 8, and heat addition at constant pressure continues for 4 % of the stroke. If the maximum cycle pressure is limited to 60 bar, make calculations for the thermal efficiency, work done per cycle and the mean effective pressure.

Assume c_p = 1.005 kJ/kg K
c_v = 0.718 kJ/kg K and γ = 1.4

Solution : Refer Fig 12.27.

Swept volume,

$$V_s = \frac{\pi}{4} d^2 l$$

or $V_S = \frac{\pi}{4}(20)^2 \times 30 = 9420 \text{ cm}^3$

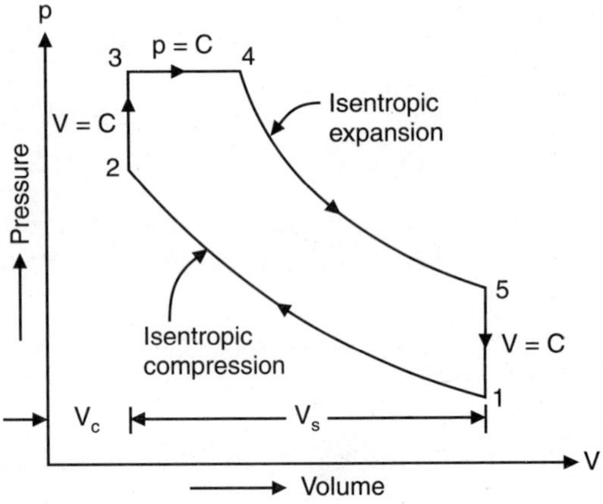

Fig. 12.27.

Compression ratio,

$$r = \frac{V_C + V_S}{V_C} = 1 + \frac{V_S}{V_C}$$

∴ Clearance volume,

$$V_C = \frac{V_S}{r-1}$$

$$= \frac{9420}{8-1} = 1346 \text{ cm}^3$$

Cylinder volume,

$$V_1 = V_C + V_S$$
$$= 1346 + 9420 = 10766 \text{ cm}^3$$

Process 1–2 :

$$T_2 = T_1 \left(\frac{V_1}{V_2}\right)^{\gamma-1}$$

$$= 300 \times (8)^{0.4} = 689.22 \text{ K}$$

$$p_2 = p_1 \left(\frac{V_1}{V_2}\right)^{\gamma}$$

$$= 1 \times (8)^{1.4} = 18.38 \text{ bar}$$

$$V_2 = V_C = 1346 \text{ cm}^3$$

Process 2–3 :

$$T_3 = \frac{p_3}{p_2} T_2$$

$$= \frac{60}{18.38} \times 689.22 = 2250 \text{ K}$$

$$V_3 = V_2 = 1346 \text{ cm}^3$$

Process 3–4 :

$$V_4 = V_3 + 0.04\ V_S$$
$$= V_2 + 0.04\ V_S$$
$$= 1346 + 0.04 \times 9420$$
$$= 1722.8 \text{ cm}^3$$

$$T_4 = \frac{V_4}{V_3} \times T_3$$

$$= \frac{1722.8}{1346} \times 2250 = 2880 \text{ K}$$

Process 4–5 :

$$T_5 = T_4 \left(\frac{V_4}{V_5}\right)^{\gamma-1}$$

$$= T_4 \left(\frac{V_4}{V_1}\right)^{\gamma-1}$$

$$= 2880 \times \left(\frac{1722.8}{10766}\right)^{0.4} = 1384 \text{ K}$$

(*a*) Heat supplied,

$$Q_1 = c_v (T_3 - T_2) + c_p (T_4 - T_3)$$
$$= 0.718 (2250 - 689.22) + 1.005 (2880 - 2250)$$
$$= 1754 \text{ kJ/kg}$$

Heat rejected,

$$Q_2 = c_v (T_5 - T_1)$$
$$= 0.718 (1384 - 300)$$
$$= 778 \text{ kJ/kg}$$

Work done = heat added – heat rejected
= 1754 – 778 = 976 kJ/kg

Efficiency,

$$\eta = \frac{\text{work done}}{\text{heat supplied}}$$

$$= \frac{976}{1754} = 0.556 \text{ to } \mathbf{55.6\ \%}$$

(*b*) The mass of air being supplied per cycle can be calculated by applying the characteristic gas equation to the fluid properties at state point 1,

$$m = \frac{p_1 V_1}{R T_1}$$

$$= \frac{1 \times 10^5 \times (10766 \times 10^{-6})}{287 \times 300}$$

$$= 0.0125 \text{ kg/cycle}$$

∴ Work done per cycle

$= 976 \times 0.0125 = \mathbf{12.2\ kJ/cycle}$

(c) Mean effective pressure, *mep*

$$= \frac{\text{work done per cycle}}{\text{swept volume}}$$

$$= \frac{12.2}{9420 \times 10^{-6}}$$

$= 1295\ \text{kN/m}^2 = \mathbf{12.95\ bar}$

EXAMPLE 12.29.

In an ideal cycle for compression ignition (diesel) engine, the cylinder is assumed to contain 1 kg of air initially at 1 bar and 320 K. This air is compressed adiabatically through a volume ratio of 16, heated at constant volume until the pressure is 70 bar, further heated at constant pressure and then expanded adiabatically to the initial volume. Finally heat is transferred from the air at constant volume to return it to the initial state. If all the processes are reversible and the heat added to the air at constant pressure is equal to that added at constant volume, determine :

(a) the pressures and temperatures at the cardinal points of the cycle,

(b) the cycle efficiency, and

(c) the mean effective pressure.

Solution : Refer Fig 12.28.

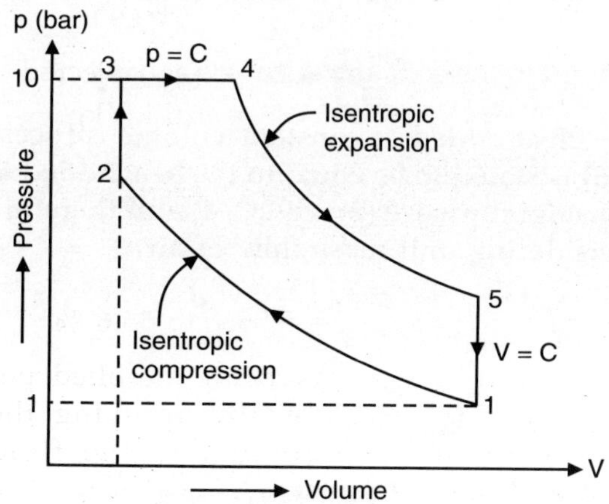

Fig. 12.28.

Process 1–2 :

$$T_2 = T_1\left(\frac{V_1}{V_2}\right)^{\gamma-1}$$

$= 320 \times (16)^{0.4} = 970\ \text{K}$

$$p_2 = p_1\left(\frac{V_1}{V_2}\right)^{\gamma}$$

$= 1 \times (16)^{1.4} = 48.5\ \text{bar}$

Process 2–3 :

$$T_3 = T_2\frac{p_3}{p_2}$$

$$= 970 \times \frac{70}{48.5} = 1400\ \text{K}$$

$Q_{2\text{-}3} = c_v\ (T_3 - T_2)$

$= 0.718 \times (1400 - 970)$

$= 308.74\ \text{kJ/kg}$

Process 3–4 :

$Q_{3\text{-}4} = Q_{2\text{-}3} = c_p\ (T_4 - T_3)$

$= 1.005\ (T_4 - 1400)$

$$\therefore \quad T_4 = \frac{308.74}{1.005} + 1400 = 1707.2\ \text{K}$$

$$\frac{V_4}{V_3} = \frac{T_4}{T_3} = \frac{1707.2}{1400} = 1.219$$

Process 4–5 :

$$T_5 = T_4\left(\frac{V_4}{V_5}\right)^{\gamma-1}$$

$$= T_4\left(\frac{V_4}{V_2} \times \frac{V_2}{V_5}\right)^{\gamma-1}$$

$$= T_4\left(\frac{V_4}{V_3} \times \frac{V_2}{V_1}\right)^{\gamma-1}$$

$= 1707.2 \times (1.219/16)^{0.4}$

$= 609.6\ \text{K}$

$$p_5 = p_1\left(\frac{T_5}{T_1}\right)$$

$$= 1 \times \frac{609.6}{320} = \mathbf{1.905\ bar}$$

(b) Heat supplied

$= Q_{2\text{-}3} + Q_{3\text{-}4}$

$= 308.74 + 308.74$

$= 617.48\ \text{kJ/kg}$

Heat rejected

$= c_v\ (T_5 - T_1)$

$= 0.718\ (609.6 - 320)$

$= 207.93\ \text{kJ/kg}$

Work done = heat supplied – heat rejected

$$= 617.48 - 207.93$$
$$= 409.55 \text{ kJ/kg}$$

Air standard efficiency,

$$\eta = \frac{\text{work done}}{\text{heat supplied}}$$
$$= \frac{409.55}{617.48} = 0.663 \text{ or } \mathbf{66.3\ \%}$$

(*c*) From characteristic gas equation,

$$v_1 = \frac{RT_1}{p_1}$$
$$= \frac{287 \times 320}{1 \times 10^5} = 0.9184 \text{ m}^3\text{/kg}$$
$$v_2 = \frac{v_1}{16} = \frac{0.9184}{16} = 0.0574 \text{ m}^3\text{/kg}$$

Swept volume

$$= v_1 - v_2$$
$$= 0.9184 - 0.0574$$
$$= 0.861 \text{ m}^3\text{/kg}$$

Mean effective pressure, *mep*

$$= \frac{\text{work done per cycle}}{\text{swept volume}}$$
$$= \frac{409.55}{0.861}$$
$$= 475.67 \text{ kN/m}^2 = \mathbf{4.7567 \text{ bar}}$$

EXAMPLE 12.30.

In an internal combustion engine operating on the dual cycle, the temperature of the working fluid at the beginning of compression is 27°C. The ratio of the maximum and minimum pressures in the cycle is 70 and the compression ratio is 15. The amounts of heat added at constant volume and constant pressure are equal. Compute the air standard thermal efficiency of the cycle. State three main reasons why the actual thermal efficiency is different from the theoretical value.

Solution : Refer Fig. 12.29. The subscripts denote the respective state points in the cycle.

Given $T_1 = 300$ K

$$\frac{p_3}{p_1} = 70 \quad \text{and} \quad \frac{V_1}{V_2} = 15$$

Process 1 – 2 is adiabatic compression process and therefore

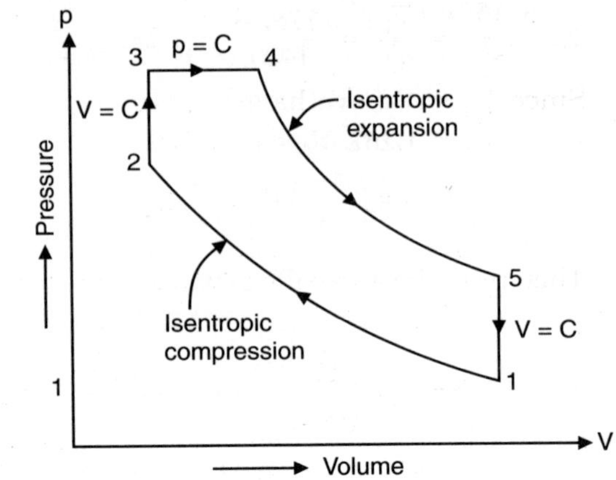

Fig. 12.29.

$$p_2 = p_1\left(\frac{V_1}{V_2}\right)^{\gamma}$$
$$= p_1 \times (15)^{1.4} = 44.3\ p_1$$
$$T_2 = T_1\left(\frac{V_1}{V_2}\right)^{\gamma-1}$$
$$= 300 \times (15)^{1.4-1} = 886.25 \text{ K}$$

Process 2–3 is constant volume process and hence

$$\frac{p_2}{T_2} = \frac{p_3}{T_3}$$
$$\therefore \quad T_3 = T_2 \times \frac{p_3}{p_2} = T_2 \times \frac{p_3}{44.3 p_1}$$
$$= \frac{886.25}{44.3} \times 70 = 1400.4 \text{ K}$$

Heat added at constant volume (process 2–3) is stated to be equal to the heat added at constant pressure (process 3–4) and therefore considering unit mass flow of air,

$$c_v\,(T_3 - T_2) = c_p\,(T_4 - T_3)$$
$$\text{or} \quad T_3 - T_2 = \gamma(T_4 - T_3)$$
$$\therefore \quad T_4 = T_3 + \frac{T_3 - T_2}{\gamma}$$
$$= 1400.4 + \frac{1400.4 - 886.25}{1.4}$$
$$= 1767.65 \text{ K}$$

Process 3–4 is constant pressure process and hence

$$\frac{V_3}{T_3} = \frac{V_4}{T_4}$$

$$\therefore \quad \frac{V_4}{V_3} = \frac{T_4}{T_3} = \frac{1767.65}{1400.4} = 1.262$$

Since $V_3 = V_2$, we have

$$V_4 = 1.262\ V_2$$

$$= 1.262 \times \left(\frac{V_1}{15}\right) = 0.0841\ V_1$$

Then from the adiabatic expansion process 4–5,

$$T_5 = T_4 \times \left(\frac{V_4}{V_5}\right)^{\gamma-1} = T_4\left(\frac{V_4}{V_1}\right)^{\gamma-1}$$

$$= 1767.65 \times \left(\frac{0.0841\ V_1}{V_1}\right)^{1.4-1}$$

$$= 655.62\ \text{K}$$

The air standard efficiency η is given by

$$\eta = \frac{\text{work done}}{\text{heat supplied}}$$

$$= \frac{\text{heat supplied} - \text{heat rejected}}{\text{heat supplied}}$$

$$= 1 - \frac{\text{heat rejected}}{\text{heat supplied}}$$

$$= 1 - \frac{c_v\,(T_5 - T_1)}{c_v(T_3 - T_2) + c_p\,(T_4 - T_3)}$$

$$= 1 - \frac{T_5 - T_1}{(T_3 - T_2) + \gamma(T_4 - T_3)}$$

$$= 1 - \frac{655.62 - 300}{(1400.4 - 886.25) + 1.4\,(1767.65 - 1400.4)}$$

$$= 1 - \frac{355.62}{514.15 + 514.15}$$

$$= \mathbf{0.654 \text{ or } 65.4\ \%}$$

The actual thermal efficiency would be different from the theoretical value due to the following reasons :

1. Air is the working substance in theoretical cycle, whereas a mixture of air and fuel constitutes the working substance in an actual cycle.

2. The theoretical cycle does not consider the effect of fuel combustion phenomenon and the associated problems like dissociation of gases, dilution of charge during suction stroke etc.

3. No account is taken of the effect of variable specific heat, effect of inlet and exhaust velocities and the heat loss through cylinder walls.

EXAMPLE 12.31.

A four stroke six cylinder engine with 10 cm bore and 15 cm length works on a dual combustion cycle with compression ratio 10.5. The pressure and temperature of air at the commencement of compression are 0.9 bar and 323 K respectively. Both compression and expansion were carried out according to polytropic law $pV^{1.33}$ = constant. While running at 1500 rpm, the fuel consumption for the engine is 0.175 kg/min. If 2/3 of heat released by the combustion of fuel is at constant volume and the remainder at constant pressure, determine

(a) heat supplied per kg of cylinder charge

(b) pressure and temperature at the corner points of the cycle

(c) power output and cycle efficiency of the engine.

Assume that calorific value of the fuel used is 44500 kJ/kg.

Solution : Refer Fig. 12.30. The subscripts denote the respective state points in the cycle.

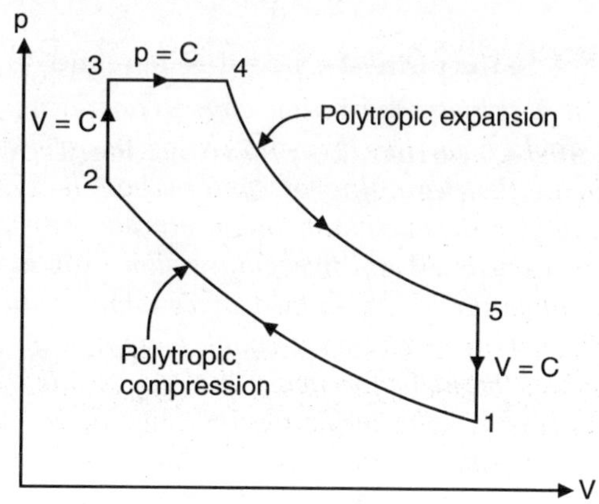

Fig. 12.30.

Given : $\frac{V_1}{V_2} = 10.5$

$p_1 = 0.9$ bar ; and $T_1 = 323$ K

(*a*) For a four stroke engine, the number of working strokes are half the rotational speed. Then

Fuel used per cycle per cylinder

$$= \frac{0.175}{6 \times (1500/2)} = 3.78 \times 10^{-5} \text{ kg}$$

Heat input per cylinder charge

$$= (3.78 \times 10^{-5}) \times 44500$$
$$= 1.6821 \text{ kJ}$$

$$\text{Swept volume per cylinder} = \frac{\pi}{4} d^2 l$$

$$= \frac{\pi}{4} \times (0.1)^2 \times 0.15$$
$$= 1.1775 \times 10^{-3} \text{ m}^3$$

The mass of the charge can be calculated by applying characteristic equation at state point 1. That is

$$m = \frac{p_1 V_1}{R T_1}$$

$$= \frac{(0.9 \times 10^5) \times (1.1775 \times 10^{-3})}{287 \times 323}$$

$$= 1.143 \times 10^{-3} \text{ kg}$$

Hence heat supplied per kg of cylinder charge

$$= \frac{1.6821}{1.143 \times 10^{-3}} = \mathbf{1471.65} \text{ kJ}$$

(*b*) **State point 1 :** $p_1 = 0.9$ bar and $T_1 = 323$ K

State point 2 : From isentropic compression process 1 – 2

$$T_2 = T_1 \left(\frac{V_1}{V_2}\right)^{n-1}$$

$$= 323 \times (10.5)^{1.33-1} = 701.88 \text{ K}$$

$$p_2 = p_1 \left(\frac{V_1}{V_2}\right)^{n}$$

$$= 0.9 \times (10.5)^{1.33} = 20.53 \text{ bar}$$

State point 3 : Heat supplied at constant volume per kg of charge

$$\frac{2}{3} \times 1471.65 = c_v (T_3 - T_2)$$
$$= 0.718 (T_3 - 701.88)$$

$$\therefore \quad T_3 = \frac{2}{3} \times \frac{1471.65}{0.718} + 701.88$$
$$= 2068.3 \text{ K}$$

$$p_3 = \frac{T_3}{T_2} \times p_2$$

$$= \frac{2068.3}{701.88} \times 20.53 = 60.5 \text{ bar}$$

State point 4 : $p_4 = p_3 = 60.5$ bar

Heat supplied at constant pressure per kg of charge,

$$\frac{1}{3} \times 1471.65 = c_p (T_4 - T_3)$$
$$= 1.005 (T_4 - 2068.3)$$

$$\therefore T_4 = \frac{1}{3} \times \frac{1471.65}{1.005} + 2068.3 = 2556.4 \text{ K}$$

$$V_4 = V_3 \times \frac{T_4}{T_3}$$

$$= 1 \times \frac{2556.4}{2068.3} = 1.236 \text{ m}^3$$

It may be noted that volume at state point 3 has been taken as unity.

State point 5 : From isentropic expansion process 4–5

$$T_5 = T_4 \times \left(\frac{V_4}{V_5}\right)^{n-1}$$

$$= 2556.4 \times \left(\frac{1.236}{10.5}\right)^{0.33} = 1261.8 \text{ K}$$

$$p_5 = p_4 \times \left(\frac{V_4}{V_5}\right)^{n}$$

$$= 60.5 \times \left(\frac{1.236}{10.5}\right)^{1.33} = 3.515 \text{ bar}$$

The pressures and temperatures as calculated above at the salient points of the cycle have been tabulated below

State point	Temperature, K	Pressure, bar
1	323	0.9
2	701.88	20.53
3	2068.3	60.5
4	2556.4	60.5
5	1261.8	3.515

(*b*) Work done during the cycle equals the area enclosed by the processes constituting the cycle.

Work done

$$= p_4(V_4 - V_3) + \frac{p_4V_4 - p_5V_5}{n-1} - \frac{p_2V_2 - p_1V_1}{n-1}$$

$$= R\left[(T_4 - T_3) + \frac{T_4 - T_5}{n-1} - \frac{T_2 - T_1}{n-1}\right]$$

$$= 0.287\left[(2556.4 - 2068.3) + \frac{2556.4 - 1261.8}{1.33-1} - \frac{701.88 - 323}{1.33-1}\right]$$

$= 0.287(488 + 3923 - 1148) = 936.48$ kJ

$$\text{Cycle, } \eta = \frac{\text{work done}}{\text{heat supplied}}$$

$$= \frac{936.48}{1471.65}\ \mathbf{0.636 \text{ or } 63.6\%}$$

Work done

$= 936.48$ kJ/kg of cylinder charge

$= 936.48 \times (1.143 \times 10^{-3})$ kJ/cycle

$$= 936.48 \times (1.143 \times 10^{-3}) \times \frac{1500}{2} \text{ kJ/min}$$

$$= 936.48 \times (1.143 \times 10^{-3}) \times \frac{750}{60} \text{ kJ/s} = 13.38 \text{ kW}$$

For 6 cylinders, the power output will be

$= \mathbf{6 \times 13.38 = 80.28}$ kW

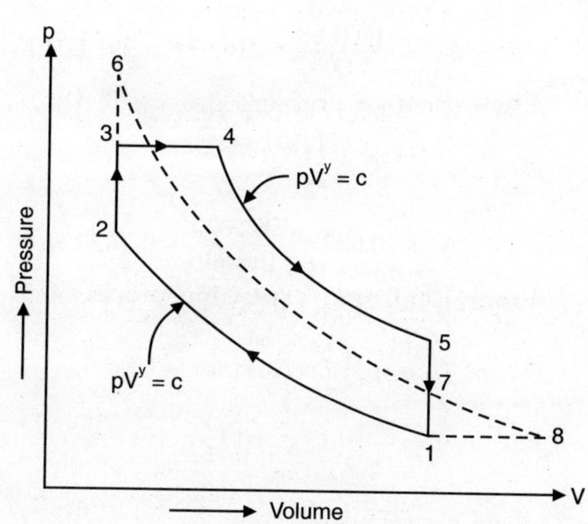

Fig. 12.31.

EXAMPLE 12.32.

In an air standard dual cycle having compression ratio 10:1, the working fluid is at 310 K temperature and 1 bar pressure at the commencement of compression. The compression and expansion processes are both frictionless adiabatic and the total heat addition equals 2800 kJ/kg of charge. Determine the mean effective pressure and thermal efficiency of the cycle if the maximum pressure limit is 70 bar.

(b) How these parameters would be effected if the total heat addition is at constant volume? Also evaluate the extra work available if the isentropic expansion proceeds down to the original pressure of 1 bar.

Solution : Refer Fig. 12.31 for the p–V plot for the engine operating on the given cycles. The subscripts denote the state points in the cycles.

For 1 kg of air within the cylinder

$$V_1 = \frac{mR\,T_1}{p_1}$$

$$= \frac{1 \times 287 \times 310}{1 \times 10^5} = 0.89 \text{ m}^3$$

$$T_2 = T_1\left(\frac{V_1}{V_2}\right)^{\gamma-1}$$

$= 310 \times (10)^{1.4-1} = 778.7$ K

$$p_2 = p_1\left(\frac{V_1}{V_2}\right)^{\gamma}$$

$= 1 \times (10)^{1.4} = 25.12$ bar

$$V_3 = V_2 = \frac{0.89}{10} = 0.089$$

Cycle (*a*) :

$$T_3 = T_2 \times \frac{p_3}{p_2}$$

$$= 778.7 \times \frac{70}{25.12} = 2169.94 \text{ K}$$

∴ Heat supplied at constant volume

$= m\, c_v\, (T_3 - T_2)$

$= 1 \times 0.718\ (2169.94 - 778.7)$

$= 998.91$ kJ

Since the total heat supplied is 2800 kJ, the heat supplied at constant pressure is equal to

$= 2800 - 998.91 = 1801.09$ kJ

∴ $1801.09 = m\, c_p\, (T_4 - T_3)$

$= 1 \times 1.005\ (T_4 - 2169.94)$

$$T_4 = \frac{1801.09}{1.005} + 2169.94 = 3962.07 \text{ K}$$

From constant pressure process 3–4 :

$$V_4 = V_3 \times \frac{T_4}{T_3}$$

$$= 0.089 \times \frac{3962.07}{2169.94} = 0.1625 \text{ m}^3$$

From isentropic expansion process 4–5:

$$T_5 = T_4 \left(\frac{V_4}{V_5}\right)^{\gamma - 1}$$

$$= 3962.07 \times \left(\frac{0.1625}{0.89}\right)^{1.4 - 1}$$

$$= 2007.13 \text{ K}$$

Heat rejected

$$= m\, c_v\, (T_5 - T_1)$$
$$= 1 \times 0.718(2007.13 - 310)$$
$$= 1218.54 \text{ kJ}$$

Work done,

$$W = \text{heat added} - \text{heat rejected}$$
$$= 2800 - 1218.54 = 1581.46 \text{ kJ}$$

Swept volume,

$$V_s = V_1 - V_2 = 0.89 - 0.089$$
$$= 0.801 \text{ m}^3$$

Mean effective pressure, *mep*

$$= \frac{\text{work done}}{\text{swept volume}}$$

$$= \frac{1581.46}{0.801}$$

$$= 1974.36 \text{ kN/m}^2 = \mathbf{19.74} \text{ bar}$$

Cycle efficiency,

$$\eta = \frac{\text{work done}}{\text{heat supplied}}$$

$$= \frac{1581.46}{2800} = \mathbf{0.565 \text{ or } 56.5\%}$$

Cycle (*b*) : Here the entire heat addition takes place during the constant volume process 2–6. Then

$$2800 = m\, c_v\, (T_6 - T_2)$$
$$= 1 \times 0.718\, (T_6 - 778.7)$$

$$T_6 = \frac{2800}{0.718} + 778.7$$
$$= 4678.42 \text{ K}$$

$$p_6 = \frac{T_6}{T_2} \times p_2$$

$$= \frac{4678.42}{778.7} \times 25.12 = 150.92 \text{ bar}$$

From isentropic expansion process 6–7,

$$T_7 = T_6 \times \left(\frac{V_6}{V_7}\right)^{\gamma-1} = T_6 \times \left(\frac{V_2}{V_1}\right)^{\gamma-1}$$

$$= 4678.42 \times \left(\frac{1}{10}\right)^{1.4-1}$$

$$= 1862.43 \text{ K}$$

Heat rejected $= m\, c_v\, (T_7 - T_1)$

$$= 1 \times 0.718\, (1862.43 - 310)$$
$$= 1114.64 \text{ kJ}$$

Work done

$$= 2800 - 1114.64 = 1685.36 \text{ kJ.}$$

Swept volume

$$V_S = V_1 - V_2$$
$$= 0.89 - 0.089 = 0.801 \text{ m}^3$$

$$mep = \frac{1685.36}{0.801}$$

$$= 2104.07 \text{ kN/m}^2 = \mathbf{21.04} \text{ bar}$$

$$\text{cycle } \eta = \frac{1685.36}{2800} = 0.6019 \quad \text{or} \quad \mathbf{60.19\%}$$

(*c*) In case the expansion continues down to the original pressure, we have

$$T_8 = T_6 \times \left(\frac{p_8}{p_6}\right)^{\frac{\gamma-1}{\gamma}}$$

$$= 4678.42 \times \left(\frac{1}{150.92}\right)^{\frac{1.4-1}{1.4}}$$

$$= 1114.25 \text{ K}$$

Then heat rejected at constant pressure process 8 – 1,

$$= m\, c_p\, (T_8 - T_1)$$
$$= 1 \times 1.005\, (1114.25 - 310)$$
$$= 808.27 \text{ K}$$

Work obtained

$$= 2800 - 808.27 = 1991.73 \text{ kJ}$$

Increase in work

$$= 1991.73 - 1685.36 = \mathbf{306.37} \text{ kJ}$$

12.8. COMPARISON OF OTTO AND DIESEL CYCLES

(a) Same compression ratio and heat supplied

The efficiency of Otto cycle depends upon compression r and is given by

$$\eta_{Otto} = 1 - \frac{1}{r^{\gamma-1}} \quad ...(i)$$

The efficiency of diesel cycle depends upon compression ratio r, the cut off ratio ρ and is prescribed by the relation

$$\eta_{Diesel} = 1 - \frac{1}{r^{\gamma-1}}\left[\frac{\rho^{\gamma}-1}{\gamma(\rho-1)}\right] \quad ...(ii)$$

Apparently for fixed compression ratio r, the efficiency of diesel cycle depends upon the factor

$$K = \frac{\rho^{\gamma}-1}{\gamma(\rho-1)}$$

For adiabatic exponent $\gamma = 1.4$, the value of K for different cut off ratios is

$\rho = 1.5$	2.0	2.5	3.0
$K = 1.092$	1.17	1.24	1.31

The value of factor K is thus always greater than unity. A close examination of expressions (*i*) and (*ii*) then reveals that the diesel cycle is always less efficient than the corresponding Otto cycle having the same compression ratio.

Mathematical analysis :

Cut off ratio,

$$\rho = \frac{V_3}{V_2} = \frac{V_3}{V_4} \times \frac{V_4}{V_2}$$

$$= \frac{V_3}{V_4} \times \frac{V_1}{V_2} = \frac{r}{r_e}$$

where r is the compression ratio and r_e is the expansion ratio.

The efficiency of diesel cycle can then be expressed as

$$\eta_{Diesel} = 1 - \frac{1}{r^{\gamma-1}} \times \frac{1}{\gamma}\left[\frac{(r/r_e)^{\gamma}-1}{(r/r_e)-1}\right] \quad ...(i)$$

It is to be noted from Fig. 12.13 that expansion V_4/V_3 ratio is less than compression ratio V_1/V_2. Let

$$r_e = r - \delta$$

where δ is a very small quantity

Then :

$$\frac{r}{r_e} = \frac{r}{r-\delta} = \left(1-\frac{\delta}{r}\right)^{-1}$$

$$= 1 + \frac{\delta}{r} + \left(\frac{\delta}{r}\right)^2 + \left(\frac{\delta}{r}\right)^3 + ...$$

Also

$$\left(\frac{r}{r_e}\right)^{\gamma} = \left(1-\frac{\delta}{r}\right)^{-\gamma}$$

$$= 1 + \frac{\delta}{r}\gamma + \frac{\gamma(\gamma+1)}{2!}\left(\frac{\delta}{r}\right)^2 + \frac{\gamma(\gamma+1)(\gamma+2)}{3!}\left(\frac{\delta}{r}\right)^3 + ...$$

When the values of $\left(\frac{r}{r_e}\right)$ and $\left(\frac{r}{r_e}\right)^{\gamma}$ are substituted in expression (*i*), we get

$$\eta_{diesel} = 1 - \frac{1}{r^{\gamma-1}}\left[\frac{\frac{\delta}{r} + \frac{\gamma+1}{2!}\left(\frac{\delta}{r}\right)^2 + \frac{(\gamma+1)(\gamma+2)}{3!}\times\left(\frac{\delta}{r}\right)^3 + ...}{\frac{\delta}{r} + \left(\frac{\delta}{r}\right)^2 + \left(\frac{\delta}{r}\right)^3 + ...}\right]$$

The ratio δ/r is small but is always positive, and the coefficients of the terms $(\delta/r)^2$, $(\delta/r)^3$ etc. in the numerator are greater than unity. As such the numerator is greater than denominator. That indicates that the efficiency of diesel cycle is lower than that of Otto cycle for the same compression ratio.

Reference may also be made to Fig. 12.32, which shows Otto, diesel and dual cycles sketched on $p - V$ and $T - s$ plots for the same compression ratio and heat input.

The salient features of these diagrams are:

– All the three cycles start at the same initial parameters corresponding to state point

1 with isentropic compression process 1—2; the compression ratio $r = V_1/V_2$ is same for all the cycles.

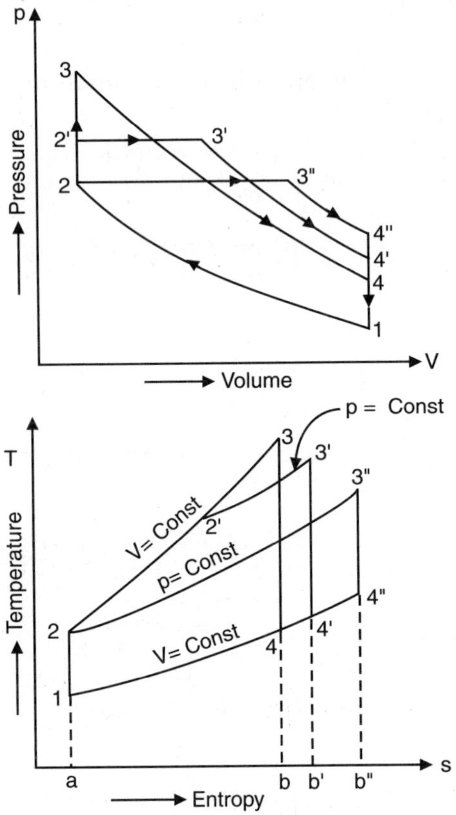

Fig. 12.32. Otto, diesel and dual cycles with same compression ratio and heat input

— Addition of heat is under varying conditions for the different cycles.

— Heat supplied is same for all the three cycles and accordingly the area of $T-s$ diagrams for the three cycles are equal

Area a – 2 – 3 – b = area a – 2 – 3″ – b''

= area a – 2 – 2′ – 3′ – b'

— The isentropic expansion for each cycle continues upto the constant volume line 1 - 4″ and the heat rejected for each cycle is represented by the appropriate area under line 1 - 4″

The thermal efficiency of any cycle is given by

$$\eta_t = \frac{Q_1 - Q_2}{Q_1} = 1 - \frac{Q_2}{Q_1}$$

For the same quantity of heat supplied, the cycle rejecting the least amount of heat will be most efficient. Fig. 12.32. shows that diesel cycle rejects more heat (area a – 1 – 4″ – b'') than the Otto cycle (area a – 1 – 4 – b). The heat rejected by dual cycle (area a – 1 – 4′ – b') lies between the heats rejected by Otto and dual cycles.

The graphical representation thus shows that the order of efficiencies of the three cycles is

Otto > dual > diesel

(*b*) Same maximum pressure and temperature : When pressure is the limiting factor in engine design, it becomes necessary to compare the air standard cycles on the basis of same maximum pressure and temperature. Fig. 12.33 shows the Otto and diesel cycles on $p-V$ and $T-s$ plots for the constant maximum pressure and temperature. Since state point 3 is common to both the cycles, the Otto cycle is limited to low compression ratio.

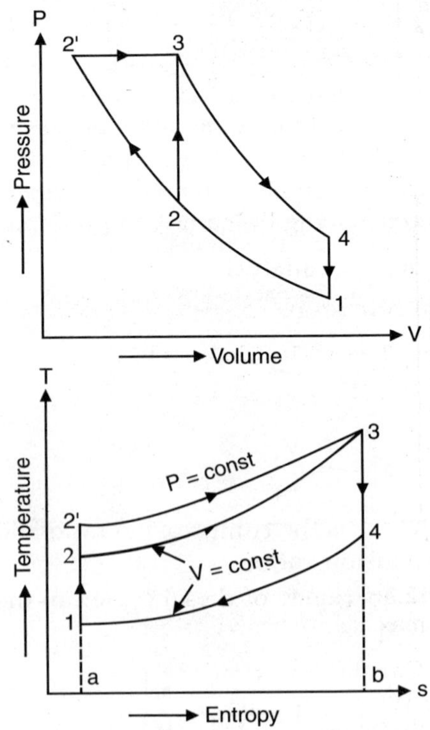

Fig. 12.33. Otto and diesel cycles with same maximum pressure and temperature

It is evident from the diagram that :

– both the cycles reject the same amount of heat given by area *b* - 4 - 1 - *a*.

– heat supplied in diesel cycle represented by area *a* - 2′ - 3 - *b* is more than the heat supplied for Otto cycle represented by area *a* - 2 - 3 - *b*.

Then from the expression $\eta = 1 - \frac{Q_2}{Q_1}$, it is evident, that for the same heat rejected Q_2, the cycle with greater heat addition Q_1 is the more efficient. Similarly it can be shown that the Dual cycle is more efficient than Otto cycle but less efficient than diesel cycle. Thus for a constant maximum pressure and temperature, the order of efficiencies for the cycles is

diesel > dual > Otto

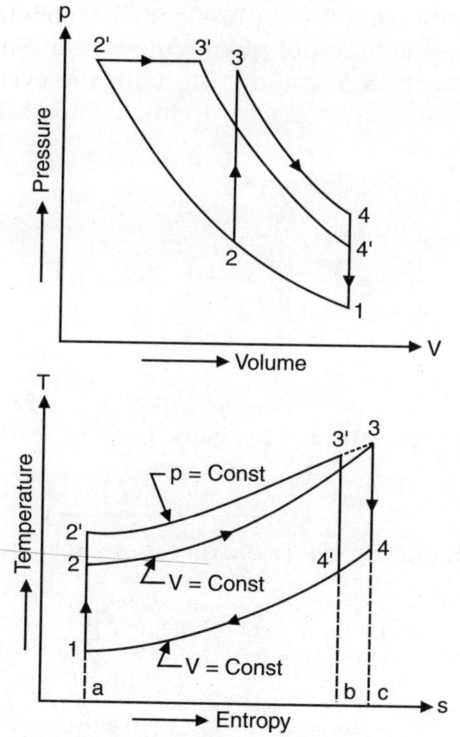

Fig. 12.34. Otto and diesel cycles with same maximum pressure and heat input

(*c*) Constant maximum pressure and heat input : When pressure is the limiting factor in engine design, it becomes necessary to compare the air standard cycles on the basis of same maximum pressure and heat supplied. Fig 12.34 shows the Otto and diesel cycles on *p*–*V* and *T* – *s* plots for the constant maximum pressure and heat input. For the same maximum pressure, points 3 and 3′ lie on the same constant pressure line and for the same heat input the areas (*a* - 2 - 3 - *c*) and (*a* - 2′ - 3′ - *b*) on *T*–*s* plot must be equal.

From visual inspection of Fig. 12.34, it is observed that heat rejected by diesel cycle represented by area (*b* - 4′ - 1 - *a*) is less than that of Otto cycle represented by area (*c* - 4 - 1 - *a*). Then from the expression $\eta = 1 - \frac{Q_2}{Q_1}$ it becomes obvious that for the same amount of heat supplied Q_1 the diesel cycle with less heat rejected Q_2 has a higher value of thermal efficiency compared to that for Otto cycle. A similar treatment would show that the efficiency of dual cycle lies between that for diesel and Otto cycles. Thus for a constant maximum pressure and temperature, the order of efficiencies for the cycle is

diesel > dual > Otto

12.9. ATKINSON CYCLE

The Atkinson cycle was conceived and developed by a British engineer, Dr J Atkinson. Fig. 12.35, shows the Atkinson cycle plotted on *p* – *V* and *T* – *s* diagram.

The sequence of operations is

Process 1–2 : Piston moves from crank end dead centre (ODC) to cover end dead centre (IDC) and air with initial state point 1 is compressed isentropically to state point 2 through compression ratio $r = V_1/V_2$

Process 2–3 : The piston is momentarily at rest at IDC and heat is supplied to the compressed air at constant volume from an external source which is brought into contact with the cylinder head. The pressure rises and the ratio $\alpha = p_3/p_2$ is called the ***explosion ratio***.

Process 3–4 : The increased high pressure exerts a greater amount of force on the piston and pushes it towards ODC. Expansion of working fluid takes place isentropically upto the lowest pressure $p_1 = p_4$ of the cycle, and work is done by the system.

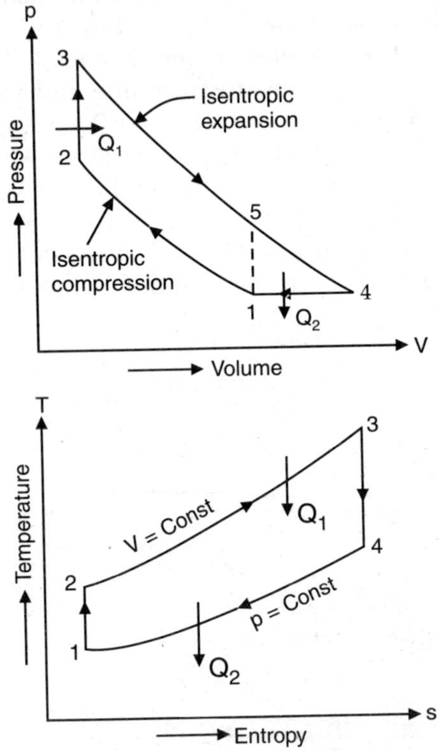

Fig. 12.35. Pressure-volume and temperature entropy diagrams for air standard Atkinson cycle

Process 4 – 1 : There occurs preliminary compression of air at constant pressure and heat is rejected to the external sink by bringing it in contact with the cylinder head. The process is so controlled that the air is brought back to the initial state 1 and the cycle is completed.

Consider unit mass of air undergoing the cyclic change :

Heat supplied during process 2–3,

$$Q_1 = c_v (T_3 - T_2)$$

Heat rejected during process 4–1,

$$Q_2 = c_p (T_4 - T_1)$$

work done = heat supplied – heat rejected

$$= c_v (T_3 - T_2) - c_p (T_4 - T_1)$$

Thermal efficiency,

$$\eta = \frac{\text{work done}}{\text{heat supplied}}$$

$$= \frac{c_v (T_3 - T_2) - c_p (T_4 - T_1)}{c_v (T_3 - T_2)}$$

$$= 1 - \gamma \frac{(T_4 - T_1)}{T_3 - T_2} \quad \text{...(12.21)}$$

The expression for thermal efficiency in terms of compression ratio r, explosion ratio α and adiabatic exponent γ can be set up by expressing all the temperatures in terms of lowest temperature T_1 in the cycle :

From isentropic compression process 1–2,

$$T_2 = T_1 \left(\frac{V_1}{V_2}\right)^{\gamma-1} = T_1 \, r^{\gamma-1}$$

From constant volume process 2–3,

$$T_3 = T_2 \frac{p_3}{p_2} = T_2 \, \alpha = \alpha \, T_1 \, r^{\gamma-1}$$

From isentropic expansion process 3–4

$$T_4 = T_3 \left(\frac{V_3}{V_4}\right)^{\gamma-1}$$

$$= T_3 \left(\frac{V_3}{V_1} \times \frac{V_1}{V_4}\right)^{\gamma-1}$$

$$= T_3 \left(\frac{V_2}{V_1} \times \frac{V_1}{V_4}\right)^{\gamma-1}$$

$$= \frac{T_3}{r^{\gamma-1}} \left(\frac{V_1}{V_4}\right)^{\gamma-1} = \alpha \, T_1 \left(\frac{V_1}{V_4}\right)^{\gamma-1}$$

From constant pressure process 4–1,

$$T_4 = \frac{V_4}{V_1} T_1$$

Equating the two expressions for T_4, we get

$$\alpha \, T_1 \left(\frac{V_1}{V_4}\right)^{\gamma-1} = \frac{V_4}{V_1} T_1 \quad \text{or} \quad \frac{V_4}{V_1} = \alpha^{1/\gamma}$$

$$\therefore \quad T_4 = \frac{V_4}{V_1} T_1 = \alpha^{1/\gamma} \, T_1$$

When these temperature values are substituted in expression 12.21, we get

$$\eta = 1 - \gamma \frac{(\alpha^{1/\gamma} T_1 - T_1)}{(\alpha \, T_1 \, r^{\gamma-1} - T_1 \, r^{\gamma-1})}$$

$$= 1 - \frac{\gamma}{r^{\gamma-1}}\left[\frac{\alpha^{1/\gamma}-1}{\alpha-1}\right] \quad ...(12.22)$$

which is the required expression for the thermal efficiency of air standard Atkinson cycle.

The operation of Atkinson cycle is achieved by a special linkage mechanism that provides alternately short and long strokes. The isentropic compression is achieved on a short stroke while the expansion is on a long stroke. The idea is to get more work than that given by Otto cycle. The area 1 - 4 - 5 represents this increased work. Further it is to be noted that heat rejection occurs at lower average temperature (T_5 being higher than T_1). These aspects make Atkinson cycle more efficient than Otto cycle. However, it has not been possible so far to make such engines commercially viable due to the complex drive mechanism and disproportionate increase in cylinder dimensions.

EXAMPLE 12.33.

In an ideal Atkinson cycle, air at 1 bar and 300 K is compressed adiabatically through a volume compression ratio of 6 : 1. The heat addition at constant volume is so regulated that a peak pressure of 18 bar is attained. The air then expands adiabatically to the original pressure and subsequently the cycle is completed by heat rejection at constant pressure. Determine (a) pressure and temperature at the cardinal points of the cycle, (b) work done per kg of air, and (c) cycle efficiency.

Solution : Refer Fig 12.36.

From isentropic compression process 1−2,

$$T_2 = T_1\left(\frac{V_1}{V_2}\right)^{\gamma-1}$$

$$= 300 \times (6)^{0.4} = 614.3 \text{ K}$$

$$p_2 = p_1\left(\frac{V_1}{V_2}\right)^{\gamma}$$

$$= 1 \times (6)^{1.4} = 12.29 \text{ bar}$$

$$p_1 = 1 \text{ bar and } T_1 = 300 \text{ K}$$

From constant volume process 2−3,

$$p_3 = 18 \text{ bar}$$

$$T_3 = \frac{p_3}{p_2}T_2 = \frac{18}{12.29}\times 614.3$$

$$= 899.7 \text{ K}$$

$$p_4 = p_1 = 1 \text{ bar}$$

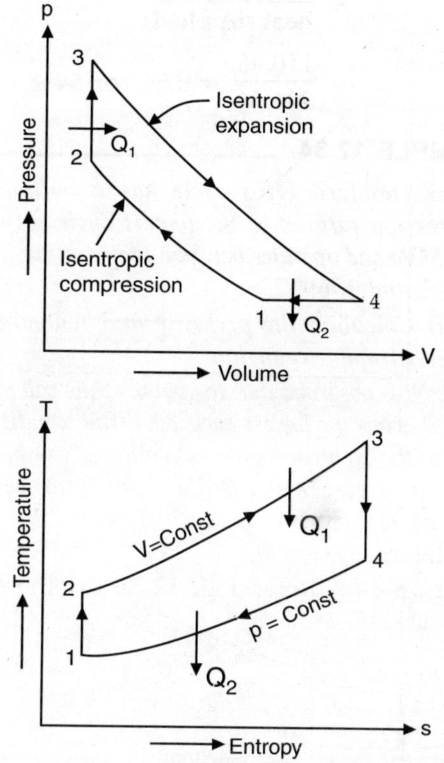

Fig. 12.36. Pressure-volume and temperature entropy diagrams for air standard Atkinson cycle

From isentropic expansion process 3−4,

$$T_4 = T_3\left(\frac{p_4}{p_3}\right)^{\frac{\gamma-1}{\gamma}} = T_3\left(\frac{p_1}{p_3}\right)^{\frac{\gamma-1}{\gamma}}$$

$$= 899.7\left(\frac{1}{18}\right)^{\frac{1.4-1}{1.4}} = 393.63 \text{ K}$$

Heat added,

$$Q_1 = c_v\,(T_3 - T_2)$$

$$= 0.718\,(899.2 - 614.3)$$

$$= 204.56 \text{ kJ/kg}$$

Heat rejected,

$$Q_2 = c_p\,(T_4 - T_1)$$

$$= 1.005\,(393.63 - 300)$$

$$= 94.1 \text{ kJ/kg}$$

∴ Work done = heat added − heat rejected

$$= 204.56 - 94.1 = 112.46 \text{ kJ/kg}$$

Thermal efficiency

$$= \frac{\text{work done}}{\text{heat supplied}}$$

$$= \frac{110.46}{204.56} = 0.54 \text{ or } \mathbf{54\%}$$

EXAMPLE 12.34.

An air standard Otto cycle has a volumetric compression ratio of 6, the lowest cycle pressure is 0.1 MPa and operates between temperature limits of 27°C and 1569°C.

(a) Calculate the pressure and temperature after isentropic expansion.

(b) It is observed that the values thus calculated are well above the lowest cycle operating conditions. As such the expansion process is allowed to continue down to pressure 0.1 MPa. Which process is required to complete the cycle ? Name the cycle so obtained.

Solution : (*a*) Refer Fig. 12.37 for the Otto cycle (1–2–3–4)

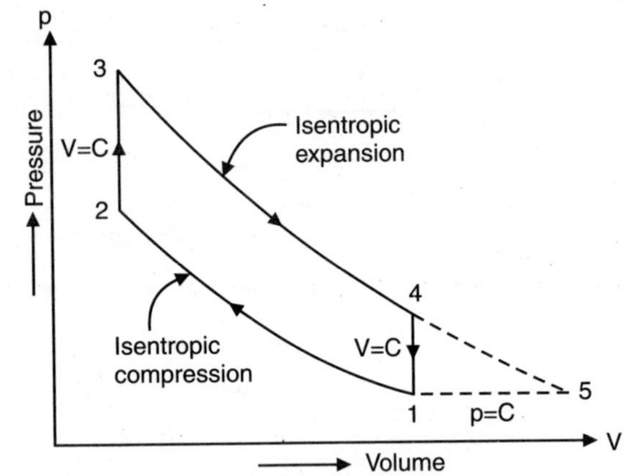

Fig. 12.37.

$$T_1 = 27 + 273 = 300 \text{ K}$$

$$T_3 = 1569 + 273 = 1842 \text{ K}$$

$$\frac{p_2}{p_1} = \left(\frac{V_1}{V_2}\right)^{\gamma}$$

$$\therefore \quad p_2 = 0.1 \times (6)^{1.4} = 1.23 \text{ MPa}$$

$$\frac{T_2}{T_1} = \left(\frac{V_1}{V_2}\right)^{\gamma-1}$$

$$\therefore \quad T_2 = 300 \times (6)^{0.4} = 614.3 \text{ K}$$

$$p_3 = p_2 \times \frac{T_3}{T_2}$$

$$= 1.23 \times \frac{1842}{614.3} = 3.69 \text{ MPa}$$

$$\frac{p_4}{p_3} = \left(\frac{V_3}{V_4}\right)^{\gamma}$$

$$\therefore \quad p_4 = 3.69\left(\frac{1}{6}\right)^{1.4} = 0.3 \text{ MPa}$$

$$T_4 = \frac{p_4}{p_1} \times T_1 = \frac{0.3}{0.1} \times 300 = 900 \text{ K}$$

$$\eta_{Otto} = 1 - \frac{1}{r^{\gamma-1}} = 1 - \frac{1}{(6)^{0.4}}$$

$$= 1 - 0.488 = 0.512 \text{ or } 51.2 \text{ \%}$$

(*b*) Isentropic expansion of the air continues upto the lowest pressure of the cycle and thereafter, there is preliminary compression of air at constant pressure. This completes the cycle (Fig. 12.37) and the cycle (1–2–3–5) so obtained is called the Atkinsion cycle.

$$\eta_{Atkinson} = 1 - \gamma \frac{(T_5 - T_1)}{(T_3 - T_2)}$$

Here $$\frac{T_5}{T_3} = \left(\frac{p_5}{p_3}\right)^{\gamma-1/\gamma}$$

$$\therefore \quad T_5 = 1842 \times \left(\frac{0.1}{3.69}\right)^{\frac{0.4}{1.4}} = 657 \text{ K}$$

$$\therefore \eta_{Atkinson} = 1 - 1.4 \frac{(657 - 300)}{(1842 - 614.3)}$$

$$= 0.593 \text{ or } 59.3\%$$

Improvement in cycle efficiency

$$= 59.3 - 51.2 = 8.1\%$$

12.10. STIRLING AND ERICSSON CYCLES

Both of these cycles are totally reversible and have the same efficiency as that of Carnot cycle when operating between the same temperature limits.

$$\eta_{Carnot} = \eta_{Stirling} = \eta_{Ericsson} = \frac{T_h - T_l}{T_h}$$

where T_h and T_l are the highest and lowest operating temperatures.

(a) Stirling cycle : Fig. 12.38 shows this thermodynamic cycle plotted on $p-V$ and $T-s$ diagrams. The cycle consists of two isothermals $1-2$ and $3-4$ connected by constant volume processes.

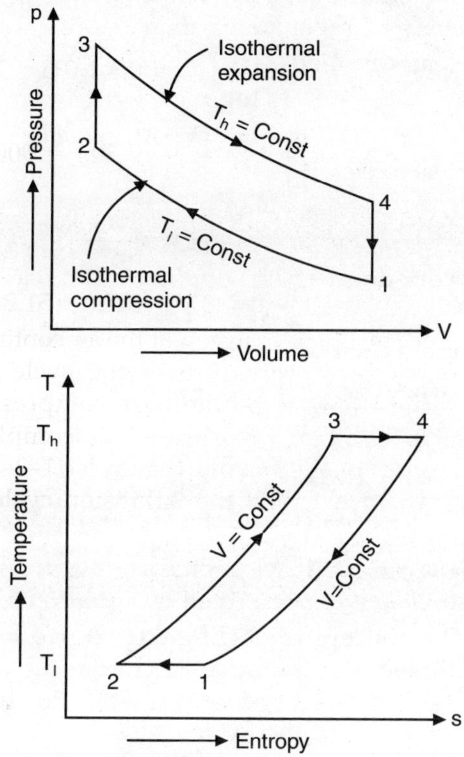

Fig. 12.38. Stirling cycle on p-V and T-s diagrams

The cycle employs the principle of regeneration; a process during which the working fluid rejects heat to a thermal storage device (called regenerator) during a constant volume heat rejection process. This heat is stored in the regenerator and is later supplied back to the working fluid during heat addition process at constant volume.

The air in the lower space of the regenerator is kept at high temperature T_h by the fuel burnt in the furnace under it. The lower temperature T_l of the air is kept constant in the upper space of the regenerator by circulating cold water. The wire gauge in the regenerator ensures the uniform temperature gradient from temperature T_h to T_l. The wire matrix is of such a capacity that during heat addition or heat rejection process, its temperature changes by an infinitesimal amount. The air is alternately passed through these two heat reservoirs at temperatures T_h and T_l.

The sequence of operation is :

Process 1 – 2 : Piston moves from crank end dead centre (ODC) to cover end dead centre (IDC) and air with initial state 1 is compressed isothermally to state point 2 through compression ratio, $r = V_1/V_2$. The cylinder head is kept in contact with heat reservoir at temperature T_l.

Process 2 – 3 : At state point 2, the air at temperature $T_1 = T_2 = T_l$ is made to enter the regenerator from top. It flows down the wire mesh and gets heated to temperature $T_3 = T_h$ at constant volume.

Process 3 – 4 : The hot air is admitted to the engine cylinder from the bottom portion of the regenerator. The air exerts a greater amount of force on the piston and pushes it towards ODC. Expansion of working fluid (air) takes place isothermally and work is produced. During expansion process, the cylinder head is kept in contact with the heat source at T_h

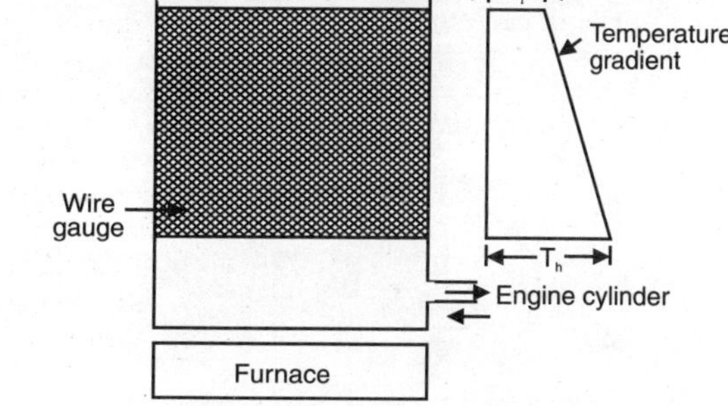

Fig. 12.39. Regenerator

Process 4—1 : At state point 4, the air at temperature $T_4 = T_h$ is made to enter the regenerator from the lower. It flows upwards and transfers its heat to the wire mesh and thus gets cooled at constant volume. The process is so controlled that ultimately the air comes to its initial state 1 and the cycle is completed.

If the regenerator is 100% efficient, then heat supplied to air during process 2—3 equals the heat rejected during process 4—1. The system exchanges a net amount of heat with energy reservoirs only during the isothermals. The heat is added during process 3—4 at temperature T_h, and heat is rejected during process 1—2 at temperature T_l. During any isothermal process, the change in internal energy is zero (temperature being constant) and then in accordance with law of energy conservation, the heat exchange equals the work transfer. Therefore, for a unit mass of air,

Heat supplied during process 3—4,

$$Q_{3-4} = w_{3-4}$$

$$= R\,T_h \log_e \frac{V_4}{V_3}$$

$$= R\,T_h \log_e \frac{V_1}{V_2} = R\,T_h \log_e r$$

Heat rejected during process 1—2,

$$Q_{1-2} = w_{1-2}$$

$$= R\,T_l \log_e \frac{V_1}{V_2} = R\,T_l \log_e r$$

Work done = heat supplied – heat rejected

$$= R\,T_h \log_e r - R\,T_l \log_e r$$

$$= R \log_e r\,(T_h - T_l)$$

Thermal efficiency,

$$\eta_h = \frac{\text{work done}}{\text{heat supplied}}$$

$$= \frac{R \log_e r\,(T_h - T_l)}{R \log_e r\, T_h}$$

$$= \frac{T_h - T_l}{T_h} \qquad \text{...(12.23)}$$

Thus the efficiency of Stirling cycle is equal to that of Carnot cycle operating between the same temperature limits.

In the above treatment, it has been presumed that the regenerator is 100 % efficient which implies that all heat stored in the regenerator during heat rejection process is available to air during the process of heat addition. But in actual practice, there is a loss of heat varying from 10 to 20% due to radiation and poor conductivity of air. If η_{reg} is the efficiency of regenerator then

Heat supplied

$$= R\,T_h \log_e r + (1 - \eta_{reg})\, c_v\,(T_h - T_l)$$

Heat rejected

$$= R\,T_l \log_e r + (1 - \eta_{reg})\, c_v\,(T_h - T_l)$$

Work done = heat supplied – heat rejected

$$= R\,(T_h - T_l) \log_e r$$

Thermal efficiency,

$$\eta = \frac{\text{work done}}{\text{heat supplied}}$$

$$= \frac{R\,(T_h - T_l)\log_e r}{R\,T_h \log_e r + (1 - \eta_{reg})\,c_v\,(T_h - T_l)} \qquad \text{...(12.24)}$$

If $\eta_{reg} = 100\%$, we get the same expression for efficiency as prescribed by equation 12.21

The concept of Stirling engine had to be abandoned due to large heat transfer area needed for the regenerator and the high temperatures of the cycle.

(b) Ericsson cycle : Fig. 12.40 shows this thermodynamic cycle plotted on $p - V$ and $T - s$ diagram. The cycle consists of two isothermals 1—2 and 3—4 connected by constant pressure process.

The cycle employs the principle of regeneration. The energy storage device (matrix) absorbs heat from the system (air) during one part of cycle (constant pressure process 4—1) and returns the same amount of heat during another part (constant pressure process 2—3) of the cycle. With the incorporation of 100% efficient regenerator, the Ericsson cycle too becomes thermodynamically reversible and so the efficiency will be the same as that for any other reversible cycle operating between the same two heat reservoirs.

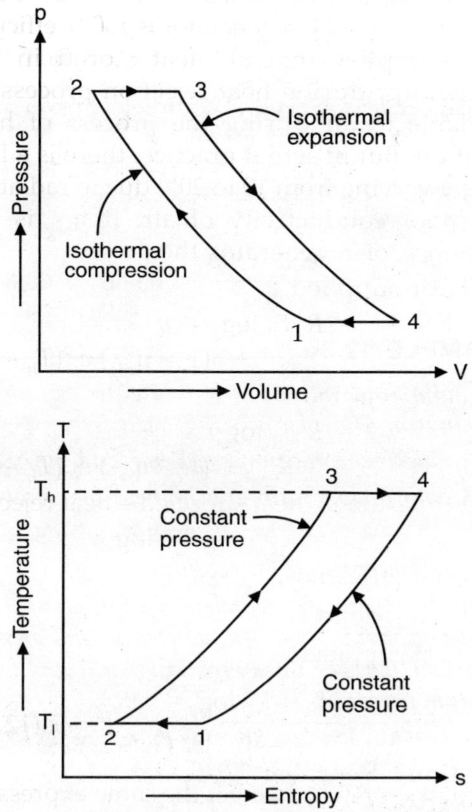

Fig. 12.40. Ericsson cycle on p-V and T-s diagrams

For a unit mass of air,

Heat supplied during process 3–4,

$$Q_{3-4} = w_{3-4}$$

$$= RT_h \log_e \frac{V_4}{V_3}$$

$$= RT_h \log_e \frac{p_3}{p_4} = RT_h \log_e r_p$$

Heat rejected during process 1–2

$$Q_{1-2} = w_{1-2} = RT_l \log_e \frac{V_1}{V_2}$$

$$= RT_l \log_e \frac{p_2}{p_1} = RT_l \log_e r_p$$

In the above expressions $r_p = \dfrac{p_2}{p_1} = \dfrac{p_3}{p_4}$ represents the pressure ratio for the isothermal compression and expansion processes.

Work done = heat added – heat rejected

$$= R\, T_h \log_e r_p - R\, T_l \log_e r_p$$

$$= R\,(T_h - T_l) \log_e r_p$$

Thermal efficiency,

$$\eta = \frac{\text{work done}}{\text{heat supplied}}$$

$$= \frac{R\,(T_h - T_l)\log_e r_p}{R\, T_h \log_e r_p}$$

$$= \frac{T_h - T_l}{T_h} \qquad ...(12.25)$$

which is the same as that for a Carnot cycle and a Stirling cycle operating between the maximum and minimum temperature limits of T_h and T_l respectively.

EXAMPLE 12.35.

Consider an air standard Stirling cycle equipped with a 100% efficient regenerator system. The isothermal compression commences from 100 kPa and 300 K, and subsequent heat addition at constant volume raises the pressure and temperature to 15 bar and 900 K respectively. The cycle is finally completed through an isothermal expansion and constant volume heat rejection. Analyse each of the four processes for work and heat transfer and determine the engine efficiency.

Solution : Refer Fig. 12.41.

$$p_1 = 100 \text{ kPa} = 100 \times 10^3 \text{ N/m}^2$$

$$v_1 = \frac{R\,T_1}{p_1} = \frac{287 \times 300}{100 \times 10^3}$$

$$= 0.861 \text{ m}^3/\text{kg}$$

$$v_2 = v_3 = \frac{RT_3}{p_3} = \frac{287 \times 900}{15 \times 10^5}$$

$$= 0.1722 \text{ m}^3/\text{kg}$$

∴ compression ratio,

$$r = \frac{v_1}{v_2} = \frac{0.861}{0.1722} = 5$$

Process 1–2 :

$$T_1 = T_2 = 300 \text{ K} = T_l$$

Heat rejected,

$$Q_{1-2} = W_{1-2} = R\, T_l \log_e r_p$$

$$= 287 \times 300 \log_e 5$$

$$= 138573 \text{ J/kg} = 138.57 \text{ kJ/kg}$$

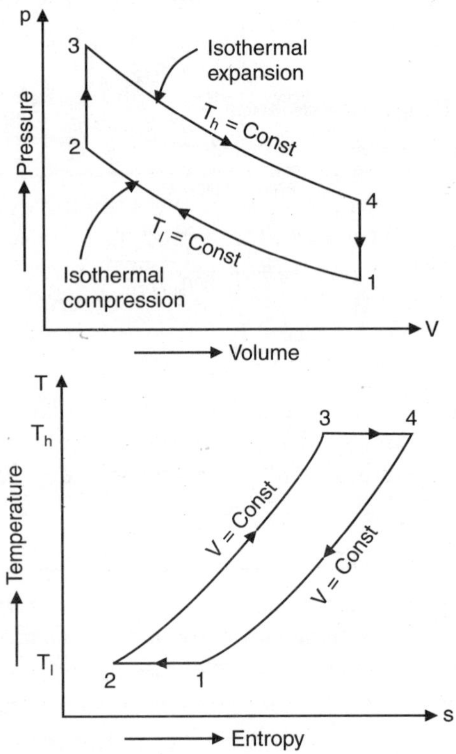

Fig. 12.41. Stirling cycle on p-V and T-s diagrams

Process 2—3 : V = constant and therefore $W_{2\text{-}3} = 0$

$$Q_{2\text{-}3} = c_v\,(T_3 - T_2)$$
$$= 0.718\ (900 - 300)$$
$$= 430.8 \text{ kJ/kg}$$

Process 3—4 :

$$T_3 = T_4 = 900 \text{ K} = T_h$$

Heat supplied,

$$Q_{3\text{-}4} = W_{3\text{-}4}$$
$$= R\,T_h \log_e r = 287 \times 900 \log_e 5$$
$$= 415718 \text{ J/kg} = 415.72 \text{ kJ/kg}$$

Process 4—1 : V = constant and therefore $W_{4\text{-}1} = 0$

$$Q_{4\text{-}1} = c_v\,(T_4 - T_1)$$

Heat supplied to air during process 2—3 equals the heat rejected during process 4—1, and therefore

work done = net heat exchange during isothermals

$$= 415.72 - 138.57$$
$$= 276.43 \text{ kJ/kg}$$

∴ Thermal efficiency,

$$\eta = \frac{276.43}{415.72} = 0.665 \quad \text{or } \mathbf{66.5\ \%}$$

Since Stirling cycle is completely reversible, its efficiency is also given by

$$\eta = \frac{T_h - T_l}{T_h}$$
$$= \frac{900 - 300}{900} = 0.666 \text{ or } \mathbf{66.6\ \%}$$

EXAMPLE 12.36.

The following particulars relate to an engine working on 4-stroke Stirling cycle :

Induction conditions = 1 bar and 300 K

Compression ratio and highest temperature = 2.5 and 850 K

Engine speed = 60 rev/min

If the regenerator efficiency is 90 % and 2000 kJ/min of heat is supplied, determine the thermal efficiency, mass of air per working cycle and mean effective pressure.

Take c_v = 0.718 kJ/kg K and R = 0.287 kJ/kg K

Solution : Refer Fig. 12.41.

$$\text{Work done} = R\,(T_h - T_l) \log_e r$$
$$= 0.287\ (850 - 300) \log_e 2.5$$
$$= 144.64 \text{ kJ/kg}$$

$$\text{Heat added} = R\,T_h \log_e r + (1 - \eta_{reg})\,c_v\,(T_h - T_l)$$
$$= 0.287 \times 850 \times \log_e 2.5 + (1 - 0.9) \times 0.718\ (850 - 300)$$
$$= 223.53 + 39.49 = 263.02 \text{ kJ/kg}$$

∴ Thermal efficiency,

$$\eta = \frac{\text{work done}}{\text{heat supplied}}$$
$$= \frac{144.64}{263.02} = 0.55 \text{ or } \mathbf{55\%}$$

(*b*) Heat supplied = 2000 kJ/min (given)
= 263.02 kJ/kg (calculated above)

$$\text{Mass of air} = \frac{2000}{263.02} = 7.60 \text{ kg/min}$$

For a 4-stroke engine, the cycle is completed in two strokes and accordingly the number of effective working cycles equal half the number of revolutions.

Number of working cycles per minute

$$= \frac{60}{2} = 30$$

∴ Mass of air per working cycle

$$= \frac{7.60}{30} = \mathbf{0.253\ kg}$$

(c) From characteristic gas equation,

$$v_1 = \frac{RT_1}{p_1} = \frac{287 \times 300}{1 \times 10^5}$$

$$= 0.861\ \text{m}^3/\text{kg}$$

$$v_2 = \frac{v_1}{2.5} = \frac{0.861}{2.5} = 0.3444\ \text{m}^3/\text{kg}$$

Swept volume, $v_1 - v_2$

$$= 0.861 - 0.3444 = 0.5166\ \text{m}^3/\text{kg}$$

$$\text{Mean effective pressure} = \frac{\text{work done}}{\text{heat supplied}}$$

$$= \frac{144.64}{0.5166}$$

$$= 279.98\ \text{kN/m}^2 = \mathbf{2.80\ bar}$$

EXAMPLE 12.37.

Consider an air standard Ericsson cycle provided with 100 % efficient regenerator. The isothermal compression commences from 100 kPa, heat supply and heat rejection in the cycle occur at 1500 K and 300 K respectively. If heat supply is 1000 kJ/kg, find the compression work, expansion work and thermal efficiency of the cycle.

Solution : Refer Fig. 12.40.

Since the regenerator is 100% efficient, heat supplied to air during constant pressure process 2–3 equals the heat rejected during constant pressure process 4–1. The cycle is then completely reversible and its efficiency equals the Carnot efficiency for the same temperature limits.

$$\eta_{Ericsson} = \eta_{Carnot} = \frac{T_h - T_l}{T_h}$$

$$= \frac{1500 - 300}{1500} = 0.8 \text{ or } 80\ \%$$

$W_{net} = \eta \times$ heat supplied at high temperature

$$= 0.8 \times 1000 = 800\ \text{kJ/kg}$$

Expansion work,

$$W_{3\text{-}4} = Q_{3\text{-}4} = \mathbf{1000\ kJ/kg}$$

Compression work,

$$W_{1\text{-}2} = W_{3\text{-}4} - W_{net}$$

$$= 1000 - 800 = \mathbf{200\ kJ/kg}$$

12.11. BRAYTON OR JOULE CYCLE

The Brayton cycle or Joule cycle is a theoretical cycle on which run the constant pressure gas turbines. A gas turbine is akin to petrol and diesel engines in its working medium and internal combustion. However, the processes of compression, combustion and expansion take place in different components in a gas turbine plant. Further, the gas turbine is a rotary machine and operates under steady and continuous supply of working medium.

For the analysis of Brayton cycle, refer to the chapter on Gas turbines and jet propulsion.

12.12. LENOIR CYCLE

The Lenoir cycle is the idealised cycle for the pulse jet engine. An ideal gas undergoes

(*i*) constant volume heating, the pressure rising from p_1 to p_2

(*ii*) reversible adiabatic expansion from pressure p_2 back to pressure p_1 and

(*iii*) isobaric compression to the volume at the start of the cycle.

Fig. 12.42. shows the Lenoir cycle on $p - V$ and $T - s$ co-ordinates.

The expansion process is isentropic and hence involves no heat interaction. Energy is absorbed as heat during process 1–2 at constant volume, and is rejected as heat during process 3–1 at constant pressure.

Consider unit mass of air undergoing the cyclic change :

Heat supplied during process 1–2

$$Q_1 = c_v (T_2 - T_1)$$

Heat rejected during process 3–1

$$Q_2 = c_p (T_3 - T_1)$$

Work done = heat supplied – heat rejected

$$= c_v (T_2 - T_1) - c_p (T_3 - T_1)$$

Thermal efficiency,

$$\eta = \frac{\text{work done}}{\text{heat supplied}}$$

$$= \frac{c_v (T_2 - T_1) - c_p (T_3 - T_1)}{c_v (T_2 - T_1)}$$

$$= 1 - \gamma \frac{(T_3 - T_1)}{(T_2 - T_1)} \quad ...(12.26)$$

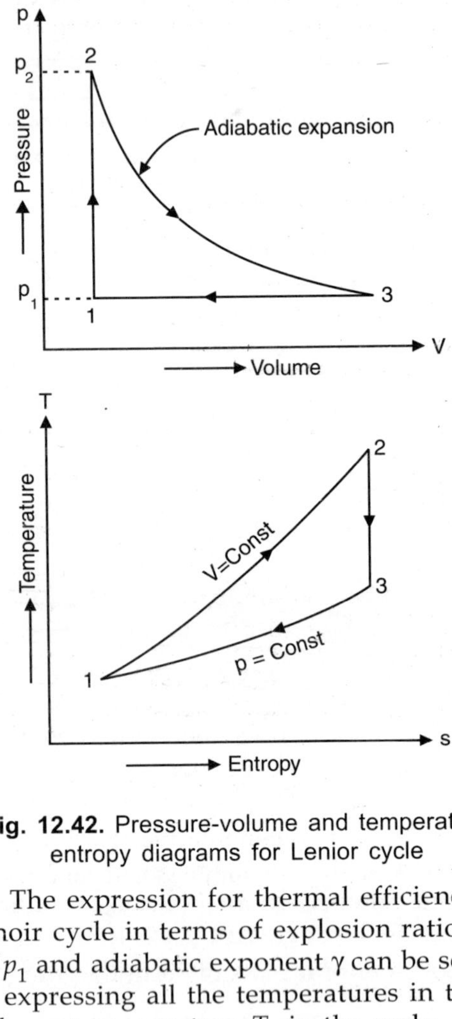

Fig. 12.42. Pressure-volume and temperature entropy diagrams for Lenior cycle

The expression for thermal efficiency of Lenoir cycle in terms of explosion ratio $\alpha = p_2/p_1$ and adiabatic exponent γ can be set up by expressing all the temperatures in terms of lowest temperature T_1 in the cycle.

From constant volume heating process 1-2,

$$T_2 = T_1 \times \frac{p_2}{p_1} = \alpha\, T_1$$

From isentropic expansion process 2–3,

$$\frac{T_3}{T_2} = \left(\frac{p_3}{p_2}\right)^{\frac{\gamma-1}{\gamma}} = \left(\frac{p_1}{p_2}\right)^{\frac{\gamma-1}{\gamma}} = \left(\frac{1}{\alpha}\right)^{\frac{\gamma-1}{\gamma}}$$

$$T_3 = T_2 \left(\frac{1}{\alpha}\right)^{\frac{\gamma-1}{\gamma}}$$

$$= \alpha T_1 \left(\frac{1}{\alpha}\right)^{\frac{\gamma-1}{\gamma}} = T_1\, \alpha^{1/\gamma}$$

When these temperature values are substituted in expression 12.28, we get

$$\eta = 1 - \gamma \left[\frac{T_1\, \alpha^{1/\gamma} - T_1}{\alpha\, T_1 - T_1}\right]$$

$$= 1 - \gamma \left[\frac{\alpha^{1/\gamma} - 1}{\alpha - 1}\right] \quad ...(12.27)$$

EXAMPLE 12.38.

At the start of energy addition in a simple Lenoir cycle, the pressure and temperature of air are 1 bar and 300 K respectively. The energy added is 1250 kJ/kg at constant volume after which isentropic expansion is followed by isobaric energy rejection and the initial state is restored. Determine the maximum cycle temperature and pressure, the cycle efficiency and mean effective pressure.

Solution : Refer Fig. 12.42.

(*a*) From constant volume heat addition process,

$$Q_1 = c_v\,(T_2 - T_1)$$

$$\text{or} \quad 1250 = 0.718\,(T_2 - 300)$$

$$\therefore \quad T_2 = \frac{1250}{0.718} + 300 = 2040.95 \text{ K}$$

$$p_2 = p_1 \frac{T_2}{T_1} = 1 \times \frac{2040.95}{300} = 6.80 \text{ bar}$$

From isentropic expansion process,

$$T_3 = T_2 \left(\frac{p_3}{p_2}\right)^{\frac{\gamma-1}{\gamma}} = T_2 \left(\frac{p_1}{p_2}\right)^{\frac{\gamma-1}{\gamma}}$$

$$= 2040.95 \left(\frac{1}{6.80}\right)^{\frac{0.4}{1.4}} = 1179.6 \text{ K}$$

From constant pressure heat rejection process,

$$Q_2 = c_p\,(T_3 - T_1)$$

$$= 1.005\,(1179.6 - 300)$$

$$= 884 \text{ kJ/kg}$$

(*b*) Work done = heat added – heat rejected

$$= 1250 - 884 = 366 \text{ kJ/kg}$$

Thermal efficiency,

$$\eta = \frac{\text{work done}}{\text{heat supplied}}$$

$$= \frac{366}{1250} = 0.293 = \mathbf{29.3\ \%}$$

(c) From characteristic gas equation,

$$v_1 = \frac{R\,T_1}{p_1} = \frac{287 \times 300}{1 \times 10^5}$$

$$= 0.861\ \text{m}^3/\text{kg}$$

From the constant pressure process 3 – 1,

$$v_3 = v_1 \frac{T_3}{T_1} = \frac{0.861 \times 1179.6}{300}$$

$$= 3.385\ \text{m}^3/\text{kg}$$

Swept volume $= v_3 - v_1$

$$= 3.385 - 0.861 = 2.524\ \text{m}^3/\text{kg}$$

$$mep = \frac{\text{work output}}{\text{swept volume}} = \frac{366}{2.524}$$

$$= 145\ \text{kN/m}^2 = \mathbf{1.45\ bar}$$

SALIENT POINTS

1. A heat engine is device which converts the chemical energy of the fuel into mechanical work.
2. A cycle is defined as a repeated series of operations occuring in a certain prescribed order.
3. The efficiency of an engine using air as a working medium is called the air standard efficiency.
4. Relative efficiency represents the ratio of actual thermal efficiency to air standard efficiency.
5. **For any cycle :** thermal efficiency

$$= \frac{\text{work done}}{\text{heat supplied}}$$

$$= \frac{\text{heat supplied} - \text{heat rejected}}{\text{heat supplied}}$$

6. **In Otto cycle** both heat addition and rejection are at constant volume, and compression and expansion are adiabatics. It is the theoretical cycle on which the petrol and gas engines work.

   In diesel cycle heat is supplied at constant pressure, heat rejection is at constant volume, compression and expansion are adiabatics. It is the basic cycle on which run the oil engines.

   The dual (mixed or composite) cycle is combination of Otto and diesel cycles. The heat addition is partly at constant volume and partly at constant pressure, heat rejection is at constant volume, compression and expansion are adiabatics.
7. Air standard efficiency for :

   Otto cycle : $\eta = 1 - \dfrac{1}{r^{\gamma-1}}$

   Diesel cycle: $\eta = 1 - \dfrac{1}{r^{\gamma-1}}\left[\dfrac{\rho^{\gamma}-1}{\gamma(\rho-1)}\right]$

   Dual cycle :

$$\eta = 1 - \frac{1}{r^{\gamma-1}}\left[\frac{\alpha\rho^{\gamma}-1}{(\alpha-1)+\gamma(\rho-1)\alpha}\right]$$

8. Mean effective pressure is that constant pressure which acting on piston through one stroke would do the same amount of work as it done by the varying pressure during the cycle

$$mep = \frac{\text{work done}}{\text{swept volume}}$$

9. For the same heat supplied and compression ratio

   $\eta_{otto} > \eta_{dual} > \eta_{diesel}$

   For the same maximum pressure and temperature

   $\eta_{diesel} > \eta_{dual} > \eta_{otto}$

   For the same maximum pressure and heat supplied

   $\eta_{diesel} > \eta_{dual} > \eta_{otto}$

REVIEW QUESTIONS

A. Conceptual and conventional questions.

1. How are the combustion and exhaust processes modelled in air standard cycles ?
2. What is the significance of an air-standard cycle?

3. Define the following terms in relation to reciprocating engines : dead centres; stroke; piston speed and compression ratio.
4. How is the mean effective pressure for a reciprocating engine defined ?
5. Name the four processes that make up the ideal Otto cycle/Diesel cycle.
6. Define compression ratio. How does it affect the air standard efficiency of an Otto cycle ?
7. What is cut off ratio ? How does it affect the air standard efficiency of a Diesel cycle?
8. List the assumptions made in the analysis of air standard cycles.
9. How do the thermal efficiencies of Otto and Diesel cycle engines compare in practice.
10. Draw a curve showing the variation of thermal efficiency with compression ratio. What limits the compression ratio for an engine working on Otto cycle ?
11. Derive an expression for the air standard efficiency and mean effective pressure of an Otto cycle. State the assumptions made.
12. Derive an expression for the air standard efficiency of the Diesel cycle in terms of the compression ratio, cut off ratio and the adiabatic index. Proceed to show that the efficiency of this cycle is less than that of the Otto cycle for the same compression ratio. Why in actual practice is a diesel engine preferred for some applications ?
13. Set up the following expression for the air standard efficiency of an engine working on dual combustion cycle

$$\eta = 1 - \frac{1}{r^{\gamma-1}}\left[\frac{\alpha\,\rho^{\gamma} - 1}{(\alpha - 1) + \gamma\,(\rho - 1)\alpha}\right]$$

where the symbols have their usual meanings. Deduce therefrom the expressions for the air standard efficiency of an Otto cycle engine and a Diesel cycle engine.
14. For the same compression ratio and heat input, which cycle is more efficient : Otto, Diesel or Dual? Explain with $p-V$ and $T-s$ diagrams
15. With the help of $p-V$ and $T-s$ diagrams, show that for the same maximum pressure and heat input,

$$\eta_{Diesel} > \eta_{Dual} > \eta_{Otto}$$

16. What are the four processes which constitute a Stirling cycle ? Show that the efficiency of an air standard Stirling cycle is identical with that of a Carnot cycle working between the same temperature limits.
17. What are the four processes which constitute an Ericsson cycle ? Show that the efficiency of an air standard Ericsson cycle is identical with that of a Carnot cycle working between the same temperature limits.
18. An engine is 75 mm bore and stroke and has a compression ratio of 5.6 to 1. What amount of metal should be machined off from the cylinder head joint face if the compression ratio is to be raised to 6 ?

[Ans. 1.30 mm**]**

19. An engine working on Otto cycle has temperatures of 300 K and 600 K at the beginning and end of compression stroke. Determine the compression ratio and air standard efficiency.

Take $\gamma = 1.4$ for the working fluid.

[Ans. 5.66; 50%**]**

20. An ideal engine operating on Otto cycle sucks air at 1.01325 bar and 288 K. The compression ratio is limited to 5 and addition of heat at constant volume amounts to 2600 kJ per kg of air induction. Determine the peak pressure of the cycle, air standard efficiency and mean effective pressure.

Take $\gamma = 1.4$ and $c_v = 0.718$ kJ/kg K.

[Ans. 7.31 MN/m^2, 47.3%, 1.91 MN/m^2**]**

21. Set up the following expression for change in efficiency of Otto cycle with variable specific heat

$$\frac{d\eta}{\eta} = \frac{(1-\eta)(\gamma-1)\log_e r}{\eta}\frac{dc_v}{c_v}$$

Proceed to determine the percentage change in efficiency if compression ratio is 7 and the specific heat at constant volume increases by 2 percent.

[Ans. -1.322%**]**

22. A gas engine working on Otto cycle operates with the following parameters :

Inlet conditions : 1 bar pressure and 320 K temperature

Compression ratio 4 and pressure ratio 4. If the working fluid is air with $R = 287$ J/kg K and $\gamma = 1.4$, make calculations for the useful work done, thermal efficiency and mean effective pressure.

[Ans. 545 kJ/kg, 0.426, 7.85 bar**]**

23. 2 kg of air at pressure 1 bar and temperature 300 K is taken through an Otto cycle of compression ratio 8 : 1 involving 5000 kJ of heat addition. Taking $\gamma = 1.4$ and $R = 287$

kJ/kg K, determine the thermal efficiency, work done and mean effective pressure.

[**Ans.** 0.565, 2825 kJ, 1875 kN/m^2]

24. The following information is available about an engine working on Otto cycle :

maximum cycle temperature = 1550 K
exhaust temperature = 720 K
ambient conditions = 1 bar, 310 K.

If the air consumption rate is 2 kg/min, determine the compression ratio, air standard efficiency and the power output.

[**Ans.** 6.8, 53.54%, 11.31 kW]

25. Consider an air standard Otto cycle having a temperature of 300 K at the beginning of compression process and a maximum temperature of 1800 K during the cycle. What compression ratio is required to develop maximum work ? Determine the correspon-ding cycle thermal efficiency and the maximum net work per cycle.

[**Ans.** 9.39, 59.9%, 452.56 kJ/kg]

26. The pressure and temperature at the beginning of compression in an ideal constant volume air standard cycle (Otto cycle) are 1 bar, 40°C and the compression ratio is 9. The heat transferred to air during the constant volume heating process is 2000 kJ/kg. Assuming that compression and expansion are isentropic, make calculations for the cycle efficiency, the maximum pressure and temperature attained during the cycle, and the mean effective pressure.

[**Ans.** 58.5%, 101.8 bar, 3540 K, 14.65 bar]

27. In an air standard Diesel cycle, the pressure and temperature at the start of compression are 1 bar, 330 K and the compression ratio is 16. The heat transferred to air during fuel injection at constant pressure is 1250 kJ/kg. Determine the cycle efficiency, the maximum pressure and temperature attained during the cycle, and the mean effective pressure.

[**Ans.** 60.2%, 48.5 bar, 2245 K, 8.47 bar]

28. An engine working on Diesel cycle has air intake conditions of 1 bar and 300 K, and it operates through compression ratio 16. After the fuel has burnt, the temperature reached is 2000 K. Make calculations for the air standard efficiency, and the net work output per cycle.

[**Ans.** 60.41%, 662 kJ/kg]

29. In a reciprocating engine operating on Diesel cycle, the compression ratio is 20, the cut off ratio is 2, and the pressure and temperature at the beginning of compression are 1 bar and 350 K respectively. If the working medium is air with γ = 1.4 and R = 287 J/kg K., determine the useful work done, thermal efficiency and mean effective pressure.

[**Ans.** 748 kJ/kg, 64.8%, 7.88 bar]

30. An engine working on Diesel cycle has a compression ratio of 14 and an expansion ratio of 7. If the air at suction has a temperature of 300 K, calculate the efficiency and work output.

[**Ans.** 59.25%, 513.3 kJ/kg]

31. The following information is available about an engine working on Diesel cycle :

Maximum cycle temperature = 1890 K
Heat supplied = 860 kJ/kg
Ambient conditions = 1 bar, 300 K

Determine the compression ratio, cut off ratio, maximum pressure and air standard efficiency.

[**Ans.** 22, 1.83, 76 bar, 66.82%]

32. The pressure and temperature at the beginning of compression of an engine working on dual cycle are 1 bar, 27°C and the compression ratio is 14. The working fluid receives equal amounts of heat during both the constant volume and constant pressure processes, and the maximum cycle temperature is limited to 1850°C. Make calculations for the explosion ratio, cut off ratio and air standard efficiency. Assume c_p = 1.005 kJ/kg K and c_v = 0.718 kJ/kg K.

[**Ans.** 1.853, 1.329, 64.12%]

33. In an engine working on dual combustion cycle, the compression ratio is 16 and the induction conditions correspond to 1 bar pressure and 50 °C temperature. The maximum pressure in the cycle is 65 bar, and equal amounts of heat are added at constant volume and constant pressure processes. Estimate the air standard efficiency and mean effective pressure.

Assume c_p = 1.005 kJ/kg and c_v = 0.718 kJ/kg K

[**Ans.** 66.3%, 3.65 bar]

34. In an air standard dual combustion cycle, the pressure and temperature at the start of compression are 1 bar and 330 K respectively. The energy added in the cycle is 1250 kJ/kg, two thirds of this being added at constant volume and the remaining at constant pressure. If the engine operates through compression ratio 16, make calculations for the maximum pressure and temperature in the cycle, the

air standard efficiency and the mean effective pressure.

[Ans. 104.8 bar, 2576 K, 66.5%, 9.38 bar**]**

35. A hot air engine works on Stirling cycle. The pressure and temperature limits on the working cycle are :

1 bar and 10 bar ; 330 K and 660 K

Determine the thermal efficiency and mean effective pressure.

[Ans. 50%, 2 bar**]**

36. Consider a regenerative air engine working on Stirling cycle between temperature limits of 15°C and 400°C. If the ratio of isothermal expansion is 3, make calculations for the ideal efficiency when (*a*) the engine is fitted with a perfect regenerator (*b*) the efficiency of the regenerator is 0.8.

[Ans. 57.2%, 45.5 %**]**

37. An engine working on Atkinson cycle takes air at 1 bar and 300 K, and compresses it isentropically through compression ratio 6. Then it receives heat at constant volume till the pressure rises to 20 bar. The cycle is completed by expanding air isentropically to intake pressure and cooling at constant pressure. Find the efficiency of the cycle. How this efficiency compares with the Carnot efficiency work under the same conditions?

[Ans. 54.6%, 69.98 %**]**

B. Fill in the blanks with appropriate word/words.

1. A thermodynamic cycle working with air as the working substance is called .... cycle.
2. The concept of .... cycle helps to evaluate one engine relative to antoher.
3. The volume of working fluid in the cylinder, when the piston is at the bottom dead centre, is referred to as ....
4. The distance travelled by the piston of an IC engine in one minute is called ....
5. A .... cycle permits the designer to arrive at maximum efficiency that can be obtained from any heat engine under standard conditions.
6. The thermal efficiency of an Otto cycle depends upon .... and ....
7. The efficiency of a diesel engine cycle approaches that of Otto engine cycle when the cut off ratio tends to ....
8. For the same compression ratio, the efficiency of dual cycle is .... than that of diesel cycle and .... than that of Otto cycle.
9. The ideal constant pressure gas turbine works on .... cycle.

Answers

1. air standard 2. air standard 3. cylinder volume 4. piston speed 5. Carnot cycle 6. compression 7. unity 8. greater, less 9. Brayton cycle

C. Multiple choice questions.

1. For a particular diesel engine cycle, the compression ratio and expansion ratio are respectively stated to be 21 : 1 and 10.5 : 1. The corresponding cut off ratio will be
   (*a*) 0.5 (*b*) 1.5 (*c*) 2 (*d*) 4
2. The ideal constant pressure gas turbine works on
   (*a*) Atkinson cycle (*b*) Brayton cycle
   (*c*) Ericsson cycle (*d*) Lenoir cycle
3. The concept of which cycle had to be abandoned due to complex drive mechanism and disproportional increase in cylinder dimensions?
   (*a*) Atkinson cycle
   (*b*) Joule cycle
   (*c*) Stirling cycle
   (*d*) Mixed or composite cycle
4. Which one of the following cycles is often referred to as limited pressure cycle?
   (*a*) Dual cycle (*b*) Ericsson cycle
   (*c*) Lenoir cycle (*d*) Joule cycle
5. Identify the cycle which is the idealised cycle for a pulse jet engine
   (*a*) Diesel cycle (*b*) Brayton cycle
   (*c*) Lenoir cycle (*d*) Stirling cycle
6. For the same compression ratio and heat supplied, the air standard efficiency of an Otto cycle compared to that of a diesel cycle is
   (*a*) less (*b*) more
   (*c*) equal (*d*) unpredictable
7. For the same compression ratio and heat input, the cycles in decreasing order of thermal efficiency are
   (*a*) Otto, dual, diesel
   (*b*) diesel, Otto, dual
   (*c*) dual, diesel, Otto
   (*d*) Otto, diesel, dual

8. The order of values of thermal efficiency of Otto, Diesel and Dual cycle, when they have equal compression ratio and heat rejection, is given by
   (a) $\eta_{otto} > \eta_{diesel} > \eta_{dual}$
   (b) $\eta_{diesel} > \eta_{dual} > \eta_{otto}$
   (c) $\eta_{dual} > \eta_{diesel} > \eta_{otto}$
   (d) $\eta_{otto} > \eta_{dual} > \eta_{diesel}$
9. Which air standard cycle consists of two isothermals connected by two constant volume processes?
   (a) Brayton cycle (b) Ericsson cycle
   (c) Stirling cycle (d) Atkinson cycle
10. Identify the gas power cycle of four processes during which work alone is transferred during two processes, and heat alone is transferred during the other two processes
   (a) Atkinson cycle (b) Carnot cycle
   (c) Diesel cycle (d) Otto cycle
11. Match the air standard cycle in column A with its heat addition and heat rejection characteristics in column B. For the given cycles, compression and expansion processes are carried out isentropically

Column A	Column B
(a) Otto cycle	(i) both heat addition and heat rejection are at constant pressure
(b) Diesel cycle	(ii) heat addition at constant volume and heat rejection at constant pressure
(c) Brayton cycle	(iii) both heat addition and heat rejection are at constant volume
(d) Atkinson cycle	(iv) heat addition at constant pressure and heat rejection at constant volume

12. The mean effective pressure of an Otto cycle is
   (a) independent of pressure ratio
   (b) directly proportional to pressure ratio
   (c) inversely proportional to pressure ratio
   (d) proportional to the square root of the pressure ratio
13. All of the following statements are correct except
   (a) For the same compression ratio, thermal efficiency of Otto cycle is greater than that of diesel cycle
   (b) For the same compression and pressure ratios, the work output of theoretical Otto cycle increases with increase in adiabatic index γ
   (c) The thermal efficiency of an ideal Otto cycle increases with compression ratio but decreases with specific heat ratio
   (d) The heat addition and rejection in a diesel cycle take place at constant pressure and constant volume respectively
14. Stirling cycle comprises
   (a) two isothermals and two adiabatics
   (b) two isothermals and two isobarics
   (c) two isothermals and two isochorics
   (d) two adiabatics, one isobaric and one isochoric

Answers

1. (a) 2. (b) 3. (a) 4. (a) 5. (c)
6. (b) 7. (a) 8. (d) 9. (c) 10. (d)
11. (a-iii), (b-iv), (c-i), (d-ii) 12. (b) 13. (c)
14. (c)

Chapter 13

Steam and its Properties

Learning objectives : A study of the subject matter included in this chapter will enable the reader to understand and explain :

- *concept of pure substance and phase*
- *process of steam generation and properties of steam*
- *analysis of heating and expansion processes of steam*
- *Mollier chart and its use in the solution of problems dealing will steam vapour*
- *throttling of steam and determination of dryness fraction of wet steam*

Steam is the most common working substance employed as working medium in steam engines, steam turbines and in atomic power plants for power generation. There occurs a change in the properties of steam during a process and that helps to attain the desired effect. Steam also finds application in process industries, and also acts as heat carrying agent in heat exchange equipment.

This chapter deals with the process of steam generation, properties of steam and the heating and expansion processes of vapours. The utility of Mollier chart in the solution of problems dealing with steam vapour has also been dealt with.

13.1. PURE SUBSTANCE AND PHASE

A ***pure substance*** is one which consists of only single species with characteristic feature of being of homogeneous and invariable composition irrespective of phase or phases in which it exists. A pure substance can exist in three different states, *e.g.*, water can be formed as solid (ice), liquid and gas (water vapour or steam). A system composed of liquid and vapour phases of water is also a pure substance. Even if some of the liquid is vaporised during a process, the system will still be chemically homogeneous and unchanged in chemical composition. However, it is not physically homogeneous because the liquid density is usually much greater than the vapour density. Air which is a mixture of 79% nitrogen and 21% oxygen by volume can also be treated as pure substance if its composition does not change when it undergoes a cooling/heating or expansion/compression process. However, when it is cooled and a part of it liquifies, it is no longer a pure substance. Other examples of pure substance are the vapours of Fr-12, Fr-22 and NH_3 etc. which are used in refrigeration and air conditioning system for producing cooling effect.

The characteristic features of a pure substance can be represented by the following thermodynamic diagrams :

$p-V$, $p-T$, $T-v$, $T-s$, $h-s$ (Mollier diagram) and $p-h$ diagrams.

The three conditions of a pure substance namely solid, liquid, vapour (or gas) are known as phases. Thus a ***phase*** is any homogeneous part of a system that is physically distinctive. It is essentially the homogeneous, chemical and physical state of aggregation of molecules comprising the substances. Each phase of a system is separated from other phases by interfaces called ***phase boundaries***. Consider a system of an ice cube in liquid water. There are two phases, solid and liquid, and the phase boundary is the surface of ice cube. The transition from one phase to another namely melting/freezing, vaporisation/condensation are known as changes of phases. Whenever a change of phase occurs, all the mass of substance does not change its phase simultaneously. For example, evaporation of water starts when its temperature is raised to the boiling point. However, all the water in the container is not simultaneously transformed to the gaseous phase. It remains in a mixed phase of gas and liquid, and the substance of this mixed phase is called ***vapour***. Unlike a gas, a vapour can be easily liquified by applying pressure at constant temperature. The simple laws of perfect gases do not apply to vapours, hence variation between their properties are depicted in charts and listed in tables.

13.2. VAPORISATION, EVAPORATION AND BOILING

Vaporisation refers to the process that involves change from the liquid phase to vapour phase. Water vapours can be obtained by evaporation and steam by boiling.

Evaporation is the process of vapour generation only from the surface of a liquid. The molecules having greater velocity, *i.e.*, kinetic energy break away from the liquid surface and escape to the surrounding atmosphere. Evaporation occurs at any temperature ; however, its intensity increases with rise in temperature of the liquid.

When heat is added to a liquid, it starts boiling and vapours get generated both from the free surface of liquid and inside it. ***Boiling*** refers to the process of vapour formation that takes place in the whole mass of the liquid. All the liquid gets converted into vapour, when vaporisation takes place in an infinite space, *i.e.*, in an open vessel. However, when the vessel is closed, vapours escaping from the liquid fill the free space above it and a fraction of these vapours return back into the liquid.

Saturation temperature and saturation pressure : Boiling takes place at a definite temperature which depends upon physical properties of the liquid and pressure. This temperature, denoted by t_s, is called the saturation temperature and it remains constant until all the liquid has transformed into vapour. At the saturation temperature, the boiling liquid is in equilibrium with its own vapour. The pressure at a specific temperature, at which the liquid boils or is in equilibrium with its own vapour, is called ***saturation pressure***. At standard atmospheric pressure (p = 1.01325 bar), water boils at 100°C. As such, at p = 1.01325 bar, the saturation temperature of water is 100°C. Likewise, the saturation pressure at 100°C boiling is 1.01325 bar. Quite often, the saturation pressure is referred to as vapour pressure.

13.3. SATURATED AND SUPERHEATED STEAM

Depending on its properties, a steam may be saturated or superheated.

The steam is called ***saturated*** when the molecules escaping from the liquid become equal to the molecules returning to it. The saturated steam has a maximum density at a given temperature and is in thermal equilibrium with the liquid. The temperature of saturated steam (vapour) and the liquid in contact with it is called the ***saturation point*** or ***boiling point***.

The liquid at its boiling point at a specified pressure is called a ***saturated liquid***. Such a liquid is in equilibrium with its own vapour at specified temperature/pressure. At standard atmospheric pressure, the water boils at 100°C and generates its own vapour. Thus corresponding to standard atmospheric pressure, the water at 100°C is a saturated liquid.

Saturated steam may be dry or wet. When the saturated vapour contains particles of liquid evenly distributed over the entire mass of vapour, it is called ***wet saturated steam.*** Wet steam is characterised by its ***dryness fraction*** which is defined as the ratio of mass of dry saturated vapour to the total mass of the mixture. That is

Dryness fraction, x

$$= \frac{\text{mass of dry saturated vapour}}{\text{mass of mixture}}$$

$$= \frac{m_g}{m} = \frac{m_g}{m_f + m_g} \qquad \text{...(13.1)}$$

where m is mass of mixture, m_g is mass of dry saturated steam and m_f is the mass of water vapour (saturated liquid) in suspension with the saturated dry steam.

The dryness fraction when expressed in percentage is called ***quality of steam.***

Wetness fraction is another term associated with wet steam. It represents the ratio of mass of water vapour or moisture in suspension with the total mass of steam which contains it.

Wetness fraction, y

$$= \frac{m_f}{m_f + m_g}$$

$$= 1 - \frac{m_g}{m_f + m_g} = 1 - x \qquad \text{...(13.2)}$$

The wetness fraction when expressed in percentage is called ***priming***. Thus,

quality + priming = 100 percent

Saturated vapour which contains no liquid particle is called ***dry saturated steam***. For dry saturated steam, $x = 1$.

Steam becomes superheated, if at a given pressure, its temperature is higher than the saturation temperature. The state/condition of superheated steam is prescribed by its pressure p and temperature t_{sup}. The temperature difference $(t_{sup} - t_s)$ is known as ***degree of superheat***

Degree of superheat

$$= t_{sup} - t_s \qquad \text{...(13.3)}$$

where t_{sup} is the temperature of superheated steam and t_s is the saturation temperature. Higher the degree of superheat, greater the volume occupied by steam. This implies greater distance between molecules and smaller the influence of molecular forces of attraction. At high superheat, the behaviour of steam becomes identical to that of ideal gases.

13.4. STEAM GENERATION

The process of steam generation can be divided into three stages namely

(*i*) heating of water upto boiling point,

(*ii*) evaporation of boiling water and its conversion into dry saturated steam

(*iii*) transformation of dry saturated steam into superheated steam.

These stages have been depicted in $T-Q$, $T-v$ and $p-v$ diagrams in Fig. 13.1. by considering 1 kg of ice at – 10°C (below freezing point). The initial state of the substance has been denoted by point 1.

(1) When heat is supplied, its temperature rises to 0°C (273 K) and thereafter, melting starts (2—3). During the process of melting (phase change from solid to liquid),

- there exists a mixed phase of ice and water
- the temperature of mixed phase remains at 0°C
- there is nominal decrease in volume ; liquid water at 0°C is heavier than ice. However, for normal substances there is increase in volume when change occurs from solid phase to liquid phase.

This transformation of phase from solid to liquid is called the ***melting*** or ***fusion of ice***, and the quantity of heat required for complete transformation of ice into water at constant temperature of 0°C is called ***latent heat of fusion***. At normal atmospheric pressure of 1.01325 bar (1 atm), it equals 335 kJ/kg.

(2) When the entire ice has melted (point 3)

- there is only one phase, *i.e.*, liquid water at 0°C.
- rise in temperature of water occurs with supply of heat and that continues till boiling point is reached. The boiling point depends upon the pressure at which heat is being supplied. The boiling temperature is called

saturation temperature and at normal atmospheric pressure its value is 100°C. The pressure corresponding to saturation temperature is called saturation pressure. Thus

At p_{sat} = 1.01325 bar, t_{sat} = 100°C

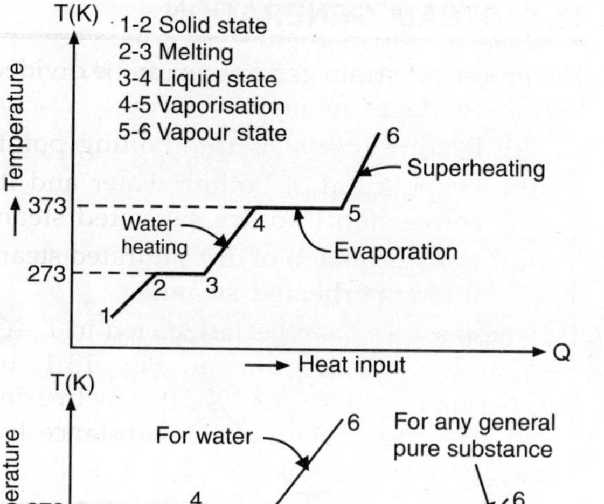

Fig. 13.1. Process of steam generation

• there is decrease in volume when temperature rises from 0°C to 4°C (process 3 – 3') and thereafter the volume increases upto the commencement of boiling (point 4).

The heat required for change from liquid state to vaporisation/boiling point is called ***sensible heat.***

(3) When more heat is added, evaporation of water starts and once again another change of phase occurs. During this phase change from state point 4 to state point 5,

• there exists a two phase mixture of water and its vapour

• the temperature remains constant till all the water gets converted into vapour. The process of vaporisation is thus both isobaric and isothermal

• there is considerable increase in volume.

At state point 5, all the water has vaporised and the state of steam at this point is called dry and saturated steam. The phase transformation from liquid to vapour is called vaporisation, and the heat required for complete transformation of water to its vapour at constant temperature is called the ***latent heat of vaporisation.*** For water evaporating at 100°C under standard atmospheric condition of 1.01325 bar, the latent heat equals 2256.9 kJ/kg.

(4) Since the saturation temperature is a function of pressure, evaporation can take place along any one of the lines 4 – 5, 4' – 5' or 4'' – 5''. The joining of points 4 – 4' – 4'' gives the ***liquid line*** and the joining of points 5 – 5' – 5'' forms a line called the ***saturated vapour line***. On the left of liquid line, there exists water in the liquid phase, and on the right of saturated vapour line the water is said to exist in the superheated phase. Between state points 4 and 5 there is a mixture of steam and water, *i.e.*, wet saturated steam with varying dryness fraction. A shift in the state point A towards the vapour line makes the steam drier and drier.

The saturated liquid curve is steeper compared to the dry saturated vapour line because the volume of liquid changes at much slower rate with increase in temperature.

(5) The horizontal distance between saturated liquid line and the dry saturated vapour line becomes less and less with increase of pressure and saturation temperature. The point at which this distance becomes zero is called the ***critical point***. Critical point represents the highest temperature and pressure above which a liquid cannot exist. At the critical point,

• the liquid is directly and fully converted into dry vapour without going into the phase of evaporation, *i.e.*, latent heat is zero

• the liquid and vapour are indistinguishable

• the properties of saturated liquid and saturated vapour are identical.

The temperature and pressure at the critical point are called ***critical temperature*** t_c and ***critical pressure*** p_c.

For water, the critical values are :

critial temperature t_c = 374.14°C ; critical pressure p_c = 220.9 bar and critical volume v_c = 0.003155 m^3/kg.

(6) Once the steam attains dry and saturated state, its temperature and volume start increasing with further supply of heat. The steam so obtained is called ***superheated steam*** and it behaves like a perfect gas.

Following advantages occur by using superheated steam in steam power plants :

(*i*) increased work capacity without having to increase its pressure.

(*ii*) greater heat content of the superheated steam leads to less steam consumption for a given output and that results in saving in the cost of fuel.

(*iii*) moisture content of steam leaving the steam turbine/engine can be kept within safe limits. Accordingly, the heat losses due to condensation of steam are avoided to a large extent.

(*iv*) superheating is done by utilizing the heat of waste furnace gases which otherwise passes uselessly to the atmosphere.

(7) Steam generation process, as discussed above, has been under standard atmospheric pressure. At different pressures below critical pressure, the nature of the various processes will be identical but the saturation temperature will have different values. In actual practice, the steam is generated at pressure varying from 10 to 200 bar.

In recent times, methods have been developed to produce steam at pressures much above the critical pressure of 220.9 bar. The steam so produced at $p > p_c$ is called ***super pressure steam*** and its formation takes place without boiling or bubbling of water.

13.5. SOLID—LIQUID—VAPOUR EQUILIBRIUM

Fig. 13.2 shows the phase (p versus T) diagram for water in which regions of solid, liquid and vapour are indicated. Its salient features are:

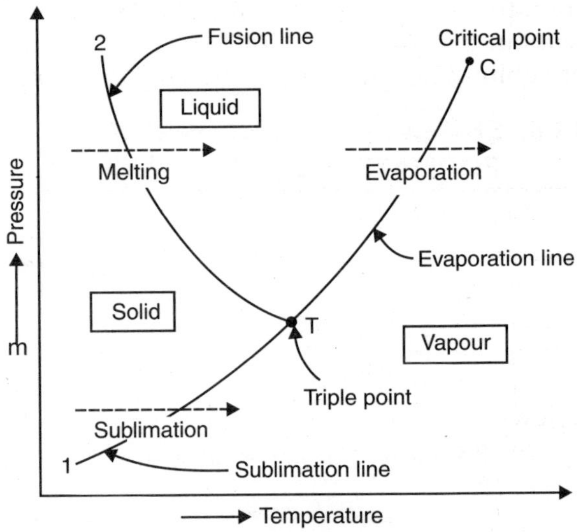

Fig. 13.2. Phase equilibrium diagram on p-T co-ordinates

1. The single phase regions are separated from each other by saturation lines. Along the sublimation line 1 – T there is equilibrium between the solid and vapour phase, *i.e.*, the solid and vapour phases co-exist. The solid and liquid phases co-exist in equilibrium along the fusion line T – 2. There is equilibrium between the liquid and vapour phase along the evaporation line T – C.

2. The fusion line for water has a negative slope ($\partial p/\partial T < 0$) indicating that melting point of ice decreases with increasing pressure. However, for most substances, the slope of fusion curve is positive.

3. The evaporation line ends at the critical point where there is no distinct change from the liquid phase to the vapour phase, *i.e.*, the liquid and vapour are indistinguishable.

4. The sublimation, fusion and evaporation lines meet at point T, called *the* ***triple point***. The triple point may thus be defined as the point at which all the three phases - solid, liquid and vapour co-exist in equilibrium. The triple point exists at a definite pressure and temperature. For water, triple point values are

t_{tp} = 0.01°C and p_{tp} = 0.611 kPa

5. Sublimation refers to the heating process where solid water or ice gets directly transformed to saturated vapour phase without first melting into water. The reverse of sublimation is referred to as ablimation.

13.6. ENTHALPY CHANGES DURING FORMATION OF STEAM

Refer Fig. 13.1 (*a*) which represents formation of steam from ice at - 10°C. Since supply of heat is at constant pressure, the heat input can be taken equal to change in enthalpy. During different stages of the process of steam formation, the enthalpy changes are as follows:

(*i*) Enthalpy changes from - 10°C to 0°C (process 1–2)

$(dh)_{1-2} = c_i\, dt$ where c_i is the specific heat of ice

$= 2.09 \{0 - (-10)\}$

$= 20.9$ kJ/kg

(*ii*) Enthalpy change during melting of ice at 0°C (process 2–3)

$(dh)_{2-3}$ = latent heat of fusion of ice

= 335 kJ/kg

(*iii*) Enthalpy change during heating of water from 0°C to the boiling/saturation temperature 100°C (process 3–4)

$(dh)_{3-4} = c_w\, dt$ where c_w is the specific heat of water

$= 4.186\ (100 - 0)$

$= 418.6$ kJ/kg

(*iv*) Enthalpy change during vaporisation at 100°C (process 4–5)

$(dh)_{4-5}$ = latent heat of vaporisation

= 2256.9 kJ/kg

(*v*) Enthalpy change during heating of dry saturated steam upto say 250°C. (process 5–6)

$(dh)_{5-6} = c_{ps}\, dt$

where c_{ps} is specific heat of superheated steam

$= 2.1\ (250 - 100) = 315$ kJ/kg

It is to be noted that enthalpy quantities are always measured from some datum.

The transformation of water from 0°C into steam involves two kinds of changes that take place on supply of heat at constant pressure.

— temperature of water changes but phase remains the same.

— phase changes but temperature remains constant.

The corresponding heat inputs are defined as :

Sensible heat (h_f) : It is the quantity of heat required to raise the temperature of 1 kg of water from 0°C to the boiling point or saturation temperature at a given pressure. Sensible heat is denoted by letter h_f and is also called ***total heat (or enthalpy) of water.*** Sensible heat increases as the pressure increases.

Latent heat of vaporisation (h_{fg}) : It is the quantity of heat required to convert 1 kg of water at saturation temperature for a given pressure into dry and saturated steam at that temperature and pressure. The value of latent heat is not constant, it decreases as the pressure increases and becomes zero when the critical pressure is reached.

Depending upon the condition of steam, the heat or enthalpy values are defined as :

(*i*) ***Total heat of wet steam*** **:** It represents the quantity of heat required to convert 1 kg of water at 0°C into wet steam at constant pressure

$$h_{wet} = h_f + x\, h_{fg} \qquad ...(13.4)$$

where x is the dryness fraction of wet steam.

The sensible heat increases with pressure.

(*ii*) ***Total heat of dry saturated steam*** **:** It represents the quantity of heat required to convert 1 kg of water at 0°C into dry saturated steam at constant pressure

$$h_s = h_f + h_{fg} = h_g \qquad ...(13.5)$$

The enthalpy of dry saturated steam increases with pressure.

(*iii*) ***Total heat of superheated steam*** : It represents the quantity of heat required to convert 1 kg of water at 0°C into superheated steam at constant pressure.

$$h_{sup} = h_f + h_{fg} + c_{ps}\,(t_{sup} - t_s) \quad ...(13.6)$$

The value of specific heat of steam at constant pressure c_{ps} depends upon the degree of superheat and the pressure of steam generation. Its average value is taken from 2.0 to 2.1 kJ/kg.

13.7. VOLUME OF DRY, WET AND SUPERHEATED STEAM

The specific volume of a fluid is the volume occupied by a unit mass of the fluid. For dry saturated steam, it is denoted by the letter v_g. Its value decreases with increase in pressure, and the values corresponding to different pressures are listed in steam tables. The Clapeyron's equation, given below, also helps to determine its value.

Reference Fig. 13.3, *abcd* represents an elementary Carnot cycle for dry saturated steam on $p - v$ diagram. The point *a* and point *b* lie on the saturated liquid line and the dry saturated steam line, respectivley. If dTs measures the difference in temperature between the isothermals *ab* and *cd*, the efficiency of this elementary cycle is

$$\eta = \frac{T_1 - T_2}{T_1} = \frac{dT_s}{T_s} \quad ...(i)$$

Efficiency of a Carnot cycle is also worked out from the relation

$$\eta = \frac{\text{work done}}{\text{heat supplied}}$$

$$= \frac{(v_g - v_f)\,dp}{h_{fg}} \quad ...(ii)$$

where dp is difference of pressure between the isothermals, v_g and v_f are the volumes of dry saturated steam and that of liquid, h_{fg} is the heat supplied which equals the latent heat (During evaporation, the heat supplied is the latent heat).

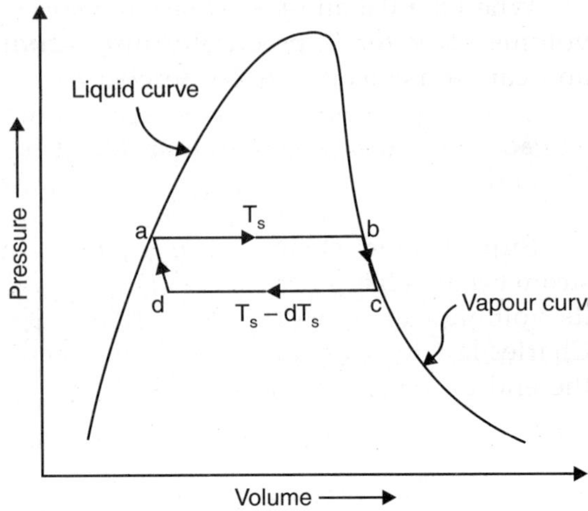

Fig. 13.3. Clapeyron's equation

From relations (*i*) and (*ii*)

$$\eta = \frac{dT_s}{T_s} = \frac{(v_g - v_f)\,dp}{h_{fg}}$$

$$\text{or} \quad (v_g - v_f) = \frac{h_{fg}}{T_s}\frac{dT_s}{dp} \quad ...(13.7)$$

Equation 13.7 is referred to as the Clapeyron's equation, and it provides a precise method for the specific volume at any required pressure or temperature. The value of $\dfrac{dT_s}{dp}$ is obtained from the slope of $T - p$ curve which is plotted from experimental data.

Wet steam : The volume of 1 kg of wet steam equals the volume of dry portion of steam plus the volume of water in suspension.

If x is the dryness fraction of wet steam, then 1 kg of wet steam consists x kg of dry steam and (1 – x) kg of water in suspension. Thus the volume of wet steam is

$$v_w = x\,v_g + (1 - x)\,v_f$$

where v_f is the specific volume of water and v_g is the specific volume of dry saturated steam.

By re-arrangement, $x = \dfrac{v_w - v_f}{v_g - v_f}$

When $x > 0.8$ and $p < 30$ bar, the specific volume of water is practically insignificant and can be neglected. Accordingly,

$$v_w = x\, v_g$$

and
$$x = \frac{v_w}{v_g} \quad ...(13.8)$$

Superheated steam : The superheated steam behaves like a perfect gas and, therefore, its volume can be worked out by applying Charles law to steam at the beginning and at the end of superheating process. Thus

$$\frac{v_g}{T_s} = \frac{v_{sup}}{T_{sup}}$$

$$v_{sup} = v_g \frac{T_{sup}}{T_s} \quad ...(13.9)$$

where v_g and T_s are respectively the specific volume and absolute temperature of dry saturated steam and, v_{sup} and T_{sup} are respectively the specific volume and absolute temperature of superheated steam.

13.8. EXTERNAL WORK AND INTERNAL LATENT HEAT

When water evaporates to form steam, its volume increases from v_f to v_g at constant pressure. The work is thus done by steam in increasing its volume. This work is called ***external work of evaporation.***

$$\text{External work} = p\,(v_g - v_f)$$

At low pressures, specific volume of water v_f is extremely small compared to v_g and hence neglected. Then

$$\text{External work} \simeq p\, v_g \quad ...(13.10)$$

In case of wet steam with dryness fraction x, the volume of steam would be xv_g and accordingly

$$\text{External work} = p\, x\, v_g$$

A certain amount of latent heat supplied to water during the evaporation process is utilized in expanding the volume and thus producing external work. The latent heat of steam minus the external work is called the ***internal latent heat.***

$$\text{Internal latent heat} = h_{fg} - p\, v_g \quad ...(13.11)$$

13.9. INTERNAL ENERGY OF STEAM

The actual heat energy above the freezing point of water stored in the steam is known as ***internal energy of steam***. The work of evaporation is utilized in doing external work and is thus not stored in the steam. Accordingly, internal energy of steam can be determined by subtracting work of evaporation from the total heat/enthalpy of steam. That is : $u = h - pv$

For wet steam :

$$u = (h_f + x\, h_{fg}) - p\, x\, v_g$$

For dry saturated steam :

$$u = (h_f + h_{fg}) - p\, v_g$$

For superheated steam :

$$u = h_f + h_{fg} + c_{ps}\,(t_{sup} - t_s) - p\, v_{sup} \quad ...(13.12)$$

13.10. ENTROPY OF STEAM

The heating and evaporation of water during the steam formation process takes place at constant pressure. Even superheating of steam is done at constant pressure. The entropy of steam can, therefore, be calculated by using the relations for entropy change at constant pressure.

Entropy of water (s_f) : Let one kg of water be heated from temperature T_1 to T_2 at constant pressure. Entropy change is then given by

$$ds = \frac{\delta Q}{T} = c_p \frac{dT}{T}$$

Upon integration from initial state 1 to final state 2,

$$s_2 - s_1 = c_p \log_e \frac{T_2}{T_1}$$

If 0°C is taken as datum : $s_1 = 0$ and $T_1 = 273$ K. That gives

$$s_f = c_p \log_e \frac{T}{273}$$

where s_f is the entropy of water at any temperature T above freezing point of water. If heating of water continues upto saturation temperature T_s, then

$$s_f = c_p \log_e \frac{T_s}{273} \quad ...(13.13)$$

Entropy of evaporation (s_e) : Evaporation of water takes place at constant pressure and constant temperature T_s. The quantity of heat supplied during evaporation equals $x\,h_{fg}$ if the steam is wet at the end of heat supply, and equals h_{fg} if the steam is dry saturated at the end of heat supply. Thus change of entropy during evaporation

$$= \frac{h_{fg}}{T_s} \text{ for dry and saturated steam}$$

$$= \frac{x h_{fg}}{T_s} \text{ for wet saturated steam}$$

with dryness fraction x

Thus entropy of wet steam reckoned above the freezing point of water (0°C) is given by

s_{wet} = entropy of water + entropy change during evaporation

$$= s_f + \frac{x h_{fg}}{T_s} = s_f + x\, s_{fg}$$

$$= s_f + x\,(s_g - s_f) \quad ...(13.14)$$

Likewise, the entropy of dry saturated steam (*i.e.*, $x = 1$) reckoned above 0°C works out

$$s_{dry} = s_f + \frac{h_{fg}}{T_s} = s_f + (s_g - s_f)$$

$$= s_g \quad ...(13.15)$$

Entropy of superheated steam : Entropy change during superheating, which takes place at constant pressure, is given by

$$ds = \frac{\text{heat of superheat}}{T}$$

$$= \frac{\delta q}{T} = c_{ps}\frac{dT}{T}$$

If T_{sup} is the temperature of superheated steam, and T_s is the saturation temperature at the considered pressure, then

$$ds = c_{ps} \log_e \frac{T_{sup}}{T_s}$$

Thus the total entropy of superheated steam reckoned above the freezing point of water (0°C) is

$$s_{sup} = s_f + s_{fg} + c_{ps} \log_e \frac{T_{sup}}{T_s} \quad ...(13.16)$$

The specific heat of superheated steam, c_{ps} varies from 2.0 to 2.3 kJ/kg K depending upon degree of superheat and superheating pressure.

13.11. STEAM TABLES

The simple laws of perfect gases donot apply to vapours, hence variations between their properties are depicted on charts and tables. The preparation of steam tables has been made possible by studying experimentally the various properties of steam at different conditions of steam formation. The experimental data was then utilized to formulate suitable equations of state in different regions, and subsequently obtain comprehensive data on steam properties. The steam tables give values of specific volume, enthalpy and entropy for saturated liquid and dry saturated vapour tabulated against pressure (p) or corresponding saturation temperature (t_s). A typical structure of such tables is given in Table 13.1.

Such tables make it possible to rapidly find the basic characteristics of steam according to the given parameters. The method of interpolation is used to find the parameters for intermediate values of pressure and temperature (not given in tables).

The properties of superheated steam are tabulated separately. The values of specific volume, enthalpy and entropy not only depend on pressure but also on temperature or degree of superheat. A typical structure of such tables is given below.

The first horizontal line gives the temperature of steam in degrees celcius and the first vertical column gives the absolute pressure in bars alongwith the saturation temperture corresponding to that pressure. The second, third and fourth horizontal lines show the specific volume, enthalpy and entropy of superheated steam arranged in increasing order of magnitude.

Table 13.1.

1. Properties of dry saturated steam

Absolute pressure, bar	Saturation temperature °C	Specific Volume m^3/kg		Specific enthalpy kJ/kg			Specific entropy kJ/kg K		
p	t_s	v_f	v_g	h_f	h_{fg}	h_g	s_f	s_{fg}	s_g
10	179.9	0.001127	0.194	762.6	2013.6	2776.2	2.1382	4.4446	6.5825
30	233.8	0.001216	0.0666	1008.4	1793.9	2802.3	2.6455	3.5382	6.1837

2. Properties of dry saturated steam

Temperature °C	Absolute pressure bar	Specific Volume m^3/kg		Specific enthalpy kJ/kg			Specific entropy kJ/kg K		
t_s	p	v_f	v_g	h_f	h_{fg}	h_g	s_f	s_{fg}	s_g
10	0.0123	0.0010004	106.4	42.0	2477.7	2519.7	0.151	8.750	8.901
50	0.1235	0.001012	12.03	209.3	2382.7	2582.7	0.704	7.372	8.076

Table 13.2. Properties of superheated steam

↓ p (bar) (t_s)	t °C →	200	250	300	350	400
5.0	v	0.425	0.474	0.523	0.570	0.617
(151.8)	h	2855.4	2960.7	3064	3167.7	3271.9
	s	7.059	7.271	7.46	7.633	7.794
15.0	v	0.132	0.152	0.169	0.187	0.203
(198.3)	h	2796.8	2923.3	3037.6	314.7.5	3255.8
	s	6.455	6.709	6.918	7.102	7.269

EXAMPLE 13.1

Calculate the dryness fraction of steam which has 1.25 kg of water in suspension with 40 kg of steam.

Solution : m_s = 40 kg and m_w = 1.25 kg

$$\text{Dryness fraction, } x = \frac{m_s}{m_s + m_w}$$

$$= \frac{40}{40 + 1.25}$$

$$= 0.9696 \simeq \mathbf{0.97}$$

EXAMPLE 13.2

Using steam tables, determine the mean specific heat for superheated steam at 1 bar and between 150°C and 200°C.

Solution : From steam tables of superheated steam and at 1 bar pressure, we have

At 150°C :

$$h_{sup} = 2776.4 \text{ kJ/kg}$$

At 200°C :

$$h_{sup} = 2875.3 \text{ kJ/kg}$$

$$\therefore \quad 2875.3 = 2776.4 + c_{ps}\,(200 - 150)$$

$$c_{ps} = \frac{2875.3 - 2776.4}{50}$$

$$= \mathbf{1.978 \text{ kJ/kg K}}$$

EXAMPLE 13.3

Determine the state of steam, i.e., whether it is wet, dry or superheated in the following cases

(i) Steam has a pressure of 10 bar and specific volume 0.175 m³/kg.

(ii) Steam has a pressure of 15 bar and a temperature of 220°C.

(iii) Steam has a pressure of 20 bar and if 2700 kJ/kg of heat is required to generate the steam from water at 0°C.

Solution : (*i*) From steam tables, specific volume of dry saturated steam at 10 bar is v_g = 0.194 m³/kg Since the specific volume of the steam whose quality is to be determined is less than the specific volume of dry saturated steam, the steam is wet and its dryness fraction is given by

$$\text{Dryness fraction, } x = \frac{\text{specific volume of given steam at 10 bar}}{\text{specific volume of dry saturated steam at 10 bar}}$$

$$= \frac{0.175}{0.194} = \mathbf{0.902}$$

(*ii*) From steam tables, saturation temperature corresponding to 15 bar pressure is 198.3°C. The given steam has a temperature of 220°C which is higher than the saturation temperature. Therefore, the given steam is superheated and its degree of superheat is

$$= 220 - 198.3 = \mathbf{21.7°C}$$

(*iii*) The dry saturated steam at 20 bar has a total heat of 2797.2 kJ/kg. The enthalpy of given steam is less than this value. As such, the given steam is wet and its dryness fraction can be worked out from the relation

$$h = h_f + x\,h_{fg}$$

$$2700 = 908.6 + x \times 1888.6$$

$$\therefore \quad x = \frac{2700 - 908.6}{1888.6} = \mathbf{0.948}$$

EXAMPLE 13.4

A spherical shell of 30 cm in radius contains saturated steam and water at 300°C. Calculate the mass of each if their volumes are equal.

Solution : Total volume of the shell $= \frac{4}{3} \pi r^3$

$$= \frac{4}{3} \pi (0.3)^3 = 0.1130 \text{ m}^3$$

$\therefore$ Volume of vapour and that of water

$$= \frac{0.1130}{2} = 0.0565 \text{ m}^3$$

From steam tables, corresponding to saturation temperature of 300°C,

$$v_f = 0.001404 \text{ m}^3/\text{kg}$$

and $v_g = 0.0216 \text{ m}^3/\text{kg}$

$\therefore$ Mass of vapour $= \frac{0.0565}{0.0216} = \mathbf{2.616 \text{ kg}}$

Mass of water $= \frac{0.0565}{0.001404} = \mathbf{40.242 \text{ kg}}$

EXAMPLE 13.5

A tank contains 100 kg of liquid water and 5 kg of water vapour under saturation conditions at 20°C. If the specific volume of saturated vapour at that temperature is taken as 57.78 m³/kg, calculate the volume of tank and the moisture content of the mixture.

Solution : At 20°C ; $v_f = 0.001002 \text{ m}^3/\text{kg}$; $v_g = 57.79 \text{ m}^3/\text{kg}$

That gives

$$V_f = m_f v_f$$
$$= 100 \times 0.001002$$
$$= 0.1002 \text{ m}^3$$
$$V_g = m_g v_g$$
$$= 5 \times 57.79 = 288.95 \text{ m}^3$$

$\therefore$ Total volume $V = V_f + V_g$

$$= 0.1002 + 288.95 = \mathbf{289 \text{ m}^3}$$

Moisture content refers to wetness fraction which is given by

$$x = \frac{m_f}{m_f + m_g}$$

$$= \frac{100}{100+5} = \mathbf{0.952} \text{ or } \mathbf{95.2\ \%}$$

EXAMPLE 13.6

A metallic drum of 1.8 m³ capacity is to be completely filled with dry saturated steam at 6 bar. For this, the drum is evacuated and the necessary amount of water is put into it. Subsequently, heat is supplied and the entire water is converted into steam. Make calculations for the amount of water to be placed in the drum and the resulting temperature of steam if (a) the steam is dry saturated, (b) the steam is wet with 0.85 dryness fraction.

Solution : The temperature of steam would correspond to the saturation temperature at 6 bar pressure

$$t = t_s \text{ at 6 bar} = \mathbf{158.8°C}$$

(*a*) Specific volume of dry and saturated steam at 6 bar pressure is

$$v = v_g = 0.315 \text{ m}^3/\text{kg}$$

$\therefore$ Mass of water = mass of steam

$$= \frac{\text{volume of drum}}{\text{specific volume of steam}}$$

$$= \frac{1.8}{0.315} = \mathbf{5.714 \text{ kg}}$$

(*b*) Specific volume of wet steam is

$$v = (1 - x)v_f + x\, v_g \simeq x\, v_g$$

(neglecting specific volume of water)

$$= 0.85 \times 0.315$$
$$= 0.2677 \text{ m}^3/\text{kg}$$

$\therefore$ Mass water $= \frac{1.8}{0.2677} = \mathbf{6.723 \text{ kg}}$

EXAMPLE 13.7.

A closed container holds 1 kg of water at 153.3°C temperature with the following composition by volume 1/3 liquid and 2/3 vapour

Determine the presure, quality, volume and enthalpy of the mixture.

Solution : Since a mixture of water and its vapour is being considered, its pressure would be the same as the saturation pressure corresponding to 153.3°C. Then from steam tables : p = **5.2 bar**

Given : $\frac{V_f}{V_g} = \frac{1/3}{2/3} = \frac{1}{2}$

or $\dfrac{m_f v_f}{m_g v_g} = \dfrac{1}{2}$

$\therefore \dfrac{m_f \times 0.001094}{m_g \times 0.361} = \dfrac{1}{2}$

That gives : $\dfrac{m_f}{m_g} = \dfrac{1}{2} \times \dfrac{0.361}{0.001094} \simeq 165$

Dryness fraction is given by,

$$x = \frac{m_g}{m_f + m_g} = \frac{m_f/165}{m_f + m_f/165}$$

$$= \frac{1/165}{1 + 1/165} = \frac{1}{166} = \mathbf{0.00602}$$

For wet steam with dryness fraction x, specific volume $v = (1 - x)v_f + x\, v_g$

$$= (1 - 0.00602) \times 0.001094 + 0.00602 \times 0.361$$

$$= 1.0874 \times 10^{-3} + 2.1732 \times 10^{-3}$$

$$= \mathbf{3.2606 \times 10^{-3}\ m^3/kg}$$

specific enthelpy

$$h = (1 - x)h_f + x\, h_g$$

$$= h_f + x(h_g - h_f)$$

$$= h_f + x\, h_{fg}$$

$$= 646.7 + 0.00602 \times 2103.8$$

$$= \mathbf{659.36}\ \text{kJ/kg}$$

EXAMPLE 13.8.

A vessel of 0.3 m³ capacity contains 1.5 kg mixture of water and steam in equilibrium at a pressure of 5 bar. Calculate (a) the volume and mass of liquid and (b) the volume and mass of water.

Solution : The specific volume of the given mixture,

$$v = \frac{0.3}{1.5} = 0.2\ \text{m}^3/\text{kg}$$

From steam tables, the specific volumes of saturated liquid and saturated vapour at 5 bar pressure are noted as

$$v_f = 0.001093\ \text{m}^3/\text{kg}$$

and $v_g = 0.375\ \text{m}^3/\text{kg}$

If x is the dryness fraction of the mixture, then

$$v = x\, v_g + (1 - x)\, v_f$$

$$0.2 = x \times 0.375 + (1 - x)\, 0.001093$$

$$= 0.375\, x + 0.001093 - 0.001093\, x$$

$$x = \frac{0.2 - 0.001093}{0.375 - 0.001093}$$

$$= \frac{0.1989}{0.3739} = 0.532$$

Now, by definition,

$$x = \frac{m_g}{m_g + m_f}$$

$\therefore$ Mass of vapour, $m_g = x\,(m_g + m_f)$

$$= 0.532 \times 1.5 = \mathbf{0.798\ kg}$$

Mass of liquid, $m_f = 1.5 - 0.798 = \mathbf{0.702\ kg}$

Volume of vapour $= m_g v_g$

$$= 0.798 \times 0.375 = \mathbf{0.2992\ m^3}$$

Volume of liquid $= m_f v_f$

$$= 0.702 \times 0.001093$$

$$= \mathbf{0.00135\ m^3}$$

EXAMPLE 13.9

Determine the amount of heat required to generate 5 kg of steam at a pressure of 10 bar and tempertature of 250°C from water at 25°C. Take specific heat for superheated steam as 2.1 kJ/kg K.

Solution : From steam tables, the steam properties at 10 bar pressure are

$$t_s = 179.9°\text{C}$$

$$h_f = 762.6\ \text{kJ/kg}$$

and $h_{fg} = 2013.6\ \text{kJ/kg}$

$\therefore$ Total heat of superheated steam,

$$h_{sup} = h_f + h_{fg} + c_{ps}\,(t_{sup} - t_{sat})$$

$$= 762.6 + 2013.6 + 2.1\,(250 - 179.9)$$

$$= 2923.41\ \text{kJ/kg}$$

Heat content of 1 kg of water at 25°C,

$$h_w = m\, c_{pw}\, \Delta t$$

$$= 1 \times 4.186 \times (25 - 0)$$

$$= 104.65\ \text{kJ/kg}$$

$\therefore$ Heat required per kg of steam generated

$$= 2923.41 - 104.65$$

$$= 2818.76\ \text{kJ/kg}$$

Total amount of heat required

$$= 5 \times 2818.76$$

$$= \mathbf{14093.8\ kJ}$$

EXAMPLE 13.10

Determine the mass of 0.25 m³ of wet steam at 5 bar pressure and 0.85 dryness fraction. Proceed to calculate the heat content of 1 m³ of this steam.

Solution : From steam tables corresponding to 5 bar pressure,

$$v_g = 0.375 \text{ m}^3/\text{kg},$$
$$h_f = 640.1 \text{ kJ/kg}$$
and $$h_{fg} = 2107.4 \text{ kJ/kg}$$

Mass of steam,

$$m = \frac{\text{volume of steam}}{\text{specific volume}}$$
$$= \frac{V}{x\,v_g} = \frac{0.25}{0.85 \times 0.375}$$
$$= \mathbf{0.784\ kg}$$

Mass of 1 m³ of steam

$$= \frac{1}{0.85 \times 0.375} = 3.137 \text{ kg}$$

Enthalpy of 1 kg of steam,

$$h = h_f + x\,h_{fg}$$
$$= 640.1 + 0.85 \times 2107.4$$
$$= 2431.39 \text{ kJ/kg}$$

∴ Total heat content of steam

$$= 3.137 \times 2431.39$$
$$= \mathbf{7627.27\ kJ}$$

EXAMPLE 13.11.

Steam is available at 8 bar and 0.9 dry. Make calculations for the final dryness fraction of steam in each of the following cases :

(i) there is a loss of 125 kJ from the steam at constant pressure.

(ii) the temperature of steam falls to 160 °C.

(iii) the steam expands to 3 bar pressure in a turbine stage and work equivalent of 200 kJ/kg is done.

Solution : (*i*) Latent heat of steam at 8 bar

$$= 2046.5 \text{ kJ/kg}$$

Increase in wetness due to removal of 125 kJ

$$= \frac{125}{2046.5} = 0.0611$$

∴ Final dryness fraction = 0.9 – 0.0611

$$= \mathbf{0.8389}$$

(*ii*) Saturation temperature at 8 bar

$$= 170.4°\text{C}$$

With fall in temperature to 160°C, the pressure also gets lowered and becomes equal to the saturation pressure corresponding to the reduced temperature of 160°C.

∴ Final pressure (corresponding to 160°C)

$$= 6.2 \text{ bar}$$

Specific volume at 8 bar = 0.24 m³/kg

Specific volume at 6.2 bar = 0.306 m³/kg

Let x be the dryness fraction of steam at 6.2 bar.

Since the volume remains constant, one may write

$$0.9 \times 0.24 = x \times 0.306 \quad \therefore \quad x = 0.706$$

(*iii*) Corresponding to 8 bar pressure,

$$h_{f1} = 720.9 \text{ kJ/kg}$$
and $$h_{fg1} = 2046.5 \text{ kJ/kg}$$

Enthalpy of steam at inlet,

$$h_1 = h_{f1} + x_1\,h_{fg1}$$
$$= 720.9 + 0.9 \times 2046.5$$
$$= 2562.75 \text{ kJ/kg}$$

Enthalpy at exit = h_1 – work done

$$= 2562.75 - 200 = 2362.75 \text{ kJ/kg}$$

Corresponding to 3 bar pressure

$$h_{f2} = 561.4 \text{ kJ/kg}$$
and $$h_{fg2} = 2163.2 \text{ kJ/kg}$$

Writing down an expression for the total heat at 3 bar, we get

$$2362.75 = 561.4 + x_2 \times 2163.2$$

∴ Dryness fraction at exit from the turbine,

$$x_2 = \frac{2362.75 - 561.4}{2163.2} = \mathbf{0.833}$$

EXAMPLE 13.12

Find the specific volume, enthalpy, internal energy and entropy of wet steam at 15 bar pressure and dryness fraction 0.8.

Solution : The specific volume of wet steam with dryness fraction x is

$$v = x\,v_g = 0.8 \times 0.132$$
$$= 0.1056 \text{ m}^3/\text{kg}$$

(*ii*) Total heat of 1 kg of wet steam with dryness fraction x is

$$h = h_f + x\, h_{fg}$$
$$= 844.7 + 0.8 \times 1945.2$$
$$= \mathbf{2400.86\ kJ/kg}$$

(*iii*) External work $= p\, v$
$$= 15 \times 10^5 \times 0.1056$$
$$= 158400\ \text{J} = 158.4\ \text{kJ}$$

Internal energy, $u = h - p\, v$
$$= 2400.86 - 158.4$$
$$= \mathbf{2242.46\ kJ/kg}$$

(*iv*) The entropy for wet steam is
$$s = s_f + x\, s_{fg}$$
$$= 2.3145 + 0.8 \times 4.1261$$
$$= \mathbf{5.6154\ kJ/kg\ K}$$

EXAMPLE 13.13

A sample of steam at 5 bar is stated to have an enthalpy of 2350 kJ/kg. Make calculations for the specific volume, internal energy and entropy of this sample of steam.

Solution : The dry saturated steam at 5 bar has a total heat of 2747.5 kJ/kg. The enthalpy of the given sample of steam is less than this value. As such the given steam is wet and its dryness fraction can be worked out from the relation

$$h = h_f + x\, h_{fg}$$
$$2350 = 640.1 + x \times 2107.4$$

i.e., dryness fraction,
$$x = \frac{2350 - 640.1}{2107.4} = 0.811$$

(*i*) Specific volume of wet steam,
$$v = x\, v_g = 0.811 \times 0.0375$$
$$= \mathbf{0.3041\ m^3/kg}$$

(*ii*) External work of evaporation,
$$= p\, v = 5 \times 10^5 \times 0.3041$$
$$= 152050\ \text{J} = 152.05\ \text{kJ/kg}$$

Internal energy, $u = h - p\, v$
$$= 2350 - 152.05$$
$$= \mathbf{2197.95\ kJ/kg}$$

(*iii*) The entropy for wet steam is
$$s = s_f + x\, s_{fg}$$
$$= 1.8604 + 0.811 \times 4.9588$$
$$= \mathbf{5.882\ kJ/kg\ K}$$

EXAMPLE 13.14

Determine the enthalpy, volume, internal energy and entropy of superheated steam at 15 bar pressure and 220°C. The volume of water may be neglected and take specific heat of superheat equal to 2.2 kJ/kg K.

Solution : Corresponding to 15 bar pressure, the following values can be read from the steam tables :

$v_g = 0.132\ \text{m}^3/\text{kg}$; $h_f = 844.7\ \text{kJ/kg}$
$h_{fg} = 1945.2\ \text{kJ/kg}$; $s_f = 2.314\ \text{kJ/kg K}$
$s_{fg} = 4.1261\ \text{kJ/kg K}$
and $t_{sat} = 198.3°\text{C}$

(*i*) Total heat of 1 kg of superheated steam is

$$h_{sup} = h_f + h_{fg} + c_{ps}\,(t_{sup} - t_{sat})$$
$$= 844.7 + 1945.2 + 2.2\,(220 - 198.3)$$
$$= \mathbf{2837.64\ kJ/kg}$$

(*ii*) The superheated steam behaves like a perfect gas and as such its volume can be worked out by applying Charles' law, *i.e.*,

$$v_{sup} = v_g \times \frac{T_{sup}}{T_{sat}}$$
$$= 0.132 \times \frac{220 + 273}{198.3 + 273}$$
$$= \mathbf{0.1381\ m^3/kg}$$

(*iii*) External work $= p\, v$
$$= 15 \times 10^5 \times 0.1381$$
$$= 207150\ \text{J} = 207.15\ \text{kJ/kg}$$

Internal energy,
$$u = h_{sup} - \text{external work}$$
$$= 2837.64 - 207.15$$
$$= \mathbf{2630.49\ kJ/kg}$$

(*iv*) The entropy of superheated steam is

$$s_{sup} = s_f + s_{fg} + c_{ps} \log_e \frac{T_{sup}}{T_{sat}}$$
$$= 2.314. + 4.1261 + 2.2 \log_e \frac{220 + 273}{198.3 + 273}$$
$$= \mathbf{6.539\ kJ/kg\ K}$$

EXAMPLE 13.15.

Determine the external work done during evaporation and internal latent heat at a pressure of 1 MPa and dryness fraction of 0.80.

Solution : From steam tables at 1 MPa, *i.e.,* 10 bar

$$v_f = 0.001127 \text{ m}^3/\text{kg}$$
and $$v_g = 0.194 \text{ m}^3/\text{kg}$$
$$h_{fg} = 2013.6 \text{ kJ/kg}$$

External work done during evaporation

$$= p\,(x\,v_g - v_f)$$

Since v_f is extremely small as compared to v_g, we may write

External work done $= p\,xv_g$

$$= 10 \times 10^5 \times 0.8 \times 0.194$$
$$= 155200 \text{ J or } 155.2 \text{ kJ}$$

Latent heat of wet steam

$$= x \times h_{fg} = 0.8 \times 2013.6$$
$$= 1610.88 \text{ kJ/kg}$$

The amount of external work is included in the latent heat of steam. The latent heat of steam minus the external work is known as internal latent heat.

$\therefore$ Internal latent heat $= 1610.88 - 155.2$

$$= \mathbf{1455.68 \text{ kJ/kg}}$$

EXAMPLE 13.16

Steam at 10 bar and 200°C is cooled till it becomes dry saturated and is then throttled to 1 bar pressure. Determine the change in enthalpy and heat transferred during each process. Also find the quality of steam at the end of throttling process.

Take c_{ps} = 2.25 kJ/kg K for superheated steam.

Solution :

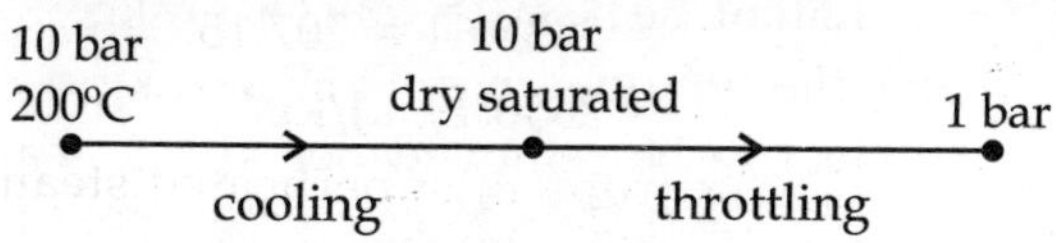

Fig. 13.4.

At 10 bar and 200°C (steam is superheated)

$$h_1 = h_{sup1} = 2827.9 \text{ kJ/kg}$$

At 10 bar and dry saturated,

$$h_2 = h_{f2} + h_{fg2}$$
$$= 762.6 + 2013.6$$
$$= 2776.2 \text{ kJ/kg}$$

$\therefore$ Change in enthalpy during cooling

$$= h_2 - h_1$$
$$= 2827.9 - 2776.2$$
$$= \mathbf{51.7 \text{ kJ/kg}}$$

(*ii*) Throttling is a constant enthalpy process, *i.e.,* $\Delta h = 0$ during process 2–3. Therefore, for the throttling process, we can write

$$h_2 = h_3$$

At 1 bar : $h_{f3} = 417.5$ kJ/kg

and $h_{fg3} = 2257.9$ kJ/kg

$$2776.2 = h_{f3} + h_{fg3} + c_{ps}\ \Delta t$$
$$= 417.5 + 2257.9 + 2.25\ \Delta t$$

$\therefore$ Degree of superheat Δt

$$= \frac{2776.2 - 417.5 - 2257.9}{2.25}$$
$$= \mathbf{44.8°C}$$

EXAMPLE 13.17.

Steam generated in a boiler at 8.5 bar and 0.95 dry is taken to a superheater where it receives 85 kJ/kg of heat at constant pressure. Subsequently the steam is throttled to a pressure of 3.5 bar. Determine the final state of steam, i.e., whether it is wet, dry or superheated.

Solution : Enthalpy of steam before entering the superheater

$$h = h_f + x\,h_{fg}$$
$$= 732.05 + 0.95 \times 2037.9$$
$$= 2668.06 \text{ kJ/kg}$$

Enthalpy of steam at exit from the superheater

$$= 2668.06 + 85 = 2753.06 \text{ kJ/kg}$$

During throttling process, the total heat of steam would remain constant. Accordingly, the steam would have a total heat of 2753.06 kJ/kg after throttling operation to 3.5 bar. From steam tables, it is found that the dry saturated steam at 3.5 bar has an enthalpy equal to 2731.6 kJ/kg. As such the steam is superheated with a heat of superheat equal to

$$= 2753.06 - 2731.6$$
$$= 21.46 \text{ kJ/kg}$$

Presuming specific heat of superheated steam equal to 2.1 kJ/kg K,

$$\text{Degree of superheat} = \frac{21.46}{2.1} = \mathbf{10.22°C}$$

EXAMPLE 13.18

Steam from a boiler is delivered at an absolute pressure of 15 bar and dryness fraction of 0.95 into a steam superheater in which the steam receives additional heat at constant pressure and its temperature increases upto 300°C. Determine the amount of heat added and the change in internal energy for unit mass of steam.

Solution : For the initial state of steam at p = 15 bar and x = 0.95, we have from steam tables

$$h_1 = h_{f1} + x_1 h_{fg1}$$
$$= 844.7 + 0.95 \times 1945.2$$
$$= 2692.64 \text{ kJ/kg}$$
$$v_1 = x_1 v_{g1}$$
$$= 0.95 \times 0.132 = 0.1254 \text{ m}^3/\text{kg}$$
$$u_1 = h_1 - p_1 v_1$$
$$= 2692.64 - \frac{15 \times 10^5 \times 0.1254}{10^3}$$
$$= 2504.54 \text{ kJ/kg}$$

The enthalpy and the specific volume of steam in the final state are found from tables for superheated steam (p = 15 bar and t = 300°C)

$$h_2 = 3037.6 \text{ kJ/kg}$$
and
$$v_2 = 0.169 \text{ m}^3/\text{kg}$$
$$u_2 = h_2 - p_2 v_2$$
$$= 3037.6 - \frac{15 \times 10^5 \times 0.169}{10^3}$$
$$= 2784.1 \text{ kJ/kg}$$

∴ Quantity of heat supplied in the superheater is

$$q = h_2 - h_1$$
$$= 3037.6 - 2692.64$$
$$= \mathbf{344.96 \text{ kJ/kg}}$$

and change in internal energy is

$$du = u_2 - u_1$$
$$= 2784.1 - 2504.54$$
$$= \mathbf{279.56 \text{ kJ/kg}}$$

EXAMPLE 13.19

Steam is supplied from the boiler at a pressure of 16 bar and 98 percent dry to a steam engine. As the steam flows through the pipeline, its pressure remains constant but it loses 25 kJ/kg of heat due to poor pipe insulation. What would be the condition (temperature and dryness fraction) of steam at the engine end of pipe line ?

Solution : Let suffix 1 and 2 refer to the state of steam at the boiler end and at the engine end of the pipe line.

From steam tables at pressure 16 bar :

$$t_s = 201.4°\text{C}$$
$$h_{f1} = h_{f2} = 858.6 \text{ kJ/kg}$$
$$h_{fg1} = h_{fg2} = 1933.2 \text{ kJ/kg}$$
$$h_1 = h_{f1} + x_1 h_{fg1}$$
$$= 858.6 + 0.98 \times 1933.2$$
$$= 2753.14. \text{ kJ/kg}$$
$$h_2 = h_{f2} + x_2 h_{fg2}$$
$$= 858.6 + x_2 \times 1933.2$$

According to the given condition :

$$h_1 - 25 = h_2$$
$$2753.14 - 25 = 858.6 + 1933.2\, x_2$$
$$\therefore \quad x_2 = \frac{2753.14 - 25 - 858.6}{1933.2} = \mathbf{0.967}$$

Thus the steam is 0.967 dry and has a temperature of 201.4°C at the engine end.

EXAMPLE 13.20.

The pressure of steam in a condenser is 10 kPa and the dryness fraction is 0.85. Calculate the number of heat units that must be abstracted so as to condense (i) 1 kg of steam (ii) 1 m³ of steam.

Solution : To condense steam, it is the latent heat that needs to be abstracted.

At 10 kPa (0.1 bar) pressure,

Latent heat, h_{fg} = 2392.8 kJ/kg

Specific volume, v_g = 14.67 m³/kg

∴ Heat to be removed per kg of steam,

$$= 0.85 \times 2392.8 = 2033.88 \text{ kJ}$$

Heat to be removed per m³ of steam,

$$\frac{2033.88}{x \times v_g} = \frac{2033.88}{0.85 \times 14.67} = \mathbf{163.11 \text{ kJ}}$$

EXAMPLE 13.21.

Exhaust steam from a turbine at 38°C and 0.95 dry enters a surface condenser at the rate of

2500 kg/hr. The condensed steam leaves the condenser with a temperature of 30°C. Make calculations for the rate at which cooling water should be supplied to the condenser if rise in its temperature is limited to 10°C.

Take specific heat of water

= 4.186 kJ/kg K

Solution : From steam tables at 38°C,

$h_{f1} = 159.2$ kJ/kg

and $h_{fg1} = 2411.5$ kJ/kg

∴ Enthalpy of steam,

$$h_1 = h_{f1} + x\, h_{fg1}$$
$$= 159.2 + 0.95 \times 2411.5$$
$$= 2450.125 \text{ kJ/kg}$$

At 30°C : $h_{f2} = 125.8$ kJ/kg

heat lost by steam = heat gained by water

$$m_s\,(h_1 - h_{f2}) = m_w \times c_{pw} \times \Delta t$$
$$2500\,(2450.125 - 125.8) = m_w \times 4.186 \times 10$$

∴ Mass of cooling water required,

$$m_w = \frac{2500\,(2450.125 - 125.8)}{4.186 \times 10}$$
$$= \mathbf{138815\ kg/hr}$$

EXAMPLE 13.22.

Steam initially at 10 bar pressure and with 2% moisture flows steadily through the jackets of a reciprocating engine. Measurements indicate that 5 kg of this wet steam condenses every minute, and the condensate leaves the jacket at the temperature of the steam. Calculate the heat that will be given up by the jacket steam per minute.

Solution : *Initial state :*

$m = 5$ kg (steady flow)

$p_1 = 10$ bar ; $x = 0.98$

$$\therefore \quad h_1 = h_f + x\, h_{fg}$$
$$= 762.6 + 0.98 \times 2013.6$$
$$= 2735.93 \text{ kJ/kg}$$

Final state : t_2 = saturation temperature of steam at 10 bar pressure = 179.9°C

$h_2 = h_f$ at 10 bar pressure

$= 762.2$ kJ/kg

From the steady flow energy equation

$$\dot{m}\left(h_1 + \frac{V_1^2}{2} + gz_1\right) + Q = \dot{m}\left(h_2 + \frac{V_2^2}{2} + gz_2\right) + w_s$$

Neglecting kinetic energy and potential energy changes and as shaft work w_s is zero,

$$Q = \dot{m}(h_2 - h_1) = 5(762.2 - 2735.93)$$
$$= \mathbf{-\ 9868.65}\ \text{kJ/min}$$

Thus the heat given up by the jacket steam per minute is 9868.65 kJ

EXAMPLE 13.23.

A vessel of 0.05 m³ capacity contains a mixture of saturated steam and water at a temperature of 300°C; the mass of liquid water being 10 kg. Determine the pressure, mass, enthalpy, specific volume, internal energy and entropy of contents present in the vessel.

Solution : As the steam in the vessel is wet saturated, the pressure corresponds to the saturated temperature of 300°C.

∴ Pressure inside the vessel = 86 bar

The vessel contains liquid and steam in saturated state

$$\therefore \quad V = m_f v_f + m_g v_g$$
$$0.05 = 10 \times 0.001404 + m_g \times 0.0216$$

∴ Mass of dry steam,

$$m_g = \frac{0.05 - 10 \times 0.001404}{0.0216}$$
$$= 1.6648 \text{ kg}$$

Dryness fraction of steam,

$$x = \frac{m_g}{m_f + m_g}$$
$$= \frac{1.6648}{10 + 1.6648} = 0.1427$$

Mass of steam in the vessel,

$$m = m_f + m_g = 10 + 1.6648$$
$$= \mathbf{11.6648\ kg}$$

Specific volume of steam,

$$v = \frac{V}{m} = \frac{0.05}{11.6648}$$
$$= \mathbf{0.004286\ m^3/kg}$$

Enthalpy of steam,

$$H = mh = m\,(h_f + x\,h_{fg})$$
$$= 11.6648\,(1345.4 + 0.1427 \times 1405.5)$$
$$= 11.6648 \times 1545.96 = \mathbf{18033\ kJ}$$

Internal energy of steam,

$$U = m\,(h_f + x\,h_{fg} - p\,x\,v_g)$$
$$= 11.6648\left(1545.96 - \frac{86\times10^5\times0.1427\times0.0216}{10^3}\right)$$
$$= \mathbf{17724\ kJ}$$

Entropy of steam,

$$S = ms = m\,(s_f + x\,s_{fg})$$
$$= 11.6648\,(3.2557 + 0.1427 \times 2.4519)$$
$$= \mathbf{42.06\ kJ/kg}$$

EXAMPLE 13.24.

Steam at 1000 kPa and 300°C enters an engine and expands to 20 kPa. If the exhaust steam has a dryness fraction of 0.9, make calculations for the drop in enthalpy and change in entropy.

Solution : At 1000 kPa, *i.e.*, 10 bar, the saturation temperature of steam is 179.9°C. As such the steam at entry to the engine is superheated and from steam tab'es for superheated steam, its enthalpy and entropy values are :

$$h_1 = 3051.2\ \text{kJ/kg}$$

and, $s_1 = 7.123$ kJ/kg K

Corresponding to 20 kPa (0.2 bar), the following values are read from the steam tables:

$h_f = 251.5$ kJ/kg ; $h_{fg} = 2358.4$ kJ/kg

$s_f = 0.8321$ kJ/kg K ; $s_{fg} = 7.0773$ kJ/kg K

$$\therefore\quad h_2 = (h_f + x\,h_{fg})_2$$
$$= 251.5 + 0.9 \times 2358.4$$
$$= 2374.06\ \text{kJ/kg}$$
$$s_2 = (s_f + x\,s_{fg})_2$$
$$= 0.8321 + 0.9 \times 7.0773$$
$$= 7.2017\ \text{kJ/kg K}$$

Hence, drop in enthalpy

$$= h_1 - h_2 = 3051.2 - 2374.06$$
$$= \mathbf{677.14.\ kJ/kg}$$

Change in entropy

$$= s_1 - s_2 = 7.123 - 7.2017$$
$$= \mathbf{-\ 0.0787\ kJ/kg\ K\ (decrease)}$$

EXAMPLE 13.25.

A vessel of 1m³ capacity contains steam at 10 bar and 0.9 dryness. The steam is allowed to blow off by opening the delivery valve and the period of blowing is so regulated that pressure drops to 5 bar. Subsequently the valve is closed and the vessel is cooled until the pressure becomes 4 bar. Make calculations for the mass of steam blown off, the dryness fraction of steam in the vessel after cooling and the heat transferred during cooling.

Assume that total heat per kg of steam in the vessel remains constant during the period of blow off.

Solution : Corresponding to the given pressures, the following parameters are read from the steam tables.

At 10 bar :

$v_g = 0.194$ m³/kg ; $h_f = 762.6$ kJ/kg

$h_{fg} = 2013.6$ kJ/kg

At 5 bar :

$v_g = 0.375$ m³/kg ; $h_f = 640.1$ kJ/kg

$h_{fg} = 2107.4$ kJ/kg

At 4 bar :

$v_g = 0.462$ m³/kg ; $h_f = 604.7$ kJ/kg

$h_{fg} = 2133.0$ kJ/kg

Initial mass of steam in the vessel,

$$= \frac{V}{v} = \frac{1}{0.9\times0.194} = 5.727\ \text{kg}$$

Since enthalpy of steam in the vessel remains constant during the period of blow off

$$h_{f1} + x_1\,h_{fg1} = h_{f2} + x_2\,h_{fg2}$$
$$762.6 + 0.9 \times 2013.6 = 640.1 + x_2 \times 2107.4$$

Solution gives : dryness fraction after blowing, $x_2 = 0.918$

Mass of steam after blowing off

$$= \frac{1}{0.918\times0.375} = 2.905\ \text{kg}$$

$\therefore$ Mass of steam blown off

$$= 5.727 - 2.905 = \mathbf{2.822\ kg}$$

(*b*) Dryness fraction of steam after cooling

$$x = \frac{V}{m\,v_g} = \frac{1}{2.905 \times 0.462} = \mathbf{0.745}$$

(*c*) From first law : $\delta Q = \delta w + du$. Since the vessel has a fixed volume, the work done is zero and accordingly the heat transferred during cooling equals the change in internal energy.

Now : $U_2 = h_2 - p_2 v_2$

$$= (h_{f_2} + x_2\, h_{fg_2}) - p_2\, x_2\, v_{g2}$$

$$= (640.1 + 0.918 \times 2107.4) - \frac{5 \times 10^5 \times 0.918 \times 0.375}{10^3}$$

$$= 2574.7 - 172.1 = 2402.6 \text{ kJ/kg}$$

and $U_3 = h_3 - p_3 v_3$

$$= (h_{f_3} + x_3\, h_{fg_3}) - p_3\, x_2\, v_{g3}$$

$$= (604.7 + 0.745 \times 2133) - \frac{4 \times 10^5 \times 0.745 \times 0.462}{10^3}$$

$$= 2193.8 - 137.7 = 2056.1 \text{ kJ/kg}$$

∴ Heat transferred during cooling,

$$= 2.822 \times (2056.1 - 2402.6)$$

$$= \mathbf{-\ 977.8 \text{ kJ}}$$

The negative sign implies that the heat is lost during the cooling process.

EXAMPLE 13.26

A closed metallic boiler drum of 0.25 m³ capacity holds steam at 10 bar pressure and 195°C. Determine the quantity of steam in the drum.

(b) At what pressure in the drum the steam would become dry saturated if the drum is cooled ?

(c) What would be the pressure and condition of steam in the drum if the cooling continues till the temperature of steam becomes 120.2°C ?

Solution : The saturation temperature corresponding to 10 bar pressure is 179.9°C. The given steam has a temperature of 195°C which is higher than the saturation temperature. Therefore the given steam is superheated and its volume is given by

$$V_{sup} = \frac{v_{g1}\, T_{sup}}{T_{sat}}$$

$$= \frac{0.194 \times (195 + 273)}{(179.9 + 273)} = 0.2 \text{ m}^3$$

Mass of water = mass of steam

$$= \frac{\text{volume of boiler drum}}{\text{specific volume of steam}}$$

$$= \frac{0.25}{0.2} = 13.25 \text{ kg}$$

(*b*) At $v_g = v_{sup} = 0.2$ m³, the saturation pressure is 9.7 bar

(*c*) After subsequent cooling the temperature becomes 120.2°C. At this temperature,

saturation pressure = 2 bar

specific volume of dry saturated steam = 0.885m³/kg

Then : $0.2 = x \times 0.885$; $x = 0.225$

Thus, when cooled to 120°C, the steam will be at 2 bar pressure and 0.225 dryness fraction.

EXAMPLE 13.27.

Two boilers A and B are delivering steam in equal proportion to a common main. Both the boilers operate at a pressure of 14 bar which is also the pressure of steam in the main. The boiler A is fitted with a superheater and supplies steam at 300°C. If the temperature of the resulting mixture in the steam main is 235°C, estimate the quality of steam supplied by the boiler B which is without a superheater

Proceed to determine the net change in the entropy of the arrangement.

Take c_{ps} = 2.25 kJ/kg K for the superheated steam.

Solution : Saturation temperature corresponding to 14 bar = 195°C. The steam in the main has a temperature of 235°C which is higher than the saturation temperature. As such the steam in the main is superheated.

Heat of steam in the main

$$= h_f + h_{fg} + c_p\ (t_{sup} - t_{sat})$$

$$= 830.1 + 1957.7 + 2.25\,(235 - 195)$$

$$= 2877.8 \text{ kJ/kg}$$

Heat supplied by boiler A

$= h_f + h_{fg} + c_{ps}\,(t_{sup} - t_{sat})$

$= 830.1 + 1957.7 + 2.25\,(300 - 195)$

$= 3024.05$ kJ/kg

If x is the dryness fraction of steam supplied by boiler B, then

Heat supplied by boiler B

$= h_f + x\,h_{fg}$

$= 830.1 + 1957.7\,x$

Assuming no losses due to radiation and condensation etc, the energy balance gives

$3024.05 + 830.1 + 1957.7\,x = 2 \times 2877.8$

The right hand side has been multiplied by 2 as the mass of steam in the main would be 2 kg.

Solution gives : $x = 0.9713$

The steam in the main has thus a dryness fraction of 0.9713 or has a quality of 97.13%

(*b*) Entropy of steam in the main

$$s_{main} = \left(s_f + s_{fg} + c_{ps}\log_e \frac{T_{sup}}{T_{sat}}\right)_{main}$$

$$= 2.2837 + 4.1814 + 2.25\log_e \frac{235+273}{195+273}$$

$= 6.65$ kJ/kg K

Entropy of steam being generated in boiler A

$$s_A = \left(s_f + s_{fg} + c_{ps}\log_e \frac{T_{sup}}{T_{sat}}\right)_A$$

$$= 2.2837 + 4.1814 + 2.25\log_e \frac{300+273}{195+273}$$

$=$ **6.9205** kJ/kg K

Entropy of steam being generated in boiler B

$s_B = (s_f + x\,s_{fg})_B$

$= 2.2837 + 0.9713 \times 4.1814$

$= 6.345$ kJ/kg K

∴ Net change in entropy

$= 2 \times 6.65 - 6.9205 - 6.345$

= 0.0345 kJ/kg K (increase)

EXAMPLE 13.28.

Two boilers A and B discharge equal amount of steam into a common steam pipe. The steam being supplied by boiler A is at 10 bar pressure and 350°C temperature. The steam from boiler B is at the same pressure but is wet with dryness fraction 0.92. Determine :

(a) equilibrium condition after mixing in the common steam pipe

(b) change in entropy of the steams from boilers A and B

(c) net increase or decrease of entropy.

Solution : For the steam (10 bar and 300°C) from boiler A (suffix *a*)

$h_a = h_{sup} = 3158$ kJ/kg

and $s_a = s_{sup} = 7.301$ kJ/kg K

For the steam (10 bar and 0.92 dry) from boiler *B*(suffix *b*)

$h_b = h_f + x\,h_{fg}$

$= 762.6 + 0.92 \times 2013.6$

$= 2615$ kJ/kg

$s_b = s_f + x\,s_{fg} = 2.138 + 0.92 \times 4.4446$

$= 6.227$ kJ/kg K

Presuming that each boiler dischanges 1 kg of steam, the enthalpy of the mixture (suffix *m*)

$$h_m = \frac{h_a + h_b}{2} = \frac{3158 + 2615}{2}$$

$= 2886$ kJ/kg

At 10 bar, the enthalpy of dry saturated steam is 2776 kJ/kg. As such, the mixed steam in the common steam pipe is superheated.

Corresponding to $p = 10$ bar and $h_m = 2886$ kJ/kg K, the following parameters are worked out by interpolation :

$$t_m = 200 + \frac{2866 - 2828}{2942 - 2828} \times (250 - 200)$$

$= 225.44$°C

$$s_m = 6.694 + \frac{225.44 - 200}{250 - 200} \times (6.925 - 6.694)$$

$= 6.811$ kJ/kg K

$$v_m = 0.206 + \frac{225.44 - 200}{250 - 200} \times (0.233 - 0.206)$$

$= 0.219$ m^3/kg

(*b*) Loss of entropy by steam from boiler A

$s_a - s_m = 7.301 - 6.811 =$ **0.49 kJ/kg K**

Gain of entropy by steam from boiler B

$s_m - s_b = 6.811 - 6.227 =$ **0.654 kJ/kg K**

(c) Net gain in entropy

$= 0.654 - 0.49 =$ **0.164 kJ/kg K**

Heating and Expansion of Vapours

Constant volume process : Consider unit mass of steam undergoing constant volume heating process from initial wet state 1 (pressure p_1 and dryness fraction x_1) to the final superheated state 2 (pressure p_2 and temperature T_{sup2})

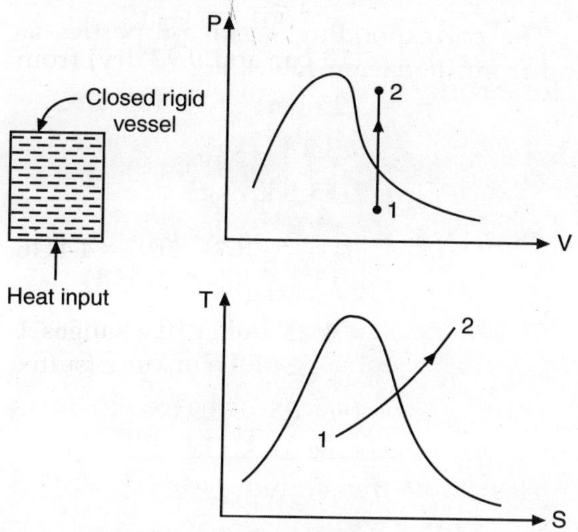

Fig. 13.5.

(i) For the constant volume process:

$$v_1 = v_2$$

$$x_1 \, v_{g1} = v_{g2} \times \frac{T_{sup}}{T_{sat}}$$

$$\therefore \quad x_1 = \frac{v_{g2}}{v_{g1}} \times \frac{T_{sup2}}{T_{sat2}}$$

(ii) Work done $w_{1-2} = \int_1^2 p\,dv = 0$

$(\because dv = 0)$

Obvious the work done is zero in a constant volume process.

(iii) Applying the first law of thermodynamics

$$\delta q = \delta w + du$$
$$= u_2 - u_1 \qquad (\because \delta w = 0)$$

The enthalpy h and internal energy u are correlated as $h = u + p\,v$. Then,

$$q_{1-2} = (h_2 - p_2 v_2) - (h_1 - p_1 v_1)$$

The enthalpy values at the end states are:

$$h_1 = h_{f_1} + x_1 h_{fg_1}$$

and $\quad h_2 = h_{g_2} + c_p (T_{sup2} - T_{sat1})$

The heating or cooling of steam in a closed rigid vessel represents a constant volume process.

EXAMPLE 13.29

Steam at 10 bar pressure and 0.9 dryness fraction is cooled at constant volume at 160°C. What will be its final condition ?

Solution : Initial volume

$$v_1 = (1 - x_1)\, v_{f1} + x_1\, v_{g1}$$

Final volume

$$v_2 = (1 - x_2)\, v_{f2} + x_2\, v_{g2}$$

Since volume remains constant,

$$v_1 = v_2, \text{ and therefore,}$$

$$(1 - x_1)\, v_{f1} + x_1\, v_{g1} = (1 - x_2)\, v_{f2} + x_2\, v_{g2}$$

Neglecting specific volume of water, we may write,

$$x_1\, v_{g1} = x_2\, v_{g2}$$

From steam tables: at 10 bar,

$$v_{g1} = 0.194 \text{ m}^3/\text{kg}$$

and at 160°C ,

$$v_{g2} = 0.306 \text{ m}^3/\text{kg}$$

$$\therefore \quad 0.9 \times 0.194 = x_2 \times 0.306$$

$$\text{or } x_2 = \frac{0.9 \times 0.194}{0.306}$$

$= \mathbf{0.5706}$ or **57.06%**

EXAMPLE 13.30.

A boiler drum contains water and steam at atmospheric pressure, all air having been expelled and the stop valve closed. Determine the heat required for generating dry saturated steam at 6 bar.

When the stop valve is opened and the boiler supplies the dry saturated steam at 6 bar, determine the additional heat input required if the feed water is at 45°C.

Solution : At atmospheric pressure of 1 bar,

$$u_1 = h_{f1} - p_1 v_{f1}$$

$$= 417.5 - \frac{1\times10^5\times0.001043}{10^3}$$

$$= 417.4 \text{ kJ/kg}$$

For dry saturated steam at 6 bar

$$u_2 = (h_{f2} + h_{fg2}) - p_2 v_{g2}$$

$$= (670.4 + 2085.0) - \frac{6\times10^5\times0.315}{10^3}$$

$$= 2566.4 \text{ kJ/kg}$$

Since the stop valve is closed, the process of steam generation takes place at constant volume, *i.e.*, the work done is zero.

∴ From principle of energy conservation.

$$\delta q = du = u_2 - u_1$$

$$= 2566.4 - 417.4 = \mathbf{2149 \text{ kJ/kg}}$$

(*b*) Heat of feed water at 45°C

$$h = c_p \,\Delta t = 4.186\,(45 - 0)$$

$$= 188.37 \text{ kJ/kg}$$

When the stop valve is open, the heat required will be

$$= (h_{f2} + h_{fg2}) - h$$

$$= (670.4 + 2085) - 188.37$$

$$= 2567 \text{ kJ/kg}$$

∴ Additional heat required

$$= 2567 - 2149 = \mathbf{418 \text{ kJ/kg}}$$

EXAMPLE 13.31

A pressure cooker contains 2 kg of steam at 5 bar pressure and 0.9 dryness fraction. Find the quantity of steam which must be transferred so as the quality of steam becomes 60% dry. Also calculate the pressure and temperature of the steam that exists in the cooker after the heat rejection.

Solution : From steam tables corresponding to 5 bar pressure,

$$v_g = 0.375 \text{ m}^3/\text{kg}$$

$$h_f = 640.1 \text{ kJ/kg}$$

and $h_{fg} = 2107.4$ kJ/kg

Specific internal energy at the initial state 1,

$$u_1 = h_1 - p_1 v_1$$

$$= (h_{f1} + x_1 h_{fg1}) - p_1 x_1 v_{g1}$$

$$= (640.1 + 0.9 \times 2107.4) - \frac{5\times10^5\times(0.9\times0.375)}{10^3}$$

$$= 2536.76 - 168.75$$

$$= 2368.01 \text{ kJ/kg}$$

As the volume remains constant during the cooling process

$$x_1 v_{g1} = x_2 v_{g2}$$

$$v_{g2} = \frac{x_1}{x_2} v_{g1}$$

$$= \frac{0.9}{0.6}\times0.375 = 0.5625 \text{ m}^3/\text{kg}$$

The corresponding steam properties as read from the steam tables are :

$$p_2 = 3.25 \text{ bar};\quad t_2 = 136.25°\text{C}$$

$$h_{f2} = 573.2 \text{ kJ/kg};$$

$$h_{fg2} = 2155.1 \text{ kJ/kg}$$

Now : $u_2 = h_{f2} + x_2 h_{fg2} - p_2 x_2 v_{g2}$

$$= (573.2 + 0.6 \times 2155.1) - \frac{3.25\times10^5\times(0.6\times0.5625)}{10^3}$$

$$= 1866.25 - 109.68$$

$$= 1756.57 \text{ kJ/kg}$$

Heat to be transferred $q = du$

$$= u_2 - u_1$$

$$= 1756.57 - 2368.01$$

$$= -611.44 \text{ kJ/kg}$$

Total heat transferred

$$= 2 \times (-611.44) = \mathbf{-1222.88 \text{ kJ}}$$

The – ve sign implies that heat is rejected from the system.

The steam remaining in the pressure cooker after rejection of heat is at :

pressure = **3.25** bar

and temperature = **136°25** C

EXAMPLE 13.32

A rigid tank of 1 m³ volume contains dry saturated steam at 0.2 MPa. Due to poor insulation, there is heat transfer to the surroundings and the pressure drops to 0.1 MPa after some time. Make calculations for the final condition of the steam and the amount of heat transferred.

Solution : At 0.2 MPa (2 bar)

$$v_1 = v_{g1} = 0.885 \text{ m}^3/\text{kg}$$

∴ Mass of steam

$$m = \frac{V}{v_1} = \frac{1}{0.885} = 1.13 \text{ kg}$$

At 0.1 MPa (1 bar)

$$v_2 = (1 - x_2)\, v_{f2} + x_2\, v_{g2}$$
$$\simeq x_2 v_{g2} = 1.694\, x_2$$

Since the tank is rigid, the volume remains constant ($v_1 = v_2$) and therefore,

$$1.694\, x_2 = 0.885\ ; \quad x_2 = 0.522$$

Thus the final condition of steam corresponds to wet steam of dryness fraction 0.522 at 0.1 MPa

(*ii*) $h_1 = h_{g1}$ (dry steam at 0.2 MPa)
$= 2706.3$ kJ/kg

$h_2 = h_{f2} + x_2\, h_{fg2}$
(wet steam at 0.1 MPa)
$= 417.5 + 0.522 \times 2257.9$
$= 1596.12$ kJ/kg

For a constant volume process work done is zero and therefore, from principle of energy conservation

$$q_{1-2} = du = (u_2 - u_1)$$
$$= (h_2 - p_2 v_2) - (h_1 - p_1 v_1)$$
$$= (h_2 - h_1) - v_1\,(p_2 - p_1)$$
$$= (1596.12 - 2706.3) - \frac{0.885\,(0.1 - 0.2) \times 10^6}{10^3}$$
$$= -\ 1021.68 \text{ kJ/kg}$$

∴ Total energy transferred as heat to the surroundings

$$Q_{1-2} = m\, q_{1-2}$$
$$= 1.13 \times (-\ 1021.68)$$
$$= \mathbf{-\ 1154.5\ kJ}$$

EXAMPLE 13.33.

A rigid vessel of 2 m³ volume is filled with superheated steam at 20 bar and 300°C. The vessel is cooled until the steam is just dry saturated. Calculate the mass of steam in the vessel, the final pressure of steam and the amount of energy transferred as heat to the surroundings. Represent the process on T – S diagram.

Solution : Refer to Fig. 13.6.

At state point 1,

$$p_1 = 20 \text{ bar}, \ t_1 = 300°\text{C}$$
$$V_1 = 2 \text{ m}^3$$

From steam tables,

$$v_1 = 0.125 \text{ m}^3/\text{kg}$$

∴ Mass of steam,

$$m = V/v_1 = \frac{2}{0.125} = \mathbf{16\ kg}$$

(*ii*) Since the tank is rigid, the volume remains constant and therefore,

$$v_2 = v_1 = 0.125 \text{ m}^3/\text{kg}$$

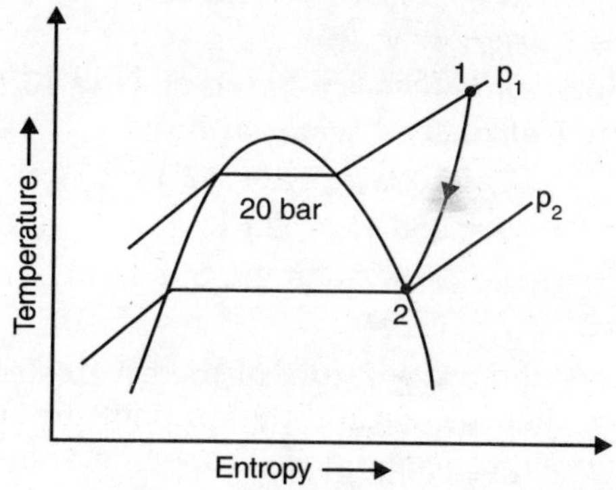

Fig. 13.6.

Further, the steam is just dry and saturated at state point 2 and so

$$v_{g2} = v_2 = 0.125 \text{ m}^3/\text{kg}$$

From steam tables, the pressure of dry saturated steam corresponding to specific volume 0.125 m³/kg is **16 bar.**

(*iii*) For a constant volume process, work done is zero and therefore, from the principle of energy conservation

$$q_{1-2} = du = (u_2 - u_1)$$
$$= (h_2 - p_2 v_2) - (h_1 - p_1 v_1)$$
$$= (h_2 - h_1) - v_1\,(p_2 - p_1)$$

At 20 bar and 300°C :

$$h_1 = 3023.5 \text{ kJ/kg}$$

At 16 bar and $x = 1$:

$$h_2 = h_g = 2791.7 \text{ kJ/kg}$$

$$\therefore \quad q_{1-2} = (2791.7 - 3023.5) - \frac{0.125\,(16 - 20) \times 10^5}{10^3}$$
$$= -\ 181.8 \text{ kJ/kg}$$

$$Q_{1\text{-}2} = m\, q_{1\text{-}2} = 16 \times (-181.8)$$
$$= \mathbf{-2908.8\ kJ}$$

EXAMPLE 13.34.

A closed vessel of 0.2 m³ capacity contains steam at a pressure of 10 bar and a temperature of 250°C. The vessel is cooled till the pressure of steam in the vessel falls to 3.5 bar. Find :

(a) final temperature and dryness fraction,

(b) change in internal energy and heat transferred,

(c) change in entropy.

Represent the cooling process on p – v, T – s and h – s diagrams.

Solution : *Initially when the steam is at 10 bar pressure (suffix 1),*

$$p_1 = 10 \text{ bar}, V_1 = 0.2 \text{ m}^3$$

and $t_1 = 250°\text{C} = 523$ K

$$t_{sat} = 179.9°\text{C} = 452.9 \text{ K}$$

and $v_{g1} = 0.194$

Since the temperature of the given steam is more than the saturation temperature, the steam is superheated and its specific volume is given by

$$v_1 = v_{g1} \times \frac{T_{sup1}}{T_{sat1}} = 0.194 \times \frac{523}{452.9}$$
$$= 0.224 \text{m}^3/\text{kg}$$

Mass of steam in the vessel,

$$m = \frac{\text{total volume}}{\text{specific volume}} = \frac{0.2}{0.224}$$
$$= 0.893 \text{ kg}$$

After the cooling process (suffix 2)

$$p_2 = 3.5 \text{ bar} ;\ v_{g2} = 0.524 \text{ m}^3/\text{kg}$$
$$h_{f2} = 584.3 \text{ kJ/kg} ;\ h_{fg2} = 2147.4 \text{ kJ/kg}$$
$$s_{f2} = 1.727 \text{ kJ/kg K} ;\ s_{fg2} = 5.212 \text{ kJ/kg K}$$

The final specific volume of dry steam v_{g2} at 3.5 bar pressure is more than the initial specific volme v_1 at 10 bar pressure. As such the final condition of steam is wet. Since the vessel is closed one, the cooling takes place at constant volume. That is

$$v_1 = v_2 = x_2\, v_{g2}$$

where x_2 is final dryness fraction

$$0.224 = x_2 \times 0.524$$

$$\therefore \quad x_2 = \frac{0.224}{0.524} = \mathbf{0.427}$$

The temperature of wet steam equals the saturation temperature at the given pressure.

At 3.5 bar : t = **138.8°C**

(*b*) At 10 bar pressure and 250°C temperature, the steam is superheated and its specific enthalpy is

$$h_1 = 2942.6 \text{ kJ/kg}$$

(from steam tables for superheated steam)

$$u_1 = h_1 - p_1 v_1$$
$$= 2942.6 - \frac{10 \times 10^5 \times 0.224}{10^3}$$
$$= 2718.6 \text{ kJ/kg}$$

u_2(internal energy of wet steam at 3.5 bar)

$$= h_2 - p_2 v_2$$
$$= \left(h_{f2} + x_2\, h_{fg2}\right) - p_2\, x_2\, v_{g2}$$
$$= (584.3 + 0.427 \times 2147.4)$$
$$- \frac{3.5 \times 10^5 \times 0.427 \times 0.524}{10^3}$$
$$= 1501.2 - 78.3 = 1422.9 \text{ kJ/kg}$$

$\therefore$ Change in internal energy $= m(u_2 - u_1)$

$= 0.893\ (1422.9 - 2718.6) =$ **– 1157** kJ

The negative sign indicates that there is decrease in internal energy.

During constant volume process, the heat transferred equals the change in internal energy.

$\therefore$ Heat transferred = – 1157 kJ

The negative sign indicates that the heat has been rejected by steam.

(*c*) From steam tables for supherheated steam, the entropy of steam at 10 bar and 250°C is

$$s_1 = 6.925 \text{ kJ/kg}$$

Entropy of steam after cooling to 3.5 bar,

$$s_2 = s_{f2} + x_2\, s_{fg2}$$
$$= 1.727 + 0.427 \times 5.212$$
$$= 3.952 \text{ kJ/kg K}$$

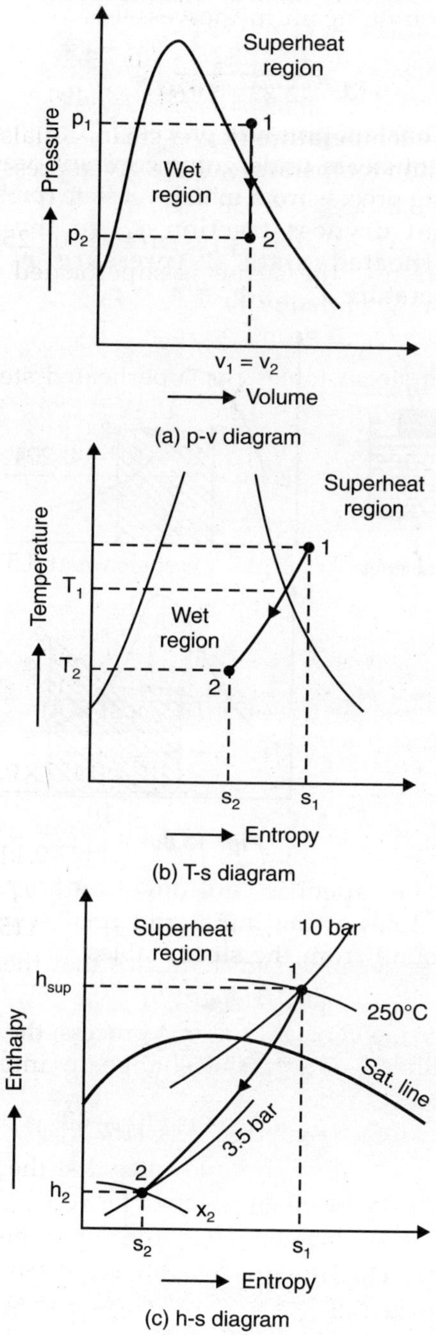

Fig. 13.7.

$\therefore$ Change in entropy $= m(s_2 - s_1)$

$= 0.893(3.952 - 6.925)$

$=$ **$-$ 2.655 kJ/K**

The negative sign indicates that entropy decreases during the cooling process.

The constant volume cooling process on p–v, T–s and h–s diagrams has been shown in Fig. 13.7.

EXAMPLE 13.35.

A closed vessel of 1.2 m³ capacity contains steam at 3 bar pressure and 0.85 dryness fraction. Steam at 10 bar and 0.96 dry is supplied to the vessel until the pressure in the vessel becomes 5 bar. Calculate the mass of steam supplied to the vessel and the dryness fraction of mixed steam in the vessel. You may neglect the volume of moisture and thermal capacity of the vessel.

Solution : Volume capacity of the vessel $= 1.2\text{ m}^3$

For the steam initially present (suffix 1) in the vessel,

$$p_1 = 3 \text{ bar} \quad \text{and} \quad x_1 = 0.85$$

$$h_{f_1} = 561.4 \text{ kJ/kg}$$

$$h_{fg_1} = 2163.2 \text{ kJ/kg and}$$

$$v_{g_1} = 0.606 \text{ m}^3/\text{kg}$$

m_1(mass of steam)

$$= \frac{\text{total volume of steam in the vessel}}{\text{specific volume of steam}}$$

$$= \frac{V}{x_1 \times v_{g_1}} = \frac{1.2}{0.85 \times 0.606} = 2.33 \text{ kg}$$

Total internal energy of steam at 3 bar pressure

$$U_1 = m_1(h_1 - p_1 v_1)$$

$$= m_1\left[(h_{f_1} + x_1 h_{fg_1}) - p_1 v_1\right]$$

$$= 2.33\left[(561.4 + 0.85 \times 2163.2) - \frac{3 \times 10^5 \times 0.85 \times 0.606}{10^3}\right]$$

$$= 2.33\,[2400.12 - 154.53] = 5232 \text{ kJ}$$

Let m_2 be the mass of steam supplied to the closed vessel at 10 bar.

At 10 bar pressure (suffix 2)

$$h_{f_2} = 762.6 \text{ kJ/kg} ;$$

$$h_{fg_2} = 2013.6 \text{ kJ/kg}$$

$$\text{and } v_{g_2} = 0.194 \text{ m}^3/\text{kg}$$

Then enthalpy of steam supplied,

$$h_2 = m_2(h_{f_2} + x_2\, h_{fg2})$$
$$= m_2(762.6 + 0.96 \times 2013.6)$$
$$= 2696\ m_2 \text{ kJ}$$

If suffix 3 denotes the characteristics of mixed steam, then

$$p_3 = 5 \text{ bar} \ ; \text{ dryness fraction} = x_3$$
$$h_{f_3} = 640.1 \text{ kJ/kg} \ ;$$
$$h_{fg_3} = 2107.4 \text{ kJ/kg and}$$
$$v_{g_3} = 0.375 \text{ m}^3\text{/kg}$$
$$m_3 = m_1 + m_2 = 2.33 + m_2$$

The mass of steam after mixing is also given by

$$m_3 = \frac{\text{total volume of steam}}{\text{specific volume of steam after mixing}}$$
$$= \frac{1.2}{x_3\, v_{g_3}}$$

$$\therefore\ 2.33 + m_2 = \frac{1.2}{x_3 \times 0.375} = \frac{3.2}{x_3}$$

That gives : $x_3 = \dfrac{3.2}{2.33 + m_2}$ ...(*i*)

Internal energy of steam after mixing,

$$U_3 = m_3(h_3 - p_3v_3)$$
$$= m_3\left[(h_{f_3} + x_3\, h_{fg_3}) - p_3v_3\right]$$
$$= (2.33 + m_2) \times$$
$$\left[(640.1 + x_3 \times 2107.4) - \frac{5 \times 10^5 \times x_3 \times 0.375}{10^3}\right]$$
$$= (2.33 + m_2)\ [640.1 + 1920\ x_3)$$
$$= (2.33 + m_2)\left[640.1 + \frac{3.2}{2.33 + m_2} \times 1920\right]$$
$$= 640.1\ m_2 + 7635.4$$

The vessel has a fixed volume and accordingly the enthalpy of steam supplied equals the change in internal energy. Therefore,

$$h_2 = U_3 - U_1$$

or $2696\ m_2 = 640.1\ m_2 + 7635.4 - 5232$

$$\therefore \quad m_2 = \frac{7635.4 - 5232}{2696 - 640.1}$$
$$= \frac{2403.4}{2055.9} = \mathbf{1.169}$$

Then from expression (*i*), the final dryness fraction of steam in the vessel is

$$x_3 = \frac{3.2}{2.33 + 1.169} = \frac{3.2}{3.499} = \mathbf{0.914}$$

Constant pressure process : Consider unit mass of steam undergoing constant pressure heating process from initial wet state (pressure, p_1 and dryness fraction x_1) to the final superheated state 2 (pressure p_2 and temperature T_{sup2}); $p_1 = p_2 = p$.

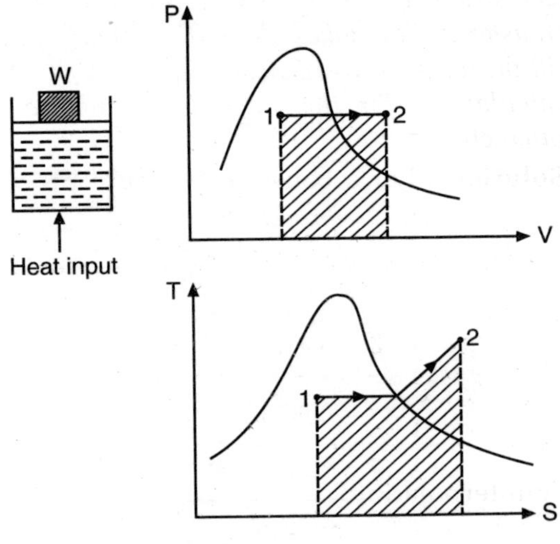

Fig. 13.8.

The specific volumes and specific enthalpies at the initial and final states can be found from the steam tables.

$$v_1 = x_1\, v_{g1} \ ; \ h_1 = h_{f1} + x_1\, h_{fg1}$$
$$v_2 = v_{g2} \times \frac{T_{sup2}}{T_{sat2}} \ ;$$
$$h_2 = h_{g_2} + c_p\,(T_{sup2} + T_{sat2})$$

(*i*) Work done $w_{1\text{-}2} = \displaystyle\int_1^2 p\, dV = p(v_2 - v_1)$

(*ii*) The enthalpy h and internal energy are correlated as $h = u + p\,v$. Then

Change in internal energy

$$(u_2 - u_1) = (h_2 - h_1) - (p_2v_2 - p_1v_1)$$
$$= (h_2 - h_1) - p(v_2 - v_1)$$

(*iii*) Applying the first law of thermodynamics

$$\delta q = \delta w + du \ ;$$

$$q_{1-2} = p(v_2 - v_1) + (h_2 - h_1) - p(v_2 - v_1)$$
$$= (h_2 - h_1)$$

Apparently the heat transfer at constant pressure is equal to changes in enthalpy.

The generation of steam in a boiler represents a constant pressure heating process.

EXAMPLE 13.36

A piston cylinder assembly contains one kg of wet steam of quality 0.8 at 0.1 MPa. Energy transfer as heat takes place at constant pressure till the temperature of steam rises to 350°C. Make calculations for the work done and the heat interaction.

Solution : For a constant pressure process

$$w_{1-2} = p\ (v_2 - v_1)$$

For steam tables at 0.1 MPa (1 bar)

$$v_1 = (1 - x_1)\ v_{f1} + x_1\ v_{g1}$$
$$\simeq x_1\ v_{g1}$$
$$= 0.8 \times 1.694$$
$$= 1.3552\ \text{m}^3/\text{kg}$$
$$v_2 = v_{sup}\ \text{(1 bar and 350°C)}\ ;$$

by interpolation

$$= \frac{2.639 + 3.103}{2}$$
$$= 2.871\ \text{m}^3/\text{kg}$$

∴ Work done

$$w_{1-2} = 0.1 \times 10^6\ (2.871 - 1.3552)$$
$$= 151580\ \text{J}$$
$$= \mathbf{151.58\ kJ/kg}$$

(*ii*) Again from steam tables at 0.1 MPa (1 bar)

$$h_1 = h_{f1} + x_1\ h_{fg1}$$
$$= 417.5 + 0.8 \times 2257.9$$
$$= 2223.82\ \text{kJ/kg}$$
$$h_2 = h_{sup}\ \text{(1 bar and 350°C)}$$
$$= \frac{3074.3 + 3278.2}{2}$$
$$= 3176.25\ \text{kJ/kg}$$

For a constant pressure process the heat interaction equals the change in enthalpy

∴ Heat interaction $= (h_2 - h_1)$

$$= 3176.25 - 2223.82$$
$$= \mathbf{952.43\ kJ/kg}$$

EXAMPLE 13.37

Wet steam at 20 bar pressure and 0.9 dryness fraction is heated reversibly at constant pressure to a temperature of 300°C. Make calculations for the work done, heat supplied and changes in internal energy and entropy. Represent the process on T – s diagram and indicate there upon the area which represents the heat interaction.

Solution : Refer to Fig 13.9.

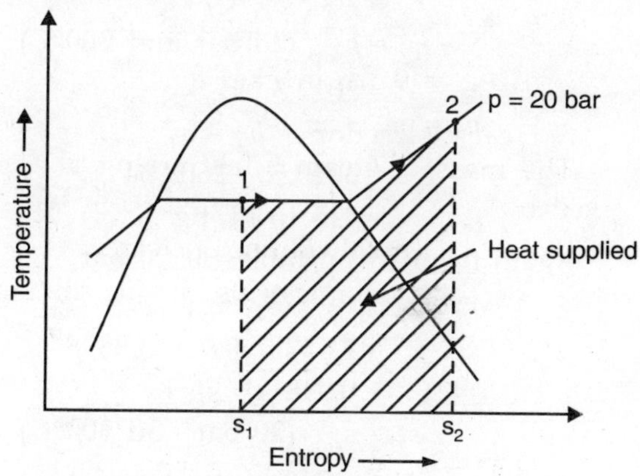

Fig. 13.9.

For a constant pressure process,

$$w_{1-2} = p\ (v_2 - v_1)$$

From steam tables at 20 bar

$$v_1 = (1 - x_1)\ v_{f1} + x_1\ v_{g1}$$
$$\simeq x_1\ v_{g1} = 0.9 \times 0.0995$$
$$= 0.08955\ \text{m}^3/\text{kg}$$
$$v_2 = v_{sup}\ \text{(20 bar and 300°C)}$$
$$= 0.125\ \text{m}^3/\text{kg}$$

∴ Work done w_{1-2}

$$= 2 \times 10^5\ (0.125 - 0.08955)$$
$$= 70900\ \text{J/kg} = \mathbf{70.9\ kJ}$$

Again from steam tables at 20 bar,

$$u_1 = u_{f1} + x_1\ u_{fg1}$$
$$= 907 + 0.9 \times 1693$$
$$= 2430.7\ \text{kJ/kg}$$
$$u_2 = u_{sup}\ \text{(20 bar and 300°C)}$$
$$= 2772.6\ \text{kJ/kg}$$

∴ Change in internal energy $du = (u_2 - u_1)$

$$= 2772.6 - 2430.7$$
$$= \mathbf{341.3\ kJ/kg}$$

From the principle of energy conservation,

$$q_{1-2} = w_{1-2} + du$$
$$= 70.9 + 341.3$$
$$= \mathbf{412.2\ kJ/kg}$$

Alternatively : For a constant pressure process, the heat interaction equals the change in enthalpy.

$$h_1 = h_{f1} + x_1\, h_{fg1}$$
$$= 908.6 + 0.9 \times 1888.6$$
$$= 2608.34 \text{ kJ/kg}$$
$$h_2 = h_{sup} \text{ (20 bar and 300°C)}$$
$$= 3023.5 \text{ kJ/kg}$$

$\therefore$ Heat interaction $= h_2 - h_1$
$$= 3023.5 - 2608.34$$
$$= 415.16 \text{ kJ/kg}$$

Again from steam tables at 20 bar,

$$s_1 = s_{f1} + x_1\, s_{fg1}$$
$$= 2.447 + 0.9 \times 3.8898$$
$$= 0.948 \text{ kJ/kg K}$$
$$s_2 = s_{sup} \text{ (20 bar and 300°C)}$$
$$= 6.766 \text{ kJ/kg K}$$

$\therefore$ Entropy change $s_2 - s_1$
$$= 6.766 - 5.948$$
$$= \mathbf{0.818\ kJ/kg\ K}$$

EXAMPLE 13.38.

A piston cylinder arrangement contains 2.5 kg of wet steam at 1.5 bar pressure and 1.875 m^3 initial volume. The piston is free to move up or down unless it reaches the stops at the top. The steam is heated until the temperature reaches 400°C; the piston is up against the stops and the volume becomes 3.87 m^3. Make calculations for the amount of work and heat interactions.

Solution : Initially the specific volume of steam is

$$v_1 = \frac{1.875}{2.5} = 1.75 \text{ m}^3/\text{kg}$$

From steam tables, specific volume of dry saturated steam at 1.5 bar pressure is 1.159 m^3/kg. Obviously then the steam is initially wet and its dryness fraction is

$$x_1 = \frac{0.75}{1.159} = 0.647$$

$\therefore$ Enthalpy of steam at the initial state is

$$h_1 = h_{f1} + x_1 h_{fg1}$$
$$= 467.1 + 0.647 \times 2226.2$$
$$= 1907.45 \text{ kJ/kg}$$
$$u_1 = h_1 - p_1 v_1$$
$$= 1907.45 - \frac{1.5 \times 10^5 \times 1.875}{10^3}$$
$$= 1666.2 \text{ kJ/kg}$$

Finally the steam is at 400°C and has a specific volume equal to

$$v_2 = \frac{3.87}{2.5} = 1.548 \text{ m}^3/\text{kg}$$

From steam tables it is found that at 400°C when specific volume of steam is 1.548 m^3/kg, the pressure of steam is 2 bar. Correspondingly,

h_2 = specific enthalpy at 2 bar pressure and 400°C temperature
$$= 3276.6 \text{ kJ/kg}$$
$$u_2 = h_2 - p_2 v_2$$
$$= 3276.6 - \frac{2 \times 10^5 \times 1.548}{10^3}$$
$$= 2967 \text{ kJ/kg}$$

$\therefore$ Heat added during the process
$$= m(h_2 - h_1)$$
$$= 2.5(3276.6 - 1907.45)$$
$$= 3423 \text{ kJ}$$

Change in the internal energy during the process $= m(u_2 - u_1)$
$$= 2.5(2967 - 1666.2) \approx 3252 \text{ kJ}$$

Then applying the first law for the given process : $\delta Q = \delta w + dU$

work done = 3423 – 3252 = **171** kJ

The + ve signs with heat and work interactions indicates that heat is supplied to steam and work is done by the expanding steam.

EXAMPLE 13.39.

A piston-cylinder assembly is placed in a vertical position and contains 1.8 kg of water at 20°C. The assembly is well insulated, the cylinder is 0.1 m^2 in cross-sectional area and the piston exerts a pressure equivalent to 8 bar of water. The water is supplied heat by 0.5 kW electric heater placed at the bottom of cylinder. How much time will elapse for the piston to rise upwards through a distance of 4 m ?

Solution : At 20°C, specific volume of water,

$v_f = 0.001002 \text{ m}^3/\text{kg}$

Total initial volume of water

$= 1.5 \times 0.001002$

$= 1.53 \times 10^{-3} \text{ m}^3/\text{kg}$

The piston moves upward when steam is generated consequent to supply of heat.

Change in volume due to rise of piston,

$= 4 \times 0.1 = 0.4 \text{ m}^3$

∴ Final volume when steam is formed,

$= 0.4 + 1.53 \times 10^{-3} \simeq 0.40153 \text{ m}^3$

Specific volume of steam,

$$v = \frac{0.40153}{1.8} = 0.2231 \text{ m}^3/\text{kg}$$

At 8 bar, specific volume of dry saturated steam,

$v_g = 0.240 \text{ m}^3/\text{kg}$

Since $v < v_g$, the steam is wet with dryness fraction

$$x = \frac{v}{v_g} = \frac{0.2231}{0.240} = 0.929$$

Now, h_1 = enthalpy of water at 20°C

$= 84 \text{ kJ/kg}$

h_2 = enthalpy of steam at 8 bar and 0.929 dry

$= h_{f2} + x_2 h_{fg2}$

$= 720.9 + 0.929 \times 2046.5$

$= 2622.1 \text{ kJ/kg}$

The steam is being raised at constant pressure, and at a constant pressure process the heat transfer equals the change in enthalpy.

∴ Heat transfer, $q_{1-2} = (h_2 - h_1)$

$= 2622.1 - 84 = 2538.1 \text{ kJ/kg}$

Total heat added by the heater

$= 1.8 \times 2538.1 = 4568.58 \text{ kJ}$

If t is the time elapsed in seconds , then

$$0.5 \times t = \frac{4568.58}{0.5} = 9137.16 \text{ s}$$

$= \mathbf{2.538 \ hours}$

EXAMPLE 13.40.

A drum of capacity 2 m³ is completely evacuated, sealed and contains saturated water at 65°C. Heat is supplied and measurements show that the steam generated is at pressure 15 bar and 250°C temperature. Make calculations for the mass of water contained in the drum and the heat supplied. Sketch the process on temperature entropy diagram.

Solution : The drum is completely filled with the steam generated

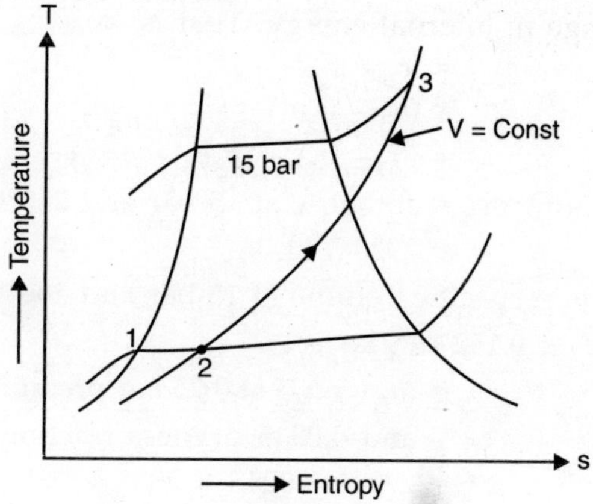

Fig. 13.10.

At 15 bar pressure and 250°C, the steam is superheated and

$v_3 = v_{sup3} = 0.152 \text{ m}^3/\text{kg}$

∴ Mass of water = mass of steam

$$= \frac{\text{volume of drum}}{\text{specific volume of steam}}$$

$$= \frac{0.2}{0.152} = \mathbf{1.316} \text{ kg}$$

(*b*) Corresponding to saturated water at 65°C

$p_1 = 0.25 \text{ bar} \,;\; h_{f_1} = h_{f_2} = 272 \text{ kJ/kg}$

$h_{fg_2} = 2346.4 \text{ kJ/kg} \;\; v_{g_1} = v_{g_2} = 6.205 \text{ m}^3$

Initially the heating takes place at constant pressure till the entire space with 2 m³ volume is filled with wet steam at 0.25 bar pressure. Then if x_2 is the dryness fraction of wet steam,

$V = 0.2 = m(x_2 v_{g2})$

$= 1.316 \times x_2 \times 6.205$

$$\therefore \quad x_2 = \frac{0.2}{1.316 \times 6.205} = 0.0245$$

Heat added during constant pressure (p = 0.25 bar) process is

$= (h_{f_2} + x_2 h_{fg_2}) - h_{f_1}$

$= x_2 h_{fg_2} = 0.0245 \times 2346.4$

$= 57.49 \text{ kJ/kg}$

Once the drum is filled with steam at 0.25 bar pressure, further addition of heat would take place at constant volume till the required condition is reached.

Heat added at constant volume equals the change in internal energy. That is

$$\begin{aligned} q_{2-3} &= u_3 - u_2 \\ &= (h_3 - p_3 v_3) - (h_2 - p_2 v_2) \\ &= (h_3 - h_2) - (p_3 v_3 - p_2 v_2) \end{aligned}$$

Now : h_3 = enthalpy at 15 bar and 250°C
= 2923.3 kJ/kg

v_3 = specific volume at 15 bar and 250°C
= 0.152 m³/kg

$h_2 = (h_f + x\,h_{fg})$ at 0.25 bar pressure and 0.0245 dryness fraction

$= 272 + 0.0245 \times 2346.4$

$= 329.5$ kJ/kg

v_2 = specific volume at 0.25 bar pressure and 0.0245 dryness fraction

$= 0.0245 \times 6.205 = 0.152$ m³/kg

$$\therefore \quad q_{2-3} = (2923.3 - 329.5) - \frac{15\times10^5\times0.152 - 0.25\times10^5\times0.152}{10^3}$$

$= 2593.8 - 224.2 = 2369.6$ kJ/kg

Heat added during generation of steam

$= 57.49 + 2369.6$

$= 2427.09$ kJ/kg

$= 2427.09 \times 1.316 =$ **3194** kJ

Reversible adiabatic or isentropic process: Consider a unit mass of steam undergoing reversible adiabatic expansion process from initial superheated state 1 (pressure p_1 and temperature T_{sup1}) to final wet state 2 (pressure p_2 and dryness fraction x_2)

(*i*) During isentropic process, entropy remains constant. Thus, $s_1 = s_2$ and so

$$s_{g_1} + c_p \log_e \frac{T_{sup1}}{T_{sat1}} = s_{f_2} + x_2\, s_{fg_2}$$

$$\therefore \quad x_2 = \frac{s_{g_1} + c_p \log_e \dfrac{T_{sup1}}{T_{sat1}} - s_{f_2}}{s_{fg_2}}$$

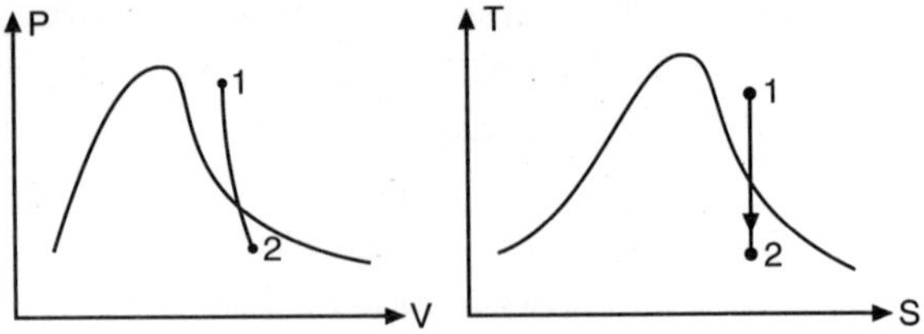

Fig. 13.11.

(*ii*) Change in internal energy,

$$\begin{aligned} du &= u_2 - u_1 \\ &= (h_2 - p_2 v_2) - (h_1 - p_1 v_1) \end{aligned}$$

where $v_1 = v_{g_1} \times \dfrac{T_{sup1}}{T_{sat1}}$

$v_2 = x_2\, v_{g_2}$

$h_1 = h_{g_1} + c_{p(T_{sup\,1} - T_{sat\,1})}$

and $h_2 = h_{g_2} + x_2\, h_{fg_2}$

(*iii*) For a reversible adiabatic process, no heat is added to or subtracted from the system. That is

$$q_{1-2} = 0$$

Then from first law of thermodynamics ($\delta q = \delta w + du$)

Work done w_{1-2}

$$\begin{aligned} &= -\,du = u_1 - u_2 \\ &= (h_1 - p_1 v_1) - (h_2 - p_2 v_2) \end{aligned}$$

EXAMPLE 13.41

Dry saturated steam initially at a pressure of 600 kPa is contained in a thermally insulated cylinder fitted with a frictionless piston. As the piston moves slowly outwards, the steam expands to a pressure of 60 kPa. Calculate the work done by steam.

Solution : At state point 1 :

$p_1 = 600$ kPa (6 bar) and $x_1 = 1$

$s_1 = s_{g1} = 6.761$ kJ/kg K

$u_1 = u_{g1} = 2568$ kJ/kg

At state point 2 :

$p_2 = 60$ kPa (0.6 bar)

$u_{f2} = 360$ kJ/kg ; $u_{fg2} = 2129$ kJ/kg

$s_{f2} = 1.145$ kJ/kg K ; $s_{fg} = 6.386$ kJ/kg K

Since the cylinder is perfectly insulated, there is no heat interaction, *i.e.*, the process is adiabatic and therefore,

$$s_1 = s_2$$
$$6.761 = 1.145 + x_2 \times 6.386$$
$$\therefore \quad x_2 = \frac{6.761 - 1.145}{6.386} = 0.879$$
and $u_2 = u_{f2} + x_2\, u_{fg2}$
$$= 360 + 0.879 \times 2129$$
$$= 2231.39 \text{ kJ/kg}$$

From the non-flow energy equation
$$q_{1-2} = w_{1-2} + (u_2 - u_1)$$
$$0 = w_{1-2} + (2231.39 - 2568)$$
$\therefore$ Work done
$$w_{1-2} = 2568 - 2231.39$$
$$= \mathbf{336.61\ kJ/kg}$$

Constant temperature (isothermal) process : Consider unit mass of steam undergoing heat addition at constant temperature from initial wet state 1 (pressure p_1 and dryness fraction x_1) to final superheated state 2 (pressure p_2 and temperature T_{sup2})

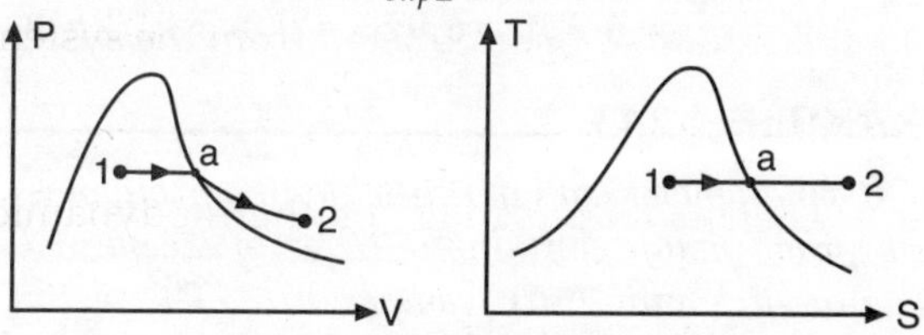

Fig. 13.12.

For an isothermal process, the temperature and pressure remain constant in the wet region, *i.e.*, $p_1 = p_a$ and $T_1 = T_2$. Once the steam is in superheated region, it behaves like a perfect gas and obeys the gas laws. Apparently then the expansion process $a - 2$ in superheated region is hyperbolic, *i.e.*, $p_a v_a = p_2 v_2$.

The entire process 1-2 can be analysed in two parts :

***(i)* Process 1-*a* :** Heating of wet steam at a constant temperature till it becomes dry saturated. This is similar to heating at constant pressure and accordingly
$$w_{1-a} = p(v_a - v_1)$$
where $\quad p_1 = p_a = p$
$$u_a - u_1 = (h_a - h_1) - p(v_a - v_1)$$
$$q_{1-a} = h_a - h_1$$

***(ii)* Process *a*-2 :** Heating of steam at constant temperature in the superheated region is hyperbolic, *i.e*, it follows the Boyle's law $p\,v$ = constant. For that
$$p_a v_a = p_2 v_2\,; \qquad v_2 = \frac{p_a}{p_2} v_a$$

Then the temperature T_{sup2} can be worked out by applying Charles' law. That is
$$\frac{v_2}{T_2} = \frac{v_a}{T_a}$$
where $\quad T_a$ = saturation temperature corresponding to pressure $p_a (= p_1)$
and $\quad v_a$ = specific volume of dry saturated steam at pressure p_a

Polytropic process (pv^n = constant) : Consider unit mass of steam undergoing polytropic process from initial state 1 (pressure p_1, dryness fraction x_1) to final state (pressure p_2, dryness fraction x_2). Then
$$p_1 v_1^n = p_2 v_2^n$$
$$p_1 (x_1 v_{g_1})^n = p_2 (x_2 v_{g_2})^n$$
$$\therefore \quad x_2 = \frac{x_1 v_{g_1}}{v_{g_2}} \left(\frac{p_1}{p_2}\right)^{1/n}$$

Depending on the value of x_2, the steam at its final state would be wet (if $x_2 < 1$); dry saturated (if $x_2 = 1$) and superheated (if $x_2 > 1$).

The volume and temperature of the superheated steam can be determined as follows:
$$p_1 (x_1 v_{g_1})^n = p_2 (v_{sup2})^n$$
$$v_{sup2} = x_1 v_{g_1} \left(\frac{p_1}{p_2}\right)^{1/n}$$
$$\text{Further,} \quad \frac{v_{sup2}}{T_{sup2}} = \frac{v_{g_2}}{T_{sat2}}$$
$$T_{sup2} = T_{sat2} \times \frac{v_{sup2}}{v_{g_2}}$$
where T_{sat2} is the saturation temperature corresponding to pressure p_2

(*i*) Work done $w_{1-2} = \dfrac{p_1 v_1 - p_2 v_2}{n-1}$

(*ii*) Change in internal energy

$$du = u_2 - u_1$$
$$= (h_2 - p_2v_2) - (h_1 - p_1v_1)$$

(*iii*) Heat absorbed or heat transferred

$$q_{1-2} = w_{1-2} + du$$
$$= w_{1-2} + (u_2 - u_1)$$

The enthalpy values are read from the steam tables corresponding to the initial and final conditions of steam. That is

$$h_1 = h_{f_1} + x_1 h_{fg_1}$$
$$h_2 = h_{f_2} + x_2 h_{fg_2}$$

for wet state

$$= h_{f_2} + h_{fg_2}$$

for dry saturated state

$$= h_{f_2} + h_{fg_2} + c_p(T_{sup2} - T_{sat2})$$

for superheated state

EXAMPLE 13.42

Steam at 10 bar and 250°C expands polytropically to a pressure of 1 bar. If the exponent n for the polytropic process is 1.2, calculate (i) specific volume and temperature after expansion, (ii) heat interaction, and (iii) change in entropy.

Solution : At 10 bar and 250°C (steam is superheated)

$v_1 = 0.2328$ m³/kg ; $u_1 = 2711$ kJ/kg
$h_1 = 2944$ kJ/kg and $s_1 = 6.926$ kJ/kg K

As $p_1 v_1^n = p_2 v_2^n$; $v_2 = v_1 (p_1/p_2)^{1/n}$

∴ $v_2 = 0.2328 \times (10)^{1/1.2}$
$= \mathbf{1.586\ m^3/kg}$

At 1 bar (state point 2)

$v_{g2} = 1.694$ m³/kg ; $u_{f2} = 417$ kJ/kg
$u_{fg2} = 2089$ kJ/kg ; $h_{f2} = 417$ kJ/kg
$h_{fg2} = 2258$ kJ/kg ; $s_{f2} = 1.303$ kJ/kg K
and $s_{fg2} = 6.056$ kJ/kg K

Since $v_2 < v_{g2}$, the steam is wet with dryness fraction

$$x_2 = \frac{v_2}{v_{g2}} = \frac{1.586}{1.694} = 0.936$$

The temperature after expansion will be the saturation temperature corresponding to 1 bar, *i.e.*, **99.6°C**

$$u_2 = u_{f2} + x_2 u_{fg2}$$
$$= 417 + 0.936 \times 2089 = 2372.8 \text{ kJ/kg}$$
$$h_2 = h_{f2} + x_2 h_{fg2}$$
$$= 417 + 0.936 \times 2258 = 2530.5 \text{ kJ/kg}$$
$$s_2 = s_{f2} + x_2 s_{fg2}$$
$$= 1.303 + 0.936 \times 6.056 = 6.971 \text{ kJ/kg}$$

$$\text{Work done } w_{1-2} = \frac{p_1 v_1 - p_2 v_2}{n - 1}$$
$$= \frac{10 \times 10^5 \times 0.2328 - 1 \times 10^5 \times 1.586}{1.2 - 1}$$
$$= 371000 \text{ J} = 371 \text{ kJ}$$

From principle of energy conservation,
Heat interaction, q_{1-2}

$$= (u_2 - u_1) + w_{1-2}$$
$$= (2372.8 - 2711) + 371$$
$$= \mathbf{32.8\ kJ/kg}$$

Entropy change $= s_2 - s_1$
$= 6.971 - 6.926 = \mathbf{0.045\ kJ/kg\ K}$

EXAMPLE 13.43.

A 30 cm diameter cylinder fitted with a frictionless leak proof piston contains 0.02 kg of steam at 6 bar pressure and 200°C temperature. The piston moves outwards steadily through a distance of 25 cm and the pressure of steam falls to 1.25 bar. During this expansion process, the pressure p and volume V are related by the polytropic law pV^n = constant. Make calculations for the value of exponent n, work done by steam and the heat interchange between steam and the surrounding walls of the cylinder.

Solution : From saturation steam tables, the saturation temperature corresponding to 6 bar pressure is

$$(ts)_p = 6 \text{ bar} = 158.9°\text{C}$$

Since temperature of the given steam (t = 200°C) is greater than the saturation temperature, the steam is initially superheated.

From the steam tables for supherheated steam, it may be noted that at p_1 = 6 bar and t_1 = 200°C,

$$v = 0.352 \text{ m}^3/\text{kg}$$
$$\therefore \quad V_1 = mv_1 = 0.02 \times 0.352$$
$$= 0.00704 \text{ m}^3$$

$$V_2 = V_1 + \text{displacement volume}$$

$$= 0.00704 + \frac{\pi}{4}(0.3)^2 \times 0.25$$

$$= 0.02470 \text{ m}^3$$

(*a*) From the polytropic law : $p_1 V_1^n = p_2 V_2^n$

$$n = \frac{\log_e (p_1/p_2)}{\log_e (V_2/V_1)}$$

$$= \frac{\log_e (6/1.25)}{\log_e (0.02470/0.00704)}$$

$$= \frac{\log_e 4.8}{\log_e 3.508} = \frac{1.5686}{1.255} = 1.25$$

Hence the exponent n for the polytropic expansion process is 1.25.

(*b*) Work done during a polytropic process is given by

$$W = \frac{p_1V_1 - p_2V_2}{n-1}$$

$$= \frac{(6 \times 10^5 \times 0.00704) - (1.25 \times 10^5 \times 0.02470)}{1.25 - 1}$$

$$= 4546 \text{ Nm (or J)} = \mathbf{4.546} \text{ kJ}$$

(*c*) From superheated steam tables, the enthalpy of steam at p = 6 bar and t = 200°C is h_1 = 2850.1 kJ/kg

Further at state point 2,

$$v_2 = \frac{V_2}{m} = \frac{0.0247}{0.02} = 1.235 \text{ m}^3/\text{kg}$$

From saturation steam tables, the specific volume of dry saturated steam at 1.25 bar is 1.376 m³/kg. Obviously the steam is wet and its dryness fraction is

$$x_2 = \frac{1.235}{1.376} = 0.897$$

Then $h_2 = (h_f + x\, h_{fg})_{\text{at } 1.25 \text{ bar}}$

$$= 444.3 + 0.897 \times 2241$$

$$= 2454.5 \text{ kJ/kg}$$

Change in internal energy

$$U_2 - U_1 = m[(h_2 - p_2v_2) - (h_1 - p_1v_1)]$$

$$= 0.02\left[\left\{2454.5 - \frac{1.25 \times 10^5 \times 1.235}{10^3}\right\} - \left\{2850.1 - \frac{6 \times 10^5 \times 0.352}{10^3}\right\}\right]$$

$$= 0.02 \,[2300 - 2639] = \mathbf{-\,6.78} \text{ kJ}$$

Applying the first law for the given process : $\delta Q + \delta w + dU$

$$Q_{1-2} = W_{1-2} + (U_2 - U_1)$$

$$= 4.546 + (-\,6.78) = \mathbf{-\,2.234} \text{ kJ}$$

The negative sign indicates that the heat is lost by the steam as it expands.

EXAMPLE 13.44

Wet steam at 1.5 bar and 0.8 dry occupies a volume of 0.05 m³ in a cylinder where compression takes place so that pv = constant (that is hyperbolically) until the volume has fallen to one-fifth of the original volume. Determine the work and heat transfers and the final condition of steam.

Solution : Specific volume of dry saturated steam at 1.5 bar = 1.159 m³/kg

∴ Mass of steam,

$$m = \frac{0.05}{1.159 \times 0.8} = 0.0539 \text{ kg}$$

Final volume,

$$V_2 = \frac{V_1}{5} = \frac{0.05}{5} = 0.01 \text{ m}^3$$

Final pressure,

$$p_2 = p_1\left(\frac{V_1}{V_2}\right) = 1.5 \times 5 = 7.5 \text{ bar}$$

If x_2 is the dryness fraction of steam after compression to 7.5 bar, then

$$x_2 = \frac{V_2}{m \times v_{g2}}$$

$$= \frac{0.01}{0.0539 \times 0.2565} = \mathbf{0.723}$$

where $v_{g2} = \dfrac{0.2728 + 0.2403}{2}$ (by interpolation)

= 0.2565 is the specific volume of dry saturated steam at 7.5 bar.

$$u_1 = u_{f1} + x_1\, u_{fg1}$$

$$= 467 + 0.8 \times 2052$$

$$= 2108.6 \text{ kJ/kg}$$

$$u_2 = u_{f2} + x_2\, u_{fg2}$$

$$= \frac{696 + 720}{2} + 0.723\left(\frac{1877 + 1857}{2}\right)$$

= **2057.8 kJ/kg**

∴ Total change in internal energy

$= m\ (u_2 - u_1)$

$= 0.0539\ (2057.8 - 2108.6)$

$= -\ 2.738$ kJ (decrease)

Work done during compression

$= p_1 V_1 \log_e V_1/V_2$

$= 1.5 \times 10^5 \times 0.05 \log_e 1/5$

$= -\ 12070.8$ J $= -\ 12.07$ kJ

The negative sign implies that work was done on the system

From non-flow energy equation,

$\delta Q = \delta W + dU$

$= -12.07 + (-\ 2.738)$

$= -\ 14.81$ kJ

The negative sign implies that heat will flow from steam to walls of cylinder.

EXAMPLE 13.45.

Steam at 15 bar and 300°C expands isentropically in a steam turbine till the temperature falls to 80°C. Find the condition of steam at the end of expansion process and the work done per kg of air. If the steam flow rate is 10 kg/s, what power will be produced by the turbine ?

Solution : At 15 bar and 300°C (steam is superheated),

$h_1 = h_{sup1} = 3037.6$ kJ/kg

$s_1 = s_{sup1} = 6.918$ kJ/kg K

At 80°C : $s_{f2} = 1.075$ kJ/kg K ;

$s_{fg2} = 6.537$ kJ/kg K

$h_{f2} = 334.9$ kJ/kg

and $h_{fg2} = 2308.8$ kJ/kg

During isentropic expansion of steam, the entropy remains constant, and therefore,

$s_1 = s_2 = s_{f2} + x_2\, s_{fg2}$

or $6.918 = 1.075 + x_2 \times 6.537$

$$\therefore \quad x_2 = \frac{6.918 - 1.075}{6.537} = \mathbf{0.8938}$$

Thus the steam is 0.8938 dry and has a temperature of 80°C after adiabatic expansion.

(*ii*) During steady flow process,

$w_{1\text{-}2} = h_1 - h_2$

$h_1 = h_{sup1} = 3037.6$ kJ/kg

$h_2 = h_{f2} + x_2\, h_{fg2}$

$= 334.9 + 0.8938 \times 2308.8$

$= 2398.5$ kJ/kg

∴ Work done per kg of steam in the turbine,

$= 3037.6 - 2398.5 =$ **639.1 kJ/kg**

Power developed by the steam turbine,

$= 639.1 \times 10 = 6391$ kJ/s = **6391 kW**

EXAMPLE 13.46.

An insulated vessel is divided into two compartments connected by a valve. Initially, one compartment contains steam at 10 bar and the other is evacuated. The valve is opened and the steam is allowed to fill the entire volume, achieving a final pressure of 1 bar. Determine the final temperature, the percentage of the vessel volume initially occupied by steam, and the amount of entropy produced.

Solution : At 10 bar and 500°C (Steam is superheated)

$u_1 = 3124.4$ kJ/kg ; $h_1 = 3478.5$ kJ/kg

$v_1 = 0.354$ m³/kg and $s = 7.762$ kJ/kg K

From non-flow energy equation,

$\delta q = \delta w + du$

For the given arrangement,

$\delta q = 0$ and $\delta w = 0$

$\therefore \quad du = 0$, *i.e.*, $u_2 = u_1$

(no change in internal energy)

At 1 bar and with $u_2 = 3124.4$ kJ/kg , interpolation gives :

$$t_2 = 400 + \frac{3131.6 - 2967.9}{3124.4 - 2967.9} \times 100$$

= **495.6 °C**

$$v_2 = 3.103 + \frac{3.565 - 3.103}{100} \times 95.6$$

= **3.5446 m³/kg**

$$s_2 = 8.544 + \frac{8.834 - 8.544}{100} \times 95.6$$

= **8.8212 kJ/kg K**

% of volume initially occupied by steam

$$= \frac{0.354}{3.5446} = 0.09987 \text{ or } \mathbf{9.987\%}$$

Entropy production $= s_2 - s_1$

$= 8.8212 - 7.762$

$= \mathbf{1.0592 \text{ kJ/kg K}}$

EXAMPLE 13.47.

A rigid vessel is initially divided into two parts A and B by a partition wall. The part A contains 2 kg of dry saturated steam at 4 bar whereas part B contains 3 kg of steam at 8 bar and dryness fraction 0.85. When the partition is removed, the two steams mix and attain a pressure of 5.5 bar. Determine the volume of vessel, condition of steam at 5.5 bar and, heat interchange between steam and surrounding walls of the vessel.

Solution : At 4 bar :

$v_g = 0.462 \text{ m}^3/\text{kg}$

∴ Volume of compartment

A $= 2 \times 0.462 = 0.924 \text{ m}^3$

At 8 bar : $v_g = 0.240 \text{ m}^3/\text{kg}$

∴ Volume of compartment

B $= 3 \times (0.85 \times 0.24) = 0.612 \text{ m}^3$

Volume of vessel $= 0.924 + 0.612$

$= \mathbf{1.536 \text{ m}^3}$

At 5.5 bar : $v_g = 0.3425 \text{ m}^3/\text{kg}$

∴ Dryness fraction of steam at 5.5 bar

$$x = \frac{1.536}{5 \times 0.3425} = \mathbf{0.897}$$

$$U_A = m_A h_A - p_A V_A$$

$$= 2\,(604.7 + 2133) - \frac{4 \times 10^5 \times 0.924}{10^3}$$

$= 5105.8$ kJ

$$U_B = m_B h_B - p_B V_B$$

$$= 3\,(720.9 + 0.85 \times 2046.5) - \frac{8 \times 10^5 \times 0.612}{10^3}$$

$= 6891.7$ kJ

Let suffix f represent the final state when the partition wall is removed.

$$U_f = m_f h_f - p_f V_f$$

$$= 5\,(655.8 + 0.897 \times 2091.5) - \frac{5.5 \times 10^5 \times 1.536}{10^3}$$

$= 11814.6$ kJ

∴ Change in internal energy dU

$= 11814.6 - 5105.8 - 6891.7$

$= -182.9$ kJ

From the law of conservation of energy,

$\delta Q = \delta W + dU$; here $\delta W = 0$

∴ Heat interchange $\delta Q = dU = \mathbf{-182.9 \text{ kJ}}$

The negative sign implies that there is a loss of heat to the surroundings.

EXAMPLE 13.48.

A rigid vessel is divided into two parts A and B by means of a frictionless, perfectly conducting piston.

Initially the part A contains 0.4 m³ of air at 10 bar and part B contains 0.4 m³ of wet steam at 10 bar. Heat is now added to both the parts until all the water in part B is evaporated and pressure becomes 15 bar. Determine the total amount of heat added to both the parts and the initial quality of steam in part B.

Solution : Let suffix 1 and 2 denote the initial and final states of the system. Then as per data given

$$p_{a1} = p_{b1} = 10 \text{ bar}$$

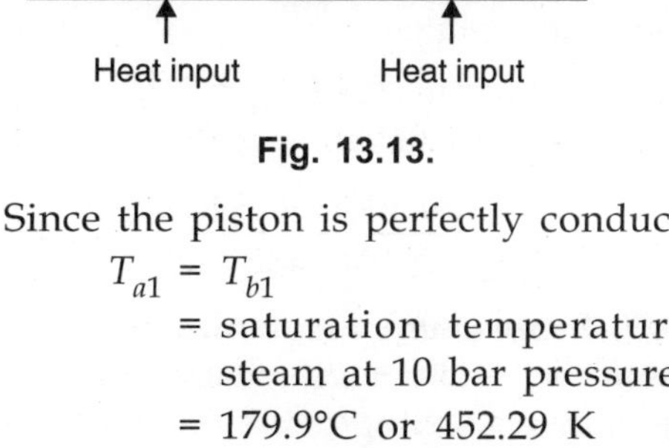

Fig. 13.13.

Since the piston is perfectly conducting,

$T_{a1} = T_{b1}$

= saturation temperature of steam at 10 bar pressure

$= 179.9°\text{C}$ or 452.29 K

Mass of air $m_a = \dfrac{p_{a_1} V_{a_1}}{RT_{a_1}}$

$$= \frac{10 \times 10^5 \times 0.4}{287 \times 452.9} = 3.077 \text{ kg}$$

At the final state :

$$p_{a_2} = p_{b_2} = 15 \text{ bar}$$

This is necessary for balancing.

$T_{a_2} = T_{b_2}$

= saturation temperature of steam at 15 bar pressure

= 198.3°C or 471.3 K

Then by applying the gas law PV/T = constant, the volume of air at the final state is

$$V_{a_2} = V_{a_1} \times \frac{T_{a_2}}{T_{a_1}} \times \frac{p_{a_1}}{p_{a_2}}$$

$$= 0.4 \times \frac{471.3}{452.9} \times \frac{10}{15} = 0.2775 \text{ m}^3$$

Then volume of steam at the final condition is

$$V_{b2} = 0.4 + 0.4 - 0.2775 = 0.5225 \text{ m}^3$$

At 15 bar, the specific volume of dry saturated steam is 0.132 m³/kg.

Then mass of steam

$$m_s = \frac{0.5225}{0.132} = 3.958 \text{ kg}$$

Initial volume of steam

$$V_1 = \frac{V_{b_1}}{m_s} = \frac{0.4}{3.958} = 0.1011 \text{ m}^3/\text{kg}$$

Then if x_1 is the dryness fraction of steam at the initial state,

$$v_1 = v_{f1} + x_1(v_{g1} - v_{f1})$$

$$\text{or } 0.1011 = 0.001127 + x_1(0.194 - 0.001127)$$

$$\therefore \quad x_1 = \frac{0.1011 - 0.001127}{0.194 - 0.001127} = \mathbf{0.518}$$

(*b*) From energy equation for a closed system : $\delta Q = \delta W + dU$. Here work done is zero and accordingly the heat interaction equals the change in internal energy. That is

$$Q = dU = (dU)_{air} + (dU)_{steam}$$

Now : $(dU)_{air} = m_a\, c_{va}\, (dT)_a$

$$= 3.077 \times 0.718\, (471.3 - 452.9)$$
$$= 40.65 \text{ kJ}$$

$$(dU)_{steam} = (h_2 - p_2v_2) - (h_1 - p_1v_1)$$
$$= (h_2 - h_1) + (p_1v_1 - p_2v_2)$$

From steam tables :

$$h_2 = h_{g2} = 2789.9 \text{ kJ/kg}$$

$$h_1 = h_{f1} + x_1 h_{fg1}$$
$$= 762.6 + 0.518 \times 2013.6$$
$$= 1805.6 \text{ kJ/kg}$$

$$v_2 = v_{g2} = 0.132 \text{ m}^3/\text{kg}$$
$$v_1 = 0.1011 \text{ m}^3/\text{kg}$$

Then,

$$(dU)_{steam} = (2789.9 - 1805.6) + \frac{10 \times 10^5 \times 0.1011}{10^3} - \frac{15 \times 10^5 \times 0.132}{10^3}$$

$$= 984.3 + 101.1 - 198$$
$$= 887.4 \text{ kJ/kg}$$
$$= 887.4 \times 3.958 = 3512.3 \text{ kJ}$$

∴ Heat interaction

$$= 40.65 + 3512.3 = \mathbf{3552.95} \text{ kJ}$$

13.12. TEMPERATURE-ENTROPY (T-s) AND ENTHALPY-ENTROPY (H-s) DIAGRAMS

The solutions to problems dealing with steam vapour as working substance are easily and rapidly obtained by referring to charts/ diagrams which have been prepared by having the data from steam tables. The salient features of two such plots which are of considerable importance are mentioned below :

(*a*) Temperature-entropy (*T-s*) diagram:

(*i*) The absolute temperature is plotted on the axis of ordinate and entropy on the abscissa.

(*ii*) The liquid boundary line originates at the axis of ordinates at a temperature of 273.16 K ; this has to be so because entropy of liquid at the triple point is equal to zero.

(*iii*) The boundary curves (liquid line *ABCDE*) and the saturated vapour line (*EFGHI*) divide the diagram into three parts ; to the

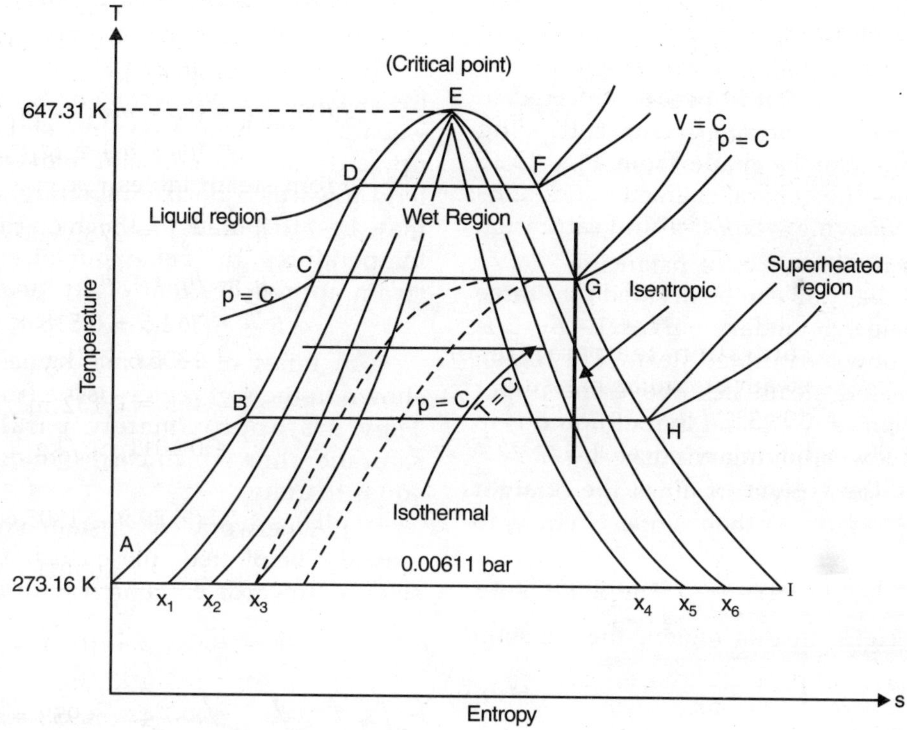

Fig. 13.14. Temperature – entropy diagram

left of AE is the liquid region, between the line AE and EI is the wet vapour region, and to the right of line EI is the region of superheated steam.

(*iv*) The boundary curves meet at point E which is the critical point of water with pressure 221.2 bar and temperature 647.31 K.

(*v*) The constant pressure lines are parallel to the constant temperature lines in the wet region. These lines become curved in the superheated region.

(*vi*) The wet region has plots of constant dryness fraction and constant volume lines.

(*vii*) The constant volume lines are steeper than constant pressure lines in the superheated zone.

(*viii*) The slopes of constant pressure line in the liquid, wet and superheated region are governed by the relations

$$\left(\frac{\partial T}{\partial s}\right)_v = \frac{T}{c_v} \quad \text{and} \quad \left(\frac{\partial T}{\partial s}\right)_p = \frac{T}{c_p}$$

(*ix*) The area confined between the line of a reversible process and the abscissa represents the quantity of heat added to 1 kg of working medium, and is equal to $q = \int T\, ds$

(*x*) An isothermal process plots as horizontal line and a reversible adiabatic process is represented by a vertical line.

The temperature-entropy diagram permits observation of change in the temperature of working medium and makes it possible to find the quantity of heat involved in the process. Accordingly it is put to an extensive use in the investigation of thermodynamic processes and cycles.

(*b*) Enthalpy-entropy ($h-s$) diagram :

Steam generation is done at constant pressure and for constant pressure heating $\delta q = dh$. Thus calculations for heat supplied can be done readily if enthalpy-entropy diargam is drawn rather than temperature-entropy diagram. In the enthalpy-entropy

$(h-s)$ diagram, enthalpy is laid along the axis of ordinate and entropy along the axis of abscissa. Such a diagram was suggested by Mollier in 1904 and it is usually referred to as the *Mollier diagram*. The following observations can be made from Fig. 13.15 which shows the general characteristics of h–s diagram showing water line and saturation line as one-curve :

(*i*) The diagram can be divided into three regions. Below boundary curves 1–A–2 is the region of wet saturated steam. The region of superheated steam lies above the upper boundary curve A–2, and that of liquid is to the left of lower boundary curve 1–A.

(*ii*) Constant pressure lines are straight in the wet region as their slope is equal to $\frac{dh}{ds} = \frac{dq}{ds} = T$, and have converging tendency towards origin. In this region, the constant temperature lines coincide with the constant pressure lines.

(*iii*) In the region of superheated steam, the isobaric and isothermal lines diverge; the isobaric lines have a positive slope and turn upwards, and the constant temperature lines tend towards a horizontal straight line. This may be attributed to the fact that at high temperatures, the behaviour of superheated steam approaches that of an ideal gas :

$$h = f(T) \text{ alone}$$

(*iv*) Lines of constant dryness fraction, shown in the wet region, issue from point A. They are approximately parallel to the saturation line which corresponds to quality equal to unity.

(*v*) A network of constant volume lines may also be plotted ; these isochors rise more steeply upwards compared with the isobars.

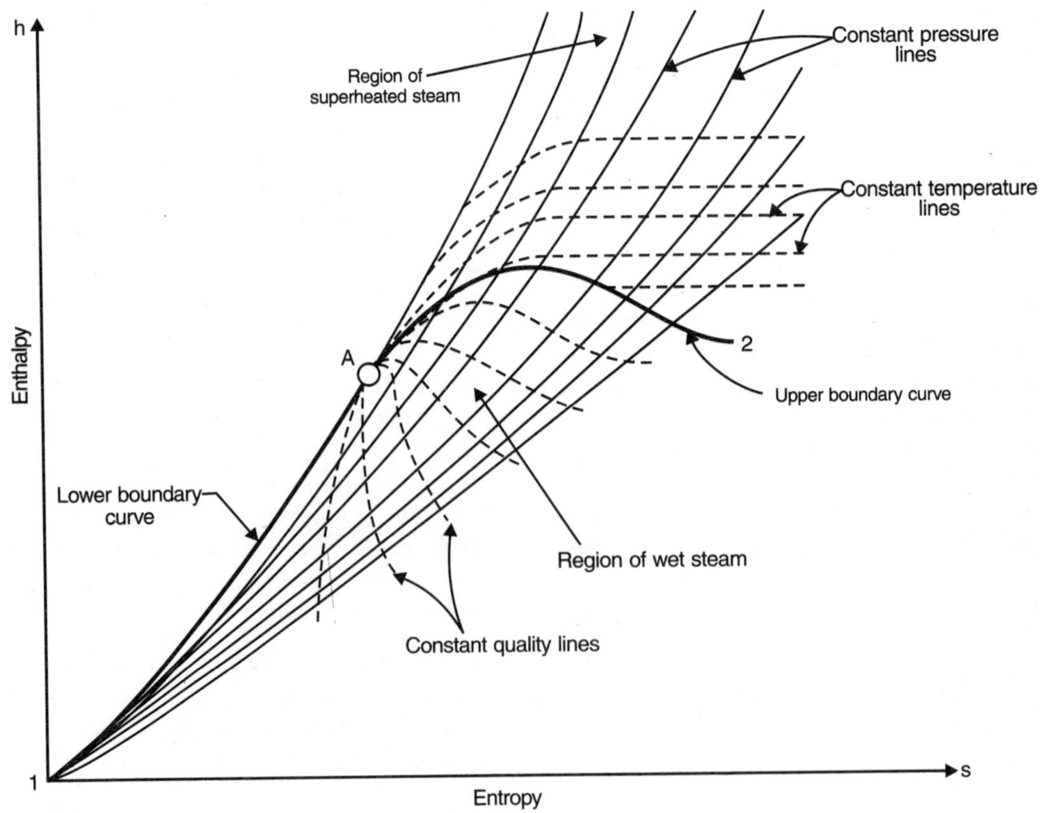

Fig. 13.15. Enthalpy – entropy diagram (Mollier Chart)

(*vi*) The throttling process, which is a constant enthalpy process, appears as a horizontal line. A reversible adiabatic process plots as a vertical line.

The saturated liquid region is not required for solving engineering problems. This makes it possible to make the working part of the diagram on a larger scale for greater convenience. Accordingly, in the commercial Mollier charts, the water line is dispensed with and only the dry steam line is drawn (Fig. 13.16). This plot has the usual constant temperature, constant pressure and constant quality lines. The throttling process has been shown represented by line 1–2–3, and the isentropic process by line $a-b-c$.

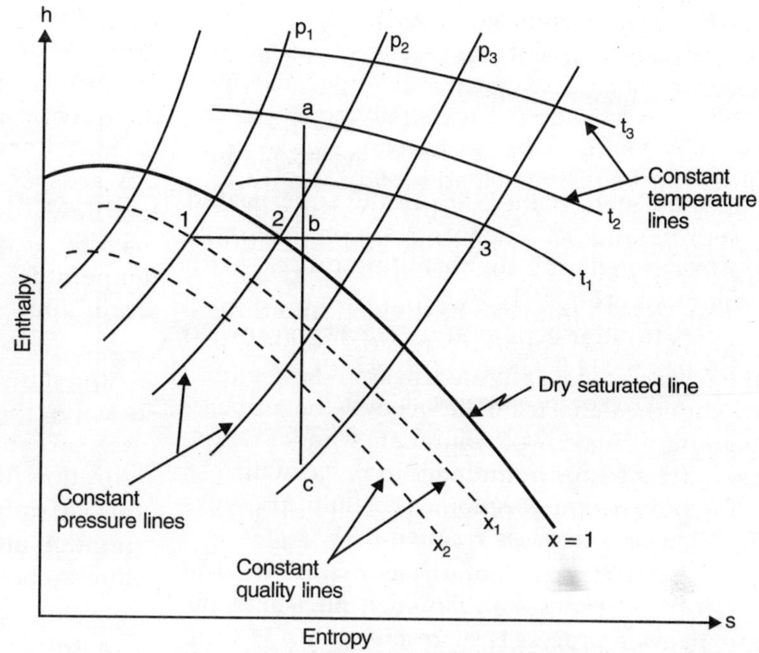

Fig. 13.16. Commercial Mollier chart

The Mollier chart is a useful tool in the analysis of steady flow devices such as nozzles and turbines. The chart makes it possible to rapidly determine the steam parameters with an accuracy sufficient for practical purposes and gives solution to many thermodynamic problems relating to a change in the state of steam.

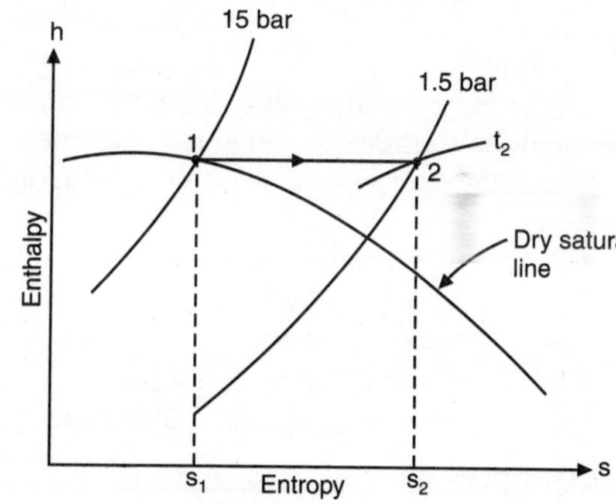

Fig. 13.17.

EXAMPLE 13.49

Represent the following processes on $h-s$ co-ordinates and use the Mollier diagram for obtaining solution to these processes for the mentioned conditions :

(a) Steam initially dry and saturated at 15 bar is throttled to a pressure of 1.5 bar. What is the final condition of steam and the entropy change during the process ?

(b) Steam initially at 0.95 dry and 12 bar expands isentropically in a non-flow process to a final dryness fraction of 0.8. What is the final pressure of steam and enthalpy change during the process ?

Solution : Refer Fig. 13.17.

(*a*) (*i*) Locate the state point 1 which is the point of intersection of constant pressure line 15 bar and the dry saturated line.

(*ii*) Enthalpy remains constant during throttling and as such the throttling process is represented by a horizontal straight line.

Draw a horizontal straight line from point 1 until it meets the constant pressure line 1.5 bar at point 2.

(*iii*) Drop perpendiculars from the initial and final state points on the axis of abscissa.

The following values are then read from the chart

$s_1 = 6.45$ kJ/kg K
$s_2 = 7.48$ kJ/kg K ; $t_2 = 162°C$

$\therefore$ Change of entropy $= s_2 - s_1$
$= 7.48 - 6.45$
$= 1.03$ kJ/kg K

The state point 2 lies in the superheated region and as such the steam becomes superheated after the throttling process with temperature $t_2 = 162°C$

Saturation temperature corresponding to 1.5 bar = 111.3°C

$\therefore$ Degree of superheat
$= 162 - 111.3 =$ **50.7°C**

(*b*) (*i*) Locate the state point *a* which is the point of intersection of constant pressure 12 bar and dryness fraction 0.95.

(*ii*) Entropy remains constant during isentropic process and as such the isentropic expansion process is represented by a vertical straight line.

Draw a vertical straight line from state point *a* till it meets the dryness fraction line 0.8 at point *b*.

(*iii*) Draw perpendiculars from the initial and final state points to the axis of ordinate.

The following values are then read from the chart :

$h_a = 2680$ kJ/kg
$h_b = 2180$ kJ/kg
$p_2 = 0.45$ bar

Change of enthalpy $= h_a - h_b$
$= 2680 - 2180 =$ **500 kJ/kg**

EXAMPLE 13.50

Steam at 15 bar and 300°C is throttled to 10 bar before supplying to steam turbine. It then undergoes isentropic expansion to 1 bar in the turbine. Determine isentropic heat drop and the condition of steam at exit from the turbine. Use enthalpy-entropy chart.

Solution : Refer Fig. 13.18.

(*i*) Locate state point 1 which is the point of intersection of 15 bar and temperature line 300 °C.

(*ii*) Enthalpy remains constant during throttling and as such throttling process is represented by a horizontal straight line.

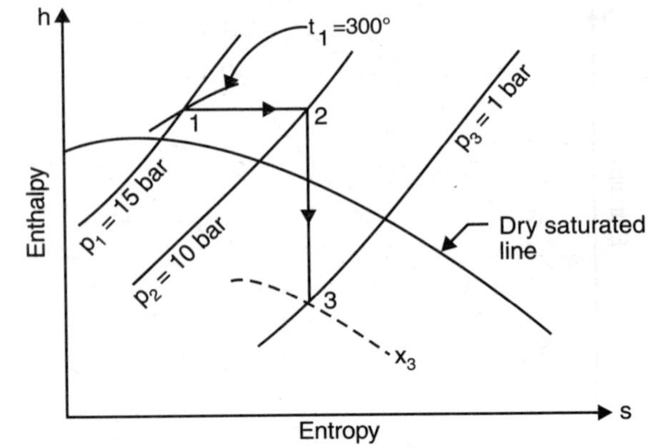

Fig. 13.18.

Draw a horizontal straight line from state point 1 until it meets the constant pressure line 10 bar at state point 2.

(*iii*) Entropy remains constant during isentropic process and as such the isentropic expansion process is represented by a vertical straight line.

Draw a vertical straight line from state point 2 until it meets the constant pressure line 1 bar at point 3.

(*iv*) Draw perpendiculars from the state points 2 and 3 onto the axis of ordinate.

The following values are then read from the chart.

$h_1 = h_2 = 3040$ kJ/kg
$h_3 = 2510$ kJ/kg ; $x_3 = 0.93$

i.e., of quality of steam leaving the turbine is 0.93 and the heat drop is

$h_2 - h_3 = 3040 - 2510$
$=$ **530 kJ/kg**

EXAMPLE 13.51

Steam at an absolute pressure of 10 bar and 250°C expands isentropically in a steam turbine till the pressure drops to 1 bar. Estimate the quality of steam leaving the turbine work developed per kg of steam. If the actual drop is 90% of isentropic drop and steam flow rate through the turbine is 500 kg/min, determine the power developed by the turbine . Use Mollier chart.

Solution : Refer Fig. 13.19.

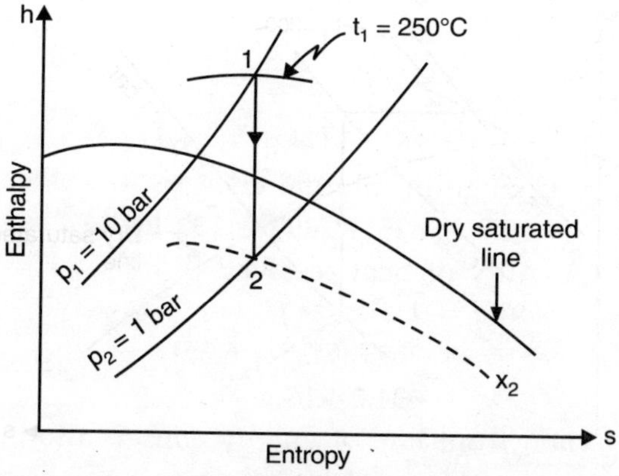

Fig. 13.19.

(*i*) Locate the state point 1 which is the point of intersection of constant pressure line 10 bar and temperature line 250°C.

(*ii*) Entropy remains constant during isentropic process and as such the isentropic expansion process is represented by a vertical straight line.

Draw a vertical straight line from state point 1 till it meets the constant pressure line 1 bar at point 2.

(*iii*) Project the initial and final state points onto the axis of ordinate.

The following values are then read from the chart

$$h_1 = 2945 \text{ kJ/kg}$$
$$h_2 = 2520 \text{ kJ/kg}, \; x_2 = 0.93$$

i.e., quality of steam leaving the turbine is **0.93**

Work done per kg of steam

$$= h_1 - h_2$$
$$= 2945 - 2520 = \mathbf{425 \text{ kJ/kg}}$$

Actual awork done

$$= 425 \times 0.9 = 382.5 \text{ kJ/kg}$$

Work done per minute

$$= 382.5 \times 500$$
$$= 191000 \text{ kJ/min}$$

Power developed

$$= \frac{191000}{60} = 3183 \text{ kJ/s}$$
$$= \mathbf{3183 \text{ kW}}$$

EXAMPLE 13.52

Consider 0.1 kg of steam at 1 bar and 150°C contained in a piston cylinder assembly. If 25 kJ of energy in the form of heat is removed from the steam while the pressure is held constant, use Mollier diagram to work out the final condition of steam.

Solution : Refer Fig. 13.20.

(*i*) Locate the state point 1 which is the point of intersection of constant pressure 1 bar and temperature 150°C.

(*ii*) Draw a perpendicular line from state point 1 to the axis of ordinate. Then from the chart h_1 = 2780 kJ/kg.

$$\text{Now, } q = \frac{25}{0.1} = 250 \text{ kJ/kg}$$

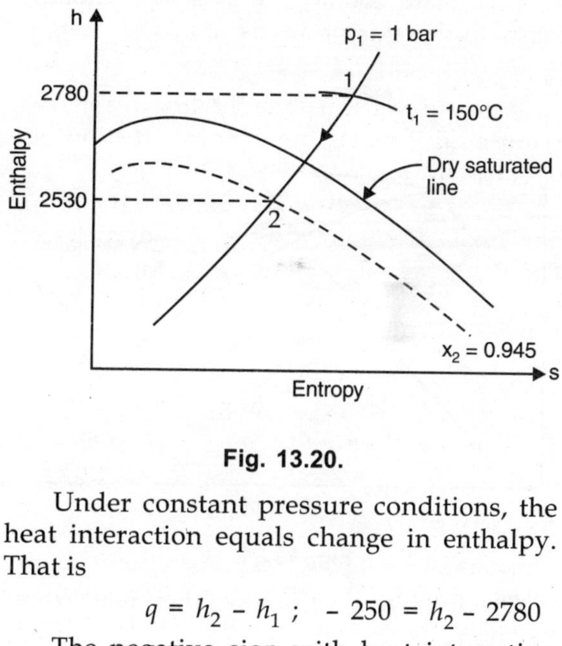

Fig. 13.20.

Under constant pressure conditions, the heat interaction equals change in enthalpy. That is

$$q = h_2 - h_1 \; ; \quad -250 = h_2 - 2780$$

The negative sign with heat interaction because it is a heat loss from the system.

$$h_2 = 2780 - 250 = 2530 \text{ kJ/kg}$$

(*iii*) Locate state point 2 which is the point of intersection of constant pressure 1 bar and enthalpy 2530 kJ/kg.

Then from the chart : x_2 = **0.945**

Hence the final state of steam is wet with dryness fraction 0.945.

EXAMPLE 13.53.

One kg of steam with initial parameters of 12 bar pressure and 0.95 dryness fraction is made to undergo isothermal expansion to 1 bar pressure. Use Mollier diagram to determine the quantity of heat received, the work performed, and the change in internal energy of steam.

Solution : Refer Fig. 13.21.

(*i*) Locate state point 1 which is the point of intersection of 12 bar pressure line and 0.95 dryness fraction line.

(*ii*) Move along the isothermal line 1—a—2 upto intersection with the given pressure p_2 = 1 bar. It is to be noted that in the wet region the constant temperature line coincides with the constant pressure line.

(*iii*) Project the initial state point 1 and the final state 2 onto the axis of ordinates and onto the axis of abscissa.

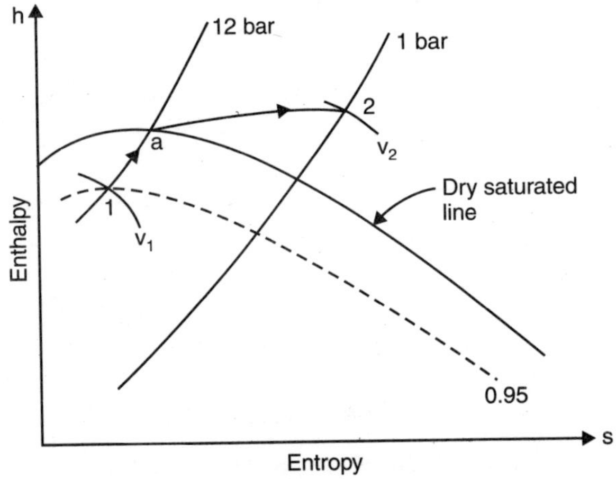

Fig. 13.21.

The following values are then read from the chart :

h_1 = 2685 kJ/kg ; v_1 = 0.16 m³/kg

s_1 = 6.35 kJ/kg K ; h_2 = 2835 kJ/kg

v_2 = 2.1 m³/kg ; s_2 = 7.75 kJ/kg

and $t_1 = t_2$ = 180°C

= 180 + 273 = 453 K

$$u_1 = h_1 - p_1 v_1$$

$$= 2685 - \frac{1 \times 10^5 \times 0.16}{10^3}$$

$$= 2493 \text{ kJ/kg}$$

$$u_2 = h_2 - p_2 v_2$$

$$= 2835 - \frac{1 \times 10^5 \times 2.1}{10^3}$$

$$= 2625 \text{ kJ/kg}$$

∴ Change in internal energy,

$$du = 625 - 2493 = \mathbf{132 \text{ kJ/kg}}$$

Quantity of heat received,

$$q = T(s_2 - s_1)$$

$$= 453 (7.75 - 6.35)$$

$$= \mathbf{634.2 \text{ kJ/kg}}$$

Then from law of energy conservation,

Work performed,

$$w = q - du = 634.2 - 132$$

$$= \mathbf{502.2 \text{ kJ/kg}}$$

EXAMPLE 13.54.

Steam from a boiler is delivered at an absolute pressure of 15 bar and dryness fraction 0.95 into a steam superheater in which the steam receives additional heat at constant pressure and its temperature increases upto 300°C. Using Mollier chart, make calculations for the amount of heat added and the change in internal energy for unit mass of steam.

Solution : Refer Fig. 13.22.

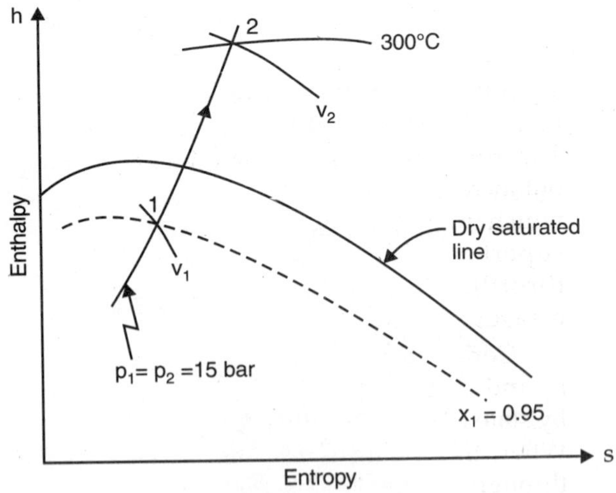

Fig. 13.22.

(*i*) Locate state point 1 which is the point of intersection of constant pressure line 15 bar and dryness fraction line 0.95

(*ii*) Move along the constant pressure line upto intersection with the given temperature t_2 = 300°C.

(*iii*) Project the initial state point 1 and the final state 2 onto the axis of ordinates.

The following values are then read from the chart

h_1 = 2695 kJ/kg ; v_1 = 0.125 m³/kg

h_2 = 3040 kJ/kg ; v_2 = 0.17 m³/kg

$$u_1 = h_1 - p_1 v_1$$

$$= 2695 - \frac{15 \times 10^5 \times 0.125}{10^3}$$

$$= 2507.5 \text{ kJ/kg}$$

$$u_2 = h_2 - p_2 v_2$$

$$= 3040 - \frac{15 \times 10^5 \times 0.17}{10^3}$$

$$= 2785 \text{ kJ/kg}$$

∴ Quantity of heat supplied in the boiler,

$$q = h_2 - h_1 = 3040 - 2695$$

$$= \mathbf{345\ kJ/kg}$$

and change in internal energy,

$$du = u_2 - u_1 = 2785 - 2507.5$$

$$= \mathbf{277.5\ kJ/kg}$$

13.13. THROTTLING AND MEASUREMENT OF DRYNESS FRACTION

Throttling is a flow process during which pressure drops but enthalpy remains constant. Throttling reduces the temperature of steam but increases the dryness fraction of wet steam which may become dry saturated or even superheated. This forms the principle of throttling calorimeter ; a device used to measure the quality of wet steam.

Consider wet saturated steam at pressure p_1 and dryness fraction x_1, and represented by state point 1 on the Mollier chart (Fig. 13.23). When this steam is throttled (made to pass through a restricted passage – a partially closed valve) its pressure drops. The pressure at exit from the valve depends upon the amount of restriction provided, and depending upon the pressure attained after throttling the steam may become dry saturated at pressure p_2 (state point 2) or become superheated (state point 3). If the values of (p_2, t_2) or (p_3, t_3) are measured, the enthalpy values corresponding to these parameters can be noted from the steam tables. Since throttling is an (constant enthalpy) process,

Enthalpy before throttling (h_1)
= enthalpy after throttling (h_2 or h_3)

$$(h_f + x h_{fg})_{\text{at } p_1} = (h_f + h_{fg})_{\text{at } p_2}$$

$$= \left[h_f + h_{f_g} + c_p (t_{sup} - t_s)\right]_{at\, p_3}$$

The above identity helps to determine the value of dryness fraction x of wet steam at pressure p_1.

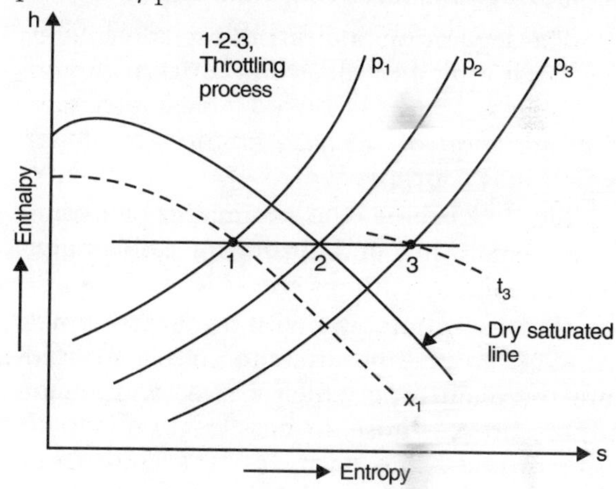

Fig. 13.23. Steam throttling process

Fig. 13.24 shows the schematic arrangement of throttling calorimeter. Pipe *A* carries the wet steam at pressure p_1 whose dryness fraction x is to be determined. A portion of this steam is taken by the sampling tube through transverse holes provided for this purpose.

The steam is then throttled to a low pressure, usually atmospheric pressure. The throttling operation is controlled by adjusting the position of throttle valve with a control screw. The throttled steam then passes through a container where its pressure and temperature are noted. Subsequently, the throttled steam is allowed to escape to atmosphere.

It is to be noted that working of throttling apparatus would be successful only if steam becomes superheated or atleast dry saturated after the throttling operation.

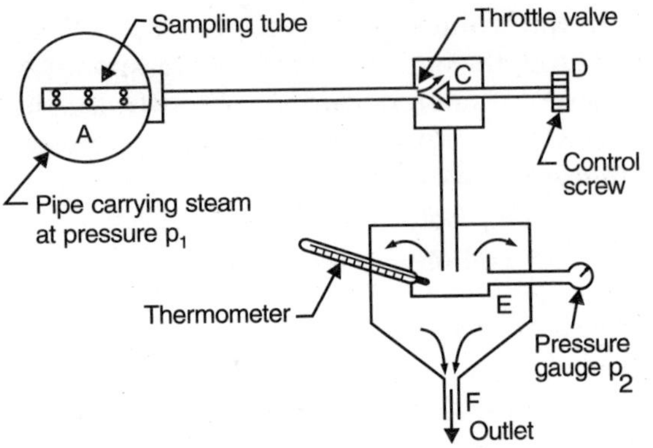

Fig. 13.24. Throttling calorimeter

Separating and throttling calorimeter

The separating and throttling calorimeter has separating and throttling units arranged in series, and is used for accurate determination of dryness fraction of steam which may be quite wet.

Fig. 13.25 shows the schematics of a combined separating and throttling calorimeter and its salient features are :

(*i*) A sampling pipe with transverse holes provided around the circumference is inserted into the main pipe which carries wet steam at pressure p_1 whose dryness fraction is to be determined. The sample of steam thus obtained is made to pass through a separating unit through a valve. This valve is kept fully open to ensure that the steam undergoes no throttling (drop in pressure) at this stage. This steam is subjected to sudden reversal of direction in its motion at a metal basket which is provided with perforations all around but not at the base. The water particles in steam have greater inertia and due to this reversal are thrown out of the mixture. The quantity of water thus separated is collected and measured.

(*ii*) The fairly dry steam is subsequently made to pass through the throttling unit where it passes through a small opening and gets throttled at constant enthalpy. Measurements are made of steam pressure before throttling, and pressure and temperature after throttled. The steam coming out of throttling calorimeter is condensed and the weight of the condensate is recorded.

Calculations

Let M = mass of wet steam as sampled from the steam main

m_1 = mass of water collected in the separating unit

m_2 = mass of steam condensed and collected from throttling unit.

Then dryness fraction of steam considering separating calorimeter alone,

$$x_1 = \frac{m_2}{M} = \frac{m_2}{m_1 + m_2} \quad ...(i)$$

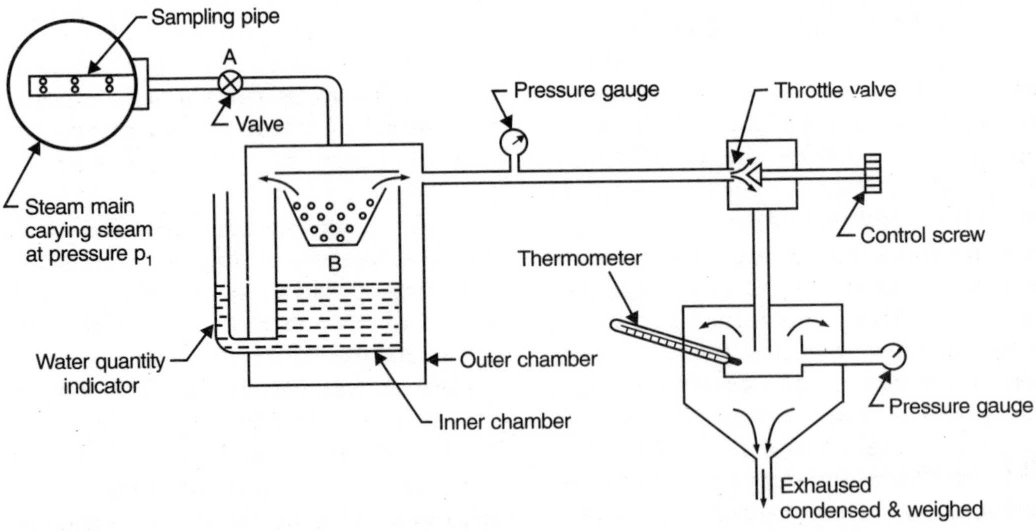

Fig. 13.25. Combined separating and throttling calorimeter

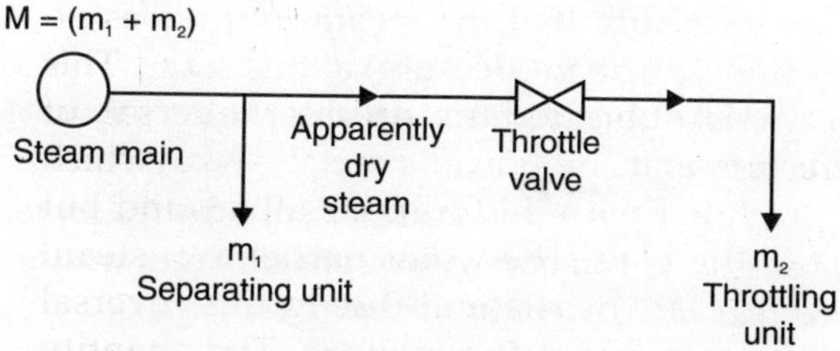

Amount of water separated by separating calorimeter,

$$(1 - x_1)(m_1 + m_2)$$

If x_2 is the dryness fraction of steam entering into throttling calorimeter, then amount of water carried by steam into the throttling calorimeter

$$= (1 - x_2)\, m_2$$

If x is the actual dryness fraction of steam in the main steam pipe, then amount of water initially in the steam

$$= (1 - x)(m_1 + m_2)$$

From principle of mass conservation,

$$(1 - x)(m_1 + m_2) = (1 - x_1)(m_1 + m_2) + (1 - x_2)\, m_2$$

or $$(1 - x) = (1 - x_1) + (1 - x_2)\frac{m_2}{m_1 + m_2}$$

Substituting the value of $\frac{m_2}{m_1 + m_2}$ from expression (*i*), we obtain

$$1 - x = 1 - x_1 + (1 - x_2)\, x_1$$
$$= 1 - x_1 + x_1 - x_1 x_2$$
$$\therefore \quad x = x_1 x_2$$

Thus the dryness fraction of steam measured with combined separating and throttling calorimeter is the product of dryness fraction x_1 and x_2 measured separately by separating and throttling units.

It may, however, be pointed out that it is not necessary to find out x_1 and x_2 separately; the net dryness fraction can also be obtained by considering the combination of two calorimeters as a system and making a heat balance.

EXAMPLE 13.55

For determining the dryness fraction of a sample of steam at 14 bar, it was passed through a throttling calorimeter attached to the steam main. The pressure and temperature after the throttling process were recorded to be 1.2 bar and 120°C respectively. Calculate the dryness fraction of steam before passing through the throttling unit. Take c_{ps} = 2.1 kJ/kg K for superheated steam.

(b) What is the minimum value of dryness fraction that can be determined by such a calorimeter if the pressure values are assumed to be same ?

Solution : Enthalpy before and after throttling process are same. That gives ;

$$h_{f1} + x_1 h_{fg1} = h_{f2} + h_{fg2} + c_{ps}(t_{sup} - t_s)$$

Substituting the values read from steam tables, we get

$$830.1 + x_1 \times 1957.7 = 439.4 + 2244.1 + 2.1\,(120 - 104.8)$$
$$= 2715.42$$

$$\therefore \quad x_1 = \frac{2715.42 - 830.1}{1957.7} = \mathbf{0.963}$$

(*b*) Minimum degree of dryness fraction that can be estimated pertains to the situation that steam after throttling must become atleast saturated.

In that case,

$$h_{f1} + x_1 h_{fg1} = h_{f2} + h_{fg2}$$
$$830.1 + x_1 \times 1957.7 = 439.4 + 2244.1 = 2683.5$$

$$\therefore \quad x_1 = \frac{2683.5 - 830.1}{1957.7} = \mathbf{0.946}$$

If the steam has a dryness fraction lower than 0.946, a separating calorimeter is used alongwith the throttling calorimeter.

EXAMPLE 13.56

A sample of steam from a boiler drum at 15 bar is passed through a throttling calorimeter in which pressure and temperature are found to be 1 bar and 120°C. Estimate the quality of steam taken from the boiler.

(b) If the calorimeter must show at least 5°C superheat for acceptable results, what is the maximum moisture that could be measured with this set up under the given pressure.

Solution : Let x be the dryness fraction of steam taken from the boiler. Then

$h_1 = (h_f + x\, h_{fg})$ at 15 bar
$= 844.7 + 1945.2\, x$

After throttling process $p_2 = 1$ bar, $t_{sat} \simeq 99.6°C$. Since the given temperature is 120°C, the steam is finally superheated.

From the tables of superheated steam, it is noted that

At $p = 1$ bar
$t = 100°C$, and $h = 2676.2$ kJ/kg

At $p = 1$ bar
$t = 150°C$, and $h = 2776.4$ kJ/kg

Then the interpolation gives :

$$h_2 = 2676.2 + \frac{2776.4 - 2676.2}{150 - 100} \times (120 - 100)$$
$$= 2716.3 \text{ kJ/kg}$$

The process is throttling during which the enthalpy remains constant

$\therefore\ 844.7 + 1945.2\, x = 2716.3$

That gives $x = \dfrac{2716.3 - 844.7}{1945.2} = \mathbf{0.962}$

(b) When $p = 1$ bar

and $t = t_{sat} + 5$
$= 99.6 + 5 = 104.6°C$

Again by interpolation,

$$h_2 = 2676.2 + \frac{2776.4 - 2676.2}{150 - 100} \times (104.6 - 100)$$
$$= 2685.4 \text{ kJ/kg}$$

Then $844.7 + 1945.2\, x = 2685.4$

and $x = \dfrac{2685.4 - 844.7}{1945.2} = \mathbf{0.946}$

Thus the maximum moisture which can be determined with the given set up only 5.4%.

EXAMPLE 13.57

The following observations were made during experiment on combined separating and throttling calorimeter for determining the dryness fraction of a sample of steam :

Mass of water trapped in the separator
= 0.5 kg

Mass of steam condensed after throttling
= 3.0 kg

Steam pressure before and after throttling = 8 bar and 15 cm of mercury above that of atmosphere

Temperature of steam after throttling
= 120°C

Make calculations of dryness fraction of steam in the main.

Take c_{ps} = 5.1 kJ/kg K for superheated steam and assume that atmospheric pressure is equivalent to that of 75.5 cm of mercury.

Solution : *At the separator unit*

$m_w = 0.5$ kg and $m_s = 3.0$ kg

Here m_w is the mass of moisture collected in the separating calorimeter and m_s is the mass of apparently dry vapour leaving the separating calorimeter and entering the throttling unit.

$\therefore\ x_1$ (dryness fraction considering only the separator unit),

$$= \frac{m_s}{m_w + m_s} = \frac{3.0}{0.5 + 3.0} = 0.857$$

At the throttling unit :

Pressure of steam after throttling
$= 75.5 + 15 = 90.5$ cm of mercury
$= \dfrac{1.01325}{76} \times 90.5 \simeq 1.20$ bar

Saturation temperature of steam corresponding to 1.20 bar = 104.8°C. The temperature of steam after throttling is given to be 120°C. Accordingly, the steam becomes superheated after throttling. Further, the total heat remains constant during throttling operation. That gives

$$h_{f2} + x_2\, h_{fg2} = h_{f3} + h_{fg3} + c_{ps}\,(t_{sup} - t_s)$$

The suffix 2 denotes the condition before throttling (8 bar pressure) and suffix 3 represents the state of steam after throttling (pressure 1.20 bar).

$720.9 + x_2 \times 2046.5$
$= 439.4 + 2244.1 + 2.1\,(120 - 104.8)$
$= 2715.42$

$\therefore\ x_2$ (dryness fraction considering throttling unit)

$$= \frac{2715.42 - 720.9}{2046.5} = 0.974$$

$\therefore$ Dryness fraction in the steam main,

$x = x_1 \times x_2$
$= 0.857 \times 0.974 = \mathbf{0.835}$

13.14. BUCKET OR BARREL CALORIMETER

The unit consists of a copper vessel which contains a measured quantity of water. This vessel is surrounded by an insulated vessel and the gap between the two vessels has air or some other insulating medium. This is to prevent any heat transfer to or from the calorimeter. The calorimeter is placed on a wooden block and the top is provided with some detachable insulating cover. The steam whose dryness fraction is to be determined is passed through a sampling tube which terminates into fine exit holes for discharging the steam into water contained in the calorimeter. Measurements are made of the weight of calorimeter before and after mixing with steam. The difference gives the mass of steam condensed. Due to condensation of steam, the temperature of water rises and the rise in temperature is measured by a mercury thermometer. The pressure of steam is noted by a pressure gauge which is fitted on the steam main.

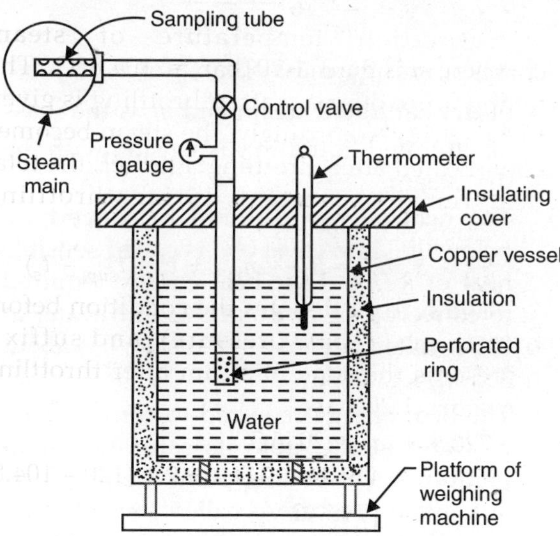

Fig. 13.26. Barrel calorimeter

Let p , t = pressure and temperature of steam

h_{fg} = latent heat of steam at pressure p

m_c , c_c = mass and specific heat of calorimeter

m_s = mass of steam condensed

m_w , c_w = mass and specific heat of water in the calorimeter

t_1 , t_2 = initial and fixed temperature of water and calorimeter

If x is the dryness fraction of steam sample, then

Heat lost by steam = $m_s[x\, h_{fg} + c_w\,(t - t_2)]$

Heat gained by water and calorimeter

$$= (m_w c_w + m_c c_c)\,(t_2 - t_1)$$

Provided there is no addition or loss of heat, the heat lost by steam equals the heat gained by water calorimeter. That is

$$m_s[x\, h_{fg} + c_w\,(t - t_2)] = (m_w\, c_w + m_c c_c)\,(t_2 - t_1)$$

With substitution of measured data, the above equation can be used to determine the dryness fraction of steam.

Note : For getting better results, the rise in temperature of water has to be sufficient enough for its accurate and easy measurement. However, excessive rise in temperature would result in conduction and radiation losses which are difficult to account for. The calculated value of dryness fraction would then be less than the actual value of dryness fraction.

EXAMPLE 13.58

Steam from a small boiler was discharged through a pipe into a barrel and the following experimental data was recorded after a few minutes of test-run.

mass of copper barrel = 1 kg

mass of barrel plus water = 3.5 kg

mass of barrel plus water + steam = 3.72 kg

initial and final temperature of water = 10°C and 50°C

pressure of steam = 6 bar

If the specific heat of the material of barrel is 0.405 kJ/kg K, determine the dryness fraction of the steam sample.

Solution : From the given data : m_c = 1 kg

$$m_c + m_w = 3.5$$

or $$m_w = 3.5 - m_c = 3.5 - 1 = 2.5 \text{ kg}$$

$m_c + m_w + m_s = 3.75$

or $m_s = 3.75 - (m_c + m_w)$

$= 3.72 - 3.5 = 0.22$ kg

$t_1 = 15°C$ and $t_2 = 50°C$

Corresponding to pressure of 6 bar,

$t = 158.9°C$ and

$h_{fg} = 2086.3$ kJ/kg

If x is the dryness fraction of the steam sample,

heat lost by steam

$= m_s[x\, h_{fg} + c_w\,(t - t_2)]$

Taking specific heat of water as 4.18 kJ/kg K

heat lost by steam

$= 0.22\ [2086.3x + 4.18(158.9 - 50)]$

$= 458.98\ x + 100.14$ kJ

heat gained by water and calorimeter

$= (m_w\, c_w + m_c\, c_c)\,(t_2 - t_1)$

$= (2.5 \times 4.18 + 1 \times 0.405)\,(50 - 10)$

$= 434.2$ kJ

Presuming that there is no addition or loss of heat, we can write

heat lost by steam = heat gained by water

$458.98\ x + 100.14 = 434.2$

∴ Dryness fraction of the steam sample,

$$x = \frac{434.2 - 100.14}{458.98} = \mathbf{0.728}$$

SALIENT POINTS

1. A pure substance consists of only single species with characteristic feature of homogeneous and invariable composition during heat transfer processes. A pure substance may exist in more than one phase depending on its energy level. Water, air, vapours of Fr – 12 and NH_3 are treated as pure substances.
2. A liquid at its boiling point at a specific pressure is called a saturated liquid. At standard atmospheric pressure, the water boils and generated its own vapours. Water in gaseous state or a two phase mixture of water and gas is called steam.
3. Saturation pressure is the pressure, for a given temperature at which a substance undergoes a phase change. Likewise saturation temperature refers to the temperature, for a given pressure, at which phase change of the substance occurs. At standard atmosphere pressure (1.01325 bar), water boils at 100°C. Therefore for water

   Saturation pressure = 1.01325 bar at 100°C

   Saturation temperature = 100°C at 1.01325 bar
4. When the saturated vapour contains particles of liquid evenly distributed over the entire mass of vapour, it is called wet saturated steam. Wet steam is characterised by its dryness fraction defined as the ratio of mass of dry saturated vapour to the total mass of mixture. That is

$$x = \frac{m_g}{m} = \frac{m_g}{m_f + m_g}$$

   where m is the mass of mixture, m_g is mass of dry saturated steam and m_f is the mass of saturated liquid in suspension with the dry steam.
5. The heat energy required to convert ice below 0°C into ice at 0°C is called sensible heat of ice, and the heat energy required to convert 1 kg of ice at 0°C into water at 0°C is called latent heat of fusion; it is about 335 kJ/kg.

   The heat energy required to convert 1 kg of water at 0°C upto its saturation temperature t_s (or boiling point temperature) at a given pressure is called sensible heat or enthalpy of water (h_f). State of steam after complete evaporation at saturation temperature is called dry saturated steam, and the heat energy required to convert 1 kg of saturated water at t_s°C into dry saturated steam is called enthalpy of steam (h_{fg}). Dry saturated steam contains no liquid particles in it. Steam becomes superheated, if at a given pressure, its temperature t_{sup} is higher than the saturation temperature. The temperature difference $(t_{sup} - t_s)$ is known as degree of superheat.

6. The simple laws of perfect gases donot apply to vapours; hence variations in their properties are depicted in tables and charts. The steam tables make it possible to rapidly find the basic characteristics of steam according to the given parameters.

• The specific enthalpy values for different conditions of steam are :

$$h = h_f + x\, h_{fg} \quad \text{for wet steam}$$
$$= h_f + h_{fg} \quad \text{for dry steam}$$
$$= h_f + h_{fg} + c_{ps}\,(t_{sup} - t_s)$$
for superheated steam

The specific heat of superheated steam c_{ps} varies from 2.0 to 2.32 kJ/kg K depending upon degree of superheat and superheating pressure.

• The change is specific entropy values are :

$$s = s_f + x\,\frac{h_{fg}}{T_s}$$

for wet satuarted steam with dryness fraction x

$$= s_f + \frac{h_{fg}}{T_s}$$

for dry saturated steam

$$= s_f + \frac{h_{fg}}{T_s} + c_{ps}\log_e\frac{T_{sup}}{T_s}$$

for superheated steam

Here s_f denotes the entropy of water reckoned above 0°C, *i.e.*, $s_f = c_p \log_e \dfrac{T_{sup}}{273}$

• The values of specific volume are :

$$v = v_f + x\, v_{fg} \quad \text{for wet steam}$$
$$= v_f + v_{fg}$$

for dry saturated steam

In the superheated segion, steam generation at constant pressure conforms to Charles' law, and accordingly,

$$v_{sup} = v_s\,\frac{T_{sup}}{T_s}$$

• The internal energy of steam represents the actual amount of energy stored, and is prescribed as

Internal energy $u = h - W_e$

where the external work of evaporation W_e equals pv.

7. Triple point corresponds to the temperature at which a pure subtance can co-exist in equilibrium state in the solid, liquid and gaseous phases of the substance. The corresponding pressure and temperature are referred to as triple point pressure (P_t) and triple point temperature (T_t). For water

$$P_t = 0.612 \text{ kPa and } T_t = 0.01°\text{C}$$

The triple point is represented by a point an $p - T$ diagram, and by a line on $p - v$ diagram.

8. Critical point refers to a particular set of pressure and temperature at which the liquid is directly and fully converted into dry vapour without going into the phase of evaporation, *i.e.*, latent heat is zero. At critical point, the density of vapour equals the density of the liquid of a substance. For water, the critical pressure p_c and critical temperature t_c are :

$$p_c = 220.9 \text{ bar and } t_c = 374.14°\text{C}$$

9. Mollier diagram (enthalphy – entropy plot) is a useful tool for easy and rapid estimation of the steam parameters and provides solution to many thermodynamic problems relating to a change in the state of steam.

10. Throttling is a flow process during which pressure drops and enthalpy remains constant. Throttling reduces the temperature of steam but increases the dryness fraction of wet steam which may become dry saturated or even superheated. This forms the principle of throttling calorimetes, a device used to measure the dryness fraction of steam.

REVIEW QUESTIONS

A. Conceptual and conventional questions.

1. What is a pure substance? Can air be treated as a pure substance?
2. What is meant by phase of a substance?
3. How do you differentiate between a saturated liquid and a saturated vapour?
4. Define the terms saturation temperature and saturation pressure.
5. What is the main feature of triple point ? State the values of pressure and temperature at the triple point of water.
6. What is critical point? What process is possible below the critical point?
7. How evaporation differs from boiling?
8. State the difference between a vapour and a gas.
9. Draw the $p-v$ diagram of water vapour (steam) and point out its peculiarities.
10. Explain the process of steam generation (change of phase of water) at constant pressure. Show the various stages on $p-v$, $T-v$ and $T-s$ diagrams.
11. What kind of steam is called wet saturated, dry saturated and superheated steam ?
12. Define quality and dryness fraction of steam.
13. Express the specific volume v of a wet vapour mixture of dryness fraction x in terms of specific volume v_f of liquid and specific volume v_g of vapour.
14. List the advantages of using superheated steam.
15. Define the terms (*i*) internal latent heat and (*ii*) external work of evaporation.
16. Sketch a Mollier chart and indicate thereupon the throttling and isentropic expansion processes.
17. How throttling helps in estimating the dryness fraction of steam ?
18. What is meant by quality of steam ? Outline the procedure followed to determine its value by using a throttling calorimeter.
19. Sketch and explain the construction and working of a separating and throttling calorimeter used for determining the dryness fraction of steam in a boiler.
20. Determine the state of steam at
    (*i*) p = 10 bar and v = 0.17 m^3/kg
    (*ii*) p = 12 bar and t = 200°C
    [**Ans.** Wet with dryness 0.875; superheated with 12.05 degree of superheat]
21. Determine the condition of steam raised at 15 bar if 2400 kJ/kg of heat has been spent to generate this steam from water at 0°C.
    [**Ans.** 0.798]
22. Two kg of steam at 200 kPa occupies a volume of (*i*) 1.04 m^3 and (*ii*) 2.08 m^3. Determine the temperature and quality of steam for any two-phase state.
    [**Ans.** Wet with dryness 0.586 ; superheated with temperature 183.5°C]
23. Determine the specific volume, density, enthalpy and entropy of wet saturated steam at 20 bar pressure and 0.9 dryness fraction.
    [**Ans.** 0.0898 m^3/kg ; 16.16 kg/m^3 ; 2610 kJ/kg and 5.95 kJ/kg K]
24. Make calculations for the mass and enthalpy of 0.5 m^3 of wet steam with degree of wetness equal to 10% and a pressure of 10 bar.
25. Steam at 10 bar pressure and 0.8 dryness fraction has been generated by a boiler. Make calculations for its enthalpy, volume, entropy and internal energy.
    [**Ans.** 2375 kJ/kg, 0.1552 m^3/kg , 5.697 kJ/kg K and 2220 kJ/kg]
26. Steam at 1 bar pressure is stated to have an entropy of 7.314. kJ/kg K. Determine its dryness fraction, specific volume and enthalpy.
    [**Ans.** 0.963, 1.631 m^3/kg , 2591.5 kJ/kg]
27. A sample of 1 kg of steam at a pressure of 20 bar exists in the superheated state with temperature equal to 250°C. Determine its enthalpy, volume, entropy and internal energy. Neglect volume of water and take specific heat of superheat equal to 2.30 kJ/kg K.
    [**Ans.** 2875.5 kJ/kg ; 0.1073 m^3/kg ; 6.512 kJ/kg K ; 2661 kJ/kg]
28. A pressure cooker has 1.0 kg of steam 0.8 dry at 3.0 bar. What amount of heat needs to be supplied to raise its pressure to 4.0 bar ?
    [**Ans.** 600 kJ]
29. Dry saturated steam at a pressure of 80 bar is expanded reversibly and hyperbolically. Determine the work and heat transfer during the process and final temperature if expansion lowers the pressure to 2 bar.
    [**Ans.** 694 kJ/kg, 689.5 kJ/kg, 142.5°C]
30. A piston-cylinder assembly contains superheated steam at 20 bar and 300°C. The steam expands adiabatically behind the fully resisted piston to 1.2 bar. If the steam initially

occupies a volume of 0.0627 m^3, determine the final condition of steam and the work and heat interactions.

[Ans. 0.5 kg, W = 209 kJ and Q = 0]

31. Steam initially at 6 MPa and 450°C is allowed to cool to 200°C at constant volume. Find the final condition and heat transfer if the mass of steam is 0.75 kg.

[Ans. x_2 = 0.4038 ; – 1423.35 kJ]

32. Two kg of dry saturated steam at 11 bar is to be generated from water available at 45°C . Determine the quantity of heat required to be supplied for raising the steam. If subsequently the steam is to be superheated by 50°C, what additional heat input is required. Take c_{ps} = 2.25 kJ/kg K for superheated steam, and c_p = 4.18 kJ/kg K for water.

[Ans. 25714.8 kJ/kg , 112.5 kJ/kg]

33. A vessel of 1.35 m^3 capacity is filled with wet steam at 10 bar pressure and 0.95 dryness fraction. The vessel is cooled until the pressure falls to 2 bar. Determine the mass of contents in the vessel and the dryness fraction of steam after cooling.

[Ans. 7.314 kg, 0.208]

34. The expansion of superheated steam at 4 MPa and 350°C takes place to 0.35 MPa by a polytropic law where pressure and volume are related by the expression $pv^{1.3}$ = constant. Make calculations for the work done, change in internal energy and heat transfer.

[Ans. 389.5 kJ/kg ; – 633.2 kJ/kg ; – 243.7 kJ/kg]

35. A throttling calorimeter sampling the steam generated by a boiler at a pressure of 20 bar gives readings of a pressure of 1 bar and temperature 130°C. Evaluate the dryness of steam taken from the boiler.

What is the maximum degree of wetness that can be measured under the given pressure ?

[Ans. 0.9675]

36. The quality of steam is to be determined by passing a sample through a separating and throttling calorimeter. The following experimental observations have been noted :

Steam main pressure = 8.4 bar

Pressure after throttling = atmospheric (1.01325 bar)

Mass of water drained from separator = 1.2 kg

Mass of steam condensed after passing through the throttle = 23.4 kg

Evaluate the dryness of the steam in the main and state, with reasons, whether the throttling calorimeter alone could have been used for the test.

[Ans. 0.9185]

B. Fill in the blanks with appropriate word/words.

1. A mixed phase of liquid and gas is called a ....
2. The dryness fraction of steam when expressed in percentage is called the .... of steam.
3. The latent heat of vaporisation is .... at the critical point.
4. A wet vapour can be completely specified by its .... and ....
5. Triple point is the point where .... exist together.
6. During evaporation at constant pressure the change in temperature due to heat addition will be ....
7. During a constant pressure heating process, the heat absorbed by steam equals the change in ....
8. Internal latent heat represents the difference between .... and .... during evaporation.
9. For successful working of throttling calorimeter to experimentally determine the value of dryness fraction of steam, the steam after throttling must be in .... or .... condition.
10. The ordinate and abscissa of Mollier chart are respectively the .... and ....

Answers

1. vapour 2. quality 3. zero 4. pressure, dryness fraction 5. three phases of solid, liquid and gas 6. zero 7. enthalpy 8. latent heat, work done 9. superheated, dry saturated 10. specific enthalpy, specific entropy

C. Multiple choice questions.

1. Vapour is a
   (*a*) pure substance
   (*b*) perfect gas
   (*c*) mixed phase of liquid and gas
   (*d*) substance homogeneous and invariable in chemical composition
2. Critical temperature for steam is close to
   (*a*) 313°C (*b*) 347°C
   (*c*) 375°C (*d*) 409°C

3. At critical pressure (225 bar)
   (*a*) entropy is zero
   (*b*) latent heat of vaporisation is zero
   (*c*) saturated vapour becomes a gas
   (*d*) sublimation occurs from a solid to vapour phase

4. Sensible heat
   (*a*) does not allow rise in temperature
   (*b*) results in temperature rise that can be sensed by a thermometer
   (*c*) brings about a change in the state of substance
   (*d*) is absorbed by a liquid at saturation temperature to get converted into a vapour at the same temperature

5. The heat absorbed by water at its saturation temperature to get conveted into dry steam at the same temperature is called
   (*a*) sensible heat (*b*) specific heat
   (*c*) total heat (*d*) latent heat

6. Mollier diagram is a plot of
   (*a*) temperature and entropy
   (*b*) enthallpy and entropy
   (*c*) pressure and enthalpy
   (*d*) pressure and volume

7. Which of the following process is represented by horizontal line parallel to abscissa of a Mollier diagram?
   (*a*) constant pressure
   (*b*) constant temperature
   (*c*) throttling (*d*) adiabatic expansion

8. The triple point of water is
   (*a*) 54.35 K (*b*) 113.84 K
   (*c*) 216.55 K (*d*) 273.16 K

9. In a separating and throttling calorimeter, the steam will be at minimum pressure
   (*a*) after separating unit
   (*b*) before throttling unit
   (*c*) after throttle valve
   (*d*) pressure remains constant throughout

10. All of the following statements are correct, except
    (*a*) the superheated vapour behaves like a perfact gas
    (*b*) the latent heat of vaporisation has a maximum value at the critical point
    (*c*) the process of leakage of steam through a narrow aperture constitutes a constant enthalpy process
    (*d*) for complete specification of superheated vapour, one needs pressure as well as temperature

11. At critical point, the enthalpy of vaporisation is
    (*a*) zero (*b*) maximum
    (*c*) minimum
    (*d*) dependent on temperature only

12. Match the two lists given below :

A. Critical point	1. All three phases–solid, liquid and vapour coexit in equilibrium
B. Sublimation	2. Phase changes from solid to liquid
C. Triple point	3. Properties of saturated liquid and saturated vapour are identical
D. Melting	4. Heating process where solid directly gets transformed to gaseous phase

13. Name the parameter that decreases with an increase in steam pressure
    (*a*) sensible heat
    (*b*) latent heat of vaporisation
    (*c*) specific entropy
    (*d*) boiling point

14. Which of the following is not a correct statement ?
    (*a*) The enthalpy of water is measured from 0°C
    (*b*) During phase transformation, both pressure and temperature remain unchanged
    (*c*) A dry saturated vapour curve is steeper than a dry saturated liquid curve
    (*d*) Sensible heat results in temperature rise that can be sensed by a thermometer

Answers

1. (*c*)	2. (*c*)	3. (*b*)	4. (*b*)	5. (*d*)
6. (*b*)	7. (*c*)	8. (*d*)	9. (*c*)	10. (*b*)
11. (*a*)	12. A-3; B-4; C-1; D-2			13. (*b*)
14. (*c*)				

Chapter 14

Steam Boilers, Mountings and Accessories

Learning objectives : A study of the subject matter presented in this chapter will enable the reader to explain the :

- *concept of fire tube and water tube boilers*
- *construction and operation of fire tube boilers : Cochran, Lancashire and locomotive boilers*
- *construction and operation of water tube boilers : Babcock-Wilcox and Stirling boilers*
- *purpose and working of boiler mountings and accessories*
- *construction and working of high pressure boilers : Lamont, Loeffler and Benson boilers*

The equipment used for producing steam is called *steam generator* or *boiler*. The American Society of Mechanical Engineers (A.S.M.E.) gives the following definition of the steam generating unit :

"A combination of apparatus for producing, furnishing or recovering heat together with the apparatus for transferring the heat so made available to the fluid being heated and vaporised."

The fluid (water) is contained in the boiler called *shell* and the thermal energy released during combustion of fuel, which may be solid, liquid or gaseous, is transferred to water and this converts water into steam at the desired temperature and pressure. The steam thus generated is used for:

(*i*) *Power generation* : Mechanical work or electric power may be generated by expanding steam in the steam engine or steam turbine.

(*ii*) *Heating* : The steam is utilized for heating the residential and industrial buildings in cold weather and for producing hot waters for hot water supply.

(*iii*) Utilization of steam for industrial processes such as for sizing and bleaching etc., in textile industries. Steam is also used in many other industries like sugar mills and chemical industries.

This chapter is devoted to the study of the constructinal and operational features of the boilers and the associated mountings and accessories.

14.1. CLASSIFICATION OF BOILERS

Boilers are mainly classified according to the following :

1. *Relative position of hot gases and water*

(*a*) *Fire tube boiler*. The hot gases pass through the tubes that are surrounded by water. The products of combustion leaving the furnace are passed through fire (smoke) tubes which are arranged within the water space. The heat energy of the flue gas is transferred to water which is converted into steam. The spent gases are then discharged to atmosphere through chimney (Fig. 14.1)

The fire tube boilers are known by certain common names such as horizontal return tubular, locomotive fire box, scotch marine and vertical tubular etc.

(*b*) *Water tube boiler*. The tubes contain water and the hot gases produced by combustion of fuel flow outside. A bank of water tubes (tubes containing water) is connected with steam-water drum through two sets of headers. The hot flue gases from the furnace are made to flow around the water tubes a sufficient number of times. The gases thus give up their heat to an appreciable extent, get cooled and are discharged to the stack. The steam formed separates from water in the drum and gets accumulated in the steam space (Fig. 14.2).

The water tube boilers are designated by the following common names :

(i) Babcock and Wilcox boiler which has usually straight but inclined tubes which connect the headers.

(ii) Stirling boiler which is a multitubular boiler having bent tubes that connect drums to headers.

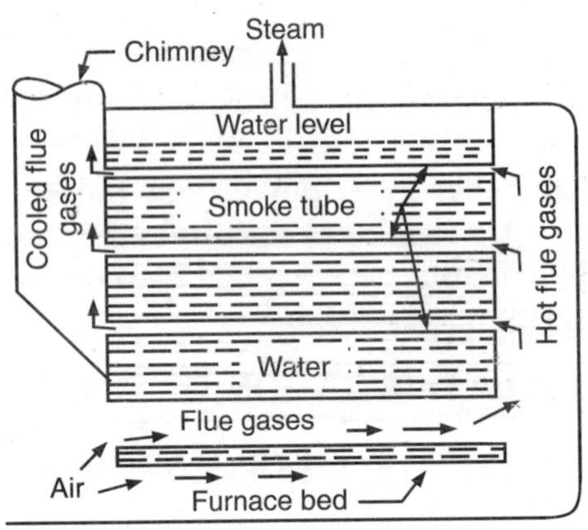

Fig. 14.1. Principle of fire tube boiler

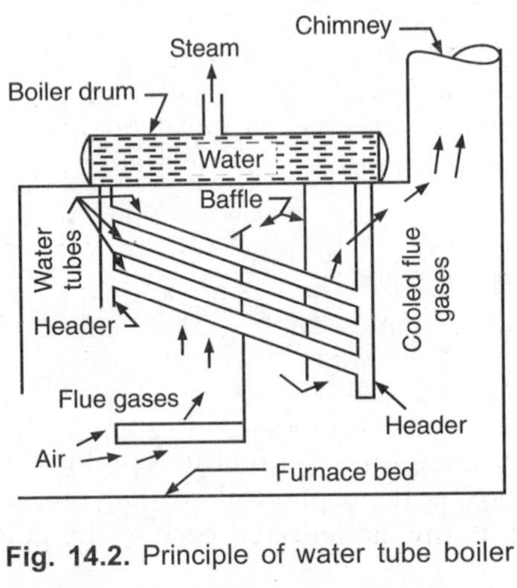

Fig. 14.2. Principle of water tube boiler

2. *Method of Firing*

(a) *Internally fired boiler* : The furnace region (space in which combustion of fuel takes place) is provided inside the boiler shell and is completely surrounded by water cooled surfaces. The method of internal firing is used in Lancashire, Locomotive and Scotch boilers.

(b) *Externally fired boiler* : The furnace region is provided outside or built under the boiler as in the case of Babcock and Wilcox boiler. The externally fired boiler has the advantage that its furnace region is simple to construct and can be easily enlarged. The distinction between internally fired and externally fired boiler is illustrated in Fig. 14.3.

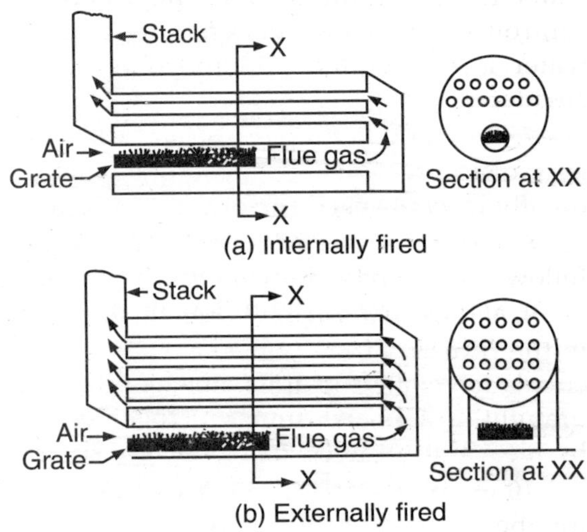

Fig. 14.3. Distinction between internally fired and externally fired boiler

3. *Pressure of steam* : Boilers producing steam at a pressure of 80 bar and above are called high pressure boilers. The high pressure boilers are Babcock and Wilcox, Lamont, Velox and Benson etc.

The boilers which produce steam at pressures lower than 80 bar are called low pressure boilers. Examples are Cochran, Cornish, Lancashire and Locomotive boilers.

4. *Method of circulatian of water* : Majority of the boilers operate with natural circulation, *i.e.*, the circulation set up by convection currents or by gravity. However at higher steam pressures, the steam becomes dense and there is very little difference in the density of steam water mixture and water alone. For effective circulation, pump (forced) circulation is employed in place of natural circulation. Forced circulation is also sometimes employed at low pressures to give freedom in tube layout.

5. ***Nature of service to be performed*** **:** Boilers which are used with stationary plants are classified as land boilers. Boilers which can be readily dismantled and easily carried from one site to another are called portable boilers. Marine and Locomotive boilers belong to another category called *mobile boilers*.

6. ***Once through boilers*** **:** (Benson) in which there is no recirculation of water, *i.e.*, the feed water leaves the tube as steam, whereas in controlled circulation boilers only a part of water is evaporated and the remainder is circulated.

7. ***Position and number of drums*** **:** Single or multidrums may be positioned longitudinally or crosswise.

8. ***Design of gas passages*** **:** The gas may follow a single pass, return pass or multipass.

9. ***Nature of draught*** **:** When the fuel burns in the furnace of the boiler, with the natural circulation of air, the draft is named as natural draught. In artificial draught, the air is forced by means of a forced fan.

10. ***Heat source*** **:** The heat energy utilized for the conversion of a fluid into a vapour may be derived from :

(*i*) combustion of solid, liquid or gaseous fuel

(*ii*) electrical and nuclear energy

(*iii*) hot waste gases of other chemical reactions

11. ***Fluid used*** **:** The boilers are classified as steam boilers using water as fluid, mercury boilers using mercury as fluid, and the boilers which are used for heating special chemicals.

12. ***Material of construction of boiler shell*** **:** Depending upon the material used for the construction of boiler shell, we classify the boilers into cast iron boilers and steel boilers. Power boilers are usually fabricated from steel plates. Low pressure heating boilers are built either of cast iron or steel. Miniature boilers have been fabricated from metals such as copper and stainless steel.

14.2. FIRE TUBE BOILERS

These boilers essentially consist of one or more fire tubes passing through a cylindrical shell which may be horizontal or vertical and is filled to at least half of its volume with water.

(*A*) Cochran Boiler

Simple vertical boilers of the fire tube type find favour in small plants requiring small quantities of steam and where the floor area is limited. The most common applications are steam rollers, pile drivers, steam shovells, portable hoisting rigs and certain other mobile applications.

Cochran boiler, illustrated in Fig. 14.4, provides an excellent example of the improved design of vertical, multitubular, internally fired natural circulation boiler.

The Cochran boiler essentially consists of :

(*i*) Boiler shell with hemispherical crown,

(*ii*) Furnace, fire box and grate,

(*iii*) Combustion chamber and flue pipes,

(*iv*) Smoke box and chimney, and

(*v*) Connections for boiler mounting and accessories.

Constructional and operational details

The unit consists of a cylindrical shell with a dome shaped top where the space is provided for steam. The shell is formed of steel plates joined together with the rivets. Both the circumferential and longitudinal joints are lap joints made steam tight by fullering or caulking operation. The fuel is burnt on grate in the furnace provided at the bottommost part of boiler. The furnace has no riveted seams exposed to flame and is pressed hydraulically from one plate to finished shape. This makes the furnace suitable to resist the intense heat produced by the combustion of fuel. The grate consists of iron bars which are arranged with spacing between them. The spacing allows the air to pass onto the fuel for combustion. The firebox is hemispherical so that the unburnt fuel, if any, is deflected back to the grate and

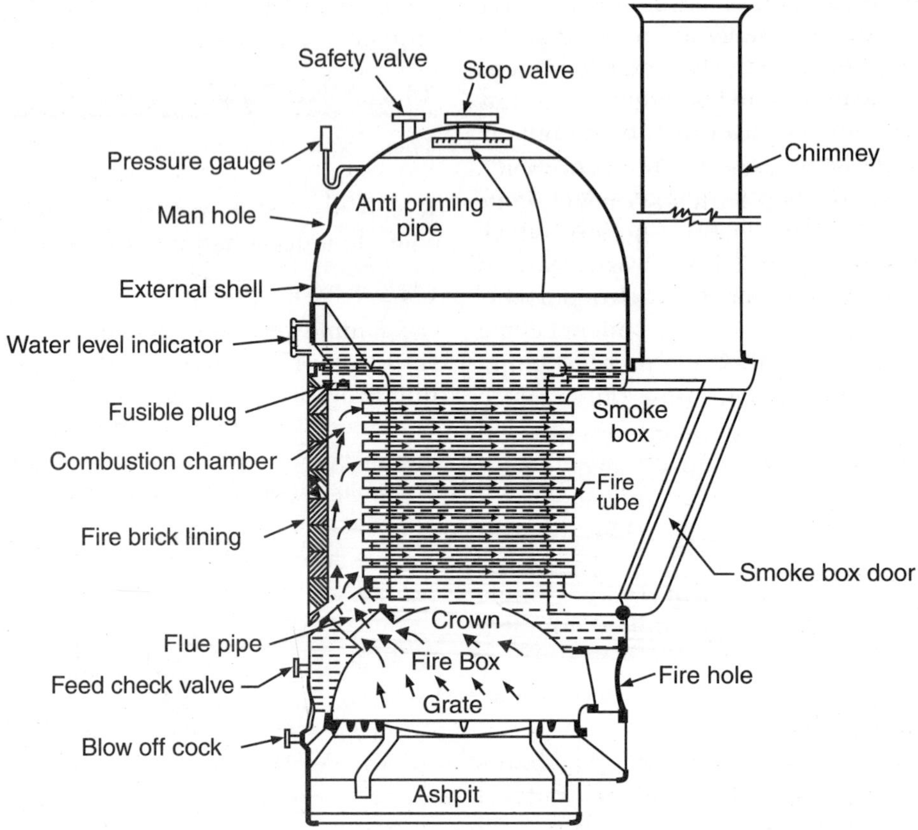

Fig. 14.4. Cochran boiler

complete combustion is achieved. An ash pit is attached beneath the furnace for collecting ash after regular intervals. The boiler can be arranged to burn almost any kind of fuel including wood, paddy husk and oil fuel. For operation as an oil fired unit, an oil burner is fitted at the fire hole. The grate is then dispensed with and a lining of fire brick is provided beneath the furnace.

The coal, on burning, produces hot flue gases and these hot products of combustion from the fire box enter through the small flue pipe into the combustion chamber which is lined with fire bricks on the outer wall of the boiler. The lining prevents the shell from being damaged due to the overheating.

The dome shaped furnace and the combustion chamber prevent the loss which could otherwise occur because of combustion being retarded and much unburnt and combustible matter leaving the furnace. The unburnt fuel is deflected back to the grate and complete combustion is achieved in combustion chamber where the high temperatures are maintained.

The hot gases passing through the horizontal smoke tubes give their heat to the water and in doing so convert water into steam which gets accumulated in the upper portion of the shell from where it can be supplied to the user. The flue tubes are generally of 65.5 mm external diameter and are 165 in number. The crown of the shell is made hemispherical in shape which gives the maximum space and strength for a certain weight of material in form of plates.

Finally the flue gases are discharged to the atmosphere through the smoke box and the chimney. The smoke box door enables the cleaning and inspection of the smoke box and fire tubes. Through a manhole provided at the crown of the shell, a man can enter the boiler for periodic cleaning and maintenance of the boiler. There are connections provided at appropriate places for fixing the usual boiler mounting such as pressure gauge, water level indicator, safety valve, steam stop valve, feed check valve and blow off cock etc.

Typical specifications of Cochran boiler are :

Size = 1 m diameter × 2 m high (evaporation 20 kg/hr)
= 3 m diameter × 6 m high (evaporation 3000 kg/hr)

Heating surface = 10 to 25 times grate area

Steam pressure = upto 20 bar

Efficiency = 70 to 75%

The Cochran boiler is compact in design and there is good external and internal accessibility.

(B) Lancashire Boiler

This is a fire tube, internally fired, horizontal, natural draft and natural circulation type of boiler.

Construction : Fig. 14.5 shows the main features of a Lancashire boiler and its brickwork setting.

(*i*) The boiler has a cylindrical horizontal shell made to suitable dimensions by several rings of steel plates which are either welded or riveted. The shell is provided with two parallel flues which run throughout the entire length of the shell. The diameter of the tubes is about 0.4 times the diameter of the shell.

(*ii*) The flue tubes built in short lengths are flanged. Further to reduce the amount of internal mechanical bracing, the flue tubes are sometimes corrugated. To accommodate a grate of sufficient area and minimum length, the flue tubes are larger in diameter at the front of the shell.

(*iii*) Each of the flue tube has its own furnace with grate of about 2 m length arranged at the front end of the shell. Through the fire door, the coal is fed to the grates where its combustion takes place. The fire bridge provided at the back of the grate prevents the fuel from falling over the end of the furnace and also restricts the space through which hot gases have to pass. This space restriction is necessary to produce proper mixture of air and gases and to give perfect combustion.

(*iv*) To regulate the gas flow and to control the amount of air entering the grate, there are dampers placed in the path of the flue gases. These dampers are operated by means of chains passing over pulleys from the front of the boiler.

(*v*) Openings are made in the boiler for inspection, steam exit and for necessary mountings and accessories.

Working : The flue gases rising from the furnace pass over the fire bridge and traverse along the horizontal path. At the rear end, the gases enter the bottom common flue, travel backward to near the front of the boiler where they bifurcate and pass into the two sides and are finally discharged to the atmosphere through chimney.

The feed water is supplied uniformly to the shell by a perforated feed pipe controlled by a feed valve. When the boiler is strongly heated, the steam generated carries a large quantity of water in the steam space. To remove these water particles, the steam is passed through an antipriming pipe before it is taken out through the stop valve and supplied to the engine when required.

Capacity and utility : Lancashire boilers are made to withstand working pressures upto 20 bar and have evaporative capacity upto 8000 kg/hr. The ratio of heating surface to grate area varies from 24 to 30. The size of the shell ranges from 2 m diameter × 6 m length to 3 m diameter × 10 m length.

The boiler is very widely used in sugar mills and chemical industries where alongwith power system, steam is also required for process work. The boiler is also commonly employed where we need large reservoir of water and steam.

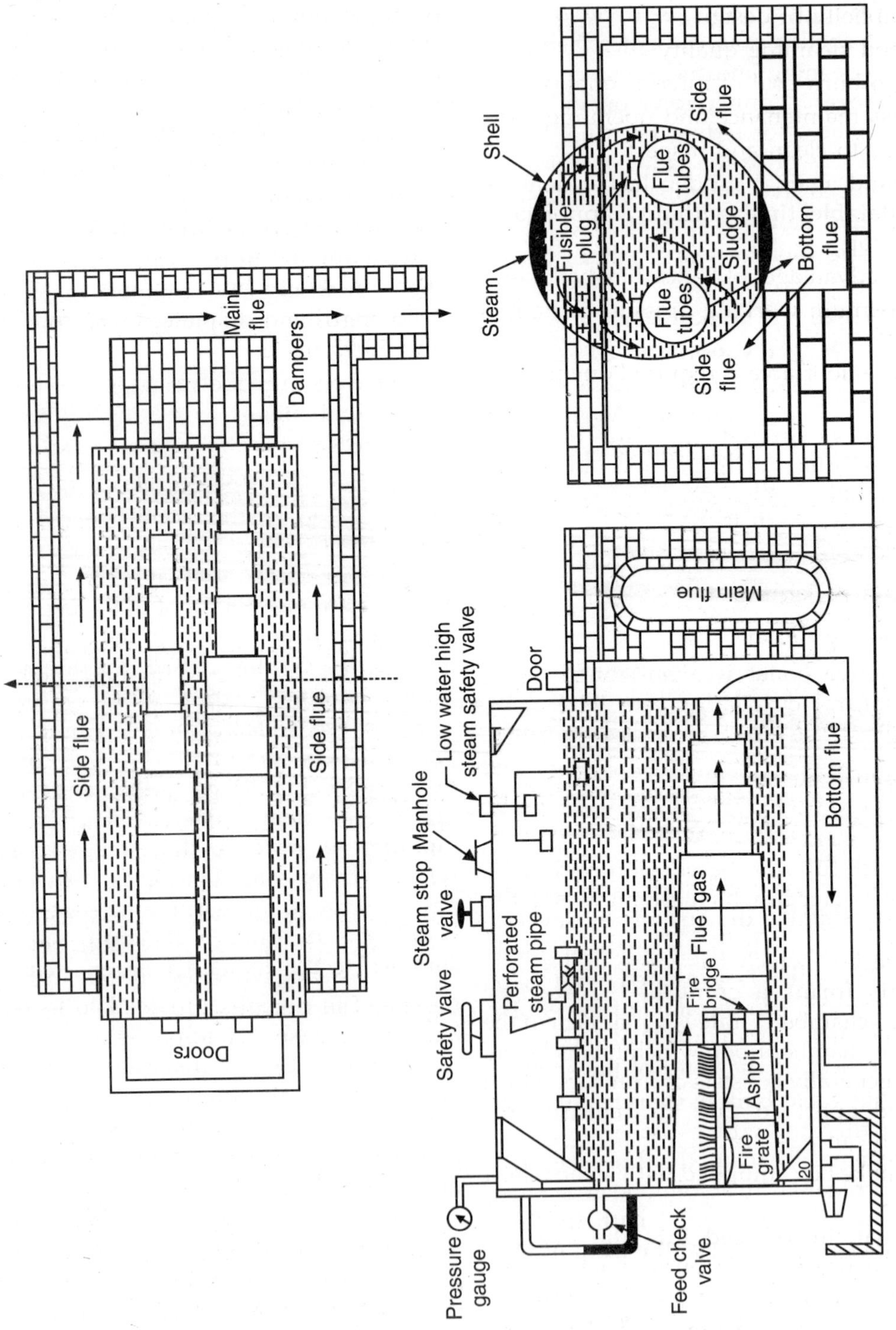

Fig. 14.5. Lancashire boiler

Advantage/disadvantages : The Lancashire boilers are very popular because of :

(*i*) simplicity of design

(*ii*) good steaming quality

(*iii*) can burn coal of inferior quality

(*iv*) less maintenance and operating cost

(*v*) easy to clean and inspect

(*vi*) stand against sudden heavy demand for considerable time without appreciable pressure drop.

The disadvantages of Lancashire boiler are :

(*i*) maximum working pressure is limited to 20 bar

(*ii*) more floor-area is required due to brick work setting

(*iii*) cracks in setting due to large temperature difference inside and outside

(*iv*) restricted grate area due to furnace being inside

(*v*) more time taken for developing steam pressure due to large water capacity.

(C) Locomotive Boiler

The locomotive boiler is a horizontal, multi tubular, natural circulation, artificial draft, internally fired, fire tube type of portable boiler.

Construction and working : The locomotive boiler shown in Fig. 14.6 comprises the followings :

(*i*) Multi tubular barrel containing an envelope of water in which are immersed fire tubes.

(*ii*) Fire box which has more or less flat sides. This forms a combined grate and combustion chamber. Grate is a platform in the combustion chamber upon which fuel is burnt. The grate generally consists of iron bars which are spaced apart so that air required for combustion can pass through them.

(*iii*) Smoke box equipped with a very short chimney.

The coal, introduced into the fire box through the fire hole, is burnt on the grate which slopes downwards towards the front.

On the shell or dome are fitted the mountings such as safety valve, stop valve, pressure gauge and water level etc. The boiler is also provided with a manhole for cleaning purposes and a large door for excess to the smoke box and tubes for cleaning. The height of the chimney has to be less so that when the locomotive is passing through a tunnel or under a bridge, the chimney does not strike against them. Moreover, a high chimney is likely to create unbalancing.

The important dimensions and specifications for the locomotive boilers which are being made at Chitranjan are :

Length and diameter of barrel : 5.203 m and 2.095 m

Size and number of super heater tubes : 14 cm and 38

Size and number of flue tubes : 5.72 cm 116

Working pressure and capacity : 14 bar and 9000 kg/hr

Heating surface and grate area : 271 m^2 and 4.2 m^2

Working : The hot flue gases rising from the grate are deflected by a fire brick arch. This is to ensure more uniform and proper heating of gases by bringing them into contact thoroughly with the entire heating source of the fire box. The dampers control the flow of air into the fire box. The hot flue gases while passing through the fire tubes impart their heat energy to surroundings water and are ultimately led to smoke box and then discharged to atmosphere through a short chimney.

The steam generated collects in the steam dome. The dome is provided to increase the steam release capacity. This incidentally increases the distance of steam from water line and reduces priming. The dome has to be centrally located so that static water level changes, (due to inclination of the rail line) do not effect the level at dome.

The driver from the cab operates the regulator by turning a lever and the dry saturated steam is led to engine for expanding and doing work. For getting super heated steam, the steam is taken to superheater header and thence to superheater tubes. These small

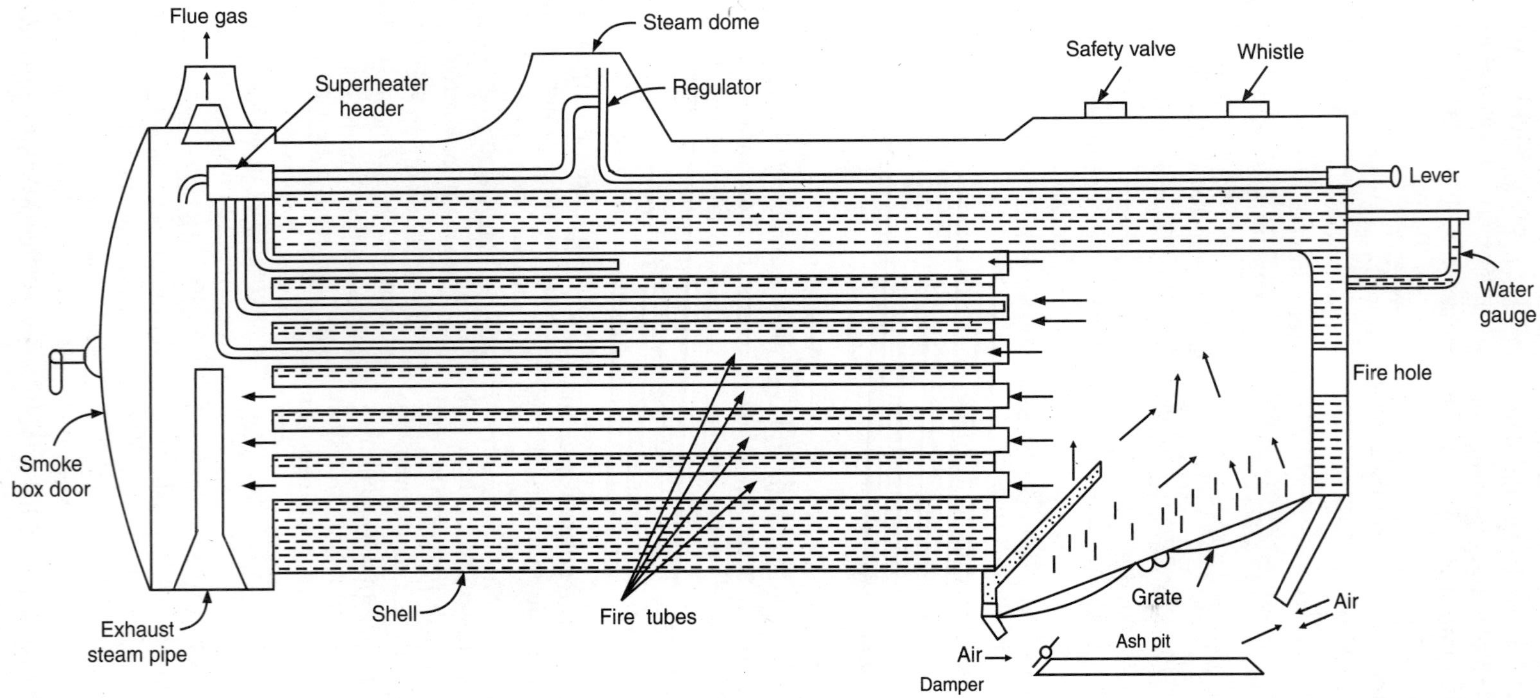

Fig. 14.6. Locomotive boiler

diameter tubes start from the superheater header and are laid inside the large diameter fire tubes which are placed at the highest level of the boiler shell. The superheated steam is then supplied to the engine.

Because of low chimney, artificial draught has to be created to drive out the burnt gases. The draught is accomplished by the following methods:

(*i*) Motion of locomotive on rails.

(*ii*) Periodic rush of spent up steam from the engine cylinder into the fire box. The vacuum thus created sucks the hot gases and helps their discharge to atmosphere.

(*iii*) A blower which is run by steam supplied directly by the boiler. The blower is operated only when supply of steam to the engine is cut off.

High draught thus produced increases the coal consumption and the steam evaporation rate. The ash of coal burnt on the grate falls into ash pit from where it can be removed at regular intervals.

Utility/applications : The locomotive boiler is so designed that it is capable of meeting the sudden and fluctuating demands of steam which may be imposed because of variation of power and speed. In addition to railways, road rollers and haulage engines, the locomotive boilers have been used in agricultural fields, saw mill plants, and stationary power service where semi – portability is desired.

Advantages/disadvantages : The Locomotive boilers are very popular because of :

(*i*) Large rate of evaporation (55 to 70 kg/sec) per square metre of heating surface. This aspect however tends to produce wet steam and loss in efficiency.

(*ii*) Freedom from brickwork, chimney and special foundations and this reduces the cost of installation.

(*iii*) Reasonable low cost and compactness makes it ideal for a portable unit. The unit can be easily moved and set up readily in different locations.

The boiler is, however, incapable of meeting very high overloads because of the danger of being damaged due to overheating. Other shortcomings are :

(*i*) maximum steam pressure is limited to 20 bar

(*ii*) leakage occurs frequently at the places where the tubes join the plates

(*iii*) large flat surface needs bracing

(*iv*) accumulation of sediments and mud particles in the water legs is liable to cause corrosion and scale formation. To safeguard against this, good feed water has to be supplied to the boiler.

14.3. WATER TUBE BOILERS

Among the water tube boilers, Babcock and Wilcox boiler and Stirling boiler are the most common type used in thermal power houses for generation of steam in large quantities.

(A) Babcock and Wilcox Boiler

This is a horizontal, externally fired, water tube, natural circulation type of stationary boiler.

Constructional features : The salient aspects of the constructional features of the Babcock and Wilcox boiler have been shown in Fig. 14.7.

(*i*) It consists of a welded steel high pressure drum mounted at the top. From each end of drum connections are made with the uptake header and a down take header. The headers are joined to each other by a large number of tubes which are kept inclined at an angle of about 15° to the horizontal. The water tubes are straight, solid drawn steel tubes about 10 cm in diameter and are expanded into the bored holes of the headers. Due to serpentine (sinusoidal) form of headers, the tubes are staggered and this exposes the complete heating surface to flue gases.

(*ii*) The furnace is arranged below the uptake header. The unit has a grate for fuel burning, provided with a chain gate stoker. The coal is fed to the chain gate stoker through the fire door. The chain speed is so adjusted that by the time coal reaches the other end of the grate, its combustion has been complete. The residual ash falls into the ashpit.

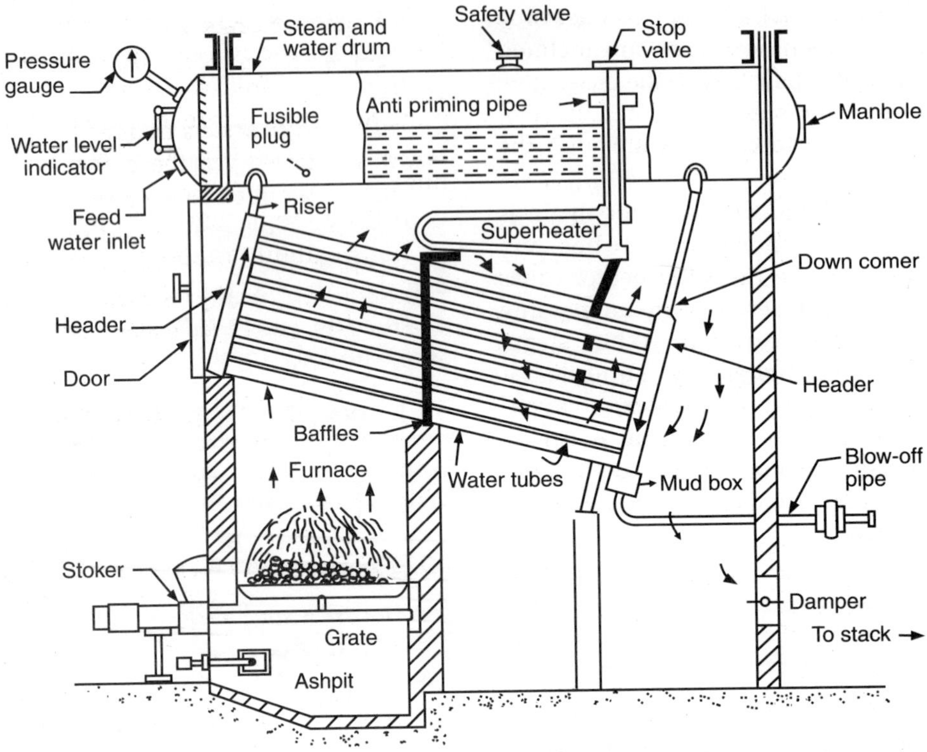

Fig. 14.7. Babcock and Wilcox boiler

(*iii*) Baffles are provided across the water tubes to act as deflectors to the flue gases and to provide them with gas passes.

(*iv*) At the bottom of the rear header is the mud box. The foreign matter held in suspension in water gets collected in it and can be blown off from time to time. The access to the interior of boiler is provided by the doors. This is necessary to clean the tubes and remove the soot. The draught is regulated by a damper which is provided in the back chamber.

(*v*) For getting the superheated steam, the unit is fitted with superheater tubes which are arranged in the combustion chamber above the water tubes.

Vents are also provided for mountings and accessories such as safety valve, pressure gauge, water level indicator, fusible plug and feed check valve etc.

The entire assembly of water tubes is hung along with the drum from steel girder frame by steel rods called slings in a room made of masonry brick lined with fire bricks.

Working : The boiler drum shell is filled with water through feed valve and a constant water level is maintained upto about $2/3^{rd}$ part in the shell. The water from the drum flows through inclined tubes via downtake header and then goes back into the shell in the form of water and steam via uptake header. Obviously the circulation of water is maintained by convective currents. The hottest water and steam rise from the tubes to the uptake header and then through riser enter the boiler drum. The steam vapours escape through water to the upper half of the drum. The cold water flows from the drum to the rear header and that completes the water circuit.

The hot combustion gases produced by burning fuel on the grate rise upwards up to the water drum and then downwards. They again move upwards and downwards between the baffles. Baffle plates make the hot gases move in such a way that a longer contact time is maintained between hot gases and inclined water tubes. During their travel, the gases give their heat to water and steam is formed. Finally, the hot gases escape to chimney through the smoke chamber.

The steam collected in the water drum is next led to the superheater tubes via antipriming pipe. Since the superheater tubes are exposed to hot gases, the steam passing through them gets superheated. When steam is raised from cold boiler, the superheater should be flooded (filled with water) to avoid its burning. The superheated steam is finally taken out through the stop valve and supplied to the engine when needed.

Capacity and utility : Evaporative capacity of such boilers ranges from 20,000 to 40,000 kg/hr and operating pressures from 11.5 to 17.5 bar are quite common. The steam from such boilers is primarily used to run steam turbines and generate electric power.

Salient aspects : The notable aspects of Babcock and Wilcox boiler in the context of merits and demerits are :

(*i*) The unit has the capability to quickly cope with high peak loads which are generally needed at thermal power stations.

(*ii*) The inspection of the boiler can be carried even when the boiler is in operation.

(*iii*) Draught loss is minimum.

(*iv*) The replacement of defective tubes can be made readily.

Further, since the boiler unit (except furnace) is suspended, any expansion or contraction of the boiler has no harmful effects on the masonry work enclosing the furnace and the boiler.

(*B*) Stirling Bent-tube Boiler

This multidrum water tube boiler, has the following distinguishing features:

(1) It consists of steam water drums, more than two in number. This gives greater steam and water storage capacity and the unit can meet peaks in load variations with less pressure drops.

(2) The drums are interlinked to each other with a bank of bent water tubes. The tubes are bent for the following reasons:

(*i*) to allow free expansion and contraction of the tubes and to prevent the undue thermal stresses which are often set up with straight tubes.

(*ii*) tube replacement becomes easier.

(*iii*) flexibility in design particularly with regard to the location of drums.

(*iv*) the tubes can enter the drums in approximately radial direction.

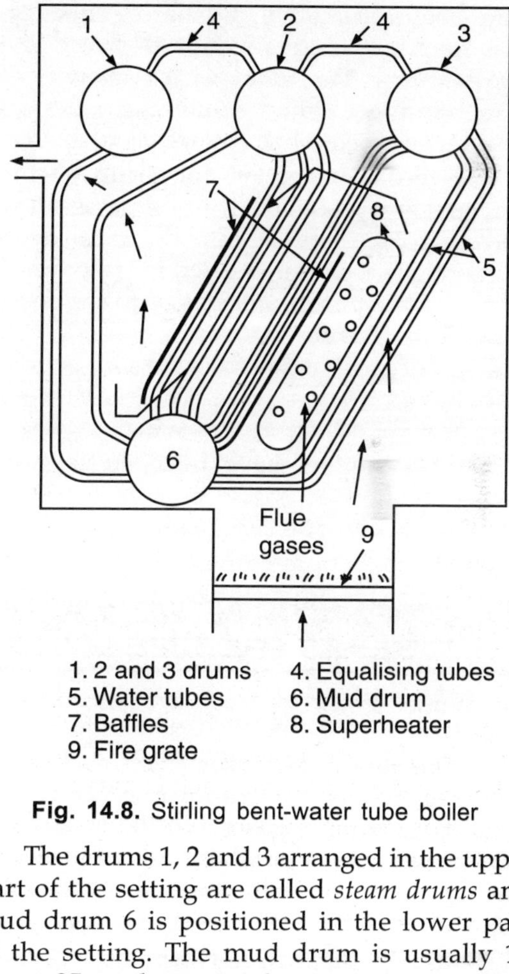

Fig. 14.8. Stirling bent-water tube boiler

The drums 1, 2 and 3 arranged in the upper part of the setting are called *steam drums* and mud drum 6 is positioned in the lower part of the setting. The mud drum is usually 10 cm to 25 cm larger in diameter than a steam drum. The steam drums are usually suspended by the steel structure and the mud drum is supported by tubes to permit free expansion and contraction of tubes. The entire unit is thus independent of the brick work setting which serves only to provide furnace space and confine heat.

The level of water in the steam drums is at the same level when the boiler is idle or operating at low ratings. However at high ratings the level in the drum 3 which contains hottest water and around which hot gases pass is often lower. This is so because the major portion of evaporation takes place in this region. The steam drums are connected by tubes above and below the water line. The tubes which connect the water space of adjacent steam drums are called circulating tubes and these help in the equalization of water level in the drums. The tubes which connect the steam space are called equalizing tubes and these serve to equalize the steam pressures.

The feed water enters the drum 1 which is farthest away from the fire and passes down the rear bank of tubes into the mud drum. Any foreign matter suspended in water settles down and the pure water passes to the drums 2 and 3. Baffles are provided to deflect the products of combustion from furnace over the tube banks. Arrangements are there to superheat the steam as and when necessary.

Stirling boilers have been successfully employed for high steam pressures (60 bar and 450°C) and for capacities up to 50,000 kg/hour.

14.4. COMPARISON BETWEEN WATER TUBE AND FIRE TUBE BOILER

The salient points of each are enumerated below:

1. The two types differ from each other with regard to the manner or order of conducting hot flue gases and the water. Fire tube boilers have products of combustion passing through tubes which are immersed in water. In water tube boilers the flue gases surround the tubes through which the fluid flows under pressure.

2. Fire tube boilers have large ratio of water content to steam capacity. Larger water capacity makes the boiler slow in reaching the operating pressure and less immune to failure in any feed water arrangement.

Comparatively small ratio of water content to steam capacity helps in reaching the steaming temperature in short time but any failure in feed water supply is liable to make the boiler overheated.

3. Because of simple and rigid construction, the fire tube boilers have greater reliability and are low in first cost. These boilers require skill for economical and efficient operation.

Complexity in design of water tube boiler requires periodic examination by skilled hands.

4. Large diameter of shell (2-4 m), limit of maximum thickness of 30 mm and stress considerations (maximum 12 kN/m^2) limit the pressure range from 17.5 bar (Lancashire boiler) and 24.5 bar (Economic boiler). Evaporative rate of these fire tube boilers is also limited to 900 kg/hour.

Because of small drum diameters, the water tube boilers can withstand high internal pressures for the same wall thickness and stress. These boilers can operate up to pressure as high as 200 bar and have wide ranges in capacity. Boiling temperature at 200 bar is about 640 K. Good carbon steel can safely stand this temperature and still keep the stresses within the limits of 8 kN/m^2.

5. With water contained in a large number of small tubes in a water tube boiler, the heating surface area is increased and evaporation rate is considerably enhanced. All large power plants use water tube boilers.

Slow rate of steam generation renders the fire tube boilers unsuitable for use in steam power plants.

6. Circulation is more positive in water tube boilers and as such there is less tendency of the deposits to settle on the heated surfaces. This positive circulation also helps in quick steam raising. However, with high pressures and high temperatures in water tube boilers, the use of pure feed water becomes essential.

7. Due to large cylindrical drums of fire tube boiler, there is ample water surface from which the steam can be quickly raised. Simple antipriming devices serve the purpose very well.

The water tube boilers with relatively small surface tend to produce wet steam and consequently superheater has to take more loads.

8. All parts of water tube boiler are easily accessible for cleaning, inspecting and repairing.

9. The explosion, if any, when it occurs becomes very serious problem in smoke tube boilers because of its large water capacity. Since water is uniformly spread in large number of water tubes in water tube boiler, no serious explosion occurs even if one of the tube fails.

10. Since the water tube boilers are mostly externally fired, the proportions of the furnace can be altered considerably to meet the fuel requirements.

11. Fire tube boilers are excellent for engines operating with rapid changes in load. When demand on the unit increases, pressure slightly goes down with consequent reduction in boiling temperature. The evaporation rate would increase and the steam in greater quantity (but at slightly lower pressure) can be supplied to the engine. With reduction in steam demand, slight increase in pressure takes place resulting in the condensation of the part of existing steam. If overpeaks exist for short duration the pressure variations are tolerable. However for large load fluctuations extending over larger durations, the water tube boilers are to be preferred.

12. Because of large size of shell, the transportation of fire tube boiler is very inconvenient. The water tube boilers however can be easily dismantled, transported conveniently and erected quickly at site.

14.5. GUIDELINES FOR THE CHOICE OF A NEW BOILER

Before making a choice for the best type of boiler for a new installation, the following factors are given due consideration :

(1) Rate of steam generation, *i.e.*, how much steam per hour is required to be produced.

(2) The pressure at which the boiler is required to operate and the quality of steam required, *i.e.*, whether we need wet, dry sor superheated steam.

(3) Whether the steam raised is to be used at a steady or fluctuating loads.

(4) Type of the fuel to be used, *i.e.*, solid, liquid or gaseous fuel.

(5) Space available for erection and the height of the building in which the boiler is to be accommodated.

(6) Probable permanency of station.

(7) Comparatively low initial, operating and maintenance costs.

(8) Easy accessability for erection and repair.

Depending upon the industrial plant where the boiler is to be installed, the following points give rough guidelines :

(*a*) Vertical boilers are used for very small factories, for waste heat, pile drivers and mobile jib cranes.

(*b*) Cylindrical smoke tube boilers for factories of medium and large size, dairies, heating installations and laundries.

(*c*) Locomotive boiler for railway engines, for agricultural and traction purposes, mobile boilers of all kinds and medium size factories.

(*d*) Water tube boilers for power stations, industrial processes, heating installations, rolling mills, chemical and wood working factories.

14.6. ESSENTIALS OF A GOOD BOILER

The essentials of a good boiler are :

(1) The boiler should be capable of generating steam at the required pressure and of the required quality quickly and with minimum fuel consumption.

(2) The initial cost, installation cost and the maintenance cost of the boiler should not be too high.

(3) The boiler should be light in weight, should need least amount of brick work construction and should occupy small floor area.

(4) The boiler should be capable of quick starting and should meet the fluctuating

demands on steam supply without being overheated.

(5) The different parts of the boiler should be easily approachable for repairs.

(6) The boiler should have only minimum joints and those too should be away from direct flames.

(7) The boiler should offer the ease of dismantling the parts and erection at site within reasonable time and labour. This is necessary for a boiler which has to be transported quite often.

(8) For efficient heat transmission rate, the water and flue gases should have maximum velocity without incurring heavy frictional losses.

(9) There should be no deposition of mud and other foreign particles on the heated surfaces.

(10) The tubes of the boiler should be sufficiently strong to allow for wear and corrosion.

(11) The boiler should conform to the safety regulations as laid down in the "Boilers Act."

14.7. BOILER MOUNTINGS AND ACCESSORIES

For efficient operation and maintenance of safety, the boilers are equipped with two categories of components or elements.

First category includes the fittings which are primarily intended for the safety of the boiler and for complete control of the process of steam generation. These units are called *boiler mountings*. The mountings form an integral part of the boiler and are mounted on the body of the boiler itself. In accordance with Indian Boiler Regulations, the following mountings are usually installed on the boiler.

(1) Two safety valves
(2) Two water level indicators
(3) Pressure gauge
(4) Fusible plug
(5) Steam stop valve
(6) Feed check valve
(7) Blow-off cock
(8) Man and mud holes

Out of the above mentioned mountings, safety valve, water level indicator and the fusible plug are called *safety fittings* and the remaining are the *control fittings*.

Second category includes the components which are installed to increase the efficiency of the steam power plant and help in the proper working of the boiler unit. These fittings are called *boiler accessories*. Usually a boiler has the following accessories attached to it.

(1) Air preheater
(2) Economiser
(3) Superheater
(4) Feed pump
(5) Injector
(6) Steam separator and steam trap

Some of the above mentioned boiler mountings and accessories are discussed in the following sections with particular reference to their function, location, construction and operation.

(A) Safety Valves

Function : The function of the safety valve is to permit the steam in the boiler to escape to atmosphere when pressure in the steam space exceeds a certain specified limit. Thus the safety valve prevents the building up of excessive pressure in the boiler.

As per Boiler Regulations, each boiler must be fitted atleast with two safety valves.

Location : The safety valve is located above the steam space in the boiler.

The safety valves operate on the principle that a valve is pressed against its seat through some agency such as strut, screw or spring by external weights or force. When the steam force due to boiler pressure acting under the valve exceeds the external force, the valve gets lifted off its seat and some of the steam rushes out until normal pressure is restored again.

The commonly used safety valves are described below :

(i) Dead weight safety valve

Construction and operation : The unit consists of a vertical steel pipe the flange of

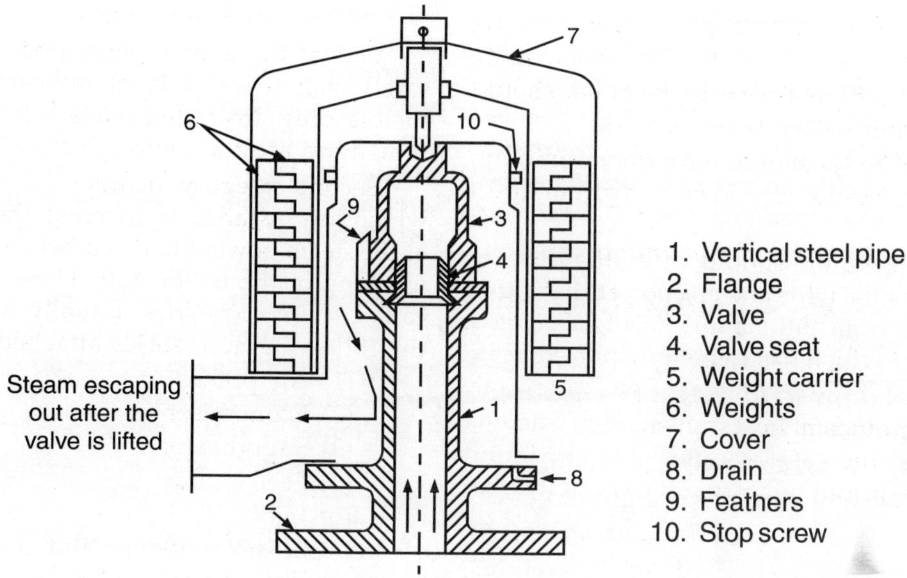

Fig. 14.9. Dead weight safety valve

which is bolted to a mounting block on the boiler shell. The gun metal valve rests on the gun metal valve seat which has been screwed to the top of the vertical pipe. A carrier on casting suspended from the top of the valve acts as a weight carrier. The dead weights consisting of cast iron rings are placed on the weight carrier. These weights are surrounded by a cast iron cover which is attached to the weight carrier. This is necessary to protect the weights from being tampered with.

Under normal working conditions the upward force exerted by steam in the boiler is balanced by the downward force equivalent to the total weights on the valve. The downward weight includes the weight of the carrier, weight of the ring, weights and the weight of the cover. With the increase in pressure beyond a certain permissible limit, the valve along with the weight carrier is lifted off its seat. The steam then escapes through the discharge pipe.

The steam which gets condensed when passing through the valve is drained by a drain pipe connected to the bottom. The feather guides the movement of the valve and the stop screw fittings on the discharge casing prevent the blowing away of the valve and the weight carrier, in case the steam pressure becomes too high.

For the valve in equilibrium position,

$$W = \frac{\pi}{4} d^2 p$$

where p is the steam pressure and d is the diameter of the valve. The total downward weight W calculated above is equivalent to

W = weight of valve + weight of cover + weight of carrier + dead weights mounted

Merits and demerits :

- Simplicity of design and construction.
- Reliable with regard to satisfactory performance during operation.
- Safe from the possibility of any tempering in the context of pressure adjustment.

However the dead weight safety valves are unsuitable for boilers which are non-stationary. It is because in portable boilers, there is the danger of weights being displaced. Moreover there occurs a reduction in the effective weights when the moving unit is negotiating a slope. Further its use is restricted only to low pressure boilers because the high pressure boilers would necessitate very heavy weights to balance the steam pressure.

EXAMPLE 14.1

The diameter of the safety valve of the dead weight type is 8 cm. The dead weights mounted on the weight carrier are equal to 3500 N. Calculate the equivalent weight of the valve, weight carrier and the cover of the safety valve in order that the valve may blow off at a steam pressure of 7.5 bar.

Solution : For the valve in equilibrium position

$$W = \frac{\pi}{4} d^2 p = \frac{\pi}{4}(0.08)^2 \times (7.5 \times 10^5)$$
$$= 3768 \text{ N}$$

The total downward weight W calculated above is equivalent to:

W = weight of valve
+ weight carrier + weight of cover
+ dead weights mounted

∴ Equivalent weight of valve, weight carrier and cover

$= 3768 - 3500 =$ **268 N**

(*ii*) Lever safety valve

Construction and operation : The lever safety valve consists of a valve, a valve seat, fulcrum, a lever and weights whose position can be adjusted on the lever.

The base of the valve body (Fig. 14.10) made of cast iron or case steel is bolted to a mounting block on the boiler. The valve rests on valve seat screwed to the valve body and this provides facility for its replacement. The valve is held by the lever which is pivoted at the fulcrum and subjected to an external weight at the other end. The hinged strut transmits the thrust of the lever to the valve. The guide prevents the lateral movement of the lever and also confines the lift of the valve. The weight can be moved along the length of the lever and the specific position of the weight, determined by the pressure of steam acting below the valve, is locked so that no unauthorised person can temper with it.

When the pressure becomes more that the specified limit, the valve rises off its seat and allows some steam to leak out until the pressure in the boiler is again balanced by the weight.

Lever safety valves are suitable for stationary boilers.

(*iii*) Spring loaded safety valve

Construction and operation : Spring loaded safety valve (Fig. 14.11) has a cast iron body which is provided with two channels for the flow of steam. The flange of the body is bolted to a mounting block on the boiler shell. The unit consists of two separate valves which are held down to their seats. The valves are either screwed or kept as force fit so that neither they get lifted by the steam pressure nor do they allow any leakage around them. The necessary thrust for keeping the valves on the valve seat against steam pressure is provided by strut (pivot) to the lever which is held down by the helical spring. The lever has two pivots

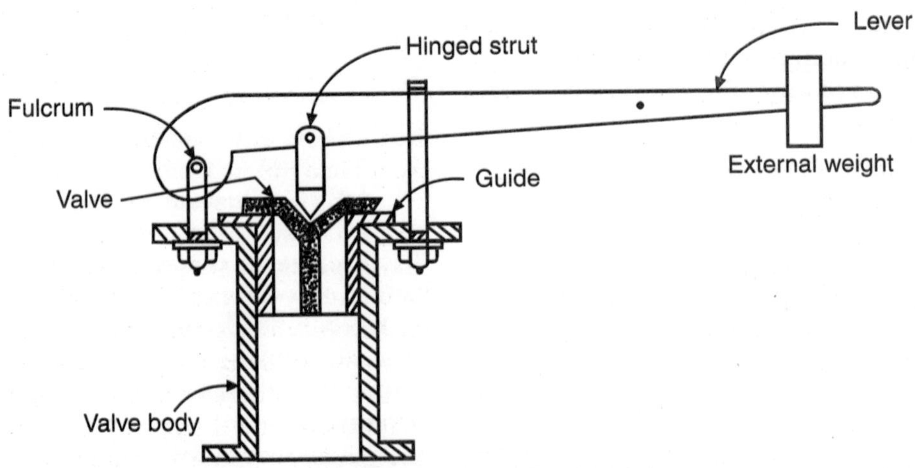

Fig. 14.10. Lever safety valve

: one is joined to the lever by a pin and the other pivot is cast integral with the lever. The helical spring has its lower end hooked in the shackle which is screwed to the valve chest by studs and nuts. The upper end of the spring is connected to the lever arm in the middle of the two valves.

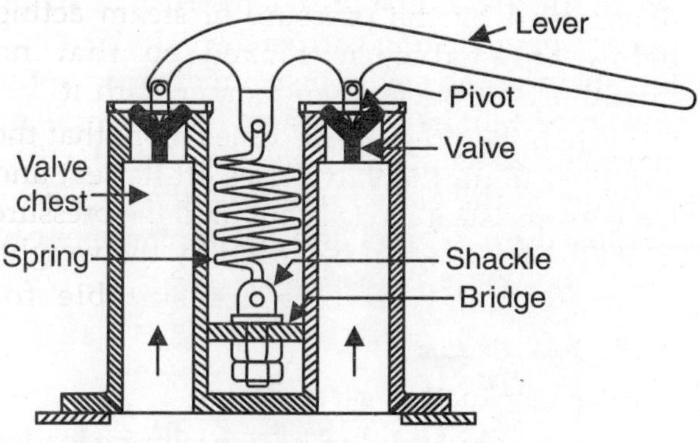

Fig. 14.11. Spring loaded safety valve

When the pressure is normal, the spring lies in the usual position and the valves touch their seats tightly. However when the boiler pressure rises above that for which the valve is set, the valve opens and releases steam from the boiler.

The tension of the spring and so the setting in the blow off pressure can be altered by loosening or screwing the nut. The valve and the valve seat are kept clean and the checking of the working of the valve is done by blowing off steam, achieved with the help of hand lifting of the lever.

The spring loaded safety valves are used with portable boilers such as locomotive and marine boilers as they are not affected by jerks and vibrations.

(*iv*) High steam-low water safety valve

Function : The boiler has not only to be safeguarded against too high a pressure but also against too low a water level in the boiler. We know that if the water level becomes too low, there is danger of overheating and softening of combustion chamber, furnace and other boiler parts. All this may ultimately lead to boiler explosion.

To safeguard against the danger likely to be caused by too low water level, we use a safety device which allows the steam to escape automatically, once the level of water falls below a certain limit. In combination with low water alarm, we use high pressure safety valve which automatically discharges the steam out when the pressure of steam rises above the normal boiler working pressure.

The above two functions are very well served by the high steam-low water safety valve (Fig. 14.12).

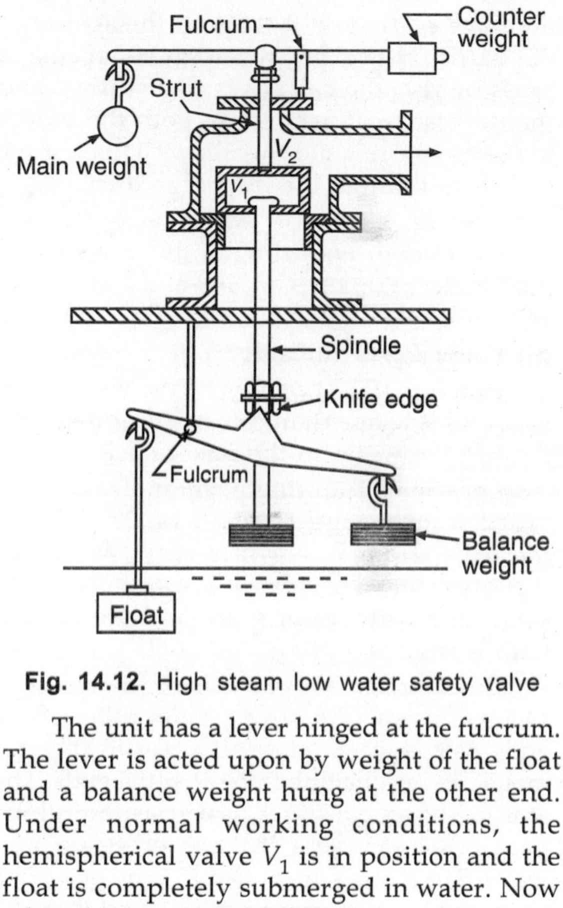

Fig. 14.12. High steam low water safety valve

The unit has a lever hinged at the fulcrum. The lever is acted upon by weight of the float and a balance weight hung at the other end. Under normal working conditions, the hemispherical valve V_1 is in position and the float is completely submerged in water. Now when the water level falls below a certain level, the float no longer remains in water and consequently its weight increases. This increase in weight is in accordance with Archimede's principle which states that objects lose weight in water. With increase in the apparent weight of the float, there occurs a swing in the lever. With the swing thus produced the knife edge presses against the spindle which gets lifted. Since the spindle is connected to the hemispherical valve, the upward movement of the spindle would throw off the valve V_1 of its position and consequently the steam would rush out producing a hissing noise which will serve as a warning to the boiler room attendant.

The arrangement of the external valve V_2 is exactly similar to that lever safety valve. It has a main weight, and the counter weight acting about the fulcrum. Under normal boiler pressure the valve V_2 is held in position by the force exerted on it through the agency of the strut. However, when there occurs a pressure rise, the balance is undisturbed and the unbalanced forces cause both the valves to rise as one unit and this allows a free escape of steam through the discharge duct.

The high steam low water safety valve is the most important fitting of the internal fire tube boilers namely Cornish and Lancashire boilers.

(*B*) Water Level Indicator

Function : The function of the water level indicator is to ascertain constantly and exactly the level of water in the boiler shell.

Location : The water level indicator, also called water gauge, is fitted in the front of the boiler from where it is easily visible to the operator. The top of the mounting is connected with steam space and the bottom with water space.

Construction and operation : The unit (Fig. 14.13) consists of a strong glass tube whose ends pass through stuffing boxes in the gun metal castings having flanges at the ends. The stuffing boxes consist of heat resisting rubber packings to prevent leakage of steam and water. The flanges are bolted to front end plate of the boiler, the upper flange being fitted to the steam space and the lower to water space in the boiler. There are two cocks namely steam cock and water cock which communicate the boiler shell spaces to the gauge glass tube. When the handle of the cocks are vertical, they are in operation and the water level in the tube corresponds to water level in the shell. A red mark on the glass tube indicates the safe water level.

The unit has an automatic arrangement to shut off the steam and water supply to the glass tube in case the tube gets broken. For this purpose a hollow metal tube is connected between the upper and lower gun metal castings. Under normal operating conditions the balls are in position as shown in Fig. 14.13. When the glass tube breaks the rush of steam in the upper casting carries the ball to fit into the opening. As such no steam escapes to the glass tube. Similarly the ball in the lower casing closes the water opening and prevents the flow of water to the glass tube. Now the cocks can be closed for repairs and the replacement of the gauge glass is made.

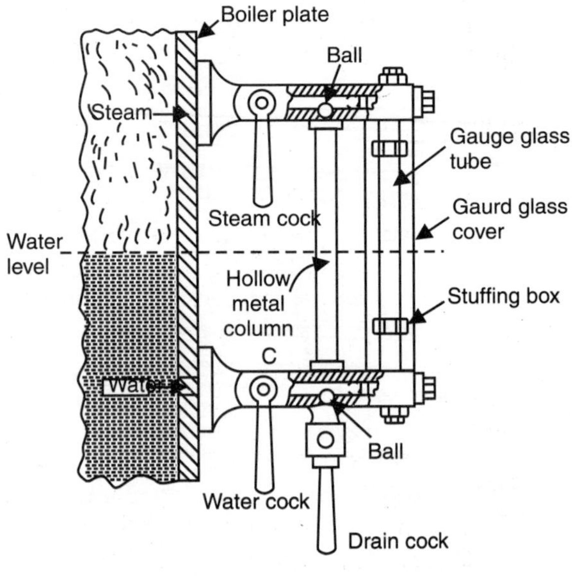

Fig. 14.13. Water level indicator

The draincock serves to blow out water at regular intervals so that there is no accumulation of sediments and foreign deposits. When closed, the handle of the cock is kept in vertical position. The various screw plugs provide access to different passages for cleaning purposes.

The protection from the flying glass pieces in the event of breakage of the tube is provided by enclosing the glass tube by a thick guard glass cover. Usually the boilers are provided with two water level indicators, one placed on each side of the boiler shell.

(C) Fusible Plug

Function : The function of the fusible plug is to extinguish the fire in the event of water level in the boiler shell falling below a certain specified limit. We know that when the water on heating transforms into steam, the level of water in the boiler falls down. If the water is not replenished and the steam generation continues then the parts which have been uncovered by water may get overheated and subsequently get melted. To safeguard against this eventuality we use fusible plug.

Location : The fusible plug is inserted at the box crown or over the combustion chamber at the lowest permissible water level.

Construction and operation : The unit (Fig. 14.14) comprises three plugs A, B and C. The hollow gun metal A having hexagonal flanges is screwed to a fire box crown plate.

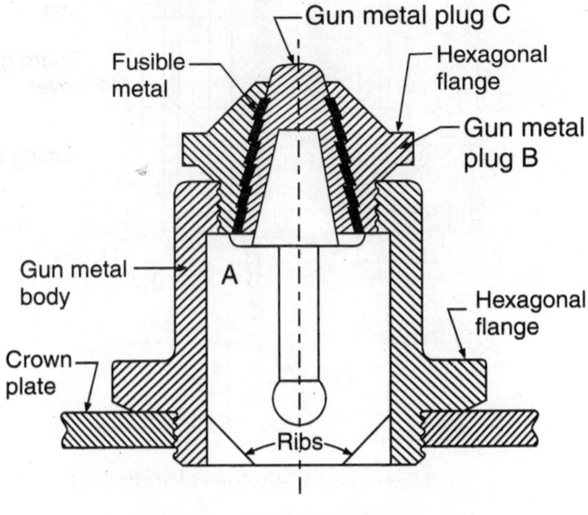

Fig. 14.14. Fusible plug

The plug B, also made of gun metal is screwed into plug A and the plug C is locked to plug B by a metal like tin or lead which has a low fusing point. The plug C is made of copper and has a conical top and round bottom.

When the boiler is operating under normal conditions, the fusible plug is covered with water and the temperature of fusible metal is below its melting point. However, when the water level falls below a certain limit, the plug gets uncovered from water and is then exposed to steam. This overheats the plug, with the result that the fusible metal melts and the plug C falls. This makes a passage and the water-steam mixture rushes into the furnace and the fire is extinguished. The ribs cast integral with gun metal body A hold the falling copper plug C and prevent its falling into the combustion chamber.

Before refiring the boiler, the operator either inserts a new plug or the same plug after doing necessary repairs.

Since the plug is subjected to heat on one side and scale deposits on the other side, it is likely to become defective. So during normal running operations, the plug is usually replaced after every two to three years service.

(D) Pressure Gauge

Function : Each boiler has to be provided with a pressure gauge which records the pressure at which the steam is being generated in the boiler.

Location : The gauge is usually mounted at the front top of the boiler shell or drum. The gauge has to be clearly visible to the attendant so that he can easily record the pressure reading.

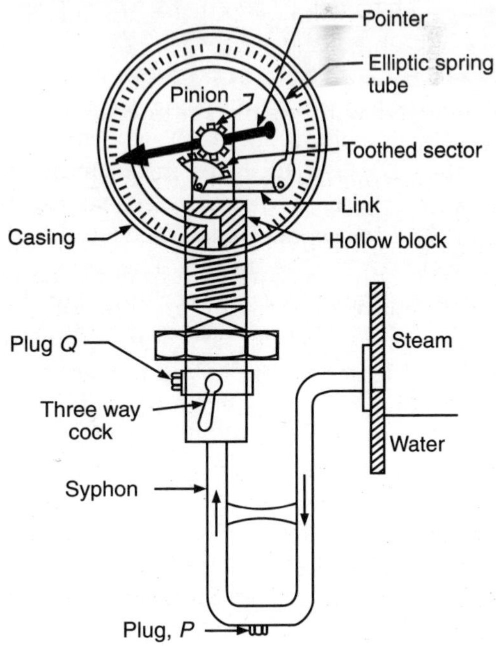

Fig. 14.15. Pressure gauge

Construction and operation : The most important pressure sensing element of Bourden's pressure gauge (Fig. 14.15) is the elliptic spring tube called Bourden tube. This tube is made of a special quality of phosphor bronze and is solid drawn. One end of this tube is closed by a plug and the other end is secured firmly to a hollow block which forms passage for steam. This block is in communication with the steam space of the boiler through a syphon. The U-shaped syphon contains condensed water and the water also fills the spring tube. When the pressure of steam is applied to the interior of the bourden tube, it has a tendency to assume a circular shape. In doing so it causes pull on the link which operates the toothed sector. The link mechanism is designed in such a way that a slight movement of the link is magnified considerably and the pointer gives a maximum deflection which can be easily read on the graduated scale.

It may be pointed out that the gauge records only the pressure difference between the steam and that of atmosphere. In order to get the absolute value the atmospheric pressure ($\simeq$ 1.033 bar) has to be added to the pressure value being shown by the gauge.

The water in the syphon prevents the steam from entering into the spring tube and thus the tube remains comparatively cool. If the steam is allowed to come into contact with the gauge tube, it would overheat the tube causing it to expand and this would affect the accuracy of the gauge. The accuracy of the pressure gauge during working of the boiler can be tested by connecting the inspector's steam pressure test gauge to the plug Q. By means of a three way cock, the pressure of steam in the boiler can be communicated either to pressure gauge or the plug Q.

A plug P is provided for cleaning the syphon.

(*E*) Blow off Cock

Function : The blow off cock serves to drain out the water from the boiler periodically for any one of the following reasons:

(*i*) to discharge mud, scale and other impurities which settle down at the bottom of the boiler.

(*ii*) to empty the boiler for internal cleaning and inspection.

(*iii*) to lower the water level rapidly if the level becomes too high.

Location : The unit is fitted at the lowest portion of the boiler. It may be mounted directly to the boiler shell or through an elbow pipe which is fitted to the boiler shell.

Construction and operation : The unit (Fig. 14.16) comprises a gun metal plug having a hole or slot in it. The casing, sometimes called shell, is bored to receive the plug.

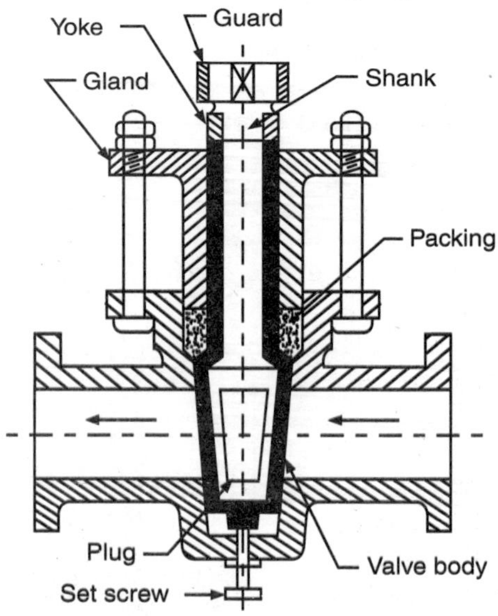

Fig. 14.16. Blow off cock

The shell is provided with two flanges, one of the flanges is connected to the boiler and the other flange is connected to a pipe which takes the blow off water out of the boiler house. The shank of the plug passes through glands and stuffing boxes. Packing and glands have to be provided so as to make the valve leak proof when the boiler is steaming.

To discharge water, the plug is rotated so that the slot made in the plug comes in line with inlet and outlet duct. In the off position the hole is at right angles to the direction of flow. This brings the solid portion of the valve in front of duct and this prevents the flow.

A set screw provided at the bottom of the plug is used to force the plug off its seat in case it gets jammed. The yoke provided at the top has two vertical slots in which the projections of the spanner, used for operating the valve, can fit. The guard prevents the spanner from being removed when the cock is in open position. This avoids the possibility of damage which may occur to the boiler by draining out water.

When it is required to take the discharge from more than one boiler through the same discharge pipe, an isolating valve has to be fitted in between the boilers. This valve is necessary to prevent the discharge of one boiler rushing into the other.

(F) Feed Check Valve

Function : The feed check valve has the following two functions to perform:

(*i*) to allow the feed water to pass into the boiler.

(*ii*) to prevent the back flow of water from the boiler in the event of the failure of the feed pump.

Location : The valve is installed between the feed pump and the water space of the boiler shell.

Construction and operation : The unit (Fig. 14.17) comprises two valves–valve V_1 called the *feed valve* and the valve V_2 known as *check valve*. The check valve allows the water to flow in one direction only and that too from the feed pump side to the boiler side. When the feed pump is working properly the pressure on the feed pump side is more than on the boiler side. Consequently the valve V_2 lifts off its seat, the delivery passage opens and the water flows to the boiler to maintain a constant water level. The lift of the valve V_2 and hence the supply of water to the boiler can be controlled by adjusting the position of the feed valve V_1. The valve V_1 can be raised or lowered by means of a hand wheel :

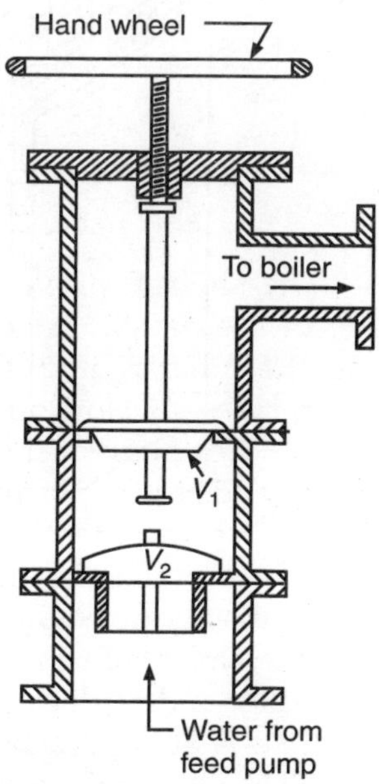

Fig. 14.17. Feed check valve

If by chance the feed pump fails, the pressure reduces on the water side. The valve V_2 then closes under the pressure of steam and this prevents the back flow of water from the holder.

In case it is felt that the check valve is not giving a satisfactory performance, sufficient water is pumped into the boiler and the feed valve is pressed closed onto its seat. The check valve is then removed and replaced after doing the necessary repair and cleaning of the valve and the valve seat.

(G) Stop Valve

Function : The function of the steam stop valve is to shut off or regulate the flow of steam from the boiler to the steam pipe or from the steam pipe to the engine. When used for the former purposes, it is called junction valve. Usually the junction valve means a regulating valve of larger size and a stop valve refers to a regulating valve of smaller size.

Location : The junction valve is mounted on the highest part of the steam space of the boiler and is connected to the steam pipe which carries the steam to the engine.

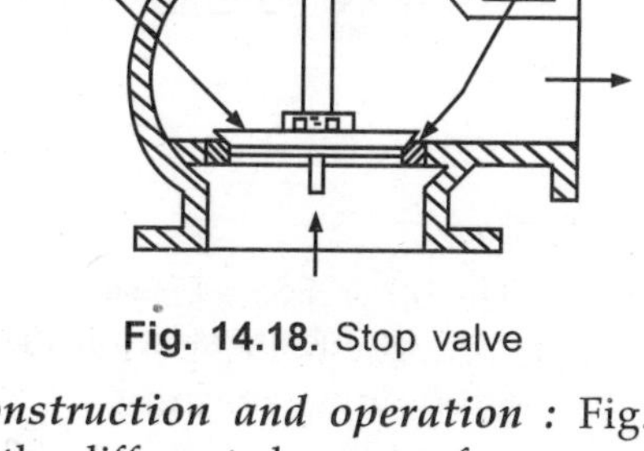

Fig. 14.18. Stop valve

Construction and operation : Fig. 14.18 shows the different elements of a common type of stop valve. The valve body usually made of cast iron or cast steel has two flanges, one flange is bolted to the boiler at the highest point of the steam space and the other flange is connected to the steam outlet pipe. There is a spindle which passes through glands and stuffing box. The glands and packing are necessary to make the valve leak proof. The upper portion of the spindle is threaded and it passes through a nut in a cross bar or yoke rests on a valve seat pillars screwed to the valve body by the lugs. The valve disc is screwed to the valve spindle by a nut. The lower edge of the nut comes in contact with the collar provided at the lower end of the spindle. It is obvious that by this arrangement the spindle is free to rotate within the valve disc. The rotation given to a hand wheel fixed to the spindle top, raises or lowers the spindle and this operation makes the valve disc to be put on and off its seat.

The usual convention to quote the size of the valve is the diameter of the steam pipe to which it is connected, *i.e.*, 50, 75 and 100 mm stop valves.

(*H*) Manholes

These are doors to allow men to enter inside the boiler for inspection and repair.

(*I*) Air heaters

Air heaters or air preheaters are waste heat recovery devices in which the air on its way to the furnace is raised in temperature by utilizing the heat of the exhaust gases.

Air preheaters are classified into the following two categories :

(*a*) *Recuperative* air heaters in which the heat transfer is through a metal wall. The flue gases pass through one side and the combustion air passes on the other side. Air is made to follow a *zig-zag* path so that it passes a number of times. The recuperative air heaters are subdivided into tubular type or plate type depending upon whether the tubes or plates are used for carrying the hot flue gases. Tubular air heaters are in the greatest favour.

(*b*) *Regenerative* air heaters in which the flue gases and the air are made to pass

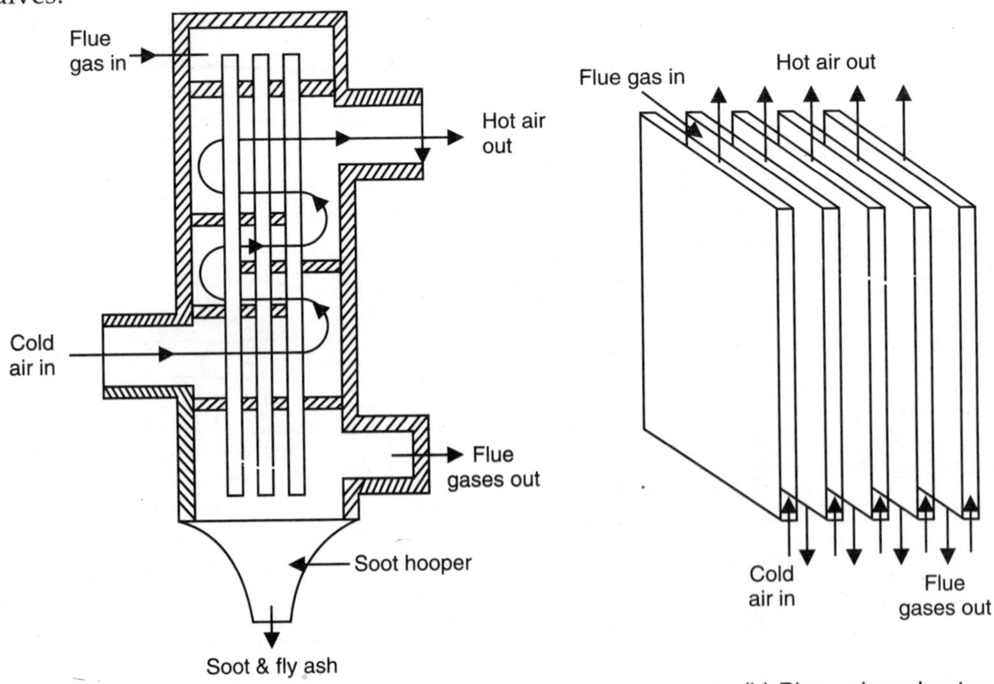

(a) Tubular air pre-heater (b) Plane air-preheater

Fig. 14.19. Air preheaters

alternatively through the same path. Initially when the hot gases pass through a certain medium called matrix, the matrix receives heat from the gases and stores in itself. Next when air is made to traverse the same path, the matrix transfers the heat stored in it to the air. This heater has a small size and is light in weight.

Roughly there is again of about 2% in the boiler efficiency for each 35-40°C rise of the temperature of combustion air. The other advantages of using preheated air are :

(*i*) increase in the evaporation rate,

(*ii*) better combustion with less soot, smoke and ash,

(*iii*) low grade and inferior fuels can be used.

(*J*) Economiser

Function : The economiser is a device which serves to recover some of the heat being carried by exhaust flue gases. The heat thus recovered is utilized in raising the temperature of feed water being supplied to the boiler. If the feed water at raised temperature is supplied to the boiler, it needs less heat for its conversion into steam and thus there is a saving in the consumption of fuel.

The other advantages claimed by installing an economiser are :

(*i*) Improvement in the thermal efficiency of the steam plant. It has been estimated that for each 5 to 6°C rise in the temperature of feed water, there is a gain of about 10% in the plant efficiency.

(*ii*) Reduction in the losses of heat with flue gases.

(*iii*) Increase in the steaming capacity of the boiler.

(*iv*) Less thermal stresses in the boiler parts and consequently long life of the boiler.

Location : The economiser unit is installed in the path of the flue gases between the boiler and the chimney.

Construction and operation : Fig. 14.20 shows the details of the Green economiser used with the boilers of medium pressure range upto 25 bar.

The system comprises a number of vertical tubes which are pressed hydraulically into the top and bottom header. The feed water from the feed pump on its way to the boiler enters the bottom header, passes through the vertical tubes and reaches the upper header. The upper header is in communication with the water

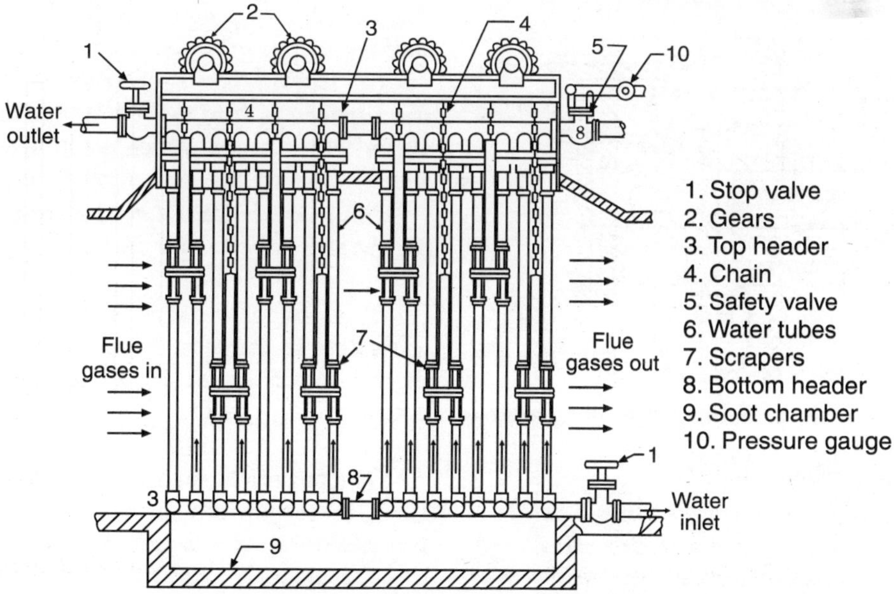

Fig. 14.20. Green's economiser

space of the boiler *via* feed check valve. When the water flows upwards inside the tubes, it gains heat from the hot flue gases which are made to pass over the external surface of the tubes.

For efficient heat transfer, the surface of the tubes has to be kept clean from soot and volatile ash deposits. This is achieved by scrapers which are kept slowly moving up and down to clean the surfaces. A pair of scrapers for adjacent tubes are coupled together by a chain which passes over pulley. The pulley is driven by mechanical drive. It is obvious that the simultaneous action of adjacent pair of scrapers is opposite in direction, *i.e.*, when one scraper is moving up, the other is going down.

The safety valve opens when the pressure of water becomes 10-15% higher than the boiler pressure.

A bypass arrangement is provided to isolate the economiser when it is not required or when it is to be cleaned and repaired. Fig. 14.21 shows such an arrangement for two Lancashire boilers fitted with an economiser. When the economiser is in service the damper A is closed and the dampers B and C are open. For isolating the economiser we close the dampers B and C and open the damper A.

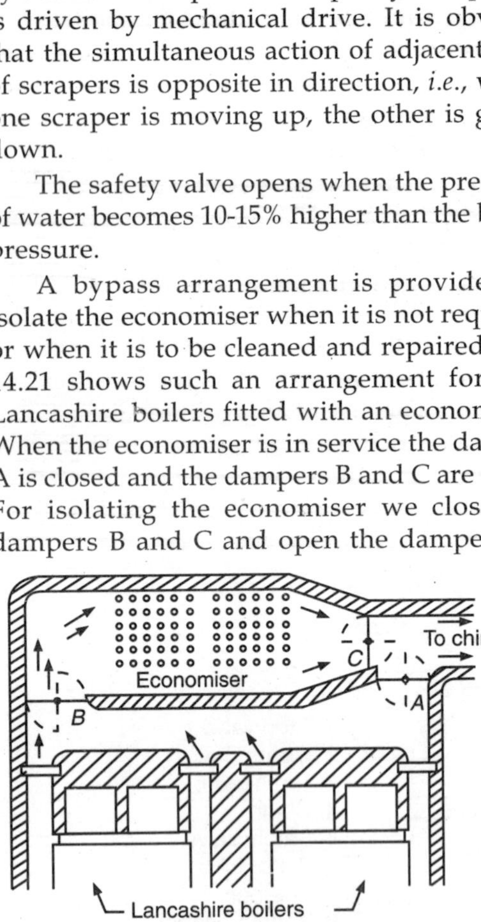

Fig. 14.21. Bypass arrangement of flue gases

(K) Steam superheater

The steam generated by a simple boiler is generally wet or at the most dry saturated. Steam superheater is a surface heat exchanger in which the wet stream is first dried at the same temperature and pressure and then raised to temperature above the saturation temperature at constant pressure. Heat of flue gases is utilized in superheating the steam and as such the superheater is placed in the path of flue gases. However in bigger installations, the superheater may be placed in an independently fired furnace. Such superheaters are known as separately fired or portable superheaters.

Since superheating results in the increased efficiency and economy of the steam plant, it has become the most important accessory of the steam power plants. The advantages of using superheated steam are :

(*i*) Increase in the amount of work output for the same weight of steam and consequently increase in cycle efficiency.

(*ii*) Loss due to condensation of steam in reciprocating steam engines is minimised to an appreciable extent.

(*iii*) No condensation of steam in the pipe line connecting the boiler to the appliance.

(*iv*) Absence of moisture in the steam reduces the friction and the erosion of steam turbine blades.

Superheaters are classified according to the mode of heat reception.

(*i*) In *convective superheaters*, the heat of the combustion gases is transferred to the surface of superheater tubes by convection.

(*ii*) In *radiation superheater*, the heat from the furnace is transferred by radiation to the superheater tubes which are placed in one or more walls of the boiler furnace.

(*iii*) In *combination type* the steam first enters the radiant superheater and then the convection one.

The superheaters are also classified according to the movement of gases and steam in the superheater. The three categories are counterflow, parallel flow and the combined flow.

The superheaters mainly follow a counter-flow arrangement in order to reduce the temperature of outgoing flue gases to the lowest possible limit. The counter flow superheater has a small size and is light in weight for the same heat transfer to the steam.

Further the superheaters are also classified according to the arrangement of superheater tubes in the boiler :

(*i*) Overdeck superheater is placed in the space over the water tubes.

(*ii*) Interdeck superheaters are placed between the water tubes which are located near the furnace.

(*iii*) Inter-tube superheaters are placed between the bank or row of water tubes.

All superheaters are essentially coil pipes made from seamless steel.

Fig. 14.22 illustrates the principle of operation of Sudgen's hair pin type superheater which is used in most of the low pressure boilers namely Cornish, Lancashire and Babcock and Wilcox boiler.

The path of flow of steam and the gases is indicated respectively by solid and dotted arrows. When the superheater is in operation, the valve V_3 is closed and the steam goes to the header H_1 *via* valve V_1. The steam when flowing inside the U shape tubes receives heat from the flue gases which are made to pass over the external surface of the tubes. The superheated steam from the header H_2 is then supplied to the steam main *via* valve V_2. For isolating the superheater when it is not needed or when it is to be cleaned and repaired, the valve V_1 is closed and the steam from the boiler is fed directly to the steam main *via* valve V_3.

The temperature control of superheated steam can be carried out by keeping the damper in some intermediate position. In that case a part of the hot flue gases passes over the superheater tubes and the remaining portion passes directly to the bottom flue.

The temperature control of superheated steam can be carried out by keeping the damper in some intermediate position. In that case a part of the hot flue gases passes over the superheater tubes and the remaining portion passes directly to the bottom flue.

Fig. 14.23. indicates the position of an economiser, an air preheater, and a superheater in a steam generating plant.

(*L*) Feed water equipment

The pressure inside a steaming boiler is high and so the feed water has to be raised in pressure before its entry can be affected in the boiler. Feed pump is a device which raises the pressure of water and forces it into the

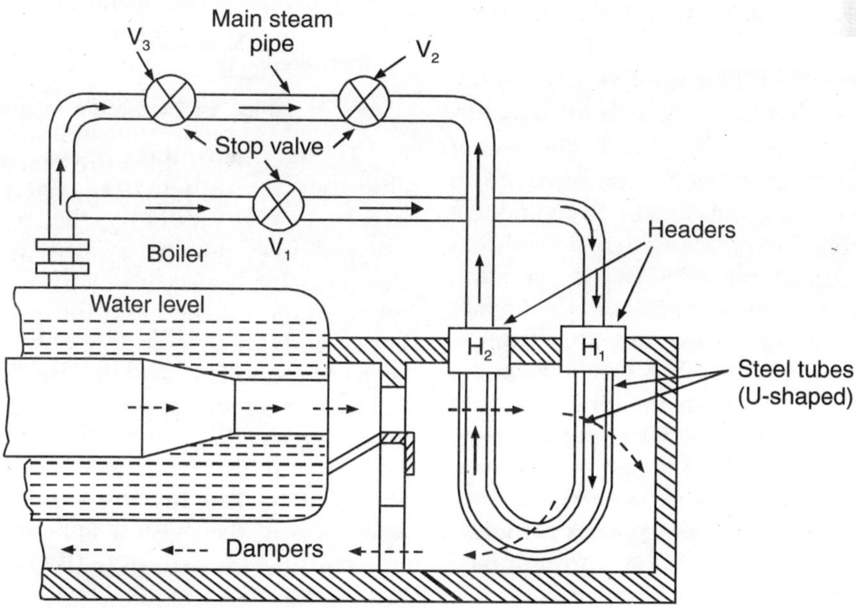

Fig. 14.22. Sudgen's superheater

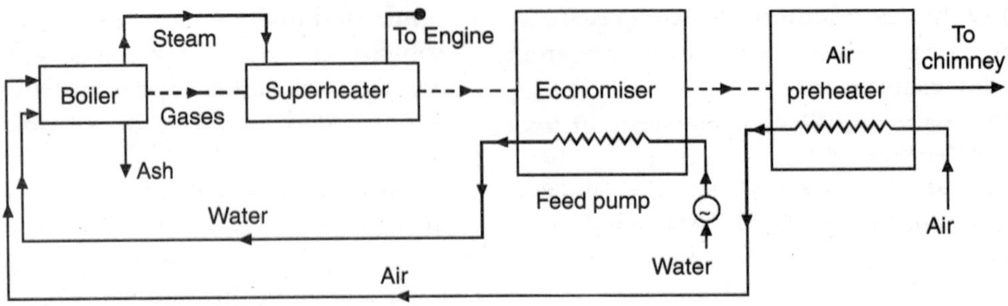

Fig. 14.23.

boiler. The commonly employed pumping devices are :

(*i*) Centrifugal pumps which run at high speeds and are driven either by electric motor or small steam turbines. The electrical driven pumps are primarily used for high capacity boilers which need a large quantity of feed water (250-300 m^3/hour). The steam turbine driven pumps too are used with medium and high steam generating capacity boilers with an output of 30 to 270 m^3/hour.

(*ii*) The reciprocating steam piston pumps are used with stationary boiler installations with a capacity of 10-15 ton/hour.

Usually a boiler house has two feed water pipe lines, one working and the other as stand-by and the output of each pump is approximately 50% more than the capacity of the boiler.

(*iii*) In vertical and locomotive boilers, the water is mainly fed with the help of injector.

With reference to Fig. 14.24, the steam enters the steam chamber whose outlet is in the form of a converging nozzle. In the nozzle section, the steam expands with a consequent gain in velocity at the expense of pressure. The resulting jet mixes with water in the mixing tube. The steam condenses and a vacuum is created in the mixing chamber. Due to vacuum, more water is drawn from the feed tank into the mixing chamber. The jet of water then enters the diverging cone where the kinetic energy of the mixture is converted into pressure. The increase of pressure is sufficient to force the water into the boiler through feed check valve.

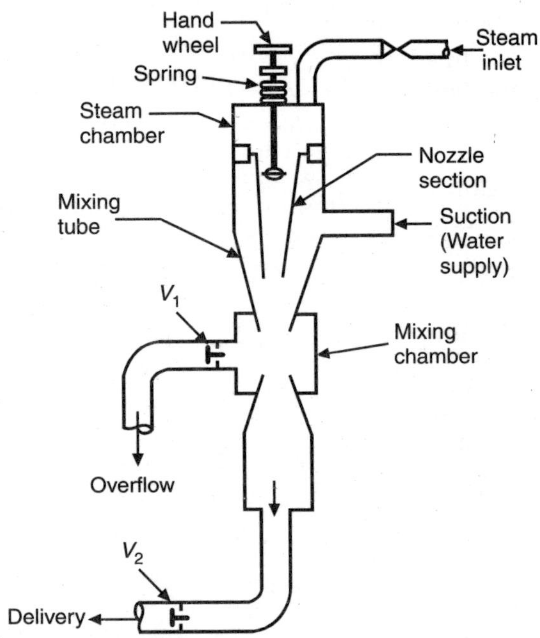

Fig. 14.24. Steam injector

To start with, the pressure in the mixing chamber is higher than the atmospheric pressure. Consequently the water and the uncondensed steam escape to atmosphere through the non-return valve V_1. The hand wheel is then turned and the annular opening between the nozzle and mixing tube is altered so as to establish a jet action. This results in the creation of rarefaction in the mixing chamber resulting in the closure of the valve V_1. The water from the feed tank rushes with high velocity to the mixing cone and regular operation of the injector is established.

The injector though limited in use to small boilers has the following advantages :

(*i*) Occupies minimum space.

(*ii*) Comparatively cheap in first cost and every day maintenance.

(*iii*) The steam consumption though more than that in case of pumps, advantage lies in the fact that entire heat of steam is recovered and returned to boiler in the feed water pumped. The injector is thus thermally much more efficient.

(*iv*) Since the injector supplies hot feed water to boiler, the boiler parts are free from thermal stresses.

In addition to the fittings described above, the boilers are also fitted with an *antipriming pipe, water separator* and the *steam trap*. We know that when a boiler is required to raise steam quickly, the steam is bound to carry water particles with it. Before being supplied to the engine, arrangements are made to separate the water particles going along with steam.

An *antipriming pipe* is a cast iron box which is fitted in the steam space of the boiler shell and under the mounting block on which the steam stop valve is to be bolted. When the steam with water particles passes through the perforations made in the upper half of the antipriming pipe, the heavier water particles separate out and are collected at the bottom of the pipe. The water thus collected is later on drained to the boiler through the holes which are made at the ends of the pipe.

The *steam separator* or drier also serves the purpose to remove the water particles being carried along with steam and the water particles resulting from a certain amount of condensation in the steam pipe between the boiler and the engine. The steam is led to pass through a chamber in which it strikes the baffle plate, gets deflected and made to change its direction. The water particles are thrown out because of greater inertia. The water collected at the bottom of the chamber is drained out at regular intervals.

Steam traps are used to collect and automatically drain away the water resulting from the partial condensation of steam without allowing any steam to escape with it. The steam condensed in the steam pipe flows to the steam trap by gravity. When the level of water in the trap becomes sufficiently high, a valve opens a passage through which the water is forced out. After the water has been discharged the passage is again closed and this prevents the escape of steam which follows water.

14.8. HIGH PRESSURE BOILERS

For high efficiency of a steam power plant, the carnot theory suggests that :

(*i*) The heat addition be at the highet possible average temperature. The average heat addition temperature is increased by increasing steam pressure and temperature.

(*ii*) The heat rejection be at the lowest possible temperature. The temperature of heat rejection is reduced by producing and maintaining high vaccum in the condenser.

Apparently, the modern trend in central power stations is to generate and use steam at higher pressures and temperatures. Boilers have been built which operate with pressure ranging from 60 bar to 200 bar and temperature upto above 600°C. The super critical boilers have pressures in excess of the critical pressure of steam (221.2 bar). Since the properties of steam in critical range are quite different, the boilers working in the supercritical range differ from the subcritical boilers both in design and operation.

The special features in the design of high pressure boilers are:

(*i*) An external pump is employed in the high pressure boilers to force water as steam through the boiler circuit. The circulation so setup is in addition to the natural circulation due to static head and convection currents. This positive circulation has been recommanded and is normally employed with boilers operating at pressure above 160 bar.

(*ii*) The high pressure boilers have parallel set of tube arrangement and this minimises resistance to flow of current. Further these boilers are characterised by the use of very small steam separating drum or by the complete absence of any drum. The supercritical pressure boilers or once through boilers, do not require any boiler drum since the water is directly converted into steam at supercritical conditions (pressure 221.2 bar and temperature 374.15°C).

(*iii*) The high pressure boilers employ certain improved arrangements of heat supply and these are:

(*a*) heating water by superheated steam

(*b*) saving of latent heat of vaporation by operating at pressures above critical range

(*c*) increasing the heat flow rate through the tube walls by using hot gases travelling at supersonic speeds.

Other special features are :

- sufficient volume of the furnace
- protection of water tubes from intense heat radiant zone
- very large combustion chamber
- sufficient area for the furnace water cooling surface

Following are the advantages of forced circulation wich is employed with boilers that operate at pressures in excess of 160 bar.

(*i*) elimination of high head needed for natural circulation. As a result, the components can be arranged horizontally giving greater accessibility and operational convenience.

(*ii*) freedom in arrangement of furnace, boiler components and tube layout.

(*iii*) due to large water velocities, size pf the tubes is small. This aspect coupled with less number of drums (or no drums as in once - through boiler) makes the unit compact and results in less weight of material for a given output.

(*iv*) reduced erection time and cost due to less weight and space requirement.

(*v*) greater possibility to operate the forced circulation boilers at other than design conditions because the circulation is not dependent on pressure and temperature.

(*vi*) less tendency of scaling due to high circulation velocity.

(*vii*) steam can be raised quickly and load changes not rapidly.

(*viii*) uniform heating of all parts and that eliminates the danger of overheating and thermal stress.

(*ix*) maintenance of uniform temperature minimises differential expansion. This eliminates leakage of gas and air, and helps in pressurising the combustion.

However, in systems with forced circulation:

- the circulating pumps consume extra power
- the safety of boiler depends on the pumping equipment
- the water has to be of highest purity

Some of the high pressure boilers are explained below regarding their construction and operation.

14.8.1. Lamont boiler

Refer Fig. 14.25 which shows the schematics of Lamont high pressure steam boiler which is of the water tube type and employs forced circulation.

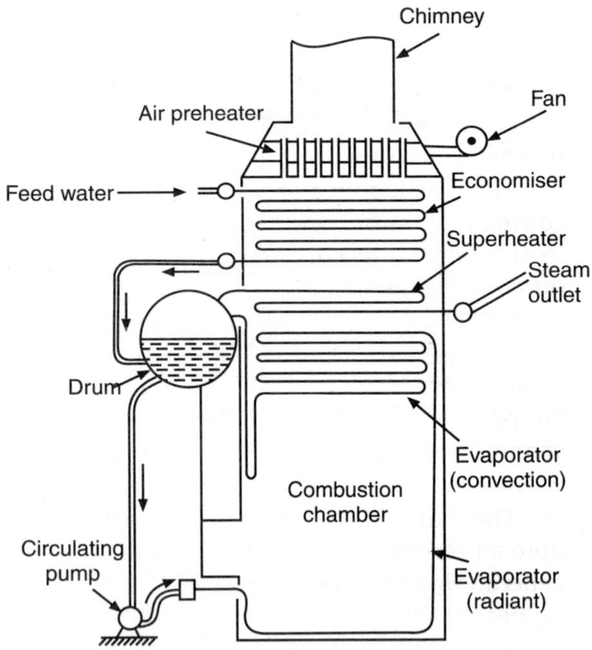

Fig. 14.25. Schematics of Lamont boiler

(*i*) The feed water from the hot well is made to pass through the economiser with the help of a feed pump. While passing through the economiser, the feed water is heated by the hot exhaust gases and is then supplied to the evaporating (steam separating) drum which is set external to the boiler.

(*ii*) The circulating pump operates at 2 to 3 bar above the boiler pressure, draws the feed water from the evaporating drum and delivers it to the header section of the evaporator. The header has orifices which distribute the feed water to the tubes in proportion to their heat absorption capacity. Generally at normal load, the water circulated in each tube is about 3.5 times the steam evaporated in the boiler. With this ratio, the overheating of tubes is kept within prescribed limits.

(*iii*) After exit from the orifices, the water is led to radiant evaporator tubes where the water absorbs the radiant heat of combustion chamber and then it is passed to convective evaporating tubes.

(*iv*) The mixture of water and steam from the evaporating tubes is taken to the boiler drum where the steam vapours are seperated. The steam vapours are drawn off to the superheater, which is of the convective type and is placed in the path of the flue gases. The superheated steam is passed to the steam prime mover where it expands and produces mechanical work.

The level of water in the drum is kept constant by supplying feed water equivalent to the steam quantity which is being drawn off continuously.

(*v*) The air drawn by the blower is heated by the flue gases before they are discharged through the chimney. The heated air from the air preheater is supplied to the combustion chamber and that improves the combustion efficiency.

The Lamont boiler can generater steam upto a pressure of about 200 bar and the steam generation rate ranges from 30,000 to 45,000 kg per hour.

14.8.2. Loeffler boiler

Refer Fig. 14.26 which shows the schematics of Loeffler high pressure boiler which is of the water tube type and employs forced circulation.

(*i*) The feed water from the hotwell is made to pass through the economiser section with the help of a feed pump. While passing through the economiser, the feed water gains some heat from the exhaust hot gases. There after the feed water is fed into the boiler drum which is housed at any convenient point outside the furnace setting.

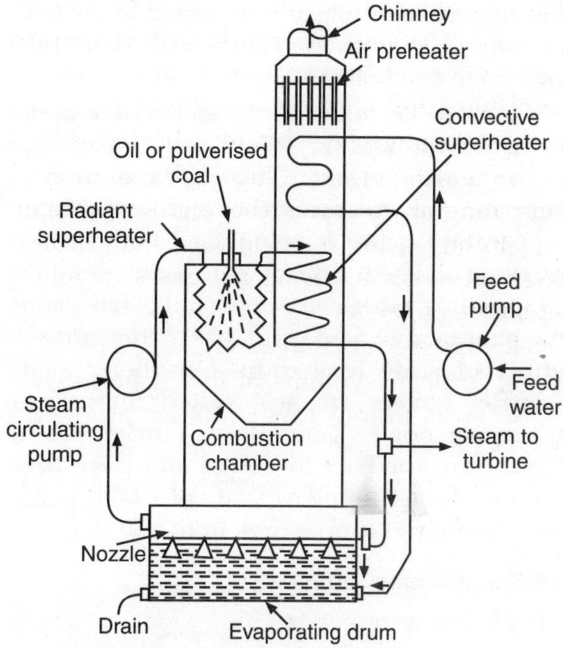

Fig. 14.26. Schematics of Loeffler boiler

(*ii*) There is a group of nozzles which receive a part of superheated steam generated in the boiler. The nozzles pass this steam to the evaporating drum where it mixes with feed water and generates steam. The rate of flow of feed water and superheated steam is so maintained that after the mixing, the water evaporates into steam which has a high dryness fraction.

(*iii*) The steam from the evaporating drum is drawn by steam circulating pump and it is passed to the superheater. The system has two superheaters : one is horizontal convection superheater located in the path of exhaust flue gases, and the other is radiant superheater which is arranged above the furnance and receives intense heat by radiation.

(*iv*) About one-third of superheater steam is supplied to the steam turbine for expansion work and the remainder two-third passes back to the evaporator through a set of nozzles. The steam gives its superheat to the water contained in the drum resulting into additional generation of saturated steam.

(*v*) A blower draws in the cold air from the surroundings and makes it flow through the air preheater which is arranged in the path of exhaust flue gases. The heated air aids in economic combustion.

It is to be noted that the Loeffler boiler evaporates water solely by means of superheated steam. The furnace heat is supplied only to economiser and superheater. Apparantly steam is being used both as heat carrying and heat absorbing medium. Since the evaporating tubes carry only supheated steam, the poor quality feed water has no detrimental effect of scale formation. For this reason, Loeffler boilers are best suited for marine transport power generation. Further, these boilers are compact in design and have been built up for steaming capacity of 1 lakh kg/hr and operating at pressures upto 150 bar

14.8.3. Benson boiler

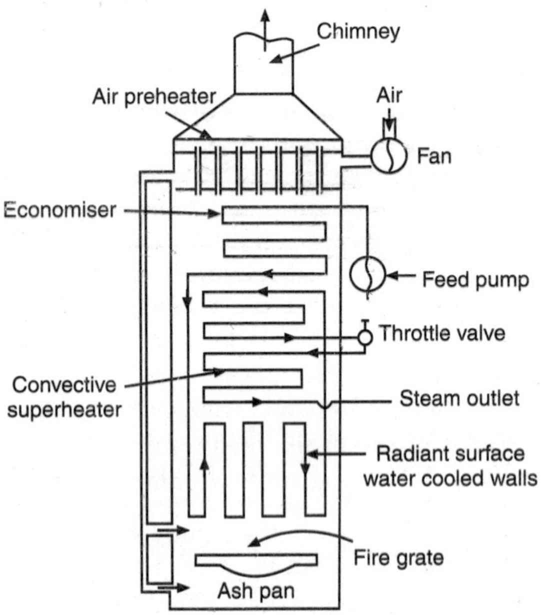

Fig. 14.27. Schematics of Benson boiler

The Benson boiler is a supercritical pressure boiler that employs forced circulation and has a unique characteristic of the absence of steam separating drum. At critical pressure, the latent heat is zero and the water transforms into steam with boiling and passes quickly from liquid to vapour phase. Apparently in such a boiler, the entire process of heating, steam formation and superheating is done in a single continuous tube. The water goes in at one end and superheated steam leaves at the other end, and due to this aspect, this type of boiler is known as mono-tube once-through forced circulation boiler.

Refer Fig. 14.27 which shows and explains disgrammaticlly the essential components of Benson boiler which are :

(*i*) economiser section

(*ii*) radiant heating section where most of the heat is transmitted and water gets heated almost to the critical temperature

(*iii*) transition section where evaporation is complete and superheating begins

(*iv*) superheating section where superheating is done to the prescribed temperature level.

The feed water is supplied by the feed pump at 225 bar pressure (higher than critical pressure) to the economiser and next to the radiant evaporator. This evaporter receives intense heat from the furnace and converts most of the water into steam. The evaporation of the remaining water occurs in the convective evaporator tubes by absorbing the heat from the exhaust flue gases. The high pressure and almost dry-saturated steam finally passes through the convective type of superheater where superheating is done to the prescribed temperature level.

Benson boilers have been built and commissioned to generate steam upto 500 bar pressure, 650°C temperature with generation rates as 1.56 lakh kg/her.

A Benson boiler presents the following advantages :

(*i*) At critical presure, steam and water co-exist at the same density and no bubble formation occurs. As such, this critical pressure boiler is free from pulsations in water circulation, and no impairment in heat transmission from flue gases to water.

(*ii*) Since the generation of steam in carried out in the evaporating tubes, the cost of drums is eliminated.

(*iii*) Circulating pumps and downcomers are dispensed with.

(iv) The boiler is self-contained unit, its parts can be easily carried and readily assembed at site.

(v) The unit is comparatively light in weight with high rate of steam generation.

(vi) The unit is compact and occupies less space.

(vii) The boiler can be put into service in 10-15 minutes depending upon its size and pressure required.

The above mentioned advantages are however, offset by the following disadvantages:

(i) The evaporation process leaves small deposits of salts in the tube and that requires frequent washing and cleaning of tubes. To obviate the problem, the requirement is of absolutely pure feed water. As such, the preparation of feed water becomes more important when compared to a conventional boiler.

(ii) The tubes are likely to get overheated when the water supply is insufficient.

(iii) The construction material for the boiler and steam turbine must be able to withstand very high pressure and temperature.

(iv) The conversion zone has to be located in the region where the flue gases have a low temperature. This is to avoid the deposition of solids and oxidation of surfaces.

(v) Small storage capacity requires close co-ordination between steam output, feed water and fuel input to maintain temperature at constant level.

SALIENT POINTS

1. A boiler is a closed vessel which generates steam at the desired pressure and temperature; the heat released by the combustion of fuel is transferred to water being heated and vaporised.
2. Boilers are generally classified on the basis of relative position of hot gases and water.

   (a) **Fire tube boiler :** The hot gases pass through the tubes that are surrounded by water, *e.g.*, Cochran, Lancashire and Locomotive boilers.

   (b) **Water tube boiler :** The tubes contain water and the hot gases produced by the combustion of fuel flow outside the tubes, *e.g.*, Babcock-Wilcox and Stirling boilers.
3. Cochran boiler is a vertical, multi-turbular fire tube, internal fired, natural circulation boiler.

   Babcock-Wilcox boiler is a horizontal drum or shell, multi-tubular, externally fired, natural circulation boiler.
4. Boiler mountings are the fittings which are primarily intended for the safety of the boiler and for complete control of the process of steam generation.

   Some boiler mountings are :

   (i) safety valves (ii) water level indicators (iii) pressure gauge (iv) fusible plug (v) steam stop valve (vi) feed check valve (vii) blow off cock and (viii) man and mud hole

   Amongst these mountings, safety valve, water level indicator and fusible plug are called safety fittings and the remaining are the control fittings.
5. Boiler accessories are the components which are installed either inside or outside the boiler to increase the efficiency of the plant.

   Some boiler accessories are :

   (i) air preheater (ii) economiser

   (iii) superheater (iv) feed pump

   (v) injector (vi) steam separator and steam trap.
6. The high pressure boilers operate with pressure ranging from 60 bar to 200 bar and temperature upto 600°C. The supercritical boilers have pressure in excess of critical pressure of steam (221.2 bar).

   Some of the high pressure boilers are :

   (i) Lamont boiler (ii) Loeffler boiler

   (iii) Benson boiler.

   Benson boiler has a unique characteristic of the absence of steam separating drum. The entire process of heating, steam formation and superheating is done in a single continuous tube; hence the name mono-tube once through boiler.

7. Natural circulation is due to static head and convection currents. Forced circulation is achieved by employing an external pump to force water or steam through the boiler circuit. This positive circulation is normally employed with boilers operating at pressures above 160 bar.

REVIEW QUESTIONS

A. Conceptual and conventional questions.

1. What is a steam boiler ? How are the boilers classified ?
2. What is the fundamental difference between internally fired and externally fired boilers ?
3. Explain the difference between a fire tube and a water tube boiler. State which type of boiler is used for power generation and why ?
4. What are the factors involved in the selection of a boiler ? Under what circumstances the water tube boilers are used in preference to fire tube boilers ? State the reasons in support of your answer.
5. Suggest a boiler for a small chemical industry which is in need of 500 kg/hour of steam at 12 atm. pressure. Draw the sketch for such a boiler showing all the mountings, and describe it.
6. Sketch and describe a Cochran vertical boiler. What are its special features ?
7. Describe the working of a locomotive boiler with the aid of a neat sketch. Show the position of various mountings. Explain the method of obtaining draught in the absence of a chimney of appreciable height in these boilers. How is the draught maintained when the locomotive engine is stationary ?
8. Draw a diagrammatic sketch to illustrate the layout of a modern water tube boiler showing the location of the superheater. Name the important parts and show on your diagram the paths taken by
   (*a*) the water and steam, (*b*) the flue gases.
   Why are the tubes of water tube boiler kept inclined ? What are the advantages of bent tubes over straight tubes in such boilers ?
9. Sketch and describe the working of a Babcock and Wilcox steam boiler.
10. Mention was made of the following in a boiler installation report :
    (*i*) Feed check valve (*ii*) Fusible plug
    (*iii*) Superheater (*iv*) Water level indicator
    (*v*) Pressure gauges (*vi*) Dead weight safety valve (*vii*) Economiser (*viii*) Steam stop valve (*ix*) Injector and (*x*) Blow off cock.
    List these separately under headings of mountings and accessories. Subdivide the mountings into control fittings and safety fittings.
11. Why mountings are essential in a boiler ? Name the different mountings and state the function of each.
12. Describe with a neat sketch, the function, location, constructional and operational details of a water level indicator.
13. Why the safety valves are needed in a boiler ? Sketch and describe the working of a spring loaded safety valve.
14. Explain the steam stop valve with reference to its function, location, construction and operation.
15. What is the function of a superheater ? Describe the construction and working of Sugden's superheater.
16. Point out the utility of an economiser in a boiler plant. Sketch and describe the working of Green's economiser.
17. Describe briefly the advantages which you would expect to be gained from incorporating an economiser, an air preheater, and a superheater in a steam generating plant.
    By a line diagram, indicate the position of these accessories in a typical boiler plant.
18. (*a*) What are high pressure boilers? How do they differ in construction and working principle from ordinary boilers?
    (*b*) Point out the advantages gained by employing forced circulation in the high pressure boilers.
19. Name the different high pressure boilers. Sketch and explain the working of any one of them.

B. Fill in the blanks with appropriate word/words.

1. Water tube boilers produce steam at a pressure which is .... than that of fire tube boilers.
2. A Cochran boiler is a ...., ... fire tube boiler.

3. A .... is a place in the combustion chamber of a boiler upon which coal is burnt.
4. Lancashire boilers are made to withstand working pressure upto .... bar.
5. The object of providing .... in between water tubes in a Bakcock-Wilcox boiler is to have better contact of flues gases with water tubes.
6. A .... boiler is best suited to meet the fluctuating demand of steam.
7. A .... is used in a boiler to put off fire in the furnace when the level of water falls to an unsafe limit.
8. The function of blow off cock in a boiler is to .... from the boiler.
9. The device which comes into operation when the pressure in the boiler exceeds the working pressure is called ....
10. A .... drains off water resulting from partial condensation of steam in pipes.
11. A unit which lifts and forces water into the boiler is called ....

Answers

1. higher 2. vertical, multi-tubular 3. grate 4. 20 5. baffles 6. locomotive 7. fusible plug 8. remove sluge 9. sefety valve 10. steam trap 11. steam injecter

C. Multiple choice questions.

1. Which aspect is not valid for fire tube locomotive boiler?
   (*a*) externally fired
   (*b*) horizontal placement
   (*c*) multi-tubular
   (*d*) mobile unit
2. Hemispherical configuration of the crown of a firebox
   (*a*) is easy to farbicate
   (*b*) provides maximum space
   (*c*) gives maximum strength
   (*d*) prevents the heat loss
3. Mark the odd one out in the context of boiler classification into fire tube and water tube boilers
   (*a*) Cochran boiler
   (*b*) Stirling boiler
   (*c*) Locomotive boiler
   (*d*) Lancashire boiler
4. Which of the following is a water tube boiler?
   (*a*) Lancashire boiler
   (*b*) Babcock-Wilcok boiler
   (*c*) Locomotive boiler
   (*d*) Cochran boiler
5. Identify the boiler mounting
   (*a*) super heater (*b*) feed check valve
   (*c*) feed pump (*d*) economiser
6. An economiser is fitted with a boiler to
   (*a*) increase the steam pressure
   (*b*) heat the fuel of low calorific value
   (*c*) heat the air entering the fire grate
   (*d*) heat the feed water by exhaust flue gases
7. All of the following are boiler mountings, except
   (*a*) blow off cock (*b*) pressure gauge
   (*c*) injector (*d*) steam stop valve
8. Identify the valve that prevents any water (after being pumped) rushing back from the boiler
   (*a*) blow off cock
   (*b*) steam stop valve
   (*c*) feed check valve
   (*d*) steam trap
9. Amongst the following fittings, which is provided more than one in number for the boiler safety
   (*a*) fusible plug
   (*b*) water level indicator
   (*c*) feed check valve
   (*d*) steam stop valve
10. Consider the following devices
    1. safety valve 2. super heater
    3. air prcheater 4. economiser
    Amongst these, boiler accessories would include
    (*a*) 1, 2 and 3 (*b*) 2, 3 and 4
    (*c*) 1 and 4 (*d*) all
11. The term priming in a boiler is associated with
    (*a*) removal of air from the boiler shell
    (*b*) water particles being carried by steam
    (*c*) firing of boilers
    (*d*) control of the rate of fuel consumption

Answers

1. (*a*)	2. (*d*)	3. (*b*)	4. (*b*)	5. (*b*)
6. (*d*)	7. (*c*)	8. (*c*)	9. (*b*)	10. (*b*)
11. (*b*)				

Chapter 15

Boiler Draught and Boiler Performance

Learning objectives : The subject matter included in this chapter will enable the reader to know about the following aspects :

- *boiler draught and its necessity*
- *natural (chimney) draught; height of chimney and condition for maximum discharge*
- *artifical draught; induced and forced fan draughts and steam jet draught*
- *boiler performance, equivalent, evaporation, boiler horse power and boiler efficiency*
- *boiler trial and heat balance sheet*

The term draught refers to the small pressure difference that causes and maintains a continuous flow of fresh air into the combustion chamber/furnace of a boiler. Essentially the draught means a force required to :

(*i*) maintain a continuous and sufficient supply of air through the fire bed to aid in the proper and complete combustion of fuel.

(*ii*) draw the resulting hot gases through the system comprising fuel bed, furnace, boiler passes and settings.

(*iii*) exhaust the flue gases to the atmosphere after they have given their energy content to the water being evaporated.

This chapter is devoted to essential aspects of natural draught produced by chimney, and the artificial draught produced by fan or steam jet. The various parameters related to performance of a boiler have also been explained.

15.1. DRAUGHT: ITS NECESSITY AND CLASSIFICATION

The amount of draught necessary for a certain boiler installation depends upon :

(*i*) nature of fuel and its depth on the grate

(*ii*) rate at which combustion is required to take place

(*iii*) design and shape of combustion chamber

(*iv*) method of burning of fuel

(*v*) resistance in the flue gas circuit offered by baffles, tubes, superheaters, economiser, air pre-heaters and dampers etc.

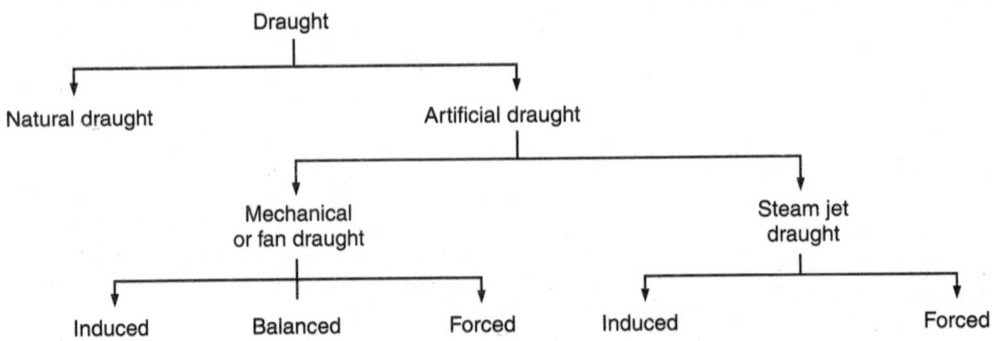

Based on the system adopted for its creation, the draught is classified as per the chart given above:

15.2. NATURAL (CHIMNEY) DRAUGHT

The natural draught is produced by the use of chimney and is accordingly referred to as chimney draught. The chimney is a vertical tubular structure made either of brick masonry, reinforced concrete or steel. The chimney carries the flue gases (combustion products) to such a height before discharging that it causes no harm to the surrounding atmosphere :

The natural draught is caused by the difference in temperature (and so density) between a column of hot gases inside the chimney and a similar column of normal atmospheric air outside the chimney.

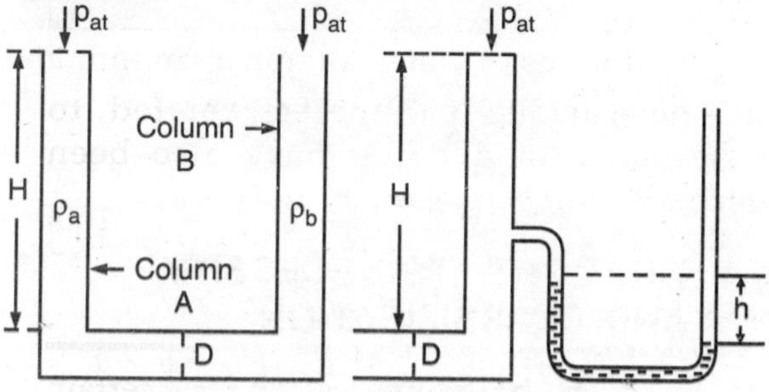

Fig. 15.1. Concept of natural draught

Consider two columns A and B containing gases of different densities ρ_a and ρ_b respectively as shown in Fig. 15.1. These densities are assumed to remain constant through the entire height H of the column. With atmospheric pressure p_{at} being exerted at the open ends of the columns, we have

Pressure at the base of column A

$$= p_{at} + \rho_a gH$$

Pressure at the base of column B

$$= p_{at} + \rho_b gH$$

Then pressure difference Δp at the base is given by :

$$\Delta p = (\rho_a - \rho_b) gH \quad ...(15.1a)$$

With cold air at atmospheric conditions filling the column A and hot fire gases filling the column B, the column A becomes unnecessary and the arrangement can be depicted as shown in Fig. 15.1*b*.

Equation 15.1 can then be rewritten as

$$\Delta p = (\rho_a - \rho_f) gH \quad ...(15.1b)$$

where ρ_a and ρ_f represents density of cold air and hot flue gases respectively.

Since the density of hot flue gases is less than the density of outside atmospheric air, a pressure difference would exist across the duct. When the duct is opened , the draught will be created and this will cause the outside air to flow into the furnace.

The draught (pressure difference between the gases at the base of chimney and the outside atmospheric air) is measured in mm of water column indicated by the manometer which is placed at the base of the chimney. In an ordinary chimney, the draught is less than 12 mm of water column which means a pressure difference

$$\Delta p = 1000 \times 9.81 \times \frac{12}{1000}$$

$$= 117.72 \text{ N/m}^2$$

between the outside air and gases at the base of chimney. This pressure difference causes the flow of air to the combustion place.

15.2.1. Height of chimney

The amount of natural draught produced by chimney depends on the height of chimney, the temperature of hot flue gases and that of atmospheric air.

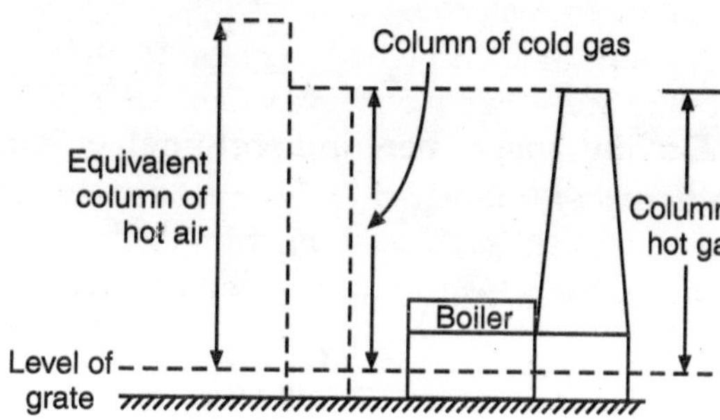

Fig. 15.2. Schematic arrangement for natural draught

Refer Fig. 15.2 for the schematic arrangement of natural draught produced by a chimney placed at a certain height above the grate. Let

T_a = absolute temperature of atmospheric air

T_f = average absolute temperature of hot flue gases

m_a = mass of air supplied per kg of fuel

$m_a + 1$ = mass of flue gases per kg of fuel

Applying characteristics gas equation ($pV = mRT$) the specific volume (volume per unit mass) of air at normal temperature pressure (T_0 = 273 K and p_0 = 1.013 bar) works out as

$$v_0 = \frac{mRT}{p_o} = \frac{1 \times 287 \times 273}{1.013 \times 10^5}$$

$$= 0.773 \text{ m}^3/\text{kg}$$

The volume of fuel burnt is small compared to the volume of air supplied and therefore can be neglected. With that assumption, the volume of flue gases (products of combustion) can be taken equal to the volume of air supplied both reduced to the same pressure and temperature conditions. Then

Volume of atmospheric air outside the chimney,

$$V_a = \frac{V_o}{T_o} \times T_a = \frac{m_a v_o}{T_o} \times T_a$$

$$= \frac{m_a \times 0.773}{273} \times T_a$$

$$= \frac{m_a T_a}{353} \text{ m}^3/\text{kg of fuel}$$

Density (mass per unit volume) of air outside the chimney

$$\rho_a = \frac{m_a}{V_a} = \frac{m_a}{m_a T_a / 353}$$

$$= \frac{353}{T_a} \text{ kg/m}^3$$

Similarly for hot flue gases inside the chimney

$$V_f = \frac{V_o}{T_o} \times T_f = \frac{m_a v_o}{T_o} \times T_f$$

$$= \frac{m_a \times 0.773}{273} \times T_f$$

$$= \frac{m_a T_f}{353} \text{ m}^3/\text{kg of fuel}$$

$$\rho_f = \frac{m_a + 1}{V_f} = \frac{m_a + 1}{m_a T_f / 353}$$

$$= \frac{m_a + 1}{m_a} \times \frac{353}{T_f}$$

Then the total static pressure difference can be obtained by invoking the relation 15.2

$$\Delta p = (\rho_a - \rho_f)\, gH$$

$$= \left[\frac{353}{T_a} - \frac{m_a + 1}{m_a} \times \frac{353}{T_f}\right] gH$$

$$= 353\, gH \left[\frac{1}{T_a} - \frac{m_a + 1}{m_a} \times \frac{1}{T_f}\right] \text{N/m}^2 \quad \text{...(15.2)}$$

• In terms of water column h_w in mm, the pressure difference can be expressed as

$$\Delta p = \rho_w\, g\, h_w = 1000 \times g \times \frac{h_w}{1000}$$

$$= g h_w \text{ N/m}^2$$

$$\therefore \quad g h_w = 353\, gH \left[\frac{1}{T_a} - \frac{m_a + 1}{m_a} \times \frac{1}{T_f}\right]$$

or

$$h_w = 353\, H \left[\frac{1}{T_a} - \frac{m_a + 1}{m_a} \times \frac{1}{T_f}\right] \text{mm of water} \quad \text{...(15.3)}$$

• In terms of hot gas column h_f in m, the pressure difference can be expressed as

$$\Delta p = \rho_f\, g\, h_f$$

$$= \frac{m_a + 1}{m_a} \times \frac{353}{T_f} \times g \times h_f$$

$$\therefore \quad \frac{m_a + 1}{m_a} \times \frac{353}{T_f} \times g \times h_f$$

$$= 353\, gH \left[\frac{1}{T_a} - \frac{m_a + 1}{m_a} \times \frac{1}{T_f}\right]$$

Simplification gives

$$h_f = H\left[\left\{\frac{m_a}{m_a+1}\right\}\times\frac{T_f}{T_a}-1\right]$$

m of hot gases ...(15.4)

Equations 15.2 to 15.4 represent the theoretical value of natural draught which depends on chimney height and weather conditions. Evidently the chimney draught

(*i*) is directly proportional to chimney draught

(*ii*) decreases with increase in temperature of atmospheric air or decrease in temperature of flue gases.

The actual draught available would be less than the theoretical draught because of friction losses offered by the passages. If the pressure loss in the chimney is equivalent to a hot gas column of h_l in meters. Then the velocity of flue gases in the chimney

$$V = \sqrt{2g(h_f - h_l)}$$

$$= \sqrt{2gh_f}\sqrt{1-\frac{h_l}{h_f}}$$

$$= K\sqrt{h_f} \quad \text{...(15.5)}$$

where $K = \sqrt{2g}\sqrt{1-\frac{h_l}{h_f}}$

It has been suggested that $K = 0.85$ for brick chimney and $K = 1.1$ for steel chimney.

If m_f is the mass flow rate of flue gases passing through the chimney, then diameter D of the chimney can be worked out from the relation

$$\rho_f = \frac{m_a+1}{V_f} = \frac{m_a+1}{m_aT_f/353}$$

$$m_f = \rho_f\left(\frac{\pi}{4}D^2\right)V_f$$

That gives

$$D = \sqrt{\frac{4m_f}{\pi\rho_f V_f}} = 1.128\sqrt{\frac{m_f}{\rho_f V_f}} \quad \text{...(15.6)}$$

15.2.2. Condition for maximum discharge

A chimney would be most effective if it discharges maximum weight of flue gases in a given time. Such a situation occurs when the absolute temperature of flue gases bears a certain relation to the absolute temperature of the outside air.

Neglecting friction losses in the passages of the chimney system, the velocity of hot flue gases passing through the chimney is

$$V = \sqrt{2gh_f}$$

Substituting the values of h_f from equation 15.4, we obtain

$$V = \sqrt{2gH\left[\frac{m_a}{m_a+1}\times\frac{T_f}{T_a}-1\right]}$$

Density of flue gas $\rho_f = \frac{p}{RT_f}$

Mass flow rate of flue gases discharged from the chimney,

$$m_f = \rho_f AV$$

$$= \frac{p}{RT_f}A\sqrt{2gH\left[\frac{m_a}{m_a+1}\times\frac{T_f}{T_a}-1\right]}$$

$$= \frac{Ap}{R}\sqrt{2gH}\sqrt{\frac{m_a}{m_a+1}\times\frac{1}{T_fT_a}-\frac{1}{T_f^2}}$$

$$= C\sqrt{\frac{m_a}{m_a+1}\times\frac{1}{T_fT_a}-\frac{1}{T_f^2}} \quad \text{...(15.7)}$$

where the parameter $\frac{Ap}{R}\sqrt{2gH}$ has been denoted by the constant C.

The condition for maximum discharge can be worked out by differentiating m_f with respect to T_f (m_a and T_a are fixed quantities) and setting the derivative equal to zero. That is

$$\frac{dm_f}{dT_f} = C\frac{d}{dT_f}\sqrt{\frac{m_a}{m_a+1}\times\frac{1}{T_fT_a}-\frac{1}{T_f^2}}$$

$$= 0$$

$$\text{or} \quad C \times \frac{1}{2}\left[\frac{m_a}{m_a+1} \times \frac{1}{T_f T_a} - \frac{1}{T_f^2}\right]^{-\frac{1}{2}}$$

$$\times \left[\frac{m_a}{m_a+1}\frac{1}{T_a}\left(-\frac{1}{T_f^2}\right) + \frac{2}{T_f^3}\right] = 0$$

$$\text{or} \quad -\frac{m_a}{m_a+1}\frac{1}{T_a T_f^2} + \frac{2}{T_f^3} = 0$$

$$\text{or} \quad \frac{m_a}{m_a+1}\frac{1}{T_a T_f^2} = \frac{2}{T_f^3}$$

Solution for T_f gives : $T_f = 2T_a \dfrac{m_a+1}{m_a}$ ...(15.8)

Evidently under conditions of maximum discharge through the chimney, the absolute temperature of flue gases should be slightly greater than twice the absolute atmospheric temperature.

Substituting the condition for maximum discharge $\dfrac{T_f}{T_a} = 2\dfrac{m_a+1}{m_a}$ in equations 15.3 and 15.4, we get

$$h_{wm} = 353\, H \left[\frac{1}{T_a} - \frac{T_f}{2T_a} \times \frac{1}{T_f}\right]$$

$$= 353\, H \left[\frac{1}{T_a} - \frac{1}{2T_a}\right] = \frac{176.5H}{T_a} \quad ...(15.9)$$

$$\text{and} \quad h_{fm} = H\left[\frac{m_a}{m_a+1} \times 2\frac{m_a+1}{m_a} - 1\right] \quad ...(15.10)$$

$$= H$$

Apparently for maximum discharge, the height of column of hot gases equals the height of chimney.

15.2.3. Chimney efficiency

With chimney draught, the temperature T_1 of flue gases (waste gases escaping to atmosphere) is higher than the corresponding temperature T_2 when the draught is by artifical means. Apparently the hot flue gases carry more heat with them when chimney is used for producing the draught.

The chimney efficiency is defined as the ratio of energy required to produce the artifical draught to the extra heat carried away by flue gases due to natural draught.

If h_f denotes the chimney draught produced in terms of column of hot gases, then the maximum energy at this head equals.

Energy utilized = $h_f\, g$ per kg of air

and

Extra heat carried by hot flue gases

$= c_{pg}\ (T_1 - T_2)$

The extra heat carried by hot flue gases represents the energy required to produce draught

∴ Efficiency of chimney

$$= \frac{h_f\, g}{c_{pg}\,(T_1 - T_2)}$$

The chimney efficiency is nearly equal to 1 percent.

15.2.4. Advantages / disadvantages of chimney draught

- The chimney is very easy to construct, has a long life, requires no maintenance and needs no auxiliay power for its operation.
- The draught produced by chimney varies with external weather conditions and the temperature of flue gases. The draught decreases with increase in outside air temperature and for producing sufficient draught, the flue gases have to be discharged at comparatively high temperatures and this results in loss of overall plant efficiency. Further, the chimney has no flexibility to create more draught to take peak loads.

15.3. ARTIFICIAL DRAUGHT

The use of chimney for producing draught of the order of 25-350 mm of water column needed for bigger power plants is neither economical nor convenient. This is because the chimney height has to be increased considerably and chimney draught depends on climatic conditions. To meet the requirements

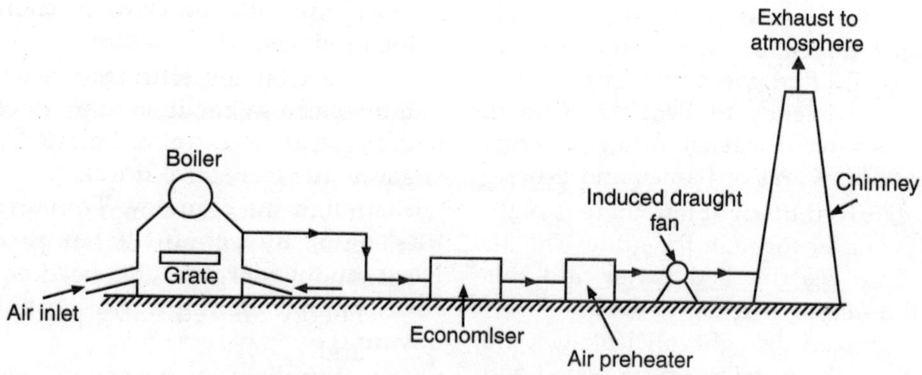

Fig. 15.3. Schematics of induced fan draught

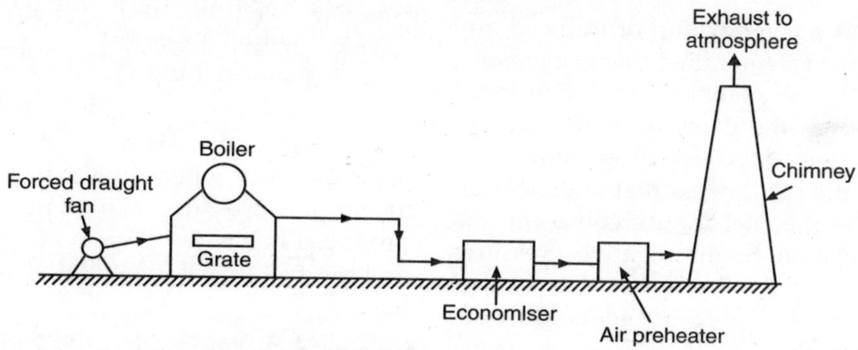

Fig. 15.4. Schematics of forced fan draught

of sufficiently high draught, some mechanical system has to be adopted. The draught so produced by a fan or steam jet is known as artificial draught.

15.3.1. Fan draught

The draught produced by a fan is called the fan draught or mechanical draught. The fan draught is of three types namely (*i*) induced (*ii*) forced (*iii*) balanced draught.

In the *induced draught* system, the fan is placed near or at the base of the chimney. The fan creates the depression which induces the flow of air through the grate. The air sucked in makes it possible to burn a large amount of fuel in a limited combustion space. The induced draught augments the draught created by the existing chimney and the system is usually employed when the boiler is provided with economisers and pre-heaters.

In the *forced fan draught*, a fan or blower installed near or at the base of boiler grate delivers air to the furnace under pressures varying from 2.5 cm to about 7.5 cm of water. The pressure helps in circulation of flue gases through components of the boiler and then through chimney to atmosphere.

The comparison between induced and forced draught systems can be made with respect to following aspects:

(*i*) Induced draught fan handles hot gases which have greater volume because of low density at elevated temperatures. As such the fan size will be greater than that of a forced fan for the same boiler. For the same reason, the induced draught system requires greater fan power.

(*ii*) Induced draught fan needs water cooled bearings since it is to withstand high temperature of gases being handled by it.

(*iii*) With induced draught, the furnace and boiler gases are at a pressure lower than that of atmosphere. Consequently when the doors are opened for firing and cleaning, there will be inrush of cold air into the furnace. This air infiltration reduces draught through the system, cools the flue gases and impairs the transmission efficiency of heating surface, However, this inward leakage of air prevents the danger of blow out of flames and ashes.

(*iv*) With forced draught fan, there is more uniform flow of air through the grate. The air penetrates into the fire bed better and this improves the rate of burning of fuel.

(*v*) With forced draught, all leakages are outward. When the fire doors are opened and the fan is operating, there will be serious danger of blow out.

Sometimes a combination of induced and forced draught system, called *balanced draught*, is employed. In this combination, the induced draught removes the gases from the furnace and the forced draught overcomes the resistance of the fuel bed so that sufficient air is supplied to the fuel for its complete and proper combustion. Slight negative pressures are maintained on the fire bed so that leakage is inwards and there is less danger of blow out of flames and ash.

15.3.2. Steamjet draught

The draught produced by steamjet is called steamjet draught and it is of two types namely forced and induced.

In forced draught, the jets of steam after having been throttled to 1.5 to 2.5 bar gauge are applied below the ashpit which is located under the fire grate of the furnace. The surrounding air is dragged along the fuel bed, furnace, fire tubes and finally to atmosphere through chimney.

In induced draught there is a blast of low pressure exhaust stream in the smoke box, *i.e*, at the base of chimney. The partial vacuum thus created forces the flue gases out through the chimney. This arrangement has been successfully employed in a locomotive boiler.

Steamjet draught has the following advantages :

- simplicity of operation
- cheapness in cost
- comparatively maintenance free ; no maintenance cost
- requires little space for installation
- capability to use low grade fuels

However the steamjet draught can be operated only when steam has been raised at proper pressure. Further, the draught produced is low (20-25 mm of water column) and the system can be installed only on small boilers.

15.3.3. Comparison between artificial and natural draughts

The mechanical or artificial draught is better and economical than the natural or chimney draught as becomes apparent from the following aspects :

(1) The artificial draught has increased rate of combustion due to high draught. Less quantity of air is then needed for complete combustion of fuel. This reduces the weight

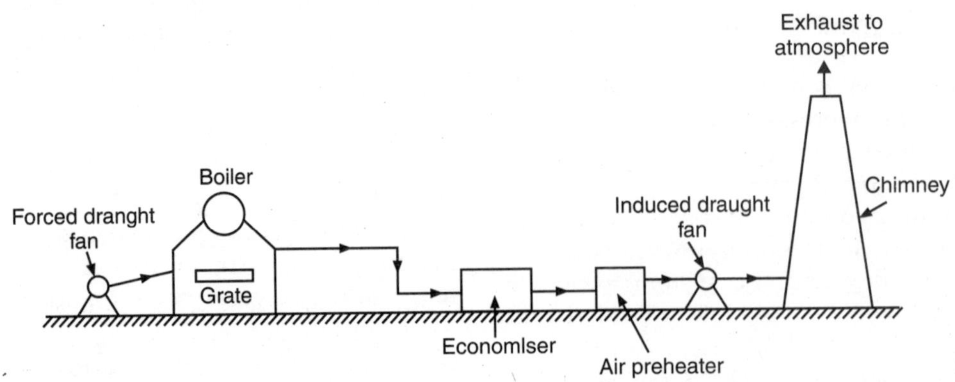

Fig. 15.5. Schematics of balanced fan draught

of flue gases per kg of fuel which means less loss of heat to exhaust gases.

(2) The artificial draught is flexible which means that the draught pressure can be controlled by regulating the flow of air or steamjets.

(3) The artificial draught is independent of climatic conditions which often seriously affect the chimney draught.

(4) Economical use of low grate fuel is possible with artificial draught due to higher intensity of draught.

(5) The function of chimney with high intensity artificial draught is only to discharge the hazardous smoke and gases high up in the atmosphere. Creation of draught is then only a secondary function and as such the chimney height can be kept small.

(6) Smoke formation and tendency for air leakage in the furnace is less with artificial draught.

(7) Efficiency of fan draught is about 7% whereas that of chimney draught is less than 1%.

EXAMPLE 15.1

Calculate the height of chimney required to generate a pressure difference of 100 mm of water column. The temperature of outside air is 30°C and the average temperature of flue gases in the chimney is 150°C. Neglect friction losses in the chimney and assume appropriate value of gas constant for air.

Solution :

$$\Delta P = 100 \text{ mm of water column}$$

$$= 100 \times g \times \frac{100}{1000}$$

$$= 100\, g \text{ N/m}^2 \qquad (\because P = \rho g h)$$

The draught in terms of presence difference in N/m² is given by

$$\Delta P = gH\,(\rho_{air} - \rho_{gases}) \qquad ...(i)$$

Density of air,

$$\rho_{air} = \left(\frac{p}{RT}\right)_{air} = \frac{1.01325 \times 10^5}{287 \times (273 + 30)}$$

$$= 1.165 \text{ kg/m}^3$$

Density of flue gases,

$$\rho_{gases} = \left(\frac{p}{RT}\right)_{gases} = \frac{1.01325 \times 10^5}{287 \times (273 + 150)}$$

$$= 0.835 \text{ kg/m}^3$$

Substituting the relevant data in expression (*i*)

$$100\, g = g \times H\,(1.165 - 0.835)$$

Then height of chimney

$$H = \frac{100}{1.165 - 0.835} = \mathbf{303 \text{ m}}$$

EXAMPLE 15.2

A thermal power plant works on natural draught and the height of chimney is restricted to 30 m. The ambient temperature is 25°C and the temperature of flue gases passing through the chimney has been measured to be 300°C. Considering the air fuel ratio to be 20 : 1, make calculations for

(a) draught produced in N/m²

(b) velocity of flue gases passing through the chimney.

It may be presumed that 50% of the theoretical discharge is lost in friction.

Solution :

$$\Delta p = 353\, gH \left[\frac{1}{T_a} - \frac{m_a + 1}{m_a} \times \frac{1}{T_f}\right]$$

$$= 353 \times 9.81 \times 30$$

$$\times \left[\frac{1}{298} - \frac{20+1}{20} \times \frac{1}{573}\right]$$

$$= 158.5 \text{ N/m}^2$$

(*b*) Draught in terms of height of flue gas column

$$h_f = H\left[\frac{m_a}{m_a + 1} \times \frac{T_f}{T_a} - 1\right]$$

$$= 30\left[\frac{20}{20+1} \times \frac{573}{298} - 1\right] = 24.94 \text{ m}$$

Considering friction loss, draught available for producing velocity,

$$h_f - h_l = 24.90 \times 0.5 = 12.45 \text{ m}$$

Then velocity of flue gases

$$V_f = \sqrt{2g(h_f - h_l)}$$

$$= \sqrt{2 \times 9.81 \times 12.45} = \mathbf{15.63\ m/s}$$

EXAMPLE 15.3

In a certain boiler installation, a steel chimney of 30 m height produces a natural draught equivalent to 17.75 mm of water column. The mean temperature of the boiler house is 298 K and that of hot gases leaving the chimney is 633 K. If the boiler uses 1350 kg of coal per hour, make calculations for

(a) air supplied per kg of coal burnt on the grate,

(b) draught in terms of column of hot flue gases,

(c) flow rate of hot gases through the chimney, and

(d) diameter of the chimney at its base.

Experimental results show that only $\frac{1}{16}$th of the total draught is available for producing velocity.

Solution : The draught in terms of mm of water column is given by

$$h_w = 353\,H\left[\frac{1}{T_a} - \frac{m_a+1}{m_a} \times \frac{1}{T_f}\right]$$

Substituting the given values, we get

$$17.75 = 353 \times 30\left[\frac{1}{298} - \frac{m_a+1}{m_a} \times \frac{1}{633}\right]$$

$$= 35.54 - \frac{m_a+1}{m_a} \times 16.73$$

That gives :

$$\frac{m_a+1}{m_a} = \frac{35.54 - 17.75}{16.73} = 1.063$$

or $m_a = \dfrac{1}{0.063} = 15.87$ kg of air/kg of fuel

(*b*) The draught in terms of column of hot flue gases is given by

$$h_f = H\left[\frac{m_a}{m_a+1} \times \frac{T_f}{T_a} - 1\right]$$

$$= 30\left[\frac{15.87}{15.87+1} \times \frac{633}{298} - 1\right]$$

$$= 30 \text{ m}$$

(*c*) Since for burning each kg of coal, we supply 15.87 kg of air, mass of flue gases per kg of coal burnt is

(15.87 + 1) = 16.87 kg/kg of coal

Mass of flue gases

$$= \frac{16.87 \times 1350}{3600} = 6.326 \text{ kg/s}$$

Density of flue gases

$$= \frac{353}{T_f} \times \frac{m_a+1}{m_a}$$

$$= \frac{353}{633} \times \frac{15.87+1}{15.87} = 0.593 \text{ kg/m}^3$$

Volume of flue gases passing through the chimney

$$= \frac{6.326}{0.593} = 10.63 \text{ m}^3\text{/s}$$

(*d*) Draught available for producing velocity

$$= \frac{h_f}{16} = \frac{30}{16} \text{ m of hot gas column}$$

If V_f is the velocity of flue gases in m/s, then

$$V_f = \sqrt{2 \times 9.81 \times \frac{30}{16}} = 6.06 \text{ m/s}$$

Area of the chimney at its base

$$= \frac{10.63}{6.06} = 1.754 \text{ m}^2$$

∴ Diameter of the chimney at its base

$$D = \sqrt{\frac{1.754 \times 4}{\pi}} = \mathbf{1.49\ m}$$

$$\left(A = \frac{\pi}{4}D^2\right)$$

EXAMPLE 15.4

The following data pertains to the natural draught produced by the chimney of a certain boiler installation :

Height of chimney = 35 m

Draught produced = 20 mm of water
Mean temperature of flue gases = 650 K
Temperature of boiler house = 310 K

Work out the amount of air supplied per kg of coal burnt on the grate and comment whether the condition for maximum discharge is satisfied by this chimney. Also determine the height of hot gas column under conditions of maximum discharge.

Solution : The draught in terms of mm of water column is given by

$$h_w = 353\,H\left[\frac{1}{T_a} - \frac{m_a+1}{m_a}\times\frac{1}{T_f}\right]$$

$$\therefore\ 20 = 353 \times 35\left[\frac{1}{310} - \frac{m_a+1}{m_a}\times\frac{1}{650}\right]$$

$$= 39.85 - \frac{m_a+1}{m_a} \times 19.0$$

That gives: $\dfrac{m_a+1}{m_a} = \dfrac{39.85-20}{19.0} = 1.0447$

∴ Mass of supplied per kg of coal,

$$m_a = \frac{1}{0.0447} = 17.57 \text{ kg}$$

(*b*) For maximum discharge : $T_f = 2T_a\,\dfrac{m_a+1}{m_a}$

$$\therefore\quad 650 = 2 \times 310 \times \frac{m_a+1}{m_a}$$

$$\frac{m_a+1}{m_a} = \frac{650}{2\times 310} = 1.048$$

$$m_a = \frac{1}{0.048} = 20.83 \text{ kg}$$

The air to be supplied per kg of fuel burnt for maximum discharge has worked out to be more than that corresponding to the draught produced. Obviously the chimney does not satisfy the condition for maximum charge.

(*c*) The draught produced in mm of water for maximum discharge is given by

$$h_{wm} = \frac{176.5\,H}{T_a} = \frac{176.5\times 35}{310}$$

$$= 19.92 \text{ mm of water}$$

Density of flue gases for maximum discharge

$$\rho_f = \frac{m_a+1}{m_a} \times \frac{353}{T_f}$$ where m_a is the mass of air to be supplied for maximum discharge

$$= \frac{20.83+1}{20.83} \times \frac{353}{650} = 0.569 \text{ kg/m}^3$$

Then column of hot gases equivalent to h_{wm} = 19.92 mm of water is given by

$$h_{fm} = \frac{\rho_w}{\rho_f} \times h_{wm}$$

$$= \frac{1000}{0.569} \times (19.92 \times 10^{-3}) = 35 \text{ m}$$

This is in conformity with the stipulation that for maximum discharge, the height of column of hot gases equals the height of chimney.

EXAMPLE 15.5.

A chimney of 1.8 m diameter fitted with a certain thermal power plant produces a draught equal to 18 mm water column. The mean temperature of flue gases is 270°C and the boiler house temperature is 32°C. If the flue gases formed per kg of fuel burnt are 22, make calculations for the mass of flue gases passing through the chimney.

Solution : Mass of air per kg of fuel m_a

$$= m_g - 1 = 22 - 1 = 21$$

$$T_a = 273 + 32 = 305$$

$$T_f = 273 + 270 = 543 \text{ K}$$

The draught produced in mm of water column is given by

$$h_w = 353\,H\left[\frac{1}{T_a} - \frac{m_a+1}{m_a}\times\frac{1}{T_f}\right]$$

$$\therefore\quad 18 = 353\,H\left[\frac{1}{305} - \frac{21+1}{21}\times\frac{1}{543}\right]$$

$$= 0.501\,H$$

That gives : height of chimney

$$= \frac{18}{0.501} = 35.93 \text{ m}$$

Draught created by hot gas column is

$$h_f = H\left[\frac{m_a}{m_a+1}\times\frac{T_f}{T_a}-1\right]$$

$$= 35.93\left[\frac{21}{21+1}\times\frac{273+270}{273+32}-1\right]$$

$$= 27.38 \text{ m}$$

Velocity of flue gases

$$V = \sqrt{2gh_f} = \sqrt{2\times 9.81\times 27.38}$$

$$= 23.18 \text{ m/s}$$

Density of flue gases

$$\rho_f = 353\ \frac{m_a+1}{m_a}\times\frac{1}{T_f}$$

$$= 353\times\frac{21+1}{21}\times\frac{1}{(273+290)}$$

$$= 0.675 \text{ kg/m}^3$$

Mass of flue gases passing through the chimney,

$$m = \rho_f AV = \rho_f\left(\frac{\pi}{4}D^2\right)V$$

$$= 0.657\times\left(\frac{\pi}{4}\times 1.8^2\right)\times 23.18$$

$$= \mathbf{38.73\ kg/s}$$

EXAMPLE 15.6.

In a boiler installation provided with a chimney for natural draught, the flue gases are at temperature of 600 K and the ambient temperature is 300 K. The air requirement is 19 kg per kg of coal which burns at the rate of 2.5 × 10³ kg/hr. Determine the diameter and height of chimney considering the following aspects :

Velocity of gases leaving the chimney is equivalent to 1.2 mm of water. Actual draught equals 90% of theoretical draught, pressure losses in furnace grate, flues, bends and chimney are 5 m, 4 m, 2.5 m and 1.5 m of water column.

Solution : Draught actually needed

= (pressure head required to overcome different losses) + velocity head

= (5 + 4 + 2.5 + 1.5) + 1.2

= 14.2 mm of water

Draught to be produced $= \dfrac{14.2}{0.9} = 15.78$ mm of water

The draught in terms of mm of water column is given by

$$h_w = 353H\left[\frac{1}{T_a}-\frac{m_a+1}{m_a}\times\frac{1}{T_f}\right]$$

Substituing the given values, we get,

$$15.78 = 353H\left[\frac{1}{300}-\frac{19+1}{19}\times\frac{1}{600}\right]$$

That gives : height of chimney $H = 28.31$ m

(*b*) Density of flue gases

$$\rho_f = \frac{353}{T_f}\times\frac{m_a+1}{m_a}$$

$$= \frac{353}{600}\times\frac{19+1}{19} = 0.619 \text{ kg/m}^3$$

For burning each kg of coal, 19 kg of air is supplied. As such mass of flue gases per kg of coal burnt is

(19 + 1) = 20 kg/kg of coal

Mass of flue gases

$$= \frac{2.5\times 10^3}{3600}\times 20 = 13.89 \text{ kg/s}$$

Volume of flue gases passing through the chimney

$$= \frac{13.89}{0.619} = 22.44 \text{ m}^3/\text{s}$$

Draught for producing velocity h_v

$= 1.2 \times 10^{-3}$ m of water

$= 1.2 \times 10^{-3} \times \dfrac{\rho_w}{\rho_f}$

$= 1.2 \times 10^{-3} \times \dfrac{1000}{0.619}$

= 1.938 m of gas column

If V_f is the velocity of flue of gases in m/s, then

$$V_f = \sqrt{2gh_v} = \sqrt{2\times 9.81\times 1.938}$$

$$= 6.166 \text{ m/s}$$

Area of the chimney at its base

$$= \frac{22.44}{6.166} = 3.639 \text{ m}^2$$

Diameter of the chimney at its base

$$D = \sqrt{\frac{3.639 \times 4}{\pi}} = \mathbf{2.15\ m}$$

$$\left(A = \frac{\pi}{D} D^2\right)$$

EXAMPLE 15.7.

The chimney provided for producing artificial draught in a boiler set up is 75 m high and is designed for maximum discharge of flue gases through it. The boiler furnace is supplied with 20 kg of air/kg of fuel and the temperature of surroundings is estimated to be 300 K. Make calculations for

(a) temperature of hot flue gases and draught produced in mm of water

(b) chimney efficiency if temperature of waste gases leaving the chimney is 365 K

(c) extra expenditure of heat on the production of the chimney draught as a percentage of the heat of fuel.

Given :

calorific value of fuel = 32800 kJ/kg

mean specific heat of waste gases = 1.004 kJ/kg K

Solution : For maximum discharge through a chimney

$$T_f = 2T_a \frac{m_a + 1}{m_a}$$

$$= 2 \times 300 \times \frac{20+1}{20} = 630 \text{ K}$$

The draught in terms of mm of water column is given by

$$h_w = 353\, H\left[\frac{1}{T_a} - \frac{m_a + 1}{m_a} \times \frac{1}{T_f}\right]$$

Substituting the given values, we get

$$h_w = 353 \times 75\left[\frac{1}{300} - \frac{20+1}{20} \times \frac{1}{630}\right]$$

$$= 44.13 \text{ mm of water column}$$

(*b*) Draught in terms of height of flue gas column

$$h_f = H\left[\frac{m_a}{m_a + 1} \times \frac{T_f}{T_a} - 1\right]$$

$$= 75\left[\frac{20}{20+1} \times \frac{630}{300} - 1\right] = 75 \text{ m}$$

Energy imparted by static head of 75 m of gas column

$$= mgh = 1 \times 9.81 \times 75$$

$$= 735.75 \text{ J/kg}$$

Extra heat carried away by the flue gases due to temperature being higher than that in artificial draught system

$$= m\, c_p\, \Delta T$$

$$= 1 \times (1.004 \times 10^3) \times (630 - 385)$$

$$= 245980 \text{ J/kg}$$

EXAMPLE 15.8.

A boiler, equipped with a chimney 30 metre high, is supplied with 20 kg of air per kg of fuel burnt. The boiler room temperature is 315 K and the mean temperature of flue gases leaving the chimney is 525 K. In an artificial draught system, the mean temperature of flue gases would have been 415 K. Taking mean specific heat of flue gases as 1.005 kJ/kg K, determine the efficiency of chimney.

Solution: Draught in terms of height of flue gas column.

$$h_f = H\left[\frac{m_a}{m_a + 1}\frac{T_f}{T_a} - 1\right]$$

$$= 30\left[\frac{20}{21} \times \frac{525}{315} - 1\right]$$

$$= 17.619 \text{ m}$$

Energy utilized per kg of chimney gases

$$= g\, h_f = 9.81 \times 17.619$$

$$= 172.84 \text{ Nm}$$

Energy required to produce draught

$$= c_{pg}\, (T_1 - T_2)$$

$$= (1.005 \times 1000) \times (525 - 415)$$

$$= 110550 \text{ Nm}$$

Efficiency of chimney

$$= \frac{172.84}{110550} \times 100 = 0.156\%$$

15.4. BOILER PERFORMANCE

The purpose of the boiler is to generate steam at the predetermined pressure and temperature by using heat produced by the combustion of fuel. The evaporation (steam raising) capacity of the boiler depends upon

(*a*) grate area, *i.e.*, the amount of fuel the boiler can consume

(*b*) heating surface, *i.e.*, the quantity of heat the boiler can absorb

The evaporation capacity may be expressed as kg/hr-m^2 of heating surface or kg/kg of fuel burnt. If the grate area and heating surface are kept constant, the amount of water which a boiler can evaporate further depends upon :

(*a*) pressure and temperature conditions under which the boiler is operating

(*b*) quality of steam being raised (wet, dry or superheated)

(*c*) temperature of feed water being supplied to boiler.

Apparently it may be realized that the number of kg of water evaporated per hour is not an exact measure of the performance of the boiler. For making comparison of the relative evaporation capacity of the boiler, the water is supposed to evaporate under the following standard conditions:

(*i*) feed water is a supplied to the boiler at 100°C

(*ii*) evaporation of water (steam raising) takes place at the same temperature of 100°C

(*iii*) the steam raised is just dry and saturated.

When water evaporates under these conditions, it requires 2256 kJ/kg of heat which equals the latent heat of steam at 100°C under atmospheric pressure of 1 bar.

Equivalent evaporation of water is then defined as "Evaporation which would be obtained if the feed water were supplied at 100°C and converted into dry saturated steam at 100°C at the standard atmospheric pressure of 1 bar."

Let W_a = weight of water actually evaporated into steam per kg of coal burnt at the boiler working pressure.

h_s = specific enthalpy of steam at the working pressure

h_w = specific enthalpy (sensible heat) of feed water.

Then : heat absorbed by water at the boiler working pressure,

$$= W_a\,(h_s - h_w) \qquad ...(a)$$

If W_e denotes the equivalent evaporation, then heat absorbed under standard conditions,

$$= W_e\,L_o \qquad ...(b)$$

where L_o is the enthalpy of evaporation (latent heat) of steam at 1 bar and 100°C, and it equals 2256 kJ/kg.

Equating expressions (*a*) and (*b*), use obtain.

$$W_e = \frac{W_a(h_s - h_w)}{L_o} \text{ kg/kg of coal burnt.} \qquad ...(15.11)$$

The factor $\left\{\dfrac{h_s - h_w}{L_o}\right\}$ in the above expression is always greater than unity and is known as *factor of evaporation* or *generation factor*. The factor of evaporation may be defined as the ratio of heat absorbed by 1 kg of feed water under working conditions to heat absorbed by 1 kg of water from and at 100°C (standard conditions).

Boiler efficiency is defined as the ratio of heat energy utilized by feed water in converting it into steam in the boiler to the heat energy produced by complete combustion of fuel during the same time. Thus

$$\eta_{\text{boiler}} = \frac{\text{energy absorbed by feed water}}{\text{energy produced by fuel}}$$

$$= \frac{W_a(h_s - h_w)}{W_f \times \text{calorific value of fuel}} \qquad ...(15.12)$$

where W_f is the weight of fuel consumed during the time the weight W_a of water is actually evaporated.

If the boiler is equipped with an economizer and a superheater, then the efficiency of the unit is expressed in terms of *overall efficiency* which takes into account the efficiency of all the three units namely the boiler, the economizer and the superheater.

Economizer efficiency is a measure of the heat absorbed by feed water from the flue gases which are on way to chimney.

$$\eta_{\text{economizer}} = \frac{\text{heat asbrobed by feedwater in the economiser}}{\text{heat content of flue gases entering the economiser}}$$

$$= \frac{W_w c_{pw}(t_{w_2} - t_{w_1})}{W_f c_{pf}(t_f - t_a)} \quad ...(15.13)$$

where

W_w and W_f = weights of water and flue gases respectively

c_{pw} and c_{pf} = specific weights heats of water and flue gases respectively

t_{w2} and t_{w1} = temperature of water entering and leaving the economizer

t_f = temperature of flue gases entering the economizer

t_a = temperature of air being supplied to boiler.

Boiler horse power is the term used for the performance unit equivalent to evaporation of 15.653 kg of water in one hour from and at 100°C. Thus

Boiler horse power

$$= \frac{\text{equivalent evaporation/hr from and at 100°C}}{15.653} \quad ...(15.14)$$

EXAMPLE 15.9.

A boiler generates 45 × 10³ kg/hr of wet steam at 10 bar pressure and 0.95 dryness fraction. The boiler receives water at 40°C and the coal burns in the furnace at the rate of 5 × 10³ kg/hr. Calculate the equivalent evaporation of boiler per kg of coal burnt.

Solution : Mass of steam generated per kg of coal burnt,

$$m = \frac{\text{mass of steam generated/hr}}{\text{mass of coal burnt/hr}}$$

$$= \frac{45 \times 10^3}{5 \times 10^3} = 9 \text{ kg /kg of fuel}$$

At 10 bar pressure, the properties of steam as read from the steam tables are :

h_f = 762.6 kJ/kg

and h_{fg} = 2013.6 kJ/kg

At 40°C feed water temperature;

h_w = 167.6 kJ/kg

Enthalpy of steam

$h = h_f + x\, h_{fg}$

= 762.6 + 0.95 + 2013.6

= 2675.52 kJ/kg

Equivalent evaporation

$$= \frac{m(h - h_w)}{\text{latent heat at 100°C}}$$

$$= \frac{9(2675.52 - 167.6)}{2257}$$

= **10** kg/kg of coal

EXAMPLE 15.10.

A boiler evaporates 5 kg of water per kg of coal burnt on the fire grate in the combustion chamber. The temperature of feed water is 30°C and the steam is raised at 10 bar absolute pressure. Calculate the equivalent evaporation from and at 100°C per kg of coal as well as the factor of evaporation if the steam produced is (a) 0.95 dry (b) superheated to 250°C.

Take specific heat of superheated steam = 2.1 kJ/kg K

Solution : From steam-tables,

At 30°C feed water temperature :

h_w = 125.77 kJ/kg

At 10 bar absolute pressure :

h_f = 762.79 kJ/kg

h_{fg} = 2015.3 kJ/kg

and T_{sat} = 179.91°C

Mass of water evaporated

m = 5 kg/kg of coal

(*a*) Enthalpy of steam

$h = h_f + xh_{fg}$

= 762.79 + 0.95 + 2015.3

= 2677.32 kJ/kg

Equivalent evaporation

$$= \frac{m(h - h_w)}{\text{latent heat at } 100°C}$$

$$= \frac{5(2677.32 - 125.77)}{2257}$$

= **5.65 kg/kg** of coal

Factor of evaporation

$$= \frac{h - h_w}{\text{latent heat at } 100°C}$$

$$= \frac{2677.32 - 125.77}{2257} = \mathbf{1.13}$$

(*b*) Enthalpy of steam h

$= h_f + h_{fg} + c_p (T_{sup} - T_{sat})$

$= 762.79 + 2015.3$

$+ 2.1 (250 - 179.91)$

$= 2925.28$/kJ/kg

Equivalent evaporation

$$= \frac{m(h - h_w)}{\text{latent heat at } 100°C}$$

$$= \frac{5(2925.28 - 125.77)}{2257}$$

= **6.20** kg/kg of coal

Factor of evaporation

$$= \frac{h - h_w}{\text{latent heat at } 100°C}$$

$$= \frac{2925.28 - 125.77}{2257} = \mathbf{1.24}$$

EXAMPLE 15.11.

The following data pertains to trial on a boiler conducted for 6 hours :

Boiler heating surface = 90.95 m^2

Total water evaporated = 22750 kg

Total fuel fired = 3520 kg

Calculate :

(a) fuel fired/m^2 *grate area and fuel fired per hour*

(b) evaporation/m^2 *heating surface and equivalent evaporation if feed water temperature is 45°C and the steam produced is dry saturated at 8 bar absolute pressure*

(c) boiler efficiency if the coal used has a calorific value of 32.5 MJ/kg

Solution : (*a*)

$$\text{Fuel fired} = \frac{3520}{90.95} = 38.70 \text{ kg/m}^2$$

$$= \frac{3520}{6} = 586.67 \text{ kg/hr}$$

(*b*) $\text{Evaporation} = \frac{22750}{90.95} = 250.14 \text{ kg/m}^2$

Mass of water evaporated/kg of coal

$$m = \frac{22750}{3520} = 6.46 \text{ kg/kg of fuel}$$

From steam tables : Corresponding to feed water temperature of 45°C

$h_w = 188.37$ kJ/kg

Corresponding to steam pressure of 8 bar

$h = 2770$ kJ/kg

Then : equivalent evaporation

$$= \frac{m(h - h_w)}{\text{latent heat at } 100°C}$$

$$= \frac{6.46(2770 - 188.37)}{2257}$$

= **7.39 kg/kg** of fuel

(*c*) Efficiency of boiler

$$\eta = \frac{m(h - h_w)}{CV}$$

$$= \frac{6.46(2770 - 188.37)}{32.5 \times 10^3} = \mathbf{0.51 \text{ or } 51\%}$$

EXAMPLE 15.12.

A boiler generates steam at 10 bar and 200°C from feed water at 37°C. The coal used in the furnace has a calorific value of 33695 kJ/kg and 10 kg of steam is generated per kg of coal burnt. Calculate:

(a) equivalent evaporation from and at 100°C

(b) generation factor,

(c) boiler horse power if 250 kg of coal is fired per hour.

If the boiler supplies steam to an engine in which the heat equivalent to brake work is 305 kJ/kg of steam supplied, what is the overall efficiency of plant ?

Solution : From steam tables,

At 37°C feed water temperature :

$h_w = 154.9$ kJ/kg

At 10 bar absolute pressure :

h_f = 762.79 kJ/kg

h_{fg} = 2015.3 kJ/kg

and T_{sat} = 179.91°C

Mass of steam generated

m = 10 kg/kg of coal

(*b*) Enthalpy of steam

$$h = h_f + h_{fg} + c_{ps}\,(T_{sup} - T_{sat})$$

$$= 762.79 + 2015.3 + 2.1\,(200 - 179.91)$$

$$= 2820.3 \text{ kJ/kg}$$

Equivalent evaporation

$$= \frac{m(h - h_w)}{\text{latent heat at } 100°\text{C}}$$

$$= \frac{5(2820.3 - 154.9)}{2257}$$

= **11.81 kg/kg** of coal

(*c*) Boiler horse power

$$= \frac{\text{equivalent evaporation per hour}}{15.653}$$

$$= \frac{11.81 \times 250}{15.653} = \mathbf{188.62}$$

(*d*) Heat equivalent of brake work

= 305 kJ/kg of steam

= (305 × 10) kJ/kg of coal

$$\text{Overall } \eta = \frac{\text{heat equivalent of brake work}}{\text{heat supplied in coal}}$$

$$= \frac{305 \times 10}{33695} = 0.09 \text{ or } \mathbf{9\%}$$

EXAMPLE 15.13.

Following observations were made during the trial of a gas fired boiler of a steam power plant:

Generator output	*: 50000 kW*
Steam conditions	*: 50 bar and 500°C*
Steam output	*: 65 kg/s*
Feed water temperature	*: 80°C*
Percentage composition of fuel (by volume)	*: CH_4 = 96.5 and C_2H_6 = 0.5*
Higher calorific value of gas	*: 38700 kJ/m³ (at 1.013 bar and 15°C)*
Gas consumption	*: 6.5 m³/s*

Calculate the boiler efficiency and overall thermal efficiency based on the lower calorific value of fuel.

Solution : Consider the combustion equations

$$\underset{1\text{ mole}}{CH_4} + 2\,O_2 = CO_2 + \underset{2\text{ moles}}{2\,H_2O}$$

$$\underset{1\text{ mole}}{C_2H_6} + 2.5\,O_2 = 2\,CO_2 + \underset{3\text{ moles}}{3\,H_2O}$$

Also 1 mole of water is equivalent to 18 kg

∴ Mass of water formed

= (2 × 0.965 × 18) + (3 × 0.005 × 18)

= 35.01kg

Assuming steam as a perfect gas and recollecting that 1 mole of gas at 1.013 bar and 0°C is equivalent to 22.41 m³, 1 mole of gas at 15°C and 1.013 bar

$$= 22.41 \times \frac{288}{273} = 23.64 \text{ m}^3$$

∴ Mass of water formed

$$= \frac{35.01}{23.64} = 1.481 \text{ kg/m}^3 \text{ of fuel}$$

The lower calorific value of fuel is given by

= higher calorific value – h_{fg} at 15°C

From steam tables ;

h_{fg} at 15°C = 2466 kJ/kg

enthalpy of steam at 50 bar and 500°C,

h_1 = 3434 kJ/kg

enthalpy of fuel water at 80°C

h_f = 335 kJ/kg

Then, LCV of fuel = 38700 – 1.481 × 2466

= 35048 kJ/m³

Boiler efficiency

$$= \frac{\text{heat utilized in raising steam}}{\text{heat supplied at the boiler grate}}$$

$$= \frac{m_w\,(h_1 - h_f)}{\text{LCV} \times \text{vol of fuel burnt}}$$

$$= \frac{65\,(3434 - 335)}{35048 \times 6.5} = 0.8842 \text{ or } 88.42\%$$

Overall thermal efficiency

$$= \frac{\text{output from generator}}{\text{heat supplied by burning fuel}}$$

$$= \frac{50000}{35048 \times 6.5} = 0.2195 \text{ or } 21.95\%$$

15.5. BOILER TRIAL AND HEAT BALANCE

For making performance study of a boiler, it is run on test conditions for 3 to 4 hours to obtain steady conditions. The steady conditions imply that the conditions of fuel, draught, temperature, pressure, water level, rate of feeding and evaporation etc., remain same throughout the test run.

The boiler trial is made

– to determine the thermal efficiency of the plant

– to check the generating capacity of the boiler when working at full load

– to prepare the heat balance sheet so that suitable measures are taken to improve the working of the plant.

The heat released by combustion of fuel on the boiler grate is always more than the heat utilized in raising steam and the difference represents the heat lost : The terms included in the heat balance are :

1. Heat supplied by fuel $Q = m_f \times CV = CV$ for unit mass of fuel burnt.

2. Heat utilized for generation of steam (Q_1)

$$Q_1 = m_w\,(h_s - h_w)$$

where h_s and h_w represent the enthalpies of steam raised and that of feed water, and m_w is the mass of water evaporated per kg of fuel.

3. Heat carried away by dry flue gases (Q_2)

$$Q_2 = m_g\,c_{pg}\,(T_g - T_a)$$

where

m_g = mass of dry flue gases per kg of fuel

c_{pg} = mean specific heat of dry flue gases

T_g = temperature of flue gases leaving the chimney

T_a = temperature of air in the boiler room

4. Heat loss to the moisture in the flue and to water vapours formed from the combustion of hydrogen (Q_3).

Mass of steam formed = $(9H_2 + m_m)$ per kg of fuel.

where H_2 is the mass of hydrogen in each kg of fuel and m_m is mass of moisture present per kg of fuel.

Further, it is presumed that the steam so formed is in superheated state at atmospheric pressure and at the flue gas temperature.

$$Q_3 = (9H_2 + m_m)\,(h'_s - h'_w)$$

$$= (9\,H_2 + m_m)\,[h_g + c_{ps}\,(T_g - T) - h'_w]$$

From steam tables :

At 1.013 bar,

$$h_g = 2676 \text{ kJ/kg}\,;\ T = 100°\text{C}$$

$$Q_3 = (9H_2 + m_m)\,[2676 + c_{ps}\,(T_g - 100) - h'_w]$$

where c_{ps} is specific heat of superheated steam and h'_w is the enthalpy of water at boiler room temperature.

5. Heat loss due to incomplete combustion (Q_4).

Incomplete combustion implies that the fuel burns to CO instead of CO_2. When 1 kg of C is burnt to

CO_2 : 33830 kJ of heat is released

CO : 10130 kJ of heat is released

As such, the heat lost due to incomplete combustion of one kg of carbon is

(33830 – 10130) = 23700 kJ

From the volumetric analysis of dry flue gases,

Mass of carbon burnt to CO

$$= \frac{CO \times C}{CO_2 + CO}$$

where C is fraction of carbon in one kg of fuel, CO and CO_2 are the percentages by volume of CO and CO_2 in the dry flue gases. Then

$$Q_4 = \frac{CO \times C}{CO_2 + CO} \times CV \text{ of CO}$$

6. Heat loss due unburnt fuel (Q_5)

This loss occurs due to unburnt carbon particles that fall through the grate bars and are lost with ash.

$$Q_5 = m_c \times CV \text{ of carbon}$$

where m_c is the unburnt mass of carbon per kg of fuel and CV is calorific value of carbon.

7. Heat unaccounted for (Q_6)

$$Q_6 = \text{heat supplied in fuel} - (\text{heat utilized in raising steam} + \text{all heat losses})$$
$$= Q - (Q_1 + Q_2 + Q_3 + Q_4 + Q_5)$$

This unaccounted heat is normally taken as the heat lost due to convection and radiation.

EXAMPLE 15.14

The following data were obtained in a boiler trial :

Mass and temperature of feed water = 680 kg/hr and 20°C

Steam pressure and its temperature = 15 bar and 300°C

Coal used and its calorific value = 98 kg/hr and 26500 kJ/kg

Flue gas formed and its temperature at chimney = 18 kg/kg of coal supplied and 300°C

Ash and unburnt coal is ash-pit = 4 kg/hr with 2200 kJ/kg calorific value

Mean specific heat of flue gases and feed water = 1 kJ/kg K and 4.187 kJ/kg K

If the ambient temperature in the boiler room is 28°C, determine the :

(a) boiler efficiency,

(b) equivalent evaporation from and at 100°C

(c) percentage heat unaccounted for, and

(d) draught produced in mm of water column if the height of chimney is 50 m

Solution : From steam tables

Enthalpy of steam at 15 bar and 300°C = 3037.5 kJ/kg

Enthalpy of feed water at 30°C = 83.74 kJ/kg

(*a*) Boiler efficiency

$$= \frac{\text{heat absorbed by feed water}}{\text{energy released by fuel}}$$

$$= \frac{m_w(h-h_w)}{m_f \times CV}$$

$$= \frac{680(3037.5-83.74)}{98 \times 26500}$$

$$= 0.7734 \text{ or } 77.34\%$$

(*b*) Mass of water evaporated per kg of fuel

$$m = \frac{680}{98} = 6.94 \text{ kg}$$

Equivalent evaporation

$$= \frac{m(h-h_w)}{\text{latent heat at } 100°\text{C}}$$

$$= \frac{6.94(3037.5-83.74)}{2257}$$

$$= 9.08 \text{ kg/kg of fuel}$$

(*c*) Heat released by coal

$$= m_f \times CV$$
$$= 98 \times 26500 = 2.597 \times 10^6 \text{ kJ/hr}$$

Heat utilized in raising steam $= m_w (h - h_w)$

$$= 680 (3037.5 - 83.74)$$
$$= 2.0086 \times 10^6 \text{ kJ/hr}$$

Heat lost to exhaust gases

$$= m_g c_{pg} (t_g - t_a)$$
$$= (18 \times 98) \times 1.0 (300 - 28)$$
$$= 0.4798 \times 10^6 \text{ kJ/hr}$$

Heat lost in ash-pit

$$= \text{mass} \times \text{calorific value of ash}$$
$$= 4 \times 2200 = 8800 \text{ kJ/hr}$$

Heat unaccounted for

$$= 2.597 \times 10^6 - 2.0086 \times 10^6 - 0.4798 \times 10^6 - 8800$$
$$= 99800 \text{ kJ/hr}$$

∴ Percentage heat unaccounted for

$$= \frac{99800}{2.597 \times 10^6} = 0.0384 \text{ or } 3.84\%$$

(*d*) Mass of air per kg of fuel

$$m_a = m_g - 1 = 18 - 1 = 17$$

The draught produced in mm of water column is given by

$$h_w = 353\, H \left[\frac{1}{T_a} - \frac{m_a+1}{m_a} \times \frac{1}{T_f}\right]$$

$$= 353 \times 50 \left[\frac{1}{301} + \frac{17+1}{17} \times \frac{1}{573}\right]$$

$= 26.02$ mm of water

EXAMPLE 15.15

The following data were recorded during test on a steam boiler :-

Pressure and condition of steam at exit from boiler = 0.95 MPa and 0.95 dry

Temperature of steam leaving the superheater = 250°C

Feed water supplied per hour = 7000 kg

Feed water temperature at entry to and at exit from economiser = 25°C and 80°C

Quantity of coal burnt/hr and its calorific value = 750 kg and 33915 kJ/kg

Determine the thermal efficiency of the plant. Also calculate the heat absorbed by feed water in various components as a percentage of total heat absorbed.

Solution : The schematic arrangement of the components in the steam plant is shown below in Fig. 15.6.

From steam tables :

Enthalpy of water at 25°C :

$h_1 = 105$ kJ/kg

Enthalpy of water at 80°C :

$h_2 = 335$ kJ/kg

Enthalpy of steam at 0.95 MPa and 0.95 dry :

$h_3 = h_f \times x\, h_{fg}$

$= 753 + 0.95 \times 2023 = 2674.8$ kJ/kg

Enthalpy of steam at 0.95 MPa and 250°C :

$h_3 = h_g + c_{ps}\,(t_{sup} - t_{sal})$

$= 2776 + 2.3\,(250 - 177) = 2943.9$ kJ/kg

Now : heat released by combustion of fuel in the boiler $= m_f \times CV = 750 \times 33915$

$= 25.436 \times 10^6$ kJ/kg

Heat utilized in raising steam from feed water at 25°C

$= 7000\,(2943.9 - 105)$

$= 19.872 \times 10^6$ kJ/kg

∴ Thermal efficiency of the plant

$$= \frac{19.872 \times 10^6}{25.436 \times 10^6} = 0.781 \text{ or } 78.1\%$$

(*b*) Heat supplied in boiler drum

$= 7000\,(2674.8 - 335)$

$= 16.378 \times 10^6$ kJ/hr

$$= \frac{16.378 \times 10^6}{19.872 \times 10^6} \times 100$$

= 82.4% of total heat absorbed in raising steam.

Heat supplied in superheater,

$= 7000(2943.9 - 2674.8)$

$= 1.884 \times 10^6$ kJ/hr

$$= \frac{1.884 \times 10^6}{19.872 \times 10^6} \times 100$$

= 9.48% of total heat absorbed in raising steam.

Heat supplied in economizer,

$= 7000\,(335 - 105) = 1.61 \times 10^6$ kJ/hr

$$= \frac{1.61 \times 10^6}{19.872 \times 10^6} \times 100$$

= 8.10% of total heat absorbed in raising steam

EXAMPLE 15.16

Draw the heat balance sheet from the data given below which pertains to the trial made on a boiler generating 500 kg/hr of steam at 10.5 bar pressure and 0.97 dryness fraction.

Fuel used and its calorific value : 75 kg/hr and 31500 kJ/kg

Moisture present in the fuel : 6% by mass

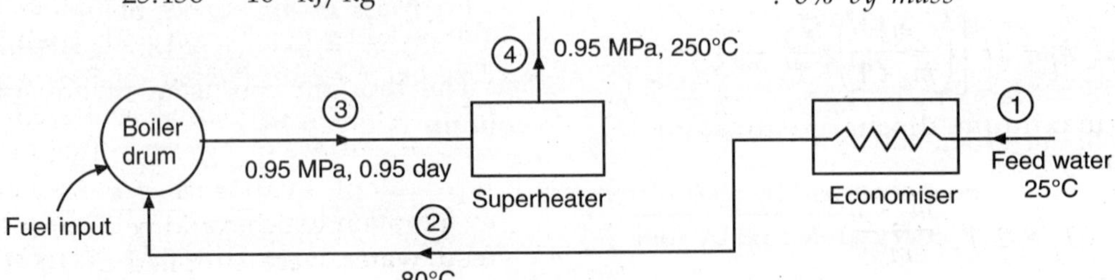

Fig. 15.6.

Mass of dry flue gases
: 10 kg/kg of fuel
Temperature of flue gases = *315°C*
Specific heat of flue gases = *1.1 kJ/kg K*
Temperature of boiler room = *38°C*
Feed water temperature = *50°C*

Solution : The fuel used has 6% of moisture present in it. Therefore

Heat supplied per kg of fuel

= (1 – 0.06) × 31500 = 29610 kJ

(*a*) Mass of water evaporated per kg of fuel

$$m_w = \frac{500}{75} = 6.67 \text{ kg}$$

From steam tables :

Enthalpy of feed water at 50°C,

h_w = 209 kJ/kg

Enthalpy of steam at 10.5 bar and 0.97 dry,

h_s = 772 + 0.97 × 2007 = 2718.8 kJ/kg

∴ Heat utilized in raising steam

$= m_w(h_s - h_w)$

= 6.67 (2718.8 – 209)

= 16740 kJ/kg of fuel

(*b*) Heat carried by flue gases

$= m_g\, c_{pg}\, (t_g - t_a)$

= 10 × 1.1 (315 – 38)

= 3047 kJ/kg of fuel

(*c*) Heat carried away by moisture

$= m_w\, [2676 + c_{ps}\, (T_g - 100) - h'_w]$

= 0.06 [2676 + 2.1 (315 – 100) – 159.1]

= 178.1 kJ/kg of fuel.

Here h'_w = sensible heat of feed water at 38°C (boiler room temperature)

= 4.186 × 38 = 159.1 kJ/kg

c_{ps} = specific heat of superheated steam assumed as 2.1 kJ/kg K

(*d*) Heat unaccounted for (lost due to convection and radiation)

= 29610 – (16740 + 3047 + 178.1)

= 9644.9 kJ/kg of fuel

SALIENT POINTS

1. Draught means the small pressure difference that causes and maintains a continuous flow of fresh air into the combustion chamber/furnace of a boiler.
2. Natural (chimney) draught is the draught produced by the use of a chimney due to the difference in densities between the hot gases inside the chimney and cold atmospheric air outside it.
3. The draught in terms of water column h_w in mm and in-terms of hot gas column h_f in m is

$$h_w = 353\, H \left[\frac{1}{T_a} - \frac{m_a+1}{m_a} \times \frac{1}{T_f}\right]$$

and $$h_f = H \left[\left(\frac{m_a}{m_a+1}\right) \times \frac{T_f}{T_a} - 1\right]$$

For maximum discharge through the chimney

$$T_f = 2\, T_a\, \frac{m_a+1}{m_a}$$

$$h_w = \frac{176.5 H}{T_a} \text{ and } h_f = H$$

where,

T_a = absolute temperature of atmospheric air

T_f = absolute temperature of hot flue gases

m_a = mass of air supplied per kg of fuel

H = height of chimney in metre

The velocity of flue gases through the chimney under a static draught of h_f metres is given by

$$V = \sqrt{2gh_f} = 4.43\sqrt{h_f}$$

4. Mechanical or artifical draught is the draught produced by fan, blower and steam jet.
5. The evaporation (steam raising) capacity of a boiler is expressed as kg/hr-m^2 of heating surface or kg/kg of fuel burnt.
6. Equivalent evaporation refers to the evaporation which could be obtained if the feed water were supplied at 100°C and

converted into dry saturated steam at 100°C at the standard atmospheric pressure of 1 bar.

$$W_e = \frac{W_e(h_s - h_w)}{L_o} \text{ kg/kg of coal burnt}$$

where W_a is weight of water actually evaporated, h_s is specific enthalpy of steam at the working pressure, h_w is specific enthalpy (sensible heat) of feed water and L_o is enthalpy of evaporation (latent heat) of steam at 1 bar and 100°C; and it equals 2256 kJ/kg.

The factor $(h_s - h_w)/L_o$ is called generation factor or factor of evaporation, and it is always greater than unity for all boilers.

7. Boiler efficiency represents the ratio of heat energy utilized by feed water in converting it into steam in the boiler to the heat energy produced by complete combustion of fuel during the same time

$$\eta_{boiler} = \frac{W_a(h_s - h_w)}{W_f \times \text{calorific value of fuel}}$$

where W_f is the weight of fuel consumed during the time the weight W_a of water is actually evaporated.

8. Boiler horse power is the term used for the performance unit equivalent to evaporation of 15.653 kg of water in one hour from and at 100°C.

Boiler horse power

$$= \frac{\text{equivalent evaporation/hr}}{15.653}$$

REVIEW QUESTIONS

A. Conceptual and conventional questions.

1. What is meant by draught ? Mention its utility in a boiler.
2. (*a*) State the functions of chimney in a boiler.
   (*b*) What arrangement of draught is made in a locomotive boiler ?
3. What are the limitations of chimney draught ?
4. Set up the following expression for the draught produced by a chimney

$$h_w = 353\,H\left[\frac{1}{T_a} - \frac{m_a + 1}{m_a} \times \frac{1}{T_f}\right] \text{ mm of water}$$

The symbols have their usual meanings.

5. Prove that for a given draught, the discharge through the chimney will be maximum when

$$T_f = 2\,T_a\,\frac{m_a + 1}{m_a}$$

where T_f is the average absolute temperature of flue gases, T_a is the absolute temperature of the air outside the chimney and m_a is the air supplied per kg of fuel burnt.

6. Obtain an expression for draught produced in mm of water column when the discharge is maximum.
7. Show that for maximum discharge from a chimney, the gas temperature is nearly twice the atmospheric temperature.
8. Prove that under maximum discharge condition, the draugth produced in terms of hot gas column is equal to the height of the chimney.
9. (*a*) Name the various methods for producing draught in a boiler.
   (*b*) Give brief description of the various types of mechanical draughts.
10. (*a*) Explain the terms: induced draught, forced draught and balanced draught.
    (*b*) State the advantages of artificial draught over natural draught.
11. Define and explain the terms :
    (*a*) Equivalent evaporation from and at 100°C
    (*b*) Boiler efficiency
12. Explain the various items that are considered in the heat balance of a boiler unit.
13. "Equivalent evaporation from and at 100°C". Explain the meaning of the term and give the reason for its use in the context of boiler performance.
14. Find the draught produced in terms of (*a*) mm of water column (*b*) metre of column of hot gases by a chimney 30 m high. The temperature of flue gases and of ambient air is 300°C and 25°C respectively and 18 kg of air is supplied for complete combustion of 1 kg of fuel.

[**Ans.** (*a*) 16 mm of water column; (*b*) 24.6 m of hot gas column]

15. A boiler is equipped with 25 m high chimney. The average temperature of flue gases in the chimney is 320°C and the ambient temperature is 25°C. The discharge of flue gases from the chimney is measured to be 21 kg of per kg of fuel burnt on fire grate of the combustion chamber. Calculate the theoretical draught created in terms of

(*a*) mm of water column (*b*) metre of hot gas column

Also make calculations for the velocity of flue gases in the chimney if 60 percent of the draught is lost in friction at grate and passages.

[**Ans.** 14 mm of water, 22.4 m of hot gas column, 13.25 m/s]

16. A steam power plant fitted with 50 m high chimney generates 1700 kW of power while consuming coal at the rate of 3 kg/kW/hr. The ambient temperature is 300 K and the amount of ambient air supplied is 18 kg of coal burnt in the combustion chamber. Neglecting friction losses and assuming the discharge through the chimney to be maximum, determine the diameter of the chimney at its base.

[**Ans.** 1.36 m]

17. The following data were taken during the test on a boiler for a period of one hour :

Steam generated = 5000 kg ; coal burnt = 700 kg

quality of steam raised = 0.92 ; calorific value of coal = 31402 kJ/kg

If the boiler pressure is 1.2 MPa and the feed water temperature is 45°C, find the boiler equivalent evaporation and efficiency.

[**Ans.** 7.71 kg/kg of coal ; 55.4%]

18. A coal fired boiler plant generates steam at a pressure of 20 bar and 250°C from feed water at 40°C. The plant consumes 1850 kg of coal per hour and the calorific value of the coal is 32400 kJ/kg. If equivalent evaporation of the boiler is estimated to be 22700 kg/hr, make calculations for actual evaporation of the boiler in kg/hr and efficiency of the boiler.

[**Ans.** 18896 kg/hr, 85.5%]

19. The following particulars refer to the test run on a steam boiler carried out for 10 hours :

Steam pressure = 14.8 bar ;

quality of steam raised = dry saturated

feed water temperature = 40°C

mass of water evaporated = 50000 kg

coal burnt and its calorific value

= 6000 kg and 29300 kJ/kg

If the total grate area of the boiler is 3 m^2, determine :

(*a*) amount of coal burnt per m^2 of grate area/hr

(*b*) amount of water evaporated per kg of coal burnt, under actual conditions.

(*c*) equivalent evaporation from and at 100°C per kg of coal burnt and

(*d*) boiler efficiency.

[**Ans.** 200 kg, 8.33 kg, 9.69 kg/kg of coal, 74.6%]

20. The following information was obtained during the trial of a boiler :

Steam evaporation rate and feed water temperature : 1050 kg/hr and 25°C

State of steam generated : 9.8 bar and 250°C

Coat used and its calorific value : 95 kg/hr and 33500 kJ/hr.

Quantity of air used per kg of coat burnt = 19 kg

Temperature of flue gases leaving the chimney = 350°C

Mean specific heat of flue gases = 1.005 kJ/kgK

Ash and unburnt coal collected in ash-pit = 9 kg/hr

Calorific value of ash and unburnt coal = 2500 kJ/kg

Temperature of boiler house = 30°C

Make calculations for the following items:

(*a*) Equivalent evaporation

(*b*) Actual evaporation per kg of coal

(*c*) Boiler efficiency

(*d*) Percentage of heat lost to flue gases and ash

(*e*) Heat loss unaccounted for.

[**Ans.** 8.55 kg/kg of coal; 6.84 kg; 57.6%; 19.2% and 0.7%; 22.47%]

B. Fill in the blanks with appropriate word/words.

1. The .... draught produced by the boiler chimney is due difference in .... of hot gases inside the chimney and the cold atmospheric air outside it.
2. The .... draught is the artificial draught created by a fan placed at the .... of the chimney or the boiler grate.
3. Induced type .... draught is used in a locomotive boiler.

4. For maximum discharge of hot gases through the chimney, the draught in terms of .... equals the height of chimney.
5. Draught is the .... which is necessary to supply required quantity of air for combustion and to remove the .... out of the system.
6. The quantity of steam generated by a boiler at full load is called .... of the boiler.
7. The evaporation capacity of a boiler may be expressed as kg/hr m^2 of .... or kg/kg of ....
8. The .... represents the ratio of heat absorbed by 1 kg of feed water under working conditions to that absorbed by 1 kg of water from and at 100°C.
9. The ratio of the heat actually used in producing the steam to the heated liberated in the furnace is called ....
10. The evaporation of 15.653 kg of water per hour from and at 100°C is called ....

Answers

1. artificial, densities 2. mechanical, base 3. steam jet 4. hot gas column 5. pressure difference, burnt gases 6. evaporative capacity 7. heating surface, coal burnt 8. generation factor 9. boiler efficiency 10. boiler horse power

C. Multiple choice questions.

1. The term heating surface in a boiler means
   (*a*) area of grate
   (*b*) volume of furnace
   (*c*) outer surface area of the boiler shell
   (*d*) surface area which is in contact with flue gases
2. Boiler rating is usually defined in terms of
   (*a*) maximum temperature of steam
   (*b*) heat transfer rate in kJ/hr
   (*c*) heating surface area
   (*d*) steam output in kg/hr
3. The capacity of a boiler is defined as
   (*a*) amount of water evaporated into steam in a period of one hour
   (*b*) volume of steam space inside the shell
   (*c*) volume of feed water inside the shell
   (*d*) maximum pressure at which steam can be raised in the boiler
4. The amount of water evaporated from and at 100°C into dry saturated steam at atmospheric pressure is called
   (*a*) evaporative capacity
   (*b*) equivalent evaporation
   (*c*) generation factor (*d*) boiler horse power
5. The ratio of heat actually used in raising steam to the heat liberated in the boiler furnace by the combustion of fuel is called
   (*a*) equivalent evaporation
   (*b*) boiler rating
   (*c*) boiler efficiency (*d*) generation factor
6. For maximum discharge of flue gas, the height of hot gas column producing draught and the height of chimney conform to the ratio
   (*a*) 1 : 2 (*b*) 1 : 1
   (*c*) 3 : 1 (*d*) 5 : 2
7. Steam jet draught is invariably used in
   (*a*) Cochran boiler (*b*) Lancashire boiler
   (*c*) Locomotive boiler
   (*d*) Babcock-Wilcox boiler
8. A draught of 1 mm of water is equivalent to
   (*a*) 9.81 N/m^2 (*b*) 98.1 N/m^2
   (*c*) 981 N/m^2 (*d*) 0.981 N/m^2
9. All of the following statements are correct, except
   (*a*) chimney needs no auxiliary power for its operation
   (*b*) In induced draught, the fan is placed near or at the base of chimney
   (*c*) balanced draught is the chimney draught supplemented by induced draught
   (*d*) compared to induced draught, the forced draught fan provides more uniform flow of air through the grate
10. Overall efficiency of a steam power plant is increased with mechanical draught due to
    (*a*) decrease in height of chimney
    (*b*) increased rate of combustion
    (*c*) use of heat of gases before being exhausted
    (*d*) complete combustion of fuel

Answers

1. (*d*)	2. (*d*)	3. (*a*)	4. (*b*)	5. (*c*)
6. (*b*)	7. (*c*)	8. (*a*)	9. (*c*)	10. (*b*)

Chapter **16**

Vapour Power Cycles

Learning objectives : A study of this chapter will enable the reader to deal with :

- *Carnot and Rankine vapour cycles*
- *modified Rankine cycle*
- *thermal refinement of Rankine cycle : reheating and regenerative heating*
- *binary vapour cycle*
- *co-generation : combined heat and power*

Vapour power cycles are external combustion systems in which the working fluid is alternately vaporised and condensed.

Water/steam is easily available, is cheap, is chemically stable and physiologically harmless and as such is the most commonly employed working fluid. Due to its use as the working substance in vapour power cycle, the cycle is often referred to as steam power cycle. The steam power plants generate a major and increasing fraction of electric power produced in the world. Mercury, though good from thermodynamic point of view, is both costly and poisonous and its use is restricted to a topping fluid in binary plants that operate at high temperatures. Ammonia, Freon and carbon dioxide find application in the operation of refrigeration systems for producing cooling effect.

The heat engine cycles which make use of steam as working fluid are Carnot cycle and Rankine cycle and these two have been discussed in the following paragraphs.

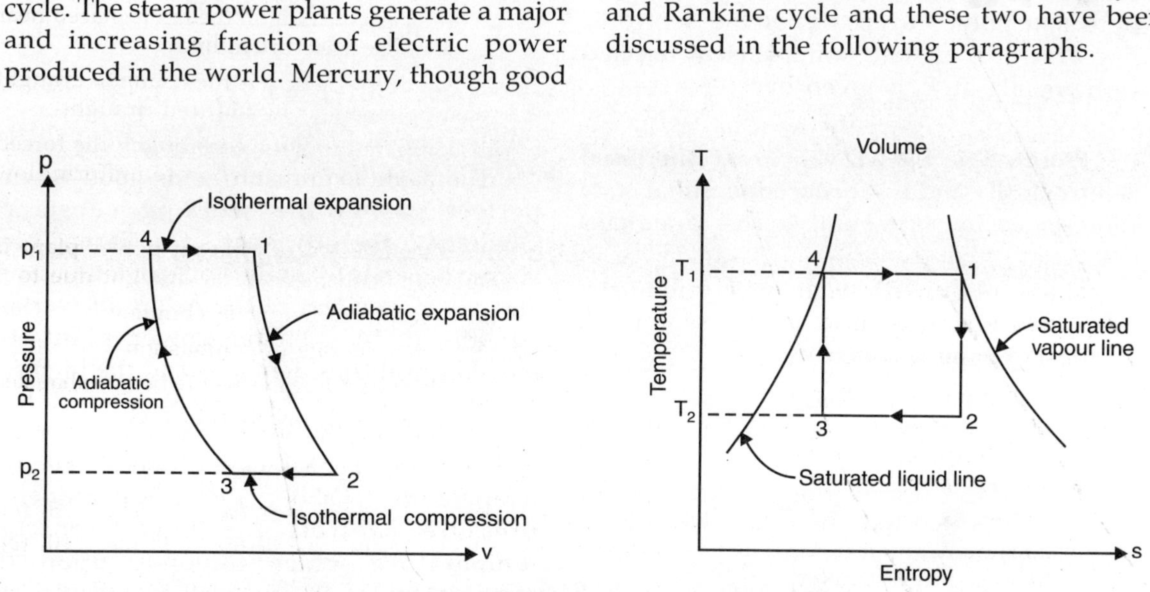

Fig. 16.1. Carnot cycle on p-v and T-s diagrams

16.1. CARNOT CYCLE

Fig 16.1. shows the pressure-volume and temperature-entropy plots of a Carnot cycle when steam is used as the working substance.

The cycle is composed of two isentropic and two isothermal processes. Since the working fluid is a vapour, the isothermal processes are also the constant pressure processes.

Process 4–1 : The boiling water contained in the steam generator (boiler) is supplied heat. Isothermal evaporation of water takes place at constant pressure until the water gets converted into dry saturated steam. Heat added isothermally at T_1 is given by

$$Q_1 = (h_1 - h_4)$$

Process 1–2 : The dry saturated steam undergoes frictionless adiabatic expansion in the steam engine/turbine and work is done by the system.

The work done by the adiabatic expansion of steam is known as shaft work or turbine work (+ve work) and equals the enthalpy drop.

$$W_t = (h_1 - h_2)$$

Process 2–3 : Heat is removed from the spent-up steam (steam leaving the prime-mover) and it condenses isothermally at constant pressure. The condensation continues upto state point 3 which lies on the isentropic line drawn from state point 4. Heat rejected isothermally at T_2 is given by

$$Q_2 = (h_2 - h_3)$$

Process 3–4 : The wet vapour is compressed isentropically and it is restored to initial state represented by state point 4. This completes the cycle.

The work done on steam during adiabatic compression is known as compressor work (- ve work) and is given by

$$W_c = (h_4 - h_3)$$

Thermal efficiency, work ratio and specific steam consumption :

The performance criterion of a steam power plant are the thermal efficiency, work ratio and specific steam consumption

Net work in the Carnot cycle,

$$W_{net} = W_t - W_c = (h_1 - h_2) - (h_4 - h_3)$$

Thermal efficiency,

$$\eta_{th} = \frac{\text{net work done}}{\text{heat supplied}}$$

$$= \frac{(h_1 - h_2) - (h_4 - h_3)}{(h_1 - h_4)}$$

$$= 1 - \frac{h_2 - h_3}{h_1 - h_4}$$

In terms of the highest and lowest temperatures under which the Carnot cycle operates, the thermal efficiency is also given by

$$\eta_{th} = \frac{T_1 - T_2}{T_1} = 1 - \frac{T_2}{T_1}$$

as worked out below :

Heat supplied to the working fluid in the boiler

= area under the isothermal line 4–1

$= T_1\, ds$

Heat rejected during condensation process

= area under isothermal line 2-3

$= T_2\, ds$

Net work done = heat supplied – heat rejected

$= T_1\, ds - T_2\, ds = (T_1 - T_2)\, ds$

Thermal efficiency,

$$\eta_{th} = \frac{\text{net work done}}{\text{heat supplied}}$$

$$= \frac{(T_1 - T_2)\, ds}{T_1\, ds} = \frac{T_1 - T_2}{T_1}$$

The same expression holds good when a perfect gas is the working substance. Obviously, the efficiency of a Carnot cycle engine depends upon the limits of temperature and is independent of the nature of working substance. In a steam power plant, the metallurgical considerations and the high cost of temperature resisting materials which have to be used for the making of the boiler and turbine components limit the highest operating temperature to about 575°C. The condenser pressure controls the lowest possible temperature which depends upon the temperature of the naturally available heat sinks (air and large quantities of water), and is normally kept at 25–30°C.

Work ratio is defined as the ratio of net work to turbine (expansion) work done in the cycle.

Work ratio,

$$r_w = \frac{\text{net work}}{\text{turbine work}}$$

$$= \frac{\text{turbine work - pump work}}{\text{turbine work}}$$

$$= 1 - \frac{\text{pump}}{\text{turbine work}}$$

A low work ratio implies large pump work. Larger the pump work, lower the work ratio.

Specific steam consumption (*ssc*) *or steam rate* is defined as the flow rate of steam per unit of power developed (kWh).

$$ssc = \frac{\text{steam flow rate}}{\text{power output}}$$

It may be expressed as reciprocal of net work as follows

$$ssc = \frac{1}{W_{net}} \frac{\text{kg}}{\text{kJ}}$$

$$= \frac{1}{W_{net}} \frac{\text{kg}}{\text{kWs}} = \frac{3600}{W_{net}} \frac{\text{kg}}{\text{kWh}}$$

The quantity of working fluid governs the size of the components of the plant. More the *ssc*, greater the plant size. More *ssc* also implies less net work output for the same quantity of heat supplied, and hence lower thermal efficiency and higher operating cost. Thus *ssc* is an index of the capital and operating cost of the steam plant.

Sometimes the cycle thermal efficiency is expressed as *heat rate* which is a measure of the rate of heat input Q_1 required to produce unit work output (1 kW).

$$\text{Heat rate} = \frac{3600\, Q_1}{W_t - W_c}$$

$$= \frac{3600}{\eta_{th}} \frac{\text{kJ}}{\text{kWh}}$$

Drawbacks of Carnot cycle :

Theoretically a Carnot cycle is most efficient; however the following difficulties are associated with it when steam is the working fluid.

(*i*) Steam condensation is not allowed to proceed to completion. The condensation process is controlled one; it has to be stopped at state point 3.

(*ii*) The working fluid at state point 3 is both in the liquid and vapour state, and these donot form a homogeneous mixture. It is rather difficult to compress the two phase mixture isentropically.

(*iii*) The vapour has a large specific volume and to accommodate greater volumes, the size of the compressor becomes quite big.

(*iv*) For running a large sized compressor, more power is required and that results in poor plant efficiency.

(*v*) Carnot cycle has high specific steam consumption, large back work ratio and low work ratio.

(*vi*) The steam at exhaust from the turbine is of low quality, *i.e.*, high moisture content. The liquid water droplets cause pitting and hence erosion of the turbine blades.

EXAMPLE 16.1

In a Carnot cyclè, heat is supplied at 625 K and is rejected at 300 K. The working fluid is water which while receiving heat evaporates from liquid at 625 K to steam at 625 K. From the steam tables, the entropy change for this process is 1.4374 KJ/kg K.

If the cycle operates with 1 kg mass of water, find the heat supplied, work done and heat rejected per cycle. What is the pressure of water during heat reception ?

Solution : Refer Fig. 16.1 for *T-s* diagram for the Carnot cycle.

Heat is supplied at 625 K and is rejected at 300 K.

$$\therefore \quad T_1 = 625 \text{ K and } T_2 = 300 \text{ K}$$

Entropy change

$$= 1.4374 \text{ kJ/kg K}$$

$$\therefore \quad s_1 - s_4 = s_2 - s_3 = 1.4374$$

Heat supplied

$$= T_1 (s_1 - s_4)$$

$$= 625 \times 1.4374 = 898.375 \text{ kJ/kg}$$

Heat rejected

$= T_2 (s_2 - s_3)$
$= 300 \times 1.4374 = 431.22$ kJ/kg

Work done

= heat supplied – heat rejected
$= 898.375 - 431.22 = 467.15$ kJ/kg

The pressure of water during heat reception is obtained from steam tables corresponding to saturation temperature of 625 K (= 352°C). This is equal to 170 bar.

EXAMPLE 16.2

A power generating plant uses steam as working fluid and operates on an ideal Carnot cycle. Dry saturated steam at 17.5 bar pressure is supplied to the engine and it expands isentropically to a condenser pressure of 0.07 bar. Assuming the liquid to be saturated at entry to the boiler, make calculations for the work ratio, thermal efficiency and specific steam consumption.

Solution : Refer Fig. 16.1 for *T-s* diagram of the Carnot cycle. Steam is supplied to the engine at state point 1 at pressure 17.5 bar and dry saturated. From steam tables

$h_1 = 2794.1$ kJ/kg

and $s_1 = 6.385$ kJ/kg K

For isentropic expansion process 1–2, entropy remains constant and therefore

$s_1 = s_2$; $s_1 = (s_{f2} + x_2\, s_{fg})_{0.07\text{ bar}}$

$6.385 = 0.559 + x_2 \times 7.717$

That gives :

$$x_2 = \frac{6.385 - 0.559}{7.717} = 0.755$$

$\therefore \quad h_2 = (h_f + x_2\, h_{fg})_{0.07\text{ bar}}$
$= 163.4 + 0.755 \times 2409.1$
$= 1982.3$ kJ/kg

Since the state point 4 lies on the liquid line,

$h_4 = (h_f)_{17.5\text{ bar}} = 878.3$ kJ/kg
$s_4 = (s_f)_{17.5\text{ bar}} = 2.385$ kJ/kg K

For the isentropic compression (pumping) process 3–4

$s_3 = s_4$
$(s_f + x_3\, s_{fg})_{0.07\text{ bar}} = s_4$
$(0.559 + x_3 \times 7.717) = 2.385$

That gives :

$$x_3 = \frac{2.385 - 0.559}{7.717} = 0.237$$

$h_3 = (h_f + x_3\, h_{fg})_{0.07\text{ bar}}$
$= 163.4 + 0.237 \times 2409.1$
$= 734.36$ kJ/kg

(*a*) Work done by the engine $= h_1 - h_2$
$= 2794.1 - 1982.3 = 811.8$ kJ/kg

Work input to pump

$= h_4 - h_3 = 878.3 - 734.36$
$= 143.94$ kJ/kg

Net work = engine work – pump work
$= 811.8 - 143.94 = 667.86$ kJ/kg

$$\therefore \text{ Work ratio} = \frac{\text{net work}}{\text{engine work}} = \frac{667.86}{811.8} = \mathbf{0.823}$$

(*b*) Heat input to the working fluid in the boiler

= heat at state point 1 – heat at state point 4
$= 2794.1 - 878.3 = 1915.4$ kJ/kg

Thermal efficiency

$$= \frac{\text{net work input}}{\text{heat supplied}} = \frac{667.86}{1915.4} = 0.3487 \text{ or } \mathbf{34.87\%}$$

Thermal efficiency of a Carnot cycle can also be worked out from the higher and lower operating temperature limits. That is

$$\eta_{carnot} = \frac{T_1 - T_2}{T_1} \quad \text{where}$$

T_1 = saturation temperature corresponding to boiler pressure of 17.5 bar = 478.7 K

T_2 = saturation temperature corresponding to condenser pressure of 0.07 bar = 312 K

$$\therefore \eta_{carnot} = \frac{478.7 - 312}{478.7} = 0.3482 \text{ or } 34.82\%$$

(*c*) Specific steam consumption

$$= \frac{3600}{W_{net}} \text{ kg/kWh}$$

$$= \frac{3600}{667.86} = \mathbf{5.39\ kg/kWh}$$

16.2. RANKINE CYCLE

Professor Rankine, who made significant contributions to the engineering and scientific development, modified the Carnot cycle and presented a technically feasible cycle, called the Rankine cycle. It is also a reversible cycle but it differs from the Carnot cycle in the following respects :

(*i*) The condensation process is allowed to proceed to completion; the exhaust steam from the engine/turbine is completely condensed. At the end of condensation process the working fluid is only liquid and not a mixture of liquid and vapour.

(*ii*) The pressure of liquid water can be easily raised to the boiler pressure (pressure at which steam is being generated in the boiler) by employing a small sized pump.

In addition, the steam may be superheated in the boiler so as to obtain exhaust steam of higher quality. That will prevent pitting and erosion of turbine blades.

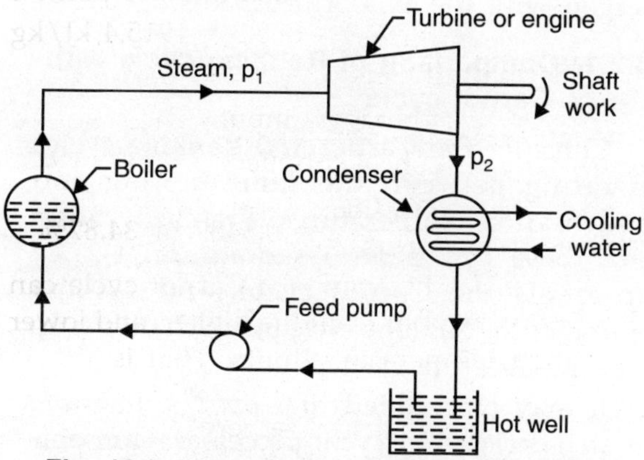

Fig. 16.2. Steam power plant operating on simple Rankine cycle

Fig. 16.2. shows the schematic line diagram of a steam power plant working on ideal Rankine cycle. The various elements are :

- A boiler which generates steam at constant pressure.
- An engine or turbine in which steam expands isentropically and work is done.
- A condenser in which heat is removed from the exhaust steam and it is completely converted into water at constant pressure.
- A hot well in which the condensate is collected.
- A pump which raises the pressure of liquid water to the boiler pressure and pumps it into the boiler for conversion into steam.

The corresponding *p-v* and *T-s* plots are shown in Fig. 16.3.

The working fluid undergoes the following series of internally reversible processes :

Process 1-2 : Isentropic expansion of steam in the engine/turbine from boiler pressure p_1 to condenser pressure p_2. The enthalpy drop occurs from h_1 to h_2 and work is obtained. From steady flow energy equation, the work done equals the enthalpy drop.

$$\text{Turbine work, } W_t = (h_1 - h_2)$$

Process 2-3 : Condensation of exhaust steam at constant pressure and temperature to liquid at the same pressure. The latent heat is rejected to cooling water in the condenser. The cooling water may be obtained from a river or by circulation from a cooling tower.

$$\text{Heat rejected, } Q_2 = (h_2 - h_3)$$

Process 3-3′ : Reversible adiabatic pumping of condensate water at state point 3 from

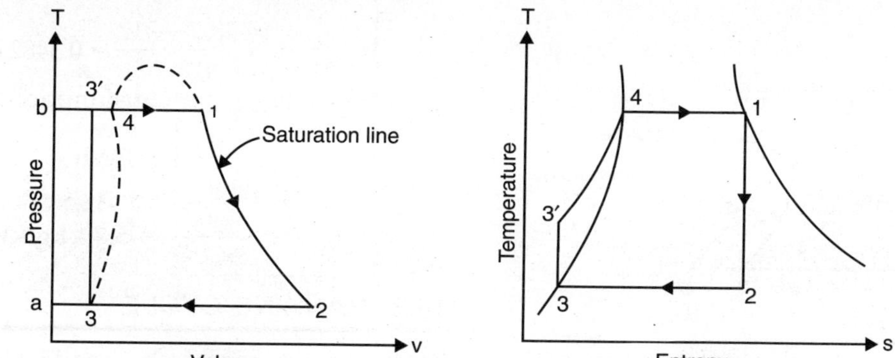

Fig. 16.3. p-v and T-s diagrams of ideal Rankine cycle

condenser pressure p_2 to boiler pressure p_1. Due to compression, the temperature of feed water is increased and this is represented by 3–3′ on *T–s* diagram. The equivalent work (pump work) is represented by area 3-a-b-3′ on the *p–v* diagram

Pump work,

$$W_p = h_{3'} - h_3 = \int v dp = v_{f_3} (p_1 - p_2)$$

Process 3′-1 : Sensible heating of water from 3′ to 4, and then evaporation from 4 to 1 in the boiler at constant pressure. The operation 3′–4 may take place either in an economiser or in a special feed heater.

Heat supplied,

$$Q_1 = h_1 - h_{3'}$$

Net work of the cycle,

$$W_n = W_t - W_p$$
$$= (h_1 - h_2) - (h_{3'} - h_3)$$

Alternatively for a cyclic process

$$\oint \delta W = \oint \delta Q$$

∴Net work,

$$W_n = (h_1 - h_{3'}) - (h_2 - h_3)$$
$$= (h_1 - h_2) - (h_{3'} - h_3)$$

Thermal efficiency of the Rankine cycle,

$$\eta_{Rankine} = \frac{\text{net work}}{\text{heat input}}$$
$$= \frac{(h_1 - h_2) - (h_{3'} - h_3)}{h_1 - h_{3'}}$$
$$= \frac{(h_1 - h_2) - (h_{3'} - h_3)}{(h_1 - h_3) - (h_{3'} - h_3)}$$

The compression process in the pump is being carried out with liquid only for which the specific volume is small. Consequently the pump work $(h_{3'} - h_3)$ is quite small compared to the turbine work and can be neglected. The Rankine cycle efficiency then approximately becomes

$$\eta_{Rankine} = \frac{h_1 - h_2}{h_1 - h_3}$$

As mentioned earlier, there is the possibility of superheating steam beyond point 1 in the boiler so as to obtain exhaust steam of higher quality at 2 and thus prevent the pitting and erosion of turbine blades. Fig 16.4 shows the *p–v* and *T–s* plots when the engine/turbine operates with superheated steam (1′–2′) and wet steam (1″–2″). In this plot, pump work has been neglected and the line 3–3′ is shown merged with the axis of ordinates.

16.2.1. Comparison of Rankine cycle with Carnot cycle

Consider the Carnot and Rankine cycles operating between the same boiling and condensation temperatures. Fig. 16.5 shows both these cycles on *T-s* diagram. 1-2-3-4 represents the Carnot cycle, and 1-2-3′-4 represents the Rankine cycle in which the pump work has been neglected.

It may be recalled that on *T-s* diagram, the area enclosed by cyclic processes represents the work done and area under the line representing heat addition gives the heat input to the cycle.

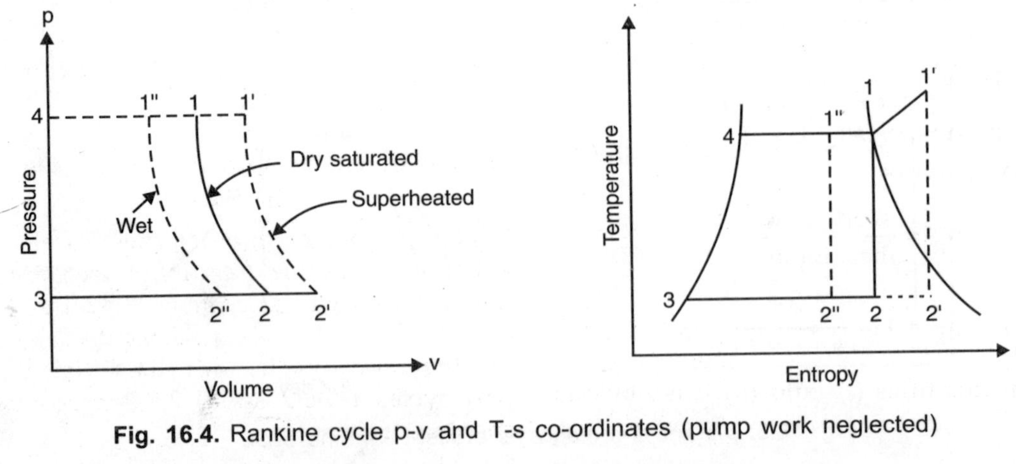

Fig. 16.4. Rankine cycle p-v and T-s co-ordinates (pump work neglected)

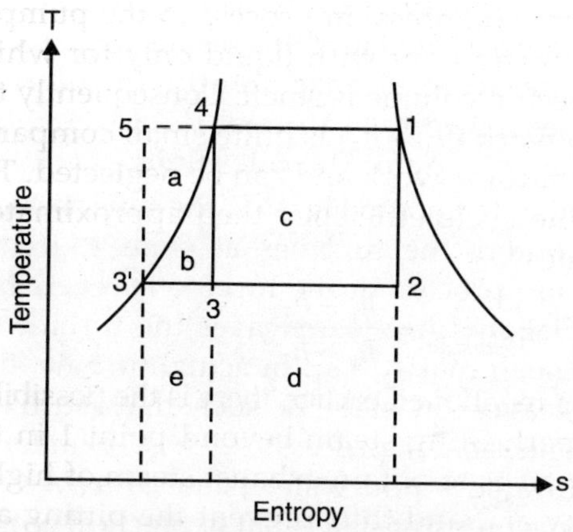

Fig. 16.5. Comparison of Carnot and Rankine cycles

Thus for the Carnot cycle,

Work done = area 1-2-3-4 = area c

Heat input = area under the line 4-1

= area c + area d

$$\eta_c = \frac{\text{work done}}{\text{heat input}} = \frac{c}{c+d}$$

Let 1-2-3′-5 be another reversible Carnot cycle, which will have thermal efficiency given by

$$= \frac{a+b+c}{a+b+c+d+e}$$

From Carnot's theorem, two reversible cycles operating within the same temperature limits have the same thermal efficiency. Accordingly efficiency of the given Carnot cycle 1-2-3-4 can be put as

$$\eta_c = \frac{a+b+c}{a+b+c+d+e}$$

$$\text{or} \quad 1 - \eta_c = \frac{d+e}{a+b+c+d+e} \quad ...(i)$$

The Rankine cycle 1-2-3′-4 has thermal efficiency given by

$$\eta_R = \frac{\text{work done}}{\text{heat input}} = \frac{b+c}{b+c+d+e}$$

$$1 - \eta_R = 1 - \frac{d+e}{b+c+d+e} \quad ...(ii)$$

From identities (*i*) and (*ii*) it is obvious that

$$1 - \eta_c < 1 - \eta_R$$

$$\text{or} \quad \eta_c > \eta_R$$

That is the Carnot cycle is more efficient than a Rankine cycle when both are operating within the same temperature limits.

The above aspect also stems from the fact that for the Carnot cycle the entire heat input is during the process 4-1 which takes place at the maximum cycle temperature. For the Rankine cycle, the heat addition does not take place at the maximum cycle temperature. The temperature during sensible heating 3'-4 is much lower and is continuously changing, and that lowers the average temperature at which heat is supplied and accordingly the efficiency of Rankine cycle goes down.

16.2.2. Rankine cycle using steam with and without superheat

Consider two Rankine cycles operating between the same boiler and condenser pressure limits. 1-2-3-4 is the basic Rankine cycle and 5-6-3-4 represents the Rankine cycle with superheat. The pump work has been neglected for both the cycles.

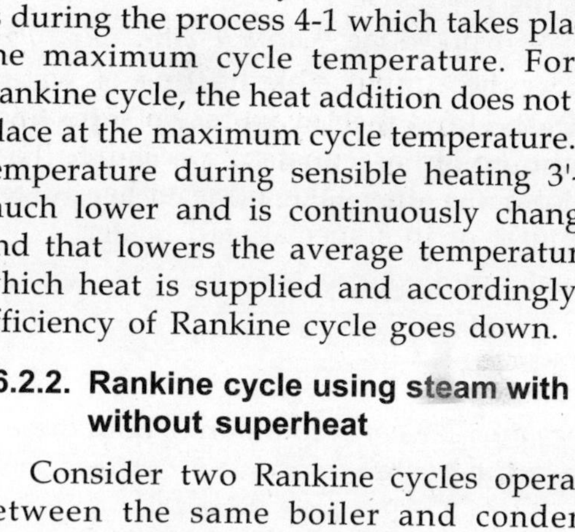

Fig. 16.6. Rankine cycle using steam with and without superheat

Let 7-8-3-4 be another theoretical cycle which has the same quantity of heat input as the superheated cycle. Since area 1-2-3-4 is common to both the superheated cycles and the theoretical cycle, only the efficiency of part cycles 1-5-6-2 and 1-7-8-2 needs to be compared.

For the part cycle 1-5-6-2,

work done = area 1-5-6-2

heat input = area 1-5-*b*-*a*

$$\therefore \quad \text{Efficiency, } \eta_1 = \frac{\text{area } 1-5-6-2}{\text{area } 1-5-b-a} \quad ...(i)$$

For the part cycle 1-7-8-2,

work done = area 1-7-8-2

heat input = area 1-7-*c*-*a*

Since both the cycles are supplied with same amount of heat, area 1-7-*c*-*a* equals the area 1-5-*b*-*a*, the efficiency of this cycle can be put as

$$\eta_2 = \frac{\text{area } 1-7-8-2}{\text{area } 1-5-b-a} \quad ...(ii)$$

Again, area 1-7-*c*-*a* = area 1-5-*b*-*a*

Since area 1-9-*b*-*a* is common to both these areas, we can write

area 1-5-9 = area 9-7-*c*-*b*

or area 1-5-9 > area 9-7-8-6

Upon adding area 1-9-6-2 to both sides, we get

area 1-5-6-2 > area 1-7-8-2

Then an examination of identities (*i*) and (*ii*) reveals that $\eta_1 > \eta_2$ which shows that thermal efficiency of a Rankine cycle using superheated steam is greater than the thermal efficiency of a corresponding Rankine cycle using steam without superheat, when both the cycles operate between the same boiling and condensation pressure limits.

The above conclusion also stems from the fact that superheating increases the average temperature at which heat is supplied and accordingly the cycle efficiency improves.

16.2.3. Effect of internal irreversibilities on cycle efficiency

In actual practice, it is impossible to find pumps and turbines operating under isentropic conditions because of inherent irreversibilities. The second law of thermodynamics for steady state, steady flow process states that whenever energy is being utilized, losses take place due to friction between working fluid and the wall and between the molecules of the fluid itself. A certain portion of work is utilized in overcoming friction and this results in increase in temperature. Consequently in all real processes, entropy of the exit fluid is greater than the entropy of the inlet fluid.

With reference to Fig. 16.7, the super heated steam enters the turbines at state 1. If the expansion process in the turbine is reversible and adiabatic the steam leaves the turbine at state 2 such that $s_2 = s_1$. In actual turbine, the steam leavers at state 2*a* such that (suffix *a* represents actual process) $s_{2a} > s_1$. Similarly the saturated liquid water (state 3) leaves an isentropic pump at state 4 but in practical pump the liquid leaves at state 4*a* such that $s_{4a} > s_4 > s_3$.

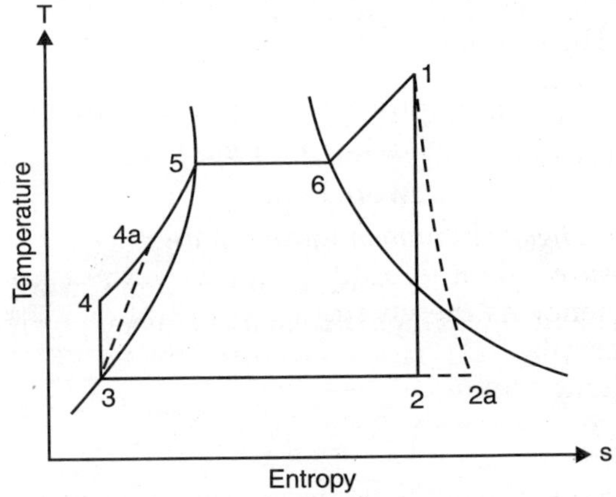

Fig. 16.7. Practical Rankine cycle

The isentropic efficiency for the turbine (η_t) and the pump (η_p) a then defined as

$$\eta_t = \frac{\text{work delivered by the actual turbine}}{\text{work delivered by the isentropic turbine}}$$

$$= \frac{h_1 - h_{2a}}{h_1 - h_2}$$

and $$\eta_p = \frac{\text{work required by the isentropic pump}}{\text{work required by the actual pump}}$$

$$= \frac{h_4 - h_3}{h_{4a} - h_3}$$

Since η_t and η_p are less than unity, we have :

(*i*) $(h_1 - h_{2a}) < (h_1 - h_2)$, *i.e.*, the actual turbine delivers less work than an isentropic turbine

(*ii*) $(h_{4a} - h_3) > (h_4 - h_3)$, *i.e.*, the actual pump consumes more work than an isentropic pump.

These aspects reduce the net work delivered by the system. There is also reduction in the heat energy supplied from $(h_1 - h_4)$ to $(h_1 - h_{4a})$. However the effect of reduction in net work predominates and accordingly the efficiency of a practical Rankine cycle is less than that of an ideal Rankine cycle.

16.2.4. Working fluid for a Rankine cycle

The working fluid for Rankine cycle should be cheap, universally available, chemically stable and harmless. Thermodynamically, an ideal working fluid for Rankine cycle has the following requirements :

• *Highest saturation temperature for a moderate pressure* : This would result in high cycle efficiency as the energy (heat) addition then takes place at the maximum temperature without having to deal with high pressure which can cause mechanical failures.

• *Low specific heat* : The sensible heat for the fluid will then be quite small in comparison with latent heat. Maximum heat addition would then occur at maximum temperature and the Rankine cycle would approach the Carnot cycle.

• *High density* : For a given output, the plant size becomes small if the fluid has high density.

• *Steep saturated vapour line* : The steam after expansion in the turbine will then have low moisture content (high dryness fraction) and the possibility of pitting and erosion of turbine blades due to liquid droplets will get minimised.

• *Higher saturation pressure than atmospheric pressure at the minimum cycle temperature* : This permits to run the engine without using a condenser and thus avoiding the undesirable condenser problems.

• *Non-toxic and non-corrosive* : Mercury has a critical temperature and pressure of 1100°C and 1.374 bar respectively whereas the corresponding values for water are 374°C and 220.72 bar, and as such mercury is good from thermodynamic point of view. However, mercury is costly, toxic and poisonous. Water being easily available and harmless is universally adopted. Mercury is used only as topping fluid in binary plants.

16.3. MODIFIED RANKINE CYCLE

The pressure-volume variation for a simple steam engine operating on Rankine cycle is shown in Fig. 16.8. The clearance is neglected and the expansion of steam continues upto the back pressure. The *p-v* diagram is very narrow at the tow and the work output during the last part of expansion stroke 1-2′ is very small. In

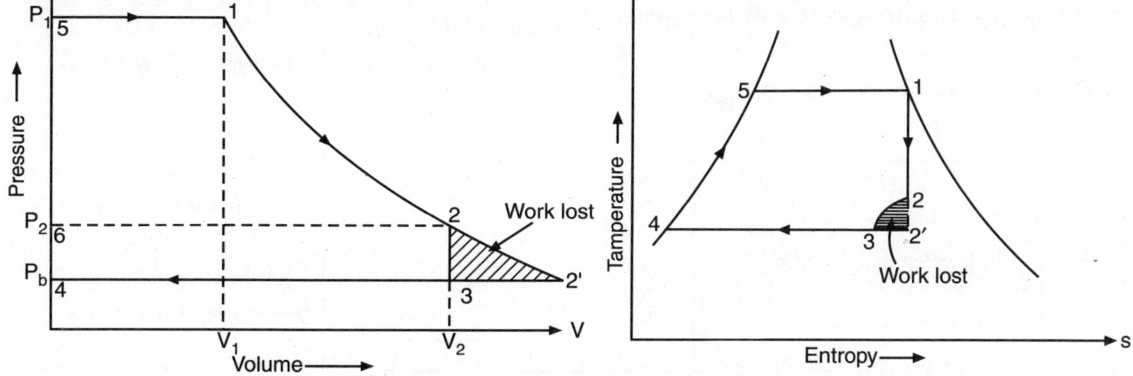

Fig. 16.8. Modified Rankine cycle

fact, this work is not even sufficient to overcome the frictional losses in the moving parts of the engine. In actual practice, the expansion process is not allowed to proceed to completion upto point 2′. Rather it is terminated at state point 2 by opening the exhaust part. The pressure then drops suddenly while the volume remains constant (process 2-3). The resulting cycle is known as the *modified Rankine cycle.*

Due to incomplete expansion, there is a small loss of work represented by the shaded area 2-2′-3. The advantage lies in the fact that there is considerable reduction in stroke length and hence cylinder size.

$$\text{Work done} = \text{area } 5 - 1 - 2 - 6 + \text{area } 2 - 3 - 4 - 6$$
$$= (h_1 - h_2) + (p_2 - p_b)\, V_2$$

$$\text{Heat supplied} = h_1 - h_4$$

∴ Efficiency of modified Rankine cycle

$$= \frac{(h_1 - h_2) + (p_2 - p_b)v_2}{h_1 - h_{f_4}}$$

EXAMPLE 16.3

Steam at 9 bar and 0.9 dry is supplied is an engine where it expands adiabatically to release pressure of 1.4 bar. Thereafter the pressure falls at constant volume to exhaust pressure of 0.3 bar. Make calculations for :

(a) *steam consumption in kg/kWh*
(b) *mean effective pressure*
(c) *thermal efficiency of the cycle*
(d) *heat to be removed in the condenser per unit mass of exhaust steam*
(e) *theoretical loss of work due to incomplete expansion.*

Solution : The cycle is represented by 1-2-3-4-5 on the *p-v* and *T-s* plots of Fig. 16.9. The area 2-2′-3 in both the diagrams represents the loss of work due to incomplete expansion.

Steam is supplied to the engine at state point 1 at pressure $p_1 = 9$ bar and 0.9 dry

$$h_1 = h_{f1} + x_1\, h_{fg1}$$
$$= 743 + 0.9 \times 2031$$
$$= 2570.9 \text{ kJ/kg}$$

Considering the isentropic expansion process 1-2 :

$$s_1 = s_2$$
$$s_{f1} + x_1\, s_{fg1} = s_{f2} + x_2\, s_{fg2}$$
$$2.094 + 0.9 \times 4.529 = 1.411 + x_2 \times 5.835$$
$$6.1701 = 1.411 + 5.835\, x_2$$
$$\therefore \quad x_2 = \frac{6.1701 - 1.411}{5.835} = 0.816$$

Then : $h_2 = h_{f2} + x_2\, h_{fg2}$
$= 458 + 0.816 \times 2232$
$= 2279.3$ kJ/kg

$$\text{Work done} = (h_1 - h_2) + (p_2 - p_3)\, x_2 v_{s2}$$

Now : $h_1 - h_2 = 2570.9 - 2279.3$
$= 291.6$ kJ/kg

$(p_2 - p_3)\, x_2\, v_{s2}$
$= (1.4 - 0.3) \times 10^5 \times 0.816 \times 1.236$
$= 110940$ J/kg $= 110.94$ kJ/kg

∴ Work done $= 291.6 + 110.94$
$= 402.54$ kJ/kg

(*a*) Specific steam consumption (steam consumption in kg/kWh)

$$= \frac{1 \times 3600}{402.54} = 8.943 \text{ kg/kWh}$$

(*b*) Mean effective pressure

$$p_m = \frac{\text{work done per kg of steam}}{\text{specific volume at state point 2}}$$

$$= \frac{402.54 \times 10^3}{0.816 \times 1.236}$$
$$= 3.99 \times 10^6 \text{ N/m}^2 = 3.99 \text{ bar}$$

(*c*) The enthalpy of steam at point 4 equals the liquid enthalpy corresponding to back pressure of 0.3 bar.

$$\text{Heat supplied} = h_1 - h_{f4}$$

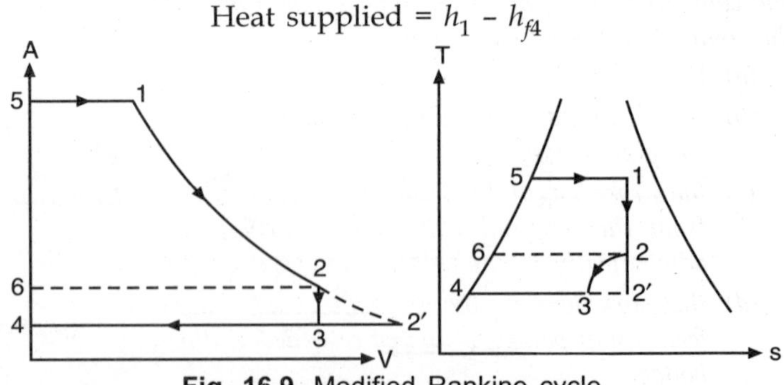

Fig. 16.9. Modified Rankine cycle

$= 2570.9 - 289$
$= 2281.9$ kJ/kg

Thermal efficiency

$$= \frac{\text{work done}}{\text{heat supplied}} = \frac{402.54}{2281.9}$$
$= 0.1764$ or 17.64%

(*d*) For a cyclic process $\oint \delta Q = \oint \delta W$

∴ Heat rejected in condenser
= heat supplied − work done
$= 2281.9 - 402.54 = 1879.36$ kJ/kg

(*e*) If the expansion is allowed to proceed to completion, then the state of steam after expansion would correspond to point 2′.

Considering the isentropic expansion process 1-2′ : $s_1 = s'_2$

$$6.1701 = 0.944 + x'_2 \times 6.822$$

$$\therefore \quad x'_2 = \frac{6.1701 - 0.944}{6.822} = 0.766$$

and $h'_2 = 289 + 0.766 \times 2336$
$= 2079.78$ kJ/kg

Work done with complete expansion
$= h_1 - h'_2$
$= 2570.9 - 2079.78 = 491.12$ kJ/kg

∴ Loss due to incomplete (restricted) expansion
$= 491.2 - 402.54 = 88.58$ kJ/kg

EXAMPLE 16.4

A steam engine of 30 cm diameter and 50 cm stroke length is supplied steam at 10 bar and 300°C. The steam expands adiabatically to 0.7 bar and then release occurs at constant volume to a condenser at 0.3 bar. Represent the cycle on p-v and T-s plots and compare it with the corresponding diagram for the complete Rankine cycle. Determine :

(*a*) *the modified Rankine cycle efficiency*
(*b*) *the Rankine efficiency corresponding to complete expansion*
(*c*) *the stroke length if the same amount of steam from the original condition is expanded adiabatically to condenser pressure*
(*d*) *the work done by the extraction and boiler feed pump per kg of water returned to the boiler.*

Solution : The cycle is represented by 1-2-3-4-5 on the *p-v* and *T-s* plots of Fig. 16.10. The area 2-2′-3 in both the diagrams represents the loss of work due to incomplete expansion.

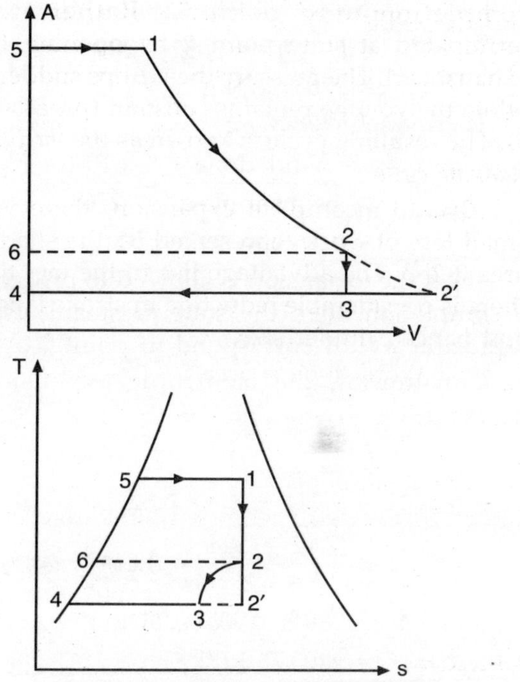

Fig. 16.10. Modified Rankine cycle

Steam is supplied to the engine at state point 1 at pressure $p_1 = 10$ bar and temperature $T_1 = 300$°C (the steam is superheated).

From steam tables :
$h_1 = 3052$ kJ/kg
and $s_1 = 7.12$ kJ/kg K

Considering the isentropic expansion 1-2 :

$$s_1 = s_2$$
$$7.12 = s_{f2} + x_2\, s_{fg2} = 1.192 + x_2 \times 6.287$$
$$\therefore \quad x_2 = \frac{7.12 - 1.192}{6.287} = 0.943$$

and $h_2 = h_{f2} + x_2\, h_{fg2}$
$= 377 + 0.943 \times 2284$
$= 2530.8$ kJ/kg

Work done $= (h_1 - h_2) + (p_2 - p_3)\, v_2\, x_2\, v_{s2}$

Now : $h_1 - h_2 = 3052 - 2530.8$
$= 521.2$ kJ/kg

$(p_2 - p_3)\, x_2\, v_{s2}$
$= (0.7 - 0.3) \times 10^5 \times 0.943 \times 2.365$
$= 89200$ J/kg $= 89.2$ kJ/kg

∴ Work done = 521.2 + 89.2 = 610.4 kJ/kg

(*a*) The enthalpy of steam at point 4 equals the liquid enthalpy corresponding to back pressure of 0.3 bar.

Heat supplied

$$= h_1 - h_{f_4} = 3052 - 289$$
$$= 2763 \text{ kJ/kg}$$

Thermal efficiency

$$= \frac{\text{work done}}{\text{heat supplied}} = \frac{610.4}{2763}$$
$$= 0.2209 \text{ or } 22.09\%$$

(*b*) If the expansion is allowed to proceed to completion, then the state of steam after expansion would correspond to point 2′.

Considering the isentropic expansion process 1-2′ :

$$s_1 = s_2'$$
$$7.12 = 0.944 + x_2' \times 6.822$$
$$\therefore \quad x_2' = \frac{7.12 - 0.944}{6.822} = 0.905$$
$$\text{and} \quad h_2' = 289 + 0.905 \times 2336$$
$$= 2403.79 \text{ kJ/kg}$$

Rankine efficiency corresponding to complete expansion

$$\eta_{th} = \frac{h_1 - h_2'}{h_1 - h_{f_4}} = \frac{3052 - 2403.79}{3052 - 289}$$
$$= 0.2346 \text{ or } 23.46\%$$

(*c*) Volume of cylinder

$$= \frac{\pi}{4}(0.3)^2 \times 0.5 = 0.03532 \text{ m}^3$$

Mass of steam at state point 2

$$= \frac{0.03532}{0.943 \times 2.365}$$
$$= 0.0158 \text{ kg/stroke}$$

Volume at point 2′

$$= m(x_2' \times v_{s2}')$$
$$= 0.0158\,(0.905 \times 5.23)$$
$$= 0.07478 \text{ m}^3$$

Let l be the new stroke length. Then

$$\frac{\pi}{4}(0.3)^2\, l = 0.07478$$

Solution gives : l = 1.0585 m = 105.85 cm

(*d*) Work is done by extraction and boiler feed pump in raising the pressure from 0.3 bar to 10 bar

1 kg water = 0.001 m^3

∴ Work done per kg of water

$$= (10 - 0.3) \times 10^5 \times 0.001$$
$$= 970 \text{ J} \quad \text{or} \quad 0.97 \text{ kJ}$$

16.4. THERMAL REFINEMENT OF RANKINE CYCLE

It may be recalled that neglecting pump work, the efficiency of a Rankine cycle (Fig. 16.3) is approximately equal to

$$\eta_{\text{Rankine}} = \frac{h_1 - h_2}{h_1 - h_3} = \frac{(\Delta h)_s}{h_1 - h_3}$$

This expression indicates that Rankine efficiency depends on the values of h_1 , Δh_s and h_3.

- The enthalpy of steam h_1 is determined by the pressure and temperature of steam at inlet to the steam turbine.
- The enthalpy of feed water h_3 is determined by the condenser pressure.
- The isentropic heat drop $(\Delta h)_s = (h_1 - h_2)$ is determined by the pressure and temperature at entrance to the stream turbine and the pressure at the end of expansion in the turbine, *i.e.*, the condenser pressure.

(*a*) Effect of raising the supply temperature

The supply pressure and condenser pressure are assumed to remain constant. The vapour gets heated above the saturation temperature giving superheated steam. Superheating increases the average temperature at which heat is transferred to steam and accordingly the cycle efficiency improves.

At supply temperature t_1 :

Enthalpy drop = $(h_1 - h_2)$

At supply temperature t_1' :

Enthalpy drop = $(h_1' - h_2')$

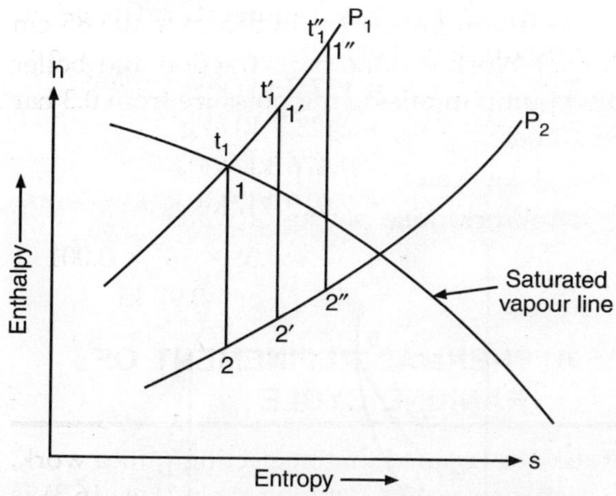

Fig. 16.11. Effect of steam supply temperature on cycle efficiency

Since the constant pressure lines on Mollier diagram diverge towards the right

$$(h_1' - h_2') > (h_1 - h_2)$$

With increase in enthalpy drop more work is developed and so the cycle efficiency should improve.

Metallurgical considerations and the high cost of temperature resisting materials which have to be used for the making of the boiler and turbine components limit the highest operating temperature of the cycle to about 575°C.

(*b*) Effect of raising the steam supply pressure

The maximum inlet temperature and exhaust (condenser) pressure are taken to remain constant.

With reference to Fig. 16.12 the maximum temperature t_1 as well as the condenser pressure P_2 are held constant. The steam supply pressures are P_1', P_1'' and P_1'''. With increase in steam supply pressure ($P_1'' > P_1'$), the isentropic heat drop has increased from $(h_1' - h_2')$ to $(h_1'' - h_2'')$. With increase in heat drop, more work is available at the turbine shaft and consequently the Rankine efficiency increases.

However, with increase in steam supply pressure, there is greater amount of moisture in the steam at the end of expansion in the turbine. When the moisture content in the low pressure stages in the turbine increases more than 10%, there is serious wear of the turbine blades and the turbine efficiency falls down.

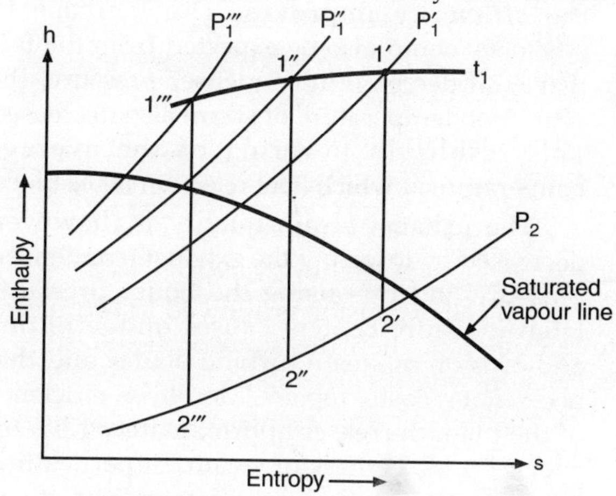

Fig. 16.12. Effect of steam supply pressure on cycle efficiency

(*c*) Effect of decrease in exhaust pressure

Fig. 16.13 shows how the Rankine cycle would operate with drop in the exhaust pressure (or temperature) at which heat is rejected. The lowest possible temperature of condenser depends upon the temperature of the naturally available heat sinks, *i.e.*, air and large mass of water. The lowest temperature is normally taken as 25-30°C with corresponding pressure being 0.032-0.042 bar.

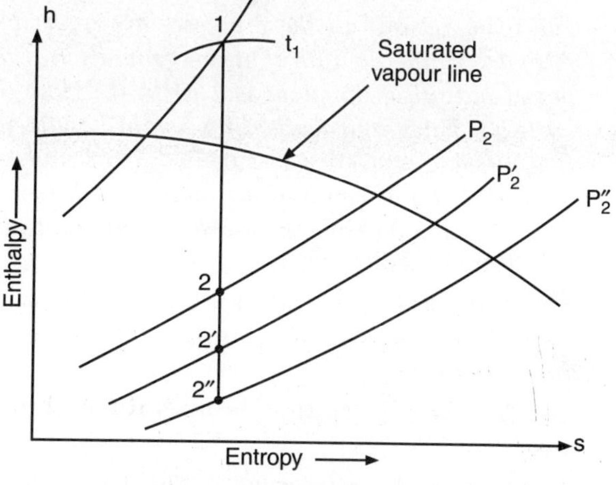

Fig. 16.13. Effect of steam exhaust pressure on cycle efficiency

From Fig. 16.13 it is clearly evident that for fixed initial state, the enthalpy drop increases with drop of exhaust pressure. Consequently, more work is developed and so the efficiency improves. The increase in efficiency could also be expected from the fact that with decrease in condenser pressure, the saturation temperature of steam also decreases. This results in lowering of the average temperature at which heat rejection takes place.

The exhaust steam quality is, however, decreased by lowering the exhaust (condenser) pressure and by raising the boiler pressure. High moisture content causes undue pitting and erosion of steam turbine blades and that necessitates costly repairs. Moreover, efficiency of the plant decreases approximately 1.5% for every 1% of wetness of steam. Superheating, however, keeps the condition of exit steam within safe limits and also provides saving in cost of fuel. This is due to more heat content of the superheated steam and consequently less steam consumption for a given output. With less steam consumption, there is reduction in work to be done in the feed and condenser pumps.

EXAMPLE 16.5

A power generating plant uses steam as a working fluid and operates on Rankine cycle between a source temperature of 311.1°C (boiler pressure 100 bar) and a sink temperature of 32.9° (condenser pressure 0.05 bar). Determine the cycle efficiency and work ratio if all the processes are reversible. Also determine the rate of steam generation if the power output of the plant is 1 MW. (b) How the cycle efficiency and work ratio would be affected if isentropic efficiency of turbine is 0.8 and the saturated liquid coming out of the condenser is compressed to the boiler pressure with an isentropic efficiency of 0.9 ?

Solution : Refer Fig 16.14 for the *T-s* plot when both the expansion and compression processes are reversible.

At 311.1°C source temperature (boiler pressure 100 bar)

$$h_1 = h_{g1} = 2727.7 \text{ kJ/kg}$$

$$s_1 = s_{g1} = 5.62 \text{ kJ/kg}$$

At 32.9°c sink temperature (condenser pressure 0.05 bar)

$$h_f = 137.8 \text{ kJ/kg}$$
$$h_{fg} = 2423.7 \text{ kJ/kg}$$
$$s_f = 0.476 \text{ kJ/kg K}$$
$$s_{fg} = 7.919 \text{ kJ/kg K}$$

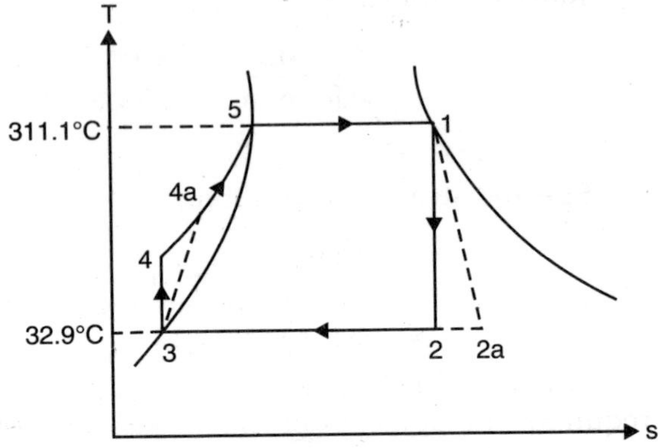

Fig. 16.14.

Considering adiabatic expansion from state 1 to state 2

$$s_1 = s_2 ;$$
$$s_1 = (s_f + x_2 s_{fg})_{0.05 \text{ bar}}$$
$$5.62 = 0.476 + x_2 \times 7.919$$
$$x_2 \approx \frac{5.62 - 0.476}{7.919} = 0.65$$
$$h_2 = (h_f + x_2 h_{fg})_{0.05 \text{ bar}}$$
$$= 137.8 + 0.65 \times 2423.7$$
$$= 1713.2 \text{ kJ/kg}$$
$$h_3 = (h_f)_{0.05 \text{ bar}} = 137.8 \text{ kJ/kg}$$

$$\text{Pump work} = v_f (p_1 - p_2)$$
$$= 0.001005 (100 - 0.05) \times 10^5$$
$$= 10045 \text{ J/kg} = 10.045 \text{ kJ/kg}$$
$$h_4 = h_3 + \text{pump work}$$
$$= 137.8 + 10.045$$
$$= 147.845 \text{ kJ/kg}$$

$$\text{Now, turbine work} = h_1 - h_2$$
$$= 2727.7 - 1713.2$$
$$= 1014.15 \text{ kJ/kg}$$

$$\text{Net work} = \text{turbine work} - \text{pump work}$$
$$= 1014.15 - 10.045$$
$$= 1004.1 \text{ kJ/kg}$$

$$\text{Heat input} = h_1 - h_4$$
$$= 2727.7 - 147.845$$

$= 2579.86$ kJ/kg

$$\text{Work ratio} = \frac{\text{net work}}{\text{turbine work}}$$

$$= \frac{1004.1}{1014.15} = 0.99$$

Rankine efficiency

$$= \frac{\text{net work}}{\text{heat input}} = \frac{1004.1}{2579.86}$$

$= 0.389$ or 38.9%

Specific steam consumption

$$= \frac{3600}{1004.1} = 3.585 \text{ kg/kWh}$$

$= 3585$ kg/MWh

i.e., for 1 MW of net power output, the required rate of steam production is 3585 kg per hour

(*b*) From the isentropic efficiency relations

$$\eta_t = \frac{h_1 - h_{2a}}{h_1 - h_2} \text{ and } \eta_p = \frac{h_4 - h_3}{h_{4a} - h_3}$$

Actual turbine work, $(h_1 - h_{2a})$

$= \eta_t \times (h_1 - h_2)$

$= 0.8 \times 1014.15 = 811.32$ kJ/kg

Actual pump work,

$$(h_{4a} - h_3) = \frac{h_4 - h_3}{\eta_p}$$

$$= \frac{10.045}{0.9} = 11.16 \text{ kJ/kg}$$

Then : $h_{4a} = h_3$ + acutal pump work

$= 137.8 + 11.16 = 148.96$ kJ/kg

Actual net work $= 811.32 - 11.16$

$= 800.16$ kJ/kg

Actual heat input $= h_1 - h_{4a}$

$= 2727.7 - 148.96$

$= 2578.74$ kJ/kg

∴ For the actual Rankine cycle

$$\eta = \frac{800.16}{2578.74} = 0.31 \text{ or } 31\%$$

$$r_w = \frac{800.16}{811.32} = 0.986$$

$$ssc = \frac{3600}{800.16} = 4.499 \text{ kg/kWh}$$

$= 4499$ kg/MWh

Comments : The effect of turbine and compressor efficiencies is to decrease the work ratio and thermal efficiency, and to increase the specific steam consumption.

EXAMPLE 16.6

A steam turbine working on Rankine cycle is supplied with dry saturated steam at 25 bar and the exhaust takes place at 0.2 bar. For a steam flow rate of 10 kg/s, determine : (a) quality steam at end of expansion, (b) turbine shaft work, (c) power required to drive the pump, (d) work ratio, (e) Rankine efficiency, and (f) heat flow in the condenser.

Solution : Refer Fig. 16.15 for the *T-s* plot

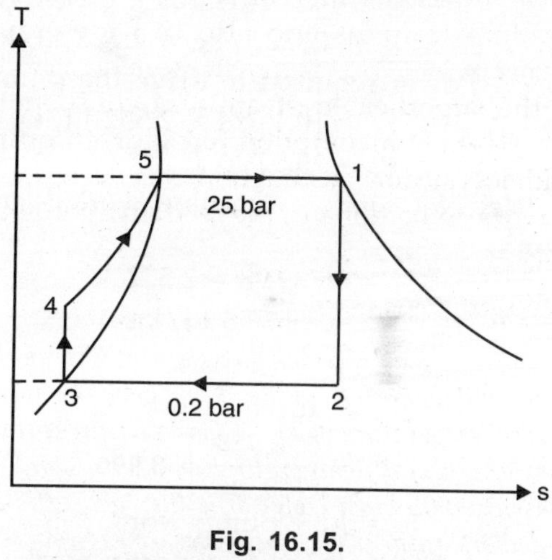

Fig. 16.15.

From steam tables : At 25 bar;

$h_1 = h_{g1} = 2800.9$ kJ/kg

$s_1 = s_{g1} = 6.253$ kJ/kg K

At 0.2 bar

$h_f = 251.5$ kJ/kg

$h_{fg} = 2358.4$ kJ/kg

$s_f = 0.8321$ kJ/kg K

$s_{fg} = 7.077$ kJ/kg K

$v_f = 0.001017$ m³/kg

(*a*) Considering reversible adiabatic expansion from state 1 to state 2.

$s_1 = s_2$; $s_1 = (s_f + x_2 s_{fg})_{0.2 \text{ bar}}$

$$6.253 = 0.8321 + x_2 \times 7.077$$

$$x_2 = \frac{6.253 - 0.8321}{7.077} = 0.776$$

The quality of steam at the end of expansion is 77.6%

(b) $h_2 = (h_f + x_2\, h_{fg})_{0.2 \text{ bar}}$
$= 251.5 + 0.776 \times 2358.4$
$= 2061.62$ kJ/kg

Turbine work
$= (h_1 - h_2) = 2800.9 - 2061.62$
$= 739.28$ kJ/kg

∴ Turbine power
$= 10$ kg/s × 739.28 kJ/kg
≈ **7393 kW**

(c) Pump work
$= v_f\,(p_1 - p_2)$
$= 0.001017\,(25 - 0.2) \times 10^5$
$= 2522$ J/kg = 2.522 kJ/kg

∴ Power required to drive the pump
$= 10$ kg/s × 2.552 kJ/kg
= **25.52 kW**

(d) Net work = turbine work – pump work
$= 739.28 - 2.522$
$= 736.758$ kJ/kg

$$\text{Work ratio} = \frac{\text{net work}}{\text{turbine work}} = \frac{736.758}{739.28} = \mathbf{0.996}$$

(e) $h_4 = h_3 + \text{pump work}$
$= (h_f)_{0.2 \text{ bar}} + \text{pump work}$
$= 251.5 + 2.522 = 254.02$ kJ/kg

Heat input $= h_1 - h_4 = 2800.9 - 254.02$
$= 2546.88$ kJ/kg

Rankine efficiency

$$= \frac{\text{net work}}{\text{heat input}} = \frac{736.758}{2546.88}$$

$= 0.289$ or 28.9%

(f) Condenser heat flow
$= (h_2 - h_3)$
$= (2061.62 - 251.5)$
$= 1810.12$ kJ/kg

With steam flow rate of 10 kg/s,

Condenser heat flow
$= 10 \times 1810.12$
$= 18100.12$ kJ/s ≈ **18100 kW**

EXAMPLE 16.7

In a thermal power plant which uses the Rankine cycle, superheated steam is produced at 1.5 MPa and 300°C and fed to the turbine where it expands to a condenser pressure of 80 kPa. The saturated liquid coming out of the condenser is pumped back to the boiler by a feed pump. Assuming ideal processes, determine :

(a) the condition of steam after isentropic expansion, (b) Rankine cycle efficiency, (c) the mean effective pressure, (d) the ideal steam consumption per kW hour, and (e) the actual steam consumption per kW hour.

Take relative efficiency = 60% and neglect pump work.

Solution : Refer Fig. 16.16 for the *T-s* plot.

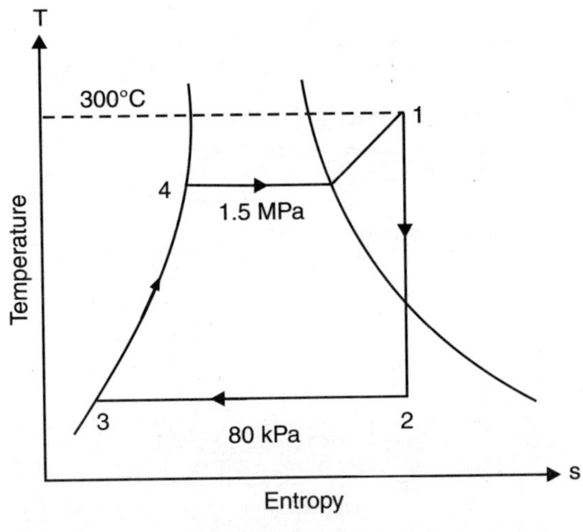

Fig. 16.16.

From steam tables :

At 1.5 MPa (15 bar) and 300°C
$h_1 = 3037.6$ kJ/kg and
$s_1 = 6.918$ kJ/kg K

At 80 kPa (0.8 bar)
$h_f = 391.7$ kJ/kg
$h_{fg} = 2274$ kJ/kg
$s_f = 1.233$ kJ/kg K
$s_{fg} = 6.202$ kJ/kg K
$v_g = 2.087$ m^3/kg

(*a*) Considering adiabatic expansion from state 1 to state 2

$$s_1 = s_2\,;\quad s_1 = (s_f + x_2\, s_{fg})_{0.8\text{ bar}}$$
$$6.918 = 1.233 + x_2 \times 6.202\,;\quad x_2 = 0.916$$

(*b*) $h_2 = (h_f + x_2\, h_{fg})_{0.8\text{ bar}}$
$= 391.7 + 0.916 \times 2274$
$= 2474.7$ kJ/kg

$h_3 = (h_f)_{0.8\text{ bar}} = 391.7$ kJ/kg

Rankine efficiency

$$= \frac{h_1 - h_2}{h_1 - h_3} \text{(neglecting pump work)}$$
$$= \frac{3037.6 - 2474.7}{3037.6 - 391.7}$$
$$= 0.2127 \text{ or } 21.27\%$$

(*c*) $V_2 = x_2 \times (v_g)_{0.8\text{ bar}}$
$= 0.916 \times 2.087 = 1.917 \text{ m}^3/\text{kg}$

Mean effective pressure

$$= \frac{\text{work done}}{\text{stroke volume}} = \frac{h_1 - h_2}{V_2}$$
$$= \frac{3037.6 - 2474.7}{1.917}$$
$$= \mathbf{293.63\ kN/m^2}$$

(*d*) Ideal specific steam consumption

$$= \frac{3600}{\text{work done}} = \frac{3600}{h_1 - h_2}$$
$$= \frac{3600}{3037.6 - 2474.7}$$
$$= 6.39 \text{ kg/kWh}$$

(*e*) Relative efficiency

$$= \frac{\text{Indicated thermal efficiency}}{\text{Rankine cycle efficiency}}$$

∴ Indicated thermal efficiency
$= 0.6 \times 0.2127 = 0.1276$

Indicated work
= indicated thermal efficiency × heat supplied
= indicated thermal efficiency × $(h_1 - h_3)$
$= 0.1276 \times (3037.6 - 391.7)$
$= 337.62$ kJ/kg

∴ Actual specific steam consumption

$$= \frac{3600}{337.62} = \mathbf{10.66\ kg/kWh}$$

EXAMPLE 16.8

In a thermal power plant operating on an ideal Rankine cycle, steam at 15 bar and 250°C enters a turbine which generates 40 kW indicated power. If the steam consumption is 300 kg/hr and condenser is maintained at 0.15 bar, determine the final condition of steam, Rankine efficiency and relative efficiency. Neglect pump work. Also determine the fuel to be supplied per hour if its calorific value is 41850 kJ/kg.

Solution : Refer Fig. 16.17 for the *T-s* plot.

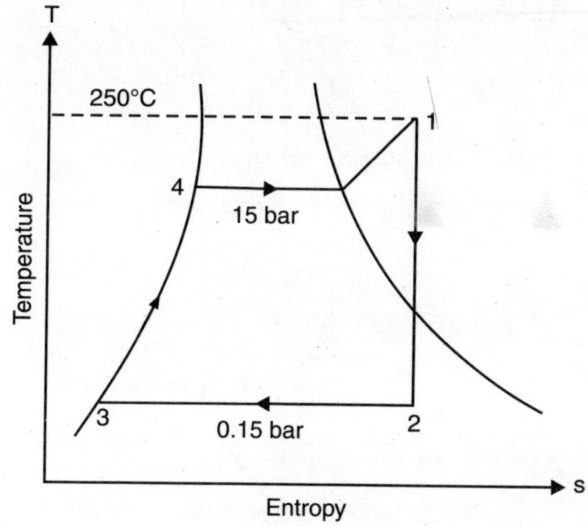

Fig. 16.17.

From steam tables :

At 15 bar, (steam is superheated)

$h_1 = 2923.5$ kJ/kg
$s_1 = 6.709$ kJ/kg K

At 0.15 bar

$h_f = 226$ kJ/kg
$h_{fg} = 2372.2$ kJ/kg
$s_f = 0.755$ kJ/kg K
$s_{fg} = 7.254$ kJ/kg K

(*a*) Considering adiabatic expansion from state 1 to state 2

$$s_1 = s_2$$
$$s_1 = (s_f + x_2\, s_{fg})_{0.15\text{ bar}}$$
$$6.709 = 0.755 + x_2 \times 7.254$$
$$x_2 = \frac{6.709 - 0.755}{7.254} = \mathbf{0.821}$$

The dryness fraction of steam at exit from the turbine is 0.821.

(*b*) $h_2 = (h_f + x_2 h_{fg})_{0.15\text{ bar}}$
$= 226 + 0.821 \times 2372.2$
$= 2174.4$ kJ/kg

$h_3 = (h_f)_{0.15\text{ bar}} = 226$ kJ/kg

Rankine efficiency

$$= \frac{h_1 - h_2}{h_1 - h_3} = \frac{2923.5 - 2174.4}{2923.5 - 226}$$

$= \mathbf{0.278}$ or **27.8%**

(*c*) Indicated power

$= 40$ kW $= 40$ kJ/s

Heat input $= (h_1 - h_3)$
$= (2923.5 - 226)$ kJ/kg

$$= \frac{300}{3600}(2923.5 - 226)$$

$= 224.79$ kJ/s

Indicated thermal efficiency

$$= \frac{40}{224.79} = 0.178$$

∴ Relative efficiency

$$= \frac{0.178}{0.278} = \mathbf{0.64} \text{ or } \mathbf{64\%}$$

EXAMPLE 16.9

In a thermal power plant operating on an ideal Rankine cycle, superheated steam produced at 5 MPa and 500°C is fed to a turbine where it expands to the condenser pressure of 10 kPa. If the net power output of the plant is to be 20 MW, determine:

(a) heat added in the boiler per kg of water, (b) thermal efficiency of the cycle, (c) mass flow rate of steam in kg/s, (d) mass flow rate of cooling water in the condenser if the cooling water enters the condenser at 25°C and leaves at 35°C.

Solution : Refer Fig. 16.18 for the *T-s* plot.

From steam tables :

At 5 MPa (50 bar) and 500°C the steam is superheated and

$h_1 = 3433.8$ kJ/kg
$s_1 = 6.976$ kJ/kg K

At 10 kPa (0.1 bar)

$h_f = 191.8$ kJ/kg
$h_{fg} = 2392.8$ kJ/kg
$s_f = 0.649$ kJ/kg K
$s_{fg} = 7.501$ kJ/kg K
$v_f = 0.00101$ m^3/kg

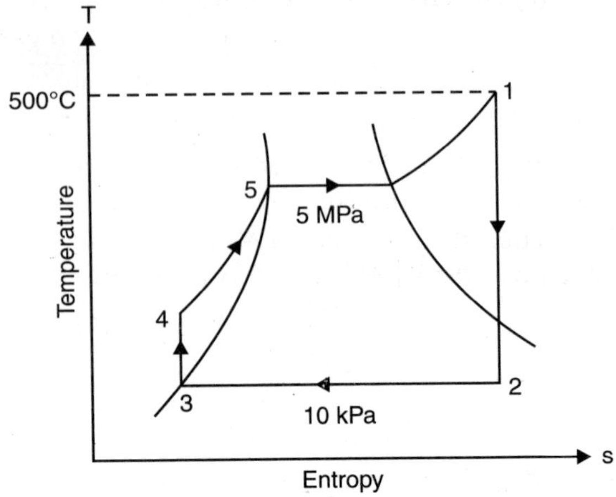

Fig. 16.18.

Considering adiabatic expansion from state 1 to state 2

$s_1 = s_2$; $s_1 = (s_f + x_2 s_{fg})$ at 10 kPa
$6.976 = 0.649 + x_2 \times 7.501$

$$x_2 = \frac{6.976 - 0.649}{7.501} = 0.843$$

$h_2 = (h_f + x_2 h_{fg})$ at 10 kPa
$= 191.8 + 0.843 \times 2392.8$
$= 2208.93$ kJ/kg

$h_3 = (h_f)$ at 10 kPa $= 191.8$ kJ/kg

Pump work $= (h_4 - h_3) = v_f (p_1 - p_2)$
$= 0.00101 (50 - 0.1) \times 10^5$
$= 5040$ J/kg $= 5.04$ kJ/kg

$h_4 = h_3 +$ pump work
$= 191.8 + 5.04 = 196.84$ kJ/kg

(*a*) Heat added in the boiler,

$h_1 - h_4 = 3433.8 - 196.84$
$= \mathbf{3236.96}$ **kJ/kg**

(*b*) Turbine work $= (h_1 - h_2)$
$= 3433.8 - 2208.93$
$= 1224.87$ kJ/kg

Net work = turbine work − pump work
$= 1224.87 - 5.04 = 1219.83$ kJ/kg

Rankine efficiency

$$= \frac{\text{net work}}{\text{heat input}} = \frac{1219.83}{3236.96}$$

$= \mathbf{0.3768}$ or **37.68%**

(*c*) Specific steam consumption,

$$ssc = \frac{3600}{\text{net work out put}} = \frac{3600}{1219.83}$$

$$= 2.951 \text{ kg/kWh}$$

i.e., for 1 kW of net power output, the required rate of steam production is 2.951 kg/hr

Therefore, for 20 MW (20×10^3 kW) of net power output, mass flow rate of steam

$$= 2.951 \times (20 \times 10^3) \text{ kg/hr}$$

$$m_s = \frac{2.951 \times (20 \times 10^3)}{3600}$$

$$= 16.39 \text{ kg/s}$$

(*d*) Heat rejected by steam in the condenser

$$= m_s\,(h_2 - h_3)$$

If m_w kg/s is the mass flow rate of cooling water in the condenser, then from energy conservation

$$m_w \times c_p\,(t_0 - t_i) = m_s\,(h_2 - h_3)$$

$$m_w = \frac{m_s\,(h_2 - h_3)}{c_p\,(t_0 - t_i)}$$

$$= \frac{16.39\,(2208.93 - 191.8)}{4.186\,(35 - 25)}$$

$$= \mathbf{789.79 \text{ kg/s}}$$

16.5. REHEAT CYCLE

In a reheat cycle, the expansion of steam is carried out in two or more stages. After partial expansion to an intermediate pressure in the high pressure turbine, the steam is withdrawn and reheated to the original temperature at constant pressure. Steam is then returned to the low pressure turbine and further expanded to the condenser pressure.

Fig. 16.19 shows the layout of a reheat Rankine cycle with two stages of expansion. The corresponding $p - V$ and $T - s$ variation has also been depicted.

With reference to Fig. 16.19 and for unit mass of steam

Turbine work

$$W_t = \text{Work done in HP turbine} + \text{work done in LP turbine}$$

$$= (h_1 - h_2) + (h_3 - h_4)$$

Pump work $W_p = (h_6 - h_5)$

Heat supplied, Q_s

$$= \text{heat supplied in boiler + heat supplied in reheater}$$

$$= (h_1 - h_6) + (h_3 - h_2)$$

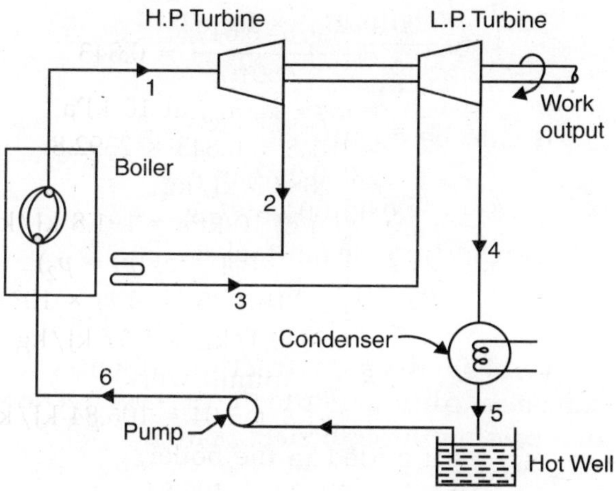

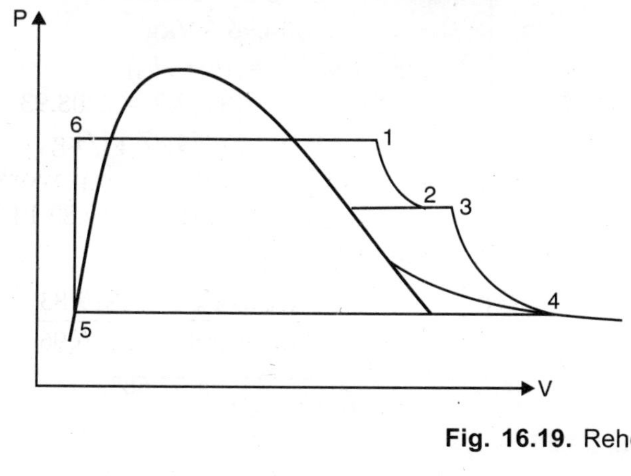

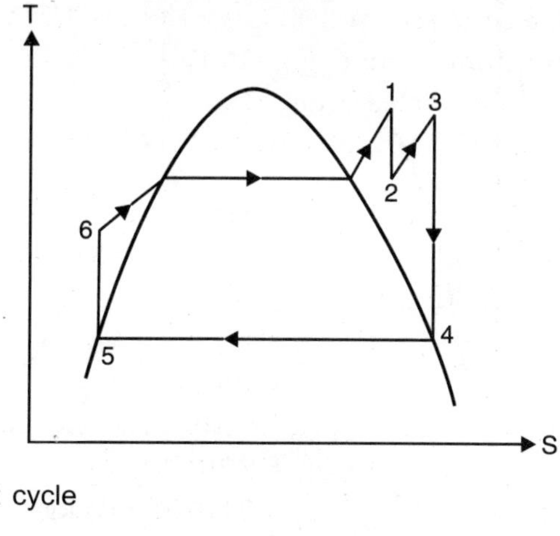

Fig. 16.19. Reheat cycle

∴ Reheat cycle efficiency,

$$\eta = \frac{\text{net work done}}{\text{heat supplied}}$$

$$= \frac{(h_1 - h_2) + (h_3 - h_4) - (h_6 - h_5)}{(h_1 - h_6) + (h_3 - h_2)}$$

The reheat cycle reduces the moisture content at low pressure turbine and accordingly the erosion and corrosion problems in the turbine are eliminated. It also increases output of the turbine and also overall efficiency of the plant. However, reheating requires more maintenance. Increase in thermal efficiency is also not appreciable in comparison to the expenditure incurred in reheating.

EXAMPLE 16.10

A steam power plant operates on a theoretical reheat cycle. Steam at 25 bar pressure and 400°C is supplied to the high pressure turbine. After its expansion to dry state, the steam is reheated at constant pressure to its original temperature. Subsequent expansion occurs in the low pressure turbine to a condenser pressure of 0.04 bar. Considering feed pump work, make calculations to determine :

(a) quality of steam at entry to condenser,

(b) thermal efficiency and

(c) specific steam consumption.

Solution : Refer Fig. 16.20 for the given reheat cycle on $T - s$ and $h - s$ diagrams.

Steam expands from state point 1 (25 bar and 400°C) to state point 2 (dry state) in the high pressure turbine, and the following enthalpy values may be read from the Mollier ($h - s$) diagram.

$$h_1 = 3240 \text{ kJ/kg}$$

and $h_2 = 2725$ kJ/kg

Corresponding to state point 2, the pressure as read from the Mollier diagram is 3 bar. Now, the reheating takes place at 3 bar to the original temperature of 400°C, and that locates the state point 3. Thereafter, the expansion of reheated steam takes place to state point 4 which intersects the condenser pressure line of 0.04 bar. From the Mollier diagram.

$$h_3 = 3275 \text{ kJ/kg}$$

and $h_4 = 2430$ kJ/kg

Now : $h_5 = h_{f4}$, *i.e.*, sensible heat at condenser pressure of 0.04 bar and from steam tables, its value is 121.5 kJ/kg.

pump work (work input to pump)

$$W_p = h_6 - h_5 = V_{f5}\,(p_6 - p_5)$$

$$= \frac{0.001004(25 \times 10^5 - 0.04 \times 10^5)}{10^3}$$

$$= 2.50 \text{ kJ/kg}$$

It is to be noted that

$$v_{f5} = v_f \text{ at } 0.04 \text{ bar}$$

$$= 0.001004 \text{ m}^3/\text{kg}$$

as read from steam tables

$$\therefore \quad h_6 = h_5 + W_p$$

$$= 121.5 + 2.50 = 124 \text{ kJ/kg}$$

(*a*) The dryness fraction at entry to condenser corresponds to the dryness fraction line passing through state point 4.

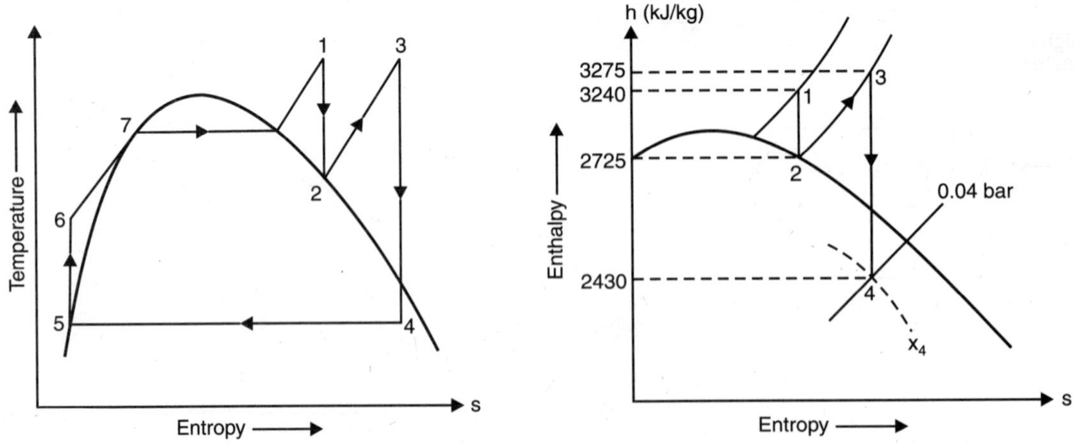

Fig. 16.20.

$x_4 = 0.945$

(value read from Mollier diagram)

(*b*) Turbine work,

$$W_t = \text{HP turbine work} + \text{LP turbine work}$$
$$= (h_1 - h_2) + (h_3 - h_4)$$
$$= (3240 - 2725) + (3275 - 2430)$$
$$= 1360 \text{ kJ/kg}$$

Pump work, $W_p = 2.50$ kJ/kg

Net work, $W_{net} = W_t - W_p = 1360 - 2.50$
$= 1357.5$ kJ/kg

Heat supplied, $Q_s = (h_1 - h_6) + (h_3 - h_2)$
$= (3240 - 124) + (3275 - 2725)$
$= 3666$ kJ/kg

∴ Thermal efficiency of reheat cycle,

$$= \frac{W_{net}}{Q_s} = \frac{1357.5}{3666}$$
$$= \mathbf{0.37} \text{ or } \mathbf{37}\ \%$$

(*c*) Specific steam consumption

$$ssc = \frac{3600}{W_{net}} = \frac{3600}{1357.5}$$
$$= \mathbf{2.652} \text{ kg/kWh}$$

16.6. REGENERATIVE HEATING

The average temperature at which heat is added to the working fluid can be increased by preheating the feed water before it enters the boiler. For this purpose, part of the steam is withdrawn or extracted at some intermediate stage during expansion in the turbine. The rest of the steam expands in the turbine to the condenser pressure.

The steam thus extracted is mixed with feed water coming from the hot well. The system of abstracting steam from any point in the turbine and subsequently using it for heating the feed water is called **Bleeding.**

Feedwater heaters

The steam extracted from the turbine at various stages of expansion is used to heat the feedwater, and the devices where this heating is done are called the regenerators or feedwater heaters. The feedwater heater is essentially a heat exchanger where heat is transferred from the steam to the feedwater either by mixing or without mixing the two fluid streams.

(*a*) Closed feedwater heaters : In the closed feedwater heaters, there is no physical contact between the bled steam and feedwater. The feedwater passes through a bank of tubes surrounded by a casing which contains the drain from bled steam.

Fig. 16.21 shows the simplest arrangement of closed heaters where the drains (bled steam after condensation) are cascaded from heater to heater and finally discharged into the hot well.

Analysis is made by assuming that in the heater

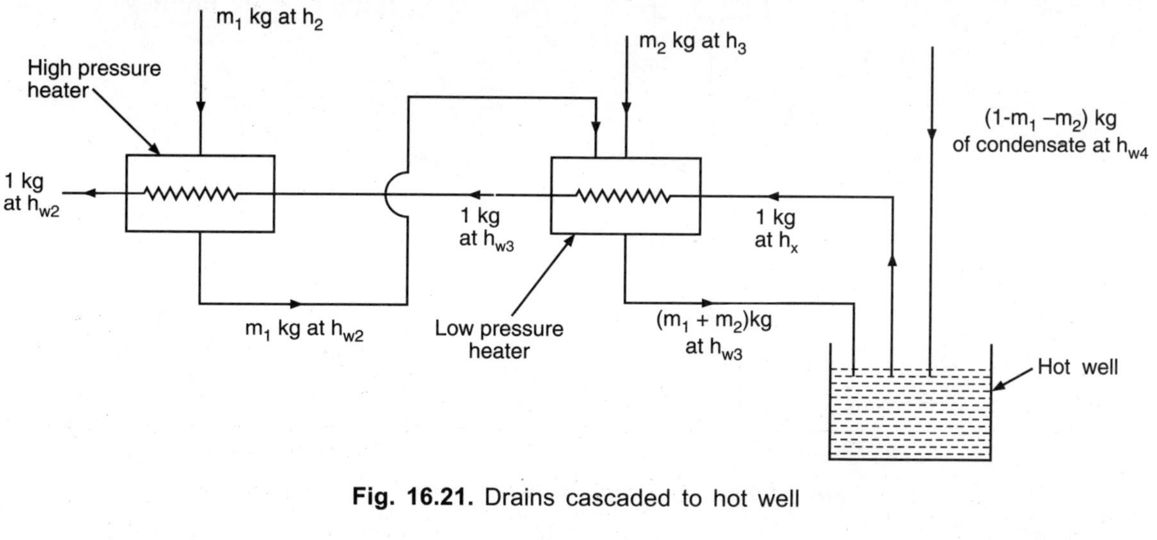

Fig. 16.21. Drains cascaded to hot well

(*i*) the bled steam just condenses by giving up its superheat and latent heat. The drain from the heater will then be at the saturation temperature corresponding to bleeding pressure.

(*ii*) the temperature of feedwater gets raised to the saturation temperature of bled steam.

From energy balance at the hot well, low pressure heater and high pressure heater, we get

$$1 \times h_x = (m_1 + m_2)\, h_{w3} + (1 - m_1 - m_2)\, h_{w4} \quad \text{...(Hot well)}$$

$$m_1 h_{w2} + m_2 h_3 + h_x = (m_1 + m_2)\, h_{w3} + h_{w3} \quad \text{...(LP heater)}$$

$$m_1 h_2 + h_{w3} = m_1 h_{w2} + h_{w2}$$

These identities can be used to determine the mass of steam bled at different sections.

In many applications, the drains are pumped ahead, the condensed steam (drain) enters the feed line immediately after the heater being considered as shown in Fig. 16.22. Energy balance for low pressure heater gives

$$m_2 h_3 + (1 - m_1 - m_2)\, h_{w4} = m_2 h_{w3} + (1 - m_1 - m_2)\, h_{w3}$$

Energy balanced for high pressure heater gives

$$m_1 h_2 + (1 - m_1)\, h_{w3} = (1 - m_1)\, h_{w2} + m_1 h_{w2}$$

These energy balance equations can be used to find out the mass of steam bled at different section of the turbine.

(*b*) Direct contact (open) feedwater heaters : In the open feedwater heaters, there is direct mixing of bled steam and the feedwater. The mixture is extracted and pumped into the next high pressure heater as shown in Fig. 16.23.

Open water heaters are less expensive and have better heat transfer characteristics. However, for such heaters the bled steam and feedwater must be at the same pressure and this necessitates the use of pump for handling the feedwater between each heater.

Fig. 16.24 shows the layout out and $T - s$ plot of a Rankine cycle employing open type feed water heating.

With reference to Fig. 16.24 and for unit mass of steam

Heat supplied in the boiler,

$$Q_s = (h_1 - h_7)$$

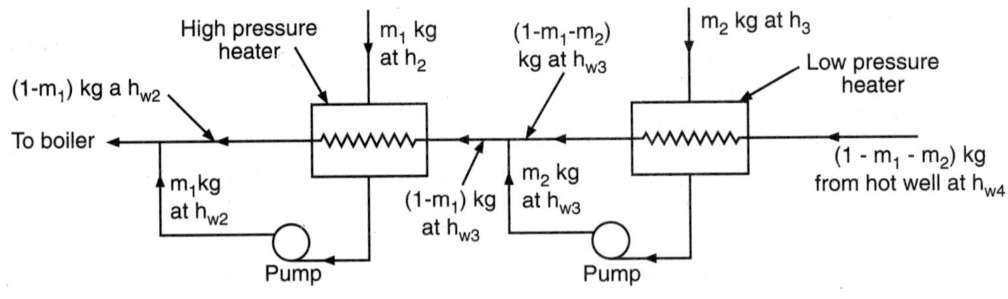

Fig. 16.22. Drains pumped ahead

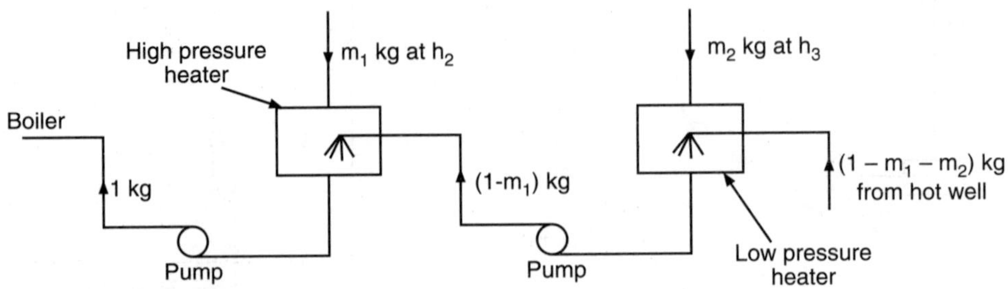

Fig. 16.23. Schematics of direct contact heaters

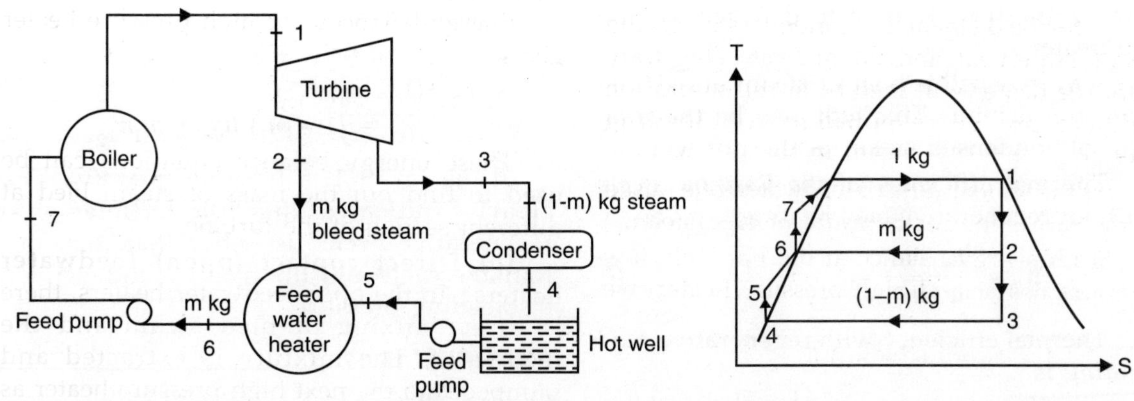

Fig. 16.24. Regenerative Rankine cycle

Work done in the turbine,

$$W_t = 1 \times (h_1 - h_2) + (1 - m)(h_2 - h_3)$$

Feed pump work Wp = work input to pump 1 and pump 2

$$= (1 - m)\, v_{f4}(p_5 - p_4) + 1 \times v_{f6}(p_7 - p_6)$$

Thermal efficiency of the regenerative cycle.

$$\eta = \frac{\text{net work done}}{\text{heat supplied}} = \frac{W_t - W_p}{Q_s}$$

In a regnerative cycle

- There is improvement in cycle economy with relatively much smaller capital expenditure.
- With an infinite number of heaters, the heating process becomes reversible and the efficiency approaches that of a Carnot cycle.
- The supply of feed water to the boiler is at increased temperature. That reduces the temperature range in the boiler and keeps the thermal stresses low.

However, work done per kg of steam decreases and as such large capacity boiler is needed for a given output. The system also becomes complicated, less flexible and involves greater maintenance and capital cost due to installation of feed water heaters.

EXAMPLE 16.11

Show that the thermal efficiency of a regnerative cycle is always greater than that of a simple Rankine cycle regardless of where steam is tapped off.

Solution : Refer Fig. 16.25 for the flow diagram and h–s diagram of a regenerative cycle with one feed heater.

Let h_{f2} = sensible heat of steam corresponding to bleeding pressure. This will

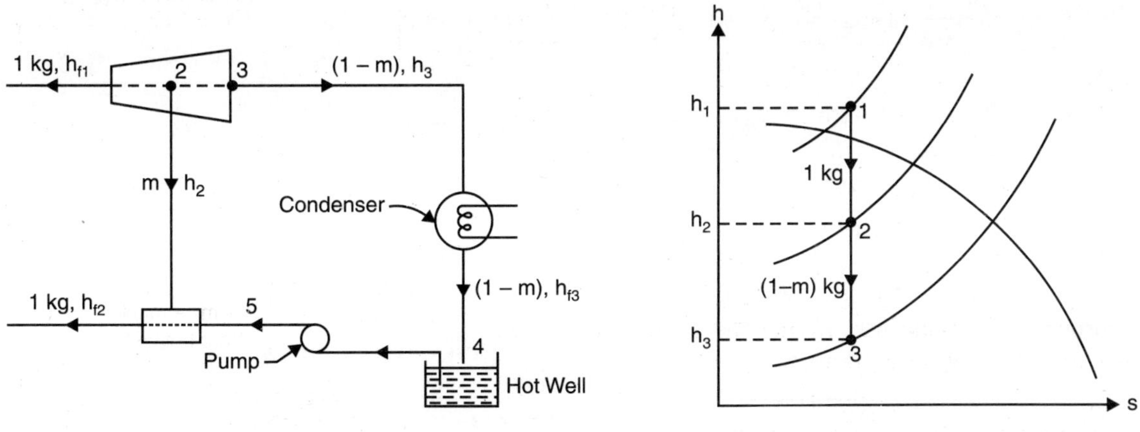

Fig. 16.25.

also be the total heat of feed water leaving the heater.

h_{f3} = sensible heat of steam at exhaust from the turbine. This will also be the total heat of condensed steam in the hot well.

Thermal efficiency of the Rankine cycle without regenerative feed heating is given by

$$\eta_R = \frac{h_1 - h_3}{h_1 - h_{f3}} \quad \text{...}(i)$$

Thermal efficiency with regenerative feed heating is

$$\eta = \frac{(h_1 - h_2) + (1 - m)(h_2 - h_3)}{h_1 - h_{f2}}$$

From energy balance for the heater, we have

$$mh_2 + (1 - m)\, h_{f3} = h_{f2}$$

$$\therefore \eta = \frac{(h_1 - h_2) + (1 - m)(h_2 - h_3)}{h_1 - mh_2 - (1 - m)\, h_{f3}}$$

$$= \frac{h_1 - h_2 + h_2 - h_3 - mh_2 + mh_3}{h_1 - mh_2 - h_{f3} + mh_{f3}}$$

$$= \frac{(h_1 - h_3) - m(h_2 - h_3)}{(h_1 - h_{f3}) - m(h_2 - h_{f3})} \quad \text{...}(ii)$$

From expression (*i*) and (*ii*)

$$\eta - \eta_R = \frac{(h_1 - h_3) - m(h_2 - h_3)}{(h_1 - h_{f3}) - m(h_2 - h_{f3})} - \frac{h_1 - h_3}{h_1 - h_{f3}}$$

$$= \frac{a - b}{c - d} - \frac{a}{c} = \frac{ad - cb}{c(c - d)}$$

Now :

$$a \times d = (h_1 - h_3) \times m(h_2 - h_{f3})$$
$$= m\,(h_1h_2 - h_1\, h_{f3} - h_2h_3 + h_3\, h_{f3})$$
$$c \times b = (h_1 - h_{f3}) \times m(h_2 - h_3)$$
$$= m\,(h_1h_2 - h_1\, h_3 - h_2h_{f3} + h_3\, h_{f3})$$
$$\therefore ad - cb = m\,(- h_1\, h_{f3} - h_2h_3 + h_1h_3 + h_2\, h_{f3})$$
$$= m\,(h_1 - h_2)\,(h_3 - h_{f3})$$

Since h_2 must be less than from h_1 in all cases and h_3 is greater than h_{f3}

$ad - cb$ is positive,

Further, $(h_1 - h_{f3})$ is greater than $m\,(h_2 - h_{f3})$ since $m < 1$

That gives

$c > d$ or $(c - d)$ is positive

$$\therefore \frac{ad - cb}{c(c - d)} \text{ is positive}$$

Thus $(\eta - \eta_R)$ is positive quantity. That clearly indicates that the efficiency of regenerative cycle is better than that of a Rankine cycle and the result is independent of bleeding pressure.

EXAMPLE 16.12

Steam at 20 bar pressure and 350° C temperature is supplied to a turbine where its expansion takes place to a condenser pressure of 0.04 bar. Calculate its thermal efficiency.

It is desired to improve the efficiency by regenerative feed heating by bleeding steam at 2 bar pressure and heating in open feed heater. Calculate the percentage improvement in thermal efficiency. Neglect pump work.

Solution : Refer Fig. 16.26 for the flow diagram and *h–s* diagram of a regnerative cycle with open feed water heater.

The following enthalpy values are noted from the Mollier (*h–s*) diagram.

$$h_1 = 3140 \text{ kJ/kg}$$
$$h_2 = 2640 \text{ kJ/kg}$$
$$h_3 = 2100 \text{ kJ/kg}$$

From steam tables

$$h_{f2} = 504.7 \text{ kJ/kg}$$
$$h_{f3} = 121.5 \text{ kJ/kg}.$$

(*a*) Thermal efficiency of the Rankine cycle without regnerative feed heating is given by

$$\eta_R = \frac{h_1 - h_3}{h_1 - h_{f3}} = \frac{3140 - 2100}{3140 - 121.5}$$

$$= \mathbf{0.3445} \text{ or } \mathbf{34.45}\ \%$$

(*b*) Thermal efficiency with regnerative feed heating is

$$\eta = \frac{(h_1 - h_2) + (1 - m)(h_2 - h_3)}{h_1 - h_{f2}}$$

From energy balance for the heater, we have

$$mh_2 + (1 - m)\, h_{f3} = h_{f2}$$

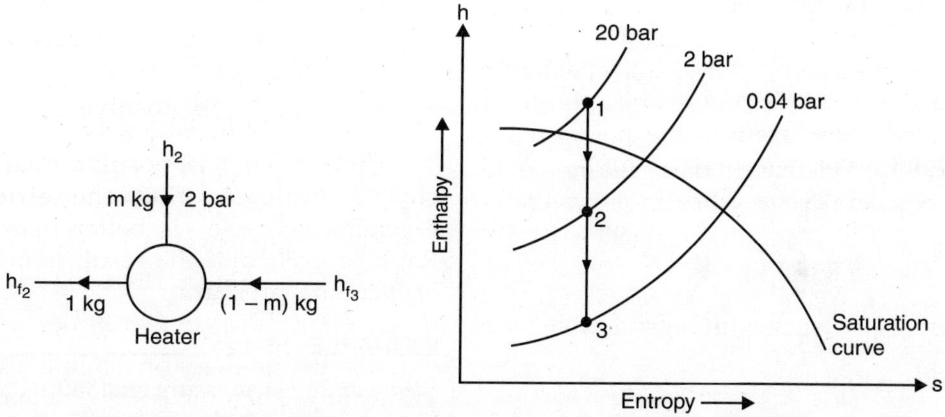

Fig. 16.26.

$$\text{or} \quad m = \frac{h_{f2} - h_{f3}}{h_2 - h_{f3}} = \frac{504.7 - 121.5}{2640 - 121.5}$$

$$= 0.152 \text{ kg}$$

Therefore, the efficiency η with regenerative feed heating is

$$= \frac{(3140 - 2640) + (1 - 0.152) \times (2640 - 2100)}{3140 - 504.7}$$

$$= \frac{500 + 457.92}{2635.3}$$

$$= \mathbf{0.3635 \text{ or } 36.35} \ \%$$

Then percentage improvement in thermal efficiency due to regnerative feed heating is

$$= \frac{0.3635 - 0.3445}{0.3445}$$

$$= \mathbf{0.05515 \text{ or } 5.515} \ \%$$

EXAMPLE 16.13

Steam leaves the boiler dry and saturated at 30 bar and is condensed at 0.035 bar. Calculate the cycle efficiency of a regnerative cycle with one feed heater.

Solution : For maximum efficiency, the optimum temperature at the tapping is

$$t_2 = \frac{t_1 + t_3}{2}$$

where t_1 is the saturation temperature at stop valve (30 bar) = 233.8°C

t_3 is saturation temperature at turbine exit (0.035 bar) = 26.7°C

$$\therefore \quad t_2 = \frac{233.8 + 26.7}{2} = 130.25°\text{C}$$

The pressure at the tapping point can be read from steam tables corresponding to saturation temperature of 130.25°C.

Tapping pressure p_2 = 2.75 bar

The enthalpy values at different points during expansion as read from the Mollier chart are :

$$h_1 = 2805 \text{ kJ/kg}$$
$$h_2 = 2390 \text{ kJ/kg}$$
$$h_3 = 1850 \text{ kJ/kg}$$

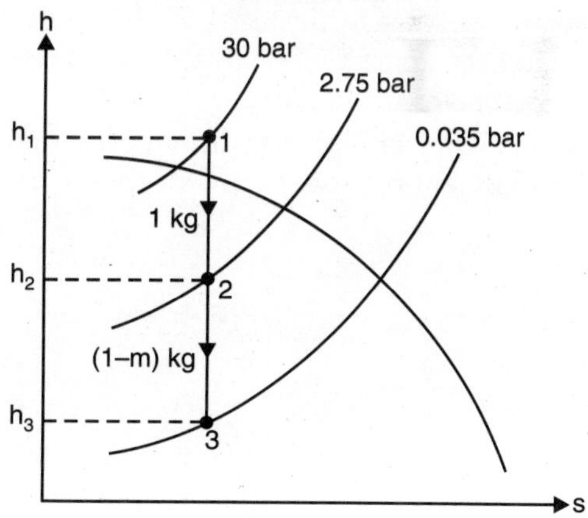

Fig. 16.27.

Further from steam tables, the liquid enthalpies at extraction pressure (2.75 bar) and condenser pressure (0.035 bar) respectively are

$$h_{f2} = 548.8 \text{ kJ/kg}$$
$$\text{and} \quad h_{f3} = 111.9 \text{ kJ/kg}$$

Let m be the mass of steams extracted from feed heating,. Then energy balance for the heater gives

$mh_2 + (1 - m)\ h_{f3} = h_{f2}$

or $\quad m = \dfrac{h_{f2} - h_{f3}}{h_2 - h_{f3}} = \dfrac{548.8 - 111.9}{2390 - 111.9}$

$= 0.192$ kg

Work done $= (h_1 - h_2) + (1 - m)\ (h_2 - h_3)$

$= (2805 - 2390) + (1 - 0.192) \times (2390 - 1850)$

$= 851.3$ kJ/kg

Heat supplied $= h_1 - h_{f2} = 2805 - 548.8$

$= 2256.2$ kJ/kg

Thermal efficiency

$= \dfrac{\text{work done}}{\text{heat supplied}} = \dfrac{851.3}{2256.2}$

$=$ **0.377** or **37.7%**

EXAMPLE 16.14

An ideal steam power cycle combines the reheat and regenerative cycles. Steam is supplied to the turbine at 17.5 bar and 300°C. The steam expands isentropically to a pressure of 8.5 bar, is reheated at constant pressure to 290°C and isentropic expansion follows to a condenser pressure of 0.04 bar. Steam is extracted at 1 bar for the purpose of feed water heating. The bled steam raises the temperature of feed water to 70°C. Assuming the temperature of hot well 26°C, determine the thermal efficiency of the steam plant.

Solution : Refer Fig. 16.28 for the representation of the expansion process on Mollier Chart.

The following values are noted from the chart :

$h_1 = 3030$ kJ/kg
$h_2 = 2880$ kJ/kg
$h_3 = 3035$ kJ/kg
$h_4 = 2610$ kJ/kg
and $h_5 = 2165$ kJ/kg

From steam tables,

At 26°C; $\quad h_{f4} = 109.1$ kJ/kg

At 70°C; $\quad h_{f5} = 293$ kJ/kg

Let m be the mass of steam extracted for feed heating. Then the energy balance for the heater gives

$m \times 2610 + (1 - m) \times 109.1 = 1 \times 293$

or $2610\ m + 109.1 - 109.1\ m = 293$

$\therefore \quad m = \dfrac{293 - 109.1}{2610 - 109.1} = 0.0735$ kg

Work done during expansion,

$=$ mass of steam $\times$ enthalpy change

$= 1(3030 - 2880) + 1(3035 - 2610) + (1 - 0.0735)\ (2610 - 2165)$

$= 150 + 425 + 412.3 = 987.3$ kJ

Heat input $= (3030 - 293) + (3035 - 2880)$

$= 2737 + 155 = 2892$ kJ

$\therefore$ Thermal efficiency of the cycle,

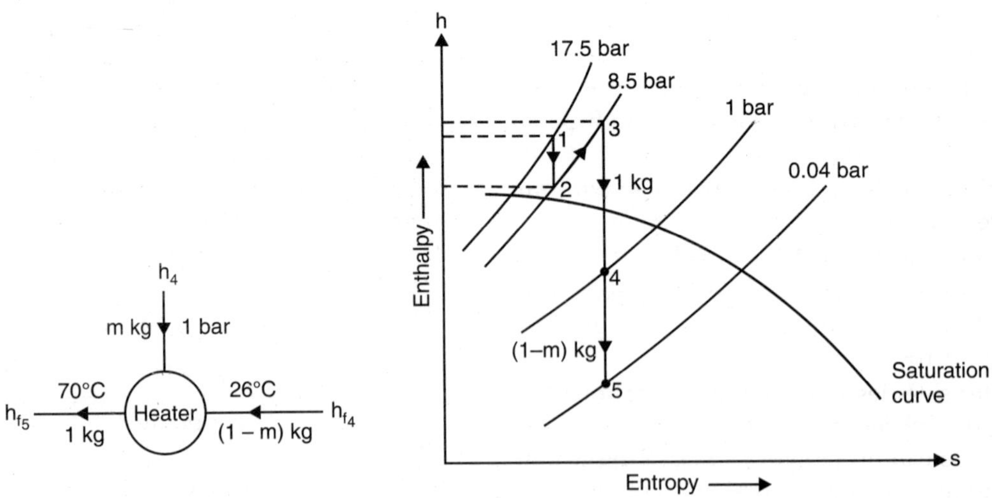

Fig. 16.28.

$$\eta = \frac{\text{work done}}{\text{heat input}} = \frac{987.3}{2892}$$

$= \mathbf{0.341}$ or $\mathbf{34.1\%}$

16.7. BINARY VAPOUR CYCLE

"A power vapour cycle that is a combination of two cycles; one in a high temperature region and the other in a low temperature region."

A binary cycle utilizes two working fluids, usually mercury and water in an endeavour to get maximum efficiency from a turbine plant. For best performance, the working fluid for vapour pressure cycles should have the following characteristics :

1. High enthalpy of vapourization and low specific heat so that less heat is required to raise the temperature of fluid to its boiling point.

2. High critical temperature; this should be well above the metallurgical limit so that latent heat can be supplied at the maximum temperature of the cycle.

3. Low pressure at the highest operating temperature.

4. Condenser pressure not too low ; the leakage problems are then minimized and the power required for the maintenance of vacuum is not much.

5. Saturated vapour line should lie close to the path of steam expansion. This keeps the moisture content at the end of expansion within safe limits.

6. Inert and easy availability at low cost.

7. Non-toxic and non-dangerous to human life.

8. Non-corrosive and stable at maximum pressure and temperature of the cycle.

9. Good heat transfer characteristics.

10. Freezing should be below the room temperature; this would facilitate the filling and draining of the equipment.

Water satisfies most of these requirements (except properties listed at 2, 3 and 5) to an appreciable extent. However, steam has a low critical temperature with which is connected a very high critical pressure.

For water :

$$T_{cri} = 374.15°\text{C and } P_{cri} = 225.65 \text{ bar}$$

This shortcoming of water is overcoming by substituting water with any one of the more suitable fluids. The critical temperature of that fluid should be well above the limit set by metallurgical considerations, and the vapour pressure be moderate at this temperature.

Two fluids which been have suggested and used to a limited extent are :

1. Diphenyl oxide which meets these requirements upto a maximum temperature of 480°C. At temperatures higher than this value, it has a tendency to decompose and also has a vapour pressure which exceeds the desirable limit. Further, it is likely to get solidified at room temperature.

2. Mercury which has a critical temperature of 588.4°C and corresponding pressure only 21 bar.

For mercury :

$$T_{cri} = 588.4°\text{C and } P_{cri} = 21 \text{ bar}$$

However, if mercury vapour expands down in a turbine to atmospheric temperature, its pressure would almost be a perfect vacuum and a tremendous volume. Such high vacuums are difficult to produce and maintain. At normal condenser pressure of 0.04 bar, it is at 216.7°C temperature and has thus a lot of thermal energy which by itself is sufficient to generate steam. This suggests that desirable properties of mercury vapour and steam be combined in one plant, leading to the development of Binary vapour cycle in which mercury is used in the high temperature range of the cycle and water is used in the low temperature range. The condenser of the high temperature cycle (topping cycle) serves as the boiler of the low temperature cycle (bottom cycle).

The schematic arrangement of the binary vapour plant and its representation on *T-s* plot is shown in Fig. 16.29.

The sequence of operation of this binary cycle is :

1-2 : Pumping of saturated mercury at state 1 to the high pressure state 2.

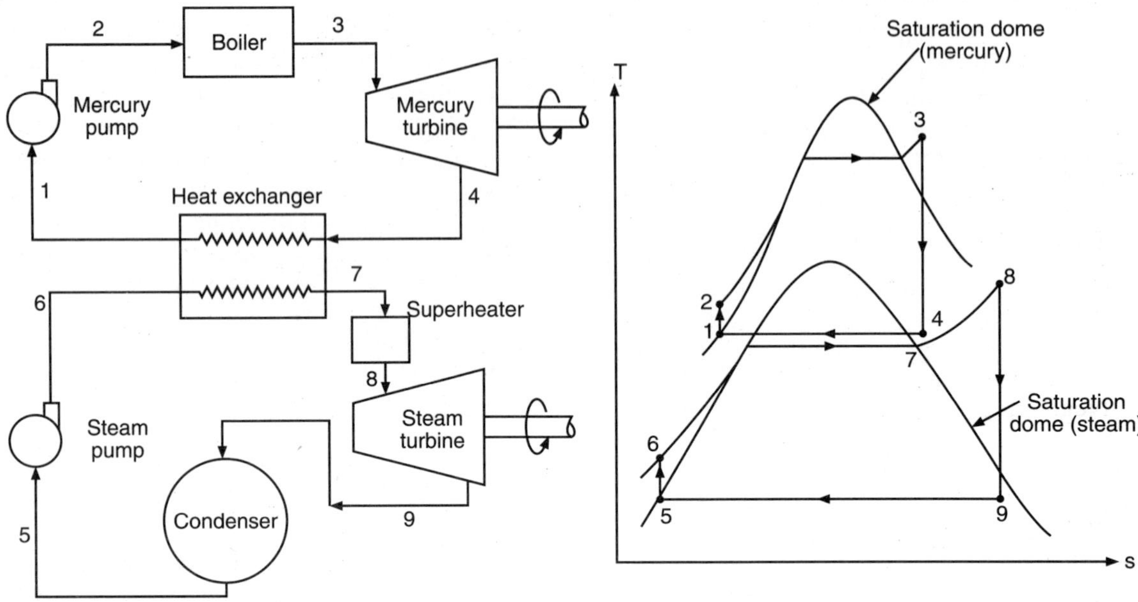

Fig. 16.29. Schematic arrangement and T-s plot of mercury-water binary vapour cycle

2-3 : Heating of liquid mercury to the saturation temperature, vaporization and superheating of mercury in a mercury boiler.

3-4 : Expansion of superheated mercury vapour in the mercury turbine.

4-1 : Condensation of the mercury vapour in the mercury condenser.

The heat removed during condensation of mercury is used for heating the liquid water to saturated vapour (Process 6-7). The saturated steam so formed is then superheated by exhaust gases leaving the mercury boiler (Process 7-8). The superheated steam at state 8 is then led to the steam turbine where it expands to normal condenser pressure (Process 8-9). The low pressure steam is then condensed (Process 9-5) to saturated liquid. The process 5-6 represents the pumping process of the feed water pump.

Apparently the binary cycle is a combination of two Rankine cycles and the two fluids do not mix with each other.

Thermal efficiencies of the order of 50% are possible with binary cycles. However, such cycles could not be attractive economically because of the following disadvantages of using mercury :

(*i*) Mercury is toxic, and also attacks most of the metals in common use.

(*ii*) Mercury does not wet boiler and condenser surfaces and this aspect gives poor heat transfer characteristics.

(*iii*) For vaporization of 1 kg of water vapour, 8 to 10 kg of mercury has to be condensed. With mercury being quite expensive, the cost of the plant becomes too high.

The thermal efficiency of a binary cycle plant can be calculated by considering the mass and energy balance of the system.

In the plant heat lost by mercury in the mercury condenser is equal to heat gained by the steam upto the dry saturated state. Let

m = mass of mercury required per kg of steam required.

The energy balance for the mercury condenser – steam boiler gives

$$m(h_4 - h_1) = 1(h_7 - h_6)$$

$$\text{or} \quad m = \frac{h_7 - h_6}{h_4 - h_1}$$

Work output from the mercury and steam turbines

$$W_t = m\,(h_3 - h_4) + 1(h_8 - h_9)$$

Work input to the pumps

$$= m\,(h_2 - h_1) + 1 \times (h_6 - h_5)$$

Network done

$$= m\,(h_3 - h_4) + (h_8 - h_9) - m(h_2 - h_1) - (h_6 - h_5)$$

Heat supplied per kg of steam

$$m(h_3 - h_2) + 1(h_8 - h_7)$$

Then thermal efficiency of the plant is

$$\eta_{th} = \frac{m(h_3 - h_4) + (h_8 - h_9) - m(h_2 - h_1) - (h_6 - h_5)}{m(h_3 - h_2) + (h_8 - h_7)}$$

Neglecting pump work

$$\eta_{th} = \frac{m(h_3 - h_4) + (h_8 - h_9)}{m(h_3 - h_2) + (h_8 - h_7)}$$

The thermal efficiency of the plant can also be evaluated by using the expression

$$\eta_{th} = \frac{Q_1 - Q_2}{Q_1} \quad \text{where}$$

Q_1 = heat supplied = $m(h_3 - h_2) + 1(h_8 - h_7)$

Q_2 = heat rejected = $1(h_9 - h_5)$

Thermal efficiencies of the order of 50% are possible with binary cycles.

16.8. COGENERATION : COMBINED HEAT AND POWER

Cogeneration or combined heat and power (CHP) is the use of heat engine or a power station to simultaneously generate both electricity and useful heat.

In the conventional method of power generation, only about 35% of primary energy fed into the power plant is actually made available to the user in the form of electricity and the remaining 65% of energy is lost. The major source of loss in the conversion process is the heat rejected into the environment through cooling towers and due to the inherent constraints of the different thermodynamic cycles employed in power generation. There will be considerable improvement in overall efficiency of the plant if this heat can be utilized for heating purposes or process work.

Cogeneration refers to the system of producing both power and process heat. The system is so designed that it meets the requirements both for thermal energy and mechanical energy from a single primary energy source. Mechanical energy may be used either to drive an alternator for producing electricity or a fan for delivering various services. Thermal energy can be used either for direct process applications or for indirectly producing steam, hot water and hot air for dryer. By product at moderate temperatures can also be used in absorption chillers for cooling. A plant producing electricity, heat and cold is called *trigeneration* or *polygeneration* plant.

Cogeneration plants are commonly found in

– Heating systems of cities and industrial plants with large heating loads

– Paper mills and textile mills

– Sugar industry, chemical industry and oil refineries

– Drying plants, rubber manufacturing plants and waste water treatment plants

Cogeneration technologies that have been widely commercialized include the back pressure and extraction steam turbines.

In the back pressure steam turbines (Fig. 16.30), the steam leaves the turbine at a pressure which is appropriate to the process steam requirement :

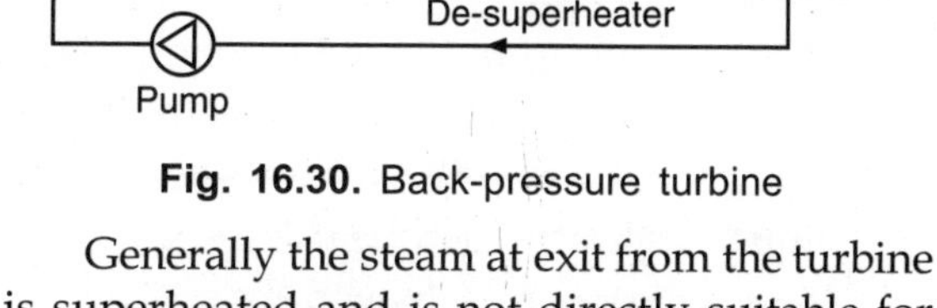

Fig. 16.30. Back-pressure turbine

Generally the steam at exit from the turbine is superheated and is not directly suitable for process work because

– the control of temperature is not convenient.

– the heat transfer rate from the superheated steam to the heating surface is lower as compared to that from saturated steam.

Accordingly the steam is made to pass through a desuperheater where it loses heat to water and gets converted into saturated steam. At the process work, the saturated steam is fully condensed and the condensate is pumped back to the boiler. The process unit replaces the condenser of the basic Rankine cycle.

The back-pressure turbines as used for process work are found to be suitable where the ratio of process heat to power is high (10 or above).

In the pass-out or extraction steam turbines (Fig. 16.31), the steam is bled at some point or points between inlet and exhaust.

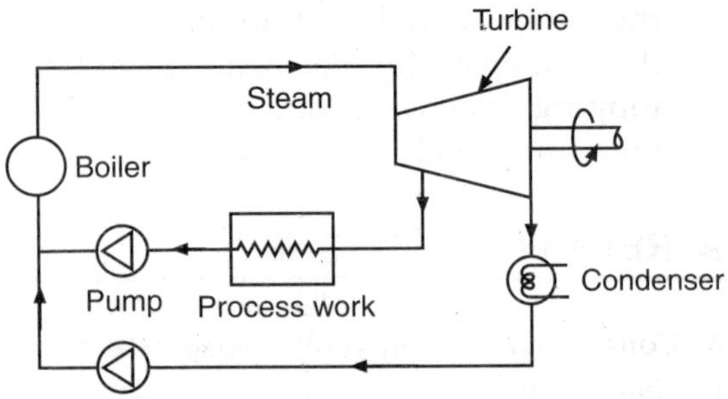

Fig. 16.31. Pass-out or extraction turbine

The steam extraction is at the pressures required for the process work. The exhaust from the processing plant and the turbine is condensed and the condensate is pumped back to the boiler. These turbines are used where the ratio of process heat to power lies in the range 4 to 10.

SALIENT POINTS

1. Carnot vapour cycle, like Carnot air cycle, is an ideal cycle; however, it can not be implemented in practice.
2. Rankine cycle is the practical vapour cycle which comprises four internally reversible stages :

   - constant pressure heat addition in the boiler
   - isentropic expansion of steam in the steam turbine
   - constant pressure heat rejection in the condenser
   - isentropic pumping of boiler feed water

   Neglecting feed pump work,

$$\eta_{Rankine} = \frac{h_1 - h_2}{h_1 - h_{f_4}}$$

3. The Rankine efficiency can be improved by raising the average temperature at which heat is supplied and decreasing the temperature at which heat is rejected.

   These aspects are achieved by :

   (*i*) increasing the boiler pressure
   (*ii*) super heating
   (*iii*) reducing the condenser pressure
4. The following methods have been suggested and are used for thermal refinement of Rankine cycle :

   (*a*) **Regeneration :** heating of feed water before it enters the boiler. For this, a part of the steam is extracted at some intermediate stage during expansion in the turbine. The steam thus extracted is mixed with feed water coming from the hot well. The system of extracting steam from any point in the turbine and subsequently using it for heating the feed water is called bleeding.

   The device where the feed water is heated is called a regenerator or feed water heater.

   (*b*) **Reheating :** expansion of steam is carried out in two or more stages. After partial expansion to an intermediate pressure, the steam is withdrawn and reheated to the original temperature at constant pressure. Steam is then sent to the low pressure turbine and further expanded to the condenser pressure.

   (*c*) **Binary vapour cycle :** a vapour cycle that is a combination of two cycles; one in high temperature cycle (topping cycle) and the other in a low temperature region (bottoming cycle). The heat rejected by the topping cycle is taken as source by the bottoming cycle.

A binary cycle utilizes two working fluids, usually mercury and water.

REVIEW QUESTIONS

A. Conceptual and conventional questions.

1. Name the four basic components of a steam power plant.
2. Why is Carnot cycle not used as a standard of reference for steam power plants ?
3. What are the difficulties experienced in carrying out Carnot cycle for a vapour ?
4. State the essential difference between Carnot and Rankine steam engine cycles.
5. How does Rankine cycle differ from the Carnot cycle for a vapour ?
6. Sketch the Rankine cycle on *p-v* and *T-s* plots.
7. What fixes the pressure and temperature limits in a Rankine cycle ?
8. Why the Rankine cycle and not the Carnot cycle is the basic cycle for a steam turbine plant ?
9. Define steam rate and heat rate in the context of a steam power plant.
10. What is the importance of work ratio in a power cycle ?
11. List the requirements of an ideal working fluid for Rankine cycle.
12. How does friction in the system affect the expansion and compression processes in the working of a Rankine cycle ?
13. Name the methods adopted for improving the performance of the Rankine cycle.
14. State the effects of following parameters in a Rankine cycle.
    (*i*) Steam pressure at inlet to turbine
    (*ii*) Steam temperature at inlet to turbine
    (*iii*) Steam pressure at exhaust
15. Why does the efficiency of a Rankine cycle increase with decreasing condenser pressure ?
16. Why does the efficiency of a Rankine cycle increase with increasing boiler pressure when the condenser is maintained at constant pressure ?
17. What is bleeding ? How does it affect the cycle efficiency ?
18. What is meant by a reheat cycle ? When is reheating of steam recommended in a steam power plant?
19. Describe a Carnot power cycle employing a condensable vapour as a working medium and mention the drawbacks associated with the design of an engine based on Carnot cycle.
20. Draw the layout of a steam power plant operating on Rankine cycle in which the boiler generates superheated steam. Depict the cycle on *p-v* and *T-s* plots and describe its working.
21. Is the thermal efficiency of a Rankine cycle equal to that of a Carnot cycle operating between the same temperature limits ? If it differs, explain the reasons for the same.
22. Explain, with a neat sketch, the working of a binary vapour cycle and represent it on temperature entropy plot.
23. (*a*) What is meant by cogeneration ?
    (*b*) How a pass-out turbine differs from a back-pressure turbine ?
24. Determine the thermal efficiency, work ratio and specific steam consumption of a Carnot cycle operating with a boiler pressure of 10 MPa and a condenser pressure of 5 kPa. Assume that the steam enters the boiler as saturated liquid and leaves as saturated vapour.
    [**Ans.** 47.6%, 0.618, 5.75 kg/kWh]
25. A hypothetical steam plant operates on the ideal Carnot cycke between pressures of 20 bar and 0.8 bar. Complete condensation occurs at the end of isentropic compression and complete vaporisation at the end of isothermal expansion. Make calculations for the thermal efficiency, work ratio and steam quality at commencement of compression.
    [**Ans.** 24.5%, 0.866, 0.196]
26. A steam power plant works on Rankine cycle between pressure limits of 20 bar and 0.04 bar. Assuming steam to be dry and saturated at inlet to turbine, make calculations for work ratio, specific steam consumption and Rankine efficiency. Account for feed pump work input.
    [**Ans.** 0.99, 4.05 kg/kWh; 33.2%]
27. A power generating plant uses steam as a working fluid and operates at a boiler pressure of 50 bar, dry saturated and a condenser pressure of 0.05 bar. Determine the cycle

efficiency, work ratio and specific steam consumption for (*i*) Carnot cycle (*ii*) Rankine cycle. For Rankine cycle, take pump work into account.

[**Ans.** (*i*) 43%, 0.724 and 5.1 kg/kWh, (*ii*) 36.5%, 0.995 and 3.715 kg/kWh]

28. A steam power plant is to operate with a boiler pressure of 50 bar and a condenser pressure of 0.07 bar. The steam leaving the boiler is at a temperature of 350°C. Compare the efficiency, specific steam consumption and work ratio of the Rankine cycle with corresponding values for a Carnot cycle working with saturated steam between the same pressure limits.

[**Ans.** (*i*) 36.7%, 3.38 kg/kWh, 0.995, (*ii*) 49.9%, 8.1 kg/kWh, 0.47]

29. A steam power plant supplies steam at pressure 88 bar and temperature 480°C to a turbine which is to develop 10 MW of power. If expansion of steam in the turbine is allowed upto a condenser pressure of 0.04 bar, find the rate of flow of steam per second to the turbine.

[**Ans.** 6.64 kg/s]

30. Superheated steam at 17.5 bar and 250°C enters the turbine of a thermal power plant operating on ideal Rankine cycle. The exhaust steam leaves the turbine at 0.07 bar. When the steam plant is supplied 300 kg/hr of fuel of calorific value 41850 kJ/kg, the turbine develops 400 kW of power. Neglecting pump work, make calculations for the Rankine efficiency, actual thermal efficiency and the relative efficiency.

[**Ans.** 31.2%, 11.46%, 36.7%]

B. Fill in the blanks with appropriate word/words.

1. Vapour power cycles are external combustion systems in which the working fluid is alternately .... and ....
2. The Carnot cycle operating with steam as the working substance is composed of two .... and two .... processes.
3. The Rankine cycle is .... efficient than Carnot cycle when both are operating within the same temperature limits.
4. The pressure of reheat steam after passing through reheater compared to initial condition is ....
5. Reheat cycle in steam power plant is used to prevent .... in last stages of turbine.
6. The efficiency of a regenerative steam cycle increases when the number of extractions are ....
7. Rankine efficiency would approach .... efficiency by providing an infinite number of feed water heaters.
8. The system of extracting steam from any point in the steam turbine and subsequently using it for heating the feed water is called ....

Answers

1. vaporised, condensed 2. isentropic, isothermal 3. less 4. less 5. excessive moisture 6. more 7. Carnot cycle 8. bleeding

C. Multiple choice questions.

1. Which aspect is not true in the context of Carnot cycle using steam as the working fluid?
   (*a*) condensation process of steam is stopped before being wholly converted into water
   (*b*) compressor has to deal with non-homogeneous mixture of water and vapour
   (*c*) high plant efficiency and high work ratio
   (*d*) bigger size of plant for a given output
2. Which of the following processes is not involved in a Rankine cycle?
   (*a*) isobaric (*b*) isothermal
   (*c*) isochoric (*d*) adiabatic
3. Consider the following processes
   1. constant pressure heat addition
   2. adiabatic compression
   3. adiabatic expansion
   4. constant pressure heat rejection

   The correct sequence of these processes in Ranking cycle is
   (*a*) 1, 2, 3, 4 (*b*) 2, 1, 4, 3
   (*c*) 2, 1, 3, 4 (*d*) 1, 3, 4, 3
4. Consider the following statements :
   The efficiency of the vapour power Rankine cycle can be increased by
   (1) increasing the temperature of the working fluid at which heat is added
   (2) increasing the pressure of the working fluid at which heat is added
   (3) decreasing the temperature of the working fluid at which heat is rejected

   Which of these statements is/are correct ?
   (*a*) 2 and 3 (*b*) 1 alone
   (*c*) 1 and 2 (*d*) 1, 2 and 3
5. The efficiency of Rankine cycle is lower than that of corresponding Carnot cycle because

(a) the average temperature at which heat is supplied in Rankine cycle is less than that corresponding to Carnot cycle
(b) the Carnot cycle has gas as working substance and Rankine cycle has steam as working substance
(c) the Rankine cycle efficiency depends upon properties of working substance, where as Carnot cycle efficiency is independent of the properties of working substance
(d) the temperature range of Carnot cycle is greater than that for Rankine cycle

6. Reheat cycle is essential in a steam power plant to :
(a) utilize the heat of flue gases
(b) heat the feed water on its way to boiler
(c) increase the plant efficiency
(d) restrict the moisture content in the low pressure stages of turbine within the limits of 10-12%

7. Which aspect is not true in the context of reheat cycle for a steam turbine ?
(a) there is increase in specific steam consumption
(b) steam has less moisture content in the low pressure stages of the turbine
(c) within certain limits, there is slight improvement in thermal efficiency
(d) use justified only in plants with high capacity

8. The reheat cycle power plant is mainly adopted to
(a) improve thermal efficiency
(b) keep the moisture content in low pressure stages to safe value
(c) decrease the capacity of condenser
(d) recover the waste heat of boiler

9. Which combination of following statements is correct ?

The incorporation of a reheater in a steam power plant always increases the
1. thermal efficiency of the plant
2. dryness fraction of steam at low pressure turbine
3. mean temperature of heat addition
4. specific work output from the turbine.

(a) 1 and 4 (b) 2 and 4
(c) 1, 3 and 4 (d) 1, 2, 3 and 4

10. In a steam power plant, feed water heater is a heat exchanger to preheat feedwater by
(a) live steam from the generator
(b) hot air from the air preheater
(c) hot flue gases coming out of the boiler furnace
(d) extracting steam from the turbine

11. The term bleeding in a steam turbine refers to :
(a) leakage of steam
(b) steam doing no useful work
(c) steam extracted for preheating feed water
(d) removal of wet steam in the low pressure stages of turbine

12. For a given set of operating pressure limits of a Rankine cycle, the highest efficiency occurs for
(a) saturated cycle (b) superheated cycle
(c) reheat cycle (d) regenerative cycle

13. In a Rankine cycle, regeneration results in higher thermal efficiency because of
(a) increase in pressure inside the boiler
(b) increase in total work delivered by the turnine
(c) additional heat added before the steam enters the low pressure turbine
(d) increase in the average temperature at which heat is added in the boiler

Answers

1. (c)	2. (c)	3. (d)	4. (d)	5. (a)
6. (d)	7. (a)	8. (b)	9. (b)	10. (d)
11. (c)	12. (d)	13. (d)		

❑❑❑

Chapter **17**

Steam Engines

Learning objectives : After having made a study of the subject matter presented in this chapter, the reader will be able to :

- *explain the constructional and operational details of simple steam engine*
- *discuss the terms diagram factor, saturation curve, missing quantity, cylinder condensation and its prevention*
- *state the reasons for compounding and explain the various methods of compounding of steam engines*
- *analyse the steam engines for work done and cylinder dimensions*
- *explain the governing methods for simple and compound steam engines*

Steam engine is an external combustion engine in which steam is used as the working substance to convert heat into work. The steam generated in the boiler is delivered to the cylinder of the reciprocating steam engine. The steam expands and performs useful work on the engine shaft. The first steam engine was developed by James Watt in 1763 and this led to the era of industrial revolution.

Attention has been focussed in this chapter to cover the topics related to the functional parts and the working of simple and compound steam engines. A brief has also been given about their governing and performance parameters.

17.1. CLASSIFICATION

Steam engine is an external combustion engine in which steam is used as the working substance to convert heat into work. The first steam engine was developed by James Watt in 1763 and that led to the era of industrial revolution.

The steam engines are generally classified as :

- *Class of service* :
  (*i*) stationary (*ii*) locomotive
  (*iii*) marine
- *Rotational speed* :
  (*i*) slow speed engine (up to 100 rpm)
  (*ii*) medium speed engine (between 100 to 250 rpm)
  (*iii*) high speed (above 250 rpm)
- *Arrangement of cylinder* :
  (*i*) horizontal steam engine – the axis of the cylinder is horizontal
  (*ii*) vertical steam engine – the axis of the cylinder is vertical

The horizontal engine is more easily accessible than a vertical engine but it occupies more floor area.

- *Acting of steam on piston* :

(*i*) single acting – steam acts on one side of the piston and the other side of the piston is open to atmosphere.

(*ii*) double acting – steam acts alternately on both sides of the piston

A double acting steam engine will theoretically develop twice the power of a single acting engine. All steam engines in practice are double acting engines.

- *Type of exhaust*

(*i*) non condensing – the steam exhausts into atmosphere

(*ii*) condensing – the steam exhausts into a condenser where the pressure maintained is much lower than atmosphere. Steam is condensed here and the condensate is sent back to boiler by means of feed pump.

- *Range of steam expansion* (number of cylinders) :

(*i*) simple steam engine – the cylinder receives steam direct from boiler and exhausts to the atmosphere or the condenser.

(*ii*) compound steam engine – the steam from the boiler partially expands in one cylinder, called high pressure (HP) cylinder. The partially expanded steam then exhausts into a larger cylinder called low pressure (LP) cylinder where expansion is completed.

The steam engines may also be classified on the type of governing (throttle or cut off), types of valves and valve gear, position of crank and on nature of expansion (expansive or non-expansive). If the steam is supplied throughout the stroke, the engine is said to be non-expansive. If the steam is supplied during part of the stroke and then allowed to expand, it is said to be expansive engine.

At present, the steam engine is receiving increasing competition from the more efficient gas or steam turbines and the internal combustion engines because of their high output, high efficiency and small bulk. However, the steam engine is still being retained for certain applications because of its

(*i*) ability to use any fuel

(*ii*) rugged and simple construction

(*iii*) low first cost

(*iv*) reliability and ease of maintenance

(*v*) easy of reversibility in the direction of rotation with suitable gear mechanism

(*vi*) excellent torque characteristics and nearly constant efficiency at part loads

Further, for units under 1500 kW, the steam economy of steam engines is as good as for steam turbines.

Steam engines are preferred for the following services :

1. Drives for pumps and compressors.

2. Laundries and hotels and in such other small or medium capacity slow speed prime movers where exhaust may be used for heating and process work.

3. Locomotive engines and for hoisting purposes which need wide flexibility of speed control and full power reversal operation.

17.2. CONSTRUCTIONAL DETAILS

Refer Fig. 17.1 which shows the schematic arrangement of a single cylinder, double acting,

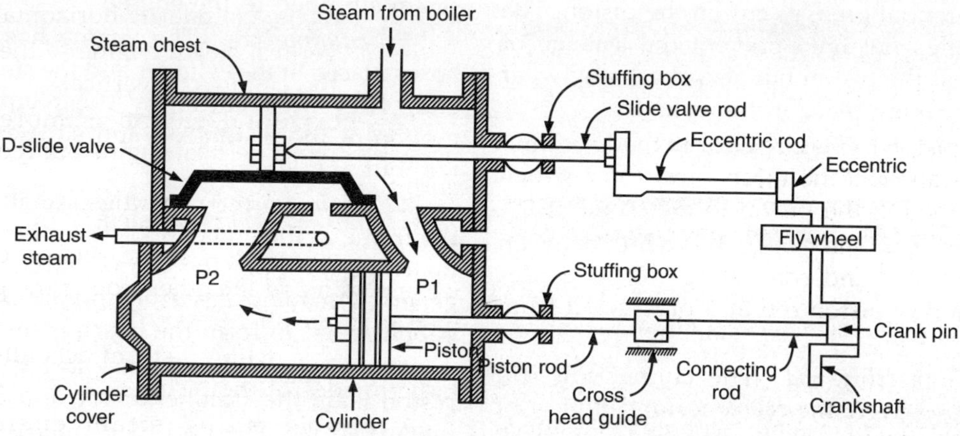

Fig. 17.1. Simple steam engine

non-condensing horizontal steam engine fitted with a slide valve.

The various parts of a simple steam engine are:

1. *Frame* : The frame supports and holds all the stationary and moving parts of the engine in position. It is essentially a heavy casting that rests upon a foundation.

2. *Cylinder* : The engine cylinder is a hollow cylindrical vessel made of cast iron and bored out perfectly true. The piston has a reciprocating motion inside it under the pressure of steam. The cylinder is bolted to the frame at one end and its ends are closed by separate covers. The end on left of the piston is called crank end and the other end through which the piston rod passes is known as crank end.

3. *Steam chest* : It is a reservoir of steam from which the steam is admitted to the engine cylinder. It is always cast integral with the cylinder and is closed by a cover known as steam chest cover which may be of circular or rectangular in shape depending on the type of valve used.

4. *Piston, piston rings and piston rod* : The piston is a cylindrical body made of cast iron, cast steel or forged steel. Under the action of steam pressure, it has a reciprocating motion from end to end of the cylinder. The piston must form a steam tight division between the two ends of the cylinder. This is accomplished by fitting atleast two piston rings in the circumferential grooves cut on the piston. The piston rings not only prevent the leakage of steam past the piston but also check any wear and tear taking place in the cylinder.

The piston rod is circular rod made of mild steel. It transmits the mechanical force which is developed at the piston to the crosshead.

5. *Crosshead and guides* : It is a link between the piston rod and connecting rod. It guides the motion of piston rod and prevents it from bending.

6. *Connecting rod* : The connecting rod serves to convert the reciprocating motion of piston into rotary motion of crank. Its one end, called the small end, is connected to the cross head by means of a pin called the gudgeon pin or wrist pin. The other end, called the big end, is connected to the crank by means of a pin known as crank pin. The connecting rod is made of steel forging of rectangular, I or H-section.

7. *Crank and crankshaft* : The crank converts the reciprocating motion of piston into rotary motion of crankshaft. The crankshaft is supported on the main bearings and is free to rotate in them. The flywheel and the eccentric are mounted on the crankshaft.

8. *Slide valve* : The slide valve is situated in the steam chest and its function is to admit the steam from the steam chest to the cylinder and exhaust it from the cylinder at the correct moment.

9. *Eccentric, eccentric rod and valve rod* : The eccentric is fitted to the crank shaft and functions to convert the rotary motion of crank shaft into reciprocating motion which is transmitted to the D-slide valve through eccentric rod and valve rod.

10. *Flywheel and governor* : The flywheel is a heavy cast iron or cast steel wheel mounted on the crank shaft. It absorbs energy when the supply is greater than demand and gives it back when the demand exceeds the supply. Thus it prevents fluctuations in engine speed and jerks to the crankshaft. Governor is a device which maintains the engine speed more or less constant at all load conditions. It is done either by controlling the quantity or the pressure of steam supplied to the engine.

11. *Stuffing box* : The engine has stuffing boxes placed at the cylinder and the steam chest end. They serve to allow the reciprocating motion of the piston rod and slide valve rod without any leakage.

12. *Main bearing* : Bearings are attached to the frame at the ends opposite to the cylinder. Bearings support the crankshaft. They are generally made of cast iron and are lined with babbit metal to form the bearing surface.

Working : With reference to Fig. 17.1, the piston is on the crank end side and ready for inward stroke. The port P1 is open and steam from the steam chest enters the cylinder on the right side of the piston. The high pressure steam pushes the piston inward and motion of the piston moves the crank, the crank shaft and the eccentric. Admission of steam continues till the eccentric moves the D-slide valve to a position that closes the port P1. At this instant, the cut off takes place and the supply of steam to the cylinder is stopped. Subsequently the steam expands, its volume increases with corresponding fall in pressure, and work is done. Towards the end of outward stroke, the D-slide valve opens the port P1 and release of steam occurs to atmosphere. Thereafter the port P1 is closed to exhaust, the outward movement of piston continues and the steam entrapped between piston and crank end is compressed till the piston reaches the extreme location on crank end side. That completes the cycle and the system gets ready for the next cycle.

A similar cycle is repeated on the other end of the cylinder. When admission and expansion take place on cover end side, exhaust and compression take place on the crank end side.

17.3. STEAM ENGINE TERMINOLOGY

(*i*) **Cylinder bore (*d*)** : It is the inner diameter of the cylinder or the liner.

(*ii*) **Stroke (*l*)** : It is the linear distance which the piston travels from the cover end of the cylinder to the crank end of the cylinder. The piston is said to make a forward (or outward) stroke when it moves from the cover end to crank end, and a backward (or in word) stroke when it moves from crank end to cover end.

(*iii*) **Crank throw or crank radius (*r*)** : The linear distance between the centre of the crank shaft and the centre of the crank pin is called the crank throw. It equals half of the piston stroke, *i.e.*, $r = \frac{l}{2}$.

Two strokes of the piston are performed in one revolution of the crankshaft.

(*iv*) **Average piston speed** : The distance travelled by the piston in one minute is called piston speed

Piston speed = $2\,l\,N$ m/min

where l is the stroke length and N is the speed of the engine in rev /min

(*v*) **Dead centre** : The extreme positions of the piston in the cylinder are called dead centres. The piston reverses its direction of the motion at the dead centres and apparently then its velocity is zero. There are two dead centres; one for each end of the stroke. For horizontal disposition of the piston cylinder assembly,

Inner dead centre (IDC) : The inner most position of the piston towards the cylinder cover end

Outer dead centre(ODC) : The outer most position of the piston towards the cylinder crank end.

For a vertical configuration, the dead centers are called the top and bottom dead centers.

(*vi*) **Swept volume (V_s)** : The volume swept by the piston as it moves from one dead centre position to another dead centre position is called the swept volume.

$$\text{Swept volume } V_s = \frac{\pi}{4} d^2 l$$

The swept volume is also called the piston displacement.

(*vii*) **Clearance space and clearance volume** : Theoretically a piston may flush with the cylinder head. However, in actual practice some clearance is always kept for wear and tear and for the reception of ports.

The small distance between the piston and cylinder cover is called piston clearance and the corresponding volume is called clearance volume (V_c).

$$V_c = \frac{\pi}{4} d^2 \times \text{clearance space}$$

The clearance volume is usually expressed as a percentage of swept volume. The clearance space is provided to serve the following objectives:

- To prevent the piston from striking the cylinder head

• To provide cushioning effect to the fast moving piston when it changes its direction of motion.

• To allow small quantity of water to be collected due to steam condensation without excessive rise of pressure.

(*viii*) **Cylinder (total) volume :** The volume of working fluid in the cylinder when the piston is at extreme position on outward stroke (*i.e.*, at ODC) is referred to as cylinder or total volume and it equals the clearance volume plus swept volume.

Cylinder volume

$= \text{clearance volume} + \text{swept volume}$

$= V_c + V_s$

17.4. HYPOTHETICAL INDICATOR DIAGRAM

Various events taking place during the operation of a steam engine can be shown on indicator diagram. In this *P-V* plot, the length of the diagram shows the volume of steam and the height represents the pressure acting on the piston at any point of stroke. The hypothetical indicator diagram shown in Fig. 17.2 presumes that:

(*i*) Steam is admitted to the engine cylinder at constant boiler pressure

(*ii*) Cut off takes place instantaneously

(*iii*) Expansion is hyperbolic ($p\, v$ = constant)

(*iv*) Release takes place instantaneously

(*v*) Exhaust takes place at constant condenser pressure

(*vi*) No compression after exhaust, and

(*vii*) Clearance is neglected

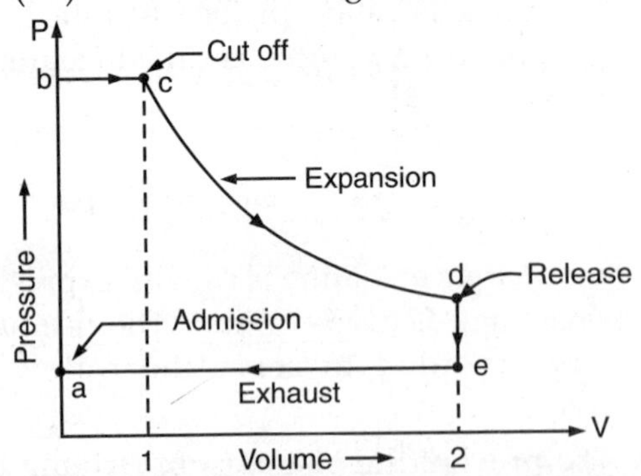

Fig. 17.2. Hypothetical indicator diagram

The events taking place during the operation of a steam engine are:

1. Admission : The inlet port opens at point *a* and some steam from the steam chest suddenly enters into the cylinder, the pressure of steam increases without change in volume. The piston occupies the extreme position on the cover end side (point *b*).

2. Cut off : With steam entering the cylinder at constant pressure, the piston is driven to the position *c* where the slide valve closes the inlet passage (port). The supply of steam is totally cut off at point *c* and this point is known as the cut off point.

3. Expansion : At cut off point *c*, the steam expands pushing the piston outward. The expansion continues till the piston reaches the dead end. The expansion takes place hyperbolically and the pressure falls considerably.

The work output is transmitted to the crankshaft by the piston rod, crosshead, connecting rod and crank in order.

4. Blow down and release : At point *d*, the slide valve opens the cylinder to exhaust and the steam gets released. The pressure falls at constant volume till it is reduced to the exhaust pressure.

The point *d* is known as release point.

5. Exhaust : At point *e*, the piston starts its inward journey and the spent up steam is exhausted at constant pressure. The exhaust of spent-up steam continues up to point *a* where the exhaust port closes, the inlet port opens and the cycle is repeated.

The steam enters the cylinder from the point of admission (*a*) to the point of cut off (*c*). Steam leaves the cylinder from the point of release (*d*) to the point of admission (*a*).

Work done and mean effective pressure : With reference to Fig. 17.3. Let

p_1 = the boiler pressure at which steam enters the cylinder (pressure at point *b*).

p_b = back pressure or condenser pressure (pressure along exhaust process *e-a*)

V_1 = volume of steam in the cylinder at the point of cut off (point *c*)

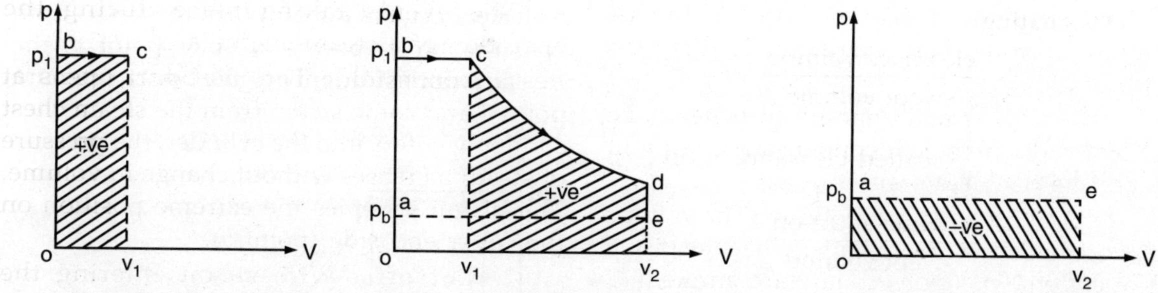

Fig. 17.3.

V_2 = volume of steam in the cylinder at the end of expansion stroke, *i.e.*, the release point *d*.

The theoretical work done per cycle by the engine equals the area of indicator (*p*-*V*) diagram.

Work done W

= area $a\ b\ c\ d\ e\ a$ of indicator plot.

$$= p_1 V_1 + p_1 V_1 \log_e \frac{V_2}{V_1} - p_b V_2$$

$$= p_1 V_1 \left(1 + \log_e \frac{V_2}{V_1}\right) - p_b V_2 \quad ...(17.1)$$

The mean effective pressure (*mep*) is the constant uniform pressure which if acting on the face of the piston throughout the stroke would have produced the same work as obtained under actual working conditions. Apparently it represents the mean height of indicator diagram and is given by

$$mep = \frac{\text{work done per cycle}}{\text{stroke volume}}$$

$$= \frac{p_1 V_1 \left(1 + \log_e \dfrac{V_2}{V_1}\right) - p_b V_2}{V_2}$$

$$= \frac{p_1}{r}(1 + \log_e r) - p_b \quad ...(17.2)$$

The volume ratio $r = \dfrac{V_2}{V_1}$ is called the expansion ratio.

Theoretical indicator diagram with clearance

Work done W

= area $a\ b\ c\ d\ e\ a$

= area $h\ b\ c\ g + g\ c\ d\ f$ – area $h\ a\ e\ f$

$$= p_1(V_1 - V_c) + p_1 V_1 \log_e \frac{V_2}{V_1} - p_b(V_2 - V_c)$$

$$= p_1(V_1 - V_c) + p_1 V_1 \log_e \frac{V_2}{V_1} - p_b V_s$$

$$= p_1 V_1[1 + \log_e r] - p_1 V_c - p_b V_s$$

where $r = \dfrac{V_2}{V_1}$ is the expansion ratio

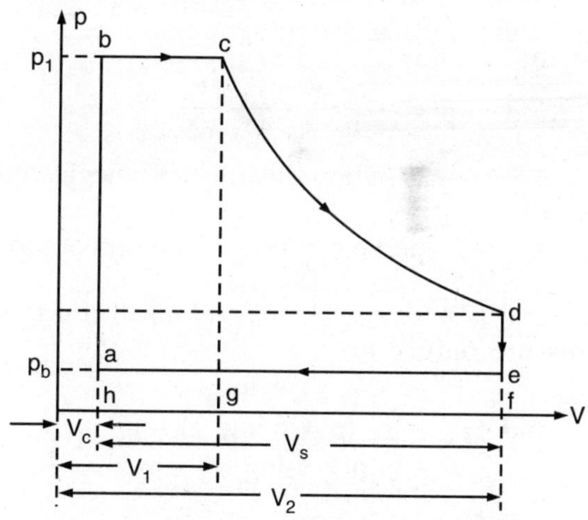

Fig. 17.4. Indicator diagram with clearance

Mean effective pressure

$$p_m = \frac{\text{work done}}{\text{stroke volume}}$$

$$= \frac{p_1 V_1 (1 + \log_e r) - p_1 V_c - p_b V_s}{V_s}$$

$$= \frac{p_1 V_1}{V_s}(1 + \log_e r) - p_1 \frac{V_c}{V_s} - p_b$$

Designating

$$c = \frac{\text{clearance volume}}{\text{swept volume}}$$

$$= \frac{V_c}{V_s} \text{ called clearance ratio}$$

$$\rho = \frac{\text{volume at cut-off}}{\text{swept volume}}$$

$$= \frac{V_1 - V_c}{V_s} \text{ called cut off ratio}$$

$$V_1 = \rho V_s + V_c \text{ or } \frac{V_1}{V_s} = \rho + c \quad ...(17.3)$$

Substituting the above results in the expression for mean effective pressure, we get

$$p_m = p_1(\rho + c)(1 + \log_e r) - p_1 c - p_b$$

$$= p_1\left[\rho + (\rho + c)\log_e r\right] - p_b \quad ...(17.4)$$

It may noted again that :

$$r = \frac{V_2}{V_1}; \quad c = \frac{V_c}{V_s}; \ V_s = V_2 - V_c$$

and $\rho = \dfrac{V_1 - V_c}{V_s} = \dfrac{V_1 - V_c}{V_2 - V_c}$

Apparently when clearance is neglected ($c = 0$, $V_s = V_2$ and $\rho = \dfrac{V_1}{V_2} = \dfrac{1}{r}$), the expressions 17.3 and 17.4 for work done and mean effective pressure reduce to

$$W = p_1 V_1 (1 + \log_e r) - p_b V_2$$

and
$$p_m = p_1 (\rho + \rho \log_e r) - p_b$$

$$= p_1\rho(1 + \log_e r) - p_b$$

$$= \frac{p_1}{r}(1 + \log_e r) - p_b$$

Theoretical indicator diagram with clearance and compression

Here : *a-b-c* is steam admission
c-d is steam expansion
d-e is steam release
e-f is steam exhaust

The process *f-a* represents the compression of steam remaining in the cylinder before the fresh steam enters into it. When the fresh steam from the boiler enters at state point *a*, the pressure rises suddenly to the boiler pressure p_1.

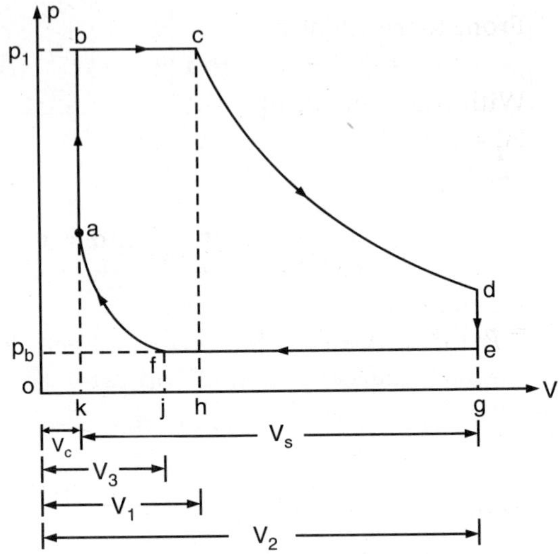

Fig. 17.5. Indicator diagram with clearance and compression

Work done

$$W = \text{area } a\,b\,c\,d\,e\,f\,a$$

$$= \text{area } k\,b\,c\,h + \text{area } h\,c\,d\,g - \text{area } j\,f\,e\,g - \text{area } k\,a\,f\,j$$

$$= p_1(V_1 - V_c) + p_1 V_1 \log_e \frac{V_2}{V_1} - p_b(V_2 - V_3) - p_b V_3 \log_e \frac{V_3}{V_c}$$

$$= p_1 V_1 \left[1 + \log_e \frac{V_2}{V_1}\right] - p_1 V_c - p_b(V_s + V_c - V_3) - p_b V_3 \log_e \frac{V_3}{V_c}$$

Let $\dfrac{V_2}{V_1} = r$ (expansion ratio)

$\dfrac{V_1 - V_c}{V_s} = \rho$ (cut off ratio)

$\dfrac{V_c}{V_s} = c$ (clearance ratio)

and

$$\frac{V_3 - V_c}{V_s} = \alpha \text{ (fraction of stroke completed at start of compression)}$$

From these identities :

$$V_1 = \rho V_s + V_c \quad \text{and} \quad V_3 = \alpha V_s + V_c$$

With these substitutions,

$$W = p_1 V_1 (1 + \log_e r) - p_1 V_c - p_b [V_s + V_c - \alpha V_s - V_c] - p_b (\alpha V_s + V_c) \log_e \left(\frac{\alpha V_s + V_c}{V_c} \right)$$

$$= p_1 V_1 (1 + \log_e r) - p_1 V_c - p_b (V_s - \alpha V_s) - p_b (\alpha V_s + V_c) \log_e \frac{\alpha + c}{c} \quad ...(17.5)$$

Mean effective pressure

$$p_m = \frac{\text{work done}}{\text{swept volume}}$$

$$= \frac{p_1 V_1}{V_s} (1 + \log_e r) - \frac{p_1 V_c}{V_s} - p_b (1 - \alpha) - p_b \left(\alpha + \frac{V_c}{V_s} \right) \log_e \frac{\alpha + c}{c}$$

$$= p_1 \left(\frac{\rho V_s + V_c}{V_s} \right) (1 + \log_e r) - p_1 c - p_b (1 - \alpha) - p_b (\alpha + c) \log_e \frac{\alpha + c}{c}$$

$$= p_1 (\rho + c) + (1 + \log_e r) - p_1 c - p_b (1 - \alpha) - p_b (\alpha + c) \log_e \frac{\alpha + c}{c}$$

Notes : (*i*) When the effect of compression is neglected, the points *f* and *a* merge, *i.e.*, $V_3 = V_c$ and α becomes zero. Substituting $\alpha = 0$, we get

$$W = p_1 V_1 (1 + \log_e r) - p_1 V_c - p_b V_s$$

and $$p_m = p_1 (\rho + c) (1 + \log_e r) - p_1 c - p_b$$

These are the expressions for work done and mean effective pressure without compression.

(*ii*) When the effects of compression and clearance are neglected, $\alpha = 0$ and $c = 0$ and upon substituting these values, we get

$$W = p_1 V_1 (1 + \log_e r) - p_b V_2 \quad (\because V_c = 0 \text{ and } V_s = V_2)$$

and $$p_m = p_1 \rho (1 + \log_e r) - p_b$$

$$= \frac{p_1}{r} (1 + \log_e r) - p_b \quad \left(\because \rho = \frac{V_1}{V_2} = \frac{1}{r} \right)$$

These are the expressions for work done and mean effective pressure without clearance and compression.

Diagram Factor

The mean effective pressure calculated from the hypothetical indicator diagram is higher than the actual mean effective pressure. This is due to the fact that the following practical losses have not been considered:

(*i*) Pressure losses occur due to throttling of steam in stop valve and friction and pipes. Further, to make the steam flow from boiler to the engine cylinder a certain pressure differential must exist. As such, the actual pressure of steam at the beginning of stroke is some what less than the boiler pressure.

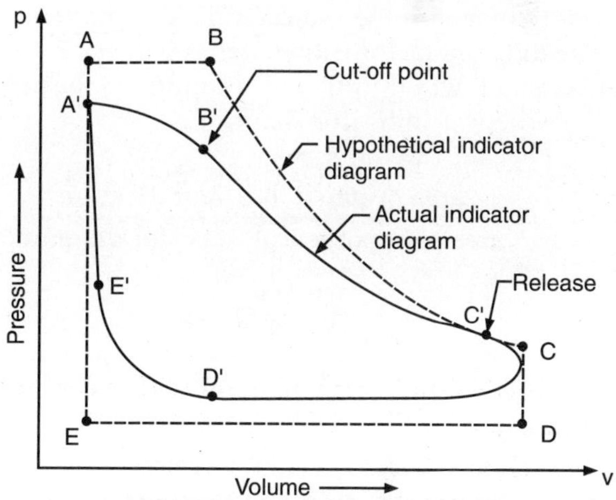

Fig. 17.6. Theoretical and actual indicator diagram of a steam engine

(*ii*) When admitted, steam comes in contact with cylinder surfaces which are at a lower temperature. This results in condensation of steam. This fact along with wire drawing in steam ports results in progressive fall in steam pressure along the admission line.

(*iii*) Due to time factor in closing the admission port, the point of cut off is rounded.

(*iv*) The expansion of steam is not truly hyperbolic. Initially the steam is hotter than the cylinder walls and is more prone to condensation. Towards the later part, the steam temperature becomes lower than the cylinder temperature. The direction of flow of heat is reversed and some of the condensed steam tends to evaporate at low pressure.

(*v*) The opening of the exhaust valve is not instantaneous resulting in the rounding off the point of release.

(*vi*) For proper blowing down of steam, the exhaust takes place at a higher back pressure due to some inevitable resistance.

(*vii*) The steam locked up in the cylinder, after the exhaust valve has been closed, is compressed before the fresh steam enters the cylinder.

Figure 17.6 shows the difference between the actual and hypothetical indicator diagrams for the same engine plotted between the same ordinates.

The area of actual indicator diagram is less than that of theoretical indicator diagram. The diagram factor is then defined as the ratio of area of actual indicator diagram to that of hypothetical indicator diagram.

Digram fractor K =

$$\frac{\text{area of actual indicator diagram}}{\text{area of hypothetical indicator diagram}}$$

$$= \frac{\text{actual } p_m}{\text{hypothetical } p_m}$$

The diagram factor varies slightly with different types of engines and its average value is about 0.7.

17.5. MASS OF STEAM IN THE CYLINDER

The total mass of steam in the cylinder at any point between cut off and release equals the sum of cushion steam (m_c) left in the clearance space and the fresh steam admitted per stroke (m_a). That is

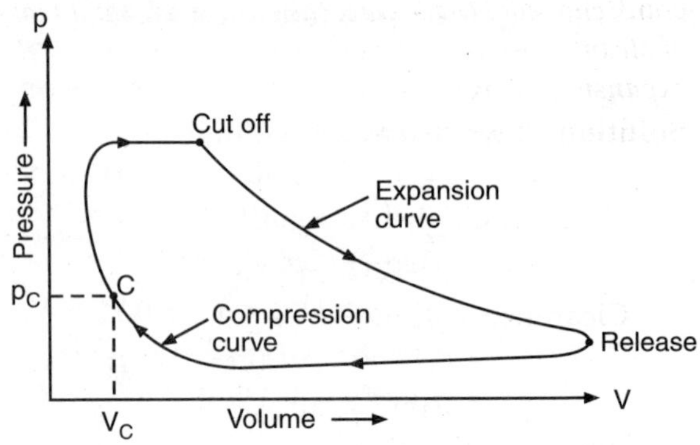

Fig. 17.7. Mass of cushion steam in the cylinder

$$m = m_c + m_a$$

• To get the mass of entrapped or cushion steam m_c, it is assumed that at any point C on the compression curve, the steam is dry and saturated. Corresponding to this point, the volume V_c and pressure p_c are noted from the indicator plot. At this pressure, the specific volume v_s of dry saturated steam is read from the steam tables. Then mass of cushion steam is given by

$$m_c = \frac{V_c}{v_s}$$

• The mass of steam admitted into the cylinder m_a is calculated from the mass of steam condensed per stroke.

$$m_a = \frac{\text{mass of steam condensed in a specified time}}{\text{number of working revolutions in the same time}}$$

$$= \frac{M}{2N}$$

where M is the mass of steam condensed per minute and N is the number of revolutions turned per minute.

EXAMPLE 17.1

A single cylinder double acting steam engine with 25 cm bore and 35 cm stroke has clearance volume 15 percent of stroke. The engine turns 180 rev/min and consumes 1500 kg of steam per hour. The indicator diagram shows that pressure of steam at any point near the end of compression stroke is 3 bar and this point lies at 82% of stroke from the crank end side. Make calculations for the total mass of steam present in the engine cylinder during the expansion stroke.

Solution : Swept (stroke) volume

$$V_s = \frac{\pi}{4} d^2 l = \frac{\pi}{4} (0.25)^2 \times 0.35$$
$$= 0.01717 \text{ m}^3$$

Clearance volume

$$V_c = 0.15 \times V_s = 0.15 \times 0.01717$$
$$= 0.00257 \text{ m}^3$$

Volume at point (say C) on the compression stroke

$$= 0.00257 + (1 - 0.82) \times 0.01717$$
$$= 0.00566 \text{ m}^3$$

If steam is assumed to be dry saturated at point C, then

$v_s = 0.6058$ m^3/kg corresponding to 3 bar pressure

Then : mass of cushion steam,

$$m_c = \frac{0.00566}{0.6058} = 0.00934 \text{ kg/stroke}$$

Mass of steam admitted

$$m_a = \frac{1500}{2 \times 60 \times 180}$$
$$= 0.06944 \text{ kg/stroke}$$

Hence, mass of steam in the engine cylinder during expansion stroke

$$= 0.00934 + 0.06944$$
$$= 0.07878 \text{ kg.}$$

17.6. SATURATION CURVE AND MISSING QUANTITY

Saturation curve is the curve that indicates the volume the steam in the cylinder would occupy during expansion stroke, if perfectly dry.

Neglecting volume of water vapours, the expansion curve gives the volume of steam at any point during expansion.

Corresponding to any point M on the expansion curve of indicator plot, let the pressure be P_M and the volume be V_M. At pressure P_M, the specific volume of dry saturated steam (v_{sM}) can be read from the steam tables. Then the volume the steam would occupy if dry and saturated

$$= m \times v_{sM}$$

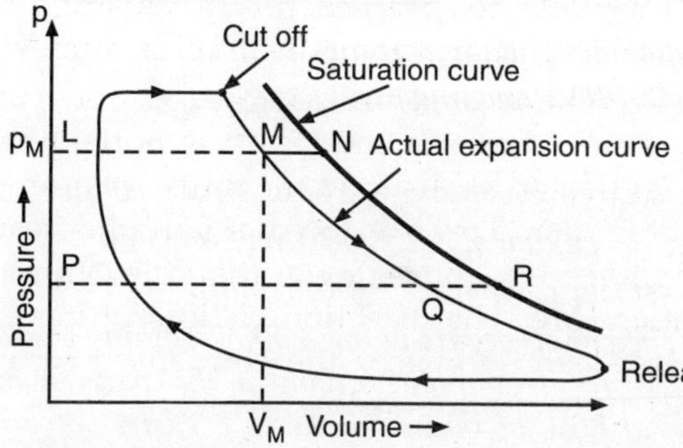

Fig. 17.8. Saturated curve and missing quantity

where m is the mass of steam in the cylinder at any point between cut off and release.

Let this volume be represented by distance LN on volume scale of the indicator plot. By adopting the same procedure, a number of points can be obtained and plotted. The curve passing through these points is called the *saturation curve.*

The *missing quantity* is the horizontal distance between the actual expansion curve and the saturation curve. For conditions at point M, the missing quantity is

$$\text{Missing quantity} = MN$$
$$= LN - LM$$
$$= m\, v_{sM} - x_M\, m\, v_{sM}$$
$$= m\, v_{sM}\,(1 - x_M)$$

where x_M is the dryness fraction of steam at point M on the actual expansion curve.

It is worthwhile to point out that :

1. The missing quantity is maximum at the point of cut off, diminishes as the expansion proceeds and becomes minimum at the point of release. This is because the steam has a tendency to evaporate at reduced pressure and get dried before release.

2. The loss of possible available work per stroke is represented by the area between the saturation curve and the expansion curve.

3. The missing quantity is mainly due to condensation of steam in the cylinder, and a

small amount may be due to leakage past the valves and the piston.

EXAMPLE 17.2

The following data pertains to trial on a single cylinder double acting steam engine :

stroke volume = 0.15 m³
clearance volume = 5% of stroke volume
engine speed = 120 rev/min and
steam condensed = 50 kg/min.

The cut off occurs at one-half of working stroke and the pressure at point A lying on the compression curve at 80% of return stroke is 1.5 bar.

Compute the quality of steam at point B lying on the expansion curve at 60% of forward stroke. The pressure recorded at this point is 4.5 bar.

Solution : Refer Fig. 17.9.

Swept (stroke) volume $V_s = 0.15$ m³

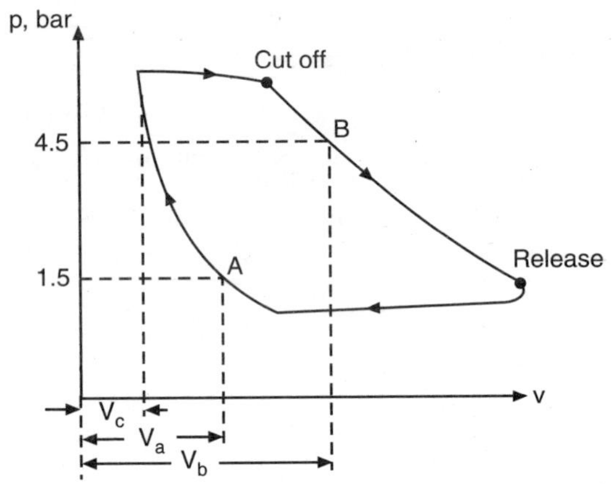

Fig. 17.9.

Clearance volume $V_c = \frac{5}{100} \times 0.15$
$= 0.0075$ m³

Volume at point A lying on the compression curve

$v_a = 0.0075 + (1 - 0.8) \times 0.15$
$= 0.0375$ m³

Volume at point B lying on expansion curve
$= 0.0075 + 0.6 \times 0.15$
$= 0.0975$ m³

If steam is assumed to be dry saturated at point A, then

$v_s = 1.164$ m³/kg corresponding to 1.5 bar pressure.

Then : mass of cushion steam

$$= \frac{0.0375}{1.164} = 0.0322 \text{ kg/stroke}$$

mass of steam admitted

$$= \frac{50}{2 \times 120} = 0.2083 \text{ kg/stroke}$$

∴ Total mass of steam at point B (a point between cut off and release)
$= 0.0322 + 0.2083$
$= 0.2405$ kg/stroke

Now, specific volume of dry saturated steam at 4.5 bar pressure = 0.414 m³/kg

Indicated mass of steam at point B

$$= \frac{\text{indicated volume at point B}}{\text{specific volume of dry steam at 4.5 bar}}$$

$$= \frac{0.0975}{0.414} = 0.2355$$

Then quality (dryness fraction) at point

$$B = \frac{0.2355}{0.2405} = 0.979$$

Missing quantity at point
B = 0.2405 – 0.2355
= 0.005 kg/stroke
= 0.005 × (2 × 120 × 60) = 72 kg/hr

EXAMPLE 17.3

A single cylinder double acting steam engine having bore 50 cm and stroke length 75 cm has clearance volume 5 percent of stroke volume. The engine runs at 60 rev/min and uses 60 kg of steam per minute. The parameters related to different events of the working cycle are as indicated below :

Event	*Fraction of stroke*	*Pressure*
Cut off	*40% of forward stroke*	*8 bar*
Release	*95% of forward stroke*	*5 bar*
Compression	*80% of return stroke*	*2 bar*

Make calculations for :

(a) mass of dry steam present in the cylinder at the points of cut off, release and compression.

(b) quality and missing quantity at the points of cut off and release.

How do you explain the difference, if any, between the missing quantities at these two points.

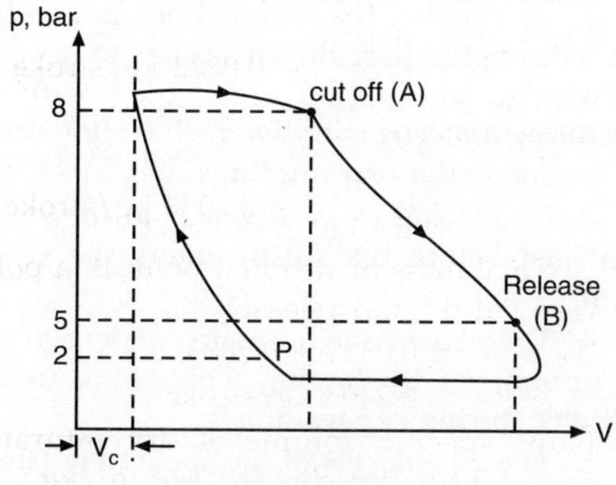

Fig. 17.10.

Solution : Refer Fig. 17.10.

Swept (stroke) volume $V_s = \frac{\pi}{4} d^2 l$

$$V_s = \frac{\pi}{4} (0.5)^2 \times 0.75 = 0.1472 \text{ m}^3$$

Clearance volume $V_c = 0.05 \times 0.1472$
$= 0.00736 \text{ m}^3$

Volume at cut off point A,
$V_a = 0.00736 + 0.4 \times 0.1472$
$= 0.06624 \text{ m}^3$

Volume at release point B,
$V_b = 0.00736 + 0.95 \times 0.1472$
$= 0.1472 \text{ m}^3$

Volume at point P on the compression curve
$V_p = 0.00736 + (1 - 0.8) \times 0.1472$
$= 0.0368 \text{ m}^3$

From tables :

At $p = 8$ bar : $v_s = 0.240 \text{ m}^3/\text{kg}$

$p = 5$ bar : $v_s = 0.375 \text{ m}^3/\text{kg}$

$p = 2$ bar : $v_s = 0.886 \text{ m}^3/\text{kg}$

Assuming the steam to be dry and saturated at any point on the compression curve,

Mass of cushion steam

$$= \frac{0.0368}{0.886} = 0.0415 \text{ kg/stroke}$$

Mass of steam admitted

$$= \frac{60}{2 \times 60} = 0.5 \text{ kg/stroke}$$

Then total mass of steam in the cylinder between cut off and release

$= 0.5 + 0.0415 = 0.5415$ kg/stroke

(*b*) Indicated mass of steam at cut off

$$= \frac{\text{indicated volume at the point of cut off}}{\text{specific volume of dry steam at 8 bar}}$$

$$= \frac{0.06624}{0.24} = 0.276 \text{ kg}$$

Then quality (dryness fraction) of steam at

$$\text{cut off} = \frac{0.276}{0.5415} = 0.51$$

Missing quantity of cut off
$= (0.5415 - 0.276)$
$= 0.2655$ kg/stroke
$= 0.2655 \times (2 \times 60 \times 60)$
$= 1911.6$ kg/hr

Indicated mass of steam at release

$$= \frac{\text{indicated volume at the point of release}}{\text{specific volume of dry steam at 5 bar}}$$

$$= \frac{0.1472}{0.375} = 0.392$$

Then quality (dryness fraction) of steam at

$$\text{release} = \frac{0.392}{0.5415} = 0.724$$

Missing quantity at release
$= (0.5415 - 0.392)$
$= 0.1495$ kg/stroke
$= 0.1495 (2 \times 60 \times 60)$
$= 1076.4$ kg/hr

From the calculations made above, it is seen that missing quantity at release is less than that at the point of cut off. This may be attributed to the fact that effects of initial condensation are more pronounced at cut off than at release. During latter part of the expansion stroke re-evaporation occurs and steam tends to become dry.

Percentage re-evaporation

$$= \frac{1911.6 - 1076.4}{1911.6} = 0.437 \text{ or } 43.7\%$$

17.7. CYLINDER CONDENSATION AND ITS PREVENTION

Consider steam admitted to a simple reciprocating steam engine which is being started from cold. The steam admitted will give its heat to warm the walls. In this process it will condense with a corresponding reduction in volume. More steam will be admitted to take its place until a sufficient pressure is built up to force the piston outwards.

Under normal working conditions, when the steam starts entering the cylinder at the point of admission it meets surface of cylinder walls, piston and parts which have been cooled by the exhaust steam from the previous stroke. The steam in the process of warming these surfaces to its own temperature is partially condensed. At cut off, the mixture of steam and water in the space behind the piston has a volume much less than that had there been no condensation. During expansion further condensation occurs due to uncovering of more cold cylinder surface and due to the work being done during expansion. During expansion the pressure of steam falls and with it drops the temperature. However, this temperature is still higher than the average temperature of cylinder walls and the condensation of steam continues although at a progressively reduced rate. Towards the end of expansion, the temperature of steam falls below that of confining walls, the direction of flow of heat reverses and the steam gains heat from the cylinder. Some of the steam does evaporate near the release point where the pressure has fallen sufficiently. The evaporated steam however passes out into exhaust without doing any useful work. Condensation is thus a loss to the engine and the loss may be as high as 10-30% of the dry incoming steam.

Condensation is likely to be augmented by the heat loss due to radiation and conduction through metal parts.

The following methods have been suggested to reduce initial condensation :

1. *Steam jacketing* : Cylinder walls are maintained at high temperature by passing fresh steam through jackets which surround the engine cylinder.

2. *Lagging* : The steam lines and other parts through which the steam has to flow on its way to the engine are lagged with an insulating material to check the heat flow from the steam to the surroundings.

3. By placing *water separator* in the steam pipe just before the steam enters the steam chest.

4. The percentage of steam condensed can be minimized by having *more steam* in the cylinder during expansion.

5. The *compression* of the spent over steam left in the clearance space raises its temperature as well as of the surrounding cylinder surface and this reduces the steam condensation to an appreciable extent.

6. *Superheating* of steam controls the cylinder condensation by the following ways :

(*i*) heat transmission rate is less from a superheated vapour than from a moist vapour,

(*ii*) moisture content in the steam at the end of expansion is less and consequently the leakage is reduced due to less density,

(*iii*) absence of droplets of water in the superheated steam which could act as nuclei for condensation to proceed,

(*iv*) In warming the cylinder walls, the superheated steam loses its heat of superheat rather than getting condensed.

7. For developing same power with steam conditions remaining unchanged, an engine running at *greater speeds* has smaller dimensions. So with increased rotational speeds, the time and the surface parameters upon which the conduction of heat depends are reduced. Heat transfer is thus lessened and so will do the condensation rate.

8. The engines are so designed that the fresh charge is not to traverse the passage which has been cooled previously by the flow of exhaust steam. This aspect has been incorporated in the design of Uniflow engine.

9. The expansion of steam is carried out in a series of cylinders, the steam exhaust from one cylinder becoming the charge for the next

cylinder. This arrangement known as *compounding* reduces the temperature range per cylinder and arrests condensation.

17.8. PERFORMANCE OF STEAM ENGINES

A regular and periodic check of the performance of an engine is essential if it is to operate at its peak efficiencies.

The objective is to determine how the engine is doing its job in relation to the energy fed to it. How the input energy is being distributed? Attempts are made to make all non-useful energy output as small as possible. Careful examination is required if it is felt that a particular form of energy loss seems excessive.

The different measurements necessary for evaluating the performance are:

1. Indicated power (IP) : The indicated power is the power developed in the engine cylinder and is reflected by the area of the indicator diagram.

For making calculations for the indicated power the following data should be known:

(*i*) mean defective pressure acting on the engine piston during the stroke

(*ii*) area of the piston and length of the stroke

(*iii*) number of working strokes

Since the steam engines are mostly double acting, power will be developed both on the cover end side and the crank end side of the cylinder

Let

p_{m1} = actual mean effective pressure acting on the piston on cover end side (bar)

L = length of stroke (m)

A = area of piston (m^2)

and N = speed of engine (rev/min)

Then : Work done per stroke for cover end side

$$= \text{force} \times \text{distance moved}$$

$$= (\text{pressure} \times \text{area}) \times \text{distance moved}$$

$$= (p_{m1} \times 10^5 \times A) \times L \text{ Nm}$$

Work done per second

$$= (p_{m1} \times 10^5 \times A) \times L \times \frac{N}{60} \text{ Nm/s or W}$$

$$= 100\, p_{m1}\, LA \frac{N}{60} \text{ kW}$$

While calculating the indicated power for the crank end side, allowance must be made for the area of piston rod which reduces the area of piston on the crank end side.

If a is the area of the piston rod, then effective piston area on crank end side will be equal to $(A - a)$. Then for mean effective pressure p_{m2} acting on the crank end side, the indicated power for crank end side of the piston will be given by

$$= 100\, p_{m2}\, L\,(A - a) \frac{N}{60} \text{ kW}$$

Then the total indicated power developed by a double acting steam engine becomes

$$\text{Total } IP = 100\, p_{m1}\, LA \frac{N}{60}$$

$$+ 100\, p_{m2}\, L(A - a) \frac{N}{60}$$

If the area of the piston rod is neglected and assuming mean effective pressure to be same on both sides (*i.e.*, $p_{m1} = p_{m2} = p_m$) then the above equation can be rewritten as

$$\text{Total } IP = 100\, p_m LA \left(\frac{2N}{60}\right) \text{ kW}$$

The indicator diagram is plotted with the help of an instrument called indicator, and the mean effective pressure is calculated from the indicator diagram by using the relation

$$p_m = \frac{\text{area of indicator diagram (mm}^2) \times \text{spring number (bar/mm)}}{\text{length of indicator diagram (mm)}}$$

2. Brake power (BP) : There is always some friction at the moving parts of the engine, and a certain portion of the indicated power developed by the engine is utilized to overcome the friction. Apparently then the actual power available from the engine for doing useful work is less than the indicated power provided inside the engine cylinder.

The power available at the shaft and which can be used for doing useful work is measured with the help of brake dynamometer is called *Brake power* (*BP*) of the engine.

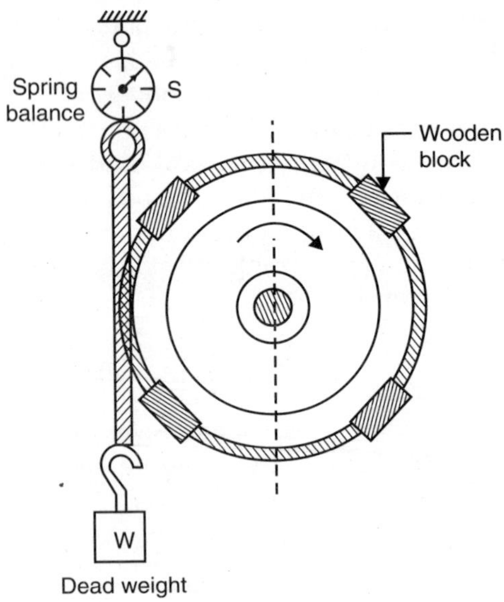

Fig. 17.11. Brake dynamometer

With reference to the brake dynamometer shown in Fig. 17.11. Let

W = weight applied on the rope (N)

S = spring balance reading (N)

D = diameter of brake wheel (m)

d = diameter of the rope (m)

N = speed of the engine (rev/min)

Then : friction or braking torque

T = effective load × effective radius

$= (W - S)R$ Nm

Work done per revolution

= torque × angle turned through one revolution

$= (W - S)R \times 2\pi$

where $R = (D + d)/2$

Since the engine turns N revolutions per minute,

Work done per second

$$= (W-S)R \times 2\pi\left(\frac{N}{60}\right) \text{ Nm/s or W}$$

∴ Brake power

$$= \frac{(W-S)R}{1000} 2\pi\left(\frac{N}{60}\right) \text{ kW}$$

$$= \frac{2\pi N(W-S)R}{60\times 10^3} \text{ kW}$$

The difference between the indicated power and the brake power is known as friction power.

$$FP = IP - BP$$

3. Steam consumption : The steam consumption is defined as the amount of steam in kg supplied to the engine per unit time (usually per hour).

4. Efficiencies : The efficiencies of a steam engine relate the power developed and heat supplied for generating steam in the boiler.

The different efficiencies of concern are:

• **Mechanical efficiency.** The ratio of brake power to indicated power is known as mechanical efficiency

$$\eta_{mech} = \frac{\text{brake power}}{\text{indicated power}}$$

The mechanical efficiency is taken to be a measure of the friction at the moving parts of the engine.

• **Thermal efficiency :** Due to restriction imposed by the second law, no engine can be constructed which can convert all the heat energy being supplied into work. The fraction which is converted into work is called thermal efficiency.

Heat supplied for raising steam in the boiler

$$= m_s(h_1 - h_w)$$

where m_s is mass of steam supplied (kg/s); h_1 is specific enthalpy of steam raised in the boiler (kJ/kg) and h_w is the specific enthalpy of water entering the boiler (kJ/kg)

Thermal efficiency may be worked on the basis of indicated power or brake power

Indicated thermal efficiency

$$\eta_{ith} = \frac{\text{heat equivalent of } IP}{\text{heal supplied in steam}}$$

$$= \frac{IP \text{ in kW}}{m_s(h_1 - h_w)}$$

Likewise : Brake thermal efficiency

$$\eta_{bth} = \frac{BP \text{ in kW}}{m_s(h_1 - h_w)}$$

• **Overall efficiency :** The overall efficiency is defined as the ratio of heat equivalent of work available at the shaft (brake power) to the energy supplied in fuel for the generation of steam in the boiler.

$$\text{overall efficiency} = \frac{BP \text{ in kW}}{m_f \times CV}$$

where m_f is fuel consumption in kg/s and CV is calorific value of fuel in kJ/kg.

EXAMPLE 17.4.

The following data pertains to a single cylinder double acting steam engine :

Boiler pressure = 14 bar
back pressure = 0.35 bar
Cut off at 0.4 stroke
diagram factor = 0.7
Cylinder diameter = 30 cm
stroke = 1.5 times cylinder bore
Mechanical efficiency = 80%

Neglecting clearance, estimate the power developed by the engine when it turns 200 rev/min.

Solution : $r = \frac{1}{\rho} = \frac{1}{0.4} = 2.5$

$$p_m \text{ (theoretical)} = \frac{p_1}{r}(1 + \log_e r) - p_b$$

$$= \frac{14}{2.5}(1 + \log_e 2.5) - 0.35$$

$$= 10.38 \text{ bar}$$

p_m (actual)

$$= \text{diagram factor} \times p_m \text{ (theoretical)}$$

$$= 0.7 \times 10.38 = 7.27 \text{ bar}$$

$$= 7.27 \times 10^5 \text{ N/m}^2$$

For a double acting steam engine, the indicated power is

$$IP = 100\, p_m\, LA\left(\frac{2N}{60}\right)$$

where p_m is the mean effective pressure in bar, L is stroke length in m, A is piston area in m^2 and N is the engine speed in rev/min.

$$\therefore \quad IP = 100 \times 7.27 \times 0.45 \times \frac{\pi}{4}(0.3)^2 \times \left(\frac{2 \times 200}{60}\right)$$

$$= 154.08 \text{ kW}$$

Mechanical efficiency

$$= \frac{\text{brake power}}{\text{indicated power}}$$

$$\therefore \text{ Brake power} = 154.08 \times 0.8$$

$$= \mathbf{132.26 \text{ kW}}$$

EXAMPLE 17.5

A double acting steam engine is supplied with dry and saturated steam at 12 bar. The cut off occurs at 40 percent of stroke and the exhaust pressure is 0.5 bar. To effect a saving in economy the cut off is timed to occur at 25 percent of stroke and the engine speed is so regulated that the power output remains constant. If the expansion of steam is hyperbolic and the diagram factor remains the same in both the situations, calculate :

(i) ratio of the changed to the original speed, and

(ii) saving in steam consumption

Solution : Since the cutoff occurs at 40% of stroke, the expansion ratio is

$$r_1 = \frac{1}{0.4} = 2.5$$

p_{m1} (theoretical)

$$= \frac{p_1}{r_1}(1 + \log_e r_1) - p_b$$

$$= \frac{12}{2.5}(1 + \log_e 2.5) - 0.5$$

$$= 8.698 - 0.5 = 8.698 \text{ bar}$$

p_{m1}(actual)

$$= \text{diagram factor} \times p_m \text{ (theoretical)} = 8.698\, k \text{ bar}$$

For a double acting steam engine, the indicated power is

$$IP = 100\, p_m\, LA\left(\frac{2N}{60}\right) \text{ kW}$$

where, p_m is the mean effective pressure in bar, L is the stroke length in m, A is the piston area in m^2 and N is the engine speed in rev/min

$$\therefore \quad (IP)_1 = 100 \times 8.698\, k\, LA\left(\frac{2N_1}{60}\right)$$

$$= 28.99\, k\, LAN_1 \text{ kW}$$

Case 2 : When cut off occurs at 20% of stroke, then

$$r_2 = \frac{1}{0.2} = 5$$

p_{m2}(theorctical)

$$= \frac{12}{5}(1 + \log_e 5) - 0.5 = 5.763 \text{ bar}$$

$$p_{m2} \text{ (actual)} = 5.763\, k \text{ bar}$$

$$(IP)_2 = 100 \times 5.763\, k\, LA\left(\frac{2N_2}{60}\right)$$

$$= 19.21\, k\, LAN_2 \text{ kW}$$

Since the power developed remains the same,

$$28.99\, k\, LAN_1 = 19.21\, k\, LAN_2$$

$\therefore$ Ratio of changed to original speed $\dfrac{N_2}{N_1}$

$$= \frac{28.99}{19.21} = 1.509$$

(*b*) When cut off takes place at 40% of stroke, the mass of steam supplied per minute is

$$m_1 = \frac{\text{volume at cut off} \times 2N_1}{\text{specific volume of steam at 12 bar}}$$

$$= \frac{\frac{\pi}{4} d^2 l \times 0.4 \times 2N_1}{v}$$

Likewise, the mass of steam supplied per minute when cut off takes place at 20% of stroke is

$$m_2 = \frac{\frac{\pi}{4} d^2 l \times 0.2 \times 2N_2}{v}$$

$\therefore$ Percentage of saving in steam consumption

$$= \frac{m_1 - m_2}{m_1} \times 100$$

$$= \frac{0.4\, N_1 - 0.2\, N_2}{0.4\, N_1} \times 100$$

$$= \frac{0.4\, N_1 - 0.2 \times 1.509\, N_1}{0.4\, N_1} \times 100$$

$$= 24.55\%$$

EXAMPLE 17.6

Dry saturated steam at 8.5 bar pressure is supplied to a steam engine having two double acting cylinders with the dimensions 20 cm bore diameter and 30 cm stroke length. The steam supply is cut off at 45% of stroke and the exhaust from the engine takes place at 0.65 bar. While turning 200 rev/min, the engine develops 50 kW brake power with 80% mechanical efficiency. Assuming clearance volume to be 8% of stroke volume, make calculations for the actual indicated mean effective pressure and the diagram factor.

Solution : Refer Figure 17.12

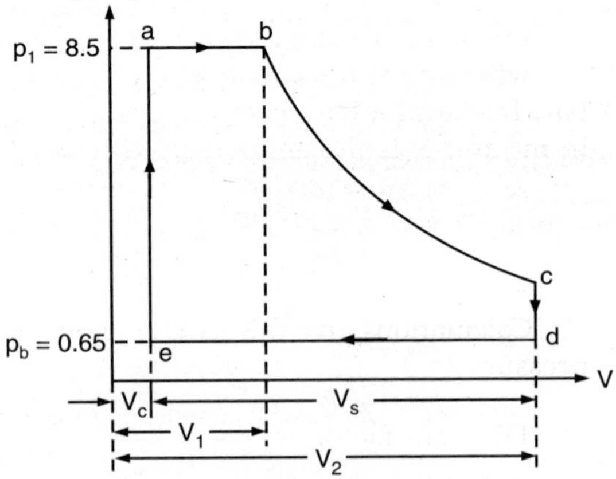

Fig. 17.12.

$$\text{Clearance ratio} \quad c = \frac{V_c}{V_s} = 0.08$$

$$\text{Cut off ratio} \quad \rho = \frac{V_1 - V_c}{V_s} = 0.45$$

Expansion ratio r

$$= \frac{V_2}{V_1} = \frac{V_c + V_s}{V_c + \rho V_s}$$

$$= \frac{0.08\,V_s + V_s}{0.08\,V_s + 0.45\,V_s} = \frac{1.08\,V_s}{0.53\,V_s}$$

$$= 2.038$$

Considering clearance, the theoretical mean effective pressure is prescribed by the relation

$$p_m = p_1(\rho + c)\,(1 + \log_e r) - p_1 c - p_b$$
$$= 8.5(0.45 + 0.08)(1 + \log_e 2.038) - 8.5 \times 0.08 - 0.65$$
$$= 7.712 - 0.68 - 0.65 = 6.382 \text{ bar}$$

Mechanical efficiency,

$$\eta_m = \frac{\text{brake power}}{\text{indicated power}}$$

$$\therefore \text{ Indicated power} = \frac{50}{0.8} = 62.5 \text{ kW}$$

Indicated power per cylinder

$$= \frac{62.5}{2} = 31.25 \text{ kW}$$

For a single cylinder double acting steam engine, the indicated power is given by

$$IP = 100\, p_m\, LA\left(\frac{2N}{60}\right)$$

where p_m is the steam supply pressure in bar, L is stroke length in m, A is piston area in m^2 and N is the engine speed in rev/min.

$$\therefore \quad 31.25 = 100 \times (p_m)_{\text{actual}} \times 0.3 \times \frac{\pi}{4}(0.2)^2 \times \left(\frac{2 \times 200}{60}\right)$$

Calculations give the actual mean effective pressure $(p_m)_{\text{acutal}}$ = 4.976 bar

$$\text{Diagram factor} = \frac{(p_m)_{\text{acutal}}}{(p_m)_{\text{theoretical}}} = \frac{4.976}{6.382}$$

$$= \mathbf{0.78}$$

EXAMPLE 17.7

A single cylinder double acting steam engine admits dry saturated steam at 12 bar. The cylinder diameter is 30 cm and stroke is 40 cm. The clearance is 8% of stroke and the cut off occurs when the total volume is equal to 0.25 of stroke volume. Assuming a diagram factor of 0.75, calculate the indicated power of the engine if it runs at 250 rev/min and the exhaust takes place to atmosphere at 1 bar.

Solution : Refer Fig. 17.13.

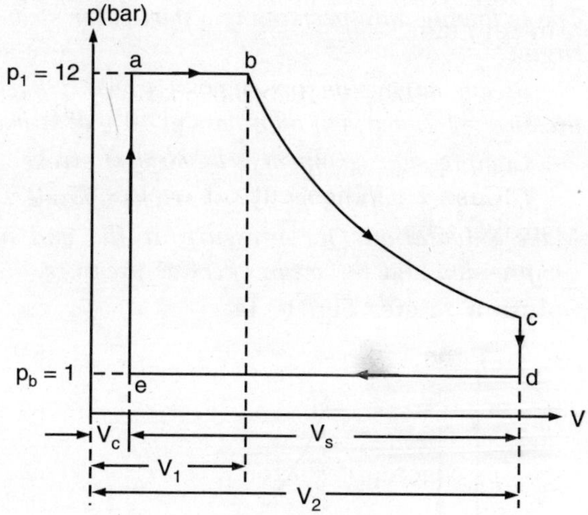

Fig. 17.13.

$$V_1 = 0.25\,V_s\,;\ V_c = 0.08\,V_s$$
$$V_2 = V_c + V_s = (0.08 + 1)V_s = 1.08\,V_s$$

$$\text{cut off ratio} \quad \rho = \frac{V_1 - V_c}{V_s}$$

$$= \frac{(0.25 - 0.08)V_s}{V_s} = 0.17$$

$$\text{expansion ratio } r = \frac{V_2}{V_1} = \frac{1.08\,V_s}{0.25\,V_s} = 4.32$$

$$\text{clearance ratio } c = \frac{V_c}{V_s} = 0.08$$

Mean effective pressure

$$p_m = p_1(\rho + c)[1 + \log_e r] - p_1 c - p_b$$
$$= 12(0.17 + 0.08)\,[1 + \log_e 4.32] - 12 \times 0.08 - 1$$
$$= 7.39 - 0.96 - 1$$
$$= 5.43 \text{ bar} = 5.43 \times 10^5 \text{ N/m}^2$$
$$p_m \text{ (actual)} = (5.43 \times 10^5) \times 0.75$$
$$= 4.07 \times 10^5 \text{ N/m}^2$$

Indicated power

$$= p_m\, LA(2N)$$

$= (4.07 \times 10^5) \times 0.4$

$$\times \frac{\pi}{4}(0.3)^2 \times (2 \times 250)$$

$= 5751 \times 10^3$ Nm/min

$= 95848$ Nm/s (W) $=$ **95.8 kW**

EXAMPLE 17.8

The following data pertains to a particular steam engine:

Steam admission pressure =14 bar ; back pressure =1.25 bar; cut off occurs at 40% of stroke

Compression commences at 30% of stroke

Clearance volume is 5% of stroke volume

Make calculations for pressure at the end of compression and the mean effective pressure.

Solution : Refer Fig. 17.14.

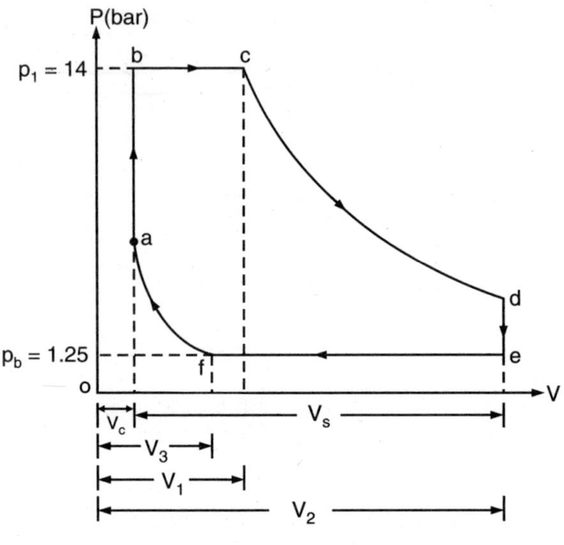

Fig. 17.14.

$$\text{clearance ratio } c = \frac{V_c}{V_s} = 0.05$$

$$\text{cut off ratio} \quad \rho = \frac{V_1 - V_c}{V_s} = 0.4$$

Fraction of stroke completed at the start of compression,

$$\alpha = \frac{V_3 - V_c}{V_s} = 0.3$$

The pressure and volume for the compression process $f\,a$ are related as

$P_f V_f = P_a V_a$ where $P_f = 1.25$ bar

and $V_f = V_3 = \alpha V_s + V_c$

$= 0.3\ V_s + 0.05\ V_s = 0.35\ V_s$

$V_a = V_c = 0.05\ V_s$

$\therefore\ 1.25 \times 0.35\ V_s = P_a \times 0.05\ V_s$

$$P_a = \frac{1.25 \times 0.35}{0.05} = 8.75 \text{ bar}$$

(*b*) Expansion ratio r

$$= \frac{V_2}{V_1} = \frac{V_c + V_s}{V_c + \rho V_s}$$

$$= \frac{0.05 V_s + V_s}{0.05 V_s + 0.4 V_s} = 2.33$$

Considering clearance and compression, the theoretical mean effective pressure is given by

$$p_m = p_1(\rho + c)(1 + \log_e r) - p_1 c - p_b(1 - \alpha) - p_b(\alpha + c) \log_e \frac{\alpha + c}{c}$$

Now :

$p_1(\rho + c)(1 + \log_e r)$

$= 14\ (0.4 + 0.05)(1 + \log_e 2.33)$

$= 11.63$

$p_1 c = 14 \times 0.05 = 0.7$

$p_b(1 - \alpha) = 1.25\ (1 - 0.3) = 0.875$

$$p_b\ (1 + c) \log_e \left(\frac{\alpha + c}{c}\right)$$

$$= 1.25\ (1 + 0.05) \log_e \left(\frac{0.3 + 0.05}{0.05}\right)$$

$= 0.85$

$\therefore\ p_m = 11.63 - 0.7 - 0.875 - 0.85 =$ **9.20 bar**

EXAMPLE 17.9

A single cylinder double acting steam engine develops 30 kW brake power when running at 120 rpm. The dry saturated steam is supplied at 1200 kN/m² and the spent up steam is exhausted at 120 kN/m². The clearance volume is 10% of displacement volume, the steam supply is cut off at 40% the stroke and compression commences at 0.85 of the return stroke.

Calculate the mean effective pressure and the cylinder dimensions.

It may be assumed that expansion and compression are hyperbolic, mechanical efficiency is 80% and the engine has mean piston speed of 75 m/min.

Solution : Refer Fig. 17.15.

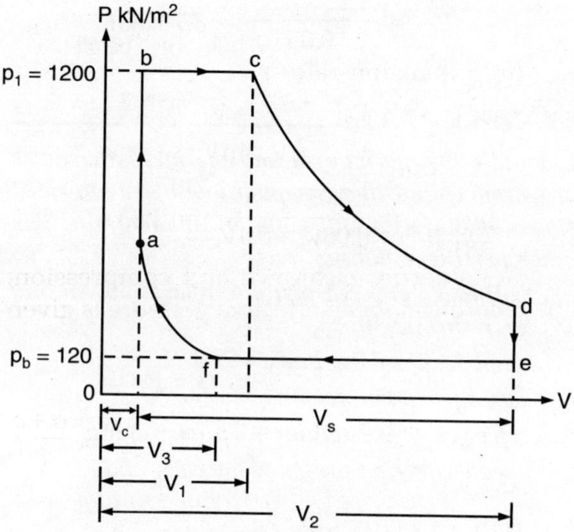

Fig. 17.15.

clearance ratio $c = \frac{V_c}{V_s} = 0.1$

cut off ratio $\rho = \frac{V_1 - V_c}{V_s} = 0.4$

Compression commences at 0.85 of the return stroke. That is

$$\frac{V_2 - V_3}{V_s} = 0.85$$

or

$$\frac{V_c + V_s - V_3}{V_s} = 0.85$$

That gives : fraction of stroke completed at the commencement of compression

$$\alpha = \frac{V_3 - V_c}{V_s} = 1 - \frac{V_c + V_s - V_3}{V_s}$$

$$= 1 - 0.85 = 0.15$$

Expansion ratio

$$r = \frac{V_2}{V_1} = \frac{V_c + V_s}{V_c + \rho V_s}$$

$$= \frac{0.1V_s + V_s}{0.1V_s + 0.4V_s} = 2.2$$

Theoretical mean effective pressure taking clearance and compression into account is given by

$$p_m = p_1(\rho + c)(1 + \log_e r) - p_1 c$$

$$- p_b(1 - \alpha) - p_b(\alpha + c)\log_e \frac{\alpha + c}{c}$$

Now :

$$p_1(\rho + c)(1 + \log_e r)$$

$$= 1200\,(0.4 + 0.1)\,(1 + \log_e 2.2)$$

$$= 1073.07$$

$$p_1 c = 1200 \times 0.1 = 120$$

$$p_b(1 - \alpha) = 120\,(1 - 0.15) = 102$$

$$p_b(1 + c)\,\text{loge}\left(\frac{\alpha + c}{c}\right)$$

$$= 120\,(1 + 0.1)\log_e\left(\frac{0.15 + 0.1}{0.1}\right)$$

$$= 27.49$$

$$\therefore p_m = 1073.07 - 120 - 102 - 27.49$$

$$= 823.58 \text{ kN/m}^2$$

(*b*) Mean piston speed $2\,l\,N = 75$ m/min

$\therefore$ Piston stroke l

$$= \frac{75}{2 \times 120} = 0.3125 \text{ m} = 31.25$$

Indicated power

$$= \frac{\text{brake power}}{\text{mechanical efficiency}} = \frac{30}{0.8}$$

$$= 37.5 \text{ kW}$$

For double acting steam engine,

Indicated power $= p_m\,LA\,(2\,N)$

$$\therefore 37.5 = 823.58 \times 0.3125 \times \frac{\pi}{4} d^2 \left(\frac{2 \times 120}{60}\right)$$

$$= 808.14\,d^2$$

Piston diameter $d = \sqrt{\frac{37.5}{808.14}}$

$= 0.215$ m $=$ **21.5 cm**

EXAMPLE 17.10.

A single cylinder double acting steam engine running at 2000 rpm receives dry saturated steam at 10 bar and develops 50 kW indicated power.

The cylinder dimensions are: diameter = 25 cm and stroke 40 cm

The cut off occurs at 20% of piston stroke and the engine operates with 0.6 bar back pressure. Assuming steam consumption to be 600 kg/hr, determine the diagram factor and the indicated thermal efficiency.

Solution : Neglecting clearance, the theoretical mean effective pressure is prescribed by the relation

$$p_m = \frac{p_1}{r}(1 + \log_e r) - p_b$$

Given:

$$r = \frac{1}{\rho} = \frac{1}{0.2} = 5$$

$$p_1 = 10 \text{ bar and } p_b = 0.6 \text{ bar}$$

$$\therefore \quad p_m = \frac{10}{5}(1 + \log_e 5) - 0.6$$

$$= 4.619 \text{ bar}$$

The actual mean effective pressure can be worked out from the correlation

$$IP = 100(p_m)_{\text{actual}} LA\left(\frac{2N}{60}\right)$$

$$50 = 100(p_m)_{\text{actual}} \times 0.4$$

$$\times \frac{\pi}{4}(0.25)^2 \times \left(\frac{2 \times 200}{60}\right)$$

Calculations give :

$$(p_m)_{\text{acutal}} = 3.82 \text{ bar}$$

Diagram factor K

$$= \frac{\text{actual mean effective pressure}}{\text{theoretical mean effective pressure}}$$

$$= \frac{3.82}{4.619} = \mathbf{0.827}$$

From steam tables :

specific enthalpy of dry saturated steam at 10 bar, h_1 = 2778.1 kJ/kg

specific enthalpy of feed water at 0.6 bar, h_w= 359.8 kJ/kg

Heat supplied for raising steam in the boiler

$$Q_s = m_s (h_1 - h_w)$$

$$= \frac{600}{3600}(2778.1 - 359.8)$$

$$= 40.3.05 \text{ kJ/s} = 403.50 \text{ kW}$$

Then : Indicated thermal efficiency

$$\eta_{ith} = \frac{50}{403.05} = 0.124 \text{ or } \mathbf{12.4\%}$$

EXAMPLE 17.11.

Calculate the mechanical and thermal efficiencies of a steam engine that is supplied with dry saturated steam at 800 kPa and runs at 100 rev/min. The other pertinent data is:

cylinder bore and piston stroke = 25 cm and 32 cm respectively

dead load on the brake= 750 N

spring balance reading = 160 N

effective brake radius = 75 cm

mean effective pressure for each end = 120 kPa

steam consumption = 65 kg/hr and

condensate temperature = 45°C

Solution : For a double acting steam engine, the indicated power is

$$IP = 100\, p_m\, LA\left(\frac{2N}{60}\right) \text{ kW}$$

where p_m is the steam supply pressure in bar, L is stroke length is m, A is piston area is m^2 and N is the engine speed is rev/min.

Given :

$$p_m = 120 \text{ kPa} = 120 \times 10^3 \text{ N/m}^2$$

$$= 1.2 \text{ bar}$$

$$\therefore IP = 100 \times 1.2 \times 0.32 \times \frac{\pi}{4}(0.25)^2 \times \frac{2 \times 100}{60}$$

$$= 6.28 \text{ kW}$$

Brake power BP

$$= \frac{2\pi N(W - S)R}{60 \times 10^3}$$

$$= \frac{2\pi \times 100(750 - 160) \times 0.75}{60 \times 10^3}$$

$$= 4.63 \text{ kW}$$

Mechanical efficiency

$$\eta_{mech} = \frac{\text{brake power}}{\text{indicated power}} = \frac{4.63}{6.28}$$

$$= 0.737 \text{ or } 73.7\%$$

From steam tables:

specific enthalpy of dry saturated steam at 800 kPa, h_1 = 2769.10 kJ/kg

specific enthalpy of feed water at 45°C, h_w = 188.42 kJ/kg

Heat supplied for raising steam in the boiler

$$Q_s = m_s(h_1 - h_w)$$

$$= \frac{65}{3600}(2769.1 - 188.42)$$

$$= 46.59 \text{ kJ/s} = 46.59 \text{ kW}$$

Then : Indicated thermal efficiency

$$\eta_{ith} = \frac{6.28}{46.59} = 0.135 \text{ or } 13.5\%$$

Brake thermal efficiency

$$\eta_{bth} = \frac{4.63}{46.59} = 0.099 \text{ or } 9.9\%$$

EXAMPLE 17.12

The following data refers to the trial run on a single cylinder double acting non-condensing steam engine running at 240 rev/min.

cylinder bore, piston rod diameter and stroke length : 25 cm, 5 cm and 35 cm

steam supply pressure and cut off =10.5 bar and 30% of stroke

area of indicator diagram for cover and crank end = 1580 mm² and 1450 mm²

length of indicator diagram and spring number = 50 mm and 0.12 bar/mm

circumference of brake wheel and brake rope = 5 m and 65 mm

dead load on brake and spring balance reading = 2000 N and 200 N

If the steam supplied is 0.9 dry, determine:

(a) indicated power, brake power and mechanical efficiency

(b) specific steam consumption on IP basis and BP basis

(c) brake thermal efficiency

Solution : The mean effective pressure is given by

$$p_m = \frac{\text{area of indicator diagram in mm}^2 \times \text{spring strength in bar/mm}}{\text{length of indicator diagram in mm}}$$

For cover end

$$p_{m1} = \frac{1580 \times 0.12}{50} = 3.79 \text{ bar}$$

For crank end

$$p_{m2} = \frac{1450 \times 0.12}{50} = 3.48 \text{ bar}$$

A (area of piston)

$$= \frac{\pi}{4}(0.25)^2 = 0.049 \text{ m}^2$$

a (area of piston rod)

$$= \frac{\pi}{4}(0.05)^2 = 0.00196 \text{ m}^2$$

∴ For cover end $(IP)_1$

$$= 100 p_{m_1} LA\left(\frac{N}{60}\right)$$

$$= 100 \times 3.79 \times 0.35 \times 0.049 \times \frac{240}{60}$$

$$= 26 \text{ kW}$$

For crank end $(IP)_2$

$$= 100\, p_{m2}\, L\, (A - a)\left(\frac{N}{60}\right)$$

$$= 100 \times 3.48 \times 0.35 \times (0.049 - 0.00196) \times \frac{240}{60}$$

$$= 22.92 \text{ kW}$$

(*a*) Total indicated power of the steam engine,

$$IP = (IP)_1 + (IP)_2$$

$$= 26 + 22.92 = 8.92 \text{ kW}$$

Brake power BP

$$= \frac{2\pi N(W-S)R_{eff}}{60\times10^3}$$

$$= \frac{N(W-S)\times\pi D_{eff}}{60\times10^3}$$

$$= \frac{240(2000-200)}{60\times10^3}\times\left(5+\frac{65}{1000}\right)$$

$= 36.47$ kW

Mechanical efficiency

$$\eta_m = \frac{BP}{IP} = \frac{36.47}{48.92} = 0.745 \text{ or } \mathbf{74.5\%}$$

(b) Steam consumption per hour

$$= \frac{\text{volume of steam consumed per revolution}}{\text{specific volume}} \times \text{rpm} \times 60$$

For cover end :

$$(m_s)_1 = \left(\frac{\pi}{4}D^2 l\times\rho\right)\times\frac{60\,\text{rpm}}{xV_g}$$

where ρ = cut off ratio and $V_g = 0.185$ m^3/kg at 10.6 bar as read from steam tables

$$(m_s)_1 = \left[\frac{\pi}{4}(0.25)^2\times0.35\times0.3\right]\times\frac{60\times240}{0.9\times0.185}$$

$= 445.54$ kg/hr

From crank end :

$$(m_s)_2 = \left[\frac{\pi}{4}(D^2-d^2)l\times\rho\right]\times\frac{60\,\text{rpm}}{xV_g}$$

$$= \left[\frac{\pi}{4}(0.25^2-0.05^2)\times0.35\times0.3\right]\times\frac{60\times240}{0.9\times0.185}$$

$= 427.17$ kg/hr

Total steam consumption

$m_s = 445.54 + 427.17$

$= 872.71$ kg/hr

Specific steam consumption (ssc) is defined as the steam consumed per hour per kW power produced.

$$\therefore\ ssc \text{ on } IP \text{ basis} = \frac{872.71}{48.92} = 17.84 \text{ kg/kWh}$$

$$ssc \text{ on } BP \text{ basis} = \frac{872.71}{36.47} = 23.93 \text{ kg/kWh}$$

(c) From steam tables, corresponding to steam pressure of 10.5 bar

$h_f = 772$ kJ/kg and $h_{fg} = 2007.85$ kJ/kg

Heat supplied in the boiler for raising steam

$h = 772 + 0.9 \times 2007.85$

$= 2579.06$ kJ/kg

The enthalpy of feed water is being neglected because the engine is non-condensing.

$$Q_s = m_s h = \frac{872.71}{3600}\times 2579.06$$

$= 625.2$ kJ/s $= 625.2$ kW

Then : Brake thermal efficiency

$$\eta_{bth} = \frac{36.47}{625.2} = 0.058 \text{ or } 5.8\%$$

17.9. GOVERNING OF STEAM ENGINES

With load variations, the engine must be capable of affecting a change in the area of indicator diagram. This is to ensure that the engine operates at fairly constant speed at all operating loads. For greater loads, more work must be produced and the area of indicator card must increase. For smaller loads, the engine should develop less power, *i.e.*, the area of the indicator diagram must decrease.

Apparently governing implies that the engine is able to adjust its power output according to the load with minimum fluctuation in speed.

Governing of steam engines is affected by the following two ways:

1. Throttle governing : Vary the throttle pressure keeping the quantity of steam admitted constant.

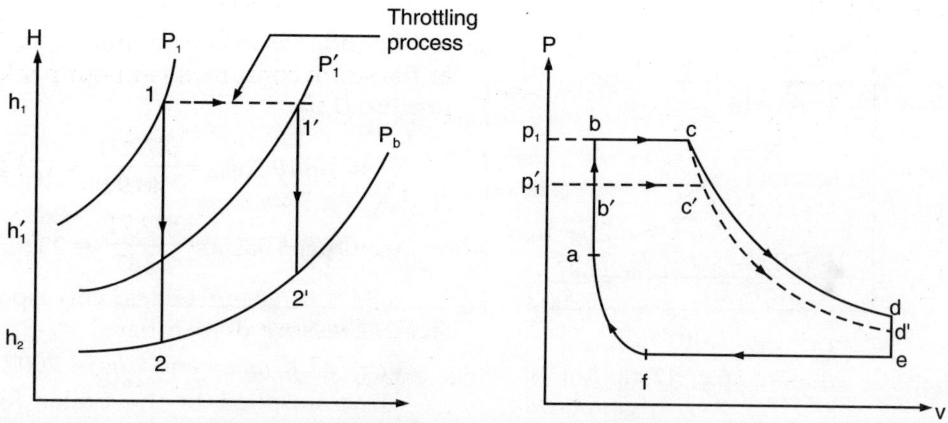

Fig. 17.16. Throttle governing of steam engine

Consider the engine to be working on a particular load and has a certain constant speed. With a decrease in load, the speed will increase if no adjustment is made with the operation of the engine. To bring the speed back to normal, the steam is throttled which means that the steam is supplied to the engine at reduced pressure. The effect of throttling has been depicted in the *h-s* and *p-v* plots of Fig. 17.18

The enthalpy drop responsible for doing work has been reduced from $(h_1 - h_2)$ to $(h_1 - h_2')$ and accordingly the work output will get reduced from area *abcdefa* to area *ab′c′d′efa*.

During throttle governing, the steam consumption (m) varies linearly in relation to indicated power (IP)

$$m = A + B\ (IP)$$

This aspect is known as Willian's law.

The system though simple and low in cost has the inherent disadvantage of low thermal efficiency at part loads. This is due to the fact that full steam pressure is not being utilized. There is degradation of energy and more quantity of steam has to be supplied to give a certain output which without throttling could be obtained with less steam charge. The system is, however, suitable for small engines where efficiency is not of great importance.

Cut off governing : Vary the steam supply keeping the supply pressure constant.

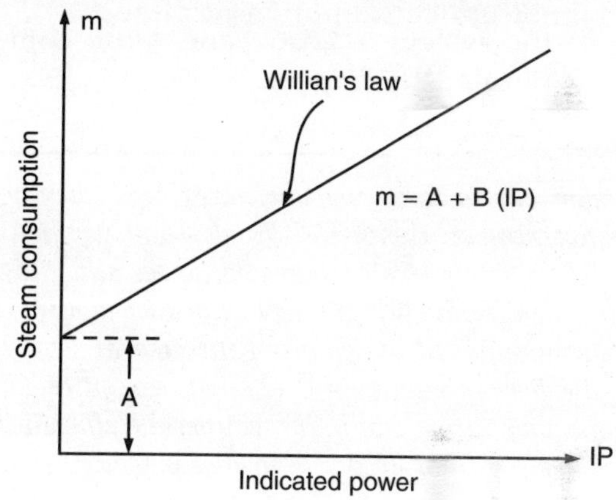

Fig. 17.17.

In cut off governing, a change in the output of the engine is affected by varying the cut off ratio, *i.e.*, by varying the amount of steam supply to the engine. With fall in load, the cut off point is shifted towards the left. (Fig. 17.20). The cut off governing is achieved with help of slide valve that operates under the control of a centrifugal governor. The effect of variable cut off on power, steam consumption and work done per kg of steam admitted is shown in Fig. 17.16.

The salient aspects of cut off governing are:

(*i*) full use is made of steam pressure and at part loads the system is more economical

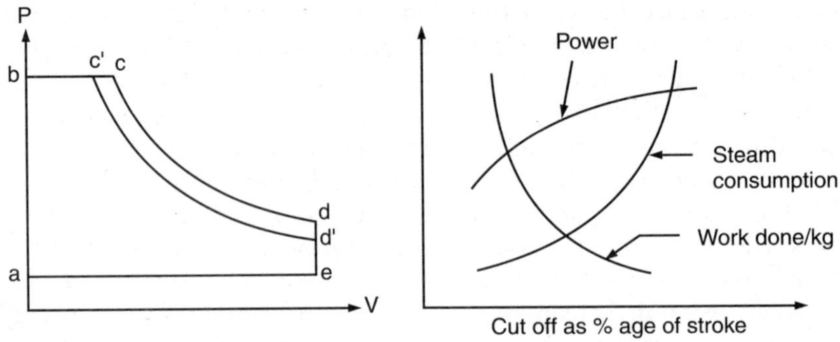

Fig. 17.18. Cut off as percentage of stroke

(*ii*) to meet very light loads, the cut off has to be very small and that promotes condensation

(*iii*) the system is costly and quite complicated to operate.

EXAMPLE 17.13

A steam engine governed by throttling (varying the inlet pressure) used 525 kg of steam per hour when developing 40 kW indicated power and 2150 kg/hr when producing 200 kW indicated power. If the steam supplied is dry and saturated at 12 bar and the back pressure is 0.125 bar, compute the steam consumption and indicated thermal efficiency when developing 150 kW indicated power.

Solution : When the power output is varied by throttling, the steam consumption in kg/hr is expressed in terms of indicated power by Willan's law which is expressed as

$$m = A + B \times \text{indicated power}$$

where A and B are constants and m is steam consumption per hour.

Then substituting of the values of steam consumption for two loads in the expression for Willan's law we have

$$525 = A + 40\,B\text{, and}$$
$$2150 = A + 200\,B$$

Upon solution, we obtain

$$B = \frac{2150 - 525}{200 - 40} = 10.156$$

and $A = 525 - 40 \times 10.156 = 118.76$

Hence for an output of 150 kW,

$$m = 118.76 + 10.156 \times 150$$
$$= 1642.16 \text{ kg/hr}$$

For dry and saturated steam at 12 bar :

$$h_1 = 2782 \text{ kJ/kg}$$

Enthalpy of water at back pressure of 0.16 bar :

$$h_f = 232 \text{ kJ/kg}$$

Indicated thermal efficiency

$$= \frac{\text{heat equivalent of indicated power}}{\text{heat supplied in raising steam}}$$

$$= \frac{150 \times 3600}{1642.16(2782 - 232)}$$

$$= 0.1289 \text{ or } 12.89\%$$

17.10. COMPOUND STEAM ENGINES

Modern tendency is to use steam at higher pressure as the high pressure steam gives a high cycle efficiency and the plant requires less space per kW of power developed. However, when the high pressure steam expands to the condenser pressure in a single cylinder, the following difficulties are encountered :

1. Expansion ratio is increased. The cylinder has to be designed not only to with-stand high initial pressure but also to accommodate greater volumes at low pressure. This means that the cylinder will be of very heavy construction.

2. The excessive initial stresses on the moving parts at the commencement of stroke ask for heavy and strong moving parts.

3. In a single cylinder engine with heavy reciprocating masses, balancing is more difficult and a large flywheel is needed to even out the turning moment which varies considerably.

4. Large pressure difference between the inlet and exhaust steam results in corresponding large temperature difference and consequently there is an increased condensation loss.

To overcome the above mentioned difficulties, the compound steam engine has been developed in which the steam is expanded in two or more cylinders. The steam from the boiler is first admitted into high pressure cylinder which then exhausts the partially expanded steam into the next cylinder where it further expands.

Individual expansion cylinders are so proportioned that each piston moves through the same stroke and the cylinders develop approximately equal powers.

Classification

The compound steam engines may be classified in following ways :

1. *Number of cylinders* : The expansion of steam may be carried out in two, three or even four cylinders in succession. A two cylinder compound steam engine is called double expansion engine and three cylinder compound engines is called triple expansion engine. A four cylinder compound steam engine is called a quadruple expansion engine.

2. *Arrangement of cylinders and cranks* : Depending upon the arrangement of cylinder and crank, the compound steam engines are divided into two categories namely Tandem compounding and Cross compounding. The cross compounding is further subdivided into Woolf type compound steam engine and the Receiver type compound steam engine.

Operational features of compounding arrangements

1. Tandem compounding

(*i*) The axis of both the high pressure and low pressure cylinders lie in the same straight line.

(*ii*) The piston in both the cylinders have a common piston rod, cross head, connecting rod and crank.

(*iii*) High pressure steam from the boiler enters the high pressure cylinder and after

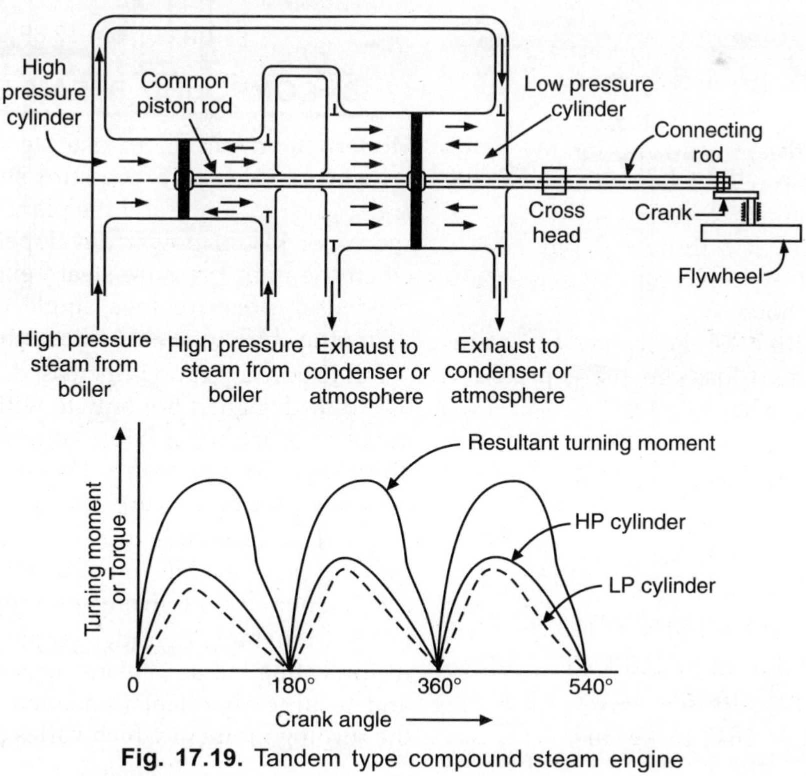

Fig. 17.19. Tandem type compound steam engine

partial expansion flows directly to the low pressure cylinder where it expands further to the back pressure.

Due to common piston rods and crank, the operation in both the cylinders is in phase with the results that maximum torques would occur at approximately the same crank angle. There would be a great variation in inertia force and resulting turning moment. This would necessitate heavy flywheel and balancing weight.

2. Woolf compounding

(*i*) The cylinders are arranged side by side and their axes are parallel to each other.

(*ii*) The cranks are 180 - degree apart, *i.e.*, when the crank of high pressure cylinder is at cover end, the crank of the low pressure cylinder will be at the crank end.

(*iii*) The time of exhaust of high pressure cylinder coincides with the time of admission to low pressure cylinder. Thus the exhaust from high pressure cylinder passes directly to the low pressure cylinder and there is continuous expansion during the stroke.

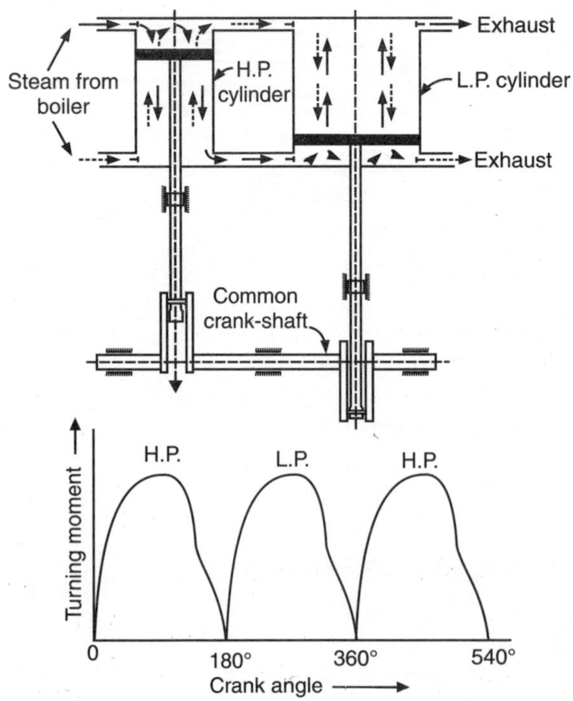

Fig. 17.20. Woolfe type compound steam engine

Since the two cranks are at 180-degree to each other, the two cycles are in phase causing large variation in the turning moment on the crankshaft. Obviously there is need of heavier flywheel to overcome the fluctuations of turning moment.

3. Receiver compounding

(*i*) The high pressure and low pressure cylinders have separate piston rods attached to different cranks set at 90-degree to each other.

(*ii*) The strokes of the two cylinders are out of phase and accordingly the steam from high pressure cylinder cannot exhaust directly to low pressure cylinder.

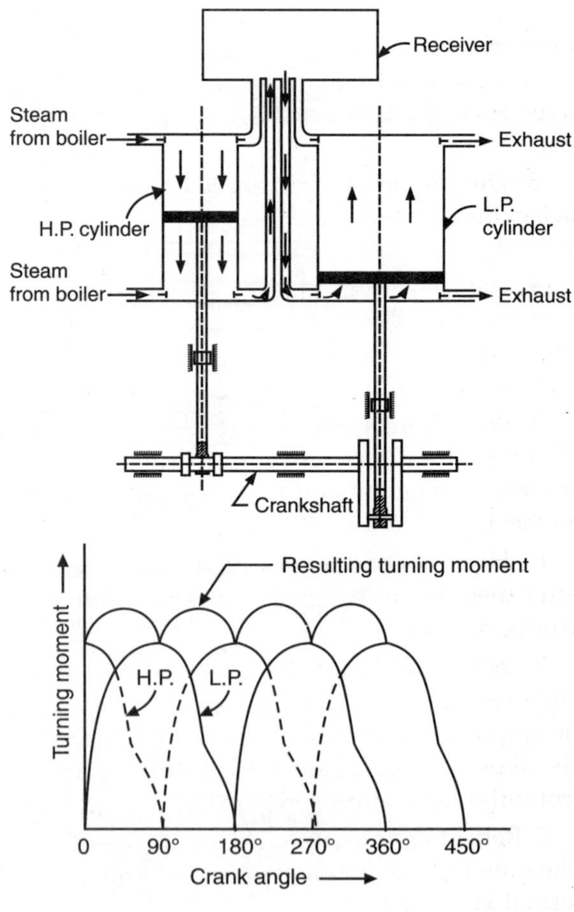

Fig. 17.21. Receiver type compound steam engine

(*iii*) The exhaust from high pressure cylinder is taken to a reservoir called receiver. The receiver holds the steam until low pressure cylinder is ready to receive it. The pressure in the receiver varies throughout the stroke according to supply and demand of steam from it. To minimize these fluctuations, the receiver is kept atleast 1 to 1.5 times the high pressure cylinder volume.

The turning moment variations in this arrangement are fairly reduced and accordingly a smaller flywheel is needed.

Advantages of compounding :

1. The temperaure range per cylinder in compound engines is less and that reduces condensation of steam.

2. Each cylinder is designed for its own pressure range. The low pressure cylinders donot have to withstand high pressures and hence need not be made of thick walls. This saves in weight and material and hence in cost.

3. Due to small difference of pressure in each cylinder there is reduced leakage of steam past valves and less stresses in the engine parts.

4. The total forces are distributed over more components and consequently each individual component is subjected to less force and there are less chances of damage.

5. With cranks of two cylinders placed at 90° to each other, more uniform turning moment is obtained and this results in lighter flywheel.

6. High speeds are possible because of better mechanical balance and more uniform turning moment.

7. With multi-cylinder construction, the cranks can be set at different angular positions. The engine can then start in any position, and this aspect is of great utility in marine, locomotive and mining services.

8. Reheating in between stages is possible. Reheating reduces condensation and improves thermal efficiency.

9. Compounding gives improvement in steam economy.

10. For the same power, the compound steam engines are cheaper than the simple steam engines. This is due to the fact that scantlings of the compound steam engine though more in number are lighter and cheaper to make.

11. When break down occurs, the compound engine can be readily modified to work on reduced load.

The above merits are offset to some extent by the following disadvantages :

(*i*) The system is more complex, needs greater attendence and maintenance.

(*ii*) There is increased wear and tear and consequently more lubrication is needed.

(*iii*) The arrangement is more bulky and needs greater floor area.

17.11. CALCULATIONS FOR CYLINDER DIMENSIONS

- **Compound engine with full expansion**

Figure 17.22 shows the hypothetical indicator diagram of a two cylinder compound steam engine in which steam has continuous full expansion upto back pressure. The diagram is drawn on the assumptions that :

(*i*) there is no clearance

(*ii*) expansion is hyperbolic in both stages

(*iii*) the expansion lines for both the high pressure (HP) and low pressure (LP) cylinders form a continuous curve

The contour *abcda* represents the *p-v* diagram for HP cylinder and contour *dcefd* represents the *p-v* diagram for LP cylinder. The cut offs for the HP cylinder and LP cylinder occur respectively at points *b* and *c* on the expansion curve *bce*.

For *high pressure cylinder* :

$$\text{Ratio of expansion } r_H = \frac{V_H}{V_1}$$

Theoretical mean effective pressure,

$$p_{mH} = \frac{p_1}{r_H}(1 + \log_e r_H) - p_2$$

For *low pressure cylinder* :

$$\text{Ratio of expansion } r_L = \frac{V_L}{V_H}$$

Theoretical mean effective pressure p_{mL}

$$= \frac{p_2}{r_L}(1 + \log_e r_L) - p_b$$

For the engine as a whole :

Ratio of expansion r

$$= \frac{V_L}{V_1} = \frac{V_L}{V_H} \times \frac{V_H}{V_1}$$

= ratio of cylinder volumes × r_H

Theoretical mean effective pressure

$$= \frac{p_1}{r}(1 + \log_e r) - p_b$$

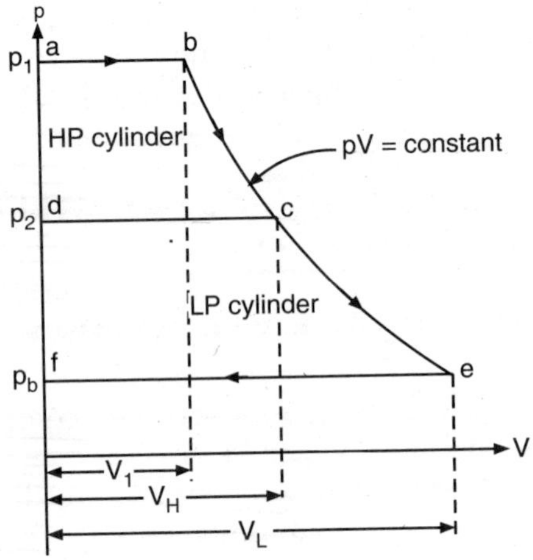

Fig. 17.22. Compound engine with full expansion

The cylinder dimensions for the compound steam engine are worked out depending on equal work done in each cylinder and equal initial force on pistons.

For equal work done in each cylinder

$$\left[\frac{p_1}{r_H}(1+\log_e r_H)-p_2\right] V_H$$

$$= \left[\frac{p_2}{r_L}(1+\log_e r_L)-p_b\right] V_L$$

or $$\left[\frac{p_1 V_1}{V_H}(1+\log_e r_H)-p_2\right] V_H$$

$$= \left[\frac{p_2 V_H}{V_L}(1+\log_e r_L)-p_b\right] V_L$$

or $p_1 V_1 (1 + \log_e r_H) - p_2 V_H$
$= P_2 V_H (1 + \log_e r_H) - p_b V_L$

From the hyperbolic expansion curve :

$$p_1 V_1 = p_2 V_H = p_b V_L$$

The above identify can then be written as

$1 + \log_e r_H = 1 + \log_e r_L$ or $r_H = r_L$

Thus the ratio of expansion must be same for both the cylinders.

Also $\frac{V_H}{V_1} = \frac{V_L}{V_H}$ or $\frac{p_1}{p_2} = \frac{p_2}{p_b}$ or $p_2 = \sqrt{p_1 p_b}$

The above relation follows from the identity $p_1 V_1 = p_2 V_H = p_b V_L$ for the hyperbolic expansion curve.

For equal initial force or *load on the pistons*

$$(p_1 - p_2) A_H = (p_2 - p_b) A_L ; \quad \frac{p_1 - p_2}{p_2 - p_b} = \frac{A_L}{A_H}$$

where A_H is area of HP piston and A_L is area of LP piston. Assuming stroke to be same for each cylinder, we get

$$\frac{p_1 - p_2}{p_2 - p_b} = \frac{V_L}{V_H} = \frac{p_2}{p_b} (\because p_2 V_H = p_b V_L)$$

$$\therefore (p_1 - p_2) p_b = (p_2 - p_b) p_2$$

$$p_1 p_b - p_2 p_b = p_2^2 - p_2 p_b$$

or $$p_2 = \sqrt{p_1 p_b}$$

Apparently the conditions of equal initial force on pistons and equal work in each cylinder are achieved theoretically if the intermediate pressure is the geometric mean of steam supply pressure and the back pressure.

The equal work criterion is for uniform turning moment developed in each cylinder. The criterion for equal inertia force is for equal loads on the pistons so that the piston rods and gears are strained to the same extent.

- **Compound engines with incomplete expansion in low pressure cylinder**

Figure 17.23 shows the hypothetical indicator diagram of a two cylinder compound engine with incomplete expansion in the low pressure cylinder. The diagram is drawn on the assumptions that :

(*i*) there is no clearance

(*ii*) expansion is hyperbolic in both stages

(*iii*) expansion is full in the HP cylinder but the toe end of the LP cylinder is cut off

The contour *abcda* represents the *p-v* diagram for HP cylinder and the contour *dcefgd* represents the *p-v* diagram for the LP cylinder. The cut offs for the HP cylinder and LP cylinder occur respectively at point *b* and *c* on the expansion curve *bce*.

Neglecting clearance, the volumes of HP cylinder and LP cylinder are

$$V_H = V_2 = A_H \times L$$

and $V_L = V_3 = A_L \times L$

where *A* denotes the piston area and *L* designates the stork length which is taken to be same for both the cylinders.

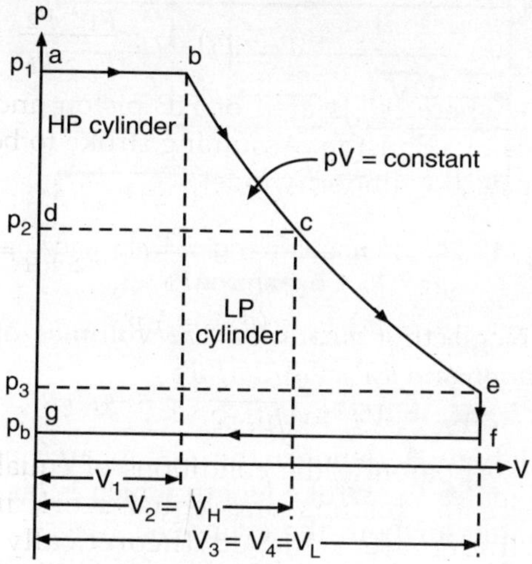

Fig. 17.23. Compound engine with incomplete expansion in LP cylinder

The expansion ratios for the HP cylinder and LP cylinder are

$$r_H = \frac{V_H}{V_1} = \frac{V_2}{V_1}$$

and $r_L = \dfrac{V_L}{V_2} = \dfrac{V_3}{V_2}$

If the diagram factor is *K* for each cylinder, then

Mean effective pressure for HP cylinder,

$$p_{mH} = K\left[\frac{p_1}{r_H}(1+\log_e r_H) - p_2\right]$$

and work done

$$W_H = p_{mH} \times \text{cylinder volume}$$

$$= K\left[\frac{p_1}{r_H}(1+\log_e r_H) - p_2\right] \times V_2$$

$$= K\,[p_1V_1 + p_1V_1 \log_e r_H - p_2V_2]$$

$$\left[\because r_H = \frac{V_2}{V_1}\right]$$

Since expansion is hyperbolic,

$p_1V_1 = p_2V_2$ and therefore,

$$W_H = K\,p_1V_1 \log_e r_H$$

Similarly, for the low pressure cylinder

$$p_{mL} = K\left[\frac{p_2}{r_L}(1+\log_e r_L) - p_b\right]$$

$$W_L = p_{mL} \times V_3$$

$$= K\left[\frac{p_2}{V_3/V_2}(1+\log_e r_L) - p_b\right] \times V_3$$

$$= K\left[p_2V_2(1+\log_e r_L) - p_bV_3\right]$$

When the work done in both the cylinders is equal

$$K\,p_1V_1 \log_e r_H = K\left[p_2V_2(1+\log_e r_L) - p_bV_3\right]$$

or $\log_e r_H = \dfrac{p_2V_2}{p_1V_1}(1 + \log_e r_L) - \dfrac{p_bV_3}{p_1V_1}$

For hyperbolic expansion

$p_1V_1 = p_2V_2$ and therefore

$$\log_e r_H = 1 + \log_e r_L - \frac{p_bV_3}{p_1V_1}$$

or $\log_e\left(\dfrac{r_H}{r_L}\right) = 1 - \dfrac{p_bV_3}{p_1V_1}$

Let r = overall ratio of expansion

$$= \frac{V_3}{V_1} = \frac{V_3}{V_2} \times \frac{V_2}{V_1} = r_H\,r_L$$

Then : $\log_e\left[\dfrac{r_H}{r/r_H}\right] = 1 - \dfrac{p_b}{p_1}r$

or $\log_e\left[\dfrac{r_H^2}{r}\right] = 1 - \dfrac{p_b}{p_1}r$

Solution gives : $r_H = \sqrt{r\,\text{antilog}\left[1 - \dfrac{p_b}{p_1}r\right]}$

Again from hyperbolic expansion :

$$p_1V_1 = p_2V_2 \quad \text{or} \quad \frac{p_1}{p_2} = \frac{V_2}{V_1} = r_H$$

That gives : $p_2 = \dfrac{p_1}{\sqrt{r \times \text{antilog}\left(1 - \dfrac{p_b}{p_1} r\right)}}$ ...(*i*)

When the initial load or thrust on both the pistons is equal

$$(p_1 - p_2) A_H = (p_2 - p_b) A_L$$

or $\dfrac{p_1 - p_2}{p_2 - p_b} = \dfrac{A_L \times L}{A_H \times L} = \dfrac{V_3}{V_2} = r_L$

Again : $r = r_H \, r_L$ or $r_L = \dfrac{r}{r_H} = r \times \dfrac{V_1}{V_2}$

$= r \dfrac{p_2}{p_1}$ $\quad (\because p_1 V_1 = p_2 V_2)$

$\therefore \dfrac{p_1 - p_2}{p_2 - p_b} = r \dfrac{p_2}{p_1}$

or $r \, p_2^2 - r \, p_2 \, p_b = p_1^2 - p_1 \, p_2$

This can be recast as

$r \, p_2^2 - p_2 \, (r \, p_b - p_1) - p_1^2 = 0$ and the solution of this quadratic equation in p_2 gives

$$p_2 = \frac{(rp_b - p_1) + \sqrt{(r \, p_b - p_1)^2 + 4rp_1^2}}{2r} \quad ...(ii)$$

Apparently the values of intermediate pressure p_2 as prescribed by expressions (*i*) and (*ii*) are different and as such it is not possible to fulfil both conditions of equal work and equal initial loads.

- **Compound engine with incomplete expansion in both cylinders**

Figure 17.24 shows the hypothetical indicator diagram for a two cylinder receiver type compound steam engine in which there is incomplete expansion in both the cylinders. This diagram is drawn on the assumptions that:

(*i*) there is no clearance

(*ii*) expansion is hyperbolic in both stages

(*iii*) expansion lines for the HP cylinder and LP cylinder form a continuous curve

The contour *abckea* represents the *p-v* diagram for HP cylinder and contour *efghje* represents the *p-v* diagram for LP cylinder. The cut offs for the HP cylinder and LP cylinder occur respectively at points *b* and *f*. The pressure loss ($p_2 - p_3$) in the receiver causes a loss of work which is shown by the shaded area.

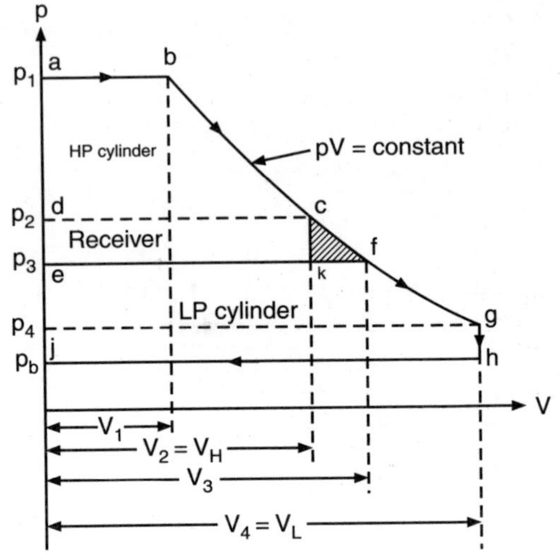

Fig. 17.24. Compound engine with imcomplete expansion

Neglecting clearance, the volumes of HP cylinder and LP cylinder are

$$V_H = V_2 = A_H \, L \text{ and } V_L = V_4 = A_L \, L$$

where *A* denotes the piston area and *L* designates the stroke length which is taken to be same for both the cylinders.

The expansion ratios for the HP cylinder and LP cylinder are

$$r_H = \frac{V_H}{V_1} = \frac{V_2}{V_1} \text{ and } r_L = \frac{V_L}{V_3} = \frac{V_4}{V_3}$$

If *K* is the diagram factor for each cylinder, then

Mean effective pressure for HP cylinder,

$$p_{mH} = K \left[\frac{p_1}{r_H}(1 + \log_e r_H) - p_3 \right]$$

and work done $W_H = p_{mH} \times$ cylinder volume

$$= K \left[\frac{p_1}{r_H}(1 + \log_e r_H) - p_3 \right] \times V_2$$

The corresponding expressions for LP cylinder are

$$p_{mL} = K \left[\frac{p_3}{r_L}(1 + \log_e r_L) - p_b \right]$$

$$W_L = p_{mL} \times V_L$$

$$= K\left[\frac{p_3}{r_L}(1+\log_e r_L) - p_b\right] V_4$$

For equal work shared by each cylinder

$$K\left[\frac{p_1}{r_H}(1+\log_e r_H) - p_3\right] V_2$$

$$= K\left[\frac{p_3}{r_L}(1+\log_e r_L) - p_b\right] V_4$$

or $\frac{p_1}{r_H}(1 + \log_e r_H) - p_3$

$$= \left[\frac{p_3}{r_L}(1+\log_e r_L) - p_b\right] \times \frac{V_4}{V_2}$$

$$= \left[\frac{p_3}{r_L}(1+\log_e r_L) - p_b\right] \times \frac{V_L}{V_H}$$

Thus : Mean effective pressure of HP cylinder = Mean effective pressure of LP cylinder × Ratio of cylinder volumes.

For equal initial loads on the pistons in both cylinders

$$(p_1 - p_3) A_H = (p_3 - p_b) A_L$$

or $\frac{p_1 - p_3}{p_3 - p_b} = \frac{A_L}{A_H} = \frac{A_L \times L}{A_H \times L} = \frac{V_L}{V_H}$

= ratio of cylinder volumes

The above expressions can be used to compute the cylinder dimensions.

EXAMPLE 17.14

A double acting compound steam engine is to develop 350 kW while running at 360 rpm. Steam is supplied at 10 bar and the back pressure is 0.3 bar. The overall expansion ratio is 10, the diagram factor is 0.8, and allowable piston speed is 432 m/min. Assuming hyperbolic expansion and neglecting clearance and receiver losses, determine the admission pressure for the low pressure cylinder and the diameter of each cylinder so that the initial load on the piston is same.

Solution : Refer Fig. 17.23.

Mean effective pressure referred to LP cylinder is

$$p_{mL} = K\left[\frac{p_1}{r}(1+\log_e r) - p_b\right]$$

$$= 0.8\left[\frac{10}{10}(1+\log_e 10) - 0.3\right]$$

$$= 2.402 \text{ bar} = 2.402 \times 10^5 \text{ N/m}^2$$

Piston speed = 2 LN = 432 m/min

∴ Stroke length

$$L = \frac{432}{2 \times 360} = 0.6 \text{ m}$$

Indicate power = $p_{mL}\, LA(2N)$

Substituting the appropriate values, we get

$$350 \times 10^3 = (2.402 \times 10^5) \times 0.6 \times \frac{\pi}{4}(d_{LP})^2 \times \left(2 \times \frac{360}{60}\right)$$

Solution gives : Diameter of LP cylinder

$$d_{LP} = 0.508 \text{ m or } 50.8 \text{ cm}$$

From the hyperbolic expansion curve *bce*,

$$p_1V_1 = p_2V_2 = p_3V_3$$
(*i*) (*ii*) (*iii*)

From identities (*i*) and (*iii*),

$$p_3 = p_1\frac{V_1}{V_3} = 10 \times \frac{1}{10} = 1 \text{ bar}$$

From identities (*ii*) and (*iii*),

$$p_2 V_H = p_3 V_L$$

That gives :

$$p_2 = p_3 \times \frac{V_L}{V_H} = p_3 \times \frac{\frac{\pi}{4}(d_{LP})^2 \times L}{\frac{\pi}{4}(d_{HP})^2 \times L}$$

$$= p_3 \times \frac{(d_{LP})^2}{(d_{HP})^2} = 1 \times \frac{(d_{LP})^2}{(d_{HP})^2}$$

$$= \frac{(d_{LP})^2}{(d_{HP})^2}$$

Since initial load on the pistons is same,

$$(p_1 - p_2)\frac{\pi}{4}(d_{HP})^2 = (p_2 - p_b)\frac{\pi}{4}(d_{LP})^2$$

or $(10 - p_2)\frac{(d_{LP})^2}{p_2} = (p_2 - 0.3) \times (d_{LP})^2$

or $\quad (10 - p_2) = (p_2 - 0.3)\, p_2$

$= p_2^2 - 0.3\, p_2$

or $\quad p_2^2 + 0.7\, p_2 - 10 = 0$

Solution of this quadratic equation gives:

$$p_2 = \frac{-0.7 \pm \sqrt{(0.7)^2 - 4\times 1(-10)}}{2}$$

$= 2.83$ bar

Therefore the admission pressure for LP cylinder is 2.83 bar

From the correlation

$$p_2 = \frac{(d_{LP})^2}{(d_{HP})^2} \text{ derived above,}$$

Diameter of HP cylinder

$$d_{HP} = \frac{d_{LP}}{\sqrt{p_2}} = \frac{50.8}{\sqrt{2.83}} = \mathbf{30.2\ cm}$$

EXAMPLE 17.15

A two cylinder, double acting Woolf compounding steam engine is required to generate 126.5 kW at a speed of 180 rev/min. The engine is supplied with dry saturated steam at 10 bar and the exhaust occurs to a condenser at 0.5 bar. The ratio of expansion for the whole engine is 6 and the stroke for each cylinder equals 1.4 times the diameter of LP cylinder. Presuming that diagram factor is 0.8, estimate.

(a) dimensions of the low pressure cylinder

(b) diameter of the high pressure cylinder when equal work is shared by each cylinder.

Neglect clearance effects.

Solution : For a Woolf compounding, the exhaust steam from HP cylinder to the LP cylinder is continuous and there is no pressure drop (complete expansion in HP cylinder). However there is pressure drop at release. Then with reference to Fig. 17.23,

Mean effective pressure referred to low pressure cylinder is

$$p_{mL} = K\left[\frac{p_1}{r}(1+\log_e r) - p_b\right]$$

$$= 0.8\left[\frac{10}{6}(1+\log_e 6) - 0.5\right]$$

$= 3.322$ bar $= 3.322 \times 10^5$ N/m²

Indicated power $= p_{mL}\, LA(2N)$

Substituting the appropriate values, we get

$$126.5 \times 10^3 = (3.322 \times 10^5) \times (1.4\, d_{LP}) \times \frac{\pi}{4}(d_{LP})^2 \times \left(2\times\frac{180}{60}\right)$$

Solution gives : Diameter of LP cylinder

$d_{LP} = 0.3869$ m or 38.69 cm

Stroke length $= 38.69 \times 1.4 = 54.16$ cm

(*b*) With complete expansion, the work done per stroke in the HP cylinder is

$$W_{HP} = K\, p_1 V_1 \log_e r_H$$

$$r = \frac{V_4}{V_1} = \frac{V_{LP}}{V_1}$$

$$V_1 = \frac{1}{6} \times \frac{\pi}{4}(0.3869)^2 \times 0.5416$$

$= 0.0106$ m³

Further, the work is shared equally by the two cylinders. As such the power developed by the HP cylinder will be equal to half the power developed by the whole engine. Thus

$$\frac{126.5}{2} \times 10^3 = K\, p_1 V_1 \log_e r_H \times \frac{2N}{60}$$

$$= 0.8 \times (10 \times 10^5) \times 0.0106 \log_e r_H \times \frac{2\times 180}{60}$$

Solution gives :

$\log_e r_H = 1.243$ or $r_H = 3.466$

$$r_H = \frac{V_2}{V_1} = \frac{V_H}{V_1};$$

$V_H = 3.446 \times 0.0106 = 0.03652$ m³

$$\therefore \frac{\pi}{4}(d_{HP})^2 \times l = 0.03652$$

or $\quad \frac{\pi}{4}(d_{HP})^2 \times 0.5416 = 0.03652$

Diameter of HP cylinder

$$= \sqrt{\frac{0.03652}{0.5416}\times\frac{4}{\pi}} = 0.2931 \text{ m or } \mathbf{29.31\ cm}$$

EXAMPLE 17.16.

In a two cylinder double acting compound steam engine, the steam is supplied at 15 bar and 0.9 dry and exhausted into a condenser at 0.4 bar. The cylinder diameter for HP and LP units are 20 cm and 30 cm and the length of stroke for both cylinders is 40 cm. The engine turns 240 revolutions per minute, the diagram factor referred to LP cylinder is 0.8 and the expansion occurs completely in HP cylinder. Considering hyperbolic expansion throughout, zero clearance and each cylinder having equal initial piston loads, determine :

(a) intermediate pressure,

(b) indicated power output, and

(c) steam consumption in kg/hr.

Solution : Refer Fig. 17.23 for *p-v* diagram of the processes involved in the working of the given compound steam engine. For equal initial loads on the pistons,

$$(p_1 - p_2)\, A_{HP} = (p_2 - p_b)\, A_{LP}$$

$$(15 - p_2)\times\frac{\pi}{4}(0.2)^2 = (p_2 - 0.4)\times\frac{\pi}{4}(0.3)^2$$

or $\quad (15 - p_2) = (p_2 - 0.4) \times 2.25$

Solution gives : Intermediate or receiver pressure,

$$p_2 = \frac{15 + 0.4\times 2.25}{3.25} = 4.89 \text{ bar}$$

(*b*) From the hyperbolic expansion process:

$$p_1 V_1 = p_2 V_2$$

Volume at cut off in the HP cylinder,

$$V_1 = \frac{p_2}{p_1}\times V_2$$

$$= \frac{4.89}{15}\times\left[\frac{\pi}{4}\times(0.2)^2\times 0.4\right]$$

$$= 4.09 \times 10^{-3} \text{ m}^3$$

Volume of LP cylinder

$$V_2 = \frac{\pi}{4}\times(0.3)^2\times 0.4 = 0.0268 \text{ m}^3$$

Overall expansion ratio

$$r = \frac{V_4}{V_1} = \frac{V_L}{V_1} = \frac{0.0268}{4.09\times 10^{-3}} = 6.55$$

$$(\because V_4 = V_L)$$

Mean effective pressure for compound engine referred to LP cylinder p_{mL}

$$= \left[\frac{p_1}{r}(1+\log_e r) - p_b\right]\times \text{diagram factor}$$

$$= \left[\frac{15}{6.55}(1+\log_e 6.55) - 0.4\right]\times 0.8$$

$$= 4.955 \text{ bar}$$

Indicated power

$$= p_{mL}\, LA(2N)$$

$$= \left[\begin{array}{l}(4.955\times 10^5)\times 0.4\\ \times\frac{\pi}{4}(0.3)^2\times\left(2\times\frac{240}{60}\right)\end{array}\right]\times\frac{1}{10^3} = 112 \text{ kW}$$

(*c*) Volume of steam admitted per hour

= (volume at cut off) × (number of working strokes per hour)

$= (4.09 \times 10^{-3}) \times (2 \times 240 \times 60)$

$= 117.79 \text{ m}^3$

Specific volume of steam being admitted

$v_1 = v_f$ at 15 bar

$= x \times v_{fg}$ at 15 bar

$= 0.9 \times 0.1318 = 0.1186 \text{ m}^3/\text{kg}$

Then, steam consumption

$$= \frac{117.79}{0.1186} = 993.17 \text{ kg/hr}$$

EXAMPLE 17.17

A double acting compound steam engine is supplied with steam at 10 bar superheated to 240°C. The cut off in the HP cylinder takes place at 50% of stroke and the overall expansion ratio is 12. The engine turns 200 rev/min, has indicated thermal efficiency of 15% with steam consumption of 5 kg/min. The stroke length equals 1.2 times the LP cylinder diameter, the back pressure for the LP cylinder is 650 mm of mercury vacuum and the overall diagram factor is 0.7. Determine :

(a) the diameters of the HP and LP cylinders

(b) the stroke length and cut off for the LP cylinders when the initial piston loads are equal.

It may assumed that the barometer reads 750 mm of mercury and expansion is incomplete in both cylinders.

Solution : Back pressure

$$p_b = \frac{750 - 650}{760} \times 1.01325$$

$$= 0.1333 \text{ bar}$$

Mean effective pressure for the combined diagram (Fig. 17.24) referred to LP cylinder is

$$p_{mL} = K\left[\frac{p_1}{r}(1+\log_e r) - p_b\right]$$

$$= 0.7\left[\frac{10}{12}(1+\log_e 12)-0.1333\right]$$

$$= 1.94 \text{ bar}$$

At 10 bar and 240°C :

$$h_1 = 2919.5 \text{ kJ/kg}$$

and $v_{s1} = 0.2274 \text{ m}^3/\text{kg}$

and at 0.1333 bar,

$$h_{f2} = 215.5 \text{ kJ/kg}$$

Indicated thermal efficiency

$$= \frac{\text{indicated power}}{\text{heat supplied through steam}}$$

∴ Indicated power

$$= \eta_{th} \times m_s (h_1 - h_{f2})$$

$$= 0.15 \times \frac{5}{60} \times (2919.5 - 215.5)$$

$$= 33.8 \text{ kJ/s (or kW)}$$

Indicated power is also given as $p_m LA (2N)$

Therefore,

$$33.8 \times 10^3 = (1.94 \times 10^5) \times (1.2\, d_{LP}) \times \frac{\pi}{4}(d_{LP})^2 \times \left(2\times\frac{200}{60}\right)$$

$$(d_{LP})^3 = 0.02774$$

or $d_{LP} = 0.3031 \text{ m} = 30.31 \text{ cm}$

and $L = 1.2 \times 30.31 = 36.37 \text{ cm}$

Now : Steam consumption

$$= 5 \times 0.2274 \text{ m}^3/\text{min}$$

$$= \frac{5\times 0.2274}{2\times 200}$$

$$= 2.84 \times 10^{-3} \text{ m}^3/\text{stroke}$$

Since cut off for the HP cylinder takes place at 50% of stroke, the volume of HP cylinder

$$= \frac{2.84\times 10^{-3}}{0.5} = 5.68 \times 10^{-3} \text{ m}^3$$

$$\therefore \quad \frac{\pi}{4}(d_{HP})^2 \times L = 5.68 \times 10^{-3}$$

$$\frac{\pi}{4}(d_{HP})^2 \times 0.3637 = 5.68 \times 10^{-3}$$

$$(d_{HP})^2 = \frac{5.68\times 10^{-3}\times 4}{\pi\times 0.3637} = 0.01989$$

That gives : diameter of HP cylinder,

$$d_{HP} = 0.141 \text{ m or } 14.1 \text{ cm}$$

(*b*) Since initial load on the pistons is same

$$(p_1 - p_3) \times \frac{\pi}{4}(d_{HP})^2 = (p_3 - p_b) \times \frac{\pi}{4}(d_{LP})^2$$

or $p_1 - p_3 = (p_3 - p_b)\left(\dfrac{d_{LP}}{d_{HP}}\right)^2$

or $10 - p_3 = (p_3 - 0.1333) \times \left(\dfrac{30.31}{14.1}\right)^2$

$$= 4.609\, p_3 - 0.6105$$

$$\therefore \quad p_3 = \frac{10+0.6105}{5.609} = 1.8917 \text{ bar}$$

Cut off for LP cylinder $\dfrac{V_3}{V_4} = \dfrac{V_3}{V_1} \times \dfrac{V_1}{V_4}$

From the hyperbolic expansion curve *bfcg*,

$$p_1V_1 = p_3V_3\ ; \quad \frac{V_3}{V_1} = \frac{p_1}{p_3}$$

∴ Cut off of LP cylinder

$$= \frac{p_1}{p_3} \times \frac{V_1}{V_4} = \frac{10}{1.8917} \times \frac{1}{12} = 0.44$$

Thus the cut off in LP cylinder takes place at 44% of stroke.

EXAMPLE 17.18

The engine of a double acting compound steam engine is supplied with steam at 8.4 bar and the exhaust takes place into a condenser at 0.28 bar. The cut off in the HP cylinder takes place at half stroke and the ratio of cylinder volumes is 4. Both the cylinders have a common stroke and have incomplete (restricted) expansions. Assuming hyperbolic expansion and neglecting clearance effects, determine

(a) cut off ratio in LP cylinder for equal initial loads on the pistons

(b) mean effective pressure referred to LP cylinder

(c) percentage loss of work due to incomplete expansion in HP cylinder

Solution : Refer Fig. 17.24 for incomplete expansion in both the cylinders

(*a*) For equal initial loads on the pistons

$$(p_1 - p_3)\, A_H = (p_3 - p_b)\, A_L$$

where p_3 is the receiver pressure.

or $$p_1 - p_3 = (p_3 - p_b) \times \frac{A_L}{A_H}$$

$$= (p_3 - p_b) \times \frac{A_L \times L}{A_L \times L}$$

$$= (p_3 - p_b) \frac{V_L}{V_H}$$

$\therefore$ $(8.4 - p_3) = (p_3 - 0.28) \times 4$

Solution gives :

$$p_3 = \frac{8.4 + 0.28 \times 4}{5} = 1.904 \text{ bar}$$

(*a*) Cut off ratio for LP cylinder $\frac{V_3}{V_4}$

$$= \frac{V_3}{V_1} \times \frac{V_1}{V_4} = \frac{V_3}{V_1} \times \frac{V_1}{V_2} \times \frac{V_2}{V_4}$$

$$= \frac{8.4}{1.904} \times 0.5 \times \frac{1}{4} = 0.55$$

Thus the cut off in the LP cylinder takes place at 55% the stroke.

(*b*) The expansion ratios for HP and LP cylinders are

$$r_H = \frac{V_2}{V_1} = \frac{1}{0.5} = 2$$

and $$r_L = \frac{V_4}{V_3} = \frac{1}{0.55} = 1.815$$

Mean effective pressure of HP cylinder

$$p_{mH} = \frac{p_1}{r_H}(1 + \log_e r_H) - p_3$$

$$= \frac{8.4}{2}(1 + \log_e 2) - 1.904$$

$$= 5.207 \text{ bar}$$

When referred to LP cylinder, the mean effective pressure of HP cylinder

$$= p_{mH} \times \frac{V_H}{V_L} = 5.207 \times \frac{1}{4}$$

$$= 1.3017 \text{ bar}$$

Mean effective pressure of LP cylinder

$$p_{mL} = \frac{p_3}{r_L}(1 + \log_e r_L) - p_b$$

$$= \frac{1.904}{1.815}(1 + \log_e 1.815) - 0.28$$

$$= 1.394 \text{ bar}$$

$\therefore$ Total mean effective pressure referred to LP cylinder

$$= 1.3017 + 1.394 = 2.696 \text{ bar}$$

(*c*) Overall expansion ratio

$$r = \frac{V_4}{V_1} = \frac{V_4}{V_2} \times \frac{V_2}{V_1} = \frac{V_L}{V_H} \times \frac{V_2}{V_1}$$

$$= 4 \times \frac{1}{0.5} = 8$$

Mean effective pressure of entire indicator diagram when the shaded area is also included, *i.e.*, when there is complete expansion in the HP cylinder.

$$= \frac{p_1}{r}(1 + \log_e r) - p_b$$

$$= \frac{8.4}{8}(1 + \log_e 8) - 0.28$$

$$= 2.953 \text{ bar}$$

Loss in mean effective pressure due to restricted expansion in HP cylinder.

$$= 2.953 - 2.696 = 0.257 \text{ bar}$$

With all other factors remaining same, the loss in work is proportional to loss in mean effective pressure.

$$\therefore \text{ Percentage loss in work} = \frac{0.257}{2.953} \times 100$$

$$= \mathbf{8.70\%}$$

EXAMPLE 17.19

A double acting compound steam engine develops 250 kW brake power at 120 rev/min when using dry saturated steam at 10 bar initial pressure against a back pressure of 0.4 bar.

The cut off in the high pressure cylinder is at ½ stroke and the cylinder volume ratio is 4. Neglecting clearance effects and assuming that initial load in the pistons is same, compute

(a) receiver steam pressure

(b) fraction of the stroke at which cut off should place in LP cylinder

(c) brake power developed in each cylinder.

Solution : Refer Fig. 17.24 for incomplete expansion in both the cylinders

(*a*) For equal initial loads on the pistons

$$(p_1 - p_3) A_H = (p_3 - p_b) A_L$$

where p_3 is the receiver pressure.

or $$p_1 - p_3 = (p_3 - p_b) \times \frac{A_L}{A_H}$$

$$= (p_3 - p_b) \times \frac{A_L \times L}{A_H \times L}$$

$$= (p_3 - p_b) \frac{V_2}{V_4}$$

$$\therefore \quad 10 - p_3 = (p_3 - 0.4) \times 4$$

Hence receiver pressure

$$p_3 = \frac{10 + 4 \times 0.4}{5} = 2.32 \text{ bar}$$

(b) Cut off ratio for LP cylinder $\frac{V_3}{V_4}$

$$= \frac{V_3}{V_1} \times \frac{V_1}{V_4} = \frac{V_3}{V_1} \times \frac{V_1}{V_2} \times \frac{V_2}{V_4}$$

From the hyperbolic expansion curve *bcfg*,

$$p_1 V_1 = p_3 V_3 \,; \quad \frac{V_3}{V_1} = \frac{p_1}{p_3}$$

∴ Cut off ratio of LP cylinder

$$= \frac{p_1}{p_3} \times \frac{V_1}{V_2} \times \frac{V_2}{V_4}$$

$$= \frac{10}{2.32} \times \frac{1}{2} \times \frac{1}{4} = 0.539$$

Thus the cut off in the LP cylinder occurs at 53.9% of the stroke.

(c) The expansion ratios for HP and LP cylinders are

$$r_H = \frac{V_2}{V_1} = \frac{1}{1/2} = 2$$

and $$r_L = \frac{V_4}{V_3} = \frac{1}{0.539} = 1.855$$

Mean effective pressure of HP cylinder

$$p_{mH} = \frac{p_1}{r_H} (1 + \log_e r_H) - p_3$$

$$= \frac{10}{2} (1 + \log_e 2) - 2.32 = 6.146 \text{ bar}$$

When referred to LP cylinder, the mean effective pressure of HP cylinder is

$$= p_{mH} \times \frac{V_H}{V_L} = 6.146 \times \frac{1}{4}$$

$$= 1.536 \text{ bar}$$

Mean effective pressure of LP cylinder

$$p_{mL} = \frac{p_3}{r_L} (1 + \log_e r_L) - p_b$$

$$= \frac{2.32}{1.855} (1 + \log_e 1.855) - 0.4$$

$$= 1.623 \text{ bar}$$

Now : $$\frac{\text{work done in HP cylinder}}{\text{work done in LP cylinder}}$$

$$= \frac{\text{mep of HP cylinder referred to LP cylinder}}{\text{mep of LP cylinder}}$$

Let P_H and P_L be the power developed in HP and LP cylinders respectively. Then

$$\frac{P_H}{P_L} = \frac{1.536}{1.623} = 0.946$$

and $$P_H + P_L = 250$$

From these two identities

$$0.946\, P_L + P_L = 250 \,;$$

$$P_L = \frac{250}{1.946} = 128.47 \text{ kW}$$

and $$P_H = 250 - 128.47 = \mathbf{121.53\ kW}$$

EXAMPLE 17.20

The following data refers to a compound steam engine which develops 460 kW when running at 150 rpm.

Steam inlet pressure = 8 bar and exhaust pressure = 0.6 bar

Cut off = 0.4 of stroke in HP cylinder and 0.5 of stroke in LP cylinder

Cylinder ratio = 4 and diagram factor = 0.8

Piston speed = 150 m/min.

Both cylinders have restricted expansion. Neglecting clearance and assuming hyperbolic expansion, compute :

(a) receiver pressure,

(b) work done per stroke,

(c) diameter of high pressure cylinder

Also compare initial loads on the pistons.

Solution : Refer Fig. 17.24 for restricted expansion in both cylinders.

The expansion ratios for HP and LP cylinders are

$$r_H = \frac{V_2}{V_1} = \frac{1}{0.4} = 2.5$$

and $$r_L = \frac{V_4}{V_3} = \frac{1}{0.5} = 2$$

(*a*) From the hyperbolic expansion curve *bcfg*,

$$p_1V_1 = p_3V_3\ ;$$

$$p_3 = p_1\frac{V_1}{V_3} = p_1 \times \frac{V_1}{V_4} \times \frac{V_4}{V_3}$$

$$= p_1 \times \frac{V_1}{V_2} \times \frac{V_2}{V_4} \times \frac{V_4}{V_3}$$

$$= 8 \times 0.4 \times \frac{1}{4} \times \frac{1}{0.5} = 1.6 \text{ bar}$$

Hence the receiver pressure is 1.6 bar

(*b*) Working strokes

$$= 2N = 2 \times \frac{150}{60} = 5 \text{ per second}$$

Work done per stroke

$$= \frac{460}{5} = 92 \text{ kW}$$

(*c*) Mean effective pressure of HP cylinder

$$p_{mH} = K\left[\frac{p_1}{r_H}(1+\log_e r_H) - p_3\right]$$

$$= 0.8\left[\frac{8}{2.5}(1+\log_e 2.5) - 1.6\right]$$

$$= 3.626 \text{ bar}$$

Work done in HP cylinder

$$= p_{mH} \times V_H$$
$$= (3.626 \times 10^5) \times V_2 \text{ Nm/stroke}$$

Mean effective pressure of LP cylinder

$$p_{mL} = K\left[\frac{p_3}{r_L}(1+\log_e r_L) - p_b\right]$$

$$= 0.8\left[\frac{1.6}{2}(1+\log_e 2) - 0.6\right]$$

$$= 0.6036 \text{ bar}$$

Work done in LP cylinder

$$= p_{mL} \times V_L = (0.6036 \times 10^5) \times V_4$$
$$= (0.6036 \times 10^5) \times 4V_2$$
$$= 2.414 \times 10^5\ V_2 \text{ Nm/stroke}$$

$$\therefore 3.626 \times 10^5\ V_2 + 2.414 \times 10^5\ V_2 = 92 \times 10^3$$

Solution gives : $$V_2 = \frac{92\times10^3}{(3.626+2.414)\times10^5} = 0.152 \text{ m}^3$$

Mean piston speed $= 2LN = 150$ m/min

$$\therefore \text{Stroke length } L = \frac{150}{2\times150} = 0.5 \text{ m}$$

$$\text{Now}: V_2 = \frac{\pi}{4}(d_{HP})^2 \times L$$

$$0.152 = \frac{\pi}{4}(d_{HP})^2 \times 0.5$$

$\therefore$ Diameter of high pressure d_{HP}

$$= \sqrt{\frac{0.152\times4}{0.5\pi}} = 0.622 \text{ m}$$

(*d*) Ratio of piston loads

$$= \frac{(p_1-p_3)A_H}{(p_3-p_b)A_L} = \frac{(p_1-p_3)V_H}{(p_3-p_b)V_L}$$

$$= \frac{8-1.6}{1.6-0.6} \times \frac{1}{4} = \mathbf{1.6}$$

EXAMPLE 17.21

A double acting compound steam engine develops 300 kW indicated power at 160 rev/min. The live steam enters the high pressure cylinder at 12 bar and leaves the low pressure cylinder at 0.3 bar. The cut off in the HP cylinder is at one-half of stroke and the diagram factor is 0.7. The ratio of cylinder volumes is 4.2 and the stroke equals two-third of the diameter of LP cylinder. Assuming hyperbolic expansion and neglecting clearance effects, compute

(a) cylinder dimensions

(b) cut off in the LP cylinder for equal initial piston loads. There occurs a pressure drop of 0.3 bar in the pipe joining the two cylinders.

Also compare the temperature drops in HP and LP cylinders.

Solution : Refer Fig. 17.25.

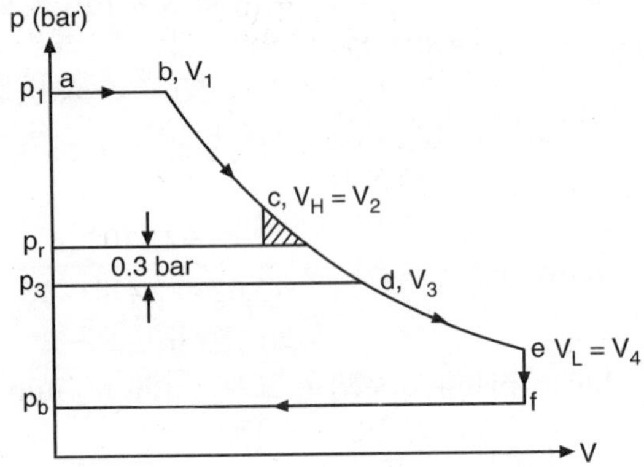

Fig. 17.25.

Overall expansion ratio

$$r = \frac{V_4}{V_1} = \frac{V_4}{V_2} \times \frac{V_2}{V_1}$$

$$= 4.2 \times 2 = 8.4$$

(*a*) Overall mean effective pressure referred to LP cylinder

$$p_{mL} = K\left[\frac{p_1}{r}(1+\log_e r)-p_b\right]$$

$$= 0.7\left[\frac{12}{8.4}(1+\log_e 8.4)-0.3\right]$$

$$= 2.918 \text{ bar}$$

Indicated power = p_{mL} LA $(2N)$

Substituting the appropriate values, we obtain

$$300 \times 10^3 = (2.918 \times 10^5) \times \frac{2}{3}\, d_{LP} \times \frac{\pi}{4}\,(d_{LP})^2 \times \left(2\times\frac{160}{60}\right)$$

Solution gives : Diameter of LP cylinder d_{LP} = 0.717 m or **71.7 cm**

$$\text{Stroke length } L = \frac{2}{3} \times 71.7 = \mathbf{47.8\ cm}$$

Ratio of cylinder volumes

$$\frac{V_4}{V_2} = \frac{V_H}{V_L} = \frac{\frac{\pi}{4}(d_{HP})^2 \times L}{\frac{\pi}{4}(d_{LP})^2 \times L}$$

∴ Diameter of high pressure cylinder

$$d_{HP} = \frac{d_{LP}}{\sqrt{\text{ratio of cylinder volumes}}}$$

$$= \frac{71.7}{\sqrt{4.2}} = \mathbf{34.98\ cm}$$

(*b*) Let p_r be the receiver pressure. Then for equal initial piston loads

$$(p_1 - p_r)\,A_H = (p_r - 0.3 - p_b)\,A_L$$

$$\text{or} \quad p_1 - p_r = (p_r - 0.3 - p_b)\,\frac{A_L}{A_H}$$

$$= (p_r - 0.3 - p_b)\,\frac{V_L}{V_H}$$

or $12 - p_r = (p_r - 0.3 - 0.3) \times 4.2$

Solution gives :

$$p_r = \frac{12+0.6\times 4.2}{5.2} = 2.792 \text{ bar}$$

$$p_3 = p_r - 0.3 = 2.792 - 0.3 = 2.492 \text{ bar}$$

$$\text{Cut off in LP cylinder} = \frac{V_3}{V_4} = \frac{V_3}{V_1} \times \frac{V_1}{V_4}$$

From the hyperbolic expansion curve

$$p_1\,V_1 = p_3\,V_3\,;\quad \frac{V_3}{V_1} = \frac{p_1}{p_3}$$

$$\therefore \text{ Cut off in LP cylinder} = \frac{p_1}{p_3} \times \frac{V_1}{V_4}$$

$$= \frac{12}{2.492} \times \frac{1}{8.4} = 0.573$$

Thus the cut off in LP cylinder occurs at 57.3% of stroke.

(*c*) Saturation temperature at 12 bar = 188°C

Saturation temperature at 2.792 bar = 131.2°C

∴ Temperature drop in HP cylinder
= 188 – 131.2 = 56.8°C

Saturation temperature at 2.492 bar = 127.4°C

Saturation temperature at 0.3 bar = 69.1°C

∴ Temperature drop in LP cylinder
= 127.4 – 69.1 = 58.3°C

$$\text{Hence :} \quad \frac{\text{Temperature drop in HP cylinder}}{\text{Temperature drop in LP cylinder}}$$

$$= \frac{56.8}{58.3} = \mathbf{0.974}$$

17.12. GOVERNING OF COMPOUND STEAM ENGINES

The area of the indicator plot (p-v diagram) represents the work done by the engine. With decrease/increase in the external load, the engine tends to run at high/low speed. For the engine to continue running at the designed speed, it becomes necessary that less/more work be developed by the engine. This stipulates that the area of indicator plot should reduce/increase. In the governing of compound steam engines, the change in the area of indicator plot (that is work output and hence speed control) is achieved by three methods namely throttle governing, cut off governing in HP cylinder and cut off governing in LP cylinder. These methods have been discussed briefly in the following paragraphs. For simplicity, it will be assumed that the engine is of non-receiver type (complete expansion in HP cylinder) and initially both the cylinders develop equal power.

1. Throttle governing : Reduction of steam supply pressure in HP cylinder

Initially at full load, the admission of steam to the high pressure cylinder is at pressure p_1 and the area *abcdefg* represents the total work done in both the cylinders. With decrease in load, the engine tends to run at a higher speed. For speed control, the work output and hence the area of indicator diagram should reduce. For that, steam needs to be admitted to the HP cylinder at the reduced pressure p_1'. The pressure of intake steam is varied by closing or opening the throttle valve under the action of a centrifugal governor.

With cut off ratios remaining unchanged for both the cylinders, the supply of steam to HP cylinder will now be along $g'a'$ and for LP cylinder along $f'b'$. The points a' and b' lie on the vertical lines drawn respectively through the initial cut off points a and b. With reference to Fig. 17.26, it is seen that :

(*i*) The back pressure of HP cylinder or the admission pressure of LP also gets reduced from p_2 to p_2'.

(*ii*) The work output for both the cylinders reduces from area *abcdefg* to area $a'b'c'def'g'$.

(*iii*) There is reduction of work output in both the cylinders. In HP cylinder, the work output decreases from area *abfg* to area $a'b'f'g'$ and in LP cylinder the reduction is from area *bcdef* to area $b'c'def'$.

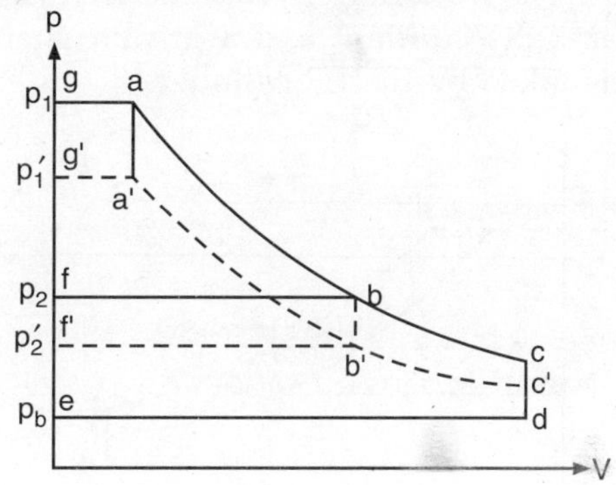

Fig. 17.26. Throttle governing of compound steam engine

It has been noted that sharing of work between the cylinders depends upon load on the engine. At loads slightly less than the designed value, there is greater reduction in LP cylinder. However, as the load goes on decreasing, there will be greater and greater reduction in HP cylinder.

The throttling method of speed control is wasteful in steam consumption as it reduces the heat potential of steam and therefore capacity of doing work per kg of steam.

2. Cut-off governing in HP cylinder : Variation of the point of cut-off in HP cylinder. This method manifests in controlling the engine output by varying the volume of intake steam in HP cylinder. The variation in the point of cut-off is under the control of a centrifugal governor :

From the indicator plot shown in Fig. 17.27, it is evident that with increased cut-off

(*i*) The expansion shifts from curve *abc* to curve $a'b'c'$.

(*ii*) The cut off point b for the LP cylinder shifts to position b' which lies vertically above the point b.

(*iii*) The back pressure of HP cylinder (or the admission pressure of LP cylinder) has been raised from p_2 to p_2'.

(*iv*) There is increase in the power output of the engine (area $a'b'c'def'g$ is greater than area $abcdefg$).

(*v*) The work output from the HP cylinder is practically constant and a greater share of load is taken by the LP cylinder.

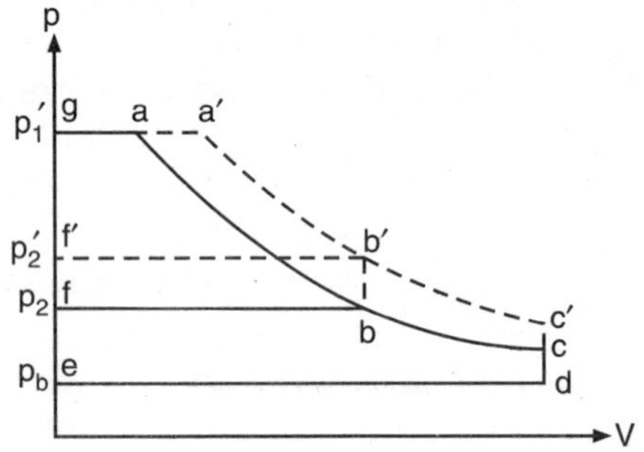

Fig. 17.27. Increased cutoff in HP cylinder

From the indicator plot shown in Fig. 17.28, it is evident that with reduced cut-off :

(*i*) The expansion shifts from curve *abc* to curve $a'b'c'$.

(*ii*) The cut off point *b* for the LP cylinder shifts to position b' which lies vertically below the point *b*.

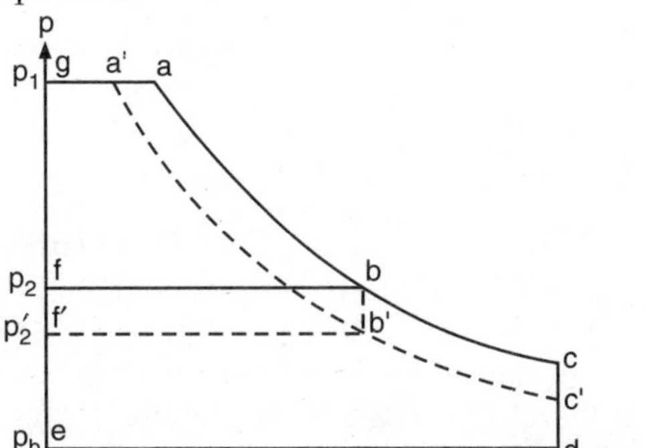

Fig. 17.28. Reduced cut off in HP cylinder

(*iii*) The back pressure of HP cylinder (or the admission pressure of LP cylinder) has decreased from p_2 to p_2'.

(*iv*) There is decrease in the power output of the engine (area $a'b'c'def'g$ is less than the area $abcdefg$)

(*v*) There is very little change in the work output from the HP cylinder and almost entire reduction takes place in the LP cylinder.

With cut off governing, work done in the HP cylinder practically remains the same. All the increase (with increased cut off) or decrease (with reduced cut off) takes place in the LP cylinder. This disturbs the balance of work done in the two cylinders. However, cut off governing is more economical as there is no drop in heat potential and hence in capacity of doing work per kg of steam.

3. Cut-off governing in LP cylinder

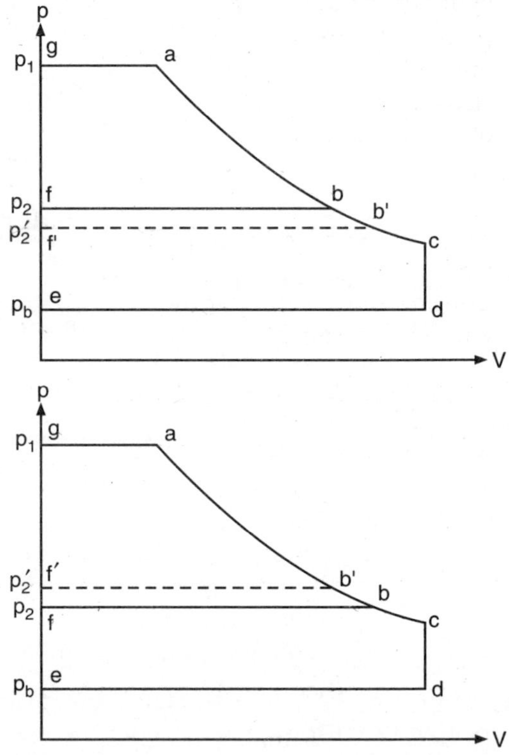

Fig. 17.29. Cut off governing in LP cylinder

From the indicator diagrams depicted in Fig. 17.29 it is to be noted that :

(*i*) When the cut off ratio is increased in LP cylinder, the back pressure of the HP cylinder falls from p_2 to p_2'. The output of the HP cylinder increases with a corresponding reduction in the work output in the LP cylinder. The increase in the work output of HP cylinder equals the decrease in the work output of LP cylinder.

(ii) By decreasing the cut off in the LP cylinder, the back pressure of HP cylinder rises from p_2 to p_2'. There is a decrease in the power output of HP cylinder and increase in the work output of LP cylinder.

Apparently the cut off governing when applied to LP cylinder does not change the total work output from the engine. It only changes the share of loads between the two cylinders; it simply shifts the load from one cylinder to the other.

SALIENT POINTS

1. A steam engine is reciprocating machine which converts heat energy of steam into mechanical energy through reciprocating (to and fro) motion of the piston in a piston-cylinder arrangement.

   (i) **Single acting** : steam is admitted on one side of the piston and one working stroke is produced during each revolution of the crankshaft.

   (ii) **Double acting** : steam is admitted, in turn on both sides of the piston and two working strokes are produced during each revolution of the crankshaft.

2. Considering theoretical indicator diagram without clearance,

$$(p_m)_{th} = \frac{p_1}{r}(1 + \log_e r) - p_b$$

   where r is the expansion ratio (ratio of swept volume to the volume at cut off)

3. Diagram factor k =

$$= \frac{\text{area of actual indicator diagram}}{\text{area of theoretical indicator diagram}}$$

$$= \frac{\text{actual work done/stroke}}{\text{theoretical work done/stroke}}$$

$$= \frac{\text{actual } p_m}{\text{theoretical } p_m}$$

   The value of diagram factor depends upon initial condition of steam, initial and back pressure of steam, speed of engine, type of engine, and its average value lies between 0.65 and 0.9.

4. Missing quantity is the volume of steam represented by the horizontal distance between the actual expansion curve and the saturation curve. Saturation curve is the curve that indicates the volume the steam in the cylinder would occupy during expansion stroke, if perfectly dry.

5. The missing quantity is mainly due to condensation of steam in the cylinder, and a small amount may be due to leakage past the valves and the piston.

6. The cylinder condensation and hence the missing quantity is reduced by :

   (i) efficienct steam jacketing and lagging

   (ii) superheating of steam supplied to engine cylinder

   (iii) making the engine run at greater speeds

   (iv) compounding

7. The governing of steam engines is carried out by :

   (a) throttle governing which controls the engine output by varying pressure of the intake steam.

   (b) cut off governing which controls the engine output by varying the quantity of intake steam.

8. For a throttle governed engine, the steam consumption (m) varies linearly in relation to indicated power (IP).

$$m = A + B\ (\text{IP})$$

   and this aspect is known as Willan's law.

9. Compounding : expansion of steam takes place in two or more cylinders. The steam from the boiler is first admitted into the high pressure cylinder which then exhausts the partially expanded steam into the next cylinder where it further expands.

   Individual expansion cylinders are so proportioned that each piston moves through the same stroke and the cylinders develop approximately equal powers.

10. The compound steam engines are classified acording to arrangement of cranks and the angles between them.

(*i*) **Tandem compounding :** the cylinders have a common piston rod working on the same crank; cranks are at 0° to each other.

(*ii*) **Woolf compounding :** the high and low pressure cylinders have different piston rods attached to two different cranks set at 180° to each other.

(*iii*) **Receiver compounding :** the high and low pressure cylinders have different piston rods attached to two different cranks set at 90° to each other.

REVIEW QUESTIONS

A. Conceptual and conventional questions.

1. Explain, with a neat sketch, the working of a single cylinder double acting steam engine.
2. List the main components of a steam engine and state the function of each
3. What do you understand by a single acting and a double acting steam engine?
4. Define the following terms as related to a steam engine:
   Dead centres; swept volume; cut off ratio; ratio of expansion and piston speed.
5. State the function of providing clearance in a steam engine.
6. (*a*) What do you understand by a hypothetical indicator diagram?
   (*b*) State the assumptions made in drawing the hypothetical indicator diagram.
   (*c*) How and why does the hypothetical indicator diagram differ from actual indicator diagram?
7. What is meant by mean effective pressure?
8. Define and explain the term diagram factor
9. Differentiate between indicated power and brake power of a steam engine
10. What is the effect of piston rod in a double acting steam engine?
11. Sketch a typical brake dynamometer and explain how it is used to measure the brake power of a steam engine.
12. Name and define the different parameters indicating the performance of a steam engine.
13. Explain with a neat sketch, the reason for employing modified Rankine cycle for reciprocating steam engines.
14. (*a*) What is a meant by saturation curve and missing quantity.
    (*b*) Suggest various methods by which missing quantity can be reduced.
15. Discuss with reference to an indicator diagram, how the missing quantity in a steam engine cylinder could be calculated and discuss the methods of reducing it.
16. What are the causes of cylinder condensation ? State the methods of reducing it?
17. (*a*) What are the methods of governing a simple steam engine?
    (*b*) How variation of cut off affects the power output and thermal efficiency?
18. (*a*) What is meant by compounding of a steam engine?
    (*b*) Give the reasons for compounding of a steam engine.
    (*c*) List the advantages of a compound steam engine over a simple steam engine.
19. How steam engines are classified?
20. (*a*) Name the various methods of compounding a steam engine
    (*b*) Explain, with a neat sketch, the working of a Woolf type compound steam engine.
21. Explain the different methods used for the governing of compound steam engines. How the different methods affect the output from individual cylinders.
22. A single cylinder double acting steam engine has a bore of 40 cm and a stroke of 60 cm. The steam is supplied at a pressure of 11 bar, cut off occurs at 50% stroke and the exhaust pressure is 0.8 bar. The net load on the brake is 4.9 kN and the brake used diameter is 4.5 m. Find the indicated power, brake power and mechanical efficiency of the engine if the diagram factor is 0.82 and the engine turns 150 rev/min.
    [**Ans.** 236 kW, 173 kW, 65.78%]
23. A single cylinder double acting steam engine admits dry saturated steam at 12 bar pressure and exhausts into a condenser at 0.9 bar.
    The cut off occurs at 25% of stroke and the engine develop 75 kW brake power while running at 250 rev/min. The stroke/bore ratio is 1.3, the diagram factor is 0.8 and the brake

thermal efficiency for the engine is 15%. Determine the cylinder dimensions and the specific steam consumption. Neglect the effects of clearance and piston rod area, and assume hyperbolic expansion.

[**Ans.** 27.5 cm bore, 35.8 cm stroke, 10.1 kg/kWh]

24. A single cylinder double acting steam engine with 15 cm bore and 20 cm stroke is to develop 20 kW at 300 rev/min with cut off occurring at 20% of stroke. The exhaust occurs at a back pressure of 0.28 bar and the diagram factor is 0.72. Neglecting the effects of clearance and area of piston rod, determine (*a*) steam supply pressure and (*b*) indicated thermal efficiency if the engine receives 225 kg of dry steam per hour.

[**Ans.** 15.6 bar, 12.75%]

25. A double acting single cylinder vertical steam engine running at 300 rpm is supplied with steam at 12 bar.

The clearance is 10 percent of stroke volume, cut off takes place at 0.4 of stroke and exhaust occurs in a condenser maintained at 0.3 bar pressure. Determine:

(*a*) mean effective pressure if diagram factor is 0.7

(*b*) indicated power if the cylinder bore is 12.5 cm and the stroke length is 20 cm

[**Ans.** 5.62 bar, 13.79 kW]

26. A throttle governed steam engine is supplied with steam at 10 bar pressure and 0.96 dryness fraction, and the condenser pressure is 0.5 bar. The engine consumed 80 kg/ hr of steam at indicated load of 7.5 kW, and 340 kg/ hr of steam at 40 kW indicated load.

Determine the steam consumption, and indicated thermal efficiency when developing 30 kW.

[**Ans.** 260 kg/ hr, 17.6%]

27. A steam engine governed by throttling has a specific steam flow rate of 6 kg/kWh when developing 60 kW and 4.5 kg/kWh when developing 150 kW. Find out the steam consumption when developing 120 kW.

Find also the indicated thermal efficiency of the engine if the steam being supplied to the engine is at 15 bar and 0.96 dry and the back pressure is 0.36 bar.

[**Ans.** 570 kg/hr, 31.5%]

28. A steam engine with 0.106 m^3 swept volume uses 45 kg of steam/min while running at a speed of 120 rev/min. The clearance is 8% of swept volume and the cut off takes place at one-half of the working stroke. From the indicator plot, it is found that pressure of steam at 0.7 stroke is 4 bar and pressure at 0.8 of return stroke on the compression curve is 1.3 bar. Determine the dryness fraction of steam in the cylinder. It may be presumed that volume of 1 kg of steam at 4 bar is 0.43 m^3 and that at 1.3 bar is 1.28 m^3.

[**Ans.** 0.4829]

29. A steam engine having cylinder diameter 45 cm, stroke length 60 cm and clearance volume 10% of swept volume runs with rotational speed of 120 rpm. From the indicator plot it is noted that the pressure of steam at 0.7 of return stroke is 1.2 bar, and the pressure of steam at 0.8 of the expansion stroke is 4 bar. If the steam condensation is 50 kg/min, calculate the dryness fraction and the missing quatity at 0.8 of the expansion stroke.

[**Ans.** 0.79, 710.8 kg/hr]

30. A double acting steam engine consumes 5 kg/min of steam when making 150 rev/min. The data corresponding to different state points is indicated in the table given below :

State point	*Pressure* (bar)	*Volume* (litres)	*Volume of dry saturated steam*
Cutoff	6.0	3.8	0.315
Release	3.0	8.25	0.605
Compression	1.103	1.15	1.67

Make calculations for :

(*a*) total mass of steam inside the cylinder

(*b*) condition of steam at cut off and release

(*c*) missing quantity at cut off and release

(*d*) percentage re-evaporation.

[**Ans.** 17.36 × 10^{-3} kg/stroke; 0.695, 0.786; 1.588 kg/min, 1.114 kg/min, 29.85%]

31. A two cylinder compound double acting steam engine develops 92 kW of power while running at 110 rev/min. The steam supply is at 7 bar and exhaust is at 0.21 bar. The cylinders develop equal power, the overall expansion ratio is 15 and the diagram factor is 0.7. Assuming hyperbolic expansion throughout, complete expansion in high pressure cylinder and neglecting clearance, determine the cylinder dimensions if the common stroke equals the diameter of low pressure cylinder.

[**Ans.** d_h = 39 cm;n d_l = L = 66.8 cm]

32. A double acting twin cylinder compound steam engine is required to give 230 kW of indicated power equally between the two cylinders. The engine turns 400 rev/min, is supplied with steam at 12 bar and the exhaust to condenser takes place at 0.17 bar. The cylinder diamters for the HP and LP cylinders are 30 cm and 60 cm respectively and the stroke length equals the diameter of HP cylinder. Neglecting clearance and any pressure drop in the pipes connecting the cylinders, estimate
   (*a*) mean effective pressure in each cylinder, and
   (*b*) initial loads on the pistons assuming them to be equal.

   [**Ans.** p_{mL} = 1.017 bar; p_{mH} = 4.068 bar; 66.8 bar]

33. A twin cylinder double acting compound steam engine is to develop 370 kW indicated power while turning 120 revolutions per minute. The engine uses steam at 8.2 bar initial pressure against a back pressure of 0.3 bar. The cut offs in HP and LP cylinders are at 0.4 and 0.5 of stroke. The ratio of cylinder volumes is 3.5, the diagram factor is 0.85 and the allowable piston speed is 150 m/min. Neglecting clearance and assuming hyperbolic expansion, determine
   (*a*) the cylinder dimensions,
   (*b*) LP receiver pressure.

   Also compare initial loads on the pistons.

   [**Ans.** d_{LP} = 91 cm, d_{HP} = 48.6 cm, L = 62.5 cm, 1.87 bar and 1.15]

34. The following data refers to a compound steam engine which develops 80 kW when running at 300 rpm.

   steam inlet pressure = 15 bar
   exhaust pressure = 0.2 bar
   common stroke = 25 cm ; diagram factor = 0.7
   cut off is at 0.5 of stroke in both cylinders

   Estimate the cylinder dimensions if equal work is to be done in both the cylinders. Both cylinders have restricted expansion and the expansion process follows the hyperbolic curve.

   [**Ans.** d_{LP} = 47.8 cm; d_{HP} = 16.05 cm]

B. Fill in the blanks with appropriate word/ words.

1. In a steam engine, the connecting rod is connected to .... by ....
2. Steam engines are generally .... acting
3. The inverse of the cut off ratio is known as ....
4. The value of diagram factor lies between .... and ....
5. The ratio of actual *mep* to the theoretical *mep* is called ....
6. For a .... steam engine, the steam is admitted through both sides of the piston.
7. For a non-condensing steam engine, the steam is exhausted into the ....
8. The stuffing box and gland are used to reduce the .... of steam.
9. The condensation of steam in the engine cylinder of a steam engine is reduced by ....
10. The hypothetical constant pressure acting on the piston which produces the same amount of work as the actual engine is called ....
11. In case of throttle governing, the straight line showing the relationship between indicated power and steam consumption is called ....
12. In cut off govering of steam engines, the engine output is controlled by varying the .... of intake steam.
13. The steam left in the clearance space from the previous stroke is called ....

Answers

1. crosshead, gudgeon pin 2. double 3. ratio of expansion 4. 0.65, 0.85 5. diagram factor 6. double acting 7. atmosphere 8. leakage 9. steam jacketing 10. mean effective pressure 11. Willan's line 12. volume 13. cushion steam

C. Multiple choice questions.

1. The reciprocating motion of the piston of a steam engine is converted into rotary motion of crank by
   (*a*) crosshead (*b*) connecting rod
   (*c*) eccentric (*d*) crankshaft
2. Which of the following is not a part of steam engine ?
   (*a*) cross head (*b*) camshaft
   (*c*) crank (*d*) eccentric
3. The steam engine part which guides motion of the piston rod and prevents it from bending is called
   (*a*) eccentric (*b*) crank shaft
   (*c*) cross head (*d*) valve rod

4. The function of D-slide valve in a steam engine is to
   (*a*) throttle the steam and reduce its pressure
   (*b*) guide the motion of piston
   (*c*) act as safety valve against high pressure of steam
   (*d*) admit and exhaust the steam at proper timings
5. What forms the connection between piston rod and connecting rod?
   (*a*) eccentric (*b*) valve rod
   (*c*) crankshaft (*d*) cross head
6. Throttle governing of steam engine is a method of controlling the engine output by varying the .... of intake steam
   (*a*) pressure (*b*) volume
   (*c*) temperature (*d*) quality
7. Willian's line for the steam engine is a straight line relationship between rate of steam consumption and
   (*a*) brake power (*b*) indicated power
   (*c*) efficiency (*d*) pressure of steam
8. In a Woolf type compound steam engine, the cranks are placed at
   (*a*) 45° (*b*) 90°
   (*c*) 180° (*d*) 270°
9. High pressure and low pressure cylinders have a common piston rod, cross head, connecting rod and crank in a
   (*a*) double acting steam engine
   (*b*) Woolf compound steam engine
   (*c*) receiver compound steam engine
   (*d*) tandem compound steam engine
10. All of the following statements are correct except
   (*a*) The temperature range per cyliner is reduced in a compound steam engine
   (*b*) The tandem type compound steam engine has small inertia force and uniform turning moment
   (*c*) The receiver type compound steam engine requires a smaller flywheel
   (*d*) The cylinders are arranged side by side in cross-compounding of steam engines

Answers

1. (*b*)	2. (*b*)	3. (*c*)	4. (*d*)	5. (*d*)
6. (*a*)	7. (*b*)	8. (*c*)	9. (*d*)	10 (*b*)

❑❑❑

Chapter 18

Flow Through Nozzles

Learning objectives : Attention has been focussed in this chapter to enable the reader to understand the concepts of :

- *wave propagation and sonic velocity*
- *one dimensional compressible flow and isentropic flow relations*
- *variation of velocity with area ratio*
- *flow of gases and vapours through nozzles, choking condition and effect of friction*
- *supersaturated flow and its effects*

A nozzle is a duct of varying cross-sectional area used for increasing the velocity of a steadily flowing stream of fluid. The fluid enters the nozzle with a high pressure and relatively small velocity. During the flow process, its pressures falls and velocity increases continuously from the entrance to the exit of the nozzle.

There arises a need for high velocity stream of fluid in the steady flow processes that occur in the following practical situations :

– In steam and gas turbines, the high speed stream of fluid is directed on the curved blades and the kinetic energy is transformed into mechanical work.

– In jet propulsion units and rocket motors, the thrust produced by the jet provides the necessary propulsive effort.

– In flow measurement, the differential pressure drop across the nozzle is correlated to velocity to find the discharge.

Other applications of nozzle are the injectors for pumping feed water into the boilers, and the ejectors for removing air from the condensers.

Attention has been focussed in this chapter to analyze the one-dimensional fluid flow through a nozzle under the following set of assumptions :

– Thermodynamic and mechanical properties change only in the direction of flow.

– The flow process is stable, *i.e.,* the velocity of the fluid anywhere along the cross-section of the nozzle does not change in the course of time.

– There are no abrupt changes in the cross-section of nozzle.

– The fluid flowing through the nozzle is ideal and the flow is isentropic in behaviour.

The effect of friction is accounted for by introducing appropriate coefficients and efficiencies.

18.1. EQUATION OF STATE

The state of a flowing liquid is described by three variables namely the velocity V, pressure p and temperature T, and the governing equations are those of mass, momentum and energy. Analysis of compressible flow of gases involves consideration of an additional variable, the fluid density ρ which is related to pressure and temperature by the characteristic gas equation :

$$pV = mRT \quad \text{or} \quad p\,v = RT$$

$$\text{or} \quad \frac{p}{\rho} = RT \qquad ...(18.1)$$

where p is the absolute pressure in N/m^2 ; V is the volume occupied by mass m of the gas ; v is the specific volume in m^3/kg ; ρ is the mass density in kg/m^3 ; T is the absolute temperature in degree kelvin, and R is the gas constant. A dimensional analysis of the state equation would reveal that the gas constant R, though constant for a particular gas, is not dimensionless but has units of J/kg K or that of energy per unit mass and unit temperature. For air, the value of gas constant may be taken as 287 J/kg K.

Taking logs on both sides :

$$\log p = \log \rho + \log R + \log T$$

Upon differentiation,

$$\frac{dp}{p} = \frac{d\rho}{\rho} + \frac{dR}{R} + \frac{dT}{T}$$

Since R is constant for a particular gas, its derivative is zero.

$$\therefore \quad \frac{dp}{p} - \frac{d\rho}{\rho} - \frac{dT}{T} = 0 \qquad ...(18.2)$$

Equation 18.2 is the differential equation of a perfect gas.

This equation shows how the changes in pressure, density and temperature are interrelated in compressible flows.

18.2. WAVE PROPAGATION AND SOUND VELOCITY

The term ***sonic*** or ***acoustic*** velocity refers to the speed at which small pressure disturbances are propagated in the form of waves through a fluid.

Consider a duct of constant area A inside which slides a piston without friction. The duct encloses a compressible fluid initially at rest with pressure p and density ρ. When a small impulse is given to the piston, a small pressure disturbance called ***sonic wave*** is created that propagates through the fluid causing the pressure of fluid to change from p to $(p + dp)$. This increase in pressure of the gas layer immediately adjacent to the piston compresses the gas and changes its density from ρ to $(\rho + d\rho)$. During a time interval dt, the pressure wave progresses a distance $dl = a\,dt$ and the gas particles move a distance $dx = dV \times dt$ from the initial position. Here a is the velocity of the pressure wave propagation and dV is the velocity of gas particles that approximately equals the velocity of piston movement.

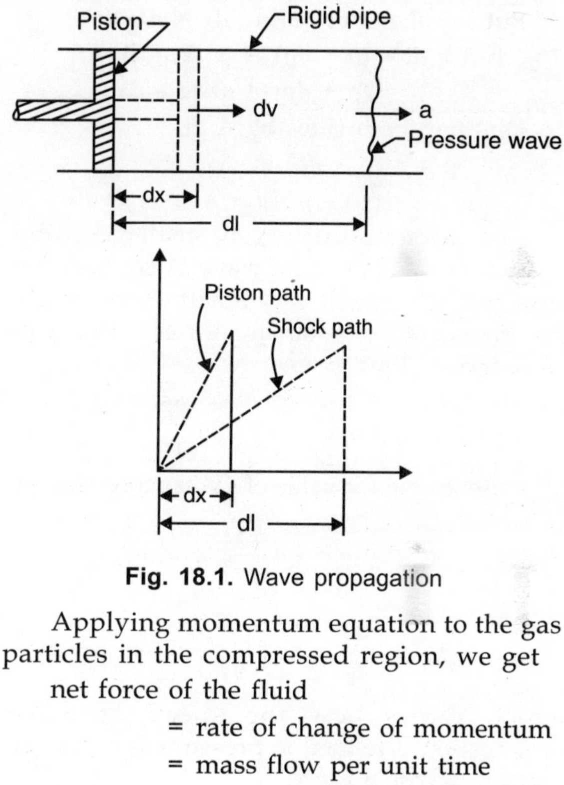

Fig. 18.1. Wave propagation

Applying momentum equation to the gas particles in the compressed region, we get

net force of the fluid

= rate of change of momentum

= mass flow per unit time × change of velocity

$$(p + dp)\,A - pA = \frac{\rho\,A\,dl}{dt} \times dV$$

$$\text{or} \quad dp = \rho \frac{dl}{dt}\,dV$$

$$= \rho\left(\frac{a\,dt}{dt}\right) dV \qquad (\because\ dl = a\,dt)$$

$$= \rho\,a\,dV \qquad ...(i)$$

Before an impulse is imparted to the piston, mass within length dl of the duct equals

= ρ × volume of fluid upto length

= $\rho\,A\,dl$

After piston movement, the gas density increases to $(\rho + d\rho)$ but the length reduces to $(dl - dx)$. Therefore, mass within the duct after piston movement equals

$$= (\rho + d\rho)\, A\, (dl - dx)$$

From continuity considerations, the gas mass should remain constant before and after the piston movement. Therefore :

$$\rho\, A\, dl = (\rho + d\rho)\, A\, (dl - dx)$$

But $\quad dl = a\, dt \quad$ and $\quad dx = dV\, dt$

$$\therefore \rho\, A\, a\, dt = (\rho + d\rho)\, A\, (a\, dt - dV\, dt)$$
$$= (\rho + d\rho)\, A\, dt\, (a - dV)$$

Dividing both sides by $A\, dt$,

$$\rho\, a = (\rho + d\rho)\, (a - dV)$$
$$= \rho\, a - \rho\, dV + a\, d\rho - d\rho\, dV$$

The piston velocity is very small compared to the velocity of pressure wave. Also the value of density change is very small. Accordingly the product of two small quantities $(d\rho\, dV)$ can be dropped. That gives :

$$a\, d\rho = \rho\, dV\, ; \quad dV = a\frac{d\rho}{\rho}$$

Substituting the value of dV in expression (i),

$$dp = \rho\, a \times \left(a\frac{d\rho}{\rho}\right) = a^2\, d\rho$$

$$a^2 = \frac{dp}{d\rho}\, ; \quad a = \sqrt{\frac{dp}{d\rho}} \qquad \text{...(18.3)}$$

which shows how the speed of wave propagation is related to pressure and density changes across a wave.

Sonic velocity for an adiabatic process : In order to evaluate the expression, $a = \sqrt{dp/d\rho}$ for sonic velocity, the manner in which pressure varies with density must be determined. With no temperature gradients at the boundaries of control volume, there is no heat transfer into or out of the control volume. Further only infinitesimal changes take place across the wave. Evidently then the departure of the fluid from thermodynamic equilibrium is negligible and that constitutes a reversible process. Accordingly, the propagation of minor disturbances through air can be thought to be a reversible adiabatic (isentropic) process which is governed by the relation,

$$\frac{p}{\rho^\gamma} = \text{constant } C\, ;$$

$$\log p = \log C + \gamma \log \rho$$

or $$\frac{dp}{p} = \frac{dC}{C} + \gamma\frac{d\rho}{\rho}$$

Since C is a constant, its derivative is zero. Therefore,

$$\frac{dp}{p} = \gamma\frac{d\rho}{\rho}\, ; \quad \frac{dp}{d\rho} = \gamma\frac{p}{\rho}$$

Equation 18.3 for the sonic velocity may then be written as

$$a = \sqrt{\frac{dp}{d\rho}} = \sqrt{\frac{\gamma p}{\rho}} = \sqrt{\gamma RT} \qquad \text{...(18.4)}$$

It may be recalled that when the pressure and density changes are infinitesimal.

$$K = -\frac{dp}{dV/V} = \frac{dp}{d\rho/\rho}$$

where K is the bulk modulus of elasticity of the flow medium. Then in terms of K, the velocity of sound in a medium is

$$a = \sqrt{\frac{K}{\rho}}$$

Thus the different expressions for the velocity of propagation of pressure wave (or sound wave) in a fluid are :

$$a = \sqrt{\frac{dp}{d\rho}} = \sqrt{\frac{K}{\rho}} = \sqrt{\frac{\gamma p}{\rho}} = \sqrt{\gamma RT} \qquad \text{...(18.5)}$$

From the expressions for sonic velocity, the following important conclusions can be drawn :

(i) Sonic velocity is dependent upon the change of density for a given change in pressure. For compressible fluids such as gases or vapours, the ratio $d\rho/dp$ is large compared to that for incompressible fluids like water. Consequently sonic velocity would be less in a gas than that in a liquid. For example, the sonic velocity in air at 15°C is about 340 m/s and that in water at the same temperature is about 1440 m/s.

(*ii*) Sonic velocity is higher in flow mediums with high bulk modulus of elasticity. Conversely, the sonic velocity would have a low value in the medium having a high value of coefficient of compressibility.

(*iii*) Sonic velocity increases with growth in temperature. Accordingly, it would decrease with height due to temperature drop at higher altitudes.

(*iv*) Sonic velocity is higher in gases with a high value of gas constant *R*. This constant depends upon molecular weight; the value being higher in a gas with low molecular weight. Accordingly, the sonic velocity will be higher in a gas which has a low molecular weight. Sonic velocity is about 1400 m/s in hydrogen (low molecular weight) and drops to about 100 m/s in some refrigerants which have a fairly high value of molecular weight.

EXAMPLE 18.1

Find the sonic velocity for the following fluids : (i) Crude oil of specific gravity 0.8 and bulk modulus 15.3×10^8 N/m²; (ii) Mercury having a bulk modulus of 2.65×10^{10} N/m².

Solution : The sonic velocity in a liquid is given by : $a = \sqrt{\dfrac{K}{\rho}}$

(*i*) Sonic velocity in the crude oil,

$$a = \sqrt{\frac{15.3\times10^8}{0.8\times1000}} = \mathbf{1383\ m/s}$$

(*ii*) Sonic velocity in mercury,

$$a = \sqrt{\frac{2.65\times10^{10}}{13.6\times1000}} = \mathbf{1396\ m/s}$$

EXAMPLE 18.2

Indicate whether the sonic velocity will increase or decrease in magnitude with increasing elevation above the earth's surface.

Find the speed of sound wave in air at sea level where the pressure and temperature are 1.01 N/m² abs and 17°C respectively.

For air : take Gas constant R = 287 J/kg K and adiabatic exponent $\gamma = 1.4$.

Solution : Temperature

$$T = 273 + 17 = 290 \text{ K}$$

The sonic velocity given by :

$$a = \sqrt{\gamma RT}$$

$$\therefore \quad a = \sqrt{1.4\times287\times290} = \mathbf{341.35\ m/s}$$

The data regarding pressure at sea level is superfluous.

18.3. BASIC EQUATIONS FOR ONE-DIMENSIONAL COMPRESSIBLE FLOW

• **Continuity Equation :** For one-dimensional flow, the continuity equation gives : $m = \rho AV$ where *m* is mass flow per unit time, *A* is flow area, *V* is fluid velocity and ρ is mass density, *i.e.*, mass per unit volume. For all applications, the mass flow is taken constant and hence the continuity equation can be written as :

$$\rho AV = \text{constant } C$$

$$\log \rho + \log A + \log V = \log C$$

Upon differentiation,

$$\frac{d\rho}{\rho} + \frac{dA}{A} + \frac{dV}{V} = \frac{dC}{C}$$

Since *C* is constant, its derivative is zero

$$\therefore \quad \frac{d\rho}{\rho} + \frac{dA}{A} + \frac{dV}{V} = 0 \quad ...(18.6)$$

The above result represents the differential form of continuity equation for one-dimensional steady, compressible flow.

Momentum Equation : Consider steady fluid flow through a pipeline whose cross-sectional area has a gradual variation along the flow direction.

We focus our attention on an arbitrary volume; the control volume. Let it be assumed that the fluid has the following properties and characteristics at inlet to control volume :

pressure *p*, mean flow velocity *V*, area *A*, average fluid density ρ and the internal energy *E*. The fluid properties change as the fluid passes through the control volume and let the corresponding parameters at exit be :

$(p + dp)$, $(V + dV)$, $(A + dA)$, $(\rho + d\rho)$ and $(E + dE)$

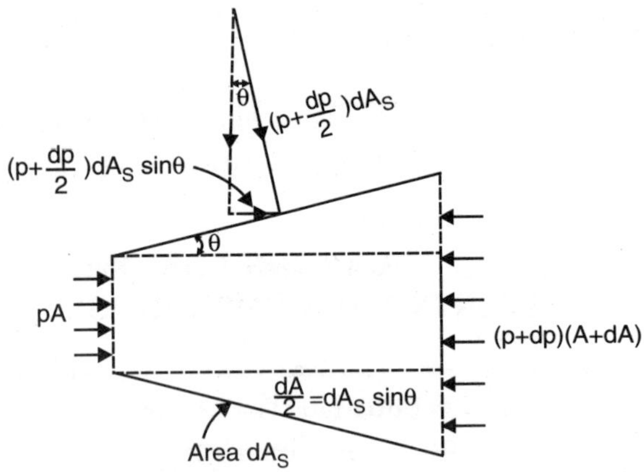

Fig. 18.2. Steady flow through a channel of varying cross-section

For the chosen control volume, the pressure forces are:

pressure force at inlet section = pA

pressure force at exit section

$= (p + dp)\ (A + dA)$

Let $(p + dp/2)$, the average of pressures acting on the inlet and exit sections, be the pressure acting on the slanted side walls of the control surface. Therefore :

average force normal to the surface of each

$$\text{wall} = \left(\frac{p+dp}{2}\right) dA_s$$

where dA_s is the side wall area. The component of this force along the x-direction equals

$$\left(\frac{p+dp}{2}\right) dA_s \sin\theta \text{ for one wall, and}$$

$$\left(\frac{p+dp}{2}\right) \times 2\, dA_s \sin\theta \text{ for both the walls.}$$

But $dA_s \sin\theta = dA/2$. Therefore x-component of pressure force acting on the side walls is $\left(\frac{p+dp}{2}\right) dA.$

Presuming an ideal flow, *i.e.*, there are no shear and friction forces and if the gravitational effects due to change in height are neglected, then net force acting on the fluid in x-direction is given by

$$= pA - (p + dp)\ (A + dA) - \left(\frac{p+dp}{2}\right) dA$$

$= -\ A\ dp$ if second order differential terms are neglected.

Now consider the change of momentum of fluid within the control volume at any instant

Momentum = mass flow rate × change of velocity

$$= (\rho\ AV)\ dV$$

According to impulse-momentum equation, time rate of change of momentum equals the force causing it. That is

$$(\rho\ AV)\ dV = -\ A\ dp$$

$$\text{or} \quad \frac{dp}{\rho} + V\ dV = 0 \qquad \text{...(18.7)}$$

This is the momentum or dynamic equation and is called the ***Euler's equation***. Integration of this equation would give the Bernoulli's equation. However, for compressible flow, the integration is possible only when the density ρ is a definite function of pressure p and this occurs only in homogeneous fluids. For isentropic flow conditions, we have :

$$\frac{p}{\rho^\gamma} = \text{constant } C$$

$$dp = C\,\gamma\,\rho^{\gamma-1}\,d\rho$$

Substituting this value of dp in equation 18.7,

$$\frac{C\gamma\rho^{\gamma-1}d\rho}{\rho} + V\ dV = 0$$

or $C\,\gamma\,\rho^{\gamma-2}\,d\rho + V\ dV = 0$

Upon integration,

$$C\frac{\gamma}{\gamma-1}\rho^{\gamma-1} + \frac{V^2}{2} = \text{constant} \qquad \text{...(18.8)}$$

Replacing constant C by $\frac{p}{\rho^\gamma}$

$$\frac{\gamma}{\gamma-1}\frac{p}{\rho} + \frac{V^2}{2} = \text{constant}$$

Had the effects of gravitational forces taken into account, the result would have been

$$\frac{\gamma}{\gamma-1}\frac{p}{\rho} + \frac{V^2}{2} + g\,y = \text{constant} \quad ...(18.9)$$

Equation 18.9 represents the Bernoulli's equation for flow of a compressible fluid. Each term has units of Nm/kg, *i.e.*, units of energy. If density ρ is replaced by specific weight w ($w = \rho g$);

$$\frac{\gamma}{\gamma-1}\frac{p}{w} + \frac{V^2}{2g} + y = \text{constant} \quad ...(18.10)$$

Each term in the above modified form of Bernoulli's equation represents energy per unit weight, *i.e.*, Nm/N and has units of length, *i.e.*, metre.

Energy Equation : Along with changes of fluid properties and characteristics are also associated the mechanical or shaft work w and the heat q which is transferred into or out of the system. Assuming that q units of heat are supplied to the system and work w_s is taken out of the system, then in accordance with the law of conservation of energy, we can write

$$E + \frac{p}{\rho} + \frac{V^2}{2} + q$$

$$= (E + dE) + \left[\frac{p}{\rho} + d\left(\frac{p}{\rho}\right)\right] + \frac{(V+dV)^2}{2} + w_s \quad ...(18.11)$$

Each term has units of heat and has been expressed per unit mass. The potential energy changes have been neglected as they are usually small for gas flow.

In the absence of heat exchange and shaft work (*e.g.*, flow through a nozzle) and neglecting the higher order differential terms, equation 18.11 gets simplified to :

$$d\left(E + \frac{p}{\rho}\right) + V\,dV = 0$$

In thermodynamics, the term $(E + p/\rho)$ is called specific enthalpy and is designated by symbol h. Then

$$dh + V\,dV = 0 \quad ...(18.12)$$

For a perfect gas,

$$E = c_v\,T \quad \text{and} \quad \frac{p}{\rho} = RT$$

$$\therefore \quad h = c_v\,T + RT = (c_v + R)\,T$$

$$= c_p\,T = \frac{\gamma}{\gamma-1}RT$$

Energy equation 18.12 may then be written as :

$$\frac{\gamma}{\gamma-1}R\,dT + V\,dV = 0$$

Upon integration :

$$\frac{\gamma}{\gamma-1}RT + \frac{V^2}{2} = \text{constant}$$

$$\text{or} \quad \frac{\gamma}{\gamma-1}\frac{p}{\rho} + \frac{V^2}{2} = \text{constant} \quad ...(18.13)$$

which is Bernoulli's equation for gases. Thus we see that the law of conservation of energy (first law of thermodynamics) for a steady, frictionless compressible flow is same as Bernoulli's equation for such a flow.

18.4. ISENTROPIC FLOW RELATIONS

Static and stagnation values **:** Consider a steady flow past a body for frictionless compressible fluid and negligible changes in elevation. The energy equation for the fluid flow yields :

$$V\,dV + dh = 0$$

$$\text{or} \quad \frac{V^2}{2} + h = \text{constant}$$

$$\text{or} \quad \frac{V^2}{2} + c_p\,T = \text{constant}$$

The kinetic energy $V^2/2$ and enthalpy $c_p\,T$ always appear in flow problems and for convenience, their summation $[c_p\,T + V^2/2]$ is considered a single quantity and is called the ***total head*** or ***stagnation enthalpy***, h_0

$$c_p\,T + \frac{V^2}{2} = h_0$$

The stagnation enthalpy represents the enthalpy value which would arise if the flow velocity were brought to zere no matter how, so long as no heat were added from outside and no external work were done. The corresponding temperature T_0 is then given by:

$$c_p T + \frac{V^2}{2} = c_p T_0$$

$$T_0 = T + \frac{V^2}{2c_p} \quad \text{...(18.14)}$$

T_0 is known as the stagnation or total head temperature of the flow steam. When a fluid flows past blunt body, there is some stagnation point P on the body where the stream divides and in which region the stream is virtually brought to rest. Consequently at the stagnation point, the temperature would reach a maximum value and would be the stagnation temperature. Whereas the static temperature T is registered by a thermometer which moves with the flowing fluid, the stagnation temperature T_0 is read on a thermometer when the flowing gas strikes it and is brought to rest. The term $V^2/2c_p$ is called the velocity or dynamic temperature.

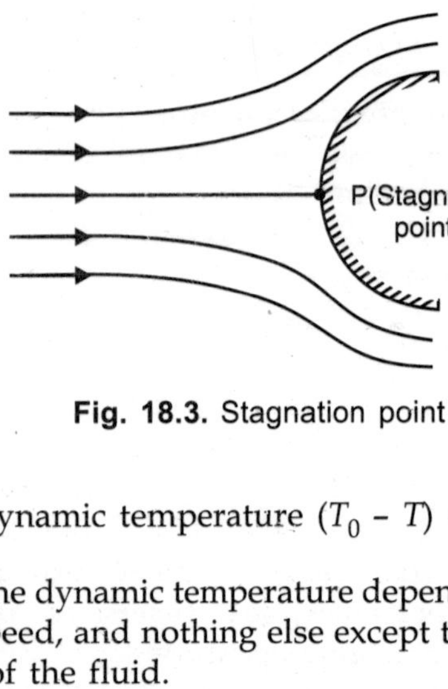

Fig. 18.3. Stagnation point

$$\text{Dynamic temperature } (T_0 - T) = \frac{V^2}{2c_p}$$

The dynamic temperature depends only on the speed, and nothing else except the specific heat of the fluid.

For ambient conditions, the static and stagnation values are same as the ambient velocity is assumed to be zero. When the fluid flow is brought to rest, the static temperature becomes the stagnation temperature.

Replacing c_p by $\frac{\gamma}{\gamma-1}R$ and $\sqrt{\gamma RT}$ by the sonic velocity a, equation 18.14 simplifies to

$$\frac{T_0}{T} = 1 + \frac{V^2}{2c_pT} = 1 + \frac{V^2}{2\times\frac{\gamma}{\gamma-1}RT}$$

$$= 1 + \frac{\gamma-1}{2}\frac{V^2}{\gamma RT}$$

$$\frac{T_0}{T} = 1 + \frac{\gamma-1}{2}\frac{V^2}{a^2} = 1 + \frac{\gamma-1}{2}M^2 \quad \text{...(18.15)}$$

For a reversible adiabatic process, the corresponding stagnation pressure p_0 and density ρ_0 are related by:

$$\frac{p_0}{p} = \left(\frac{T_0}{T}\right)^{\frac{\gamma}{\gamma-1}}; \quad \frac{\rho_0}{\rho} = \left(\frac{T_0}{T}\right)^{\frac{1}{\gamma-1}}$$

Accordingly, the following relationships hold good for temperature, pressure and density in terms of local Mach number :

$$\frac{T_0}{T} = 1 + \frac{\gamma-1}{2}M^2$$

$$\frac{p_0}{p} = \left(1+\frac{\gamma-1}{2}M^2\right)^{\frac{\gamma}{\gamma-1}}$$

$$\text{and} \quad \frac{\rho_0}{\rho} = \left(1+\frac{\gamma-1}{2}M^2\right)^{\frac{1}{\gamma-1}} \quad \text{...(18.16)}$$

For given stagnation conditions, the values of pressure, temperature and density decrease as the Mach number is increased.

For isentropic behaviour of a fluid, the stagnation fluid properties remain constant. Then it follows from equation 18.16 that the following relations hold good for the fluid property ratios between any two cross-sections of an isentropic fluid flow through a duct.

$$T_1\left(1+\frac{\gamma-1}{2}M_1^2\right) = T_2\left(1+\frac{\gamma-1}{2}M_2^2\right) = T_0 = \text{constant}$$

Hence :

$$\frac{T_2}{T_1} = \frac{1+\frac{\gamma-1}{2}M_1^2}{1+\frac{\gamma-1}{2}M_2^2}$$

Likewise :

$$\frac{p_2}{p_1} = \left(\frac{1+\frac{\gamma-1}{2}M_1^2}{1+\frac{\gamma-1}{2}M_2^2}\right)^{\frac{\gamma}{\gamma-1}}$$

and $$\frac{\rho_2}{\rho_1} = \left(\frac{1+\frac{\gamma-1}{2}M_1^2}{1+\frac{\gamma-1}{2}M_2^2}\right)^{\frac{1}{\gamma-1}} \quad ...(18.17)$$

18.5. VARIATION OF VELOCITY WITH AREA RATIO

For a steady flow, the differential form of continuity equation as obtained in equation 18.6 is :

$$\frac{d\rho}{\rho} + \frac{dA}{A} + \frac{dV}{V} = 0$$

For incompressible flow, there is no change in density and so $\frac{d\rho}{\rho} = 0$

$$\therefore \quad \frac{dA}{A} = -\frac{dV}{V} \quad ...(18.18)$$

i.e., a small fractional decrease in area implies a corresponding fractional increase in velocity and vice versa. Thus a liquid would speed up in a contracting nozzle and slow down in an expanding diffuser. This result, however, cannot be applied for a gas flow in which density changes.

For compressible flow, we recall the momentum equation

$$\frac{dp}{\rho} + V\,dV = 0\ ; \quad \frac{dp}{d\rho} \times \frac{d\rho}{\rho} + V\,dV = 0$$

or $$a^2 \frac{d\rho}{\rho} + V\,dV = 0 \qquad \left(\because\ a^2 = \frac{dp}{d\rho}\right)$$

Substituting for $\frac{d\rho}{\rho}$ from continuity equation, we obtain

$$a^2\left(\frac{dA}{A} + \frac{dV}{V}\right) = V\,dV$$

or $$\frac{dA}{A} = \frac{V\,dV}{a^2} - \frac{dV}{V} = \frac{dV}{V}\left(\frac{V^2}{a^2} - 1\right)$$

$$= \frac{dV}{V}(M^2 - 1) \quad ...(18.19)$$

where M is the local Mach number $= \frac{V}{a}$

The equation 18.19 shows that the ratio dA/A can be negative or positive depending on the Mach number of the flow; hence the importance of Mach number at high speeds. The following conclusions can be drawn from equation 18.19.

(*i*) For subsonic velocities ($M < 1$): dA and dV are opposite in sign, *i.e.*, an increase of cross-sectional area causes a reduction of velocity and vice versa. This statement is also true for an incompressible flow.

(*ii*) For supersonic velocities ($M > 1$): dA and dV are of same sign. An increase of cross-sectional area then causes an increase of velocity and vice versa.

(*iii*) When $M = 1$: dA must be zero and since the second derivative is positive, A must be minimum. Thus if the velocity of flow equals the sonic velocity any where; it must do so where the cross-section is of minimum area.

For acceleration of fluid in the subsonic region, the cross-section of the flow region converges down from entry to minimum area which is exit; such a nozzle is called ***converging*** or ***subsonic nozzle*** (Fig. 18.4*a*). For acceleration of fluid in the supersonic velocity range, the flow passage diverges from entry to exit. Such a nozzle is called ***diverging*** or ***supersonic nozzle*** (Fig. 18.4*b*). In a convergent-divergent nozzle (Fig. 18.4*c*), the cross-section of the passage converges down from entry area called ***mouth*** to minimum area called ***throat***, and then diverges from throat to exit. The fluid accelerates from subsonic in the convergent portion to supersonic in the divergent portion. The flow velocity at the throat equals the sonic velocity.

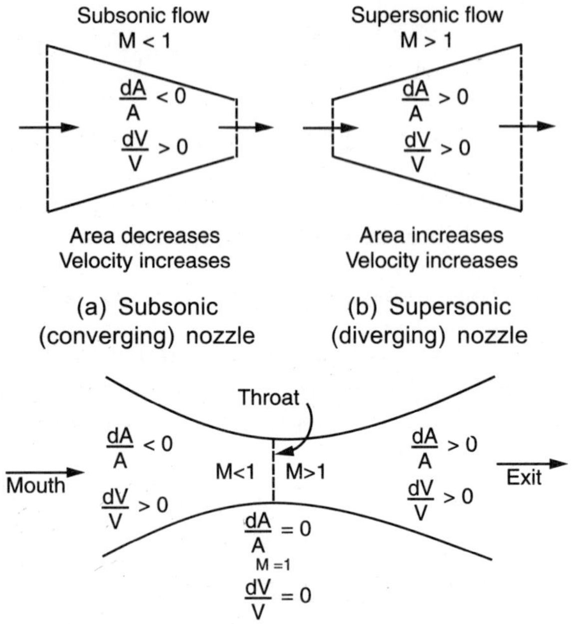

Fig. 18.4. Flow through nozzle

A diffuser is a passage of varying cross-section that serves to achieve reduction in velocity of the flowing fluid. When the flow is subsonic ($M < 1$), the cross-section of the diffusing passage diverges from entry to exit. However, when the flowing fluid is in the supersonic regime ($M > 1$), the cross-section of the diffusing duct decreases down from entry to exit. The area-velocity relationship for a diffuser has been shown in Fig. 18.5.

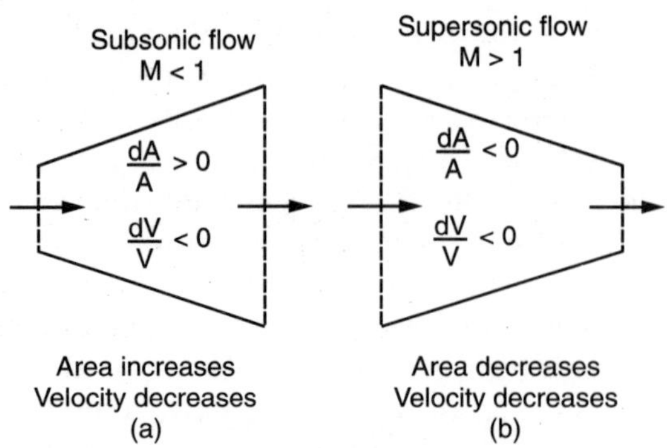

Fig. 18.5. Flow through diffusers

Some additional relationships between changes in velocity, Mach number, temperature, pressure and area can be established by combining the differential forms of the continuity, perfect gas, isentropic, Mach number and energy equations.

Continuity : ρAV = constant

$$\frac{d\rho}{\rho} + \frac{dA}{A} + \frac{dV}{V} = 0$$

Perfect gas : $p = \rho RT$

$$\frac{dp}{p} = \frac{d\rho}{\rho} + \frac{dT}{T}$$

$$\left(\because \frac{dR}{R} \text{ is zero for a particular gas}\right)$$

Isentropic : $\dfrac{p}{\rho^\gamma}$ = constant

$$\frac{dp}{p} - \gamma\frac{d\rho}{\rho} = 0$$

Mach number : $M = \dfrac{V}{a} = \dfrac{V}{\sqrt{\gamma RT}}$

$$\frac{dM}{M} = \frac{dV}{V} - \frac{1}{2}\frac{dT}{T}$$

Energy : $c_p T + \dfrac{V^2}{2}$ = constant

$$c_p\, dT + V\, dV = 0$$

or $$\frac{\gamma}{\gamma-1} R\, dT + V\, dV = 0$$

or $$\frac{\gamma}{\gamma-1}\frac{R\, dT}{V^2} + \frac{dV}{V} = 0$$

or $$\frac{\gamma}{\gamma-1}\frac{RT}{V^2}\frac{dT}{T} + \frac{dV}{V} = 0$$

$$\therefore \quad \frac{1}{(\gamma-1)M^2}\frac{dT}{T} + \frac{dV}{V} = 0$$

$(\because \gamma RT = a^2)$

From these relations, we obtain :

$$\frac{dV}{V} = \underbrace{\left(\frac{1}{1+\frac{\gamma-1}{2}M^2}\right)}_{+\text{ve}}\frac{dM}{M} \quad ...(18.20)$$

i.e., both velocity and Mach number either increase or decrease together.

$$\frac{dT}{T} = \underbrace{\left(\frac{-(\gamma-1)M^2}{1+\frac{\gamma-1}{2}M^2}\right)}_{-\text{ve}}\frac{dM}{M} \quad ...(18.21)$$

i.e., temperature changes are opposite to Mach number changes

$$\frac{dp}{p} = \underbrace{\left(\frac{\gamma - M^2}{1+\frac{\gamma-1}{2}M^2}\right)}_{-\text{ve}}\frac{dM}{M} \quad ...(18.22)$$

i.e., pressure changes are opposite to Mach number changes

$$\frac{d\rho}{\rho} = \underbrace{\left(\frac{M^2}{1+\frac{\gamma-1}{2}M^2}\right)}_{-\text{ve}}\frac{dM}{M} \quad ...(18.23)$$

i.e., density changes are opposite to Mach number changes

$$\frac{dA}{A} = \left(\frac{-(1-M^2)}{1+\frac{\gamma-1}{2}M^2}\right)\frac{dM}{M} \quad ...(18.24)$$

The term inside the brackets may be positive or negative depending upon the magnitude of the Mach number.

Equation 18.24 can be integrated to obtain the following relationship between the critical throat area A_c where the Mach number is unity and the area A at any other section where $M \gtrless 1$

$$\frac{A}{A_c} = \frac{1}{M}\left(\frac{2+(\gamma-1)M^2}{\gamma+1}\right)^{\frac{\gamma+1}{2(\gamma-1)}} \quad ...(18.25)$$

EXAMPLE 18.3

A supersonic flow at Mach number 1.5 has been observed in a given passage. Workout the percent of the original passage area needed if a 20 percent increase in velocity is to be attained. Also workout the corresponding passage area required had the initial flow been subsonic at Mach number 0.5.

Solution : For one-dimensional compressible flow through a duct,

$$\frac{dA}{A} = \frac{dV}{V}(M^2 - 1) = 0.2\ (1.5^2 - 1)$$
$$= 0.25$$

Therefore, the area will be 25 percent greater than the original area or the area required will be 1.25 times the original area.

When $M = 0.5$;

$$\frac{dA}{A} = 0.2\ (0.5^2 - 1) = -\ 0.15$$

Therefore, the area will be 15 percent smaller than the original area or the required area will be 0.85 times the original area.

EXAMPLE 18.4

Establish the following relations for one-dimensional compressible flow through ducts of varying area :

(i) $\dfrac{dA}{A} = \dfrac{dp}{\rho V^2}\ (1 - M^2)$

(ii) $\dfrac{dA}{A} = \dfrac{1}{\gamma}\dfrac{dp}{p}\left(\dfrac{1-M^2}{M^2}\right)$

The symbols have their usual meanings.

Solution : Recall the momentum equation for compressible flow,

$$\frac{dp}{\rho} + V\,dV = 0\ ; \quad \frac{dp}{\rho V^2} + \frac{dV}{V} = 0$$

Substituting for $\frac{dV}{V}$ from continuity equation $\frac{d\rho}{\rho} + \frac{dA}{A} + \frac{dV}{V} = 0$, we obtain

$$\frac{dp}{\rho V^2} - \frac{d\rho}{\rho} - \frac{dA}{A} = 0$$

or $$\frac{dA}{A} = \frac{dp}{\rho V^2} - \frac{d\rho}{\rho} = \frac{dp}{\rho V^2} - \frac{dp}{\rho a^2}$$

$$\left(\because\ a^2 = \frac{dp}{d\rho}\right)$$

$$= \frac{dp}{\rho V^2}\left(1 - \frac{V^2}{a^2}\right)$$

$\therefore \quad \dfrac{dA}{A} = \dfrac{dp}{\rho V^2}\ (1 - M^2)$ which is the required result.

(*ii*) Now,

$$\rho V^2 = \rho \frac{V^2}{a^2} \times a^2 = \rho M^2 \gamma \frac{p}{\rho} = \gamma p M^2$$

$$\left(\because \; a^2 = \frac{\gamma p}{\rho}\right)$$

Accordingly the area relation derived above can also be rewritten as

$$\frac{dA}{A} = \frac{1}{\gamma} \frac{dp}{p} \left(\frac{1-M^2}{M^2}\right)$$

EXAMPLE 18.5

(a) In the case of an isentropic flow of a compressible fluid through a variable area duct, show that :

$$\frac{A}{A^*} = \frac{1}{M} \left(\frac{1+\frac{\gamma-1}{2}M^2}{\frac{\gamma+1}{2}}\right)^{\frac{\gamma-1}{2(\gamma-1)}}$$

where γ is the ratio of specific heats, M is the Mach number at a section where area is A and A^* is the critical area of flow.

(b) A supersonic nozzle is to be designed for flow with Mach number 3 at the exist section which is 20 cm in diameter. The pressure and temperature of air at the nozzle exit are to be 8 kPa and 200 K respectively. Determine the reservoir pressure, the temperature and the throat area. Take γ = 1.4.

Solution : The reservoir temperature and pressure are the stagnation conditions which can be determined from the relations :

$$T_0 = T\left(1+\frac{\gamma-1}{2}M^2\right)$$

$$p_0 = p\left(1+\frac{\gamma-1}{2}M^2\right)^{\frac{\gamma}{\gamma-1}}$$

$$\text{hence } T_0 = 200\left(1+\frac{1.4-1}{2}\times 3^2\right) = \mathbf{560\ K}$$

$$\text{and} \quad p_0 = 8\left(1+\frac{1.4-1}{2}\times 3^2\right)^{\frac{1.4}{1.4-1}}$$

$$= \mathbf{293.86\ kPa}$$

By employing the equation given in part (*a*) :

$$\frac{A}{A^*} = \frac{1}{3}\left(\frac{1+\frac{1.4-1}{2}\times 3^2}{\frac{1.4+1}{2}}\right)^{\frac{1.4-1}{2(1.4-1)}}$$

$$= 4.23$$

Hence the critical area of flow,

$$A^* = \frac{A}{4.23} = \frac{\pi/4 \times 0.2^2}{4.23} = \mathbf{0.00742\ m^2}$$

18.6. DISCHARGE THROUGH A CONVERGENT NOZZLE

Flow through a converging nozzle is identical to that from a reservoir in which the gas is initially at rest. The gas flows through the minimum cross-sectional area of the nozzle at exit where pressure p equals the atmospheric pressure. The Bernoulli's equation

$$\frac{\gamma}{\gamma-1}\frac{p_0}{\rho_0} = \frac{\gamma}{\gamma-1}\frac{p}{\rho} + \frac{V^2}{2}$$

can be used to obtain the velocity V at the nozzle exit.

$$V = \left[2\frac{\gamma}{\gamma-1}\left(\frac{p_0}{\rho_0} - \frac{p}{\rho}\right)\right]^{1/2}$$

$$= \left[2\frac{\gamma}{\gamma-1}\frac{p_0}{\rho_0}\left(1 - \frac{p}{\rho}\times\frac{\rho_0}{p_0}\right)\right]^{1/2}$$

For isentropic flow,

$$\frac{\rho_0}{\rho} = \left(\frac{p_0}{p}\right)^{1/\gamma} = \left(\frac{p}{p_0}\right)^{-1/\gamma}$$

$$\therefore \quad V = \left[2\frac{\gamma}{\gamma-1}\frac{p_0}{\rho_0}\left(1-\left\{\frac{p}{p_0}\right\}^{(\gamma-1)/\gamma}\right)\right]^{1/2} \qquad ...(18.26)$$

The mass flow rate for compressible fluid is

$$m = \rho AV$$

$$= \rho A\left[2\frac{\gamma}{\gamma-1}\frac{p_0}{\rho_0}\left(1-\left\{\frac{p}{p_0}\right\}^{(\gamma-1)/\gamma}\right)\right]^{1/2}$$

$$= A\left[2\frac{\gamma}{\gamma-1}\frac{p_0}{\rho_0}\rho^2\left\{1-\left(\frac{p}{p_0}\right)^{(\gamma-1)/\gamma}\right\}\right]^{1/2}$$

Making the substitution $\rho = \rho_0\left(\frac{p_0}{p}\right)^{1/\gamma}$

mass flow rate in

$$= A\left[2\frac{\gamma}{\gamma-1}\frac{p_0}{\rho_0}\times\rho_0^2\left(\frac{p}{p_0}\right)^{2/\gamma}\left\{1-\left(\frac{p}{p_0}\right)^{(\gamma-1)/\gamma}\right\}\right]^{1/2}$$

$$= A\left[\frac{2\gamma}{\gamma-1}p_0\rho_0\left\{\left(\frac{p}{p_0}\right)^{2/\gamma}-\left(\frac{p}{p_0}\right)^{(\gamma+1)/\gamma}\right\}\right]^{1/2} \quad ...(18.27)$$

Pressure ratio (p/p_0) for maximum mass flow rate : A nozzle is always designed to discharge the maximum flow rate for the given pressure drop. There is only one value of ratio (p/p_0) which will provide the maximum flow for the nozzle. This value can be found by differentiating $[(p/p_0)^{2/\gamma} - (p/p_0)^{(\gamma+1)/\gamma}]$ with respect to (p/p_0) and equating the result to zero. That is,

$$\frac{2}{\gamma}\left(\frac{p}{p_0}\right)^{\frac{2}{\gamma}-1} - \frac{\gamma+1}{\gamma}\left(\frac{p}{p_0}\right)^{\frac{\gamma+1}{\gamma}-1} = 0$$

or $$\frac{2}{\gamma}\left(\frac{p}{p_0}\right)^{\frac{2-\gamma}{\gamma}} = \frac{\gamma+1}{\gamma}\left(\frac{p}{p_0}\right)^{\frac{1}{\gamma}}$$

or $$\frac{2/\gamma}{(\gamma+1)/\gamma} = \frac{(p/p_0)^{1/\gamma}}{(p/p_0)^{(2-\gamma)/\gamma}}$$

or $$\frac{2}{\gamma+1} = \left(\frac{p}{p_0}\right)^{\frac{1}{\gamma}-\frac{2-\gamma}{\gamma}} = \left(\frac{p}{p_0}\right)^{\frac{\gamma-1}{\gamma}}$$

$$\therefore \quad \frac{p}{p_0} = \left(\frac{2}{\gamma+1}\right)^{\frac{\gamma}{\gamma-1}}$$

The exit pressure p for maximum discharge is called the **critical pressure** p_c and the ratio of critical pressure to initial pressure p_0 is called the **critical pressure** ratio :

$$\frac{p_c}{p_0} = \left(\frac{2}{\gamma+1}\right)^{\frac{\gamma}{\gamma-1}} \quad ...(18.28)$$

It may be pointed out that a convergent nozzle is employed when the exit pressure is equal to or more than the critical pressure and a convergent-divergent nozzle is used when the discharge pressure is less than the critical pressure.

Relevant relations for critical density and temperature are :

$$\frac{\rho_c}{\rho_0} = \left(\frac{2}{\gamma+1}\right)^{\frac{1}{\gamma-1}} ; \quad \frac{T_c}{T_0} = \frac{2}{\gamma+1} \quad ...(18.29)$$

For air as the fluid medium, $\gamma = 1.4$ and so the above results reduce to

$$\frac{p_c}{p_o} = 0.528 ; \quad \frac{\rho_c}{\rho_o} = 0.634 ; \quad \frac{T_c}{T_o} = 0.833$$

Maximum mass flow rate : The critical velocity for maximum mass flow rate can be obtained by substituting for the critical pressure ratio in the following expression for flow velocity :

$$V^2 = 2\frac{\gamma}{\gamma-1}\frac{p_0}{\rho_0}\left[1-\left(\frac{p}{p_0}\right)^{\frac{\gamma-1}{\gamma}}\right]$$

$$V_c^2 = 2\frac{\gamma}{\gamma-1}\frac{p_0}{\rho_0}\left[1-\left(\frac{2}{\gamma+1}\right)^{\frac{\gamma}{\gamma-1}\times\frac{\gamma-1}{\gamma}}\right]$$

$$= 2\frac{\gamma}{\gamma-1}\frac{p_0}{\rho_0}\left(1-\frac{2}{\gamma+1}\right)$$

$$= 2\frac{\gamma}{\gamma-1}\frac{p_0}{\rho_0}\times\frac{\gamma-1}{\gamma+1} = 2\frac{\gamma}{\gamma+1}\frac{p_0}{\rho_0}$$

$$\therefore \quad V_c = \left(2\frac{\gamma}{\gamma+1}\frac{p_0}{\rho_0}\right)^{1/2} = \left(\frac{2}{\gamma+1}\right)^{1/2} a_0$$

$$\left(\because \ a_0^2 = \gamma\frac{p_0}{\rho_0}\right)$$

The mass flow rate through a nozzle is :

$$m = A\left[2\frac{\gamma}{\gamma-1}p_0\rho_0\left(\left(\frac{p}{p_0}\right)^{2/\gamma} - \left(\frac{p}{p_0}\right)^{(\gamma+1)/\gamma}\right)\right]^2$$

The maximum mass rate can be obtained by substituting the critical pressure ratio $\frac{p_c}{p_0} = \left(\frac{2}{\gamma+1}\right)^{\frac{\gamma}{\gamma-1}}$ in the above expression:

∴ Maximum mass flow:

$$=A\left[2\frac{\gamma}{\gamma-1}p_0\rho_0\left\{\left(\frac{2}{\gamma+1}\right)^{\frac{\gamma}{\gamma-1}\times\frac{2}{\gamma}} - \left(\frac{2}{\gamma+1}\right)^{\frac{\gamma}{\gamma-1}\times\frac{\gamma+1}{\gamma}}\right\}\right]^{1/2}$$

$$=A\left[2\frac{\gamma}{\gamma-1}p_0\rho_0\left\{\left(\frac{2}{\gamma+1}\right)^{\frac{2}{\gamma-1}} - \left(\frac{2}{\gamma+1}\right)^{\frac{\gamma+1}{\gamma-1}}\right\}\right]^{1/2}$$

$$= A\left[2\frac{\gamma}{\gamma-1}p_0\rho_0\left(\frac{2}{\gamma+1}\right)^{\frac{\gamma+1}{\gamma-1}}\times\left\{\left(\frac{2}{\gamma+1}\right)^{\frac{1-\gamma}{\gamma-1}} - 1\right\}\right]^{1/21/2}$$

$$=A\left[2\frac{\gamma}{\gamma-1}p_0\rho_0\left(\frac{2}{\gamma+1}\right)^{\frac{\gamma+1}{\gamma-1}}\times\left(\frac{\gamma+1}{2} - 1\right)\right]^{1/2}$$

$$=A\left[2\frac{\gamma}{\gamma-1}p_0\rho_0\left(\frac{2}{\gamma+1}\right)^{\frac{\gamma+1}{\gamma-1}}\times\frac{\gamma-1}{2}\right]^{1/2}$$

$$=A\left[\gamma\, p_0\rho_0\left(\frac{2}{\gamma+1}\right)^{\frac{\gamma+1}{\gamma-1}}\right]^{1/2} \qquad ...(18.30)$$

The above result could also be obtained by making substitution for the critical values in the expression

$$m_{max} = A\,\rho_c\,V_c$$

$$= A\,\rho_0\left(\frac{2}{\gamma+1}\right)^{\frac{1}{\gamma-1}} \times \left(2\frac{\gamma}{\gamma+1}\frac{p_0}{\rho_0}\right)^{1/2}$$

$$= A\left[\rho_0^2\left(\frac{2}{\gamma+1}\right)^{\frac{2}{\gamma-1}}\times 2\frac{\gamma}{\gamma+1}\frac{p_0}{\rho_0}\right]^{1/2}$$

$$= A\left[\gamma\, p_0\rho_0\left(\frac{2}{\gamma+1}\right)^{\frac{2}{\gamma-1}+1}\right]^{1/2}$$

$$= A\left[\gamma\, p_0\rho_0\left(\frac{2}{\gamma+1}\right)^{\frac{\gamma+1}{\gamma-1}}\right]^{1/2}$$

Taking $\gamma = 1.4$, the above expression becomes :

$$m_{max} = A\sqrt{p_0\rho_0} \times \left[1.4\times\left(\frac{2}{1.4+1}\right)^{\frac{1.4+1}{1.4-1}}\right]^{1/2}$$

$$= 0.685\, A\,\sqrt{p_0\rho_0} \qquad ...(18.31)$$

A passage in which the sonic velocity has been reached and thus in which the flow rate is maximum is often said to be ***choked*** or in ***choking conditions.*** It is evident from equation 18.27 that for a fixed value of inlet pressure the mass flow depends on the nozzle exit pressure. Fig. 18.6 depicts the variation of actual and theoretical mass flow rate versus p/p_0. From both the curves it is found that the flow rate increases with a decrease in the pressure ratio p/p_0 and attains the maximum value at the critical pressure ratio p_c/p_0. = 0.528 for air. Thereafter the theoretical mass flow rate decreases with a further decrease in the exit pressure below the critical value. This is contrary to the actual results where the mass flow rate remains constant after the attainment of maximum value. Physical interpretation of the choking may be given by the fact that the sonic velocity represents the maximum speed at which pressure effects can be propagated. Thus once the velocity at throat reaches the sonic value, no message can be transmitted upstream and so the fluid there is quite unware of any further reduction of pressure at the downstream end of the nozzle. The convergent part of the nozzle continues to behave independently of conditions beyond the throat until these become subsonic again.

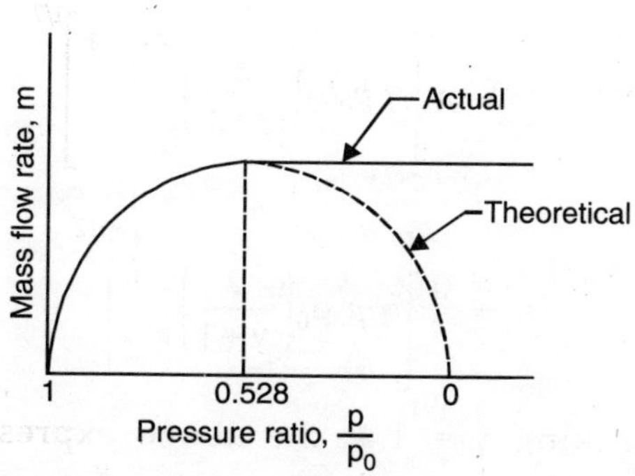

Fig. 18.6. Mass flow rate versus pressure ratio

EXAMPLE 18.6

A tank contains air at a pressure 135 kPa and temperature 27°C. The local barometric pressure is 100 kPa. Air discharges out of the tank and into atmosphere through a convergent nozzle. Determine the outflow velocity and the mass rate of air. The cross-sectional area at the nozzle outlet is 500 mm².

How the mass flow rate would change if the pressure in the tank is 450 kPa ? The other data remains same.

Solution : Assuming isentropic flow, the reservoir would be at the stagnation conditions represented by the suffix 0.

$$p_0 = 135 \text{ kPa}$$
$$T_0 = 273 + 27 = 300 \text{ K}$$
$$\rho_0 = \frac{p_0}{RT_0} = \frac{135\times10^3}{287\times300}$$
$$= 1.568 \text{ kg/m}^3$$

Case (*a*) : $p/p_0 = 100/135 = 0.741$ which is greater than the critical pressure ratio. The outflow velocity, therefore, does not reach the critical velocity and is determined from the relation :

$$V = \left[\frac{2\gamma}{\gamma-1}\frac{p_0}{\rho_0}\left\{1-\left(\frac{p}{p_0}\right)^{\frac{\gamma-1}{\gamma}}\right\}\right]^{1/2}$$

$$= \left[2\times\frac{1.4}{1.4-1}\times\frac{135\times10^3}{1.568}\left(1-(0.741)^{(1.4-1)/1.4}\right)\right]^{1/2}$$

$$= 222.52 \text{ m/s}$$

From isentropic relation, the density of air at nozzle exit is given by

$$\rho = \rho_0\left(\frac{p}{p_0}\right)^{\frac{1}{\gamma}} = 1.568 \times (0.741)^{1/1.4}$$
$$= 1.266 \text{ kg/m}^3$$

∴ Mass flow

$$= \rho AV \text{ at the nozzle exit}$$
$$= 1.266 \times (500 \times 10^{-6}) \times 222.52$$
$$= \mathbf{0.1408 \text{ kg/s}}$$

Case (*b*) : $p/p_0 = 100/450 = 0.22$ which is less than the critical pressure ratio. The nozzle exit therefore, becomes a critical section.

$$\frac{T_c}{T_0} = \frac{2}{\gamma+1} = 0.833$$
$$T_c = 300 \times 0.833 = 250 \text{ K}$$
$$V_c = a_c = \sqrt{\gamma RT_c}$$
$$= \sqrt{1.4\times287\times250} = 316.94 \text{ m/s}$$
$$\rho_c = \frac{p_c}{RT_c} = \frac{0.528\times450\times10^3}{287\times250}$$
$$= 3.31 \text{ kg/m}^3$$

∴ Mass flow

$$= \rho AV \text{ at the critical conditions}$$
$$= 3.31 \times (500 \times 10^{-6}) \times 316.94$$
$$= 0.524 \text{ kg/s}$$

This could have alternatively been found by using the relation :

$$\text{mass flow} = 0.685\, A\sqrt{p_0\rho_0}$$

$$\rho_0 = \frac{p_0}{RT_0} = \frac{450\times10^3}{287\times300} = 5.23 \text{ kg/m}^3$$

$$\therefore \quad m = 0.685 \times (500 \times 10^{-6}) \times \sqrt{450\times10^3\times5.23}$$

$$= \mathbf{0.525 \text{ kg/s}}$$

EXAMPLE 18.7

A supersonic nozzle is to be designed for air flow with Mach number 3 at the exit section which is 250 mm in diameter. The pressure and temperature of air at the nozzle exit are 8.5 kN/m² and 215 K. Make calculations for (a) reservoir pressure and temperature and (b) throat area.

Solution : Since the velocity in the reservoir is zero, the temperature and pressure correspond to stagnation conditions represented by the suffix 0.

The Mach number is related to the static and stagnation pressures by the relation :

$$p_0 = p\left[1+\frac{\gamma-1}{2}M^2\right]^{\frac{\gamma}{\gamma-1}}$$

$$= 8.5\left[1+\frac{1.4-1}{2}\times 3^2\right]^{\frac{1.4}{1.4-1}}$$

$$= 8.5 \times (2.8)^{3.5} = \mathbf{312.23\ kN/m^2}$$

The Mach number is related to the static and stagnation temperatures by the relation,

$$T_0 = T\left[1+\frac{\gamma-1}{2}M^2\right]$$

$$= 215\left[1+\frac{1.4-1}{2}\times 3^2\right]$$

$$= 215 \times 2.8 = \mathbf{602\ K}$$

The following correlation exists between the critical throat area A_c (where the Mach number is unity) and the area A at any other section where $M \gtrless 1$

$$\frac{A}{A_c} = \frac{1}{M}\left[\frac{2+(\gamma-1)M^2}{\gamma+1}\right]^{\frac{\gamma+1}{2(\gamma-1)}}$$

$$= \frac{1}{3}\left[\frac{2+(1.4-1)\times 3^2}{1.4+1}\right]^{\frac{2.4}{0.8}}$$

$$= \frac{1}{3}(2.33)^3 \doteq 4.216$$

$$\therefore\ A_c = \frac{A}{4.216} = \frac{\frac{\pi}{4}\times(0.25)^2}{4.126} = \mathbf{0.0116\ m^2}$$

EXAMPLE 18.8

Air flows through a nozzle with velocity 75 m/s, temperature 300 K and pressure 650 kPa at the inlet section which has an area of 10 cm². If pressure at the nozzle exit is 250 kPa, calculate mass flow rate through the nozzle and the flow velocity at the exit section.

Assume one-dimensional isentropic flow.

Solution : Let suffix 1 and 2 represent the conditions at the inlet and exit sections of the nozzle.

$$\text{Mass flow rate } M = \rho_1 A_1 V_1 = \frac{p_1}{RT_1} A_1 V_1$$

$$= \frac{650\times 10^3}{287\times 300} \times (10\times 10^{-4}) \times 75$$

$$= 0.566 \text{ kg/s}$$

$$M_1 = \frac{V_1}{a_1} = \frac{V_1}{\sqrt{\gamma R T_1}}$$

$$= \frac{7.5}{\sqrt{1.4\times 287\times 300}} = 0.216$$

The following correlation holds good for pressure ratio between any two cross-sections of an isentropic fluid flow through a duct.

$$\frac{p_2}{p_1} = \left[\frac{1+\frac{\gamma-1}{2}M_1^2}{1+\frac{\gamma-1}{2}M_2^2}\right]^{\frac{\gamma}{\gamma-1}}$$

$$\therefore\ \frac{250}{650} = \left[\frac{1+\frac{1.4-1}{2}\times 0.216^2}{1+\frac{1.4-1}{2}M_2^2}\right]^{\frac{1.4}{1.4-1}}$$

$$= \left[\frac{1.00933}{1+0.2M_2^2}\right]^{3.5}$$

$$\text{or}\ \left(\frac{250}{650}\right)^{1/3.5} = \frac{1.00933}{1+0.2M_2^2}$$

$$\text{or}\ 0.7608 = \frac{1.00933}{1+0.2M_2^2}$$

$$\text{or}\ 1 + 0.2\,M_2^2 = \frac{1.00933}{0.7608} = 1.3266$$

$$\therefore\ M_2 = \left[\frac{1.3266-1}{0.2}\right]^{1/2} = 1.278$$

Further, for isentropic flow

$$\frac{T_2}{T_1} = \left(\frac{p_2}{p_1}\right)^{\frac{\gamma-1}{\gamma}} = \left(\frac{250}{650}\right)^{\frac{1.4-1}{1.4}}$$

$$= 0.7608$$

$$\therefore\ T_2 = 0.7608\ T_1$$

$$= 0.7608 \times 300 = 228.24 \text{ K}$$

Flow velocity at exit

$$V_2 = M_2 a_2 = M_2\sqrt{\gamma R T_2}$$
$$= 1.278\ \sqrt{1.4 \times 287 \times 228.24}$$
$$= \mathbf{387\ m/s}$$

EXAMPLE 18.9

Air flows through a convergent-divergent nozzle. At some section in the nozzle pressure = 2 bar, velocity = 170 m/s and temperature = 200°C and cross sectional area = 1000 mm². Assuming isentropic flow conditions determine : (a) stagnation temperature and stagnation pressure, (b) sonic velocity and Mach number at this section (c) velocity, Mach number and flow area at outlet section where pressure is 1.1 bar (d) pressure, temperature, velocity and flow area at throat of the nozzle.

Take R = 287 J/kg K, c_p = 1000 J/kg K and γ = 1.4

Solution : Stagnation temperature

$$T_0 = T + \frac{V^2}{2c_p}$$

$$= 473 + \frac{170^2}{2 \times 1000} = 487.45\ \text{K}$$

$$\frac{p_0}{p} = \left(\frac{T_0}{T}\right)^{\frac{\gamma}{\gamma-1}}$$

$$= \left(\frac{487.45}{473}\right)^{\frac{1.4}{1.4-1}} = 1.11$$

∴ Stagnation pressure,

$$p_0 = 1.11\ p = 1.11 \times 2 = 2.22\ \text{bar}$$

(*b*) Sonic velocity,

$$a = \sqrt{\gamma R T} = \sqrt{1.4 \times 287 \times 473}$$
$$= 435.95\ \text{m/s}$$

Mach number

$$M = \frac{V}{a} = \frac{170}{435.95} = 0.39$$

(*c*) Let subscript *e* refer to the conditions at the outlet section. Then

$$\frac{p_0}{p_e} = \left(1 + \frac{\gamma - 1}{2} M_e^2\right)^{\frac{\gamma}{\gamma-1}}$$

$$\frac{2.22}{1.1} = \left(1 + \frac{1.4-1}{2} M_e^2\right)^{\frac{1.4}{1.4-1}}$$

$$= (1 + 0.2\ M_e^2)^{3.5}\ ; \quad M_e^2 = 1.11$$

Mach number at exit,

$$M_e = \sqrt{1.11} = 1.054$$

Temperature at exit,

$$T_e = T_0 \left(\frac{p_e}{p_0}\right)^{\frac{\gamma-1}{\gamma}}$$

$$= 487.45 \left(\frac{1.1}{2.22}\right)^{\frac{1.4-1}{1.4}} = 398.7\ \text{K}$$

Sonic velocity at exit,

$$a_e = \sqrt{\gamma R T_e} = \sqrt{1.4 \times 287 \times 398.7}$$
$$= 400.25\ \text{m/s}$$

∴ Velocity at exit,

$$V_e = M_e a_e = 1.054 \times 400.25$$
$$= 421.86\ \text{m/s}$$

From continuity considerations :

$$\rho A V = \rho_e A_e V_e$$

$$\frac{p}{RT} A V = \frac{p_e}{R T_e} A_e V_e$$

∴ Flow area at exit,

$$A_e = A \times \frac{p}{p_e} \times \frac{V}{V_e} \times \frac{T_e}{T}$$

$$= 1000 \times \frac{2}{1.1} \times \frac{170}{421.86} \times \frac{398.7}{473}$$

$$= 617.45\ \text{mm}^2$$

(*d*) Let subscript *t* refer to the conditions at the throat section of the nozzle. At throat, critical conditions prevail, *i.e.*, the flow velocity becomes equal to the sonic velocity and Mach number attains a unit value.

$$\frac{T_0}{T_t} = 1 + \frac{\gamma - 1}{2} M^2 = 1 + \frac{\gamma - 1}{2} \times 1^2$$

$$= \frac{\gamma + 1}{2}$$

∴ Temperature at throat,

$$T_t = T_0 \times \frac{2}{\gamma + 1} = 487.45 \times \frac{2}{1.4 + 1}$$

$$= 406.21\ \text{K}$$

∴ Pressure at throat,

$$p_t = p_0 \times \left(\frac{T_t}{T_0}\right)^{\frac{\gamma}{\gamma-1}}$$

$$= 2.22 \left(\frac{406.21}{487.45}\right)^{\frac{1.4}{1.4-1}} = 1.173 \text{ bar}$$

Sonic velocity (corresponding to throat conditions)

$$= \sqrt{\gamma R T_t} = \sqrt{1.4 \times 287 \times 406.21}$$
$$= 404 \text{ m/s}$$

∴ Flow velocity at throat, V_t = 404 m/s

From continuity considerations,

$$\rho AV = \rho_t A_t V_t ; \quad \frac{p}{RT} AV = \frac{p_t}{RT_t} A_t V_t$$

∴ Flow area at throat,

$$A_t = A \times \frac{p}{p_t} \times \frac{V}{V_t} \times \frac{T_t}{T}$$

$$= 1000 \times \frac{2}{1.173} \times \frac{170}{404} \times \frac{406.21}{473}$$

$$= \mathbf{616 \ mm^2}$$

18.7. VELOCITY AND HEAT DROP

Consider steady flow of fluid flowing from section 1 to section 2 of any flow passage. The steady flow equation on unit mass basis is

$$h_1 + \frac{V_1^2}{2} + gz_1 + Q = h_2 + \frac{V_2^2}{2} + gz_2 + W_s \quad ...(18.32)$$

The only purpose of nozzle being the conversion of heat energy (pressure energy) into kinetic energy, the flow through a nozzle is characterized by the following features :

- no shaft work ; $W_s = 0$
- walls are thermally insulated ; the system does not exchange any heat with the surroundings, *i.e.*, the flow is adiabatic ; $Q = 0$
- the nozzle is placed with its axis horizontal, change in elevation between the entry and exit point is negligible ; $(z_1 - z_2) = 0$

With these stipulations SFEE reduces to

$$\frac{V_2^2 - V_1^2}{2} = (h_1 - h_2)$$

Thus an increase in kinetic energy equals the decrease in enthalpy.

Theoretical velocity at exit V_2

$$= \sqrt{2(h_1 - h_2) + V_1^2}$$

The velocity at inlet V_1, called the *approach velocity*, is negligible in comparison with exit velocity V_2 and can be neglected . Then

$$V_2 = \sqrt{2(h_1 - h_2)}$$

If h_1 and h_2 are enthalpies of the working fluid in kJ/kg and the velocity is in m/s, then

$$\frac{V_2^2}{2 \times 10^3} = (h_1 - h_2)$$

or $$V_2 = 44.7\sqrt{(h_1 - h_2)} \quad ...(18.33)$$

For a perfect gas :

$h_1 - h_2 = c_p (T_1 - T_2)$ and therefore

$$V_2 = 44.7\sqrt{c_p(T_1 - T_2)} \quad ...(18.34)$$

18.8. FLOW OF VAPOURS

The equations for critical pressure and temperature set up in previous sections apply to perfect gases only and not to vapours. This is due to the fact that in case of vapours, the exponent of expansion is no longer constant but depends upon the initial and final states of the vapour. Sufficient close approximation is however obtained for a steam nozzle if the expansion is assumed to follow the law pv^n = constant.

For steam initially superheated : $n = 1.3$

For steam initially dry saturated : $n = 1.135$

For steam having initially dryness fraction x, the value of exponent n can be found by using the Zeuner's equation given below :

$$n = 1.035 + x \times 0.1 \quad ...(18.35)$$

This value of n is used in place of γ in determining the critical pressure for steam.

$$\therefore \text{ For steam, } \frac{p_c}{p_1} = \left(\frac{2}{n+1}\right)^{\frac{n}{n-1}}$$

∴ Critical pressure ratio for superheated steam

$$= \left(\frac{2}{1.3+1}\right)^{\frac{1.3}{0.3}} = 0.546$$

and critical pressure ratio for dry saturated steam

$$= \left(\frac{2}{1.135+1}\right)^{\frac{1.135}{0.135}} = 0.578$$

18.9. EFFECT OF FRICTION AND NOZZLE EFFICIENCY

Assuming no friction through the nozzle, the adiabatic expansion can be represented by a vertical line 1-2s on *h-s* chart (Fig. 18.7). However in actual expansion of fluid through a nozzle there are frictional losses due to :

(*a*) friction between fluid and the nozzle walls.

(*b*) friction within the fluid itself ; and

(*c*) energy dissipation due to formation of eddies and shocks.

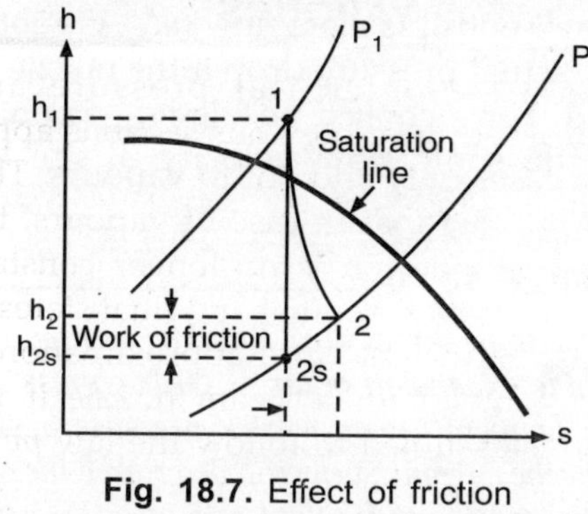

Fig. 18.7. Effect of friction

The energy lost in friction is converted into heat and that tends to dry the steam and increase its enthalpy. The expansion is no longer isentropic and there is progressive increase in entropy. The actual expansion is along the curve (1 – 2), the point 2 being on the constant pressure line p_2.

The effect of friction losses are :

(*i*) Reduction in the actual enthalpy drop from $(h_1 - h_{2s})$ to $(h_1 - h_2)$

(*ii*) Reduction in the exit velocity

(*iii*) Increase in the dryness fraction of steam leaving the nozzle

(*iv*) Increase in the specific volume of fluid due to increase in dryness fraction of steam

Due to increased specific volume and reduced velocity, the mass flow rate decreases

$$\left(m = \frac{AV}{v}\right)$$

Most of the friction in the convergent divergent nozzle is assumed to occur between the throat and exit (diverging section) and expansion from mouth to throat (converging section) is taken to be truly isentropic. This is due to the fact that

(*i*) the diverging flow is more susceptible to losses due to higher velocities and turbulence in that portion.

(*ii*) the losses in the convergent portion of the nozzle are much less owing to small inlet velocity and minimum turbulence.

Nozzle efficiency : Nozzle efficiency is a measure of the degree of perfection of nozzle in fulfilling its sole purpose of delivering the maximum possible amount of kinetic energy for the given pressure drop.

The nozzle efficiency is usually expressed as the ratio of actual enthalpy drop to the isentropic enthalpy drop between the same pressure limits.

$$\eta_{nozzle} = \frac{\text{actual enthalpy drop}}{\text{isentropic enthalpy drop}}$$

$$= \frac{h_1 - h_2}{h_1 - h_{2s}} \qquad ...(18.36)$$

The exit velocity V_2 considering friction is then given by

$$V_2 = 44.7\sqrt{(h_1 - h_{2s}) \times \eta_{nozzle}}$$

For a perfect gas, the expression for nozzle efficiency can be written as

$$\eta_{nozzle} = \frac{c_p(T_1 - T_2)}{c_p(T_1 - T_{2s})} = \frac{T_1 - T_2}{T_1 - T_{2s}} \qquad ...(18.37)$$

Applying the steady flow energy equation between the entry (velocity V_1) and exit (velocity V_2 in actual and V_{2s} in isentropic) sections, we get

$$\text{isentropic : } h_1 + \frac{V_1^2}{2} = h_{2s} + \frac{V_{2s}^2}{2}$$

$$(h_1 - h_{2s}) = \frac{V_{2s}^2 - V_1^2}{2}$$

$$\text{actual : } \quad h_1 + \frac{V_1^2}{2} = h_2 + \frac{V_2^2}{2}$$

$$(h_1 - h_2) = \frac{V_2^2 - V_1^2}{2}$$

The nozzle efficiency may then be written as

$$\eta_{nozzle} = \frac{V_2^2 - V_1^2}{V_{2s}^2 - V_1^2} \quad \text{...(18.38)}$$

The velocity V_1 at entry to nozzle is negligibly small as compared to the exit velocity V_2 or V_{2s}. Then

$$\eta_{nozzle} = \left(\frac{V_2}{V_{2s}}\right)^2$$

The nozzle efficiency depends upon the following factors :

(*i*) size and shape of the nozzle
(*ii*) angle of nozzle divergence
(*iii*) material of which the nozzle is made
(*iv*) workmanship in the making of nozzle
(*v*) fluid velocity, the state and the nature of fluid flowing
(*vi*) friction and turbulence in the nozzle flow passage

The nozzle efficiency ranges from 90 to 95%; the larger nozzles with straight line axis have the highest efficiency.

Velocity coefficient : The velocity coefficient is defined as the ratio of actual velocity at the nozzle exit to the velocity at exit for isentropic flow from the same inlet state to the same exit pressure.

Velocity coefficient

$$C_v = \frac{\text{actual velocity}}{\text{isentropic velocity}}$$

$$= \frac{\text{actual enthalpy drop}}{\text{isentropic enthalpy drop}}$$

$$= \sqrt{\frac{h_1 - h_2}{h_1 - h_{2s}}}$$

$$= \sqrt{\text{nozzle efficiency}} \quad \text{...(18.39)}$$

Thus the velocity coefficient is equal to the square root of the nozzle efficiency (when inlet velocity is assumed negligible).

Experimental investigations have shown that the velocity coefficient depends upon nozzle dimensions, roughness of nozzle walls and the flow velocity, and its value ranges from 0.93 to 0.97.

Coefficient of discharge : The coefficient of discharge is associated with the mass flow rate and is defined as the ratio of actual mass flow rate through the nozzle to the mass flow which would be passed if the flow were isentropic from the same inlet state to the same exit pressure.

$$\text{Coefficient of discharge } C_d = \frac{m_{act}}{m_{isen}} \quad \text{...(18.40)}$$

The isentropic mass flow (m_{isen}) is obtained by the actual pressure drop if the nozzle is not choked. For a choked nozzle m_{isen} is obtained from the sonic flow velocity at the throat section.

EXAMPLE 18.10

A nozzle with throat area 200 mm² is used to have isentropic expansion of air at 60 bar pressure and 310 K temperature to a back pressure of 40 bar. Neglect the entrance velocity, determine the outflow velocity and the mass flow rate of air.

Solution : Critical pressure

$$p_c = p_1 \left(\frac{2}{\gamma+1}\right)^{\frac{\gamma}{\gamma-1}}$$

$$= 60 \left(\frac{2}{1.4+1}\right)^{\frac{1.4}{1.4-1}} = 31.7 \text{ bar}$$

Since the nozzle exit pressure (40 bar) is more than the critical pressure (31.7 bar), the exit velocity does not reach the critical velocity and the nozzle is only of convergent section. The outflow velocity is then given by

$$V_2 = \left[2\frac{\gamma}{\gamma-1} p_1 v_1 \left\{ 1 - \left(\frac{p_2}{p_1} \right)^{\frac{\gamma-1}{\gamma}} \right\} \right]^{1/2}$$

Now :

$$v_1 = \frac{mRT_1}{p_1} = \frac{1 \times 287 \times 310}{60 \times 10^5}$$

$$= 0.01483 \text{ m}^3/\text{kg}$$

$$2\frac{\gamma}{\gamma-1} p_1 v_1 = 2 \times \frac{1.4}{1.4-1} \times 60 \times 10^5 \times 0.01483$$

$$= 622860$$

$$1 - \left(\frac{p_2}{p_1} \right)^{\frac{\gamma-1}{\gamma}} = 1 - \left(\frac{40}{60} \right)^{\frac{1.4-1}{1.4}} = 0.1095$$

$$\therefore V_2 = (622860 \times 0.1095)^{1/2} = 261.16 \text{ m/s}$$

The mass flow rate of air is worked out from the relation

$$m = A_2 \left[2\frac{\gamma}{\gamma-1} \frac{p_1}{v_1} \left\{ \left(\frac{p_2}{p_1} \right)^{\frac{2}{\gamma}} - \left(\frac{p_2}{p_1} \right)^{\frac{\gamma+1}{\gamma}} \right\} \right]^{1/2}$$

Now :

$$\frac{\gamma}{\gamma-1} \frac{p_1}{v_1} = \frac{1.4}{1.4-1} \times \frac{60 \times 10^5}{0.01483} = 2.83 \times 10^9$$

$$\left(\frac{p_2}{p_1} \right)^{\frac{2}{\gamma}} - \left(\frac{p_2}{p_1} \right)^{\frac{\gamma+1}{\gamma}} = \left(\frac{40}{60} \right)^{\frac{2}{1.4}} - \left(\frac{40}{60} \right)^{\frac{1.4+1}{1.4}}$$

$$= 0.5603 - 0.499$$

$$= 0.0613$$

$$\therefore m = 200 \times 10^{-6} \, (2.83 \times 10^9 \times 0.0613)^{1/2}$$

$$= 2.6 \text{ kg/s}$$

EXAMPLE 18.11

In a propulsion nozzle, gases expand from 4 bar and 650 K to a back pressure of 1 bar at the rate of 15 kg/s. The coefficient of discharge is 0.98 and the nozzle efficiency is 95%. Assuming inlet velocity to be negligible, compute the throat and exit areas of the nozzle.

For gases $\gamma = 1.3$ and $c_p = 1.1$ kJ/kg K

Solution : Critical pressure

$$p_c = p_1 \left(\frac{2}{\gamma+1} \right)^{\frac{\gamma}{\gamma-1}} = 4 \left(\frac{2}{1.3+1} \right)^{\frac{1.3}{1.3-1}}$$

$$= 2.183 \text{ bar}$$

Since the nozzle exit pressure (1 bar) is less than the critical pressure (2.813 bar), the nozzle must be of convergent-divergent type. The mass flow is determined from the critical throat velocity.

Critical temperature at throat

$$T_c = \left(\frac{2}{\gamma+1} \right) T_1 = \left(\frac{2}{1.3+1} \right) \times 650$$

$$= 565.22 \text{ K}$$

Critical throat velocity

$$V_c = \sqrt{2(h_1 - h_c)} = \sqrt{2c_p(T_1 - T_c)}$$

$$= \sqrt{2 \times (1.1 \times 10^3)(650 - 565.22)}$$

$$= 431.87 \text{ m/s}$$

Specific gas constant

$$R = \frac{c_p(\gamma-1)}{\gamma} = \frac{1.1 \times 10^3 \times (1.3-1)}{1.3}$$

$$= 253.85 \text{ J/kg K}$$

Specific volume

$$v_{sc} = \frac{RT_c}{p_c} = \frac{253.85 \times 565.22}{2.183 \times 10^5}$$

$$= 0.657 \text{ m}^3/\text{kg}$$

Mass flow rate for isentropic flow

$$m_s = \frac{15}{0.98} = 15.306 \text{ kg/s}$$

Throat area

$$A_c = \frac{m_s v_{sc}}{V_c} \qquad \left(\because m = \frac{AV}{v} \right)$$

$$= \frac{15.306 \times 0.657}{431.87} = 0.0233 \text{ m}^2$$

$$= 233 \text{ cm}^2$$

For an isentropic expansion from the inlet conditions down to the back pressure, the temperature at nozzle exit is T_{2s} and

$$\frac{T_1}{T_{2s}} = \left(\frac{p_1}{p_2} \right)^{\frac{\gamma-1}{\gamma}}$$

$$\frac{T_1}{T_{2s}} = \left(\frac{4}{1}\right)^{\frac{1.3-1}{1.3}} = 1.38$$

$$T_{2s} = \frac{T_1}{1.38} = \frac{650}{1.38} = 471 \text{ K}$$

Nozzle efficiency

$$= 0.95 = \frac{T_1 - T_2}{T_1 - T_{2s}}$$

$\therefore$ Nozzle exit temperature

$$T_2 = T_1 - 0.95\,(T_1 - T_{2s})$$
$$= 650 - 0.95\,(650 - 471) = 479.95 \text{ K}$$

Specific volume

$$v_{s_2} = \frac{RT_2}{p_2} = \frac{253.85 \times 479.95}{1 \times 10^5}$$
$$= 1.218 \text{ m}^3/\text{kg}$$

Exit velocity

$$V_2 = \sqrt{2(h_1 - h_2)} = \sqrt{2c_p(T_1 - T_2)}$$
$$= \sqrt{2 \times (1.1 \times 10^3) \times (650 - 479.95)}$$
$$= 6.11.64 \text{ m/s}$$

Throat area

$$A_2 = \frac{m_2\, v_{s_2}}{V_2} = \frac{15.306 \times 1.218}{611.64}$$
$$= 0.03048 \text{ m}^2 = 304.8 \text{ cm}^2$$

EXAMPLE 18.12

(a) Explain what do you mean by critical pressure ratio in connection with reversible adiabatic flow through a nozzle.

(b) When a nozzle is subjected to this pressure ratio, show that the velocity of steam at throat for maximum discharge is given by

$$(i)\ V_2 = \sqrt{2\frac{n}{n+1} p_1 v_1} \qquad (ii)\ V_2 = \sqrt{n\, p_2 v_2}$$

where n is index of expansion, p_1 is initial pressure of steam, and v_1 is specific volume of steam.

Assume frictionless adiabatic flow through a convergent-divergent nozzle and neglect entering steam velocity.

Solution : For frictionless adiabatic flow and neglecting entering velocity, the velocity of steam at throat of nozzle is given by

$$\frac{V_2^2}{2} = (h_1 - h_2)$$

The isentropic steam flow through the nozzle approximately follows the law : pv^n = constant

and the enthalpy drop $(h_1 - h_2)$ represents work done during Rankine cycle. That is

$$h_1 - h_2 = \frac{n}{n-1}(p_1 v_1 - p_2 v_2)$$

$$\therefore \quad \frac{V_2^2}{2} = \frac{n}{n-1}(p_1 v_1 - p_2 v_2)$$

$$= \frac{n}{n-1} p_1 v_1 \left(1 - \frac{p_2}{p_1} \times \frac{v_2}{v_1}\right) \quad ...(i)$$

From the relation $p_1 v_1^n = p_2 v_2^n$ we have :

$$\frac{v_2}{v_1} = \left(\frac{p_2}{p_1}\right)^{-1/n} \text{ and as such}$$

$$\frac{V_2^2}{2} = \frac{n}{n-1} p_1 v_1 \left[1 - \frac{p_2}{p_1} \times \left(\frac{p_2}{p_1}\right)^{-\frac{1}{n}}\right]$$

$$= \frac{n}{n-1} p_1 v_1 \left[1 - \left(\frac{p_2}{p_1}\right)^{\frac{n-1}{n}}\right]$$

For maximum discharge

$$\frac{p_2}{p_1} = \left(\frac{2}{n+1}\right)^{\frac{n}{n-1}} \text{ and then}$$

$$\frac{V_2^2}{2} = \frac{n}{n-1} p_1 v_1 \left[1 - \left(\frac{2}{n+1}\right)\right]$$

$$= \frac{n}{n-1} p_1 v_1 \times \frac{n-1}{n+1} = \frac{n}{n+1} p_1 v_1$$

$$\text{Hence } V_2 = \sqrt{2\frac{n}{n+1} p_1 v_1}$$

(*b*) The expansion (*i*) can also be written as

$$\frac{V_2^2}{2} = \frac{n}{n-1} p_2 v_2 \left[\frac{p_1}{p_2} \times \frac{v_1}{v_2} - 1\right]$$

$$= \frac{n}{n-1} p_2 v_2 \left[\left(\frac{p_2}{p_1}\right)^{-1} \times \left(\frac{p_2}{p_1}\right)^{\frac{1}{n}} - 1\right]$$

$$= \frac{n}{n-1} p_2 v_2 \left[\left(\frac{p_2}{p_1} \right)^{\frac{1-n}{n}} - 1 \right]$$

For maximum discharge

$$\frac{p_2}{p_1} = \left(\frac{2}{n+1} \right)^{\frac{n}{n-1}} \text{ and then}$$

$$\frac{V_2^2}{2} = \frac{n}{n-1} p_2 v_2 \left[\left(\frac{2}{n+1} \right)^{\frac{n}{n-1} \times \frac{1-n}{n}} - 1 \right]$$

$$= \frac{n}{n-1} p_2 v_2 \left[\frac{n+1}{2} - 1 \right]$$

$$= \frac{n}{n-1} p_2 v_2 \times \frac{n-1}{2} = \frac{n}{2} p_2 v_2$$

That gives :

$$V_2 = \sqrt{n\, p_2 v_2}$$

Upon differentiation of pv^n = const, we have

$$p\, n\, v^{n-1}\, dv + v^n\, dp = 0$$

or $$-\frac{dp}{dv} = \frac{p\, n\, v^{n-1}}{v^n} = \frac{pn}{v}$$

and $$-v^2 \frac{dp}{dv} = n\, p\, v$$

But $-v^2 \dfrac{dp}{dv} = a^2$ where a is the sonic velocity

Thus : $a^2 = n\, pv$ and $V_2 = \sqrt{n\, p_2 v_2} = a$

This indicates that under conditions of maximum discharge, the flow velocity at throat equals the sonic velocity at the throat conditions.

EXAMPLE 18.13

Determine the throat and exit area for a steam nozzle designed to discharge steam at the rate of 0.25 kg/s. The steam enters the nozzle at 10 bar and 250°C and the exit pressure is 2 bar. Neglect inlet velocity and assume isentropic expansion which follows the law $pv^{1.3}$ = const. Calculations may be made using Mollier chart or otherwise.

Solution : Let points 1, 2 and 3 represent the conditions at entrance, throat and exit of the nozzle.

From steam tables

v_1 = 0.233 m³/kg at p_1 = 10 bar

and T_1 = 250°C

(*a*) For optimum conditions

$$\frac{p_2}{p_1} = \left(\frac{2}{n+1} \right)^{\frac{n}{n-1}}$$

$$= \left(\frac{2}{1.3+1} \right)^{\frac{1.3}{1.3-1}} = 0.5457$$

Velocity at throat

$$V_2 = \left[\frac{2n}{n-1} p_1 v_1 \left\{ 1 - \left(\frac{p_2}{p_1} \right)^{\frac{n-1}{n}} \right\} \right]^{1/2}$$

$$= \left[\frac{2 \times 1.3}{1.3-1} \times (10 \times 10^5) \times 0.233 \times \left\{ 1 - (0.5457)^{\frac{1.3-1}{1.3}} \right\} \right]^{1/2}$$

$$= 513.15 \text{ m/s}$$

Specific volume at throat $v_2 = v_1 \left(\dfrac{p_1}{p_2} \right)^{\frac{1}{n}}$

$$= 0.233 \left(\frac{1}{0.5457} \right)^{\frac{1}{1.3}}$$

$$= 0.371 \text{ m}^3/\text{kg}$$

Mass flow rate

$$m = \frac{A_2 V_2}{v_2} \text{ and therefore}$$

Throat area

$$A_2 = \frac{m v_2}{V_2} = \frac{0.25 \times 0.371}{513.15}$$

$$= 1.807 \times 10^{-4} \text{ m}^2 = 1.807 \text{ cm}^2$$

(*b*) Velocity at nozzle exit

$$V_3 = \left[\frac{2n}{n-1} p_1 v_1 \left\{ 1 - \left(\frac{p_3}{p_1} \right)^{\frac{n-1}{n}} \right\} \right]^{1/2}$$

$$= \left[\frac{2 \times 1.3}{1.3 - 1} \times (10 \times 10^5) \times 0.233 \left\{ 1 - \left(\frac{2}{10} \right)^{\frac{1.3-1}{1.3}} \right\} \right]^{1/2}$$

= 791.45 m/s

Specific volume at exit $v_3 = v_1 \left(\frac{p_1}{p_3} \right)^{\frac{1}{n}}$

$$= 0.233 \left(\frac{10}{2} \right)^{\frac{1}{1.3}} = 0.8035 \text{ m}^3/\text{kg}$$

∴ Exit area

$$A_3 = \frac{m\, v_3}{V_3} = \frac{0.25 \times 0.8035}{791.45}$$

$= 2.538 \times 10^{-4}$ m² = **2.538 cm²**

EXAMPLE 18.14

Steam at a pressure of 7 bar and 180°C is discharged through a convergent-divergent nozzle at 1 bar. The mass flow rate of steam is estimated to 2 kg/s. The pressure at the nozzle throat is 4 bar and the steam enters the nozzle with velocity of 75 m/s. Expansion upto nozzle throat is isentropic and the frictional resistance equivalent to 60 kJ/kg of steam reheats the steam in the divergent section. Make calculations for :

(a) suitable areas for the throat and exit.

(b) overall efficiency of nozzle based on enthalpy drop between the actual state of steam and the back pressure.

Solution : Refer Fig. 18.8 for the expansion process represented on the Mollier chart :

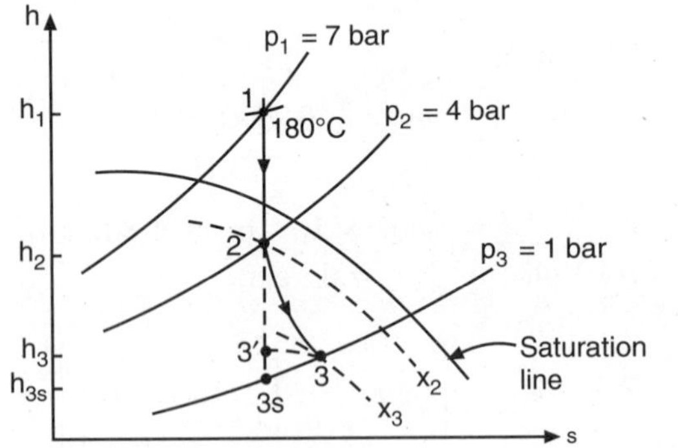

Fig. 18.8.

The enthalpy and dryness fraction values as read from the Mollier chart are :

h_1 = 2795 kJ/kg

h_2 = 2680 kJ/kg

and x_2 = 0.98

h_{3s} = 2460 kJ/kg

$h_3 = h'_3$ = 2460 + 60 = 2520 kJ/kg

and x_3 = 0.93

The specific volumes of dry saturated steam corresponding to state points 2(p_2 = 4 bar) and state point 3(p_3 = 1 bar) as read from the steam tables are :

v_{s2} = 0.462 m³/kg

and v_{s3} = 1.695 m³/kg

The velocity of steam V_2 at the throat of nozzle is given by

$$\frac{V_2^2}{2} = (h_1 - h_2) + \frac{V_1^2}{2}$$

$$\frac{V_2^2}{2 \times 1000} = (2795 - 2600) + \frac{75^2}{2 \times 1000}$$

or $V_2^2 = 2 \times 1000 \times (2795 - 2680) + 75^2$

= 235625

V_2 = 485.4 m/s

Area at nozzle throat

$$A_2 = \frac{m \times v_2}{V_2} = \frac{2 \times (0.98 \times 0.462)}{485.4}$$

$= 1.8655 \times 10^{-3}$ m² = 18.65 cm²

The velocity of steam V_3 at exit from the nozzle is given

$$\frac{V_3^2}{2} = (h_1 - h_3) + \frac{V_1^2}{2}$$

$$\frac{V_3^2}{2 \times 1000} = (2795 - 2520) + \frac{75^2}{2 \times 1000}$$

or $V_3^2 = 2 \times 1000 \times (2795 - 2520) + 75^2$

= 555625

V_3 = 745.4 m/s

Area at exit from nozzle

$$A_3 = \frac{m \times v_3}{V_3} = \frac{2 \times (0.93 \times 1.695)}{745.4}$$

$= 4.229 \times 10^{-3}$ m² = 42.29 cm²

(*b*) Overall nozzle efficiency

$$= \frac{\text{actual heat drop}}{\text{isentropic heat drop}}$$

$$= \frac{h_1 - h_3}{h_1 - h_{3s}}$$

$$= \frac{2795 - 2520}{2795 - 2460} = 0.821 \text{ or } 82.1\%$$

EXAMPLE 18.15

Steam at 10 bar and 0.97 after expansion in a group of convergent nozzles to a pressure of 0.15 bar flows in an impulse turbine which develops 150 kW of power with a steam consumption of 8 kg/kWh. Determine

(a) the pressure at throat,

(b) the number of nozzles required each of which is 6 mm diameter at the throat.

(c) suitable diameter at exit from each nozzle.

It may be assumed that 10% of the overall isentropic enthalpy drop reheats the steam in the divergent portion of the nozzle.

Solution : For the wet steams, Zeuner's equation can be used for determining the value of adiabatic expansion through the nozzle.

$$n = 1.035 + 0.1\,x$$

$$= 1.035 + 0.1 \times 0.97 = 1.132$$

Let points 1, 2 and 3 represent the conditions at entrance, throat and exit of the nozzle.

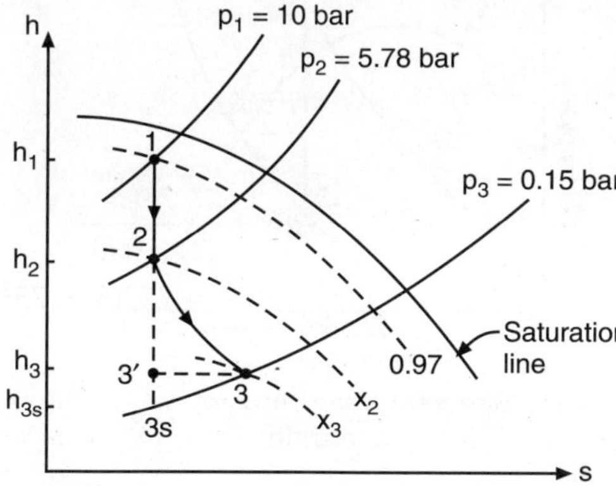

Fig. 18.9.

(*a*) Pressure at throat

$$P_2 = P_1 \left(\frac{2}{n+1}\right)^{\frac{n}{n-1}}$$

$$= 10 \left(\frac{2}{1.132+1}\right)^{\frac{1.132}{1.132-1}} = 5.78 \text{ bar}$$

(*b*) The process of expansion may be represented on the Mollier diagram (*h-s* plot) as shown in Fig. 18.9.

From *h-s* chart :

$$h_1 = 2718 \text{ kJ/kg} \quad \text{and}$$

$$h_{3s} = 2085 \text{ kJ/kg}$$

10% of isentropic enthalpy drop from entrance to exit

$$= \frac{10}{100}(2718 - 2085) = 63.3 \text{ kJ/kg}$$

Locate point 3′ at a distance corresponding to 63.3 kJ from 3*s* on the vertical line 1-2-3*s*. From 3′, draw a horizontal line till it meets the 0.15 bar pressure line at point 3. The path 1-2-3 then represents the actual expansion process in the nozzle from inlet to outlet.

The following enthalpy and dryness fraction values at throat (point 2) and exit (point 3) of the nozzle are then read from the *h-s* chart.

$$h_2 = 2618 \text{ kJ/kg} \; ; \quad x_2 \cong 0.935$$

$$h_3 = 2148 \text{ kJ/kg} \; ; \quad x_3 \cong 0.815$$

From steam tables :

$$v_{s_2} = 0.3275 \text{ m}^3\text{/kg}$$

at 5.78 bar pressure

and $v_{s_3} = 10.06$ m³/kg at 0.15 bar pressure

Steam consumption

$$m = \frac{8 \times 150}{3600} = 0.333 \text{ kg/s}$$

Velocity at throat

$$V_2 = 44.7\sqrt{(h_1 - h_2)}$$

$$= 44.7\sqrt{(2718 - 2618)} = 447 \text{ m/s}$$

Mass flow through a nozzle

$$= \frac{A_2 V_2}{v_2} = \frac{\frac{\pi}{4}(0.006)^2 \times 447}{0.935 \times 0.3275}$$

$$= 0.0412 \text{ kg/s}$$

$\therefore$ Number of nozzles required $= \dfrac{0.333}{0.0412} \cong 8$

(*c*) Velocity at exit

$$= 44.7\sqrt{(h_1 - h_3)}$$

$$= 44.7\sqrt{(2718 - 2148)} = 1067.2 \text{ m/s}$$

Total exit area

$$= \frac{m v_3}{V_3} = \frac{0.333 \times (10.06 \times 0.815)}{1067.2}$$

$$= 2.558 \times 10^{-3} \text{ m}^2 = 25.58 \text{ cm}^2$$

Area for one nozzle

$$= \frac{25.58}{8} = 3.1975 \text{ cm}^2$$

$\therefore$ Nozzle diameter at exit

$$= \sqrt{\frac{3.1975 \times 4}{\pi}} = \mathbf{2.018 \text{ cm}}$$

18.10. SUPERSATURATED FLOW AND ITS EFFECTS

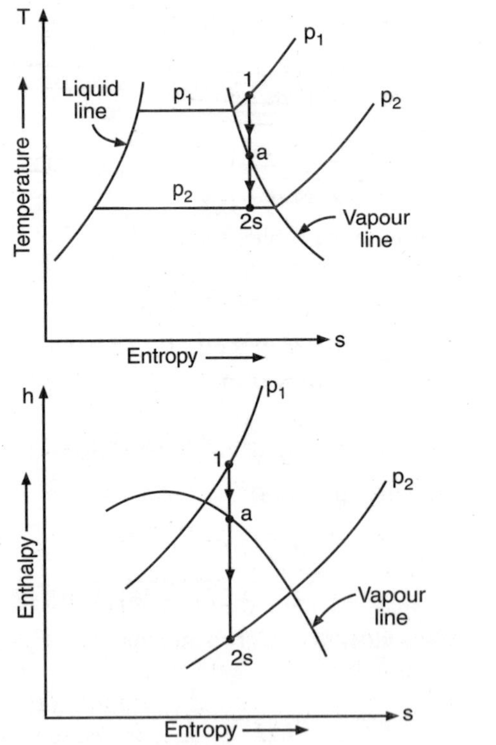

Fig. 18.10. Expansion of steam under thermal equilibrium

Generally it is assumed that during isentropic expansion of steam vapour in a nozzle, the condensation proceeds at correct rate. This means that the steam vapour begins to condense as soon as the vapour line is reached. With reference to *T-s* and *h-s* plots of Fig. 18.10 the initial state of steam in the superheated region is at pressure p_1 and temperature T_1 $(= T_{sup1})$. This steam expands isentropically through the nozzle and condensation within the vapour begins immediately at point *a* where the isentropic line 1-2 meets the saturated vapour line. There is a stable mixture of steam and liquid at any point during expansion. If the expansion is stopped at any instant, no subsequent changes take place in the condition of steam. Under such conditions, the steam is said to be expanding in *thermal equilibrium*.

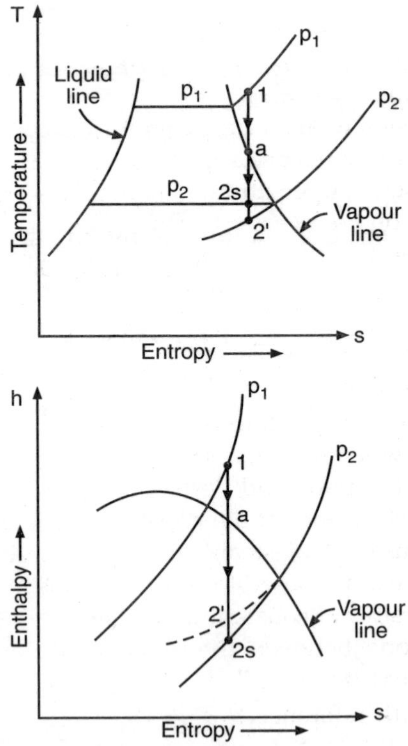

Fig. 18.11. Supersaturated flow of steam through a nozzle

In actual practice, the condensation of steam does not take place as rapidly as does the change in pressure and temperature. For condensation to take place, two or more molecules must come together to act as a nucleus which should attract other molecules. Condensation would take place when a recognizable droplet of water has been formed around the nucleus. There is a strong tendency for immediate evaporation of a minute water droplet formed during rapid expansion. Because of sonic and supersonic speeds in the nozzle, the steam takes only a very small fraction of second (0.001 second or even less) for expansion. The condensation cannot keep pace with expansion and the steam continues to expand in dry state. The steam remains in a state of homogeneous vapour and does not consist of vapour and liquid phases which the condition for thermal equilibrium demands. The steam in such conditions behaves more or less like a perfect gas, and is set to be in *metastable, supersaturated* or *non-equilibrium* state.

The supersaturated flow of steam through a nozzle has been represented in *T-s* and *h-s* plots of Fig. 18.11. The line 1-2s represents expansion with thermal equilibrium throughout the entire expansion process. Line 1-2′ represents supersaturated expansion. The point 2′ lies on the constant pressure line p_2 when it is extended into the saturated region with the same slope as it had in the superheated region.

Wilson line : The limit of supersaturated is shown on *h-s* diagram by a curve known as Wilson line.

The Wilson line represents the saturated steam when its density is about 8 times that of normal vapour and corresponds to about 4% of moisture line. At the Wilson line, the ultimate condensation of steam to proper equilibrium mixture is supposed to take place suddenly and irreversibly at constant total energy (enthalpy). The zone between the Wilson line and dry saturated line is called *supersaturated field*.

With reference to Fig. 18.12.

(*i*) line 1-2*s*-3*s* represents the isentropic expansion of steam under thermal equilibrium condition.

(*ii*) The curvature of constant pressure line p_2 in the superheated area is extended to wet area, and this curve intersects the Wilson line at 2′.

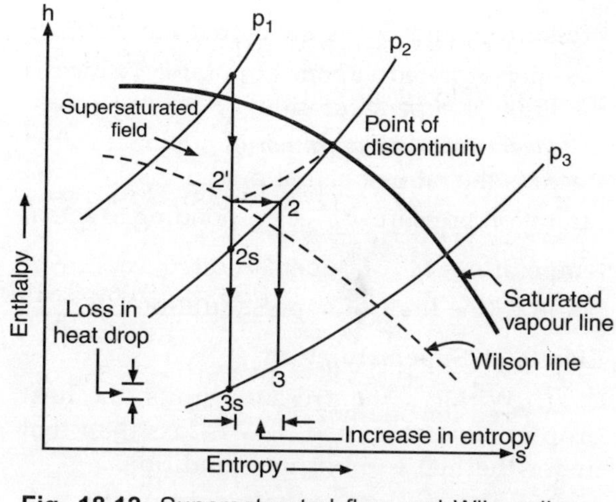

Fig. 18.12. Supersaturated flow and Wilson line

(*iii*) At point 2′ the limit of supersaturation is reached and steam condenses to normal state at the pressure p_2. The condensation takes place at constant total energy along 2′– 2 where point 2 lies on the normal constant pressure line of p_2.

(*iv*) The line 2-3 represents the subsequent normal stable expansion of steam to the exit pressure p_3.

(*v*) The entire expansion process is shown by 1-2′-2-3; the vertical distance between 1 and 3 represents the total enthalpy drop.

Degree of undercooling and supersaturation

During supersaturated flow (process 1-2′), the steam vapour expands with index of expansion n = 1.3 and accordingly.

$$pV^{1.3} = \text{constant and } T_2' = T_1\left(\frac{p_2'}{p_1}\right)^{\frac{0.3}{1.3}}$$

where p_2' and T_2' are saturation pressure and temperature at state 2′ which is the limit of supersaturation. The temperature T_2' is less than the saturation temperature T_{2s} for the pressure P_2. This is due to the fact that gain in kinetic energy of steam has been at the expense of sensible heat. The vapour is said to be *super-cooled* or *under-cooled*. The amount of under cooling is known as *degree of under cooling* and it is defined as $(T_{2s} - T_2')$ where T_{2s} is the saturation temperature corresponding to pressure p_2 and T_2' is the actual temperature of supersaturated vapour at point 2′ (which is the limit of supersaturation).

Degree of supersaturation of supersaturated steam is the ratio of actual pressure p_2 and the saturation pressure p_{2s}' corresponding to actual temperature T_2' of supersaturated vapour at point 2′ (the limit of supersaturation).

Effects of Supersaturation

(*i*) Within same pressure limits, the heat drop in supersaturated flow is less than that under thermal equilibrium conditions.

(*ii*) Due to less enthalpy drop, the exit velocity for supersaturated flow is less than that for equilibrium flow.

(*iii*) Stabilization process during which condensation takes place suddenly is irreversible process and accordingly there is increase in entropy immediately after supersaturated flow.

(*iv*) The specific volume at exit with supersaturated flow is considerably less than the specific volume at exit with equilibrium flow.

Now the mass flow through a given exit area is given by continuity equation, *i.e.*,

$$m = \frac{\text{area} \times \text{velocity}}{\text{specific volume}}$$

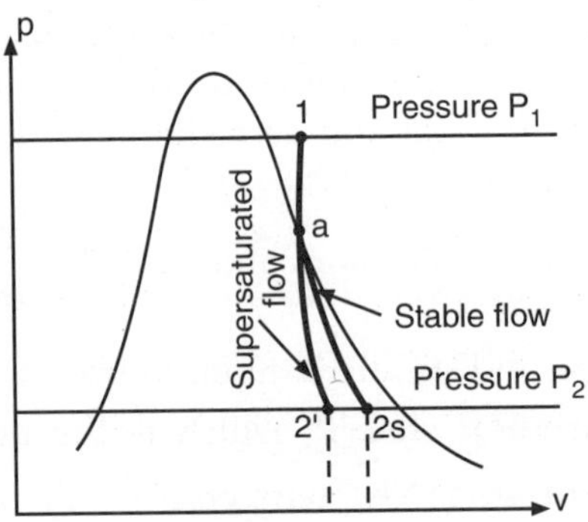

Fig. 18.13. Supersaturated flow on pressure volume diagram

Since the difference in enthalpy drop is small and we take square of enthalpy drop while calculating velocity, the effect of supersatured flow on the exit velocity becomes still smaller. Consequently the effect of decreased specific volume (increased density which is about 8 times the normal in supersaturated (zone) predominates over the small decrease in exit velocity. This affords the explanation to the well established fact that discharge with supersaturated flow is about 2-5% greater than that with equilibrium flow.

EXAMPLE 18.16

Dry saturated steam expands in a nozzle from 10 bar to 4 bar. If the steam remains in dry state during expansion, compute the degree of undercooling and degree of supersaturation.

Solution : From steam tables, the saturation temperature at 10 bar pressure is 179.9°C

$$\therefore \quad T_1 = 273 + 179.9 = 452.9 \text{ K}$$

Let T_2' be the temperature at point 2′ where saturation occurs. For supersaturated flow, the correlation between pressure and temperature is

$$p = T^{\frac{1.3}{0.3}} \times \text{constant}$$

$$\text{Then : } T_2' = T_1 \times \left(\frac{p_2}{p_1}\right)^{\frac{0.3}{1.3}}$$

$$= 452.9 \times \left(\frac{4}{10}\right)^{\frac{0.3}{1.3}} = 366.6 \text{ K}$$

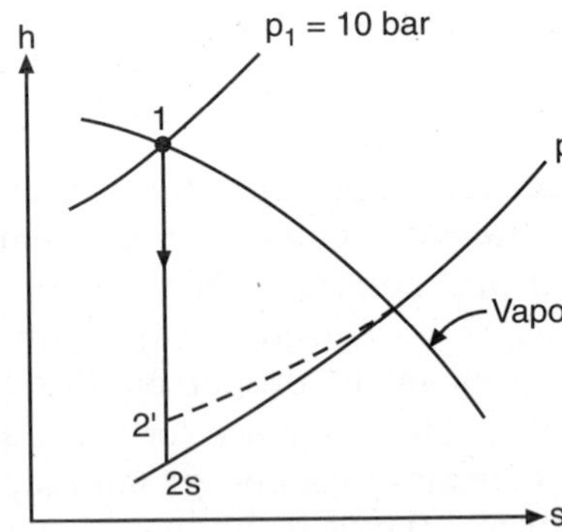

Fig. 18.14.

At 4 bar pressure, the saturation temperature is 143.6°C or 416.6 K

Then : degree of undercooling

$$= 416.6 - 366.6 = 50°\text{C}$$

Saturation pressure corresponding to T_2' (366.6 – 273 = 93.6°C) is

$$= 0.7011 + \frac{0.8453 - 0.7011}{5} \times 3.6 = 0.8049 \text{ bar}$$

$\therefore$ Degree of supersaturation $= \dfrac{4}{0.8049} = 4.97$

EXAMPLE 18.17

A convergent-divergent nozzle receives steam at 10 bar and 200°C and exhausts to 5 bar. The mass flow rate of steam is 0.1 kg/s and the flow is supersaturated with $pv^{1.3}$ = constant. Neglecting inlet velocity, make calculations for the nozzle exit area, degree of undercooling and the degree of supersaturation.

If the nozzle coefficient is 0.92 and mean value of c_p = 1.925 kJ/kg K, determine the exit area.

Use the relation :

$$pv \times 10^3 = 2.308\ (h - 1943)$$

where v is in m³/kg, p is in bar and h is in kJ/kg.

Solution : Refer Fig. 18.15.

With supersaturated flow, the isentropic enthalpy drop is given by

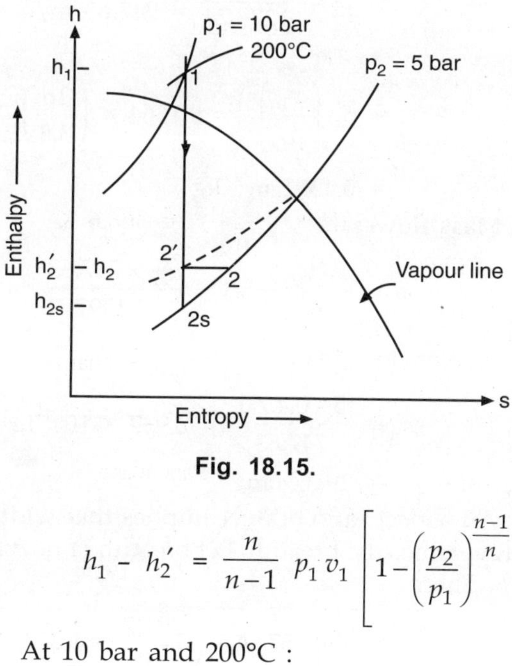

Fig. 18.15.

$$h_1 - h_2' = \frac{n}{n-1}\, p_1 v_1 \left[1 - \left(\frac{p_2}{p_1}\right)^{\frac{n-1}{n}}\right]$$

At 10 bar and 200°C :

h_1 = 2827 kJ/kg and v_1 = 0.206 m³/kg

$$h_1 - h_2' = \frac{1.3}{1.3-1} \times (10 \times 10^5) \times 0.206 \times \left[1 - \left(\frac{5}{10}\right)^{\frac{1.3-1}{1.3}}\right]$$

$$= 131917 \text{ J/kg} \quad \text{or} \quad 131.92 \text{ kJ/kg}$$

Then velocity at point 2,

$$V_2 = 44.7\sqrt{(h_1 - h_2)}$$
$$= 44.7\sqrt{131.92} = 513.41 \text{ m/s}$$

$$h_2 = h_2' = h_1 - 131.92$$
$$= 2827 - 131.92 = 2695.08 \text{ kJ/kg}$$

$$v_2 = \frac{2.308}{p_2 \times 10^3} \times (h_2 - 1943)$$
$$= \frac{2.308}{5 \times 10^3}(2695.08 - 1943)$$
$$= 0.3472 \text{ m}^3/\text{kg}$$

Mass flow rate

$$m = \frac{A_2 V_2}{v_2}\ ; \quad 0.1 = \frac{A_2 \times 513.41}{0.3472}$$

$\therefore$ Nozzle exit area

$$A_2 = \frac{0.1 \times 0.3472}{513.41}$$
$$= 18.76 \times 10^{-5} \text{ m}^2 = 67.626 \text{ mm}^2$$

Let T_2' be the temperature at which saturation occurs. Then

$$T_2' = T_1 \left(\frac{p_2}{p_1}\right)^{\frac{n-1}{n}}$$
$$= (200 + 273) \times \left(\frac{5}{10}\right)^{\frac{1.3-1}{1.3}}$$
$$= 403.1 \text{ K or } 130.1°\text{C}$$

At 5 bar, the saturation temperature = 151.8°C

Then : degree of undercooling

$$= 151.8 - 130.1 = 21.7°\text{C}$$

Saturation pressure corresponding to temperature T_2' (130.1°C) is 2.701 bar.

Then : degree of supersaturation

$$= \left(\frac{5}{2.701}\right) = 1.851$$

(b) When nozzle coefficient is 0.92 :

$$h_1 = 2827 \text{ kJ/kg}$$
$$h_1 - h_2 = 131.92 \times 0.92 = 121.37 \text{ kJ/kg}$$
$$V_2 = 44.7\sqrt{(h_1 - h_2)}$$
$$= 44.7\sqrt{121.37} = 492.44 \text{ m/s}$$
$$h_2 = (h_1 - 121.37)$$
$$= 2827 - 121.37 = 2705.6 \text{ kJ/kg}$$

$$v_2 = \frac{2.308}{5\times10^3}(2705.6 - 1943)$$
$$= 0.352 \text{ m}^3/\text{kg}$$

Nozzle exit area

$$A_2 = \frac{m\,v_2}{V_2} = \frac{0.1\times0.352}{492.44}$$
$$= 7.148 \times 10^{-5} \text{ m}^2 = 71.48 \text{ mm}^2$$

EXAMPLE 18.18

A group of six nozzles is supplied 5 kg/s of steam at 16 bar and 250°C and expansion of steam takes place against a back pressure of 3.5 bar. The expansion is in supersaturated or metastable state. Neglecting friction, make calculations for

(a) the dimensions of the nozzle. The nozzle is of rectangular cross-section with aspect ratio 3 : 1

(b) degree of undercooling and supersaturation.

(c) loss in available heat drop due to irreversibility

(d) increase in entropy.

Also compare the mass flow rate with metastable expansion to that under conditions of thermal equilibrium.

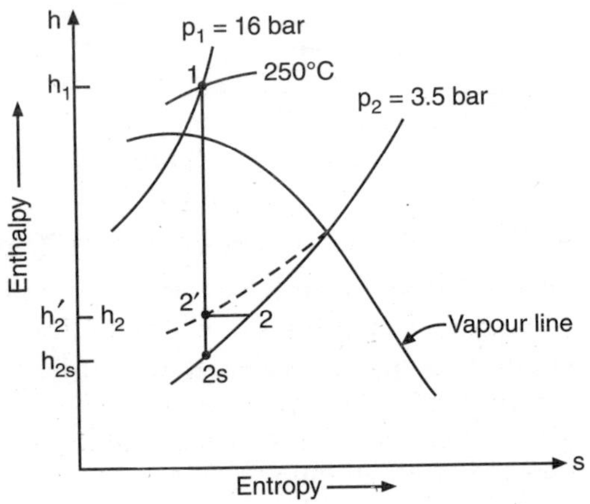

Fig. 18.16.

Solution :

(*a*) With supersaturated flow, the index expansion is taken as n = 1.3 and isentropic enthalpy drop is given by

$$h_1 - h_2' = \frac{n}{n-1}\,p_1 v_1\left[1-\left(\frac{p_2}{p_1}\right)^{\frac{n-1}{n}}\right]$$

At 16 bar and 250°C :

$$v_1 = \frac{T_{sup}}{T_s} \times v_s$$
$$= \frac{250+273}{201.4+273} \times 0.1237 = 0.1364 \text{ m}^3/\text{kg}$$

Then :

$$h_1 - h_2' = \frac{1.3}{1.3-1}(16 \times 10^8) \times 0.1364\left[1-\left(\frac{3.5}{16}\right)^{\frac{1.3-1}{1.3}}\right]$$
$$= 945707\,(1 - 0.7042)$$
$$= 279740 \text{ J/kg or } 279.74 \text{ kJ/kg}$$

Then velocity at point 2,

$$V_2 = 44.7\sqrt{(h_1 - h_2')}$$
$$= 44.7\sqrt{279.74} = 747.63 \text{ m/s}$$

$$v_2 = v_1\left(\frac{p_1}{p_2}\right)^{\frac{1}{n}} = 0.1364 \times \left(\frac{16}{3.5}\right)^{\frac{1}{1.3}}$$
$$= 0.4391 \text{ m}^3/\text{kg}$$

Mass flow rate

$$m = \frac{A_2V_2}{v_2}\ ; \quad 5 = \frac{A_2\times747.63}{0.4391}$$

∴ Nozzle area

$$A_2 = \frac{5\times0.4391}{747.63} = 2.937 \times 10^{-3} \text{ m}^2$$
$$= 29.37 \text{ cm}^2$$

An aspect ratio of 3 : 1 implies that width is three times the breadth. Let breadth of nozzle be b. Then

$$6 \times b \times 3b = 29.37\ ;\ b = \sqrt{\frac{29.37}{18}} = 1.277 \text{ cm}$$

∴ breadth = 1.277 cm

and width = 3 × 1.277 = 3.831 cm

(*b*) Let T_2' be the temperature at which saturation occurs.

$$T_2' = T_1\left(\frac{p_2}{p_1}\right)^{\frac{n-1}{n}}$$

$$= (250 + 273) \times \left(\frac{3.5}{16}\right)^{\frac{1.3-1}{1.3}} = 368.32 \text{ K}$$

At 3.5 bar pressure, the saturation temperature is 138.9°C or 411.9 K

Then : degree of undercooling

$$= 411.9 - 368.32 = 43.58°\text{C}$$

Saturation pressure corresponding to temperature T_2' (368.32 – 273 = 95.32°C) is 0.8457 bar.

Then : degree of supersaturation

$$= \frac{3.5}{0.8457} = 4.138$$

(*c*) For supersaturated flow :

Enthalpy drop $(h_1 - h_2)$ = 279.74 kJ/kg as calculated above.

For expansion under thermal equilibrium condition

enthalpy drop $(h_1 - h_{2s})$ = 290 kJ/kg (read from Mollier chart)

∴ Loss in available heat drop

$$= 290 - 279.74 = 10.26 \text{ kJ/kg}$$

(*d*) Increase in entropy

$$= \frac{\text{loss in available heat}}{\text{absolute temperature at 3.5 bar}}$$

$$= \frac{10.26}{138.9 + 273} = 0.0249 \text{ kJ/kg K}$$

(*e*) The actual velocity with flow under thermal equilibrium conditions

$$= 44.7\sqrt{h_1 - h_{2s}} = 44.7\sqrt{290} = 761.2 \text{ m/s}$$

Dryness fraction at point 2*s* (read from Mollier chart) = 0.955

Specific volume of dry saturated steam at 3.5 bar = 0.524 m³/kg

Then specific volume of steam at point 2*s*

$$= 0.955 \times 0.524 = 0.500 \text{ m}^3\text{/kg}$$

For exit area *A*, the mass flow under metastable conditions,

$$= \frac{A \times 747.63}{0.4391} = 1702.64\ A$$

For the same area, the mass flow under thermal equilibrium conditions

$$= \frac{A \times 761.2}{0.5} = 1522.4\ A$$

∴ Percentage increase in discharge due to supersaturation

$$= \frac{1702.64A - 1522.4A}{1522.4A}$$

$$= 0.1184 \text{ or } 11.84\%$$

18.11. NOZZLES OFF THE DESIGN PRESSURE RATIO

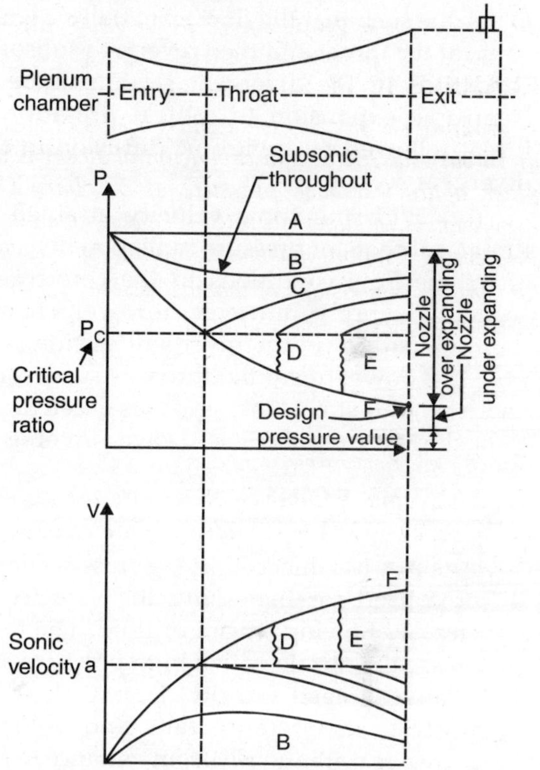

Fig. 18.17. Possible flow conditions in a nozzle

With the experimental set up shown in Fig. 18.17 and assuming the fluid to come into the nozzle from a large plenum chamber (reference stagnation pressure p_0) the distribution of pressure and velocity for various values of exit pressure can be studied.

When the back pressure equals the intake pressure, there will be no flow and the pressure will remain constant throughout the nozzle. The flow starts as the back pressure is reduced; and depending upon the value of downstream pressure p_d prevailing beyond the nozzle exit, the following are the possible conditions of flow in the nozzle:

(*i*) When the downstream pressure p_d is reduced to such a value that the pressure at the throat of the nozzle is well above the critical pressure, the flow is subsonic throughout the nozzle. The convergent-divergent nozzle performs like a venturi; the flow accelerates in the convergent portion and decelerates isentropically in the divergent portion (curve B).

(*ii*) When the discharge pressure reduces to a value such that the flow just attains a sonic value at the throat and then reverts to subsonic conditions in the divergent section, there is isentropic expansion to critical pressure at throat followed by isentropic diffusion in the divergent section (curve C).

(*iii*) With the sonic velocity attained at throat, subsequent pressure reduction does not affect the flow conditions in the convergent section. The pressure waves, however, change the flow pattern in the divergent section. The flow just downstream the throat is supersonic and when this supersonic stream is decelerated, a shock wave results. Shock waves involve an abrupt rise of pressure, an increase of entropy and a deceleration from supersonic to subsonic velocity. The flow across the shock is non-isentropic. As the back pressure is lowered, the shock wave progresses down the nozzle (curve *D* and *E*).

(*iv*) When the discharge pressure equals the nozzle design outlet pressure, there is steady acceleration in the entire passage; being subsonic in convergent section, sonic at throat and supersonic in the divergent section.

(*v*) With further reduction of back pressure below the design value, the expansion occurs outside the nozzle. The process is violent and irreversible and continues until back pressure is reached.

A nozzle when operating below the design value is said to be ***under-expanding*** and when operating above the design pressure is called ***over-expanding.***

SALIENT POINTS

1. A nozzle is a duct of varying cross-sectional area used for increasing the velocity of steadily flowing stream of fluid. The fluid enters the nozzle with a high pressure and relatively small velocity. During the flow process, its pressure falls and velocity increases continuously from entrance to exit of the nozzle.
2. The different expressions for the velocity of propagation of wave in a fluid are :
$$a = \sqrt{dp/d\rho} = \sqrt{K/\rho} = \sqrt{\gamma P/\rho} = \sqrt{\gamma RT}$$
3. The variation of velocity with area ratio is
$$\frac{dA}{A} = \frac{dV}{V}(M^2 - 1)$$
where M is the Local Mach number;
$$M = V/a$$
4. Types of nozzles are :
(*a*) converging or subsonic nozzle; the cross-section of flow region decreases continuously from entry to exit and the acceleration of fluid is in the subsonic region ($M < 1$).
(*b*) diverging or supersonic nozzle; the flow passage diverges from entry to exit and the acceleration of fluid is in the supersonic velocity range ($M > 1$).
(*c*) convergent-divergent nozzle; the cross-section of the flow passage converges down from entry area (called mouth) to minimum area (called throat), and then diverges from throat to exit. The flow velocity at the throat equals the sonic velocity.
5. For steady flow through nozzle, the gain in kinetic energy equals the enthalphy drop. That is
$$\frac{V_2^2 - V_1^2}{2} = (h_1 - h_2)$$
For enthalphy values in kJ/kg and flow velocity in m/s
$$V_2 = \sqrt{V_1^2 + 2000(h_1 - h_2)}$$
Neglecting the approach velocity V_1 which is usually very small,
$$V_2 = 44.72\sqrt{(h_1 - h_2)}$$
For a perfect gas : $h_1 - h_2 = c_p(T_1 - T_2)$ and so
$$V_2 = 44.72\sqrt{c_p(T_1 - T_2)}$$

6. The exit pressure p for maximum discharge is called the critical pressure p_c and the ratio of critical pressure to initial pressure p_0 is called the critical pressure ratio.

$$\frac{p_c}{p_0} = \left(\frac{2}{\gamma+1}\right)^{\frac{\gamma}{\gamma-1}} \quad \text{for air}$$

$$= \left(\frac{2}{n+1}\right)^{\frac{n}{n-1}} \quad \text{for vapours}$$

The exponent n = 1.3 for superheated steam, 1.135 for dry saturated steam and $n = 1.035 + 0.1\,x$ for wet steam with dryness fraction x.

7. In actual expansion of fluid through a nozzle, there are losses due to friction and energy dissipation resulting from formation of eddies and shocks. As a result of friction losses :

   - the expansion is no longer isentropic and there is progressive increase in entropy
   - energy lost in friction is converted into heat and that tends to dry the steam and increase its enthalpy.
   - reduction in actual enthalpy drop and the exit velocity.
   - increase in specific volume and decrease in mass flow rate.

   The nozzle efficiency is expressed as

$$\eta_{nozzle} = \frac{\text{actual enthalpy drop}}{\text{isentropic enthalpy drop}}$$

   Considering friction,

$$V_2 = 44.7\sqrt{(h_1 - h_2) \times \eta_{nozzle}}$$

8. The condensation of steam during expansion does not take place as rapidly as does the change in pressure and temperature. The steam remains in a state of homogenous vapour and continues to expand in dry state. The steam is then said to be in meta stable or super saturated state.
   The steam mass flow rate with super saturated flow is about 2 – 5% greater than that with equilibrium flow.
9. The limit of supersaturation is shown on $h - s$ plot by a curve known as Wilson line; it represents the saturated steam when its density is about 8 times that of normal vapour and corresponds to about 4% of moisture line.
10. A nozzle operating with back pressure below the design value is said to be under expanding and when operating above the design pressure is called over-expanding.

REVIEW QUESTIONS

A. Conceptual and conventional questions.

1. Differentiate between the compressible and incompressible flow. Mention some of the flow situations where compressibility of the fluid has to be considered.
2. Prove that the velocity of propagation of an infinitesimal pressure wave is given by

$$\sqrt{\frac{\partial p}{\partial \rho}} = \sqrt{\frac{\gamma p}{\rho}} = 20.1\sqrt{T} \text{ m/s for air}$$

   where p is pressure, ρ is mass density, T is temperature in K and γ is the ratio of specific heats.
3. Define the function of a nozzle and name some applications where it is used?
4. Derive an expression relating the critical pressure ratio to the index of expansion n for expansion in a nozzle.
5. Show that maximum discharge of steam per unit area through a nozzle takes places when the ratio of steam pressure at the throat to the inlet pressure is $\left(\frac{2}{n+1)}\right)^{\frac{n}{n-1}}$ where n is the index of adiabatic expansion.
6. (*a*) Explain, with the help of h-s diagram, the effect of friction on flow through a steam nozzle.
   (*b*) Define nozzle efficiency and state the factors on which it depends.
7. What is superstaurated flow? State the effects of supersaturation in a nozzle.
8. Explain the super saturated or metastable flow of steam through a nozzle and state the significance of Wilson's line.

9. Explain the phenomenon of supersaturation expansion of steam, and its effects upon discharge of steam from a nozzle as compared with expansion in thermal equilibrium?
10. The actual discharge of steam through a nozzle is observed to be 2 to 5% more than the calculated value. Comment on the validity of this statement.
11. A convergent-divergent nozzle receives air at a fixed pressure and exhausts into a chamber where the pressure can be controlled. Describe the flow behaviour in the nozzle when the pressure in the chamber is reduced from the no flow condition.
12. What is meant by over-expansion and under-expansion as applied to fluid flow through a nozzle.
13. Air expands under reversible adiabatic conditions in a nozzle from 4 bar and 725°C to a final pressure of 1 bar. Determine the shape of the nozzle required and the air flow rate if the minimum nozzle diameter is 1.25 cm. If the nozzle requires a divergent section, determine the maximum velocity developed in the nozzle and the suitable exit diameter.
[**Ans.** convergent-divergent, 0.063 kg/s, 1.375 cm]
14. Dry saturated steam at a pressure of 8 bar enters a convergent divergent nozzle and leaves it at a pressure of 1.5 bar. If the flow is isentropic and if the corresponding expansion index is 1.133, find the ratio of cross-sectional area at the exit to the throat for maximum discharge.
[**Ans.** 1.595]
15. Dry saturated steam expands in a group of convergent-divergent nozzles from 11 bar to 1.5 bar. The mass flow rate is 5 kg/s, the nozzle efficiency is 90% and the index of expansion is 1.2. Make calculations for the pressure at throat, number of nozzles and the exit area. At throat, the cross-sectional area of each nozzle is estimated to be 1 cm^2.
16. Dry saturated steam is supplied to a convergent nozzle at 11 bar where it expands and gets discharged into space at 6 bar. Neglecting the effects of friction and inlet velocity to nozzle, compute the exit velocity and mass flow when
(*a*) the flow is isentropic and in thermal equilibrium
(*b*) the flow is supersturated.
Comment on the mass flow rate in the two conditions
[**Ans.** 423.2 m/s and 0.083 kg/s ; 471 m/s and 0.1 kg/s]
17. (*a*) How does the velocity and pressure vary with area for (*i*) subsonic flow and (*ii*) supersonic flow.
(*b*) Show that for steady one-dimensional isentropic compressible flow through a duct,

$$\frac{dA}{A} = \frac{dV}{V}(M^2 - 1)$$

and explain its implication in the design of nozzles and diffusers.
Determine the Mach number of flow if 25 percent increase in velocity is produced by 25 percent increase in area.
[**Ans.** 1.414]
18. Gas is flowing form a tank where it is at stagnation state through a converging nozzle to the surrounding air. By measuring the pressure difference between the tank and the surrounding air and knowing the nozzle throat area, the nozzle can be used as a means of determining the mass flow rate of the air for steady state conditions. Show that this is so and specify the data to be recorded if : (*a*) $M < 1$ at exit, (*b*) $M = 1$ at exit.

B. Fill in the blanks with appropriate word/words.

1. The flow through a nozzle is regarded as ....
2. The steam enters the nozzle at .... pressure and .... velocity.
3. The ratio of useful heat drop to the isentropic heat drop in a nozzle is called ....
4. The critical pressure ratio for initially dry saturated steam is .... as compared to that for initially wet steam.
5. When the nozzle operates with the maximum mass flow, the nozzle is said to be ....
6. The density of supersaturated steam is about .... that of saturated steam at the corresponding pressure.
7. When the fluid expands in a convergent-divergent nozzle, the maximum friction losses are expeceted to occur in ....
8. The isentropic expansion of steam through a nozzle for the steam initially dry saturated at inlet is approximated by the relation pv^n = constant where n ....
9. When the back pressure of a nozzle is below the designed value of pressure at the nozzle exit, the nozzle is said to be ....
10. Supersaturated in the flow of steam through nozzles is due to .... in the condension of steam during expansion.

11. For supersonic flow, a convergent passage is a .... and the divergent passage is a ....
12. The unit value of mach number (M = 1) occurs only at the .... and no where else, and this happens only when the discharge is ....

Answers

1. adiabatic 2. high, low 3. nozzle efficiency 4. less 5. choked 6. eight times 7. divergent portion 8. 1.135 9. under expanding 10. time lag 11. diffuser nozzle 12. throat, maximum

C. State true or false. If false, write the correct statement.

1. Nozzles are more efficient than diffusers.
2. The Mach number becomes greater than unity when the nozzle operates with maximum mass flow rate.
3. For supersonic flow, the converging duct is a diffuser.
4. A convergent nozzle is used to obtain supersonic velocity.
5. When steam flows with maximum mass per unit area, the velocity at the exit of the convergent nozzle becomes sonic.
6. A nozzle is a flow passage of varying cross-sectional area in which the velocity of fluid increases and pressure drops in the direction of flow.
7. The cross-sectional area of throat decides the mass flow through nozzle given inlet conditions.
8. Nozzle is called over expanding if the back pressure of nozzle is below the designed value of pressure at exit of nozzle.
9. Phenomenon of supersaturation causes decrease in discharge by 2-5%
10. Degree of undercooling refers to difference in saturation temperature at state in stable equilibrium and temperature of the state in unstable equilibrium.

Answers

1. True 2, False; Mach number becomes unity 3. True 4. False; sonic velocity 5. True 6. True 7. True 8. False, under expanding 9. False; increase 10. True

D. Multiple choice questions.

1. A steam nozzle is a passage of varying cross-section used in steam turbines. It converts
   (*a*) enthalpy of steam into pressure energy
   (*b*) enthalpy of steam into kinetic energy
   (*c*) enthalpy of steam into work
   (*d*) pressure energy of steam into work
2. Steam at exit from the nozzle has
   (*a*) high pressure and low velocity
   (*b*) high pressure and high velocity
   (*c*) low pressure and low velocity
   (*d*) low pressure and high velocity
3. The expansion process of steam through a nozzle can be approximated to
   (*a*) throttling process
   (*b*) free expansion process
   (*c*) adiabatic process
   (*d*) polytropic process
4. In a varying area passage, the area *A* and velocity *C* are correlated by the expression
   (*a*) $\frac{dA}{A} = \frac{dC}{C}(M^2 - 1)$
   (*b*) $\frac{dC}{C} = \frac{dA}{A}(M^2 - 1)$
   (*c*) $\frac{dA}{A} = \frac{dC}{C}(1 - M)$
   (*d*) $\frac{dC}{C} = \frac{dA}{A}(1 - M)$
   where *M* denotes the Mach number; ratio of flow velocity to sonic velocity
5. For subsonic flow through a nozzle, the following changes occur in velocity and pressure along the flow direction
   (*a*) pressure decreases; velocity increases
   (*b*) pressure increases; velocity decreases
   (*c*) increase in both pressure and velocity
   (*d*) velocity increases and pressure remains constant
6. Under ideal conditions, the velocity of steam at the outlet of a nozzle for a heat drop of 400 kJ/kg will be approximately
   (*a*) 1200 m/s (*b*) 900 m/s
   (*c*) 600 m/s
   (*d*) the same as sonic speed
7. The cross-section of a nozzle is
   (1) circular (2) elliptical
   (3) square (4) rectangular
   Which of these statements is/are correct ?
   (*a*) 1 and 2 (*b*) 2 and 3
   (*c*) 1 and 3 (*d*) 1, 2 and 3

8. For steam having inintial dryness fraction x, the value of exponent n for the adiabatic expansion through a nozzle can be worked out by using the Zeuner's equation prescribed as
(a) $n = 1.035 + 0.1x$ (b) $n = 1.35 + 0.1x$
(c) $n = 1.3 + 0.15x$ (d) $n = 1.15 + 0.1x$

9. Which relation is valid for flow of superheated flow through a nozzle?
(a) $pv^{1.135} = C$ (b) $pv^{1.3} = C$
(c) $pv^{1.035} = C$ (d) $pv^{1.55} = C$

10. The value of critical pressure ratio for dry saturated steam is
(a) 0.525 (b) 0.546
(c) 0.578 (d) 0.627

11. A steam nozzle is said to be choked when
(a) there is stoppage of flow through the nozzle
(b) flow rate cannot be increased by increasing the supply pressure
(c) critical conditions prevail and mass flow rate is maximum
(d) shock effect is felt in the nozzle

12. In a convergent nozzle, the mass flow rate is not a function of
(a) initial pressure of steam
(b) initial condition of steam; whether wet, dry or superheated
(c) throat diameter
(d) exit condition of steam

13. In a nozzle designed for maximum discharge conditions, the flow velocity in the convergent section of a nozzle is
(a) subsonic (b) sonic
(c) supersonic
(d) depends upon the initial pressure and condition of steam.

14. Which of the following statements is/are relevant to critical flow through a steam nozzle?
1. flow rate through the nozzle is minimum
2. flow rate through the nozzle is maximum
3. velocity at the throat is supersonic
4. velocity at the throat is sonic
Select the correct answer using the codes given below:
Codes :
(a) 1 above (b) 1 and 3
(c) 2 and 4 (d) 4 alone

15. Which is not true regarding effect of friction loss in a nozzle ?
(a) reduction in exit velocity
(b) increase in final dryness friction
(c) increase in specific volume of steam
(d) increase in mass flow rate

16. The effect of friction in a steam nozzle is to .... velocity and .... dryness fraction
(a) increase, increase (b) increase, decrease
(c) decrease, increase (d) decrease, decrease

17. The expansion of steam through a nozzle is so rapid that steam may not get time to condense and may remain in a state of homogeneous vapour .The expansion process is then stated to be
(a) critical
(b) thermally equilibrium
(c) metastable (d) supersaturated

18. Wilson line is associated with
(a) total steam consumption with respect to power output
(b) supersonic flow of steam through a nozzle
(c) nozzle flow with friction
(d) supersaturated flow of steam through a nozzle

19. Which is not true regarding effect of supersaturated flow in a steam nozzle ?
(a) reduction in heat drop
(b) reduction in exit velocity
(c) decrease in mass flow rate
(d) increase in entropy

20. A convergent divergent nozzle is said to be choked when
1. critical pressure is attained at the throat
2. velocity at the throat becomes sonic
3. exit velocity becomes supersonic.
Which of these statements are correct ?
(a) 1, 2 and 3 (b) 1 and 2
(c) 2 and 3 (d) 1 and 3

Answers

1. (b)	2. (d)	3. (c)	4. (a)	5. (a)
6. (b)	7. (d)	8. (a)	9. (b)	10. (c)
11. (c)	12. (d)	13. (a)	14. (c)	15. (d)
16. (c)	17. (d)	18. (d)	19. (c)	20. (b)

Chapter **19**

Steam Turbines

Learning objectives : The subject matter included in this chapter will enable the reader to understand the concepts of :

- *impulse and reaction principle of steam turbines*
- *efficiency of impulse turbine blading*
- *steam turbine staging*
- *degree of reaction and efficiency of reaction turbine blading*
- *stage efficiency, overall efficiency and reheat factor*
- *governing of steam turbine*

Steam turbine is a contrivance which is designed to convert a part of the energy of steam, evidenced by high pressure and temperature, into mechanical power.

The turbine mainly consists of nozzle (or set of nozzles) and a rotary blade wheel. The steam is expanded from high pressure to low pressure either in nozzles or in the blading and the kinetic energy thus obtained is supplied to the rotary blade where it is transformed into mechanical work. The power is made available at the turbine shaft directly or with the help of reduction gears. Power so produced may be utilized in the various diverse fields of industry such as for power generation and for transport. The power from smaller units is equally adaptable to run fans, pumps and compressors directly.

The first turbine historically recorded and working on the reaction principle, that of Hero some 2000 years ago is shown diagramatically in Fig. 19.1. The steam generated in the boiler, flowing through hollow trunnions enters a hollow spherical receiver. With the increase of pressure in the receiver, the steam issues tangentially from nozzles at the ends of two opposite arms. The reactive force of the steam leaving the nozzles makes the sphere turn about its axis.

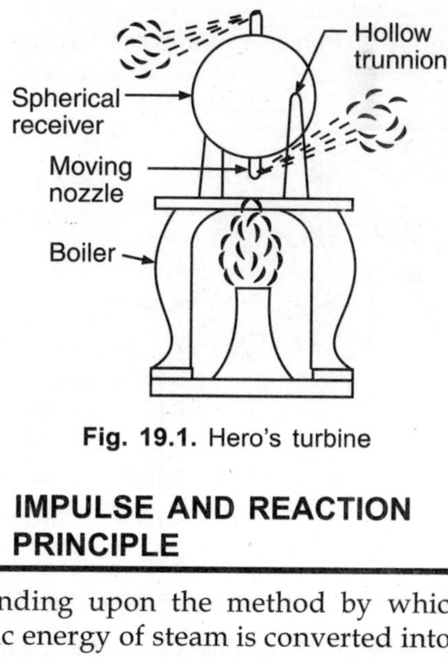

Fig. 19.1. Hero's turbine

19.1. IMPULSE AND REACTION PRINCIPLE

Depending upon the method by which the kinetic energy of steam is converted into shaft work, the steam turbines are classified into *impulse turbine* and *reaction turbine*. In the steam turbine embodying the impulse principle (Fig. 19.2), the steam at high pressure and temperature but at low velocity expands through the nozzles to exhaust pressure, thereby gaining a high velocity. The nozzles are stationary and secured either in a diaphragm or directly in the casing. The high velocity jet issuing from the nozzle impinges on the blades fixed on the periphery of a rotor. The blade changes the direction of steam without changing its pressure. The resulting change of momentum gives the motive force to the turbine shaft.

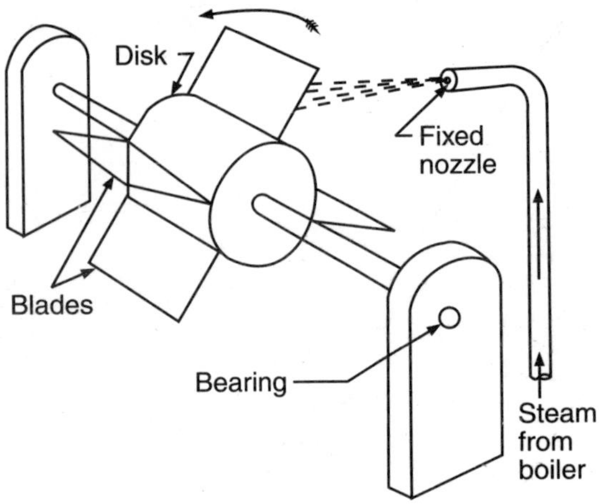

Fig. 19.2. Principle of impulse turbine

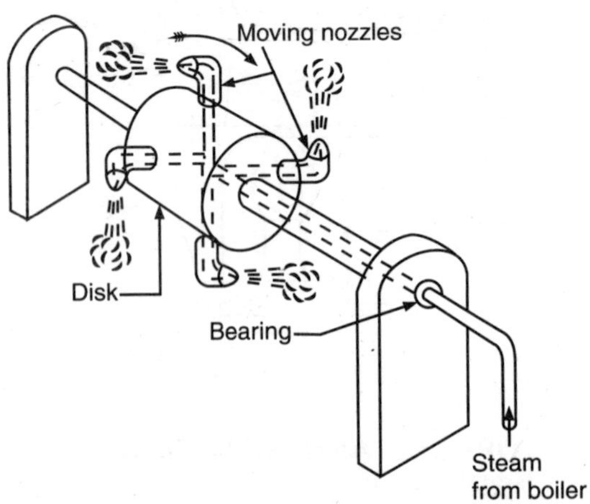

Fig. 19.3. Principle of reaction turbine

The principle underlying the reaction turbine is illustrated in Fig. 19.3. In such a turbine the steam expands through the nozzles which are moving. The steam from the boiler at high pressure and temperature is piped through a hollow shaft to a hollow disc. The disc has four radial openings through tubes, the ends of which are shaped as nozzles. When the steam escapes through these tubes, it expands and there is increase in steam velocity relative to the rotating disc. The resulting reaction force sets the disc in rotation. The disc and the shaft rotate in a direction opposite to the direction of steam jet.

19.2. FLOW THROUGH NOZZLES

A **nozzle** is a device for increasing the velocity of a steadily flowing stream of fluid. The fluid enters the nozzle with a high pressure and relatively small velocity. As it flows through the nozzle, it expands and in the process the pressure falls and the velocity increases continuously from the entrance to the exit of the nozzle.

The following thermodynamic relation applies for flow of gas through a duct

$$\frac{dA}{A} = \frac{dV}{V}(M^2 - 1) \qquad \text{...(19.1)}$$

where A denotes area of duct and V represents the flow velocity. The Mach number M is defined as the ratio of flow velocity to sonic velocity in the flowing medium.

The following conclusions can be drawn from this relation :

(*i*) For subsonic velocities ($M < 1$), dA and dV must be opposite in sign, *i.e.*, an increase of cross-sectional area causes a reduction of velocity and *vice versa*.

For acceleration of fluid in the subsonic region, the cross-section of the flow passage converges down from entry to minimum area which is exit. Such a nozzle is called *converging* or *subsonic nozzle* (Fig. 19.4(*a*)).

(*ii*) For supersonic velocities ($M > 1$), dA and dV are of same sign. An increase of cross-section then causes an increase of velocity and *vice versa*.

For acceleration of fluid in the supersonic velocity range, the flow passage diverges from entry to exit. Such a nozzle is called *diverging* or *supersonic nozzle* (Fig. 19.4(*b*)).

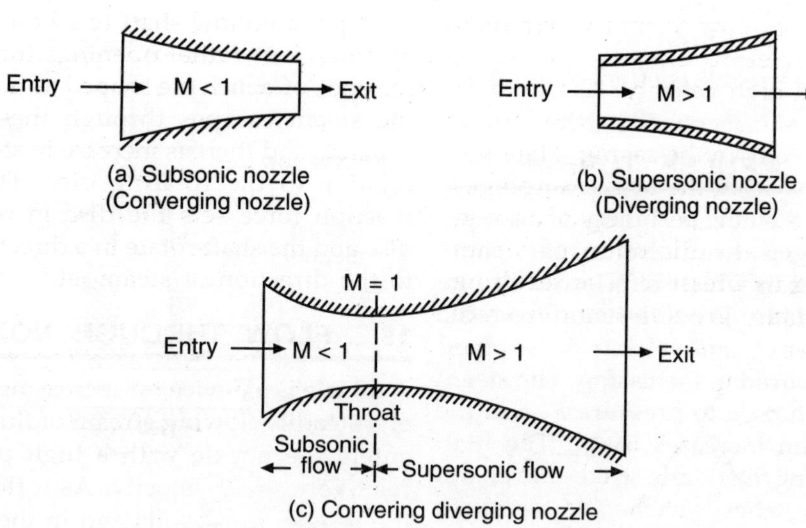

Fig. 19.4. Types of steam nozzles

(*iii*) When $M = 1$, dA must be zero and since the second derivative is positive, A must be minimum. Thus if the velocity of flow equals the sonic velocity anywhere, it must do where the cross-section is of minimum area.

In a convergent-divergent nozzle (Fig. 19.4.(*b*)), the cross-section of the passage converges down from entry area called **mouth** to minimum area called **throat** and then diverges up from throat to exit. The fluid accelerates from subsonic in the convergent portion to supersonic in the divergent portion. The velocity of flow at the throat equals sonic velocity ($M = 1$).

The steady flow energy equation for a unit mass flow is

$$h_1 + \frac{V_1^2}{2} + gz_1 + q = h_2 + \frac{V_2^2}{2} + gz_2 + w_s \qquad ...(19.2)$$

For flow through a nozzle :

(*i*) no external work is done on or by the fluid ($w_s = 0$),

(*ii*) changes in potential energy are negligible ($gz_1 = gz_2$),

(*iii*) the velocity of gas flow is so high that there is no time available for the heat exchange with the surroundings. The expansion process through the nozzle is then considered adiabatic ($q = 0$).

The steady flow energy equation for the nozzle then takes the form

$$h_1 + \frac{V_1^2}{2} = h_2 + \frac{V_2}{2}$$

$$\text{or} \quad \frac{V_2^2 - V_1^2}{2} = (h_1 - h_2) \qquad ...(19.3)$$

The gain in kinetic energy thus equals the enthalpy drop.

If enthalpy of steam is in kJ/kg and the flow velocity is in m/s, then equation 19.3 may be written as

$$\frac{V_2^2 - V_1^2}{2 \times 10^3} = (h_1 - h_2)$$

$$\text{or} \quad V_2 = \sqrt{V_1^2 + 2000\,(h_1 - h_2)} \qquad ...(19.4)$$

The initial velocity V_1 called the approach velocity is usually very small and can be neglected. Then

$$V_2 = \sqrt{2000\,(h_1 - h_2)}$$

$$= 44.72\sqrt{(h_1 - h_2)} \qquad ...(19.5)$$

A certain amount of friction is always present between steam and the nozzle surfaces. This reduces the heat drop by 10 to 15 per cent and a corresponding reduction occurs in the exit velocity of steam. In that case

$$V_2 = 44.72\sqrt{K(h_1 - h_2)} \qquad ...(19.6)$$

where K is the nozzle coefficient or nozzle efficiency.

19.3. SIMPLE IMPULSE TURBINE

Fig. 19.5. shows the nozzle and blade arrangement of a single stage impulse turbine. There is one set of nozzles which is followed by one set of moving blade. The live steam from the boiler at pressure P_0 and velocity V_0 is piped to the nozzle secured in the casing. The steam expands in the nozzle to pressure P_1 and the velocity of steam increases to V_1. The high velocity jet leaving the nozzle strikes the blades mounted on the wheel attached to the shaft. The jet causes the wheel and the shaft to rotate and thus useful work is made available at the shaft.

The state of affairs with respect to change of pressure and velocity variation in the single stage impulse wheel is also indicated in Fig. 19.5.

(*i*) The pressure drops during the passage of steam through the nozzles and then practically remains constant in moving blades.

(*ii*) Velocity increases due to pressure drop in the nozzles. Some of the kinetic energy is imparted to moving blades to produce motion of the shaft and as such the velocity decreases when flowing over moving blades.

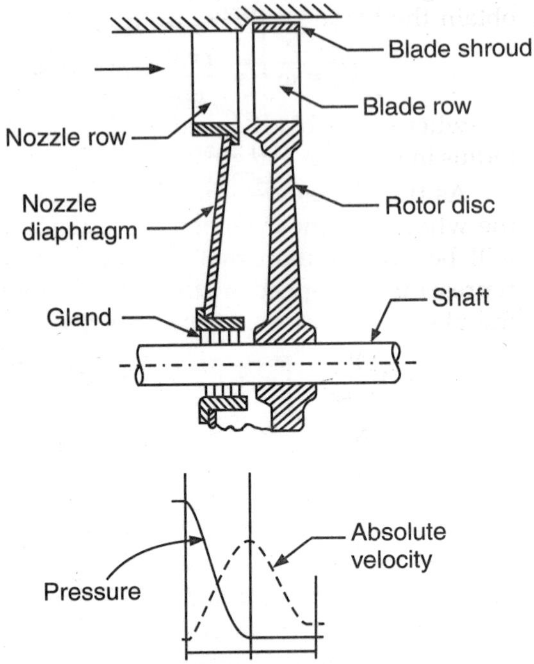

Fig. 19.5. Simple impulse turbine

With reference to Fig. 19.6, let us assume that the axis of the nozzle is inclined at an angle a to the negative direction of motion of blade. The angle a is called the **nozzle angle.** The steam issuing from the nozzle has an absolute velocity V_{a1}. The blade is also moving with speed V_b called **blade speed.** If the wheel on which the blades are mounted (Fig. 19.7.) is

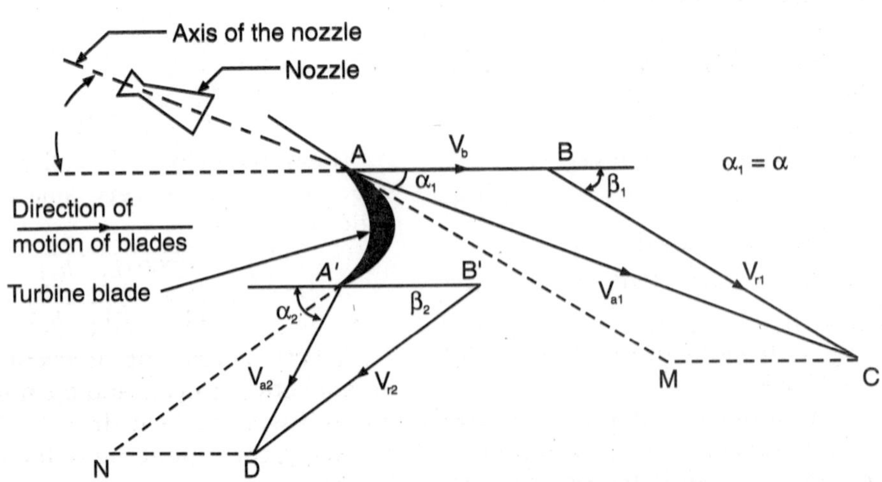

Fig. 19.6. Velocity triangles for impulse turbine

rotating at N revolutions per minute, we can obtain the blade speed V_b from the relation

$$V_b = \omega r = \frac{2\pi N}{60} \times r \qquad ...(19.7)$$

where V_b = blade speed in m/s, r = blade radius in m and ω = angular velocity of rotation

As the blades are situated radially round the wheel the linear velocity of outside edge will be greater than that of inside edge. V_b represents the speed at the mean height of the blade.

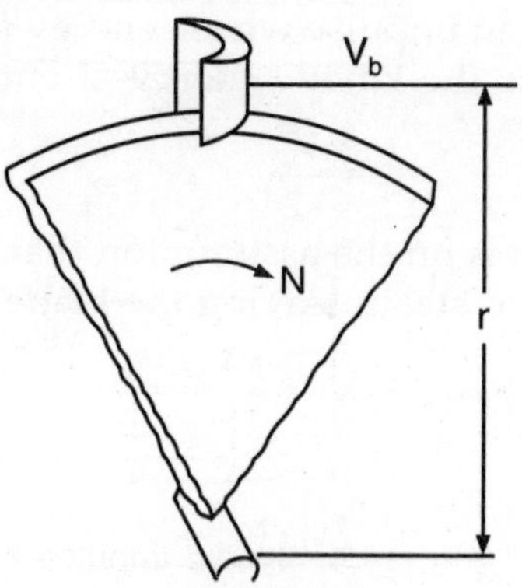

Fig. 19.7

Vectorial subtraction of the blade speed V_b from the absolute velocity V_{a1} gives the relative velocity V_{r1} with which the steam from the stationary nozzle enters the moving blade. Thus, if in Fig. 19.7., AB represents the blade speed V_b both in direction and magnitude and vector AC represents the absolute steam velocity V_{a1}, then vector BC would give the velocity of steam relative to blade (V_{r1}). The relative velocity V_{r1} is inclined at an angle β_1 to the plane of rotation (negative direction of motion) of the blade. The angle β_1 is called the **inlet angle** to the moving blade. For smooth flow without any shock and disturbance in the blade passages, V_{r1} should be tangential to the blade profile at inlet to the blade.

The direction of the steam jet changes as it passes over the blade. The relative velocity at outlet V_{r2} will be inclined at an angle β_2 with the plane of rotation. The angle β_2 is called the **exit angle** of the moving blade. If there is no friction in flow passages, the magnitude of the relative velocity leaving (V_{r2}) is equal to the relative velocity entering (V_{r1}). However, in practice, the steam experiences friction whilst flowing across the blade passages. The effect of friction is usually expressed by means of ratio :

$$k = \frac{V_{r2}}{V_{r1}} \qquad ...(19.8)$$

The factor k is called the *blade factor* or *blade velocity coefficient*.

Vectorial subtraction of blade speed from the relative velocity V_{r2} gives the absolute velocity at exit V_{a2} with its direction α_2 with the plane of rotation. Thus in Fig. 19.6., the diagonal A′D of the parallelogram A′B′D N represents the absolute velocity at exit from the moving blade.

The triangle ABC which represents the conditions when the steam enters the moving blade is called *inlet velocity triangle*. Similarly, the triangle A′B′D which represents the condition when the steam leaves the moving blade is called the *exit velocity triangle*.

Since inlet and outlet is assumed to take place at same radius, the velocity triangles have a common base and for convenience the velocity triangles are combined in one diagram as shown in Fig. 19.8.

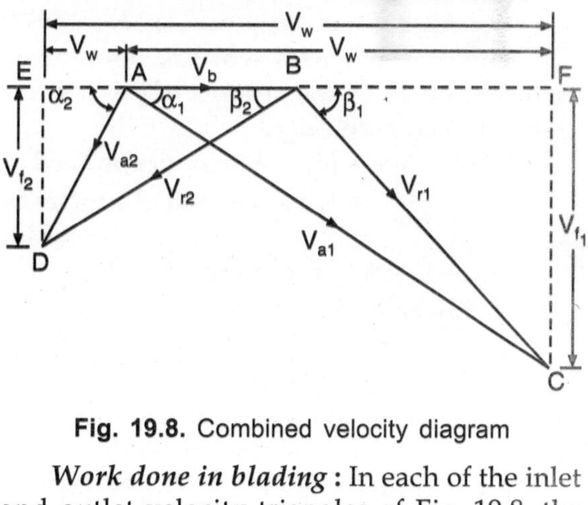

Fig. 19.8. Combined velocity diagram

Work done in blading : In each of the inlet and outlet velocity triangles of Fig. 19.8, the absolute velocity has two components :

(*i*) Component of absolute velocity in the direction of blade motion. This component (designated by V_w) is called the whirl component or the tangential component.

(*ii*) Component of absolute velocity in a direction at right angles to the direction of blade motion. This component (designated by V_f) is called the flow component or the axial component in an axial flow turbine.

By Newton's second law of motion, force in the tangential direction is equal to time rate of momentum in that direction.

Force = rate of change of momentum
= rate of change of (mass × velocity)
= mass flow × change of velocity

The change of velocity which produces force on the blade in the direction of motion of blade is the change in the whirl velocity.

Change in whirl velocity

$$V_w = -V_{w2} - (V_{w1}) = -(V_{w2} + V_{w1})$$

The value of V_{w2} is negative because it is measured in a direction opposite to that of the blade motion.

∴ Tangential force

$$= -m\,(V_{w1} + V_{w2}) \quad ...(19.9)$$

where m is the mass of steam flowing through the blades in kg/s.

The reaction of this force provides the driving thrust on the wheel.

Driving force on the wheel

$$= m\,(V_{w1} + V_{w2}) \text{ N} \quad ...(19.10)$$

The rate at which work is done on the wheel is given by the product of driving force and the blade velocity.

Work done = force × distance moved

$$= m\,(V_{w1} + V_{w2})\,V_b \text{ Nm/s} \quad ...(19.11)$$

Since 1 Nm/s represents 1 watt,

∴ Power of blading

$$= m\,(V_{w1} + V_{w2})\,V_b \text{ watts}$$

Axial Thrust : A change in the velocity of flow across the blades of an axial flow turbine gives an end thrust along the shaft.

$$\text{Axial thrust} = m\,(V_{f2} - V_{f1}) \text{ N} \quad ...(19.12)$$

If $V_{f2} < V_{f1}$, the end thrust is along the turbine shaft in the direction of velocity of flow.

If $V_{f2} > V_{f1}$, the end thrust is in opposite direction to the velocity of flow.

19.4. EFFICIENCY OF IMPULSE TURBINE BLADING

The effectiveness of conversion of steam energy into work is expressed in terms of diagram efficiency η_d which is defined as the ratio of work done to energy supplied.

$$\eta_d = \frac{\text{work done by blades}}{\text{energy supplied to blades}}$$

This efficiency is also called the *utilization factor* or the *blading efficiency*.

For an impulse wheel, energy supplied to blades is the kinetic energy at entrance

$$= \frac{V_{a1}^2}{2}$$

This is on the assumption that the kinetic energy of steam leaving the blade is wasted.

$$\therefore \quad \eta_d = \frac{(V_{w1} + V_{w2})\,V_b}{\frac{V_{a1}^2}{2}}$$

$$= \frac{2V_b\,(V_{w1} + V_{w2})}{V_{a1}^2} \quad ...(19.13)$$

The following transformations are carried out to determine the different parameters which affect the diagram efficiency of the steam turnine blading.

From the velocity vector diagrams (Fig. 19.8),

$$V_{w1} = V_{a1}\cos\alpha_1 = V_{r1}\cos\beta_1 + V_b$$

$$V_{w2} = V_{a2}\cos\alpha_2 = V_{r2}\cos\beta_2 - V_b$$

∴ Change in tangential component

$$\Delta V_w = V_{w1} + V_{w2}$$
$$= V_{r1}\cos\beta_1 + V_{r2}\cos\beta_2$$
$$= V_{r1}\cos\beta_1\left[1 + \frac{V_{r2}\cos\beta_2}{V_{r1}\cos\beta_1}\right]$$

The ratio $\frac{V_{r2}}{V_{r1}}$ represents the blade velocity coefficient k and let $C = \frac{\cos\beta_2}{\cos\beta_1}$. Then

$$\Delta V_w = V_{r1}\cos\beta_1\,(1 + kC)$$
$$= (V_{a1}\cos\alpha_1 - V_b)(1 + kC)$$

The change in tangential component provides the driving thrust on the rotor, and the rate of work done (or power) on the rotor is given by the product of the driving thrust and the blade velocity. Then for unit mass of steam,

Rate of work

$$= (V_{a1} \cos \alpha_1 - V_b)(1 + kC)\, V_b \quad ...(19.14)$$

and diagram efficiency

$$\eta_d = \frac{(V_{a1} \cos \alpha_1 - V_b)(1 + kC)V_b}{V_{a1}^2/2}$$

$$= \frac{2(V_{a1} \cos \alpha_1 - V_b)(1 + kC)V_b}{V_{a1}^2} \quad ...(19.15)$$

The ratio between the blade speed and the absolute steam velocity at inlet is called the *blade speed ratio,* and it forms an important factor in the steam turbine blading. By substituting $\rho = V_b / V_{a1}$ in the expression for diagram efficiency, we get

$$\eta_d = 2\,(\rho \cos \alpha_1 - \rho^2)(1 + kC)$$

Apparently the value of diagram efficiency depends upon

(*i*) nozzle angle α_1

(*ii*) blade speed ratio ρ

(*iii*) blade angles β_1 and β_2

(*iv*) blade velocity coefficient k

It may be noted from equation 19.14 that for given set of conditions, the work done increases with reduction in nozzle angle α_1 and it becomes maximum when $\cos \alpha_1 = 1$, *i.e.*, $\alpha_1 = 0$. In practice α_1 cannot be kept zero as the turbine wheel has many blades and each has to receive the steam in succession. Moreoever, a reduction in nozzle angle implies a reduced axial or flow velocity ($V_{f1} = V_{a1} \cos \alpha_1$). Lower flow velocity means larger annular area for the same mass flow of steam. If the area is increased, friction losses in the blade passages become more. As a best compromise, the nozzle angle is normally kept between the limits of 14 to 25 degree.

Assumping α_1, k and C to be constant, the diagram efficiency depends on the value of blade speed ratio ρ. The optimum value of ρ for maximum diagram efficiency can be worked out by differentiating the expression

$$\eta_d = 2\,(\rho \cos \alpha_1 - \rho^2)(1 + kC) \quad ...(19.16)$$

with respect to ρ and setting the derivative equal to zero. That gives :

$$2(1 + kC)\,(\cos \alpha_1 - 2\rho) = 0$$

Since $2(1 + kC)$ is not equal to zero, we have

$$\rho = \frac{\cos \alpha_1}{2} \quad ...(19.17)$$

From the analyis made above, the following observations can be made :

1. Substituting $\rho = \frac{\cos \alpha_1}{2}$ in the expression 19.16 for diagram efficiency,

$$(\eta_d)_{max} = 2(1 + kC)\left[\frac{\cos \alpha_1}{2} \times \cos \alpha_1 - \frac{\cos^2 \alpha_1}{4}\right]$$

$$= (1 + kC)\frac{\cos^2 \alpha_1}{2}$$

Considering the blades to be symmetrical ($\beta_1 = \beta_2$ and so $C = 1$) and neglecting friction in blade passages ($k = 1$),

$$(\eta_d)_{max} = \cos^2 \alpha_1 \quad ...(19.17a)$$

2. Substituting $\rho = \frac{V_b}{V_{a1}} = \frac{\cos \alpha_1}{2}$ in equation 19.14, we get the rate of work done corresponding to maximum diagram efficiency

$$= \left(\frac{2V_b}{\cos \alpha_1} \times \cos \alpha_1 - V_b\right) \times 2\, V_b$$

(k=1 and C=1)

$$= 2\, V_b^2 \quad ...(19.18)$$

3. The identity $\rho = \frac{V_b}{V_{a1}} = \frac{\cos \alpha_1}{2}$ clearly indicates that for fixed value of nozzle angle, the blade speed be half of the absolute velocity of steam jet if maximum work is to be produced. For any other value of blade speed, the absolute velocity at outlet from the blade will increase. Consequently more energy will be carried away by exhaust steam and efficiency will decrease. The optimum value of V_b / V_{a1} corresponds to the case when the outlet velocity is axial.

EXAMPLE 19.1.

How is the heat of steam converted into work in steam turbines ?

The data pertaining to an impulse turbine is as follows :

Steam velocity = 500 m/s
Blade speed = 200 m/s
Exit angle of moving blade = 25°
Nozzle angle = 20°

Neglecting the effect of friction when passing through blade passages, calculate :

(a) inlet angle of moving blade, (b) exit velocity and direction, (c) work done per kg of steam, (d) axial thrust and power for a steam flow rate of 5 kg/s, (e) diagram or blade efficiency.

Solution : Always draw a rough sketch of the velocity diagram first. Then decide on a suitable scale. The vector diagram is shown in Fig. 19.9 and the steps in its construction are :

(*i*) Draw AB equal to blade speed V_b = 200 m/s to a suitable scale.

(*ii*) Draw AC with its inclination with AB equal to nozzle angle = 20° and set off AC = V_{a1} = 500 m/s on the same scale.

(*iii*) Join BC to complete the inlet velocity triangle ABC.

(*iv*) Set off BD with inclination β_2 = 25° and cut BD = *BC*. This is because in the given problem, the effect of friction has been neglected and $V_{r2} = V_{r1}$.

(*v*) Join AD to complete the outlet velocity triangle ABD.

The following results are obtained from the vector diagram :

(*a*) β_1 = inlet angle of moving blade = 33°

(*b*) Exit velocity V_{a2} = 162.5 m/s

α_2= Direction of steam at exit = 56°

(*c*) Change in whirl component,

$V_w = EF$ = 555 m/s

Work done/kg of steam

$= V_b \times V_w$

= 555 × 200 = 111000 Nm/kg

(*d*) Work done/s

= 111000 × 5 = 555000 Nm/s

∴ Power output

= 555000 J/s = 555000 W

= 555 kW

Change in flow velocity

$V_f = V_{f1} - V_{f2}$ = 45 m/sec

Axial thrust

$= m\,(V_{f1} - V_{f2})$

= 5 × 45 = 225 N

(*e*) Kinetic energy supplied to the blade

$$= \frac{V_{a1}^2}{2}$$

$$= \frac{(500)^2}{2} = 125000 \text{ Nm/kg}$$

Diagram efficiency

$$= \frac{\text{work done}}{\text{kinetic energy supplied}}$$

$$= \frac{111000}{125000} = 0.888 \quad \text{or} \quad \mathbf{88.8\%}$$

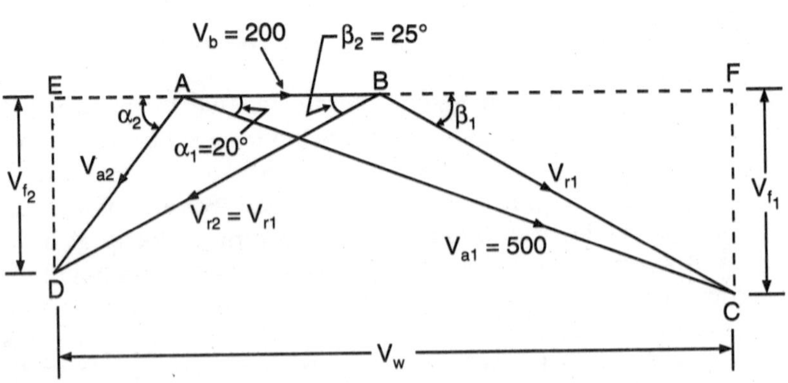

Fig. 19.9.

EXAMPLE 19.2.

The blade speed of a single ring of impulse blading is 300 m/s and the nozzle angle is 20°. The isentropic heat drop is 475 kJ/kg and the nozzle efficiency is 0.85. Given that the blade velocity coefficient is 0.7 and the blades are symmetrical, draw the vector diagrams and calculate for a mass flow of 1 kg/s :

(a) Axial thrust on the blading (b) Work done (c) Blade (diagram) efficiency, and maximum blade efficiency (d) Heat equivalent of the friction of blading.

Solution : Useful heat drop

$$= 475 \times 0.85 = 403.75 \text{ kJ/kg}$$

Velocity of steam at exit from the nozzle

$$V_{a1} = 44.72 \times \sqrt{403.75} \simeq 900 \text{ m/s}$$

Further, we are given that :

$$\alpha_1 = 20° \; ; \quad V_b = 300 \text{ m/s}$$

$$\beta_1 = \beta_2 \text{ (blades being symmetrical)}$$

and $V_{r2} = 0.7\, V_{r1}$

From this data, the vector diagram (Fig. 19.10) can be drawn and the following results are obtained :

$$\beta_1 = \beta_2 = 29.5°$$

$$V_w = 927.5 \text{ m/s}$$

$$V_{f1} - V_{f2} = 92.5 \text{ m/s}$$

$$V_{r1} = 630.0 \text{ m/s}$$

and $V_{r2} = 441.0$ m/s

(*a*) Axial thrust

$$= m\,(V_{f1} - V_{f2})$$

$$= 1 \times 92.5 = \mathbf{92.5\ N}$$

(*b*) Work done

$$= m\, V_b\, V_w = 1 \times 300 \times 927.5$$

$$= 278250 \text{ Nm/s}$$

$$= \mathbf{278.25\ kW}$$

(*c*) Blade efficiency

$$= \frac{2 V_b V_w}{V_{a1}^2} = \frac{2 \times 300 \times 927.5}{(900)^2}$$

$$= 0.688 \text{ or } \mathbf{68.8\%}$$

Maximum blade efficiency under optimum conditions $= \cos^2 \alpha$

$$= \cos^2 20 = \mathbf{0.88 \text{ or } 88\%}$$

(*d*) Energy loss in blade friction

$$= m \frac{V_{r1}^2 - V_{r2}^2}{2} = 1 \times \frac{630^2 - 441^2}{2}$$

$$= 101209 \text{ J/kg}$$

$$= \mathbf{101.21 kJ/kg}$$

EXAMPLE 19.3.

The following data refers to one stage of an impulse turbine :

Isentropic nozzle heat drop 185 kJ/kg ; reheat of steam due to blade friction 10% of isentropic heat drop ; nozzle angle 20° ; ratio of blade speed to whirl component of steam speed 0.5 ; velocity coefficient for blades 0.95. You may also assume that steam at entry to nozzle has a velocity of 30 m/s.

Find the blade angles so that steam may enter them without shock and leave them in an axial direction.

Make an energy balance sheet for the stage per kg of steam.

Solution : Useful heat drop which is converted into kinetic energy

$$= 185 \times 0.9 = 166.5 \text{ kJ}$$

If V_{a1} is the steam velocity issuing from the nozzle,

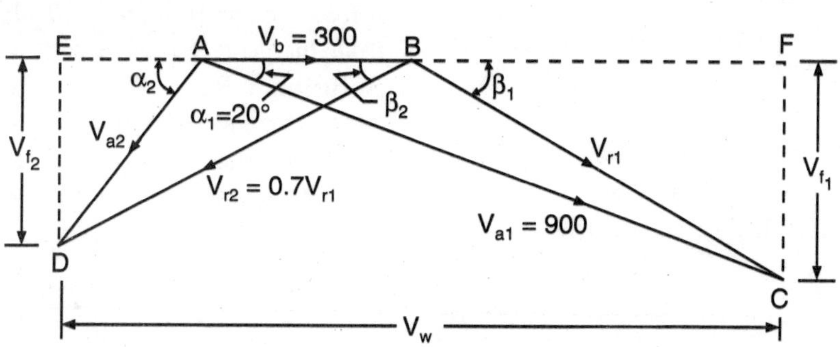

Fig. 19.10.

Then, $\frac{V_{a1}^2 - 30^2}{2 \times 10^3} = 166.5$

$\therefore \quad V_{a1} = 577.8$ m/s

Refer to Fig. 19.11, set off any horizontal line AB which represents the direction of motion of blades. From A set off AC = V_{a1} = 577.8 m/s and inclined at 20° to blade motion. The tangential component of V_{w1} scales 542 m/s.

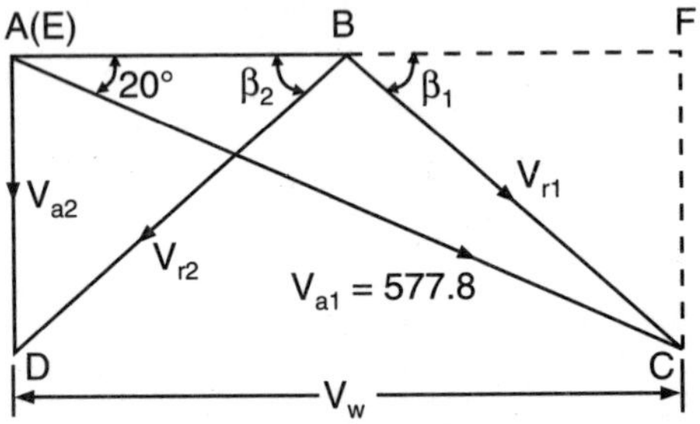

Fig. 19.11.

Blade velocity V_b = 542 × 0.5 = 271 m/s (cut off length AB equal to 271 m/s on the chosen scale and join BC. ABC is then the vector diagram of steam conditions at inlet.

Since the steam is to leave the blades in axial direction, the direction of absolute exit velocity is along AD which has been drawn at 90° to the line AB. From B swing an arc BD = 0.95 BC and let this meet the line AD at D. ABD is the exit vector diagram.

From the velocity triangles

(*a*) Angle β_1

= entry angle of the blade = 35.6°

Angle β_2

= exit angle of the blade = 30.8°

V_{a2} = 167 m/s

V_{r1} = 337.5 m/s

V_{r2} = 320.75 m/s

Work done per kg

= $V_w \times V_b$ = 542 × 271

= 146.9 × 10^3 Nm/kg

= 146.9 kJ/kg of steam

Friction loss

$$= \frac{V_{r1}^2 - V_{r2}^2}{2} = \frac{(337.5)^2 - (320.75)^2}{2}$$

= 5510 Nm/kg of steam

= 5.51 kJ/kg of steam

Kinetic energy loss at exit

$$= \frac{V_{a2}^2}{2} = \frac{167^2}{2}$$

= 13950 Nm/kg of steam

= 13.95 kJ/kg of steam

KE supplied at inlet

$$= \frac{V_{a1}^2}{2} = \frac{30^2}{2} = 450 \text{ Nm/kg of steam}$$

= 0.45 kJ/kg of steam

From the above calculations we can prepare the heat balance sheet as given below :

Credit	kJ	Debit	kJ
Isentropic heat drop	185	Useful work	146.9
KE entering	0.45	Nozzle friction	18.5
Total	185.45	Blade friction	5.51
		K.E. loss at exist	13.95
			184.86

EXAMPLE 19.4.

The velocity of steam at inlet to a single row impulse turbine is 400 m/s and the nozzle angle is 20°. The mean blade speed is 150 m/s and the axial thrust on the blade is estimated to be zero. Make calculations for :

(a) inlet and outlet angles of moving blades,

(b) power developed for a steam flow rate of 1.5 kg/s, and

(c) magnitude and direction of velocity of steam at exit.

Neglect the effect of friction when passing through blade passages and obtain your solution from the geometry of velocity diagrams.

Solution : Refer Figure 19.12 for the rough proportionate sketch of the velocity diagrams.

Since there is no axial thrust on the blade

$m(V_{f1} - V_{f2}) = 0 ; \quad V_{f1} = V_{f2}$

$V_{r1} \sin \beta_1 = V_{r2} \sin \beta_2$

When the effect of friction is neglected,

$V_{r2} = V_{r1}$ and accordingly

$\beta_1 = \beta_2$

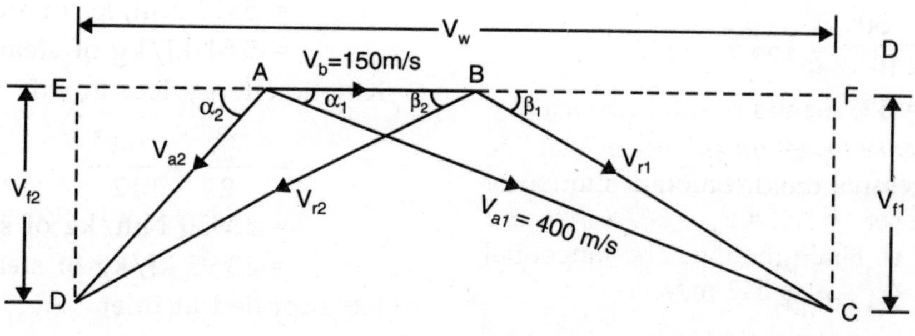

Fig. 19.12.

Obviously inlet and outlet angles of moving blade are equal.

(a) $\quad FC = V_{r1} \sin \beta_1 = V_{a1} \sin \alpha_1 \quad ...(i)$

$BF = V_{r1} \cos \beta_1$

$= V_{a1} \cos \beta_1 - V_b \quad ...(ii)$

From these identities

$$\tan \beta_1 = \frac{V_{a_1} \sin\alpha_1}{V_{a_1} \cos\beta_1 - V_b}$$

$$= \frac{400 \sin 20°}{400 \cos 20° - 150}$$

$$= \frac{136.81}{375.88 - 150} = 0.6056$$

$$\beta_1 = \tan^{-1} (0.6056) = 31.20° = \beta_2$$

(b) $\quad V_{f1} = V_{r1} \sin \beta_1 = V_{a1} \sin \alpha_1$

$$\therefore \quad V_{r1} = \frac{V_{a_1} \sin\alpha_1}{\sin\beta_1}$$

$$= \frac{400 \times \sin 20°}{\sin 31.20} = 264.11 \text{ m/s}$$

Change in whirl velocity component

$V_w = EF$

$= EB + BF$

$= V_{r2} \cos \beta_2 + V_{r1} \cos \beta_1$

$= 2\, V_{r1} \cos \beta_1$

$(\because V_{r2} = V_{r1} \text{ and } \beta_2 = \beta_1)$

$= 2 \times 264.11 \cos 31.20$

$= 451.82$ m/s

Power developed

$P = m\, V_w\, V_b$

$= 1.5 \times 451.82 \times 150$

$= 101659$ Nm/s $\simeq$ **101.66 kW**

(c) $\quad DE = V_{a2} \sin \alpha_2 = V_{r2} \sin \beta_2$

$EA = V_{a2} \cos \alpha_2 = V_{r2} \cos \beta_2 - V_b$

From these identities

$$\tan \alpha_2 = \frac{V_{r_2} \sin\beta_2}{V_{r_2} \cos\beta_2 - V_b}$$

$$= \frac{264.11 \sin 31.20°}{264.11 \cos 31.20° - 150} = 1.802$$

$\therefore \quad \alpha_2 = \tan^{-1} (1.802) = $ **60.97°**

Since $V_{f2} = V_{f1}$, we have

$V_{a2} \sin \alpha_2 = V_{a1} \sin \alpha_1$

$$\therefore \quad V_{a2} = V_{a1} \frac{\sin\alpha_1}{\sin\alpha_2}$$

$$= 400 \times \frac{\sin 20}{\sin 60.97} = \mathbf{156.52\ m/s}$$

EXAMPLE 19.5.

The following particulars refer to a single row impulse turbine :

Enthalpy drop in the nozzle and nozzle angle = 45 kJ/kg and 16°

Mean diameter of blade ring = 0.3 m

Rotational speed of wheel = 10,000 rev/min

Blade exit angle equals the blade inlet angle.

Determine : (a) Blade inlet angle for shockless entry, and (b) Work done and axial thrust for steam flow rate of 1 kg/s.

Neglect the effect of friction when passing through blade passages and obtain your solution form the geometry of velocity diagrams.

Solution : Enthalpy drop = 45 kJ/kg

Velocity of steam at exist from the nozzle

$$V_{a1} = 44.72 \times \sqrt{45} = 300 \text{ m/s}$$

$$\text{Blade speed} = \frac{\pi d_1 N}{60}$$

$$= \frac{\pi \times 0.3 \times 10000}{60} = 157 \text{ m/s}$$

Refer Fig. 19.13, for the rough proportionate sketch of the velocity diagrams.

Velocity of flow across the blades at inlet and outlet are :

$$V_{f1} = V_{r1} \cos \beta_1$$

and $V_{f2} = V_{r2} \cos \beta_2$

Further, we are given that

$$\beta_1 = \beta_2 \quad \text{and} \quad V_{r2} = V_{r1}$$

($\because$ there is no friction in blade passages)

Accordingly

$$V_{f1} = V_{f2}$$

(*a*) $\quad FC = V_{r1} \sin \beta_1 = V_{a1} \sin \alpha_1 \quad \ldots(i)$

$BF = V_{r1} \cos \beta_1$

$= V_{a1} \cos \beta_1 - V_b \quad \ldots(ii)$

From these identities

$$\tan \beta_1 = \frac{V_{a_1} \sin \alpha_1}{V_{a_1} \cos \beta_1 - V_b}$$

$$= \frac{300 \sin 16°}{300 \cos 16° - 157}$$

$$= \frac{82.69}{288.38 - 157} = 0.629$$

$\beta_1 = \tan^{-1} (0.629) = \mathbf{32.19°} = \beta_2$

(*b*) Axial thrust on the blade

$m(V_{f1} - V_{f2}) = 0 \quad (\because V_{f1} = V_{f2})$

Obviously the blades will not be subjected to any axial thrust.

From the identity :

$V_{f1} = V_{r1} \sin \beta_1 = V_{a1} \sin \alpha_1$, we have

$$V_{r1} = V_{a1} \frac{\sin \alpha_1}{\sin \beta_1}$$

$$= 300 \frac{\sin 16°}{\sin 32.19} = 155.22 \text{ m/s}$$

Change in whirl velocity component

$V_w = EF$

$= EB + BF$

$= V_{r2} \cos \beta_2 + V_{r1} \cos \beta_1$

$= 2 V_{r1} \cos \beta_1$

$(\because V_{r2} = V_{r1} \text{ and } \beta_2 = \beta_1)$

$= 2 \times 155.22 \cos 32.19°$

$= 262.72$ m/s

Power developed

$P = m V_w V_b$

$= 1 \times 262.72 \times 157$

$= 41247$ Nm/s $= \mathbf{41.25\ kW}$

19.5. STEAM TURBINE STAGING

If the entire expansion of steam from the boiler pressure to condenser pressure is carried out in a single stage, the absolute velocity of steam would be very high. Since for maximum efficiency the blade speed and the absolute steam velocity are related to each other by the relation $\frac{V_b}{V_{a1}} = \frac{\cos \alpha_1}{2}$, with increase in steam speed the blade velocity will be correspondingly high. We generally prefer small diameter rotors because they give more economic construction. But the high value of blade speed in small machines would give high speed of the order of 20,000 to 25,000 rev/min. The high rotational speeds are not directly applicable and a reduction gear box has to be installed between the turbine disc and the external equipment to give speeds which are in the useful range. Moreover at high rotational speeds, the

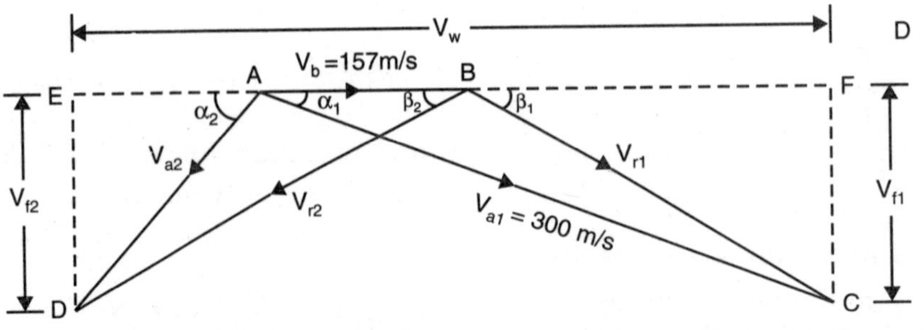

Fig. 19.13.

centrifugal stresses become excessive and there is danger of structural failure. Usually these considerations limit the blade speed to a maximum value of 400 m/s.

Moreoever with single stage, the velocity of steam at exit is sufficiently high and there is a considerable (about 10-12%) loss of kinetic energy.

The one wheel turbine is therefore not very useful. Further development of steam turbines led to the emergence of compounding whereby speed of rotation is reduced and at the same full use is made of the energy in the steam. This allows production of turbines of larger size and high power output.

This is achieved by :

(*a*) Converting the total enthalpy fall into kinetic energy in one stage but dividing the conversion of kinetic energy of steam into mechanical energy of the wheel in portions: *Velocity compounding*.

(*b*) Dividing the fall of total pressure drop in portions. Instead of one set of nozzles, there are several sets separated from each other by a turbine wheel. Each set provides a limited drop in enthalpy and pressure and deals with its own turbine wheel (the wheel following the nozzle) : *Pressure compounding.*

(*c*) Combination of pressure and velocity compounding.

Velocity compounding : (The Curtis turbine)

The velocity compounded stage, called the Curtis stage after its designer, is used to employ lower blade speed ratio and higher utilization of the kinetic energy of steam. The physical arrangement of velocity compounding is shown in Fig. 19.14. The velocity staging has the following main features :

(1) Like a single stage impulse wheel, velocity compounding allows a larger pressure drop in one set of nozzles. The number of nozzles which are in actual operation will be a function of the load on the turbine. The high load requires more steam necessitating that more nozzles are to be put into service.

(2) According to the velocity with which the steam issues from the nozzle group, two or more rings of moving blades separated by rings of fixed blades (guide blades) are keyed in series on the common shaft. Thus a two row Curtis stage consists of :

(*i*) convergent divergent nozzle, (*ii*) rotor, (*iii*) stator (guide blade), and (*iv*) another rotor.

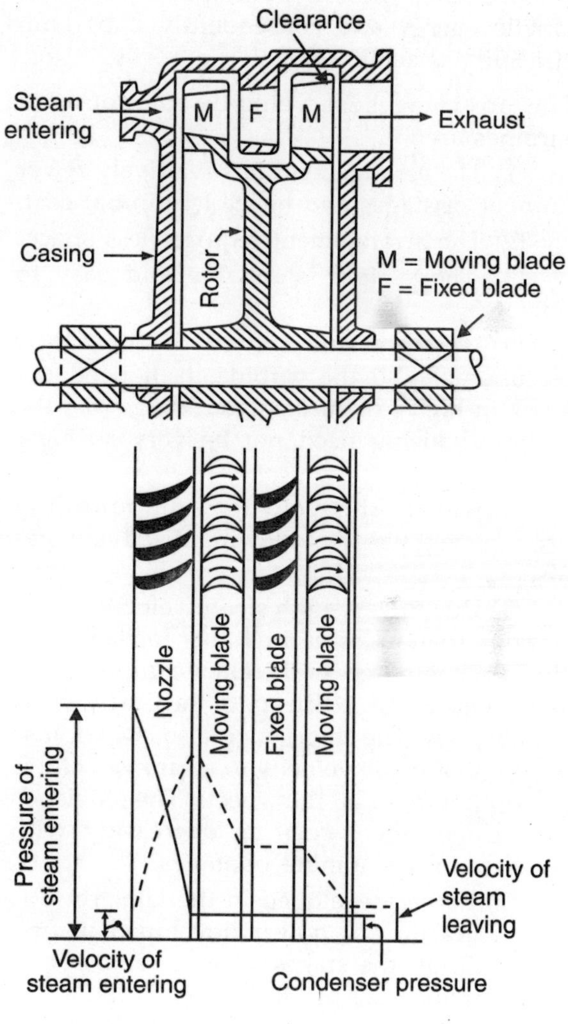

Fig. 19.14. Curtis or velocity staging

The function of the stator blades is merely to turn the steam into the direction required for entry into the next row of rotor blades without altering the pressure and velocity. However, there may be a slight reduction in velocity as a result of friction losses occurring in the guide blades.

(3) A fall in the velocity occurs every time when the steam passes over rings of moving blades.

(4) Since the rows of blades are connected to the same shaft, there is no loss of power.

(5) The speeds and angles are such that final velocity of steam from the second wheel is axial. Thus there is minimum rejection of kinetic energy and consequently maximum efficiency is achieved.

The advantages of velocity compounded turbines are :

(1) The arrangement has relatively fewer number of stages and hence less initial cost.

(2) The arrangement requires less space.

(3) The system is reliable and easy to operate

(4) The pressure fall in the nozzle is considerable, so the turbine itself need not work in high pressure surroundings and the turbine housing need not be very strongly made.

However, the velocity compounding arrangement has the following disadvantages also :

(1) Due to very high steam velocity in the nozzles, friction losses are rather too large and thus as compared with other compounding arrangements, the efficiency of the Curtis wheel is pretty low. The efficiency is also low because the ratio of blade velocity to steam velocity is not optimum for all the wheels. The efficiency of turbine wheel goes on decreasing with increase in the number of stages.

(2) Power developed in the later rows is only a fraction of power developed in first row. Still all the stages require same space, material and cost of fabrication.

The velocity compounded turbines are mainly used as drives for centrifugal compressors, pumps, small generators and for small units working with high pressure steam, *i.e.*, for driving feed pumps in big power units.

Pressure compounding. (The Rateau turbine).

"A number of simple impulse turbines arranged in series" :

The physical arrangement of the different units of pressure compounding is shown in Fig. 19.15. The pressure compounding arrangement, also called the Rateau compounding, has the following main features :

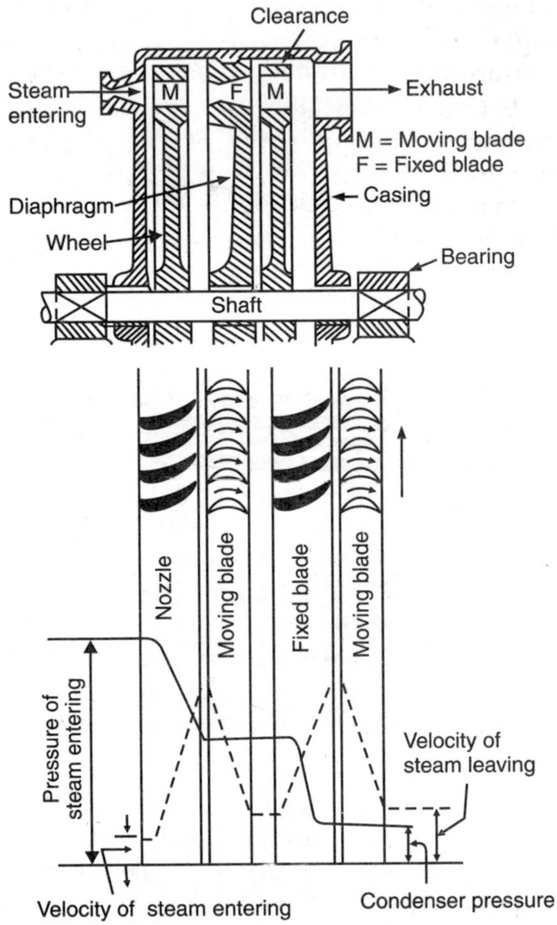

Fig. 19.15. Pressure compounding

(1) In the nozzles only a small pressure drop (and so enthalpy drop) is provided, giving limited increase of kinetic energy.

(2) As in velocity compounding, there are rotating rows of blades (wheels) and fixed rows of blades, all stages being keyed to the shaft in series.

(3) The rotating blades have the typical symmetrical shape of impulse turbine. The fixed blades will not only change the direction of steam speed, but will increase the speed

also. So they are in fact nozzles. The kinetic energy at exit from the first wheel is not lost but carried over to the next row.

(4) For all rotating blades the velocity diagrams and power output may be same.

(5) Since pressure gradually goes down, volume will increase and therefore the blade height has to be increased towards the low pressure side.

In Curtis wheel, although volume remains constant, the second and third row of blades need greater height as well. This is due to the fact that the steam speed while flowing across them is smaller than when flowing through the first row and so for the same mass flow the area has to be increased in accordance with continuity equation

$$m = \frac{AV}{v}$$

Another difference between velocity and pressure compounding is that in Curtis, all rows of rotating blades are working under the same low pressure. Leakage along the blade tips will hardly happen. It is even common practice to drill holes through the stationary wheels for being sure of equal pressure. In pressure compounding turbines, however, each stationary wheel is a pressure barrier between two stages and therefore should be securely locked against the moving parts.

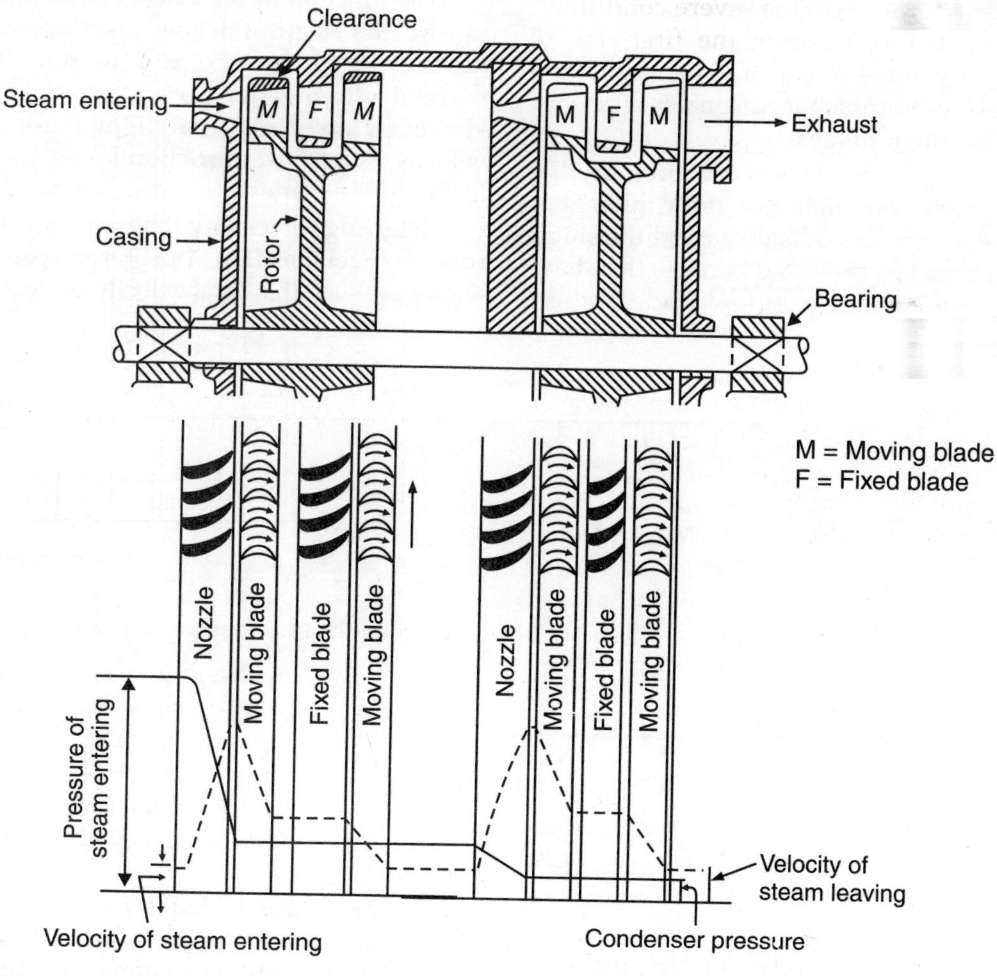

Fig. 19.16. Pressure velocity compounded turbine

Pressure and velocity compounding. The total pressure drop of steam is divided into stages and the velocity obtained in each stage is also compounded. This has the advantage of allowing a bigger pressure drop in each stage. Consequently less stages are necessary and a compact turbine will suffice for a given pressure drop.

The usual practice in multistage impulse turbines is however to have the first stage compounded for velocity and have single row wheels for the remaining stages. We know that modern power plants usually operate on high pressures and temperatures. The usage of velocity compounding in the first stage helps to achieve a large change in enthalpy and naturally the turbine casing and the rotor will be protected from extreme severe conditions. Moreover, before entering the first row of blades the volume of steam is increased and this fact is responsible for reasonable efficiency even with small blades.

In the remaining stages, the pressure has a value much lower than that at the inlet. Thus there are no chances of leakage and the single wheels should be preferred because they have the inherent advantage of low blade friction.

19.5.1. Two stage velocity compounding

The velocity compounding manifests when the entire enthalpy fall into kinetic energy takes place in one stage and the conversion of kinetic energy into mechanical energy of the wheel is achieved in portions. Depending on the magnitude with which the steam issues from the nozzle group, two or more rings of moving blades separated by rings of fixed blades (guide blades) are keyed in series on the common shaft. A two row velocity compounded stage (called the Curtis stage) would consist of :

(*i*) convergent-divergent nozzle
(*ii*) rotor
(*iii*) stator (guide blade)
(*iv*) another rotor

The function of the stator blades is merely to turn the steam into the direction required for entry into the next row of rotor blades without altering the pressure and velocity. However, there may be slight reduction in velocity as a result of friction losses occurring in the guide blades.

Referring to velocity diagrams for the two row arrangement (Fig. 19.17), the speeds and angles are such that final velocity of steam from

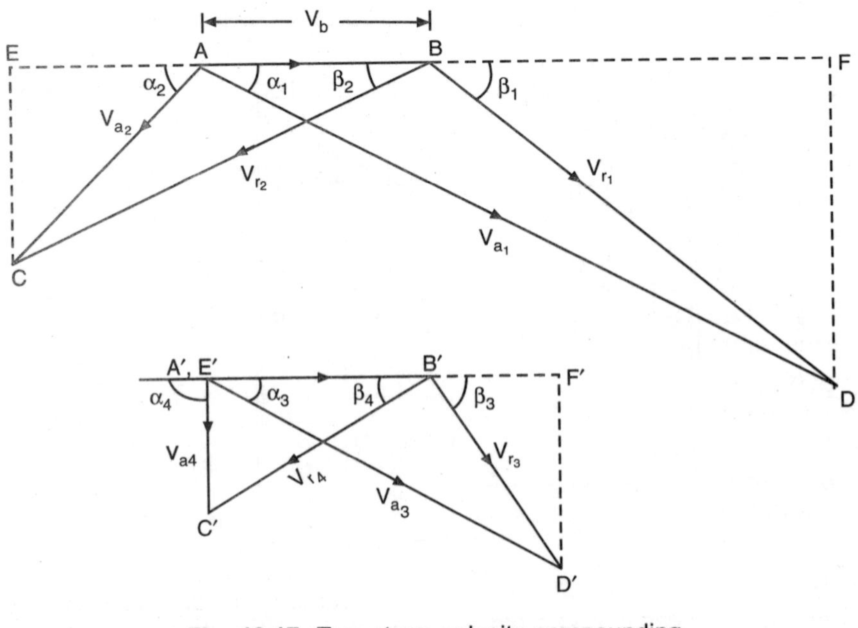

Fig. 19.17. Two stage velocity compounding

the second wheel is axial. Then there is minimum rejection of kinetic energy and consequently maximum efficiency is achieved.

Energy transfer from 1st row of moving blades,

E_1 = blade speed × change in whirl component

$= V_b(V_{a1} \cos \alpha_1 + V_{a2} \cos \alpha_2)$

$= V_b[(V_{r1} \cos\beta_1 + V_b) + (V_{r2} \cos\beta_2 - V_b)]$

Assuming ideal case of no friction loss and symmetrical blading :

$$V_{r1} = V_{r2} \quad \text{and} \quad \beta_1 = \beta_2$$

$$\therefore \quad E_1 = 2V_b V_{r1} \cos \beta_1$$

$$= 2V_b [V_{a1} \cos \alpha_1 - V_b] \quad ...(i)$$

When flowing through the first rotor, the fluid is turned but the magnitude of its velocity is unchanged. Accordingly

$$V_{a3} = V_{a2}$$

Energy transfer from 2nd row of moving blades is given by :

$$E_2 = V_b [V_{a3} \cos \alpha_3 + V_{a4} \cos \alpha_4]$$

$$= V_b [V_{a3} \cos \alpha_3] \quad (\because \alpha_4 = 90°)$$

$$= V_b [V_{r3} \cos \beta_3 + V_{r4} \cos \beta_4]$$

Taking $\beta_3 = \beta_4$ and $V_{r3} = V_{r4}$, we get

$$E_2 = 2 V_b V_{r3} \cos \beta_3 \quad ...(ii)$$

If α_3 is equal to α_2, then

$$V_{a3} \cos \alpha_3 = V_{a2} \cos \alpha_2$$

$$= V_{r2} \cos \beta_2 - V_b$$

Since $V_{r1} = V_{r2}$ and $\beta_2 = \beta_1$, the above identity can be written as

$$V_{a3} \cos \alpha_3 = V_{r1} \cos \beta_1 - V_b$$

From the velocity triangle :

$V_{r1} \cos \beta_1 = V_{a1} \cos \alpha_1 - V_b$ and then

$V_{a3} \cos \alpha_3 = V_{a1} \cos \alpha_1 - 2V_b$

Now :

$$V_{r3} \cos \beta_3 = V_{a3} \cos \alpha_3 - V_b$$

Substituting the value of $V_{a3} \cos \alpha_3$ as determined above, we get

$$V_{r3} \cos \beta_3 = (V_{a1} \cos \alpha_1 - 2V_b) - V_b$$

$$= V_{a1} \cos \alpha_1 - 3V_b$$

The expression (*ii*) for the energy transfer from 2nd row of moving blades may then be written as

$$E_2 = 2V_b (V_{a1} \cos \alpha_1 - 3V_b)$$

Hence total energy transfer is given by :

$$E = E_1 + E_2 = 2V_b [2V_{a1} \cos \alpha_1 - 4V_b]$$

∴ Diagram efficiency

$$\eta_d = \frac{2V_b[2V_{a1} \cos\alpha_1 - 4V_b]}{V_{a1}^2/2}$$

$$= \frac{8V_b}{V_{a1}} \cos \alpha_1 - 16 \left(\frac{V_b}{V_{a1}}\right)^2$$

$$= 8\rho \cos \alpha_1 - 16\rho^2$$

For maximum efficiency, $\frac{d\eta_d}{d\rho} = 0$

$$\therefore \quad 8 \cos \alpha_1 - 32\rho = 0 \quad \text{or} \quad \rho = \frac{\cos\alpha_1}{4}$$

$$\eta_d \max = 8 \frac{\cos\alpha_1}{4} \times \cos \alpha_1 - 16 \frac{\cos^2 \alpha_1}{16}$$

$$= \cos^2 \alpha_1$$

The rate of doing work corresponding to maximum diagram efficiency can be obtained by substituting $\rho = \frac{\cos\alpha_1}{4} = \frac{V_b}{V_{a1}}$ in the expression for work done or energy transfer.

∴ Maximum rate of doing work per kg of steam/s

$$= 2V_b \left[2V_{a1} \times \frac{4V_b}{V_{a1}} - 4V_b\right]$$

$$= 2V_b (4 V_b) = 8V_b^2$$

The analysis has been done for two rows of moving blades, but of course it is possible to make use of 3 or even 4 wheels following the same pattern. In general, we may write :

Optimum blade speed ratio : $\rho = \frac{\cos\alpha_1}{2n}$

Work done in last row : $= \frac{1}{2^n}$ of total work

where n is the number of rows.

Thus for a three row velocity compounded wheels the ideal $\rho = \frac{\cos\alpha_1}{6}$ and the last row of blades would do only $\frac{1}{8}$th of the total work.

Because of the decreasing utility in last row with the increase in the number of rows, it is seldom now-a-days that more than two rows are used.

Comparison of relations $\rho = \frac{\cos \alpha_1}{2}$ for a single stage impulse wheel and $\rho = \frac{\cos \alpha_1}{4}$ for a two row impulse wheel reveals that for the same blade speed and the same nozzle angle, the steam velocity at exit from the nozzle of two row velocity compounded turbine is twice that for a simple impulse turbine. Since blade work is proportional to kinetic energy, theoretically the work of a two row Curtis wheel is four times that of a simple wheel, blade speeds being equal in both the cases.

The advantages of velocity compounded turbine are :

(1) The arrangement has relatively fewer number of stages and hence less initial cost.

(2) The arrangement requires little space.

(3) The system is reliable and easy to operate.

(4) The pressure fall in the nozzle is considerable, so the turbine itself need not work in high pressure surroundings and the turbine housing need not be very strongly made.

However the velocity compounding arrangement has the following disadvantages also :

(*i*) Due to very high steam velocity in the nozzles, friction losses are rather too large.

(*ii*) The efficiency of turbine wheel decreases with increase in the number of stages.

(*iii*) Power developed in the later rows is only a fraction of power developed in first row. Still all stages require same space, material and cost of fabrication.

The velocity compounded turbines are mainly used as drives for centrifugal compressors, pumps and small generators and for small units working with high pressure steam, *i.e.*, for driving feed pumps in big power units.

EXAMPLE 19.6.

Steam issues from the nozzle of an impulse turbine at 600 m/s and this velocity is compounded in two rings of moving blades separated by fixed blades. The nozzle angle is 16° while the discharge angles are 18° for the first moving ring, 21° for the fixed ring and 35° for the second moving ring, all measured relative to the plane of rotation :

Designing for a blade velocity of 120 m/s and a 10% drop in velocity during passage through each ring of blades, calculate :

(a) blade inlet angle for each row

(b) driving force and axial thrust for each row of moving blades

(c) diagram power per kg/s flow of steam and diagram efficiency of the stage.

What would be the maximum possible diagram efficiency for the given inlet velocity and nozzle angle.

Solution : The velocity diagrams are drawn as shown in Fig. 19.18 and the steps in its construction are as follows :

(*i*) Draw a horizontal line AB representing V_b = 120 m/s according to some chosen scale.

(*ii*) Set of the direction of V_{a1} at 16° to AB and take AC = 600 m/s.

(*iii*) Measure BC representing V_{r1} and calculate V_{r2}.

V_{r1} = 485 m/s

and V_{r2} = 0.9 × 485 = 436.5 m/s

(*iv*) Set off BD = V_{r2} at 18° to the horizontal and join A with D.

This completes the velocity vector diagram for the first row of moving blades.

(*v*) Measure AD representing V_{a2} and calculate V_{a3}. V_{a3} is the steam velocity with which it enters the second row of moving blades.

V_{a2} = 328 m/s

and V_{a3} = 0.9 × 328 = 295.2 m/s

(*vi*) The blade speed for each moving blade is same and so take A′B′ = 120 m/s

(*vii*) Set off the direction of V_{a3} at 21° to A′B′ and take A′C′ = 295.2 m/s

(*viii*) Measure B′C′ representing V_{r3} and calculate V_{r4}.

V_{r3} = 186 m/s

and V_{r4} = 0.9 × 186 = 167.4 m/s

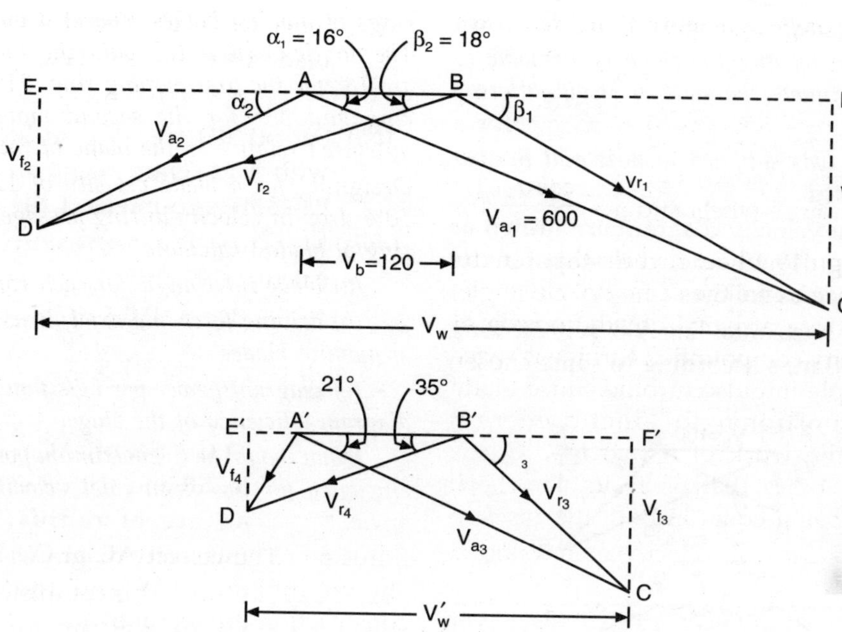

Fig. 19.18.

(*ix*) Set off B′D′ = V_{r4} at 35° to the horizontal and join A′ with D′.

This completes the velocity vector diagram for the 2nd row of moving blade.

By measurement from the velocity diagrams :

Inlet blade angle, first row of moving blades β_1 = 20°

Inlet blade angle, second row of moving blade β_3 = 34.5°

(*b*) V_w = EF = 870 m/s

V'_w = E′F′ = 285 m/s

$V_f = (V_{f1} - V_{f2})$ = 30 m/s

$V'_f = (V_{f3} - V_{f4})$ = 9 m/s

Driving force

$= mV_w$

= 1 × 870 = 870 N for first row

= 1 × 285 = 285 N for second row

Total driving force

= 870 + 285 = 1155 N

Axial thrust

$= mV_f$

= 1 × 30 = 30 N for first row

= 1 × 9 = 9 N for second row

Total axial thrust

= 30 + 9 = 39 N

(*c*) Power output = 1155 × 120

= 138600 Nm/s

= 138.6 kW

(*d*) Diagram efficiency

$$= \frac{2V_b \times V_w}{V_{a1}^2}$$

$$= \frac{2 \times 120 \times (870 + 285)}{(600)^2}$$

= 0.77 or 77%

EXAMPLE 19.7.

The nozzle angle in a velocity compounded stage of an impulse turbine is 20° to the plane of rotation and the mean blade speed is 100 m/s for each moving blade row. There are two rows of moving blades with exit angles 26° and 30°. Between these rows there is a row of fixed blades with exit angle 28°. The relative velocity of steam drops by 10% during passage through ring of blades and the final discharge is axial. Determine the velocity of steam leaving the nozzles and the blade efficiency.

If the nozzle efficiency is 95% and the kinetic energy of steam leaving the stage is available as heat energy, estimate the state of steam entering the next stage.

The steam was supplied to nozzle at 8.5 bar and dry saturated.

Solution : The velocity diagram are drawn as shown in Fig. 19.19 and the steps in its construction are as follows :

(*i*) Draw a horizontal line A′B′ representing V_b = 100 m/s according to some chosen scale.

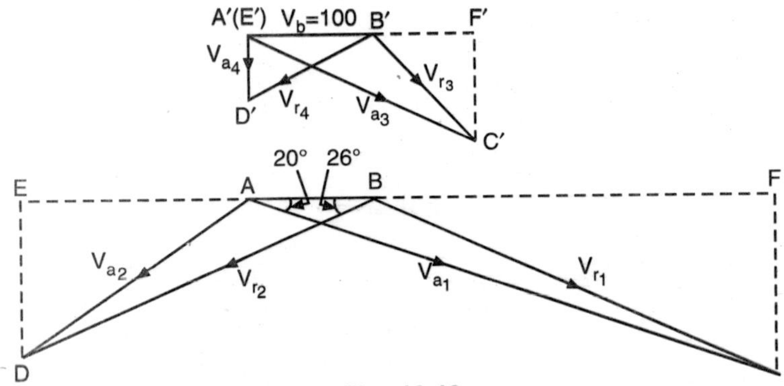

Fig. 19.19.

(*ii*) Set off the direction of V_{r4} (B′D′) at 30° to V_b and draw V_{a4} (A′D′) at 90° to V_b. It will intersect B′D′ at D′.

(*iii*) Measure V_{r4} and calculate V_{r3}

$$V_{r4} = 115 \text{ m/s}$$

and $$V_{r3} = \frac{115}{0.9} = 128 \text{ m/s}$$

(*iv*) Set off A′C′, *i.e.*, V_{a3} at 28° to A′B′, and with B′ as centre swing an arc of radius 128 m/s to intersect A′C′ at C′.

This completes the velocity vector diagram for the second row of moving blade.

(*v*) Measure A′C′ representing V_{a3} and calculate V_{a2}

$$V_{a3} = 210 \text{ m/s}$$

and $$V_{a2} = \frac{210}{0.9} \; 233.3 \text{ m/s}$$

(*vi*) Since the blade speed for each moving blade row is same, take

$$AB = 100 \text{ m/s}$$

(*vii*) Set off BD (*i.e.*, V_{r2}) at 26° to AB and with A as centre swing an arc of radius 233.3 m/s to intersect BD at D.

(*viii*) Measure BD representing V_{r2} and calculate V_{r1}

$$V_{r2} = 317.5 \text{ m/s}$$

and $$V_{r1} = \frac{317.5}{0.9} \; 353 \text{ m/s}$$

(*ix*) Set off AC (*i.e.*, V_{a1}) at 20° to AB and with *B* as centre swing an arc of radius 353 m/s to intersect AC at C.

This completes the velocity vector diagram for first row of moving blade.

AC (*i.e.*, V_{a1}) represents the velocity of steam leaving the nozzle.

By measurement,

$$V_{a1} = 445 \text{ m/s}$$

$$V_w = EF + E'F' = 596 + 18 = 614 \text{ m/s}$$

Blade efficiency

$$= \frac{2V_bV_w}{V_{a1}^2}$$

$$= \frac{2 \times 100 \times 614}{(445)^2} = 0.62 \text{ or } 62\%$$

Now :

$$V_{a1} = 44.7 \sqrt{\Delta H_{isen} \times \eta_n}$$

Isentropic heat drop

$$= \left(\frac{445}{44.7}\right)^2 \times \frac{1}{0.95}$$

$$= 104.32 \text{ kJ/kg}$$

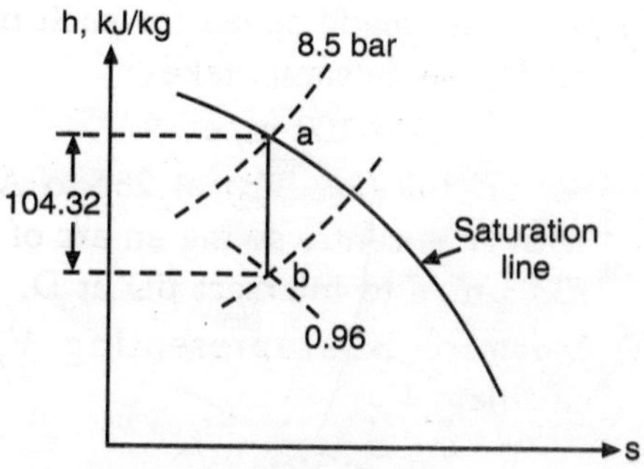

Fig. 19.20.

The expansion process *ab* on Mollier chart is shown in Fig. 19.20. From state point *a*, which represents the initial steam condition (8.5 bar, dry saturated), a vertical line *ab* is set off such that *ab* equals the isentropic enthalpy drop (104.32 kJ/kg). The point *b* then represents the steam condition at exit from nozzle.

Corresponding to state point *b*, dryness fraction x_b = 0.96 and pressure P_b = 5 bar.

19.6. REACTION TURBINE

The pure reaction turbine based on the principle of Hero was built in 1883 by Gustaf De Laval. The turbine demonstrated by De Laval had a tip speed of 180 m/s with speed of rotation 42,000 rpm. The turbine was inefficient and did not become of commercial interest.

The *impulse reaction turbine* (hereafter mentioned as reaction turbine) receives its driving force partly as an impulse force and partly as a reaction force. The turbine may be looked upon as a development of pressure compounding. The fixed blades in the pressure compounding, as mentioned earlier, are asymmetrical having a decreasing area towards the exit side resulting in expansion and increase of speed. In a reaction turbine the same thing is done with the moving blades giving them exactly the same space as the stationary ones. So not only the absolute velocity increases in the fixed blades, the relative velocity will also increase in the moving blade.

The pressure and velocity variations, and the velocity diagram for one reaction stage is shown in Fig. 19.21. A stage constitutes one set of fixed blades followed by one set of moving blades. Since expansion of steam takes place in the moving blades also, the relative velocity leaving V_{r2} is greater than the velocity entering.

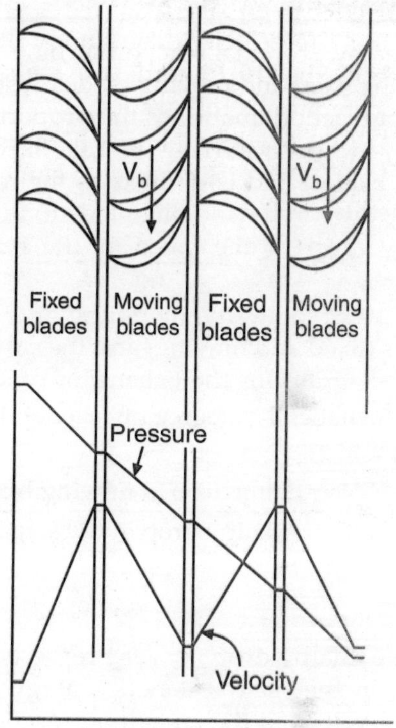

Fig. 19.21. (a) Pressure and velocity variations for a reaction turbine

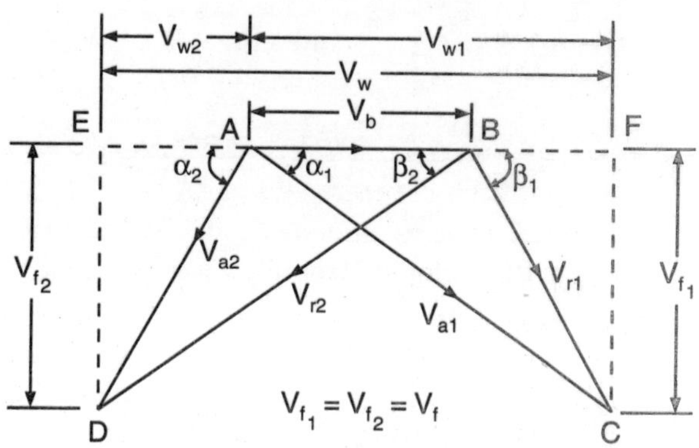

Fig. 19.21. (b) Velocity vector diagram for a reaction turbine

The following expressions for work done and diagram efficiency for impulse turbines are applicable to reaction turbines also

Work done = $m\,(V_{w1} + V_{w2})\,V_b$ Nm/s

Diagram efficiency

$$= \frac{2V_b\,(V_{w1}+V_{w2})}{V_{a1}^2}$$

19.7. DEGREE OF REACTION

The enthalpy drop in a reaction turbine occurs both in the fixed and moving blades. The term degree of reaction quantifies the proportion of enthalpy drops occurring in the fixed and moving blades, and is defined as the ratio of enthalpy drop in the moving blade to the total enthalpy drop in the stage of the reaction turbine.

Fig. 19.22 shows the mounting of two sets of blades (fixed and moving) and the enthalpy-entropy diagram for the expansion process.

Mathematically, degree of reaction can be expressed as

$$R = \frac{\text{enthalpy drop in moving blade}}{\text{enthalpy drop in the stage}}$$

$$= \frac{h_2 - h_3}{h_1 - h_3}$$

The enthalpy drop $(h_2 - h_3)$ represents the increase in the relative kinetic energy in the moving blade, and the enthalpy drop $(h_1 - h_3)$ in the stage is a measure of work output. That is

$$(h_2 - h_3) = \frac{V_{r_2}^2 - V_{r_1}^2}{2} \quad \text{and} \quad (h_1 - h_3) = V_b\,V_w$$

$$\therefore \quad R = \frac{\left(V_{r_2}^2 - V_{r_1}^2\right)/2}{V_b V_w} = \frac{V_{r_2}^2 - V_{r_1}^2}{2\,V_b\,V_w} \qquad \text{...(19.19)}$$

From the velocity triangles for a reaction turbine (Fig. 19.21),

$$V_{r_2} = V_f \operatorname{cosec}\beta_2\,; \quad V_{r_1} = V_f \operatorname{cosec}\beta_1$$

and $V_w = V_f\,(\cot\beta_2 + \cot\beta_1)$

where V_f is the flow velocity which remains constant throughout a stage.

Substituting the above results in identity 19.19, we get

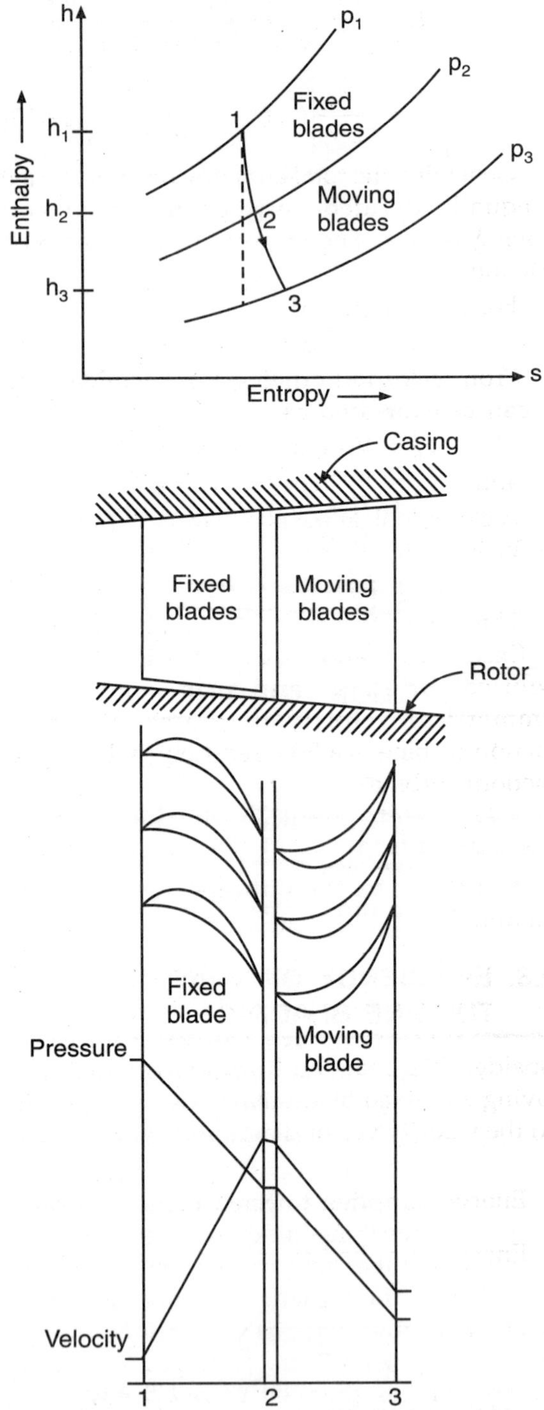

Fig. 19.22. Reaction turbine stage

$$R = \frac{V_f^2 (\text{cosec}^2 \beta_2 - \text{cosec}^2 \beta_1)}{2 V_f (\cot \beta_2 + \cot \beta_1) V_b}$$

$$= \frac{V_f}{2 V_b} (\cot \beta_2 - \cot \beta_1) \quad ...(19.20)$$

Generally, the total enthalpy drop in a stage is equally divided between the fixed and moving blades. The stage is then called 50% reaction stage.

For 50% degree of reaction,

$$V_b = V_f (\cot \beta_2 - \cot \beta_1)$$

From velocity triangles, the blade velocity V_b can be expressed as

$$V_b = V_f (\cot \beta_2 - \cot \alpha_2)$$

and $V_b = V_f (\cot \alpha_1 - \cot \beta_1)$

A closer look at the above three expressions for V_b indicates that

$$\alpha_1 = \beta_2 \text{ and } \beta_1 = \alpha_2$$

This shows that for a 50% reaction turbine, the fixed (stator) and moving (rotor) blades are identical in shape and that would give symmetrical velocity diagrams. Parson's reaction turbine is a 50% reaction (half-degree reaction) turbine.

- For simple impulse turbine, the degree of reaction $R = 0$
- For pure reaction turbine, the degree of reaction $R = 1$

19.8. EFFICIENCY OF REACTION TURBINE BLADING

Consider 50% reaction turbine for which the moving and fixed blades have the same profile and the velocity vector diagram is symmetrical.

$$\text{Energy supplied to fixed blade} = \frac{V_{a1}^2}{2}$$

Energy supplied to moving blade

$$= \frac{V_{r2}^2 - V_{r1}^2}{2}$$

$$\text{Total energy input} = \frac{V_{a1}^2}{2} + \frac{V_{r2}^2 - V_{r1}^2}{2}$$

From the symmetry of velocity vector diagrams, $V_{r2} = V_{a1}$ and therefore,

Total energy input

$$= V_{a1}^2 - \frac{V_{r1}^2}{2}$$

Again from the velocity vector diagram :

$$V_{r1}^2 = V_{a1}^2 + V_b^2 - 2V_b V_{a1} \cos \alpha_1$$

∴ Total energy input

$$= V_{a1}^2 - \left[\frac{V_{a1}^2 + V_b^2 - 2V_b V_{a1} \cos \alpha_1}{2} \right]$$

$$= \frac{V_{a1}^2 - V_b^2 + 2V_b V_{a1} \cos \alpha_1}{2}$$

Rate of doing work

$$= V_w V_b$$

$$= (V_{a1} \cos \alpha_1 + V_{r1} \cos \beta_2 - V_b) V_b$$

Since, $\alpha_1 = \beta_2$ and $V_{a1} = V_{r1}$

Rate of doing work

$$= (2 V_{a1} \cos \alpha_1 - V_b) V_b \quad ...(19.21)$$

Diagram efficiency

$$\eta_d = \frac{\text{rate of doing work}}{\text{energy input}}$$

$$= \frac{2V_b (2V_{a1} \cos \alpha_1 - V_b)}{V_{a1}^2 - V_b^2 + 2V_b V_{a1} \cos \alpha_1}$$

$$= \frac{\frac{2V_b}{V_{a1}} \left[2 \cos \alpha_1 - \frac{V_b}{V_{a1}} \right]}{1 - \left(\frac{V_b}{V_{a1}} \right)^2 + 2\frac{V_b}{V_{a1}} \cos \alpha_1}$$

$$= \frac{2\rho (2 \cos \alpha_1 - \rho)}{1 - \rho^2 + 2\rho \cos \alpha_1} \quad ...(19.22)$$

where $\rho = \frac{V_b}{V_{a1}}$ is the blade speed ratio.

The efficiency would be maximum when the denominator $(1 - \rho^2 + 2\rho \cos \alpha_1)$ is maximum or when its differential with respect to ρ is zero. Therefore, for maximum diagram efficiency

$$- 2\rho + 2 \cos \alpha_1 = 0$$

or $$\rho = \cos \alpha_1 \quad ...(19.23)$$

Substituting $\rho = \cos \alpha_1$ in equation 19.22, the maximum diagram efficiency for 50% reaction turbine is

$$(\eta_d)_{max} = \frac{2\cos\alpha_1\,(2\cos\alpha_1 - \cos\alpha_1)}{1-\cos^2\alpha_1 + 2\cos\alpha_1\cos\alpha_1}$$

$$= \frac{2\cos^2\alpha_1}{1+\cos^2\alpha_1} \qquad ...(19.24)$$

Corresponding to maximum diagram efficiency,

Rate of doing work

$$= (2\,V_{a1}\cos\alpha_1 - V_b)\,V_b$$

$$= \left(2V_{a1}\frac{V_b}{V_{a1}} - V_b\right)V_b = V_b^2 \qquad ...(19.25)$$

EXAMPLE 19.8

In the stage of a reaction turbine the steam leaves the fixed blades at a pressure of 2 bar with a dryness fraction of 0.95. For each pair of fixed and moving blades, the exit angle is 20 degree. When the rotor turns 300 rev/min, the turbine develops 2.95 kW. The steam supply rate is 250 kg/min and the tip leakage of steam is estimated to be 10 percent. Assuming that axial velocity equals 0.75 times the blade speed, make calculations for the drum diameter and the height of blades.

Solution : Blade speed

$$V_b = \frac{\pi dN}{60} = \frac{\pi d \times 300}{60} = 15.7\,d \text{ m/s}$$

where d is the mean drum diameter in metres.

∴ Flow velocity

$$V_f = 0.75 \times 15.7\,d = 11.77\,d \text{ m/s}$$

Steam flow rate through the blades

$$m = 0.9 \times \frac{250}{60} = 3.75 \text{ kg/s}$$

Power output = $m\,V_w\,V_b$

$2.95 \times 10^3 = 3.75 \times V_w \times 15.7\,d$

∴ Change in whirl component

$$V_w = \frac{2.95\times10^3}{3.75\times15.7\,d} = \frac{50.1}{d} \text{ m/s}$$

With reference to velocity vector diagram (Fig. 19.17) the change in whirl component is also given by

$$V_w = 2\,V_f\cot\alpha_1 - V_b$$

$$= 2 \times 11.77\,d\cot 20 - 15.7\,d$$

$$= 48.97\,d \text{ m/s}$$

Equating the two expression for whirl component, we have

$$\frac{50.1}{d} = 48.97\,d\,; \quad d^2 = 1.023$$

∴ Drum diameter

$= d =$ **1.011 m**

(*b*) Flow velocity

$$V_f = 11.77\,d$$

$$= 11.77 \times 1.011 = 11.9 \text{ m/s}$$

Mass flow of steam

$$m = \frac{AV_f}{v} = \frac{\pi d h\,V_f}{v}$$

For steam at 2 bar and 0.95 dryness fraction, the specific volume of steam as noted from steam tables is

$$v = 0.95 \times 0.92 = 0.8569 \text{ m}^3\text{/kg}$$

$$\therefore \quad 3.75 = \frac{\pi\times1.011\times h\times11.9}{0.8569}$$

Solution gives : $h = 0.0851$ m

Thus the mean blade height is 0.0851 m

EXAMPLE 19.9

Dry saturated steam at 3 bar is admitted to one stage of a 50% reaction turbine having identical blading. The mean blade height is 4 cm, the exit angle of moving blade is 20° and the velocity of steam is 100 m/s. At the mean radius, the axial flow velocity equals three-fourth of blade speed. For a steam flow rate of 10,000 kg/hr, calculate : (a) the rotor speed in rev/min, (b) the power output of stage, (c) the diagram efficiency, (d) the percentage increase in relative velocity in the moving blades, and (e) the enthalpy drop of steam in this stage.

Solution : Refer Fig. 19.23 for the velocity vector diagram :

Flow velocity

$$V_f = V_1\sin\alpha_1$$

$$= 100\sin 20 = 34.2 \text{ m/s}$$

Mean blade speed

$$V_b = \frac{4}{3}\,V_f = \frac{4}{3} \times 34.2 = 45.6 \text{ m/s}$$

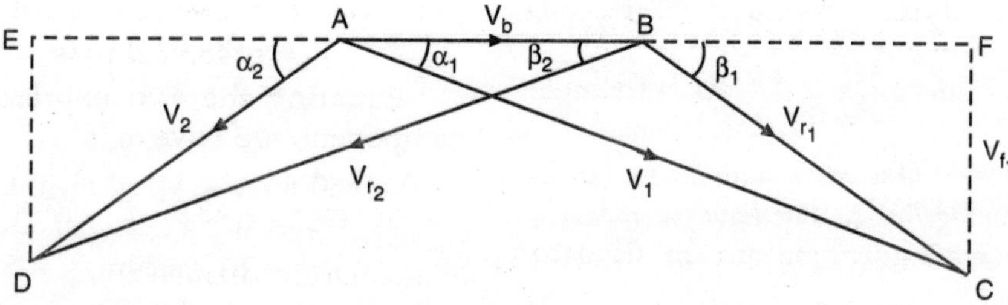

Fig. 19.23

Mess flow of stream

$$m = \frac{AV_f}{v} = \frac{\pi dh\, V_f}{v}$$

For dry saturated steam at 3 bar, the specific volume of steam as noted from steam tabels is 0.6055 m³/kg. Then

$$\frac{10000}{3600} = \pi \times d \times \frac{4}{100} \times \frac{34.2}{0.6055}$$

Solution gives : Mean blade diameter

$$d = 0.391 \text{ m}$$

(*a*) Now : $V_b = \frac{\pi dN}{60}$; $45.6 = \frac{\pi \times 0.391 \times N}{60}$

$$\therefore \text{ Rotor speed } N = \frac{45.6 \times 60}{\pi \times 0.391}$$

$$= \mathbf{2228.5 \text{ rev/min.}}$$

(*b*) The velocity diagram can now be plotted to scale. As the blades are identical and the degree of reaction is 50%, the diagram would be symmetrical

Alternatively, the calculations may be made as follows :

$$BF = AF - AB = V_1 \cos \alpha_1 - V_b$$
$$= 100 \cos 20 - 45.6 = 48.37 \text{ m/s}$$

EA = BF = 48.37 m/s as the diagram is symmetrical

$$\beta_1 = \tan^{-1} \frac{FC}{BF}$$

$$= \tan^{-1} \left(\frac{34.2}{48.37}\right) = 35.26°$$

$$V_{r1} = \frac{V_{f_1}}{\sin \beta_1} = \frac{34.2}{\sin 35.26} = 59.27 \text{ m/s}$$

Whirl velocity

$$EF = EA + AB + BF$$
$$= 48.37 + 45.6 + 48.37 = 142.34 \text{ m/s}$$

Power output = $m\, V_w\, V_b$

$$= \frac{10000}{3600} \times 142.34 \times 45.6$$

= 18029 W or **18.029 kW**

(*c*) Work done / kg of steam = $V_w\, V_b$

Energy input/kg of steam

$$= \frac{V_1^2}{2} + \frac{V_{r_2}^2 - V_{r_1}^2}{2} = \frac{2V_1^2 - V_{r_1}^2}{2}$$

$$(\because V_{r_2} = V_1)$$

∴ Diagram efficiency η_d

$$= V_w\, V_b \times \frac{2}{2V_1^2 - V_{r_1}^2}$$

$$= (142.34 \times 45.6) \times \frac{2}{2(100)^2 - (59.27)^2}$$

= 0.787 or **78.7%**

(*d*) Percentage increase in relative velocity

$$= \frac{V_{r_2} - V_{r_1}}{V_{r_1}} - \frac{100 - 59.27}{59.27}$$

= 0.687 or 68.7%

(*e*) Enthalpy drop in the moving blades

$$= \frac{V_{r_2}^2 - V_{r_1}^2}{2}$$

$$= \frac{(100)^2 - (59.27)^2}{2} = 3243 \text{ J/kg}$$

Total enthalpy drop per stage

= 2 × 3243 = 6486 J/kg

= **6.486 kJ/kg**

EXAMPLE 19.10

At one stage of a multi-stage Parson's reaction turbine, the mean ring diameter is 1.4 m and the speed ratio is 0.7. If the rotor turns 3000 rev/min, determine

(a) the blade inlet angle if the blade exit angle is 20°,

(b) the diagram efficiency, and

(c) the percentage increase in diagram efficiency if the turbine is designed to run at the best theoretical speed.

Solution : Refer Fig. 19.24 for a rough sketch of the velocity vector diagram :

Blade speed

$$V_b = \frac{\pi d N}{60}$$

$$= \frac{\pi \times 1.4 \times 3000}{60} = 219.8 \text{ m/s}$$

Velocity of steam

$$V_1 = \frac{219.8}{0.7} = 314 \text{ m/s}$$

$$\left(\because \rho = \frac{V_b}{V_1} = 0.7\right)$$

From the inlet velocity triangle *ABC*,

$$V_{r_1} = \sqrt{V_1^2 + V_b^2 - 2V_1V_b\cos\alpha_1}$$

$$= \sqrt{(314)^2 + (219.8)^2 - 2 \times 314 \times 219.8 \cos 20}$$

$$= 131.2 \text{ m/s}$$

$$\frac{V_{r_1}}{\sin\alpha_1} = \frac{V_b}{\sin C};$$

$$\sin C = \frac{219.8}{131.2}\sin 20 = 0.573$$

$$\angle C = 34.96°$$

∴ Blade exit

$$\beta_1 = 20 + 34.96 = 54.96°$$

(*b*) Whirl velocity

$$V_w = EF = (2V_1 \cos\alpha_1 - V_b)$$

$$= 2 \times 314 \cos 20 - 219.8$$

$$= 370.32 \text{ m/s}$$

Work done per kg of steam

$$= V_w V_b = 370.32 \times 219.8$$

$$= 81396 \text{ Nm/s} = 81.396$$

Energy supplied per kg of steam

$$= \frac{V_1^2}{2} + \frac{V_{r_2}^2 - V_{r_1}^2}{2} = \frac{2V_1^2 - V_{r_1}^2}{2}$$

$$(\because V_{r2} = V_1)$$

$$= \frac{2(314)^2 - (131.2)^2}{2}$$

$$= 89989 \text{ Nm/s} = 89.989 \text{ kJ/s}$$

Diagram efficiency

$$\eta_d = \frac{\text{work done}}{\text{energy supplied}}$$

$$= \frac{81.396}{89.989} = 0.9045 \text{ or } \mathbf{90.45\%}$$

(*c*) The best theoretical speed is

$$V_b = V_1 \cos\alpha_1$$

$$= 314 \cos 20 = 295.06 \text{ m/s}$$

$$V_{r_1} = \sqrt{V_1^2 + V_b^2 - 2V_1V_b\cos\alpha_1}$$

$$= \sqrt{(314)^2 + (295.06)^2 - 2 \times 314 \times 295.06 \cos 20}$$

$$= 107 \text{ m/s}$$

$$V_w = 2V_1 \cos\alpha_1 - V_b$$

$$= 2 \times 314 \cos 20 - 295.06 = 295.06 \text{ m/s}$$

Workdone/kg of steam

$$= V_w V_b$$

$$= 295.06 \times 295.06$$

$$= 87060 \text{ Nm/s} = 87.06 \text{ kJ/s}$$

Energy supplied per kg of steam

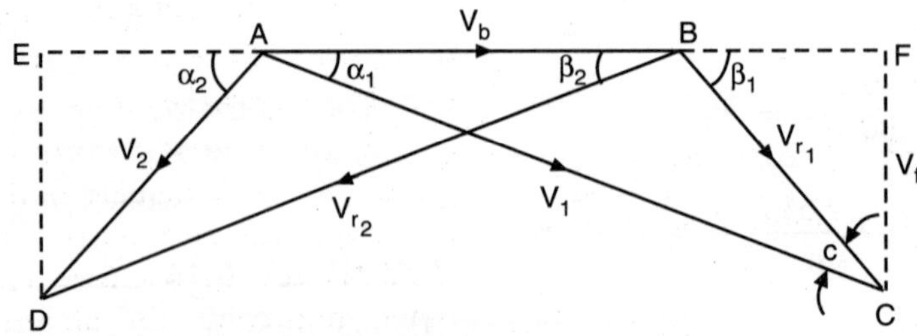

Fig. 19.24

$$= \frac{2V_1^2 - V_{r_1}^2}{2}$$

$$= \frac{2(314)^2 - (107)^2}{2}$$

$$= 92872 \text{ Nm/s} = 92.87 \text{ kJ/s}$$

Diagram efficiency

$$\eta_d = \frac{87.06}{92.87} = 0.937 \text{ or } 93.7\%$$

Percentage increase in diagram efficiency

$$= \frac{93.7 - 90.45}{90.45} = 0.0359 \text{ or } \mathbf{3.59\%}$$

EXAMPLE 19.11

A 50% reaction turbine (with symmetrical velocity triangles) running at 400 rev/min has the exit angle of the blades as 20° and the velocity of steam relative to the blades is 1.35 times the mean blade speed. The steam mass flow rate is 8.25 kg/s and at a particular stage the specific volume is 1.382 m³/kg. For this stage, make calculations for

(a) a suitable blade height if the rotor mean diameter is 10 times the blade height.

(b) the diagram work.

Solution : Refer Fig. 19.25 for the velocity vector diagram

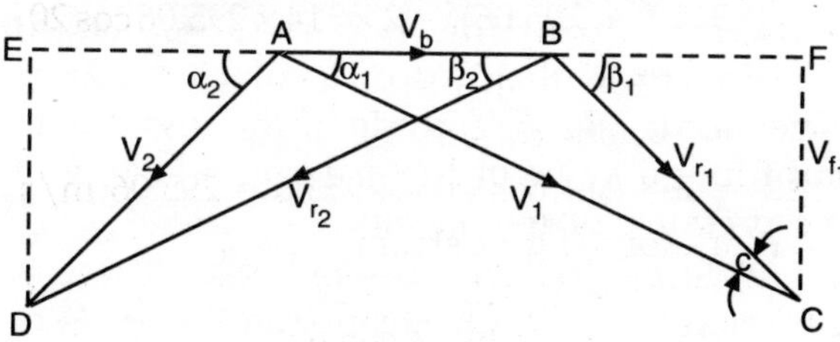

Fig. 19.25.

$$\alpha_1 = \beta_2 = 20°$$

Blade speed

$$V_b = \frac{\pi d N}{60}$$

$$= \frac{\pi d \times 400}{60} = 20.93\, d \text{ m/s}$$

From inlet velocity triangle ABC,

$$\frac{V_{r_1}}{\sin \alpha_1} = \frac{V_b}{\sin C}$$

$$\text{or } \sin C = \frac{V_b}{V_{r_1}} \times \sin \alpha_1$$

$$= \frac{V_b}{1.35 V_b} \times \sin 20 = 0.2533$$

$$\therefore \quad \angle C = 14.68°$$

Blade angle $\beta_1 = 20 + 14.68$

$$= 34.68°$$

Flow velocity

$$V_{f1} = V_{r1} \sin \beta_1 = 1.35\, V_b \sin 34.68$$

$$= 0.768\, V_b = 0.768 \times 20.93\, d$$

$$= 16.07\, d$$

Area of flow $A = \pi\, d\, h$

$$= \pi\, d \times \frac{d}{10} = \frac{\pi d^2}{10}$$

$$\text{Mass flow } m = \frac{A_f V_f}{v} = \frac{\pi d^2}{10} \times \frac{16.07\, d}{1.382}$$

$$= 3.65\, d^3$$

$\therefore$ Mean diameter of rotor

$$d = \left(\frac{m}{3.65}\right)^{1/3} = \left(\frac{8.25}{3.65}\right)^{1/3}$$

$$= 1.312 \text{ m}$$

$$\text{Blade height } h = \frac{1.312}{10} = \mathbf{0.1312 \text{ m}}$$

19.9. IMPULSE TURBINES VERSUS REACTION TURBINES

The salient points of difference between an impulse and reaction steam turbine are restated below:

(1) In impulse turbine the fluid is expanded completely in the nozzle and it remains at constant pressure during its passage through the rotor blades.

In reaction turbine the fluid is only partially expanded in the nozzle, the remaining expansions taking place in the rotor blades.

(2) In impulse turbines, when the steam glides over moving blades, the relative velocity of steam either remains constant or reduces slightly due to friction. However, since in reaction turbine the steam is continuously expanding, relative velocity does increase.

(3) Impulse blades are of the plate or profile type and are symmetrical as shown in Fig. 19.26(*a*). Reaction turbine blades use aerofoil section and are asymmetrical. The blade is thicker at one end (Fig. 19.26(*b*)) and this provides a suitable shaped passage for the steam to expand.

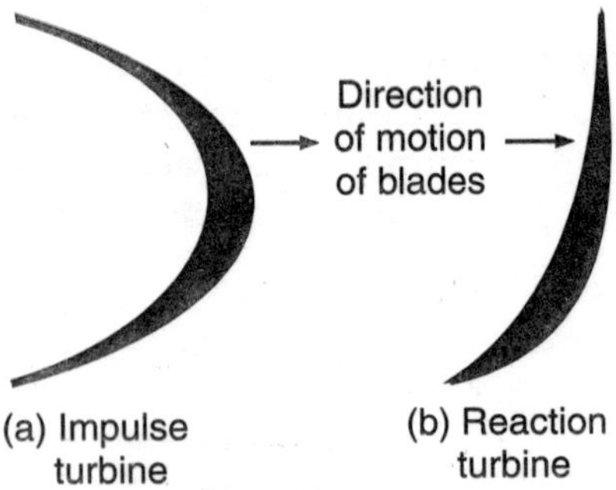

Fig. 19.26. Turbine blading

(4) The blades of an impulse turbine are only in action when they are in front of the nozzles, whereas the blades of the reaction turbine are in action all the time.

(5) Impulse turbines have the same pressure on the two sides of the rotor blades, whereas different pressures exist on the two sides of moving blades of a reaction turbine.

(6) Because of small pressure drop in each stage, the number of stages required for a reaction turbine are much greater than those for an impulse turbine of the same power.

(7) The steam velocity and the blade speed for a reaction turbine are low as compared with those of impulse turbine.

(8) The variation of diagram efficiency with blade speed ratio is shown in Fig. 19.27, both for an impulse turbine and a reaction turbine. From this figure, it is clear that for a reaction turbine the efficiency curve is reasonably flat in the region of maximum value of diagram efficiency. This point is of great significance as the small variations in the blade speed ratio, from that for optimum conditions, can be accepted without having much variation in the value of diagram efficiency.

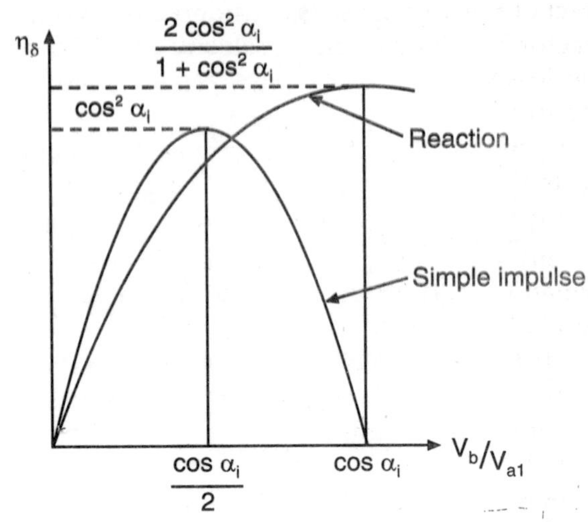

Fig. 19.27. Variation of diagram efficiency for impulse and reaction turbines

19.10. LOSSES IN STEAM TURBINES

Some of the losses which occur within any turbine are as follows :

(1) *Losses in regulating valves.* The steam generated in the boiler has to flow through stop and regulating valves before entering the steam turbine. At these valves, the steam gets throttled and the pressure of steam at entry to the turbine is less than the boiler pressure. Although throttling occurs at constant enthalpy, the enthalpy drop available for work is reduced. The magnitude of the loss due to throttling may be of the order of 5 to 10% of the pressure of live steam.

(2) *Steam friction losses in nozzle.* There is a friction between the steam and the walls of the nozzle and internal friction between the fluid layers itself due to turbulence and eddies formation. The friction results in less available enthalpy drop and consequently the velocity of steam leaving the nozzle is less. The effect of friction is taken into consideration by nozzle efficiency which is defined as the ratio of actual enthalpy drop to isentropic enthalpy drop.

(3) *Steam friction losses in blade passages.* While plotting the velocity triangles for impulse turbine, mention was made of the fact that the steam experiences resistance to its motion when flowing through the blade passages. The

effect of this resistance is considered by taking a factor k defined as blade velocity coefficient. The losses in the blade passages are mainly due to :

(*i*) energy loss which occurs when the steam meets the leading edges of the blade profile,

(*ii*) friction losses in the blades passages,

(*iii*) leaving or wake losses due to the formation of eddies as soon as the steam leaves the blade passage.

(4) *Residual velocity loss*. The steam leaves the turbine with a certain absolute velocity, the equivalent kinetic energy of which is wasted.

In a single stage impulse wheel this loss may be as high as 10-12%. However, in multistage turbines the loss is much less because a part of this energy is utilized when the steam passes to the next stage. When this kinetic energy is passed over to the next stage, it is termed *carry over*.

(5) *Losses due to disc friction and windage.* When the turbine disc rotates, the steam is thrown radially towards the casing due to centrifugal action. A part of kinetic energy is utilized to overcome the friction of steam particles which are being dragged along the surface of blades. In impulse turbines due to partial admission, there occurs turbulence at wheel sections where steam is not being supplied. Moreover, when the blades in turn come in front of the nozzle to receive steam, some work has to be done to clear away the spent steam filling the blade passages. Loss due to these factors is about 10%.

(6) *Loss due to leakage.* The steam leakage losses occur at the following points :

(*i*) between the turbine shaft and the bearings

(*ii*) at the clearance space provided between the diaphragm and the stationary disc in the case of impulse turbines

(*iii*) around the tips of reaction turbine blades.

(7) *Loss due to mechanical friction and bearing.* The friction in the bearings can be kept to a minimum value by a proper lubrication system.

(8) *Losses due to wetness of steam.* Towards the last stages of a multistage condensing turbine, the steam is usually wet and is associated with minute droplets of water. The water and steam have different velocities and donot form a homogeneous mixture. The water particles have to be dragged along with steam and consequently a part of the kinetic energy of steam is lost.

(9) *Losses due to radiation.* Usually the turbines are heavily insulated to reduce the heat loss to surroundings by radiation and as such these losses are negligible.

19.11. STAGE EFFICIENCY, OVERALL EFFICIENCY AND REHEAT FACTOR

With in a turbine, the various losses occur due to:

(*i*) nozzle friction caused by imperfections in nozzles

(*ii*) friction and turbulence in blade passages

(*iii*) disc friction and blade windage.

The energy utilized in overcoming the friction reheats the steam and the quality (dryness fraction) of steam improves. This contributes towards an increase in entropy and reduction in enthalpy drop. The work available from the turbine then becomes less than that due to isentropic expansion.

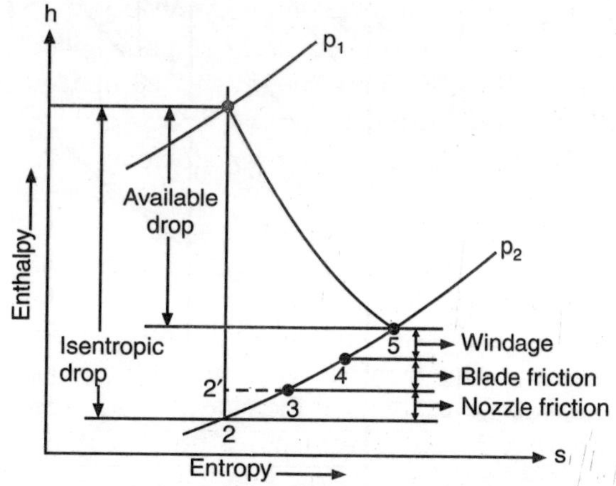

Fig. 19.28. Losses in a steam turbine

Figure 19.28 shows the expansion of steam when passing through the turbine.

(*i*) The expansion of steams takes place in the nozzle. This path 1-2 represents the isentropic expansion in the nozzle, but the actual path of steam in the nozzle is 1-3.

(*ii*) 3-4 shows the loss due blade friction and 4-5 represents the loss due to disc friction and blade windage. The point 5 denotes the actual condition of steams at the outlet from the first stage of the turbine. The stage consists of a set of nozzle and a set of moving blades.

For nozzle: isentropic heat drop = $h_1 - h_2$

Actual heat drop = $h_1 - h_3$

$$\text{Nozzle efficiency} = \frac{h_1 - h_3}{h_1 - h_2} = \frac{V_{a_1}^2 / 2}{h_1 - h_2}$$

Moving blade: work done = $h_1 - h_4$

$$\text{Blade efficiency} = \frac{h_1 - h_4}{h_1 - h_3} = \frac{V_w \times V_b}{V_{a_1}^2 / 2}$$

The gross stage efficiency is defined as the ratio of work done by the moving blades to the isentropic enthalpy drop.

Gross stage efficiency

$$= \frac{h_1 - h_4}{h_1 - h_2} = \frac{h_1 - h_3}{h_1 - h_2} \times \frac{h_1 - h_4}{h_1 - h_3}$$

= nozzle efficiency × blade efficiency

If the losses due to windage, leakage and throttling etc. are also taken into account, then

Net work supplied to turbine shaft = $h_1 - h_5$

Net stage efficiency

$$= \frac{\text{work delivered to shaft}}{\text{isentropic heat drop}}$$

$$= \frac{h_1 - h_5}{h_1 - h_2}$$

Figure 19.29 shows the expansion of steam through a multistage turbine consisting of four stages. The initial pressure is p_1 and the back pressure is p_5.

(*i*) State point 1 represents the initial condition of steam.

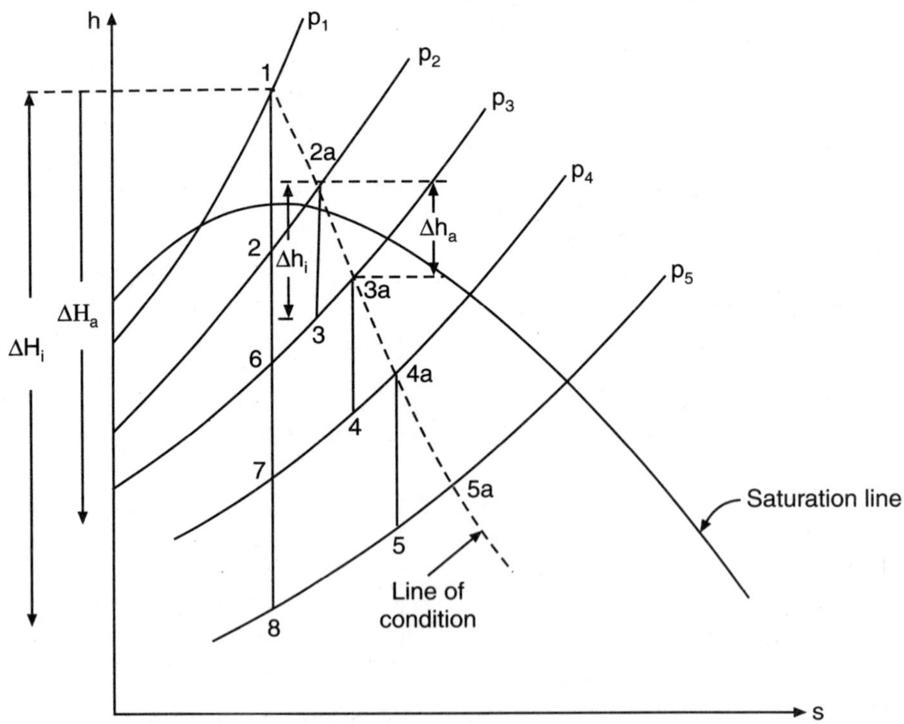

Fig. 19.29.

(ii) The process 1-2 represents the isentropic expansion of steam in the first stage. However, due to various losses, the actual state point of steam after expansion in the first stage is 2*a* and not 2. As such the enthalpy drop available for work is $(h_1 - h_{2a})$ and not $(h_1 - h_2)$.

(iii) The expansion process for the second stage starts from state point 2*a* and not from 2. The path 2*a* – 3 represents the isentropic expansion in the second stage while the actual expansion lies along the path 2*a* – 3*a*.

(iv) The isentropic and actual expansions in the succeeding stages are represented by

3*a* – 4 and 3*a* – 4*a* for third stage

4*a* – 5 and 4*a* – 5*a* fourth stage

(v) The points 1, 2*a*, 3*a*, 4*a* and 5a represent the condition of steam at the blade outlets of different stages. The dotted line passing through these points is called the *line of condition*.

(vi) The sum of isentropic heat drops

$$(h_1 - h_2) + (h_{2a} - h_3) + (h_{3a} - h_4) + (h_{4a} - h_5)$$

is called the cumulative heat drops.

The heat drop $(h_1 - h_8)$ represents the isentropic (Rankine) heat drop for the initial and final pressure limits.

(vii) Overall efficiency is a measure of the degree of perfection of a turbine as a whole in transforming the enthalpy of steam into shaft work.

Overall efficiency

$$\eta_0 = \frac{\text{actual enthalpy drop in the turbine}}{\text{isentropic enthalpy drop}}$$

$$= \frac{h_1 - h_{5a}}{h_1 - h_8} = \frac{\Delta H_a}{\Delta H_i}$$

Considering only one stage (say second stage)

stage efficiency

$$\eta_s = \frac{h_{2a} - h_{3a}}{h_{2a} - h_3} = \frac{\Delta h_a}{\Delta h_i}$$

(viii) The available enthalpy drop can be expressed either in terms of the isentropic enthalpy drop as

$$\Delta H_a = \eta_0 \times \Delta H_i$$

or in terms of isentropic stage enthalpy drop as

$$\Delta H_a = \eta_s (\Delta h_{i1} + \Delta h_{i2} + \Delta h_{i3} + ...)$$
$$= \eta_s \Sigma \Delta h_i$$

From the above two identities for ΔH_a

$$\eta_s \Delta H_i = \eta_s \Sigma \Delta h_i \text{ or } \eta_0 = \eta_s \frac{\Sigma \Delta h_i}{\Delta H_i}$$

The ratio of cumulative heat drop $\Sigma \Delta h_i$ to direct isentropic or Rankine heat drop ΔH_i is called the *reheat factor* (RF)

$$\therefore \quad \eta_0 = \eta_s \times \text{RF}$$

With reference to Fig. 19.29

$$\Delta H_i = (h_1 - h_2) + (h_2 - h_6) + (h_6 - h_7) + (h_7 - h_8)$$

$$\Sigma \Delta h_i = (h_1 - h_2) + (h_{2a} - h_3) + (h_{3a} - h_4) + (h_{4a} - h_5)$$

On the enthalpy-entropy plot, the constant pressure lines diverge as we move from left to right. As such

$$(h_{2a} - h_3) > (h_2 - h_6) \; ; \; (h_{3a} - h_4) > (h_6 - h_7)$$

and $(h_{4a} - h_5) > (h_7 - h_8)$

Consequently the cumulative heat drop is greater than the Rankine heat drop. Conclusion is then made that reheat factor RF has a value greater than unity. Normally RH is taken as 1.04.

Reheat factor depends upon the turbine stage efficiency, the initial pressure and superheat, final pressure and the number of stages. If the number of stages are large for a given pressure range and the stage efficiency is low, the reheat factor becomes large. A lower value of reheat factor is generally desirable.

EXAMPLE 19.12

Steam is supplied to a 50% reaction turbine at 14 bar and 315°C, and it expands to a pressure of 0.14 bar. The turbine has 20 successive stages, all stages develop equal power and the total power output is 11760 kW. Calculate the steam flow rate presuming that the turbine has a stage efficiency of 75% for each stage and the reheat factor is 1.04.

(b) At a certain place in the turbine, the steam has a pressure of 1.05 bar and is dry saturated. The exit angle of the blade is 20° and the blade speed ratio is 0.7 : Determine the rotor speed and the

mean diameter of the annulus at this point in the turbine. It is given that the height of blade is 1/12 of blade mean diameter.

Solution : From Mollier chart,

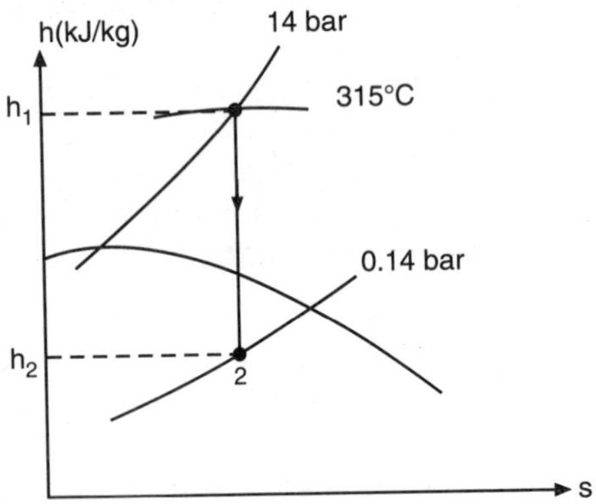

Fig. 19.30.

Isentropic enthalpy drop

$= h_1 - h_2 = 3077 - 2269 = 808$ kJ/kg

Overall efficiency h_0

= stage efficiency × reheat factor

= 0.75 × 1.04 = 0.78

∴ Actual enthalpy drop

= 0.78 × 808 = 630.24 kJ/kg

Power output

= m × enthalpy drop

where m is the steam flow rate in kg/s

$$\therefore \quad m = \frac{11760}{630.24} = \mathbf{18.66} \text{ kg/s}$$

(*b*) With reference to Fig. 19.21.

Whirl velocity

V_w = AF + BF – AB

For 50% reaction turbines, the blades are identical and the velocity triangles are similar and so BF = AF

$$\therefore \quad V_w = 2AF - AB$$

$$= (2V_1 \cos 20 - V_b)$$

$$= \left(2\times\frac{V_b}{0.7}\cos 20 - V_b\right) \qquad \left(\because \frac{V_b}{V_1} = 0.7\right)$$

$$= 1.685\ V_b \text{ m/s}$$

workdone/kg of steam

$= V_w V_b = 1.685\ V_b \times V_b = 1.685\ V_b^2$

This work done equals the enthalpy drop per stage.

$$\therefore\ 1.685\ V_b^2 = \frac{630.24\times 10^3}{20}$$

Solution gives : $V_b = 136.75$ m/s

Flow velocity $V_f = V_1 \sin 20$

$$= \frac{V_b}{0.7} \times \sin 20$$

$$= \frac{136.75}{0.7} \times \sin 20 = 66.82 \text{ m/s}$$

At 1.05 bar, specific volume of dry steam = 1.653 m³/kg

Mass flow rate

$$= \frac{AV_f}{v} = \frac{\pi dh \times V_f}{v}$$

$$\text{or} \quad 18.66 = \frac{\pi d(d/12)\times 66.82}{1.653}$$

∴ Mean blade diameter

$$d = \left[\frac{18.66\times 1.653\times 12}{66.82\times\pi}\right]^{1/2}$$

$$= \mathbf{1.328\ m}$$

Blade speed

$$V_b = \frac{\pi dN}{60}\ ;\ 136.75 = \frac{\pi\times 1.328\times N}{60}$$

∴ Rotor speed

$$N = \frac{136.75\times 60}{\pi\times 1.328} = \mathbf{1967.67} \text{ rev/min}$$

EXAMPLE 19.13

A four-stage steam turbine receives steam at 35 bar and 435°C and exhausts at 0.04 bar. Each stage has an efficiency ratio of 76 percent. If the pressure at the end of stages are 5, 1.2 and 0.25 bar respectively, determine :

(a) Rankine enthalpy drop,

(b) work done,

(c) reheat factor, and

(d) mass flow rate of steam if the turbine is to generate one kilowatt hour of power.

Solution : Refer Fig. 19.31 for the 4-stage expansion process as indicated on the Mollier chart.

1-6 is the isentropic expansion from initial state 1 (35 bar, 435°C) to the exhaust state 2(0.04 bar)

1-2′, 2-3′, 3-4′ and 4-5′ are the isentropic expansions in first second, third and four stage respectively.

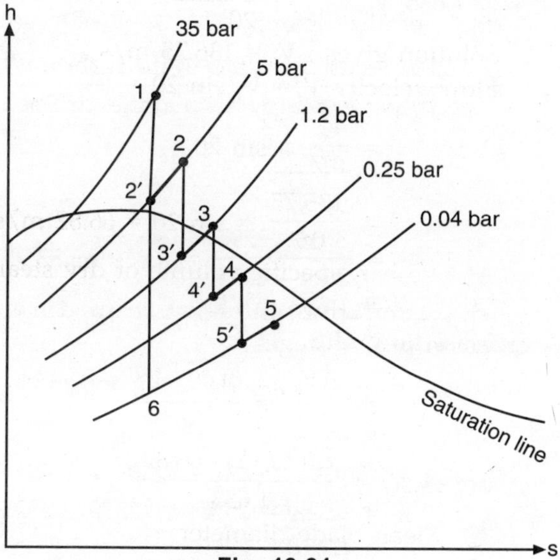

Fig. 19.31.

The enthalpy values of the steam at different points are :

$$h_1 = 3320 \text{ kJ/kg}$$
$$h_6 = 2110 \text{ kJ/kg}$$
$$h_2' = 2820 \text{ kJ/kg}$$
$$h_2 = h_1 - (h_1 - h_2')\,\eta_s$$
$$= 3320 - (3320 - 2820) \times 0.76$$
$$= 2940 \text{ kJ/kg}$$
$$h_3' = 2660 \text{ kJ/kg}$$

and $h_3 = h_2 - (h_2 - h_3') \times \eta_s$
$$= 2940 - (2940 - 2660) \times 0.76$$
$$= 2727 \text{ kJ/kg}$$
$$h_4' = 2486 \text{ kJ/kg}$$

and $h_4 = h_3 - (h_3 - h_4') \times \eta_s$
$$= 2727 - (2727 - 2486) \times 0.76$$
$$= 2544 \text{ kJ/kg}$$
$$h_5' = 2298 \text{ kJ/kg}$$

and $h_5 = h_4 - (h_4 - h_5') \times \eta_s$
$$= 2544 - (2544 - 2298) \times 0.76$$
$$= 2357 \text{ kJ/kg}$$

The calculations made above help to locate the points 1, 2, 3, 4 and 5 on the $h - s$ plot.

(*a*) Rankine enthalpy drop (isentropic heat drop)

$$h_1 - h_6 = 3320 - 2110 = 1210 \text{ kJ/kg}$$

(*b*) Workdone

$$= h_1 - h_5 = 3320 - 2357 = 963 \text{ kJ/kg}$$

(*c*) Isentropic enthalpy drop in each stage is :

$$h_1 - h_2' = 3320 - 2820 = 500 \text{ kJ/kg}$$
$$h_2 - h_3' = 2940 - 2660 = 280 \text{ kJ/kg}$$
$$h_3 - h_4' = 2727 - 2486 = 241 \text{ kJ/kg}$$
$$h_4 - h_5' = 2544 - 2298 = 246 \text{ kJ/kg}$$

Reheat factor

$$= \frac{\text{commulative heat drop}}{\text{total isentropic heat drop}}$$
$$= \frac{500+280+241+246}{1210} = \mathbf{1.047}$$

(*d*) 1 kWh = 3600 kJ/s

∴ Steam required

$$= \frac{3600}{963} = \mathbf{3.738} \text{ kg/s}$$

EXAMPLE 19.14

In a five-stage turbine, the steam is supplied at 17.5 bar and 300°C, and the condition of exhaust steam is 0.04 bar and 0.87 dry. The pressure values and the steam conditions at intermediate stages are as given below :

Stage	*Pressure*	*Condition*
Ist intermediate	*7.5 bar*	*230°C*
2nd intermediate	*2.8 bar*	*145°C*
3rd intermediate	*0.85 bar*	*0.97 dry*
4th intermediate	*0.2 bar*	*0.92 dry*

Trace the condition curve on Mollier chart and determine :

(a) efficiency of each stage (b) over all turbine efficiency and (c) reheat factor

Solution : The expansion process is represented on Mollier diagram in Fig. 19.32. The enthalpy values at different points have also been indicated on the diagram.

(*a*) Stage efficiency is defined as

$$\eta_s = \frac{\text{useful enthalpy change in stage}}{\text{isentropic enthalpy drop in that stage}}$$

$$\therefore \eta_s \text{ (first stage)} = \frac{h_1 - h_2}{h_1 - h_2'} = \frac{3068 - 2905}{3068 - 2845} = 0.731 \text{ or } 73.1\%$$

$$\eta_s \text{ (second stage)} = \frac{h_2 - h_3}{h_2 - h_3'} = \frac{2905 - 2750}{2905 - 2705} = 0.775 \text{ or } 77.5\%$$

$$\eta_s \text{ (third stage)} = \frac{h_3 - h_4}{h_3 - h_4'} = \frac{2750 - 2590}{2750 - 2555} = 0.80 \text{ or } 80\%$$

$$\eta_s \text{ (fourth stage)} = \frac{h_4 - h_5}{h_4 - h_5'} = \frac{2590 - 2420}{2590 - 2375} = 0.791 \text{ or } 79.1\%$$

$$\eta_s \text{ (fifth stage)} = \frac{h_5 - h_6}{h_5 - h_6'} = \frac{2420 - 2235}{2420 - 2210} = 0.881 \text{ or } 88.1\%$$

(*b*) Overall efficiency

$$= \frac{\text{useful enthalpy drop in the turbine}}{\text{isentropic enthalpy drop in the turbine}} = \frac{h_1 - h_6}{h_1 - h_7'} = \frac{3068 - 2235}{3068 - 2075} = 0.839 \text{ or } 83.9\%$$

(*c*) Cumulative enthalpy drop during expansion in five-stages

$$= (h_1 - h_2') + (h_2 - h_3') + (h_3 - h_4') + (h_4 - h_5') + (h_5 - h_6')$$

$$= (3068 - 2845) + (2905 - 2705) + (2750 - 2555) + (2590 - 2375) + (2420 - 2210)$$

$$= 223 + 200 + 195 + 215 + 210$$

$$= 1043 \text{ kJ/kg}$$

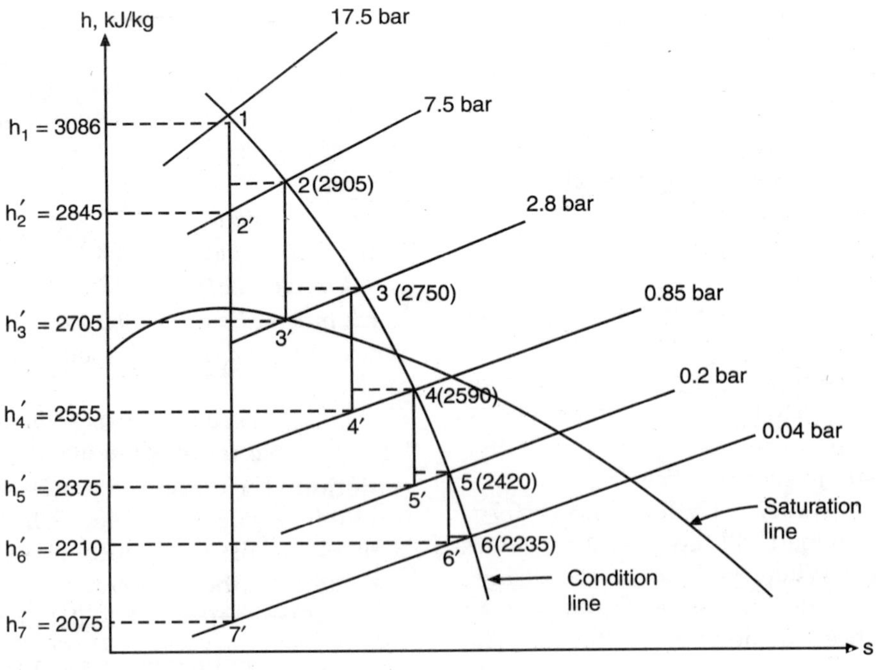

Fig. 19.32.

Rankine (isentropic) enthalpy drop in the turbine,

$$(h_1 - h_7') = 3068 - 2075 = 993 \text{ kJ/kg}$$

∴ Reheat factor

$$= \frac{\text{cummulative enthalpy drop}}{\text{isentropic enthalpy drop}}$$

$$= \frac{1043}{993} = 1.05$$

19.12. STEAM TURBINE GOVERNING

A turbine is designed to operate at the most economic capacity (the capacity at which the turbine would run with minimum of steam consumption). The capacity of the turbine, while under operation may vary between the wide limits of no load to full load. The methods of governing commonly employed for keeping the speed fairly constant are discussed below :

1. Throttle governing. The flow of steam entering the turbine is restricted by a double beat valve which is controlled by a centrifugal governor. As the speed of the turbine rises with a reduction in turbine load, the fly bar moves out-wards. The outward movement of the fly bar is due to the increase in centrifugal force which is there at high speeds. The governor lever arm swings about a pivot and adjusts the valve setting to throttle the steam. Although the enthalpy remains constant, the availability is decreased and original speed is restored. With increase in turbine load the governor operates in the reverse direction.

Because of less initial cost and simple mechanism, throttle governing is most widely used particularly on small units.

2. Nozzle control. Diagrammatic arrangement of such a system in shown in Fig. 19.34.

The nozzles are made up in groups, the supply of steam to each group being controlled by a regulating valve. Under conditions of operation at full load, all the regulating valves are fully open. When the load on the turbine is less than the design value, the supply of steam to a group of nozzles is shut off. The nozzle governing has the advantage of using steam at essentially full boiler pressure with throttling occuring only in the valves which are partly open.

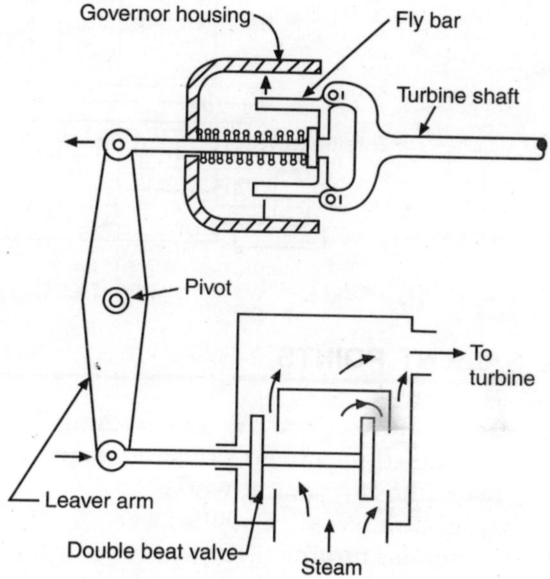

Fig. 19.33. Throttle governing

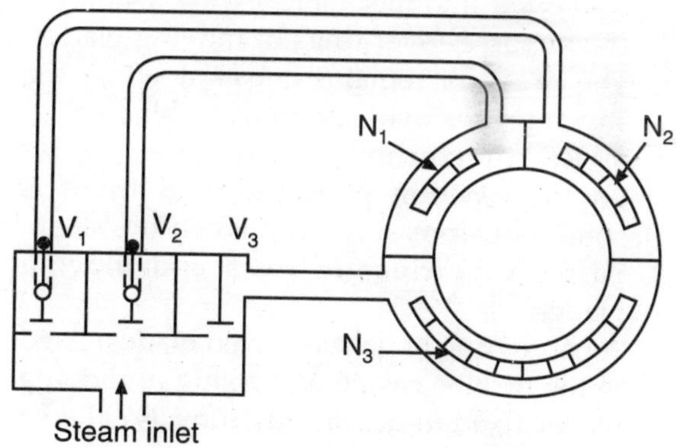

Fig. 19.34. Nozzle governing

3. By-pass governing. This arrangement (Fig. 19.35) is used in conjunction with throttle governing mostly in the case of reaction turbines. Main throttle valve delivers steam to the nozzles of the high pressure stages. To cope with loads greater than those of designed capacity, fresh steam is fed directly to one more intermediate stage of the turbine.

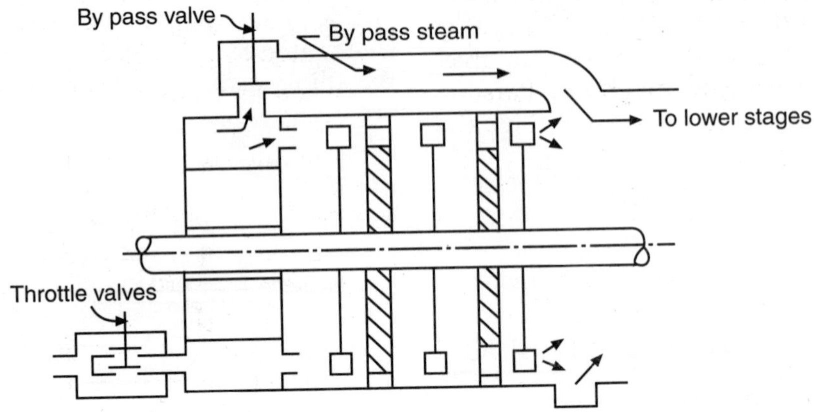

Fig. 19.35. By-pass governing

SALIENT POINTS

1. A steam turbine is a rotating machine that is designed to convert pressure energy of steam into mechanical work.
2. Steam turbines are classifies as :
   (*a*) **Impulse turbine :** The pressure of steam is converted to kinetic energy in a nozzle, and subsequently kinetic energy is converted into mechanical work when the steam passes over rings of moving blades. The pressure remains constant when the steam passes over the moving blades.
   (*b*) **Reaction turbine :** The turbine operates on the reaction principle, and there is continuous pressure drop when the steam passes over rings of fixed and moving blades.
3. A pair of ring of nozzles (fixed blades) fixed to the turbine casing and a ring of moving blades fixed to the turbine rotor is called a stage or turbine pair.
4. For an impulse turbine
   (*i*) When there is friction, V_{r2} is less than V_{r1} and the ratio V_{r2}/V_{r1} is called blade velocity coefficienct K. The value of K varies from 0.73 to 0.85 depending upon the shape of blades.
   (*ii*) Power produced = $m\,(V_{w1} \pm V_{w2})\,V_b$
   (*iii*) Axial thrust = $m\,(V_{f2} - V_{f1})$
   (*iv*) Diagram efficiency =

$$= \frac{\text{Work done by blades}}{\text{energy supplied to blades}}$$

$$= \frac{(V_{w1}+V_{w2})V_b}{V_{a1}^2/2} = \frac{2V_b\,(V_{w1}+V_{w2})}{V_{a1}^2}$$

$$= 2\,(\rho \cos \alpha_1 - \rho^2)\,(1 + KC)$$

   where ρ (blade speed ratio) is V_b/V_{a1}, K is V_{r2}/V_{r1} and C is $\cos \beta_2/\cos \beta_1$.

$$(\eta_d)_{max} = (1 + KC)\,\frac{\cos \alpha_1}{2}$$

   Neglecting friction in blade passages $K = 1$ and considering blades to be symmetrical ($\beta_1 = \beta_2$) and so $C = 1$,

$$(\eta_d)_{max} = \cos^2\alpha_1$$

5. Compounding of impulse turbines is the arrangement done to reduce the rotor speed within practical limits. Methods used are : (*i*) Velocity compounding (*ii*) Pressure compounding (*iii*) Pressure-velocity compounding.
   All these methods consist of a multiple series system of rotors keyed to a common shaft and the steam pressure or jet velocity is absorbed in stages as steam flows over rotor blades.
6. For a reaction turbine
   (*i*) The expansion of steam takes place in the moving blades also, and therefore the relative velocity V_{r2} is greater than V_{r1}.
   (*ii*) The enthalpy drop occurs both in the fixed and moving blades

$$\text{Degree of reaction} = \frac{(\Delta h)\,\text{in moving blade}}{(\Delta h)\,\text{in stage}}$$

For 50% reaction turbine the fixed (stator) and moving (rotor) blades are identical in shape ($\alpha_1 = \beta_2$ and $\beta_1 = \alpha_2$) and that gives symmetrical velocity diagrams. Parson's reaction turbine is 50% reaction turbine and for such a turbine

$$\eta_d = \frac{2\rho(2\cos\alpha_1 - \rho)}{1-\rho^2+2\rho\cos\alpha_1}$$

The efficiency would be maximum when $\rho = \cos\alpha_1$, and

$$(\eta_d)_{max} = \frac{2\cos^2\alpha_1}{1+\cos^2\alpha_1}$$

7. Reheat factor is the performance parameter in multi-stage turbines and is defined as

$$\text{Reheat factor} = \frac{\text{cummulative heat drop}}{\text{isentropic heat drop}}$$

The reheat factor depends upon stage efficiency of the turbine, initial pressure and temperature of steam, exit pressure of steam and number of stages (larger the number of stages, more is the reheat factor). The value of reheat factor varies from 1.02 to 1.06.

8. The governing of steam turbine (maintaining constant speed respective of the load) is done by : (*i*) throttle governing (*ii*) nozzle control governing (*iii*) by-pass governing All these methods control the supply of steam to the turbine.
9. Compared to steam engines, the steam turbines are of smaller size, occupy less space, lesser weight to power ratio, higher output and efficiency, perfect balance and vibration free.

REVIEW QUESTIONS

A. Conceptual and conventional questions.

1. What is the principle of operation of a steam turbine ? State the fundamental difference between an impulse and a reaction turbine.
2. Explain the working of a single stage impulse turbine. Sketch the pressure and velocity variations along the axis of the turbine.
3. Draw the velocity vector diagram of a single stage impulse turbine and set up expressions for the axial thrust, tangential force, work done and diagram efficiency.
4. What is the necessity of compounding a steam turbine? Discuss the salient features of the various methods used for compounding of steam turbines.
5. Sketch and explain the velocity compounded impulse turbine. Indicate the pressure and velocity variations along the turbine axis. Where this system of compounding is used?
6. Make a comparison between the impulse and reaction type steam turbines.
7. Discuss the various losses occuring in a steam turbine.
8. (*a*) Define degree of reaction.
   (*b*) Show that with 50% degree of reaction turbine, the fixed and the moving blades have the same shape and the velocity diagram is symmetrical.
9. (*a*) Define reheat factor. Mention the factors on which it depends.
   (*b*) Set up the relationship between overall efficiency, stage efficiency and reheat factor.
10. (*a*) What is meant by governing ?
    (*b*) Describe briefly the important methods used for the governing of steam turbines.
11. In a single stage impulse turbine, steam leaves the nozzle with a velocity of 1000 m/s inclined at an angle of 20° to the plane of rotation. The mean blade speed is 400 m/s and the blades are symmetrical. Neglecting the effect of friction in the blade passages and taking mass flow rate of steam as 0.75 kg/s, determine : (*a*) blade angles, (*b*) axial thrust, (*c*) tangential force on the blades, (*d*) diagram power, and (*e*) diagram efficiency.
    [Ans. $\beta_1 = \beta_2 = 32.5°$, zero, 810 N, 325 kW and 86.5%]
12. In a single stage impulse turbine, the nozzle is inclined at 20° to the plane of the wheel. The isentropic heat drop for the stage is 450 kJ/kg and the nozzle efficiency is 0.9. Taking blade speed = 300m/s, blade velocity coefficient = 0.85 and blade exit angle = 25°, make calculations for the following parameters for a mass flow of 1 kg/s :

(*i*) inlet angle of moving blade,
(*ii*) axial thrust,
(*iii*) driving force on the wheel and diagram power,
(*iv*) energy lost in the blade passages due to frictions, and
(*v*) the diagram efficiency.

[**Ans.** β_1 = 29.5°, 82.8 N, 1030 N and 308.5 kW, 54.5 kW, 76.2%]

13. For a stage of impulse turbine with single acting wheel and equiangular blades, the nozzle angle is 20°. The velocity coefficient for the blades is 0.83. What is the maximum blade efficiency ?

If the blade efficiency is 90% of maximum value, what are the possible ratios of blade speed to steam speed.

[**Ans.** 81.3%, 0.6165 or 0.3235]

14. Explain the essential differences in the functions of moving blades in the impulse and reaction types of turbines.

The speed of a single stage impulse turbine is 3000 rev/min and the mean blade diameter is 1 meter. The nozzle of the turbine is inclined at 20° to the plane of the wheel and the moving blade inlet and exit angles are 35° and 30° respectively. Assuming a friction factor (blade velocity coefficient) of 0.8, determine :

(*a*) the power developed for a steam consumption rate of 9000 kg/hour,
(*b*) the blade or diagram efficiency ,
(*c*) the axial thrust.

Assume that the steam enters the blades without shock.

[**Ans.** 120 kW, 81.5%, 892 N]

15. At a particular stage of a reaction turbine, the mean blade speed is 60 m/s and the steam pressure is 3.5 bar with a temperature of 175°C. The fixed and moving blades are identical with inlet and outlet angles of 30° and 20° respectively. Determine.

(*a*) the blade height if it is one-tenth of the blade ring diameter and the mass flow rate of stream is 800 kg/min.
(*b*) the power developed by the pair, and
(*c*) the specific enthalpy drop if the stage efficiency is 85%.

[**Ans.** 64 mm, 218 kW, 19.2 kJ/kg]

16. A Parson's steam turbine delivers dry saturated steam at 2.8 bar from the fixed blades at 85 m/s. The mean height of blade is 50 mm and the moving blade exit angle is 20°. The axial-flow velocity of the steam is 75 percent of the blade velocity at the mean radius and the flow rate of steam is 3 kg/s. Neglecting the effect of the blade tip thickness on the annulus area, compute for this stage the rotational speed, power and efficiency, and the enthalpy drop of the steam.

[**Ans.** 1685 rpm, 14.1 kW and 78.6%; 4.68 kJ/kg]

B. Fill in the blanks with appropriate word/words.

1. In a steam turbine embodying impulse principle, the expansion of steam takes place in ....
2. The effect of friction when steam flows across the blade passages is expressed by the ratio of relative velocities at outlet and inlet : $K = V_{r_2}/V_{r_1}$. The factor K is called ....
3. .... is defined as the ratio of work done by blades and the energy supplied to the blades.
4. Neglecting the effect of friction and assuming blades to be symmetrical, the maximum blading efficiency of a single stage impulse turbine in terms of nozzle angle α is ....
5. A velocity compounding turbine has number of .... blade rings separated by rings of .... blades.
6. When steam flows over the blades of a reaction turbine, its pressure .... and velocity ....
7. In a reaction turbine, the steam is only partially expanded in the .... and the remaining expansion takes place in the ....
8. Degree of reaction is defined as the ratio of enthalpy drop in .... to the enthalpy drop ....
9. With .... design, the nozzles and the blades have the same shape and the velocity diagram is symmetrical.
10. For each stage of a Parson's reaction turbine, the velocity ratio for a fixed blade .... the velocity ratio for moving blade.
11. De Laval turbine is .... with single ....
12. The value of reheat factor is of the order of .... to ....

Answers

1. stationary nozzle 2. blade velocity coefficient 3. blading efficiency or diagram efficiency 4. $\cos^2\alpha$ 5. moving, guide 6. decreases, increases 7. nozzle, rotor blades 8. moving blades, per stage 9. 50% degree of reaction 10. equals 11. impulse turbine, single rotor 12. 1.02 to 1.06

C. Multiple choice questions.

1. When steam flows over moving blades of an impulse turbine
   (*a*) pressure drops and velocity increases
   (*b*) pressure remains constant and velocity decreases
   (*c*) both pressure and velocity remain constant
   (*d*) both pressure and velocity decrease
2. The blade speed ratio defines the relation between
   (*a*) relative velocity at outlet and inlet
   (*b*) absolute steam velocity at inlet and outlet
   (*c*) blade speed and absolute steam velocity at inlet
   (*d*) whirl speed and absolute steam velocity at outlet
3. If the nozzle angle is 30 degree, the efficiency of an impulse turbine would be maximum when the blade speed equals
   (*a*) 0.433 V_{a_1} (*b*) 0.5 V_{a_1}
   (*c*) 0.75 V_{a_1} (*d*) 0.866 V_{a_1}
4. A single stage impulse turbine is generally not used in practice because
   (*a*) it cannot generater more power
   (*b*) heat energy is not utilized fully
   (*c*) it has very low efficiency
   (*d*) it needs large reduction gearing due to high speeds
5. Compounding of steam turbine is done to
   (*a*) balance the rotor
   (*b*) reduce the blade friction
   (*c*) reduce the rotor speed
   (*d*) reduce axial thrust
6. Curtis turbine is
   (*a*) single rotor impulse turbine
   (*b*) impulse-reaction turbine
   (*c*) velocity compounded impulse turbine
   (*d*) pressure compounded impulse turbine
7. Theoretically the work of two row Curtis wheel is .... times that of a simple wheel; blade speeds being equal in both the cases
   (*a*) 2 (*b*) 4 (*c*) 6 (*d*) 8
8. What changes occur in the magnitude of pressure and velocity when steam flows through the fixed blades of a reaction turbine?
   (*a*) velocity decreases and pressure remains constant
   (*b*) both pressure and velocity decrease
   (*c*) both pressure and velocity increase
   (*d*) pressure falls and velocity increases
9. Which is the false statement in connection with Parson's reaction turbine ?
   (*a*) the moving and the fixed blades are identical
   (*b*) the velocity diagram is symmetrical about a vertical centre line
   (*c*) the relative velocity of steam either remains constant or reduces slightly when the steam glides over moving blades
   (*d*) the turbine has 50% degree of reaction
10. In Parson's reaction turbine, the velocity diagram triangles at inlet and outlet are
    (*a*) asymmetrical (*b*) isosceles
    (*c*) right angled (*d*) congruent
11. In Parson's reaction turbine, the maximum blading efficiency is obtained when
    (*a*) $\rho = \dfrac{\cos\alpha}{2}$ (*b*) $\rho = \cos\alpha$
    (*c*) $\rho = \cos^2\alpha$ (*d*) $\rho = \dfrac{1+\cos\alpha}{2}$
    where ρ is the blade speed ratio and α is the nozzle angle.
12. Match List I (turbine) with List II (classification)
    A. Parson's 1. Pressure compounded
    B. De Laval 2. Reaction
    C. Rateau 3. Simple impulse
    D. Curtis 4. Velocity compounded
13. Throttle governing in stream turbines
    (*a*) leads to significant pressure loss
    (*b*) increases the efficiency
    (*c*) increases heat losses
    (*d*) decreases steam temperature
14. In an impulse-reaction turbine, the heat drops in the fixed and moving blades are 15 kJ/kg and 30 kJ/kg respectively. The degree of reaction for this stage will be
    (*a*) 1/3 (*b*) 1/2 (*c*) 2/3 (*d*) 3/4
15. The ratio of cumulative enthalpy drop to isentropic enthalpy drop is known as
    (*a*) overall efficiency
    (*b*) degree of reaction
    (*c*) gross stage efficiency
    (*d*) reheat factor

Answers

1. (*b*)	2. (*c*)	3. (*a*)	4. (*d*)	5. (*c*)
6. (*c*)	7. (*b*)	8. (*d*)	9. (*c*)	10. (*d*)
11. (*b*)	12. (A-2, B-3, C-1, D-4)			13. (*a*)
14. (*c*)	15. (*d*)			

❑❑❑

Chapter **20**

Steam Condensers

Learning objectives : Attention has been focussed in this chapter to make the reader conversant with :

- *steam condenser and its utility*
- *elements of a condensing plant and types of condensers*
- *condenser vacuum, vaccum efficiency and condenser efficiency*
- *air leakage and loss of vacuum, construction and working of Edward air pump*
- *cooling towers and cooling ponds*

20.1. CONDENSER AND ITS UTILITY

The efficiency of a steam power plant working on Carnot cycle is given by

$$\eta = \frac{T_1 - T_2}{T_1} = 1 - \frac{T_2}{T_1}$$

where T_1 is the temperature at which heat is supplied and T_2 is the temperature at which heat is rejected. This expression reveals that for obtaining maximum efficiency, it is necessary that the difference between the temperature T_1 and the temperature T_2 should be as large as possible. The temperature T_1 can have the maximum optimum value consistent with metallurgical considerations and so for maximum efficiency, the temperature T_2 should have the minimum value. There is a definite relation between steam temperature and pressure. Low exhaust temperature means low exhaust pressure. Steam cannot be exhausted to atmosphere if it is expanded in the steam turbine or engine to a back pressure which is lower than the atmospheric pressure. However, the steam by exchanging heat with water in a vessel can be condensed resulting in a fall in its temperature and pressure. However, the condensation of steam will cause reduction in pressure only when water is contained in a closed vessel. With the vessel open to atmosphere the condensation will of course be there but the pressure will not drop below the atmospheric pressure.

The closed vessel in which steam is condensed by abstraction of heat and in which vacuum is maintained is called *condenser*.

Condensation of steam enables expansion of steam to a lower back pressure. The available enthaply drop increases. Consequently more work is done and the plant efficiency improves.

Due to increased cost involved in the creation and maintenance of higher vacuum, a limit is imposed beyond which the reduction of back pressure does not prove economical. For steam engines the lowest practical exhaust pressure is 65 cm of mercury vacuum. The steam engines being of positive displacement have to provide piston displacement equivalent to volume of exhaust steam and as such the cylinder dimensions impose a limit on the engine condenser pressure. Steam turbines being steady flow machines may be designed to operate at 73.25 cm of mercury vacuum or even more depending upon the capacity of the plant and cooling water available.

The hatched areas in Fig. 20.1(*a*) and 20.1(*b*) represent the increase in work done by a steam engine and a steam turbine by exhausting the steam into a condenser.

The following are the advantages obtained by incorporating a condenser in a steam engine or steam turbine plant :

(1) Improvement in the efficiency of the plant due to increased available enthalpy drop.

(2) Reduction in steam consumption per kW/hour. Increase in vacuum from 71 to 73.5 cm. of Hg gives about 45% reduction in steam consumption.

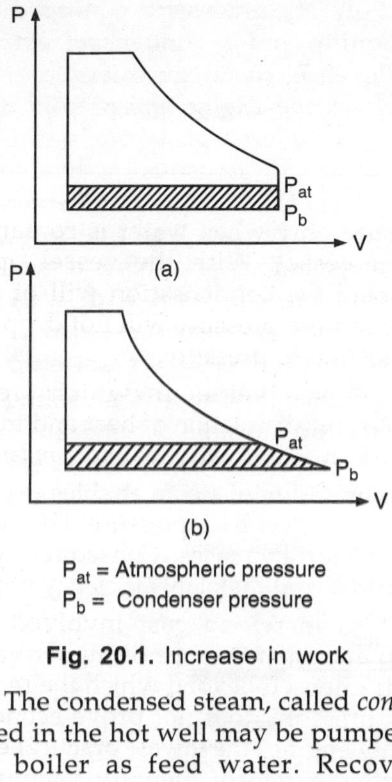

P_{at} = Atmospheric pressure
P_b = Condenser pressure

Fig. 20.1. Increase in work

(3) The condensed steam, called *condensate* collected in the hot well may be pumped back to the boiler as feed water. Recovery of condensate reduces the make up water that must be added to the system from 100% when non-condensing to 1.5% when condensing. For steam power plants where sufficient quantity of good quality boiler feed water is not available, recovery of condensate is very important. For marine practice where sea water is treated before being used in the boiler, recovery is a necessity.

(4) To prevent the encrustation of boiler, the feed water if not available in pure form has to be treated first in the water softening plant. The recovered condensate reduces the capital and running cost of the water softening plant.

(5) Provision for the supply of hot water to the boiler results in fuel economy and safety from thermal stresses.

20.2. ELEMENTS OF A CONDENSING PLANT

Fig. 20.2 shows the essential elements which comprise a condensing unit:

(1) A closed vessel in which steam is condensed.

(2) A condensate pump to extract the condensed steam from the condenser and feed it to hot well.

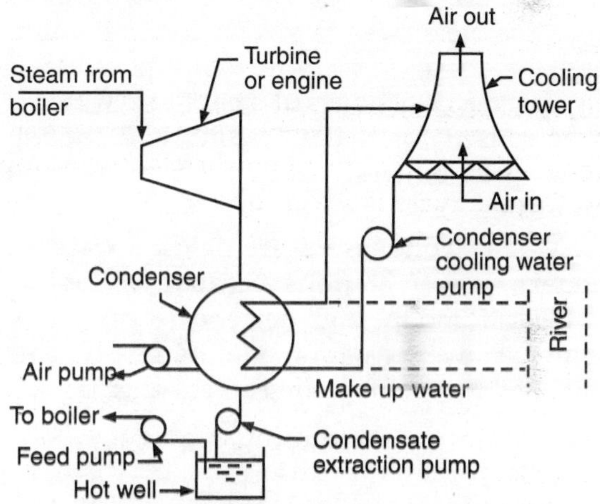

Fig. 20.2. Elements of a condensing unit

(3) A dry air pump to remove air and non-condensable gases. Sometimes a single pump known as wet air pump serves to remove both air and condensate.

(4) A feed water pump to force the condensate from the hot well to the boiler.

(5) A cooling water pump for circulating cooling water.

(6) An arrangement such as cooling tower or spray pond to recool the circulating water after it gets heated in the condenser. This is necessary when the supply of cooling water is scarce and same water has to be used over and over again for cooling purposes.

(7) An atmospheric relief valve for relieving the pressure in the condenser when the condenser does not function properly. The steam then escapes through the valve and engine operates as non-condensing.

20.3. TYPES OF CONDENSERS

The condensers can be classified into two groups namely jet condensers and surface condensers. These two types are discussed briefly in the following sections.

(A) *Mixing* or *jet condensers*. There is a direct contact between the steam and cooling water and the heat exchange is by direct conduction. The steam quickly condensers in water introduced in the form of spray or jet and the stream of warm water (condensed steam + water) is continuously withdrawn. Since the steam escapes with cooling water, the recovery of condensate is not possible. If the mixture of condensed steam and water is to be reused for boiler feeding, the cooling water must be fresh and free from harmful impurities.

Mixing condensers are seldom used in modern power stations because of loss of condensate and the high power of jet condenser pumps. However with moderate size turbine units and for reciprocating steam engines, the jet condensers are used especially where an abundant supply of good feed water is available.

The jet condensers are divided into:

(*a*) Parallel flow in which the steam and the cooling water flow are in the same direction.

(*b*) Counter flow in which steam flows in the opposite direction to the cooling water.

Depending upon the arrangement of the removal of condensate, the jet condensers are further subdivided into the following three categories:

(*i*) *Low level jet condenser*. A low water level parallel flow jet condenser is equipped with a dry air pump and a condensate extraction pump (Fig. 20.3). The mixture of exhaust steam and air from the engine enters from the top of the condenser and while descending downwards comes in contact with a spray of cooling water. The cooling water enters from the top and is broken into fine streams by suitable arranged baffles. This is done to have intimate mixing of steam with water. The condensate gets progressively cooled as it goes down and this reduces the plant efficiency.

In a counter flow type, the cooling water enters at the top, cascades downwards through a series of perforated trays. The spent steam enters near the bottom. In this arrangement

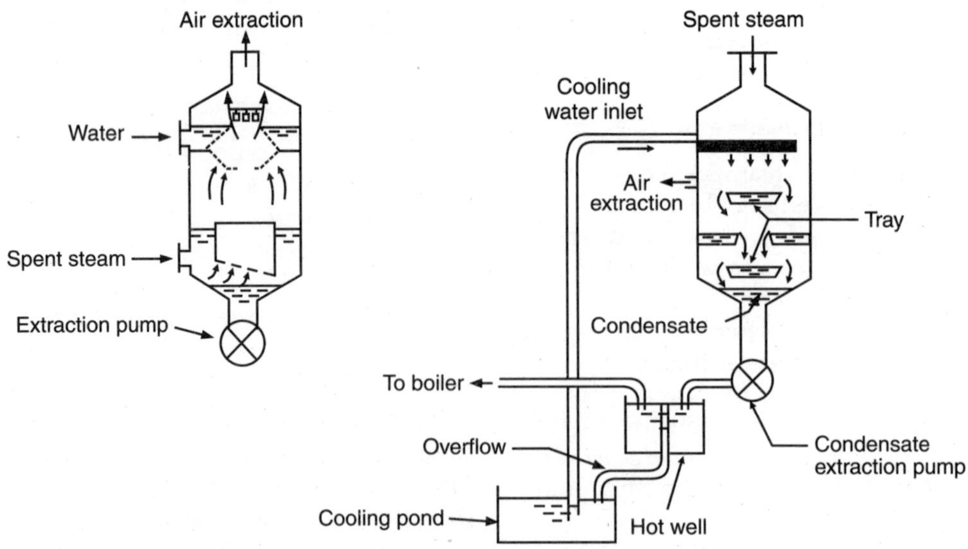

Fig. 20.3. Jet condensers (Low-Level)

there is no under-cooling of condensate and the temperature of the condensate may approach that of the incoming steam. To bring about condensation of steam less quantity of cooling water will be required. With air extraction at the top, the air while ascending up increases in density due to progressive cooling and this reduces the required capacity of air extraction pump.

The vacuum created by air pump is sufficient to draw the cold water into the condenser shell and as such there is no need of cooling water pump.

(*ii*) *Barometric* or *high level jet condenser*. This high level jet condenser (Fig. 20.4) is similar to low level jet condenser except that a column of water in the tail pipe about 10.36 m high forces the condensate to drain away by gravity into the hot well.

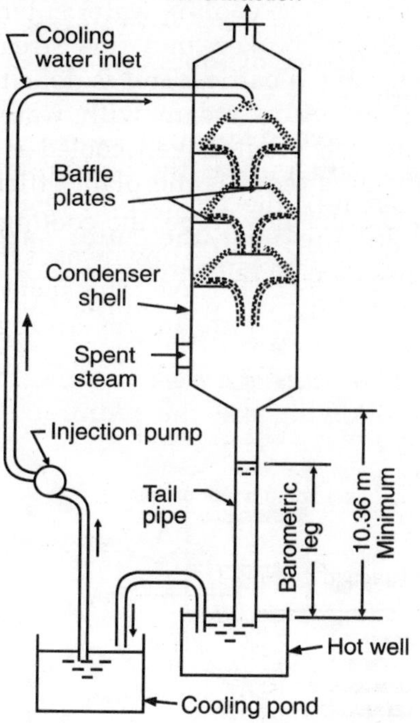

Fig. 20.4. Barometric jet condenser

Theoretically the height *H* of the tail pipe is given by :

H = (Atmospheric pressure
– condenser pressure)
+ friction loss in tail pipe
+ water velocity head

Unless a supply of fresh water under pressure is available, an injection pump has to be installed to pump the cooling water into the condenser shell. Its installation is appropriate only where enough head required for tail pipe is available.

(*iii*) *Ejector condenser*. The distinctive feature of ejector condenser (Fig. 20.5) is that a momentum of flowing water removes the mixture of condensate and cooling water against the atmospheric pressure.

The cooling water under a head of about 6 m enters the condenser from the top and passes through a series of convergent nozzles. At the nozzle throat the velocity increases with a corresponding reduction in pressure. This reduction in pressure draws the spent steam into the nozzle through ports and provides intimate mixing. The steam enters through non-return valve so that in the event of failure in the supply of injection water, there is no sudden rush of water from the hot well into the engine.

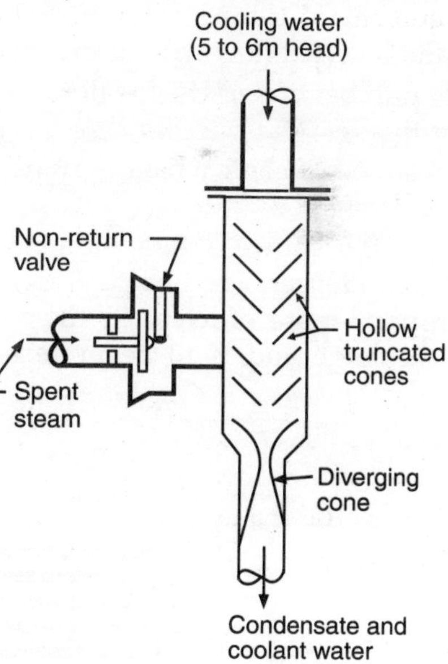

Fig. 20.5. Ejector condenser

In the diverging cone, the kinetic energy is partially transformed into pressure energy. This pressure is enough to overcome the resistance of the atmospheric pressure and so the water discharges automatically into the hot well.

The condenser is used for moderate vacuum and dispenses with extraction pump.

(B) *Surface condensers*. The heat is convectively transferred through a wall interposed between steam and water. The steam is drawn across a nest of tubes which are arranged in certain pattern and are maintained at a temperature lower than that of steam by a flow of cooling water through them.

Because the steam and circulating water do not mix, the condensate can be recovered. Moreover cooling water need not be up to a high standard of purity. This fact is of great advantage for ships which can carry only a limited amount of pure treated water for steam raising and use sea water for cooling purposes. All the marine installations are equipped with surface condensers.

The surface condensers may be classified according to:

(*a*) direction of flow of condensate and arrangement of tubing : down flow, central flow and inverted flow type surface condensers

(*b*) number of passes of water : single pass or multipass

(*c*) shape of shell which may be circular, oval or *U* shaped

Fig. 20.6 shows a two pass down flow surface condenser. The two pass arrangement is compact, more efficient in the process of heat exchange and is to be preferred when the supply of cooling water is limited. The suction of the extraction pump, installed at the bottom, causes the spent steam entering from the top to flow downwards over a nest of tubes. The cooling water enters at one end of the bottom-set of tubes, flows through them till it reaches the other end of the shell. The water then rises up and flows in the opposite direction through the upper half of the tubes and finally leaves through the outlet.

A section of tubes near the air pump suction is screened off by providing a baffle [Fig. 20.7(*a*)] This is done to reduce the amount of water vapour going along with air. Moreover the lowest temperature maintained in this section increases the density of air and so we need an extraction pump of a small capacity.

In the central flow type surface condenser [Fig. 20.7(*b*)] the air extraction pump is placed in the centre of the tube nest. The steam flowing radially towards the centre passes over the entire periphery of tubes and is extracted at the bottom.

In the inverted flow type surface condenser, the steam enters near the bottom and flows upwards since the air suction is at the top. After flowing near the outer surface the condensed steam falls downwards and leaves from the bottom where the condensate extraction pump is installed.

Requirements of a modern surface condenser : The following are the requirements of a

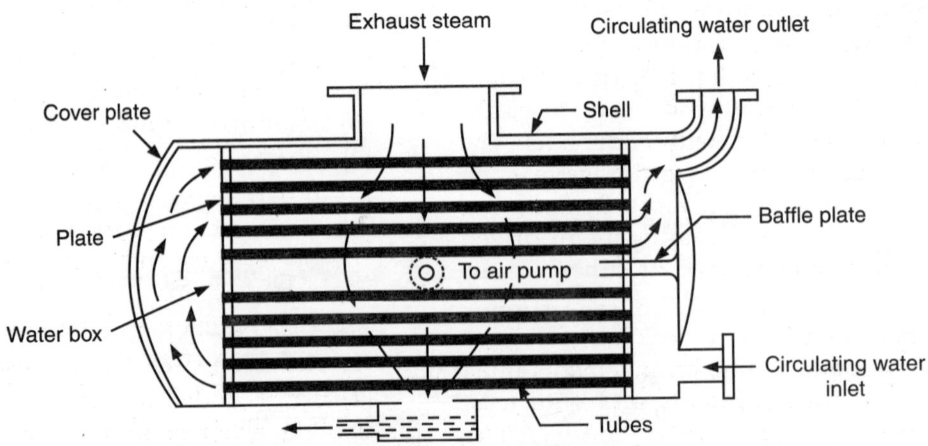

Fig. 20.6. Two-pass downflow surface condenser

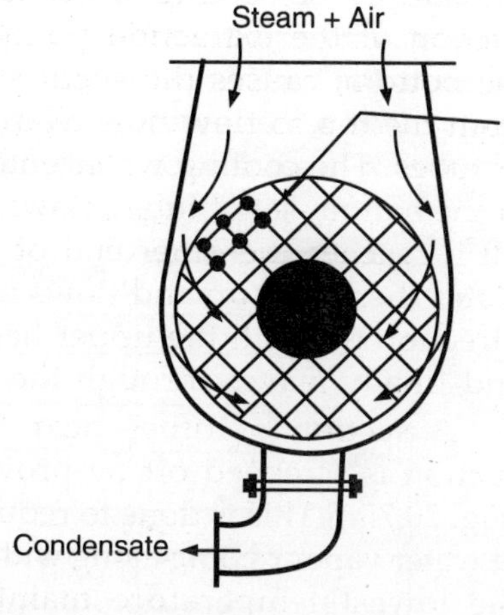

Fig. 20.7. (a) Central flow surface condender

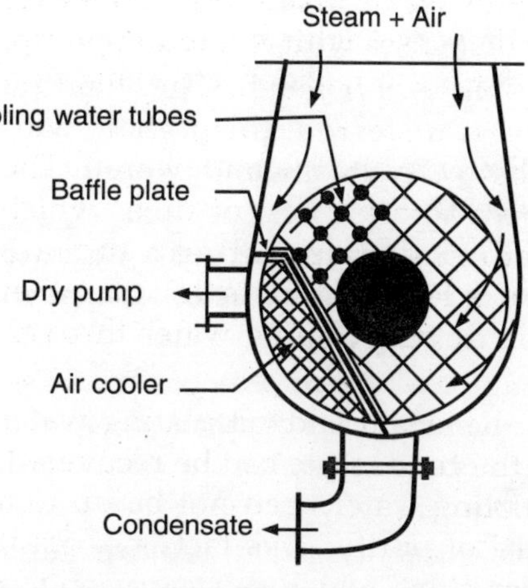

Fig. 20.7. (b) Down flow surface condender

modern surface condenser, sometimes referred to as an ideal condenser :

(1) The steam should be well distributed in the vessel and the pressure drop should be minimum.

(2) The steam should enter the condenser with least possible resistance.

(3) Circulating water should pass through the surface condenser with least friction but at a velocity consistent with high efficiency.

(4) The steam should lose only its latent heat and there should be no condensate depression (undercooling of condensate). To achieve this the quantity of water circulating through the tubes should be so regulated that temperature of leaving water is the same as the saturation temperature of steam.

(5) The water is to flow inside the tubes and vapour outside so that outside surface of the tubes (which is rather difficult to clean) does not get deposited with sediments. The cooling water, if dirty, would leave such deposits on the inside of the tubes. By removing the end cover plates and passing motor driven brushes, the inside surface of tubes can be cleaned.

(6) There should be no leakage of air into the condenser and if any, arrangements must be made to extract it rapidly and with least expenditure of mechanical energy. The presence of air in the condenser would destroy the vacuum and will hamper the rate of heat transfer from the steam to the coolant because of its poor conductivity.

(7) The air extraction should be at the coolest section of the tubes and the air exit should be shielded from the down flowing condensate by means of a baffle. This is to extract the air with a comparatively much smaller amount of water vapour, *i.e.*, without entailing much loss of potential condensate.

20.4. COMPARISON OF JET AND SURFACE CONDENSERS

Jet condenser : A jet condenser has the following advantages :

(1) More intimate mixing of steam and cooling water.

(2) Requires less quantity of circulating water to affect the steam condensation.

(3) Equipment simple and low in cost.

(4) Less building space needed.

(5) In barometric and ejecter condensers, condensate extraction pump is dispensed with and the low level jet condenser does not require cooling water pump.

The disadvantage of jet condenser are :

(1) There is waste of condensate.

(2) If the condensate has to be salvaged, the cooling water should be clean and free from harmful impurities.

(3) With low level jet condenser, there is greater possibility of engine being flooded in the event of failure of condensate extraction pump.

(4) The piping to and from the barometric condenser is costly.

(5) With barometric condenser a vacuum loss (1–1.5 cm of Hg) occurs due to leakage in the long exhaust pipe line.

(6) Vacuum seldom exceeds 66 cm of Hg as the cooling water always contains dissolved air which gets liberated under the condenser vacuum conditions.

(7) The air extraction pump requires high power which may be about two times of that required for surface condenser.

Surface condenser : A surface condenser has the following advantages :

(1) High vacuum can be attained and greater plant efficiency is achieved.

(2) Any kind of cooling water can be used and so the cost of water softening plant is considerably reduced.

(3) Chances of losing vacuum are minimum because water supply is not affected by drop in vacuum.

(4) The condensate can be salvaged and used for boiler feed.

(5) The arrangement can be conveniently adapted to weigh the condensate for tests in laboratory.

(6) Suitable for high capacity units.

The surface condenser is, however, bulky and so requires considerable floor space for its erection. Moreover, the unit has a high capital and maintenance costs.

20.5. EVAPORATIVE CONDENSER

The steam to be condensed enters a coiled finned pipe system at A (Fig. 20.8). The water from a cooling pond is pumped to a horizontal header C which is fitted with spray nozzles. The cooling water sprayed over the finned tubes gets evaporated. The steam loses its heat both to the cooling water and the current of air circulating over the water film. The heated air moves upwards carrying along with it a portion of cooling water evaporated into vapour. The remainder of the cooling water falls into the cooling pond and the loss of water evaporated is replenished by the addition of a requisite quantity of cold make up water.

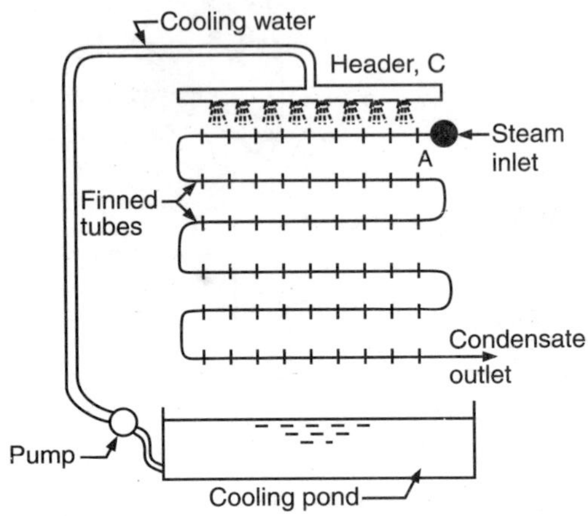

Fig. 20.8. Evaporative condenser

The arrangement is simple, cheap, does not require large quantity of cooling water and so needs a cooling water pump of small capacity.

The vacuum created is, however, not so high as in the surface condenser. Moreover the cloud of evaporated water is also a nuisance to the surroundings.

20.6. CONDENSER VACUUM

The vacuum obtainable from a condenser depends upon tightness of valves and joints, amount of air infiltration and the temperature of steam after condensation. The vacuum is not uniform throughout the condenser, being least at the air pump suction, high in the body of the condenser and still higher at the engine exhaust valve.

The degree of vacuum measured by means of a vacuum gauge (Fig. 20.9) can be expressed in the following different ways :

(*a*) Excess of atmospheric pressure over observed vacuum. A 65 cm of vacuum means that the pressure of atmosphere is 65 cm of Hg above the condenser pressure.

(*b*) Percentage of vacuum, *i.e.*, ratio of observed vacuum to atmospheric pressure. If the gauge reads 65 cm of Hg with the barometer reading standing at 75 cm, then percentage vacuum

$$= \frac{65}{75} \times 100$$

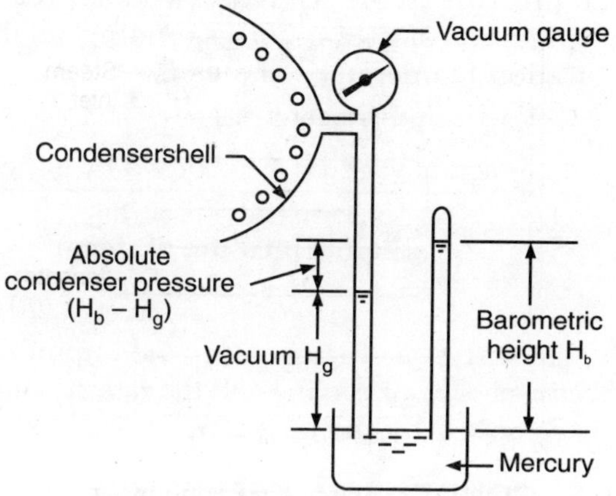

Fig. 20.9. Vacuum measurement

(*c*) Absolute pressure

Let H_b = barometric height in cm of Hg

H_g = vacuum gauge reading in cm of Hg

Then absolute pressure in the condenser

$$= (H_b - H_g) \text{ cm of Hg}$$

The vacuum in the condenser is thus a function of both absolute pressure as well as the barometric pressure. The barometric height is a variable quantity, changing from place to place. Accordingly the gauge reading would also vary, if the absolute pressure is to remain constant. For purposes of comparison it is more convenient to refer the vacuum readings to a standard barometric height of 76 cm of Hg.

$$\therefore \text{ Corrected vacuum} = 76 - (H_b - H_g) \quad ...(20.1)$$

Standard atmospheric pressure

$$= 76 \text{ cm of Hg} = 1.01325 \text{ bar}$$

$\therefore$ Pressure equivalent of 1 cm of Hg

$$= \frac{1.01325}{76} = 0.01333 \text{ bar}$$

20.7. LOSS OF VACUUM AND AIR LEAKAGE

The loss of vacuum in a condenser may be caused by :

(*i*) Air infiltration

(*ii*) Reduced circulation of cooling water

(*iii*) Scale or lime accumulation on the surface of tubes

(*iv*) Plugging of ejector jet

Because of low pressure in the condenser, the air infiltrates into the system and the condenser is thus always filled with a mixture of water, steam and air. The sources of air in a condenser are:

(*i*) Leakage through packing glands and microscopic holes in the shell and the joints

(*ii*) Leakage through vents from atmospheric relief valve and other accessories

(*iii*) In jet condenser, the air enters with injection water and gets liberated at low pressure

(*iv*) The air contained in spent up steam is also liberated at low pressure.

In a well designed and properly maintained surface condenser, the amount of air leakage is about :

= 5 kg per 10,000 kg of steam condensed for a steam turbine plant.

= 15 kg per 10,000 kg of steam condensed for a steam engine plant.

The presence of air in the condenser affects its performance in the following ways :

(1) Back pressure in the stream power plant increases and correspondingly the work output decreases.

(2) For the same absolute pressure the partial pressure of steam decreases with air infiltration. From steam tables, we see that at lower pressures the steam has more latent heat. To remove this greater quantity of heat, more cooling water has to be supplied.

(3) Because of poor thermal conductivity of air, the rate of heat transfer from the vapour is reduced and the surface area of the tubes has to be increased for a condenser duty.

(4) The abstraction of heat by the cooling water is partly from steam and partly from air. This aspect also reduces the rate of steam condensation.

(5) For the maintenance of proper vacuum, we have to install an air extraction pump which would remove the air continuously from the shell. Inspite of shielding the air extraction section, some steam does escape with air resulting in loss of high temperature feed water to the boiler. Moreover, the condensate is under-cooled and more heat has to be supplied to water in the boiler. Loss of condensate and its undercooling lowers the overall plant efficiency.

The following procedure is adopted to check whether there is air leakage in the condenser.

(*i*) Run the plant until the temperature and pressure conditions are steady in the condenser.

(*ii*) Isolate the condenser by shutting off steam supply and simultaneously closing the condensate and air extraction pumps.

If there is a leakage, the readings of vacuum gauge and thermometer will record a fall.

Source of air leakage can be checked by adopting the following methods:

(*i*) The condenser is put under air pressure and its effect on soap water is noted at the points where infiltration is likely to occur.

(*ii*) When the condenser is operating, peppRament oil is put on the suspected joint and a check is made on the peppRament odour in the air ejector discharge.

(*iii*) With condenser under vacuum, large leakages can be detected by passing a candle flame over possible openings.

20.8. VACUUM EFFICIENCY

The vacuum efficiency is a measure of the degree of perfection in achieving the aim of maintaining a desired vacuum in condenser. The vacuum efficiency may be defined as the ratio of actual vacuum as recorded by the vacuum gauge to the ideal vacuum.

$$\text{Vacuum efficiency} = \eta_{vac} = \frac{\text{actual vacuum as recorded by gauge}}{\text{ideal vacuum}} \quad ...(20.2)$$

The ideal vacuum means the vacuum due to steam alone when air is absent. In that case total pressure in the condenser will approach the pressure of steam corresponding to the saturation temperature of steam.

$$\therefore \text{ Vacuum efficiency } \eta_{vac} = \frac{\text{actual vacuum as recorded by gauge}}{\text{barometric pressure} - \text{absolute pressure of steam}} \quad ...(20.3)$$

The vacuum efficiency depends upon the effectiveness of air cooling and the rate at which it is removed by the air pump.

20.9. CONDENSER EFFICIENCY

The purpose of an ideal condenser is to remove only the latent heat so that temperature of condensate equals the saturation temperature corresponding to condenser pressure. In other words there is no under cooling of condensate. Further, the maximum temperature to which the cooling water can be raised is the condensate temperature. The condenser efficiency is then defined as the ratio of actual rise in the temperature of cooling water to the maximum possible rise.

$$\text{Condenser efficiency} = \frac{t_2 - t_1}{t_3 - t_1} \quad ...(20.4)$$

where t_3 = saturation temperature corresponding to condenser pressure, t_2 and t_1 are the outlet and inlet temperatures of cooling water.

20.10. DALTON'S LAW AND AIR-VAPOUR MIXTURE

Dalton's law forms the basis for the analytical treatment of the problems dealing with mixture of gases or of gas and vapour. The law states:

"The total pressure exerted by a mixture of gases or a mixture of gas and vapour (which have

no chemical action on each other) is equal to the sum of partial pressure of the constituents".

The partial pressure of each constituent of the mixture is the pressure exerted by the constituent taken separately if the quantity of the constituent occupies alone the same volume as that of the mixture and at the same temperature. Thus each constituent behaves as if it occupies the volume alone and is independent of the presence of other constituent.

In a condenser, there is a mixture of steam and air leaking into the condenser.

∴ Pressure of mixture = partial pressure of steam + partial pressure of air

$$p = p_s + p_a \qquad ...(20.5)$$

The total pressure p is measured by a vacuum gauge, *i.e.*, p = (barometric reading– gauge reading). If the condenser temperature is known then p_s can be read from the steam tables. By difference the partial pressure of air p_a can be obtained.

EXAMPLE 20.1

The following observations were recorded during a test on a steam condenser :

Recorded condenser vacuum
= 71 cm of Hg

Barometric reading = 76.5 cm of Hg

Mean condenser temperature
= 34 °C

Temperature of hot well
= 28.5°C

Condensate collected = 1800 kg/hour

Flow rate of cooling water
= 57500 kg/hour

Inlet temperature of cooling water
= 8.5°C

Outlet temperature of cooling water
= 26°C

Calculate: (a) vacuum corrected to standard barometer of 76 cm of Hg, (b) vacuum efficiency, (c) under cooling of condensate, (d) condenser efficiency, (e) state of steam entering the condenser and (f) mass of air present per cubic meter of condenser volume and the mass of air present per kg of uncondensed steam.

Solution : (*a*) Corrected vacuum

$$= 76 - (H_b - H_g)$$
$$= 76 - (76.5 - 71) = \mathbf{70.5\ cm\ of\ Hg}$$

(*b*) From steam tables, the absolute pressure of steam corresponding to condenser temperature of 34°C

$$= 0.053 \text{ bar}$$
$$= \frac{0.0532}{0.01333} = 3.99 \text{ cm of Hg}$$

Vacuum efficiency η_{vac}

$$= \frac{\text{actual vacuum as recorded by gauge}}{\text{barometric pressure} - \text{absolute pressure of steam}}$$

$$= \frac{71}{76.5 - 3.99}$$

$$= \frac{71}{72.51} = 0.979 \simeq \mathbf{98\%}$$

(*c*) Condensate under-cooling

= condenser temperature – temperature of hot well

$= 34 - 28.5 = \mathbf{5.5°C}$

(*d*) Absolute condenser pressure

= barometer reading – vacuum reading

$= 76.5 - 71 = 5.5$ cm of Hg

$= 5.5 \times 0.01333 = 0.0733$ bar

Saturation temperature corresponding to 0.0733 bar = 39.95°C

Therefore the maximum temperature to which cooling water can be raised is 39.95°C.

Condenser efficiency

$$= \frac{\text{actual rise in temperature of cooling water}}{\text{maximum permissible temperature rise}}$$

$$= \frac{26 - 8.5}{39.95 - 8.5} = \mathbf{55.7\%}$$

(*e*) From steam tables for a pressure of 0.0733 bar

$$h_f = 167.23 \text{ kJ/kg}$$
and $$h_{fg} = 2406 \text{ kJ/kg}$$

The enthaply of condensate (h_c) corresponding to hot well temperature of 28.5°C is 119.4 kJ/kg.

Heat absorbed by cooling water

= Heat given up by steam

$$m_w\, c_w(t_2 - t_1) = m_s\left[(h_f + h_{fg}) - h_c\right]$$

$$57500 \times 4.186\ (26.0 - 8.5)$$

$$= 1800\ \left[(167.23 + x \times 2406) - 119.4\right]$$

or $2340 = 2406\ x + 47.83$

$$\therefore \quad x = \frac{2340 - 47.83}{2406} = 0.953$$

Thus steam at entry to the condenser has a dryness fraction of 0.953.

(*f*) At condenser temperature of 34°C,

Partial pressure of steam,

$$p_s = 0.0532 \text{ bar}$$

∴ From Dalton's law, partial pressure of air

$$p_a = 0.0733 - 0.0532 = 0.0201 \text{ bar}$$

From characteristic gas equation,

$$pV = mRT$$

∴ Mass of air present per m^3 of condenser volume

$$= \frac{0.0201 \times 10^5 \times 1}{287 \times 307} = 0.0228 \text{ kg}$$

Volume of 1 kg of steam at 34°C = 26.61 m^3

Air associated with 1 kg of steam will have the same volume.

∴ Mass of air present per kg of uncondensed steam

$$= \frac{0.0201 \times 10^5 \times 26.61}{287 \times 307} = 0.607 \text{ kg}$$

EXAMPLE 20.2

A condenser deals with 1000 kg of steam per hour with a dryness fraction of 0.9. The mean condenser temperature is 45°C. The air associated with the steam in the condenser is 250 kg/hour. What would be the vacuum reading? Barometer reading is 75cm of Hg. Correct this vacuum to a standard barometer reading of 76 cm of mercury.

Solution : From steam tables :

Partial pressure of steam at 45°C,

$$p_s = 0.0958 \text{ bar}$$

Volume of 1 kg of steam at 45°C,

$$v_s = 15.28 \text{ m}^3$$

∴ Total volume of steam

$$= m \times (x \times v_s)$$
$$= 1000(0.9 \times 15.28)$$
$$= 13752 \text{ m}^3/\text{hr}$$

According to Dalton's law this is also the volume of 250 kg of air at 45°C.

From characteristic gas equation,

$$pV = mRT$$

partial pressure of air,

$$p_a = \frac{250 \times 287 \times (45 + 273)}{13752}$$
$$= 1659 \text{ N/m}^2 = 0.01659 \text{ bar}$$

∴ Total condenser pressure

$$= 0.0958 + 0.01659 = 0.11239 \text{ bar}$$
$$= \frac{0.11239}{0.01333} = 8.43 \text{ cm of Hg}$$

∴ Vacuum reading

$$= 75 - 8.43 = 66.57 \text{ cm of Hg}$$

Corrected vacuum

$$= 76 - (75 - 66.57)$$

= **67.57** cm of Hg

EXAMPLE 20.3

In a surface condenser test, the vacuum gauge reading was 65 cm of Hg when the barometer read 75 cm of Hg. The temperature in the condenser was 50°C. If the quantity of air entering the condenser was reduced to 80kg/hr by suitable means, find the resultant alteration in vacuum. If the condensate weighs 1000kg/hour, calculate quantities of air and vapour which the pump has to deal with.

Solution : Absolute condenser pressure

$$= (75 - 65) \times 0.01333 = 0.1333 \text{ bar}$$

From steam tables, at 50°C

partial pressure of steam

$$= 0.1233 \text{ bar}$$

and volume of 1 kg of steam

$$= 12.04 \text{ m}^3$$

∴ Partial pressure of air

$$= 0.1333 - 0.1233 = 0.01 \text{ bar}$$

volume of steam/hr

$$= 12.04 \times 1000 = 12040 \text{ m}^3$$

This is also the volume of air

∴ Mass of air,

$$m = \frac{pV}{RT} = \frac{0.01 \times 10^5 \times 12040}{287 \times 327}$$

$$= 128.29 \text{ kg/hr}$$

When the mass of air is reduced to 80 kg/hr, the partial pressure of air will become

$$= \frac{80}{128.29} \times 0.01 = 0.00623 \text{ bar}$$

∴ Absolute pressure of condenser will change to

$$0.1233 + 0.00623 = 0.12953 \text{ bar}$$

$$= \frac{0.12953}{0.01333}$$

$$= 9.717 \text{ cm of Hg}$$

New vacuum reading

$$= 75 - 9.717$$

$$= 65.283 \text{ cm of Hg}$$

∴ Alteration of vacuum

$$= 65.283 - 65$$

$$= 0.283 \text{ cm of Hg}$$

Capcaity of pump

$$= 1000 + 80 = 1080 \text{ kg/hr}$$

EXAMPLE 20.4

A steam turbine discharges 5000 kg/hr of steam at 40°C and 0.85 dry. The air leakage in the condenser is estimated to be 15 kg/hr. The temperature at the suction of air pump is 32°C and the temperature of condensate is 35°C. Determine : (a) vacuum gauge reading, (b) capacity of air pump, (c) loss of condensate in kg/hr and (d) quantity of cooling water if its temperature rise is limited to 10°C.

Solution : At 40°C temperature

$$v_s = 19.54 \text{ m}^3/\text{kg}$$

$$p_s = 0.0737 \text{ bar}$$

$$h_f = 167.6 \text{ kJ/kg}$$

and $h_{fg} = 2406.7$ kJ/kg

Then enthalpy of steam being discharged by the turbine is

$$h_1 = h_f + x\,h_{fg}$$

$$= 167.6 + 0.85 \times 2406.7$$

$$= 2213.29 \text{ kJ/kg}$$

For 1 kg of steam in the condenser,

Specific volume

$$v_1 = 0.85 \times 19.54 = 16.609 \text{ m}^3/\text{kg}$$

This is also the volume occupied by

$$\frac{15}{5000} = 3 \times 10^{-3} \text{ kg of air}$$

Applying characteristic gas equation

$pv = mRT$, we get

Partial pressure of air in the condenser

$$p_a = \frac{3 \times 10^{-3} \times 287 \times (273 + 40)}{16.609}$$

$$= 16.22 \text{ N/m}^2 = 1.622 \times 10^{-4} \text{ bar}$$

Then absolute pressure in the condenser

$$p = p_a + p_s$$

$$= 1.622 \times 10^{-4} + 0.0737$$

$$= 0.07386 \text{ bar}$$

$$= \frac{760}{1.01325} \times 0.07386$$

$$= 55.4 \text{ mm of Hg}$$

(*a*) Taking standard atmospheric pressure

$$= 760 \text{ mm of Hg}$$

Vacuum gauge reading

$$= 760 - 55.4 = 704.6 \text{ mm of Hg}$$

(*b*) At the air pump suction where temperature is 32°C

Steam vapour pressure $p_s = 0.0476$ bar

∴ Partial pressure of air

$$= p - p_s$$

$$= 0.07386 - 0.0476 = 0.02626 \text{ bar}$$

Capacity air pump

$$V = \frac{mRT}{p}$$

$$= \frac{15 \times 287 \times (273 + 32)}{0.02626 \times 10^5} = 500 \text{ kg/hr}$$

This is also the volume of vapour mixed with air.

(*c*) The loss of condensate in air pump equals the mass of water for the volume (capacity) of air pump. Then

$$m_w = \frac{p_s V}{R_w T}$$

$$= \frac{(0.0476 \times 10^5) \times 500}{\left(\frac{8314}{18}\right) \times (273 + 32)} = 16.89 \text{ kg/hr}$$

(*d*) From heat energy balance

heat lost by steam = heat gained by water

Corresponding to 35°C temperature, enthalpy of condensate = 146.7 kJ/kg

Then : 5000(2213.29 –146.7)

$$= m_w \times 4.186 \times 10$$

∴ Quantity of cooling water

$$= \frac{5000(2213.29 - 146.7)}{4.186 \times 10}$$

$$= 246845 \text{ kg/hr}$$

EXAMPLE 20.5

The following procedure was adopted to check the leakage of air into a condenser.

After running the plant to reach steady state conditions, the condenser was completely isolated by shutting down the steam supply and also the condensate and air extraction pumps. The temperature and vacuum readings were noted to be 39°C and 68.5 cm mercury at shut down, and 28°C and 27 cm mercury respectively 5 minutes after the shut down. If effective volume of the condenser is 1 m³, calculate

(a) the quantity of air leakage into the condenser during the period of observation and

(b) the quantity of water vapour condensed during the period.

The barometer is stated to read 75 cm of mercury.

Solution : (*i*) *At shut down :*

Absolute pressure in the condenser

$$= (75 - 68.5) = 6.5 \text{ cm of Hg}$$

$$= \frac{6.5}{760} \times 1.01325 = 0.08665 \text{ bar}$$

Steam vapour pressure corresponding to 39°C = 0.06999 bar

∴ Partial pressure of air

$$p_a = 0.08665 - 0.06999$$

$$= 0.01666 \text{ bar}$$

Mass of air present in the condenser volume of 1 m³

$$m_{a_1} = \frac{p_a \times V}{RT}$$

$$= \frac{0.01666 \times 10^5 \times 1}{287(273+39)} = 0.0186 \text{ kg}$$

Specific volume of steam at 39°C

$$= 20.54 \text{ m}^3/\text{kg}$$

∴ Mass of vapour (steam) present

$$m_{s_1} = \frac{1}{20.54} = 0.0487 \text{ kg}$$

(*ii*) *Five minutes after shut down :*

Absolute pressure

$$= (75 - 27) = 48 \text{ cm of Hg}$$

$$= \frac{48}{760} \times 1.01325 = 0.63984 \text{ bar}$$

Steam vapour pressure corresponding 28°C

$$= 0.03782 \text{ bar}$$

∴ Partial pressure of air

$$= 0.63984 - 0.03782 = 0.6020 \text{ bar}$$

Mass of air present in the condenser volume of 1 m³

$$m_{a_2} = \frac{0.6020 \times 10^5 \times 1}{287(273+28)} = 0.697 \text{ kg}$$

Specific volume of steam at 28°C

$$= 36.69 \text{ m}^3/\text{kg}$$

∴ Mass of vapour present

$$m_{s_2} = \frac{1}{36.69} = 0.0272 \text{ kg}$$

That gives :

Mass of air leaked in 5 minutes

$$= 0.697 - 0.0186 = 0.6784 \text{ kg}$$

Mass of steam condensed in 5 minutes

$$= 0.0487 - 0.0272 = 0.0215 \text{ kg}$$

EXAMPLE 20.6

A steam turbine develops 2500 kW using 15000 kg of steam per hour which is supplied at a pressure of 30 bar and 300°C. The exhaust from the turbine is condensed at a vacuum of 725 mm of mercury while the barometer reading is 758 mm of mercury. The condensate is removed from the condenser at a temperature of 28°C and the temperature of circulating water increases from 7 to 27°C. Assuming that there is no heat loss by convection and radiation, determine

(a) dryness fraction of steam entering the condenser

(b) mass circulating water/hr and the cooling ratio

(c) minimum amount of cooling water required per kg of steam

(d) overall thermal efficiency

Solution : Enthalpy of steam at 30 bar and 300°C,

h_1 = 2995 kJ/kg

Enthalpy drop

$(h_1 - h_2)$ = heat equivalent of work done
= 2500 × 3600
= 90 × 10^5 kJ/hr ...(*i*)

Condenser pressure

p = (758 – 725) = 33 mm of mercury

$= \frac{33}{760} \times 1.01325 = 0.04399$ bar

Corresponding to 0.04399 bar, it is noted from steam tables that :

$h_f \cong 128$ kJ/kg and $h_{fg} \cong 2430$ kJ/kg

(*a*) If x is the dryness fraction of steam entering the condenser, then enthalpy drop in the turbine

$15000\left[2995-(128+2430x)\right]$ kJ/hr ...(*ii*)

From identities (*i*) and (*ii*), we get that

$15000\left[2995-(128+2430x)\right] = 90 \times 10^5$

or $2430x = 2995 - \frac{90\times10^5}{1500} - 120 = 2067$

∴ Dryness fraction of steam

$= \frac{2275}{2430} = \mathbf{0.936}$

(*b*) From heat energy balance,

heat lost by steam

= heat gained by water

$m_s\left[h_f + xh_{fg} - h_w\right] = m_w \times c_{pw} \times \Delta T$

where h_w is the enthalpy of condensate at 28°C = 117.2 kJ/kg

$15000\left[(128+0.936\times2430)-117.2\right]$

$= m_w \times 4.186 \times 20$

Solution gives : cooling water flow rate

m_w = **409450** kg/hr

Cooling ratio represents the amount of cooling water required per kg of steam

∴ Cooling ratio $= \frac{409450}{15000}$ = **27.297**

(*c*) Maximum possible temperature of cooling water at outlet may correspond to saturation temperature corresponding to condenser pressure of 0.04399 bar which is approximately 30.7°C.

Then : maximum rise in temperature of circulating water

= 30.7 – 7 = 23.7°C

Cooling water flow rate

$= 409450 \times \frac{20}{23.7} = 345527$ kg/hr

$= \frac{345527}{15000}$ = **23.03** kg/kg of steam

(*d*) Work output per kg of steam

$= \frac{2500\times3600}{15000} = 600$ kJ

Heat added per kg of steam

= 2995 – 117.2 = 2877.8 kJ

∴ Overall efficiency

$= \frac{600}{2877.8}$ = 0.2085 or **20.85%**

EXAMPLE 20.7

A surface condenser, fitted with separate air and condensate extraction pump deals with 12000 kg of steam per hour at a pressure of 7.5 kN/m². The steam entering the condenser is 0.95 dry and the temperature at the air and condensate extraction pumps is 30°C. If the air leakage is estimated to be 10 kg/hr, make calculations for

(a) the surface area required if the average rate of heat transfer is 4 kJ/cm²/s.

(b) the cylinder diameter of the single acting reciprocating dry pump if it runs at 75 rpm. The volumetric efficiency of the pump is 0.9 and the stroke equals 1.25 times the cylinder bore.

Solution : Mass of steam dealt with

$= \frac{12000}{3600} = 3.33$ kg/s

From steam tables : At 7.5 kN/m² or 0.075 bar pressure

h_f = 169 kJ/kg

and h_{fg} = 2405 kJ/kg

Further, at 30°C temperature :

h_{f_1} = 126 kJ/kg

Then : heat lost by steam

$$= m_s \left[(h_f + x f_{fg}) - h_{f1}\right]$$

$$= 3.33\ [(169 + 0.95 \times 2405) - 126]$$

$$= 7751.4 \text{ kJ/s}$$

∴ Surface area required

$$= \frac{7751.4}{4} = \mathbf{1937.85} \text{ cm}^2$$

(*b*) Total pressure of air and steam in the condenser = 0.075 bar

Partial pressure of steam at 30°C temperature,

$$p_s = 0.0425$$

∴ Partial pressure of air,

$$p_a = 0.075 - 0.0425 = 0.0325 \text{ bar}$$

Mass of air leakage

$$= \frac{10}{60} = 0.1667 \text{ kg/min.}$$

Then from characteristic gas equation, the volume of air dealt with

$$V_a = \frac{m_a R_a T_a}{p_a}$$

$$= \frac{0.1667 \times 287 \times (273 + 30)}{0.0325 \times 10^5}$$

$$= 4.46 \text{ m}^3\text{/min}$$

Air volume dealt per stroke

$$= \frac{4.46}{0.9 \times 75}$$

$$= 0.0661 \text{ m}^3 = 66.1 \times 10^3 \text{ cm}^3$$

The volume of air dealt per stroke of pump is also given by $\frac{\pi}{4}\ d^2 l$

$$\therefore \frac{\pi}{4}\ d^2 \times 1.25\ d = 66.1 \times 10^3\ ;\ d^3 = 67.34 \times 10^3$$

Hence the diameter of dry air pump,

$$d = \left[67.34 \times 10^3\right]^{\frac{1}{3}} = 40.53 \text{ cm}$$

EXAMPLE 20.8

At a thermal power station, a surface condenser is required to deal with 15000 kg of steam per hour, and the air leakage is estimated at 0.6 kg per 1500 kg of steam. At the air pump suction, the vacuum reading is 705 mm of water and the temperature is 35°C. Determine :

(a) the discharge capacity of wet air pump which removes both air and condensate. The volumetric efficiency of the pump is 92 percent.

(b) the dimensions of the pump. The pump is single acting, runs at 60 rev/min and the piston stroke equals 1.25 times the pump diameter.

Solution : Total pressure of steam and air in the condenser, *i.e.*, the condenser pressure is

$$p = 760 - 705$$

$$= 55 \text{ mm of mercury}$$

$$= \frac{55}{760} \times 1.01325 = 0.0733 \text{ bar}$$

Corresponding to 35°C, the partial pressure of steam

$$p_s = 0.0562 \text{ bar} \quad \text{(from steam tables)}$$

∴ Partial pressure of air

$$p_a = 0.0733 - 0.0562 = 0.0171 \text{ bar}$$

Mass of air leakage into the condenser

$$m_a = \frac{0.6}{1500} \times \frac{15000}{60} = 0.1 \text{ kg/min}$$

Then from characteristic gas equation, the volume of air leakage is

$$V_a = \frac{m_a RT}{p_a}$$

$$= \frac{0.1 \times 287 \times (35 + 273)}{0.0171 \times 10^5}$$

$$= 5.169 \text{ m}^3\text{/min}$$

Mass of steam condensed

$$m_s = \frac{15000}{60} = 250 \text{ kg/min}$$

Taking density of water as 1000 kg/m³, the volume of condensate is

$$V_c = \frac{250}{1000} = 0.25 \text{ m}^3\text{/min}$$

∴ Volume of air-condensate mixture to be dealt with

$$= 5.169 + 0.25 = 5.419 \text{ m}^3\text{/min}$$

Mixture volume dealt per stroke

$$= \frac{5.419}{0.9 \times 60} = 0.1 \text{ m}^3 = 100000 \text{ cm}^3$$

The discharge capacity of the wet air pump is also given by $\frac{\pi}{4} d^2 l$

$$\therefore \frac{\pi}{4} d^2 \times 1.25\, d = 100000 \; ; \quad d^3 = 101910$$

Hence : diameter of pump,

$$d = 46.53 \text{ cm}$$

Stroke length,

$$l = 46.56 \times 1.25 = 58.16 \text{ cm}$$

EXAMPLE 20.9

A surface condenser is designed to handle 2500 kg of steam per hour and the air leakage is estimated as 0.8 per 1000 kg of steam. The steam enters into the condenser dry saturated at 38°C and the condensate is extracted at the lowest point of the condenser at a temperature of 37°C. The system is provided with a separate air extraction pump which draws air over an air cooler, and the air leaves the cooling section at 31°C. Assuming that condenser pressure remains practically constant, compute :

(a) mass of steam condensed in the air cooler section

(b) volume of air to be handled by the dry air pump

(c) reduction in air pump capacity resulting from the cooling of air.

The following data is taken from the steam tables :

Temperature in °C	*38*	*37*	*31*
Pressure in bar	*0.0676*	*0.0640*	*0.0459*
Specific volume in m³/kg	*21.63*	*22.77*	*31.20*

Solution : Air infiltration

$$= \frac{0.8}{1000} \times 2500 = 2 \text{ kg/hr}$$

Total volume of dry steam that enters the condenser at 38°C

$$= 25000 \times 21.63 = 54075 \text{ m}^3\text{/hr}$$

This is also the volume of 2 kg of air and it exerts its own partial pressure.

Partial pressure of air

$$p_a = \frac{m_a R T_a}{V_a} = \frac{2 \times 287 \times (273+38)}{54075}$$

$$= 3.30 \text{ N/m}^2 = 0.33 \times 10^{-4} \text{ bar}$$

This is negligibly small and can be neglected. Obviously then the total pressure in the condenser is the saturation pressure corresponding to 38°C (the temperature at which steam enters the condenser. Thus

$$p = 0.0676 \text{ bar}$$

At condensate extraction (37°C)

Partial pressure of steam

$$p_s = 0.0640 \text{ bar}$$

Partial pressure of air

$$p_a = p - p_s$$

$$= 0.0676 - 0.0640 = 0.0036 \text{ bar}$$

Applying the characteristic gas equation to 2 kg of air that is to be extracted per hour at partial pressure of 0.0036 bar, we have

Volume of air removed

$$= \frac{2 \times 287 \times (273+37)}{0.0036 \times 10^5}$$

$$= 494.28 \text{ m}^3\text{/hr}$$

By Dalton's law this will also be the volume of steam at 37°C

∴ Mass of steam going with air

$$= \frac{494.28}{22.77} = 21.71 \text{ kg/hr}$$

At air cooler section (31°C)

Partial pressure of steam

$$p_s = 0.0459 \text{ bar}$$

Partial pressure of air

$$p_a = p - p_s$$

$$= 0.0676 - 0.0459 = 0.0217 \text{ bar}$$

Applying characteristic gas equation,

Volume of air removed

$$V_a = \frac{2 \times 287 \times (273+31)}{0.0217 \times 10^5}$$

$$= 80.4 \text{ m}^3\text{/hr}$$

Mass of steam associated with this air and which is partially condensed in the air cooler

$$= \frac{80.4}{31.2} = 2.577 \text{ kg/hr}$$

∴ Mass of steam condensed in air cooler section (or saving in condensate)

$$= 21.71 - 2.577 = 19.133 \text{ kg/hr}$$

Air pump capacity (volume of air handled by dry air pump) = 494.28 m³/hr

Air pump capacity with air cooler

$= 80.4\ m^3/hr$

$\therefore$ Percentage reduction in air pump capacity due to air cooling

$$= \frac{494.28 - 80.4}{494.28}\ 0.837 \text{ or } \mathbf{83.7\%}$$

EXAMPLE 20.10.

A two-pass surfaces condenser handles 20 × 10³ kg per hour of steam which is 0.87 dry. The condenser maintains a vacuum of 70 cm of mercury when the barometer reads 76 cm of mercury. The condensate is removed from the condenser at 35°C and the rise in the temperature of cooling water is limited to 22°C. If pressure head equivalent to 8 m of water is required to force water through the condenser and pipe, calculate :

(a) flow area required for a water velocity of 65 m/min, heat transfer rate is

15 × 10⁵ kJ/m²/hr, and

(b) power required to run the circulating water pump if it is 70 percent efficient.

Solution : Absolute condenser pressure

$= 76 - 70 = 6$ cm of mercury

$$= \frac{1.01325}{76} \times 6 = 0.08 \text{ bar}$$

Corresponding to condenser pressure of 0.08 bar, it is noted from steam tables that

$h_f = 173.38$ kJ/kg

and $h_{fg} = 2403.4$ kJ/kg

Also at condensate temperature of 35°C,

$h_w = 146.66$ kJ/kg

From heat energy balance, heat lost by steam

= heat gained by water

Now : heat lost by steam

$$= m_s \left[(h_f + h_{fg}) - h_w\right]$$

$$= 20 \times 10^3 \left[(173.38 + 0.87 \times 2403.34) - 146.66\right]$$

$= 42.35 \times 10^6$ kJ/hr

heat gained by water $= m_w\, c_{pw}\, \Delta T$

$= m_w \times 4.18 \times 22$

$= 91.96\, m_w$ kJ/hr

$\therefore 91.96\, m_w = 42.35 \times 10^6$

Solution gives : cooling water flow rate

$m_w = 4.605 \times 10^5$ kg/hour

Volume flow of water

$$= (4.605 \times 10^5) \times \frac{1}{1000}$$

$= 460.5\ m^3/hr$

where density of water has been taken as 1000 kg/m³

(*a*) Flow area required

$$= \frac{\text{volume flow of water}}{\text{flow velocity of water}}$$

$$= \frac{460.5}{60} \times \frac{1}{65} = 0.1181\ m^2$$

(*b*) Cooling surface area required

$$= \frac{\text{heat lost by steam}}{\text{average heat transfer rate}}$$

$$= \frac{42.35 \times 10^6}{15 \times 10^5} = 28.23\ m^2$$

(*c*) Velocity head $= \dfrac{V^2}{2g}$

$$= \frac{\left(\frac{65}{60}\right)^2}{2 \times 9.81} = 0.0598 \text{ m}$$

Total head $h = 8 + 0.0598 = 8.059$ m

Power required to run the circulating water pump

$$= (m_w\, gh) \times \frac{1}{\eta}$$

$$= \frac{4.605 \times 10^5}{3600} \times 9.81 \times 8.059 \times \frac{1}{0.7}$$

$= 14447$ Nm/s $= 14.447$ kW

EXAMPLE 20.11.

A barometric jet condenser handles 4500 kg of steam per hour which is 0.96 dry. The condenser maintains 65 cm mercury vecuum when the barometer indicates 76 cm mercury. The cooling water enters the condenser at 15°C temperature and the mixture of condensate and cooling water leaves at 40°C. Assuming no undercooling, make calculations for

(a) *the minimum height of the tail pipe above the level of hot well, and*

(b) *the amount of cooling water required.*

Solution : Absolute condenser pressure

$= (76 - 65) = 11$ cm of marcury

$= \frac{11}{100} \times 13.6 = 1.496$ m of water

Barometric height

$= \frac{76}{100} \times 13.6 = 10.336$ m of water

∴ Length of tail pipe

$= 10.336 - 1.496 = 8.84$ m

(*b*) As there is no undercooling of condensate, the saturation temperature of steam entering the condenser will be the same as the temperature of the mixture of condensate and cooling water. Corresponding to this 40°C temperature.

$h_w = h_f = 167.54$ kJ/kg

and $h_{fg} = 2406.7$ kJ/kg

From energy balance, heat gained by cooling water = heat lost by steam

$$m_w \, c_{pw} \, (t_2 - t_1) = m_s \left[(h_f + x h_{fg}) - h_w\right]$$

or $m_w \times 4.18\ (40 - 15)$

$= 4500 \left[(167.54 + 0.96 \times 2406.7) - 167.54\right]$

∴ Amount of cooling water required

$$m_w = \frac{4500 \times (0.96 \times 2406.7)}{4.18\,(40 - 15)}$$

$= 99492$ kg/hr

EXAMPLE 20.12.

The steam at exit from a steam turbine enters a jet condenser at the rate of 325 kg/minute and the volume of cooling required for condensation of steam is 0.02 m³/kg of steam. The cooling water enters the condenser at 20°C and the condensate (mixture of condensed steam and water) leaves at 30°C. The condenser maintains a vacuum of 680 mm of Hg whilst the borometer reads 760 mm Hg. The air leakage into the condenser is 0.05% of the steam mass that enters the condenser. If the volume of air dissolved in the water injected is 5% of the volume at atmospheric pressure, calculate the volume handling capacity of air pump for removing air and condensate. Take volumetric efficiency of air pump as 88%.

Solution : Absolate condenser pressure

$= (760 - 680) \times 0.001333$

$= 0.1066$ bar

From steam tables : At 30°C partial pressure of steam

$p_s = 0.0426$ bar

∴ partial pressure of air

$p_a = 0.1066 - 0.0426 = 0.064$ bar

Volume of cooling water required

$= 325 \times 0.02 = 6.5$ m³/min

Volume of air entering the condenser with cooling water $= 6.5 \times 0.05 = 0.325$ m³/min

This air enters the condenser at atmospheric pressure (p = 1.0132 bar) and 20°C temperature.

Then, mass of air entering the condenser with cooling water

$$= \frac{pv}{RT} = \frac{(1.0132 \times 10^5) \times 0.325}{287 \times (273 + 20)}$$

$= 0.3916$ kg/min

Mass of air leakage into the condenser

$= \frac{0.05}{100} \times 325 = 0.1625$ kg/min

As such, the total mass of air inside the condenser

$= 0.3916 + 0.1625$

$= 0.5541$ kg/min

The pressure and temperature corresponding to this mass of air are 0.064 bar (partial pressure of air) and 30°C (temperature of mixture of steam and water)

$$\text{Volume of air} = \frac{mRT}{p}$$

$$= \frac{0.5541 \times 287 \times (273 + 30)}{0.064 \times 10^5}$$

$= 7.529$ m³/min

Volume of steam condensed

= mass of steam × V_f at 30°C

$= 325 \times 0.001004 = 0.3263$ m³/min

Then, total capacity of wet air pump

= volume of condensate
+ volume of air
+ volume of cooling water

$= 0.3263 + 7.529 + 6.5$

$= 14.325 \text{ m}^3/\text{min}$

∴ Actual capacity of air pump

$$= \frac{14.355}{0.88} = 16.31 \text{ m}^3/\text{min}$$

20.11. EDWARD AIR PUMP

The pressure of air and other non-condensable gases, which enter the condenser along with steam or infiltrate through the joints which are not absolutely tight, adversely affect the performance of the condenser by making the condenser air bound, *i.e.*, by increasing the condenser pressure. An increase in the condenser pressure results into a decrease in pressure drop through the turbine and that adversely affects the turbine performance. Further, the non-condensable gases are highly corrosive and their removal from the condenser is essential.

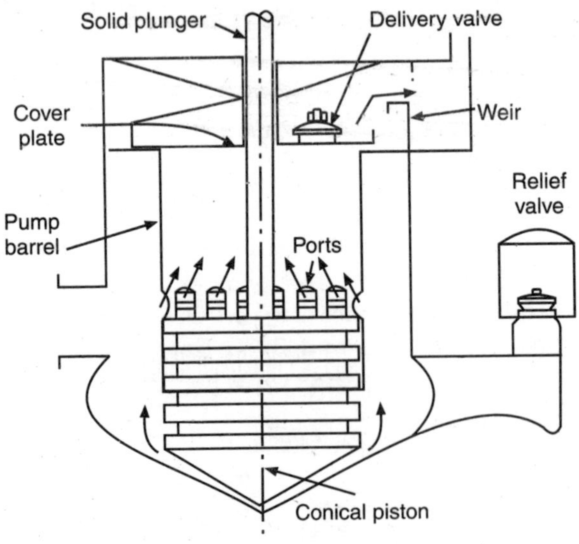

Fig. 20.10. Edrward air pump

Refer Fig. 20.10 for the construction and operation of Edward air. It is a wet air pump that is provided to maintain a desired vacuum in the condenser by extracting a mixture of condensate (in the form of water) and non-condensable gases (particularly air) from the condenser.

Construction: The unit consists of a conical piston that reciprocates inside a barrel. At the top of the barrel is a cover plate that carries a number of delivery valves for discharging the charge (moist air and water in fluid state) to the hot well. The bottom of the barrel follows the contour of the conical end of the piston. The upper surface of the piston has circumferential grooves which hold water and form a labyrinth packing. There are ports at the lower end of the barrel and these connect the pump with the condenser through flange.

Working: Consider the piston to be at the top and begining to execute the downward stroke. A partial vacuum is created in the barrel above the piston and that closes the delivery valve. When the piston uncovers the ports during its downward movement, the air and water vapour from the condenser rush into the space above the piston. With further downward movement, the condensate entrapped in the conical bottom of the pump is forced above the piston through ports. At this instant, pump is said to be fully charged of with water, air and steam above the piston.

During upward motion of the piston, the charge is compressed to a pressure higher than the atmospheric pressure. The delivery valve opens and the charge flows over the weir to hotwell. The weir maintains a sufficient constant head of water above the cover plate and that seals the delivery valve against air leakage (preventing the air to enter the barrel.)

There is a water sealed relief valve at the barrel of cylinder. If pressure below the piston exceeds the atmospheric pressure, the relief valve opens to atmosphere and that allows the condensate and air to escape through it.

Note : (1) This reciprocating pump is quite reliable and satisfactory as long as relatively low vacuum is required as is the case with land and marine condensing steam engines. For large power plants operating with high vacuum and dealing with large quantities of steam, such reciprocating pumps have limitations of size, speed and the resistances affered by passage to flow of air. Further, in central power stations the practice of removing both water and gases together is considered objectionable. The presence of oxygen in feed water is likely to cause damage to high pressure boilers. This has led to the development of steam jet ejector which is a dry air pump, a pump that removes only the air.

20.12. COOLING TOWERS AND COOLING PONDS

The cooling towers and cooling ponds are employed for cooling the hot water coming out of the condenser so that the resulting cooled water can be used again for circulation in the condenser. The cooling is affected by exposing the hot water to an atmosphere of air. The cooling tower becomes a necessity when the supply of cooling water is scarce and the same water has to be used over and over again for cooling purposes.

The cooling process in the cooling tower is normally affected by the following factors :

- temperature and humility of air,
- velocity with which the air enters the cooling tower,
- reach of air to different sections of the cooling tower,
- arrangement of plates/baffles inside the tower, and
- size and height of the tower.

The cooling towers operate on draught which may be natural, induced or forced.

Natural draught cooling tower : Water from the condenser is supplied to troughs and nozzles with the help of a pump installed at a height of about 8-10 m from the bottom. Water falls in the form of spray and meets the air which enters from the bottom. Water loses its heat to air and gets cooled. The cooled water is collected at the bottom and is returned to condenser for circulation. There are no moving parts in the natural draught cooling tower. Air flow through the shell is created by the density difference between the atmospheric air and the air inside the tower which has been warmed by the hot circulating water.

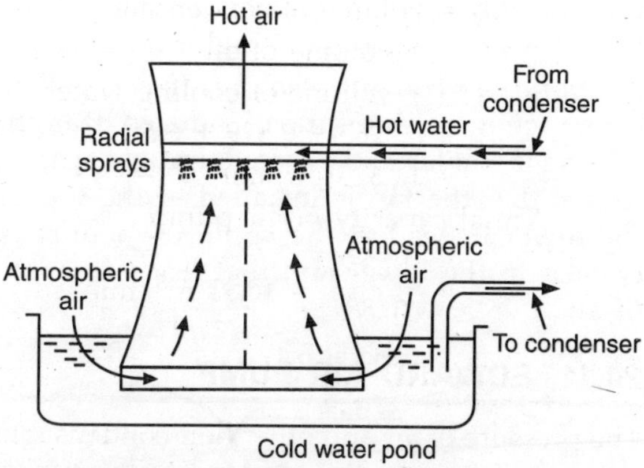

Fig. 20.11. Natural draught cooling tower

Some of the water during its fall is evaporated and is carried along by the heated air . The loss of water due to evaporation is made up by fresh water called *make up water.* The reduction in temperature of water is called *range.*

The performance of natural draught tower is affected by variation of atmospheric conditions and wind velocity.

Forced draught cooling tower : Hot water coming from the condenser enters the unit from the top and gets sprayed through nozzles. Water droplets falling downwards meet the air going upward. A good supply of air is made by a fan installed at the bottom of the tower. Spray eliminators are provided at the top and these entrap the water which is likely to go along with hot air leaving from the top.

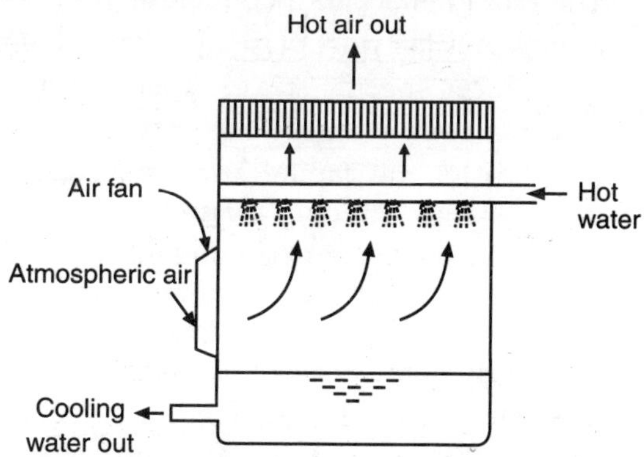

Fig. 20.12. Forced draught cooling tower

Compared to natural draught cooling tower, the forced draught system is considerably small in size of tower and pipes, has high efficiency but has high running cost.

Induced draught cooling tower : The construction and working of induced draught cooling is similar to the forced draught system except that the fan is installed at the top of the tower instead at the bottom. Vacuum is created in the tower and that leads to entry of air.

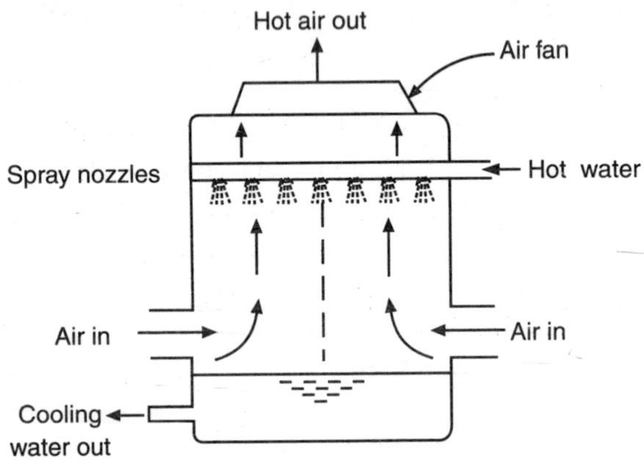

Fig. 20.13. Induced draught cooling tower

Cooling towers are generally of hyperbolic shape and are made of timber, steel or concrete. The timber towers have a short life, a limited cooling capacity, high maintenance cost and rot easily due to exposure to sun, wind, water etc. In contrast, the advantages of concrete cooling towers are: low maintenance, large capacity, increased stability, and improved draught and air circulation.

Cooling pond : The hot water coming from the condenser is sprayed into the cooling pond through nozzles. These nozzles are kept at a distance from each other so that there is no interference from between two sprays. Further, for effective cooling, the nozzles are fitted at a height of 1 to 2 m above the ground level. The evaporation from the water surface produces the cooling effect.

For the same duty, the surface area required in a cooling pond is about 30 times the size of cooling tower. Further, a good amount of water is lost by evaporation and by the wind blowing across the cooling pond.

SALIENT POINTS

1. A steam condenser is a closed vessel into which the exhaust steam from an engine or a turbine is exhausted and condensed by means of cooling water at a pressure below atmospheric.

   The condensed steam is called condensate.
2. The employment of a condenser
   (*i*) improves the work output per kg of steam
   (*ii*) reduces specific steam consumption
   (*iii*) improves thermal efficiency
   (*iv*) reduces the capital and running cost of the water softening plant
3. The basic elements of a condensing plant are : condenser, condensate and air extraction pumps, circulating pump and cooling tower.
4. Condensers are classified as :

   (A) jet (direct or mixing) condenser in which there is a direct contact between the exhaust steam and the cooling water. The jet condensers are further classified as counter or parallel flow, high or low level and ejector condensers.

   (*B*) surface condensers in which water flows through a network of tubes over which passes the exhaust steam. The surface condensers are further classified according to direction of flow of condensate (central flow, down flow), number of passes of water (single pass or multi pass) and shape of shell (circular, oval or V-shaped).
5. The vacuum gauge reading corrected to standard atmosphere is

   $$= 760 - (\text{barometer reading} - \text{vacuum gauge reading})$$

   The vacuum efficiency is prescribed as

   $$\eta_v = \frac{\text{actual vacuum}}{\text{ideal vaccum}}$$

   $$= \frac{\text{barometric pressure} - \text{actual pressure}}{\text{barometric pressure} - \text{ideal pressure}}$$

The condenser efficiency is

$$\eta_{con} = \frac{t_2 - t_1}{t_3 - t_1}$$

where t_3 is saturation temperature corresponding to condenser pressure, t_2 and t_1 are the outlet and inlet temperature of cooling water.

6. In a condenser, there is a mixture of steam and air leaking into the condenser, and according to Dalton's law :

$$p = p_s + p_a$$

where p is the total pressure of the mixture p_s and p_a are the partial pressures of steam and air respectively.

7. The air leakage into the conderser is due to :

(*i*) leakage through packing glands and microscopic holes in the shell and joints.

(*ii*) dissolved air in spent up steam or in cooling water (jet condensers)

Due to air leakage, there is loss of vaccum (increase in the condenser pressure) and that reduces the expansion work.

8. Cooling towers and cooling ponds are employed for cooling the hot water coming out of the condenser so that it can be recirculated for cooling purposes. The cooling towers may be of natural draught, induced draught or forced draught type.

REVIEW QUESTIONS

A. Conceptual and conventional questions.

1. Enumerate the advantages of incorporating a condenser in a steam power plant.
2. Make a neat sketch of barometric jet condenser and explain its working.
3. (*a*) State the functions of condenser in a steam power plant.

   (*b*) Describe, with a neat sketch, the construction and working of a surface condenser.
4. State the relative advantages and disadvantages of surface condenser and a jet condenser.
5. (*a*) State Dalton's law of partial pressure. How does it apply to the condenser of a steam plant.

   (*b*) Define and explain the terms vacuum efficiency and condenser efficiency of a condensing plant.
6. (*a*) State briefly the sources and effects of air leakage into a condenser.

   (*b*) How do you proceed to check the leakage of air into a condenser ?
7. (*a*) What purpose is served by a cooling tower in a condensing plant ?

   (*b*) Name the different types of cooling towers and give a brief description of each one of them.
8. The vacuum in a surface condenser is noted to be 711 mm of mercury and the barometer stands at 760 mm of mercury. The hot well temperature is 32°C, and the inlet and outlet temperatures of cooling water are 12°C and 28°C respectively. Calculate the condenser efficiency and the vacuum efficiency.

   [**Ans.** 80% ; 98.15%]
9. The following observations pertain to a condenser test :

   Vacuum = 700 mm of Hg and barometer reading = 760 mm of Hg

   Inlet and outlet temperature of cooling water = 16.5° C and 31°C.

   Mean temperature of condensation = 29°C

   Mass of condensate = 1200 kg/hr, and

   Mass of cooling water = 45500 kg/hr

   Determine :

   (*a*) Mass of air present per unit of condenser volume

   (*b*) state of steam at entry to condenser

   (*c*) condensate under cooling, condenser efficiency and vacuum efficiency.

   [**Ans.** 0.026 kg ; 094 ; 6°C, 78.5% and 97.5%]
10. The following observations were made on a condensing plant in which the mean temperature of condensation was 35°C. The recorded condenser vacuum was 68.73 cm of Hg and the barometer read 74.0 cm of Hg. The condenser dealt with 1170 kg/hour of steam. 45000 kg/hour of cooling water was supplied and there was a rise of 14°C in its temperature.

    Determine : (*a*) corrected vacuum corresponding to barometric reading of 76cm of Hg

    (*b*) vacuum efficiency (*c*) dryness fraction of

steam as it enters the condenser (*d*) mass of air present per cubic meter of condenser volume.

[**Ans.** 70.73cm of Hg, 98.5%, 0.923, 0.01505 kg/m^3]

11. The vacuum in a surface condenser is 70.5 cm of mercury when the barometer reads 75 cm. At inlet to air cooling section, the temperature is 33°C and at the outlet 29°C. If air leakage is estimated to be 2.3 kg/hr, make calculations for the mass of steam condensed in the cooling section.

[**Ans.** 4.6 kg/hr]

12. The following observations refer to a test on a steam condensate plant :

Recorded condenser vacuum
= 710 mm of mercury

Barometer reading = 760 mm of mercury

Mean condenser temperature
= 29°C

Condensate collected
= 22700 kg/hr

The air infiltration into the condenser is prescribed by

$$\left(\frac{\text{steam condensed in kg/hr}}{2000}+2\right)$$

Compute the discharging capacity of wet air pump which removes both air and condensate. Take volumetric efficiency of air equal to 80 percent. If the air pump is single acting and runs at 60 rev/min and the stroke equals 1.5 times the piston diameter, compute the dimensions of the pump.

[**Ans.** 574 m^3/hr, 51.4 cm and 77.1 cm]

B. Fill in the blanks with appropriate word/words.

1. Condenser is a .... vessel where condensation of steam takes place at constant pressure which is generally .... than atmospheric pressure.
2. .... pump is used to extract out the air and dissolved gases from the condenser.
3. High level jet condenser is also called as .... condenser.
4. In a high level jet condenser, the tail pipe that discharges condensate from the condenser into the hot well is about .... long.
5. The absolute pressure inside a condenser is the sum of partial pressures of steam and air inside it. This statement is referred to as ....
6. In an .... condenser, the mixture of condensate and water is discharged automatically in the hotwell and there is no need of extraction pump.
7. In an ideal condenser, the temperature of condensate equals the .... corresponding to condenser pressure.
8. The vacuum obtainable in a condenser is dependent on the .... of circulating water.
9. The thermal efficiency of a steam power plant is increased with the employment of surface condenser because the .... of heat refection is reduced.

Answers

1. closed, less 2. air extraction 3. barometric
4. 10.36 m 5. Dalton's law of partial pressure
6. ejector 7. saturation temperature
8. temperature 9. average temperature

C. Multiple choice questions.

1. Which aspect may or may not be true regarding the advantage gained by using a condenser in a steam power plant ?
   (*a*) expansion of steam to a lower back pressure
   (*b*) increase in available enthalpy drop and consequently more work output
   (*c*) reduction in steam consumption per kW-hour
   (*d*) fuel economy and safety from thermal stresses in the boiler shell
2. Which one of the followings is a wrong statement?
   (*a*) the employment of condenser permits the expansion of steam to a pressure lower than atmospheric pressure
   (*b*) the thermodynamic efficiency of a condensing steam turbine unit is higher than that of a non-condensing one
   (*c*) the vacuum in a condenser is a function of both absolute pressure and the barometric pressure
   (*d*) the ratio of actual vacuum in the condenser to ideal vacuum is called condenser efficiency
3. The inlet and outlet temperatures of cooling water being supplied to a condenser are t_1 and t_2 respectively. If t_s is the saturation

temperature corresponding to condenser pressure, the condenser efficiency will be

(a) $\frac{t_2 - t_1}{t_s - t_2}$ (b) $\frac{t_2 - t_1}{t_s - t_1}$

(c) $\frac{t_2 - t_1}{t_s}$ (d) $\frac{t_1 + t_2}{2\,t_s}$

4. In a steam condenser, the barometric reading and the recorded condenser vacuum are stated to be 76.5 cm and 71 cm of mercury. The vacuum corrected to standard barometer of 76 cm of mercury would be
   (a) 70.5 cm of Hg (b) 71.5 cm of Hg
   (c) 69.5 cm of Hg (d) 72.5 cm of Hg
5. The length of tail pipe fitted to the barometric or high level jet condenser is about
   (a) 6.2 m (b) 7.6 m
   (c) 8.5 m (d) 10.4 m
6. Which aspect is not true in the context of a jet condenser ?
   (a) more intimate mixing of steam and cooling water
   (b) air extraction pump requires high power
   (c) requires considerable space for erection
   (d) equipment simple and low in cost
7. The maximum limit of vacuum attained in a jet condenser is about
   (a) 55 cm of Hg (b) 60 cm of Hg
   (c) 65 cm of Hg (d) 70 cm of Hg
8. All of the following statements about surface condenser are correct, except
   (a) high vacuum can be attained and greater plant efficiency is achieved
   (b) condensate can be salvaged and used for boiler feed
   (c) high capital and maintenance cost
   (d) maximum chances of losing vacuum
   (e) suitable for high capacity units
9. In a surface condenser used in steam power station, under cooling of condensate is undersirable as this would
   (a) not absorb the gases in steam
   (b) reduce efficiency of the plant
   (c) increase the cooling water requirements
   (d) increase the thermal stresses in the condenser.
10. With the removal of air from a surface condenser, there is
    (a) no change in absolute pressure
    (b) rise in absolute pressure
    (c) fall in absolute pressure
    (d) rise in temperature of condensed steam
11. Edwards air pump removes .... from the condenser
    (a) only air
    (b) only un-condensed vapour
    (c) moist air (air plus vapour)
    (d) moist air and also the condensate
12. Consider the following statements
    1. In a shell and tube condenser surface condenser, cooling water passes through the tubes and steam surrounds them
    2. Evaporative type of condenser has steam in pipes surrounded by water
    3. The cooling section in a surface condenser allows greater quantity of water to be extracted along with air

    Which of the statements made above are correct?
    (a) 1 and 2 (b) 2 and 3
    (c) 1 and 3 (d) 1, 2 and 3

Answers

1. (d)	2. (d)	3. (b)	4. (a)	5. (d)
6. (c)	7. (c)	8. (d)	9. (c)	10. (c)
11. (d)	12. (a)			

❑❑❑

Chapter 21

Compressed Air: Positive Displacement and Dynamic Compressors

Learning objectives : After having made a study of the subject matter presented in this chapter, the reader will be able to :

- *mention the different applications of compressed air and state how compressors are classified*
- *explain the working of single stage reciprocating compressor; isothermal efficiency, clearance volume and volumetric efficiency*
- *state the conditions under which the use of multi-stage compression is adopted, advantages of multi-stage compression*
- *explain the working of roots blower and vane blower*
- *make an analysis of centrifugal compressors; construction and operation; slip factor and power input factor*
- *make an analysis of axial flow compressors; construction and operation; degree of reaction*
- *explain the phenomenon of surging and choking in rotary compressors*

A compressor is a machine which takes in a definite quantity of atmospheric air and raises its pressure with an appreciable increase in density. After compression, the air at higher pressure is delivered to a storage vessel from where it is conveyed through pipelines to the site where it is required to be used. In contrast to the engine, the compressor does not produce work but is a machine that consumes mechanical/electrical energy supplied by an external power source.

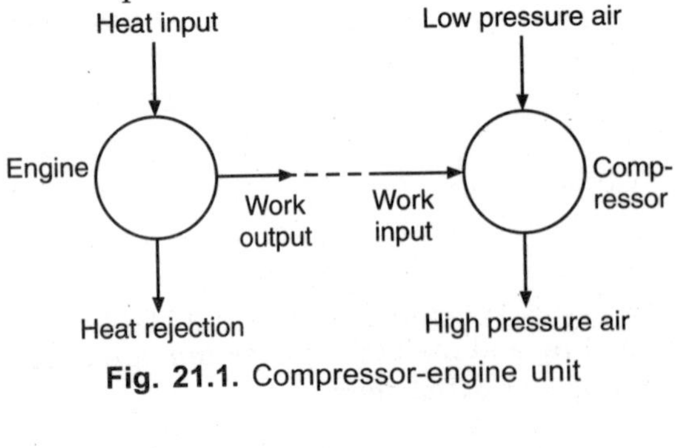

Fig. 21.1. Compressor-engine unit

Fig. 21.1 shows the schematic arrangement of a compressor-prime mover unit.

21.1. UTILITY OF COMPRESSED AIR

Compressed air is a precious commodity and it finds an extensive use in industries, workshops, automobiles and mines etc.

A : Low compressed air (P < 1.0 MN/m^2) is used for

(*i*) spray painting, car servicing and filling air in tyres

(*ii*) creating blast in a blast furnace

(*iii*) conveying materials like sand, concrete and powdered fuel along a pipeline

(*iv*) boosting (supercharging) of internal combustion engines

B : Medium compressed air (P = 1.0 to 8.0 MN/m^2) in large quantities is used in gas turbines that drive aircraft.

C : High compressed air (P = 8.0 to 10.0 MN/m^2) finds application in

(*i*) driving a wide variety of pneumatic tools like drills and hammers

(*ii*) pneumatic brakes for locomotives and pneumatic steering of heavy vehicles

(*iii*) pneumatic presses, pneumatic rivetting and cleaning of surfaces

(*iv*) Injecting fuel in diesel engines

The compressed air is also used in

- refrigeration and cooling of large buildings
- transmitting control system pressures to remote locations
- pneumatic drilling of hardrocks and blasting them
- driving vibrators for mixing cement and concrete with water to make slurry
- operation of doors, rams, cage gates and conveyors.

An important application of compressed air is in running the mining machines where the use of internal combustion engines and electrical machines is precluded because of fire risks due to the pressure of inflammable gases.

21.2. CLASSIFICATION OF COMPRESSORS

The compressors are classified on the basis of :

1. Design and principle of operation

(*i*) *Positive displacement* units wherein the fluid (air or gas) is confined within solid boundaries. The pressure is developed by change in volume or by displacement action.

In reciprocating compressors, there is a piston-cylinder arrangement. The interaction between the sliding piston and the fluid involves a change in volume and consequently pressure rise. The reciprocating units are best suited for small mass flow rates and large pressure ratios.

In the rotary displacement compressors, the air is compressed by trapping it in a reducing space formed by a set of engaging surfaces. The change in pressure is either by the backflow of air (roots blower) or both by squeezing action and backflow of air (vane blower). The rotary machines can operate at high speeds and can handle large mass flow rates than their reciprocating counterparts.

(*ii*) *Non-positive compressors* : The fluid is not positively contained within specified boundaries but is in a continuous steady flow throughout the machine. There is transfer of energy from the machine rotor to the fluid and the rise in pressure is primarily due to dynamic effects. For such machines, the delivery pressure is not very high but continuous delivery of air at uniform rate is maintained. Examples are centrifugal and axial flow compressors.

Dynamic compressors and blade compressors are the other two names given to the non-positive type of compressors.

2. Number of stages

The compressors may be single stage or multi-stage. Multi-stage compression is generally adopted when compression ratio exceeds 5.

3. Pressure limits

Based on delivery pressure, the compressors are classified as :

Low pressure	: delivery pressure upto 1 MN/m^2
Medium pressure	: delivery pressure 1 to 8 MN/m^2
High pressure	: delivery pressure 8 to 10 MN/m^2
Super high pressure	: delivery pressure is above 10 MN/m^2

4. Pressure ratio

According to pressure ratio, the compressors are named as

Fans	: pressure ratio is 1 to 1.1
Blowers	: pressure ratio is 1.1 to 2.5
Compressors	: pressure ratio is above 2.5

Further, boosters are the machines where suction side is at a pressure greater than atmospheric pressure. Compressors employed for creating and maintaining vacuum are called air pumps or exhausters.

5. Capacity

Classification of compressors is also done on the basis of volume of air delivered per unit time.

Low capacity having volume delivery
$= 0.15\ m^3/s$ or less

Medium capacity having volume delivery
$= 0.15\ m^3/s$ to $5\ m^3/s$

High capacity having volume delivery
$= 5\ m^3/s$ and above

21.3. SINGLE ACTING RECIPROCATING OF COMPRESSOR

A reciprocating compressor has a piston-cylinder arrangement where the piston is fitted with piston rings and has reciprocating motion inside the cylinder. The prime mover (an engine or electric motor) drives the crank shaft, the crank rotates and converts the rotary motion into reciprocating motion with the help of connecting rod. The inlet and outlet valves are provided in the cylinder head. The valves are mostly of thin steel plates, and are provided with light springs for reducing their inertia. The valves operate due to small pressure difference across them; their smaller inertia gives rapid opening and closing action. The cylinder is surrounded by water jacket or metallic fins for proper cooling of air during compression.

To understand the working of compressor, let the piston be at inner dead centre (IDC) and the clearance volume be neglected.

When the piston moves outward, the vacuum is created and the inlet valve opens. The atmospheric air is sucked into the cylinder and the suction continues during movement of piston from cover end (IDC) to crank end (ODC). The volume of air handled equals the stroke volume of the cylinder. After the completion of suction, the inward movement of the piston commences, the inlet valve closes and the air is compressed (the pressure starts increasing continuously). The compression continues until the air pressure inside the cylinder becomes slightly more than the pressure on delivery side which is connected to the receiver. The delivery valve then opens and the air is delivered to the receiver at constant pressure. After exhaust (delivery), the air pressure in the cylinder falls below atmospheric pressure, inlet valve opens and the cycle is repeated.

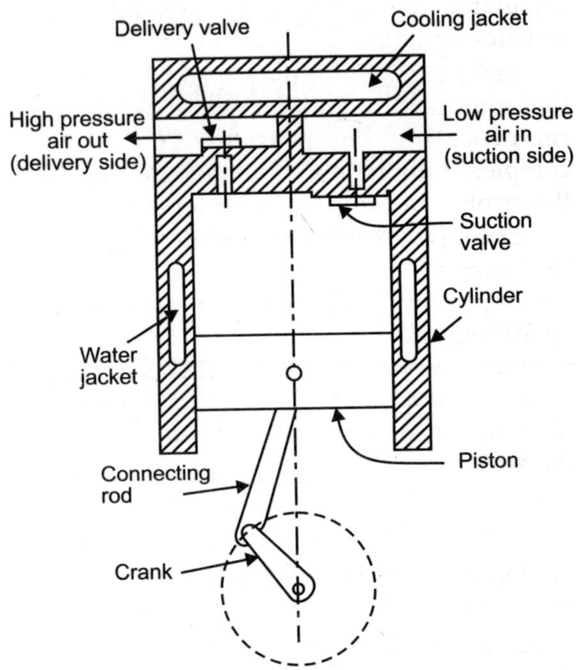

Fig. 21.2. Reciprocating air compressor

All these operations (suction, compression and delivery) take place in two strokes of the piston, *i.e.*, is one revolution of the crank. The suction takes place in one stroke (piston movement from cover end to crank end), on one side of the piston and compression and delivery take place during second stroke (piston movement from crank end to cover end). When these operations occur alternatively on both sides (cover end side and crank end side) the unit is called *double acting compressor.*

21.4. POWER REQUIRED TO DRIVE THE COMPRESSOR

To study the operation of a single stage reciprocating compressor and to establish a relation for the work required to drive it, let it be presumed that (*i*) there is no clearances

(*ii*) there are no losses due to friction

(*iii*) the air does not undergo any pressure loss when passing through the pipes and valves

(*iv*) the physical properties at air of any instant during the working cycle can be calculated by applying the general characteristics equation; $pV = mRT$

Refer Fig. 21.3 for the *p-v* and *T - s* plots of the compression process for a single stage reciprocating compressor. The cycle is completed in one revolution of the crank and the strokes in order of operation are :

Stroke 0 - 1 : The piston moves outward, the suction valve opens and the air at atmospheric pressure and temperature conditions is sucked into the cylinder. The volume of air handled is equal to stroke volume of the cylinder. The stroke volume can be determined from the cylinder diameter *d* and the stroke length *l* as

$$V_s = \frac{\pi}{4} d^2 l = V_1$$

During suction, the flow work is done by air.

Stroke 1 - 2 - 3 : The piston moves inward, the inlet valve closes and the air is compressed. The compression process has the following three possibilities :

(*i*) isentropic (adiabatic) compression along 1 - 2′

(*ii*) polytropic compression 1 - 2

(*iii*) isothermal compression 1 - 2″

When the air has been compressed to delivery pressure p_2, the delivery valve opens and the air is delivered to the receiver at constant pressure p_2. During compression, the work is done on air and the temperature of air at the end of compression depends on the law of compression and heat exchanged with the surrounding. During delivery 2 - 3,

(*i*) pressure and temperature remain constant

(*ii*) there is no exchange of heat with surroundings

(*iii*) the mass of air in the cylinder decreases

(*iv*) flow work is done on air in pushing it out of cylinder.

After exhaust, the pressure in the cylinder drops from p_2 to p_1 and the cycle is repeated.

Area 0 - 1 - 2 - 3 of *p - V* plot represents the network when the compression follows the polytropic law.

Work input per cycle = area 0 - 1 - 2 - 3

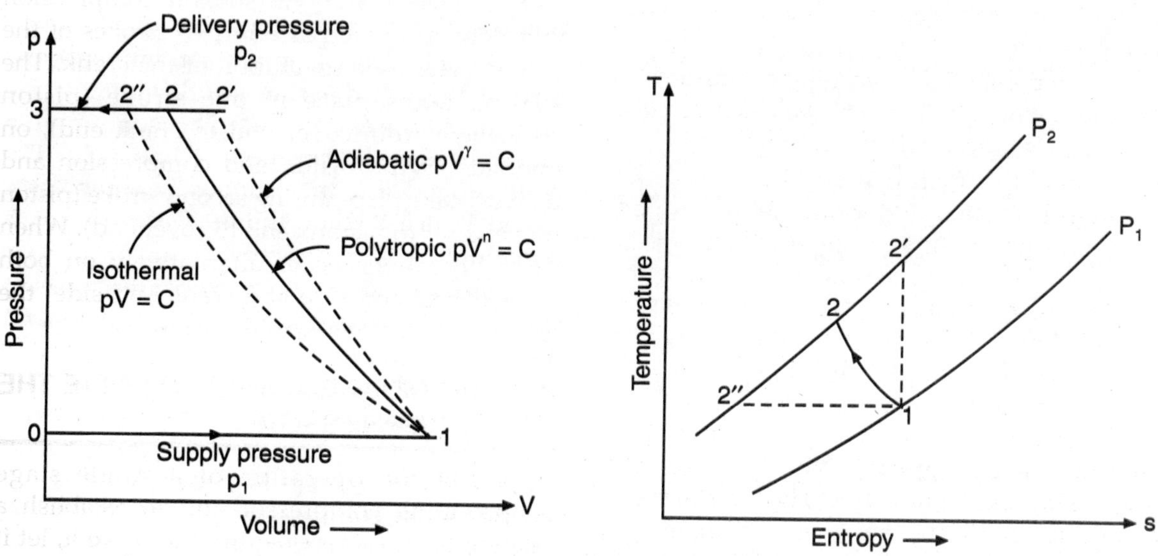

Fig. 21.3.

= work done on air during compression 1 - 2 + work done on air during delivery 2 - 3 - work done by air during suction 0 - 1

$$= \frac{p_2V_2 - p_1V_1}{n-1} + p_2V_2 - p_1V_1$$

$$= \frac{n}{n-1}(p_2V_2 - p_1V_1)$$

$$= \frac{n}{n-1} mR(T_2 - T_1)$$

Invoking the relation $\frac{T_2}{T_1} = \left(\frac{p_2}{p_1}\right)^{\frac{n-1}{n}}$, the above equation can be recast as

Working input/cycle (W_a) in running the compressor

$$= \frac{n}{n-1} mRT_1 \left[\left(\frac{p_2}{p_1}\right)^{\frac{n-1}{n}} - 1\right]$$

$$= \frac{n}{n-1} p_1V_1 \left[\left(\frac{p_2}{p_1}\right)^{\frac{n-1}{n}} - 1\right] \quad ...(21.1)$$

With adiabatic compression :

Work input/cycle

$$W_a = \frac{\gamma}{\gamma-1} mRT_1 \left[\left(\frac{p_2}{p_1}\right)^{\frac{\gamma-1}{\gamma}} - 1\right]$$

$$= \frac{\gamma}{\gamma-1} p_1V_1 \left[\left(\frac{p_2}{p_1}\right)^{\frac{\gamma-1}{\gamma}} - 1\right] \quad ...(21.2)$$

With isothermal compression :

Work input/cycle

W_a = area 0 - 1 - 2″ - 3

$$= p_2V_2'' + p_1V_1 \log_e \frac{V_1}{V_2''} - p_1V_1$$

$$= p_1V_1 \log_e \frac{V_1}{V_2''} \quad (\because p_1V_1 = p_2V_2'')$$

$$= p_1V_1 \log_e \frac{p_2}{p_1} = mRT_1 \log_e \frac{p_2}{p_1} \quad ...(21.3)$$

For a single acting air compressor, air power P_a is given by

$$P_a = W_a \times \frac{N}{60}$$

For double acting air compressor :

$$P_a = W_a \times \frac{2N}{60} \quad ...(21.4)$$

where N is the speed of the compressor in revolutions per minute

The power input to the compressor can also be determined if there is an arrangement for taking the indicator diagram.

$$P_a = k\, p_m\, LA \frac{N}{60} \text{ kJ/s} \quad ...(21.5)$$

where p_m is the mean effective pressure in kPa, L is the stroke length in meter, A is the area of the piston in m^2, N is the speed in rev/min and k is a factor : 1 for single acting and 2 for double acting.

Best value for compression index

For a given compressor,

(*i*) The pressure ratio $\frac{p_2}{p_1}$ is fixed which means a fixed height (line 0–3) of p-V diagram.

(*ii*) The capacity given by the cylinder volume fixes the stroke length which means line 0 - 1 of the p-V plot has a fixed value.

The only process which can influence the area of the p-V diagram (work input) is the path of compression 1-2. The compression index decides the position of line 1–2. The work expended in the running the compressor would be least when the area of p-V plot is minimum

Comparison of the expressions derived above for the work required to compress and deliver unit mass of air for a given pressure ratio would show that isothermal compression requires least amount of energy. However, isothermal compression cannot be achieved in practice but an attempt is made to approach it by cooling the compressor either by providing fins or a water jacket to the compressor cylinder.

The term *isothermal efficiency* of a compressor is a measure of the degree to which isothermal compression has been achieved. It is expressed as the ratio of isothermal work to actual work.

$$\text{Isothermal efficiency} = \frac{\text{isothermal work input}}{\text{actual work input}}$$

$$= \frac{p_1 V_1 \log_e \frac{p_2}{p_1}}{\frac{n}{n-1} p_1 V_1 \left[\left(\frac{p_2}{p_1}\right)^{\frac{n-1}{n}} - 1\right]} \quad \text{...(21.6)}$$

21.5. EFFECTIVE OF CLEARANCE AND VOLUMETRIC EFFICIENCY

When the piston is at its extreme left position it does not flush with the cylinder head. A dead space remains between the cylinder head and the piston. The dead space is called clearance space. The clearance ensures mechanical freedom to the moving parts without much wear and tear and provides the necessary space for valve operation.

The volume occupied by the air when the piston is at its extreme left position is known as clearance volume V_c and the volume swept by the piston as it moves in the cylinder from its extreme position on the left to right is referred to as swept volume or piston displacement V_s.

Swept volume or piston displacement volume

$$V_s = \text{area of piston} \times \text{stroke length}$$

$$= \frac{\pi}{4} d^2 l$$

The pressure-volume diagram for a single stage compressor when clearance is considered is shown in Fig. 21.4. The small quantity of air at pressure p_2 remaining in the dead space at the end of delivery line 2-3 expands polytropically along 3–4. Theoretically the expansion continues until pressure p_1 is reached at point 4. The inlet valve then opens and the induction of fresh air into the cylinder begins. The intake continues for the remaining period of stroke. The volume sucked in during suction is $(V_1 - V_4)$. At point 1, the inlet valve closes and the charge is compressed polytropically along the path 1–2 until delivery pressure p_2 is reached at point 2. The delivery valve then drops and the air is discharged to the receiver. The delivery continues till the piston reaches the extreme position on the left. The cycle is then repeated.

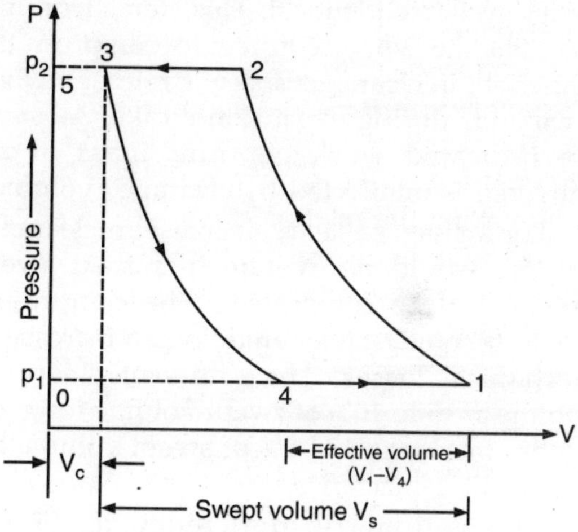

Fig. 21.4. Effect of clearance and volumetric efficiency

Work input/cycle = area 1-2-3-4

= area (0-1-2-5) – area (3-4-0-5)

$$= \frac{n}{n-1} p_1 V_1 \left[\left(\frac{p_2}{p_1}\right)^{\frac{n-1}{n}} - 1\right]$$

$$- \frac{n}{n-1} p_1 V_4 \left[\left(\frac{p_2}{p_1}\right)^{\frac{n-1}{n}} - 1\right]$$

$$= \frac{n}{n-1} p_1 (V_1 - V_4) \left[\left(\frac{p_2}{p_1}\right)^{\frac{n-1}{n}} - 1\right] \quad \text{...(21.7)}$$

where $(V_1 - V_4)$ is the effective volume sucked in during the suction stroke. Work input thus depends on effective volume and not on the swept volume.

Equation 21.7 can be rewritten as

Work input cycle

$$= \frac{n}{n-1} (m_1 - m_4) R (T_2 - T_1)$$

$$= \frac{n}{n-1} m_e R (T_2 - T_1) \quad ...(21.7a)$$

where m_e is the effective mass of air induced per cycle.

The expressions 21.1 and 21.7 *a* for work input/cycle are identical. This stems from the fact that the work required to compress the mass held in clearance space is released when it expands during expansion stroke. As such, the indicated work per unit mass of air delivered is unaffected by clearance volume.

For a given capacity, increase in clearance volume would necessitate increased swept capacity of the compressor. The compressor would become heavier entailing an increase in mechanical losses. The limits of clearance volume are 5 to 10% of swept volume for good quality machines and 2% of swept volume for sleeve valve machines.

The volumetric deficiency η_v of the reciprocating compressor is defined as the ratio of actual volume sucked in to the piston displacement or swept volume of the compressor. Thus

$$\eta_v = \frac{V_1 - V_4}{V_s} = \frac{V_s + V_c - V_4}{V_s} = 1 + \frac{V_c}{V_s} - \frac{V_4}{V_s} \quad ...(i)$$

Considering polytropic expansion process 3–4, we have

$$p_3 V_3^n = p_4 V_4^n; \; V_4 = V_3 \left(\frac{p_3}{p_4}\right)^{1/n}$$

But $p_3 = p_2$; $p_4 = p_1$ and $V_3 = V_c$

$$\therefore \quad V_4 = V_c \left(\frac{p_2}{p_1}\right)^{1/n} \quad ...(ii)$$

Substituting the value of V_4 as given by expression (*ii*) in expression (*i*), we get

$$\eta_v = 1 + \frac{V_c}{V_s} - \frac{V_c}{V_s}\left(\frac{p_2}{p_1}\right)^{1/n}$$

$$= 1 - \frac{V_c}{V_s}\left[\left(\frac{p_2}{p_1}\right)^{1/n} - 1\right] \quad ...(21.8)$$

Volumetric efficiency of a compressor referred to ambient conditions is given by

$$\eta_v = \frac{(V_1 - V_4) \text{ referred to ambient conditions}}{V_s}$$

$$= \frac{p_1 T_0}{p_0 T_1} \frac{V_1 - V_4}{V_s}$$

Here, the suffix 1 represents the suction and suffix 0 represents the ambient conditions

$\therefore$ η_v (at ambient suction condition)

$$= \frac{p_1 T_0}{p_0 T_1}\left[1 - \frac{V_c}{V_s}\left\{\left(\frac{p_2}{p_1}\right)^{1/n} - 1\right\}\right] \quad ...(21.9)$$

When free air or ambient conditions are not given, they are assumed to be 1 bar and 15°C.

The expression of volumetric efficiency as given by equations 21.8 and 21.9 become identical when the suction conditions are the same as ambient conditions. The dimensions of the compressor cylinder are determined by using the volumetric efficiency based on suction conditions.

From expressions 21.8 and 21.9, it is evident that factors influencing the volumetric efficiency of a reciprocating compressor are :

clearance factor $C = \frac{V_c}{V_s}$; pressure ratio $\frac{p_2}{p_1}$

and the polytropic index of compression n.

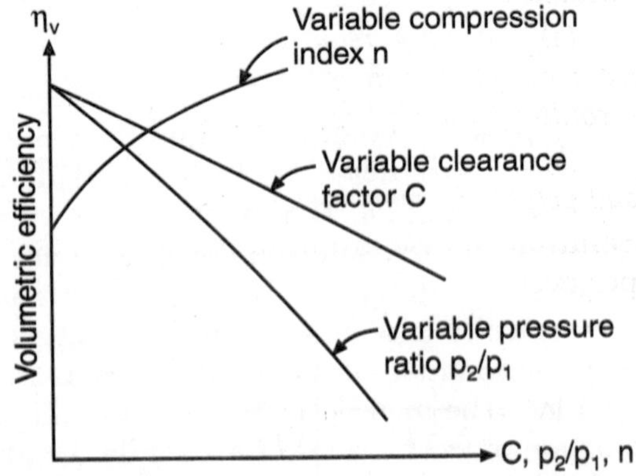

Fig. 21.5. Variation in volumetric efficiency

Figure 21.5 shows the variation in volumetric efficiency when each of these parameters is changed, keeping the other two factors constant. From this plot, it is seen that volumetric efficiency

- decreases as the clearance factor C increases
- decreases with increase of p_2/p_1 and
- increase with increase of index n.

The volumetric efficiency as expressed by equations 21.8 and 21.9 is further reduced due to

1. increase in temperature of fresh (sucked) air when it comes in contact with heated cylinder walls
2. wire drawing or throttling (pressure drops) at suction and delivery valves
3. leakage in the system, *i.e.*, the valve and piston leakage
4. inertia effects in opening the inlet valve
5. the air does not get enough time to fill the cylinder fully during suction strokes if the compressor runs at high speeds

Further, in actual operation, the inlet valve opens only when the high pressure air left in the dead space expands to a pressure below the suction pressure. Some portion of the suction stroke may thus be wasted.

EXAMPLE 21.1

A single cylinder reciprocating compressor has a piston displacement of 0.1 m³. The suction pressure and temperature are 1 bar and 298 K respectively. If the delivery pressure after compression is 7 bar, calculate :

(a) the work required to compress the air isentropically with $\gamma = 1.4$ and polytropically according to law $pV^{1.25}$ = constant

(b) the isothermal efficiency for the isentropic and polytropic compression processes.

Solution : (*a*) Isentropic (adiabatic) work done per cycle

$$W_a = \frac{\gamma}{\gamma - 1} p_1 V_1 \left[\left(\frac{p_2}{p_1} \right)^{\frac{\gamma - 1}{\gamma}} - 1 \right]$$

$$= \frac{1.4}{1.4 - 1} (1 \times 10^5) \times 0.1 \left[\left(\frac{7}{1} \right)^{\frac{1.4 - 1}{1.4}} - 1 \right]$$

$$= \mathbf{26061\ Nm}$$

Polytropic work done per cycle

$$W_p = \frac{n}{n - 1} p_1 V_1 \left[\left(\frac{p_2}{p_1} \right)^{\frac{n - 1}{n}} - 1 \right]$$

$$= \frac{1.25}{1.25 - 1} (1 \times 10^5) \times 0.1 \left[\left(\frac{7}{1} \right)^{\frac{1.25 - 1}{1.25}} - 1 \right]$$

$$= \mathbf{23788\ Nm}$$

(*b*) Isothermal work done per cycle

$$W_i = p_1 V_1 \log_e \left(\frac{p_2}{p_1} \right)$$

$$= (1 \times 10^5) \times 0.1 \times \log_e \left(\frac{7}{1} \right)$$

$$= \mathbf{19459\ Nm}$$

Isothermal efficiency for isentropic compression

$$= \frac{W_i}{W_a} = \frac{19459}{26061} = 0.7467 = \mathbf{74.67\%}$$

Isothermal efficiency for polytropic compression

$$= \frac{W_i}{W_p} = \frac{19459}{23788} = \mathbf{0.818} = \mathbf{81.8\%}$$

EXAMPLE 21.2

A two-cylinder double acting reciprocating compressor sucks in air at pressure 0.98 bar and temperature 300 K. The delivery of the compressed air to the receiver is at 6 bar. The breathing capacity of the compressor is stated to be 2.5 m³/min when measured at 1 bar and 288 K. The mean speed of the compressor is limited to 120 m/minute and the stroke is 0.75 times the cylinder diameter. If the law of compression is $pV^{1.3}$, make calculations for

(a) cylinder dimensions, i.e., the cylinder diameter and stroke length

(b) compressor speed in rev/min

(c) shaft power if the mechanical efficiency is 85%

(d) isothermal efficiency

Neglect the effect of clearance and piston rod diameter.

Solution : Designating the free air conditions by suffix 0 and the suction conditions by suffix 1, then

$$V_1 = \frac{p_0 V_0}{T_0} \times \frac{T_1}{p_1}$$

$$= \frac{1 \times 2.5}{288} \times \frac{300}{0.98} = 2.66 \text{ m}^3/\text{min}$$

For a double acting compressor, the swept volume per minute is

$$= \text{number of cylinders} \times \frac{\pi}{4} d^2 l \times (2N)$$

The compressor speed N has been taken twice since the compressor is double acting.

Further, mean piston speed 2 lN = 120 (given)

$$\therefore\ 2.66 = 2 \times \frac{\pi}{4} d^2 \times 120$$

Solution gives : cylinder diameter

$$d = \sqrt{\frac{2.66 \times 4}{2\pi \times 120}} = 0.1188 \text{ m} = \mathbf{11.84 \text{ cm}}$$

Stroke length

$$l = 0.75 \times 11.84 = \mathbf{8.88 \text{ cm}}$$

(*b*) Invoking the relation : 2 $l\,N$ = mean piston speed

Compressor speed

$$N = \frac{120}{2 \times 0.088} = 681.82 \text{ rev/min}$$

$$(c)\ \text{Work input} = \frac{n}{n-1} p_1 V_1 \left[\left(\frac{p_2}{p_1}\right)^{\frac{n-1}{n}} - 1\right]$$

$$= \frac{1.3}{1.3-1}(0.98 \times 10^5) \times \left(\frac{2.66}{60}\right) \left[\left(\frac{6}{0.98}\right)^{\frac{1.3-1}{1.3}} - 1\right]$$

$$= 9770 \text{ Nm/s} = 9770 \text{ W} = 9.77 \text{ kW}$$

Shaft power

$$= \frac{\text{indicated power}}{\text{mechanical efficiency}}$$

$$= \frac{9.77}{0.85} = \mathbf{11.49 \text{ kW}}$$

$$(d)\ \text{Isothermal work} = p_1 V_1 \log_e \left(\frac{p_2}{p_1}\right)$$

$$= 0.98 \times 10^5 \times \left(\frac{2.66}{60}\right) \log_e \left(\frac{6}{0.98}\right)$$

$$= 7872 \text{ Nm/s}$$

$\therefore$ Isothermal efficiency

$$= \frac{\text{isothermal work}}{\text{polytropic work}}$$

$$= \frac{7872}{9770} = 0.806 \text{ or } \mathbf{80.6\%}$$

EXAMPLE 21.3

A reciprocating air compressor designed to delivery air at a gauge pressure of 7 bar has a stroke of 75 cm length and the clearance volume is 3 percent of swept volume. To study the effect of clearance on free air delivery and work expended, the compressor was over hauled and a distance piece 0.5 cm thick was inserted between the cylinder head and cylinder. The compressor was then commissioned under the changed clearance. Make calculations for the percentage change in (a) the volume of free air delivered, and (b) the power necessary to drive the compressor.

Before and after overhauling, the suction is at atmospheric pressure of 1 bar, and the compression and expansion processes follow the polytropic law $pV^{1.3}$ = *constant.*

Solution : Delivery pressure

$$p_2 = 7 + 1 = 8 \text{ bar}$$

Before overhauling

Stroke volume V_s = (piston area A in cm²) × (stroke length l in cm) = $A \times 75 = 75\,A$ cm³

The volume remaining in the dead space at the end of delivery line 2–3 equals the clearance volume. Therefore

$$V_3 = V_c = 0.03 \times 75\,A = 2.25\,A \text{ cm}^3$$

Cylinder volume

$$V_1 = V_s + V_c = 75\,A + 2.25\,A = 77.25\,A \text{ cm}^3$$

From the expansion curve 3 – 4 :

$$p_3 V_3^n = p_4 V_4^n$$

$$V_4 = V_3 \left(\frac{p_3}{p_4}\right)^{1/n} = 2.25\,A\left(\frac{8}{1}\right)^{1/1.3} = 11.13\,A \text{ cm}^3$$

∴ Volume of free air sucked in

$$= V_1 - V_4 = (77.25\,A - 11.13\,A) = 66.12\,A \text{ cm}^3$$

After Overhauling :

When the distance piece is inserted, the clearance space is increased. Consequently for the same stroke volume, the cylinder volume increases

$$V_3 = V_c = 2.25\,A + 0.5\,A = 2.75\,A \text{ cm}^3$$

$$V_1 = 75\,A + 2.75\,A = 77.75\,A \text{ cm}^3$$

$$V_4 = V_3 \left(\frac{p_3}{p_4}\right)^{1/n} = 2.75\,A\left(\frac{8}{1}\right)^{1/1.3} = 13.61\,A \text{ cm}^3$$

Volume of free air sucked in

$$= 77.75\,A - 13.61\,A = 64.14\,A \text{ cm}^3$$

∴ Change in free air delivery

$$= \frac{66.12 - 64.14}{66.12} = 0.0299 = \mathbf{2.99\%} \text{ (decrease)}$$

The power expended in running the compressor is prescribed as

$$= \frac{n}{n-1}\, p_1 \text{ (volume of free air sucked in)} \times \left[\left(\frac{p_2}{p_1}\right)^{\frac{n-1}{n}} - 1\right]$$

With index n and pressures p_1 and p_2 remaining unchanged the power expended in running the compressor is directly proportional to free air sucked in. Therefore, the change in power expended will be same as change in free air taken in. That is

Change in power expended = **2.99%** (decrease)

EXAMPLE 21.4

A single acting reciprocating compressor having 8 cm bore and 10 cm stroke is required to compress air from 1 bar to delivery pressure of 6 bar. The air after compression is stored in a receiver having a capacity of 1.2 m³. The compressor is driven at 500 rev/min and it consumes 1.2 kW brake power. Initially the pressure and temperature in the reservoir were noted to be 1 bar and 15°C. After 25 minutes of pumping, the pressure in the reservoir rises to 7 bar and the temperature becomes 50°C. Determine :

(a) free air delivered at 15°C

(b) volumetric efficiency at atmospheric conditions and

(c) mechanical and isothermal efficiency for the compressor

It may be pressured that indicated mean effective pressure is 2.25 bar.

Solution : Mass of air initially in the reservoir,

$$m_1 = \frac{p_1 V_1}{R T_1} = \frac{(1 \times 10^5) \times 1.2}{287 \times 288} = 1.452 \text{ kg}$$

Mass of air finally in the reservoir,

$$m_2 = \frac{p_2 V_2}{R T_2} = \frac{(6 \times 10^5) \times 1.2}{287 \times 323} = 7.767 \text{ kg}$$

Then : mass of air delivered to reservoir in 25 minutes of pumping

$$= 7.767 - 1.452 = 6.315 \text{ kg}$$

(*a*) Volume of free air delivered per minute at 1 bar and 15°C

$$V = \frac{mRT}{p} = \left(\frac{6.315}{25}\right) \times \frac{287 \times 288}{1 \times 10^5} = 0.209 \text{ m}^3$$

(*b*) Volume delivered per stroke

$$= \frac{0.209}{500} = 0.418 \times 10^{-3} \text{ m}^3$$

Swept volume

$$V_s = \frac{\pi}{4}(0.08)^2 \times 0.1 = 0.502 \times 10^{-3} \text{ m}^3$$

Then volumetric efficiency

$$= \frac{0.418 \times 10^{-3}}{0.502 \times 10^{-3}} = 0.832 \text{ or } \mathbf{83.2\%}$$

(c) Indicated power

$$= p_{mi}\, l\, A\left(\frac{N}{60}\right)$$

$$= (2.25 \times 10^5) \times 0.1 \times \frac{\pi}{4}(0.08)^2 \times \frac{500}{60}$$

$$= 942 \text{ Nm/s} = 942 \text{ W} = 0.942 \text{ kW}$$

Compressor mechanical efficiency

$$= \frac{\text{indicated power}}{\text{brake power}} = \frac{0.942}{1.2}$$

= 0.785 or **78.5%**

Isothermal work

$$= p_1 V_1 \log_e\left(\frac{p_2}{p_1}\right)$$

$$= (1 \times 10^5) \times \left(\frac{0.209}{60}\right) \log_e\left(\frac{7}{1}\right)$$

= 677.82 Nm/s

= 677.82 W = 0.6778 kW

Isothermal efficiency

$$= \frac{\text{isothermal work}}{\text{indicated work}} = \frac{0.6778}{0.942}$$

= 0.719 or **71.9%**

EXAMPLE 21.5

The following data pertains to the performance test of single acting reciprocating compressor with bore 15 cm and stroke 12 cm.

Suction pressure and temperature
= 1 bar and 20°C

Discharge pressure and temperature
= 6 bar and 177°C

Speed of machine and shaft power
= 800 rev/min and 6 kW

Mass of air delivered = 1.6 kg/min

Make calculations for the volumetric efficiency, indicated power, isothermal efficiency, mechanical efficiency and overall efficiency.

Solution : Displacement volume

$$= \frac{\pi}{4}\, d^2 l \times \text{speed of machine}$$

$$= \frac{\pi}{4}(0.151)^2 \times 0.12 \times 800$$

$$= 1.696 \text{ m}^3/\text{min}$$

Volume of air delivered at suction conditions (*FAD*)

$$= \frac{mRT_1}{p_1} = \frac{1.6 \times 287 \times 293}{1 \times 10^5}$$

$$= 1.345 \text{ m}^3/\text{min}$$

∴ Volumetric efficiency

$$\eta_v = \frac{\text{volume of air delivered}}{\text{displacement volume}}$$

$$= \frac{1.345}{1.696} = 0.793 \text{ or } \mathbf{79.3\%}$$

Indicated power

$$= \frac{n}{n-1} mRT_1\left[\left(\frac{p_2}{p_1}\right)^{\frac{n-1}{n}} - 1\right]$$

The exponent n can be worked out from the

relation : $\frac{T_2}{T_1} = \left(\frac{p_2}{p_1}\right)^{\frac{n-1}{n}}$

$$\frac{n-1}{n} = \frac{\log_e T_2/T_1}{\log_e p_2/p_1} \text{ or } \frac{1}{n} = 1 - \frac{\log_e T_2/T_1}{\log_e p_2/p_1}$$

$$= 1 - \frac{\log_e (450/293)}{\log_3 (6/1)} = 1 - 0.239$$

$$n = \frac{1}{1-0.239} = \frac{1}{0.761} = 1.314$$

Then : Indicated power

$$= \frac{1.314}{1.314-1} \times \frac{1.6}{60} \times 287$$

$$\times 293\left[\left(\frac{6}{1}\right)^{\frac{1.314-1}{1.314}} - 1\right]$$

= 5016 N m/s = 5016 W = **5.016 kW**

EXAMPLE 21.6

A single-stage single acting reciprocating compressor sucks air at 1 bar and 30°C and delivers 0.6 kg of air per minute at 6 bar. The bore and stroke of the compressor are 10 cm and 15 cm respectively and clearance is 3% of stroke. Taking the index and compression and expansion as 1.3, work out

(a) the volumetric efficiency of the compressor,

(b) the power required if the mechanical efficiency is 85% and

(c) speed of the machine in rev/minute.

Solution : (*a*) Volumetric efficiency

$$\eta_v = 1 - \frac{V_c}{V_s}\left[\left(\frac{p_2}{p_1}\right)^{1/n} - 1\right]$$

$$= 1 - 0.03\left[\left(\frac{6}{1}\right)^{1/1.3} - 1\right]$$

$$= 0.911 \text{ or } \mathbf{91.1\%}$$

(*b*) Indicated power

$$= \frac{n}{n-1} mRT_1\left[\left(\frac{p_2}{p_1}\right)^{\frac{n-1}{n}} - 1\right]$$

$$= \frac{1.3}{1.3-1} \times \left(\frac{0.6}{60}\right) \times 287 \times 303 \left[\left(\frac{6}{1}\right)^{\frac{1.3-1}{1.3}} - 1\right]$$

$$= 1928.9 \text{ N m/s}$$

$$= 1928.9 \text{ W} \simeq 1.929 \text{ kW}$$

Power required =

$$\frac{\text{indicated power}}{\text{mechnanical efficiency}} = \frac{1.929}{0.85} = \mathbf{2.269 \text{ kW}}$$

(*c*) Volume of air delivered at suction conditions,

$$= \frac{mRT_1}{p_1} = \frac{0.6 \times 287 \times 303}{1 \times 10^5} = 0.5217 \text{ m}^3/\text{min}$$

$$\text{Displacement volume} = \frac{\text{volume delivered}}{\text{volumetric efficiency}}$$

$$= \frac{0.5217}{0.961} = 0.5727 \text{ m}^3/\text{min}$$

For a single acting compressor, the displacement volume equals :

$$\frac{\pi}{4} d^2 l \times \text{speed in rev/min}$$

$$\therefore \text{Speed} = \frac{\text{displacement volume}}{\frac{\pi}{4} d^2 l}$$

$$= \frac{0.5727}{\frac{\pi}{4}(0.1)^2 \times 0.15} = \mathbf{486.4 \text{ rev/min}}$$

EXAMPLE 21.7

Consider a single stage reciprocating compressor with intake conditions 0.95 bar pressure, 300 K temperature and delivery pressure 7 bar. The clearance volume is 5% of the stroke volume which is 0.06 m³/s. Determine :

(a) volumetric efficiency referred to intake conditions and atmospheric conditions of 1 bar and 290 K,

(b) free air delivery,

(c) power required to drive the compressor if the ratio of actual to indicated power is 1.25.

The compression and expansion follow the polytropic law $pV^{1.3}$ = constant.

Solution : (*a*) Volumetric efficiency with respect to inlet conditions (suffix 1) is

$$\eta_{v1} = 1 - \frac{V_c}{V_s}\left[\left(\frac{p_2}{p_1}\right)^{1/n} - 1\right]$$

$$= 1 - 0.05\left[\left(\frac{7}{0.95}\right)^{1/1.3} - 1\right] = \mathbf{0.8177}$$

Volumetric efficiency with respect to ambient/atmospheric conditions (suffix 0) is

$$\eta_{v0} = \frac{p_1}{p_0} \times \frac{T_0}{T_1} \times \eta_{v1}$$

$$= \frac{0.95}{1} \times \frac{290}{300} \times 0.8177 = \mathbf{0.751}$$

(*b*) Free air delivery = stroke volume × η_{v0}

$= 0.06 \times 0.751 = \mathbf{0.04506}$ m³/s

(*c*) Work input

$$W = \frac{n}{n-1}(V_1 - V_4)\left[\left(\frac{p_2}{p_1}\right)^{\frac{n-1}{n}} - 1\right]$$

where $(V_1 - V_4)$ is the effective volume sucked in during the suction stroke.

$(V_1 - V_4)$ = stroke volume × η_{v1}

$= 0.06 \times 0.8177 = 0.04906$ m³/sec

Then :

$$W = \frac{1.3}{1.3-1} \times (0.95 \times 10^5) \times 0.04906 \left[\left(\frac{7}{0.95} \right)^{\frac{1.3-1}{1.3}} - 1 \right]$$

$= 11819$ N m/s

$= 11819$ W $= 11.819$ kW

∴ Power required to drive the compressor

$= 11.819 \times 1.25 =$ **14.77 kW**

EXAMPLE 21.8

A single cylinder reciprocating compressor in which the clearance is 5 percent of active stroke sucks air at 1 bar and 300 K. While running at 150 rev/min, the machine compresses air to 6 bar pressure and delivers 5 m³ of free air per second to a receiver. Presuming that the compression and expansion curves follow the polytropic law $pV^{1.3}$ = constant, make calculations for

(a) volumetric efficiency and volume of air taken in per stroke

(b) cylinder dimensions if stroke equals 1.25 times diameter

(c) compressor power

Solution : (*a*) Volumetric efficiency with respect to inlet conditions (suffix 1) is

$$\eta_{v1} = 1 - \frac{V_c}{V_s} \left[\left(\frac{p_2}{p_1} \right)^{\frac{1}{n}} - 1 \right]$$

$$= 1 - 0.05 \left[\left(\frac{6}{1} \right)^{\frac{1}{1.3}} - 1 \right]$$

$= 0.852$ or 85.2%

Volumetric efficiency with respect to ambient/atmospheric conditions (suffix 0) is

$$\eta_{v0} = \frac{p_1}{p_0} \times \frac{T_0}{T_1} \times \eta_{v1}$$

$$= \frac{1}{1} \times \frac{288}{300} \times 0.852 = 0.8179$$

The free air/ambient conditions have been assumed to be 1 bar and 15°C

(*b*) Free air delivery = stroke volume × η_{v0}

$5 = V_s \times 0.8179$; $V_s = 6.113$ m³/min

Swept volume per stroke

$$= \frac{6.113}{150} = 0.04075 \text{ m}^3 = 4075 \text{ cm}^3$$

If d and l are the piston diameter and stroke length respectively in cm, then

$$\frac{\pi}{4} d^2 l = 4075 \quad \text{or} \quad \frac{\pi}{4} d^2 (1.25d) = 4075$$

Solution gives :

$$d = \sqrt[3]{\frac{4075 \times 4}{1.25\,\pi}} = \mathbf{15.63\ cm}$$

and $l = 1.25 \times 15.63 =$ **19.54 cm**

(*c*) Work input

$$W = \frac{n}{n-1} p_1 (V_1 - V_4) \left[\left(\frac{p_2}{p_1} \right)^{\frac{n-1}{n}} - 1 \right]$$

where $(V_1 - V_4)$ is the effective volume sucked in during the suction stroke

$(V_1 - V_4)$ = stroke volume × η_{v1}

$$= \left(\frac{6.113}{60} \right) \times 0.852 = 0.868 \text{ m}^3/\text{s}$$

Then :

$$W = \frac{1.3}{1.3-1} \times (1 \times 10^5) \times 0.068 \left[\left(\frac{6}{1} \right)^{\frac{1.3-1}{1.3}} - 1 \right]$$

$= 19253$ N m/s

$= 19253$ W $=$ **19.25 kW**

EXAMPLE 21.9

A single acting single cylinder reciprocating compressor of 20 cm bore and 25 cm stroke is constructed so that its clearance can be altered by moving the cylinder head; the stroke being unaffected.

(a) With the clearance volume set at 700 cm³, the compressor sucks air at 1 bar and 32°C. The compressor turns 300 rev/min and delivers the compressed air to a receiver at 5 bar. Taking the index of compression and expansion as 1.25,

calculate the free air delivered at 1.013 bar and 15°C and also the power required to run the machine. Assume mechanical efficiency of 80%.

(b) To what minimum value may the clearance volume be reduced when the delivery pressure is 4.2 bar. It may be assumed that the same driving power is available and that the suction pressure, speed, value of exponent n and the mechanical efficiency remain unaltered.

Solution : (*a*) Refer Fig. 21.6.

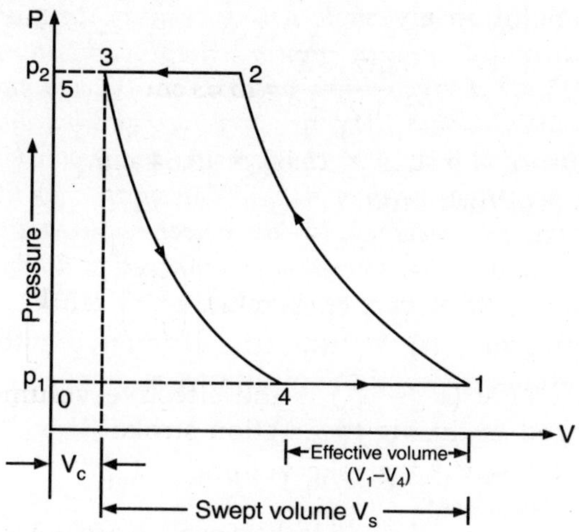

Fig. 21.6.

Swept volume

$$V_S = \frac{\pi}{4}d^2 l = \frac{\pi}{4}(20)^2 \times 25 = 7850 \text{ cm}^3$$

Volumetric efficiency at suction conditions,

$$\eta_{vs} = 1 - \frac{V_c}{V_s}\left[\left(\frac{p_2}{p_1}\right)^{1/n} - 1\right]$$

$$= 1 - \frac{700}{7850}\left[\left(\frac{5}{1}\right)^{1/1.25} - 1\right] = 0.766$$

Volumetric efficiency at free air conditions (suffix 0)

$$\eta_{v0} = \frac{T_0}{T_s} \times \frac{p_s}{p_0} \times \eta_{vs}$$

$$= \frac{288}{305} \times \frac{1}{1.013} \times 0.766 = 0.714$$

Free air delivery (*FAD*)

$= V_s \times \eta_{v0} \times$ speed of machine

$= 7850 \times 0.714 \times 300$

$= 1681470 \text{ cm}^3/\text{min}$

$\simeq 1.681 \text{ m}^3/\text{min}$

Mass of air delivered

$$m = \frac{p_0 \,(FAD)}{RT_0} = \frac{1.013 \times 10^5 \times 1.681}{287 \times 288}$$

$= 2.06$

Indicated power

$$= \frac{n}{n-1} mRT_1 \left[\left(\frac{p_2}{p_1}\right)^{\frac{n-1}{n}} - 1\right]$$

$$= \frac{1.25}{1.25-1} \times \frac{2.06}{60} \times 287 \times 305 \left[\left(\frac{5}{1}\right)^{\frac{1.25-1}{1.25}} - 1\right]$$

$= 5706$ N m/s $= 5706$ W $= 5.706$ kW

Power required to drive the compressor

$$= \frac{\text{indicated power}}{\text{mechanical efficiency}}$$

$$= \frac{5.706}{0.8} = \mathbf{7.132 \text{ kW}}$$

(*b*) When the delivery pressure is 4.2 bar and all other parameters remain unaltered, then the mass of air delivered can be worked out from the relation for indicated power. That is

$$\text{Indicated power} = \frac{n}{n-1} m_1 RT_1 \left[\left(\frac{p_2}{p_1}\right)^{\frac{n-1}{n}} - 1\right]$$

$$5.706 = \frac{1.25}{1.25-1} \times \frac{m_1}{60} \times 287 \times 305 \left[\left(\frac{4.2}{1}\right)^{\frac{1.25-1}{1.25}} - 1\right]$$

Solution gives : $m_1 = 2.353$ kg/min

Apparently the mass flow rate increases and accordingly the volumetric efficiency will take the value

$$\eta_{vs} = \frac{m_1}{m} \times \text{initial } \eta_{vs}$$

$$= \frac{2.353}{2.06} \times 0.766 = 0.875$$

$$\therefore\ 0.875 = 1-\frac{V_c}{V_s}\left[\left(\frac{p_2}{p_1}\right)^{1/n}-1\right]$$

$$= 1-\frac{V_c}{7850}\left[\left(\frac{4.2}{1}\right)^{1/1.25}-1\right]$$

$$= 1-\frac{V_c}{7850}\times 2.152$$

Solution gives the requisite minimum clearance

$$V_c = (1-0.875)\times\frac{7850}{2.152} = \mathbf{455.97\ cm^3}$$

EXAMPLE 21.10

Set up the following relation for heat rejected per stage per kg of air in a reciprocating compressor with perfect inter cooling :

$$= \left[c_p + c_v\left(\frac{\gamma - n}{n-1}\right)\right][T_2 - T_1]$$

where $(T_2 - T_1)$ is the temperature rise during compression. The other symbols have their usual meanings.

Solution : The total heat rejected equals the summation of heat rejected during compression process and the heat rejected in the intercooler.

When the air is compressed polytropically from state (p_1, v_1) to state (p_2, v_2), the specific work input is

$$w = \frac{p_2v_2 - p_1v_1}{n-1} = \frac{R\,(T_2-T_1)}{n-1}$$

Then heat rejected during compression process is

$$q_1 = \frac{\gamma-n}{\gamma-1}\times\text{work done}$$

$$= \frac{\gamma-n}{\gamma-1}\times\frac{R\,(T_2-T_1)}{n-\text{i}} \qquad ...(i)$$

Heat rejected in the intercooler at constant pressure is

$$q_2 = c_p\,(T_2 - T_1) \qquad ...(ii)$$

Summing (*i*) and (*ii*), the total heat rejected per unit mass of air is

$$q = \frac{\gamma-n}{\gamma-1}\,\frac{R\,(T_2-T_1)}{n-1} + c_p\,(T_2-T_1)$$

Replacing $\frac{R}{\gamma-1}$ by c_v, we get

$$q = \frac{\gamma-n}{n-1}\,c_v\,(T_2-T_1) + c_p\,(T_2-T_1)$$

$$= \left[c_p + c_v\left(\frac{\gamma-n}{n-1}\right)\right][T_2-T_1]$$

which is the required result.

EXAMPLE 21.11

A double acting single acting reciprocating air compressor has a piston displacement of 0.015 m³ per revolution, operates at 500 rpm and has 5% clearance. The air is received at 1 bar and delivered at 6 bar. The compression and expansion are polytropic with n = 1.3. Determine : (i) the volumetric efficiency, (ii) the power required and (iii) the heat transferred and its direction during compression if inlet temperature of air is 20°C.

Solution : (*i*) Volumetric efficiency with reference to suction conditions is

$$\eta_v = 1-\frac{V_c}{V_s}\left[\left(\frac{p_2}{p_1}\right)^{1/n}-1\right]$$

$$= 1-0.05\left[\left(\frac{6}{1}\right)^{1/1.3}-1\right]$$

$= 0.8517$ or **85.17%**

(*ii*) Power required to drive the compressor

$$= \frac{n}{n-1}\,p_1V_1\left[\left(\frac{p_2}{p_1}\right)^{\frac{n-1}{n}}-1\right]$$

where V_1 = piston displacement per revolution × $2N$ × η_v

The compressor speed is taken twice as the compressor is double acting.

$$V_1 = 0.015 \times (2\times 500)\times 0.8517$$
$$= 12.77\ \text{m}^3/\text{min}$$

∴ Power required

$$= \frac{1.3}{1.3-1}\times(1\times10^5)\times\left(\frac{12.77}{60}\right)\times\left[\left(\frac{6}{1}\right)^{\frac{1.3-1}{1.3}}-1\right]$$

$= 47228$ N m/s = **47.23 kW**

(*iii*) Work input during compression process

$$= \frac{p_2V_2 - p_1V_1}{n-1} = \frac{m\,R\,(T_2 - T_1)}{n-1}$$

Heat rejected during compression process

$$= \frac{\gamma - n}{\gamma - 1} \times \text{work done}$$

$$= \frac{\gamma - n}{\gamma - 1} \times \frac{m\,R\,(T_2 - T_1)}{n-1}$$

Mass of air

$$m = \frac{p_1V_1}{RT_1} = \frac{1\times10^5 \times 12.77}{287\times293}$$

$= 15.186$ kg/min $= 0.2531$ kg/sec

$$T_2 = T_1\left(\frac{p_2}{p_1}\right)^{\frac{n-1}{n}}$$

$$= 293\left(\frac{6}{1}\right)^{\frac{1.3-1}{1.3}} = 443 \text{ K}$$

Then : Heat rejected

$$= \frac{1.4-1.3}{1.4-1} \times 0.2531 \times \frac{287(443-293)}{1.3-1}$$

$= 9080$ J/s $=$ **9.08 kW**

EXAMPLE 21.12

An air compressor cylinder has 12 cm bore and 15 cm stroke and the clearance is 5% of stroke volume. The machine operates between 1 bar, 300 K and 5 bar. Taking exponent for polytronic compression as 1.3, determine :

(a) the mass flow of air and the flow capacity in m³/min at 750 rev/minute,

(b) the mean effective pressure and

(c) the heat transferred as a fraction of the indicated work.

Solution : Refer Fig. 21.7.

V_1 = stroke volume V_s + clearance volume V_c

$$= \frac{\pi}{4}(0.12)^2 \times 0.15 + \left[\frac{\pi}{4}(0.12)^2 \times 0.15\right] \times 0.05$$

$= 16.95 \times 10^{-4} + 0.85 \times 10^{-4}$

$= 17.81 \times 10^{-4}$ m³

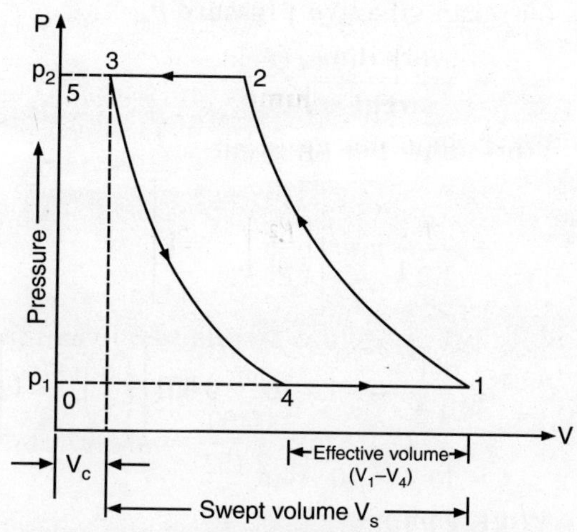

Fig. 21.7.

$$V_4 = V_3 \times \left(\frac{p_3}{p_4}\right)^{1/n} = V_c \times \left(\frac{p_2}{p_1}\right)^{1/n}$$

$$= 0.85\times10^{-4} \times \left(\frac{5}{1}\right)^{1/1.3} = 2.93\times10^{-4} \text{ m}^3$$

(*a*) Mass flow of air/cycle $= \dfrac{V_1 - V_4}{v_1}$

where v_1 = specific volume of air that enters the compressor

$$= \frac{RT_1}{p_1} = \frac{287\times300}{1\times10^5} = 0.861 \text{ m}^3/\text{kg}$$

$$\therefore \text{ Mass flow} = \frac{17.81\times10^{-4} - 2.93\times10^{-4}}{0.861}$$

$=$ **17.28 × 10⁻⁴ kg**

Flow capacity

$= (V_1 - V_4) \times$ speed of the machine

$= (17.81 \times 10^{-4} - 2.93 \times 10^{-4}) \times 750$

$=$ **1.117** m³/min

(*b*) Volumetric efficiency

$$\eta_v = \frac{V_1 - V_4}{V_s}$$

$$= \frac{17.81\times10^{-4} - 2.93\times10^{-4}}{16.96\times10^{-4}}$$

$= 0.877$ or **87.7%**

(*c*) Mean effective pressure p_m

$$= \frac{\text{work done/cycle}}{\text{swept volume}}$$

Work done per kg of air

$$= \frac{n}{n-1} p_1 v_1 \left[\left(\frac{p_2}{p_1} \right)^{\frac{n-1}{n}} - 1 \right]$$

$$= \frac{1.3}{1.3-1} \times 1 \times 10^5 \times 0.861 \left[\left(\frac{5}{1} \right)^{\frac{0.3}{1.3}} - 1 \right]$$

$= 1.67 \times 10^5$ Nm

Work done per cycle

$= 1.67 \times 10^5 \times 17.28 \times 10^{-4} = 288.6$ Nm

∴ Mean effective pressure

$$p_m = \frac{288.6}{16.96 \times 10^{-4}}$$

$= 1.7 \times 1065$ N/m^2 = **1.7 bar**

21.6. MULTISTAGE COMPRESSION

Multistaging means compression of air from the suction pressure to delivery pressure in more than one cylinder. The cylinders are arranged in series one after the other. The air after compression in the first cylinder and before entering the next cylinder is cooled at constant pressure after passing it through an intercooler/heat exchanger.

The process of carrying out the compression of gas above 0.6 to 0.8 MN/m^2 in a single compressor causes the following adverse effects:

(*i*) higher compression ratios result into high delivery temperature and that may damage the lubrication system.

(*ii*) at high pressure ratios, there is considerable fall in volumetric efficiency. This aspect is obvious from the expression

$$\eta_v = 1 - \frac{V_c}{V_s} \left[\left(\frac{p_2}{p_1} \right)^{\frac{\gamma-1}{\gamma}} - 1 \right]$$

and from the plot 21.8. With increase in pressure ratio from p_2/p_1 to p_3/p_1, the effective volume sucked in reduces from $(V_1 - V_4)$ to $(V_1 - V_7)$. At a certain limit of pressure ratio, the volumetric efficiency may even approach zero which means no delivery. Before this extreme condition is reached, the compression process is split up into two or more stages; the following arrangement is normally used with regard to the number of stages :

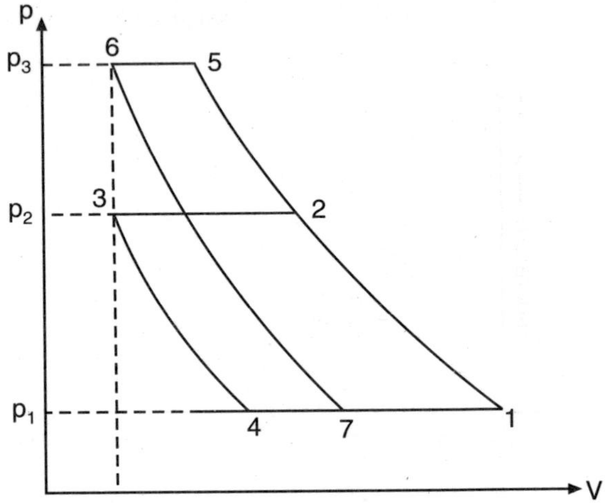

Fig. 21.8. Effect of pressure ratio on volumetric efficiency

(*a*) Single stage compression for delivery pressures upto 5.6 bar

(*b*) Two stage compression for delivery pressures from 5.6 to 35 bar

(*c*) Three stage compression for delivery pressures from 35 bar to 84 bar

(*d*) Four stage compression for delivery pressures above 84 bar

Fig. 21.9. shows the schematic arrangement of a two-stage compressor with intercooler, and the pressure-volume and temperature-entropy plots of the compression cycle.

The air enters the low pressures cylinder at a point 1 with volume V_1 and pressure p_1. The air is compressed polytropically to point 2. This compressed air at intermediate pressure p_2 and temperature T_2 is then cooled at constant pressure by passing it through the intercooler. Cooling is represented by line 2-*a*. The air then enters the high pressure cylinder where it is further compressed polytropically to pressure p_3 along the path *a*-3.

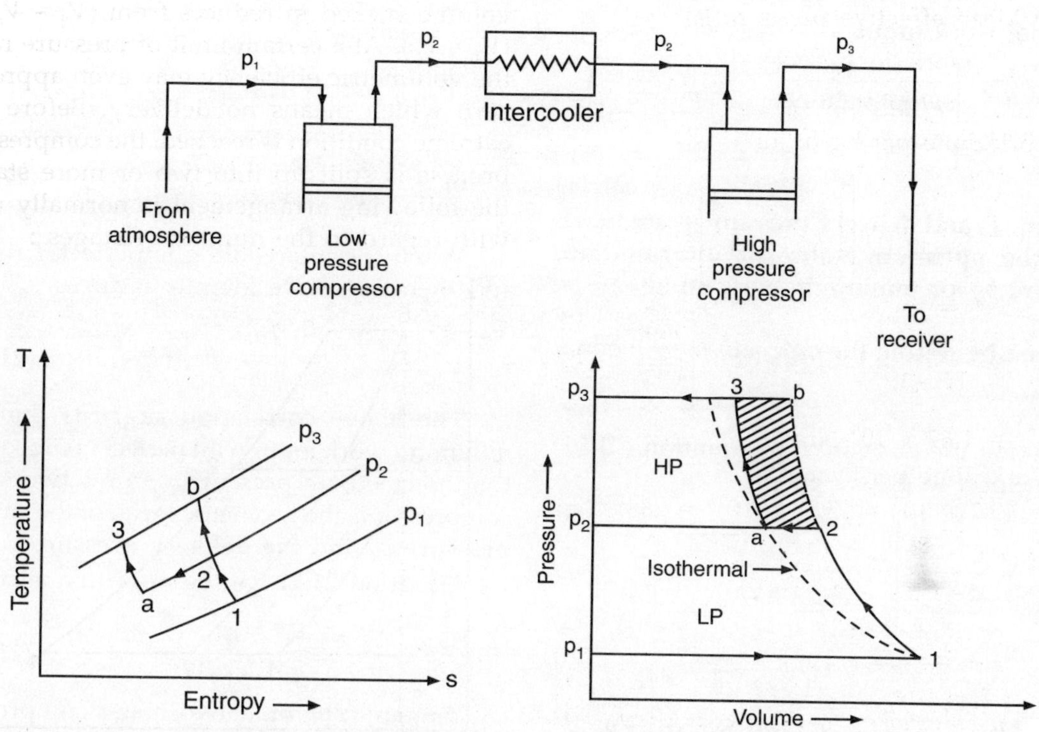

Fig. 21.9. Multi-stage compression with intercooling

The cooling is said to be perfect if the air after compression in the low pressure cylinder is cooled to the original temperature T_1. For perfect cooling, the state points 1 and a would lie on the same isothermal curve. Without intercooling, the compression would have followed the path 1-2-b. The shaded area a-2-b-3 represents the saving in work input due to intercooling.

Work input to low pressure cylinder

$$= \frac{n}{n-1} p_1 V_1 \left[\left(\frac{p_2}{p_1}\right)^{\frac{n-1}{n}} - 1\right]$$

Work input to high pressure cylinder

$$= \frac{n}{n-1} p_2 V_a \left[\left(\frac{p_3}{p_2}\right)^{\frac{n-1}{n}} - 1\right]$$

∴ Total work input to compressor

$$= \frac{n}{n-1}\left[p_1 V_1 \left\{\left(\frac{p_2}{p_1}\right)^{\frac{n-1}{n}} - 1\right\} + p_2 V_a \left\{\left(\frac{p_3}{p_2}\right)^{\frac{n-1}{n}} - 1\right\}\right] \quad ...(21.10)$$

These expressions have been written with the assumptions that :

(*i*) The index n for polytropic compression is same for each stage.

(*ii*) Intercooling is at constant pressure.

(*iii*) There is no pressure drop between the two stages, *i.e.*, exhaust pressure of low pressure stage equals the suction pressure of the high pressure stage.

(*iv*) Low pressure and high pressure cylinders have the same mass of air.

For unit mass of air

$p_1 V_1 = RT_1$ and $p_2 V_a = RT_a$

Making these substitutions, equation 21.10 can be rewritten as

Total work input

$$= \frac{n}{n-1} R\left[T_1\left\{\left(\frac{p_2}{p_1}\right)^{\frac{n-1}{n}} - 1\right\} + T_a\left\{\left(\frac{p_3}{p_2}\right)^{\frac{n-1}{n}} - 1\right\}\right] \quad ...(21.11)$$

If p_1, T_1 and delivery pressure p_3 are fixed, then the optimum value of intermediate pressure p_2 for minimum work input can be obtained by setting the derivative $\frac{dW}{dp_2}$ equal to zero.

Taking the derivative of equation (21.11) with respect to p_2, we get

$$\frac{dW}{dp_2} = \frac{n}{n-1} R\left[T_1 \frac{n-1}{n}\left(\frac{p_2}{p_1}\right)^{\frac{n-1}{n}-1} \times \frac{1}{p_1}\right.$$

$$\left. + T_a \frac{n-1}{n}\left(\frac{p_3}{p_2}\right)^{\frac{n-1}{n}-1} \times \left(-\frac{p_3}{p_2^2}\right)\right]$$

$$= \frac{n}{n-1} R \frac{n-1}{n}\left[\frac{T_1}{p_1}\left(\frac{p_2}{p_1}\right)^{-\frac{1}{n}} - \frac{p_3 T_a}{p_2^2}\left(\frac{p_3}{p_2}\right)^{-\frac{1}{n}}\right]$$

$$= R\left[\frac{T_1}{p_1}\left(\frac{p_2}{p_1}\right)^{-\frac{1}{n}} - \frac{p_3\, T_a}{p_2^2}\left(\frac{p_3}{p_2}\right)^{-\frac{1}{n}}\right]$$

For optimum value of pressure p_2, the above identify is set equal to zero. Then

$$\frac{T_1}{p_1}\left(\frac{p_2}{p_1}\right)^{-\frac{1}{n}} = \frac{p_3\, T_a}{p_2^2}\left(\frac{p_3}{p_2}\right)^{-\frac{1}{n}}$$

or
$$\frac{p_2^{-\frac{1}{n}}}{p_1 p_1^{-\frac{1}{n}}} \times p_2^2\, p_2^{-\frac{1}{n}} = \frac{T_a}{T_1} p_3\, p_3^{-\frac{1}{n}}$$

or
$$\frac{p_2^{\frac{2(n-1)}{n}}}{p_1^{\frac{n-1}{n}}} = \frac{T_a}{T_1} p_3^{\frac{n-1}{n}}$$

Dividing both sides by $p_1^{\frac{n-1}{n}}$, we get

$$\left(\frac{p_2}{p_1}\right)^{\frac{2(n-1)}{n}} = \frac{T_a}{T_1}\left(\frac{p_3}{p_1}\right)^{\frac{n-1}{n}}$$

or
$$\frac{p_2}{p_1} = \sqrt{\left(\frac{T_a}{T_1}\right)^{\frac{n}{n-1}} \frac{p_3}{p_1}} \quad ...(21.12)$$

When the intercooling is perfect $T_a = T_1$, and then the above identity reduces to

$$\frac{p_2}{p_1} = \sqrt{\frac{p_3}{p_1}} \text{ or } p_2 = \sqrt{p_1 p_3} \quad ...(21.13)$$

The above correlation suggests that for minimum work input with perfect intercooling, the intermediate pressure p_2 for a two stage compressor is the geometric mean of the supply pressure p_1 and the delivery pressure p_3.

Equation 21.12 can be rewritten as

$$\frac{p_2}{p_1} = \frac{P_3}{P_2} = \sqrt{\frac{P_3}{P_1}} \quad ...(21.14)$$

Apparently for a two stage compressor with perfect intercooling, the pressure ratios for each stage are equal and are given by the square root of overall pressure ratio.

In general, if there are N stages, the pressure ratio for each stage is

$$\frac{p_2}{p_1} = \frac{p_3}{p_2} = \frac{p_4}{p_3} = \ldots\ldots\ldots\ldots \frac{p_{N+1}}{p_N} \quad ...(21.15)$$

With perfect intercooling ($T_a = T_1$) and therefore the expression 21.11 for input can be rewritten as

Work input

$$= \frac{n}{n-1} RT_1\left[\left(\frac{p_2}{p_1}\right)^{\frac{n-1}{n}} + \left(\frac{p_3}{p_2}\right)^{\frac{n-1}{n}} - 2\right] \quad ...(21.16)$$

The expression for minimum work input can be determined by substituting $\frac{p_2}{p_1} = \frac{p_3}{p_2} = \sqrt{\frac{p_3}{p_1}}$ in the above identity.

Minimum work input

$$= \frac{n}{n-1} RT_1\left[\left(\frac{p_3}{p_1}\right)^{\frac{n-1}{2n}} + \left(\frac{p_3}{p_1}\right)^{\frac{n-1}{2n}} - 2\right]$$

$$= \frac{2n}{n-1} RT_1 \left[\left(\frac{p_3}{p_1} \right)^{\frac{n-1}{2n}} - 1 \right] \quad ...(21.17)$$

If the expansion process is carried out in a three stage compressor with perfect cooling in between, then

Work expended

$$= \frac{n}{n-1} RT_1 \left[\left(\frac{p_2}{p_1} \right)^{\frac{n-1}{n}} + \left(\frac{p_3}{p_2} \right)^{\frac{n-1}{n}} + \left(\frac{p_4}{p_3} \right)^{\frac{n-1}{n}} - 3 \right]$$

and optimum pressure ratio for each stage is

$$\frac{P_2}{P_1} = \frac{p_3}{p_2} = \frac{p_4}{p_3} = \sqrt[3]{\frac{p_4}{p_1}}$$

∴ Minimum work input

$$= \frac{n}{n-1} RT_1 \left[\left(\frac{p_4}{p_1} \right)^{\frac{n-1}{3n}} + \left(\frac{p_4}{p_1} \right)^{\frac{n-1}{3n}} + \left(\frac{p_4}{p_1} \right)^{\frac{n-1}{n}} - 3 \right]$$

$$= \frac{3n}{n-1} RT_1 \left[\left(\frac{p_4}{p_1} \right)^{\frac{n-1}{3n}} - 1 \right] \quad ...(21.18)$$

Equations 21.17 and 21.18 clearly indicate that minimum work required per kg mass of air in a *N*-stage compressor with perfect intercooling at all stages shall be given by

$$W = \frac{Nn}{n-1} RT_1 \left[\left(\frac{p_{N+1}}{p_1} \right)^{\frac{n-1}{Nn}} - 1 \right] \quad ...(21.19)$$

where p_{N+1} is final delivery pressure and p_1 is inlet suction pressure.

Compared to single state compression, the multistage compression provides the following main advantages :

1. Improvement in volumetric efficiency of the compressor.

2. Reduction in work input to compressor. This is achieved by cooling the compressed air before its compression in the succeeding stage.

3. Low operating temperatures and consequently there is little chance of lubrication problems.

4. Leakage loss is minimum due to better cylinder lubrication and reduced pressure difference across the two sides of the piston and valve.

5. When air is being compressed in one cylinder, there is suction in the other cylinder. This provides more uniform torque with consequent reduction in the size of flywheel. For the same reasons, the mechanical balance is more perfect.

6. The compressing machine has a lighter construction. Further, since the operating temperatures are lower, cheaper material can be used for construction. These aspects result in saving the material cost of construction.

The multi-stage compression has increased number of parts and the unit becomes complicated. That requires more initial investment and maintenance costs. However, this gets neutralized by considerable saving in operation cost in terms of energy consumed in running the compressor.

EXAMPLE 21.13

A two stage single acting air compressor compresses 10 m³/min of air from 100 kPa and 20°C to 1400 kPa with perfect intercooling. What will be the pressure in the intercooler ? If index of compression is 1.3, find the percentage saving in work by compressing in two stages compared to single stage compression.

Solution : Intermediate pressure, $p_2 = \sqrt{p_1 p_3}$ where p_1 and p_3 are the suction and discharge pressures respectively.

$$p_2 = \sqrt{100 \times 1400} = \mathbf{374\ kPa}$$

Work input in two stage compression is

$$= \frac{2n}{n-1} p_1 v_1 \left[\left(\frac{p_3}{p_1} \right)^{\frac{n-1}{2n}} - 1 \right]$$

$$= \frac{2 \times 1.3}{1.3 - 1} (100 \times 1000) \times \frac{10}{60}$$

$$\times \left[\left(\frac{1400}{100} \right)^{\frac{1.3-1}{2 \times 1.3}} - 1 \right] = 51413 \text{ N m/s}$$

For single stage compression, the work input is

$$= \frac{n}{n-1} p_1 v_1 \left[\left(\frac{p_3}{p_1} \right)^{\frac{n-1}{n}} - 1 \right]$$

$$= \frac{1.3}{1.3-1} (100 \times 1000) \times \frac{10}{60}$$

$$\times \left[\left(\frac{1400}{100} \right)^{\frac{1.3-1}{1.3}} - 1 \right] = 60567 \text{ N m/s}$$

∴ Percentage saving in work

$$= \frac{60567 - 51413}{60567} = 0.151 \text{ or } \mathbf{15.1\%}$$

EXAMPLE 21.14

A multi-stage reciprocating air compressor admits air at 1 bar, 300 K and delivers the compressed air at 160 bar. The intercooling between stages is at optimum intercooler pressure and perfect intercooling, and the maximum discharge temperature at any stage is limited to 425 K. For all the stages, the compression and expansion index is 1.25. Neglecting the effect of clearance, determine the minimum number of stages required.

Solution : Let N be the number of stages. When perfect and optimum intercooling is considered, then pressure ratio r_p for each stage is same and is given by

$$r_p = \left(\frac{\text{delivery pressure}}{\text{suction pressure}} \right)^{1/N} = (160)^{1/N}$$

For any stage :

$$\frac{T_2}{T_1} = (r_p)^{\frac{n-1}{n}} \quad \therefore \quad \frac{425}{300} = (160)^{\frac{1.25-1}{1.25 N}}$$

Taking logs, we get

$$\log_e \left(\frac{425}{300} \right) = \frac{0.25}{1.25 N} \log_e 160$$

$$\text{or} \quad 0.3485 = \frac{0.25}{1.25 N} \times 5.075$$

∴ Minimum number of stages required is

$$N = \frac{0.25 \times 5.075}{1.25 \times 0.3485} = 2.91 \text{ say 3 stages}$$

EXAMPLE 21.15

A two stage air compressor takes in air at 1 bar and 300 K, and delivers the compressed air to a receiver at 9 bar. After compression in the first stage, the air is cooled to initial temperature in the intercooler and the intercooler pressure is so chosen that work is divided equally between the two stages. Neglecting clearance and taking compression index n = 1.3 for both the stages, determine :

(a) work expended in running the compressor,

(b) heat abstracted in the intercooler and the mass of cooling water if rise in its temperature is limited to 20°C.

Solution : With perfect intercooling

Intermediate pressure

$$p_2 = \sqrt{p_1 p_3} = \sqrt{1 \times 9} = 3 \text{ bar}$$

Temperature at end of compression,

$$T_2 = T_1 \left(\frac{p_2}{p_1} \right)^{\frac{n-1}{n}} = 300 \times \left(\frac{3}{1} \right)^{\frac{1.3-1}{1.3}} = 386 \text{ K}$$

(*a*) Work done/kg of air,

$$W = \frac{2n}{n-1} R T_1 \left[\left(\frac{p_3}{p_1} \right)^{\frac{n-1}{2n}} - 1 \right]$$

$$= \frac{2 \times 1.3}{1.3-1} \times 287 \times 300 \left[\left(\frac{9}{1} \right)^{\frac{1.3-1}{2 \times 1.3}} - 1 \right]$$

$$= \mathbf{215312 \text{ Nm}}$$

(*b*) Heat abstracted in the intercooler per kg of air cooled

$$= c_{pa} (T_2 - T_1)$$
$$= 1.005 (386 - 300) = \mathbf{86.4 \text{ kJ}}$$

If m is the mass of cooling water, then

mc_{pw} × rise in temperature of water

= heat abstracted in the intercooler

$$\therefore m = \frac{86.4}{4.186 \times 20} = \mathbf{1.032 \text{ kg/kg of air}}$$

EXAMPLE 21.16

A two stage single acting reciprocating compressor takes in air at the rate of 0.2 m³/s. The intake pressure and temperature of air are 0.1 MPa and

16°C. The air is compressed to a final pressure of 0.7 MPa. The intercooling is perfect and the intermediate pressure is for minimum work input conditions.

The compression index in both the stages is 1.25 and the compressor runs at 600 rev per min. Neglecting clearance, determine

(a) the intermediate pressure,

(b) the total volume of each cylinder,

(c) the power required to drive the compressor and

(d) the rate of heat rejection in the intercooler.

Take c_p = 1.005 kJ/kg K and
R = 0.287 kJ/kg K

Solution : Refer Fig. 21.10.

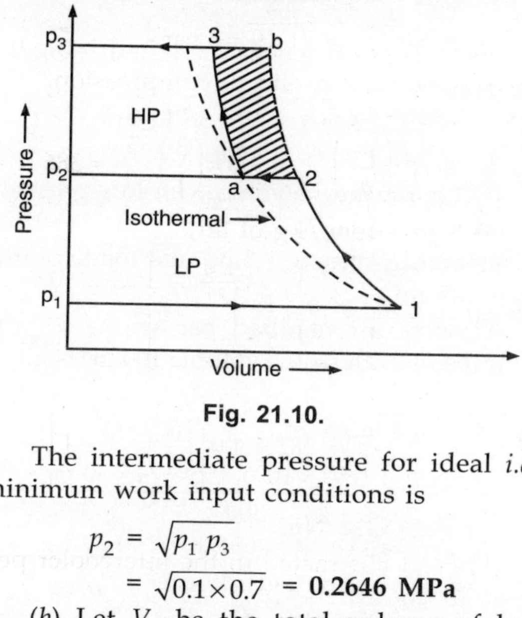

Fig. 21.10.

The intermediate pressure for ideal *i.e.*, minimum work input conditions is

$$p_2 = \sqrt{p_1 p_3}$$
$$= \sqrt{0.1 \times 0.7} = \mathbf{0.2646\ MPa}$$

(*b*) Let V_1 be the total volume of low pressure cylinder. Then,

$$V_1 \times \frac{N}{60} = 0.2\,;\ V_1 = \frac{60 \times 0.2}{600} = \mathbf{0.02\ m^3}$$

Since the intercooling is perfect, the points 1 and *a* lie on the isothermal curve and accordingly

$$p_2 V_a = p_1 V_1$$

$$V_a = \frac{p_1}{p_2} V_1$$

$$= \frac{0.1}{0.2646} \times 0.02 = 0.756 \times 10^{-2}\ \text{m}^3$$

$V_a = 0.756 \times 10^{-2}$ m^3 represents the total volume of high pressure cylinder.

(*c*) For ideal conditions, the work input to drive the compressor is

$$= \frac{2n}{n-1} p_1 V_1 \left[\left(\frac{p_3}{p_1} \right)^{\frac{n-1}{2n}} - 1 \right]$$

Here V_1 is the volume flow rate of air.

$$W = \frac{2 \times 1.25}{1.25 - 1} \times (0.1 \times 10^6) \times 0.2 \left[\left(\frac{0.7}{0.1} \right)^{\frac{1.25-1}{2 \times 1.25}} - 1 \right]$$

$$= 42963\ \text{N m/s} = 42963\ \text{J/s} = \mathbf{42.963\ kW}$$

(*d*) Mass flow rate of air,

$$m = \frac{pV}{RT}$$

$$= \frac{0.1 \times 10^6 \times 0.2}{287 \times 289} = 0.241\ \text{kg/s}$$

$$T_2 = T_1 \left(\frac{p_2}{p_1} \right)^{\frac{n-1}{n}}$$

$$= 289 \left(\frac{0.2646}{0.1} \right)^{\frac{1.25-1}{1.25}} = 351\ \text{K}$$

Then : heat rejected in the intercooler

$$= mc_p (T_2 - T_a)$$
$$= mc_p (T_2 - T_1)$$
$$= 0.241 \times 1.005\ (351 - 289)$$
$$= \mathbf{15.02\ kJ/s}$$

EXAMPLE 21.17

Consider a two-stage single acting reciprocating compressor with the following data :

Pressure and temperature at commencement of compression = 0.97 bar and 300 K

Clearance volume for low pressure cylinder = 5% of its swept volume

Delivery pressure and compressor speed = 18 bar and 200 rev/min

Compression and expansion index for each stage = 1.25

Capacity of compressor = 2.5 m³/min under free conditions of 1 bar and 290 K

If the compressor has perfect intercooling and it is to operate with intermediate (intercooler) pressure of 4 bar, calculate :

(a) theoretical power spent in driving the compressor,

(b) dimensions of low pressure cylinder if the bore is equal to stroke,

(c) volumetric efficiency and isothermal efficiency,

(d) mass of cooling water required for abstracting heat in the intercooling if the temperature rise of cooling water is limited to 20°C.

Solution : Refer Fig. 21.11 for the *p-v* plot for the two stage compressor.

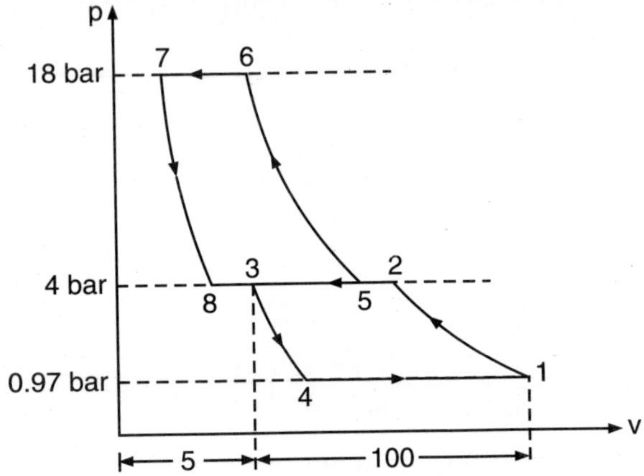

Fig. 21.11.

$$T_2 = T_1\left(\frac{p_2}{p_1}\right)^{\frac{n-1}{n}}$$

$$= 300 \times \left(\frac{4}{0.97}\right)^{\frac{1.25-1}{1.25}} = 398.27 \text{ K}$$

$$T_6 = T_5\left(\frac{p_6}{p_5}\right)^{\frac{n-1}{n}}$$

$$= 300 \times \left(\frac{18}{4}\right)^{\frac{1.25-1}{1.25}} = 405.3 \text{ K}$$

$$(\because T_5 = T_1 = 300 \text{ K})$$

Work input in low pressure cylinder per kg of air,

$$W_{lp} = \frac{n}{n-1} R\,(T_2 - T_1)$$

$$= \frac{1.25}{1.25-1} \times 0.287\,(398.27 - 300)$$

$$= 141.02 \text{ kJ}$$

Work input in high pressure cylinder per kg of air

$$W_{hp} = \frac{n}{n-1} R\,(T_6 - T_5)$$

$$= \frac{1.25}{1.25-1} \times 0.287\,(405.3 - 300)$$

$$= 151.10 \text{ kJ}$$

Total work input per kg of air

$$= 141.02 + 151.10 = 292.12 \text{ kJ}$$

Mass of air handled per minute,

$$m = \frac{pV}{RT} = \frac{(1 \times 10^5) \times 2.5}{287 \times 290} = 3 \text{ kg}$$

(*a*) Theoretical power spent in driving the compressor

$$= 292.12 \times 3 = 876.36 \text{ kJ/min}$$

$$= 14.61 \text{ kJ/s} = \mathbf{14.61} \text{ kW}$$

(*b*) Let the stroke volume for low pressure cylinder be equal to 100 m^3. Then

clearance volume = 5 m^3 and total volume V_1 = 105 m^3

Mass of air handled per stroke for the compressor with stroke volume 100 m^3 is given by

$$m = m_1 - m_3\,.$$

$$= \frac{0.97 \times 10^5 \times 105}{287 \times 300} - \frac{4 \times 10^5 \times 5}{287 \times 398.27}$$

$$= 118.29 - 17.50 = 100.79 \text{ kg}$$

Actual volume per minute

$$= \frac{100}{100.79} \times 3 = 2.976 \text{ m}^3$$

$$\therefore\ 2.976 = \frac{\pi}{4} d^2 l\, N$$

where d = bore and l = stroke

$$= \frac{\pi}{4} d^2 \times d \times 200$$

Solution gives : $d = l = 0.267$ m = **26.7 cm**

(*c*) The volumetric efficiency is always referred to low pressure cylinder

$$V_2 = V_1\left(\frac{p_1}{p_2}\right)^{1/n}$$

$$= 10^5 \times \left(\frac{0.97}{4}\right)^{1/1.25} = 33.80 \text{ m}^3$$

Then volume delivered by low pressure cylinder at 4 bar and 398.27 K

$$= V_2 - V_3 = 33.80 - 5 = 28.8 \text{ m}^3$$

The corresponding volume delivered under free air conditions

$$= 28.8 \times \frac{4}{1} \times \frac{298}{398.27} = 86.2 \text{ m}^3$$

∴ Volumetric efficiency

$$= \frac{86.2}{100} = 0.862 \text{ or } \mathbf{86.2\%}$$

When the compressor is working within the same pressure limits, the isothermal work done per kg of air is

$$= R\,T_i \log_e \left(\frac{p_d}{p_i}\right)$$

$$= 287 \times 300 \log_e \left(\frac{18}{0.97}\right)$$

$$= 251480 \; J = 251.48 \text{ kJ}$$

∴ Isothermal efficiency

$$= \frac{251.48}{292.12} = 0.861 \text{ or } \mathbf{86.1\%}$$

(*d*) Heat abstracted in the intercooler per kg of air cooled

$$= c_{pa}\,(T_2 - T_5)$$
$$= 1.005\,(398.27 - 300)$$
$$(\because T_5 = T_1 = 300 \text{ K})$$
$$= 98.76 \text{ kJ/kg of air}$$

Then : Mass of cooling water

$$= \frac{98.76}{c_{pw} \times \Delta t} = \frac{98.76}{4.186 \times 20}$$

$$= 1.18 \text{ kg of water/kg of air}$$

EXAMPLE 21.18

A two stage reciprocating air compressor has air entering at 1 bar, 290 K and leaving high pressure stage at 5 bar. An intercooler working at 2 bar pressure is provided between the high pressure and low pressure stages which cools the air at constant pressure to 295 K. The air is then taken to the high pressure stage where further compression continues upto the delivery pressure. The compression process follows the law $pV^{1.3}$ in both the stages and the compressor turns 200 rev/min. Calculate the power needed in compressing 0.2 m^3 of air at suction conditions to low pressure cylinder.

(a) What should be the intermediate pressure if the machine is required to operate with minimum work input ?

(b) Both the cylinders have the same stroke and their piston diameters are 10 cm and 6 cm and the volumetric efficiency is 0.85 for low pressure cylinder. Comment whether the intercooler pressure will be steady or will it rise or fall as the compressor continues working ? Assume perfect intercooling.

Solution : Refer Fig. 21.12.

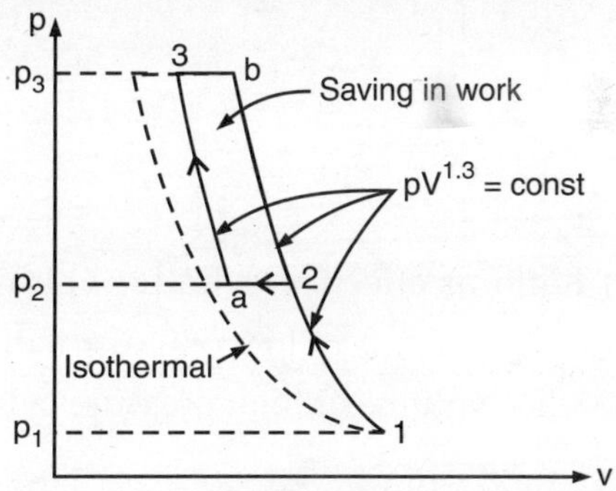

Fig. 21.12.

Work input per kg of air

$$= \frac{n}{n-1} R\,T_1 \left[\left(\frac{p_2}{p_1}\right)^{\frac{n-1}{n}} - 1\right]$$

$$+ \frac{n}{n-1} R\,T_a \left[\left(\frac{p_3}{p_2}\right)^{\frac{n-1}{n}} - 1\right]$$

$$= \frac{1.3}{1.3-1} \times 287 \times 290 \left[\left(\frac{2}{1}\right)^{\frac{1.3-1}{1.3}} - 1\right]$$

$$+ \frac{1.3}{1.3-1} \times 287 \times 295 \left[\left(\frac{5}{2}\right)^{\frac{1.3-1}{1.3}} - 1\right]$$

$$= 62561 + 86390$$
$$= 148951 \text{ Nm/kg of air}$$

Applying characteristic gas equation,

$$m = \frac{pV}{RT} = \frac{1\times10^5\times0.2}{287\times290}$$
$$= 0.2403 \text{ kg/stroke}$$

$$= 0.2403\times\frac{200}{60} = 0.801 \text{ kg/s}$$

∴ Power needed to run the compressor

$$= 148951 \times 0.801$$
$$= 119310 \text{ Nm/s} = \mathbf{119.31\ kW}$$

(*a*) For minimum work input or maximum efficiency, the intermediate pressure is

$$p_2 = \left[\left(\frac{T_a}{T_1}\right)^{\frac{n}{n-1}}\times\frac{p_3}{p_1}\right]^{1/2}$$

$$= \left[\left(\frac{295}{290}\right)^{\frac{1.3}{1.3-1}}\times\frac{5}{1}\right]^{1/2} = \mathbf{2.32\ bar}$$

(*c*) Ratio of effective cylinder volumes

$$= \eta_v\times\frac{\text{volume of low pressure cylinder}}{\text{volume of high pressure cylinder}}$$

$$= 0.85\times\left(\frac{10}{6}\right)^2 = 2.36$$

The ratio of cylinder volumes can also be obtained by isothermal law to the conditions at the commencement of compression in low pressure cylinder and at entry to high pressure cylinder. That gives :

$$\frac{V_{lp}}{V_{hp}} = \frac{p_2}{p_1} = 2$$

Apparently the cylinder in the second stage is small to suck in all the air compressed in the first stage. This leads to accumulation of air in the intercooler resulting in rise of pressure.

EXAMPLE 21.19

A two-stage single acting reciprocating air compressor takes atmospheric air at 15°C, compresses it adiabatically to intermediate pressure, cools to initial temperature, and then compresses adiabatically in high pressure cylinder. For both the stages, the clearance volumes and pressure ratios are 10 percent and 2 respectively. Make calculations for the swept volume of high pressure cylinder corresponding to 50 litre swept volume for the low pressure cylinder.

Solution : Refer Fig. 21.13.

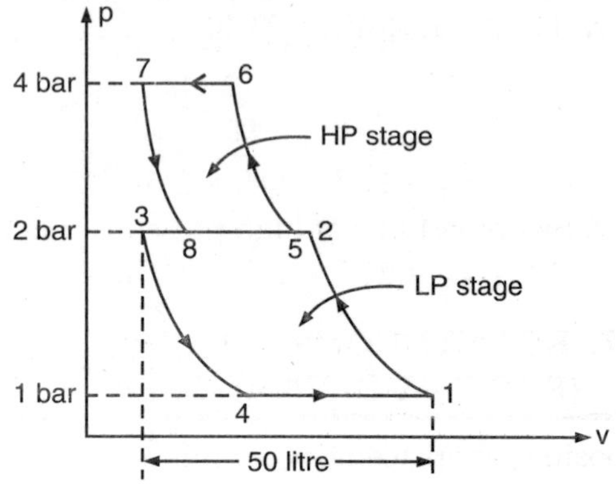

Fig. 21.13.

$$V_1 = 50+\frac{10}{100}\times50 = 55 \text{ litre}$$

From the compression process 1 – 2,

$$T_2 = T_1\left(\frac{p_2}{p_1}\right)^{\frac{\gamma-1}{\gamma}}$$

$$= 288\times\left(\frac{2}{1}\right)^{\frac{1.4-1}{1.4}} = 351 \text{ K}$$

$$V_2 = \frac{V_1}{(p_2/p_1)^{1/\gamma}} = \frac{55}{(2)^{1/1.4}}$$
$$= 33.52 \text{ litre}$$

Corresponding to high pressure stage,

$$V_5 = V_7 + 10\ V_7 = 11\ V_7$$

$$V_8 = V_7\left(\frac{p_7}{p_8}\right)^{1/\gamma}$$

$$= V_7\left(\frac{4}{2}\right)^{1/1.4} = 1.64\ V_7$$

When the air at 351 K temperature and intermediate pressure is cooled to initial temperature 288 K, its volume reduces to $(V_5 - V_8)$ from $(V_2 - V_3)$.

At constant pressure Charles' law applies and therefore

$$\frac{V_2 - V_3}{T_2} = \frac{V_5 - V_8}{T_5}$$

$$V_5 - V_8 = \frac{288}{351}(33.52 - 5) = 23.40$$

$$\therefore\ 11V_7 - 1.64\ V_7 = 23.40$$

$$V_7 = \frac{23.40}{11 - 1.64} = 2.5$$

$$V_5 = 11V_7 = 11 \times 2.5 = 27.5$$

∴ Swept volume of high pressure cylinder

$V_5 - V_7 = 27.5 - 2.5 =$ **25 litre**

21.7. ROTARY COMPRESSORS (POSITIVE DISPLACEMENT TYPE)

In positive displacement compressors, the air is trapped in between two sets of engaging surfaces and the change in pressure is either by the back flow of air (roots blower) or both by squeezing action and back flow of air (vane blower).

• ***Roots blower*** **:** The unit consists of two rotors which are on separate parallel axes, are lobed and revolve in opposite directions. One of the rotor is connected to an external drive and the second rotor is gear driven from the first. The lobes of the rotor are of cycloidal or involute form and this ensures their correct mating. The high pressure delivery side is sealed from the low pressure suction side at all angular positions.

To reduce wear, a small clearance (0.1 mm to 0.2 mm) is provided between the lobes and between the lobe and casing. The clearance, however, follows a leakage path and has an adverse effect on the performance of the unit.

The air trapped between the lobes and casing moves along a part of the casing and is finally discharged to the receiver. As the flow area does not decrease from entry to exit, theoretically the air should not undergo any pressure rise. However, when the outlet port is uncovered, the high pressure air in the receiver offers resistance to the delivery of blower discharge and there occurs an instantaneous pressure rise. In Fig. 21.14, the air in spaces A and C is at atmospheric pressure p_1. The air in space B will be increased in pressure to the value of receiver pressure p_2.

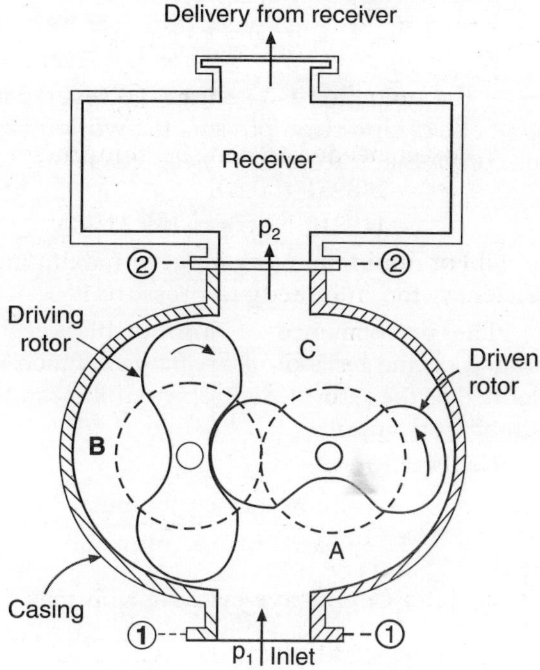

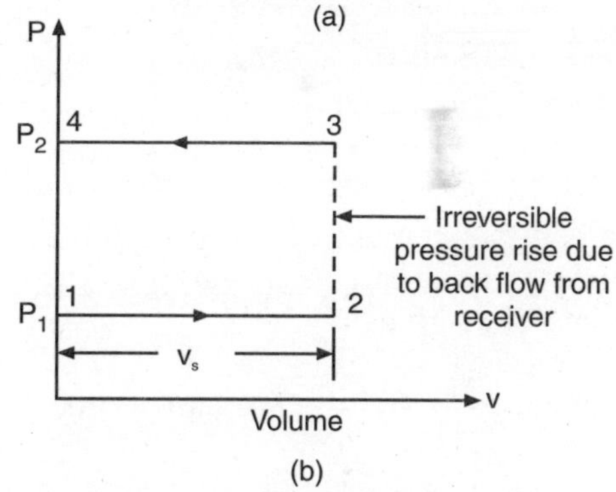

Fig. 21.14. Roots blower

The cycle of operation has been shown in the *p-v* plot wherein the pressure rise from p_1 to p_2 is shown dotted as it occurs by irreversible constant volume process 2-3.

Work done per cycle = $(p_2 - p_1)\ V$

where V is the volume between a lobe face and the casing, *i.e.*, in space A. The free air delivery for a two lobe rotor will be 4V per revolution, and it will be 6V per revolution for a three lobe rotor. Then for a two lobe rotor.

Work done per revolution of rotor

$$= (p_2 - p_1) \times 4V$$

If the volume dealt per minute at p_1 and T_1 is denoted by V_s , then work done per minute

$$W_{\text{roots}} = (p_2 - p_1)\ V_s \quad \text{...(21.20)}$$

If the operation is executed by reversible adiabatic compression process, the work input for compression from p_1 to p_2 is given by

$$W_{\text{ideal}} = \frac{\gamma}{\gamma - 1}\ p_1\ V_s \left[\left(\frac{p_2}{p_1} \right)^{\frac{\gamma-1}{\gamma}} - 1 \right] \quad \text{...(21.21)}$$

The performance of a roots blower is judged on the basis of *roots efficiency* which is defined as the ratio of isentropic work input to actual work input.

Roots efficiency

$$\eta_{\text{roots}} = \frac{\text{isentropic work input}}{\text{actual work input}}$$

$$= \frac{\frac{\gamma}{\gamma-1}\left[p_1 V_s (r_p)^{\frac{\gamma-1}{\gamma}} - 1 \right]}{p_1 V_s (r_p - 1)}$$

$$= \frac{\frac{\gamma}{\gamma-1}\left[(r_p)^{\frac{\gamma-1}{\gamma}} - 1 \right]}{r_p - 1} \quad \text{...(21.22)}$$

where r_p is the pressure ratio $\frac{p_2}{p_1}$.

Replacing $\frac{\gamma}{\gamma - 1}$ by $\frac{c_p}{R}$, we can write

$$\eta_{\text{roots}} = \frac{c_p}{R} \left[\frac{r_p^{\frac{\gamma-1}{\gamma}} - 1}{r_p - 1} \right] \quad \text{...(21.23)}$$

An insight into the above expression reveals that the roots efficiency decreases with increase in pressure ratio.

The roots blowers have been designed for delivery capacities ranging from 0.14 m^3/min to 1400 m^3/min. The pressure ratios are of the order of 1 to 2 for a single stage machine and from 1 to 3 for a two stage machine.

• ***Vane blower :*** A vane blower consists of the following elements :

(*i*) A rotor which is located eccentrically in a cylindrical outer casing.

(*ii*) A set of vanes mounted on the rotor concentrically in the outer casing. The vanes are of non-metallic material usually fibre or carbon.

(*iii*) Intake and delivery ports. The suction side opens to a large area of vanes and the delivery opens to a much smaller area.

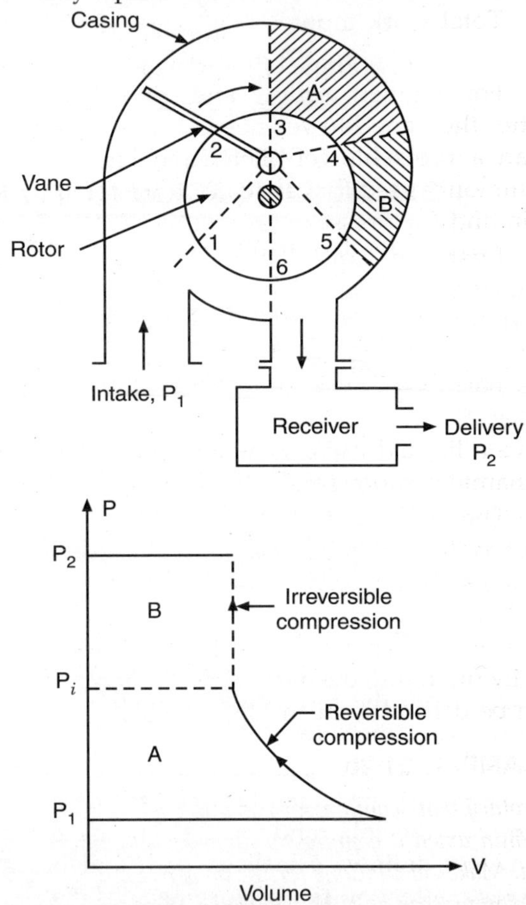

Fig. 21.15. Vane blower

Travelling in the clockwise direction, each vane will start collecting air after it leaves the position marked 1 and no more air will be sucked after the vane moves to position 2. Thereafter the compression of air entrapped between the vanes commences due to decreasing space volume between the rotor and the casing. When the vane travels from position 3 to 4, the air occupying volume A is compressed into a smaller volume B. The unit is so designed that half of the total pressure rise takes place by internal reversible compression before the entrapped air is open to the delivery port. The air is further compressed in an irreversible manner due to the momentarily back flow of high pressure air from the receiver.

The pressure-volume diagram for the compression process is shown in Fig. 21.15.

Total work input

$$= \text{area } A + \text{area } B$$

For a given airflow and given pressure ratio, the vane blower needs less work input than a roots blower. The vane blowers are commonly used for capacities upto 150 m^3 per min and for pressure ratios upto 8.5 : 1.

Lysholm and screw type positive displacement compressors have helical lobes that intermesh. The entrapped air is progressively compressed as it moves through the narrowing passage ways formed by the lobes. Such machines though quite efficient have a limited capacity when compared with dynamic compressors.

The vane and roots blowers are being superceded by centrifugal blowers in respect of their use in superchanging aero-engines. This is because the centrifugal blowers are comparatively much more efficient, can be more easily fitted into the design of aero-engines and can be driven at much higher speeds.

EXAMPLE 21.20

A rotary vane compressor has a free air delivery of 5 m^3/min when it compressor air from 100 kPa to 200 kPa. Make calculations for the power required to run the compressor and the isentropic efficiency when

(a) the ports are so placed that there is no internal compression, and

(b) the placement of ports is such that 50% of pressure rise is due to internal adiabatic compression and there after the back flow occurs.

Solution : Since there is no internal compression, the entire pressure rise will be due to back flow. The machine is then of roots type for which

$$\text{Power required} = (p_2 - p_1)\, V$$

$$= (200 - 100) \times \frac{5}{60}$$

$$= 8.33 \text{ kN m/s} = 8.33 \text{ kW}$$

(*b*) When there is 50% pressure rise due to internal compression, pressure p_i after internal compression is

$$p_i = 100 + (200 - 100) \times \frac{50}{100} = 150 \text{ kPa}$$

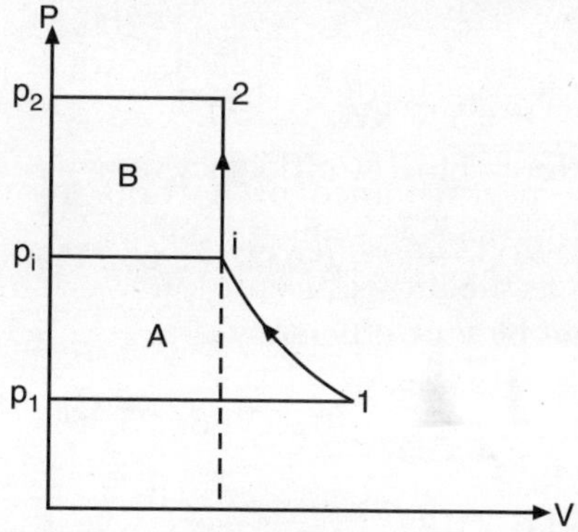

Fig. 21.16.

Workdone due to internal compression is given by area A

$$= \frac{\gamma}{\gamma - 1} p_1 V_1 \left[\left(\frac{p_i}{p_1}\right)^{\frac{\gamma-1}{\gamma}} - 1\right]$$

$$= \frac{1.4}{1.4-1} \times 100 \times \frac{5}{60} \left[\left(\frac{150}{100}\right)^{\frac{1.4-1}{1.4}} - 1\right]$$

$$= 3.586 \text{ kW}$$

Work done due to back flow is given by area B

$$= (p_2 - p_i)\, V_i$$

$$= (200 - 150) \times \left[\frac{5}{60} \times \left(\frac{100}{150}\right)^{1/1.4}\right]$$

$$\left\{\because V_i = V_1 \left(\frac{p_i}{p_1}\right)^{1/\gamma}\right\}$$

$$= 3.12 \text{ kW}$$

Total power required = 3.586 + 3.12
= 6.706 kW

Work required if the entire compression had been reversible adiabatic, *i.e.*, isentropic

$$= \frac{\gamma}{\gamma - 1} p_1 V_1 \left[\left(\frac{p_2}{p_1} \right)^{\frac{\gamma-1}{\gamma}} - 1 \right]$$

$$= \frac{1.4}{1.4 - 1} \times 100 \times \frac{5}{60} \times \left[\left(\frac{200}{100} \right)^{\frac{1.4-1}{1.4}} - 1 \right]$$

$= 6.395$ kW

∴ Roots blower efficiency

$$= \frac{6.395}{8.33} = 0.6377 \text{ or } \mathbf{63.77\%}$$

Vane blower efficiency

$$= \frac{6.395}{6.706} = 0.9536 \text{ or } \mathbf{95.36\%}$$

EXAMPLE 21.21

The supercharger of an aircraft is supplied a combustible mixture having air-fuel ratio of 16 : 1. The mixture is initially at 0.56 bar and 270 K, and is compressed by the supercharger to 1 bar with 85% isentropic efficiency. What power will be spent in driving the superchanger if fuel consumption is 4.5 kg/min. The adiabatic exponent γ for the mixture is 1.38, and the combustible mixture formed by 1 kg of fuel has a volume of 12.8 m³ at 1 bar and 273 K.

Solution : Mass of mixture/kg of fuel

$= 1 + 16 = 17$ kg

1 kg of fuel or 17 kg of mixture at 1 bar and 273 K has a volume of 12.8 m³. Then from characteristic equation,

$$R_{mix} = \frac{pV}{mT}$$

$$= \frac{(1 \times 10^5) \times 12.8}{17 \times 273} = 275.8 \text{ J/kg K}$$

$$c_p = \frac{\gamma}{\gamma - 1} R$$

$$= \frac{1.38}{1.38 - 1} \times 275.8$$

$= 1001.59$ J/kg K $\simeq 1.001$ kJ/kg K

If T_2' denotes the temperature after isentropic compression, than

$$T_2' = T_1 \left(\frac{p_2}{p_1} \right)^{\frac{\gamma-1}{\gamma}}$$

$$= 270 \left(\frac{1}{0.56} \right)^{\frac{1.38-1}{1.38}} = 316.67 \text{ K}$$

$$\eta_c = \frac{T_2' - T_1}{T_2 - T_1}$$

$$T_2 = T_1 + \frac{T_2' - T_1}{\eta_c}$$

$$= 270 + \frac{316.67 - 270}{0.85} = 324.90 \text{ K}$$

The work input or power required to drive the supercharger is

$$= m\, c_p\, (T_2 - T_1)$$

$$= \left(\frac{17 \times 4.5}{60} \right) \times 1.001\,(324.90 - 270)$$

$= \mathbf{70.06}$ **kW**

21.8. DYNAMIC COMPRESSORS

In dynamic compressors, the fluid is not positively contained within specified boundaries but is in continuous steady flow throughout the machine. There is transfer of energy from the machine rotor to the fluid and the rise in pressure is primarily due to dynamic effects. These rotary compressors are also called the blade compressors. The bladed type rotary compressors are of two types namely the radial or centrifugal compressor and the axial flow compressor.

The salient points of difference between the reciprocating and bladed type rotary compressors are:

1. The reciprocating compressors are suitable for low discharge (maximum limit 300 m³/min) at high delivery pressures (as high as 1000 bar). The blade compressors are used where high efficiency, medium pressure rise (upto 10 bar) and large flow rate (300 m³/min) are the prime considerations.

The displacement compressors of the root and vane type are mostly employed for supercharging of IC engines and for producing scavenging action in two stroke diesel engines.

2. The reciprocating compressors provide intermittent delivery of high pressure air and therefore require receiver for maintaining continuous delivery. In blade compressors, a continuous delivery of air at uniform rate is provided and therefore no receiver is required.

The rotary displacement compressors too give an intermittent supply like the reciprocating compressors.

3. The rotary compressors are compact in design. The reciprocating compressor have more bulk and weight.

4. The rotary compressors run at much higher speeds (30,000 to 40,000 rpm) and can be coupled directly to high speed prime movers such as steam turbines, electric motors and high speed engines without intermediate gearing. This aspect is not possible in slow speed running reciprocating compressors.

5. The lubrication requirements are quite severe in reciprocating compressors. Consequently the air delivered from recipro-cating compressors is of poor quality due to its contamination with lubricant. In contrast, there are only few sliding parts in rotary compressors which require lubrication. As such, the rotary compressors will give more clean supply of air.

6. The reciprocating compressors have higher maintenance expenses, lesser continuity of service are require greater operating attention.

7. The rotary air compressors have better balancing. As such, their running is vibration less and only small foundations are required.

8. Reciprocating compressors have high volumetric efficiency in comparison to rotating ones; thereby they consume less power.

9. The compression process in a reciprocating compressor may approach isothermal compression due to slow speed, cylinder and interstage cooling. In bladed compressors running at high speed, there is lot of friction and eddies formation between the blade passages. This causes internal heating of air to a temperature which is much higher than obtained by adiabatic compression. If no cooling devices are used, the index n in the compression law pv^n = constant may become as high as 1.7. The high value of compression index demands a large amount of compression work. The compression index can be reduced by intercooling and surrounding the air passages by cold water jackets. However, it is generally impossible to provide sufficient cooling that could bring the compression curve to the left of isentropic line. It is precisely due to this reason that the *thermodynamic efficiency of bladed compressors is referred to adiabatic conditions where as that of reciprocating compressors is referred to isothermal conditions.*

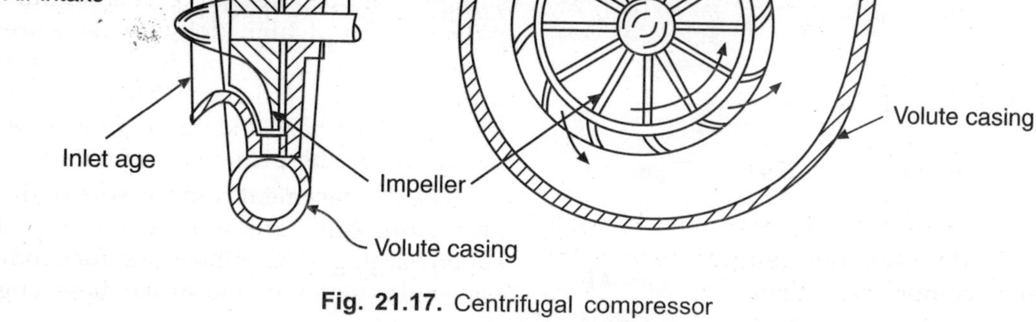

Fig. 21.17. Centrifugal compressor

21.9. CENTRIFUGAL COMPRESSOR

A centrifugal compressor consists of the following elements :

(i) *Series of inlet buckets* : These buckets are attached to and rotate with the shaft.

(ii) *Rotating impeller* : The impeller is a disc with radial or approximately radial vanes. The air is supplied to these vanes by the inlet buckets.

(iii) *Casing* : The casing surrounds the rotating impeller.

(iv) *Diffuser* : The diffuser is housed in radial portion of the casing.

The air enters the eye of the impeller with a low velocity V_1 and atmospheric pressure p_1. The impeller is driven at high speed 20,000 to 30,000 rev/min. Depending upon the centrifugal action of the impeller, the air moves radially outwards and is guided by the impeller vanes during its movement. The impeller transfers its energy to the air causing a rise both in static pressure and temperature, and an increase in velocity. The air now enters the diverging passage called diffuser where it is efficiently slowed down. The kinetic energy is converted into pressure energy with the result that there is further rise in static pressure. If the kinetic energy is not fully recovered as pressure energy, it gets dissipated into thermal energy which is not desired function. Thus the efficiency of the compressor increases by the use of diffuser ring and volute casing.

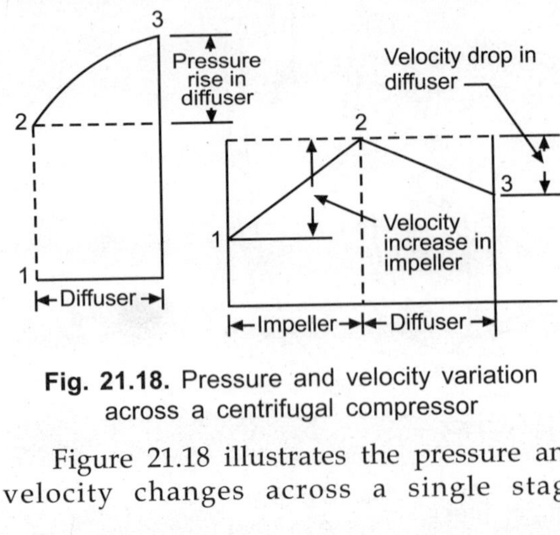

Fig. 21.18. Pressure and velocity variation across a centrifugal compressor

Figure 21.18 illustrates the pressure and velocity changes across a single stage compressor. About half of the total pressure rise per stage is usually achieved in the impeller and the other half in the diffuser.

The impeller may be single sided or double sided. In double sided impeller, there is an eye on either side of the impeller and the air is taken in from both sides. The double side has the advantage that the impeller is subjected to approximately equal stresses in the axial direction.

Centrifugal compressors are designed to supply large quantities of air with pressure ratios 4 to 5. For higher pressure ratios, multistage is adopted..

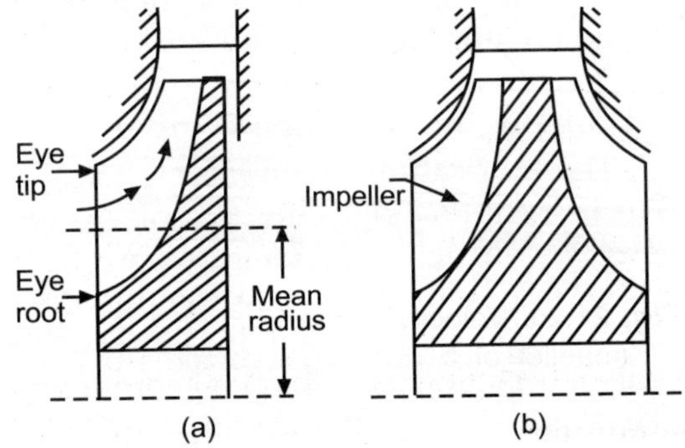

Fig. 21.19. Single and double sided impellers

Centrifugal compressors have been successfully used as blowing engines in steel mills, compressors for low pressure refrigeration and for supercharging internal combustion engines. They are also employed for compressing air in small gas turbine aircrafts.

21.9.1. Velocity Vector Diagrams for Centrifugal Compressor

The velocity vector diagrams at inlet to and at outlet from the impeller of a centrifugal compressor are shown in Fig. 21.20. Following nomenclature has been used while drawing these diagrams :

V = absolute velocity of fluid ;

V_b = blade or peripheral velocity ;

V_r = relative fluid velocity ;

V_f = flow component of velocity and

V_w = whirl or tangential component of velocity.

Further, suffix 1 and 2 represent the conditions at inlet and outlet of impeller.

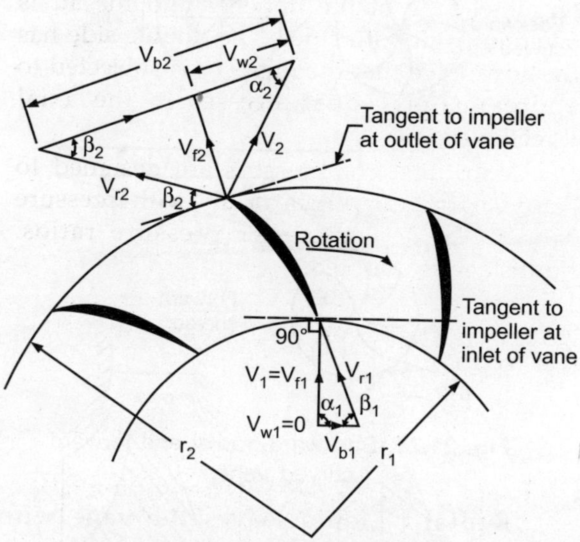

Fig. 21.20. Velocity vector diagrams for the impeller of a centrifugal compressor

Assumptions

- infinite number of vanes; the air flowing in the passages follows the path outlined by the vanes.
- no energy loss in the impeller due to friction and eddy formation.
- even/uniform velocity distribution in the narrow passages formed between two adjacent vanes.
- air enters the impeller eye in an axial/radial direction, *i.e.*, the whirl component V_{w1} of the inlet absolute velocity V_1 is zero and the flow component V_{f1} equals the absolute velocity itself.
- no loss due to shock at entry, *i.e.*, inlet edge of the impeller blades is parallel to the relative velocity.

In general, when we consider the flow of a fluid through a rotor of any shape, the rate of change of angular momentum is given by :

$$= m(V_{w2}\, r_2 - V_{w1}\, r_1) \qquad \text{...(21.24)}$$

where m is the mass of air entering and leaving the impeller in a unit time,

r_2, r_1 = impeller radius at outlet and inlet respectively.

From the momentum theorem, the change of angular momentum equals the torque exerted by the impeller on air.

$$\text{Torque} = m(V_{w2}\, r_2 - V_{w1}\, r_1)$$

Now, energy transfer

$$= \text{torque} \times \text{rotational speed}$$
$$= m\,(V_{w2}\, r_2 - V_{w1}\, r_1) \times \omega$$
$$= m(V_{w2}\, V_{b2} - V_{w1}\, V_{b1}) \qquad \text{...(21.25)}$$

For an axial or radial fluid entry (air entering the impeller has no whirl component), term $V_{w1}\, V_{b1}$ vanishes and therefore,

$$E = m\, V_{w2}\, V_{b2}$$

The energy transfer per unit mass is called the Euler head and is designed by H_e. That is

$$H_e = V_{w2}\, V_{b2} \qquad \text{...(21.26)}$$

Evidently, it is only the discharge conditions that determine the energy transfer. Equation 6.26, stipulates that for delivering high heads the peripheral velocity V_{b2} must be high and the vector V_{w2} must be large enough, *i.e.*, sufficient whirl should be imparted to the fluid. Increase in V_{b2} is achieved by increasing the speed of rotation and the impeller diameter. Provision of sufficient number of vanes of suitable size and shape augments the whirl component V_{w2}.

Each of the velocity triangles is a combination of two right-angled triangles. For example, the outlet triangle gives

$$V_{f2}^2 = V_2^2 - V_{w2}^2$$

Also $V_{f2}^2 = V_{r2}^2 - (V_{b2} - V_{w2})^2$

Combining these two expressions, we obtain

$$V_2^2 - V_{w2}^2 = V_{r2}^2 - (V_{b2} - V_{w2})^2$$
$$= V_{r2}^2 - V_{b2}^2 - V_{w2}^2 + 2V_{b2}\, V_{w2}$$
$$V_{b2}\, V_{w2} = \frac{1}{2}(V_2^2 + V_{b2}^2 - V_{r2}^2)$$

Likewise, the inlet triangle would yield :

$$V_{b1}\, V_{w1} = \frac{1}{2}(V_1^2 + V_{b1}^2 - V_{r1}^2)$$

Upon substitution in equation 21.24, we get

$$H_e = \frac{V_2^2 - V_1^2}{2g} + \frac{V_{b2}^2 - V_{b1}^2}{2g} + \frac{V_{r1}^2 - V_{r2}^2}{2g} \quad ...(21.27)$$

Evidently, the head developed in a rotating bladed impeller is a combination of three distinct effects.

- The term $\frac{(V_2^2 - V_1^2)}{2g}$ represents an increase in kinetic energy or dynamic head. Subsequent conversion of dynamic head into static pressure is achieved through retardation in the volute casing/diffuser following the impeller.
- The term $\frac{(V_{b2}^2 - V_{b1}^2)}{2g}$ represents an increase in static pressure accomplished within the impeller due to centrifugal force acting on the fluid; due to the movement of rotating fluid from one radius of rotation to another.
- The term $\frac{(V_{r1}^2 - V_{r2}^2)}{2g}$ represents a change in kinetic energy due to retardation of fluid flow relative to impeller; this term, therefore, represents a conversion of kinetic energy within the impeller itself.

Relative proportions of these terms vary considerably in different designs.

21.9.2. Vane shape and head capacity relationship

The shape/form of impeller vanes/blades is fundamentally changed depending on the blade angle β_2 which has a significant influence on the conversion of energy. The different blade configurations and the respective velocity vectors are schematically represented in Fig. 21.21.

Backward curved : Outlet tip of blade curves in a direction opposite to that of motion, and the angle between the blade tip and the tangent to rotor at exit is acute ($\beta_2 < 90°$).

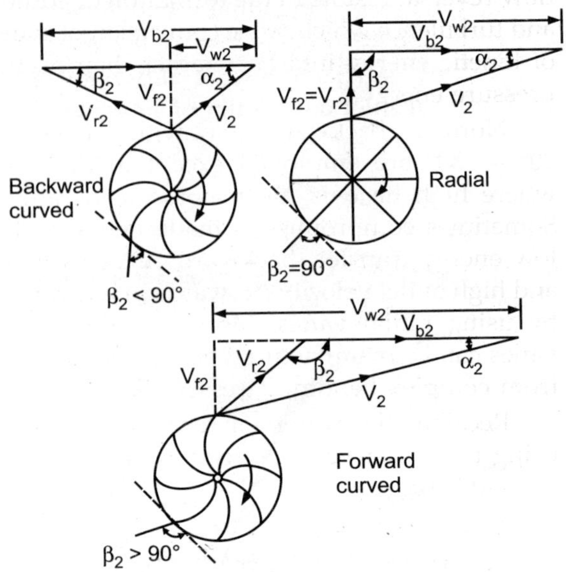

Fig. 21.21. Backward, radial and forward curved vanes.

Radial : Liquid leaves the vane with relative velocity in a radial direction and angle $\beta_2 = 90°$.

Forward curved : Outlet tip of blade curves in the direction of motion, and the angle between the blade tip and the tangent to rotor at exit is obtuse ($\beta_2 > 90°$)

Direction of exit relative velocity is fixed by angle β_2 and obviously the velocity vector changes depending upon vane configuration. For backward curved vanes the tangential component V_{w2} is much reduced and consequently for a given impeller speed the impeller will have a low energy transfer.

In the case of forward curved vanes, V_{w2} is increased and consequently the energy transfer is maximum for forward-curved vanes. However, the absolute velocity V_2 at impeller outlet is also increased. A high value of V_2 is not desirable as its conversion into static pressure cannot be very efficiently carried out in the diffuser section. During the process of diffusion, there is always a tendency for the liquid to break away from the walls of diverging passages. If the diffusion is too rapid, *i.e.*, if it is carried out in a small diffuser section, the liquid may reverse its direction and flow back in the direction of pressure gradient. The flow reversal results in the formation of eddies and turbulence which cause conversion of some of kinetic energy into heat rather than useful pressure energy.

Normally backward vanes with β_2 between 20° - 30° are employed except in the case where high head is the major consideration. Sometimes compromise is made between the low energy transfer (backward curved vanes) and high outlet velocity (forward curved vanes) by using radial vanes. Moreover, the radial vanes can be manufactured easily and are free from complex bending stresses.

Recalling the expression for Euler head and using the outlet velocity vector diagram.

Euler head H_e

$$= V_{w2} V_{b2}$$

$$= V_{b2} [V_{b2} - V_{f2} \cot \beta_2]$$

$$= V_{b2}\left[V_{b2} - \frac{Q}{A_2}\cot\beta_2\right] \quad ...(21.28)$$

For a particular compressor running at constant speed, the blade velocity V_{b2} is fixed and so are the parameters β_2 and area A_2. Thus

$$H_e = K_1 - K_2 Q \quad ...(21.29)$$

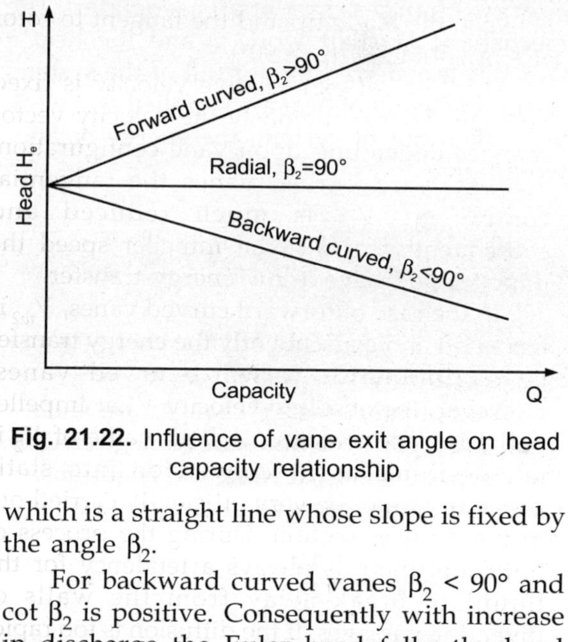

Fig. 21.22. Influence of vane exit angle on head capacity relationship

which is a straight line whose slope is fixed by the angle β_2.

For backward curved vanes $\beta_2 < 90°$ and $\cot \beta_2$ is positive. Consequently with increase in discharge the Euler head falls; the head capacity characteristics has a negative slope. For radial vanes $\beta_2 = 90°$ and $\cot \beta_2 = 0$. Thus the head remains constant with variation in flow rate. For forward curved vanes $\beta_2 > 90°$ and $\cot \beta_2$ is negative. Obviously with increase in mass flow rate the Euler head rises: the head capacity characteristic has a positive slope.

21.9.3. Slip and slip factor

Due to inertial effects, the liquid which is trapped between the impeller vanes is reluctant to move round with the impeller. This results in a difference of pressure force across the vane; there being a high pressure (+ve sign) on the leading face and a low pressure (–ve sign) on the trailing face. This difference of pressure, called ***vane loading***, increases with an increase in the number of vanes. Because of pressure variation, a velocity gradient exists across the channel. On the high pressure side the air follows the blade contour; it leaves the vane tangentially. On the low pressure side, the air leaves the vanes with a certain circumferential component (Fig. 21.23*a*). The air will thus be discharged at a certain average angle β'_2 which is less than the geometric blade angle β_2. This aspect is shown illustrated in Fig. 21.23*b* for an impeller with radial vanes.

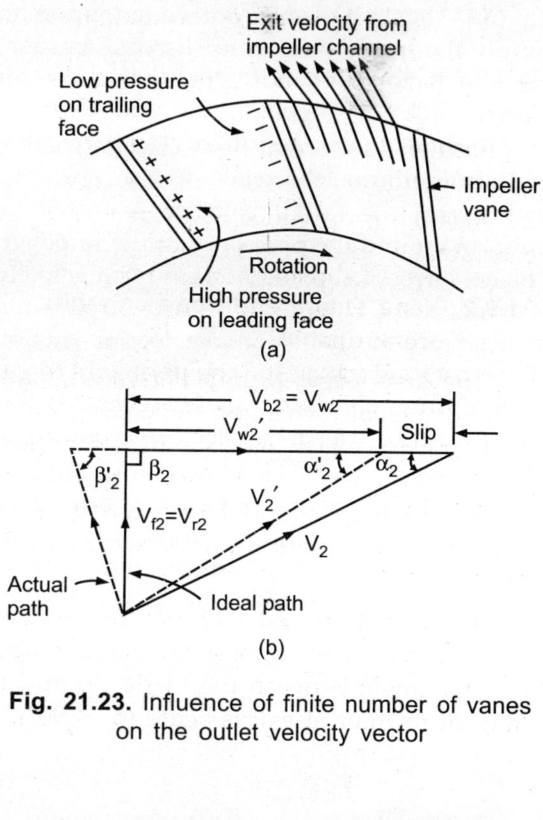

Fig. 21.23. Influence of finite number of vanes on the outlet velocity vector

Due to this deviation in the flow path, the tangential component gets reduced from V_{w2} to V'_{w2}. The difference or deviation (V_{w2} – V'_{w2}) is ***slip*** of the impeller. The ideal slip coefficient is then defined as the ratio of whirl component with flow path deviation to the whirl component without flow path deviation.

$$\text{Slip factor } \sigma = \frac{V'_{w2}}{V_{w2}} \qquad ...(21.30)$$

The slip factor depends upon the number of vanes n ; greater the number of vanes, smaller is the slip. An empirical relation for slip factor is

$$\sigma = 1 - \frac{0.63}{n} \text{ where } n \text{ is the number of vanes.} \qquad ...(21.31)$$

(*i*) With increase in slip factor, V_{w2} increases and when it approaches V_{b2}, the greatest amount of work can be usefully put into the compressor of given size.

(*ii*) Greater the number of vanes, higher is the value of slip factor. However, with an increase in the number of vanes, the effective area of flow decreases.

(*iii*) The impeller is not called upon to supply the head loss due to slip and as such, the slip factor does not represent a loss of energy.

Further due to real fluid effects (friction and separation on the walls of disc, shrouds, and vanes) the radial velocity may not be uniform around the periphery of the impeller. The net effect of slip and non-uniform velocity is to reduce the Euler head. A further reduction in head occurs due to intake losses, friction and separation losses and the discharge losses in the diffuser. Obviously there are three heads which describe the performance of a centrifugal compressor.

- the Euler, virtual or theoretical head H_e which assumes the fluid exit angle equal to the vane angle (no fluid deviation) and uniform radial velocity around the periphery or impeller.
- the ideal head H_i which considers the effects of slip and non-uniform velocity distribution, but ignores the friction and turbulence losses.
- The actual head H that considers the inlet losses, friction and separation losses etc.

Fig. 21.24 depicts these three heads for a backward curved vane.

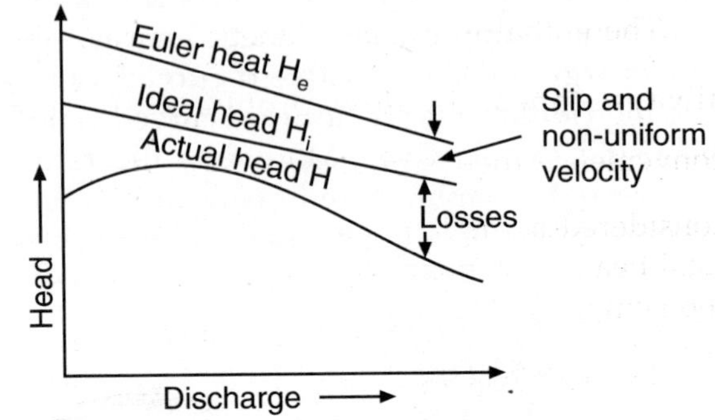

Fig. 21.24. Head-capacity relationship

21.9.4. Static and total head values

Consider a steady flow of unit mass of fluid through a certain system. The total energy at any section consists of internal energy u, flow work $p\, v$, kinetic energy $V^2/2$ and the potential energy due to position z. Along with these energy changes are also associated the mechanical or shaft work w and the heat q which is transferred into or out of the system. Assuming q unit of heat are supplied and work w is taken out of system, the energy balance would give:

$$u_1 + \frac{p_1 v_1}{J} + \frac{V_1^2}{2} + z_1\, g + q = u_2 \frac{p_2 v_2}{J} + \frac{V_2^2}{2} + z_2\, g + w$$

The potential energy changes are usually small for a gas flow and can be neglected. In the absence of external work and heat transfer, the equation for energy balance reduces to

$$u_1 + p_1 v_1 + \frac{V_1^2}{2} = u_2 + p_2 v_2 + \frac{V_2^2}{2}$$

For a perfect gas: $u + p\, v = h = c_p T$ where h is specific enthalpy and c_p is specific heat at constant pressure

$$\therefore \quad c_p T_1 + \frac{V_1^2}{2} = c_p T_2 + \frac{V_2^2}{2}$$

$$\text{or} \quad c_p T + \frac{V^2}{2} = \text{constant}$$

The enthalpy c_pT and kinetic energy $\frac{V^2}{2}$ always appear in flow problems and for convenience their summation $\left(c_pT + V^2/2\right)$ is considered a single quantity. It is then called total head or stagnation enthalpy and is given the symbol h_0.

$$\therefore \quad c_p T + \frac{V^2}{2} = h_o = \text{constant}$$

Thus for adiabatic flow with no work crossing the boundaries of the system, the stagnation enthalpy is constant. Further h_0 can be put as c_pT_0

$$\therefore \quad c_p T + \frac{V^2}{2} = c_p T_0$$

Considering constant specific heat :

$$T + \frac{V^2}{2c_p} = T_0 \quad \text{...(21.32)}$$

T_0 is called *stagnation* or *total temperature.* Thus we can say that sum of static temperature T and temperature equivalent of kinetic energy is constant in an adiabatic flow and is equal to total temperature. The stagnation temperature T_0 may be defined as the temperature attained by a gas originally at static temperature and flowing with a velocity when brought to rest. Whereas the static temperature (actual temperature) is registered by a thermometer moving with gas with the speed equal to that of gas, the stagnation temperature would be read on a thermometer when the flowing gas strikes it and is brought to rest. The term $V^2/2c_p$ is usually given the symbol θ_c and is called the *velocity temperature.*

For ambient conditions, the static and stagnation temperatures are same, as the ambient velocity is assumed to be zero.

When the fluid is brought to rest, the static temperature becomes the stagnation temperature. The process is adiabatic but need not be reversible, *i.e.*, friction and turbulence etc. remain in the system and appear as temperature.

Total head or *stagnation pressure* is defined as the pressure which would be attained if the fluid is brought to rest adiabatically and reversibly. Thus conversion of static pressure p to stagnation pressure p_0 can be made from the isentropic relation between pressure and temperature.

$$\therefore \quad \frac{p_0}{p} = \left(\frac{T_0}{T}\right)^{\frac{\gamma}{\gamma-1}}$$

Adiabatic process and *isentropic process.* In adiabatic process the system does not exchange any heat with the surroundings, *i.e.*, no heat enters or leaves the working fluid externally. The ideal reversible adiabatic process is called isentropic process and in this process the entropy remains constant. During adiabatic compression in a rotary compressor, there is friction between molecules of air and between air and blade passages, eddies formation and shocks at entry and exit. These factors cause internal generation of heat and consequently the maximum temperature reached would be more than that for adiabatic compression. This results in a progressive increase in entropy. Such a process though adiabatic is not isentropic. The heat generated by friction etc may be removed continuously with the result that the process might not involve any entropy change. The process would then be isentropic but not adiabatic as heat has been transferred.

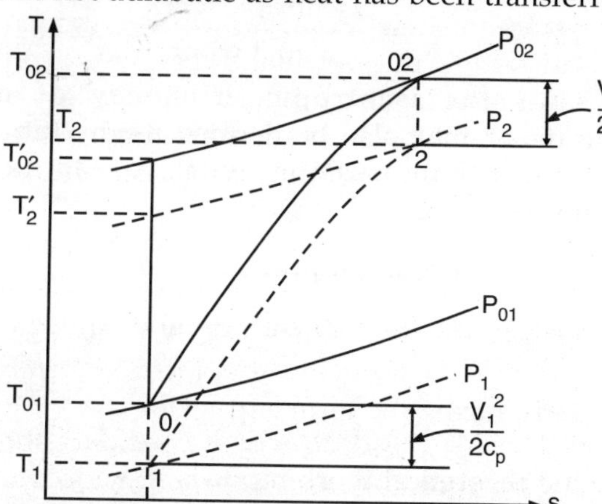

Fig. 21.25. Concept of static and total head values

Isentropic efficiency of rotary compressor may be defined as the ratio of isentropic temperature rise to actual temperature rise.

With reference to Fig. 21.25 which represents a combined diagram for static and stagnation (total head) values, we have

Isentropic efficiency

$$= \frac{\text{isentropic temperature rise}}{\text{actual temperature rise}}$$

$$\eta_c = \frac{T'_{02} - T_{01}}{T_{02} - T_{01}} \text{ (based on total values)}$$

$$= \frac{T_2' - T_1}{T_2 - T_1} \text{ (based on static values)} \quad ...(21.33)$$

During compression process, work has to be imparted to the impeller. The energy balance equation would then yield:

$$c_p T_1 + \frac{V_1^2}{2} = c_p T_2 + \frac{V_2^2}{2} - W$$

or $\quad c_p T_{01} = c_p T_{02} - W$

or $\quad W = c_p (T_{02} - T_{01})$

Thus the work input is the product of specific heat at constant pressure and temperature rise. This relation is true both for adiabatic and isentropic processes.

$$\text{Isentropic efficiency} = \frac{T'_{02} - T_{01}}{T_{02} - T_{01}}$$

$$\eta_c = \frac{c_p (T'_{02} - T_{01})}{c_p (T_{02} - T_{01})}$$

$$= \frac{\text{isentropic work}}{\text{actual work}} \quad ...(21.34)$$

Thus the isentropic efficiency of a compressor may also be defined as the ratio isentropic compression work to actual compression work.

21.9.5. Compression pressure ratio

Energy transfer or work done $= V'_{w2} V_{b2}$

Introducing the term slip factor $\sigma = \frac{V'_{w2}}{V_{w2}}$,

the theoretical work input is

$= \sigma V_{w2} V_{b2}$

For impeller with radial vanes, $V_{w2} = V_{b2}$ and then theoretical work required $= \sigma V_{b2}^2$

The actual work required is obtained by multiplying the above quantity by a parameter P called the *work factor* or *power input factor*. The power factor accounts for the extra work needed due to fluid friction and windage loss.

The increase in work input resulting from power input factor is entirely absorbed in overcoming friction. The additional work then reappears in the form of heat and that leads to rise in temperature. As such the high grade mechanical energy gets conveted into low grade heat energy. This suggests that power input factor should be as low as possible so that the compresion process follows the isentropic law. A low value of power input factor implies high value of isentropic efficiency.

$$\text{Actual work required} = P\,\sigma V_{b2}^2 \quad ...(21.35)$$

The values of slip factor and the power factor are usually taken as

$$\sigma = 0.9 \text{ and } P = 1.04$$

Neglecting changes in potential energy the steady flow energy equation between sections 1 and 2 for adiabatic (no heat interaction) compression process is

$$h_1 + \frac{V_1^2}{2} + W = h_2 + \frac{V_2^2}{2}$$

$$\text{or } c_p T_1 + \frac{V_1^2}{2} + W = c_p T_2 + \frac{V_2^2}{2}$$

$$\text{or } W = c_p \left(T_2 + \frac{V_2^2}{2c_p}\right) - c_p \left(T_1 + \frac{V_1^2}{2c_p}\right)$$

From definition of stagnation temperature,

$$T_1 + \frac{V_1^2}{2c_p} = T_{01} \text{ and } T_2 + \frac{V_2^2}{2c_p} = T_{02}$$

$$\therefore \quad W = c_p (T_{02} - T_{01}) \quad ...(21.36)$$

Thus the impeller work (actual work required) is given by the product of specific heat at constant pressure and changes in stagnation temperature from inlet to exit.

Equating the expressions 21.31 and 21.32 for work done, the stagnation temperature rise through the impeller is

$$(T_{02} - T_{01}) = \frac{P\sigma V_{b2}^2}{c_p}$$

Let T'_{02} denote the stagnation temperature at exit from the compressor when the compression process is reversible adiabatic (isentropic). Then in terms of stagnation pressure ratio p_{02}/p_{01}, we have

$$\frac{T'_{02}}{T_{01}} = \left(\frac{p_{02}}{p_{01}}\right)^{\frac{\gamma-1}{\gamma}}$$

The temperature ratio T'_{02}/T_{01} can also be written as

$$\frac{T'_{02}}{T_{01}} = 1 + \frac{T'_{02} - T_{01}}{T_{01}}$$

$$= 1 + \frac{T'_{02} - T_{01}}{T_{02} - T_{01}} \times \frac{T_{02} - T_{01}}{T_{01}}$$

$$= 1 + \eta_c \times \frac{T_{02} - T_{01}}{T_{01}}$$

$$= 1 + \eta_c \times \frac{P\sigma V_{b2}^2}{c_p T_{01}}$$

Thus

$$\left(\frac{p_{02}}{p_{01}}\right)^{\frac{\gamma-1}{\gamma}} = 1 + \eta_c \times \frac{P\sigma V_{b2}^2}{c_p T_{01}}$$

$$\therefore \quad \frac{p_{02}}{p_{01}} = \left[1 + \eta_c \frac{P\sigma V_{b2}^2}{c_p T_{01}}\right]^{\frac{\gamma}{\gamma-1}} \quad ...(21.37)$$

Apparently the stagnation pressure ratio depends on the square of blade tip speed.

21.9.6. Pressure coefficient

The pressure coefficient represents the ratio of isentropic work to energy transfer, *i.e.*, Euler work.

Pressure coefficient

$$\psi = \frac{\text{Isentropic work}}{\text{Euler work}} = \frac{c_p (T'_{02} - T_{01})}{V_{w2} V_{b2}}$$

For a radial vaned impeller : $V_{w2} = V_{b2}$ and therefore

$$\psi = \frac{c_p (T'_{02} - T_{01})}{V_{b2} V_{b2}}$$

$$= \frac{c_p}{V_{b2}^2} \times \frac{T'_{02} - T_{01}}{T_{02} - T_{01}} \times (T_{02} - T_{01})$$

$$= \frac{c_p}{V_{b2}^2} \eta_c (T_{02} - T_{01})$$

The product $c_p (T_{02} - T_{01})$ represents the actual work required to run the compressor. That is

$$c_p (T_{02} - T_{01}) = P \sigma V_{b2}^2$$

The pressure coefficient may then be written as

$$\psi = \frac{\eta_c}{V_{b2}^2} \times P\sigma V_{b_2}^2 = \eta_c P \sigma \quad ...(21.38)$$

EXAMPLE 21.22

Air is sucked into centrifugal compressor at static values of 1 bar and 300 K with inlet velocity of 60 m/s. The compressor operates with total head pressure ratio 4 and isentropic efficiency 75 percent. If the compressor delivers 20 kg/min of free air, determine :

(a) total head temperature of air at exit from compressor, and

(b) brake power required to drive the compressor. The motor attached to the compressor has mechanical efficiency of 90%.

Solution : Stagnation temperature inlet

$$T_{01} = T_1 + \frac{V_1^2}{2c_p}$$

$$= 300 + \frac{60^2}{2 \times (1.005 \times 10^3)} = 301.79 \text{ K}$$

For the isentropic process 1 – 2′

$$\frac{T'_{02}}{T_{01}} = \left(\frac{P_{02}}{P_{01}}\right)^{\frac{\gamma-1}{\gamma}}$$

$$T'_{02} = 301.79 \times (4)^{\frac{1.4-1}{1.4}} = 448.63 \text{ K}$$

$$\eta_c = \frac{T'_{02} - T_{01}}{T_{02} - T_{01}}$$

$$T_{02} = T_{01} + \frac{T'_{02} - T_{01}}{\eta_c}$$

$$= 301.79 + \frac{448.63 - 301.79}{0.75} = 497.58 \text{ K}$$

Hence the total head temperature at exit from the compressor is 497.58 K

(*b*) Work input = $m\, c_p\, (T_{02} - T_{01})$

$$= \frac{20}{60} \times 1.005\,(497.58 - 301.79)$$

$$= 65.59 \text{ kJ/s} = 65.59 \text{ kW}$$

$$\therefore \quad \text{Brake power required} = \frac{65.59}{0.9} = \mathbf{72.88} \text{ kW}$$

EXAMPLE 21.23

An uncooled centrifugal compressor running at 15000 rev/min sucks in air at 290 K axially into the eye of its impeller. The air leaves the impeller with a whirl velocity which equals 0.85 times the velocity of the blade tip. The blade tip has an outer diameter of 40 cm. If the isentropic efficiency of the compressor is 0.75, calculate :

(a) The rise in temperature of air compressed, and

(b) The static head pressure ratio of the compressor. The kinetic energy of air at outlet from the compressor is equal to that at inlet to the impeller.

Solution : Velocity of blade tip

$$V_{b2} = \frac{\pi d N}{60} = 314 \text{ m/s}$$

Work input $W = (V_{w2} V_{b2} - V_{w1} V_{b1})$

Since air enters axially

$V_{w1} = 0$. Further $V_{w2} = 0.85\, V_{b2}$

Therefore

$$W = 0.85\, V_{b2}^2$$

$$= 0.85 \times (314)^2 = 83807 \text{ J/kg}$$

$$= 83.81 \text{ kJ/kg}$$

The work done is also given by :

$$W = c_p\, \Delta T$$

where ΔT is the actual rise in temperature.

∴ Actual rise in temperature of air

$$\Delta T = \frac{83.81}{1.005} = \mathbf{83.39°C}$$

(*b*) Isentropic efficiency

$$\eta_c = \frac{\text{isentropic temperature rise}}{\text{actual temperature rise}}$$

$$= \frac{T'_2 - T_1}{T_2 - T_1}$$

Then : $T'_2 - T_1 = 0.75 \times 83.39 = 62.54$

$$T'_2 = 290 + 62.54 = 352.54 \text{ K}$$

$(\because\ T_1 = 290 \text{ K})$

Since the kinetic energy is same at inlet to and at exit from the compressor, the entire work done is utilized in increasing the pressure.

∴ Static head pressure rise

$$\frac{p_2}{p_1} = \left(\frac{T'_2}{T_1}\right)^{\frac{\gamma}{\gamma-1}} = \left(\frac{352.54}{290}\right)^{\frac{1.4}{1.4-1}} = \mathbf{1.98}$$

EXAMPLE 21.24

A centrifugal compressor running at 16000 rev/min takes in air at 295 K and compresses it through a pressure ratio of 4 : 1 with an isentropic efficiency of 80 percent. The mean diameter of the eye and the impeller tip diameter are 25 cm and 60 cm respectively, and the absolute air velocity at inlet is 160 m/s. The blades are radially inclined at exit, and the guide vanes at inlet give the air an angle of pre-whirl of 20° to the axial direction at all radii. Calculate the slip factor for the impeller.

Solution : With isentropic compression,

$$T'_2 = T_1 \left(\frac{p_2}{p_1}\right)^{\frac{\gamma-1}{\gamma}}$$

$$= 295\left(\frac{4}{1}\right)^{\frac{1.4-1}{1.4}} = 438.54 \text{ K}$$

$$\eta_c = \frac{T'_2 - T_1}{T_2 - T_1}$$

$$T_2 = T_1 + \frac{T'_2 - T_1}{\eta_c}$$

$$= 295 + \frac{438.54 - 295}{0.8} = 474.42 \text{ K}$$

Work input per kg of air

$$W = c_p (T_2 - T_1)$$
$$= 1.005 (474.42 - 295) = 180.32 \text{ kJ}$$

Speed at blade tip

$$V_{b2} = \frac{\pi d_t N}{60}$$
$$= \frac{\pi \times 0.6 \times 16000}{60} = 502.4 \text{ m/s}$$

Speed at impeller eye

$$V_{b1} = \frac{\pi d_e N}{60}$$
$$= \frac{\pi \times 0.25 \times 16000}{60} = 209.3 \text{ m/s}$$

Whirl velocity at inlet

$$V_{w1} = 160 \times \sin 20 = 54.72 \text{ m/s}$$

Now : $W = V_{w2} V_{b2} - V_{w1} V_{b1}$

or $180.32 \times 10^3 = V_{w2} \times 502.4 - 54.72 \times 209.3$

$$\text{or } V_{w2} = \frac{180.32 \times 10^3 + 54.72 \times 209.3}{502.4}$$
$$= 381.71 \text{ m/s}$$

∴ Slip factor

$$s = \frac{V_{w2}}{V_{b2}} = \frac{381.71}{502.4} = \mathbf{0.76}$$

EXAMPLE 21.25

A single-sided centrifugal compressor for a gas turbine is required to deliver 10 kg/s of air while operating with a total pressure ratio of 4.5 while turning 18000 rev/min. Initial conditions of air are 1.013 bar pressure and 300 K temperature. The air enters the inlet eye axially with a velocity of 140 m/s with no prewhile. Assuming isentropic efficiency for the compressor as 80% and slip factor as 0.92, make calculations for:

(a) rise in total temperature, (b) tip speed of the impeller and tip diameter, (c) annuals area of inlet eye, and (d) power required to drive the compressor.

Solution : $T_{01} = T_1 + \dfrac{V_1^2}{2c_p}$

$$= 300 + \frac{140^2}{2 \times (1.005 \times 10^3)}$$
$$= 309.75 \text{ K}$$

$$T_{02}' = T_{01} \left(\frac{p_{02}}{p_{01}}\right)^{\frac{\gamma - 1}{\gamma}}$$
$$= 309.75 \times (4.5)^{\frac{1.4-1}{1.4}} = 476.24 \text{ K}$$
$$\eta_c = \frac{T_{02}' - T_{01}}{T_{02} - T_{01}}$$
$$T_{02} = T_{01} + \frac{T_{02}' - T_{01}}{\eta_c}$$
$$= 309.75 + \frac{476.24 - 309.75}{0.8}$$
$$= 517.86 \text{ K}$$

∴ Rise in total temperature

$$= (T_{02} - T_{01})$$
$$= 517.86 - 309.75 = 208.11 \text{ K}$$

(*b*) Specific work input (work input per kg of air)

$$w = c_p (T_{02} - T_{01})$$
$$= 1.005 \times 208.11 = 209.15 \text{ kJ}$$

The specific work input is also given by

$$w = V_{w2} V_{b2} - V_{w1} V_{b1}$$

Since there is no prewhere at inlet, V_{w1} = 0 and therefore

$$w = V_{w2} V_{b2} = \sigma V_{b_2}^2 \left(\because \sigma = \frac{V_{w2}}{V_{b2}}\right)$$

∴ Tip speed of impeller

$$V_{b_2} = \left[\frac{209.15 \times 10^3}{0.92}\right]^{\frac{1}{2}} = 476.8 \text{ m/s}$$

Tip diameter

$$d = \frac{V_{b2}}{\pi N/60}$$
$$= \frac{476.8 \times 60}{\pi \times 18000} = 0.506 \text{ m}$$

(*c*) From characteristic gas equation

$$\rho = \frac{p}{RT}$$
$$= \frac{1.013 \times 10^5}{287 \times 300} = 1.1765 \text{ kg/m}^3$$

$m = \rho A V$; $10 = 1.1765 \times A \times 140$

∴ Annulus area of inlet eye

$$A = \frac{10}{1.1765 \times 140} = 0.0607 \text{ m}^2$$

(*d*) Power required to drive the compressor

$$= m\, c_p\, (T_{02} - T_{01})$$
$$= 10 \times 209.15$$
$$= 2091.5 \text{ kJ/s} = \mathbf{2091.5 \text{ kW}}$$

EXAMPLE 21.26

A centrifugal compressor is to be designed for compressing 20 m³/s of air from 1 bar and 1.5°C through a pressure ratio 1.5 : 1. The compression follows the law $pV^{1.5}$ = constant, the flow velocity at inlet and outlet remains the same and equal to 60 m/s. If the inlet and outlet impeller diameters are 0.6 m and 1.2 m respectively and the compressor turns 5000 rev/min, calculate:

(a) the blade angles at inlet and outlet of impeller,

(b) the angle at which the air from the impeller enters the casing, and

(c) the breadth of impeller blade at inlet and outlet.

It may be assumed that no diffuser is employed and the entire pressure increase occurs in the impeller and that the blades have a negligible thickness.

Solution : Refer Fig. 22.16.

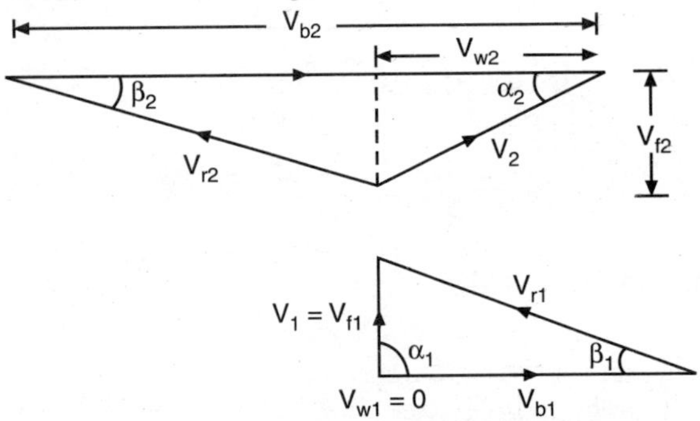

Fig. 21.26.

The temperature T_2 at outlet from the compressor is given by

$$T_2 = T_1 \left(\frac{p_2}{p_1}\right)^{\frac{n-1}{n}}$$

$$= 288\,(1.5)^{\frac{1.5-1}{1.5}} = 329.63 \text{ K}$$

The peripheral velocies of impeller tip at inlet and outlet are

$$V_{b1} = \frac{\pi d_1 N}{60}$$

$$= \frac{\pi \times 0.6 \times 5000}{60} = 157 \text{ m/s}$$

$$V_{b2} = \frac{\pi d_2 N}{60}$$

$$= \frac{\pi \times 1.2 \times 5000}{60} = 314 \text{ m/s}$$

Work input/kg of air is

$$c_p\,(T_2 - T_1) = \frac{V_{b2}\, V_{w2}}{1000}$$

∴ Whirl component

$$V_{w2} = \frac{1.005\,(329.63 - 288) \times 1000}{314}$$

$$= 133.24 \text{ m/s}$$

(*a*) Blade angle at impeller inlet

$$\beta_1 = \tan^{-1} \frac{V_{f_1}}{V_{b_1}}$$

$$= \tan^{-1} \frac{60}{157} = 20.91°$$

Blade angle at impeller outlet,

$$\beta_2 = \tan^{-1} \frac{V_{f_2}}{V_{b_2} - V_{w_2}}$$

$$= \tan^{-1} \frac{60}{314 - 133.24} = 18.36°$$

(*b*) Blade angle at inlet to casing

$$\alpha_2 = \tan^{-1} \frac{V_{f_2}}{V_{w_2}}$$

$$= \tan^{-1} \frac{60}{133.24} = 24.24°$$

(*c*) If *V* is discharge at inlet, then

$$V_1 = \pi\, d_1\, b_1\, V_{f1}$$

where b_1 is breadth of impeller at inlet

$$b_1 = \frac{V_1}{\pi\, d_1 V_{f_1}} = \frac{20}{\pi \times 0.6 \times 60}$$

$$= 0.1769 \text{ m} = 17.69 \text{ cm}$$

From the relation

$\dfrac{p_1 V_1}{T_1} = \dfrac{p_2 V_2}{T_2}$, the discharge at outlet from the impeller is

$$V_2 = \frac{p_1 V_1}{T_1} \times \frac{T_2}{p_2}$$

$$= \frac{1 \times 10^5 \times 20}{288} \times \frac{329.63}{1.5 \times 10^5} = 15.26 \text{ m}^3$$

Then breadth of impeller at output b_2

$$= \frac{V_2}{\pi \, d_2 \, V_{f_2}}$$

$$= \frac{15.26}{\pi \times 1.2 \times 60} = 0.0675 \text{ m} = 6.75 \text{ cm}$$

21.10. AXIAL FLOW COMPRESSOR

In an axial flow compressor, the flow proceeds throughout the compressor in a direction essentially parallel to the axis of the machine.

The unit essentially consists of adjacent rows of rotor blades and stator blades. The rotor blades are mounted on the rotating drum and the stator blades are fixed to the casing. One stage of the machine comprises a row of rotor blades followed by a row of stator blades. For efficient operations, the blades are of airfoil section. The blades are so designed that wasteful losses due to shock and turbulence are prevented and the blades are free from stalling troubles. The blades are said to be stalled when the air stream fails to follow the

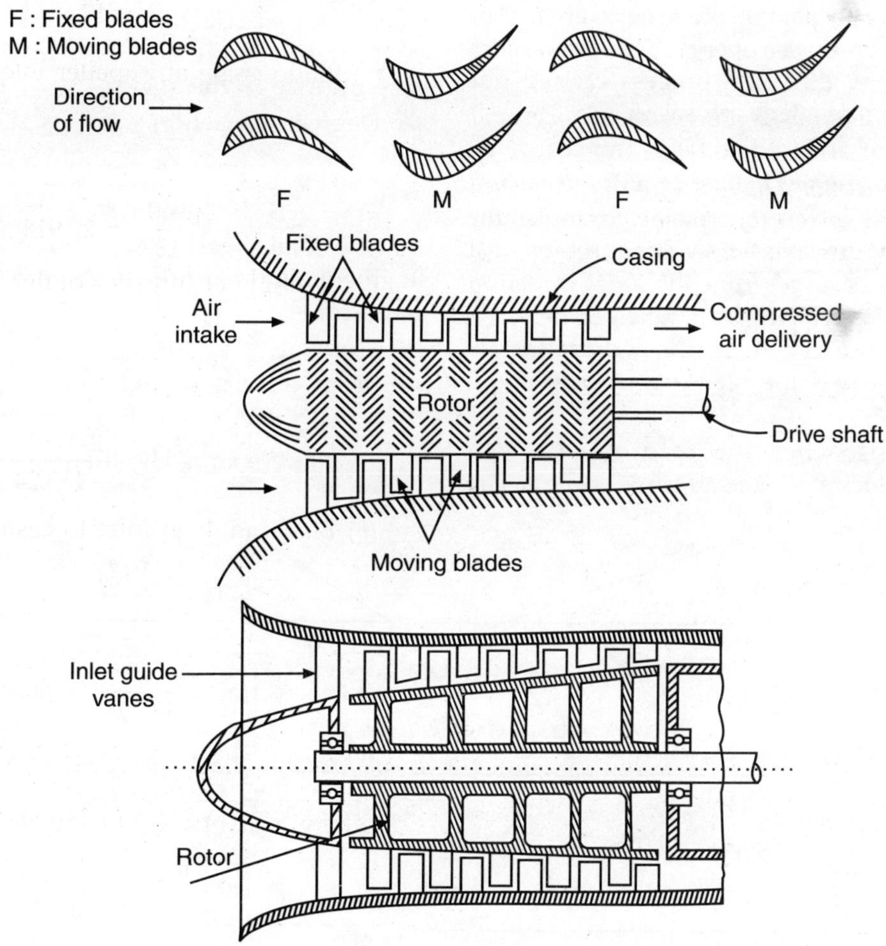

Fig. 21.27. Axial flow compressor

blade contour. Whereas the compressor blades have airfoil section, the turbine blades have profiles formed by a number of circular arcs. This is so because the acceleration process being carried out in the converging blade passages of a reaction turbine is much more efficient and stable process as compared with the diffusing or decelerating process being carried out in the diverging passage between the blades of an axial flow compressor.

The annulus area is usually reduced from inlet to outlet of the compressor. This is to keep the flow velocity constant throughout the compressor length. In the diverging passages of the moving blades, there is rise in pressure due to diffusion. The absolute velocity is also increased due to work input. The fixed blades serve the following two purposes :

(*i*) convert a part of the kinetic energy of the fluid into pressure energy. This conversion is achieved by diffusion process carried out in the diverging blade passages.

(*ii*) guide and redirect the fluid flow so that entry to the next stage is without shock.

Fig. 21.28 shows the velocity triangles for one stage of an axial flow compressor. All angles are measured from the axial direction and the blade velocity V_b is taken to be same at blade entry and exit. This is because the air enters and leaves the blades at almost equal radii.

Air approaches the rotor blade with absolute velocity V_1 and at an angle α_1. The relative velocity V_{r1}, obtained by the vectorial addition of absolute velocity V_1 and blade velocity V_b has the inclination β_1 with the axial direction. Due to diffusion in the diverging passages formed by rotor blades, there is some pressure rise. This is at the expense of relative velocity and so the relative velocity decreases from V_{r1} to V_{r2}. Since work is being done on the air by rotor blades, the air would ultimately leave the rotor blades and the diffusion and declaration take place in the diverging passage of the stator blades. Finally the air leaves the stator blades with velocity V_3 at an angle α_3 and is redirected to the next stage. Generally it is assumed that absolute velocity V_3 leaving the compressor stage equals the approach velocity V_1.

Degree of reaction. The term degree of reaction for an axial flow compressor is defined as the ratio of pressure rise in the rotor to pressure rise in the stage.

Degree of reaction

$$R = \frac{\text{pressure rise in the rotor}}{\text{pressure rise in the stage}}$$

The total pressure rise in the compressor stage equals the work input per stage and is given by

$$(\Delta p)_{\text{stage}} = (V_{w2} - V_{w1})\, V_b$$

From velocity triangles:

$$V_{w2} = V_b - V_f \tan \beta_2$$

and $$V_{w1} = V_b - V_f \tan \beta_1$$

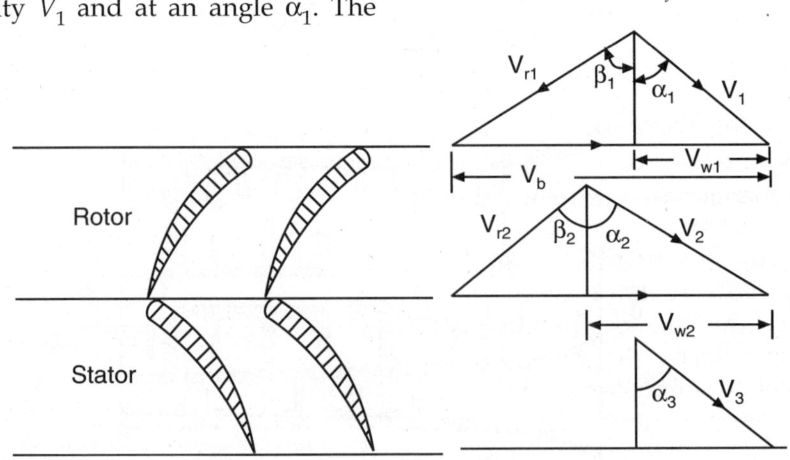

Fig. 21.28. Velocity diagrams for axial flow compressor

$$\therefore (\Delta p)_{stage} = V_b V_f (\tan \beta_1 - \tan \beta_2) \quad ...(21.39)$$

The pressure rise in the rotor blades is at the expense of change in relative kinetic energy and it equals

$$(\Delta p)_{rotor} = \frac{V_{r_1}^2 - V_{r_2}^2}{2}$$

From velocity triangles:

$$V_{r_1}^2 = V_f^2 + (V_f \tan \beta_1)^2$$

and $V_{r_2}^2 = V_f^2 + (V_f \tan \beta_2)^2$

$$\therefore (\Delta p)_{rotor} = \frac{V_f^2}{2} (\tan^2 \beta_1 - \tan^2 \beta_2) \quad ...(21.40)$$

$$\text{Then : } R = \frac{\frac{V_f^2}{2}(\tan^2 \beta_1 - \tan^2 \beta_2)}{V_b V_f (\tan \beta_1 - \tan \beta_2)}$$

$$= \frac{V_f}{2V_b} (\tan \beta_1 + \tan \beta_2) \quad ...(21.41)$$

The usual practice is to design the compressors for 50% reaction. Then

$$\frac{V_b}{V_f} = \tan \beta_1 + \tan \beta_2$$

From velocity triangles,

$$\frac{V_b}{V_f} = \tan \alpha_1 + \tan \beta_1$$

and $\frac{V_b}{V_f} = \tan \alpha_2 + \tan \beta_2$

It follows from the above identities that

$$\alpha_1 = \beta_2 \text{ and } \beta_1 = \alpha_2$$

The equality of angles (the exit angle β_2 of moving blade is equal to the inlet angle α_1 of the fixed blades) suggests that the blades are identical for the 50% reaction axial flow compressor. Since the blades are identical (symmetrical), the velocity diagrams would also be symmetrical. In symmetrical blades the losses due to tip clearance and fluid friction are minimum.

EXAMPLE 21.27

An eight-stage axial flow compressor of 50% reaction design has a mean blade velocity of 185 m/s and a constant axial velocity of 100 m/s through out the compressor. The work is divided equally between the stages and the machine provides an overall pressure ratio 6 : 1 with an overall isentropic efficiency 90%. Assuming an air temperature of 290 K, estimate the power required and the angles of the impeller blade.

Solution : Refer Fig. 21.28 for the velocity vector diagrams. For 50% reaction design, the blades are symmetrical and velocity vector diagrams are identical. Apparently then

$$\alpha_1 = \beta_2 \text{ and } \alpha_2 = \beta_1$$

Considering the compression process to be isentropic, the temperature of air at exit from the compressor stages is

$$T'_{0N} = T_{01} \left(\frac{P_{0N}}{P_{01}}\right)^{\frac{\gamma - 1}{\gamma}}$$

$$= 290\,(6)^{\frac{1.4-1}{1.4}} = 484.1 \text{ K}$$

$$\eta_c = \frac{T'_{0N} - T_{01}}{T_{0N} - T_{01}}$$

$$T_{0N} = T_{01} + \frac{T'_{0N} - T_{01}}{\eta_c}$$

$$= 290 + \frac{484.1 - 290}{0.9} = 505.67 \text{ K}$$

Work done/kg of air for all the stages is

$$c_p (T_{0N} - T_{01}) = 1.005\,(505.67 - 290)$$
$$= 216.75 \text{ kJ/kg}$$

The work done per stage can also be worked out as

$$= (V_{w_2} - V_{w_1}) V_b$$
$$= V_f (\tan \alpha_2 - \tan \alpha_1) V_b$$

Since work input to each stage is same, total work input

$$= V_f (\tan \alpha_2 - \tan \alpha_1) V_b \times N$$
$$= 100 (\tan \alpha_2 - \tan \alpha_1) \times 185 \times 8$$

$$\therefore \tan \alpha_2 - \tan \alpha_1 = \frac{216.75}{100 \times 185 \times 8} = 1.464$$

or $\tan\beta_1 - \tan\alpha_1 = 1.464$ $(\because \alpha_2 = \beta_1)$

From velocity diagram :

$$\tan\alpha_1 + \tan\beta_1 = \frac{V_b}{V_f} = \frac{185}{100} = 1.85$$

From the above two identities, we get

$$\tan\beta_1 = \frac{1.464 + 1.85}{2} = 1.657$$

$\therefore$ $\beta_1 = 58.89° = \alpha_2$

Then from the first identity ;

$$\tan\alpha_1 = \tan\beta_1 - 1.464$$
$$= 1.657 - 1.464 = 0.193$$

$\therefore$ $\alpha_1 = 10.92° = \beta_2$

(*b*) Power input for 1 kg/s of air supply,

$$= m\,c_p\,(T_{ON} - T_{01})$$
$$= 1 \times 1.005(505.67 - 290)$$
$$= \mathbf{216.75\ kW}$$

EXAMPLE 21.28

A multi-stage axial flow compressor takes in air from stagnation conditions of 1 bar and 290 K. When delivering 20 kg/s of air, the compressor absorbs 4500 kW. The polytropic efficiency of compression is 0.9 and the stage stagnation pressure ratio is constant. Calculate the pressure at compressor outlet, the number of stages and the overall isentropic efficiency of compressor. Assume that the temperature rise in the first stage is 20°C.

Solution : $T_{02} = T_{01} + 20 = 290 + 20 = 310$ K

In terms of stagnation pressure and stagnation temperature ratio, the polytropic efficiency is given as

$$\eta_p = \frac{\log_e (p_{02}/p_{01})^{\frac{\gamma-1}{\gamma}}}{\log_e (T_{02}/T_{01})}$$

$$0.9 = \frac{\log_e (p_{02}/p_{01})^{\frac{1.4-1}{1.4}}}{\log_e (310/290)}$$

$$= \frac{\log_e (p_{02}/p_{01})^{0.286}}{0.0667}$$

$$\text{or } \log_e\left(\frac{p_{02}}{p_{01}}\right) = \frac{0.9 \times 0.0667}{0.286} = 0.201$$

$\therefore$ Stagnation pressure ratio $\dfrac{p_{02}}{p_{01}} = 1.223$

$$\text{Power input} = m\,c_p\,(T_{ON} - T_{01})$$
$$4500 = 20 \times 1.005\,(T_{ON} - T_{01})$$

$$\text{or } T_{ON} = \frac{4500}{20 \times 1.005} + T_{01}$$
$$= 223.88 + 290 = 513.88 \text{ K}$$

$\therefore$ Stagnation temperature ratio

$$\frac{T_{ON}}{T_{01}} = \frac{513.88}{290} = 1.772$$

For a compressing unit,

$$\eta_p = \frac{\gamma-1}{\gamma} \times \frac{n}{n-1} \quad \text{(See Page 768)}$$

$$\frac{n}{n-1} = \frac{0.9 \times 1.4}{1.4 - 1} = 3.15$$

$$\frac{p_{ON}}{p_{01}} = \left(\frac{T_{ON}}{T_{01}}\right)^{\frac{n}{n-1}} = (1.772)^{3.15} = 6.06$$

$\therefore$ Stagnation pressure at outlet

$$p_{ON} = 6.06 \times 1 = 6.06 \text{ bar}$$

(*b*) The stage stagnation pressure is stated to be constant, *i.e.*, same for each stage. Then

$$\frac{p_{02}}{p_{01}} = \frac{p_{03}}{p_{02}} = \frac{p_{04}}{p_{03}} = \ldots \frac{p_{ON}}{p_{0(N-1)}}$$

$$\text{As such } \left(\frac{p_{02}}{p_{01}}\right)^N = \frac{p_{ON}}{p_{01}}$$

$$\text{or } N = \frac{\log_e (p_{ON}/p_{01})}{\log_e (p_{02}/p_{01})}$$

$$= \frac{\log_e 6.06}{\log_e 1.223} = \frac{1.802}{0.2013} = 8.95$$

$\therefore$ Number of stages $N = 9$

(*c*) Considering the compression process to be isentropic, the temperature of air leaving the compressor would be

$$T'_{ON} = T_{01}\left(\frac{p_{ON}}{p_{01}}\right)^{\frac{\gamma-1}{\gamma}}$$

$$= 290\,(6.06)^{\frac{1.4-1}{1.4}} = 485.5 \text{ K}$$

∴ Overall isentropic efficiency

$$\eta = \frac{T'_{0N} - T_{01}}{T_{0N} - T_{01}}$$

$$= \frac{485.5 - 290}{513.88 - 290} = 0.873 \text{ or } \mathbf{87.3\%}$$

EXAMPLE 21.29

A multi-stage axial flow compressor of 50% reaction design at the mean radius operates with overall stagnation pressure ratio 3.5, stagnation isentropic efficiency 85 percent and inlet stagnation temperature 320 K. The relative air angles at rotor inlet and outlet are 130° and 100°C respectively, and these angles have been measured in the same direction from the blade velocity direction. The blade velocity is constant at 200 m/s, and flow velocity too remains constant through out the compressor. Taking work done factor for each stage as 0.85, calculate the stagnation polytropic efficiency and the number of stages.

Solution : Refer Fig. 21.29 for the velocity vector diagram.

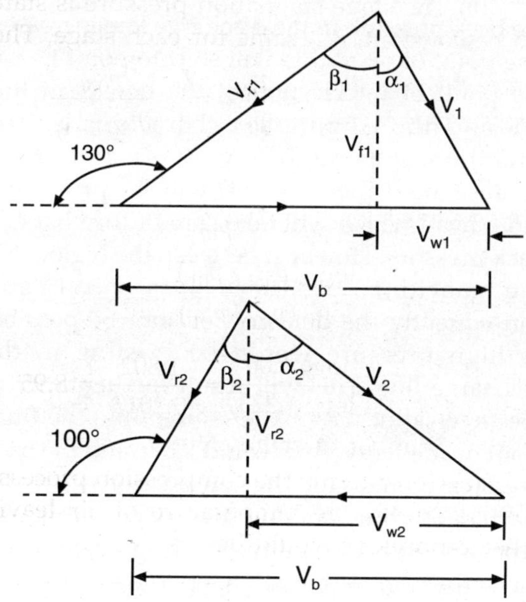

Fig. 21.29.

For a 50% design, the inlet and outlet velocity vector diagrams for the compressor are similar.

$$\therefore \quad \beta_1 = \alpha_2 = 90 - (180 - 130) = 40°$$

$$\beta_2 = \alpha_1 = 90 - (180 - 100) = 10°$$

Considering the compression processes to be isentropic, the temperature at outlet from the compressor stages would be

$$T'_{0N} = T_{01}\left(\frac{p_{0N}}{p_{01}}\right)^{\frac{\gamma-1}{\gamma}}$$

$$= 320\,(3.5)^{\frac{1.4-1}{1.4}} = 457.88$$

$$\eta_c = \frac{T'_{0N} - T_{01}}{T_{0N} - T_{01}}$$

$$T_{0N} = T_{01} + \frac{T'_{0N} - T_{01}}{\eta_c}$$

$$= 320 + \frac{457.88 - 320}{0.85} = 482.21 \text{ K}$$

In terms of stagnation pressure and stagnation temperature ratios, the polytropic efficiency is given as

$$\eta_p = \frac{\log_e (p_{0N}/p_{01})^{\frac{\gamma-1}{\gamma}}}{\log_e (T_{0N}/T_{01})}$$

$$= \frac{\log_e (3.5)^{\frac{1.4-1}{1.4}}}{\log_e (482.21/320)}$$

$$= \frac{0.3584}{0.410} = 0.874 \text{ or } 87.4\%$$

(*b*) From the velocity vector diagrams

$$\frac{V_b}{V_f} = \tan \alpha_1 + \tan \beta_1$$

$$= \tan 10° + \tan 40° = 1.016$$

$$\therefore \quad V_f = \frac{200}{1.016} = 196.85 \text{ m/s}$$

$$V_{w_2} = V_f \tan \alpha_2$$

$$= 196.85 \tan 40° = 165.18 \text{ m/s}$$

$$V_{w_1} = V_f \tan \alpha_1$$

$$= 196.85 \tan 10° = 34.71 \text{ m/s}$$

Work done per stage

$= (V_{w2} - V_{w1})\, V_b \times P$

$= (165.18 - 34.71) \times 200 \times 0.85$

$= 22180 \text{ Nm/s} = 22.18 \text{ kW}$

Total work done

$= c_p\,(T_{0N} - T_{01})$

$= 1.005\,(482.21 - 320) = 163.02 \text{ kW}$

$$\therefore \text{ Number of stages } N = \frac{163.02}{22.18} = 7.35$$

Hence 8 stages are required for the given multi-stage compressor to perform the assigned task.

21.11. SURGING, CHOKING AND STALLING

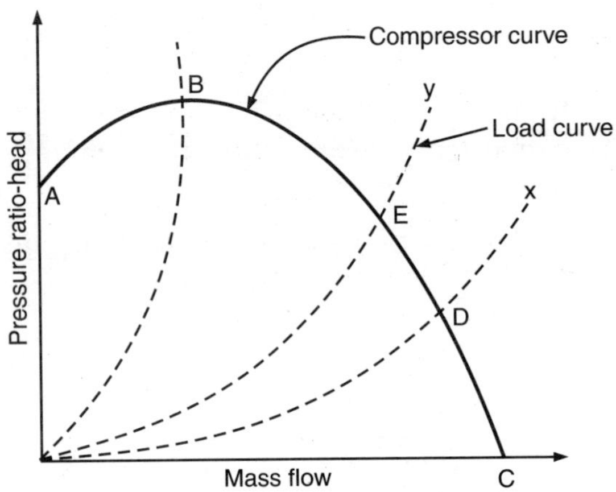

Fig. 21.30. Surging

The head capacity characteristic of a compressor consists of two portions, portion *AB* having a positive slope and portion BC having a negative slope (Fig. 21.30) A and C are the two extreme points of this curve. Point *A* represents the flow conditions when discharge valve is fully closed and C represents the other extreme condition when discharge valve is fully open. When the discharge valve is fully closed, the flow rate is zero and the static pressure developed at this instant is called shut off head. This head is partially due to centrifugal action and partially due to kinetic energy component. At full discharge valve opening, the discharge is maximum but pressure rise is zero. Consequently compression efficiency is zero.

When the discharge valve is opened slightly, delivery of mass flow commences. The diffuser contributes towards pressure rise and consequently the static pressure rises. With more opening, the pressure goes on increasing till the maximum pressure rise represented by point B is attained. This is the point of maximum efficiency for given conditions of rotational speed, inlet temperature and suction pressure. The increased mass flow rate with further opening of the valve, beyond point *B* is with less pressure rise and consequently with reduced efficiency. Finally when the pressure ratio approaches unity, the mass flow would be maximum, but the compression efficiency will be zero. The opening and closing of the valve can be linked with external load on the compressor. The compressor has to operate in conjunction with this external load. The point of intersection of the compressor curve and the load curve gives the operating point of the system comprising compressor and external load. D is one such point. With the increase in external load due to closing of delivery valve, the back pressure in the discharge line increases. The point of operation will shift to point E. E is the point of intersection of the new load line OY and the compressor characteristic. The operation at point E is possible and stable because now the machine will be producing more head which will take care of the changed back pressure. However between the region AB, the machine produces less head and consequently the delivery will not be possible as high pressure would be existing in the discharge line. This will result momentarily in the reversal of flow. After some time the fluid from the delivery pipe would leave and the back pressure would fall. The compressor will again start delivering air and the cycle would be repeated with instability occuring again. Thus when the compressor has to operate at less mass flow rates, the air surges and pulsates throughout the compressor and the compressor does not give a steady flow of air.

This unsteady, periodic and reversal of flow manifests the phenomenon of **surging**. Prolonged surging results in overheating and the stress reversal in the blades may lead to fatigue failure.

Theoretically the mass flow rate becomes maximum when pressure ratio is unity. But at this instant the compression efficiency is zero *i.e.*, there is no compression at all. An upper limit is imposed on mass flow which is practically possible in a compressor and this mass flow is known as choking mass flow. **Choking** means fixed mass flow rate regardless of pressure ratio, *i.e.*, the characteristic becomes vertical.

Because of surging and choking, the ideal head-capacity characteristic is much modified (Fig. 21.31) and only a part of the curve is useful.

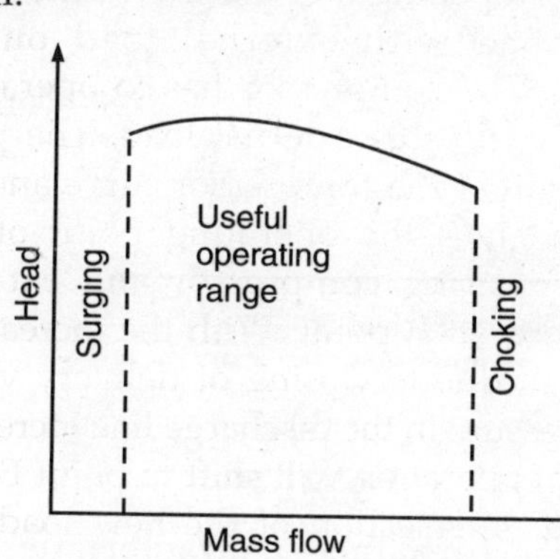

Fig. 21.31. Modified head-capacity characteristic

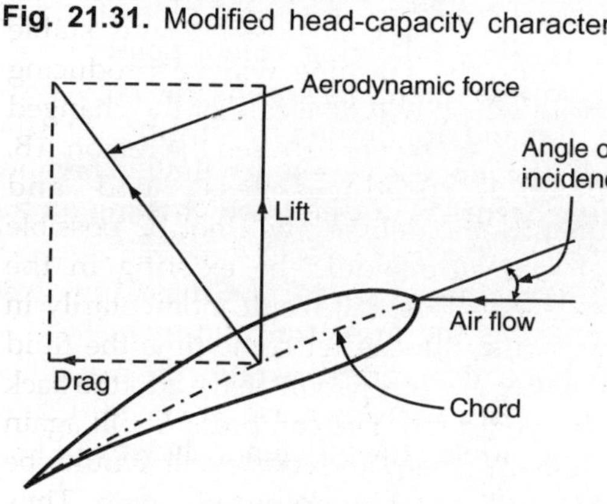

Fig. 21.32. Aerodynamic force

Stalling of the blades of axial flow compressor can be explained in terms of airfoil theory. It is known that when a moving stream of air meets the leading edge of the airfoil (compressor blade) at a certain angle of attack, lift is produced. Angle of attack is the inclination of the chord (line joining the leading edge and trailing edge) with air stream (Fig. 21.32).

Aerodynamic force ∝ kinetic energy of stream × surface area of airfoil

$$\propto \frac{1}{2}\rho V^2 \times \text{area}$$

This force acts through the centre of pressure and is normal to the chord line. The aerodynamic force has two components:

(*i*) one parallel to the direction of air stream called drag

(*ii*) one perpendicular to the direction of air stream called lift

$$\text{Drag } D = C_d \times \frac{1}{2}\rho V^2 A$$

$$\text{Lift } L = C_l \times \frac{1}{2}\rho V^2 A$$

where C_d and C_l are respectively drag and lift coefficients. The variation of these coefficients with change in angle incidence for a typical cascade test is plotted in Fig. 21.33. Cascade means a row of airfoils. The results on a cascade resemble compressor performance.

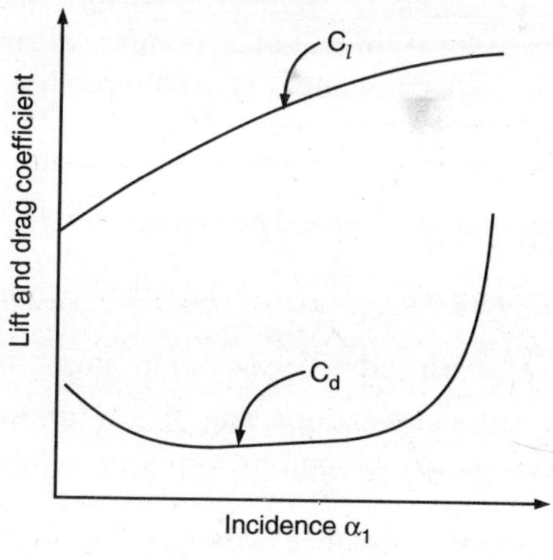

Fig. 21.33. Variation of C_l and C_d with angle of incidence

It may be seen that C_l (or lift force) increase with increasing incidence but only upto a certain maximum value occuring at a small positive value of incidence. With further increases in angle of incidence a critical value is reached beyond which C_l decrease and C_d (or drag force) increases rapidly. The reason is that air stream breaks away from the surface and there is separation of flow over large part of the blade. The separation causes violent eddies and distortion in flow. There is loss of lift and increase in drag. With an increase in angle of attack, the area of separated flow goes on increasing.

The phenomenon of reduction in lift force at higher angles of incidence is known as ***stalling***.

21.12. PERFORMANCE CHARACTERISTICS OF ROTARY COMPRESSORS

The common method for plotting the performance characteristics of a compressor is to show how the power, efficiency and pressure ratio vary with change in mass flow rate (Fig. 21.34 (*a*)). Such a plot makes no consideration of varying inlet temperature and pressure. Moreover the comparison of the performance of similar compressors of different sizes cannot be predicted from such plots. To account for the effect of inlet conditions on performance and to make it possible to predict the performance of a compressor from that of a geometrical similar model, the performance curves are plotted with dimensionless parameters. The dimensionless parameters for a compressor are ; pressure ratio parameter p_2/p_1, speed parameter $N/\sqrt{T_1}$ and flow parameter $m\sqrt{T_1}/p_1$. The speed parameter $N/\sqrt{T_1}$ is varied in reasonable steps and a family of curves depicting the relationship between pressure ratio p_2/p_1 and mass flow $m\sqrt{T_1}/p_1$ are drawn as shown in Fig. 21.34 (*b*).

These curves are limited by surge line at low mass flow rate and choking at low pressure ratio. The operating line is the locus of points of maximum efficiency. This line is close to surge line near design point but at lower speed, it moves farther away from the surge line. Surge line gives the limit of stability. Beyond this curve on the left hand side, there are pulsations in the flow rate.

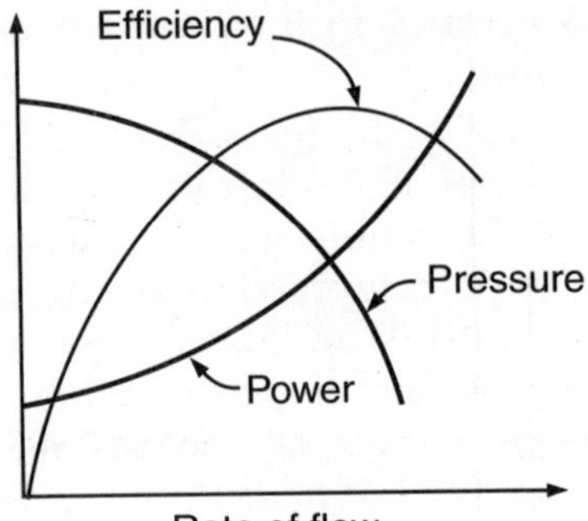

(a)

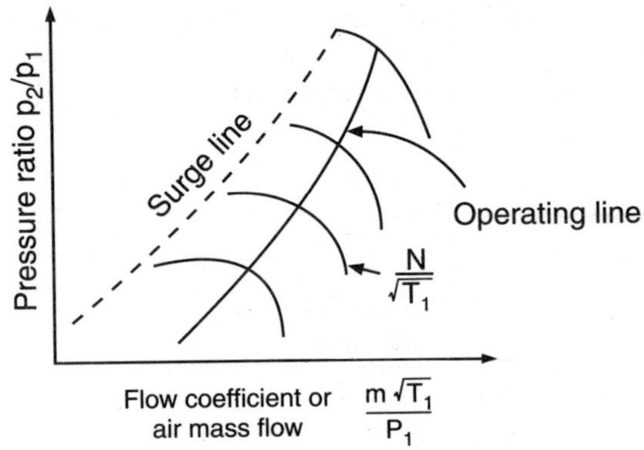

(b)

Fig. 21.34.

The performance characteristics of a centrifugal compressor, shown plotted in Fig. 21.35 have the following salient points:

1. At low rotational speeds, the curves are fairly flat and are limited by surging. At high speeds, the range is very much limited because surging occurs at one end and choking on the other.

2. For a given speed, the mass flow rate decreases as the pressure ratio increases.

3. For a given pressure ratio, the mass flow rate increases with increase in speed. Under such conditions, the efficiency, however, falls rapidly.

With respect to Fig. 21.36 which represents the performance curves for an axial flow compressor, the following salient points can be noticed :

1. There is fall in efficiency when there is decrease in pressure on mass flow:

2. Near the design point, the volume change is not apreciable with increase of pressure.

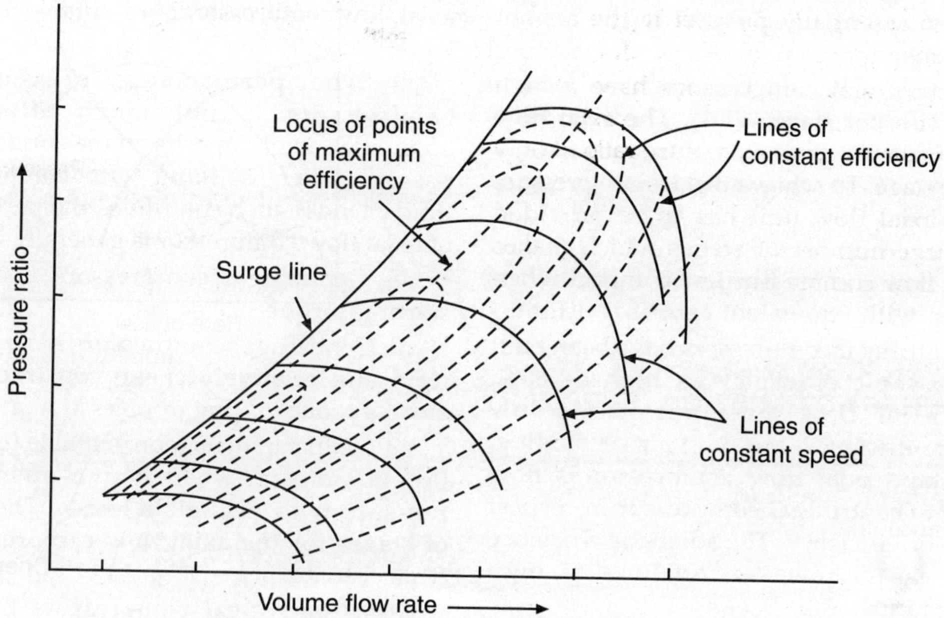

Fig. 21.35. Performance curves of centrifugal compressor

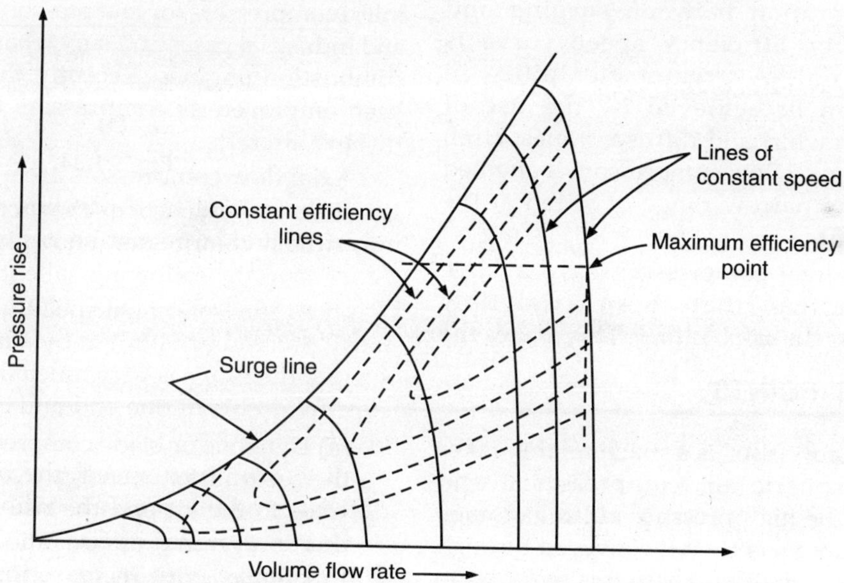

Fig. 21.36. Performance curves of an axial flow compressor

3. The compressor gives optimum performance at large volume flows and at high speed.

21.13. CENTRIFUGAL COMPRESSOR VERSUS AXIAL FLOW COMPRESSOR

Below is given the comparison between some of the main features of centrifugal and axial flow compressors. The comparison helps to form some criterion for the choice between blade compressors for a particular application.

1. In centrifugal compressors, the flow through the impeller takes place largely in a plane which is perpendicular to the axis of the compressor. In axial flow compressors, the flow proceeds throughout the compressor in

a direction essentially parallel to the axis of the machine.

2. Centrifugal compressors have a high pressure rise per stage (4 : 1). The axial flow machine however gives a pressure ratio of only 1.2 : 1 per stage. To achieve the desired pressure rise, the axial flow unit has to be provided with a large number of stages which makes the axial flow compressor less compact when compared with equivalent centrifugal unit.

3. Centrifugal compressors have been built with isentropic efficiency as high as 80%. Earlier axial flow machines had a low isentropic efficiency, but with airfoil blading, a multistage axial flow compressor is now superior to centrifugal compressor in respect of adiabatic efficiency. The adiabatic efficiency in both cases, however, declines at high pressure ratios.

4. Centrifugal compressors have a wide range of operation between surging and choking limits. Efficiency speed curve is reasonably flat. A greater flexibility in operation can be achieved by the use of adjustable prewhirl and diffuser vanes. High efficiency for axial flow compressors is attained only with in a narrow range of speed at the design pressure.

5. Centrifugal compressors have a large frontal area than that of an axial flow compressor of the same rating. This makes the axial flow compressor more suitable for aircraft work.

6. The performance of centrifugal compressors is not much affected by accumulation of deposits on the surface of flow passage when working with contaminating fluids. Under such conditions the performance of axial flow compressor is generally impaired.

7. Centrifugal compressors need a low starting torque.

8. Centrifugal compressors have a simple, rigid and relatively cheap construction and are less prone to icing troubles at high altitudes.

9. Multistaging is more suitable for an axial flow compressor where it gives an increase in pressure with negligible losses. The number of stages for the axial flow compressor vary from 5 to 14.

10. Centrifugal compressors have been successfully used as blowing engines in steel mills, compressors for low pressure refrigeration and industrial gases and supercharging internal combustion engines. Recently they have also been employed as compressors for small gas turbine aircrafts.

Axial flow compressors are mostly used in gas turbines and high pressure units, *i.e.*, for industrial and large marine gas turbine plants.

SALIENT POINTS

1. An air compressor is a machine that takes in atmospheric air, compresses it and delivers the high pressure air to a storage vessel. Subsequently it is conveyed through pipelines to the site where it is required to be used.

2. Compressors are classified as :

   (*i*) Positive displacement where in the air is confined within solid boundaries, and the pressure is developed by change in volume (*e.g.*, piston cylinder arrangement in reciprocating compressors) or by traping it in a reducing space formed by a set of engaging surfaces (*e.g.*, roots blower and vane blower.)

   (*ii*) Dynamic or blade compressors in which there is transfer of energy from the machine rotor to the air and the rise in pressure is due to dynamic effects.

3. In a single acting reciprocating compressor, the suction, compression and delivery take place in two strokes of the piston *i.e.*, in one revolution of crank.

   Neglecting clearance and friction losses, the work input per cycle is

$$(i)\ W_a = \frac{\gamma}{\gamma-1}\, p_1 v_1 \left[\left(\frac{p_2}{p_1}\right)^{\frac{\gamma-1}{\gamma}} - 1\right]$$

(ii) $W_p = \frac{n}{n-1} p_1 v_1 \left[\left(\frac{p_2}{p_1}\right)^{\frac{n-1}{n}} - 1\right]$

(iii) $W_i = p_1 v_1 \log_e \frac{p_2}{p_1}$

where W_a, W_p and W_i are the work inputs when the compression process follows the adiabatic, polytropic and isothermal laws.

4. The isothermal efficiency represents the ratio of isothermal work to actual work.
5. Taking clearance into consideration, work input per cycle

$$= \frac{n}{n-1} p_1 (V_1 - V_4) \left[\left(\frac{p_2}{p_1}\right)^{\frac{n-1}{n}} - 1\right]$$

where $(V_1 - V_4)$ is the effective volume sucked in during the suction stroke.
Work input depends on effective volume and not an the swept volume. Apparently the indicated work per unit mass of air delivered is unaffected by clearance volume.
6. The volumetric efficiency represents the ratio of actual volume sucked in to the piston displacement or swept volume of the compressor.

$$\eta_v = 1 - \frac{V_c}{V_s}\left[\left(\frac{p_2}{p_1}\right)^{1/n} - 1\right]$$

The volumetric efficiency decreases as the clearance factor V_c/V_s increases, decreases with increase of p_2/p_1 and increases with increase of index n.
7. Multistaging means compression of air from the suction pressure in more than one cylinder. The air after compression in the first cylinder and before entering the next cylinder is cooled at constant pressure by passing it through an inter-cooler.
For a two stage compressor or with perfect intercooling
(i) the pressure ratios for each stage are equal and are given by the square root of overall pressure ratio. That is

$$\frac{p_2}{p_1} = \frac{p_3}{p_2} = \sqrt{\frac{p_3}{p_1}} \text{ and } p_2 \sqrt{p_1 p_3}$$

(ii) the work input in each stage is same and the total work input is

$$W = \frac{2n}{n-1} RT_1 \left[\left(\frac{p_3}{p_1}\right)^{\frac{n-1}{2n}} - 1\right]$$

(iii) When extended to N stages

$$\frac{p_2}{p_1} = \frac{p_3}{p_2} = \frac{p_4}{p_3} = \ldots = \frac{p_{N+1}}{p_N}$$

$$W = \frac{Nn}{n-1} RT_1 \left[\left(\frac{p_{N+1}}{p_1}\right)^{\frac{n-1}{Nn}} - 1\right]$$

8. When multistage compression is employed with intercooling between the stages, the volumetric efficiency improves, the work input becomes less and there is reduction in flywheel size and lubrication and leakage problems.
9. The centrifugal compressor consists of rotor (or impeller) to which a number of curved vanes are attached symmetrically. The rotor rotates in an air tight volute casing. When the impeller rotates, it transfers its energy to the air causing a rise in pressure.
(i) The flow through the impeller takes place in a plane which is perpendicular to the axis of the compressor.
(ii) For an axial or radial entry, the transfer per unit mass (called the Euler heat) is

$$H_e = V_{w2} V_{b2}$$

(iii) Due to inertial effects, there is deviation in the flow path and the tangential component gets reduced from V_{w2} to V'_{w2}.
The ratio V'_{w2}/V_{w2} is called slip factor σ.
(iv) For an impeller with radial vanes ($V_{w2} = V_{b2}$). The theoretical work

$$= V'_{w2} V_{b2} = \sigma V_{w2} V_{b2} = \sigma V_{b2}^2$$

Actual work required $= P \sigma V_{b2}^2$
The factor P accounts for the extra work needed due to fluid friction and windage loss, and is called the work factor or power input factor.

(v) The pressure coefficienct ψ represents the ratio of isentropic work to Euler work.

$$\psi = \eta_c P\sigma$$

where η_c denotes the isentropic efficiency of the compressor.

10. The axial flow compressor consists of adjacent rows of rotor blades and stator blades. The rotor blades are mounted on the rotating drum and the stator blades are fixed to the casing.

The flow proceeds throughout the compressor in a direction essentially parallel to the axis of the machine.

One stage of the machine comprises a row of rotor blades followed by a row of stator blades. The degree of reaction is defined as

$$R = \frac{\text{pressure rise in the rotor}}{\text{pressure rise in the stage}}$$

The blades would be identical for the 50% axial flow compressor.

11. Surging manifests in unsteady, periodical and reversal of flow. Prolonged surging results in overheating and the stress reversal in the blades may lead to fatigue failure.
12. Choking means fixed mass flow rate regardless of pressure ratio, *i.e.*, the head-capacity characteristic become vertical.
13. For efficient operation, the blades of an axial flow compressor are of air foil section. The phenomenon of reduction in lift at higher angles of incidence of blades is known as stalling.
14. Centrifugal compressors have been successfully used as blowing engines in sugar mills, compressors for low pressure refrigeration and industrial gases, and for supercharging IC engines.

Axial flow compressors are mostly used for aircraft work, and industrial and large marine gas turbine plants.

REVIEW QUESTIONS

A. Conceptual and conventional questions.

1. Mention the different applications of compressed air.
2. "Compressed air is a useful commodity." Comment on the validity of this statement by citing a few applications of compressed air.
3. (*a*) How compressors are classified ?

   (*b*) Distinguish between positive and non-positive displacement type compressors.
4. Explain, with a neat sketch, the working of a single stage reciprocating compressor.
5. Define the following terms in relation to reciprocating compressors :

   (*i*) isothermal efficiency (*ii*) free air delivery and (*iii*) volumetric efficiency
6. Define volumetric efficiency of a single stage reciprocating compressor and set up an expression for it in terms of pressure ratio r and clearance ratio C. The compression and expansion curves follow the law pv^n = constant.
7. Set up the following expression for volumetric efficiency with respect to free air delivery,

$$\eta_{vol} = \frac{p_1 T_0}{p_0 T_1}\left[1 + C - C\left(\frac{p_2}{p_1}\right)^{1/n}\right]$$

where the symbols in the expression have their usual meanings.

8. (*a*) State the conditions under which the use of multistage compression is adopted. Explain the underlying principle by means of pressure-volume and temperature-entropy plots.

   (*b*) Enumerate the advantages of multi-stage compression.
9. (*a*) Describe, with a neat sketch, the working of a roots blower. Represent the processes involved on *p-v* diagram.

   (*b*) Set up the expression for roots efficiency.
10. (*a*) Draw the neat sketch of a vane type compressor and explain its working.

    (*b*) Set up an expression for the efficiency of a vane blower.
11. Comment on the validity of the statement :

    The thermodynamic efficiency of a centrifugal compressor is referred to adiabatic conditions where as that of reciprocating compressor is referred to isothermal conditions.
12. (*a*) Draw neat sketch of a centrifugal compressor and explain its working.

(*b*) How pressure changes take place in the impeller and diffuser section of a centrifugal compressor ?

13. (*a*) Sketch the ideal velocity diagrams for a centrifugal compressor with backward, radial and forward curved blades.

(*b*) Set up the Euler equation for work input to a centrifugal compressor. Transfer this equation to give the head-capacity relationship for one-dimensional ideal flow conditions.

14. Explain the terms slip factor and power input factor in a centrifugal compressor.

15. (*a*) Describe, with a neat sketch, the working of an axial flow compressor.

(*b*) State the difference between the blading of a reaction turbine and that of an axial flow compressor.

16. Explain the term degree of reaction for an axial flow compressor. Why is the degree of reaction generally 50% ?

17. Explain the phenomenon of surging and choking in rotary compressors.

18. Sketch the salient characteristics of centrifugal and axial flow compressors and mention their salient features.

19. Describe briefly the relative merits and demerits of centrifugal and axial flow compressors.

20. A single stage double acting reciprocating air compressor receives air at 100 kPa and 290 K, and compresses it to 600 kPa according to the law $pv^{1.25}$ = constant. While running at 100 rev/min, the average piston speed of the compressor is 150 m/min. Taking the cylinder diameter as 30 cm, determine the power required to drive the compressor.

[**Ans.** 30 kW]

21. A single stage single acting reciprocating compressor delivers 150 m³ of free air per minute, compressing it from 1 bar to 8 bar. The clearance volume is one-sixteenth of swept volume and the compressor turns 300 rev/min. The stroke length equals 1.5 times the cylinder bore. Taking the exponent n for compression and expansion as 1.3, determine the diameter and stroke of the compressor.

[**Ans.**82.5 cm and 123.75 cm]

22. A single stage double acting reciprocating air compressor is designed to deliver 14 m³/min of air measured at 1.013 bar and 15°C. The unit has clearance volume of 5% of the swept volume, runs at 300 rev/min and operates with delivery pressure of 7 bar. Taking the exponent n for compression and expansion as 1.3, calculate the swept volume of the cylinder, the delivery temperature and indicated power.

[**Ans.** 0.028 m³/cycle, 450 K, 57.5 kW]

23. Consider a two-stage compression with intercooling between pressure limits P_1 and P_2. The air enters the first stage at P_1, T_1 and leaves at an intermediate pressure P. Then the air is cooled at constant pressure to the original temperature T_1. The air is then compressed in the second stage to P_2. For fixed values of P_1, P_2 and T_1, determine the optimum value of the intermediate pressure P, which will minimize the work input.

24. An ideal four-stage reciprocating compressor is required to compressor air from 0.95 bar and 300 K to 25 bar. Assuming equal stage pressure ratios and perfect intercooling, calculate the interstage pressure, air delivery temperature and the isothermal efficiency. The law of compressor is $pv^{1.2}$ = constant for each stage.

[**Ans.** 3.83 bar, 15.41 bar, 62.08 bar; 378.6 K; 88.6%]

25. A single-acting, two-stage reciprocating air compressor is to be designed to compressor 4.5 kg of air per minute from 1.013 bar and 298 K through an overall pressure ratio of 9. The compressor runs at 300 rev/min and the clearance volumes of both stages are 5% of their respectives swept volumes. The intercooling is complete and both stages have the same pressure ratio. Presuming the law of compression and expansion to be $pv^{1.3}$ = constant for both stages, make calculations for the indicated power and the swept volumes required for each cylinder.

[**Ans.** 15.5 kW; 0.0131 m³ ; 0.00436 m³]

26. A centrifugal compressor takes in air at 1 bar pressure and 288 K. The corresponding values at exit from the compressor are 2 bar and 370 K. Determine (*a*) the index of compression, (*b*) the isentropic efficiency, and (*c*) the power required to drive the compressor.

It may be assumed that the compressor is uncooled and no heat is lost to surroundings.

[**Ans.** n = 1.57, 76.9%; 164.8 kW]

27. A single sided centrifugal compressor has radial impeller blades of tip diameter 55 cm and runs at 18000 rev/min. The other pertinent data is : mean diameter of eye = 20 cm
absolute velocity of air at inlet = 150 m/s
pre-whirl to the axial direction at the radii = 20°
slip factor and isentropic efficiency of compressor = 0.8 and 0.82

Assuming that air is sucked in at 290 K and 1 bar, make calculations for the pressure ratio and the power input required to drive the compressor.

[Ans. 4.95, 205.5 kW]

28. Air is compressed in a single sided impeller type centrifugal compressor running at 20,000 rev/min with an isentropic efficiency of 80 percent. The compression of air is from 1 bar static pressure to 4.5 bar total pressure. The mass flow rate of air is 10 kg/s, and the ambient air at 1 bar and 20°C enters the impeller eye with a velocity of 150 m/s with no pre-whirl. Assuming that the ratio of whirl speed to tip speed is 0.95, calculate:
    (*a*) rise in total temperature during compression if the change in kinetic energy is negligible,
    (*b*) the impeller tip speed and tip diameter,
    (*c*) the power expended in running the compressor, and
    (*d*) the external diameter of the eye for which the internal diameter is 15 cm.

[Ans. 183°C, 440.2 m/s and 0.42 m, 180 kW, 30.6 cm

B. Fill in the blanks with appropriate word/words.

1. The pressure radio in the case of reciprocating compressor is .... than that of dynamic compressor.
2. The interstage pressure for minimum work input in a two stage compressor is .... of the suction and delivery pressures.
3. The criterion of thermodynamic efficiency of reciprocating compressor is .... while that of rotary compressor is ....
4. A .... compressor provides a continuous and uniform stream of compressed air.
5. Reciprocating compressors are usually preferred for .... pressure and .... discharge.
6. The capacity of a reciprocating air compressor expressed in m^3/min refers to the volume of .... delivered.
7. In a reciprocating compressor, the ratio of clearance volume to swept volume is called the ....
8. For the same pressure ratio, the efficiency of vane blower is .... than that of roots blower.
9. .... refers to the instability in the flow occurring due to flow separation from the blade surfaces of a compressor.
10. The state of operation of compressor at maximum mass flow rate is called ....
11. For a multi-stage compressor, the .... is constant through out for all stages.
12. Degree of reaction in an axial compressor is defined as the ratio of static enthalpy rise in the .... to static enthalpy rise in the ....

Answers

1. higher 2. geometric mean 3. isothermal efficiency, adiabatic efficiency 4. centrifugal compressor 5. high, low 6. free air 7. clearance ratio 8. higher 9. stalling 10. choking 11. polytropic efficiency 12. rotor, stage

C. State true of false. If false, write the correct statement.

1. The work input for compressing air is minimum when the compression is done isentropically.
2. The volumetric efficiency of a compressor decreases with increase in pressure ratio.
3. The clearance volume has no effect on work input to the compressor for delivering unit kg of air.
4. The positive displacement compressors may be of axial flow type or centrifugal type depending upon nature of flow in the machine.
5. Pressure ratio per stage in axial flow compressor is more than that of centrifugal compressor.
6. In a centrifugal compressor, the air enters the impeller eye radially and flows axially outwards.
7. Rotary compressors of the root blower type and vane blower type belong to the category of positive displacement compressors.
8. A greater portion of the pressure rise in a centrifugal compressor occurs primarily in the impeller.

Answers

1. False; isothermally 2. True 3. True 4. False; dynamic compressors 5. False; less 6. False; axially or radially 7. True 8. False

D. Multiple choice questions.

1. Identify the non-positive displacement compressor
    (*a*) axial flow compressor
    (*b*) reciprocating compressor

(c) roots blower
(d) vane blower

2. The work input to air compressor will be least if the exponent n in the process pv^n = constant equals
(a) $n = 1$ (b) $n = 1.2$
(c) $n = \gamma = 1.4$ (adiabatic compression)
(d) n takes the infinite value

3. With increase in clearance volume, the specific work input (work input per kg of air)
(a) increases
(b) decreases
(c) remains constant
(d) unpredictable ; may increase or decrease depending on capacity of compressor.

4. The volumetric efficiency of a compressor is the ratio of
(a) displacement volume to clearance volume
(b) displacement volume to volume of air actually sucked inside the cylinder
(c) actual free air delivered to the displacement volume
(d) volume of air before compression to volume of air after compression.

5. Consider the following factors
(1) cylinder size
(2) clearance ratio
(3) delivery pressure
(4) compressor shaft power
The factors which affect the volumetric efficiency of a single stage compressor will include.
(a) 1 and 2 (b) 3 and 4
(c) 2 and 3 (d) 1 and 4

6. An inter-cooler is used with multi-stage reciprocating compressor to cool.
(a) the air before it is sucked inside the compressor cylinder
(b) the air before it is discharged to the receiver
(c) the air after one stage of compression before it is sent to the next stage
(d) the compressor cylinder

7. A four-stage compressor, with perfect intercooling between stages, compresses air from 1 bar to 16 bar. The optimum pressure in the last intercooler will be
(a) 6 bar (b) 8 bar
(c) 12 bar (d) 16 bar

8. The performance of dynamic compressors is compared on the basis of
(a) isothermal efficiency
(b) adiabatic (isentropic) efficiency
(c) pressure coefficient
(d) power input factor

9. In a rotary bladed compressor, the ratio of isentropic work to Euler work is called
(a) slip factor (b) pressure efficient
(c) work factor (d) degree of reaction

10. Which one of the following factors accounts for the extra work needed due to friction and windage losses in a centrifugal compressor ?
(a) whirl factor (b) slip factor
(c) work factor (d) pressure coefficient

11. Which of the following blades gives the maximum pressure rise in a centrifugal compressor ?
(a) forward curved
(b) backward curved
(c) radial
(d) twisted blades ; blades with mixed curvature

12. Stalling of blades in axial flow compressors is the phenomenon of
(a) air stream blocking the passage
(b) motion of air at sonic velocity
(c) unsteady, periodic and reversed flow
(d) air stream not able to follow the blade contour

13. Which one of the following types of compressor is mostly used for supercharging of internal combustion engines ?
(a) radial flow
(b) axial flow
(c) roots blower
(d) reciprocating compressor

Answers

1. (a)	2. (a)	3. (c)	4. (c)	5. (c)
6. (c)	7. (b)	8. (b)	9. (b)	10. (c)
11. (a)	12. (d)	13. (c)		

❑❑❑

Chapter **22**

Internal Combustion Engines

Learning objectives : The reader will be able to explain the following aspects after having made a study of the subject matter presented in this chapter :

- ❑ *internal combustion engines and their classification*
- ❑ *engine parts and their functions*
- ❑ *working of two stroke/four stroke petrol and diesel engines*
- ❑ *combustion phenomenon in petrol and diesel engines, pre-ignition, detonation and diesel knock*
- ❑ *engine systems; fuel supply, engine cooling and lubrication*
- ❑ *performance parameters and heat balance calculations*

22.1. HEAT ENGINES

A machine or device which derives heat from the combustion of fuel and converts part of this energy into mechanical work is called a heat engine. Heat engines are broadly classified into internal combustion engines and external combustion engines.

An internal combustion engine is a reciprocating heat engine in which fuel mixed with correct amount of air is burnt inside a cylinder. The gaseous products of combustion form the working substance which make the piston move and produce mechanical work at the engine crankshaft. In contrast, the combustion of fuel in steam engines is external. Combustion takes place on the fire grate of the boiler, and the heat energy of fuel thus released is used to convert water into steam. The steam is then led to the steam engine or steam turbine where work is produced. Obviously, the working substance (steam) is generated in a boiler which is outside the power producing device.

Compared to steam engines, the IC engines are noted for :

– high overall efficiency. The efficiency of IC engines ranges from 30 to 35% whereas efficiency of steam engines lies between 15–20%

– compact and small size

– low weight to power ratio

– easy and quick starting. In steam engines, firing of the boiler and generation of steam takes sufficient time

– less maintenance and operating cost.

The important applications of IC engines are:

• Road vehicles, locomotives, ships and aircraft. As such IC engines enable passengers and cargos to cross lands, oceans and skies.

• Portable stand by units for power generation in case of scarcity of electric power.

• Extensively used in farm tractors, lawn movers, concrete mixing devices and motor boats.

22.2. CLASSIFICATION OF IC ENGINES

The IC engines are classified on the basis of following systems and their variations :

• *Number of strokes required for the completion of one cycle*

(*i*) Two stroke engines in which the engine cycle is completed in two stokes of the piston, *i.e.*, in one revolution of crank shaft.

(*ii*) Four stroke engines in which the engine cycle is completed in four strokes of the piston, *i.e.*, in two revolutions of crank shaft.

• *Thermodynamic cycle* : The thermodynamic cycles commonly used are :

(*i*) Constant volume combustion (Otto) cycle : Most of the petrol and gas engines work on this cycle.

(*ii*) Constant pressure combustion (Diesel) cycle : Low speed diesel engines work on this cycle.

(*iii*) Mixed or limited pressure (Dual) cycle: The high speed diesel engines work on this cycle.

• *Ignition system* : The following two methods are used for the ignition of fuel.

(*i*) Spark ignition : Petrol engines use a spark for the ignition of compressed charge (mixture of air and petrol) and the spark may be produced by magneto or battery.

(*ii*) Compression ignition : Diesel engines have a high compression ratio. The resulting high temperature is utilized to burn the fuel.

• *Kind of fuel used* :

(*i*) Light oil engines using kerosene or petrol. Petrol engines fall under this category.

(*ii*) Heavy oil or diesel oil engines: The oil used may be crude oil or mineral oil.

(*iii*) Gas engines : The gas used may be coal gas, producer gas, blast furnace gas or coke oven gas.

(*iv*) By-fuel engines : The gas is used as the main fuel and liquid fuel is used for starting purposes.

• *Number and arrangement of cylinders*

(*i*) In-line engines : all the cylinders are arranged with their axes parallel and they transmit power to a single crank shaft.

(*ii*) V-engines : the engines contain two banks of cylinders connected to the same crank and crank shaft. The crank shaft length for V-type engines is half of that for in-line engines.

(*iii*) Radial engines : the cylinders are arranged radially and are connected to a single crank shaft. Radial engines occupy less floor area and have little balancing problem.

• *Fuel supply system*

(*i*) Carburettor engines : mixture of petrol and air is prepared in the carburettor and is supplied to the engine during suction stroke.

(*ii*) Solid injection or airless injection : a fuel pump is used to inject the fuel in diesel engines.

(*iii*) Air injection : fuel is supplied under pressure to the cylinder of diesel engines by using compressed air.

• *Cooling system*

(*i*) Water cooled engines in which the heat from the cylinder walls is transferred to cooling water which is kept circulating in the water jackets provided in the cylinder block. The water picks up heat and is taken to the radiator where the heat is transferred to the surrounding air. The water is repeatedly returned to the engine after being cooled in the radiator. Medium and large sized engines and the automobile engines use the water cooling system.

(*ii*) Air cooling in which the heat from the cylinder walls is directly transferred to surrounding air. Air cooling is generally employed for small capacity engines like scooter and motor cycle engines.

• *Lubrication system* : Lubrication system refers to the act of reducing friction by introducing a substance (called lubricant) between the mating parts of the engine.

(*i*) Splash lubrication system suitable for small capacity engines with moderate speed and bearing loads.

(*ii*) Pressure lubrication system used for heavy duty engines.

• *Governing system* (speed control under variable load)

(*i*) Quality control engines in which composition of mixture (air-fuel ratio) is changed by admitting more or less fuel in accordance with variation in load on the engine. This method is used in diesel engines.

(*ii*) Quantity control engines in which the air-fuel mixture has a constant composition. However, the quantity of the mixture supplied is changed in accordance with load on the engine. This is used in petrol and gas engines.

• *Valve location :*

(*i*) overhead valve engine

(*ii*) side valve engine

• *Speed* : Engines having speeds above 900 rpm are called high speed engines, and less than 400 rpm are called slow speed engines.

• *Field of application :*

(*i*) Stationary engines used for small and medium capacity electric power plants, concrete mixers and pumping units

(*ii*) Mobile engines installed in motor vehicles, air planes and ships.

22.3. ENGINE PARTS AND THEIR FUNCTIONS

• ***Cylinder and cylinder head :*** The cylinder is the main body of the engine wherein direct combustion of fuel takes place. The cylinder is a stationary component and the piston reciprocates inside it. The cylinder head closes one end of the cylinder, and it is usually casted as one piece and is bolted to the top of the cylinder. It contains the valve seats and ports, and supports the valves and valve - actuated mechanism. Cylinders are usually made of ordinary cast iron. However, for heavy duty engines, alloy steels are used. In the interest of weight saving, particularly in aeroplanes, use is made of aluminium and magnesium alloys.

• ***Piston and piston rings :*** A piston is a metal cup with its crown facing the combustion space. The function of the piston, together with the rings, is to confine the gases in the combustion space and thus transmit the full force of expansion to the connecting rod and crank shaft. The piston also acts as a bearing for the small end of connecting rod. Pistons are usually made of grey iron or of aluminium alloys for high speed engines. Aluminium has the advantage of low density (the density of aluminium is about two-fifths that of cast iron).

The leakage of gases between the walls of piston and cylinder is prevented by means of three to six cast iron rings which may be square or rectangular in cross-section. These rings are inserted into the grooves provided on the piston. There are usually two sets of piston rings:

(*i*) upper piston rings (called the compression rings) provide gas tight seal and prevent the leakage of high pressure gas.

(*ii*) lower piston rings (called oil rings) provide effective seal and prevent the leakage of oil into the cylinder head.

• ***Connecting rod :*** The connecting rod transmits the force given by the piston to the crank, causing it to turn and thus convert the reciprocating motion of the piston into rotary motion of the crankshaft. The rotary motion is required to make wheels turn, a cutting blade spin or a pulley rotate.

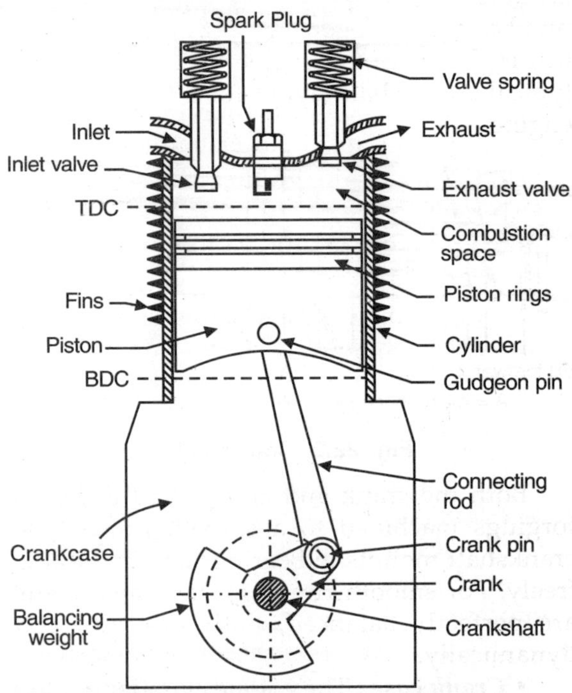

Fig. 22.1. Nomenclature of engine components

The connecting rod connects the piston at one end and the crank at the other end. The piston end is called the small end and the crank end is called the big end. The connection at the small end is made by a pin called the gudgeon pin, wrist pin or piston pin. At the big end, the connecting rod embraces the crank arm by a pin named crank pin.

The connecting rod is usually a small forging of I-section which provides the maximum stiffness with minimum weight. It is normally tapered along its length so as to provide smaller cross-sectional area towards the small end. It has also a passage for the transfer of lubricating oil from the big end bearing to the gudgeon pin.

• ***Crank and crankshaft :*** The reciprocating motion of the piston is converted into rotary motion by the connecting rod and crank mechanism. All the auxiliary mechanism of the engine having mechanical transmission are geared in one way or the other to the crank shaft and obtain their motive power from it. The shape of the crank shaft, *i.e.*, the mutual arrangement of the cranks depends on the number and arrangement of cylinders and the turning order of the engine. Fig. 22.2 shows a typical crank shaft layout for a four-cylinder engine.

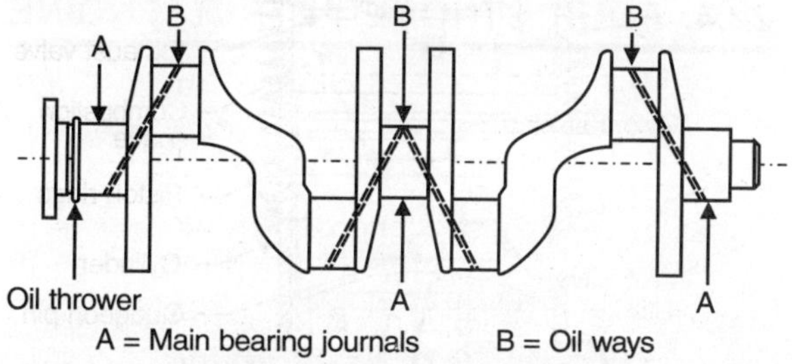

Fig. 22.2. Crankshaft

Both the crank and crank shaft are steel forgings machined to a smooth finish. The crankshaft mounts in bearings and can rotate freely. For smooth running, the crank shafts are perfectly balanced, both statically and dynamically.

• ***Crank case :*** The engine cylinder, piston and crankshaft are housed in the crank case which also serves as an oil sump for the storage of the lubricating oil. The oil level is checked with the help of an oil stick or dip stick.

The crank case is generally made of cast iron.

• ***Camshaft and valve mechanism :*** The camshaft operates the intake and exhaust valves through the cams, cam followers, push rods and rocker arms. On a four stroke engine, the inlet and exhaust valves operate once per cycle, *i.e.*, in two revolutions of the crankshaft. Consequently, the cam shaft is driven by the crankshaft at exactly half its rotational speed. The crankshaft material is commonly a steel forging with journals and cam faces case hardened.

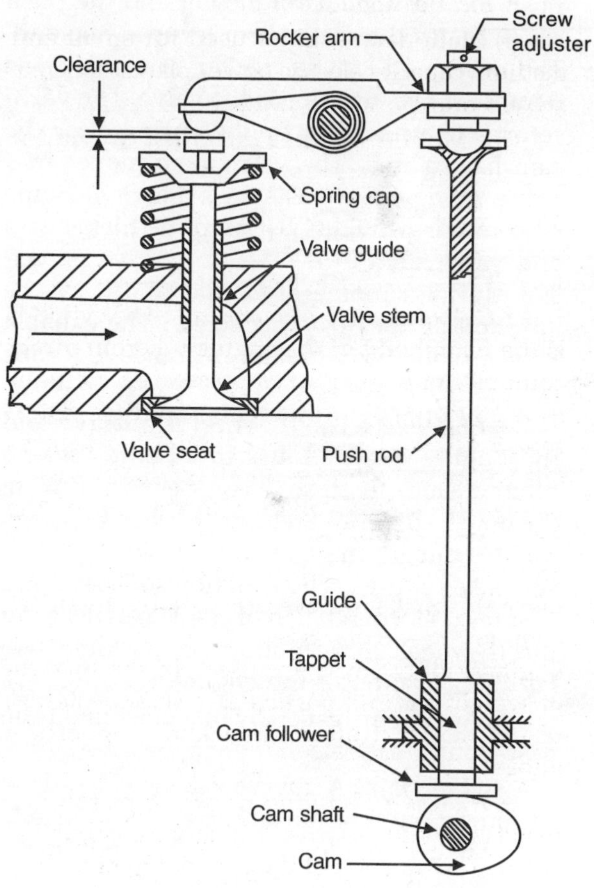

Fig. 22.3. Overhead valve mechanism

The valves are usually mushroom shaped (known as poppet valves) with conical seating surfaces. The poppet valve is practically universal on all modern car and commercial vehicle engines. The face of the valve and its seat on the cylinder are very accurately ground at an angle of 30 degree or 45 degree. Steel containing a small percentage of nickel and chromium is the usual valve material. Such an alloy has good heat-resisting qualities, and is considered good enough to withstand high temperatures, mechanical forces, corrosive and erosive effects of the high velocity cylinder gases.

The valves may be provided at the top or on the side of the engine cylinder. Fig. 13.3 shows a typical overhead-valve assembly with the principle parts named. The cam lifts the push rod through cam follower and the push rod actuates the rocker arm lever at one end. The other end of the rocker arm then gets depressed and that opens the valve. The valve returns to its seating by the spring after the cam has rotated. The valve stem moves in a valve guide which acts as a bearing. Some clearance is provided between the rocker arm and valve stem to take care of valve expansion during running of engine, and it can be adjusted by the abjusting screw. The exhaust valve usually has a greater clearance as it runs hotter.

• *Flywheel :* The flywheel is a heavy and thoroughly balanced disk fitted onto the end of crankshaft. It stores excess energy during the power stroke and returns this stored energy for use during the auxiliary strokes. Thus it serves to reduce cyclic variations of speed and ensures uniform rotation of crankshaft. Heavier the flywheel, more uniform and stable will be the operation of the engine.

Flywheels are generally manufactured from cast iron or cast steel.

• *Governor :* A governor is used to adjust the power output from an engine in conformity with the external load and accordingly make the engine operate at constant speed. The task is accomplished by regulating the quantity of charge in petrol engines, and the amount of fuel in diesel engines.

All types of engines generally have the components described above. Given below are the components used either only for diesel engines or for petrol engines.

Fuel pump and injector : In diesel engines, fuel pumps are used to deliver the correct quantity of fuel at the precise instant required for a wide range of loads and speeds. The nozzle atomises the fuel and distributes it into the combustion chamber of the engine. Atomisation of fuel means breaking of the fuel stream into mist like spray. Atomisation ensures that each particle of fuel is surrounded by air needed for combustion and that assists in rapid and successful burning of fuel. Atomisation is made possible by the high velocity of fuel through the nozzle which is due to high pressure created by the pump.

Carburettor and spark plug : The carburettor delivers to the gas and petrol engines a combustible mixture of air and fuel in a condition that can be easily and efficiently burnt in the engine cylinder. The process of mixture formation is called carburation.

Towards the end of compression stroke, the combustible mixture is ignited by a spark plug which has to spark several thousand times a minute under a wide range of temperatures. Each cylinder is provided with its own spark plug screwed into the lid.

22.4. FOUR STROKE PETROL ENGINE

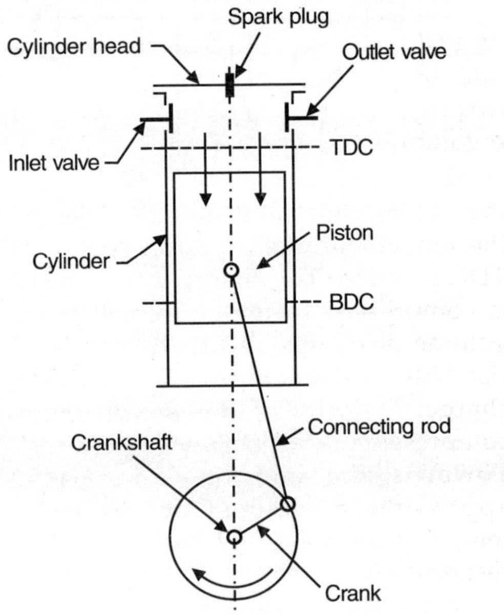

Fig. 22.4. Four stroke cycle system

A cycle is a sequence of operations constantly repeated, and *four-stroke* refers to the number of strokes of the piston required to complete one cycle. Refer Fig. 22.4 for the arrangement of different parts of a four stroke cycle system. The piston reciprocating inside a cylinder, is connected to the crankshaft through connecting rod and the crank. The inlet (suction) and outlet (exhaust) valves are housed in the cylinder head. The cylinder head is also provided with an electric spark plug.

All events of the cycle namely suction, compression, combustion and expansion, and exhaust are completed in two revolutions of the crankshaft.

The salient features of the four strokes in a petrol engine are as given below :

1. *Intake or suction stroke* : (Fig. 22.5a) Initially the piston is at top dead centre (TDC) position, the inlet valve is open and the outlet valve is closed. The piston moves downwards towards bottom dead centre (BDC) position and the pressure inside the cylinder is reduced to a value below the atmospheric pressure. The vacuum thus created causes the charge to rush in and fill the space vacated by the piston. The charge consists of a mixture of air and petrol prepared by the carburettor.

The suction continues till the piston reaches its BDC position. The piston has now made one stroke and the crankshaft has turned through 180°C, *i.e.*, has made half the revolution.

2. *Compression stroke* : (Fig. 22.5b) Both the valves (inlet and outlet) are closed and the movement of the piston is from BDC to TDC position. The charge inside the cylinder is compressed to the clearance volume; the volume decreases and there is a continuous rise both in temperature and pressure of the charge. Majority of the petrol engines use compression ratios between 5 to 1 and 8 to 1. Towards the end of compression, the approximate values of pressure and temperature are 6 – 12 bar and 250 – 300°C respectively.

3. *Working, expansion or power stroke* : (Fig. 22.5c) When the piston reaches TDC position, the charge is ignited by causing an electric spark between the electrodes of a spark plug which is located in the cylinder head. During combustion, the chemical energy of the fuel is released and there is rise both in pressure and temperature of the gases at almost constant volume. The temperature of the gases increases to about 1800 – 200°C and the pressure reaches 30 – 40 bar.

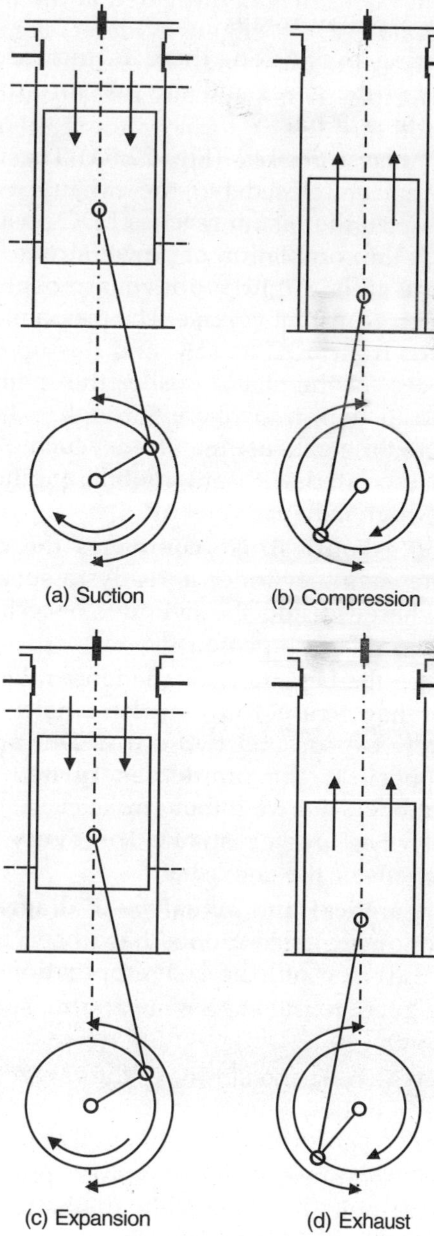

Fig. 22.5. Operation of a four stroke cycle petrol engine

With both valves closed, the gases at increased pressure and temperature expand, push the piston down the cylinder and work is done by the system. The reciprocating motion of the piston is subsequently converted into rotary motion of the crankshaft by connecting rod and crank. It is the rotary motion which is required to make wheels run, a cutting blade spin or a pulley rotate.

During expansion, there is increase in volume of the gases and the pressure drops to as low as 3 bar.

4. *Exhaust Stroke* : (Fig. 22.5d) The inlet valve remains closed but the exhaust valve opens when the piston reaches BDC position towards the completion of power stroke. The pressure falls slightly above atmospheric pressure at constant volume. The piston moves upwards from BDC to TDC and this upward movement of the piston pushes the spent up gases into the atmosphere through exhaust valve and the exhaust manifold. Much of the noise associated with automobile engines is due to high exhaust velocity.

The exhaust stroke completes the cycle and the engine cylinder is ready to suck the fresh charge inside the cylinder once again and the cycle is repeated.

Since the beginning of suction stroke, the piston has made four strokes inside the cylinder : two up and two down. During the same period, the crank has turned two revolutions. Thus for a four stroke cycle, there is only one power stroke for every two revolutions of the crankshaft.

Theoretical and actual $p-V$ diagrams : The following assumptions have been made while carrying out the above operations :

(*i*) Suction and exhaust is at atmospheric pressure.

(*ii*) Opening and closing of the valves (both inlet and outlet) is instantaneous and at dead centres.

(*iii*) Compression and expansion processes are isentropic, *i.e.*, reversible adiabatic.

(*iv*) The combustion of fuel takes place instantaneously at constant volume.

(*v*) There is sudden drop in gas pressure to the atmospheric pressure at the end of expansion stroke.

Such a theoretical operation of the cycle can be represented by the $p-V$ plot as shown in Fig. 22.6a.

However, an actual cycle deviates considerably form the hypothetical one because of the following reasons :

(*i*) For efficient suction, the pressure inside the cylinder at suction is less than the atmospheric pressure. This pressure difference, called intake depression, is needed to overcome the resistance to the flow of charge through the restricted area of the inlet passages.

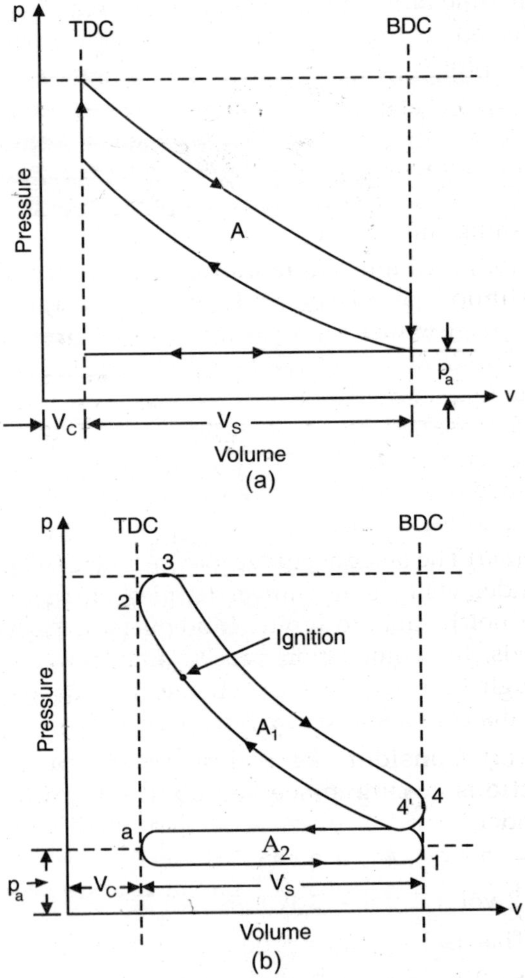

Fig. 22.6. Theoretical and actual p-V diagram for four stroke petrol engine

(*ii*) For proper exhaust of burnt gases, there has to be a net positive pressure from inside of cylinder to outside. Accordingly the exhaust line does not coincide with atmospheric pressure but is slightly above it.

(*iii*) Compression and expansion are not isentropic. Friction is always present, and there is considerable interchange of heat between the charge and the cylinder walls.

(*iv*) There is always a time lag between the ignition of charge and its actual combustion. Consequently, the combustion does not take place at constant volume and the pressure rise is not along a straight line.

(*v*) Opening and closing of the valves takes sometime and is never instantaneous. This is reflected in rounding off the corners of the $p-V$ plot.

(*vi*) Ignition of charge, and opening of the valves is never at dead centres. These events occur at some degree on either side of dead centres to get better charging and scavenging (pushing out of burnt gases) performance.

(*vii*) At high temperature, there is every likelihood of dissociation of products of combustion such as CO_2 and H_2O (steam). This splitting is an endothermic process and requires heat. Further, the value of exponent γ decreases with rise in temperature. Due to these aspects, the pressure and temperature attained during the actual cycle are lower than the theoretical values.

(*viii*) The layer of charge just in contact will cylinder walls is at a lower temperature and may not be able to ignite. Moreover at high speeds, the combustion process may not get enough time required for chemical reactions and the combustion may remain incomplete.

(*ix*) Consider the following chemical reactions taking place inside the engine cylinder,

- $2\,CO + O_2 = 2\,CO_2$

2 vol + 1 vol = 2 vol

That is 3 volumes produce 2 volumes

- $2\,H_2 + O_2 = 2\,H_2O$

2 vol + 1 vol = 2 vol

Again 3 volumes produce 2 volumes.

Thus we see that due to chemical reaction, the volume decreases. THis apparant contraction in volume causes a reduction in pressure.

(*x*) Thermodynamic properties are different for fresh charge and the products of combustion. The specific heats increase with temperature. For the same difference between c_p and c_v, the value of adiabatic exponent γ decreases with rise in temperature.

Because of dissociation and lower value of γ at high temperatures, the pressure and temperature attained after compression are lower than the theoretical values.

When these modifications are taken into account, the $p-V$ plot takes the form as shown in Fig. 22.6(*b*). The suction line $a-1$ lies below the atmospheric pressure line, and the exhaust $1-a$ lies slightly above the atmospheric line. The area enclosed by the exhaust and suction lines is called negative loop or pumping loop.

This represents the work required for the admission of fresh charge and for the removal of burnt gases. Net work is obtained by subtracting the pump loss from the gross output

Network per cycle = $(A_1 - A_2)$

The area $(A_1 - A_2)$ is always less than the area A of theoretical $p-V$ diagram.

22.5. FOUR STROKE DIESEL ENGINE

The engine completes its working cycle in four strokes of the piston, uses diesel oil as fuel and, therefore, is known as four-stroke diesel engine. Fig. 22.7 shows the different parts of the engine and illustrates its principle of operation. The piston reciprocates inside a cylinder, is connected to the crankshaft through connecting rod and the crank. The inlet (suction) and outlet (exhaust) valves are housed in the cylinder head. The cylinder head is also provided with a nozzle for injecting the fuel.

The sequence of individual strokes is as follows :

1. *Intake or suction stroke* : Initially the piston is at top dead centre (TDC) position, the inlet valve is open and the outlet valve is closed. The piston moves downwards towards bottom dead centre position (BDC) and the pressure

inside the cylinder is reduced to a value below the atmospheric pressure. The vacuum thus created causes the air from atmosphere to rush in and fill the space vacated by the piston. The suction stroke is completed when the piston reaches the BDC position. The piston has then made one stroke and the crankshaft has turned through 180°, *i.e.*, has made half the revolution.

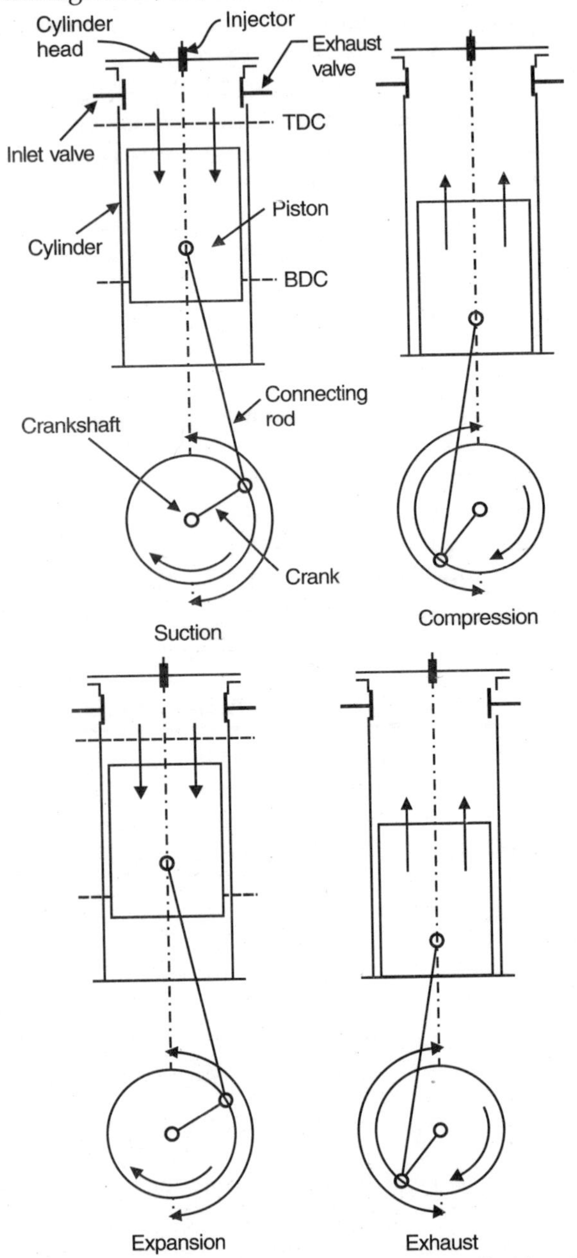

Fig. 22.7. Operation of a four stroke cycle diesel engine

2. *Compression stroke* : Both the inlet and outlet valves are closed and the movement of the piston is from BDC to TDC position (Fig. 22.7 b). The air inside the cylinder is compressed to clearance volume; the volume decreases and there is a continuous rise both in temperature and pressure of the air. Majority of the diesel engines use compression ratios between 15 : 1 and 20 : 1. Towards the end of compression, the approximate values of pressure and temperature are 60 bar and 600°C respectively.

3. *Working, expansion* or *power stroke* : When the piston reaches TDC position, a fine spray of diesel is injected into the combustion space containing the high temperature compressed air. The fuel vapours are raised to self ignition temperature and combustion occurs at approximately constant pressure.

The atomisation of fuel and its supply to the combustion space can also be accomplished by compressed air supplied from compressed air bottles. The air entering the combustion space is so regulated that the pressure theoretically remains constant during the burning period.

With both valves closed, the combustion products at increased pressure and temperature push the piston down the cylinder with a large force. Expansion of the gases takes place and work is done by the system. The reciprocating motion of the piston is subsequently converted into rotary motion of the crankshaft by connecting rod and crank.

The expansion stroke gets completed as the piston reaches its BDC position. During expansion, there is increase in volume of the gases and the pressure drops.

4. *Exhaust stroke* : The inlet valve remains closed while the exhaust valve opens. The piston moves upwards from BDC to TDC position and this upward movement of the piston pushes the spent up gases into the atmosphere through exhaust valve and the exhaust manifold.

The exhaust stroke completes the cycle and the engine cylinder is ready to suck fresh air inside the cylinder once again and the cycle is repeated. Much of the noise associated with automobile engines is due to high velocity of exhaust gases.

Since the beginning of suction stroke, the piston has made four strokes inside the cylinder : two up and two down. During the same period, the crank has turned two revolutions. Thus for a four stroke cycle, there is only one power stroke for every revolutions of the crankshaft.

The following assumptions have been made while carrying out the above operations.

(*i*) Suction and exhaust is at atmospheric pressure.

(*ii*) Opening and closing of the valves (both inlet and outlet) is instantaneous and at dead centres.

(*iii*) Compression and expansion processes are isentropic, *i.e.*, reversible adiabatic.

(*iv*) The combustion of fuel takes place at constant pressure during a small part of expansion stroke.

(*v*) There is a sudden drop in gas pressure to atmospheric pressure at the end of expansion stroke.

Such a theoretical operation of the cycle can be represented by the $p - V$ plot as shown in Fig. 22.8(*a*).

However, the actual cycle differs considerably from the hypothetical one due to the reasons mentioned in section 22.4 above. When those modifications are taken into account, then the plot takes the form as shown in Fig. 22.8 (b). The actual pressure inside the cylinder is slightly less than the atmospheric pressure during suction stroke, and slightly higher than the atmospheric pressure during exhaust stroke. Also the corners are rounded off in the actual $p - V$ plot. The area exclosed by the exhaust and suction lines is called negative loop or pumping loop and it represents the work required for the admission of fresh charge and for the removal of burnt gases. Net work is obtained by subtracting the pump loss from the gross output

Net work per cycle = $(A_1 - A_2)$

The area $(A_1 - A_2)$ is always less than the area A of the theoretical p–V diagram.

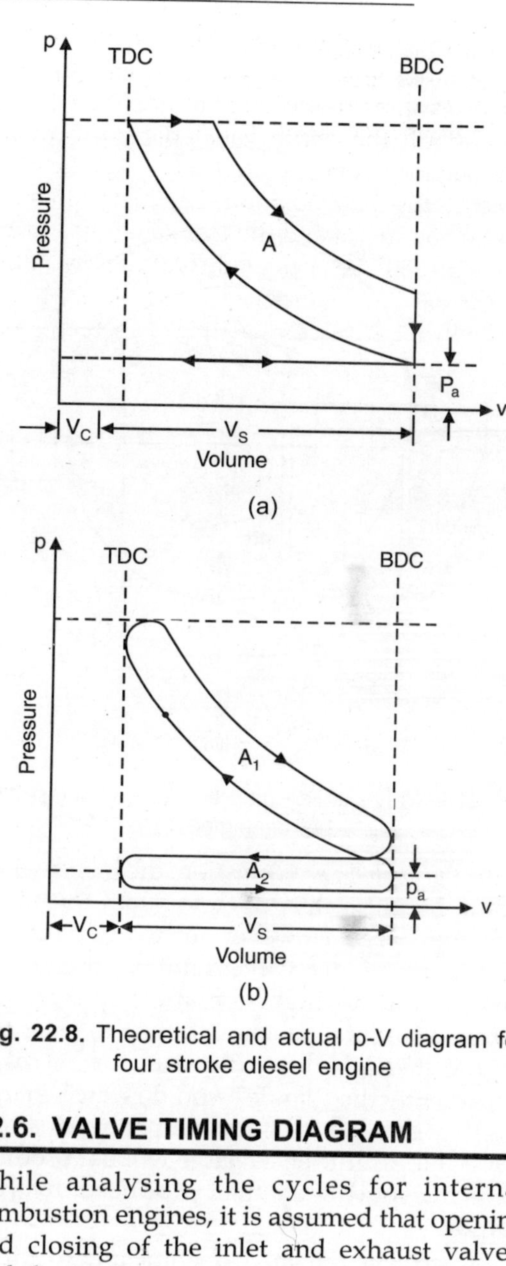

Fig. 22.8. Theoretical and actual p-V diagram for four stroke diesel engine

22.6. VALVE TIMING DIAGRAM

While analysing the cycles for internal combustion engines, it is assumed that opening and closing of the inlet and exhaust valves, and the ignition of charge (or injection of fuel occurs at the engine dead centres). However, in actual operation, these events occur at some degrees on either side of dead centres to get better charging and scavenging performance.

Valve-timing diagram shows the correct positions pertaining to opening and closing of the valves. Refer Fig. 22.9. which shows the typical valve timing diagrant of vertical 4-stroke petrol engine :

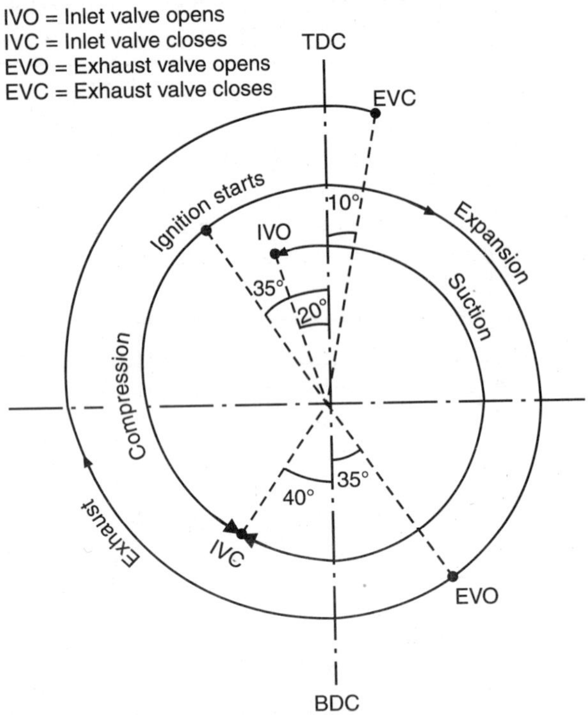

Fig. 22.9. Valve timing diagram four a four stroke petrol engine

1. *Inlet valve* : At admission, the intake valve should be sufficiently open to avoid throttling of intake charge. However due to inertia effect, the opening of inlet valve is not instantaneous. Further, due to high operating speeds, the charge would not get enough time to enter the engine cylinder during suction stroke. Apparaently the cylinder would receive charge less than its capacity if the inlet valve closes at BDC. This insufficienct charge will not produce the same power as that produced by full charge.

To achieve full valve opening and to admit full amount of charge,

(*i*) a lead of 20 – 30° to given to the intake valve, *i.e.*, intake valve opens 20 – 30° before TDC.

(*ii*) there is a delay of 30 – 50° to exhaust valve, *i.e.*, the exhaust valve closes 30 – 50° after BDC.

For the crank angle positions shown for the opening and closure of inlet valve, it remains open for (20 + 180 + 40) = 240°.

2. *Ignition* : The top dead centre would be the proper time to produce spark if the charge could burn instantaneously. However, there is always a time lag between the timing of spark and that of actual ignition. For best results with regard to power and economy, and to avoid explosion knock, the ignition of charge is timed to occur as early as engine permits. This shifting of ignition timing ahead of TDC is called ignition advance.

Depending upon the speed of the engine, the ignition of charge is times to occur 20 – 35° before TDC.

3. *Exhaust valve* : Exhaust valve opens before BDC and closes after TDC. This is to ensure that full amount of burnt gases is expelled out from the engine cylinder.

(*i*) Earlier opening helps in driving out the exhaust gases due to their pressure being higher than the atmosphere.

(*ii*) During late closure, the kinetic energy of fresh charge is utilized to assist in the maximum exhausting of the cylinder.

It needs to be pointed out that early opening of exhaust valve entails loss of power due to shorting of power stroke, but it is well compensated by better scavenging and less negative work that needs to be done by piston during return stroke.

Generally, the exhaust valve opens at about 35° before BDC and closes at 10° after TDC. Thus the exhaust valve remains open for (35 + 180 + 10) = 225°.

The noteable aspects of the valve timing diagram (Fig. 22.10) for a four stroke diesel engine are :

(*i*) The inlet valve opens 10 – 25° in advance of top dead cenbtre position, and closes 25 – 40° after the bottom dead centre position.

(*ii*) Exhaust valve opens 30 – 50° ahead of bottom dead centre positions and its closure occurs 10 – 15° after top dead centre position.

(*iii*) There is a brief time interval for the fuel oil to mix with the hot compressed air in the cylinder and ignite. The injection of fuel is timed to occur 10 – 15° before TDC so that by the time piston reaches TDC, the actual combustion of fuel starts. The fuel valve remains open for some time as per design.

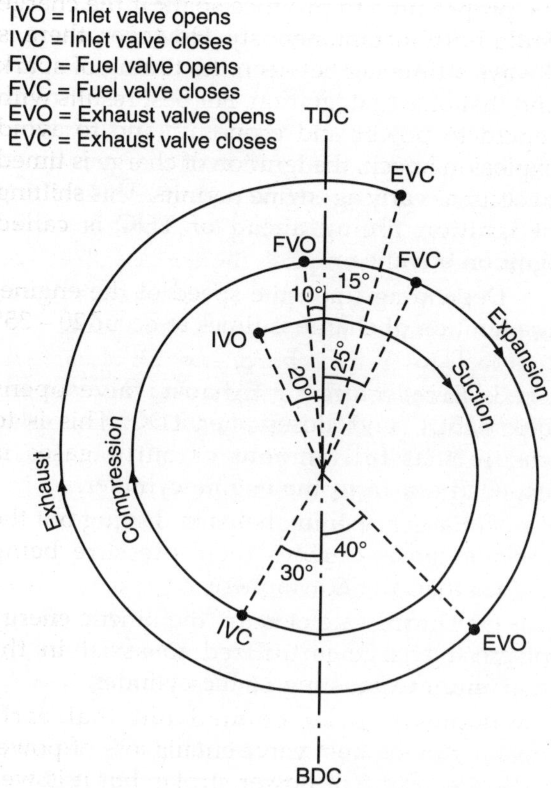

Fig. 22.10. Valve timing diagram for four stroke diesel engine

Note:

(*i*) The value of crank positions as mentioned above are only average and considerable difference exists with different engines.

(*ii*) The opening and closing timings for the valves are function of engine speed. For high speed engines, the inlet valve is made to close later, the exhaust valve opens earlier and the ignition (or injection) is timed to occur earlier.

(*iii*) A valve is said to lead when it opens before TDC, and lags if the closure is after BDC.

(*iv*) For some part of the cycle near TDC, both the inlet and exhaust valve remain open, and this period is called **overlap**. During overlap the incoming charge helps to drive out the traces of exhaust gases, and this phenomenon is called **scavenging**.

22.7. COMPARISON BETWEEN PETROL AND DIESEL ENGINES

Below are given the differences between the construction and operation of a diesel engine and a petrol engine of similar capacity and number of cylinders.

(*i*) *Basic cycles* : The petrol engine works on Otto cycle whereas a diesel engine works on Diesel cycle.

(*ii*) *Fuel used* : Petrol engine is a light oil engine and uses gasoline or petrol as fuel. Diesel engine is a heavy oil engine and uses diesel as fuel. The diesel has a high self ignition temperature compared to petrol.

(*iii*) *Induction of fuel* : Mixture of petrol and air in required strength is prepared in the carburettor and inducted into the engine cylinder during the suction stroke of a petrol engine.

During suction stroke of a diesel engine, only air from the atmosphere is sucked in the engine cylinder. A fuel pump is used to inject fuel directly into the combustion space where it meets the air which has been compressed.

(*iv*) *Compression ratio* : Majority of petrol engines use compression ratio between 5 to 1 and 8 to 1. The upper limit is fixed by anti-knock rating of fuels.

The compression ratio in diesel engines lies between 15 : 1 and 20 : 1 ; the upper limit is fixed because of an increase in the weight of engine with increase in compression ratio.

(*v*) *Thermal efficiency* : For the same compression ratio, the efficiency of a diesel engine is lower than that of a petrol engine. However, this aspect is of not much practical significance since the petrol engines work with compression ratio not exceeding 8 whereas diesel engines can safely have compression ratio as high as 20. High compression ratio for a diesel engine is a must not only for high efficiency but also to prevent diesel knock – phenomenon of uncontrolled and rapid combustion.

(*vi*) *Ignition* : Petrol engines use a spark plug to ignite the charge (mixture of air and petrol) after it has been compressed. The combustion of fuel in diesel engines is due to high temperature of compressed air.

(*vii*) *Weight* : The cylinder walls of diesel engines have to be made thicker to sustain the high pressures attained due to higher compression ratios. The weight of a diesel engine amounts to between 30 to 50 % more than that of a petrol engine giving the same power output.

(*viii*) *Speed* : The petrol engines are high speed engines due to light weight, and the diesel engines run at comparatively low speeds due to heavy weight.

(*ix*) *Vibration and noise* : The vibration and noise level is higher with a diesel engine because of higher maximum pressure.

(*x*) *Load control* : The petrol engines are quantity control engines. The air-fuel mixture prepared by the mixture being inducted is controlled by the throttle valve in accordance with load on the engine.

The diesel engines are quality control engines. The composition of the mixture (air-fuel ratio) is changed by admitting more or less fuel with variation in engine load. The fuel pump regulates the supply of fuel injected into the combustion space.

(*xi*) *Power* : A petrol engine can be designated to give a better mean effective pressure mainly due to its improved combustion efficiency.

(*xii*) *Cost* : The initial cost of a diesel engine is higher, mainly due to the cost of the fuel pump. Since diesel is cheaper than petrol, the running cost of a diesel engine works out to be low. Further, the experience of most operators is that a diesel engine requires less maintenance than a petrol engine. However, a larger battery is necessary for a diesel engine.

(*xiii*) *Fire risk* : This is minimised in diesel engines owing to the higher ignition point of the fuel used.

(*xiv*) *Applications* : Petrol engines are used in cars, scooters and motor cycles.

Diesel engines are used in heavy duty vehicles like trucks, buses and locomotive engines.

22.8. TWO STROKE SYSTEM

The working cycle is completed in two strokes of the piston or in one revolution of the crankshaft as against two crankshaft revolutions in a four stroke cycle engine. The preparatory strokes (suction and exhaust) are combined with the working strokes (compression and expansion). The following two methods have been used to accomplish the desired objective.

(*i*) Providing a separate pump outside the engine cylinder to compress the charge (air-fuel mixture from carburettor or air alone from atmosphere) before forcing it into the cylinder. The pump is an integral part of the engine and gets its motive power from the engine itself. The arrangement is referred to as *two channel system* and is used for large capacity multi-cylinder engines.

(*ii*) Crank-case compression system where the crankcase works as an air pump as the piston moves up and down. The charge (air-fuel mixture or air alone) is compressed by the pumping action of the underside of the piston before being supplied to the engine cylinder. The arrangement is referred to as *three-channel system* and is commonly used for single cylinder small power engines such as scooter and motor cycle engines.

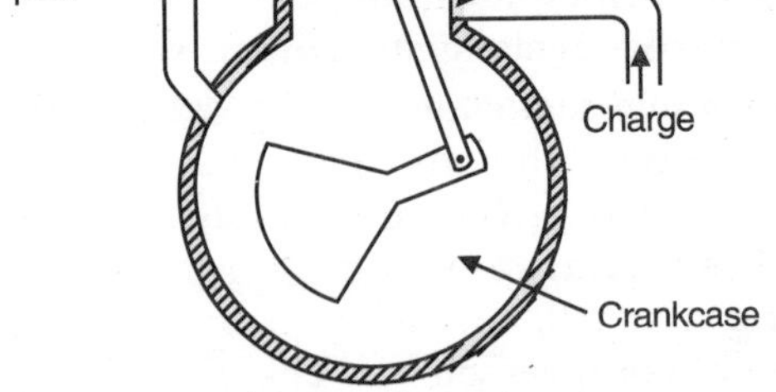

Fig. 22.11. Two-stroke cycle system

Construction : Fig. 22.11 shows the arrangement of a typical three-port engine employing crank case compression. The piston which is closely fitted in the cylinder is connected to the crankshaft through connecting rod and crank. The top of the piston is usually crown-shaped and that assists in sweeping the spent-up gases towards the exhaust port with the help of fresh charge. The engine employs ports as against valves as provided in a four-stroke system. These ports are cut in the cylinder walls and are three in number : the transfer port, inlet or induction port and the exhaust port. The inlet and exhaust ports are located on one side, and the transfer port is provided on the other side. The cylinder top is provided with an electric spark plug in a petrol engine, or a nozzle for injecting the fuel in a diesel engine.

Working : The charge is led to the crankcase through the inlet port. The charge consists of a mixture of air and petrol prepared by the carburettor in case of petrol engine. The diesel engine admits only fresh air through the atmosphere. The transfer port takes the compressed charge from the crankcase to the engine cylinder. The spent up gases are discharged to the atmosphere through the exhaust port. The closing and opening of the ports is controlled by up and down motion of the piston inside the cylinder. The piston crown helps to prevent the loss of incoming fresh charge (charge being carried with the spent up gases) and uses its momentum for exhausting only the burnt gases. *Scavenging* is the term applied to the process of forcing the burnt gases through exhaust port by deflecting fresh charge across the cylinder.

Sequence of events : Fig. 22.12 explains the working principle and sequence of events in a two-stroke cycle system.

Ignition and induction : In Fig. 22.12*a*, the piston occupies the almost TDC position towards the end of compression stroke. The compressed charge is being ignited by providing a spark, or fuel is being injected into the hot compressed air. The combustion of fuel occurs and thermal energy is released. There occurs a rise both in the pressure and temperature of combustion products.

At the same time, a partial vacuum (pressure lower than atmosphere) exists in the crank case and fresh charge is being inducted into the crank case through the inlet port which is uncovered by the piston.

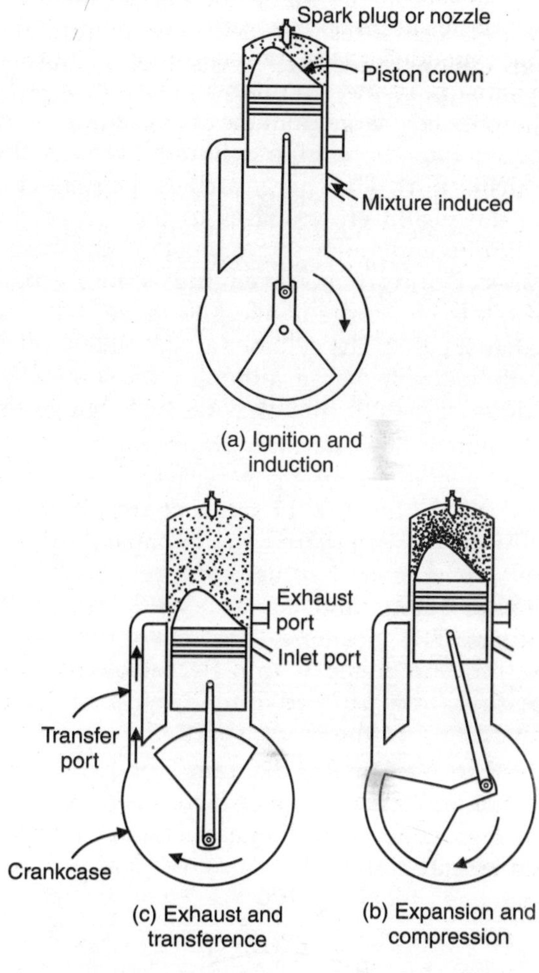

Fig. 22.12. Working of two-stroke cycle engine

Expansion and compression : (Fig. 22.12*b*). The high pressure gases push the piston down, expansion takes place and power is developed. With downward movement of the piston, the charge in the crank case gets compressed by the underside of the piston to a pressure of about 1.4 bar absolute.

After completion of about 80% of expansion stroke, the piston uncovers the exhaust port. Some of the combustion products which are still above atmospheric pressure escape to the atmosphere. On its further downward motion, the piston uncovers the transfer port and allows the slightly compressed charge from the crank case to be admitted into the cylinder via the transfer port.

Exhaust and transference : (Fig. 22.12*c*). The piston lies at its bottom dead centre position. The expanded gases are escaping through the exhaust port and simultaneously the slightly compressed charge from the crank case is being forced into the engine cylinder through the transfer port. The charge strikes the deflector on the piston crown, rises to the top of the cylinder and pushes out most of the burnt gases. During this scavenging action, a part of the fresh charge is likely to leave with the exhaust gases. The cylinder is completely filled with the fresh charge, although it is somewhat diluted due to its mixing with the burnt gases.

When the piston moves upward from its BDC position, it first covers the transfer port and stops the flow of fresh charge into the cylinder. A little later, the exhaust port too gets covered and actual compression of the charge begins and continues till the piston reaches TDC position. The cycle of the engine is thus completed within two strokes of the piston (one up and one down) and one revolution of the crank shaft.

The port timing diagrams for two stroke petrol and diesel engines have been depicted in Fig. 22.13, and these diagrams are self explanatary.

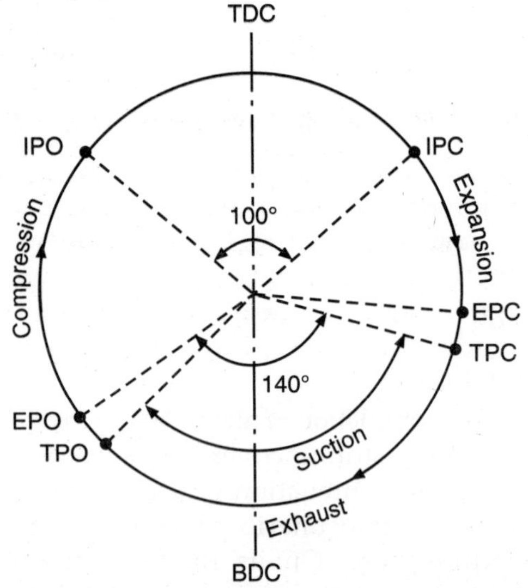

(a) Petrol engine

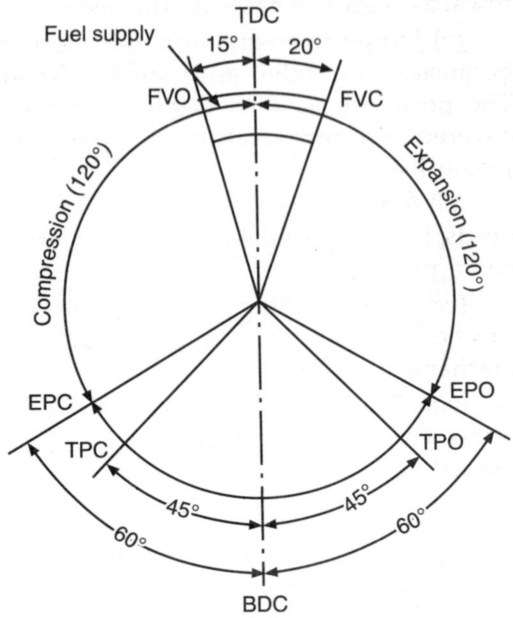

(b) Diesel engine

Fig. 22.13. Port timing diagrams for two stroke petrol and diesel engines

22.9. COMPARISON BETWEEN TWO-STROKE AND FOUR-STROKE ENGINES

Merits of two-stroke engines

(*i*) For the same power output, a two stroke engine is simple in design, easy to manufacture and operate.

(*ii*) A two stroke cycle engine gives one working stroke for each revolutions of the crankshaft. The four stroke cycle engine gives one working stroke for every two revolutions of the crankshaft. As such, a two stroke engine develops theoretically twice the power developed by four stroke engine for the same engine speed and cylinder volume.

(*iii*) The number of working strokes is twice than that in a four stroke engine. Consequently, the turning moment is uniform and hence the need only for a lighter flywheel.

(*vi*) Less friction loss due to the absence of suction and exhaust strokes; consequently high mechanical efficiency. Absence of cams, cam shaft and rockers etc. also contributes towards high mechanical efficiency.

(*v*) Simpler construction and mechanism because of no valve and valve mechanism. The ports are easy to design and they are covered and uncovered by the movement of piston itself.

(*vi*) A two stroke engine occupies less space, needs lighter foundations and requires few spare parts.

(*vii*) The reversing of a two stroke engine can be achieved by a simple reversing gear mechanism.

(*viii*) The initial cost of a two stroke engine is less due to light weight and absence of valve mechanism.

(*ix*) The two stroke engines are much easier to start.

Demerits of two-stroke engines :

(*i*) Scavenging (driving out of burnt gases) is not complete due to short time available for exhaust. This results in the dilution of fresh charge.

(*ii*) Exhaust and inlet parts are uncovered (open) simultaneously during a certain period. Some fresh charge is likely to escape without giving any work output.

(*iii*) Thermal efficiency of a two-stroke engine is likely to be lower due to some charge escaping without burning and poor scavenging. Consequently fresh charge is diluted which results not only decrease in performance but also slow running , low combustion pressure and poor efficiency.

(*iv*) For the same stroke and clearance volume, the effective compression ratio is lower in a two stroke engine. This too lowers the engine efficiency.

(*v*) More wear and tear of moving parts due to double the number of power strokes.

(*vi*) The piston gets overheated due to firing in each revolution of crankshaft. Higher temperatures make the cooling and lubrication requirements quite severe.

(*vii*) Greater consumption of lubricating oil due to high operating temperatures.

(*viii*) Noisy exhaust due to sudden release of burnt gases.

Applications of two-stroke engines

Two stroke engines are generally used where low cost, compactness and light weight are the major considerations such as in scooters, motor cycles, mopeds and other light vehicles. The two-stroke opposed piston diesel engines are quite suitable for marine installations (ship propulsion) where the engine room is small.

22.10. CARBURATION AND CARBURETTOR

Carburation refers to the entire process of vaporising and atomising the petrol fuel and mixing it with the air in varying proportions, to suit the changing operating conditions of the engine.

* *Atomisation* : breaking up the fuel into very small particles so that it gets properly mixed with air

* *Vaporisation* : change in state of the fuel from liquid to vapour

Carburettor is the device which supplies the metred spray of petrol fuel mixed with correct amount of air for efficient combustion in cylinder at all operating conditional namely :

- starting, idling and low load running
- acceleration and deceleration
- maximum power and maximum economy (minimum specific fuel consumption)

The spark ignition engines use octane ($C_8 H_{18}$) as fuel for which the chemically correct or stoichiometric mixture of air-fuel ratio is approximately 15 : 1 by mass. With this mixture, there is most rapid combustion of fuel, almost the greatest power and reasonable economy of fuel. Richer mixtures (11 to 15 : 1) given power and better fuel economy results when the mixture is weak (16 to 18 : 1).

A carburettor is designed to meet the following main requirements :

(*i*) maintain a constant level of petrol in the mixing chamber

(*ii*) prepare the homogenous air-fuel mixture by atomising and vaporising the fuel

(*iii*) meter the liquid fuel so as to produce the required air-fuel ratios at all operating conditions like idling, low load running, cruising and maximum power range.

(*iv*) provide rich mixtures for easy and reliable starting of engine from cold conditions.

(*v*) simplicity of maintence and repairs, durability and reliability of operation.

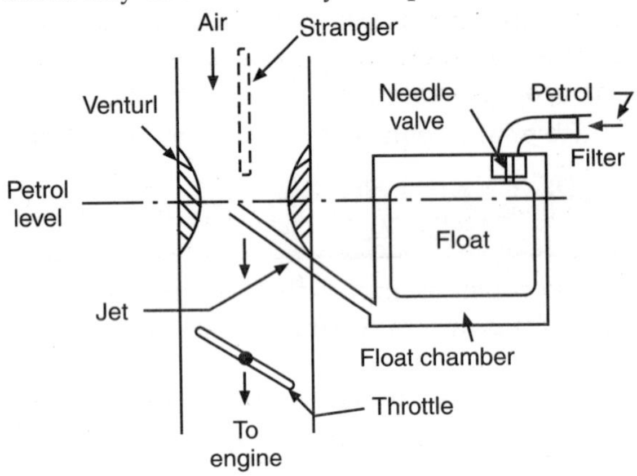

Fig. 22.14. Simple plain tube carburettor

Refer Fig. 22.14 which shows the schematics of a simple plane tube carburettor. The main components of the unit with their functions are :

(*i*) *Float and float chamber* : The purpose of the float chamber is to reserve/maintain a small volume of petrol at atmospheric pressure and at a constant level for supplying to various jets and passages. The float chamber has a float (a hollow brass pressing) which controls the position of a needle valve in the top of the chamber. When the petrol in the chamber is at the prescribed level, the needle valve blocks the inlet passage and that cuts off the petrol supply. During running of engine, the fuel is withdrawn from the chamber and there is fall in the petrol level. Consequently the float descends allowing the needle valve to open petrol passage and make the petrol to flow again to the chamber. The float then rises and keeps the petrol at constant level at all operating conditions of the engine.

The fuel in the float chamber is always in communication with atmospheric pressure through a small hole made in the cover. A filter is usually fitted at entrance to the float chamber and this serves to prevent the foreign matter from choking or clogging the needle valve or jets.

(*ii*) *Venturi* : The venture is restriction (a narrower passage) with smooth walls which allow a frictionless flow.

When the air passes through the venturi during suction stroke of the piston, its velocity increases at the throat (the section where cross-sectional area is minimum) with corresponding reduction in pressure. There is a fuel jet (nozzle) which is placed inside the venturi throat and is connected with the float chamber. Since pressure is atmospheric on the petrol in the float chamber, pressure would be lowest and vaccum or suction develops at the throat. This vaccum (called carburettor depression) is utilized in sucking the petrol from the float chamber through the fuel jet. The sucked in petrol vaporises, under low pressure conditions at the throat, and mixes with incoming air stream forming a homogenous air-fuel mixture. The vaporisation of fuel depends upon the nature of fuel, temperature and velocity of incoming air and the pressure at the throat.

The tip of the jet is kept slightly higher than the level of fuel in the float chamber. This is to prevent the fuel leaking into the carburettor when the engine is not running. The difference of level between the tip of the nozzle jet and the fuel level in the float chamber is called **nozzle lip**, and it is usually kept as 1.5 mm.

(*iii*) *Throttle* : The throttle valve is of butterfly type hinged and placed between the mixing chamber and the intake manifold of the engine. The purpose of the valve is to control the quantity of air-fuel mixture being supplied to the engine cylinder, and hence the power output from the engine speed.

The control of the throttle valve is normally with the accelerator pedal through certain linkages. When the throttle is fully open, the engine sucks the maximum amount of mixture and the engine speed goes up till equilibrium is reached between the engine power and the load on the engine. When the throttle is partially closed, the engine slows down until new equilibrium is reached. Apparently, the position of throttle helps to run the engine at widely changing speeds.

(*iv*) *Choke valve* : The following difficulties are encountered when a cold engine is being cranked for starting :

- only a small vaccum is built up at the venturi throat due to low air speed through the venturi tube. The small vaccum allows only a small quantity of fuel to be sucked into the fuel nozzle.

– the mixture may also become weak due to condensation of the petrol in the air-fuel mixture on its way to the induction pipe.

The engine starting becomes quite difficult due to th air-fuel mixture being weak and non-homogenous. For getting rich mixture needed for cold starting, a choke or strangler is fitted on the air intake side. When the choke is partially closed, the pressure near the throttle due to engine suction becomes very low. Since the pressure acting on the petrol in the float chamber is atmospheric, a greater pressure differential is obtained. Due to greater pressure depression, enough petrol is forced into the air passage and a rich mixture needed for cold starting is obtained. After the engine starts and gets warm, the choke is closed and normal air-fuel ratio is restored.

This simple carburettor entails the following limitations :

- Working is affacted by changes in environmental conditions of pressure and temperature.
- The unit gives proper mixture at only one engine speed and load, and as such is suitable only for engines operating under constant load and speed conditions.
- No arrangement exists for providing extra fuel during acceleration and overload.

For pracise control of the mixture composition at varying operating conditions, this simple plain tube carburettor has been modified by adding idling jets, compensating jets and the valves for increasing the richness of air-fuel mixture.

22.11. FUEL INJECTION IN COMPRESSION IGNITION ENGINES

In diesel engines, only air with a small amount of residual gases is compressed through a high compression ratio (16 : 1 to 20 : 1) and the diesel oil (chemical formula C_{12} H_{26} and specific gravity 0.82 – 0.88) is injected into the engine cylinder almost towards the end of compression stroke. The temperature of air attained after compression is high than the ignition temperature of fuel. The heterogenous fuel air mixture then ignites due to heat of compression; there is no need of any out side means for ignition.

The main requirements of fuel injection system are :

(*i*) correct metering of fuel to be injected per cycle as per load and speed of the engine.

(*ii*) proper timing of fuel injection so that maximum power is obtained at minimum fuel consumption.

(*iii*) suitable rate of fuel injection to have desired heat release during the combustion process.

(*iv*) proper atomisation of fuel so that each particle of fuel is surrounded by oxygen needed for combustion; this will assist in rapid and successful burning of fuel.

(*v*) uniform distribution of sprayed fuel into the cylinder so there is penetration of fuel in all parts of the combustion chamber.

(*vi*) instantaneous start and termination of fuel injection to avoid any dribling of fuel droplets into the cylinder.

The engine performance, *i.e.*, its power output, fuel economy and efficiency etc are greatly affected by the correct working of the fuel injection system. There are two distinct methods for fuel injection namely the air injection and the airless injection. In *air injection* system, air is compressed to about 60 – 80 bar and is stored in bottles. The kinetic energy of a blast from this air is utilized to feed fuel spray into the engine cylinder. The nozzle is fed with fuel from a pump, the air atomises the fuel by friction and creates turbulence in the combustion chamber for its thorough mixing with fuel. The system has the merit in :

– very good atomisation and distribution,

– requirement of no precision fuel pump,

– no choking of fuel valve, and

– slight effect of supercharging due to compressed air supply.

However, the system has become obsolete due to its cumbersome design which renders it unsuitable for portable engines, requirement of more space, poor combustion efficiency and considerable amount of power needed to run the compressor. Their use is restricted to high powered engines using heavy viscous fuels.

The *airless injection system* (also known by different names such as mechanical, solid and hydraulic injection) is commonly employed on diesel engines. The system supplies fuel at a very high pressure (15 bar), from the fuel pump to the fuel injector from where it is injected to the combustion chamber at appropriate time.

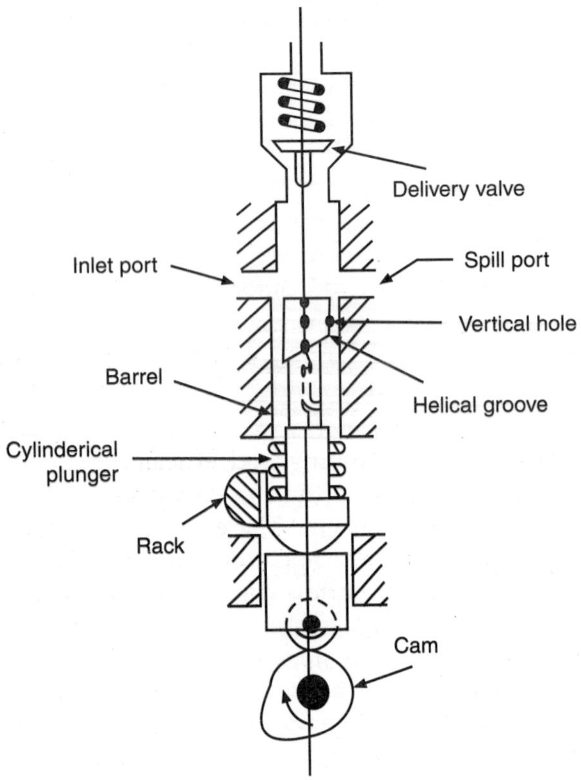

Fig. 22.15. Diesel fuel pump

The constructional and operational details of the diesel fuel pump are :

Construction

(*i*) The pump has a cylinderical plunger which operates inside a closely fitting barrel. The barrel is closed at the top by a spring loaded delivery valve from where a pipe leads the fuel to the engine cylinder via injector.

(*ii*) The plunger has a constant stroke and reciprocating motion to it is given by a cam on a camshaft. The plunger can also be rotated by a rack and pinion arrangement.

(*iii*) The plunger has a vertical slot and a helical groove which supplies varying quantity of fuel as per demand. Further, in the upper portion of the pump cylinder there are two ports. One port communicates with the fuel tank by an intake pipe through the filter, and the other is connected to spill port which spills to the fuel tank.

Operation

(*i*) At the starting position the plunger is at the bottom of its stroke and the two parts (inlet port and spill port) leading to the suction space are open and the cylinder is filled with fuel oil.

(*ii*) With starts of delivery stroke, the plunger moves up and displaces some fuel back from the port in the barrel. This act continues until the top edge of plunger covers the ports and the remainder of the fuel oil is forced out through the delivery pipe to the engine cylinder.

(*iii*) The supply of fuel to the injector continues until the helical edge uncovers the port and releases fuel back to suction lines. At this position of the plunger, the pump space communicates with the spill port through the vertical groove and is thus connected to the atmosphere. That releases the pressure on the spring, the delivery valve falls back on its seat and the supply of fuel to the injector is stopped. At this instant, the fuel in the barrel escapes through the by-pass, *i.e.*, spill port.

During downward stroke of the plunger, suction is created and fuel is drawn into the barrel through the supply port.

(*iv*) The quantity of fuel being supplied to the injector depends upon the position of the helical groove. When the plunger is turned to the right by means of tooth rack, the effective pump stroke is increased. This implies that the instant at which the space communicates with the spill port can be delayed and so more fuel can be supplied to the engine at higher loads. The rack is operated by a governor or by accelerator pedal.

The function of an injector is to :

(*i*) split up the fuel into a spray, and

(*ii*) deliver the atomised fuel directly into the combustion chamber where its proper penetration and distribution is achieved.

Refer Fig. 22.16 which shows the outlines of a spring load injector which is used in conjunction with the fuel pump.

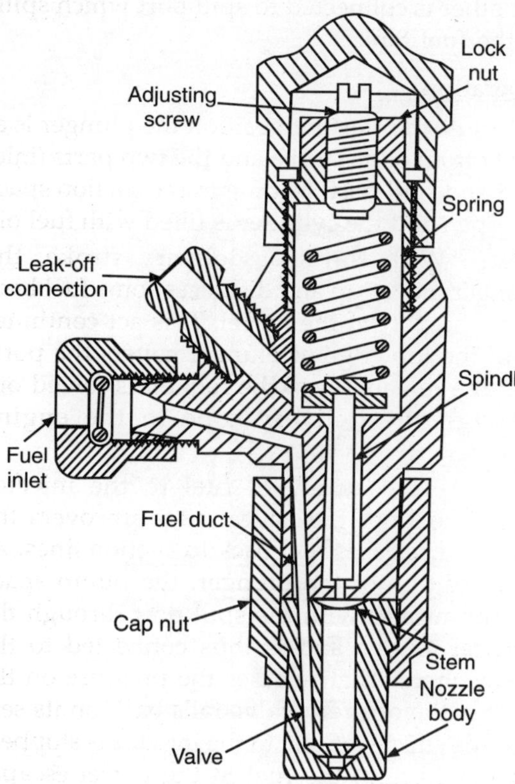

Fig. 22.16. Spring loaded fuel injector

The unit consists of :

(*i*) A housing and a detachable nozzle (atomiser) body which are connected by a cap or union nut.

(*ii*) A spray nozzle which is held on its seat by means of a spindle and helical spring. The compression of the spring can be adjusted by means of an adjusting nut. This permits to adjust the pressure at which the nozzle valve is to lift against the spring force.

(*iii*) The nozzle is provided with small sized holes 0.15 mm to 0.4 mm in diameter. The number of holes depends upon the type of combustion chamber and varies from 1 to 8 in a multi-hole nozzle.

High pressure oil from the pump is led to the injector through the fuel duct and ultimately reaches the valve seat. The valve gets lifted off its seat against the spring force due to fuel pressure, and the fuel is injected into the combustion chamber. A small quantity of oil which leaks into the body of the injector provides the necessary lubrication and is finally removed through the leak off connection.

22.12. IGNITION SYSTEM FOR PETROL ENGINE

The combustible air-fuel mixture in a petrol engine is ignited by an electric spark produced between the electrodes of a spark plug.

The battery ignition system consists of a storage battery of 6-12 volt; an ignition coil; a circuit breaker and spark plugs.

(*i*) The ignition coil consists of primary and secondary coils wound on a soft iron core. The primary winding has a small number of turns of a thick wire and the secondary winding has a large number of turns of thin wire.

(*ii*) The contact breaker is a spring controlled switch and is operated by means of a cam. The lobes of the cam correspond to the number of cylinders and for a four stroke cycle engine, the cam is motored at half the engine speed.

(*iii*) Distributor and the spark plugs for a multi-cylinder engine.

When the primary circuit is closed, current starts flowing and magnetic field is setup. With opening of breaker points, this magnetic field collapses and a high voltage of the order of 10,000 to 15,000 volts is set up at the secondary coil. The high voltage thus generated is directed to the rotor of the distributor which in turn supplies spark to each of the spark plug in

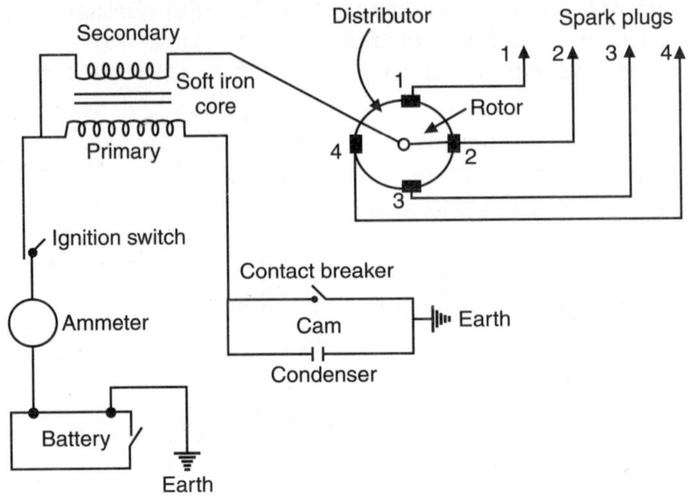

Fig. 22.17. Battery ignition system

accordance with the firing order of the engine. Because of high voltage, spark jumps across the spark plug gap (0.45 to 0.6 mm) and ignites the combustible air-fuel mixture.

The condenser connected in parallel with the circuit breaker prevents sparking across the gap between the points, and causes more rapid collapse of the magnetic field (primary current) and hence giving a higher voltage in the secondary circuit.

The coil ignition system is employed with medium and heavy spark engines such as automotive cars. The scooters, motor cycles and small motor boat engines operate with magneto ignition system where magneto acts its own generator and so no battery is required.

Spark plug : Towards the end of compression stroke, the combustle mixture in a petrol engine is ignited by an electric spark produced between the electrodes of a spark plug. The spark plug receives high voltage ignition current from secondary coil of the ignition system and supplies spark which jumps across its electrodes.

Refer Fig. 22.18 which shows the cross-section of a spark plug. The unit essentially consists of a central electrode, a threaded metallic body with a ground electrode, and an insulator that separates the two electrodes.

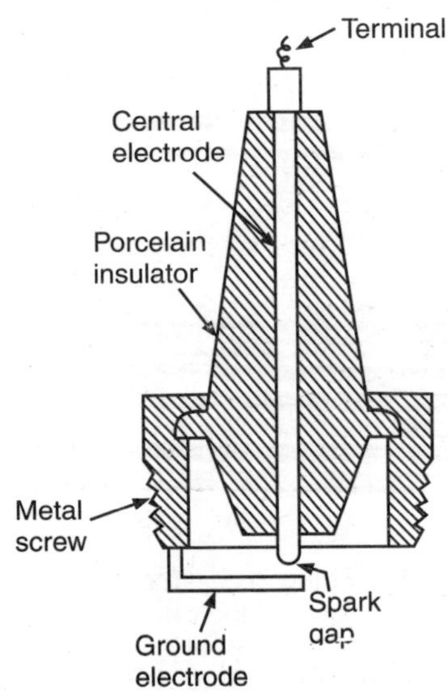

Fig. 22.18. Spark plug

The central electrode passes through the parcelain insulator. The electrode extends for a short length through the bottom of insulator and it has external contact at the top of wire from ignition coil. There is a metal screw that surrounds the bottom part of the insulator and fastened into the cylinder head. The lower part of the metal screw has a ground electrode welded to it. The ground electrode is bent towards the central electrode. The small gap between the central electrode and the ground electrode is called *spark gap*. When high tension current is given to the terminal of central electrode, it jumps across the gap and spark is produced.

Note : (*i*) The spark gap is generally kept between 0.6 mm to 1 mm. Two large or two small gap has an adverse effect on the efficiency of ignition system.

(*ii*) The sparking voltage at the spark plug is 10000 to 15000 volt and the spark lasts for about 0.001 second.

(*iii*) When the spark plug is short and has thick central electrodes and insulator, it runs cooler and is called cold plug. The plug would run hotter when it has long and thin exposed electrode, and is then known as hot plug.

The firing order in a multi-cylinder engine depends upon the number of cylinders, the arrangement of cylinders and the design of crankshaft. There are two possible firing orders of the four stroke cycle engine as 1-3-4-2 or 1-2-4-3 (Fig. 22.19). Although the first ordre is more frequently used, both satisfy the condition as regards eveness of torque.

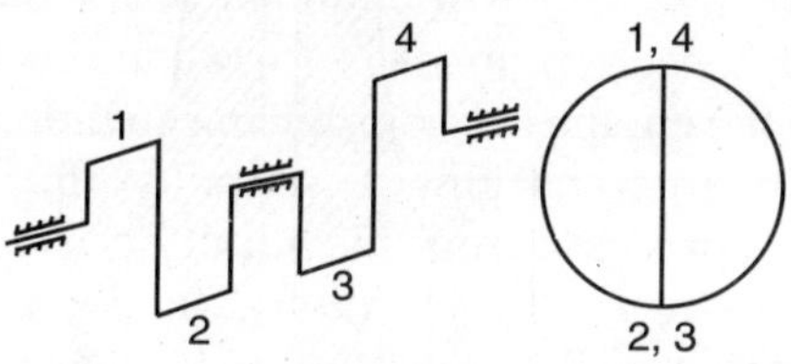

Fig. 22.19. Firing order for four stroke cycle engine

The firing interval of any engine is equal to $720/n$ where n is the number of cylinders. In a six cylinder engine, the firing interval is $720/6 = 120°$ and the corresponding spacing of crank is shown below.

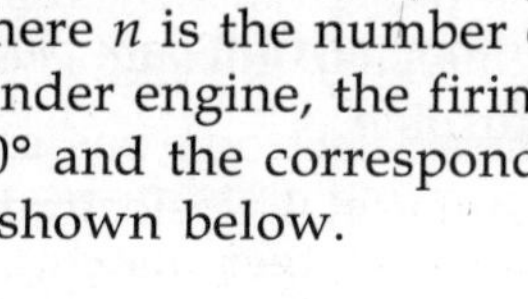

Fig. 22.20. Spacing of cranks

22.13. GOVERNING (SPEED CONTROL) SYSTEM

The term governing implies keeping the engine speed constant regardless of the charges in load on the engine. Speed control is essential is the stationary engines used for driving electrical generators, pumps and compressors that are to operate at constant speed irrespective of load.

The methods commonly used for governing IC engines are quantity governing, quality governing and, the hit and miss governing.

In *quantity governing,* the quantity of the mixture supplied to the engine is changed in accordance with load on the engine; the mixture strength (air-fuel ratio) remains same. The control is accomplished by means of a throttle valve which is placed in the intake manifold immediately after the carburettor. In automobiles, the throttle valve is operated by the foot or accelerator pedal through certain link rods. In stationary engines which must run at constant spped the throttle is positioned as a function of load and the movement of throttle valve is regulated by the lift of governor. With a decrease in load on the engine, the speed increases and the fly balls move outwards. That makes the governer sleeve rise up and partly close the throttle valve. The quantity of mixture entering the engine cylinder then decreases. The converse happens when there is increases in load on the engine. With a change in the quantity of mixture, there will be change in the mean effective pressure and accordingly the power developed by the engine.

The quantity governing is used in many gas engines and is invariably employed with all petrol engines.

In *quality governing,* the air flow rate remains constant but the composition of mixture is changed by admitting more or less fuel in accordance with variation in load. The variation in the quantity of fuel admitted is affected by one the following methods :

– changing the angular position of the helical groove of the fuel pump plunger relative to the suction port. This would alter the amount of fuel delivered due to change in effective stroke of the pump.

- delaying the closure of the suction valve of the pump.

- inserting a control valve to the delivery side of the fuel pump, this by-passes a part of the fuel back to the reservoir.

The quality governing is mostly used in high speed diesel engines.

The speed control system becomes flexible and most economical when both the quantity and quality of the mixture are changed depending on the load.

In *hit and miss governing*, every charge is of normal strength but an explosion of charge is occasionally omitted when the engine speed has risen above the mean speed. The governor causes the inlet valve to remain closed (in gas engine) and puts the fuel pump out of action in diesel engines. Since no fuel is supplied to the engine, it will perform an idle stroke during which no power will be developed.

22.14. SUPERCHARGING

The following methods have been suggested and are usually employed for improving the performance of an engine of a given size and capacity:

(*i*) *Run the engine at greater speeds* : An increase in speed results into an increase in charge induced per unit time and that increases the power output. However, the system has to encounter increased friction losses and a drop in volumetric efficiency (breathing capacity) beyond a certain limit of speed. Further, the increased engine speeds necessitate a rigid and robust engine due to increased bearing loads and inertia forces.

(*ii*) *Use of higher compression ratios* : Engine output and thermal efficiency generally increase with increase in compression ratio. However, too high a value of compression ratio results into an enormous increase in cylinder pressure. The engine cylinder has to be made strong enough to resist the high pressure and consequently the increased structural weight of the engine offsets the advantage of increased efficiency. Further, the combustion process takes place in an abnormal manner at high compression ratio. There is almost an instantaneous pressure rise which results in excessive noise and engine detonation. If detonation becomes severe, the engine over heats and the output starts declining. The value of compression ratio at which the engine power starts decreasing is called the highest useful compression ratio (HUCR).

(*iii*) *Employment of two stroke cycle systems* : For an engine working on two-stroke cycle, the power output is theoretically two-times the power output from a four-stroke cycle engine at the same speed and capacity. However, this approach is not considered appropriate due to certain practical difficulties of cooling, low thermal efficiency and low volumetric efficiency.

(*iv*) *Increasing the charge density* : If the density of charge is more, then for a given capacity, the engine would suck a greater mass of charge and generate more power.

The process where by a mixture of air and fuel (in case of petrol engines) or air only (in case of diesel engines) is forced under pressure to the engine cylinder during suction stroke is called supercharging. Essentially, the supercharging process means the admittance of dense charge to the engine cylinder. The device used for increasing the density, *i.e.*, increasing the pressure of charge above atmospheric pressure is called supercharger. Basically the superchargers are air compressors which may be of the reciprocating or rotary type. The units are driven either by engine shaft through gearing or by exhaust gas turbines.

The reciprocating units are used with large, slow speed engines – their short coming being: need large space, are bulky and heavy inconstructions, and have greater lubrication problems. The supercharging of aero- engines is done by rotary blowers (roots and vane blowers). These units can be run at much greater speeds by coupling them directly to the shaft of the exhaust gas turbines.

The engines which are supplied charge at pressure higher than atmospheric pressure are called supercharged engines and the engines which suck air at atmospheric pressure are called naturally aspirated engines.

The effects and purpose of supercharging an engine are :

(*i*) To increase the power output from an engine by supplying charge at pressure and density higher than atmospheric.

(*ii*) To reduce the weight to power ratio. This aspect attains significance in case of aircraft and racing cars.

(*iii*) To reduce the bulk of the engine to fit into a limited space as in case of marine engines and locomotives.

(*iv*) To compensate for the loss of power in aircraft applications at high attitudes where the atmosphere is rarified.

Further

- the mechanical efficiency of a supercharged engine is more than that of a naturally aspirated engine when both are made to run at the same speed.
- the specific fuel consumption of a super-charged engine is less due to proper combustion achieved by better turbulence and proper formation of air-fuel mixture and due to increased mechanical efficiency.

A supercharged engine has the following limitations:

(*i*) the engine components have to be designed to sustain the high pressure, and accordingly the engine structure becomes heavy.

(*ii*) an increase in *mep* is likely to increase the bearing pressure and that results into increased friction losses.

(*iii*) power is required to drive the supercharger (usually driven by the engine) and it partly offsets the power gain due to supercharging.

(*iv*) the engine parts have to be made of materials which can sustain high temperatures encountered in supercharged engines. For keeping the temperature levels with in safe limits, a greater load would be put on the engine cooling system.

In a supercharged spark ignition (SI) engine, the pressure and temperature are high and the delay period in short. These aspects increase the possibility of detonation. For preventing detonation either the engine has to operate with low compression ratio or a fuel with better octane rating has to be employed. With reduced compression ratio, the thermal efficiency of the engine is adversely affected.

The higher inlet pressure and temperature achieved by supercharging a compression ignition (CI) engine shorten the delay period and that lessons the possibility of knock. The rate of pressure rise during combustion is quite moderate and the engine gives a smoother and quieter operation. Apparently, the use of supercharging is more favorable with diesel engines than with patrol engines.

22.15. COMBUSTION IN SPARK IGNITION ENGINES

With reference to Fig. 22.21 given below :

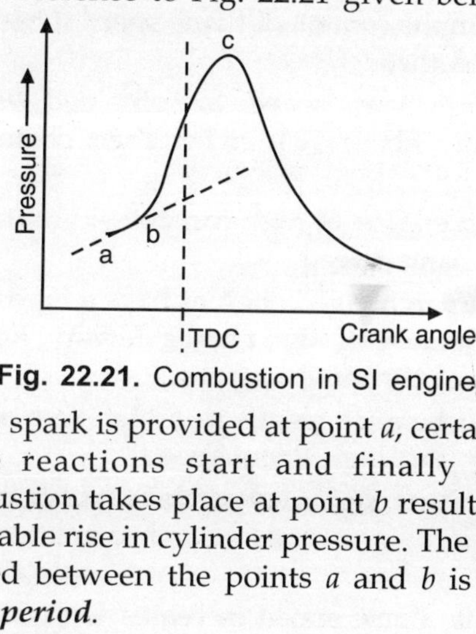

Fig. 22.21. Combustion in SI engines

(*i*) spark is provided at point *a*, certain pre-flame reactions start and finally actual combustion takes place at point *b* resulting in noticeable rise in cylinder pressure. The period elapsed between the points *a* and *b* is called ***delay period.***

(*ii*) During delay period, the molecules of mixture prepare to ignite. The number of degrees before *tdc* position at which ignition occurs is called *angle of advance*. The delay period is very short, about 0.002 second. But in high speed engines, this will correspond to appreciable movement of the crank. If an engine runs at 3000 rev/min, then during a delay period of 0.002 seconds, the crank will turn through,

$$= \frac{300 \times 360}{60} \times 0.002 \rightarrow 36°$$

(*iii*) The delay period is reduced with increase in compression ratio, intake temperatures and pressure. A higher value of self ignition temperature of fuel increases the delay period.

(*iv*) The charge adjacent to the spark plug ignites first and a luminous flame front develops which imparts its heat to the unburnt charge lying adjacent to it. When the unburnt charge attains the self-ignition temperature, it starts burning and a new wavefront develops. The flame front thus spreads through out the charge, heat is released and consequently there occurs a rise in pressure and maximum pressure rise is attained at point *c*; a few degree after tdc. The combustion stage corresponding to the flame propagation in the combustion chamber is represented by the line *b-c*.

Sometimes, the burning of fuel continues even during the expansion stroke, and this phenomenon is called *after burning*.

(*v*) In the context of flame speed it has been observed that :

- high flame speeds are obtained with an A/F ratio about 10% richer than chemically correct.
- an engine of high compression ratio has a high flame speed.
- superchanged engines have a high flame speed because of high charge density. Reverse is true for throttled engines.
- turbulence results in a high rate of heat transfer and high flame speed.
- at high engine speed, there will be greater inlet velocity or turbulence and so greater flame speed.
- the flame speed increases with increase in engine load and intake temperature.

(*vi*) The mechanism of *detonation* (sharp metallic sound) in spark ignition engines manifests in setting up pressure waves in the gases. These shock waves travel at very high velocity (900 – 2700 m/s) and are repeatly reflected from the walls of combustion space. The waves by their impact against the walls of the combustion chamber set them in vibration, and this gives rise to high pitched noises. The region in which detonation occurs is fasthest from the spark plugs and is known as *detonation zone*. Even with most severe detonation, it is rarely more than 1/4 th of clearance space.

The following factors promote detonation in a petrol engine:

- increase in inlet pressure and temperature.
- increase in compression ratio.
- advanced spark time.
- increase in coolant temperature and the engine load.

Further, it has been noted that detonation is

- maximum for slightly rich mixtures.
- much relieved in small cylinders with engine running at the same speed.
- much less in engines having compact combustion chamber.
- less with fuels of aromatic series

Due to detonation :

- the engine parts are set in vibration and the vibrating engine parts may get damaged due to fatigue failure.
- the operating temperatures are about 300°C higher and that increases the heat loss to the chamber walls.
- the temperature of the spark plug is greatly increased when the high pressure gases pass through its cavity time and again.

The high temperature may adversely affect the electrical and thermal properties of the plug and then render it useless. Overheating of the spark plug may even lead to pre-ignition. Pre-ignition is the spontaneous auto-ignition of the fresh combustible charge before the end of compression stroke. The high pressure wave set up in the cylinder due to pre-ignition opposes the piston movement during the last part of compression stroke and there occurs a serious loss in efficiency and power output of engine.

Detonation follows the spark while preignition precedes it. Further, pre-ignition can occur with any type of fuel and is due to the presence of some hot spots in any portion of combustion space. The hot spots may be an overheated spark plug, exhaust valve or the presence of some carbon deposits on the cylinder walls. Detonation, however, occurs with fuels which are volatile in nature and is affected by the composition and molecular structure of the fuel. The high temperatures and pressures resulting from pre-ignition, however, promote detonation.

The following methods have been suggested to reduce detonation :

(*i*) Compact design of combustion chamber and central location of spark plug. This is to reduce the path of the flame front travel from the spark plug to the remotest part of combustion chamber.

(*ii*) Better cooling of the end charge for getting rid of the excess heat.

(*iii*) Keep the inside temperature limits within the reasonable limits by injecting water into the cylinder before igniting the charge.

(*iv*) Use salt or sodium cooled exhaust valves particularly when the engine is operating at high compression ratios.

(*v*) Provide the combustion chamber with squish recesses. This will facilitate the cooling of the last portion of the charge and accelerate the flame front speed by providing the necessary turbulence for proper mixing.

(*vi*) Retard the spark timings so that the peak pressures are reached only during the power stroke and are also of lower magnitude.

(*vii*) Operate the engine with extremely lean or rich mixtures so that flame temperatures are kept low.

The anti-knock value of a gasoline engine is rated in terms of its *octane number*. The octane rating is a measure of the ability of fuel to resist detonation during combustion. The fuel under test is used to run the engine under specific conditions and the octane rating is measured with a specially designed knock meter. The engine under the same conditions is then operated on a mixture of iso-octane and *n*-heptane.

The iso-octane resists detonation very powerfully and is arbitrarily assigned a rating of 100 octane number. The *n*-heptane detonates very easily even at every low compression ratios and represents zero octane number. The two hydrocarbons may be mixed in varying proportions by volume and number of blends of fuel can be obtained. Such blends will have anti-knock ratings ranging from 0 to 100 octane numbers.

The percentage of iso-octane in such a blend which gives the same detonating results as that recorded by using the gasoline under test is called the octane rating of the fuel. Thus if a similar knock intensity is observed with a blend consisting of 75% iso-octane and 25% *n*-heptane, then we will say that the commercial fuel has a octane rating of 75%. Higher the octane number, better is the quality of fuel.

For understanding the progress of flame front and its effect on burning of the end charge in the combustion chamber, refer to Fig. 22.22 given below :

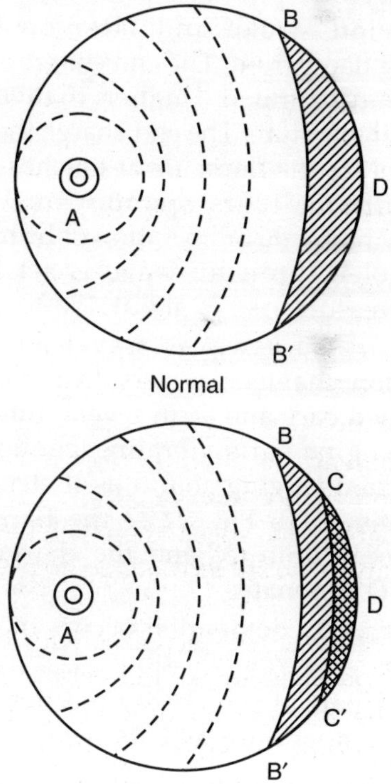

Fig. 22.22. Flame travel for normal combustion and knocking combustion

There is communication of flame from molecule to molecule. The charge near the flame is heated, attains the temperature of self-ignition, burns and forms a new wave front.

This flame front passes its heat to the next section which in turn ignites and so on.

1. Normal combustion : Consider the normal flame front which travels across the combustion chamber from *A* towards *D*. As the flame advances, it compresses the unburned charge *BB′D* raising both its pressure and temperature. However if the temperature of this unburned charge is below the self ignition temperature, it will not ignite automatically and the flame from *BB′* will be able to travel to the end gas upto the end point *D*. Under such conditions, the burning of the entire charge proceeds smoothly, the release of heat occurs at uniform rate and the pressure is within limits.

2. Knocking combustion : The end gas experiences rapid compression during compression stroke and later on by the advancing flame front. The end charge acquires a temperature much higher than the self ignition temperature. The end charge then auto-ignites before the flame front reaches it. This auto ignition occurs spontaneously with extremely high rate of liberation of heat energy and a violent pressure wave is set up; the abrupt pressure rise is about 150 to 20 bar.

The pressure waves travel across the combustion chamber at very fast speed and are reflected back and forth several times. That sets the engine parts vibrating giving rise to high pitched ringing sound as if struck by a light hammer. In Fig. 22.22 the flame front reaches only upto *CC* and the charge in the area *CC′D* detonates.

Apparently detonation occurs in a petrol engine if

- end part of the charge reaches its self-ignition temperature and its delay period is over

- the flame does not reach the end part of the charge

Combustion chambers for SI engines :

The main requirements for the design of combustion chambers for SI engines are :

- compact so that the flame has to travel a short distance to reach the end gas

- low surface to volume ratio for keeping the heat losses to minimum

- adequate turbulence which will ensure quick propagation of flame front.

- proper location of spark plug ; the spark plug is to be so located that the distance of the farthest point of combustion space from the spark plug is small.

- location of exhaust valves away from the end gas; this is to reduce the possibility of detonation as the exhaust valves are the hottest spot in the combustion chamber.

- large area for the inlet and exhaust valves; this will permit the free flow of charge into and products out of the cylinder.

- location of the valves should be such that they can be actuated with relatively simple mechanism.

The commonly used valve arrangements for SI engines are :

1. *T-head type* : Both the inlet and exhaust valves are located in the engine block on opposite sides.

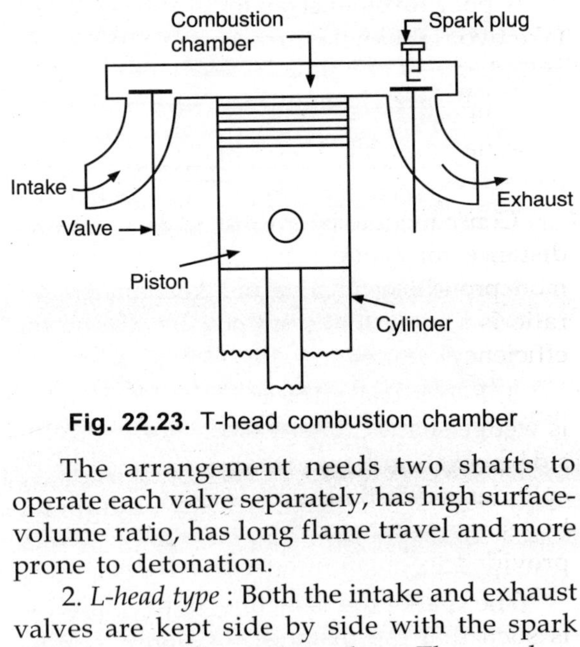

Fig. 22.23. T-head combustion chamber

The arrangement needs two shafts to operate each valve separately, has high surface-volume ratio, has long flame travel and more prone to detonation.

2. *L-head type* : Both the intake and exhaust valves are kept side by side with the spark plug located above the valves. These valves are operated through tappet directly from a single cam shaft. The top of the combustion chamber has a flat surface.

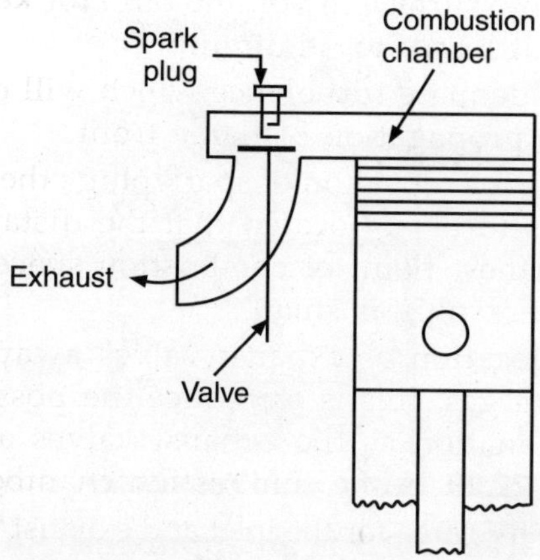

Fig. 22.24. L-head combustion chamber

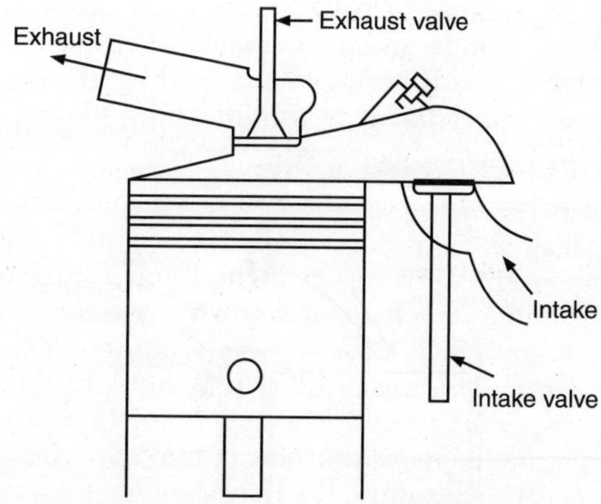

Fig. 22.25. F-head combustion chamber

The merits/demerits of this arrangement are :

- easy casting, maintenance and lubrication of the valve mechanism.

- easy removal of cylinder head, this permits decarburization without disturbing the valve gear mechanism.

- poor turbulence as the intake air has to take two right angle turns before reaching the cylinder.

- unsatisfactory mixing of air and fuel.

- flame front has to travel more distance before reaching the end charge.

Consequent to poor turbulence and greater distance for flame travel, such systems are more prone to detonation and their compression ratio is limited to 4 (low power and thermal efficiency).

3. *F-head type* : The combustion chamber is wedge shaped, inlet valve is located in the side pocket and the exhaust valve is in the cylinder head. Both the valves are operated by a single camshaft, with a rocker arrangement provided to operate the valve in the head.

The spark plug is inclined and its location is such that the distance for flame travel is less and the detonation tendency is arrested.

Fig. 22.26, refers to the combustion chamber arrangement used in aircraft and automobile racing car engines.

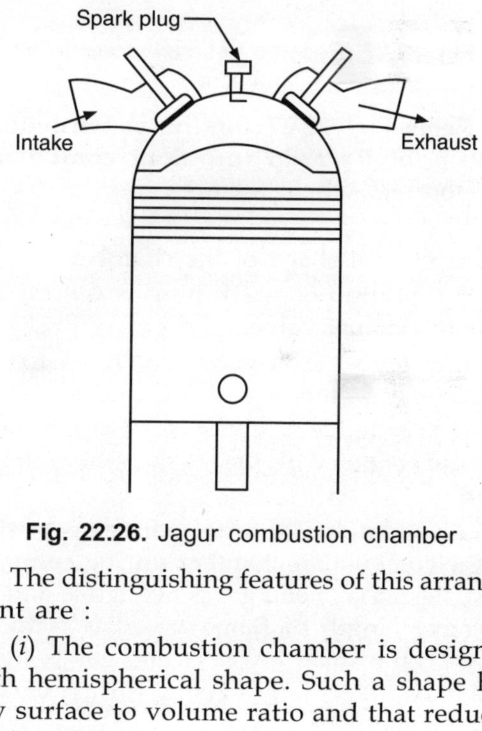

Fig. 22.26. Jagur combustion chamber

The distinguishing features of this arrangement are :

(*i*) The combustion chamber is designed with hemispherical shape. Such a shape has low surface to volume ratio and that reduces the heat losses.

(*ii*) Both the valves are in cylinder head and are inclined at a certain angle to the cylinder axes.

(*iii*) The piston crown is so shaped that reguiste amount of turbulence is created.

(*iv*) The spark plug is located central.

Due to these features, the flame speeds are quite high, the flame travel distance is small, and that helps in reducing detonation.

4. *Recardo turbulent combustion chamber* : The system is an improvement over the side valve (L-head) arrangement in which the heat losses are reduced to minimum by having a hemi-spherical combustion space

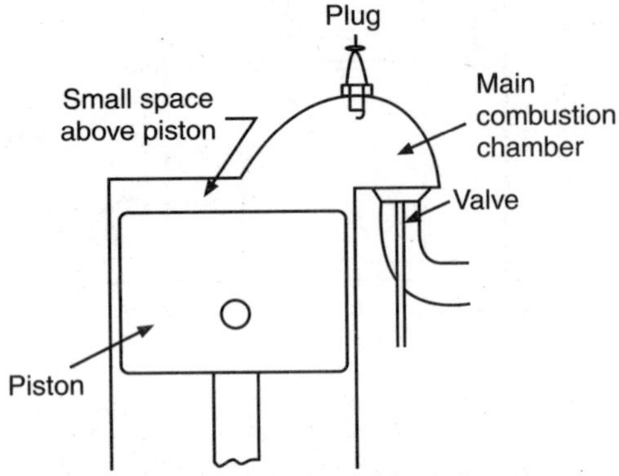

Fig. 22.27. Recardo turbulent combustion chamber

Refer Fig. 22.27 for the constructional aspects of Recardo turbulent combustion chamber; its salient features are:

(*i*) minimum surface-volume ratio due to hemispherical shape of the chamber.

(*ii*) location of spark plug is central and near to exhaust valve.

(*iii*) high turbulence which ensures a homogeneous mixture of fuel and air.

(*iv*) negligible clearance as the piston comes in close contact with the combustion chamber head.

Accordingly, the whole charge is pushed in the combustion chamber during compression, the mass of end gas is negligible and the effective length of flame travel is reduced. Further, the small lower of end gas is coded due to relatively large cooling surface of combustion chamber.

The reduced distance for flame travel, cooling of end gas, central location of spark plug and its nearness to exhaust valve reduce the tendency to detonate.

5. *Rover head combustion chamber* : The design of Rover head combustion chamber processes all the requirements of a good combustion chamber, and its main distinguishing characteristics are :

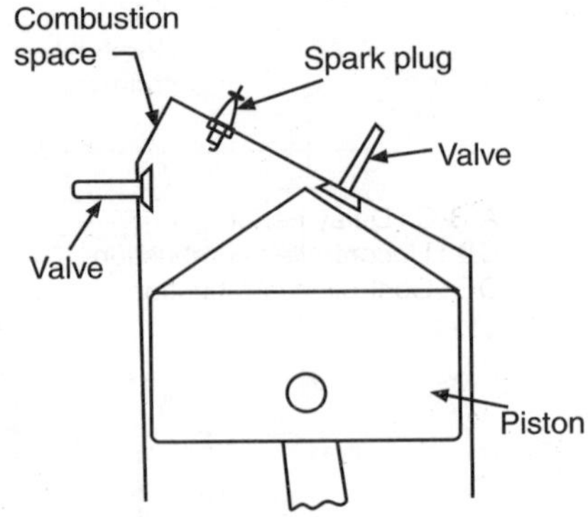

Fig. 22.28. Rover head combustion chamber

(*i*) The unit has an approximately hemispherical shape for its head. That gives a small surface to volume ratio, and hence better thermal efficiency and power output.

(*ii*) The combustion space is compact and the spark plug is centrally located. Obviously the length of flame travel is short.

(*iii*) The relatively cool piston and cool inlet valve keep the thin layer of charge (end gas) between the piston and cylinder head at a low temperature. With reduction in temperature, the delay period gets prolonged and that ensures normal combustion with least tendency to detonate.

However, the system has complicated valve mechanism and an uncommon shape for the piston. Further, the cylinder head cannot be removed (for decarbunising) without disturbing the valve gear mechanism.

22.16. COMBUSTION IN COMPRESSION IGNITION ENGINES

(*i*) Fuel injection starts at point *A* and actual combustion takes at point *C*.

(*ii*) Period *ABC* denotes the delay period or ignition delay.

AB → physical delay during which fuel particles are atomised and vaporised.

BC → chemical delay during which fuel thoroughly mixes with air and is raised to self-ignition temperature.

(*iii*) The actual combustion starts at point *C*; the fuel gets burnt in an uncontrolled manner and the pressure rise in the cylinder takes place

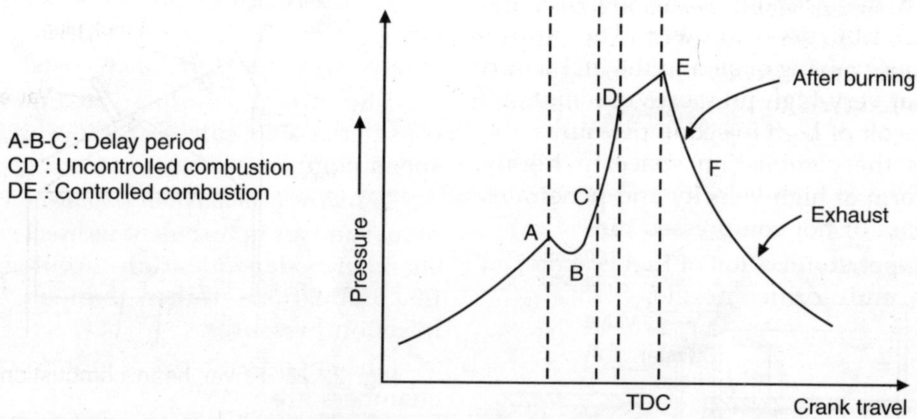

Fig. 22.29. Combustion in CI engines

at an enormous rate. Due to rapid increase in pressure rise, a sudden load is applied to the engine structure and this may result in fatigue failure of the engine components. The engine has a rough running and produces a violent noise called *diesel knock*. The knock results in drop in engine power and efficiency.

(*iv*) After the combustion of main bulk of fuel, the flame spreads throughout the whole combustion space. During the later part of combustion (*DE*), the rate of burning and the pressure rise remains controlled to some extent.

(*v*) Where as detonation in SI engines is due to auto-ignition of last part of charge to burn, the knock in diesel engines is due to instantaneous and rapid burning of the first part of charge. Due to this inherent difference, the factors tending to increase detonation in SI engines, tend to reduce knocking in CI engines.

(*vi*) The anti-knock value of the diesel fuel is rated in terms of the *cetane number*. Cetane is a straight chain paraffin, has very good ignitability and is arbitrarily assigned a rating of 100 cetane number. α-methyl naphthalene (a hydro carbon) has a very poor ignitability and is arbitrarily assigned a cetane number of zero.

These two are mixed in varying proportions and a number of fuel blends can be obtained. The cetane number of fuel is defined as the percentage by volume of cetane in such blends which gives the same knocking results as the fuel in question, when tested in a special engine under certain specified conditions. Thus if a similar knock intensity is observed with a blend consisting of 60% cetane and 40% α-methyl naphthalene, then we will say that the commercial fuel has a cetane rating of 60. Higher the centane number, better is the quality of diesel fuel.

Combustion chambers for CI engines : The main requirements for the design of combustion chambers for CI engines are :

(*i*) fine atomisation of fuel and its proper mixing with air

(*ii*) requiste turbulence and short delay periods for smooth running of the engine.

(*iii*) complete burning of fuel; this is necessary for obtaining peak pressures.

(*iv*) low surface to volume ratio, this is for reducing heat losses and achieving high thermal efficiency.

To combustion chambers for CI engines essentially belong to two categories namely (*i*) the direct or open type and (*ii*) the indirect or swirl chambers

1. Direct combustion chambers

The non-turbulent open or direct type combustion chambers are formed into the piston and the fuel in atomised form is injected directly into the main combustion space. The mixing of fuel and air depends on the induction swirl provided by directing the flow of air in desired direction by surrounding the valve head or squish. Squish is the rapid movement of air from the cylinder walls towards the centre of combustion chamber. The squish ensures proper mixing of air and the fuel which is injected at very high pressures of about 300 bar. As a result of high injection pressures, the fuel enters the combustion space in highly atomised form at high velocity and penetrates into the mass of hot compressed air.

The proper atomisation of fuel is achieved by using a multi-orifice nozzle.

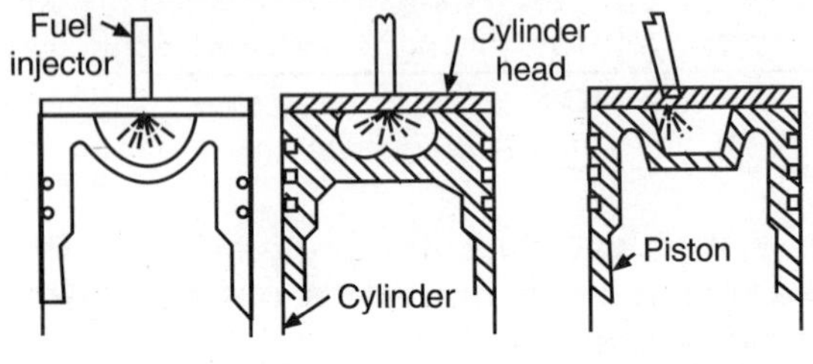

Fig. 22.30. Direct or open combustion chambers

The different shapes of the commonly used combustion chambers are shown in Fig. 22.30. The cylindrical chamber provides a better squish compared to that of hemispherical chamber. Further, the squish of cylindrical chamber can be varied by changing the depth to diameter ratio of the cavity. In the truncated cone design of the chamber, the injection nozzle is inclined to the cylinder axis, and the air swirl is obtained by surrounding the inlet valve.

The advantages and short comings of direct injection systems are :

(*i*) simple design and shape for combustion chamber

(*ii*) small heat loss to cylinder walls because of less turbulence and that ensures easier starting from cold.

(*iii*) less turbulence requires considerably excess supply of air for better utilization of fuel. With the increased charge, the thermal efficiency improves.

(*iv*) supply of excess air results into lower maximum temperatures and that adversely affects the power output.

(*v*) need of high injection pressures and accordingly the engine parts need to be manufactured from high quality materials.

(*vi*) there are greater chances of frequent clogging of small nozzle openings and this entails higher maintenance costs.

The direct combustion chambers are considered quite suitable for low and moderate speed engines.

2. *Indirect combustion chamber* : The mixing of fuel and air in turbulent indirect combustion chamber is dependent on turbulence created in combustion rather than on very high injection pressures.

The important types of indirect combustion chambers are :

(*a*) *Pre-combustion chamber* : The arrangement consists of two chambers namely the pre-combustion (ante) chamber and the main chamber. The pre-combustion chamber is a small heated chamber which is located in the cylinder head, and it communicates with the main chamber through one or several restricted passages.

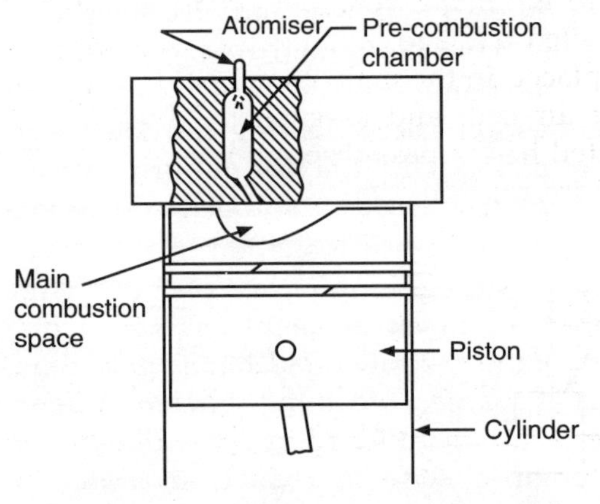

Fig. 22.31. Pre-combustion chamber

Towards the end compression stroke, the entire fuel is injected into the pre-combustion chamber where it mixes with air and combustion is initiated. Vaporization and partial combustion (5 to 10%) of fuel takes place in the ante-chamber. This chamber is kept hot by an electrically heated plug. Consequently the delay period is reduced and the combustion is normal.

The pressure in the pre-chamber is higher than that in the main chamber and this pressure differential forces the burning and unburnt mixture into the main chamber at very high velocity. The high velocity aids in rapid mixing and burning of charge in the main chamber.

The advantages/disadvantages of pre-combustion chamber design are:

- no need of high injection pressures
- single orifice nozzles can be used and these have less changes of clogging
- smooth running of the engine due to reduced delay period as the pre-combustion chamber is designed to run hot
- rapid combustion of fuel in the main chamber gives high mean effective pressure.
- cold starting is difficult, heat losses are high (due to high rate of burning and turbulence) and consequently thermal efficiency is low.

(*b*) *Air-cell combustion chamber*. The combustion space of the unit has two portions namely a minor chamber (called air cell) and a major chamber. The air cell communicates with the major chamber through a small restricted passage called neck. The fuel injector is placed in the main chamber and in front of the air cell, and as such the fuel when injected has to pass through the main chamber before reaching the air cell.

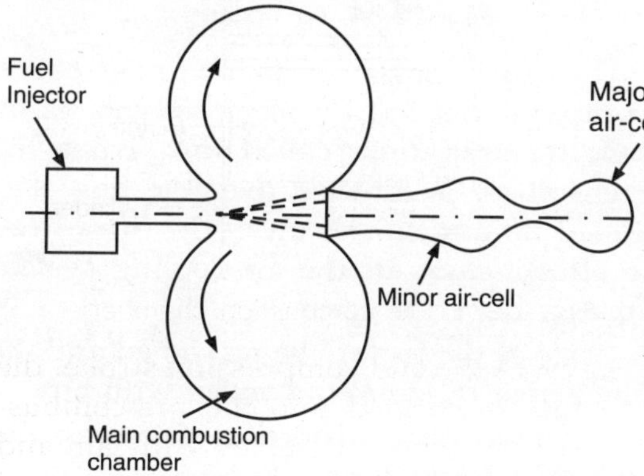

Fig. 22.32. Air cell combustion chamber

The combustion process, which initiates in the main chamber, has a low burning rate due to insufficient mixing of fuel and air. Subsequently combustion in the air-cell progresses at a rapid rate. The increase in pressure in the air-cell forces the burning gases into the main chamber. The design of the main chamber is such that the gas stream from the air-cell splits into two vortices and high swirl is created. High turbulence and high temperatures of gases reduces the delay period. Apparently the rate of pressure rise is controlled and the engine runs smooth.

With this arrangement the engine runs smooth and its cold starting is possible. However, a portion of air which remains in their cell continues burning even when the piston is on the expansion stroke. This late burning reduces both power and efficiency of the engine.

(*c*) *Swirl or turbulent chamber*

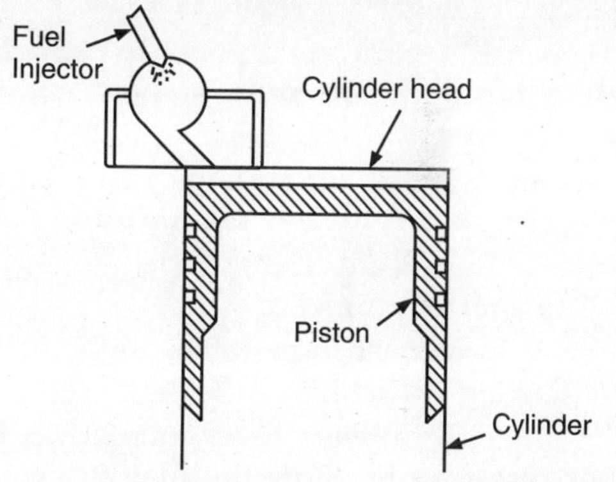

Fig. 22.33. Swirl or turbulent chamber

The salient operational features of the swirl or turbulent combustion chamber are :

(*i*) During compression stroke, about 75 to 85 percent of compressed air is forced into the combustion chamber which is separated from the engine cylinder.

(*ii*) The entry of air into the chamber is in tangential direction and that gives high rotational movement.

(*iii*) The fuel is injected into the swirl chamber a few degrees before top dead centre; there is partial mixing of injected fuel with the rotating air.

(*iv*) The combustion starts and there is development of high pressures in the swirl chamber. After a bulk of combustion has taken place in this chamber, the combustion products at high pressure are discharged into the main chamber. The resistance to gas flow in the communicating passage imparts high turbulence and assists in further combustion.

The swirl chamber is kept hot by providing a thick insulation. Accordingly high temperatures are maintained. That reduces the delay period resulting into controlled pressure rise and smooth running of the engine.

Main advantages of swirl chamber are :

- mixing of fuel and air is dependent on air swirl and not on high injection pressures and as such the system needs low fuel injection pressures.

- due to requirement of low pressure, single orifice nozzles can be used. These nozzles have less tendency to clog and accordingly maintenance problems are reduced.

- there is control on the pressure rise during combustion and that gives smooth running of the engine.

- there is better utilization of air resulting into higher mean effective pressure and power output.

The short comings are :

• air loses heat to cylinder walls making cold starting difficult.

• high turbulence level increasing heat losses resulting into low thermal efficiency.

• high specific fuel consumption

The swirl combustion chambers are generally used with high speed engines.

22.17. COOLING SYSTEM

The engine components such as cylinder walls, cylinder head, piston and valves absorb a considerable amount of heat that is liberated during combustion of air-fuel mixture in the engine cylinder. With over heating

(*i*) The valves may warp and the components may expand excessively leading to seized piston and piston rings.

(*ii*) The lubricating oil may decompose giving gummy and carbon deposits.

(*iii*) Due to large variations in the temperature of different engine parts, thermal stresses are set up and these may cause distortion and cracking of certain parts.

(*iv*) In spark ignition engines, the overheating of spark plug may cause pre-ignition which further promotes detonation and leads to loss in engine power and efficiency.

The engine has to be provided with an efficient cooling system so that temperature level does not exceed beyond a certain prescribed limit. The amount of heat to be removed is 20 to 30% of the heat released during combustion. However, too much removal of heat is undesirable as that leads to drop in thermal efficiency.

The cooling systems in common use are :

1. Direct (air) cooling in which heat from the cylinder walls is transferred to the surrounding air

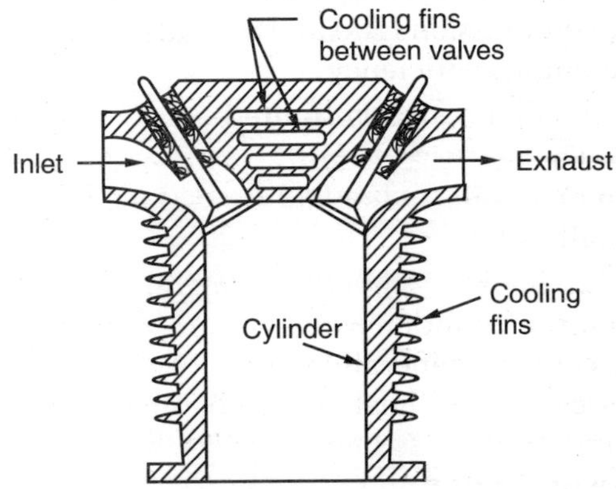

Fig. 22.34. Air cooling

The outer surface of the cylinder barrel and the cylinder head is increased in area by providing projections called fins. When the current of air is directed over the fins, it is warmed up and carries away the extra heat. The effectiveness of the air cooling system primarily depends upon :

- size of the surface being cooled and the surface area of metal in contact with air

- temperature difference between the heated surface and air

- air velocity *i.e.,* rate of air flow

- thermal conductivity of the metal of the surface.

In motor cycles and aeroplanes, the high velocity air is obtained by the motion of unit itself. The stationary units have to be provided with a fan to circulate air at a high velocity.

The air cooling system has the merits of :

- simple design, light in weight.
- no requirement of cooling water and anti-freeze.
- less starting trouble; can be operated in cold climate where water may freeze.
- more freedom in the positioning of engine as there is no radiator cooling jacket etc.
- reduced maintenance and capital cost.

The short comings are :

- noisy operation
- cooling is not uniform all around the cylinders
- high operating temperatures which limit the compression ratio and that adversely effects the engine efficiency
- engine size is limited

The use of air cooling system is restricted to motor cycles, aeroplanes and small stationary installations.

2. *Indirect (liquid) cooling* in which heat from the cylinder walls is first transferred to the coolant, generally water. The coolent next passes on the heat to atmospheric air. The liquid cooling systems are either of thermo-syphon or of pump circulating type.

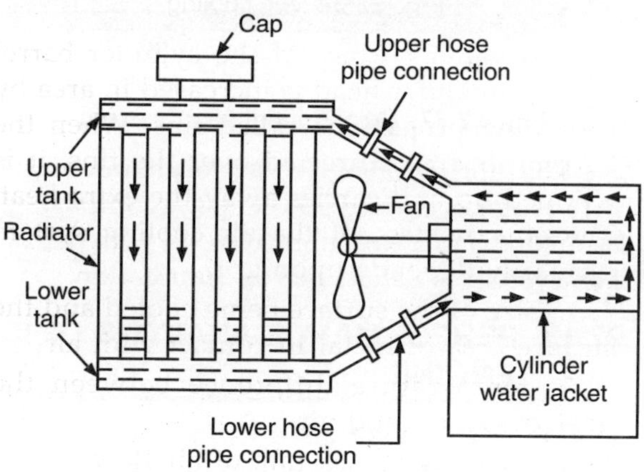

Fig. 22.35. Thermo-syphon system of cooling

Fig. 11.24 shows the schematic arrangement of an engine cooled on thermo-syphon principle. The circulation of water through the system is accomplished by convection currents which are set up because of the density difference of hot and cold waters. The system consists of a radiator having upper and lower tanks which are connected through pipes respectively to upper and lower water jackets. The cylinder water becomes lighter when heated, rises up and goes to the upper part of the radiator. In the radiator, water fralls through a large number of vertical tubes which are exposed to air. Consequently, the water reaching the bottom part of radiator is cool and is ready to take the place of water which earlier left the engine cylinder. Usually a fan is installed between the radiator and engine block to increase the rate of air flowing over the radiator tubes.

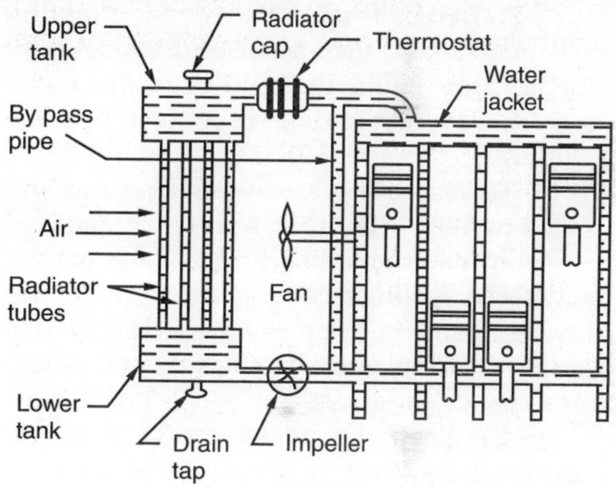

Fig. 22.36. Pump circulating water cooling

In large engines and those of higher output, an impeller pump is installed to assist in the circulation of water through the water jacket and the radiator. The system is then called pump or pressure cooling. The merits of this system are:

- the flow of water is at an increased rate
- possibility of uniform cooling; the coolant can be directed towards hottest parts of the engine such as exhaust valves;
- compact engine design with minimum frontal area; also installation of a small radiator.

In the liquid cooling systems, a thermostat is provided in the water circuit. The thermostat allows the water to go to the radiator when the water has been raised to a pre-determined temperature of about 80–85 °C. If the temperature of heated water is less, it is by-passed to the pump. Further, when operating under extreme cold conditions, the freezing of water in the radiator tubes is prevented by adding anti-freeze solution containing ethylene glycol.

22.18. LUBRICATION SYSTEM

Lubrication system manifests in providing a lubricating oil at the rotating and sliding parts of an engine, and this is necessary for the proper running and maintenance of the engine.

The objectives of the lubrication system are :

(*i*) To reduce friction between the mating parts and reduce their wear and tear. An oil film is formed and that prevents the metal parts from coming into direct contact with one another.

(*ii*) To absorb shock between bearings and other engine parts. This reduces noise and increases the engine life.

(*iii*) To act as a cooling medium for removing heat.

(*iv*) To form a good seal between piston rings and cylinder walls.

(*v*) To keep the engine parts clean; to prevent deposition of carbon and soot; to absorb and carry away harmful substances that result when the combustion is incomplete.

The different systems for lubricating the engine parts are the splash system, pressure system (wet and dry) and petrol system.

In ***splash lubrication***, the crankshaft is kept immerbed in crankcase which is full of lubricating oil. When the crankshaft rotates, the oil is splashed in different directions and the engine parts get lubricated.

Arrangement is provided through a system of specially drilled holes and galleries to take the oil to crankpin, connecting rod bearings, main bearings and the camshaft bearings. The oil after lubricating different parts drains back to the pump. The splash lubrication is mostly used on small one cylinder stationary engines.

In ***pressure lubrication*** (Fig. 22.38), the lubricating oil under pressure is circulated through drilled passages and oil lines to various engine parts which require lubrication. In the wet sump system, the pump that supplies oil is either submerged in the oil or is placed slightly above the oil level. The system is used in the small automobile engines. In the dry sump system the pump is outside the sump and it forces oil into a small reservoir from where it is delivered to the main gallery by another pump. The oil reservoir is usually placed behind the radiator and thus takes the advantage of air blast for cooling the lubricating oil. The dry sump system is generally used in large stationary and marine engines.

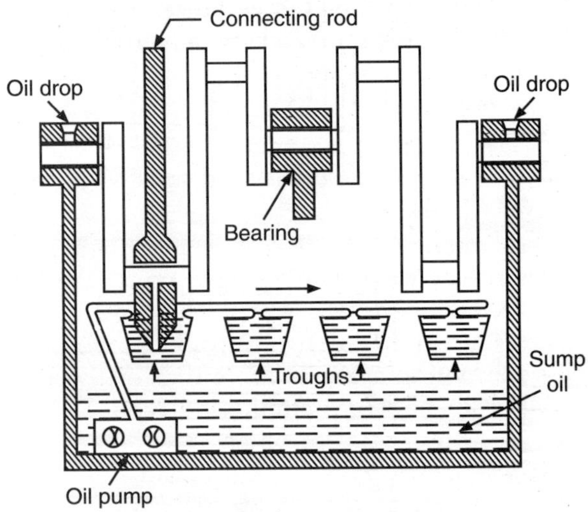

Fig. 22.37. Splash lubrication system

The lubrication of two stroke engines fitted with motor cycle and scooters is accomplished by mixing 3 to 6% of lubricating oil with petrol. The system is called petroil lubrication.

22.19. PERFORMANCE ANALYSIS OF IC ENGINE

Engine performance is an indication of the degree of success with which the engine does the job assigned to it, *i.e.*, the conversion of chemical energy contained in the fuel into useful mechanical work. The engine performance is expressed in terms of certain parameters and a study needs to be made how those parameters are affected by the operating conditions, design concepts and modifi-

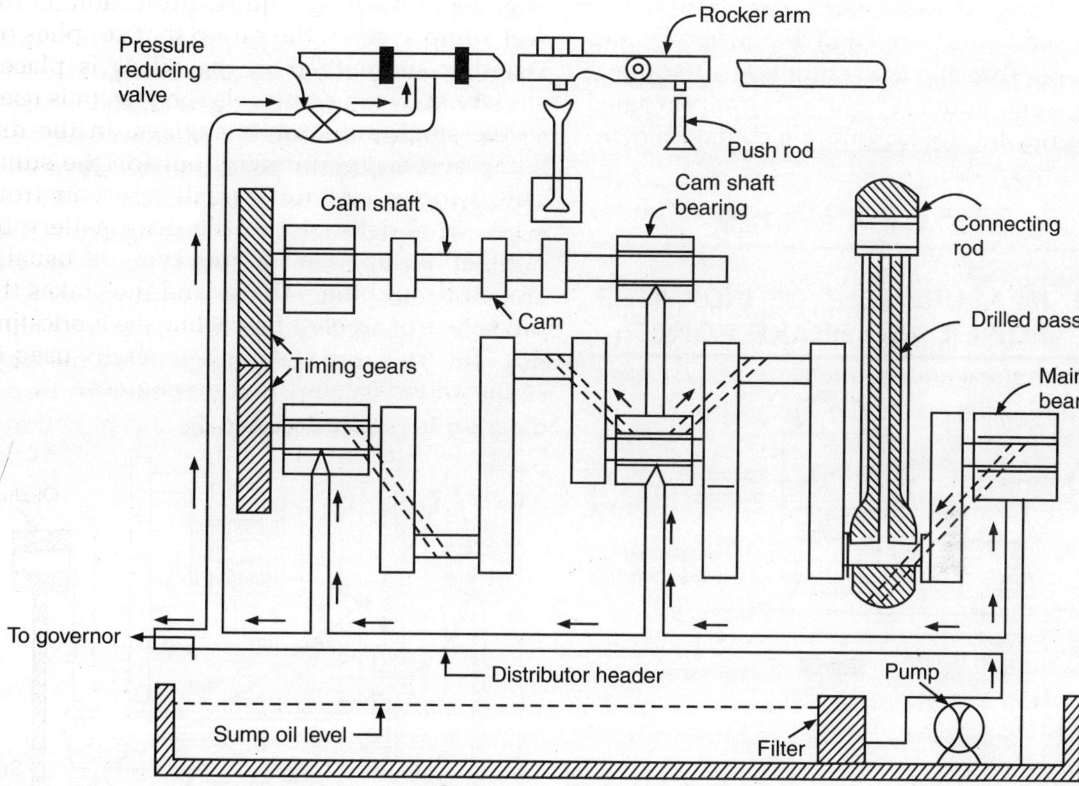

Fig. 22.38. Pressure lubrication system

cations etc. The desired engine performance depends on power requirements of the apparatus or vehicle to which the engine has to be connected. A test run on the engine may be made.

(*i*) to find the power output under certain conditions

(*ii*) to determine the fuel economy

(*iii*) to obtain detailed information about the various losses and for preparing heat balance.

Refer Fig. 22.39 for the energy flow through an IC engine.

When the mixture of air and fuel supplied is burnt inside the engine cylinder, the chemical energy gets converted into heat energy. The total heat thus generated cannot be converted into work and a substantial part is

(*a*) lost to cooling water flowing through the cylinder jackets,

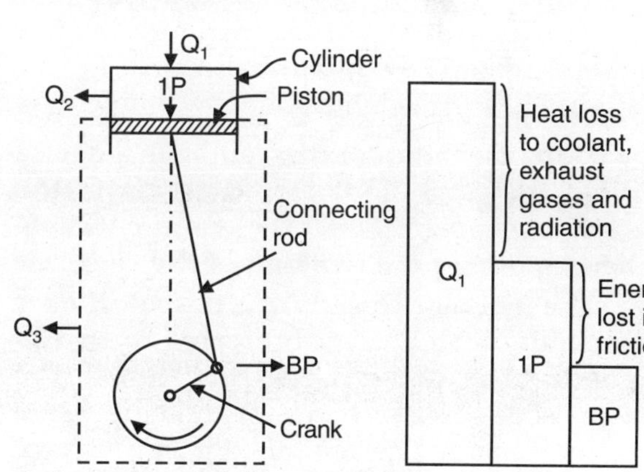

Fig. 22.39. Energy flow in an IC engine

(*b*) carried by exhaust gases, and

(*c*) lost to surroundings by way of radiation.

The remaining part of the energy exerts a force on the piston which moves and mechanical work is done. The power developed inside the engine cylinder due to combustion of fuel is called ***indicated power (IP)***.

A part of the indicated power is lost due to friction between the piston and cylinder as well as in the bearings. The net output available at the output shaft is called ***brake power*** **(*BP*)**.

The difference between indicated power and brake power is called the ***friction power*** **(*FP*)**.

22.20. MEASUREMENT OF INDICATED POWER AND BRAKE POWER

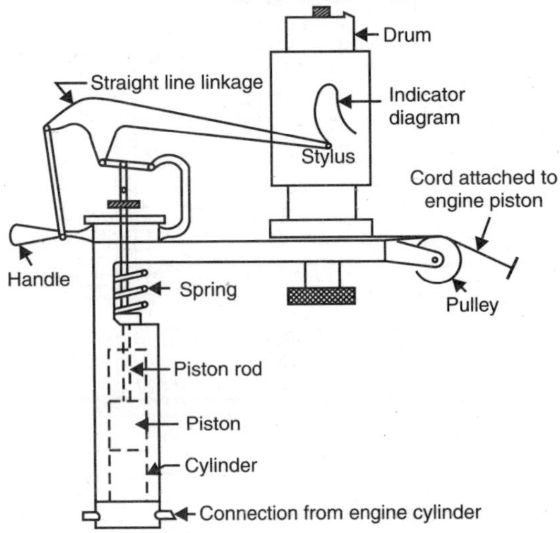

Fig. 22.40. Engine indicator

The *indicated power* is usually measured by taking the indicator diagram with a device called indicator. Indicator diagram is the graphical representation of the pressure-volume changes during the working of the cycle.

The indicator consists of a piston-cylinder assembly which communicates with the cylinder head of the engine being tested. The piston slides inside the cylinder and the piston rod is connected to the straight line linkage through a spring of proper stiffness. The spring controls the piston movement which is caused by any variation in pressure inside the engine cylinder. Higher the spring stiffness, less the magnification and lower the spring constant, greater the magnification. The movement of the indicator piston is transferred to a drum by the linkage at the end of which is attached a stylus (pencil). The stylus moves in the vertical direction in proportion to the piston movement and records the gas pressure. The drum is made to rotate about its axis by a pulley and a cord connected to the engine piston through a reducing mechanism. The motion of the cord is proportional to the piston stroke. Obviously a diagram would be obtained on a strip of paper wrapped round the rotating drum. The vertical movement of the stylus and the horizontal movement of the cord combine to produce a closed figure known as indicator diagram. The area of the diagram is a measure of the work developed and the length of the diagram is proportional to the swept volume.

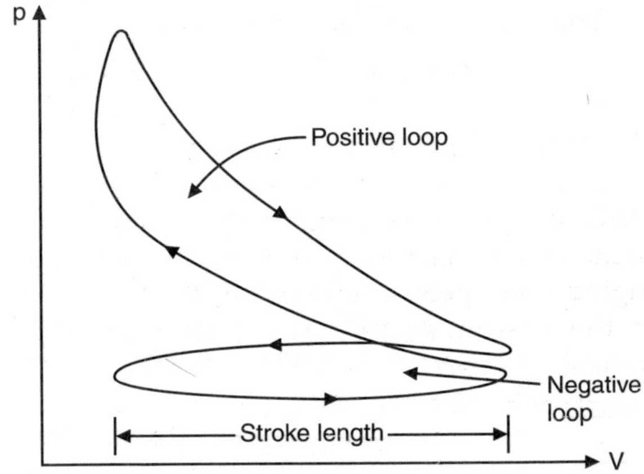

Fig. 22.41.

The indicator diagram has a positive loop (area between the compression and expansion lines) and a negative loop (area between the suction and exhaust lines). The positive loop gives the gross work done by the piston during the cycle and the negative loop represents the pump loss due to admission of fresh charge and removal of exhaust gases.

Net work done per cycle

$$= \text{area of positive loop} - \text{area of negative loop}$$

The area of the indicator card can be measured either with a planimeter or by the ordinate method. The indicated mean effective pressure (p_{mi}) is then given by

$$p_{mi} = \frac{\text{area of indicator diagram}}{\text{base length of indicator diagram}} \times \text{spring constant}$$

The mean effective pressure represents the constant pressure if it is acted over the full length of the stroke would produce the same amount of work as is actually developed by the engine during a cycle.

If area is in mm^2, base length is in mm and the spring constant is in N/mm^2 per mm of vertical movement of the indicator stylus, then mean effective pressure works out in N/mm^2.

Further, if p_{mi} = indicated mean effective pressure, N/m^2

A = area of piston, m^2

L = length of stroke, m

and N = speed of the engine, rpm

Then : force on the piston

$= p_{mi} \times A$ (N)

work done per stroke $= p_{mi}\ LA$ (Nm)

work done per minute $= p_{mi}\ LAN$ (Nm/min)

In a two stroke cycle, the cycle is completed in two strokes of the piston or in one revolution of the crank shaft. In a four stroke cycle the engine cycle gets completed in four strokes of the piston or two revolutions of the crankshaft. Then

work done/minute $= p_{mi}\ LAN$

for two stroke engine

$$= p_{mi}\ LA \frac{N}{2}$$

for four stroke engine

Indicated power,

$$IP = \frac{p_{mi}\ LA\,(N \text{ or } N/2)}{60 \times 10^3} \times n \text{ kW}$$

where n is the number of cylinders in the engine.

The ***brake power*** (the useful power available at the engine crankshaft) is determined with the help of a dynamometer coupled to the engine crankshaft. A rope brake dynamometer consists of one or more ropes wrapped around the flywheel of the engine whose brake power is to be measured. The ropes are spaced evenly across the width of the rim by means of U-shaped wooden blocks located at different points of the rim of the flywheel. The upward ends of the ropes are connected together and attached to a spring balance, and the downward ends are kept in place by a dead weight. The rotation of the flywheel produces frictional force and the rope tightens. Consequently a force is induced in the spring balance. Generation of heat is enormous and that necessitates a cooling arrangement for the brake.

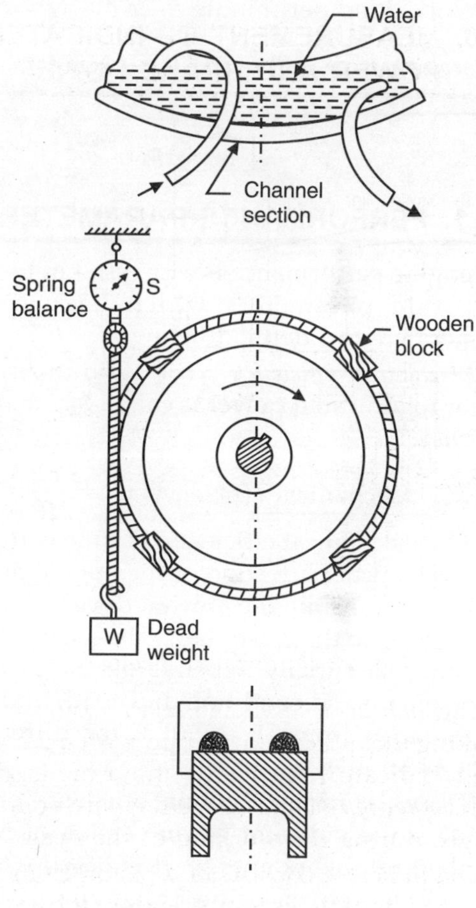

Fig. 22.42. Rope brake dynamometer

The power of the engine is absorbed as frictional heat between the rope and pulley surface.

Let W = dead weight at the end of the rope, N

S = spring balance reading, N

N = engine speed, rpm

D = diameter of the brake wheel, m

d = diameter of the rope, m

Then : effective radius of brake drum

$$= \frac{D+d}{2}$$

Torque, $T = (W - S) \times \frac{D+d}{2}$

Work done per revolution

= torque × angle turned per revolution

$= 2\pi T$

Work done per minute $= 2\pi T \times N$

Brake power $(BP) = 2\pi T \times \frac{N}{60}$ Nm/s or W

$$= \frac{2\pi NT}{60 \times 10^3} \text{ kW}$$

22.21. PERFORMANCE PARAMETERS

The engine performance is expressed in terms of certain parameters which have been enumerated and defined below :

Mechanical efficiency : The ratio of brake power to indicated power is called *mechanical efficiency*.

$$\text{Mechanical efficiency } \eta_m = \frac{BP}{IP}$$

The mechanical efficiency is a measure of the mechanical perfection of the engine or its ability to transmit the power developed in the cylinder to the drive shaft. The mechanical efficiency essentially depends on the design of the engine, its condition and load, and its value varies from 75 to 90 % for different types of engines.

Thermal efficiency: Thermal efficiency of an engine is the indicator of conversion of heat supplied into work energy. It is based either on indicated power or on brake power. The indicated thermal efficiency of an IC engine is the ratio of the heat converted into indicated work to the heat energy in fuel supplied to the engine.

Indicated thermal efficiency

$$\eta_{ith} = \frac{\text{indicated power in watt}}{m_f \times CV}$$

where m_f = mass of fuel supplied

CV = lower calorific value of fuel in kJ/kg

Brake thermal efficiency is the ratio of the heat converted into useful shaft work to the heat supplied to the engine in fuel.

Brake thermal efficiency,

$$\eta_{bth} = \frac{\text{brake power in watt}}{m_f \times CV}$$

From the expression for η_{ith} and η_{bth} as given above, we can write

$$\frac{\eta_{bth}}{\eta_{ith}} = \frac{BP}{IP} = \eta_m$$

$$\therefore \quad \eta_{bth} = \eta_m \times \eta_{ith}$$

The value of brake thermal efficiency usually varies from 25 – 30 % for spark ignition engines and 30 – 40 % for compression ignition engines.

Relative efficiency : The relative efficiency or efficiency ratio is defined as the ratio of indicated thermal efficiency to the thermal efficiency of a theoretical reversible cycle operating with the same compression ratio. The relative efficiency for various cycles usually ranges from 60 – 80 %.

Volumetric efficiency : The engine output depends upon the maximum amount of charge taken in during the suction stroke. Volumetric efficiency is essentially an indication of the breathing capacity of the engine and it is defined as the ratio of actual volume of the charge (reduced to NTP) drawn in during the suction stroke to the swept volume of the engine.

The average value of the volumetric efficiency of an engine ranges from 70 to 80 percent.

Specific output : The specific output of an engine is defined as the brake power per unit of piston displacement.

$$\text{specific output} = \frac{BP}{A \times L}$$

$$= \text{constant} \times p_{mb} \times \text{rotational speed}$$

Thus for a given displacement of the piston, the brake power can either be increased by increasing the speed of the engine or by increasing the brake mean effective pressure.

Specific fuel consumption : The fuel consumption and the evaluation of fuel cost is an important factor considered to have an indication of the relative economy of the engine under test when compared with test results on other engines. The criterion of economical power production is the specific fuel consumption (sfc) which is defined as *the mass of fuel required to be supplied to an engine to develop unit* kW *power per hour*. That is

$$sfc = \frac{m_f}{BP} \text{ kg/kWh}$$

The value of *sfc* is approximately

= 1.47 kg/kWh for diesel engines

= 1.85 kg/kWh for petrol engines

Air-fuel ratio : It is defined as the ratio of the mass of air to the mass of fuel in the air-fuel mixture. This ratio remains practically constant for petrol engines over a wide range of operation. However, for diesel engines the air-fuel ratio varies with load as the mass of air remains constant but the mass of fuel changes with load on the engine.

Specific weight : It is an indication of the engine bulk and is defined as the weight of engine in kg per each unit of brake power developed.

22.22. MORSE TEST

The Morse test is applicable to multi-cylinder engines and is used for determining their indicated power without any elaborate arrangement in the context of indicators employed to plot the pressure-volume diagram. The test consists in rendering inoperative, in turn, each cylinder of the engine by shorting the spark plug if petrol engine is under test or by disconnecting an injector in case of diesel engine. The speed falls because of the loss of power with one cylinder cut out, but is restored by reducing the load.

The procedure for conducting the morse test on a multi-cylinder petrol engine is outlined below :

(*i*) Start the engine with throttle fully open and with all cylinders working, take the dynamometer readings and note the engine speed.

(*ii*) Keeping the throttle opening fixed, render the first cylinder inoperative by shorting its spark plug. This will cause the engine speed to decrease. Restore the speed back to original value by adjusting the dynamometer load and note down the new dynamometer reading.

(*iii*) Make the first cylinder operative and cut out the second, third and fourth cylinder in turn.

(*iv*) Repeat the procedure taking care that speed remains constant and the throttle opening is undisturbed while the test is being conducted.

When all the four cylinders are working,

$$(BP)_4 = (IP)_4 - (FP)_4 \qquad ...(i)$$

When one cylinder is cut out, its contribution towards power development is lost and the speed of the engine falls. The brake load is reduced, so that the speed of the engine is restored to its original value. This is to main frictional power (*FP*) constant which is supposed to be independent of load and directly proportional to speed. Now only three cylinders will deliver power but still all the four cylinders will consume friction power. That is

$$(BP)_3 = (IP)_3 - (FP)_4 \qquad ...(ii)$$

From expressions (*i*) and (*ii*),

$$(BP)_4 - (BP)_3 = (IP)_4 - (IP)_3$$

Thus the difference between the measured brake power with four and three plugs will be the *IP* of the cylinder rendered inoperative. Similarly, *IP* of each cylinder can be found out separately. Addition of *IP* of different cylinders would then give the *IP* of the engine as a whole.

The following errors are associated in the determination of indicated power by the morse test :

(*i*) The cutting off a cylinder affects the distribution of charge to the cylinders and accordingly the breathing capacity (volumetric efficiency) of the engine becomes different.

(*ii*) The cutting off a cylinder causes pulsations in the exhaust manifold and that affects the performance of the engine.

(*iii*) The pumping and mechanical friction losses are assumed to be same whether a particular cylinder is working or not; and this is not essentially true as became evident from the test runs made by Recardo.

22.23. HEAT BALANCE SHEET

The engine is supplied with a certain quantity of energy in fuel form. Only a fraction of this energy is converted into useful mechanical work which is available at the engine crankshaft and the remainder is lost. It is the purpose of energy balance to trace the energy distribution, *i.e.*, to determine how the input energy is distributed, where the energy leaves and how much. If during the test it is revealed that at certain loading conditions, a certain form of energy loss is excessive, then a careful examination of that form is needed.

A heat balance test of an engine indicates the following items; each expressed in heat units.

Credit

Heat released by fuel $Q_f = m_f \times CV$ where m_f is the fuel consumption and CV is the lower calorific value of fuel

Debit

1. Heat equivalent of useful work or power available at the crankshaft (Q_b)

2. Heat loss to cooling water

$$Q_w = m_w \times c_w \times \Delta T$$

where m_w is mass of cooling water, c_w = specific heat of water and ΔT is the rise in temperature of cooling water

3. Heat carried by exhaust gases

$$Q_g = m_g \times c_g \times (t_g - t_a)$$

where m_g is the mass of exhaust gases, c_g is the mean specific heat of exhaust gases, t_g is temperature of exhaust gases and t_a is the temperature of ambient air.

The mass of exhaust gases equals the sum of air consumption of the engine and the fuel burnt in the same time.

4. Unaccounted and radiation loss

This loss cannot be measured directly and is calculated by subtracting the items Q_b, Q_w and Q_g from the item Q_f. That is

$$Q_{rad} = Q_f - (Q_b + Q_w + Q_g)$$

It may be pointed out that the heat equivalent of friction power is dissipated in cooling loss, exhaust loss and the radiation loss. Friction energy is a hidden part of these measurements and accordingly the separate inclusion of friction loss is omitted from the energy balance of internal combustion engines.

Heat in fuel is approximately distributed as follows :

1/3 in brake power produced, 1/3 in cooling water and 1/3 in exhaust gases.

EXAMPLE 22.1

A 4-cylinder, two stroke cycle petrol engine develops 30 kW at 2500 rev/min. The mean effective pressure on each piston side is 8 bar and mechanical efficiency is 80%. Calculate (a) the diameter and stroke of each cylinder if the stroke to bore ratio is 1.5 (b) the fuel consumption if the brake thermal efficiency is 28% and the fuel used has a calorific value 43900 kJ/kg.

Solution : Indicated power IP

$$= \frac{\text{brake power}}{\text{mechanical efficiency}} = \frac{30}{0.8} = 37.5 \text{ kW}$$

Indicated power per cylinder

$$= \frac{37.5}{4} = 9.375 \text{ kW}$$

The indicated power per cylinder is also given by

$$IP = 100\ \frac{p_m LAN}{60}$$

where p_m is the mep in bar, L and A are the stroke length in m and piston area in m^2, n is the number of working strokes per minute.

For a two stroke engine, $n = N$ (the engine speed in rev/min)

$$\therefore\ 9.375 = \frac{100 \times 8 \times 1.5\, D \times \frac{\pi}{4} D^2 \times 2500}{60}$$

$$(\because\ L = 1.5\ D)$$

Solution gives :

$$D^3 = 2.388 \times 10^{-4} \text{ or } D = 0.0638 \text{ m}$$

$\therefore$ D = **63.8** mm

and $L = 63.8 \times 1.5 =$ **95.7** mm

(*b*) Brake thermal efficiency $\eta_{bth} = \dfrac{BP}{m_f \times CV}$

That is : $0.28 = \dfrac{30}{m_f \times 43900}$

$$m_f = \frac{30}{0.28 \times 43900}$$

$= 2.44 \times 10^{-3}$ kg/s = 8.786 kg/hr

The brake specific fuel consumption is then

$$bsfc = \frac{m_f \text{ in kg/hr}}{BP \text{ in kW}}$$

$$= \frac{8.786}{30} = \mathbf{0.293} \text{ kg/kW/hr}$$

EXAMPLE 22.2

The following particulars refer to a 4-cylinder, 4-stroke petrol engine;

bore = 65 mm

storke length = 95 mm

torque developed = 64 Nm when engine turns 3000 rev/min

clearance volume = 63 cm³ for each cylinder

relative efficiency = 0.5 and calorific value of petrol = 42000 kJ/kg

Make calculations for brake mean effective pressure and the fuel consumption in kg/hr.

Solution : Swept volume per cylinder

$$V_s = \frac{\pi}{4} d^2 l$$

Compression ratio

$$r = \frac{V_s + V_c}{V_c} = \frac{315.08 + 63}{63} = 6$$

Air standard efficiency

$$\eta = 1 - \frac{1}{r^{r-1}} = 1 - \frac{1}{(6)^{1.4-1}} = 0.512$$

Then: Brake thermal efficiency = relative efficiency × air standard efficiency

$$\eta_{bth} = 0.5 \times 0.512 = 0.2562$$

Brake power

$$BP = 2\pi NT$$
$$= 2\pi \times 3000 \times 64$$
$$= 1205760 \text{ Nm/min}$$
$$= 20096 \text{ Nm/s} \simeq 21.1 \text{ kW}$$

$$\text{Brake power per cylinder} = \frac{21.1}{4} = 5.025 \text{ kW}$$

The brake power per cylinder is also given as

$$BP/\text{cylinder} = \frac{100 p_m L\, A\, n}{60}$$

where p_m is mep in bar, L and A are the stroke length (m) piston area (m²) respectively, n is the number of working strokes per minute.

For a 4-stroke engine $n = \frac{N}{2}$ (N is the engine speed in rev/min)

$$\therefore\ 5.025 = \frac{100\, p_m \times 0.095 \times \frac{\pi}{4}(0.065)^2 \times \frac{3000}{2}}{60}$$

Solution gives : Brake mean effective pressure p_m = 6.38 bar

(*b*) Brake thermal efficiency

$$\eta_{bth} = \frac{BP}{m_f \times CV}$$

That is :

$$0.2562 = \frac{21.1}{m_f \times 42000};$$

$$m_f = \frac{21.1}{0.2562 \times 42000}$$

$= 1.96 \times 10^{-3}$ kg/s = **7.06** kg/hr

EXAMPLE 22.3

The following data was recorded during a test run made on a single cylinder, four stroke engine having a compression ratio of 6.

Bore and stroke = 10 cm and 12.5 cm respectively

Dead load and spring balance reading = 60 N and 20 N

Effective radius of brake drum = 40 cm

Fuel consumption = 1.2 kg/hr

Calorific value of fuel = 42500 kJ/kg

If the engine turns 2000 rev/min and the indicated mean effective pressure is 0.25 MPa, determine its

(a) indicated power and brake power

(b) mechanical, overall and relative efficiencies.

Solution : For a four stroke engine, the indicated power is given by

$$IP = p_{mi}\, LA\left(\frac{N}{2}\right)$$

$$= \frac{(0.25 \times 10^6) \times 0.125 \times \frac{\pi}{4}(0.1)^2 \times \frac{2000}{2}}{60 \times 10^3}$$

$= 4.088$ kW

Brake power

$$BP = 2\pi NT$$

$$= \frac{[2\pi \times 2000] \times [(60-20) \times 0.4]}{60 \times 10^3}$$

$= \mathbf{3.349}$ kW

(*b*) Mechanical efficiency

$$\eta_m = \frac{BP}{IP} = \frac{3.349}{4.088} = \mathbf{0.879} \text{ or } \mathbf{81.9}\ \%$$

Overall efficiency

$$\eta_0 = \frac{\text{heat equivalent of brake power}}{\text{heat supplied in fuel}}$$

$$= \frac{BP}{m_f \times CV} = \frac{3.349 \times 10^3}{\left(\frac{1.2}{3600}\right) \times 42500}$$

$= \mathbf{0.2364}$ or $\mathbf{23.64}$ %

Air standard efficiency,

$$\eta_{air} = 1 - \frac{1}{r^{\gamma-1}}$$

$$= 1 - \frac{1}{(6)^{1.4-1}} = 0.5116$$

∴ Relative efficiency,

$$\eta_r = \frac{\eta_0}{\eta_{air}} = \frac{0.2364}{0.5116} = \mathbf{0.462} \text{ or } \mathbf{46.2}\ \%$$

EXAMPLE 22.4

A four cylinder, four-stroke car engine with bore 80 mm and stroke length 100 mm has engine to rear axle ratio of 5 : 1. The car moves at a speed of 60 km/hr, travels a distance of 4 km and the fuel consumption during this period is 0.3225 kg. When the tyres are fully inflated, the wheel diameter is 650 mm and the mean effective pressure as estimated from the indicator diagram is 5.5 bar. If the fuel used has a calorific value of 43500 kJ/kg, determine the indicated power and thermal efficiency.

Solution : Speed of the car

$$V = 60 \text{ km/hr} = \frac{60 \times 1000}{60}$$

$= 1000$ m/min

Then revolutions made by the wheel (tyre rim),

$$N_i = \frac{V}{\pi_1 D} = \frac{1000}{\pi_1 \times 0.65}$$

$= 489.95$ rev/min

As the engine to rear axle ratio is 5 : 1, the speed of engine shaft is

$$N = 489.95 \times 5 = 2449.75 \text{ rev/min}$$

Indicated power *IP*

$$= p_{mi} LA \frac{N}{2} \text{ for a four stroke cycle}$$

$$= \frac{5.5 \times 10^5 \times 0.1 \times \frac{\pi}{4}(0.08)^2 \times 2449.75/2}{60 \times 10^3}$$

$= \mathbf{22.56}$ kW

(*b*) Speed of the car = 60 km/hr

Time taken by the car to cover a distance of 4 km $= \frac{4}{60}$ hr

Fuel consumed in $\frac{4}{60}$ hr $= 0.3225$ kg

That gives : Fuel consumption per hour

$$= 0.3225 \times \frac{60}{4} = 4.8375 \text{ kg}$$

Indicated thermal efficiency,

$$\eta_{ith} = \frac{IP}{m_f \times CV} = \frac{22.56}{\left(\frac{4.8375}{3600}\right) \times 43500}$$

$= \mathbf{0.385}$ or $\mathbf{38.5}$ %

EXAMPLE 22.5

A four stroke petrol engine having six cylinders is to operate with compression ratio 6 and deliver 300 kW brake power when running at 2400 rev/min. Determine :

(a) bore and stroke of the engine,
(b) fuel consumption in kg/hr, and
(c) specific fuel consumption in kg/kWh

The following data may be assumed :
stroke = 1.25 times the bore;
mechanical efficiency = 0.8
indicated mean effective pressure = 10 bar,
relative efficiency = 0.5, and

calorific value of fuel used = 44 MJ/kg

Solution : Indicated power

$$= \frac{\text{brake power}}{\text{mechanical efficiency}}$$

$$= \frac{300}{0.8} = 375 \text{ kW}$$

Indicated power per cylinder

$$= \frac{375}{6} = 62.5 \text{ kW}$$

In terms of indicated mean effective pressure p_{mi}, the indicated power per cylinder is given by

$$IP = p_{mi}\ LA\frac{N}{2} \text{ for a four-stroke engine}$$

$$\therefore\ 62.5 = \frac{10\times10^5\times1.25\,D\times\frac{\pi}{4}D^2\times2400/2}{60\times10^3}$$

$$= 19625\ D^3$$

That gives :

$$D^3 = \frac{62.5}{19625} = 0.003185$$

or $D = (0.003185)^{1/3}$

$= 0.1474$ m $= 14.74$ cm

and $L = 1.25 \times 14.74 = 18.425$ cm

Thus cylinder bore = **14.74** cm and stroke = **18.425** cm

(*b*) Air standard efficiency

$$\eta_{air} = 1-\frac{1}{r^{\gamma-1}} = 1-\frac{1}{6^{1.4-1}} = 0.512$$

On the basis of indicated thermal efficiency,

$$\eta_r = \frac{\eta_{ith}}{\eta_{air}}$$

$$\therefore\ \eta_{ith} = 0.5 \times 0.512 = 0.256$$

Using the relation :

$$\eta_{ith} = \frac{IP}{m_f \times CV}, \text{ we have}$$

$$10.256 = \frac{375\times10^3}{\left(\frac{m_f}{3600}\right)(44\times10^6)}$$

$\therefore$ Fuel consumed/hr,

$$m_f = \frac{375\times10^3\times3600}{0.256\times44\times10^6}$$

$= \mathbf{119.85}$ kg/hr

(*c*) Specific fuel consumption

$$sfc = \frac{\text{fuel consumed / hr}}{\text{brake power}}$$

$$= \frac{119.85}{300} \simeq \mathbf{0.4} \text{ kg/kWh}$$

EXAMPLE 22.6

A 4-stroke six cylinder engine has a bore of 80 mm and stroke of 100 mm. While running at a mean speed of 12.5 m/s, its fuel consumption is 20 kg/hr and develops a torque of 150 Nm. Assuming a clearance volume of 75 cm³ per cylinder, determine,

(a) brake power and brake mean effective pressure

(b) brake thermal efficiency if the calorific value of fuel used is 42.5 MJ/kg

(c) relative efficiency on the basis of brake thermal efficiency.

Solution : Mean piston speed $S = 2\ LN$

$$\therefore\ 12.5 = 2\times0.1\times\frac{N}{60};\ N = 3750 \text{ rev/min}$$

(*a*) Brake power,

$$BP = 2\pi NT$$

$$= \frac{2\pi\times3750\times150}{60\times10^3} = \mathbf{58.875} \text{ kW}$$

In terms of brake mean effective pressure (p_{mb}), the brake power for an engine working an four-stroke cycle is

$$BP = p_{mb}\ LA\left(\frac{N}{2}\right) \times \text{number of cylinders}$$

$$58.875 = \frac{p_{mb}\times0.1\times\frac{\pi}{4}(0.08)^2\times\frac{3750}{2}}{60\times10^3}\times6$$

Solution gives :

$p_{mb} = 625000$ N/m² $= \mathbf{6.25}$ bar

(*b*) Brake thermal efficiency,

$$\eta_{bth} = \frac{\text{heat equivalent of brake power}}{\text{heat supplied in fuel}}$$

$$= \frac{\text{brake power}}{m_f \times CV}$$

$$= \frac{58.875 \times 10^3}{\left(\frac{20}{3600}\right) \times (42.5 \times 10^6)} = \mathbf{0.249} \text{ or } \mathbf{24.9}\ \%$$

(*c*) Swept volume

$$V_S = \frac{\pi}{4} d^2 l = \frac{\pi}{4} (8)^2 \times 10 = 502.4 \text{ cm}^2$$

Clearance volume, $V_C = 75$ cm^3

Compression ratio,

$$r = \frac{V_S + V_C}{V_C} = \frac{502.4 + 75}{75} \simeq 7.7$$

Air standard efficiency,

$$\eta_{air} = 1 - \frac{1}{r^{\gamma - 1}} = 1 - \frac{1}{(7.7)^{1.4-1}}$$

$$= 0.558$$

$$\therefore \quad \eta_r = \frac{\eta_{bth}}{\eta_{air}} = \frac{0.249}{0.558}$$

$$= \mathbf{0.446} \text{ or } \mathbf{44.6\%}$$

EXAMPLE 22.7

Following data was obtained during the trial of a two cylinder, two-stroke engine :

Bore and stroke = 100 mm and 150 mm

Area of positive and negative loops of indicator diagram = 6 cm^2 and 0.25 cm^2

Length of the indicator = 6 cm

Spring constant = 3.8 bar per cm

Net brake load and effective drum radius = 235 N and 0.42 m

Fuel consumption = 4.5 kg/hr

If the engine turns 1600 rev/min and the fuel used has a calorific value of 43.5 MJ/kg, determine;

(a) indicated power and brake power

(b) mechanical and thermal efficiencies.

Solution : Net area of indicator diagram

$$= 6 - 0.25 = 5.75 \text{ cm}^2$$

Indicated mean effective pressure,

$$p_{mi} = \frac{\text{area of indicator diagram}}{\text{length of indicator diagram}} \times \text{spring constant}$$

$$= \frac{5.75}{6} \times 3.8 = 3.642 \text{ bar}$$

For a two-stroke engine, the indicated power is given by

$$= (p_{mi}\ LAN) \times \text{number of cylinders}$$

$$= \frac{3.642 \times 10^5 \times 0.15 \times \frac{\pi}{4} (0.1)^2 \times 1600}{60 \times 10^3} \times 2$$

$$= \mathbf{22.87} \text{ kW}$$

Brake power, $BP = 2\pi NT$

$$= \frac{2\pi \times 1600 \times (235 \times 0.42)}{60 \times 10^3} = \mathbf{16.43} \text{ kW}$$

(*b*) Mechanical efficiency,

$$\eta_m = \frac{BP}{IP} = \frac{16.43}{22.87} = \mathbf{0.722} \text{ or } \mathbf{72.2}\ \%$$

Indicated thermal efficiency,

$$\eta_{ith} = \frac{IP}{m_f \times CV}$$

$$= \frac{22.87 \times 10^3}{\left(\frac{4.5}{3600}\right) \times (43.5 \times 10^6)} = \mathbf{0.42} \text{ or } \mathbf{42}\ \%$$

Brake thermal efficiency,

$$\eta_{bth} = \frac{BP}{m_f \times CV}$$

$$= \frac{16.52}{\left(\frac{4.5}{3600}\right) \times (4.35 \times 10^6)}$$

$$= 0.303 \text{ or } 30.3\ \%$$

η_{bth} could also be worked out by using the relation

$$\eta_{bth} = \eta_m \times \eta_{ith}$$

$$= 0.722 \times 0.42 = 0.303 \text{ or } 30.3\ \%$$

EXAMPLE 22.8

A four-cylinder four-stroke petrol engine with bore 60 mm and stroke 90 mm was tested at full throttle at constant speed. The fuel supply was fixed at 0.082 kg/min and the spark plugs of the four cylinders were successively short circuited without change of speed; the brake torque being correspondingly adjusted. The brake power was measured with a dynamometer having a torque arm of 35 cm and the measurements of brake loads at rated speed of 3000 rev/min were as follows :

With all cylinders working -- 160 N
With cylinder No. 1 cut out -- 113.5 N
With cylinder No. 2 cut out -- 109 N
With cylinder No. 3 cut out -- 106.5 N
With cylinder No. 4 cut out -- 120 N

Taking calorific value of fuel used as 44 MJ/kg, determine the engine torque; brake mean effective pressure and bake thermal efficiency; specific fuel consumption; mechanical efficiency and indicated mean effective pressure.

(b) If the engine was tested in an atmosphere of 1 bar and 15°C and the air-fuel ratio was 15 : 1, determine the volumetric efficiency of the engine.

Solution : Torque $T = W \times R = 160 \times 0.35$

$= \mathbf{56}$ Nm

Brake power

$$BP = 2\pi NT = \frac{2\pi \times 3000 \times 56}{60 \times 10^3}$$

$= 17.584$ kW

BP per cylinder

$$= \frac{17.584}{4} = 4.396 \text{ kW}$$

In terms of brake mean effective pressure p_{mb}, the brake power per cylinder can be expressed as

$$BP = p_{mb}\, LA \frac{N}{2}$$

$$\therefore\ 4.396 = \frac{p_{mb} \times 0.09 \times \frac{\pi}{4}(0.06)^2 \times \frac{3000}{2}}{60 \times 10^3}$$

$= 6.36 \times 10^{-6}\, p_{mb}$

That gives :

$$p_{mb} = \frac{4.396}{6.36 \times 10^{-6}}$$

$= 6.91 \times 10^5$ N/m^2 = **6.91** bar

Brake thermal efficiency,

$$\eta_{bth} = \frac{\text{brake power}}{m_f \times CV}$$

$$= \frac{17.584 \times 10^3}{\left(\frac{0.082}{60}\right) \times (44 \times 10^6)}$$

$= \mathbf{0.292}$ or **29.2** %

Specific fuel consumption,

$$sfc = \frac{\text{fuel consumption per hr}}{\text{brake power}}$$

$$= \frac{0.082 \times 60}{17.584} = \mathbf{0.279} \text{ kg/kWh}$$

Indicated load = (160 − 113.5) + (160 − 109) + (160 − 106.5) + (160 − 120)

= 46.5 + 51 + 53.5 + 40 = 191 N

Mechanical efficiency,

$$\eta_m = \frac{\text{brake load}}{\text{indicated load}}$$

$$= \frac{160}{191} = \mathbf{0.838} \text{ or } \mathbf{83.8}\ \%$$

Indicated mean effective pressure,

$$p_{mi} = \frac{p_{mb}}{\eta_m} = \frac{6.91}{0.838} = \mathbf{8.246} \text{ bar}$$

(*b*) Mass flow rate of air

= 0.082 × 15 = 1.23 kg/m

Then from characteristic equation $pV = mRT$,

Volume of air drawn,

$$V = \frac{1.23 \times 287 \times (273 + 15)}{1 \times 10^5}$$

$= 1.017$ m^3/min

Swept volume of the engine,

$$V_S = \frac{\pi}{4} \times (0.06)^2 \times 0.09 \times \frac{3000}{2}$$

$= 1.526$ m^3/min

∴ Volumetric efficiency,

$$\eta_v = \frac{1.017}{1.526} = \mathbf{0.666} \text{ or } \mathbf{66.6}\%$$

EXAMPLE 22.9

In a test on a single cylinder oil engine with bore 30 cm, stroke 45 cm and working on the four stroke cycle, the following data were obtained :

Net brake load and effective drum diameter
= 1650 N and 150 cm

Engine speed and duration of test
= 200 rpm and 1 hour

Fuel consumption = 7.5 kg with calorific value 42 MJ/kg

Inlet and outlet temperatures of cooling water
= 15°C and 60°C

Total mass of jacket cooling water = 450 kg

Total air consumption = 350 kg
Inlet air and exhaust gas temperatures
= 20°C and 300°C
Mean specific heat of exhaust gases
= 1.025 kJ/kg K

If the indicator diagram gives a mean effective pressure equal to 6 bar, calculate the indicated and brake power, the mechanical efficiency, the thermal efficiencies and prepare a heat balance chart for the engine trial.

Solution : Heat supplied in fuel

$$Q = \frac{7.5}{3600} \times (42 \times 10^3) = 88.67 \text{ kJ/s}$$

(*a*) Indicated power,

$$IP = p_{mi} \, LA \frac{N}{2} \times \text{number of cylinders}$$

$$= \frac{6 \times 10^5 \times 0.45 \times \frac{\pi}{4}(0.3)^2 \times 200/2}{60 \times 10^3} \times 1 = 31.79 \text{ kW}$$

Brake power,

$$BP = 2\pi NT$$

$$= \frac{2\pi \times 200 \times (1650 \times 0.75)}{60 \times 10^3} = 25.9$$

(*b*) Mechanical efficiency,

$$\eta_m = \frac{BP}{IP} = \frac{25.9}{31.79} = 0.815$$

Indicated thermal efficiency,

$$\eta_{ith} = \frac{31.79 \times 10^3}{\left(\frac{7.5}{3600}\right) \times (42 \times 10^5)} = \mathbf{0.36}$$

Brake thermal efficiency

$$\eta_{bth} = \frac{25.9 \times 10^3}{\left(\frac{7.5}{3600}\right) \times (42 \times 10^6)} = \mathbf{0.296}$$

(*c*) 1. Heat equivalent of brake power,

$$Q_b = 27.63 \text{ kJ/s}$$

2. Heat lost to cooling water,

$$Q_w = m \, c_p \, dT$$

$$= \frac{450}{3600} \times 4.186 \times (60 - 15) = 23.55 \text{ kJ/s}$$

3. Heat lost in exhaust gases

$$Q_g = m \, c_p \, dT$$

$$= \frac{350 + 7.5}{3600} \times 1.025 \times (300 - 20)$$

$$= 28.5 \text{kJ/s}$$

4. Heat lost by radiation and unaccounted for (by difference)

$$Q_{rad} = 88.67 - (25.9 + 23.55 + 28.50)$$

$$= 10.72 \text{ kJ/s}$$

The heat balance account on second basis may then be tabulated as indicated below:

Credit	kJ/s	percent	Debit	kJ/s	percent
Heat supplied in fuel	88.67	100	Brake power	25.90	29.21
			Cooling water	23.55	26.56
			Exhaust gases	28.50	32.14
			Radiation and unaccounted for	10.72	12.09
	88.67	100		88.67	100

22.24. PERFORMANCE CHARACTERISTICS

Towards closure, it will appropriate to draw the performance curve for the IC engines and comment on their general nature.

A. The curves indicated in Fig. 22.43 pertain to a spark ignition engine operating at full throttle and variable load. The difference parameters have been plotted against speed throughout the operating range of the engine. Such tests are conducted at constant throttle setting in the petrol engines, and at constant full pump setting in diesel engines.

The salient aspects of these plots are:

(*i*) At a certain speed in the operating range, the engine sucks maximum quantity of charge per cylinder per cycle. This occurs when there is sufficient time for the cylinder to get filled up with charge. Greater pressures are developed and the torque is high. The torque developed has the maximum value when the air charge per cycle is greatest. At high speeds the quantity of charge decreases (there is fall in volumetric efficiency), there is less push on the piston and the engine develops low torque.

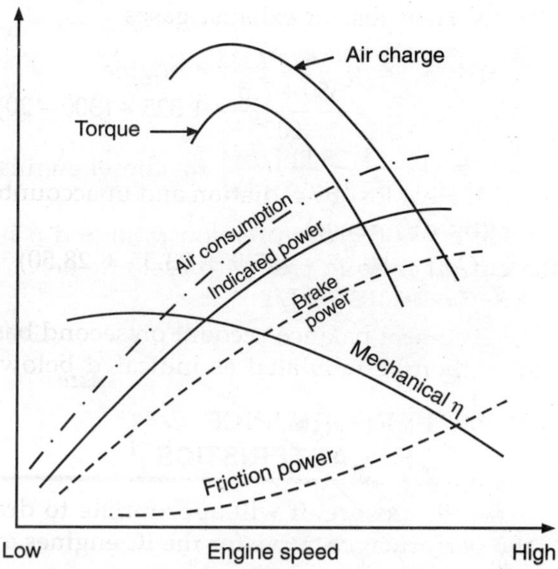

Fig. 22.43. Performance curves for an SI engine

(*ii*) There is progressive increase in air consumption (quantity of air sucked by the engine per unit time) with increase in engine speed. The may be attributed to the fact that the effect of speed is more predominant in the relation.

Air consumption
$$= \text{air induced/cycle} \times \text{speed}$$

However, after a certain speed, the fall in air charge per cycle is more rapid than an increase in the number of cycles per unit time. Obviously then the air consumption curve also starts falling.

Since the indicated power produced in the cylinder is directly proportional to air consumption, the curve for indicated power is identical with that for air consumption. Further, the brake power is lower than indicated power and the difference between these curves at any speed gives the friction power.

The friction power increases approximately at a rate proportional to the square of speed.

(*iii*) The maximum values of brake power and indicated power occur at different speeds; the brake power reaching the maximum value at a lower speed. This is due to the fact that:

- fall in indicated power after a maximum value is due to drop in volumetric efficiency, and

- reduction in brake power is both due to reduced volumetric efficiency and increase in friction power at higher speeds.

This is also the reason for drop in mechanical efficiency at higher speeds.

(*iv*) Refer to Fig. 22.44 for the plot of fuel consumption and brake specific fuel consumption (bsfc) against speed.

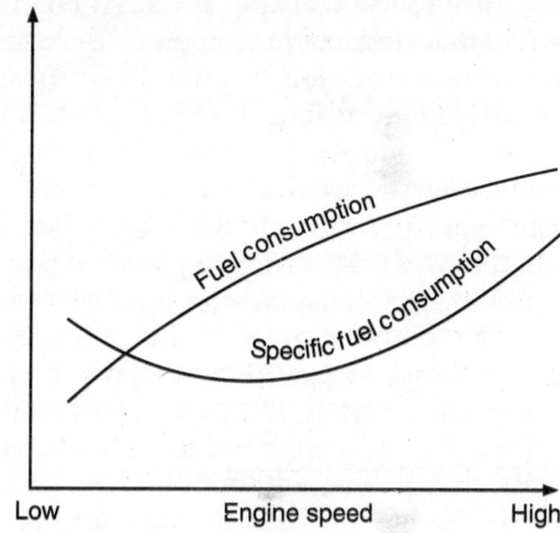

Fig. 22.44. Variation of fuel consumption and bsfc with speed

It may be noted that:

- the quantity of fuel consumed per unit time increases with speed,

- the *bsfc* drops as engine speed increases in the low speed range, reaches a minimum value and then starts increasing in the high speed range. The speed where *bsfc* is minimum is called the optimum engine speed.

At low speeds, the engine has a poor combustion efficiency and there is proportionately more heat loss to the chamber, and apparently higher fuel consumption for the power output. At optimum speed, the engine has a fairly high volumetric efficiency and low mechanical losses. Consequently, there is less fuel consumption per brake power. There occurs a great fall in volumetric efficiency and rapid increase in mechanical losses at very high speeds. The resulting small increase in brake power cannot keep pace with increase in fuel consumption. Consequently the fuel consumption per brake power increases at higher speeds.

Willan's line : We can test given engine at constant speed and several loads. For a good diesel engine, the fuel consumption versus indicated power is a straight line passing through the origin.

The mechanical losses depend upon engine speed, and maximum pressure and minimum temperature of the working fluid. At change of load, both pressure and temperature change. However, this variation is quite insignificant for a diesel engine as long as the engine works on the same compression ratio. As such the mechanical losses are independent of load at constant speed. As the brake power equals indicated power minus friction power, a graph is drawn of fuel consumption against brake power. This graph is called Willan's line and is extended back to cut the brake power axis. This helps to estimate the power loss of the engine at a particular engine speed, and thereby the indicated power without the use of an indicator. Such a test run at constant speed and variable load enables us to draw the graphs showing the variation of mechanical efficiency, specific fuel consumption and brake or indicated thermal efficiency against brake speed.

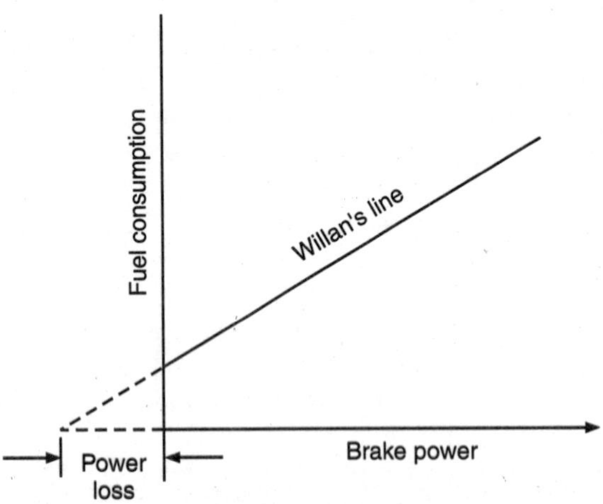

Fig. 22.45. Willan's line

Effect of compression ratio : Engine output and thermal efficiency generally increase with increase in compression ratio as is apparent from the following expressions for air standard efficiency.

$$\eta = 1 - \frac{1}{\gamma^{\gamma-1}} \quad \text{for petrol engine}$$

$$= 1 - \frac{1}{\gamma^{\gamma-1}}\left[\frac{\rho^{\gamma}-1}{\gamma(\rho-1)}\right] \quad \text{for diesel engine}$$

Here γ is the compression ratio and ρ is the cut off ratio in case of diesel engine.

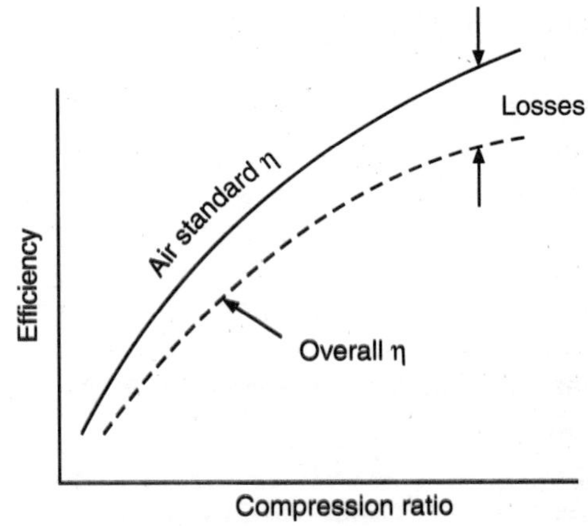

Fig. 22.46.

However, too high a value of compression ratio results into an enormous increase in cylinder pressure. The engine cylinder has to be made strong enough to resist the high pressure and consequently the increased structural weight of the engine offsets the advantage of increased efficiency. Further, the combustion process takes place in an abnormal manner at high compression ratio. There is almost an instantaneous pressure rise which results in excessive noise and engine detonation. If detonation becomes severe, the engine overheats and the output starts declining. The value of compression ratio at which the engine power starts decreasing is called the highest useful compression ratio (HUCR). Further, the gain in efficiency at high values of compression ratio follows the law of diminishing returns. An increase in compression ratio from 5 to 9 results in 29% increase in efficiency. Subsequent increase in compression ratio from 9 to 13 yields a gain of only 9 percent.

Also, the following losses need to be taken into account while dealing with actual (overall) efficiency.

- heat loss to exhaust gases
- incomplete combustion
- heat loss through water jackets and finned surfaces
- friction losses and power spent in driving the auxiliaries.

The friction and cooling losses go on increasing with compression ratio. Apparently the actual efficiency will be lower than theoretical and the difference between the two will go on increasing with compression ratio.

The ability to operate the engine with high compression ratio depends on the provision of better quality of fuel and improved of better quality of fuel and improved design of combustion chamber. The average compression ratio for petrol engine is 5 to 9, and the diesel engines work on compression ratio in the range 14 to 20.

Effect of air-fuel ratio : The quantity of air supplied is an important parameter that aids in the proper combustion of fuel and accordingly affects the engine performance. The petrol engines generally operate with air-fuel ratio of 10 to 20. Refer Fig. which shows the typical variation of specific fuel consumption (sfc) against brake mean effective pressure (bmep) for a spark ignition engine.

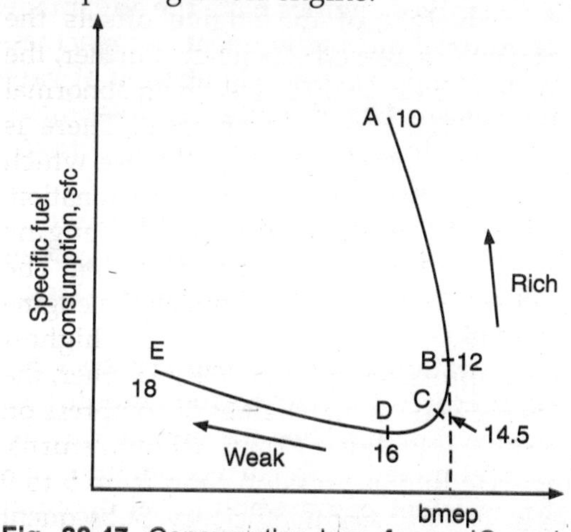

Fig. 22.47. Consumption loop for an IC engine

The following information can be gleaned from this consumption loop or hook curve:

(*i*) The mixture is richest at point A where the air-fuel ratio has a minimum value of 10. With increase in air-fuel ratio the *bmep* increases and its maximum value pertains to point *B* which corresponds to A/F ratio of 13. During the span *A* to *B*, the *sfc* also drops.

(*ii*) With further increase in A/F ratio, the *bmep* starts decreasing. The engine however operates economically (*i.e*, with less *sfc*) and maximum economy is reached when the A/F ratio is about 16. The point *C* lying between the point of maximum *bmef* (*B*) and maximum economy(*D*) corresponds to the condition when the air supplied is just sufficient for chemically correct combustion. For most hydrocarbon fuels, the stoichiometric A/F ratio is 14.5.

(*iii*) Increasing the A/F ratio beyond point *D*, makes the mixture too lean. The operating of the engine is then adversely affected both with regard to *bmep* and *sfc*.

(*iv*) At the two extreme points *A* and *E*, the performance of the engine is not satisfactory. For very rich mixture (*A*), the engine would run unsteadily with great possibility of combustion continuing in the exhaust system. Again at the point *E* where the mixture is weakest (very lean) the engine will have unsteady run. The combustion process will be so slow that the gases may continue burning in the clearance space till the beginning of next induction stroke.

Refer Fig. 22.48 which shows the variation of *bsfc* and *bmep* plotted on a base representing A/F ratio for a petrol engine.

An examination of this plot reveals that

(*i*) The maximum *bmep* (engine output) is obtained when the A/F ratio is 12:1. This stems from the fact that with rich mixture the combustion is incomplete which means presence of CO in the burnt gases. The presence of CO suppresses dissociation and this leads to high temperatures and pressures during the expansion stroke. Consequently, the engine develops more power. However, with very rich mixtures, the combustion will be excessively incomplete resulting into loss in power output. Also when the engine is made to operate with very weak mixtures, the energy released will be less and the power produced will be low.

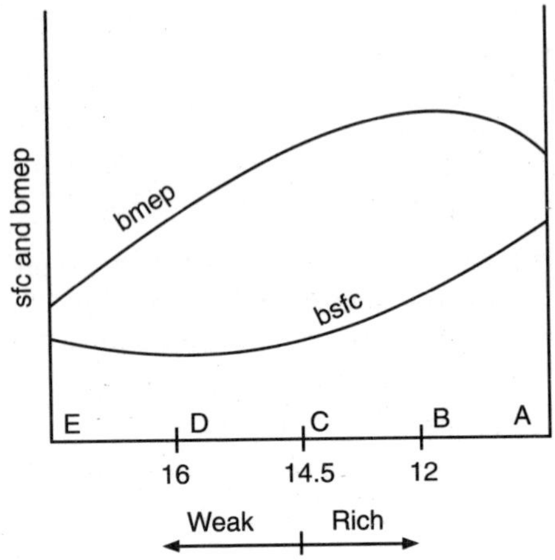

Fig. 22.48.

(*ii*) The *bsfc* is minimum and maximum fuel economy is obtained when the A/F ratio is about 17:1. The is due to the fact that with excess air, every particle of fuel is in contact with air and this ensures proper and full combustion. Further with lean mixtures, the maximum temperatures produced are low and consequently heat loss to the cylinder walls is minimum. A high value of *bsfc* with very lean mixtures is due to low power developed and increased losses. Also when the mixtures are two rich, the *bsfc* would be high due to low power developed.

Refer Fig. 22.49 which shows the typical consumption loop for a compression ignition (CI) engine.

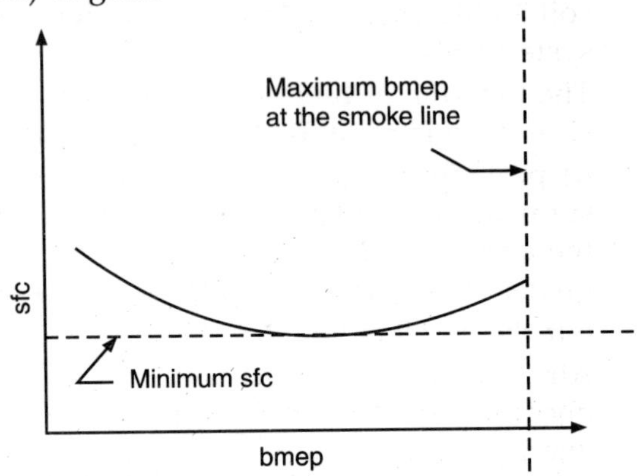

Fig. 22.49. Consumption loop for CI engine

The curve is reasonably flat which means low *sfc* over a large range of *bmep* (load). This aspect makes the CI engines more suitable for road vehicles.

The limit to the maximum *bmep* is indicated by the appearance of maximum smoke in the exhaust and this corresponds to an A/F ratio of about 16:1. The diesel engines tend to be dirty when operated with rich mixtures.

SALIENT POINTS

1. A heat engine is a device which converts the chemical energy of the fuel into mechanical work.
2. IC engines are classified as spark ignition and compression ignition engines.
3. Four stroke engines have suction, compression, expansion or power and the exhaust strokes completed in two revolutions of the cranskshaft.
4. In two stroke engines, one cycle is completed in one revolution of crankshaft by eliminating the suction and exhaust strokes. These processes are carried out simultaneously during compression and expansion strokes.
5. Two strokes engines have the advantages of uniform turning moment, compactness, twice the power and elimination of valve mechanism. However, they are less efficient due to loss of fresh charge during scavenging process.
6. Valve timing diagrams represent the opening and closing of inlet and exhaust valves (or ports) in the actual working of IC engines.
7. Carburation is the process of preparation of mixture of atomised fuel and air. The device that supplies the metred quantity of fuel mixed with correct amount of air for efficient combustion at all operating conditions is called carburettor.

Towards the end of compression stroke, the combustible mixture is ignited by a spark plug which receives a high voltage (10,000 to 12,000 volts) from an ignition system and produces a spark across the gap between its electrodes.

The main ignition systems are the battery coil ignition system and magneto ignition system.

8. The function of fuel injection system in a CI engine is to :

   (*i*) provide atomised fuel in the form of spray according to the load and speed of the engine, and

   (*ii*) distribute the charge uniformly through out the combustion chamber

   Air injection and the solid or airless injection are the two methods of fuel injection in CI engines.

9. The basic cooling systems provided for IC engines are the air cooling and water cooling.

   In the air cooling method, the heat is carried away by forcing the air around the engine cylinder. The heat transfer rate is increased by providing fins on the cylinder head and cylinder barrel. Air cooling is used in scooters, motor cycles and small capacity aircraft engines.

   For medium and high capacity engines, cooling is done by themo-syphon water cooling and pressurised water cooling. The thermo-syphon arrangement has water jackets surrounding the cylinder and connected to the top and bottom of radiator tubes. The circulation of water through the system is accomplished by convection currents which are set up due to density difference of hot and cold waters. Air flow over the radiator takes place due to vehicle motion or is provided by a fan.

   In pressurised water cooling system, a pump is employed which speeds up the flow of water and that

   (*i*) enables a smaller radiator to be fitted,

   (*ii*) permits cool water to be directed towards the hotter parts of the engine such as exhaust valve seatings.

10. The lubrication system needs to be provided on IC engines to reduce the friction (by introducing the substance called lubricant) between the mating parts.

    Lubrication systems may be of spalsh or of pressure (wet sump or dry sump) type. The lubrication of two stroke engines fitted with motor cycle and scooters is accomplished by mixing 3 to 6% of lubricating oil with petrol; the system is called petrol lubrication system.

11. Supercharging is the process where by charge at pressure and density higher than atmospheric is supplied to engine cylinder during suction stroke.

    Supercharging is used to :

    (*i*) increase the power output from an engine of given capacity,

    (*ii*) compensate for the loss of power in aircraft applications at high altitudes where atmosphere is rarified.

12. Governing implies keeping the engine speed constant regardless of the changes in load on the engine.

    The methods commonly used for governing IC engines are quantity governing, quality governing and, the hit and miss governing.

13. The combustion stages in SI engies are : ignition lag, flame propagation and after burning.

    (*i*) Ignition lag or delay period refers to the time that elapses between the act of actual combustion and the provision of spark. The number of degrees before *tdc* position at which spark is provided is called the angle of advance.

    The delay period is reduced with high intake pressure and temperature, increase in compression ratio, slightly rich mixtures, lesser residual gases, and low self ignition temperature.

    (*ii*) The flame refers to the gas rendered luminous by liberation of chemical energy.

    An increase in the following engine variables increases the speed of flame propagation :

Compression ratio, intake pressure and temperature, engine speed turbulence, load on engine and about 10% rich mixture.

(*iii*) Detonation refers to the pinging sound due to gas vibrations caused by abnormal combustion resulting from auto-ignition of the end charge of spark ignition engines. Detonation occurs when the end charge reaches its self ignition temperature without flame front reaching it.

(*iv*) The factors that promote detonation are : advanced spark time, increase in compression ratio, increase in intake pressure and temperature, increase in coolant temperature and load on the engine.

(*v*) The percentage of octane in a mixture of iso-octane and *n*-heptane by volume which gives the same detonating results as given by the petrol fuel under test is called the octane rating of the fuel. Higher the octane number, better is the anti-detonation quality of fuel.

(*vi*) Pre-ignition is the premature combustion which starts before the application of spark. Detonation follows the spark while pre-ignition precedes it.

14. The stages of combustion in compression ignition engine are : delay period, uncontrolled combustion, controlled combustion and after burning.

(*i*) Knocking refers to the thudding sound due to direct blow on engine parts caused by sudden pressure rise due to abnormal combustion of initial part of the charge in compression ignition engine.

(*ii*) Factors which tend to increase detonation tendency in SI engines, will reduce the knocking tendency in CI engines.

(*iii*) The percentage of cetane in a mixture of cetane and α-methyl naphthalene by volume which knocks under the same conditions as diesel fuel is called the cetane number of fuel. High the cetane number, better is the anti-knock quality of fuel.

(*iv*) Knocking and detonation lead to local melting of piston and rings, mechanical failure of engine parts, loss of power and engine efficiency.

(*v*) Dopes are the compounds that when added in small quantities have pronounced effect on anti-knock characteristics of the engine.

For SI engines, these compounds are tetra ethyl lead and tetra methyl lead.

Amyl nitrate and amyl nitrite are the suitable dopes/additives for CI engines.

15. Engine testing is done to verify the engine performance as per design specifications and justify the engine rating.

(*i*) Indicated power (*IP*) of the engine is a measure of the power developed within the engine cylinder.

(*ii*) Brake power represents the useful power available at the engine crankshaft.

(*iii*) Mechanical efficiency, $\eta_m = \dfrac{BP}{IP}$

$$\text{Thermal efficiency (indicated)} = \frac{IP}{m_f \times CV}$$

$$\text{Thermal efficiency (brake)} = \frac{BP}{m_f \times CV}$$

where power (*IP* or *BP*) is in kW, m_f is mass of fuel in kg/s and *CV* is calorific value of fuel in kJ/kg.

Relative efficiency or efficiency ratio

$$= \frac{\text{thermal efficiency}}{\text{air standard efficiency}}$$

Specific fuel consumption *sfc*

$$= \frac{m_f}{BP} \text{ kg/kWh}$$

where m_f is the fuel consumed per hour. In general *sfc* for diesel engines is about 0.27 kg/kWh and is about 0.3 kg/kWh for petrol engines.

(*iv*) Heat balance sheet represents the account of energy supplied in fuel and its corresponding expenditure. The energy balance should include brake power and not indicated power. Friction power is not to form a separate item of energy balance because part of its value reappears in cooling water and exhaust gases.

16. $IP = \frac{p_m LA N}{60 \times 10^3}$ kW

Here N stands for working cycles per minute which are half the rev/min in four stroke engines and equal to rev/min in two stroke engines.

$$BP = \frac{2\pi NT}{60 \times 10^3} = \frac{\pi N (W - S)(D + d)}{60 \times 10^3} \text{ kW}$$

Here N stand for the rev/min of brake drum, irrespective of whether it is a four stroke or a two stroke engine.

REVIEW QUESTIONS

A. Conceptual and conventional questions.

1. Differentiate between internal and external combustion engines.
2. List the various advantages of IC engines over external combustion (steam) engines.
3. Mention the various applications of IC engines.
4. How IC engines are classified ?
5. State the purpose of the following parts of an IC engine :
   spark plug, piston rings, crank and crank shaft, cam shaft & valve mechanism, fly wheel.
6. What is the function of compression and oil control rings provided on the piston of an IC engine ?
7. Mention the material used for the following components of an IC engine :
   cylinder block, piston, piston rings, connecting rod, crank and crank shaft, valves, flywheel.
8. Sketch the overhead valve mechanism of an IC engine, and label its different parts. How it operates and what function does it perform ?
9. Describe, with neat sketches, the sequence of events in the working of a four-stroke petrol/ diesel engine.
10. Give a neat sketch of the theoretical and actual $p - V$ diagrams for a four-stroke petrol/diesel engine. Describe briefly the factors which account for deviations between these two plots.
11. Give as many differences as you can between the construction and operation of a compression ignition engine and a petrol engine.
12. Compare diesel engine with a petrol engine with special reference to maximum pressure, efficiency, power to weight ratio, cost, and load control. Give explanations.
13. On a two-stroke engine, is there a power stroke in each revolution of the crank shaft ?
14. How does a two-stroke engine differ from a four stroke engine ?
15. Explain, with neat sketches, the sequence of events in the working of a two-stroke petrol/ diesel engine.
16. Mention the relative merits and demerits of two stroke engines when compared with four stroke engines.
17. What is a valve timing diagram? Draw the typical valve timing diagram of a four stroke petrol engine. Give the reasons for early opening of the exhaust valve and late closing the inlet valve.
18. What is meant by overlap in a valve timing diagram? Give reasons for its use and mention any possible disadvantage.
19. (*a*) List the requirements of a good carburettor
    (*b*) Explain, with a heat sketch, the working of a simple plain tube carburettor. Mention, the limitation of this carburettor.
20. (*a*) What are the basic requirements of the fuel injection system of a diesel engine.
    (*b*) Explain, with sketches, the construction and working of a fuel injection pump and a injector.
21. (*a*) Draw the schematic arrangement of the different elements of battery–coil ignition system and state the purpose served by each element.
    (*b*) Give a brief description of the construction of a spark plug.
22. (*a*) State the necessity of cooling of an IC engine
    (*b*) Describe water cooling of an IC engine. Why a thermostat and a pump are usually provided with such a cooling system?
23. (*a*) State the objectives of lubrication in an IC engine
    (*b*) Explain, with the help of a neat sketch, the splash method of lubrication
24. (*a*) Point out the different stages of combustion in a CI engine
    (*b*) What is meant by the term delay period? How does it affect the knocking tendency of CI engine.
25. (*a*) State the difference between the terms pre-ignition and detonation in SI engines
    (*b*) Give a brief description of the various factors which influence the tendency of an SI engine

to detonate. What are the effects of detonation and suggest the methods used to suppress it.

26. "Factors tending to increase detonation in SI engines decrease knocking in CI engines." Justify the validity of this statement of by taking examples of few factors.

27. (*a*) What are the requirements of a good combustion chamber for a spark ignition engine?

    (*b*) State the salient features of L-head (side valve) combustion chamber and Recardo turbulent combustion chamber

    (*c*) What is the most favourable position for spark plug location and why?

28. State the purpose, effects and limitations of supercharging an IC engine.

29. What is meant by governing as applied to an IC engine. Explain briefly the different methods used for this purpose.

30. Define indicated power, brake power and various efficiencies as applied to spark ignition and compression ignition engines.

31. Define specific fuel consumption and state its importance.

32. Sketch and explain the rope brake arrangement for determining the brake power of an engine.

33. Describe the Morse test for determining the indicated power of a multi cylinder engine. State the assumptions made.

34. (*a*) Sketch a typical consumption loop (curve showing the variation of specific fuel consumption with mean affective pressure) and point out its salient features.

    (*b*) Why oil engines are preferred to petrol engine for road transport.

35. A four-cylinder four-stroke petrol engine develops 15 kW while running at 1000 rev/min. The mean effective pressure as estimated from the indicator card is 5.5 bar. Determine the bore and stroke of the engine, if the length of stroke is 1.5 times the bore.

    [Ans. 88.5 mm, 132.75 mm**]**

36. The following data pertains to a test rum on a four-stroke, four cylinder spark ignition engine.

Bore and stroke	: 6 cm and 9.5 cm
Clearance volume	: 53.5 cm^3
Heating value of fuel	: 44.1 MJ/kg
Relative efficiency based on brake thermal efficiency	: 50 %

When tested on load and running at 2800 rev/min, the engine develops 60 Nm torque. Taking mechanical efficiency equal to 80 %, determine the specific fuel consumption and brake mean effective pressure.

[Ans. 0.4 kg/kWh, 7.02 bar**]**

37. The following data refers to a test on a single cylinder oil engine working on four stroke cycle:

Diameter of brake wheel and rope diameter	: 60 cm and 3 cm
Dead load and spring balance reading	: 250 N and 50 N
Bore and piston stroke	: 10 cm and 15 cm
Speed of engine	: 400 rev/min
Length and area of indicator diagram	: 6 cm and 4 cm^2
Spring stiffness	: 12 N/cm^2 per cm
Fuel consumption	: 0.32 kg/kWh
Calorific value of fuel	: 43950 kJ/kg

Determine brake power, indicated power, mechanical efficiency and indicated thermal efficiency of the engine.

[Ans. 2.64 kW, 3.14 kW, 84%, 30.4%**]**

38. A two stroke diesel engine has a bore 100 mm and a stroke of 150 mm. Running at a mean speed of 5m/s, it develops a torque of 60 Nm. The engine has a mechanical efficiency of 80 percent and the indicated thermal efficiency is 40 percent.

    Assuming a calorific value of fuel used as 45000 kJ/kg, determine the indicated power, the indicated mean effective pressure and the fuel consumption per kWh of brake power.

    [Ans.7.85 kW, 4 bar, 0.25 kg/kWh**]**

39. A four-stroke petrol engine with four cylinders, coupled to a hydraulic dynamometer was tested at full throttle at constant speed. For each cylinder, the bore and stroke length were 80 mm and 100 mm respectively. Fuel was supplied at the rate of 5.5 kg/hr and the plugs of the four cylinders were successively short-circuited without the change of speed. The power measurements were as follows.

With all cylinders working	---	14.8 kW
With cylinder 1 cut-off	---	10.2 kW
With cylinder 2 cut-off	---	10.4 kW
With cylinder 3 cut-off	---	10.5 kW
With cylinder 4 cut of	---	10.3 kW

The clearance volume for each cylinder was 100 cm^3 and the fuel used had a calorific value of 42000 kJ/kg. You are required to determine the mechanical efficiency, the indicated thermal efficiency, the air standard efficiency and the relative efficiency.

[Ans. 83.1%, 27.7%; 51.2%; 54%**]**

40. During the trial of a single cylinder, four stroke oil engine, the following results were obtained:

Cylinder bore and piston stroke	: 20 cm and 40 cm
Torque and speed	: 407 Nm and 250 rpm
Indicated mean effective pressure	: 6 bar
Oil consumption and its heating value	: 4 kg/hr and 43 MJ/kg
Cooling water flow rate	: 4.5 kg/min
Air used per kg of fuel	: 30 kg
Rise in temperature of cooling water	: 45°C
Temperature of exhaust gases	: 420°C
Room temperature	: 20°C
Mean specific heat of water and hot gases	: 4.18 and 1 kJ/kg K

Determine the indicated power, brake power and draw up the heat balance sheet for the test in kJ/hr.

[Ans. 15.71 kW, 10.65 kW**]**

B. Fill in the blanks with appropriate word/words.

1. A four stroke engine theoretically operates on .... cycle.
2. The compression ratio of an IC engine is the ratio of .... to ....
3. The extreme positions of the piston in the cylinder of an IC engine are called ....
4. The compression ratio varies from .... in SI engines and from .... in CI engines.
5. A two stroke engine employs .... cut in the cylinder walls instead of ....
6. In a four stroke cycle engine the cycle is completed in .... crank revolutions.
7. The top of the piston in two stroke engine is ....
8. A .... provides the connection between small end of connecting rod and piston
9. The specific fuel comsumption is defined as the mass of fuel required to be supplied to an engine to develope ....
10. An engine indicator is used to measure the .... of an engine.

Answers

1. Otto 2. total cylinder volume, clearance volume 3. dead centres 4. 6-8, 16-20 5. ports, valves 6. two 7. crown shaped 8. gudgeon 9. unit kW power per hour 10. indicated power

C. Multiple choice questions.

1. Which is a false statement in the context of the parts of an IC engine ?
   (*a*) the piston is connected to the small end of connecting rod by a gudgeon pin
   (*b*) crank pin connects the big end of connecting rod to crank arm
   (*c*) the reciprocating motion of the piston is converted to the rotary motion by the connecting rod and crank mechanism
   (*d*) the crankshaft provides the power and direction to the suction and exhaust valves
2. What is the main shaft of an engine that controls the movement of piston ?
   (*a*) axle (*b*) drive shaft
   (*c*) crankshaft (*d*) camshaft
3. What is mounted on the crankshaft to transform rotary motion into back-and-forth motion for actuating the valves?
   (*a*) crank (*b*) crank and slider
   (*c*) camshaft (*d*) flywheel
4. In four stroke engines, the camshaft is connected to the crankshaft by gears or chain and rotates at
   (*a*) half the crankshaft speed
   (*b*) three-fourth the crankshaft speed
   (*c*) equal the crankshaft speed
   (*d*) double the crankshaft speed
5. The slotted grooves on oil rings for pistons help to
   (*a*) seal the cylinder
   (*b*) minimise friction
   (*c*) prevent the piston from wear
   (*d*) provide an escape for the oil that the slots edge out from the cylinder walls

6. The number of working strokes per minute in case of four stroke IC engine will be equal to
   (*a*) $N/2$ (*b*) N
   (*c*) $2N$ (*d*) $4N$
   where N is represents the revolutions turned by the crankshaft in one minute

7. For the same compression ratio and heat supplied, the air standard efficicency of an Otto cycle compared to that of a Diesel cycle is
   (*a*) less (*b*) more
   (*c*) equal (*d*) unpredictable

8. Which is not the common component between a petrol and a diesel engine ?
   (*a*) camshaft (*b*) exhaust silencer
   (*c*) spray nozzle (*d*) dynamo

9. Which is not a part of petrol engine ?
   (*a*) valve mechanism (*b*) fuel injector
   (*c*) induction coil (*d*) air filter

10. Which is the correct statement in the context of two-stroke engines ?
    (*a*) compression ratio is always lower than that of a four stroke cycle engine
    (*b*) there is only one valve for inlet and exhaust
    (*c*) charge enters the engine cylinder through ports only
    (*d*) a Diesel engine cannot operate on two stroke cycle

11. Compared to four-stroke cycle engine, a two-stroke cycle engine
    (*a*) can be easily started
    (*b*) has lesser shocks and vibrations
    (*c*) greater thermal efficiency
    (*d*) less wear and tear of moving parts

12. Which aspect is not true for a two-stroke cycle engine when compared with a four stroke cycle engine ?
    (*a*) cycle is completed in one stroke of the piston
    (*b*) uniform turning moment and hence a lighter flywheel
    (*c*) theoretically develops twice the power
    (*d*) noisy exhaust
    (*e*) more consumption of lubricating oil

13. A two stroke engine has a speed of 750 rpm. A four stroke engine having an identical cylinder size runs at 1500 rpm. The theoretical output of the two-stroke engine will
    (*a*) be twice that of the four stroke engine
    (*b*) be half that of the four stroke engine
    (*c*) be the same as that of the four stroke engine
    (*d*) depends upon whether it is SI or CI engine.

14. Scavenging air means
    (*a*) air used under compression
    (*b*) air used for forcing the burnt gases out of the cylinder during the exhaust period
    (*c*) forced air for cooling the engine cylinder
    (*d*) burnt air containing combustion products

15. All of the following statements are correct, except
    (*a*) For the same power output, a petrol engine is of light construction and low initial cost when compared to a diesel engine.
    (*b*) Morse test is used to determine the brake power of a multicylinder engine
    (*c*) The thermal efficiency of a well-designed and properly maintained diesel engine usually lies in the range 35-40%
    (*d*) A governor is used to adjust the power output from an engine in conformity with the external load.

16. The preference of diesel engines for road transport is due to
    (*a*) easy starting
    (*b*) low operating cost
    (*c*) low specific fuel consumption over a large range of load
    (*d*) complete combustion of charge
    (*e*) harmless exhaust

17. Morse test in multi-cylinder engines is done to determine
    (*a*) air-flow to the engine
    (*b*) volumetric efficiency
    (*c*) indicated power
    (*d*) mechanical efficiency

18. Which one of the followings is a wrong statement?
    (*a*) valve timing diagram shows the correct crank positions corresponding to opening and closing of the valves
    (*b*) valve overlap represents the angle during which both the inlet and exhaust valves remain open
    (*c*) closing of inlet valve before bottom dead centre would reduce volumetric efficiency of the engine

(*d*) ignition of charge in a petrol engine takes place very close to top dead centre

19. All of the following statements are correct, except
(*a*) a carburettor prepares a homogeneous air-fuel mixture by atomising and vaporising the fuel
(*b*) vacuum at the throat of venturi sucks the fuel through the fuel jet
(*c*) throttle valve controls the supply of air-fuel mixture
(*d*) choke is kept open when cranking a cold engine for starting

20. Consider the following statements :
1. Petrol engines are quality governed
2. Combustion in spark ignition engines is homogeneous
3. Lean mixture is required during idling
4. Choke supplies rich mixture for cold starting.

Which of these are correct statements ?
(*a*) 1 and 3 (*b*) 2 and 4
(*c*) 1, 3 and 4 (*d*) 1 and 3

21. The purpose of ignition coil in the ignition system of a petrol engine is to
(*a*) increase the voltage
(*b*) reduce sparking in the system
(*c*) distribute high currents to the engine cylinder
(*d*) regulate the current produced by the dynamo

22. What material is used for the insulating body of a spark plug ?
(*a*) silica (*b*) dolomite
(*c*) alumina (*d*) glass

23. A particular firing order is necessary in a multi-cylinder engine so as to
(*a*) have better balance of the engine
(*b*) give good distribution of fuel to all the cyclinders
(*c*) obtain uniform toque
(*d*) operate the ignition system smoothly

24. In diesel engines, the injection pressure is of the order of
(*a*) 5-8 MN/m^2 (*b*) 10-15 MN/m^2
(*c*) 17.5-25 MN/m^2 (*d*) 40-50 MN/m^2

25. Supercharging is the process of
(*a*) providing a draft of air for cooling purposes
(*b*) supplying air to remove the spent up gases
(*c*) raising the pressure of charge fed to the cylinder
(*d*) supplying more fuel for getting greater output

26. Velocity of flame propagation in the SI engine is maximum for air-fuel mixture which is .... stoichiometric
(*a*) 10% richer than
(*b*) equal to
(*c*) more than 10% richer than
(*d*) 10% leaner than

27. Delay period refers to the period that elapses between providing a spark and the commencement of actual combustion of fuel. An increase in which of the following factors increases the delay period ?
(*a*) self ignition temperature of fuel
(*b*) intake temperature
(*c*) inlet pressure
(*d*) compression ratio

28. All of the following factors promote detonation in spark ignition engines, except
(*a*) increase in inlet pressure and temperature of charge
(*b*) higher compression ratio
(*c*) high self ignition temperature of fuel
(*d*) advanced spark timing

29. Reference fuels for knock rating of SI engines fuels would include
(*a*) iso-octane and α-methyl napthalene
(*b*) normal octane and anciline
(*c*) iso-octane and *n*-hexane
(*d*) *n*-heptane and iso-octane

30. Which is not the characteristic of diesel knock in a compression ignition engine ?
(*a*) instantaneous and rapid burning of the last part of the charge
(*b*) pressure rise in the cylinder occurs at an enormous rate
(*c*) rough running and fatigue failure of engine components
(*d*) drop in engine power and efficiency

31. Consider the following measures:
1. Increasing the compression ratio
2. Increasing the intake air temperature

3. Increasing the length to diameter ratio of the cylinder
4. Increasing the engine speed

The measures necessary to reduce the tendency to knock in CI engines would include

(*a*) 1, 2 and 3 (*b*) 1, 2 and 4
(*c*) 1, 3 and 4 (*d*) 2, 3 and 4

32. All of the following statements are correct, except
   (*a*) the detonation follows the spark while pre-ignition precedes it
   (*b*) doping refers to the addition of certain additives to petrol to improve its resistance to detonation
   (*c*) the two reference fuels used for cetane rating are cetane and *n*-heptane
   (*d*) factors tending to increase detonation in SI engines tend to reduce knocking in CI engines

33. All of the following statements are correct except
   (*a*) Petrol engines normally operate in the compression ratio range of 6 – 10
   (*b*) A two stroke engine has one inlet port, one exhaust port and one transfer port
   (*c*) Two stroke IC engines are less efficient than their four-stroke counterparts due to low compression ratio and shorter stroke length
   (*d*) The direction of spray is not critical in indirect combustion chambers

34. Operation of a spark ignition engine at high compression ratios is not desirable as it may
   (*a*) lead to incomplete combustion and poor efficiency
   (*b*) necessinate robust construction for the engine cylinder
   (*c*) promote engine knocking causing damage to the engine
   (*d*) cause an excessive temperature rise in the cylinder resulting into back fire

35. In spark ignition engines, the flame speed increases with increasing
   1. fuel-air ratio 2. pressure ratio
   3. turbulence 4. fuel intake rate

   Which of the statements made above are correct?

   (*a*) 1 and 2 (*b*) 2 and 3
   (*c*) 1 and 4 (*d*) 3 only

36. Consider the following factors :
   1. compression ratio
   2. engine speed
   3. engine output
   4. fuel viscosity

   Which of these affect the ignition delay in a CI engine.

   (*a*) 1 and 2 (*b*) 2 and 4
   (*c*) 1, 2 and 3 (*d*) 1, 2 and 4

37. Which of the followings is identified with octane rating of 100 ?
   (*a*) iso-octane (*b*) *n*-heptane
   (*c*) tetraethyl lead (*d*) α-methylnapthalene

38. Identify the aspect an increase in which promotes the tendency of knocking in CI engines
   (*a*) compression ratio
   (*b*) inlet air temperature
   (*c*) coolant water temperature
   (*d*) rate of fuel atomisation

Answers

1. (*d*)	2. (*c*)	3. (*c*)	4. (*a*)	5. (*d*)
6. (*a*)	7. (*b*)	8. (*d*)	9. (*b*)	10. (*c*)
11. (*a*)	12. (*a*)	13. (*c*)	14. (*b*)	15. (*b*)
16. (*c*)	17. (*c*)	18. (*d*)	19. (*d*)	20. (*b*)
21. (*a*)	22. (*c*)	23. (*c*)	24. (*b*)	25. (*c*)
26. (*a*)	27. (*a*)	28. (*c*)	29. (*d*)	30. (*a*)
31. (*b*)	32. (*c*)	33. (*c*)	34. (*c*)	35. (*d*)
36. (*c*)	37. (*a*)	38. (*c*)		

❑❑❑

Chapter **23**

Gas Turbines and Jet Propulsion

Learning objectives : A study of the subject matter included in this chapter will enable the reader to explain and discuss :

- *the system of operation of constant pressure and constant volume combustion gas turbine*
- *the operation of Brayton cycle and its thermal refinement*
- *the advantages of gas turbines over IC engines and steam turbines*
- *the difference between the jet propulsion and rocket propulsion systems*
- *the working difference between turbo jet, turbo prop, and pulse jet propulsion systems*

The introduction of gas turbine as a contrivance for producing mechanical power is the realisation of a long cherished dream of the engineers. The turbine is capable of performing a wide range of functions :

(*i*) In the field of aeronautics, the gas turbine is widely used as a propulsion power plant in turboprop and turbojet aircraft. With employment in avitation, high flying speeds approaching sonic and supersonic values have been attained and flights at high altitudes have become possible.

(*ii*) In the marine field, the gas turbine is used as main propulsion engine, as an emergency high power boost engine and to drive auxillaries and emergency devices.

(*iii*) On land the applications include : supply of energy for chemical heating processes ; supply of compressed air for process work, for peak load and stand by power plant in electric power stations and for locomotive propulsion.

The gas turbine competes favourably with other prime-movers because of its simplicity and reliable working, few auxiliaries and high rotating speed, freedom from vibration and ability to generate large power from units of relatively small size and weight.

The gas turbine is akin to petrol and diesel engines in its working medium and internal combustion, and is similar to steam turbine with regard to continuous and steady flow of the working medium.

A rotary compressor, either of centrifugal or axial flow type, sucks air from atmosphere and compresses it. The pressurised air is then introduced into the combustion chamber where heat is added to air and its temperature is raised. The addition of heat to air or any other working medium is accomplished by injection of fuel under high pressure in the combustor or by employing an air heater. The high temperature air is then directed into the gas turbine which may be either of impulse or reaction type, analogous to the corresponding steam turbine types. The high velocity jet produced by expansion in the turbine nozzle impinges on the turbine moving blades and the kinetic energy is transformed into mechanical work of turbine rotor.

Majority of the gas turbines use liquid fuels. Gaseous fuels (natural gas, generator gas, or gas produced by the underground gassification of solid fuels) too are used in many gas turbine installations. Recently, the use of nuclear fuel is assuming special importance.

23.1. SYSTEM OF OPERATION

The gas turbine may work on open or closed system.

In the *open system*, fresh air is drawn into the compressor from the atmosphere, heat is added by combustion of fuel in the air itself and the spent up combustion products are rejected to the atmosphere. The arrangement thus necessitates continuous replacement of the working medium. The open system is not a true heat engine as the system is not restored to its original state but undergoes a chemical combustion. The combustion takes place between air and fuel and the working fluid is obviously formed by gaseous combustion products.

In a ***closed system***, the compressed working fluid which may be air, hydrogen, helium or carbon dioxide is delivered to a device called heat exchanger or simply a heater. In the heater, the fluid is heated to the required temperature and the hot fluid then expands in the gas turbine. The exhaust from the turbine is cooled and again fed to the compressor, thus closing the cycle. The working agent thus does not leave the plant and is used repeatedly in successive cycles.

Besides the above distinction the gas turbine installations operate with the combustion of fuel either at constant pressure or at constant volume.

(a) Constant pressure gas turbine : In constant pressure gas turbines, the combustion process in the combustor goes on at constant pressure and the turbine works on Joule or Brayton cycle. Fig. 23.1 shows the schematic diagram of a constant pressure gas turbine installation with an axial flow compressor and reaction turbine mounted on a common shaft.

Fresh atmospheric air drawn into the circuit is compressed in the axial flow compressor to a combustion pressure of the order of 1.5 to 4 bar. The pressurised air then enters the combustion chamber into which hydrocarbon fuel is sprayed by means of injection nozzles. With chemically correct combustion, the burning of a hydrocarbon fuel occurs at temperatures around 2000°C. The temperature of the resulting combustion products has to be reduced to a value compatible with the creep strength of the gas turbine blading. This can be done by addition of large quantity of

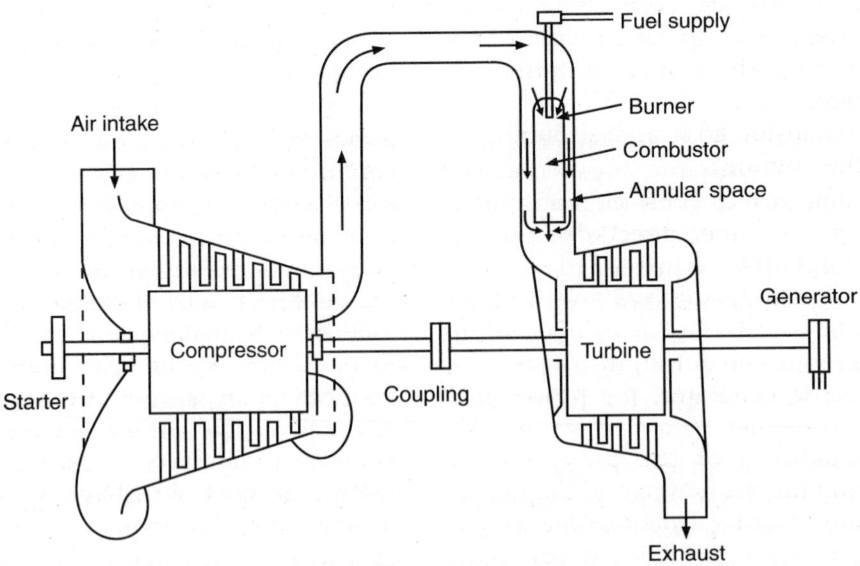

Fig. 23.1. Constant pressure gas turbine

air or by injection of water or by partial abstraction of heat by water cooled surfaces. The plant shown in Fig. 23.1. is based on the use of excess air as a cooling medium. Before entering the combustor, the pressurised air is split up into two circuits :

(*i*) a small part of air called *primary air* or *combustion air* enters the burner and is consumed during the combustion of the fuel oil.

(*ii*) the remainder large part of air called *secondary air* flows through the annular space between the walls of the combustor and the burner jackets. The secondary air takes heat from and incidently cools the burner.

Furtherdown the combustor, the secondary air mixes with the products of combustion issuing from the burner and this reduces their temperature to that admissible for the gas turbine blading. Metallurgical considerations and the high cost of heat resisting materials limit the maximum temperature to about 1000-1200K for reliable continuous operation. For efficient operation at such high temperatures either turbine blades are given a ceramic coating or the turbine blades, discs and shafts are cooled.

The diluted combustion products now enter the various rings of the moving blades of the turbine. The gases expand continuously and the pressure energy is converted into kinetic energy which in turn is transformed into mechanical work of the turbine rotor. The gases after expansion go straight from the gas turbine to the chimney.

The turbine drives both the compressor and the generator either directly or through gearing. The work required to run the compressor is called *back work* and it equals about 65 to 80% of the total turbine output. The rest of the turbine output may be used to drive an electric generator, the propeller of ship or the propeller of an aeroplane. The auxiliary machines of the plant are the lubricating and fuel oil pumps and the starting motor. While starting, the turbine is first motored up to some minimum speed called *coming in speed*. Starting motor brings the turbine to coming in speed. The coming in speed is approximately 30 to 40% of the rated operating speed. The fuel is then turned on, ignited and the turbine picks up speed.

Theoretically, the combustion takes place at constant pressure. In actual gas turbine engine, there occurs a small pressure drop due to friction of the working fluid and due to a change in the momentum of the fluid caused by change in density during combustion. Further the pressure also varies under load changes and so these turbines are better named as *continuous combustion turbines*. The continuous combustion turbine has the following disadvantages :

(*i*) For cooling the combustion products, about 3 to 4 times air excess of stochiometric air has to be provided. This increases the compressor size and the power losses in the compression process

(*ii*) The system necessitates several stages if high pressures for high thermal efficiency are to be attained. The frictional losses then increase and the compressor efficiency drops.

(*b*) ***Constant volume combustion gas turbine*** : In constant volume combustion gas turbine, the combustion process in the combustor goes on at constant volume and the turbine works on Atkinson cycle. Fig. 23.2 shows the schematic lay out of the most successful turbine of this category–the Holzwarth explosion turbine.

Air taken from the surrounding atmosphere, through the compressor inlet, is compressed to a pressure of about 3 bar. The pressurised air then enters the combustor through valves V_1 and V_2. These valves close after the combustor has been filled. A requisite quantity of fuel, from the fuel tank, is injected into the combustor through valve V_3 which is then closed. The fuel air mixture formed in the combustor is ignited by means of spark plugs. The combustion takes place at constant volume since the fuel and air valves are closed. During the explosion process, there occurs an instantaneous pressure rise to about 10-14 bar and the non-return valve fitted at the combustor top is forced open outwardly. This

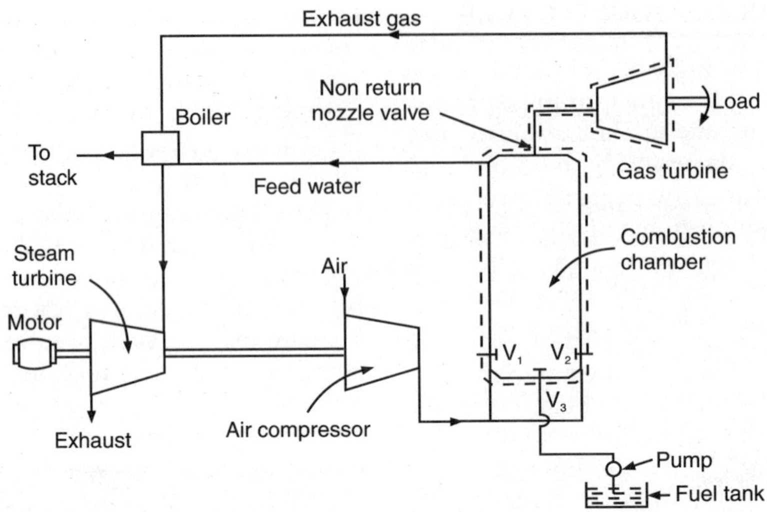

Fig. 23.2. Constant volume gas turbine plant (Holzwarth turbine)

permits the combustion products to escape from the combustor and enter the turbine wheel. The turbine is fitted with expansion nozzles in which the pressure drops and the heat energy of the gases is converted into kinetic energy. The non-return valve closes when the pressure inside the combustor falls due to the escaping combustion products and their subsequent expansion in the turbine. The combustion process takes about 0.8 to 1.4 seconds and thus about 60 fuel combustion cycles are completed per minute.

Since only a small quantity of excess air is supplied, high temperatures are developed. For efficient operation and to enable the metal of the turbine rotor blades to withstand the centrifugal and bending stresses at the high working temperatures, the combustor, nozzles, impeller and the rotor blades are cooled by circulating water in jackets. The heat given to the cooling water is recovered by taking the hot jacket water to the steam boiler where it is evaporated into steam. The heat needed for steam raising is supplied by the exhaust gases leaving the gas turbine. The steam thus generated actuates the steam turbine which drives the compressor. Thus the total output of the gas turbine is utilized to run the generator which is mechanically coupled to the gas turbine. The starting motor is operated only at the time of starting to bring the turbine to coming in speed.

The constant volume combustion system gives the advantages namely :

(*i*) A relatively small size of the compressor unit is needed since the air supplied to the combustor is only slightly more than that needed for chemically correct combustion. This reduces the power consumption of the compressor unit.

(*ii*) The air is compressed only to about 1/4 of the final explosion pressure.

The short comings of the system are :

(*i*) A number of automatically operated valves are necessary for the combustor and the complicated system increases the cost of plant

(*ii*) The components of the unit have to be strong enough to withstand the high operating pressures

(*iii*) Recovery of the heat given up to the cooling water by evaporating it and using the steam generated to actuate a steam turbine for driving the compressor, renders the system more elaborate and complicated

(*iv*) Combustion is intermittent and this impairs the smooth running of the unit.

23.2. BRAYTON OR JOULE CYCLE

The Brayton cycle or Joule cycle is a theoretical cycle on which run the constant pressure gas turbines. A gas turbine is akin to petrol and diesel engines in its working medium and internal combustion. However, the processes of compression, combustion and expansion take place in different components in a gas turbine plant. Further, the gas turbine is a rotary machine and operates under steady and continuous supply of working medium.

The schematic arrangement of the different components of a gas turbine plant and, $p-v$ and $T-s$ plots for an ideal Brayton cycle are shown in Fig. 23.3. The cycle consists of two isentropic processes (compression and expansion) and two constant pressure processes (combustion and exhaust). The sequence of operations is :

1–2 : Isentropic compression air in a rotary compressor which raises the pressure and temperature of air from p_1, T_1 to p_2, T_2.

2–3 : The air at condition 2 is passed through a heat exchanger where heat is externally supplied to it at constant pressure. This raises the temperature from T_2 to T_3.

3–4 : Isentropic expansion of high temperature and high pressure air in the turbine during which work is done by the system.

4–1 : The air at state point 4 is passed through a heat exchanger and heat is rejected at constant pressure. This cools the air to initial condition 1 and the cycle is completed.

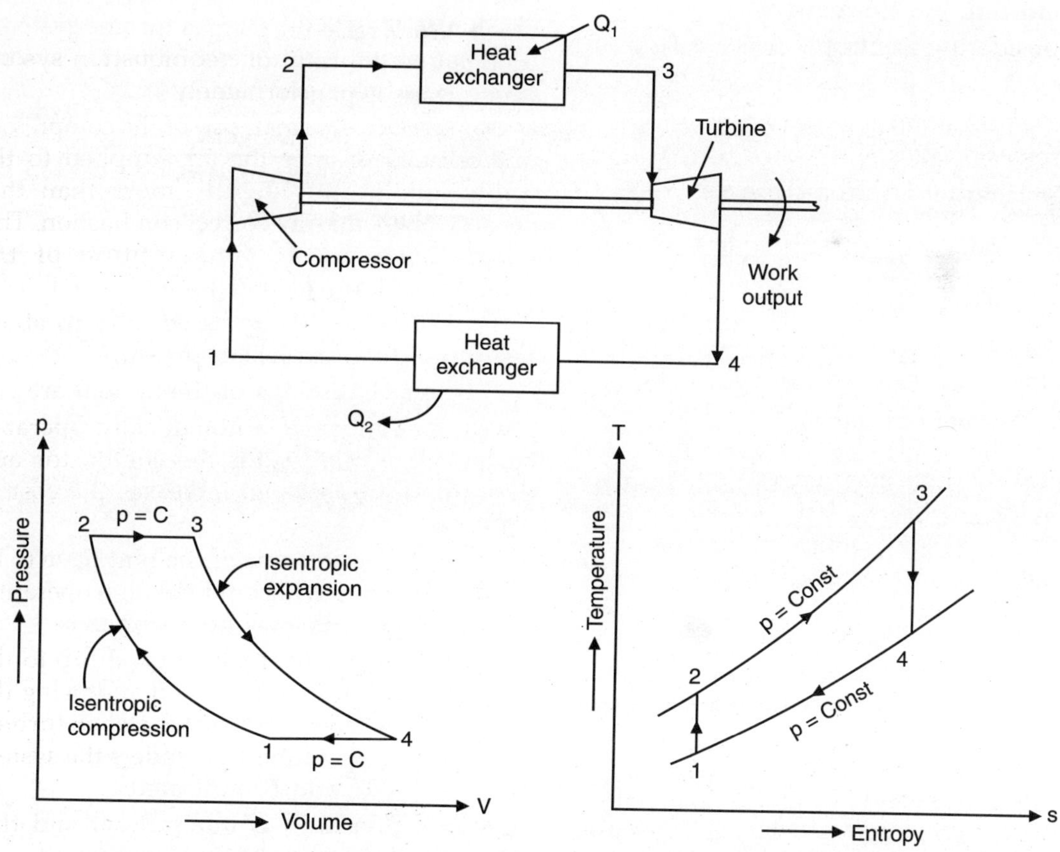

Fig. 23.3. Schematic diagram of closed cycle gas turbine and, p-V and T-s plot for an ideal Brayton cycle

For unit mass of air,

Heat supplied, $Q_1 = c_p (T_3 - T_2)$

Heat rejected, $Q_2 = c_p (T_4 - T_1)$

Work done = heat supplied − heat rejected

$$= c_p (T_3 - T_2) - c_p (T_4 - T_1)$$

$$\text{Thermal efficiency}, \eta = \frac{\text{work done}}{\text{heat supplied}}$$

$$= \frac{c_p (T_3 - T_2) - c_p (T_4 - T_1)}{c_p (T_3 - T_2)}$$

$$= 1 - \frac{T_4 - T_1}{T_3 - T_2} \quad \text{...(23.1)}$$

The efficiency of Brayton cycle in terms of pressure ratio $r_p = \frac{p_2}{p_1} = \frac{p_3}{p_4}$ and adiabatic exponent γ can be obtained by carrying out the following transformations :

Considering isentropic process 1–2,

$$\frac{T_2}{T_1} = \left(\frac{p_2}{p_1}\right)^{\frac{\gamma-1}{\gamma}} = (r_p)^{\frac{\gamma-1}{\gamma}}$$

Considering isentropic process 3–4,

$$\frac{T_3}{T_4} = \left(\frac{p_3}{p_4}\right)^{\frac{\gamma-1}{\gamma}} = (r_p)^{\frac{\gamma-1}{\gamma}}$$

$$\text{Thus, } \frac{T_2}{T_1} = \frac{T_3}{T_4} \quad \text{or} \quad \frac{T_4}{T_1} = \frac{T_3}{T_2}$$

∴ Thermal efficiency,

$$\eta = 1 - \frac{T_1\left(\frac{T_4}{T_1} - 1\right)}{T_2\left(\frac{T_3}{T_2} - 1\right)}$$

$$= 1 - \frac{T_1}{T_2} = 1 - \left(\frac{p_1}{p_2}\right)^{\frac{\gamma-1}{\gamma}}$$

$$= 1 - \left(\frac{1}{r_p}\right)^{\frac{\gamma-1}{\gamma}} \quad \text{...(23.2)}$$

It follows from equation 23.2 that the thermal efficiency of Brayton cycle is a function of pressure ratio and that the efficiency increases with increase in pressure ratio. The variation of thermal efficiency with pressure ratio is shown in Fig. 10.4. The curve tends to become flat at higher pressure ratios which implies that though the efficiency is increasing, the rate of increase starts diminishing at higher pressures.

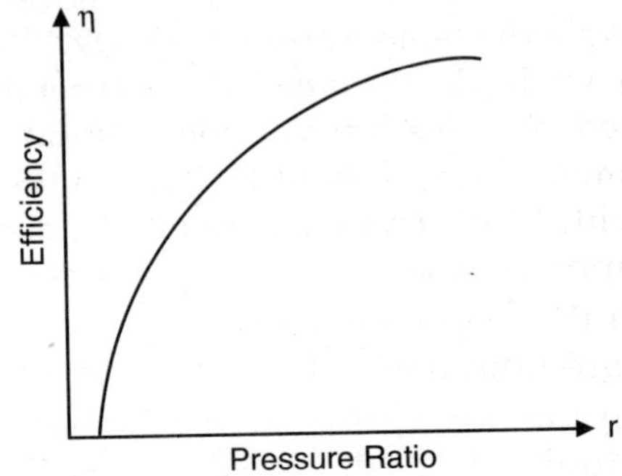

Fig. 23.4. Efficiency of Brayton cycle versus pressure ratio

• **Work ratio** for the gas turbine system is defined as the ratio of net or useful work to the turbine expansion work.

$$\text{Work ratio} = \frac{\text{net work}}{\text{turbine work}}$$

Net work = turbine work − compressor work

$$= c_p(T_3 - T_4) - c_p(T_2 - T_1)$$

$$= c_p T_3 \left[1 - \frac{T_4}{T_3}\right] - c_p T_1 \left[\frac{T_2}{T_1} - 1\right]$$

$$= c_p T_3 \left[1 - \frac{1}{(r_p)^m}\right] - c_p T_1 \left[(r_p)^m - 1\right]$$

where r_p is the pressure ratio and $m = \frac{\gamma - 1}{\gamma}$

Work ratio

$$= \frac{c_p T_3 \left[1 - \frac{1}{(r_p)^m}\right] - c_p T_1 \left[(r_p)^m - 1\right]}{c_p T_3 \left[1 - \frac{1}{(r_p)^m}\right]}$$

$$= 1 - \frac{T_1}{T_3}(r_p)^m \quad \text{...(23.3)}$$

For maximum work ratio, the temperature T_1 at compressor entry should be as low as possible and the temperature T_3 at entry to the turbine should be as high as possible. The compressor inlet temperature is normally the atmospheric temperature (say 288 K at sea level) while the turbine inlet temperature is decided by metallurgical considerations (the maximum value of about 1000 K that the metal can withstand). From equations 23.2 and 23.3, it is apparent that

(*i*) the thermal efficiency is a function of pressure ratio only

(*ii*) the net work and the work ratio are functions of both pressure ratio and temperature ratio

Apparently the performance of an actual plant depends upon both the pressure ratio and the temperature ratio.

- **Pressure ratio for optimum specific work output :** The specific work output (work output per unit mass) is

$$W = c_p T_3 \left[1 - \frac{1}{(r_p)^m}\right] - c_p T_1 \left[(r_p)^m - 1\right]$$

The optimum value of pressure ratio r_p that gives maximum specific work output can be obtained by differentiating the above expression with respect to r_p and equating the result to zero. Treating T_1 and T_3 as constant, we have

$$0 = c_p T_3 \left[-m(r_p)^{-m-1}\right] - c_p T_1 \left[m(r_p)^{m-1}\right]$$

$$\frac{(r_p)^{m-1}}{(r_p)^{-m-1}} = \frac{T_3}{T_1} \quad \text{or} \quad (r_p)^{2m} = \frac{T_3}{T_1}$$

That gives:

$$r_p = \left(\frac{T_3}{T_1}\right)^{\frac{1}{2m}} = \left(\frac{T_3}{T_1}\right)^{\frac{\gamma}{2(\gamma-1)}} \quad \text{...(23.4)}$$

- **Isentropic compression and expansion efficiencies :** In practice, it is not possible to achieve either isentropic compression or isentropic expansion. These processes tend to become irreversible due to losses caused by internal friction, turbulence and leakage. Consequently in actual cycle, the temperatures at the end of compression and at the end of expansion are higher than in an ideal case for the same pressure ratio (Fig. 23.5).

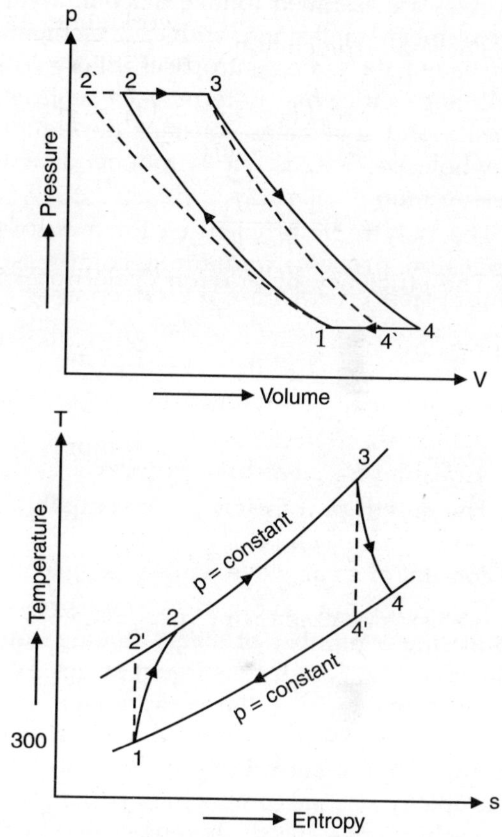

Fig. 23.5. Isentropic compression and expansion efficiencies

Efficiency of compression process (compressor)

$$\eta_c = \frac{\text{isentropic temperature size}}{\text{actual temperature rise}} = \frac{T_2' - T_1}{T_2 - T_1} \quad \text{...(23.5)}$$

Efficiency of expansion process (turbine)

$$\eta_t = \frac{\text{actual decrease in temperature}}{\text{isentropic decrease in temperature}}$$

$$= \frac{T_3 - T_4}{T_3 - T_4'} \quad \text{...(23.6)}$$

- **Polytropic efficiency :** The isentropic efficiency of a complete machine (compressor or turbine) varies with pressure ratio. The turbine isentropic efficiency increases with pressure ratio whereas the compressor isentropic efficiency decreases with pressure ratio.

Quite often, the compression and expansion processes are assumed to take place in a very large number small stages with each elemental stage having the same isentropic efficiency. This small stage efficiency is termed as *polytropic efficiency* and it remains constant throughout the whole process, *i.e.*, it is independent of pressure ratio.

The polytropic efficiencies for a turbine (expansion process) and for a compressor (compression process) are prescribed as

$$\eta_{p_t} = \frac{n-1}{n} \times \frac{\gamma}{\gamma - 1} \quad \text{... turbine}$$

$$\eta_{p_c} = \frac{\gamma-1}{\gamma} \times \frac{n}{n - 1} \quad \text{... compressor}$$

The polytropic efficiency is essentially the true aerodynamic efficiency of the machine on the concept of small stage within the machine.

Consider an axial flow compressor comprising a number of stages having equal stage efficiency for all constituent stages. With reference to Fig 23.6, the compression takes place from pressure p_1 to p_2 in 4-stages with intermediate pressures being p_a, p_b and p_c. The isentropic and actual compression processes for the machine are respectively represented by the paths 1-2′ and 1-2.

The small stage or polytropic efficiency is

$$\eta_{p_c} = \frac{\text{isentropic enthalpy rise per stage}}{\text{actual enthalpy rise per stage}}$$

$$= \frac{(\Delta h_i)_{\text{stage}}}{(\Delta h_a)_{\text{stage}}} = \frac{(dT_i)_{\text{stage}}}{(dT_a)_{\text{stage}}}$$

Let it be assumed that the end states 1 and 2 of the actual irreversible compression lie on the path that obeys the law $p = K_1 \rho^n$. Upon differentiation, we get

$$dp = K_1 n \rho^{n-1} d\rho$$

$$= \frac{p}{\rho^n} \times n\rho^{n-1} d\rho = \frac{np}{\rho} d\rho$$

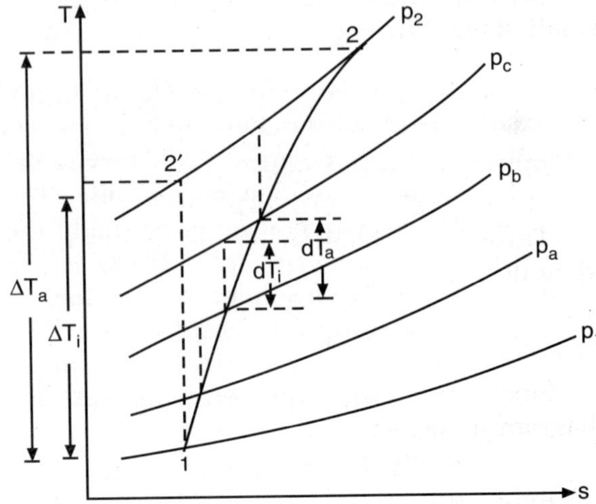

Fig. 23.6. Multistage compression

Further from characteristic equation:

$$\rho = \frac{p}{RT}$$

This gives

$$d\rho = d\left(\frac{p}{RT}\right) = \frac{RT\,dp - p\,RdT}{(RT)^2}$$

Combining the above two identities, we can write

$$dp = \frac{np}{p/RT}\left[\frac{RT\,dp - pR\,dT}{(RT)^2}\right]$$

$$= n\,dp - \frac{np\,dT}{T}$$

or $$dT_a = \left(\frac{n-1}{n}\right)\frac{T}{p}dp \quad \text{...(i)}$$

Similarly for isentropic path 1-2′ which follows the law $p = K_2 \rho^\gamma$

$$dT_i = \frac{\gamma - 1}{\gamma}\frac{T}{p}dp \quad \text{...(ii)}$$

From expressions (*i*) and (*ii*), the polytropic efficiency for the compression process can be written as

$$\eta_{p_c} = \frac{dT_i}{dT_a} = \frac{\gamma-1}{\gamma} \times \frac{n}{n-1} \quad \text{...(23.7)}$$

It may be noted that:

(*i*) for the entire machine: $\eta_c = \dfrac{\Delta T_i}{\Delta T_a}$

(*ii*) from the expression for polytropic or small stage efficiency

$$\Sigma\, dT_a = \Sigma\left(\frac{dT_i}{\eta_{p_c}}\right) = \frac{\Sigma\, dT_i}{\eta_{p_c}}$$

(*iii*) for the multistage compressor:

$$\Sigma\, dT_a = \Delta T_a$$

Through correlation of the above three identities,

$$\eta_c = \eta_{p_c} \times \frac{\Delta T_i}{\Sigma\, dT_i}$$

Since the constant pressure lines on *T-s* diagram diverge towards right,

$$\Delta T_i < \Sigma\, dT_i \text{ and hence } \eta_c < \eta_{pc}$$

Evidently the isentropic efficiency of the whole machine is less than the polytropic (small stage) efficiency.

Consider expansion in the gas turbine taking place from pressure p_1 to p_2 in 4-stages with intermediate pressures being p_a, p_b and p_c as shown in Fig. 23.7. The actual and isentropic expansion processes are respectively represented by the paths 1-2 and 1-2′

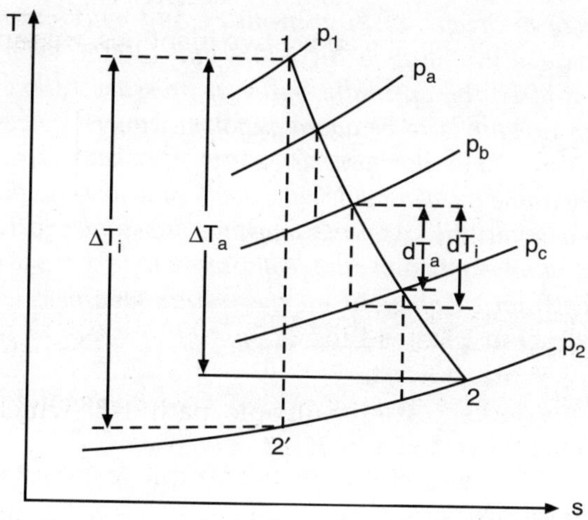

Fig. 23.7. Multistage expansion

By following the procedure as outlined above, it can be shown that

η_{p_t} (polytropic efficiency of expansion in turbine)

$$= \frac{n-1}{n} \times \frac{\gamma}{\gamma - 1} \text{ and } \eta_t < \eta_{pt} \quad ...(23.8)$$

Conclusion can thus be made that:

(*i*) polytropic efficiency takes care of variation of isentropic efficiency with pressure ratio and can be assumed constant.

(*ii*) polytropic efficiency is greater than the isentropic efficiency of the whole machine.

EXAMPLE 23.1

A Brayton cycle operates with air entering the compressor at 1 bar and 25 °C. The pressure ratio across the compressor is 3 to 1, and the maximum temperature in the cycle is 650 °C. Determine the compressor work, turbine work, thermal efficiency and work ratio of the cycle.

Solution : Refer Fig. 23.8.

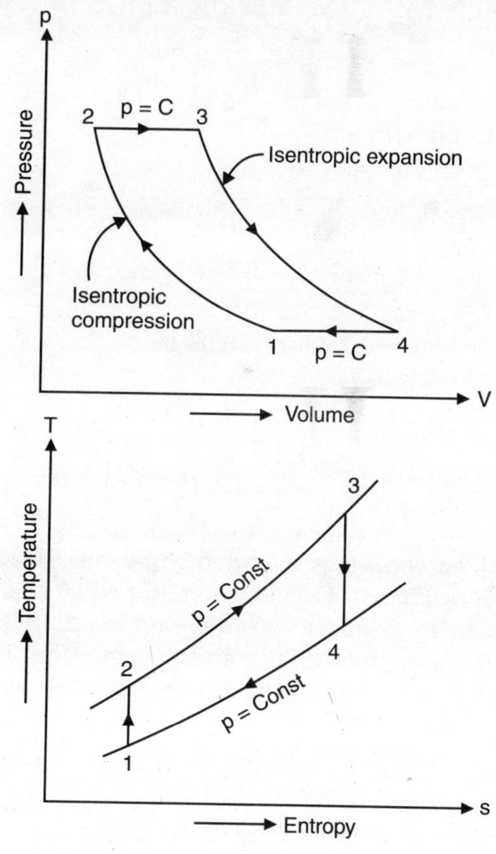

Fig. 23.8.

p_1 =1 bar, T_1 = (25 + 273) = 298 K

and T_3 = 650 + 273 = 923 K

Process 1 – 2 is reversible adiabatic

$$\frac{T_2}{T_1} = \left(\frac{p_2}{p_1}\right)^{\frac{\gamma-1}{\gamma}} = (3)^{\frac{0.4}{1.4}} = 1.369$$

$\therefore \quad T_2 = 298 \times 1.369 = 407.96$ K

Process 3—4 is reversible adiabatic

$$\frac{T_3}{T_4} = \left(\frac{p_3}{p_4}\right)^{\frac{\gamma-1}{\gamma}} = (3)^{\frac{0.4}{1.4}} = 1.369$$

$$\therefore \quad T_4 = \frac{T_3}{1.369} = \frac{923}{1.369} = 674.21 \text{ K}$$

Heat added to air,

$$Q_1 = c_p (T_3 - T_2) = 1.005 (923 - 407.96) = 517.61 \text{ kJ/kg}$$

Heat rejected from air,

$$Q_2 = c_p (T_4 - T_1) = 1.005 (674.21 - 298) = 378.09 \text{ kJ/kg}$$

Cycle efficiency,

$$\eta = \frac{Q_1 - Q_2}{Q_1} = \frac{517.61 - 378.09}{517.61} = 0.2695 \text{ or } \mathbf{26.95\ \%}$$

Alternatively :

$$\eta = 1 - \left[\frac{1}{r_p}\right]^{\frac{\gamma-1}{\gamma}} = 1 - \left[\frac{1}{3}\right]^{\frac{0.4}{1.4}}$$

$$= 1 - 0.73 = 0.27 \text{ or } 27\ \%$$

The compressor and turbine are steady flow machines. Further the compression and expansion processes are reversible adiabatic. Then from steady flow energy equation, the work interactions through these machines equal the enthalpy drops.

Compression work,

$$W_c = h_2 - h_1 = c_p (T_2 - T_1) = 1.005 (407.96 - 298) = 112.51 \text{ kJ/kg}$$

Turbine work,

$$W_t = h_3 - h_4 = c_p (T_3 - T_4) = 1.005 (923 - 674.21) = 250.03 \text{ kJ/kg}$$

Net work output,

$$W_n = W_t - W_c = 250.03 - 112.51 = 139.52 \text{ kJ/kg}$$

Thermal efficiency is also given by,

$$\eta = \frac{\text{net work output}}{\text{heat supplied}} = \frac{139.52}{517.61} = 0.2695 \text{ or } \mathbf{26.95\ \%}$$

Work ratio is defined as the ratio of net work output to turbine expansion work

$$\text{Work ratio} = \frac{139.52}{250.03} = 0.558 \text{ or } \mathbf{55.8\ \%}$$

EXAMPLE 23.2

In an air-standard Brayton cycle, the minimum temperature T_1 is governed by the ambient atmosphere and the maximum temperature T_3 is dictated by the material of construction of the turbine blades. For fixed values of T_1 and T_3, determine the pressure ratio r_p for obtaining maximum net work per unit mass of air undergoing the cyclic change.

(b) Consider a gas turbine plant operating on Brayton cycle with minimum and maximum temperature limits being 300 K and 1000 K. What would be the optimum value of pressure ratio if the turbine is to be operated for maximum power output ? For the pressure ratio thus calculated, determine the plant efficiency, work ratio and specific power output. Assume working substance to be air and expansion and compression to be truly isentropic. Comment on the results obtained.

Solution : Refer Fig. 23.9.

Work output,

$$W = \text{heat supplied} - \text{heat rejected} = c_p (T_3 - T_2) - c_p (T_4 - T_1)$$

1—2 and 3—4 are isentropic processes. Therefore,

$$\frac{T_2}{T_1} = \left(\frac{p_2}{p_1}\right)^{\frac{\gamma-1}{\gamma}} \text{ and } \frac{T_3}{T_4} = \left(\frac{p_3}{p_4}\right)^{\frac{\gamma-1}{\gamma}}$$

$$\text{But, } \frac{p_2}{p_1} = \frac{p_3}{p_4} = r_p\text{, the pressure ratio}$$

$$\therefore \quad W = c_p \left(T_3 - T_1\, r_p^{\frac{\gamma-1}{\gamma}} \right) - c_p \left(T_3\, r_p^{\frac{1-\gamma}{\gamma}} - T_1 \right)$$

$$= c_p\, T_3 \left(1 - r_p^{\frac{1-\gamma}{\gamma}}\right) - c_p T_1 \left(r_p^{\frac{\gamma-1}{\gamma}} - 1\right)$$

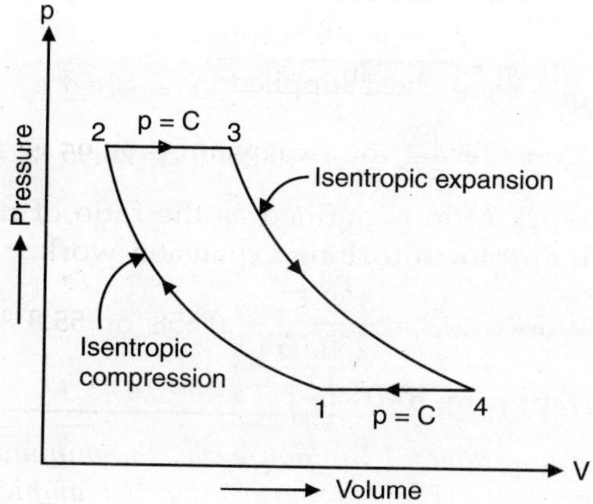

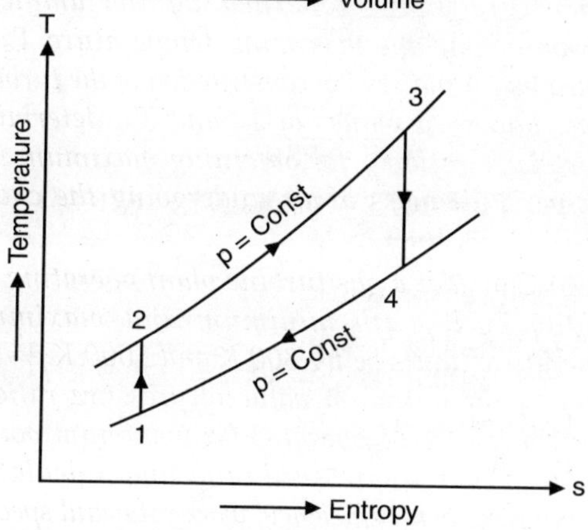

Fig. 23.9.

The pressure ratio for maximum work output can be obtained by differentiating the above expression with respect to r_p and setting the derivative equal to zero.

That is

$$\frac{dW}{dr_p} = -c_p\, T_3\, \frac{1-\gamma}{\gamma}(r_p)^{\frac{1-2\gamma}{\gamma}} - c_p\, T_1\, \frac{\gamma-1}{\gamma}(r_p)^{\frac{-1}{\gamma}}$$

= 0 for maximum work output

$$\text{or } T_3\, \frac{\gamma-1}{\gamma}(r_p)^{\frac{1-2\gamma}{\gamma}} = T_1\, \frac{\gamma-1}{\gamma}(r_p)^{\frac{-1}{\gamma}}$$

$$\text{or} \quad \frac{T_3}{T_1} = (r_p)^{\frac{2(\gamma-1)}{\gamma}} \quad \text{or} \quad r_p = \left(\frac{T_3}{T_1}\right)^{\frac{\gamma}{2(\gamma-1)}}$$

which is the required expression.

(*b*) For maximum power output, pressure ratio is given by

$$r_p = \left(\frac{T_3}{T_1}\right)^{\frac{\gamma}{2(\gamma-1)}}$$

$$= \left(\frac{1000}{300}\right)^{\frac{1.4}{2(1.4-1)}} = 8.22$$

1–2 and 3–4 are isentropic processes. Therefore,

$$T_2 = T_1\left(r_p\right)^{\frac{\gamma-1}{\gamma}}$$

$$= 300\,(8.22)^{\frac{1.4-1}{1.4}} = 548 \text{ K}$$

$$T_4 = \frac{T_3}{\left(r_p\right)^{\frac{\gamma-1}{\gamma}}} = \frac{1000}{(8.22)^{\frac{1.4-1}{1.4}}} = 548 \text{ K}$$

The compressor and turbine are steady flow machines. Further the compression and expansion processes are reversible adiabatic. Then from steady flow energy equation, the work interactions through these machines equal the enthalpy drops . Therefore,

Compressor work,

$$W_c = c_p(T_2 - T_1)$$
$$= 1.005\ (548 - 300) = 249.24 \text{ kJ/kg}$$

Turbine work,

$$W_t = c_p\ (T_3 - T_4)$$
$$= 1.005\ (1000 - 548) = 454.26 \text{ kJ/kg}$$

Net work output,

$$W_n = W_t - W_c$$
$$= 454.26 - 249.24 = 205.02 \text{ kJ/kg}$$

Heat input,

$$Q_1 = c_p\ (T_3 - T_2)$$
$$= 1.005\ (1000 - 548)$$
$$= 454.26 \text{ kJ/kg}$$

Plant efficiency,

$$\eta = \frac{\text{net work output}}{\text{heat input}}$$

$$= \frac{205.02}{454.26} = 0.4513 \text{ or } \mathbf{45.13\ \%}$$

Work ratio is defined as the ratio of net work output to turbine expansion work

$$\text{Work ratio} = \frac{205.02}{454.26} = 0.4513$$

Specific power output is the power available if 1 kg of air is supplied per second.

Specific power output

$$= 205.02\frac{\text{kJ}}{\text{kg}}\times 1\frac{\text{kg}}{\text{s}}$$

$$= 205.02 \text{ kJ/s} = \mathbf{205.02 \text{ kW}}$$

Comments : Under conditions of maximum power output, the temperature at compressor exit (T_2) equals the temperature at exit from the turbine (T_4).

EXAMPLE 23.3

Air enters the compressor of a gas turbine plant operating on Brayton cycle at 1 bar pressure and 300 K temperature. The pressure ratio is 5 and the maximum cycle temperature is limited to 1075 K. If the compressor and turbine efficiencies are 80% and 85% respectively make calculations for the network output, cycle efficiency and the work ratio.

Solution : Refer Fig. 23.10 for the *T-s* diagram. The subscripts denote the state points in the cycle.

Considering the compression process 1–2,

Given : $p_1 = 1$ bar and $T_1 = 300$ K

$r_p = 5$ and $T_3 = 1075$ K

$\eta_c = 0.80$ and $\eta_t = 0.85$

$$T_2' = T_1\left(\frac{p_2}{p_1}\right)^{\frac{\gamma-1}{\gamma}}$$

$$= 300\times(5)^{\frac{1.4-1}{1.4}} = 475.36 \text{ K}$$

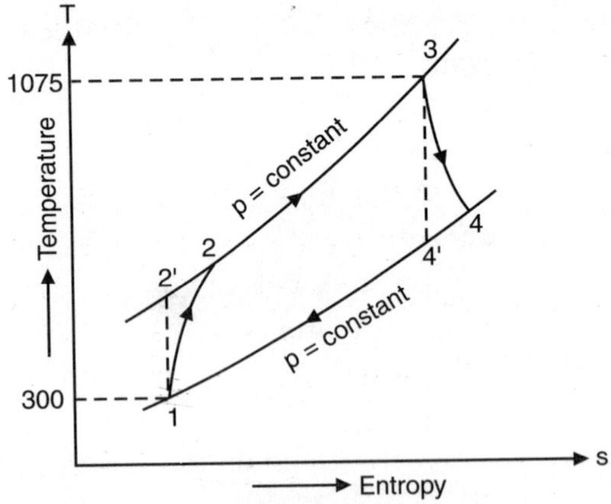

Fig. 23.10.

$$\eta_c = \frac{T_2'-T_1}{T_2-T_1}$$

$$\text{or} \quad 0.8 = \frac{475.36-300}{T_2-300}$$

$$\text{or} \quad T_2 = \frac{475.36-300}{0.8} + 300 = 519.2 \text{ K}$$

Considering the expansion process 3–4,

$$T_4' = T_3\times\left(\frac{p_4}{p_3}\right)^{\frac{\gamma-1}{\gamma}}$$

$$= 1075\times\left(\frac{1}{5}\right)^{\frac{1.4-1}{1.4}} = 678.45 \text{ K}$$

$$\eta_c = \frac{T_3-T_4}{T_3-T_4'}$$

$$\text{or } 0.85 = \frac{1075-T_4}{1075-678.45}$$

$$\text{or} \quad T_4 = 1075 - 0.85(1075 - 678.45)$$

$$= 737.93 \text{ K}$$

Compressor work,

$$W_c = c_p\,(T_2 - T_1)$$

$$= 1.005(519.2 - 300) = 220.3 \text{ kJ/kg}$$

Turbine work,

$$W_t = c_p\,(T_3 - T_4)$$

$$= 1.005(1075 - 737.93)$$

$$= 338.76 \text{ kJ/kg}$$

Heat supplied,

$$Q = c_p\,(T_3 - T_2)$$

$$= 1.005\,(1075 - 519.2) = 558.58 \text{ kJ}$$

Net work output,

$$W_n = W_t - W_c$$

$$= 338.76 - 220.3 = \mathbf{118.46} \text{ kJ/kg}$$

Work ratio is defined as the ratio of net work output to turbine expansion work.

$$\text{Work ratio} = \frac{118.46}{338.76} = \mathbf{0.349}$$

Plant efficiency

$$\eta = \frac{\text{net work output}}{\text{heat supplied}}$$

$$= \frac{118.46}{558.58} = \mathbf{0.212} \text{ or } \mathbf{21.2\%}$$

EXAMPLE 23.4

In a simple constant pressure turbine, the temperatures at inlet to compressor and turbine are stated to be 290 K and 1000 K respectively. For the turbine to operate under conditions of maximum power output, what should be the value of pressure ratio?

For this optimum value of pressure ratio, make calculations for the plant efficiency, work ratio and specific power output. It may be assumed that the working fluid is air and the expansion/compression processes follow the isentropic law $pV^{1.4}$ = constant.

Solution : The pressure ratio for maximum power output is given by

$$r_p = \left(\frac{T_3}{T_1}\right)^{\frac{\gamma}{2(\gamma-1)}} = \left(\frac{1000}{290}\right)^{\frac{1.4}{2(1.4-1)}} = 8.72$$

Further :

$$T_2 = T_1 (r_p)^{\frac{\gamma-1}{\gamma}}$$

$$= 290\,(8.72)^{\frac{1.4-1}{1.4}} = 538.4 \text{ K}$$

$$T_4 = \frac{T_3}{(r_p)^{\frac{\gamma-1}{\gamma}}} = \frac{1000}{(8.72)^{\frac{1.4-1}{1.4}}} = 538.6 \text{ K}$$

With in calculation errors, $T_2 = T_4 = 538.5$ K. Apparently for maximum power output, the temperature at the end of compression (T_2) equals the temperature at the end of expansion (T_4).

Compressor work

$W_c = c_p(T_2 - T_1)$

$= 1.005\,(538.5 - 290) = 249.74$ kJ/kg

Turbine work

$W_t = c_p\,(T_3 - T_4)$

$= 1.005\,(1000 - 538.5) = 463.81$ kJ/kg

Network

$W_n = W_t - W_c$

$= 463.81 - 249.74 = 214.07$ kJ/kg

Heat supplied

$= c_p(T_3 - T_2)$

$= 1.005\,(1000 - 538.5) = 463.81$ kJ/kg

Then:

(*a*) Plant efficiency

$$= \frac{\text{network}}{\text{heat supplied}}$$

$$= \frac{214.07}{463.81} = 0.461 \text{ or } \mathbf{46.1\%}$$

(*b*) Work ratio

$$= \frac{\text{network}}{\text{turbine work}} = \frac{214.07}{463.81} = \mathbf{0.461}$$

(*c*) The specific power output refers to the power available if air is supplied at the rate of 1 kg/sec

∴ Specific power output

$$= 214.07 \frac{\text{kJ}}{\text{kg}} \times 1 \frac{\text{kg}}{\text{s}}$$

$= 214.07$ kJ/s $=$ **214.07 kW**

EXAMPLE 23.5

A gas turbine unit receives air at 1 bar, 300 K and compresses it adiabatically to 4 bar with 80% compressor efficiency. The air leaving the combustion chamber is heated up 850 K and expands upto 1 bar with turbine expansion efficiency 85%. The fuel used has a heating value of 43950 kJ/kg. If the air flow rate is 1.35 kg/sec, determine:

(a) air-fuel ratio for the set-up

(b) power output from the plant,

(c) overall efficiency of the plant, and

(d) back work ratio

It may be presumed that for air

c_p = 1 kJ/kg K and c_v = 0.714 kJ/kg K

and for the products of combustion

c_p = 1.045 kJ/kg K and c_v = 0.774 kJ/kg K

Solution : Refer Fig. 23.11 for *T-s* diagram of the gas turbine cycle.

$$\gamma_a = \frac{1.00}{0.714} = 1.4$$

$$\gamma_g = \frac{1.045}{0.774} = 1.35$$

From compression process

$$T_2' = T_1 \left(\frac{p_2}{p_1}\right)^{\frac{\gamma-1}{\gamma}}$$

$$= 300\left(\frac{4}{1}\right)^{\frac{1.4-1}{1.4}} = 445.8 \text{ K}$$

$$\eta_c = \frac{T_2' - T_1}{T_2 - T_1}$$

$$T_2 = T_1 + \frac{T_2' - T_1}{\eta_c}$$

$$= 300 + \frac{445.8 - 300}{0.8} = 482.25 \text{ K}$$

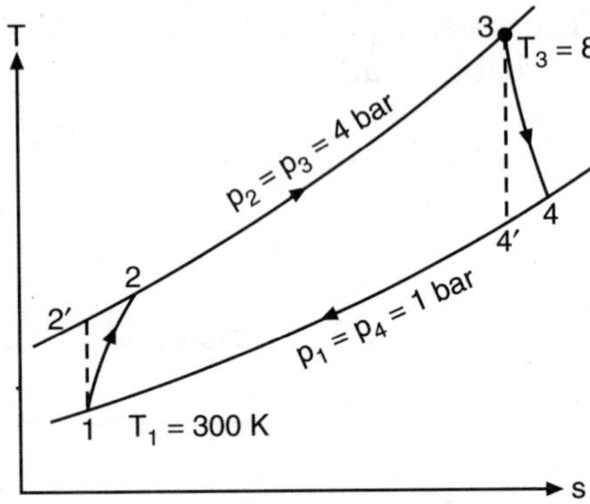

Fig. 23.11.

From expansion process

$$T_4' = \frac{T_3}{\left(\frac{p_3}{p_4}\right)^{\frac{\gamma-1}{\gamma}}}$$

$$= \frac{850}{\left(\frac{4}{1}\right)^{\frac{1.35-1}{1.35}}} = 593.6 \text{ K}$$

$$\eta_t = \frac{T_3 - T_4}{T_3 - T_4'} ;$$

$$T_4 = T_3 - (T_3 - T_4')\,\eta_t$$

$$= 850 - (850 - 593.6) \times 0.85$$

$$= 632.1 \text{ K}$$

(*a*) Enthalpy of air at entry to combustion chamber

$$= m_a\, c_{p_a}\, T_2$$

Enthalpy of gases at exit from combustion chamber

$$= (m_a + m_f)\, c_{p_g}\, T_3$$

The gain in enthalpy is due to thermal energy released by the combustion of fuel. Therefore

$$m_f \times CV = (m_a + m_f)\, c_{p_g}\, T_3 - m_a\, c_{p_a}\, T_2$$

or $$m_f \times 43950 = (m_a \times m_f) \times 1.045 \times 850 - m_a \times 1.0 \times 482.25$$

$$= 888.25\, m_a + 888.25\, m_f - 482.25\, m_a$$

or $$\frac{m_a}{m_f} = \frac{43950 - 888.25}{888.25 - 482.25}$$

$$= 106 \text{ or air-fuel ratio} = \mathbf{106 : 1}$$

(*b*) Compressor work

$$W_c = m_a\, c_{pa}\, (T_2 - T_1)$$
$$= 1 \times 1.0\, (482.25 - 300)$$
$$= 182.25 \text{ kJ/kg of air}$$

Turbine work

$$W_t = (m_a + m_f)\, c_{pg}\, (T_3 - T_4)$$

$$= \left(1 + \frac{1}{106}\right) \times 1.045\,(850 - 632.1)$$

$$= 229.85 \text{ kJ/kg of air}$$

Net work

$$W_n = W_t - W_c$$
$$= 229.85 - 182.25$$
$$= 47.6 \text{ kJ/kg of air}$$

∴ Power output

$$= 1.35 \frac{\text{kg}}{\text{s}} \times 47.6 \frac{\text{kJ}}{\text{kg}}$$

$$= 64.26 \text{ kJ/s} = \mathbf{64.26} \text{ kW}$$

(*c*) Heat supplied/kg of air

$$= \frac{\text{calorific value of fuel}}{\text{air-fuel ratio}}$$

$$= \frac{43950}{106} = 414.62 \text{ kJ}$$

Overall efficiency

$$= \frac{\text{net work}}{\text{heat supplied}}$$

$$= \frac{47.6}{414.62} = 0.114 \text{ or } \mathbf{11.4\%}$$

(*d*) Even though the turbine develops great power, the thermal efficiency of a gas turbine plant is less because a big amount of power is consumed by the compressor. The compressor work is called back work and the ratio of back work to turbine work is called the back work ratio.

Back work ratio

$$= \frac{\text{compressor work}}{\text{turbine work}}$$

$$= \frac{182.25}{229.85} = 0.793 \text{ or } \mathbf{79.3\%}$$

EXAMPLE 23.6

In a gas turbine plant, the air at 1 bar and 288 K is compressed to 4 bar with isentropic compression efficiency of 80%. After passing through regenerator and combustion chamber, the temperature of air is raised to 975 K. The combustion gases then expand in the turbine with 85% isentropic efficiency. The exhaust from the turbine is made to pass through the regenerator which has 75% effectiveness. Determine the thermal efficiency of the plant.

It may be assumed that there is a pressure loss of 0.1 bar in the combustion chamber and a pressure loss of 0.04 bar on each side of regenerator. Further $\gamma = 1.4$ and $c_p = 1$ kJ/kg K both for air and gases.

Solution : Refer Fig. 23.12 from compression process

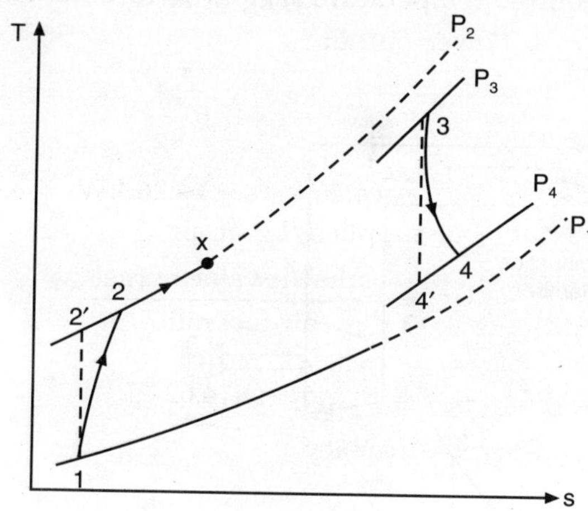

Fig. 23.12.

$$T_2' = T_1 \left(\frac{p_2}{p_1}\right)^{\frac{\gamma-1}{\gamma}}$$

$$= 288 \left(\frac{4}{1}\right)^{\frac{1.4-1}{1.4}} = 428.13 \text{ K}$$

$$\eta_c = \frac{T_2' - T_1}{T_2 - T_1}$$

$$T_2 = T_1 + \frac{T_2' - T_1}{\eta_c}$$

$$= 288 + \frac{428.13 - 288}{0.8} = 463.16 \text{ K}$$

From expansion process

$p_3 = p_2$ – pressure loss in (regenerator + combustion chamber)

$= 4 - (0.04 + 0.1) = 3.86$ bar

$p_4 = p_1$ + pressure loss in regenerator

$= 1 + 0.04 = 1.04$ bar

$$\frac{T_4'}{T_3} = \left(\frac{p_4}{p_3}\right)^{\frac{\gamma-1}{\gamma}} = \left(\frac{1.04}{3.86}\right)^{\frac{1.4-1}{1.4}} = 0.6872$$

$$T_4' = 0.6872 \times T_3 = 0.6872 \times 975 = 670 \text{ K}$$

$$\eta_t = \frac{T_3 - T_4}{T_3 - T_4'}$$

$$T_4 = T_3 - (T_3 - T_4')\,\eta_t$$

$$= 975 - (975 - 670) \times 0.85 = 715.75 \text{ K}$$

Work input to compressor

$$W_c = c_p(T_2 - T_1) = 1.005\,(463.16 - 288)$$

$$= 176.03 \text{ kJ/kg}$$

Work output from turbine

$$W_t = c_p(T_3 - T_4)$$

$$= 1.005(975 - 715.75) = 260.55 \text{ kJ/kg}$$

∴ Net work output

$$W_n = W_t - W_c$$

$$= 260.55 - 176.03 = 84.52 \text{ kJ/kg}$$

The regenerative effectiveness is given by

$$\varepsilon = \frac{T_x - T_2}{T_4 - T_2} = \frac{T_x - 463.16}{715.75 - 463.16}$$

or $T_x = 463.16 + (715.75 - 463.16) \times 0.75$

$= 652.58$ K

Heat supplied in the combustion chamber

$$c_p(T_3 - T_x) = 1.005(975 - 652.58)$$

$$= 324.03 \text{ kJ/kg}$$

∴ Thermal efficiency of the plant

$$\eta_{th} = \frac{\text{net work output}}{\text{heat supplied}}$$

$$= \frac{84.52}{324.03} = 0.2608 \text{ or } \mathbf{26.08\%}$$

23.3. THERMAL REFINEMENT OF A GAS TURBINE CYCLE

The following systems are practised for improving the power output and thermal efficiency of a gas turbine plant:

(*i*) Regeneration
(*ii*) Intercooling
(*iii*) Reheating

The improvement is however, at the sacrifice of simplicity weight, space and cost.

23.3.1. Regeneration

The regeneration manifests in preheating the air with turbine exhaust and that results in decrease of heat supplied in the combustion chamber and thereby gaining of thermal efficiency. The temperature of exhaust gases leaving the turbine is quite high. A part of this heat can be recovered by utilizing the hot gases to preheat the compressed air before it enters the combustion chamber. The heat exchange is done in a counter flow heat exchanger called regenerator.

Fig. 23.13 shows the schematic arrangement of gas turbine plant incorporating a regenerator, along with the corresponding *T-s* plot. The temperature of the compressed air has been raised from T_2 to T_a in the regenerator with a fall in the exhaust gas temperature from T_4 to T_b.

Net work output = turbine work − compressor work

$$= c_p\,(T_3 - T_4) - c_p\,(T_2 - T_1)$$

Heat supplied in the combustor

$$= c_p\,(T_3 - T_a)$$

The thermal efficiency of the Brayton cycle employing regeneration is then given by

$$\eta_{th} = \frac{c_p(T_3 - T_4) - c_p\,(T_2 - T_1)}{c_p\,(T_3 - T_a)}$$

With introduction of regenerator in the circuit,

(*i*) there is no change in compressor work, turbine work and so the net workoutput.

(*ii*) there is substantial reduction in the quantity of fuel required to bring the air to required temperature at entry to turbine. This

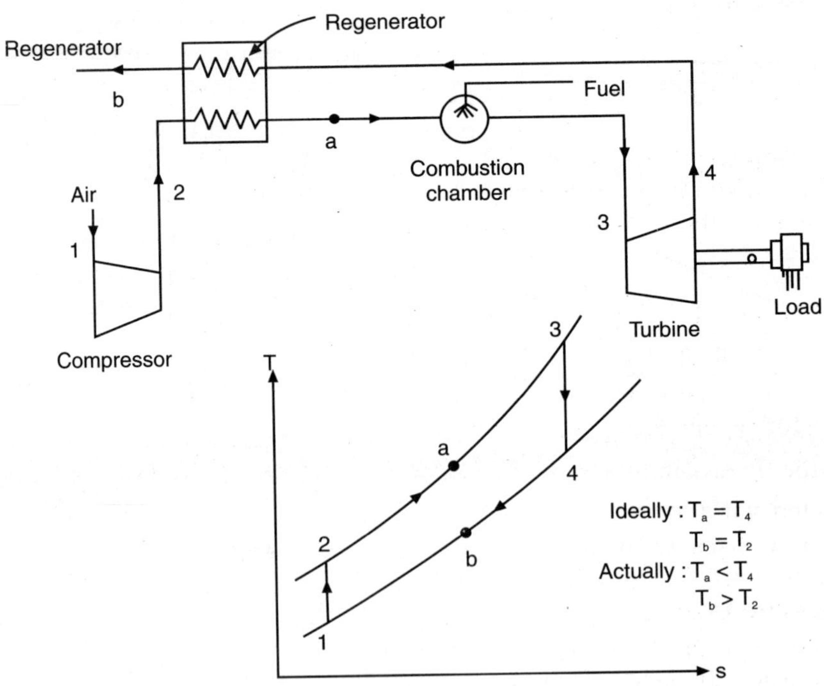

Fig. 23.13. Gas turbine cycle with regeneration

stems from the fact that the temperature of air entering the combustion chamber has increased.

These aspects result in an increase in thermal efficiency of the cycle.

(*iii*) it is theoretically possible to raise the temperature of compressed air leaving the compressor from T_2 to $T_a = T_4$ and lower the temperature of exhaust gases from T_4 to $T_b = T_2$. The system is then called ideal regenerator and with that the thermal efficiency becomes

$$\eta_{th} = \frac{(T_3 - T_4) - (T_2 - T_1)}{T_3 - T_4}$$

$$= 1 - \frac{T_2 - T_1}{T_3 - T_4}$$

$$= 1 - \frac{T_1}{T_3}\left[\frac{T_2/T_1 - 1}{1 - T_4/T_3}\right]$$

From the isentropic compression and expansion processes,

$$\frac{T_2}{T_1} = \frac{T_3}{T_4} = (r_p)^{\frac{\gamma-1}{\gamma}}$$

where r_p = pressure ratio

With this substitution

$$\eta_{th} = 1 - \frac{T_1}{T_3}(r_p)^{\frac{\gamma-1}{\gamma}} \qquad \text{...(23.9)}$$

When the compressor and turbine efficiencies η_c and η_t are taken into consideration, the above expression for thermal efficiency would take the form

$$\eta_{th} = 1 - \frac{1}{\eta_c\,\eta_t}\frac{T_1}{T_3}(r_p)^{\frac{\gamma-1}{\gamma}} \qquad \text{...(23.10)}$$

Variation of thermal efficiency with cycle pressure ratio for regenerative cycle is shown in Fig. 23.14 for fixed values of $t = T_3/T_1$.

The depicted variation suggests that :

(*i*) Efficiency rises with an increase in temperature ratio $t = T_3/T_1$, and decreases with increasing pressure ratio.

(*ii*) When r_p = 1, the efficiency becomes equal to the Carnot efficiency. That is

$$\eta = \frac{T_3 - T_1}{T_3} = \frac{T_{max} - T_{min}}{T_{max}}$$

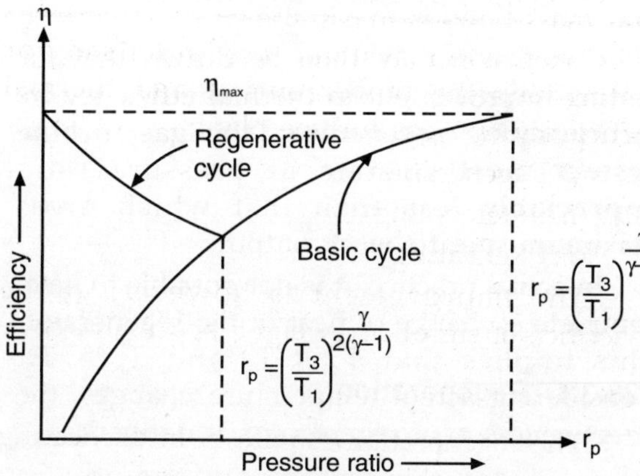

Fig. 23.14.

(*iii*) With regeneration, the peak efficiency occurs at much lower pressure ratio. This may be attributed to the fact that with increase in pressure ratio, the temperature T_2 of compressed air increases while the turbine exhaust temperature T_4 reduces. Consequently the heat available for regeneration is reduced. The point at which the regenerative curve meets the basic cycle curve corresponds to the pressure ratio when $T_2 = T_4$. In that case, the regenerator becomes superfluous. It is precisely due to this reason, that the constant $t = T_3/T_1$ line is not extended beyond the point where it meets the efficiency curve for the basic cycle.

Apparently then:

$$c_p(T_3 - T_4) = c_p(T_3 - T_2)$$

or $$T_3\left[1 - \frac{T_4}{T_3}\right] = T_3\left[1 - \frac{T_2}{T_3}\right]$$

$$= T_3\left[1 - \frac{T_2}{T_1} \times \frac{T_1}{T_3}\right]$$

or $$1 - \frac{1}{(r_p)^{\frac{\gamma-1}{\gamma}}} = 1 - \frac{T_1}{T_3}(r_p)^{\frac{\gamma-1}{\gamma}}$$

Simplification gives:

$$r_p = \left(\frac{T_3}{T_1}\right)^{\frac{\gamma}{2(\gamma-1)}} \qquad \text{...(23.11)}$$

As shown earlier, (equation 23.4), this is the optimum value of pressure ratio at which the specific work output (the net output per unit mass flow) is maximum.

Conclusion may then be drawn that for getting improvement in thermal efficiency by incorporating regenerator, the gas turbine system must operate at pressure ratio appreciably less than that which gives maximum specific work output.

In actual practice, it is not possible to have complete exchange of heat in the regenerator. This implies that $T_a < T_4$ and $T_b > T_2$. Considering actual temperature changes, the effectiveness $\in$ of regenerator is defined as

$$\in = \frac{\text{actual heat received by compressed air}}{\text{maximum possible heat which could be transferred from gases}}$$

$$= \frac{m_a\, c_{pa}\,(T_a - T_2)}{mg\, c_{pg}\,(T_4 - T_2)} \quad \text{...(23.12)}$$

The performance of the regenerator is also expressed by thermal ratio R which is prescribed as

$$R = \frac{\text{actual temperature rise of air}}{\text{theoretical temperature rise with counter flow heat exchanges of infinite size}}$$

$$= \frac{T_a - T_2}{T_4 - T_2} \quad \text{...(23.13)}$$

Apparently, the thermal ratio equals the regenerator effectiveness when $m_a\, c_{pa} = m_g\, c_{pg}$

A regenerator would give the highest thermal efficiency if its surface for heat transfer is of infinite area. However, the employment of such a regenerator gets restricted due to:

(*i*) a drop in pressure of compressed air due to resistance in flow of gases through piping

(*ii*) an increase of back pressure of the turbine

(*iii*) weight and space limitations, and cost

These aspects need to be weighed against the saving in fuel consumption that would be affected by using a regenerator.

23.3.2. Intercooling

The compression process in the compressor unit is achieved in two or more stages, and intercooling is done at constant pressure after every stage. The cooled air has a lesser volume enabling it to be compressed in a small compressor with less expenditure of energy.

Fig. 23.15 shows the schematic arrangement of a gas turbine plant involving two stage compression process with complete intercooling at the intermediate pressure p_i. For simplicity the compression and expansion processes have been assumed isentropic while sketching the corresponding *T-s* plot.

Consider the compression in two stages 1 – *a* and *b* – 2 and perfect cooling, *i.e.*, the temperature T_a at exit from the low pressure compressor is reduced to its original temperature $T_b = T_1$ before entry to the high pressure compressor. Then the compressor work done on the air is

Compressor work

$$W_c = c_p\,[(T_a - T_1) + (T_2 - T_b)]$$

$$= c_p\left[T_1\left\{\frac{T_a}{T_1} - 1\right\} + T_b\left\{\frac{T_2}{T_b} - 1\right\}\right]$$

$$= c_p\, T_1\left[\frac{T_a}{T_1} + \frac{T_2}{T_b} - 2\right] \quad (\because T_b = T_1)$$

$$= c_p\, T_1\left[\left(\frac{p_i}{p_1}\right)^m + \left(\frac{p_2}{p_i}\right)^m - 2\right]$$

where $m = \dfrac{\gamma - 1}{\gamma}$

The optimum value of p_i for minimum compressor work can be determined by differentiating the above expression with respect to p_i and equating the result to zero. That is

$$\frac{dW_c}{dp_i} = c_p\, T_1\left[\frac{m}{p_1^m}\, p_i^{m-1} + p_2^m\,(-m) \times \frac{1}{p_i^{m+1}}\right] = 0$$

or $$\frac{p_i^{m-1}}{p_1^m} = \frac{p_2^m}{p_i^{m+1}}\ ;\ p_i^{2m} = (p_1\, p_2)^m$$

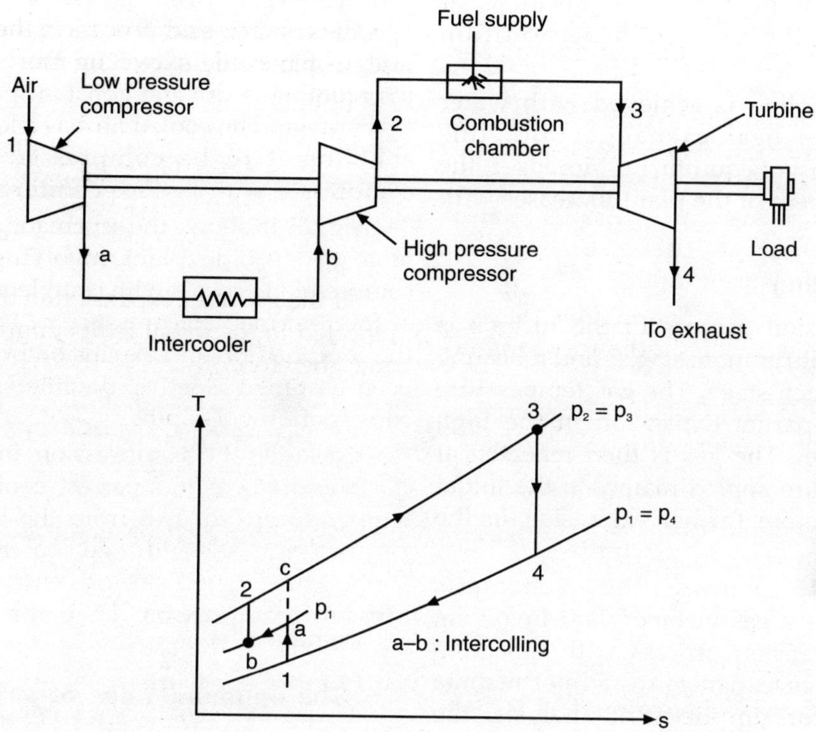

Fig. 23.15. Two stage compression with intercooling

That gives:

$$p_i = \sqrt{p_1 p_2} \quad \text{or} \quad \frac{p_i}{p_1} = \frac{p_2}{p_i} \qquad \text{...(23.14)}$$

Thus for minimum compression work, the pressure ratio in each stage is equal. With introduction of intercooler in the circuit,

(*i*) The temperatures after compression in each stage are equal and the work input for each compression stage is same.

Compressor work

$$\begin{aligned} W_c &= c_p (T_a - T_1) + c_p (T_2 - T_b) \qquad \text{...}(i) \\ &= c_p (T_a - T_1) + c_p (T_a - T_1) \\ &= 2\, c_p (T_a - T_1) \end{aligned}$$

With single stage compressor, the compressor work is

$$\begin{aligned} &= c_p(T_c - T_1) \\ &= c_p(T_a - T_1) + c_p(T_c - T_a) \qquad \text{...}(ii) \end{aligned}$$

The curvature of constant pressure lines on *T-s* diagram is such that the vertical distances between them increases towards the right. Accordingly $(T_c - T_a) > (T_2 - T_b)$. From the comparison of identities. (*i*) and (*ii*), it becomes evident that work supplied to compressor gets reduced if compression process is carried out in two stages with intercooling after the first stage.

(*ii*) The turbine work is the same as that for the basic cycle.

$$\text{Work ratio } r_w = \frac{W_t - W_c}{W_t}$$

Since intercooling reduces the compressor work without affecting the turbine work, an improvement results in work ratio with the installation of an intercooler.

(*iii*) The delivery of air after the compression process in two stages is at a lower temperature than that in single stage compressor, *i.e.*, $T_2 < T_c$. The lower temperature leads to additional heat requirement in the combustion chamber. For that more amount of fuel is to be burnt for attaining the required turbine inlet temperature. Although the net output is increased by intercooling, the thermal efficiency may not increase because of simultaneous increase in heat addition requirement.

(*iv*) Intercooling is achieved with water cooled surface heat exchanger and this increases the cost of equipment. Moreover, the water requirement of the plant increases with intercooling.

23.3.3. Reheating

The expansion process in the turbine is carried out in two or more stages, and reheating is done after each stage. The gas temperature drops due to partial expansion in the high pressure section. The gas is then reheated at constant pressure approximately to the initial temperature before further expansion in the next stage.

Fig. 23.16 shows the schematic arrangement of a gas turbine plant involving two stage expansion process with reheating (after first stage expansion) at intermediate pressure p_i. For simplicity of analysis, the compression process 1-2 and expansion processes 3-*a* and *b*-4 have been assumed isentropic while sketching the corresponding *T-s* plot.

Turbine work

$$W_t = c_p(T_3 - T_a) + c_p(T_b - T_4)$$

$$= c_p\, T_3\left[1 - \frac{T_a}{T_3}\right] + c_p\, T_b\left[1 - \frac{T_4}{T_b}\right]$$

If the gas after expansion in the first stage is reheated to the initial temperature, then $T_b = T_3$ and, therefore

$$W_t = c_p\, T_3\left[\left(1 - \frac{T_a}{T_3}\right) + 1 - \frac{T_4}{T_b}\right]$$

$$= c_p\, T_3\left[2 - \frac{1}{(p_2/p_i)^m} - \frac{1}{(p_i/p_1)^m}\right]$$

where $m = \dfrac{\gamma - 1}{\gamma}$

The optimum value of p_i for maximum

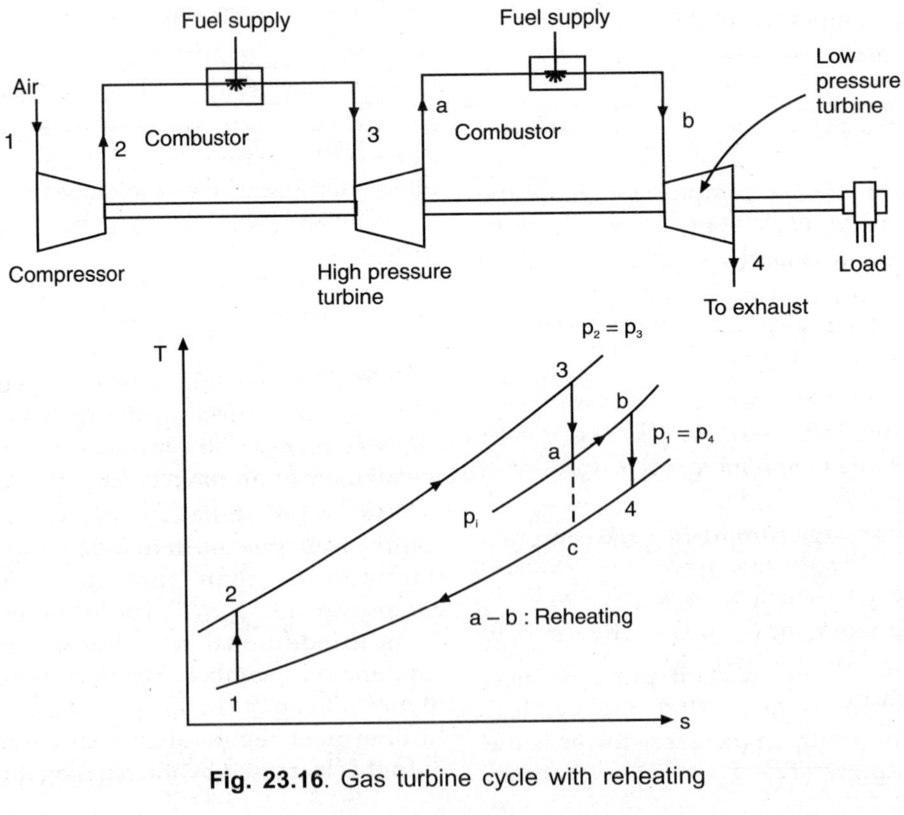

Fig. 23.16. Gas turbine cycle with reheating

turbine work can be determined by differentiating the above expression with respect to p_i and equating the result to zero. That is

$$\frac{dW_t}{dp_i} = \left[-\frac{1}{(p_2)^m} m\,(p_i)^{m-1} - (p_1)^m (-m)\,(p_i)^{-m-1} = 0 \right.$$

or

$$\frac{1}{(p_2)^m}(p_i)^{m-1} = (p_1)^m\,(p_i)^{-m-1}$$

$$(p_i)^{2m} = (p_1 p_2)^m$$

That gives:

$$p_i = \sqrt{p_1\,p_2}$$

or $$\frac{p_2}{p_i} = \frac{p_i}{p_1} \qquad ...(23.15)$$

Thus for maximum turbine work the pressure ratio for each expansion stage is equal.

With introduction of reheating system in the circuit.

(*i*) The temperatures after expansion in each stage are equal and the work output for each expansion stage is same.

Turbine work

$$= c_p(T_3 - T_a) + c_p(T_b - T_4) \qquad ...(i)$$
$$= c_p(T_3 - T_a) + c_p(T_3 - T_a)$$
$$= 2c_p\,(T_3 - T_a)$$

With single stage expansion, the turbine work is

$$c_p(T_3 - T_c) = c_p(T_3 - T_a) + c_p(T_a - T_c) \qquad ...(ii)$$

The constant pressure lines on *T-s* diagram diverge from left to right, *i.e.*, the vertical distance between them increases towards right. Accordingly $(T_b - T_4) > (T_a - T_c)$. From the comparison of identities (*i*) and (*ii*), it becomes evident that the turbine work of a reheating cycle is greater than that of a basic cycle.

(*ii*) The work input to compressor is the same as that for the basic cycle.

Work ratio

$$r_w = \frac{W_t - W_c}{W_t}$$

Since reheating increases the turbine work without affecting the compressor work, there is improvement in work ratio due to reheating.

(*iii*) There is greater fuel consumption or heat supply because of additional heat supplied for reheating. Although the net output is also increased by reheating, the effect of greater fuel consumption generally predominates and there occurs a small drop in thermal efficiency.

(*iv*) With reheating, the temperature of gases leaving the low pressure turbine is much higher. As such the reheat cycle offers potential for use of regenerator which would recover some of the energy going with exhaust gases.

When two turbines are used, the high pressure turbine usually feeds the compressor and the low pressure turbine provides the net output. This arrangement improves the plant efficiency because the high pressure turbine (called the compressor turbine) can be operated at a speed most suitable for the compressor. This is quite independent of the speed of the low pressure turbine (called the power turbine) that supplies the useful power for external load.

EXAMPLE 23.7

In a gas turbine plant operating on Brayton cycle the air at 15°C and 1 bar atmospheric pressure enters the compressor. After compression to 5 bar, the air enters the combustion chamber and is heated to maximum temperature of 1075 K. There after the heated air enters the turbine and expands to atmospheric pressure. The polytropic efficiency for the compressor and turbine is 0.87 and the fuel used has a calorific value of 41800 kJ/kg. Make calculations for the specific output, over all efficiency, air-fuel ratio and specific fuel consumption for the plant.

Take C_p = 1.005 kJ/kg K both for air and gases.

Solution : Refer Fig. 23.17.

For compressor unit :

$$\eta_{pc} = \frac{n}{n-1} \times \frac{\gamma - 1}{\gamma}$$

$$0.87 = \frac{n}{n-1} \times \frac{1.4-1}{1.4}$$

$$\frac{n}{n-1} = \frac{1.4 \times 0.87}{1.4-1} = 3.045$$

$n = 1.489$

Then $T_2 = \left(\frac{p_2}{p_1}\right)^{\frac{n-1}{n}}$

$$= 288\left(\frac{5}{1}\right)^{\frac{1.489-1}{1.489}} = 488.58 \text{ K}$$

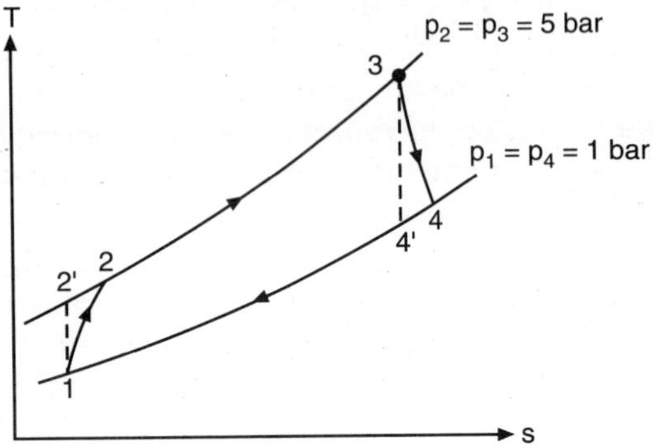

Fig. 23.17

For turbine unit:

$$\eta_{p_t} = \frac{n-1}{n} \times \frac{\gamma}{\gamma - 1};$$

$$0.87 = \frac{n-1}{n} \times \frac{1.4}{1.4-1}$$

$$\frac{n-1}{n} = 0.2486 \text{ or } n = 1.33$$

Then $\frac{T_3}{T_4} = \left(\frac{p_3}{p_4}\right)^{\frac{n-1}{n}} = \left(\frac{5}{1}\right)^{\frac{1.33-1}{1.33}} = 1.492$

$$T_4 = \frac{T_3}{1.492} = \frac{1075}{1.492} = 720.51 \text{ K}$$

(*a*) Turbine work $= c_p\,(T_3 - T_4)$
$= 1.005\,(1075 - 720.51)$
$= 356.26$ kJ/kg

Compressor work $= c_p\,(T_2 - T_1)$
$= 1.005\,(488.58 - 288)$
$= 201.58$

Net work output
= turbine work − compressor work
= 356.26 − 201.58 = 154.68 kJ/kg

The term specific out implies the power out for 1 kg of air supplied per second. That is $m = 1$ kg/s

∴ Specific power output

$$= 154.68 \frac{\text{kJ}}{\text{kg}} \times 1 \frac{\text{kg}}{\text{s}}$$

= 154.68 kJ/s = **154.68 kW**

(*b*) Heat supplied $= c_p(T_3 - T_2)$
$= 1.005(1075 - 488.58)$
$= 589.35$ kJ/kg

overall thermal efficiency

$$\eta_{th} = \frac{\text{net work output}}{\text{heat supplied}}$$

$$= \frac{154.68}{589.35} = 0.262 \text{ or } \mathbf{26.2\%}$$

(*c*) Enthalpy of air at entry to combustion chamber $= m_a\, c_{pa}\, T_2$

Enthalpy of gases at exit from combustion chamber $= (m_a + m_f)\, c_{pg}\, T_3$

Gain in enthalpy

$$= (m_a + m_f)\, c_{pg}\, T_3 - m_a\, c_{pa}\, T_2$$

This gain in enthalpy is due to thermal energy released by the combustion of fuel. Heat balance then gives

$$m_f \times 41800 = (m_a + m_f) \times 1.005 \times 1075 - m_a \times 1.005 \times 488.58$$
$$= 1080.37\, m_a + 1080.37\, m_f - 491.02\, m_a$$

∴ Air fuel ratio

$$\frac{m_a}{m_f} = \frac{41800 - 1080.37}{1080.37 - 491.02} = 69.09$$

(*d*) Fuel consumption per hour for 1 kg/s of air $= \frac{3600}{69.09} = \frac{3600}{69.09}$

Then specific fuel consumption (fuel consumption per kW of power output) is

$$sfc = \frac{3600}{69.09 \times 154.68}$$

= **0.3369 kg/kWh**

EXAMPLE 23.8

An open cycle gas turbine power plant takes in air at 1 bar and 300 K, and operates with pressure ratio 5.5. The compression is carried out in two stages with equal pressure ratio for each stage and perfect intercooling in between. The maximum temperature of the cycle is limited to 975 K. The isentropic efficiency for each compressor stage is 85% and that of turbine is 90%. For air flow rate of 1.2 kg/s, determine the efficiency of power plant and the power developed.

Take c_p = 1.02 kJ/kgK and γ = 1.4, ignore combustion and ducting losses, and assume that the mass of fuel is small compares with that of air.

Solution : Refer Fig. 23.18 with equal pressure ratio in each stage.

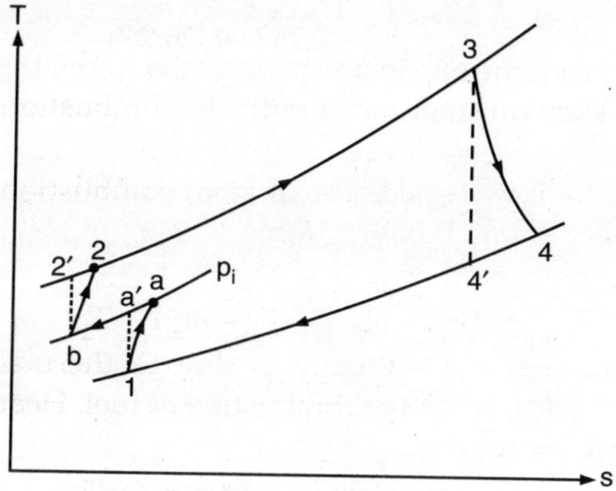

Fig. 23.18.

$$\frac{p_i}{p_1} = \frac{p_2}{p_i} = \sqrt{\frac{p_2}{p_1}} = \sqrt{5.5} = 2.345$$

$$T_a' = T_1\left(\frac{p_i}{p_1}\right)^{\frac{\gamma-1}{\gamma}}$$

$$= 300 \times (2.345)^{\frac{1.4-1}{1.4}} = 382.8 \text{ K}$$

Compressor efficiency

$$\eta_c = \frac{T_a' - T_1}{T_a - T_1}$$

$$\therefore \quad T_a = T_1 + \frac{T_a' - T_1}{\eta_c}$$

$$= 300 + \frac{382.8 - 300}{0.85} = 397.4 \text{ K}$$

With perfect intercooling, equal pressure ratio, same isentropic efficiency for both compressor stages, the work input in each stage would be same.

$$T_a - T_1 = T_2 - T_b \; ; \; T_b = T_1 \text{ and } T_2 = T_a$$

$$W_c = mc_p\,(T_a - T_1) + mc_p\,(T_2 - T_b)$$
$$= 2\,mc_p\,(T_a - T_1)$$
$$= 2 \times 1.2 \times 1.02(397.4 - 300)$$
$$= 238.43 \text{ kJ/s}$$

From expansion process 3-4 in the turbine

$$\frac{T_3}{T_4'} = \left(\frac{p_3}{p_4}\right)^{\frac{\gamma-1}{\gamma}}$$

$$= \left(\frac{5.5}{1}\right)^{\frac{1.4-1}{1.4}} = 1.628$$

$$T_4' = \frac{T_3}{1.628} = \frac{975}{1.628} = 598.9 \text{ K}$$

$$\eta_t = \frac{T_3 - T_4}{T_3 - T_4'}$$

$$\therefore \; T_4 = T_3 - (T_3 - T_4')\,\eta_t$$
$$= 975 - (975 - 598.9) \times 0.9 = 636.5$$

Turbine work

$$W_t = m\,c_p\,(T_3 - T_4)$$
$$= 1.2 \times 1.02\,(975 - 636.5)$$
$$= 414.32 \text{ kJ/s}$$

Network

$$W_n - W_c = 414.32 - 238.43$$
$$= \mathbf{175.89 \text{ kJ/s (kW)}}$$

The net work represents the power developed by the gas turbine plant.

(*b*) Heat supplied in the combustion chamber,

$$= mc_p(T_3 - T_2)$$
$$= 1.2 \times 1.02(975 - 397.4)$$
$$= 706.98 \text{ kJ/s}$$

Thermal efficiency

$$= \frac{\text{net work output}}{\text{heat supplied}}$$

$$= \frac{175.89}{706.98} = 0.2488 \text{ or } \mathbf{24.88\%}$$

EXAMPLE 23.9

An open cycle gas turbine plant sucks in air at 1 bar and 288 K and compresses it to 6 bar. After passing through regenerator and combustion chamber, the working medium expands in the high pressure turbine which drives the compressor. Further expansion continues in the low pressure turbine. Find the specific output and thermal efficiency. Consider the following data:

Polytropic efficiency of compression and expansion = 85%

Thermal ratio of regenerator = 0.75

Heat exchanger pressure loss = 0.07 bar on each side

Combustion pressure loss = 0.1 bar

Maximum cycle temperature = 900 K

For air: $\gamma = 1.4$ and $c_{pa} = 1.005$ kJ/kg K

For gases : $\gamma = 1.33$ and $c_{pg} = 1.13$ kJ/kg K

Neglect all mechanical losses.

Solution : Refer Fig. 23.19.

Compressor polytropic efficiency

$$= \frac{\gamma - 1}{\gamma} \times \frac{n}{n-1}$$

$$0.85 = \frac{1.4 - 1}{1.4} \times \frac{n}{n-1} \; ; n = 1.5$$

$$T_2 = T_1 \left(\frac{p_2}{p_1}\right)^{\frac{n-1}{n}}$$

$$= 288 \left(\frac{6}{1}\right)^{\frac{1.5-1}{1.5}} = 523 \text{ K}$$

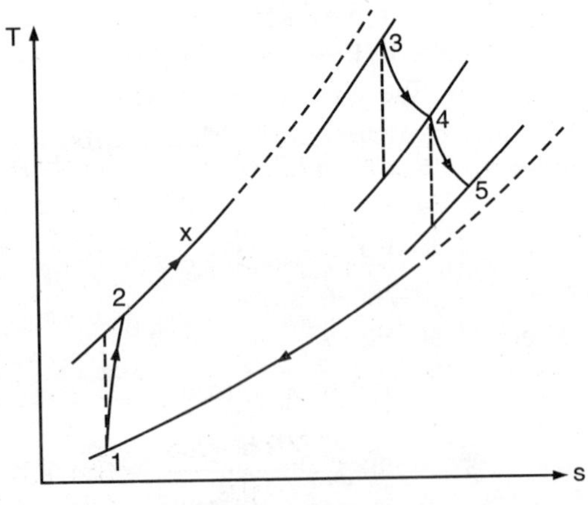

Fig. 23.19.

The high pressure turbine has only to run the compressor and hence its output must equal the work requirement of the compressor.

$$\therefore \quad c_{pg}(T_3 - T_4) = c_{pa}(T_2 - T_1)$$

$$T_4 = T_3 - \frac{c_{pa}}{c_{pg}}(T_2 - T_1)$$

$$= 900 - \frac{1.005}{1.13}(523 - 288)$$

$$= 691 \text{ K}$$

Turbine polytropic efficiency

$$= \frac{\gamma}{\gamma - 1} \times \frac{n-1}{n}$$

$$0.85 = \frac{1.33}{1.33 - 1} \times \frac{n-1}{n} \; ; n = 1.27$$

From the expansion process 3-4 in the high pressure turbine,

$$\frac{T_3}{T_4} = \left(\frac{p_3}{p_4}\right)^{\frac{n-1}{n}}$$

$$\frac{900}{691} = \left(\frac{p_3}{p_4}\right)^{\frac{1.27-1}{1.27}} \; ; \frac{p_3}{p_4} = 3.47$$

$p_3 = p_2 -$ (pressure loss in regenerator and combustion chamber)

$= 6 - (0.07 + 0.1) = 5.83$ bar

$\therefore$ Intermediate pressure $p_4 = \dfrac{5.83}{3.47}$

$= 1.68$ bar

Pressure at exit from low pressure turbine,

$p_5 = p_1$ + pressure loss in regenerator

$= 1 + 0.07 = 1.07$ bar

From the expansion process 4 - 5 in the low pressure turbine,

$$\frac{T_4}{T_5} = \left(\frac{p_4}{p_5}\right)^{\frac{n-1}{n}}$$

$$\frac{691}{T_5} = \left(\frac{1.68}{1.07}\right)^{\frac{1.27-1}{1.27}} \; ; T_5 = 628 \text{ K}$$

Power output from the low pressure turbine,

$$= c_{pg}(T_4 - T_5)$$
$$= 1.13(691 - 628)$$
$$= 71.19 \text{ kJ/kg}$$

If T_x is the temperature of air after it passes out through heat exchanger, then

$$\text{Thermal ratio} = \frac{T_x - T_2}{T_5 - T_2} ; 0.75 = \frac{T_x - 523}{628 - 523}$$

$$\text{or} \quad T_x = 523 + (628 - 523)\ 0.75$$
$$= 601.75 \text{ K}$$

Heat supplied in combustion chamber,

$$= c_{pg}\ (T_3 - T_x)$$
$$= 1.13\ (900 - 601.75)$$
$$= 337.02 \text{ kJ/kg}$$

∴ Thermal efficiency of the plant

$$= \frac{\text{output from power turbine (LP turbine)}}{\text{heat supplied in combustion chamber}}$$

$$= \frac{71.19}{337.02} = 0.211 \text{ or } \mathbf{21.1\%}$$

Specific power output

= power output per kg of air supplied per second

$$= 71.19 \frac{\text{kJ}}{\text{kg}} \times 1 \frac{\text{kg}}{\text{s}} = \mathbf{71.19\ kW}$$

EXAMPLE 23.10

In a gas turbine installations, two stages of compression are used with intercooling between stages; the overall pressure ratio is 5 : 1. The maximum temperature in the cycle is 1050 K and the expansion to the original pressure is carried out in two stages with reheating to 1050 K between the stages. The intermediate pressure between both the compression and expansion stages is for optimum conditions.

The isentropic efficiencies of the compressors and turbines are 0.8 and 0.85 respectively. Determine –

(a) power output for a mass flow of 15 kg/s of air,

(b) overall thermal efficiency and

(c) air-fuel ratio.

It may be assumed that temperature of air at suction is 15°C, c_p = 1.005 kJ/kg K throughout the cycle and calorific value of fuel is 42000 kJ/kg.

Solution : Refer Fig. 23.20.

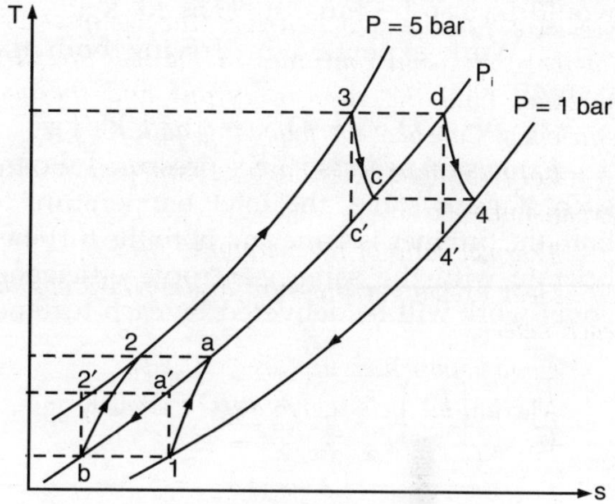

Fig. 23.20.

For optimum conditions, the intermediate pressure p_i between the compression and expansion stages is

$$p_i = \sqrt{p_1 p_2}$$
$$= \sqrt{5} = 2.236 \text{ bar}$$

Compression process : Since intercooling is perfect, pressure ratio for each stage is same and each compressor operates with same isentropic efficiency, equal work will be expended in driving in each compressor. From isentropic compression process in the low pressure compressor.

$$T_a' = T_1 \left(\frac{p_i}{p_1}\right)^{\frac{\gamma - 1}{\gamma}}$$

$$= 288 \times (2.236)^{\frac{1.4-1}{1.4}} = 362.5 \text{ K}$$

$$\eta_c = \frac{T_a' - T_1}{T_a - T_1} ;$$

$$T_a = T_1 + \frac{T_a' - T_1}{\eta_c}$$

$$= 288 + \frac{362.5 - 288}{0.8} = 381.1 \text{ K}$$

Work input to low pressure compressor

$= c_p (T_a - T_1)$

$= 1.005(381.1 - 288) = 93.56$ kJ/kg

The work input to the high compressor would be also be equal to 93.56 kJ/kg.

∴ Work expended in driving both the compressors

$$W_c = 93.56 + 93.56 = 187.1 \text{ kJ/kg}$$

Expansion process : Since pressure ratio for each stage is same, the inlet temperature to both the turbines is same and both the turbines operate with the same isentropic efficiency, equal work will be delivered by each turbine.

$$\frac{T_3}{T_c'} = \left(\frac{p_3}{p_i}\right)^{\frac{\gamma-1}{\gamma}} = (2.236)^{\frac{1.4-1}{1.4}} = 1.2588$$

$$T_c' = \frac{T_3}{1.2588} = \frac{1050}{1.2588} = 834 \text{ K}$$

$$\eta_t = \frac{T_3 - T_c}{T_3 - T_c'}$$

$$\therefore \quad T_c = T_3 - (T_3 - T_c')\,\eta_t$$

$$= 1050 - (1050 - 834) \times 0.85$$

$$= 866.4 \text{ K}$$

Work delivered by high pressure turbine

$= c_p (T_3 - T_c)$

$= 1.005(1050 - 866.4) = 184.52$ kJ/kg

The work output from the low pressure turbine wil also be equal to 184.52 kJ/kg

∴ Work delivered by both the turbines

$W_t = 184.52 + 184.52 = 369.04$ kJ/kg

Net work W_n

= turbine work – compressor work

$= 369.04 - 187.1 = 181.94$ kJ/kg

Power output

$$= 181.94 \frac{\text{kJ}}{\text{kg}} \times 15 \frac{\text{kg}}{\text{s}} = \mathbf{2729 \text{ kW}}$$

(*b*) Heat supplied $= c_p(T_3 - T_2) + c_p(T_d - T_c)$

$= 1.005(1050 - 381.1) + 1.005(1050 - 866.4)$

$= 856.76$ kJ/kg

Thermal efficiency

$$\eta = \frac{\text{network output}}{\text{heat supplied}}$$

$$= \frac{181.94}{856.76} = 0.212 \text{ or } \mathbf{21.2\%}$$

(*c*) From heat balance,

$$m_f \times CV = (m_a + m_f)\, c_p\, T_3 - m_a\, c_p\, T_2 + (m_a + m_f)\, c_p\, (T_d - T_c)$$

$$\simeq m_a[c_p (T_3 - T_2) + c_p (T_d - T_c)]$$

$$(\because m_f << m_a)$$

∴ Air fuel ratio

$$\frac{m_a}{m_f} = \frac{CV}{c_p (T_3 - T_2) + c_p (T_d - T_c)}$$

$$= \frac{42000}{856.76} = \mathbf{49.02 : 1}$$

EXAMPLE 23.11

Consider an open gas turbine installation working on Brayton cycle and provided with intercooring, regenerator and reheating systems. 15 kg/s of air enters the low pressure compressor at 1 bar, 290 K and after compression to 4 bar is cooled upto 295 K. After subsequent compression to 8 bar in the high pressure compressor, the air passes through a regenerator which has an effectiveness of 80 percent. The air is next led to combustion chamber where it is heated upto 1250 K. The heated air enters the high pressure turbine where it expands to 4 bar. The air is reheated thereafter to 1200 K before being expand to 1 bar in the low pressure turbine. The exhaust from the turbine is passed through regenerator before being discharged out of cycle. Calculate the power output, thermal efficiency, and fuel-air ratio in each combustion chamber. It may be assumed that air is the working fluid through out, compression and expansion processes are isentropic, the fuel used has a heating value of 41800 kJ/kg, mechanical efficiencies of the compressor and turbine are 0.96 and 0.98 respectively and the efficiency of combustion is 0.97.

Solution : Refer Fig. 23.21.

From compression prcesses

$$T_a = T_1 \left(\frac{p_a}{p_1}\right)^{\frac{\gamma-1}{\gamma}}$$

$$= 290 \left(\frac{4}{1}\right)^{\frac{1.4-1}{1.4}} = 431.1 \text{ K}$$

$$T_2 = T_b \left(\frac{p_2}{p_b}\right)^{\frac{\gamma-1}{\gamma}}$$

$$= 295 \left(\frac{8}{4}\right)^{\frac{1.4-1}{1.4}} = 359.68 \text{ K}$$

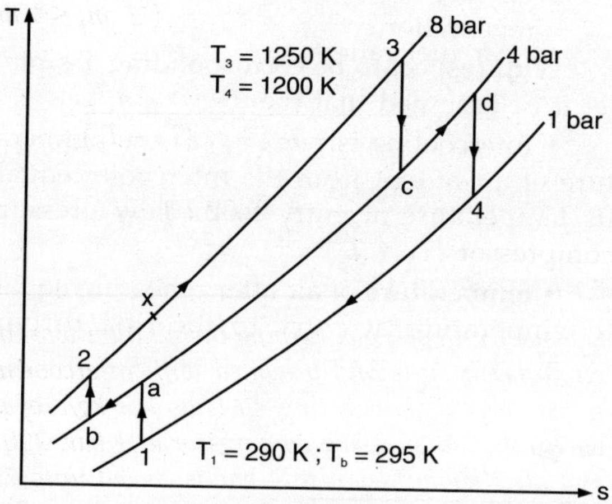

Fig. 23.21.

From expansion processes

$$T_c = T_3 \left(\frac{p_c}{p_3}\right)^{\frac{\gamma-1}{\gamma}}$$

$$= 1250 \left(\frac{4}{8}\right)^{\frac{1.4-1}{1.4}} = 1025.2 \text{ K}$$

$$T_4 = T_d \left(\frac{p_4}{p_d}\right)^{\frac{\gamma-1}{\gamma}}$$

$$= 1200 \left(\frac{1}{4}\right)^{\frac{1.4-1}{1.4}} = 807.2 \text{ K}$$

(*a*) Compressor work

$$W_c = \left[c_p (T_a - T_1) + c_p (T_2 - T_b)\right] \times \frac{1}{\eta_{mc}}$$

$$= [1.005(431.1-290) + 1.005(359.68-295)] \times \frac{1}{0.96}$$

$$= 215.42 \text{ kJ/kg}$$

Turbine work

$$W_t = \left[c_p (T_3 - T_c) + c_p (T_d - T_4)\right] \times \eta_{mt}$$

$$= [1.005(1250-1025.2) + 1.005(1200-807.2)] \times 0.98$$

$$= 608.27 \text{ kJ/kg}$$

Network output

$$W_{net} = W_t - W_c = 608.27 - 215.42 = 392.85 \text{ kJ/kg}$$

∴ Power output

$$P = 392.85 \frac{\text{kJ}}{\text{kg}} \times 15 \frac{\text{kg}}{\text{s}} = 5892.75 \text{ kW}$$

(*b*) In the regenerator,
Effectiveness

$$\epsilon = \frac{T_x - T_2}{T_4 - T_2} \; ; \; T_x = T_2 + \epsilon (T_4 - T_2)$$

$$T_x = 359.68 + 0.8 (807.2 - 359.68) = 717.7 \text{ K}$$

Heat added in the combustion chambers

$$= \left[c_p (T_3 - T_x) + c_p (T_d - T_c)\right] \times \frac{1}{\eta_{com}}$$

$$= [1.005(1250-717.7) + 1.005(1200-1025.2)] \frac{1}{0.97}$$

$$= 732.61 \text{ kJ/kg}$$

∴ cycle thermal efficiency

$$\eta = \frac{\text{network output}}{\text{heat added}}$$

$$= \frac{392.85}{732.61} = 0.536 \text{ or } 53.6\%$$

(*c*) Fuel-air ratio

$$= \frac{c_p (T_3 - T_x)}{CV}$$

$$= \frac{1.005 (1250 - 717.7)}{41800}$$

= 0.0128 for combustion chamber 1

$$= \frac{c_p (T_d - T_c)}{CV}$$

$$= \frac{1.005 (1200 - 1025.2)}{41800}$$

= 0.0042 for combustion chamber 2

23.4. CLOSED AND SEMI-CLOSED GAS TURBINE CYCLES

In a closed gas turbine system :

(*i*) the compressed working fluid which may be air, hydrogen, helium or carbon dioxide is delivered to a heater where it is heated to the required temperature.

(*ii*) the heated high pressure working fluid expands in the turbine and work is obtained.

(*iii*) the exhaust from the turbine is passed through a pre-coolar where cooling is done to a suitable temperature.

(*iv*) the cooled working fluid next enters the compressor and the cycle is completed.

Apparently the working agent does not leave the plant and is used repeatedly in successive cycles.

Fig. 23.22, shows the layout of a typical closed cycle gas turbine plant that incorporates :

- multi-stage compression with intercooling in between
- multi-stage expansion with reeheating between stages, and
- regenration

With respect to be corresponding T-s plot, let it be assumed that the

- intercooling is perfect, *i.e.*, the temperature of air at exit from the intercooler equals its temperature at entry to the low pressure compressor ($T_3 = T_1$)
- temperature of air after reheating equals it temperature at entry to the first turbine ($T_8 = T_6$)

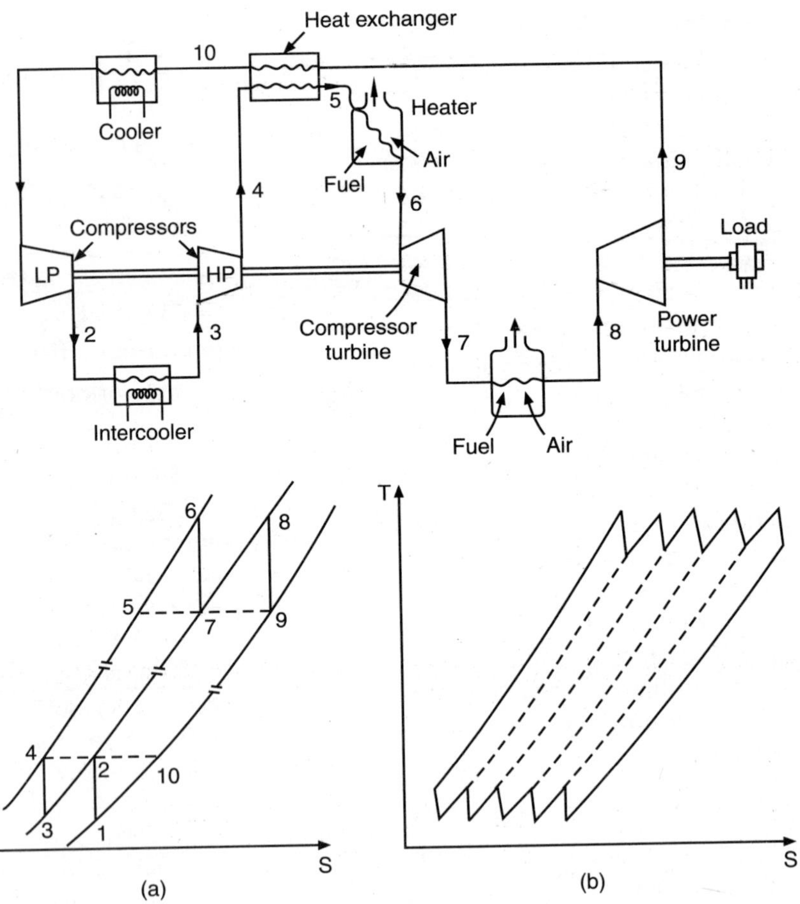

Fig. 23.22. Closed cycle turbine plant employing intercooling, reheating and generator

• temperature of air after leaving the regenerator equals its temperature at exit from the low pressure turbine ($T_5 = T_9$)

Further, the system approaches the Ericsson cycle if the number of compression and expansion stages are infinite. Both the compression and expansion processes become isothermal. The heat supply at higher temperature T_{max}, the heat rejection is at low temperature T_{min} and the efficiency of the plant becomes equal to the Carnot efficiency $(T_{max} - T_{min})/T_{min}$.

The advantages of closed gas turbine plant are :

1. The system can be made to operate at high pressure throughout the cycle. Consequently, there is reduction both in the volume of air being circulated and the size of components.

2. The cycle operates with a limited amount of air which is used continuously over and over again. The compressor is free from the deposits of atmospheric dirt and dust.

3. The combustion is arranged to take place in a separate unit external to the cycle. That permits the use of any type of fuel.

4. The working medium remains free from pollution of combustion products. Apparently the interior of the plant remains clean, and there is reduction in damage to turbine blades due to corrosion and erosion.

5. The starting of the plant can be made easier with reduction of inlet pressure at entry to compressor.

6. The fractional loads can be handled efficiently with a change in the density of the gas system.

7. A closed system operating under high pressure has a high value of Reynolds number at which heat conduction and convection rates are quite high. That permits the use of small size heat exchanges and intercoolers that operate efficiently.

8. For the same temperature limits, the closed systems have been found to operate with appreciably high efficiency.

9. A closed system can operate with gases other than air as the working medium. A gas with high value of adiabatic exponent and a greater density are recommend for this purpose.

These advantages, to some extent, are offset by the following disadvantages :

1. The system is complex and entails higher costs.

2. The performance of the unit is adversely affected due to combustion being external

3. The regenerators and other heat exchangers employed with the closed system need to be sufficiently strong to resist the high pressure involved.

4. The system needs a greater supply of cooling water, and this aspects eliminates its use in the field of aeronautics.

Sometimes use is made of a semi-closed cycle in which an attempt is made to obtain the best characteristics of both the open and closed cycles. The basic pressure can be varied in accordance with load (closed system) and combustion takes place in a compact combustion chamber (open system).

23.5. GAS TURBINES versus RECIPROCATING IC ENGINES

The principal advantages likely to be shown by a gas turbine installation are :

1. *Mechanical efficiency.* A reciprocating engine has a larger number of pairs of sliding or bearing members and each of these members is a source of friction loss. The gas turbine is less prone to friction loss as the only source of friction are the high speed main bearings. Consequently a gas turbine unit has a high value of mechanical efficiency ranging from 95 to 97 per cent. The mechanical efficiency of an IC engine under full load conditions lies between 85 to 95%.

2. *Complete expansion.* In an IC engine, the full adiabatic expansion of the products of combustion cannot be affected inside the cylinder. This involves a loss of work. Since the gas turbine is a rotary machine, full expansion can be attained and consequently the gas turbine gives greater output.

3. *Flexibility.* The processes of compression, combustion and expansion take place in

different components in a gas turbine and as such these units can be designed, tested and developed individually. This also allows a great variety in the arrangement of the system. The arrangement usually depends upon the duty, performance desired and the space allocation of the plant.

4. *Balancing*. The gas turbine with its purely rotational parts can be very accurately balanced at all speeds. The operation of a dynamically balanced gas turbine is smooth and without much noise. Further the torque of the gas turbine is more uniform and so the unit does not require stronger driven members and mountings. Again the blades of a gas turbine are continuously under the effect of steady flow of combustion products and this renders them free from torsional vibrations.

5. *Compact and low weight unit*. The gas turbines are noted for their ability to produce large power from units of comparatively small size and weight. The gas turbine has a specific weight of 10kg/kW as compared with 55 kg/kW for a diesel engine.

6. *High-operational speed*. Due to the absence of reciprocating parts, the gas turbine unit can be driven at much higher speeds which are not possible in an IC engine of similar output.

7. *Simplicity*. The unit has an easy and inexpensive lubrication. The ignition system is also simple as the spark is needed only for a short duration and the subsequent combustion of fuel is self sustained. Further, the combustor has a light weight, is small and inexpensive and gives a high rate of heat release.

8. *Fuels*. Cheap and readily available fuels such as benzene, powdered coal and heavy graded hydrocarbons can be used with gas turbines without impairing their efficiency to an appreciable extent.

9. *Low operating pressures*. The gas turbines operate at comparatively much lower pressures than those of an IC engine. Consequently the pressure joints and the piping system can be designed without much difficulty. Further, the components of the gas turbine subjected to these pressures can be made light.

10. *Maintenance and installation costs*. The gas turbine is a simple plant with few accessories and consequently the maintenance cost is low and infrequent. Further, since the unit is compact, it needs smaller floor foundations and the building to house the plant and as such the installation cost is less.

11. *Operation at high altitudes*. The output of the gas turbine increases with a decrease in the compressor inlet temperature of the working fluid. Since at higher altitudes the atmospheric air has a low temperature the avitation gas turbine is capable of well maintaining its output. However, the performance of an IC engine is adversely affected at higher altitudes because of rarified atmosphere.

Among the shortcomings of the gas turbine are :

(*i*) The gas turbine unit has a lower value of thermal efficiency because a considerable portion of the power produced in the turbine is spent up in running the compressor.

(*ii*) The starting of the gas turbine is rather difficult.

(*iii*) Metallurgical considerations limit the cycle temperature to about 800°C. In an I.C engine, temperatures as high as 2000°C are possible. The piston and the cylinder head of the IC engine can withstand the high temperature since it exists only for a fraction of a second.

(*iv*) The cooling system of a gas turbine engine is much more elaborate than that of an IC engine.

23.6. GAS TURBINES versus STEAM TURBINES

The principal advantages of a gas turbine plant in relation to steam plant are enumerated below:

(*i*) The gas turbine plant can be brought into service with great rapidity (8 to 12 minutes) from cold to full load. In steam plant, the firing of the boiler and generation of steam take a considerable longer time.

(*ii*) Gas turbines have a low weight to power ratio due to their compact design. Whereas the specific weight of a gas turbine is 10 kg/kW, it is 25 kg/kW for a steam turbine.

(*iii*) The capital and maintenance costs of a gas turbine unit are less than those of an equivalent steam plant.

(*iv*) Omission of boiler, condenser and their auxiliaries leads to simplification of a gas turbine plant.

(*v*) Due to fewer parts and controls, the operation of a gas turbine is rather simpler.

(*vi*) The gas turbine plant stops and becomes cool within a few seconds of cut-off of fuel supply. The stored energy is quickly absorbed by the compressor in pumping cooled air through and thus cooling the combustor and gas turbine.

(*vii*) The gas turbines have a higher efficiency because they operate at higher temperatures ($\simeq$ 1000 K) as compared with steam turbine where temperatures are about 750 K.

(*viii*) The gas turbine unit is not much dependent on water supply and can be used quite effectively in the regions where an abundant supply of water is not available for steam raising.

23.7. JET PROPULSION

The propulsive work needed for the propulsion of aircrafts, missiles and spaceships is generally achieved by

(*i*) using an engine or a gas turbine to drive the air screw or the propeller.

(*ii*) expanding high temperature and high pressure gases that are discharged in the rearward direction as high velocity jet (jet propulsion).

In both cases, the accelerating mass of gas produces a force and its reaction called the thrust or propolsive force is used to propel the aircraft in the forward direction. The screw propeller differs from jet propulsion in the following two aspects :

(*a*) The efficiency of screw propeller is high (about 85%) at low altitudes but drops with rise in altitude. However, the efficiency of jet propulsion systems is quite low at low altitudes and becomes high at greater heights.

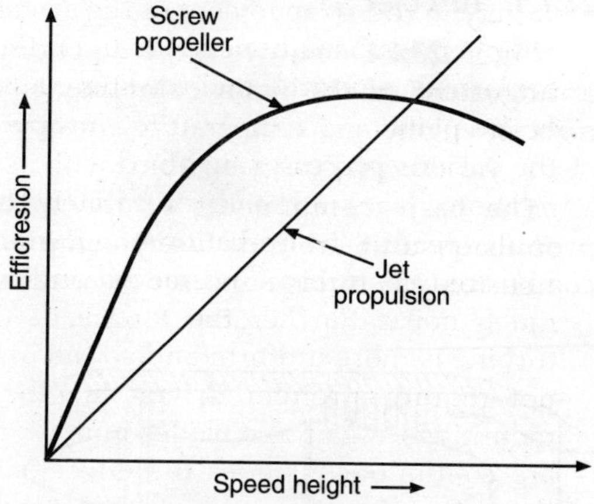

Fig. 12.23. Effect of speed and height an efficiency

(*b*) When the speed approaches the sonic value, there is rapid fall in the lift coefficient of the airfoil blading. Consequently, there occurs a loss in the tractive effort of the screw propeller. With jet propulsion, there is no loss in the tractive effort and so propulsion is possible even at high speeds.

The jet propulsion systems are classified as :

1. Air breathing engines : The oxygen necessary for combustion is taken from the surrounding atmosphere.

2. Rocket engines : The fuel and the oxidizer are contained in the body of the unit which is to be propelled.

Further sub-divisions of the jet propulsion systems are as indicated below :

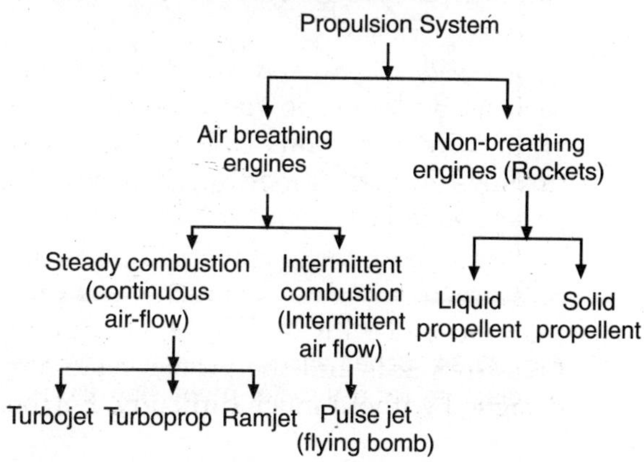

These propulsion systems have been discussed briefly in the following section.

23.7.1. Turbojet

Fig. 23.24 shows the schematic arrangement of the main components of a turbojet engine and temperature-entropy plot of the various processes involved.

The basic components of the turbojet propulsion unit are : diffuser, compressor, combustor, gas turbine and set of nozzles.

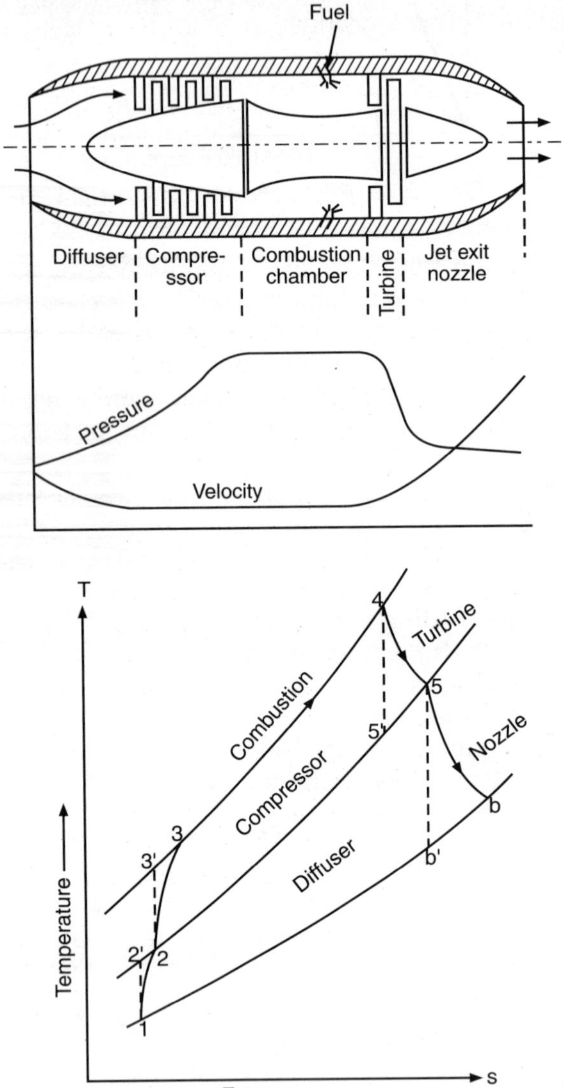

Fig. 23.24. Schematic arrangement and h-s diagram for a turbojet propulsion unit

• **Diffuser :** The diffuser is fitted at the entrance of turbojet and ambient air enters it at a velocity equal to the speed of aircraft. In the diffusing section, the entering air slows down and a part of its kinetic energy is converted into pressure. This type of air compression is called the *ram compression.*

• **Rotary compressor :** The air after ram compression gets further compressed to a pressure of 3.2 to 4 bar in a rotary compressor which is usually of the axial flow type.

• **Combustion chamber :** The pressurized air from the compressor is passed on to the combustion chamber where fuel is injected and burnt. During combustion, the pressure remains fairly constant.

• **Gas turbine :** The hot combustion products then enter the gas turbine where they are expanded partially, *i.e.,* undergo limited expansion. The turbine is so designed that the power output from it is just sufficient to drive the compressor and auxiliary equipment like fuel and lubricating pumps.

• **Exit nozzle :** The exhaust from the turbine is at a pressure which is considerably in excess of atmospheric pressure. These gases expand in the exit nozzle and there occurs conversion of hot gases into kinetic energy. The gases leave the nozzle as a jet with high velocity; the exit velocity is much higher than that at the entry. Due to increase in velocity, there is change in momentum which provides a thrust that propels the aircraft in a direction opposite to the exiting jet.

This method of propulsion is best suited to aircrafts travelling at or above 800 km/hr.

Compared to multicylinder piston engines, a turbojet engine has the following advantages :

(*i*) Simple construction, few pairs of mating surfaces which wear and consequently inexpensive maintenance cost.

(*ii*) Low specific weight : 1/4 to 1/2 of that of reciprocating engine

(*iii*) Small frontal area and better aerodynamic shape

(*iv*) No loss of power in transmission as the thrust is applied directly (*v*) Uninterrupted and smooth power supply and consequently engine vibrations are practically absent

(*vi*) Low working pressures and so the structure and casing can be of lighter construction

(*vii*) Cheap and readily available fuels such as paraffin and kerosene can be used

(*viii*) No high speed restrictions, speeds as high as 3000 km/hr can be achieved

The shortcomings of a turbojet propulsion unit are:

- shorter life, noisy operation and requirement for quite expensive materials
- inefficient operation at speeds below 550 km/hr
- take-off length is too much and as such longer strip is required

Further, during take-off and climb, the thrust produced is relatively low. This effect can be overcome by thrust augmentation which is achieved by

(*i*) water or liquid injection into the ram air

(*ii*) incorporating after burner or tail pipe burning

The liquid injection serves to cool the ram air by which the mass flow rate of air can be increased. Quite often, additional fuel is burnt in the turbine exhaust before it moves on to the exit nozzle. This tail pipe burning provides an additional thrust due to increased mass flow of gases in the nozzle. This increase in thrust is, however, at the expense of fuel economy and is to be preferred during short periods of take off where reduction in efficiency is less important than the increased thrust.

23.7.2. Turbo-prop

Fig. 23.25 shows the schematic arrangement of a turbo-prop propulsion unit; the system has a propeller geared to turbojet engine.

The various actions performed in turbo-prop unit are :

(*i*) The diffuser serves to slow down the air that enters the unit at the velocity of aircraft. There is conversion of kinetic energy of incoming air into pressure energy (ram compression).

(*ii*) At exit from the diffuser is located the rotary compressor which compresses the air to 3 to 4 bar pressure.

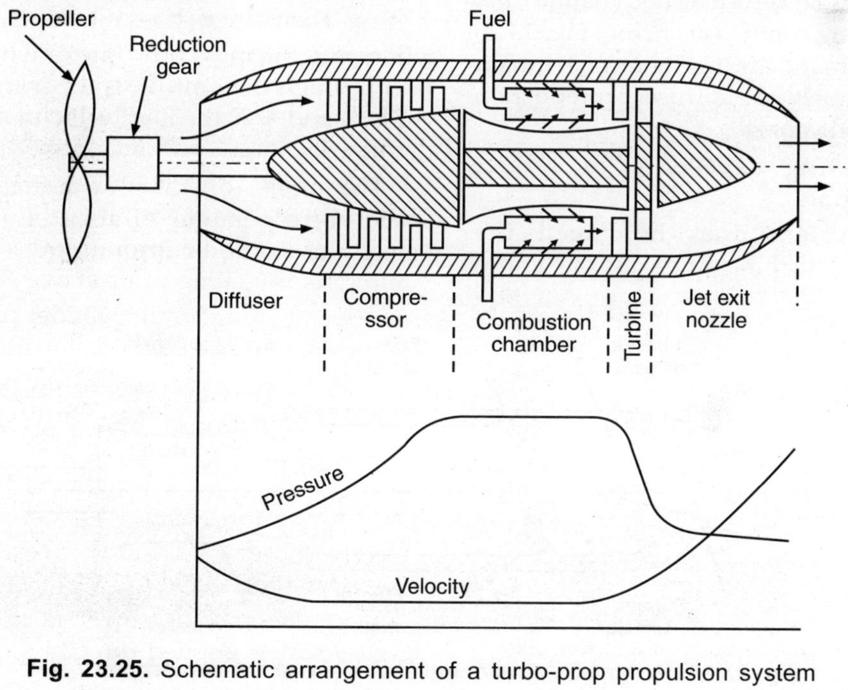

Fig. 23.25. Schematic arrangement of a turbo-prop propulsion system

(*iii*) The compressed air is taken to the combustion chamber where fuel is injected. Combustion takes place, heat is released causing rise in temperature.

(*iv*) The combustion products expand partly (80 to 90%) in the turbine, and the power developed by the turbine is utilized in running the compressor and the propeller. The power transmission to the propeller is through a reduction gear.

(*v*) The remaining (10 to 20%) of the expansion takes place in the nozzle and that produces high velocity jet.

The aircraft gets its forward motion from the propeller and the high velocity jet produced by the nozzle. Quite often the thrust output of the unit is increased by heating the turbine exhaust before expansion in the nozzle.

A turbo-prop combines in it the merits of turbojet engine (low specific weight, small frontal area, simplicity of construction, smooth running and freedom from vibrations) and that of a propeller (high power for take off and climb, and high propulsive efficiency at speeds below 800 km/hr).

23.7.3. Ramjet (Athodyd)

The ramjet is an aero-thermodynamic duct (Athodyd) having continuous firing. Fig. 23.26 shows the schematic arrangement of its main components namely the diffuser, combustion chamber and exit nozzle.

The operating aspects of a ramjet propulsion unit are :

(*i*) The system does not need the compressor and turbine units as the entire compression process is achieved through ram compression.

(*ii*) The atmospheric air enters the ramjet with supersonic speed and gets slowed down to sonic speed in the supersonic diffuser. A shock wave is formed accompanied by an increase in pressure. Further increase in pressure occurs in the subsonic diffuser. The pressure increase causes the temperature of air to rise above the ignition temperature.

(*iii*) The unit has fuel injectors which inject the fuel in the combustion chamber. The fuel-air mixture is then ignited by means of a spark plug and temperatures of the order of 2000 K are attained due to heat released by combustion.

(*iv*) Due to pressure barrier (the hot gases are at a pressure which is higher than that at the after end of the diffuser) the hot gases are constrained to move towards the nozzle and undergo expansion. The pressure energy is converted into the kinetic energy and the high velocity gases leaving the nozzle provide forward thrust to the unit.

The net thrust developed by the engine is given by the change in momentum expressed as

$$\text{Thrust} = m(V_j - V_a)$$

where m is the mass flow rate of working fluid (air) through the unit, V_a is the incoming flow velocity and V_j is the high velocity at exit from the nozzle.

The ramjet operates at sufficiently high speed corresponding to about 1.5 times the velocity of sound or even more.

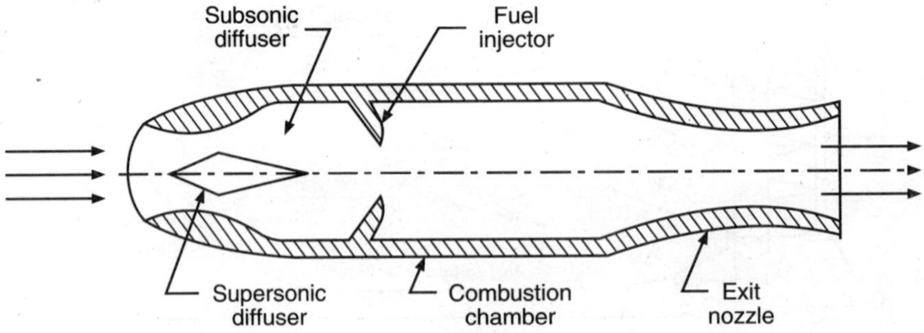

Fig. 23.26. Schematic arrangement of a ramjet propulsion unit

The notable aspects of a ramjet engine are :

(*i*) No moving parts and light in weight as compared with turbojet engine.

(*ii*) Wide range of fuels can be burnt in the combustion chamber.

(*iii*) High fuel consumption at low and moderate speeds.

(*iv*) Unit is not self-operating, it needs to be accelerated to a sufficiently high speed by an external power source (turbojet or rocket) where ram pressure is sufficient for initiating the combustion of fuel.

(*v*) Need for careful and proper design of diffusers; this is essential for efficient conversion of kinetic energy associated with high incoming velocities into pressure.

(*vi*) Steady combustion requires certain elaborate devices in the form of flame holders or pilot flame.

Further, since sufficient pressure rise for good efficiency can only be obtained at very high speeds, the ramjet applications are generally at high speeds.

23.7.4. Pulse jet (Flying bomb)

In contrast to turbo-jet and ram-jet engines the pulse jet has intermittent combustion and its operation is akin to a reciprocating engine. However like an athodyd, the pulse jet develops thrust by a high velocity jet of exhaust gases without the aid of a turbine or a compressor.

Fig. 23.27 shows the schematic arrangement of a pulse jet engine; its operating characteristics are :

(*i*) The atmospheric air entering the pulse jet is compressed by ram effect in the diffuser section and the grid passages. The opening and closing of the grid passages is achieved by V-shaped non-return valves.

(*ii*) The fuel is injected into the combustion chamber through fuel injection nozzles. The air pressure from compressed air bottles is the motive force for the working of nozzles. The fuel-air mixture is then ignited by means of a spark plug. When the engine starts operating normally, the spark is turned off and the subsequent combustion takes place because of residual flame and the hot ignition bars which retain heat from previous explosion.

(*iii*) Heat is liberated /evolved during the combustion of air-fuel mixture and that raises the pressure and temperature of combustion products. The hot gases are at a pressure which is higher than that of ram pressure and consequently the non-return valves get closed. The hot gases are constrained to move towards the tail pipe and undergo expansion. The pressure energy is converted into kinetic energy and the high velocity gases leaving the tail pipe provide forward thrust to the unit.

When the gases escape to the atmosphere, the static pressure in the diffuser section forces the valves to open. The fresh air from the atmosphere is then admitted into the unit and the cycle is repeated.

The advantages/disadvantages of a pulse jet are :

– simplicity of construction

– cost less compared to that of a turbo jet engine

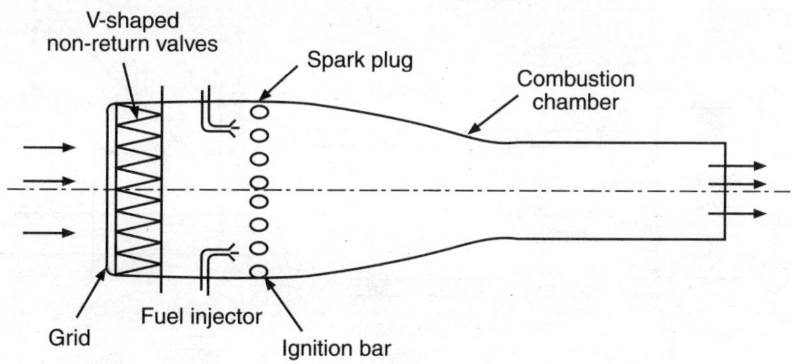

Fig. 23.27. Schematic arrangement of a pulse jet propulsion unit

– well adopted to pilot-less aircraft

– high noise level and severe vibrations

– high rate of fuel consumption and low thermodynamic efficiency

– limitations of mechanical valve arrangement

– intermittent combustion as compared to the continuous combustion in a turbo jet engine.

Further, the maximum operating altitude is limited by air density considerations. However, unlike the ram jet engines, it is capable of producing static thrust and thrust in excess of drag at much lower speeds.

23.8. ROCKET ENGINES

In a rocket propulsion system, the entire mass of the fuel and the oxidizing agent is carried by the rocket body itself. As such, the rocket does not depend on any surrounding air for burning the fuel. Essentially it operates better in vacuum (space) since no part of the thrust developed will be wasted in overcoming the friction.

The fuel and the oxidizer are commonly known as *propellant*. The requirements of an ideal propellant are :

– high heat value

– low toxicity and corrosiveness

– stability and ease of handling and storing

– reliable and smooth burning

– high density so that it occupies less space

The rockets find utility in the following fields :

• satellites and space ships

• long range artillery and lethal weapons

• jet assisted take off

• signalling and fire work display

The rocket motors can be operated either with liquid propellants or solid propellants. Fig. 23.28 shows the elements of a liquid propellant rocket using liquid oxygen and hydrocarbon fuel.

The liquid propellants (fuel and oxidizer) at low pressure are stored separately and are supplied by pumps to the combustion chamber where ignition of fuel takes place. The pumps are driven by:

(*i*) high pressure stored gases. This is done where only small amount of propellants is needed.

(*ii*) a steam or a gas turbine. For the steam turbine, the steam is raised by the action of a catalyst on hydrogen peroxide. The working fluid for the gas turbine unit is forced by the rocket fuel and the oxidizer. The turbine actuated pumps are used when large quantity of propellants are needed to produce greater thrust for longer duration.

When the atomized propellants react in the combustion chamber, combustion occurs at constant pressures and high temperatures are generated. The hot combustion gases then enter the nozzle unit where they undergo adiabatic

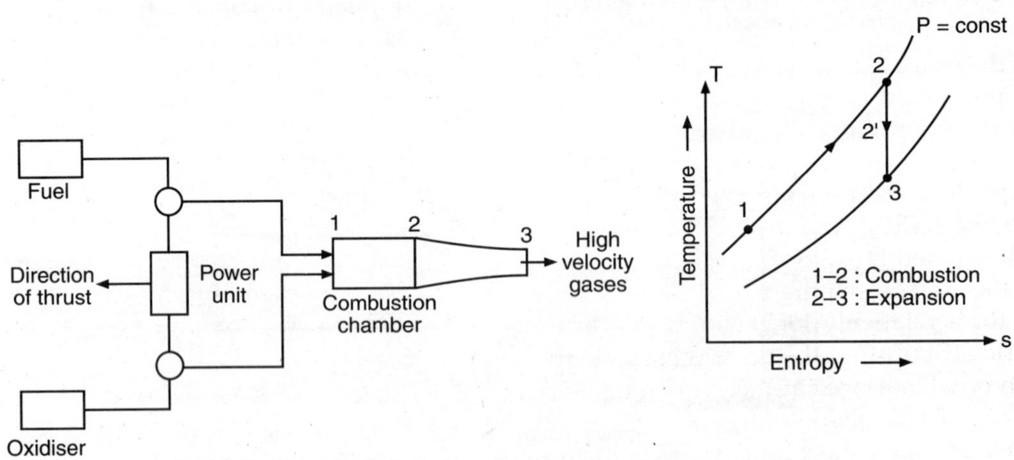

Fig. 23.28. Schematics of a liquid propellant rocket engine

expansion down to atmospheric pressure. The gases leave the nozzle at high velocities that produce the thrust necessary for the propulsion of the rocket. Quite often the liquid propellant is circulated around the walls of the combustion chamber and nozzle, and this provides the necessary cooling action. Liquid propellant rockets are used where high specific impulse, on-off operation and longer firing times are desired.

The fuel and the oxidizer in the *solid propellant* rockets are premixed in required proportion. The mixture is formed into a solid grain which is placed in the chamber where combustion is to take place. Fig. 23.29 shows the component parts of a solid propellant rocket using (*a*) restricted burning and (*b*) unrestricted burning.

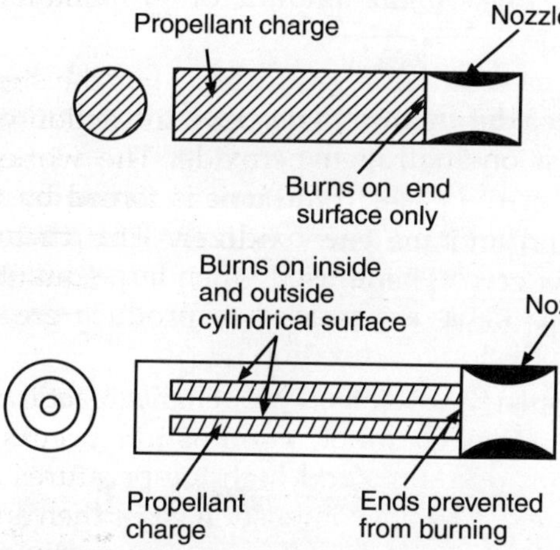

Fig. 23.29. Component parts of a solid propellant rocket

In the restricted burning rocket, the burning of the propellant is restricted so that in burns in one direction only. The restricted burning, also called the cigarette burning, is accomplished by pouring the charge when liquid. On solidification, it fits the chamber tight. Restricted burning rockets are preferred when the unit is required to deliver a small thrust for a relatively longer duration. In the unrestricted burning, the entire charge is free to burn on all surfaces at the same time. Such rockets are used wherein large thrust need to be developed for a short time, *i.e.*, in launching rockets called boosters.

The solid propellant rockets have the advantage of simplicity of design; no valves, pumps and control mechanism. However, their combustion chamber has to be large enough to store the entire propellant and also strong enough to withstand the high combustion pressure. Being solid, the propellant cannot be used to cool the nozzle and combustion chamber and as such there is danger of overheating. These rockets are used when firing time is short and thrust is required only for shorter durations.

23.9. THRUST AND PROPULSIVE EFFICIENCY

Let V_a = velocity of aircraft (moving forward)

V_j = velocity of jet relative to the aircraft (flowing backwards)

m_a = mass flow rate of air entering the aircraft

m_f = mass flow rate of fuel injected into the combustion chamber.

Since atmospheric air is assumed to be still, V_a also represents the velocity of air entering the aircraft. Further, the air-fuel ratio for the engine is very high ($m_a >> m_f$). The mass of fuel can be considered as negligible and with that assumption, the mass of exiting jet is equal to mass of entering air.

• **Thrust :** The force produced due to difference between the momentum of high velocity exhaust gases leaving the nozzle and low velocity air entering the aircraft is called thrust.

Now, absolute velocity of gases leaving the aircraft

$$= (V_j - V_a)$$

absolute velocity of air entering the aircraft

$$= V_a - V_a = 0$$

Thrust = rate of change of momentum of working fluid

= mass flow rate of air × change of velocity

$$= m_a \left[(V_j - V_a) - 0\right] = m_a (V_j - V_a)$$

Thrust work = forward thrust × speed of aircraft

$$= m_a (V_j - V_a) \times V_a$$

If the mass flow is in kg/s and the velocities are in m/s, then thrust works out in Nm/s or W.

The *thrust power* is the rate at which work must be developed by the engine if the aircraft is to be kept moving at a constant velocity against friction force or drag.

$$\text{Thrust power} = \frac{m_a(V_j - V_a)\times V_a}{1000}\ \text{kW}$$

• **Propulsive power :** The energy supplied to the system, called the propulsive work, is required to produce thrust work and to give kinetic energy to the gases. That is

Propulsive work

$$= m_a\,(V_j - V_a)\,V_a + \frac{m_a(V_j - V_a)^2}{2}$$

$$= \frac{m_a(V_j^2 - V_a^2)}{2}$$

The propulsive work is the net work output from the engine and equals the increase in kinetic energy of the working fluid.

$$\text{Propulsive power} = \frac{m_a(V_j^2 - V_a^2)}{2\times 1000}\ \text{kW}$$

• **Propulsive efficiency :** The ratio of thrust work to propulsive work is called the propulsive efficiency of the propulsion unit.

$$\eta_p = \frac{\text{thrust work}}{\text{propulsive work}}$$

$$= \frac{m_a(V_j - V_a)V_a}{m_a\,(V_j^2 - V_a^2)/2}$$

$$= \frac{2(V_j - V_a)V_a}{(V_j - V_a)(V_j + V_a)}$$

$$= \frac{2V_a}{V_j + V_a} = \frac{2}{\dfrac{V_j}{V_a} + 1}$$

Evidently the propulsive efficiency increases with an increase in the aircraft velocity V_a. An efficiency of 100% is attained when the aircraft velocity V_a approaches the jet velocity V_j. But, then the thrust power becomes zero. Accordingly a lower value (about 60%) of propulsion efficiency is used to obtain reasonable thrust.

• **Thermal efficiency :** The ratio of the propulsive work to the energy released by the combustion of fuel is called the thermal efficiency of the propulsion unit.

$$\eta_{th} = \frac{\text{propulsive work}}{\text{heat released by the combustion of fuel}}$$

$$= \frac{m_a(V_j^2 - V_a^2)/2}{m_f\times CV} = \frac{m_a(V_j^2 - V_a^2)}{2(m_f\times CV)}$$

where CV is the calorific value of fuel.

The product of thermal efficiency and the propulsive efficiency defines the overall efficiency. That is

$$\eta_o = \eta_p\times\eta_{th} = \frac{2V_a}{V_j + V_a}\times\frac{m_a(V_j^2 - V_a^2)}{2(m_f\times CV)}$$

$$= \frac{m_a(V_j - V_a)V_a}{m_f\times CV}$$

For maximum overall efficiency, the aircraft velocity is one-half of the jet velocity.

The air is self-contained in rocket propulsion and the entry velocity relative to aircraft is zero. Therefore,

$$\text{Thrust work} = mV_j\,V_a$$

$$\text{Propulsive work} = mV_j\,V_a + \frac{1}{2}\,m(V_j - V_a)^2$$

$$= \frac{1}{2}\,m(V_j^2 + V_a^2)$$

Rocket propulsive efficiency

$$= \frac{mV_j V_a}{\frac{1}{2}m(V_j^2 + V_a^2)}$$

$$= \frac{2V_a/V_j}{1+(V_a/V_j)^2} = \frac{2\sigma}{1+\sigma^2}$$

where σ is the ratio of rocket velocity to jet velocity.

EXAMPLE 23.12

An aeroplane flies at an altitute with a velocity of 800 km/hr. The pressure and temperature at that altitude are 1.206 × 10⁴ N/m² and 217 K respectively. Calculate

(a) the maximum possible temperature and pressure intensity on the airplane body

(b) the critial and maximum possible velocity of air relative to the aeroplane.

Solution : Velocity of plane V

$$= 800 \text{ km/hr} = \frac{800 \times 10^3}{3600} = 222.2 \text{ m/s}$$

The maximum possible temperature on the airplane body will be the stagnation temperature.

$$T_0 = T + \frac{V^2}{2c_p}$$

$$= 217 + \frac{222.2^2}{2 \times 1005} = 241.55 \text{ K}$$

The specific heat of air at constant pressure has been assumed to be 1005 J/kg K

From isentropic correlation between pressure and temperature, the pressure intensity on the airplane body is

$$p_0 = p\left(\frac{T_0}{T}\right)^{\frac{\gamma}{\gamma-1}}$$

$$= 1.206 \times 10^4 \times \left(\frac{241.55}{217}\right)^{\frac{1.4}{1.4-1}}$$

$$= 1.755 \times 10^4 \text{ N/m}^2$$

(ii) Critical velocity of air

$$= \sqrt{\gamma R T}$$

$$= \sqrt{1.4 \times 287 \times 217} = 295.28 \text{ m/s}$$

Maximum possible velocity of air relative to plane

$$= \sqrt{\gamma R T_0}$$

$$= \sqrt{1.4 \times 287 \times 241.55}$$

$$= 311.54 \text{ m/s}$$

EXAMPLE 23.13

A jet propulsion system has to create a thrust of 90 tonnes to move the system at a velocity of 650 km/hr. Find the exit gas velocity and propulsion efficiency if the gas flow rate through the system is restricted to 25 kg/s.

Solution : Thrust = 90 tonnes

$$= \frac{90 \times 10^3}{9.81} = 9174.3 \text{ N}$$

Forward velocity of machine

$$V_a = \frac{650 \times 10^3}{3600} = 180.5 \text{ m/s}$$

Now : Thrust

$$= m(V_j - V_a)\,; \quad 9174.3 = 25(V_j - 180.5)$$

∴ Velocity of jet (exit gas velocity)

$$V_j = \frac{9174.3}{25} + 180.5 = 547.5 \text{ m/s}$$

$$\text{Propulsion efficiency } \eta_p = \frac{2V_a}{V_j + V_a}$$

$$= \frac{2 \times 180.5}{547.5 + 180.5} = 0.496 \text{ or } 49.6\%$$

EXAMPLE 23.14.

The engine of a turbojet system has two jets and while at an altitude of 11000 m, it has a velocity of 225 m/s. The density of air at the flight altitude is 0.173 kg/m³. The resistance or drag to the motion of turbojet is 6375 N and the propulsion efficiency of jets is 52 percent. Make calculations for : (a) absolute velocity of jet (b) volume of air compressed (c) diameter of jet (d) power output of the unit (e) fuel consumption /thrust power/hr (f) air-fuel ratio for the engine.

It may be presumed that overall efficiency of the plant is 16 percent and the fuel used has a calorific value of 49200 kJ/kg.

Solution : $\eta_{propulsion} = \frac{2V_a}{V_j + V_a}\,; 0.52 = \frac{2 \times 225}{V_j + 225}$

$$\therefore \text{ Jet velocity } V_j = \frac{2 \times 225}{0.52} - 225$$

$$= 640.4 \text{ m/s}$$

(a) Absolute velocity of jet

$$= (V_j - V_a)$$

$$= 640.4 - 225 = 415.4 \text{ m/s}$$

(b) Propulsive force (thrust) $= m(V_j - V_a)$

$$\therefore 6375 = m(640.4 - 225)$$

$$\text{and} \quad m = \frac{6375}{640.4 - 225} = 15.35 \text{ kg/s}$$

Volume of air,

$$V = \frac{15.35}{0.173} = 88.73 \text{ m}^3/\text{s}$$

(*c*) Neglecting mass of fuel,

$$V = 2a\ V_j$$

where a is the area of one jet.

$$\therefore \quad 88.73 = 2a \times 640.4\ ; \quad a = 0.0693 \text{ m}^2$$

$$\frac{\pi}{4} d^2 = 0.0693\ ;\ d = 0.297 \text{ m or } 29.7 \text{ cm}$$

(*d*) Thrust power

$$= \text{thrust (drag)} \times \text{plane velocity}$$

$$= \frac{6375 \times 225}{10^3} = 1434.37 \text{ kW}$$

Propulsive efficiency is the ratio of thrust power to propulsive (turbine) power.

∴ Power output of the turbine

$$= \frac{1434.37}{0.52} = 2758.4 \text{ kW}$$

(*e*) Overall efficiency

$$= \frac{\text{thrust power}}{\text{heat supplied in fuel}}$$

Let m_f be the fuel consumption/kW thrust power /hr

$$0.16 = \frac{1}{m_f \times 49200} \times 3600$$

$$m_f = \frac{1}{0.16 \times 49200} \times 3600$$

$$m_f = 0.4573 \text{ kg/kW/hr}$$

(*f*) Air-fuel ratio $= \dfrac{\text{air used per hour}}{\text{fuel used per hour}}$

$$= \frac{15.35 \times 3600}{0.4573 \times 1434.37}$$

$$= 84.24$$

EXAMPLE 23.15.

A turbojet engine consumes 45 kg/s of air while flying at a speed of 800 km/hr. Make calculations for :

(a) jet exit velocity if the enthalpy change is 186 kJ/kg and the velocity coefficient for the nozzle is 0.95.

(b) mass flow rate of fuel and the thrust specific fuel consumption if the air-fuel ratio is 80.

(c) thermal efficiency of the system if combustion efficiency is 95% and the fuel used has a calorific value of 43950 kJ/kg.

(d) propulsive power, propulsive efficiency and overall efficiency.

Solution : Velocity of plane

$$V_a = \frac{800 \times 10^3}{3600} = 222.2 \text{ m/s}$$

(*a*) Jet velocity

$$V_j = C_v \times 44.7 \sqrt{\text{enthalpy drop in kJ/kg}}$$

$$= 0.95 \times 44.7 \sqrt{186} = 579.14 \text{ m/s}$$

(*b*) Fuel consumption

$$= \frac{45}{80} \times 3600 = 2025 \text{ kg/hr}$$

$$\text{Thrust} = m(V_j - V_a)$$

$$= 45(579.14 - 222.2) = 16062 \text{ N}$$

Thrust specific fuel consumption

$$= \frac{2025}{16062} = 0.126 \text{ kg/hr/N thrust}$$

(*c*) Thermal efficiency

$$= \frac{\text{work output}}{\text{heat input}}$$

$$= \frac{\text{gain in kinetic energy}}{\text{heat input}}$$

$$\therefore \eta_{thermal} = \frac{\frac{1}{2}(V_j^2 - V_a^2)}{m_f \times CV \times \eta_{comb}}$$

$$= \frac{\frac{1}{2}(579.14^2 - 222.2^2) \times 10^{-3}}{\frac{1}{80} \times 43950 \times 0.95}$$

$$= 0.274 \text{ or } \mathbf{27.4\%}$$

(*d*) Propulsive power

$$= \frac{1}{2} m(V_j^2 - V_a^2)$$

$$= \frac{1}{2} \times \frac{45(579.14^2 - 222.2^2)}{10^3}$$

$$= 6436 \text{ kW}$$

Thrust power = thrust × speed of plane

$$= \frac{16062 \times 222.2}{10^3} = 3569 \text{ kW}$$

Propulsive efficiency is the ratio of thrust power to propulsive power

$$\eta_p = \frac{3569}{6436} = 0.554 \quad \text{or} \quad 55.4\%$$

Alternatively :

$$\eta_p = \frac{2V_a}{V_j + V_a} = \frac{2 \times 222.2}{579.14 \times 222.2}$$

$$= 0.554 \quad \text{or} \quad 55.4\%$$

Overall efficiency

$$\eta_0 = \frac{\text{thrust work}}{\text{heat supplied in fuel}}$$

$$= \frac{(V_j - V_a)V_a}{m_f \times CV \times \eta_{comb}}$$

$$= \frac{(579.14 - 222.2) \times 222.2 \times 10^{-3}}{\frac{1}{80} \times 43950 \times 0.95}$$

$$= 0.152 \quad \text{or} \quad 15.2\%.$$

Alternatively :

$$\eta_0 = \eta_{thermal} \times \eta_p = 0.274 \times 0.554$$

$$= 0.152 \quad \text{or} \quad 15.2\%$$

EXAMPLE 23.16.

An ideal jet propulsion engine draws air at 0.9 bar, 290 K with a velocity of 200 m/s. The pressure ratio across the compressor is 4.5 : 1 and the maximum temperature existing in the operating cycle is 1150 K. The output from the turbine unit is just equal to the work needed to run the compressor. After leaving the turbine, the air expands adiabatically to the ambient pressure of 0.9 bar. For air mass flow rate of 1 kg/s, determine the propulsive power developed by the engine.

Solution : With reference to Fig. 23.29, 1 – 1′ represents the process in the diffuser section where the incoming velocity gets converted to pressure. For the diffusing process,

$$T_1' = T_1 + \frac{V_1^2}{2c_p}$$

$$= 290 + \frac{200^2}{2 \times (1.005 \times 1000)} = 309.9 \text{ K}$$

The c_p has been assumed to be 1.005 kJ/kg K.

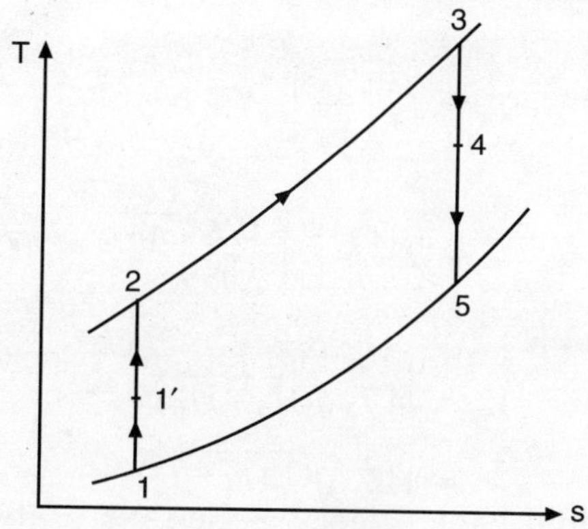

Fig. 23.30.

For the adiabatic process 1 – 1′.

$$\frac{p_1'}{p_1} = \left(\frac{T_1'}{T_1}\right)^{\frac{\gamma}{\gamma-1}}$$

$$= \left(\frac{309.9}{290}\right)^{\frac{1.4-1}{1.4}} = 1.261$$

$$\therefore p_1' = 1.261 \times 0.9 = 1.135 \text{ bar}$$

and $p_2 = 4.5 \times 1.135 = 5.1$ bar

Now :

$$\frac{T_2}{T_1'} = \left(\frac{p_2}{p_1'}\right)^{\frac{\gamma-1}{\gamma}} = (4.5)^{\frac{1.4-1}{1.4}} = 1.537$$

$$\therefore T_2 = 1.537\, T_1' = 1.537 \times 309.9 = 476.3 \text{ K}$$

The work required for running the compressor equals the output from the turbine and therefore

$$c_p (T_2 - T_1') = c_p (T_3 - T_4)$$

$$\therefore T_4 = T_3 - (T_2 - T_1')$$

$$= 1150 - (476.3 - 309.9) = 983.6 \text{ K}$$

For the expansion process 3-4,

$$\frac{T_3}{T_4} = \left(\frac{p_3}{p_4}\right)^{\frac{\gamma-1}{\gamma}}$$

$$\therefore \quad \frac{p_3}{p_4} = \left(\frac{T_3}{T_4}\right)^{\frac{\gamma}{\gamma-1}} = \left(\frac{1150}{983.6}\right)^{\frac{1.4}{1.4-1}} = 1.728$$

and $p_4 = \frac{5.1}{1.728} = 2.95$ bar

$(\because p_3 = p_2 = 5.1 \text{ bar})$

From the process 4-5 :

$$T_5 = T_4 \left(\frac{p_5}{p_4}\right)^{\frac{\gamma-1}{\gamma}}$$

$$= 983.6 \times \left(\frac{0.9}{2.95}\right)^{\frac{1.4-1}{1.4}} = 700.4 \text{ K}$$

Now :

$$V_5 = 44.7\sqrt{(H_4 - H_5)}$$

$$= 44.7\sqrt{C_p(T_4 - T_5)}$$

$$= 44.7\sqrt{1.005(983.6 - 700.4)}$$

$$= 754.1 \text{ m/s}$$

Thrust $F = m(V_5 - V_1) = 1 \times (754.1 - 225)$
$= 529.1$ N

$\therefore$ Propulsive power of the engines
$= F \times V_1 = 529.1 \times 225$
$= 11904.5$ Nm/s or J/s
$= 119.05$ kW

EXAMPLE 23.17

A flying bomb flies with a velocity of 600 km/hr at an altitude where the barometric pressure is 0.865 bar. The bomb has a range of 250 km, consumes 425 kg of petrol and produces 2675 N of thrust. The maximum temperature rise in the combustion chamber is 800°C and the discharge nozzle has a diameter of 30 cm. Make calculations for :

(a) the air-fuel ratio (b) the approximate temperature of exhaust gases (c) the velocity of exhaust gases relative to bomb, and (d) the propulsive efficiency and overall thermal efficiency of the power unit.

Assume calorific value of fuel as 42700 kJ/kg, specific heat of exhaust gases as 1.05 kJ/kg K and gas constant R = 300 J/kg K.

Solution : Let m_a be the mass of air supplied per kg of fuel. Then from energy balance in the combustion chamber, we have

$$m_f \times CV = (m_a + m_f)\, c_{pg}\, T_3 - m_a\, c_{pa}\, T_2$$

Asuming $c_{pg} = c_{pa}$, the above identity can be approximately written as

$$m_f \times CV = (m_a + m_f)\, c_p\, \Delta T$$

where ΔT is the temperature rise $(T_3 - T_2)$ in the combustion chamber. Further,

$$\frac{m_f}{m_a} \times CV = \left(1 + \frac{m_f}{m_a}\right) c_p\, \Delta T$$

or

$$\frac{m_f}{m_a} \times 42000 = \left(1 + \frac{m_f}{m_a}\right) \times 1.05 \times 800$$

$$= 920 + 920\frac{m_f}{m_a}$$

Then: fuel-air ratio

$$\frac{m_f}{m_a} = \frac{920}{42000 - 920} = 0.0224$$

and air-fuel ratio $\frac{m_a}{m_f} = \frac{1}{0.0224} = 44.64$

(*b*) Velocity of machine

$$= \frac{600 \times 10^3}{3600} = 166.67 \text{ m/s}$$

Duration of flight $= \frac{250 \times 10^3}{166.7} = 1500$ s

Fuel consumption

$$m_f = \frac{425}{1500} = 0.2833 \text{ kg/s}$$

Air consumption

$$m_a = 0.2833 \times 44.64$$
$$= 12.646 \text{ kg/s}$$

Propulsive force (thrust)

$$= (m_a + m_f)\, V_j - m_a\, V_a$$

$$= m_a\left[\left(1 + \frac{m_a}{m_f}\right) V_j - V_a\right]$$

where V_a is forward velocity of the flying bomb and V_j is the velocity of jet relative to the bomb. Substituting the appropriate values, we obtain

$$2650 = 1.646\left[(1 + 0.0224)V_j - 166.67\right]$$

That gives, velocity of jet relative to bomb

$$V_j = \frac{2650 + 12.646 \times 166.67}{1.0224 \times 12.646}$$

$$= 368 \text{ m/s}$$

Specific volume of exhaust gases

$$v = \frac{A\,V_j}{m} = \frac{A\,V_j}{m_a + m_f}$$

$$= \frac{\frac{\pi}{4}(0.3)^2 \times 368}{12.646 + 0.2833} = 2.011 \text{ m}^3/\text{kg}$$

Using characteristic gas equation $pv = RT$, the temperature of exhaust gases is

$$T = \frac{pv}{R}$$

$$= \frac{(0.95 \times 10^5) \times 2.011}{300} = 636.82 \text{ K}$$

(c) Propulsive efficiency

$$\eta_p = \frac{\text{thrust power}}{\text{propulsive power}}$$

$$= \frac{\text{thrust power}}{\text{gain in kinetic energy}}$$

$$= \frac{\left[(m_a + m_f)V_j - m_a V_a\right]V_a}{\frac{1}{2}\left[(m_a + m_f)V_j^2 - m_a V_a^2\right]}$$

$$= \frac{[(12.646 + 0.2833) \times 368 - 12.646 \times 166.67] \times 166.67}{\frac{1}{2}\left[(12.646 + 0.2833) \times 368^2 - 0.2833 \times (166.67)^2\right]}$$

$= 0.631$ or 63.1%

Neglectging the mass of fuel, the propulsive efficiency would be

$$\eta_p = \frac{2V_a}{V_j + V_a}$$

$$= \frac{2 \times 166.6T}{368 + 166.67} = 0.623 \text{ or } 62.3\%$$

Thermal efficiency

$$\eta_{th} = \frac{\text{propulsive work; } i.e., \text{ gain in kinetic energy}}{\text{heat released by combustion of fuel}}$$

$$= \frac{\frac{1}{2}\left[(m_a + m_f)V_j^2 - m_a V_a^2\right]}{m_f \times CV}$$

$$= \frac{\frac{1}{2}\left[(12.646 + 0.2833) \times 368^2 - 12.646 \times 166.67^2\right]}{0.2833 \times (42000 \times 1000)}$$

$= 0.0588$ or 5.88%

Overall thermal efficiency $\eta_0 = \eta_p \times \eta_{th}$

$= 0.631 \times 0.0588$

$= 0.0371$ or $3.71\ \%$

EXAMPLE 23.18

A turbojet plane flying with a speed of 900 km/hr inducts air at rate of 45 kg/s. The air-fuel ratio is 60 and the expansion of gases occurs upto ambient pressure. Calculate the following parameters:

(a) jet velocity, maximum thrust, specific thrust and thrust power.

(b) specific fuel consumption

It may be assumed that calorific value of fuel is 43000 kJ/kg and for maximum thrust, the flight to jet velocity ratio is 0.5.

Solution : Forward velocity of the turbojet V_a

$= 900$ km/hr

$$= \frac{900 \times 10^3}{3600} = 250 \text{m/s}$$

Jet velocity

$$V_j = \frac{250}{0.5} = 500 \text{ m/s}$$

Propulsive force (thrust)

$= (m_a + m_f)\,V_j - m_a\,V_a$

$$= \left(45 + \frac{45}{60}\right)500 - 45 \times 250$$

$= 11625$ N $= 11.625$ kN

Specific thrust (*i.e.*, thrust per kg of air supplied per second)

$$= \frac{11625}{45} = 258.33 \text{ N/kg/s of air}$$

Thrust power = thrust × speed of plane

$= 116.25 \times 250$

$= 2906.25$ kJ/s $= 2906.25$ kW

(b) Propulsive efficiency

$$\eta_p = \frac{\text{thrust power}}{\text{propulsive power}}$$

$$= \frac{\text{thrust power}}{\text{gain in kinetic energy}}$$

$$= \frac{[(m_a + m_f)V_j - m_a V_a]V_a}{\frac{1}{2}[(m_a + m_f)V_j^2 - m_a V_a^2]}$$

$$= \frac{\left[\left(45 + \frac{45}{60}\right) \times 500 - 45 \times 250\right] \times 250}{\frac{1}{2}\left[\left(45 + \frac{45}{60}\right) \times 500^2 - 45 \times 250^2\right]}$$

$= 0.6739$ or 67.39%

Neglecting the mass of fuel, the propulsive efficiency would be

$$\eta_p = \frac{2V_a}{V_j + V_a}$$

$$= \frac{2 \times 250}{500 + 250} = 0.6667 \text{ or } 66.67\%$$

Thermal efficiency

$$\eta_{th} = \frac{\text{work output or propulsive work}}{\text{heat input}}$$

$$= \frac{\text{gain in kinetic energy}}{\text{heat released by combustion of fuel}}$$

$$= \frac{\frac{1}{2}[(m_a + m_f)V_j^2 - m_a V_a^2]}{m_f \times \text{calorific value of fuel}}$$

$$= \frac{\frac{1}{2}\left[\left(45 + \frac{45}{60}\right) \times 500^2 - 45 \times 250^2\right]}{\frac{45}{60} \times (43000 \times 1000)}$$

$= 0.1337$ or 13.37%

Overall efficiency $\eta_0 = \eta_p \times \eta_{th}$

$= 0.6739 \times 0.1337 = 0.0910$ or 9.01%

Specific fuel consumption

$$= \frac{\text{fuel consumed per hour}}{\text{propulsive force}}$$

$$= \frac{\frac{45}{60} \times 3600}{11625} = \mathbf{0.2322} \text{ kg/hr N}$$

SALIENT POINTS

1. The gas turbine is a contrivance for producing mechanical power and its application lies in the field of aviation, marine, supercharging, railways and road transport, electric power generation as peak load and stand by power plant.
2. The gas turbine is akin to petrol and diesel engines in its working medium and internal combustion and is similar to steam turbine with regard to continuous and steady flow of the working medium.
3. The gas turbine competes favourably with other prime movers because of its simplicity and reliable working, few auxiliaries and high rotative speed, freedom from vibration, low installation and running cost and ability to generate large power from units of relatively small size.
4. The gas turbines may be of constant volume or explosion type, and of constant pressure type working on open cycle or closed cycle.

The constant pressure turbines operate on Brayton cycle comprising two adiabatics and two constant pressure processes; the addition and rejection of heat are at constant pressure. In terms of pressure ratio r_p, the performance parameters of Brayton cycle are :

$$\eta = 1 - \left(\frac{1}{r_p}\right)^{\frac{\gamma-1}{\gamma}}$$

$$\text{Work ratio} = \frac{\text{net work}}{\text{turbine work}}$$

$$= 1 - \frac{T_1}{T_3}(r_p)^{\frac{\gamma-1}{\gamma}}$$

For maximum work ratio, the temperature T_1 at compressor entry should be as low as possible and temperature T_3 at entry to turbine should be as large as possible. Further, the pressure ratio for optimum specific work output is

$$r_p = \left(\frac{T_3}{T_1}\right)^{\frac{\gamma}{2(\gamma-1)}}$$

5. The isentropic efficiency of compressor (η_c) and turbine (η_t) are :

$$\eta_c = \frac{T_2' - T_1}{T_2 - T_1} \text{ and } \eta_t = \frac{T_3 - T_4}{T_3 - T_4'}$$

where T_2 and T_4 are the temperatures at exit from the compressor and turbine when the compression and expansion processes are isentropic; T_2' and T_4' represent the corresponding actual values.

Polytropic efficiency is the small stage efficiency that remains constant throughout the entire process, *i.e.*, it is independent of pressure ratio.

$$\eta_{pt} = \frac{n-1}{n} \times \frac{\gamma}{\gamma-1} \quad \text{...turbine}$$

$$\eta_{pc} = \frac{\gamma-1}{\gamma} \times \frac{n}{n-1} \quad \text{...compressor}$$

6. The following systems have been suggested for improving the performance of a gas turbine plant :
   - **Regeneration :** no change in compressor work, turbine work and so the network; reduction in the quantity of fuel input and improvement in thermal efficiency.
   - **Intercooling :** reduction in work supplied to compressor, no change in turbine work an improvement in network and work ratio; thermal efficiency may not increase because of simultaneous increase in requirement of heat addition.
   - **Reheating :** increase in turbine work without affecting the compressor work and so improvement in work ratio; greater fuel consumption and there occurs a small drop in thermal efficiency.
7. Jet propulsion devices are used to propel aircrafts, missiles and rockets. The jet propulsion systems are classified as :
   - Air breathing engines which take the air necessary for combustion from the atmosphere. For example turbojet, turbo prop, ram jet and pulse jet.
   - Rocket engines where the fuel and oxidiser are contained in the body of the unit which is to be propelled.
8. The turbojet uses ram compressor in addition to a compressor for compressing the air. Hot gases produced by combustion of fuel in the combustor expand in the turbine and further in the nozzle. The resulting high velocity jet propels the aircraft.

   In a turbo prop engine, the aircraft gets its forward motion from the propeller and the high velocity jet produced by the nozzle.

   Compression air in ram jet depends entirely on ram compression at supersonic speeds.
9. Thrust augmentation refers to the method of providing extra thrust at the time of take off or climb for short durations. This is achieved by (*i*) water or liquid injection into the ram air or by (*ii*) employing after burner or tail pipe burning, *i.e.*, burning of additional fuel in the turbine exhaust before it moves on to the exit nozzle.
10. The rocket motors can be operated either with liquid propellants or solid propellants. The liquid propellants (fuel and oxidiser) at low pressure are stored separately and are supplied by pumps to the combustion chamber where ignition of fuel takes place. The fuel and oxidiser in the solid propellant rockets are premixed in required proportion. The mixture is formed into a solid grain which is placed in the chamber where combustion is to take place.
11. For a jet propulsion system

    Thrust = $m_a (V_j - V_a)$

    Thrust work = $m_a (V_j - V_a) V_a$

    Propulsive work = $\frac{1}{2} m_a (V^2_j - V^2_a)$

$$\text{Propulsion efficiency} = \frac{\text{thrust work}}{\text{propulsive work}}$$

$$= \frac{2V_a}{V_j + V_a}$$

where m_a is mass flow rate of air entering the aircraft, V_a is velocity of aircraft (moving forward) and V_j is velocity of jet relative to the aircraft (flowing backwards)

For a rocket propulsion system

Thrust work = $m\ V_j V_a$

Propulsive work = $\frac{1}{2}\ m\ (V_j^2 + V_a^2)$

$$\text{Propulsive efficiency} = \frac{2V_a/V_j}{1+(V_a/V_j)^2} = \frac{2\sigma}{1+\sigma^2}$$

REVIEW QUESTIONS

A. Conceptual and conventional questions.

1. Mention the applications of gas turbine.
2. Describe, with a neat sketch, the working of a simple constant pressure gas turbine.
3. (*a*) Name the main components of a gas turbine and state the function of each component.
   (*b*) State the difference in the working of an open cycle and a closed cycle gas turbine.
4. Sketch Brayton cycle on *p-v* and *T-s* plots and derive a relation for its thermal efficiency in terms of pressure ratio.
5. In a Brayton cycle, the ambient conditions limit the minimum temperature T_1 and the metallurgical aspects for the turbine blades limit the maximum temperature T_3. For fixed values of T_1 and T_3, determine the optimum pressure ratio r_p which gives the maximum work in the Brayton cycle.
6. Discuss the influence of reheating, regeneration and intercooling on the performance of a gas turbine cycle.
7. Set up the condition for optimum pressure ratio for minimum compressor requirement in two-stage perfect intercooled compression.
8. Discuss the means for improving the efficiency and specific output of a gas turbine working on simple Brayton cycle.
9. An open gas turbine cycle with compressor isentropic efficiency η_c and turbine isentropic efficiency η_t uses air as ideal working substance. If r_p is the pressure ratio for both compression and expansion and γ is index of adiabatic compression or expansion, show that the condition of
   (*a*) possible work output is : $s < \eta_c\ \eta_t\ t$
   (*b*) maximum output is : $s^2 = \eta_c\ \eta_t\ t$

   where $s = (r_p)^{\frac{\gamma-1}{\gamma}}$ and $t = \frac{T_{max}}{T_{min}}$; T_{max} and T_{min} are the maximum and minimum temperature limits in the cycle.
10. List the advantages of gas turbine over (*a*) IC engines (*b*) steam turbines.
11. State the difference between the jet propulsion and rocket propulsion systems.
12. Explain the working difference between propellor-jet, turbo-jet and turbo-prop propulsion systems.
13. (*a*) Explain the working of a turbo jet propulsion system with the help of schematic arrangement of its different components and the *T-s* diagram of its basic cycle.
    (*b*) List the advantages and disadvantages of a turbo-jet engine.
14. Define the following terms in the context of propulsion systems :
    (*i*) thrust (*ii*) thrust power
    (*iii*) propulsive efficiency
    (*iv*) thermal efficiency.
15. A gas turbine plant operates between minimum and maximum temperature limits of 305 K and 1080 K respectively. Find the maximum work done per kg of air and the corresponding cycle efficiency. How does this efficiency compare with the Carnot cycle efficiency when operating between the same temperature limits ?
    [**Ans.** 9.14, 46.9%, 238.36 kJ/kg, 71.76%]
16. In a gas turbine installation, air is compressed adiabatically from 1 bar and 15°C in an axial flow compressor to 4 bar; the isentropic efficiency of compressor being 80 percent. The air is then heated to 650°C in a combustion chamber and is divided into two streams: one stream going to compressor turbine and the other to power turbine; the efficiency of each turbine is 85 percent. Find the ratio of heated air which passes through compressor turbine.
    [**Ans.** 67.8%]
17. A gas turbine power plant operates with air entering the compressor at 1 bar and 15°C, and getting compressed adiabatically through

compression ratio 5 : 1. The air after compression passes through a heat exchanger which allows heating of compressed air through 70 percent of the maximum possible range. Subsequent heating raises the air to turbine inlet temperature of 800°C and thereafter the hot air is expanded back to atmospheric pressure of 1 bar. Assuming compressor and turbine efficiencies to be 0.85 each, find the thermal efficiency of the plant.

[Ans. 33.5%]

18. A gas turbine set draws in atmospheric air at 1 bar and 298 K; there are two pressure stages with intercooler and the total pressure ratio is 8 : 1. The compression in both the stages occurs adiabatically with compressor efficiency 83%. The compressed air passes through a regenerator which recovers 60% of the available heat. The maximum gas temperature at entry to turbine unit is 865 K where expansion occurs to 1 bar pressure with 86% efficiency.

    Take c_p = 1.00 kJ/kg K and γ = 1.4 for both air and gases.

19. A gas turbine plant delivers 1500 kW and operates with compressor inlet pressure and temperature of 1 bar and 15°C, and a turbine inlet pressure and temperature of 4 bar and 900°C. If the isentropic efficiency of compressor is 85% and the overall plant efficiency is 21%, determine the isentropic efficiency of the turbine and the mass flow of air in kg/s.

    [Ans. 82.38%, 9.87 kg/s]

20. Air enters the compressor of a simple gas turbine at 1 bar and 25°C and is compressed to 7 bar with isentropic efficiency 82 percent. After attaining a temperature of 1050 K in the combustion chamber, the gases expand down to ambient pressure with 85 percent isentropic efficiency. The combustion efficiency is 0.89 and the pressure drop through the combustion chamber is 0.11 bar. If the heating value of the fuel is 42 MJ/kg, determine.

    (*a*) thermal efficiency,

    (*b*) work ratio, and

    (*c*) specific fuel consumption

    Take c_p = 1.005 kJ/kg K for air and

    c_p = 1.11 kJ/kg K for gases.

    [Ans. 22.5% ; 0.36, 0.38 kg/kWh]

21. A turbojet engine experiences 6100 N of drag when flying at a speed of 800 km/hr at 9500 m attitude where the density of the ambient air is 0.17 kg/m^3. The propulsive efficiency and overall efficiency of the unit are stated to be 55% and 17% respectively. If calorific value of the fuel used is 46000 kJ/kg, make calculations for the following parameters :

    (*a*) absolute velocity of jet

    (*b*) volume of air compressed

    (*c*) diameter of jet

    (*d*) power output of the unit and

    (*e*) air-fuel; ratio.

    [Ans. 82.38%, 9.87 kg/min, 462 mm, 2465 kW, asd 53 : 1]

22. A turbo jet engine operates between the pressure limits of 35 kPa and 350 kPa. The inlet air temperature is – 40°C and the upper temperature limit is 1370 K. Assuming isentropic compression and expansion and an inlet velocity of 100 m/s, make calculations for

    (*a*) the specific momentum thrust of the engine,

    (*b*) the heat input and power delivered by the engine per unit mass

    Take the gas to be equivalent to air and the velocity at the nozzle inlet to be negligible.

    [Ans. 845 N; 706.5 kW, 84.5 kW]

B. Fill in the blanks with appropriate word/words.

1. A gas turbine is the most useful plant for .... and .... load.
2. The air standard Brayton cycle consists of two isentropic and two .... processes.
3. The thermal efficiency of an ideal gas turbine cycle depends only on the ....
4. A constant volume combustion gas turbine works on .... cycle in which heat supply is at .... and heat rejection is at ....
5. The isentropic efficiency of a turbine is defined as the ratio of the .... output to the .... output between the same pressure limits.
6. The net work output of a Brayton cycle depends on the .... and the maximum ....
7. In a two-stage gas turbine plant, reheating after the first stage .... thermal efficiency.
8. For minimum compressor work in a gas turbine plant, the .... and .... for the two stages are equal.
9. The efficiency of a gas turbine plant increases with .... in ambient air temperature at .... altitudes.

10. The thermal efficiency of a gas turbine plant is improved by .... as this decreases the quantity of heat supplied.
11. The performance of a heat exchanger is often assessed by the term .... defined as the ratio of temperature difference available.

Answers

1. stand by and peak load 2. isobaric 3. pressure 4. Atkinson, constant volume, constant pressure 5. actual work, isentropic work 6. pressure ratio, cycle temperature 7. reduces 8. pressure ratio, work input 9. decrease, high 10. regeneration 11. thermal ratio

C. Multiple choice questions.

1. The ideal constant pressure gas turbine works on
   (*a*) Atkinson cycle (*b*) Ericsson cycle
   (*c*) Brayton cycle (*d*) Bell coleman cycle
2. Which aspect is not true in the context of constant volume gas turbine?
   (*a*) works on Atkinson cycle
   (*b*) intermittent combustion
   (*c*) high operating pressures are involved
   (*d*) thermal efficiency is higher than that of a constant pressure gas turbine
3. While starting a gas turbine, it is first motored to some minimum speed called
   (*a*) no load speed (*b*) motoring speed
   (*c*) coming in speed (*d*) run away speed
4. What factors limit the maximum temperature in a gas turbine cycle?
   (*a*) quality of fuel
   (*b*) combustion efficiency
   (*c*) turbine blade material
   (*d*) rotational speed of turbine blades
5. The thermal efficiency of a air standard Brayton cycle is a function of pressure ratio r_p and is given by
   (*a*) $\eta = 1 - (r_p)^m$ (*b*) $\eta = 1-\left(1/r_p\right)^m$
   (*c*) $\eta = 1-\left(1/r_p\right)^{\frac{1}{2m}}$ (*d*) $\eta = 1-\left(1/r_p\right)^{\frac{2}{m}}$
   where $m = \dfrac{\gamma-1}{\gamma}$
6. The optimum value of pressure ratio that gives maximum output per unit mass (specific work output) is given by $(r_p)_{opt}$ =
   (*a*) $(T_3/T_1)^{\frac{\gamma}{\gamma-1}}$ (*b*) $(T_3/T_1)^{\frac{\gamma}{2(\gamma-1)}}$
   (*c*) $(T_3/T_1)^{\frac{\gamma-1}{2\gamma}}$ (*d*) $(T_3/T_1)^{\gamma-1}$
   where T_3 is the turbine inlet temperature and T_1 is the temperature of air at inlet to compressor.
7. Work ratio is defined as the ratio of
   (*a*) actual heat drop and isentropic heat drop in a turbine
   (*b*) compressor work and turbine work
   (*c*) net work output and turbine work
   (*d*) net work output and compressor work
8. Which of the following units increase the work ratio in a gas turbine plant?
   1. Regenerator 2. Intercooling 3. Reheating
   Select the correct answer using the codes given be loss:
   (*a*) 1 and 2 (*b*) 2 and 3
   (*c*) 1 and 3 (*d*) 1, 2 and 3
9. In a gas turbine cycle, regenerator is used to
   (*a*) heat the fuel being supplied to the combustion chamber
   (*b*) heat the compressed air on its way to combustion chamber
   (*c*) heat the gases leaving the combustion chamber
   (*d*) heat the turbine exhaust before it enters the low pressure stage
10. The thermal efficiency of a simple open cycle gas turbine plant is improved by regeneration as this
    (*a*) decreases the temperature of gases at the turbine inlet
    (*b*) decreases the quantity of heat supplied in the combustion chamber
    (*c*) increases the turbine output
    (*d*) lowers the work input to compressor
11. Consider the following statements :
    1. The speed of rotation of the moving elements of gas turbines is much higher than those of steam turbines.
    2. Gas turbine plants are heavier and larger in size than steam turbine plants
    3. Gas turbine requires cooling water for its operation
    4. Almost any kind of fuel can be used with gas turbines

Which of the statement given above are correct?

(a) 1 and 2 (b) 1 and 3

(c) 1 and 4 (d) 3 and 4

12. A jet engine works on the principle of conservation of

(a) mass (b) discharge

(c) energy (d) momentum

13. The compressor does not form a part of

(a) ramjet (b) turbo-jet

(c) turbo-prop (d) aircraft gas turbine

14. The surrounding air is not used as oxidiser in

(a) flying bomb (b) rocket propulsion

(c) turbo-prop (d) ramjet

15. In a turbo-jet engine, thrust may be increased by

1. increasing the jet velocity
2. increasing the mass flow rate of air
3. after buring of fuel

Which of these statements are correct?

(a) 1 and 2 (b) 2 and 3

(c) 1 and 3 (d) 1, 2 and 3

16. The propulsion efficiency of a turbo-jet engine is

(a) $\eta_p = \frac{\sigma}{1+2\sigma}$ (b) $\eta_p = \frac{2\sigma}{1+\sigma}$

(c) $\eta_p = \frac{\sigma}{2(1+\sigma)}$ (d) $\eta_p = \frac{2\sigma}{1+\sigma^2}$

17. For maximum overall efficiency of a turbo-jet engine, the aircraft velocity and jet velocity conform to the ratio

(a) 1 : 2 (b) 2 : 3

(c) 2 : 1 (d) 1 : 3

18. Consider the following statements indicating a comparison between rocket and jet propulsion systems:

1. Both the rocket and jet engines carry the fuel and oxidant
2. Rockets donot employ compressor or propeller
3. Rockets can operate in vaccum also
4. Rockets can use solid fuels and oxidants

Which of these statements are correct?

(a) 1, 2, 3 and 4 (b) 1 and 2

(c) 2, 3 and 4 (d) 1, 3 and 4

19. For a gas turbine operating between temperature limits of T_1 and T_3 ($T_3 > T_1$), the optimum value of pressure ratio that gives maximum specific work output is given by :

(a) $r_p = \left(\frac{T_3}{T_1}\right)^{1/m}$ (b) $r_p = \left(\frac{T_3}{T_1}\right)^{1/2m}$

(c) $r_p = \left(\frac{T_3}{T_1}\right)^{m/2}$ (d) $r_p = \left(\frac{T_3}{T_1}\right)^{2/m}$

where $m = \frac{\gamma-1}{\gamma}$

20. Which of the following combinations leads to higher efficiency of an ideal Brayton cycle with regeneration?

(a) high temperature ratio and high pressure ratio

(b) high temperature ratio and low pressure to ratio

(c) low temperature ratio and high pressure ratio

(d) low temperature ratio and low pressure ratio

where temperature ratio refers to ratio of maximum to minimum operating temperatures of the cycle.

21. Consider the following statements in the context of a gas turbine plant

1. With regeneration, the thermal efficiency becomes equal to Carnot efficiency when pressure ratio r_p is unity
2. With reheating, there is greater fuel consumption and there occurs a small decrease in thermal efficiency
3. The work ratio improves with the installation of an intercooler

Which of these statements are correct?

(a) 1 and 2 (b) 2 and 3

(c) 1 and 3 (d) 1, 2 and 3

Answers

1. (c)	2. (d)	3. (c)	4. (c)	5. (b)
6. (b)	7. (c)	8. (b)	9. (b)	10. (b)
11. (c)	12. (d)	13. (a)	14. (b)	15. (b)
16. (b)	17. (a)	18. (c)	19. (b)	20. (b)
21. (d)				

Chapter **24**

Fuels and Combustion

Learning objectives : Attention has been focussed in this chapter on the following aspects of fuels and combustion :

- *characteristics of an ideal fuel, types of fuels and their sources*
- *combustion reaction and estimation of air required for combustion*
- *calorific value of fuels, Bomb calorimeter and Junker's gas calorimeter*
- *analysis of the products of combustion, Orsat apparatus*
- *chemical thermodynamics; stoichiometric coefficiencts, formation reaction, adiabatic flame temperature etc*

Corresponding to the three natural laws of thermodynamics, the study of combustion thermodynamics is dealt under three aspects namely :

- Study of relationship between the compositions of the reactants and the products (law of conservation of matter)
- Study of the interrelation between heat, work and energy changes in processes undergone by systems which are not pure substances (1st law of thermodynamics)
- Study whether a reaction will proceed forward or backward (2nd law of thermodynamics)

Thus the thermodynamic analysis of a chemically reactive system essentially provides information regarding the feasibility of a reaction, the equilibrium degree of reaction and the effect of operating conditions on the degree of conversion. These aspects will be discussed briefly with major emphasis on combustion of fuels which plays a significant role in many power producing devices like IC engines and boilers. Attention is focussed on the types of fuels and their calorific values, estimation of the quantity of air required whether combustion is complete or incomplete, and an analysis of the products of combustion.

24.1. CHARACTERISTICS OF AN IDEAL FUEL

Fuel represents a combustible substance which, once raised to ignition temperature, continues to burn without any external support provided a sufficient quantity of oxygen is available for combustion. Wood, coal, kerosene, petrol, diesel, producer gas, oil gas etc. are some of the fuels.

The requirements of an ideal fuel are:

(*i*) high heating or calorific value

(*ii*) low content of moisture and non-combustible matter

(*iii*) readily available and low in cost

(*iv*) easy to store and transport

(*v*) high combustion efficiency without much smoke and harmful products like CO, SO_2, H_2S, etc.

(*vi*) moderate ignition temperature. *Ignition temperature* is the lowest temperature to which the fuel must be pre-heated so that it starts burning smoothly. Fire hazards are caused at low ignition temperature and there is difficulty in combustion at high ignition temperature.

(*vii*) combustion should be at moderate speeds and easily controllable

(*viii*) size of solid fuel should be uniform so that combustion is regular

24.2. CLASSIFICATION OF FUELS

A fuel is a complex substance which generally belongs to the hydrocarbon family of which the major constituents are carbon and hydrogen. There may be traces of sulphur and non-combustible substances like nitrogen, water vapours and ash etc. Sulphur is an unwanted element because its combustion products destroy the metal parts of boiler units and poison the atmosphere.

According to its physical state, fuel may be solid, liquid or gaseous, and to the method by which it is obtained-natural and artificial.

24.2.1. Solid fuels

The solid fuels may be (i) natural or primary which are found in nature as such or (ii) artificial or secondary which are prepared from primary fuels.

The common examples of natural solid fuels are wood and variety of coals (peat, lignite, bituminous, anthracite). Coal is high carbonaneous matter that has been formed as a result of alteration of vegetable matter under certain favourable conditions. The various forms of coal are generally recognised on the basis of rank or degree of alteration or coalification from the parent material.

The progressive transformation of wood to anthracite results in:

- decrease in moisture content
- decrease in hydrogen, oxygen, nitrogen and sulphur contents with a corresponding increase in carbon content.
- decrease in volatile matter
- increase in the heating or calorific value
- increase in hardness

Anthracite is the highest rank coal which is used for steam raising, in metallurgical processes and for household purposes.

Wood charcoal, coke, pulverised coal and briquette coal belong to the class of prepared solid fuels.

The analysis of the composition of coal is generally done in two ways namely proximate analysis and ultimate analysis. In ***proximate analysis***, no attempt is made to determine the percentage of individual elements comprising the fuel. Rather, the percentages of moisture, volatile matter, fixed carbon and ash (incombustible matter) are determined and specified. Such an analysis would normally be indicated as

$$\text{moisture} = 5 \text{ to } 25\%$$
$$\text{volatile matter} = 10 \text{ to } 55\%$$
$$\text{fixed carbon} = 60 \text{ to } 85\%$$
$$\text{and} \quad \text{ash} = 2 \text{ to } 8\%$$

In ***ultimate analysis***, the percentages of various elements are specified such as

$$\text{moisture} = 5 \text{ to } 25\%$$
$$\text{carbon} = 50 \text{ to } 90\%$$
$$\text{hydrogen} = 3 \text{ to } 10\%$$
$$\text{oxygen} = 2 \text{ to } 5\%$$
$$\text{nitrogen} = 0.5 \text{ to } 3\%$$
$$\text{sulphur} = 0.8 \text{ to } 3.8\%$$

and remaining incombustible matter.

The ultimate analysis is helpful to determine the amount of air required for combustion, the calorific value of fuel, and the constituents of products of combustion. The proximate analysis is easy, quick and is adopted for commercial purposes.

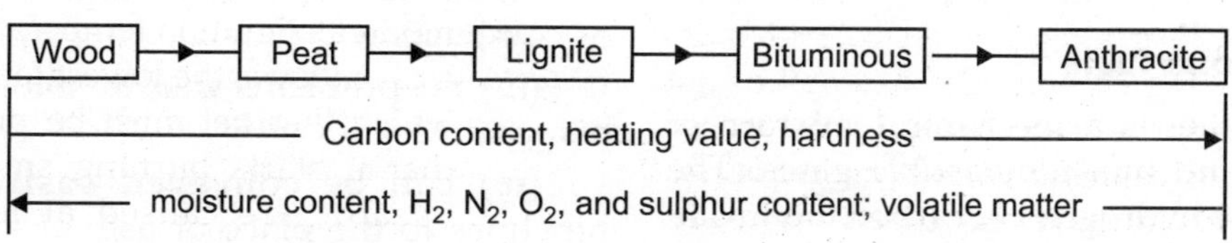

24.2.2. Liquid fuels

Most of the liquid fuels are obtained from petroleum which is a mixture of many hydrocarbons (paraffins, olefins, napthalene and aromatics) together with small amounts of organic compounds containing oxygen, nitrogen and sulphur. When crude petroleum is refined by boiling at different temperatures, it gives the liquid fuels namely petrol, diesel and kerosene oil.

Coal tar, tar oil, shale oil and natural gas oil etc. are distilled artificial oils.

Petrol is highly volatile, inflammable and is used as fuel for internal combustion engines and aeroplanes. Diesel oil is used as a diesel engine fuel for road and rail transport. Kerosene is used as domestic fuel, as jet engine fuel and for making oil gas. Alcohol is another liquid fuel formed by fermentation of vegetable matter. Its use as commercial fuel in IC engines has been to a limited extent.

The advantages of liquid fuel over solid fuel are:

(*i*) higher calorific (heating) value

(*ii*) no loss in heating value on storage

(*iii*) better control of combustion through adjustment of fuel supply

(*iv*) can be easily pumped over great distance along pipes

(*v*) easy to store from space and safety point of view

(*vi*) clean in use and economic in labour

(*vii*) require less furnace space for combustion

(*viii*) undergo complete combustion with slight excess of air

The disadvantages of liquid fuels are:

(*i*) less availability and as such high cost

(*ii*) costly containers required for storage and transport

(*iii*) efficient burning requires good design of burners

24.2.3. Gaseous fuels

Gaseous fuel is a mechanical mixture of combustible and non-combustible gases. The former include hydrogen H_2, carbon monoxide CO, methane CH_4, various hydro carbons and hydrogen sulphide. The hydrocarbons are the most valuable components in a gaseous fuel.

The gaseous fuels of engineering importance are:

• **Natural gas** is obtained from the wells dug in the petroleum-bearing fields. It essentially consists of methane (CH_4) together with small amounts of other gases such as ethane (C_2H_6), CO_2 and CO. It is an excellent domestic fuel and can be conveyed over very large distances in pipelines.

• **Coke oven gas** is obtained by the carbonisation of bituminous coals and is used for industrial heating and power generation.

• **Coal gas** is obtained when coal is carbonised or heated in the absence of air at about 1300° C. The coal gas consists mainly of H_2 , CH_4, CO and traces of various other hydro-carbons. The coal gas is very rich amongst combustible gases and is used for street and domestic lighting, heating and in metallurgical operations for providing reducing atmosphere.

• **Blast furnace gas** is the by-product in the production of pig iron in a blast furnace. The gas finds extensive use as fuel in steel works, for power generation in gas engines, for steam raising in boilers and for preheating the blast for furnace.

• **Producer gas** is obtained by passing air-water blast over a bed of red hot coal or coke at about 1100°C in a special reactor called gas producer.

It is a cheap, clean and easily preparable gas. It is used mostly for heating open hearth furnaces (for steel and glass manufacture) muffle furnaces (for coke and coal gas manufacture) and as a reducing agent in metallurgical operations.

The advantages of gaseous fuels are:

(*i*) have high heat content

(*ii*) possibility of remote and instant control of combustion

(*iii*) no problems due to smoke and ash disposal

(*iv*) can be conveyed easily through pipelines to the place of use (*v*) free from solid and liquid impurities.

(*vi*) possibility of complete combustion with slight excess of air due to uniform mixing of air and fuel

(*vii*) clean in use; burns without soot and smoke

The disadvantages of gaseous fuels are:

(*i*) need large storage tanks

(*ii*) highly inflammable; high chances of fire hazards

(*iii*) high in cost

Gaseous fuels are used in industrial applications where uniform heating is required and the product is not to be contaminated.

24.3. COMBUSTION REACTION AND COMBUSTION ANALYSIS

Combustion involves a chemical reaction during which the interatomic bonds of the molecules of the fuel and oxygen are broken and there is rearrangement of atoms in new molecular combination. Before considering combustion problems, it is necessary to understand the following terms associated with combustion :

The combustion of a fuel involves the oxidation of its constituents and the process is represented by a chemical reaction. Consider complete combustion of methane (CH_4) in the presence of oxygen (O_2) forming carbon dioxide (CO_2) and water (H_2O). The water may be a liquid or a vapour depending on whether the product has been cooled sufficiently to cause condensation. This combustion reaction can be represented as

$$\underset{\text{Reactants}}{CH_4 + 2\,O_2} \longrightarrow \underset{\text{Products}}{CO_2 + 2\,H_2O}$$

With regard to chemical reaction,

(*i*) The chemical equation is obtained by balancing the atoms of each of the atomic species involved in the reaction.

(*ii*) The initial constituents (methane and oxygen) which start the reaction are called *reactants,* and the final constituents (carbon dioxide and water) which are formed due to arrangement of atoms and electrons are called the *products*. The reactants disappear and the products are formed.

(*iii*) The complete combustion means that all the carbon in the fuel will burn to CO_2 and all hydrogen to H_2O.

(*iv*) 1 mole of CH_4 reacts with 2 moles of oxygen yielding 1 mole of CO_2 and 2 moles of water. The relative masses of carbon, oxygen and hydrogen are 12 kg, 32 kg and 2 kg. Accordingly *on mole basis* :

$$1 \text{ mole of } CH_4 + 2 \text{ moles of } O_2 = 1 \text{ mole of } CO_2 + 2 \text{ moles of } H_2O$$

on mass basis

$$1(12 + 4) \text{ kg of } CH_4 + 2(32) \text{ kg of } O_2 = 1(12 + 32) \text{ kg of } CO_2 + 2(2 + 16) \text{ kg of } H_2O$$

or

$$16 \text{ kg of } CH_4 + 64 \text{ kg of } O_2 = 44 \text{ kg of } CO_2 + 36 \text{ kg of } H_2O$$

The mass representation clearly stipulates that the total mass of products is equal to the total mass of reactants. Further, during the chemical reaction, the mass of each element is conserved *i.e.,* remains the same. Apparently the chemical equation involves the principle of conservation of mass of each element. However, the total number of moles is not necessarily conserved during a chemical reaction.

(*v*) Combustion is an exothermic reaction and the energy released during combustion equals the difference in the energy of reactants and that of products formed.

$$\text{Fuel + Oxygen} = CO_2 + H_2O + \text{Heat}$$

Products : more energy content	Reactants : lesser energy content

The combustion energy is released to the surroundings in the form of heat which can be used economically for industrial and domestic purposes.

(*vi*) The oxygen required for combustion of fuel is supplied in the form of air which contains oxygen. Presuming that CO_2, argon and other gases to be present in negligible quantities, air can be treated as a mixture containing 21% oxygen and 79% nitrogen by volumes (or 23% oxygen and 77% nitrogen by mass), the combustion equation can be written as

$$CH_4 + 2\,O_2 + 2\left(\frac{79}{21}\right)N_2 \longrightarrow CO_2 + 2\,H_2O + 2\left(\frac{79}{21}\right)N_2$$

Apparently, there is no free oxygen in the combustion products and nitrogen does not undergo any oxidation. Nitrogen is an inert gas as far as chemical reaction is concerned and appears on both sides of the equation.

(*vii*) The minimum amount of air which supplies the requisite amount of oxygen for complete combustion of fuel (carbon, hydrogen and traces of other elements that may oxidize) is called *theoretical* or *stoichiometric* air. The term complete combustion implies that all the carbon in fuel burns to CO_2 and all hydrogen to H_2O, and the products do not contain any oxygen.

(*viii*) The reaction is, however, not likely to proceed to completion if the fuel is supplied just with stoichiometric air. The objective of complete combustion is achieved by supplying excess air; excess air means the amount of air supplied in excess to that theoretically needed to burn fuel completely. The quantity of excess air depends on the type of fuel and the method of combustion used and may vary from 25 to 100 percent. Excess air ensures better combustion but at the same time cools the furnace and increases heat loss in the exhaust gases. When more air is supplied than that required by the fuel for its complete combustion, the mixture is said to be *weak* or *lean*. However, when insufficient air is supplied, we call the air-fuel mixture *rich*. When the air supplied is just enough to burn the fuel completely, the mixture is called *stoichiometric* or *chemically correct.*

The 150% theoretical air means that air supplied is 1.5 times the stoichiometric air, and with that the combustion equation for methane fuel is written as

$$CH_4 + 1.5 \times 2O_2 + 1.5 \times 2\left(\frac{79}{21}\right)N_2 \longrightarrow CO_2 + 2\,H_2O + O_2 + 1.5 \times 2\left(\frac{79}{21}\right)N_2$$

(*ix*) The air-fuel (AF) ratio is defined as the ratio of the quantity of air supplied for combustion to the quantity of fuel. This ratio maybe expressed on mass basis (for solid and liquid fuels) or on molar basis (for gaseous fuels).

$$\text{AF ratio} = \frac{\text{mass of air}}{\text{mass of fuel}} \quad \text{(on mass basis)}$$

$$\text{AF ratio} = \frac{\text{moles of air}}{\text{moles of fuel}} \quad \text{(on molar basis)}$$

The inverse of air-fuel ratio is called fuel-air (FA) ratio.

Now, consider 1 kg sample of a fuel having the mass composition as C kg of carbon, H_2 kg of hydrogen, S kg of sulphur and O_2 kg of oxygen and remaining incombustible matter.

• *Chemical reaction* when carbon (C) combines with oxygen (O_2) to form carbon dioxide (CO_2) is given by

$$\begin{array}{cccc} C & + \; O_2 & = & CO_2 \\ 12 & 32 & & 44 \\ 1 & \frac{8}{3} & = & \frac{11}{3} \end{array}$$

That is to say that 1 kg of carbon requires $\frac{8}{3}$ kg of oxygen for its complete combustion and forms $\frac{11}{3}$ kg of carbon dioxide.

$\therefore$ C kg of carbon requires $\frac{8}{3}$C kg of oxygen.

• *Chemical reaction* when hydrogen (H_2) combines with oxygen (O_2) to form water vapour (H_2O) is given by

$$\begin{array}{cccc} 2\,H_2 & + \; O_2 & = & 2\,H_2O \\ 4 & 32 & & 18 \\ 1 & 8 & = & 9 \end{array}$$

Thus, 1 kg of H_2 requires 8 kg of oxygen and forms 9 kg of water vapour.

$\therefore$ H_2 kg of hydrogen requires 8 H_2 kg of oxygen.

• *Chemical reaction* when sulphur (S) combines with oxygen (O_2) to form sulphur dioxide is given by

$$\begin{array}{cccc} S & + \; O_2 & = & SO_2 \\ 32 & 32 & & 64 \\ 1 & 1 & = & 2 \end{array}$$

Thus 1 kg of S requires 1 kg of oxygen and forms 2 kg of sulphur dioxide.

∴ *S* kg of sulphur requires S kg of oxygen.

There is O_2 kg of oxygen already present in the fuel.

∴ Net amount of oxygen required for complete combustion of 1 kg of fuel

$$= \left(\frac{8}{3}C + 8H_2 + S\right) - O_2$$

$$= \frac{8}{3}C + 8\left(H_2 - \frac{O_2}{8}\right) + S$$

The oxygen required for combustion is supplied in the form of air which contains oxygen. Ignoring the presence of CO_2, argon and other gases (which are present in negligible amount) air can be treated as a mixture containing 23% oxygen and 77% of nitrogen by mass (or 21% oxygen and 79% nitrogen by volume). Then minimum air required for complete combustion of the given fuel sample is

$$= \left[\frac{8}{3}C + 8\left(H_2 - \frac{O_2}{8}\right) + S\right] \times \frac{1}{0.23}$$

The minimum amount of air which supplies the requisite amount of oxygen for complete combustion of fuel is called ***theoretical*** or ***stoichiometric*** air.

The reaction is, however, not likely to proceed to completion if the fuel is supplied just with stochiometric air. The objective of complete combustion is achieved by supplying excess air; excess air means the amount of air supplied in excess to that theoretically needed to burn fuel completely. The quantity of excess air depends on the type of fuel and the method of combustion used and may vary from 25 to 100 percent. Excess air insures better combustion but at the same time cools the furnace and increases heat loss in the exhaust gases. When more air is supplied than required by the fuel for its complete combustion, the mixture is said to be ***weak*** or ***lean***. However, when insufficient air is supplied, we call the air-fuel mixture ***rich***. When the air supplied is just enough to burn the fuel completely, the mixture is called ***stochiometric*** or ***chemically*** *correct*.

The air-fuel ratio (AF) is defined as the mass of air to the mass of fuel. If theoretically or chemically correct air is supplied, the ratio is called ***stoichiometric air-fuel ratio.*** The inverse of air-fuel ratio is called fuel-air (FA) ratio.

24.4. CALORIFIC VALUE OF FUELS

A fuel generally has number of constituents. Some constituents react exothermally and thereby generate heat and some others do not react chemically at all. The heat generated due to combustion of unit mass of solid (or liquid) fuel is called its ***calorific value*** and is expressed in terms of kJ/kg. For gaseous fuels, the calorific value is expressed as heat liberated by combustion of unit cubic metre of gas at standard temperature and pressure (15°C and 760 mm of mercury).

Hydrogen which is one of the important constituent of most of the fuels burns to form water vapours. The formation of water vapours consumes some of the heat generated. If the combustion products are cooled back to the initial temperature (usually specified as 15°C), the water vapours get condensed and that liberates the latent heat. Obviously, then the maximum amount of heat is produced and this maximum heat is called the ***higher calorific value*** *(HCV)* of fuel.

Consider unit mass of solid fuel consisting of C kg of carbon, H kg of hydrogen, S kg of sulphur and O kg of oxygen. Further let the individual calorific value of carbon be 33900 kJ/kg, of hydrogen be 144000 kJ/kg and of sulphur be 9290 kJ/kg. Then higher calorific value of fuel will be

$$HCV = (33900\ C + 144000\ H + 9290\ S)\ kJ/kg$$

when no oxygen is present.

In case some oxygen is also present in fuel, then

$$HCV = \left[\begin{array}{c} 33900\,C + 144000\left(H - \frac{O}{8}\right) \\ +9290\,S \end{array}\right] kJ/kg$$

The parameter $\left(H - \frac{O}{8}\right)$ represents the quantity of hydrogen available for combustion.

If the cooling of combustion products to room temperature is not possible, then there is no condensation of water vapour and the associated heat is carried to the atmosphere. The heat then available for heating purposes and commercial use is less than the total heat evolved by the combustion of fuel. This heat is called the lower calorific value (*LCV*) of fuel.

$LCV = HCV$ – heat carried by water vapour formed per kg of fuel burnt

From the chemical reaction $2\,H_2 + O_2 = 2\,H_2O$, it is seen than 4 kg of hydrogen combine with 32 kg of oxygen to form 36 kg of water vapour. That is the water vapour formed is 9 times the mass of hydrogen in fuel.

$$\therefore \quad LCV = HCV - 9H \times h_{fg}$$

where h_{fg} = 2460 kJ/kg is the latent heat of steam at saturation temperature of 15°C

Note: Dulong has suggested the following empirical formula for calculating the *HCV* of fuel

$$HCV = \left[33800\,C + 144000\left(H - \frac{O}{8}\right) + 9270\,S\right] \text{kJ/kg}$$

EXAMPLE 24.1

Estimate the higher and lower calorific values of a fuel having the following composition by mass:

C = 88% ; H_2 = 10% ;
O_2 = 3% ; S = 2%; N_2 = 20%

and the rest is incombustible matter

Solution : Using formula suggested by Dulong for the calculation for the calorific value of a solid fuel, we have:

$$HCV = 33800\,C + 144000\left(H - \frac{O}{8}\right) + 9270\,S$$

where *C*, *H*, *O* and *S* are carbon, hydrogen, oxygen and S in 1 kg of fuel. Then

$$HCV = 33800 \times 0.8 + 144000\left(0.1 - \frac{0.03}{8}\right) + 9270 \times 0.02$$

$$= 27040 + 13860 + 185.4$$

$$= \mathbf{41085.4\ kJ/kg}$$

Amount of steam (water vapour) formed

$$= 0.1 \times 9 = 0.9 \text{ kg/kg fuel}$$

Heat absorbed in formation of steam

$$= 0.9 \times 2460 = 2214 \text{ kJ}$$

$$\therefore \quad LCV = 41085.4 - 2214 = \mathbf{38871.4\ kJ/kg}$$

24.4.1. Bomb calorimeter

The bomb calorimeter is used for finding out the heating (calorific) value of solid or liquid fuel.

Construction : The bomb is a thick walled constant volume vessel made of stainless steel which provides considerable resistance to corrosion and can withstand a pressure of about 150 atmospheres. The unit has a silica crucible in which is placed a weighted quantity of the fuel sample. The bomb is provided with an electric connection for producing spark, a connection for oxygen supply and a release valve for burnt gases. The bomb is surrounded by jacket containing water. The temperature of this water is kept uniform with the help of a motor driven stirrer. The water temperature is measured by a special thermometer reading upto 0.01 degree. The bottom of the bomb also contains about 5 gm of water to absorb vapours of sulphur or nitrogen in fuel.

Procedure: The steps necessary for the conduct of test run are outlined below:

(*i*) A fuel pallet (briquette) of 0.75 to 1 gm is accurately weighted and placed in the crucible.

(*ii*) A fuse wire of known mass is stretched between the electrodes and it is ensured that the fuse wire is in contact with the fuel.

(*iii*) The oxygen required for burning of fuel is admitted in the bomb through oxygen valve to a pressure of about 25-30 bar.

(*iv*) The bomb is placed in the calorimeter (water jacket) which contains a measured quantity of distilled water.

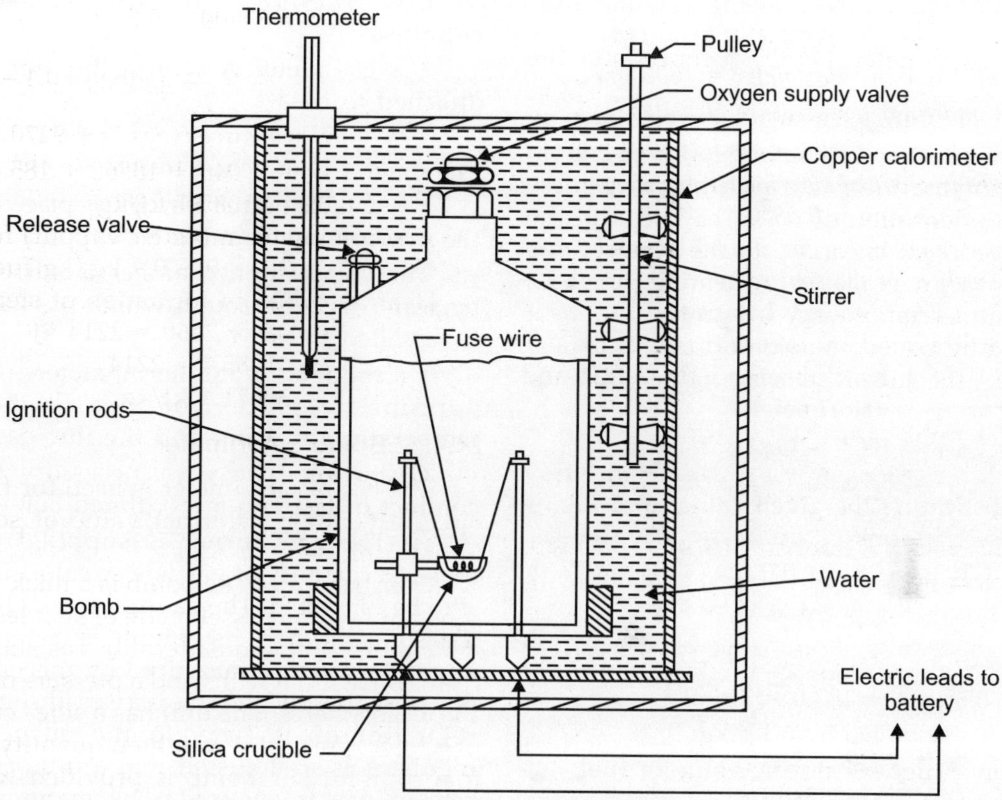

Fig. 24.1. Bomb calorimeter

(*v*) The water in the calorimeter is continuously stirred at moderate rate and temperature readings are noted. When the temperature becomes constant, the electric circuit is closed and the fuel gets ignited. Due to combustion of fuel, heat is liberated and temperature of water starts rising. Temperature readings are noted at regular intervals until constantancy is observed.

(*vi*) Having completed the experiment, the bomb is removed from the calorimeter. The pressure is slowly released through the exhaust valve and the contents of the bomb are weighed.

Calculations: The heat generated by the combustion of fuel and fuse wire is partly absorbed by water and partly by the metal of bomb and calorimeter. That is

$$m_f \times CV + m_{fw}\, CV_{fw} = (m_w + m_c)\, c_{pw}\, dT$$

where

m_f, CV = mass and calorific value of fuel

m_{fw}, CV_{fw} = mass and calorific value of fuse wire

m_w = mass of water in the calorimeter

m_c = water equivalent of bomb and calorimeter

c_{pw} = specific heat of water

dT = corrected temperature rise of water in the calorimeter

The corrected temperature rise refers to observed rise in temperature plus correction for radiation loss.

EXAMPLE 24.2

The following data pertains to a test run made to determine the calorific value of a sample of coal having hydrogen content of 5 percent.

mass of coal burnt = 0.85 gm

mass of fuse wire burnt and its calorific value
= 0.028 gm; 6700 kJ/kg
mass of water in calorimeter = 1800 gm
initial and final temperature of water
= 16.5°C and 20.25°C
water equivalent of calorimeter = 350 gm
cooling correction = 0.08°C

Make calculations for the higher and lower calorific values of the coal sample.

Solution : From energy balance

heat liberated by fuel and fuse wire = heat absorbed by water and calorimeter

$$m_f \times CV + (m \times CV)_{fuse} = (m_w + m_c)\, c_{pw}\, (dT + a)$$

Substituting the given values and taking all masses in gms

$$0.85 \times CV + 0.028 \times 6700 = (1800 + 350) \times 4.186\,[(20.25 - 16.5) + 0.08]$$

or $0.85\ CV + 187.6 = 34469.6$

$CV = 40331.76$ kJ/kg

Hence higher calorific value of fuel,

$HCV =$ **40331.76 kJ/kg**

The burning equation for hydrogen gives

$2\ H_2 + O_2 = 2\ H_2O$

4 kg of hydrogen + 32 kg of oxygen = 36 kg of water vapour

The fuel contains 0.05 kg of hydrogen which will form

$$= 0.05 \times \frac{36}{4}$$

= 0.45 kg of water vapour

Heat absorbed in formation of steam

$= 0.45 \times 2460 = 1107$ kJ/kg

∴ Lower calorific value

$LCV = 40331.76 - 1107$

= **39224.76 kJ/kg**

24.4.2. Junker's gas calorimeter

Refer Fig. 24.2. for the constructional details of Junker's gas calorimeter which is used for the determination of the calorific value of gaseous fuels.

Construction: The unit essentially comprises:

- a gas meter A with a thermometer B attached to it
- a gas pressure regulator C with a manometer D attached to it
- a suitable burner E which is placed inside the combustion chamber F

The combustion chamber is surrounded by water jacket through which pass a set of tubescalled flues.

- a set of accurate thermometers fitted at appropriate places for measurement of temperature of water and the flue gases.

Operation: The steps necessary for the conduct of test run are outlined below:

(*i*) The gas burner is supplied with a metered quantity of gas whose calorific value is to be estimated. The gas first passes through a gas meter where its volume is metered and its temperature is measured by thermometer B. Subsequently the gas passes through the regulator on its way to the burner. The regulator is so adjusted that a uniform flow rate of gas is ensured. The manometer D measures the pressure of gas being taken to burner.

(*ii*) When the gas burns, the hot products of combustion travel upwards through the centre of the chamber. The chamber is so designed that the gases after having reached the top get deflected downwards through the flues and finally escape to the atmosphere through outlet. The temperature of the combustion products at exit is measured by thermometer fitted into the exit passage.

(*iii*) The water enters the jacket, surrounding the combustion chamber, at the bottom and leaves from the top. The water which flows through the jacket is collected, weighed and its temperatures at inlet and outlet are noted.

(*iv*) Most of the fuels contain some percentage of hydrogen which burns to form water vapours. The amount of condensate formed due to condensation of water vapours is also collected and weighed.

Calculations: The calorific value of gas is normally expressed in kJ/m^3 at standard atmospheric pressure conditions (15°C and 760 mm of mercury)

Let p_1, V_1 and T_1 = pressure (bar), volume (m^3) and absolute temperature (K) of gas supplied to burner

Then equivalent volume under standard conditions (designated by symbol 0) works out as

$$V_0 = \frac{p_1 V_1}{T_1} \times \frac{T_0}{p_0}$$

From energy balance:

heat given by fuel = heat taken by water

$$V_0 \times CV = m_w \times c_{pw} \times \Delta T$$

where CV is the calorific value of fuel, m_w is mass of cooling water, c_{pw} is specific heat of water and ΔT is rise in temperature of cooling water. Thus

Highest calorific value (HCV) of fuel

$$= \frac{m_w \times c_{pw} \times \Delta T}{V_0}$$

The lower calorific value (LCV) can be obtained by substracting the heat consumed in the formation of steam. That is

Lower calorific value (LCV)

$$= HCV - m_c \times h_{fg}$$

where m_c is the mass of condensate and h_{fg} is the latent heat of steam corresponding to 15°C temperature.

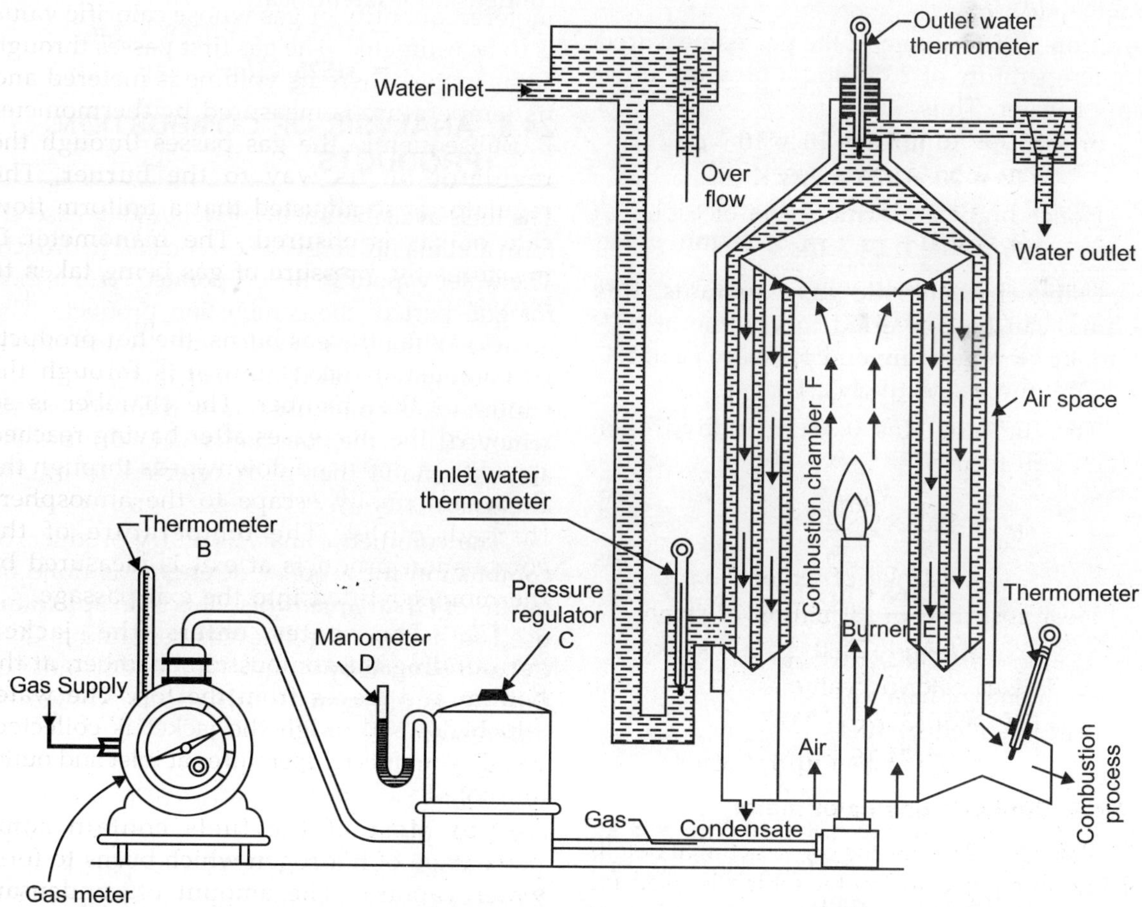

Fig. 24.2. Junker's gas calorimeter

EXAMPLE 24.3

The following data pertains to the test run made in the laboratory to determine the calorific value of a gaseous fuel by Junker's gas calorimeter:

Gas burnt in the calorimeter = 10 litres

Gas temperature and pressure at inlet = 20°C and 50 mm of water above atmospheric

Cooling water passing through the calorimeter = 8 litre

Inlet and outlet temperature of water = 15°C and 25°C

Steam condensed during test run = 10 cc

Determine the higher and lower calorific value of the fuel sample.

You may take barometric pressure as 750 mm of mercury and latent heat of vaporisation of water as 2465 kJ/kg.

Solution : The volume of the gas is measured at a temperature of 20°C and a pressure of 50 mm of water. Thus

$$V_1 = 10 \text{ litres} = 10 \times 10^{-3} \text{ m}^3$$

$$T_1 = 20 + 273 = 293 \text{ K}$$

$$p_1 = 750 + \frac{50}{13.6} = 753.68 \text{ mm of Hg}$$

Using characteristic gas equations, this volume can be converted to volume at STP (standard temperature and pressure *i.e.*, 15°C and 760 mm respectively). That is

$$\frac{p_1V_1}{T_1} = \frac{p_0V_0}{T_0}$$

$$V_0 = \frac{p_1V_1}{T_1} \times \frac{T_0}{p_0}$$

$$= \frac{753.68 \times 10 \times 10^{-3}}{293} \times \frac{288}{760}$$

$$= 9.748 \times 10^{-3} \text{ m}^3$$

From energy balance,
heat liberated by fuel
= heat taken up by water

$$V_0 \times CV = m_w \times c_{pw} \times \Delta T$$

Given : $m_w = 8 \text{ litre} = 8 \times 10^{-3} \text{ m}^3 = 8 \text{ kg}$ $(\because 1 \text{ kg} = 10^3 \text{ m}^3)$

∴ Calorific value of fuel,

$$CV = \frac{m_w \times c_{pw} \times \Delta T}{V_0}$$

$$= \frac{8 \times 4.186 \times (25-15)}{9.748 \times 10^{-3}}$$

$$= \mathbf{34353.7 \text{ kJ/m}^3}$$

Thus the higher specific value (*HCV*) of fuel is 34353.7 kJ/m^3. The lower calorific value (*LCV*) can be found by subtracting the heat consumed in the formation of steam.

Condensate collected

$$= 10 \text{ cc}$$

$$= 10 \times 10^{-6} \text{ m}^3 = 10 \times 10^{-3} \text{ kg}$$

Amount of water formed, *i.e.*, steam condensed per m^3 of gas burnt

$$= \frac{10 \times 10^{-3}}{9.748 \times 10^{-3}} = 1.0258 \text{ kg}$$

∴ Lower calorific value = *HCV* – (mass of condensate × latent heat)

$$= 34353.7 - 1.0258 \times 2465$$

$$= \mathbf{31825 \text{ kJ/m}^3}$$

24.5. ANALYSIS OF COMBUSTION PRODUCTS

The fuels available are in solid, liquid or gaseous form and mainly consist of carbon and hydrogen. The water vapour in the evaporated form is then integral part of the combustion products. The analysis of flue gases made taking water vapour into account is called the ***wet analysis.*** If the vapours are condensed and the condensate removed, then the flue gases will be dry and the analysis made thereof is referred to as ***dry analysis.***

The volumetric analysis of dry products of combustion for a boiler or engine is done by means of ***Orsat apparatus.*** This is an apparatus used to analyse the flue gas containing CO_2, CO and oxygen by volume. If the remainder of flue gas is nitrogen, it will also be known by difference.

Essentially the unit consists of three flasks containing chemicals to absorb different gases. For example :

- Flask 1 contains KOH (one part KOH + 2 parts water by weight) and this absorbs CO_2.
- Flask 2 contains alkaline solution of pyrogallic acid (5 gm pyrogallic acid in 15 cc

of water mixed with 120 gm KOH in 80 cc of water) to absorb O_2.

• Flask 3 contains cuprous chloride (Cuo dissolved in 20 times its weight of concentrated HCl acid till it becomes colourless) to absorb CO

Each flask is really made of two flasks joined together with a glass tubing. Three such flasks are arranged in such a way that they form two rows called front row and back row. The top of each flask in a row is connected to a common header called front header and back header. The front header is further connected to eudiometer generally calibrated from 0 to 100 cc. The eudiometer is surrounded by water jacket to keep its temperature constant. On the otherside, the front header is provided with a 3-way valve which can connect it to flue gas supply or atmosphere or may remain closed. The back header is connected to rubber bladder.

The front part of each flask is provided with one way glass tap at the neck joining the front header. It is also provided with a number of glass tubes inside it to accelerate the absorbing action of solution by providing more wet surface. The back part of each flask has oil film which prevents the solution from any reaction with atmosphere. Eudiometer is connected by flexible rubber pipe to aspirator bottle whose position can be changed by lifting or lowering it. This consequently changes the level of liquid (pure water) contained in eudiometer. The aspirator bottle contains salt water which cannot dissolve any flue gas.

Operation: The following procedure is adopted for successful working of the experimental test run :

Step 1. The whole apparatus is thoroughly cleaned, stoppers greased and then tested for air tightness. The absorption bulbs (flasks) are filled with their respective solutions to level just below their rubber connections and their stop cocks A, B and C are then closed. The jacket and aspirator bottle are filled with water. The 3-way cock is opened to atmosphere and the aspirator bottle is raised up to expel any gas or air in the eudiometer or header. The 3-way valve is now closed and the aspirator bottle is brought down so as to read zero on eudiometer scale.

Step 2. The 3-way cock is now connected to flue gas supply and the aspirator bottle is

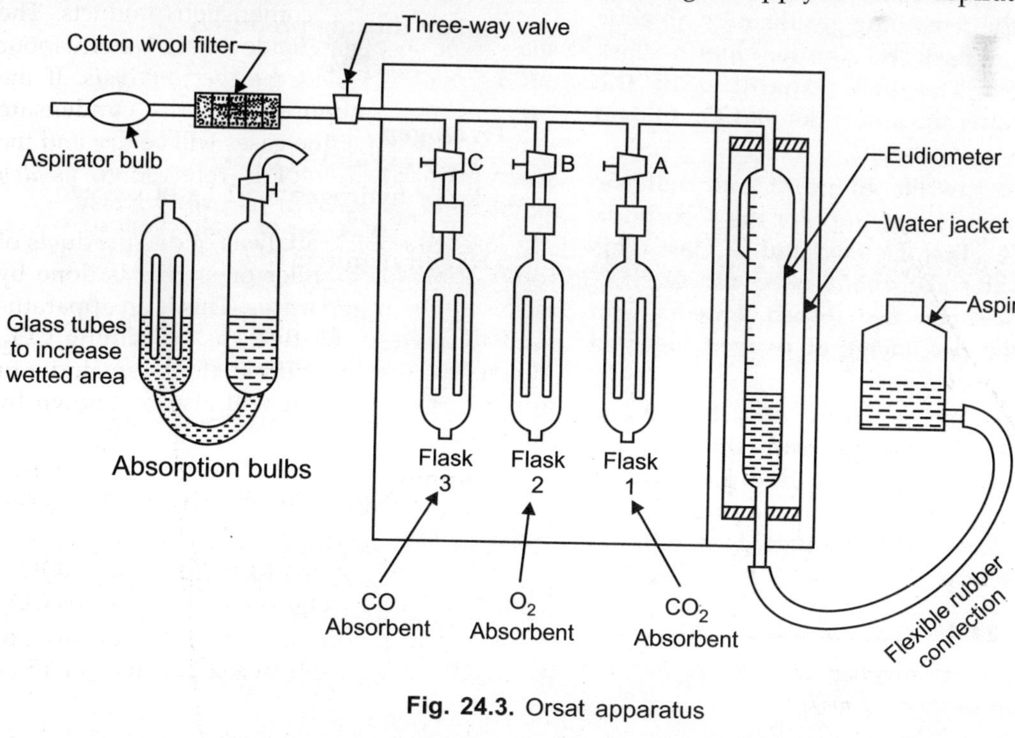

Fig. 24.3. Orsat apparatus

lowered to draw in the gas. The gas mixes with the air that may be present in the apparatus. Accordingly the 3-way cock is opened to atmosphere and the gas expelled by raising the aspirator bottle. The process of sucking and exhausting the gas is repeated 3-4 times so as to expel the air from the apparatus. Finally, the gas is sucked in the eudiometer and its volume is adjusted to 100 cc. The 3-way stop cock is then closed.

Step 3. The stopper *A* of flask 1 containing caustic potash solution is opened and the aspirator bottle is raised to push the gas in the flask. The aspirator bottle is moved up and down several times to ensure that all the CO_2 has been absorbed by the chemical solution in flask 1. The unabsorbed gas is then taken back to the eudiometer till the level of solution in flask 1 stands at the constant mark (original level) and then its stopper is closed. The levels of water in the eudiometer and aspirator bottle are equalised and the volume of residual gas is noted. The decrease in volume gives the volume of CO_2 in 100 cc of the flue gas sample.

Step 4. The procedure is repeated and the volumes of O_2 and CO are similarly determined by passing the remaining gas through alkaline pyrogallic acid flask and cuprous chloride flask respectively. The gas remaining in the eudiometer after the absorption of CO_2, O_2 and CO is taken as nitrogen.

It is worthwhile to point out that the following order is necessary for the absorption of gases: CO_2 first, O_2 next and CO last. This is because the absorbent used for O_2, *i.e.*, pyrogallic acid can also absorb some CO_2. In that case, the percentage of oxygen detected would be reported more.

The apparatus is handy, easy to use and fairly accurate. In order to ensure that only dry gas is being analysed, the flue gas may be passed through a U-tube containing calcium chloride or some other drying agent before it enters the set-up.

EXAMPLE 24.4

A sample of coal supplied to a boiler has the following composition by mass :

Carbon = 88 percent
Hydrogen = 5 percent
Oxygen = 3 percent
Nitrogen = 1 percent
Sulphur = 0.5 percent
Incombustible matter = 2.5 percent

Calculate : (a) mass of air required for complete combustion of 1 kg of coal, (b) dry analysis both by mass and volume of the products of combustion when 15 percent excess air is supplied.

Solution : Each constituent is taken separately and the amount of oxygen required for complete combustion is worked out by writing the relevant chemical equation. Complete combustion means that all the carbon in the fuel is converted to CO_2, all the hydrogen is converted to H_2O and all the sulphur gets transformed to SO_2.

Carbon

$C + O_2 = CO_2$

12 kg of C + 32 kg of oxygen
= 44 kg of carbon dioxide

Oxygen required

$$= \frac{32}{12} \times 0.88 = 2.347 \text{ kg}$$

Carbon dioxide produced

$$= \frac{44}{12} \times 0.88 = 3.227 \text{ kg}$$

Hydrogen

$2\,H_2 + O_2 = 2\,H_2O$

4 kg of hydrogen + 32 kg of oxygen
= 36 kg of water vapour

Oxygen required

$$= \frac{32}{4} \times 0.05 = 0.4 \text{ kg}$$

Water vapour formed

$$= \frac{36}{4} \times 0.05 = 0.45 \text{ kg}$$

Sulphur

$S + O_2 = SO_2$

32 kg of sulphur + 32 kg of oxygen
= 64 kg of sulphur dioxide

Oxygen required

$$= \frac{32}{32} \times 0.005 = 0.005 \text{ kg}$$

Sulphur dioxide formed

$$= \frac{64}{32} \times 0.005 = 0.01 \text{ kg}$$

∴ Total oxygen required for combustion of 1 kg of coal

$$= (2.347 + 0.4 + 0.005) = 2.752 \text{ kg}$$

But there is 0.03 kg of oxygen already present in 1 kg of coal. Then net oxygen required for combustion of 1 kg of coal

$$= 2.752 - 0.03 = 2.722 \text{ kg}$$

Since air contains 23 percent of oxygen by mass,

Air required for complete combustion of 1 kg of coal $= \frac{2.722}{0.23} = \mathbf{11.965\ kg}$

This implies that stochiometric air-fuel ratio is 11.965

(*b*) Actual air supplied

$$= 11.965 \times 1.15 = 13.76 \text{ kg}$$

This air contains 0.77 × 13.76 = 10.472 kg of nitrogen

and (13.76 – 10.472) = 3.288 kg of oxygen

Hence in the products we have :

CO_2 = 3.227 kg

H_2O = 0.45 kg ; SO_2 = 0.01 kg

N_2 = 10.472 + 0.01 = 10.482 kg

and excess

O_2 = 3.288 – 2.722 = 0.566 kg

The dry flue gases will not contain any water. The products are then entered in the table given below and percentage composition of dry products by mass and volume is obtained.

Product	*Mass*	*% by mass*	*Molecular weight*	*Proportional volume*	*% by volume*
CO_2	3.227	22.59	44	$\frac{3.227}{44} = 0.0733$	15.76
SO_2	0.01	0.07	44	$\frac{0.01}{44} = 0.000227$	0.049
N_2	10.482	73.38	28	$\frac{10.482}{28} = 0.374$	80.39
O_2	0.566	3.96	32	$\frac{0.566}{32} = 0.01769$	3.80
	4.285			0.4652	

EXAMPLE 24.5

Calculate the minimum quantity of air required for complete combustion of 1 m³ of gaseous fuel which has the following composition by volume.

H_2 = 30%; CH_4 = 40%; CO = 15%

CO_2 = 5%; O_2 = 2% and N_2 = 8%

What will be the volumes of individual dry gases if 40 percent excess air is used ? Proceed to calculate the percentage contraction in volume after the combustion products have been cooled.

Solution : Each constituent is taken separately and the amount of oxygen required for complete combustion is worked out by writing the relevant chemical equation :

Hydrogen

$$2\,H_2 + O_2 = 2\,H_2O$$
$$2 \text{ vol} + 1 \text{ vol} = 2 \text{ vol}$$

$$\text{Oxygen required} = \frac{1}{2} \times 0.3 = 0.15 \text{ m}^3$$

$$\text{Water vapour formed} = \frac{2}{2} \times 0.3 = 0.3 \text{ m}^3$$

Methane

$$CH_4 + 2\,O_2 = CO_2 + 2\,H_2O$$
$$1 \text{ vol} + 2 \text{ vol} = 1 \text{ vol} + 2 \text{ vol}$$

$$\text{Oxygen required} = \frac{2}{1} \times 0.4 = 0.8 \text{ m}^3$$

$$\text{Carbon dioxide formed} = \frac{1}{1} \times 0.4 = 0.4 \text{ m}^3$$

$$\text{Water vapour formed} = \frac{2}{1} \times 0.4 = 0.8 \text{ m}^3$$

Carbon monoxide

$$2\,CO + O_2 = 2\,CO_2$$

$$2 \text{ vol} + 1 \text{ vol} = 2 \text{ vol}$$

$$\text{Oxygen required} = \frac{1}{2} \times 0.15 = 0.075 \text{ kg}$$

$$\text{Carbon dioxide formed} = \frac{2}{2} \times 0.15 = 0.15 \text{ kg}$$

It may be noted that other constituents namely CO_2, O_2 and N_2 donot undergo any reaction.

∴ Amount of oxygen required for complete combustion of fuel

$$= 0.15 + 0.8 + 0.075 = 1.025 \text{ m}^3$$

But there is 0.02 m^3 of oxygen already present in 1 m^3 of fuel. Then net oxygen required for combustion of 1 m^3 of fuel,

$$= 1.025 - 0.02 = 1.005 \text{ m}^3$$

Since air contains 21 percent of oxygen by volume, air required for complete combustion of 1 m^3 of gas

$$= \frac{1.005}{0.21} = \mathbf{4.786 \text{ m}^3}$$

(*b*) Actual air supplied $= 4.786 \times 1.4 = 6.7 \text{ m}^3$

$$\text{Excess air} = 6.7 - 4.786 = 1.914 \text{ m}^3$$

CO_2 in flue gas = CO_2 formed during combustion of CH_4 and CO + CO_2 in fuel

$$= 0.4 + 0.15 + 0.05 = \mathbf{0.06 \text{ m}^3}$$

N_2 in flue gas = N_2 in fuel + N_2 in air supplied

$$= 0.08 + 6.7 \times 0.79 = \mathbf{5.373 \text{ m}^3}$$

O_2 in flue gas = Oxygen associated with excess air supplied

$$= 1.914 \times 0.21 = \mathbf{0.402 \text{ m}^3}$$

(*c*) From combustion of hydrogen, it is apparent that 2 volume of hydrogen combine with 1 volume of oxygen to give 2 volume of H_2O. When the products of combustion are cooled, the water vapour will get condensed and the condensate will have negligible volume. Thus for 2 volume of hydrogen, there is a reduction of 3 volume.

Similarly for 1 volume of CH_4 there is reduction of 2 volume and for 2 volume of CO there is reduction of 1 volume

∴ Total reduction in volume

$$= \frac{3}{2} \times 0.3 + \frac{2}{1} \times 0.4 + \frac{1}{2} \times 0.15$$

$$= 0.45 + 0.8 + 0.075 = 1.325 \text{ m}^3$$

Total volume

$$= 1 + 1.4 \times 4.786 = 7.7 \text{ m}^3$$

Hence percentage reduction in volume

$$= \frac{1.325}{7.7} \times 100 = \mathbf{17.21\%}$$

EXAMPLE 24.6

The fuel supplied to an internal combustion engine has the following composition by mass :

Carbon = 87 percent and hydrogen = 13 per cent

The air fuel ratio is 13 : 1. If all the carbon is burnt either to CO or CO_2 and if there is no free oxygen in the combustion products, determine :

(a) volumetric analysis of the products of combustion, and

(b) heat lost by incomplete combustion expressed as a percentage of gross calorific value :

You may assume the following heat values :

carbon burning to CO_2 =34320 kJ/kg

carbon burning to CO = 10360 kJ/kg and

hydrogen burning to H_2O =144420 kJ/kg

Solution : The combustion equations alongwith proportion by mass are :

- $C + O_2 = CO_2$

  12 kg of carbon + 32 kg of oxygen = 44 kg of carbon dioxide

- $2\,C + O_2 = 2\,CO$

  24 kg of carbon + 32 kg of oxygen = 56 kg of carbon dioxide

- $2\,H_2 + O_2 = 2\,H_2O$

  4 kg of hydrogen + 32 kg of oxygen = 36 kg of water vapour

Air supplied per kg of fuel = 13 kg

Since air contains 23 percent of oxygen by mass,

Oxygen supplied = 13 × 0.23 = 2.99 kg

Let it be presumed that x kg of carbon burns to CO_2 and the remainder to CO. Then

Oxygen required for burning x kg of carbon to CO_2

$$= \frac{32}{12} \times x = 2.667x \text{ kg}$$

Oxygen required for burning $(0.87 - x)$ kg of carbon to CO

$$= \frac{32}{24}(0.87 - x) = (1.16 - 1.33x) \text{ kg}$$

Oxygen required for burning hydrogen to

$$H_2O = \frac{32}{4} \times 0.13 = 1.04 \text{ kg}$$

Since there is no free oxygen left in the exhaust gases

$$2.667x + (1.16 - 1.333x) + 1.04 = 2.99$$

$$\text{That gives : } x = \frac{2.99 - 1.04 - 1.16}{2.667 - 1.333} = 0.592$$

The products of combustion are :

$$CO_2 = \frac{44}{12} \times 0.592 = 2.17 \text{ kg}$$

$$CO = \frac{56}{24}(0.87 - 0.592) = 0.648 \text{ kg}$$

N_2 (contained in the air supplied)

$$= 13 \times 0.77 = 10.01 \text{ kg}$$

The products are then entered in the table given below and percentage composition of products by mass and volume is obtained.

(*b*) Gross calorific value

$$= 0.87 \times 34320 + 0.13 \times 144420$$
$$= 29858 + 18775 = 48633 \text{ kJ}$$

Heat released for complete combustion of carbon to CO_2

$$= 0.87 \times 34320 = 29858 \text{ kJ}$$

Heat released during incomplete combustion

$$= 0.592 \times 34320 + (0.87 - 0.592) \times 10360$$
$$= 20317 + 2880 = 23197 \text{ kJ}$$

Heat loss due to incomplete combustion of carbon

$$= 29858 - 23197 = 6661 \text{ kJ}$$

Loss expressed as percentage of gross heat value

$$= \frac{6661}{48663} \times 100 = \mathbf{13.69\%}$$

EXAMPLE 24.7

The fuel supplied for a boiler had the following gravimetric analysis :

Carbon = 80%
Hydrogen = 5%
Oxygen = 6%

and the remaining incombustible matter

The percentage volumetric analysis of dry fuel gases was :

CO_2 = 10%, CO = 1.5%; O_2 = 8.5% and N_2 = 80%.

If the atmospheric and flue gas temperatures were recorded as 20°C and 400°C respectively, make calculations for :

(a) mass of air supplied per kg of fuel
(b) heat carried by excess air and the products of combustion

Product	*Mass*	*% by mass*	*Molecular weight*	*Proportional volume*	*% by volume*
CO_2	2.17	16.92	44	$\frac{2.17}{44} = 0.0493$	11.48
CO	0.648	5.05	28	$\frac{0.648}{28} = 0.0231$	5.38
N_2	10.01	78.03	28	$\frac{10.01}{28} = 0.357$	83.14
	12.828			0.4294	

(c) *heat lost by incomplete combustion*

(d) *density of dry gases at 1 bar and 15°C*

(e) *partial pressure of each gas if the total pressure of dry gases is 1 bar*

Following data may be assumed :

Calorific value of C to CO_2 = 33900 kJ/kg and for C to CO = 10045 kJ/kg

Further, within the given temperature range, c_p for CO_2, O_2, N_2, CO and air = 1.13, 1.025, 1.088, 1.021 and 0.994 kJ/kg K respectively.

Solution : To calculate the weight of dry flue gases per kg of fuel, there is need to convert the volumetric analysis to analysis by weight. (See table given at the bottom)

Mass of carbon per kg of dry flue gas

= amount of carbon in 0.147 kg of CO_2 + amount of carbon in 0.014 kg of CO

$$= \frac{12}{44} \times 0.147 + \frac{12}{28} \times 0.014$$

= 0.046 kg

Mass of carbon per kg of fuel = 0.8

∴ Mass of dry flue gases per kg of fuel

$$= \frac{\text{mass of carbon in 1 kg of fuel}}{\text{mass of carbon in 1 kg of flue gas}}$$

$$= \frac{0.8}{0.046} = 17.40 \text{ kg}$$

The burning equation for hydrogen gives

$2H_2 + O_2 = 2H_2O$

4 kg of hydrogen + 32 kg of oxygen = 36 kg of water vapour

The fuel contains 0.05 kg of hydrogen which will form

$$= 0.05 \times \frac{36}{4} = 0.45 \text{ kg of water}$$

∴ Total mass of products

= 17.40 + 0.45 = 17.85 kg

From the mass conservation principle, the mass of reactants equals the mass of products. Then

Combustible matter + air supplied per kg of fuel = 17.85

(0.8 + 0.05 + 0.06) + x = 17.85

∴ Air supplied per kg of fuel,

x = 17.85 – (0.8 + 0.05 + 0.06)

= **16.94 kg**

(*b*) $C + O_2 = CO_2$

12 kg of carbon + 32 kg of oxygen = 44 kg of CO_2

Oxygen required for burning of 0.8 kg of carbon

$$= \frac{32}{12} \times 0.8 = 2.13 \text{ kg}$$

$2H_2 + O_2 = 2H_2O$

4 kg of hydrogen + 32 kg of oxygen = 36 kg of H_2O

Oxygen required for burning of 0.05 kg of hydrogen

$$= \frac{32}{4} \times 0.05 = 0.45 \text{ kg}$$

Oxygen available in fuel = 0.06 kg

∴ Theoretical oxygen required for complete combustion

= 2.13 + 0.45 – 0.06 = 2.52 kg

Thus theoretical air required for complete combustion

$$= \frac{2.52}{0.23} = 10.96 \text{ kg}$$

Excess air supplied = 16.94 – 10.96 = 5.98 kg

Heat carried by excess air = $m_a \, c_{pa} \, \Delta T$

= 5.98 × 0.994 × (400 – 20) = **2258.76 kJ**

Constituent	*% by volume*	*Molecular weight*	*Proportional weight*	*% by mass*
CO_2	10.0	44	440	14.7
CO	1.5	28	42	1.4
O_2	8.5	32	272	9.08
N_2	80.0	28	2240	74.8
			2994	

Average specific heat of flue glass

$$= \frac{\Sigma c_p \times m}{\Sigma m}$$

where m is the mass proportion of each gas.

∴ Average specific heat

$$= \frac{1.13 \times 14.7 + 1.021 \times 1.4 + 1.025 \times 9.08 + 1.088 \times 74.8}{14.7 + 1.4 + 9.08 + 74.8}$$

$$= \frac{16.61 + 1.43 + 9.31 + 81.38}{100} = 1.087 \text{ kJ/kg K}$$

∴ Heat carried by products of combustion

$= m\ c_p\ \Delta T$

$= 17.85 \times 1.087 \times (400 - 20) =$ **7373.12 kJ**

(*c*) From mass balance

Carbon in 0.147 kg of CO_2

$$= \frac{12}{44} \times 0.147 = 0.04$$

Carbon in 0.014 kg of CO

$$= \frac{12}{28} \times 0.014 = 0.006$$

Total carbon = 0.04 + 0.006 = 0.046

∴ Carbon burnt to CO

$$= \frac{0.006}{0.046} \times 0.8 = 0.1043$$

The carbon burnt to CO per kg of fuel can also be calculated by using the relation

$$= \frac{CO \times C}{CO_2 + CO}$$

where C = carbon content per kg of fuel; CO_2 and CO are percentages of carbon dioxide and carbon monoxide by volume in the dry flue gas

$$\therefore \text{Carbon burnt to CO} = \frac{1.5 \times 0.8}{10 + 1.5} = 0.1043 \text{ kg}$$

The burning of 1 kg of carbon to CO instead of CO_2 involves a loss of (33900 – 10045) = 23855 kJ in heat produced.

∴ Heat lost due to incomplete combustion

= 0.1043 × 23855 = **2488 kJ**

(*d*) Mean molecular weight of the flue gases,

$$M_{mix} = \frac{\Sigma nM}{\Sigma n}$$

where M is the molecular weight of individual gases and n stands for number of moles or volumes.

$$M_{mix} = \frac{10 \times 44 + 1.5 \times 28 + 8.5 \times 32 + 80 \times 28}{10 + 1.5 + 8.5 + 80}$$

$= 29.94$

Gas constant for the mixture,

$$R_{mix} = \frac{R_{mal}}{M_{mix}} = \frac{8314}{29.94} = 277.7 \text{ kJ/kg K}$$

Density of the mixture can now be obtained by applying the characteristic gas equation

$$\rho_{mix} = \frac{1}{v_{mix}} = \frac{p_{mix}}{R_{mix} \times T_{mix}}$$

$$= \frac{1 \times 10^5}{277.7 \times 288} = \mathbf{1.25\ m^3/kg}$$

(*e*) Partial pressure of the constituent of a gas mixture is given by the relation partial pressure

$$= \frac{\text{number of moles}}{\text{total moles}} \times \text{total pressure}$$

$$\therefore \quad p_{CO_2} = \frac{10}{100} \times 1 = \mathbf{0.1\ bar}$$

$$p_{CO} = \frac{1.5}{100} \times 1 = \mathbf{0.015\ bar}$$

$$p_{O_2} = \frac{8.5}{100} \times 1 = \mathbf{0.085\ bar}$$

$$p_{N_2} = \frac{80}{100} \times 1 = \mathbf{0.8\ bar}$$

EXAMPLE 24.8

The following is the composition of a sample of coal

Carbon 85% ; hydrogen 7% ; sulphur 3% ; oxygen 2% and remainder incombustible matter

For an air-fuel mixture of 11.5 : 1, determine

(a) mixture strength as a percentage weak or rich

(b) volumetric analysis of the dry products of combustion

Solution : The combustion equations alongwith proportion by mass are :

• $C + O_2 = CO_2$

12 kg of C + 32 kg of oxygen

= 44 kg of carbon dioxide

$$\text{oxygen required} = \frac{32}{12} \times 0.85 = 2.27 \text{ kg}$$

• $2\,H_2 + O_2 = 2\,H_2O$

4 kg of hydrogen + 32 kg of oxygen = 36 kg of water vapour

$$\text{oxygen required} = \frac{32}{4} \times 0.07 = 0.56 \text{ kg}$$

• $S + O_2 = SO_2$

32 kg of sulphur + 32 kg of oxygen = 64 kg of sulphur dioxide

$$\text{oxygen required} = \frac{32}{32} \times 0.03 = 0.03 \text{ kg}$$

Total oxygen required for complete combustion of 1 kg of coal

$$= 2.27 + 0.56 + 0.03 = 2.86 \text{ kg}$$

But there is 0.02 kg of oxygen already present in 1 kg of coal. Then weight of net oxygen required for combustion of 1 kg of coal

$$= 2.86 - 0.02 = 2.84 \text{ kg}$$

Since air contains 23 percent of oxygen by mass,

$$\text{Minimum air required} = \frac{2.84}{0.23} = 12.348$$

Hence theoretical or stochiometer air-fuel ratio is 12.348 : 1

(*a*) Mixture strength is defined as the ratio of stochiometric air-fuel ratio to actual air fuel ratio. That is

$$\text{Mixture strength} = \frac{12.348}{11.5} \times 100 = 107.37\%$$

Obviously the mixture is 7.37% rich.

(*b*) Deficient amount of air

$$= 12.348 - 11.5 = 0.848 \text{ kg}$$

In the absence of proper amount of air, a portion of carbon will burn to carbon monoxide and the corresponding chemical reaction is

$2\,C + O_2 = 2\,CO$

24 kg of carbon + 32 kg of oxygen = 56 kg of carbon monoxide.

Amount of air saved by burning 1 kg of C to CO instead of CO_2

$$= \frac{\text{oxygen saved}}{0.23}$$

$$= \left(\frac{32}{12} - \frac{32}{24}\right) \times \frac{1}{0.23} = 5.797$$

$$\therefore \text{ Carbon burnt to CO} = \frac{0.848}{5.797} = 0.146$$

Carbon burnt to $CO_2 = 0.85 - 0.146 = 0.704$ kg

Thus with deficient air, *i.e.*, with rich mixture, the products of combustion would be

$$CO_2 = \frac{44}{12} \times 0.704 = 2.581 \text{ kg}$$

$$CO = \frac{56}{24} \times 0.146 = 0.341 \text{ kg}$$

$$SO_2 = \frac{64}{32} \times 0.03 = 0.06 \text{ kg}$$

N_2 (supplied as part of air)

$$= 11.5 \times 0.77 = 8.855 \text{ kg}$$

The percentage of dry flue gases by volume will be as given below :

Constituent	*% by mass*	*Molecular mass*	*Proportional volume*	*% by volume*
CO_2	2.581	44	$\frac{2.581}{44} = 0.0586$	15.10
CO	0.341	28	$\frac{0.341}{28} = 0.0122$	3.14
SO_2	0.06	64	$\frac{0.06}{64} = 0.000937$	0.24
N_2	8.855	28	$\frac{8.855}{28} = 0.3162$	81.5
			$\Sigma = 0.38798$	

EXAMPLE 24.9

A hydrocarbon fuel with chemical formula $C_{10}H_{22}$ is burnt with an air-fuel ratio of 13 : 1 by mass. The carbon burns to CO_2 and CO and the whole amount of hydrogen burns to form water vapour. There is no evidence of any free oxygen or any free carbon in the exhaust. Determine the complete volumetric analysis of the combustion products.

Solution : The combustion equation for the hydrocarbon is

$$\underset{284}{2\,C_{10}H_{22}} + \underset{992}{31\,O_2} = \underset{880}{20\,CO_2} + \underset{396}{22\,H_2O}$$

Oxygen required for combustion of 1 kg of fuel

$$= \frac{992}{284} \times 1 = 3.493 \text{ kg}$$

Since air contains 23 percent of oxygen by mass,

$$\text{Minimum air required} = \frac{3.493}{0.23} = 15.187 \text{ kg}$$

Hence theoretical or stochiometric air-fuel ratio is 15.187 : 1

Deficiency of air = 15.187 – 13 = 2.187 kg/kg of fuel

In the absence of proper amount of air, the combustion of fuel will not be complete. A portion of carbon will burn to CO_2 and the remainder to CO. The corresponding chemical reactions for the burning of carbon element are :

$$\underset{12}{C} + \underset{32}{O_2} = \underset{44}{CO_2};\quad \underset{24}{2\,C} + \underset{32}{O_2} = \underset{56}{2\,CO}$$

Amount of less air required for burning 1 kg of C to CO instead of CO_2

$$= \frac{\text{less oxygen}}{0.23}$$

$$= \left(\frac{32}{12} - \frac{32}{24}\right) \times \frac{1}{0.23} = 5.797$$

Hence $\frac{2.187}{5.797} = 0.379$ kg of carbon is burnt to CO

Carbon present in 1 kg of $C_{10}H_{22}$

$$= \frac{120}{120+22} = 0.845$$

∴ Carbon burnt to CO_2

$$= 0.845 - 0.379 = 0.466 \text{ kg}$$

Hence in the products, we will have :

$$CO_2 = \frac{44}{12} \times 0.466 = 1.709 \text{ kg}$$

$$CO = \frac{56}{24} \times 0.379 = 0.884 \text{ kg}$$

$$H_2O = \frac{396}{284} \times 1 = 1.394$$

N_2 (supplied as a part of air)

$$= 13 \times 0.77 = 10.01$$

The percentage composition of products of composition by volume will be as given below :

Constituent	*Mass*	*Molecular mass*	*Proportional volume*	*% by volume*
CO_2	1.709	44	$\frac{1.709}{44} = 0.0388$	7.68
CO	0.884	28	$\frac{0.884}{28} = 0.0316$	6.25
H_2O	1.394	18	$\frac{1.394}{18} = 0.0774$	15.32
N_2	10.01	28	$\frac{10.01}{28} = 0.3575$	70.75
			$\Sigma = 0.5053$	

EXAMPLE 24.10

A hydrocarbon fuel of unknown composition is burned with air and an Orsat analysis of the composition products gives

$CO_2 = 7.26\%$; $CO = 2.42\%$ *and*

oxygen $= 7.5\%$

Determine : (a) the empirical formula of the fuel (b) composition of fuel on mass basis (c) percent theoretical air used for combustion.

Solution : The Orsat analysis gives the composition of CO_2, CO, O_2 and N_2 and the analysis is on dry basis.

Therefore

$$N_2 = 100 - (7.26 + 2.42 + 7.5) = 82.82$$

Let the hydrocarbon fuel be represented by the empirical formula $C_x H_y$. Considering 100 moles of combustion products, the combustion equation can be written as

$$C_x H_y + a\, O_2 + \frac{79}{21} a\, N_2 = 7.26\, CO_2 + 2.42\, CO + 7.5\, O_2 + 82.82\, N_2 + b\, H_2O$$

Then from mass balance of the constituent elements we have :

Carbon balance :

$$x = 7.26 + 2.42 \quad \text{or} \quad x = 9.68$$

Nitrogen balance :

$$\frac{79}{21} a = 82.82 \quad \text{or} \quad a = 22.01$$

Oxygen balance :

$$a = 7.26 + \frac{2.42}{2} + 7.5 + \frac{b}{2}$$

$$= 15.97 + \frac{b}{2}$$

$$\therefore \quad b = (a - 15.97) \times 2$$

$$= (22.01 - 15.97) \times 2 = 12.08$$

Hydrogen balance :

$$y = 2b = 2 \times 12.08 = 24.16$$

Thus the empirical formula of the fuel is

$$C_x H_y = C_{9.68} H_{24.16}$$

(*b*) Composition of fuel on mass basis :

$$\text{Carbon} = \frac{9.68 \times 12}{9.68 \times 12 + 24.16 \times 1} \times 100$$

$$= 82.78\%$$

$$\text{Hydrogen} = 100 - 82.78 = 17.22\%$$

(*c*) Actual air-fuel ratio $= \dfrac{\text{moles of air}}{\text{moles of fuel}}$

$$= \frac{a + \frac{79}{21} a}{1}$$

$$= 22.01 + \frac{79}{21} \times 22.01 = 104.81$$

The following chemical reaction can be used to obtain the requirement of theoretical oxygen

$$C_{9.68} H_{24.16} + 15.72\, O_2 = 9.68\, CO_2 + 12.08\, H_2O$$

That is 15.72 mol oxygen is needed per mole of air.

∴ Theoretical air-fuel ratio

$$= \frac{15.72 + 15.72 \times \frac{79}{21}}{1} = 74.86$$

Percent theoretical air

$$= \frac{104.81}{74.86} \times 100 = \mathbf{140\%}$$

EXAMPLE 24.11

The composition by weight of fuel is 65% hexane ($C_6 H_{14}$) having a net calorific value of 45200 kJ/kg and 35% benzene ($C_6 H_6$) having a net calorific value of 40980 kJ/kg. Determine :

(a) the air-fuel ratio by mass of correct mixture of this blend.

(b) the calorific value of one cubic metre of this blend at NTP, and

(c) the percentage by volume of CO in the dry exhaust gases from a 30% rich mixture assuming complete combustion of hydrogen.

Further, it may be assumed that air contains 21% of oxygen by volume and 1 kg mole of the fuel occupies 22.42 m³ at NTP.

Solution : For complete combustion of fuel, the combustion equations with proportion by masses are :

Hexane :

$2\, C_6 H_{14}$	+	$19\, O_2$	=	$12\, CO_2$	+	$14\, H_2O$
2 (72 + 14)		19 × 32	=	12 (12 + 32)		14 (2 + 16)
172		608		528		252

Oxygen required $= 0.65 \times \dfrac{608}{172} = 2.3$ kg

Benzene

$$2\,C_6H_6 \quad + \quad 15\,O_2 = \quad 12\,CO_2 + 6\,H_2O$$
$$2\,(72 + 6) \quad 15 \times 32 = 12\,(12 + 32) \quad 6\,(2 + 6)$$
$$156 \qquad 480 \qquad 528 \qquad 108$$

Oxygen required $= 0.35 \times \dfrac{480}{156}$

$= 1.08$ kg

$\therefore$ Theoretical oxygen required for complete combustion of 1 kg of fuel,

$= 2.3 + 1.08 = 3.38$ kg

Since air contain 23% of oxygen by mass, Theoretical air required for complete combustion

$$= \frac{3.38}{0.23} = 14.695 \text{ kg} \simeq 14.7 \text{ kg}$$

(*a*) Air-fuel ratio by mass = **14.7 : 1**

(*b*) Calorific value per kg of coal

$= 0.65 \times 45200 + 0.35 \times 40980$

$= 43723$ kJ

Expressing constituents of the mixture on mole basis (as shown in table given below), we have :

$\therefore$ Calorific value of 1 kg mole of blend

$$= \frac{43723}{0.5219} = 83776 \text{ kJ/kg mol}$$

At NTP, 1 kg mole of fuel occupies 22.42 m^3

$\therefore$ Calorific value of fuel blend

$$= \frac{83776}{22.42} = \mathbf{3776.67\ kJ/m^3}$$

(*c*) Quantity of air supplied for 30 percent rich mixture

$$= \frac{14.7}{1.3} = 11.3 \text{ kg/kg of fuel}$$

We may now write the combustion equation in terms of number of moles ensuring that the number of moles of each constituent before combustion are equal to the number of moles after combustion.

$$\frac{0.65}{86}(C_6H_{14}) + \frac{0.35}{78}(C_6H_6) + \frac{11.3\times0.23}{32}(O_2) + \frac{11.3\times0.77}{28}(N_2)$$
$$= a(CO_2) + b(CO) + d(H_2O) + e(N_2)$$

Carbon monoxide would appear in the products since insufficient amount of air has been supplied.

From carbon balance :

$$\frac{0.65}{86} \times 6 + \frac{0.35}{78} \times 6 = a + b$$

or $\quad a + b = 0.0723 \quad \ldots(i)$

From hydrogen (H_2) balance :

$$\frac{0.65}{86} \times 7 + \frac{0.35}{78} \times 3 = d\,;$$

$$d = 0.06645$$

From oxygen (O_2) balance :

$$\frac{11.3\times0.23}{32} = a + \frac{b}{2} + \frac{d}{2}$$

or $\quad 0.1624 = 2a + b + d$

or $\quad 2a + b = 0.1624 - d$

$= 0.1624 - 0.06645 = 0.09595 \quad \ldots(ii)$

From identities (*i*) and (*ii*)

$2a + (0.0723 - a) = 0.09595$

$a = 0.02365$

and $\quad b = 0.0723 - 0.02365 = 0.04865$

From nitrogen (N_2) balance :

$$\frac{11.3\times0.77}{28} = e\,; \quad e = 0.3107$$

Constituent	*Mass* *m*	*Molecular mass* M	*Number of moles* *n*
C_6H_{14}	0.65	86	0.00756
C_6H_6	0.35	78	0.00448
O_2	3.38	32	0.1056
N_2	(14.7 − 3.38) = 11.32	28	0.4043
	$\Sigma m = 15.7$		$\Sigma n = 0.5219$

Then for a 30% rich mixture, the percentage of CO by volume in the dry exhaust is

$$= \frac{b}{a+b+e}$$

$$= \frac{0.04865}{0.02365+0.04865+0.3107}$$

$= 0.127$ or **12.7%**

24.6. CHEMICAL THERMODYNAMICS

24.6.1. Stoichiometric coefficients

Consider the following chemical equation which pertains to chemical reaction between ethylene and oxygen

$$C_2H_4 + 3\ O_2 \longrightarrow 2\ CO_2 + 2\ H_2O$$

This equation tells that 1 mole of C_2H_4 reacts with 3 moles of O_2 to produce 2 moles of CO_2 and 2 moles of H_2O. That is

$$1\ (C_2H_4) + 3\ (O_2) \longrightarrow 2\ (CO_2) + 2\ (H_2O)$$

The coefficients 1, 3, 2 and 2 associated with the chemical species C_2H_4, O_2, CO_2 and H_2O are called *stoichiometric coefficients.*

During a chemical reaction, the reactants are consumed and the products are formed. As such the following sign convention is adopted for writing the stoichiometric coefficients :

Assign a positive value for the stoichiometric coefficients associated with products, and a negative value for the stoichiometric coefficients associated with reactants.

The stoichiometric coefficients along with their signs are called stoichiometric numbers.

With this sign convention, the stoichiometric numbers associated with C_2H_4 and O_2 are – 1 and – 3 respectively where as the stoichiometric coefficients associated with CO_2 and H_2O are + 2 and + 2 respectively. The chemical reaction can then be expressed as

$$2\ CO_2 + 2\ H_2O + (-\ 1)\ C_2H_4 + (-\ 3)\ O_2 = 0$$

or $\Sigma\ s_i\ A_i = 0$

where s_i is the stoichiometric coefficient associated with species A_i.

Likewise, in the chemical equation representing the synthesis of ammonia

$$N_2 + 3\ H_2 \longrightarrow 2\ NH_3$$

the stoichiometric coefficients for N_2, H_2 and NH_3 are 1, 3 and 2; and the corresponding stoichiometric numbers are – 1, – 3 and 2.

24.6.2. Standard reference state

The energy liberated or consumed during a chemical reaction depends on the temperature of reactants and products. For the analysis of a combustion process to be convenient, the reactants and products are taken to be in their standard states and at the same temperature. The standard state for liquid and gaseous fuels is taken to be their purest form at 1 bar pressure., The standard state for a pure solid fuel is the most stable allotropic form of the fuel. The standard form of carbon is, however, graphite as it is more stable than diamond.

The combustion analysis is concerned with changes in the energy of a system during a process and not with the absolute value of energy at a particular state. As such, any state can be chosen as the reference state and assigned a value of zero to internal energy (U) and enthalpy (H) of a substance at that state. For convenience of tabulating the data, the common practice is however to choose 25°C and 1 atm as the reference state. This condition is referred to as the standard reference state, and the properly values at this state are indicated by a superscript 0.

The change of enthalpy during a chemical reaction is designated as Δh_T^o. Here superscript o denotes that all the reactants and products are in their respective standard states and the subscript T denotes the temperature at which the reaction is carried out.

For example, the enthalpy values of NH_3 vapour at 1 atm pressure and 100°C and 25°C temperature are stated as

$$h_T = h_{373\ K} = 1692.6\ kJ/kg$$

$$h^\circ = h_{298\ K} = 1528.1\ kJ/kg$$

Then the enthalpy of NH_3 vapour at 1 atm and 100°C relative to the standard state of 25°C is found out as

$$Dh_T^\circ = 1692.6 - 1528.1 = 164.5\ kJ/kg$$

24.6.3. Formation reaction

A formation reaction is a chemical reaction in which only one mole of a single compound is formed from its constituent elements.

(*i*) Consider the chemical reaction,

$$C + 2\,H_2 + \frac{1}{2}\,O_2 = CH_3OH$$

Here one mole of a single compound CH_3OH has been produced starting from C, H_2 and O_2 which are elements. As such it is a formation reaction.

(*ii*) The reaction $N_2 + 3\,H_2 \longrightarrow 2\,NH_3$ is not a formation reaction as 2 moles of NH_3 have been formed even though the reactants are elements. However, this reaction becomes a formation reaction if written as

$$\frac{1}{2}\,N_2 + \frac{3}{2}\,H_2 \longrightarrow NH_3$$

(*iii*) The reaction $CO + 2\,H_2 \longrightarrow CH_3\,OH$ is not a formation reaction because here 1 mole of CH_3OH has been formed beginning from CO and H_2 in which CO is not an element. Likewise, the reactions

$$CH_3\,CHO + H_2 \longrightarrow C_2H_5OH$$

$$C_2H_4 + H_2O \longrightarrow C_2H_5OH$$

are not formation reactions.

24.6.4. Internal energy

The experimental investigations have shown that when a unit mass of fuel undergoes complete combustion, the quantum of energy that is released depends on the temperature at which the combustion process is carried out. Further, during the combustion process, the state of reactants is identified by the reference temperature T_0 and another property (pressure or volume) and the products are taken to be formed at the same state.

• **Combustion at constant volume :** For a chemically reacting closed system (the combustion process being carried out at constant volume), the conservation of energy relation $\delta Q = \delta W + dU$ can be expressed as

$$Q = U_2 - U_1$$

($\because \delta W = 0$ for constant volume combustion)

Taking $U_1 = U_{Ro}$; the internal energy of reactants at temperature T_0

$U_2 = U_{Po}$; the internal energy of products at temperature T_0

We can write

$Q = U_{Po} - U_{Ro}$; the subscripts R and P refer to reactants an products respectively.

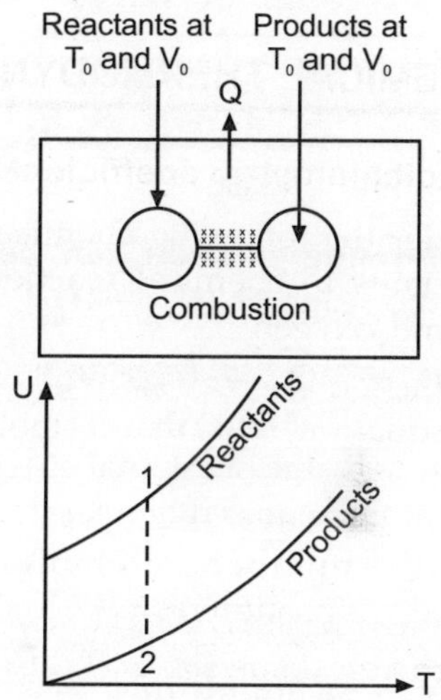

Fig. 24.4.

Thus the heat transferred during the constant volume combustion process equals the change in internal energy. For combustion, ΔU_0 (change in internal energy) is a negative quantity and as such the heat is transferred from the system. These aspects are illustrated above in Fig. 24.4.

The heat so transferred is called the internal energy of combustion.

It is appropriate to mention that in case of actual constant volume combustion process, the initial and final temperatures will not be same as T_0 (their reference temperature). For analysis purposes, the change in internal energy between reactants at state 1 and the products at state 2 can be considered to occur in the following three stages :

(*i*) Change of reactants from state 1 to T_0

(*ii*) Constant volume combustion process from reactants to products at T_0

(*iii*) Change of products from T_0 to state 2.

Then the change in internal energy between states 1 and 2 can be written as

$$U_{P_2} - U_{R_1} = (U_{P_2} - U_{P_0}) + (U_{P_0} - U_{R_0}) + (U_{R_0} - U_{R_1})$$

$$= (U_{P_2} - U_{P_0}) + \Delta U_0 + (U_{R_0} - U_{R_1})$$

Products Reactants

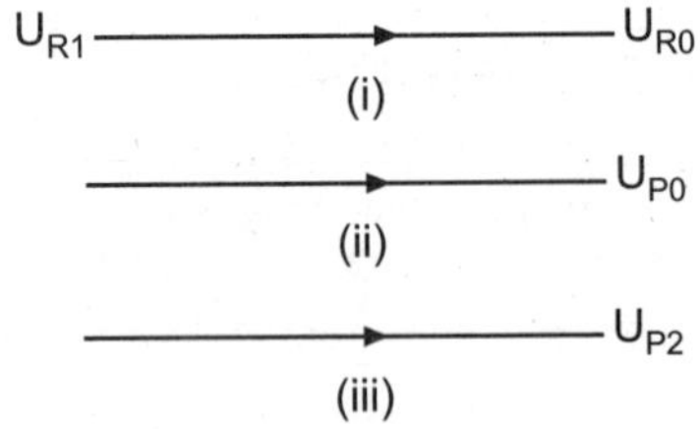

The following relation can be used to determine the value of $(U_{R_0} - U_{R_1})$:

$$U_{R_0} - U_{R_1} = \sum_R n_i (u_{i_0} - u_{i_1})$$

where u_i = value of internal energy for any constituent at the required temperature T_0 or T_1

n_i = number of moles of the constituent.

$\sum_R$ = summation of all the constituents of the reactants

If calculations are made on mass basis, then

$$U_{R_0} - U_{R_1} = \sum_R m_i (u_{i_0} - u_{i_1})$$

where u_i is the internal energy per unit mass.

In terms of average values of the specific heats for the required temperature range, the above expression can be written as

$$U_{R_0} - U_{R_1} = \sum_R m_i c_{v_i}(T_0 - T_1)$$

$$= (T_0 - T_1) \sum_R m_i c_{v_i}$$

The similar expressions for the products can be written as

$$U_{P_2} - U_{P_0} = \sum_P n_i (u_{i_2} - u_{i_0}) \quad \text{... on mole basis}$$

$$= \sum_P m_i (u_{i_2} - u_{i_0}) \quad \text{... on mass basis}$$

$$= (T_2 - T_0) \sum_P m_i c_{vi}$$

24.6.5 . Standard enthalpy change

When a chemical reaction is carried out at constant pressure the conservation of energy relation

$\delta Q = \delta W + dU$ can be recast as

$$Q = P(V_2 - V_1) + (U_2 - U_1)$$
$$= (U_2 + P_2V_2) + (U_1 + P_1V_1)$$
$$= H_2 - H_1$$

Likewise for steady-state and steady-flow (SSSF) chemical process in an open system (in the absence of kinetic energy and potential energy changes and without any shaft work), we have

$$Q = H_2 - H_1$$

When the initial state 1 is that of reactants and the final state 2 is that of products, we can write

$$Q = \Sigma H_P - \Sigma H_R$$

That is the heat transfer in a chemical process equals the difference in the enthalpies of the products and reactants.

Let it be assumed that during the chemical reaction, the reactants as well as the products are in their respective standard state and at the same temperature. The enthalpy change associated with such a reaction is called the standard heat (or standard enthalpy) change and is designated by Δh_T° where

— subscript T denotes the temperature at which reaction is carried out

— superscript 0 denotes that the reactant and products are in their standard states.

Further, the state of species in the chemically reacting systems are denoted as gas (*g*), liquid (*l*) or solid (*s*).

Consider the following chemical reaction which corresponds to the oxidation of glucose.

$$C_6H_{12}O_6\ (s) + 6\ O_2(g) \longrightarrow 6\ CO_2\ (g) + 6\ H_2O\ (l)$$

If the reactants and products are all at the reference temperature T, then

$$(H_{CO_2} + H_{H_2O}) - (H_{glucose} + H_{O_2}) = \Delta H_T^\circ$$

The enthalpy change associated with such chemical reactions is measured in the laboratory with the help of a flow calorimeter which is surrounded by water jackets. (Fig. 24.5).

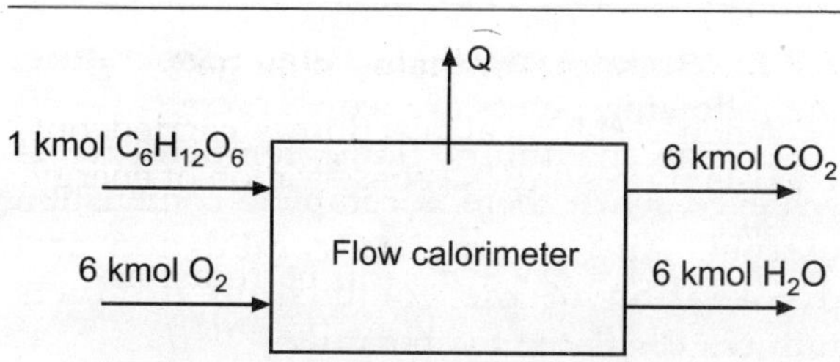

Fig. 24.5. Oxidation reaction for glucose

The reactants (glucose and oxygen at 1 atm and 25°C) enter into the calorimeter, reaction takes place, and the products (CO_2 and H_2O) are made to leave at 1 atm and 25°C. The symbol Q represents the energy in the form of heat which leaves the calorimeter.

The following remarks apply to the identity $Q = \Delta H_T^\circ$

(*i*) ΔH_T° (or Q) > 0 ; Energy in the form of heat needs to be supplied for the reaction to take place. This heat is absorbed from the surroundings, *i.e.*, heat is added to system. Such a reaction is called an *endothermic* reaction.

(*ii*) ΔH_T° (or Q) < 0 : Energy in the form of heat is liberated due to chemical reaction. This heat must be transferred to the surroundings implying heat rejection by the system. Such a reaction is called an *exothermic* reaction.

(*iii*) ΔH_T° (or Q) = 0 : The reaction is *thermo-neutral*, *i.e.*, energy is neither absorbed nor evolved due to chemical reaction.

Obviously, the sign of ΔH_T° indicates whether energy is liberated or consumed during a chemical reaction.

The glucose-oxidation reaction is exothermic in nature. During this reaction, heat is liberated and temperature of the products rises. To bring this temperature back to the initial temperature of the reactants, the heat must be rejected by the systems.

Apparently then ΔH_T° is negative. For the glucose-oxidation reaction, the enthalpy of reaction (or standard heat) per K mol of glucose is stated as

$$\Delta H_T^\circ = -\ 28170 \text{ kJ/k mol glucose}$$

24.6.6. Enthalpy of formation

The enthalpy of a formation of a compound is its enthalpy at constant pressure isothermal reaction in a steady flow process in which

(*i*) the reactants are chemical elements

(*ii*) the products and reactants are both at the standard state (1 atm and 25°C).

Consider the reaction : $C + O_2 = CO_2$

Here the product CO_2 is the compound and the reactants C and O_2 are the chemical elements. Further, the reaction stipulates that in a steady flow process 1 kg-mole of carbon reacts with 1 kg-mole of oxygen (both at the standard state of 1 atm and 25°C) to produce 1 kg-mole of carbon dioxide. Also, through carefully conducted experiments it is noted that 393520 kJ of heat is transferred from the system so that the product CO_2 leaves at the same pressure and temperature values. Neglecting changes in kinetic and potential energies and applying the first law to the process, we have

$$Q = h_p - h_R$$

where h_P and h_R refer to the enthalpy values of the products and reactants respectively.

At the reference state (1 atm and 25°C), the enthalpy of reacting elements (C and O_2) is taken to be 0. Then enthalpy of formation of 1 kg mole of CO_2 at 1 atm 25°C is then

$$h_P = Q = -\ 393\ 520 \text{ kJ}$$

The negative sign indicates that the enthalpy of CO_2 is less than the enthalpy of the carbon and the oxygen. In other words, chemical energy in the form of heat is released when C and O_2 combine to form CO_2. Apparently a negative enthalpy for a compound implies that heat is released for the formation of a compound from its reacting element.

The enthalpy of formation of 1 kg made of CO_2 is generally stated as

$$\Delta h_f^\circ = -\ 393\ 520 \text{ kJ/kg mol } CO_2$$

As usual, superscript 0 indicates that the value of standard state and the subscript f indicates that it is the enthalpy of formation.

Likewise corresponding to the formation reaction for the synthesis of ammonia

$$\frac{1}{2}\ N_2 + \frac{3}{2}\ H_2 \longrightarrow NH_3 + 46100 \text{ kJ/kg mol}$$

the enthalpy of formation of kg mole of ammonia would be designated as

$$\Delta h_f^\circ = -46100 \text{ kJ/kg mole ammonia.}$$

24.6.7. Adiabatic flame temperature

The maximum flame temperature that can be attained in any combustion reaction for the given products is called the adiabatic flame temperature if

(*i*) the combustion process takes place adiabatically

(*ii*) the burner where the combustion of fuel takes place operates under steady-flow, steady-state condition

(*iii*) there is no work transfer

(*iv*) no changes in kinetic energy and potential energy are involved

With these stipulations, all the heat of combustion goes to increase the enthalpy and hence temperature of the products.

In relation to adiabatic flame temperature, the following points are worth noting :

- The maximum flame temperature is achieved when there is complete combustion of fuel. Any incomplete combustion and dissociation of the combustion products reduces the flame temperature.
- For a given fuel and given pressure and temperature of the reactants, the maximum adiabatic flame temperature corresponds to complete combustion with stoichiometric air-fuel mixture. Any increase in air-fuel ratio results into a decrease in the maximum temperature.
- The adiabatic flame temperature can be controlled by adjusting the air-fuel ratio. This is done so in gas turbines by supplying excess air to the system with a view to keep the flame temperature lower than the maximum permissible temperature limit set by metallurgical considerations in the design and construction of combustion chamber and turbine blades.

SALIENT POINTS

1. A fuel is the material that can be burnt to release heat energy, *e.g.*, coal, gasoline, fuel oils and natural gases.
2. Combustion is a chemical reaction between the fuel and oxidiser (oxygen supplied by air). With regard to the chemical reaction.

   $$CH_4 + 2\,O_2 \rightarrow CO_2 + 2\,H_2O$$

   (*i*) The chemical equation is obtained by balancing the atoms of each of the species in the recation.

   (*ii*) The initial constituents (methane and oxygen) which start the reaction are called reactants, and the final constituents (carbon dioxide and water) which are formed due to arrangement of atoms and electrons are called products.

   (*iii*) Complete combustion means complete conversion of all combustible elements of fuel into stable oxidies. That is all the carbon in the fuel will burn to CO_2 and all hydrogen to H_2O.

   (*iv*) The oxygen required for combustion of fuel is supplied in the form of air. Air contains 21% oxygen and 79% nitrogen by volume (or 23% oxygen and 77% nitrogen by mass).

   The minimum amount of air which supplies the requiste amount of oxygen for complete combustion of fuel is called theoretical or stoichiometric air, and the corresponding air-fuel ratio is called stoichiometric air-fuel ratio.

   $$\text{A/F ratio} = \frac{\text{mass of air}}{\text{mass of fuel}}$$

   (*v*) An incomplete combustion refers to the condition when some unburnt fuel remains in the products. This may happen due to insufficienct supply of oxygen or due to inadequate air-fuel mixing.
3. Combustion is an exothermic reaction, *i.e.*, combustion converts chemical energy to thermal energy. The heat generated due to combustion of unit mass of solid (or liquid) fuel is called its calorific value, and it is determined by using bomb calorimeter.

For gaseous fuel, calorific value refers to the heat liberated by combustion of unit cubic metre of gases, and Junker's gas calorimeter is used for finding its value.

4. An analysis of the dry combustion products is made by using Orsat apparatus.
5. Enthalpy of formation refers to the increase in enthalpy when a compound is formed from its constituent elements in their natural form and in a standard state (1 atm and 15°C).
6. The temperature of the combustion products is referred to as the adiabatic flame temperature when the combustion process takes place adiabatically with no work and changes in kinetic or potential energy.

   The adiabatic flame temperature depends on the state of reactants, degree of completion of reaction, and the amount of air used.

REVIEW QUESTIONS

A. Conceptual and conventional questions.

1. What are fuels ? How are they classified ? List examples of various types of fuels.
2. Enumerate the characteristics of a good fuel.
3. What is meant by ignition temperature of fuel ? Why should an ideal fuel have moderate value for ignition temperature ?
4. Name the different varieties of coal formed inside the earth. What is proximate and ultimate analysis of coal ?
5. Define the terms weak (lean) mixture, rich mixture and stoichiometric (chemically correct) mixture.
6. Define calorific value of fuel. Differentiate between higher and lower calorific value of fuel.
7. What is meant by dry and wet analysis of the products of combustion ?
8. Describe briefly the constructional and operational aspects of a system used to experimentally determine the calorific value of a sample of coal. How the result is calculated from the data obtained?
9. Describe, with a neat sketch, the Orsat apparatus used for analysing the flue gases from a boiler trial.
10. What are stoichiometric coefficients ?
11. Point out the standard states normally used while analyzing chemical reacting systems
12. (*a*) What is meant by standard enthalpy change of a chemical reaction.

    (*b*) How the sign adopted for enthalpy change makes it possible to identify whether energy is liberated or absorbed during a chemical reaction.
13. What is formation reaction and what is meant by standard enthalpy of formation ?
14. Define adiabatic flame temperature and state its importance.
15. The ultimate analysis of coal used in a boiler is as given below :

    C = 82%; H_2 = 6%; O_2 =4%; moisture = 2% and ash = 6%.

    Calculate the minimum air required for complete combustion of this coal sample.

    If the actual air supply is 18 kg/kg of coal, hydrogen is completely burnt, 80% of carbon burns to CO_2 and the remainder to CO, work out the volumetric analysis of the dry products of combustion.

    [**Ans.** 11.41, CO_2 = 8.84%, CO = 2.22%, O_2 = 8.74% and N_2 = 80.2%]
16. A sample of fuel oil contains 85% carbon and 15% hydrogen. It was found that 10% carbon in the fuel is burnt to CO even when 25% excess air is supplied. Determine :

    (*i*) minimum amount of air required per kg of fuel oil

    (*ii*) oxygen in the flue gas per kg of fuel burnt

    (*iii*) mass analysis of the flue gas

    [**Ans.** 15.07 kg, 0.98 kg, CO_2 = 15.17%, CO = 1.07%, O_2 = 5.3%, N_2 = 78.46%]
17. Consider a sample of producer gas having the following analysis on volume basis :

    H_2 = 14%, CH_4 = 3%, CO = 27%, CO = 4.5%, O_2 = 1% and N_2 = 50.5%

    If the gas is burnt with 10% excess air, estimate the air-fuel ratio both on volumetric and mass basis.

    [**Ans.** 1.335 k mol/k mol fuel, 1.56 kg/kg fuel]

18. During trial in a boiler, the volumetric analysis of a sample of flue gas as estimated by Orsat apparatus was

CO_2 = 10.4%, CO = 0.2%, O_2 = 78%

and N_2 = 81.6% (by difference)

The gravimetric analysis of the coal was reported as

C = 78%, H_2 = 6%, O_2 = 3% and incombustible matter = 13%

Make an estimate of (*a*) weight of dry flue gases per kg of fuel and (*b*) weight of excess air per kg of fuel.

[**Ans.** 18.4, 6.56]

19. Ethyl alcohol (C_2H_6O) has been used as fuel in a petrol engine. Determine :

(*a*) the stoichiometric air-fuel ratio

(*b*) the air-fuel ratio, dry analysis of the exhaust gas for a mixture strength of 90%.

[**Ans.** 8.96, 9.95, CO_2 = 13.45%, N_2 = 84.31%, O_2 = 2.24%]

20. The products of combustion of an unknown hydrocarbon C_xH_y has the following composition as measured by an Orsat apparatus

CO_2 = 8%, CO = 0.9%, O_2 = 8.8% and

N_2 = 82.3% (by difference)

Determine : (*a*) the empiricial formula and the composition of the fuel, (*b*) the air-fuel ratio and (*c*) the percentage excess air used.

[**Ans.** $C_{8.9}$ $H_{18.5}$, C= 58.23% and H_2 = 14.77%; (*b*) 24 : 1 and (*c*) 67.22%]

21. The following data pertains to volumetric analysis made on dry basis of the exhaust gas from a diesel engine :

CO_2 = 9.98%, CO = 2%, O_2 = 4.69% and

N_2 = 83.33%.

Presuming that the diesel fuel has the empirical formula C_{12} H_{26}, make calculations for the actual air-fuel ratio and the percent excess air used for combustion.

[**Ans.** 17.98, 19.78%]

22. Motor gasoline with empirical formula C_8H_{18} has been burnt with air which is 150 percent of stoichiometric air. Determine the air-fuel ratio and work out the volumetric analysis of the combustion products. Proceed to calculate the partial pressure of water vapour formed and check whether these would condense or not if the products are cooled to 20°C. Take atmospheric pressure as 1.01325 bar.

[**Ans.** 22.3 kg/kg of fuel, CO_2 = 8.53%, O_2 = 6.67%, H_2O = 9.6% and N_2 = 75.2%, p_{H_2O} = 9.727, kPa vapours will condense]

B. Fill in the blanks with appropriate word/words.

1. The combustion of fuel involves the .... of its constituents.
2. With regard to a chemical reaction, the constituents which start the reaction are called ....
3. The minimum amount of air which supplies the requiste amount of oxygen for complete combustion of fuel is called .... air
4. The volumetric analysis of dry products of combustion for a boiler or an engine is done by means of ....
5. A reaction is said to be .... when energy is neither absorbed nor evolved during the chemical reaction.
6. The standard enthalpy change during chemical reaction is designated as Δh_T^o. Here, subscript ° denotes that the reactants and products are in their ....
7. The maximum flame temperature that can be attained in any chemical reaction for the given products is called the ....
8. Consider the following chemical reaction

$$1(C_2H_4) + 3\ O_2 \longrightarrow 2\ (CO_2) + 2\ (H_2O)$$

The coefficient's 1, 3, 2 and 2 associated with the chemical species C_2 H_4, O_2, CO_2 and H_2O are called ....

Answers

1. oxidation 2. reactants 3. stoichiometric 4. Orsat apparatus 5. thermo-neutral 6. standard reference states 7. adiabatic flame temperature 8. stoichiometric coefficients

C. State true or false. If false, write the correct statement.

1. During a chemical reaction, the mass of each element is conserved.
2. When more air is supplied than that required by the fuel for its complete combustion, the air-fuel mixture is said to be rich.
3. In proximate analysis of the composition of coal, an attempt is made to determine the percentage of individual elements comprising the coal.
4. Ignition temperature is the lowest temperature to which the fuel must be preheated to initiate its smooth burning.
5. The term complete combustion implies that all the carbon burns to CO_2 and all hydrogen to H_2O, and the products donot contain any oxygen.

6. The air can be treated as a mixture containing 21% oxygen and 79% nitrogen by mass.
7. The bomb calorimeter is used for the determination of the calorific value of gaseous fuels.
8. The Orsat apparatus is used for the gravimetric analysis of the dry products of combustion for a boiler or an engine.
9. The stoichometric coefficients along with their sign are called stoichiometric numbers.
10. The reaction $N_2 + 3\ H_2 \longrightarrow 2\ NH_3$ represents a formation reaction.

Answers

1. True 2. False; the mixture is said to be lean or weak 3. False ; in ultimate analysis 4. True 5. True 6. False; by volume. The composition by mass is 23% oxygen and 77% nitrogen 7. False; the bomb calorimeter is used for finding out the heating value of solid or liquid fuel. The calorific value of gaseous fuels is determined by using Junker's gas calorimeter 8. False; volumetric analysis 9. True 10. False; here two moles of NH_3 have been formed even though the reactants are elements. The given reaction becomes a formation reaction if it is written as $\frac{1}{2}N_2 + \frac{3}{2}H_2 \rightarrow NH_3$ (one mole of single compound NH_3 is formed)

D. Multiple choice questions.

1. The percentage by weight of oxygen in atmospheric air is
   (*a*) 19 (*b*) 21
   (*c*) 23 (*d*) 27
2. For complete burning of 1 kg of carbon, the air required will be about
   (*a*) 9.67 kg (*b*) 11.6 kg
   (*c*) 12.7 kg (*d*) 14.5 kg
3. Identify the false statement
   (*a*) liquid fuels have a higher calorific value than solid fuels
   (*b*) coal gas consists mainly of hydrogen, monoxide and various hydrocarbons
   (*c*) a good fuel has a high ignition point
   (*d*) calorific value of gaseous fuel is expressed in kJ/m^3
4. Which of the following constituents of a fuel does not contribute to its calorific value of combustion?
   (*a*) carbon (*b*) hydrogen
   (*c*) sulphur (*d*) nitrogen
5. Bomb calorimeter is used to determine
   (*a*) calorific value of solid or liquid fuels
   (*b*) calorific value of gaseous fuels
   (*c*) ash content of solid fuels
   (*d*) incombustible matter in solid fuel
6. Orsat apparatus is meant for
   (*a*) volumetric analysis of the flue gas
   (*b*) direct determination of nitrogen in the flue gas
   (*c*) determining the combustion efficiency
   (*d*) finding out the calorific value of gaseous fuels
7. In Orsat apparatus, oxygen is absorbed in
   (*a*) cupurous chloride
   (*b*) potassium cabonate
   (*c*) dilute potassium sulphate
   (*d*) alkaline solution of pyrogallic acid
8. The sequence of absorption in the flue gas analysis by Orsat's apparatus is
   (*a*) CO_2, CO and O_2
   (*b*) CO_2, O_2 and CO
   (*c*) O_2, CO_2 and CO
   (*d*) CO, O_2 and CO_2

Answers

1. (*c*)	2. (*b*)	3. (*c*)	4. (*d*)	5. (*a*)
6. (*a*)	7. (*d*)	8. (*b*)		

□□□

Chapter **25**

Refrigeration Systems and Psychrometry

Learning objectives : The reader will be able to explain the following aspects after making a study of the subject matter presented in this chapter :

- *refrigeration and its applications*
- *air standard refrigeration cycles*
- *vapour compression refrigeration; refrigerants and their properties*
- *vapour absorption systems and electrolux refrigerator*
- *properties of moist air, psychrometric processes and human comfort*

A : REFRIGERATION SYSTEMS

25.1. REFRIGERATION: WHAT IS IT?

Science of providing and maintaining temperatures below that of surroundings

Refrigeration and air conditioning is the fascinating branch of science which deals with the chilling or freezing of a substance by removing some of its heat. This artificial withdrawl of heat produces within the substance or within a space a temperature below the general temperature of its surroundings. Refrigeration essentially means continued abstraction of heat from a substance (perishable foods, drinks and medicines etc.) at low temperature level and then transfer this heat to another system at high potential of temperature. To accomplish this, mechanical work must be performed to satisfy the second law of thermodynamics.

Air conditioning refers to the simultaneous control of temperature, humidity, cleanliness and air motion within a confined region or space.

25.2. REFRIGERATOR AND HEAT PUMP

Refrigerators and heat pumps are ***reversed*** heat engines. The adjective reversed means operating backwards. The direction of heat and work interactions are opposite to that of a heat engine, *i.e.*, work input and heat output. These machines (refrigerators and heat pumps) are used to remove heat from a body at low temperature level and then transfer this heat to another body at high potential of temperature. When the main purpose of the machine is to remove heat from the cooled space, it is called a ***refrigerator***. A refrigerator operates between the temperature of surroundings and a temperature below that of the surroundings. Refrigerators are essentially used to preserve food items and drugs at low temperature.

The term ***heat pump*** is applied to a machine whose objective is to heat a medium which may already be warmer than its surroundings. A heat pump thus operates between the temperature of the surroundings and a temperature above that of the surroundings. Heat pumps are generally used to keep the rooms warm in winter.

The transfer of heat against a reverse temperature gradient in a refrigerator and heat pump is accomplished by supplying energy to the machine. A schematic representation of heat pump and a refrigerator has been shown in Fig. 25.1.

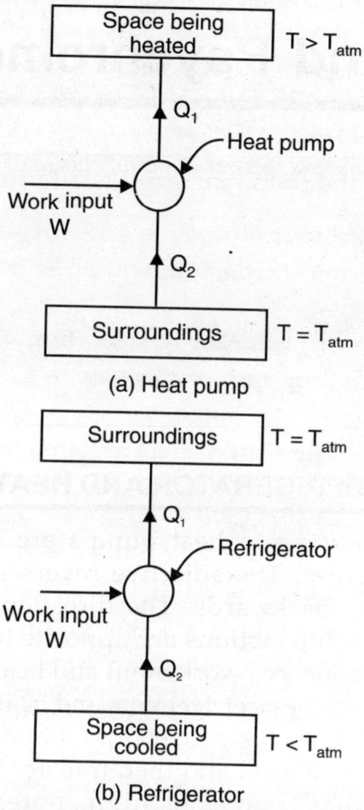

Fig. 25.1. Functional difference between a heat pump and a refrigerator

In the context of refrigerators and heat pumps, the performance is expressed in terms of ***coefficient of performance*** (*COP*) which represents the ratio of desired effect to work input

$$COP = \frac{\text{desired effect}}{\text{work input}}$$

In a ***refrigerator***, the desired effect is the amount of heat Q_2 extracted from the space being cooled, *i.e.*, the space at low temperature.

$$(COP)_{ref} = \frac{\text{heat extracted at low temperature}}{\text{work input}}$$

$$= \frac{Q_2}{W}$$

From the principle of energy conservation :

$W = Q_1 - Q_2$

$$\therefore\ (COP)_{ref} = \frac{Q_2}{Q_1 - Q_2}$$

For most of the refrigerating machines, the values of *COP* lie between 3 and 4, and the *COP* are greatest when temperature differences are least.

In a ***heat pump***, the desired effect is the amount of heat Q_1 supplied to the space being heated.

$$(COP)_{ref} = \frac{\text{heat rejected at high temperature}}{\text{work input}}$$

$$= \frac{Q_1}{Q_1 - Q_2}$$

$$= 1 + \frac{Q_2}{Q_1 - Q_2} = 1 + (COP)_{ref}$$

Thus the *COP* of a machine operating as a heat pump is higher than the *COP* of the same machine when operating as a refrigerator by unity.

Note : In the context of a refrigerator system :

1. The amount of heat extracted from the body at low temperature, *i.e.*, the space being cooled is called *refrigerating effect* and is designated by *N*.

2. The *COP* of a refrigerator based on the theoretical values of refrigerating effect and work input is termed as *theoretical COP*. The theoretical refrigerating effect and work input are calculated by applying the laws of thermodynamics to the refrigeration cycle.

3. The *COP* of a refrigerator based on actual values of refrigerating effect and work input is termed as *actual COP*. The actual refrigerating effect and work input are obtained during test run on a refrigerating plant.

4. The ratio of actual *COP* to theoretical *COP* is known as *relative COP*.

$$\text{Relative } COP = \frac{\text{actual } COP}{\text{theroretical } COP}$$

5. *Refrigeration efficiency* is defined as the ratio of *COP* of a cycle to the *COP* of a Carnot cycle operating in the same temperature range.

6. A single machine can fulfil both the functions of cooling and heating simultaneously. For example, cool a food storage space as a refrigerator and heat a water system as a heat pump.

25.2.1. Rating or capacity of a refrigerating unit

The refrigerating machines are usually rated in terms of their cooling capacity; the standard unit being *ton of refrigeration.*

One ton of refrigeration is defined as the refrigerating effect that freezes one ton (2000 pound mass) of liquid water during a period of 24 hours. The water is to be liquid at 0°C before and ice at 0°C after the process.

$$2000 \text{ pound mass} = \frac{2000}{2.205} = 907 \text{ kg}$$

Enthalpy of fusion of water at 0°C

$$= 333.43 \text{ kJ/kg}$$

$$\therefore \text{ 1 ton of refrigeration} = \frac{907 \times 333.43}{24 \times 60}$$

$$= 210 \text{ kJ/min} = 3.5 \text{ kJ/s} = 3.5 \text{ kW}$$

A ton of refrigeration is thus not a unit of mass but a measure of the rate of cooling. The refrigerating capacity decides the mass flow rate of a given working substance (refrigerant) working under specified conditions.

Mass flow rate of refrigerant

$$= \frac{\text{refrigeration capacity}}{\text{refrigerating effect per unit mass}}$$

Quite often, power needed to produce a refrigeration effect equivalent of 1 ton of refrigeration is used as a measure to calculate the cost of operation or motor size of the refrigerating unit. Then

$$\text{kW per ton of refrigeration} = \frac{3.5}{COP}$$

Further, the ratio of heat removal rating (kJ/hr) of a refrigeration system to energy input (kWhr) of the machine is called energy efficiency ratio (EER).

EXAMPLE 25.1

(i) The capacity of a refrigeration system is specified to be 12 tons. What is then the cooling rate of the machine ?

(ii) 250 litres of drinking water is required per hour at 10°C. Would the use of 1.5 ton refrigerating system be justified if the available water is at 30°C ?

(iii) A refrigerating machine takes 1.25 kW and produces 25 kg/hr of ice at 0°C from water available at 30°C. Determine refrigerating effect, tonnage and coefficient of performance of machine. Take

Specific heat of water = 4.18 kJ/kg K

Enthalpy of solidification of water from and at 0°C = 335 kJ/kg

Solution : (*i*) 1 ton of refrigeration ≡ 3.5 kJ/s

$$\therefore \text{ Cooling rate of machine} = 12 \times 3.5 = 42 \text{ kJ/s}$$

(*ii*) Refrigeration effect required for cooling the water

$$= mc_p\, \Delta T$$

$$= 250 \times 4.18 \times (30\text{-}10) = 20900 \text{ kJ/hr}$$

1 ton of refrigeration

$$\equiv 3.5 \text{ kJ/s} = 12600 \text{ kJ/hr}$$

∴ Tonnage required

$$= \frac{20900}{12600} = 1.658$$

As such, the use of 1.5 ton machine will not serve the purpose.

(*iii*) Refrigeration effect

≡ removal of heat from water at 30°C to convert into ice at 0°C

$$= m\left[c_{pw}\, \Delta T + L\right]$$

$$= 25\,\left[4.18(30-0)+335\right] = 11510 \text{ kJ/hr}$$

1 ton of refrigeration

$$\equiv 3.5 \text{ kJ/s} = 12600 \text{ kJ/hr}$$

∴ Tonnage required

$$= \frac{11510}{12600} = 0.913$$

$$COP = \frac{\text{refrigerating effect}}{\text{work input}}$$

$$= \frac{11510/3600}{1.25} = 2.558$$

25.2.2. Methods of refrigeration

Refrigeration is the technique of producing cooling effect by abstraction/withdrawl of heat so that temperature below that of surroundings is produced in a substance or within a space. The desired cooling effect can be produced by :

1. Evaporation

When a liquid evaporates, it absorbs heat from the surroundings equivalent to its latent heat of vaporisation, and that results in lowering the temperature of surroundings. For example, we feel cooling effect when there is evaporation of a drop of spirit placed on the palm of hand. Likewise, the evaporation of moisture from the skin of a human body helps to keep it cool. There is a common practice to cool the water for drinking purposes by keeping it in the porous earthen pots. The water evaporates through the pores and that produces cooling effect. The army people keep small water containers made of metal and covered with water soaked namada; the walls of the metallic container get cooled and that cools the water kept inside it.

The principle of evaporative refrigeration is employed in desert (room) coolers. The dry atmospheric air is made to pass through water soaked packings. When this water evaporates, it takes heat from the air causing it to cool.

2. Dissolution of salts in water

When certain salts are dissolved in water, they absorb heat and lower the temperature of water and create a sort of refrigeration bath for cooling substance. Sodium chloride lowers the water temperature upto – 20°C while calcium chloride upto – 50°C. The salt can be recovered by evaporation of water from the solution.

3. Ice refrigeration (change of phase)

The use of ice to refrigerate and thus preserve food goes back to the prehistoric times and the ancient cultures of Chinese, Greeks, Romans and Persians. Ice and snow were stored in caves or dugouts lined with straw or other insulating materials. This practice worked well down through the centuries, with ice houses remaining in use. Greater work was done on developing better insulation products for long distance shipment of ice and the ice harvesting became a big business.

The natural ice or artificially produced ice is brought into contact with the substance to be cooled. The ice melts and the heat required for melting of ice is supplied by the substance being cooled. The cooling effect produced by ice is

$$Q = \dot{m} \times h_{sf}$$

where $\dot{m}$ and h_{sf} are the rate of fusion of ice and enthalpy of fusion respectively. At normal atmospheric pressure of 1.01325 bar (1 atm), h_{sf} equals 335 kJ/kg.

Ice refrigeration prevents dehydration and preserves the fresh appearance of eatable products like fruit and vegetables. However, ice has to be fed to the refrigerator of and on and that is neither convenient nor economical. Further, water coming out of the refrigerator poses a problem in its disposal.

4. Dry ice refrigeration (sublimation)

Solid carbon dioxide (called dry ice) has a peculiar characteristics that it changes from solid state to vapour state without passing through liquid state. During change of phase, it absorbs heat equivalent to its latent heat of vaporisation and produces cooling effect.

At one atmospheric pressure, solid CO_2 produces 573 kJ/kg of refrigeration maintaining a temperature of – 78.5°C.

Dry ice is used to preserve food stuff during transportation. The slabs of ice are usually packed on either side or on top of food packages in cartons. When dry ice evaporates, it absorbs heat from the food stuff and preserves it in frozen state.

The refrigeration methods (1) to (4) as mentioned above are called the ***natural methods.*** These methods are non-cyclic and the temperatures attainable are limited. Further, there is continuous consumption of the refrigerating substance and that necessitates replenishment. However, these methods are sometimes convenient forms of cooling where small refrigeration is required such as in the laboratory and workshop.

5. Chemical methods

Here the heat required for the completion of chemical reaction is taken from the substance being cooled. The chemical method for producing cooling effect cannot be followed on a commercial scale.

6. Air or gas refrigeration

Expansion of gas lowers its pressure and temperature. This cooling effect results without change in the phase of gas. The different air refrigeration systems use this thermodynamic principle for producing low temperatures.

7. Throttling process

Throttling is the expansion of fluid from high pressure to low pressure. This process occurs when fluid passes through an obstruction (partially opened valve or a small orifice) placed in the fluid flow passage. Under certain conditions, a gas gets cooled during throttling and that produces cooling effect.

8. Mechanical refrigeration

The natural and chemical methods have been successfully replaced by mechanical or heat energy refrigeration techniques. In these methods, the heat is abstracted from the substance or space (which is to be cooled) and is pumped to a system (which is at high temperature level) by taking energy from an external source as input to refrigerating machine. The refrigeration system consists of a cycle of processes with the same quantity of working fluid (refrigerant) in continuous circulation.

Mechanical refrigeration systems are broadly classified into :

- air or gas refrigeration
- vapour compression system
- vapour absorption system, and
- steam jet refrigeration

Energy needed for the mechanical systems is essentially in the form of mechanical, electrical or thermal. Due to world energy crisis, concerted efforts are being made by various agencies to develop refrigeration systems which utilize waste heat, solar energy, wind energy and bio-energy etc for their functioning. A lot of research is being done to devise ways and means which make the refrigeration systems more energy efficient.

9. Non-conventional refrigeration systems

- ***Thermo-electric*** cooling uses the Peltier effect to create a heat flux between the junction of two different types of materials.

When a direct current is made to pass through electrical junctions between unlike metals, heating or cooling effect is produced depending on the direction of flow of current. One junction gets heated and the other gets cooled. The cold junction is located inside the space desired to be cooled and the warmed junction lies outside in the surroundings. The hot and cold junctions get reversed when a change occurs in the direction of flow of current. Apparently the same setup can be used for heating and cooling of an insulated space.

This effect is commonly used in camping and portable coolers, and for cooling electronic components and small instruments.

- ***Magnetic refrigeration*** (adiabatic demagnetisation) is a cooling technology based on magnet caloric effect, an intrinsic property of magnetic solids. A strong magnetic field is applied to the refrigerant, forcing its various magnetic dipoles to align and putting these degrees of freedom of the refrigerant into a state of lowered entropy. A heat sink then absorbs the heat released by the refrigerant due to its loss of entropy. Thermal contact with the heat sink is then broken so that the system is insulated and the magnetic field is switched off. This increases the heat capacity of the refrigerant, thus decreasing its temperature below the temperature of heat sink.

- ***Thermo acoustic*** refrigeration which uses sound waves in a pressurised gas to drive heat exchange.

- ***Vortex tube*** that operates with compressed air and is used for spot cooling. A high pressure gas is allowed to expand through a nozzle fitted tangentially to a pipe. This causes simultaneous discharge of cool air at the core and hot air at the periphery.

25.2.3. Applications of refrigeration

Refrigeration has played an important role in the growth and attainment of the present-day standard of living. Its various applications can be essentially grouped into the following categories :

1. Domestic refrigeration deals with providing a low temperature place for preserving eatables, drinks and medicines. The use of refrigerators in our kitchens for the storage of fruits and vegetables has allowed us to add fresh salads to our diets year round, and to store fish and meats for long periods.

2. Commercial refrigeration is related to refrigeration fixtures of the type used by restaurants and hotels, retail stores and institutions for storing, displaying, processing and dispensing of perishable commodities of different types.

3. Industrial refrigeration is concerned with systems which are much larger in size and cooling capacity that those for commercial applications. Typical industrial applications are:

- Ice and large food plants
- Chilling and storage of food stuffs including beverages, meat, poultry products, dairy products, vegetables, fruits and fruit juices etc
- Processing of food products and farm crops
- Processing of textiles, printing work and photographic materials
- Oil refineries and synthetic rubber manufacturing. In oil refineries, chemical manufacturing and petrochemical plants, refrigeration is used to maintain processes at their required low temperatures.

Refrigeration is also used to liquefy gases like oxygen, nitrogen, propane and methane. Further, in compressed air purification, it is used to condense water vapour from compressed air to reduce its moisture air content.

4. Transport refrigeration applies to refrigerated railway cars and trucks for local delivery and long distance transport of temperature sensitive food stuffs and other materials.

5. Air-conditioning refers to the simultaneous control of temperature, humidity, cleanliness and air motion. The conditioning of a space done to provide comfortable and healthy conditions for the occupants is called *comfort air-conditioning*. Residences, offices, theatres and hospitals are the spaces air-conditioned for this purposes.

There are some manufacturing processes which need to be done under controlled environmental conditions. Even some of the sophisticated and precision instruments need controlled conditions for their effective working and upkeep. The conditioning done for this purpose is called *industrial air-conditioning* and this is concerned with production of environment suitable for :

- computer centres
- pharmaceutical units
- printing works and photographic products
- precision devices and production shop laboratories etc.

The applications cited above clearly indicate that refrigeration and air-conditioning which was considered luxury in the society a few decades ago has become the necessity of the present society and a tool for higher productivity.

25.3. AIR REFRIGERATION CYCLES

The air refrigeration cycles use air as the working fluid. The piston-cylinder arrangement is filled with a constant amount of air which remains gaseous (no phase change) and homogeneous throughout and at all times. Air refrigeration was one of the earliest methods of producing cold. However, it became out of date (obsolete) for several years due to its low performance and high power requirement. Presently the system is being preferred for aircraft refrigeration where low weight and volume of the associated equipment are the main considerations.

The air refrigeration systems may work on reversed Carnot cycle or reversed Brayton cycle.

25.3.1. Reversed Carnot cycle

Refrigerators and heat pumps are reversed heat engines. They operate backwards and have heat and work interactions opposite to that of a heat engine. Heat at low temperature is removed and rejected at higher temperature and the task is accomplished by supplying work to the system.

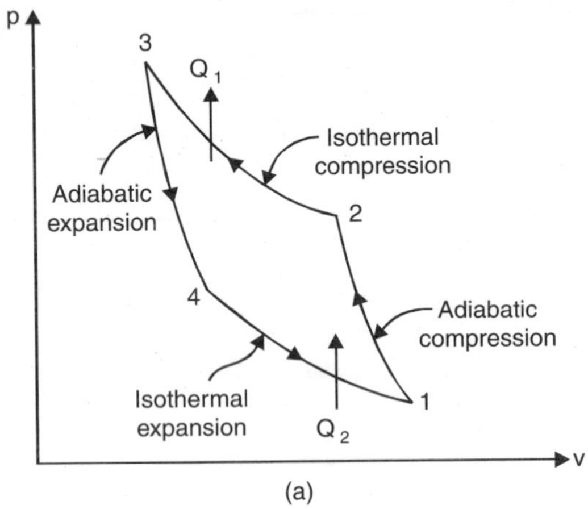

(a)

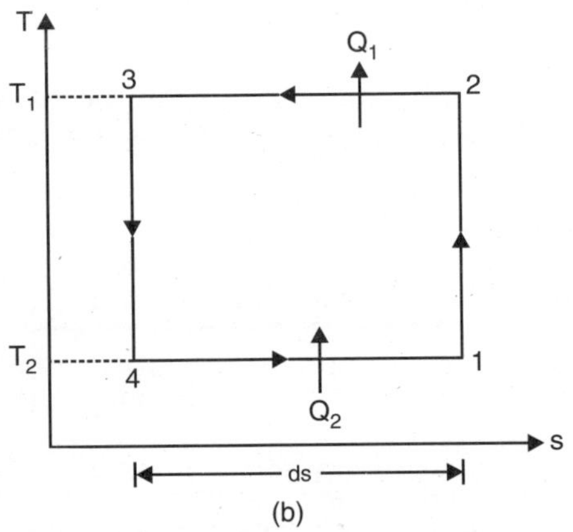

(b)

Fig. 25.2. Reversed Carnot cycle

The *p-v* and *T-s* diagrams of a reversed Carnot cycle are shown in Fig. 25.2 and the sequence of operation is

1-2 : adiabatic compression of air with the aid of an external power. The temperature of air rises from T_2 to T_1.

2-3 : isothermal compression of air during which heat is rejected to another space at temperature T_1.

3-4 : adiabatic expansion of air in the clearance space of the cylinder and the temperature of air falls from T_1 to T_2.

4-1 : isothermal expansion at which heat is absorbed at low temperature T_2 from the space being cooled.

Heat abstracted $Q_2 = T_2\, ds$

Heat rejected $Q_1 = T_1\, ds$

Work input W

$$= \text{heat rejected} - \text{heat abstracted}$$

$$= (T_1 - T_2)\, ds$$

$$\therefore \; (COP)_{ref} = \frac{\text{heat abstracted at low temperature}}{\text{work input}}$$

$$= \frac{T_2\, ds}{(T_1 - T_2)\, ds} = \frac{T_2}{T_1 - T_2}$$

$$(COP)_{heat\ pump} = \frac{\text{heat rejected at high temperature}}{\text{work input}}$$

$$= \frac{T_1\, ds}{(T_1 - T_2)\, ds} = \frac{T_1}{T_1 - T_2}$$

Although theoretically more efficient, Carnot heat pumps and refrigerators like Carnot heat engines have not been used in practice. This is because the machine has to run at high speed during the portions of the cycle when adiabatic processes are being performed and run very slowly during other portions when isothermal processes are being carried out. The variation of speed during different processes constituting the cycle is not possible.

EXAMPLE 25.2

A refrigerating machine of 1 ton rating works on reversed Carnot cycle. It requires 1.5 kW to maintain low evaporator temperature of 250 K. Calculate coefficient of performance of the machine when working as a (i) refrigerator (ii) heat pump. Also determine the temperature at which heat is rejected.

Solution : 1 ton of refrigeration = 210 kJ/min = 3.5 kJ/s = 3.5 kW

(*i*) $(COP)_{refrigerator} = \dfrac{\text{refrigerating effect}}{\text{work input}}$

$= \dfrac{3.5}{1.5} = \mathbf{2.33}$

(*ii*) $(COP)_{heat\ pump} = 1 + (COP)_{refrigerator}$

$= 1 + 2.33 = \mathbf{3.33}$

(*b*) In terms of operating temperatures, the *COP* of a refrigerator is given as

$$COP = \frac{T_2}{T_1 - T_2}; \quad 2.33 = \frac{250}{T_1 - 250}$$

or $T_1 = \dfrac{250 + 250 \times 2.33}{2.33} = \mathbf{357.3\ K}$

EXAMPLE 25.3

A refrigerator system operating on a reversed Carnot cycle produces 400 kg/hr of ice at – 5°C from water at 30°C. Make calculations for

(a) the power required to drive the machine,

(b) the heat rejected from the system.

Take latent heat of freezing = 335 kJ/kg and specific heat of ice = 2.1 kJ/kgK

Solution : T_1 = 25°C = 298 K and

T_2 = – 5°C = 268 K

In terms of operating temperatures, the *COP* of the machine is given as

$$COP = \frac{T_2}{T_1 - T_2} = \frac{268}{298 - 268} = 8.93$$

Heat absorbed or net refrigerating effect

= (5 × 2.1 + 335 + 4.186 × 30) × 400

= 188432 kJ/hr = 52.34 kJ/s

Now, $COP = \dfrac{\text{refrigerating effect}}{\text{work input}}$

$$8.93 = \frac{52.34}{W}$$

∴ Work required to run the machine,

$$W = \frac{52.34}{8.93} = 5.86 \text{ kJ/s} = \mathbf{5.86\ kW}$$

(*b*) From the principle of energy conservation (1st law)

heat rejected = heat absorbed + work input

= 52.34 + 5.86 = **58.2 kJ/s**

EXAMPLE 25.4

A cold storage is to be maintained at – 5°C while the surroundings are at 35°C. The heat leakage from the surroundings into the cold storage is estimated to be 29 kJ/s and the actual coefficient of performance of the refrigeration plant is one-third of an ideal plant working between the same temperature limits. Find the power required in kW to drive the plant.

Solution : Refer Fig. 25.3 for the schematic diagram for the given system.

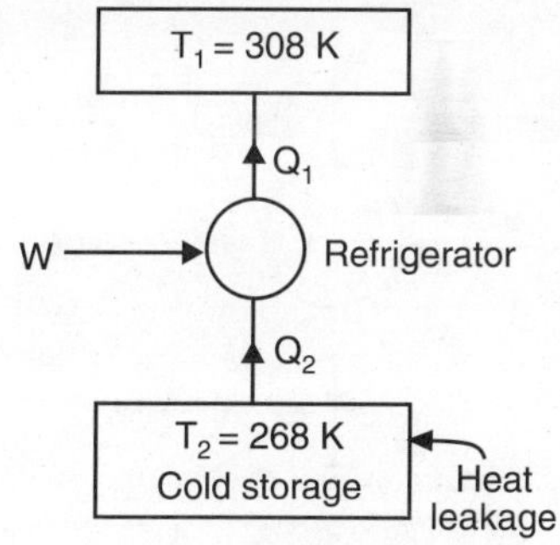

Fig. 25.3.

If the cold storage is to be maintained at a constant temperature of 268 K, it must reject heat at the same rate at which heat is leaking into it from the surroundings.

Q_2 = 29 kJ/s

COP for ideal refrigerator

$$= \frac{T_2}{T_1 - T_2} = \frac{268}{308 - 268} = 6.7$$

$$\text{Actual } COP = \frac{1}{3} \times 6.7 = 2.233$$

Also : $COP = \dfrac{\text{refrigerating effect}}{\text{work input}} = \dfrac{Q_2}{W}$

∴ Work input required

$$= \frac{Q_2}{COP} = \frac{29}{2.233} = 12.987 \text{ kJ/s}$$

$= \mathbf{12.987\ kW}$

EXAMPLE 25.5

A refrigerator working on Carnot cycle has a coefficient of performance of 5.0 when it extracts

8350 kJ/min from a space to be cooled. Determine (i) the heat rejected to the surroundings and (ii) the power required to run the compressor.

If the temperature maintained in the refrigerator is 253°C, determine the temperature at which heat is rejected.

Solution : Refer Fig. 25.4 for the schematic arrangement of the system.

$$COP = \frac{\text{heat absorbed from the system being cooled}}{\text{work input}}$$

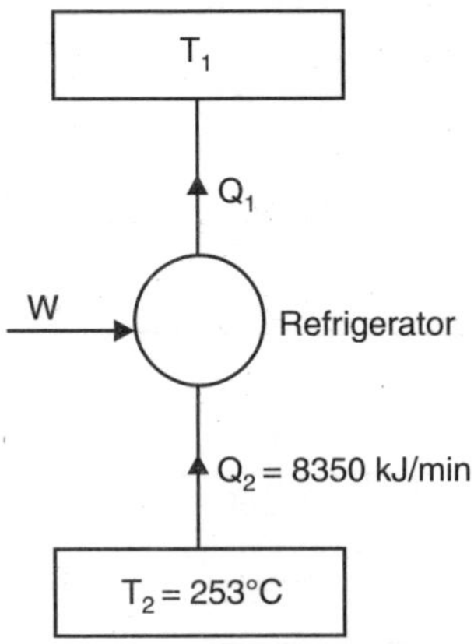

Fig. 25.4.

For a refrigerator,

∴ Work input to compressor

$$= \frac{8350}{5}$$

$$= 1670 \text{ kJ/min} = 27.83 \text{ kJ/s}$$

Thus the compressor requires 27.83 kW power for its operation.

(*ii*) From the principle of energy conservation (1st law),

$$Q_1 = Q_2 + W$$
$$= 8350 + 1670 = 10020 \text{ kJ/min}$$

Thus 10020 kJ/min of heat is rejected to the surroundings.

(*iii*) For a refrigerator operating on Carnot cycle, the *COP* is given by

$$COP = \frac{T_2}{T_1 - T_2} \; ; \; 5 = \frac{253 + 273}{T_1 - (253 + 273)}$$

Solution gives :

$$T_1 = \mathbf{631.2 \text{ K}} \text{ or } \mathbf{358.2°C}$$

EXAMPLE 25.6

A refrigerator operates on a reversed Carnot cycle whose coefficient of performance is 5.0. The evaporator is maintained at a temperature of – 6°C and the power required to run the refrigerator is 3.5 kW. Determine the refrigerating effect and the condenser temperature of the refrigerator.

Solution : Refer Fig 25.5.

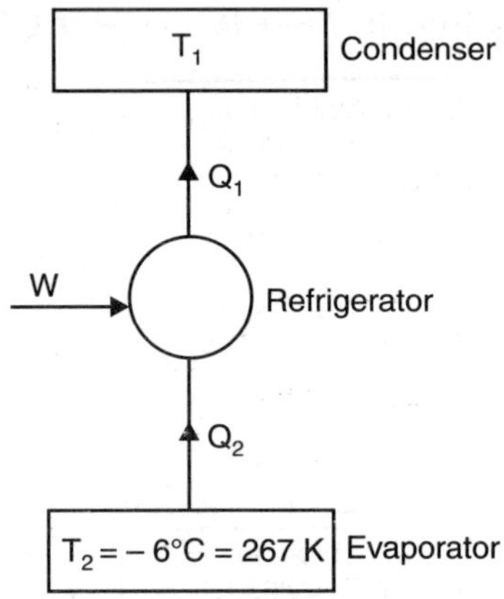

Fig. 25.5.

In a refrigerator system, heat is removed from a body at low temperature (evaporator) and this heat is transferred to another body (condenser) at a high potential of temperature. Further, refrigerating effect is the amount of heat extracted from the body at low temperature (evaporator).

$$T_2 = -6°C = 273 + (-6) = 267 \text{ K}$$

For a refrigerator working on reversed Carnot cycle,

$$(COP)_{ref} = \frac{T_2}{T_1 - T_2} \; ; \; 5 = \frac{267}{T_1 - 267}$$

∴ Condenser temperature,

$$T_1 = 267 + \frac{267}{5} = \mathbf{320.4 \text{ K}}$$

In terms of heat and work interactions,

$$(COP)_{ref} = \frac{Q_2}{Q_1 - Q_2} = \frac{Q_2}{W}$$

∴ Heat extracted from the space being cooled (refrigeration effect),

$$Q_2 = W \times (COP)_{ref}$$
$$= 3.5 \times 5 = \mathbf{17.5 \text{ kW}}$$

EXAMPLE 25.7

A domestic food refrigerator maintains a temperature of – 10°C whilst the ambient air temperature is – 30°C. The heat leakage into the freezer is estimated to be at the continuous rate of 2 kJ/s. Determine the least power needed to pump out this heat continuously.

Solution : The refrigerator is required to remove heat from the freezer at the same rate at which heat leaks into it. That is : $Q_2 = 2$ kJ/s

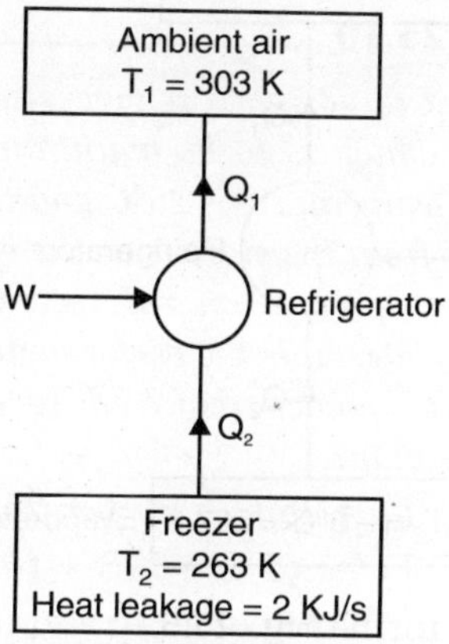

Fig. 25.6.

For minimum power requirement, the refrigerator operates on reversed Carnot cycle. Accordingly

$$COP = \frac{T_2}{T_1 - T_2} \quad ...(i)$$

In terms of heat and work interactions

$$COP = \frac{Q_2}{W} = \frac{Q_2}{Q_1 - Q_2} \quad ...(ii)$$

From expressions (*i*) and (*ii*)

$$\frac{T_1 - T_2}{T_2} = \frac{Q_1 - Q_2}{Q_2} \quad \text{or} \quad \frac{T_1}{T_2} = \frac{Q_1}{Q_2}$$

$$\therefore \quad Q_1 = \frac{T_1}{T_2} \times Q_2$$

$$= \frac{303}{263} \times 2 = 2.304 \text{ kJ/s}$$

and $W = Q_1 - Q_2$

$= 2.304 - 2 =$ **0.304** kJ/s

Hence the least power required to pump the heat continuosly is 0.304 kW

EXAMPLE 25.8

A heat pump is used to meet the heating requirements of a house and maintain it at 300 K. On a day when the outdoor air temperature drops to 275 K, it is estimated that the house loses heat at the rate of 75 MJ/hr. If the coefficient of performance of the heat pump under these conditions is 2.5, determine:

(a) the power consumed by the heat pump, and

(b) the rate at which heat is absorbed from the cold outside air.

Solution : Refer Fig. 25.7 for the schematic diagram for the given system.

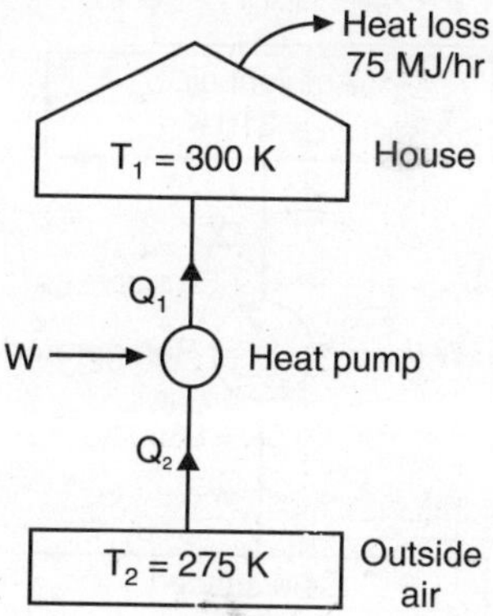

Fig. 25.7.

If the house is to be maintained at a constant temperature of 300 K, it must receive heat at the same rate at which it is losing heat. That is

$$Q_1 = 75 \text{ MJ/hr}$$

For the heat pump; $COP = \dfrac{Q_1}{W}$

∴ Power consumed by the pump,

$$W = \frac{75}{2.5} = 30 \text{ MJ/hr}$$

$$W = \frac{30 \times 10^3}{3600} = 8.33 \text{ kJ/s} = \mathbf{8.33\ kW}$$

(*b*) From the principle of energy conservation; $Q_1 = W + Q_2$

∴ Rate at which heat is absorbed from the cold outside air is

$$Q_2 = Q_1 - W = 75 - 30 = \mathbf{45\ MJ/hr}$$

EXAMPLE 25.9

A refrigerator operating on a reversed Carnot cycle consumes 150 W power in summer when the ambient atmosphere is at 310 K. The heat leakage into the refrigerator through the doors is estimated to be at the continuous rate of 15 W per degree temperature difference between the ambient air and the cold space of the refrigerator. For continuous operation of the refrigerator, determine the temperature at which the cold space is maintained.

Solution : The refrigerator is requried to remove heat from the cold space at the same rate at which heat leaks into it. That is :

$$Q_2 = 15(T_1 - T_2)$$

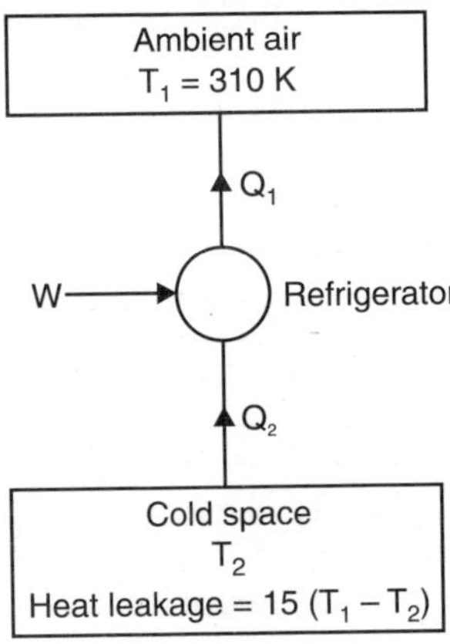

Fig. 25.8.

Since the refrigerator operates on the reversed Carnot cycle, its coefficient of performances can be written as

$$COP = \frac{T_2}{T_1 - T_2} \quad ...(i)$$

In terms of heat and work interactions

$$COP = \frac{Q_2}{W} \quad ...(ii)$$

From expressions (*i*) and (*ii*),

$$\frac{T_2}{T_1 - T_2} = \frac{15(T_1 - T_2)}{150}$$

$$\frac{T_2}{310 - T_2} = \frac{310 - T_2}{10}$$

or $\quad (310 - T_2)^2 = 10\,T_2$

or $\quad 96100 + T_2^2 - 620\,T_2 = 10\,T_2$

or $\quad T_2^2 - 630\,T_2 + 96100 = 0$

$$T_2 = \frac{630 \pm \sqrt{396900 - 384400}}{2} = \frac{630 \pm 111.80}{2}$$

$$= 370.9 \text{ K} \quad \text{or} \quad 259.1 \text{ K}$$

The temperature T_2 = 370.9 K is greater than the temperature of ambient air and as such is inadmissible. Therefore the cold space has to be maintained at a temperature of **259.1 K**.

EXAMPLE 25.10

It is proposed to design a refrigeration plant for a food store which is to be maintained at –5°C. The ambient temperature is 25°C and the estimated heat transfer from the store is at the rate of 5 kW. If the system operates on reversed Carnot heat engine cycle, determine the performance index and the minimum power required to operate the refrigeration plant.

Solution : T_1 = 25°C = 273 + 25 = 298 K

T_2 = – 5°C = 273 + (– 5) = 268 K

The performance of a refrigeration system is expressed in terms of coefficient of performance, and the *COP* of the refrigerator working between two thermal reservoirs

T_1 (298 K) and T_2 (268 K) is given by

$$(COP)_{ref} = \frac{T_2}{T_1 - T_2} = \frac{268}{298 - 268} = 8.93$$

In terms of heat interactions,

$$(COP)_{ref} = \frac{Q_2}{Q_1 - Q_2}\,; \quad 8.93 = \frac{5}{Q_1 - 5}$$

or $\quad 8.93\,Q_1 - 44.65 = 5$

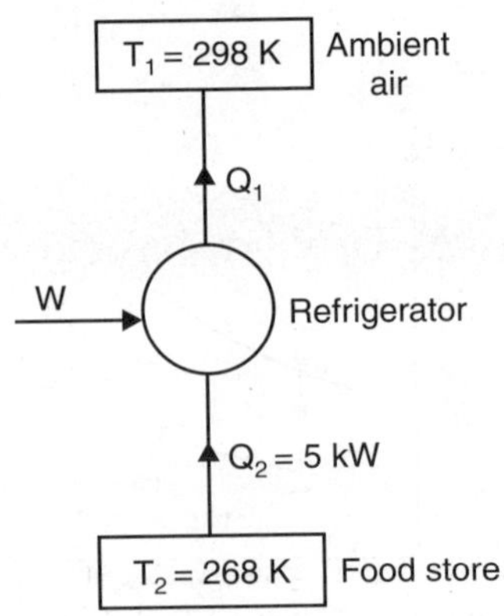

Fig. 25.9.

$$Q_1 = \frac{49.65}{8.93} = 5.56 \text{ kW}$$

From principle of energy conservation (1st law): $Q_1 = W + Q_2$

∴ Power required to operate the refrigerator,

$$W = Q_1 - Q_2 = 5.56 - 5 = \mathbf{0.56\ kW}$$

A reversible refrigerator requires the minimum power to operate it. Therefore, the minimum power required is 0.56 kW.

EXAMPLE 25.11

A perfect reversed heat engine is used for making ice at– 5°C from water available at 25°C. The temperature of freezing mixture is – 10°C. Calculate the quantity of ice formed per kWh.

For ice : specific heat = 2.1 kJ/kgK and latent heat = 335 kJ/kg

Solution : The temperature limits for the reversed heat engine are:

$$T_1 = 25°\text{C} = 298 \text{ K}$$

and $$T_2 = -10°\text{C} = 263 \text{ K}$$

A perfect reversed heat engine works on Carnot cycle for which *COP* in terms of temperatures is given by

$$COP = \frac{T_2}{T_1 - T_2} = \frac{263}{298 - 263} = 7.514$$

Work input = 1 kWh = 3600 kJ

In terms of heat and work interactions,

$$COP = \frac{Q_2}{Q_1 - Q_2} = \frac{Q_2}{W}$$

∴ Total heat extracted from ice,

$$Q_2 = W \times COP$$
$$= 3600 \times 7.514 = 27050 \text{ kJ/hr}$$

Removal of heat from 1 kg of water at 25°C so that it becomes ice at – 5°C is given by

$$= m\,c_{pw}\,\Delta T + mL + m\,c_{pi}\,\Delta T$$
$$= 1\times 4.186\,(25 - 0) + 1\times 335 + 1 \times 2.1\,[0 - (-5)]$$
$$= 104.65 + 335 + 10.5$$
$$= 450.15 \text{ kJ}$$

∴ Ice formed per kWh

$$= \frac{27050}{450.15} = \mathbf{60.09\ kg}$$

25.3.2. Bell-coleman cycle

Fig. 25.10 shows the schematic arrangement, *p-v* and *T-s* plots of the Bell-Coleman air

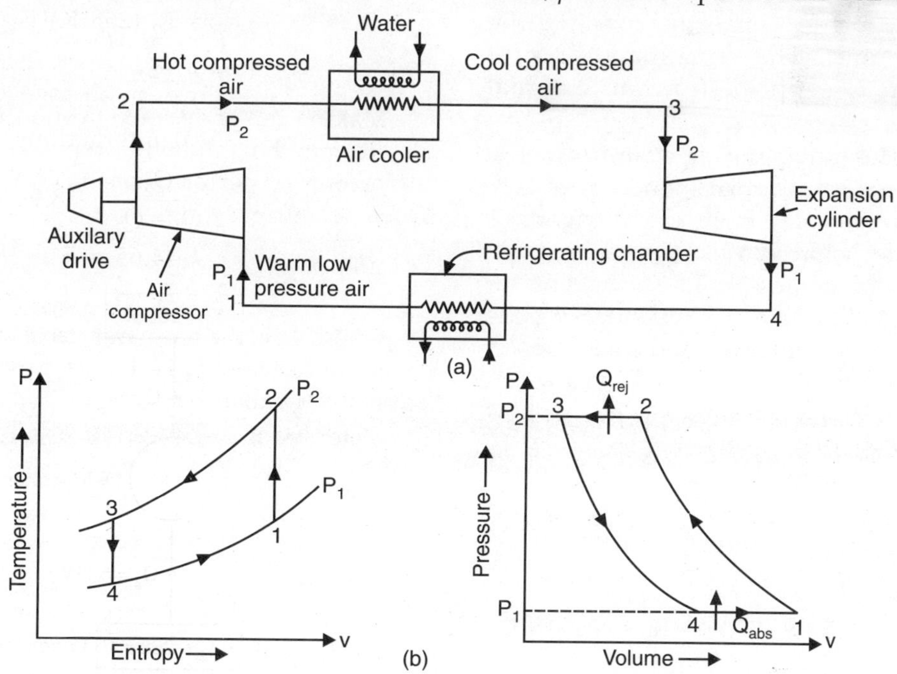

Fig. 25.10. Bell Coleman cycle

refrigerator working on the reversed Brayton heat engine cycle.

The machine essentially consists of a refrigerating (cold) chamber, a compressor, an air cooler and an expander. The four stages of operation are :

Process 1-2 (compressor) : The warm air leaving the refrigerating chamber at state point 1 enters the compressor where it is compressed isentropically. During compression, there occurs an increase both in temperature and pressure.

Process 2-3 (air cooler) : The hot compressed air leaving the compressor passes through pipes which are immersed in circulating water. The heat is rejected to water at constant pressure $p_2 = p_3$ and that cools the air to temperature T_3.

Process 3-4 (expander) : The air at high pressure ($p_3 = p_2$) and moderate temperature T_3 is taken to expansion cylinder where it expands isentropically to lower pressure ($p_4 = p_1$) and lowest temperature T_4 existing in the cycle. The temperature T_4 is sufficiently below the temperature of refrigerating chamber.

The work made available during the expansion process assists in running the compressor unit. The remaining work of compression is supplied from an external source.

Process 4-1 (refrigerating chamber) : The cold low pressure air corresponding to state point 4 absorbs heat from the refrigerator which may be brine solution (for making ice) or space to be cooled (cold storage or air conditioning). This heat absorption takes place at constant pressure ($p_4 = p_1$) and consequently the temperature increases from T_4 to T_1.

The air next enters the compressor and the cycle is repeated.

For unit mass flow of air,

Heat absorbed from the cold chamber, *i.e.*, refrigerating effect

$$Q_{abs} = c_p\,(T_1 - T_4)$$

Heat rejected to circulating water in the cooler

$$Q_{rej} = c_p\,(T_2 - T_3)$$

Work input in driving the compressor

$$= \text{heat rejected} - \text{heat absorbed}$$

$$= c_p\,(T_2 - T_3) - c_p\,(T_1 - T_4)$$

Coefficient of performance

$$(COP) = \frac{\text{refrigerating effect}}{\text{work input}}$$

$$= \frac{c_p(T_1 - T_4)}{c_p(T_2 - T_3) - c_p(T_1 - T_4)}$$

$$= \frac{T_1 - T_4}{(T_2 - T_3) - (T_1 - T_4)}$$

$$= \frac{T_4\left[\dfrac{T_1}{T_4} - 1\right]}{T_3\left[\dfrac{T_2}{T_3} - 1\right] - T_4\left[\dfrac{T_1}{T_4} - 1\right]} \quad ...(i)$$

From the isentropic compression (1 - 2) and expansion (3 - 4) processes

$$\frac{T_2}{T_1} = \left(\frac{p_2}{p_1}\right)^{\frac{\gamma - 1}{\gamma}}$$

and $$\frac{T_3}{T_4} = \left(\frac{p_3}{p_4}\right)^{\frac{\gamma - 1}{\gamma}} = \left(\frac{p_2}{p_1}\right)^{\frac{\gamma - 1}{\gamma}}$$

$$\therefore \quad \frac{T_2}{T_1} = \frac{T_3}{T_4} \quad \text{or} \quad \frac{T_2}{T_3} = \frac{T_1}{T_4}$$

Making the above substitution in identity (*i*), we get the following expression for the coefficient of performance (*COP*) of a Bell-Coleman refrigerating cycle.

$$COP = \frac{T_4}{T_3 - T_4}$$

where T_3 is the temperature of air entering the expander and T_4 is the temperature of air leaving the expander.

In terms of pressure ratio,

$$COP = \frac{1}{\dfrac{T_3}{T_4} - 1} = \frac{1}{\left(\dfrac{p_2}{p_1}\right)^{\frac{\gamma - 1}{\gamma}} - 1}$$

$$= \frac{1}{(r_p)^{\frac{\gamma - 1}{\gamma}} - 1} \quad ...(25.1)$$

Remarks :

(1) The *COP* of the ideal Bell-Coleman cycle depends upon the ratio of upper and lower pressure limit. The lower the pressure ratio, the higher is the *COP*.

(2) At a pressure ratio equal to unity, the *COP* is infinity but the refrigerating effect is zero. The effect of increasing pressure ratio is to increase the refrigerating effect. The optimum pressure is arrived at by striking a compromise between recurring cost of power consumption (governed by *COP*) and the capital cost of equipment (governed by refrigerating effect and hence capacity).

(3) The Bell-Coleman refrigerating systems had been built and operated extensively on ships for carrying frozen meat because the refrigerant used (air) is non-toxic, non-poisonous, easily available and non-expensive. The system still finds application in aircraft air-conditioning because of safety.

Due to low specific heat and high specific volume, a large amount of air has to be circulated for a given cooling load (refrigerating effect). The unit has a cumbersome design, low *COP* and high power consumption per ton of refrigeration.

(4) In modern air refrigeration cycles, the large displacement volumes can be handled best by the rotary type compressor and expander, as these are more compact for a given flow than the reciprocating machines.

(5) Quite often, the compression and expansion processes take place polytropically where pressure and volume are related as

$$pv^n = \text{constant}$$

25.3.3. Open and closed systems

The Bell-Coleman refrigerator may work on the open system or closed system. In ***open system***, the air comes in direct contact with the refrigerating chamber or the substances being cooled. In ***closed system***, the air is passed through coils placed in the chamber. Quite often, the coils carrying cold air are kept in the brine tank. The air extracts heat from the brine and the resulting chilled brine is then circulated around the cold chamber.

The brine is an aqueous solution of sodium chloride (NaCl) or calcium chloride ($CaCl_2$) in water. The brine does not generally freeze at low temperature, and its freezing point depends on salt water ratio.

The closed system is preferred over the open system on account of the following reasons :

(*i*) In the open system the air collects moisture from the stuff being cooled in the refrigerating chamber. The moisture may freeze during expansion and the snow deposits may choke up the valves.

(*ii*) In the open system, the expansion process can be continued only upto atmospheric pressure prevailing in the refrigerated chamber. No such restriction exists in the closed system.

(*iii*) In the closed system, the suction to the compressor may be at high pressure. The size of the compressor and expander can be kept reasonably small by using dense air.

EXAMPLE 25.12

An air refrigeration open system operating between 1 MPa and 100 kPa is required to produce a cooling effect of 2000 kJ/min. The temperature of air leaving the cold chamber is – 5°C and at leaving the cooler is 30°C. Neglecting losses and clearance in the compressor and expander, determine :

(a) mass of air circulated per hour,

(b) compressor work, expander work and the cycle work,

(c) coefficient of performance and the power required to run the machine.

Solution : Refer Fig. 25.11 for the *p-v* diagram for the cycle.

Given : $T_1 = -5°C = 268$ K,

$T_3 = 30°C = 303$ K

$p_4 = p_1 = 100 \text{ kPa} = 10^5 \text{ Pa} = 1 \text{ bar}$

$p_2 = p_3 = 1 \text{ MPa} = 10^6 \text{ Pa} = 10 \text{ bar}$

The compression process 1-2 follows the law pv^γ = constant and accordingly

$$\frac{T_2}{T_1} = \left(\frac{p_2}{p_1}\right)^{\frac{\gamma-1}{\gamma}} = \left(\frac{10}{1}\right)^{\frac{1.4-1}{1.4}} = 1.932$$

$$\therefore \quad T_2 = 1.932 \times 268 = 517.78 \text{ K}$$

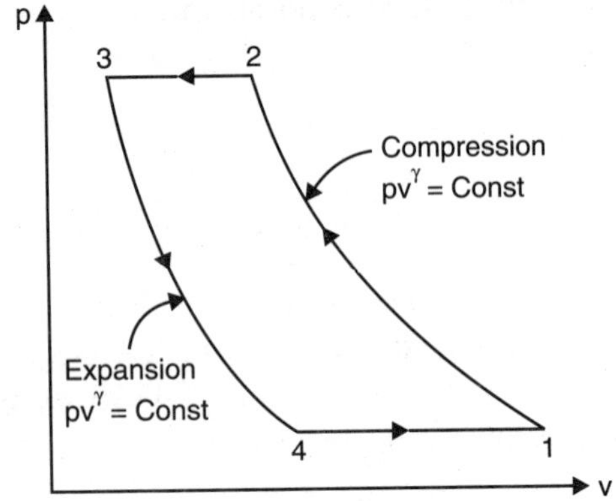

Fig. 25.11.

For the adiabatic expansion process 3-4,

$$\frac{T_4}{T_3} = \left(\frac{p_4}{p_3}\right)^{\frac{\gamma-1}{\gamma}}$$

$$= \left(\frac{1}{10}\right)^{\frac{1.4-1}{1.4}} = \frac{1}{1.932}$$

$$\therefore \quad T_4 = \frac{T_3}{1.932} = \frac{303}{1.932} = 156.83 \text{ K}$$

(*a*) Refrigerating effect per kg of air flow

$$= c_p (T_1 - T_4)$$
$$= 1.005 (268 - 156.83)$$
$$= 111.73 \text{ kJ/kg}$$

If m is the mass of air circulated per minute, then

$$111.73 \times m = 2000$$

$$m = \frac{2000}{111.73} = \mathbf{17.9 \text{ kg/min}}$$

(*b*) Compression work,

$$W_{comp} = \frac{\gamma}{\gamma-1} mR (T_2 - T_1)$$

$$= \frac{1.4}{1.4-1} \times 17.9 \times 287 (517.78 - 268)$$

$$= 4491182 \text{ J/min} = \mathbf{4491.18 \text{ kJ/min}}$$

Expander work

$$W_{exp} = \frac{\gamma}{\gamma-1} mR (T_3 - T_4)$$

$$= \frac{1.4}{1.4-1} \times 17.9 \times 287 (303 - 156.83)$$

$$= 2628217 \text{ J/min} = \mathbf{2628.22 \text{ kJ/min}}$$

Cycle work $W_{cycle} = W_{comp} - W_{exp}$

$$= 4491.18 - 2628.22 = \mathbf{1862.96 \text{ kJ/min}}$$

(*c*) $$COP = \frac{\text{refrigeration effect}}{\text{network expended}}$$

$$= \frac{2000}{1862.96} = \mathbf{1.0736}$$

Power required P = cycle work, *i.e.*, network expended per second

$$= \frac{1862.96}{60} = 31.05 \text{ kJ/s} = \mathbf{31.05 \text{ kW}}$$

EXAMPLE 25.13.

In a refrigerator working on Bell-Coleman cycle, air is drawn into the cylinder of the compressor from the cold chamber at 1 bar and – 3°C. After reversible adiabatic compression to 5 bar, the air is cooled at constant pressure to a temperature of 17°C. After subsequent polytropic expansion ($pv^{1.25}$ = C) to 1 bar in the expansion cylinder, the air is passed to the cold chamber. Sketch the p-v and T-s diagrams of the cycle, determine for unit mass flow of air :

(a) refrigerating effect

(b) work expended and

(c) coefficient of performance.

Proceed to calculate the capacity of the plant in tons of refrigeration for a mass flow rate of 100 kg/hr.

Solution : Refer Fig. 25.12 for the *p-v* and *T-s* diagram for the cycle.

Given : $p_a = p_b = 1$ bar

$p_c = p_d = 5$ bar

$T_b = -3 + 273 = 270$ K

and $T_d = 17 + 273 = 290$ K

The compression process *b-c* follows the law $pv^{1.4}$ = constant

$$\frac{T_c}{T_b} = \left(\frac{p_c}{p_b}\right)^{\frac{\gamma-1}{\gamma}} = \left(\frac{5}{1}\right)^{\frac{1.4-1}{1.4}} = 1.584$$

$$\therefore \quad T_c = 1.584\, T_b = 1.584 \times 270 = 427.68 \text{ K}$$

The expansion process *d-a* follows the law $pv^{1.25}$ = constant.

$$\frac{T_a}{T_d} = \left(\frac{p_a}{p_d}\right)^{\frac{n-1}{n}} = \left(\frac{1}{5}\right)^{\frac{1.25-1}{1.25}} = 0.725$$

$$\therefore \quad T_a = 0.725\, T_d$$
$$= 0.725 \times 290 = 210.25 \text{ K}$$

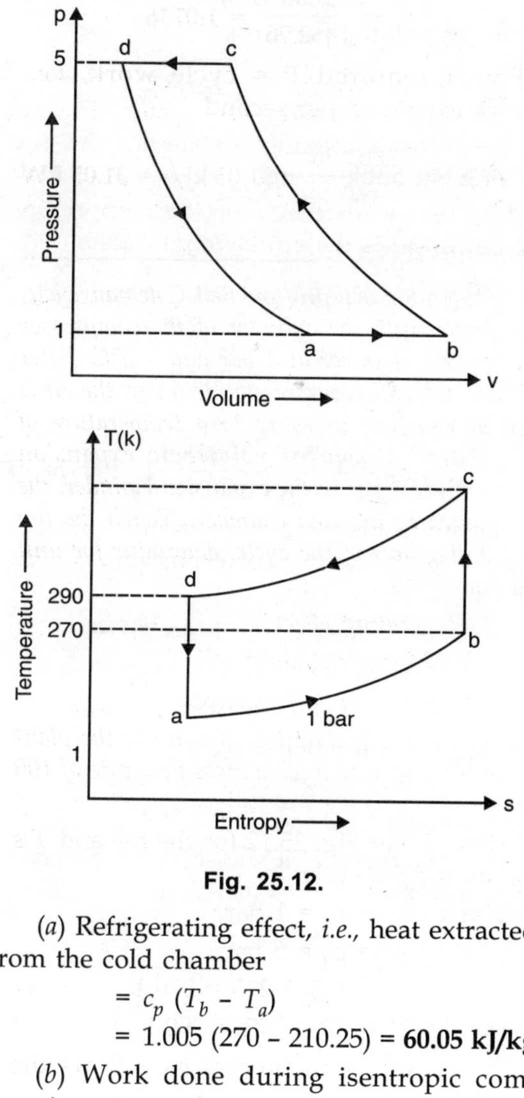

Fig. 25.12.

(*a*) Refrigerating effect, *i.e.*, heat extracted from the cold chamber

$$= c_p\,(T_b - T_a)$$
$$= 1.005\,(270 - 210.25) = \mathbf{60.05\ kJ/kg}$$

(*b*) Work done during isentropic compression process *b-c*

$$W_{comp} = \frac{\gamma}{\gamma - 1} \times R\,(T_c - T_b)$$

$$= \frac{1.4}{1.4-1} \times 287\,(427.68 - 270)$$

$$= 158390 \text{ J/kg} = 158.39 \text{ kJ/kg}$$

Work done during polytropic expansion process *d-a*

$$W_{exp} = \frac{n}{n-1} R\,(T_d - T_a)$$

$$= \frac{1.25}{1.25-1} \times 287\,(290 - 210.25)$$

$$= 114440 \text{ J/kg} = 114.44 \text{ kJ/kg.}$$

$\therefore$ Net work expended $= W_{comp} - W_{exp}$

$= 158.39 - 114.44 = \mathbf{43.95\ kJ/kg}$

(*c*) Coefficient of performance,

$$COP = \frac{\text{refrigerating effect}}{\text{network expended}}$$

$$= \frac{60.05}{43.95} = \mathbf{1.366}$$

(*d*) Refrigerating effect per second

$$= 60.05 \times \frac{100}{3600} = 1.668 \text{ kJ/s}$$

1 ton of refrigeration = 3.5 kJ/s

$$\therefore \text{ Capacity of plant} = \frac{1.668}{3.5} = \mathbf{0.476\ tons}$$

EXAMPLE 25.14

An air refrigerating machine of 10 ton capacity operating on Bell-Coleman cycle has an upper pressure limit of 4 bar. The pressure and temperature at the start of compression are 1 bar and – 20°C respectively. The compressed air cooled at constant pressure to a temperature of 40°C enters the expansion cylinder. Assuming both compression and expansion processes to be adiabatic with $\gamma = 1.4$, *make calculations for :*

(a) coefficient of performance,

(b) quantity of air in circulation per minute,

(c) piston displacement of compressor and expander,

(d) bore of compressor and expansion cylinders,

(e) power required to drive the unit.

The unit turns 200 revolutions per minute, is double acting and the stroke length is 200 mm.

Solution : Refer Fig. 25.13 for the *p-v* plot of the operating cycle.

Given : $p_4 = p_1 = 1$ bar; $p_2 = p_3 = 4$ bar

$T_1 = -\,20°\text{C} = 253$ K

and $T_3 = 40°\text{C} = 313$ K

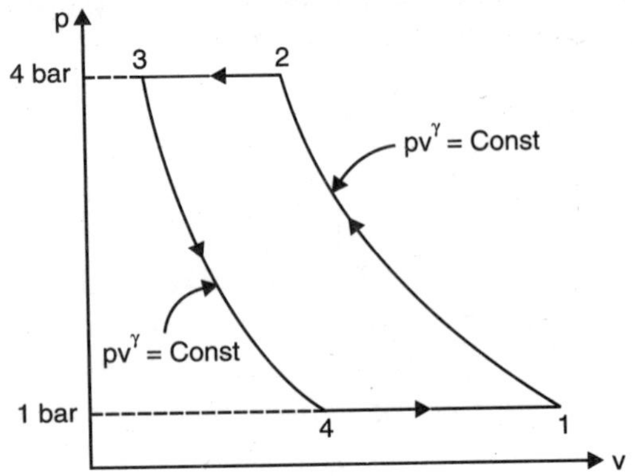

Fig. 25.13.

For the adiabatic compression process 1-2

$$\frac{T_2}{T_1} = \left(\frac{p_2}{p_1}\right)^{\frac{\gamma-1}{\gamma}} = \left(\frac{4}{1}\right)^{\frac{1.4-1}{1.4}} = 1.486$$

$\therefore \quad T_2 = 1.486 \times T_1$

$= 1.486 \times 253 = 376 \text{ K}$

For the adiabatic expansion process 3-4

$$\frac{T_4}{T_3} = \left(\frac{p_4}{p_3}\right)^{\frac{\gamma-1}{\gamma}} = \left(\frac{1}{4}\right)^{\frac{1.4-1}{1.4}} = \frac{1}{1.486}$$

$$\therefore \quad T_4 = \frac{T_3}{1.486} = \frac{313}{1.486} = 210.7 \text{ K}$$

(*a*) Since both the compression and expansion processes are isentropic

$$COP = \frac{T_4}{T_3 - T_4} = \frac{210.7}{313 - 210.7} = \mathbf{2.06}$$

(*b*) Refrigerating effect, *i.e.*, heat extracted from the cold chamber per kg of air

$= c_p (T_1 - T_4)$

$= 1.005 (253 - 210.7)$

$= 42.51$ kJ/kg

If m is the mass of air circulated per minute, then

Refrigerating effect = 42.51 m kJ/min

The absorbing capacity of the system is stated to be 10 ton or 10 × 210 kJ/min and therefore

$42.51\, m = 10 \times 210$

or $$m = \frac{10 \times 210}{42.51} = 49.4 \text{ kJ/min}$$

(*c*) and (*d*) Piston displacement of compressor is given by the volume corresponding to state point 1.

$$V_1 = \frac{m R T_1}{p_1} = \frac{49.4 \times 287 \times 253}{1 \times 10^5}$$

$= \mathbf{35.87\ m^3/min}$

Swept volume per stroke

$$= \frac{35.87}{2 \times 200} = 0.0897 \text{ m}^3$$

The factor 2 accounts for the fact that the machine is double acting.

If d_c is the diameter of the compressor cylinder and l is the stroke length, then

$$\frac{\pi}{4} d_c^2\, l = 0.0897$$

or $$\frac{\pi}{4} d_c^2 \times \frac{200}{1000} = 0.0897$$

$$\therefore \quad d_c = \left(\frac{0.0897 \times 1000 \times 4}{\pi \times 200}\right)^{\frac{1}{2}} = \mathbf{0.756\ m}$$

Piston displacement of expansion cylinder is given by the volume corresponding to state point 4.

$$V_4 = \frac{m R T_4}{p_4} = \frac{49.4 \times 287 \times 210.7}{1 \times 10^5}$$

$= \mathbf{29.87\ m^3/min}$

Swept volume per stroke

$$= \frac{29.87}{2 \times 200} = 0.0747 \text{ m}^3$$

If d_e is the diameter of the expansion cylinder, then

$$\frac{\pi}{4} d_e^2\, l = 0.0747$$

or $$\frac{\pi}{4} d_e^2 \times \frac{200}{1000} = 0.0747$$

$$\therefore \quad d_e = \left(\frac{0.0747 \times 1000 \times 4}{\pi \times 200}\right)^{\frac{1}{2}} = \mathbf{0.69\ m}$$

(*e*) Work done $= \dfrac{\text{refrigerating effect}}{\text{coefficient of performance}}$

$$= \frac{2100}{2.06} = 1019.42 \text{ kJ/min}$$

$= 16.99$ kJ/s

∴ Power required to drive the unit = **16.99 kW**

25.4. VAPOUR COMPRESSION REFRIGERATION

Vapour compression refrigeration is the most important and practical form of refrigeration system for domestic and commercial utility. The working fluids (refrigerants) used are NH_3, CO_2, Freons etc., and these fluids have the characteristic ability to change alternately between the vapour and liquid phases without leaving the plant. The alternate condensation and evaporation take place at temperature and pressure close to atmospheric conditions. During evaporation, the liquid refrigerant absorbs its latent heat from the space/product to be cooled and changes from the liquid to gaseous state. The vapour is next compressed (raised in temperature and pressure) and subsequently condensed. While condensing, the vapour liquefies by rejecting heat to an external system (typically available cooling water or cooling air). As such, this refrigeration system may be likened to a latent heat pump that pumps heat from the space being refrigerated to either water or air in the condenser. The system is called vapour compression as it is the compression of refrigerant that permits pumping of heat.

25.4.1. Carnot vapour-compression refrigeration cycle

Fig. 25.14 shows the schematic representation and *T-s* diagram of a Carnot vapour-compression refrigeration cycle.

The sequence of operation of different processes constituting the cycle is :

1-2 : The refrigerant as a mixture of liquid and vapour corresponding to state point 1 enters the compressor where isentropic compression takes place. The compression process increases the temperature of refrigerant from lower limit T_2 to the upper limit T_1. Work is supplied to the system and after compression, the vapour is wet or saturated but not superheated.

2-3 : The refrigerant in the form of vapour enters the condenser at state 2 and heat is rejected at constant pressure and temperature. At exit from the condenser, the refrigerant becomes saturated liquid at state point 3.

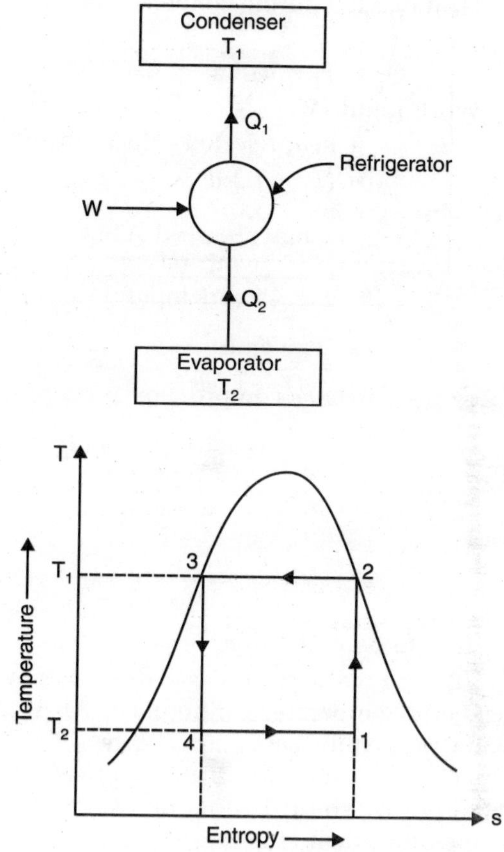

Fig. 25.14. Carnot vapour-compression refrigeration cycle

3-4 : The refrigerant at state point 3 enters the expansion cylinder, expands isentropically and its temperature drops to lower temperature T_2 at the end of the expansion process. Work is obtained during the expansion process.

4-1 : The liquid refrigerant at point 4 enters the evaporator and extracts heat at constant pressure and temperature from the space or substance being cooled and thus produces refrigerating effect.

The cycle consists of two isothermals and two adiabatics and obviously is a reversible cycle.

From the temperature-entropy plot,

Heat absorbed (refrigerating effect) during

evaporation process (4 – 1)

$$Q_2 = T_2 \times ds$$

Heat rejected during condensation process (2 – 3)

$$Q_1 = T_1 \times ds$$

Work input W

$$= \text{heat rejected} - \text{heat absorbed}$$
$$= (T_1 - T_2)\, ds$$

$$\therefore\ (COP)_{ref} = \frac{\text{heat absorbed at low temperature}}{\text{work input}}$$

$$= \frac{T_2\, ds}{(T_1 - T_2)ds} = \frac{T_2}{T_1 - T_2}$$

$$(COP)_{heat\ pump} = \frac{\text{heat rejected at high temperature}}{\text{work input}}$$

$$= \frac{T_1\, ds}{(T_1 - T_2)ds} = \frac{T_1}{T_1 - T_2} \quad ...(25.2)$$

It is to be noted that:

(*i*) The values of *COP* decrease with increased temperature difference. Also *COP* tends to infinity as T_2 approaches T_1. This aspect suggests that the temperature of refrigerant should not be at a lower temperature than is necessary.

(*ii*) A Carnot refrigeration cycle consists of all reversible processes. As such no refrigerator can have a *COP* greater than that of a Carnot refrigerator when operating between the same temperature limits. As such a Carnot refrigerator is considered as a standard of perfection against which the performance of a practical refrigerator can be compared.

(*iii*) A Carnot refrigerator has the practical difficulty due to isentropic expansion of saturated liquid to a low pressure. At low pressure, the liquid vaporises and the expander has to deal with a two phase mixture of relatively low quality.

EXAMPLE 25.15

A Carnot refrigerator has working temperatures of – 15°C and 40°C. If it operates with F-12 as working substance, calculate :

(a) the work of isentropic compression and expansion,

(b) refrigerating effect and heat rejected per kg of the refrigerant, and

(c) coefficient of performance of the cycle.

Use the following data in respect of properties of F-12

Tempe-rature	*Enthalpy*	*Entropy*	*(kJ/kg)*	*(kJ/kg K)*
°C	*Liquid*	*Vapour*	*Liquid*	*Vapour*
– 15	186.28	344.92	0.9487	1.5632
40	238.53	367.15	1.1298	1.5405

Solution : Refer Fig. 25.15 for the Carnot cycle represented on *T-s* plane.

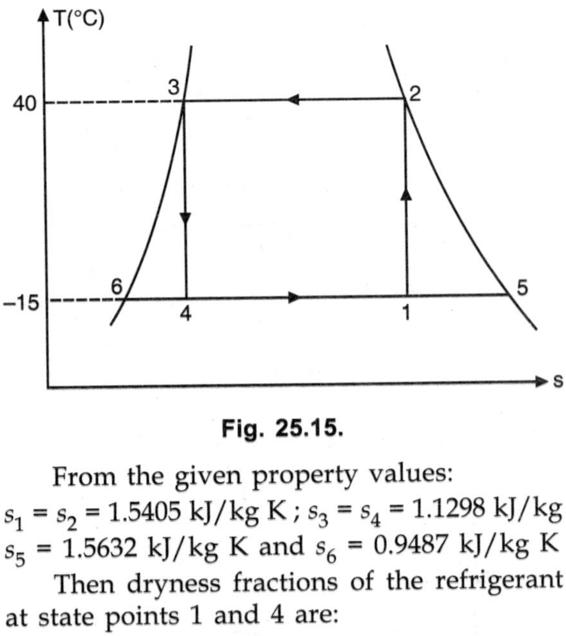

Fig. 25.15.

From the given property values:

$s_1 = s_2 = 1.5405$ kJ/kg K ; $s_3 = s_4 = 1.1298$ kJ/kg

$s_5 = 1.5632$ kJ/kg K and $s_6 = 0.9487$ kJ/kg K

Then dryness fractions of the refrigerant at state points 1 and 4 are:

$$x_1 = \frac{s_1 - s_6}{s_5 - s_6}$$

$$= \frac{1.5405 - 0.9487}{1.5632 - 0.9487} = \frac{0.5918}{0.6145} = 0.963$$

$$x_4 = \frac{s_4 - s_6}{s_5 - s_6} = \frac{1.1298 - 0.9487}{1.5632 - 0.9487}$$

$$= \frac{0.1811}{0.6145} = 0.295$$

The corresponding enthalpies,

$h = h_f + x\, h_{fg}$, at these state points are :

$h_1 = 186.28 + 0.963\ (344.92 - 186.28)$

$= 339.05$ kJ/kg

$h_4 = 186.28 + 0.295\ (344.92 - 186.28)$

$= 233.08$ kJ/kg

(*a*) Work of compression

$W_c = h_2 - h_1 = 367.15 - 339.05$

$=$ **28.1** kJ/kg

Work of expansion

$W_e = h_3 - h_4$

$= 238.53 - 233.08 =$ **5.45** kJ/kg

(*b*) Refrigerating effect

$N = h_1 - h_4$

$= 339.05 - 233.08 =$ **105.97** kJ/kg

Heat rejected $= h_2 - h_3$

$= 367.15 - 238.53 =$ **128.62** kJ/kg

(*c*) Net work done $= W_c - W_e$

$= 28.1 - 5.45 = 22.65$ kJ/kg

$$COP = \frac{\text{refrigerant effect}}{\text{net work done}} = \frac{105.97}{22.65} = \mathbf{4.68}$$

25.4.2. Vapour compression refrigeration cycle

Fig. 25.16 shows the schematic arrangement and *T-s* diagram of a simple vapour-compression refrigeration cycle.

The system comprises a number of flow processes carried out in separate components namely compressor, condenser, expansion valve or capillary tube and evaporator. The working medium is always a liquid gas, *i.e.*, CO_2, NH_3 and Freon etc. Liquid gas is a fluid which at normal temperature may exist either as a liquid or as a gas depending upon pressure. The refrigerant is not allowed to leave the system but is circulated again and again, alternately condensing and evaporating.

The sequence of operation of the cycle is :

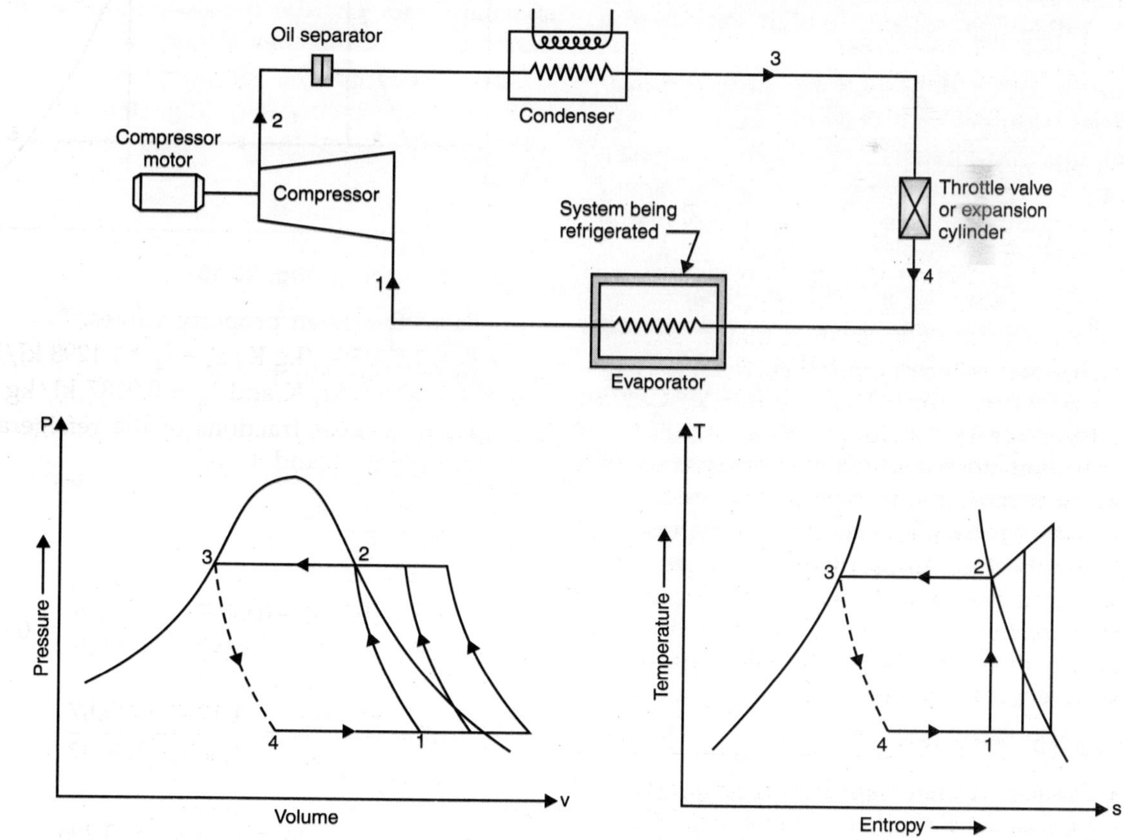

Fig. 25.16. Simple vapour compression cycle

1. Reversible adiabatic compression (1-2) : The vapour at a low pressure p_1, temperature T_1 and preferably dry (state 1) is drawn from the evaporator during suction stroke of the compressor. During compression, the vapour is compressed isentropically to a higher pressure p_2 and temperature T_2. At compressor exit, the vapour may be wet, dry or superheated. In addition to raising the pressure of vapour, the compressor creates the pressure difference between the evaporator and condenser and that maintains a continuous flow of refrigerant through the system.

The compressor is of rotary type for domestic refrigerators and of reciprocating type for large plants and machines, *viz.*, ice plants and cold storages. For plants with capacity over 50 tons, centrifugal compressors are used and these are directly coupled with high speed gas turbines or electrical controls.

2. Constant pressure condensation (2-3) : The vapour refrigerant at high temperature and pressure (state 2) coming from the compressor is delivered to the condenser where its condensation takes place at constant pressure. The condenser first absorbs the heat of superheat and the temperature of the vapour falls to saturation temperature. Subsequently the refrigerant loses its latent heat and is changed into a high pressure liquid (state 3).

For small machines (*i.e.* domestic refrigerators), the cooling medium is atmosphere and for larger units it may be a stream of water.

3. Throttling (3-4) : The expansion of the high pressure liquid refrigerant is done by the throttling process. During throttling pressure drops but the enthalpy remains constant. Due to drop in pressure, the liquid starts boiling and the required heat is provided by the refrigerant itself. Consequently the temperature falls and at throttle exit, the refrigerant is a mixture of liquid and vapour (very wet vapour of 10 to 20% quality). Throttling is an irreversible process and has been shown by dotted line on *T-s* plot.

The throttling device may be a capillary tube, a thermostatic expansion valve, an automatic expansion valve and float valve etc.

4. Constant pressure evaporation (4-1) : The wet vapour after throttling passes through evaporator coils placed in the medium (usually brine) which is to be cooled. Since the temperature of the vapour is below that of brine, it absorbs latent heat from the brine and gets evaporated at constant pressure. The vaporisation continues upto state point 1 and this stage of the refrigerant depends upon the heat absorbed. The refrigerant may remain wet, become dry saturated or get slightly superheated.

To complete the refrigerant cycle, the refrigerant vapour from the evaporator is routed back into the compressor.

The components of the refrigeration system are connected by pipes such as :

– Suction line that conveys the low pressure vapour from the evaporator to the suction inlet of the compressor.

– Discharge line (or hot gas line) that delivers the high pressure, high temperature refrigerant from the compressor to the condenser.

– Liquid line that carries the liquid refrigerant from the receiver tank to the expansion/throttle valve.

The receiver tank installed between the condenser and expansion device provides storage for the condensed liquid so that a constant supply of liquid is assured to the evaporator as per requirement.

COP from *T-s* diagram : Refer Fig. 25.17, which depicts the *T-s* diagram for an ideal vapour compression cycle.

The following assumptions are made with reference to this diagram:

(*i*) compression is isentropic;

(*ii*) there is no undercooling of refrigerant, *i.e.*, the refrigerant after leaving the condenser is a saturated liquid;

(*iii*) there is no change of enthalpy during throttling process;

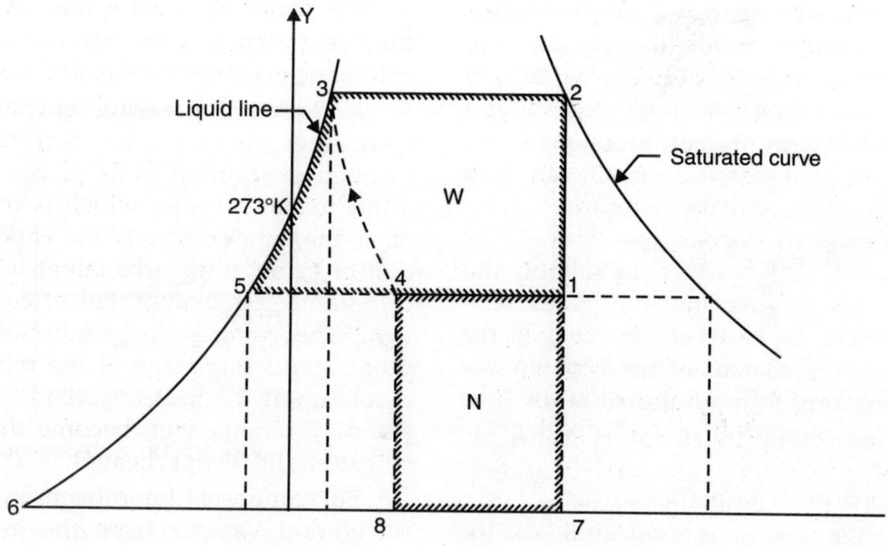

Fig. 25.17. COP from T-s diagram

(*iv*) no leakage of refrigerant;

(*v*) no pressure loss due to friction and heat exchange in the pipelines.

Work input during compression, W

= total heat at point 2 − total heat at point 1

= (area 7-2-3-5-6) − (area 7-1-5-6)

= area 1-2-3-5

Thus the compressor work is represented by the area 1-2-3-5.

Refrigerating effect (heat extracted by the refrigerant), N

= heat at point 1 − heat at point 4

= (area 7-1-5-6) − area (8-4-5-6)

= area 7-1-4-8

Thus the refrigerating effect is shown by the area 7-1-4-8. Knowing refrigerating effect N and work input to compressor W, the coefficient of performance can be determined.

$$COP = \frac{N}{W} = \frac{\text{area } 1\text{-}2\text{-}3\text{-}5}{\text{area } 7\text{-}1\text{-}4\text{-}8} = \frac{h_2 - h_1}{h_1 - h_4}$$

Wet and dry compression : For maximum refrigerating effect, the abstraction of heat in the evaporator should be continued until the vapour is dry saturated or slightly superheated. If the vapour is perfectly dry and saturated at the beginning of compression, the compression is called ***dry compression.*** In dry compression, vapour throughout the compression process lies in the superheated region. In ***wet compression,*** the vapour is wet or just dry at the end of compression process. In wet compression, the vapour remains wet throughout the compression process. Fig. 25.18 shows *T-s* diagrams of wet and dry compression processes.

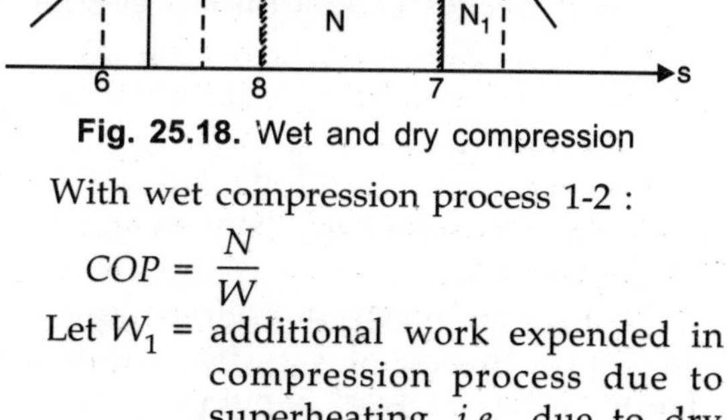

Fig. 25.18. Wet and dry compression

With wet compression process 1-2 :

$$COP = \frac{N}{W}$$

Let W_1 = additional work expended in compression process due to superheating, *i.e.*, due to dry compression and N_1 = additional refrigerating effect as heat extracted from the medium being refrigerated with dry compression.

Then, COP with dry compression

$$= \frac{N+N_1}{W+W_1} = \frac{N}{W}\left[\frac{1+\frac{N_1}{N}}{1+\frac{W_1}{W}}\right]$$

The ratio of W_1 to W is greater than the ratio of N_1 to N. The coefficient of performance is thus lowered with dry compression and thus the amount of superheat needs to be kept to minimum.

It is worthwhile to point out that :

(*i*) Wet compression is not desirable as the liquid droplets in the refrigerant would enter the compressor and damage the valves and other moving parts. Further, the liquid droplets would gather the lubricating oil from the cylinder walls of a reciprocating compressor. The oil would then pass to the condenser and adversely affect the heat transfer rate from the refrigerant to the condenser cooling water.

(*ii*) Dry compression is safeguard to the compressor cylinder and is almost universally employed. Volumetric efficiency is also high if dry compression is used. During suction stroke with wet compression, there occurs evaporation of liquid particles and that reduces the volumetric efficiency.

Sub-cooling : Sometimes the condensed liquid refrigerant is sub-cooled (under-cooled) at constant pressure before entering the throttle valve. Sub-cooling means reducing the temperature of the liquid below saturation temperature corresponding to condenser pressure.

With reference to Fig. 25.19, the process 3-3′ represents under-cooling of the liquid refrigerant before throttling operation. For most of the refrigerants, under-cooling follows the liquid line.

With under-cooling, the refrigerating effect has been increased by the amount N'. Theoretically the work input to compressor remains unchanged and accordingly under-cooling tends to increase COP and improve the performance of the unit.

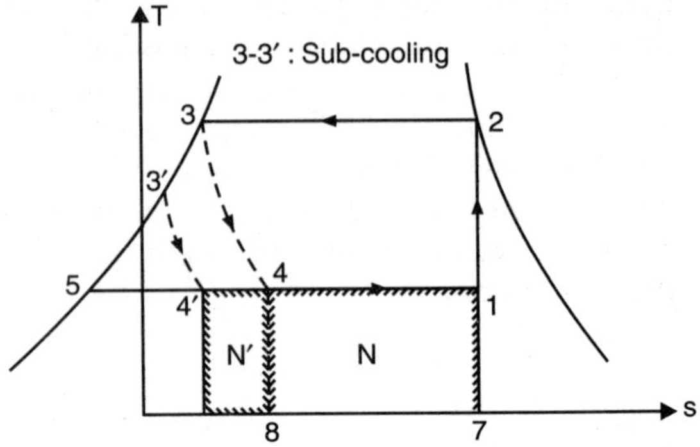

Fig. 25.19. Sub-cooling

Sub-cooling is achieved by

(*i*) circulating greater quantity of cooling water in the condenser,

(*ii*) using low temperature water as sink for rejection of heat from the refrigerant,

(*iii*) incorporating a separate sub-cooler

It is worthwhile to point out that in addition to compressor work, under-cooling needs some extra energy to obtain the extra coolant. There is also some increase in financial aspects due to separate sub-cooler.

Throttle valve *versus* expansion cylinder

The throttling process is highly irreversible and that makes the vapour compression cycle irreversible. Further, no work is obtained during throttling and the entire amount of work required for compression has to be provided by an external source. If the refrigerant after condensation process is expanded adiabatically in an expansion cylinder, the work is obtained and that can be fed to the compressing unit. That reduces the external mechanical work required in running the compressor. Also, the end state 4′ with expansion cylinder is at lower enthalpy when compared with enthalpy corresponding to end state 4 achieved with throttle valve. This aspect increases the refrigerating effect. The net result is that with expansion cylinder, the coefficient of performance of the machine can be improved.

The expansion cylinder system is, however,

mechanically more complicated and involves greater first cost which is not justified for small gains in cooling capacity. Moreover, with throttle valve the flow of refrigerant can be controlled. The expansion cylinder has fixed cylinder volume which cannot accommodate varying flow. Modern vapour compression plants prefer throttle valve to simplify the design of installation and to improve the process of control.

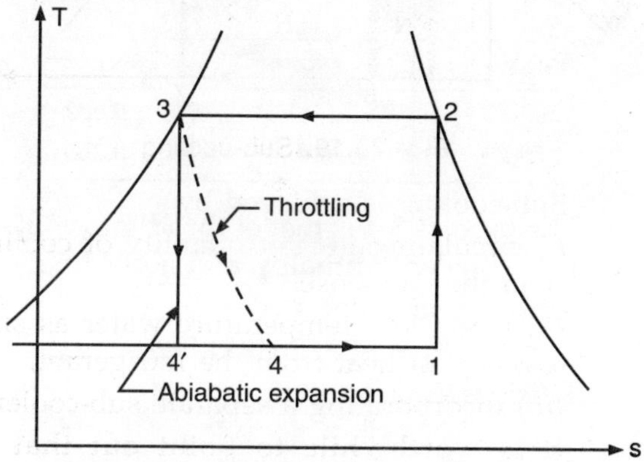

Fig. 25.20. Throttle valve versus expansion cylinder

Air-refrigeration system *versus* **vapour compression system**

Compared to the air-refrigeration system, the vapour-compression system has the following merits and demerits :

(1) high *COP*,

(2) small size for the given refrigerating effect,

(3) less running cost (only 1/5th that of air-refrigeration system) when operated on ground level,

(4) simple adjustment of the throttle valve helps to achieve required temperature at the evaporator,

(5) can be employed over large range of temperatures,

(6) initial cost is high,

(7) suffers from problems associated with toxicity, inflammability and leakage of the refrigerant.

Reverse Carnot cycle *versus* **vapour compression refrigeration**

A brief compression of vapour compression cycle with the reverse Carnot cycle is given in Table 25.1.

Applications : Vapour compression refrigeration has been and is the most widely used method for air-conditioning large public buildings, private residences, hotels, hospitals, theatres, restaurants and automobiles. It is also used in domestic and commercial refrigerators, large-scale warehouses for storage of foods and meats, refrigerated trucks and railroad cars, and a host of other commercial and industrial services.

Table 25.1.

Reverse Carnot cycle	*Vapour compression refrigeration*
(*i*) ideal cycle with highest *COP*	(*i*) practical cycle with lower *COP*
(*ii*) independent of working substance	(*ii*) performance depends on the type of vapour used
(*iii*) heat absorption and heat rejection are at constant temperature	(*iii*) heat absorption and heat rejection are at constant pressure as well as at constant temperature
(*iv*) involves an expansion process which is isentropic (reversible adiabatic)	(*iv*) expansion is by throttling which is irreversible and at constant enthalpy
(*v*) there is no superheating and no under-cooling of the refrigerant	(*v*) the refrigerant may be superheated in the compressor and undercooled in the condenser so as to improve the efficiency of the system
(*vi*) canot be practically implemented	(*vi*) most popular and widely used refrigeration system for commercial, industrial and domestic services

EXAMPLE 25.16

A standard vapour compression refrigerator using F-12 as the refrigerant operates between the condenser pressure of 10 bar and the evaporator pressure of 1.5 bar. The evaporator absorbs 75 kJ/min of energy as heat and the vapour is dry saturated at exit from the compressor.

Represent the cycle on T-s plane and calculate :

(a) flow rate of refrigerant

(b) power consumed and

(c) COP of the cycle

Also calculate COP of the Carnot refrigerator operating between the same temperature limits.

The relevant properties of F-12 are given below :

Pressure bar	Saturation temperature, °C	Enthalpy kJ/kg		Entropy kJ/kg K	
		Liquid	Vapour	Liquid	Vapour
10	41.7	76.8	203.65		0.682
1.5	- 20.1	17.82	178.84	0.073	0.709

Solution : Refer Fig. 25.21 for *T-s* plot of the given cycle

Since the compression process 1-2 is isentropic,

entropy at point 1 = entropy at point 2

$$s_{f1} + x_1 s_{fg1} = s_{g2}$$

$$0.073 + x_1 (0.709 - 0.073) = 0.682$$

$$\text{or} \quad x_1 = \frac{0.682 - 0.073}{0.709 - 0.073} = 0.9575$$

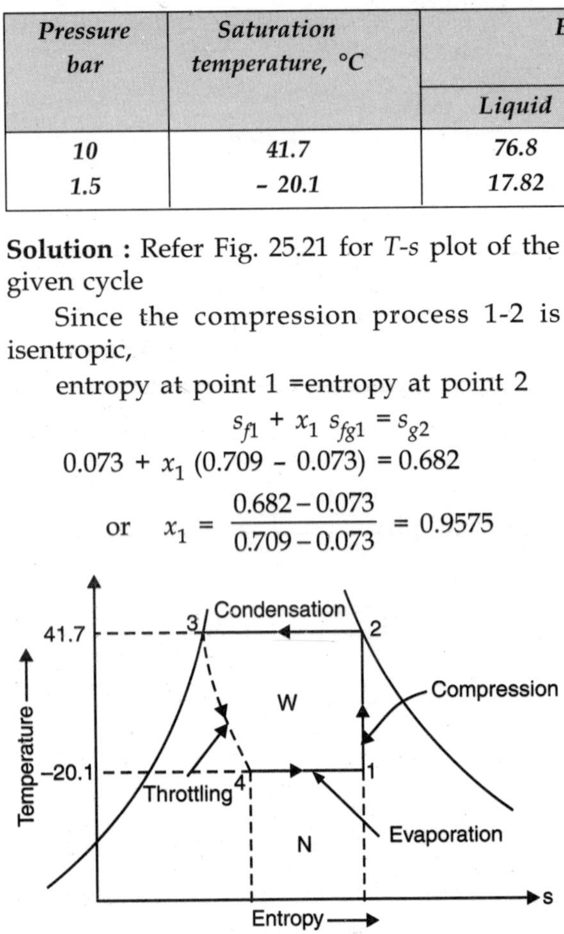

Fig. 25.21.

Enthalpy at point 1 :

$$h_1 = h_{f1} + x_1 h_{fg1}$$
$$= 17.82 + 0.9575 (178.84 - 17.82)$$
$$= 172 \text{ kJ/kg}$$

Enthalpy at point 2 :

$$h_2 = h_{g2} = 203.65 \text{ kJ/kg}$$

Enthalpy at point 3 :

$$h_3 = h_{f2} = 76.8 \text{ kJ/kg}$$

The process 3-4 is that of throttling, it takes place at constant enthalpy and accordingly

Enthalpy at point 4 :

$$h_4 = h_3 = 76.8 \text{ kJ/kg}$$

Refrigerating effect

$$N = h_1 - h_4$$
$$= 172 - 76.8 = 95.2 \text{ kJ/kg}$$

Work done

$$W = h_2 - h_1$$
$$= 203.65 - 172 = 31.65 \text{ kJ/kg}$$

(*a*) If *m* is the mass flow of refrigerant/min, then energy balance in the evaporator gives

$m \times 95.2 = 75$; m = **0.788 kg/min**

(*b*) Power consumed

$$= \text{work done per second}$$
$$= 31.65 \times \frac{0.788}{60}$$
$$= 0.416 \text{ kJ/s} = \mathbf{0.416 \text{ kW}}$$

(*c*) Coefficient of performance of the cycle,

$$COP = \frac{\text{refrigerating effect}}{\text{work done}}$$
$$= \frac{95.2}{31.65} = \mathbf{3.0}$$

The coefficient of performance of a refrigerator working on reversed Carnot cycle is given by

$$COP = \frac{T_2}{T_1 - T_2}$$

where T_2 = temperature in the evaporator
= 273 + (– 20.1) = 252.9 K

T_1 = temperature in the condenser
= 273 + 41.7 = 314.7 K

$$\therefore \quad COP = \frac{252.9}{314.7 - 252.9} = \mathbf{4.09}$$

EXAMPLE 25.17

A refrigerating plant running on CO_2 operates between saturation temperatures of 268 K (in the evaporator) and 298 K (in the condenser coils). The vapour has a dryness fraction of 0.6 at entry to the compressor, isentropic compression takes place in the compressor and there is no under-cooling of the working fluid in the condenser. The machine has a relative efficiency of 60% and 5.5 kg of CO_2 is circulated per minute.

Calculate the amount of ice produced per day at 0°C from water at 15°C.

Take latent heat of ice as 335 kJ/kg and use the following table for the properties of CO_2.

Temperature °C	*Liquid heat* kJ/kg	*Latent heat* kJ/kg	*Entropy of liquid* kJ/kg K
268	– 7.53	245.8	– 0.04187
298	81.25	121.5	0.2513

Solution : Refer Fig. 25.22 for *T-s* plot of the given cycle. It has been presumed that the refrigerant remains wet even after compression.

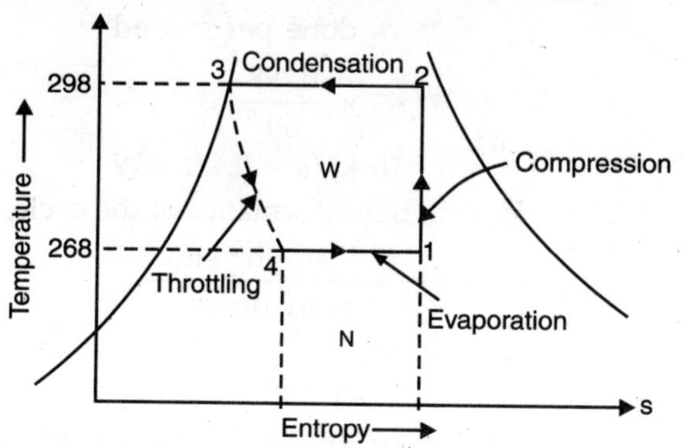

Fig. 25.22.

Since the compression process 1-2 is adiabatic,

entropy at point 1 = entropy at point 2

$$s_{f1} + x_1 \frac{h_{fg1}}{T_1} = s_{f2} + x_2 \frac{h_{fg2}}{T_1}$$

or $\quad - 0.04187 + 0.6 \times \dfrac{245.8}{268} = 0.2513 + x_2 \times \dfrac{121.5}{298}$

or $\quad - 0.04187 + 0.5503 = 0.2513 + 0.4077\, x_2$

∴ Dryness fraction of vapour at exit from the compressor,

$$x_2 = \frac{-0.04187 + 0.5503 - 0.2513}{0.4077} = 0.631$$

Total heat at point 1 :

$h_1 = h_{f1} + x_1 h_{fg1}$
= – 7.53 + 0.6 × 245.8 = 139.95 kJ/kg

Total heat at point 2 :

$h_2 = h_{f2} + x_2 h_{fg2}$
= 81.25 + 0.631 × 121.5 = 157.92 kJ/kg

Total heat at point 3 :

h_3 = liquid heat corresponding to condenser temperature of 298 K
= 81.25 kJ/kg

The expansion process 3-4 is of throttling during which enthalpy remains constant. That is

$h_4 = h_3$ = 81.25 kJ/kg

Work done, $W = h_2 - h_1$
= 157.92 – 139.95 = 17.97 kJ/kg

Refrigerating effect, $N = h_1 - h_4$
= 139.95 – 81.25 = 58.7 kJ/kg

$$COP = \frac{N}{W} = \frac{58.7}{17.97} = 3.266$$

The ratio of actual *COP* to the theoretical *COP* is known as relative coefficient of performance.

∴ Actual *COP* = 0.6 × 3.266 = 1.96

With 5.5 kg of CO_2 circulated per minute,

$$\text{Work done} = \frac{5.5}{60} \times 17.97 = 1.647 \text{ kJ/s}$$

∴ Actual refrigerating or cooling effect
= 1.647 × 1.96 = 3.228 kJ/s

Heat to be extracted during one day of 24 hours

$= 24 \times 3600 \times 3.228$

$= 278899$ kJ

Heat extracted from 1 kg of water at 15°C for formation of 1 kg of ice at 0°C

$= 1 \times 4.187 \times (15 - 0) + 335$

$= 397.80$ kJ

$\therefore$ Mass of ice formed $= \dfrac{278899}{397.80} = \mathbf{701\ kg}$

EXAMPLE 25.18.

In an ammonia vapour refrigeration plant, the pressure range is from 3.15 bar in the evaporator to 10.5 bar in the condenser, the compression is isentropic and before entering the throttle valve the refrigerant ammonia is sub-cooled. The temperature of refrigerant at entry and exit from the condenser is 50°C and 20°C respectively and the water being circulated in the condenser at the rate of 10.5 kg/min has a temperature rise 10°C. The compressor unit is single cylinder, single acting with bore 10 cm, stroke 15 cm and while running at 200 rev/min has an indicated mean effective pressure equal to 3.5 bar.

If the plant forms 50 kg/hr of ice at 0°C from water at 15°C, determine :

(a) COP (b) flow rate of refrigerant and

(c) condition of vapour at entry to compressor.

Take latent heat of ice as 335 kJ/kg. The relevant properties of NH_3 are as given below :

Pressure bar	*Saturation temperature, K*	*Enthalpy kJ/kg*		*Specific heat kJ/kg K*	
		Liquid	*Vapour*	*Liquid*	*Vapour*
3.5	264	– 35.6	1264	–	–
10.5	304	134	1294	4.6	2.8

Solution : Refer Fig. 25.23 for *T-s* plot of the cycle. Since the temperature of refrigerant at compressor exit (condenser inlet) $T_2 = 273 + 50 = 323$ K is higher than the saturation temperature of 304 K corresponding to compressor exit pressure of 10.5 bar, the refrigerant at compressor exit would be superheated.

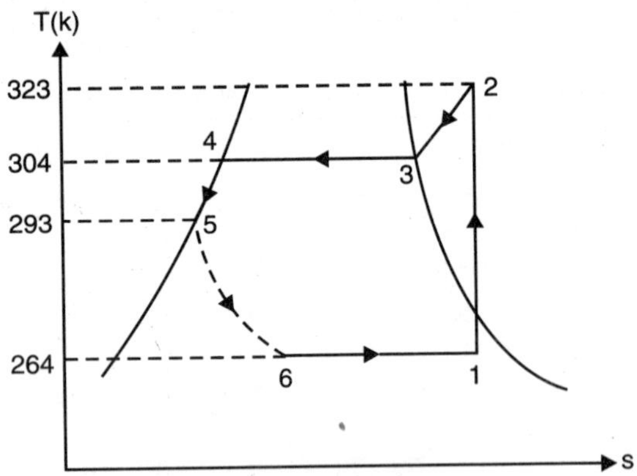

Fig. 25.23.

Compressor work $= p_m\ LAN$

$$= \frac{(3.5\times10^5)\times0.15\times\frac{\pi}{4}(0.1)^2\times200}{10^3}$$

$= \mathbf{82.42\ kJ/min}$

Heat extracted for ice formation (refrigerating effect)

$= (4.186 \times 15) + 335$

$= 397.8$ kJ/kg of ice formed

$= 397.8 \times \dfrac{50}{60} = 331.5$ kJ/min

(*a*) Coefficient of performance, *COP*

$$= \frac{\text{refrigerating effect}}{\text{work input}}$$

$$= \frac{331.5}{82.42} = \mathbf{4.02}$$

(*b*) Let *m* be the mass flow of refrigerant per minute. Then energy balance in the condenser gives :

heat lost by refrigerant

= heat gained by condenser water

$m\ (h_2 - h_5) = m_w\ c_{pw}\ \Delta t$

From the given data :

$$h_4 - h_5 = (c_p)_l \times (T_4 - T_5)$$
$$= 4.6(304 - 293) = 50.6 \text{ kJ/kg}$$
$$h_2 - h_3 = (c_p)_v \times (T_2 - T_3)$$
$$= 2.8(323 - 304) = 52.2 \text{ kJ/kg}$$
$$h_3 - h_4 = 1294 - 134 = 1160 \text{ kJ/kg}$$

Upon adding :

$$h_2 - h_5 = 50.6 + 53.2 + 1160$$
$$= 1263.8 \text{ kJ/kg}$$

∴ Mass flow rate of refrigerant m

$$= \frac{m_w cp_w \Delta t}{h_2 - h_5} = \frac{10.5 \times 4.186 \times 10}{1263.8}$$
$$= \mathbf{0.348 \text{ kg/min}}$$

(c) Refrigerating effect $(h_1 - h_6)$

$$= 331.5 \text{ kJ/min} = \frac{331.5}{0.348}$$
$$= 952.6 \text{ kJ/kg of refrigerant}$$

During throttling operation 5-6, the enthalpy remains constant

$$\therefore \quad h_6 = h_5 = h_4 - (h_4 - h_5)$$
$$= 134 - 50.6 = 83.4 \text{ kJ/kg}$$

That gives :

$$h_1 = 952.6 + 83.4 = 1036 \text{ kJ/kg}$$

Further $h_1 = h_{f1} + x_1 h_{fg1}$

$$= -35.6 + x_1 [1264 - (-35.6)]$$
$$= -35.6 + 1229.6 x_1$$
$$\therefore \quad 1036 = -35.6 + 1229.6 x_1$$
$$x_1 = \frac{1036 + 35.6}{1229.6} = \mathbf{0.87}$$

Thus the refrigerant at entry to the compressor has a dryness fraction of 0.87.

EXAMPLE 25.19.

A food storage locker requires a refrigeration system of 50 kW capacity at an evaporator temperature of – 10°C and a condenser temperature of 35°C. The refrigerant used is ammonia which is subcooled by 5°C before entering the expansion valve and the vapour is dry saturated before leaving the evaporator. The two cylinder compressor with stroke equal to 1.2 times the bore operates at 1000 rev/min. Determine :

(a) the coefficient of performance,

(b) the mass of refrigerant to be circulated per minute,

(c) the power required,

(d) the heat removed through condenser, and

(e) the cylinder dimensions.

The volumetric efficiency of the compressor is 80 percent. You may use the extract of ammonia property table given below :

Saturation temperature°C	*Pressure bar*	*Enthalpy kJ/kg*		*Entropy kJ/kg K*		*Specific volume m³/kg*		*Specific heat kJ/kg K*	
		Liquid	*Vapour*	*Liquid*	*Vapour*	*Liquid*	*Vapour*	*Liquid*	*Vapour*
– 10	*2.91*	*154*	*1450*	*0.83*	*5.75*	*–*	*0.42*	*–*	*2.49*
35	*13.5*	*366*	*1489*	*1.56*	*5.21*	*1.7*	*0.096*	*4.56*	*2.90*

Solution : Refer Fig. 25.24, for *T-s* plot of the given cycle.

$$T_1 = T_6 = -10 + 273 = 263 \text{ K}$$
$$T_3 = T_4 = 35 + 273 = 308 \text{ K}$$
$$T_5 = 308 - 5 = 303 \text{ K}$$
$$h_1 = 1450 \text{ kJ/kg}$$
$$h_3 = 1489 \text{ kJ/kg}$$
$$h_4 = 366 \text{ kJ/kg}$$
$$h_5 (= h_6) = h_4 - c_{pl} \times (T_4 - T_5)$$
$$= 366 - 4.56 (308 - 303)$$
$$= 366 - 22.8 = 343.2 \text{ kJ/kg}$$

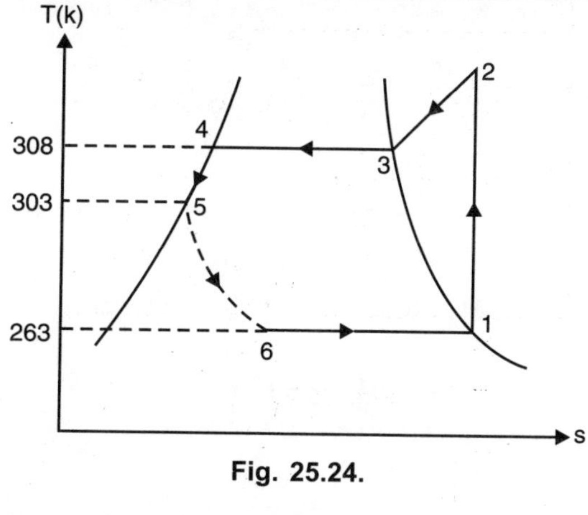

Fig. 25.24.

Since the compression process 1-2 is isentropic, we have $s_1 = s_2$

$$\text{or } 5.75 = s_3 + c_p \log_e \left(\frac{T_2}{T_3}\right)$$

$$= 5.21 + 2.9 \log_e \frac{T_2}{308}$$

$$\text{or } \log_e \left(\frac{T_2}{308}\right) = \frac{5.75 - 5.21}{2.9} = 0.1862$$

$$\text{or } \frac{T_2}{308} = e^{0.1862} = 1.2047$$

∴ Temperature at exit from the compressor, $T_2 = 308 \times 1.2047 = 371$ K

Now, $h_2 = h_3 + c_{pv} \times (T_2 - T_3)$
$= 1489 + 2.90 (371 - 308)$
$= 1671.7$ kJ/kg

(*a*) Refrigerating effect,
$N = h_1 - h_6 = 1450 - 343.2$
$= 1106.8$ kJ/kg

Work input,
$W = h_2 - h_1 = 1671.7 - 1450$
$= 221.7$ kJ/kg

Coefficient of performance,

$$COP = \frac{N}{W} = \frac{1106.8}{221.7} = \mathbf{4.99}$$

(*b*) Mass of refrigerant m

$$= \frac{\text{total refrigeration effect}}{\text{refrigeration effect per kg of refrigerant}}$$

$$= \frac{50}{1106.8} = \mathbf{0.04517\ kg/s}$$

(*c*) Power required
$= 221.7$ kJ/kg $= 221.7 \times 0.04517$
$= 10.01$ kJ/s $=$ **10.01 kW**

(*d*) Heat removed through condenser
$= m\,(h_2 - h_5)$
$= 0.04517\,(1671.7 - 343.2)$
$=$ **60 kJ/s**

(*e*) Volume flow rate
$= m \times$ specific volume of vapour at entry to compressor
$= 0.04517 \times 0.42 = 0.01897$ m³/s

Compressor displacement

$$= \frac{0.01897}{0.8} = 0.0237 \text{ m}^3/\text{s}$$

$$= 1.422 \text{ m}^3/\text{min}$$

This is equal to $\frac{\pi}{4} d^2\, l\, N\, n$ where d is the diameter, l is the stroke, N is rev/min and n is the number of cylinders of the compressor. As such

$$\frac{\pi}{4} d^2 \times 1.2d \times 1000 \times 2 = 1.422$$

$$d^3 = 7.548 \times 10^{-4}$$

∴ Diameter of cylinder,
$d = 0.09127$ m $=$ **91.27 mm**

and length of cylinder, $l = 1.2 \times 91.27$
$=$ **109.52 mm**

EXAMPLE 25.20.

The following data were obtained from test on a 20 ton ice making refrigerating plant with NH_3 as refrigerant using chilled brine as an intermediate circuit.

temperature of water entering and leaving the condenser = 20°C and 27°C

temperature of condensation of NH_3 = 25°C
temperature of brine in the evaporator = – 15°C
temperature of NH_3 at entry to expansion valve = 20°C

Condition of refrigerant at entry to compressor = dry saturated

For one ton of refrigeration, determine
(a) power expended in running the compressor
(b) amount of cooling water in the condenser,
(c) COP of the plant.

How does COP of this cycle compare with that of a reversed Carnot cycle operating between the same evaporator and condenser temperatures.

The relevant properties of NH_3 are given below :

Unit	Saturation temperature °C	Enthalpy kJ/kg		Entropy kJ/kg K		Specific heat kJ/kg K	
		Liquid	Vapour	Liquid	Vapour	Liquid	Vapour
Evaporator	– 15	350	1663	3.82	9.01	4.39	2.30
Condenser	25	536	1703	4.59	8.51	4.65	2.80

Solution : Refer Fig. 25.25, for *T-s* plot of the given cycle.

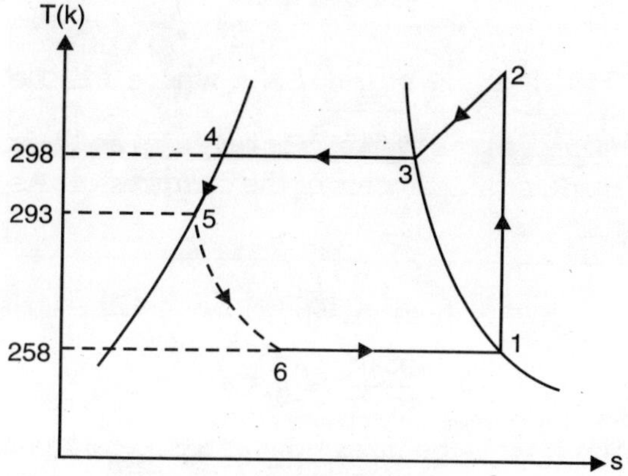

Fig. 25.25.

$$T_1 = T_6 = -15 + 273 = 258 \text{ K}$$
$$T_3 = T_4 = 25 + 273 = 298 \text{ K}$$
$$T_5 = 20 + 273 = 293 \text{ K}$$
$$h_1 = 1663 \text{ kJ/kg}$$
$$h_3 = 1703 \text{ kJ/kg}$$
$$h_4 = 536 \text{ kJ/kg}$$
$$h_5 (= h_6) = h_4 - c_{pl} \times (T_4 - T_5)$$
$$= 536 - 4.65 \times (298 - 293)$$
$$= 512.75 \text{ kJ/kg}$$

Since the compression process 1-2 is adiabatic

entropy at 1 = entropy at 2

$$9.01 = 8.51 + 2.80 \log_e \frac{T_2}{298}$$

$$\text{or } \log_e \frac{T_2}{298} = \frac{9.01 - 8.51}{2.80} = 0.17857$$

$$\text{or } \frac{T_2}{2.98} = e^{0.17857} = 1.1955$$

∴ Temperature at exit from the compressor, $T_2 = 298 \times 1.1955 = 356$ K

Now, $h_2 = h_3 + c_{pv} (T_2 - T_3)$
$= 1703 + 2.8 (356 - 298)$
$= 1865.4$ kJ/kg

(*a*) Work done

$$W = h_2 - h_1 = 1865.4 - 1663$$
$$= 202.4 \text{ kJ/kg}$$

Refrigerating effect

$$N = h_1 - h_6 = 1663 - 512.75$$
$$= 1150.25 \text{ kJ/kg}$$

Coefficient of performance,

$$COP = \frac{N}{W} = \frac{1150.25}{202.4} = \mathbf{5.68}$$

(*b*) Let m_r be the refrigerant flow rate per ton of refrigeration. Then

$$m_r \times 1150.25 = 210 ;$$
$$m_r = 0.1826 \text{ kg/min}$$

∴ Power input per ton of refrigeration

$$= 202.4 \times 0.1826 = 36.95 \text{ kJ/min}$$
$$= \frac{36.95}{60} = 0.616 \text{ kJ/s} = \mathbf{0.616 \text{ kW}}$$

(*c*) Energy balance in the condenser gives heat lost by refrigerant

= heat gained by condenser water

$$m_r (h_2 - h_5) = m_w c_p \Delta t$$
$$0.1826 (1865.4 - 512.75) = m_w \times 4.186 \times (27 - 20)$$

∴ Mass flow rate of cooling water in the condenser

$$= \frac{0.1826(1865.4 - 512.75)}{4.186 \times (27 - 20)}$$
$$= 8.43 \text{ kg/min}$$

Thus the amount of cooling water is 8.43 kg/min or 8.43 kg per ton of refrigeration.

(*d*) *COP* of reversed Carnot cycle

$$= \frac{T_e}{T_h - T_e} = \frac{258}{298 - 258} = \mathbf{6.45}$$

EXAMPLE 25.21.

A heat pump using Freon as the working fluid is used to carry a part of the heating load of a factory.

The heat pump raises the temperature of 6000 litres of circulating water from 35°C to 46°C. The unit operates between evaporator and condenser pressure limits of 0.14 bar and 2.1 bar. The vapour is in the superheated state at entry to the compressor and has an enthalpy 205 kJ/kg and entropy equal to 0.75 kJ/kg K. After adiabatic compression, the refrigerant Freon attains a temperature of 73.5°C. After condensation, refrigerant enters the throttle valve at 68°C and the vapour gets throttled to evaporator pressure of 0.14. Sketch the cycle on T-s plot and calculate the coefficient of performance and the power input required to run the unit.

Take specific heat of Freon liquid = 5.02 kJ/kg K and specific heat of superheated vapour at 2.1 bar = 1.3 kJ/kg K. Further the relevant properties of freon are as given below :

Unit	Pressure bar	Saturation temperature °C	Enthalpy kJ/kg		Entropy of vapour kJ/kg K
			Liquid	Vapour	
Condenser	2.1	71.1	102	237	0747
Evaporator	0.14	– 1.15	–	–	

Solution : Refer Fig. 25.26, for *T-s* plot of the given cycle.

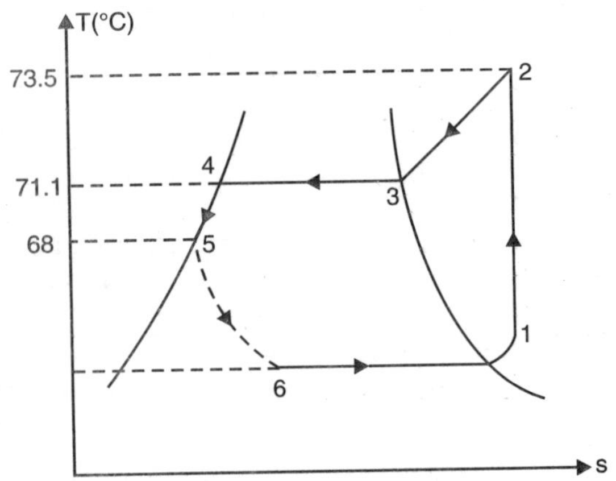

Fig. 25.26.

$$h_1 = 205 \text{ kJ/kg}$$

$$h_6 = h_5 = h_4 - c_{pl}(t_4 - t_5) = 102 - 5.02(71.1 - 68) = 86.44 \text{ kJ/kg}$$

$$h_2 = h_3 + c_{pv}(t_2 - t_3) = 237 + 1.3(73.5 - 71.1) = 240.12 \text{ kJ/kg}$$

(*a*) Work done

$$= h_2 - h_1 = 240.12 - 205 = 35.12 \text{ kJ/kg}$$

Heat given by refrigerant to water in the condenser,

$$(h_2 - h_5) = 240.12 - 86.44 = 153.68 \text{ kJ/kg}$$

∴ *COP* of heat pump

$$= \frac{153.68}{35.12} = \mathbf{4.376}$$

(*b*) If m_r is the mass flow of refrigerant per second, then energy balance in the condenser gives :

heat given by refrigerant = heat taken by water

$$m_r \times 153.68 = \frac{6000}{3600} \times 4.186 \times (46 - 35) ;$$

$$m_r = 0.4988 \text{ kg/s}$$

∴ Power input to compressor

$$= 35.12 \times 0.4988 = 17.518 \text{ kJ/s} = \mathbf{17.518 \text{ kW}}$$

25.5. REFRIGERANTS

A refrigerant is the working substance or heat carrying medium used in a refrigeration cycle to absorb heat from a low temperature system and discard the heat so absorbed to another system at higher temperature. During heat absorption, the refrigerant evaporates (changes from liquid to vapour state) and during heat rejection it condenses (changes from vapour to liquid state). Obviously, the refrigerant undergoes a phase change from gas to liquid and back. These processes occur in the vapour compression system as well as in the vapour absorption system. A wide variety of refrigerants are available, each having typical properties which limit its efficient use in a particular refrigerating machine.

The selection of refrigerant for an application is essentially based on certain parameters such as pressure-temperature characteristics, density, viscosity, flammability, toxicity, miscibility with oil, and heat absorbing capacity etc.

Early mechanical systems employed sulphur dioxide gas or anhydrous ammonia with small home refrigerators. Being toxic, sulphur dioxide disappeared from the market with the introduction of Freon. Ammonia is still used in large commercial plants, well away from residential areas, where a leak will not cause widespread injuries. Until concerns about depletion of ozone layer arose in 1980s, the most widely used refrigerants were the halomethanes R-12 and R-22, with R-12 being more common in automotive air conditioning and small refrigerators and R-22 being used for residential and light commerical air conditioning, refrigerators and freezers. Some very early systems used R-11 because its low boiling point allows low pressure systems for which low mechanical strength components can be used.

25.5.1. An ideal refrigerant—desirable properties

The properties of an ideal refrigerant include :

1. Non-flammable, non-explosive, non-viscous, non-toxic and non-corrosive to metals
2. Easy availability, low cost and easy storage and disposal
3. Low specific heat of liquid and high total heat of vaporization
4. Low freezing point, low boiling point and high critical temperature
5. Moderate density in liquid form and a relatively high density in gaseous form
6. High thermal conductivity and electrical insulation
7. Easy to liquefy at moderate pressure
8. Ability to operate on a positive pressure
9. Long life chemical stability
10. Odorous easy detection of leakage
11. Mixes well with oil
12. Eco-friendly, and
13. High COP and low power consumption

No single refrigerant satisfies all the qualities desired of a refrigerant and which can be used for all types of applications. Obviously no ideal refrigerant exists.

25.5.2. Classification of refrigerants

The refrigerants are usually classified as follows on the basis of :

- **Working principle :**

– *Primary refrigerant :* These are the heat carrier substances that take part directly in the refrigeration system and cool the product/ space by absorption of latent heat. The refrigerant undergoes a cyclic process involving compression, condensation, expansion and evaporation with change in phase.

– *Secondary refrigerant :* These are the heat carrier substances that are first cooled by primary refrigerants, and are subsequently used as a medium for producing cooling effect in the product or space being refrigerated. The secondary refrigerants do not undergo any cyclic process; no change in phase and produce the cooling affect by absorption of sensible heat.

- **Chemical composition :**

The generic classification of refrigerants is shown below :

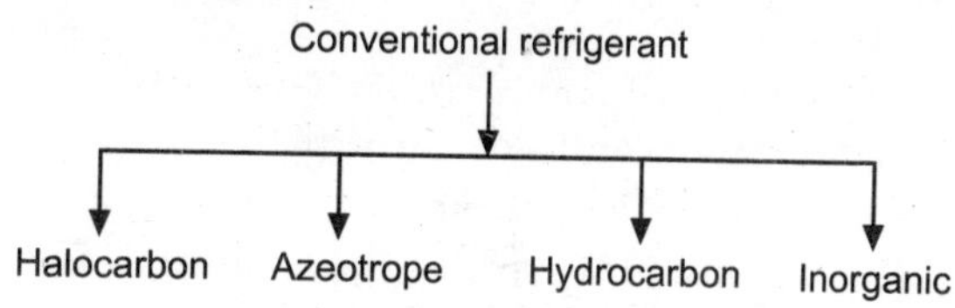

(*i*) ***Halocarbon refrigerants*** : These are synthetically produced and were developed as Freon family of refrigerants. These are either methane-based or ethane-based, where the hydrogen atoms are replaced by chlorine or fluorine atoms.

The halocarbon refrigerants have wide application in domestic, commerical and industrial installations.

(*ii*) ***Azeotropes :*** An azeotrope is a stable mixture of different substances. This mixed refrigerant has fixed thermodynamic properties and it cannot be separated into its constituents by distillation. The vapour and liquid phases of this blend retain identical composition over a wide range of temperatures.

(*iii*) ***Hydrocarbon refrigerants*** : These refrigerants are essentially compounds of hydrogen and carbon. These refrigerants have satisfactory thermodynamic properties, but are flammable and thus unsafe.

Examples: ethane, propane, iso-butane and ethylene

(*iv*) ***Inorganic refrigerants :*** These are non-hydrocarbon refrigerants and are mostly available in nature. These refrigerants were exclusively used before the advent of halocarbon refrigerants, and are still in use due to their inherent physical and thermodynamic properties.

Examples : air, ammonia, carbon dioxide, sulphur dioxide and water.

• **Nature of refrigerants :**

(*i*) Natural which exist in nature such as water, ammonia etc.

(*ii*) Artificial which are man-made and are synthetically produced such as azetropes and halocarbon refrigerants.

• **Safety :**

(*i*) Safe refrigerants which are non-toxic and non-flammable.

Examples : CO_2 , water and Freon family refrigerants

(*ii*) Toxic refrigerants which have irritating, obnoxious smell and are harmful.

Examples : SO_2 and NH_3

(*iii*) Flammable refrigerants which are strongly prone to catch fire.

Examples : Butane and isopropane etc

25.5.3. Selection of a refrigerant

The following factors need consideration when choosing a refrigerant for the vapour compression refrigeration system. The selection primarily depends on satisfying the essential requirements for a particular application.

• **Thermodynamic requirements :**

(*i*) *Evaporating and condensing pressures* : Pressure of a refrigerant corresponds to saturation pressures during heat abstraction (evaporation) and heat rejection (condensation) processes.

The evaporating pressure should be positive and as near atmospheric as possible. A positive pressure is required to restrict the possibility of the entry of air and moisture into the system.

The condensing pressure should be moderately low. This will permit the use of light weight equipment and piping on the high pressure side of the system.

(*ii*) *Critical temperature and pressure* : The critical temperature and pressure of the refrigerant must be well above the maximum operating temperature and pressure limits. The refrigerant has better heat transfer rate below the critical points.

(*iii*) *Freezing point* : The freezing point of the refrigerant must be below the lower temperature of the cycle. Except in the case of water for which the freezing point is 0°C, all other refrigerants have reasonably low value for the freezing point.

(*iv*) *Latent heat of vaporisation* : A high latent heat of vaporization is desirable since it permits greater amount of heat to be absorbed from the space being refrigerated. Then for the same amount of refrigerating effect, less mass of the refrigerant needs to be circulated in the system. That reduces the initial cost of the refrigerant. The size of the system also becomes small and hence low initial cost.

• **Chemical requirements**

(*i*) non-flammable and non-explosive

(*ii*) odourless, non-irritating and non-toxic

(*iii*) non-corrosive to metal

(*iv*) chemical stability

(*v*) high electrical insulation

The refrigerant should neither decompose nor react chemically with metals and lubricating oil used in the compressor. The refrigerant should not cause the oil to be carried over into the condenser and evaporator.

The presence of oil in these units reduces the heat transfer rate and causes fouling.

Further, the refrigerant should have no effect on the quality of stored food and other products. The dielectric strength of a refrigerant is primarily important in hermetically sealed units in which the electric motor is exposed to the refrigerant

- **Physical requirements**

(*i*) *Low viscosity and high thermal conductivity* : The refrigerant should have a low viscosity and high thermal conductivity. This would ensure better heat transfer rates in the condenser and evaporator.

(*ii*) *Low specific heat* : This would minimise the amount of vapour formed during throttling or expansion and consequently improve the refrigerating capacity.

(*iii*) *Low specific volume* : The power input to the compressor depends on the flow rate of the refrigerant. For the volume of the refrigerant dealt to be small, the refrigerant must have a low value of specific volume.

Reciprocating compressors are used with refrigerants having high pressures and small volumes of the suction vapour. Centrifugal compressors are adopted when large volumes of refrigerant at low pressures have to be dealt with.

- **General requirements**

(*i*) leakage tendency should be low and leaks readily detectable

(*ii*) cheap in cost and easy availability

(*iii*) high coefficient of performance

(*iv*) low power consumption per ton of refrigeration.

The properties pertinent to the most commonly used refrigerants are given below :

- **Ammonia (NH_3)**

(*i*) excellent thermal properties : high latent heat and high heat transfer coefficient

(*ii*) large refrigerating effects with relatively small sized machinery

(*iii*) lighter than oil and can be easily separated. However in the presence of moisture, ammonia forms an emulsion with oil and that creates certain operating difficulties

(*iv*) low specific volume and so low volumetric displacement

(*v*) low mass of liquid to be circulated per ton of refrigeration

(*vi*) cheap refrigerant and low running cost

(*vii*) leaks easily detectable due to pungent odour. The leak is detected by bringing a sulphur candle near the leak : dense white smoke is given off when ammonia vapour is present.

(*viii*) highly volatile and explosive when in contact with air

(*ix*) toxic, irritating to eyes, nose and throat

(*x*) non-corrosive to ferrous metals but attacks copper and its alloys (brass and bronze)

Ammonia is an ideal refrigerant and is invariably used in ice factories, breweries and cold storage plants

- **Sulphur dioxide (SO_2)**

(*i*) colourles, non-explosive and non-flammable

(*ii*) extremely toxic and pungent irritating odour

(*iii*) works at low pressure and has small latent heat of vaporisation

(*iv*) corrosive when in contact with moisture due to formation of H_2SO_4

(*v*) leaks easily detectable on account of its pungent odour. The leakage is detected by bringing aqueous ammonia near the leak; white smoke is given off

(*vi*) low refrigerating effect and high specific volume. Therefore, for the same refrigerating effect with this refrigerant, we need larger compressors with higher speeds

(*vii*) not readily miscible with oil, liquid SO_2 is heavier than oil and accordingly it floats on the surface of oil. This aspect simplifies the problem of oil removal and return

In earlier days, sulphur dioxide was widely used in household and small commerical units.

- **Carbon dioxide (CO_2)**

(*i*) colourless and odourless

(*ii*) non-toxic, non-corrosive and non-flammable

(*iii*) higher concentration causes suffocation

(*iv*) leaks difficult to detect ; leak detection done by soap solution

(*v*) low specific volume makes the plant more compact; size of the compressor is small

(*vi*) heavier working parts needed because of high operating pressures

(*vii*) non-miscible with oil ; there is no dilution of oil in the compressor crankcase

(*viii*) low critical temperature and accordingly low temperature cooling medium is required in the condenser

(*ix*) chemically stable ; no reaction with oil and metals in the presence of moisture

The use of this refrigerant has been limited to marine application and to the manufacture of dry ice (solid CO_2).

- **Methyl chloride (CH_3Cl)**

(*i*) Neither flammable nor toxic

(*ii*) Sweet and non-irritating odour; however, higher concentrations may produce unconsciousness

(*iii*) Tendency to dissolve oil and so provision has to be made for their separation

(*iv*) Acroline is added to CH_3Cl as an irritant to make its detection easier on leakage from plant

(*v*) Because of reasonably low operating pressures, it is suitable for small scale refrigeration and domestic refrigerators

- **Freon-F_{12}**

(*i*) Non-toxic, non-flammable and non-explosive

(*ii*) Non-reactive with lubricating oil

(*iii*) Highly inert and stable; does not break even under extreme operating conditions

(*iv*) Most suitable refrigerant in high, medium and low temperature applications

(*v*) Completely safe refrigerant, does not affect the food products and is also non-irritant to human body

(*vi*) Fully oil miscible and that simplifies the problem of oil return from condenser and evaporator

(*vii*) Low latent heat requires large mass flow rate and as such large swept volume of the compressor

(*viii*) Leaks difficult to detect

(*ix*) Freon-12 is superceding NH_3 in domestic and commercial plants where its low toxicity is of special value. Home window air-conditioners, package units and small central air-conditioners also use F_{12} as refrigerant. Being an excellent insulator, it is universally employed in sealed compressing units.

- **Water (H_2O)**

(*i*) non-toxic, non-flammable

(*ii*) thermally stable and chemically inert even at high temperatures

(*iii*) very high enthalpy of vaporization

(*iv*) very safe refrigerant ; no leakage problem on health or environment

(*v*) high level transfer coefficient allows small dimensions for condenser and evaporators

(*vi*) available in abundance at no cost

(*vii*) light system components due to very low operating pressures with water

Water is essentially used as ice to produce the cooling effect. Its high freezing point limits its use in vapour compression systems. Water is used as a solvent in vapour absorption system, and also in steam jet refrigeration.

Freon v/s Ammonia

Freon-12 is superceding NH_3 in domestic and commercial plants because of its low toxicity; completely safe refrigerant as it does not affect the food products and is also non-irritant to human body. Home window air conditioners, package units and small central air-conditioners also use F-12 as refrigerant. Being an excellent insulator, it is universally employed in sealed compressor units.

Before the advent of Freon refrigerants, ammonia has been used universally for many applications due to low running cost, easy maintenance (no evacuation and drying), low volumetric displacement, low weight of liquid refrigerant per ton of refrigeration etc. Even today, ammonia is used in cold storages, ice manufacturing plants, skating rinks and breweries without exception.

Since Freon-12 is associated with ozone depletion, its production and use will be completely phased out in the coming years. With its replacement being both costly and time consuming, ammonia will remain ever green and popular due to its ODP (Ozone Depletion Potential) and GWP (Global Warming Potential) being practically zero.

25.5.4. Secondary refrigerants

There are several applications where the direct use of a refrigerant is not desirable and has to be avoided. For example :

• toxic refrigerants (NH_3 and SO_2) cannot be used for the air-conditioning of schools and residential buildings.

• big cold storages need large quantity of refrigerant, and the cost of refrigerant may become higher than that of the refrigeration system itself.

Under such situations, indirect cooling is adopted and use is made of a liquid (water/brine) which is much cheaper than the refrigerant. The liquid is first cooled by the refrigerant and subsequently this chilled liquid is circulated by pumps to the space/products which are to be refrigerated. The liquid becomes a carrier of refrigeration and is known as secondary refrigerant.

The use of secondary refrigerant has the following advantages :

(*i*) easy to handle and easy to control

(*ii*) different temperatures can be maintained in different parts of the building by controlling the flow rate of chilled liquid

(*iii*) reduction in size and overall cost of the system

(*iv*) toxic refrigerants can be kept away from the space and products being refrigerated

(*v*) network of pipeline is reduced in size and length, and low grades materials can be used for piping and heat exchangers

The commonly used secondary refrigerants are :

Water, brines, glycols and halocarbons

The following aspects need consideration while selecting a secondary refrigerant for a particular application :

– low freezing point and high specific heat. Higher the specific heat, lower will be the rate of flow required

– non-flammable and non-toxic

– compatibility with the materials of the equipment

– neutral (neither acidic nor basic) so as to minimize corrosion

– low viscosity, it should not rise fast with fall in temperatures otherwise pumping power will become high and uneconomical

(A) Water : Water is used as the secondary refrigerant for temperature applications above its freezing point of 0°C such as air-conditioning plants and industrial cooling applications. In air-conditioning plants, air is cooled and dehumidified by using chilled water in water spraying unit.

(B) Brines : For sub-zero applications, brines are used as secondary refrigerant. Brine is a solution formed by dissolving a soluble substance in water ; the soluble substance could be salt (NaCl ; $CaCl_2$) or a glycol. When any salt is mixed with water, the solution so formed has a freezing point lower than that of water and it further decreases with increases in salt concentration The strength or concentration of the brine solution is a measure of the mass of salt or glycol in the solution expressed as a percentage of the total mass of the solution.

(C) Glycols : Ethylene and propylene glycols are often mixed with water in varying proportions to depress the freezing point of water. The freezing point decreases to – 9.3°C with 15% ethylene glycol (by volume) and further drops to – 41.5°C with 65% strength. Standard steel and copper pipings are used for glycol lines.

These glycols are being extensively used as anti-freeze elements in brewing and dairy industry due to their following advantages over brines :

(*i*) non-corrosive and non-electrolytic even in the presence of water (*ii*) extremely stable ; no evaporation during service

(*iii*) not much increase in the density of water solutions

(D) Halocarbons : The commonly used halocarbons used as secondary refrigerants are listed below along with their freezing points. R-11 (–110°C) ; R-12 (– 136°C) and R-30 (– 96°C)

Since these halocarbons are volatile, they need a sealed system.

25.6. VAPOUR ABSORPTION SYSTEM

The vapour absorption system is a heat operated system in which the refrigerant (NH_3) is alternately absorbed and liberated from the absorbent (water). The system operates on the principle that NH_3 and water though completely soluble in each other have different boiling points at the same pressure. When the high pressure solution is heated, the NH_3 vapours are driven out because the boiling point of NH_3 is lower than that of water.

Fig. 25.27 shows the schematic arrangement of different components in the water-ammonia (also called aqua-ammonia) absorption system. The various components of the system have been arranged according to their pressures and temperatures.

A strong aqua-ammonia solution is formed by dissolving fairly dry ammonia in cold water. The process of absorption is exothermic *i.e.,* heat is generated. The resulting rise in temperature of the solution adversely affects the capacity of water to absorb ammonia. To have a continuous and much more absorption, the absorber is kept cooled by removing the heat generated to water which is circulated through the cooling coils kept immersed in the absorber.

The high concentrated NH_3 solution known as strong liquor is then circulated through the system by the pump which raises the pressure to about 10.5 bar. For a steady flow process, compression work $= \int_{p_1}^{p_2} V\, dp$

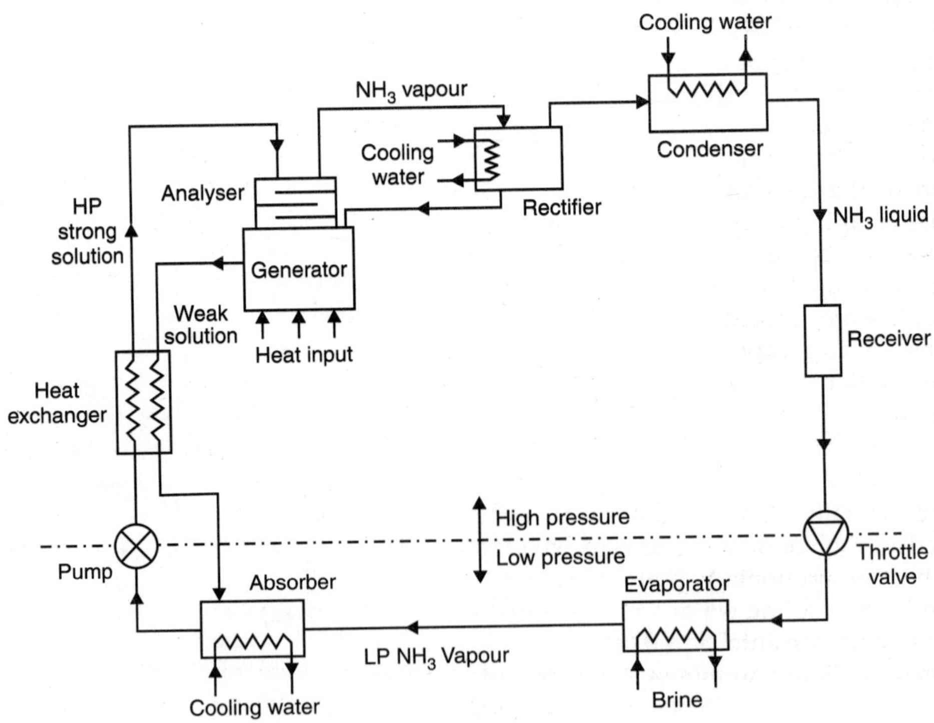

Fig. 25.27. Vapour absorption refrigeration system

where V is the volume of the fluid to be handled and p_1, p_2 are the lower and higher pressures of the cycle.

The pump requires a very small input due to low volume of the liquid solution.

The solution next passes through heat exchanger where the strong ammonia solution is heated by the hot weak ammonia solution returning from generator to absorber. The warm high pressure solution then enters the generator where it is further heated by steam coils, gas or electricity. At high pressure and temperature existing in the generator, the solubility of ammonia in water is appreciably reduced and consequently NH_3 vapours are driven out of solution. Since the boiling point of NH_3 is less than that of water, the vapour leaving the generator are mainly those of NH_3. The weak NH_3 solution left in the generator called weak aqua is throttled and returned to the absorber via heat exchanger.

Ammonia vapours leaving the generator contain some water vapours. These water vapours if allowed to go to condenser and expansion valve, are likely to freeze resulting in choking of flow. Analyser and rectifier are the devices incorporated to remove water vapours escaping with ammonia vapour. Analyser is essentially an open type of cooler that consists of a series of trays mounted on the top of the generator. Most of the water particles escaping with the hot rising ammonia vapours get cooled, condense into liquid stream and fall back into the generator. A rectifier is water cooled heat exchanger which condenses water vapour and some ammonia and the condensed mixture is drained back to the generator. Thus final reduction or elimination of the water vapour takes place in the rectifier.

The virtually pure ammonia vapours are then led to the condensing unit in which the latent heat of ammonia vapours is rejected to the condenser cooling water and the ammonia vapours condense into liquid ammonia. The high pressure liquid ammonia is then throttled to a pressure of about 1.75 bar. The wet vapour at low pressure and temperature finally enter the evaporator, absorb heat from the brine and get partially dried up. The low pressure ammonia vapours leaving the evaporator are passed on the absorber and the cycle is repeated.

It is to be noted that the absorption system consists of a condenser, throttling or expansion valve and an evaporator similar to vapour compression refrigerator. The combination of absorber, pump, heat exchanger and generator replaces the compressor and this combination can be considered equivalent to a simple heat engine and refrigerator inter-connected for work transfer (Fig. 25.28)

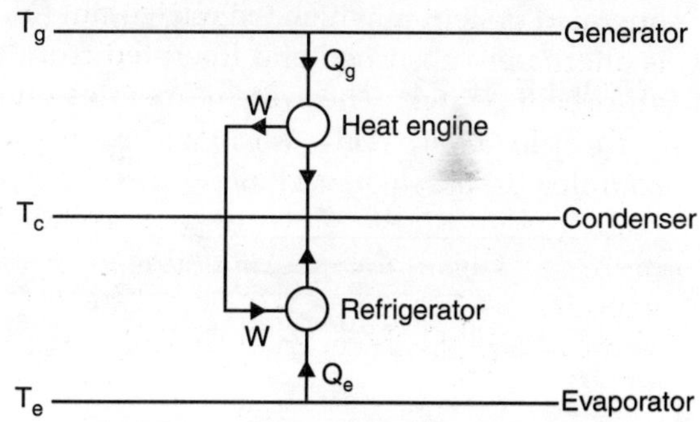

Fig. 25.28.

Work equivalent of heat Q_g supplied from generator

$$= \frac{T_g - T_c}{T_g} \times Q_g$$

Work required by the refrigerator in pumping Q_e units of heat from temperature T_e to T_c

$$= \frac{T_c - T_e}{T_e} \times Q_e$$

If it is presumed that the refrigerator is being run by the power output of the heat engine, then

$$\frac{T_g - T_c}{T_g} \times Q_g = \frac{T_c - T_e}{T_e} \times Q_e$$

∴ Ideal COP

$$= \frac{\text{heat extracted from the evaporator}\,(Q_e)}{\text{heat supplied in the generator}\,(Q_g)}$$

$$= \frac{T_g - T_c}{T_c - T_e} \times \frac{T_e}{T_g}$$

The above expression shows that for high value of *COP*,

(*i*) temperature of generator T_g be as high as possible

(*ii*) temperature of condenser T_c be as low as possible and

(*iii*) temperature of evaporator T_e be as high as possible.

The absorption system is used at places where ample cooling water and low pressure heating streams are available. The system is simple in design and low in operating cost.

EXAMPLE 25.22.

In a vapour absorption refrigeration system, the generator is operated by solar heat and the temperature reaches 385 K. The refrigeration temperature is 260 K and the environmental conditions are stated to be at 325 K. Determine the maximum possible COP of the system.

Solution : T_g = 385 K; T_c = 325 K

and T_e = 260 K

$$\therefore \text{ Ideal } COP = \frac{T_g - T_c}{T_c - T_e} \times \frac{T_e}{T_g}$$

$$= \frac{385 - 325}{325 - 260} \times \frac{260}{385} = \mathbf{0.623}$$

Vapour Compression *Versus* Vapour Absorption Systems

The salient points of difference between vapour compression and vapour absorption systems of refrigeration have been given below :

Vapour Compression System	*Vapour Absorption System*
1. Energy supplied for running the compressor has to be mechanical or electrical *i.e.*, high grade energy.	The energy consumption of the system is mainly in the generator in the form of heat which is a low grade energy.
2. Energy supply is low being only one-third of the refrigerating effect.	Energy input is high being about one and half times the refrigerating effect.
3. Due to moving parts of the compressor unit, the system is quite noisy in operation and has appreciable wear and tear.	The moving parts are only in the pump which is a small element of the system. As such, the operation of the system is quiet and subjected to little wear.
4. The system has a poor performance at partial loads.	The performance of the system is not affected by load variations. The load variations can be easily met by controlling the quantity of aqua circulated and heat input to generator.
5. Due to reciprocating parts, there are chances of leakage of the refrigerant.	No chances of leakage as there are no reciprocating components.
6. The charging of refrigerant is quite simple.	The charging of refrigerant is rather difficult.
7. The valves and other compressor parts get damaged by the presence of traces of water vapour in the suction lines.	There is no danger from the liquid traces at the exit of the evaporator.
8. The system is used when no cheap source of energy is available.	The system is useful at places where waste heat is available in abundance, *i.e.*, in electric power houses, gas turbines and chemical industries.
9. Equipment is bulky and initial cost is high.	More precise equipment.
10. The space requirements, automatic control needs and high capacity (above 1000 tons) favour the use of vapour absorption systems.	

25.7. ELECTROLUX REFRIGERATOR

The electrolux refrigerator is of ammonia absorption type that uses ammonia as a refrigerant, water as an absorbent and hydrogen gas which reacts neither with ammonia nor with water. The main features of this system are :

(*i*) The system operates at a constant total pressure of about 14 bar throughout the circuit.

(*ii*) When liquid NH_3 is exposed to an atmosphere of hydrogen, it readily evaporates resulting in a fall in its temperature.

(*iii*) The circulation is accompanied by gravity and by a small amount of heat energy supplied by an electric heater or a gas. Hydrogen, being light, also facilitates circulation due to large difference in densities between the generator and evaporator units.

(*iv*) The unit does not incorporate any pump or compressor. Instead, it has a generator, a separator and an absorber.

Fig. 25.29, shows the schematic arrangement of the different components of the electrolux refrigerator.

The absorber absorbs ammonia vapour contained in ($NH_3 + H_2$) mixture coming from the evaporator and forms a strong solution of ammonia dissolved in distilled water. Ammonia is much more soluble in water than in hydrogen. As such NH_3 vapour gets dissolved in water while hydrogen is driven upward. Further, the absorption of NH_3 vapour is more and quick in cold water and accordingly the absorber is kept cooled by cooling fins or coils around it.

The strong ammonia solution flows down from the absorber to the generator via heat exchanger *P*. In the heat exchanger, the strong solution of ammonia is heated by the weak solution of ammonia returning from the generator.

Small heat energy from gas or electric heater is supplied to the generator and this causes the ammonia vapour to be driven out of the solution. The particles of water clinging to ammonia vapour are retained in the rectifier unit and flow by gravity to the absorber.

The pure ammonia vapours rise upwards to the air cooled condensing unit, condense to liquid ammonia which flows by gravity to the evaporator via a liquid trap. The evaporator contains hydrogen at 12 bar pressure. The partial pressure of ammonia vapour in the

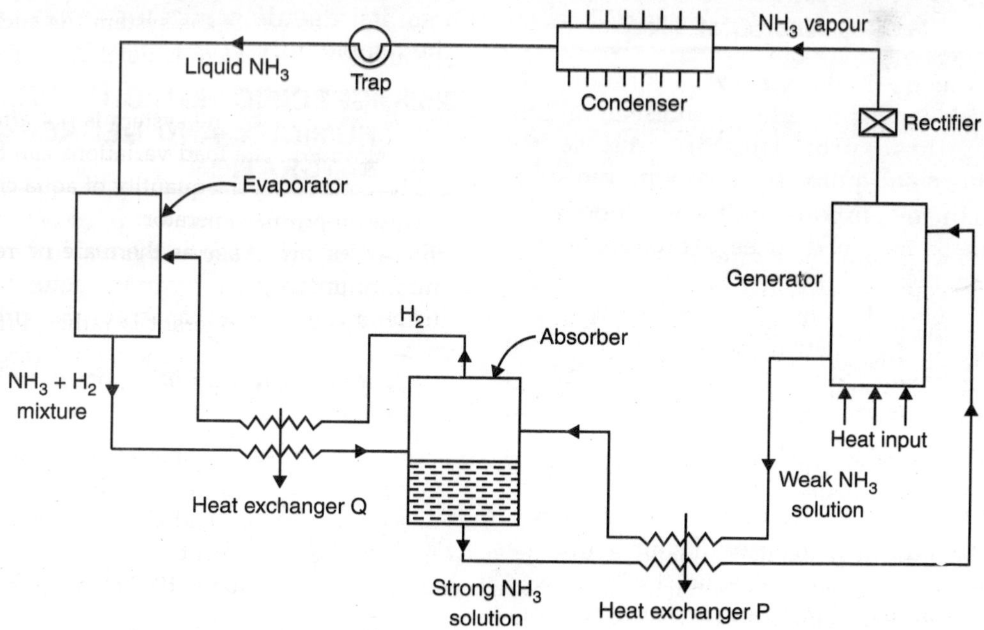

Fig. 25.29. Schematic arrangement of an electrolux refrigerator

evaporator gets reduced to 2 bar according to Dalton's law of partial pressure, the corresponding temperature being - 18°C. Ammonia then evaporates at low temperature and the heat required for evaporation is taken from the system being refrigerated.

The mixture of ammonia and hydrogen flows to the absorber where ammonia is absorbed and the separated hydrogen rises to the top. The hydrogen is taken back to the evaporator and on its way is cooled in heat exchanger Q by the ammonia hydrogen mixture flowing from the evaporator to the absorber.

The coefficient of performance of the electrolux refrigerator is given by

$$COP = \frac{\text{heat absorbed in the evaporator}}{\text{heat supplied in the generator}}$$

The electrolux refrigerator is mostly used for household refrigeration in remote place, where there is no electricity. The system does not incorporate any machinery (pump or compressor) and is, therefore, quiet in operation and free from wear and tear. Further, the unit is compact, durable and easy to regulate.

B : PSYCHROMETRY

Psychrometry is the science of studying the thermodynamic properties of moist air and the use of these properties to analyse the conditions and processes involving moist air.

For many purposes, the composition of real air can be assumed to be a mixture of two components:

Standard dry air : Dry air is a mixture of number of gases such as oxygen, nitrogen, carbon dioxide, hydrogen, argon etc. Oxygen and nitrogen are the main constituents with the following composition

21% oxygen and 79% of nitrogen ..... by volume

23% oxygen and 77% nitrogen ..... by mass

The molecular mass of dry air is taken as 20.966 and gas constant equal to 287 J/kgK. For psychrometric purpose, dry air is assumed to be a pure substance and not a mixture.

Water vapour : Air has affinity for water and consequently the atmospheric air always contains some water; water vapour content varies from 0 to 3% by mass.

The moist air is essentially a mixture of dry air and water vapour. The amount of water vapour present depends on the absolute pressure and temperature of the mixture.

For the **moist air**, which is a mixture of dry air and water vapour

$$p_t = p_a + p_v$$

where p_t = total pressure of moist air

p_a = partial pressure of dry air

p_v = partial pressure of water vapour

The ***saturated air-mixture*** is the mixture of dry air and water vapour in which the partial pressure of the water vapour is equal to its saturation pressure corresponding to the temperature of the mixture.

The ***unsaturated air mixture*** is the mixture of dry air and water vapour in which the partial pressure of the water vapour is less than its saturation pressure corresponding to the temperature of the mixture.

The ***super saturated air mixture*** is mixture of dry air and water vapour in which the partial pressure of the water vapour is greater than its saturation pressure corresponding to the temperature of the mixture.

25.8. SPECIFIC HUMIDITY, RELATIVE HUMIDITY AND DEGREE OF SATURATION

Humidity refers to the dampness, *i.e.*, the water content of air. ***Absolute humidity*** represents the amount of water vapour actually present in the air, expressed as gram per cubic meter of air.

The ***specific humidity*** or ***humidity ratio*** or ***moisture content*** is the ratio of mass of water vapour to the mass of dry air in a given volume of the mixture. Consider a mixture consisting of m_a kg of dry air and m_v kg of water vapour contained in a vessel of volume V at total pressure p_t and temperature T. Then

$$\text{Specific humidity, } \omega = \frac{m_v}{m_a}$$

Since both masses occupy volume V,

$$\omega = \frac{m_v/V}{m_a/V} = \frac{\rho_v}{\rho_a} = \frac{v_a}{v_v}$$

where ρ is the density and v is the specific volume

If both the vapour and dry air are considered as perfect gases, then from the characteristic gas equation,

$$m_a = \frac{p_a V}{R_a T} \quad \text{and} \quad m_v = \frac{p_v V}{R_v T}$$

That gives : $\omega = \dfrac{m_v}{m_a}$

$$= \frac{p_v V}{R_v T} \times \frac{R_a T}{p_a V} = \frac{R_a}{R_v} \times \frac{p_v}{p_a}$$

Taking, R_a = 287 J/kgK and R_v = 461 J/kgK

$$\omega = \frac{287}{461}\frac{p_v}{p_a}$$

$$= 0.622\frac{p_v}{p_a} = 0.622\frac{p_v}{p_t - p_v} \quad ...(25.3)$$

The above relation shows that if the total pressure remains constant, the specific humidity is a function of partial pressure of water vapour only. This relationship has been obtained by assuming that behaviour of water vapour is identical with that of an ideal gas. Such an assumption is quite valid at low pressure and normal humidity conditions.

The ***relative humidity*** is the ratio (expressed as a percentage) of the amount of water vapour actually present in a given volume of air to the maximum amount that the air could hold under the same pressure and temperature conditions.

Relative humidity

$$\phi = \frac{\text{mass of water vapour in a given volume}}{\text{mass of water vapour in the same volume if saturated at the same temperature}}$$

With the assumption that vapours behave as perfect gases, we have

$$m_v = \frac{p_v V_v}{R_v T_v} \quad \text{and} \quad m_{vs} = \frac{p_{vs} V_{vs}}{R_{vs} T_{vs}}$$

where vs subscript is used for saturated vapour.

$$\therefore \quad \phi = \frac{m_v}{m_{vs}} = \frac{p_v V_v}{R_v T_v} \times \frac{R_{vs} T_{vs}}{p_{vs} V_{vs}}$$

Also $V_v = V_{vs}$; $R_v = R_{vs}$ and $T_v = T_{vs}$.

That gives

$$\phi = \frac{p_v}{p_{vs}} \quad ...(25.4)$$

where p_{vs} is the saturation pressure at the temperature of the mixture. This saturation pressure is obtained from the steam tables corresponding to the given temperature.

Apparently, the relative humidity can also be defined as the ratio of the partial pressure of water vapour in a given volume of mixture to the partial pressure of water vapour when the same volume of mixture is saturated at the same temperature. This implies that ϕ equals unity for saturated air; 100 percent relative humidity means that air contains the maximum moisture it can hold.

Essentially, the term relative humidity compares the humidity of the given air with the humidity of saturated air at the same pressure and temperature. For saturated air, the relative humidity is 100 percent.

The ***degree of saturation*** μ represents the ratio of mass of water vapour associated with unit mass of dry air to the mass of water vapour associated with unit mass of dry air saturated at the same temperature.

$$\mu = \frac{\text{mass of water vapour with unit mass of dry air}}{\text{mass of water vapour with unit mass of dry saturated air}}$$

This relation implies that μ represents the ratio of specific humidity of moist air to specific humidity of saturated air at the same temperature. That is

$$\mu = \frac{\omega}{\omega_s} = \frac{0.622\dfrac{p_v}{p_t - p_v}}{0.622\dfrac{p_{vs}}{p_t - p_{vs}}}$$

$$= \frac{p_v}{p_{vs}} \frac{p_t - p_{vs}}{p_t - p_v}$$

$$= \frac{p_v}{p_{vs}} \left[\frac{1 - \frac{p_{vs}}{p_t}}{1 - \frac{p_v}{p_t}} \right] \quad ...(25.5)$$

The following observations can be made from the above relation:

(*i*) If the air is dry, then $p_v = 0$ and therefore $\mu = 0$.

If the relative humidity $\phi = p_v/p_{vs} = 1$, then $p_v = p_{vs}$ and accordingly $\mu = 1$.

Thus the degree of saturation varies between 0 and 1.

(*ii*) The degree of saturation is a measure of the capacity of the air to absorb moisture.

When $\mu = 1$, then $\omega = \omega_s$

which implies that air is holding the maximum amount of water vapour.

(*iii*) The expression for the degree of saturation can be recast as

$$\mu = \frac{p_v}{p_{vs}} \left[\frac{1 - \frac{p_{vs}}{p_t}}{1 - \frac{p_v}{p_{vs}} \times \frac{p_{vs}}{p_t}} \right] = \phi \left[\frac{1 - \frac{p_{vs}}{p_t}}{1 - \phi \frac{p_{vs}}{p_t}} \right]$$

$$\text{or} \quad \phi - \phi \frac{p_{vs}}{p_t} = \mu - \mu \phi \frac{p_{vs}}{p_t}$$

$$\text{or} \quad \phi \left[1 - \frac{p_{vs}}{p_t} + \mu \frac{p_{vs}}{p_t} \right] = \mu$$

$$\text{or} \quad \phi \left[1 - (1 - \mu) \frac{p_{vs}}{p_t} \right] = \mu$$

$$\therefore \quad \phi = \frac{\mu}{1 - (1 - \mu) \frac{p_{vs}}{p_t}} \quad ...(25.6)$$

The difference between relative humidity and degree of saturation is usually less than 2%.

25.9. DRY BULB TEMPERATURE AND WET BULB TEMPERATURE

The ***dry-bulb temperature*** (*dbt*) is the normal temperature of an air-vapour mixture as indicated or recorded by any temperature measuring device placed in the mixture. This temperature is not affected by the moisture content in the mixture.

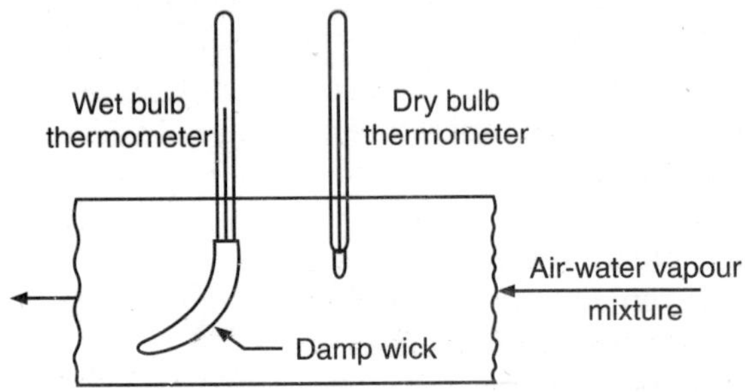

Fig. 25.30. Dry and wet bulb temperature

The ***wet-bulb temperature*** (*wbt*) of an air-vapour mixture is the temperature measured by a thermometer whose bulb is covered by a wick soaked in water. When the air passes over the wet wick, the moisture contained in the wick tends to evaporate. That produces cooling effect at the bulb and an equilibrium temperature lower than that of the air stream is recorded.

It is worthwhile to note that:

(*i*) The wet-bulb temperature is lower than the dry-bulb temperature and the difference is known as the ***wet-bulb depression.***

(*ii*) The wet-bulb depression is greatest when the air is initially completely dry, *i.e.*, capable of absorbing a maximum amount of moisture.

(*iii*) When the air is initially saturated, there will be no evaporation of water and hence the two thermometers will record equal temperatures. This implies that the depression is zero with 100 percent relative humidity.

(*iv*) The dry and wet-bulb temperatures are simultaneously measured by instruments called ***psychrometers.***

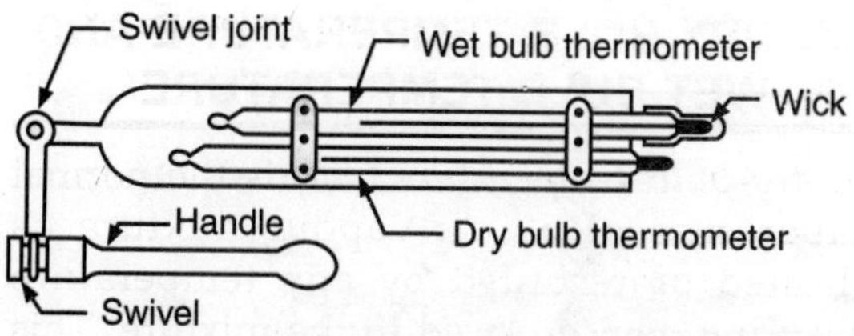

Fig. 25.31. Sling psychrometer

A sling psychrometer consists of two identical mercury-in-glass themometers mounted on a suitable frame and arranged with a swivel-mounted handle as shown in Fig. 25.32. The temperature sensing bulb of one of the thermometers is covered with a knitted or woven cotton wick which is wetted with pure clean water. For better and accurate measurements of the wet bulb temperature, a fast movement of air past the moistened wick is necessary. This is to ensure that the surrounding air does not cling to the moistened wick and that the air at the wet-bulb thermometer is always in immediate contact with the wet wick. The necessary air motion, 5 m/s to 10 m/s, is provided by rotating the psychrometer frame with the swivel mounted handle. The readings are taken after swinging the psychrometer in a smooth circular path for 15 to 20 seconds. With a too short duration, the temperature will not be depressed to its proper value. If the swinging period is too large, the wick will dry and the bulb temperature will not remain at its minimum value.

(*v*) When dry and wet bulb temperatures are known, the other psychrometric properties can be determined by calculations.

(*vi*) Many investigators have suggested different expressions to determine the partial pressure of water vapour in air from the wet and dry bulb temperature readings. Carrier's equation, as given below, is most widely used.

$$p_v = (p_{vs})_{wb} - \frac{(p_t - p_{vs})(dbt - wbt)}{1544 - 1.44\, wbt} \quad ...(25.7)$$

where, p_v = partial pressure of water vapour

p_{vs} = partial pressure of water vapour when air is fully saturated

p_t = total (barometric pressure of moist air)

dbt and *wbt* = dry bulb and wet bulb temperature respectively in °C.

25.10. DEW POINT AND ADIABATIC SATURATION TEMPERATURE

Dew point refers to the temperature at which the mixture becomes saturated, *i.e.*, the moisture (water vapour) present in the mixture begins to condense consequent to continuous cooling at constant pressure.

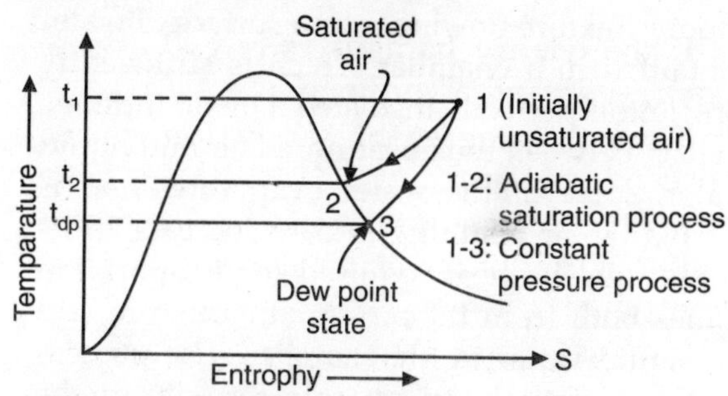

Fig. 25.32. Concept of dew point temperature

With reference to Fig. 25.33, let point 1 represent the initial state of air-water vapour mixture. The vapour is superheated, the pressure equals the partial pressure of the superheated vapour and temperature corresponds to the temperature of the mixture. When the mixture is cooled at constant pressure along the path 1-3, the cooling continues until the vapour attains the saturated state at point 3. With further cooling, the vapour condenses, *i.e.*, the moisture is released. The state point 3 represents the dew point temperature (*dpt*) of the air-water vapour mixture. The dew point temperature thus corresponds to the saturation temperature of steam at the partial pressure of water vapour in the mixture. *For saturated air, the dry bulb temperature, the wet bulb temperature and the dew point temperature is same.*

The adiabatic saturation temperature (or the thermodynamic wet bulb temperature) refers to the temperatures at which the air can be brought adiabatically to saturation state by the evaporation of water into a flowing stream.

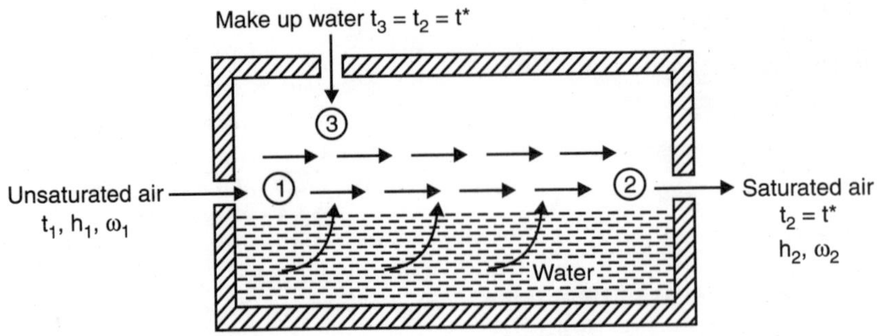

Fig. 25.33. Concept of adiabatic saturation process

Consider a stream of unsaturated air-vapour mixture flowing over a surface of water contained in a chamber which is sufficiently long and is perfectly insulated. Due to intimate contact between the unsaturated air and liquid water, some of the water evaporates and is carried by air resulting into an increase in its humidity. The heat required for evaporation comes both from the air-vapour mixture and the liquid water in the chamber. The process continues until the energy transferred from the air to the water is equal to the energy required to vaporise the water. By the time, air reaches the exit section, it becomes saturated and an equilibrium is established. The temperature of the saturated air at the exit section is known as ***adiabatic saturation temperature*** or the ***thermodynamic wet bulb temperature.*** The steady state thermal equilibrium conditions are maintained by adding make up water steadily at the rate of evaporation. This make up water has to be at the adiabatic saturation temperature, *i.e.*, the temperature of the mixture at the exit section.

With reference to Fig. 25.32, the adiabatic saturation process is represented by path 1-2. During the adiabatic saturation process, the partial pressure of vapour increases, although the total pressure of the air-vapour mixture remains constant. The unsaturated air at dry bulb temperature t_1 is cooled adiabatically to dry bulb temperature t_2 which is equal to adiabatic saturation temperature t^*. For all practical purposes, the adiabatic saturation temperature is taken equal to wet bulb temperature.

EXAMPLE 25.22 (a)

A metal beaker contains water initially at room temperature and the water is cooled by gradually adding ice water to it. When the water temperature reaches 13°C, the moisture from room air begins to condense on the beaker. Make calculations for the specific humidity and parts by mass of water vapour in the room air. Take:

room air temperature = 22°C and barometric pressure = 1.01325 bar

Solution : From steam tables, the partial pressure of water vapour at *dpt* of 13°C

$$p_v = 1500 \text{ N/m}^2$$

partial pressure of dry air

$$p_a = 101325 - 1500 = 99825 \text{ N/m}^2$$

(*a*) Specific humidity or humidity ratio,

$$\omega = \frac{m_v}{m_a} = 0.622\frac{p_v}{p_a}$$

$$= 0.622 \times \frac{1500}{99825}$$

$$= \mathbf{0.009346} \text{ kg/kg of dry air}$$

(*b*) Parts of mass of water vapour,

$$\frac{m_v}{m} = \frac{\omega}{1+\omega} = \frac{0.009346}{1+0.009346}$$

$$= \mathbf{0.00926} \text{ kg/kg of mixture}$$

EXAMPLE 25.23

The air supplied to an air-conditioned room is noted to be at temperature 20°C and specific humidity 0.0085. Corresponding to these conditions, detetrmine the partial pressure of vapour, relative humdiity and dew point temperature.

Take barometric or total pressure = 1.0132 bar

Solution : Specific humidity

$$\omega = 0.622\frac{p_v}{p_a} = 0.622\frac{p_v}{p_t - p_v}$$

Thus : $0.0085 = 0.622\dfrac{p_v}{1.0132 - p_v}$

∴ Partial pressure of vapour

$$p_v = \frac{1.0132 \times 0.0085}{0.622 + 0.0085} = \mathbf{0.01366} \text{ bar}$$

(*b*) The relative humidity is defined as $\phi = p_v/p_{vs}$. From steam tables, the saturation vapour pressure at 20°C = 0.0234 bar.

Then: $\phi = \dfrac{0.01366}{0.0234} = 0.5837$ or **58.38%**

(*c*) The dew point temperature is the saturation temperature of water at a pressure of 0.01366 bar

DPT (from steam tables by interpolation)

$$= 11 + (12 - 11) \times \frac{(0.01366 - 0.01312)}{0.01401 - 0.01312}$$

$= 11 + 0.607 =$ **11.607°C**

EXAMPLE 25.24

10 gm of water vapour was removed from a given sample of one kg of atmospheric air at 40°C and 60% relative humidity. The temperature of air after the removal of moisture reduced to 30°C. Determine the humidity ratio, partial pressure of vapour, relative humidity and dew point temperature of this air (air after removal of moisture).

Take total atmospheric pressure as 101.325 kPa.

Solution : The relative humidity is defined as $\phi = p_v/p_{vs}$. From steam tables, the saturation vapour pressure at 40°C = 7.384 kPa

That gives :

$$p_v = \phi\, p_{vs} = 0.6 \times 7.384 = 4.43 \text{ kPa}$$

Specific humidity or humidity ratio

$$\omega = 0.622\frac{p_v}{p_t - p_v}$$

$$= 0.622 \times \frac{4.43}{101.325 - 4.43}$$

$= 0.0284$ kg/kg of dry air

$= 28.4$ gm/kg of dry air

(*a*) Since 10 gm of water vapour per kg has been removed,

Specific humdity = 28.4 – 10 = 18.4 gm/kg

= **0.0184** kg/kg of dry air

(*b*) At this state, the air is at 30°C with specific humidity 0.0184 kg/kg of dry air. Then

$$0.0184 = 0.622\frac{p_v}{101.325 - p_v}$$

That gives: partial pressure of water vapour,

$$p_v = \frac{101.325 \times 0.0184}{0.0184 + 0.622} = \mathbf{2.92} \text{ kPa}$$

(*c*) At 30°C,

$p_{vs} = 4.246$ kPa (from steam tables)

∴ Relative humidity

$$\phi = \frac{p_v}{p_{vs}} = \frac{2.92}{4.246} = 0.688 \text{ or } 68.6\%$$

(*d*) The dew point temperature is the saturation temperature at the pressure of 2.92 kPa. Then from steam tables

DPT = 24°C

EXAMPLE 25.25

A mixture of air and water-vapour occupies a volume of 650 m³ at 1 bar pressure and 35°C temperature. If its relative humidity is 78 percent, calculate the specific humidity, the dew point and the masses of air and vapour in the mixture.

Solution : Relative humidity

$$\phi = \frac{m_v}{m_s} = \frac{p_v}{p_s}$$

From steam tables at 35°C : $p_s = 0.0563$ bar

∴ partial pressure of vapour

$p_v = \phi \times p_s = 0.78 \times 0.0563 = 0.0439$ bar

(*a*) Specific humidity or humidity ratio,

$$\omega = 0.622\frac{p_v}{p_a} = 0.622\frac{p_v}{p_t - p_v}$$

Taking total (barometric) pressure equal to 1.01325 bar,

$$\omega = 0.622 \times \frac{0.0439}{1.01325 - 0.0439}$$

$=$ **0.0282** kg/kg of dry air

(b) The dew point refers to the saturation temperature corresponding to the vapour pressure

Dew point = t_{sat} at 0.0439 bar

From steam tables: t_{sat} =29°C at 0.04 bar

and t_{sat} = 31°C at 0.045 bar

Then by interpolation, the saturation temperature at 0.0439 bar is

$$(t_{sat})_{0.0439\ bar} = 29 + (31 - 29) \times \frac{0.0439 - 0.04}{0.045 - 0.04}$$

$$= \mathbf{30.56°C}$$

(c) Partial pressure of air

$p_a = p_t - p_v$

$= 1.01325 - 0.0439 = 0.9693$ bar

Applying characteristic gas equation, $pV = mRT$, we have

$$\text{Mass of air } m_a = \frac{p_a V}{R_a T} = \frac{0.9693 \times 10^5 \times 650}{287 \times (273 + 35)}$$

$$= 712.75 \text{ kg}$$

Since $\omega = \frac{m_v}{m_a}$, we get

Mass of water vapour

$m_v = \omega \times m_a = 0.0282 \times 712.75 = \mathbf{20.1\ kg}$

EXAMPLE 25.26

An air-water vapour mixture at 20°C and 760 mm mercury has a relative humidity of 70 percent. Determine (a) partial pressures of vapour and air (b) humidity ratio (c) saturation ratio (d) dew point and (e) density of mixture.

Solution : Relative humidity $\phi = \frac{m_v}{m_s} = \frac{p_v}{p_s}$

From steam tables at 20°C : p_s = 0.0234 bar

∴ partial pressure of vapour $p_v = \phi \times p_s$

$= 0.70 \times 0.0234 = 0.01638$ bar

Corresponding to 760 mm mercury, the barometric pressure of the mixture is 1.01325 bar

∴ p_t = 1.01325 bar

and partial pressure of air

$p_a = p_t - p_v$

$= 1.01325 - 0.01638 = 0.99687$ bar

(b) Humidity ratio or specific humidity,

$$\omega = 0.622 \frac{p_v}{p_a} = 0.622 \times \frac{0.01638}{0.99687}$$

$= 0.01022$ kg/kg of dry air

(c) Saturation ratio, $\mu = \phi \frac{p_t - p_s}{p_b - p_v}$

$$= 0.7 \times \frac{1.01325 - 0.0234}{1.01325 - 0.01638} = \mathbf{0.695}$$

(d) The dew point refers to the saturation temperature corresponding to the vapour pressure.

Dew temperature = t_{sat} at 0.01638 bar

From steam tables:

t_{sat} = 13°C at 0.015 bar

t_{sat} = 17°C at 0.02 bar

Then by interpolation, the saturation temperature at 0.01638 bar is

$$(t_{sat})_{0.01638\ bar} = 13 + (17 - 13) \times \frac{0.01638 - 0.015}{0.02 - 0.015}$$

$= 14.104°C$

∴ Dew point = **14.104°C**

(e) Density of mixture = $\rho_a + \rho_v$

$= \rho_a + \omega\, \rho_a = \rho_a (1 + \omega)$

$$= \frac{p_a}{R_a T_a} (1 + \omega)$$

$$= \frac{0.99687 \times 10^5}{287 \times 293} (1 + 0.01022)$$

$= \mathbf{1.197\ kg/m^3}$

EXAMPLE 25.27

An air-water vapour mixture at 25°C and 1 bar has a relative humidity of 75 percent. Determine:

(a) the partial pressures of water vapour and dry air

(b) the specific volume of each constituent

(c) the dew point, the specific humidity and the saturation ratio

(d) the density of the mixture.

Also find the amount of water vapour condensed, if the mixture is cooled at constant pressure to a temperature of 10°C.

Solution : Corresponding to 25°C, the saturated vapour pressure p_s as read from the steam tables is

p_s = 0.0317 bar

∴ partial pressure of vapour

$p_v = \phi\, p_s = 0.75 \times 0.0317 = \mathbf{0.0237\ bar}$

and partial pressure of air

$$p_a = p_t - p_v$$
$$= 1 - 0.0237 = \mathbf{0.9762\ bar}$$

(*b*) Applying the perfect gas equation $pv = RT$ to both air and water vapour,

For air :

$$v_a = \frac{R_a T_a}{p_a} = \frac{287 \times 298}{0.9762 \times 10^5}$$
$$= \mathbf{0.876\ m^3/kg}$$

For vapour :

$$v_v = \frac{R_v T_v}{p_v} = \frac{(8314/18) \times 298}{0.0237 \times 10^5}$$
$$= \mathbf{57.91\ m^3/kg}$$

(*c*) The dew point refers to the saturation temperature corresponding to vapour pressure

Dew point = t_{sat} at 0.02377 bar

From steam tables:

t_{sat} = 17°C at 0.020 bar

t_{sat} = 21.1°C at 0.025 bar

Then by interpolation, the saturation temperature at 0.02377 bar is

$$= 17 + (21.1 - 17) \times \frac{0.02377 - 0.020}{0.025 - 0.20}$$
$$= \mathbf{20.09°C}$$

$$\text{Specific humidity } \omega = 0.622\ \frac{p_v}{p_a}$$
$$= 0.622 \times \frac{0.02377}{0.9762}$$
$$= \mathbf{0.0151\ kg/kg\ of\ dry\ air}$$

$$\text{Saturation ratio } \mu = \phi\ \frac{p_t - p_s}{p_t - p_v}$$
$$= 0.75 \times \frac{1 - 0.0317}{1 - 0.02377} = \mathbf{0.744}$$

(*d*) Density of the mixture,

$$\rho = \rho_a + \rho_v = \frac{1}{v_a} + \frac{1}{v_v}$$
$$= \frac{1}{0.876} + \frac{1}{57.91}$$
$$= 1.1415 + 0.01727 = 1.1588\ \text{kg/m}^3$$

Alternatively:

$$\rho = \rho_a\,(1 + \omega) = \frac{1}{0.876}\,(1 + 0.0151)$$
$$= 1.1588\ \text{kg/m}^3$$

(*e*) Since 10°C is below the dew point temperature of 20.09°C, there will be condensation of some of the vapour and the mixture will have a relative humidity of 100 percent.

At 10°C : p_s = 0.01227 bar

$$p_v = \phi\, p_s = 1 \times 0.01227 \text{ bar}$$
$$= 0.01227 \text{ bar}$$

$$\omega = 0.622\ \frac{p_v}{p_a}$$
$$= 0.622 \times \frac{0.01227}{1 - 0.01227}$$
$$= 0.007727\ \text{kg/kg of dry air}$$

∴ Amount of vapour condensed per kg of dry air

$$= 0.0151 - 0.007727 = \mathbf{0.007373\ kg}$$

EXAMPLE 25.28

Moist air at a total pressure of 760 mm of mercury occupies 3.2 m³ at 35°C; the relative humidity being 0.75. It is subsequently cooled at constant pressure to 5°C. Using the property values given below, determine the mass of moisture that gets deposited during the process.

Saturation temperature °C	*Pressure bar*	*Specific volume m³/kg*
35	0.0563	25.22
5	0.0087	147.1

Solution : Barometric pressure p_b

= 760 mm of Hg = 1.01325 bar

For initial state at 35°C,

$$\text{mass of water vapour present} = \frac{m}{v}\phi$$
$$= \frac{3.2}{25.22} \times 0.75 = 0.09516 \text{ kg}$$

partial pressure of water vapour

$$p_v = \phi \times p_{vs}$$
$$= 0.75 \times 0.0563 = 0.04222 \text{ bar}$$

partial pressure of air

$$p_a = p_b - p_v$$
$$= 1.01325 - 0.04222 = 0.97103 \text{ bar}$$

For final state at 5°C

Here the air would be saturated and therefore partial pressure of air

$$p_a = 1.01325 - 0.0087 = 1.00455 \text{ bar}$$

Applying characteristic gas equation to initial and final states, we have

$$\frac{p_1V_1}{T_1} = \frac{p_2V_2}{T_2}$$

(mass and gas constant same)

∴ Volume of air

$$V_2 = \frac{p_1V_1}{T_1} \times \frac{T_2}{p_2}$$

$$= \frac{0.97103 \times 3.2}{273+35} \times \frac{273+5}{1.00455} = 2.792 \text{ m}^3$$

$$\text{Mass of water vapour} = \frac{2.792}{147.1} = 0.01898 \text{ kg}$$

∴ Mass of moisture deposited

$$= 0.09516 - 0.01898 = \mathbf{0.07618 \text{ kg}}$$

EXAMPLE 25.29.

For a given sample of moist air, the wet bulb and dry bulb temperatures have been measured as 30°C and 45° respectively. Assuming total pressure of the mixture to be 1 bar, determine the following parameters for this air sample:

(a) vapour pressure

(b) dew point temperature

(c) specific humidity, relative humidity and the degree of saturation

(d) vapour density of air and

(e) enthalpy of the mixture.

Solution : The following data can be read from the steam tables :

At 45°C : $p_s = 0.0958$ bar;

$v_v = 15.28 \text{ m}^3/\text{kg}$

and $h_v = 2582$ kJ/kg

At 30°C : $p_s = 0.0425$ bar;

$v_v = 32.93 \text{ m}^3/\text{kg}$

and $h_v = 126$ kJ/kg.

(*a*) We may use the Carrier equation to determine the partial pressure of water vapour in air

$$p_v = (p_s)_{wbt} - \frac{[p-(p_s)_{wbt}][dbt - wbt]}{1547 - 1.44\, wbt}$$

$$= 0.0425 - \frac{(1-0.0425)(45-30)}{1547 - 1.44 \times 30}$$

$$= 0.0425 - 0.00955 = \mathbf{0.03285 \text{ bar}}$$

(*b*) The dew point refers to the saturation temperature corresponding to the vapour pressure.

Dew point temperature

$$= t_{sat} \text{ at } 0.03285 \text{ bar}$$

From steam tables :

$t_{sat} = 24.1$°C at 0.030 bar

$t_{sat} = 26.7$°C at 0.035 bar

Then by interpolation, the saturation temperature at 0.03285 bar is

$$= 24.1 + (26.7 - 24.1) \times \frac{0.03285 - 0.030}{0.035 - 0.030}$$

$$= \mathbf{25.582°C}$$

(*c*) Specific humidity or humidity ratio

$$\omega = 0.622\, \frac{p_v}{p_t - p_v}$$

$$= 0.622 \times \frac{0.03285}{1 - 0.03285}$$

$$= \mathbf{0.02113} \text{ kg/kg of dry air}$$

$$\text{Relative humidity } \phi = \frac{p_v}{p_s}$$

$$= \frac{0.03285}{0.0958} = \mathbf{0.3429}$$

$$\text{Saturated ratio } \mu = \phi \frac{p_t - p_s}{p_t - p_v}$$

$$= 0.3429 \times \frac{1-0.0958}{1-0.03285} = \mathbf{0.32}$$

(*d*) Vapour density of saturated air is the reciprocal of specific volume of saturated air

$$= \frac{1}{15.28} = 0.0654 \text{ kg/m}^3$$

For the moist air with relative humidity 0.3429, the vapour density would be

$$= 0.0654 \times 0.3429$$

$$= \mathbf{0.02243 \text{ kg/m}^3}$$

(e) Enthalpy of mixture,

= enthalpy of dry air + enthalpy of vapour associated with dry air

$= c_{pa}\ dbt + \omega \times (h_v)_{dbt}$

$= 1.005 \times 45 + 0.02113 \times 2582$

= **99.78** kJ/kg of dry air

25.11. PSYCHROMETRIC CHART AND PROCESSES

The subject which deals with the behaviour of moist air is known as psychrometry, and the properties of moist air are called ***psychrometric properties***.

Humidity calculations can be made by using the equations relating the dry and wet bulb temperatures to the humidity. The method, however, tends to be tedious, cumbersome and time consuming. The key to humidity calculations is then provided by the ***Psychrometric*** or ***Hygrometric chart*** which graphically describes the relationship between the properties of moist air, *i.e.*, the dry bulb, the wet bulb and dew point temperatures of the mixture and its humidity. Fig. 25.34 shows how these parameters are laid out on a typical psychrometric chart.

The psychrometic chart has the number of details and its salient aspects are:

1. The dry bulb temperature is taken as abscissa and specific humidity (*i.e.* moisture content) as ordinate.

The dry bulb temperature lines are vertical and uniformly spaced. The specific humidity lines are horizontal and also uniformly spaced. The saturation curve is drawn by plotting the various saturation points at corresponding dry bulb temperatures. The saturation curve

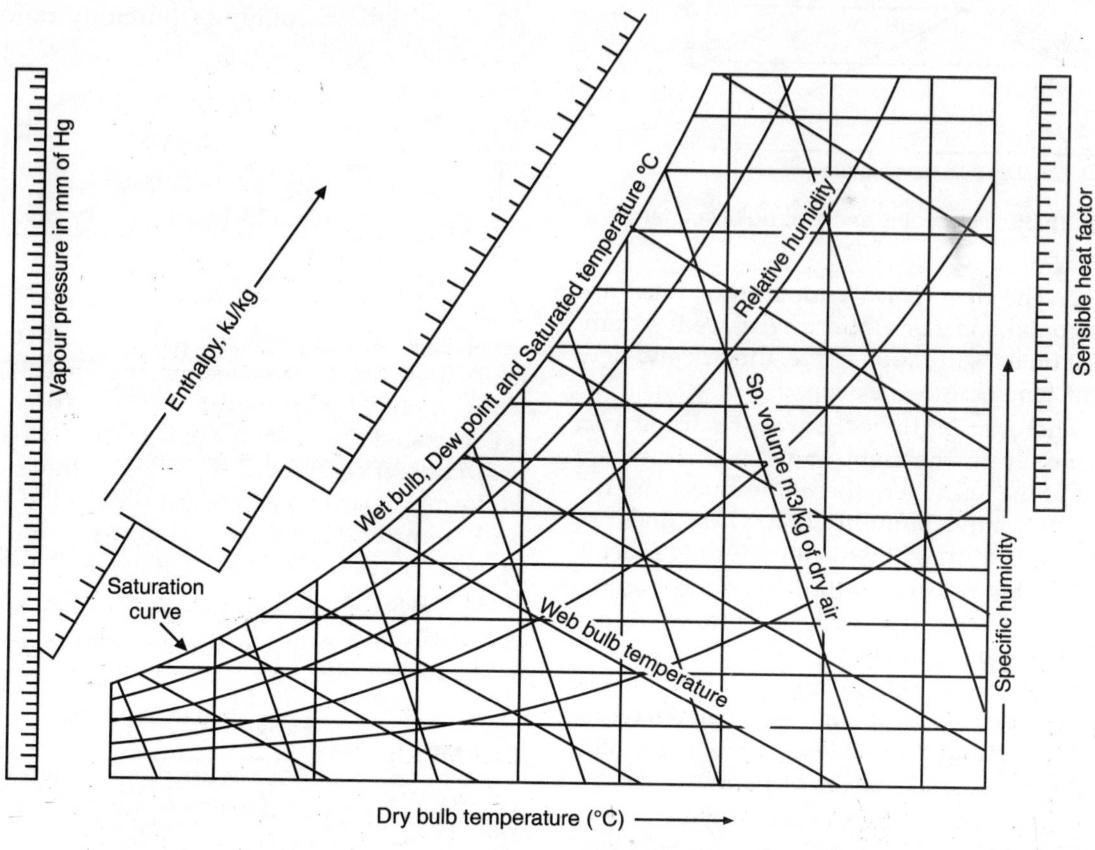

Fig. 25.34. Psychrometric chart

represents 100 percent relative humidity at various dry bulb temperatures. It also indicates the wet bulb and dew point temperatures.

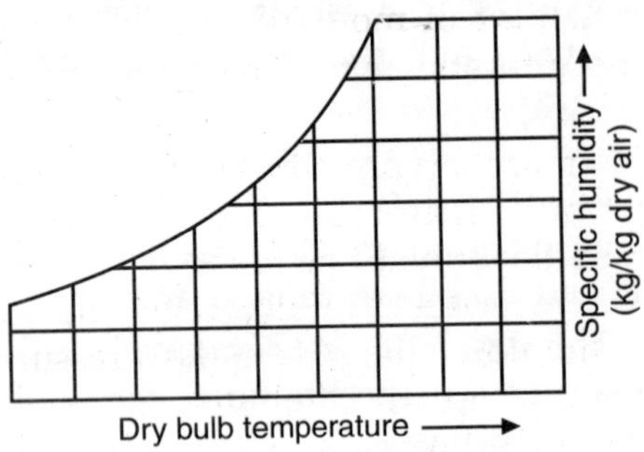

Fig. 25.35. Dry bulb and specific humidity lines

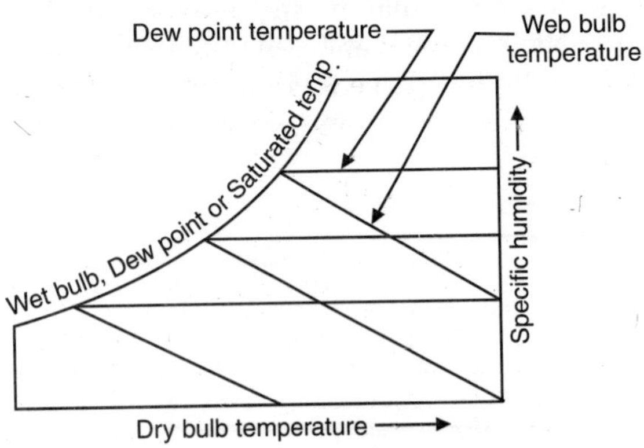

Fig. 25.36. Dew point and wet bulb temperature lines

2. The dew point temperature lines are horizontal and non-uniformly spaced. At any point on the saturation curve, the dry and dew point temperatures are equal.

The wet bulb temperature lines run diagonally to the right and their values are read at the left where these lines meet the 100 percent relative humidity line. These lines are inclined and straight but not uniformly spaced.

3. The relative humidity lines curve upwards to the right with the percent values indicated on the lines themselves. The relative humidity curve depicts quantity of moisture actually present in the air as a percentage of the total amount possible at various dry bulb temperatures and masses of vapour.

The specific volume (volume of air-vapour mixture per kg of dry air) lines are indicated by obliquely inclined straight lines. These lines are uniformly spaced and are drawn upto the saturation curve.

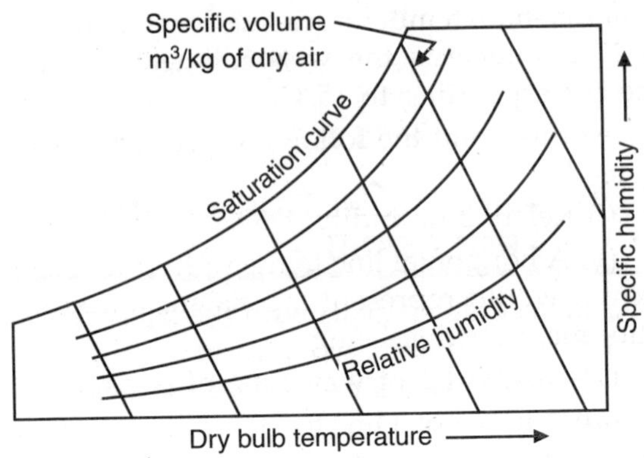

Fig. 25.37. Relative humidity and specific volume lines

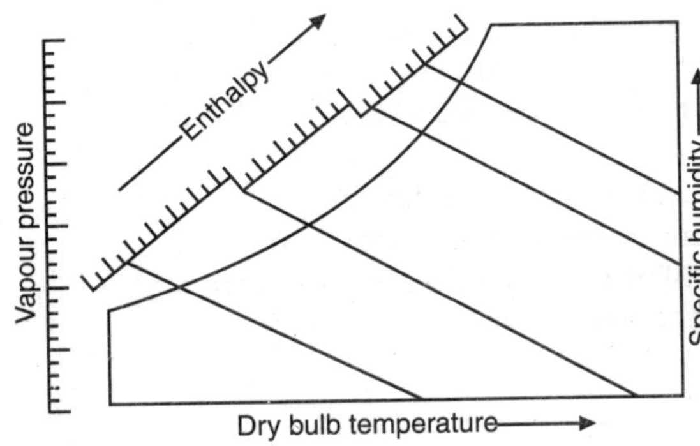

Fig. 25.38. Enthalpy and vapour pressure lines

4. The vapour pressure and enthalpy (total heat) lines are also scaled on the chart. The total heat at saturation temperature is represented by a diagonal system of co-ordinates. These inclined straight lines are uniformly spaced and are parallel to the wet bulb temperature lines. The scale on the diagonal lines is separate from the body of the chart and is indicated above the saturation line.

Pressure of water vapour is shown in the scale on left and is the absolute pressure of steam in mm of mercury.

EXAMPLE 25.30

Atmospheric air at 1 bar pressure has 15°C wet-bulb temperature and 25°C dry-bulb temperature. With the help of a psychrometric chart, determine the salient psychrometric properties of the air.

Solution : Refer Fig. 25.39.

Locate the state point *S* on the psychrometric chart. This is the point where a constant dry-bulb temperature line corresponding to 25°C intersects the wet-bulb temperature line corresponding to 15°C.

At this point the following properties can be read:

(*i*) Relative humidity = 33.5%

(*ii*) A horizontal line to the right from state point *S* would represent a constant humidity ratio line; its value being

$\omega = 0.0067$ kg of water/kg of dry air

(*iii*) A horizontal line to the left drawn from the state point *S* would intersect the saturation line; the temperature at the intersecting point gives the value of dew point temperature equal to *dpt* = 8°C.

(*iv*) The constant wet-bulb temperature line is produced to meet the enthalpy line ; the enthalpy value approximates to 43 kJ/kg of dry air.

(*v*) A horizontal line drawn from the state point *S* meets the vapour pressure scale at 0.0105 bar. This represents the partial pressure of the vapour.

(*vi*) Saturation pressure corresponding to *dbt* = 25°C can be obtained by drawing a horizontal line to the left from the point where *dbt* = 25°C line cuts the 100% relative humidity line. The saturation pressure value is 0.0316 bar.

(*vii*) Saturation pressure corresponding to the wet-bulb temperature is obtained by drawing a horizontal line to the left from the point where *wbt* = 15°C line cuts the 100% relative humidity line. Here the pressure is scaled as 0.0166 bar.

(*viii*) Humidity ratio corresponding 25°C dry-bulb temperature can be obtained by drawing a horizontal line to the right from the point where *dbt* = 25°C line cuts the 100% relative humidity line. The corresponding value is scaled as 0.0202 kg of water/kg of dry air.

The degree of saturation is then calculated as:

$$\mu = \frac{\omega}{\omega_s} = \frac{0.0067}{0.0202} = 33\%$$

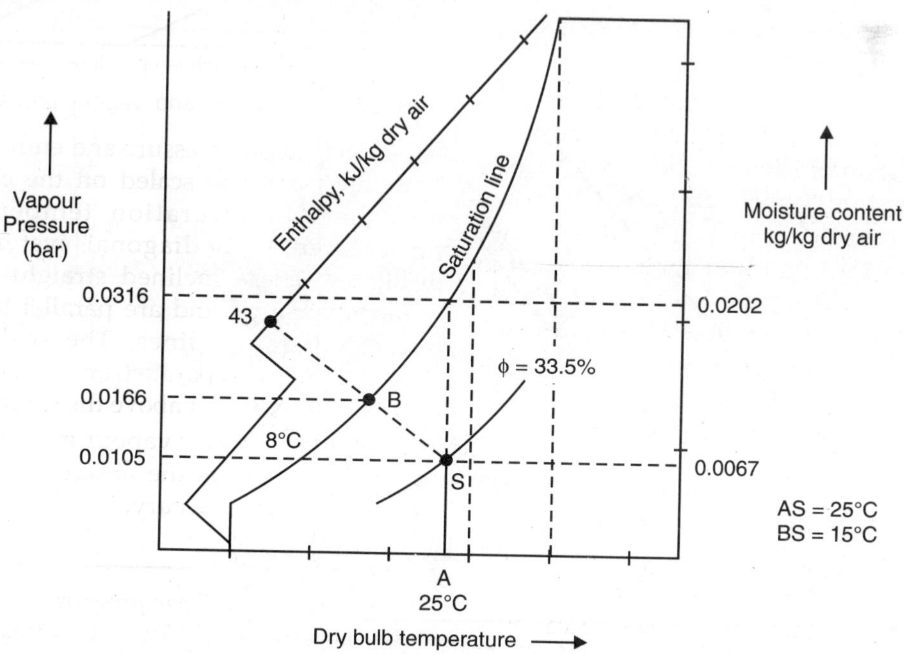

Fig. 25.39.

EXAMPLE 25.31

The atmospheric air at 35°C dbt and 25°C wbt is cooled to 20°C. Using psychrometric chart, find (i) humidity ratio, (ii) relative humidity and (iii) dew point temperature. Comment whether or not there will be any condensation on the duct that supplies this cooled air to the conditioned space.

Solution : Locate the condition of air at 35°C *dbt* and 25°C *wbt* (Point 1) on the psychrometric chart and read therefrom :

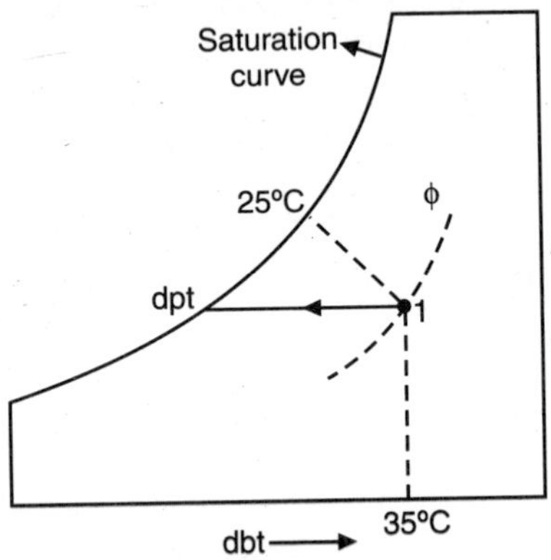

Fig. 25.40.

Humidity ratio ω = 0.016 kg/kg of dry air

Relative humidity ϕ = 0.45 or 45%

Dew point temperature *dpt* = 18.4°C

The condensation of moisture begins to take place when the dew point temperature has been reached. Since the air is cooled only to 20°C which is higher than the dew point temperature, there will be no condensation of moisture on the duct surface.

25.12. PSYCHROMETRIC PROCESSES

The most common processes in air-conditioning which involve the air-water vapour mixture are mentioned below and indicated on the psychrometric chart in Fig. 25.41.

The initial state is represented by point 0. Heating and cooling processes are represented by horizontal lines while humidification and dehumidification plot as vertical lines. Humidification represents the process wherein the moisture is added to air but its dry bulb temperature is maintained constant. In dehumification process, the moisture is removed from air without changing its dry bulb temperature. The combination of these processes are set between them.

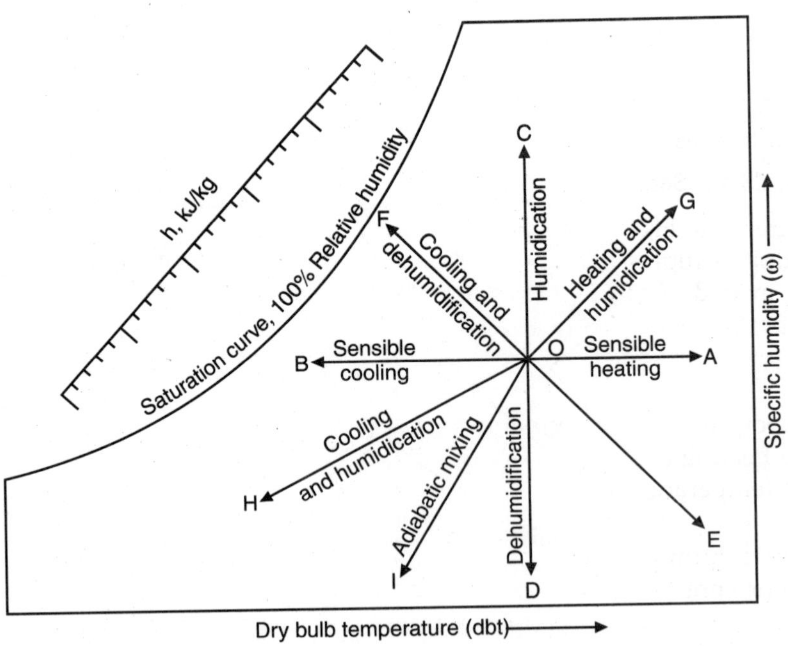

Fig. 25.41. Psychrometric processes

25.12.1. Sensible heating

The mixture is heated without any change in its moisture content. The process results when the mixture is made to pass over a surface whose temperature is above the dry bulb-temperature of the mixture. The heating surface may be the electric resistance heating coils or steam passed through the coils or hot water passed through the coils.

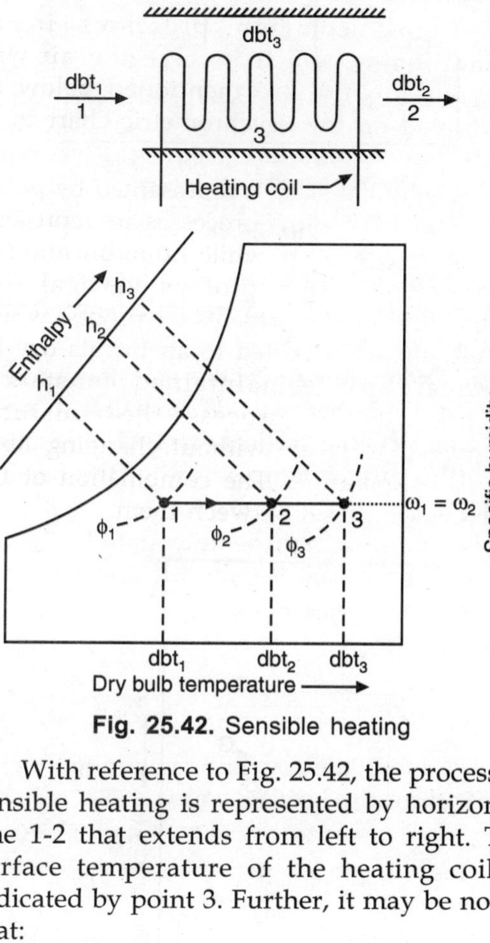

Fig. 25.42. Sensible heating

With reference to Fig. 25.42, the process of sensible heating is represented by horizontal line 1-2 that extends from left to right. The surface temperature of the heating coil is indicated by point 3. Further, it may be noted that:

(*i*) Air-vapour mixture at temperature dbt_1 passes over a heating coil of temperature dbt_3 and acquires temperature dbt_2 at exit; dbt_2 is less than dbt_3.

(*ii*) Specific humidity during sensible heating remains constant, *i.e.*, $\omega_1 = \omega_2$

(*iii*) Dry bulb temperature increases from dbt_1 to dbt_2 and the relative humidity reduces from ϕ_1 to ϕ_2.

(*iv*) The amount of heat absorbed by the mixture can be obtained from the psychrometric chart by the enthalpy difference $(h_2 - h_1)$. Alternatively, it can be calculated from the relation:

$$\begin{aligned} h &= h_2 - h_1 \\ &= c_{pa}\,(dbt_2 - dbt_1) + \omega\, c_{pv}\,(dbt_2 - dbt_1) \\ &= (c_{pa} + \omega\, c_{pv})\,(dbt_2 - dbt_1) \\ &= c_{pm}\,(dbt_2 - dbt_1) \end{aligned}$$

The factor $c_{pm} = (c_{pa} + \omega\, c_{pv})$ is called ***humid specific heat*** and its value is generally taken as 1.022 kJ/kg K.

(*v*) At exit, the air-vapour mixture ideally should attain temperature dbt_3 of the heating coil but it leaves at dbt_2. The difference is due to inefficiency of the heating coil and this is a prescribed by the terms ***by-pass factor*** and ***coil-efficiency*** defined as

$$\text{By-pass factor } (BPF) = \frac{dbt_3 - dbt_2}{dbt_3 - dbt_1}$$

and Coil efficiency $\eta_c = 1 - BPF$

The coil efficiency is also called ***contact factor.***

The factor $(1 - BPF)$ is the mass actually coming in perfect contact with the coil. The by-pass factor is an important tool for measuring the performance of the heating coils. The coil gives better performances with lower by-pass factor.

EXAMPLE 25.32

The moist air at 10°C and 50% relative humidity enters a steam heating coil at the rate of 50 kg/sec and the temperature at exit is noted to be 30°C. Determine :

(a) sensible heat transfer,

(b) mass flow rate of steam if it enters saturated at 100°C and the condensate leaves at 65°C.

Solution : Refer Fig. 25.43.

(*i*) Locate the initial state point 1 on psychrometric chart as per dbt_1 = 10°C and ϕ_1 = 50%.

(*ii*) The state point 2 is located by drawing a constant specific humidity line from point 1 to intersect the vertical line drawn through 30°C *dbt* at point 2.

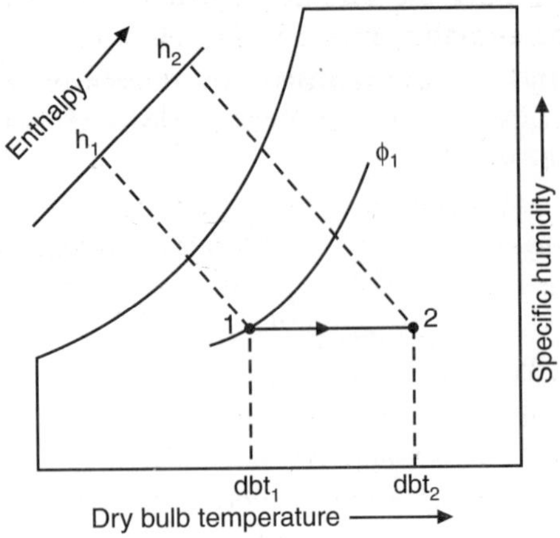

Fig. 25.43.

The line 1-2 represents sensible heating of air.

From the chart :

h_1 = 19.5 kJ/kg of dry air and

h_2 = 40 kJ/kg of dry air

Then :

Sensible heat transfer $Q = m_a (h_2 - h_1)$

$= 50(40 - 19.5) = \mathbf{1025}$ kJ/s

(*b*) From steam tables :

Enthalpy of saturated steam at 100°C,

h_g = 2676 kJ/kg

Enthalpy of condensate at 65°C,

h_f = 272 kJ/kg

Then : steam mass flow rate

$$m_s = \frac{Q}{h_g - h_f} = \frac{1025}{2676 - 272}$$

$$= \mathbf{0.426} \text{ kg/s}$$

EXAMPLE 25.33

A sample of atmospheric air having 15°C dry bulb temperature and 10°C wet bulb temperature is made to enter a heating coil whose surface temperature is 40°C. Using psychrometric chart, determine :

(a) dry bulb temperature, wet bulb temperature and relative humidity of air at exit from the heating coil, and

(b) heat added to the air.

It may be assumed that the heating coil has a by-pass factor of 0.5.

Solution : Refer Fig. 25.44. The situation corresponds to sensible heating, *i.e.*, heating of air without any change in specific humidity.

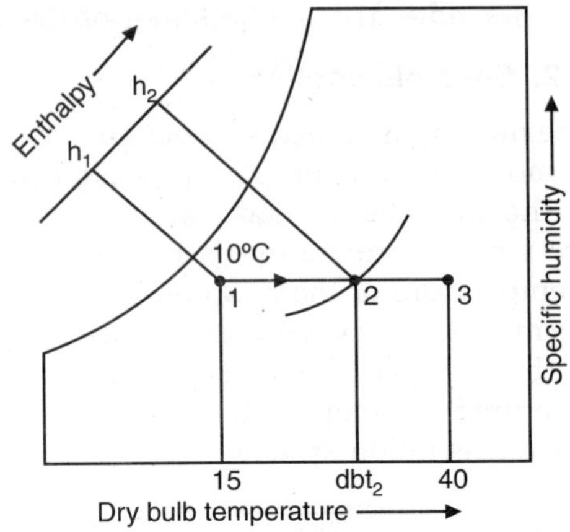

Fig. 25.44.

(*i*) Mark point 1 corresponding to dry bulb temperature of 15°C and wet bulb temperature of 10°C.

(*ii*) Draw a horizontal line (constant specific humidity) 1-3 that extends from left to right and intersects the vertical line drawn through 40°C (the surface temperature of the heating coil).

The point 2, representing state of the air on leaving the coil, lies on this line.

We know that by-pass factor in sensible heating is

$$BPF = \frac{dbt_3 - dbt_2}{dbt_3 - dbt_1} \quad \text{or} \quad 0.5 = \frac{40 - dbt_2}{40 - 15}$$

Therefore, the dry bulb temperature of air at exit from the heating coil is

$$dbt_2 = 40 - 0.5\,(40 - 15) = 27.5°\text{C}$$

This helps to locate point 2 on the sensible heating line 1-3.

The following parameters can then be read from the psychrometric chart :

wet bulb temperature of air leaving the coil at point 2, wbt_2 = 16.5°C

relative humidity of air leaving the coil at point 2, ϕ_2 = 29.5%

enthalpy of air when leaving the coil at point 2, h_2 = 45 kJ/kg dry air enthalpy of air when entering the coil at point 1, h_1 = 31.5 kJ/kg dry air

Then, heat added to air $h = h_2 - h_1$

$= 45 - 31.5 = 13.5$ kJ/kg of dry air

25.12.2. Sensible cooling

The mixture is cooled without any change in its moisture content. The process results when the mixture is made to pass over a surface whose temperature is below the dry bulb temperature of the mixture. The cooling surface may be the cooled water or gas flowing through coils or the refrigerant at low temperature in the coils of the evaporator of a vapour refrigeration system.

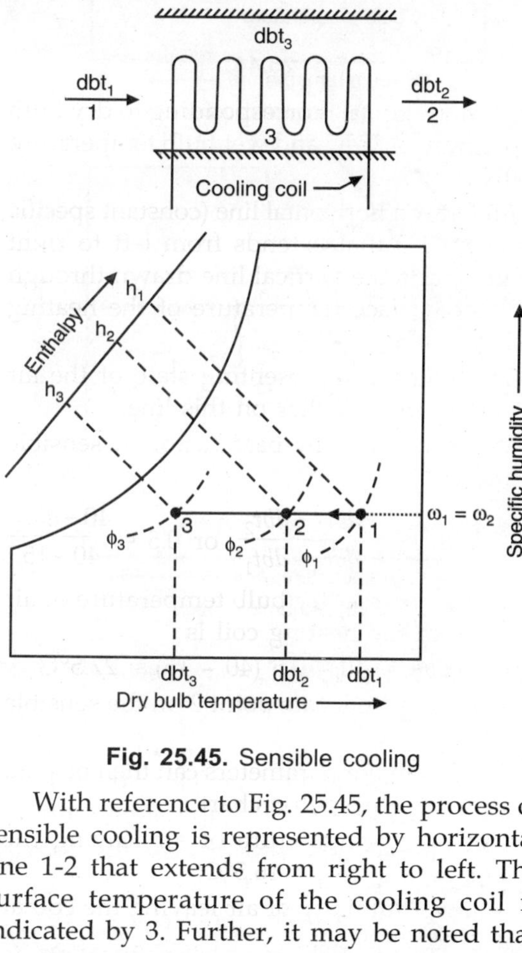

Fig. 25.45. Sensible cooling

With reference to Fig. 25.45, the process of sensible cooling is represented by horizontal line 1-2 that extends from right to left. The surface temperature of the cooling coil is indicated by 3. Further, it may be noted that

(*i*) Air-vapour mixture at temperature dbt_1 passes other a cooling coil of temperature dbt_3 and acquires temperature dbt_2 at exit; dbt_2 is more than dbt_3.

(*ii*) Specific humidity during sensible cooling remains constant, *i.e.*, $\omega_1 = \omega_2$.

(*iii*) Dry bulb temperature decreases from dbt_1 to dbt_2 and the relative humidity increases from ϕ_1 to ϕ_2.

(*iv*) The amount of heat rejected by the mixture can by obtained from the psychrometric chart by the enthalpy difference $(h_1 - h_2)$. Alternatively, it can be calculated from the relation:

$$\begin{aligned} h &= h_1 - h_2 \\ &= c_{pa}\,(dbt_1 - dbt_2) + \omega\, c_{pv}\,(dbt_1 - dbt_2) \\ &= (c_{pa} + \omega\, c_{pv})\,(dbt_1 - dbt_2) \\ &= c_{pm}\,(dbt_1 - dbt_2) \end{aligned}$$

The factor $c_{pm} = (c_{pa} + \omega\, c_{pv})$ is called ***humid specific heat*** and its value is generally taken as 1.022 kJ/kg K.

(*v*) At exit, the air-vapour mixture should ideally attain temperature dbt_3 of the cooling coil but it leaves at dbt_2. The difference is due to inefficiency of the cooling coil and this is prescribed by the terms ***by-pass factor*** and ***coil efficiency*** defined as

$$\text{By-pass factor } (BPF) = \frac{dbt_2 - dbt_3}{dbt_1 - dbt_3}$$

and $\quad$ Coil efficiency $\eta_c = 1 - BPF$

EXAMPLE 25.34

Atmospheric air at 35°C dbt and 55% relative humidity has been cooled to a state corresponding to 25°C dbt and 50% relative humidity. Determine the capacity of the refrigerating machine if it is to handle 125 m³ of air per minute at initial state.

Solution : Refer Fig. 25.46.

(*i*) Set the initial state 1 according to

$$dbt_1 = 35°\text{C} \quad \text{and} \quad \phi_1 = 55°\text{C}$$

(*ii*) Locate point 2 by drawing a horizontal line through point 1 such that dbt_2 = 25°C and ϕ_2 = 50°C

The following values are then read from the chart :

$$h_1 = 85.8 \text{ kJ/kg}\ ; \quad v_{s_1} = 0.9 \text{ m}^3\text{/kg}$$
$$h_2 = 49.8 \text{ kJ/kg}\ ; \quad v_{s_2} = 0.85 \text{ m}^3\text{/kg}$$

Then : mass of air cooled

$$m_a = \frac{V}{v_{s_1}} = \frac{125}{0.9} = 138.38 \text{ kg/min}$$

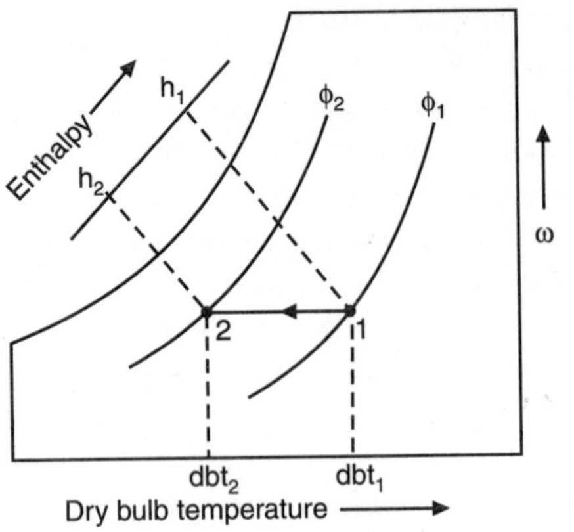

Fig. 25.46.

Heat removed by the refrigerating machine (refrigerating effect)

$$= m_a\,(h_1 - h_2)$$

$$= 138.38\,(85.8 - 49.8) = 4981.68 \text{ kJ/min}$$

Since 1 ton of refrigeration is equivalent to 210 kJ/min,

Capacity of refrigerating machine

$$= \frac{4981.68}{210} = 23.72 \text{ ton}$$

25.12.3. Humidification and dehumidification

Humidification represents the process wherein the moisture is added but its dry bulb temperature is maintained constant. In dehumidification process, the moisture is removed from air without changing its dry bulb temperature. These processes are obviously represented as vertical lines on the psychrometric chart.

With reference to Fig. 25.47, it may be noted that

(*i*) In humidification the specific humidity increases from ω_1 to ω_2 and the relative humidity also increases from ϕ_1 to ϕ_2.

However in dehumidification, there is decrease in specific humidity from ω_1 to ω_2 and also a decrease in relative humidity from ϕ_1 to ϕ_2.

(*ii*) The increase in enthalpy in humidification is represented by the intercept $(h_2 - h_1)$. Since there is no change in dry bulb temperature during humidification, the sensible heat would remain constant. As such the enthalpy change $(h_2 - h_1)$ equals the latent heat of vaporisation of the increased moisture content $(\omega_2 - \omega_1)$.

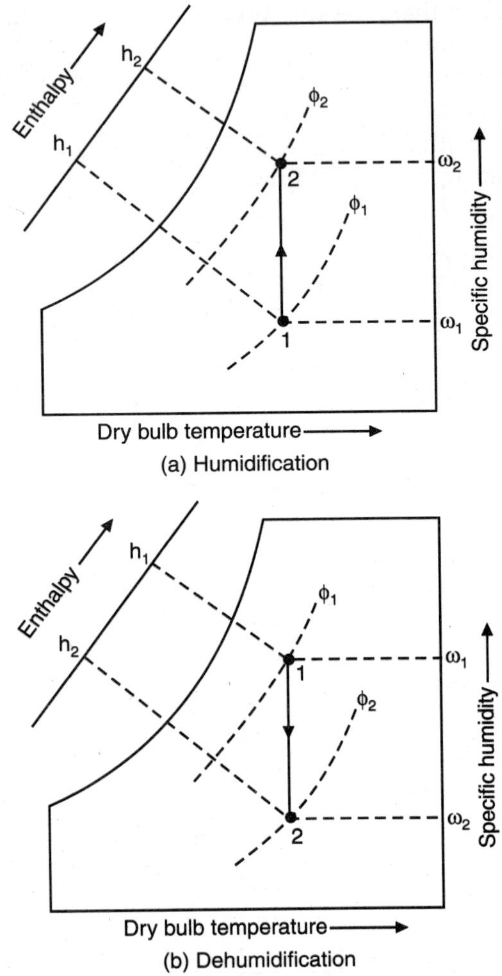

Fig. 25.47. Humidification and dehumidificaion processes

$$\therefore \quad h_2 - h_1 = (h_{fg})_{db}\,(\omega_2 - \omega_1)$$

where $(h_{fg})_{db}$ is the latent heat of vaporisation at the dry bulb temperature *dbt*

For dehumidification process, the decrease in enthalpy can be written as

$$h_1 - h_2 = (h_{fg})_{db}\,(\omega_1 - \omega_2)$$

(*iii*) In practice, pure humidification and dehumidificaton processes are not possible. These are always accompanied by heating or cooling.

25.12.4. Heating and humidification

The process is achieved when the moist air is made to pass through spray water whose temperature is maintained at a temperature higher than the dry bulb temperature of the air. The unsaturated air tends to become saturated and the heat of vaporisation is absorbed from the spray water.

With reference to Fig. 25.48, the process of heating and humidification is represented by line 1-2.

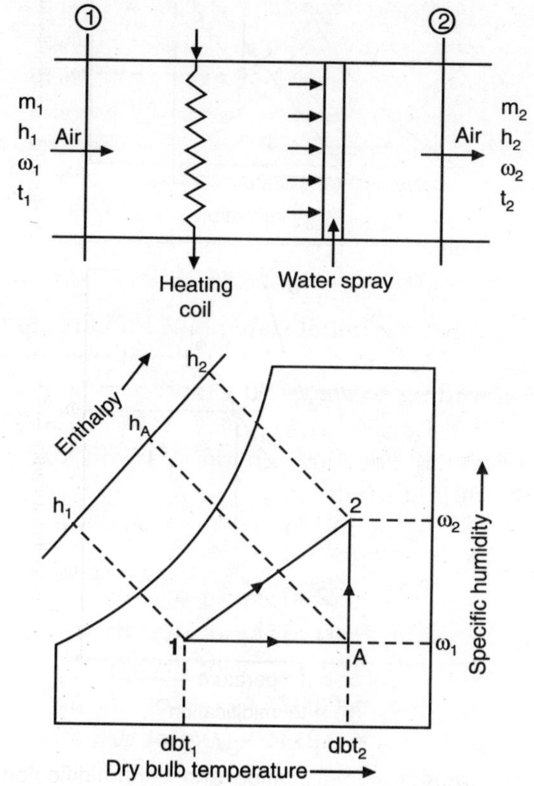

Fig. 25.48. Heating and humidification

The air-vapour mixture enters at condition 1 and leaves at condition 2. It is to be noted that during the heating and humidification process,

(*i*) There is an increase in specific humidity, dry and wet bulb temperatures, dew point temperature and enthalpy.

(*ii*) The relative humidity may either increase or decrease.

(*iii*) The process may be assumed to follow the paths 1-A and A-2. Path 1-A represents sensible heating during which the dry bulb temperature increases. Path A-2 represents humidification during which, there is no change in dry bulb temperature and latent heat of vaporisation is supplied with increased moisture content $(\omega_2 - \omega_1)$. Then

$$\text{Total heat added } h = h_2 - h_1$$
$$= (h_2 - h_A) + (h_A - h_1)$$
$$= \text{latent heat + sensible heat}$$

The ratio of sensible heat to total heat is known as ***sensible heat factor*** (*SHF*) or ***sensible heat ratio*** (*SHR*)

SHF can be read from a scale on the psychrometric chart located in the right corner of the chart

$$SHF = \frac{\text{sensible heat}}{\text{total heat}}$$

$$= \frac{\text{sensible heat}}{\text{sensible heat + latent heat}}$$

The process of heating and humidification has practical application in winter air conditioning.

25.12.5. Cooling and dehumidification

The process takes place when the moist air is passed through a cooling coil whose effective surface temperature is lower than the dew point temperature of mixture. The surface temperature of the cooling coil is called the ***apparatus dew point*** (**ADP**)

With reference to Fig. 25.49, the process of cooling and dehumidification is represented by line 1-2. The moist air enters the cooling coil at condition 1 and leaves at condition 2. It is to be noted that during the cooling and dehumidification process,

(*i*) the dry bulb temperature decreases from dbt_1 to dbt_2

(*ii*) the air is cooled and condensation of moisture takes place, *i.e.*, it is dehumidified

(*iii*) the specific humidity decreases from ω_1 to ω_2

(*iv*) the relative humidity at outlet is generally higher than that at inlet (*v*) the process may be assumed to follow the path 1-A and A-2. Path 1-A represents sensible cooling during which the dry bulb temperature decreases. Path A-2 represents dehumidification during which there is no change in the dry bulb temperature and latent heat condensation is removed with decreased moisture content $(\omega_1 - \omega_2)$

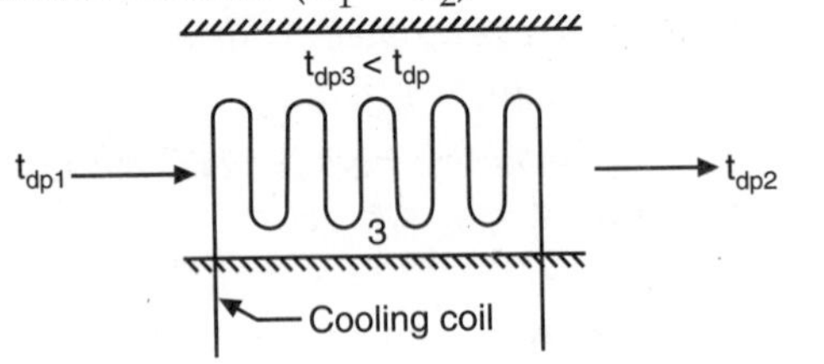

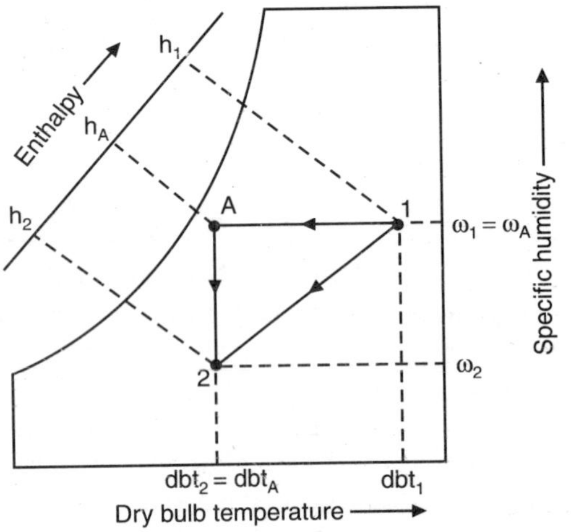

Fig. 25.49. Cooling and dehumidification

Total heat removed $h = h_1 - h_2$

$= (h_1 - h_A) + (h_A - h_2)$

= latent heat + sensible heat

The ratio of sensible heat to total heat is known as ***sensible heat factor*** (*SHF*)

$$SHF = \frac{\text{sensible heat}}{\text{total heat}}$$

$$= \frac{\text{sensible heat}}{\text{sensible heat + latent heat}}$$

The cooling and dehumidification process has practical application in summer air-conditioning.

EXAMPLE 25.35

Air at 15°C dry bulb temperature and 25% relative humidity is heated and humidified at 30°C dry bulb temperature and 50% relative humidity. Using psychrometric chart, calculate the heat and moisture added to air and the sensible heat factor for the process.

Solution : Refer Fig. 25.50 which shows the given process laid on psychrometric chart.

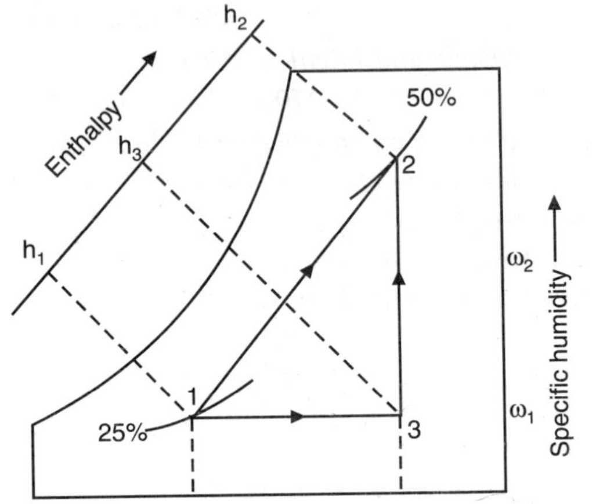

Fig. 25.50.

(*i*) Set the initial state point 1 according to dbt_1 = 15°C and ϕ_1 = 25% and final state point 2 according to dbt_2 = 30°C and ϕ_2 = 50%.

(*ii*) Locate the point 3 by drawing horizontal line through point 1 and vertical line through point 2.

The following values are then read from the chart :

h_1 = 22 kJ/kg of dry air

h_2 = 65 kJ/kg of dry air

h_3 = 38 kJ/kg of dry air

ω_1 = 0.0027 kg/kg of dry air

ω_2 = 0.0134 kg/kg of dry air

Then :

(*a*) Heat added to air = $h_2 - h_1$

= 65 – 22 = **43** kJ/kg of dry air

(*b*) Moisture added to air = $\omega_2 - \omega_1$

= 0.0134 – 0.0027

= **0.0107** kg/kg of dry air

(*c*) Sensible heat factor

$$SHF = \frac{\text{sensitble heat}}{\text{total heat}}$$

$$= \frac{h_3 - h_1}{h_2 - h_1} = \frac{38 - 22}{65 - 22} = \mathbf{0.372}$$

EXAMPLE 25.36

The following data refers to summer air-conditioning of a small office hall with capacity for 30 persons :

Outside conditions : 34°C dbt and 28°C wbt

Inside design conditions :

24°C dbt and 50% RH

Sensible and latent heat load in the room

2100 kJ/min and 700 kJ/min

Volume of air supplied : 0.4 m³/min per person.

Make calculations for sensible heat factor for the air-conditioning plant.

Solution : Refer Fig. 25.51 which shows the given process laid on psychrometric chart.

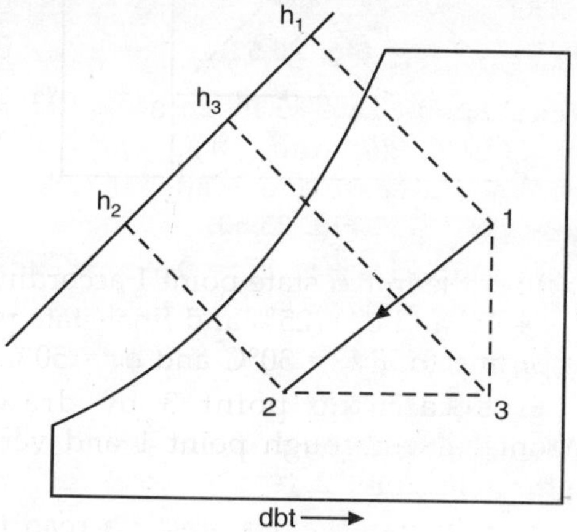

Fig. 25.51.

(*i*) Locate points 1 and 2 representing outdoor air (34°C *dbt* and 28°C *wbt*) and room air (24°C dbt and 50% RH) conditions respectively.

(*ii*) Draw a horizontal line towards right from point 2 and a vertical line downwards from point 1 and mark point 3 where these lines intersect.

The horizontal and vertical lines represent cooling and dehumidification respectively.

From the chart :

h_1 = 90 kJ/kg of dry air

v_1 = 0.9 m³/kg of dry air

h_2 = 48 kJ/kg of dry air

h_3 = 58.8 kJ/kg of dry air

Then, mass of air supplied

$$m_a = \frac{0.4 \times 30}{0.9} = 13.33 \text{ kg/min}$$

Sensible heat removed from air

$= m_a (h_3 - h_2)$

$= 13.33 (58.8 - 48) = 143.96$ kJ/min

∴ Total sensible heat

$SH = 2100 + 143.96 = 2243.96$ kJ/min

Latent heat removed from air

$= m_a (h_1 - h_3)$

$= 13.33 (90 - 58.8) = 415.9$ kJ/min

∴ Total latent heat $LH = 700 + 415.9$

$= 1115.9$ kJ/min

Thus sensible heat factor

$$SHF = \frac{SH}{SH + LT} = \frac{2243.96}{2243.9 + 1115.9} = \mathbf{0.668}$$

25.12.6. Adiabatic mixing

Let two streams of moist air having different specific humidities and enthalpies be allowed to mix without the addition or rejection of either heat or moisture, *i.e.*, adiabatically and at constant total moisture content.

With reference to Fig. 25.52, the state of the resultant mixture lies on the straight line that joins the state of two streams on psychrometric chart. The location of final state on the straight line depends on the masses involved, and on the enthalpy and specific humidity of each stream.

The fundamental equations of the process are:

Mass balance: $m_1 + m_2 = m_3$ ...(*i*)

Energy balance: $m_1h_1 + m_2h_2 = m_3h_3$ ...(*ii*)

Mass balance of water vapour:

$m_1 \omega_1 + m_2 \omega_2 = m_3 \omega_3$ ...(*iii*)

From expressions (*i*) and (*ii*),

$$m_1 h_1 + m_2 h_2 = (m_1 + m_2) h_3 = m_1 h_3 + m_2 h_3$$

or $$m_1 (h_1 - h_3) = m_2 (h_3 - h_2)$$

where *m*, *h*, ω represent the mass, enthalpy and specific humidity of moist air at a particular state point.

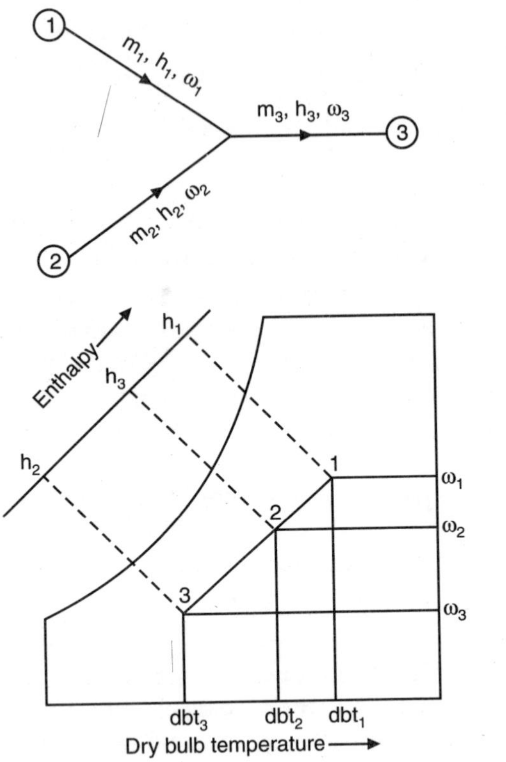

Fig. 25.52. Adiabatic mixing of two air streams

That gives:

$$\frac{m_1}{m_2} = \frac{h_3 - h_2}{h_1 - h_3} \qquad ...(a)$$

Likewise from expressions (*i*) and (*iii*)

$$\frac{m_1}{m_2} = \frac{\omega_3 - \omega_2}{\omega_1 - \omega_3} \qquad ...(b)$$

Thus from identities (*a*) and (*b*), we have

$$\frac{m_1}{m_2} = \frac{h_3 - h_2}{h_1 - h_3} = \frac{\omega_3 - \omega_2}{\omega_1 - \omega_3}$$

Obviously, the final state point 3 divides the line 1-2 in the inverse ratio of the mixing masses.

EXAMPLE 25.37

120 m³/min of air at 35°C dry bulb temperature and 45% relative humidity is mixed with 325 m³/min of recirculated air at 20°C dry bulb temperature and 10°C dew point temperature. Determine the enthalpy, specific volume, humidity ratio and dew point temperature of the mixed steam.

Solution : Refer Fig. 25.53 which shows the processes involved drawn on psychrometric chart as per given data :

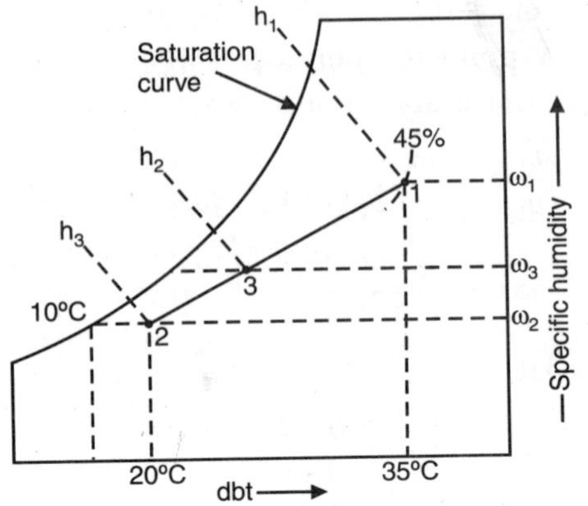

Fig. 25.53.

Locate point 1 (35°C *dbt* and 45% RH) and point 2(20°C *dbt* and 10°C *dpt*) on the psychrometric chart and read the following values :

$$h_1 = 76 \text{ kJ/kg} \; ; \quad v_{s_1} = 0.895 \text{ m}^3\text{/kg}$$
$$h_2 = 39.2 \text{ kJ/kg} \; ; \quad v_{s_2} = 0.842 \text{ m}^3\text{/kg}$$
$$\omega_1 = 0.0158 \text{ kg/kg of dry air}$$
$$\omega_2 = 0.0076 \text{ kg/kg of dry air}$$

Then : mass of fresh air at state 1,

$$m_1 = \frac{V_1}{v_{s_1}} = \frac{325}{0.895} = 134.23 \text{ kg/min}$$

Mass of recirculated air at state 2,

$$m_2 = \frac{V_2}{v_{s_2}} = \frac{325}{0.842} = 385.98 \text{ kg/min}$$

For adiabatic mixing :

$$\frac{m_1}{m_2} = \frac{h_3 - h_2}{h_1 - h_3} \; ; \quad \frac{134.23}{385.98} = \frac{h_3 - 39.2}{76 - h_3}$$

or $0.348 \times 76 - 0.348\, h_3 = h_3 - 39.2$

$$\therefore \quad h_3 = \frac{0.348 \times 76 + 39.2}{1 + 0.348}$$

$$= 48.7 \text{ kJ/kg of dry air}$$

Thus the specific enthalpy of the resulting mixture is 48.7 kJ/kg of dry air. Now locate point 3 on the line 1-2 corresponding to h_3 = 48.7 kJ/kg of dry air. Then the following values may be read from psychrometric chart :

Specific volume

$v_{s3} = 0.854$ m^3/kg of dry air

Humidity ratio

$\omega_3 = 0.0097$ kg/kg of dry air

Dew point temperature $dpt_3 = 13.8$°C

Alternatively : For adiabatic mixing

$$\frac{m_1}{m_2} = \frac{\omega_3 - \omega_2}{\omega_1 - \omega_3}\ ; \quad \frac{134.23}{385.98} = \frac{\omega_3 - 0.0076}{0.0158 - \omega_3}$$

or $0.348 \times 0.0158 - 0.348\ \omega_3 = \omega_3 - 0.0076$

$$\therefore \quad \omega_3 = \frac{0.348 \times 0.0158 + 0.0076}{1 + 0.348}$$

$= 0.0097$ kg/kg of dry air

The point 3 can then be located on line 1-2 and the relevant data read from the chart.

EXAMPLE 25.38

A stream of moist air at 10°C dry bulb and 90 percent relative humidity mixes with another stream of moist air at 40°C dry bulb and 30 percent relative humidity. The flow rates of these streams are in the ratio of one part of first to two parts of the second stream. Presuming barometric pressure equal to 1 bar and considering adiabatic mixing, determine the temperature, specific enthalpy, humidity ratio and relative humidity of the air after mixing.

Solution : Locate state 1 corresponding to $dbt_1 = 10$°C and $\phi_1 = 90\%$ and state 2 corresponding to $dbt_2 = 40$°C and $\phi_2 = 30\%$. Draw the line joining states 1 and 2 on a psychromatic chart. Then, from the chart

$\omega_1 = 0.007$ kg/kg of dry air and

$\omega_2 = 0.014$ kg/kg of dry air

For adiabatic mixing

$$\frac{m_1}{m_2} = \frac{\omega_3 - \omega_2}{\omega_1 - \omega_3}\ ; \quad \frac{1}{2} = \frac{\omega_3 - 0.014}{0.007 - \omega_3}$$

or $0.007 - \omega_3 = 2\omega_3 - 2 \times 0.014$

$$\therefore \quad \omega_3 = \frac{0.007 + 2 \times 0.014}{1 + 2}$$

= 0.0117 kg/kg of dry air

Thus the specific humidity of the resulting mixture is 0.0117 kg/kg of dry air. Now locate point 3 on line 1-2 corresponding to $\omega_3 = 0.0117$ kg/kg of dry air. Then we may find that at point 3,

dry bulb temperature dbt_3 = **29.5°C**

relative humidity ϕ_3 = **45%**

specific enthalpy h_3 = **55** kJ/kg of dry air

25.13. AIRCONDITIONING AND HUMAN COMFORT

Air-conditioning is an artificial process that involves cooling as well as heating coupled with ventilation, filtration and air circulation. It is essentially the process of treating air to control simultaneously its temperature, humidity, cleanliness and distribution to meet the comfort requirements of the occupants of the conditioned space. The functioning of an air-conditioning system can be conceived as depicted in Fig 25.54.

Apart from the creation of an acceptable thermal environment (controlled temperature), control of humidity is of great importance both

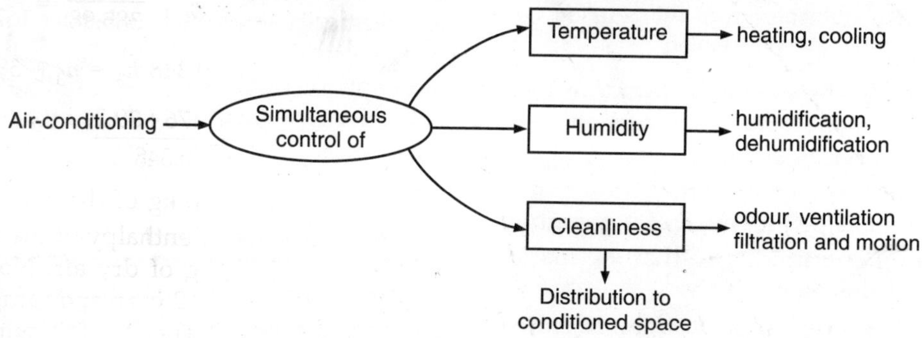

Fig. 25.54. Functioning of air-conditioning

in humid and arid climates. Further, the air inside the conditioned space gets fouled due to absorption of pollutants from different sources and for human comfort, the indoor air has be purified.

25.14. APPLICATIONS OF AIR-CONDITIONING

Air-conditioning which was once considered as luxury, has now become a necessity in our day-to-day life. The air-conditioning has applications in diverse fields such as

(*i*) Residential and office buildings

(*ii*) Hospitals, cinema halls and departmental stores

(*iii*) Libraries, museums, computer centres and research laboratories

(*iv*) Transport vehicles :

(*a*) cars, buses and rail coaches

(*b*) aircrafts, space shuttles and rockets

(*c*) submarines

(*v*) Printing, textile and photographic products

(*vi*) Food and process industries

(*vii*) Production shop laboratories, manufacture of materials and precision devices.

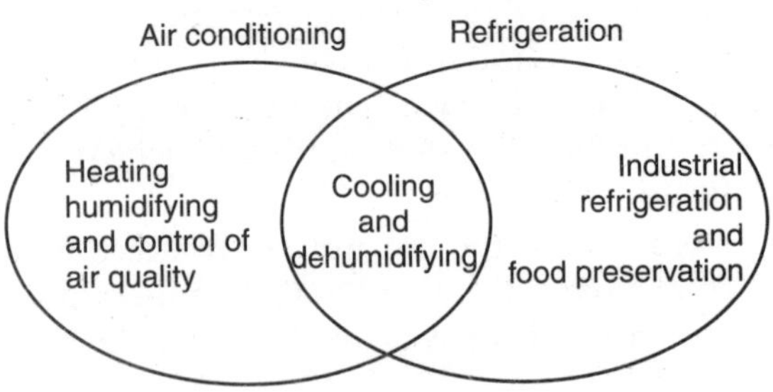

Fig. 25.55. Relationship of refrigeration and air-conditioning

Air-conditioning essentially performs three services in the manufacture of precision metal parts. These services are :

(*a*) maintenance of uniform temperatures so that the metals neither expand nor contract

(*b*) control of humidity so that the rusting of metals is prevented

(*c*) filtration of air so as to minimize dust. Cleanliness of air conditioned space is absolutely essential where electronic components are being manufactured.

The fields of refrigeration and air-conditioning are very closely inter-related as indicated in Fig. 25.55.

25.15. COMFORT AIR-CONDITIONING AND ITS TYPES

Comfort air-conditioning deals with the creation of an optimum environmental conditions conducive to human health, comfort and efficiency. Air-conditioning systems in homes, offices, stores, restaurants, theatres, schools and hospitals etc. are of this type.

The comfort air-conditioning systems are generally classified into the following three categeries :

• **Summer air-conditioning** : These systems when properly designed and installed maintain the temperature and humidity of indoor air to a level at which persons feel comfortable. Essentially it involves reducing the air temperature and humidity (in humid tropics) by the process of cooling and dehumidification.

• **Winter air-conditioning :** These systems are meant for the control of environmental conditions of indoor air so as to provide comfort in winter. Essentially it involves an increase in sensible heat and water content of air by the process of heating and humidification. The heating is done by furnaces or boilers fired with solid, liquid or gaseous fuels.

• **Year-round air-conditioning** : This system manifests in the control of temperature and humidity in an enclosed space for all times of the year ; this is despite a change in the atmospheric conditions. Essentially the system comprises the heating and cooling equipment with associated components and automatic controls.

25.16. HUMAN COMFORT

Thermal comfort is a condition of mind which expresses satisfaction with thermal environment. It is the state where the person is entirely unaware of his surroundings; no consideration whether the space is too hot or too cold. Dissatisfaction with the thermal environment may be caused by the body as a whole (being too hot or cold) or by the unwanted heating or cooling of a particular part of the body (local discomfort).

Human comfort refers to the control of temperature and humidity of air and its circulation so that the resulting environment becomes human friendly; the state of environment where persons feel comfortable. Comfort is however a subjective quality; it is dependent on the preferences of an individual and varies with the age, sex, state of health and clothing etc. of a person.

25.17. THERMAL INTERCHANGE WITH ENVIRONMENT

The human body is continually gaining and producing heat as well as losing heat to its surroundings to maintain temperature equilibrium. Human body can be likened to a machine which receives fuel in the form of food, converts a fraction of energy in the fuel into work and rejects the remainder as heat to the surroundings. It is the continuous process of heat rejection which requires a thermal balance.

The various mechanisms by which temperature control is accomplished and a person feels comfortable are shown in Fig. 25.56.

The generation of heat occurs in the body cells, circulatory system carries this heat to skin where it is released to environment. The physical basis of comfort lies in the thermal balance of the body, *i.e.*, the heat produced by the body's metabolism must be dissipated to the environment, otherwise the body would overheat.

The thermal balance of the body can be expressed as

$$Q_b = W \pm Q_c \pm Q_r \pm Q_e \pm Q_s$$

where :

(*i*) Q_b is the heat produced within the body itself as a result of metabolic processes. The rate at which the body produces heat is referred to as metabolic rate. The metabolic heat produced depends on the rate of food consumption which itself is dependent on individual's health, his physical activity and environment. The metabolic heat production will be low for a person who is sick or weak or is fasting.

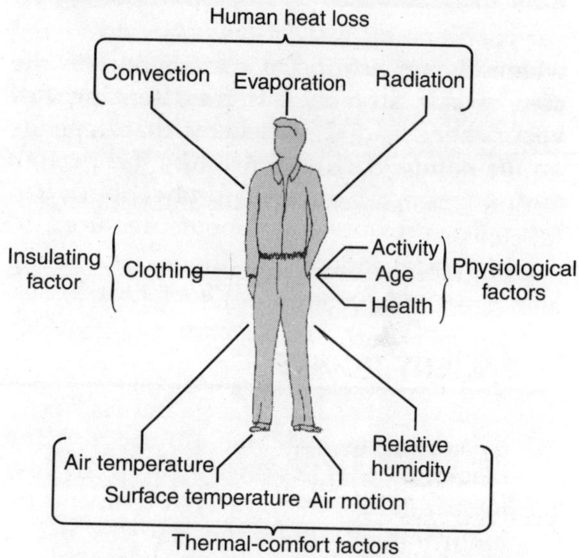

Fig. 25.56. Thermal interchange with environment

The heat generated is transported from the warm body core to the body surface partly by conduction through the tissues and partly by blood flow to the skin.

(*ii*) Q_c is the heat gain or loss by convection; it is prescribed by the relation

$$Q_c = UA\ (t_b - t_s)$$

where U is overall heat transfer coefficient the body, A is area of body surface, t_b is body temperature and t_s is temperature of surroundings. Q_c is +ve when heat is lost and –ve when heat is gained from the surroundings.

(*iii*) Q_r is the heat gain or loss by radiation; it depends on the mean radiant temperature which refers to the average surface temperature of the surrounding objects in the conditioned space. The body loses heat if the radiant temperature is more than the dry bulb temperature of the room. Conversely it gains heat.

(*iv*) Q_e is the evaporative heat loss due to water evaporating at skin surface; the evaporation rate depends on the humidity of air and the moisture available for evaporation. Essentially the heat loss by evaporation depends on difference of pressure between skin surface and surrounding air. It is worked out from the relation

$$Q_e = C_d \, A \, L \, C_c \, (p_s - p_v)$$

where C_d is the diffusion coefficient, A is the area of skin surface, L is the latent heat of vaporization and C_c is a factor that depends on the nature of clothes worn by the person, and $(p_s - p_v)$ is the difference between saturation pressure and vapour pressure.

In typical indoor conditions, convecting losses account for about 35% of heat losses, radiative losses are about 45% and evaporative losses are of the order of 20%.

(*v*) Q_s is the heat stored in the body reflected by rise in temperature of body tissues.

When Q_c, Q_r and Q_e are high and positive and Q_b is low, Q_s will become –ve. Hence a weak or fasting person will feel more cold when exposed to low temperature, high draft air.

The body's thermo-regulatory system tries to maintain energy balance and accordingly the temperature of the body remains fairly constant at about 36.9°C for tissues at the surface or the skin, and about 37.2°C for the deep tissues or core. A value of 40.5°C is considered high and 43.5°C is certainly fatal.

SALIENT POINTS

1. Refrigerators and heat pumps are the machines that transfer heat from low temperature region to high temperature environment. Both the devices use a working find (called refrigerant) and require work input for their operation.
2. The duty of a refrigerator is to extract heat Q_2 from the space to be cooled and that of heat pump is to deliver heat Q_1 to the environment to be heated (temperature T_1)
3. The performance of refrigerators and heat pumps is expressed in terms of coefficient of performance (COP) defined as

$$COP = \frac{\text{desired objective}}{\text{work input}}$$

$$= \frac{Q_2}{Q_1 - Q_2} \text{ for refrigerator}$$

$$= \frac{Q_1}{Q_1 - Q_2} \text{ for heat pump}$$

4. For a refigerator and a heat pump working on reversed Carnot cycle, the above expressions can be rewritten as

$$\text{COP} = \frac{T_2}{T_1 - T_2} \text{ for refrigerator}$$

$$= \frac{T_1}{T_1 - T_2} \text{ for heat pump}$$

From these identities, it follows that

$$(COP)_{\text{heat pump}} = 1 + (COP)_{\text{refrigerator}}$$

5. The cooling capacity of a refrigeration system refers to its rate of heat removal from the refrigerated space (space to be cooled). The unit of cooling capacity is ton of refrigerator and 1 ton is equivalent to 210 kJ/min or 3.5 kJ/s of heat removal.
6. The Bell Coleman cycle is the gas refrigeration cycle consisting of adiabatic compression, isobaric cooling, adiabatic expansion and heat absorption (refrigeration effect) at constant pressure and its performance coefficient is

$$COP = \frac{T_4}{T_3 - T_4}$$

where T_3 and T_4 denote the temperatures of the working gas at entry to and at exit from the expander

(*i*) The Bell-Coleman refrigerator may work on open system or closed system.

(*ii*) The gas refrigeration cycle is simple in operation, has lighter components and that makes it ideally suitable for air craft cooling.

7. Vapour compression refrigeration is the most important and practical form of refrigeration system for domestic and commercial utility.

(*i*) The flow processes of the system are carried out sequence wise in separate components namely compressor, condenser, expansion valve or capillary and evaporator.

(*ii*) The working medium (called refrigerant) is a gas that alternately evaporates and condenses and its performance coefficient is

$$COP = \frac{h_2 - h_1}{h_1 - h_4}$$

where h_1 and h_2 are the enthalpy values at inlet to and at exit from the compressor and h_4 is the enthalby at entry to the evaporator

(*iii*) Wet compression: the vapour is wet or just dry at the end of compression process

Dry compression : the vapour lies in the superheated zone throughout the compression process.

The COP of the machine decreases with dry compression.

(*iv*) Sub-cooling means reducing the temperature of condensed liquid refrigerant below saturation temperature corresponding to condenser pressure.

With undercooling, there is improvement in the *COP* of the unit.

8. The basic characteristics of a good refrigerant are :
   - Easy availability and low cost
   - Non-toxic, non-corresive and non-flammable
   - Chemical stability
   - High heat of vaporisation and high *COP*
   - Easy detectability and environment friendly (no impact on ozone layer)

9. The commonly used refrigerants are:
   - CFC'S (Chloro fluoro carbons)
   - NH_3, CO_2, air and water
   - Hydrocarbons (ethane, ethylene, propane)

10. The vapour absorption system is a heat operated system in which the refrigerant is alternately absorbed and liberated from the absorbent (transport medium).

(*i*) In NH_3 - H_2O refrigeration system, NH_3 is the refrigerant and H_2O is the absorbent. In Li Br - H_2O refrigeration system, H_2O is the refrigerant and Li Br is the absorbent

(*ii*) The absorption system consists of a condenser, throttling or expansion valve and an evaporator similar to vapour compression refrigeration. The compressor unit is replaced by a combination of absorber, pump, heat exchanger and generator.

(*iii*) The absorption system has more precise equipment, its operation is quiet and is useful at places where waste heat is available in abundance *i.e.*, in effective power houses, gas turbines and chemical industries.

11. The electrolux refrigerator is of ammonia absorption type that uses ammonia as refrigerant, water as an absorbent and hydrogen. In the presence of hydrogen, liquid ammonia readily evaporates resulting in fall in its temperature.

12. Psychrometry is the science of studying the properties of moist air and the use of these properties to analyse the conditions and processes involving moist air (mixture of dry air and water vapour). For moist air

$$p_t = p_a + p_v$$

where p_t is total pressure of moist air, p_a is partial pressure of dry air and p_v is partial pressure of water vapour.

(*i*) Specific humidity (or humidity ratio) defines the ratio of the mass of water vapour (m_v) to the mass of dry air (m_a) in a given volume of air sample:

$$\omega = \frac{m_v}{m_a} = 0.622 \frac{p_v}{p_t - p_v}$$

(*ii*) Relative humidity ϕ represents the ratio of actual mass of water vapour (m_v) in a given volume to that it could have if it were saturated (m_s) at the same temperature.

$$\phi = \frac{m_v}{m_s} = \frac{p_v}{p_s}$$

where p_s is the saturation pressure at the temperature of the given air sample.

(*iii*) Degree of saturation μ represents the ratio of the specific humidity of a given air sample to the specific humidity of the sample when saturated at the same temperature

$$\mu = \frac{\omega}{\omega_s} = \phi \frac{p_t - p_{vs}}{p - p_v}$$

(*iv*) The dry bulb and wet bulb temperatures are simultaneously measured by instruments called psychrometers. The wet bulb temperature is lower than dry bulb temperature and the difference is known as wet bulb depression.

(*v*) Dew point temperature is the temperature at which water vapour present in air starts condensing

(*vi*) For saturated air the dry bulb temperature, the wet bulb temperature and the dew point temperature is same.

(*vii*) The enthalpy of moist air is the sum of the enthalpy of dry air and the enthalpy of water vapour associated with it.

$$m_h = m_a h_a + m_v h_v$$

(*viii*) The Carrier's equation as indicated below is used to find the partial pressure of water vapour present in the moist air

$$p_v = p_w - \frac{(p - p_w)(dbt - wbt)}{1544 - 1.44\, wbt}$$

where the temperatures (*dbt* and *wbt*) are in °C and p_w is saturation pressure corresponding to wet bulb temperature.

(*ix*) The psychrometric charts depicts the graphical relationship of the properties of moise air varying with dry bulb temperature.

(*x*) The most common processes which involve the air-water vapour mixture are:

- Sensible heating: raising the temperature without change in moisture content.
- Sensible cooling: lowering the temperature without change in moisture content.
- Humidification: adding of moisture without change in dry bulb temperature
- Dehumidification: removing of moisture without change in dry bulb temperature
- Heating and humidification
- Cooling and dehumidification
- Adiabatic mixing of air streams.

(xi) The dry bulb temperature, relative humidity and air motion are the primary factors that determine human comfort.

REVIEW QUESTIONS

A. Conceptual and conventional questions.

1. What is a reversed heat engine? Point out the purpose for which it is used.
2. State the difference between a refrigerator and a heat pump. How do these machines satisfy the second law of thermodynamics?
3. Define the following terms:
   (*i*) refrigeration
   (*ii*) refrigerating effect
   (*iii*) ton of refrigeration
4. What is meant by *COP*? What value of *COP* is desirable, large or small and why?
5. Set up a relation for the *COP* of a heat-pump and that of a refrigeration. Proceeed to show that

   $(COP)_{heat\ pump} = 1 + (COP)_{refrigeration}$
6. State and briefly explain the various techniques of producing cooling.
7. Name and explain the commonly used unit of refrigeration.
8. Define refrigeration. How ice and dry ice are used for the purpose of refrigeration?
9. Mention the various applications of refrigeration.
10. Comment on the validity of following statements:
    (*i*) *COP* of heat pump is always greater than unity.
    (*ii*) One ton of refrigeration plant produces one ton of ice.

11. Sketch the schematics of refrigeration system operating on Bell-Coleman cycle and explain its working.
12. Distinguish between open air and dense air refrigeration cycles.
13. Describe, with a neat schematic arrangement, the working of a simple vapour compression refrigeration cycle. Represent the cycle on *p-v* and *T-s* plots.
14. Differentiate between wet and dry compression. Why is wet compression not employed though it is thermodynamically better?
15. For a vapour compression refrigeration machine, explain the effect of undercooling and superheating on the coefficient of performance.
16. State the functions of the following components of a simple vapour compression refrigeration system
    (*i*) compressor (*ii*) condenser
    (*iii*) expansion valve
    (*iv*) evaporator
17. (*a*) Define a refrigerant and name a few widely used refrigerants.
    (*b*) List the characteristics of an ideal refrigerant.
18. State the properties and applications of the following refrigerants :
    Ammonia, carbon dioxide, Freon-12
19. Distinguish between primary and secondary refrigerants. Give examples of each category.
20. What is moist air and saturated air
21. Define and explain the following terms in relation to psychrometry
    (*a*) dry bulb, wet bulb and dew point temperatures
    (*b*) relative humidity and specific humidity.
22. Establish the following expression for air-vapour mixture

$$\text{specific humidity } \omega = 0.622 \times \frac{p_v}{p_b - p_v}$$

    where p_v is partial pressure of water vapour and p_b is the barometric pressure?
23. What is a sling psychrometer? Draw its neat sketch and explain its use.
24. What is a psychrometric chart? What information does it provide?
25. Explain the concept of adiabatic saturation temperature and adiabatic dew point temperature.
26. Explain the following processes and represent these on a psychrometric chart:
    (*a*) adiabatic mixing of two streams
    (*b*) heating and humidification
    (*c*) cooling and dehumidification
27. Define the terms by-pass factor and coil efficiency in relation to the processes of sensible heating and sensible cooling.
28. Name any five psychrometric processes and represent them on the psychrometric chart.

B. Fill in the blanks with appropriate word/words.

1. A refrigeration system removes heat from a system at .... temperature and transfers the same to a system at .... temperature.
2. The COP of a Carnot refrigeration cycle decreases on .... the upper temperature and .... the lower temperature.
3. One ton of refrigeration is equivalent to .... kW.
4. The Bell-coleman refrigeration cycle uses .... as the working fluid.
5. The bank of tubes at the back of a domestic refrigerator of vapour compression type are the .... tubes.
6. In a vapour compression refrigeration system, the capillary tube is located between the .... and ....
7. The brine is an aqueous solution of .... in water.
8. .... is the refrigerant commonly used for commercial ice plants.
9. The refrigerant Freon-12 is a compound consisting of carbon, chlorine and ....
10. .... is the secondary refrigerant normally used in milk chilling plants.
11. The correct sequece of the components of a vapour compression refrigerator is evaportor, ..., ... and ....
12. An .... is a mixture of refrigeration withut phase separation.
13. A vertical downward line on a psychromatric chart represents ....
14. The wet bulb temperature would be zero when the relative humidity is .... percent.
15. The effective temperature is a measure of ...., ... and ....
16. The desirable air velocity in the occupied zone for summer air conditioners is in the range of .... metre/min.

17. A human body feels more comfort when the heat produced due to metabolism of human body is equal to ....
18. The normally desired comfort conditions in an air conditioning system are .... dry bulb temperature and .... relative humidity.

Answers

1. Low, high 2. increasing, decreasing 3. 3.5 4. air 5. condenser 6. condenser, evaporator 7. sodium chloride 8. ammonia 9. flourine 10. brine 11. compressor, condenser, throttle valve 12. azeotrope 13. dehumidification 14. 100% 15. dry bulb temperature, relative humidity, air motion 16. 4-5 m/s 17. heat dessipation to surroundings 18. 25°C, 50%

C. Multiple choice questions.

1. The capacity of a refrigerating machine is expressed as
   (*a*) lowest temperature attainable
   (*b*) rate of abstraction of heat from the space being cooled
   (*c*) inside volume of the cabinet
   (*d*) gross weight of the machine in tons
2. One ton of refrigeration implies that the machine has a refrigerating effect (capacity of heat extraction from the system being cooled) equal to
   (*a*) 1 kJ/s (*b*) 2.5 kJ/s
   (*c*) 3.5 kJ/s (*d*) 5 kJ/s
3. A Bell-Coleman cycle is a reversed
   (*a*) Brayton cycle (*b*) Rankine cyucle
   (*c*) Atkinson cycle (*d*) Ericsson cycle
4. Which part of the vapour compression refrigeration cycle produces the refrigeration effect
   (*a*) condenser (*b*) throttle valve
   (*c*) evaporator (*d*) compressor
5. In a refrigeration system expansion valve (or capillary tube) is incorporated between
   (*a*) evaporator and compressor
   (*b*) condenser and evaporator
   (*c*) compressor and condenser
   (*d*) superheater and subcooler
6. Which is usually the costliest item in a refrigeration system
   (*a*) condenser (*b*) capillary tube
   (*c*) evaporator (*d*) compressor
7. The refrigerant commonly used for commercial ice plants is
   (*a*) Freon - 12 (*b*) NH_3
   (*c*) CO_2 (*d*) SO_2
8. Which of the following pair of refrigerants is both non-toxic and non-flammable
   (*a*) NH_3 and SO_2 (*b*) CO_2 and SO_2
   (*c*) CO_2 and F-12 (*d*) F-12 and NH_3
9. The refrigerant obtained by mixing two refrigerants in proper proportion is called
   (*a*) secondary refrigerant
   (*b*) auxiliary refrigeraint
   (*c*) synthetic refrigerant
   (*d*) azetrope
10. The expression $\dfrac{0.622\, p_v}{p_t - p_v}$ is used to determine
   (*a*) relative humidity
   (*b*) specific humidity
   (*c*) degree of saturation
   (*d*) partial pressure
11. Consider the statements : A psychrometer measures
   1. Wet bulb temperature
   2. Dew point temperature
   3. dry bulb temperature
   Which of these statements is correct?
   (*a*) 1 and 2 (*b*) 2 and 3
   (*c*) 1 and 3 (*d*) 1, 2 and 3
12. Match List-I with List II

List-I	List-II
A. Degree of saturation	1. latent enthalpy of moist air
B. Dry bulb temperature	2. total enthalpyof moist air
C. Wet bulb temperature	3. capacity of air to absorb moisture
D. Dew point temperature	4. sensible enthalpy of moist air

13. On a psychrometric chart, the constant wet bulb temperature lines coincide with lines of constant
   (*a*) relative humidity
   (*b*) enthalpy
   (*c*) volume
   (*d*) dew point temperature

14. Consider the following statements : In a psychrometric chart
    1. vertical lines indicate wet bulb temperature
    2. horizontal lines indicate specific humidity
    3. sensible heating/cooling is represented by an inclined line.

    Which of these statemets is/are correct?
    (*a*) 1 only (*b*) 2 only
    (c) 3 only (*d*) 1, 2 and 3
15. In sensible heating or cooling process .... remains constant
    (*a*) dry bulb temperaure
    (*b*) wet bulb temperature
    (*c*) humidity ratio
    (*d*) relative humidity
16. Tick the statement which is not correct
    (*a*) For unsaturated air, the dew point is more than wet bulb temperature
    (*b*) Saturated air is the dry air with maximum amount of water vapour diffused into it for a particular dry bulb temperature
    (*c*) The wet bulb depression is zero when the relative humidity is 100 percent
    (*d*) Dry bulb temperature is the air temperature recorded by a thermometer when it is not affected by the moisture in air
17. By pass factor depends on the
    1. velocity of air flow
    2. pitch of the heating/cooling coil fins
    3. number of rows in the coil in the flow direction

    Which of these statements is/are true?
    (*a*) 1 and 2 (*b*) 1 and 3
    (*c*) 2 and 3 (*d*) 1, 2 and 3
18. Identify the process which is generally practiced in winter air conditioning
    (*a*) sensible heating
    (*b*) chemical dehumidification
    (*c*) humidification
    (*d*) heating with dehumidification
19. Consider the following factors :
    1. air motion
    2. dry bulb temperature
    3. relative humidity
    4. wet bulb temperature
    5. elevation of the place and its latitude

    Which of these affect the human comfort?
    (*a*) 1, 2 and 3 (*b*) 1, 2, 3 and 4
    (*c*) 2, 3 and 5 (*d*) 1, 2, 3 and 5

Answers

1. (*b*)	2. (*c*)	3. (*a*)	4. (*c*)	5. (*b*)
6. (*d*)	7. (*b*)	8. (*c*)	9. (*d*)	10. (*b*)
11. (*c*)	12.A–3, B–4, C–2, D–1			13. (*b*)
14. (*b*)	15. (*c*)	16. (*a*)	17. (*d*)	18. (*d*)
19. (*a*)				

Chapter 26

Heat Transfer

Learning objectives : After having made a study of the subject matter presented in this chapter, the reader will be able to :

- *explain the modes and basic laws of heat transfer*
- *analyse the conduction through walls, cylinders and spheres*
- *explain the effect of variable thermal conductivity and the concept of critical thickness of insulation*
- *state the significance of laws of radiation; Planck's law, Stefan Boltzman law, Kirchoff's law and Lambert's cosine law*
- *list the variables specifying the system with free and forced convection, and formulate the relevant dimensionless numbers*
- *classify the heat exchangers and make their performance analysis based on LMTD and NTU approach*

26.1. MODES AND BASIC LAWS OF HEAT TRANSFER

The literature on heat transfer generally recognises three distinct modes of heat transmission; ***conduction***, ***convection*** and ***radiation***. These three modes are similar in that a temperature differential must exist and the heat exchange is in the direction of decreasing temperature. Each method has, however, different physical picture and different controlling laws.

26.1.1. Conduction

Thermal conduction is a mechanism of heat propagation from a region of higher temperature to a region of lower temperature within a medium (solid, liquid or gaseous) or between different mediums in direct physical contact. Conduction does not involve any movement of macroscopic portions of matter relative to one another. The thermal energy may be transferred by means of electrons which are free to move through the lattice structure of the material. In addition, or alternatively, it may be transferred as vibrational energy in the lattice structure. Irrespective of the exact mechanism, the observable effect of conduction is an equalization of temperature.

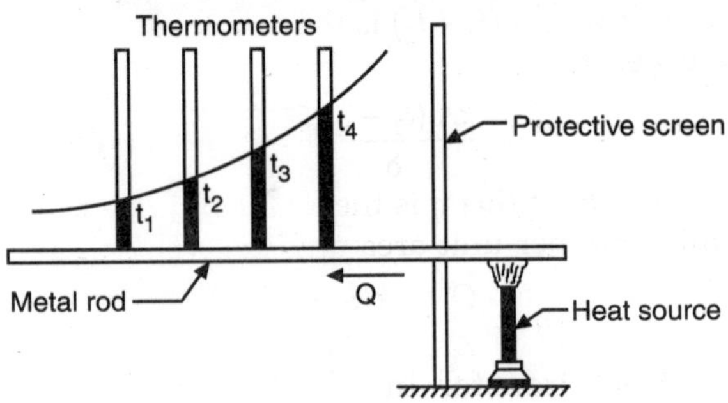

Fig. 26.1. Conduction heat flow along a rod

Consider the flow of heat along a metal rod, one end of which is placed adjacent to a flame. The elementary particles (molecules, atoms, electrons) composing the rod, and which are in immediate vicinity of the flame, get heated. Because of the resulting temperature growth, their kinetic energy increases and this puts them in a violent state of agitation, and they start vibrating about their mean positions. Consequently, these more active particles collide with less active molecules lying next to them. During collision, the less active particles also get excited, *i.e.*, thermal energy is imparted to them. The process is repeated for layer after layer of molecules until the other end of the rod is reached. Each layer of molecules is at a slightly higher temperature than the preceding one, *i.e.*, a temperature gradient exists along the length of the rod. The rate of heat flow between the two ends depends upon the length of the rod, temperature difference between the two ends, and the physical and chemical composition of the bar material.

The rate equation for one-dimensional steady flow of heat by conduction is prescribed by the *Fourier law* :

$$Q = -kA\frac{dt}{dx} \quad ...(26.1)$$

where Q is the heat transfer rate, A is the area of heat transfer surface, dt is the temperature difference for a short perpendicular distance dx, and the thermal conductivity k is a characteristic of the surface material. Since the temperature gradient is negative in the positive x-direction, the minus sign in the equation gives positive heat flow.

If δ is the path length in the direction of heat flow and $(t_1 - t_2)$ is the temperature difference, then

$$Q = \frac{kA\,(t_1 - t_2)}{\delta} \quad ...(26.2)$$

The ***heat flux*** q is the heat conducted per unit time per unit area and is given by

$$q = \frac{Q}{A} = \frac{k(t_1 - t_2)}{\delta} \quad ...(26.3)$$

Heat transfer in metal rods, in heat treatment of steel forgings and through the walls of heat exchange equipment are some practical examples of heat conduction.

26.1.2. Convection

Thermal convection is a process of energy transport affected by the circulation or mixing of a fluid medium (gas, liquid or a powdery substance). Convection is possible only in a fluid medium and is directly linked with the transport of medium itself. The effectiveness of heat transfer by convection depends largely upon the mixing motion of the fluid.

With respect to origin, two types of convection are distinguished; forced and natural or free convection.

In ***natural*** or ***free convection***, the circulation of the fluid medium is caused by buoyancy effects, *i.e.*, by the difference in the densities of the cold and heated particles. Consider heat flow from a hot plate to atmosphere. The stagnant layer of air in the immediate vicinity of the plate gets thermal energy by conduction. The energy thus transferred serves to increase the temperature and internal energy of the air particles. Because of temperature rise these particles become less dense (and therefore lighter) than the surrounding air. The lighter air particles move upwards to a region of low temperature where they mix with and transfer a part of their energy to the cold particles. Simultaneously the cold air particles descend downwards to fill the space vacated by the hot air particles. The circulation pattern, upward movement of the warm air and the downward movement of the cold air, is called the ***convection currents***.

Some examples of free convection are :

- chilling effect of a cold wind on a warm body,
- heat flow from a hot pavement to surrounding atmosphere and heating of air in a room by a stove,
- cooling of billets in the atmosphere ,
- heat exchange on the outside of cold and warm pipes,
- hot-water heating system.

In ***forced convection***, the flow of fluid is caused by a pump, fan or by atmospheric winds. These mechanical devices provide a definite circuit for the circulating currents and that speeds up the heat transfer rate. Examples of forced convection are :

- flow of water in condenser tubes,
- fluid passing through the tubes of a heat exchanger,

• cooling of internal combustion engines,
• air conditioning installation and nuclear reactors.

Regardless of the particular nature, the appropriate rate equation for the convective heat transfer between a surface and an adjacent fluid is prescribed by ***Newton's law of cooling***.

$$Q = hA\,(t_s - t_f) \quad ...(26.4)$$

where, Q is the convective heat flow rate, A is the area exposed to heat transfer, t_s and t_f are the surface and fluid temperatures respectively. The heat transfer co-efficient h depends upon the thermodynamic and transport properties (*e.g.* density, viscosity, specific heat and thermal conductivity of the fluid), the geometry of the surface, the nature of fluid flow, and the prevailing thermal conditions.

Convection mechanism involving phase changes leads to the important fields of boiling (evaporation) and condensation.

26.1.3. Radiation

Thermal radiation is the transmission of heat in the form of radiant energy or wave motion from one body to another across an intervening space. Unlike heat transfer by conduction and convection, transport of thermal radiation does not necessarily affect the material medium between the heat source and the receiver. An intervening medium is not even necessary and the radiation can be affected through vacuum or a space devoid of any matter. Radiation exchange, in fact, occurs most effectively in vacuum. A material present between the heat source and the receiver would either reduce or eliminate entirely the propagation of radiation energy.

The mechanism of the heat flow by radiation consists of three distinct phases:

(*i*) *Conversion of thermal energy of the hot source into electromagnetic waves* : All bodies above absolute zero temperature are capable of emitting radiant energy. Energy released by a radiating surface is not continuous but is in the form of successive and separate (discrete) packets or quanta of energy called ***photons***. The photons are propagated through the space as rays; the movement of swarm of photons is described as the electromagnetic waves.

(*ii*) *Passage of wave motion through intervening space*. The photons, as carriers of energy, travel with unchanged frequency in straight paths with speed equal to that of light.

(*iii*) *Transformation of waves into heat*: When the photons approach the cold receiving surface, there occurs reconversion of wave motion into thermal energy which is partly absorbed, reflected or transmitted through the receiving surface.

The most vivid evidence of radiation heat transfer is that represented by solar energy which passes through inter-stellar space (conditions close to that for perfect vacuum) on its way to the earth surface. Solar radiation plays an important part in the design of heating and ventilating systems. Heat transfer by radiation is encountered in boiler furnaces, billet reheating furnaces and other types of heat exchange apparatus. The design and construction of engines, gas turbines, nuclear reactors and solar collectors is also significantly influenced by the radiation heat transfer.

The basic rate equations for radiation heat transfer are based on ***Stefan-Boltzman law*** :

$$E_b = \sigma_b\, A\, T^4 \quad ...(26.5)$$

where, E_b is the energy radiated per unit time, T is the absolute temperature of the surface, and σ_b is the Stefan-Boltzman constant.

$$\sigma_b = 5.67 \times 10^{-8}\ \text{W/m}^2\ \text{K}^4$$

Equation 26.5 is essentially valid for an ideal radiator or a black body-suffix b designates a black surface. The radiant energy emitted by a real surface is less than that for an ideal emitter and is given by

$$E = \varepsilon\, \sigma_b\, A\, T^4 \quad ...(26.6)$$

where, ε is a radiative property of the surface and is called ***emissivity***; its value depends upon surface characteristics and temperature. It indicates how effectively the surface emits radiations compared to an ideal or black body radiator

26.2. STEADY AND UNSTEADY HEAT TRANSFER

Heat exchange between two systems may take place under steady (stable) thermal conditions or under unsteady (unstable) thermal conditions. Steady state implies that temperature at each point of the system remains constant in the course of time, and it is a function only of space co-ordinates

$$t = f(x, y, z) \; ; \; \frac{dt}{d\tau} = 0 \qquad ...(26.7)$$

Steady state results in a constant rate of heat exchange (heat influx equals heat efflux), and there is no change in the internal energy of the system during such a process. Typical examples of steady state heat transfer are :

- cooling of an electric bulb by the surrounding atmosphere,
- heat flow from the products of combustion to water in the tubes of a boiler, from the hot to cold fluid in a heat exchanger, and from a refrigerated space to cooling surface of the evaporator.

Under unsteady thermal conditions, temperature of the system changes continuously with time. Temperature is obviously a function of space and time co-ordinates.

$$t = f(x, y, z, \tau) \; ; \; \frac{dt}{d\tau} \neq 0 \qquad ...(26.8)$$

Unsteady state results in heat transfer rate which changes with time. Further, a change in temperature indicates a change of internal energy of the system. Energy storage is thus a part and parcel of unsteady heat flow. Typical examples of unsteady heat transfer are :

- warm-up periods of furnaces
- boilers and turbines
- cooling of castings in a foundry
- heat treatment and stress relieving of metal castings.

A special kind of unsteady process is the transient state wherein the system is subjected to cyclic variations in the temperature of its environment. The temperature at a particular point of the system returns periodically to the same value; the rate of heat flow and energy storage also undergo periodic variations. Examples are : heating or cooling of the water of an IC engine; heating or cooling of the walls of a building during the 24-hours cycle of the day.

Further, the heat transfer in a system may be in one, two or more directions. In a one dimensional heat flow, there is a single predominant direction in which temperature differential exists and obviously the heat flow takes place; heat flow in the other two directions can be safely neglected. When the temperature is a function of two co-ordinates, heat flow is two-dimensional. A three-dimensional heat flow stipulates that temperature is a function of three co-ordinates, and consequently heat flow occurs in all three directions.

26.3. SIGNIFICANCE OF HEAT TRANSFER

The discipline of the heat transfer encompasses a great many fascinating areas like :

- Design of steam generators, condensers, and other heat exchange equipments in power plant engineering; solar energy conversion for space heating and for electric power production.
- IC engines, refrigeration and air-conditioning units, superheaters and condensers and many other cooling and heating appliances in mechanical engineering. The operation of refrigeration and air-conditioning units depends greatly on the effective transfer of heat in condensers and evaporators.
- Design of cooling systems for electric motors, generators and transformers in electrical engineering so that the heat generated during the flow of current through the windings of these machines can be effectively dissipated. This is to avoid the conditions which will cause overheating and damage of the equipment.
- Evaporation, condensation, heating and cooling of fluids in chemical operations.
- Construction of dams and other heavy structures, calculation of thermal expansion of suspension bridges and railway tracks, minimisation of building-heat losses by means of improved insulation techniques.

• Proper functioning of valves and other controls operated by temperature changes, thermal control of space vehicles.

• Heat treatment of metals where diffusion rate of carbon in steel is required to be made to estimate the period for which the steel component must be exposed to carburizing atmosphere.

An engineer utilises his knowledge of heat transfer either to transmit heat in the most effective/economic way, or to protect his equipment against excessive heat gains or losses. The various engineering problems involving heat transfer can be categorised into two groups.

(*i*) Heat flow situations where maximum heat transfer is desirable with minimum possible heat exchange area. Gas turbine blades, walls of IC engines and combustion chambers, outer surface of a space vehicle all depend for their durability on rapid removal of heat from their surfaces. The design of heat exchangers is considered to be optimum under specified temperature conditions when maximum heat transfer occurs with minimum surface area.

(*ii*) Heat flow situations where heat transfer is undesirable and its flow is to be prevented. The walls of centrally heated buildings, and the steam pipes in a steam power plant are properly insulated to restrict heat losses.

With few exceptions, engineering problems involve more than one of the three modes of heat transfer and this aspect results into a complicated heat exchange pattern. The significance of heat transfer and the simultaneous occurrence of different modes of heat transfer can be well-judged by citing the following examples :

(*i*) Closed container filled with hot coffee and kept in a room whose air and walls are at a fixed temperature. Fig. (26.2).

All three models of heat transfer contribute towards cooling of coffee, and different paths for energy transfer from coffee are :

– free convection from the coffee to the flask, q_1

– heat conduction through the flask, q_2

– free convection from the flask to the air, q_3

– radiation exchange between the outer surface of the flask and the inner surface of the cover, q_4

– free convection from air to the cover, q_5

– heat conduction through the cover, q_6

– free convection from the cover to the room air, q_7

– radiation exchange between the outer surface of the cover and the surroundings, q_8

(*ii*) Automobile engine with thermosyphon cooling system. Here the relevant heat transfer processes are :

– free convection and radiation from hot combustion gases to cylinder walls

– conduction through cylinder walls

– free convection from cylinder walls to water and from water to radiator tubes

– conduction through walls of radiator tubes

– convection from radiator tubes to surrounding air

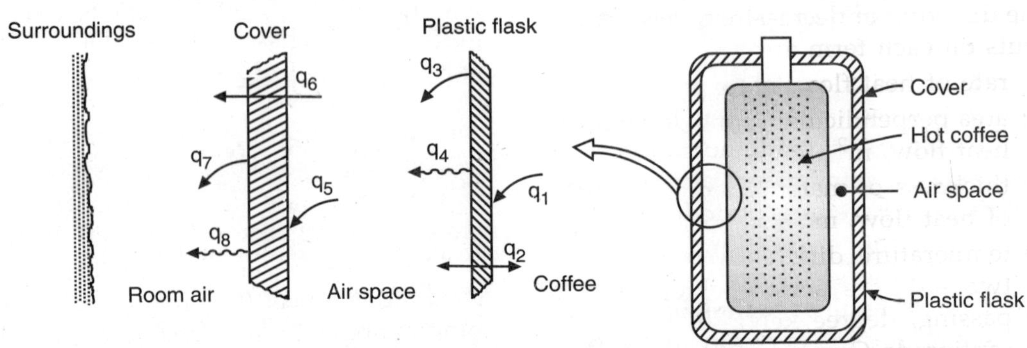

Fig. 26.2. Cooling of hot coffee

(*iii*) Heat flow through a wall consisting of two plates separated by vacuum (Fig. 26.3).

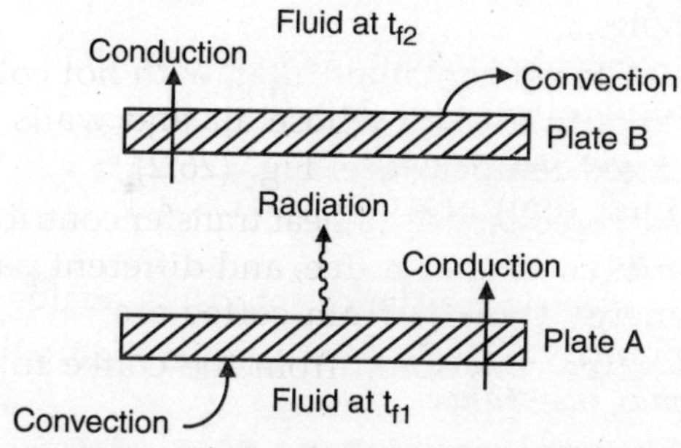

Fig. 26.3. Heat transfer through a composite wall

Heat is convected from the fluid at temperature t_{f1} to plate A, conducted through plate A, radiated from plate A to plate B, conducted through plate B, and finally convected from plate B to the fluid at temperature t_{f2}.

Likewise in the process of steam generation, the boiler tubes receive heat from the products of combustion by all the three modes of heat transfer.

26.4. FOURIER EQUATION

Conduction is primarily a molecular phenomenon requiring temperature gradient as the driving force. Experimental evidence does indicate that steady-state one-dimensional flow of heat by conduction through a homogeneous material is given by the ***Fourier law***

$$Q = -k A \frac{dt}{dx}$$

$$q = \frac{Q}{A} = -k \frac{dt}{dx} \qquad ...(26.9)$$

The heat flux q (heat conducted per unit time per unit area) flows along normal to area A in the direction of decreasing temperature. The units on each term are :

Q : rate of heat flow, kJ/hr

A : area perpendicular to the direction of heat flow, m^2

dx : thickness of material along the path of heat flow, m

dt : temperature difference between the two surfaces across which heat is passing, degree kelvin K or degree centigrade C.

The ratio dt/dx represents the change in temperature per unit thickness, *i.e.*, the temperature gradient. The negative sign indicates that the heat flow is in the direction of negative temperature gradient, and that serves to make the heat flow positive. The proportionality factor k is called the ***heat conductivity*** or ***thermal conductivity*** of the material through which the heat propagates. Thermal conductivity provides an indication of the rate at which heat energy is transferred through a medium by the diffusion (conduction) process. For a prescribed temperature gradient and geometric parameters, the heat flow rate increase with increasing thermal conductivity.

The Fourier law is essentially based on the following assumptions :

- steady state conduction which implies that the time rate of heat flow between any two selected points is constant with time. This also means that the temperature of the fixed points within a heat conducting body does not change with time : $t \neq f(\tau)$.
- one-directional heat flow; only one space co-ordinate is required to describe the temperature distribution within the heat conducting body; $t = f(x)$. The surfaces in the y-and z-directions are perfectly insulated.
- bounding surfaces are isothermal in character, *i.e.*, constant and uniform temperatures are maintained at the two faces.
- isotropic and homogeneous material, *i.e.*, thermal conductivity has a constant value in all the directions.
- constant temperature gradient and a linear temperature profile.
- no internal heat generation.

Some essential features of the Fourier relation are enumerated below :

- Fourier law predicts how heat is conducted through a medium from a region of high temperature to a region of low temperature
- Fourier law is valid for all matter regardless of its state: solid, liquid or gas.
- Fourier law is a vector expression indicating that heat flow rate is normal to an isotherm and is in the direction of decreasing temperature.

• Fourier law cannot be derived from first principles; it is a generalization based on experimental evidence.

• Fourier law helps to define the transport property k, *i.e.*, the thermal conductivity of the heat conducting medium.

Assuming $dx = 1\text{m}$; $A = 1\text{m}^2$ and $dt = 1°$, we obtain

$$Q = k$$

Hence thermal conductivity may be defined as the amount of heat conducted per unit time across unit area and through unit thickness, when a temperature difference of unit degree is maintained across the bounding surface.

The units of thermal conductivity are worked out from the Fourier law written in the form:

$$k = -\frac{Q}{A}\frac{dx}{dt}$$

$$\text{Thus } [k] = \frac{\text{kJ}}{\text{hr}}\frac{1}{\text{m}^2}\frac{\text{m}}{\text{deg}} = \text{kJ/m-hr-deg}$$

The unit kJ/m-hr-deg could also be specified as J/m-s-deg or W/m-deg.

26.5. THERMAL RESISTANCE

Observations indicate that in systems involving flow of fluid, heat and electricity, the flow quantity is directly proportional to the driving potential and inversely proportional to the flow resistance. In a hydraulic circuit, the pressure along the path is the driving potential and roughness of the pipe is the flow resistance. The current flow in a conductor is governed by the voltage potential and electrical resistance of the material. Likewise, temperature difference constitutes the driving force for heat conduction through a medium.

From Ohm's law

$$\text{current } (i) = \frac{\text{voltage potential } (dV)}{\text{electrical resistance } (Re)}$$

and from Fourier's law

heat flow rate (Q)

$$= \frac{\text{temperature potential } (dt)}{\text{thermal resistance } (dx/kA)}$$

Obviously there is a one-one correspondence between the flow of electric current and heat, *i.e.*,

– electric current (amperes) is analogous to thermal heat flow rate (kJ/hr).

– electric voltage (volts) corresponds to thermal temperature difference (degree Kelvin).

– electric resistance (ohms) is analogous to quantity dx/kA. This quantity is called ***thermal resistance***.

Thermal resistance, $R_t = (dx/kA)$, is expressed in the units s-deg/J or deg/W. The reciprocal of thermal resistance is called ***thermal conductance*** and it represents the amount of heat conducted through a solid wall of area A and thickness dx when a temperature difference of unit degree is maintained across the bounding surfaces.

26.6. CONDUCTION THROUGH A PLANE WALL

Consider one-dimensional heat conduction through a homogeneous, isotropic wall of thickness δ with constant thermal conductivity k and constant cross-sectional area A. The wall is insulated on its lateral faces and constant but different temperatures t_1 and t_2 are maintained at its boundary surfaces. Obviously temperature varies only in the direction normal to the wall and the temperature potential causes heat transfer in the positive x-direction.

Let attention is focussed on an elementary strip of thickness dx located at a distance x from the reference plane. The temperature difference acorss the strip is dt and obviously the temperature gradient is dt/dx. Since Q is constant in a steady state, the terms of the Fourier rate equations $Q = -kA\,(dt/dx)$ may be separated and integrated directly between the limits $t = t_1$ at $x = 0$ and $t = t_2$ at $x = \delta$.

Thus,

$$Q\int_0^{\delta} dx = -kA\int_{t_1}^{t_2} dt$$

$$Q\,\delta = kA\,(t_1 - t_2)$$

$$Q = \frac{kA\,(t_1 - t_2)}{\delta} \quad ...(26.10)$$

Similarly between the limits $t = t_1$ at $x = 0$ and $t = t$ at $x = x$,

$$Q = \frac{kA(t_1 - t)}{x}$$

Comparing this identity with Equation 26.10, we get

$$t = t_1 + \left(\frac{t_2 - t_1}{\delta}\right) x$$

or
$$\frac{t - t_1}{t_2 - t_1} = \frac{x}{\delta} \quad ...(26.11)$$

The expression for heat flow rate, equation 26.10, can be written as

$$Q = \frac{t_1 - t_2}{\delta/kA} = \frac{t_1 - t_2}{R_t} \quad ...(26.12)$$

where $R_t = \delta/kA$ is the thermal resistance to heat flow. Equivalent thermal circuit for heat flow through a plane wall has been included in Fig. 26.4.

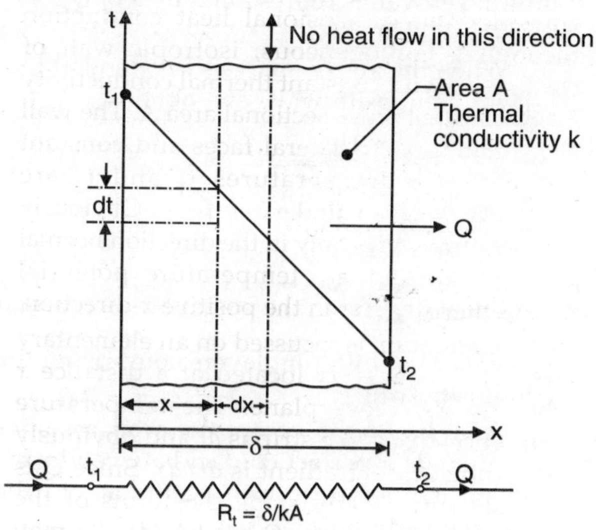

Fig. 26.4. Stready state conduction through a plane wall

EXAMPLE 26.1

The interior of an oven is maintained at a temperature of 850°C by means of a suitable control apparatus. The oven walls are 500 mm thick and are fabricated from a material of thermal conductivity 0.3 W/m-deg. For an outside wall temperature of 250°C, workout the resistance to heat flow and the heat flow per square metre of wall surface. Also calculate the temperature at a point 200 mm from the interior side.

Solution : For a plane wall of thickness δ, conductivity k and area A, the thermal resistance R_t is given by

$$R_t = \frac{\delta}{kA} = \frac{0.5}{0.3 \times 1} = \mathbf{1.667\ deg/W}$$

For one-dimensional steady state, the conduction heat flow through the wall is

$$Q = \frac{t_1 - t_2}{R_t} = \frac{850 - 250}{1.667}$$
$$= \mathbf{359.92\ W}$$

The temperature at distance x from the interior side is

$$t = t_1 + \left(\frac{t_2 - t_1}{\delta}\right) x$$
$$= 850 + \left(\frac{250 - 850}{0.5}\right) \times 0.2$$
$$= \mathbf{610°C}$$

26.7. CONDUCTION THROUGH A COMPOSITE WALL

A composite wall refers to a wall of a several heterogeneous layers, *e.g.*, walls of dwelling houses where bricks are given a layer of plaster on either side. Likewise walls of furnaces, boilers and other heat exchange devices consist of several layers; a layer for mechanical strength or for high temperature characteristics (fire brick), a layer of low thermal conductivity material to restrict the flow of heat (insulating brick) and another layer for structural requirements for good appearnce (ordinary brick).

Fig. 26.5 shows one such composite wall having three layers of different materials tightly fitted to one another. The layers have thicknesses δ_1, δ_2, δ_3 and their thermal conductivities correspond to the average temperature conditions. The surface temperature of the wall are t_1 and t_4 and the temperatures at the interfaces are t_2 and t_3.

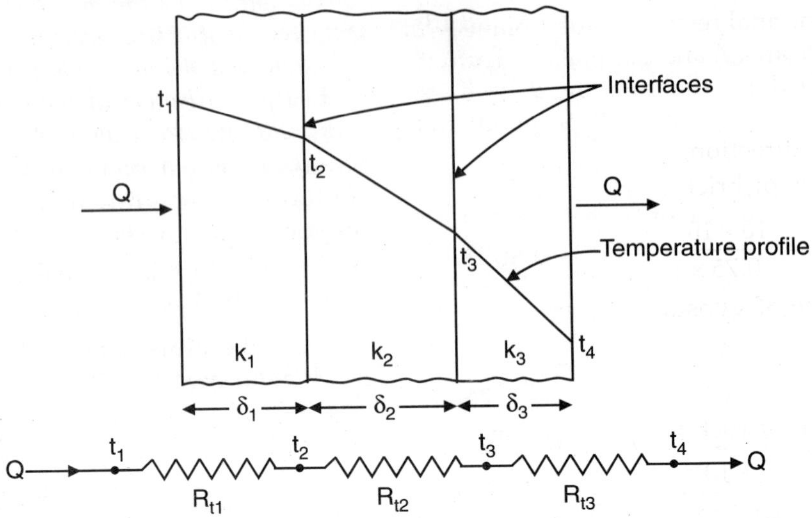

$$R_{t1} = \frac{\delta_1}{k_1 A} \quad ; \quad R_{t2} = \frac{\delta_2}{k_2 A} \quad ; \quad R_{t3} = \frac{\delta_3}{k_3 A}$$

Fig. 26.5. Steady state conduction through a composite wall

Under steady state conditions, heat flow does not vary across the wall, *i.e.*, it is same for every layer. Thus

$$Q = \frac{k_1 A}{\delta_1}(t_1 - t_2) = \frac{k_2 A}{\delta_2}(t_2 - t_3)$$
$$= \frac{k_3 A}{\delta_3}(t_3 - t_4)$$

Rewriting the above expression in terms of temperature drop across each layer,

$$t_1 - t_2 = \frac{Q\delta_1}{k_1 A} \; ; \; t_2 - t_3 = \frac{Q\delta_2}{k_2 A} \text{ and}$$
$$t_3 - t_4 = \frac{Q\delta_3}{k_3 A}$$

Summation gives the overall temperature difference across the wall

$$t_1 - t_4 = Q\left(\frac{\delta_1}{k_1 A} + \frac{\delta_2}{k_2 A} + \frac{\delta_3}{k_3 A}\right)$$

Then

$$Q = \frac{(t_1 - t_4)}{\delta_1/k_1 A + \delta_2/k_2 A + \delta_3/k_3 A}$$
$$= \frac{(t_1 - t_4)}{R_{t_1} + R_{t_2} + R_{t_3}} \qquad \text{...(26.13)}$$

The distribution of temperature in a plane multilayer wall is represented by a polygonal line (Fig. 26.5).

When the above analysis is extended to a n-layer composite wall, one obtains :

$$Q = \frac{(t_1 - t_{n+1})}{\sum_1^n \delta/kA} \qquad \text{...(26.14)}$$

where $\sum_1^n \delta/kA$ is sum of the thermal resistances of different layers comprising the composite wall.

Anaysis of the composite wall assumes that there is a perfect contact between layers and no temperature drop occurs across the interface between materials.

EXAMPLE 26.2

An exterior wall of a house may be approximated by 10 cm layer of common brick (k = 0.75 W/m-deg) followed by 4 cm layer of gypsum plaster (k = 0.5 W/m-deg). What thickness of loosely packed rock wool insulation (k = 0.065 W/m-deg) should be added to reduce the heat loss or gain through the wall by 75% ?

Solution : Thermal resistance for a plane wall of thickness δ, area A and thermal condcutivity k is prescribed by the relation $R_t = \delta/kA$. Considering unit area of wall perpendicular to heat flow direction,

Resistance of brick work

$$= \frac{10 \times 10^{-2}}{0.75 \times 1} = 0.133 \text{ deg/W}$$

Resistance of gypsum plaster

$$= \frac{4 \times 10^{-2}}{0.5 \times 1} = 0.08 \text{ deg/W}$$

Resistance of rock wool insulation

$$= \frac{x \times 10^{-2}}{0.065} = 0.1538\, x \text{ deg/W}$$

where x is thickness of rock wool insulation in cm

Case I : *Rockwool insulation is not used :*

$$\text{Heat loss or gain} = \frac{\Delta t}{\sum R_t}$$

$$= \frac{\Delta t}{0.133 + 0.08} = 4.695\, \Delta t$$

Case II : *Rockwool insulation is used :*

Heat loss or gain = (1 – 0.75) × 4.695 Δ*t*
= 1.174 Δ*t*

$$\therefore\ 1.174\, \Delta t = \frac{\Delta t}{0.133 + 0.08 + 0.1538x}$$

$$\text{or } (0.133 + 0.08 + 0.1538\, x) = \frac{1}{1.174} = 0.8518$$

∴ Thickness of rock wool,

$$x = \frac{0.8518 - 0.133 - 0.08}{0.1538}$$

= 4.153 cm

EXAMPLE 26.3

A storage chamber of interior dimensions 10 m × 8 m × 2.5 m high has its inside maintained at a temperature of – 20°C whilst the outside is at 25°C. The walls and ceiling of the chamber have three layers made of

60 mm thick board (k = 0.2 W/m-deg) on the inside

90 mm thick insulation (k = 0.04 W/m-deg) at mid

240 mm thick concrete (k = 1.8 W/m-deg) on the outside

Neglecting flow of heat through the floor, determine the rate at which heat can flow towards inside of the chamber.

Solution : Neglecting corners and edges, the area of heat flow is

$$A = 2\,(10 \times 2.5) + 2\,(8 \times 2.5) + (10 \times 8) = 170 \text{ m}^2$$

Recalling that thermal resistance for a slab of thickness δ, conductivity k and area A equals δ/kA, we have

$$R_t = \frac{1}{A}\left[\frac{\delta_1}{k_1} + \frac{\delta_2}{k_2} + \frac{\delta_3}{k_3}\right]$$

$$= \frac{1}{170}\left[\frac{60 \times 10^{-3}}{0.2} + \frac{90 \times 10^{-3}}{0.04} + \frac{240 \times 10^{-3}}{1.8}\right]$$

$$= \frac{1}{170}(0.3 + 2.25 + 0.133)$$

$$= 0.01578 \text{ deg/W}$$

$$\therefore \text{ Heat flow rate, } Q = \frac{\Delta t}{R_t}$$

$$= \frac{25 - (-20)}{0.01578} = \mathbf{2851.7\ W}$$

EXAMPLE 26.4

The walls of house in cold region comprise three layers

15 cm outer brick work (k = 0.75 W/m-deg)

1.25 cm inner wooden paneling

k = 0.2 W/m-deg

7.5 cm intermediate layer of insulating material

The insulation layer is stated to offer resistance twice the thermal resistance of brick work. If the inside and outside temperatures of the composite wall are 20°C and –15°C respectively, determine the rate of heat loss per unit area of the wall and the thermal conductivity of the insulating material.

Solution : Thermal resistance for a plane wall of thickness δ, area A and thermal conductivity k is prescribed by the relation $R_t = \delta/kA$

Resistance of brick work,

$$R_{t_1} = \frac{15 \times 10^{-2}}{0.75 \times 1} = 0.2 \text{ deg/W}$$

Resistance of wooden paneling,

$$R_{t_2} = \frac{1.25 \times 10^{-2}}{0.2 \times 1} = 0.0625 \text{ deg/W}$$

Resistance of insulating material,

$$R_{t_3} = 2 \times 0.2 = 0.4 \text{ deg/W}$$

Total resistance,

$$R_t = \sum R_t = 0.2 + 0.0625 + 0.4$$
$$= 0.6625 \text{ deg/W}$$

$$\therefore \text{ Heat loss, } Q = \frac{\Delta t}{\sum R_t}$$

$$= \frac{20 - (-15)}{0.6625} = \mathbf{52.83\ W}$$

(*b*) Thermal conductivity of insulating material,

$$k = \frac{\delta}{A\, R_t} = \frac{7.5 \times 10^{-2}}{1 \times 0.4}$$

$$= \mathbf{0.1875\ W/m\text{-}deg}$$

EXAMPLE 26.5

A furnace wall is a made up of steel plate 10 mm thick (k = 62.8 kJ/m-hr-deg) lined on inside with silica bricks 150 mm thick (k = 7.32 kJ/m-hr-deg) and on the outside with magnesia bricks 200 mm thick (k = 18.84 kJ/m-hr-deg). The inside and outside surfaces of the wall are at temperautre 650°C and 125°C respectively. Make calculations for the heat loss from unit area of the wall.

It is required that the heat loss be reduced to 10 MJ/hour by means of air gap between steel and magnesia bricks. Estimate the necessary width of air gap if thermal conductivity for air is = 0.126 kJ/m-hr-deg.

Solution : Thermal resistance for a plane wall of thickness δ, area *A* and thermal conductivity *k* is prescribed by the relation $R_t = \delta/kA$.

∴ Resistance of steel plate,

$$R_{t_1} = \frac{0.01}{62.8 \times 1} = 0.000159 \text{ deg hr/kJ}$$

Resistance of silica bricks,

$$R_{t_2} = \frac{0.15}{7.32 \times 1} = 0.02049 \text{ deg hr/kJ}$$

Resistance of magnesia bricks,

$$R_{t_3} = \frac{0.20}{18.84 \times 1} = 0.01061 \text{ deg hr/kJ}$$

Total resistance of the composite wall,

$$\sum R_t = R_{t_1} + R_{t_2} + R_{t_3}$$
$$= 0.03126 \text{ deg hr/kJ}$$

Heat loss from the wall

$$= \frac{\Delta t}{\sum R_t} = \frac{650 - 125}{0.03126}$$
$$= 16795 \text{ kJ/hr}$$

To reduce the heat loss to 10 MJ/hr, the thermal resistance should be increased to :

$$\frac{650 - 125}{10 \times 10^3} = 0.0525 \text{ deg hr/kJ}$$

∴ Thermal resistance for the air gap,

$$\frac{\delta}{kA} = 0.0525 - 0.03126$$
$$= 0.02124 \text{ deg hr/kJ}$$

∴ Thickness of air gap,

$$\delta = 0.02124 \times 0.126 \times 1$$
$$= 2.676 \times 10^{-3} \text{ m} = \mathbf{2.676\ mm}$$

EXAMPLE 26.6

A furnace wall comprises three layers: 13.5 cm thick inside layer of fire brick, 7.5 cm thick middle layer of insulating brick and 11.5 cm thick outside layer of red brick. The furnace operates at 870°C and it is anticipated that the outside of this composite wall can be maintained at 40°C by the circulation of air. Assuming close bonding of layers at their interfaces, find the rate of heat loss from the furnace and the wall interface temperature. The wall measures 5 m × 2 m and the data on thermal conductivities is :

Fire brick k_1 = 1.2 W/m-deg
Insulating brick k_2 = 0.14 W/m-deg
Red brick k_3 = 0.85 W/m-deg

Solution : The wall area (5 m × 2 m) = 10 m² is constant for all layers of the composite wall.

The thermal resistance R_t of a slab (thickness δ, conductivity *k* and area *A*) is given by :

$$R_t = \frac{\delta}{kA}$$

∴ Resistance of fire brick

$$R_{t_1} = \frac{0.135}{1.2 \times 10} = 0.01125 \text{ deg/W}$$

Resistance of insulating brick

$$R_{t_2} = \frac{0.075}{0.14 \times 10} = 0.05357 \text{ deg/W}$$

Resistance of red brick

$$R_{t_3} = \frac{0.115}{0.85 \times 10} = 0.01353 \text{ deg/W}$$

Total resistance of the composite wall,

$$\sum R_t = R_{t_3} + R_{t_2} + R_{t_3}$$
$$= 0.07835 \text{ deg/W}$$

The heat will flow from the inside of the furnace (temperature t_1 = 870°C) to outside of the composite wall (temperature t_4 = 40°C)

$$\text{Heat loss} = \frac{t_1 - t_4}{\sum R_t} = \frac{870 - 40}{0.07835}$$
$$= \mathbf{10593.5 \text{ W}}$$

(*b*) Since heat flowing through each layer is same, then for inside layer of fire brick

$$10593.5 = \frac{t_1 - t_2}{0.01125} = \frac{870 - t_2}{0.01125}$$

where t_2 is the temperature at the interface of fire brick and insulating brick.

$$t_2 = 870 - 10593.5 \times 0.01125$$
$$= 750.82°C$$

Similarlty for the mid layer of insulating brick,

$$10593.5 = \frac{t_2 - t_3}{0.05357} = \frac{750.82 - t_3}{0.05357}$$

where t_3 is the temperature at the interface of insulating brick and the red brick

$$t_3 = 750.82 - 10593.5 \times 0.05357$$
$$= 183.33°C$$

Check : The temperature t_3 could also be worked out by considering the heat flow through the outside layer of red brick.

$$10593.5 = \frac{t_3 - t_4}{0.01357} = \frac{t_3 - 40}{0.01357}$$
$$t_3 = 40 + 10593.5 \times 0.01357$$
$$= 183.75°C$$

This is same as calculated above.

EXAMPLE 26.7

Explain the concept of thermal contact resistance.

A furnace wall consists of an inside layer of silica brick 10 cm thick. (k = 6.28 kJ/m-hr°C) followed by a 20 cm layer of magnesite brick (k = 20.95 kJ/m-hr°C) on the outside. The inside surface of the silica brick wall is maintained at 750°C whilst the outside surface of magnesite is at 125°C. The conact thermal resistance between the two walls at the interface is 0.000716 hr-°C/kJ per unit wall area. What is the rate of heat loss per unit area of the wall? Also calculate the temperature drop at the interface.

Solution : Calculation of heat flow through a multi-layer composite wall are made on the presumption that

(*i*) there is perfect contact between adjacent layers

(*ii*) the temperature is continuous at the interface although there is discontinuty in temperature gradient

(*iii*) there is no fall of temperature at the interface, *i.e.*, the temperature of the layers are same at the plane of contact

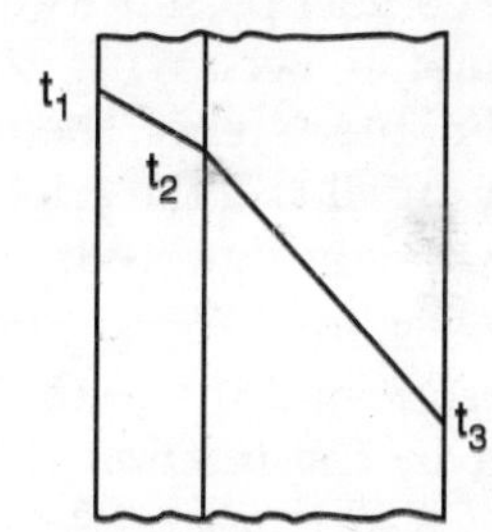

(i) No temperature drop at interface

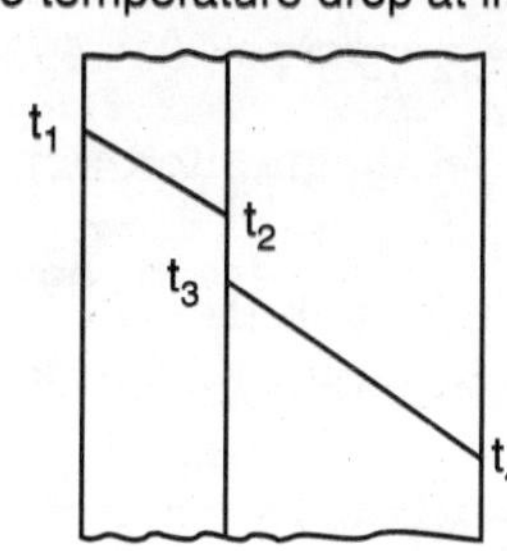

(ii) Temperature drop at interface

Fig. 26.6. Concept of contact resistance

However in real systems, the contact surfaces touch only at discrete locations due to surface roughness, interspersed with void spaces which are usually filled with air. Obviously, there is not a single place of contact. This implies that the area for heat flow at the interface will be small compared to the geometric area of the face. Due to this apparent decrease in the heat flow area and also due to the presence of air voids, there occurs a large resistance to heat flow at the interface. This resistance is referred to as ***thermal contact resistance*** and it causes temperature drop between two materials at the interface.

Let t_2 and t_3 represent the temperature at the theoretical plane interface obtained by consideration of heat flow in materials on either side. Then for a given wall area A, the thermal contact resistance R_c is defined as

$$R_c = \frac{t_2 - t_3}{Q/A} \frac{\text{m}^2\text{-deg-hr}}{\text{kJ}}$$

$$= \frac{t_2 - t_3}{Q} \frac{\text{deg-hr}}{\text{kJ}}$$

The value of metallic contact resistance depends on the metals involved, the surface roughness, the contact pressure and temperature and the matter occupying void spaces. For solid to solid interference, the temperature drop is usually small and the effect of contact resistance is generally ignored.

Thermal resistance for a plane wall of thickness δ, area A and thermal conductivity k is prescribed by the relation $R_t = \delta/kA$

∴ Resistance of silica brick

$$R_{t_1} = \frac{0.1}{6.28 \times 1} = 0.01592 \frac{\text{hr-deg}}{\text{kJ}}$$

Resistance of magnesite brick

$$R_{t_2} = \frac{0.2}{20.95 \times 1} = 0.009546 \frac{\text{hr-deg}}{\text{kJ}}$$

Contact thermal rsistance

$$R_c = 0.000716 \frac{\text{hr-deg}}{\text{kJ}}$$

Total resistance of the composite wall,

$$\Sigma R_t = (R_{t_1} + R_{t_2} + R_c)$$

$$= (0.01592 + 0.009546 + 0.000716)$$

$$= 0.02618 \frac{\text{hr-deg}}{\text{kJ}}$$

Heat loss form the wall $= \dfrac{\Delta t}{\Sigma R_t}$

$$= \frac{750 - 125}{0.02618} = \mathbf{23873\ kJ/hr}$$

The same heat flows through each layer of composite system. Accordingly for the silica brick

$$Q = \frac{t_1 - t_2}{R_{t_1}}; \quad 23873 = \frac{750 - t_2}{0.01592}$$

$$\therefore \quad t_2 = 750 - 23873 \times 0.01592$$

$$= 369.95°\text{C}$$

For the magnesite brick,

$$Q = \frac{t_3 - t_4}{R_{t_2}}; \quad 23873 = \frac{t_3 - 125}{0.009546}$$

$$\therefore \quad t_3 = 125 + 23873 \times 0.009546$$

$$= 352.89°\text{C}$$

Hence, temperature drop at the interface

$$= (t_2 - t_3) = 369.95 - 352.89$$

$$= \mathbf{17.06°C}$$

EXAMPLE 26.8

Determine the heat transfer rate across a composite slab which is made of different materials with top and bottom as shown in Fig. 26.7. The entire left-hand face is held at the temperature T_1 while the entire right hand face is at the temperature T_2. The conductivities of the two different materials are stated as k_a and k_b, and their areas as viewed in the direction of slab thickness δ are A_a and A_b respectively. Steady state exists, there is no heat generation and the slab is so thin that any edge effects can be neglected. Interpret the result in terms of an electrical circuit.

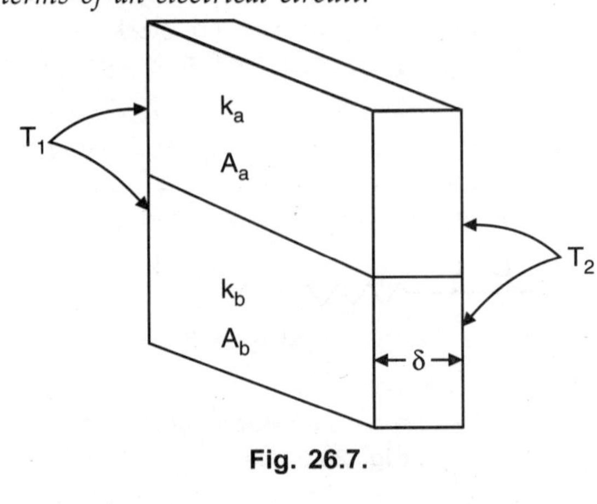

Fig. 26.7.

Solution : Applying Fourier law of heat conduction separately to materials a and b, we obtain

$$Q_a = \frac{k_a A_a (T_1 - T_2)}{\delta}$$

$$Q_b = \frac{k_b A_b (T_1 - T_2)}{\delta}$$

The desired quantity of heat transfer through the slab equals the sum of Q_a and Q_b

$$Q = Q_a + Q_b$$

$$= \frac{k_a A_a (T_1 - T_2)}{\delta} + \frac{k_b A_b (T_1 - T_2)}{\delta}$$

$$= \left\{\frac{1}{\delta/(k_a A_a)} + \frac{1}{\delta/(k_b A_b)}\right\}(T_1 - T_2)$$

$$= \left(\frac{1}{R_{t_a}} + \frac{1}{R_{t_b}}\right)(T_1 - T_2)$$

$$= \frac{1}{R_t}\ (T_1 - T_2)$$

Apparently, the two thermal resistances

$$R_{t_a} = \frac{\delta}{k_a A_a} \quad \text{and} \quad R_{t_b} = \frac{\delta}{k_b A_b}$$

appear in the same way as two electrical resistors in parallel. Accordingly the electrical circuit for the heat transfer through the given composite slab will be as indicated below :

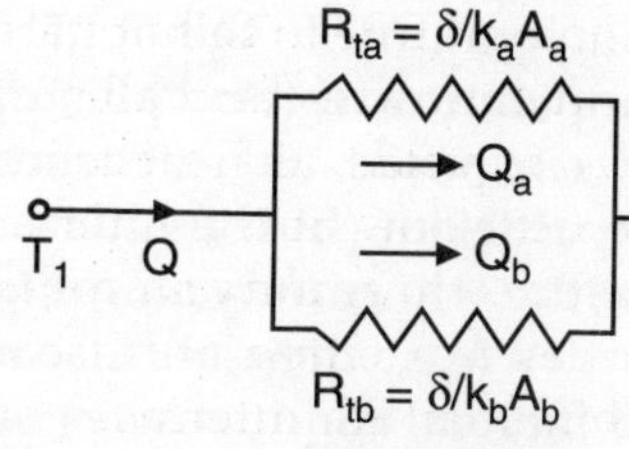

Fig. 26.8.

26.8. HEAT FLOW BETWEEN SURFACE AND SURROUNDINGS

When a moving fluid comes into contact with a stationary surface, a thin boundary layer develops adjacent to the wall and in this layer there is no relative velocity with respect to surface. In a heat exchange process, this layer is called ***stagnant film*** and heat flow in the layer is covered both by conduction and convection processes. Since thermal conductivity of fluids is low, the heat flow from the moving fluid of the wall is mainly due to convection.

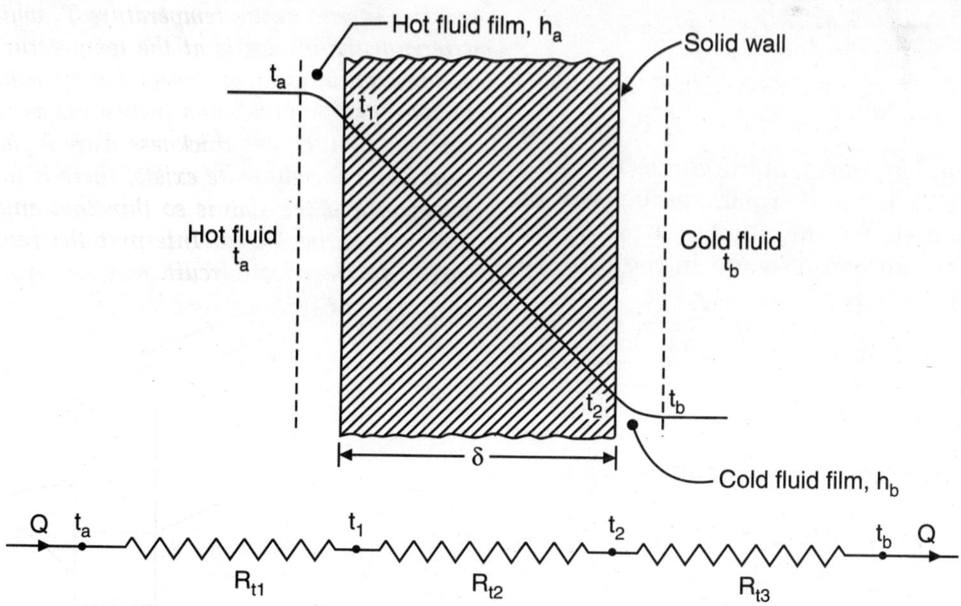

$$R_{t1} = \frac{1}{h_a A} \quad ; \quad R_{t2} = \frac{\delta}{kA} \quad ; \quad R_{t3} = \frac{1}{h_b A}$$

Fig. 26.9. Heat conduction through a wall separating two fluids

The rate of convective heat transfer between a solid boundary and adjacent fluid is given by the ***Newton-Rikhman law*** :

$$Q = hA(t_s - t_f) \quad ...(26.15)$$

where t_f is the temperature of the hot moving fluid, t_s is the temperature of the wall surface and A is the area exposed to heat transfer. The convective coefficient (film coefficient) h depends upon the thermodynamic and transport properties (*e.g.*, density, viscosity, specific heat and thermal conductivity) of the fluid, the geomeny of the surface, the nature of fluid flow and the prevailing thermal conditions. The resistance of the stagnant boundary layer is included in the convection coefficient. The dimensions of h are W/m^2-deg. The factor $1/hA$ represents the thermal resistance of the film heat transfer process.

The heat transfer through a wall separating two moving fluids involves: (*i*) flow of heat from the fluid of high temperature to the wall, (*ii*) heat conduction through the wall and (*iii*) transport of heat from the wall to the cold fluid (Fig. 26.9)

Under steady state conditions, the heat flow can be expressed by the equations :

$$Q = h_a A\,(t_a - t_1) = \frac{kA}{\delta}\,(t_1 - t_2) = h_b A\,(t_2 - t_b) \quad ...(26.16)$$

where h_a and h_b represent the convective film coefficients, k is the thermal conductivity of the solid wall having thickness δ. These expressions can be presented in the form :

$$t_a - t_1 = \frac{Q}{h_a A}\ ;\ t_1 - t_2 = \frac{Q\delta}{kA} \text{ and}$$

$$t_2 - t_b = \frac{Q}{h_b A}$$

Summation of these equalities yields

$$(t_a - t_b) = Q\left[\frac{1}{h_a A} + \frac{\delta}{kA} + \frac{1}{h_b A}\right]$$

$$\therefore \quad Q = \frac{(t_a - t_b)}{1/h_a A + \delta/kA + 1/h_b A} \quad ...(26.17)$$

The denominator $(1/h_a A + \delta/kA + 1/h_b A)$ is the sum of thermal resistance of different sections through which heat has to flow.

Quite often, the heat flow through a composite section is written in the form

$$Q = UA\,(t_a - t_b) \quad ...(26.18)$$

where U is the ***overall heat transfer coefficient***. It represents the intensity of heat transfer from one fluid to another through a wall separating them. Numerically it equals the quantity of heat passing through unit area of wall surface in unit time at a temperature difference of unit degree. The coefficient U has dimensions of W/m^2-deg.

Comparing equations 26.17 and 26.18.

$$\frac{1}{U} = \frac{1}{h_a} + \frac{\delta}{k} + \frac{1}{h_b} \quad ...(26.19)$$

Apparently the overall heat transfer coefficient is reciprocal of unit thermal resistance to heat flow.

EXAMPLE 26.9

A lake surface is covered by a 8 cm thick layer of ice (k = 8 kJ/m-hr-deg) when the ambient air temperature is – 12.5°C. A thermo-couple embedded on the upper surface of the layer indicates a temperature of – 5°C. Assuming steady state conduction in ice and no liquid subcooling at the bottom surface of the ice layer, find the heat transfer coefficient at the upper surface. Also work out the heat loss per square kilometer of area.

Solution : Since water at the bottom surface of ice layer is to remain in liquid state (no sub-cooling), the minimum temperature required at that surface is 0°C.

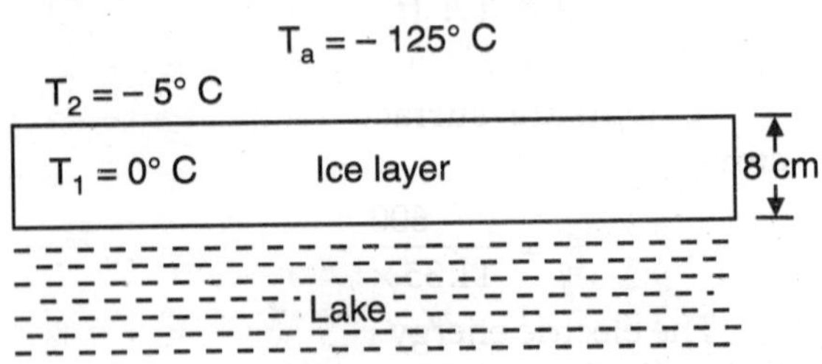

Fig. 26.10.

Invoking Fourier's law, the conduction heat transfer rate across the ice layer becomes

$$Q = \frac{k A \Delta t}{\delta}$$

$$= \frac{8 \times (1000 \times 1000) \times \{0 - (-5)\}}{0.08}$$

$$= 5 \times 10^8 \text{ kJ/hr}$$

Since steady state conditions prevail, the conduction heat flow across the ice slab equals the convective heat transfer between the top surface of ice slab and surrounding atmosphere. Accordingly

$$Q = h A \Delta t$$

$$5 \times 10^8 = h \times (1000 \times 1000) \times [-5 - (-12.5)]$$

∴ Heat transfer coefficient at the upper surface,

$$h = \frac{5 \times 10^8}{(1000 \times 1000) \times 7.5}$$

$$= \mathbf{66.67\ m^2\text{-hr-deg}}$$

EXAMPLE 26.10

The oven of an electric stove, of total outside surface area 2.9 m², dissipates electric energy at the rate of 600 watt. The surrounding room air is at 20°C and the surface coefficient of heat transfer between the room air and the surface of the oven is estimated to be 11.35 W/m²-deg. Determine the average steady-state temperature of the outside surface of the stove. What would be the inside surface temperature if wall thickness of stove is 3.8 cm and thermal conductivity of the stove material is 0.069 W/m-deg ?

Solution : The electric energy is dissipated as convective heat flow from the outside surface of stove to the ambient air. Therefore

$$Q = h A (t_o - t_a)$$

$$600 = 11.35 \times 2.9 \times (t_o - 20)$$

∴ Outside surface temperature of the stove,

$$t_o = \frac{600}{11.35 \times 2.9} + 20 = \mathbf{38.22°C}$$

The electric energy is first conducted across the wall of the oven. Invoking Fourier's law of heat conduction

$$Q = \frac{kA(t_i - t_o)}{\delta}$$

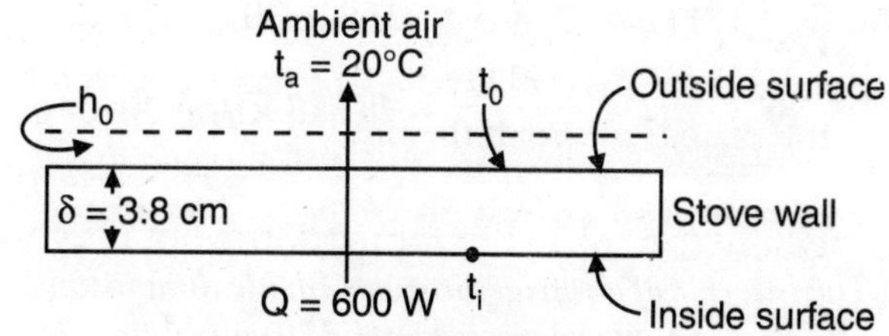

Fig. 26.11.

$$600 = \frac{0.069 \times 2.9 \times (t_i - 38.22)}{0.038}$$

$$= 5.26\ (t_i - 38.22)$$

∴ Inside surface temperature of the stove

$$t_i = \frac{600}{5.26} + 38.22 = \mathbf{152.29°C}$$

EXAMPLE 26.11

Consider a plane composite wall that is made of two materials of thermal conductivities k_a = 735 kJ/m-hr-deg and k_b = 165 kJ/m-hr-deg and thickness δ_a = 5 cm and δ_b = 2.5 cm. Material A adjoins a hot fluid at 150°C for which h_a = 42 kJ/m²-hr-deg and the material B is in contact with a cold fluid at 30°C and h_b = 85 kJ/m²-hr-deg. Calculate (a) the rate of heat transfer through a wall which is 2 m high and 2.5 m wide (b) the overall coefficient of heat transfer.

Solution : The wall area (2 m × 2.5 m) = 5 m² is constant for all the layers. Total thermal resistance of the composite system is given by

$$\Sigma R_t = \frac{1}{A}\left[\frac{1}{h_a} + \frac{\delta_a}{k_a} + \frac{\delta_b}{k_b} + \frac{1}{h_b}\right]$$

$$= \frac{1}{5}\left[\frac{1}{42} + \frac{0.05}{735} + \frac{0.025}{165} + \frac{1}{85}\right]$$

$$= 0.007013 \ \frac{\text{hr-deg}}{\text{kJ}}$$

$$\text{Rate of heat transfer} = \frac{\Delta t}{\sum R_t}$$

$$= \frac{150 - 30}{0.007013} = 17111 \text{ kJ/hr}$$

(*b*) Heat transfer from the hot to cold fluid can also be expressed as :

$$Q = U A \Delta t$$

where U is the overall coefficient of heat transfer

$\therefore\ 17111 = U \times 5 \times (150 - 30)$

or $\quad U = \dfrac{17111}{5 \times 120} = \mathbf{28.518\ kJ/m^2\text{-}hr\text{-}deg}$

EXAMPLE 26.12

The interior of a refrigerator has inside dimensions 60 cm × 45 cm base area and 120 cm high. The composite wall is made of two 3 mm mild steel sheets (k = 145 kJ/m-hr-deg) with 6 cm of glass wool (k = 0.188 kJ/m-hr-deg) insulation sandwiched between them. The average values of convective heat transfer coefficients at the interior and exterior wall are 40.8 and 52.3 kJ/m²-hr-deg respectively.

(a) Calculate the individual resistance of this composite wall and the resistances at the surfaces, and the overall conductance.

(b) Draw the thermal circuit

(c) For the air temperature inside the refrigerator at 6.5°C and outside of 25°C, determine the rate at which heat must be removed from the refrigerator. Also, calculate the temperature on the outer surface of the metal sheet.

Solution : Neglecting corner effects, the area for heat flow is

$$A = 2(1.2 \times 0.6 + 0.6 \times 0.45 + 0.45 \times 1.2) = 3.06\ \text{m}^2$$

This is constant for all layers of the composite wall and the various thermal resistances to flow of heat are offered by :

(*i*) outside air film $= \dfrac{1}{h_o A} = \dfrac{1}{52.3 \times 3.06}$
$= 6.248 \times 10^{-3}$ deg-hr/kJ

(*ii*) mild sheet sheet $= \dfrac{\delta_1}{k_1 A} = \dfrac{0.003}{145 \times 3.06}$
$= 6.761 \times 10^{-6}$ deg-hr/kJ

(*iii*) glass wool insulation $= \dfrac{\delta_2}{k_2 A}$
$= \dfrac{0.06}{0.188 \times 3.06}$
$= 0.1043$ deg-hr/kJ

(*iv*) mild steel sheet $= \dfrac{\delta_3}{k_3 A} = \dfrac{0.003}{145 \times 3.06}$
$= 6.761 \times 10^{-6}$ deg-hr/kJ

(*v*) inside air film $= \dfrac{1}{h_i A} = \dfrac{1}{40.8 \times 3.06}$
$= 8.0097 \times 10^{-3}$ deg-hr/kJ

The total resistance equals the sum of individual resistances and the summation gives

$$\Sigma R_t = 0.11857 \text{ deg-hr/kJ}$$

$$\text{Overall conductance} = \frac{1}{\sum R_t} = \frac{1}{0.11857} = 8.434 \text{ kJ/hr-deg}$$

$$\text{Heat loss} = \frac{\Delta t}{\sum R_t} = \frac{25 - 6.5}{0.11857} = \mathbf{156\ kJ/hr}$$

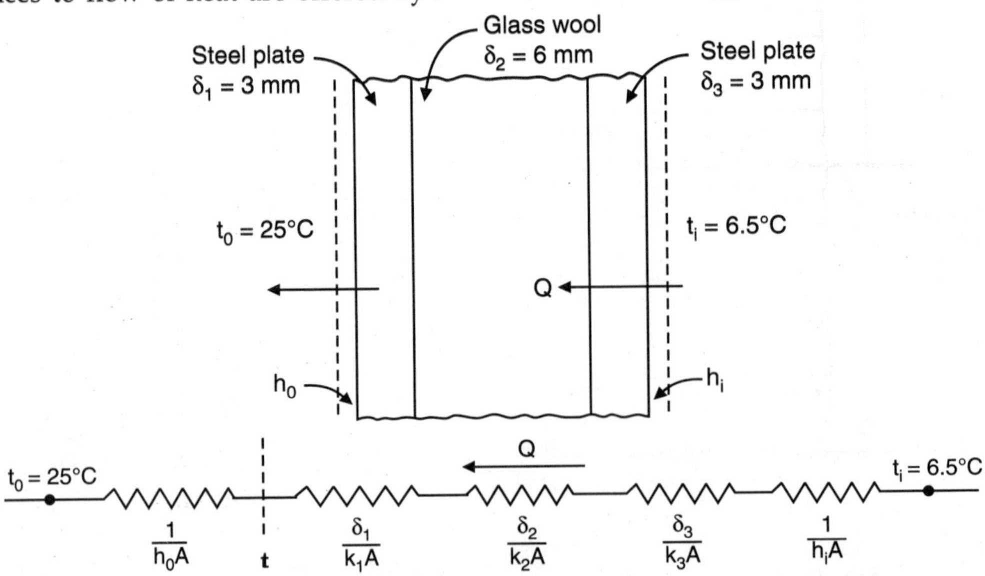

Fig. 26.12.

Since heat flowing through each layer is same; then for the outside air film

$$156 = h_o A (t_o - t)$$
$$= 52.3 \times 3.06(25 - t)$$

where t is the temperature at the outer surface of metal sheet. Solution gives

$$t = 25 - \frac{156}{52.3 \times 3.06} = 24.02°C$$

26.9. CONDUCTION THROUGH A CYLINDRICAL WALL

Cylindrical metal tubes constitute an essential element of power stations, oil refineries and most process industries. The boilers have tubes in them, the condensers contain bank of tubes, the heat exchangers are tubulars and all these units are connected by tubes. Evidently then the radial heat transfer rate through a tube or any insulation which may surround it is quite important.

Consider heat conduction through a cylindrical tube of inner radius r_1, outer radius r_2 and length l. The inside and outside surfaces of the tube are at constant temperatures t_1 and t_2 and thermal conductivity k of the tube material is constant within the given temperature range. If both ends are perfectly insulated, the heat flow is limited to radial direction only. Further if temperature t_1 at the inner surface is greater than temperature t_2 at the outer surface, the heat flows radially outwards.

Consider an infinitesimally thin cylindrical element at radius r. Let thickness of this elementary ring be dr and the change of temperature across it be dt. Then invoking Fourier law of heat conduction

$$Q = -kA\frac{dt}{dr} = -k(2\pi rl)\frac{dt}{dr}$$

Separating the variables and integrating within the prescribed boundary conditions, we obtain :

$$\frac{Q}{2\pi kl}\int_{r_1}^{r_2}\frac{dr}{r} = -\int_{t_1}^{t_2} dt$$

$$\frac{Q}{2\pi kl}\log_e\frac{r_2}{r_1} = (t_1 - t_2)$$

or $$Q = 2\pi kl\,\frac{(t_1 - t_2)}{\log_e(r_2/r_1)} = \frac{(t_1 - t_2)}{R_t} \quad \text{...(26.20)}$$

Evidently for conduction in a hollow cylinder, the thermal resistance takes the form

$$R_t = \frac{\log_e(r_2/r_1)}{2\pi kl}$$

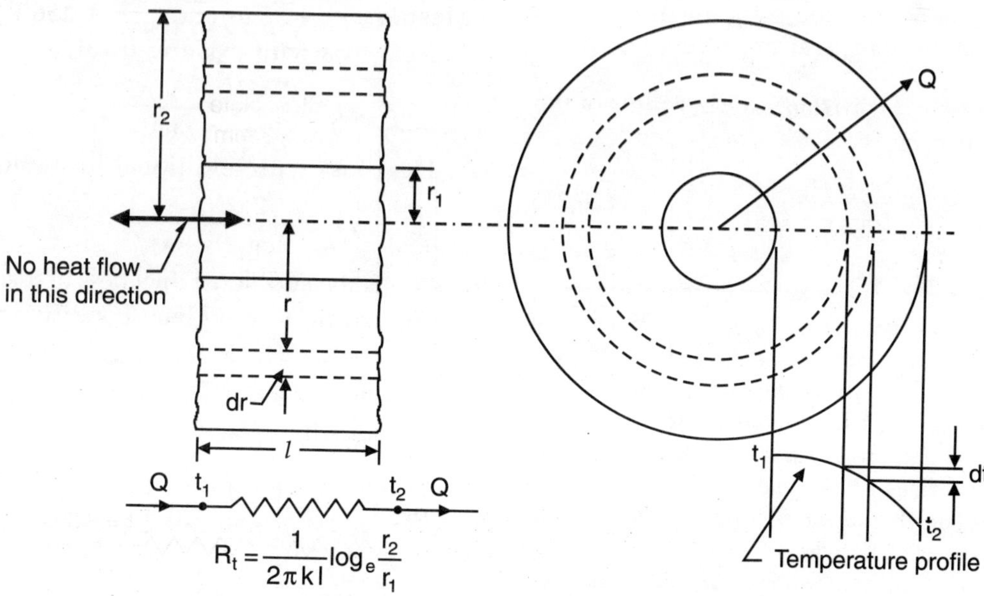

Fig. 26.13. Steady state conduction through a cylindrical wall

Similarly between the limits $t = t_1$ at $r = 0$ and $t = t$ at $r = r$,

$$Q = \frac{-2\pi kl(t-t_1)}{\log_e(r/r_1)}$$

Comparing this identity with Equation 26.20, we get

$$\frac{t-t_1}{t_2-t_1} = \frac{\log_e(r/r_1)}{\log_e(r_2/r_1)} \quad ...(26.21)$$

(*iii*) Quite often it is considered advantageous to write the heat flow equation through a cylinder in the same form as that for heat flow through a plane wall. Then thickness δ will be equal to $(r_2 - r_1)$ and the area A will be an equivalent area A_m. Thus

$$Q = \frac{kA}{\delta}(t_1 - t_2) = kA_m \frac{(t_1 - t_2)}{r_2 - r_1} \quad ...(26.22)$$

Comparing equations 26.20 and 26.22,

$$A_m = \frac{2\pi(r_2 - r_1)l}{\log_e r_2 / r_1} = \frac{A_2 - A_1}{\log_e A_2 / A_1}$$

where A_1 and A_2 are the inner and outer surface areas of the cylindrical tube. The equivalent area A_m is called the ***logarithmic mean area*** of the tube. Further,

$$A_m = 2\pi r_m l = \frac{2\pi(r_2 - r_1)l}{\log_e r_2 / r_1}$$

Obviously, ***logarithmic mean radius*** of the cylindrical tube is :

$$r_m = \frac{r_2 - r_1}{\log_e r_2 / r_1} \quad ...(26.23)$$

EXAMPLE 26.13

A cylindrical cement tube of radii 0.05 cm and 1.0 cm has a wire embedded into it along its axis. To maintain a steady temperature difference of 120°C between the inner and outer surfaces, a current of 5 ampere is made to flow in the wire. Make calculations for the amount of heat generated per metre length and the thermal conductivity of cement.

Take resistance of wire equal to 0.1 ohm per cm of length.

Solution : Resistance of wire,

R = 0.1 Ω per cm length
= 10 Ω per m length

Heat generated = $I^2 R$
= $5^2 \times 10$ = **250 W/m** length

Under steady state conditions, the heat generated equals the heat transfer through the cylindrical element.

$$\therefore \quad Q = \frac{2\pi kl(t_1 - t_2)}{\log_e(r_2/r_1)}$$

$$250 = \frac{2\pi \times k \times 1 \times 120}{\log_e(1.0/0.05)}$$

∴ Thermal conductivity of cement,

$$k = \frac{250 \log_e(1.0/0.05)}{2\pi \times 120}$$

= **0.994 W/m-deg**

EXAMPLE 26.14

A hot fluid is being conveyed through a long pipe of 4 cm outer diameter and covered with 2 cm thick insulation. It is proposed to reduce the conduction heat loss to the surroundings to one-third of the present rate by further covering with same insulation. Calculate the additional thickness of insulation.

Solution : r_1 = 2 cm and r_2 = 4 cm

Heat loss with existing insulation,

$$Q_1 = \frac{2\pi kl(t_1 - t_2)}{\log_e(r_2/r_1)}$$

Heat loss with additional insulation,

$$Q_2 = \frac{2\pi kl(t_1 - t_2)}{\log_e(r_2 + x)/r_1}$$

where x is the additional thickness of insulation

According to the given condition :

$$Q_2 = Q_1/3$$

$$\therefore \quad \frac{2\pi kl(t_1 - t_2)}{\log_e(r_2 + x)/r_1} = \frac{1}{3}\frac{2\pi kl(t_1 - t_2)}{\log_e(r_2/r_1)}$$

$$\text{or} \quad \frac{r_2 + x}{r_1} = \left(\frac{r_2}{r_1}\right)^3 = \left(\frac{4}{2}\right)^3 = 8$$

$$\therefore \quad x = 8r_1 - r_2$$

= 8 × 2 − 4 = **12 cm**

EXAMPLE 26.15

A steam pipe, 7.5 cm external diameter and 25 m long, conveys 1000 kg of steam per hour at a pressure of 20 bar. The steam enters the pipe with dryness fraction 0.98 and it is required that the steam at exit from the pipe must have a minimum dryness fraction of 0.96. The task is accomplished by suitably lagging the pipe; the thermal conductivity of lagging being 0.7 kJ/m-hr-deg. Neglecting thermal resistance of the steam pipe (no temperature drop across the steam pipe), determine the minimum thickness of lagging required to meet the necessary conditions. The temperature at the outside surface of lagging be taken as 25°C.

Also determine the temperature at the point half way between the inner and outer surfaces.

Solution : A reference to steam tables indicates that at 20 bar pressure

Saturation temperature of steam = 212.4°C

Latent heat of steam = 1890 kJ/kg

∴ Heat loss per kg of steam passing through the pipe,

$$= (0.98 - 0.96) \times 1890 = 37.8 \text{ kJ}$$

Total heat loss

$$Q = 37.8 \times 1000 = 37800 \text{ kJ/hr}$$

In terms of thermal resistance and temperature drop, the heat loss is given by :

$$Q = \Delta t / R_t$$

$$\therefore\ 37800 = \frac{212.4 - 25}{R_t};$$

$$\text{or}\quad R_t = \frac{212.4 - 25}{37800} = 0.00496 \text{ deg-hr/kJ}$$

For a cylindrical wall, the thermal resistance is

$$R_t = \frac{\log_e (r_2/r_1)}{2\pi k l};$$

$$0.00496 = \frac{\log_e (r_2/r_1)}{2\pi \times 0.7 \times 25}$$

$$\log_e (r_2/r_1) = 0.545;\quad (r_2/r_1) = 1.725$$

r_1 = 3.75 cm (given), and hence

$$r_2 = 3.75 \times 1.725 = 6.47$$

∴ Thickness of lagging = 6.47 − 3.75

= 2.72 cm

Radius at mid-plane of the pipe,

$$r = \frac{r_1 + r_2}{2} = \frac{3.75 + 6.47}{2} = 5.11 \text{ cm}$$

The temperature at a point of the cylindrical pipe wall can be obtained by using the relation

$$\frac{t - t_1}{t_2 - t_1} = \frac{\log_e (r/r_1)}{\log_e (r_2/r_1)}$$

$$\frac{t - 212.4}{25 - 212.4} = \frac{\log_e (5.11/3.75)}{\log_e (6.47/3.75)} = 0.5673$$

$$\therefore\quad t = 212.4 - (212.4 - 25) \times 0.5673$$

$$= \mathbf{106.1°C}$$

EXAMPLE 26.16

The hot combustion gases at 150°C flow through a hollow cylindrical pipe of 10 cm inner diameter and 12 cm outer diameter. The pipe is located in a space at 30°C and the thermal conductivity of the pipe material is 200 W/mK. Neglecting surface heat transfer coefficients, calculate the heat loss through the pipe per unit length and the temperature at a point halfway between the inner and outer surface. What should be the surface area normal to the direction of heat flow so that the heat transfer through the pipe can be determined by considering material of the pipe as a plane wall of the same thickness ?

Solution : In terms of geometrical parameters, thermal resistance of a cylindrical pipe is

$$R_t = \frac{1}{2\pi k l} \log_e r_2/r_1$$

$$= \frac{1}{2\pi \times 200 \times 1} \log_e (12/10)$$

$$= 1.4516 \times 10^{-4} \text{ deg/W}$$

Heat loss Q

$$= \frac{\Delta t}{R_t} = \frac{150 - 30}{1.4516 \times 10^{-4}}$$

$$= \mathbf{826674\ W}$$

(*ii*) Radius at halfway through the pipe wall,

$$r = \frac{r_1 + r_2}{2} = \frac{10 + 12}{2} = 11 \text{cm}$$

Thermal resistnace of cylindrical pipe upto its mid-plane

$$= \frac{1}{2\pi k l} \log_e (r/r_1)$$

$$= \frac{1}{2\pi \times 200 \times 1} \log_e (11/10)$$

$$= 7.5884 \times 10^{-5} \text{ deg/W}$$

Since heat flow through each section is same ;

$$826674 = \frac{t_1 - t}{7.5884 \times 10^{-5}}$$

∴ Temperature at the mid plane,

$$t = 150 - 826674 \times 7.5884 \times 10^{-5}$$
$$= \mathbf{87.27°C}$$

(*iii*) The equivalent logarithmic mean area is

$$A_m = \frac{A_2 - A_1}{\log_e (A_2 / A_1)} = \frac{2\pi (r_2 - r_1) l}{\log_e (r_2 / r_1)}$$

$$= \frac{2\pi(0.12 - 0.10) \times 1}{\log_e (12/10)} = 0.689 \text{ m}^2$$

26.10. CONDUCTION THROUGH A MULTI-LAYER CYLINDRICAL WALL

Multi-layer cylindrical walls are frequently employed to reduce heat losses from metallic pipes meant for handling a hot fluid. The pipe is generally wrapped in one or more layers of heat insulation, *e.g.*, a steam pipe used for conveying high pressure steam in a steam power plant may have cylindrical metal wall, a layer of insulating material and then a layer of protecting plaster. The arrangement is called ***lagging*** of the pipe system.

Fig. 16.14 shows conduction of heat through a composite cylindrical wall having three layers of different materials. There is a perfect contact between the layers and so an equal interface temperature for any two neighbouring layers. For steady state conduction, the heat flow through each layer is same and it can be described by the following set of equations :

$$Q = 2\pi k_1 l \frac{(t_1 - t_2)}{\log_e (r_2 / r_1)}$$

$$= 2\pi k_2 l \frac{(t_2 - t_3)}{\log_e (r_3 / r_2)}$$

$$= 2\pi k_3 l \frac{(t_3 - t_4)}{\log_e (r_4 / r_3)}$$

These equations help to determine the temperature difference for each layer of the composite cylinder, *i.e.*,

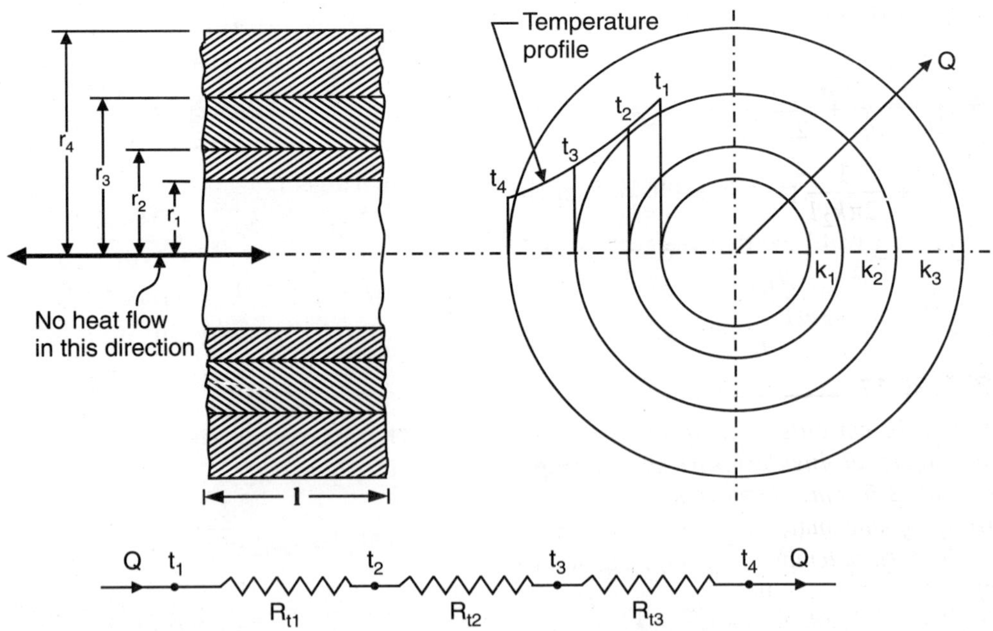

Fig. 26.14. Conduction through a composite cylindrical wall

$$t_1 - t_2 = \frac{Q}{2\pi k_1 l} \log_e \frac{r_2}{r_1}$$

$$t_2 - t_3 = \frac{Q}{2\pi k_2 l} \log_e \frac{r_3}{r_2}$$

$$t_3 - t_4 = \frac{Q}{2\pi k_3 l} \log_e \frac{r_4}{r_3}$$

From summation of these equalities, one obtains

$$(t_1 - t_4) = Q\left[\frac{1}{2\pi k_1 l} \log_e \frac{r_2}{r_1} + \frac{1}{2\pi k_2 l} \log_e \frac{r_3}{r_2} + \frac{1}{2\pi k_3 l} \log_e \frac{r_4}{r_3}\right]$$

Thus the heat flow rate through a composite cylindrical wall is

$$Q = \frac{(t_1 - t_4)}{\frac{1}{2\pi k_1 l} \log_e \frac{r_2}{r_1} + \frac{1}{2\pi k_2 l} \log_e \frac{r_3}{r_2} + \frac{1}{2\pi k_3 l} \log_e \frac{r_4}{r_3}} \quad ...(26.24)$$

The quantity in the denominator is the sum of thermal resistances of different layers comprising the composite cylinder.

If the internal and external heat transfer coefficients for the composite cylinder are h_i and h_0 respectively, then the total thermal resistance to heat flow would be

$$R_t = \frac{1}{2\pi r_1 l h_i} + \frac{1}{2\pi k_1 l} \log_e \frac{r_2}{r_1} + \frac{1}{2\pi k_2 l} \log_e \frac{r_3}{r_2} + \frac{1}{2\pi k_3 l} \log_e \frac{r_4}{r_3} + \frac{1}{2\pi r_4 l h_0} \quad ...(26.25)$$

EXAMPLE 26.17

A steam pipe 10 cm outside diameter is covered with two layers of insulation, each having a thickness of 2.5 cm. The average thermal conductivity of one material is 3 times that of other and the surface temperatures of the insulated steam pipe are fixed. Examine the position of better insulating layer relative to the steam pipe if heat dissipation from steam is to be minimum. What % age saving in heat dissipation results from that arrangement?

Solution : The geometrical dimensions of the lagged steam pipe are :

r_1 = 5 cm ; r_2 = 7,5 cm ; r_3 = 10 cm

Let k and $3k$ be the thermal conductivities of the two insulating materials. Obviously the material with low thermal conductivity is a better insulator.

(*i*) Better insulator (material with low thermal conductivity) is placed inside, *i.e.*, next to the steam pipe.

Thermal resistance

$$= \frac{1}{2\pi k_1 l} \log_e \frac{r_2}{r_1} + \frac{1}{2\pi k_2 l} \log_e \frac{r_3}{r_2}$$

$$= \frac{1}{2\pi k l} \log_e \frac{r_2}{r_1} + \frac{1}{2\pi (3k) l} \log_e \frac{r_3}{r_2}$$

$$= \frac{1}{2\pi k l}\left[\log_e \frac{7.5}{5.0} + \frac{1}{3} \log_e \frac{10}{7.5}\right]$$

$$= \frac{0.2506}{\pi k l}$$

Heat dissipation

$$Q_1 = \frac{\Delta T}{R_t} = \frac{\Delta T \times \pi k l}{0.2506} = 3.99\ \pi\, k\, l\, \Delta T$$

(*ii*) Better insulator is outside

Thermal resistance

$$= \frac{1}{2\pi k_2 l} \log_e \frac{r_2}{r_1} + \frac{1}{2\pi k_1 l} \log_e \frac{r_3}{r_2}$$

$$= \frac{1}{2\pi (3k) l} \log_e \frac{r_2}{r_1} + \frac{1}{2\pi k l} \log_e \frac{r_3}{r_2}$$

$$= \frac{1}{2\pi k l}\left[\frac{1}{3} \log_e \frac{r_2}{r_1} + \log_e \frac{r_3}{r_2}\right]$$

$$= \frac{1}{2\pi k l}\left[\frac{1}{3} \log_e \frac{7.5}{5.0} + \log_e \frac{10}{7.5}\right]$$

$$= \frac{0.2114}{\pi k l}$$

Heat dissipation

$$Q_2 = \frac{\Delta T}{R_t} = \frac{\Delta T}{0.2114} \times \pi k l$$

$$= 4.7303\ \pi\, k\, l\, \Delta T$$

$$\therefore \quad \frac{Q_1}{Q_2} = \frac{3.99}{4.7303} = 0.843$$

Obviously the heat dissipation is small when the material with low thermal conductivity is placed next to the steam pipe

Saving in heat dissipation

$$= \frac{4.7303 - 3.99}{3.99} \times 100 = \mathbf{18.55\%}$$

EXAMPLE 26.18

An insulated steam pipe of 16 cm diameter is covered with 4 cm thick layer of insulation (k = 0.9 W/m-deg) and carries process steam. Determine the percentage change in the rate of heat loss if an extra 2 cm thick layer of lagging (k = 1.25 W/m-deg) is provided. Given that surrounding temperature remains constant and the heat transfer coefficient for both the configurations is 12 W/m²-deg.

Solution : Since the thickness and conductivity of the pipe material are not given, it will be assumed that resistance of the pipe material to heat flow is negligible.

r_1 = 8 cm ; r_2 = 8 + 4 = 12 cm and r_3 = 12 + 2 = 14 cm

Case I : Resistance to heat flow,

$$\Sigma R_t = R_{t_1} + R_{t_2}$$

$$= \frac{1}{2\pi k_1 l}\log_e\frac{r_2}{r_1} + \frac{1}{2\pi r_2 l h_o}$$

$$= \frac{1}{2\pi l}\left[\frac{1}{0.9}\log_e\frac{12}{8} + \frac{1}{0.12 \times 12}\right]$$

$$= \frac{1}{2\pi l} \times 1.1449 \text{ deg/W}$$

Heat loss,

$$Q_1 = \frac{\Delta t}{\Sigma R_t} = \frac{2\pi l \Delta t}{1.1449} = 1.747\ \pi\, l\, \Delta t$$

Case II : Resistance to heat flow,

$$\Sigma R_t = R_{t_1} + R_{t_2} + R_{t_3}$$

$$= \frac{1}{2\pi k_1 l}\log_e\frac{r_2}{r_1} + \frac{1}{2\pi k_2 l}\log_e\frac{r_3}{r_2} + \frac{1}{2\pi r_3 l h_o}$$

$$= \frac{1}{2\pi l}\left[\frac{1}{0.9}\log_e\frac{12}{8} + \frac{1}{1.25}\log_e\frac{14}{12} + \frac{1}{0.14 \times 12}\right]$$

$$= \frac{1}{2\pi l}\left[0.4505 + 0.1233 + 0.5952\right]$$

$$= \frac{1}{2\pi l} \times 1.169 \text{ deg/W}$$

Heat loss,

$$Q_2 = \frac{\Delta t}{\Sigma Rt} = \frac{2\pi l \Delta t}{1.169} = 1.711\ \pi\, l\, \Delta t$$

∴ Decrease in heat loss

$$= \frac{Q_1 - Q_2}{Q_1} = \frac{1,747 - 1.711}{1.747}$$

$= 0.0206$ or **2.06 %**

EXAMPLE 26.19

A steam main 75 mm inside diameter and 90 mm outside diameter is lagged with two successive layers of insulation. The layer in contact with the pipe is 38 mm asbestos and the asbestos layer is covered with 25 mm thick magnesia insulation. The surface coefficients for inside and outside surfaces are 227 W/m²K and 6.8 W/m²K respectively. If the steam temperature is 375°C and the ambient temperature is 35°C, calculate the steady state loss of heat from steam for 60 m length of pipe. Also workout the overall coefficient of heat transfer based on the inside and outside surfaces of the lagged steam main. Comment upon the important generalisation of your result.

Thermal conductivity values are :

Pipe material : 45 W/mK

Asbestos : 0.14 W/mK

Magnesia Insulation : 0.07 W/mK

Solution : r_1 = 37.5 mm = 0.0375 m ;
r_2 = 0.045 m
r_3 = 0.045 + 0.038 = 0.083 m
r_4 = 0.083 + 0.025 = 0.108 m

The various thermal resistances to flow of heat are offered by:

(*i*) Inner steam film

$$= \frac{1}{h_i A_1} = \frac{1}{227 \times 2\pi \times 0.0375 \times 60}$$

$= 3.117 \times 10^{-4}$

(*ii*) Pipe material $= \dfrac{1}{2\pi k_1 l}\log_e\dfrac{r_2}{r_1}$

$$= \frac{1}{2\pi \times 45 \times 60} \log_e \left(\frac{0.045}{0.0375}\right)$$

$= 0.1075 \times 10^{-4}$

(*iii*) Asbestos insulation

$$= \frac{1}{2\pi k_2 l} \log_e \frac{r_3}{r_2}$$

$$= \frac{1}{2\pi \times 0.14 \times 60} \log_e \left(\frac{0.083}{0.045}\right)$$

$= 116.048 \times 10^{-4}$

(*iv*) Magnesia insulation

$$= \frac{1}{2\pi k_3 l} \log_e \frac{r_4}{r_3}$$

$$= \frac{1}{2\pi \times 0.07 \times 60} \log_e \left(\frac{0.108}{0.083}\right)$$

$= 99.822 \times 10^{-4}$

(*v*) Outside air film

$$= \frac{1}{h_o A_o} = \frac{1}{6.8 \times 2\pi \times 0.108 \times 60}$$

$= 36.137 \times 10^{-4}$

The resistance of series path is equal to the sum of individual resistances and the summation gives

$$\Sigma R_t = 255.2315 \times 10^{-4} \text{deg/W}$$

Heat lost through conduction,

$$Q = \frac{\Delta t}{\Sigma R_t} = \frac{375 - 35}{255.2315 \times 10^{-4}}$$

$= 13321$ W

(*b*) The heat loss can also be expressed as

$$Q = U_o A_o \Delta T = U_i A_i \Delta T$$

where U_o and U_i are the overall coefficients of heat transfer based on the outside area A_o and inside area A_i respectively.

$$U_o = \frac{13321}{(2\pi \times 0.108 \times 60)(375 - 35)}$$

$= 0.962$ W/m^{2}K

$$U_i = \frac{13321}{(2\pi \times 0.0375 \times 60)(375 - 35)}$$

$= 62.19$ W/m^{2}K

26.11. CONDUCTION THROUGH A SPHERE

Consider heat conduction through a hollow sphere of inner radius r_1, outer radius r_2 and made of a material of constant thermal conductivity. The inner and outer surfaces are maintained at constant but different temperatures t_1 and t_2 respectively. Geometrical symmetry indicates that the heat flow is limited to radial direction only. Further if temperature t_1 at the inner surface is greater than temperature t_2 at the outer surface, the heat flows radially outwards.

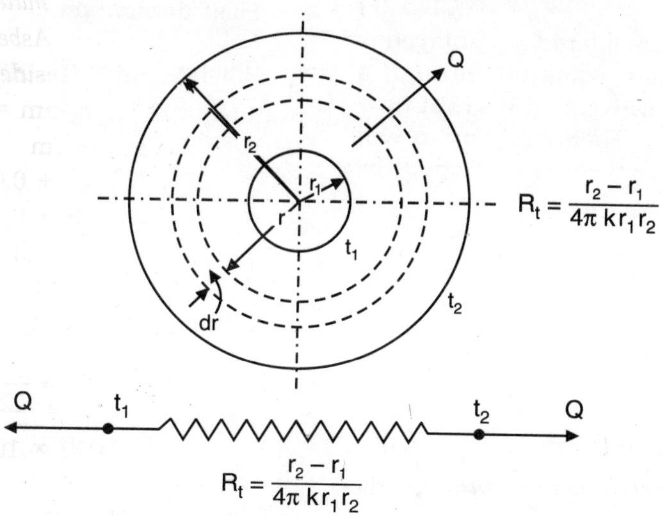

Fig. 26.15. Steady state conduction through a sphere

Consider an infinitesimal thin spherical element at radius r. Let thickness of this elementary ring be dr and the change of temperature across it be dt. Then invoking Fourier law of heat conduction.

$$Q = -kA\frac{dt}{dr} = -k\,(4\pi r^2)\frac{dt}{dr}$$

Separating the variables and integrating within the prescribed boundary conditions

$$\frac{Q}{4\pi k}\int_{r_1}^{r_2}\frac{dr}{r^2} = -\int_{t_1}^{t_2} dt$$

$$\frac{Q}{4\pi k}\left(\frac{1}{r_1}-\frac{1}{r_2}\right) = (t_1 - t_2)$$

or $$Q = \frac{4\pi\, k(t_1-t_2)r_1r_2}{r_2-r_1}$$

$$= \frac{(t_1-t_2)}{(r_2-r_1)/4\pi k r_1 r_2} \qquad ...(26.26)$$

Similarly between the limits $t = t_1$ at $r = 0$ and $t = t$ at $r = r$,

$$Q = \frac{4\pi k(t_1-t)r_1 r}{r-r_1}$$

Comparing this identity with equation 26.26 we get

$$\frac{t-t_1}{t_2-t_1} = \frac{1/r_1-1/r}{1/r_1-1/r_2} = \frac{r_2}{r}\left(\frac{r-r_1}{r_2-r_1}\right) \qquad ...(26.27)$$

Quite often it is considered advantageous to write the heat flow equation through a sphere in the same form as that for heat flow through a plane wall. Then thickness δ will be equal to $(r_2 - r_1)$ and the areas A will be an equivalent area A_m. Thus

$$Q = \frac{kA\,(t_1-t_2)}{\delta} = \frac{kA_m\,(t_1-t_2)}{(r_2-r_1)} \qquad ...(26.28)$$

Comparing equations 26.26 and 26.28,

$$A_m = 4\pi r_1 r_2$$

Further, $A_m = 4\pi r_m^2 = 4\pi r_1 r_2$

Obviously ***geometric mean radius*** of the spherical shell is,

$$r_m = \sqrt{r_1 r_2}$$

Spherical composites are dealt with in the same way as composite walls and cylinders. The heat flow through a three-layer composite spherical shall will be:

$$Q = \frac{(t_1-t_4)}{\dfrac{(r_2-r_1)}{4\pi k_1 r_1 r_2}+\dfrac{(r_3-r_2)}{4\pi k_2 r_2 r_3}+\dfrac{(r_4-r_3)}{4\pi k_3 r_3 r_4}} \qquad ...(26.29)$$

EXAMPLE 26.20

A hollow sphere of inner radius 30 mm and outer radius 50 mm is electrically heated at the inner surface at a rate of 10^5 W/m². At the outer surface, it dissipates heat by convection into a fluid at 100°C and a heat transfer coefficient of 400 W/m² K. Determine the temperature at the inside and outside surfaces of the sphere.

It may be presumed that there is no energy generation and the thermal conductivity of the material is constant at 15 W/mK.

Solution : $r_1 = 0.03$ m and $r_2 = 0.05$ m

Rate of heat dissipation $Q = q \times 4\pi\, r_1^2$

$= 10^5 \times (4\pi \times 0.03^2) = 1130.4$ W

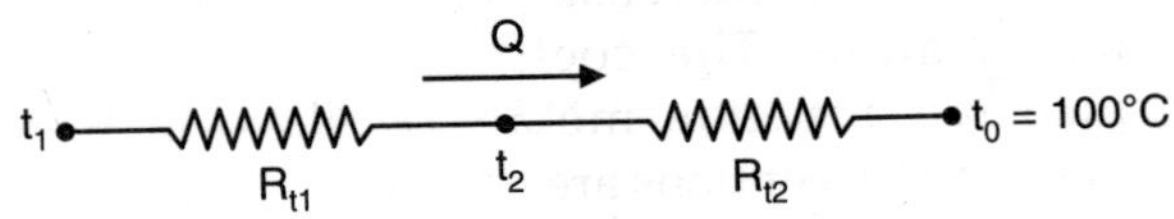

Fig. 26.16.

Heat dissipation $Q = \dfrac{t_1 - t_o}{R_{t1} + R_{t2}}$

where R_{t1}(resistance to conduction)

$$= \frac{r_2-r_1}{4\pi k\, r_1 r_2}$$

$$= \frac{0.05-0.03}{4\pi\times 15\times 0.05\times 0.03}$$

$= 0.0708$ k/W

and R_{t2}(resistance to convection)

$$= \frac{1}{4\pi r_2^2\, h} = \frac{1}{4\pi(0.05)^2\times 400}$$

$= 0.0796$ k/W

Then : $1130.4 = \dfrac{t_1-100}{0.0708+0.0796} = \dfrac{t_1-100}{0.1504}$

∴ Temperature at inner surface of the sphere,

$$t_1 = 1130.4 \times 0.1504 + 100 = 270°C$$

Under steady state conditions, heat flow through each section is same. Therefore

$$1130.4 = \frac{t_2 - t_0}{R_2} = \frac{t_2 - 100}{0.0796}$$

∴ Temperature at outer surface of the sphere,

$$t_2 = 1130.4 \times 0.0796 + 100$$
$$= \mathbf{189.98°C}$$

26.12. EFFECT OF VARIABLE CONDUCTIVITY

For most materials, the dependence of thermal conductivity on temperature is almost linear

$$k = k_o (1 + \beta t) \qquad ...(26.30)$$

where k_0 is the thermal conductivity at zero degree centigrade temperature and β is a constant whose value depends upon the material. This constant may be positive or negative depending upon whether thermal conductivity increases or decreases with temperature. The coefficient β is usually positive for non-metals and insulation materials (exceptions are magnesite bricks) and negative for metallic conductors (exceptions are aluminium and certain non-ferrous alloys).

With variable thermal conductivity, Fourier law of heat conduction through a plane wall can be expressed as

$$Q = -k_o (1 + \beta t) A \frac{dt}{dx}$$

Separating the variables and integrating within the limits $t = t_1$ at $x = 0$ and $t = t_2$ at $x = \delta$

$$Q \int_0^{\delta} dx = -k_0 A \int_{t_1}^{t_2} (1 + \beta t)\, dt$$

$$Q\,\delta = -k_0 A \left[\left(t_2 + \frac{\beta}{2} t_2^2 \right) - \left(t_1 + \frac{\beta}{2} t_1^2 \right) \right]$$

$$= k_0 A (t_1 - t_2) \left\{ 1 + \frac{\beta}{2} (t_1 + t_2) \right\}$$

Thus, the heat flow rate,

$$Q = \frac{k_m A}{\delta} (t_1 - t_2) \qquad ...(26.31)$$

The quantity $k_m = k_0 \left(1 + \frac{\beta}{2} (t_1 + t_2) \right)$ represents the mean thermal conductivity evaluated at the arithmetic mean temperature $\frac{1}{2}(t_1 + t_2)$.

Temperature variation

Fourier law of heat conduction through a plane wall is

$$Q = -k A \frac{dt}{dx} \; ; \quad \frac{Q}{A} = -\left(k \frac{dt}{dx} \right)$$

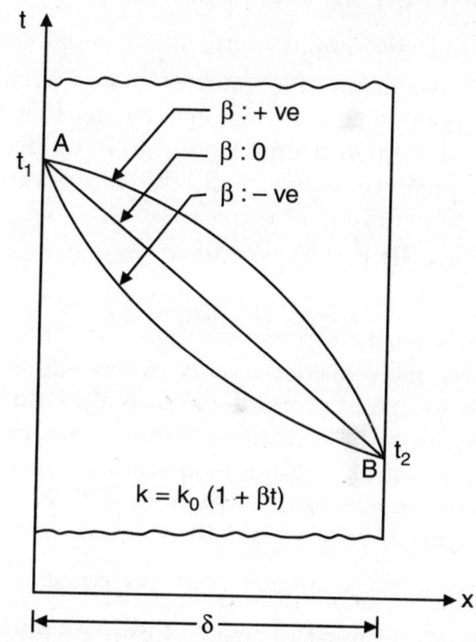

Fig. 26.17. Temperature profile with variable thermal conductivity

For steady state heat conduction in the infinite wall, Q/A is constant and so has to be the parameter $(k\,dt/dx)$. Obviously then, the temperature variation in the wall is governed by the condition that parameter $(k\,dt/dx)$ is constant.

Case I β = 0 : We have,

$$k = k_0 (1 + \beta t) = k_0 = \text{constant}$$

The thermal conductivity does not vary with temperature and equals the constant value k_0. Accordingly for the parameter

$(k\,dt/dx)$ to be constant, the term dt/dx must be constant. Therefore, the slope of temperature curve is constant and the temperature profile is linear.

Case II $\beta > 0$: We have,

$$k = k_0(1 + \beta t)\ ;\ i.e.,\ k \propto t$$

The thermal conductivity of the wall material is directly proportional to temperature; k increases with increasing temperature or decreases with decreasing temperature. Since the temperature decreases in x-direction, the thermal conductivity would also decrease. Accordingly to maintain the parameter $(k\,dt/dx)$ constant, the term dt/dx must increase. Consequently the value of slope increases from point A and B and that means that the curve should go steeper from A to B. Evidently with positive value of β, the temperature variation curve is of convex nature.

Case III $\beta < 0$: We have,

$$k = k_0\,(1 - \beta t),\ i.e.,\ k \propto \frac{1}{t}$$

The thermal conductivity of the wall material is inversely proportional to temperature; k decreases with increasing temperature or increases with decreasing temperature. Since the temperature decreases in x-direction, the thermal conductivity would increase. In order to maintain the parameter $(k\,dt/dx)$ constant, the term dt/dx must decrease. Consequently the value of the slope of temperature profile decreases. Evidently the temperature profile will be concave for negative value of coefficient β.

EXAMPLE 26.21

The inner and outer surfaces of a furnace wall, 25 cm thick, are at 300°C and 30°C respectively. The thermal conductivity of the wall material varies with temperature and is prescribed by the relation

$$k = (1.45 + 0.5 \times 10^{-5}\,t^2)\ \text{kJ/m-hr-deg}$$

where t is the temperature in degree centigrade. Proceed from the basic principles to calculate the heat loss per square metre of the wall surface area.

Solution : Invoking Fourier's law of heat conduction,

$$Q = -kA\frac{dt}{dx}$$

$$Q\,dx = -\,kA dt$$

$$= -(1.45 + 0.5 \times 10^{-5}\,t^2)A\,dt$$

Integrating over the wall thickness δ,

$$Q\,\delta = -\left\{1.45\,(t_2 - t_1) + \frac{0.5 \times 10^{-5}}{3}\left(t_2^3 - t_1^3\right)\right\}A$$

$$= \left\{1.45\,(t_1 - t_2) + \frac{0.5 \times 10^{-5}}{3}\left(t_1^3 - t_2^3\right)\right\}A$$

Inserting the appropriate values,

$$Q \times 0.25 = \left\{\begin{aligned}&1.45(300 - 30)\\ &+\frac{0.5\times 10^{-5}}{3}(300^3 - 30^3)\end{aligned}\right\}\times 1$$

$$= 436.45$$

$\therefore$ Heat loss from the furnace,

$$Q = \frac{436.45}{0.25} = \mathbf{1745.8\ kJ/m^2\text{-}hr}$$

EXAMPLE 26.22

A plane wall of thickness δ has its surfaces maintained at temperatures T_1 and T_2. The wall is made of a material whose thermal conductivity varies with temperature according to the relation $k = k_0T^2$. Set up an expression to workout the steady state heat conduction through the wall. Further proceed to calculate the temperature at which mean thermal conductivity be evaluated so as to get the same heat flow by its substitution in the simplified Fourier relation.

Solution : Invoking Fourier's law of heat conduction.

$$Q = -kA\frac{dt}{dx} = -\,k_0T^2A\frac{dt}{dx}$$

Separating the variables and integrating within the prescribed boundary conditions, we obtain :

$$Q\int_0^{\delta} dx = -Ak_0\int_{T_1}^{T_2} T^2 dx$$

$$\text{or}\quad Q\,\delta = -\frac{Ak_0}{3}\left(T_2^3 - T_1^3\right)$$

$$\therefore\quad Q = \frac{Ak_0}{3\delta}\left(T_1^3 - T_2^3\right)$$

which is the required relation.

If this heat flow is to be obtained by substituting an average value of thermal conductivity in the simplified Fourier relation, we have

$$\frac{Ak_0}{3\delta}\left(T_1^3 - T_2^3\right) = \frac{k_m A(T_1 - T_2)}{\delta}$$

$$= \frac{k_0 T_m^2 A(T_1 - T_2)}{\delta}$$

Upon simplification, the desired temperature works out as

$$T_m = \sqrt{\frac{T_1^2 + T_2^2 + T_1 T_2}{3}}$$

EXAMPLE 26.23

Derive an expression for the heat flow rate through a hollow sphere of inside radius r_1 and outside radius r_2, whose internal and external surfaces are maintained at temperatures t_1 and t_2 respectively. The thermal conductivity of the sphere material has a quadratic variation with temperature :

$$k = k_0\,(1 + \alpha t + \beta t^2)$$

Solution : Invoking Fourier law of heat condition,

$$Q = -kA\frac{dt}{dr}$$

where A is the area normal to radial direction r.

Substituting $A = 4\pi r^2$ and

$k = k_0(1 + \alpha t + \beta t^2)$, we obtain

$$Q = -\,k_0\,(1 + \alpha t + \beta t^2) \times 4\pi r^2 \times \frac{dt}{dr}$$

$$Q\frac{dr}{r^2} = -\,4\pi k_0(1 + \alpha t + \beta t^2)\,dt$$

In steady state, Q is constant and hence integration through the spherical walls yields

$$Q\int_1^{r_2}\frac{dr}{r^2} = -\,4\pi k_0\int_{t_1}^{t_2}\left(1 + \alpha t + \beta t^2\right)dt$$

$$\text{or}\quad -\,Q\left[\frac{1}{r}\right]_{r_1}^{r_2} = -\,4\pi k_0\left[\left(t + \frac{\alpha t^2}{2} + \frac{\beta t^3}{3}\right)\right]_{t_1}^{t_2}$$

$$\text{or}\quad Q\frac{r_2 - r_1}{r_1 r_2} = 4\pi k_0\left[(t_1 - t_2) + \frac{\alpha}{2}\left(t_1^2 - t_2^2\right) + \frac{\beta}{3}\left(t_1^3 - t_2^3\right)\right]$$

$$\text{or}\quad Q = \frac{4\pi k_0\, r_1 r_2}{r_2 - r_1}(t_1 - t_2) \times \left[1 + \frac{\alpha}{2}(t_1 + t_2) + \frac{\beta}{3}\left(t_1^2 + t_2^2 + t_1 t_2\right)\right]$$

which is the required expression.

26.13. CRITICAL THICKNESS OF INSULATION

Contrary to the common belief that addition of insulating material on a surface always brings about a decrease in the heat transfer rate, there are instances when the addition of insulation to the outside surfaces of cylindrical or spherical walls (geometries which have non-constant cross-sectional areas) does not reduce the heat loss. In fact, under certain circumstances, it actually increases the heat flow upto a certain thickness of insulation. To establish this fact, consider a thin-walled metallic cylinder of length l, radius r_i and transporting a fluid at temperature t_i which is higher than the ambient temperature t_o. Surrounding this cylinder is an annular section of insulating sheathing of thickness $(r - r_i)$ and thermal conductivity k.

With stipulations of :

(*i*) steady state conditions

(*ii*) one-dimensional heat flow only in the radial direction

(*iii*) negligible thermal resistance due to cylinder wall

(*iv*) negligible radiation exchange between outer surface of insulation and surroundings, the heat transmission can be expressed as

$$Q = \frac{(t_i - t_o)}{1/2\pi r_i l h_i + 1/2\pi k l \times \log_e(r/r_i) + 1/2\pi r l h_o} \quad ...(26.32)$$

where h_i and h_o are the film coefficients at the inner and outer surface respectively. The denominator represents the sum of thermal

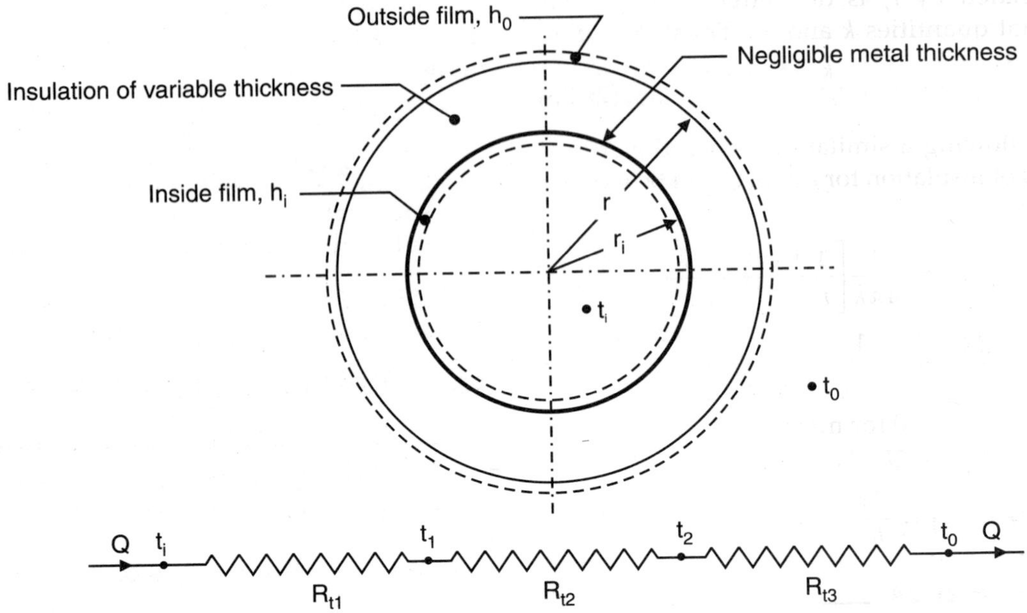

$$R_{t_1} = \frac{1}{2\pi r_i l h_i}\ ,\ R_{t_2} = \frac{1}{2\pi k l}\log_e \frac{r}{r_i}\ ,\ R_{t_3} = \frac{1}{2\pi r l h_0}$$

Fig. 26.18. Critical thickness of pipe insulation

resistance to heat flow. The values of k, r_i, h_i and h_o are constant; therefore the total thermal resistance will depend upon thickness of insulation which specifies the outer radius r of the arrangement. An examination of equation 26.32 would reveal that with increase in r (*i.e.*, thickness of insulation), the thermal resistance of insulating increases but the resistance due to convection coefficient at the outer surface drops. The thermal resistance due to inside film coefficient remains unaffected with change in radius r. Obviously, addition of insulation material can either increase or decrease the rate of heat transmission depending upon a change in the total resistance with outer radius r.

The effect of insulation thickness can be studied by differentiating the total resistance R_t with respect to r and setting the derivative equal to zero.

$$\frac{dR_t}{dr} = \frac{d}{dr}\left[\frac{1}{2\pi r_i l h_i} + \frac{1}{2\pi k l}\log_e \frac{r}{r_i} + \frac{1}{2\pi r l h_o}\right]$$

$$= \frac{1}{2\pi k l} \times \frac{1}{r} - \frac{1}{2\pi r^2 l h_o}$$

$$\therefore \quad \frac{1}{2\pi k l} \times \frac{1}{r} - \frac{1}{2\pi r^2 l h_o} = 0$$

which gives $r = \dfrac{k}{h_o}$

To determine whether the foregoing result maximises or minimises the total resistance, the second derivative needs to be calculated

$$\frac{d^2 R_t}{dr^2} = -\frac{1}{2\pi k l}\frac{1}{r^2} + \frac{1}{\pi r^3 l h_o}$$

At $r = \dfrac{k}{h_o}$

$$\frac{d^2 R_t}{dr^2} = -\frac{1}{2\pi k l}\left(\frac{h_o^2}{k^2}\right) + \frac{1}{\pi l h_o}\left(\frac{h_o^3}{k^3}\right)$$

$$= \frac{h_o^2}{2\pi k^3 l}$$

which is indeed positive.

Then $r = k/h_o$ represents the condition for minimum resistance and consequently maximum heat flow rate. The insulation radius at which resistance to heat flow is minimum is called the ***critical radius***. The critical radius, designated by r_c is dependent only on the thermal quantities k and h_o. Thus

$$r = r_c = \frac{k}{h_o} \quad ...(26.33)$$

Following a similar approach the critical radius of insulation for a sphere can be worked out as :

$$R_t = \frac{1}{4\pi k}\left[\frac{1}{r_1} - \frac{1}{r}\right] + \frac{1}{4\pi r^2 h_o}$$

$$\frac{dR_t}{dr} = \frac{1}{4\pi k r^2} - \frac{2}{4\pi r^3 h_o}$$

$= 0$ for maximum heat flow rate

or $r^3 h_o = 2k\, r^2$

$$\text{or } r\ (= r_c) = \frac{2k}{h_o} \quad ...(26.34)$$

EXAMPLE 26.24

"Addition of insulating material does not always bring about a decrease in the heat transfer rate for geometries with non-constant cross-section area." Comment upon the validity of this statement.

A pipe of outside diameter 20 mm is to be insulated with asbestos which has a mean thermal conductivity of 0.1 W/m-deg. The local coefficient of convective heat to the surroundings is 5 W/m²-deg. Comment upon the utility of asbestos as the insulating material.

What should be the minimum value of thermal conductivity of insulating material to reduce heat transfer ?

Solution : The critical radius of insulation for optimum heat transfer from pipe is given by :

$$r_c = \frac{k}{h_o} = \frac{0.1}{5}$$

$= 0.02$ m or 20 mm

$r_o = 10$ mm

For insulation to be properly effective in restricting heat transmission, the pipe radius r_o must be greater than or equal to the critical radius r_c. Here $r_o < r_c$ and as such there is no point in using absestos as the insulating material. Addition of asbestos insulation will increase the heat transfer rate and that is not desirable. An insulating material with smaller thermal conductivity needs to be employed.

For insulation to be effective, the pipe radius should be greater than the critical radius, *i.e.*,

$$r_o > r_c \ : \ 0.01 \geq \frac{k}{5}$$

or $k \leq 5 \times 0.01 = 0.05$ W/m-deg

Apparently, the maximum conductivity of insulation premitted is 0.05 W/m-deg

EXAMPLE 26.25

A cable of 10 mm outside is to be laid in an atmosphere of 25°C (h_o = 12.5 W/m²-deg) and its surface temperature is likely to be 75°C due to heat generated within it. How would the heat flow from the cable be affected if it is insulated with rubber having thermal conductivity k = 0.15 W/m-deg ?

Solution : The critical radius of insulation for the given cable is,

$$r_c = \frac{k}{h_o} = \frac{0.15}{12.5} = 0.012 \text{ m} = 12 \text{ mm}$$

Radius of the cable : $r_o = 5$ mm

Since $r_c > r_o$, the heat flow from the cable would be increased by insulating the cable.

The heat flow from the cable when provided with critical insulation is given by

$$Q = \frac{2\pi \Delta t}{\frac{1}{k}\log_e\left(\frac{r_c}{r_o}\right)}$$

for unit length of cable

$$= \frac{2\pi(75-25)}{\frac{1}{0.15}\log_e\left(\frac{12}{5}\right)} = \frac{2\pi \times 50}{5.836}$$

$=$ **53.80 W per metre length**

26.14. SALIENT FEATURES AND CHARACTERISTICS OF RADIATION

Some salient features and characteristics of radiation are enumerated below :

(*i*) The emission of thermal radiations is associated with thermally excited conditions which depend upon the nature of surface and its absolute temperature.

(*ii*) The distinction between one form of radiation and another lies only in its frequency and wavelength which are related by

$$c \text{ (speed of light)} = \lambda \text{ (wavelength)} \times f \text{ (frequency)}$$

Consequently longer wavelengths correspond to lower frequencies and a shorter wavelengths to higher frequencies. Again, a high temperature body will have a high frequency quantum and so shorter wavelengths.

(*iii*) Thermal radiation is limited to range of wavelength between 0.1 and 100 μm; it thus includes the entire visible and infrared, and a part of the ultraviolet spectrum.

The sun with an effective surface temperature of 5600°C emits most of its radiations at the extreme lower end of the spectrum 0.1 to 4 μm. The radiations from a lamp filament are in the frequency range 1 to 10 μm. Most solids and liquids have a continuous spectrum; they emit radiations of all the wavelengths. Gases and vapours radiate energy only at certain bands of wavelength and hence are called the **selective emitters**. The emission of thermal radiation depends upon the nature, temperature and state of the emitting surface . However with gases the dependence is also upon the thickness of emitting layer and the gas pressure.

(*iv*) Thermal radiations exhibit characteristics similar to those of visible light, and follow optical laws. They can be reflected, refracted, and are subject to scattering and absorption when they pass through a media. They get polarised and weakened in strength with the inverse square of radial distance from the radiating surface.

(*v*) All matter emits radiant energy and bodies at high temperature emit at a greater rate than bodies at low temperature. Normally a body radiating heat is simultaneously receiving heat from other bodies as incident radiation. The net exchange of heat between two radiating surfaces is due to the fact that one at higher temperature radiates more and receives less energy for its absorption. An isolated body which remains at constant temperature emits just as much energy by radiation as it receives. The entire system is then in a state of mobile thermal equilibrium.

(*vi*) Heat transfer by conduction and convection depends primarily on the temperature difference gradient and is little affected by temperature level. With other factors (k, h and A) remaining constant, the heat transfer due to conduction and convection from a hot source at 1000°C to the surroundings at 200°C would practically remain same if the hot source and the surroundings take up the temperature values as 900°C and 100°C respectively. However, even with the same temperature difference, the heat exchange by radiation gets enhanced at elevated temperature of the source and the surroundings.

The most vivid evidence of radiation transfer is that represented by the solar energy which passes through interstellar space (conditions close to that for perfect vacuum) on its way to the earth surface. Solar radiation plays an important part in the design of heating and ventilating systems. Heat transfer by radiation is encountered in boiler furnaces, billet reheating furnaces and other types of heat exchangers. The design and construction of engines, gas turbines, nuclear reactors and solar collectors is also significantly influenced by the radiation heat transfer.

26.15. ABSORPTIVITY, REFLECTIVITY AND TRANSMISSIVITY

The total radiant energy (Q_0) impinging upon a body would be partially or totally absorbed by it (Q_a), reflected from its surface (Q_r), or transmitted through it (Q_t) in accordance with the characteristics of the body (Fig. 26.19). By the conservation of energy principle, the total sum must be equal to the incident radiation.

That is :

$$Q_a + Q_r + Q_t = Q_0$$

Dividing throughout by Q_0, we have

$$\frac{Q_a}{Q_0} + \frac{Q_r}{Q_0} + \frac{Q_t}{Q_0} = \frac{Q_0}{Q_0}$$

$$\alpha + \rho + \tau = 1 \qquad ...(26.35)$$

where

α = absorptivity or fraction of total energy absorbed by the body

ρ = reflectivity or fraction of total energy reflected from the body

τ = transmissivity or fraction of total energy transmitted through the body

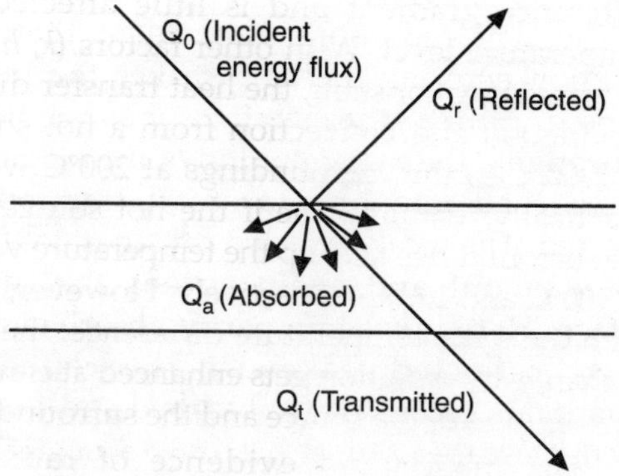

Fig. 26.19. Absorption, reflection and transmission of radiation

The factors α, ρ and τ are dimensionless and vary from 0 to 1. The value depends upon the nature of the surface of the body, its temperature and wavelength of incident rays. The response of the body to incident radiations is, however, completely independent of and unaffected by the simulatneous emission from the body.

Black surfaces are effective absorbers of radiation in the wavelengths that are encountered in heat transfer. Accordingly the name ***black body*** is assigned to a perfect absorber of radiation. The thermal radiations impinging upon a black body are totally absorbed by it; the radiations are neither reflected from the surface nor transmitted through it. For a black body $\alpha = 1$ and $\rho = \tau = 0$. Incidently this implies that a black body is a perfectly non-reflecting and non transmitting surface. Snow, with its absorptivity 0.985, is nearly black to thermal radiations. The absorptivity of surfaces can be increased to 90-95% by coating their surfaces with lamp black or a dark rought paint. In actual practice there does not exist a perfectly black body which will absorb all the incident radiations. The absorptivity of a surface depends upon the direction of incident radiation, temperature of the surface, composition and structure of the irradiated surface and the spectral distribution of incident radiation. When a surface absorbs a certain fixed percentage of impinging radiations, the surface is called the ***gray body***. The absorptivity of a gray body is necessarily below unity, but it remains constant over the entire range of temperature and wavelength of incident radiation.

A body that reflects all the incident thermal radiations is called a ***specular body*** (if the reflection is regular) or an ***absolutely white body*** (if the reflection is diffused). For such bodies $\rho = 1$, and $\alpha = \tau = 0$. The specular and diffused type of reflections have been indicated in Fig. 26.20.

Regular (specular) reflection implies that angle between the reflected beam and the normal to the surface equals the angle made by the incident radiation with the same normal. Reflection from highly polished and smooth surfaces approaches specular characteristics. In a diffused reflection, the incident beam is reflected in all directions, *i.e.*, there is directional independence of the reflected beam. Most of the engineering materials have rough surfaces, and these rough surfaces give diffused reflections. Diffused reflection is

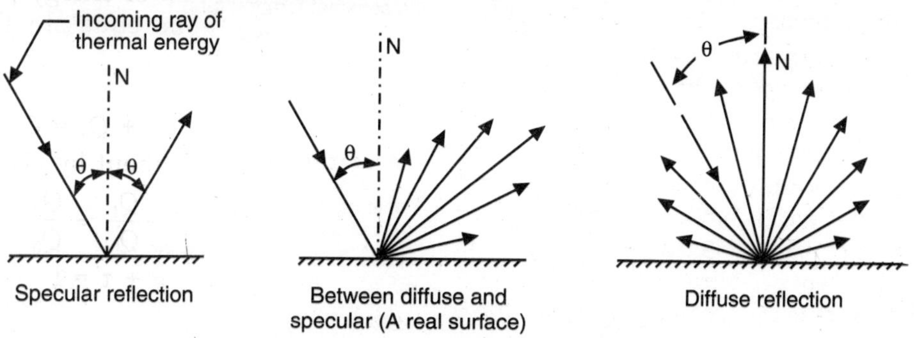

Fig. 26.20. Specular and diffused reflections

sometimes likened to the situation in which incident energy is absorbed near the surface and then re-emitted.

A body that allows all the incident radiations to pass through it is called **transparent** or **diathermaneous**. For such bodies $\tau = 1$, and $\alpha = \rho = 0$. Transmissivity varies with wavelength of incident radiation. A material may be non-transparent for a certain wavelength band and transparent for another. A thin glass plate transmits most of the thermal radiations from the sun, but absorbs in equally great measure the thermal radiations emitted from the low temperature interior of a building.

The absorption of a radiation is a surface phenomenon; it occurs in a very thin layer (approx 1 μm thick) of material near the surface. Since most of the solids and liquids encountered in engineering are thick enough to cover this layer, they can be considered non-transparent (opaque, diathermanous) to thermal radiations. Exceptions are few solid substances like glass, quartz, rock salt and most liquids in the visible and near infrared range. For opaque bodies $\tau = 0$ and so $\alpha + \rho = 1$. This result does suggest that good absorbs are bad reflectors and vice versa. However gases have relatively high transmissivity, they transmit an appreciable portion of radiation even in layers of fairly large thickness. Further, the gases are known to reflect very little of the radiation impinging on their interface. Therefore, for gases reflectivity can be neglected and so $\alpha + \tau = 1$. In a transparent medium, the absorption occurs throughout the material.

Consider a large hollow sphere or cylinder provided with only one small opening and let it be maintained at a uniform temperature. The inner surface of the cylinder is coated with lamp black which absorbs about 95% of the incident radiation. A beam of thermal radiation entering the hole strikes the inner surface. Since the surface has a high absorptivity, the major portion of the radiation is absorbed and only a small fraction is reflected. The weak reflected ray does not find any way out and again strikes the inner surface. Here it is again partly absorbed and partly reflected. Likewise the reflected radiation is successively absorbed and finally when it escapes out, it has only a negligible amount of energy associated with it.

Let Q represent the radiant energy that enters the hole. This radiation gets reduced to $(1 - \alpha) Q = \rho Q$ after first internal reflection, $\rho^2 Q$ after second internal reflection, ... $\rho^n Q$ after the nth reflection and this approaches zero at $n \to \infty$. The few rays that emerge from the hole will have suffered many reflections; the emergent flux then becomes essentially zero and that give absorptivity $\alpha = 1$ for the hole. A small hole leading into a cavity (Hohlraum) thus acts very nearly as a black body because all the radiant energy entering through it gets absorbed. The smaller the opening, better the approximation to black body behaviour. For most experimentation, a hole of 2.5 cm diameter in the end of a hollow cylinder 25 cm long and 7.5 cm diameter would suffice. Isothermal furnaces, with small

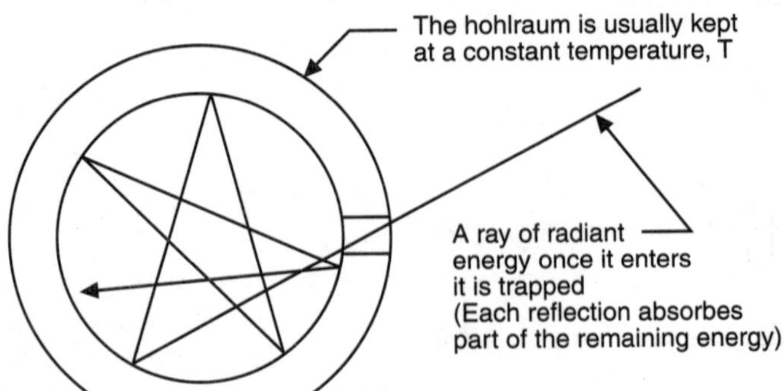

Fig. 26.21. Concept of block body

apertures, approximate a black body and are frequently used to calibrate heat flux gauges, thermometers and other radiometric devices.

The values pertaining to black body are commonly designated by the suffix '*b*'.

EXAMPLE 26.26

Why does a cavity with a small hole behave as a black body ?

Thermal radiation strikes a surface which has a reflectivity of 0.55 and a transmissivity of 0.032. The absorbed flux as measured indirectly by heating effect works out to be 95 W/m². Determine the rate of incident flux.

Solution : From an energy balance,

$$\alpha + \rho + \tau = 1$$

or $$\frac{Q_a}{Q_0} + \rho + \tau = 1$$

or $$\frac{Q_a}{Q_0} + 0.032 + 0.55 = 1$$

∴ Incident flux Q_0

$$= \frac{Q_a}{1 - 0.032 - 0.55} = \frac{95}{0.418}$$

$$= \mathbf{227.27\ W/m^2}$$

EXAMPLE 26.27

A thin metal plate of 4 cm diameter is suspended in atmospheric air whose temperature is 290 K. The plate attains a temperature of 295 K when one of its face receives radiant energy from a heat source at the rate of 2 W. If heat transfer coefficient on both surfaces of the plate is stated to be 87.5 W/m²-deg, workout the reflectivity of the plates.

Solution : Heat lost by convection from both sides of the plate

$$= 2h\ A\ \Delta t$$

The factors 2 accounts for two sides of the plate

$$= 2 \times 87.5 \times \left\{\frac{\pi}{4}(0.04)^2\right\} \times (295 - 290)$$

$$= 1.1\ \text{W}$$

For most of solids, the transmissivity is zero.

∴ Energy lost by reflection

$$= 2.0 - 1.1 = 0.9\ \text{W}$$

Reflectivity,

$$\rho = \frac{Q_r}{Q} = \frac{0.9}{2.0} = \mathbf{0.45}$$

26.16. WAVELENGTH DISTRIBUTION OF BLACK BODY RADIATION : PLANCK'S LAW

The laws governing the distribution of radiant energy over wavelength for a black body at a fixed temperature were formulated by Planck. Based upon extensive experimental evidence, Planck suggested the following law for the spectral distribution of emissive power :

$$(E_\lambda)_b = 2\pi c^2 h \frac{\lambda^{-5}}{\exp[ch/k\lambda T] - 1} \quad \text{...(26.36)}$$

The symbols used have the following meanings :

h = Planck constant, 6.6236 × 10⁻³⁴ Js

c = Velocity of light in vacuum (2.998 × 10⁸ m/s)

k = Boltzman constant, 13.802 × 10⁻⁴ J/K

λ = Wavelength of radiation waves, m

T = Absolute temperature of the black body, K

Quite often the above expression is written as

$$(E_\lambda)_b = \frac{C_1 \lambda^{-5}}{\exp[C_2/\lambda T] - 1} \quad \text{...(26.37)}$$

where

$$C_1 = 2\pi c^2 h = 0.374 \times 10^{-15}\ \text{Jm}^2/\text{s}$$

$$C_2 = \frac{ch}{k} = 1.4385 \times 10^{-2}\ \text{mK}$$

The quantity $(E_\lambda)_b$ denotes the monochromatic (single wavelength) emissive power, and is defined as the energy emitted by the black surface (in all directions) at a given wavelength λ per unit wavelength interval around λ. That is, the rate of energy emission in the interval $d\lambda$ is equal to $(E_\lambda)_b\, d\lambda$.

The variation of distribution of the monochromatic emissive power with wavelength is called the **spectral energy distribution**, and this has been depicted in Fig. 26.22 for a number of selected temperatures.

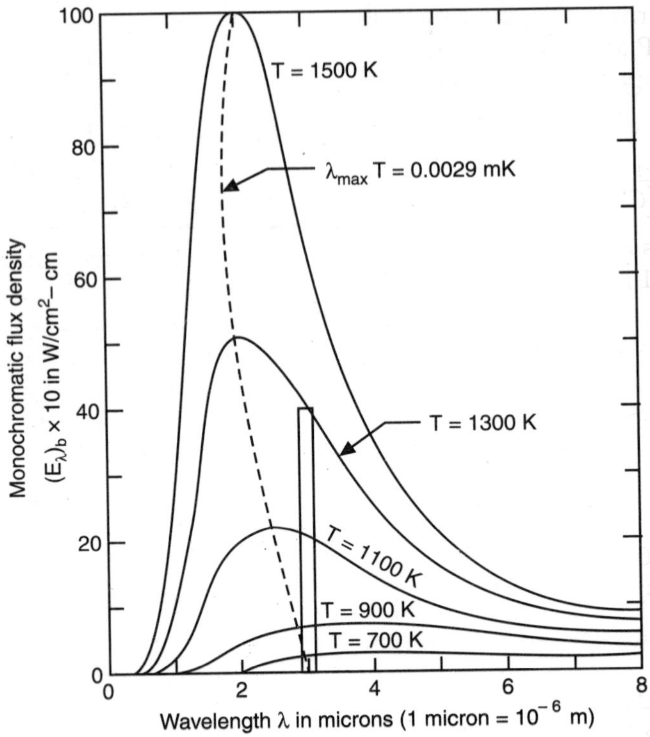

Fig. 26.22. Radiation of black body as a function of wavelength and temperature

The following important features can be noted from this plot :

(*i*) The monochromatic emissive power varies across the wavelength spectrum; the distribution is continuous but non-uniform. The emitted radiation is practically zero at the zero wavelength. With increase in wavelength, the monochromatic emissive power increases and attains a certain maximum value. With further increase in wavelength, the emissive power diminishes and drops again to almost zero value at infinite wavelength.

(*ii*) At any wavelength the magnitude of the emitted radiation increases with increasing temperature.

(*iii*) The wavelength at which the monochromatic emissive power is maximum shifts in the direction of shorter wavelengths as the temperature increases. This shift signifies that at elevated temperature, much of the energy is emitted in a narrow band ranging on both sides of wavelength at which the monochromatic emissive power is maximum. For example, the sun with its surface temperature of about 5600°C emits 90% of its radiations between 0.1 and 3μ.

(*iv*) At any temperature, the area under the monochromatic emissive power versus wavelength gives the rate of radiant energy emitted within the wavelength interval $d\lambda$. Thus $dE_b = (E\lambda)_b\, d\lambda$. Upon integration over the entire range of wavelength,

$$E_b = \int_{\lambda=0}^{\lambda=\infty} (E_\lambda)_b \, d\lambda \qquad ...(26.38)$$

The integral measures the total area under the monochromatic emissive power versus wavelength curve for the black body, and it represents the total emissive power per unit area (radiant energy flux density) radiated from a black body.

(*a*) For shorter wavelengths, the factor $C_2/\lambda T$ becomes very large, In that case

$$\exp\left[\frac{C_2}{\lambda T}\right] > 1$$

Obviously the term (– 1) appearing in the denominator of the Planck's distribution law can be neglected compared to this large value. The Planck's law then reduces to

$$(E_\lambda)_b = \frac{C_1 \lambda^{-5}}{\exp[C_2/\lambda T]} \quad ...(26.39)$$

Equation 26.39 is called **Wien's law**, and it is accurate within 1 percent for λT less than 3000 μK.

(*b*) For longer wavelengths, the factor $C_2/\lambda T$ is small. In that case exp $(C_2/\lambda T)$ can be expanded in series to give

$$\exp[C_2/\lambda T] = 1 + \frac{C_2}{\lambda T} + \frac{1}{2!}\left(\frac{C_2}{\lambda T}\right)^2 + ...$$

$$= 1 + \frac{C_2}{\lambda T}$$

The Planck's distribution law then becomes

$$(E_\lambda)_b = \frac{C_1 \lambda^{-5}}{1 + C_2/\lambda T - 1} = \frac{C_1 T}{C_2 \lambda^4} \quad ...(26.40)$$

The above identity is known as **Rayleigh-Jean's Law**. It is accurate within 1 percent for $\lambda T > 8 \times 10^5$ μK.

26.17. TOTAL EMISSIVE POWER ; STEFAN-BOLTZMAN LAW

The ***total emissive power*** *E* of a surface is defined as the total radiant energy emitted by the surface in all directions over the entire wavelength range per unit surface area per unit time.

The basic rate equation for radiation transfer is based on Stefan-Boltzman law which states that the amount of radiant energy emitted per unit time from unit area of black surface is proportional to the fourth power of its absolute temperature.

$$E_b = \sigma_b T^4 \quad ...(26.41)$$

where σ_b is the radiation coefficient of a black body. This rate equation can be set-up by the integration of monochromatic emissive power over the entire band width of wavelength for $\lambda = 0$ to $\lambda = \infty$.

$$E_b = \int_0^\infty (E_\lambda)_b \, d\lambda$$

$$= \int_0^\infty \frac{C_1 \lambda^{-5}}{\exp[C_2/\lambda T] - 1} d\lambda$$

Let $\frac{C_2}{\lambda T} = y$; $\lambda = \frac{C_2}{yT}$; $d\lambda = \frac{-C_2}{y^2 T} dy$

With this substitution, the new integration limits are :

At $\lambda = 0$, $y = \infty$ and at $\lambda = \infty$, $y = 0$

$$\therefore \quad E_b = -C_1 \int_\infty^0 \frac{y^5 T^5 C_2}{C_2^5 [\exp(y) - 1] y^2 T} dy$$

$$= \frac{C_1 T^4}{C_2^4} \int_0^\infty y^3 [\exp(y) - 1]^{-1} dy$$

Expanding $[\exp(y) - 1]^{-1}$ by a series, we obtain

$$E_b = \frac{C_1 T^4}{C_2^4} \times \int_0^\infty y^3 [\exp(-y) + \exp(-2y) + \exp(-3y) + ...] dy$$

The integral is of the form

$$\int_0^\infty y^n [\exp(-ay)] dy = \frac{n!}{a^{n+1}}$$

$$\therefore \quad E_b = \frac{C_1 T^4}{C_2^4} \left[\frac{3!}{1^4} + \frac{3!}{2^4} + \frac{3!}{3^4} + ...\right]$$

$$= \frac{C_1 T^4}{C_2^4} \times 6.48$$

Substituting the value for constants C_1 and C_2 which are :

$C_1 = 0.374 \times 10^{-15}$ Jm²/s ;
$C_2 = 1.4388 \times 10^{-2}$ mK

we get :

$$E_b = \frac{0.374 \times 10^{-15}}{1.4388 \times 10^{-2}} \times 6.48 \times T^4$$

$$= 5.67 \times 10^{-8} T^4 = \sigma_b T^4 \quad ...(26.42)$$

where

$$\sigma_b = 5.67 \times 10^{-8} \text{ W/m}^2\text{K}^4$$

is the ***radiation coefficient*** or the Stefan-Boltzman constant.

Undoubtedly, the Stefan Boltzman law helps us to determine the amount of radiations emitted in all the directions and over the entire wavelength spectrum from a simple knowledge of the temperature of the black body.

Normally a body radiating heat is simultaneously receiving heat from other bodies as radiation. Consider that surface 1 at temperature T_1 is completely enclosed by another black surface at temperature T_2. The net radiant heat flux is then given by

$$Q_{net} = \sigma_b \, (T_1^4 - T_2^4) \qquad ...(26.43)$$

EXAMPLE 26.28

List the salient features of a black body radiation.

Calculate the radiant flux density from a black body at 400°C ? If the emitted radiant energy is to be doubled, to what temperature surface of the black body needs to be raised ?

Solution : A black body is an ideal or hypothetical surface having the following radiation heat transfer characteristics :

(*i*) A black body absorbs all the incident radiation regardless of wavelength and direction.

(*ii*) A black body neither reflects nor transmits any amount of incident radiation.

(*iii*) For a prescribed wavelength a black body radiates the maximum energy possible at the temperature of the body.

(*iv*) The black body is a diffused emitter. This implies that the radiation emitted by a black surface is a function of wavelength and temperature but is independent of direction.

From Stefan-Boltzman law, the rate of energy transmission from a black body is

$$E = \sigma_b T^4$$
$$= 5.67 \times 10^{-8} \times (400 + 273)^4$$
$$= 11631.7 \text{ W/m}^2$$

Let T be the absolute temperature at which the radiant flux gets doubled. Then

$$2\,(11631.7) = 5.67 \times 10^{-8}\, T^4$$

$$T^4 = \frac{2 \times 11631.7}{5.67 \times 10^{-8}} \;;$$

$$T = 800.3 \text{ K or } \mathbf{527.3°C}$$

26.18. WIEN'S DISPLACEMENT LAW

From the spectral distribution of black body emissive power, it is apparent that the wavelength associated with maximum rate of emission depends upon the absolute temperature of the radiating surface. The nature of this dependence can be obtained by differentiating the Planck's expression

$$E_\lambda = \frac{C_1 \lambda^{-5}}{\exp\left[\dfrac{C_2}{\lambda T}\right] - 1}$$

with respect to λ and setting the derivative equal to zero. That is

$$\frac{d}{d\lambda}(E_\lambda) = \frac{d}{d\lambda}\left(\frac{C_1 \lambda^{-5}}{\exp\left(\dfrac{C_2}{\lambda T}\right) - 1}\right) = 0$$

$$\therefore \quad \frac{\left[\exp\left(\dfrac{C_2}{\lambda T}\right) - 1\right] C_1 (-5)\lambda^{-6} - C_1 \lambda^{-5} \times \left[\exp\left(\dfrac{C_2}{\lambda T}\right)\right] \times (C_2/T)(-1)\lambda^{-2}}{\left[\exp\left(\dfrac{C_2}{\lambda T}\right) - 1\right]^2} = 0$$

Since the denominator ≠ 0, we have

$$-\,5C_1 \left[\exp\left(\frac{C_2}{\lambda T}\right) - 1\right]\lambda^{-6} + \frac{C_1 C_2}{T}\left[\exp\left(\frac{C_2}{\lambda T}\right)\right]\lambda^{-7} = 0$$

Simplification gives the following transcendental equation

$$\frac{C_2}{5\lambda T} + \left[\exp\left(-\frac{C_2}{\lambda T}\right) - 1\right] = 0$$

whose solution by hit and trial method gives

$$\frac{C_2}{\lambda T} = 4.965$$

$$\therefore \lambda_{max}\, T = \frac{C_2}{4.965} = \frac{1.4388 \times 10^{-2}}{4.965}$$
$$= 2.898 \times 10^{-3} = 0.0029 \text{ mK} \qquad ...(26.44)$$

where λ_{max} denotes the wavelength at which emissive power is maximum. The Wien's displacement law may be stated as "***the product of absolute temperature and the wavelength at which the emissive power is maximum, is***

constant". The law suggests that λ_{max} is inversely proportional to the absolute temperature and accordingly the maximum spectral intensity of radiation shifts towards the shorter wavelength with rising temperature. The locus of points described by Wien's law has been plotted as the dashed curve in Fig. 26.22.

The Wien's displacement law finds application in the prediction of a very high temperature through measurement of wavelength.

A combination of Planck's law and the Wien's displacement law yields the correlation for maximum monochromatic emissive power for a black body.

$$(E_\lambda)_{max} = \frac{C_1 \lambda^{-5}}{\exp\left[\frac{C_2}{\lambda T}\right] - 1}$$

$$= \frac{0.374 \times 10^{-15} \times (2.898 \times 10^{-3}/T)^{-5}}{\exp\left[\frac{1.4388 \times 10^{-2}}{2.898 \times 10^{-3}}\right] - 1}$$

$$= \frac{1.829 \times 10^{-3} T^5}{142.30}$$

$= 1.285 \times 10^{-5} T^5$ W/m^2 per metre wavelength ...(26.45)

Thus the magnitude of the maximum monochromatic emissive power varies proportionally with the fifth power of the absolute temperature of the black surface.

EXAMPLE 26.29

What are the ranges of wavelength of electromagnetic waves covering ultra-violets visible, infrared and thermal radiation.

A small black body has a total emissive power of 4.5 kW/m^2. Determine its surface temperature and the wavelength of emission maximum. In which range of the spectrum does this wavelength fall ?

Solution : From Stefan Boltzman law, the rate of energy transmission from a black body is

$$E = \sigma_b T^4 ;$$

$$4.5 \times 1000 = 5.67 \times 10^{-8} T^4$$

$$\therefore \quad T = \left[\frac{4.5 \times 1000}{5.67 \times 10^{-8}}\right]^{\frac{1}{4}} = 530.77 \text{ K}$$

The wavelength of emission maximum is given by Wien's law. That is

$$\lambda_{max} T = 2.898 \times 10^{-3}$$

$$\lambda_{max} = \frac{2.898 \times 10^{-3}}{530.77}$$

$$= 5.46 \times 10^{-6} \text{ m} = 5.46 \text{ μm}$$

This wavelength falls in the infrared region of the spectrum.

EXAMPLE 26.30

Making use of Planck's law of distribution, establish the relation for the Wien's displacement law.

The sun emits maximum radiation at λ = 0.52 μm. Assuming the sun to be a black body, calculate the surface temperature of the sun and the emissive ability of the sun's surface at that temperature. Also determine the maximum monochromatic emissive power of the sun's surface.

Solution : From Wien's displacement law

$$T = \frac{2.898 \times 10^{-3}}{\lambda_{max}} = \frac{2.898 \times 10^{-3}}{0.52 \times 10^{-6}}$$

$$= \mathbf{5573 \text{ K}}$$

From Stefan's Boltzman law,

$$E = \sigma_b T^4 = 5.67 \times 10^{-8} (5573)^4$$

$$= \mathbf{5.47 \times 10^7 \text{ W/m}^2}$$

Maximum monochromatic emissive power can be worked out from the relation

$$(E_\lambda)_{max} = 1.285 \times 10^{-5} T^5$$

$$= 1.285 \times 10^{-5} (5573)^5$$

$= \mathbf{6.908 \times 10^{13}}$ **W/m^2** per metre wavelength

EXAMPLE 26.31

A furnace emits radiation at 2000 K. Treating it as a black body radiation, calculate the

(a) monochromatic radiant flux density at 1 μm wavelength.

(b) wavelength at which emission is maximum and the corresponding radiant flux density

(c) total emissive power, and

Solution : (*a*) From Planck's law of distribution,

$$(E_\lambda)_b = \frac{C_1 \lambda^{-5}}{\exp[C_2/\lambda T]-1}$$

$$= \frac{0.374\times10^{-15}\times(1\times10^{-6})^{-5}}{\exp[1.4388\times10^{-2}/1\times10^{-6}\times2000]-1}$$

$$= \frac{0.374\times10^{15}}{1331.4-1}$$

= **2.81 × 10⁷ W/m² per meter wavelength**

(*b*) From Wien's displacement law ;

$$\lambda_{max}\, T = 2.898 \times 10^{-3}$$

$$\therefore \quad \lambda_{max} = \frac{2.898\times10^{-3}}{2000} = \mathbf{1.449 \times 10^{-6}\ m}$$

Maximum radiant flux density,

$1.285 \times 10^{-5}\, T^5 = 1.285 \times 10^{-5} \times (2000)^5$

= **4.11 × 10¹¹ W/m²** per metre wavelength

(*c*) From Stefan-Boltzman law,

$$E = \sigma_b T^4$$
$$= 5.67 \times 10^{-8} \times (2000)^4$$
$$= \mathbf{907200\ W/m^2}$$

26.19. KIRCHOFF'S LAW

Consider two surfaces, one absolutely black at temperature T_b and the other non-black at temperature T (Fig. 26.23). The surfaces are arranged parallel to each other and so close that the radiation of one falls totally on the other. The radiant energy E emitted by the non-black surface impinges on the black surface and gets fully absorbed. Likewise the radiant energy E_b emitted by the black surface strikes the non-black surface. If the non-black surface has absorptivity α, it will absorb αE_b radiations and the remainder $(1 - \alpha)\, E_b$ will be reflected back for full absorption at the black surface. Radiant interchange for the non-black surface equals $(E - \alpha E_b)$. If both the surfaces are at the same temperature, $T = T_b$, then the conditions correspond to mobile thermal equilibrium for which the resultant interchange of heat is zero.

Under these conditions :

$$E - \alpha E_b = 0 \quad \text{or} \quad \frac{E}{\alpha} = E_b$$

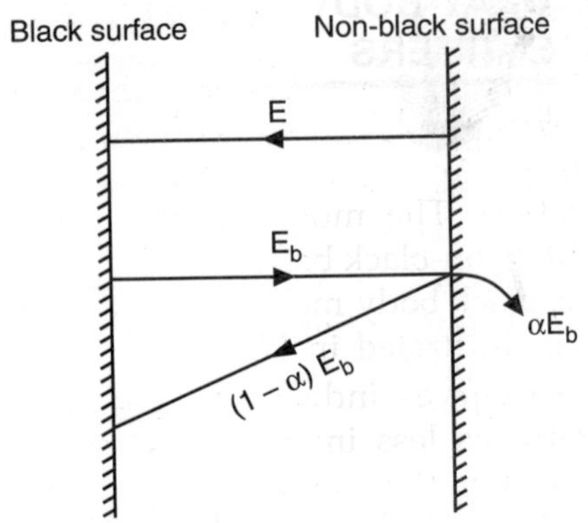

Fig. 26.23. Exchange of heat between a black and a non-black surface

This relationship can be extended by considering different surfaces in turn, Thus in general :

$$\frac{E_1}{\alpha_1} = \frac{E_2}{\alpha_2} = \frac{E_3}{\alpha_3} = \ldots\ldots = \frac{E_b}{\alpha_b}$$
$$= E_b = f(T) \qquad \ldots(26.46)$$

This is because the absorptivity α_b of a black body equals unity.

Equations 26.46 shows that the ratio of the emissive power E to absorptivity α is same for all bodies, and is equal to the emissive power of a black body at the same temperature. This relationship is known as the Kirchoff's law.

The ratio of the emissive power of a certain non-black body E to the emissive power of a black body E_b, both bodies being at the same temperature, is called the **emissivity** of the body. Emissivity of a body is a function of its physical and chemical properties and the state of its surface-whether rough or smooth.

From equation 26.46 :

$$E/E_b = \alpha \quad \text{or} \quad \epsilon = \alpha \qquad \ldots(26.47)$$

where, $\epsilon = E/E_b$ is emissivity.

Kirchoff's law can also be stated as : "*The* ***emissivity*** ϵ ***and absorptivity*** α ***of a real surface are equal for radiation with identical temperature and wavelength***". The equivalence of ϵ and α does suggest that a perfect absorber (the perfect black body) is also a perfect radiator.

26.20. GRAY BODY AND SELECTIVE EMITTERS

Consider two bodies, one absolutely black and the other non-black and let these be at the same temperature. The monochromatic emissive power of a non-black body varies significantly from the black body monochromatic emissive power as illustrated in Fig. 26.24.

These curves indicate that a non-black body radiates less intensively than a black body. Further the radiation spectrum for a non-black body may be similar or radically different from that of a black body.

When the emissivity of non-black surface is constant at all temperatures and through out the entire range of wavelength, the surface is called a **gray body**. The radiation spectrum for a gray body, though reduced in vertical scale, is continuous and identical to the corresponding curve for a perfectly black surface; there is no shift in the peak of the curves. However for many materials the emissivity is different for the various wavelengths of the emitted energy. The radiating bodies exhibiting this behaviour are called **selective emitters**. The radiation spectrum for a selective emitter does not follow any definite pattern, and it varies entirely from that of a black body.

Stefan-Boltzman law when applied to a gray body takes the form

$$E = \sigma T^4 \quad ...(26.48)$$

The constant σ is different for different bodies and its value depends upon the nature of the body, the state of its surface and temperature. It is always less than the radiation coefficient σ_b for a black body; it's value ranges from 0.0 to 5.67×10^{-8} W/m^2 K^4.

The emissivity of the gray surface may be expressed as :

$$\in = \frac{E}{E_b}$$

$$= \frac{\sigma T^4}{\sigma_b T^4} = \frac{\sigma}{\sigma_b} \quad ...(26.49)$$

Values of emissivities range from 0.0 to 1.0.

The emitted radiant energy flux density for non-black body, as prescribed by equation 26.42 may be rewritten as :

$$E = \in \sigma_b T^4 \text{ W/m}^2 \quad ...(26.50)$$

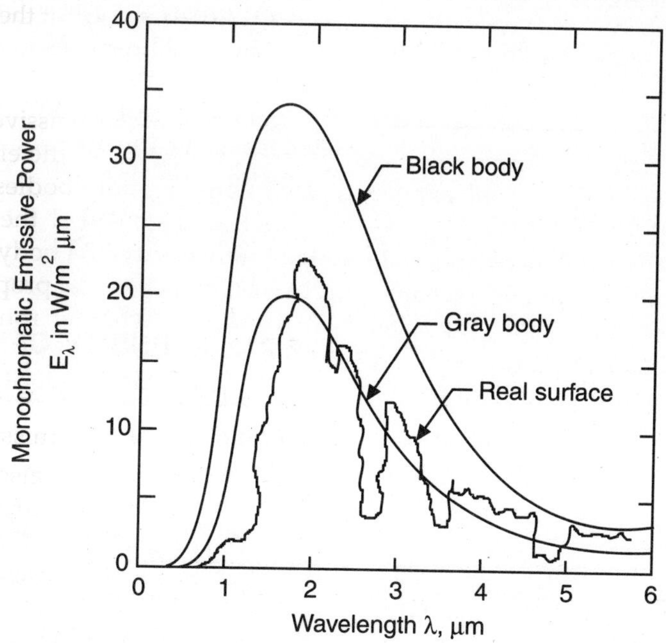

Fig. 26.24. Emission from black, gray and selective emitters

EXAMPLE 26.32

Can a surface behave essentially like a black surface and still look practically white to the eye.

(b) A 100 watt light bulb has a tungsten filament (emissivity $\in = 0.30$) which is required to operate at 2780 K. If the bulb is completely evacuated, calculate the minimum surface area of the tungsten filament. Assume that no radiant energy strikes the bulb surface and that steady state conditions prevail.

Solution : The operating conditions stipulate that :

(*i*) The bulb is completely evacuated, and the vacuum precludes any heat transfer by convection.

(*ii*) No radiant energy strikes the bulb surface,

(*iii*) Steady state conditions prevail, *i.e.*, there is no storage of energy.

Accordingly the entire power supplied by the tungsten filament is dissipated by the radiation process. That is

$$E = \in \sigma_b A T^4$$

$$100 = 0.3 \times (5.67 \times 10^{-8}) \times A \times 2780^4$$

$$= 101.6 \times 10^4 A$$

∴ Surface area of the bulb filament,

$$A = \frac{100}{101.46 \times 10^4} = \mathbf{0.98 \times 10^{-4}\ m^2}$$

EXAMPLE 26.33

A gray body ($\in = 0.8$) emits the same amount of heat as a black body at 1075 K. Find out the required temperature of the gray body.

(b) If a black body at 1000 K and a gray body at 1250 K emit the same amount of radiation, what should be the emissivity of the gray body.

Solution : Since the black and gray body are emitting the same amount of radiation,

$$\sigma_b A T_b^4 = \sigma_b \in A T_g^4$$

Presuming the two bodies to be of same area, we have

$$T_b^4 = \in T_g^4$$

$$(a) \quad T_g = \frac{T_b}{(\in)^{0.25}} = \frac{1075}{(0.8)^{0.25}} = \frac{1075}{0.9457}$$

$$= \mathbf{1136.72\ K}$$

$$(b) \quad \in = \left(\frac{T_b}{T_g}\right)^4 = \left(\frac{1000}{1250}\right)^4 = \mathbf{0.4096}$$

EXAMPLE 26.34

A polished metal pipe 5 cm outside diameter and 370 K temperature at the outer surface is exposed to ambient conditions at 295 K temperature. The emissivity of the surface is 0.2 and the convection coefficient of heat transfer is 11.35 W/m²-deg. Calcualte the heat transfer by radiation and natural convection per metre length of the pipe. Take thermal radiation constant

$$\sigma_b = 5.67 \times 10^{-8}\ W/m^2K^4$$

What would be the overall coefficient of heat transfer by the combined mode of convection and radiation ?

Solution : Surface area A of the pipe per metre run is equal to

$$\pi dl = \pi \times 0.05 \times 1 = 0.157\ m^2$$

$$Q_r \text{ (radiation)} = \in \sigma_b A (T_1^4 - T_2^4)$$

$$= 0.2 \times 5.67 \times 10^{-8} \times 0.157\ (370^4 - 295^4)$$

$$= 19.88\ W$$

$$Q_c \text{ (convection)} = h A \Delta t$$

$$= 11.35 \times 0.157\ (370 - 295)$$

$$= 133.64\ W$$

$$Q_t \text{ (total)} = 19.88 + 133.64 = 153.52\ W$$

The total heat exchange can be expressed as :

$$Q_t = U A \Delta T$$

where U is the overall coefficient of heat transfer

$$\therefore \quad U = \frac{Q_t}{A \Delta T} = \frac{153.52}{0.157(370 - 295)}$$

$$= \mathbf{13.04\ W/m^2\text{-}deg}$$

26.21. INTENSITY OF RADIATION AND LAMBERT'S COSINE LAW

Subtended plane and solid angles : The plane angle α is defined by a region by the rays of a circle, and is measured as the ratio of the element of arc of length l on the circle to the radius r of the circle: $\alpha = l/r$

The solid angle ω is defined by a region by the rays of a sphere, and is measured as :

$$\omega = \frac{A_n}{r^2} = \frac{A\cos\theta}{r^2} \qquad ...(26.51)$$

where

A_n = projection of the incident surface normal to the line of propagation

A = area of incident surface

θ = angle between the normal to the incident surface and the line of propagation

r = length of the line of propagation between the radiating and the incident surfaces

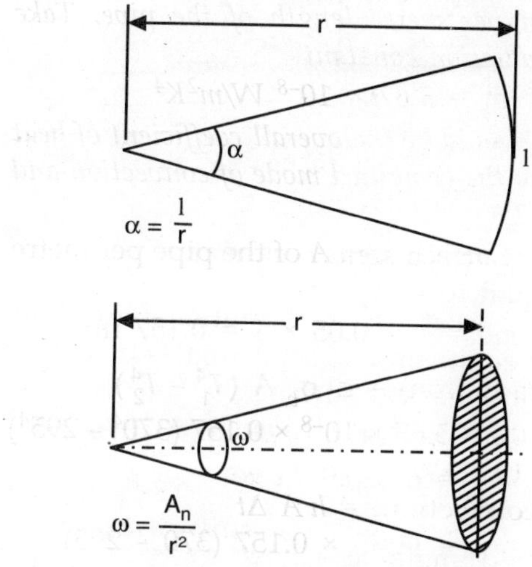

Fig. 26.25. Plane and solid angles

The relationship between A, A_n and θ has been illustrated in Fig. 26.26. When the incident surface is a sphere, the projection of surface normal to the line of propagation is the silhoutte disk of the sphere, which is a circle of the diameter of the sphere. The unit of measure of solid angle is the steradian (sr).

Intensity of radiation : Consider a small black surface dA (emitter) arbitrarily located at a point in the space under consideration and emitting radiations in different directions. A black body radiation collector through which the radiations pass is located at an angular position characterised by zenith angle θ towards the surface normal and the azimuth angle ϕ of a spherical coordinate system. Further, the collector subtends a solid angle $d\omega$ when viewed from a point on the emitter.

The intensity of radiation I is the energy emitted (of all wavelengths) in a particular direction per unit surface area and through a unit solid angle. The area is the projected area of the surface on a plane perpendicular to the direction of radiation. The collector or the incidence surface measures a variation of the emitted radiations depending upon its angular position. Maximum amount of radiation is measured (received) by the collector when it is at the position normal to the emitter, and the intensity in a direction θ from normal to the emitter follows the Lambert's cosine law :

"The intensity of radiation in a direction θ from the normal to a black emitter is proportional to cosine of the angle θ."

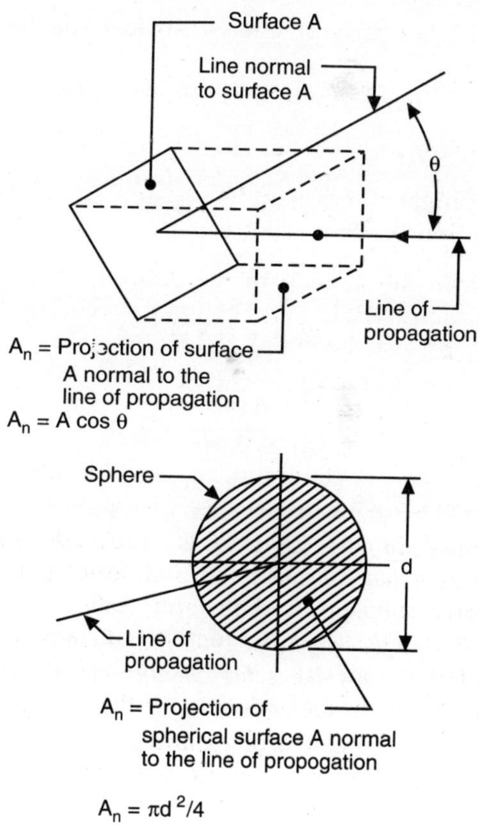

Fig. 26.26. Projection of an incident surface normal to the line of propagation

If I_n denotes the normal intensity and I_θ represents the intensity at angle θ from the normal, then

$$I_\theta = I_n \cos\theta \qquad ...(26.52)$$

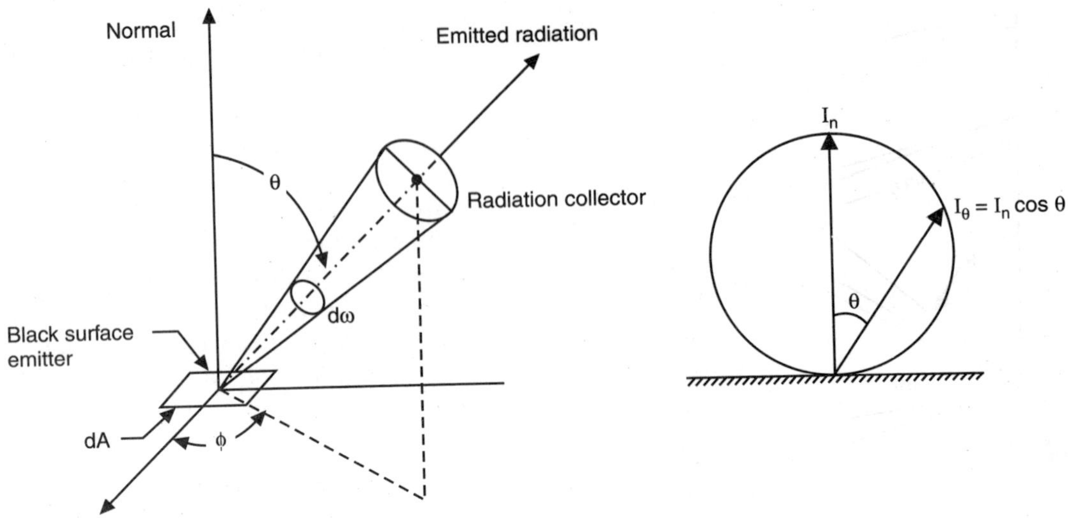

Fig. 26.27. Spatial distribution of radiations emitted from a surface element

Apparently the energy radiated out decreases with increase in θ and becmes zero at θ = 90°.

When the collector is oriented at an angle θ_1 from the normal to the emitter, then the radiations striking and being absorbed by the collector can be expressed as :

$$(dE_b)_{\theta_1} = I_{\theta_1} d\omega_1 \, dA$$

$$= I_n \cos\theta_1 \, d\omega_1 \, dA \quad ...(26.53)$$

where $d\omega_1$ is the solid angle subtended by the collector at the surface of the emitter.

The collector could be located at different angular positions and still maintain the same radial distance from the emitter. Let it subtend a solid angle $d\omega_2$ at the emitter surface when located in a direction θ_2 from the normal. Then the rate of flow of energy through it will be:

$$(dE_b)_{\theta_2} = I_{\theta_2} d\omega_2 \, dA$$

$$= I_n \cos\theta_2 \, d\omega_2 \, dA \quad ...(26.54)$$

It follows from equations 26.53 and 26.54 that for any surface located at an angle θ from the normal and subtending a solid angle $d\omega$ at the emitter dA,

$$(dE_b)_\theta = I_n \cos\theta \, d\omega \, dA \quad ...(26.55)$$

Relation between the normal intensity and emissive power : To establish a relation between the normal intensity and the emissive power, we relate the differential solid angle $d\omega$ to the zenith and azimuth angles by noting that for a spherical surface :

$$\text{area of collector} = (r\,d\theta) \times (r \sin\theta \, d\phi)$$

$$= r^2 \sin\theta \, d\theta \, d\phi$$

$$\therefore \text{ solid angle } d\omega = \frac{r^2 \sin\theta \, d\theta \, d\phi}{r^2}$$

$$= \sin\theta \, d\theta \, d\phi$$

Then the radiations leaving the emitter and striking the collector is :

$$dE_b = I_n \cos\theta \sin\theta \, d\theta \, d\phi \, dA \quad ...(26.56)$$

The total energy E_b radiated by the emitter and passing through a spherical region can be worked out by integrating equation 26.56 over the limits.

$$\theta = 0 \text{ to } \theta = \frac{\pi}{2}$$

and $\quad \phi = 0$ to $\phi = 2\pi$

Thus,

$$\int_{sphere} dE_b = I_n \, dA \int_0^{\pi/2} \sin\theta \cos\theta \, d\theta \int_0^{2\pi} d\phi$$

$$E_b = I_n \, dA \times \frac{1}{2}(2\pi) = I_n \, \pi \, dA \quad ...(26.57)$$

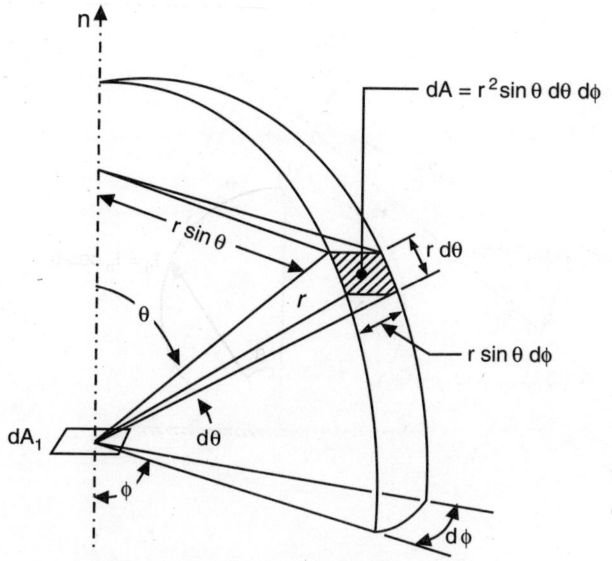

Fig. 26.28. Differential solid angle in terms of zenith and azimuthal angles

But the total emissive power of the emitter with area dA and temperature T is also given by:

$$E_b = \sigma_b T^4 dA$$

Combining this with equation 26.57, we get :

$$I_n = \frac{\sigma_b T^4}{\pi dA} \times dA = \frac{\sigma_b T^4}{\pi} \quad ...(26.58)$$

Thus for a unit surface, the ***intensity of normal radiation I_n is the 1/π times the emissive power*** E_b.

EXAMPLE 26.35.

Define intensity of radiation and prove that the intensity of normal radiation is 1/π times the total emissive power.

A black body of 0.2 m² area has an effective temperature of 800 K. Calcualte (a) the total rate of energy emission (b) the intensity of normal radiation (c) the intensity of radiation along a direction 60° to the normal, and (d) the wavelength of maximum monochromatic emissive power.

Solution : From Stefan-Boltzman law, the rate of energy emission E_b is :

$$E_b = \sigma_b AT^4$$

$$= (5.67 \times 10^{-8}) \times 0.2 \times 800^4$$

$$= \mathbf{4644.86\ W}$$

(*ii*) The intensity of normal radiation is

$$I_n = \frac{\sigma_b T^4}{\pi} = \frac{5.67\times10^{-8}\times800^4}{\pi}$$

$$= 7396.28\ \text{W/m}^2\ \text{steradian}$$

(*iii*) From Lambert's cosine law :

$$I_\phi = I_n \cos\phi$$

$$= 7396.28 \times \cos 60^\circ$$

$$= 3698.14\ \text{W/m}^2\ \text{steradian}$$

(*iv*) From Wien's displacement law, the wavelength λ_{max} for maximum monochromatic emissive power is :

$$\lambda_{max} = \frac{2.898\times10^{-3}}{T} = \frac{2.898\times10^{-3}}{800}$$

$$= \mathbf{36.25 \times 10^{-6}\ m}$$

26.22. HEAT EXCHANGE BETWEEN BLACK BODIES

Consider heat exchange between elementary areas dA_1 and dA_2 of two black radiating bodies having areas A_1 and A_2 respectively. The elementary areas are at a distance r apart and the normals to these areas make angles θ_1 and θ_2 with the line joining them. The surface dA_1 is at temperature T_1 and the surface dA_2 is at temperature T_2.

If the surface dA_2 subtends a solid angle $d\omega_1$ at the centre of the surface dA_1, then radiant energy emitted by dA_1 and impinging on (and absorbed by) the surface dA_2 is :

$$dQ_{12} = I_{0_1}\ d\omega_1\ dA_1$$

$$= I_{n1} \cos\theta_1\ d\omega_1\ dA_1$$

Projected area of dA_2 normal to the line joining dA_1 and dA_2 = $dA_2 \cos\theta_2$

$$\text{solid angle } d\omega_1 = \frac{dA_2 \cos\theta_2}{r^2}$$

$$\therefore\ dQ_{12} = I_{n1} \cos\theta_1 \frac{dA_2 \cos\theta_2}{r^2} dA_1$$

$$= I_{n1} \frac{\cos\theta_1 \cos\theta_2\, dA_1 dA_2}{r^2}$$

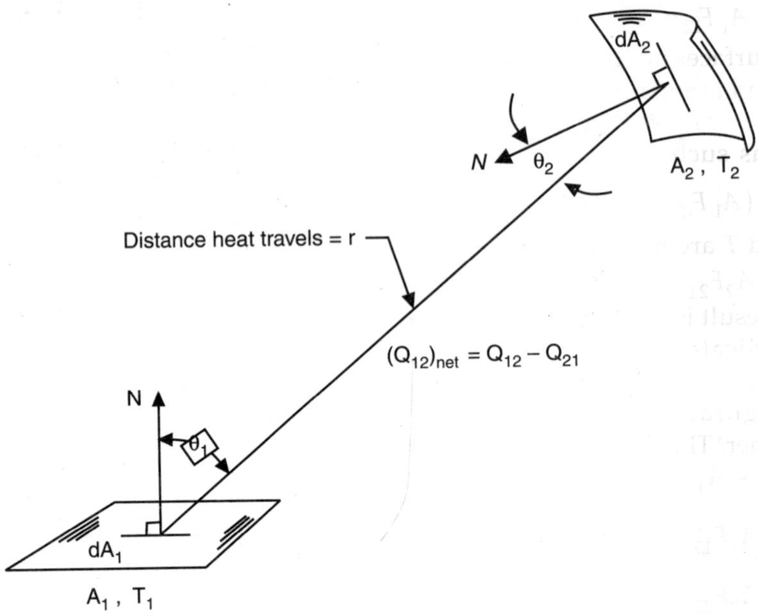

Fig. 26.29. Radiant heat exchange between two black surfaces

But $I_{n1} = \dfrac{\sigma_1 T_1^4}{\pi}$ and therefore

$$dQ_{12} = \frac{\sigma_1 T_1^4}{\pi} \frac{\cos\theta_1 \cos\theta_2\, dA_1\, dA_2}{r^2} \quad ...(26.59)$$

Integration of equation 26.59 over finite areas A_1 and A_2 gives :

$$Q_{12} = \frac{\sigma_1 T_1^4}{\pi} \int_{A_1} \int_{A_2} \cos\theta_1 \cos\theta_2 \frac{dA_1\, dA_2}{r^2} \quad ...(26.60)$$

The solution to this equation is simplified by introducting a term called ***radiation shape factor, geometrical factor, configuration factor or view factor***. The configuration factor depends only on the specific geometry of the emitter and collection surfaces, and is defined as :

"The fraction of the radiative energy that is diffused from one surface element and strikes the other surface directly with no intervening reflections."

The radiation shape factor is represented by the symbol F_{ij} which means the shape factor form a surface A_i to another surface A_j. Thus the radiation shape factor F_{12} of surface A_1 to surface A_2 is

$$F_{12} = \frac{\text{direct radiation from surface 1 incident upon surface 2}}{\text{total radiation from emitting surface 1}}$$

$$= \frac{Q_{12}}{\sigma_1 A_1 T_1^4}$$

$$= \frac{1}{\sigma_1 A_1 T_1^4} \times \frac{\sigma_1 T_1^4}{\pi} \int_{A_1} \int_{A_2} \cos\theta_1 \cos\theta_2 \frac{dA_1\, dA_2}{r^2}$$

$$= \frac{1}{A_1 \pi} \int_{A_1} \int_{A_2} \cos\theta_1 \cos\theta_2 \frac{dA_1\, dA_2}{r^2} \quad ...(26.61)$$

Thus the amount of radiation leaving A_1 and striking A_2 may be written as :

$$Q_{12} = A_1 F_{12} \sigma_1 T_1^4 \quad ...(26.62)$$

Similarly the energy leaving A_2 and arriving A_1 is :

$$Q_{21} = A_2 F_{21} \sigma_2 T_2^4 \quad ...(26.63)$$

and the net energy exchange from A_1 to A_2 is :

$$(Q_{12})_{net} = A_1 F_{12} \sigma_1 T_1^4 - A_2 F_{21} \sigma_2 T_2^4$$

When the surfaces are thought to be black ($\sigma_1 = \sigma_2 = \sigma_b$) and are maintained at the same temperature ($T_1 = T_2 = T$), there is no heat exchange and as such

$$0 = (A_1 F_{12} - A_2 F_{21}) \sigma_b T_1^4$$

Since σ_b and T are both non-zero entities,

$$A_1 F_{12} = A_2 F_{21} \quad ...(26.64)$$

The above result is known as a ***reciprocity theorem***. It indicates that the net radiant interchange may be evaluated by computing one way configuration factor from either surface to the other. Thus the net heat exchange between surfaces A_1 and A_2 is

$$(Q_{12})_{net} = A_1 F_{12}\, \sigma_b \left(T_1^4 - T_2^4\right)$$

$$= A_2 F_{21}\, \sigma_b \left(T_1^4 - T_2^4\right) \quad ...(26.65)$$

Equation 26.65 applies only to black surfaces and must not be used for surfaces having emissivities very different from unity.

The unknown configuration factor is worked out by adding and subtracting known factors of related geometries. The method is based on the definition of shape factor, the reciprocity principle and the energy conservation law.

The inter-relation between various shape factors is called shape factor algebra

The following facts and properties will be useful for the calculation of shape factors of specific geometries and for the analysis of radiant heat exchange between surfaces :

(*i*) The value of shape factor depends only on the geometry and orientation of surfaces with respect to each other. Once the shape factor between two surfaces is known, it can be used for calculating the radiant heat exchange between the surfaces at any temperature.

(*ii*) All the radiation streaming out from a convex surface 1 is intercepted by the enclosing surface 2. As such the shape factor of convex surface with respect to the enclosure F_{12} is unity. Then in conformity with reciprocity theorem, the other shape factor F_{21} is merely the ratio of areas.

(*iii*) The radiant energy emitted by one part of concave surface is intercepted by another part of the same surface. Accordingly a concave surface has a shape factor with respect to itself. The shape factor of a surface with respect ot itself is denoted by F_{11}.

For a flat or convex surface, the shape factor with respect to itself is zero. This aspect stems from the fact that for any part of flat or convex surface, one cannot see any other part of the same surface.

(*iv*) If one of the two surfaces (say A_i) is divided into sub areas A_{i1}, A_{i2}, ...A_{in}, then

$$A_i F_{ij} = \Sigma A_{in} F_{in\,j} \quad ...(26.66)$$

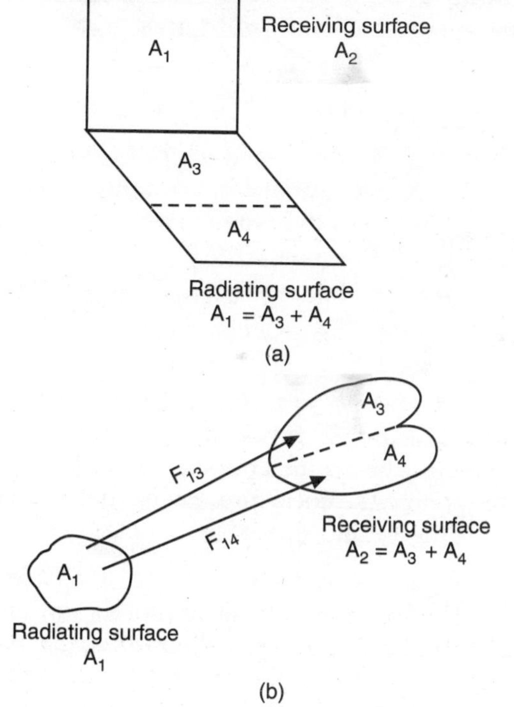

Fig. 26.30. Relation between shape factors

Thus with respect to Fig. 26.30 (*a*), wherein the radiating surface A_1 has been split up into areas A_3 and A_4,

$$A_1 F_{12} = A_3 F_{32} + A_4 F_{42}$$

Obviously $F_{12} \neq F_{32} + F_{42}$

Thus if the transmitting surface is sub divided, the shape factor for that surface with respect to the receiving surface is not equal to the sum of the individual shape factors.

If the receiving surface A_2 [Fig. 26.30 (*b*)] is divided into subareas A_3 and A_4, we have

$A_1 F_{12} = A_1 F_{13} + A_1 F_{14}$ or $F_{12} = F_{13} + F_{14}$

Apparently the shape factor from a radiating surface to a subdivided receiving surface is simply the sum of the individual shape factors.

(*v*) Any radiating surface will have finite area and therefore will be enclosed by many surfaces. The total unit radiation being emitted by the radiating surface will be received and absorbed by each of the confining surfaces. Since a shape factor is the fraction of total radiation leaving the radiating surface and falling upon a particular receiving surface, the energy balance would give :

$$\sum_{j=1}^{n} F_{ij} = 1 \; ; \; i = 1, 2, \ldots n \qquad \ldots(26.67)$$

The interior surface of a complete enclosed space has been subdivided in *n*-parts-each part having a finite area $A_1, A_2 \ldots\ldots\ldots\ldots A_n$. Thus

$$F_{11} + F_{12} + F_{13} \ldots + F_{1n} = 1$$
$$F_{21} + F_{22} + F_{23} \ldots + F_{2n} = 1$$
$$\cdot \quad \cdot \quad \cdot \quad \cdot \quad \cdot \quad \cdot \quad \cdot$$
$$F_{n1} + F_{n2} + F_{n3} \ldots + F_{nn} = 1$$

Accordingly when a radiating surface exchanges heat with a number of black surfaces comprising the enclosure, the net heat transfer from the radiating surface will be

$$(Q_{12})_{net} = A_1 F_{12} \sigma_b \left(T_1^4 - T_2^4\right) + A_1 F_{13} \sigma_b \left(T_1^4 - T_3^4\right) + A_1 F_{14} \sigma_b \left(T_1^4 - T_4^4\right) \qquad \ldots(26.68)$$

EXAMPLE 26.36

The sun is nearly spherical radiation source that is approximately 1.385 × 10⁹ m diameter and located at 1.50 × 10¹¹ m from the earth. On a clear day, the energy flux associated with solar radiation incident on the outer surface of earth has been accurately measured and known to be 1135 W/m² with an additional 285 W/m² absorbed by the earth atmosphere. Assuming that the sun emits as black body, estimate its surface temperature.

Solution : The radiant heat exchange between the sun and the earth may be written as :

$$Q_{12} = \sigma_b \left(T_1^4 - T_2^4\right) \frac{\cos\theta_1 \cos\theta_2 \, dA_1 \, dA_2}{\pi r^2}$$

where the suffex 1 and 2 pertain to sun and earth respectively. From the given data :

(*i*) $\theta_1 = \theta_2 = 90°$; hence $\cos\theta_1 = \cos\theta_2 = 1$

(*ii*) Temperature T_2 of the earth can be neglected in comparison to temperature T_1 of the sun.

(*iii*) The sun, emitting diffusely, appears as a disk of area $dA_1 = (\pi/4)d_1^2$ and

(*iv*) The heat interchange corresponds to unit area of the earth surface and so $dA_2 = 1$.

$$\therefore \; Q_{12} = \frac{\sigma_b T_1^4}{\pi r^2} \frac{\pi}{4} d_1^2 = \frac{\sigma_b T_1^4}{4\pi r^2} d_1^2$$

or (1135 + 285) =

$$\frac{5.67 \times 10^{-8} \times T_1^4}{4\pi \times \left(1.50 \times 10^{11}\right)^2} \times \left(1.385 \times 10^9\right)^2$$

Solution gives :

$$T_1 = \mathbf{5854.8 \; K}$$

The sun's temperature is generally accepted to be of this approximate magnitude.

EXAMPLE 26.37

Calculate the shape factors for the configurations shown in the figure given below :

(a) long tube with cross-section of an equilateral triangle

(b) black body inside a black enclosure

(c) diagonal partition within a long square duct.

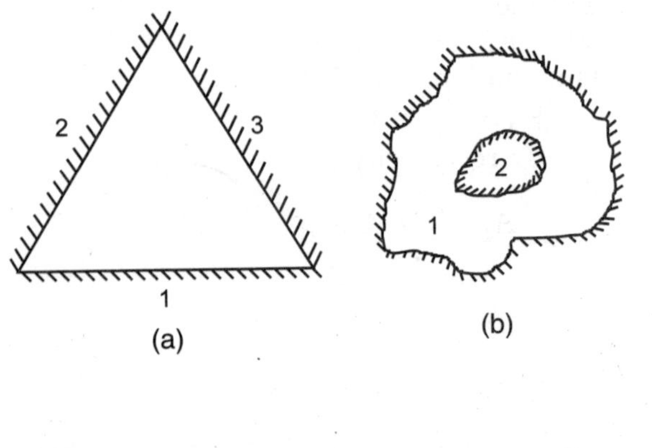

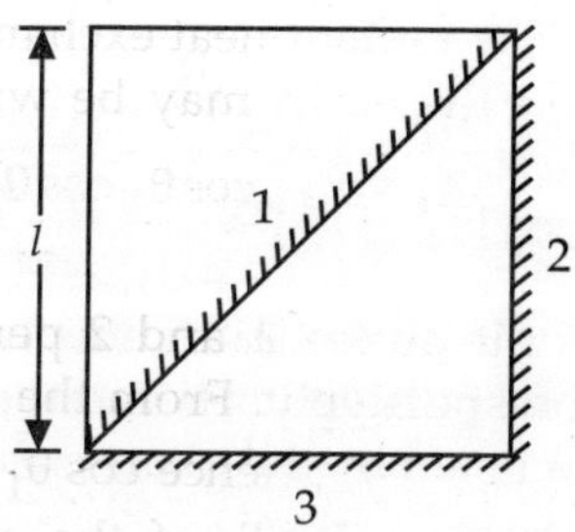

(c)

Fig. 26.31.

Solution : The desired shape factors can be worked out by invoking the summation rule, the reciprocity theorem and from inspection of the geometry.

(*a*) By summation rule for radiation from surface 1.

$$F_{11} + F_{12} + F_{13} = 1$$

The flat surface 1 cannot see itself and so $F_{11} = 0$. That gives

$$F_{12} + F_{13} = 1$$

Due to symmetry, the radiation from surface 1 is equally divided between surfaces 2 and 3, and therefore

$$F_{12} = F_{13} = 0.5$$

Likewise, considering radiation from surface 2 :

$$F_{21} + F_{23} = 1 \qquad \text{(because } F_{22} = 0\text{)}$$

$$F_{23} = 1 - F_{21}$$

By reciprocity :

$$F_{21} = \frac{A_1}{A_2} F_{12} = F_{12}$$

(because $A_1 = A_2$)

$$\therefore \quad F_{23} = 1 - F_{12} = 1 - 0.5 = 0.5$$

(*b*) The concave surface 1 can see itself and the rest of radiation falls on the enclosed surface 2. Invoking the conservation principle (summation rule) for surface 1

$$F_{11} + F_{12} = 1$$

By reciprocity :

$$A_1 F_{12} = A_2 F_{21} \quad \text{or} \quad F_{12} = \frac{A_2}{A_1} F_{21}$$

$$\therefore \quad F_{11} = 1 - F_{12} = 1 - \frac{A_2}{A_1} F_{21}$$

Further, all the radiations coming out from the black surface 2 are intercepted by the enclosing surface, Therefore $F_{21} = 1$, and so

$$F_{11} = 1 - \frac{A_2}{A_1}$$

(*c*) From summation rule :

$$F_{11} + F_{12} + F_{13} = 1$$

By inspection $F_{11} = 0$ as the flat surface cannot see itself

$$\therefore F_{12} + F_{13} = 1$$

Due to symmetry, the radiation form surface 1 is equally divided between the surfaces 2 and 3 and therefore

$$F_{12} = F_{13} = 0.5$$

By reciprocity :

$$F_{21} = \frac{A_1}{A_2} F_{12} = \frac{\sqrt{2}\, l}{l} \times 0.5 = 0.71$$

EXAMPLE 26.38

Establish a relation for the shape factor of a cavity with respect to itself. The cavity is closed on its outer surface with a flat surface.

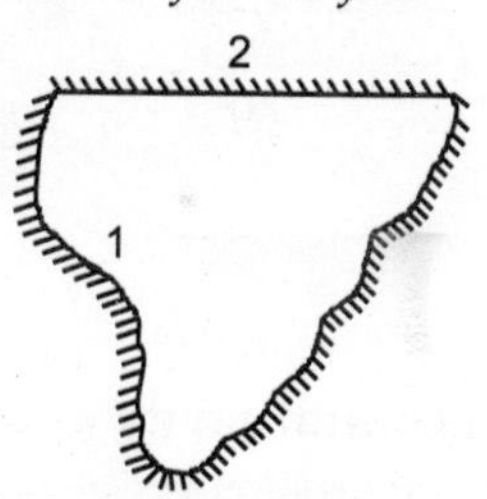

Solution : Invoking the conservation principle (summation rule) for surface 1 of the cavity,

$$F_{11} + F_{12} = 1 \qquad \text{...(}i\text{)}$$

For the closing surface 2 :

$$F_{21} + F_{22} = 1$$

But $F_{22} = 0$ as the surface is flat and cannot see itself.

$$\therefore \quad F_{21} = 1 - F_{22} = 1 - 0 = 1$$

By reciprocity :

$$F_{12} = \frac{A_2}{A_1} F_{21} = \frac{A_2}{A_1}$$

∴ From expression (*i*)

$$F_{11} = 1 - F_{12} = 1 - \frac{A_2}{A_1}$$

The above relation is valid fol all types of cavities as shown in Fig. 26.32.

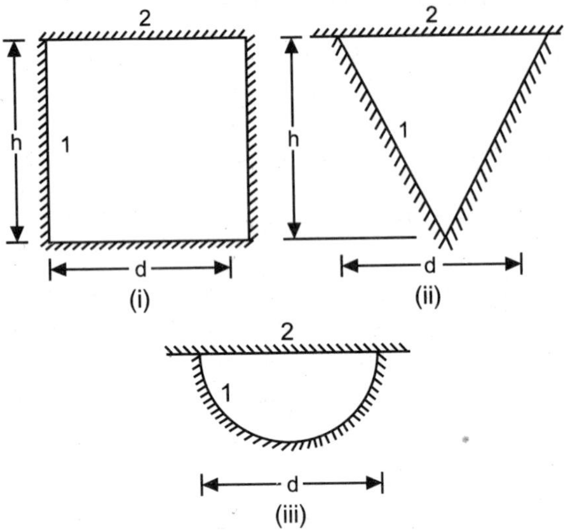

Fig. 26.32.

(*i*) For a cylindrical cavity of depth h and diameter d,

$$F_{11} = 1 - \frac{A_2}{A_1} = 1 - \frac{\frac{\pi}{4}d^2}{\frac{\pi}{4}d^2 + \pi d h}$$

$$= 1 - \frac{d}{d + 4h} = \frac{4h}{4h + d}$$

(*ii*) For a conical cavity of diameter d and depth h,

$$F_{11} = 1 - \frac{A_2}{A_1} = 1 - \frac{\frac{\pi}{4}d^2}{\frac{\pi d l}{2}} = 1 - \frac{d}{2l}$$

$$= 1 - 2 \sin \alpha$$

where l is the slant height of the cone and α is half vertex angle.

In terms of depth h

$$F_{11} = 1 - \frac{d}{2\sqrt{h^2 + \frac{d^2}{4}}} = 1 - \frac{d}{\sqrt{4h^2 + d^2}}$$

(*iii*) For a hemispherical bowl of diameter d

$$F_{11} = 1 - \frac{A_2}{A_1} = 1 - \frac{\frac{\pi}{4}d^2}{\frac{\pi d^2}{2}} = 1 - \frac{1}{2} = 0.5$$

Thus, half of the radiation from hemispherical bowl falls on itself and the remaining half is intercepted by the plane closing surface.

EXAMPLE 26.39

A small sphere with a surface temperature of 550 K is located at the geometric centre of a large sphere with an inner surface temperature of 280 K. The outside diameter of the small sphere is 5 cm, and the inside diameter of the large sphere is 25 cm. Assuming that both sides approach black body behaviour, determine how much of the emission from the inner surface of the large sphere is incident upon the outer surface of the small sphere.

Solution : Let suffix 1 designate the small sphere and suffix 2 denote the large sphere. We are asked to determine the configfuration factor F_{21}.

All the radiation being emitted by the small sphere is incident upon and absorbed by the inner surface of the large sphere. Therefore configuration factor between 1 and 2 is $F_{12} = 1$

By reciprocity theorem: $A_1 F_{12} = A_2 F_{21}$

$$4\pi r_1^2 F_{12} = 4\pi r_2^2 F_{21}$$

$$\therefore \quad F_{21} = F_{12} \times \frac{r_1^2}{r_2^2} = \frac{1 \times (2.5)^2}{(12.5)^2} = 0.04$$

Thus 4% of the emission from the inner surface of the large sphere is incident upon the small sphere and absorbed by it.

From energy balance for the large sphere,

$$F_{21} + F_{22} = 1$$

or $\quad F_{22} = 1 - F_{21} = 1 - 0.04 = 0.96$

Thus 96% of emission from the large sphere is absorbed by the inner surface of the sphere itself. The net interchange of heat between the two sphere is :

$$(Q_{12})_{net} = F_{12} A_1 \sigma_b \left(T_1^4 - T_2^4\right)$$

$$= 1 \times (4\pi \times 0.025^2) \times 5.67 \times 10^{-8} (550^4 - 280^4)$$

$$= \mathbf{37.99\ W}$$

26.23. HEAT EXCHANGE BETWEEN NON-BLACK BODIES

Infinite parallel planes

The analysis of radiant heat exchange between two non-black parallel surfaces shall be based on the following assumptions :

(*i*) The surfaces are arranged at small distance from each other and are of equal area so that practically all radiation emitted by one surface falls on the other. The configuration factor of either surface is, therefore, unity

(*ii*) The surfaces are diffuse and uniform in temperature, and that the reflective and emissive properties are constant over all the surface.

(*iii*) The surfaces are separated by a non-absorbing medium such as air.

The surface 1 emits radiant energy E_1 which strikes the surface 2. From it a part $\alpha_2 E_1$ is absorbed by the surface 2 and the remainder $(1 - \alpha_2) E_1$ is reflected back to surface 1. On reaching surface 1, a part $\alpha_1 (1 - \alpha_2) E_1$ is absorbed and the remainder $(1 - \alpha_1)(1 - \alpha_2) E_1$ is reflected and so on. The amount of radiant energy which left surface 1 per unit time is :

$$Q_1 = E_1 - \left[\alpha_1 (1-\alpha_2) E_1 + \alpha_1 (1-\alpha_1)(1-\alpha_2)^2 E_1 + \alpha_1 (1-\alpha_1)^2 (1-\alpha_2)^2 E_1 + ...\right]$$

$$= E_1 - \alpha_1 (1 - \alpha_2) E_1 \times \left[1 + (1-\alpha_1)(1-\alpha_2) + (1-\alpha_1)^2 (1-\alpha_2)^2 + ...\right]$$

$$= E_1 - \alpha_1 (1 - \alpha_2) E_1 [1 + P + P^2 + ...]$$

where $P = (1 - \alpha_1)(1 - \alpha_2)$

Since P is less than unity, the series $1 + P + P^2 +$ ...., when extended to infinity gives $1/(1 - P)$

$$\therefore \quad Q_1 = E_1 - \frac{\alpha_1 (1-\alpha_2) E_1}{1-P}$$

$$= E_1 \left[1 - \frac{\alpha_1 (1-\alpha_2)}{1-(1-\alpha_1)(1-\alpha_2)}\right]$$

From Kirchoff's law, emissivity and absorptivity of a surface are equal and so $\alpha_1 = \in_1$

$$\therefore \quad Q_1 = E_1 \left[1 - \frac{\in_1 (1-\in_2)}{1-(1-\in_1)(1-\in_2)}\right]$$

$$= E_1 \left[\frac{\in_2}{\in_1 + \in_2 - \in_1 \in_2}\right] \quad ...(26.69)$$

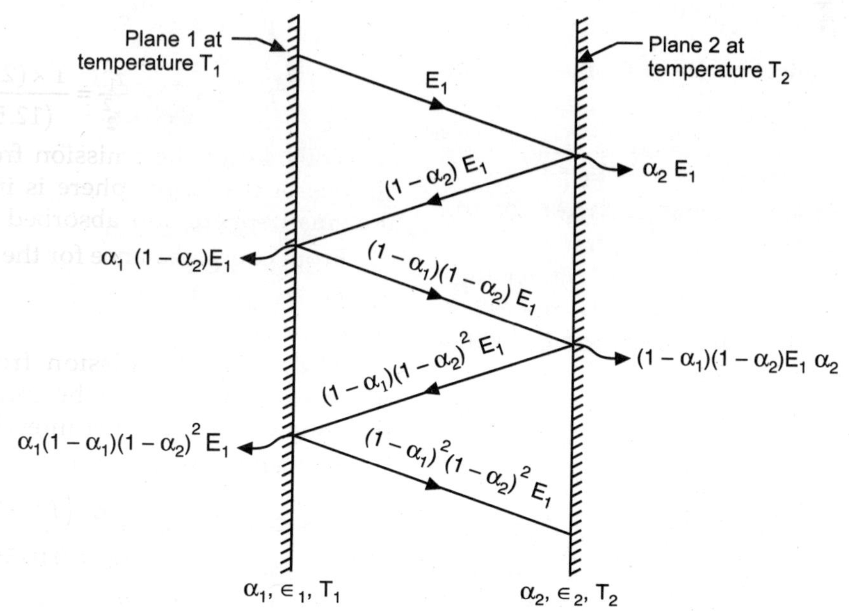

Fig. 26.33. Radiant heat exchange between two non-black parallel surfaces

Similarly the surface 2 emits radiation of emissive power E_2. From it a part $\alpha_1 E_2$ is absorbed by surface 1 and the remainder $(1 - \alpha_1) E_2$ is reflected back to it. On reaching surface 2, a part $\alpha_2 (1 - \alpha_1) E_2$ is absorbed and the rest $(1 - \alpha_1)(1 - \alpha_2) E_2$ is reflected and so on. Proceeding exactly in the same way, we can determine the amount of heat which leaves surface 2 per unit time.

$$Q_2 = E_2 \left[\frac{\epsilon_1}{\epsilon_1 + \epsilon_2 - \epsilon_1 \epsilon_2}\right] \quad ...(26.70)$$

The net heat flow from surface 1 to surface 2 per unit time is then given by

$$Q_{12} = Q_1 - Q_2$$

$$= \frac{E_1 \epsilon_2}{\epsilon_1 + \epsilon_2 - \epsilon_1 \epsilon_2} - \frac{E_2 \epsilon_1}{\epsilon_1 + \epsilon_2 - \epsilon_1 \epsilon_2}$$

$$= \frac{E_1 \epsilon_2 - E_2 \epsilon_1}{\epsilon_1 + \epsilon_2 - \epsilon_1 \epsilon_2} \quad ...(26.71)$$

From Stefan-Blotzman law for non-black surfaces,

$$E_1 = \epsilon_1 \sigma_b T_1^4 \quad \text{and} \quad E_2 = \epsilon_2 \sigma_b T_2^4$$

$$\therefore \quad Q_{12} = \frac{\epsilon_1 \sigma_b T_1^4 \epsilon_2 - \epsilon_2 \sigma_b T_2^4 \epsilon_1}{\epsilon_1 + \epsilon_2 - \epsilon_1 \epsilon_2}$$

$$= \frac{\epsilon_1 \epsilon_2}{\epsilon_1 + \epsilon_2 - \epsilon_1 \epsilon_2} \sigma_b \left(T_1^4 - T_2^4\right)$$

$$= f_{12} \sigma_b \left(T_1^4 - T_2^4\right) \quad ...(26.72)$$

where,

$$f_{12} = \frac{\epsilon_1 \epsilon_2}{\epsilon_1 + \epsilon_2 - \epsilon_1 \epsilon_2} = \frac{1}{1/\epsilon_1 + 1/\epsilon_2 - 1}$$

and is called the **interchange factor** for the radiation from surface 1 to surface 2.

While calculating the radiant interchange between two gray surfaces, both the interchange factor f_{12} and the geometric factor F_{12} are considered and net heat interchange is computed from the relation

$$Q_{net} = f_{12} F_{12} A_1 \sigma_b \left(T_1^4 - T_2^4\right)$$

$$= (F_g)_{12} A_1 \sigma_b \left(T_1^4 - T_2^4\right) \quad ...(26.73)$$

The factor $(F_g)_{12} = f_{12} F_{12}$ is called the *gray body factor* and for two surfaces exchanging heat, it is prescribed by the relation :

$$(F_g)_{12} = \frac{1}{\frac{1-\epsilon_1}{\epsilon_1} + \frac{1}{F_{12}} + \frac{1-\epsilon_2}{\epsilon_2} \times \frac{A_1}{A_2}} \quad ...(26.74)$$

EXAMPLE 26.40

Two opposed, parallel, infinite planes are maintained at 420 K and 480 K respectively. Calculate the net heat flux between these planes if one has an emissivity of 0.8 and other an emissivity of 0.7. Does it matter which plate has which emissivity ? How this heat flux will be affected if

(i) the temperature difference is doubled by raising the temperature 480 K to 540 K.

(ii) the planes are assumed to be black ?

Solution : The rate of heat interchange between the two plates is given by :

$$Q_{12} = (F_g)_{12} A_1 \sigma_b \left(T_1^2 - T_2^4\right)$$

The gray body factor $(F_g)_{12}$ is

$$(F_g)_{12} = \frac{1}{\frac{1-\epsilon_1}{\epsilon_1} + \frac{1}{F_{12}} + \frac{1-\epsilon_2}{\epsilon_2} \times \frac{A_1}{A_2}}$$

For infinite long parallel planes which see each other and nothing else,

$$F_{12} = 1 \quad \text{and} \quad A_1 = A_2$$

$$\therefore \quad (F_g)_{12} = \frac{1}{\frac{1}{\epsilon_1} + \frac{1}{\epsilon_2} - 1} = \frac{1}{\frac{1}{0.87} + \frac{1}{0.7} - 1}$$

$$= 0.59$$

$$Q_{12} = 0.59 \times 1 \times 5.67 \times 10^{-8} (480^4 - 420^4)$$

$$= 734.86 \text{ W/m2}$$

It is immaterial which plane has which emissivity since the emissivities are independent of temperature.

If T_2 = 540 K and the surfaces have the given emissivities, then

$$Q_{12} = 0.59 \times 1 \times (5.67 \times 10^{-8}) \times (540^4 - 420^4)$$

$$= \mathbf{1803.55\ W/m^2}$$

Obviously the heat flux increases by a factor of 1803.55/734.86 = 2.45 when the temperature is doubled.

If the surfaces are black,

$$Q_{12} = A_1 \sigma_b (T_1^4 - T_2^4)$$

$= 1 \times (5.67 \times 10^{-8}) \times (480^4 - 420^4)$
$= \mathbf{1245.52\ W/m^2}$

EXAMPLE 26.41

A 250 × 250 mm ingot casting, 1.5 m high and at 1225 K temperature, is stripped from its mold. The casting is made to stand on end on the floor of a large foundry whose wall, floor and roof can be assumed to be at 300 K temperature. Make calculation for the rate of radiant heat interchange between the casting and the room. The casting material has an emissivity of 0.85. Take Stefan Boltzman constant $\sigma_b = 5.67 \times 10^{-8}\ W/m^2\ K^4$.

Solution : The rate of heat interchange between the ingot and the room is given by :

$$Q_{12} = (F_g)_{12}\, A_1\, \sigma_b \left(T_1^4 - T_2^4\right)$$

The configuration corresponds to a completely enclosed body, and small compared with the enclosing body, *i.e.*, $A_1 <<< A_2$ and $F_{12} = 1$. Hence

$$(F_g)_{12} = \frac{1}{\dfrac{1-\epsilon_1}{\epsilon_1} + \dfrac{1}{F_{12}} + \dfrac{1-\epsilon_2}{\epsilon_2} \times \dfrac{A_1}{A_2}}$$

$$= \frac{1}{\dfrac{1-\epsilon_1}{\epsilon_1} + 1 + 0} = \epsilon_1 = 0.85$$

Area A_1 of the ingot radiating energy,

$A_1 = (0.25 \times 0.25) + (4 \times 0.25 \times 1.5)$
$= 0.0625 + 1.5 = 1.5625\ m^2$

$\therefore\ Q_{12} = 0.85 \times 1.5625 \times 5.67 \times 10^{-8} \times (1225^4 - 300^4)$
$= \mathbf{171.12 \times 10^3\ W}$

26.24. RADIATION SHIELDS

Many situations are encountered where it is desired to reduce the overall heat transfer between two radiating surfaces. The task is accomplished by placing radiation shields between the emitting surfaces.

The shields are thin opaque partitions arranged in the direction perpendicular to the propagation of radiated heat, and made of materials of very low absorptivity and high reflectivity (thin sheets of aluminium, copper etc.) The shields introduce a sort of additional resistance in the heat flow path and accordingly the net heat flux is reduced.

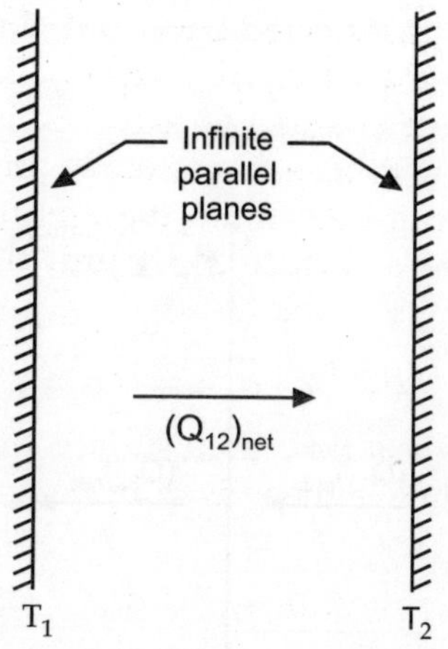

Fig. 26.34. Heat exchange between two infinite parallel planes without radiation shields

(*i*) With no radiation shields, the net heat exchange between the infinite parallel planes is given by :

$$Q_{12} = \left(F_g\right)_{12} A_1\, \sigma_b \left(T_1^4 - T_2^4\right)$$

For the given cofiguration of parallel planes which see each other and nothing else,

$$F_{12} = 1 \text{ and } A_1 = A_2 = A$$

$$(F_g)_{12} = \frac{1}{\dfrac{1-\epsilon_1}{\epsilon_1} + \dfrac{1}{F_{12}} + \dfrac{1-\epsilon_2}{\epsilon_2} \times \dfrac{A_1}{A_2}}$$

$$= \frac{1}{\left(\dfrac{1}{\epsilon_1} - 1\right) + 1 + \left(\dfrac{1}{\epsilon_2} - 1\right)} = \frac{1}{\dfrac{1}{\epsilon_1} + \dfrac{1}{\epsilon_2} - 1}$$

$$\therefore\quad Q_{12} = \frac{A\,\sigma_b \left(T_1^4 - T_2^4\right)}{\dfrac{1}{\epsilon_1} + \dfrac{1}{\epsilon_2} - 1}$$

(*ii*) The placement of a radiation shield between these two planes would neither remove nor add any heat to the system. Under steady state conditions, the screens attain a uniform temperature of T_3, and heat transfer between plane 1 and the shield $(Q_{13})_{net}$

= heat transfer between the shield and the plane 2 $(Q_{32})_{net}$

$$\frac{A\sigma_b\left(T_1^4 - T_3^4\right)}{\frac{1}{\epsilon_1}+\frac{1}{\epsilon_3}-1} = \frac{A\sigma_b\left(T_3^4 - T_2^4\right)}{\frac{1}{\epsilon_3}+\frac{1}{\epsilon_2}-1} \quad ...(26.75)$$

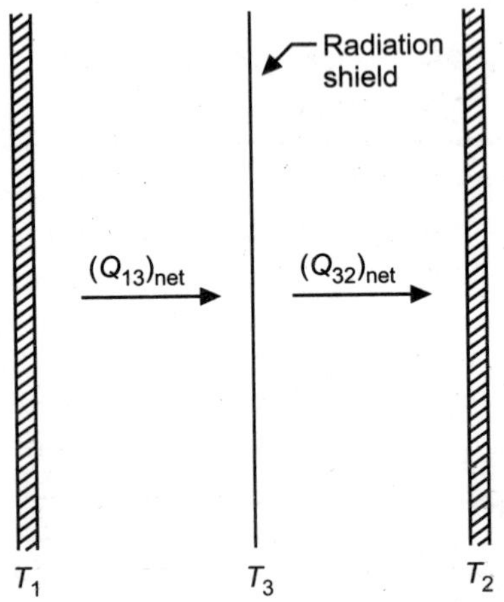

Fig. 26.35. Heat exchange between two infinite parallel planes with radiation shield

Both faces of the radiation shield have been assumed to have the same emissivity. Simplification of expression 26.75 yields :

$$T_3^4 = \frac{T_1^4\left(\frac{1}{\epsilon_3}+\frac{1}{\epsilon_2}-1\right)+T_2^4\left(\frac{1}{\epsilon_1}+\frac{1}{\epsilon_3}-1\right)}{\left(\frac{1}{\epsilon_3}+\frac{1}{\epsilon_2}-1\right)+\left(\frac{1}{\epsilon_1}+\frac{1}{\epsilon_3}-1\right)} \quad ...(26.76)$$

Each side of expression 26.75 represents the heat flow through the system. Substituting the value of T_3 in left hand side of expression 26.75, we obtain

$$(Q_{12})_{net} = \frac{A\sigma_b\left(T_1^4 - T_2^4\right)}{\left(\frac{1}{\epsilon_3}+\frac{1}{\epsilon_2}-1\right)+\left(\frac{1}{\epsilon_1}+\frac{1}{\epsilon_3}-1\right)} \quad ...(26.77)$$

The ratio of radiant energy transfer with one shield to that without any shield is obtained from expressions 26.74 and 26.77. That gives,

Radiant energy transfer

$$\frac{\text{With shield}}{\text{Without shield}} = \frac{\frac{1}{\epsilon_1}+\frac{1}{\epsilon_2}-1}{\left(\frac{1}{\epsilon_1}+\frac{1}{\epsilon_3}-1\right)+\left(\frac{1}{\epsilon_3}+\frac{1}{\epsilon_2}-1\right)} \quad ...(26.78)$$

When $\epsilon_1 = \epsilon_2 = \epsilon_3$, the above fraction takes the value 1/2. Thus by inserting one shield, the heat transfer rate is reduced to one-half of the original value. The corresponding temperature T_3 of the shield attains the value :

$$T_3^4 = \frac{1}{2}\left(T_1^4 + T_2^4\right) \quad ...(26.79)$$

EXAMPLE 26.42

Two parallel square plates, each 4 m² area, are large compared to a gap of 5 mm separating them. One plate has a temperature of 800 K and surface emissivity of 0.6, while the other has a temperature of 300 K and a surface emissivity of 0.9. Find the net energy exchange by radiation between the plates.

If a thin polished metal sheet of surface emissivity 0.1 on both sides is now located centrally between the two plates, what will be its steady state temperature ? How the heat transfer would be altered ? Neglect the convection and edge effects if any. Comment upon the significance of this exercise.

Solution : The rate of heat interchange between the two plates is given by :

$$Q_{12} = \left(F_g\right)_{12} A_1 \sigma_b \left(T_1^4 - T_2^4\right)$$

For infinite long parallel planes which see each other and nothing else, $F_{12} = 1$ and $A_1 = A_2$.

$$(F_g)_{12} = \frac{1}{\frac{1-\epsilon_1}{\epsilon_1}+\frac{1}{F_{12}}+\frac{1-\epsilon_2}{\epsilon_2}\times\frac{A_1}{A_2}}$$

$$= \frac{1}{\frac{1}{\epsilon_1}+\frac{1}{\epsilon_2}-1}$$

$$= \frac{1}{\frac{1}{0.6}+\frac{1}{0.9}-1} = 0.562$$

$\therefore$ $Q_{12} = 0.562 \times 4 \times 5.67 \times 10^{-8} (800^4 - 300^4)$
= 51175.8 Watts or **51.176 kW**

(*ii*) Let suffix 3 designate the sheet which has been inserted between the two plates.

Heat flow from plate 1 to sheet,

$$Q_{13} = (F_g)_{13} A_1 \sigma_b \left(T_1^4 - T_3^4\right)$$

$$\text{Now } (F_g)_{13} = \frac{1}{\frac{1}{\epsilon_1} + \frac{1}{\epsilon_3} - 1}$$

$$= \frac{1}{\frac{1}{0.06} + \frac{1}{0.1} - 1} = 0.09374$$

$$\therefore Q_{13} = 0.09374 A_1 \sigma_b \left(800^4 - T_3^4\right) \quad ...(a)$$

Heat flow from sheet to plate 2,

$$Q_{32} = (F_g)_{32} A_3 \sigma_b \left(T_3^4 - T_2^4\right)$$

$$\text{Now } (F_g)_{32} = \frac{1}{\frac{1}{\epsilon_3} + \frac{1}{\epsilon_2} - 1}$$

$$= \frac{1}{\frac{1}{0.1} + \frac{1}{0.9} - 1} = 0.0989$$

$$\therefore Q_{32} = 0.0989 A_3 \sigma_b \left(T_3^4 - 300^4\right)$$

Under steady state conditions,

$$0.09374 A_1 \sigma_b\left(800^4 - T_3^4\right) = 0.0989 A_3 \sigma_b\left(T_3^4 - 300^4\right)$$

Recognising that $A_1 = A_3$, we get

$$0.09374 \left(800^4 - T_3^4\right) = 0.0989 \left(T_3^4 - 300^4\right)$$

$$T_3^4 = \frac{0.09374 \times 800^4 + 0.0989 \times 300^4}{0.09374 + 0.0989}$$

T_3 = **671.65 K**

Therefore equilibrium temperature T_3 of the shield is **671.65 K**

Any one of the expressions (*a*) and (*b*) can now be used to workout the heat interchange between the plates:

∴ Heat flow through the system,

$$= 0.09374 \times 4 \times (5.67 \times 10^{-8}) \times (800^4 - 671.65^4)$$

= 4381.66 Watts or **4.382 kW**

The placement of a radiation shield reduces the radiant heat transfer by a factor of 51.176/4.382 = 11.678 times. Screens are thus placed between surfaces to cut down the radiation loss in heat insulation structure members. For example, the bulb of a thermometer or a thermocouple junction is often shielded in order to reduce radiation effects to a minimum in the act of temperature measurement of a fluid.

26.25. FREE AND FORCED CONVECTION

With respect to the cause of fluid circulation or flow, two types of convection are distinguished:

Free convection : Circulation of bulk fluid motion is caused by changes in fluid density resulting from temperature gradients between the solid surface and the main mass of fluid. The stagnant layer of fluid in the immediate vicinity of the hot body gets thermal energy by conduction. The energy thus transferred serves to increase the temperature and internal energy of fluid particles. Because of temperature rise, these particles become less dense and hence lighter than the surrounding fluid particles. The lighter fluid particles move upwards to a region of low temperature where they mix with and transfer a part of their energy to the cold particles. Simultaneously the cool heavier particles descend downwards to fill the space vacated by the warm fluid particles. The circulation pattern, upward movement of the warm fluid and downward movement of cool fluid, is called ***convection currents***. These currents are setup naturally due to gravity alone and are responsible for heat convection.

Designers of furnaces, house heating systems, architectural projects, roads and concrete structures will be concerned with free convection.

Forced convection : Flow of fluid is caused by a pump, a fan or by the atmospheric winds. These mechanical devices provide a definite circuit for the circulating currents and that speeds up the heat transfer rate. Examples of forced convection are cooling of internal combustion engines, air conditioning installations and nuclear reactors, condenser tubes and other heat exchange equipment.

Regardless of the particular nature (free or forced), the appropriate rate equation for the convective heat transfer between a surface and an adjacent fluid is prescribed by *Newton's law* of *cooling*.

$$Q = hA\,(t_s - t_f)$$

where Q is the convective heat flow rate, A is the surface area exposed to heat transfer, t_s is the surface temperature of the solid, and t_f is the undisturbed temperature of the fluid. The constant of proportionality h relates the heat transfer per unit time and unit area to the overall temperature difference. The unit of h are W/m^2-deg, and it is referred to as convective heat transfer coefficient, the surface conductance or the film coefficient.

26.25.1. Dimensionless group for free convection

The different variables specifying the system behaviour have been indicated in Fig. 26.36. which represents the free convectiopn of fluid flow over a flat plate. The physical quantites with their dimensions in M-L-T-θ-H system of units are :

Variable	Symbol	Dimension
Fluid viscosity	μ	$ML^{-1}\,T^{-1}$
Fluid density	ρ	ML^{-3}
Fluid thermal conductivity	k	$HL^{-1}\,T^{-1}\theta^{-1}$
Fluid heat capacity	c_p	$HM^{-1}\,\theta^{-1}$
Fluid coefficient of thermal expansion	β	θ^{-1}
Temperature difference	$\Delta t = (t_a - t_f)$	θ
Significant length	l	L
Heat transfer coefficient	h	$HL^{-2}\,T^{-1}\,\theta^{-1}$

The coefficient of thermal expansion, β is prescribed by the relation;

$$\rho_1 = \rho_2\,\{1 + \beta\,(t_2 - t_1)\}$$

where ρ_1 and ρ_2 are the fluid densities at temperature t_1 and t_2 respectively,

Buoyant force

$$= (\rho_1 - \rho_2)g$$

$$= \left[\rho_2\left\{1+\beta\,(t_2-t_1)\right\}-\rho_2\right]g = \rho_2\,(\beta g\,\Delta t)$$

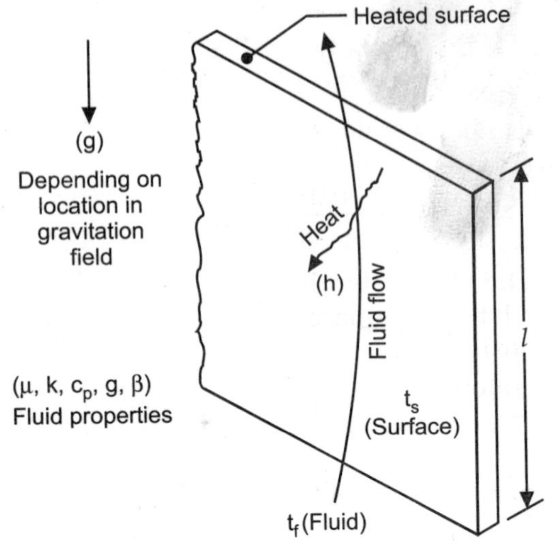

Fig. 26.36. Dimensional analysis variables for free convection

This suggests the inclusion of variables β, g, ΔT into the list of those important to the natural convection situation. The parameter $(\beta g\,\Delta T)$ represents the buoyant force and has the dimensions of (LT^{-2}).

Buckingham's π-method : It can be premised that the fuctional relationship is,

$$f\,(\mu, \rho, k, c_p, \beta g\,\Delta T, l, h) = 0$$

There are 8 physical quantities (βg and ΔT are counted separately) and 5 fundamental units; hence (8 - 5) or 3 π-terms. We choose fluid viscosity μ, thermal conductivity k, product $(\beta g\,\Delta t)$ and the characteristic length l as the core group (repeated variables) with unknown exponents, and establish the π terms as follows :

$$\pi_1 = \mu^a\,k^b\,(\beta g\,\Delta T)^c\,l^d\,\rho$$

$$1 = (ML^{-1}\,T^{-1})^a \times (HL^{-1}\,T^{-1}\,\theta^{-1})^b \times (LT^{-2})^c \times (L)^d \times ML^{-3} \quad ...(i)$$

Equating the exponents of fundamental dimensions on both sides :

$M : 0 = a + 1$

$L : 0 = -a - b + c + d - 3$

$T : 0 = -a - b - 2c$

$\theta : 0 = -b$

$H : 0 = b$

Solution gives :

$a = -1;\ \ b = 0;\ \ c = 1/2$ and $d = 3/2$

Substituting these values in relation (*i*)

$$\pi_1 = (\mu)^{-1}\,(k)^0\,(\beta g\,\Delta T)^{1/2}\,(l)^{3/2}\,\rho$$

$$= \frac{l^{3/2}\,\rho\,(\beta g\,\Delta T)^{1/2}}{\mu} = \frac{l^3\,\rho^2\,(\beta g\,\Delta T)}{\mu^2}$$

The above step has been worked on the postulate that square of a π-term is also non-dimensional.

Following the same procedure, one would obtain ;

$$\pi_2 = \mu^a\,k^b\,(\beta g\,\Delta T)^c\,l^d\,c_p\,; \qquad \pi_2 = \frac{\mu c_p}{k}$$

$$\pi_3 = \mu^a\,k^b\,(\beta g\,\Delta T)^c\,l^d\,h\,; \qquad \pi_3 = \frac{hl}{k}$$

Thus the functional relationship becomes :

$$\phi\left(\frac{l^3\rho^2\beta g\,\Delta T}{\mu^2}, \frac{\mu c_p}{k}, \frac{hl}{k}\right) = 0$$

or

$$\frac{hl}{k} = \phi\left(\frac{l^3\rho^2\beta g\,\Delta T}{\mu^2}, \frac{\mu c_p}{k}\right)$$

$$Nu = \phi\,(Gr,\ Pr) \qquad \text{...(26.80)}$$

where

$Nu = h\,l/k$, a dimensionless group called Nusselt number

$Gr = \left(l^3\,\rho^2\,\beta g\,\Delta T/\mu^2\right)$, a dimensionless group called Grashof number

$Pr = \mu\,c_p/k$, a dimensionless group called Prandtl number

It is a usual practice to rewrite the above correlation in the form

$$Nu = C\,(Gr)^a\,(Pr)^b \qquad \text{...(26.81)}$$

The constant C and the exponents a and b are evaluated through experiments.

26.25.2. Dimensionaless groups for forced convection

The different variables specifying the system behaviour have been indicated in Fig. 26.37 which represents the forced convection of fluid flow over a flat plate. The physical quantities with their dimensions in M-L-T-θ-H system of units are :

Variable	Symbol	Dimension
Fluid viscosity	μ	$ML^{-1}\,T^{-1}$
Fluid density	ρ	ML^{-3}
Fluid thermal conductivity	k	$HL^{-1}\,T^{-1}\theta^{-1}$
Fluid heat capacity	c_p	$HM^{-1}\,\theta^{-1}$
Temperature difference	$\Delta t = (t_a - t_f)$	θ
Flow velocity	V	LT^{-1}
Plate length	l	L
Heat transfer coefficient	h	$HL^{-2}\,T^{-1}\,\theta^{-1}$

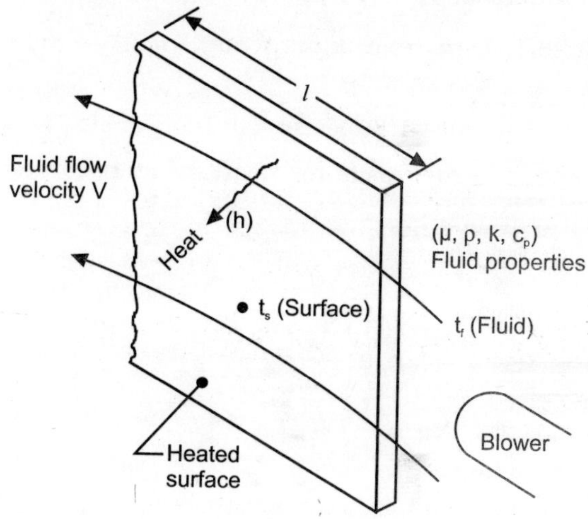

Fig. 26.37. Dimensional analysis variables for forced convection

Buckingham's π-Method. It can be premised that the functional relationship is

$$f\,(\mu,\ \rho,\ k,\ c_p,\ \Delta t,\ V,\ l,\ h) = 0$$

There are 8 physical quantities and 5 fundamental units, hence (8 - 5) or 3 π-terms. We choose fluid viscosity μ, thermal conductivity k, velocity V and the characteristic length l as the core group (repeated variables) with unknown exponents and establish the π-terms as follows:

$$\pi_1 = \mu^a\,k^b\,V^c\,l^d\,\rho$$

$$1 = (ML^{-1}\,T^{-1})^a \times (HL^{-1}\,T^{-1}\,\theta^{-1})^b \times (LT^{-1})^c \times (L)^d \times (ML^{-3}) \qquad \text{...(i)}$$

Equating the exponents of fundamental dimensions on both sides :

$M : 0 = a + 1$
$L : 0 = -a - b + c + d - 3$
$T : 0 = -a - b - c$
$\theta : 0 = -b$
$H : 0 = b$

Solution gives :

$a = -1$; $b = 0$; $c = 1$; $d = 1$.

Substituting these values in relation (*i*),

$$\pi_1 = \mu^{-1} \times k^0 \times V^1 \times l^1 \times \rho = \frac{\rho V l}{\mu}$$

Following the same procedure, one would obtain :

$$\pi_2 = \mu^a k^b V^c l^d c_p ; \qquad \pi_2 = \frac{\mu c_p}{k}$$

$$\pi_3 = \mu^a k^b V^c l^d h ; \qquad \pi_3 = \frac{hl}{k}$$

Thus the functional relationship becomes :

$$\phi\left(\frac{\rho V l}{\mu}, \frac{\mu c_p}{k}, \frac{hl}{k}\right) = 0$$

or $$\frac{hl}{k} = \phi\left(\frac{\rho V l}{\mu}, \frac{\mu c_p}{k}\right)$$

$$Nu = \phi\,(Re, Pr) \qquad \ldots(26.82)$$

where

$Nu = (h\,l/k)$, a dimensionless group called Nusselt number

$Re = (\rho\, V\, l/\mu)$, a dimensionless group called Reynolds number

$Pr = (\mu\, c_p/k)$, a dimensionless number called Prandtl nunber

It is usual practice to rewrite the above correlation in the form

$$Nu = C\,(Re)^a\,(Pr)^b \qquad \ldots(26.83)$$

The constant C and the exponents a and b are evaluated through experiments.

Note : If μ, ρ, c_p and V were chosen as the core group (repeated variables), then the analysis would have yielded the following non-dimentional groups :

$$Re = \frac{\rho V l}{\mu} ; \quad Pr = \frac{\mu c_p}{k} ; \quad St = \frac{h}{\rho V c_p}$$

where St is called the Stanton number.

Accordingly another correlation for forced convection would be of the form :

$$St = \phi\,(Re, Pr)$$

EXAMPLE 26.42 (a)

A horizontal heated plate at 200°C and facing upwards has been placed in still air at 20°C. If the plate measures 1.25 m × 1 m, make calculations for the heat loss by natural convection.

The convective film coefficient for free convection is given by the following empirical relation :

$$h = 0.32\,(\theta)^{0.25}\ W/m^2K$$

where θ is the mean film temperature in degrees kelvin.

Solution : Mean film temperature,

$$\theta = 273 + (200 + 20)/2 = 383°C$$

∴ Convective film coefficient,

$$h = 3.02 \times (383)^{0.25}$$
$$= 13.36\ \text{W/m}^2\text{-deg}$$

Rate of heat loss,

$$Q = h\,A\,\Delta t$$
$$= 13.36 \times (1.25 \times 1) \times (200 - 20)$$
$$= \mathbf{3006\ W}$$

EXAMPLE 26.43

Two horizontal steam mains with diameters 5 cm and 15 cm are so laid in a boiler house that any mutual heat effect is precluded. The mains are at the same surface temperature of 500°C whilst the ambient air is at 50°C. Work out the ratios of the heat transfer coefficients, and of the heat losses from one metre length of the mains.

Solution : The steam mains are located in the boiler house where the ambient air is stationary. The situation then corresponds to that of free convection for which the following correlation applies :

$$Nu = C\,(Gr \times Pr)^{0.25}$$

$$Nu = C\left[\frac{d^3 \rho^2 \beta g \Delta t}{\mu^2} \times \frac{\mu c_p}{k}\right]^{0.25}$$

Since both the mains are at the same temperature and are exposed to the same ambient conditions, the relevant fluid properties are same for both.

$\therefore \quad Nu \propto (d^3)^{0.25} \propto (d)^{0.75}$

$$\frac{hd}{k} \propto (d)^{0.75} \quad \text{or} \quad h \propto \frac{1}{(d)^{0.25}}$$

$$\therefore \quad \frac{h_1}{h_2} = \left(\frac{d_2}{d_1}\right)^{0.25} = \left(\frac{15}{5}\right)^{0.25} = \mathbf{1.316}$$

Heat loss

$$Q = h\,A\,\Delta t$$

$$\therefore \quad \frac{Q_1}{Q_2} = \frac{h_1 \times \pi d_1 l}{h_2 \times \pi d_2 l}$$

$$= \frac{h_1 d_1}{h_2 d_2} = 1.316 \times \frac{5}{15} = \mathbf{0.438}$$

EXAMPLE 26.44

Air flows at 1.8 m/s past a 15 cm diameter pipe in a direction normal to the axis. The pipe surface is at 30°C and it receives 4.95 kW/m² of heat from the air at 250°C temperature. The convective heat transfer coefficient is prescribed by the relation

$$\left[\frac{hd}{k}\right] = \left[\frac{V\,d\,c_p}{k}\right]^{0.7}$$

Under similar operating conditions, make calculations for the heat loss from carbon dioxide at 200°C when it flows at 1.5 m/s normal to a 25 cm diameter pipe at 25°C.

The relevant thermo-physical properties for air and carbon dioxide are :

For air at 250°C :

c_p = 1.038 kJ/kgK ;

$k = 4.270 \times 10^{-2}$ W/mK

For CO_2 at 200°C :

c_p = 0.99 kJ/kgK ;

$k = 3.12 \times 10^{-2}$ W/mK

Solution : Let suffix 1 denote air and suffix 2 denote carbon dioxide. Then,

$$\frac{h_1 d_1}{k_1} = C\left[\frac{V_1 d_1 c_{p1}}{k_1}\right]^{0.7}$$

$$\frac{h_2 d_2}{k_2} = C\left[\frac{V_2 d_2 c_{p2}}{k_1}\right]^{0.7}$$

$$\therefore \quad \frac{h_1}{h_2} \times \frac{d_1}{d_2} \times \frac{k_2}{k_1} = \left[\frac{V_1}{V_2} \times \frac{d_1}{d_2} \times \frac{c_{p1}}{c_{p2}} \times \frac{k_2}{k_1}\right]^{0.7}$$

$$\frac{h_1}{h_2} \times \frac{0.15}{0.25} \times \frac{3.12 \times 10^{-2}}{4.27 \times 10^{-2}} = \left[\frac{1.8}{1.6} \times \frac{0.15}{0.25} \times \frac{1.038}{0.99} \times \frac{3.12 \times 10^{-2}}{4.27 \times 10^{-2}}\right]^{0.7}$$

$$0.438\,\frac{h_1}{h_2} = 0.659 \quad \text{or} \quad \frac{h_1}{h_2} = \frac{0.659}{0.438} = 1.505$$

From the convective heat equation,

$$Q = h\,A\,\Delta t$$

Therefore for the same heat flow area,

$$\frac{Q_1}{Q_2} = \frac{h_1\,\Delta t}{h_2\,\Delta t_2} = 1.505 \times \frac{(250-30)}{(200-25)}$$

$$= 1.892$$

$\therefore$ Heat loss from carbon dioxide,

$$Q_2 = \frac{4.95}{1.892} = \mathbf{2616\ kW/m^2}$$

EXAMPLE 26.45

Investigate the effect of following conditions on the average value of heat transfer coefficient in flow through a tube :

(i) Two-fold increase in flow velocity by varying mass flow rate;

(ii) Two-fold increase in the diameter of tube, the flow velocity is maintained constant by a change in the rate of liquid flow.

It may be presumed that there is no change in the temperatures of the liquid and the tube wall, and that the flow through the tube is turbulent in character.

Solution : In case of turbulent flow through a tube, the average value of heat transfer coefficient is prescribed by the relation

$$Nu = \frac{h\,d}{k} = 0.023\,(Re)^{0.8}\,(Pr)^{0.33}$$

$$\text{or} \qquad h = \frac{k}{d} \times 0.023\left(\frac{\rho\,V\,d}{\mu}\right)^{0.8}\left(\frac{\mu\,c_p}{k}\right)^{0.33}$$

$$= 0.023\,k\left(\frac{\rho}{\mu}\right)^{0.8}\left(\frac{\mu c_p}{k}\right)^{0.33}\frac{(V)^{0.8}}{(d)^{0.2}}$$

Further as there is no change in the temperatures, the fluid properties remain constant.

(*i*) When the tube diameter and the fluid properties remain constant

$$h \propto (V)^{0.8}$$

$$\frac{h_2}{h_1} = \left(\frac{V_2}{V_1}\right)^{0.8} = (2)^{0.8} = 1.741$$

That is, a two-fold increase in the flow velocity results into 74.1% increase in heat transfer.

(*ii*) When the flow velocity and the fluid properties remain unchanged

$$h \propto \frac{1}{(d)^{0.2}}$$

$$\frac{h_2}{h_1} = \left(\frac{d_1}{d_2}\right)^{0.2} = \left(\frac{1}{2}\right)^{0.2} = 0.87$$

That is, with two-fold increase in tube diameter, the heat transfer decreases by 13%.

EXAMPLE 26.46

The oil pan of an IC engine approximates a flat plate 0.3 m wide by 0.45 m long and protrudes below the framework of the automobile. The engine oil is at 95°C and the ambient air temperature is 35°C. If the automobile runs at 36 km/hr, make calculations for the rate of heat transfer from the oil-pan surface. Assume negligible resistance to conduction through the oil pan.

Solution : At the mean film temperature, $t_f = (95 + 35)/2 = 65°C$, the thermo-physical properties of air are :

$$\nu = 18.46 \times 10^{-6}\ \text{m}^2/\text{s}\ ;$$
$$k = 0.0293\ \text{W/m-deg}\ ;$$
$$Pr = 0.695$$

Velocity of automobile $V = 36$ km/hr

$$= \frac{36 \times 1000}{3600} = 10\ \text{m/s}$$

$$Re = \frac{\rho V l}{\mu} = \frac{V l}{\nu} = \frac{10 \times 0.45}{18.46 \times 10^{-6}}$$

$$= 2.438 \times 10^5 \quad \text{(laminar)}$$

$$Nu = 0.664\,(Re)^{0.5}\,(Pr)^{0.33}$$
$$= 0.664\,(2.438 \times 10^5)^{0.5} \times (0.695)^{0.33}$$
$$= 290.81$$

$$h = Nu \times \frac{k}{l} = 290.81 \times \frac{0.0293}{0.45}$$
$$= 18.93\ \text{W/m-deg}$$

Convective heat transfer,

$$Q = h\,A\,\Delta t$$
$$= 18.93 \times (0.3 \times 0.45) \times (95 - 35)$$
$$= \mathbf{170.37\ W}$$

26.26. HEAT EXCHANGERS

Heat exchanger is process equipment designed for the effective transfer of heat energy between two fluids ; a hot fluid and a coolant. The purpose may be either to remove heat from a fluid or to add heat to a fluid. Notable examples are :

(*i*) boilers (evaporators), superheaters and condensers of a power plant

(*ii*) automobile radiators and oil coolers of heat engines

(*iii*) evaporator of an ice plant and milk-chiller of a pasteurising plant

(*iv*) condensers and evaporators in refrigeration units

(*v*) water and air heaters or coolers

The heat transferred in the heat exchanger may be in the form of latent heat (*e.g.* in boilers and condensers) or sensible heat (*e.g.* in heaters and coolers).

26.27. CLASSIFICATION OF HEAT EXCHANGERS

Many types of heat exchangers have been developed to meet the widely varying applications. Based upon their

– operating principle
– arrangement of flow path
– design and certain constructional features,

the heat exchangers can be classified into the following categories :

26.27.1. Nature of heat exchange process

Based upon the nature of heat exchange process, the heat exchangers are classified into direct contact, regenerators and recuperators.

In **direct contact or open heat exchangers,** the energy transfer between the hot and cold fluids is brought about by their complete physical mixing ; there is simultaneous transfer of heat and mass. Use of such units is restricted to the situations where mixing between the fluids is either harmless or is desirable. Examples are water cooling towers and jet condensers in steam power plants. Figure 26.38. represents a direct contact heat exchanger. Steam is being bubbled into water; steam gets condensed and releases heat that warms up the water.

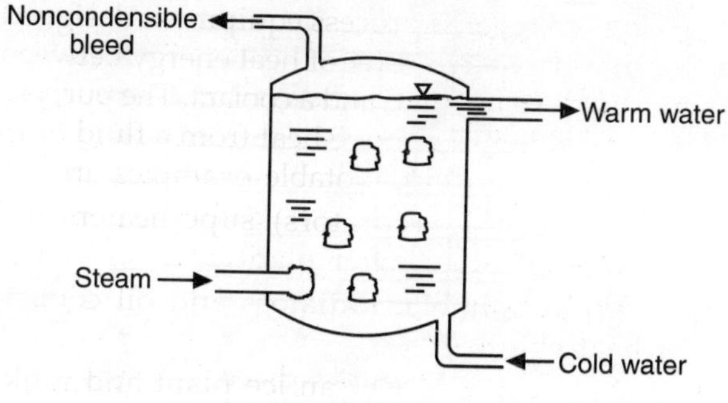

Fig. 26.38. Direct contact or open heat exchanger

In a **regenerator**, the hot fluid is passed through a certain medium called matrix. The heat is transferred to the solid matrix and accumulates there; the operation is called heating period. The heat thus stored in the matrix is subsequently transferred to the cold fluid by allowing it to pass over the heated matrix. The regenerators are quite often used in connection with engines and gas turbines. Other applications are : regenerators of open hearth and glass melting furnaces and air heaters of blast furnaces. The operation of a regenerator is intermittent; the matrix alternately stores heat extracted from the hot fluid and then delivers it to the cold fluid. However in some of the regenerators the matrix is made to rotate through the fluid passages arranged side by side and that renders the heat exchange process continuous. The effectiveness of a regenerator depends upon the heat capacity of the regenerating material and the rate of absorption and release of heat.

In a **recuperator**, the fluids flow simultaneously on either side of a separating wall; the heat transfer occurs between the fluid streams without mixing or physical contact with each other. The wall provides an element of thermal resistance between the fluids and the heat transfer consists of:

- convection between the hot fluid and the wall
- conduction through the wall
- convection between wall and the cold fluid

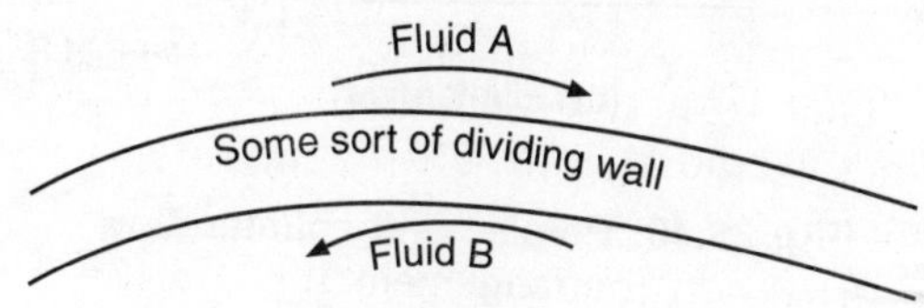

Fig. 26.39. Heat exchange in a recuperator

Such exchangers are used when the two fluids cannot be allowed to mix, *i.e.*, when the mixing is undesirable. Majority of the industrial applications have exchangers of the recuperator type. Notable examples are :

(*i*) boilers, superheaters and condensers; economisers and the air preheaters in steam power plants

(*ii*) automobile radiators

(*iii*) condensers and evaporators in refrigeration units

(*iv*) oil heaters for an airplane

(*v*) heat exchanger inside a gas furnace etc.

The open-type (direct contact) heat exchangers and the recuperators operate under steady state conditions; and the transfer of heat inside a regenerator takes place essentially under transient conditions.

26.27.2. Relative direction of motion of fluids

According to the direction of flow of fluids, the heat exchangers are classified into three categories: parallel flow, counter flow and the cross flow.

In the ***co-current*** or ***parallel*** flow arrangement, the fluids (hot and cold) enter the unit from the same side, flow in the same direction and subsequently leave from the same side

(Fig. 26.40 *a*). Obviously the flow of fluids is uni-directional and parallel to each other.

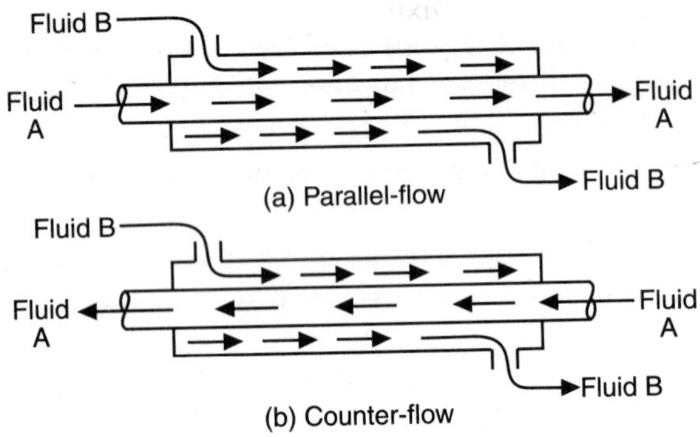

Fig. 26.40. Parallel and counter-flow arrangement

In the ***counter-current*** or ***counter-flow*** arrangement, the fluids (hot and cold) enter the unit from opposite ends, travel in opposite directions and subsequently leave from opposite ends (Fig. 26.40 *b*). Obviously the flow of fluids is opposite in direction to each other. For a given surface area, the counter-flow arrangement gives the maximum heat transfer rate and is naturally preferred for the heating and cooling of fluids.

In the ***cross-flow*** arrangement, the two fluids (hot and cold) are directed at right angles to each other. Figure 26.41 shows two common arrangements of cross-flow heat exchangers. In Figure 26.41 (*a*) the fluid A flows inside the separate tubes and its different streams do not mix. The fluid B flows over the tube banks and gets perfectly mixed. In Figure 13.5 (*b*), each of the fluid stays in prescribed paths and are not allowed to mix as they flow through the heat exchanger. When mixing occurs, the temperature variations are primarily in the flow direction. When unmixed, there is temperature gradient along the stream as well as in the direction perpendicular to it. Apparently, temperatures of the fluids leaving the unit are not uniform for the unmixed streams. The cross flow heat exchangers are commonly employed in air or gas heating and cooling applications, *e.g.*, the automobile radiator and the cooling unit of a air-conditioning duct.

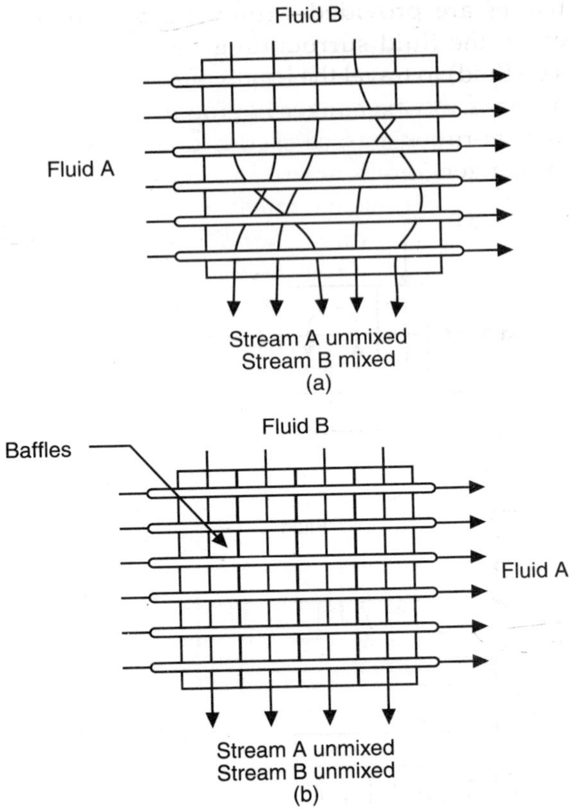

Fig. 26.41. Cross-flow arrangement

26.27.3. Mechanical design of heat exchange surface

(*i*) *Concentric tubes* : Two concentric pipes are used, each carrying one of the fluids. The direction of flow may correspond to unidirectional or counter flow arrangement (Fig. 26.40).

(*ii*) *Shell and tube* : One of the fluids is carried through a bundle of tubes enclosed by a shell. The other fluid is forced through the shell and flows over the outside surface of tubes. The direction of flow for either or both fluids may change during its passage through the heat exchanger.

(*iii*) *Multiple shell and tube passes* : The two fluids may flow through the exchanger only once (single pass), one or both fluids may traverse the exchanger more than once (multi-pass). By suitable header design, the fluid within the tubes (tube side fluid) can be made to traverse back and forth from one end of the shell to the other. Quite often longitudinal baffles are provided within the shell which cause the fluid surrounding the tubes (shell side fluid) to travel the length of shell a number of times. An exchanger having *n*-shell passes and *m*-tubes passes is designated as *n-m* exchanger.

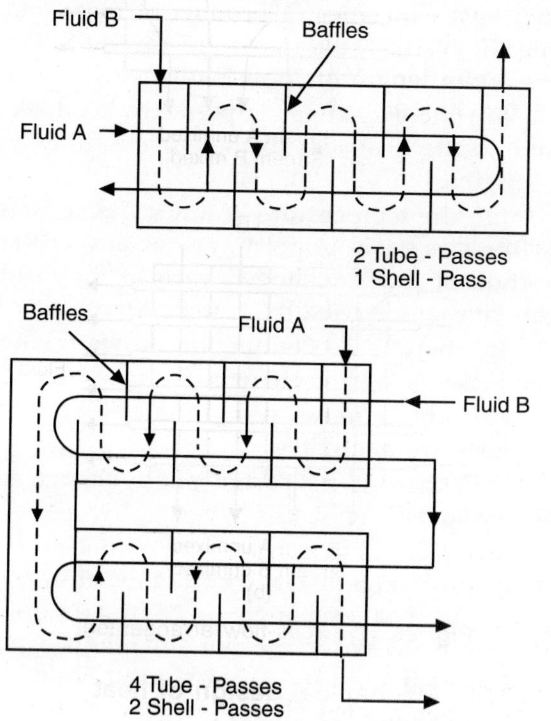

Fig. 26.42. Multiple shell and tube passes

A multiple shell and tube exchanger is preferred to ordinary counter-flow design due to its low cost of manufacture, easy dismantling for cleaning and repair and reduced thermal stresses due to expansion.

26.27.4. Physical state of heat exchanging fluids (condensation and evaporation)

(*i*) *Condenser* : The hot fluid (condensing steam) remains at constant temperature all along its passage through the heat exchanger, whilst the temperature of the other fluid (coolant water) gradually increases from inlet to outlet. Obviously the hot fluid loses only part of its heat.

(*ii*) *Evaporator* : During heat exchange in the evaporation of water into steam, the cold-fluid (boiling water) evaporates at constant temperature whilst the temperature of hot gases continuously decreases from inlet to outlet.

The heat exchangers can be further classified on the basis of following design parameters :

(*i*) temperature and pressure levels of the fluid

(*ii*) corrosiveness, toxicity and scale forming tendency of the fluids.

(*iii*) economic considerations such as cost, ease of manufacture, necessary space and required life etc. The cost considerations may, however, be subordinate to weight and size limitation in space and aeronautical limitations. Exchangers of compact design are employed where weight, space and cost limitations are severe.

26.28. LOGARITHMIC MEAN TEMPERATURE DIFFERENCE

During heat exchange between two fluids, the temperature of the fluids change in the direction of flow and consequently there occurs a change in the thermal head causing the flow of heat. Fig. 26.43(*a*) represents the temperature conditions existing in surface condenser or feed water heater. The hot fluid is steam and the cold fluid is water. Here the temperature of steam remains constant but the temperature of water is progressively rising. During heat exchange in evaporation of water into steam (Fig. 26.43 *b*), the water evaporates at constant temperature and the temperature of hot gases continuously decreases in flowing from inlet to outlet.

In many instances, both the fluids experience a change in temperature while flowing through the heat exchange equipment. In a parallel flow system, the thermal head (temperature potential) causing the flow of heat is maximum at the inlet. However, the thermal head goes on diminishing along the flow path and is minimum at the outlet. In a counter flow system, both the fluids are in their coldest state at the exit.

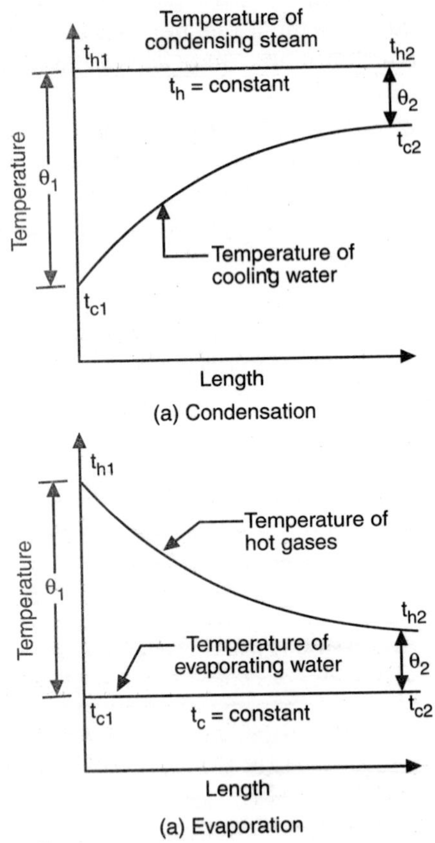

Fig. 26.43. Temperature changes of mediums during (a) condensation (b) evaporation

To calculate the rate of heat transfer by the expression, $Q = U\,A\,\Delta t$, an average value of the temperature difference between the fluids has to be determined. The calculations become simplified if it is assumed that :

(*i*) the overall heat transfer coefficient U is constant throughout the heat exchanger

(*ii*) the specific heats and mass flow rates of both the fluids are constant. This also implies that heat capacities (a product of mass and specific heat) of the fluids are constant over the entire length of flow path

(*iii*) the exchanger is perfectly insulated and so the heat loss to the surroundings is negligible

(*iv*) the temperature at any cross-section of the stream is uniform, *i.e.*, at any cross-section of heat exchanger, each of the fluid can be characterised by a single temperature

(*v*) there is no conduction of heat along the tubes of heat exchanger

(*vi*) the kinetic and potential energy changes are negligible

(*vii*) there is no partial phase change in the systems.

Consider heat transfer across an element of length dx at a distance x from the entrance side of the heat exchanger. Let at this section, the temperature of the hot fluid be t_h and that of cold fluid be t_c. Within the limits of this elementary length, the temperature t_h and t_c of the heat exchanging fluids can be considered to remain constant. Then heat flow dQ through this elementary length is given by :

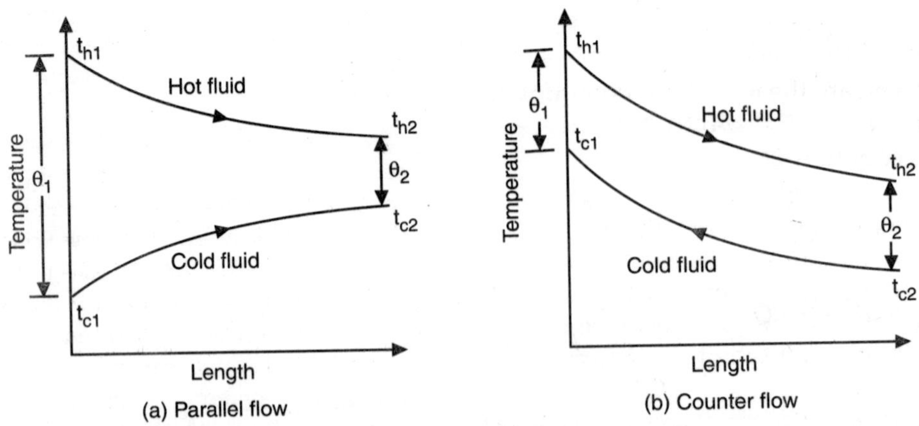

Fig. 26.44. Temperature changes of mediums (fluids) during (a) parallel flow (b) counter flow

$$dQ = U\,dA\,(t_h - t_c) = U\,dA\,\theta \qquad ...(26.84)$$

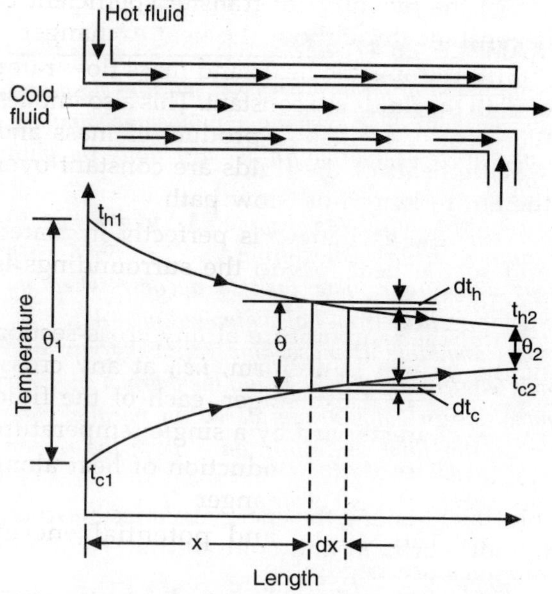

Fig. 26.45. Temperature changes of mediums (fluids) during parallel flow arrangement

where $\theta = (t_h - t_c)$, the temperature difference between the fluids. Due to heat exchange, the temperature of the hot fluid decreases by dt_h and the temperature of cold fluid increases by dt_c. Then for heat exchange between the fluids, we can write

$$dQ = -\,m_h\,c_h\,dt_h = m_c\,c_c\,dt_c$$
$$= -\,C_h\,dt_h = C_c\,dt_c \qquad ...(26.85)$$

where $C_h = m_h\,c_h$ = heat capacity or water equivalent of the hot fluid

and $C_c = m_c\,c_c$ = heat capacity or water equivalent of the cold fluid.

m_h and m_c are the mass flow rates of fluids and c_h and c_c are the respective specific heats.

From equation 26.85,

$$dt_h = \frac{-dQ}{C_h} \quad \text{and} \quad dt_c = \frac{dQ}{C_c}$$

$$\therefore\ dt_h - dt_c = -\,dQ\left[\frac{1}{C_h} + \frac{1}{C_c}\right]$$

$$d\theta = -\,dQ\left[\frac{1}{C_h} + \frac{1}{C_c}\right] \qquad ...(26.86)$$

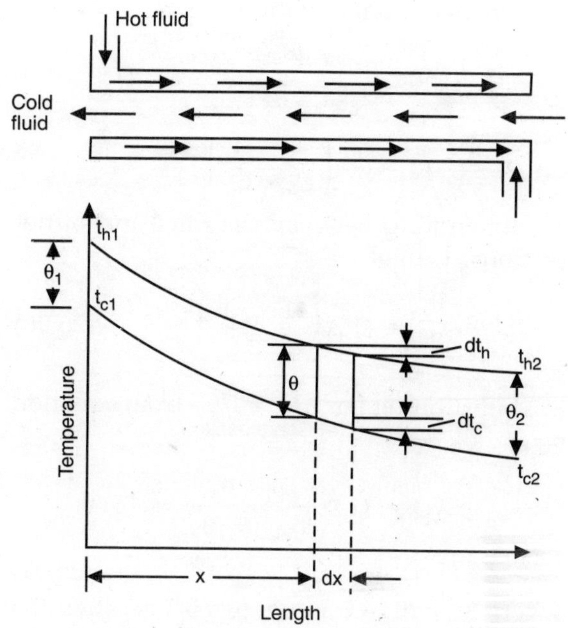

Fig. 26.46. Temperature changes of mediums (fluids) during counterflow arrangements

In a counter flow system, the temperatures of both the fluids decrease in the direction of heat exchanger length. In that case

$$dQ = -\,C_h\,dt_h = -\,C_c\,dt_c$$

$$\therefore \quad dt_h = -\frac{dQ}{C_h} \quad \text{and} \quad dt_c = -\frac{dQ}{C_c}$$

$$\text{or } (dt_h - dt_c) = d\theta = -\,dQ\left[\frac{1}{C_h} - \frac{1}{C_c}\right] \qquad ...(26.87)$$

From equations 26.86 and 26.87, we can write

$$d\theta = -\,dQ\left[\frac{1}{C_h} \pm \frac{1}{C_c}\right] \qquad ...(26.88)$$

The +ve sign refers to the parallel flow heat exchanger and –ve sign refers to the counter flow heat exchanger.

Integration of equation 26.88 between the inlet 1 and outlet 2 gives :

$$\theta_1 - \theta_2 = Q\left[\frac{1}{C_h} \pm \frac{1}{C_c}\right] \qquad ...(26.89)$$

Further, from equations 26.84 and 26.89,

$$U\,dA\,\theta = \frac{-d\theta}{1/C_h \pm 1/C_c}$$

or $$\frac{d\theta}{\theta} = -U\,dA\left(\frac{1}{C_h} \pm \frac{1}{C_c}\right)$$

Integrating between the inlet and outlet sections, we get :

$$\log_e \frac{\theta_1}{\theta_2} = U\,A\left(\frac{1}{C_h} \pm \frac{1}{C_c}\right) \quad ...(26.90)$$

Substituting for $1/C_h \pm 1/C_c$ from equation 26.89, we get :

$$Q = U\,A\,\frac{\theta_1 - \theta_2}{\log_e(\theta_1/\theta_2)} = U\,A\,\theta_m \quad ...(26.91)$$

where $\theta_m = (\theta_1 - \theta_2)/\log_e(\theta_1/\theta_2)$ is called the **logarithmic mean temperature difference** (LMTD)

If the heat capacity of the two fluids are equal in a counter flow arrangement, then it follows from equation 26.89, that $\theta_1 = \theta_2$

Then θ_m =

$$\frac{\theta_1 - \theta_2}{\log_e(\theta_1/\theta_2)} = \frac{0}{\log_e 1} = \frac{0}{0} = \text{indeterminate}$$

Applying L' Hospital's rule :

$$\lim_{\theta_1 \to \theta_2} = \frac{\theta_1 - \theta_2}{\log_e(\theta_1/\theta_2)} = \frac{\partial}{\partial \theta_1}(\theta_1 - \theta_2)_{\theta_1 = \theta_2}$$

$$\div \frac{\partial}{\partial \theta_1}\log_e(\theta_1/\theta_2)_{\theta_1 = \theta_2}$$

$$= \frac{1}{(1/\theta_1)_{\theta_1 = \theta_2}} = \theta_1 = \theta_2 \quad ...(26.92)$$

Clearly the effective temperature difference must equal θ_1 and θ_2 in case the temperature differences on either side of a heat exchanger are equal. Further, the logarithmic mean temperature difference for a counter flow unit is greater than that of a parallel flow system and accordingly the counter flow heat exchanger can transfer more heat than a similar parallel flow heat exchanger. Conversely, a counter flow exchanger needs a smaller heating surface for the same rate of heat transfer.

EXAMPLE 26.47.

Explain the difference between parallel flow and counter flow heat exchanger and show how the temperature of the two fluids vary along the path of flow.

A cold liquid (sp. heat 2.95 kJ/kg K) at 10 kg/min is to be heated from 25°C to 55°C in a heat exchanger. The task is accomplished by extracting heat from hot water (sp heat 4.186 kJ/kg K) available at mass flow rate 5 kg/min and inlet temperature 85°C. Should the thermal engineer make design calculations based on parallel flow or counter flow configuration ? State the reason thereof.

Solution : Making the energy balance between the hot (suffix h) and cold (suffix c) fluids,

$$m_h\,c_h\,(t_{h1} - t_{h2}) = m_c\,c_c\,(t_{c2} - t_{c1})$$

or

$$5 \times 4.186 \times (85 - t_{h2}) = 10 \times 2.95 \times (55 - 25) = 855$$

$$\therefore \quad t_{h2} = 85 - \frac{855}{5 \times 4.186} = \mathbf{42.72°C}$$

The hot water exit temperature (t_{h2} = 42.72°C), is less than the cold fluid exit temperature (t_{c2} = 55°C). The parallel flow configuration is not possible with such temperature distribution. Hence the design calculations need to be made for counter flow configuration of the heat exchanger.

EXAMPLE 26.48.

In a food processing plant, a brine solution is heated from – 12°C to – 65°C in a double pipe parallel flow heat exchanger by water entering at 35°C and leaving at 20.5°C at the rate of 9 kg/min. Determine the heat exchanger area for an overall heat transfer coefficient of 860 W/m² K. For water c_p = 4.186 × 10³ J/kgK

Solution : Heat transfer from the water,

$$Q = m\,c_p\,\Delta t$$
$$= 9 \times 4.186 \times 10^3\,(35 - 20.5)$$
$$= 546.27 \times 10^3 \text{ J/min}$$
$$= 9104.5 \text{ J/s}$$

Log-mean temperature difference,

$$\theta_m = \frac{\theta_1 - \theta_2}{\log_e(\theta_1/\theta_2)}$$

For parallel flow arrangement (Fig. 26.40) :

$$\theta_1 = t_{h1} - t_{c1} = 35 - (-12) = 47°C$$
$$\theta_2 = t_{h2} - t_{c2} = 20.5 - (-6.5) = 27°C$$

The subscripts h and c represent the hot and cold fluids respectively :

$$\therefore \quad \theta_m = \frac{47 - 27}{\log_e(47/27)} = 36.10°C$$

Heat exchange, $Q = U\,A\,\theta_m$

∴ Heating surface area,

$$A = \frac{Q}{U\theta_m} = \frac{9104.5}{860 \times 36.10}$$

$$= \mathbf{0.293\ m^2}$$

EXAMPLE 26.49.

A tubular heat exchanger is to be designed for cooling oil from a temperature of 80°C to 30°C by a large of stagnant water which may be assumed to remain constant at a temperature of 20°C. The heat transfer surface consists of 30 m long straight tube of 20 mm inside diameter. The oil (specific heat = 2.5 kJ/kg K and specific gravity = 0.8) flows through the cylindrical tube with an average velocity of 50 cm/s. Calculate the overall heat transfer coefficient for the oil cooler.

Solution : The mass rate of flow of oil (hot fluid) through the heat transfer tube is

$$m_h = V\,A\,\rho$$

$$= 0.5 \times \frac{\pi}{4}(0.02)^2 \times (0.8 \times 1000)$$

$$= 0.1256\ \text{kg/s}$$

An energy balance on the hot fluid then gives the total energy transferred

$$Q = m_h\,c_h\,(t_{h1} - t_{h2})$$
$$= 0.1256 \times 2.5 \times (80 - 30)$$
$$= 15.7\ \text{kJ/s}$$

The energy transferred is also given by

$$Q = U\,A\,\theta_m = UA\frac{\theta_1 - \theta_2}{\log_e(\theta_1/\theta_2)}$$

Since temperature of one of the fluid remains constant during the flow passage, it is immaterial whether the calculations are made for parallel flow or counter flow arrangement. Both arrangements would give the same values for log mean temperature difference and heating surface area for a specified load.

For parallel flow arrangement,

$$\theta_1 = t_{h1} - t_{c1} = 80 - 20 = 60°C$$
$$\theta_2 = t_{h2} - t_{c2} = 30 - 20 = 10°C$$

$$\theta_m = \frac{\theta_1 - \theta_2}{\log_e(\theta_1/\theta_2)} = \frac{60 - 10}{\log_e(60/10)}$$

$$= 27.90°C$$

$$A = \pi\,d\,l = \pi \times 0.02 \times 30$$
$$= 1.884\ \text{m}^2$$

∴ Overall heat transfer coefficient for the oil cooler,

$$U = \frac{Q}{A\theta_m} = \frac{15.7}{1.884 \times 27.90}$$

$$= \mathbf{0.2987\ kJ/m^2\text{-}s\text{-}deg}$$

EXAMPLE 26.50.

Exhaust gases (c_p = 1.12 kJ/kg-deg) flowing through a tubular heat exchanger at the rate of 1200 kg/hr are cooled from 400°C to 120°C. The cooling is affected by water (c_p = 4.18 kJ/kg K) that enters the system at 10°C at the rate of 1500 kg/hr. If the overall heat transfer coefficient is 500 kJ/m²-hr-deg, what heat exchanger area is required to handle the load for (a) parallel flow and (b) counter flow arrangement ?

Solution : The unknown exit temperature of the cooling water may be found from an energy balance on the two fluids, *i.e.*,

heat gained by water

= heat lost by exhaust gases

$$m_c\,c_c\,(t_{c2} - t_{c1}) = m_h\,c_h\,(t_{h1} - t_{h2})$$

$$t_{c2} = t_{c1} + \frac{m_h\,c_h}{m_c\,c_c}(t_{h1} - t_{h2})$$

$$= 10 + \frac{1200 \times 1.12}{1500 \times 4.18}(400 - 120) \cong 70°C$$

From an energy balance on the hot fluid, the heat transfer rate is,

$$Q = m_h\,c_h\,(t_{h1} - t_{h2})$$
$$= 1200 \times 1.12\,(400 - 120)$$
$$= 376320\ \text{kJ/hr}$$

(*a*) **Parallel flow arrangement** (Fig. 26.40). The log mean temperature difference is,

$$\theta_m = \frac{\theta_1 - \theta_2}{\log_e(\theta_1/\theta_2)}$$

where $\theta_1 = t_{h1} - t_{c1} = 400 - 10 = 390°C$
and $\theta_2 = t_{h2} - t_{c2} = 120 - 70 = 50°C$

$$\therefore \quad \theta_m = \frac{390 - 50}{\log_e(390/50)} = 165.53°C$$

Now, heat exchange $Q = U\,A\,\theta_m$

$\therefore$ Heating surface area,

$$A = \frac{Q}{U\theta_m} = \frac{376320}{500 \times 165.53} = \mathbf{4.547\ m^2}$$

(*b*) **Counter flow arrangement** (Fig. 26.40). The log mean temperature difference is,

$$\theta_m = \frac{\theta_1 - \theta_2}{\log_e(\theta_1/\theta_2)}$$

$$\theta_1 = t_{h1} - t_{c2} = 400 - 70 = 330°C$$
$$\theta_2 = t_{h2} - t_{c1} = 120 - 10 = 110°C$$

$$\theta_m = \frac{320 - 110}{\log_e(330/110)} = 200.25°C$$

Heat exchange, $Q = U\,A\,\theta_m$

$\therefore$ Heating surface area,

$$A = \frac{Q}{U\theta_m} = \frac{376320}{500 \times 200.25} = \mathbf{3.758\ m^2}$$

26.29. EFFECTIVENESS AND NUMBER OF TRANSFER UNITS (NTU)

The concept of log-mean temperature difference for estimating/analysing the performance of a heat exchanger unit is quite useful only when the inlet and outlet temperatures of the fluids are either known or can be easily determined from the relevant data. The usual design is, however, based on known fluid inlet temperature and estimated heat transfer coefficients. The unknown parameters may be the outlet conditions and heat transfer or the surface area required for a specified heat transfer. From energy balance, the heat given up by the hot fluid is picked up by the coolant on being transferred through the heat exchanger. Thus

$$Q = m_c\, c_c\,(t_{c2} - t_{c1})$$
$$= m_h\, c_h\,(t_{h1} - t_{h2}) = U\,A\,\theta_m$$

where,

$$\theta_m = \frac{\theta_1 - \theta_2}{\log_e(\theta_1/\theta_2)}$$
$$= \frac{(t_{h1} - t_{c2}) - (t_{h2} - t_{c1})}{\log_e[(t_{h1} - t_{c2})/(t_{h2} - t_{c1})]}$$

The outlet conditions for each fluid can be worked out by eliminating Q between the above equations. The resulting equation however becomes unwieldly requiring a trial-and-error iteration approach owing to the logarithmic function in θ_m.

An estimate/analysis of the heat exchanger can be made more conveniently by the NTU approach which is based on the concept of capacity ratio, effectiveness and number of transfer units.

Capacity ratio* (C) :** The product $m\,c$ (mass × specific heat) of a fluid flowing in a heat exchanger is repeatedly encountered and is termed as the ***capacity rate. It indicates the capacity of the fluid to store energy at a given rate

Capacity rate of the hot fluid, $C_h = m_h\, c_h$

Capacity rate of the cold fluid, $C_c = m_c\, c_c$

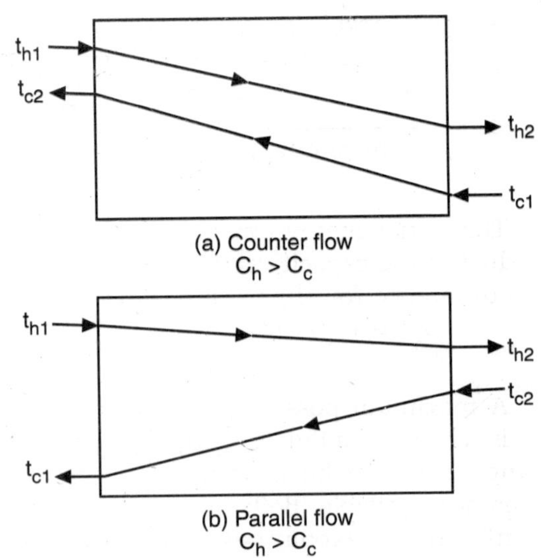

Fig. 26.47. Typical temperature profiles for counter-flow and parallel flow heat exchangers

The ***capacity ratio*** C is defined as the ratio of the minimum to maximum capacity rate. In parallel or counter flow heat exchangers, the hot or cold fluid may have the minimum value.

$$\text{if } m_h\, c_h > m_c\, c_c : C = \frac{m_c\, c_c}{m_h\, c_h} \quad ...(26.93)$$

$$\text{if } m_h\, c_h < m_c\, c_h : C = \frac{m_h\, c_h}{m_c\, c_c} \quad ...(26.94)$$

The temperature distribution along the exchanger length would be different depending on capacity ratio. The relative temperature change of the two fluids is inversely related to their capacity rates; the one with a smaller value of capacity rate experiencing the greater change in temperature. When $m_h\ c_h > m_c\ c_c$ in the counter-flow arrangement, the temperature tends to converge at the inlet. However, the temperature would be diverging when $m_c\ c_c > m_h\ c_h$. Further, in a parallel flow arrangement, t_{c2} approaches t_{h2} for an infinitely long heat exchanger ; consequently the counter-flow configuration is more effective in its operation.

Heat exchanger effectiveness **(∈)** = The effectiveness of a heat exchanger is defined as the ratio of the energy actually transferred to the maximum theoretical energy transfer.

$$\in = \frac{Q_{act}}{Q_{max}}$$

$$= \frac{\text{actual heat transfer}}{\text{maximum possible heat transfer}} \quad ...(26.95)$$

The actual energy transfer is given by the product of the capacity rate and the temperature difference for either fluid, *i.e.*,

$$Q_{act} = m_h\, c_h\, (t_{h1} - t_{h2})$$
$$= m_c\, c_c\, (t_{c2} - t_{c1})$$

A maximum possible heat transfer (Q_{max}) rate is achieved if a fluid undergoes temperature change equal to he maximum temperature difference available. Both for the parallel and counter-flow exchanger, the maximum temperature difference equals the inlet temperature of hot fluid minus the inlet temperature of the cold fluid, *i.e.*,

maximum available temperature difference

$$= (t_{h1} - t_{c1})$$

Further, the maximum possible heat transfer occurs when the fluid of a small heat capacity rate undergoes the maximum temperature difference available. The maximum possible energy transfer, therefore, becomes :

$$Q_{max} = C_{min}\, (t_{h1} - t_{c1}) \quad ...(29.96)$$

The effectiveness of heat exchanger is then :

$$\in = \frac{m_h\, c_h (t_{h1} - t_{h2})}{C_{min}(t_{h1} - t_{c1})} = \frac{C_h (t_{h1} - t_{h2})}{C_{min}(t_{h1} - t_{c1})}$$

$$= \frac{m_c\, c_c (t_{c2} - t_{c1})}{C_{min}(t_{h1} - t_{c1})} = \frac{C_c (t_{c2} - t_{c1})}{C_{min}(t_{h1} - t_{c1})} \quad ...(26.97)$$

Since either the hot or cold fluid may have the minimum value of capacity heat rate, there are two possible values of effectiveness

$$C_h < C_c : \in_h = \frac{(t_{h1} - t_{h2})}{(t_{h1} - t_{c1})}$$

$$C_c < C_h : \in_c = \frac{(t_{c2} - t_{c1})}{(t_{h1} - t_{c1})} \quad ...(26.98)$$

The subscript on ∈ designates the fluid which has the minimum heat capacity rate. Apparently, the effectiveness is simply a ratio of the temperature change of the fluid with the smaller heat capacity to the maximum temperature difference available in the heat exchanger.

Number of transfer units **(*NTU*)** : The number of transfer units (*NTU*) is a measure of the size of heat exchanger; it provides some indication of the size of the heat exchanger. It is defined as :

$$NTU = \frac{UA}{m_c c_c} \quad \text{when } m_h\, c_h > m_c\, c_c$$

$$NTU = \frac{UA}{m_h c_h} \quad \text{when } m_h\, c_h < m_c\, c_c \quad ...(26.99)$$

The denominator is always the smaller thermal capacity rate and therefore

$$NTU = \frac{UA}{(mc)_{min}} = \frac{UA}{C_{min}} \quad ...(26.100)$$

The following relations have been set up for the effectiveness of parallel and counter flow heat exchangers in terms of capacity ratio C and number of transfer units (*NTU*) :

$$\in = \frac{1-\exp[-NTU(1+C)]}{1+C} \quad \text{...parallel flow}$$

$$= \frac{1-\exp[-NTU(1-C)]}{1-C\exp[-NTU(1-C)]} \quad \text{...counter flow}$$

...(26.101)

EXAMPLE 26.51.

Hot water having specific heat 4200 J/kgK flows through a heat exchanger at the rate of 4 kg/min with an inlet temperature of 100°C. A cold fluid having a specific heat 2400 J/kg K flows in at a rate of 8 kg/min and with inlet temperature 20°C. Make calculations for the maximum possible effectiveness if the fluid flow conforms to (a) parallel flow arrangement (b) counter flow arrangement.

Solution : Thermal capacity rates of the hot and cold fluids are :

$$C_h = m_h c_h = \frac{4}{60} \times 4200 = 280 \text{ W/K}$$

$$C_c = m_c c_c = \frac{8}{60} \times 2400 = 320 \text{ W/K}$$

Obviously,

$$C_{min} = 280 \text{ W/K and } C_{max} = 320 \text{ W/K}$$

Capacity ratio,

$$C = \frac{C_{min}}{C_{max}} = \frac{280}{320} = 0.875$$

The effectiveness for a parallel flow heat exchanger is

$$\in = \frac{1-\exp[-NTU(1+C)]}{1+C}$$

Effectiveness will be maximum when $NTU = \infty$ for the given value of capacity ratio C.

∴ Maximum possible effectiveness

$$= \frac{1-\exp[-\infty]}{1+C}$$

$$= \frac{1}{1+C} = \frac{1}{1+0.875} = \mathbf{0.533}$$

(*b*) The effectiveness for a counter flow heat exchanger is

$$\in = \frac{1-\exp[-NTU(1-C)]}{1-C\exp[-NTU(1-C)]}$$

Effectivenss will be maximum when $NTU = \infty$ for the given value of capacity ratio C

∴ Maximum possible effectiveness

$$= \frac{1-\exp(-\infty)}{1-C\exp(-\infty)} = \frac{1-0}{1-0} = \mathbf{1}$$

EXAMPLE 26.52.

A counter flow heat exchanger is used to cool 2000 kg/hr of oil (c_p = 2.5 kJ/kg K) from 105°C to 30°C by the use of water entering at 15°C. If the overall heat transfer coefficient is expected to be 1.5 kW/m²K, make calculations for the water flow rate, the surface area required and the effectiveness of heat exchanger. Presume that the exit temperature of the water is not to exceed 80°C. Use NTU-effectiveness approach.

Solution : The mass flow rate of water can be determined from an energy balance on the two fluids, *i.e.*,

heat lost by oil (hot fluid)

= heat gained by water (coolant)

$$m_h c_h (t_{h1} - t_{h2}) = m_c c_c (t_{c2} - t_{c1})$$

$$2000 \times 2.5 (105 - 30) = m_c \times 4.18 (80 - 15)$$

∴ mass flow rate of coolant (water) m_c

$$= \frac{2000 \times 2.5 \times (105-30)}{4.18 \times (80-15)}$$

$$= 1380.2 \text{ kg/hr}$$

(*b*) Thermal capacity of the water stream (coolant)

$$C_c = m_c c_c = 1380.2 \times 4.18$$

$$= 5769.24 \text{ kJ/hrK}$$

Thermal capacity of the oil stream (hot fluid)

$$C_h = m_h c_h = 2000 \times 2.5$$

$$= 5000 \text{ kJ/hrK}$$

Obviously C_{min} = 5000 kJ/hrK

and C_{max} = 5769.24 kJ/hrK

$$\text{Capacity ratio } C = \frac{C_{min}}{C_{max}} = \frac{5000}{5769.24} = 0.867$$

When hot fluid has the minimum heat capacity, then

$$\text{Effectiveness } \in = \frac{t_{h1} - t_{h2}}{t_{h1} - t_{c1}} = \frac{105-300}{105-15}$$

$$= 0.833$$

The number of transfer units (NTU) can be computed from the following expression for effectiveness of a counter flow exchanger

$$\in = \frac{1-\exp[-NTU(1-C)]}{1-C\exp[-NTU(1-C)]}$$

Rearranging

$$\frac{\in -1}{\in C-1} = \exp[-NTU(1-C)]$$

$$\frac{0.833-1}{0.833\times 0.867-1} = \exp[-NTU(1-0.867)]$$

or $\log_e \frac{0.167}{0.278} = 0.133\ NTU$

or $NTU = 3.84$

But $NTU = \frac{UA}{C_{min}}$

Therefore the heat transfer area is

$$A = \frac{NTU \times C_{min}}{U} = \frac{3.84\times 5000}{1.5\times 3600} = 3.55 \text{ m}^2$$

SALIENT POINTS

1. Heat transfer is the propagation of heat energy between two systems due to temperature differential and takes place in the direction of decreasing temperature.
2. There are three distinct modes for heat transmission namely conduction, convection and radiation.
3. For a plane slab, the conduction heat flow is given by Fourier law :

$$Q = -kA\frac{dt}{dx} = \frac{kA(t_1-t_2)}{\delta}$$

$$= \frac{(t_1-t_2)}{\delta/kA} = \frac{t_1-t_2}{R_t}$$

The thermal conductivity k is a characteristic of the surface material, its unit are kJ/m-hr-deg or W/m-deg. The parameter $h_t = \delta/kA$ is called thermal resistance, it is expressed as s-deg/J or deg/W.

4. The rate of convective heat transfer between a solid boundary and adjacent fluid is given by the Newton-Rikhman law:

$$Q = hA(t_s - t_f) = \frac{t_s - t_f}{\frac{1}{hA}}$$

Here hA represents thermal resistance. The heat transfer coefficienct has units of W/m^2-deg.

5. For a two layer composite wall with heat transfer coefficients on inside (h_i) and outside (h_o)

$$Q = \frac{\Delta t}{\Sigma R_t} = \frac{\Delta t}{\frac{1}{h_i A}+\frac{\delta_1}{k_1 A}+\frac{\delta_2}{k_2 A}+\frac{1}{h_o A}}$$

Quite often, the heat flow through a composite section is written in the form

$$Q = UA\,\Delta T$$

where U is the overall heat transfer coefficienct. Then :

$$\frac{1}{U} = \frac{1}{h_i} + \frac{\delta_1}{k_1} + \frac{\delta_2}{k_2} + \frac{1}{h_o}$$

6. The conduction heat flow through a cylindrical metal tube and multi-layer cylindrical wall are :

$$Q = \frac{(t_1-t_2)}{\frac{1}{2\pi kl}\log_e\frac{r_2}{r_1}} = \frac{(t_1-t_2)}{R_t}$$

and

$$Q = \frac{\Delta T}{\frac{1}{2\pi k_1 l}\log_e\frac{r_2}{r_1}+\frac{1}{2\pi k_2 l}\log_e\frac{r_3}{r_2}}$$

The thermal resistance due to internal and external heat transfer coefficients would respectively be

$$\frac{1}{2\pi r_1 l h_i} \text{ and } \frac{1}{2\pi r_3 l h_o}$$

7. For a hollow sphere, the conduction heat transmission is

$$Q = \frac{t_1 - t_2}{\dfrac{r_2 - r_1}{4\pi k r_1 r_2}}$$

8. Addition of insulation material does not always bring about a decrease in heat transfer rate for geometries with non-constant cross-sectional area.

The critical radius of insulation for minimum resistance and consequently maximum heat flow rate is

$$r_c = \frac{k}{h_o} \text{ for a cylinder}$$

$$= \frac{2k}{h_o} \text{ for a sphere}$$

9. Thermal convection is the process of energy transport affected by the circulation or mixing of a fluid medium

(*i*) In natural or free convection, the circulation of the free medium is caused by buoyancy effects. For free convection over a flat plate

$$N_u = \phi(G_r, P_r)$$

$$\text{where } N_u \text{ (Nusselt number)} = \frac{hl}{k}$$

$$G_r \text{ (Grashof number)} = \frac{l^3 \rho^2 \beta g \Delta T}{\mu^2}$$

$$P_r \text{ (Prandtl number)} = \frac{\mu c_p}{k}$$

(*ii*) In forced convection, the flow of fluid is caused by a pump, fan or by winds. For forced convection over a flat plate,

$$N_u = \phi (R_e, P_r) \text{ and } St = \phi (R_e, P_r)$$

$$\text{where } R_e \text{ (Reynolds number)} = \frac{\rho Vl}{\mu}$$

$$St \text{ (Stanton number)} = \frac{h}{\rho V c_p}$$

10. Thermal radiation is the transmission of heat in the form of radiant energy or wave motion. The laws governing the mechanism of heat flow by radiation are :

(*i*) **Stefan Boltzman law :** The amount of radiant energy emitted per unit time from a black surface is proportional to fourth power of its absolute temperature

$$E_b = \sigma_b AT^4 \text{ Watt}$$

where the radiant coefficient

$$\sigma_b = 5.67 \times 10^8 \text{ W/m}^2 \text{ K}^4$$

(*ii*) **Wien's displacement law :** The product of absolute temperature and the wavelength at which emissive power is maximum is constant

$$\lambda_{max} T = 0.0029 \text{ mK}$$

(*iii*) **Kirchoff's law :** The emissivity ϵ and absorptivity α of a real surface are equal for radiation with identical temperature and wavelength.

$$\epsilon = \alpha \text{ where emissivity } \epsilon = \frac{E}{E_b}$$

(*iv*) **Lambert's cosine law :** The intensity of radiation in a direction θ from the normal to the black emitter is proportional to cosine of angle θ

$$I_\theta = I_n \cos\theta = \frac{\sigma_b T^4}{\pi} \cos\theta$$

where for a unit surface, the intensity of normal radiation I_n is $1/\pi$ times the emissive power E_b.

(*v*) The rate of heat exchange between two surfaces with emissivities ϵ_1 and ϵ_2 is given by

$$Q_{12} = (F_g)_{12} A_1 \sigma_b \left(T_1^4 - T_2^4\right)$$

where

$$(F_g)_{12} = f_{12} \times F_{12}$$

$$= \frac{1}{\dfrac{1-\epsilon_1}{\epsilon_1} + \dfrac{1}{F_{12}} + \dfrac{1-\epsilon_1}{\epsilon_2} \times \dfrac{A_1}{A_2}}$$

The configuration factor F_{12} considers the orientation and geometry of the black radiating surfaces; how the two surfaces view each other and to what extent the two surfaces radiate solely to each other. The interchange factor f_{12} allows for the departure of the two surfaces from complete blackness; a function of the emissivities of two surfaces.

For infinite long parallel planes ($A_1 = A_2$) which see each other and nothing else, and for the configuration corresponding to a completely enclosed body which is small compared with the enclosing body ($A_1 <<< A_2$), the configuration factor $F_{12} = 1$.

(*vi*) The radiation shields are thin obaque partitions that introduce additional resistance in the heat flow path and accordingly the net heat flux is reduced.

Radiant energy transfer

$$\frac{\text{with shield}}{\text{without shield}} = \frac{\frac{1}{\epsilon_1}+\frac{1}{\epsilon_2}-1}{\left(\frac{1}{\epsilon_1}+\frac{1}{\epsilon_3}-1\right)+\left(\frac{1}{\epsilon_2}+\frac{1}{\epsilon_2}-1\right)}$$

This is on the presumption that both faces of the radiation shield have the same emissivity ϵ_3.

Further, when $\epsilon_1 = \epsilon_2 = \epsilon_3$, the above fraction takes the value 1/2. Thus by inserting one shield, the heat transfer is reduced to one-half of the original value. The corresponding temperature T_3 of the shield will attain the value :

$$T_3^4 = \frac{1}{2}\left(T_1^4 + T_2^4\right)$$

11. Heat exchanger is a process equipment designed for the effective transfer of heat energy between two fluids, a hot fluid and a coolant.

In the performance analysis of heat exchanger,

heat loss by hot fluid

= heat gained by cold fluid

$$m_h\, c_h\left(t_{h_1} - t_{h_2}\right) = m_c\, c_c\left(t_{c_2} - t_{c_1}\right)$$

Further, the structure of the heat exchanger transfers this heat and is written as

$$= UA\,\theta_m$$

where U is the overall heat transfer coefficient between the two fluids, A is the effective heat transfer area and θ_m is the logaritmic mean temperature difference across the heat exchanger structure

$$\theta_m = \frac{\theta_1 - \theta_2}{\log_e \frac{\theta_1}{\theta_2}}$$

where θ_1 and θ_2 are the temperature differences between the fluids at entry to and at exit from the heat exchanger.

(*ii*) The LMTD for a counter flow unit is greater than that for a parallel flow system. As such, the counter flow exchanger needs a smaller heating surface for the same rate of heat transfer.

(*iii*) When one of the fluid remains at constant temperature during the flow passage (evaporator or condenser), both the parallel and counter flow arrangements give the same values of LMTD and heating surface area for a specified load.

12. The number of transfer units (*NTU*) is a measure of the size of heat exchanger.

$$NTU = \frac{UA}{C_{min}}$$

where the capacity ratio C represents the ratio of the minimum to the maximum capacity rate ($m_c\, c_c$ or $m_h\, c_h$).

Further, the effectiveness of heat exchanger is prescribed as :

$$\epsilon = \frac{1-\exp[-NTU(1+C)]}{1+C}$$

...parallel flow

$$= \frac{1-\exp[-NTU(1-C)]}{1-C\exp[-NTU(1-C)]}$$

...counter flow

REVIEW QUESTIONS

A. Conceptual and conventional questions.

1. Write the Fourier rate equation for heat transfer by conduction. Give the units and physical significance of each term appearing in this equation.
2. Define thermal conductivity, thermal resistance and thermal conductance.

3. How are Fourier's law and Ohm's law similar?
4. Compute the heat flux through a 10 cm thick brick wall (k = 0.69 W/m-deg) which has temperatures of 105°C and 10°C maintained on its two faces. How this heat flux would vary when 3 cm thick layers of magnesia insulation are added both on the inside and outside faces of the wall ? Also determine the interface temperature for the composite wall.
(**Ans.** 972 W/m^2, 140.5 W/m^2, 93.1°C and 72.8°C)
5. It is necessary to insulate a plane wall 2.5 m long and 2 m high so that amount of heat flowing per hour through this wall will not exceed 8500 kJ. The temperature of wall underneath the insulation is 400°C and the outer surface of the insulation is at a temperature of 50°C.
Determine: (*a*) thickness of insulation layer made of asbestos material of thermal conductivity k = 0.537 W/m-deg; (*b*) temperature in the insulation layer at a distance of 0.05 m from the inner surface; and (*c*) temperature gradient of the insulation.
(**Ans.** 0.1139 m, 246.5°C, 3072.87 deg/m)
6. The door of a domestic refrigerate has an area of 0.7 m^2 and it basically consists of a thin metal sheet with a 25 mm thick layer of insulation on the inside. The thermal conductivity of this insulation is 0.25 W/m-deg and heat transfer on each side of the door is 10 W/m^2-deg. Determine the heat flow rate through the door and the temperature of the metal sheet. The refrigerated chamber and the room are at 0°C and 20°C respectively. Neglect thermal resistance due to the sheet metal.
(**Ans.** 46.62 W, 13.32°C)
7. A refrigerant at – 30°C flows inside a copper tube (k = 400 W/m-deg) of inside diameter 1 cm and wall thickness 2 mm. To reduce losses, a 3 mm thick shell of thermocole type of material (k = 0.03 W/m-deg) is lagged onto the pipe. Calculate the heat leakage to the refrigerant per metre length of pipe. The internal and external film coefficient are 450 and 5 W/m^2-deg respectively and the outside air temperature is 30°C.
8. A 20 cm thick plane wall is built from fire brick material whose thermal conductivity varies linearly with temperature : $k = k_o (1 + \beta t)$ where k = 0.815 W/m-deg and β = 0.0005/°C. The temperature of wall surface are 1000°C and 200°C. Determine the heat flux from the wall and temperature at 50 mm away from the higher temperature surface. What would be the error in temperature calculated on the basis of an average thermal conductivity ?
(**Ans.** 6.13 W)
9. Explain the terms absorptivity, reflectivity and transmissivity of radiant energy. How are they related to each other for a black body and an opaque body ?
10. (*a*) What do you understand by a 'black body' and a 'gray body' as applied to radiation problems. Explain how radiation streaming out of a small hole in a large hollow body can be considered as black body radiation.
(*b*) A black body may be approximated by means of a large hollow container with a small opening. Exactly which surface of the arrangement can be considered as black body ?
11. (*a*) Define monochromatic and total emissive power. How is the later related to the absolute temperature ? Describe how the mono-chromatic emissive power varies with the wavelength for emissions from a black body ? At what wavelength is the black body monochromatic emissive power the maximum ?
(*b*) State and explain the following laws relating to thermal radiation and temperature of a radiating body :
Planck's law : Stefan Boltzman law and Wien's displacement law.
12. (*a*) State and prove Kirchoff's law of radiation. What restrictive conditions are inherent in the derivation of Kirchoff's law ?
(*b*) Define Lambert's cosine law of radiation and prove that the intensity of radiation is always constant at any angle of emission for a diffused surface.
13. A polished metal pipe 5 cm outside diameter and 370 K temperature at the outer surface is exposed to ambient conditions at 295 K temperature. The emissivity of the surface is 0.2 and the convection coefficient of heat transfer is 11.35 W/m^2-deg. Calcualte the heat transfer by radiation and natural convection per metre length of the pipe. Take thermal radiation constant $\sigma_b = 5.67 \times 10^{-8}$ W/m^{2}K^4.
What would be the overall coefficient of heat transfer by the combined mode of convection and radiation ?
14. Show that the quantity of radiant heat interchange between two gray surfaces can be expressed by the relation

$$(Q_{ij})_{net} = F_{ij} f_{ij} A_i \sigma_b \left(T_i^4 - T_j^4\right)$$

The notations have their usual meanings. State all the assumptions made in the derivation of this relation.

How the relation gets modified when the surfaces are parallel to each other ?

15. A thermos flask has a double walled bottle and the space between the walls is evacuated so as to reduce the heat flow. The bottle surfaces are silver plated and the emissivity of each surface is 0.025. If the contents of the bottle are at 375 K, find the rate of heat loss from the thermos bottle to the ambient air at 300 K. What thickness of cork (k = 0.03 W/m-deg) would be required if the same insulating effect is to be achieved by the use of cork ?

16. Give a general equation for the rate of heat transfer by convection, and hence define the coefficient of heat transfer. List the various factors on which the value of this coefficient depends.

17. What thermo-physical variables are involved in the equations describing the phenomenon of free convection ?

    Apply dimesional analysis to these variables and develop a generalised correlation between certain non-dimensional parameters.

18. List the variables that affect the forced heat transfer coefficient.

    Using dimensional analysis, demonstrate that the following dimensionless parameters are possible combinations of the appropriate varibles describing forced convection :

$$\frac{V x \rho}{\mu}, \frac{\mu c_p}{k}, \frac{h x}{k} \text{ and } \frac{h}{\rho c_p V}$$

    What names have been given to these non-dimensional groups?

19. (a) Discuss the importance of heat exchangers for industrial use.

    (b) What are the basics for the classification of heat exchangers?

    (c) Sketch a two-shell pass, four tube pass, reversed current heat exchanger. Label the different parts.

20. Show the temperature variation along the length of a heat exchanger when

    (i) hot and cold fluids flow in parallel and counter flow fashion

    (ii) steam condenses on the outside of a condenser tube with water flowing inside the tube as coolant.

    (iii) hot fluid as used for evaporating another liquid.

21. Working in terms of inlet and outlet temperatures of the fluids and overall heat transfer coefficient, develop an expression for the heat transfer from one fluid to another in a conventional (i) parallel flow and (ii) counter flow heat exchanger.

22. A counter flow lubricating oil cooler of surface area 24 m^2 has been employed to cool 2.75 × 10^5 kg/hr of oil (c_p = 2.1 kJ/kg-deg) from 62.5°C at inlet to 48.5°C at discharge. The inlet and outlet temperatures of cooling water are 24°C and 32°C respectively. Make calculations for the overall heat transfer coefficient.

    What will be the required surface area for a parallel flow device having the same capacity under identical conditions?

    (Ans. 34 W/m^2-deg, 25.4 ^{2})

23. Explain the meaning of terms : heat capacity ratio, effectiveness and number of heat transfer units as applied to heat exchangers.

B. Fill in the blanks with appropriate word/words.

1. Heat transfer takes place according to .... law of thermodynamics.
2. The material medium between the heat source and receiver is not affected during the heat transmission by ....
3. Upto critical radius of insulation, the heat loss .... by adding insulation.
4. Thermal conductivity is defined as the heat flow per unit time across .... when .... is unity.
5. The Stefan Boltzman constant has units of ....
6. A surface for which emissivity is constant at all temperature and throughout the entire range of wavelength is called ....
7. The relationship, λ_{max} T = constant, between the temperature of a black body and the wavelength at which maximum value of monochromatic emissive power occurs is known as ....
8. The value of shape factor for two infinite parallel surfaces separated by distance x is ....
9. The .... number has a significant role in free convection.

10. For forced convection, Nusselt number is a function of Prandtl number and .... number.

Answers

1. First 2. radiation 3. increases 4. unit area, temperature gradient 5. $kJ/m^2 - hr - K^4$ 6. gray 7. Wien's displacement 8. unity 9. Grashof 10. Reynolds number

C. Multiple choice questions.

1. The essential condition for the transfer of heat from one body to another is
   (*a*) both bodies must be in physical contact
   (*b*) heat content of one body must be more than that of the other
   (*c*) one of the bodies must have a high value of thermal conductivity
   (*d*) there must exist a temperature difference between the bodies

2. Consider the following statements :
   The Fourier heat conduction equation
   $$Q = -k A \frac{dt}{dx}$$
   presumes
   1. Steady state conditions
   2. Constant value of thermal conductivity
   3. Uniform temperature at the wall surfaces
   4. One-dimensional heat flow.

   Which of these statements are correct ?
   (*a*) 1, 2 and 3 (*b*) 1, 2 and 4
   (*c*) 2, 3 and 4 (*d*) 1, 3 and 4

3. In M-L-T-θ system (*T* being time and θ temperature), what is the dimension of thermal conductivity?
   (*a*) $ML^{-1}T^{-1}\theta^{-3}$ (*b*) $MLT^{-1}\theta^{-1}$
   (*c*) $ML\theta^{-1}T^{-3}$ (*d*) $ML\theta^{-1}T^{-2}$

4. A wall of thickness 0.6 m has normal area 1.5 m² and is made up of a material of thermal conductivity 0.4 W/mK. If the temperatures on the two sides are 800°C and 100°C, the thermal resistance of the wall could be
   (*a*) 1 W/K (*b*) 1.8 W/K
   (*c*) 1 K/W (*d*) 1.8 K/W

5. A composite slab has two layers of different materials having thermal conductivities k_1 and k_2. If each layer has same thickness, the equivalent thermal conductivity of the slab would be
   (*a*) $\frac{k_1 k_2}{k_1 + k_2}$ (*b*) $\frac{k_1 k_2}{2(k_1 + k_2)}$
   (*c*) $\frac{2k_1}{k_1 + k_2}$ (*d*) $\frac{2k_1 k_2}{k_1 + k_2}$

6. A gas turbine blade (idealised as a flat plate of surface area *A*, thickness δ and thermal conductivity *k*) has hot gases at temperature T_1 on one side and cooling air at temperature T_2 on the other side. If h_1 and h_2 are the corresponding surface coefficients of heat transfer, then the overall heat transfer coefficient *U* is given by
   (*a*) $\frac{1}{U} = \frac{1}{h_1} + \frac{\delta}{k} + \frac{1}{h_2}$
   (*b*) $\frac{1}{U} = \frac{1}{h_1} + \frac{k}{\delta} + \frac{1}{h_2}$
   (*c*) $U = h_1 + \frac{\delta}{k} + h_2$
   (*d*) $U = \frac{1}{h_1} + \frac{\delta}{k} + \frac{1}{h_2}$

7. A hollow cylinder of inner radius r_1 and outer radius r_2 is subjected to steady state heat transfer which results in constant surface temperature t_1 and t_2 at radii r_1 and r_2 respectively. For constant thermal conductivity *k*, the radial heat flow per unit length of cylinder is given by
   (*a*) $\frac{2\pi k(t_1 - t_2)}{\log_e (r_2/r_1)}$ (*b*) $\frac{2\pi(t_1 - t_2)}{k \log_e (r_2/r_1)}$
   (*c*) $\frac{(t_1 - t_2)}{2\pi k \log_e (r_2/r_1)}$ (*d*) $\frac{k(t_1 - t_2)}{2\pi \log_e (r_2/r_1)}$

8. For steady state and constant value of thermal conductivity, the temperature distribution associated with radial conduction through a cylinder is
   (*a*) linear (*b*) logarithmic
   (*c*) parabolic (*d*) exponential

9. The thermal resistance for heat conduction through a hollow sphere of inner radius r_1 and outer radius r_2 is
   (*a*) $\frac{r_2 - r_1}{4\pi k r_1 r_2}$ (*b*) $\frac{(r_2 - r_1) r_1 r_2}{4\pi k}$
   (*c*) $\frac{4\pi k (r_2 - r_1)}{r_1 r_2}$ (*d*) $\frac{k(r_2 - r_1)}{4\pi r_1 r_2}$

where k is the thermal conductivity of the material of sphere

10. The heat flow equation through a sphere of inner radius r_1 and outer radius r_2 is to be written in the same form as that for heat flow through a plane wall. For wall thickness $(r_2 - r_1)$, the equivalent mean radius for the spherical shell is
(a) $\frac{r_1 + r_2}{2}$ (b) $r_1 r_2$
(c) $\sqrt{r_1 r_2}$ (d) $\frac{r_1 + r_2}{\log_e (r_2/r_1)}$

11. With an increase in the thickness of insulation around a circular pipe, heat loss to surroundings due to
(a) convection increases while that due to conduction decreases
(b) convection decreases while that due to conduction increases
(c) convection and conduction decreases
(d) convection and conduction increases

12. A thin cylinder of radius r is lagged to an outer radius r_0 with an insulating layer of thermal conductivity k. If h_0 is the film coefficient at the outer surface of lagging, then minimum resistance and consequently maximum heat flow rate occurs when r_0 equals
(a) $\sqrt{k h_0}$ (b) k/h_0
(c) $2k/h_0$ (d) h_0/k

13. The thermal conductivity of thermal insulation applied to a hollow spherical vessel containing very hot material is 0.5 W/mK. The convective heat transfer at the outer surface of insulation is 10 W/m^2 K. What is the critical radius of sphere?
(a) 0.1 m (b) 0.2 m
(c) 1.0 m (d) 2.0 m

14. A body which partly absorbs and partly reflects but does not allow any radiation to pass through it ($\alpha + \rho = 1$ and $\tau = 0$) is called
(a) diathermanous (b) opaque
(c) grey (d) specular

15. The ratio of total emissive power of body to the total emissive power of a black body at the same temperature is called
(a) absorptivity (b) transmissivity
(c) reflectivity (d) emissivity

16. The law governing the distribution of radiant energy over wavelength for a black body at fixed temperature is referred to as
(a) Planck's law (b) Wien's formula
(c) Kirchoff's law (d) Lambert's law

17. If the temperature of a hot body is increased by 50%, the amount of radiation emitted by it would increase by nearly
(a) 50% (b) 100%
(c) 200% (d) 500%

18. The emissive power is multiplied with the factor .... to obtain the intensity of normal radiation for a unit surface
(a) $1/\sqrt{\pi}$ (b) $1/\pi$
(c) $1/2\pi$ (d) $\sqrt{\pi}$

19. Two radiating surface $A_1 = 6$ m^2 and $A_2 = 4$ m^2 have shape factor $F_{1\text{-}2} = 0.1$. Then the shape factor $F_{2\text{-}1}$ will be
(a) 0.18 (b) 0.15
(c) 0.12 (d) 0.10

20. A body cooling from 80°C to 70°C takes 10 minutes when left exposed to environmental conditions. If the body is to cool further from 70°C to 60°C under the same external conditions, it will take
(a) same time of 10 minutes
(b) more than 10 minutes
(c) less than 10 minutes
(d) time will depend upon the environmental conditions

21. The Nusselt number in natural convection is a function of fluid Prandtl number and
(a) Stanton number
(b) Biot number
(c) Grashoff number
(d) Reynolds number

22. For forced convection, Nusselt number is a function of
(a) Prandtl and Grashoff number
(b) Reynolds and Prandtl number
(c) Reynolds and Grashoff number
(d) Reynolds number only

23. Choose the correct statement with respect to a counter flow heat exchanger :
(a) both the fluids at inlet are in their coldest state

(*b*) both the fluids at exit are in their hottest state

(*c*) both the fluids at inlet are in their hottest state

(*d*) one fluid is hottest and the other is coldest at inlet

24. In a heat exchanger with one fluid evaporating or condensing, the surface area required is least in
(*a*) parallel flow
(*b*) counter flow
(*c*) cross flow
(*d*) same in all the above arrangements

25. In a heat exchanger, the hot liquid enters with a temperature of 180°C and leaves at 160°C. The cooling fluid enters at 30°C and leaves at 110°C. The capacity ratio of the heat exchanger is
(*a*) 0.25 (*b*) 1.5
(*c*) 0.33 (*d*) 0.2

26. For an opaque surface, the absorptivity (α), transmissivity (τ) and reflectivity (ρ) are related by the equation
(*a*) $\alpha + \rho + \tau = 1$ (*b*) $\alpha + \rho = \tau$
(*c*) $\alpha + \rho = 1$ (*d*) $\alpha + \rho = 0$

27. Consider two infinitely long thin concentric tubes of circular cross-section as shown below:

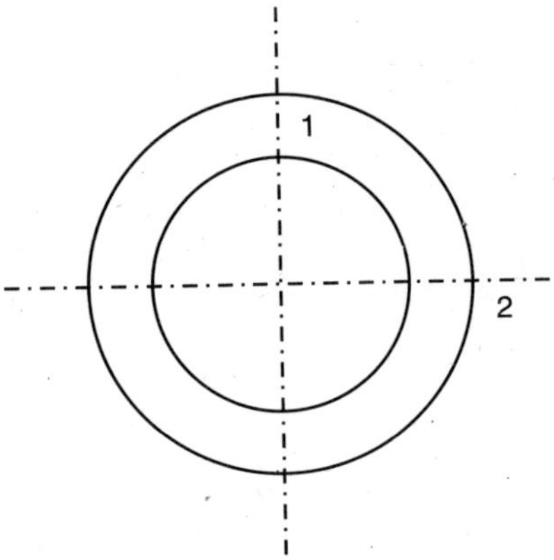

If D_1 and D_2 are the diameters of the inner and outer tubes respectively, then the view factor F_{22} is given by
(*a*) Zero (*b*) $\frac{D_2}{D_1} - 1$
(*c*) $1 - \frac{D_1}{D_2}$ (*d*) $\frac{D_1}{D_2}$

28. A pipe of 25 mm outer diameter carries steam. The heat transfer coefficient between the cylinders and surroundings is 5 W/m^2 K. It is proposed to reduce the heat loss from the pipe by adding insulation having a thermal conductivity of 0.05 W/mK. Which one of the following statements is true?
(*a*) the outer radius of the pipe is equal to critical radius
(*b*) the outer radius of the pipe is less than the critical radius
(*c*) adding the insulation will reduce the heat loss
(*d*) adding the insulation will increase the heat loss

29. The quantity of heat radiation from a body is dependent on its
1. area 2. temperature
3. shape
Which of the statements is correct?
(*a*) 1 and 2 (*b*) 2 and 3
(*c*) 2 only (*d*) 1, 2 and 3

Answers

1. (*d*)	2. (*d*)	3. (*c*)	4. (*c*)	5. (*d*)
6. (*a*)	7. (*a*)	8. (*b*)	9. (*a*)	10. (*c*)
11. (*a*)	12. (*b*)	13. (*a*)	14. (*b*)	15. (*d*)
16. (*a*)	17. (*d*)	18. (*b*)	19. (*b*)	20. (*b*)
21. (*c*)	22. (*b*)	23. (*c*)	24. (*d*)	25. (*a*)
26. (*c*)	27. (*c*)	28. (*c*)	29. (*d*)	

INDEX

INDEX

Win Prizes

ATTENTION READERS

Win Prizes

We request for your frank assessment regarding various aspects of the book as given below :

ENGINEERING THERMODYNAMICS

(Principles and Practices)

By: Dr. D.S. Kumar

Please Fill up the Sheet in Neat and Clean Writing

(i) What topics of your syllabus, important for your examination point of view, are not covered in this book ?

....

....

....

....

(ii) What are the chapters wherein the treatment of the subject matter is not systematic or organised or updated ?

....

....

....

....

(iii) Have you come across misprints/mistakes/factual inaccuracies in the book? Please specify the chapters and the page numbers.

....

....

....

....

(iv) What other books on the same subject have you found/heard better than the present book? Please specify in terms of price and quality

....

....

....

....

(v) Further suggestions and comments for the improvement of this book.

....

....

....

....

(PS: If need be please attach a separate sheet to write your views)

OTHER DETAILS

(i) Who recommended you this book ? (Please tick (✓) in the box near the option relevant to you.

☐ Teacher ☐ Friends ☐ Bookseller

....

....

....

....

(ii) Name of the recommending teacher, his designation and address :

....

....

....

....

(iii) Name and address of the bookseller you purchased the book from :

....

....

....

....

(iv) Name and address of your institution (Please mention the University or Board as the case may be)

....

....

....

....

(v) Your name and complete postal address :

....

....

....

....

(vi) Course you are enrolled in :

....

....

....

....

The best assessment will be awarded each month: The award will be in the form of our publications as decided by the Editorial Board.

Please mail the filled up coupon at your earliest to :

S.K. KATARIA & SONS®

4885/109, Prakash Mahal, Dr. Subhash Bhargav Lane,

Opposite Delhi Medical Association, Daryaganj, New Delhi-110002 (India)

Phone: +91-11-23243489, +91-11-23269324, +91-11-43551243 Telefax: 011-23243489

Email: katariabooks@yahoo.com; katariabook@gmail.com

Website : www.skkatariaandsons.com, www.katsonbooks.com